**Using *The Practice of Statistics*, Fifth Edition, for A**

(The percents in parentheses reflect cove

| Topic Outline for AP® Statistics<br>from the College Board's *AP® Statistics Course Descriptio..* | Chapter and Section references |
|---|---|
| **I. Exploring data: describing patterns and departures from patterns (20%–30%)** | |
| **A. Constructing and interpreting graphical displays of distributions of univariate data (dotplot, stemplot, histogram, cumulative frequency plot)** | **Dotplot, stemplot, histogram 1.2; Cumulative frequency plot 2.1** |
| 1. Center and spread | 1.2 |
| 2. Clusters and gaps | 1.2 |
| 3. Outliers and unusual features | 1.2 |
| 4. Shape | 1.2 |
| **B. Summarizing distributions of univariate data** | **1.3 and 2.1** |
| 1. Measuring center: median, mean | 1.3 |
| 2. Measuring spread: range, interquartile range, standard deviation | 1.3 |
| 3. Measuring position: quartiles, percentiles, standardized scores ($z$-scores) | Quartiles 1.3; percentiles and $z$-scores 2.1 |
| 4. Using boxplots | 1.3 |
| 5. The effect of changing units on summary measures | 2.1 |
| **C. Comparing distributions of univariate data (dotplots, back-to-back stemplots, parallel boxplots)** | **Dotplots and stemplots 1.2; boxplots 1.3** |
| 1. Comparing center and spread | 1.2 and 1.3 |
| 2. Comparing clusters and gaps | 1.2 and 1.3 |
| 3. Comparing outliers and unusual features | 1.2 and 1.3 |
| 4. Comparing shape | 1.2 and 1.3 |
| **D. Exploring bivariate data** | **Chapter 3 and Section 12.2** |
| 1. Analyzing patterns in scatterplots | 3.1 |
| 2. Correlation and linearity | 3.1 |
| 3. Least-squares regression line | 3.2 |
| 4. Residual plots, outliers, and influential points | 3.2 |
| 5. Transformations to achieve linearity: logarithmic and power transformations | 12.2 |
| **E. Exploring categorical data** | **Sections 1.1, 5.2, 5.3** |
| 1. Frequency tables and bar charts | 1.1 (we call them bar graphs) |
| 2. Marginal and joint frequencies for two-way tables | Marginal 1.1; joint 5.2 |
| 3. Conditional relative frequencies and association | 1.1 and 5.3 |
| 4. Comparing distributions using bar charts | 1.1 |
| **II. Sampling and experimentation: planning and conducting a study (10%–15%)** | |
| **A. Overview of methods of data collection** | **Sections 4.1 and 4.2** |
| 1. Census | 4.1 |
| 2. Sample survey | 4.1 |
| 3. Experiment | 4.2 |
| 4. Observational study | 4.2 |
| **B. Planning and conducting surveys** | **Section 4.1** |
| 1. Characteristics of a well-designed and well-conducted survey | 4.1 |
| 2. Populations, samples, and random selection | 4.1 |
| 3. Sources of bias in sampling and surveys | 4.1 |
| 4. Sampling methods, including simple random sampling, stratified random sampling, and cluster sampling | 4.1 |
| **C. Planning and conducting experiments** | **Section 4.2** |
| 1. Characteristics of a well-designed and well-conducted experiment | 4.2 |
| 2. Treatments, control groups, experimental units, random assignments, and replication | 4.2 |
| 3. Sources of bias and confounding, including placebo effect and blinding | 4.2 |
| 4. Completely randomized design | 4.2 |
| 5. Randomized block design, including matched pairs design | 4.2 |
| **D. Generalizability of results and types of conclusions that can be drawn from observational studies, experiments, and surveys** | **Section 4.3** |

| Using *The Practice of Statistics*, Fifth Edition, for Advanced Placement (AP®) Statistics<br>(The percents in parentheses reflect coverage on the AP® exam.) | |
|---|---|
| **Topic Outline for AP® Statistics**<br>from the College Board's *AP® Statistics Course Description* | ***The Practice of Statistics*, 5th ed.**<br>Chapter and Section references |
| **III. Anticipating patterns: exploring random phenomena using probability and simulation (20%–30%)** | |
| **A. Probability** | **Chapters 5 and 6** |
| 1. Interpreting probability, including long-run relative frequency interpretation | 5.1 |
| 2. "Law of large numbers" concept | 5.1 |
| 3. Addition rule, multiplication rule, conditional probability, and independence | Addition rule 5.2; other three topics 5.3 |
| 4. Discrete random variables and their probability distributions, including binomial and geometric | Discrete 6.1; Binomial and geometric 6.3 |
| 5. Simulation of random behavior and probability distributions | 5.1 |
| 6. Mean (expected value) and standard deviation of a random variable, and linear transformation of a random variable | Mean and standard deviation 6.1; Linear transformation 6.2 |
| **B. Combining independent random variables** | **Section 6.2** |
| 1. Notion of independence versus dependence | 6.2 |
| 2. Mean and standard deviation for sums and differences of independent random variables | 6.2 |
| **C. The Normal distribution** | **Section 2.2** |
| 1. Properties of the Normal distribution | 2.2 |
| 2. Using tables of the Normal distribution | 2.2 |
| 3. The Normal distribution as a model for measurements | 2.2 |
| **D. Sampling distributions** | **Chapter 7; Sections 8.3, 10.1, 10.2, 11.1** |
| 1. Sampling distribution of a sample proportion | 7.2 |
| 2. Sampling distribution of a sample mean | 7.3 |
| 3. Central limit theorem | 7.3 |
| 4. Sampling distribution of a difference between two independent sample proportions | 10.1 |
| 5. Sampling distribution of a difference between two independent sample means | 10.2 |
| 6. Simulation of sampling distributions | 7.1 |
| 7. *t* distribution | 8.3 |
| 8. Chi-square distribution | 11.1 |
| **IV. Statistical inference: estimating population parameters and testing hypotheses (30%–40%)** | |
| **A. Estimation (point estimators and confidence intervals)** | **Chapter 8 plus parts of Sections 9.3, 10.1, 10.2, 12.1** |
| 1. Estimating population parameters and margins of error | 8.1 |
| 2. Properties of point estimators, including unbiasedness and variability | 8.1 |
| 3. Logic of confidence intervals, meaning of confidence level and confidence intervals, and properties of confidence intervals | 8.1 |
| 4. Large-sample confidence interval for a proportion | 8.2 |
| 5. Large-sample confidence interval for a difference between two proportions | 10.1 |
| 6. Confidence interval for a mean | 8.3 |
| 7. Confidence interval for a difference between two means (unpaired and paired) | Paired 9.3; unpaired 10.2 |
| 8. Confidence interval for the slope of a least-squares regression line | 12.1 |
| **B. Tests of significance** | **Chapters 9 and 11 plus parts of Sections 10.1, 10.2, 12.1** |
| 1. Logic of significance testing, null and alternative hypotheses; *P*-values; one-and two-sided tests; concepts of Type I and Type II errors; concept of power | 9.1; power in 9.2 |
| 2. Large-sample test for a proportion | 9.2 |
| 3. Large-sample test for a difference between two proportions | 10.1 |
| 4. Test for a mean | 9.3 |
| 5. Test for a difference between two means (unpaired and paired) | Paired 9.3; unpaired 10.2 |
| 6. Chi-square test for goodness of fit, homogeneity of proportions, and independence (one- and two-way tables) | Chapter 11 |
| 7. Test for the slope of a least-squares regression line | 12.1 |

**Annotated Teacher's Edition**

For the **AP**® Exam

# The Practice of Statistics

FIFTH EDITION

Publisher: Ann Heath
Editorial Assistant: Matt Belford
Development Editor: Donald Gecewicz
Senior Marketing Manager: Janie Pierce-Bratcher
Photo Editor: Cecilia Varas
Photo Researcher: Julie Tesser
Art Director: Diana Blume
Cover Designers: Diana Blume, Rae Grant
Text Designer: Patrice Sheridan
Cover Image: Joseph Devenney/Getty Images
Senior Project Editor: Vivien Weiss
Illustrations: Network Graphics
Production Manager: Susan Wein
Composition: codeMantra
Printing and Binding: RR Donnelley



Library of Congress Control Number: 2013958418
ISBN-13: 978-1-4641-5401-0
ISBN-10: 1-4641-5401-5


First printing 2014



Printed in the United States of America

W. H. Freeman and Company
41 Madison Avenue
New York, NY 10010
Houndmills, Basingstoke RG21 6XS, England
www.whfreeman.com

**Annotated Teacher's Edition**

**Josh Tabor**
*Canyon del Oro High School*

For the AP® Exam

FIFTH EDITION

# The Practice of Statistics

**Daren S. Starnes**
*The Lawrenceville School*

**Josh Tabor**
*Canyon del Oro High School*

**Daniel S. Yates**
*Statistics Consultant*

**David S. Moore**
*Purdue University*

W. H. Freeman and Company/BFW
New York
A Macmillan Higher Education Company

## A Word from the Author

*Josh Tabor has enjoyed teaching general and AP® Statistics to high school students for more than 18 years, most recently at his alma mater, Canyon del Oro High School in Oro Valley, Arizona. In recognition of his outstanding work as an educator, Josh was named one of the five finalists for Arizona Teacher of the Year in 2011. He is a past member of the AP® Statistics Development Committee (2005–2009), as well as an experienced Table Leader and Question Leader at the AP® Statistics Reading. Each year, Josh leads one-week AP® Summer Institutes and one-day College Board workshops around the country and frequently speaks at local, national, and international conferences.*

My involvement with *The Practice of Statistics* started with a relatively brief and casual email from Daren Starnes in 2009. Who knew that it would lead to hundreds and hundreds of hours of challenging—but very rewarding—work, helping to revise the student edition and writing the *Annotated Teacher's Edition?* I am quite certain that neither I nor my wife knew exactly what we were getting into. As we began planning for the fifth edition, I had a better sense of the massive amount of work that lay ahead. But the excitement of making a great book even better was more than enough motivation to dive in.

My primary goal for the *Annotated Teacher's Edition* (ATE) was to provide every teacher with the experience of being in one of my summer workshops, including sharing insights from my years in the classroom; my time as a Reader, Table Leader, and Question Leader at the AP® Reading; and my four years on the AP® Statistics Development Committee. You will see these insights throughout the ATE in the form of Teaching Tips, AP® Exam Common Errors, Alternate Activities, Alternate Examples, and other features.

When I started teaching AP® Statistics in 1996, there were very few resources for teachers, and it wasn't too difficult to consult each and every one of them as I was planning my course. Teachers today have a different problem—too *many* resources. While this is a good problem to have, it doesn't make it any easier to deal with.

Thus, another goal for the *Annotated Teacher's Edition* was to collect the best resources and put them all together in one place. No longer will teachers need to refer to four or five different books to plan the next chapter. Pacing guides and suggested assignments, teaching and technology tips, and alternate activities and examples are all included in a single volume.

Writing the *Annotated Teacher's Edition* has also made me a better teacher. I carefully considered each sentence that I wrote, knowing that many teachers depend on the supplementary materials provided with a textbook. For each teaching tip, alternate example, or alternate activity, I asked myself, "Is this something I would use in my own class? Is this something I would recommend for other teachers?" If I couldn't answer "yes" to both of these questions, the material didn't make it into the ATE. Consequently, I had to spend long hours scouring the Internet for applets, activities, and, especially, real data. Artificial or made-up data would not suffice!

I hope you find the ATE to be a helpful resource, whether you are a rookie or a veteran teacher of AP® Statistics. Creating this resource has been a great experience for me, but I certainly didn't do it alone. Many thanks go to the following people:


- Daren Starnes, for providing wise advice and thoughtful feedback. I am very proud of what we accomplished in the Student Edition and Annotated Teacher's Edition. Most of all, as a teacher, I am thrilled that I get to use *TPS* 5e with my own students!
- Ann Heath, Publisher, for entrusting me with this role and helping guide me through the world of textbook publishing. Thanks especially for all of your encouragement when you could tell I was feeling the pressure.
- Enrico Bruno and Matt Belford, Editorial Assistant, for all the behind-the-scenes work on the manuscripts and other tasks.
- Patti Brecht, copyeditor, for reading each chapter so carefully and keeping me consistent from page to page.
- Jon Stockdale and Andrew Roberts at Math Made Visible, for your attention to detail and thoughtful comments about the alternate examples and solutions to exercises.
- Michael Legacy, for reviewing the teaching tips and other advice to teachers. Your comments and suggestions were very helpful.
- Most importantly, a huge "Thank you!" to my wife, Anne, and my daughter, Jordan, for allowing me to spend seemingly endless hours working on this project. You were always supportive, even though it made our lives a little crazy at times. I love you both and always will.

# About the *Annotated Teacher's Edition*

Since the first edition of *The Practice of Statistics* (*TPS*) was published, teachers have lauded the ancillary materials that accompany the student textbook. We have been through four editions of the famous Teacher's Resource Binder, starting with the Yellow Binder, then moving to the Gold, Platinum, and Titanium Binders.

With the third edition of *TPS*, the support package was expanded to include an *Annotated Teacher's Edition* (ATE), which provided teachers with an additional resource for just-in-time teaching tips, advice for the AP® exam, and answers placed on the same page as the exercises.

For the fifth edition, we have consolidated all of the best material into the ATE and no longer offer a cumbersome binder of printed material. Instead, all of the great material that used to be found in the Teacher's Resource Binder can now be accessed in three ways: by clicking on the link in the Teacher's e-Book (TE-book), opening the Teacher's Resource CD (TRCD), or logging onto the book's Web site www.whfreeman.com/tps5e (teacher log-in required). Look for the Teacher's Resource Materials TRM icon throughout the ATE for reminders about the available resources.

The *Annotated Teacher's Edition* includes:

- The complete student text.
- An introduction with general advice for teaching AP® Statistics.
- "Blue Pages" that precede the wrap-around student pages at the beginning of each chapter. These provide an in-depth guide to the chapter content and support materials, including:
  - An overview of the chapter, featuring a discussion of the most important ideas and how the content relates to the AP® Statistics Topic Outline.
  - A summary of each section in the chapter, highlighting the big ideas.
  - A list of resources, including a comprehensive list of AP® Free-Response questions appropriate for that chapter.
  - A pacing guide for the chapter, featuring learning objectives and suggested homework assignments.
- Answers (including graphs) to all Exercises, Check Your Understanding problems, Data Explorations, Case Closed problems, and FRAPPYs. Care has been taken to make sure that the answer appears on the same page as the corresponding exercise. *Note:* In many cases, answers in the ATE are abbreviated to fit on the same page as the exercise. Please see the Solutions Manual or click the link to Full Solutions in the Teacher's e-book for complete solutions to these exercises.
- Teaching Tips that are placed strategically throughout the chapter just when you need them.
- Alternate Examples for each example in the student text. These examples are also available in Microsoft Word format for easy export into PowerPoint or class notes.
- AP® Exam Common Errors that highlight mistakes that are often made by students—and how your students can avoid them.
- AP® Exam Formula Sheet References that let you know which formulas in the text are provided on the AP® exam.
- Alternate Activities for many of the activities in the student text.
- Technology Tips to provide guidance when working with graphing calculators and other technology.
- Links to Applets from the *TPS* Web site and many others.
- Links to valuable resources in the Teacher's Resource Materials, including video overviews of each activity, section, and chapter, three quizzes for each section, four tests for every chapter, printable Learning Objective grids, and FRAPPY materials.

# Other Resources Available for *The Practice of Statistics 5e*

In addition to the support provided in the *Annotated Teacher's Edition*, there are many resources available for use with *The Practice of Statistics* 5e. To see a sample (Chapter 4) of most of these supplements, check out the following preview site: www.highschool.bfwpub.com/Preview/TPS5e.

## For Teachers

### Teacher's Resource CD (TRCD)

- The content formerly found in the Yellow, Gold, Platinum, and Titanium Binder has been moved into the digital age for easier accessibility through the TRCD, the book's companion Web site, and through direct links embedded in the TE-Book. The teacher's resources on the CD include:
  - Additional resources that work in tandem with the ATE, including additional problems, materials for activities, rubrics for projects, and review materials for the AP® exam.
  - Three prepared **Quizzes** for each Section and four **Tests** per chapter in PDF format. These questions are also included in ExamView®.
  - Word files for the Alternate Examples in the Teacher's Edition.
  - Jason Molesky's **Lecture PowerPoint Presentations**.
  - The Image PowerPoints and JPG Image Bank.
  - The Teacher's **Solutions Manual** in PDF format.
  - **Statistical applets** referenced in the Activities and Data Explorations in the textbook.
  - **Supplemental Chapters** in PDF format:
    - Chapter 13 Analysis of Variance
    - Chapter 14 Multiple Linear Regression
    - Chapter 15 Logistic Regression
  - **Data files** for the larger data sets in a variety of formats, including: TI-83/84, TI-89, TI-Nspire, HP Prime, Fathom, Minitab, Excel, ASCII.

### Test Bank in ExamView® Assessment Suite

- The test bank, written by experienced AP® Reader and Table Leader Tim Brown (The Lawrenceville School), includes more than 600 multiple-choice questions, in AP® style, arranged by chapter. Each chapter also includes a table that identifies the content of each question. The test items have been checked for correlation with the text content and notation, overall usability, and accuracy. All questions can be found in the ExamView® test generator and may be printed, as desired.

**ExamView® Assessment Suite**

- ExamView® Test Generator guides teachers through the process of creating online or paper tests and quizzes quickly and easily. Users may select from the bank of more than 1000 test questions or opt for the step-by-step tutorial to write their own questions. Questions may be scrambled to create different versions of tests. Tests may be printed in many different types of formats to provide maximum flexibility or may be administered online using the ExamView® Player. Student results flow to the ExamView® Test Manager to provide a comprehensive assessment management system for the teacher.

**Teacher's Solutions Manual**

- Written by author Josh Tabor and accuracy checked by Ann Cannon, this manual includes complete, worked solutions to each exercise in the student textbook. Care has been taken to make sure that the language, methodology, and notation are consistent with that used by AP® readers.

**Lecture PowerPoint Presentations**

- Created by Jason Molesky of Lakeville Area Public Schools, the PowerPoint presentation slides are a great tool for both new and experienced teachers alike. Written by section, the slides highlight the important ideas in each section and include many examples from the book. For teachers who prefer to build their own materials, we offer the **IMAGE PowerPoints and JPG Image Bank**. All of these resources are accessible through the Teacher's e-book, the Teacher's Resource CD (TRCD), and on the Teacher's side of the book's Web site.

**Teacher's Companion Web site** www.whfreeman.com/tps5e

- The password-protected Teacher's Web site requires user registration as an instructor and gives teachers access to the book's print and video resources. It also features all of the online student Web materials, plus a gradebook to view students' online quiz results, state standards correlation guides, a chapter pacing guide, an AP® audit approved sample syllabus, and many more teaching resources.

## For Students

***The Practice of Statistics* 5e Student e-book**

- Optimized for most devices from laptops to iPads to smartphones, the new e-reader may be viewed through a browser or downloaded quickly and easily to your own device. The e-Book offers page fidelity so that it matches the print page exactly. Packed with features, the e-Book allows students to highlight and annotate the text; keep notes; take quizzes; link directly to the tutorial videos, applets, and other tools; communicate with teachers and peers; and much more.

***Strive for a 5: Preparing for the AP® Statistics Examination***

- Written by Jason Molesky and Michael Legacy, the Strive for a 5 Guide is designed for use with *The Practice of Statistics*, Fifth Edition. It is designed to help students evaluate their understanding of the material covered in the textbook, reinforce the key concepts, develop conceptual understanding and communication skills, and prepare students to succeed on the AP® Statistics Exam. This book is divided into two sections: a study guide and a test preparation section. The study guide is written to be used throughout the course, while the prep section offers additional AP® test strategies and includes two full-length AP® style practice exams—each with 40 multiple-choice questions, 5 free-response questions, and an investigative task. Each exam has a full answer key with page references to relevant content in *TPS* 5e.

**Companion Web site**

- The textbook's Web site, www.whfreeman.com/tps5e, includes the worked example, exercise, and technology corner video tutorials, applets, online quizzes, data sets, extra chapters, and other resources.

## Tutorial Videos for Students and Teachers

These videos are keyed to the student text with the play button and are accessible through a click in the student e-Book (and TE-Book) or by link from the book's Web site. They feature the following:

- **Worked examples:** Experienced AP® teacher Jason Molesky walks through key examples to give step-by-step instructions, just as teachers do in the classroom.
- **Worked exercises:** The "For Practice, Try . . ." exercises are featured in step-by-step instructional videos by Doug Tyson and Paul Buckley to give students help when they need it.
- **Technology Corners:** Detailed, key-stroke-level instructional videos by Leigh Nataro show students how to use their TI-83/84, TI-89, or TI-Nspire graphing calculators.

## For Teachers Only

- PD **PD videos:** Brief chapter and section overviews by Daren Starnes, Josh Tabor, and Doug Tyson offer advice on what to emphasize in a given section, how to anticipate and overcome common pitfalls, as well as tips for helping students realize success in the course and on the AP® exam.
- **Resource videos:** Beth Benzing's *Activity Overview* videos introduce each Activity in the text and provide guidance on how to use them most effectively in class. Also featured is Jason Molesky's "Mastering the FRQ with the FRAPPY" video.

# Goals for the Fifth Edition

After getting overwhelmingly positive feedback about the changes we made to the fourth edition of *The Practice of Statistics*, we set several new and continuing goals for the fifth edition.

- **Maintain a clear focus on the AP® exam.** As the only textbook written exclusively for AP® Statistics, we wrote every sentence with an eye towards success on the AP® exam. You'll see this throughout the book, from the AP® Topic Outline on the first page to the AP®-rubric based, four-step inference process on the last page.
- **Incorporate media.** With the growth of instructional videos on the Internet, we wanted to make sure that teachers and students using *TPS* would have a reliable source for video content that was aligned to the textbook and the AP® exam.

- **Emphasize learning objectives.** At the beginning of each section, we identify the relevant learning objectives so students know what they are expected to learn. At the end of each chapter, these learning objectives are summarized in a grid, with references to examples and chapter review exercises that address each objective so students can assess their progress.
- **Increase opportunities for review.** In addition to the review exercises and AP® Practice Test for each chapter, we have added several new features to help students review the chapter content. First, there is a series of Case Closed questions, which relate to the chapter-opening Case Study. Second, we have added a FRAPPY! to each chapter so students can practice doing AP®-style free-response questions and evaluate other students' responses. Finally, we have added a much more in-depth narrative review at the end of each chapter to help students keep their eyes on the big ideas.
- **Update examples, exercises, and activities.** In each chapter, we have enhanced many examples, exercises, and activities with new contexts or updated data. In particular, we rewrote many of the multiple-choice questions to focus more on conceptual understanding and less on basic skills.
- **Preserve an appealing design.** Today's students are accustomed to operating in a visually stimulating environment. A textbook with an engaging look and feel is more likely to get used regularly. We have once again attempted to create visual appeal without making the pages too busy or distracting.
- **Keep the focus on readability.** Explanations that are too technical or formal discourage students from reading. Narrative that's excessively casual can lead to sloppy statistics. Finding a comfortable middle ground, where students hear a consistent voice and tone throughout, was a major priority for the new edition. Readability tests and reviewer comments suggest that the fifth edition has further improved accessibility without sacrificing accuracy.

## Content and Structure

We made very few structural changes from the fourth edition to the fifth edition. Only two topics moved from one section to another:

- Density curves moved from Section 2.1 to Section 2.2, alongside Normal distributions.
- Power moved from Section 9.1 to Section 9.2, so students would have more experience with significance testing before learning this challenging topic.

Many other sections were reorganized to make them more efficient for teachers and students, including Sections 3.2, 4.2, 5.3, 8.3, 11.2, and 12.2.

The optional material on Simpson's Paradox was removed from Section 1.1 and now appears in the Teacher's Resource Materials. We have also eliminated the term "lurking variable" from the fifth edition. This term doesn't have a universally accepted definition and we have found that it causes trouble for students without any obvious benefits.

In the fourth edition, we introduced the four-step State–Plan–Do–Conclude process to help students answer inference questions on the AP® exam. In the fifth edition, we continue to emphasize this process in Chapters 8–12, but scale back its use in earlier chapters. We have also introduced a three-step framework for doing Normal distribution and binomial distribution calculations that matches recent AP® rubrics.

In Section 4.2, we have revised the material on experiments to include comparison as the first of four principles of experimental design. We have also rewritten the subsections that cover control and blocking, using graphs to illustrate difficult concepts.

Finally, we have revised the way we identify and check the conditions for inference. For inference about proportions, we have renamed the "Normal" condition as the "Large Counts" condition. This name helps students remember what they need to check and thus avoid potential misstatements about Normality of the population or sample. We also use the "Large Counts" name for all three chi-square tests. For inference about means, we renamed the "Normal" condition as the "Normal/Large Sample Size" condition, to emphasize that there are two ways in which this condition can be satisfied. Finally, to address independence of observations in a sample when sampling without replacement, we now explicitly identify this requirement as the "10%" condition.

## Why Did You Do That?

There is no single best way to organize a first course in statistics. That said, our choices reflect thinking about both content and pedagogy. Here are comments on several "frequently asked questions" about the order and selection of material in *The Practice of Statistics*, Fifth Edition.

### *Why not begin with data production?*

It is certainly reasonable to do so—the natural flow of a planned study is from study design to data analysis to inference. In fact, the book is structured so that you can start with Chapter 4 on designing studies if you wish. We put data analysis first because the graphical and numerical techniques introduced there are useful even for data that are not produced according to a careful plan.

### *Why do Normal distributions appear so early in the book?*

Density curves such as the Normal curves are just another tool to describe the distribution of a quantitative variable, along with dotplots, stemplots, histograms, and boxplots. We prefer not to suggest that Normal distributions are tied to probability, as the traditional order does. And we find it very helpful to break up the indigestible lump of probability that troubles students so much. Encountering Normal distributions early helps to do so and strengthens the "probability distributions are like data distributions" way of approaching probability.

### *What happened to inference about a population mean $\mu$ when $\sigma$ is known?*

Previous editions of the text introduced confidence intervals and significance tests about a population mean $\mu$ in an "idealized" setting where the population standard deviation $\sigma$ is known. Doing so allows use of the standard Normal curve and $z$ procedures. Unfortunately, this approach requires considerable time, is rarely used in practice, and sometimes confuses students later when we expect them to use $z$ for proportions and $t$ for means.

### *Why did you choose this inference sequence?*

Students find the ideas and methods of estimating an unknown parameter much more straightforward than the logic of testing a claim about a parameter. That's a pedagogical argument for introducing confidence intervals first. Mixing confidence intervals and tests in the opening chapter on inference (from experience) leads to more confusion than clarity. Students also struggle to distinguish settings that involve proportions and means. We attack this issue by including inference methods for both proportions and means in the same chapter. This allows us to say "$z$ is for proportions and $t$ is for means" in Chapter 8 and again in Chapters 9 and 10. Finally, we ask students to put all these ideas together by constructing confidence intervals and performing tests in settings that involve comparing two proportions or two means. For consistency, we tackle inference for proportions first and inference for means second in every case.

# Your AP® Statistics Primer

## Before You Begin Teaching AP® Statistics

Teaching AP® Statistics can be a tremendously rewarding experience, but it takes a lot of preparation. If you are like most teachers of AP® Statistics, you probably took few, if any, statistics courses in college. And even if you have some statistics courses in your background, they were probably very different from the AP® course.

Don't worry! The *Annotated Teacher's Edition* includes an abundance of resources to help you teach AP® Statistics. There are also many other resources available from the College Board and other groups to make your life easier.

Here is a list of things you should do before you start teaching the course, if possible:

1. Attend a College Board approved workshop. Ideally, you should attend a one-week summer institute. These institutes are held all over the country and are led by College Board endorsed consultants, including Daren and myself. There are also one- and two-day workshops offered by the College Board during the school year.
2. Download and read the AP® Statistics Course Description, available at apcentral.collegeboard.com/stats. This document contains the most important information you need to know, including an introduction to the course, an overview of course content, information about the AP® exam, and sample questions.
3. Download and read the AP® Statistics Teacher's Guide, written by our own Michael Legacy. Designed specifically to help teachers who are new to AP® Statistics, this document (found at apcentral.collegeboard.com/stats) includes several sample syllabi and advice for the AP® Audit.
4. Participate in the AP® Teacher Community, an online forum for AP® Statistics teachers (apcommunity.collegeboard.org). Here you can ask questions, get advice, give suggestions, and share resources. If you are like most other AP® Statistics teachers, you are the only one at your school who teaches statistics. The community is a great remedy for this problem. Plus, the authors are regular contributors.
5. Complete the AP® Course Audit. See the AP® Coordinator at your school or go to www.collegeboard.com/html/apcourseaudit for more information. A sample syllabus and pacing guide are included in the Teacher's Resource Materials.

**TRM Sample Syllabus**

## Sequencing and Pacing

With one possible exception (see below), we strongly advise that you go through the chapters and sections in order. The decisions about the arrangement of topics in the fifth edition were made after careful thought and discussion, and they reflect the preferences of not only the authors, but also of many other current users of *The Practice of Statistics*.

### Alternate Organization

Some experienced teachers, including two of the authors, choose to teach Chapter 4: Designing Studies at the beginning of the school year. Because data must be collected before it can be analyzed, it might make sense to learn how to properly collect data before learning about methods of analysis. This fits with the four-step statistical process suggested by the *Guidelines for Assessment and Instruction in Statistics Education* (www.amstat.org/GAISE): Formulate questions, collect data, analyze data, interpret results.

Starting with Chapter 4 also sends a clear message to students that this isn't a typical math class—there is more reading, writing, and thinking than they might be used to. On the other hand, some of the material in this chapter is challenging and many teachers choose to keep Chapter 4 in its place so students can be properly "warmed up" before tackling these difficult topics. Also, if you typically have students adding courses during the first couple weeks of school, starting with Chapter 1 might make more sense as this material should be easier to make up for students who join the course late.

## Pacing

Pacing is often one of the biggest challenges for new (and experienced) teachers of AP® Statistics. Depending on your schedule and when the school year starts, you may not have much time to spare before the AP® exam in early May. The table below shows a suggested pacing guide based on a schedule with 110 class sessions before the AP® exam.

| Chapter | Class sessions |
|---|---|
| 1 | 9 |
| 2 | 7 |
| 3 | 8 |
| 4 | 11 |
| 5 | 8 |
| 6 | 9 |
| 7 | 7 |
| 8 | 8 |
| 9 | 8 |
| 10 | 8 |
| 11 | 6 |
| 12 | 7 |
| Midterms/Finals | 4 |
| Review for AP® exam | 10 |
| Total | 110 |

If you have more (or fewer) class sessions, you will need to adjust your schedule accordingly. Specific pacing guides with learning objectives and suggested assignments are provided at the beginning of each chapter and in the Sample Syllabus and Pacing Guide found in the Teacher's Resource Materials. The Teacher's Resource Materials also include a pacing guide for a "semesterized" block schedule, courtesy of AP® Statistics teacher Doug Tyson.

**TRM Pacing Guide**

**TRM Pacing Guide (Block Schedule)**

# Preparing Students for the AP® Exam

Unfortunately, there is no foolproof way to prepare students for success on the AP® exam. By choosing *The Practice of Statistics,* however, you are off to a good start; it was written by high school teachers specifically for high school students in an AP® Statistics class. Here are some of the features of *TPS* that are designed specifically to help students excel on the AP® exam:

- Every topic on the official AP® Statistics Topic Outline is covered in detail in the textbook.
- For each chapter and section, teachers may access a video overview presented by the authors or by an experienced AP® Reader. Look for the PD icon throughout the *Annotated Teacher's Edition.* PD
- For every activity in the text, teachers may access a video clip by an experienced AP® teacher and Reader offering insights and advice on how to use the activity effectively. Look for the red "play" icon throughout the *Annotated Teacher's Edition.*
- Students can access short videos presented by veteran AP® Readers for selected examples and exercises, plus all technology corners. Great care was taken to ensure that the solutions provided in the videos meet (and even exceed!) the standards on recent AP® rubrics. Look for the blue "play" icon throughout the student edition.
- Throughout the student text and *Annotated Teacher's Edition,* you will find AP® Exam Tips, including a cumulative set of tips in Appendix A.
- At the end of each chapter in the student textbook, there is an AP® Statistics Practice Test with multiple-choice and free-response questions like those found on the real AP® exam.
- At the end of each chapter in the student textbook, there is a FRAPPY! (Free-Response AP® Problem, Yay!), which asks students to complete free-response questions similar to those found on the AP® exam. After completing the questions, students can look at sample responses on the book's Web site and discuss the good and bad parts of each response. A model solution, scoring rubric, and commentary on the student samples are available for teachers in the Teacher's Resource Materials. For a more detailed explanation of the FRAPPY! process, watch the video from Jason Molesky.

  **PD FRAPPY!**

- At the end of each set of exercises, we have provided "recycle and review" questions that integrate content from earlier chapters to reinforce what students learned previously and to help them make connections from chapter to chapter.
- At the end of Chapters 4, 7, 10, and 12, there is a *cumulative* AP® Statistics Practice Test with multiple-choice and free-response questions like those found on the real AP® exam.
- Additional helpful tools in the Teacher's Resource Materials include a worksheet that asks students to choose the correct inference procedure in a given setting and a set of flashcards that reviews important topics from throughout the course. Make sure to use these!

  **TRM Inference Review**

  **TRM Flashcards**

Here are some additional strategies for preparing students for the AP® exam:

- Make reading and writing an integral part of your course. On the AP® exam, students are often surprised at how little they actually need to calculate. The majority of questions involve quite a bit of reading and writing. Make sure your students get plenty of practice throughout the year and show them previous AP® questions so they understand the high expectations of AP® Readers. Also, emphasize correct use of statistical vocabulary—incorrect use of statistical terms can lower scores.
- Make sure your students understand the structure of the AP® exam.

- Section I consists of 40 multiple-choice questions to be answered in 90 minutes. This section accounts for 50% of the exam score. There is no penalty for guessing, so make sure your students answer every question.
- Section II consists of six free-response questions to be answered in 90 minutes. This section accounts for 50% of the exam score. The first five questions should take about 13 minutes each and count for 75% of the Section II score. The last question, called the investigative task, should take about 25 minutes and is worth 25% of the Section II score. According to the AP® Statistics Course Description, "The purpose of the investigative task is not only to evaluate the student's understanding in several content areas but also to assess his or her ability to integrate statistical ideas and apply them in a new context or in a nonroutine way."

- Shortly before taking the actual AP® exam, have students come in after school or on the weekend to take a full-length AP® exam for practice. It is very beneficial for students to experience a real exam, given under real conditions, so they can get a sense of how fast they need to work and what topics they need to review. If possible, administer the practice exam in the same room that students will take the AP® exam. Currently, the complete 2002 and 2007 AP® exams are available for purchase from the College Board. Also, the international versions of the exam are available for recent years through your AP® audit account (look for the link to "Secure Documents" after logging in). These forms are secure, so students can't find them on AP® Central. Please help keep them secure by only using them in class and collecting them when you are done! *Note:* The international version of the exam is different from the version administered to the majority of students in the United States and Canada. However, it is comparable both in quality and difficulty to other versions. There is also a "Practice Exam" available through your audit account, but this exam was never used with actual students and is no longer secure.
- Use as many actual AP® questions as possible. There are a number of ways to incorporate previous free-response questions into your course. Some teachers include one or two AP® free-response questions on each test. Other teachers assign them for homework or in cooperative groups during class. If you assign them for homework, be aware that students can access the solutions online. However, because there are more than 100 released free-response questions, it's probably safe to include one or two on an exam. To help choose which questions to use, we have included a list of released AP® questions that are appropriate for each chapter in the *Annotated Teacher's Edition*. You can find the lists in the blue pages that precede each chapter. Also, the Teacher's Resource Materials contain two different sortable AP® Free-Response Question Indices with all the questions listed in one place. *Note:* Questions from the secure versions of the AP® exam are not included in these indices.

**TRM Free-response question index (sortable by chapter)**

**TRM Free-response question index (sortable by topic)**

- Make sure students know how the AP® exam is graded. Show them the scoring guidelines (rubrics) and examples of student responses after you answer old AP® questions in class. The rubrics and examples of student responses (with commentary) are available at apcentral.collegeboard.com/stats.
- Early in the year, give students a copy of the AP® Topic Outline and get in the habit of alerting students when you have finished one topic and are starting a new one. We have also placed the AP® Topic Outline in the front endpapers of the student text and of this ATE, including chapter and section references.
- Early in the year, give students a copy of the formula sheets and tables that are provided on the AP® exam so they can familiarize themselves with these resources. You can find these materials in the AP® Statistics Course Description or in any released set of questions on AP® Central. Many teachers also copy these on colored card stock for students to use on chapter tests. It is better for students to spend their time learning which formula to use and how to use it rather than spending time to memorize the formulas. We have also placed the formula sheets and tables in the back matter of the student text for easy reference.
- Give challenging assessments that mimic the AP® exam. Make sure to include multiple-choice and free-response questions and don't be afraid to ask difficult questions. Also, don't be afraid to be a strict grader—the rubrics on the real AP® exam are often quite strict. To compensate for these practices, you will probably need to adjust the scores on your assessments. Experience will help you know how much to adjust them, but remember that getting a raw score of 70 out of 100 will almost certainly earn a student a 5 on the AP® exam.
- Give cumulative assessments. One of the hardest things for students to do on the AP® exam is to identify which method is appropriate for a specific problem. Giving students cumulative assessments will give them practice making these tough decisions and will keep the material fresh in their minds. Some teachers include cumulative questions on each exam, while others give cumulative exams every couple of chapters. The review questions at the end of each set of exercises and the cumulative AP® Practice Tests in the text should help as well.
- Have students use an AP® preparation book, such as *Strive for a 5*, which was written specifically for *The Practice of Statistics* by Jason Molesky and Michael Legacy.

## Assessments

Although we have provided quizzes for every section and tests for every chapter, you probably won't have time to use all of them. And while some teachers find time for quizzes, many teachers only formally assess their students at the end of each chapter. If you are new to teaching AP® Statistics, it is probably better to err on the side of fewer days of assessment so you have more time to cover the content. It will also save you time grading! Furthermore, don't forget our advice about being a strict grader and scaling the test scores like they do on the real AP® exam.

The tests and quizzes in the Teacher's Resource Materials are designed to be ready to go. We include multiple versions so that you can offer make-up exams for students who haven't achieved mastery of the content. Versions A and B are roughly parallel and version C is typically a little more challenging. Starting in Chapter 2, there is a fourth version of the chapter test that is cumulative (version D).

We have also included the questions from the TRM tests and quizzes as extra question banks in ExamView® Assessment Suite to make it easy to develop your own assessments. ExamView® is an easy-to-use resource for creating multiple-choice tests or cumulative assessments to help students prepare for the AP® exam.

*IMPORTANT: Please do not post any of the questions or solutions online, as some students search online for questions and answers to cheat. We receive many emails and complaints from teachers who have discovered that their students found answers online. Once posted, they are readily available to all AP® students with access to the Internet.*

## Homework

The number of exercises available in the textbook should greatly exceed the number you would need to assign in any given school year. In fact, as teachers get more experience with AP® Statistics, they tend to assign fewer and fewer homework problems. It is better to have students spend time doing a few problems in depth rather than giving shallow answers to a large set of problems. This philosophy is reflected in the suggested homework assignments provided at the beginning of each chapter.

The exercise sets are designed so that the problems generally come in pairs. For example, problems 1 and 2 will cover the same concepts, but in different contexts. This way, teachers have the flexibility to assign odd-numbered problems (with the solutions in the back of the student text) or to assign even-numbered problems. The suggested assignments generally include odd-numbered problems because we want students to get immediate feedback about their work and make corrections before the next class. *Note*: In addition to answers to the odd-numbered section exercises, the back-of-book solutions include answers to *all* Check Your Understanding questions, Chapter Review exercises, and AP® Practice Test exercises.

Finally, there is no consensus on how best to grade and count homework. Some teachers collect homework daily while others collect it only occasionally and at random. Some teachers grade homework problems for correctness while others grade it only for effort. Often, this decision is based on your own preferences and school culture.

## Activities and Projects

One of the best parts about teaching AP® Statistics is that it is much easier to incorporate activities and projects than it is in other math classes. Activities and projects provide one of the best ways to help students learn, hands-on, the important concepts in AP® Statistics.

To emphasize the importance of activities, we start each chapter with a classroom-tested activity designed to introduce students to some of the most important ideas in that chapter. There are also additional activities included throughout the book (and alternate activities in the *Annotated Teacher's Edition*). Although you may feel overwhelmed trying to get through the content, experienced teachers agree that getting students engaged in activities is time well spent.

Projects are a great way to have students put their knowledge to work. Suggestions for projects are included throughout the *Annotated Teacher's Edition*, with rubrics available in the Teacher's Resource Materials. Although it is probably not possible to assign a project each chapter, many teachers try to assign one project per semester.

## Current Events

Another great part of teaching AP® Statistics is that there is no shortage of real-life applications. Almost every day, newspaper articles describe current polls, studies, or experiments that you can discuss with your students. When students see that the ideas they are learning about actually get used in the real world, their motivation is sure to increase! If you don't have time to read the newspaper cover to cover looking for articles, join the AP® Statistics Teacher Community (see Before You Begin Teaching AP® Statistics, above). Several times a week, someone posts a link to an article that can be used in class. Alternatively, you can offer extra credit to students for bringing in articles related to statistics.

## Technology

Technology is a vital part of AP® Statistics. Gone are the days when students in a statistics class spent most of their time calculating means, standard deviations, and correlations by hand. Instead, students in an AP® Statistics class are expected to use technology to do these calculations so they can focus on understanding and interpreting what they are calculating.

### Graphing Calculators

All students are expected to have a graphing calculator with the capability of making statistical graphs and performing calculations, such as finding the standard deviation of a set of data or the equation of a least-squares regression line. Students are allowed to use a graphing calculator on all parts of the AP® Statistics exam and should definitely be familiar with its uses by the time they take the exam.

Because the TI-83/84 and TI-89 are the most prominently used calculators in the high school market, *The Practice of Statistics* includes Technology Corners throughout the student text, with step-by-step instructions for using these specific calculators. Also, Appendix B contains similar material for the TI-Nspire. Instructions for the HP Prime can be found on the book's Web site, www.whfreeman.com/tps5e.

Although using technology is essential, teachers have different opinions about when to introduce students to the technology. For example, some teachers want students to have the experience of calculating things by hand before they show students how to calculate it with technology. In some cases, like calculating the standard deviation, having students do it by hand once or twice may be useful because the calculations reinforce what the standard deviation is all about (deviations from the mean).

Use of the Normal table is another subject of debate about how much to use technology. Some teachers strongly prefer to have students calculate $z$-scores and use the Normal table whereas other teachers quickly move to the `normalcdf` function on the calculator. You will need to make this decision for yourself based on your own preferences and how tech savvy your students are. Make sure to pay attention to the AP® Exam Tips that often accompany the Technology Corners so that your students know how to properly show their work.

### Computers

In many college statistics courses, students are asked to use statistical software, such as Minitab, rather than graphing calculators. Unlike high schools, colleges often have computer labs open for students at all times or have classes that meet in computer labs. Also unlike most high schools, they have the money to afford statistical software for their labs!

To help ensure that colleges continue to give credit to students who pass the AP® exam, questions on the AP® exam often include output from statistical software. In *The Practice of Statistics,* we have included examples of computer output throughout the book so students will be familiar with these types of questions.

Ideally, students in AP® Statistics will have the opportunity to use computer software on a regular basis. However, this is not a possibility at many schools. At a minimum, you should have a copy of at least one program that you can use for demonstrations in class. We recommend a copy of *Fathom* and the *Fathom Guide* that was written specifically for *The Practice of Statistics* 4e. *Fathom* was written to be a tool for teaching statistics and is great for performing simulations and doing analysis on large data sets. There are also some free statistical software packages and applets available on the Internet that you can use to create graphs and generate computer output.

## What to Do after the AP® Exam

If you still have time with your students after the AP® exam, there are several options:

1. Have your students do a project. This is an ideal time for students to bring together all of the topics from the course. Have them design a study, conduct the study, analyze the data, and make the appropriate inferences. A suggested end-of-year project assignment and rubric are included in the Teacher's Resource Materials.

   **TRM Final Project**

2. Learn some additional statistical topics. The Teacher's Resource CD and the textbook's Web site include three additional chapters:
   - Chapter 13: ANOVA (comparing more than two means)
   - Chapter 14: Multiple Regression (using more than one explanatory variable to predict a response variable)
   - Chapter 15: Logistic Regression (predicting a categorical response variable)

## Recruiting Students for Next Year

In many cases, students have no idea what a statistics class is all about. It is your job to promote the hands-on learning and real-world applications that make AP® Statistics a class worth taking! One way to do this is to visit upper-level math classes like Algebra II and above and give them a short presentation about the class. When I talk to potential students, I try to hit three main themes:

1. It is a fresh start: This isn't a typical math class. You don't need to remember how to factor a polynomial, graph a sine curve, or prove that triangles are congruent. However, this doesn't mean the class will be easy! You will still need good critical thinking and communication skills.
2. It is great preparation for college: Many majors require students to take a statistics course and be familiar with the methods of analysis in AP® Statistics. Also, AP® Statistics is taught over a full year (instead of a semester in college), so you have more time to learn the material. It helps if you can gather a list of majors at your local university that require a statistics course.
3. You will *never* ask "When will we ever use this?" Problems are almost always based on real-world studies in a variety of fields. We will read articles from current newspapers and magazines and learn how to think critically about claims made in the media.

I also try to give prospective students an example or two of the types of studies we design and analyze so they have a more concrete understanding of what actually happens in class. For example: Is there an association between gender and superpower preference? How does the wording of a survey question affect the responses? Is yawning contagious?

Many students will ask you how AP® Statistics compares to AP® Calculus and which class they should take. Here are some questions to consider discussing with these students:

1. *What credit or placement will you get for my AP® math class?* Each college/university has a policy about credit and/or placement awarded for specific AP® exams. At most schools that give credit, earning a "passing" score on the AP® Statistics or AP® Calculus exams results in one semester of college credit. The key might be which math course you'll need for your chosen major.
2. *What college major do you plan to pursue?* If you plan to pursue a degree in mathematics, physics, or chemistry, you will be expected to take calculus in college. If you select an English, history, fine arts, foreign language, or social science major, then you are more likely to need a statistics course. Business majors will probably need to take both calculus and statistics. You want to choose a course that sets you up well for your college math requirements.
3. *What is the difference between AP® Calculus and AP® Statistics?*

| **AP® Calculus:** | **AP® Statistics:** |
|---|---|
| Graphical, numerical, and algebraic | Collecting and analyzing data |
| Builds on precalculus concepts | Computation de-emphasized |
| Computational proficiency helps | Focus on communication and interpretation |
| Emphasizes techniques, applications | Writing is critical |

4. *Still can't decide?* Take both courses!

## Additional Internet Resources

In addition to the book's Web site www.whfreeman.com/tps5e, AP® Central apcentral.collegeboard.com/stats, and the AP® Teacher Community apcommunity.collegeboard.org, there are many other Web sites that AP® Statistics teachers have found useful. Here is a short list of some of the best:

1. The listserv archives mathforum.org/kb/forum.jspa?forumID=67. Before there was the AP® Teacher Community,

# Contents

**Additional topics on theTRCD and at www.whfreeman.com/tps5e**

# About the Authors

**DAREN S. STARNES** is Mathematics Department Chair and holds the Robert S. and Christina Seix Dow Distinguished Master Teacher Chair in Mathematics at The Lawrenceville School near Princeton, New Jersey. He earned his MA in Mathematics from the University of Michigan and his BS in Mathematics from the University of North Carolina at Charlotte. Daren is also an alumnus of the North Carolina School of Science and Mathematics. Daren has led numerous one-day and weeklong AP® Statistics institutes for new and experienced AP® teachers, and he has been a Reader, Table Leader, and Question Leader for the AP® Statistics exam since 1998. Daren is a frequent speaker at local, state, regional, national, and international conferences. For two years, he served as coeditor of the Technology Tips column in the NCTM journal *The Mathematics Teacher.* From 2004 to 2009, Daren served on the ASA/NCTM Joint Committee on the Curriculum in Statistics and Probability (which he chaired in 2009). While on the committee, he edited the *Guidelines for Assessment and Instruction in Statistics Education* (GAISE) pre-K–12 report and coauthored (with Roxy Peck) *Making Sense of Statistical Studies,* a capstone module in statistical thinking for high school students. Daren is also coauthor of the popular text *Statistics Through Applications,* First and Second Editions.

**JOSH TABOR** has enjoyed teaching general and AP® Statistics to high school students for more than 18 years, most recently at his alma mater, Canyon del Oro High School in Oro Valley, Arizona. He received a BS in Mathematics from Biola University, in La Mirada, California. In recognition of his outstanding work as an educator, Josh was named one of the five finalists for Arizona Teacher of the Year in 2011. He is a past member of the AP® Statistics Development Committee (2005–2009), as well as an experienced Table Leader and Question Leader at the AP® Statistics Reading. Each year, Josh leads one-week AP® Summer Institutes and one-day College Board workshops around the country and frequently speaks at local, national, and international conferences. In addition to teaching and speaking, Josh has authored articles in *The Mathematics Teacher, STATS Magazine,* and *The Journal of Statistics Education.* He is the author of the *Annotated Teacher's Edition* and *Teacher's Resource Materials* for *The Practice of Statistics* 4e and 5e, along with the *Solutions Manual* for *The Practice of Statistics* 5e. Combining his love of statistics and love of sports, Josh teamed with Christine Franklin to write *Statistical Reasoning in Sports,* an innovative textbook for on-level statistics courses.

**DANIEL S. YATES** taught AP® Statistics in the Electronic Classroom (a distance-learning facility) affiliated with Henrico County Public Schools in Richmond, Virginia. Prior to high school teaching, he was on the mathematics faculty at Virginia Tech and Randolph-Macon College. He has a PhD in Mathematics Education from Florida State University. Dan received a College Board/Siemens Foundation Advanced Placement Teaching Award in 2000.

**DAVID S. MOORE** is Shanti S. Gupta Distinguished Professor of Statistics (Emeritus) at Purdue University and was 1998 President of the American Statistical Association. David is an elected fellow of the American Statistical Association and of the Institute of Mathematical Statistics and an elected member of the International Statistical Institute. He has served as program director for statistics and probability at the National Science Foundation. He is the author of influential articles on statistics education and of several leading textbooks.

# Content Advisory Board and Supplements Team

**Jason Molesky,** Lakeville Area Public School, Lakeville, MN
*Media Coordinator—Worked Examples, PPT lectures, Strive for a 5 Guide*

Jason has served as an AP® Statistics Reader and Table Leader since 2006. After teaching AP® Statistics for eight years and developing the FRAPPY system for AP® exam preparation, Jason moved into administration. He now serves as the Director of Program Evaluation and Accountability, overseeing the district's research and evaluation, continuous improvement efforts, and assessment programs. Jason also provides professional development to statistics teachers across the United States and maintains the "Stats Monkey" Web site, a clearinghouse for AP® Statistics resources.

**Tim Brown,** The Lawrenceville School, Lawrenceville, NJ
*Content Advisor, Test Bank, TRM Tests and Quizzes*

Tim first piloted an AP® Statistics course the year before the first exam was administered. He has been an AP® Reader since 1997 and a Table Leader since 2004. He has taught math and statistics at The Lawrenceville School since 1982 and currently holds the Bruce McClellan Distinguished Teaching Chair.

**Doug Tyson,** Central York High School, York, PA
*Exercise Videos, Learning Curve*

Doug has taught mathematics and statistics to high school and undergraduate students for 22 years. He has taught AP® Statistics for 7 years and served as an AP® Reader for 4 years. Doug is the co-author of a curriculum module for the College Board, conducts student review sessions around the country, and gives workshops on teaching statistics.

**Paul Buckley,** Gonzaga College High School, Washington, DC
*Exercise Videos*

Paul has taught high school math for 20 years and AP® Statistics for 12 years. He has been an AP® Statistics Reader for six years and helps to coordinate the integration of new Readers (Acorns) into the Reading process. Paul has presented at Conferences for AP®, NCTM, NCEA (National Catholic Education Association) and JSEA (Jesuit Secondary Education Association).

**Leigh Nataro,** Moravian Academy, Bethlehem, PA
*Technology Corner Videos*

Leigh has taught AP® Statistics for nine years and has served as an AP® Statistics Reader for the past four years. She enjoys the challenge of writing multiple-choice questions for the College Board for use on the AP® Statistics exam. Leigh is a National Board Certified Teacher in Adolescence and Young Adulthood Mathematics and was previously named a finalist for the Presidential Award for Excellence in Mathematics and Science Teaching in New Jersey.

**Ann Cannon,** Cornell College, Mount Vernon, IA
*Content Advisor, Accuracy Checker*

Ann has served as Reader, Table Leader, and Question Leader for the AP® Statistics exam for the past 13 years. She has taught introductory statistics at the college level for 20 years and is very active in the Statistics Education Section of the American Statistical Association, serving on the Executive Committee for two 3-year terms. She is co-author of *STAT2: Building Models for a World of Data* (W. H. Freeman and Company).

**Michael Legacy,** Greenhill School, Dallas, TX
*Content Advisor, Strive for a 5 Guide*

Michael is a past member of the AP® Statistics Development Committee (2001–2005) and a former Table Leader at the Reading. He currently reads the Alternate Exam and is a lead teacher at many AP® Summer Institutes. Michael is the author of the 2007 College Board AP® Statistics Teacher's Guide and was named the Texas 2009–2010 AP® Math/Science Teacher of the Year by the Siemens Corporation.

**James Bush,** Waynesburg University, Waynesburg, PA
*Learning Curve, Media Reviewer*

James has taught introductory and advanced courses in Statistics for over 25 years. He is currently a Professor of Mathematics at Waynesburg University and is the recipient of the Lucas Hathaway Teaching Excellence Award. James has served as an AP® Statistics Reader for the past seven years and conducts many AP® Statistics preparation workshops.

**Beth Benzing,** Strath Haven High School, Wallingford/ Swarthmore School District, Wallingford, PA
*Activities Videos*

Beth has taught AP® Statistics for 14 years and has served as a Reader for the AP® Statistics exam for the past four years. She serves as Vice President on the board for the regional affiliate for NCTM in the Philadelphia area and is a moderator for an on-line course, *Teaching Statistics with Fathom*.

**Heather Overstreet,** Franklin County High School, Rocky Mount, VA
*TI-Nspire Technology Corners*

Heather has taught AP® Statistics for nine years and has served as an AP® Statistics Reader for the past six years. While working with Virginia Advanced Study Strategies, a program for promoting AP® math, science, and English courses in Virginia High Schools, she led many AP® Statistics Review Sessions and served as a Laying the Foundation trainer of teachers of pre-AP® math classes.

# Acknowledgments

*Teamwork*: It has been the secret to the successful evolution of *The Practice of Statistics* (*TPS*) through its fourth and fifth editions. We are indebted to each and every member of the team for the time, energy, and passion that they have invested in making our collective vision for *TPS* a reality.

To our team captain, Ann Heath, we offer our utmost gratitude. Managing a revision project of this scope is a Herculean task! Ann has a knack for troubleshooting thorny issues with an uncanny blend of forthrightness and finesse. She assembled an all-star cast to collaborate on the fifth edition of *TPS* and trusted each of us to deliver an excellent finished product. We hope you'll agree that the results speak for themselves. Thank you, Ann, for your unwavering support, patience, and friendship throughout the production of these past two editions.

Truth be told, we had some initial reservations about adding a Development Editor to our team starting with the fourth edition of *TPS*. Don Gecewicz quickly erased our doubts. His keen mind and sharp eye were evident in the many probing questions and insightful suggestions he offered at all stages of the project. Thanks, Don, for your willingness to push the boundaries of our thinking.

Working behind the scenes, Enrico Bruno busily prepared the manuscript chapters for production. He did yeoman's work in ensuring that the changes we intended were made as planned. For the countless hours that Enrico invested sweating the small stuff, we offer him our sincere appreciation.

We are deeply grateful to Patrice Sheridan and to Diana Blume for their aesthetic contributions to the eye-catching design of *TPS* 5e. Our heartfelt thanks also go to Vivien Weiss and Susan Wein at W. H. Freeman for their skillful oversight of the production process. Patti Brecht did a superb job copyediting a very complex manuscript.

A special thank you goes to our good friends on the high school sales and marketing staff at Bedford, Freeman, and Worth (BFW) Publishers. We feel blessed to have such enthusiastic professionals on our extended team. In particular, we want to thank our chief cheerleader, Cindi Weiss, for her willingness to promote *The Practice of Statistics* at every opportunity.

We cannot say enough about the members of our Content Advisory Board and Supplements Team. This remarkable group is a veritable who's who of the AP® Statistics community. We'll start with Ann Cannon, who once again reviewed the statistical content of every chapter. Ann also checked the solutions to every exercise in the book. What a task! More than that, Ann offered us sage advice about virtually everything between the covers of *TPS* 5e. We are so grateful to Ann for all that she has done to enhance the quality of *The Practice of Statistics* over these past two editions.

Jason Molesky, aka "Stats Monkey," greatly expanded his involvement in the fifth edition by becoming our Media Coordinator in addition to his ongoing roles as author of the *Strive for a 5 Guide* and creator of the PowerPoint presentations for the book. Jason also graciously loaned us his Free Response AP® Problem, Yay! (FRAPPY!) concept for use in *TPS* 5e. We feel incredibly fortunate to have such a creative, energetic, and deeply thoughtful person at the helm as the media side of our project explodes in many new directions at once.

Jason is surrounded by a talented media team. Doug Tyson and Paul Buckley have expertly recorded the worked exercise videos. Leigh Nataro has produced screencasts for the Technology Corners on the TI-83/84, TI-89, and TI-Nspire. Beth Benzing followed through on her creative suggestion to produce "how to" screencasts for teachers to accompany the Activities in the book. James Bush capably served as our expert reviewer for all of the media elements and is now partnering with Doug Tyson on a new Learning Curve media component. Heather Overstreet once again compiled the TI-Nspire Technology Corners in Appendix B. We wish to thank the entire media team for their many contributions.

Tim Brown, my Lawrenceville colleague, has been busy creating many new, high-quality assessment items for the fifth edition. The fruits of his labors are contained in the revised *Test Bank* and in the quizzes and tests that are part of the *Teacher's Resource Materials.* We are especially thankful that Tim was willing to compose an additional cumulative test for Chapters 2 through 12.

We offer our continued thanks to Michael Legacy for composing the superb questions in the Cumulative AP® Practice Tests and the *Strive for a 5 Guide.* Michael's expertise as a two-time former member of the AP® Statistics Test Development Committee is invaluable to us.

Although Dan Yates and David Moore both retired several years ago, their influence lives on in *TPS* 5e. They both had a dramatic impact on our thinking about how statistics is best taught and learned through their pioneering work as textbook authors. Without their early efforts, *The Practice of Statistics* would not exist!

Thanks to all of you who reviewed chapters of the fourth and fifth editions of *TPS* and offered your constructive suggestions. The book is better as a result of your input.

—Daren Starnes and Josh Tabor

*A final note from Daren*: It has been a privilege for me to work so closely with Josh Tabor on *TPS* 5e. He is a gifted statistics educator; a successful author in his own right; and a caring parent, colleague, and friend. Josh's influence is present on virtually every page of the book. He also took on the thankless task of revising all of the solutions for the fifth edition in addition to updating his exceptional *Annotated Teacher's Edition.* I don't know how he finds the time and energy to do all that he does!

The most vital member of the *TPS* 5e team for me is my wonderful wife, Judy. She has read page proofs, typed in data sets, and endured countless statistical conversations in the car, in airports, on planes, in restaurants, and on our frequent strolls. If writing is a labor of love, then I am truly blessed to share my labor with the person I love more than anyone else in the world. Judy, thank you so much for making the seemingly impossible become reality time and time again. And to our three sons, Simon, Nick, and Ben—thanks for the inspiration, love, and support that you provide even in the toughest of times.

*A final note from Josh*: When Daren asked me to join the *TPS* team for the fourth edition, I didn't know what I was getting into. Having now completed another full cycle with *TPS* 5e, I couldn't have imagined the challenge of producing a textbook—from the initial brainstorming sessions to the final edits on a wide array of supplementary materials. In all honesty, I still don't know the full story. For taking on all sorts of additional tasks—managing the word-by-word revisions to the text, reviewing copyedits and page proofs, encouraging his co-author, and overseeing just about everything—I owe my sincere thanks to Daren. He has been a great colleague and friend throughout this process.

I especially want to thank the two most important people in my life. To my wife Anne, your patience while I spent countless hours working on this project is greatly appreciated. I couldn't have survived without your consistent support and encouragement. To my daughter Jordan, I look forward to being home more often and spending less time on my computer when I am there. I also look forward to when you get to use *TPS* 7e in about 10 years. For now, we have a lot of fun and games to catch up on. I love you both very much.

# Acknowledgments

## Fifth Edition Survey Participants and Reviewers

Blake Abbott, Bishop Kelley High School, Tulsa, OK
Maureen Bailey, Millcreek Township School District, Erie, PA
Kevin Bandura, Lincoln County High School, Stanford, KY
Elissa Belli, Highland High School, Highland, IN
Jeffrey Betlan, Yough School District, Herminie, PA
Nancy Cantrell, Macon County Schools, Franklin, NC
Julie Coyne, Center Grove HS, Greenwood, IN
Mary Cuba, Linden Hall, Lititz, PA
Tina Fox, Porter-Gaud School, Charleston, SC
Ann Hankinson, Pine View, Osprey, FL
Bill Harrington, State College Area School District, State College, PA
Ronald Hinton, Pendleton Heights High School, Pendleton, IN
Kara Immonen, Norton High School, Norton, MA
Linda Jayne, Kent Island High School, Stevensville, MD
Earl Johnson, Chicago Public Schools, Chicago, IL
Christine Kashiwabara, Mid-Pacific Institute, Honolulu, HI
Melissa Kennedy, Holy Names Academy, Seattle, WA
Casey Koopmans, Bridgman Public Schools, Bridgman, MI
David Lee, SPHS, Sun Prairie, WI
Carolyn Leggert, Hanford High School, Richland, WA
Jeri Madrigal, Ontario High School, Ontario, CA
Tom Marshall, Kents Hill School, Kents Hill, ME
Allen Martin, Loyola High School, Los Angeles, CA
Andre Mathurin, Bellarmine College Preparatory, San Jose, CA
Brett Mertens, Crean Lutheran High School, Irvine, CA
Sara Moneypenny, East High School, Denver, CO
Mary Mortlock, The Harker School, San Jose, CA
Mary Ann Moyer, Hollidaysburg Area School District, Hollidaysburg, PA
Howie Nelson, Vista Murrieta HS, Murrieta, CA
Shawnee Patry, Goddard High School, Wichita, KS
Sue Pedrick, University High School, Hartford, CT
Shannon Pridgeon, The Overlake School, Redmond, WA
Sean Rivera, Folsom High, Folsom, CA
Alyssa Rodriguez, Munster High School, Munster, IN
Sheryl Rodwin, West Broward High School, Pembroke Pines, FL
Sandra Rojas, Americas HS, El Paso, TX
Christine Schneider, Columbia Independent School, Boonville, MO
Amanda Schneider, Battle Creek Public Schools, Charlotte, MI
Steve Schramm, West Linn High School, West Linn, OR
Katie Sinnott, Revere High School, Revere, MA
Amanda Spina, Valor Christian High School, Highlands Ranch, CO
Julie Venne, Pine Crest School, Fort Lauderdale, FL
Dana Wells, Sarasota High School, Sarasota, FL
Luke Wilcox, East Kentwood High School, Grand Rapids, MI
Thomas Young, Woodstock Academy, Putnam, CT

## Fourth Edition Focus Group Participants and Reviewers

Gloria Barrett, Virginia Advanced Study Strategies, Richmond, VA
David Bernklau, Long Island University, Brookville, NY
Patricia Busso, Shrewsbury High School, Shrewsbury, MA
Lynn Church, Caldwell Academy, Greensboro, NC
Steven Dafilou, Springside High School, Philadelphia, PA
Sandra Daire, Felix Varela High School, Miami, FL
Roger Day, Pontiac High School, Pontiac, IL
Jared Derksen, Rancho Cucamonga High School, Rancho Cucamonga, CA
Michael Drozin, Munroe Falls High School, Stow, OH
Therese Ferrell, I. H. Kempner High School, Sugar Land, TX
Sharon Friedman, Newport High School, Bellevue, WA
Jennifer Gregor, Central High School, Omaha, NE
Julia Guggenheimer, Greenwich Academy, Greenwich, CT
Dorinda Hewitt, Diamond Bar High School, Diamond Bar, CA
Dorothy Klausner, Bronx High School of Science, Bronx, NY
Robert Lochel, Hatboro-Horsham High School, Horsham, PA
Lynn Luton, Duchesne Academy of the Sacred Heart, Houston, TX
Jim Mariani, Woodland Hills High School, Greensburgh, PA
Stephen Miller, Winchester Thurston High School, Pittsburgh, PA
Jason Molesky, Lakeville Area Public Schools, Lakeville, MN
Mary Mortlock, Harker School, San Jose, CA
Heather Nichols, Oak Creek High School, Oak Creek, WI
Jamis Perrett, Texas A&M University, College Station, TX
Heather Pessy, Mount Lebanon High School, Pittsburgh, PA
Kathleen Petko, Palatine High School, Palatine, IL
Todd Phillips, Mills Godwin High School, Richmond, VA
Paula Schute, Mount Notre Dame High School, Cincinnati, OH
Susan Stauffer, Boise High School, Boise, ID
Doug Tyson, Central York High School, York, PA
Bill Van Leer, Flint High School, Oakton, VA
Julie Verne, Pine Crest High School, Fort Lauderdale, FL

Steve Willot, Francis Howell North High School, St. Charles, MO
Jay C. Windley, A. B. Miller High School, Fontana, CA

## Reviewers of previous editions:

Christopher E. Barat, Villa Julie College, Stevenson, MD
Jason Bell, Canal Winchester High School, Canal Winchester, OH
Zack Bigner, Elkins High School, Missouri City, TX
Naomi Bjork, University High School, Irvine, CA
Robert Blaschke, Lynbrook High School, San Jose, CA
Alla Bogomolnaya, Orange High School, Pepper Pike, OH
Andrew Bowen, Grand Island Central School District, Grand Island, NY
Jacqueline Briant, Bishop Feehan High School, Attleboro, MA
Marlys Jean Brimmer, Ridgeview High School, Bakersfield, CA
Floyd E. Brown, Science Hill High School, Johnson City, TN
James Cannestra, Germantown High School, Germantown, WI
Joseph T. Champine, King High School, Corpus Christi, TX
Jared Derksen, Rancho Cucamonga High School, Rancho Cucamonga, CA
George J. DiMundo, Coast Union High School, Cambria, CA
Jeffrey S. Dinkelmann, Novi High School, Novi, MI
Ronald S. Dirkse, American School in Japan, Tokyo, Japan
Cynthia L. Dishburger, Whitewater High School, Fayetteville, GA
Michael Drake, Springfield High School, Erdenheim, PA
Mark A. Fahling, Gaffney High School, Gaffney, SC
David Ferris, Noblesville High School, Noblesville, IN
David Fong, University High School, Irvine, CA
Terry C. French, Lake Braddock Secondary School, Burke, VA
Glenn Gabanski, Oak Park and River Forest High School, Oak Park, IL
Jason Gould, Eaglecrest High School, Centennial, CO
Dr. Gene Vernon Hair, West Orange High School, Winter Garden, FL
Stephen Hansen, Napa High School, Napa, CA
Katherine Hawks, Meadowcreek High School, Norcross, GA
Gregory D. Henry, Hanford West High School, Hanford, CA
Duane C. Hinders, Foothill College, Los Altos Hills, CA
Beth Howard, Saint Edwards, Sebastian, FL
Michael Irvin, Legacy High School, Broomfield, CO
Beverly A. Johnson, Fort Worth Country Day School, Fort Worth, TX
Matthew L. Knupp, Danville High School, Danville, KY
Kenneth Kravetz, Westford Academy, Westford, MA
Lee E. Kucera, Capistrano Valley High School, Mission Viejo, CA
Christina Lepi, Farmington High School, Farmington, CT
Jean E. Lorenson, Stone Ridge School of the Sacred Heart, Bethesda, MD
Thedora R. Lund, Millard North High School, Omaha, NE
Philip Mallinson, Phillips Exeter Academy, Exeter, NH
Dru Martin, Ramstein American High School, Ramstein, Germany
Richard L. McClintock, Ticonderoga High School, Ticonderoga, NY
Louise McGuire, Pattonville High School, Maryland Heights, MO
Jennifer Michaelis, Collins Hill High School, Suwanee, GA
Dr. Jackie Miller, Ohio State University
Jason M. Molesky, Lakeville South High School, Lakeville, MN
Wayne Nirode, Troy High School, Troy, OH
Heather Pessy, Mount Lebanon High School, Pittsburgh, PA
Sarah Peterson, University Preparatory Academy, Seattle, WA
Kathleen Petko, Palatine High School, Palatine, IL
German J. Pliego, University of St. Thomas
Stoney Pryor, A&M Consolidated High School, College Station, TX
Judy Quan, Alameda High School, Alameda, CA
Stephanie Ragucci, Andover High School, Andover, MA
James M. Reeder, University School, Hunting Valley, OH
Joseph Reiken, Bishop Garcia Diego High School, Santa Barbara, CA
Roger V. Rioux, Cape Elizabeth High School, Cape Elizabeth, ME
Tom Robinson, Kentridge Senior High School, Kent, WA
Albert Roos, Lexington High School, Lexington, MA
Linda C. Schrader, Cuyahoga Heights High School, Cuyahoga Heights, OH
Daniel R. Shuster, Royal High School, Simi Valley, CA
David Stein, Paint Branch High School, Burtonsville, MD
Vivian Annette Stephens, Dalton High School, Dalton, GA
Charles E. Taylor, Flowing Wells High School, Tucson, AZ
Reba Taylor, Blacksburg High School, Blacksburg, VA
Shelli Temple, Jenks High School, Jenks, OK
David Thiel, Math/Science Institute, Las Vegas, NV
William Thill, Harvard-Westlake School, North Hollywood, CA
Richard Van Gilst, Westminster Christian Academy, St. Louis, MO
Joseph Robert Vignolini, Glen Cove High School, Glen Cove, NY
Ira Wallin, Elmwood Park Memorial High School, Elmwood Park, NJ
Linda L. Wohlever, Hathaway Brown School, Shaker Heights, OH

# To the Student

## Statistical Thinking and You

The purpose of this book is to give you a working knowledge of the big ideas of statistics and of the methods used in solving statistical problems. Because data always come from a real-world context, doing statistics means more than just manipulating data. *The Practice of Statistics* (*TPS*), Fifth Edition, is full of data. Each set of data has some brief background to help you understand what the data say. We deliberately chose contexts and data sets in the examples and exercises to pique your interest.

*TPS* 5e is designed to be easy to read and easy to use. This book is written by current high school AP® Statistics teachers, for high school students. We aimed for clear, concise explanations and a conversational approach that would encourage you to read the book. We also tried to enhance both the visual appeal and the book's clear organization in the layout of the pages.

Be sure to take advantage of all that *TPS* 5e has to offer. You can learn a lot by reading the text, but you will develop deeper understanding by doing Activities and Data Explorations and answering the Check Your Understanding questions along the way. The walkthrough guide on pages xxv–xxxii gives you an inside look at the important features of the text.

You learn statistics best by doing statistical problems. This book offers many different types of problems for you to tackle.

- **Section Exercises** include paired odd- and even-numbered problems that test the same skill or concept from that section. There are also some multiple-choice questions to help prepare you for the AP® exam. Recycle and Review exercises at the end of each exercise set involve material you studied in previous sections.
- **Chapter Review Exercises** consist of free-response questions aligned to specific learning objectives from the chapter. Go through the list of learning objectives summarized in the Chapter Review and be sure you can say "I can do that" to each item. Then prove it by solving some problems.
- The **AP® Statistics Practice Test** at the end of each chapter will help you prepare for in-class exams. Each test has 10 to 12 multiple-choice questions and three free-response problems, very much in the style of the AP® exam.
- Finally, the **Cumulative AP® Practice Tests** after Chapters 4, 7, 10, and 12 provide challenging, cumulative multiple-choice and free-response questions like ones you might find on a midterm, final, or the AP® Statistics exam.

The main ideas of statistics, like the main ideas of any important subject, took a long time to discover and take some time to master. The basic principle of learning them is to be persistent. Once you put it all together, statistics will help you make informed decisions based on data in your daily life.

# *TPS* and AP® Statistics

*The Practice of Statistics* (*TPS*) was the first book written specifically for the Advanced Placement (AP®) Statistics course. Like the previous four editions, *TPS* 5e is organized to closely follow the AP® Statistics Course Description. Every item on the College Board's "Topic Outline" is covered thoroughly in the text. Look inside the front cover for a detailed alignment guide. The few topics in the book that go beyond the AP® syllabus are marked with an asterisk (*).

Most importantly, *TPS* 5e is designed to prepare you for the AP® Statistics exam. The entire author team has been involved in the AP® Statistics program since its early days. We have more than 80 years' combined experience teaching introductory statistics and more than 30 years' combined experience grading the AP® exam! Two of us (Starnes and Tabor) have served as Question Leaders for several years, helping to write scoring rubrics for free-response questions. Including our Content Advisory Board and Supplements Team (page xix), we have two former Test Development Committee members and 11 AP® exam Readers.

*TPS* 5e will help you get ready for the AP® Statistics exam throughout the course by:

- **Using terms, notation, formulas, and tables consistent with those found on the AP® exam.** Key terms are shown in bold in the text, and they are defined in the Glossary. Key terms also are cross-referenced in the Index. See page F-1 to find "Formulas for the AP® Statistics Exam" as well as Tables A, B, and C in the back of the book for reference.
- **Following accepted conventions from AP® exam rubrics when presenting model solutions.** Over the years, the scoring guidelines for free-response questions have become fairly consistent. We kept these guidelines in mind when writing the solutions that appear throughout *TPS* 5e. For example, the four-step State–Plan–Do–Conclude process that we use to complete inference problems in Chapters 8 through 12 closely matches the four-point AP® scoring rubrics.
- **Including AP® Exam Tips in the margin where appropriate.** We place exam tips in the margins and in some Technology Corners as "on-the-spot" reminders of common mistakes and how to avoid them. These tips are collected and summarized in Appendix A.
- **Providing hundreds of AP®-style exercises throughout the book.** We even added a new kind of problem just prior to each Chapter Review, called a FRAPPY (Free Response AP® Problem, Yay!). Each FRAPPY gives you the chance to solve an AP®-style free-response problem based on the material in the chapter. After you finish, you can view and critique two example solutions from the book's Web site (www.whfreeman.com/tps5e). Then you can score your own response using a rubric provided by your teacher.

Turn the page for a tour of the text. See how to use the book to realize success in the course and on the AP® exam.

# READ THE TEXT and use the book's features to help you grasp the big ideas.

Read the LEARNING OBJECTIVES at the beginning of each section. Focus on mastering these skills and concepts as you work through the chapter.

## 3.1 Scatterplots and Correlation

**WHAT YOU WILL LEARN** By the end of the section, you should be able to:

- Identify explanatory and response variables in situations where one variable helps to explain or influences the other.
- Make a scatterplot to display the relationship between two quantitative variables.
- Describe the direction, form, and strength of a relationship displayed in a scatterplot and identify outliers in a scatterplot.
- Interpret the correlation.
- Understand the basic properties of correlation, including how the correlation is influenced by outliers.
- Use technology to calculate correlation.
- Explain why association does not imply causation.

Scan the margins for the purple notes, which represent the "voice of the teacher" giving helpful hints for being successful in the course.

Often, using the regression line to make a prediction for $x = 0$ is an extrapolation. That's why the $y$ intercept isn't always statistically meaningful.

**DEFINITION: Extrapolation**

**Extrapolation** is the use of a regression line for prediction far outside the interval of values of the explanatory variable $x$ used to obtain the line. Such predictions are often not accurate.

Few relationships are linear for all values of the explanatory variable. *Don't make predictions using values of x that are much larger or much smaller than those that actually appear in your data.*

Take note of the green DEFINITION boxes that explain important vocabulary. Flip back to them to review key terms and their definitions.

Watch for CAUTION ICONS. They alert you to common mistakes that students make.

Look for the boxes with the blue bands. Some explain how to make graphs or set up calculations while others recap important concepts.

**HOW TO MAKE A SCATTERPLOT**

1. Decide which variable should go on each axis.
2. Label and scale your axes.
3. Plot individual data values.

Make connections and deepen your understanding by reflecting on the questions asked in THINK ABOUT IT passages.

**THINK ABOUT IT**

**What does correlation measure?** The Fathom screen shots below provide more detail. At the left is a scatterplot of the SEC football data with two lines added—a vertical line at the group's mean points per game and a horizontal line at the mean number of wins of the group. Most of the points fall in the upper-right or lower-left "quadrants" of the graph. That is, teams with above-average points per game tend to have above-average numbers of wins, and teams with below-average points per game tend to have numbers of wins that are below average. This confirms the positive association between the variables.

Below on the right is a scatterplot of the standardized scores. To get this graph, we transformed both the $x$- and the $y$-values by subtracting their mean and dividing by their standard deviation. As we saw in Chapter 2, standardizing a data set converts the mean to 0 and the standard deviation to 1. That's why the vertical and horizontal lines in the right-hand graph are both at 0.

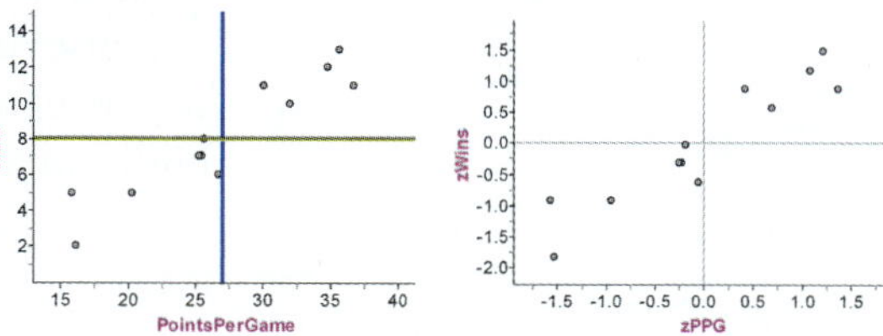


Notice that all the products of the standardized values will be positive—not surprising, considering the strong positive association between the variables. What if there was a negative association between two variables? Most of the points would be in the upper-left and lower-right "quadrants" and their $z$-score products would be negative, resulting in a negative correlation.

Read the AP® EXAM TIPS. They give advice on how to be successful on the AP® exam.

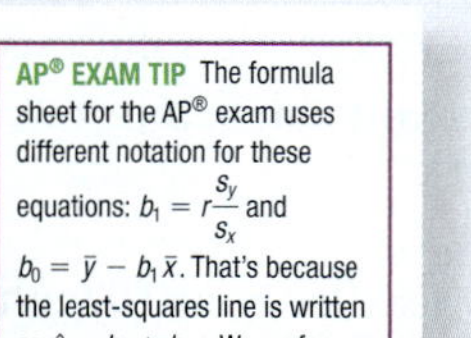

**AP® EXAM TIP** The formula sheet for the AP® exam uses different notation for these equations: $b_1 = r\frac{s_y}{s_x}$ and $b_0 = \bar{y} - b_1\bar{x}$. That's because the least-squares line is written as $\hat{y} = b_0 + b_1x$. We prefer our simpler versions without the subscripts!

# LEARN STATISTICS BY *DOING* STATISTICS

## ACTIVITY | Reaching for Chips

**MATERIALS:**
200 colored chips, including 100 of the same color; large bag or other container

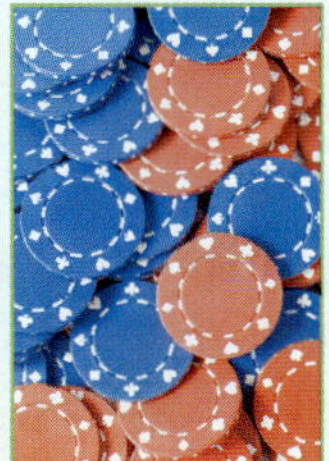

Before class, your teacher will prepare a population o
having the same color (say, red). The parameter is th
chips in the population: $p = 0.50$. In this Activity, y
variability by taking repeated random samples of size

1. After your teacher has mixed the chips thoroughly
should take a sample of 20 chips and note the sampl
When finished, the student should return all the chi
and pass the bag to the next student.

*Note:* If your class has fewer than 25 students, have s
samples.

2. Each student should record the $\hat{p}$-value in a chart
value on a class dotplot. Label the graph scale from (
spaced 0.05 units apart.

3. Describe what you see: shape, center, spread, and
usual features.

## ACTIVITY | I'm a Great Free-Throw Shooter!

**MATERIALS:**
Computer with Internet access and projection capability

A basketball player claims to make 80% of the free throws that he attempts. We think he might be exaggerating. To test this claim, we'll ask him to shoot some free throws—virtually—using *The Reasoning of a Statistical Test* applet at the book's Web site.

1. Go to www.whfreeman.com/tps5e and launch the applet.

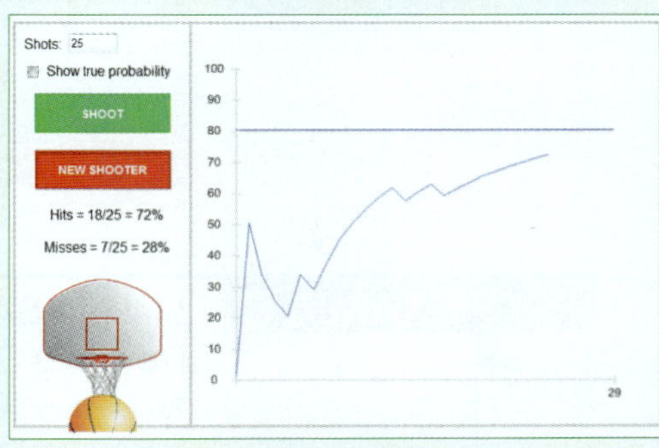


2. Set the applet to take 25 shots. Click "Shoot." How many of the 25 shots did the player make? Do you have enough data to decide whether the player's claim is valid?

3. Click "Shoot" again for 25 more shots. Keep doing this until you are convinced *either* that the player makes less than 80% of his shots *or* that the player's claim is true. How large a sample of shots did you need to make your decision?

4. Click "Show true probability" to reveal the truth. Was your conclusion correct?

5. If time permits, choose a new shooter and repeat Steps 2 through 4. Is it easier to tell that the player is exaggerating when his actual proportion of free throws made is closer to 0.8 or farther from 0.8?

Every chapter begins with a hands-on ACTIVITY that introduces the content of the chapter. Many of these activities involve collecting data and drawing conclusions from the data. In other activities, you'll use dynamic applets to explore statistical concepts.

DATA EXPLORATIONS ask you to play the role of data detective. Your goal is to answer a puzzling, real-world question by examining data graphically and numerically.

## DATA EXPLORATION The SAT essay: Is longer better?

Following the debut of the new SAT Writing test in March 2005, Dr. Les Perelman from the Massachusetts Institute of Technology stirred controversy by reporting, "It appeared to me that regardless of what a student wrote, the longer the essay, the higher the score." He went on to say, "I have never found a quantifiable predictor in 25 years of grading that was anywhere as strong as this one. If you just graded them based on length without ever reading them, you'd be right over 90 percent of the time."[3] The table below shows the data that Dr. Perelman used to draw his conclusions.[4]

**Length of essay and score for a sample of SAT essays**

| | | | | | | | | | | | |
|---|---|---|---|---|---|---|---|---|---|---|---|
| Words: | 460 | 422 | 402 | 365 | 357 | 278 | 236 | 201 | 168 | 156 | 133 |
| Score: | 6 | 6 | 5 | 5 | 6 | 5 | 4 | 4 | 4 | 3 | 2 |
| Words: | 114 | 108 | 100 | 403 | 401 | 388 | 320 | 258 | 236 | 189 | 128 |
| Score: | 2 | 1 | 1 | 5 | 6 | 6 | 5 | 4 | 4 | 3 | 2 |
| Words: | 67 | 697 | 387 | 355 | 337 | 325 | 272 | 150 | 135 | | |
| Score: | 1 | 6 | 6 | 5 | 5 | 4 | 4 | 2 | 3 | | |

Does this mean that if students write a lot, they are guaranteed high scores? Carry out your own analysis of the data. How would you respond to each of Dr. Perelman's claims?

CHECK YOUR UNDERSTANDING questions appear throughout the section. They help you to clarify definitions, concepts, and procedures. Be sure to check your answers in the back of the book.

## CHECK YOUR UNDERSTANDING

Identify the explanatory and response variables in each setting.

1. How does drinking beer affect the level of alcohol in people's blood? The legal limit for driving in all states is 0.08%. In a study, adult volunteers drank different numbers of cans of beer. Thirty minutes later, a police officer measured their blood alcohol levels.

2. The National Student Loan Survey provides data on the amount of debt for recent college graduates, their current income, and how stressed they feel about college debt. A sociologist looks at the data with the goal of using amount of debt and income to explain the stress caused by college debt.

# EXAMPLES: Model statistical problems and how to solve them



You will often see explanatory variables called *independent variables* and response variables called *dependent variables*. Because the words "independent" and "dependent" have other meanings in statistics, we won't use them here.

It is easiest to identify explanatory and response variables when we actually specify values of one variable to see how it affects another variable. For instance, to study the effect of alcohol on body temperature, researchers gave several different amounts of alcohol to mice. Then they measured the change in each mouse's body temperature 15 minutes later. In this case, amount of alcohol is the explanatory variable, and change in body temperature is the response variable. When we don't specify the values of either variable but just observe both variables, there may or may not be explanatory and response variables. Whether there are depends on how you plan to use the data.

Read through each EXAMPLE, and then try out the concept yourself by working the FOR PRACTICE exercise in the Section Exercises.

**EXAMPLE** Linking SAT Math and Critical Reading Scores

*Explanatory or response?*

Julie asks, "Can I predict a state's mean SAT Math score if I know its mean SAT Critical Reading score?" Jim wants to know how the mean SAT Math and Critical Reading scores this year in the 50 states are related to each other.

PROBLEM: For each student, identify the explanatory variable and the response variable if possible.

SOLUTION: Julie is treating the mean SAT Critical Reading score as the explanatory variable and the mean SAT Math score as the response variable. Jim is simply interested in exploring the relationship between the two variables. For him, there is no clear explanatory or response variable.

**For Practice** *Try Exercise* 1

Need extra help? Examples and exercises marked with the PLAY ICON are supported by short video clips prepared by experienced AP® teachers. The video guides you through each step in the example and solution and gives you extra help when you need it.

Example: Explanatory or Response?

Julie asks, "Can I predict a state's mean SAT Math score if I know its mean SAT Critical Reading score?"

Julie is treating the mean SAT Critical Reading score as the explanatory variable and the mean SAT Math score as the response variable.

Jim wants to know how the mean SAT Math and Critical Reading scores this year in the 50 states are related to each other.

The red number box next to the exercise directs you back to the page in the section where the model example appears.

1. **Coral reefs** How sensitive to changes in water temperature are coral reefs? To find out, measure the growth of corals in aquariums where the water temperature is controlled at different levels. Growth is measured by weighing the coral before and after the experiment. What are the explanatory and response variables? Are they categorical or quantitative?

**EXAMPLE** Gesell Scores

STEP 4 *Putting it all together*

Does the age at which a child begins to talk predict a later score on a test of mental ability? A study of the development of young children recorded the age in months at which each of 21 children spoke their first word and their Gesell Adaptive Score, the result of an aptitude test taken much later.[16] The data appear in the table below, along with a scatterplot, residual plot, and computer output. Should we use a linear model to predict a child's Gesell score from his or her age at first word? If so, how accurate will our predictions be?

4-STEP EXAMPLES: By reading the 4-Step Examples and mastering the special "State–Plan–Do–Conclude" framework, you can develop good problem-solving skills and your ability to tackle more complex problems like those on the AP® exam.

Age (months) at first word and Gesell score

| CHILD | AGE | SCORE | CHILD | AGE | SCORE | CHILD | AGE | SCORE |
|---|---|---|---|---|---|---|---|---|
| 1 | 15 | 95 | 8 | 11 | 100 | 15 | 11 | 102 |
| 2 | 26 | 71 | 9 | 8 | 104 | 16 | 10 | 100 |
| 3 | 10 | 83 | 10 | 20 | 94 | 17 | 12 | 105 |
| 4 | 9 | 91 | 11 | 7 | 113 | 18 | 42 | 57 |
| 5 | 15 | 102 | 12 | 9 | 96 | 19 | 17 | 121 |
| 6 | 20 | 87 | 13 | 10 | 83 | 20 | 11 | 86 |
| 7 | 18 | 93 | 14 | 11 | 84 | 21 | 10 | 100 |

# EXERCISES: Practice makes perfect!

Start by reading the SECTION SUMMARY to be sure that you understand the key concepts.

## Section 3.2 Summary

- A **regression line** is a straight line that describes how a response variable $y$ changes as an explanatory variable $x$ changes. You can use a regression line to **predict** the value of $y$ for any value of $x$ by substituting this $x$ into the equation of the line.
- The **slope** $b$ of a regression line $\hat{y} = a + bx$ is the rate at which the predicted response $\hat{y}$ changes along the line as the explanatory variable $x$ changes. Specifically, $b$ is the *predicted* change in $y$ when $x$ increases by 1 unit.
- The **$y$ intercept** $a$ of a regression line $\hat{y} = a + bx$ is the predicted response $\hat{y}$ when the explanatory variable $x$ equals 0. This prediction is of no statistical use unless $x$ can actually take values near 0.

Practice! Work the EXERCISES assigned by your teacher. Compare your answers to those in the Solutions Appendix at the back of the book. Short solutions to the exercises numbered in red are found in the appendix.

Most of the exercises are paired, meaning that odd- and even-numbered problems test the same skill or concept. If you answer an assigned problem incorrectly, try to figure out your mistake. Then see if you can solve the paired exercise.

Look for icons that appear next to selected problems. They will guide you to

- an Example that models the problem.
- videos that provide step-by-step instructions for solving the problem.
- earlier sections on which the problem draws (here, Section 2.2).
- examples with the 4-Step State–Plan–Do–Conclude way of solving problems.

## Section 3.2 Exercises

35. **What's my line?** You use the same bar of soap to shower each morning. The bar weighs 80 grams when it is new. Its weight goes down by 6 grams per day on average. What is the equation of the regression line for predicting weight from days of use?

36. **What's my line?** An eccentric professor believes that a child with IQ 100 should have a reading test score of 50 and predicts that reading score should increase by 1 point for every additional point of IQ. What is the equation of the professor's regression line for predicting reading score from IQ?

37. **Gas mileage** We expect a car's highway gas mileage to be related to its city gas mileage. Data for all 1198 vehicles in the government's recent *Fuel Economy Guide* give the regression line: predicted highway mpg = 4.62 + 1.109 (city mpg).
   (a) What's the slope of this line? Interpret this value in context.
   (b) What's the $y$ intercept? Explain why the value of the intercept is not statistically meaningful.
   (c) Find the predicted highway mileage for a car that gets 16 miles per gallon in the city.

38. **IQ and reading scores** Data on the IQ test scores and reading test scores for a group of fifth-grade children give the following regression line: predicted reading score = −33.4 + 0.882(IQ sco
   (a) What's the slope of this line? Interpret thi context.
   (b) What's the $y$ intercept? Explain why the intercept is not statistically meaningful.
   (c) Find the predicted reading score for a cl IQ score of 90.

39. **Acid rain** Researchers studying acid rai the acidity of precipitation in a Colorad area for 150 consecutive weeks. Acidity i by pH. Lower pH values show higher ac researchers observed a linear pattern ov They reported that the regression line pl 0.0053(weeks) fit the data well.[19]

pg 166

in Joan's midwestern home. The figure below shows the original scatterplot with the least-squares line added. The equation of the least-squares line is $\hat{y} = 1425 - 19.87x$.

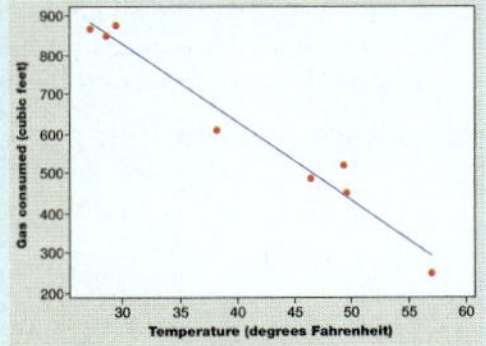


(a) Identify the slope of the line and explain what it means in this setting.
(b) Identify the $y$ intercept of the line. Explain why it's risky to use this value as a prediction.
(c) Use the regression line to predict the amount of natural gas Joan will use in a month with an average temperature of 30°F.

41. **Acid rain** Refer to Exercise 39. Would it be appropriate to use the regression line to predict pH after 1000 months? Justify your answer.

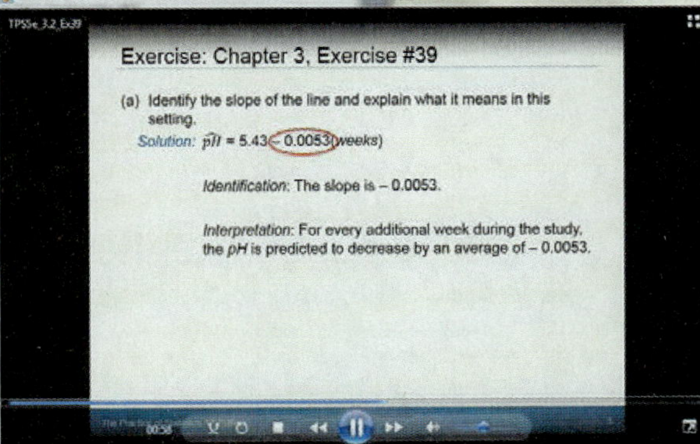


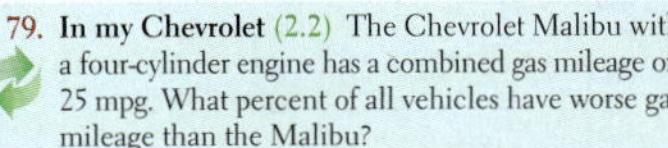

79. **In my Chevrolet** (2.2) The Chevrolet Malibu with a four-cylinder engine has a combined gas mileage of 25 mpg. What percent of all vehicles have worse gas mileage than the Malibu?

67. **Beavers and beetles** Do beavers benefit beetles? Researchers laid out 23 circular plots, each 4 meters in diameter, in an area where beavers were cutting down cottonwood trees. In each plot, they counted the number of stumps from trees cut by beavers and the number of clusters of beetle larvae. Ecologists think that the new sprouts from stumps are more tender than other cottonwood growth, so that beetles prefer them.

STEP 4 pg 185

# REVIEW and PRACTICE for quizzes and tests

## Chapter Review

**Section 3.1: Scatterplots and Correlation**

In this section, you learned how to explore the relationship between two quantitative variables. As with distributions of a single variable, the first step is always to make a graph. A scatterplot is the appropriate type of graph to investigate associations between two quantitative variables. To describe a scatterplot, be sure to discuss four characteristics: direction, form, strength, and outliers. The direction of an association might be positive, negative, or neither. The form of an association can be linear or nonlinear. An association is strong if it closely follows a specific form. Finally, outliers are any points that clearly fall outside the pattern of the rest of the data.

The correlation *r* is a numerical summary that describes the direction and strength of a linear association. When $r > 0$, the association is positive, and when $r < 0$, the association is negative. The correlation will always take values between $-1$ and 1, with $r = -1$ and $r = 1$ indic ing a perfectly linear relationship. Strong linear assoc tions have correlations near 1 or $-1$, while weak lin relationships have correlations near 0. However, it is possible to determine the form of an association from only the correlation. Strong nonlinear relationships can have a correlation close to 1 or a correlation close to 0, depending on the association. You also learned that outliers can greatly affect the value of the correlation and that correlation does not imply causation. That is, we can't assume that changes in one variable cause changes in the other variable, just because they have a correlation close to 1 or –1.

**Section 3.2: Least-Squares Regression**

In this section, you learned how to use least-squares regression lines as models for relationships between variables that have a linear association. It is important to understand the difference between the actual data and the model used to describe the data. For example, when you are interpreting the slope of a least-squares regression

Review the CHAPTER SUMMARY to be sure that you understand the key concepts in each section.

### What Did You Learn?

| Learning Objective | Section | Related Example on Page(s) | Relevant Chapter Review Exercise(s) |
|---|---|---|---|
| Identify explanatory and response variables in situations where one variable helps to explain or influences the other. | 3.1 | 144 | R3.4 |
| Make a scatterplot to display the relationship between two quantitative variables. | 3.1 | 145, 148 | R3.4 |
| Describe the direction, form, and strength of a relationship displayed in a scatterplot and recognize outliers in a scatterplot. | 3.1 | 147, 148 | R3.1 |
| Interpret the correlation. | 3.1 | 152 | R3.3, R3.4 |
| Understand the basic properties of correlation, including how the correlation is influenced by outliers. | 3.1 | 152, 156, 157 | R3.1, R3.2 |
| Use technology to calculate correlation. | 3.1 | Activity on 152, 171 | R3.4 |
| Explain why association does not imply causation. | 3.1 | Discussion on 156, 190 | R3.6 |
| Interpret the slope and *y* intercept of a least-squares regression line. | 3.2 | 166 | R3.2, R3.4 |
| Use the least-squares regression line to predict *y* for a given *x*. Explain the dangers of extrapolation. | 3.2 | 167, Discussion on 168 (for extrapolation) | R3.2, R3.4, R3.5 |
| Calculate and interpret residuals. | 3.2 | 169 | R3.3, R3.4 |
| Explain the concept of least squares. | 3.2 | Discussion on 169 | R3.5 |
| Determine the equation of a least-squares regression line using technology or computer output. | 3.2 | Technology Corner on 171, 181 | R3.3, R3.4 |
| Construct and interpret residual plots to assess whether a linear model is appropriate. | 3.2 | Discussion on 175, 180 | R3.3, R3.4 |
| Interpret the standard deviation of the residuals and $r^2$ and use these values to assess how well the least-squares regression line models the relationship between two variables. | 3.2 | 180 | R3.3, R3.5 |
| | | Discussion on 188 | R3.1 |
| | | 183 | R3.5 |

Use the WHAT DID YOU LEARN? table to guide you to model examples and exercises to verify your mastery of each LEARNING OBJECTIVE.

## Chapter 3 Chapter Review Exercises

*These exercises are designed to help you review the important ideas and methods of the chapter.*

**R3.1 Born to be old?** Is there a relationship between the gestational period (time from conception to birth) of an animal and its average life span? The figure shows a scatterplot of the gestational period and average life span for 43 species of animals.[30]

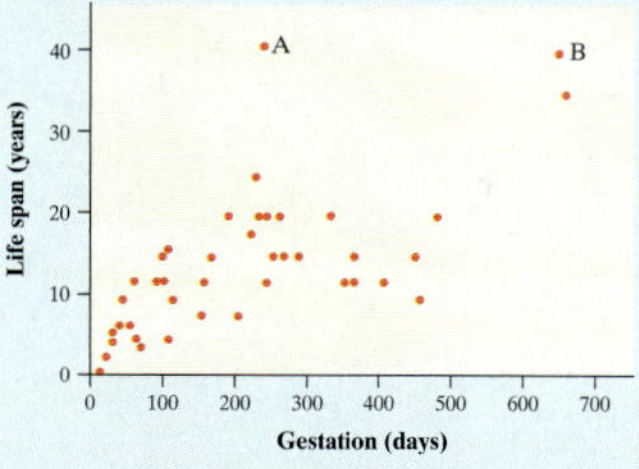


(a) Describe the association shown in the scatterplot.

**R3.3 Stats teachers' cars** A random sample of AP® Statistics teachers was asked to report the age (in years) and mileage of their primary vehicles. A scatterplot of the data, a least-squares regression printout, and a residual plot are provided below.

```
Predictor   Coef    SE Coef   T      P
Constant    3704    8268      0.45   0.662
Age         12188   1492      8.17   0.000

S = 20870.5   R-Sq = 83.7%   R-Sq(adj) = 82.4%
```

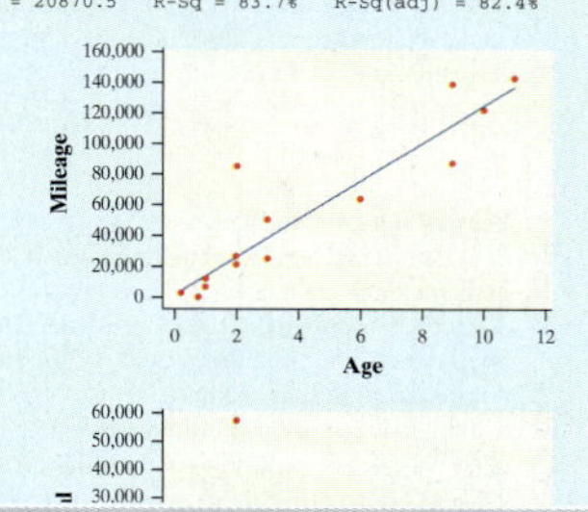


Tackle the CHAPTER REVIEW EXERCISES for practice in solving problems that test concepts from throughout the chapter.

# and the AP® Exam

Each chapter concludes with an AP® STATISTICS PRACTICE TEST. This test includes about 10 AP®-style multiple-choice questions and 3 free-response questions.

## Chapter 3 AP® Statistics Practice Test

**Section I: Multiple Choice** *Select the best answer for each question.*

T3.1 A school guidance counselor examines the number of extracurricular activities that students do and their grade point average. The guidance counselor says, "The evidence indicates that the correlation between the number of extracurricular activities a student participates in and his or her grade point average is close to zero." A correct interpretation of this statement would be that

(a) active students tend to be students with poor grades, and vice versa.

(b) students with good grades tend to be students who are not involved in many extracurricular activities, and vice versa.

(c) students involved in many extracurricular activities are just as likely to get good grades as bad grades; the same is true for students involved in few extracurricular activities.

(d) there is no linear relationship between number of activ-

alcoholic beverages for each of 11 regions in Great Britain was recorded. A scatterplot of spending on alcohol versus spending on tobacco is shown below. Which of the following statements is true?

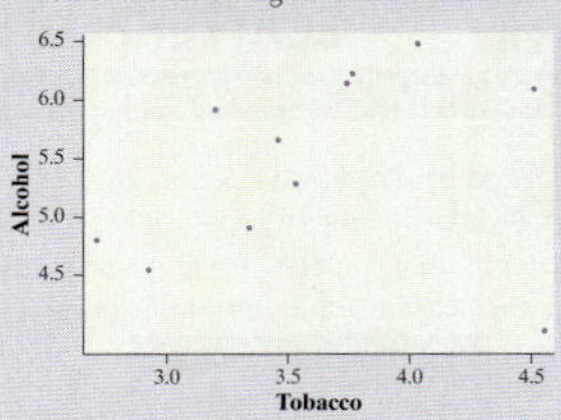


(a) The observation (4.5, 6.0) is an outlier.

(b) There is clear evidence of a negative association be-
...ending on alcohol and tobacco.
...uation of the least-squares line for this plot
...approximately $\hat{y} = 10 - 2x$.
...elation for these data is $r = 0.99$.
...ervation in the lower-right corner of the plot is
...al for the least-squares line.

## Cumulative AP® Practice Test 1

**Section I: Multiple Choice** *Choose the best answer for Questions AP1.1 to AP1.14.*

AP1.1 You look at real estate ads for houses in Sarasota, Florida. Many houses range from \$200,000 to \$400,000 in price. The few houses on the water, however, have prices up to \$15 million. Which of the following statements best describes the distribution of home prices in Sarasota?

(a) The distribution is most likely skewed to the left, and the mean is greater than the median.

AP1.4 For a certain experiment, the available experimental units are eight rats, of which four are female (F1, F2, F3, F4) and four are male (M1, M2, M3, M4). There are to be four treatment groups, A, B, C, and D. If a randomized block design is used, with the experimental units blocked by gender, which of the following assignments of treatments is impossible?

**Section II: Free Response** Show all your work. Indicate clearly the methods you use, because you will be graded on the correctness of your methods as well as on the accuracy and completeness of your results and explanations.

AP1.15 The manufacturer of exercise machines for fitness centers has designed two new elliptical machines that are meant to increase cardiovascular fitness. The two machines are being tested on 30 volunteers at a fitness center near the company's headquarters. The volunteers are randomly assigned to one of the machines and use it daily for two months. A measure of cardiovascular fitness is administered at the start of the experiment and

the two machines. Note that higher scores indicate larger gains in fitness.

| Machine A | | Machine B |
|---|---|---|
| | 0 | 2 |
| 54 | 1 | 0 |
| 876320 | 2 | 159 |
| 97411 | 3 | 2489 |

Four CUMULATIVE AP® TESTS simulate the real exam. They are placed after Chapters 4, 7, 10, and 12. The tests expand in length and content coverage from the first through the fourth.

## FRAPPY! Free Response AP® Problem, Yay!

The following problem is modeled after actual AP® Statistics exam free response questions. Your task is to generate a complete, concise response in 15 minutes.

*Directions: Show all your work. Indicate clearly the methods you use, because you will be scored on the correctness of your methods as well as on the accuracy and completeness of your results and explanations.*

Two statistics students went to a flower shop and randomly selected 12 carnations. When they got home, the students prepared 12 identical vases with exactly the same amount of water in each vase. They put one tablespoon of sugar in 3 vases, two tablespoons of sugar in 3 vases, and three tablespoons of sugar in 3 vases. In the remaining 3 vases, they put no sugar. After the vases were prepared, the students randomly assigned 1 carnation to each vase and observed how many hours each flower continued to look fresh. A scatterplot of the data is shown below.

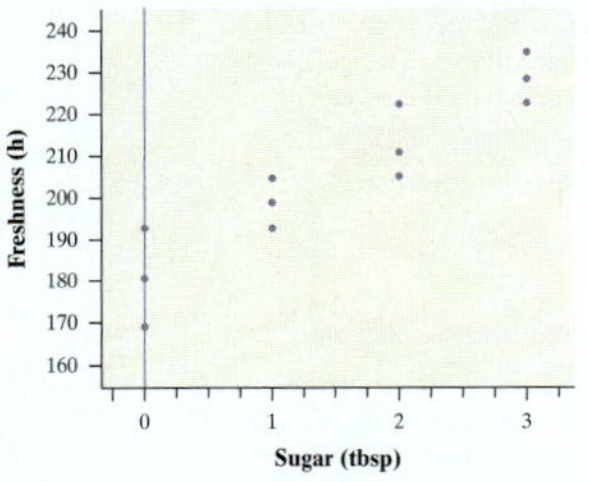


(a) Briefly describe the association shown in the scatterplot.

(b) The equation of the least-squares regression line for these data is $\hat{y} = 180.8 + 15.8x$. Interpret the slope of the line in the context of the study.

(c) Calculate and interpret the residual for the flower that had 2 tablespoons of sugar and looked fresh for 204 hours.

(d) Suppose that another group of students conducted a similar experiment using 12 flowers, but included different varieties in addition to carnations. Would you expect the value of $r^2$ for the second group's data to be greater than, less than, or about the same as the value of $r^2$ for the first group's data? Explain.

After you finish, you can view two example solutions on the book's Web site (www.whfreeman.com/tps5e). Determine whether you think each solution is "complete," "substantial," "developing," or "minimal." If the solution is not complete, what improvements would you suggest to the student who wrote it? Finally, your teacher will provide you with a scoring rubric. Score your response and note what, if anything, you would do differently to improve your own score.

Learn how to answer free-response questions successfully by working the FRAPPY! THE FREE RESPONSE AP® PROBLEM, YAY! that comes just before the Chapter Review in every chapter.

# Use TECHNOLOGY to discover and analyze

Use technology as a tool for discovery and analysis. TECHNOLOGY CORNERS give step-by-step instructions for using the TI-83/84 and TI-89 calculator. Instructions for the TI-Nspire are in an end-of-book appendix. HP Prime instructions are on the book's Web site and in the e-Book.

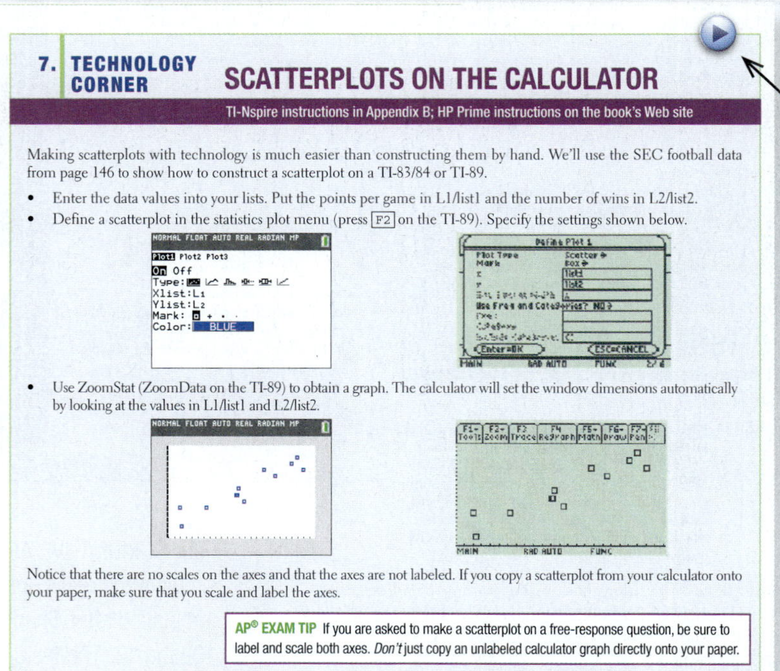

**7. TECHNOLOGY CORNER** **SCATTERPLOTS ON THE CALCULATOR**

TI-Nspire instructions in Appendix B; HP Prime instructions on the book's Web site

Making scatterplots with technology is much easier than constructing them by hand. We'll use the SEC football data from page 146 to show how to construct a scatterplot on a TI-83/84 or TI-89.

- Enter the data values into your lists. Put the points per game in L1/list1 and the number of wins in L2/list2.
- Define a scatterplot in the statistics plot menu (press F2 on the TI-89). Specify the settings shown below.
- Use ZoomStat (ZoomData on the TI-89) to obtain a graph. The calculator will set the window dimensions automatically by looking at the values in L1/list1 and L2/list2.

Notice that there are no scales on the axes and that the axes are not labeled. If you copy a scatterplot from your calculator onto your paper, make sure that you scale and label the axes.

**AP® EXAM TIP** If you are asked to make a scatterplot on a free-response question, be sure to label and scale both axes. *Don't* just copy an unlabeled calculator graph directly onto your paper.

You can access video instructions for the Technology Corners through the e-Book or on the book's Web site.

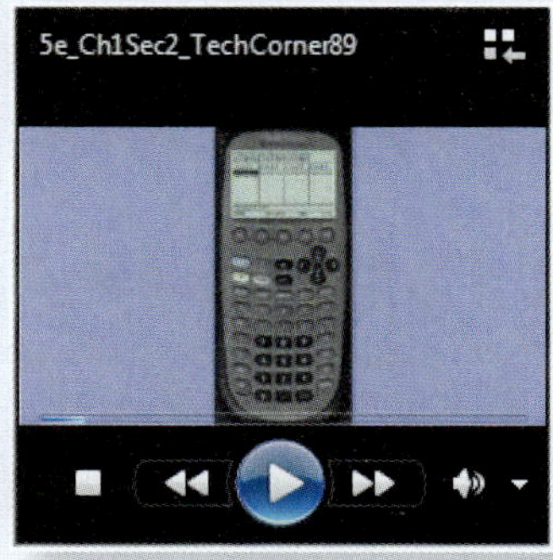


Find the Technology Corners easily by consulting the summary table at the end of each section or the complete table inside the back cover of the book.

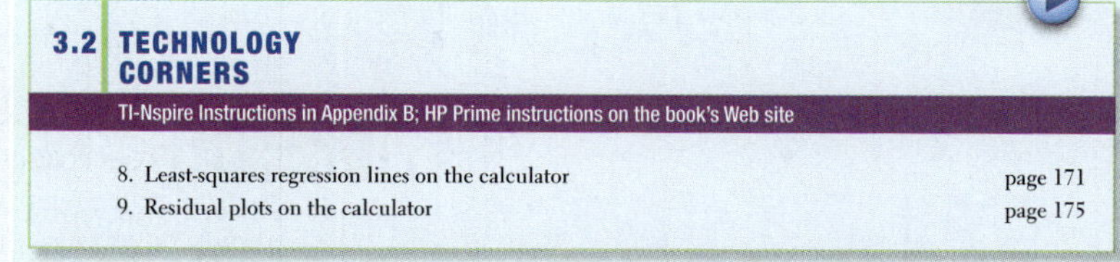

**3.2 TECHNOLOGY CORNERS**

TI-Nspire Instructions in Appendix B; HP Prime instructions on the book's Web site



59. **Merlins breeding** Exercise 13 (page 160) gives data on the number of breeding pairs of merlins in an isolated area in each of seven years and the percent of males who returned the next year. The data show that the percent returning is lower after successful breeding seasons and that the relationship is roughly linear. The figure below shows Minitab regression output for these data.

pg 181

**Regression Analysis: Percent return versus Breeding pairs**

| Predictor | Coef | SE Coef | T | P |
|---|---|---|---|---|
| Constant | 266.07 | 52.15 | 5.10 | 0.004 |
| Breeding pairs | -6.650 | 1.736 | -3.83 | 0.012 |

S = 7.76227 R-Sq = 74.6% R-Sq(adj) = 69.5%

(a) What is the equation of the least-squares regression line for predicting the percent of males that return from the number of breeding pairs? Use the equation to predict the percent of returning males after a season with 30 breeding pairs.

(b) What percent of the year-to-year variation in percent of returning males is accounted for by the straight-line relationship with number of breeding pairs the previous year?

Other types of software displays, including Minitab, Fathom, and applet screen captures, appear throughout the book to help you learn to read and interpret many different kinds of output.

# Overview What Is Statistics?

Does listening to music while studying help or hinder learning? If an athlete fails a drug test, how sure can we be that she took a banned substance? Does having a pet help people live longer? How well do SAT scores predict college success? Do most people recycle? Which of two diets will help obese children lose more weight and keep it off? Should a poker player go "all in" with pocket aces? Can a new drug help people quit smoking? How strong is the evidence for global warming?

These are just a few of the questions that statistics can help answer. But what is statistics? And why should you study it?

## Statistics Is the Science of Learning from Data

Data are usually numbers, but they are not "just numbers." *Data are numbers with a context.* The number 10.5, for example, carries no information by itself. But if we hear that a family friend's new baby weighed 10.5 pounds at birth, we congratulate her on the healthy size of the child. The context engages our knowledge about the world and allows us to make judgments. We know that a baby weighing 10.5 pounds is quite large, and that a human baby is unlikely to weigh 10.5 ounces or 10.5 kilograms. The context makes the number meaningful.

In your lifetime, you will be bombarded with data and statistical information. Poll results, television ratings, music sales, gas prices, unemployment rates, medical study outcomes, and standardized test scores are discussed daily in the media. Using data effectively is a large and growing part of most professions. A solid understanding of statistics will enable you to make sound, data-based decisions in your career and everyday life.

## Data Beat Personal Experiences

It is tempting to base conclusions on your own experiences or the experiences of those you know. But our experiences may not be typical. In fact, the incidents that stick in our memory are often the unusual ones.

### Do cell phones cause brain cancer?

Italian businessman Innocente Marcolini developed a brain tumor at age 60. He also talked on a cellular phone up to 6 hours per day for 12 years as part of his job. Mr. Marcolini's physician suggested that the brain tumor may have been caused by cell-phone use. So Mr. Marcolini decided to file suit in the Italian court system. A court ruled in his favor in October 2012.

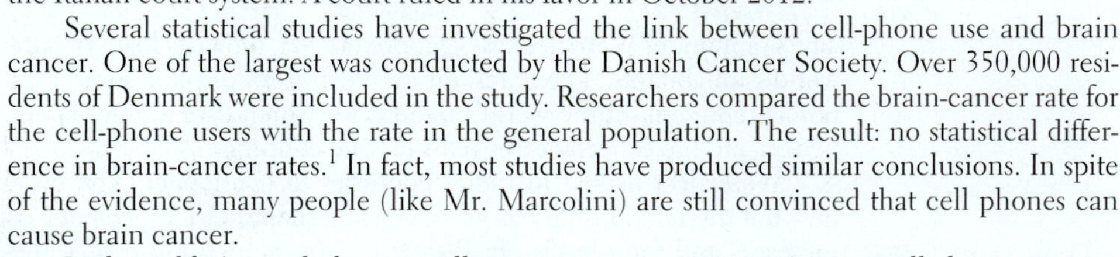

Several statistical studies have investigated the link between cell-phone use and brain cancer. One of the largest was conducted by the Danish Cancer Society. Over 350,000 residents of Denmark were included in the study. Researchers compared the brain-cancer rate for the cell-phone users with the rate in the general population. The result: no statistical difference in brain-cancer rates.[1] In fact, most studies have produced similar conclusions. In spite of the evidence, many people (like Mr. Marcolini) are still convinced that cell phones can cause brain cancer.

In the public's mind, the compelling story wins every time. A statistically literate person knows better. ***Data are more reliable than personal experiences because they systematically describe an overall picture rather than focus on a few incidents.***

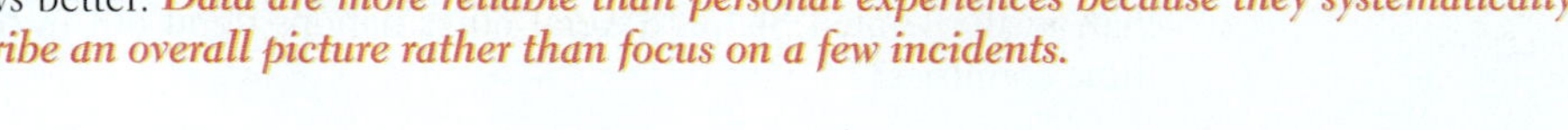

## Where the Data Come from Matters

### Are you kidding me?

The famous advice columnist Ann Landers once asked her readers, "If you had it to do over again, would you have children?" A few weeks later, her column was headlined "70% OF PARENTS SAY KIDS NOT WORTH IT." Indeed, 70% of the nearly 10,000 parents who wrote in said they would not have children if they could make the choice again. Do you believe that 70% of all parents regret having children?

You shouldn't. The people who took the trouble to write Ann Landers are not representative of all parents. Their letters showed that many of them were angry with their children. All we know from these data is that there are some unhappy parents out there. A statistically designed poll, unlike Ann Landers's appeal, targets specific people chosen in a way that gives all parents the same chance to be asked. Such a poll showed that 91% of parents *would* have children again.

Where data come from matters a lot. If you are careless about how you get your data, you may announce 70% "No" when the truth is close to 90% "Yes."

### Who talks more—women or men?

According to Louann Brizendine, author of *The Female Brain*, women say nearly three times as many words per day as men. Skeptical researchers devised a study to test this claim. They used electronic devices to record the talking patterns of 396 university students from Texas, Arizona, and Mexico. The device was programmed to record 30 seconds of sound every 12.5 minutes without the carrier's knowledge. What were the results?

According to a published report of the study in *Scientific American*, "Men showed a slightly wider variability in words uttered. . . . But in the end, the sexes came out just about even in the daily averages: women at 16,215 words and men at 15,669."[2] When asked where she got her figures, Brizendine admitted that she used unreliable sources.[3]

***The most important information about any statistical study is how the data were produced.*** Only carefully designed studies produce results that can be trusted.

## Always Plot Your Data

Yogi Berra, a famous New York Yankees baseball player known for his unusual quotes, had this to say: "You can observe a lot just by watching." That's a motto for learning from data. ***A carefully chosen graph is often more instructive than a bunch of numbers.***

### Do people live longer in wealthier countries?

The Gapminder Web site, www.gapminder.org, provides loads of data on the health and well-being of the world's inhabitants. The graph on the next pages displays some data from Gapminder.[4] The individual points represent all the world's nations for which data are available. Each point shows the income per person and life expectancy in years for one country.

We expect people in richer countries to live longer. The overall pattern of the graph does show this, but the relationship has an interesting shape. Life expectancy rises very quickly as personal income increases and then levels off. People in very rich countries like the United States live no longer than people in poorer but not extremely poor nations. In some less wealthy countries, people live longer than in the United States. Several other nations stand out in the graph. What's special about each of these countries?

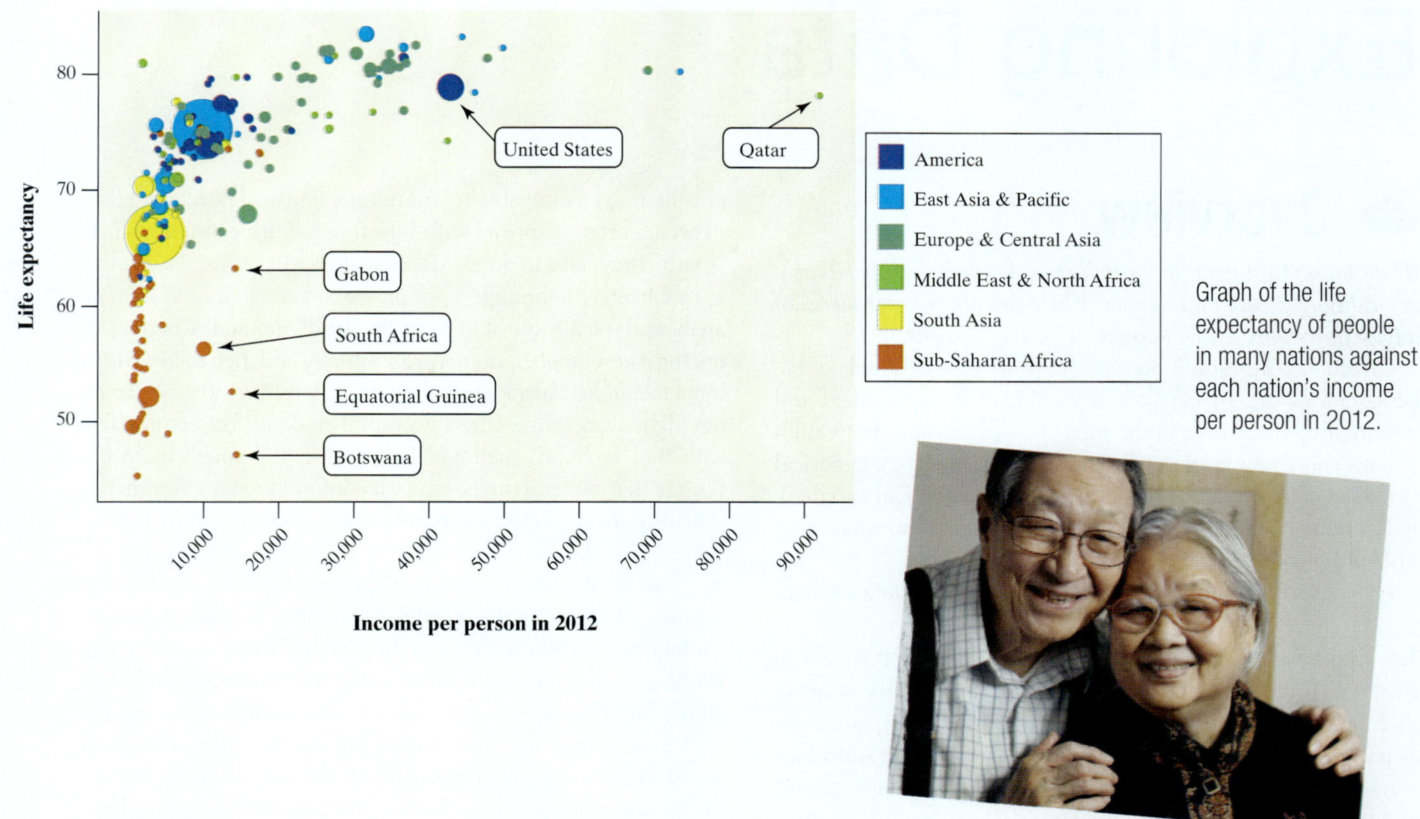


Graph of the life expectancy of people in many nations against each nation's income per person in 2012.

## Variation Is Everywhere

Individuals vary. Repeated measurements on the same individual vary. Chance outcomes—like spins of a roulette wheel or tosses of a coin—vary. Almost everything varies over time. Statistics provides tools for understanding variation.

### Have most students cheated on a test?

Researchers from the Josephson Institute were determined to find out. So they surveyed about 23,000 students from 100 randomly selected schools (both public and private) nationwide. The question they asked was "How many times have you cheated during a test at school in the past year?" Fifty-one percent said they had cheated at least once.[5]

If the researchers had asked the same question of *all* high school students, would exactly 51% have answered "Yes"? Probably not. If the Josephson Institute had selected a different sample of about 23,000 students to respond to the survey, they would probably have gotten a different estimate. *Variation is everywhere!*

Fortunately, statistics provides a description of how the sample results will vary in relation to the actual population percent. Based on the sampling method that this study used, we can say that the estimate of 51% is very likely to be within 1% of the true population value. That is, we can be quite confident that between 50% and 52% of *all* high school students would say that they have cheated on a test.

***Because variation is everywhere, conclusions are uncertain. Statistics gives us a language for talking about uncertainty that is understood by statistically literate people everywhere.***

Chapter

1

# Exploring Data

## PD Overview

*If you haven't yet read the Introduction to the Annotated Teacher's Edition, please do this now. There are many great pieces of advice that apply to all chapters, including this one!*

Chapters 1 through 3 present the principles and tools of basic data analysis. In Chapter 1, we introduce categorical and quantitative variables and the idea of a distribution. In Section 1.1, we show how to describe the distribution of a categorical variable and the relationship between two categorical variables. In Sections 1.2 and 1.3, we focus on graphing and summarizing distributions of a single quantitative variable.

Here are three principles for effectively teaching data analysis.

**1. Emphasize the Strategy, not Just the Skills:** It is easy to treat data analysis as simply a collection of descriptive statistics such as the mean, median, and standard deviation. However, as part of a larger strategy for exploring data, we also introduce ways to visualize the data using dotplots, stemplots, histograms, and boxplots. We emphasize this strategy by following a consistent approach to data analysis:

- Begin with a graph, move to numerical descriptions of specific aspects of the data, and (sometimes) to a compact mathematical model. Determining which graphs, numerical summaries, and models are appropriate depends on the context and type of data.
- Look for an overall pattern and for striking departures from that pattern. Departures such as outliers may influence the choice of descriptive summaries, and the presence and clarity of the overall pattern suggest what models may be useful.

**2. Focus on Understanding, not Just Mechanics:** Statistics exams would be much easier to grade if correctly graphing a histogram or calculating a standard deviation by hand showed mastery of data analysis. Unfortunately, students can possess these skills and still have no idea what data analysis is all about. Modern statistics courses use technology to quickly graph the data and calculate the associated numerical summaries so the focus of the course is on what we can learn from these summaries. This includes knowing which summaries are appropriate in different situations and how to interpret the summaries. It also includes being able to communicate this knowledge. Students are often surprised with how much writing and how little "math" they have to do in AP® Statistics!

In Chapters 1 through 3, we present a variety of tools for data analysis. Typically, we start with "by hand" methods when introducing a new graph or numerical summary and then follow with a more technological approach. Some teachers move to the technology right away, while others wait until students feel comfortable with the "by hand" methods. In any case, it is much more important that students understand what information the summaries provide rather than the mechanics of how they are calculated.

**3. Use Real Data:** Remember the mantra: data are not just numbers, they are numbers with a context. The context enables students to communicate conclusions in words and to judge whether their conclusions are sensible. Data usually come with a little bit of background, though for an introductory course the background may not fully reflect the complexities of the real world. We're willing to oversimplify for the sake of clarity, but not to ask students to perform empty operations with mere numbers.

We have tried to provide small- and moderate-size data sets in adequate number for basic instruction, both in the student text and in the *Annotated Teacher's Edition*. All of the medium to large data sets used in examples and exercises are available in a variety of formats in the Student's e-Book, on the Instructor's Resource CD, and on the book's companion Web site (www.whfreeman.com/tps5e) for easy downloading. The Resources section in the Introduction to the *Annotated Teacher's Edition* also highlights additional places to look for good data sets. Finally, nothing is more interesting to students than data about themselves. The more data you can collect in class the better. Consider using Mr. Starnes's Infamous AP® Statistics Survey in the *Teacher's Resource Materials* to build up your collection of student data.

## Relationship to the AP® Statistics Course Description

Chapter 1 covers the following content as described in the Topic Outline from the AP® Statistics Course Description. Topics marked with a superscript are covered in other chapters as indicated.

I. Exploring Data: describing patterns and departures from patterns

A. Constructing and interpreting graphical displays of distributions of univariate data (dotplot, stemplot, histogram, cumulative frequency plot[ch2])

1. Center and spread
2. Clusters and gaps
3. Outliers and unusual features
4. Shape

B. Summarizing distributions of univariate data

1. Measuring center: median, mean
2. Measuring spread: range, interquartile range, standard deviation
3. Measuring position: quartiles, percentiles[ch2], standardized scores (*z*-scores)[ch2]
4. Using boxplots
5. The effect of changing units on summary measures[ch2]

C. Comparing distributions of univariate data (dotplots, back-to-back stemplots, parallel boxplots)

1. Comparing center and spread
2. Comparing clusters and gaps
3. Comparing outliers and unusual features
4. Comparing shape

E. Exploring Categorical Data

1. Frequency tables and bar charts
2. Marginal and joint[ch5] frequencies for two-way tables
3. Conditional relative frequencies and association[ch5]
4. Comparing distributions using bar charts

# The Big Ideas

One of the challenges in teaching the AP® Statistics course is keeping students focused on the big picture and not just the details of each section. We outline the big ideas for the chapter here.

## Chapter 1 Introduction Data Analysis: Making Sense of Data

In this brief section, we introduce several fundamental concepts that will be important throughout the course: the idea of a distribution and the distinction between quantitative and categorical variables. We also introduce a strategy for exploring data:

- Begin by examining each variable by itself. Then move on to study relationships between variables.
- Start with a graph or graphs. Then add numerical summaries.

Also included in this section is an excellent activity for the first day of class. We strongly encourage you to use it with your students. The activity introduces students to the major themes in AP® Statistics and gives them a taste of what the rest of the course will be like. In less than one class period, you can have your students identify the evidence for a claim, use simulation to create a sampling distribution, and analyze the distribution to make a decision using the same inferential reasoning that they will be learning later in the course.

## Section 1.1 Analyzing Categorical Data

In Section 1.1, we use our data analysis strategy to explore categorical data. Starting with the distribution of a single categorical variable, we show students how to display the data with pie charts and bar graphs and what to look for when describing these displays. Make sure students get in the habit of properly labeling their graphs. Failure to do this is a very common error on the AP® exam for which students can lose points. Students should also be able to recognize misleading graphs and be careful to avoid making misleading graphs themselves.

Next, we show students how to analyze the relationship between two categorical variables. After looking at the marginal distributions of each variable separately, students should create conditional distributions of one variable for each category of the other variable. Graphing and comparing these conditional distributions will allow students to look for an association between the variables. If there is no association between the two variables, then graphs of the conditional distributions will look the same. However, if there are differences in the conditional distributions, an association exists between the variables. In previous editions, we included optional material on Simpson's paradox in this section. For this edition, we have moved this content to the *Teacher's Resource Materials.*

## Section 1.2 Displaying Quantitative Data With Graphs

In this section, we begin our study of quantitative variables by introducing three types of graphs: dotplots, stemplots, and histograms. Each of the graphs has its own distinct benefits, but all of them are good tools for examining the distribution of a quantitative variable. On the AP® exam, students will be expected to create these types of graphs, label them properly, and comment on their characteristics.

Dotplots are typically the easiest graph to construct, and many students may already be familiar with this type of display. A dotplot is much less abstract than a histogram or boxplot, making it easier for students to focus on interpreting the plot rather than trying to understand what it represents. For this reason, we will continue to use dotplots throughout the book as one of our primary ways to display distributions of quantitative data.

Stemplots are similar to dotplots in that they give a quick picture of the distribution and show each individual observation. They also have the added benefit of including the actual numerical values of the observations. However, this means that stemplots of large data sets can be cumbersome.

Histograms group numerical observations into classes and then use bars to represent the number of observations in each class. Histograms are the best choice when there are a large number of observations. Make sure to read the Think About It on page 35 with your students and emphasize the connection between dotplots and histograms.

To examine the distribution of a quantitative variable, students should look at a graph for the overall pattern (shape, center, spread) and striking departures from that pattern (outliers). We introduce the acronym SOCS (shape, outliers, center, spread) to help students remember these four characteristics. Likewise, when comparing distributions, students should include explicit comparison words such as "the center of the distribution of male heights is *greater than* the center of the distribution of female heights." A very common error on the AP® exam is describing the characteristics of each distribution separately without making these explicit comparisons.

In this section, we only briefly address center and spread and focus on shape. Because there will be a much more thorough discussion of center and spread in Section 1.3, we recommend that you spend most of your time in this section discussing the various shapes distributions can take and save the details of center and spread for later.

## Section 1.3 Describing Quantitative Data with Numbers

In this section, we introduce numerical summaries to describe the center and spread of a distribution of quantitative data and show how to identify outliers in a distribution. Although it can be helpful if students know how to calculate summary statistics by hand, it is more important that students know when they are appropriate to use and how to interpret them. With this in mind, we also present instructions for using technology to compute these numerical summaries. By the end of this section, students should feel very comfortable comparing several distributions of quantitative data by explicitly discussing the similarities in their shape, center, spread, and outliers.

To measure the center of a distribution of quantitative data, students should use the mean or median of the distribution. Students should understand that the median is a more resistant measure of center than the mean and should be able to choose which measure of center is more appropriate for various distribution shapes.

To measure the spread (variability) of a distribution of quantitative data, students can use the range, interquartile range, or standard deviation. The interquartile range ($IQR$) is a more resistant measure of spread than the range because it ignores the upper 25% and lower 25% of the distribution. The standard deviation is the most commonly used measure of spread and approximates the typical distance of the values in a distribution from the mean. The standard deviation is less resistant than the $IQR$ because it uses every value in the distribution rather than just the middle 50%. Having students develop a solid understanding of what the standard deviation measures will definitely help them be successful in later chapters.

To identify outliers in a distribution of quantitative data, students are introduced to a rule using the quartiles and interquartile range of a distribution. We also introduce boxplots as a graphical way to summarize the distribution of quantitative data. Boxplots are great for comparing distributions because they identify the five-number summary, allowing students to compare center (median) and spread (range, $IQR$). However, boxplots do not reveal modes, clusters, or gaps in a distribution. As always, it is imperative that students use proper labels when creating a visual display of data!

Throughout the book, there will be many different styles of examples. Some are narrative, some are in a problem/solution format and others illustrate the "4-Step" process. The purpose of the 4-Step examples is to model the process for answering an open-ended statistical problem. We suggest that you get students accustomed to the State–Plan–Do–Conclude format now, as it is a great template for answering all kinds of statistical questions throughout the course.

Chapter

1

# Chapter 1 Resources

## TRM Additional *TPS* Teacher's Resource Materials

The following resources can be found by clicking on the links in the new Teacher's e-Book (TE-book), opening the Teacher's Resource CD (TRCD), or logging onto the book's companion Web site www.whfreeman.com/tps5e (teacher log-in required).

- Alternate First-Day Activities (page 1)
    - Mr. Starnes's Infamous AP® Statistics Survey
    - Getting to Know Your Textbook
- Chapter 1 Learning Objective Grid (page 2)
- Simpson's paradox optional material (page 18)
- Activity: Motivating SOCS (page 26)
- Activity: Do you know your geography? (page 59)
- Activity: The memory game (page 59)
- Chapter 1 Project Idea with Rubric (page 66)
- FRAPPY! materials: Question, Model Solution, Scoring Rubric, Sample Responses with commentary (page 74)
- Quizzes: 3 versions for each section; versions A and B are roughly parallel
    - Section 1.1 (page 19)
    - Section 1.2 (page 40)
    - Section 1.3 (page 68)
- Tests: 3 versions; versions A and B are roughly parallel (page 74)

## Free-Response Questions from Previous AP® Exams

Questions can be found on the AP® Central Web site: http://apcentral.collegeboard.com/apc/members/exam/exam_questions/8357.html.

Students should be able to answer all the free-response questions listed with material learned in this chapter. Questions that contain content from this chapter but also require content from later chapters are listed in the last chapter required to complete the entire question. This list will be updated after each AP® exam and will be posted on the book's Web site: http://www.whfreeman.com/tps5e.

Questions marked with an asterisk are from exams with released multiple-choice questions. You may want to save these questions until the end of the year so you can give your students a complete released exam for practice.

| Year | # | Content |
|---|---|---|
| 2011B | 1 | • Estimating medians of histograms<br>• Comparing histograms<br>• Relationship between mean and median |
| 2010B | 1 | • Comparing distributions (boxplots)<br>• Stemplots<br>• Comparing boxplots and stemplots |
| 2007B | 1 | • Constructing a stemplot<br>• Summarizing a distribution of univariate data (stemplot)<br>• Bimodal distribution |
| 2006 | 1 | • Comparing distributions of univariate data (dotplots)<br>• Comparing variability<br>• Measuring center |
| 2005B | 1 | • Describing shape of a stemplot<br>• Mean vs. median<br>• Midrange |
| 2004 | 1 | • Constructing parallel boxplots<br>• Outliers<br>• Properties of boxplots<br>• Mean vs. median |
| 2002B | 5 | • Constructing parallel boxplots<br>• Comparing distributions of univariate data (boxplots) |
| 2001 | 1 | • Identifying outliers and unusual values |
| 2000 | 3 | • Graphing and comparing two frequency distributions |

## Applets

- The *Mean and Median* applet allows students to add, delete, and drag points to investigate the relationship between the mean and the median. www.whfreeman.com/tps5e (click on Statistical Applets and choose Mean and Median).
- The *One-Variable Statistical Calculator* applet allows students to use existing data sets or enter their own data set and calculate summary statistics and graph histograms or stemplots. www.whfreeman.com/tps5e (click on Statistical Applets and choose One-Variable Statistical Calculator).

- The *Histogram Bin-Width* applet allows students to explore how changing the bin width in a histogram can greatly alter the apparent shape of a distribution. www.rossmanchance.com/applets (choose Histogram Bin-Width).
- The *Dotplot Summaries* applet allows students to guess the positions of the mean and median on dotplots of various shapes. It also allows students to guess the size of the standard deviation and the interquartile range. www.rossmanchance.com/applets (choose Dotplot Summaries).
- The *Descriptive Statistics and Graphs* applets allow you to calculate summary statistics and make graphical displays of categorical and quantitative data, including dotplots, histograms, and boxplots. www.lock5stat.com/statkey (look on the left-hand side for these applets; press "Edit Data" to enter your own data).

## TI-Nspire Activities

These Activities are designed to illustrate important statistical concepts using the TI-Nspire. The Activities can be found at www.whfreeman.com/tps5e.

- *Exploring Bar Graphs and Pie Charts*: Introduction to graphing categorical variables.
- *Two-Way Tables—Data Exploration*: Using the Nspire to create and analyze marginal and conditional distributions of categorical data.
- *Exploring Quantitative Data: Part I and Part II*: In Part I, students will create a histogram from a data set and describe it using SOCS (Shape, Outliers, Center, Spread). In Part II, students will use the same data set to create a boxplot and compare the features of the boxplot with those of the histogram.
- *Exploring How Shape Affects Measures of Center*: Students will enter a data set, create a graphical display, and calculate numerical summaries to examine how skewness and outliers affect the mean and median.

# Chapter 1: Pacing Guide, Learning Objectives, and Suggested Assignments

This pacing guide is based on a schedule with 110, 50-minute sessions before the AP® exam. The 110 days include 4 days for midterms and 10 days for AP® exam review. If you have a different number of sessions before the AP® exam, you can modify the pacing guide to suit your needs. If you have additional time, consider incorporating quizzes, released AP® free-response questions, or additional activities. See the Resources section above for suggestions.

The suggested homework assignments list odd-numbered exercises, whenever possible, so students can check their answers against the back-of-the-book answers. If you would rather students not have access to the answers while doing homework, adding 1 to the exercise numbers usually will do the trick, because the homework exercises typically are paired. For example, Exercises 1 and 2 will generally cover the same topics, but in different contexts. You may also choose to include the Recycle and Review questions at the end of each section, which review topics from previous sections or chapters. These questions are denoted with a symbol.

# Chapter 1

| Day | Topics | **Learning Objectives** Students will be able to... | Suggested Assignment |
|---|---|---|---|
| 1 | Chapter 1 Introduction | • Identify the individuals and variables in a set of data.<br>• Classify variables as categorical or quantitative. | 1, 3, 5, 7, 8 |
| 2 | 1.1 Bar Graphs and Pie Charts, Graphs: Good and Bad | • Display categorical data with a bar graph. Decide whether it would be appropriate to make a pie chart.<br>• Identify what makes some graphs of categorical data deceptive. | 11, 13, 15, 17 |
| 3 | 1.1 Two-Way Tables and Marginal Distributions, Relationships between Categorical Variables: Conditional Distributions | • Calculate and display the marginal distribution of a categorical variable from a two-way table.<br>• Calculate and display the conditional distribution of a categorical variable for a particular value of the other categorical variable in a two-way table.<br>• Describe the association between two categorical variables by comparing appropriate conditional distributions. | 19, 21, 23, 25, 27–32 |
| 4 | 1.2 Dotplots, Describing Shape, Comparing Distributions, Stemplots | • Make and interpret dotplots and stemplots of quantitative data.<br>• Describe the overall pattern (shape, center, and spread) of a distribution and identify any major departures from the pattern (outliers).<br>• Identify the shape of a distribution from a graph as roughly symmetric or skewed.<br>• Compare distributions of quantitative data using dotplots or stemplots. | 37, 39, 41, 43, 45, 47 |
| 5 | 1.2 Histograms, Using Histograms Wisely | • Make and interpret histograms of quantitative data.<br>• Compare distributions of quantitative data using histograms. | 53, 55, 59, 60, 65, 69–74 |
| 6 | 1.3 Measuring Center: Mean and Median, Comparing Mean and Median, Measuring Spread: $IQR$, Identifying Outliers, Five-Number Summary and Boxplots | • Calculate measures of center (mean, median).<br>• Calculate and interpret measures of spread (range, $IQR$).<br>• Choose the most appropriate measure of center and spread in a given setting.<br>• Identify outliers using the $1.5 \times IQR$ rule.<br>• Make and interpret boxplots of quantitative data. | 79, 81, 83, 87, 89, 91, 93 |
| 7 | 1.3 Measuring Spread: Standard Deviation, Choosing Measures of Center and Spread, Organizing a Statistics Problem | • Calculate and interpret measures of spread (standard deviation).<br>• Choose the most appropriate measure of center and spread in a given setting.<br>• Use appropriate graphs and numerical summaries to compare distributions of quantitative variables. | 95, 97, 99, 103, 105, 107–110 |
| 8 | Chapter 1 Review/FRAPPY! | | Chapter 1 Review Exercises |
| 9 | Chapter 1 Test | | |

**PD Chapter 1 Overview**

To watch the video overview of Chapter 1 (for teachers), click on the link in the TE-book, or look on the TRCD or the book's companion Web site (teacher log-in required).

# Chapter 1

# Exploring Data

case study

## Do Pets or Friends Help Reduce Stress?

If you are a dog lover, having your dog with you may reduce your stress level. Does having a friend with you reduce stress? To examine the effect of pets and friends in stressful situations, researchers recruited 45 women who said they were dog lovers. Fifteen women were assigned at random to each of three groups: to do a stressful task alone, with a good friend present, or with their dogs present. The stressful task was to count backward by 13s or 17s. The woman's average heart rate during the task was one measure of the effect of stress. The table below shows the data.[1]

Average heart rates during stress with a pet (P), with a friend (F), and for the control group (C)

| GROUP | RATE | GROUP | RATE | GROUP | RATE | GROUP | RATE |
|---|---|---|---|---|---|---|---|
| P | 69.169 | P | 68.862 | C | 84.738 | C | 75.477 |
| F | 99.692 | C | 87.231 | C | 84.877 | C | 62.646 |
| P | 70.169 | P | 64.169 | P | 58.692 | P | 70.077 |
| C | 80.369 | C | 91.754 | P | 79.662 | F | 88.015 |
| C | 87.446 | C | 87.785 | P | 69.231 | F | 81.600 |
| P | 75.985 | F | 91.354 | C | 73.277 | F | 86.985 |
| F | 83.400 | F | 100.877 | C | 84.523 | F | 92.492 |
| F | 102.154 | C | 77.800 | C | 70.877 | P | 72.262 |
| P | 86.446 | P | 97.538 | F | 89.815 | P | 65.446 |
| F | 80.277 | P | 85.000 | F | 98.200 | | |
| C | 90.015 | F | 101.062 | F | 76.908 | | |
| C | 99.046 | F | 97.046 | P | 69.538 | | |

Based on the data, does it appear that the presence of a pet or friend reduces heart rate during a stressful task? In this chapter, you'll develop the tools to help answer this question.

1

### Teaching Tip

Try to start each example by asking a question, such as "Do Pets or Friends Help Reduce Stress?" According to the *Guidelines for Assessment and Instruction in Statistics Education*, endorsed by the American Statistical Association, "Statistical problem solving is an investigative process that involves four components: Formulate Questions, Collect Data, Analyze Data, and Interpret Results" (see www.amstat.org/education/gaise/ for more information about the four components of the statistical process). Too often, our students' experiences with statistics have been limited to analyzing data. It's no wonder that many students come into AP® Statistics thinking that statistics is just boring number crunching. By providing a motivating question at the beginning of each example, students will have a better sense of why we are analyzing the data and what to look for.

### TRM Additional First-Day Activities

At the beginning of the school year, we recommend that you start with an engaging activity, such as the hiring discrimination Activity on page 5. We also recommend that you have students read the Overview "What Is Statistics?" on page xxi and complete "Mr. Starnes's Infamous AP® Statistics Survey." Use this survey to gather data about your students that you can use all year long. Students are usually more interested in themselves than other people!

Finally, to help students take advantage of all *TPS* has to offer, have them read the Preface for Students on page xii and complete the "Getting to Know Your Textbook" activity. The survey and activity may be found in the Teacher's Resource Materials located in the TE-book, or on the TRCD or the book's companion Web site (teacher log-in required).

**TRM Learning Objectives Grid**

At the beginning of each section, we present the relevant learning objectives. Point these out to your students and refer back to them when you cover them in class. At the end of each chapter, there is a "What Did You Learn?" grid that lists all the objectives for the chapter. For each objective, the grid shows the section in which it was covered, page references for examples that illustrate the objective, and relevant chapter review exercises that students can use to assess their understanding of that objective. Encourage your students to use this grid as part of their preparations for the chapter test. There is also a PDF version of the grid with an additional column that students can use to keep track of their progress. Find it in the Teacher's Resource Materials located in the TE-book, or on the TRCD or the book's companion Web site (teacher log-in required).

### Teaching Tip

Emphasize that "individuals" don't have to be people. For example, the individuals in a study could be the trees in an orchard or the M&M'S® chocolate candies in a bag. Individuals are also called *cases*.

### Teaching Tip

Throughout the course, the distinction between categorical and quantitative variables will be very important. However, do not worry about the distinction between *discrete* quantitative variables and *continuous* quantitative variables for now. We will make this distinction in later chapters. Also, there is no need to distinguish between variables measured on ordinal, interval, or ratio scales.

# Introduction Data Analysis: Making Sense of Data

**WHAT YOU WILL LEARN** By the end of the section, you should be able to:

- Identify the individuals and variables in a set of data.
- Classify variables as categorical or quantitative.

Statistics is the science of data. The volume of data available to us is overwhelming. For example, the Census Bureau's American Community Survey collects data from 3,000,000 housing units each year. Astronomers work with data on tens of millions of galaxies. The checkout scanners at Walmart's 10,000 stores in 27 countries record hundreds of millions of transactions every week.

In all these cases, the data are trying to tell us a story—about U.S. households, objects in space, or Walmart shoppers. To hear what the data are saying, we need to help them speak by organizing, displaying, summarizing, and asking questions. That's **data analysis**.

## Individuals and Variables

Any set of data contains information about some group of **individuals**. The characteristics we measure on each individual are called **variables**.

**DEFINITION: Individuals and variables**

**Individuals** are the objects described by a set of data. Individuals may be people, animals, or things.

A **variable** is any characteristic of an individual. A variable can take different values for different individuals.

A high school's student data base, for example, includes data about every currently enrolled student. The students are the *individuals* described by the data set. For each individual, the data contain the values of *variables* such as age, gender, grade point average, homeroom, and grade level. In practice, any set of data is accompanied by background information that helps us understand the data. When you first meet a new data set, ask yourself the following questions:

1. *Who* are the individuals described by the data? How many individuals are there?
2. *What* are the variables? In what *units* are the variables recorded? Weights, for example, might be recorded in grams, pounds, thousands of pounds, or kilograms.

We could follow a newspaper reporter's lead and extend our list of questions to include *Why*, *When*, *Where*, and *How* were the data produced? For now, we'll focus on the first two questions.

Some variables, like gender and grade level, assign labels to individuals that place them into categories. Others, like age and grade point average (GPA), take numerical values for which we can do arithmetic. It makes sense to give an average GPA for a group of students, but it doesn't make sense to give an "average" gender.

**TRM Introduction Alternate Examples**

You can find the Alternate Examples for this section in Microsoft Word format by clicking on the link in the TE-book, opening the TRCD, or logging onto the book's companion Web site www.whfreeman.com/tps5e (teacher log-in required).

**DEFINITION: Categorical variable and quantitative variable**

A **categorical variable** places an individual into one of several groups or categories.

A **quantitative variable** takes numerical values for which it makes sense to find an average.

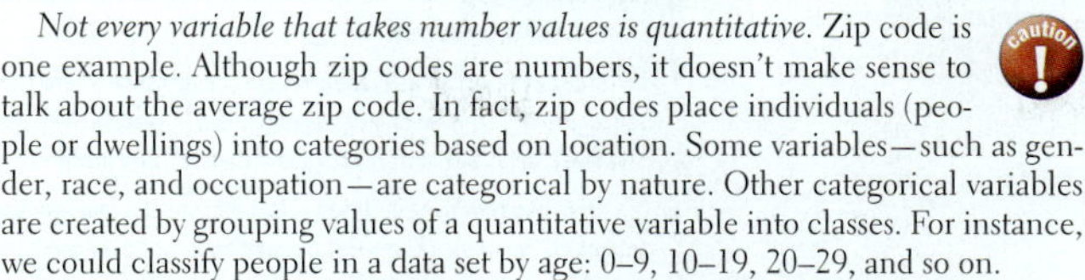

**AP® EXAM TIP** If you learn to distinguish categorical from quantitative variables now, it will pay big rewards later. You will be expected to analyze categorical and quantitative variables correctly on the AP® exam.

*Not every variable that takes number values is quantitative.* Zip code is one example. Although zip codes are numbers, it doesn't make sense to talk about the average zip code. In fact, zip codes place individuals (people or dwellings) into categories based on location. Some variables—such as gender, race, and occupation—are categorical by nature. Other categorical variables are created by grouping values of a quantitative variable into classes. For instance, we could classify people in a data set by age: 0–9, 10–19, 20–29, and so on.

The proper method of analysis for a variable depends on whether it is categorical or quantitative. As a result, it is important to be able to distinguish these two types of variables. The type of data determines what kinds of graphs and which numerical summaries are appropriate.

## EXAMPLE Census at School

### *Data, individuals, and variables*

CensusAtSchool is an international project that collects data about primary and secondary school students using surveys. Hundreds of thousands of students from Australia, Canada, New Zealand, South Africa, and the United Kingdom have taken part in the project since 2000. Data from the surveys are available at the project's Web site (www.censusatschool.com). We used the site's "Random Data Selector" to choose 10 Canadian students who completed the survey in a recent year. The table below displays the data.

| Province | Gender | Language spoken | Handed | Height (cm) | Wrist circum. (mm) | Preferred communication |
|---|---|---|---|---|---|---|
| Saskatchewan | Male | 1 | Right | 175 | 180 | In person |
| Ontario | Female | 1 | Right | 162.5 | 160 | In person |
| Alberta | Male | 1 | Right | 178 | 174 | Facebook |
| Ontario | Male | 2 | Right | 169 | 160 | Cell phone |
| Ontario | Female | 2 | Right | 166 | 65 | In person |
| Nunavut | Male | 1 | Right | 168.5 | 160 | Text messaging |
| Ontario | Female | 1 | Right | 166 | 165 | Cell phone |
| Ontario | Male | 4 | Left | 157.5 | 147 | Text Messaging |
| Ontario | Female | 2 | Right | 150.5 | 187 | Text Messaging |
| Ontario | Female | 1 | Right | 171 | 180 | Text Messaging |

There is at least one suspicious value in the data table. We doubt that the girl who is 166 cm tall really has a wrist circumference of 65 mm (about 2.6 inches). Always look to be sure the values make sense!

## Teaching Tip

Other examples of categorical variables that take number values include area codes and the floor of a building on which employees work.

## Teaching Tip

Emphasize that it is fairly common to transform a quantitative variable into a categorical variable. Some additional examples include giving students a letter grade (categorical) based on their scores on a test (quantitative) or reporting that a student exceeds the standard, meets the standard, or doesn't meet the standard (categorical) based on the results of a standardized test. This often occurs when we want to know how many individuals are in each category rather than wanting to know an average value.

## Teaching Tip

The "play" icon at the top of this example (and many others) indicates that students can watch a video presentation of the example by clicking on the link in the e-book or viewing the videos at www.whfreeman.com/tps5e. If you have a projector and Internet access, we recommend that you show your students how to access the videos.

## WEB SITE

The American Statistical Association hosted several Webinars about the CensusAtSchool project that may be of interest. See http://www.amstat.org/education/webinars/.

## ALTERNATE EXAMPLE

### *U.S. Census data*

Here is information about 10 randomly selected U.S. residents from a recent census, imported using Fathom software:

| State | Number of family members | Age | Gender | Marital status | Total income | Travel time to work |
|---|---|---|---|---|---|---|
| Kentucky | 2 | 61 | Female | Married | 21,000 | 20 |
| Florida | 6 | 27 | Female | Married | 21,300 | 20 |
| Wisconsin | 2 | 27 | Male | Married | 30,000 | 5 |
| California | 4 | 33 | Female | Married | 26,000 | 10 |
| Michigan | 3 | 49 | Female | Married | 15,100 | 25 |
| Virginia | 3 | 26 | Female | Married | 25,000 | 15 |
| Pennsylvania | 4 | 44 | Male | Married | 43,000 | 10 |
| Virginia | 4 | 22 | Male | Never married/ single | 3,000 | 0 |
| California | 1 | 30 | Male | Never married/ single | 40,000 | 15 |
| New York | 4 | 34 | Female | Separated | 30,000 | 40 |

## ALTERNATE EXAMPLE (continued)

PROBLEM:

**(a)** Who are the individuals in this data set?
**(b)** What variables are measured? Identify each as categorical or quantitative.
**(c)** Describe the individual in the first row.

SOLUTION:

**(a)** The 10 randomly selected U.S. residents from the recent census.
**(b)** Categorical: state, gender, marital status. Quantitative: number of family members, age, total income, travel time to work.
**(c)** This person is a 61-year-old married female from Kentucky who drives 20 minutes to work and has a total income of $21,000. She has two family members in her household.

PROBLEM:

(a) Who are the individuals in this data set?
(b) What variables were measured? Identify each as categorical or quantitative.
(c) Describe the individual in the highlighted row.

We'll see in Chapter 4 why choosing at random, as we did in this example, is a good idea.

SOLUTION:

(a) The individuals are the 10 randomly selected Canadian students who participated in the CensusAtSchool survey.

(b) The seven variables measured are the province where the student lives (categorical), gender (categorical), number of languages spoken (quantitative), dominant hand (categorical), height (quantitative), wrist circumference (quantitative), and preferred communication method (categorical).

(c) This student lives in Ontario, is male, speaks four languages, is left-handed, is 157.5 cm tall (about 62 inches), has a wrist circumference of 147 mm (about 5.8 inches), and prefers to communicate via text messaging.

**For Practice** *Try Exercise* **3**

Most data tables follow the format shown in the example—each row is an individual, and each column is a variable. Sometimes the individuals are called *cases*.

To make life simpler, we sometimes refer to "categorical data" or "quantitative data" instead of identifying the variable as categorical or quantitative.

A variable generally takes values that vary (hence the name "variable"!). Categorical variables sometimes have similar counts in each category and sometimes don't. For instance, we might have expected similar numbers of males and females in the CensusAtSchool data set. But we aren't surprised to see that most students are right-handed. Quantitative variables may take values that are very close together or values that are quite spread out. We call the pattern of variation of a variable its **distribution**.

**DEFINITION: Distribution**

The **distribution** of a variable tells us what values the variable takes and how often it takes these values.

Section 1.1 begins by looking at how to describe the distribution of a single categorical variable and then examines relationships between categorical variables. Sections 1.2 and 1.3 and all of Chapter 2 focus on describing the distribution of a quantitative variable. Chapter 3 investigates relationships between two quantitative variables. In each case, we begin with graphical displays, then add numerical summaries for a more complete description.

**HOW TO EXPLORE DATA**

- Begin by examining each variable by itself. Then move on to study relationships among the variables.
- Start with a graph or graphs. Then add numerical summaries.

## CHECK YOUR UNDERSTANDING

Jake is a car buff who wants to find out more about the vehicles that students at his school drive. He gets permission to go to the student parking lot and record some data. Later, he does some research about each model of car on the Internet. Finally, Jake

## Teaching Tip

If your students are curious about why it is important to randomly select individuals, tell them that random selection gives each individual an equal chance to be chosen.

## Teaching Tip

The icon in the lower-right corner of this example (and many others) refers to an exercise that is very similar to the example. Students can use these exercises to assess whether they understand the material covered in the example. Also, when students encounter the referenced exercise, there will be an icon directing them back to this example and a "play" icon reminding students that a video is available for the exercise.

### TECHNOLOGY TIP

When using computer software or applets, data must be entered in the format used in these examples: individuals in the rows and variables in the columns.

## Teaching Tip

Whenever you use the word "distribution" in this chapter, it is a good idea to remind students that a distribution tells us (1) what values a variable takes and (2) how often it takes these values. This applies to distributions of both categorical and quantitative variables. Also, a "value" doesn't have to be a number. The values of the variable *color* include blue, red, and so on.

## Teaching Tip

When introducing distributions of data, it might help to "plant a seed" by telling students that there are many different types of distributions. In this chapter, we will focus on describing the characteristics of distributions of sample data, both categorical and quantitative. However, in Chapter 2, we will introduce the Normal distribution, a theoretical probability distribution that we will use to model distributions of sample data that are roughly symmetric and bell-shaped. In Chapters 5 through 7, we will learn about other probability distributions, such as the binomial distribution. In every case, a distribution identifies the values the variable can take and how often it takes these values.

## Teaching Tip

Do your best to encourage students to genuinely explore data rather than report the first thing they see. Tell them they are on a treasure hunt for interesting information.

makes a spreadsheet that includes each car's model, year, color, number of cylinders, gas mileage, weight, and whether it has a navigation system.

1. Who are the individuals in Jake's study?
2. What variables did Jake measure? Identify each as categorical or quantitative.

## From Data Analysis to Inference

Sometimes, we're interested in drawing conclusions that go beyond the data at hand. That's the idea of **inference**. In the CensusAtSchool example, 9 of the 10 randomly selected Canadian students are right-handed. That's 90% of the *sample*. Can we conclude that 90% of the *population* of Canadian students who participated in CensusAtSchool are right-handed? No.

If another random sample of 10 students was selected, the percent who are right-handed might not be exactly 90%. Can we at least say that the actual population value is "close" to 90%? That depends on what we mean by "close."

The following Activity gives you an idea of how statistical inference works.

### ACTIVITY | Hiring discrimination—it just won't fly!

**MATERIALS:**
Bag with 25 beads (15 of one color and 10 of another) or 25 identical slips of paper (15 labeled "M" and 10 labeled "F") for each student or pair of students

An airline has just finished training 25 pilots—15 male and 10 female—to become captains. Unfortunately, only eight captain positions are available right now. Airline managers announce that they will use a lottery to determine which pilots will fill the available positions. The names of all 25 pilots will be written on identical slips of paper. The slips will be placed in a hat, mixed thoroughly, and drawn out one at a time until all eight captains have been identified.

A day later, managers announce the results of the lottery. Of the 8 captains chosen, 5 are female and 3 are male. Some of the male pilots who weren't selected suspect that the lottery was not carried out fairly. One of these pilots asks your statistics class for advice about whether to file a grievance with the pilots' union.

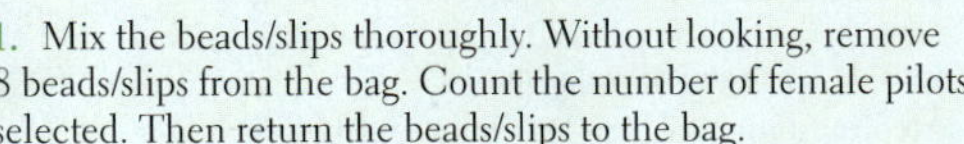

The key question in this possible discrimination case seems to be: *Is it plausible (believable) that these results happened just by chance?* To find out, you and your classmates will *simulate* the lottery process that airline managers said they used.

1. Mix the beads/slips thoroughly. Without looking, remove 8 beads/slips from the bag. Count the number of female pilots selected. Then return the beads/slips to the bag.
2. Your teacher will draw and label a number line for a class *dotplot*. On the graph, plot the number of females you got in Step 1.
3. Repeat Steps 1 and 2 if needed to get a total of at least 40 simulated lottery results for your class.
4. Discuss the results with your classmates. Does it seem believable that airline managers carried out a fair lottery? What advice would you give the male pilot who contacted you?
5. Would your advice change if the lottery had chosen 6 female (and 2 male) pilots? What about 7 female pilots? Explain.

### ✓ ANSWERS TO CYU

**1.** The cars in the student parking lot.

**2.** He measured the car's model (categorical), year (quantitative), color (categorical), number of cylinders (quantitative), gas mileage (quantitative), weight (quantitative), and whether it has a navigation system (categorical).

### Activity Overview

Please do this Activity! It is a great way to introduce students to the major themes of the course. Watch the video overview of this Activity available for teachers only in the TE-book, or on the TRCD or the book's companion Web site (teacher log-in required). Here is an example of a dotplot showing the number of females selected in 50 simulated lotteries. As you can see, selecting 5 or more female pilots is not that surprising.

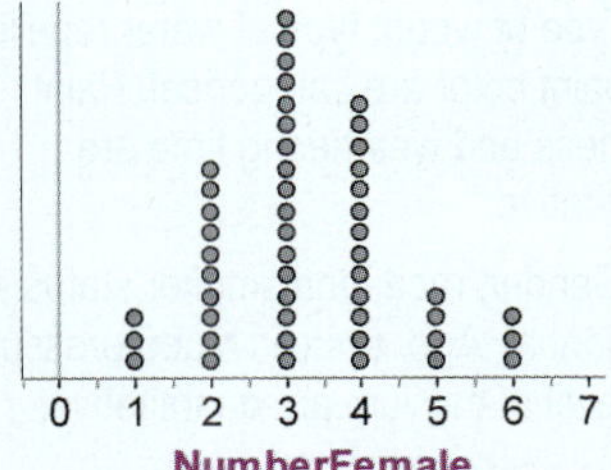

## Teaching Tip

Make sure to point out the two icons next to Exercise 3. The red square reminds students that the example on page 3 is very similar to this exercise. The "play" icon reminds students that there is a video solution available in the student e-book or at the book's Web site www.whfreeman.com/tps5e.

## Teaching Tip

The solutions provided in the *Annotated Teacher's Edition* have been shortened to fit in the margin. Complete solutions are available in the *Solutions Manual* or Teacher's e-Book (TE-book).

**TRM Full Solutions to Chapter 1 Introduction Exercises**

### Answers to Chapter 1 Introduction Exercises

**1.1** Type of wood, type of water repellent, and paint color are categorical. Paint thickness and weathering time are quantitative.

**1.2** Gender, race, and smoker status are categorical. Age, systolic blood pressure, and level of calcium are quantitative.

**1.3** **(a)** AP® Statistics students who completed a questionnaire on the first day of class. **(b)** Categorical: gender, handedness, and favorite type of music. Quantitative: height, homework time, and the total value of coins in a student's pocket. **(c)** The individual is a female who is right-handed. She is 58 inches tall, spends 60 minutes on homework, prefers Alternative music, and had 76 cents in her pocket.

Our ability to do inference is determined by how the data are produced. Chapter 4 discusses the two main methods of data production—sampling and experiments—and the types of conclusions that can be drawn from each. As the Activity illustrates, the logic of inference rests on asking, "What are the chances?" *Probability*, the study of chance behavior, is the topic of Chapters 5 through 7. We'll introduce the most common inference techniques in Chapters 8 through 12.

## Introduction Summary

- A data set contains information about a number of **individuals.** Individuals may be people, animals, or things. For each individual, the data give values for one or more **variables.** A variable describes some characteristic of an individual, such as a person's height, gender, or salary.
- Some variables are **categorical** and others are **quantitative.** A categorical variable assigns a label that places each individual into one of several groups, such as male or female. A quantitative variable has numerical values that measure some characteristic of each individual, such as height in centimeters or salary in dollars.
- The **distribution** of a variable describes what values the variable takes and how often it takes them.

## Introduction Exercises

The solutions to all exercises numbered in red are found in the Solutions Appendix, starting on page S-1.

1. **Protecting wood** How can we help wood surfaces resist weathering, especially when restoring historic wooden buildings? In a study of this question, researchers prepared wooden panels and then exposed them to the weather. Here are some of the variables recorded: type of wood (yellow poplar, pine, cedar); type of water repellent (solvent-based, water-based); paint thickness (millimeters); paint color (white, gray, light blue); weathering time (months). Identify each variable as categorical or quantitative.
2. **Medical study variables** Data from a medical study contain values of many variables for each of the people who were the subjects of the study. Here are some of the variables recorded: gender (female or male); age (years); race (Asian, black, white, or other); smoker (yes or no); systolic blood pressure (millimeters of mercury); level of calcium in the blood (micrograms per milliliter). Identify each as categorical or quantitative.
3. **A class survey** Here is a small part of the data set that describes the students in an AP® Statistics class. The data come from anonymous responses to a questionnaire filled out on the first day of class.

pg 3 

| Gender | Hand | Height (in.) | Homework time (min) | Favorite music | Pocket change (cents) |
|---|---|---|---|---|---|
| F | L | 65 | 200 | Hip-hop | 50 |
| M | L | 72 | 30 | Country | 35 |
| M | R | 62 | 95 | Rock | 35 |
| F | L | 64 | 120 | Alternative | 0 |
| M | R | 63 | 220 | Hip-hop | 0 |
| F | R | 58 | 60 | Alternative | 76 |
| F | R | 67 | 150 | Rock | 215 |

(a) What individuals does this data set describe?

(b) What variables were measured? Identify each as categorical or quantitative.

(c) Describe the individual in the highlighted row.

4. **Coaster craze** Many people like to ride roller coasters. Amusement parks try to increase attendance by building exciting new coasters. The following table displays data on several roller coasters that were opened in a recent year.[2]

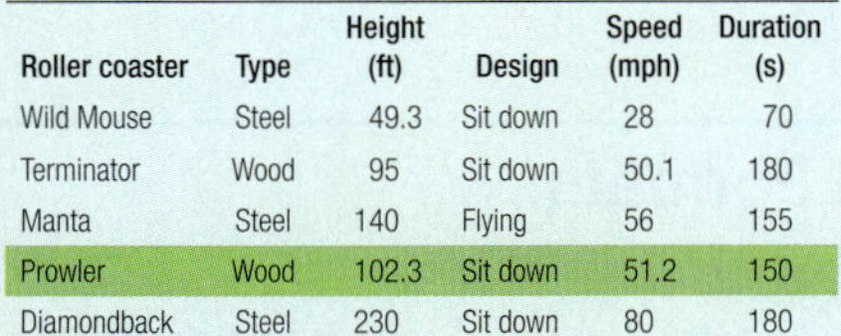

| Roller coaster | Type | Height (ft) | Design | Speed (mph) | Duration (s) |
|---|---|---|---|---|---|
| Wild Mouse | Steel | 49.3 | Sit down | 28 | 70 |
| Terminator | Wood | 95 | Sit down | 50.1 | 180 |
| Manta | Steel | 140 | Flying | 56 | 155 |
| Prowler | Wood | 102.3 | Sit down | 51.2 | 150 |
| Diamondback | Steel | 230 | Sit down | 80 | 180 |

(a) What individuals does this data set describe?

(b) What variables were measured? Identify each as categorical or quantitative.

(c) Describe the individual in the highlighted row.

5. **Ranking colleges** Popular magazines rank colleges and universities on their "academic quality" in serving undergraduate students. Describe two categorical variables and two quantitative variables that you might record for each institution.

6. **Students and TV** You are preparing to study the television-viewing habits of high school students. Describe two categorical variables and two quantitative variables that you might record for each student.

***Multiple choice: Select the best answer.***

*Exercises 7 and 8 refer to the following setting.* At the Census Bureau Web site www.census.gov, you can view detailed data collected by the American Community Survey. The following table includes data for 10 people chosen at random from the more than 1 million people in households contacted by the survey. "School" gives the highest level of education completed.

| Weight (lb) | Age (yr) | Travel to work (min) | School | Gender | Income last year ($) |
|---|---|---|---|---|---|
| 187 | 66 | 0 | Ninth grade | 1 | 24,000 |
| 158 | 66 | n/a | High school grad | 2 | 0 |
| 176 | 54 | 10 | Assoc. degree | 2 | 11,900 |
| 339 | 37 | 10 | Assoc. degree | 1 | 6000 |
| 91 | 27 | 10 | Some college | 2 | 30,000 |
| 155 | 18 | n/a | High school grad | 2 | 0 |
| 213 | 38 | 15 | Master's degree | 2 | 125,000 |
| 194 | 40 | 0 | High school grad | 1 | 800 |
| 221 | 18 | 20 | High school grad | 1 | 2500 |
| 193 | 11 | n/a | Fifth grade | 1 | 0 |

7. The individuals in this data set are

(a) households.
(b) people.
(c) adults.
(d) 120 variables.
(e) columns.

8. This data set contains

(a) 7 variables, 2 of which are categorical.
(b) 7 variables, 1 of which is categorical.
(c) 6 variables, 2 of which are categorical.
(d) 6 variables, 1 of which is categorical.
(e) None of these.

# 1.1 Analyzing Categorical Data

**WHAT YOU WILL LEARN** **By the end of the section, you should be able to:**

- Display categorical data with a bar graph. Decide if it would be appropriate to make a pie chart.
- Identify what makes some graphs of categorical data deceptive.
- Calculate and display the marginal distribution of a categorical variable from a two-way table.
- Calculate and display the conditional distribution of a categorical variable for a particular value of the other categorical variable in a two-way table.
- Describe the association between two categorical variables by comparing appropriate conditional distributions.

The values of a categorical variable are labels for the categories, such as "male" and "female." The distribution of a categorical variable lists the categories and gives either the *count* or the *percent* of individuals who fall within each category. Here's an example.

**1.4 (a)** Roller coasters that were opened in a recent year. **(b)** Categorical: type and design. Quantitative: height, speed, and duration. **(c)** The roller coaster is the Prowler, a wood, sit-down-type coaster. Its height is 102.3 feet, its speed is 51.2 mph, and the duration of the ride is 150 seconds.

**1.5** Student answers will vary. For example, quantitative variables could be graduation rate and student-faculty ratio and categorical variables could be region of the country and type of institution (2-year college, 4-year college, university).

**1.6** Student answers will vary. For example, categorical variables could be whether or not a student watches reality shows and how the student receives the television programming (antenna, cable, satellite, Internet), and quantitative variables could be the amount of time spent watching TV per week and the number of different channels viewed during a week.

**1.7** b

**1.8** c

### PD Section 1.1 Overview

There is a video overview of Section 1.1 (for teachers) available in the TE-book, or on the TRCD or the book's companion Web site (teacher log-in required).

### TRM Section 1.1 Alternate Examples

You can find the Alternate Examples for this section in Microsoft Word format by clicking on the link in the TE-book, opening the TRCD, or logging onto the book's companion Web site www.whfreeman.com/tps5e (teacher log-in required).

## Teaching Tip

The tables in this example are not data tables—they are summaries of a distribution of categorical variable. In other words, the rows do not contain individuals but rather the values of the variable, station format, and how often it takes each value. To emphasize this point, ask your students what the data table would look like. Here is one possibility. The completed table would have 13,828 rows!

| Station | Type of station |
|---|---|
| KRQQ | Contemporary hit |
| KFFN | News/talk/info |
| WFRM | Country |
| . . . | . . . |

EXAMPLE

## Radio Station Formats

*Distribution of a categorical variable*

The radio audience rating service Arbitron places U.S radio stations into categories that describe the kinds of programs they broadcast. Here are two different tables showing the distribution of station formats in a recent year:[3]

**Frequency table**

| Format | Count of stations |
|---|---|
| Adult contemporary | 1556 |
| Adult standards | 1196 |
| Contemporary hit | 569 |
| Country | 2066 |
| News/Talk/Information | 2179 |
| Oldies | 1060 |
| Religious | 2014 |
| Rock | 869 |
| Spanish language | 750 |
| Other formats | 1579 |
| **Total** | **13,838** |

**Relative frequency table**

| Format | Percent of stations |
|---|---|
| Adult contemporary | 11.2 |
| Adult standards | 8.6 |
| Contemporary hit | 4.1 |
| Country | 14.9 |
| News/Talk/Information | 15.7 |
| Oldies | 7.7 |
| Religious | 14.6 |
| Rock | 6.3 |
| Spanish language | 5.4 |
| Other formats | 11.4 |
| **Total** | **99.9** |

In this case, the *individuals* are the radio stations and the *variable* being measured is the kind of programming that each station broadcasts. The table on the left, which we call a **frequency table**, displays the counts (*frequencies*) of stations in each format category. On the right, we see a **relative frequency table** of the data that shows the percents (*relative frequencies*) of stations in each format category.

It's a good idea to check data for consistency. The counts should add to 13,838, the total number of stations. They do. The percents should add to 100%. In fact, they add to 99.9%. What happened? Each percent is rounded to the nearest tenth. The exact percents would add to 100, but the rounded percents only come close. This is **roundoff error.** Roundoff errors don't point to mistakes in our work, just to the effect of rounding off results.

## Bar Graphs and Pie Charts

Columns of numbers take time to read. You can use a **pie chart** or a **bar graph** to display the distribution of a categorical variable more vividly. Figure 1.1 illustrates both displays for the distribution of radio stations by format.

Pie charts show the distribution of a categorical variable as a "pie" whose slices are sized by the counts or percents for the categories. A pie chart must include all the categories that make up a whole. In the radio station example, we needed the "Other formats" category to complete the whole (all radio stations) and allow us to make a pie chart. Use a pie chart only when you want to emphasize each

## Teaching Tip

Pie charts are not on the AP® Topic Outline, but we include them here because they are so common in the media. Statisticians prefer bar graphs to pie charts because they are easier to read and compare.

ALTERNATE EXAMPLE

### *College majors*

Here is the distribution of bachelor's degrees awarded in 2010, according to http://nces.ed.gov/programs/digest/d11/tables/dt11_286.asp:

| Major | Number of degrees | Percent of degrees |
|---|---|---|
| Business | 358,293 | 21.7 |
| Social sciences/history | 172,780 | 10.5 |
| Health professions | 129,634 | 7.9 |
| Education | 101,265 | 6.1 |
| Psychology | 97,216 | 5.9 |
| Visual and performing arts | 91,802 | 5.6 |
| Biological and biomedical sciences | 86,400 | 5.2 |
| Communication and related programs | 81,266 | 4.9 |
| Engineering | 72,654 | 4.4 |
| English language and literature | 53,231 | 3.2 |
| Other | 405,473 | 24.6 |
| **Total** | **1,650,014** | **100.0** |

In this case, the individuals are the degrees, and the variable being measured is the field in which the degree was granted. The middle column displays the counts (frequencies) of degrees in each major and the column on the right shows the percents (relative frequencies) of degrees in each major.

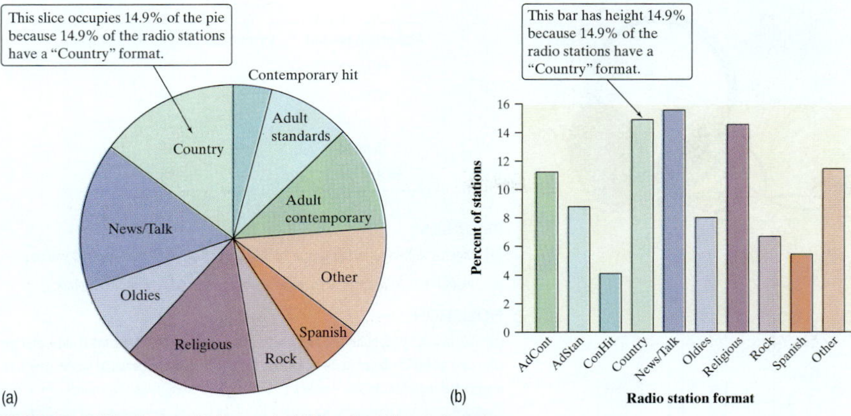


**FIGURE 1.1** (a) Pie chart and (b) bar graph of U.S. radio stations by format.

category's relation to the whole. Pie charts are awkward to make by hand, but technology will do the job for you.

Bar graphs are also called *bar charts.*

Bar graphs represent each category as a bar. The bar heights show the category counts or percents. Bar graphs are easier to make than pie charts and are also easier to read. To convince yourself, try to use the pie chart in Figure 1.1 to estimate the percent of radio stations that have an "Oldies" format. Now look at the bar graph—it's easy to see that the answer is about 8%.

Bar graphs are also more flexible than pie charts. Both graphs can display the distribution of a categorical variable, but a bar graph can also compare any set of quantities that are measured in the same units.

## EXAMPLE Who Owns an MP3 Player?

### *Choosing the best graph to display the data*

Portable MP3 music players, such as the Apple iPod, are popular—but not equally popular with people of all ages. Here are the percents of people in various age groups who own a portable MP3 player, according to an Arbitron survey of 1112 randomly selected people.[4]

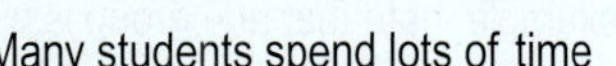

### AP® Exam Common Error

Many students spend lots of time constructing graphs only to forget the labels. This will always cause a deduction in the student's score. It is very important for students to make sure their graph diplays the data with the proper labels and scaling.

### ALTERNATE EXAMPLE

### *What do teens post?*

Here are the percentages of 12- to 17-year-olds who post various types of personal information on their social media profiles, according to the Pew Internet Parent/Teen Privacy Survey in 2012:

| Type of information | Percent who post |
|---|---|
| Photo of themselves | 91 |
| School name | 71 |
| City or town where they live | 71 |
| Email address | 53 |
| Cell phone number | 20 |

**PROBLEM:**

**(a)** Make a well-labeled bar graph to display the data. Describe what you see.
**(b)** Would it be appropriate to make a pie chart for these data? Explain.

**SOLUTION:**

**(a)**

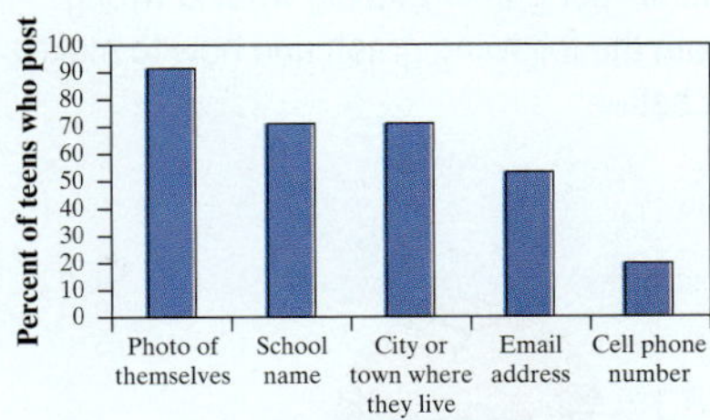


**(b)** A pie chart would not be appropriate for these data because each percent in the table refers to a different type of information, not to parts of a single whole.

## Teaching Tip

In this example, note that age group is a categorical variable, even though age is often used as a quantitative variable.

| Age group (years) | Percent owning an MP3 player |
|---|---|
| 12 to 17 | 54 |
| 18 to 24 | 30 |
| 25 to 34 | 30 |
| 35 to 54 | 13 |
| 55 and older | 5 |

**PROBLEM:**

(a) Make a well-labeled bar graph to display the data. Describe what you see.

(b) Would it be appropriate to make a pie chart for these data? Explain.

**SOLUTION:**

(a) We start by labeling the axes: age group goes on the horizontal axis, and percent who own an MP3 player goes on the vertical axis. For the vertical scale, which is measured in percents, we'll start at 0 and go up to 60, with tick marks for every 10. Then for each age category, we draw a bar with height corresponding to the percent of survey respondents who said they have an MP3 player. Figure 1.2 shows the completed bar graph. It appears that MP3 players are more popular among young people and that their popularity generally decreases as the age category increases.

(b) Making a pie chart to display these data is not appropriate because each percent in the table refers to a different age group, not to parts of a single whole.

**FIGURE 1.2** Bar graph comparing the percents of several age groups who own portable MP3 players.

**For Practice** *Try Exercise*  15

## Graphs: Good and Bad

Bar graphs compare several quantities by comparing the heights of bars that represent the quantities. Our eyes, however, react to the *area* of the bars as well as to their height. When all bars have the same width, the area (width × height) varies in proportion to the height, and our eyes receive the right impression. When you draw a bar graph, make the bars equally wide.

Artistically speaking, bar graphs are a bit dull. It is tempting to replace the bars with pictures for greater eye appeal. Don't do it! The following example shows why.

### WEB SITE

There are many good Web sites that illustrate deceptive graphs. For example, try http://www.datavis.ca/gallery/index.php or do an Internet search for "deceptive graphs." The classic book *How to Lie with Statistics* by Darrell Huff also includes some good examples.

**EXAMPLE**

## Who Buys iMacs?

### *Beware the pictograph!*

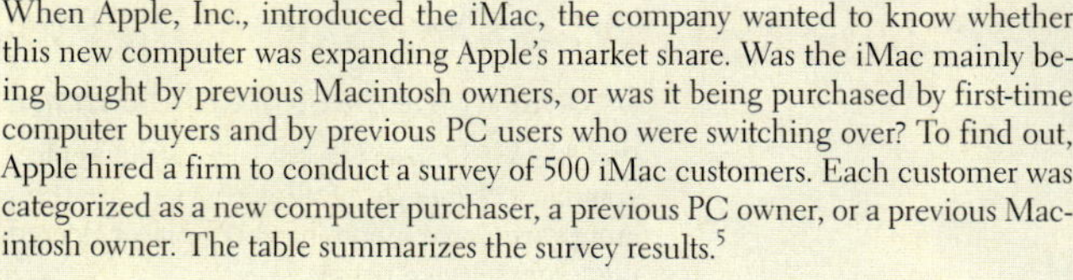

When Apple, Inc., introduced the iMac, the company wanted to know whether this new computer was expanding Apple's market share. Was the iMac mainly being bought by previous Macintosh owners, or was it being purchased by first-time computer buyers and by previous PC users who were switching over? To find out, Apple hired a firm to conduct a survey of 500 iMac customers. Each customer was categorized as a new computer purchaser, a previous PC owner, or a previous Macintosh owner. The table summarizes the survey results.[5]

**ALTERNATE EXAMPLE**

### *A misleading graph?*

**PROBLEM:** In the 2008–2009 regular season, basketball player Dwayne Wade averaged 7.5 assists per game and 2.2 steals per game. Explain what is wrong with the following graph and how to make it better.

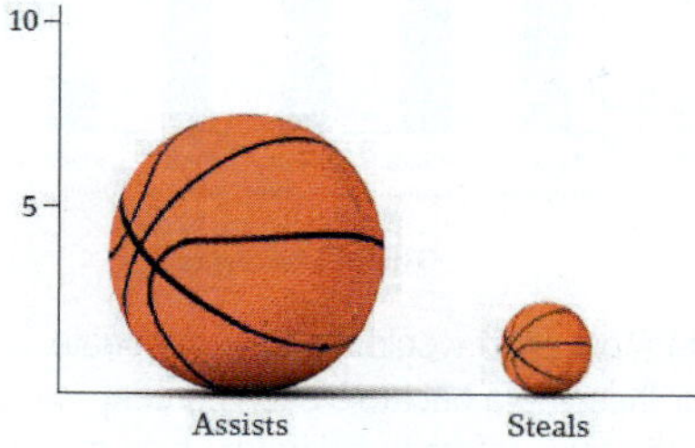


**SOLUTION:** Although the heights of the basketballs are correct, our eyes respond to the areas of the basketballs. Wade averaged about 3 times as many assists as steals, but the area of the basketball representing assists is over 9 times as big. It would be better to use equally wide bars for assists and steals so that only the height of the bars determines the area.

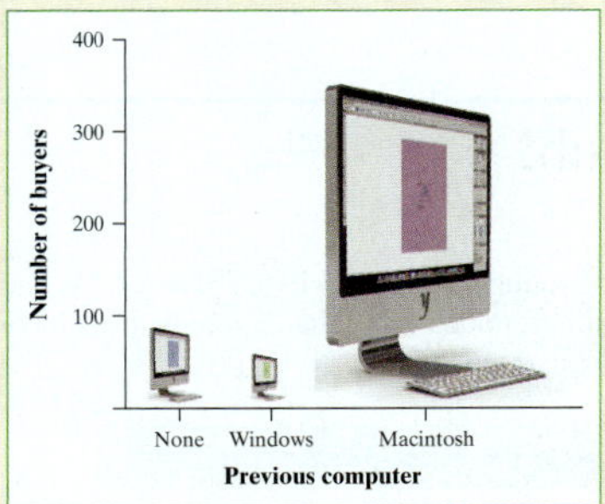


| Previous ownership | Count | Percent (%) |
|---|---|---|
| None | 85 | 17.0 |
| PC | 60 | 12.0 |
| Macintosh | 355 | 71.0 |
| **Total** | **500** | **100.0** |

PROBLEM:

(a) Here's a clever graph of the data that uses pictures instead of the more traditional bars. How is this graph misleading?

(b) Two possible bar graphs of the data are shown below. Which one could be considered deceptive? Why?

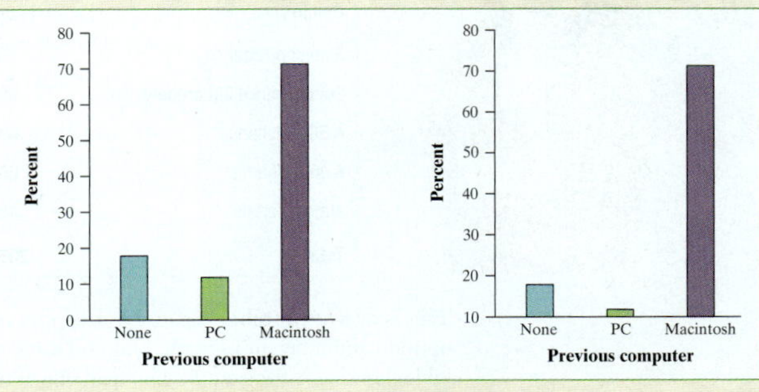


SOLUTION:

(a) Although the heights of the pictures are accurate, our eyes respond to the area of the pictures. The pictograph makes it seem like the percent of iMac buyers who are former Mac owners is at least ten times higher than either of the other two categories, which isn't the case.

(b) The bar graph on the right is misleading. By starting the vertical scale at 10 instead of 0, it looks like the percent of iMac buyers who previously owned a PC is less than half the percent who are first-time computer buyers. We get a distorted impression of the relative percents in the three categories.

**For Practice** *Try Exercise* 17

There are two important lessons to be learned from this example: *(1) beware the pictograph, and (2) watch those scales.*

## Two-Way Tables and Marginal Distributions

We have learned some techniques for analyzing the distribution of a single categorical variable. What do we do when a data set involves two categorical variables? We begin by examining the counts or percents in various categories for one of the variables. Here's an example to show what we mean.

### Teaching Tip

We will encounter two-way tables again in Chapter 5 (probability) and Chapter 11 (chi-square tests).

## EXAMPLE I'm Gonna Be Rich!

### *Two-way tables*

A survey of 4826 randomly selected young adults (aged 19 to 25) asked, "What do you think the chances are you will have much more than a middle-class income at age 30?" The table below shows the responses.[6]

**Young adults by gender and chance of getting rich**

| | Gender | | |
|---|---|---|---|
| Opinion | Female | Male | Total |
| Almost no chance | 96 | 98 | **194** |
| Some chance but probably not | 426 | 286 | **712** |
| A 50-50 chance | 696 | 720 | **1416** |
| A good chance | 663 | 758 | **1421** |
| Almost certain | 486 | 597 | **1083** |
| **Total** | **2367** | **2459** | **4826** |

This is a **two-way table** because it describes two categorical variables, gender and opinion about becoming rich. Opinion is the *row variable* because each row in the table describes young adults who held one of the five opinions about their chances. Because the opinions have a natural order from "Almost no chance" to "Almost certain," the rows are also in this order. Gender is the *column variable*. The entries in the table are the counts of individuals in each opinion-by-gender class.

How can we best grasp the information contained in the two-way table above? First, *look at the distribution of each variable separately*. The distribution of a categorical variable says how often each outcome occurred. The "Total" column at the right of the table contains the totals for each of the rows. These row totals give the distribution of opinions about becoming rich in the entire group of 4826 young adults: 194 thought that they had almost no chance, 712 thought they had just some chance, and so on. (If the row and column totals are missing, the first thing to do in studying a two-way table is to calculate them.) The distributions of opinion alone and gender alone are called **marginal distributions** because they appear at the right and bottom margins of the two-way table.

**DEFINITION: Marginal distribution**

The **marginal distribution** of one of the categorical variables in a two-way table of counts is the distribution of values of that variable among all individuals described by the table.

## Teaching Tip

In *TPS* 5e, we follow a convention for two-way tables that places the values of the explanatory variable in the columns and the values of the response variable in the rows. (A response variable measures an outcome of a study, and an explanatory variable may help explain or predict changes in the response variable.) This matches how we will construct scatterplots in Chapter 3, plotting the explanatory variable on the horizontal axis and the response variable on the vertical axis. Not everyone follows this convention, however, so it isn't wrong to place the values of the explanatory variable in the rows and the values of the response variable in the columns.

## ALTERNATE EXAMPLE

### *Cell phones*

The Pew Research Center asked a random sample of 2024 adult cell phone owners from the United States which type of cell phone they own: iPhone, Android, or other (including non-smart phones). Here are the results, broken down by age category http://pewinternet.org/Reports/2013/Smartphone-Ownership-2013.aspx:

| | 18–34 | 35–54 | 55+ | Total |
|---|---|---|---|---|
| iPhone | 169 | 171 | 127 | **467** |
| Android | 214 | 189 | 100 | **503** |
| Other | 134 | 277 | 643 | **1054** |
| **Total** | **517** | **637** | **870** | **2024** |

Percents are often more informative than counts, especially when we are comparing groups of different sizes. We can display the marginal distribution of opinions in percents by dividing each row total by the table total and converting to a percent. For instance, the percent of these young adults who think they are almost certain to be rich by age 30 is

$$\frac{\text{almost certain total}}{\text{table total}} = \frac{1083}{4826} = 0.224 = 22.4\%$$

## EXAMPLE I'm Gonna Be Rich!

### *Examining a marginal distribution*

PROBLEM:

(a) Use the data in the two-way table to calculate the marginal distribution (in percents) of opinions.

(b) Make a graph to display the marginal distribution. Describe what you see.

SOLUTION:

(a) We can do four more calculations like the one shown above to obtain the marginal distribution of opinions in percents. Here is the complete distribution.

| Response | Percent |
|---|---|
| Almost no chance | $\frac{194}{4826} = 4.0\%$ |
| Some chance | $\frac{712}{4826} = 14.8\%$ |
| A 50–50 chance | $\frac{1416}{4826} = 29.3\%$ |
| A good chance | $\frac{1421}{4826} = 29.4\%$ |
| Almost certain | $\frac{1083}{4826} = 22.4\%$ |

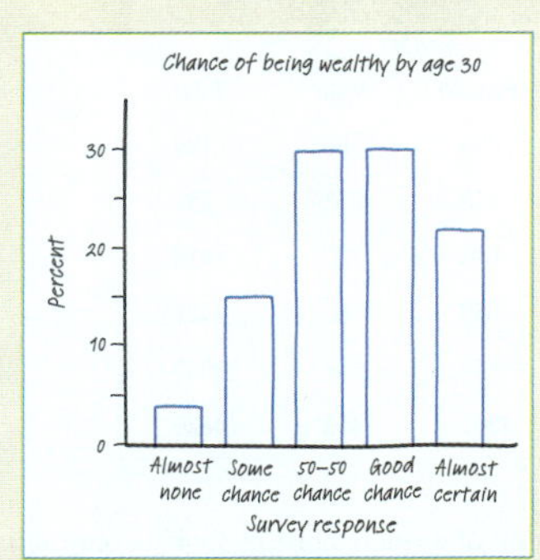


**FIGURE 1.3** Bar graph showing the marginal distribution of opinion about chance of being rich by age 30.

(b) Figure 1.3 is a bar graph of the distribution of opinion among these young adults. It seems that many young adults are optimistic about their future income. Over 50% of those who responded to the survey felt that they had "a good chance" or were "almost certain" to be rich by age 30.

**For Practice** *Try Exercise*  19

Each marginal distribution from a two-way table is a distribution for a single categorical variable. As we saw earlier, we can use a bar graph or a pie chart to display such a distribution.

**ALTERNATE EXAMPLE**

### *Cell phones*

PROBLEM:

**(a)** Use the cell phone data to calculate the marginal distribution (in percents) of type of cell phone.

**(b)** Make a graph to display the marginal distribution. Describe what you see.

SOLUTION:

**(a)** iPhone: 467/2024 = 23.1%, Android: 503/2024 = 24.9%, Other: 1054/2024 = 52.1%. *Note:* The percents do not add to 100% because of round-off error.

**(b)** A bar graph is shown below. The percent who own iPhones and Androids are about the same, but more than half of cell phone owners do not own either an iPhone or an Android.

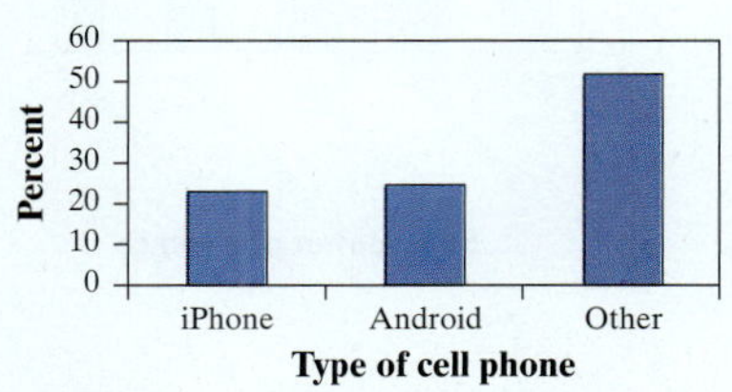

## CHECK YOUR UNDERSTANDING

A random sample of 415 children aged 9 to 17 from the United Kingdom and the United States who completed a CensusAtSchool survey in a recent year was selected. Each student's country of origin was recorded along with which superpower they would most like to have: the ability to fly, ability to freeze time, invisibility, superstrength, or telepathy (ability to read minds). The data are summarized in the table.[7]

| | Country | |
|---|---|---|
| Superpower | U.K. | U.S. |
| Fly | 54 | 45 |
| Freeze time | 52 | 44 |
| Invisibility | 30 | 37 |
| Superstrength | 20 | 23 |
| Telepathy | 44 | 66 |

1. Use the two-way table to calculate the marginal distribution (in percents) of superpower preferences.
2. Make a graph to display the marginal distribution. Describe what you see.

### ✓ ANSWERS TO CYU

**1.** Fly: 99/415 = 23.9%, Freeze time: 96/415 = 23.1%, Invisibility: 67/415 = 16.1%, Superstrength: 43/415 = 10.4%, Telepathy: 110/415 = 26.5%.

**2.** A bar graph is shown below. It appears that telepathy, ability to fly, and ability to freeze time were the most popular choices, with about 25% of students choosing each one. Invisibility was the 4th most popular and superstrength was the least popular.

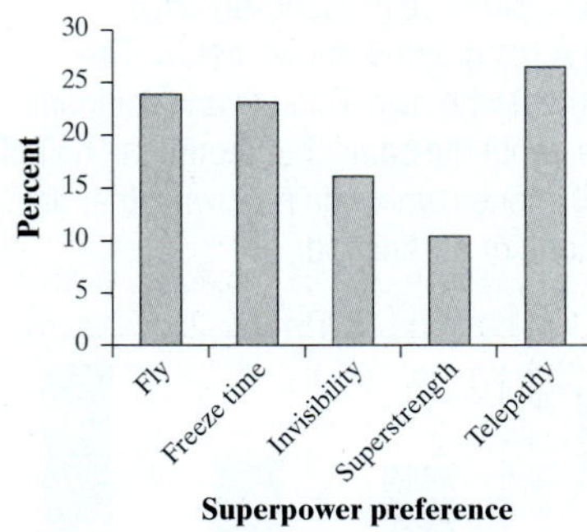


## Relationships between Categorical Variables: Conditional Distributions

The two-way table contains much more information than the two marginal distributions of opinion alone and gender alone. *Marginal distributions tell us nothing about the relationship between two variables.* To describe a relationship between two categorical variables, we must calculate some well-chosen percents from the counts given in the body of the table.

**Young adults by gender and chance of getting rich**

| | Gender | | |
|---|---|---|---|
| Opinion | Female | Male | Total |
| Almost no chance | 96 | 98 | **194** |
| Some chance but probably not | 426 | 286 | **712** |
| A 50-50 chance | 696 | 720 | **1416** |
| A good chance | 663 | 758 | **1421** |
| Almost certain | 486 | 597 | **1083** |
| **Total** | **2367** | **2459** | **4826** |

### Teaching Tip

If you are displaying the table with a projector or on the board, physically cover up the male column to emphasize that you are concerned with only the females for now.

**Conditional distribution of opinion among women**

| Response | Percent |
|---|---|
| Almost no chance | $\frac{96}{2367} = 4.1\%$ |
| Some chance | $\frac{426}{2367} = 18.0\%$ |
| A 50-50 chance | $\frac{696}{2367} = 29.4\%$ |
| A good chance | $\frac{663}{2367} = 28.0\%$ |
| Almost certain | $\frac{486}{2367} = 20.5\%$ |

We can study the opinions of women alone by looking only at the "Female" column in the two-way table. To find the percent of *young women* who think they are almost certain to be rich by age 30, divide the count of such women by the total number of women, the column total:

$$\frac{\text{women who are almost certain}}{\text{column total}} = \frac{486}{2367} = 0.205 = 20.5\%$$

Doing this for all five entries in the "Female" column gives the **conditional distribution** of opinion among women. See the table in the margin. We use the term "conditional" because this distribution describes only young adults who satisfy the condition that they are female.

**DEFINITION: Conditional distribution**

A **conditional distribution** of a variable describes the values of that variable among individuals who have a specific value of another variable. There is a separate conditional distribution for each value of the other variable.

Now let's examine the men's opinions.

## EXAMPLE I'm Gonna Be Rich!

### *Calculating a conditional distribution*

PROBLEM: Calculate the conditional distribution of opinion among the young men.

SOLUTION: To find the percent of *young men* who think they are almost certain to be rich by age 30, divide the count of such men by the total number of men, the column total:

$$\frac{\text{men who are almost certain}}{\text{column total}} = \frac{597}{2459} = 24.3\%$$

If we do this for all five entries in the "Male" column, we get the conditional distribution shown in the table.

**Conditional distribution of opinion among men**

| Response | Percent |
|---|---|
| Almost no chance | $\frac{98}{2459} = 4.0\%$ |
| Some chance | $\frac{286}{2459} = 11.6\%$ |
| A 50-50 chance | $\frac{720}{2459} = 29.3\%$ |
| A good chance | $\frac{758}{2459} = 30.8\%$ |
| Almost certain | $\frac{597}{2459} = 24.3\%$ |

**For Practice** *Try Exercise* 21

There are *two sets* of conditional distributions for any two-way table: one for the column variable and one for the row variable. So far, we have looked at the conditional distributions of opinion for the two genders. We could also examine the five conditional distributions of gender, one for each of the five opinions, by looking separately at the rows in the original two-way table. For instance, the conditional distribution of gender among those who responded "Almost certain" is

| Female | Male |
|---|---|
| $\frac{486}{1083} = 44.9\%$ | $\frac{597}{1083} = 55.1\%$ |

**ALTERNATE EXAMPLE**

### *Cell phones*

PROBLEM: Use the cell phone data to calculate the conditional distribution of cell phone type for the 18- to 34-year-olds.

SOLUTION: There were 517 people in the sample who were between 18 and 34 years old. Of these 517, 169/517 = 32.7% own an iPhone, 214/517 = 41.4% own an Android, and 134/517 = 25.9% own another type of cell phone.

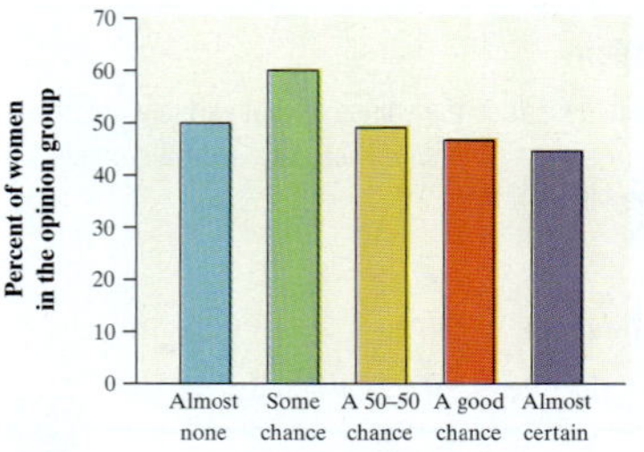


**FIGURE 1.4** Bar graph comparing the percents of females among those who hold each opinion about their chance of being rich by age 30.

That is, of the young adults who said they were almost certain to be rich by age 30, 44.9% were female and 55.1% were male.

Because the variable "gender" has only two categories, comparing the five conditional distributions amounts to comparing the percents of women among young adults who hold each opinion. Figure 1.4 makes this comparison in a bar graph. The bar heights do *not* add to 100%, because each bar represents a different group of people.

**Which conditional distributions should we compare?** Our goal all along has been to analyze the relationship between gender and opinion about chances of becoming rich for these young adults. We started by examining the conditional distributions of opinion for males and females. Then we looked at the conditional distributions of gender for each of the five opinion categories. Which of these two gives us the information we want? Here's a hint: think about whether changes in one variable might help explain changes in the other. In this case, it seems reasonable to think that gender might influence young adults' opinions about their chances of getting rich. To see whether the data support this idea, we should compare the conditional distributions of opinion for women and men.

Software will calculate conditional distributions for you. Most programs allow you to choose which conditional distributions you want to compute.

## 1. TECHNOLOGY CORNER ANALYZING TWO-WAY TABLES

Figure 1.5 presents the two conditional distributions of opinion, for women and for men, and also the marginal distribution of opinion for all of the young adults. The distributions agree (up to rounding) with the results in the last two examples.

Session

| | Female | Male | All |
|---|---|---|---|
| A: Almost no chance | 96 | 98 | 194 |
| | 4.06 | 3.99 | 4.02 |
| B: Some chance but probably not | 426 | 286 | 712 |
| | 18.00 | 11.63 | 14.75 |
| C: A 50-50 chance | 696 | 720 | 1416 |
| | 29.40 | 29.28 | 29.34 |
| D: A good chance | 663 | 758 | 1421 |
| | 28.01 | 30.83 | 29.44 |
| E: Almost certain | 486 | 597 | 1083 |
| | 20.53 | 24.28 | 22.44 |
| All | 2367 | 2459 | 4826 |
| | 100.00 | 100.00 | 100.00 |

Cell Contents: Count
% of Column

**FIGURE 1.5** Minitab output for the two-way table of young adults by gender and chance of being rich, along with each entry as a percent of its column total. The "Female" and "Male" columns give the conditional distributions of opinion for women and men, and the "All" column shows the marginal distribution of opinion for all these young adults.

### Teaching Tip

When each value of a categorical variable has only two outcomes (e.g., success/failure), it is sufficient to show bars for just the percentage of successes (or failures) for each value of the categorical variable. The figure below shows segmented bar charts for the same data. No additional information is gained by including the segments for the males.

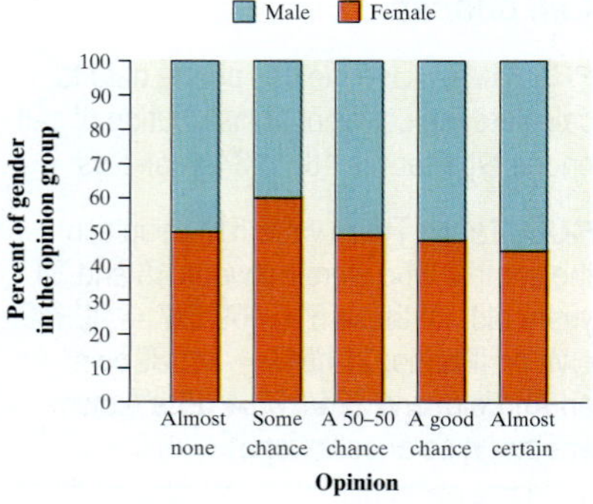


### Teaching Tip

Although we haven't introduced the ideas of an explanatory variable and a response variable in a formal way, it can be helpful to think about the role each variable plays when deciding which conditional distributions to compare. For example, in the cell phone ownership example, it seems like age group would help explain or predict which type of cell phone a person owns.

### TECHNOLOGY TIP

Remind students that they can watch a video presentation of this technology corner on the e-book or at the Web site www.whfreeman.com/tps5e.

### TECHNOLOGY TIP

Make sure to point out the "key" at the bottom of the computer output. Almost all computer output showing two-way tables will have a key of some kind to explain what each value in the table represents.

## Putting It All Together: Relationships Between Categorical Variables

Now it's time to complete our analysis of the relationship between gender and opinion about chances of becoming rich later in life.

### EXAMPLE Women's and Men's Opinions

*Conditional distributions and relationships*

PROBLEM: *Based on the survey data, can we conclude that young men and women differ in their opinions about the likelihood of future wealth? Give appropriate evidence to support your answer.*

SOLUTION: *We suspect that gender might influence a young adult's opinion about the chance of getting rich. So we'll compare the conditional distributions of response for men alone and for women alone.*

| Response | Percent of Females | Percent of Males |
|---|---|---|
| Almost no chance | $\frac{96}{2367} = 4.1\%$ | $\frac{98}{2459} = 4.0\%$ |
| Some chance | $\frac{426}{2367} = 18.0\%$ | $\frac{286}{2459} = 11.6\%$ |
| A 50-50 chance | $\frac{696}{2367} = 29.4\%$ | $\frac{720}{2459} = 29.3\%$ |
| A good chance | $\frac{663}{2367} = 28.0\%$ | $\frac{758}{2459} = 30.8\%$ |
| Almost certain | $\frac{486}{2367} = 20.5\%$ | $\frac{597}{2459} = 24.3\%$ |

*We'll make a side-by-side bar graph to compare the opinions of males and females. Figure 1.6 displays the completed graph.*

*Based on the sample data, men seem somewhat more optimistic about their future income than women. Men were less likely to say that they have "some chance but probably not" than women (11.6% vs. 18.0%). Men were more likely to say that they have "a good chance" (30.8% vs. 28.0%) or are "almost certain" (24.3% vs. 20.5%) to have much more than a middle-class income by age 30 than women were.*

Chance of being wealthy by age 30

Female
Male

Percent

Almost none, Some chance, 50-50 chance, Good chance, Almost certain

Opinion

**FIGURE 1.6** Side-by-side bar graph comparing the opinions of males and females.

**For Practice** *Try Exercise* 

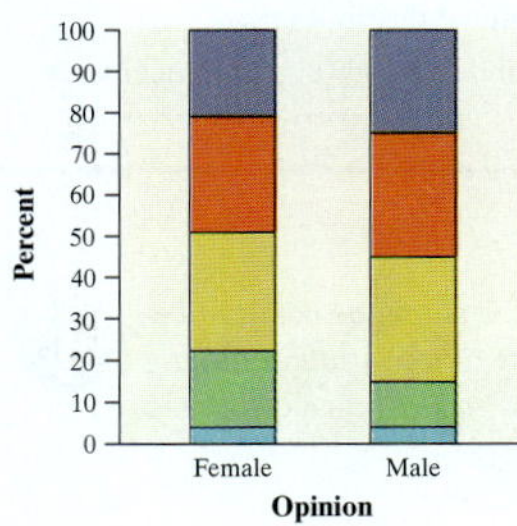


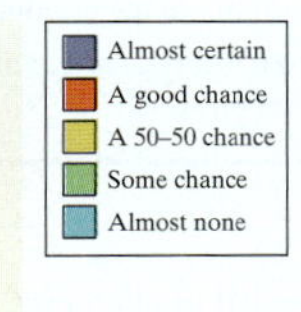


**FIGURE 1.7** Segmented bar graph comparing the opinions of males and females.

We could have used a **segmented bar graph** to compare the distributions of male and female responses in the previous example. Figure 1.7 shows the completed graph. Each bar has five segments—one for each of the opinion categories. It's fairly difficult to compare the percents of males and females in each category because the "middle" segments in the two bars start at different locations on the vertical axis. The side-by-side bar graph in Figure 1.6 makes comparison easier.

### ALTERNATE EXAMPLE

*Cell phones*

PROBLEM: Based on the cell phone data, is there evidence of a difference in the distribution of cell phone ownership for the three age groups? Justify your answer.

SOLUTION: To investigate, we will compare the conditional distributions of cell phone ownership for each of the three age groups.

- From earlier, in the 18–34 category: 32.7% own an iPhone, 41.4% own an Android, and 25.9% own another type of cell phone.
- For the 35–54 category: 171/637 = 26.8% own an iPhone, 189/637 = 29.7% own an Android, and 277/637 = 43.5% own another type of cell phone.
- For the 55+ category: 127/870 = 14.6% own an iPhone, 100/870 = 11.5% own an Android, and 643/870 = 73.9% own another type of cell phone.

Here is a side-by-side bar graph of these distributions:

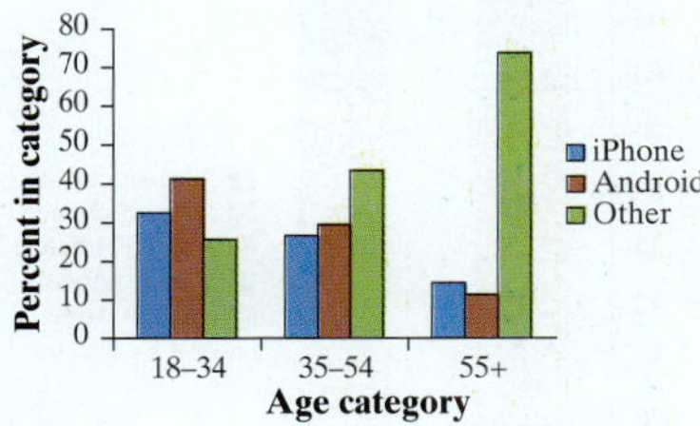


Based on the sample data, more than half of the 18- to 34-year-olds and 35- to 54-year-olds own an iPhone or an Android, while much less than half of the 55-and-older group own one of these types. Furthermore, younger people tend to prefer the Android to the iPhone, while older people show a slight preference for the iPhone over the Android.

### Teaching Tip

Segmented bar graphs are not specifically listed on the AP® Topic Outline, but they have shown up on the AP® exam (2011 #2) and are popular in the media. They also closely match the structure of the two-way table, with the values of the explanatory variable represented by the vertical bars and the values of the response variable as horizontal segments within each bar.

Both graphs provide evidence of an **association** between gender and opinion about future wealth in this sample of young adults. Men more often rated their chances of becoming rich in the two highest categories; women said "some chance but probably not" much more frequently.

> **DEFINITION: Association**
>
> We say that there is an **association** between two variables if knowing the value of one variable helps predict the value of the other. If knowing the value of one variable does not help you predict the value of the other, then there is no association between the variables.

Can we say that there is an association between gender and opinion in the *population* of young adults? Making this determination requires formal inference, which will have to wait a few chapters.

**What does "no association" mean?** Figure 1.6 (page 17) suggests an association between gender and opinion about future wealth for young adults. Knowing that a young adult is male helps us predict his opinion: he is more likely than a female to say "a good chance" or "almost certain." What would the graph look like if there was *no* association between the two variables? In that case, knowing that a young adult is male would not help us predict his opinion. He would be no more or less likely than a female to say "a good chance" or "almost certain" or any of the other possible responses. That is, the conditional distributions of opinion about becoming rich would be the *same* for males and females. The segmented bar graphs for the two genders would look the same, too.

## CHECK YOUR UNDERSTANDING

Let's complete our analysis of the data on superpower preferences from the previous Check Your Understanding (page 14). Here is the two-way table of counts once again.

| | Country | |
|---|---|---|
| **Superpower** | **U.K.** | **U.S.** |
| Fly | 54 | 45 |
| Freeze time | 52 | 44 |
| Invisibility | 30 | 37 |
| Superstrength | 20 | 23 |
| Telepathy | 44 | 66 |

1. Find the conditional distributions of superpower preference among students from the United Kingdom and the United States.
2. Make an appropriate graph to compare the conditional distributions.
3. Is there an association between country of origin and superpower preference? Give appropriate evidence to support your answer.

There's one caution that we need to offer: *even a strong association between two categorical variables can be influenced by other variables lurking in the background.* The Data Exploration that follows gives you a chance to explore this idea using a famous (or infamous) data set.

### Teaching Tip

Emphasize the concept of association between two variables. This will be a focus throughout the course. It is OK to use the word "relationship" as a synonym for "association," but do not use the word "correlation." It is reserved for a specific measure of the strength of a linear relationship between two quantitative variables.

### Teaching Tip

Formal inference for an association between categorical variables requires a chi-square test for independence, covered in Chapter 11.

### Teaching Tip

Here is what the segmented bar graphs might look like if there was no association between gender and opinion. Notice that the female bar looks exactly the same as the male bar.

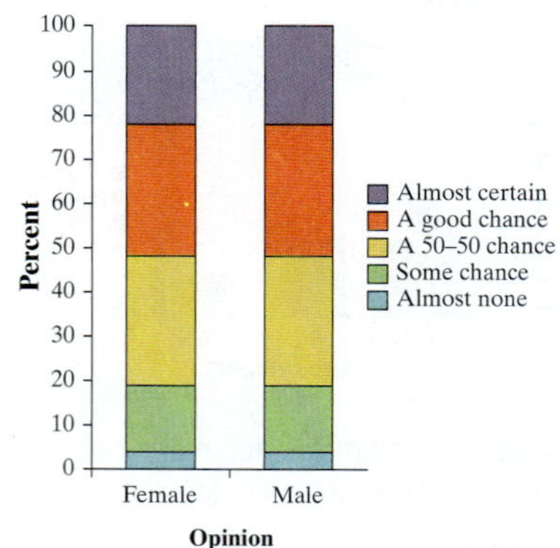


### ANSWERS TO CYU

1. For the U.K. students: 27% said fly, 26% said freeze time, 15% said invisibility, 10% said superstrength, and 22% said telepathy. For the U.S. students: 20.9% said fly, 20.5% said freeze time, 17.2% said invisibility, 10.7% said superstrength, and 30.7% said telepathy.

2. A bar graph is shown below.

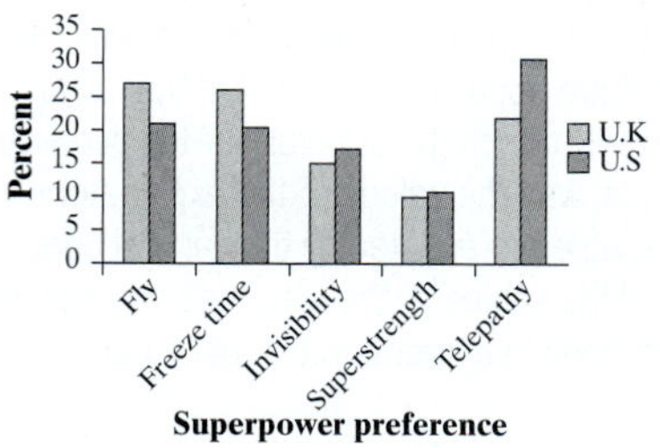


3. There is an association between country of origin and superpower preference. Students in the U.K. are more likely to choose flying and freezing time, while students in the U.S. are more likely to choose invisibility or telepathy. Superstrength is about equally unpopular in both countries.

### TRM Simpson's Paradox Materials

Optional material, including exercises, on Simpson's paradox can be found in the TE-book, on the TRCD or the book's companion Web site (teacher log-in required). Although this is an interesting topic, it is not on the AP® Statistics Topic Outline.

## DATA EXPLORATION A Titanic disaster

In 1912 the luxury liner *Titanic*, on its first voyage across the Atlantic, struck an iceberg and sank. Some passengers got off the ship in lifeboats, but many died. The two-way table below gives information about adult passengers who lived and who died, by class of travel.

| | Class of Travel | | |
|---|---|---|---|
| Survival status | First class | Second class | Third class |
| Lived | 197 | 94 | 151 |
| Died | 122 | 167 | 476 |

Here's another table that displays data on survival status by gender and class of travel.

| | Class of Travel | | | | | |
|---|---|---|---|---|---|---|
| | First class | | Second class | | Third class | |
| Survival status | Female | Male | Female | Male | Female | Male |
| Lived | 140 | 57 | 80 | 14 | 76 | 75 |
| Died | 4 | 118 | 13 | 154 | 89 | 387 |

The movie *Titanic*, starring Leonardo DiCaprio and Kate Winslet, suggested the following:

- First-class passengers received special treatment in boarding the lifeboats, while some other passengers were prevented from doing so (especially third-class passengers).
- Women and children boarded the lifeboats first, followed by the men.

1. What do the data tell us about these two suggestions? Give appropriate graphical and numerical evidence to support your answer.
2. How does gender affect the relationship between class of travel and survival status? Explain.

## Section 1.1 Summary

- The distribution of a categorical variable lists the categories and gives the count (**frequency**) or percent (**relative frequency**) of individuals that fall within each category.
- **Pie charts** and **bar graphs** display the distribution of a categorical variable. Bar graphs can also compare any set of quantities measured in the same units. When examining any graph, ask yourself, "What do I see?"
- A **two-way table** of counts organizes data about two categorical variables measured for the same set of individuals. Two-way tables are often used to summarize large amounts of information by grouping outcomes into categories.

### Answers to Data Exploration

**1.** Overall, 62% of first-class passengers survived, but only 36% of second-class passengers and 24% of third-class passengers survived. Also, 74% of females survived, but only 18% of males. See graphs below.

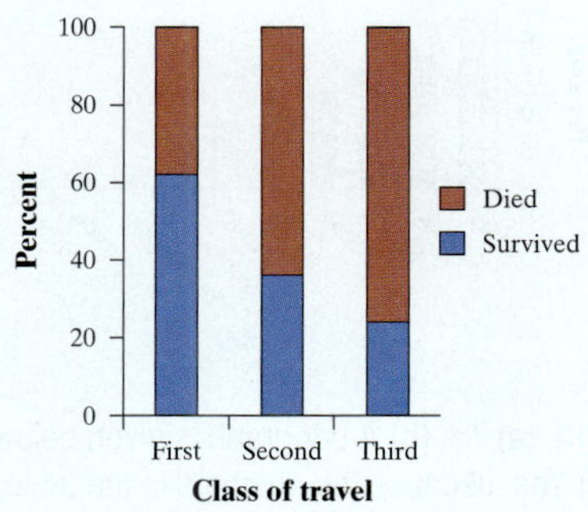


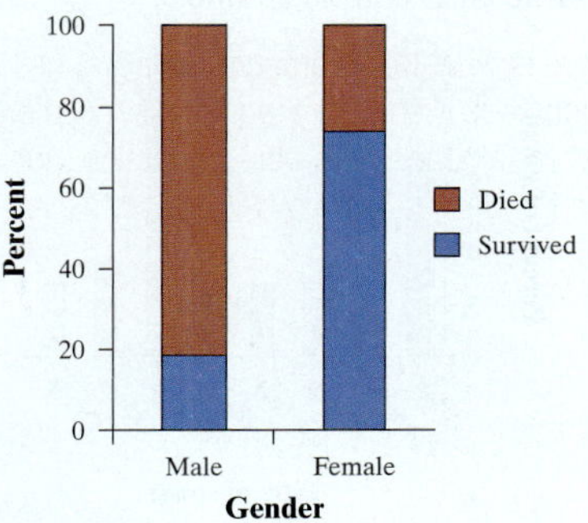


**2.** Among first-class passengers, 97% of females survived, while only 33% of males survived. Among second-class passengers, 86% of females survived, while only 8% of males survived. Among third-class passengers, 46% of females survived, while only 16% of males survived. In all cases, females were much more likely to survive than males. Interestingly, a higher proportion of third-class males survived than second-class males.

### TRM Section 1.1 Quizzes

There are three different quizzes available for this section in the Teacher's Resource Materials. Versions A and B are roughly parallel. Find them by clicking on the link in the TE-book, opening the TRCD, or logging onto the book's companion Web site www.whfreeman.com/tps5e (teacher log-in required).

**TRM Full Solutions to Section 1.1 Exercises**

## Answers to Section 1.1 Exercises

**1.9 (a)** 1% **(b)** A bar graph is given below. **(c)** Yes, because the numbers in the table refer to parts of a single whole.

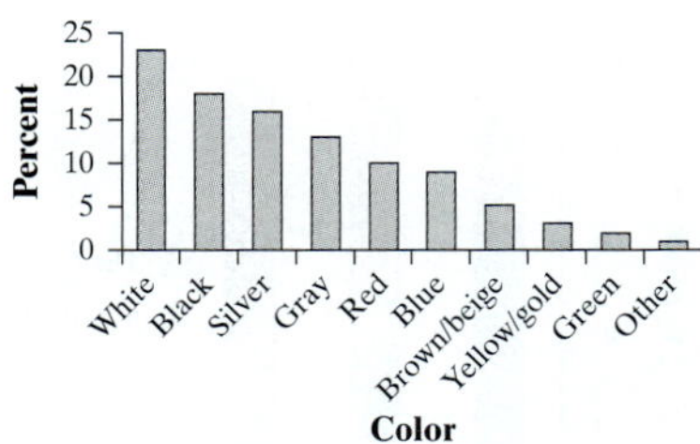


**1.10 (a)** 7% **(b)** A bar graph is given below. **(c)** Yes, because the numbers in the table refer to parts of a single whole.

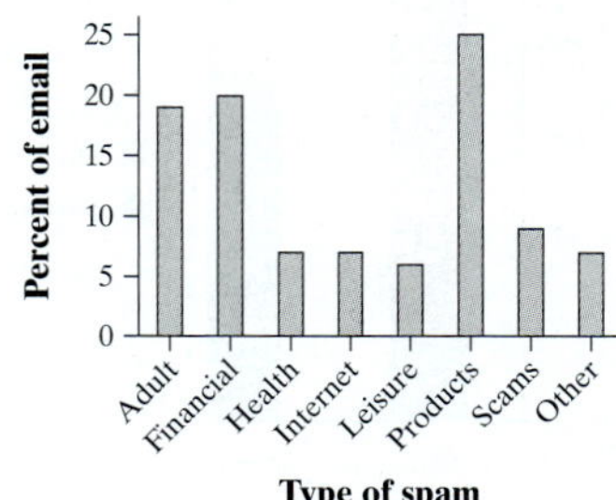


- The row totals and column totals in a two-way table give the **marginal distributions** of the two individual variables. It is clearer to present these distributions as percents of the table total. Marginal distributions tell us nothing about the relationship between the variables.
- There are two sets of **conditional distributions** for a two-way table: the distributions of the row variable for each value of the column variable, and the distributions of the column variable for each value of the row variable. You may want to use a **side-by-side bar graph** (or possibly a **segmented bar graph**) to display conditional distributions.
- There is an **association** between two variables if knowing the value of one variable helps predict the value of the other. To see whether there is an association between two categorical variables, compare an appropriate set of conditional distributions. Remember that even a strong association between two categorical variables can be influenced by other variables.

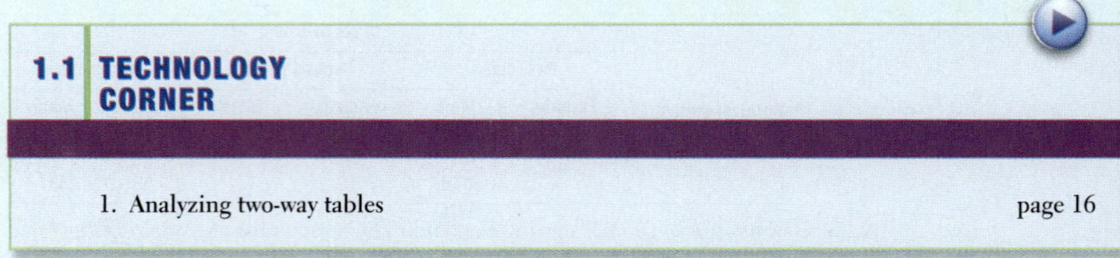


# Section 1.1 Exercises

**9. Cool car colors** The most popular colors for cars and light trucks change over time. Silver passed green in 2000 to become the most popular color worldwide, then gave way to shades of white in 2007. Here is the distribution of colors for vehicles sold in North America in 2011.[8]

| Color | Percent of vehicles |
|---|---|
| White | 23 |
| Black | 18 |
| Silver | 16 |
| Gray | 13 |
| Red | 10 |
| Blue | 9 |
| Brown/beige | 5 |
| Yellow/gold | 3 |
| Green | 2 |

(a) What percent of vehicles had colors other than those listed?

(b) Display these data in a bar graph. Be sure to label your axes.

(c) Would it be appropriate to make a pie chart of these data? Explain.

**10. Spam** Email spam is the curse of the Internet. Here is a compilation of the most common types of spam:[9]

| Type of spam | Percent |
|---|---|
| Adult | 19 |
| Financial | 20 |
| Health | 7 |
| Internet | 7 |
| Leisure | 6 |
| Products | 25 |
| Scams | 9 |
| Other | ?? |

(a) What percent of spam would fall in the "Other" category?

(b) Display these data in a bar graph. Be sure to label your axes.

(c) Would it be appropriate to make a pie chart of these data? Explain.

**11. Birth days** Births are not evenly distributed across the days of the week. Here are the average numbers of babies born on each day of the week in the United States in a recent year:[10]

| Day | Births |
|---|---|
| Sunday | 7374 |
| Monday | 11,704 |
| Tuesday | 13,169 |
| Wednesday | 13,038 |
| Thursday | 13,013 |
| Friday | 12,664 |
| Saturday | 8459 |

(a) Present these data in a well-labeled bar graph. Would it also be correct to make a pie chart?

(b) Suggest some possible reasons why there are fewer births on weekends.

**12. Deaths among young people** Among persons aged 15 to 24 years in the United States, the leading causes of death and number of deaths in a recent year were as follows: accidents, 12,015; homicide, 4651; suicide, 4559; cancer, 1594; heart disease, 984; congenital defects, 401.[11]

(a) Make a bar graph to display these data.

(b) To make a pie chart, you need one additional piece of information. What is it?

**13. Hispanic origins** Below is a pie chart prepared by the Census Bureau to show the origin of the more than 50 million Hispanics in the United States in 2010.[12] About what percent of Hispanics are Mexican? Puerto Rican?

**Percent Distribution of Hispanics by Type: 2010**

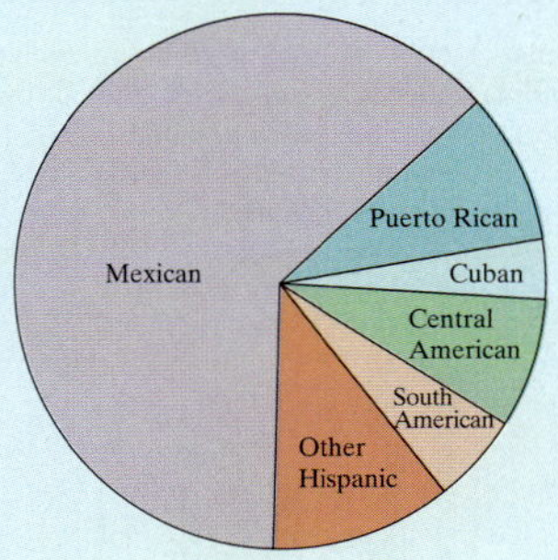


*Comment:* You see that it is hard to determine numbers from a pie chart. Bar graphs are much easier to use. (The Census Bureau did include the percents in its pie chart.)

**14. Which major?** About 1.6 million first-year students enroll in colleges and universities each year. What do they plan to study? The pie chart displays data on the percents of first-year students who plan to major in several discipline areas.[13] About what percent of first-year students plan to major in business? In social science?

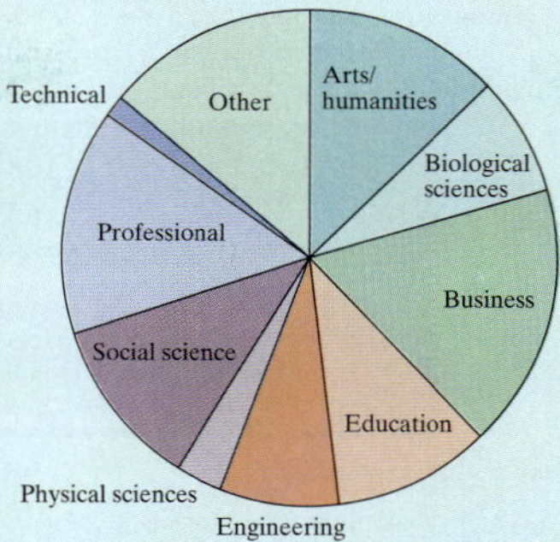


**15. Buying music online** Young people are more likely than older folk to buy music online. Here are the percents of people in several age groups who bought music online in a recent year:[14]

| Age group | Bought music online |
|---|---|
| 12 to 17 years | 24% |
| 18 to 24 years | 21% |
| 25 to 34 years | 20% |
| 35 to 44 years | 16% |
| 45 to 54 years | 10% |
| 55 to 64 years | 3% |
| 65 years and over | 1% |

(a) Explain why it is *not* correct to use a pie chart to display these data.

(b) Make a bar graph of the data. Be sure to label your axes.

**16. The audience for movies** Here are data on the percent of people in several age groups who attended a movie in the past 12 months:[15]

| Age group | Movie attendance |
|---|---|
| 18 to 24 years | 83% |
| 25 to 34 years | 73% |
| 35 to 44 years | 68% |
| 45 to 54 years | 60% |
| 55 to 64 years | 47% |
| 65 to 74 years | 32% |
| 75 years and over | 20% |

(a) Display these data in a bar graph. Describe what you see.

**1.11 (a)** A bar graph is given below. A pie chart would also be appropriate because the numbers in the table refer to parts of a single whole. **(b)** Perhaps induced or C-section births are scheduled for weekdays so doctors don't have to work as much on the weekend.

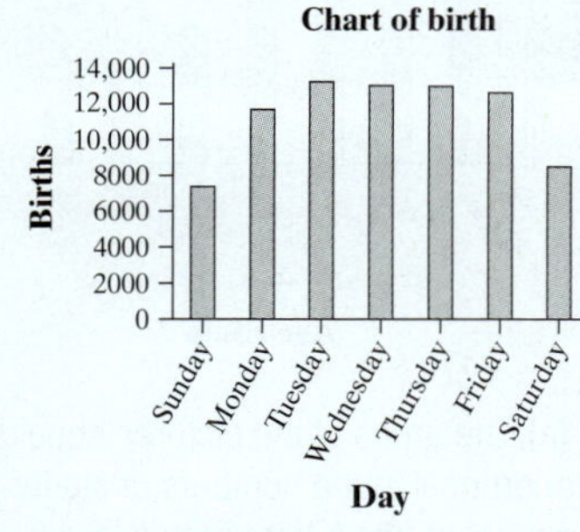


**1.12 (a)** A bar graph is given below. **(b)** You need to know the total number of deaths among people aged 15–24 in the U.S. so that you can figure out how many deaths would be categorized as "other."

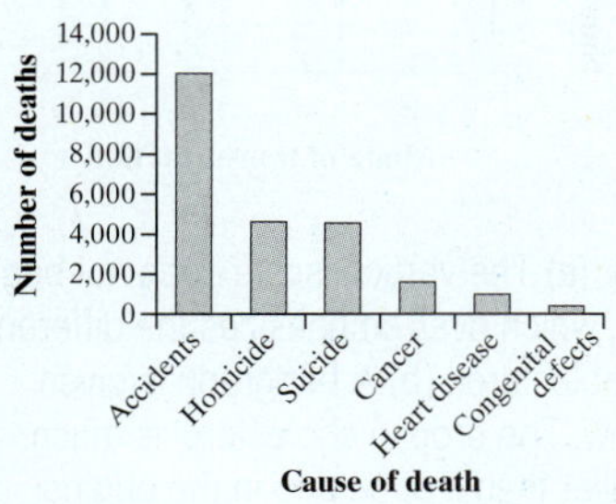


**1.13** About 63% are Mexican and 9% are Puerto Rican.

**1.14** About 20% for Business and 12% for Social Science.

**1.15 (a)** The given percents represent fractions of different age groups, rather than parts of a single whole. **(b)** A bar graph is given below.

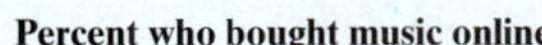


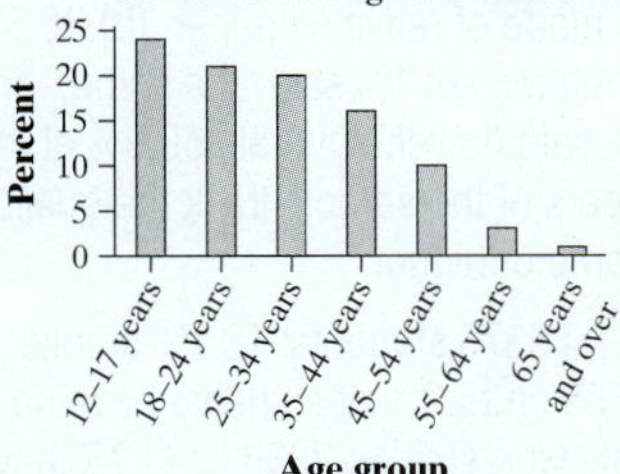


**1.16 (a)** A bar graph is given on the following page. Movie attendance seems to drop off as people get older.

**1.16 (b)** No. The given percents represent fractions of different age groups, rather than parts of a single whole. **(c)** We do not know how many people are in each age group.

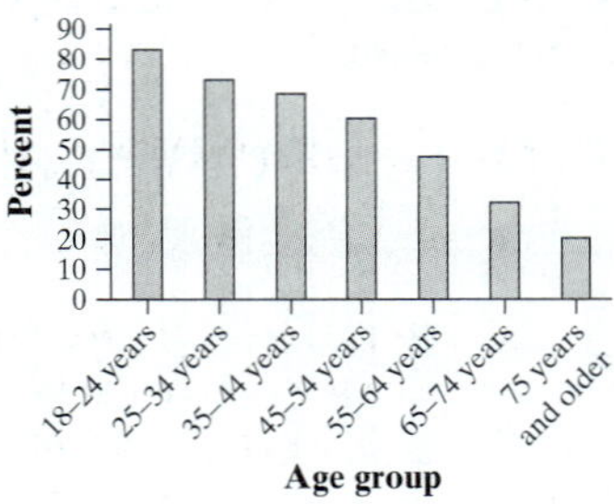


**1.17 (a)** The areas of the pictures should be proportional to the numbers of students they represent. **(b)** A bar graph is given below.

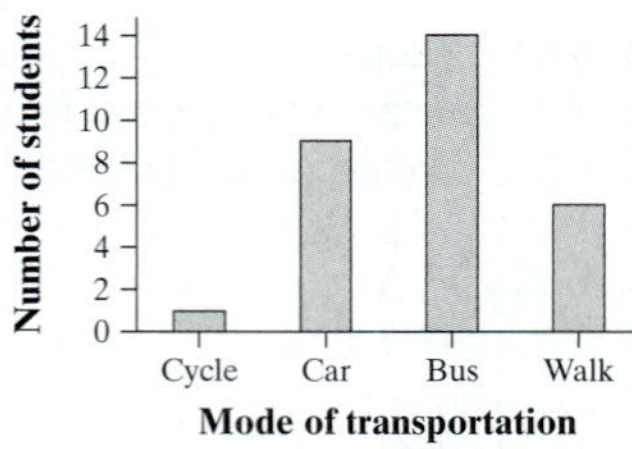


**1.18 (a)** The vertical scale does not begin at 0, which over-emphasizes the difference in cholesterol. **(b)** A bar graph is given below. The drop in cholesterol is much smaller than it appeared in the original graph.

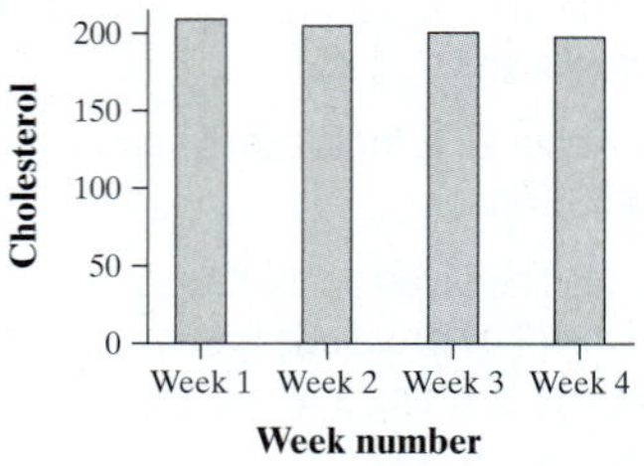


**1.19 (a)** 133 people; 36 buyers of coffee filters made of recycled paper. **(b)** 36.8% said "higher," 24.1% said "the same," and 39.1% said "lower." Overall, 60.9% of the members of the sample think the quality is the same or higher.

**1.20 (a)** 5375 students; 18.7% smoke. **(b)** 1356 or 25.2% of students have no parents who smoke; 2239 or 41.7% have one parent who smokes; 1780 or 33.1% have both parents who smoke.

(b) Would it be correct to make a pie chart of these data? Why or why not?

(c) A movie studio wants to know what percent of the total audience for movies is 18 to 24 years old. Explain why these data do not answer this question.

**17. Going to school** Students in a high school statistics class were given data about the main method of transportation to school for a group of 30 students. They produced the pictograph shown. pg 10

(a) How is this graph misleading?

(b) Make a new graph that isn't misleading.

**18. Oatmeal and cholesterol** Does eating oatmeal reduce cholesterol? An advertisement included the following graph as evidence that the answer is "Yes."

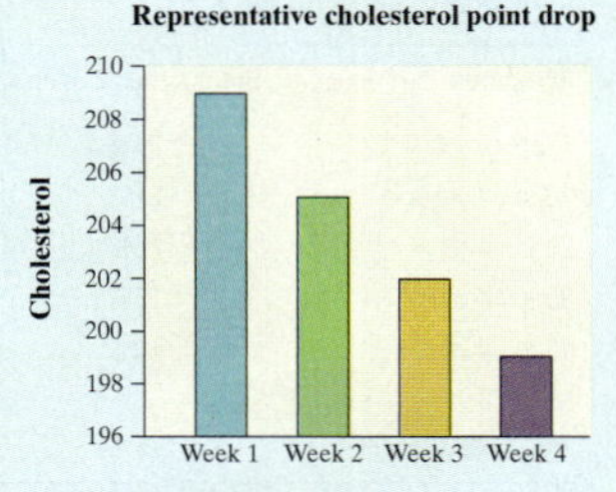


(a) How is this graph misleading?

(b) Make a new graph that isn't misleading. What do you conclude about the relationship between eating oatmeal and cholesterol reduction?

**19. Attitudes toward recycled products** Recycling is supposed to save resources. Some people think recycled products are lower in quality than other products, a fact that makes recycling less practical. People who use a recycled product may have different opinions from those who don't use it. Here are data on attitudes toward coffee filters made of recycled paper from a sample of people who do and don't buy these filters:[16] pg 13

| | Buy recycled filters? | |
|---|---|---|
| Think quality is | Yes | No |
| Higher | 20 | 29 |
| The same | 7 | 25 |
| Lower | 9 | 43 |

(a) How many people does this table describe? How many of these were buyers of coffee filters made of recycled paper?

(b) Give the marginal distribution (in percents) of opinion about the quality of recycled filters. What percent of the people in the sample think the quality of the recycled product is the same or higher than the quality of other filters?

**20. Smoking by students and parents** Here are data from a survey conducted at eight high schools on smoking among students and their parents:[17]

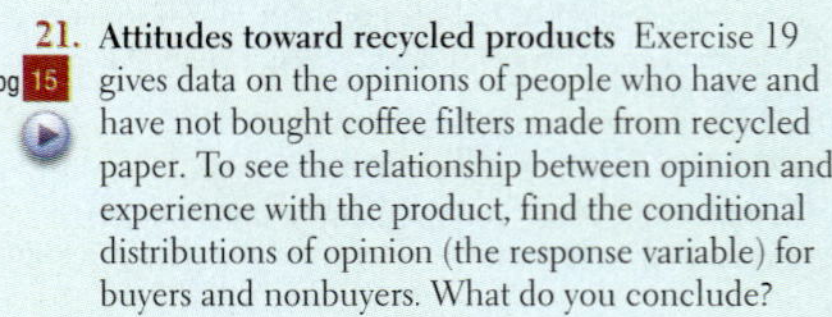

| | Neither parent smokes | One parent smokes | Both parents smoke |
|---|---|---|---|
| Student does not smoke | 1168 | 1823 | 1380 |
| Student smokes | 188 | 416 | 400 |

(a) How many students are described in the two-way table? What percent of these students smoke?

(b) Give the marginal distribution (in percents) of parents' smoking behavior, both in counts and in percents.

**21. Attitudes toward recycled products** Exercise 19 gives data on the opinions of people who have and have not bought coffee filters made from recycled paper. To see the relationship between opinion and experience with the product, find the conditional distributions of opinion (the response variable) for buyers and nonbuyers. What do you conclude? pg 15

**22. Smoking by students and parents** Refer to Exercise 20. Calculate three conditional distributions of students' smoking behavior: one for each of the three parental smoking categories. Describe the relationship between the smoking behaviors of students and their parents in a few sentences.

**23. Popular colors—here and there** Favorite vehicle colors may differ among countries. The side-by-side bar graph shows data on the most popular colors of cars in a recent year for the United States and Europe. Write a few sentences comparing the two distributions.

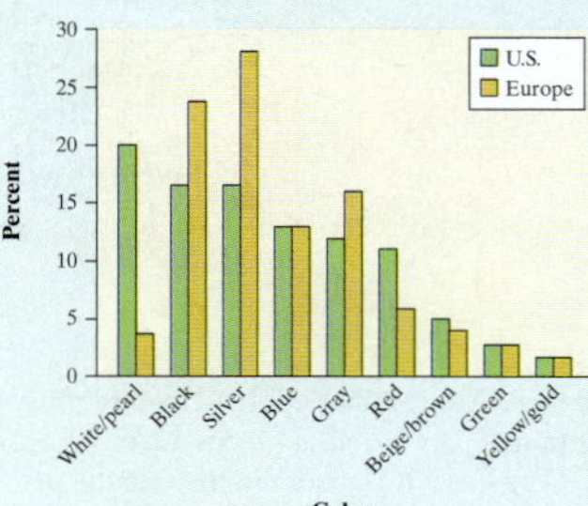


**1.21** For buyers, 55.6% said higher, 19.4% said the same, and 25% said lower. For the nonbuyers, 29.9% said higher, 25.8% said the same, and 44.3% said lower. We see that buyers are much more likely to consider recycled filters higher in quality and much less likely to consider them lower in quality than nonbuyers.

**1.22** For students with no parents who smoke, 86.1% don't smoke and 13.9% smoke. For students with one parent who smokes, 81.4% don't smoke and 18.6% smoke. For students with both parents who smoke, 77.5% don't smoke and 22.5% smoke. When the number of parents smoking goes from 0 to 1 to 2, the percentage of students who smoke increases as well.

**1.23** Americans are much more likely to choose white/pearl and red, while Europeans are much more likely to choose silver, black, or gray. Preferences for blue, beige/brown, green, and yellow/gold are about the same for both groups.

24. **Comparing car colors** Favorite vehicle colors may differ among types of vehicle. Here are data on the most popular colors in a recent year for luxury cars and for SUVs, trucks, and vans.

| Color | Luxury cars (%) | SUVs, trucks, vans (%) |
|---|---|---|
| Black | 22 | 13 |
| Silver | 16 | 16 |
| White pearl | 14 | 1 |
| Gray | 12 | 13 |
| White | 11 | 25 |
| Blue | 7 | 10 |
| Red | 7 | 11 |
| Yellow/gold | 6 | 1 |
| Green | 3 | 4 |
| Beige/brown | 2 | 6 |

(a) Make a graph to compare colors by vehicle type.

(b) Write a few sentences describing what you see.

25. **Snowmobiles in the park** Yellowstone National Park surveyed a random sample of 1526 winter visitors to the park. They asked each person whether they owned, rented, or had never used a snowmobile. Respondents were also asked whether they belonged to an environmental organization (like the Sierra Club). The two-way table summarizes the survey responses.

pg 17

| | Environmental Club | | |
|---|---|---|---|
| | No | Yes | Total |
| Never used | 445 | 212 | **657** |
| Snowmobile renter | 497 | 77 | **574** |
| Snowmobile owner | 279 | 16 | **295** |
| **Total** | **1221** | **305** | **1526** |

Do these data suggest that there is an association between environmental club membership and snowmobile use among visitors to Yellowstone National Park? Give appropriate evidence to support your answer.

26. **Angry people and heart disease** People who get angry easily tend to have more heart disease. That's the conclusion of a study that followed a random sample of 12,986 people from three locations for about four years. All subjects were free of heart disease at the beginning of the study. The subjects took the Spielberger Trait Anger Scale test, which measures how prone a person is to sudden anger. Here are data for the 8474 people in the sample who had normal blood pressure. CHD stands for "coronary heart disease." This includes people who had heart attacks and those who needed medical treatment for heart disease.

| | Low anger | Moderate anger | High anger | Total |
|---|---|---|---|---|
| CHD | 53 | 110 | 27 | **190** |
| No CHD | 3057 | 4621 | 606 | **8284** |
| **Total** | **3110** | **4731** | **633** | **8474** |

Do these data support the study's conclusion about the relationship between anger and heart disease? Give appropriate evidence to support your answer.

***Multiple choice: Select the best answer for Exercises 27 to 34.***

*Exercises 27 to 30 refer to the following setting.* The National Survey of Adolescent Health interviewed several thousand teens (grades 7 to 12). One question asked was "What do you think are the chances you will be married in the next ten years?" Here is a two-way table of the responses by gender:[18]

| | Female | Male |
|---|---|---|
| Almost no chance | 119 | 103 |
| Some chance, but probably not | 150 | 171 |
| A 50-50 chance | 447 | 512 |
| A good chance | 735 | 710 |
| Almost certain | 1174 | 756 |

27. The percent of females among the respondents was

(a) 2625. (b) 4877. (c) about 46%. (d) about 54%. (e) None of these.

28. Your percent from the previous exercise is part of

(a) the marginal distribution of females.

(b) the marginal distribution of gender.

(c) the marginal distribution of opinion about marriage.

(d) the conditional distribution of gender among adolescents with a given opinion.

(e) the conditional distribution of opinion among adolescents of a given gender.

29. What percent of females thought that they were almost certain to be married in the next ten years?

(a) About 16% (b) About 24% (c) About 40% (d) About 45% (e) About 61%

30. Your percent from the previous exercise is part of

(a) the marginal distribution of gender.

(b) the marginal distribution of opinion about marriage.

(c) the conditional distribution of gender among adolescents with a given opinion.

(d) the conditional distribution of opinion among adolescents of a given gender.

(e) the conditional distribution of "Almost certain" among females.

**1.25** A table and a side-by-side bar graph comparing the distributions of snowmobile use for environmental club members and nonmembers are shown below. There appears to be an association between environmental club membership and snowmobile use. The visitors who are members of an environmental club are much more likely to have never used a snowmobile and less likely to have rented or owned a snowmobile than visitors who are not in an environmental club.

| | Not a member | Member |
|---|---|---|
| Never used | 445/1221 = 36.4% | 212/305 = 69.5% |
| Snowmobile renter | 497/1221 = 40.7% | 77/305 = 25.2% |
| Snowmobile owner | 279/1221 = 22.9% | 16/305 = 5.2% |

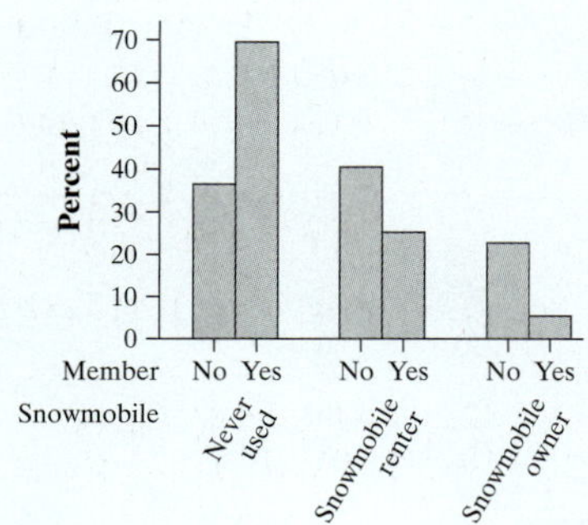


**1.26** A table and a side-by-side bar graph comparing the distributions of CHD rate for different anger levels are shown below. The data do support the study's conclusion because the percent of people in the study with CHD increases as the anger level increases.

**1.24** **(a)** A side-by-side bar graph is shown below. **(b)** Luxury car owners are more likely to choose black, white pearl, and yellow/gold. SUV, truck, and van owners are more likely to choose white, blue, red, and beige/brown. Preferences for silver, gray, and green are about the same for both groups.

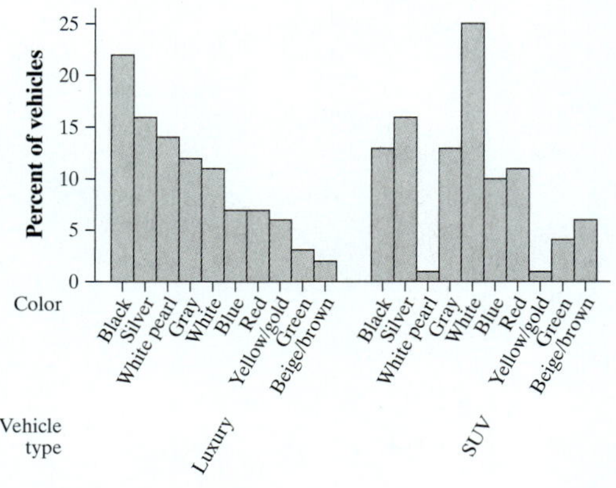


| | Low anger | Moderate anger | High anger |
|---|---|---|---|
| CHD | 53/3110 = 1.7% | 110/4731 = 2.3% | 27/633 = 4.3% |
| No CHD | 3057/3110 = 98.3% | 4621/4731 = 97.7% | 606/633 = 95.7% |

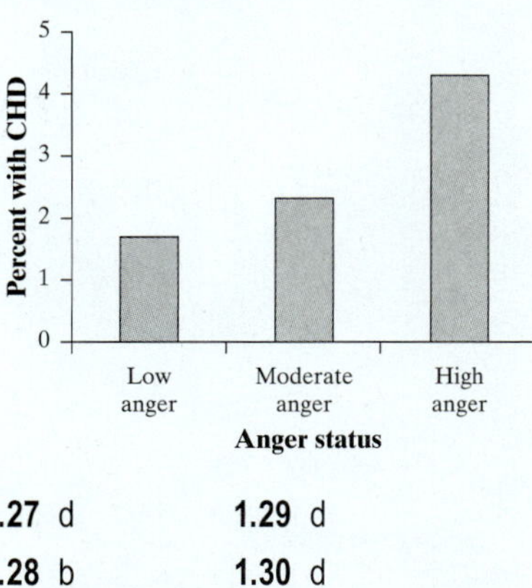


**1.27** d

**1.28** b

**1.29** d

**1.30** d

1.31 b

1.32 d

1.33 d

1.34 c

**1.35** Answers will vary. Two possible tables are given below.

| 10 | 40 |
|---|---|
| 50 | 0 |

| 30 | 20 |
|---|---|
| 30 | 20 |

**1.36 (a)** The individuals are vehicles. **(b)** The variables are make/model (categorical), vehicle type (categorical), transmission type (categorical), number of cylinders (quantitative), city MPG (quantitative), and highway MPG (quantitative).



31. For which of the following would it be inappropriate to display the data with a single pie chart?
(a) The distribution of car colors for vehicles purchased in the last month.
(b) The distribution of unemployment percentages for each of the 50 states.
(c) The distribution of favorite sport for a sample of 30 middle school students.
(d) The distribution of shoe type worn by shoppers at a local mall.
(e) The distribution of presidential candidate preference for voters in a state.

32. The following bar graph shows the distribution of favorite subject for a sample of 1000 students. What is the most serious problem with the graph?

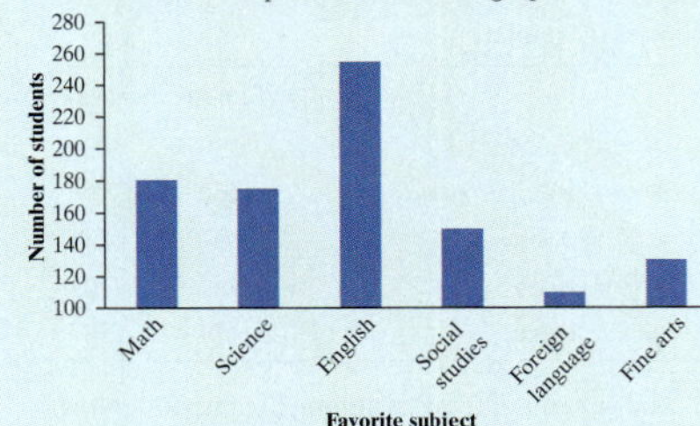


(a) The subjects are not listed in the correct order.
(b) This distribution should be displayed with a pie chart.
(c) The vertical axis should show the percent of students.
(d) The vertical axis should start at 0 rather than 100.
(e) The foreign language bar should be broken up by language.

33. In the 2010–2011 season, the Dallas Mavericks won the NBA championship. The two-way table below displays the relationship between the outcome of each game in the regular season and whether the Mavericks scored at least 100 points.

| | 100 or more points | Fewer than 100 points | Total |
|---|---|---|---|
| Win | 43 | 14 | 57 |
| Loss | 4 | 21 | 25 |
| **Total** | **47** | **35** | **82** |

Which of the following is the best evidence that there is an association between the outcome of a game and whether or not the Mavericks scored at least 100 points?
(a) The Mavericks won 57 games and lost only 25 games.
(b) The Mavericks scored at least 100 points in 47 games and fewer than 100 points in only 35 games.
(c) The Mavericks won 43 games when scoring at least 100 points and only 14 games when scoring fewer than 100 points.
(d) The Mavericks won a higher proportion of games when scoring at least 100 points (43/47) than when they scored fewer than 100 points (14/35).
(e) The combination of scoring 100 or more points and winning the game occurred more often (43 times) than any other combination of outcomes.

34. The following partially complete two-way table shows the marginal distributions of gender and handedness for a sample of 100 high school students.

| | Male | Female | Total |
|---|---|---|---|
| Right | $x$ | | 90 |
| Left | | | 10 |
| **Total** | **40** | **60** | **100** |

If there is no association between gender and handedness for the members of the sample, which of the following is the correct value of $x$?
(a) 20.
(b) 30.
(c) 36.
(d) 45.
(e) Impossible to determine without more information.

35. **Marginal distributions aren't the whole story** Here are the row and column totals for a two-way table with two rows and two columns:

| | | |
|---|---|---|
| $a$ | $b$ | 50 |
| $c$ | $d$ | 50 |
| 60 | 40 | 100 |

Find *two different* sets of counts $a$, $b$, $c$, and $d$ for the body of the table that give these same totals. This shows that the relationship between two variables cannot be obtained from the two individual distributions of the variables.

36. **Fuel economy (Introduction)** Here is a small part of a data set that describes the fuel economy (in miles per gallon) of model year 2012 motor vehicles:

| Make and model | Vehicle type | Transmission type | Number of cylinders | City mpg | Highway mpg |
|---|---|---|---|---|---|
| Aston Martin Vantage | Two-seater | Manual | 8 | 14 | 20 |
| Honda Civic Hybrid | Subcompact | Automatic | 4 | 44 | 44 |
| Toyota Prius | Midsize | Automatic | 4 | 51 | 48 |
| Chevrolet Impala | Large | Automatic | 6 | 18 | 30 |

(a) What are the individuals in this data set?
(b) What variables were measured? Identify each as categorical or quantitative.

# 1.2 Displaying Quantitative Data with Graphs

**WHAT YOU WILL LEARN** By the end of the section, you should be able to:

- Make and interpret dotplots and stemplots of quantitative data.
- Describe the overall pattern (shape, center, and spread) of a distribution and identify any major departures from the pattern (outliers).
- Identify the shape of a distribution from a graph as roughly symmetric or skewed.
- Make and interpret histograms of quantitative data.
- Compare distributions of quantitative data using dotplots, stemplots, or histograms.

To display the distribution of a categorical variable, use a bar graph or a pie chart. How can we picture the distribution of a quantitative variable? In this section, we present several types of graphs that can be used to display quantitative data.

## Dotplots

One of the simplest graphs to construct and interpret is a **dotplot**. Each data value is shown as a dot above its location on a number line. We'll show how to make a dotplot using some sports data.

### EXAMPLE Gooooaaaaallllll!

#### *How to make a dotplot*

How good was the 2012 U.S. women's soccer team? With players like Abby Wambach, Megan Rapinoe, and Hope Solo, the team put on an impressive showing en route to winning the gold medal at the 2012 Olympics in London. Here are data on the number of goals scored by the team in the 12 months prior to the 2012 Olympics.[19]

| | | | | | | | | | | | | |
|---|---|---|---|---|---|---|---|---|---|---|---|---|
| 1 | 3 | 1 | 14 | 13 | 4 | 3 | 4 | 2 | 5 | 2 | 0 | 4 |
| 1 | 3 | 4 | 3 | 4 | 2 | 4 | 3 | 1 | 2 | 4 | 2 | |

Here are the steps in making a dotplot:

- *Draw a horizontal axis (a number line) and label it with the variable name.* In this case, the variable is number of goals scored.
- *Scale the axis.* Start by looking at the minimum and maximum values of the variable. For these data, the minimum number of goals scored was 0, and the maximum was 14. So we mark our scale from 0 to 14, with tick marks at every whole-number value.
- *Mark a dot above the location on the horizontal axis corresponding to each data value.* Figure 1.8 displays a completed dotplot for the soccer data.

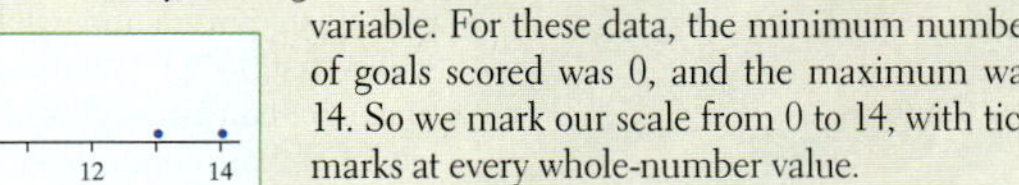


**FIGURE 1.8** A dotplot of goals scored by the U.S. women's soccer team in 2012.

**PD Section 1.2 Overview**

There is a video overview of Section 1.2 (for teachers) available in the TE-book, or on the TRCD or the book's companion Web site (teacher log-in required).

### Teaching Tip

Make a big deal about the switch from categorical variables to quantitative variables. It is important that students begin to use this distinction to classify the different methods of data analysis.

### Teaching Tip

Don't minimize the importance of dotplots! They are much less abstract than histograms or boxplots and make it easy for students to recognize key features of a distribution of quantitative data.

### ALTERNATE EXAMPLE

#### *Frozen pizza*

*Consumer Reports* magazine rated frozen pizza in its January 2011 issue. Here are the amounts of sodium (in milligrams) in a single serving of 16 different brands of cheese pizza:

| | | | | | | | |
|---|---|---|---|---|---|---|---|
| 580 | 740 | 850 | 850 | 870 | 850 | 670 | 670 |
| 630 | 690 | 780 | 610 | 790 | 570 | 700 | 640 |

To make a dotplot, draw a horizontal axis and scale it from 550 to 900 by 50s, because the values go from 570 to 870. Then place a dot for each brand of pizza at the appropriate location. Here is the completed dotplot:

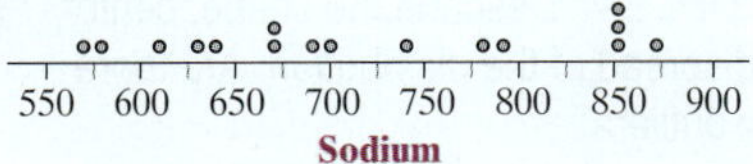


**TRM Section 1.2 Alternate Examples**

You can find the Alternate Examples for this section in Microsoft Word format by clicking on the link in the TE-book, opening the TRCD, or logging onto the book's companion Web site www.whfreeman.com/tps5e (teacher log-in required).

**APPLET**

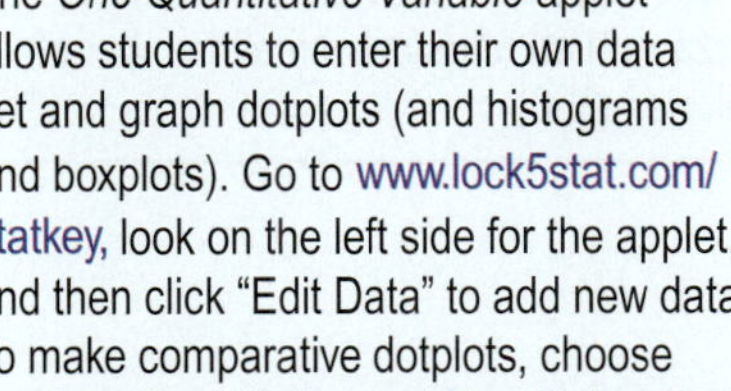

The *One Quantitative Variable* applet allows students to enter their own data set and graph dotplots (and histograms and boxplots). Go to www.lock5stat.com/statkey, look on the left side for the applet, and then click "Edit Data" to add new data. To make comparative dotplots, choose the *One Quantitative and One Categorical Variable* applet, click on "Edit Data," and enter data in the format shown.

### TRM Motivating SOCS

Consider using the "Motivating SOCS" activity found in the TE-book, or on the TRCD or the book's companion Web site (teacher log-in required). It presents several different distributions and leads students to discover that the most important characteristics to identify are shape, center, spread, and outliers.

### AP® Exam Common Error

When students are asked to describe a distribution on the AP® exam, many students only comment on the shape. To get full credit, they need to address shape, center, spread, and any outliers. Emphasizing the SOCS acronym will help.

### Teaching Tip

Don't spend much time discussing center and spread at this point. We will discuss these characteristics in much more detail in Section 1.3.

### ALTERNATE EXAMPLE

#### *Frozen pizza*

Here are the number of calories per serving for 16 brands of frozen cheese pizza, along with a dotplot of the data:

| 340 | 340 | 310 | 320 | 310 | 360 | 350 | 330 |
|---|---|---|---|---|---|---|---|
| 260 | 380 | 340 | 320 | 310 | 360 | 350 | 330 |

260 280 300 320 340 360 380 400

**Calories**

PROBLEM: Describe the shape, center, and spread of the distribution. Are there any outliers?

SOLUTION: *Shape*: There are peaks at 320 and 340 calories and a main cluster of values between 310 and 360 calories. *Center*: The middle value is 330 calories. *Spread*: The values vary from 260 calories to 380 calories. *Outliers*: There is one pizza with an unusually small number of calories (260).



Making a graph is not an end in itself. The purpose of a graph is to help us understand the data. After you make a graph, always ask, "What do I see?" Here is a general strategy for interpreting graphs of quantitative data.

**HOW TO EXAMINE THE DISTRIBUTION OF A QUANTITATIVE VARIABLE**

In any graph, look for the **overall pattern** and for striking **departures** from that pattern.

- You can describe the overall pattern of a distribution by its **shape, center,** and **spread**.
- An important kind of departure is an **outlier,** an individual value that falls outside the overall pattern.

You'll learn more formal ways of describing shape, center, and spread and identifying outliers soon. For now, let's use our informal understanding of these ideas to examine the graph of the U.S. women's soccer team data.

*Shape:* The dotplot has a peak at 4, a single main cluster of dots between 0 and 5, and a large gap between 5 and 13. The main cluster has a longer tail to the left of the peak than to the right. What does the shape tell us? The U.S. women's soccer team scored between 0 and 5 goals in most of its games, with 4 being the most common value (known as the **mode**).

*Center:* The "midpoint" of the 25 values shown in the graph is the 13th value if we count in from either end. You can confirm that the midpoint is at 3. What does this number tell us? In a typical game during the 2012 season, the U.S. women's soccer team scored about 3 goals.

When describing a distribution of quantitative data, don't forget your SOCS (shape, outliers, center, spread)!

*Spread:* The data vary from 0 goals scored to 14 goals scored.

*Outliers:* The games in which the women's team scored 13 goals and 14 goals clearly stand out from the overall pattern of the distribution. So we label them as possible outliers. (In Section 1.3, we'll establish a procedure for determining whether a particular value is an outlier.)

## EXAMPLE Are You Driving a Gas Guzzler?

### *Interpreting a dotplot*

The Environmental Protection Agency (EPA) is in charge of determining and reporting fuel economy ratings for cars (think of those large window stickers on a new car). For years, consumers complained that their actual gas mileages were noticeably lower than the values reported by the EPA. It seems that the EPA's tests—all of which are done on computerized devices to ensure consistency—did not consider things like outdoor temperature, use of the air conditioner, or realistic acceleration and braking by drivers. In 2008 the EPA changed the method for measuring a vehicle's fuel economy to try to give more accurate estimates.

The following table displays the EPA estimates of highway gas mileage in miles per gallon (mpg) for a sample of 24 model year 2012 midsize cars.[20]

| Model | mpg | Model | mpg | Model | mpg |
|---|---|---|---|---|---|
| Acura RL | 24 | Dodge Avenger | 30 | Mercedes-Benz E350 | 30 |
| Audi A8 | 28 | Ford Fusion | 25 | Mitsubishi Galant | 30 |
| Bentley Mulsanne | 18 | Hyundai Elantra | 40 | Nissan Maxima | 26 |
| BMW 550I | 23 | Jaguar XF | 23 | Saab 9-5 Sedan | 28 |
| Buick Lacrosse | 27 | Kia Optima | 34 | Subaru Legacy | 31 |
| Cadillac CTS | 27 | Lexus ES 350 | 28 | Toyota Prius | 48 |
| Chevrolet Malibu | 33 | Lincoln MKZ | 27 | Volkswagen Passat | 31 |
| Chrysler 200 | 30 | Mazda 6 | 31 | Volvo S80 | 26 |

Figure 1.9 shows a dotplot of the data:

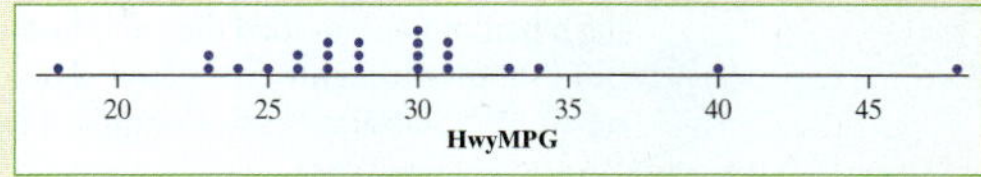


**FIGURE 1.9** Dotplot displaying EPA estimates of highway gas mileage for model year 2012 midsize cars.

PROBLEM: Describe the shape, center, and spread of the distribution. Are there any outliers?

SOLUTION:

*Shape:* The dotplot has a peak at 30 mpg and a main cluster of values from 23 to 34 mpg. There are large gaps between 18 and 23, 34 and 40, 40 and 48 mpg.

*Center:* The midpoint of the 24 values shown in the graph is 28. So a typical model year 2012 midsize car in the sample got about 28 miles per gallon on the highway.

*Spread:* The data vary from 18 mpg to 48 mpg.

*Outliers:* We see two midsize cars with unusually high gas mileage ratings: the Hyundai Elantra (40 mpg) and the Toyota Prius (48 mpg). The Bentley Mulsanne stands out for its low gas mileage rating (18 mpg). All three of these values seem like clear outliers.

The 2012 Nissan Leaf, an electric car, got an EPA estimated 92 miles per gallon on the highway. With the U.S. government's plan to raise the fuel economy standard to an average of 54.5 mpg by 2025, even more alternative-fuel vehicles like the Leaf will have to be developed.

**For Practice** *Try Exercise* 39

## Describing Shape

When you describe a distribution's shape, concentrate on the main features. Look for major peaks, not for minor ups and downs in the graph. Look for clusters of values and obvious gaps. Look for potential outliers, not just for the smallest and largest observations. Look for rough **symmetry** or clear **skewness**.

For his own safety, which way should Mr. Starnes go "skewing"?

**DEFINITION: Symmetric and skewed distributions**

A distribution is roughly **symmetric** if the right and left sides of the graph are approximately mirror images of each other.

A distribution is **skewed to the right** if the right side of the graph (containing the half of the observations with larger values) is much longer than the left side. It is **skewed to the left** if the left side of the graph is much longer than the right side.

### Teaching Tip

Emphasize that in the real world, distributions of data are almost never perfectly symmetric. If distributions are close to symmetric, then have your students describe the shape as "roughly" or "approximately" symmetric.

### Teaching Tip

Here is a fun way to help students remember how to describe a skewed distribution. If a distribution is skewed to the left, it looks like a left foot (big toes on the right and shorter toes to the left). Likewise, a distribution that is skewed to the right will look like a right foot (big toe on the left and shorter toes on the right). You can also say that a distribution that is skewed to the right is "stretched out to the right."

ALTERNATE EXAMPLE

### *Die rolls and frozen pizza*

Here are dotplots for two different sets of quantitative data:

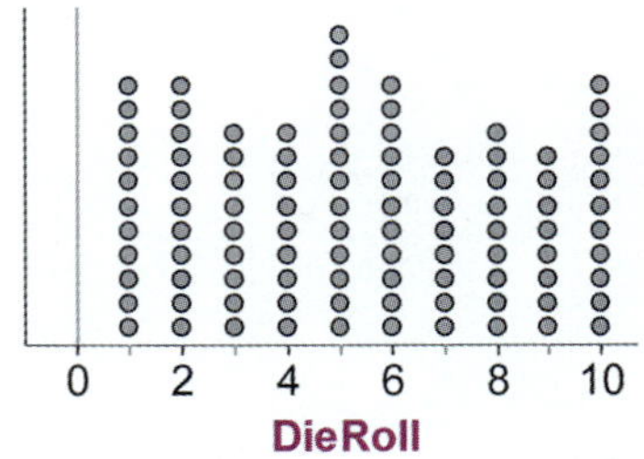


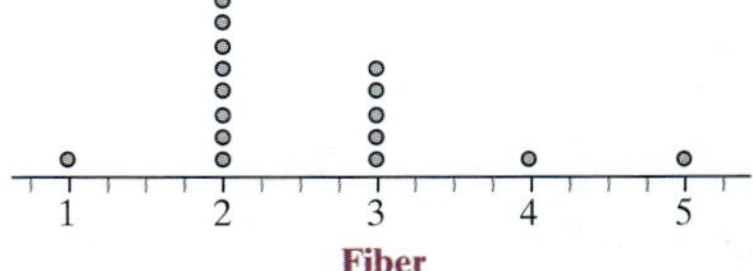


The first dotplot shows the outcomes of 100 rolls of a 10-sided die. This distribution is roughly symmetric with no obvious modes. Don't worry about the small differences in the number of dots for each die roll—this is bound to happen just by chance even if the frequencies should be the same. A distribution with this shape can be called "approximately uniform." The second dotplot shows the number of grams of fiber in one serving of frozen cheese pizza. This distribution is skewed to the right with a peak at 2 grams of fiber.

### Teaching Tip

Emphasize that students should not worry about little ups and downs in a distribution. Little irregularities are to be expected when working with real data.

### Teaching Tip

A nice way to generate a bimodal distribution is to ask each student to measure the length of the longest hair on his or her head. Typically, there will be a big peak on the left representing the short-haired students and a smaller peak on the right representing the long-haired students.

For brevity, we sometimes say "left-skewed" instead of "skewed to the left" and "right-skewed" instead of "skewed to the right." We could also describe a distribution with a long tail to the left as "skewed toward negative values" or "negatively skewed" and a distribution with a long right tail as "positively skewed."

*The direction of skewness is the direction of the long tail, not the direction where most observations are clustered.* See the drawing in the margin on page 27 for a cute but corny way to help you keep this straight.

## EXAMPLE Die Rolls and Quiz Scores

### *Describing shape*

Figure 1.10 displays dotplots for two different sets of quantitative data. Let's practice describing the shapes of these distributions. Figure 1.10(a) shows the results of rolling a pair of fair, six-sided dice and finding the sum of the up-faces 100 times. This distribution is roughly symmetric. The dotplot in Figure 1.10(b) shows the scores on an AP® Statistics class's first quiz. This distribution is skewed to the left.

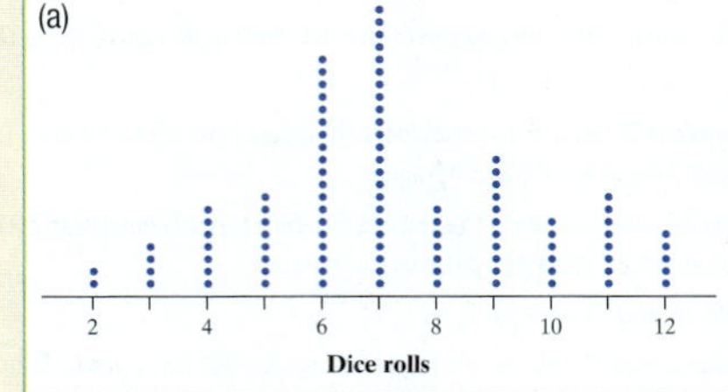


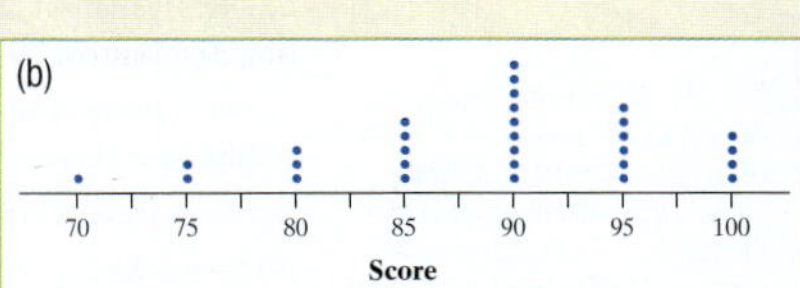


**FIGURE 1.10** Dotplots displaying different shapes: (a) roughly symmetric; (b) skewed to the left.

Although the dotplots in the previous example have different shapes, they do have something in common. Both are **unimodal**, that is, they have a single peak: the graph of dice rolls at 7 and the graph of quiz scores at 90. (We don't count minor ups and downs in a graph, like the "bumps" at 9 and 11 in the dice rolls dotplot, as "peaks.") Figure 1.11 is a dotplot of the duration (in minutes) of 220

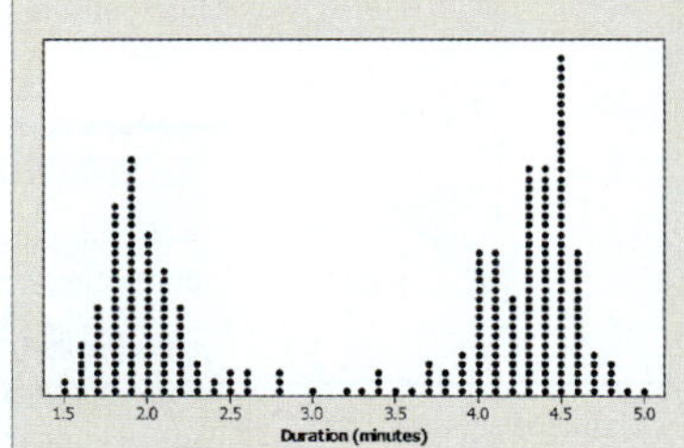


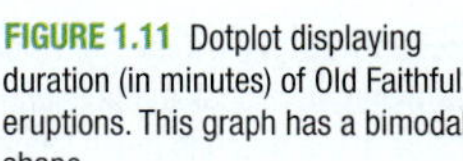

**FIGURE 1.11** Dotplot displaying duration (in minutes) of Old Faithful eruptions. This graph has a bimodal shape.

eruptions of the Old Faithful geyser. We would describe this distribution's shape as roughly symmetric and **bimodal** because it has two clear peaks: one near 2 minutes and the other near 4.5 minutes. (Although we could continue the pattern with "trimodal" for three peaks and so on, it's more common to refer to distributions with more than two clear peaks as **multimodal**.)

**What shape will the graph have?** Some variables have distributions with predictable shapes. Many biological measurements on individuals from the same species and gender—lengths of bird bills, heights of young women—have roughly symmetric distributions. Salaries and home prices, on the other hand, usually have right-skewed distributions. There are many moderately priced houses, for example, but the few very expensive mansions give the distribution of house prices a strong right skew. Many distributions have irregular shapes that are neither symmetric nor skewed. Some data show other patterns, such as the two peaks in Figure 1.11. Use your eyes, describe the pattern you see, and then try to explain the pattern.

## CHECK YOUR UNDERSTANDING

The Fathom dotplot displays data on the number of siblings reported by each student in a statistics class.

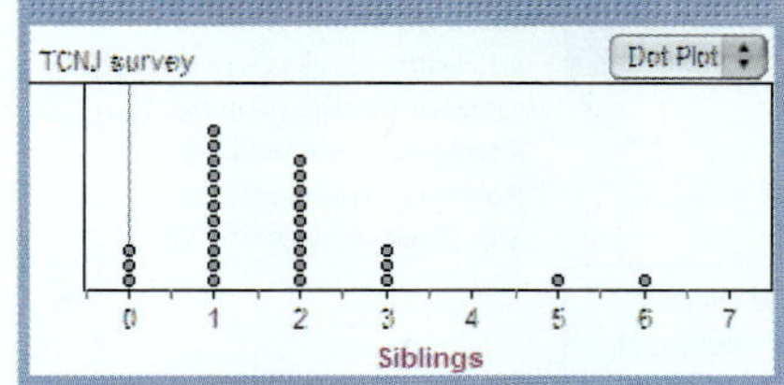


1. Describe the shape of the distribution.
2. Describe the center of the distribution.
3. Describe the spread of the distribution.
4. Identify any potential outliers.

## Comparing Distributions

Some of the most interesting statistics questions involve comparing two or more groups. Which of two popular diets leads to greater long-term weight loss? Who texts more—males or females? Does the number of people living in a household differ among countries? As the following example suggests, you should always discuss shape, center, spread, and possible outliers whenever you compare distributions of a quantitative variable.

### ANSWERS TO CYU

1. This distribution is skewed to the right and unimodal.
2. The midpoint of the 28 values is between 1 and 2.
3. The number of siblings varies from 0 to 6.
4. There are two potential outliers at 5 and 6 siblings.

## Teaching Tip

Notice that the dotplots in this example are stacked vertically and are on the same scale, making it much easier to compare the distributions.

**ALTERNATE EXAMPLE**

### *Energy cost: Top versus bottom freezers*

How do the annual energy costs (in dollars) compare for refrigerators with top freezers and refrigerators with bottom freezers? The data below are from the May 2010 issue of *Consumer Reports*.

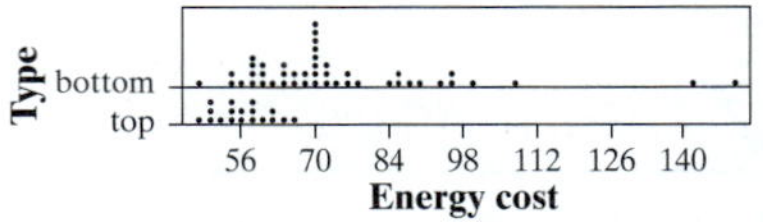


PROBLEM: Compare the distributions of energy cost for these two types of refrigerators.

SOLUTION: *Shape*: The distribution for bottom freezers looks skewed to the right and *possibly* bimodal, with modes near $58 and $70 per year. The distribution for top freezers looks roughly symmetric, with its main peak centered near $55. *Center*: The typical energy cost for the bottom freezers is greater than the typical cost for the top freezers (midpoint of $69 vs. midpoint of $56). *Spread*: There is much more variability in the energy costs for bottom freezers. *Outliers*: There are a couple of bottom freezers with unusually high energy costs (over $140 per year). There are no outliers for the top freezers.

## EXAMPLE Household Size: U.K. versus South Africa

### *Comparing distributions*

How do the numbers of people living in households in the United Kingdom (U.K.) and South Africa compare? To help answer this question, we used CensusAtSchool's "Random Data Selector" to choose 50 students from each country. Figure 1.12 is a dotplot of the household sizes reported by the survey respondents.

PROBLEM: Compare the distributions of household size for these two countries.

SOLUTION:

*Shape:* The distribution of household size for the U.K. sample is roughly symmetric and unimodal, while the distribution for the South Africa sample is skewed to the right and unimodal.

*Center:* Household sizes for the South African students tended to be larger than for the U.K. students. The midpoints of the household sizes for the two groups are 6 people and 4 people, respectively.

*Spread:* The household sizes for the South African students vary more (from 3 to 26 people) than for the U.K. students (from 2 to 6 people).

*Outliers:* There don't appear to be any outliers in the U.K. distribution. The South African distribution seems to have two outliers in the right tail of the distribution—students who reported living in households with 15 and 26 people.

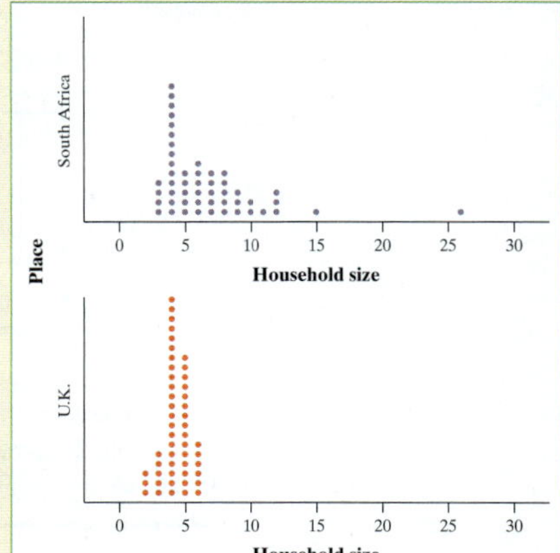


**FIGURE 1.12** Dotplots of household size for random samples of 50 students from the United Kingdom and South Africa.

**AP® EXAM TIP** When comparing distributions of quantitative data, it's not enough just to list values for the center and spread of each distribution. You have to explicitly *compare* these values, using words like "greater than," "less than," or "about the same as."

**For Practice** *Try Exercise* 43

Notice that we discussed the distributions of household size only for the two *samples* of 50 students in the previous example. We might be interested in whether the sample data give us convincing evidence of a difference in the *population* distributions of household size for South Africa and the United Kingdom. We'll have to wait a few chapters to decide whether we can reach such a conclusion, but our

## Teaching Tip

Remind students to be careful with their language when describing a distribution. For example, it isn't correct to say that the "center of South Africa" is higher than the "center of the United Kingdom." After all, the middle of South Africa is much further south than the middle of the U.K.! It is better to say that the "center of the distribution of household size in South Africa is much greater than the center of the distribution of household size in the U.K."

## AP® Exam Common Error

When students are asked to compare distributions, there are two very common errors that students make. First, they forget to address all four characteristics of the distribution (think SOCS). Second, they don't explicitly compare each characteristic. Discussing the SOCS for each distribution separately will not receive credit! Students must use phrases like "about the same as" or "is much greater than" when comparing center and spread to receive full credit.

## Teaching Tip

We will learn how to perform inference for a difference between two means in Chapter 10.

ability to make such an inference later will be helped by the fact that the students in our samples were chosen at random.

## Stemplots

Another simple graphical display for fairly small data sets is a **stemplot** (also called a stem-and-leaf plot). Stemplots give us a quick picture of the shape of a distribution while including the actual numerical values in the graph. Here's an example that shows how to make a stemplot.

**EXAMPLE**

### How Many Shoes?

*Making a stemplot*

How many pairs of shoes does a typical teenager have? To find out, a group of AP® Statistics students conducted a survey. They selected a random sample of 20 female students from their school. Then they recorded the number of pairs of shoes that each respondent reported having. Here are the data:

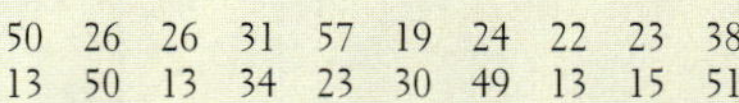

| | | | | | | | | | |
|---|---|---|---|---|---|---|---|---|---|
| 50 | 26 | 26 | 31 | 57 | 19 | 24 | 22 | 23 | 38 |
| 13 | 50 | 13 | 34 | 23 | 30 | 49 | 13 | 15 | 51 |

Here are the steps in making a stemplot. Figure 1.13 displays the process.

- *Separate each observation into a* **stem**, *consisting of all but the final digit, and a* **leaf**, *the final digit. Write the stems in a vertical column with the smallest at the top, and draw a vertical line at the right of this column. Do not skip any stems, even if there is no data value for a particular stem.* For these data, the tens digits are the stems, and the ones digits are the leaves. The stems run from 1 to 5.
- *Write each leaf in the row to the right of its stem.* For example, the female student with 50 pairs of shoes would have stem 5 and leaf 0, while the student with 31 pairs of shoes would have stem 3 and leaf 1.
- *Arrange the leaves in increasing order out from the stem.*
- *Provide a key that explains in context what the stems and leaves represent.*

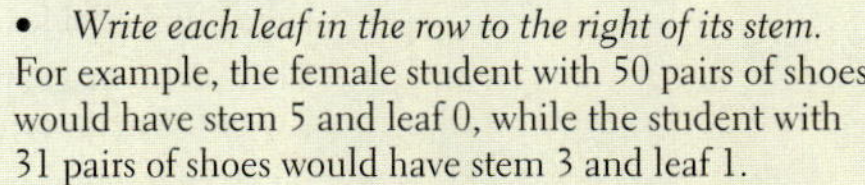

| Stems | Add leaves | Order leaves |
|---|---|---|
| 1 | 1 \| 93335 | 1 \| 33359 |
| 2 | 2 \| 664233 | 2 \| 233466 |
| 3 | 3 \| 1840 | 3 \| 0148 |
| 4 | 4 \| 9 | 4 \| 9 |
| 5 | 5 \| 0701 | 5 \| 0017 |

Add a key: Key: 4|9 represents a female student who reported having 49 pairs of shoes.

**FIGURE 1.13** Making a stemplot of the shoe data. (1) Write the stems. (2) Go through the data and write each leaf on the proper stem. (3) Arrange the leaves on each stem in order out from the stem. (4) Add a key.

The AP® Statistics students in the previous example also collected data from a random sample of 20 male students at their school. Here are the numbers of pairs of shoes reported by each male in the sample:

| | | | | | | | | | |
|---|---|---|---|---|---|---|---|---|---|
| 14 | 7 | 6 | 5 | 12 | 38 | 8 | 7 | 10 | 10 |
| 10 | 11 | 4 | 5 | 22 | 7 | 5 | 10 | 35 | 7 |

**ALTERNATE EXAMPLE**

***Who's taller?***

Who is taller, males or females? A sample of 14-year-olds from the United Kingdom was randomly selected using the CensusAtSchool Web site. Here are the heights of the students (in centimeters):

| | |
|---|---|
| **Male:** | **154, 157, 187, 163, 167, 159, 169, 162, 176, 177, 151, 175, 174, 165, 165, 183, 180** |
| **Female:** | **160, 169, 152, 167, 164, 163, 160, 163, 169, 157, 158, 153, 161, 165, 165, 159, 168, 153, 166, 158, 158, 166** |

Here are two stemplots for male heights, one with split stems:

| Stem | Leaves |
|---|---|
| 15 | 1479 |
| 16 | 235579 |
| 17 | 4567 |
| 18 | 037 |

Key: 15|1 represents a male who is 151 cm tall.

| Stem | Leaves |
|---|---|
| 15 | 14 |
| 15 | 79 |
| 16 | 23 |
| 16 | 5579 |
| 17 | 4 |
| 17 | 567 |
| 18 | 03 |
| 18 | 7 |

Key: 15|1 represents a male who is 151 cm tall.

Here is a back-to-back stemplot comparing male and female heights:

| Female | | Male |
|---|---|---|
| 332 | 15 | 14 |
| 98887 | 15 | 79 |
| 433100 | 16 | 23 |
| 99876655 | 16 | 5579 |
| | 17 | 4 |
| | 17 | 567 |
| | 18 | 03 |
| | 18 | 7 |

Key: 15|1 represents a student who is 151 cm tall.

## AP® Exam Common Error

When making a stemplot, a common error is to forget the key or labels. Make sure to deduct points for forgetting these features now so that your students won't lose points on the AP® exam.

The *One-Variable Statistical Calculator* applet allows students to use existing data sets or enter their own data set and graph stemplots. It also allows you to split stems. Go to www.whfreeman.com/tps5e (click on Statistical Applets and choose *One-Variable Statistical Calculator*).

## ANSWERS TO CYU

1. Both males and females have distributions that are skewed to the right, though the distribution for the males is more heavily skewed. The midpoint for the males (9 pairs) is less than the midpoint for the females (26 pairs). The number of shoes owned by females varies more (from 13 to 57) than for males (from 4 to 38). The male distribution has three likely outliers at 22, 35, and 38. The females do not have any likely outliers.

2. b

3. e

4. c

What would happen if we tried the same approach as before: using the first digits as stems and the last digits as leaves? The completed stemplot is shown in Figure 1.14(a). What shape does this distribution have? It is difficult to tell with so few stems. We can get a better picture of male shoe ownership by **splitting stems**.

In Figure 1.14(a), the values from 0 to 9 are placed on the "0" stem. Figure 1.14(b) shows another stemplot of the same data. This time, values having leaves 0 through 4 are placed on one stem, while values ending in 5 through 9 are placed on another stem. Now we can see the single peak, the cluster of values between 4 and 14, and the large gap between 22 and 35 more clearly.

What if we want to compare the number of pairs of shoes that males and females have? That calls for a **back-to-back stemplot** with common stems. The leaves on each side are ordered out from the common stem. Figure 1.15 is a back-to-back stemplot for the male and female shoe data. Note that we have used the split stems from Figure 1.14(b) as the common stems. The values on the right are the male data from Figure 1.14(b). The values on the left are the female data, ordered out from the stem from right to left. We'll ask you to compare these two distributions shortly.

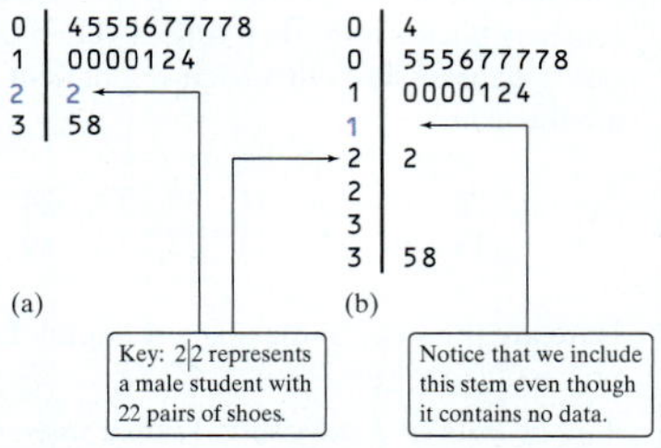


**FIGURE 1.14** Two stemplots showing the male shoe data. Figure 1.14(b) improves on the stemplot of Figure 1.14(a) by splitting stems.

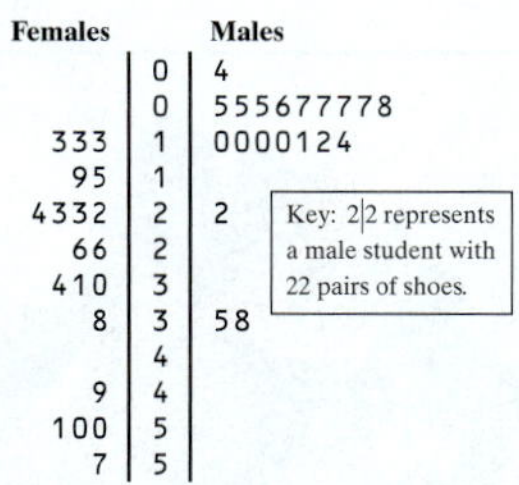


**FIGURE 1.15** Back-to-back stemplot comparing numbers of pairs of shoes for male and female students at a school.

Here are a few tips to consider before making a stemplot:

- Stemplots do not work well for large data sets, where each stem must hold a large number of leaves.
- There is no magic number of stems to use, but five is a good minimum. Too few or too many stems will make it difficult to see the distribution's shape.
- If you split stems, be sure that each stem is assigned an equal number of possible leaf digits (two stems, each with five possible leaves; or five stems, each with two possible leaves).
- You can get more flexibility by rounding the data so that the final digit after rounding is suitable as a leaf. Do this when the data have too many digits. For example, in reporting teachers' salaries, using all five digits (for example, $42,549) would be unreasonable. It would be better to round to the nearest thousand and use 4 as a stem and 3 as a leaf.

Instead of rounding, you can also *truncate* (remove one or more digits) when data have too many digits. The teacher's salary of $42,549 would truncate to $42,000.

## CHECK YOUR UNDERSTANDING

1. Use the back-to-back stemplot in Figure 1.15 to write a few sentences comparing the number of pairs of shoes owned by males and females. Be sure to address shape, center, spread, and outliers.

## Teaching Tip

Make sure to look at the Think About It on pages 35–36 before you introduce histograms. Consider using this approach as a nice transition from dotplots to histograms. Then, to make histograms less abstract, have students use their imagination to fill each bar full of dots each time they encounter a histogram.

```
 6 | 8
 7 |
 8 | 8
 9 | 79
10 | 08
11 | 15566
12 | 01222344445788899
13 | 01233333444899
14 | 02666
15 | 23
16 | 8
```

Key: 8|8 represents a state in which 8.8% of residents are 65 and older.

*Multiple choice: Select the best answer for Questions 2 through 4.*

Here is a stemplot of the percents of residents aged 65 and older in the 50 states and the District of Columbia. The stems are whole percents and the leaves are tenths of a percent.

2. The low outlier is Alaska. What percent of Alaska residents are 65 or older?
(a) 0.68 (b) 6.8 (c) 8.8 (d) 16.8 (e) 68

3. Ignoring the outlier, the shape of the distribution is
(a) skewed to the right. (c) skewed to the middle. (e) roughly symmetric.
(b) skewed to the left. (d) bimodal.

4. The center of the distribution is close to
(a) 11.6%. (b) 12.0%. (c) 12.8%. (d) 13.3%. (e) 6.8% to 16.8%.

## Histograms

Quantitative variables often take many values. A graph of the distribution is clearer if nearby values are grouped together. One very common graph of the distribution of a quantitative variable is a **histogram**. Let's look at how to make a histogram using data on foreign-born residents in the United States.

EXAMPLE

### Foreign-Born Residents

#### *Making a histogram*

What percent of your home state's residents were born outside the United States? A few years ago, the country as a whole had 12.5% foreign-born residents, but the states varied from 1.2% in West Virginia to 27.2% in California. The following table presents the data for all 50 states.[21] The *individuals* in this data set are the states. The *variable* is the percent of a state's residents who are foreign-born. It's much easier to see from a graph than from the table how your state compared with other states.

| State | Percent | State | Percent | State | Percent |
|---|---|---|---|---|---|
| Alabama | 2.8 | Louisiana | 2.9 | Ohio | 3.6 |
| Alaska | 7.0 | Maine | 3.2 | Oklahoma | 4.9 |
| Arizona | 15.1 | Maryland | 12.2 | Oregon | 9.7 |
| Arkansas | 3.8 | Massachusetts | 14.1 | Pennsylvania | 5.1 |
| California | 27.2 | Michigan | 5.9 | Rhode Island | 12.6 |
| Colorado | 10.3 | Minnesota | 6.6 | South Carolina | 4.1 |
| Connecticut | 12.9 | Mississippi | 1.8 | South Dakota | 2.2 |
| Delaware | 8.1 | Missouri | 3.3 | Tennessee | 3.9 |
| Florida | 18.9 | Montana | 1.9 | Texas | 15.9 |
| Georgia | 9.2 | Nebraska | 5.6 | Utah | 8.3 |
| Hawaii | 16.3 | Nevada | 19.1 | Vermont | 3.9 |
| Idaho | 5.6 | New Hampshire | 5.4 | Virginia | 10.1 |
| Illinois | 13.8 | New Jersey | 20.1 | Washington | 12.4 |
| Indiana | 4.2 | New Mexico | 10.1 | West Virginia | 1.2 |
| Iowa | 3.8 | New York | 21.6 | Wisconsin | 4.4 |
| Kansas | 6.3 | North Carolina | 6.9 | Wyoming | 2.7 |
| Kentucky | 2.7 | North Dakota | 2.1 | | |

ALTERNATE EXAMPLE

#### *NBA scoring averages*

The table presents the average points scored per game (PPG) for the 30 NBA teams in the 2012–2013 regular season:

| Team | PPG | Team | PPG | Team | PPG |
|---|---|---|---|---|---|
| Atlanta Hawks | 98.0 | Houston Rockets | 106.0 | Oklahoma City Thunder | 105.7 |
| Boston Celtics | 96.5 | Indiana Pacers | 94.7 | Orlando Magic | 94.1 |
| Brooklyn Nets | 96.9 | Los Angeles Clippers | 101.1 | Philadelphia 76ers | 93.2 |
| Charlotte Bobcats | 93.4 | Los Angeles Lakers | 102.2 | Phoenix Suns | 95.2 |
| Chicago Bulls | 93.2 | Memphis Grizzlies | 93.4 | Portland Trail Blazers | 97.5 |
| Cleveland Cavaliers | 96.5 | Miami Heat | 102.9 | Sacramento Kings | 100.2 |
| Dallas Mavericks | 101.1 | Milwaukee Bucks | 98.9 | San Antonio Spurs | 103.0 |
| Denver Nuggets | 106.1 | Minnesota Timberwolves | 95.7 | Toronto Raptors | 97.2 |
| Detroit Pistons | 94.9 | New Orleans Hornets | 94.1 | Utah Jazz | 98.0 |
| Golden State Warriors | 101.2 | New York Knicks | 100.0 | Washington Wizards | 93.2 |

Because the smallest value is 93.2 and the largest value is 106.1, we choose classes of width 2 starting at 93 (93 to <95, 95 to <97, 97 to <99, 99 to <101, 101 to <103, 103 to <105, and 105 to <107). Here are a frequency histogram and a relative frequency histogram for these data. The relative frequency histogram looks exactly the same as the frequency histogram except for the vertical scale.

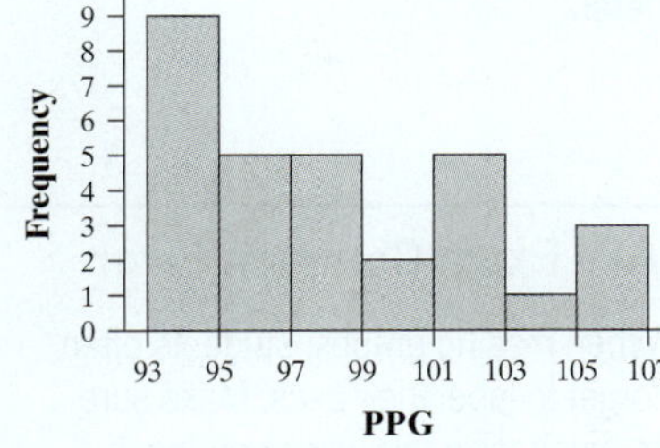


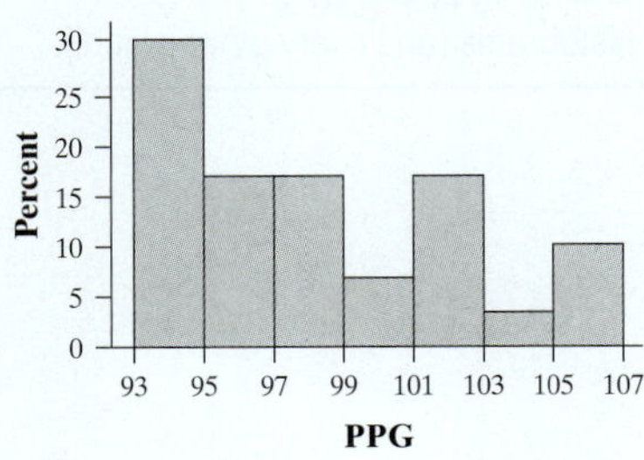


Now here are histograms using class widths of 1 starting at 93. The choice of class width can influence the appearance of a histogram. In the histograms below, we can see gaps from 99 to 100 and from 104 to 105.

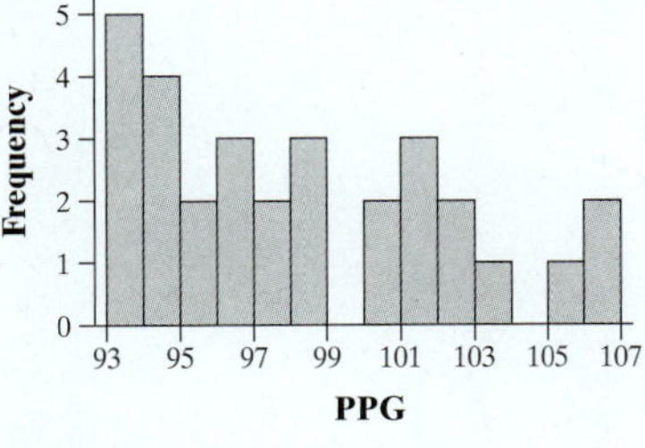


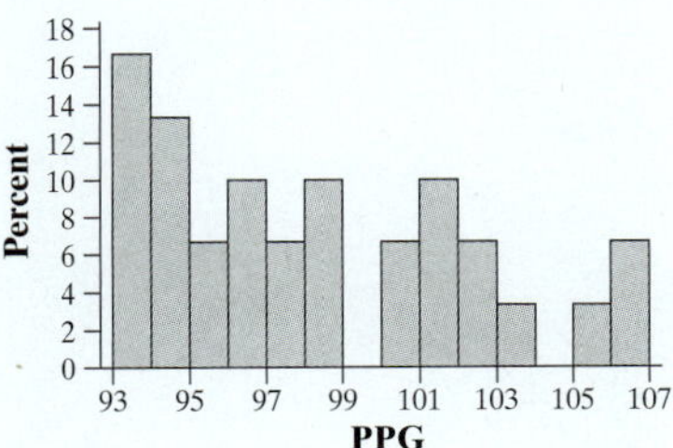

## Teaching Tip

Relative frequency histograms are typically more useful than frequency histograms. They make it easier to compare two distributions, especially when the number of individuals is very different.

## AP® Exam Common Error

When making graphs, students often forget to label their axes. Make sure to emphasize this when you teach histograms (and every other graph!).

Here are the steps in making a histogram:

- *Divide the data into classes of equal width.* The data in the table vary from 1.2 to 27.2, so we might choose to use classes of width 5, beginning at 0:

0–5   5–10   10–15   15–20   20–25   25–30

But we need to specify the classes so that each individual falls into exactly one class. For instance, what if exactly 5.0% of the residents in a state were born outside the United States? Because a value of 0.0% would go in the 0–5 class, we'll agree to place a value of 5.0% in the 5–10 class, a value of 10.0% in the 10–15 class, and so on. In reality, then, our classes for the percent of foreign-born residents in the states are

0 to <5   5 to <10   10 to <15   15 to <20   20 to <25   25 to <30

- *Find the count (frequency) or percent (relative frequency) of individuals in each class.* Here are a frequency table and a relative frequency table for these data:

Notice that the frequencies add to 50, the number of individuals (states) in the data, and that the relative frequencies add to 100%.

**Frequency table**

| Class | Count |
|---|---|
| 0 to <5 | 20 |
| 5 to <10 | 13 |
| 10 to <15 | 9 |
| 15 to <20 | 5 |
| 20 to <25 | 2 |
| 25 to <30 | 1 |
| **Total** | **50** |

**Relative frequency table**

| Class | Percent |
|---|---|
| 0 to <5 | 40 |
| 5 to <10 | 26 |
| 10 to <15 | 18 |
| 15 to <20 | 10 |
| 20 to <25 | 4 |
| 25 to <30 | 2 |
| **Total** | **100** |

- *Label and scale your axes and draw the histogram.* Label the horizontal axis with the variable whose distribution you are displaying. That's the percent of a state's residents who are foreign-born. The scale on the horizontal axis runs from 0 to 30 because that is the span of the classes we chose. The vertical axis contains the scale of counts or percents. Each bar represents a class. The base of the bar covers the class, and the bar height is the class frequency or relative frequency. Draw the bars with no horizontal space between them unless a class is empty, so that its bar has height zero.

Figure 1.16(a) shows a completed frequency histogram; Figure 1.16(b) shows a completed relative frequency histogram. The two graphs look identical except for the vertical scales.

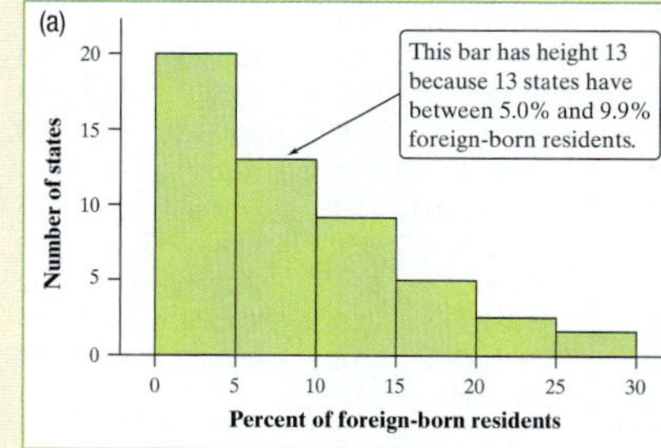


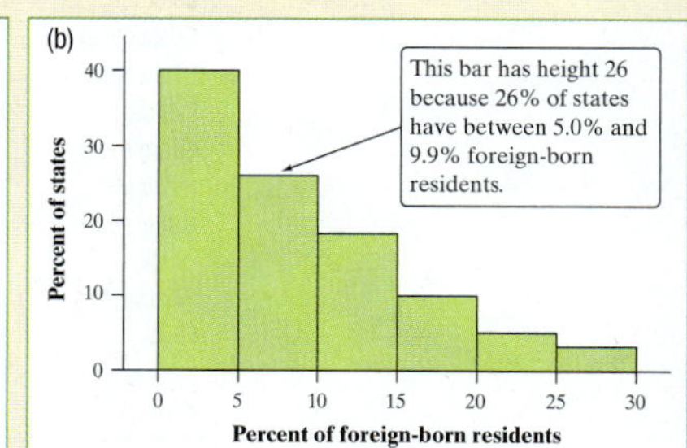


**FIGURE 1.16** (a) Frequency histogram and (b) relative frequency histogram of the distribution of the percent of foreign-born residents in the 50 states.

What do the histograms in Figure 1.16 tell us about the percent of foreign-born residents in the states? To find out, we follow our familiar routine: describe the pattern and look for any departures from the pattern.

*Shape:* The distribution is skewed to the right and unimodal. Most states have fewer than 10% foreign-born residents, but several states have much higher percents.

To find the center, remember that we're looking for the value having 25 states with smaller percents foreign-born and 25 with larger.

*Center:* From the graph, we see that the midpoint would fall somewhere in the 5.0% to 9.9% class. (Arranging the values in the table in order of size shows that the midpoint is 6.1%.)

*Spread:* The percent of foreign-born residents in the states varies from less than 5% to over 25%.

*Outliers:* We don't see any observations outside the overall pattern of the distribution.

Figure 1.17 shows (a) a frequency histogram and (b) a relative frequency histogram of the same distribution, with classes half as wide. The new classes are 0–2.4, 2.5–4.9, and so on. Now California, at 27.2%, stands out as a potential outlier in the right tail. The choice of classes in a histogram can influence the appearance of a distribution. Histograms with more classes show more detail but may have a less clear pattern.

**APPLET** The *Histogram Bin-Width* applet allows students to explore how changing the classes in a histogram can greatly alter the apparent shape of a distribution. Go to www.rossmanchance.com/applets and choose "Histogram Bin-Width."

(a)

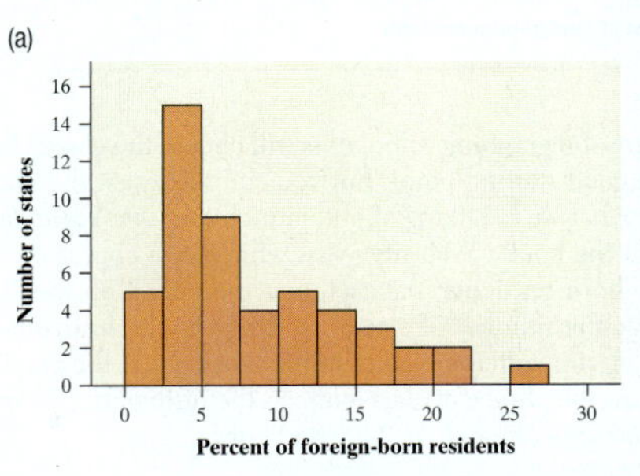


(b)

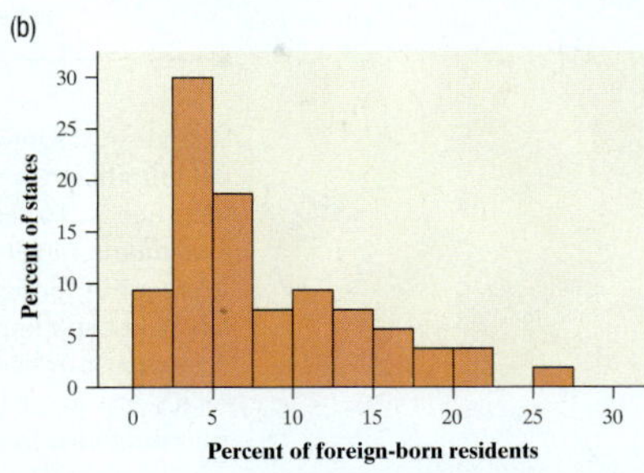


**FIGURE 1.17** (a) Frequency histogram and (b) relative frequency histogram of the distribution of the percent of foreign-born residents in the 50 states, with classes half as wide as in Figure 1.16.

Here are some important things to consider when you are constructing a histogram:

- Our eyes respond to the area of the bars in a histogram, so *be sure to choose classes that are all the same width.* Then area is determined by height, and all classes are fairly represented.
- There is no one right choice of the classes in a histogram. Too few classes will give a "skyscraper" graph, with all values in a few classes with tall bars. Too many will produce a "pancake" graph, with most classes having one or no observations. Neither choice will give a good picture of the shape of the distribution. Five classes is a good minimum.

**What are we actually doing when we make a histogram?** The dotplot on the left below shows the foreign-born resident data. We grouped the data values into classes of width 5, beginning with 0 to <5, as indicated by the dashed lines. Then we tallied the number of values in each class. The dotplot on the right shows the results of that process. Finally, we drew bars of the appropriate height for each class to get the completed histogram shown.

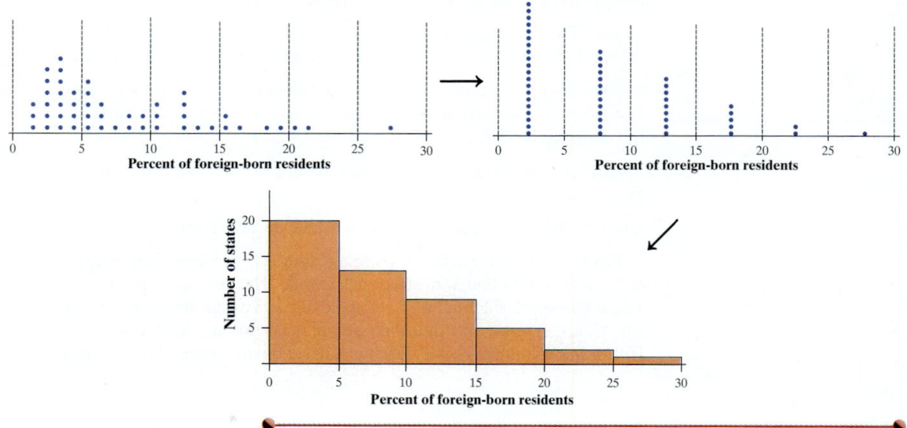

Statistical software and graphing calculators will choose the classes for you. The default choice is a good starting point, but you should adjust the classes to suit your needs. To see what we're talking about, launch the *One-Variable Statistical Calculator* applet at the book's Web site, www.whfreeman.com/tps5e. Select the "Percent of foreign-born residents" data set, and then click on the "Histogram" tab. You can change the number of classes by dragging the horizontal axis with your mouse or by entering different values in the boxes above the graph. By doing so, it's easy to see how the choice of classes affects the histogram. *Bottom line: Use your judgment in choosing classes to display the shape.*

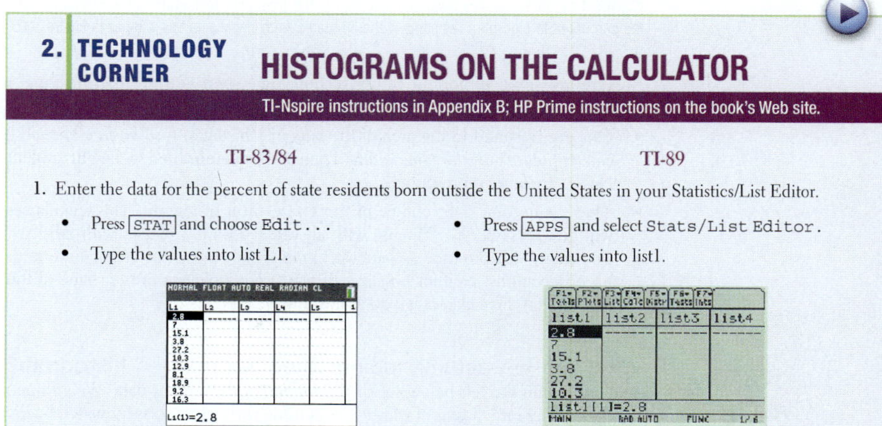

2. TECHNOLOGY CORNER

## HISTOGRAMS ON THE CALCULATOR

TI-Nspire instructions in Appendix B; HP Prime instructions on the book's Web site.

TI-83/84 | TI-89

1. Enter the data for the percent of state residents born outside the United States in your Statistics/List Editor.

TI-83/84:
- Press STAT and choose Edit...
- Type the values into list L1.

TI-89:
- Press APPS and select Stats/List Editor.
- Type the values into list1.

### Teaching Tip

From now on, each time you see a histogram, ask your students to "populate" each bar with dots—it will make histograms less abstract.

### APPLET

The *One-Variable Statistical Calculator* applet allows students to use existing data sets or enter their own data set and graph histograms. It also allows you to change the class size to see what happens to the shape. Go to www.whfreeman.com/tps5e (click on Statistical Applets and choose One-Variable Statistical Calculator). You can also make histograms at www.lock5stat.com/statkey by choosing the *One Quantitative Variable* or *One Quantitative and One Categorical Variable* (for comparative histograms) applets.

### TECHNOLOGY TIP

The screen shots we show for the TI-83/84 are from the TI-84 Plus C and use operating system 2.55MP. In many cases, the instructions are the same for older and newer operating systems. When they are not, we will provide separate instructions.

### TECHNOLOGY TIP

Remind students that they can watch a video presentation of this technology corner on the e-book or at the Web site www.whfreeman.com/tps5e.

### TECHNOLOGY TIP

Remind students that there are instructions for using the TI-Nspire in Appendix B. Also, if some of your students use a Casio graphing calculator, an AP® Statistics Resource Guide is available at http://education.casio.com/resource/PDFs/ap_statistics_resource_guide.PDF.

2. Set up a histogram in the Statistics Plots menu.

**TI-83/84**

- Press [2nd] [Y=] (STAT PLOT).
- Press [ENTER] or [1] to go into Plot1.

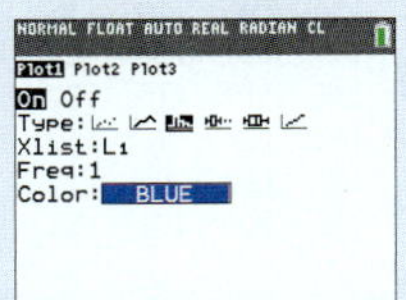

- Adjust the settings as shown.

**TI-89**

- Press [F2] and choose `Plot Setup...`
- With Plot1 highlighted, press [F1] to define.

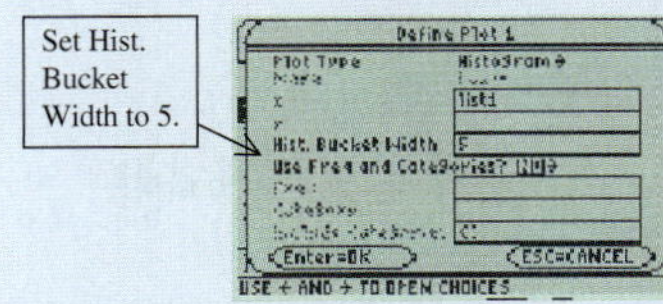


- Adjust the settings as shown.

3. Use ZoomStat (ZoomData on the TI-89) to let the calculator choose classes and make a histogram.

**TI-83/84**

- Press [ZOOM] and choose `ZoomStat`.
- Press [TRACE] and [◄] [►] to examine the classes.

**TI-89**

- Press [F5] (ZoomData).
- Press [F3] (Trace) and [◄] [►] to examine the classes.

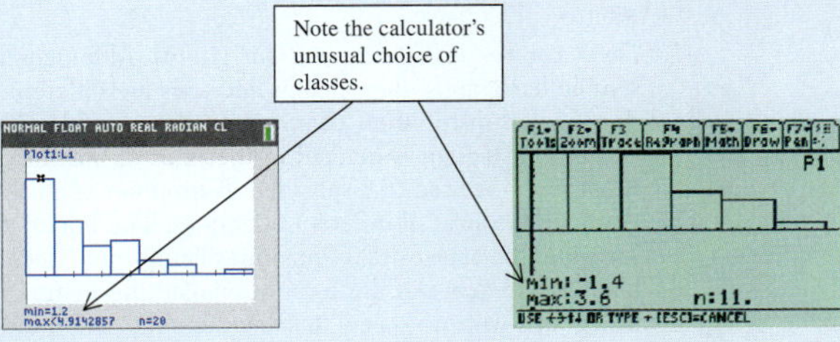


4. Adjust the classes to match those in Figure 1.16, and then graph the histogram.

**TI-83/84**

- Press [WINDOW] and enter the values shown below.
- Press [GRAPH].
- Press [TRACE] and [◄] [►] to examine the classes.

**TI-89**

- Press [◆] [F2] (WINDOW) and enter the values shown below.
- Press [◆] [F3] (GRAPH).
- Press [F3] (Trace) and [◄] [►] to examine the classes.

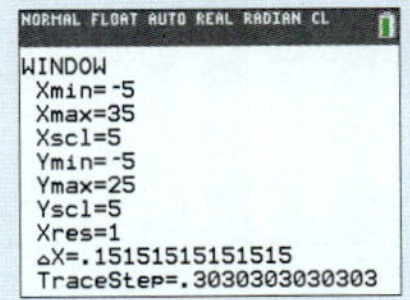


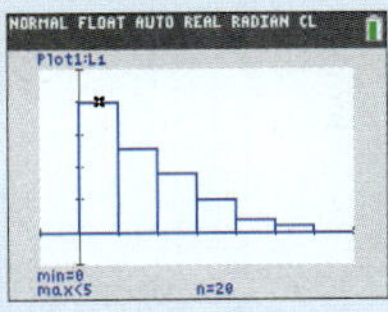

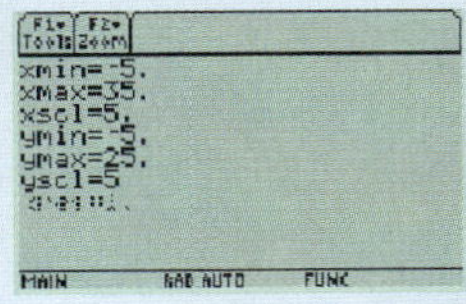


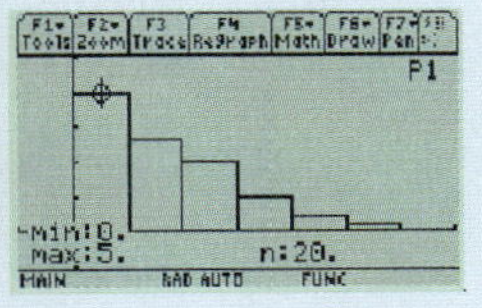

5. See if you can match the histogram in Figure 1.17.

> **AP® EXAM TIP** If you're asked to make a graph on a free-response question, be sure to label and scale your axes. Unless your calculator shows labels and scaling, don't just transfer a calculator screen shot to your paper.

## TECHNOLOGY TIP

Students may ask what the "Freq" setting is for. "Freq" stands for frequency, so when Freq = 1, each observation in L1 is counted once. The only time you might use something other than 1 is if you have data already in a frequency table. For example, suppose that the distribution of scores for your students on last year's AP® exam was

| Score | Frequency |
|---|---|
| 1 | 3 |
| 2 | 5 |
| 3 | 14 |
| 4 | 6 |
| 5 | 4 |

If you enter the values 1 to 5 in L1 and the frequencies in L2, then you should use Freq = L2 when making a graph.

## ANSWERS TO CYU

**1.** One possible histogram is shown below.

**2.** The distribution is roughly symmetric and bell-shaped. The typical IQ appears to be between 110 and 120 and the IQ's vary from 80 to 150. There do not appear to be any outliers.

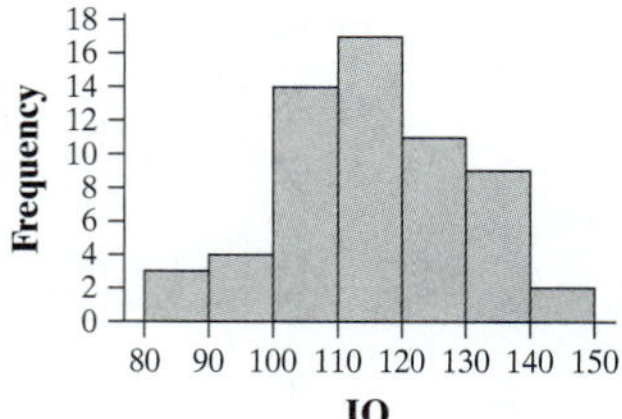

### CHECK YOUR UNDERSTANDING

Many people believe that the distribution of IQ scores follows a "bell curve," like the one shown in the margin. But is this really how such scores are distributed? The IQ scores of 60 fifth-grade students chosen at random from one school are shown below.[22]

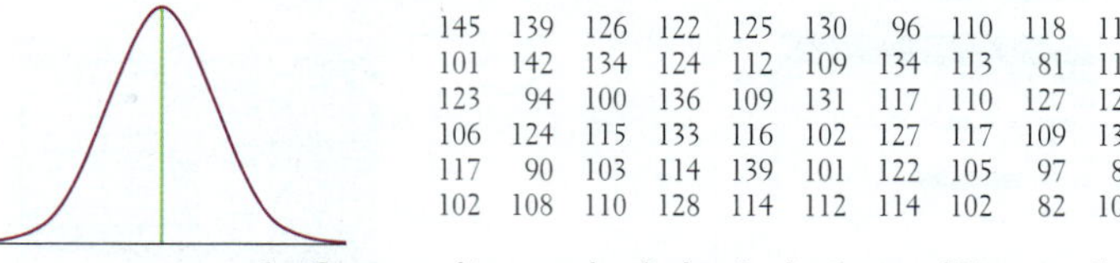

| | | | | | | | | | |
|---|---|---|---|---|---|---|---|---|---|
| 145 | 139 | 126 | 122 | 125 | 130 | 96 | 110 | 118 | 118 |
| 101 | 142 | 134 | 124 | 112 | 109 | 134 | 113 | 81 | 113 |
| 123 | 94 | 100 | 136 | 109 | 131 | 117 | 110 | 127 | 124 |
| 106 | 124 | 115 | 133 | 116 | 102 | 127 | 117 | 109 | 137 |
| 117 | 90 | 103 | 114 | 139 | 101 | 122 | 105 | 97 | 89 |
| 102 | 108 | 110 | 128 | 114 | 112 | 114 | 102 | 82 | 101 |

1. Construct a histogram that displays the distribution of IQ scores effectively.
2. Describe what you see. Is the distribution bell-shaped?

## Using Histograms Wisely

We offer several cautions based on common mistakes students make when using histograms.

1. *Don't confuse histograms and bar graphs.* Although histograms resemble bar graphs, their details and uses are different. A histogram displays the distribution of a quantitative variable. The horizontal axis of a histogram is marked in the units of measurement for the variable. A bar graph is used to display the distribution of a categorical variable or to compare the sizes of different quantities. The horizontal axis of a bar graph identifies the categories or quantities being compared. Draw bar graphs with blank space between the bars to separate the items being compared. Draw histograms with no space, to show the equal-width classes. For comparison, here is one of each type of graph from previous examples.

Histogram

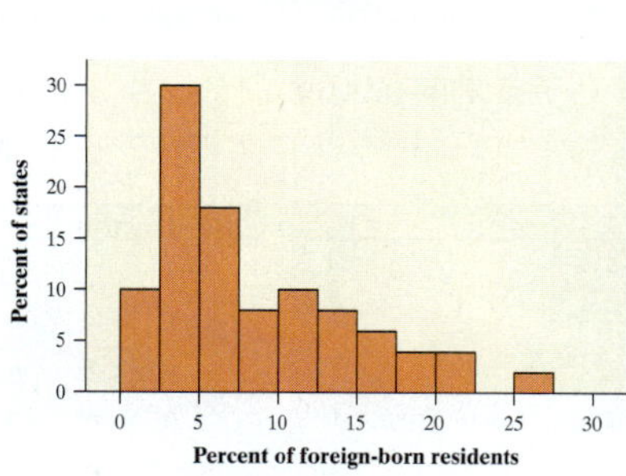


Bar graph

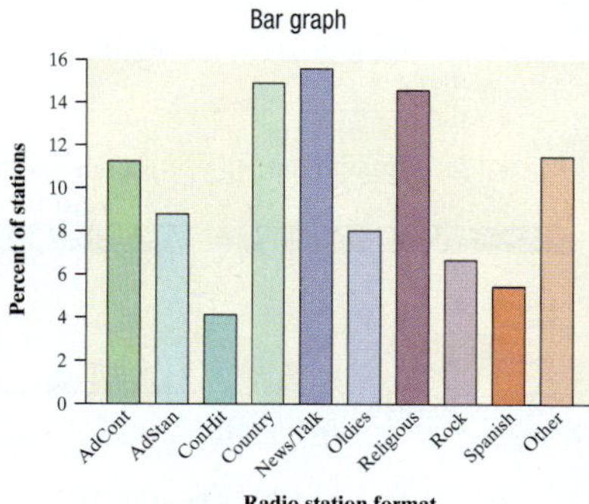


2. *Use percents instead of counts on the vertical axis when comparing distributions with different numbers of observations.* Mary was interested in comparing the reading levels of a medical journal and an

airline magazine. She counted the number of letters in the first 400 words of an article in the medical journal and of the first 100 words of an article in the airline magazine. Mary then used Minitab statistical software to produce the histograms shown in Figure 1.18(a). This figure is misleading—it compares frequencies, but the two samples were of very different sizes (100 and 400). Using the same data, Mary's teacher produced the histograms in Figure 1.18(b). By using relative frequencies, this figure provides an accurate comparison of word lengths in the two samples.

(a)

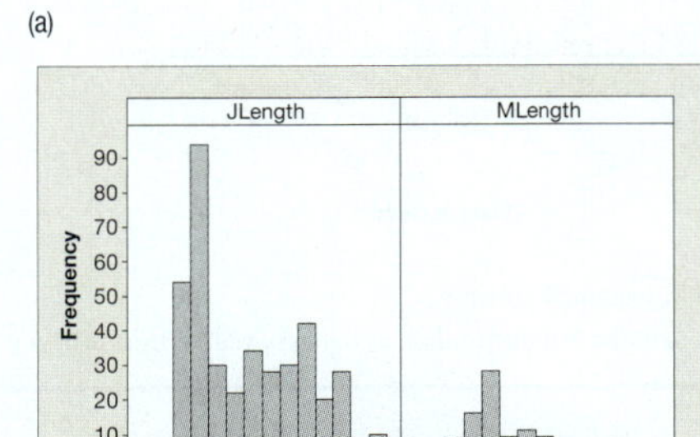


(b)

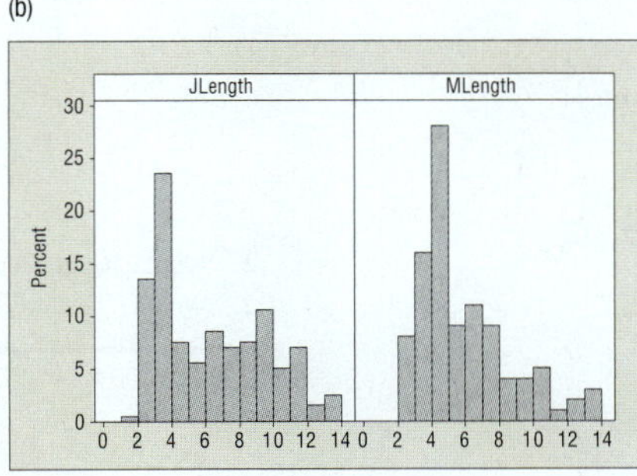


**FIGURE 1.18** Two sets of histograms comparing word lengths in articles from a journal and from an airline magazine. In (a), the vertical scale uses frequencies. The graph in (b) fixes this problem by using percents on the vertical scale.

3. *Just because a graph looks nice doesn't make it a meaningful display of data.* The students in a small statistics class recorded the number of letters in their first names. One student entered the data into an Excel spreadsheet and then used Excel's "chart maker" to produce the graph shown below left. What kind of graph is this? It's a bar graph that compares the raw data values. But first-name length is a quantitative variable, so a bar graph is not an appropriate way to display its distribution. The dotplot on the right is a much better choice.

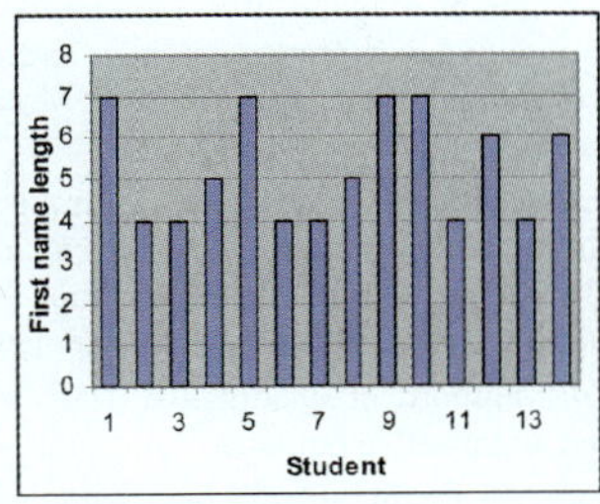


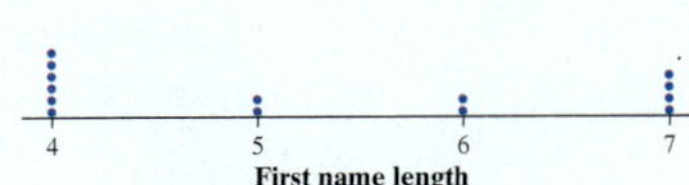


### TECHNOLOGY TIP

The bar graph on the left is not helpful because it doesn't allow us to identify the shape, center, or spread of the distribution of name length. To be a proper histogram, the variable should be on the horizontal axis and frequency (or relative frequency) should be on the vertical axis. Warn your students against using Excel for graphing quantitative variables unless they have a statistics add-on (which they probably don't). It is much better to use statistical software that is designed for such a purpose.

## CHECK YOUR UNDERSTANDING

About 1.6 million first-year students enroll in colleges and universities each year. What do they plan to study? The graph on the next page displays data on the percents of first-year students who plan to major in several discipline areas.[23]

## ✓ ANSWERS TO CYU

**1.** This is a bar graph because field of study is a categorical variable.

**2.** No, because the variable is categorical and the categories could be listed in any order on the horizontal axis.

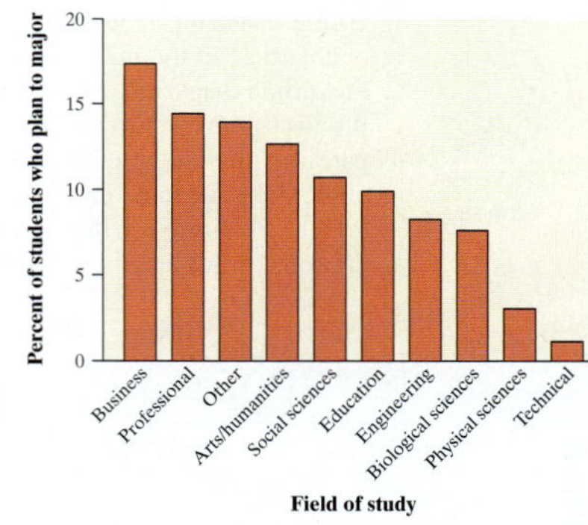


1. Is this a bar graph or a histogram? Explain.
2. Would it be correct to describe this distribution as right-skewed? Why or why not?

## Section 1.2 Summary

- You can use a **dotplot, stemplot,** or **histogram** to show the distribution of a quantitative variable. A dotplot displays individual values on a number line. Stemplots separate each observation into a stem and a one-digit leaf. Histograms plot the counts (frequencies) or percents (relative frequencies) of values in equal-width classes.
- When examining any graph, look for an **overall pattern** and for notable **departures** from that pattern. **Shape, center,** and **spread** describe the overall pattern of the distribution of a quantitative variable. **Outliers** are observations that lie outside the overall pattern of a distribution. Always look for outliers and try to explain them. Don't forget your SOCS!
- Some distributions have simple shapes, such as **symmetric, skewed to the left,** or **skewed to the right**. The number of **modes** (major peaks) is another aspect of overall shape. So are distinct clusters and gaps. Not all distributions have a simple overall shape, especially when there are few observations.
- When comparing distributions of quantitative data, be sure to compare shape, center, spread, and possible outliers.
- Remember: histograms are for quantitative data; bar graphs are for categorical data. Also, be sure to use relative frequency histograms when comparing data sets of different sizes.

### TRM Section 1.2 Quizzes

There are three different quizzes available for this section in the Teacher's Resource Materials. Versions A and B are roughly parallel. Find them by clicking on the link in the TE-book, opening the TRCD, or logging onto the book's companion Web site www.whfreeman.com/tps5e (teacher log-in required).

**1.2 TECHNOLOGY CORNER**

TI-Nspire instructions in Appendix B; HP Prime instructions on the book's Web site.

2. Histograms on the calculator page 36

# Section 1.2 Exercises

37. **Feeling sleepy?** Students in a college statistics class responded to a survey designed by their teacher. One of the survey questions was "How much sleep did you get last night?" Here are the data (in hours):

| 9 | 6 | 8 | 6 | 8 | 8 | 6 | 6.5 | 6 | 7 | 9 | 4 | 3 | 4 |
|---|---|---|---|---|---|---|---|---|---|---|---|---|---|
| 5 | 6 | 11 | 6 | 3 | 6 | 6 | 10 | 7 | 8 | 4.5 | 9 | 7 | 7 |

(a) Make a dotplot to display the data.

(b) Describe the overall pattern of the distribution and any departures from that pattern.

38. **Olympic gold!** The following table displays the total number of gold medals won by a sample of countries in the 2012 Summer Olympic Games in London.

| Country | Gold medals | Country | Gold medals |
|---|---|---|---|
| Sri Lanka | 0 | Thailand | 0 |
| China | 38 | Kuwait | 0 |
| Vietnam | 0 | Bahamas | 1 |
| Great Britain | 29 | Kenya | 2 |
| Norway | 2 | Trinidad and Tobago | 1 |
| Romania | 2 | Greece | 0 |
| Switzerland | 2 | Mozambique | 0 |
| Armenia | 0 | Kazakhstan | 7 |
| Netherlands | 6 | Denmark | 2 |
| India | 0 | Latvia | 1 |
| Georgia | 1 | Czech Republic | 4 |
| Kyrgyzstan | 0 | Hungary | 8 |
| Costa Rica | 0 | Sweden | 1 |
| Brazil | 3 | Uruguay | 0 |
| Uzbekistan | 1 | United States | 46 |

(a) Make a dotplot to display these data. Describe the overall pattern of the distribution and any departures from that pattern.

(b) Overall, 205 countries participated in the 2012 Summer Olympics, of which 54 won at least one gold medal. Do you believe that the sample of countries listed in the table is representative of this larger population? Why or why not?

39. **U.S. women's soccer—2012** Earlier, we examined data on the number of goals scored by the U.S. women's soccer team in games played in the 12 months prior to the 2012 Olympics. The dotplot below displays the goal differential for those same games, computed as U.S. score minus opponent's score. pg 26

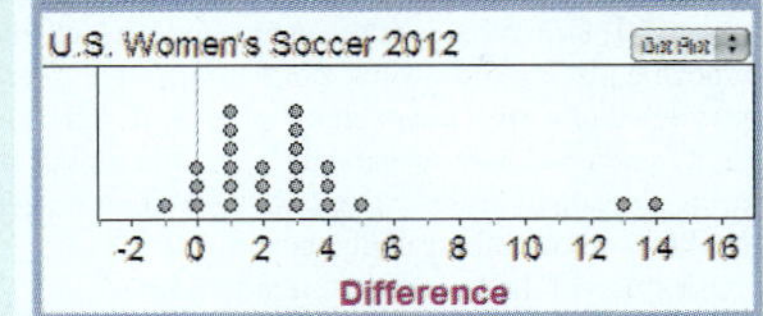


(a) Explain what the dot above −1 represents.

(b) What does the graph tell us about how well the team did in 2012? Be specific.

40. **Fuel efficiency** In an earlier example, we examined data on highway gas mileages of model year 2012 midsize cars. The following dotplot shows the difference (highway − city) in EPA mileage ratings for each of the 24 car models from the earlier example.

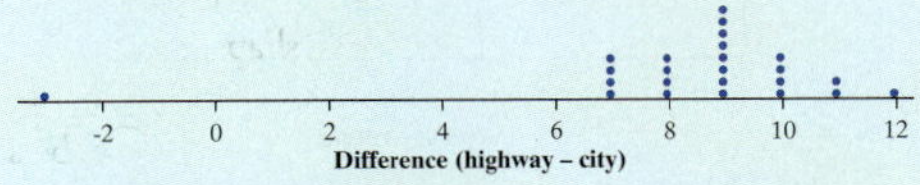


(a) Explain what the dot above 12 represents.

(b) What does the graph tell us about fuel economy in the city versus on the highway for these car models? Be specific.

**TRM Full Solutions to Section 1.2 Exercises**

## Answers to Section 1.2 Exercises

**1.37 (a)** The graph is shown below. **(b)** The distribution is roughly symmetric with a midpoint of 6 hours. The hours of sleep vary from 3 to 11. There do not appear to be any outliers.

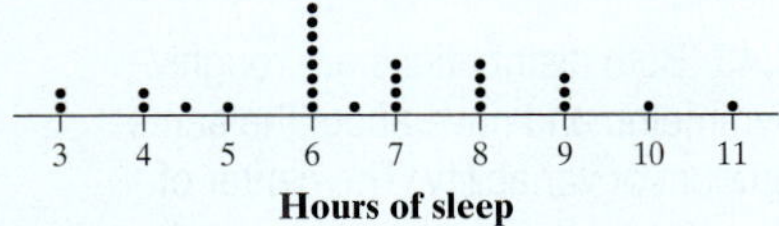


**1.38 (a)** The graph is shown below. The distribution is strongly skewed to the right with a peak at 0. The midpoint is 1 gold medal and the number of gold medals varies from 0 to 46. There are outliers at 29 (Great Britain), 38 (China), and 46 (United States). **(b)** No, because 19 out of the 30 countries in the sample (or about 63%) won gold medals. Overall, only about 54/205 = 26% won gold medals.

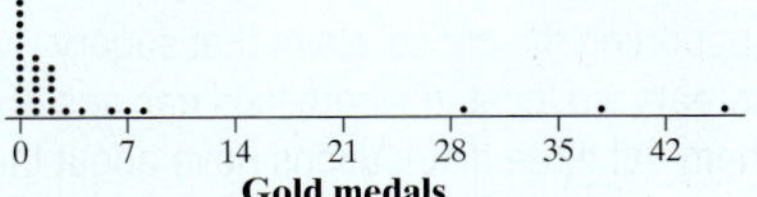


**1.39 (a)** This dot represents a game where the opposing team won by 1 goal. **(b)** All but 4 of the 25 values are positive, which indicates that the U.S. women's soccer team had a very good season. They won 21/25 = 84% of their games.

**1.40 (a)** This dot represents a car that got 12 mpg more on the highway than it did in city driving. **(b)** The EPA mpg rating is higher on the highway than in the city for all but one of the cars. The other 23 cars averaged at least 7 mpg more on the highway.

**1.41** As coins get older, they get taken out of circulation and new coins are introduced, meaning that most coins will be from recent years with a few from previous years.

**1.42** The shape of this distribution is roughly symmetric and somewhat uniform. This isn't surprising because we expect that the last digit in a phone number is equally likely to be any digit from 0 to 9.

**1.43** Both distributions are roughly symmetric and have about the same amount of variability. The center of the internal distribution is greater than the center of the external distribution, indicating that external rewards do not promote creativity. Neither distribution appears to have outliers.

**1.44** The distributions of sugar content on shelves 1 and 3 are both skewed to the right, while the distribution of sugar content on shelf 2 is skewed to the left. The center for the distribution of sugar on shelf 2 is higher than the center for the distributions of sugar on shelves 1 or 3, partially supporting the critics' claim that sugary cereals are located where kids can see them. All three distributions have about the same amount of variability and none of the distributions appear to have any outliers.

**1.45** **(a)** Otherwise, most of the data would appear on just a few stems, making it hard to identify the shape of the distribution. **(b)** Key: 12|1 means that 12.1% of that state's residents are aged 25 to 34. **(c)** The distribution of percent of residents aged 25–34 is roughly symmetric with a possible outlier at 16.0%. The center is around 13%. Other than the outlier at 16.0%, the values vary from 11.4% to 15.1%.

**1.46** **(a)** Otherwise, most of the data would appear on just a few stems, making it hard to identify the shape of the distribution. **(b)** Key: 2|3 means that an 8-ounce serving of that soft drink has 23 mg of caffeine. **(c)** This distribution is slightly skewed to the right. The center is around 28 mg and the values vary from 15 mg to 47 mg. There do not appear to be any outliers.



**41. Dates on coins** Suppose that you and your friends emptied your pockets of coins and recorded the year marked on each coin. The distribution of dates would be skewed to the left. Explain why.

**42. Phone numbers** The dotplot below displays the last digit of 100 phone numbers chosen at random from a phone book. Describe the shape of the distribution. Does this shape make sense to you? Explain.

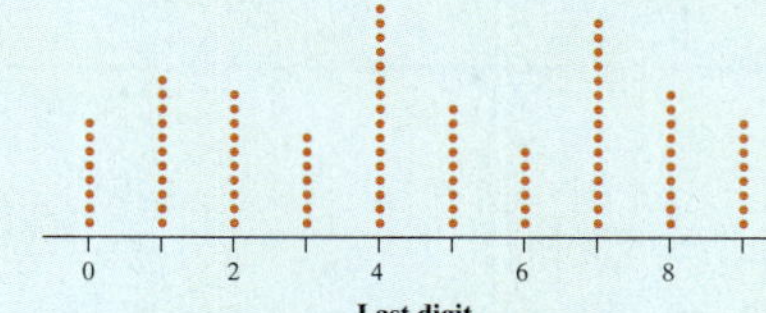


**43. Creative writing** Do external rewards—things like money, praise, fame, and grades—promote creativity? Researcher Teresa Amabile designed an experiment to find out. She recruited 47 experienced creative writers who were college students and divided them into two groups using a chance process (like drawing names from a hat). The students in one group were given a list of statements about external reasons (E) for writing, such as public recognition, making money, or pleasing their parents. Students in the other group were given a list of statements about internal reasons (I) for writing, such as expressing yourself and enjoying playing with words. Both groups were then instructed to write a poem about laughter. Each student's poem was rated separately by 12 different poets using a creativity scale.[24] These ratings were averaged to obtain an overall creativity score for each poem.

pg 30

Dotplots of the two groups' creativity scores are shown below. Compare the two distributions. What do you conclude about whether external rewards promote creativity?

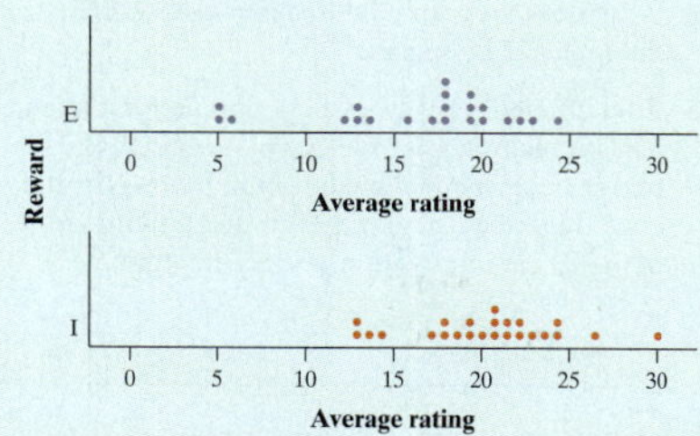


**44. Healthy cereal?** Researchers collected data on 77 brands of cereal at a local supermarket.[25] For each brand, the sugar content (grams per serving) and the shelf in the store on which the cereal was located (1 = bottom, 2 = middle, 3 = top) were recorded. A dotplot of the data is shown below. Compare the three distributions. Critics claim that supermarkets tend to put sugary kids' cereals on lower shelves, where the kids can see them. Do the data from this study support this claim?

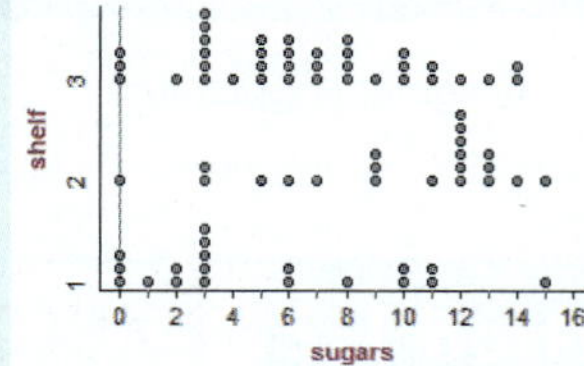


**45. Where do the young live?** Below is a stemplot of the percent of residents aged 25 to 34 in each of the 50 states. As in the stemplot for older residents (page 33), the stems are whole percents, and the leaves are tenths of a percent. This time, each stem has been split in two, with values having leaves 0 through 4 placed on one stem, and values ending in 5 through 9 placed on another stem.

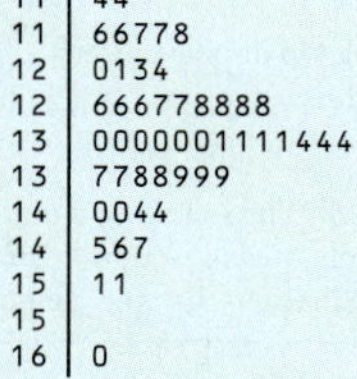

11 | 44
11 | 66778
12 | 0134
12 | 666778888
13 | 0000001111444
13 | 7788999
14 | 0044
14 | 567
15 | 11
15 |
16 | 0

(a) Why did we split stems?

(b) Give an appropriate key for this stemplot.

(c) Describe the shape, center, and spread of the distribution. Are there any outliers?

**46. Watch that caffeine!** The U.S. Food and Drug Administration (USFDA) limits the amount of caffeine in a 12-ounce can of carbonated beverage to 72 milligrams. That translates to a maximum of 48 milligrams of caffeine per 8-ounce serving. Data on the caffeine content of popular soft drinks (in milligrams per 8-ounce serving) are displayed in the stemplot below.

1 | 556
2 | 033344
2 | 55667778888899
3 | 113
3 | 55567778
4 | 33
4 | 77

(a) Why did we split stems?

(b) Give an appropriate key for this graph.

(c) Describe the shape, center, and spread of the distribution. Are there any outliers?

**1.47** **(a)** The stemplots are given at right. The stemplot with split stems makes it easier to see the shape of the distribution. **(b)** The distribution is slightly skewed to the right with a center near 780 mm, and values that vary from around 600 mm to 960 mm. There do not appear to be any outliers. **(c)** In El Niño years there is typically less rain than in other years (18 of 23 years).

**Without splitting stems**

6|0 3 5 5 7
7|0 1 2 4 4 8 8 9 9 9
8|1 1 3 6 6 7
9|0 6

Key: 6|3 = 630 mm of rain

**With splitting stems**

6|0 3
6|5 5 7
7|0 1 2 4 4
7|8 8 9 9 9
8|1 1 3
8|6 6 7
9|0
9|6

**47. El Niño and the monsoon** It appears that El Niño, the periodic warming of the Pacific Ocean west of South America, affects the monsoon rains that are essential for agriculture in India. Here are the monsoon rains (in millimeters) for the 23 strong El Niño years between 1871 and 2004:[26]

| | | | | | | | | | | | |
|---|---|---|---|---|---|---|---|---|---|---|---|
| 628 | 669 | 740 | 651 | 710 | 736 | 717 | 698 | 653 | 604 | 781 | 784 |
| 790 | 811 | 830 | 858 | 858 | 896 | 806 | 790 | 792 | 957 | 872 | |

(a) To make a stemplot of these rainfall amounts, round the data to the nearest 10, so that stems are hundreds of millimeters and leaves are tens of millimeters. Make two stemplots, with and without splitting the stems. Which plot do you prefer? Why?

(b) Describe the shape, center, and spread of the distribution.

(c) The average monsoon rainfall for all years from 1871 to 2004 is about 850 millimeters. What effect does El Niño appear to have on monsoon rains?

**48. Shopping spree** A marketing consultant observed 50 consecutive shoppers at a supermarket. One variable of interest was how much each shopper spent in the store. Here are the data (in dollars), arranged in increasing order:

| | | | | | | | | | |
|---|---|---|---|---|---|---|---|---|---|
| 3.11 | 8.88 | 9.26 | 10.81 | 12.69 | 13.78 | 15.23 | 15.62 | 17.00 | 17.39 |
| 18.36 | 18.43 | 19.27 | 19.50 | 19.54 | 20.16 | 20.59 | 22.22 | 23.04 | 24.47 |
| 24.58 | 25.13 | 26.24 | 26.26 | 27.65 | 28.06 | 28.08 | 28.38 | 32.03 | 34.98 |
| 36.37 | 38.64 | 39.16 | 41.02 | 42.97 | 44.08 | 44.67 | 45.40 | 46.69 | 48.65 |
| 50.39 | 52.75 | 54.80 | 59.07 | 61.22 | 70.32 | 82.70 | 85.76 | 86.37 | 93.34 |

(a) Round each amount to the nearest dollar. Then make a stemplot using tens of dollars as the stems and dollars as the leaves.

(b) Make another stemplot of the data by splitting stems. Which of the plots shows the shape of the distribution better?

(c) Write a few sentences describing the amount of money spent by shoppers at this supermarket.

**49. Do women study more than men?** We asked the students in a large first-year college class how many minutes they studied on a typical weeknight. Here are the responses of random samples of 30 women and 30 men from the class:

| Women | | | | | Men | | | | |
|---|---|---|---|---|---|---|---|---|---|
| 180 | 120 | 180 | 360 | 240 | 90 | 120 | 30 | 90 | 200 |
| 120 | 180 | 120 | 240 | 170 | 90 | 45 | 30 | 120 | 75 |
| 150 | 120 | 180 | 180 | 150 | 150 | 120 | 60 | 240 | 300 |
| 200 | 150 | 180 | 150 | 180 | 240 | 60 | 120 | 60 | 30 |
| 120 | 60 | 120 | 180 | 180 | 30 | 230 | 120 | 95 | 150 |
| 90 | 240 | 180 | 115 | 120 | 0 | 200 | 120 | 120 | 180 |

(a) Examine the data. Why are you not surprised that most responses are multiples of 10 minutes? Are there any responses you consider suspicious?

(b) Make a back-to-back stemplot to compare the two samples. Does it appear that women study more than men (or at least claim that they do)? Justify your answer.

**50. Basketball playoffs** Here are the numbers of points scored by teams in the California Division I-AAA high school basketball playoffs in a single day's games:[27]

| | | | | | | | | | | | | |
|---|---|---|---|---|---|---|---|---|---|---|---|---|
| 71 | 38 | 52 | 47 | 55 | 53 | 76 | 65 | 77 | 63 | 65 | 63 | 68 |
| 54 | 64 | 62 | 87 | 47 | 64 | 56 | 78 | 64 | 58 | 51 | 91 | 74 |
| 71 | 41 | 67 | 62 | 106 | 46 | | | | | | | |

On the same day, the final scores of games in Division V-AA were

| | | | | | | | | | | | |
|---|---|---|---|---|---|---|---|---|---|---|---|
| 98 | 45 | 67 | 44 | 74 | 60 | 96 | 54 | 92 | 72 | 93 | 46 |
| 98 | 67 | 62 | 37 | 37 | 36 | 69 | 44 | 86 | 66 | 66 | 58 |

(a) Construct a back-to-back stemplot to compare the points scored by the 32 teams in the Division I-AAA playoffs and the 24 teams in the Division V-AA playoffs.

(b) Write a few sentences comparing the two distributions.

**51. Returns on common stocks** The return on a stock is the change in its market price plus any dividend payments made. Total return is usually expressed as a percent of the beginning price. The figure below shows a histogram of the distribution of the monthly returns for all common stocks listed on U.S. markets over a 273-month period.[28] The extreme low outlier represents the market crash of October 1987, when stocks lost 23% of their value in one month.

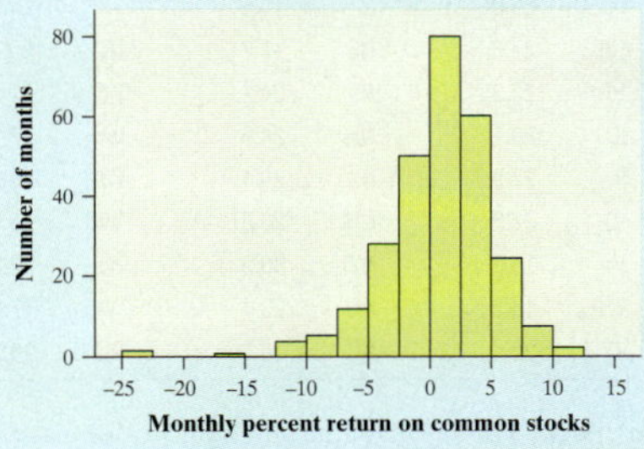


(a) Describe the overall shape of the distribution of monthly returns.

(b) What is the approximate center of this distribution?

(c) Approximately what were the smallest and largest monthly returns, leaving out the outliers?

(d) A return less than zero means that stocks lost value in that month. About what percent of all months had returns less than zero?

**1.48 (a)** and **(b)** The stemplots are shown at right and above. The stemplot with the split stems shows the skewness, gaps, and outliers more clearly. **(c)** The distribution of the amount of money spent is skewed to the right, with a center around $28, and values that vary from $3 to $93. There are some outliers on the high end ($83, $86, and $93).

**(a)**

| Stem | Leaves |
|---|---|
| 0 | 399 |
| 1 | 1345677889 |
| 2 | 000123455668888 |
| 3 | 25699 |
| 4 | 1345579 |
| 5 | 0359 |
| 6 | 1 |
| 7 | 0 |
| 8 | 366 |
| 9 | 3 |

Key: 9 | 3 = $93 spent

**(b)**

| Stem | Leaves |
|---|---|
| 0 | 3 |
| 0 | 99 |
| 1 | 134 |
| 1 | 5677889 |
| 2 | 0001234 |
| 2 | 55668888 |
| 3 | 2 |
| 3 | 5699 |
| 4 | 134 |
| 4 | 5579 |
| 5 | 03 |
| 5 | 59 |
| 6 | 1 |
| 6 | |
| 7 | 0 |
| 7 | |
| 8 | 3 |
| 8 | 66 |
| 9 | 3 |

Key: 9 | 3 = $93 spent

**1.49 (a)** Most people will round their answers to the nearest 10 minutes (or 30 or 60). The students who claimed 300 and 360 minutes of studying on a typical weeknight may have been exaggerating. **(b)** The stemplots suggest that women (claim to) study more than men. The center for women (about 175 minutes) is greater than the center for men (about 120 minutes).

| Women | Stem | Men |
|---|---|---|
| | 0 | 03333 |
| 96 | 0 | 56668999 |
| 22222222 | 1 | 02222222 |
| 88888888875555 | 1 | 558 |
| 4440 | 2 | 00344 |
| | 2 | |
| | 3 | 0 |
| 6 | 3 | |

Key: 2 | 3 = 230 minutes

**1.50 (a)** The stemplots are given below. **(b)** Both distributions are roughly symmetric, although the distribution in Division I-AAA is unimodal and the distribution in Division V-AA has more than one peak. While the Division V-AA center is slightly larger than the Division I-AAA center, the scores in both divisions are about equally variable. The team that scored 106 points in Division I-AAA might be an outlier.

| Division I-AAA | Stem | Division V-AA |
|---|---|---|
| 8 | 3 | 677 |
| 7761 | 4 | 4456 |
| 8654321 | 5 | 48 |
| 87554443322 | 6 | 0266779 |
| 876411 | 7 | 24 |
| 7 | 8 | 6 |
| 1 | 9 | 23688 |
| 6 | 10 | |

Key: 3 | 6 = 36 points scored

**1.51 (a)** The distribution is slightly skewed to the left and unimodal. **(b)** The center is between 0% and 2.5%. **(c)** The highest return was between 10% and 12.5%. Ignoring the low outliers, the lowest return was between −12.5% and −10% **(d)** About 37% of these months (102 out of 273) had negative returns.

**1.52** The distribution of lengths of words in Shakespeare's plays is skewed to the right and unimodal. The center is around 4 letters. The lengths of the words vary from 1 to 12 letters.

**1.53** **(a)** The histogram is given below. **(b)** The distribution of travel times is roughly symmetric. The center is near 23 minutes and the values vary from 15.5 to 30.9 minutes. There do not appear to be any outliers.

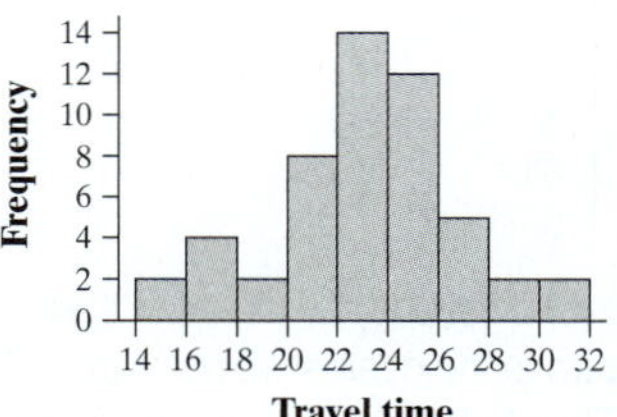


**1.54** **(a)** The histogram is given below. **(b)** The distribution of the emissions is skewed to the right with center near 4 metric tons per person. The values of $CO_2$ vary from 0.1 to 19.6 metric tons per person. There appear to be three outliers: Canada, Australia, and the United States.

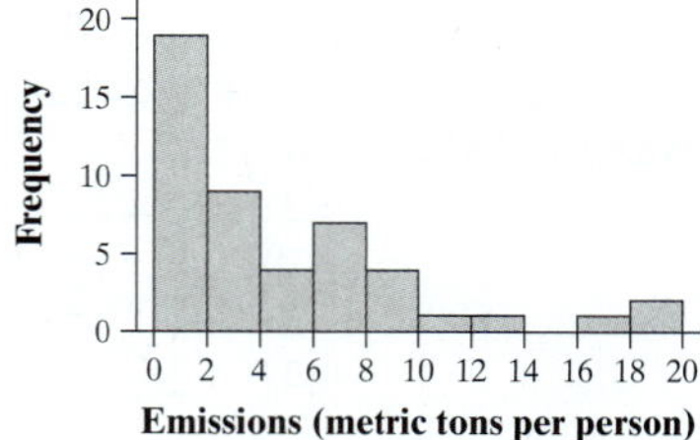


**1.55** The histogram is given below. The distribution of DRP scores is roughly symmetric with center around 35. The DRP scores vary from 14 to 54. There don't appear to be any outliers.

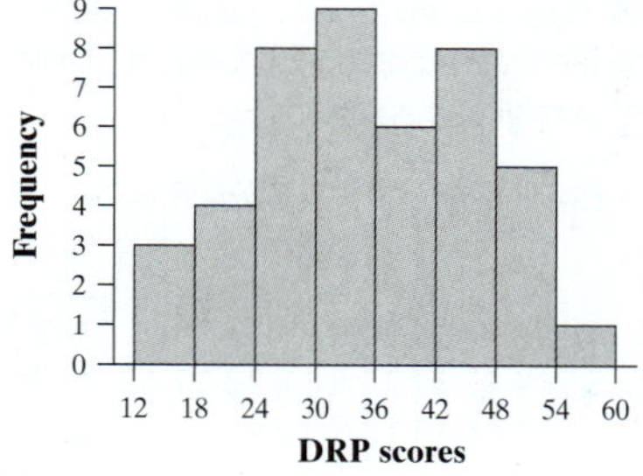




52. **Shakespeare** The histogram below shows the distribution of lengths of words used in Shakespeare's plays.[29] Describe the shape, center, and spread of this distribution.

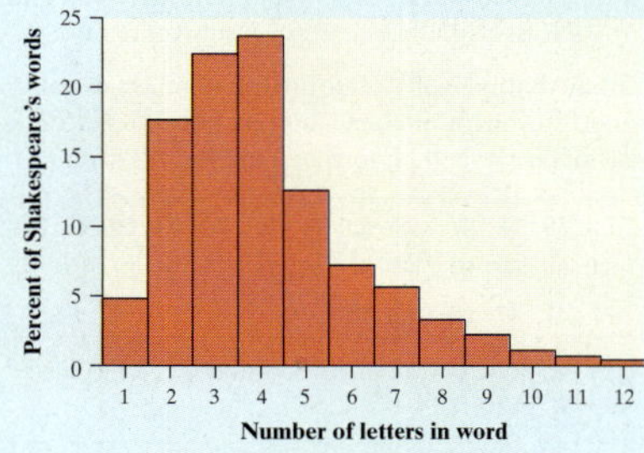


53. **Traveling to work** How long do people travel each day to get to work? The following table gives the average travel times to work (in minutes) for workers in each state and the District of Columbia who are at least 16 years old and don't work at home.[30]

| | | | | | |
|---|---|---|---|---|---|
| AL | 23.6 | LA | 25.1 | OH | 22.1 |
| AK | 17.7 | ME | 22.3 | OK | 20.0 |
| AZ | 25.0 | MD | 30.6 | OR | 21.8 |
| AR | 20.7 | MA | 26.6 | PA | 25.0 |
| CA | 26.8 | MI | 23.4 | RI | 22.3 |
| CO | 23.9 | MN | 22.0 | SC | 22.9 |
| CT | 24.1 | MS | 24.0 | SD | 15.9 |
| DE | 23.6 | MO | 22.9 | TN | 23.5 |
| FL | 25.9 | MT | 17.6 | TX | 24.6 |
| GA | 27.3 | NE | 17.7 | UT | 20.8 |
| HI | 25.5 | NV | 24.2 | VT | 21.2 |
| ID | 20.1 | NH | 24.6 | VA | 26.9 |
| IL | 27.9 | NJ | 29.1 | WA | 25.2 |
| IN | 22.3 | NM | 20.9 | WV | 25.6 |
| IA | 18.2 | NY | 30.9 | WI | 20.8 |
| KS | 18.5 | NC | 23.4 | WY | 17.9 |
| KY | 22.4 | ND | 15.5 | DC | 29.2 |

(a) Make a histogram of the travel times using classes of width 2 minutes, starting at 14 minutes. That is, the first class is 14 to 16 minutes, the second is 16 to 18 minutes, and so on.

(b) The shape of the distribution is a bit irregular. Is it closer to symmetric or skewed? Describe the center and spread of the distribution. Are there any outliers?

54. **Carbon dioxide emissions** Burning fuels in power plants and motor vehicles emits carbon dioxide ($CO_2$), which contributes to global warming. The table below displays $CO_2$ emissions per person from countries with populations of at least 20 million.[31]

(a) Make a histogram of the data using classes of width 2, starting at 0.

(b) Describe the shape, center, and spread of the distribution. Which countries are outliers?

Carbon dioxide emissions (metric tons per person)

| Country | $CO_2$ | Country | $CO_2$ |
|---|---|---|---|
| Algeria | 2.6 | Mexico | 3.7 |
| Argentina | 3.6 | Morocco | 1.4 |
| Australia | 18.4 | Myanmar | 0.2 |
| Bangladesh | 0.3 | Nepal | 0.1 |
| Brazil | 1.8 | Nigeria | 0.4 |
| Canada | 17.0 | Pakistan | 0.8 |
| China | 3.9 | Peru | 1.0 |
| Colombia | 1.3 | Philippines | 0.9 |
| Congo | 0.2 | Poland | 7.8 |
| Egypt | 2.0 | Romania | 4.2 |
| Ethiopia | 0.1 | Russia | 10.8 |
| France | 6.2 | Saudi Arabia | 13.8 |
| Germany | 9.9 | South Africa | 7.0 |
| Ghana | 0.3 | Spain | 7.9 |
| India | 1.1 | Sudan | 0.3 |
| Indonesia | 1.6 | Tanzania | 0.1 |
| Iran | 6.0 | Thailand | 3.3 |
| Iraq | 2.9 | Turkey | 3.0 |
| Italy | 7.8 | Ukraine | 6.3 |
| Japan | 9.5 | United Kingdom | 8.8 |
| Kenya | 0.3 | United States | 19.6 |
| Korea, North | 3.3 | Uzbekistan | 4.2 |
| Korea, South | 9.3 | Venezuela | 5.4 |
| Malaysia | 5.5 | Vietnam | 1.0 |

55. **DRP test scores** There are many ways to measure the reading ability of children. One frequently used test is the Degree of Reading Power (DRP). In a research study on third-grade students, the DRP was administered to 44 students.[32] Their scores were:

| | | | | | | | | | | |
|---|---|---|---|---|---|---|---|---|---|---|
| 40 | 26 | 39 | 14 | 42 | 18 | 25 | 43 | 46 | 27 | 19 |
| 47 | 19 | 26 | 35 | 34 | 15 | 44 | 40 | 38 | 31 | 46 |
| 52 | 25 | 35 | 35 | 33 | 29 | 34 | 41 | 49 | 28 | 52 |
| 47 | 35 | 48 | 22 | 33 | 41 | 51 | 27 | 14 | 54 | 45 |

Make a histogram to display the data. Write a paragraph describing the distribution of DRP scores.

**56. Drive time** Professor Moore, who lives a few miles outside a college town, records the time he takes to drive to the college each morning. Here are the times (in minutes) for 42 consecutive weekdays:

| | | | | | | | | | | |
|---|---|---|---|---|---|---|---|---|---|---|
| 8.25 | 7.83 | 8.30 | 8.42 | 8.50 | 8.67 | 8.17 | 9.00 | 9.00 | 8.17 | 7.92 |
| 9.00 | 8.50 | 9.00 | 7.75 | 7.92 | 8.00 | 8.08 | 8.42 | 8.75 | 8.08 | 9.75 |
| 8.33 | 7.83 | 7.92 | 8.58 | 7.83 | 8.42 | 7.75 | 7.42 | 6.75 | 7.42 | 8.50 |
| 8.67 | 10.17 | 8.75 | 8.58 | 8.67 | 9.17 | 9.08 | 8.83 | 8.67 | | |

Make a histogram to display the data. Write a paragraph describing the distribution of Professor Moore's drive times.

**57. The statistics of writing style** Numerical data can distinguish different types of writing and, sometimes, even individual authors. Here are data on the percent of words of 1 to 15 letters used in articles in *Popular Science* magazine:[33]

| Length: | 1 | 2 | 3 | 4 | 5 | 6 | 7 | 8 | 9 | 10 | 11 | 12 | 13 | 14 | 15 |
|---|---|---|---|---|---|---|---|---|---|---|---|---|---|---|---|
| Percent: | 3.6 | 14.8 | 18.7 | 16.0 | 12.5 | 8.2 | 8.1 | 5.9 | 4.4 | 3.6 | 2.1 | 0.9 | 0.6 | 0.4 | 0.2 |

(a) Make a histogram of this distribution. Describe its shape, center, and spread.

(b) How does the distribution of lengths of words used in *Popular Science* compare with the similar distribution for Shakespeare's plays in Exercise 52? Look in particular at short words (2, 3, and 4 letters) and very long words (more than 10 letters).

**58. Chest out, Soldier!** In 1846, a published paper provided chest measurements (in inches) of 5738 Scottish militiamen. The table below summarizes the data.[34]

| Chest size | Count | Chest size | Count |
|---|---|---|---|
| 33 | 3 | 41 | 934 |
| 34 | 18 | 42 | 658 |
| 35 | 81 | 43 | 370 |
| 36 | 185 | 44 | 92 |
| 37 | 420 | 45 | 50 |
| 38 | 749 | 46 | 21 |
| 39 | 1073 | 47 | 4 |
| 40 | 1079 | 48 | 1 |

(a) Make a histogram of this distribution.

(b) Describe the shape, center, and spread of the chest measurements distribution. Why might this information be useful?

**59. Paying for championships** Does paying high salaries lead to more victories in professional sports? The New York Yankees have long been known for having Major League Baseball's highest team payroll. And over the years, the team has won many championships. This strategy didn't pay off in 2008, when the Philadelphia Phillies won the World Series. Maybe the Yankees didn't spend enough money that year. The graph below shows histograms of the salary distributions for the two teams during the 2008 season. Why can't you use this graph to effectively compare the team payrolls?

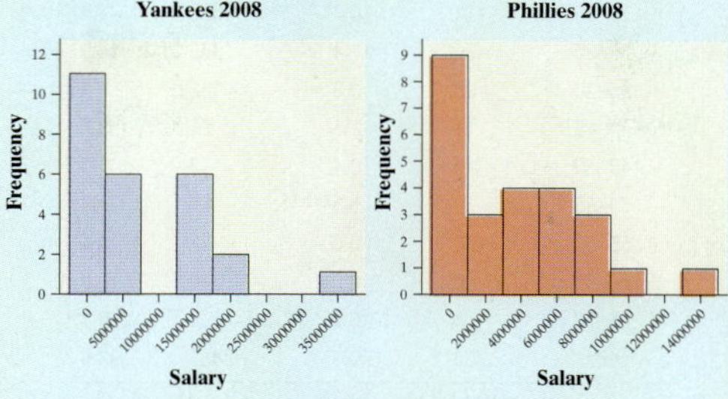


**60. Paying for championships** Refer to Exercise 59. Here is another graph of the 2008 salary distributions for the Yankees and the Phillies. Write a few sentences comparing these two distributions.

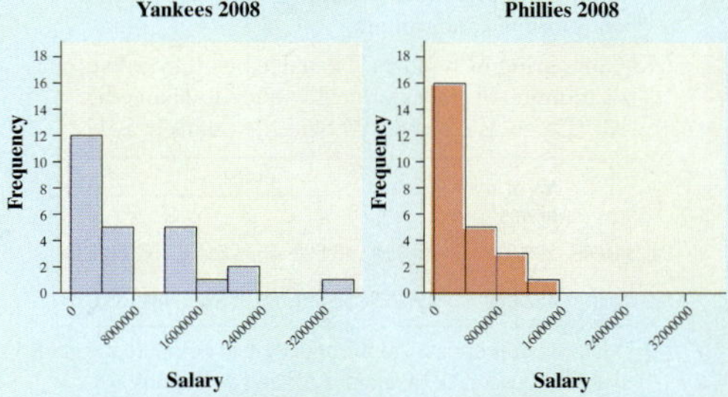


**61. Birth months** Imagine asking a random sample of 60 students from your school about their birth months. Draw a plausible graph of the distribution of birth months. Should you use a bar graph or a histogram to display the data?

**62. Die rolls** Imagine rolling a fair, six-sided die 60 times. Draw a plausible graph of the distribution of die rolls. Should you use a bar graph or a histogram to display the data?

**63. Who makes more?** A manufacturing company is reviewing the salaries of its full-time employees below the executive level at a large plant. The clerical staff is almost entirely female, while a majority of the production workers and technical staff is male. As a result, the distributions of salaries for male and female employees may be quite different. The following table gives the frequencies and relative frequencies for women and men.

**1.58 (a)** The histogram is given below. **(b)** The distribution of chest sizes is roughly symmetric with center around 40 inches and values that vary from 33 to 48 inches. This information is important so that the military can order the correct distribution of uniform sizes.

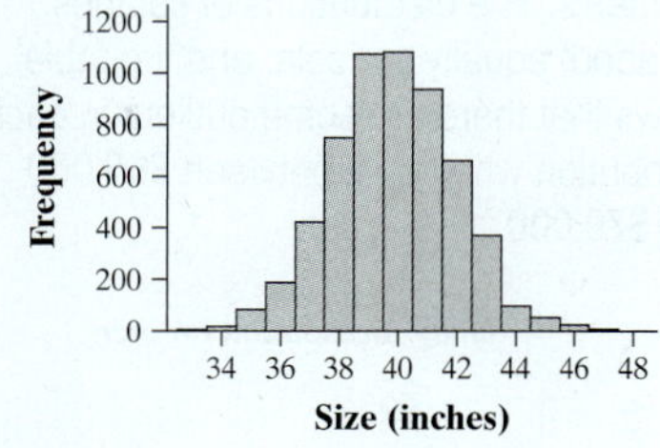


**1.59** The scale on the horizontal axis is very different from one graph to the other.

**1.60** Both distributions are skewed to the right. The salaries for the Yankees have a higher center and more variability. The Yankees have an outlier somewhere between \$32 and \$36 million.

**1.61** A bar graph should be used because birth month is a categorical variable. A possible bar graph is given below.

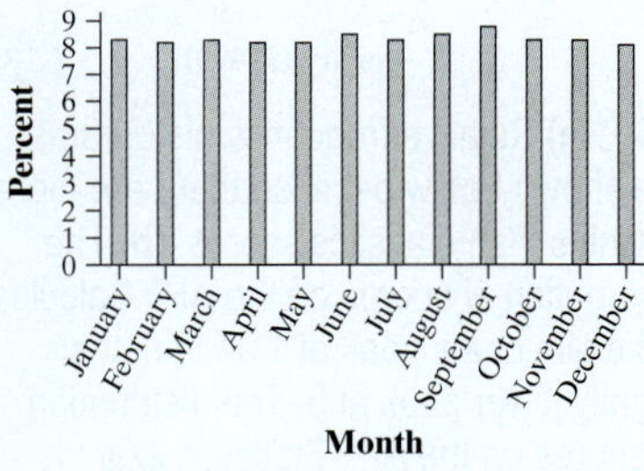


**1.62** A histogram should be used because die roll is a quantitative variable. A possible histogram is given below.

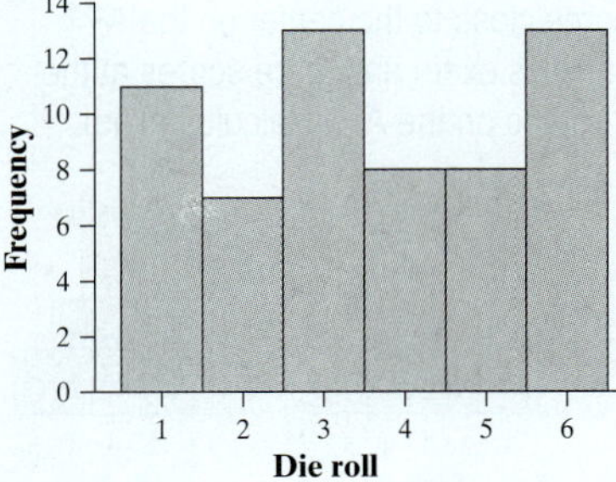


**1.56** The histogram is given below. This distribution of drive times is roughly symmetric with a center around 8.5 minutes and values that vary from 6.75 to 10.17 minutes. There are no clear outliers, although times around 10 minutes might be considered outliers.

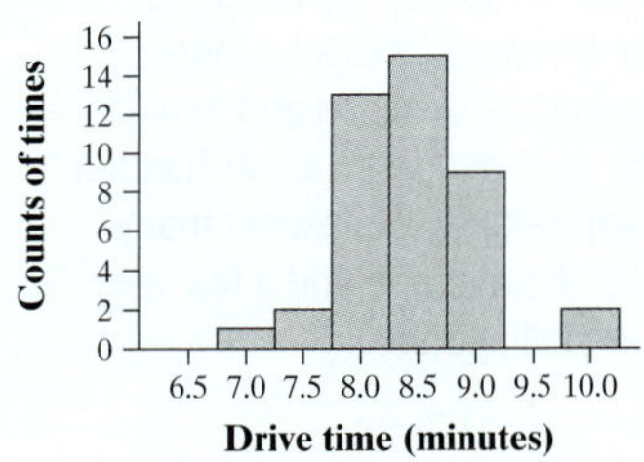


**1.57 (a)** The histogram is given below. The distribution of word lengths is skewed to the right and single-peaked. The center is around 4 letters, with words that vary from 1 to 15 letters. There don't appear to be any outliers. **(b)** There are more short words in Shakespeare's plays and more very long words in *Popular Science* articles.

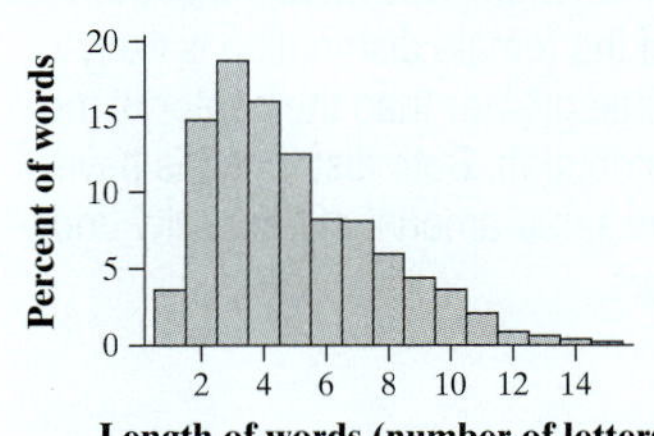

**1.63** **(a)** The percents for women sum to 100.1% due to rounding errors. **(b)** Relative frequency histograms are shown below because there are considerably more men than women. **(c)** Both histograms are skewed to the right. The center of the women's distribution of salaries is less than the men's. The distributions of salaries are about equally variable, and the table shows that there are some outliers in each distribution who make between \$65,000 and \$70,000.

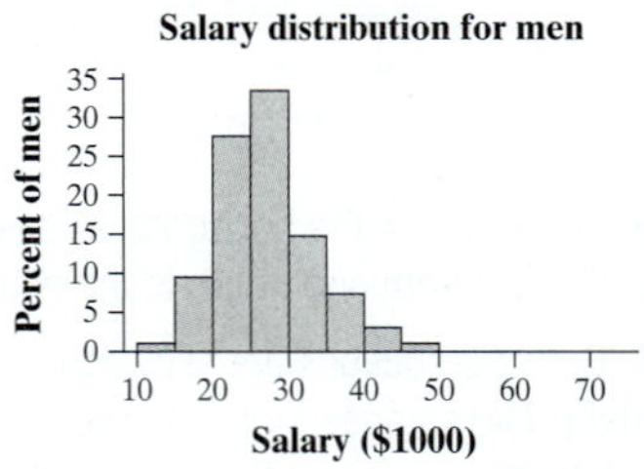


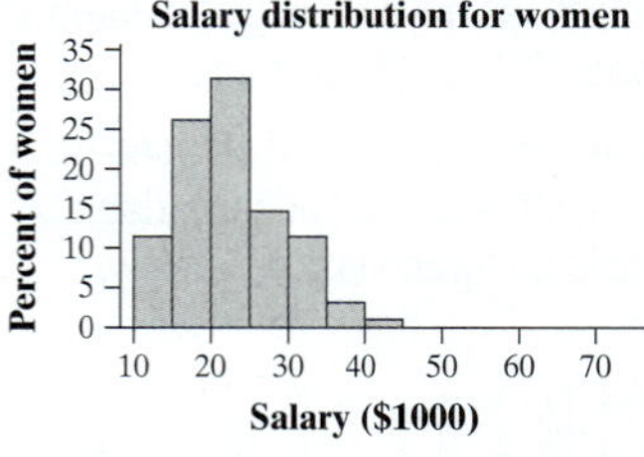


**1.64** **(a)** Relative frequency histograms are shown below because there are more calculus than statistics scores. **(b)** The distribution of scores on the AP® Calculus AB exam has a peak at 1 and another slightly lower peak at 5. The distribution of scores on the AP® Statistics exam is more uniform with scores of 1 and 3 being the most frequent and scores of 5 being the least frequent. The center of both distributions is 3. Although scores on both exams vary from 1 to 5, there are more scores close to the center on the AP® Statistics exam and more scores at the extremes on the AP® Calculus exam.

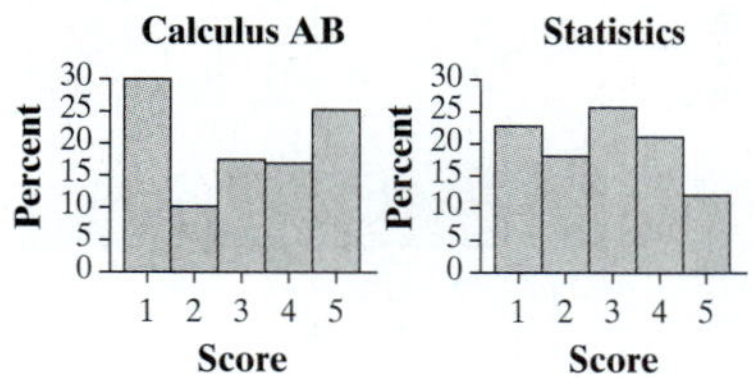


**1.65** The distribution of age is skewed to the right for both males and females, meaning that younger people outnumber older people. Among the younger Vietnamese, there are more males than females. After age 35, however, females seem to outnumber the males, making the center of the female distribution a little greater than the male distribution. Both distributions have about the same amount of variability and no outliers.

| | Women | | Men | |
|---|---|---|---|---|
| Salary (\$1000) | Number | % | Number | % |
| 10–15 | 89 | 11.8 | 26 | 1.1 |
| 15–20 | 192 | 25.4 | 221 | 9.0 |
| 20–25 | 236 | 31.2 | 677 | 27.6 |
| 25–30 | 111 | 14.7 | 823 | 33.6 |
| 30–35 | 86 | 11.4 | 365 | 14.9 |
| 35–40 | 25 | 3.3 | 182 | 7.4 |
| 40–45 | 11 | 1.5 | 91 | 3.7 |
| 45–50 | 3 | 0.4 | 33 | 1.3 |
| 50–55 | 2 | 0.3 | 19 | 0.8 |
| 55–60 | 0 | 0.0 | 11 | 0.4 |
| 60–65 | 0 | 0.0 | 0 | 0.0 |
| 65–70 | 1 | 0.1 | 3 | 0.1 |
| Total | 756 | 100.1 | 2451 | 99.9 |

(a) Explain why the total for women is greater than 100%.

(b) Make histograms for these data, choosing the vertical scale that is most appropriate for comparing the two distributions.

(c) Write a few sentences comparing the salary distributions for men and women.

64. **Comparing AP® scores** The table below gives the distribution of grades earned by students taking the AP® Calculus AB and AP® Statistics exams in 2012.[35]

| | No. of exams | Grade | | | | |
|---|---|---|---|---|---|---|
| | | 5 | 4 | 3 | 2 | 1 |
| Calculus AB | 266,994 | 67,394 | 45,523 | 46,526 | 27,216 | 80,335 |
| Statistics | 153,859 | 19,267 | 32,521 | 39,355 | 27,684 | 35,032 |

(a) Make an appropriate graphical display to compare the grade distributions for AP® Calculus AB and AP® Statistics.

(b) Write a few sentences comparing the two distributions of exam grades.

65. **Population pyramids** A population pyramid is a helpful graph for examining the distribution of a country's population. Here is a population pyramid for Vietnam in the year 2010. Describe what the graph tells you about Vietnam's population that year. Be specific.

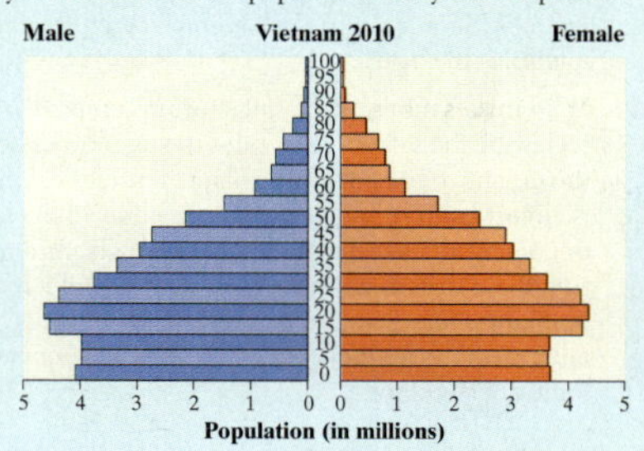


66. **Population pyramids** Refer to Exercise 65. Here is a graph of the projected population distribution for China in the year 2050. Describe what the graph suggests about China's future population. Be specific.

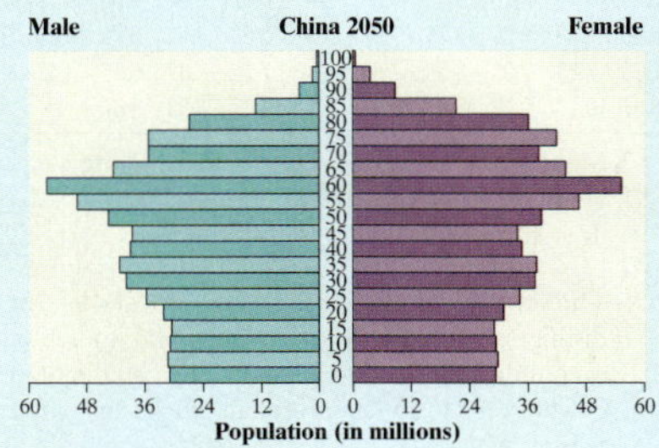


67. **Student survey** A survey of a large high school class asked the following questions:

(i) Are you female or male? (In the data, male = 0, female = 1.)

(ii) Are you right-handed or left-handed? (In the data, right = 0, left = 1.)

(iii) What is your height in inches?

(iv) How many minutes do you study on a typical weeknight?

The figure below shows graphs of the student responses, in scrambled order and without scale markings. Which graph goes with each variable? Explain your reasoning.

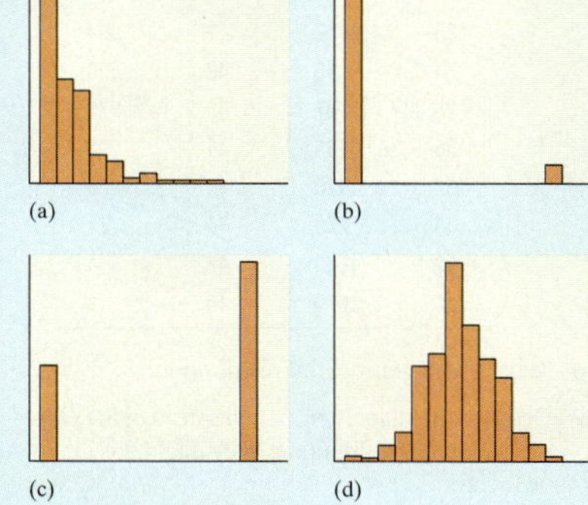


68. **Choose a graph** What type of graph or graphs would you make in a study of each of the following issues at your school? Explain your choices.

(a) Which radio stations are most popular with students?

(b) How many hours per week do students study?

(c) How many calories do students consume per day?

**1.66** The distribution of age is slightly skewed to the left for both males and females, meaning that older people will outnumber younger people. Each age group until age 65 will have slightly more men than women. After age 65, it appears that each age group will have slightly more women than men. This means that the center of the female distribution is likely to be a little greater than the center of the male distribution. Both distributions have about the same amount of variability and no outliers.

**1.67** **(a)** Amount of studying. We would expect some students to study very little, but most students to study a moderate amount. Any outliers would likely be high outliers, leading to a right-skewed distribution. **(b)** Right- vs. left-handed. About 90% of the population is right-handed (represented by the bar at 0). **(c)** Gender. We would expect a more similar percentage of males and females than for the right-handed and left-handed students. **(d)** Heights. We expect many heights near the average and a few very short or very tall people.

*Multiple choice: Select the best answer for Exercises 69 to 74.*

**69.** Here are the amounts of money (cents) in coins carried by 10 students in a statistics class: 50, 35, 0, 97, 76, 0, 0, 87, 23, 65. To make a stemplot of these data, you would use stems

(a) 0, 1, 2, 3, 4, 5, 6, 7, 8, 9.
(b) 0, 2, 3, 5, 6, 7, 8, 9.
(c) 0, 3, 5, 6, 7.
(d) 00, 10, 20, 30, 40, 50, 60, 70, 80, 90.
(e) None of these.

**70.** The histogram below shows the heights of 300 randomly selected high school students. Which of the following is the best description of the shape of the distribution of heights?

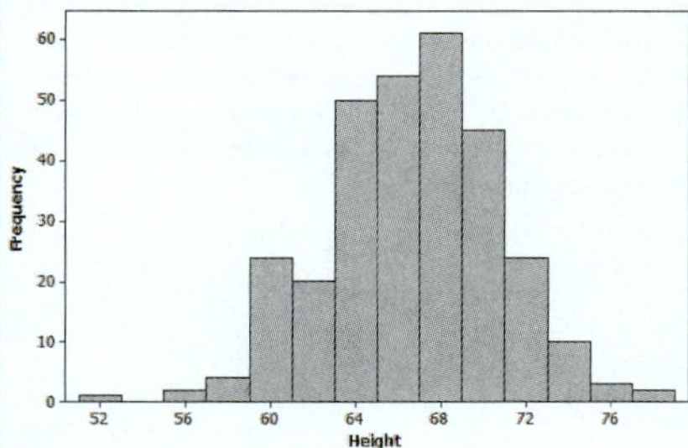


(a) Roughly symmetric and unimodal
(b) Roughly symmetric and bimodal
(c) Roughly symmetric and multimodal
(d) Skewed to the left
(e) Skewed to the right

**71.** You look at real estate ads for houses in Naples, Florida. There are many houses ranging from \$200,000 to \$500,000 in price. The few houses on the water, however, have prices up to \$15 million. The distribution of house prices will be

(a) skewed to the left.
(b) roughly symmetric.
(c) skewed to the right.
(d) unimodal.
(e) too high.

**72.** The following histogram shows the distribution of the percents of women aged 15 and over who have never married in each of the 50 states and the District of Columbia. Which of the following statements about the histogram is correct?

(a) The center of the distribution is about 36%.
(b) There are more states with percents above 32 than there are states with percents less than 24.
(c) It would be better if the values from 34 to 50 were deleted on the horizontal axis so there wouldn't be a large gap.
(d) There was one state with a value of exactly 33%.
(e) About half of the states had percents between 24% and 28%.

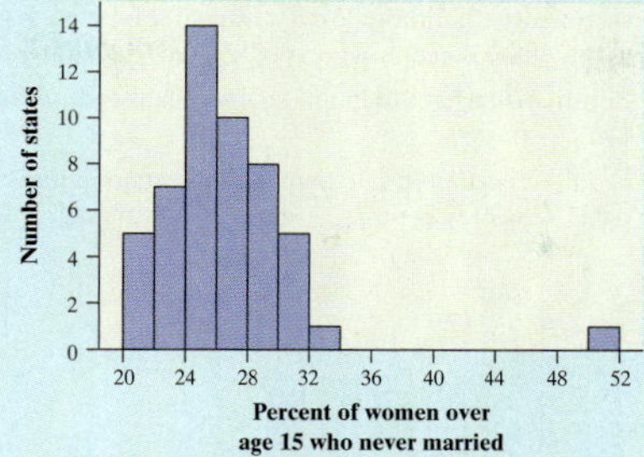


**Percent of women over age 15 who never married**

**73.** When comparing two distributions, it would be best to use relative frequency histograms rather than frequency histograms when

(a) the distributions have different shapes.
(b) the distributions have different spreads.
(c) the distributions have different centers.
(d) the distributions have different numbers of observations.
(e) at least one of the distributions has outliers.

**74.** Which of the following is the best reason for choosing a stemplot rather than a histogram to display the distribution of a quantitative variable?

(a) Stemplots allow you to split stems; histograms don't.
(b) Stemplots allow you to see the values of individual observations.
(c) Stemplots are better for displaying very large sets of data.
(d) Stemplots never require rounding of values.
(e) Stemplots make it easier to determine the shape of a distribution.

**75. Baseball players (Introduction)** Here is a small part of a data set that describes Major League Baseball players as of opening day of the 2012 season:

| Player | Team | Position | Age | Height | Weight | Salary |
|---|---|---|---|---|---|---|
| Rodriguez, Alex | Yankees | Infielder | 37 | 6-3 | 225 | 29,000,000 |
| Gonzalez, Adrian | Dodgers | Infielder | 30 | 6-2 | 225 | 21,000,000 |
| Cruz, Nelson | Rangers | Outfielder | 32 | 6-2 | 240 | 5,000,000 |
| Lester, Jon | Red Sox | Pitcher | 28 | 6-4 | 240 | 7,625,000 |
| Strasburg, Stephen | Nationals | Pitcher | 24 | 6-4 | 220 | 3,000,000 |

**1.68** **(a)** Because radio station is a categorical variable, use a bar graph. **(b)** Because hours studied per week is quantitative, either use a dotplot, stemplot, or a histogram. **(c)** Because calories consumed is quantitative, either use a dotplot, stemplot, or a histogram.

**1.69** a

**1.70** a

**1.71** c

**1.72** e

**1.73** d

**1.74** b

**1.75** **(a)** Major League Baseball players who were on the roster on opening day of the 2012 season. **(b)** 6. Two variables are categorical (team, position) and the other 4 are quantitative (age, height, weight, and salary).

**1.76** **(a)** More people "love" the newer devices such as the iPod, Broadband, and HDTV than older technologies like cable TV and pay TV. **(b)** No, because the categories are not dividing up a whole into pieces. Individuals could be represented in more than one bar.

**1.77** **(a)** 71/858 = 8.3% were elite soccer players and 43/858 = 5.0% of the people had arthritis. **(b)** 10/71 = 14.1% of elite soccer players had arthritis and 10/43 = 23.3% of those with arthritis were elite soccer players.

**1.78** Yes. The percent of each group who have arthritis is 10/71 = 14.1% for the elite soccer players, 9/215 = 4.2% for the non-elite soccer players, and 24/572 = 4.2% for the people who did not play. This can also be seen in the following bar graph.

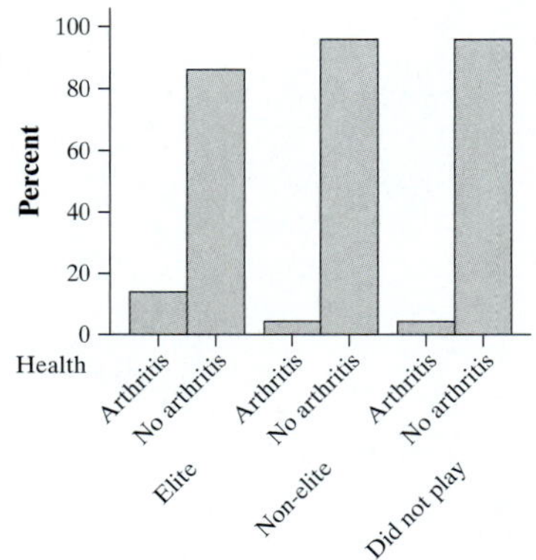


(a) What individuals does this data set describe?

(b) In addition to the player's name, how many variables does the data set contain? Which of these variables are categorical and which are quantitative?

76. **I love my iPod!** (1.1) The rating service Arbitron asked adults who used several high-tech devices and services whether they "loved" using them. Below is a graph of the percents who said they did.[36]

(a) Summarize what this graph tells you in a sentence or two.

(b) Would it be appropriate to make a pie chart of these data? Why or why not?

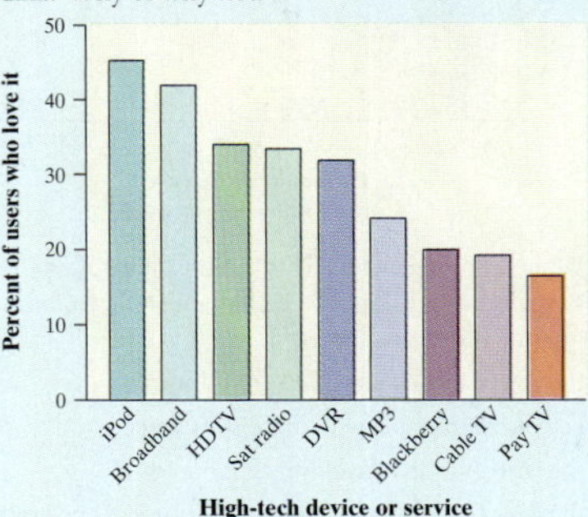


77. **Risks of playing soccer** (1.1) A study in Sweden looked at former elite soccer players, people who had played soccer but not at the elite level, and people of the same age who did not play soccer. Here is a two-way table that classifies these individuals by whether or not they had arthritis of the hip or knee by their mid-fifties:[37]

| | Elite | Non-Elite | Did not play |
|---|---|---|---|
| Arthritis | 10 | 9 | 24 |
| No arthritis | 61 | 206 | 548 |

(a) What percent of the people in this study were elite soccer players? What percent had arthritis?

(b) What percent of the elite soccer players had arthritis? What percent of those who had arthritis were elite soccer players?

78. **Risks of playing soccer** (1.1) Refer to Exercise 77. We suspect that the more serious soccer players have more arthritis later in life. Do the data confirm this suspicion? Give graphical and numerical evidence to support your answer.

# 1.3 Describing Quantitative Data with Numbers

**WHAT YOU WILL LEARN** By the end of the section, you should be able to:

- Calculate measures of center (mean, median).
- Calculate and interpret measures of spread (range, *IQR*, standard deviation).
- Choose the most appropriate measure of center and spread in a given setting.
- Identify outliers using the 1.5 × *IQR* rule.
- Make and interpret boxplots of quantitative data.
- Use appropriate graphs and numerical summaries to compare distributions of quantitative variables.

How long do people spend traveling to work? The answer may depend on where they live. Here are the travel times in minutes for 15 workers in North Carolina, chosen at random by the Census Bureau:[38]

30 20 10 40 25 20 10 60 15 40 5 30 12 10 10

### PD Section 1.3 Overview

There is a video overview of Section 1.3 (for teachers) available in the TE-book, or on the TRCD or the book's companion Web site (teacher log-in required).

### Teaching Tip

Remember to mention the Learning Objectives so students know what they will be expected to learn in this section.

### TRM Section 1.3 Alternate Examples

The Alternate Examples for this section in Microsoft Word format may be accessed by clicking on the link in the TE-book, opening the TRCD, or logging onto the book's companion Web site www.whfreeman.com/tps5e (teacher log-in required).

We aren't surprised that most people estimate their travel time in multiples of 5 minutes. Here is a stemplot of these data:

```
0 | 5
1 | 000025
2 | 005      Key: 2|5 is a NC
3 | 00       worker who travels 25
4 | 00       minutes to work.
5 |
6 | 0
```

The distribution is single-peaked and right-skewed. The longest travel time (60 minutes) may be an outlier. Our main goal in this section is to describe the center and spread of this and other distributions of quantitative data with numbers.

## Measuring Center: The Mean

The most common measure of center is the ordinary arithmetic average, or **mean**.

**DEFINITION: The mean $\bar{x}$**

To find the **mean** $\bar{x}$ (pronounced "x-bar") of a set of observations, add their values and divide by the number of observations. If the $n$ observations are $x_1, x_2, \ldots, x_n$, their mean is

$$\bar{x} = \frac{\text{sum of observations}}{n} = \frac{x_1 + x_2 + \cdots + x_n}{n}$$

or, in more compact notation,

$$\bar{x} = \frac{\sum x_i}{n}$$

The $\Sigma$ (capital Greek letter sigma) in the formula for the mean is short for "add them all up." The subscripts on the observations $x_i$ are just a way of keeping the $n$ observations distinct. They do not necessarily indicate order or any other special facts about the data.

Actually, the notation $\bar{x}$ refers to the mean of a *sample*. Most of the time, the data we'll encounter can be thought of as a sample from some larger population. When we need to refer to a *population* mean, we'll use the symbol $\mu$ (Greek letter mu, pronounced "mew"). If you have the entire population of data available, then you calculate $\mu$ in just the way you'd expect: add the values of all the observations, and divide by the number of observations.

**EXAMPLE**

### Travel Times to Work in North Carolina

*Calculating the mean*

Here is a stemplot of the travel times to work for the sample of 15 North Carolinians.

```
0 | 5
1 | 000025
2 | 005      Key: 2|5 is a NC
3 | 00       worker who travels 25
4 | 00       minutes to work.
5 |
6 | 0
```

### AP® Exam Formula Sheet

The formula for the sample mean $\bar{x}$ is included on the AP® formula sheet that can be used on both parts of the AP® exam. We recommend that you give each student a copy of the formula sheet right away so they can get accustomed to referring to it. Also, consider making a laminated copy for use on chapter quizzes and tests.

### Teaching Tip

Although the distinction between a sample mean $\bar{x}$ and a population mean $\mu$ isn't as important now, this distinction will be extremely important as we get further into the course. Make sure you are very careful to use the proper notation at all times.

**ALTERNATE EXAMPLE**

### *McDonald's fish and chicken sandwiches*

Here are data on the amount of fat (in grams) in 9 different McDonald's fish and chicken sandwiches, along with a stemplot:

| Sandwich | Fat (g) |
|---|---|
| Filet-O-Fish® | 19 |
| McChicken® | 16 |
| Premium Crispy Chicken Classic Sandwich | 22 |
| Premium Crispy Chicken Club Sandwich | 33 |
| Premium Crispy Chicken Ranch Sandwich | 27 |
| Premium Grilled Chicken Classic Sandwich | 9 |
| Premium Grilled Chicken Club Sandwich | 20 |
| Premium Grilled Chicken Ranch Sandwich | 14 |
| Southern Style Crispy Chicken Sandwich | 19 |

**ALTERNATE EXAMPLE (continued)**

| | |
|---|---|
| 0 | 9 |
| 1 | 4 |
| 1 | 699 |
| 2 | 02 |
| 2 | 7 |
| 3 | 3 |

Key: 3|3 represents 33 grams of fat in a McDonald's fish or chicken sandwich.

PROBLEM:
**(a)** Find the mean amount of fat for fish and chicken sandwiches.
**(b)** The Premium Crispy Chicken Club Sandwich is a potential outlier. How much does this one sandwich increase the mean?

SOLUTION:
**(a)** The mean amount of fat in McDonald's fish and chicken sandwiches is

$$\bar{x} = \frac{9 + 14 + \cdots + 33}{9} = 19.89 \text{ grams}$$

**(b)** The mean without the Premium Crispy Chicken Club Sandwich is

$$\bar{x} = \frac{9 + 14 + \cdots + 27}{8} = 18.25 \text{ grams}$$

This sandwich increases the mean amount of fat by about 1.6 grams.

## Teaching Tip

Students need to understand and be able to explain how unusual values (both large and small) affect numerical summaries such as the mean.

PROBLEM:
(a) Find the mean travel time for all 15 workers.
(b) Calculate the mean again, this time excluding the person who reported a 60-minute travel time to work. What do you notice?

SOLUTION:
(a) The mean travel time for the sample of 15 North Carolina workers is

$$\bar{x} = \frac{\sum x_i}{n} = \frac{x_1 + x_2 + \cdots + x_n}{n} = \frac{30 + 20 + \cdots + 10}{15} = \frac{337}{15} = 22.5 \text{ minutes}$$

(b) If we leave out the longest travel time, 60 minutes, the mean for the remaining 14 people is

$$\bar{x} = \frac{\sum x_i}{n} = \frac{x_1 + x_2 + \cdots + x_n}{n} = \frac{277}{14} = 19.8 \text{ minutes}$$

This one observation raises the mean by 2.7 minutes.

**For Practice** *Try Exercise* 79

The previous example illustrates an important weakness of the mean as a measure of center: *the mean is sensitive to the influence of extreme observations*. These may be outliers, but a skewed distribution that has no outliers will also pull the mean toward its long tail. Because the mean cannot resist the influence of extreme observations, we say that it is *not* a **resistant measure** of center.

**What does the mean mean?** A group of elementary schoolchildren was asked how many pets they have. Here are their responses, arranged from lowest to highest:[39]

1 3 4 4 4 5 7 8 9

What's the mean number of pets for this group of children? It's

$$\bar{x} = \frac{\text{sum of observations}}{n} = \frac{1+3+4+4+4+5+7+8+9}{9} = 5 \text{ pets}$$

But what does that number tell us? Here's one way to look at it: if every child in the group had the same number of pets, each would have 5 pets. In other words, the mean is the "fair share" value.

The mean tells us how large each data value would be if the total were split equally among all the observations. The mean of a distribution also has a physical interpretation, as the following Activity shows.

## ACTIVITY | Mean as a "balance point"

**MATERIALS:**
Foot-long ruler, pencil, and 5 pennies per group of 3 to 4 students

In this Activity, you'll investigate an interesting property of the mean.

1. Stack all 5 pennies above the 6-inch mark on your ruler. Place your pencil under the ruler to make a "seesaw" on a desk or table. Move the pencil until the ruler balances. What is the relationship between the location of the pencil and the mean of the five data values: 6, 6, 6, 6, 6?

2. Move one penny off the stack to the 8-inch mark on your ruler. Now move one other penny so that the ruler balances again without moving the pencil. Where did you put the other penny? What is the mean of the five data values represented by the pennies now?
3. Move one more penny off the stack to the 2-inch mark on your ruler. Now move both remaining pennies from the 6-inch mark so that the ruler still balances with the pencil in the same location. Is the mean of the data values still 6?
4. Do you see why the mean is sometimes called the "balance point" of a distribution?

## Measuring Center: The Median

In Section 1.2, we introduced the median as an informal measure of center that describes the "midpoint" of a distribution. Now it's time to offer an official "rule" for calculating the median.

**DEFINITION: The median**

The **median** is the midpoint of a distribution, the number such that about half the observations are smaller and about half are larger. To find the median of a distribution:

1. Arrange all observations in order of size, from smallest to largest.
2. If the number of observations $n$ is odd, the median is the center observation in the ordered list.
3. If the number of observations $n$ is even, the median is the average of the two center observations in the ordered list.

Medians require little arithmetic, so they are easy to find by hand for small sets of data. Arranging even a moderate number of values in order is tedious, however, so finding the median by hand for larger sets of data is unpleasant.

### Teaching Tip

Emphasize the idea that the mean is the balance point of a distribution. This will help students understand why unusually large or small values have such a big influence on the mean. It also allows students to do a "reality check" for their calculations: "Does the distribution balance at the value I calculated for the mean?"

**ALTERNATE ACTIVITY**

### *Mean as a "balance point"*

If you don't have the time to do the ruler Activity, you can still illustrate the basic concepts quickly on the board, using dotplots. Instead of placing 5 pennies on the 6-inch mark, place a stack of 5 dots above 6 on a dotplot running from 0 to 12. Then "move" one of the dots from 6 to 8 and ask students what needs to happen for the mean to remain equal to 6. When they suggest "moving" a dot from 6 to 4, point out that the deviations from 6 (the mean) are in balance. The dot above 4 has a deviation of $-2$ and the dot above 8 has a deviation of $+2$. Together, these add to 0. Now, "move" another dot from 6 to 2 and have students discuss how to keep the deviations balanced at 6.

### Teaching Tip

In *TPS* 5e, we do not use a symbol to represent the median like we do for the mean ($\bar{x}$ or $\mu$). Although some people use the symbol $M$ (e.g., $M = 20$), we have found it is easier for students to simply write "median = 20."

## EXAMPLE Travel Times to Work in North Carolina

*Finding the median when n is odd*

What is the median travel time for our 15 North Carolina workers? Here are the data arranged in order:

5 10 10 10 10 12 15 **20** 20 25 30 30 40 40 60

The count of observations $n = 15$ is odd. The bold **20** is the center observation in the ordered list, with 7 observations to its left and 7 to its right. This is the median, 20 minutes.

The next example shows you how to find the median when there is an even number of data values.

## EXAMPLE Stuck in Traffic

*Finding the median when n is even*

People say that it takes a long time to get to work in New York State due to the heavy traffic near big cities. What do the data say? Here are the travel times in minutes of 20 randomly chosen New York workers:

10 30 5 25 40 20 10 15 30 20
15 20 85 15 65 15 60 60 40 45

**PROBLEM:**

(a) Make a stemplot of the data. Be sure to include a key.

(b) Find the median by hand. Show your work.

**SOLUTION:**

(a) Here is a stemplot of the data. The stems indicate 10 minutes and the leaves indicate minutes.

```
0 | 5
1 | 005555
2 | 0005
3 | 00
4 | 005
5 |
6 | 005
7 |
8 | 5
```

Key: 4|5 is a New York worker who reported a 45-minute travel time to work.

(b) Because there is an even number of data values, there is no center observation. There is a center pair—the bold 20 and 25 in the stemplot—which have 9 observations before them and 9 after them in the ordered list. The median is the average of these two observations:

$$\frac{20 + 25}{2} = 22.5 \text{ minutes}$$

**For Practice** *Try Exercise* **81**

```
0 | 5
1 | 000025
2 | 005
3 | 00
4 | 00
5 |
6 | 0
```

Key: 2|5 is a NC worker who travels 25 minutes to work.

## Comparing the Mean and the Median

Our discussion of travel times to work in North Carolina illustrates an important difference between the mean and the median. The median travel time (the midpoint of the distribution) is 20 minutes. The mean travel time is higher, 22.5 minutes. The mean is pulled toward the right tail of this right-skewed distribution. The median,

### ALTERNATE EXAMPLE

**McDonald's fish and chicken sandwiches**

Here are the amounts of fat (in grams) in the 9 McDonald's fish and chicken sandwiches, sorted from smallest to largest:

9 14 16 19 **19** 20 22 27 33

The bold 19 in the middle of the list is the median. There are 4 values below the median and 4 values above the median.

### ALTERNATE EXAMPLE

**McDonald's beef sandwiches**

Here are data for the amount of fat (in grams) for McDonald's beef sandwiches:

| Sandwich | Fat |
|---|---|
| Big Mac® | 29 |
| Cheeseburger | 12 |
| Daily Double | 24 |
| Double Cheeseburger | 23 |
| Double Quarter Pounder® with cheese | 43 |
| Hamburger | 9 |
| McDouble | 19 |
| McRib® | 26 |
| Quarter Pounder® Bacon and Cheese | 29 |
| Quarter Pounder® Bacon Habanero Ranch | 31 |
| Quarter Pounder® Deluxe | 27 |
| Quarter Pounder® with Cheese | 26 |

**PROBLEM:**

**(a)** Make a stemplot of the data. Be sure to include a key.

**(b)** Find the median by hand. Show your work.

**SOLUTION:**

**(a)** Here is a stemplot of the data:

```
0 | 9
1 | 2
1 | 9
2 | 34
2 | 66799
3 | 1
3 |
4 | 3
```

Key: 2|3 represents 23 grams of fat in a McDonald's beef sandwich.

**(b)** Because there is an even number of values (12), there is no middle value. The center pair of values, the bold **26**s in the stemplot, have an average of 26. Thus, median = 26. About half of the beef sandwiches at McDonald's have less than 26 grams of fat and about half have more than 26 grams of fat.

unlike the mean, is *resistant.* If the longest travel time were 600 minutes rather than 60 minutes, the mean would increase to more than 58 minutes but the median would not change at all. The outlier just counts as one observation above the center, no matter how far above the center it lies. The mean uses the actual value of each observation and so will chase a single large observation upward.

You can compare the behavior of the mean and median by using the *Mean and Median* applet at the book's Web site, www.whfreeman.com/tps5e.

**COMPARING THE MEAN AND MEDIAN**

The mean and median of a roughly symmetric distribution are close together. If the distribution is exactly symmetric, the mean and median are exactly the same. In a skewed distribution, the mean is usually farther out in the long tail than is the median.[40]

The mean and median measure center in different ways, and both are useful.

**THINK ABOUT IT**

**Should we choose the mean or the median?** Many economic variables have distributions that are skewed to the right. College tuitions, home prices, and personal incomes are all right-skewed. In Major League Baseball (MLB), for instance, most players earn close to the minimum salary (which was $480,000 in 2012), while a few earn more than $10 million. The median salary for MLB players in 2012 was about $1.08 million—but the mean salary was about $3.44 million. Alex Rodriguez, Prince Fielder, Joe Mauer, and several other highly paid superstars pull the mean up but do not affect the median.

Reports about incomes and other strongly skewed distributions usually give the median ("midpoint") rather than the mean ("arithmetic average"). However, a county that is about to impose a tax of 1% on the incomes of its residents cares about the mean income, not the median. The tax revenue will be 1% of total income, and the total is the mean times the number of residents.

## CHECK YOUR UNDERSTANDING

Here, once again, is the stemplot of travel times to work for 20 randomly selected New Yorkers. Earlier, we found that the median was 22.5 minutes.

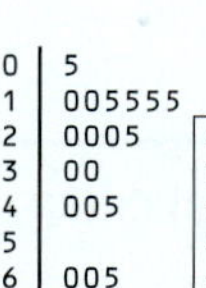


1. Based only on the stemplot, would you expect the mean travel time to be less than, about the same as, or larger than the median? Why?
2. Use your calculator to find the mean travel time. Was your answer to Question 1 correct?
3. Would the mean or the median be a more appropriate summary of the center of this distribution of drive times? Justify your answer.

## Measuring Spread: Range and Interquartile Range (*IQR*)

A measure of center alone can be misleading. The mean annual temperature in San Francisco, California, is 57°F—the same as in Springfield, Missouri. But the wardrobe needed to live in these two cities is very different! That's because daily

### APPLET

Take the time to show students the *Mean and Median* applet at the book's Web site www.whfreeman.com/tps5e. Start by clicking below the line to add points on the left side. Then grab the biggest point and drag it out to the right to see what happens to the mean and median. Next, repeat with a cluster of points on the right and drag one point to the left.

### APPLET

*The Dotplot Summaries* applet allows students to guess the positions of the mean and median on dotplots of various shapes. It also allows students to guess the size of the standard deviation and the interquartile range. Go to www.rossmanchance.com/applets and choose "Dotplot Summaries."

### ANSWERS TO CYU

**1.** Because the distribution is skewed to the right, we would expect the mean to be larger than the median.

**2.** Yes. The mean is 31.25 minutes, which is greater than the median of 22.5 minutes.

**3.** Because the distribution is skewed, the median would be a better measure of the center of the distribution.

### Teaching Tip

To introduce measures of spread, draw two dotplots with the same center and the same shape but very different amounts of variability. Hopefully, this will convince students that simply describing the shape and center does not give a complete picture of a distribution.

## Teaching Tip

Point out that the range is not a resistant measure of spread because the minimum and maximum values could be outliers. The interquartile range is a resistant measure of spread because it ignores the smallest 25% and largest 25% of the values in a distribution.

## Teaching Tip

When discussing measures of center and spread, you can make a connection with coordinate geometry: the mean is like the midpoint, because to find the midpoint between two points, you average the *x* and *y* coordinates of the points. Calculating the range is like finding the distance between two points, because we are trying to find out how spread out they are.

## Teaching Tip

Point out that the definition for quartiles in this book implies that the median is never included when computing quartiles.

**ALTERNATE EXAMPLE**

### *McDonald's fish and chicken sandwiches*

Here are the amounts of fat in the 9 McDonald's fish and chicken sandwiches, in order:

9 14 16 19 **19** 20 22 27 33

Because there is an odd number of observations, the median is the middle one, the bold **19** in the list. The first quartile is the median of the 4 observations to the left of the median, which is 15 (the average of 14 and 16). The third quartile is the median of the 4 observations to the right of the median, which is 24.5 (the average of 22 and 27). Thus, the interquartile range is $IQR = 24.5 - 15 = 9.5$ grams of fat.



Note that the range of a data set is a single number that represents the distance between the maximum and the minimum value. In everyday language, people sometimes say things like, "The data values range from 5 to 85." Be sure to use the term *range* correctly, now that you know its statistical definition.

temperatures vary a lot more in Springfield than in San Francisco. *A useful numerical description of a distribution requires both a measure of center and a measure of spread.*

The simplest measure of variability is the **range**. To compute the range of a quantitative data set, subtract the smallest value from the largest value. For the New York travel time data, the range is $85 - 5 = 80$ minutes. The range shows the full spread of the data. But it depends on only the maximum and minimum values, which may be outliers.

We can improve our description of spread by also looking at the spread of the middle half of the data. Here's the idea. Count up the ordered list of observations, starting from the minimum. The **first quartile** $Q_1$ lies one-quarter of the way up the list. The second quartile is the median, which is halfway up the list. The **third quartile** $Q_3$ lies three-quarters of the way up the list. These **quartiles** mark out the middle half of the distribution. The **interquartile range** (*IQR*) measures the range of the middle 50% of the data. We need a rule to make this idea exact. The process for calculating the quartiles and the *IQR* uses the rule for finding the median.

**HOW TO CALCULATE THE QUARTILES $Q_1$ AND $Q_3$ AND THE INTERQUARTILE RANGE (*IQR*)**

To calculate the **quartiles**:

1. Arrange the observations in increasing order and locate the median in the ordered list of observations.
2. The **first quartile** $Q_1$ is the median of the observations that are to the left of the median in the ordered list.
3. The **third quartile** $Q_3$ is the median of the observations that are to the right of the median in the ordered list.

The **interquartile range** (*IQR*) is defined as

$$IQR = Q_3 - Q_1$$

Be careful in locating the quartiles when several observations take the same numerical value. Write down all the observations, arrange them in order, and apply the rules just as if they all had distinct values.

Let's look at how this process works using a familiar set of data.

**EXAMPLE**

## Travel Times to Work in North Carolina

### *Calculating quartiles*

Our North Carolina sample of 15 workers' travel times, arranged in increasing order, is

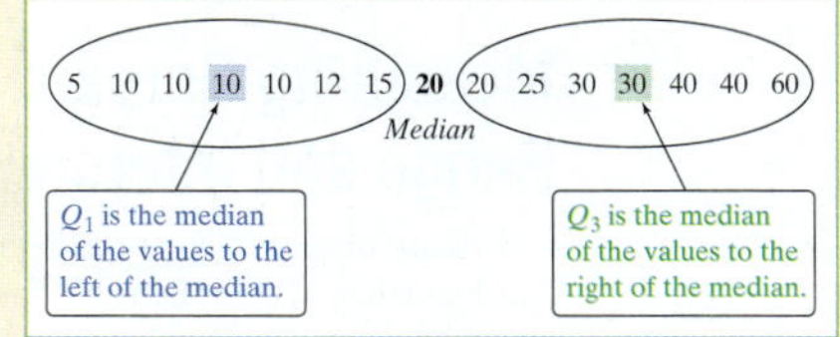

There is an odd number of observations, so the median is the middle one, the bold **20** in the list. The first quartile is the median of the 7 observations to the left of the median. This is the 4th of these 7 observations, so $Q_1 = 10$ minutes (shown in blue). The third quartile is the median of the 7 observations to the right of the median, $Q_3 = 30$ minutes (shown in green). So the spread of the middle 50% of the travel times is $IQR = Q_3 - Q_1 = 30 - 10 = 20$ minutes. *Be sure to leave out the overall median when you locate the quartiles.*

The quartiles and the interquartile range are *resistant* because they are not affected by a few extreme observations. For example, $Q_3$ would still be 30 and the *IQR* would still be 20 if the maximum were 600 rather than 60.

## EXAMPLE Stuck in Traffic Again

### *Finding and interpreting the* IQR

In an earlier example, we looked at data on travel times to work for 20 randomly selected New Yorkers. Here is the stemplot once again:

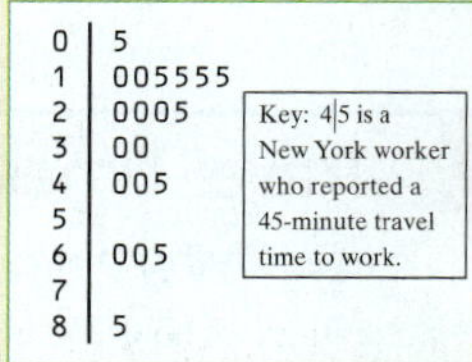


PROBLEM: Find and interpret the interquartile range (*IQR*).

SOLUTION: We begin by writing the travel times arranged in increasing order:

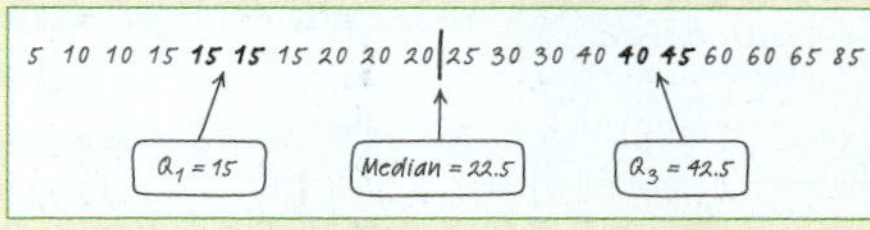


There is an even number of observations, so the median lies halfway between the middle pair. Its value is 22.5 minutes. (We marked the location of the median by |.) The first quartile is the median of the 10 observations to the left of 22.5. So it's the average of the two bold 15s: $Q_1 = 15$ minutes. The third quartile is the median of the 10 observations to the right of 22.5. It's the average of the bold numbers 40 and 45: $Q_3 = 42.5$ minutes. The interquartile range is

$$IQR = Q_3 - Q_1 = 42.5 - 15 = 27.5 \text{ minutes}$$

*Interpretation:* The range of the middle half of travel times for the New Yorkers in the sample is 27.5 minutes.

**For Practice** *Try Exercise* 89a

### TECHNOLOGY TIP

Some statistical software packages have different methods for calculating quartiles. However, the differences are usually small, especially when there are a large number of individuals in the data set.

### ALTERNATE EXAMPLE

#### *McDonald's beef sandwiches*

PROBLEM: Find and interpret the *IQR* for the distribution of fat in McDonald's beef sandwiches.

SOLUTION: Here are the 12 amounts of fat in order:

9 12 19 23 24 26 | 26 27 29 29 31 43

$Q_1 = 21$ (between 19 and 23), median (at |), $Q_3 = 29$ (between 29 and 29)

Because there is no middle value, the median falls between the two middle values (the two 26s). The first quartile is the median of the 6 values to the left of the median ($Q_1 = 21$), and the third quartile is the median of the 6 values to the right of the median ($Q_3 = 29$). Thus, $IQR = 29 - 21 = 8$ grams of fat. The range of the middle 50% of amounts of fat in McDonald's beef sandwiches is 8 grams.

### Teaching Tip

There are other rules for determining outliers, such as any value that is more than 2 standard deviations from the mean. However, we prefer the 1.5 × *IQR* rule because it is calculated using measures that are resistant to outliers. The mean and standard deviation, on the other hand, are calculated using all the values, including any outliers.

**ALTERNATE EXAMPLE**

### *McDonald's beef sandwiches*

**PROBLEM:** Determine whether there are any outliers in the distribution of fat for McDonald's beef sandwiches.

**SOLUTION:** Here are the 12 amounts of fat in order:

9 12 19 23 24 26 26 27 29 29 31 43

Earlier we found that $Q_1 = 21$, median = 26, and $Q_3 = 29$. Outliers are any values below $Q_1 - 1.5 \times IQR = 21 - 1.5(8) = 9$ and any values above $Q_3 + 1.5 \times QR = 29 + 1.5(8) = 41$. The hamburger, with just 9 grams of fat, is not quite an outlier because it is exactly on the boundary. However, the Double Quarter Pounder® with cheese is an outlier because 43 is greater than 41.

### Teaching Tip

Outliers often reveal interesting information, provided they are not mistakes. An example is the fascinating story about the discovery of the "ozone hole" in the atmosphere in Antarctica. *Nimbus-7* satellite data were collected in the 1970s that would have shown unusually low ozone levels over Antarctica, but a NASA computer program included a filter that was designed to discard sudden, large drops in ozone concentration as "errors." The phenomenon went unnoticed until an article in the May 1985 issue of *Nature* published summarized data collected by the British Antarctic Survey that showed the spring ozone levels had dropped to 10% below normal January levels for Antarctica. After the article appeared, NASA reran the *Nimbus-7* data without the filter program. Evidence of the ozone hole was then seen as far back as 1976. So here is a vivid example in which outliers were not the result of instrument error, computation error, or recording error but were, in fact, the message. It also dramatizes the need for the repetition of scientific experiments.

## Identifying Outliers

In addition to serving as a measure of spread, the interquartile range ($IQR$) is used as part of a rule of thumb for identifying outliers.

**DEFINITION: The 1.5 × *IQR* rule for outliers**

Call an observation an outlier if it falls more than 1.5 × *IQR* above the third quartile or below the first quartile.

| Stem | Leaves |
|---|---|
| 0 | 5 |
| 1 | 005555 |
| 2 | 0005 |
| 3 | 00 |
| 4 | 005 |
| 5 | |
| 6 | 005 |
| 7 | |
| 8 | 5 |

Key: 4|5 is a New York worker who reported a 45-minute travel time to work.

Does the 1.5 × $IQR$ rule identify any outliers for the New York travel time data? In the previous example, we found that $Q_1 = 15$ minutes, $Q_3 = 42.5$ minutes, and $IQR = 27.5$ minutes. For these data,

$$1.5 \times IQR = 1.5(27.5) = 41.25$$

Any values not falling between

$$Q_1 - 1.5 \times IQR = 15 - 41.25 = -26.25 \quad \text{and}$$
$$Q_3 + 1.5 \times IQR = 42.5 + 41.25 = 83.75$$

are flagged as outliers. Look again at the stemplot: the only outlier is the longest travel time, 85 minutes. The 1.5 × $IQR$ rule suggests that the three next-longest travel times (60 and 65 minutes) are just part of the long right tail of this skewed distribution.

**EXAMPLE**

### Travel Times to Work in North Carolina

#### *Identifying outliers*

Earlier, we noted the influence of one long travel time of 60 minutes in our sample of 15 North Carolina workers.

| Stem | Leaves |
|---|---|
| 0 | 5 |
| 1 | 000025 |
| 2 | 005 |
| 3 | 00 |
| 4 | 00 |
| 5 | |
| 6 | 0 |

Key: 2|5 is a NC worker who travels 25 minutes to work.

**PROBLEM:** Determine whether this value is an outlier.

**SOLUTION:** Earlier, we found that $Q_1 = 10$ minutes, $Q_3 = 30$ minutes, and $IQR = 20$ minutes. To check for outliers, we first calculate

$$1.5 \times IQR = 1.5(20) = 30$$

By the 1.5 × $IQR$ rule, any value *greater than*

$$Q_3 + 1.5 \times IQR = 30 + 30 = 60$$

*or less than*

$$Q_1 - 1.5 \times IQR = 10 - 30 = -20$$

would be classified as an outlier. The maximum value of 60 minutes is not quite large enough to be an outlier because it falls right on the upper cutoff value.

**For Practice** *Try Exercise* **89b**

Whenever you find outliers in your data, try to find an explanation for them. Sometimes the explanation is as simple as a typing error, like typing 10.1 as 101. Sometimes a measuring device broke down or someone gave a silly response, like the student in a class survey who claimed to study 30,000 minutes per night. (Yes,

On the other hand, an outlier may be an outlier simply due to a mistake by the person gathering the data or because of a measuring-instrument malfunction. Manhattan, Kansas, is sometimes called the "Little Apple," to distinguish it from that other Manhattan. A few years ago, a house there appeared in the county appraiser's records valued at $200,059,500. That would be quite a house even on Manhattan Island. As you might guess, the entry was wrong: the true value was $59,500. But before the error was discovered, the county, the city, and the school board had based their budgets on the total appraised value of real estate, which the one outlier jacked up by 6.5%. It can pay to spot outliers before you trust your data.

**AP® EXAM TIP** You may be asked to determine whether a quantitative data set has any outliers. Be prepared to state and use the rule for identifying outliers.

that really happened.) In all these cases, you can simply remove the outlier from your data. When outliers are "real data," like the long travel times of some New York workers, you should choose measures of center and spread that are not greatly affected by the outliers.

## The Five-Number Summary and Boxplots

The smallest and largest observations tell us little about the distribution as a whole, but they give information about the tails of the distribution that is missing if we know only the median and the quartiles. To get a quick summary of both center and spread, use all five numbers.

**DEFINITION: The five-number summary**

The **five-number summary** of a distribution consists of the smallest observation, the first quartile, the median, the third quartile, and the largest observation, written in order from smallest to largest. That is, the five-number summary is

Minimum $Q_1$ Median $Q_3$ Maximum

These five numbers divide each distribution roughly into quarters. About 25% of the data values fall between the minimum and $Q_1$, about 25% are between $Q_1$ and the median, about 25% are between the median and $Q_3$, and about 25% are between $Q_3$ and the maximum.

The five-number summary of a distribution leads to a new graph, the **boxplot** (sometimes called a box-and-whisker plot).

**HOW TO MAKE A BOXPLOT**

- A central box is drawn from the first quartile ($Q_1$) to the third quartile ($Q_3$).
- A line in the box marks the median.
- Lines (called whiskers) extend from the box out to the smallest and largest observations that are not outliers.
- Outliers are marked with a special symbol such as an asterisk (*).

Here's an example that shows how to make a boxplot.

**EXAMPLE**

### Home Run King

*Making a boxplot*

Barry Bonds set the major league record by hitting 73 home runs in a single season in 2001. On August 7, 2007, Bonds hit his 756th career home run, which broke Hank Aaron's longstanding record of 755. By the end of the 2007 season when Bonds retired, he had increased the total to 762. Here are data on the number of home runs that Bonds hit in each of his 21 complete seasons:

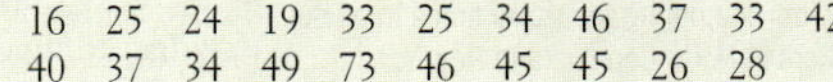

| | | | | | | | | | | |
|---|---|---|---|---|---|---|---|---|---|---|
| 16 | 25 | 24 | 19 | 33 | 25 | 34 | 46 | 37 | 33 | 42 |
| 40 | 37 | 34 | 49 | 73 | 46 | 45 | 45 | 26 | 28 | |

### Teaching Tip

Try to reinforce the fact that the five-number summary divides the data roughly into quarters by asking students to approximate what percent of the data fall above or below certain values. For example, pick a value above the median but below $Q_3$ and ask students what percent of the values in the distribution fall above the chosen value. They should say, "Somewhere between 25% and 50% of the values," unless they have the raw data.

### Teaching Tip

To help students remember how to make a boxplot and check for outliers, teach them the *IQR* Dance. Search for "*IQR* Dance" on YouTube to see Josh Tabor's class show their moves.

**ALTERNATE EXAMPLE**

*The previous home run king*

Here are the number of home runs that Hank Aaron hit in each of his 23 seasons:

| | | | | | | | |
|---|---|---|---|---|---|---|---|
| 13 | 27 | 26 | 44 | 30 | 39 | 40 | 34 |
| 45 | 44 | 24 | 32 | 44 | 39 | 29 | 44 |
| 38 | 47 | 34 | 40 | 20 | 12 | 10 | |

**PROBLEM:** Make a boxplot for these data.

**SOLUTION:** Here are the data in order, with the five-number summary identified:

**10** 12 13 20 24 **26** 27 29 30 32 34 **34** 38 39 39 40 **40** 44 44 44 44 45 **47**

*Min* (10), $Q_1$ (26), *Med* (34), $Q_3$ (40), *Max* (47)

The boundaries for outliers are 26 − 1.5 × (44 − 26) = −1 and 44 + 1.5 × (44 − 26) = 71, so there are no outliers. Here is a boxplot of the data:

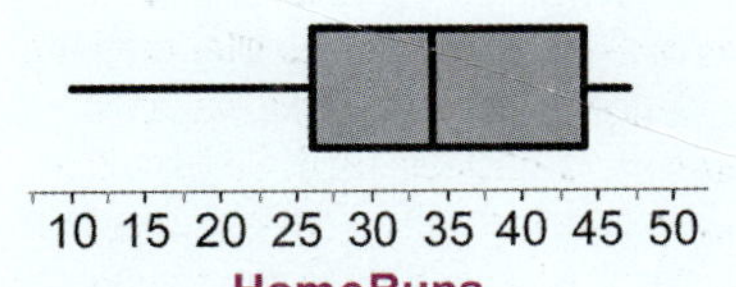

PROBLEM: Make a boxplot for these data.

SOLUTION: Let's start by ordering the data values so that we can find the five-number summary.

16 19 24 25 (25 26) 28 33 33 34 34 37 37 40 42 (45 45) 46 46 49 73

*Min* $Q_1 = 25.5$ *Median* $Q_3 = 45$ *Max*

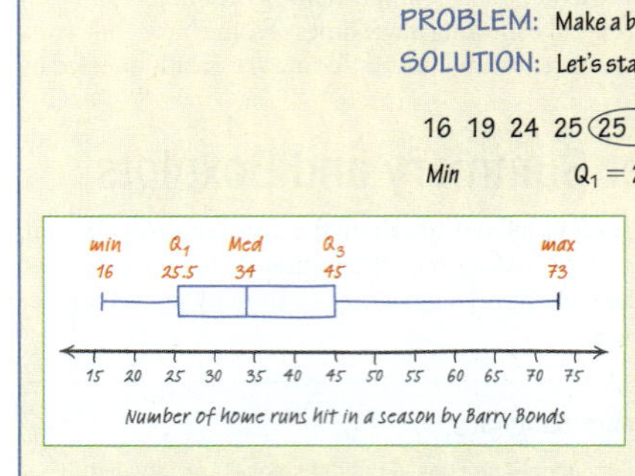


Now we check for outliers. Because $IQR = 45 - 25.5 = 19.5$, by the $1.5 \times IQR$ rule, any value *greater than* $Q_3 + 1.5 \times IQR = 45 + 1.5 \times 19.5 = 74.25$ or *less than* $Q_1 - 1.5 \times IQR = 25.5 - 1.5 \times 19.5 = -3.75$ would be classified as an outlier. So there are no outliers in this data set. Now we are ready to draw the boxplot. See the finished graph at left.

**For Practice** *Try Exercise* **91**

**THINK ABOUT IT**

**What are we actually doing when we make a boxplot?** The top dotplot shows Barry Bonds's home run data. We have marked the first quartile, the median, and the third quartile with blue lines. The process of testing for outliers with the $1.5 \times IQR$ rule is shown in visual form. Because there are no outliers, we draw the whiskers to the maximum and minimum data values, as shown in the finished boxplot at right.

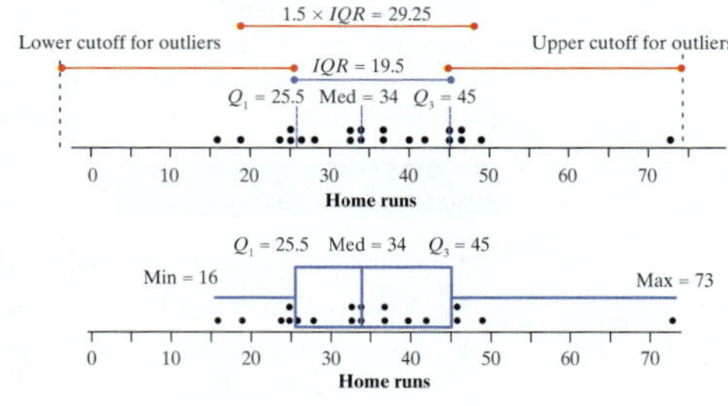


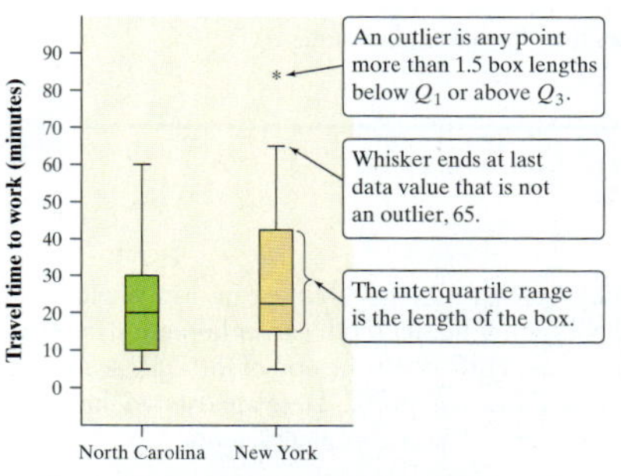


**FIGURE 1.19** Boxplots comparing the travel times to work of samples of workers in North Carolina and New York.

Figure 1.19 shows boxplots (this time, they are oriented vertically) comparing travel times to work for the samples of workers from North Carolina and New York. We will identify outliers as isolated points in the graph (like the * for the maximum value in the New York data set).

Boxplots show less detail than histograms or stemplots, so they are best used for side-by-side comparison of more than one distribution, as in Figure 1.19. As always, be sure to discuss shape, center, spread, and outliers as part of your comparison. For the travel time to work data:

*Shape:* We see from the graph that both distributions are right-skewed. For both states, the distance from the minimum to the median is much smaller than the distance from the median to the maximum.

*Center:* It appears that travel times to work are generally a bit longer in New York than in North Carolina. The median, both quartiles, and the maximum are all larger in New York.

## Teaching Tip

After reading the Think About It on this page, remind students that they can do a quick visual check for outliers. Simply add 1½ box lengths to either end of the box to find the boundary for outliers.

## AP® Exam Common Error

Many students refer to the box in a boxplot as the *IQR*. For example, they will say things like "the median is in the middle of the *IQR*." This is incorrect because the *IQR* is a number (the length of the box), not the box itself.

## AP® Exam Common Error

When describing the shape of a boxplot, many students will describe a symmetric boxplot as "approximately Normal." This is incorrect because a boxplot does not reveal where the modes of a distribution are. Emphasize this with your students whenever you are using boxplots to compare distributions.

## Teaching Tip

Point out to your students how we are using explicit comparison words in our description of these two distributions. For example, when comparing center, we say times are "a bit longer" in New York than North Carolina, and when comparing spread, we say that travel times are "more variable" in New York.

## Teaching Tip

Some teachers prefer to use a vertical orientation for boxplots, especially when comparing distributions. Then the explanatory variable is on the horizontal axis and the response variable is on the vertical axis, as on a scatterplot.

*Spread:* Travel times are also more variable in New York, as shown by the lengths of the boxes (the *IQR*) and the range.

*Outliers:* Earlier, we showed that the maximum travel time of 85 minutes is an outlier for the New York data. There are no outliers in the North Carolina sample.

## CHECK YOUR UNDERSTANDING

The 2011 roster of the Dallas Cowboys professional football team included 8 offensive linemen. Their weights (in pounds) were

310 307 345 324 305 301 290 307

1. Find the five-number summary for these data by hand. Show your work.
2. Calculate the *IQR*. Interpret this value in context.
3. Determine whether there are any outliers using the $1.5 \times IQR$ rule.
4. Draw a boxplot of the data.

## 3. TECHNOLOGY CORNER — MAKING CALCULATOR BOXPLOTS

TI-Nspire instructions in Appendix B; HP Prime instructions on the book's Web site.

The TI-83/84 and TI-89 can plot up to three boxplots in the same viewing window. Let's use the calculator to make parallel boxplots of the travel time to work data for the samples from North Carolina and New York.

1. Enter the travel time data for North Carolina in L1/list1 and for New York in L2/list2.
2. Set up two statistics plots: Plot1 to show a boxplot of the North Carolina data and Plot2 to show a boxplot of the New York data. The setup for Plot1 is shown below. When you define Plot2, be sure to change L1/list1 to L2/list2.

TI-83/84

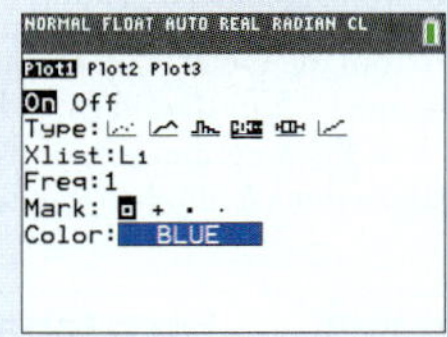


TI-89

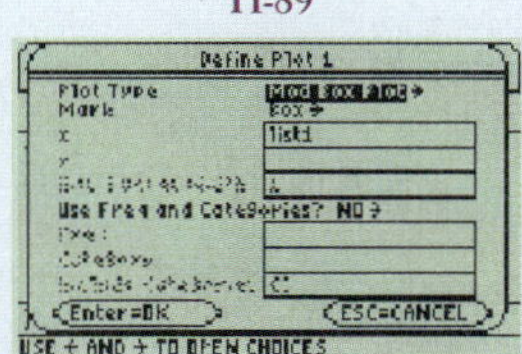


*Note:* The calculator offers two types of boxplots: one that shows outliers and one that doesn't. We'll always use the type that identifies outliers.

3. Use the calculator's Zoom feature to display the parallel boxplots. Then Trace to view the five-number summary.

TI-83/84

- Press ZOOM and select ZoomStat.
- Press TRACE.

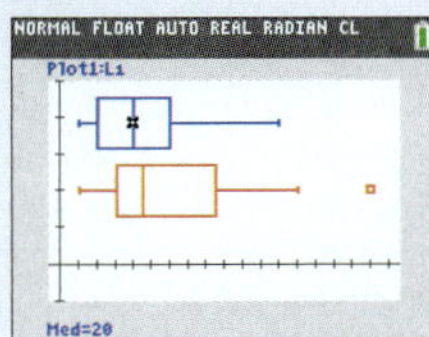


TI-89

- Press F5 (ZoomData).
- Press F3 (Trace).

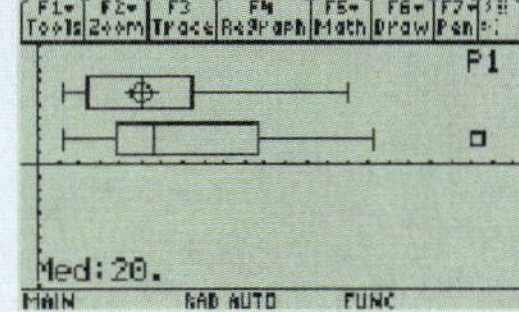


## ANSWERS TO CYU

1. The data in order are: 290, 301, 305, 307, 307, 310, 324, 345. The 5-number summary is 290, 303, 307, 317, 345.
2. The *IQR* is 14 pounds. The range of the middle half of the data is 14 pounds.
3. Any outliers occur below $303 - 1.5(14) = 282$ or above $317 + 1.5(14) = 338$, so 345 pounds is an outlier.
4. The boxplot is given below.

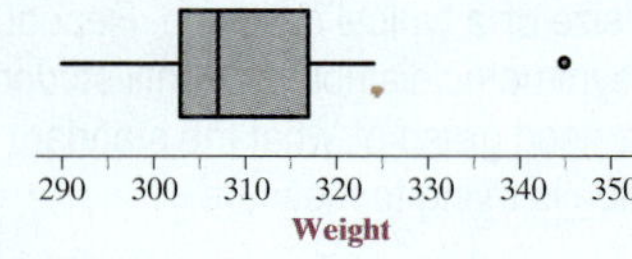


## TECHNOLOGY TIP

Always have your students use the boxplot option that marks the outliers. The TI-Nspire automatically graphs boxplots this way.

## TRM Do You Know Your Geography?

Consider doing the "Do you know your geography?" Activity found in the Teacher's Resource Materials. The activity has each student answer a couple of questions about geography. However, there are two versions of the "geography quiz." Comparing the results with parallel boxplots demonstrates how the results of a survey can be manipulated by changing the way the questions are asked. Thanks to Sanderson Smith for sharing this great activity.

## TRM The Memory Game

Consider doing "The memory game" Activity found in the Teacher's Resource Materials. The activity has each student listen to some statements and then take a quiz about how much they remember. However, one version of the quiz provides additional instructions. Comparing the results with parallel boxplots demonstrates how the results of a study can be different when given additional directions. Thanks to Adam Shrager for sharing this great activity.

APPLET

Before getting into the details of how to calculate the standard deviation, spend some time using the Dotplot Summaries applet, which allows students to guess the size of the standard deviation. Go to www.rossmanchance.com/applets and choose "Dotplot Summaries." Press Draw Sample until you get a relatively symmetric distribution and choose "Guess Std. Deviation." Move the slider so that nearly all the points are outside the boundaries and ask—"Does this seem like a typical deviation from the mean?" Then, move the slider so that nearly all the points are inside the boundary and ask the same question. Finally, have students guide you to the size of a typical deviation. Repeat for more symmetric distributions until students have a good grasp of what the standard deviation is trying to measure.

**ALTERNATE EXAMPLE**

### *Foot lengths*

Here are the foot lengths (in centimeters) for a random sample of seven 14-year-olds from the United Kingdom:

25 22 20 25 24 24 28

The mean foot length is 24 cm. Here are the deviations from the mean $(x_i - \bar{x})$ and the squared deviations from the mean $(x_i - \bar{x})^2$:

| $x$ | $x_i - \bar{x}$ | $(x_i - \bar{x})^2$ |
|---|---|---|
| 25 | 25 − 24 = 1 | $(1)^2 = 1$ |
| 22 | 22 − 24 = −2 | $(-2)^2 = 4$ |
| 20 | 20 − 24 = −4 | $(-4)^2 = 16$ |
| 25 | 25 − 24 = 1 | $(1)^2 = 1$ |
| 24 | 24 − 24 = 0 | $(0)^2 = 0$ |
| 24 | 24 − 24 = 0 | $(0)^2 = 0$ |
| 28 | 28 − 24 = 4 | $(4)^2 = 16$ |
| | Sum = 0 | Sum = 38 |

The "average" squared deviation is $\frac{38}{7-1} = 6.33$, so the standard deviation $\sqrt{6.33} = 2.52$ cm. This 2.52 cm is the typical distance each foot length is from the mean.

## Measuring Spread: The Standard Deviation

The five-number summary is not the most common numerical description of a distribution. That distinction belongs to the combination of the mean to measure center and the *standard deviation* to measure spread. The standard deviation and its close relative, the *variance*, measure spread by looking at how far the observations are from their mean. Let's explore this idea using a simple set of data.

**EXAMPLE**

### How Many Pets?

#### *Investigating spread around the mean*

In the Think About It on page 50, we examined data on the number of pets owned by a group of 9 children. Here are the data again, arranged from lowest to highest:

1 3 4 4 4 5 7 8 9

Earlier, we found the mean number of pets to be $\bar{x} = 5$. Let's look at where the observations in the data set are relative to the mean.

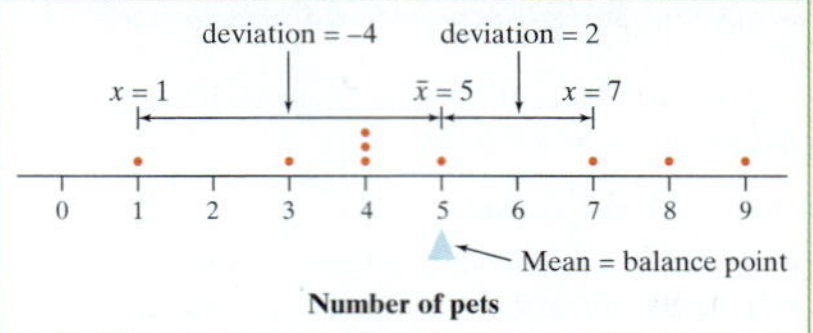


**FIGURE 1.20** Dotplot of the pet data with the mean and two of the deviations marked.

Figure 1.20 displays the data in a dotplot, with the mean clearly marked. The data value 1 is 4 units below the mean. We say that its *deviation* from the mean is −4. What about the data value 7? Its deviation is 7 − 5 = 2 (it is 2 units above the mean). The arrows in the figure mark these two deviations from the mean. The deviations show how much the data vary about their mean. They are the starting point for calculating the variance and standard deviation.

The table below shows the deviation from the mean $(x_i - \bar{x})$ for each value in the data set. Sum the deviations from the mean. You should get 0, because the mean is the balance point of the distribution. Because the sum of the deviations from the mean will be 0 for *any* set of data, we need another way to calculate spread around the mean.

| Observations $x_i$ | Deviations $x_i - \bar{x}$ | Squared deviations $(x_i - \bar{x})^2$ |
|---|---|---|
| 1 | 1 − 5 = −4 | $(-4)^2 = 16$ |
| 3 | 3 − 5 = −2 | $(-2)^2 = 4$ |
| 4 | 4 − 5 = −1 | $(-1)^2 = 1$ |
| 4 | 4 − 5 = −1 | $(-1)^2 = 1$ |
| 4 | 4 − 5 = −1 | $(-1)^2 = 1$ |
| 5 | 5 − 5 = 0 | $0^2 = 0$ |
| 7 | 7 − 5 = 2 | $2^2 = 4$ |
| 8 | 8 − 5 = 3 | $3^2 = 9$ |
| 9 | 9 − 5 = 4 | $4^2 = 16$ |
| | sum = 0 | sum = 52 |

How can we fix the problem of the positive and negative deviations canceling out? We could take the absolute value of each deviation. Or we could square the deviations. For mathematical reasons beyond the scope of this book, statisticians choose to square rather than to use absolute values.

We have added a column to the table that shows the square of each deviation $(x_i - \bar{x})^2$. Add up the squared deviations. Did you get 52? Now we compute the average squared deviation—sort of. Instead of dividing by the number of observations $n$, we divide by $n - 1$:

$$\text{"average" squared deviation} = \frac{16+4+1+1+1+0+4+9+16}{9-1} = \frac{52}{8} = 6.5$$

This value, 6.5, is called the **variance**.

Because we squared all the deviations, our units are in "squared pets." That's no good. We'll take the square root to get back to the correct units—pets. The resulting value is the **standard deviation**:

$$\text{standard deviation} = \sqrt{\text{variance}} = \sqrt{6.5} = 2.55 \text{ pets}$$

This 2.55 is the "typical" distance of the values in the data set from the mean. In this case, the number of pets typically varies from the mean by about 2.55 pets.

As you can see, the "average" in the standard deviation calculation is found in a rather unexpected way. Why do we divide by $n - 1$ instead of $n$ when calculating the variance and standard deviation? The answer is complicated but will be revealed in Chapter 7.

**DEFINITION: The standard deviation $s_x$ and variance $s_x^2$**

The **standard deviation** $s_x$ measures the typical distance of the values in a distribution from the mean. It is calculated by finding an average of the squared deviations and then taking the square root. This average squared deviation is called the **variance**. In symbols, the variance $s_x^2$ is given by

$$s_x^2 = \frac{(x_1 - \bar{x})^2 + (x_2 - \bar{x})^2 + \cdots + (x_n - \bar{x})^2}{n-1} = \frac{1}{n-1}\sum(x_i - \bar{x})^2$$

and the standard deviation is given by

$$s_x = \sqrt{\frac{1}{n-1}\sum(x_i - \bar{x})^2}$$

Here's a brief summary of the process for calculating the standard deviation.

## Teaching Tip

The average of the absolute values of the deviations is called the mean absolute deviation, or MAD. This is not on the AP® exam, but you can tell students that it is a legitimate way to measure the spread of a distribution. However, it is not nearly as useful as the standard deviation for many reasons. For example, the Normal distributions are determined by their mean and standard deviation, not the mean absolute deviation.

## Teaching Tip

If your students really want to know why they are dividing by $n - 1$, there are a number of explanations. The key is to recognize that the standard deviation we are calculating is really just an estimate of the true standard deviation—the "average" distance all the values in the population are from the mean of the population. Because we don't know the mean of the population ($\mu$), we can't calculate the deviations from $\mu$. So we use $\bar{x}$ instead. However, because $\bar{x}$ is the exact balancing point of the data, the data will almost always be closer to $\bar{x}$, on average, than they will be to $\mu$. Thus, the sum of the squared deviations from $\bar{x}$ will underestimate the sum of the squared deviations from $\mu$. To correct for this, we divide by a slightly smaller number when finding the average ($n - 1$ instead of $n$). We will demonstrate this with a simulation in Chapter 7.

## Teaching Tip

Insist that your students interpret the standard deviation. We will be using standard deviations all year, so the sooner they understand what a standard deviation measures, the better. It is the typical distance the values in a distribution are from the mean.

### AP® Exam Formula Sheet

Make sure to point out that the equation for the standard deviation is on the formula sheet that is provided on both parts of the AP® exam.

**TECHNOLOGY TIP**

After at most one day of calculating the standard deviation by hand, teach your students how to use their calculators to find the standard deviation (see next page). Understanding what the standard deviation measures is much more important than knowing how to calculate it by hand. When using the calculator, tell your students that they should always use $s_x$ rather than $\sigma_x$ unless they know that the data are from the entire population (which is rare).

## Teaching Tip

Ask your students how the formula in the cartoon is different than the formula we have just introduced. In AP® Statistics, we always divide by $n - 1$ when using data from a sample.

## Teaching Tip

Ask students to rank the standard deviation, *IQR*, and range from the most resistant (*IQR*) to the least resistant (range). Make sure that they explain their reasoning!

**HOW TO FIND THE STANDARD DEVIATION**

To find the standard deviation of $n$ observations:

1. Find the distance of each observation from the mean and square each of these distances.
2. Average the distances by dividing their sum by $n - 1$.
3. The standard deviation $s_x$ is the square root of this average squared distance:

$$s_x = \sqrt{\frac{1}{n-1}\sum(x_i - \bar{x})^2}$$

Many calculators report two standard deviations. One is usually labeled $\sigma_x$, the symbol for the standard deviation of a population. This standard deviation is calculated by dividing the sum of squared deviations by $n$ instead of $n - 1$ before taking the square root. If your data set consists of the entire population, then it's appropriate to use $\sigma_x$. Most often, the data we're examining come from a sample. In that case, we should use $s_x$.

More important than the details of calculating $s_x$ are the properties that describe the usefulness of the standard deviation:

- $s_x$ measures *spread about the mean* and should be used only when the mean is chosen as the measure of center.
- $s_x$ is *always greater than or equal to 0*. $s_x = 0$ only when there is no variability. This happens only when all observations have the same value. Otherwise, $s_x > 0$. As the observations become more spread out about their mean, $s_x$ gets larger.
- $s_x$ has the *same units of measurement as the original observations*. For example, if you measure metabolic rates in calories, both the mean $\bar{x}$ and the standard deviation $s_x$ are also in calories. This is one reason to prefer $s_x$ to the variance $s_x^2$, which is in squared calories.
- Like the mean $\bar{x}$, $s_x$ is *not resistant*. A few outliers can make $s_x$ very large. *The use of squared deviations makes $s_x$ even more sensitive than $\bar{x}$ to a few extreme observations.* For example, the standard deviation of the travel times for the 15 North Carolina workers is 15.23 minutes. If we omit the maximum value of 60 minutes, the standard deviation drops to 11.56 minutes.

caution

## CHECK YOUR UNDERSTANDING

The heights (in inches) of the five starters on a basketball team are 67, 72, 76, 76, and 84.

1. Find the mean. Show your work.
2. Make a table that shows, for each value, its deviation from the mean and its squared deviation from the mean.
3. Show how to calculate the variance and standard deviation from the values in your table.
4. Interpret the standard deviation in this setting.

## Numerical Summaries with Technology

Graphing calculators and computer software will calculate numerical summaries for you. That will free you up to concentrate on choosing the right methods and interpreting your results.

### 4. TECHNOLOGY CORNER — COMPUTING NUMERICAL SUMMARIES WITH TECHNOLOGY

TI-Nspire instructions in Appendix B; HP Prime instructions on the book's Web site.

Let's find numerical summaries for the travel times of North Carolina and New York workers from the previous Technology Corner (page 59). We'll start by showing you the necessary calculator techniques and then look at output from computer software.

**I. One-variable statistics on the calculator** If you haven't done so already, enter the North Carolina data in L1/list1 and the New York data in L2/list2.

1. Find the summary statistics for the North Carolina travel times.

**TI-83/84**

- Press [STAT] [▶] (CALC); choose 1-VarStats. **OS 2.55 or later:** In the dialog box, press [2nd] [1] (L1) and [ENTER] to specify L1 as the List. Leave FreqList blank. Arrow down to Calculate and press [ENTER]. **Older OS:** Press [2nd] [1] (L1) and [ENTER].

**TI-89**

- Press [F4] (Calc); choose 1-Var Stats.
- Type list1 in the list box. Press [ENTER].

Press [▼] to see the rest of the one-variable statistics for North Carolina.

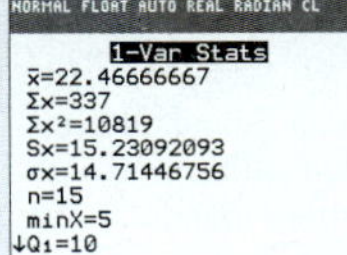


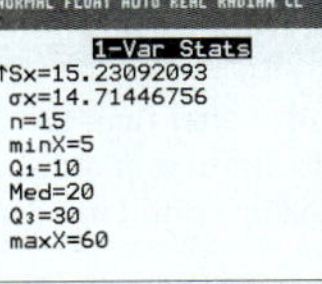


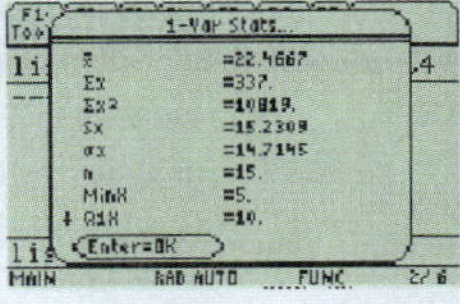


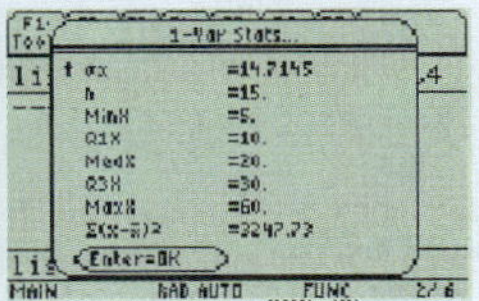


### ANSWERS TO CYU

1. The mean is 75.
2. The table is given below.

| Observation | Deviation | Squared Deviation |
|---|---|---|
| 67 | 67 − 75 = −8 | $(-8)^2 = 64$ |
| 72 | 72 − 75 = −3 | $(-3)^2 = 9$ |
| 76 | 76 − 75 = 1 | $1^2 = 1$ |
| 76 | 76 − 75 = 1 | $1^2 = 1$ |
| 84 | 84 − 75 = 9 | $9^2 = 81$ |
| Total | 0 | 156 |

3. The variance is $s_x^2 = \dfrac{156}{5-1} = 39$ inches squared and the standard deviation is $s_x = \sqrt{39} = 6.24$ inches.
4. The players' heights typically vary by about 6.24 inches from the mean height of 75 inches.

### Teaching Tip

When working with a distribution that is roughly symmetric, unimodal and bell-shaped, you can foreshadow the 68–95–99.7 rule from Section 2.2 by asking students to calculate the percent of values within 1, 2, and 3 standard deviations of the mean.

### TECHNOLOGY TIP

We recommend that you start allowing students to calculate numerical summaries with technology at this point. Students will be expected to use their calculators on the AP® exam, which allows the writers of the AP® exam to ask more conceptual questions about the application and interpretation of the numerical summaries. Make sure your students get plenty of practice with these types of questions.

APPLET

In addition to graphing calculators or statistical software, students can calculate numerical summaries using applets. Try the *One-Variable Statistical Calculator* applet at www.whfreeman.com/tps5e or the *One Quantitative Variable* applet at www.lock5stat.com/statkey.

## AP Exam Tip

On the AP® exam, students will be allowed to use any reasonable definition of quartiles on the free-response section. On the multiple-choice section, the answer choices will be designed so that any reasonable method of finding quartiles will give the correct answer.

## Teaching Tip

Don't mix and match the numerical summaries by pairing the median with the standard deviation or pairing the mean with the *IQR*.

## Teaching Tip

A boxplot is a visual representation of the five-number summary that makes it a graph/numerical summary hybrid. Unlike other types of graphs (dotplots, stemplots, histograms), boxplots don't show the modes of a distribution. Also unlike other types of graphs, boxplots identify a measure of center (median) and two measures of spread (range, *IQR*).

2. Repeat Step 1 using L2/list2 to find the summary statistics for the New York travel times.

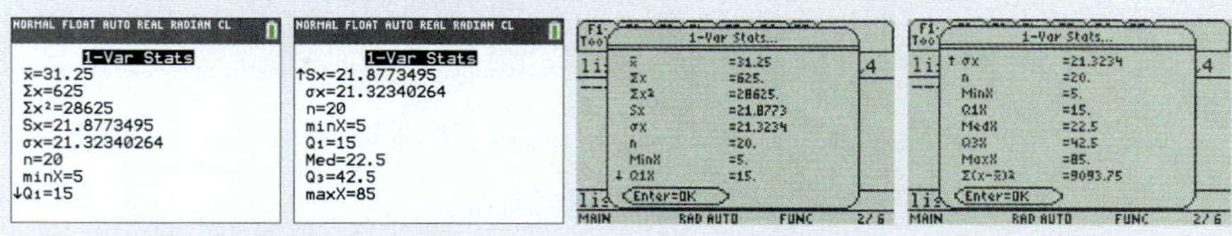


**II. Output from statistical software** We used Minitab statistical software to produce descriptive statistics for the New York and North Carolina travel time data. Minitab allows you to choose which numerical summaries are included in the output.

**Descriptive Statistics: Travel time to work**

| Variable | N | Mean | StDev | Minimum | $Q_1$ | Median | $Q_3$ | Maximum |
|---|---|---|---|---|---|---|---|---|
| NY Time | 20 | 31.25 | 21.88 | 5.00 | 15.00 | 22.50 | 43.75 | 85.00 |
| NC Time | 15 | 22.47 | 15.23 | 5.00 | 10.00 | 20.00 | 30.00 | 60.00 |

**What's with that third quartile?** Earlier, we saw that the quartiles of the New York travel times are $Q_1 = 15$ and $Q_3 = 42.5$. Look at the Minitab output in the Technology Corner. Minitab says that $Q_3 = 43.75$. What happened? Minitab and some other software use different rules for locating quartiles. Results from the various rules are always close to each other, so the differences are rarely important in practice. But because of the slight difference, Minitab wouldn't identify the maximum value of 85 as an outlier by the $1.5 \times IQR$ rule.

## Choosing Measures of Center and Spread

We now have a choice between two descriptions of the center and spread of a distribution: the median and *IQR*, or $\bar{x}$ and $s_x$. Because $\bar{x}$ and $s_x$ are sensitive to extreme observations, they can be misleading when a distribution is strongly skewed or has outliers. In these cases, the median and *IQR*, which are both resistant to extreme values, provide a better summary. We'll see in the next chapter that the mean and standard deviation are the natural measures of center and spread for a very important class of symmetric distributions, the Normal distributions.

**CHOOSING MEASURES OF CENTER AND SPREAD**

The median and *IQR* are usually better than the mean and standard deviation for describing a skewed distribution or a distribution with strong outliers. Use $\bar{x}$ and $s_x$ only for reasonably symmetric distributions that don't have outliers.

*Remember that a graph gives the best overall picture of a distribution. Numerical measures of center and spread report specific facts about a distribution, but they do not describe its entire shape.* Numerical summaries do not highlight the presence of multiple peaks or clusters, for example. *Always plot your data.*

## Organizing a Statistics Problem

As you learn more about statistics, you will be asked to solve more complex problems. Although no single strategy will work on every problem, it can be helpful to have a general framework for organizing your thinking. Here is a four-step process you can follow.

**HOW TO ORGANIZE A STATISTICS PROBLEM: A FOUR-STEP PROCESS**

**State:** What's the question that you're trying to answer?

**Plan:** How will you go about answering the question? What statistical techniques does this problem call for?

**Do:** Make graphs and carry out needed calculations.

**Conclude:** Give your conclusion in the setting of the real-world problem.

To keep the four steps straight, just remember: Statistics Problems Demand Consistency!

Many examples and exercises in this book will tell you what to do—construct a graph, perform a calculation, interpret a result, and so on. Real statistics problems don't come with such detailed instructions. From now on, you will encounter some examples and exercises that are more realistic. They are marked with the four-step icon. Use the four-step process as a guide to solving these problems, as the following example illustrates.

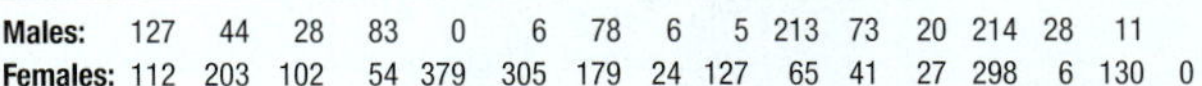

### EXAMPLE Who Texts More—Males or Females?

*Putting it all together*

For their final project, a group of AP® Statistics students wanted to compare the texting habits of males and females. They asked a random sample of students from their school to record the number of text messages sent and received over a two-day period. Here are their data:

| | | | | | | | | | | | | | | | |
|---|---|---|---|---|---|---|---|---|---|---|---|---|---|---|---|
| **Males:** | 127 | 44 | 28 | 83 | 0 | 6 | 78 | 6 | 5 | 213 | 73 | 20 | 214 | 28 | 11 |
| **Females:** | 112 | 203 | 102 | 54 | 379 | 305 | 179 | 24 | 127 | 65 | 41 | 27 | 298 | 6 | 130 | 0 |

What conclusion should the students draw? Give appropriate evidence to support your answer.

STATE: **Do males and females at the school differ in their texting habits?**

PLAN: **We'll begin by making parallel boxplots of the data about males and females. Then we'll calculate one-variable statistics. Finally, we'll compare shape, center, spread, and outliers for the two distributions.**

### Teaching Tip

We begin to utilize the four-step State–Plan–Do–Conclude process in Chapters 1, 3, and 5. Starting in Chapter 8, however, the State–Plan–Do–Conclude process becomes the way we organize nearly every problem. This is because the four steps closely match the rubrics for inference questions on the AP® exam.

### Teaching Tip

For the following alternate example, it will probably be more interesting for your students if you collect and use their number of contacts.

### ALTERNATE EXAMPLE

*Who has more contacts—males or females?*

The following data show the number of contacts that a sample of high school students had in their cell phones. What conclusion can we draw? Give appropriate evidence to support your answer.

| | |
|---|---|
| **Male:** | 124 41 29 27 44 87 85 260 290 31 168 169 167 214 135 114 105 103 96 144 |
| **Female:** | 30 83 116 22 173 155 134 180 124 33 213 218 183 110 |

STATE: Do males and females differ in the number of contacts they have in their phones? PLAN: We will compute numerical summaries, graph parallel boxplots, and compare shape, center, spread, and outliers. DO: Here are the numerical summaries and boxplots for each distribution:

| | $n$ | $\bar{x}$ | $s_x$ | Min | $Q_1$ | Med | $Q_3$ | Max |
|---|---|---|---|---|---|---|---|---|
| Male | 20 | 122.15 | 73.83 | 27 | 64.5 | 109.5 | 167.5 | 290 |
| Female | 14 | 126.71 | 65.75 | 22 | 83 | 129 | 180 | 218 |

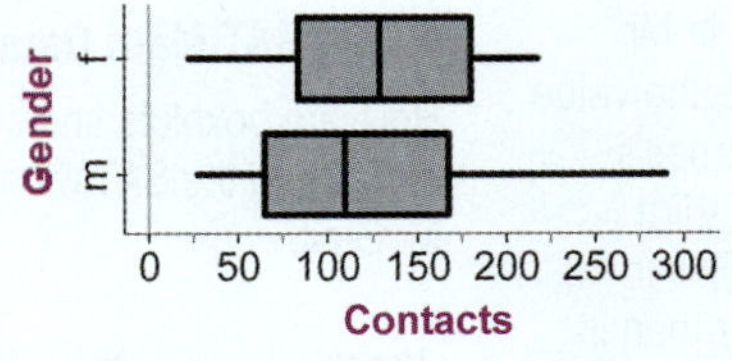


*Shape*: The female distribution is approximately symmetric, but the male distribution is slightly skewed to the right.
*Center*: The median number of contacts for the females is a little larger than the median number of contacts for the males.
*Spread*: The distribution of contacts for males is more spread out than the distribution of females because both the *IQR* and range are larger.
*Outliers*: Neither of the distributions has any outliers.

CONCLUDE: In the samples, females typically have more contacts than males, because both the mean and median values are slightly larger. However, the differences are small, so this is not convincing evidence that one gender has more contacts than the other.

## Teaching Tip

Try not to say things like the "average female sent 128 texts." Instead, say "a typical female sent an average of 128 texts." Averages are for quantitative variables, not categorical ones.

### TRM Chapter 1 Project and Rubric

Consider using the project described at the beginning of this chapter in the Additional *TPS* Resources. It is a nice way for students to put their new knowledge to work! There is a rubric included as well.

DO: Figure 1.21 is a sketch of the boxplots we got from our calculator. The table below shows numerical summaries for males and females.

| | $\bar{x}$ | $s_x$ | Min | $Q_1$ | Med | $Q_3$ | Max | *IQR* |
|---|---|---|---|---|---|---|---|---|
| **Male** | 62.4 | 71.4 | 0 | 6 | 28 | 83 | 214 | 77 |
| **Female** | 128.3 | 116.0 | 0 | 34 | 107 | 191 | 379 | 157 |

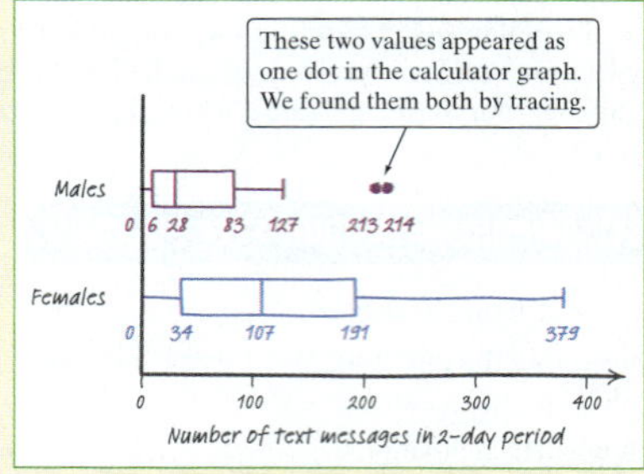


**FIGURE 1.21** Parallel boxplots of the texting data.

Due to the strong skewness and outliers, we'll use the median and *IQR* instead of the mean and standard deviation when comparing center and spread.

*Shape:* Both distributions are strongly right-skewed.

*Center:* Females typically text more than males. The median number of texts for females (107) is about four times as high as for males (28). In fact, the median for the females is above the third quartile for the males. This indicates that over 75% of the males texted less than the "typical" (median) female.

*Spread:* There is much more variation in texting among the females than the males. The *IQR* for females (157) is about twice the *IQR* for males (77).

*Outliers:* There are two outliers in the male distribution: students who reported 213 and 214 texts in two days. The female distribution has no outliers.

CONCLUDE: The data from this survey project give very strong evidence that male and female texting habits differ considerably at the school. A typical female sends and receives about 79 more text messages in a two-day period than a typical male. The males as a group are also much more consistent in their texting frequency than the females.

**For Practice** *Try Exercise* **105**

Now it's time for you to put what you have learned into practice in the following Data Exploration.

## DATA EXPLORATION Did Mr. Starnes stack his class?

Mr. Starnes teaches AP® Statistics, but he also does the class scheduling for the high school. There are two AP® Statistics classes—one taught by Mr. Starnes and one taught by Ms. McGrail. The two teachers give the same first test to their classes and grade the test together. Mr. Starnes's students earned an average score that was 8 points higher than the average for Ms. McGrail's class. Ms. McGrail wonders whether Mr. Starnes might have "adjusted" the class rosters from the computer scheduling program. In other words, she thinks he might have "stacked" his class. He denies this, of course.

To help resolve the dispute, the teachers collect data on the cumulative grade point averages and SAT Math scores of their students. Mr. Starnes provides the GPA data from his computer. The students report their SAT Math scores. The following table shows the data for each student in the two classes. Note that the two data values in each row come from a single student.

## Answers to Data Exploration

### I. The GPA Data:

Here are boxplots and summary statistics comparing the GPA data for McGrail and Starnes.

| Variable | N | Mean | Median | StDev |
|---|---|---|---|---|
| Starnes GPA | 15 | 3.2129 | 3.2000 | 0.3642 |
| McGrail GPA | 18 | 3.1347 | 3.0000 | 0.3580 |

| Variable | Minimum | Maximum | Q1 | Q3 |
|---|---|---|---|---|
| Starnes GPA | 2.6000 | 3.8000 | 2.9000 | 3.5600 |
| McGrail GPA | 2.8000 | 3.9800 | 2.9000 | 3.2588 |

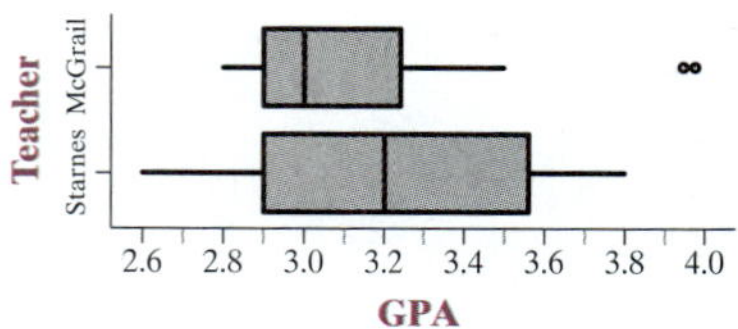


The GPA distribution in Mr. Starnes's class is roughly symmetric, whereas the distribution for Ms. McGrail's class is right-skewed with two high outliers. The median GPA is higher in Mr. Starnes's class. Even though the range of GPAs is about the same for students in both classes, the *IQR* is much larger for students in Mr. Starnes's class, mostly because the value of $Q_3$ is much higher for Mr. Starnes's class. Because the median and third quartile of the distribution of GPAs in Mr. Starnes's class are much higher than in Ms. McGrail's class, there is evidence that he stacked his class.

### II. The SAT Math Data:

Here are boxplots and summary statistics comparing the SATM data for Starnes and McGrail.

| Variable | N | Mean | Median | StDev |
|---|---|---|---|---|
| Starnes GPA | 15 | 622.0 | 630.0 | 61.3 |
| McGrail GPA | 18 | 600.1 | 600.0 | 44.1 |

| Variable | Minimum | Maximum | Q1 | Q3 |
|---|---|---|---|---|
| Starnes GPA | 510.0 | 710.0 | 570.0 | 670.0 |
| McGrail GPA | 500.0 | 680.0 | 587.5 | 622.5 |

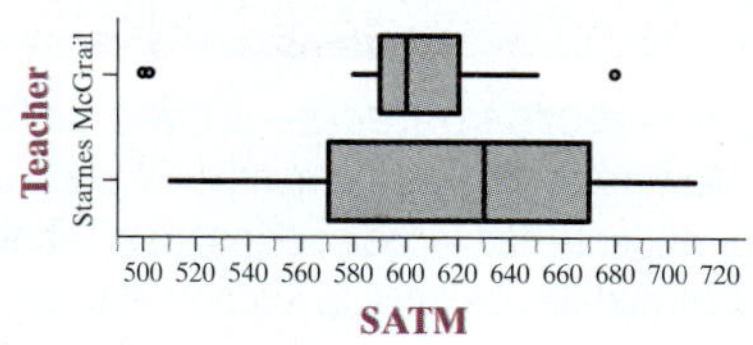


McGrail's distribution is somewhat symmetric, though there are two low outliers and one high outlier. Starnes's distribution of SAT Math scores is slightly skewed to the left. The median SATM score is higher for Starnes's class. The SATM scores vary more in Starnes's class, mostly due to the larger values for $Q_3$ and the maximum in Starnes's distribution. Again, because the median, $Q_3$, and maximum are all higher for students in Mr. Starnes's class, it appears that Mr. Starnes stacked his class.

| Starnes GPA | Starnes SAT-M | McGrail GPA | McGrail SAT-M |
|---|---|---|---|
| 2.9 | 670 | 2.9 | 620 |
| 2.86 | 520 | 3.3 | 590 |
| 2.6 | 570 | 3.98 | 650 |
| 3.6 | 710 | 2.9 | 600 |
| 3.2 | 600 | 3.2 | 620 |
| 2.7 | 590 | 3.5 | 680 |
| 3.1 | 640 | 2.8 | 500 |
| 3.085 | 570 | 2.9 | 502.5 |
| 3.75 | 710 | 3.95 | 640 |
| 3.4 | 630 | 3.1 | 630 |
| 3.338 | 630 | 2.85 | 580 |
| 3.56 | 670 | 2.9 | 590 |
| 3.8 | 650 | 3.245 | 600 |
| 3.2 | 660 | 3.0 | 600 |
| 3.1 | 510 | 3.0 | 620 |
| | | 2.8 | 580 |
| | | 2.9 | 600 |
| | | 3.2 | 600 |

Did Mr. Starnes stack his class? Give appropriate graphical and numerical evidence to support your conclusion.

**AP® EXAM TIP** Use statistical terms carefully and correctly on the AP® exam. Don't say "mean" if you really mean "median." Range is a single number; so are $Q_1$, $Q_3$, and *IQR*. Avoid colloquial use of language, like "the outlier *skews* the mean." Skewed is a shape. If you misuse a term, expect to lose some credit.

**case closed**

## Do pets or friends help reduce stress?

Refer to the chapter-opening Case Study (page 1). You will now use what you have learned in this chapter to analyze the data.

1. Construct an appropriate graph for comparing the heart rates of the women in the three groups.
2. Calculate numerical summaries for each group's data. Which measures of center and spread would you choose to compare? Why?
3. Determine if there are any outliers in each of the three groups. Show your work.
4. Write a few sentences comparing the distributions of heart rates for the women in the three groups.
5. Based on the data, does it appear that the presence of a pet or friend reduces heart rate during a stressful task? Justify your answer.

### Answers to Case Closed

**1.** Parallel boxplots for the three groups are shown here.

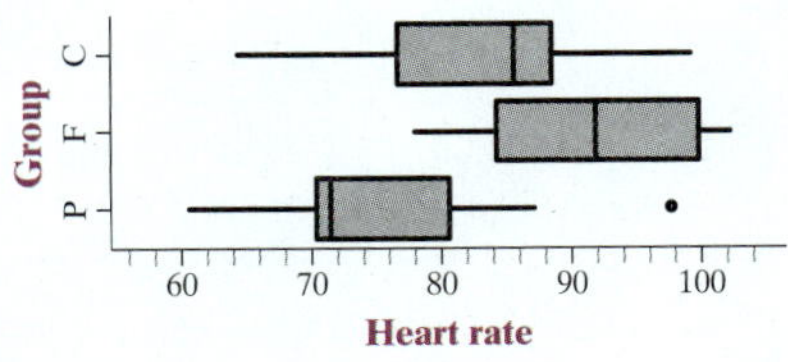


**2.** Summary statistics are shown below. Because of the skewness in the Control distribution and the outlier in the Pet distribution, it would be best to use the median and *IQR* to compare center and spread.

| | *n* | Mean | St. Dev. | Min. | $Q_1$ | Med. | $Q_3$ | Max. |
|---|---|---|---|---|---|---|---|---|
| Control | 15 | 82.5 | 9.24 | 62.6 | 75.5 | 84.7 | 87.8 | 99.0 |
| Friend | 15 | 91.3 | 8.34 | 76.9 | 83.4 | 91.4 | 99.7 | 102.2 |
| Pet | 15 | 73.5 | 9.97 | 58.7 | 68.9 | 70.1 | 79.7 | 97.5 |

**3.** Control: outliers are less than $75.5 - 1.5(87.8 - 75.5) = 57.05$ or greater than $87.8 + 1.5(87.8 - 75.5) = 106.25$, so there are no outliers in the control group. Friend: outliers are less than $83.4 - 1.5(99.7 - 83.4) = 58.95$ or greater than $99.7 + 1.5(99.7 - 83.4) = 124.15$, so there are no outliers in the friend group. Pet: outliers are less than $68.9 - 1.5(79.7 - 68.9) = 52.7$ or greater than $79.7 + 1.5(79.7 - 68.9) = 95.9$, so the heart rate of 97.5 is a high outlier.

**4.** The distributions of heart rate for the control group and the friend group are both slightly skewed to the left, while the distribution of heart rate for the pet group is skewed to the right with a high outlier at 97.5. The median heart rate for the pet group is much lower than the medians of the other two groups. In fact, the median heart rate for the pet group is less than $Q_1$ for the other two groups! The median heart rate for the friend group is the largest. The *IQR*s for the three groups are fairly similar, with the pet group having a slightly smaller *IQR* than the other two.

**5.** It appears that the presence of a pet helps reduce heart rate because the Pet distribution has the smallest median. However, the presence of a friend seems to increase heart rate as the Friend distribution has the largest median.

## TRM Section 1.3 Quizzes

There are three different quizzes available for this section in the Teacher's Resource Materials. Versions A and B are roughly parallel. Find them by clicking on the link in the TE-book, opening the TRCD, or logging onto the book's companion Web site www.whfreeman.com/tps5e (teacher log-in required).



# Section 1.3 Summary

- A numerical summary of a distribution should report at least its **center** and its **spread,** or **variability**.
- The **mean** $\bar{x}$ and the **median** describe the center of a distribution in different ways. The mean is the average of the observations, and the median is the midpoint of the values.
- When you use the median to indicate the center of a distribution, describe its spread using the **quartiles.** The **first quartile** $Q_1$ has about one-fourth of the observations below it, and the **third quartile** $Q_3$ has about three-fourths of the observations below it. The **interquartile range (*IQR*)** is the range of the middle 50% of the observations and is found by $IQR = Q_3 - Q_1$.
- An extreme observation is an **outlier** if it is smaller than $Q_1 - (1.5 \times IQR)$ or larger than $Q_3 + (1.5 \times IQR)$.
- The **five-number summary** consisting of the median, the quartiles, and the maximum and minimum values provides a quick overall description of a distribution. The median describes the center, and the *IQR* and **range** describe the spread.
- **Boxplots** based on the five-number summary are useful for comparing distributions. The box spans the quartiles and shows the spread of the middle half of the distribution. The median is marked within the box. Lines extend from the box to the smallest and the largest observations that are not outliers. Outliers are plotted as isolated points.
- The **variance** $s_x^2$ and especially its square root, the **standard deviation** $s_x$, are common measures of spread about the mean. The standard deviation $s_x$ is zero when there is no variability and gets larger as the spread increases.
- The median is a **resistant** measure of center because it is relatively unaffected by extreme observations. The mean is nonresistant. Among measures of spread, the *IQR* is resistant, but the standard deviation and range are not.
- The mean and standard deviation are good descriptions for roughly symmetric distributions without outliers. They are most useful for the Normal distributions introduced in the next chapter. The median and *IQR* are a better description for skewed distributions.
- Numerical summaries do not fully describe the shape of a distribution. *Always plot your data.*

## 1.3 TECHNOLOGY CORNERS

TI-Nspire Instructions in Appendix B; HP Prime instructions on the book's Web site.

# Section 1.3 Exercises

79. **Quiz grades** Joey's first 14 quiz grades in a marking period were

pg 49

| 86 | 84 | 91 | 75 | 78 | 80 | 74 |
|---|---|---|---|---|---|---|
| 87 | 76 | 96 | 82 | 90 | 98 | 93 |

Calculate the mean. Show your work.

80. **Cowboys** The 2011 roster of the Dallas Cowboys professional football team included 7 defensive linemen. Their weights (in pounds) were 321, 285, 300, 285, 286, 293, and 298. Calculate the mean. Show your work.

81. **Quiz grades** Refer to Exercise 79.
(a) Find the median by hand. Show your work.
(b) Suppose Joey has an unexcused absence for the 15th quiz, and he receives a score of zero. Recalculate the mean and the median. What property of measures of center does this illustrate?

pg 52

82. **Cowboys** Refer to Exercise 80.
(a) Find the median by hand. Show your work.
(b) Suppose the heaviest lineman had weighed 341 pounds instead of 321 pounds. How would this change affect the mean and the median? What property of measures of center does this illustrate?

83. **Incomes of college grads** According to the Census Bureau, the mean and median income in a recent year of people at least 25 years old who had a bachelor's degree but no higher degree were \$48,097 and \$60,954. Which of these numbers is the mean and which is the median? Explain your reasoning.

84. **House prices** The mean and median selling prices of existing single-family homes sold in July 2012 were \$263,200 and \$224,200.[41] Which of these numbers is the mean and which is the median? Explain how you know.

85. **Baseball salaries** Suppose that a Major League Baseball team's mean yearly salary for its players is \$1.2 million and that the team has 25 players on its active roster. What is the team's total annual payroll? If you knew only the median salary, would you be able to answer this question? Why or why not?

86. **Mean salary?** Last year a small accounting firm paid each of its five clerks \$22,000, two junior accountants \$50,000 each, and the firm's owner \$270,000. What is the mean salary paid at this firm? How many of the employees earn less than the mean? What is the median salary? Write a sentence to describe how an unethical recruiter could use statistics to mislead prospective employees.

87. **Domain names** When it comes to Internet domain names, is shorter better? According to one ranking of Web sites in 2012, the top 8 sites (by number of "hits") were google.com, youtube.com, wikipedia.org, yahoo.com, amazon.com, ebay.com, craigslist.org, and facebook.com. These familiar sites certainly have short domain names. The histogram below shows the domain name lengths (in number of letters in the name, not including the extensions .com and .org) for the 500 most popular Web sites.

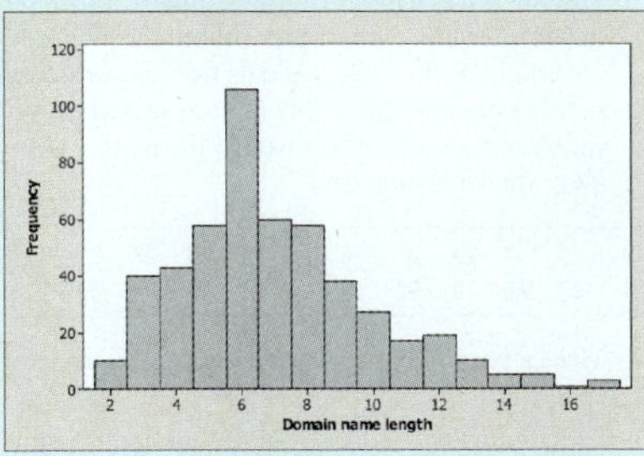


(a) Estimate the mean and median of the distribution. Explain your method clearly.
(b) If you wanted to argue that shorter domain names were more popular, which measure of center would you choose—the mean or the median? Justify your answer.

88. **Do adolescent girls eat fruit?** We all know that fruit is good for us. Below is a histogram of the number of servings of fruit per day claimed by 74 seventeen-year-old girls in a study in Pennsylvania.[42]

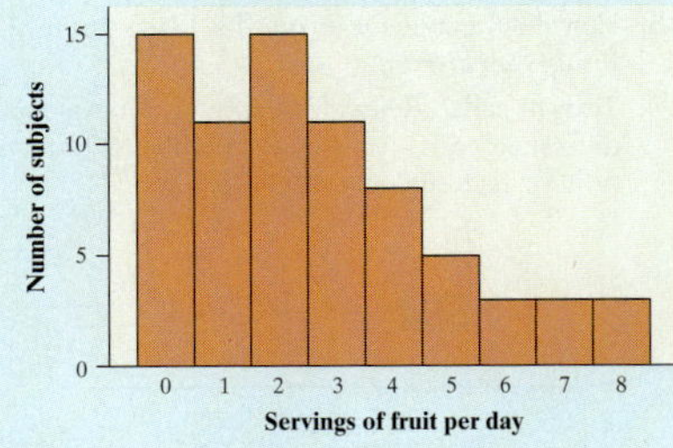


(a) With a little care, you can find the median and the quartiles from the histogram. What are these numbers? How did you find them?
(b) Estimate the mean of the distribution. Explain your method clearly.

**TRM Full Solutions to Section 1.3 Exercises**

## Answers to Section 1.3 Exercises

**1.79** $\bar{x} = 85$

**1.80** $\bar{x} = 295.4$ pounds

**1.81** **(a)** median = 85. **(b)** $\bar{x} = 79.33$ and median = 84. The median did not change much but the mean did, showing that the median is more resistant to outliers than the mean.

**1.82** **(a)** median = 293 pounds. **(b)** $\bar{x} =$ 298.3 pounds and median is unchanged (293 pounds). This is because the median is more resistant to outliers than the mean.

**1.83** The mean is \$60,954 and the median is \$48,097. The distribution of salaries is likely to be quite right skewed because of a few people who have a very large income, making the mean larger than the median.

**1.84** The mean house price is \$263,200 and the median is \$224,200. The distribution of house prices is likely to be quite skewed to the right because of a few very expensive homes, making the mean larger than the median.

**1.85** The team's annual payroll is $1.2 \times 25 = 30$ or \$30 million. No, because the median only describes the middle value in the distribution. It doesn't provide specific information about any of the other values.

**1.86** The mean salary is \$60,000. Seven of the eight employees earned less than the mean. The median is \$22,000. An unethical recruiter would report the mean salary as the "typical" salary because it is much larger than the median salary.

**1.87** **(a)** Estimating the frequencies of the bars (from left to right) as 10, 40, 42, 58, 105, 60, 58, 38, 27, 18, 20, 10, 5, 5, 1, and 3, the mean is $\bar{x} = \frac{3504}{500} = 7.01$.
The median is the average of the 250th and 251st values, which is 6. **(b)** Because the median is less than the mean, we would use the median to argue that shorter domain names are more popular.

**1.88** **(a)** Estimating the frequencies of the bars (from left to right) as 15, 11, 15, 11, 8, 5, 3, 3, and 3, the median is the average of the 37th and 38th values, which is 2. $Q_1$ is the 19th observation, which is 1. $Q_3$ is the 56th observation, which is 4. **(b)** Using the frequencies from part (a),

$$\bar{x} = \frac{194}{74} = 2.62$$

**1.89** **(a)** $IQR = 91 - 78 = 13$. The middle 50% of the data have a range of 13 points. **(b)** Any outliers are below $78 - 1.5(13) = 58.5$ or above $91 + 1.5(13) = 110.5$. There are no outliers.

**1.90** **(a)** $IQR = 300 - 285 = 15$ pounds. The middle 50% of the data have a range of 15 pounds. **(b)** Any outliers are below $285 - 1.5(15) = 262.5$ or above $300 + 1.5(15) = 322.5$. There are no outliers.

**1.91** **(a)** Outliers are anything below $3 - 1.5(40) = -57$ or above $43 + 1.5(40) = 103$, so 118 is an outlier. The boxplot is shown below. **(b)** The article claims that teens send 1742 texts a month, which is about 58 texts a day. Nearly all of the members of the class (21 of 25) sent fewer than 58 texts per day, which seems to contradict the claim in the article.

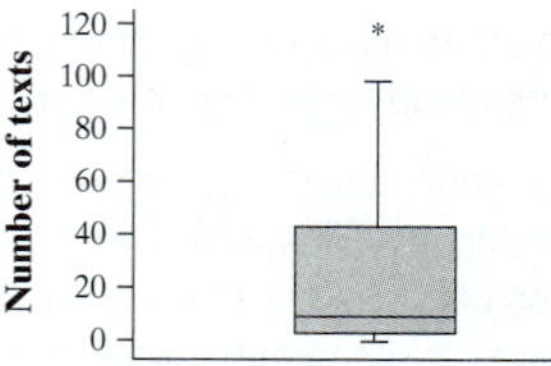


**1.92** **(a)** Outliers are anything below $82 - 1.5(11) = 65.5$ or above $93 + 1.5(11) = 109.5$, so 43 and 45 are outliers. The boxplot is shown below. **(b)** Most students did quite well. In fact, more than 75% of the class scored higher than 80. Only two of the students did very poorly.

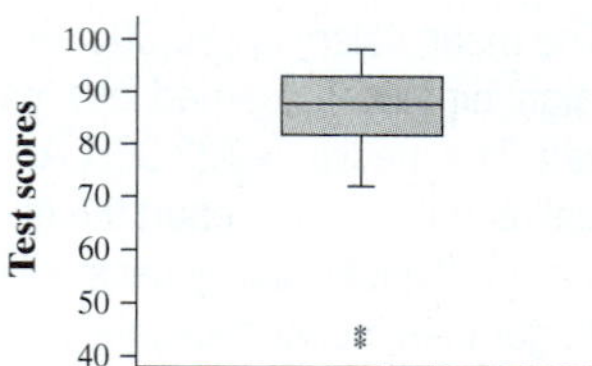


**1.93** **(a)** Positive numbers indicate students who had more text messages than calls. Because the first quartile is about 0, roughly 75% of the students had more texts than calls, which supports the article's conclusion. **(b)** No. Students in statistics classes tend to be upperclassmen and their responses might differ from underclassmen.



**89. Quiz grades** Refer to Exercise 79.

pg 55 (a) Find and interpret the interquartile range ($IQR$).

pg 56 (b) Determine whether there are any outliers. Show your work.

**90. Cowboys** Refer to Exercise 80.

(a) Find and interpret the interquartile range ($IQR$).

(b) Determine whether there are any outliers. Show your work.

pg 57

**91. Don't call me** In a September 28, 2008, article titled "Letting Our Fingers Do the Talking," the *New York Times* reported that Americans now send more text messages than they make phone calls. According to a study by Nielsen Mobile, "Teenagers ages 13 to 17 are by far the most prolific texters, sending or receiving 1742 messages a month." Mr. Williams, a high school statistics teacher, was skeptical about the claims in the article. So he collected data from his first-period statistics class on the number of text messages and calls they had sent or received in the past 24 hours. Here are the texting data:

| | | | | | | | | | | | | |
|---|---|---|---|---|---|---|---|---|---|---|---|---|
| 0 | 7 | 1 | 29 | 25 | 8 | 5 | 1 | 25 | 98 | 9 | 0 | 26 |
| 8 | 118 | 72 | 0 | 92 | 52 | 14 | 3 | 3 | 44 | 5 | 42 | |

(a) Make a boxplot of these data by hand. Be sure to check for outliers.

(b) Explain how these data seem to contradict the claim in the article.

**92. Acing the first test** Here are the scores of Mrs. Liao's students on their first statistics test:

| | | | | | | | | | | |
|---|---|---|---|---|---|---|---|---|---|---|
| 93 | 93 | 87.5 | 91 | 94.5 | 72 | 96 | 95 | 93.5 | 93.5 | 73 |
| 82 | 45 | 88 | 80 | 86 | 85.5 | 87.5 | 81 | 78 | 86 | 89 |
| 92 | 91 | 98 | 85 | 82.5 | 88 | 94.5 | 43 | | | |

(a) Make a boxplot of the test score data by hand. Be sure to check for outliers.

(b) How did the students do on Mrs. Liao's first test? Justify your answer.

**93. Texts or calls?** Refer to Exercise 91. A boxplot of the difference (texts – calls) in the number of texts and calls for each student is shown below.

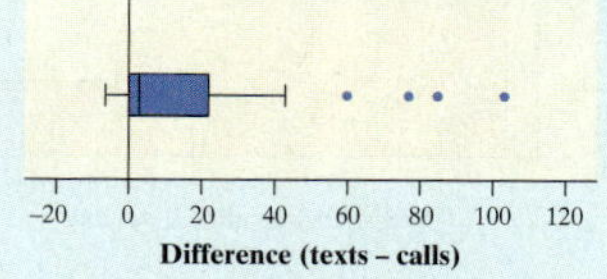


(a) Do these data support the claim in the article about texting versus calling? Justify your answer with appropriate evidence.

(b) Can we draw any conclusion about the preferences of all students in the school based on the data from Mr. Williams's statistics class? Why or why not?

**94. Electoral votes** To become president of the United States, a candidate does not have to receive a majority of the popular vote. The candidate does have to win a majority of the 538 electoral votes that are cast in the Electoral College. Here is a stemplot of the number of electoral votes for each of the 50 states and the District of Columbia.

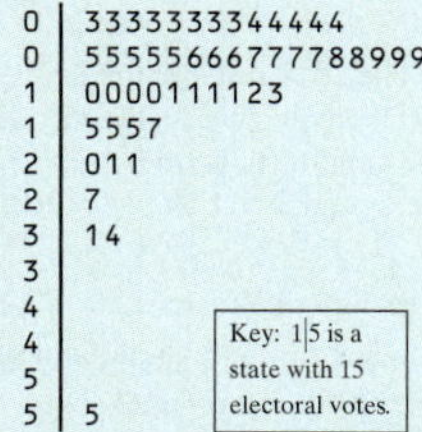

| Stem | Leaves |
|---|---|
| 0 | 3333333344444 |
| 0 | 55555666777788999 |
| 1 | 0000111123 |
| 1 | 5557 |
| 2 | 011 |
| 2 | 7 |
| 3 | 14 |
| 3 | |
| 4 | |
| 4 | |
| 5 | |
| 5 | 5 |

Key: 1|5 is a state with 15 electoral votes.

(a) Make a boxplot of these data by hand. Be sure to check for outliers.

(b) Which measure of center and spread would you use to summarize the distribution—the mean and standard deviation or the median and $IQR$? Justify your answer.

**95. Comparing investments** Should you put your money into a fund that buys stocks or a fund that invests in real estate? The boxplots compare the daily returns (in percent) on a "total stock market" fund and a real estate fund over a one-year period.[43]

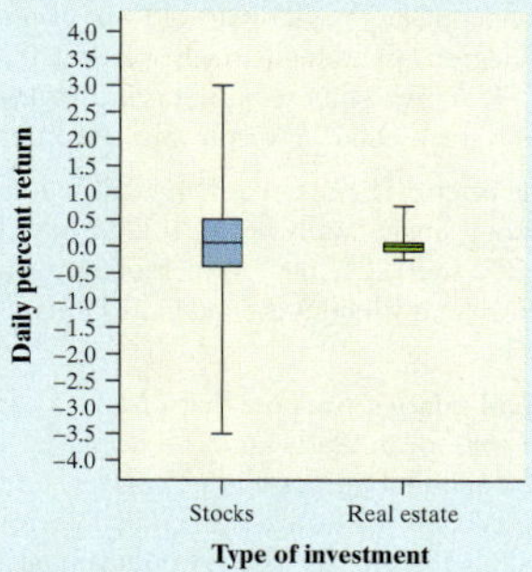


(a) Read the graph: about what were the highest and lowest daily returns on the stock fund?

(b) Read the graph: the median return was about the same on both investments. About what was the median return?

(c) What is the most important difference between the two distributions?

**1.94** **(a)** Outliers are anything below $4 - 1.5(8) = -8$ or greater than $12 + 1.5(8) = 24$ so 27, 31, 34, and 55 are outliers. The boxplot is shown below. **(b)** Use the median and $IQR$ because the distribution is skewed with several high outliers.

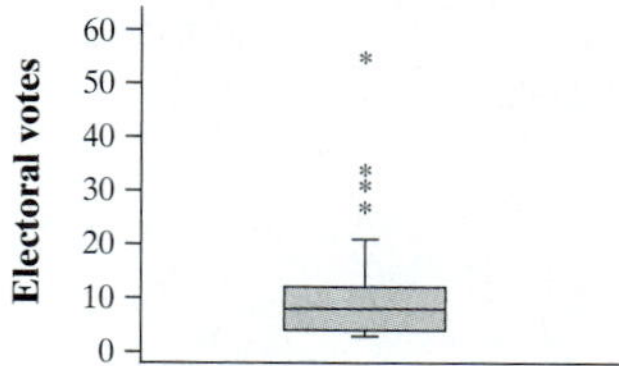


**1.95** **(a)** About 3% and −3.5%. **(b)** About 0.1%. **(c)** The stock fund is much more variable. It has higher positive returns, but also higher negative returns.

96. **Income and education level** Each March, the Bureau of Labor Statistics compiles an Annual Demographic Supplement to its monthly Current Population Survey.[44] Data on about 71,067 individuals between the ages of 25 and 64 who were employed full-time were collected in one of these surveys. The boxplots below compare the distributions of income for people with five levels of education. This figure is a variation of the boxplot idea: because large data sets often contain very extreme observations, we omitted the individuals in each category with the top 5% and bottom 5% of incomes. Write a brief description of how the distribution of income changes with the highest level of education reached. Give specifics from the graphs to support your statements.

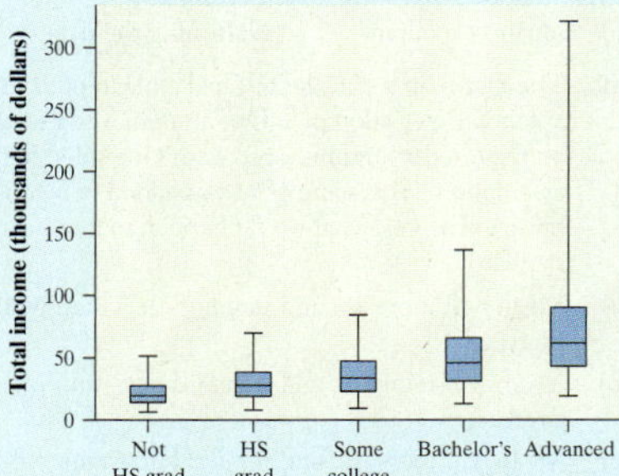


97. **Phosphate levels** The level of various substances in the blood influences our health. Here are measurements of the level of phosphate in the blood of a patient, in milligrams of phosphate per deciliter of blood, made on 6 consecutive visits to a clinic: 5.6, 5.2, 4.6, 4.9, 5.7, 6.4. A graph of only 6 observations gives little information, so we proceed to compute the mean and standard deviation.
    (a) Find the standard deviation from its definition. That is, find the deviations of each observation from the mean, square the deviations, then obtain the variance and the standard deviation.
    (b) Interpret the value of $s_x$ you obtained in part (a).

98. **Feeling sleepy?** The first four students to arrive for a first-period statistics class were asked how much sleep (to the nearest hour) they got last night. Their responses were 7, 7, 9, and 9.
    (a) Find the standard deviation from its definition. That is, find the deviations of each observation from the mean, square the deviations, then obtain the variance and the standard deviation.
    (b) Interpret the value of $s_x$ you obtained in part (a).
    (c) Do you think it's safe to conclude that the mean amount of sleep for all 30 students in this class is close to 8 hours? Why or why not?

99. **Shopping spree** The figure displays computer output for data on the amount spent by 50 grocery shoppers.

**Descriptive Statistics: Amount spent**

| Variable | Total Count | Mean | StDev | Minimum | Q1 | Median | Q3 | Maximum |
|---|---|---|---|---|---|---|---|---|
| Amount spent | 50 | 34.70 | 21.70 | 3.11 | 19.06 | 27.85 | 45.72 | 93.34 |

    (a) What would you guess is the shape of the distribution based only on the computer output? Explain.
    (b) Interpret the value of the standard deviation.
    (c) Are there any outliers? Justify your answer.

100. **C-sections** Do male doctors perform more cesarean sections (C-sections) than female doctors? A study in Switzerland examined the number of cesarean sections (surgical deliveries of babies) performed in a year by samples of male and female doctors. Here are summary statistics for the two distributions:

| | $\bar{x}$ | $s_x$ | Min | $Q_1$ | Med | $Q_3$ | Max | *IQR* |
|---|---|---|---|---|---|---|---|---|
| **Male doctors** | 41.333 | 20.607 | 20 | 27 | 34 | 50 | 86 | 23 |
| **Female doctors** | 19.1 | 10.126 | 5 | 10 | 18.5 | 29 | 33 | 19 |

    (a) Based on the computer output, which distribution would you guess has a more symmetrical shape? Explain.
    (b) Explain how the *IQR*s of these two distributions can be so similar even though the standard deviations are quite different.
    (c) Does it appear that male doctors perform more C-sections? Justify your answer.

101. **The *IQR*** Is the interquartile range a resistant measure of spread? Give an example of a small data set that supports your answer.

102. **What do they measure?** For each of the following summary statistics, decide (i) whether it could be used to measure center or spread and (ii) whether it is resistant.
    (a) $\dfrac{Q_1 + Q_3}{2}$ (b) $\dfrac{\text{Max} - \text{Min}}{2}$

103. **SD contest** This is a standard deviation contest. You must choose four numbers from the whole numbers 0 to 10, with repeats allowed.
    (a) Choose four numbers that have the smallest possible standard deviation.
    (b) Choose four numbers that have the largest possible standard deviation.
    (c) Is more than one choice possible in either part (a) or (b)? Explain.

**1.96** All five income distributions are skewed to the right. As education level rises, the median, quartiles, and extremes all rise. Also, the variability in income increases as the education level increases.

**1.97 (a)** $s_x = \sqrt{\dfrac{2.06}{6-1}} = 0.6419$ mg/dl. **(b)** The phosphate level typically varies from the mean by about 0.6419 mg/dl.

**1.98 (a)** $s_x = \sqrt{\dfrac{4}{4-1}} = 1.15$ hours. **(b)** The hours of sleep typically varies from the mean by about 1.15 hours. **(c)** No. The first 4 students to arrive in the classroom are not likely to be representative of the entire class in terms of the amount of sleep they got last night.

**1.99 (a)** Skewed to the right, because the mean is much larger than the median and $Q_3$ is much further from the median than $Q_1$. **(b)** The amount of money spent typically varies from the mean by $21.70. **(c)** Any points below $19.06 - 1.5(26.66) = -20.93$ or above $45.72 + 1.5(26.66) = 85.71$ are outliers. Because the maximum of 93.34 is greater than 85.71, there is at least one outlier.

**1.100 (a)** The distribution for the female doctors, because the mean and median are relatively close together. The mean and median for the male doctors are much farther apart. **(b)** The *IQR* measures only the range of the middle 50% of the data, while the standard deviation uses every observation and is not resistant to outliers. The middle 50% of the data sets may have similar spreads, but if the lower and upper 25% of one distribution are more spread out than in the other distribution, it will have a larger standard deviation. **(c)** It does appear that males perform more C-sections. Each of the numbers in the 5-number summary was larger for the males than for the females.

**1.101** Yes. For example, in data set 1, 2, 3, 4, 5, 6, 7, 8 the *IQR* is 4. If 8 is changed to 88, the *IQR* will still be 4.

**1.102 (a)** Center, because we are finding a middle point between two positions in the distribution. It would be resistant to outliers, because any outliers would occur below $Q_1$ or above $Q_3$. **(b)** Spread, because it finds half of the range (which is a measure of spread). This measure would not be resistant to outliers because if outliers exist, they would include the max, the min, or both.

**1.103 (a)** One possible answer is 1, 1, 1, 1. **(b)** 0, 0, 10, 10. **(c)** For part (a), any set of four identical numbers will have $s_x = 0$. For part (b), however, there is only 1 possible answer. We want the values to be as far from the mean as possible, so our best choice is two values at each extreme.

**1.104** Variable A because there are many observations far from the mean. Because of the bell shape to the distribution of variable B, many of the observations are quite close to the mean, making the typical distance from the mean smaller than for variable A.

**1.105** STATE: Do the data indicate that men and women differ in their study habits and attitudes toward learning? PLAN: We will draw side-by-side boxplots of the data about men and women, compute summary statistics, and compare the shape, center, and spread of both distributions. DO: The boxplots are given below, as is a table of summary statistics.

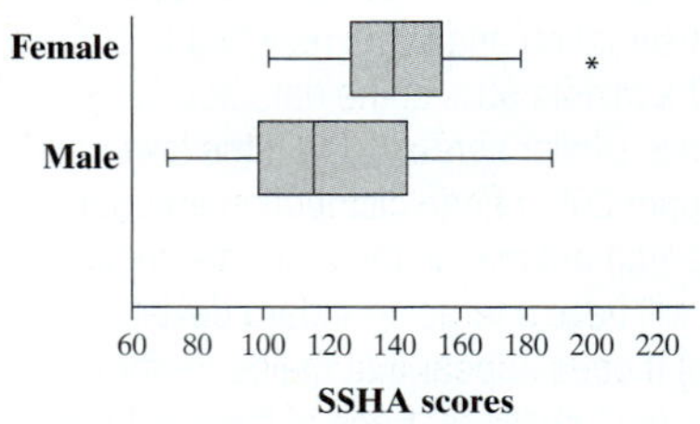


| Variable | N | Mean | StDev | Minimum |
|---|---|---|---|---|
| Women | 18 | 141.06 | 26.44 | 101.00 |
| Men | 20 | 121.25 | 32.85 | 70.00 |

| Variable | Q1 | Median | Q3 | Maximum |
|---|---|---|---|---|
| Women | 126.00 | 138.50 | 154.00 | 200.00 |
| Men | 98.00 | 114.50 | 143.00 | 187.00 |

Both distributions are slightly skewed to the right. Both the mean and median are higher for women than for men. The scores for men are more variable than the scores for women. There is no outlier in the male distribution, and a single outlier appears at 200 in the female distribution. CONCLUDE: Men and women differ in their study habits and attitudes towards learning. The typical score for females is about 24 greater than the typical score for males. Female scores are also more consistent than male scores.

**1.106** STATE: Do the different types of flowers have different distributions of length? PLAN: We will draw side-by-side boxplots of the data about each type of flower, compute summary statistics, and compare the shape, center, and spread of the distributions. DO: The boxplots are given here, as is a table of summary statistics.

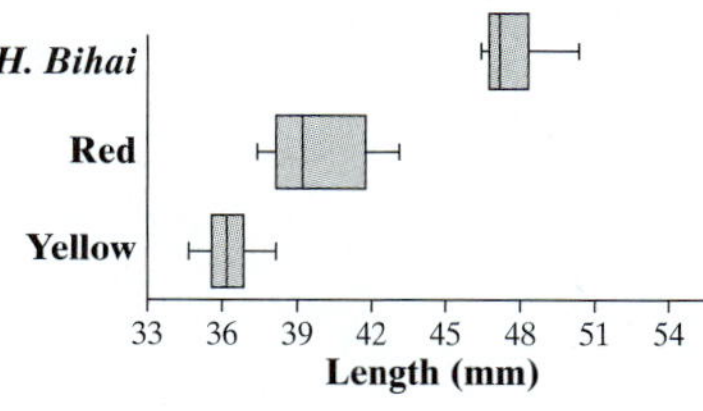


| Variable | N | Mean | StDev | Minimum |
|---|---|---|---|---|
| *H. bihai* | 16 | 47.597 | 1.213 | 46.340 |
| Red | 23 | 39.711 | 1.799 | 37.400 |
| Yellow | 25 | 36.180 | 0.975 | 34.570 |

| Variable | Q1 | Median | Q3 | Maximum |
|---|---|---|---|---|
| *H. bihai* | 46.710 | 47.120 | 48.245 | 50.260 |
| Red | 38.070 | 39.160 | 41.690 | 43.090 |
| Yellow | 35.430 | 36.110 | 36.820 | 38.130 |

The distributions for *H. bihai* and *H. caribaea* red are skewed to the right, while the distribution of *H. caribaea* yellow is roughly symmetric. Both the mean and median length are greatest for *H. bihai*, followed by *H. caribaea* red and then *H. caribaea* yellow. The lengths for *H. caribaea* red are the most variable, while the lengths of *H. caribaea* yellow are the least variable. There are no outliers in any of the distributions.

CONCLUDE: The three flower varieties have distinct distributions of lengths. *H. bihai* is clearly the longest variety—the shortest *bihai* was over 3 mm longer than the tallest red. Red is generally longer than yellow, with a few exceptions.

**1.107** d

**1.108** b

**1.109** e



**104. Measuring spread** Which of the distributions shown has a larger standard deviation? Justify your answer.

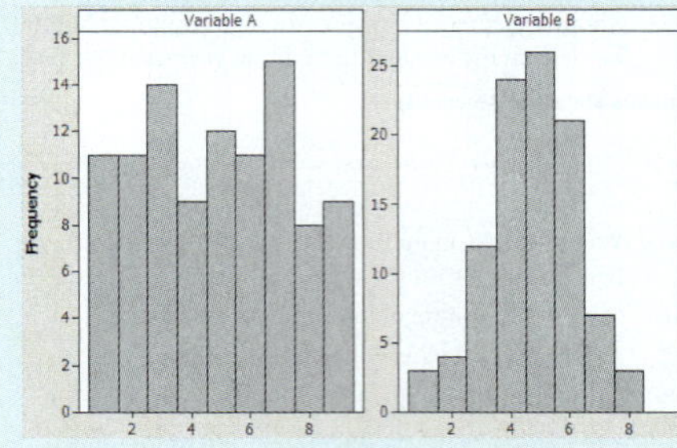


**105. SSHA scores** Here are the scores on the Survey of Study Habits and Attitudes (SSHA) for 18 first-year college women:

pg 65

STEP 4

| | | | | | | | | |
|---|---|---|---|---|---|---|---|---|
| 154 | 109 | 137 | 115 | 152 | 140 | 154 | 178 | 101 |
| 103 | 126 | 126 | 137 | 165 | 165 | 129 | 200 | 148 |

and for 20 first-year college men:

| | | | | | | |
|---|---|---|---|---|---|---|
| 108 | 140 | 114 | 91 | 180 | 115 | 126 |
| 92 | 169 | 146 | 109 | 132 | 75 | 88 |
| 113 | 151 | 70 | 115 | 187 | 104 | |

Do these data support the belief that men and women differ in their study habits and attitudes toward learning? (Note that high scores indicate good study habits and attitudes toward learning.) Follow the four-step process.

**106. Hummingbirds and tropical flower** Researchers from Amherst College studied the relationship between varieties of the tropical flower *Heliconia* on the island of Dominica and the different species of hummingbirds that fertilize the flowers.[45] Over time, the researchers believe, the lengths of the flowers and the forms of the hummingbirds' beaks have evolved to match each other. If that is true, flower varieties fertilized by different hummingbird species should have distinct distributions of length.

STEP 4

The table below gives length measurements (in millimeters) for samples of three varieties of *Heliconia*, each fertilized by a different species of hummingbird. Do these data support the researchers' belief? Follow the four-step process.

***H. bihai***

| | | | | | | | |
|---|---|---|---|---|---|---|---|
| 47.12 | 46.75 | 46.80 | 47.12 | 46.67 | 47.43 | 46.44 | 46.64 |
| 48.07 | 48.34 | 48.15 | 50.26 | 50.12 | 46.34 | 46.94 | 48.36 |

***H. caribaea* red**

| | | | | | | | |
|---|---|---|---|---|---|---|---|
| 41.90 | 42.01 | 41.93 | 43.09 | 41.47 | 41.69 | 39.78 | 40.57 |
| 39.63 | 42.18 | 40.66 | 37.87 | 39.16 | 37.40 | 38.20 | 38.07 |
| 38.10 | 37.97 | 38.79 | 38.23 | 38.87 | 37.78 | 38.01 | |

***H. caribaea* yellow**

| | | | | | | | |
|---|---|---|---|---|---|---|---|
| 36.78 | 37.02 | 36.52 | 36.11 | 36.03 | 35.45 | 38.13 | 37.10 |
| 35.17 | 36.82 | 36.66 | 35.68 | 36.03 | 34.57 | 34.63 | |

*Multiple choice: Select the best answer for Exercises 107 to 110.*

107. If a distribution is skewed to the right with no outliers,

(a) mean < median.
(b) mean ≈ median.
(c) mean = median.
(d) mean > median.
(e) We can't tell without examining the data.

108. The scores on a statistics test had a mean of 81 and a standard deviation of 9. One student was absent on the test day, and his score wasn't included in the calculation. If his score of 84 was added to the distribution of scores, what would happen to the mean and standard deviation?

(a) Mean will increase, and standard deviation will increase.
(b) Mean will increase, and standard deviation will decrease.
(c) Mean will increase, and standard deviation will stay the same.
(d) Mean will decrease, and standard deviation will increase.
(e) Mean will decrease, and standard deviation will decrease.

109. The stemplot shows the number of home runs hit by each of the 30 Major League Baseball teams in 2011. Home run totals above what value should be considered outliers?

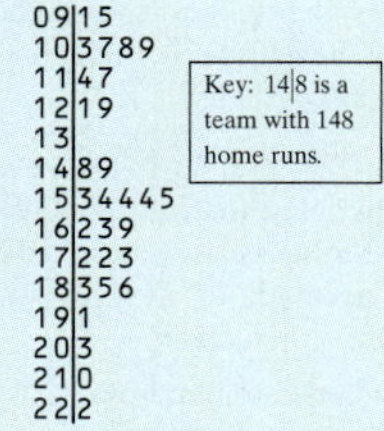


(a) 173 (b) 210 (c) 222 (d) 229 (e) 257

**110.** Which of the following boxplots best matches the distribution shown in the histogram?

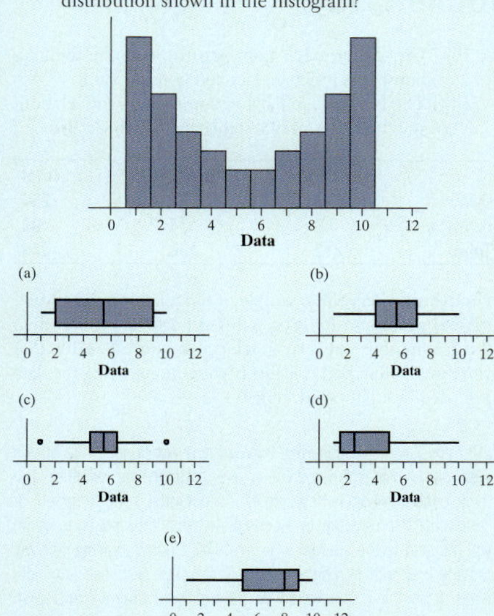


*Exercises 111 and 112 refer to the following setting.*

We used CensusAtSchool's "Random Data Selector" to choose a sample of 50 Canadian students who completed a survey in a recent year.

**111. How tall are you?** (1.2) Here are the students' heights (in centimeters).

| | | | | | | | | | |
|---|---|---|---|---|---|---|---|---|---|
| 166.5 | 170 | 178 | 163 | 150.5 | 169 | 173 | 169 | 171 | 166 |
| 190 | 183 | 178 | 161 | 171 | 170 | 191 | 168.5 | 178.5 | 173 |
| 175 | 160.5 | 166 | 164 | 163 | 174 | 160 | 174 | 182 | 167 |
| 166 | 170 | 170 | 181 | 171.5 | 160 | 178 | 157 | 165 | 187 |
| 168 | 157.5 | 145.5 | 156 | 182 | 168.5 | 177 | 162.5 | 160.5 | 185.5 |

Make an appropriate graph to display these data. Describe the shape, center, and spread of the distribution. Are there any outliers?

**112. Let's chat** (1.1) The bar graph displays data on students' responses to the question "Which of these methods do you most often use to communicate with your friends?"

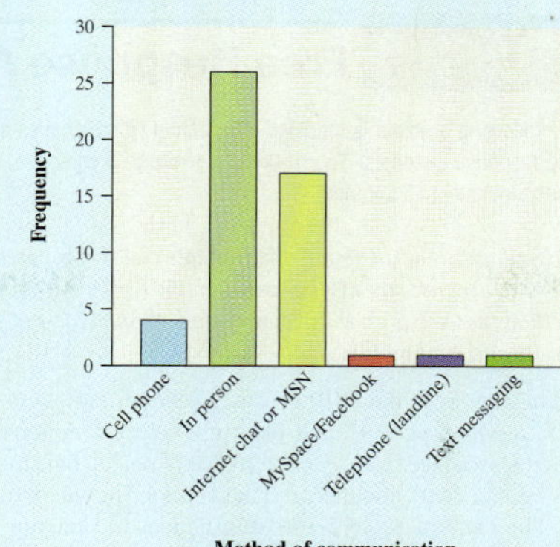


(a) Would it be appropriate to make a pie chart for these data? Why or why not?

(b) Jerry says that he would describe this bar graph as skewed to the right. Explain why Jerry is wrong.

**113. Success in college** (1.1) The 2007 Freshman Survey asked first-year college students about their "habits of mind"—specific behaviors that college faculty have identified as being important for student success. One question asked students, "How often in the past year did you revise your papers to improve your writing?" Another asked, "How often in the past year did you seek feedback on your academic work?" The figure is a bar graph comparing male and female responses to these two questions.[46]

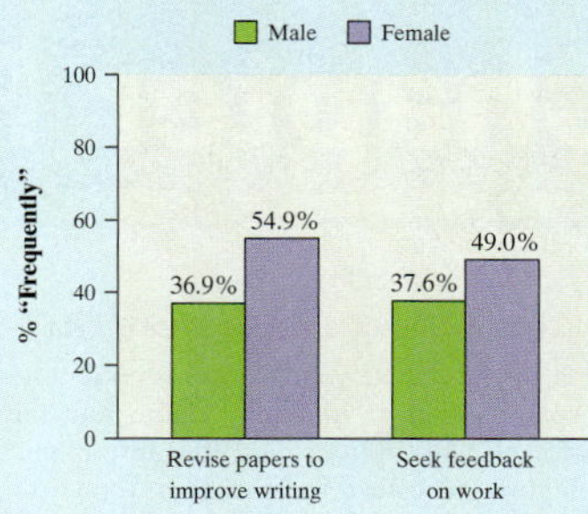


What does the graph tell us about the habits of mind of male and female college freshmen?

**1.110** a

**1.111** A histogram is given below. This distribution is roughly symmetric with a center around 170 cm and values that vary from 145.5 cm to 191 cm. There do not appear to be any outliers.

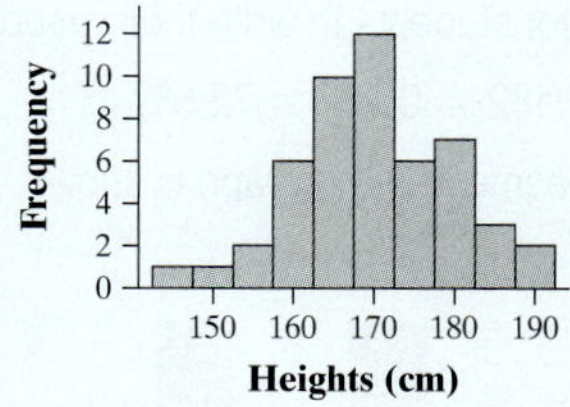


**1.112** **(a)** Yes, because the categories form parts of a whole. **(b)** The graph should not be described as skewed to the right because this is not a distribution of quantitative data. The categories could be graphed in any order.

**1.113** Women appear to be more likely to engage in behaviors that are indicative of good "habits of mind." They are especially more likely to revise papers to improve their writing. The difference is a little smaller for seeking feedback on their work, although the percentage is still higher for females.

TRM FRAPPY!

We present a model solution here. Please consult the TE-book, or Teacher's Resource Materials for sample student responses, a scoring rubric, and a black line master of the original question with space for students to write their responses.

**(a)** 132/182 = 0.725 = 72.5%

**(b)** A segmented bar graph is shown below.

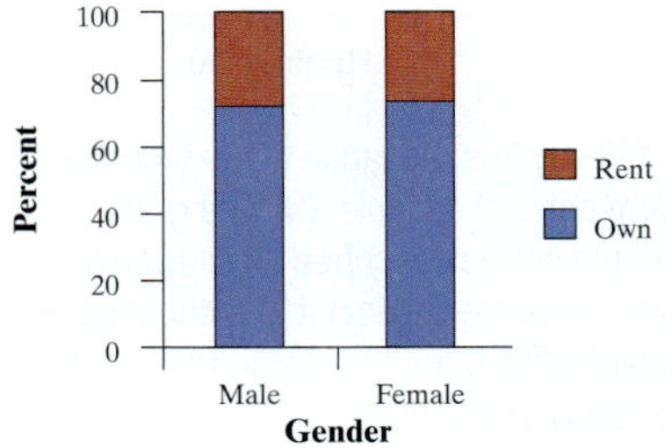


**(c)** There doesn't seem to be much of an association between gender and housing status. The percent of each gender that owns their home is roughly the same, although it is slightly higher for females (73.5% vs. 72.5%).

**(d)** A graph of housing status for married and not married respondents is shown below. Because the percent of married respondents who own their home (81.1%) is quite a bit larger than the percent of not married respondents (60.3%), there seems to be an association between marital status and housing status. Also, this association is stronger than the association between gender and housing status because the difference in percents who own a home is greater for marital status (81.1 − 60.3 = 20.8%) than gender (73.5 − 72.5 = 1%).

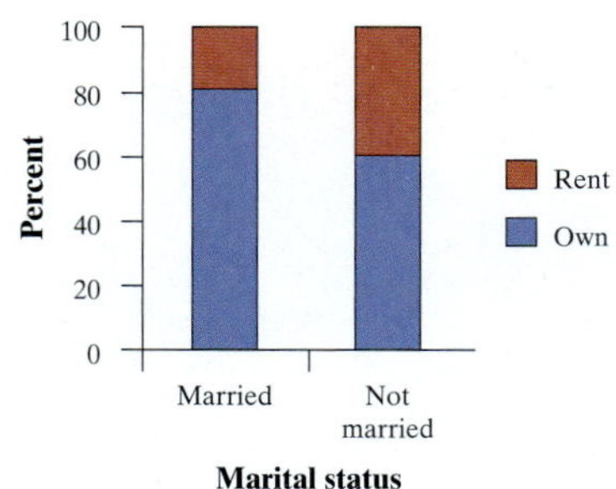


## FRAPPY! Free Response AP® Problem, Yay!

The following problem is modeled after actual AP® Statistics exam free response questions. Your task is to generate a complete, concise response in 15 minutes.

*Directions: Show all your work. Indicate clearly the methods you use, because you will be scored on the correctness of your methods as well as on the accuracy and completeness of your results and explanations.*

Using data from the 2010 census, a random sample of 348 U.S. residents aged 18 and older was selected. Among the variables recorded were gender (male or female), housing status (rent or own), and marital status (married or not married). The two-way table below summarizes the relationship between gender and housing status.

| | Male | Female | Total |
|---|---|---|---|
| Own | 132 | 122 | **254** |
| Rent | 50 | 44 | **94** |
| **Total** | **182** | **166** | **348** |

(a) What percent of males in the sample own their home?

(b) Make a graph to compare the distribution of housing status for males and females.

(c) Using your graph from part (b), describe the relationship between gender and housing status.

(d) The two-way table below summarizes the relationship between marital status and housing status.

| | Married | Not Married | Total |
|---|---|---|---|
| Own | 172 | 82 | **254** |
| Rent | 40 | 54 | **94** |
| **Total** | **212** | **136** | **348** |

For the members of the sample, is the relationship between marital status and housing status stronger or weaker than the relationship between gender and housing status that you described in part (c)? Justify your choice using the data provided in the two-way tables.

After you finish, you can view two example solutions on the book's Web site (www.whfreeman.com/tps5e). Determine whether you think each solution is "complete," "substantial," "developing," or "minimal." If the solution is not complete, what improvements would you suggest to the student who wrote it? Finally, your teacher will provide you with a scoring rubric. Score your response and note what, if anything, you would do differently to improve your own score.

# Chapter Review

### Introduction: Data Analysis: Making Sense of Data

In this brief section, you learned several fundamental concepts that will be important throughout the course: the idea of a distribution and the distinction between quantitative and categorical variables. You also learned a strategy for exploring data:

- Begin by examining each variable by itself. Then move on to study relationships between variables.
- Start with a graph or graphs. Then add numerical summaries.

### Section 1.1: Analyzing Categorical Data

In this section, you learned how to display the distribution of a single categorical variable with pie charts and bar graphs and what to look for when describing these displays. Remember to properly label your graphs! Poor labeling is an easy way to lose points on the AP® exam. You should also be able to recognize misleading graphs and be careful to avoid making misleading graphs yourself.

Next, you learned how to investigate the association between two categorical variables. Using a two-way table, you learned how to calculate and display marginal and conditional distributions. Graphing and comparing conditional distributions allow you to look for an association between the variables. If there is no association between the two variables, graphs of the conditional distributions will look the same. However, if differences in the conditional distributions do exist, there is an association between the variables.

TRM **Chapter 1 Tests**

There are three different tests available for this chapter. Versions A and B are roughly parallel. Find them in the TE-book, or on the TRCD or the book's companion Web site (teacher log-in required).

### Section 1.2: Displaying Quantitative Data with Graphs

In this section, you learned how to create three different types of graphs for a quantitative variable: dotplots, stemplots, and histograms. Each of the graphs has distinct benefits, but all of them are good tools for examining the distribution of a quantitative variable. Dotplots and stemplots are handy for small sets of data. Histograms are the best choice when there are a large number of observations. On the AP® exam, you will be expected to create each of these types of graphs, label them properly, and comment on their characteristics.

When you are describing the distribution of a quantitative variable, you should look at its graph for the overall pattern (shape, center, spread) and striking departures from that pattern (outliers). Use the acronym SOCS (shape, outliers, center, spread) to help remember these four characteristics. Likewise, when comparing distributions, you should include explicit comparison words such as "is greater than" or "is approximately the same as." When asked to compare distributions, a very common mistake on the AP® exam is describing the characteristics of each distribution separately without making these explicit comparisons.

### Section 1.3: Describing Quantitative Data with Numbers

To measure the center of a distribution of quantitative data, you learned how to calculate the mean and the median of a distribution. You also learned that the median is a resistant measure of center but the mean isn't resistant because it can be greatly affected by skewness or outliers.

To measure the spread of a distribution of quantitative data, you learned how to calculate the range, interquartile range, and standard deviation. The interquartile range (*IQR*) is a resistant measure of spread because it ignores the upper 25% and lower 25% of the distribution, but the range isn't resistant because it uses only the minimum and maximum value. The standard deviation is the most commonly used measure of spread and approximates the typical distance of a value in the data set from the mean. The standard deviation is not resistant—it is heavily affected by extreme values.

To identify outliers in a distribution of quantitative data, you learned the $1.5 \times IQR$ rule. You also learned that boxplots are a great way to visually summarize the distribution of quantitative data. Boxplots are helpful for comparing distributions because they make it easy to compare both center (median) and spread (range, *IQR*). Yet boxplots aren't as useful for displaying the shape of a distribution because they do not display modes, clusters, gaps, and other interesting features.

## What Did You Learn?

| Learning Objective | Section | Related Example on Page | Relevant Chapter Review Exercise(s) |
|---|---|---|---|
| Identify the individuals and variables in a set of data. | Intro | 3 | R1.1 |
| Classify variables as categorical or quantitative. | Intro | 3 | R1.1 |
| Display categorical data with a bar graph. Decide whether it would be appropriate to make a pie chart. | 1.1 | 9 | R1.2, R1.3 |
| Identify what makes some graphs of categorical data deceptive. | 1.1 | 10 | R1.3 |
| Calculate and display the marginal distribution of a categorical variable from a two-way table. | 1.1 | 13 | R1.4 |
| Calculate and display the conditional distribution of a categorical variable for a particular value of the other categorical variable in a two-way table. | 1.1 | 15 | R1.4 |
| Describe the association between two categorical variables by comparing appropriate conditional distributions. | 1.1 | 17 | R1.5 |
| Make and interpret dotplots and stemplots of quantitative data. | 1.2 | Dotplots: 25<br>Stemplots: 31 | R1.6 |
| Describe the overall pattern (shape, center, and spread) of a distribution and identify any major departures from the pattern (outliers). | 1.2 | Dotplots: 26 | R1.6, R1.9 |
| Identify the shape of a distribution from a graph as roughly symmetric or skewed. | 1.2 | 28 | R1.6, R1.7, R1.8, R1.9 |



**TRM Learning Objectives Grid**

Make sure to discuss this grid with your students. Ask them to read each learning objective and self-assess whether or not they can do each one. Then, point out that each learning objective has at least one related example (with page number) and is assessed at least once in the chapter review exercises (with exercise number).

## What Did You Learn? *(continued)*

| Learning Objective | Section | Related Example on Page(s) | Relevant Chapter Review Exercise(s) |
|---|---|---|---|
| Make and interpret histograms of quantitative data. | 1.2 | 33 | R1.7, R1.8 |
| Compare distributions of quantitative data using dotplots, stemplots, or histograms. | 1.2 | 30 | R1.8, R1.10 |
| Calculate measures of center (mean, median). | 1.3 | Mean: 49<br>Median: 52 | R1.6 |
| Calculate and interpret measures of spread (range, *IQR*, standard deviation). | 1.3 | *IQR*: 55<br>Std. dev.: 60 | R1.9 |
| Choose the most appropriate measure of center and spread in a given setting. | 1.3 | 65 | R1.7 |
| Identify outliers using the 1.5 × *IQR* rule. | 1.3 | 56 | R1.6, R1.7, R1.9 |
| Make and interpret boxplots of quantitative data. | 1.3 | 57 | R1.7 |
| Use appropriate graphs and numerical summaries to compare distributions of quantitative variables. | 1.3 | 65 | R1.8, R1.10 |

# Chapter 1 Chapter Review Exercises

*These exercises are designed to help you review the important ideas and methods of the chapter.*

**R1.1 Hit movies** According to the Internet Movie Database, *Avatar* is tops based on box office sales worldwide. The following table displays data on several popular movies.[47]

| Movie | Year | Rating | Time (minutes) | Genre | Box office (dollars) |
|---|---|---|---|---|---|
| *Avatar* | 2009 | PG-13 | 162 | Action | 2,781,505,847 |
| *Titanic* | 1997 | PG-13 | 194 | Drama | 1,835,300,000 |
| *Harry Potter and the Deathly Hallows: Part 2* | 2011 | PG-13 | 130 | Fantasy | 1,327,655,619 |
| *Transformers: Dark of the Moon* | 2011 | PG-13 | 154 | Action | 1,123,146,996 |
| *The Lord of the Rings: The Return of the King* | 2003 | PG-13 | 201 | Action | 1,119,929,521 |
| *Pirates of the Caribbean: Dead Man's Chest* | 2006 | PG-13 | 151 | Action | 1,065,896,541 |
| *Toy Story 3* | 2010 | G | 103 | Animation | 1,062,984,497 |

(a) What individuals does this data set describe?

(b) Clearly identify each of the variables. Which are quantitative?

(c) Describe the individual in the highlighted row.

**R1.2 Movie ratings** The movie rating system we use today was first established on November 1, 1968. Back then, the possible ratings were G, PG, R, and X. In 1984, the PG-13 rating was created. And in 1990, NC-17 replaced the X rating. Here is a summary of the ratings assigned to movies between 1968 and 2000: 8% rated G, 24% rated PG, 10% rated PG-13, 55% rated R, and 3% rated NC-17.[48] Make an appropriate graph for displaying these data.

**R1.3 I'd die without my phone!** In a survey of over 2000 U.S. teenagers by Harris Interactive, 47% said that "their social life would end or be worsened without their cell phone."[49] One survey question asked the teens how important it is for their phone to have certain features. The figure below displays data on the percent who indicated that a particular feature is vital.

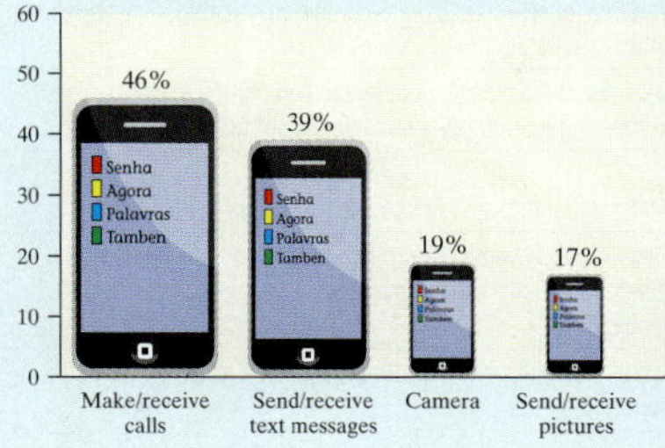


**TRM Full Solutions to Chapter 1 Review Exercises**

## Answers to Chapter 1 Review Exercises

**R1.1 (a)** Movies **(b)** Quantitative: Year, time, box office sales. Categorical: Rating, genre. *Note*: Year might be considered categorical if we want to know how many of these movies were made each year rather than the average year. **(c)** This movie is Avatar, released in 2009. It was rated PG-13, runs 162 minutes, is an action film, and had box office sales of $2,781,505,847.

**R1.2** A bar chart is given below.

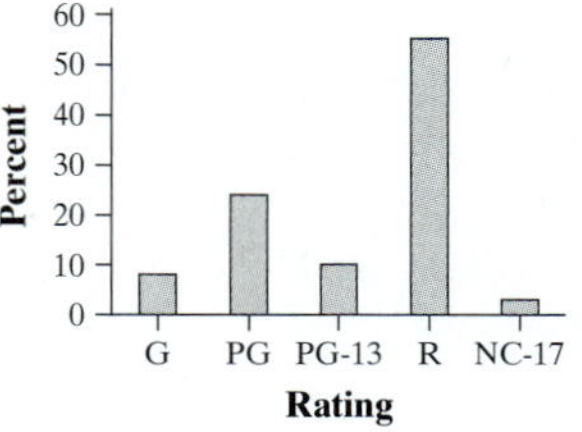


**R1.3 (a)** The "bars" are different widths. For example, the bar for "send/receive text messages" should be roughly twice the size of the bar for "camera" when it is actually about 4 times as large. **(b)** No, because they do not describe parts of a whole. Students were free to answer in more than one category. **(c)** A bar graph is given below.

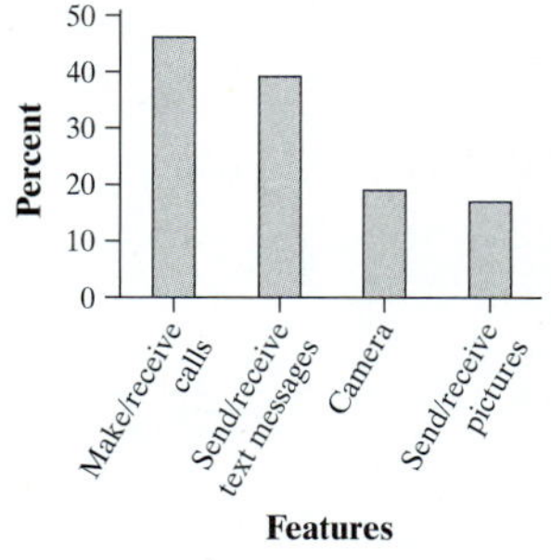


**R1.4 (a)** 148/219 = 67.6%. Marginal distribution, because it is part of the distribution of one variable for all categories of the other variable. **(b)** 78/82 = 95.1% of the younger students were Facebook users. 78/148 = 52.7% of the Facebook users were younger.

**R1.5** There does appear to be an association between age and Facebook status. From both the table and the graph given at right, we can see that as age increases, the percent of Facebook users decreases. For younger students, about 95% are members. That drops to 70% for middle students and drops even further to 31.3% for older students.

| | Facebook user? | |
|---|---|---|
| Age | Yes | No |
| Younger (18–22) | 95.1% | 4.9% |
| Middle (23–27) | 70.0% | 30.0% |
| Older (28 and up) | 31.3% | 68.7% |

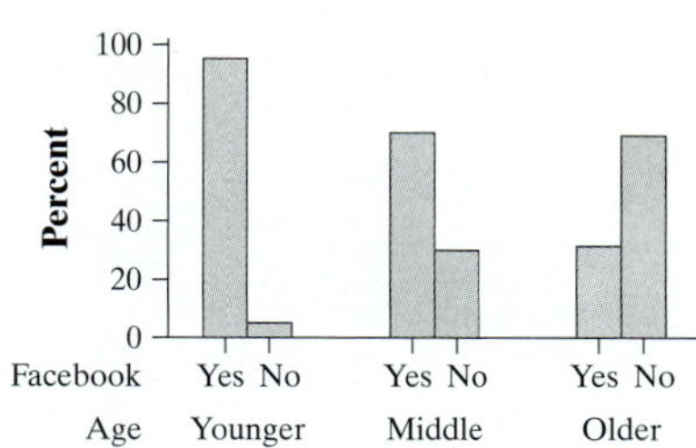

(a) Explain how the graph gives a misleading impression.
(b) Would it be appropriate to make a pie chart to display these data? Why or why not?
(c) Make a graph of the data that isn't misleading.

**R1.4 Facebook and age** Is there a relationship between Facebook use and age among college students? The following two-way table displays data for the 219 students who responded to the survey.[50]

| | Age | | |
|---|---|---|---|
| **Facebook user?** | **Younger (18–22)** | **Middle (23–27)** | **Older (28 and up)** |
| Yes | 78 | 49 | 21 |
| No | 4 | 21 | 46 |

(a) What percent of the students who responded were Facebook users? Is this percent part of a marginal distribution or a conditional distribution? Explain.
(b) What percent of the younger students in the sample were Facebook users? What percent of the Facebook users in the sample were younger students?

**R1.5 Facebook and age** Use the data in the previous exercise to determine whether there is an association between Facebook use and age. Give appropriate graphical and numerical evidence to support your answer.

**R1.6 Density of the earth** In 1798, the English scientist Henry Cavendish measured the density of the earth several times by careful work with a torsion balance. The variable recorded was the density of the earth as a multiple of the density of water. Here are Cavendish's 29 measurements:[51]

| | | | | | | | | | |
|---|---|---|---|---|---|---|---|---|---|
| 5.50 | 5.61 | 4.88 | 5.07 | 5.26 | 5.55 | 5.36 | 5.29 | 5.58 | 5.65 |
| 5.57 | 5.53 | 5.62 | 5.29 | 5.44 | 5.34 | 5.79 | 5.10 | 5.27 | 5.39 |
| 5.42 | 5.47 | 5.63 | 5.34 | 5.46 | 5.30 | 5.75 | 5.68 | 5.85 | |

(a) Present these measurements graphically in a stemplot.
(b) Discuss the shape, center, and spread of the distribution. Are there any outliers?
(c) What is your estimate of the density of the earth based on these measurements? Explain.

**R1.7 Guinea pig survival times** Here are the survival times in days of 72 guinea pigs after they were injected with infectious bacteria in a medical experiment.[52] Survival times, whether of machines under stress or cancer patients after treatment, usually have distributions that are skewed to the right.

| | | | | | | | | | | | |
|---|---|---|---|---|---|---|---|---|---|---|---|
| 43 | 45 | 53 | 56 | 56 | 57 | 58 | 66 | 67 | 73 | 74 | 79 |
| 80 | 80 | 81 | 81 | 81 | 82 | 83 | 83 | 84 | 88 | 89 | 91 |
| 91 | 92 | 92 | 97 | 99 | 99 | 100 | 100 | 101 | 102 | 102 | 102 |
| 103 | 104 | 107 | 108 | 109 | 113 | 114 | 118 | 121 | 123 | 126 | 128 |
| 137 | 138 | 139 | 144 | 145 | 147 | 156 | 162 | 174 | 178 | 179 | 184 |
| 191 | 198 | 211 | 214 | 243 | 249 | 329 | 380 | 403 | 511 | 522 | 598 |

(a) Make a histogram of the data and describe its main features. Does it show the expected right skew?
(b) Now make a boxplot of the data. Be sure to check for outliers.
(c) Which measure of center and spread would you use to summarize the distribution—the mean and standard deviation or the median and *IQR*? Justify your answer.

**R1.8 Household incomes** Rich and poor households differ in ways that go beyond income. Following are histograms that compare the distributions of household size (number of people) for low-income and high-income households.[53] Low-income households had annual incomes less than \$15,000, and high-income households had annual incomes of at least \$100,000.

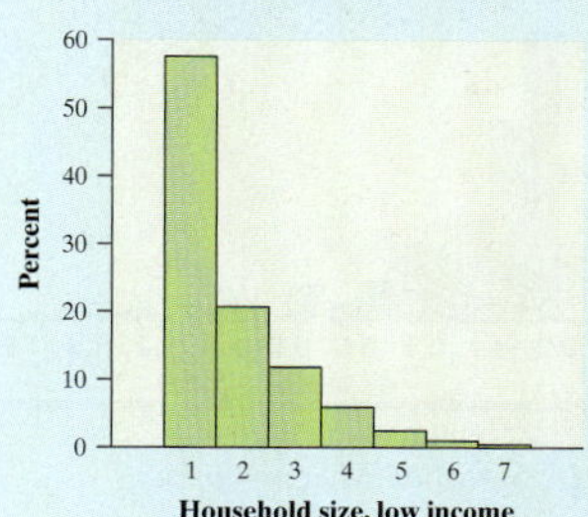


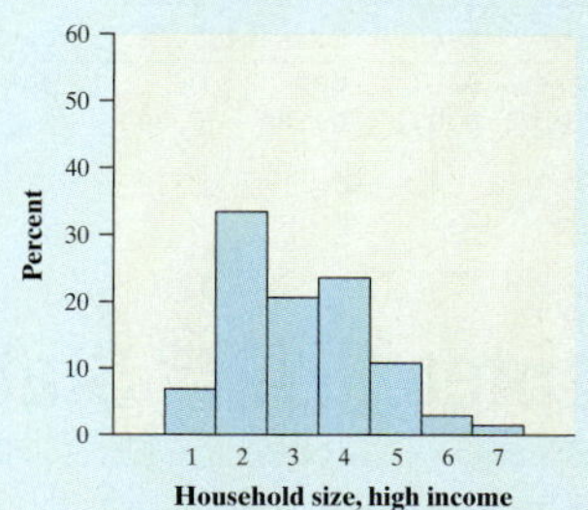


(a) About what percent of each group of households consisted of two people?
(b) What are the important differences between these two distributions? What do you think explains these differences?

*Exercises R1.9 and R1.10 refer to the following setting.* Do you like to eat tuna? Many people do. Unfortunately, some of the tuna that people eat may contain high levels of mercury. Exposure to mercury can be especially hazardous for pregnant women and small children. How much mercury is safe to consume? The Food and Drug Administration will take action (like removing the product from store shelves) if the mercury concentration in a six-ounce can of tuna is 1.00 ppm (parts per million) or higher.

**R1.7 (a)** A histogram is given below. The survival times are right skewed, as expected. The median survival time is 102.5 days and the range of survival times is $598 - 43 = 555$ days. There are several high outliers with survival times above 500.

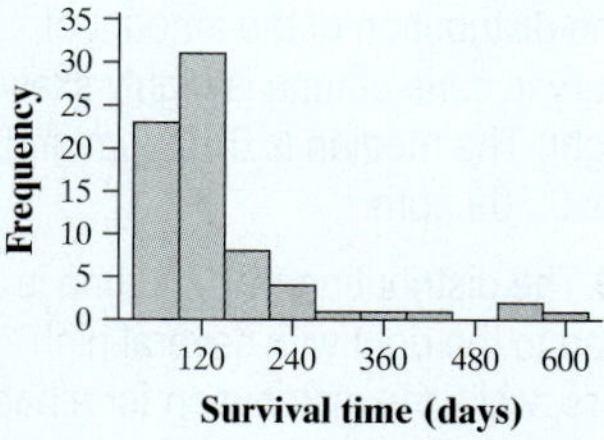


**(b)** The boxplot is given below.

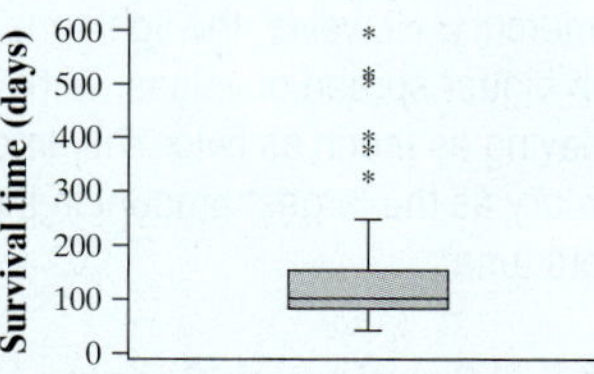


**(c)** Use the median and *IQR* to summarize the distribution because the outliers will have a big effect on the mean and standard deviation.

**R1.8 (a)** About 20% of low-income and 33% of high-income households. **(b)** The shapes of both distributions are skewed to the right; however, the skewness is much stronger in the distribution for low-income households. On average, household size is larger for high-income households. One-person households might have less income because they would include many young single people who have no job or retired single people with a fixed income.

**R1.6 (a)** A stemplot is given at right. **(b)** The distribution is roughly symmetric with one possible outlier at 4.88. The center of the distribution is between 5.4 and 5.5. The densities vary from 4.88 to 5.85. **(c)** Because the distribution is roughly symmetric, we can use the mean to estimate the Earth's density to be about 5.45 times the density of water.

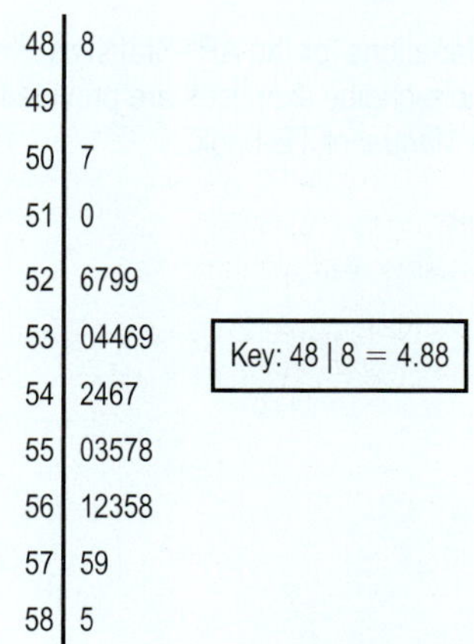

R1.9 **(a)** The amount of mercury per can of tuna will typically vary from the mean by about 0.3 ppm. **(b)** Any point below $0.071 - 1.5(0.309) = -0.393$ or above $0.38 + 1.5(0.309) = 0.8435$ would be considered an outlier. There are no low outliers, but there are several high outliers. **(c)** The distribution of the amount of mercury in cans of tuna is highly skewed to the right. The median is 0.18 ppm and the *IQR* is 0.309 ppm.

**R1.10** The distribution for light tuna is skewed to the right with several high outliers, while the distribution for albacore tuna is more symmetric with just a couple of high outliers. Because it has a greater center, the albacore tuna generally has more mercury. However, the light tuna has a much bigger spread of values, with some cans having as much as twice the amount of mercury as the largest amount in the albacore tuna.

**TRM Full Solutions to Chapter 1 AP® Statistics Practice Test**

## Answers to Chapter 1 AP® Statistics Practice Test

**T1.1** d

**T1.2** e



What is the typical mercury concentration in cans of tuna sold in stores? A study conducted by Defenders of Wildlife set out to answer this question. Defenders collected a sample of 164 cans of tuna from stores across the United States. They sent the selected cans to a laboratory that is often used by the Environmental Protection Agency for mercury testing.[54]

R1.9 **Mercury in tuna** A histogram and some computer output provide information about the mercury concentration in the sampled cans (in parts per million, ppm).

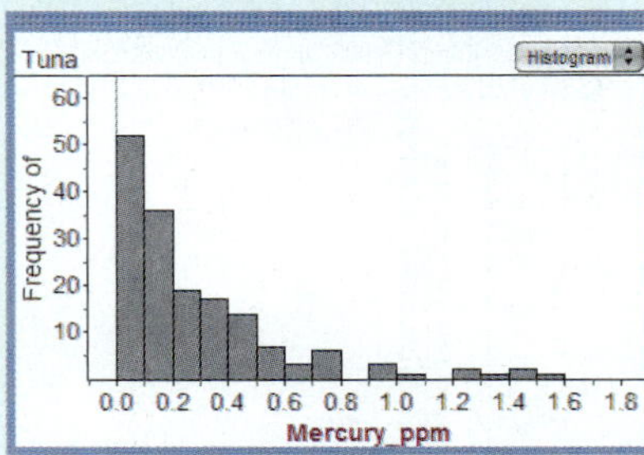


*Descriptive Statistics: Mercury_ppm*

| Variable | N | Mean | StDev | Min |
|---|---|---|---|---|
| Mercury | 164 | 0.285 | 0.300 | 0.012 |
| **Variable** | **$Q_1$** | **Med** | **$Q_3$** | **Max** |
| Mercury | 0.071 | 0.180 | 0.380 | 1.500 |

(a) Interpret the standard deviation in context.
(b) Determine whether there are any outliers.
(c) Describe the shape, center, and spread of the distribution.

R1.10 **Mercury in tuna** Is there a difference in the mercury concentration of light tuna and albacore tuna? Use the parallel boxplots and the computer output to write a few sentences comparing the two distributions.

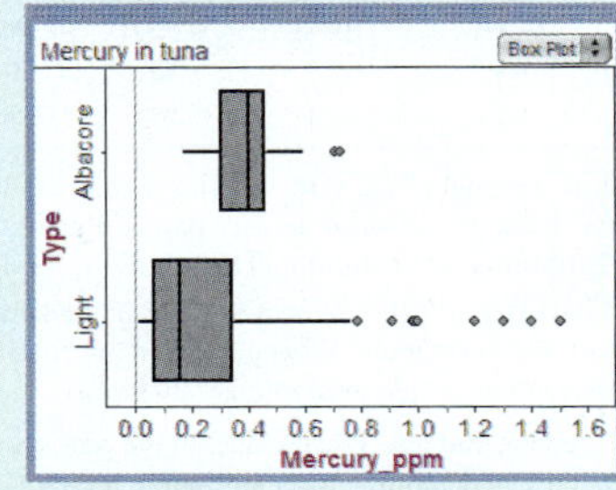


*Descriptive Statistics: Mercury_ppm*

| Type | N | Mean | StDev | Min |
|---|---|---|---|---|
| Albacore | 20 | 0.401 | 0.152 | 0.170 |
| Light | 144 | 0.269 | 0.312 | 0.012 |
| **Type** | **$Q_1$** | **Med** | **$Q_3$** | **Max** |
| Albacore | 0.293 | 0.400 | 0.460 | 0.730 |
| Light | 0.059 | 0.160 | 0.347 | 1.500 |

# Chapter 1 AP® Statistics Practice Test

**Section I: Multiple Choice** *Select the best answer for each question.*

T1.1 You record the age, marital status, and earned income of a sample of 1463 women. The number and type of variables you have recorded is

(a) 3 quantitative, 0 categorical.
(b) 4 quantitative, 0 categorical.
(c) 3 quantitative, 1 categorical.
(d) 2 quantitative, 1 categorical.
(e) 2 quantitative, 2 categorical.

T1.2 Consumers Union measured the gas mileage in miles per gallon of 38 vehicles from the same model year on a special test track. The pie chart provides information about the country of manufacture of the model cars tested by Consumers Union. Based on the pie chart, we conclude that

(a) Japanese cars get significantly lower gas mileage than cars from other countries.
(b) U.S. cars get significantly higher gas mileage than cars from other countries.
(c) Swedish cars get gas mileages that are between those of Japanese and U.S. cars.
(d) cars from France have the lowest gas mileage.
(e) more than half of the cars in the study were from the United States.

## Teaching Tip

Brief explanations for the AP® Statistics Practice Test multiple-choice exercises are provided in the *Solutions Manual* or TE-book.

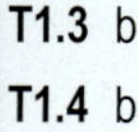

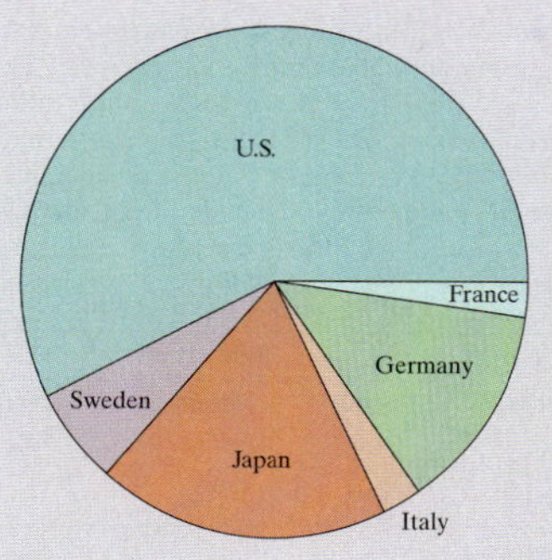


**T1.3** Which of the following bar graphs is equivalent to the pie chart in Question T1.2?

(a)

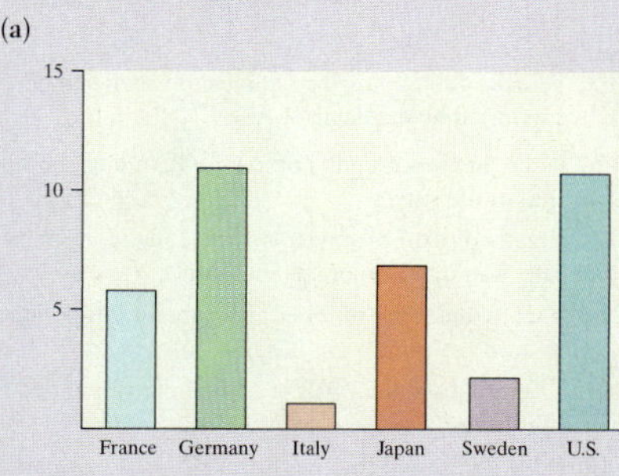


(b)

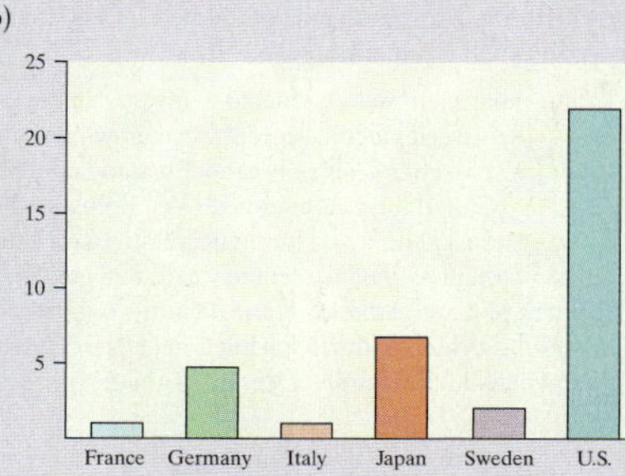


(c)

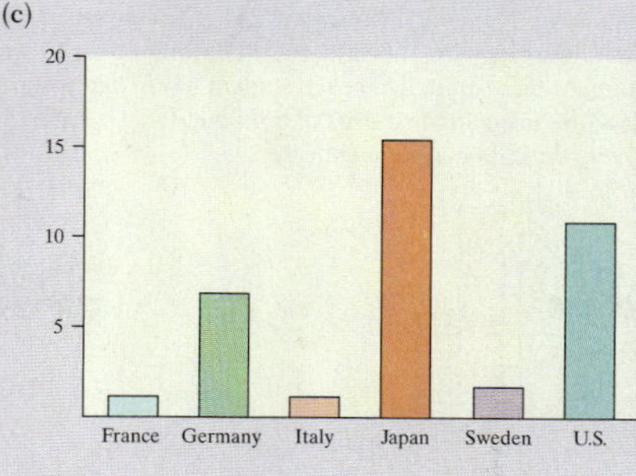


(d)

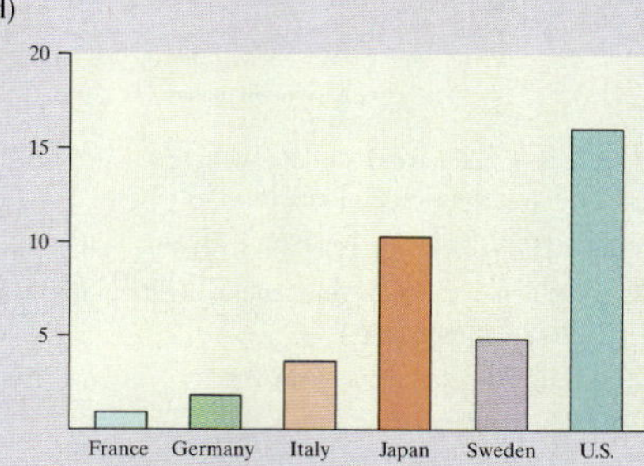


(e) None of these.

**T1.4** Earthquake intensities are measured using a device called a seismograph, which is designed to be most sensitive to earthquakes with intensities between 4.0 and 9.0 on the Richter scale. Measurements of nine earthquakes gave the following readings:

| 4.5 | L | 5.5 | H | 8.7 | 8.9 | 6.0 | H | 5.2 |
|---|---|---|---|---|---|---|---|---|

where L indicates that the earthquake had an intensity below 4.0 and an H indicates that the earthquake had an intensity above 9.0. The median earthquake intensity of the sample is

(a) 5.75.
(b) 6.00.
(c) 6.47.
(d) 8.70.
(e) Cannot be determined.

**T1.3** b

**T1.4** b

T1.5 c

T1.6 c

T1.7 b

T1.8 c

T1.9 e

T1.10 b



*Questions T1.5 and T1.6 refer to the following setting.* In a statistics class with 136 students, the professor records how much money (in dollars) each student has in his or her possession during the first class of the semester. The histogram shows the data that were collected.

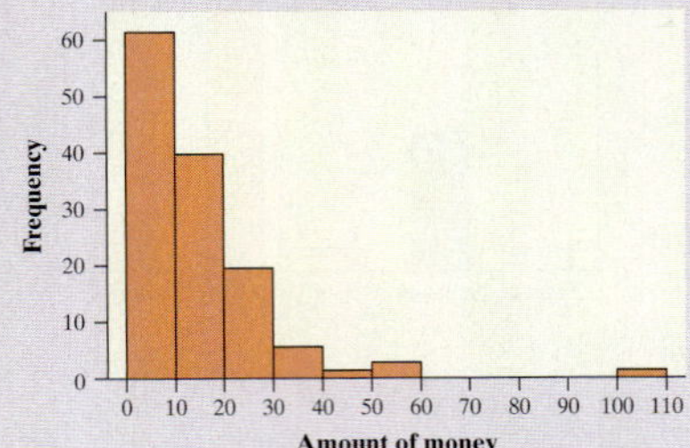


**T1.5** The percentage of students with less than $10 in their possession is closest to

(a) 30%. (b) 35%. (c) 45%. (d) 60%. (e) 70%.

**T1.6** Which of the following statements about this distribution is *not* correct?

(a) The histogram is right-skewed.
(b) The median is less than $20.
(c) The *IQR* is $35.
(d) The mean is greater than the median.
(e) The histogram is unimodal.

**T1.7** Forty students took a statistics examination having a maximum of 50 points. The score distribution is given in the following stem-and-leaf plot:

```
0 | 28
1 | 2245
2 | 01333358889
3 | 001356679
4 | 22444466788
5 | 000
```

The third quartile of the score distribution is equal to

(a) 45. (b) 44. (c) 43. (d) 32. (e) 23.

**T1.8** The mean salary of all female workers is $35,000. The mean salary of all male workers is $41,000. What must be true about the mean salary of all workers?

(a) It must be $38,000.
(b) It must be larger than the median salary.
(c) It could be any number between $35,000 and $41,000.
(d) It must be larger than $38,000.
(e) It cannot be larger than $40,000.

*Questions T1.9 and T1.10 refer to the following setting.* A survey was designed to study how business operations vary according to their size. Companies were classified as small, medium, or large. Questionnaires were sent to 200 randomly selected businesses of each size. Because not all questionnaires in a survey of this type are returned, researchers decided to investigate the relationship between the response rate and the size of the business. The data are given in the following two-way table:

| | Business size | | |
|---|---|---|---|
| **Response?** | **Small** | **Medium** | **Large** |
| Yes | 125 | 81 | 40 |
| No | 75 | 119 | 160 |

**T1.9** What percent of all small companies receiving questionnaires responded?

(a) 12.5% (b) 20.8% (c) 33.3% (d) 50.8% (e) 62.5%

**T1.10** Which of the following conclusions seems to be supported by the data?

(a) There are more small companies than large companies in the survey.
(b) Small companies appear to have a higher response rate than medium or big companies.
(c) Exactly the same number of companies responded as didn't respond.
(d) Overall, more than half of companies responded to the survey.
(e) If we combined the medium and large companies, then their response rate would be equal to that of the small companies.

**T1.11** An experiment was conducted to investigate the effect of a new weed killer to prevent weed growth in onion crops. Two chemicals were used: the standard weed killer (C) and the new chemical (W). Both chemicals were tested at high and low concentrations on a total of 50 test plots. The percent of weeds that grew in each plot was recorded. Here are some boxplots of the results. Which of the following is *not* a correct statement about the results of this experiment?

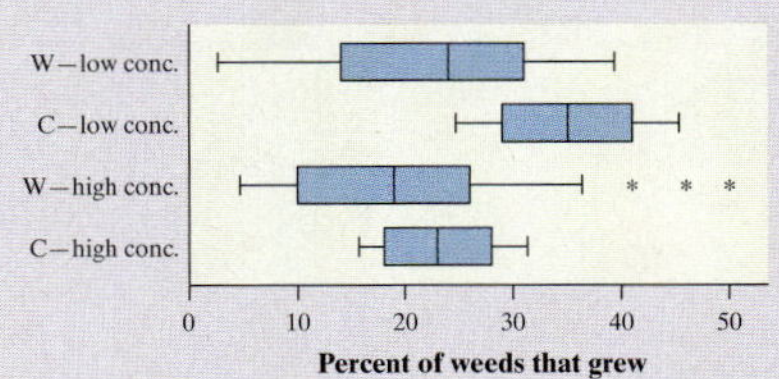

(a) At both high and low concentrations, the new chemical (W) gives better weed control than the standard weed killer (C).

(b) Fewer weeds grew at higher concentrations of both chemicals.

(c) The results for the standard weed killer (C) are less variable than those for the new chemical (W).

(d) High and low concentrations of either chemical have approximately the same effects on weed growth.

(e) Some of the results for the low concentration of weed killer W show fewer weeds growing than some of the results for the high concentration of W.

**Section II: Free Response** *Show all your work. Indicate clearly the methods you use, because you will be graded on the correctness of your methods as well as on the accuracy and completeness of your results and explanations.*

T1.12 You are interested in how much time students spend on the Internet each day. Here are data on the time spent on the Internet (in minutes) for a particular day reported by a random sample of 30 students at a large high school:

| | | | | | | | | | |
|---|---|---|---|---|---|---|---|---|---|
| 7 | 20 | 24 | 25 | 25 | 28 | 28 | 30 | 32 | 35 |
| 42 | 43 | 44 | 45 | 46 | 47 | 48 | 48 | 50 | 51 |
| 72 | 75 | 77 | 78 | 79 | 83 | 87 | 88 | 135 | 151 |

(a) Construct a histogram of these data.

(b) Are there any outliers? Justify your answer.

(c) Would it be better to use the mean and standard deviation or the median and *IQR* to describe the center and spread of this distribution? Why?

T1.13 A study among the Pima Indians of Arizona investigated the relationship between a mother's diabetic status and the appearance of birth defects in her children. The results appear in the two-way table below.

| | Diabetic Status | | | |
|---|---|---|---|---|
| **Birth Defects** | **Nondiabetic** | **Prediabetic** | **Diabetic** | **Total** |
| None | 754 | 362 | 38 | |
| One or more | 31 | 13 | 9 | |
| **Total** | | | | |

(a) Fill in the row and column totals in the margins of the table.

(b) Compute (in percents) the conditional distributions of birth defects for each diabetic status.

(c) Display the conditional distributions in a graph. Don't forget to label your graph completely.

(d) Do these data give evidence of an association between diabetic status and birth defects? Justify your answer.

T1.14 The back-to-back stemplot shows the lifetimes of several Brand X and Brand Y batteries.

| Brand X | | Brand Y |
|---|---|---|
| | 1 | |
| | 1 | 7 |
| | 2 | 2 |
| | 2 | 6 |
| 2110 | 3 | |
| 99775 | 3 | |
| 3221 | 4 | 223334 |
| | 4 | 56889 |
| 4 | 5 | 0 |
| 5 | 5 | 5 |

Key: 4|2 represents 420–429 hours.

(a) What is the longest that any battery lasted?

(b) Give a reason someone might prefer a Brand X battery.

(c) Give a reason someone might prefer a Brand Y battery.

T1.15 During the early part of the 1994 baseball season, many fans and players noticed that the number of home runs being hit seemed unusually large. Here are the data on the number of home runs hit by American League and National League teams in the early part of the 1994 season:

| | | | | | | | | | | | | | | |
|---|---|---|---|---|---|---|---|---|---|---|---|---|---|---|
| **American League:** | 35 | 40 | 43 | 49 | 51 | 54 | 57 | 58 | 58 | 64 | 68 | 68 | 75 | 77 |
| **National League:** | 29 | 31 | 42 | 46 | 47 | 48 | 48 | 53 | 55 | 55 | 55 | 63 | 63 | 67 |

Compare the distributions of home runs for the two leagues graphically and numerically. Write a few sentences summarizing your findings.

**T1.11** d

**T1.12** **(a)** A histogram is given below. **(b)** Any point below $30 - 1.5(47) = -40.5$ or above $77 + 1.5(47) = 147.5$ is an outlier. So, 151 minutes is an outlier. **(c)** Median and *IQR*, because the distribution is skewed and has a high outlier.

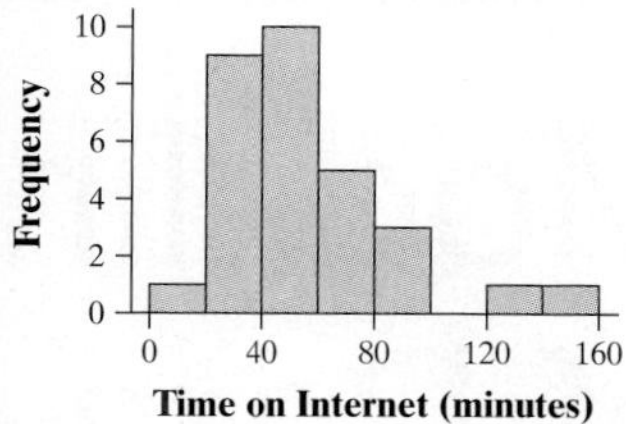


**T1.13** **(a)** Row totals are 1154, 53, and 1207. Column totals are 785, 375, 47, and 1207. **(b)** Nondiabetic: 96.1% none and 3.9% one or more. Prediabetic: 96.5% none and 3.5% one or more. Diabetic: 80.9% none and 19.1% one or more. **(c)** The graph is given below.

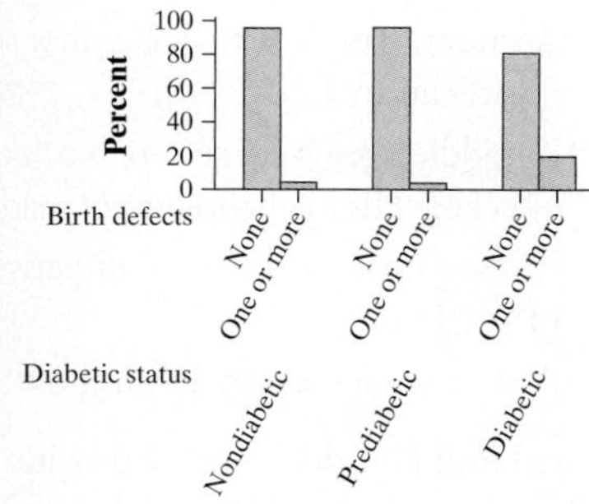


**(d)** Yes. Nondiabetics and prediabetics appear to have babies with birth defects at about the same rate. However, those with diabetes have a much higher rate of babies with birth defects.

**T1.14** **(a)** Between 550 and 559 hours. **(b)** Because it has a higher minimum lifetime or because its lifetimes are more consistent (less variable). **(c)** Because it has a higher median lifetime.

**T1.15** Side-by-side boxplots and descriptive statistics for both leagues are given below. Both distributions are roughly symmetric, although there are two low outliers in the NL. The data suggest that the number of home runs is somewhat less in the NL. All 5 numbers in the 5-number summary are less for the NL teams than for the AL teams. However, there is more variability among the AL teams.

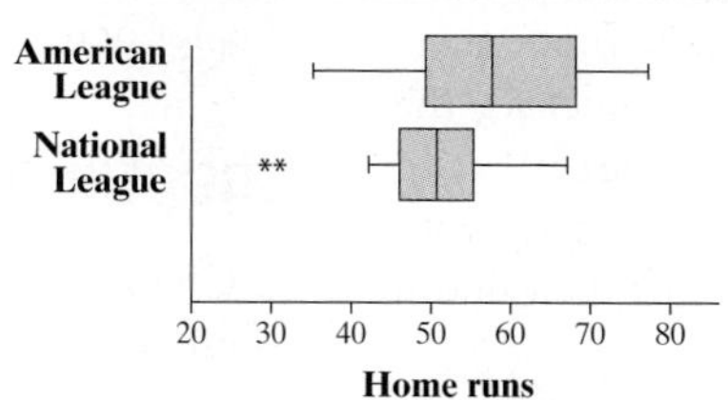


| Variable | N | Mean | StDev | Minimum |
|---|---|---|---|---|
| American League | 14 | 56.93 | 12.69 | 35.00 |
| National League | 14 | 50.14 | 11.13 | 29.00 |

| Variable | $Q_1$ | Median | $Q_3$ | Maximum |
|---|---|---|---|---|
| American League | 49.00 | 57.50 | 68.00 | 77.00 |
| National League | 46.00 | 50.50 | 55.00 | 67.00 |

Chapter

2

# Modeling Distributions of Data

## PD Overview

In Chapter 2, we continue exploring distributions of quantitative variables. Consider each of the following situations:

- A student gets a test back with a score of 78 marked clearly at the top.
- A middle-aged man goes to his doctor to have his cholesterol checked. His total cholesterol reading is 210 mg/dl.
- An employee in a large company earns an annual salary of $42,000.
- A 10th-grader scores 46 on the PSAT Writing test.

Isolated numbers don't always provide enough information. It is human nature to ask, "Where do I stand?" when presented with numbers like these. This chapter focuses on ways to describe an individual's position within a distribution.

We begin by learning two different ways to measure an individual's location in a distribution: percentiles and standardized scores ($z$-scores). This leads to a discussion of how various transformations affect the characteristics of a distribution.

Then, we extend our strategy for data analysis by suggesting that when a distribution has a clear overall pattern, we can model that pattern with a smooth curve, called a density curve. The most commonly used density curve is the Normal curve, a symmetric, bell-shaped curve that is a good model for the distribution of many different quantitative variables. We will learn how to use Normal curves to identify percentiles in a symmetric, bell-shaped distribution and how to identify the value of a variable at a given percentile.

*Note:* If you used the 4th edition of *The Practice of Statistics*, notice that the material about density curves has been moved from Section 2.1 to Section 2.2.

### Relationship to the AP® Statistics Course Description

Chapter 2 covers the following content as described in the Topic Outline from the AP® Statistics Course Description. Topics marked with a superscript are covered in other chapters as indicated.

I: Exploring data: describing patterns and departures from patterns

A. Constructing and interpreting graphical displays of distributions of univariate data (dotplot$^{ch1}$, stemplot$^{ch1}$, histogram$^{ch1}$, cumulative frequency plot)

B. Summarizing distributions of univariate data

3. Measuring position: quartiles$^{ch1}$, percentiles, standardized scores ($z$-scores)
5. The effect of changing units on summary measures

III: Anticipating patterns: exploring random phenomena using probability and simulation

C. The Normal distribution

1. Properties of the Normal distribution
2. Using tables of the Normal distribution
3. The Normal distribution as a model for measurements

## The Big Ideas

One of the challenges in teaching the AP® Statistics course is keeping students focused on the big picture and not just the details of each section. We outline the big ideas for the chapter here.

### Chapter 2 Introduction

In Chapter 1, we learned to describe characteristics of distributions such as shape, center, spread, and unusual values. In this chapter, we focus on identifying the location of individuals within a distribution. The Activity on page 84 is a great way to get students thinking about where they stand in a distribution.

### Section 2.1 Describing Location in a Distribution

In this section, we focus on describing the location of individuals in a distribution. Percentiles are perhaps the most common way to identify where a value is located in a distribution.

Students are probably already familiar with this concept, as almost all standardized tests report a student's performance with a percentile. For example, if Anne scored in the 97th percentile, that means she scored better than 97% of people who took the test.

A cumulative relative frequency graph is a handy tool for identifying percentiles in a distribution. You can use it to estimate the percentile for a particular value of a variable or estimate the value of a variable at a particular percentile.

Another very common way to measure the positions of individuals in a distribution is with a standardized score, also called a *z*-score. A standardized score measures how many standard deviations an observation is above or below the mean of the distribution. Using *z*-scores is also a great way to compare observations from different distributions because they are on a standardized scale.

To find the standardized score for a particular observation, we transform the value by subtracting the mean and dividing the difference by the standard deviation. This is just one example where we transform data from one scale to another. Another common example is converting measurements from one unit to another, such as inches to centimeters. Adding a positive constant to (or subtracting it from) each value in a data set changes the measures of location, but not the shape or spread of the distribution. Multiplying each value in a data set by a positive constant changes the measures of location and measures of spread, but not the shape of the distribution.

## Section 2.2 Density Curves and Normal Distributions

We begin this section by introducing density curves as a way to model distributions of data with a smooth curve. In many cases, using a density curve is more convenient than working with the actual data. The most common type of density curve is the Normal curve, which will be the focus of most of this section.

Some people may find it a bit unusual to present density curves and Normal distributions as part of data analysis. However, Normal curves have long been used as approximations of distributions of data that are not necessarily connected with probability theory. And giving students a brief taste of "probability" here will make it easier for them when we formally introduce random variables in Chapter 6. We think of data analysis as progressing from graphs to numerical summaries of specific aspects of the data to (when appropriate) a mathematical model for the overall pattern. That's the role Normal distributions play in data analysis for a single quantitative variable.

After introducing the basic properties of the Normal distributions and the reasons we should study them in a statistics course, we present the 68–95–99.7 rule, sometimes called the empirical rule. This handy rule allows students to tie together the ideas of standardized scores and Normal distributions by providing approximations for what proportion of the observations should fall within 1, 2, and 3 standard deviations of the mean.

Unfortunately, the 68–95–99.7 rule doesn't tell us much about what proportion of the observations fall within 1.5 standard deviations of the mean or what proportion of the observations are more than 2.5 standard deviations above the mean. The standard Normal distribution and Table A come to the rescue, providing the proportion of observations below each standardized score from −3.49 to 3.49.

After students are comfortable using the standard Normal distribution, we move to questions involving other Normal distributions, such as the distance a golf ball travels after being "crushed" by Tiger Woods. We also learn how to "go backward" and find the distance that corresponds to a specific percentile of the distribution.

In this section, we also present methods for performing Normal calculations on the TI-83/84/89. Instructions for the TI-Nspire are given in Appendix B. Regardless of whether your students use Table A or their calculators, make sure they follow the three-step process that we detail on pages 118 and 120. It is very important that students always state the distribution (including the parameters) and the values of interest, show work, and answer the question. If students choose to use technology, make sure they pay close attention to the AP® Tips that provide guidelines for receiving full credit.

Finally, we spend time showing students how to investigate whether a distribution of data is approximately Normal. Knowing how to decide if a sample could have come from a Normal population will be a very important skill when we learn about inference in Chapters 8 to 12. We introduce it here because it nicely ties together our data analysis strategy:

- Graph the data.
- Look for an overall pattern and departures from this pattern.
- Calculate numerical summaries.
- If the data follow a regular overall pattern, describe it with a smooth curve.

We also introduce a special type of graph designed specifically for assessing Normality: the Normal probability plot. Although it is not on the AP® Statistics Topic Outline, it is a useful tool and easy to create with technology. The more linear the Normal probability plot, the more Normal the data.

# Chapter 2: Resources

## TRM Additional *TPS* Teacher's Resource Materials

The following resources can be found by clicking on the links in the new Teacher's e-Book (TE-book), opening the Teacher's Resource CD (TRCD), or logging onto the book's companion Website www.whfreeman.com/tps5e (teacher log-in required).

- Printable Chapter 2 Learning Objective Grid (page 85)
- Alternate Activity: Investigating Normal distributions (page 109)
- Extra practice with the Normal distribution (page 121)
- FRAPPY! materials: Question, Model Solution, Scoring Rubric, Sample Responses with commentary (page 134)
- Quizzes: 3 versions for each section; versions A and B are roughly parallel
    - Section 2.1 (page 99)
    - Section 2.2 (page 127)
- Tests: 4 versions; versions A and B are roughly parallel, version D is cumulative (page 135)

## Free-Response Questions from Previous AP® Exams

Questions can be found on the AP® Central Web site: http://apcentral.collegeboard.com/apc/members/exam/exam_questions/8357.html.

Students should be able to answer all the free-response questions listed below with material learned in this chapter. Questions that contain content from this chapter but also require content from later chapters are listed in the last chapter required to complete the entire question. This list will be updated after each AP® exam and will be posted on the book's Web site: www.whfreeman.com/tps5e.

| Year | # | Content |
|---|---|---|
| 2011 | 1 | • Assessing normality from summary statistics<br>• Calculating and interpreting a *z*-score<br>• Using *z*-scores to make a comparison |
| 2009B | 1 | • Estimating median and *IQR* from a boxplot<br>• Linear transformations of data |
| 2008 | 1 | • Comparing distributions with boxplots<br>• Linear transformations of data<br>• Effect of shape on the relationship between mean and median |
| 2006B | 1 | • Interpreting cumulative relative frequency graphs |
| 1997 | 1* | • Interpreting cumulative relative frequency graphs<br>• Finding the median and *IQR* from a cumulative relative frequency graph<br>• Comparing center and spread |

Questions marked with an asterisk are from exams with released multiple-choice questions. You may want to save these questions until the end of the year so you can give your students a complete released exam for practice.

## Applets

- The *Normal Density Curve* applet allows students to find the area under a Normal curve. Students can choose the mean, standard deviation, and boundaries. www.whfreeman.com/tps5e (click on Statistical Applets and choose Normal Density Curve).
- The *z-Table Normal Distribution* applet allows students to find areas under a Normal curve and to work backward to find percentiles. Students can enter the boundaries, mean, and standard deviation. http://davidmlane.com/hyperstat/z_table.html.
- The *Normal* applet allows students to find areas under a Normal curve and to work backward to find percentiles. Students can enter the mean and standard deviation, choose which direction to shade, and specify the boundary value or area. www.lock5stat.com/statkey (click on Normal in the Theoretical Distributions menu).

## TI-Nspire Activities

These Activities are designed to illustrate important statistical concepts using the TI-Nspire. The Activities can be found at www.whfreeman.com/tps5e.

- *Transformations of Distributions*: This Activity helps students understand how a linear transformation affects the shape, center, and spread of a distribution.
- *Exploring Normality*: This Activity helps students determine if sample data could have come from a Normal population.

Chapter

# 2

## Chapter 2: Pacing Guide, Learning Objectives, and Suggested Assignments

This pacing guide is based on a schedule with 110, 50-minute sessions before the AP® exam. The 110 days include 4 days for midterms and 10 days for AP® exam review. If you have a different number of sessions before the AP® exam, you can modify the pacing guide to suit your needs. If you have additional time, consider incorporating quizzes, released AP® free-response questions, or additional activities. See the Resources section above for suggestions.

The suggested homework assignments list odd-numbered exercises, whenever possible, so students can check their answers against the back-of-the-book answers. If you would rather students not have access to the answers while doing homework, adding 1 to the exercise numbers usually will do the trick, because the homework exercises typically are paired. For example, Exercises 1 and 2 will generally cover the same topics, but in different contexts. You may also choose to include the Recycle and Review questions at the end of each section, which review topics from previous sections or chapters. These questions are denoted with a symbol.

| Day | Topics | **Learning Objectives** Students will be able to. . . | Suggested Assignment |
|---|---|---|---|
| 1 | 2.1 Measuring Position: Percentiles; Cumulative Relative Frequency Graphs; Measuring Position: z-scores | • Find and interpret the percentile of an individual value within a distribution of data.<br>• Estimate percentiles and individual values using a cumulative relative frequency graph.<br>• Find and interpret the standardized score (z-score) of an individual value within a distribution of data. | 1, 3, 5, 9, 11, 13, 15 |
| 2 | 2.1 Transforming Data | • Describe the effect of adding, subtracting, multiplying by, or dividing by a constant on the shape, center, and spread of a distribution of data. | 17, 19, 21, 23, 25–30 |
| 3 | 2.2 Density Curves; The 68–95–99.7 Rule;The Standard Normal Distribution | • Estimate the relative locations of the median and mean on a density curve.<br>• Use the 68–95–99.7 rule to estimate areas (proportions of values) in a Normal distribution.<br>• Use Table A or technology to find (i) the proportion of z-values in a specified interval, or (ii) a z-score from a percentile in the standard Normal distribution. | 33, 35, 39, 41, 43, 45, 47, 49, 51 |
| 4 | 2.2 Normal Distribution Calculations | • Use Table A or technology to find (i) the proportion of values in a specified interval, or (ii) the value that corresponds to a given percentile in any Normal distribution. | 53, 55, 57, 59 |
| 5 | 2.2 Assessing Normality | • Determine if a distribution of data is approximately Normal from graphical and numerical evidence. | 54, 63, 65, 66, 67, 69–74 |
| 6 | Chapter 2 Review/FRAPPY! | | Chapter 2 Review Exercises |
| 7 | Chapter 2 Test | | |

**PD** **Chapter 2 Overview**

Watch the video overview of Chapter 2 (for teachers) by clicking on the link in the TE-book, opening the TRCD, or logging onto the book's companion Web site www.whfreeman.com/tps5e (teacher log-in required).

# Chapter 2

# Modeling Distributions of Data

case study

## Do You Sudoku?

The sudoku craze has officially swept the globe. Here's what Will Shortz, crossword puzzle editor for the *New York Times*, said about sudoku:

> *As humans we seem to have an innate desire to fill up empty spaces. This might explain part of the appeal of sudoku, the new international craze, with its empty squares to be filled with digits. Since April 2005, when sudoku was introduced to the United States in* The New York Post, *more than half the leading American newspapers have begun printing one or more sudoku a day. No puzzle has had such a fast introduction in newspapers since the crossword craze of 1924–25.*[1]

Since then, millions of people have made sudoku part of their daily routines.

One of the authors played an online game of sudoku at www.websudoku.com. The graph provides information about how well he did. (His time is marked with an arrow.)

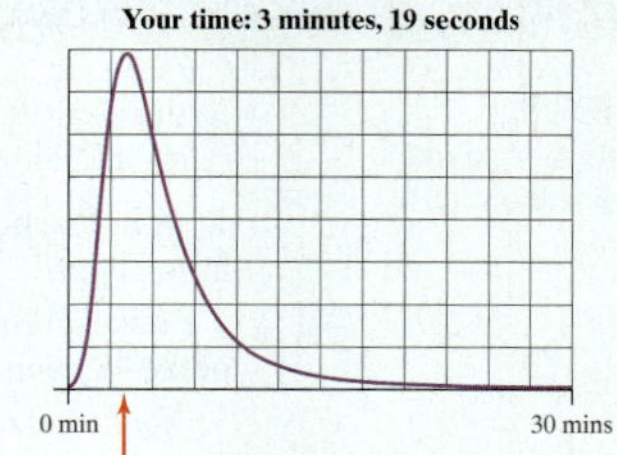


In this chapter, you'll learn more about how to describe the location of an individual observation—like the author's sudoku time—within a distribution.

83

## Teaching Tip

Remind students that the "percent correct" on a test does not measure the same thing as a "percentile."

**Activity Overview**

You can watch an overview of this Activity (for teachers) by clicking on the link in the TE-book, or looking on the TRCD or the book's companion Web site www.whfreeman.com/tps5e (teacher log-in required).

# Introduction

Suppose Jenny earns an 86 (out of 100) on her next statistics test. Should she be satisfied or disappointed with her performance? That depends on how her score compares with the scores of the other students who took the test. If 86 is the highest score, Jenny might be very pleased. Maybe her teacher will "curve" the grades so that Jenny's 86 becomes an "A." But if Jenny's 86 falls below the "average" in the class, she may not be so happy.

Section 2.1 focuses on describing the location of an individual within a distribution. We begin by discussing a familiar measure of position: *percentiles*. Next, we introduce a new type of graph that is useful for displaying percentiles. Then we consider another way to describe an individual's position that is based on the mean and standard deviation. In the process, we examine the effects of transforming data on the shape, center, and spread of a distribution.

Sometimes it is helpful to use graphical models called *density curves* to describe the location of individuals within a distribution, rather than relying on actual data values. Such models are especially helpful when data fall in a bell-shaped pattern called a *Normal distribution*. Section 2.2 examines the properties of Normal distributions and shows you how to perform useful calculations with them.

## ACTIVITY | Where do I stand?

**MATERIALS:**
Masking tape to mark number line scale

In this Activity, you and your classmates will explore ways to describe where you stand (literally!) within a distribution.

1. Your teacher will mark out a number line on the floor with a scale running from about 58 to 78 inches.
2. Make a human dotplot. Each member of the class should stand at the appropriate location along the number line scale based on height (to the nearest inch).
3. Your teacher will make a copy of the dotplot on the board for your reference.
4. What percent of the students in the class have heights less than yours? This is your *percentile* in the distribution of heights.

5. Work with a partner to calculate the mean and standard deviation of the class's height distribution from the dotplot. Confirm these values with your classmates.
6. Where does your height fall in relation to the mean: above or below? How far above or below the mean is it? How many standard deviations above or below the mean is it? This last number is the *z-score* corresponding to your height.
7. *Class discussion:* What would happen to the class's height distribution if you converted each data value from inches to centimeters? (There are 2.54 centimeters in 1 inch.) How would this change of units affect the measures of center, spread, and location (percentile and *z*-score) that you calculated?

Want to know more about where you stand—in terms of height, weight, or even body mass index? Do a Web search for "Clinical Growth Charts" at the National Center for Health Statistics site, www.cdc.gov/nchs.

# 2.1 Describing Location in a Distribution

**WHAT YOU WILL LEARN** By the end of the section, you should be able to:

- Find and interpret the percentile of an individual value within a distribution of data.
- Estimate percentiles and individual values using a cumulative relative frequency graph.
- Find and interpret the standardized score (*z*-score) of an individual value within a distribution of data.
- Describe the effect of adding, subtracting, multiplying by, or dividing by a constant on the shape, center, and spread of a distribution of data.

Here are the scores of all 25 students in Mr. Pryor's statistics class on their first test:

| 79 | 81 | 80 | 77 | 73 | 83 | 74 | 93 | 78 | 80 | 75 | 67 | 73 |
|---|---|---|---|---|---|---|---|---|---|---|---|---|
| 77 | 83 | **86** | 90 | 79 | 85 | 83 | 89 | 84 | 82 | 77 | 72 | |

The bold score is Jenny's **86**. How did she perform on this test relative to her classmates?

The stemplot displays this distribution of test scores. Notice that the distribution is roughly symmetric with no apparent outliers. From the stemplot, we can see that Jenny did better than all but three students in the class.

```
6 | 7
7 | 2334
7 | 5777899
8 | 00123334
8 | 569
9 | 03
```

Key: 7|2 is a student who scored 72 on the test

## Measuring Position: Percentiles

One way to describe Jenny's location in the distribution of test scores is to tell what percent of students in the class earned scores that were below Jenny's score. That is, we can calculate Jenny's **percentile**.

**DEFINITION: Percentile**

The ***p*th percentile** of a distribution is the value with *p* percent of the observations less than it.

Using the stemplot, we see that Jenny's 86 places her fourth from the top of the class. Because 21 of the 25 observations (84%) are below her score, Jenny is at the 84th percentile in the class's test score distribution.

### PD Section 2.1 Overview

Watch the video overview of Section 2.1 (for teachers) by clicking on the link in the TE-book, opening the TRCD, or logging onto the book's companion Web site www.whfreeman.com/tps5e (teacher log-in required).

### TRM Learning Objectives Grid

Remind students about the importance of the Learning Objectives. We include them at the beginning of each section so students know what they are expected to learn. At the end of the chapter, the Learning Objectives are included in a grid, along with references to examples and chapter review exercises. A printable version of this grid is included in the Teacher's Resource Materials.

### Teaching Tip

Be careful with your language when describing percentiles. Percentiles are specific locations in a distribution, so an observation isn't "in" the 84th percentile. Rather, it is "at" the 84th percentile.

### TRM Section 2.1 Alternate Examples

You can find the Alternate Examples for this section in Microsoft Word format by clicking on the link in the TE-book, opening the TRCD, or logging onto the book's companion Web site www.whfreeman.com/tps5e (teacher log-in required).

## Teaching Tip

Remind students that the "play" icon at the top of this example (and many others) indicates that they can watch a video presentation of the example. Find the videos at www.whfreeman.com/tps5e. Also, remind them that the icon in the lower right corner indicates that there is a link between this example and Exercise 1. If students are having trouble with the exercise, they can look back to this example for help.

## Teaching Tip

If two observations have the same value, as in part (c) of this example, they will be at the same percentile. To find the percentile, calculate the percent of the values in the distribution that are below *both* values.

### ALTERNATE EXAMPLE

*Wins in Major League Baseball*

The stemplot below shows the number of wins for each of the 30 Major League Baseball teams in 2012.

```
5 | 5
6 | 14
6 | 6899
7 | 234
7 | 569
8 | 113
8 | 56889
9 | 033444
9 | 578
```

*Key*: 6|1 represents a team with 61 wins.

PROBLEM: Find the percentiles for the following teams:

**(a)** The Minnesota Twins, who won 66 games.

**(b)** The Washington Nationals, who won 98 games.

**(c)** The Texas Rangers and Baltimore Orioles, who both won 93 games.

SOLUTION:

**(a)** The Minnesota Twins, who won 66 games, are at the 10th percentile because 3/30 teams had fewer wins than they did.

**(b)** The Washington Nationals, who won 98 games, are at about the 97th percentile because 29/30 teams had fewer wins than they did.

**(c)** The two teams with 93 wins, the Texas Rangers and the Baltimore Orioles, are at the 73rd percentile because only 22/30 teams had fewer wins than they did.

## EXAMPLE Mr. Pryor's First Test

*Finding percentiles*

PROBLEM: Use the scores on Mr. Pryor's first statistics test to find the percentiles for the following students:

(a) Norman, who earned a 72.

(b) Katie, who scored 93.

(c) The two students who earned scores of 80.

SOLUTION:

(a) Only 1 of the 25 scores in the class is below Norman's 72. His percentile is computed as follows: 1/25 = 0.04, or 4%. So Norman scored at the 4th percentile on this test.

(b) Katie's 93 puts her at the 96th percentile, because 24 out of 25 test scores fall below her result.

(c) Two students scored an 80 on Mr. Pryor's first test. Because 12 of the 25 scores in the class were less than 80, these two students are at the 48th percentile.

**For Practice** *Try Exercise* **1**

*Note:* Some people define the $p$th percentile of a distribution as the value with $p$ percent of observations *less than or equal to* it. Using this alternative definition of percentile, it is possible for an individual to fall at the 100th percentile. If we used this definition, the two students in part (c) of the example would fall at the 56th percentile (14 of 25 scores were less than or equal to 80). Of course, because 80 is the median score, it is also possible to think of it as being the 50th percentile. Calculating percentiles is not an exact science, especially with small data sets! We'll stick with the definition of percentile we gave earlier for consistency.

## Cumulative Relative Frequency Graphs

| Age | Frequency |
|---|---|
| 40–44 | 2 |
| 45–49 | 7 |
| 50–54 | 13 |
| 55–59 | 12 |
| 60–64 | 7 |
| 65–69 | 3 |

There are some interesting graphs that can be made with percentiles. One of the most common graphs starts with a frequency table for a quantitative variable. For instance, the frequency table in the margin summarizes the ages of the first 44 U.S. presidents when they took office.

Let's expand this table to include columns for relative frequency, cumulative frequency, and cumulative relative frequency.

- To get the values in the *relative frequency* column, divide the count in each class by 44, the total number of presidents. Multiply by 100 to convert to a percent.
- To fill in the *cumulative frequency* column, add the counts in the frequency column for the current class and all classes with smaller values of the variable.
- For the *cumulative relative frequency* column, divide the entries in the cumulative frequency column by 44, the total number of individuals. Multiply by 100 to convert to a percent.

## Teaching Tip

Tell students that percentiles should be whole numbers, so they should round their answer to the nearest integer if they get a decimal. For example, in a class of 13 people, the student with the highest score would be at about the 92nd percentile (12/13 = 0.923).

## AP® Exam Tip

On the AP® exam, students will be allowed to use any reasonable definition of "percentile" on the free-response section. On the multiple-choice section, the answer choices will be designed so that any reasonable method of finding percentiles will give the correct answer. When working with large data sets, the different definitions will give very similar values for the percentiles of a distribution.

Here is the original frequency table with the relative frequency, cumulative frequency, and cumulative relative frequency columns added.

| Age | Frequency | Relative frequency | Cumulative frequency | Cumulative relative frequency |
|---|---|---|---|---|
| 40–44 | 2 | 2/44 = 0.045, or 4.5% | 2 | 2/44 = 0.045, or 4.5% |
| 45–49 | 7 | 7/44 = 0.159, or 15.9% | 9 | 9/44 = 0.205, or 20.5% |
| 50–54 | 13 | 13/44 = 0.295, or 29.5% | 22 | 22/44 = 0.500, or 50.0% |
| 55–59 | 12 | 12/44 = 0.273, or 27.3% | 34 | 34/44 = 0.773, or 77.3% |
| 60–64 | 7 | 7/44 = 0.159, or 15.9% | 41 | 41/44 = 0.932, or 93.2% |
| 65–69 | 3 | 3/44 = 0.068, or 6.8% | 44 | 44/44 = 1.000, or 100% |

Some people refer to cumulative relative frequency graphs as "ogives" (pronounced "o-jives").

To make a **cumulative relative frequency graph,** we plot a point corresponding to the cumulative relative frequency in each class at the smallest value of the *next* class. For example, for the 40 to 44 class, we plot a point at a height of 4.5% above the age value of 45. This means that 4.5% of presidents were inaugurated *before* they were 45 years old. (In other words, age 45 is the 4.5th percentile of the inauguration age distribution.)

It is customary to start a cumulative relative frequency graph with a point at a height of 0% at the smallest value of the first class (in this case, 40). The last point we plot should be at a height of 100%. We connect consecutive points with a line segment to form the graph. Figure 2.1 shows the completed cumulative relative frequency graph.

Here's an example that shows how to interpret a cumulative relative frequency graph.

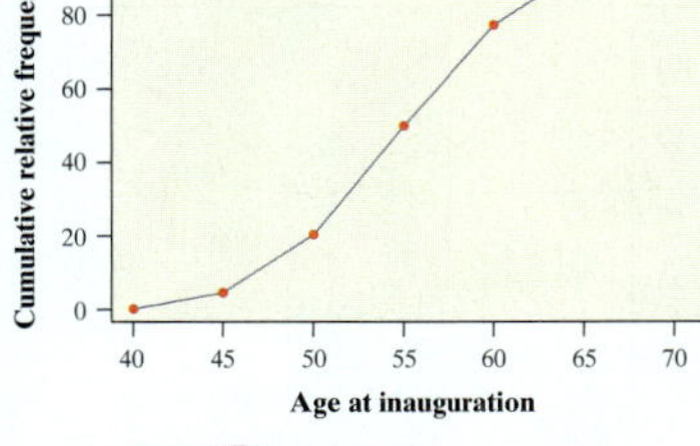


**FIGURE 2.1** Cumulative relative frequency graph for the ages of U.S. presidents at inauguration.

## EXAMPLE Age at Inauguration

### *Interpreting a cumulative relative frequency graph*

What can we learn from Figure 2.1? The graph grows very gradually at first because few presidents were inaugurated when they were in their 40s. Then the graph gets very steep beginning at age 50. Why? Because most U.S. presidents were in their 50s when they were inaugurated. The rapid growth in the graph slows at age 60.

Suppose we had started with only the graph in Figure 2.1, without any of the information in our original frequency table. Could we figure out what percent of presidents were between 55 and 59 years old at their inaugurations? Sure. Because the point at age 60 has a cumulative relative frequency of about 77%, we know that about 77% of presidents were inaugurated before they were 60 years old. Similarly, the point at age 55 tells us that about 50% of presidents were younger than 55 at inauguration. As a result, we'd estimate that about 77% − 50% = 27% of U.S. presidents were between 55 and 59 when they were inaugurated.

## ALTERNATE EXAMPLE

### *State median household incomes*

The table and cumulative relative frequency graph show the distribution of median household incomes for the 50 states and the District of Columbia in a recent year.

| Median income ($1000s) | Frequency | Relative frequency | Cumulative frequency | Cumulative relative frequency |
|---|---|---|---|---|
| 35 to < 40 | 1 | 1/51 = 0.020 | 1 | 1/51 = 0.020 |
| 40 to < 45 | 10 | 10/51 = 0.196 | 11 | 11/51 = 0.216 |
| 45 to < 50 | 14 | 14/51 = 0.275 | 25 | 25/51 = 0.490 |
| 50 to < 55 | 12 | 12/51 = 0.236 | 37 | 37/51 = 0.725 |
| 55 to < 60 | 5 | 5/51 = 0.098 | 42 | 42/51 = 0.824 |
| 60 to < 65 | 6 | 6/51 = 0.118 | 48 | 48/51 = 0.941 |
| 65 to < 70 | 3 | 3/51 = 0.059 | 51 | 51/51 = 1.000 |

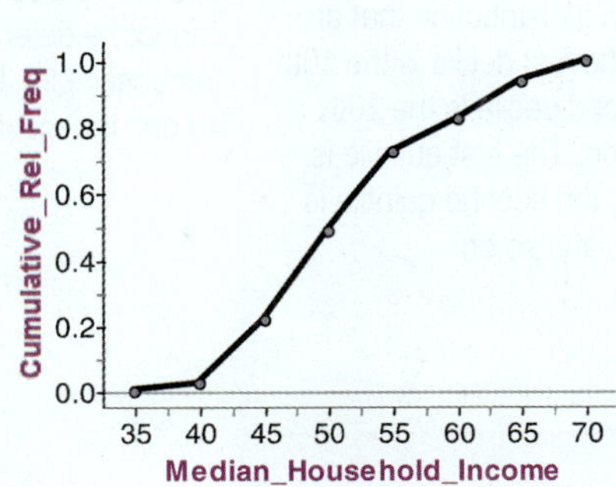


## ALTERNATE EXAMPLE *(continued)*

The point at (50, 0.49) means 49% of the states had median household incomes less than $50,000. The point at (55, 0.725) means that 72.5% of the states had median household incomes less than $55,000. Thus, 72.5% − 49% = 23.5% of the states had median household incomes between $50,000 and $55,000 because the cumulative relative frequency increased by 0.235. Due to rounding error, this value is slightly different from the relative frequency for the 50 to <55 category.

## Teaching Tip

Ask your students to think about what the steepness of the graph tells them about the distribution of median household incomes. This graph is steepest between 40 and 55, meaning that the cumulative relative frequency is increasing quickly. This means that there are lots of states with median household incomes between $40,000 and $55,000. The graph is less steep between 35 and 40 and above 55, meaning there are fewer states with these incomes.

As a challenge, ask your students to sketch a cumulative relative frequency graph for the following distribution shapes: skewed right, skewed left, uniform, and bimodal.

**ALTERNATE EXAMPLE**

*State median household incomes*

PROBLEM: Use the cumulative relative frequency graph for the state income data to answer each question.
**(a)** At what percentile is California, with a median household income of $57,445?
**(b)** Estimate and interpret the first quartile of this distribution.

SOLUTION:
**(a)** California is at about the 78th percentile.

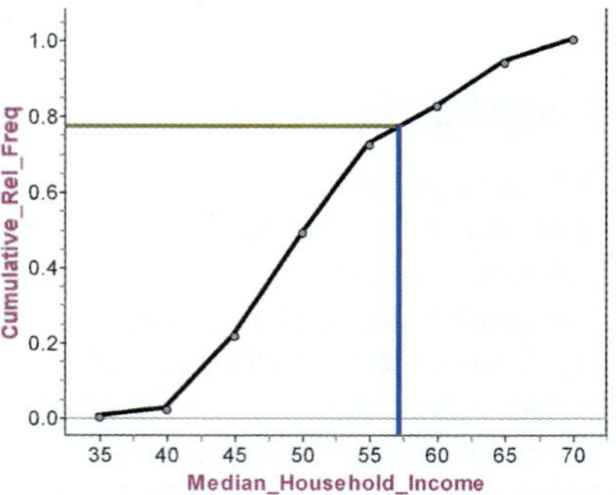


**(b)** The first quartile of this distribution is the 25th percentile. About 25% of states have median incomes less than $45,000.

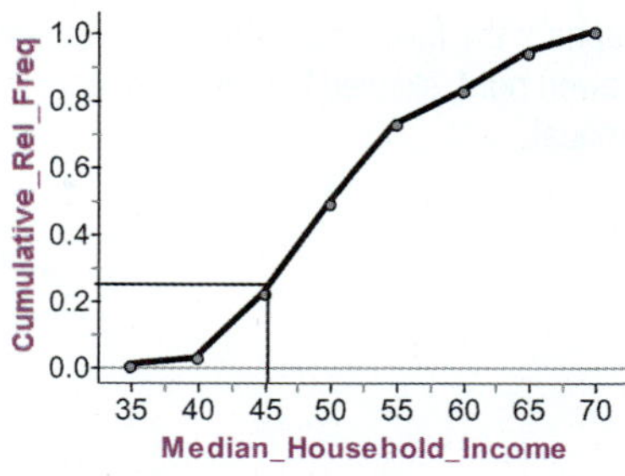

A cumulative relative frequency graph can be used to describe the position of an individual within a distribution or to locate a specified percentile of the distribution, as the following example illustrates.

## EXAMPLE Ages of U.S. Presidents

*Interpreting cumulative relative frequency graphs*

PROBLEM: Use the graph in Figure 2.1 on the previous page to help you answer each question.

(a) Was Barack Obama, who was first inaugurated at age 47, unusually young?

(b) Estimate and interpret the 65th percentile of the distribution.

SOLUTION:

(a) To find President Obama's location in the distribution, we draw a vertical line up from his age (47) on the horizontal axis until it meets the graphed line. Then we draw a horizontal line from this point of intersection to the vertical axis. Based on Figure 2.2(a), we would estimate that Barack Obama's inauguration age places him at the 11% cumulative relative frequency mark. That is, he's at the 11th percentile of the distribution. In other words, about 11% of all U.S. presidents were younger than Barack Obama when they were inaugurated and about 89% were older.

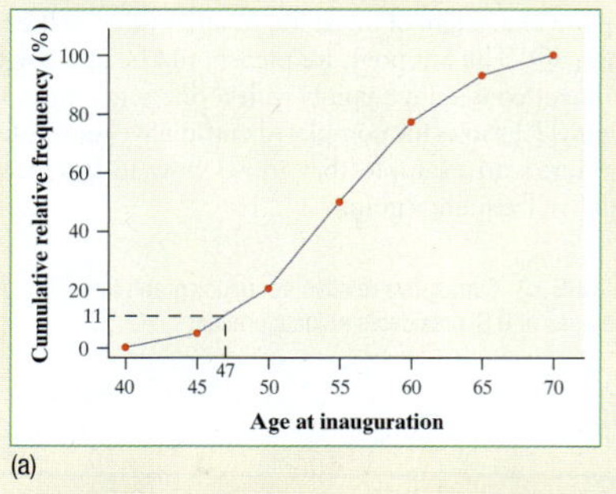


(a)

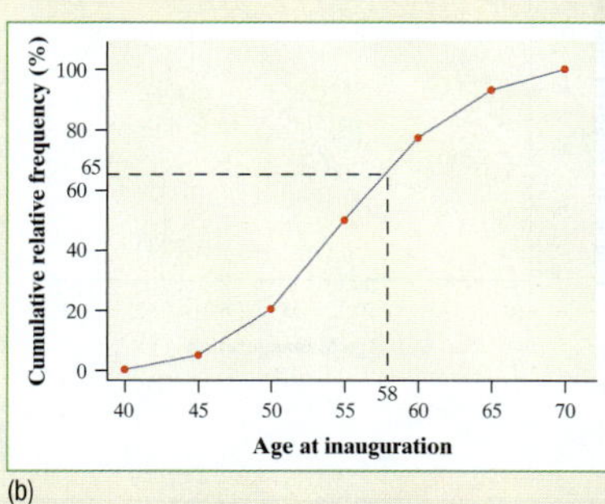


(b)

FIGURE 2.2 The cumulative relative frequency graph of presidents' ages at inauguration is used to (a) locate President Obama within the distribution and (b) determine the 65th percentile, which is about 58 years.

(b) The 65th percentile of the distribution is the age with cumulative relative frequency 65%. To find this value, draw a horizontal line across from the vertical axis at a height of 65% until it meets the graphed line. From the point of intersection, draw a vertical line down to the horizontal axis. In Figure 2.2(b), the value on the horizontal axis is about 58. So about 65% of all U.S. presidents were younger than 58 when they took office.

**For Practice** *Try Exercise* 9

**THINK ABOUT IT**

**Percentiles and quartiles:** Have you made the connection between percentiles and the quartiles from Chapter 1? Earlier, we noted that the median (second quartile) corresponds to the 50th percentile. What about the first quartile, $Q_1$? It's at the median of the lower half of the ordered data, which puts it about one-fourth of the way through the distribution. In other words, $Q_1$ is roughly the 25th percentile. By similar reasoning, $Q_3$ is approximately the 75th percentile of the distribution.

### Teaching Tip

Remind students that the values we get from the graph are just estimates. However, without the actual data, this is the best we can do.

### Teaching Tip

Deciles and quintiles are other ways to measure location in a distribution that are commonly used. The first decile is the 10th percentile, the second decile is the 20th percentile, and so on. The first quintile is the 20th percentile, the second quintile is the 40th percentile, and so on.

### Teaching Tip

To check for understanding, ask your students to make a boxplot of the data using the age at inauguration example. They should be able to estimate the first quartile, median, and third quartile by finding the values of the 25th, 50th, and 75th percentiles. The min and max values cannot be determined exactly from the cumulative frequency plot, but the minimum can't be less than 40 and the maximum can't be more than 70.

## CHECK YOUR UNDERSTANDING

1. *Multiple choice: Select the best answer.* Mark receives a score report detailing his performance on a statewide test. On the math section, Mark earned a raw score of 39, which placed him at the 68th percentile. This means that

(a) Mark did better than about 39% of the students who took the test.

(b) Mark did worse than about 39% of the students who took the test.

(c) Mark did better than about 68% of the students who took the test.

(d) Mark did worse than about 68% of the students who took the test.

(e) Mark got fewer than half of the questions correct on this test.

2. Mrs. Munson is concerned about how her daughter's height and weight compare with those of other girls of the same age. She uses an online calculator to determine that her daughter is at the 87th percentile for weight and the 67th percentile for height. Explain to Mrs. Munson what this means.

*Questions 3 and 4 relate to the following setting.* The graph displays the cumulative relative frequency of the lengths of phone calls made from the mathematics department office at Gabalot High last month.

3. About what percent of calls lasted less than 30 minutes? 30 minutes or more?

4. Estimate $Q_1$, $Q_3$, and the $IQR$ of the distribution.

### Teaching Tip

Make sure to have students do the Check Your Understanding because the questions differ from those in the examples.

### ✓ ANSWERS TO CYU

**1.** c

**2.** Her daughter weighs more than 87% of girls her age and she is taller than 67% of girls her age.

**3.** About 65% of calls lasted less than 30 minutes, which means that about 35% of calls lasted 30 minutes or longer.

**4.** $Q_1 = 13$ minutes, $Q_3 = 32$ minutes, and $IQR = 19$ minutes.

## Measuring Position: *z*-Scores

```
6 | 7
7 | 2334
7 | 5777899
8 | 00123334
8 | 569
9 | 03
```

Key: 7|2 is a student who scored 72 on the test

Let's return to the data from Mr. Pryor's first statistics test, which are shown in the stemplot. Figure 2.3 provides numerical summaries from Minitab for these data. Where does Jenny's score of 86 fall relative to the mean of this distribution? Because the mean score for the class is 80, we can see that Jenny's score is "above average." But how much above average is it?

We can describe Jenny's location in the distribution of her class's test scores by telling how many standard deviations above or below the mean her score is. Because the mean is 80 and the standard deviation is about 6, Jenny's score of 86 is about one standard deviation above the mean.

Converting observations like this from original values to standard deviation units is known as **standardizing**. To standardize a value, subtract the mean of the distribution and then divide the difference by the standard deviation.

The relationship between the mean and the median is about what you'd expect in this fairly symmetric distribution.

Minitab

Session

**Descriptive Statistics: Test 1 scores**

| Variable | N | Mean | Median | TrMean | StDev | SE Mean |
|---|---|---|---|---|---|---|
| Test 1 scores | 25 | 80.00 | 80.00 | 80.00 | 6.07 | 1.21 |

| Variable | Minimum | Maximum | Q1 | Q3 |
|---|---|---|---|---|
| Test 1 scores | 67.00 | 93.00 | 76.00 | 83.50 |

**FIGURE 2.3** Minitab output for the scores of Mr. Pryor's students on their first statistics test.

### Teaching Tip

Make sure to discuss why knowing that Jenny is 6 points above average doesn't tell you much about her location in the distribution. Depending on the spread of the distribution, Jenny might be just barely above average or really far above average. This is why we need to incorporate a measure of spread (standard deviation) to have a good understanding of how far above the mean she really is.

### TECHNOLOGY TIP

In the Minitab output, "TrMean" represents the trimmed mean. To calculate the trimmed mean, Minitab removes the smallest 5% of values and the largest 5% of values and calculates the mean of the middle 90%. This makes the trimmed mean more resistant to outliers than the regular mean. This trimmed-mean method is used in some sporting events where subjective judging exists, such as in diving and gymnastics. Also, "SE Mean" is the standard error of the mean, which is equal to the sample standard deviation divided by the square root of the sample size. We will learn much more about the standard error of the mean and how to use it in Chapters 7 to 10.

**DEFINITION: Standardized score (*z*-score)**

If $x$ is an observation from a distribution that has known mean and standard deviation, the **standardized score** for $x$ is

$$z = \frac{x - \text{mean}}{\text{standard deviation}}$$

A standardized score is often called a ***z*-score.**

A *z*-score tells us how many standard deviations from the mean an observation falls, and in what direction. Observations larger than the mean have positive *z*-scores. Observations smaller than the mean have negative *z*-scores. For example, Jenny's score on the test was $x = 86$. Her *standardized* score (*z*-score) is

$$z = \frac{x - \text{mean}}{\text{standard deviation}} = \frac{86 - 80}{6.07} = 0.99$$

That is, Jenny's test score is 0.99 standard deviations *above* the mean score of the class.

## EXAMPLE Mr. Pryor's First Test, Again

### *Finding and interpreting z-scores*

PROBLEM: Use Figure 2.3 on the previous page to find the standardized scores (*z*-scores) for each of the following students in Mr. Pryor's class. Interpret each value in context.

(a) Katie, who scored 93.

(b) Norman, who earned a 72.

SOLUTION:

(a) Katie's 93 was the highest score in the class. Her corresponding *z*-score is

$$z = \frac{93 - 80}{6.07} = 2.14$$

In other words, Katie's result is 2.14 standard deviations above the mean score for this test.

(b) For Norman's 72, his standardized score is

$$z = \frac{72 - 80}{6.07} = -1.32$$

Norman's score is 1.32 standard deviations below the class mean of 80.

**For Practice** *Try Exercise* **15(b)**

We can also use *z*-scores to compare the position of individuals in different distributions, as the following example illustrates.

### Teaching Tip

Point out that a *z*-score is not measured in the same units as the variable. It would be incorrect to say Jenny was 0.99 points above average. Instead, we say she is 0.99 standard deviations above average.

### Teaching Tip

Emphasize that a *z*-score is directional. If the *z*-score is positive, it means the observation is above average. If the *z*-score is negative, it means the observation is below average.

### Teaching Tip

Later in the course, many students assume that all *z*-scores follow a Normal distribution, which isn't true. A *z*-score simply reveals how many standard deviations a value is above or below the mean, no matter what the shape of the distribution. To illustrate this, calculate and graph the *z*-scores for a distribution of data that is non-Normal. The distribution of *z*-scores should have the same shape as the original distribution, but with a mean of 0 and a standard deviation of 1. See the Think About It on page 97 for an example.

### ALTERNATE EXAMPLE

*Wins in Major League Baseball*

In 2012 the mean number of wins for teams in Major League Baseball was 81 with a standard deviation of 11.9 wins.

PROBLEM: Find and interpret the *z*-scores for the following teams.

**(a)** The New York Yankees, with 95 wins.

**(b)** The New York Mets, with 74 wins.

SOLUTION:

**(a)** $z = \frac{95 - 81}{11.9} = 1.18$. The Yankees were 1.18 standard deviations above the mean number of wins.

**(b)** $z = \frac{74 - 81}{11.9} = -0.59$. The Mets were 0.59 standard deviations *below* the mean number of wins.

## EXAMPLE Jenny Takes Another Test

### *Using z-scores for comparisons*

The day after receiving her statistics test result of 86 from Mr. Pryor, Jenny earned an 82 on Mr. Goldstone's chemistry test. At first, she was disappointed. Then Mr. Goldstone told the class that the distribution of scores was fairly symmetric with a mean of 76 and a standard deviation of 4.

PROBLEM: **On which test did Jenny perform better relative to the class? Justify your answer.**

SOLUTION: **Jenny's z-score for her chemistry test result is**

$$z = \frac{82 - 76}{4} = 1.50$$

**Her 82 in chemistry was 1.5 standard deviations above the mean score for the class. Because she scored only 0.99 standard deviations above the mean on the statistics test, Jenny did better relative to the class in chemistry.**

**For Practice** *Try Exercise* **11**

We often standardize observations to express them on a common scale. We might, for example, compare the heights of two children of different ages by calculating their *z*-scores. At age 2, Jordan is 89 centimeters (cm) tall. Her height puts her at a *z*-score of 0.5; that is, she is one-half standard deviation above the mean height of 2-year-old girls. Zayne's height at age 3 is 101 cm, which yields a *z*-score of 1. In other words, he is one standard deviation above the mean height of 3-year-old boys. So Zayne is taller relative to boys his age than Jordan is relative to girls her age. The standardized heights tell us where each child stands (pun intended!) in the distribution for his or her age group.

### CHECK YOUR UNDERSTANDING

Mrs. Navard's statistics class has just completed the first three steps of the "Where Do I Stand?" Activity (page 84). The figure below shows a dotplot of the class's height distribution, along with summary statistics from computer output.

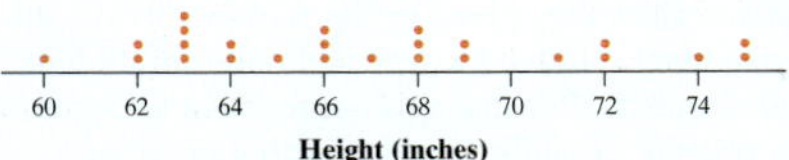


| Variable | $n$ | $\bar{x}$ | $s_x$ | Min | $Q_1$ | Med | $Q_3$ | Max |
|---|---|---|---|---|---|---|---|---|
| Height | 25 | 67 | 4.29 | 60 | 63 | 66 | 69 | 75 |

1. Lynette, a student in the class, is 65 inches tall. Find and interpret her *z*-score.
2. Another student in the class, Brent, is 74 inches tall. How tall is Brent compared with the rest of the class? Give appropriate numerical evidence to support your answer.
3. Brent is a member of the school's basketball team. The mean height of the players on the team is 76 inches. Brent's height translates to a *z*-score of −0.85 in the team's height distribution. What is the standard deviation of the team members' heights?

## ALTERNATE EXAMPLE

### *Home run kings*

The single-season home run record for Major League Baseball has been set just three times since Babe Ruth hit 60 home runs in 1927. Roger Maris hit 61 in 1961, Mark McGwire hit 70 in 1998, and Barry Bonds hit 73 in 2001. In an absolute sense, Barry Bonds had the best performance of these four players, because he hit the most home runs in a single season. However, in a relative sense, this may not be true. Baseball historians suggest that hitting a home run has been easier in some eras than others. This is due to many factors, including the quality of batters, quality of pitchers, hardness of the baseball, dimensions of ballparks, and possible use of performance-enhancing drugs. To make a fair comparison, we should see how these performances rate relative to those of other hitters during the same year.

PROBLEM: Compute the standardized scores for each performance using the information in the table. Which player had the most outstanding performance relative to his peers?

| Year | Player | HR | Mean | SD |
|---|---|---|---|---|
| 1927 | Babe Ruth | 60 | 7.2 | 9.7 |
| 1961 | Roger Maris | 61 | 18.8 | 13.4 |
| 1998 | Mark McGwire | 70 | 20.7 | 12.7 |
| 2001 | Barry Bonds | 73 | 21.4 | 13.2 |

SOLUTION: The table below shows the *z*-scores for each athlete. Although all four performances were outstanding, Babe Ruth can still lay claim to being the single-season home run champ, *relatively speaking*.

| Player | z-score |
|---|---|
| Babe Ruth | $z = \frac{60 - 7.2}{9.7} = 5.44$ |
| Roger Maris | $z = \frac{61 - 18.8}{13.4} = 3.15$ |
| Mark McGwire | $z = \frac{70 - 20.7}{12.7} = 3.88$ |
| Barry Bonds | $z = \frac{73 - 21.4}{13.2} = 3.91$ |

### Teaching Tip

When comparing *z*-scores from different distributions, it is important that the distributions be roughly the same shape. For example, a *z*-score of +2 is very unusual in a uniform distribution but not that uncommon in a distribution that is strongly skewed to the right.

### ANSWERS TO CYU

1. $z = -0.466$. Lynette's height is 0.466 standard deviations below the mean height of the class.
2. $z = 1.63$. Brent's height is 1.63 standard deviations above the mean height of the class.
3. $-0.85 = \frac{74 - 76}{\sigma}$, so $\sigma = 2.35$ inches.

**ALTERNATE EXAMPLE**

*Estimating the width of your classroom*

Instead of using the data provided, have your students estimate the width of your classroom in meters and then do the same transformations we complete in the text. That is, subtract the true width from each value to find the errors and then convert the errors to feet.

## Transforming Data

To find the standardized score ($z$-score) for an individual observation, we transform this data value by subtracting the mean and dividing the difference by the standard deviation. Transforming converts the observation from the original units of measurement (inches, for example) to a standardized scale. What effect do these kinds of transformations—adding or subtracting; multiplying or dividing—have on the shape, center, and spread of the entire distribution? Let's investigate using an interesting data set from "down under."

Soon after the metric system was introduced in Australia, a group of students was asked to guess the width of their classroom to the nearest meter. Here are their guesses in order from lowest to highest:[2]

| | | | | | | | | | | | | |
|---|---|---|---|---|---|---|---|---|---|---|---|---|
| 8 | 9 | 10 | 10 | 10 | 10 | 10 | 10 | 11 | 11 | 11 | 11 | 12 |
| 12 | 13 | 13 | 13 | 14 | 14 | 14 | 15 | 15 | 15 | 15 | 15 | 15 |
| 15 | 15 | 16 | 16 | 16 | 17 | 17 | 17 | 17 | 18 | 18 | 20 | 22 |
| 25 | 27 | 35 | 38 | 40 | | | | | | | | |

Figure 2.4 includes a dotplot of the data and some numerical summaries.

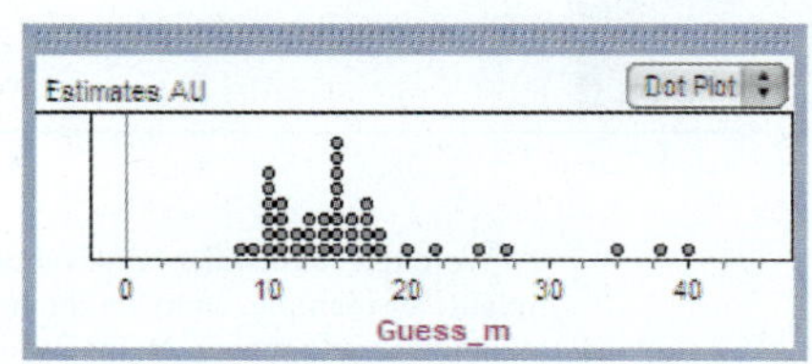


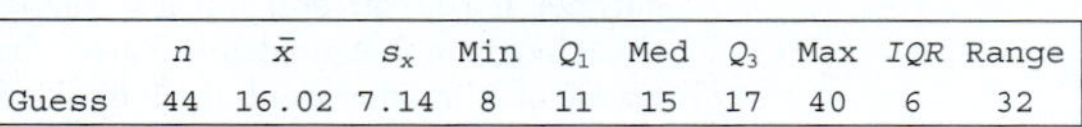

| | $n$ | $\bar{x}$ | $s_x$ | Min | $Q_1$ | Med | $Q_3$ | Max | $IQR$ | Range |
|---|---|---|---|---|---|---|---|---|---|---|
| Guess | 44 | 16.02 | 7.14 | 8 | 11 | 15 | 17 | 40 | 6 | 32 |

**FIGURE 2.4** Fathom dotplot and summary statistics for Australian students' guesses of the classroom width.

Let's practice what we learned in Chapter 1 and describe what we see.

*Shape:* The distribution of guesses appears skewed to the right and bimodal, with peaks at 10 and 15 meters.

*Center:* The median guess was 15 meters and the mean guess was about 16 meters. Due to the clear skewness and potential outliers, the median is a better choice for summarizing the "typical" guess.

*Spread:* Because $Q_1 = 11$, about 25% of the students estimated the width of the room to be fewer than 11 meters. The 75th percentile of the distribution is at about $Q_3 = 17$. The $IQR$ of 6 meters describes the spread of the middle 50% of students' guesses. The standard deviation tells us that the typical distance of students' guesses from the mean was about 7 meters. Because $s_x$ is not resistant to extreme values, we prefer the $IQR$ to describe the variability of this distribution.

*Outliers:* By the $1.5 \times IQR$ rule, values greater than $17 + 9 = 26$ meters or less than $11 - 9 = 2$ meters are identified as outliers. So the four highest guesses—which are 27, 35, 38, and 40 meters—are outliers.

**ALTERNATE EXAMPLE**

*Test scores*

Here are a graph and table of summary statistics for a sample of 30 test scores. The maximum possible score on the test was 50 points.

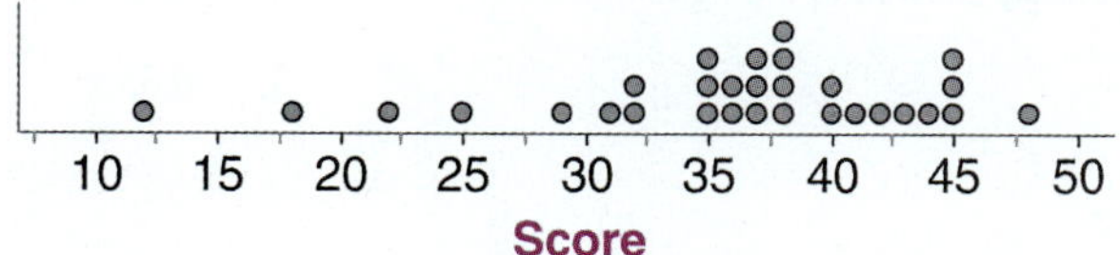


| | $n$ | $\bar{x}$ | $s_x$ | Min | $Q_1$ | Med | $Q_3$ | Max | $IQR$ | Range |
|---|---|---|---|---|---|---|---|---|---|---|
| Score | 30 | 35.8 | 8.17 | 12 | 32 | 37 | 41 | 48 | 9 | 36 |

**Effect of adding or subtracting a constant:** By now, you're probably wondering what the actual width of the room was. In fact, it was 13 meters wide. How close were students' guesses? The student who guessed 8 meters was too low by 5 meters. The student who guessed 40 meters was too high by 27 meters (and probably needs to study the metric system more carefully). We can examine the distribution of students' guessing errors by defining a new variable as follows:

$$\text{error} = \text{guess} - 13$$

That is, we'll subtract 13 from each observation in the data set. Try to predict what the shape, center, and spread of this new distribution will be. Refer to Figure 2.4 as needed.

## EXAMPLE Estimating Room Width

### *Effect of subtracting a constant*

Let's see how accurate your predictions were (you did make predictions, right?). Figure 2.5 shows dotplots of students' original guesses and their errors on the same scale. We can see that the original distribution of guesses has been shifted to the left. By how much? Because the peak at 15 meters in the original graph is located at 2 meters in the error distribution, the original data values have been translated 13 units to the left. That should make sense: we calculated the errors by subtracting the actual room width, 13 meters, from each student's guess.

**FIGURE 2.5** Fathom dotplots of students' original guesses of classroom width and the errors in their guesses.

From Figure 2.5, it seems clear that subtracting 13 from each observation did not affect the shape or spread of the distribution. But this transformation appears to have decreased the center of the distribution by 13 meters. The summary statistics in the table below confirm our beliefs.

| | $n$ | $\bar{x}$ | $s_x$ | Min | $Q_1$ | Med | $Q_3$ | Max | $IQR$ | Range |
|---|---|---|---|---|---|---|---|---|---|---|
| Guess (m) | 44 | 16.02 | 7.14 | 8 | 11 | 15 | 17 | 40 | 6 | 32 |
| Error (m) | 44 | 3.02 | 7.14 | -5 | -2 | 2 | 4 | 27 | 6 | 32 |

The error distribution is centered at a value that is clearly positive—the median error is 2 meters and the mean error is about 3 meters. So the students generally tended to overestimate the width of the room.

As the example shows, subtracting the same positive number from each value in a data set shifts the distribution to the left by that number. Adding a positive constant to each data value would shift the distribution to the right by that constant.

### Teaching Tip

Be careful when discussing measures of location. In the alternate example, the $z$-score of the student with the top score of 48 won't change, because this student is still in the same location in the distribution after the 5 points are added. Likewise, this student will still be at the 97th percentile because she still ranks ahead of 29 of the 30 students. However, the *value* of each percentile will increase by 5. So if an original test score of 25 was at the 10th percentile, then the 10th percentile in the new distribution will be 30.

Suppose that the teacher was nice and added 5 points to each test score. How would this change the shape, center, and spread of the distribution? Here are graphs and summary statistics for the original scores and the +5 scores:

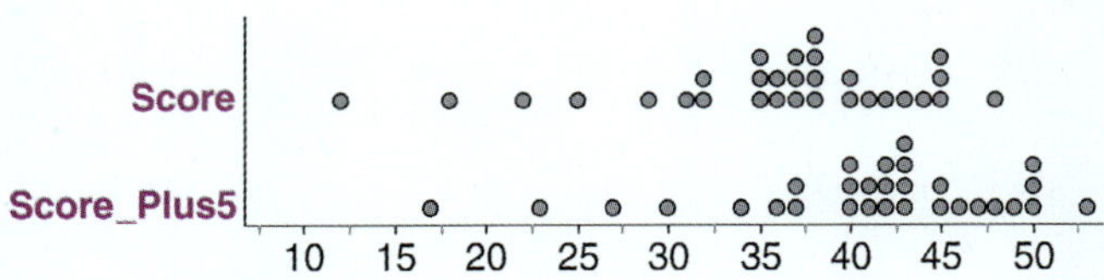


| | $n$ | $\bar{x}$ | $s_x$ | Min | $Q_1$ | Med | $Q_3$ | Max | $IQR$ | Range |
|---|---|---|---|---|---|---|---|---|---|---|
| Score | 30 | 35.8 | 8.17 | 12 | 32 | 37 | 41 | 48 | 9 | 36 |
| Score +5 | 30 | 40.8 | 8.17 | 17 | 37 | 42 | 46 | 53 | 9 | 36 |

From both the graph and summary statistics, we can see that the measures of center and measures of location all increased by 5. However, the shape of the distribution did not change nor did the spread of the distribution.

## Teaching Tip

You might want to show your students this short "proof" for why the *IQR* doesn't change when adding a number *a* to each observation:

$$\text{Original } IQR = Q_3 - Q_1$$

$$\begin{aligned} \text{New } IQR &= (Q_3 + a) - (Q_1 + a) \\ &= Q_3 + a - Q_1 - a \\ &= Q_3 - Q_1 = \text{original } IQR \end{aligned}$$



Let's summarize what we've learned so far about transforming data.

### EFFECT OF ADDING (OR SUBTRACTING) A CONSTANT

Adding the same positive number *a* to (subtracting *a* from) each observation

- adds *a* to (subtracts *a* from) measures of center and location (mean, median, quartiles, percentiles), but
- does not change the shape of the distribution or measures of spread (range, *IQR*, standard deviation).

**Effect of multiplying or dividing by a constant:** Because our group of Australian students is having some difficulty with the metric system, it may not be helpful to tell them that their guesses tended to be about 2 to 3 meters too high. Let's convert the error data to feet before we report back to them. There are roughly 3.28 feet in a meter. So for the student whose error was −5 meters, that translates to

$$-5 \text{ meters} \times \frac{3.28 \text{ feet}}{1 \text{ meter}} = -16.4 \text{ feet}$$

To change the units of measurement from meters to feet, we multiply each of the error values by 3.28. What effect will this have on the shape, center, and spread of the distribution? (Go ahead, make some predictions!)

## EXAMPLE Estimating Room Width

### *Effect of multiplying by a constant*

Figure 2.6 includes dotplots of the students' guessing errors in meters and feet, along with summary statistics from computer software. The shape of the two distributions is the same—right-skewed and bimodal. However, the centers and spreads

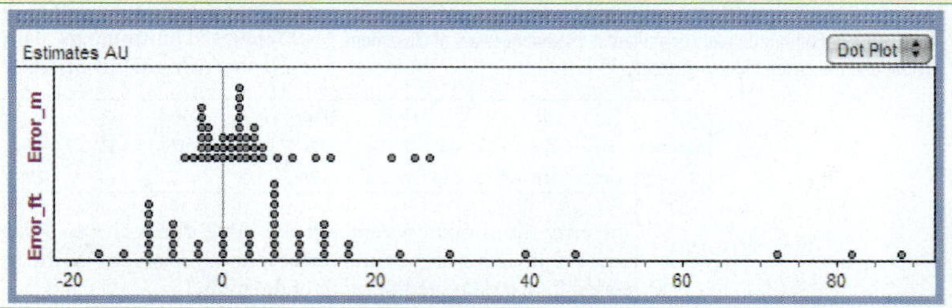


| | $n$ | $\bar{x}$ | $s_x$ | Min | $Q_1$ | Med | $Q_3$ | Max | $IQR$ | Range |
|---|---|---|---|---|---|---|---|---|---|---|
| Error (m) | 44 | 3.02 | 7.14 | −5 | −2 | 2 | 4 | 27 | 6 | 32 |
| Error (ft) | 44 | 9.91 | 23.43 | −16.4 | −6.56 | 6.56 | 13.12 | 88.56 | 19.68 | 104.96 |

**FIGURE 2.6** Fathom dotplots and numerical summaries of students' errors guessing the width of their classroom in meters and feet.

### ALTERNATE EXAMPLE

*Test scores*

Suppose that the teacher in the previous alternate example wanted to convert the *original* test scores to percents. Because the test was out of 50 points, he should multiply each score by 2 to make them out of 100. Here are graphs and summary statistics for the original scores and the doubled scores:

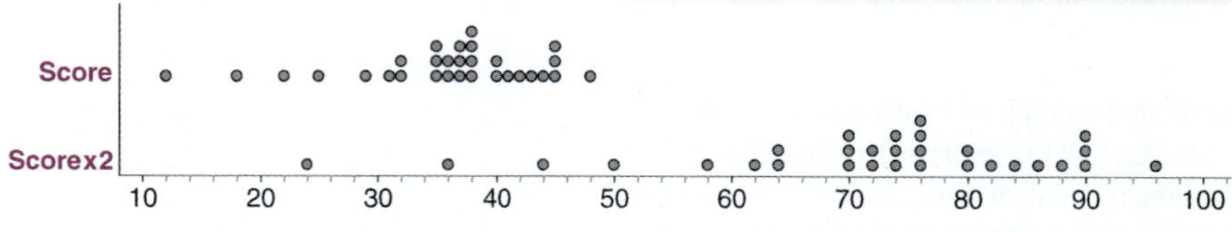


| | $n$ | $\bar{x}$ | $s_x$ | Min | $Q_1$ | Med | $Q_3$ | Max | $IQR$ | Range |
|---|---|---|---|---|---|---|---|---|---|---|
| Score | 30 | 35.8 | 8.17 | 12 | 32 | 37 | 41 | 48 | 9 | 36 |
| Score × 2 | 30 | 71.6 | 16.34 | 24 | 64 | 74 | 82 | 96 | 18 | 72 |

From the graphs and summary statistics, we can see that the measures of center, location, and *spread* all have doubled, just like the individual observations. But even though the distribution is more spread out, the shape hasn't changed. It is still skewed to the left with the same clusters and gaps.

of the two distributions are quite different. The bottom dotplot is centered at a value that is to the right of the top dotplot's center. Also, the bottom dotplot shows much greater spread than the top dotplot.

When the errors were measured in meters, the median was 2 and the mean was 3.02. For the transformed error data in feet, the median is 6.56 and the mean is 9.91. Can you see that the measures of center were multiplied by 3.28? That makes sense. If we multiply all the observations by 3.28, then the mean and median should also be multiplied by 3.28.

What about the spread? Multiplying each observation by 3.28 increases the variability of the distribution. By how much? You guessed it—by a factor of 3.28. The numerical summaries in Figure 2.6 show that the standard deviation, the interquartile range, and the range have been multiplied by 3.28.

We can safely tell our group of Australian students that their estimates of the classroom's width tended to be too high by about 6.5 feet. (Notice that we choose not to report the mean error, which is affected by the strong skewness and the three high outliers.)

As before, let's recap what we discovered about the effects of transforming data.

### EFFECT OF MULTIPLYING (OR DIVIDING) BY A CONSTANT

Multiplying (or dividing) each observation by the same positive number $b$

- multiplies (divides) measures of center and location (mean, median, quartiles, percentiles) by $b$,
- multiplies (divides) measures of spread (range, $IQR$, standard deviation) by $b$, but
- does not change the shape of the distribution.

It is not common to multiply (or divide) each observation in a data set by a *negative* number $b$. Doing so would multiply (or divide) the measures of spread by the *absolute value* of $b$. We can't have a negative amount of variability! Multiplying or dividing by a negative number would also affect the shape of the distribution as all values would be reflected over the $y$ axis.

**Putting it all together: Adding/subtracting and multiplying/dividing:** What happens if we transform a data set by both adding or subtracting a constant and multiplying or dividing by a constant? For instance, if we need to convert temperature data from Celsius to Fahrenheit, we have to use the formula °F = 9/5(°C) + 32. That is, we would multiply each of the observations by 9/5 and then add 32. As the following example shows, we just use the facts about transforming data that we've already established.

## EXAMPLE Too Cool at the Cabin?

### *Analyzing the effects of transformations*

During the winter months, the temperatures at the Starnes's Colorado cabin can stay well below freezing (32°F or 0°C) for weeks at a time. To prevent the pipes from freezing, Mrs. Starnes sets the thermostat at 50°F. She also buys a digital thermometer that records the indoor temperature each night at midnight.

### Teaching Tip

You might want to show your students this short "proof" for how the $IQR$ changes when multiplying each observation by the same number $b$:

Original $IQR = Q_3 - Q_1$

New $IQR = Q_3 \cdot b - Q_1 \cdot b = b(Q_3 - Q_1)$
$= b(\text{original } IQR)$

### Teaching Tip

Note that variance ($s_x^2$) is not included among the measures of spread listed in the summary box. When each observation in a distribution is multiplied by $b$, the variance is multiplied by $b^2$. This fact will be important in Chapter 6.

### Teaching Tip

Your students may wonder whether there are any transformations that change the shape of a distribution. As we will discover in Chapter 8 when learning about the $t$ distributions, multiplying (or dividing) the values in a distribution by a *variable* changes the shape. Also, multiplying (or dividing) the values by a negative constant reflects the distribution around the line $x = 0$. If the original distribution was skewed to the right, the new distribution will be skewed to the left.

Unfortunately, the thermometer is programmed to measure the temperature in degrees Celsius. A dotplot and numerical summaries of the midnight temperature readings for a 30-day period are shown below.

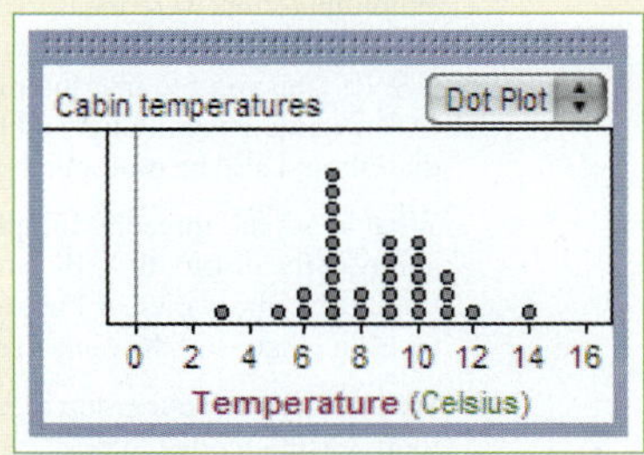


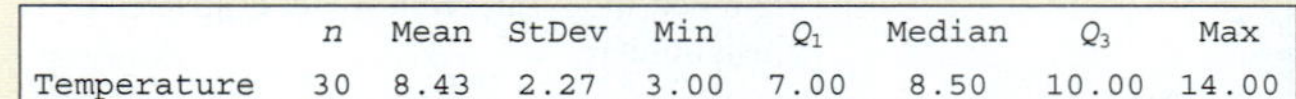

| | $n$ | Mean | StDev | Min | $Q_1$ | Median | $Q_3$ | Max |
|---|---|---|---|---|---|---|---|---|
| Temperature | 30 | 8.43 | 2.27 | 3.00 | 7.00 | 8.50 | 10.00 | 14.00 |

PROBLEM: Use the fact that °F = (9/5)°C + 32 to help you answer the following questions.

(a) Find the mean temperature in degrees Fahrenheit. Does the thermostat setting seem accurate?

(b) Calculate the standard deviation of the temperature readings in degrees Fahrenheit. Interpret this value in context.

(c) The 93rd percentile of the temperature readings was 12°C. What is the 93rd percentile temperature in degrees Fahrenheit?

SOLUTION:

(a) To convert the temperature measurements from Celsius to Fahrenheit, we multiply each value by 9/5 and then add 32. Multiplying the observations by 9/5 also multiplies the mean by 9/5. Adding 32 to each observation increases the mean by 32. So the mean temperature in degrees Fahrenheit is (9/5)(8.43) + 32 = 47.17°F. The thermostat doesn't seem to be very accurate. It is set at 50°F, but the mean temperature over the 30-day period is about 47°F.

(b) Multiplying each observation by 9/5 multiplies the standard deviation by 9/5. However, adding 32 to each observation doesn't affect the spread. So the standard deviation of the temperature measurements in degrees Fahrenheit is (9/5)(2.27) = 4.09°F. This means that the typical distance of the temperature readings from the mean is about 4°F. That's a lot of variation!

(c) Both multiplying by a constant and adding a constant affect the value of the 93rd percentile. To find the 93rd percentile in degrees Fahrenheit, we multiply the 93rd percentile in degrees Celsius by 9/5 and then add 32: (9/5)(12) + 32 = 53.6°F.

**For Practice** *Try Exercise* **19**

Let's look at part (c) of the example more closely. The data value of 12°C is at the 93rd percentile of the distribution, meaning that 28 of the 30 temperature readings are less than 12°C. When we transform the data, 12°C becomes 53.6°F. The value of 53.6°F is at the 93rd percentile of the transformed distribution because 28 of the 30 temperature readings are less than 53.6°F. What have we learned? Adding (or subtracting) a constant does not change an individual data

## ALTERNATE EXAMPLE

*Taxicabs*

In 2010 taxicabs in New York City charged an initial fee of \$2.50 plus \$2 per mile. In equation form, *fare* = 2.50 + 2(*miles*). At the end of a month, a businessman collects all his taxicab receipts and calculates some numerical summaries. The mean fare he paid was \$15.45, with a standard deviation of \$10.20.

PROBLEM: What are the mean and standard deviation of the lengths of his cab rides in miles?

SOLUTION: Solving the equation *fare* = 2.50 + 2(*miles*) for *miles*, we get $miles = \frac{fare - 2.50}{2}$. Thus, to find the mean number of miles, subtract 2.50 from the mean fare to get 12.95 and divide by 2 to get 6.475 miles. To find the standard deviation, we divide the standard deviation of the fares by 2 to get 5.10 miles. We don't subtract 2.50 because subtracting a constant from each observation does not change the spread of a distribution.

## Teaching Tip

Likewise, if you are 2 standard deviations below the mean temperature in degrees Celsius, you will be 2 standard deviations below the mean temperature in degrees Fahrenheit.

value's location within a distribution. Neither does multiplying or dividing by a positive constant.

**THINK ABOUT IT**

**Connecting transformations and z-scores:** What does all this transformation business have to do with $z$-scores? To standardize an observation, you subtract the mean of the distribution and then divide by the standard deviation. What if we standardized *every* observation in a distribution?

Returning to Mr. Pryor's statistics test scores, we recall that the distribution was roughly symmetric with a mean of 80 and a standard deviation of 6.07. To convert the entire class's test results to $z$-scores, we would subtract 80 from each observation and then divide by 6.07. What effect would these transformations have on the shape, center, and spread of the distribution?

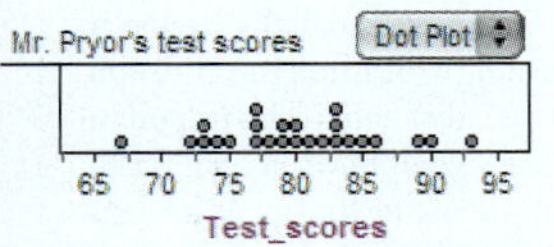


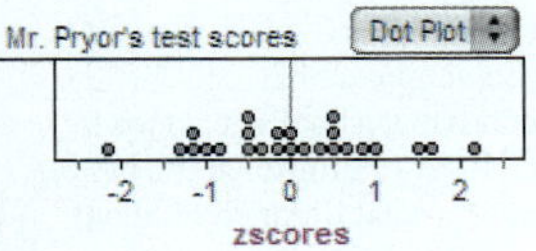


- **Shape:** The shape of the distribution of $z$-scores would be the same as the shape of the original distribution of test scores. Neither subtracting a constant nor dividing by a constant would change the shape of the graph. The dotplots confirm that the combination of these two transformations does not affect the shape.
- **Center:** Subtracting 80 from each data value would also reduce the mean by 80. Because the mean of the original distribution was 80, the mean of the transformed data would be 0. Dividing each of these new data values by 6.07 would also divide the mean by 6.07. But because the mean is now 0, dividing by 6.07 would leave the mean at 0. That is, the mean of the $z$-score distribution would be 0.
- **Spread:** The spread of the distribution would not be affected by subtracting 80 from each observation. However, dividing each data value by 6.07 would also divide our common measures of spread by 6.07. The standard deviation of the distribution of $z$-scores would therefore be $6.07/6.07 = 1$.

This is a result worth noting! If you start with *any* set of quantitative data and convert the values to standardized scores ($z$-scores), the transformed data set will have a mean of 0 and a standard deviation of 1. The shape of the two distributions will be the same. We will use this result to our advantage in Section 2.2.

The Minitab computer output below confirms the result: *If we standardize every observation in a distribution, the resulting set of z-scores has mean 0 and standard deviation 1.*

**Descriptive Statistics: Test scores, z-scores**

| Variable | $n$ | Mean | StDev | Minimum | $Q_1$ | Median | $Q_3$ | Maximum |
|---|---|---|---|---|---|---|---|---|
| Test scores | 25 | 80.00 | 6.07 | 67.00 | 76.00 | 80.00 | 83.50 | 93.00 |
| z-scores | 25 | 0.00 | 1.00 | −2.14 | −0.66 | 0.00 | 0.58 | 2.14 |

Many other types of transformations can be very useful in analyzing data. We have only studied what happens when you transform data through addition, subtraction, multiplication, or division.

## CHECK YOUR UNDERSTANDING

The figure on the next page shows a dotplot of the height distribution for Mrs. Navard's class, along with summary statistics from computer output.

1. Suppose that you convert the class's heights from inches to centimeters (1 inch = 2.54 cm). Describe the effect this will have on the shape, center, and spread of the distribution.

### Teaching Tip

In Chapter 12, we will learn about transforming nonlinear data using logarithms and other mathematical operations.

### ANSWERS TO CYU

**1.** Shape will not change. However, it will multiply the center (mean, median) and spread (range, *IQR*, standard deviation) by 2.54.

## ✓ ANSWERS TO CYU

**2.** Shape and spread will not change. It will, however, add 6 inches to the center (mean, median).

**3.** Shape will not change. However, it will change the mean to 0 and the standard deviation to 1.

## Answers to Data Exploration

**1.** The shape and spread of the distribution won't change at all. However, the center of the distribution (along with other measures of location) will be 24,800 higher. So the new mean will be $27.75 + 24{,}800 = 24{,}827.75$ nanoseconds (24,826.21 with the outliers).

**2.** The shape of the distribution won't change at all. However, both the measures of location and the measures of spread will be divided by $10^9$. So the new mean will be

$$\frac{24{,}827.75}{10^9} = 0.00002482775 \text{ seconds,}$$

and the new standard deviation will be

$$\frac{5.08}{10^9} = 0.00000000508 \text{ seconds}$$

(0.00002482621 seconds and 0.00000001075 seconds with the outliers).

**3.** If distance = (rate)(time), then the speed of light = (distance)/(time). Ignoring the two low outliers, Newcomb's estimate for the speed of light is *speed* =

$$\frac{7400}{0.00002482775} = 290{,}053{,}589.2$$

meters/second. *Note:* The actual speed of light is approximately 299,792,000 meters per second.

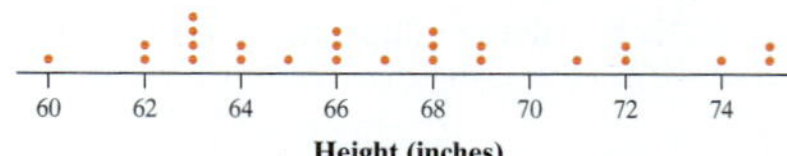

| Variable | $n$ | $\bar{x}$ | $s_x$ | Min | $Q_1$ | Med | $Q_3$ | Max |
|---|---|---|---|---|---|---|---|---|
| Height | 25 | 67 | 4.29 | 60 | 63 | 66 | 69 | 75 |

2. If Mrs. Navard had the entire class stand on a 6-inch-high platform and then had the students measure the distance from the top of their heads to the ground, how would the shape, center, and spread of this distribution compare with the original height distribution?

3. Now suppose that you convert the class's heights to $z$-scores. What would be the shape, center, and spread of this distribution? Explain.

## DATA EXPLORATION The speed of light

Light travels fast, but it is not transmitted instantly. Light takes over a second to reach us from the moon and over 10 billion years to reach us from the most distant objects in the universe. Because radio waves and radar also travel at the speed of light, having an accurate value for that speed is important in communicating with astronauts and orbiting satellites. An accurate value for the speed of light is also important to computer designers because electrical signals travel at light speed.

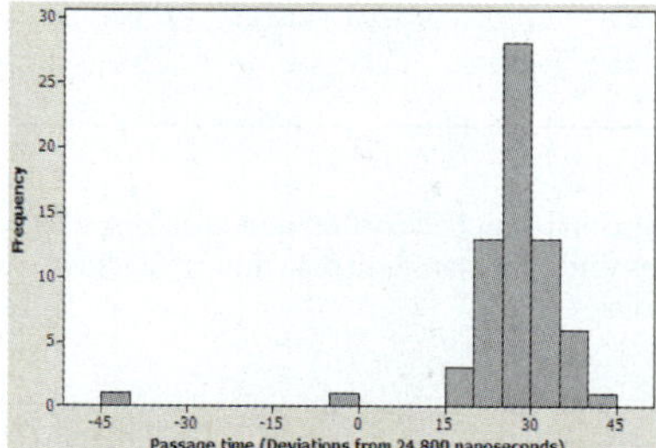

The first reasonably accurate measurements of the speed of light were made more than a hundred years ago by A. A. Michelson and Simon Newcomb. The table below contains 66 measurements made by Newcomb between July and September 1882.[3]

Newcomb measured the time in seconds that a light signal took to pass from his laboratory on the Potomac River to a mirror at the base of the Washington Monument and back, a total distance of about 7400 meters. Newcomb's first measurement of the passage time of light was 0.000024828 second, or 24,828 nanoseconds. (There are $10^9$ nanoseconds in a second.) The entries in the table record only the deviations from 24,800 nanoseconds.

| | | | | | | | | | | | | | |
|---|---|---|---|---|---|---|---|---|---|---|---|---|---|
| 28 | 26 | 33 | 24 | 34 | −44 | 27 | 16 | 40 | −2 | 29 | 22 | 24 | 21 |
| 25 | 30 | 23 | 29 | 31 | 19 | 24 | 20 | 36 | 32 | 36 | 28 | 25 | 21 |
| 28 | 29 | 37 | 25 | 28 | 26 | 30 | 32 | 36 | 26 | 30 | 22 | 36 | 23 |
| 27 | 27 | 28 | 27 | 31 | 27 | 26 | 33 | 26 | 32 | 32 | 24 | 39 | 28 |
| 24 | 25 | 32 | 25 | 29 | 27 | 28 | 29 | 16 | 23 | | | | |

Passage time (Deviations from 24,800 nanoseconds)

Descriptive Statistics: Passage time

| Variable | $n$ | Mean | Stdev | Min | $Q_1$ | Med | $Q_3$ | Max |
|---|---|---|---|---|---|---|---|---|
| P.Time | 66 | 26.21 | 10.75 | −44 | 24 | 27 | 31 | 40 |
| P.Time* | 64 | 27.75 | 5.08 | 16 | 24.5 | 27.5 | 31 | 40 |

The figure provides a histogram and numerical summaries (computed with and without the two outliers) from Minitab for these data.

1. We could convert the passage time measurements to nanoseconds by adding 24,800 to each of the data values in the table. What effect would this have on the shape, center, and spread of the distribution? Be specific.

2. After performing the transformation to nanoseconds, we could convert the measurements from nanoseconds to seconds by dividing each value by $10^9$. What effect would this have on the shape, center, and spread of the distribution? Be specific.

3. Use the information provided to estimate the speed of light in meters per second. Be prepared to justify the method you used.

## Section 2.1 Summary

- Two ways of describing an individual's location within a distribution are **percentiles** and **z-scores.** An observation's percentile is the percent of the distribution that is below the value of that observation. To standardize any observation $x$, subtract the mean of the distribution and then divide the difference by the standard deviation. The resulting $z$-score

$$z = \frac{x - \text{mean}}{\text{standard deviation}}$$

says how many standard deviations $x$ lies above or below the distribution mean. We can also use percentiles and $z$-scores to compare the location of individuals in different distributions.
- A **cumulative relative frequency graph** allows us to examine location within a distribution. Cumulative relative frequency graphs begin by grouping the observations into equal-width classes (much like the process of making a histogram). The completed graph shows the accumulating percent of observations as you move through the classes in increasing order.
- It is common to **transform data**, especially when changing units of measurement. When you add a constant $a$ to all the values in a data set, measures of center (median and mean) and location (quartiles and percentiles) increase by $a$. Measures of spread do not change. When you multiply all the values in a data set by a positive constant $b$, measures of center, location, and spread are multiplied by $b$. Neither of these transformations changes the shape of the distribution.

## Section 2.1 Exercises

pg 86
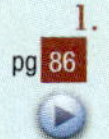

1. **Shoes** How many pairs of shoes do students have? Do girls have more shoes than boys? Here are data from a random sample of 20 female and 20 male students at a large high school:

| | | | | | | | | | | |
|---|---|---|---|---|---|---|---|---|---|---|
| **Female:** | 50 | 26 | 26 | 31 | 57 | 19 | 24 | 22 | 23 | 38 |
| | 13 | 50 | 13 | 34 | 23 | 30 | 49 | 13 | 15 | 51 |
| **Male:** | 14 | 7 | 6 | 5 | 12 | 38 | 8 | 7 | 10 | 10 |
| | 10 | 11 | 4 | 5 | 22 | 7 | 5 | 10 | 35 | 7 |

(a) Find and interpret the percentile in the female distribution for the girl with 22 pairs of shoes.

(b) Find and interpret the percentile in the male distribution for the boy with 22 pairs of shoes.

(c) Who is more unusual: the girl with 22 pairs of shoes or the boy with 22 pairs of shoes? Explain.

2. **Old folks** Here is a stemplot of the percents of residents aged 65 and older in the 50 states:

```
 7 | 0
 8 | 8
 9 | 8
10 | 019
11 | 16777
12 | 01122456778999
13 | 0001223344455689
14 | 023568
15 | 24
16 | 9
```

Key: 15|2 means 15.2% of this state's residents are 65 or older

(a) Find and interpret the percentile for Colorado, where 10.1% of the residents are aged 65 and older.

(b) Find and interpret the percentile for Rhode Island, where 13.9% of the residents are aged 65 and older.

(c) Which of these two states is more unusual? Explain.

### TRM Section 2.1 Quizzes

There are three different quizzes available for this section in the Teacher's Resource Materials. Versions A and B are roughly parallel. Find them by clicking on the link in the new TE-book, opening the TRCD, or logging onto the book's companion Web site www.whfreeman.com/tps5e (teacher log-in required).

### Teaching Tip

Make sure to point out the two icons next to Exercise 1. The red square reminds students that the example on page 86 is very similar to this exercise. The "play" icon reminds students that there is a video solution available in the e-Book or at www.whfreeman.com/tps5e.

### TRM Full Solutions to Section 2.1 Exercises

### Answers to Section 2.1 Exercises

**2.1** **(a)** She is at the 25th percentile, meaning that 25% of the girls had fewer pairs of shoes than she did. **(b)** He is at the 85th percentile, meaning 85% of the boys had fewer pairs of shoes than he did. **(c)** The boy is more unusual because only 15% of the boys have as many or more than he has. The girl has a value that is closer to the center of the distribution.

**2.2** **(a)** Colorado is at the 8th percentile, meaning 8% of states have a smaller percent of residents aged 65 and older. **(b)** Rhode Island is at the 80th percentile, meaning 80% of states have a smaller percent of residents aged 65 and older. **(c)** Colorado is the more unusual state, having only 8% of the states with smaller percentages of residents aged 65 and older. Rhode Island has a value that is closer to the center of the distribution.

3. **Math test** Josh just got the results of the statewide Algebra 2 test: his score is at the 60th percentile. When Josh gets home, he tells his parents that he got 60 percent of the questions correct on the state test. Explain what's wrong with Josh's interpretation.

4. **Blood pressure** Larry came home very excited after a visit to his doctor. He announced proudly to his wife, "My doctor says my blood pressure is at the 90th percentile among men like me. That means I'm better off than about 90% of similar men." How should his wife, who is a statistician, respond to Larry's statement?

5. **Growth charts** We used an online growth chart to find percentiles for the height and weight of a 16-year-old girl who is 66 inches tall and weighs 118 pounds. According to the chart, this girl is at the 48th percentile for weight and the 78th percentile for height. Explain what these values mean in plain English.

6. **Run fast** Peter is a star runner on the track team. In the league championship meet, Peter records a time that would fall at the 80th percentile of all his race times that season. But his performance places him at the 50th percentile in the league championship meet. Explain how this is possible. (Remember that lower times are better in this case!)

*Exercises 7 and 8 involve a new type of graph called a* ***percentile plot****. Each point gives the value of the variable being measured and the corresponding percentile for one individual in the data set.*

7. **Text me** The percentile plot below shows the distribution of text messages sent and received in a two-day period by a random sample of 16 females from a large high school.

(a) Describe the student represented by the highlighted point.

(b) Use the graph to estimate the median number of texts. Explain your method.

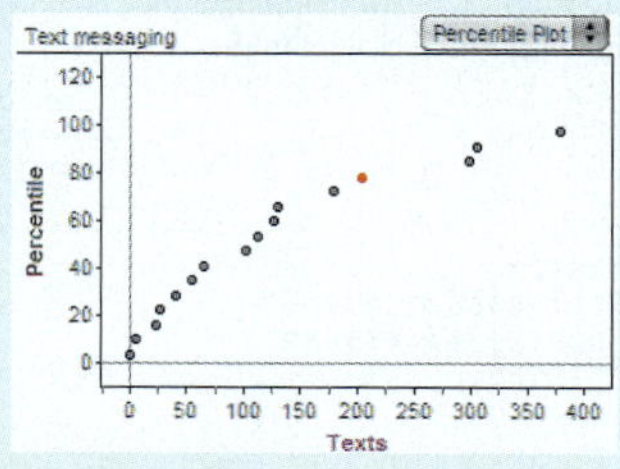


8. **Foreign-born residents** The following percentile plot shows the distribution of the percent of foreign-born residents in the 50 states.

(a) The highlighted point is for Maryland. Describe what the graph tells you about this state.

(b) Use the graph to estimate the 30th percentile of the distribution. Explain your method.

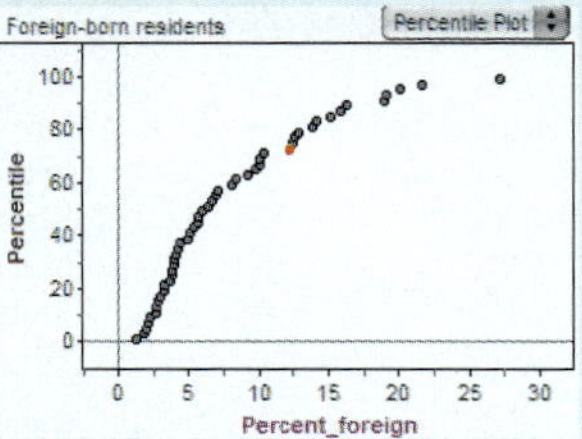


9. **Shopping spree** The figure below is a cumulative relative frequency graph of the amount spent by 50 consecutive grocery shoppers at a store.

pg 88

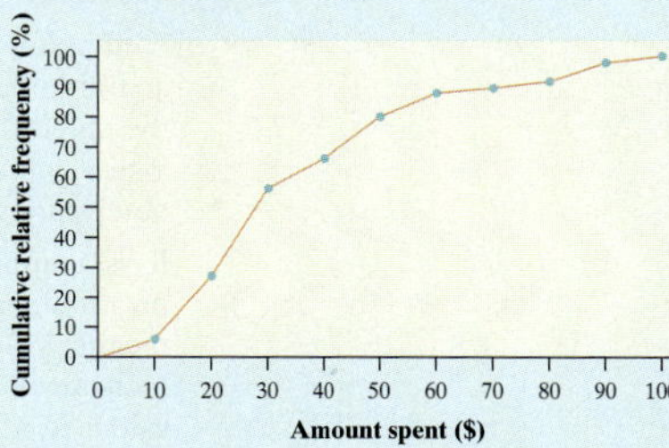


(a) Estimate the interquartile range of this distribution. Show your method.

(b) What is the percentile for the shopper who spent $19.50?

(c) Draw the histogram that corresponds to this graph.

10. **Light it up!** The graph below is a cumulative relative frequency graph showing the lifetimes (in hours) of 200 lamps.[4]

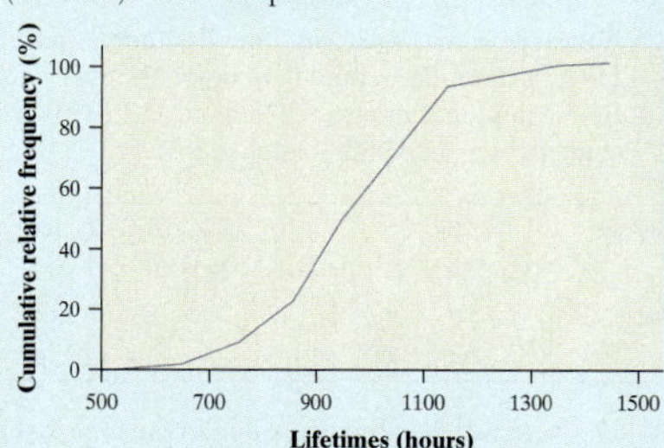


(a) Estimate the 60th percentile of this distribution. Show your method.

(b) What is the percentile for a lamp that lasted 900 hours?

(c) Draw a histogram that corresponds to this graph.

**2.3** A percentile only describes the relative location of a value in a distribution. Scoring at the 60th percentile means that Josh's score is better than 60% of the students taking this test. His percentage of correct answers could be greater than 60% or less than 60%, depending on the difficulty of the test.

**2.4** About 90% of men similar to Larry have lower (better) blood pressures. Larry is only better off than 10% of men.

**2.5** The girl weighs more than 48% of girls her age, but is taller than 78% of the girls her age.

**2.6** Peter's time was longer (slower) than 80% of his previous race times that season, but it was longer (slower) than only 50% of the racers at the league championship meet. Because this time was relatively slow for Peter but at the median for the runners in the league championship, Peter must be a good runner.

**2.7** **(a)** The student sent about 205 text messages in the 2-day period and sent more texts than about 78% of the students in the sample. **(b)** Locate 50% on the $y$-axis, read over to the points, and then go down to the $x$-axis. The median is approximately 115 text messages.

**2.8** **(a)** Maryland has about 12% foreign-born residents, which is greater than about 72% of the other states. **(b)** Locate 30% on the $y$-axis, read over to the points, and go down to the $x$-axis. The 30th percentile is approximately 4.5% foreign-born.

**2.9** **(a)** $IQR \approx \$46 - \$19 = \$27$ **(b)** About the 26th percentile. **(c)** The histogram is below.

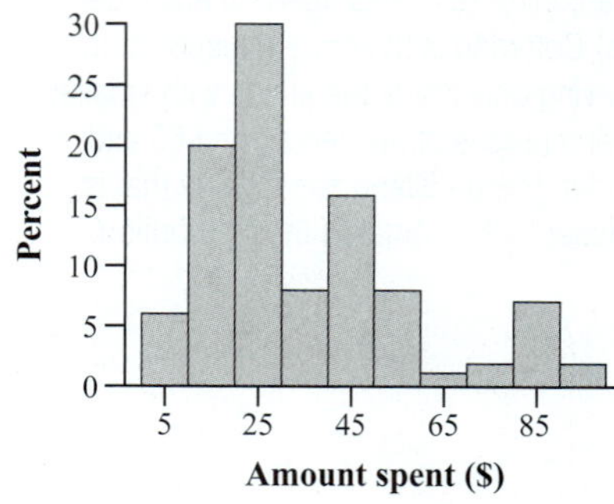


**2.10** **(a)** About 1000 hours. **(b)** About the 35th percentile. **(c)** The histogram is below.

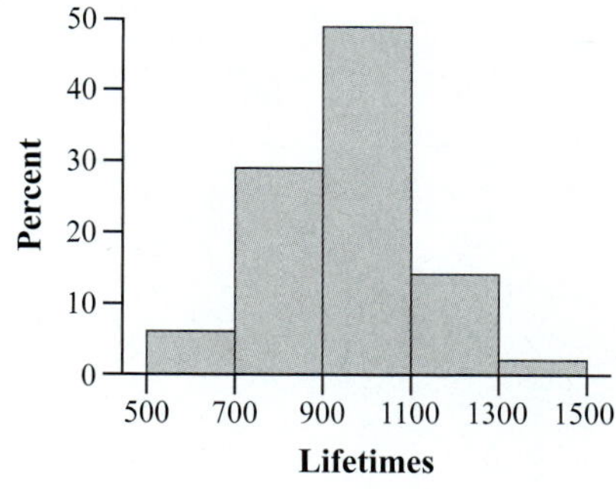

**11. SAT versus ACT** Eleanor scores 680 on the SAT Mathematics test. The distribution of SAT scores is symmetric and single-peaked, with mean 500 and standard deviation 100. Gerald takes the American College Testing (ACT) Mathematics test and scores 27. ACT scores also follow a symmetric, single-peaked distribution—but with mean 18 and standard deviation 6. Find the standardized scores for both students. Assuming that both tests measure the same kind of ability, who has the higher score?

pg 91
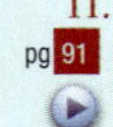

**12. Comparing batting averages** Three landmarks of baseball achievement are Ty Cobb's batting average of 0.420 in 1911, Ted Williams's 0.406 in 1941, and George Brett's 0.390 in 1980. These batting averages cannot be compared directly because the distribution of major league batting averages has changed over the years. The distributions are quite symmetric, except for outliers such as Cobb, Williams, and Brett. While the mean batting average has been held roughly constant by rule changes and the balance between hitting and pitching, the standard deviation has dropped over time. Here are the facts:

| Decade | Mean | Standard deviation |
|---|---|---|
| 1910s | 0.266 | 0.0371 |
| 1940s | 0.267 | 0.0326 |
| 1970s | 0.261 | 0.0317 |

Find the standardized scores for Cobb, Williams, and Brett. Who was the best hitter?[5]

**13. Measuring bone density** Individuals with low bone density have a high risk of broken bones (fractures). Physicians who are concerned about low bone density (osteoporosis) in patients can refer them for specialized testing. Currently, the most common method for testing bone density is dual-energy X-ray absorptiometry (DEXA). A patient who undergoes a DEXA test usually gets bone density results in grams per square centimeter (g/cm$^2$) and in standardized units.

Judy, who is 25 years old, has her bone density measured using DEXA. Her results indicate a bone density in the hip of 948 g/cm$^2$ and a standardized score of $z = -1.45$. In the reference population of 25-year-old women like Judy, the mean bone density in the hip is 956 g/cm$^2$.[6]

(a) Judy has not taken a statistics class in a few years. Explain to her in simple language what the standardized score tells her about her bone density.

(b) Use the information provided to calculate the standard deviation of bone density in the reference population.

**14. Comparing bone density** Refer to the previous exercise. One of Judy's friends, Mary, has the bone density in her hip measured using DEXA. Mary is 35 years old. Her bone density is also reported as 948 g/cm$^2$, but her standardized score is $z = 0.50$. The mean bone density in the hip for the reference population of 35-year-old women is 944 grams/cm$^2$.

(a) Whose bones are healthier—Judy's or Mary's? Justify your answer.

(b) Calculate the standard deviation of the bone density in Mary's reference population. How does this compare with your answer to Exercise 13(b)? Are you surprised?

*Exercises 15 and 16 refer to the dotplot and summary statistics of salaries for players on the World Champion 2008 Philadelphia Phillies baseball team.*[7]

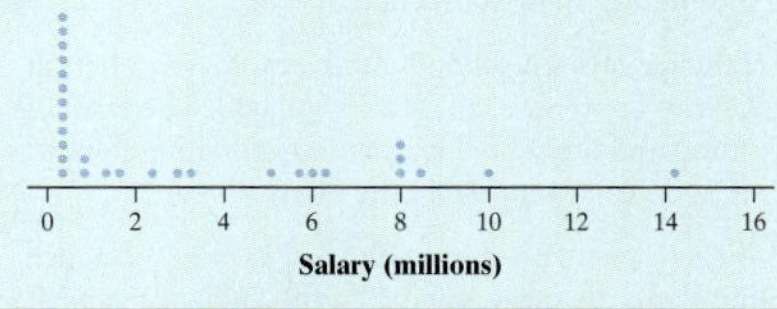


| Variable | $n$ | Mean | Std. dev. | Min | $Q_1$ | Med | $Q_3$ | Max |
|---|---|---|---|---|---|---|---|---|
| Salary | 29 | 3388617 | 3767484 | 390000 | 440000 | 1400000 | 6000000 | 14250000 |

**15. Baseball salaries** Brad Lidge played a crucial role as the Phillies' "closer," pitching the end of many games throughout the season. Lidge's salary for the 2008 season was $6,350,000.

pg 90

(a) Find the percentile corresponding to Lidge's salary. Explain what this value means.

(b) Find the $z$-score corresponding to Lidge's salary. Explain what this value means.

**16. Baseball salaries** Did Ryan Madson, who was paid $1,400,000, have a high salary or a low salary compared with the rest of the team? Justify your answer by calculating and interpreting Madson's percentile and $z$-score.

17. The scores on Ms. Martin's statistics quiz had a mean of 12 and a standard deviation of 3. Ms. Martin wants to transform the scores to have a mean of 75 and a standard deviation of 12. What transformations should she apply to each test score? Explain.

18. Mr. Olsen uses an unusual grading system in his class. After each test, he transforms the scores to have a mean of 0 and a standard deviation of 1. Mr. Olsen then assigns a grade to each student based on the transformed score. On his most recent test, the class's scores had a mean of 68 and a standard deviation of 15. What transformations should he apply to each test score? Explain.

**2.11** Eleanor. Her standardized score ($z = 1.8$) is higher than Gerald's ($z = 1.5$).

**2.12** Cobb: $z = 4.15$, Williams: $z = 4.26$, Brett: $z = 4.07$. All three hitters were at least 4 standard deviations above their peers, but Williams's $z$-score is the highest.

**2.13 (a)** Your bone density is far below average—about 1.5 times farther below average than a typical below-average density. **(b)** Solving $-1.45 = \dfrac{948 - 956}{\sigma}$ gives $\sigma = 5.52$ g/cm$^2$.

**2.14 (a)** Because Mary's $z$-score (0.5) is higher than Judy's ($z = -1.45$), Mary's bones are healthier when comparisons are made to other women in their age groups. **(b)** Solving $0.5 = \dfrac{948 - 944}{\sigma}$ gives $\sigma = 8$ g/cm$^2$. It isn't surprising that older women have a larger standard deviation because there is more time for their good or bad health habits to have an effect, creating a wider range of bone densities.

**2.15 (a)** He is at the 76th percentile, meaning his salary is higher than 76% of his teammates. **(b)** $z = 0.79$. Lidge's salary was 0.79 standard deviations above the mean salary.

**2.16** Madson is at the 48th percentile and his $z$-score is $z = -0.53$. Madson had a typical salary compared to the rest of the team because it was around the 50th percentile, but his $z$-score was negative, indicating that his salary was below the mean, probably due to some very large salaries for other players.

**2.17** Multiply each score by 4 and add 27.

**2.18** Subtract 68 from each score (the mean) and divide the difference by 15 (the standard deviation).

**2.19** **(a)** mean = 87.188 inches and median = 87.5 inches. **(b)** The standard deviation (3.20 inches) and *IQR* (3.25 inches) do not change because adding a constant to each value in a distribution does not change the spread.

**2.20** **(a)** The mean and median salaries will each increase by $1000 because adding a constant to each value in a distribution increases measures of location by the same amount. **(b)** The extremes and quartiles will also each increase by $1000 because they are measures of location. The standard deviation will not change because adding a constant to each value in a distribution does not change the spread.

**2.21** **(a)** Mean = 5.77 feet and median = 5.79 feet. **(b)** Standard deviation = 0.267 feet and *IQR* = 0.271 feet.

**2.22** **(a)** The mean and median will each increase by 5% because multiplying each value in a distribution by a constant (1.05) multiplies measures of center by the same amount. **(b)** Yes. Multiplying each value in a distribution by a constant (1.05) multiplies measures of spread by the same amount.

**2.23** Mean $= \frac{9}{5}(25) + 32 = 77$ °F and standard deviation $= \frac{9}{5}(2) = 3.6$ °F.

**2.24** Mean = (3.2 − 0.2) (2.54) = 7.62 cm and standard deviation = 0.1 (2.54) = 0.254 cm.

**2.25** c

**2.26** a

**2.27** c

**2.28** b



19. **Tall or short?** Mr. Walker measures the heights (in inches) of the students in one of his classes. He uses a computer to calculate the following numerical summaries:

pg 95

| Mean | Std. dev. | Min | $Q_1$ | Med | $Q_3$ | Max |
|---|---|---|---|---|---|---|
| 69.188 | 3.20 | 61.5 | 67.75 | 69.5 | 71 | 74.5 |

Next, Mr. Walker has his entire class stand on their chairs, which are 18 inches off the ground. Then he measures the distance from the top of each student's head to the floor.

(a) Find the mean and median of these measurements. Show your work.

(b) Find the standard deviation and *IQR* of these measurements. Show your work.

20. **Teacher raises** A school system employs teachers at salaries between $28,000 and $60,000. The teachers' union and the school board are negotiating the form of next year's increase in the salary schedule.

(a) If every teacher is given a flat $1000 raise, what will this do to the mean salary? To the median salary? Explain your answers.

(b) What would a flat $1000 raise do to the extremes and quartiles of the salary distribution? To the standard deviation of teachers' salaries? Explain your answers.

21. **Tall or short?** Refer to Exercise 19. Mr. Walker converts his students' original heights from inches to feet.

(a) Find the mean and median of the students' heights in feet. Show your work.

(b) Find the standard deviation and *IQR* of the students' heights in feet. Show your work.

22. **Teacher raises** Refer to Exercise 20. If each teacher receives a 5% raise instead of a flat $1000 raise, the amount of the raise will vary from $1400 to $3000, depending on the present salary.

(a) What will this do to the mean salary? To the median salary? Explain your answers.

(b) Will a 5% raise increase the *IQR*? Will it increase the standard deviation? Explain your answers.

23. **Cool pool?** Coach Ferguson uses a thermometer to measure the temperature (in degrees Celsius) at 20 different locations in the school swimming pool. An analysis of the data yields a mean of 25°C and a standard deviation of 2°C. Find the mean and standard deviation of the temperature readings in degrees Fahrenheit (recall that °F = (9/5)°C + 32).

24. **Measure up** Clarence measures the diameter of each tennis ball in a bag with a standard ruler. Unfortunately, he uses the ruler incorrectly so that each of his measurements is 0.2 inches too large. Clarence's data had a mean of 3.2 inches and a standard deviation of 0.1 inches. Find the mean and standard deviation of the corrected measurements in centimeters (recall that 1 inch = 2.54 cm).

*Multiple choice: Select the best answer for Exercises 25 to 30.*

25. Jorge's score on Exam 1 in his statistics class was at the 64th percentile of the scores for all students. His score falls

(a) between the minimum and the first quartile.
(b) between the first quartile and the median.
(c) between the median and the third quartile.
(d) between the third quartile and the maximum.
(e) at the mean score for all students.

26. When Sam goes to a restaurant, he always tips the server $2 plus 10% of the cost of the meal. If Sam's distribution of meal costs has a mean of $9 and a standard deviation of $3, what are the mean and standard deviation of the distribution of his tips?

(a) $2.90, $0.30
(b) $2.90, $2.30
(c) $9.00, $3.00
(d) $11.00, $2.00
(e) $2.00, $0.90

27. Scores on the ACT college entrance exam follow a bell-shaped distribution with mean 18 and standard deviation 6. Wayne's standardized score on the ACT was −0.5. What was Wayne's actual ACT score?

(a) 5.5 (b) 12 (c) 15 (d) 17.5 (e) 21

28. George has an average bowling score of 180 and bowls in a league where the average for all bowlers is 150 and the standard deviation is 20. Bill has an average bowling score of 190 and bowls in a league where the average is 160 and the standard deviation is 15. Who ranks higher in his own league, George or Bill?

(a) Bill, because his 190 is higher than George's 180.
(b) Bill, because his standardized score is higher than George's.
(c) Bill and George have the same rank in their leagues, because both are 30 pins above the mean.
(d) George, because his standardized score is higher than Bill's.
(e) George, because the standard deviation of bowling scores is higher in his league.

*Exercises 29 and 30 refer to the following setting.* The number of absences during the fall semester was recorded for each student in a large elementary school. The distribution of absences is displayed in the following cumulative relative frequency graph.

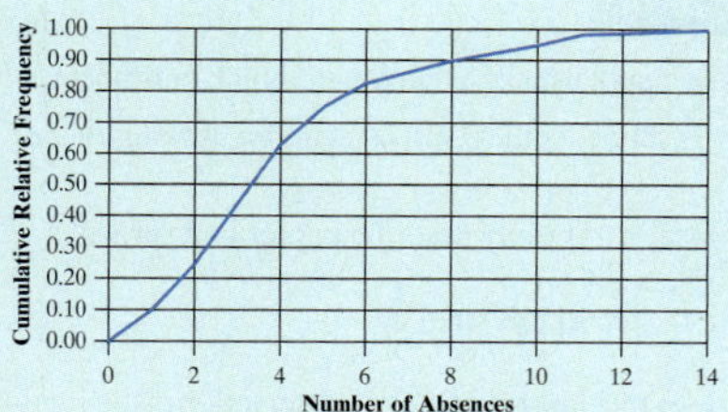


29. What is the interquartile range (*IQR*) for the distribution of absences?

(a) 1 (c) 3 (e) 14

(b) 2 (d) 5

30. If the distribution of absences was displayed in a histogram, what would be the best description of the histogram's shape?

(a) Symmetric

(b) Uniform

(c) Skewed left

(d) Skewed right

(e) Cannot be determined

*Exercises 31 and 32 refer to the following setting.* We used CensusAtSchool's Random Data Selector to choose a sample of 50 Canadian students who completed a survey in a recent year.

31. **Travel time** (1.2) The dotplot below displays data on students' responses to the question "How long does it usually take you to travel to school?" Describe the shape, center, and spread of the distribution. Are there any outliers?

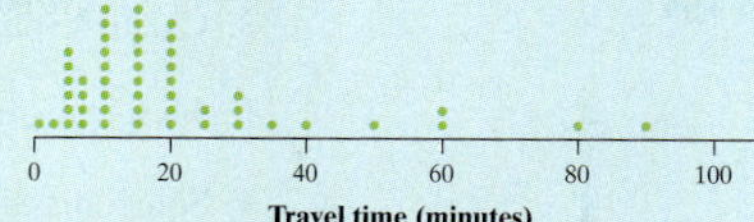


32. **Lefties** (1.1) Students were asked, "Are you right-handed, left-handed, or ambidextrous?" The responses are shown below (R = right-handed; L = left-handed; A = ambidextrous).

| | | | | | | | | | | | | | |
|---|---|---|---|---|---|---|---|---|---|---|---|---|---|
| R | R | R | R | R | R | R | R | R | R | R | L | R | R |
| R | R | R | R | R | R | R | R | R | R | R | R | R | A |
| R | R | R | R | A | R | R | L | R | R | R | R | L | A |
| R | R | R | R | R | R | R | R | | | | | | |

(a) Make an appropriate graph to display these data.

(b) Over 10,000 Canadian high school students took the CensusAtSchool survey that year. What percent of this population would you estimate is left-handed? Justify your answer.

**2.29** c

**2.30** d

**2.31** The distribution is skewed to the right with a center around 20 minutes and the range close to 90 minutes. The two largest values appear to be outliers.

**2.32** **(a)** A bar graph is given below.
**(b)** Because 3/50 = 6% of the sample was left-handed, our best estimate is 6% for the population percentage.

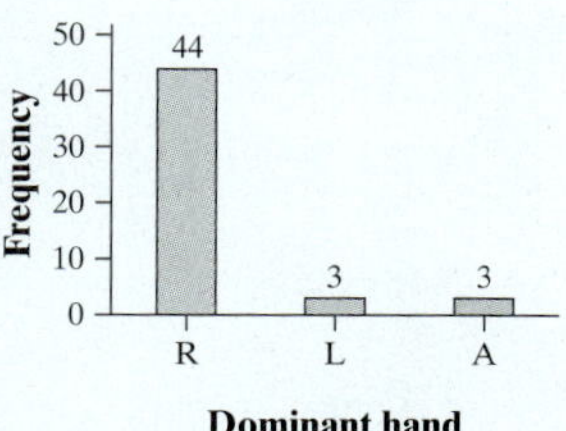


# 2.2 Density Curves and Normal Distributions

**WHAT YOU WILL LEARN** By the end of the section, you should be able to:

- Estimate the relative locations of the median and mean on a density curve.
- Use the 68–95–99.7 rule to estimate areas (proportions of values) in a Normal distribution.
- Use Table A or technology to find (i) the proportion of $z$-values in a specified interval, or (ii) a $z$-score from a percentile in the standard Normal distribution.
- Use Table A or technology to find (i) the proportion of values in a specified interval, or (ii) the value that corresponds to a given percentile in any Normal distribution.
- Determine whether a distribution of data is approximately Normal from graphical and numerical evidence.

**PD Section 2.2 Overview**

Watch the video overview of Section 2.2 (for teachers) by clicking on the link in the TE-book, opening the TRCD, or logging onto the book's companion Web site www.whfreeman.com/tps5e (teacher log-in required).

**TRM Section 2.2 Alternate Examples**

You can find the Alternate Examples for this section in Microsoft Word format by clicking on the link in the TE-book, opening the TRCD, or logging onto the book's companion Web site www.whfreeman.com/tps5e (teacher log-in required).

In Chapter 1, we developed a kit of graphical and numerical tools for describing distributions. Our work gave us a clear strategy for exploring data from a single quantitative variable.

**EXPLORING QUANTITATIVE DATA**

1. Always plot your data: make a graph, usually a dotplot, stemplot, or histogram.
2. Look for the overall pattern (shape, center, spread) and for striking departures such as outliers.
3. Calculate numerical summaries to briefly describe center and spread.

In this section, we add one more step to this strategy.

4. Sometimes the overall pattern of a large number of observations is so regular that we can describe it by a smooth curve.

## Density Curves

Figure 2.7 is a histogram of the scores of all 947 seventh-grade students in Gary, Indiana, on the vocabulary part of the Iowa Test of Basic Skills (ITBS).[8] Scores on this national test have a very regular distribution. The histogram is symmetric, and both tails fall off smoothly from a single center peak. There are no large gaps or obvious outliers. The smooth curve drawn through the tops of the histogram bars in Figure 2.7 is a good description of the overall pattern of the data.

**FIGURE 2.7** Histogram of the Iowa Test of Basic Skills (ITBS) vocabulary scores of all seventh-grade students in Gary, Indiana. The smooth curve shows the overall shape of the distribution.

### EXAMPLE Seventh-Grade Vocabulary Scores

#### *From histogram to density curve*

Our eyes respond to the areas of the bars in a histogram. The bar areas represent relative frequencies (proportions) of the observations. Figure 2.8(a) is a copy of Figure 2.7 with the leftmost bars shaded. The area of the shaded bars in Figure 2.8(a) represents the proportion of students with vocabulary scores less than 6.0. There are 287 such students, who make up the proportion 287/947 = 0.303 of all Gary seventh-graders. In other words, a score of 6.0 corresponds to about the 30th percentile.

The total area of the bars in the histogram is 100% (a proportion of 1), because all of the observations are represented. Now look at the curve drawn through the tops of the bars. In Figure 2.8(b), the area under the curve to the left of 6.0 is shaded. In moving from histogram bars to a smooth curve, we make a specific choice: adjust the scale of the graph so that *the total area under the curve is exactly 1*. Now

### Teaching Tip

In Chapter 3, we will continue using this strategy when analyzing relationships between quantitative variables. Instead of describing a distribution with a smooth curve, we will describe a linear relationship between two variables with a line.

### ALTERNATE EXAMPLE

#### *Batting averages*

The histogram below shows the distribution of batting average (proportion of hits) for the 432 Major League Baseball players with at least 100 plate appearances in a recent season. The smooth curve shows the overall shape of the distribution.

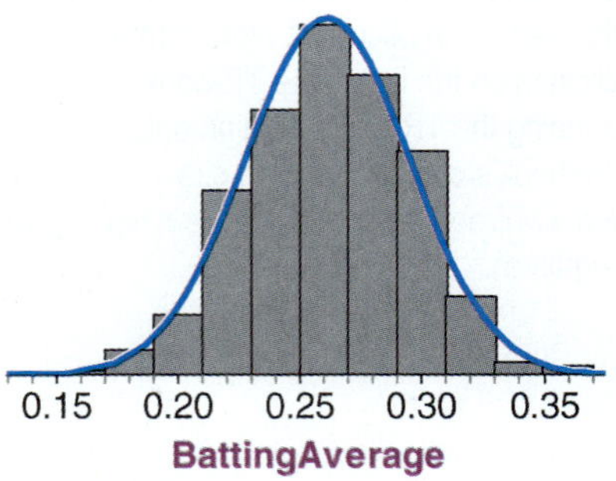


In the top graph at right, the bars in red represent the proportion of players who had batting averages of at least 0.270. There are 177 such players out of a total of 432, for a proportion of 0.410. In the bottom graph at right, the area under the curve to the right of 0.270 is shaded. This area is 0.391, only 0.019 away from the actual proportion of 0.410.

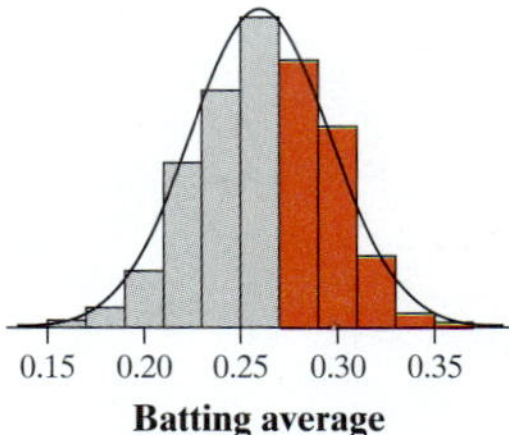


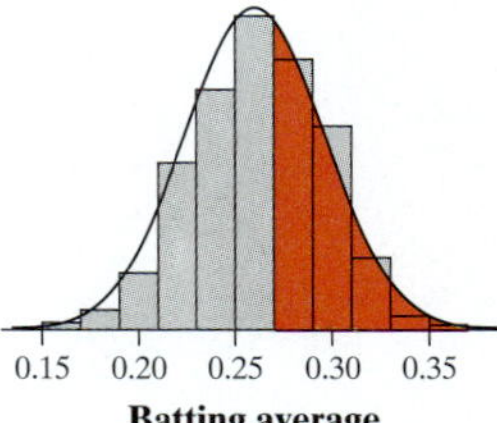

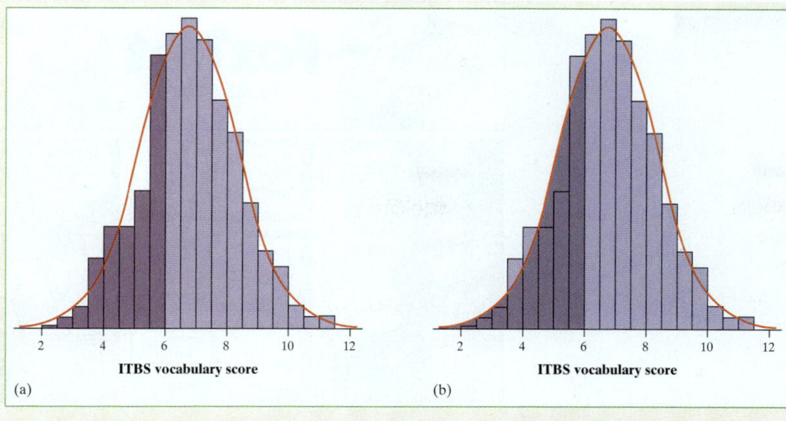


**FIGURE 2.8** (a) The proportion of scores less than or equal to 6.0 in the actual data is 0.303. (b) The proportion of scores less than or equal to 6.0 from the density curve is 0.293.

the total area represents all the observations, just like with the histogram. We can then interpret areas under the curve as proportions of the observations.

The shaded area under the curve in Figure 2.8(b) represents the proportion of students with scores lower than 6.0. This area is 0.293, only 0.010 away from the actual proportion 0.303. So our estimate based on the curve is that a score of 6.0 falls at about the 29th percentile. You can see that areas under the curve give good approximations to the actual distribution of the 947 test scores. In practice, it might be easier to use this curve to estimate relative frequencies than to determine the actual proportion of students by counting data values.

A curve like the one in the previous example is called a **density curve.**

**DEFINITION: Density curve**

A **density curve** is a curve that

- is always on or above the horizontal axis, and
- has area exactly 1 underneath it.

A density curve describes the overall pattern of a distribution. The area under the curve and above any interval of values on the horizontal axis is the proportion of all observations that fall in that interval.

Density curves, like distributions, come in many shapes. A density curve is often a good description of the overall pattern of a distribution. Outliers, which are departures from the overall pattern, are not described by the curve. *No set of real data is exactly described by a density curve. The curve is an approximation that is easy to use and accurate enough for practical use.*

caution !

## Teaching Tip

The most commonly used density curve is the Normal curve, which we will learn about in a few pages. We will use other density curves, such as *t* curves and $\chi^2$ curves, to model the distributions of the *t* and $\chi^2$ test statistics in later chapters.

## Describing Density Curves

Our measures of center and spread apply to density curves as well as to actual sets of observations. Areas under a density curve represent proportions of the total number of observations. The median of a data set is the point with half the observations on either side. So the **median of a density curve** is the "equal-areas point," the point with half the area under the curve to its left and the remaining half of the area to its right.

Because density curves are idealized patterns, a symmetric density curve is exactly symmetric. The median of a symmetric density curve is therefore at its center. Figure 2.9(a) shows a symmetric density curve with the median marked. It isn't so easy to spot the equal-areas point on a skewed curve. There are mathematical ways of finding the median for any density curve. That's how we marked the median on the skewed curve in Figure 2.9(b).

What about the mean? The mean of a set of observations is their arithmetic average. As we saw in Chapter 1, the mean is also the "balance point" of a distribution. That is, if we think of the observations as weights strung out along a thin rod, the mean is the point at which the rod would balance. This fact is also true of

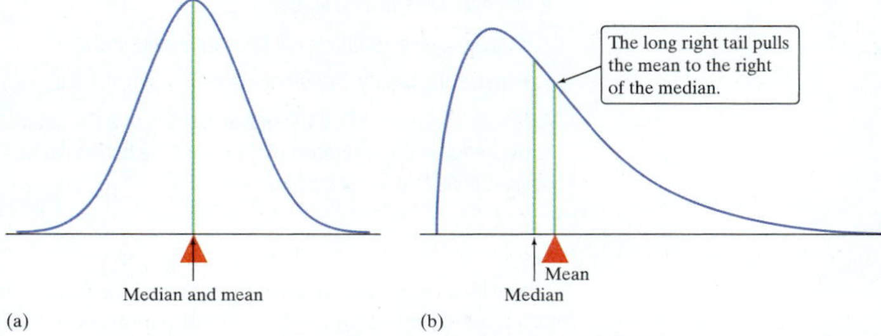


**FIGURE 2.9** (a) The median and mean of a symmetric density curve both lie at the center of symmetry. (b) The median and mean of a right-skewed density curve. The mean is pulled away from the median toward the long tail.

### Teaching Tip

Whenever your students look at a density curve, tell them to imagine that it is tracing the distribution of dots on a dotplot. Dotplots are much less abstract than density curves. This should help them understand what the density curve represents and how to approximate the median and mean.

### Teaching Tip

Remind your students about the balancing-point Activity with the pennies from Chapter 1 (page 51).

### Teaching Tip

If some of your students have taken calculus, ask them how they can find the median if they know the equation of the density curve. To do this, solve the following for $M$, assuming $f(x)$ is the equation of the density curve: $\int_{-\infty}^{M} f(x)\,dx = 0.5$. Likewise, they can find the mean of the density curve by evaluating $\int_{-\infty}^{\infty} x \cdot f(x)\,dx$.

### Teaching Tip

This would be a good time to quickly review that the median is much more resistant to outliers and skewness than the mean.

density curves. The **mean of a density curve** is the point at which the curve would balance if made of solid material. Figure 2.10 illustrates this fact about the mean.

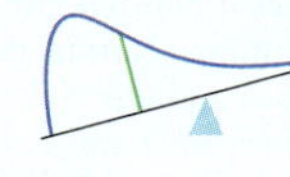
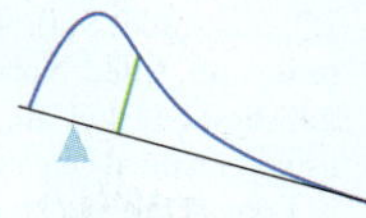
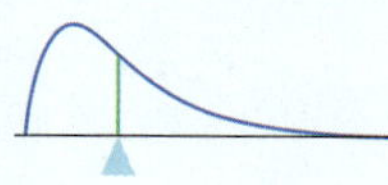

**FIGURE 2.10** The mean is the balance point of a density curve.

A symmetric curve balances at its center because the two sides are identical. *The mean and median of a symmetric density curve are equal,* as in Figure 2.9(a). We know that the mean of a skewed distribution is pulled toward the long tail. Figure 2.9(b) shows how the mean of a skewed density curve is pulled toward the long tail more than the median is.

**DISTINGUISHING THE MEDIAN AND MEAN OF A DENSITY CURVE**

The **median** of a density curve is the equal-areas point, the point that divides the area under the curve in half.

The **mean** of a density curve is the balance point, at which the curve would balance if made of solid material.

The median and mean are the same for a symmetric density curve. They both lie at the center of the curve. The mean of a skewed curve is pulled away from the median in the direction of the long tail.

Because a density curve is an idealized description of a distribution of data, we distinguish between the mean and standard deviation of the density curve and the mean $\bar{x}$ and standard deviation $s_x$ computed from the actual observations. The usual notation for the mean of a density curve is $\mu$ (the Greek letter mu). We write the standard deviation of a density curve as $\sigma$ (the Greek letter sigma). We can roughly locate the mean $\mu$ of any density curve by eye, as the balance point. There is no easy way to locate the standard deviation $\sigma$ by eye for density curves in general.

You probably noticed that we used the same notation for the mean and standard deviation of a population in Chapter 1, $\mu$ and $\sigma$, as we do here for the mean and standard deviation of a density curve.

## CHECK YOUR UNDERSTANDING

Use the figure shown to answer the following questions.

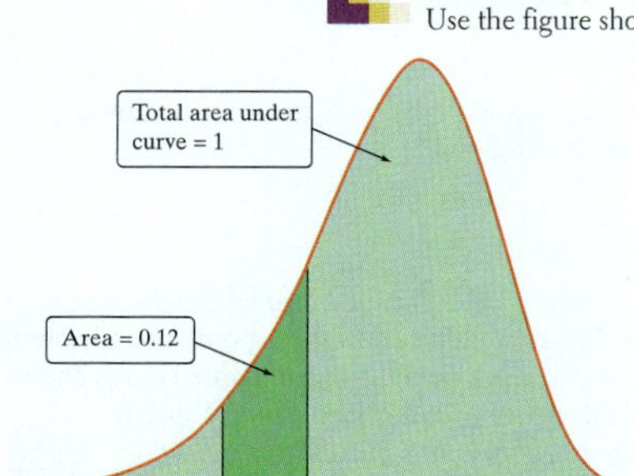


1. Explain why this is a legitimate density curve.
2. About what proportion of observations lie between 7 and 8?
3. Trace the density curve onto your paper. Mark the approximate location of the median.
4. Now mark the approximate location of the mean. Explain why the mean and median have the relationship that they do in this case.

### Teaching Tip

Have students look back at the chapter opener about Sudoku times (page 83) to see another example of a density curve. Ask your students to describe the relationship between the mean completion time and the median completion time. The mean will be higher because the distribution of times is skewed to the right.

### Teaching Tip

In general, we will use Greek letters to represent the "true" value of a numerical summary for a population (e.g., $\mu$ = population mean) and non-Greek letters to represent the *estimated* value of the numerical summary for a sample (e.g., $\bar{x}$ = sample mean).

### ANSWERS TO CYU

1. It is legitimate because it is positive everywhere and it has total area under the curve = 1.
2. 12%
3. Point A in the graph below is the approximate median. About half of the area is to the left of A and half of the area is to the right of A.
4. Point B in the graph below is the approximate mean (balance point). The mean is less than the median in this case because the distribution is skewed to the left.

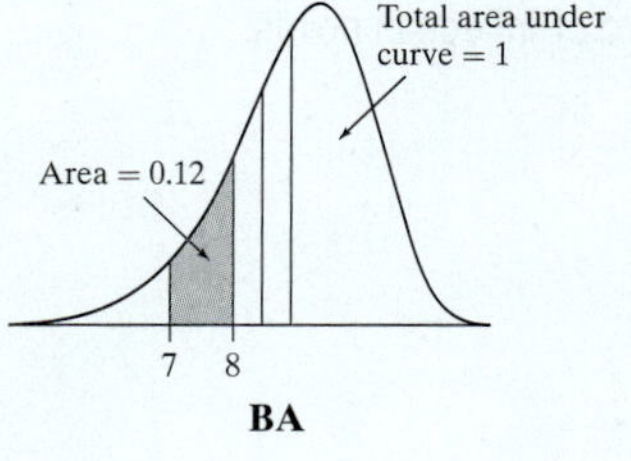

## Normal Distributions

One particularly important class of density curves has already appeared in Figures 2.7, 2.8, and 2.9(a). They are called **Normal curves.** The distributions they describe are called **Normal distributions.** Normal distributions play a large role in statistics, but they are rather special and not at all "normal" in the sense of being usual or typical. We capitalize Normal to remind you that these curves are special.

Look at the two Normal curves in Figure 2.11. They illustrate several important facts:

- All Normal curves have the same overall shape: symmetric, single-peaked, and bell-shaped.
- Any specific Normal curve is completely described by giving its mean $\mu$ and its standard deviation $\sigma$.

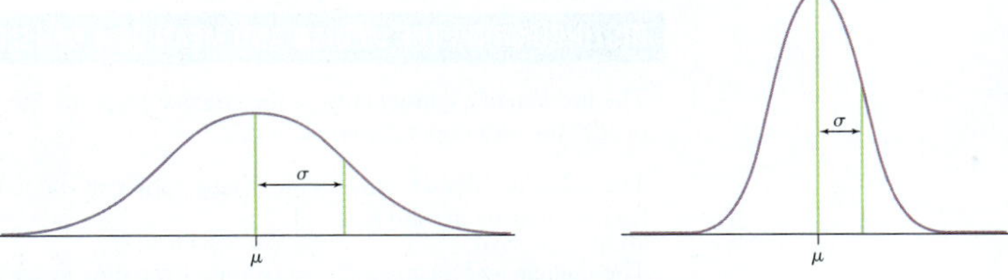


**FIGURE 2.11** Two Normal curves, showing the mean $\mu$ and standard deviation $\sigma$.

- The mean is located at the center of the symmetric curve and is the same as the median. Changing $\mu$ without changing $\sigma$ moves the Normal curve along the horizontal axis without changing its spread.
- The standard deviation $\sigma$ controls the spread of a Normal curve. Curves with larger standard deviations are more spread out.

The standard deviation $\sigma$ is the natural measure of spread for Normal distributions. Not only do $\mu$ and $\sigma$ completely determine the shape of a Normal curve, but we can locate $\sigma$ by eye on a Normal curve. Here's how. Imagine that you are skiing down a mountain that has the shape of a Normal curve. At first, you descend at an ever-steeper angle as you go out from the peak:

Fortunately, before you find yourself going straight down, the slope begins to grow flatter rather than steeper as you go out and down:

*The points at which this change of curvature takes place are located at a distance $\sigma$ on either side of the mean $\mu$.* (Advanced math students know these points as "inflection points.") You can feel the change as you run a pencil along a Normal curve and so find the standard deviation. *Remember that $\mu$ and $\sigma$ alone do not specify the shape of most distributions.* The shape of density curves in general does not reveal $\sigma$. These are special properties of Normal distributions.

### Teaching Tip

When the standard deviation $\sigma$ increases, the Normal curve is more spread out. This also means that the Normal curve will be shorter, because the area under the curve must remain equal to 1.

### Teaching Tip

Speaking of advanced math, some of your students may be interested in the equation of a Normal curve. It includes two of the most interesting numbers in mathematics: $\pi$ and $e$. If students who have taken calculus complain about the lack of math in AP® Statistics, challenge them to find areas under the Normal curve by integrating the function for the Normal curve. They will find it fairly difficult!

$$f(x) = \frac{1}{\sigma\sqrt{2\pi}} e^{-\frac{(x-\mu)^2}{2\sigma^2}}$$

### Teaching Tip

The fact that the standard deviation $\sigma$ is the natural measure of spread for Normal distributions is a good explanation for why we square the deviations when calculating the standard deviation instead of taking the absolute values of the deviations. If we calculated the mean absolute deviation (MAD) as our measure of spread, the inflection points would not be a distance of one MAD from the mean.

### Teaching Tip

Some teachers refer to the concave-up portions of the Normal curve as "happy" faces and the concave-down portion as a "sad" face. This might help the students who haven't taken calculus and have never heard of inflection points.

**DEFINITION: Normal distribution and Normal curve**

A **Normal distribution** is described by a Normal density curve. Any particular Normal distribution is completely specified by two numbers: its mean $\mu$ and standard deviation $\sigma$. The mean of a Normal distribution is at the center of the symmetric **Normal curve.** The standard deviation is the distance from the center to the change-of-curvature points on either side.

We abbreviate the Normal distribution with mean $\mu$ and standard deviation $\sigma$ as $N(\mu, \sigma)$.

Normal curves were first applied to data by the great mathematician Carl Friedrich Gauss (1777–1855). He used them to describe the small errors made by astronomers and surveyors in repeated careful measurements of the same quantity. You will sometimes see Normal distributions labeled "Gaussian" in honor of Gauss.

Why are the Normal distributions important in statistics? Here are three reasons. First, Normal distributions are good descriptions for some distributions of *real data*. Distributions that are often close to Normal include

- scores on tests taken by many people (such as SAT exams and IQ tests),
- repeated careful measurements of the same quantity (like the diameter of a tennis ball), and
- characteristics of biological populations (such as lengths of crickets and yields of corn).

Second, Normal distributions are good approximations to the results of many kinds of *chance outcomes*, like the number of heads in many tosses of a fair coin. Third, and most important, we will see that many *statistical inference* procedures are based on Normal distributions.

*Even though many sets of data follow a Normal distribution, many do not*. Most income distributions, for example, are skewed to the right and so are not Normal. Some distributions are symmetric but not Normal or even close to Normal. The uniform distribution of Exercise 35 (page 128) is one such example. Non-Normal data, like non-normal people, not only are common but are sometimes more interesting than their Normal counterparts.

caution

## The 68–95–99.7 Rule

Earlier, we saw that the distribution of Iowa Test of Basic Skills (ITBS) vocabulary scores for seventh-grade students in Gary, Indiana, is symmetric, single-peaked, and bell-shaped. Suppose that the distribution of scores over time is exactly Normal with mean $\mu = 6.84$ and standard deviation $\sigma = 1.55$. (These are the mean and standard deviation of the 947 actual scores.) The figure shows the Normal density curve for this distribution with the points 1, 2, and 3 standard deviations from the mean labeled on the horizontal axis.

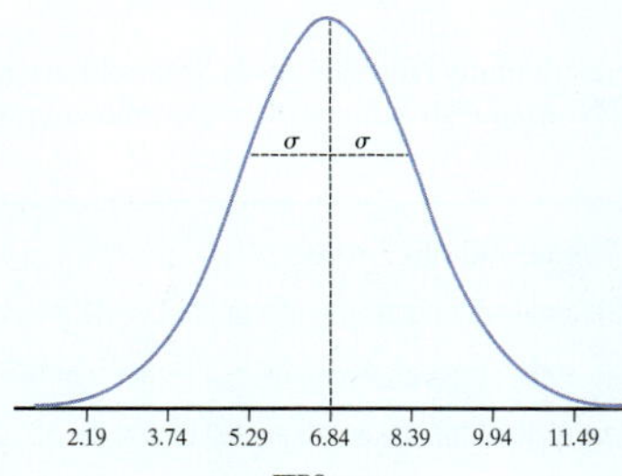


How unusual is it for a Gary seventh-grader to get an ITBS score above 9.94? As the following activity shows, the answer to this question is surprisingly simple.

### Teaching Tip

Before the euro was the official currency of Germany, Gauss and the Normal curve were featured on the 10 deutsche mark bill. To show a picture of the Gauss on the bill, go to http://upload.wikimedia.org/wikipedia/commons/0/0d/10_DM_Serie4_Vorderseite.jpg.

### ALTERNATE ACTIVITY

In addition to the Activity on the next page, you might also consider using the *Investigating Normal Distributions* Alternate Activity. In this Activity, students are given several distributions of data that are approximately Normal. Then, they are asked to calculate what percent of the values are within 1, 2, and 3 standard deviations from the mean and summarize the patterns that they discover.

**TRM Investigating Normal Distributions**

### Activity Overview

You can watch an overview of this Activity (for teachers) by clicking on the link in the TE-book, opening the TRCD, or logging onto the book's companion Web site www.whfreeman.com/tps5e (teacher log-in required).

### TECHNOLOGY TIP

If you want to find the area between two values when using the *Normal Curve* applet, drag the left boundary to the right of the other boundary and make sure the 2-tail box is checked.

### APPLET

If you would rather type in the boundaries than move the boundaries with your mouse, try the *Z-Table Normal Curve* applet at http://davidmlane.com/hyperstat/z_table.html or the *Normal* applet at www.lock5stat.com/statkey.

### Teaching Tip

Students often have a hard time sketching a Normal curve. Ask them to follow these instructions until they can draw a reasonably accurate curve freehand:

1. Draw a number line with seven evenly spaced tick marks. Label the one in the middle with the mean value $\mu$ and those to each side with $\mu \pm \sigma$, $\mu \pm 2\sigma$, and $\mu \pm 3\sigma$.
2. Draw a dot above the mean value where the peak of the curve will be.
3. Above the marks 1 standard deviation from the mean, draw dots about 60% as high as the dot above the mean.
4. Above the marks 2 standard deviations from the mean, draw dots about 15% as high as the dot above the mean (or about one-fourth as high as the dots above 1 standard deviation from the mean).
5. Above the marks 3 standard deviations from the mean, draw dots just above the horizontal axis.
6. Connect the seven dots with a smooth curve.

## ACTIVITY | The Normal density curve applet

**MATERIALS:** Computer with Internet access

In this Activity, you will use the *Normal Density Curve* applet at the book's Web site (www.whfreeman.com/tps5e) to explore an interesting property of Normal distributions. A graph similar to what you will see when you launch the applet is shown below. The applet finds the area under the curve in the region indicated by the green flags. If you drag one flag past the other, the applet will show the area under the curve between the two flags. When the "2-Tail" box is checked, the applet calculates symmetric areas around the mean.

Use the applet to help you answer the following questions.

1. If you put one flag at the extreme left of the curve and the second flag exactly in the middle, what proportion is reported by the applet? Why does this value make sense?
2. If you place the two flags exactly one standard deviation on either side of the mean, what does the applet say is the area between them?

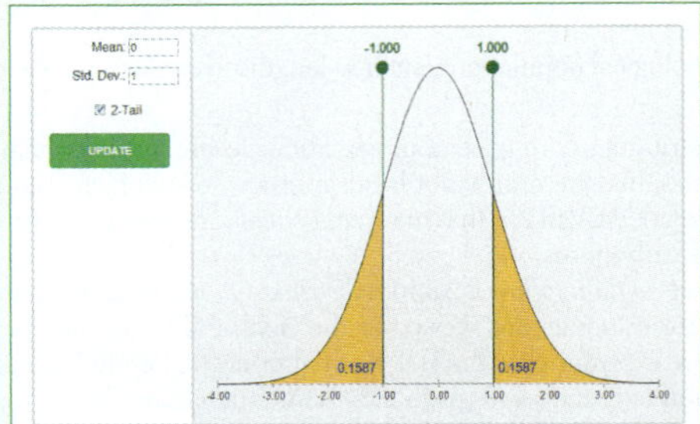


3. What percent of the area under the Normal curve lies within 2 standard deviations of the mean?
4. Use the applet to show that about 99.7% of the area under the Normal density curve lies within three standard deviations of the mean. Does this mean that about 99.7%/2 = 49.85% will lie within one and a half standard deviations? Explain.
5. Change the mean to 100 and the standard deviation to 15. Then click "Update." What percent of the area under this Normal density curve lies within one, two, and three standard deviations of the mean?
6. Change the mean to 6.84 and the standard deviation to 1.55. (These values are from the ITBS vocabulary scores in Gary, Indiana.) Answer the question from Step 5.
7. *Summarize:* Complete the following sentence: "For any Normal density curve, the area under the curve within one, two, and three standard deviations of the mean is about ___%, ___%, and ___%."

Although there are many Normal curves, they all have properties in common. In particular, all Normal distributions obey the following rule.

**DEFINITION: The 68–95–99.7 rule**

In a Normal distribution with mean $\mu$ and standard deviation $\sigma$:

- Approximately **68%** of the observations fall within $\sigma$ of the mean $\mu$.
- Approximately **95%** of the observations fall within $2\sigma$ of the mean $\mu$.
- Approximately **99.7%** of the observations fall within $3\sigma$ of the mean $\mu$.

### Teaching Tip

You may notice that we ask questions about the "proportion" or "percent" of observations that are between two boundaries, rather than the "probability" that a randomly selected observation is between two boundaries. While we will spend plenty of time using Normal distributions to calculate probabilities in Chapter 6 and beyond, for now we prefer to use the Normal distribution as a model for data.

### Teaching Tip

There are often several ways to solve problems like those in the example on the following page. For instance, in part (b), a student might add the area to the left of the mean (34%) to the area to the right of the mean (50% − 2.5% = 47.5%) and get 34% + 47.5% = 81.5%.

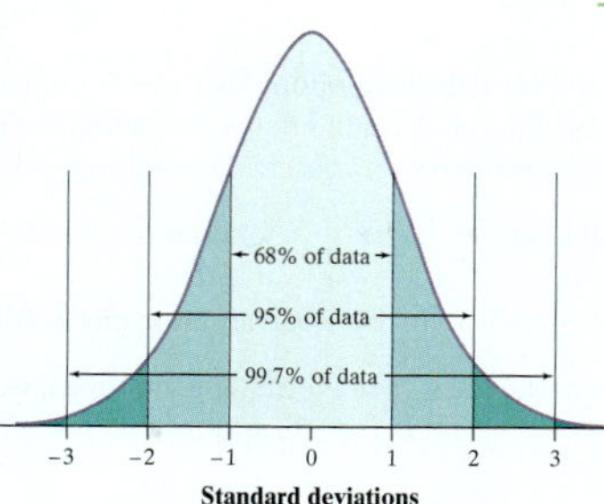


FIGURE 2.12 The 68–95–99.7 rule for Normal distributions.

Figure 2.12 illustrates the 68–95–99.7 rule. (Some people refer to this result as the "empirical rule.") By remembering these three numbers, you can think about Normal distributions without constantly making detailed calculations.

Here's an example that shows how we can use the 68–95–99.7 rule to estimate the percent of observations in a specified interval.

## EXAMPLE ITBS Vocabulary Scores

### *Using the 68–95–99.7 rule*

PROBLEM: The distribution of ITBS vocabulary scores for seventh-graders in Gary, Indiana, is $N(6.84, 1.55)$.

(a) What percent of the ITBS vocabulary scores are less than 3.74? Show your work.

(b) What percent of the scores are between 5.29 and 9.94? Show your work.

SOLUTION:

(a) Notice that a score of 3.74 is exactly two standard deviations below the mean. By the 68–95–99.7 rule, about 95% of all scores are between

$$\mu - 2\sigma = 6.84 - (2)(1.55) = 6.84 - 3.10 = 3.74$$

and

$$\mu + 2\sigma = 6.84 + (2)(1.55) = 6.84 + 3.10 = 9.94$$

The other 5% of scores are outside this range. Because Normal distributions are symmetric, half of these scores are lower than 3.74 and half are higher than 9.94. That is, about 2.5% of the ITBS scores are below 3.74. Figure 2.13(a) shows this reasoning in picture form.

(b) Let's start with a picture. Figure 2.13(b) shows the area under the Normal density curve between 5.29 and 9.94. We can see that about 68% + 13.5% = 81.5% of ITBS scores are between 5.29 and 9.94.

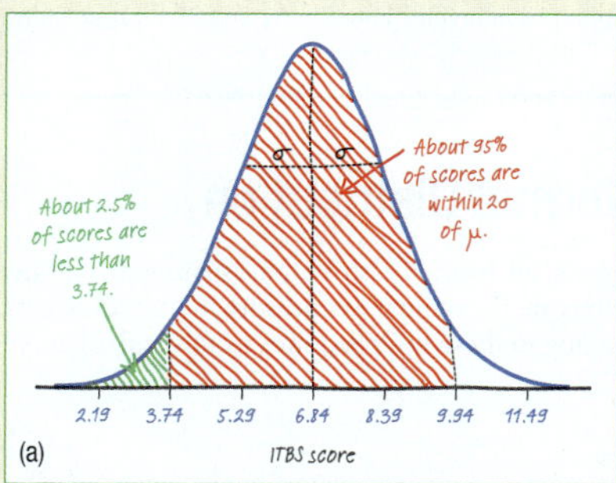


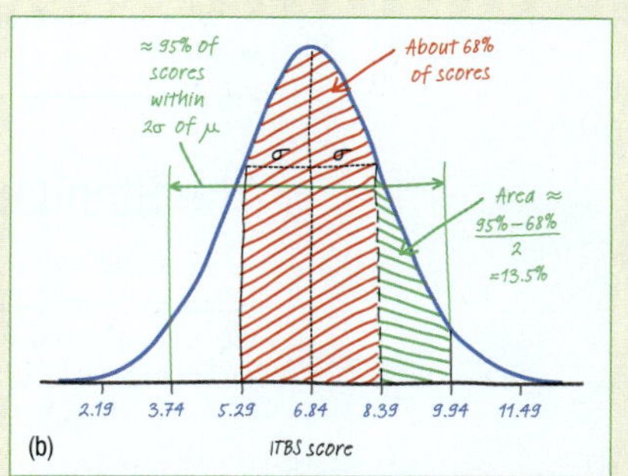


FIGURE 2.13 (a) Finding the percent of Iowa Test scores less than 3.74. (b) Finding the percent of Iowa Test scores between 5.29 and 9.94.

**For Practice** *Try Exercise* 43

## ALTERNATE EXAMPLE

### *Batting averages*

In the previous alternate example about batting averages for Major League Baseball players, the mean of the 432 batting averages was 0.261 with a standard deviation of 0.034. Suppose that the distribution is exactly Normal with $\mu = 0.261$ and $\sigma = 0.034$.

PROBLEM:

**(a)** Sketch a Normal density curve for this distribution of batting averages. Label the points that are 1, 2, and 3 standard deviations from the mean.

**(b)** What percent of the batting averages are greater than 0.329? Show your work.

**(c)** What percent of the batting averages are between 0.193 and 0.295? Show your work.

SOLUTION:

**(a)** Here is a Normal curve for the distribution of batting averages. The mean and the points 1, 2, and 3 standard deviations from the mean are labeled.

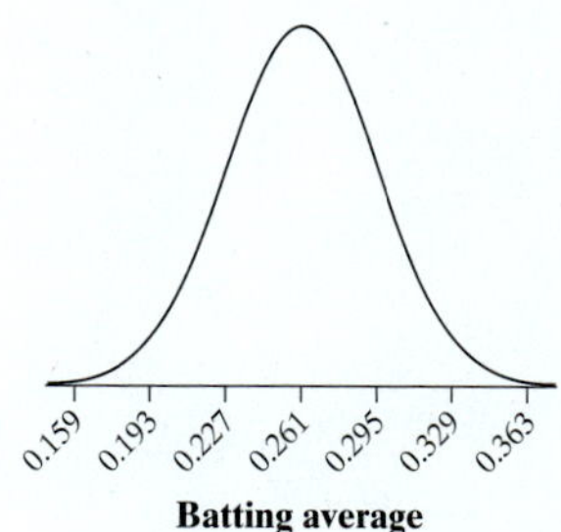


## ALTERNATE EXAMPLE *(continued)*

**(b)** Here is a curve showing the proportion of batting averages above 0.329. Because 0.329 is exactly 2 standard deviations above the mean, we know that about 95% of batting averages will be between 0.193 and 0.329. Because the curve is symmetric, half of the remaining 5%, or 2.5%, should be greater than 0.329.

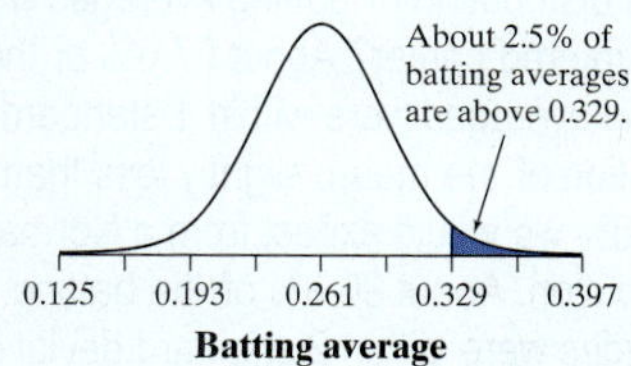


**(c)** Here is a curve showing the proportion of batting averages between 0.193 and 0.295. We know that about 68% of batting averages are between 0.227 and 0.295. Also, we know that half of the difference between 95% and 68% (27%) should be between 0.193 and 0.227. Therefore, about 13.5% + 68% = 81.5% of batting averages will be between 0.193 and 0.295.

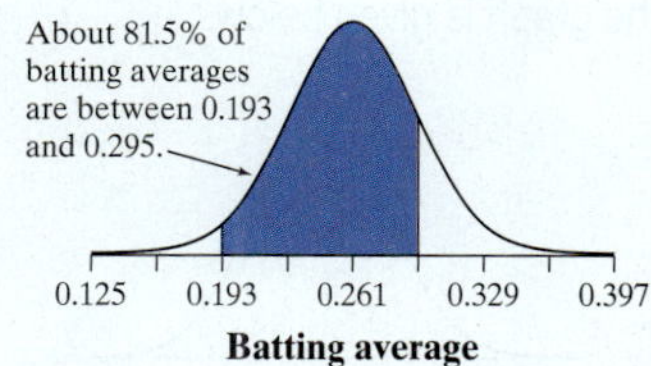

Chebyshev's inequality is an interesting result, but it is not required for the AP® Statistics exam.

The 68–95–99.7 rule applies *only* to Normal distributions. Is there a similar rule that would apply to *any* distribution? Sort of. A result known as **Chebyshev's inequality** says that in any distribution, the proportion of observations falling within $k$ standard deviations of the mean is *at least* $1 - \frac{1}{k^2}$. If $k = 2$, for example, Chebyshev's inequality tells us that at least $1 - \frac{1}{2^2} = 0.75$ of the observations in *any* distribution are within 2 standard deviations of the mean. For Normal distributions, we know that this proportion is much higher than 0.75. In fact, it's approximately 0.95.

THINK ABOUT IT

**All models are wrong, but some are useful!** The 68–95–99.7 rule describes distributions that are exactly Normal. Real data such as the actual ITBS scores are never exactly Normal. For one thing, ITBS scores are reported only to the nearest tenth. A score can be 9.9 or 10.0 but not 9.94. We use a Normal distribution because it's a good approximation, and because we think of the knowledge that the test measures as continuous rather than stopping at tenths.

How well does the 68–95–99.7 rule describe the actual ITBS scores? Well, 900 of the 947 scores are between 3.74 and 9.94. That's 95.04%, very accurate indeed. Of the remaining 47 scores, 20 are below 3.74 and 27 are above 9.94. The tails of the actual data are not quite equal, as they would be in an exactly Normal distribution. Normal distributions often describe real data better in the center of the distribution than in the extreme high and low tails.

As famous statistician George Box once noted, "All models are wrong, but some are useful!"

## CHECK YOUR UNDERSTANDING

The distribution of heights of young women aged 18 to 24 is approximately $N(64.5, 2.5)$.

1. Sketch a Normal density curve for the distribution of young women's heights. Label the points one, two, and three standard deviations from the mean.
2. What percent of young women have heights greater than 67 inches? Show your work.
3. What percent of young women have heights between 62 and 72 inches? Show your work.

## The Standard Normal Distribution

As the 68–95–99.7 rule suggests, all Normal distributions share many properties. In fact, all Normal distributions are the same if we measure in units of size $\sigma$ from the mean $\mu$ as center. Changing to these units requires us to standardize, just as we did in Section 2.1:

$$z = \frac{x - \mu}{\sigma}$$

If the variable we standardize has a Normal distribution, then so does the new variable $z$. (Recall that subtracting a constant and dividing by a constant don't

### Teaching Tip

Chebyshev's inequality is not a topic on the AP® exam. We mention it here as a contrast to the 68–95–99.7 rule, which applies only to approximately Normal distributions.

### AP® Exam Common Error

Students sometimes try to apply the 68–95–99.7 rule to non-Normal distributions. Make sure to emphasize that this rule works reasonably well only for distributions that are approximately Normal.

### ALTERNATE EXAMPLE

*Batting averages*

How well does the 68–95–99.7 rule apply to the distribution of batting averages we encountered earlier? About 67.6% of the batting averages were within 1 standard deviation of the mean, slightly less than the 68% we would expect from a Normal distribution. About 96.3% of the batting averages were within 2 standard deviations of the mean, slightly more than the 95% we would expect from a Normal distribution. Finally, about 99.8% of the batting averages were within 3 standard deviations of the mean—quite close to the 99.7% we expected.

### ANSWERS TO CYU

1. The graph is given below.

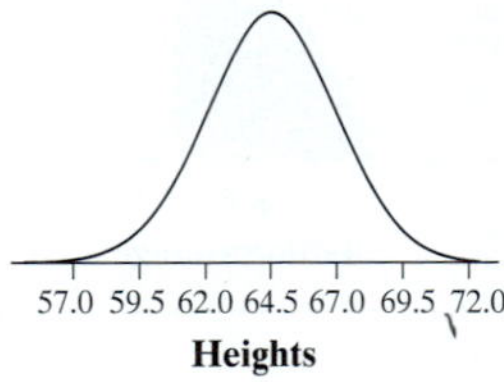


2. Approximately $\frac{100\% - 68\%}{2} = 16\%$

3. Approximately $\frac{100\% - 68\%}{2} = 16\%$ have heights below 62 inches and approximately $\frac{100\% - 99.7\%}{2} = 0.15\%$ of young women have heights above 72 inches, so the remaining 83.85% have heights between 62 and 72 inches.

change the shape of a distribution.) This new distribution with mean $\mu = 0$ and standard deviation $\sigma = 1$ is called the **standard Normal distribution**.

**DEFINITION: Standard Normal distribution**

The **standard Normal distribution** is the Normal distribution with mean 0 and standard deviation 1 (Figure 2.14).

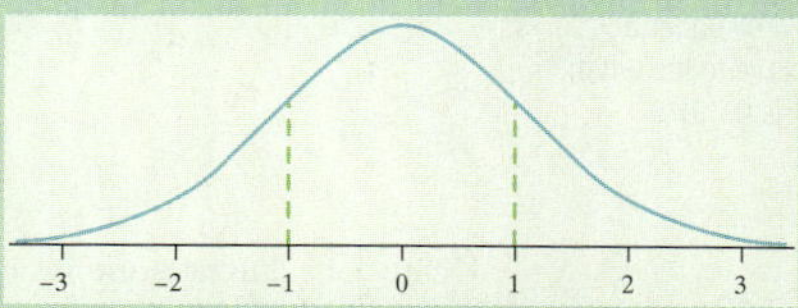


FIGURE 2.14 The standard Normal distribution.

If a variable $x$ has any Normal distribution $N(\mu, \sigma)$ with mean $\mu$ and standard deviation $\sigma$, then the standardized variable

$$z = \frac{x - \mu}{\sigma}$$

has the standard Normal distribution $N(0,1)$.

An area under a density curve is a proportion of the observations in a distribution. Any question about what proportion of observations lies in some range of values can be answered by finding an area under the curve. In a standard Normal distribution, the 68–95–99.7 rule tells us that about 68% of the observations fall between $z = -1$ and $z = 1$ (that is, within one standard deviation of the mean). What if we want to find the percent of observations that fall between $z = -1.25$ and $z = 1.25$? The 68–95–99.7 rule can't help us.

Because all Normal distributions are the same when we standardize, we can find areas under any Normal curve from a single table, a table that gives areas under the curve for the standard Normal distribution. Table A, the **standard Normal table,** gives areas under the standard Normal curve. You can find Table A in the back of the book.

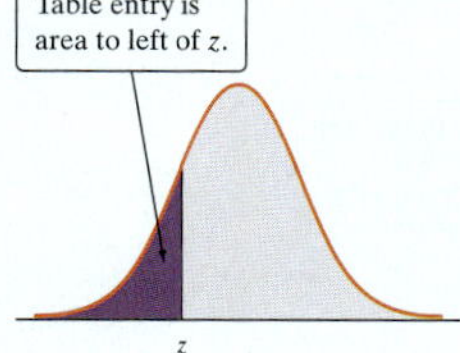


**DEFINITION: The standard Normal table**

**Table A** is a table of areas under the standard Normal curve. The table entry for each value $z$ is the area under the curve to the left of $z$.

| z | .00 | .01 | .02 |
|---|---|---|---|
| 0.7 | .7580 | .7611 | .7642 |
| 0.8 | .7881 | .7910 | .7939 |
| 0.9 | .8159 | .8186 | .8212 |

For instance, suppose we wanted to find the proportion of observations from the standard Normal distribution that are less than 0.81. To find the area to the left of $z = 0.81$, locate 0.8 in the left-hand column of Table A, then locate the remaining digit 1 as .01 in the top row. The entry opposite 0.8 and under .01 is .7910. This is the area we seek. A reproduction of the relevant portion of Table A is shown in the margin.

## Teaching Tip

Emphasize that standardized scores (z-scores) have a Normal distribution only if the original variable has a Normal distribution. Standardizing a variable doesn't change the shape of a distribution, so the distribution of $z$-scores will have the same shape as the original distribution.

## Teaching Tip

In Section 2.1, we used percentiles and z-scores to identify where an observation was located in its distribution. If the distribution is Normal, the standard Normal table allows us to convert from a percentile to a z-score and vice versa. If the distribution is non-Normal, converting from a percentile to a z-score can be much more difficult, depending on the shape.

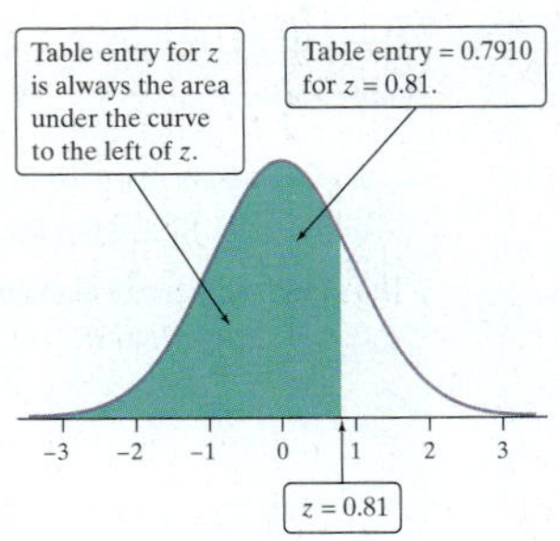


**FIGURE 2.15** The area under a standard Normal curve to the left of the point $z = 0.81$ is 0.7910.

Figure 2.15 illustrates the relationship between the value $z = 0.81$ and the area 0.7910. *Note that we have made a connection between z-scores and percentiles when the shape of a distribution is Normal.*

## EXAMPLE Standard Normal Distribution

### *Finding area to the right*

| $z$ | .07 | .08 | .09 |
|---|---|---|---|
| −1.8 | .0307 | .0301 | .0294 |
| −1.7 | .0384 | .0375 | .0367 |
| −1.6 | .0475 | .0465 | .0455 |

What if we wanted to find the proportion of observations from the standard Normal distribution that are *greater than* −1.78? To find the area to the right of $z = -1.78$, locate −1.7 in the left-hand column of Table A, then locate the remaining digit 8 as .08 in the top row. The corresponding entry is .0375. (See the excerpt from Table A in the margin.)

This is the area *to the left* of $z = -1.78$. To find the area *to the right* of $z = -1.78$, we use the fact that the total area under the standard Normal density curve is 1. So the desired proportion is $1 - 0.0375 = 0.9625$.

Figure 2.16 illustrates the relationship between the value $z = -1.78$ and the area 0.9625.

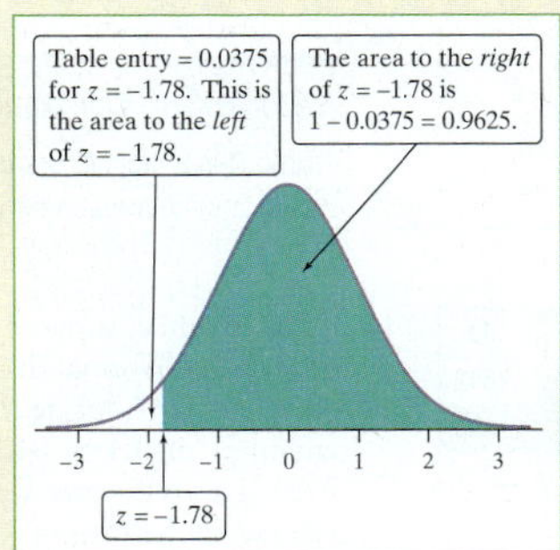


**FIGURE 2.16** The area under a standard Normal curve to the right of the point $z = -1.78$ is 0.9625.

### ALTERNATE EXAMPLE

### *Finding area to the right*

Suppose we wanted to find the proportion of observations in a Normal distribution that were more than 1.53 standard deviations above the mean. That is, we want to know what proportion of observations in the standard Normal distribution are greater than $z = 1.53$. To find this proportion, locate the value 1.5 in the left-hand column of Table A, then locate the remaining digit 3 as .03 in the top row. The corresponding entry is 0.9370. This is the area to the *left* of $z = 1.53$. To find the area above $z = 1.53$, subtract 0.9370 from 1 to get 0.0630.

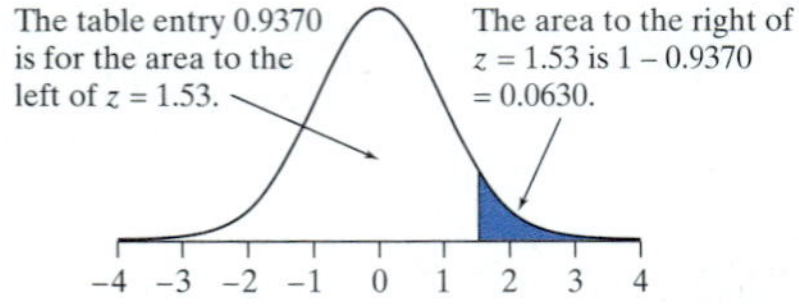

*A common student mistake is to look up a z-value in Table A and report the entry corresponding to that z-value, regardless of whether the problem asks for the area to the left or to the right of that z-value.* To prevent making this mistake, always sketch the standard Normal curve, mark the $z$-value, and shade the area of interest. And before you finish, make sure your answer is reasonable in the context of the problem.

## EXAMPLE Catching Some "z"s

### *Finding areas under the standard Normal curve*

**PROBLEM:** Find the proportion of observations from the standard Normal distribution that are between −1.25 and 0.81.

**SOLUTION:** From Table A, the area to the left of $z = 0.81$ is 0.7910 and the area to the left of $z = -1.25$ is 0.1056. So the area under the standard Normal curve between these two $z$-scores is $0.7910 - 0.1056 = 0.6854$. Figure 2.17 shows why this approach works.

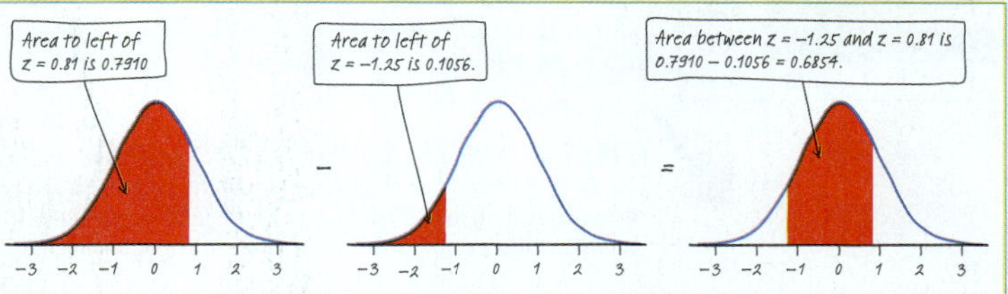


**FIGURE 2.17** One way to find the area between $z = -1.25$ and $z = 0.81$ under the standard Normal curve.

Here's another way to find the desired area. The area to the left of $z = -1.25$ under the standard Normal curve is 0.1056. The area to the *right* of $z = 0.81$ is $1 - 0.7910 = 0.2090$. So the area *between* these two $z$-scores is

$$1 - (0.1056 + 0.2090) = 1 - 0.3146 = 0.6854$$

Figure 2.18 shows this approach in picture form.

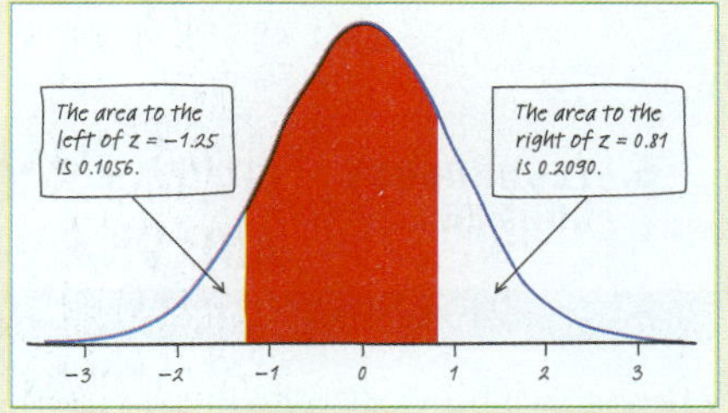


**FIGURE 2.18** The area under the standard Normal curve between $z = -1.25$ and $z = 0.81$ is 0.6854.

**For Practice** *Try Exercise* 49

## AP® Exam Common Error

When doing Normal distribution calculations, students often report the area to the left of a boundary when they are asked about the area to the right of the boundary. Insisting that your students do a quick sketch with proper shading will prevent many of these mistakes.

## ALTERNATE EXAMPLE

### *Finding areas under the standard Normal curve*

**PROBLEM:** Find the proportion of observations from the standard Normal distribution that are between −0.58 and 1.79.

**SOLUTION:** To find this proportion, we must find the proportion of values that are less than $z = 1.79$ and then subtract the proportion of values that are less than $z = -0.58$. The difference in these proportions is the proportion of observations that are between $z = -0.58$ and $z = 1.79$.

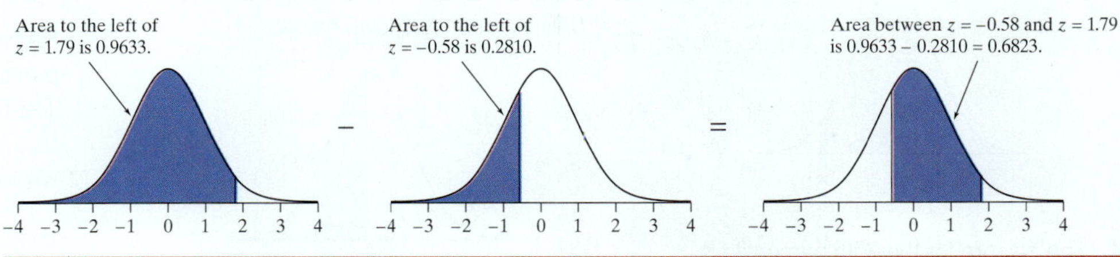

## ALTERNATE EXAMPLE

### *Working backward*

In a standard Normal distribution, 20% of the observations are greater than what value? Using Table A, we should look up an area of 0.8000 because the table always lists area to the left of a boundary. The closest area to 0.8000 is 0.7995, which corresponds to a $z$-score of $z = 0.84$. Thus, approximately 20% of the observations in a standard Normal distribution are greater than $z = 0.84$.

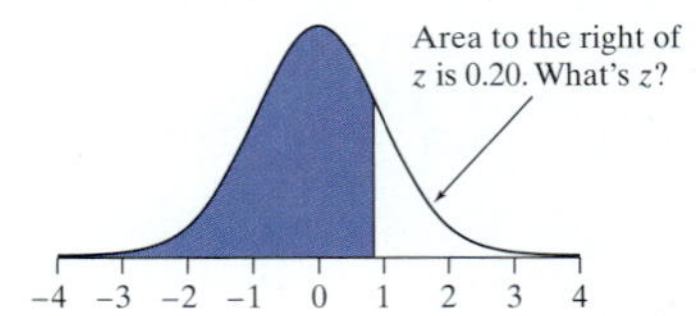


## ANSWERS TO CYU

**1.** The proportion is 0.9177. A graph is shown below.

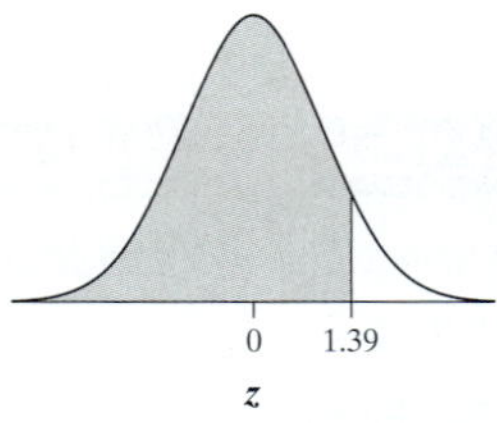


**2.** The proportion is 0.9842. A graph is shown below.

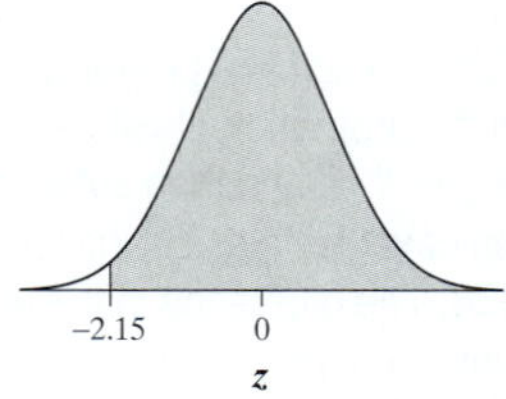


**3.** The proportion is $0.9649 - 0.2877 = 0.6772$. A graph is shown below.

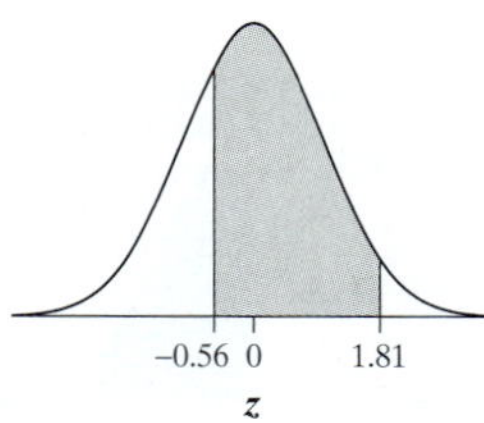


**4.** The $z$-score for the 20th percentile is $z = -0.84$. A graph is shown below.

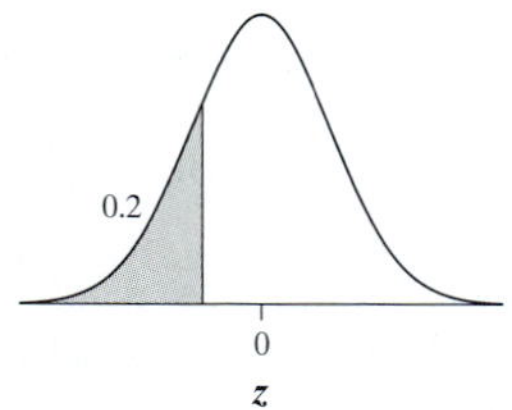


**Working backward: From areas to $z$-scores:** So far, we have used Table A to find areas under the standard Normal curve from $z$-scores. What if we want to find the $z$-score that corresponds to a particular area? For example, let's find the 90th percentile of the standard Normal curve. We're looking for the $z$-score that has 90% of the area to its left, as shown in Figure 2.19.

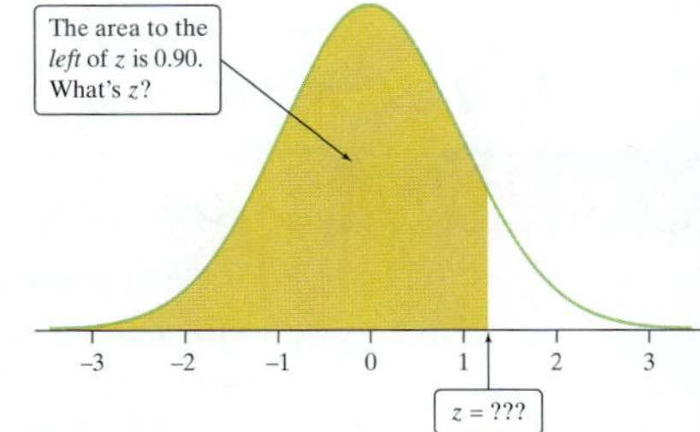


**FIGURE 2.19** The $z$-score with area 0.90 to its left under the standard Normal curve.

| $z$ | .07 | .08 | .09 |
|---|---|---|---|
| 1.1 | .8790 | .8810 | .8830 |
| 1.2 | .8980 | .8997 | .9015 |
| 1.3 | .9147 | .9162 | .9177 |

Because Table A gives areas to the left of a specified $z$-score, all we need to do is find the value closest to 0.90 in the middle of the table. From the reproduced portion of Table A, you can see that the desired $z$-score is $z = 1.28$. That is, the area to the left of $z = 1.28$ is approximately 0.90.

## CHECK YOUR UNDERSTANDING

Use Table A in the back of the book to find the proportion of observations from a standard Normal distribution that fall in each of the following regions. In each case, sketch a standard Normal curve and shade the area representing the region.

1. $z < 1.39$
2. $z > -2.15$
3. $-0.56 < z < 1.81$

Use Table A to find the value $z$ from the standard Normal distribution that satisfies each of the following conditions. In each case, sketch a standard Normal curve with your value of $z$ marked on the axis.

4. The 20th percentile
5. 45% of all observations are greater than $z$

You can use the *Normal Density Curve* applet at www.whfreeman.com/tps5e to confirm areas under the standard Normal curve. Just enter mean 0 and standard deviation 1, and then drag the flags to the appropriate locations. Of course, you can also confirm Normal curve areas with your calculator.

## 5. TECHNOLOGY CORNER — FROM $z$-SCORES TO AREAS, AND VICE VERSA

TI-Nspire instructions in Appendix B; HP Prime instructions on the book's Web site.

*Finding areas:* The `normalcdf` command on the TI-83/84 (`normCdf` on the TI-89) can be used to find areas under a Normal curve. The syntax is `normalcdf(lower bound,upper bound,mean,standard deviation)`. Let's use this command to confirm our answers to the previous two examples.

**5.** 45% of the observations are greater than $z = 0.13$. A graph is shown below.

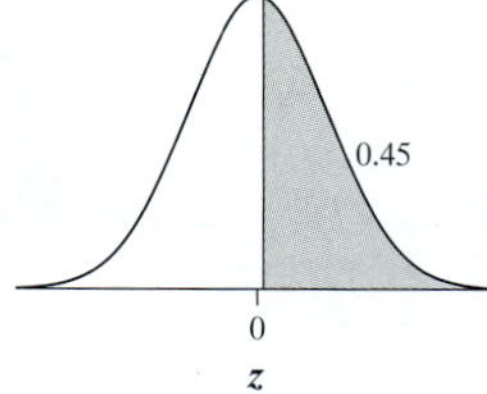


## TECHNOLOGY TIP

Remind students that they can watch a video presentation of this technology corner (for the TI-83/84, TI-89, or TI-Nspire) by clicking on the blue play button in the e-Book or at www.whfreeman.com/tps5e.

1. What proportion of observations from the standard Normal distribution are greater than –1.78?

   Recall that the standard Normal distribution has mean 0 and standard deviation 1.

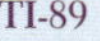

**TI-83/84**

- Press 2nd VARS (`Distr`) and choose `normalcdf(`. **OS 2.55 or later:** In the dialog box, enter these values: `lower:-1.78`, `upper:100000`, $\mu$`:0`, $\sigma$`:1`, choose `Paste`, and then press ENTER. **Older OS:** Complete the command `normalcdf(-1.78,100000,0,1)` and press ENTER.

**TI-89**

- In the Stats/List Editor, Press F5 (`Distr`) and choose `Normal Cdf(`.
- In the dialog box, enter these values: `lower:-1.78`, `upper:100000`, $\mu$`:0`, $\sigma$`:1`, and then choose ENTER.

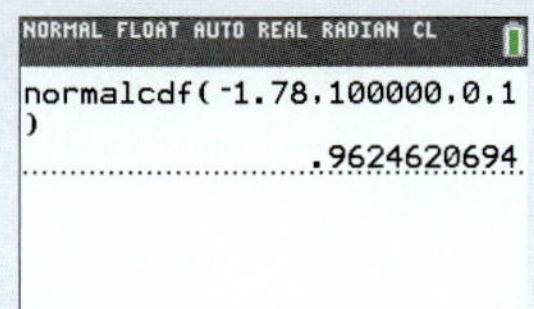


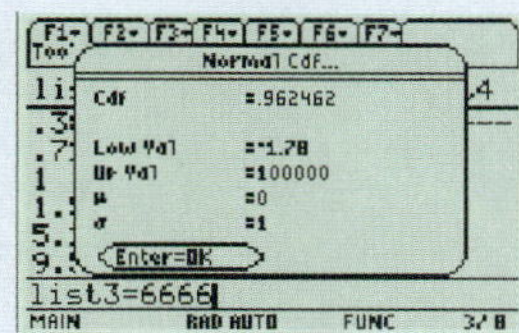


*Note:* We chose 100000 as the upper bound because it's many, many standard deviations above the mean.

These results agree with our previous answer using Table A: 0.9625.

2. What proportion of observations from the standard Normal distribution are between –1.25 and 0.81?

   The screen shots below confirm our earlier result of 0.6854 using Table A.

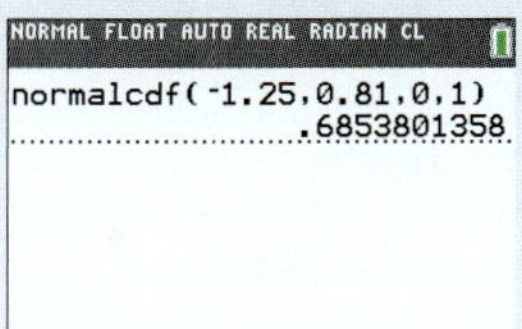


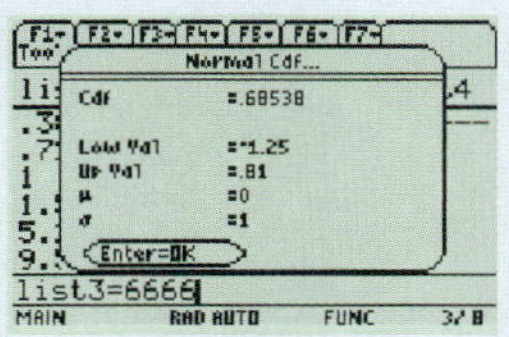


*Working backward:* The TI-83/84 and TI-89 `invNorm` function calculates the value corresponding to a given percentile in a Normal distribution. For this command, the syntax is `invNorm(area to the left, mean, standard deviation)`.

3. What is the 90th percentile of the standard Normal distribution?

**TI-83/84**

- Press 2nd VARS (`Distr`) and choose `invNorm(`. **OS 2.55 or later:** In the dialog box, enter these values: `area:.90`, $\mu$`:0`, $\sigma$`:1`, choose `Paste`, and then press ENTER. **Older OS:** Complete the command `invNorm(.90,0,1)` and press ENTER.

**TI-89**

- In the Stats/List Editor, Press F5 (`Distr`), choose `Inverse`, and `Inverse Normal...`.
- In the dialog box, enter these values: `area:.90`, $\mu$`:0`, $\sigma$`:1`, and then choose ENTER.

These results match what we got using Table A.

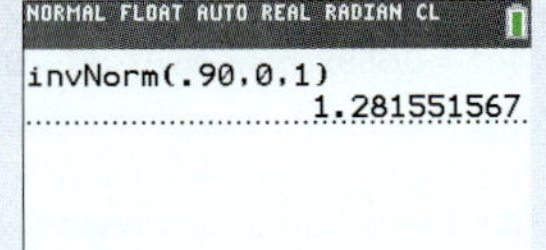


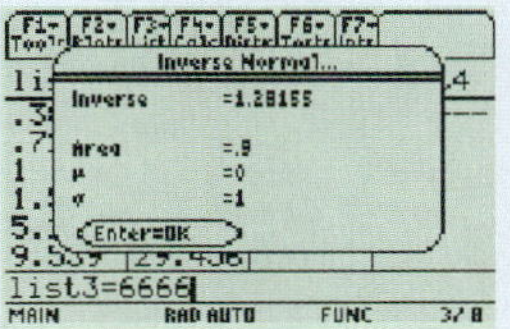


### TECHNOLOGY TIP

If your students are using a TI-84 with OS 2.55 or later, make sure that they turn on the Stat Wizards by pressing Mode and scrolling down to the second page. If students are using a TI-83 Plus or TI-84 with an older operating system, they can run an app called *Catalog Help* (ctlghelp) that will remind them what values they need to enter for certain commands. Press the APPS button to see if it is already loaded. If it is, press Enter and read the directions. If it isn't, download it at education.ti.com or copy it from another calculator.

### TECHNOLOGY TIP

We use 100,000 as an upper bound when finding areas to the right in a standard Normal distribution because it is virtually impossible to be 100,000 standard deviations above the mean in a Normal distribution. In theory, the upper bound should be infinity, but many calculators do not have an infinity button. When doing Normal calculations for Normal distributions other than the standard Normal distribution, you may need to pick a different upper (or lower) boundary, depending on the scale of the numbers involved.

### TECHNOLOGY TIP

If you are using a TI-83/84 with an older operating system (or do not have the Stat Wizards turned on), the calculator assumes that $\mu = 0$ and $\sigma = 1$ unless told otherwise. For example, to find the proportion of values that are greater than $z = -1.78$, you can enter the command `normalcdf(-1.78,100000)`.

### TECHNOLOGY TIP

When using the `invNorm` command, emphasize that students must always enter the area to the *left* of the boundary. This is consistent with Table A, which always lists the area to the left of a boundary value.

### TECHNOLOGY TIP

In addition to the `normalcdf` and `invNorm` commands, the calculator also lists a `normalpdf` command. It is not important that your students know how to use this command, but some students may ask what `normalpdf` does. Tell them that it gives the height of a Normal curve for a particular value of *x*, mean $\mu$, and standard deviation $\sigma$. For example, `normalpdf(x value:0,μ:0,μ:1)` = 0.3989 gives the height of the standard Normal curve at its peak ($x = 0$).

## Normal Distribution Calculations

We can answer a question about areas in *any* Normal distribution by standardizing and using Table A or by using technology. Here is an outline of the method for finding the proportion of the distribution in any region.

**HOW TO FIND AREAS IN ANY NORMAL DISTRIBUTION**

**Step 1: State the distribution and the values of interest.** Draw a Normal curve with the area of interest shaded and the mean, standard deviation, and boundary value(s) clearly identified.

**Step 2: Perform calculations—show your work!** Do one of the following: (i) Compute a $z$-score for each boundary value and use Table A or technology to find the desired area under the standard Normal curve; or (ii) use the normalcdf command and label each of the inputs.

**Step 3: Answer the question.**

Here's an example of the method at work.

**EXAMPLE**

### Tiger on the Range

*Normal calculations*

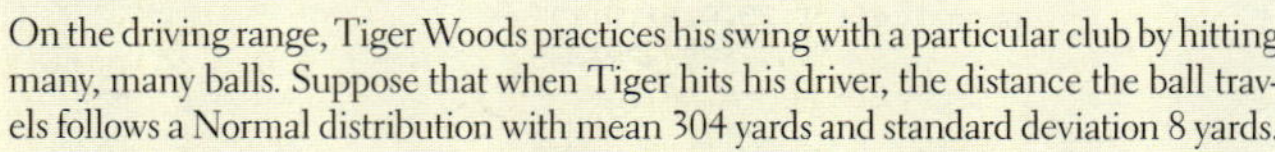

On the driving range, Tiger Woods practices his swing with a particular club by hitting many, many balls. Suppose that when Tiger hits his driver, the distance the ball travels follows a Normal distribution with mean 304 yards and standard deviation 8 yards.

PROBLEM: What percent of Tiger's drives travel at least 290 yards?

SOLUTION:

Step 1: State the distribution and the values of interest. The distance that Tiger's ball travels follows a Normal distribution with $\mu = 304$ and $\sigma = 8$. We want to find the percent of Tiger's drives that travel 290 yards or more. Figure 2.20 shows the distribution with the area of interest shaded and the mean, standard deviation, and boundary value labeled.

N (304, 8)

272 280 288 296 304 312 320 328 336

x = 290
z = −1.75

**FIGURE 2.20** Distance traveled by Tiger Woods's drives on the range.

Step 2: Perform calculations—show your work! For the boundary value $x = 290$, we have

$$z = \frac{x - \mu}{\sigma} = \frac{290 - 304}{8} = -1.75$$

So drives of 290 yards or more correspond to $z \geq -1.75$ under the standard Normal curve. From Table A, we see that the proportion of observations less than $-1.75$ is 0.0401. The area to the *right* of $-1.75$ is therefore $1 - 0.0401 = 0.9599$. This is about 0.96, or 96%.

*Using technology:* The command `normalcdf(lower:290,upper:100000,μ:304,σ:8)` also gives an area of 0.9599.

Step 3: Answer the question. About 96% of Tiger Woods's drives on the range travel at least 290 yards.

**For Practice** *Try Exercise* 53(a)

### Teaching Tip

Emphasize the three-step process described in the yellow summary box with your students. Insist that your students perform these three steps for each Normal distribution calculation. Make sure that you complete all three steps as well when doing examples and reviewing homework exercises.

### AP® Exam Common Error

Students often do not get full credit on the AP® exam because they *only* use "calculator-speak" to show their work on Normal calculation questions. For example, `normalcdf(290, 100000,304,8)`. This is not considered clear communication because some graders may not be familiar with the syntax for each calculator. To get full credit when using a calculator, students should follow the three-step process, making sure to carefully label each of the inputs in the calculator command.

### Teaching Tip

To make Normal curves less abstract, have your students mentally "fill" each Normal curve with dots. Explain that the Normal curve is tracing what a dotplot would look like after many observations. In this example, you can even suggest that students fill the Normal curve with little golf balls, each one representing the distance that Tiger Woods drives a ball.

**ALTERNATE EXAMPLE**

*Serving speed*

In a recent tournament, tennis player Rafael Nadal averaged 115 miles per hour (mph) on his serves. Assume that the distribution of his serve speeds is Normal with a standard deviation of 6 mph.

PROBLEM: About what percent of Nadal's serves would you expect to exceed 120 mph?

SOLUTION: Step 1: State the distribution and the values of interest. Rafael Nadal's serve speed follows a Normal distribution with mean 115 and standard deviation 6. We want to find the percent of his serves that exceed 120 mph (see graph at right).

Step 2: Perform calculations—show your work. The standardized score for the boundary value is $z = \frac{120 - 115}{6} = 0.83$.

From Table A, the proportion of $z$-scores above 0.83 is $1 - 0.7967 = 0.2033$. *Using technology:* The command `normalcdf(lower:120, upper:1000,μ:115,σ:6)` gives an area of 0.2023.

Step 3: Answer the question. About 20% of Nadal's serves will exceed 120 mph.

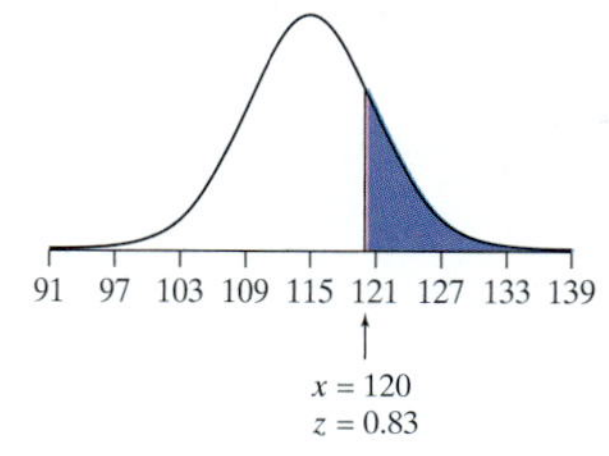

**What proportion of Tiger Woods's drives go exactly 290 yards?** There is *no* area under the Normal density curve in Figure 2.20 exactly over the point 290. So the answer to our question based on the Normal model is 0. Tiger Woods's actual data may contain a drive that went exactly 290 yards (up to the precision of the measuring device). The Normal distribution is just an easy-to-use approximation, not a description of every detail in the data. One more thing: the areas under the curve with $x \geq 290$ and $x > 290$ are the same. According to the Normal model, the proportion of Tiger's drives that go at least 290 yards is the same as the proportion that go more than 290 yards.

The key to doing a Normal calculation is to sketch the area you want, then match that area with the area that the table (or technology) gives you. Here's another example.

## EXAMPLE Tiger on the Range (Continued)

### *More complicated calculations*

PROBLEM: What percent of Tiger's drives travel between 305 and 325 yards?

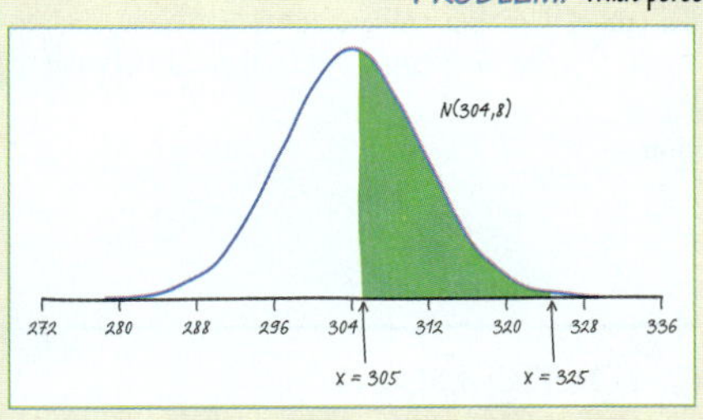


**FIGURE 2.21** Distance traveled by Tiger Woods's drives on the range.

SOLUTION:

Step 1: State the distribution and the values of interest. As in the previous example, the distance that Tiger's ball travels follows a Normal distribution with $\mu = 304$ and $\sigma = 8$. We want to find the percent of Tiger's drives that travel between 305 and 325 yards. Figure 2.21 shows the distribution with the area of interest shaded and the mean, standard deviation, and boundary values labeled.

Step 2: Perform calculations—show your work! For the boundary value $x = 305$, $z = \dfrac{305 - 304}{8} = 0.13$. The standardized score for $x = 325$ is $z = \dfrac{325 - 304}{8} = 2.63$.

From Table A, we see that the area between $z = 0.13$ and $z = 2.63$ under the standard Normal curve is the area to the left of 2.63 minus the area to the left of 0.13. Look at the picture below to check this. From Table A, area between 0.13 and 2.63 = area to the left of 2.63 − area to the left of 0.13 = 0.9957 − 0.5517 = 0.4440.

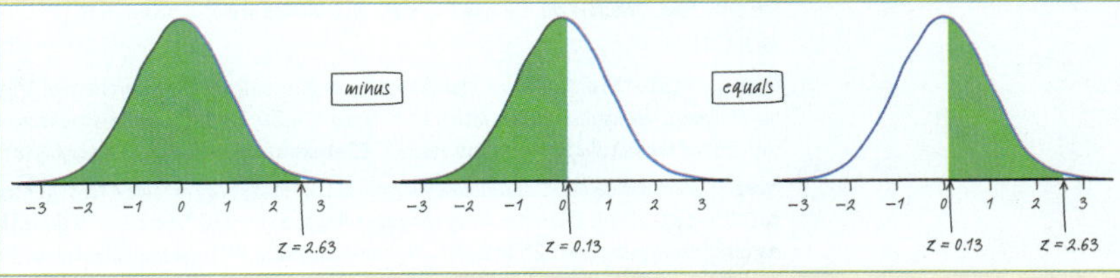


*Using technology:* The command `normalcdf(lower:305,upper:325,μ:304,σ:8)` gives an area of 0.4459.

Step 3: Answer the question. About 45% of Tiger's drives travel between 305 and 325 yards.

**For Practice** *Try Exercise* 53(b)

### Teaching Tip

The fact that the proportion of observations with $x \geq 290$ is the same as the proportion of observations with $x > 290$ is true when using any density curve, not just a Normal curve. Students will need to remember this fact later in the course when we discuss the differences between discrete random variables and continuous random variables.

### ALTERNATE EXAMPLE

*Serving speed (continued)*

PROBLEM: What percent of Rafael Nadal's serves are between 100 and 110 mph?

SOLUTION: Step 1: State the distribution and the values of interest. Rafael Nadal's serve speed follows a Normal distribution with mean 115 and standard deviation 6. We want to find the percent of his serves that are between 100 and 110 mph (see graph below).

Step 2: Perform calculations—show your work. The standardized scores for the boundary values are $z = \dfrac{100 - 115}{6} = -2.50$ and $z = \dfrac{110 - 115}{6} = -0.83$. From Table A, the proportion of *z*-scores below $z = -2.50$ is 0.0062 and the proportion of *z*-scores below −0.83 is 0.2033. Thus, the proportion of *z*-scores between −2.50 and −0.83 is 0.2033 − 0.0062 = 0.1971.

*Using technology:* The command `normalcdf(lower:100,upper:110,μ:115,σ:6)` gives an area of 0.1961.

Step 3: Answer the question. About 20% of Nadal's serves are between 100 and 110 mph.

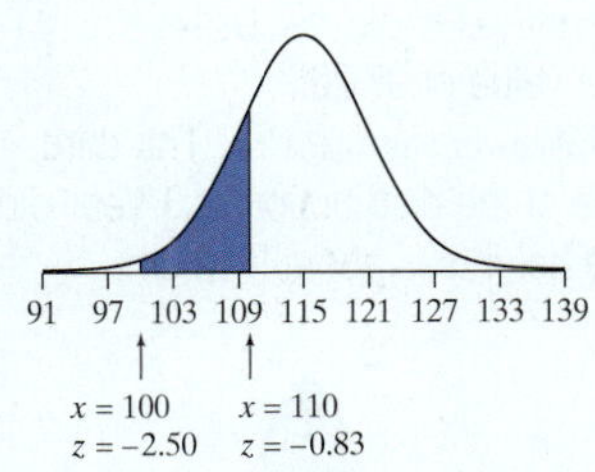

## Teaching Tip

Emphasize that Table A does not need to go beyond $z = -3.50$ and $z = 3.50$ because it is extremely rare to be more than 3.5 standard deviations from the mean in a Normal distribution.

## Teaching Tip

Again, make sure you insist that students follow this three-step process when doing Normal distribution questions that "go backward." This will help students avoid costly errors and omissions on the AP® exam.

### ALTERNATE EXAMPLE

*Heights of 3-year-old females*

According to http://www.cdc.gov/growthcharts/, the heights of 3-year-old females are approximately Normally distributed with a mean of 94.5 cm and a standard deviation of 4 cm.

PROBLEM: What is the third quartile of this distribution?

SOLUTION: Step 1: State the distribution and the values of interest. For 3-year-old females, height follows a Normal distribution with mean 94.5 cm and standard deviation 4 cm. $Q_3$ is the 75th percentile, so we want the boundary value $x$ with 75% of the distribution to its left (see graph below).

Step 2: Perform calculations—show your work. Look in the body of Table A for the value closest to 0.75. A z-score of 0.67 gives the closest value (0.7486). Solving $0.67 = \frac{x - 94.5}{4}$ gives $x = 97.18$ cm.

*Using technology:* The command `invNorm(area:0.75,μ:94.5,σ:4)` gives a value of 97.20.

Step 3: Answer the question. The third quartile of the distribution of 3-year-old female heights is about 97.2 cm.

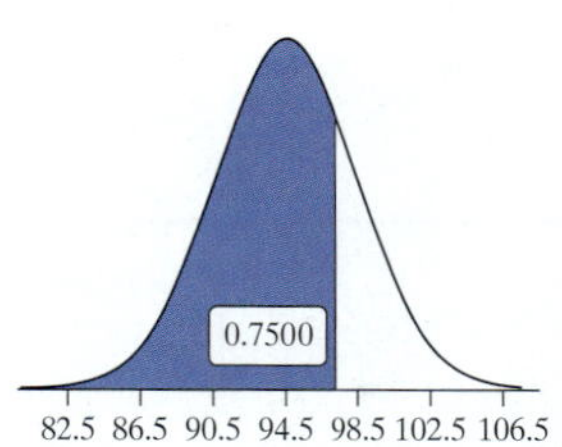


Table A sometimes yields a slightly different answer from technology. That's because we have to round z-scores to two decimal places before using Table A.

Sometimes we encounter a value of $z$ more extreme than those appearing in Table A. For example, the area to the left of $z = -4$ is not given directly in the table. The $z$-values in Table A leave only area 0.0002 in each tail unaccounted for. For practical purposes, we can act as if there is approximately zero area outside the range of Table A.

**Working backwards: From areas to values:** The previous two examples illustrated the use of Table A to find what proportion of the observations satisfies some condition, such as "Tiger's drive travels between 305 and 325 yards." Sometimes, we may want to find the observed value that corresponds to a given percentile. There are again three steps.

### HOW TO FIND VALUES FROM AREAS IN ANY NORMAL DISTRIBUTION

**Step 1: State the distribution and the values of interest.** Draw a Normal curve with the area of interest shaded and the mean, standard deviation, and unknown boundary value clearly identified.

**Step 2: Perform calculations—show your work!** Do one of the following: (i) Use Table A or technology to find the value of $z$ with the indicated area under the standard Normal curve, then "unstandardize" to transform back to the original distribution; or (ii) Use the invNorm command and label each of the inputs.

**Step 3: Answer the question.**

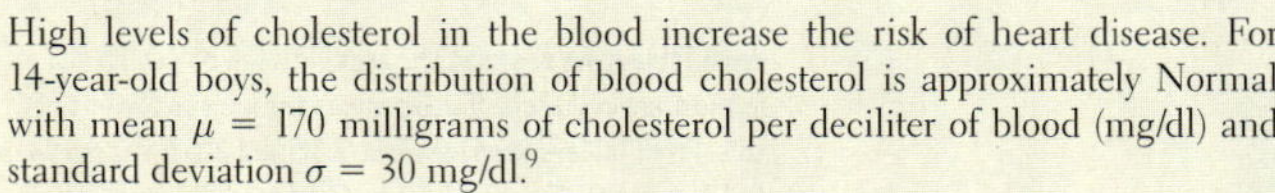

## EXAMPLE Cholesterol in Young Boys

*Using Table A in reverse*

High levels of cholesterol in the blood increase the risk of heart disease. For 14-year-old boys, the distribution of blood cholesterol is approximately Normal with mean $\mu = 170$ milligrams of cholesterol per deciliter of blood (mg/dl) and standard deviation $\sigma = 30$ mg/dl.[9]

PROBLEM: What is the 1st quartile of the distribution of blood cholesterol?

SOLUTION:

Step 1: State the distribution and the values of interest. The cholesterol level of 14-year-old boys follows a Normal distribution with $\mu = 170$ and $\sigma = 30$. The 1st quartile is the boundary value $x$ with 25% of the distribution to its left. Figure 2.22 shows a picture of what we are trying to find.

Step 2: Perform calculations–show your work! Look in the body of Table A for the entry closest to 0.25. It's 0.2514. This is the entry corresponding to $z = -0.67$. So $z = -0.67$ is the standardized score with area 0.25 to its left. Now unstandardize. We know that the standardized score for the unknown cholesterol level $x$ is $z = -0.67$. So $x$ satisfies the equation $\frac{x - 170}{30} = -0.67$.

Solving for $x$ gives

$$x = 170 + (-0.67)(30) = 149.9$$

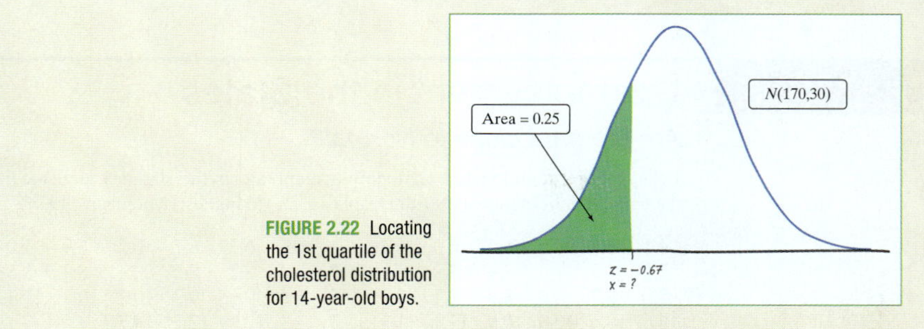


**FIGURE 2.22** Locating the 1st quartile of the cholesterol distribution for 14-year-old boys.

*Using technology:* The command `invNorm(area:0.25,μ:170,σ:30)` gives $x = 149.8$.

**Step 3: Answer the question.** The 1st quartile of blood cholesterol levels in 14-year-old boys is about 150 mg/dl.

**For Practice** *Try Exercise* **53(c)**

## CHECK YOUR UNDERSTANDING

Follow the method shown in the examples to answer each of the following questions. Use your calculator or the *Normal Curve* applet to check your answers.

1. Cholesterol levels above 240 mg/dl may require medical attention. What percent of 14-year-old boys have more than 240 mg/dl of cholesterol?
2. People with cholesterol levels between 200 and 240 mg/dl are at considerable risk for heart disease. What percent of 14-year-old boys have blood cholesterol between 200 and 240 mg/dl?
3. What distance would a ball have to travel to be at the 80th percentile of Tiger Woods's drive lengths?

## Assessing Normality

The Normal distributions provide good models for some distributions of real data. Examples include SAT and IQ test scores, the highway gas mileage of 2014 Corvette convertibles, state unemployment rates, and weights of 9-ounce bags of potato chips. The distributions of some other common variables are usually skewed and therefore distinctly non-Normal. Examples include economic variables such as personal income and total sales of business firms, the survival times of cancer patients after treatment, and the lifetime of electronic devices. While experience can suggest whether or not a Normal distribution is a reasonable model in a particular case, it is risky to assume that a distribution is Normal without actually inspecting the data.

In the latter part of this course, we will use various statistical inference procedures to try to answer questions that are important to us. These tests involve sampling individuals and recording data to gain insights about the populations from which they come. Many of these procedures are based on the assumption that the population is approximately Normally distributed. Consequently, we need to develop a strategy for assessing Normality.

### TRM Normal Distribution Worksheets

Find additional worksheets about the Normal distribution by clicking on the link in the TE-book, opening the TRCD, or logging onto the book's companion Web site www.whfreeman.com/tps5e (teacher log-in required).

### ANSWERS TO CYU

**1.** For 14-year-old boys, the amount of cholesterol follows an $N(170, 30)$ distribution and we want to find the percent of boys with cholesterol of more than 240 (see the following graph). $z = \frac{240 - 170}{30} = 2.33$. From Table A, the proportion of $z$-scores above 2.33 is $1 - 0.9901 = 0.0099$. *Using technology:* `normalcdf(lower:240,upper:1000,μ:170,σ:30)` $= 0.0098$. About 1% of 14-year-old boys have cholesterol above 240 mg/dl.

### ANSWERS TO CYU *(continued)*

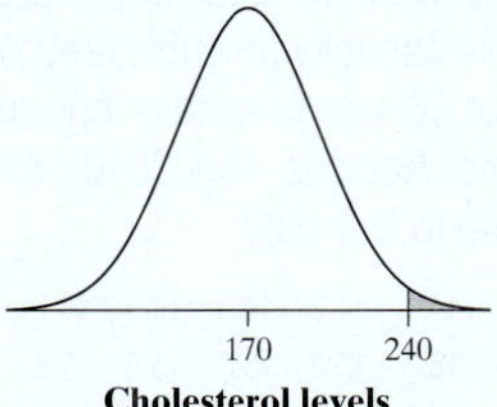


**2.** For 14-year-old boys, the amount of cholesterol follows an $N(170, 30)$ distribution and we want to find the percent of boys with cholesterol between 200 and 240 (see graph below). $z = \frac{200 - 170}{30} = 1$ and $z = \frac{240 - 170}{30} = 2.33$. From Table A, the proportion of $z$-scores between 1 and 2.33 is $0.9901 - 0.8413 = 0.1488$. *Using technology:* `normalcdf(lower:200,upper:240,μ:170,σ:30)` $= 0.1488$. About 15% of 14-year-old boys have cholesterol between 200 and 240 mg/dl.

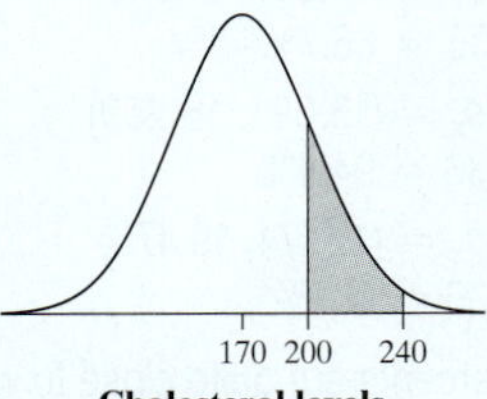


**3.** For Tiger Woods, the distance his drives travel follows an $N(304, 8)$ distribution and the 80th percentile is the boundary value $x$ with 80% of the distribution to its left (see graph below). A $z$-score of 0.84 gives the area closest to 0.80 (0.7995). Solving $0.84 = \frac{x - 304}{8}$ gives $x = 310.7$. *Using technology:* `invNorm(area:0.8,μ:304,σ:8)` $= 310.7$. The 80th percentile of Tiger Woods's drive lengths is about 310.7 yards.

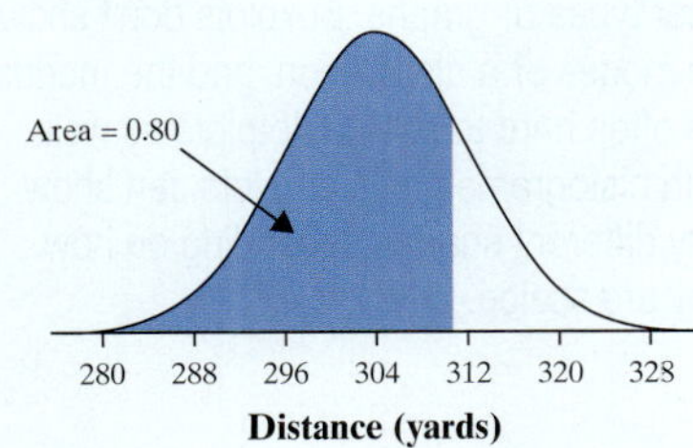

ALTERNATE EXAMPLE

*No space in the fridge?*

The measurements listed below describe the usable capacity (in cubic feet) of a sample of 36 side-by-side refrigerators (*Consumer Reports,* May 2010). Are the data close to Normal?

| | | | | | | | | |
|---|---|---|---|---|---|---|---|---|
| 12.9 | 13.7 | 14.1 | 14.2 | 14.5 | 14.5 | 14.6 | 14.7 | 15.1 |
| 15.2 | 15.3 | 15.3 | 15.3 | 15.3 | 15.5 | 15.6 | 15.6 | 15.8 |
| 16.0 | 16.0 | 16.2 | 16.2 | 16.3 | 16.4 | 16.5 | 16.6 | 16.6 |
| 16.6 | 16.8 | 17.0 | 17.0 | 17.2 | 17.4 | 17.4 | 17.9 | 18.4 |

Here is a histogram of these data. It seems roughly symmetric and bell-shaped.

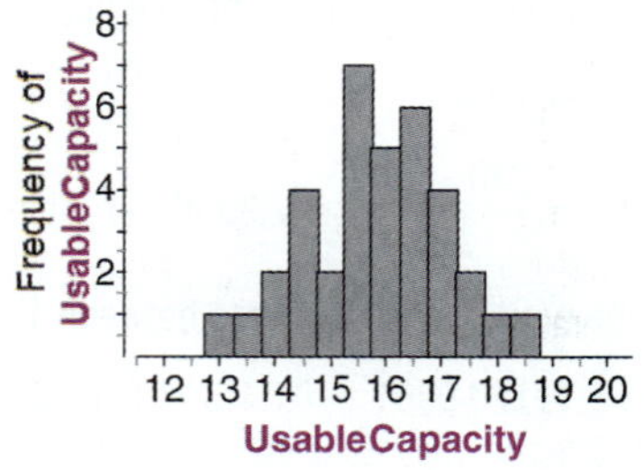


The mean and standard deviation of these data are $\bar{x} = 15.825$ and $s_x = 1.217$.

- $\bar{x} \pm 1s_x = (14.608, 17.042)$
  24 of 36 = 66.7%
- $\bar{x} \pm 2s_x = (13.391, 18.259)$
  34 of 36 = 94.4%
- $\bar{x} \pm 3s_x = (12.174, 19.476)$
  36 of 36 = 100%

These percents are quite close to what we would expect based on the 68–95–99.7 rule. Combined with the graph, this gives good evidence that this distribution is close to Normal.

## Teaching Tip

Students will not be tested on Normal probability plots on the AP® exam; however, students are welcome to use them. We include Normal probability plots because they were designed for assessing Normality and don't suffer from the drawbacks of the other types of graphs. Boxplots don't show the modes of a distribution, and the modes are often hard to see on dotplots as well. Both histograms and stemplots can show very different shapes depending on how they are scaled.

EXAMPLE

## Unemployment in the States

### *Are the data close to Normal?*

Let's start by examining data on unemployment rates in the 50 states. Here are the data arranged from lowest (North Dakota's 4.1%) to highest (Michigan's 14.7%).[10]

| | | | | | | | | | | | | |
|---|---|---|---|---|---|---|---|---|---|---|---|---|
| 4.1 | 4.5 | 5.0 | 6.3 | 6.3 | 6.4 | 6.4 | 6.6 | 6.7 | 6.7 | 6.7 | 6.9 | 7.0 |
| 7.0 | 7.2 | 7.4 | 7.4 | 7.4 | 7.8 | 8.0 | 8.0 | 8.2 | 8.2 | 8.4 | 8.5 | 8.5 |
| 8.6 | 8.7 | 8.8 | 8.9 | 9.1 | 9.2 | 9.5 | 9.6 | 9.6 | 9.7 | 10.2 | 10.3 | 10.5 |
| 10.6 | 10.6 | 10.8 | 10.9 | 11.1 | 11.5 | 12.3 | 12.3 | 12.3 | 12.7 | 14.7 | | |

- *Plot the data.* Make a dotplot, stemplot, or histogram. See if the graph is approximately symmetric and bell-shaped.

Figure 2.23 is a histogram of the state unemployment rates. The graph is roughly symmetric, single-peaked, and somewhat bell-shaped.

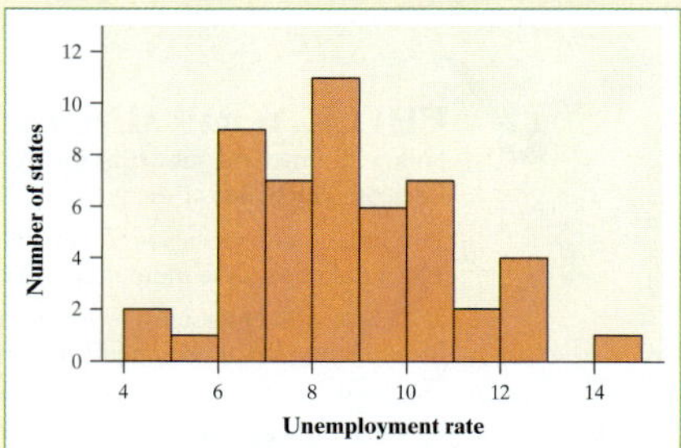


**FIGURE 2.23** Histogram of state unemployment rates.

- *Check whether the data follow the 68–95–99.7 rule.*

We entered the unemployment rates into computer software and requested summary statistics. Here's what we got:

```
Mean = 8.682 Standard deviation = 2.225.
```

Now we can count the number of observations within one, two, and three standard deviations of the mean.

| | | |
|---|---|---|
| Mean ± 1 SD: | 6.457 to 10.907 | 36 out of 50 = 72% |
| Mean ± 2 SD: | 4.232 to 13.132 | 48 out of 50 = 96% |
| Mean ± 3 SD: | 2.007 to 15.357 | 50 out of 50 = 100% |

These percents are quite close to the 68%, 95%, and 99.7% targets for a Normal distribution.

If a graph of the data is clearly skewed, has multiple peaks, or isn't bell-shaped, that's evidence that the distribution is *not* Normal. However, just because a plot of the data *looks* Normal, we can't say that the distribution *is* Normal. The 68–95–99.7 rule can give additional evidence in favor of or against Normality. A **Normal probability plot** also provides a good assessment of whether a data set follows a Normal distribution.

## EXAMPLE Unemployment in the States

### *Making a Normal probability plot*

Most software packages, including Minitab, Fathom, and JMP, can construct Normal probability plots (sometimes called Normal quantile plots) from entered data. The TI-83/84 and TI-89 will also make these graphs. Here's how a Normal probability plot is constructed.

1. *Arrange the observed data values from smallest to largest.* Record the percentile corresponding to each observation (but remember that there are several definitions of "percentile"). For example, the smallest observation in a set of 50 values is at either the 0th percentile (because 0 out of 50 values are below this observation) or the 2nd percentile (because 1 out of 50 values are at or below this observation). Technology usually "splits the difference," declaring this minimum value to be at the $(0 + 2)/2 =$ 1st percentile. By similar reasoning, the second-smallest value is at the 3rd percentile, the third-smallest value is at the 5th percentile, and so on. The maximum value is at the $(98 + 100)/2 =$ 99th percentile.

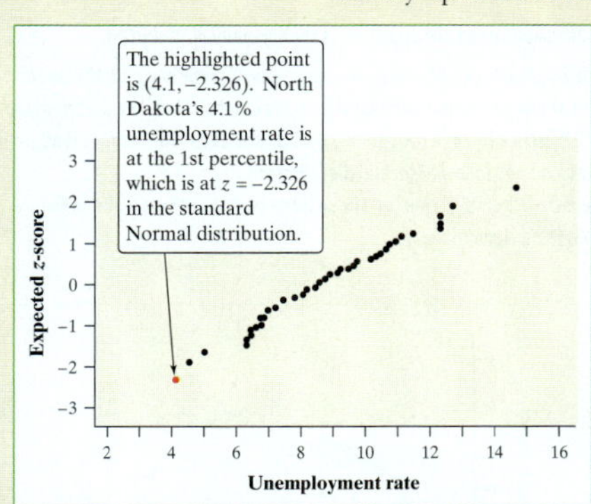


**FIGURE 2.24** Normal probability plot of the percent of unemployed individuals in each of the 50 states.

2. *Use the standard Normal distribution (Table A or* `invNorm`*) to find the z-scores at these same percentiles.* For example, the 1st percentile of the standard Normal distribution is $z = -2.326$. The 3rd percentile is $z = -1.881$; the 5th percentile is $z = -1.645$; . . . ; the 99th percentile is $z = 2.326$.

3. *Plot each observation x against its expected z-score from Step 2.* If the data distribution is close to Normal, the plotted points will lie close to some straight line. Figure 2.24 shows a Normal probability plot for the state unemployment data. There is a strong linear pattern, which suggests that the distribution of unemployment rates is close to Normal.

Some software plots the data values on the horizontal axis and the $z$-scores on the vertical axis, while other software does just the reverse. The TI-83/84 and TI-89 give you both options. We prefer the data values on the horizontal axis, which is consistent with other types of graphs we have made.

As Figure 2.24 indicates, real data almost always show some departure from Normality. *When you examine a Normal probability plot, look for shapes that show clear departures from Normality. Don't overreact to minor wiggles in the plot.* When we discuss statistical methods that are based on the Normal model, we will pay attention to the sensitivity of each method to departures from Normality. Many common methods work well as long as the data are approximately Normal.

#### INTERPRETING NORMAL PROBABILITY PLOTS

If the points on a **Normal probability plot** lie close to a straight line, the data are approximately Normally distributed. Systematic deviations from a straight line indicate a non-Normal distribution. Outliers appear as points that are far away from the overall pattern of the plot.

### Teaching Tip

Here is another way to explain what the $y$ coordinates represent in a Normal probability plot. Imagine taking many random samples of size 50 from the standard Normal distribution and sorting them from lowest to highest. The average of the minimum value from each random sample is the "expected" $z$-score for the minimum value when taking samples of size 50 from the standard Normal distribution. This "expected" $z$-score is what is paired with the actual minimum value in the data set.

### ALTERNATE EXAMPLE

*No space in the fridge? (continued)*

Here is a Normal probability plot (also called a Normal quantile plot) of the refrigerator data from the previous page. It is quite linear, supporting our earlier decision that the distribution is close to Normal.

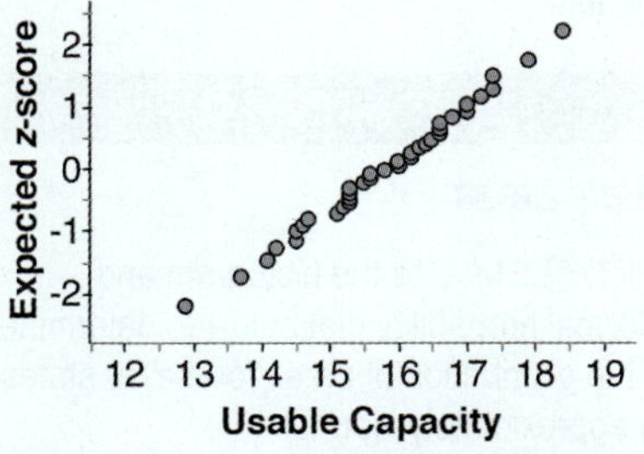


### Teaching Tip

We prefer putting the $x$-values on the horizontal axis so that when there are values to the right of what we expect, we know the distribution is skewed to the right. You can also point out that if all the points on a Normal probability plot suddenly fell down to the horizontal axis, the result would be a dotplot of the data.

## Teaching Tip

When interpreting a Normal probability plot, do not worry too much about little wiggles or small deviations from a linear pattern at the ends of the distribution. When a distribution is skewed, as in the example on this page, there will be a noticeable bend in the plot.

### TECHNOLOGY TIP

If you have Fathom, here is a nice way to help students understand how a Normal probability plot works. Start with a distribution of about 20 values that is approximately Normal. Create two graphs for this distribution, one a dotplot and the other a Normal probability plot (Fathom calls it a Normal quantile plot). Making sure to keep the graphs on the same scale, drag dots on the dotplot to change the shape from approximately Normal to skewed right and watch what happens on the Normal probability plot. Then, drag dots on the dotplot to make the distribution skewed to the left.

### ALTERNATE EXAMPLE

*State areas*

PROBLEM: Use the histogram and Normal probability plot below to determine if the distribution of areas for the 50 states is approximately Normal.

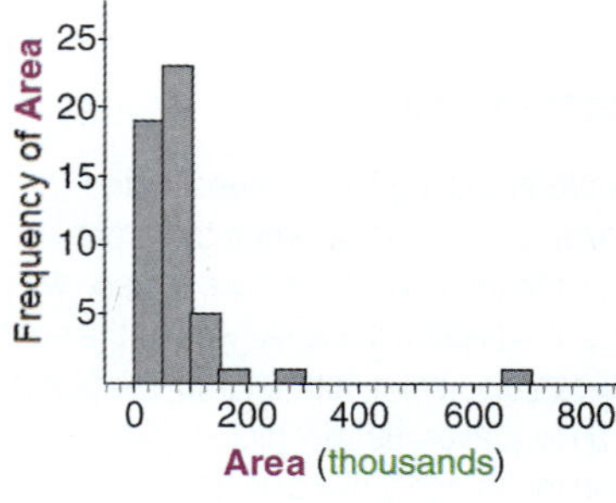


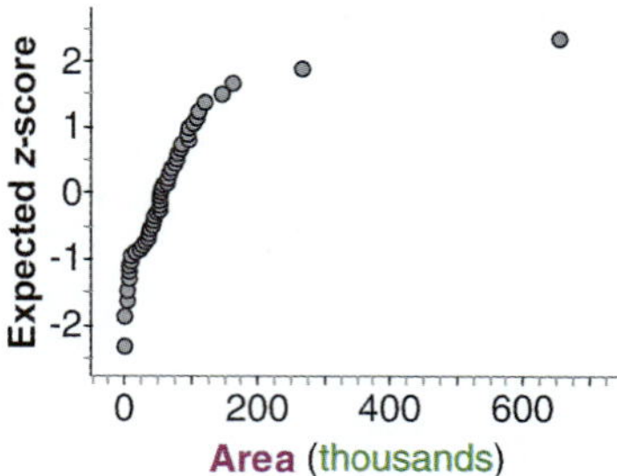


SOLUTION: Both the histogram and the Normal probability plot indicate that this distribution is strongly skewed to the right. In particular, there is one state (Alaska) whose area is much larger than we would expect if the distribution was approximately Normal.

**AP® EXAM TIP** Normal probability plots are not included on the AP® Statistics topic outline. However, these graphs are very useful for assessing Normality. You may use them on the AP® exam if you wish—just be sure that you know what you're looking for (a linear pattern).

Let's look at an example of some data that are *not* Normally distributed.

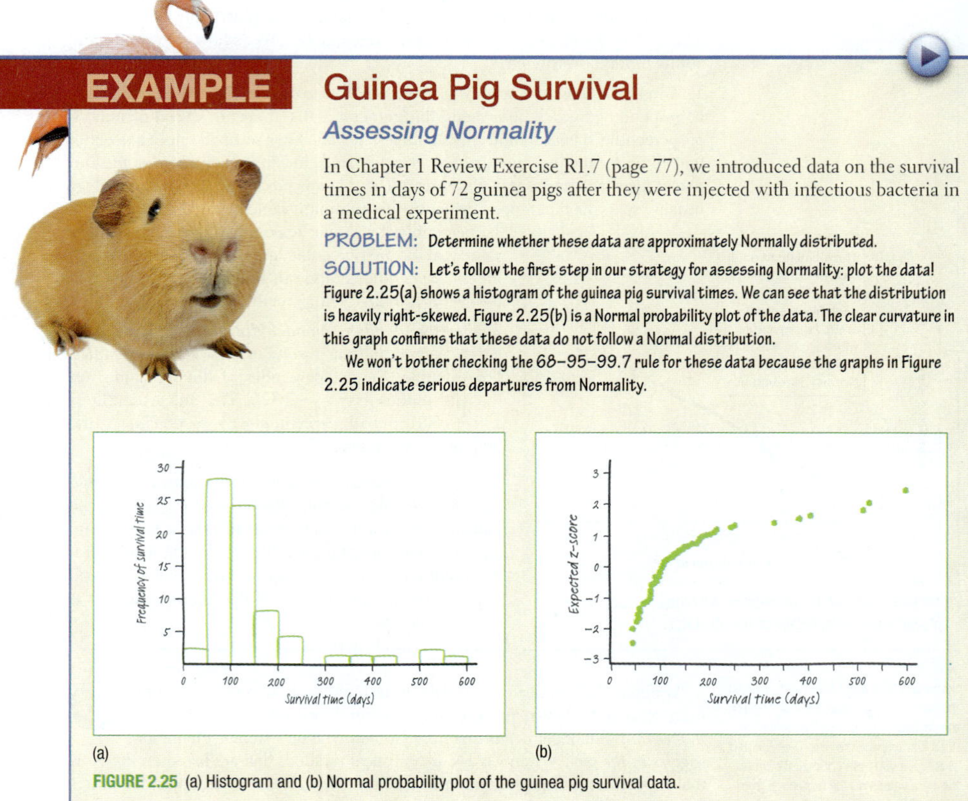

## EXAMPLE Guinea Pig Survival

*Assessing Normality*

In Chapter 1 Review Exercise R1.7 (page 77), we introduced data on the survival times in days of 72 guinea pigs after they were injected with infectious bacteria in a medical experiment.

PROBLEM: Determine whether these data are approximately Normally distributed.

SOLUTION: Let's follow the first step in our strategy for assessing Normality: plot the data! Figure 2.25(a) shows a histogram of the guinea pig survival times. We can see that the distribution is heavily right-skewed. Figure 2.25(b) is a Normal probability plot of the data. The clear curvature in this graph confirms that these data do not follow a Normal distribution.

We won't bother checking the 68–95–99.7 rule for these data because the graphs in Figure 2.25 indicate serious departures from Normality.



**FIGURE 2.25** (a) Histogram and (b) Normal probability plot of the guinea pig survival data.

For Practice Try Exercise 63

**How can we determine shape from a Normal probability plot?** Look at the Normal probability plot of the guinea pig survival data in Figure 2.25(b). Imagine drawing a line through the leftmost points, which correspond to the smaller observations. The larger observations fall systematically

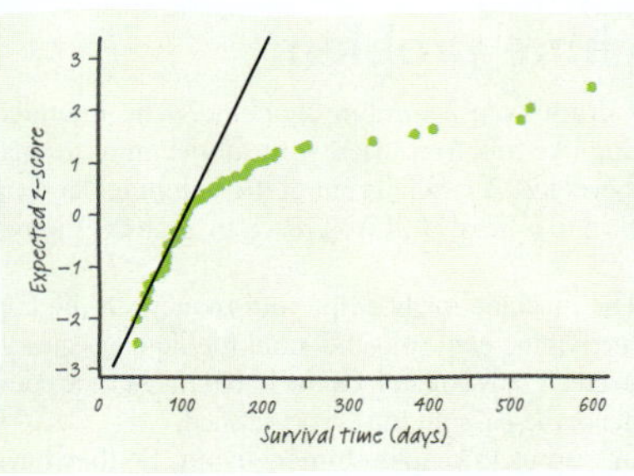


to the right of this line. That is, the right-of-center observations have much larger values than expected based on their percentiles and the corresponding $z$-scores from the standard Normal distribution.

This Normal probability plot indicates that the guinea pig survival data are strongly right-skewed. *In a right-skewed distribution, the largest observations fall distinctly to the right of a line drawn through the main body of points.* Similarly, left skewness is evident when the smallest observations fall to the left the line.

If you're wondering how to make a Normal probability plot on your calculator, the following Technology Corner shows you the process.

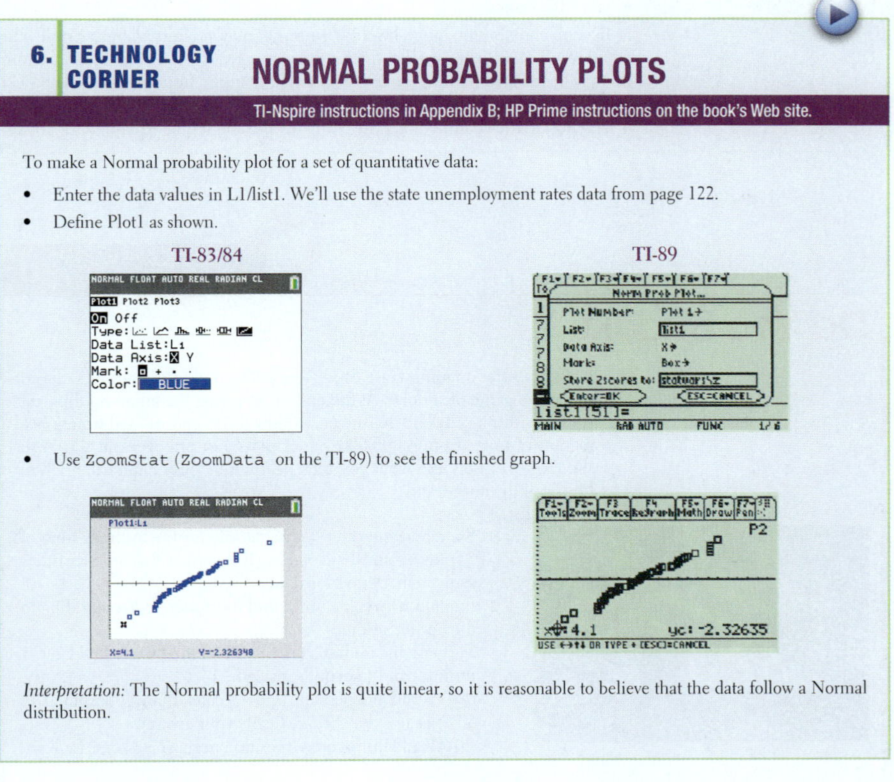

## 6. TECHNOLOGY CORNER NORMAL PROBABILITY PLOTS

**TI-Nspire instructions in Appendix B; HP Prime instructions on the book's Web site.**

To make a Normal probability plot for a set of quantitative data:

- Enter the data values in L1/list1. We'll use the state unemployment rates data from page 122.
- Define Plot1 as shown.

TI-83/84 TI-89

- Use `ZoomStat` (`ZoomData` on the TI-89) to see the finished graph.

*Interpretation:* The Normal probability plot is quite linear, so it is reasonable to believe that the data follow a Normal distribution.

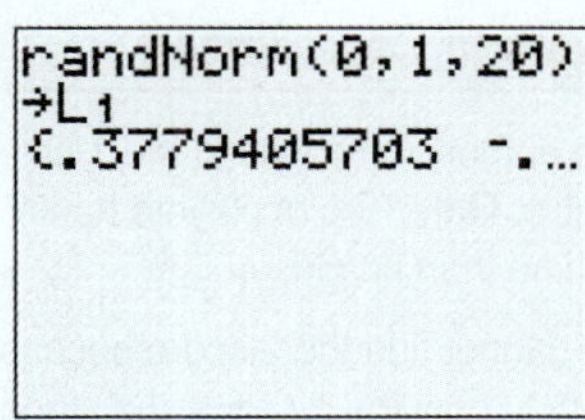

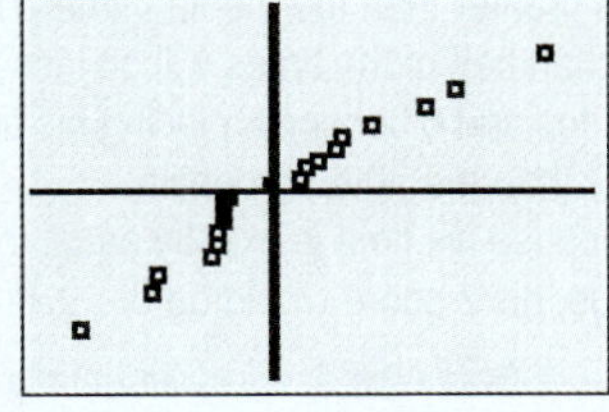

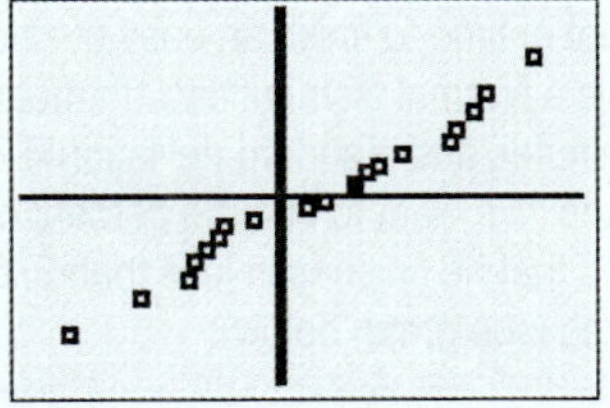

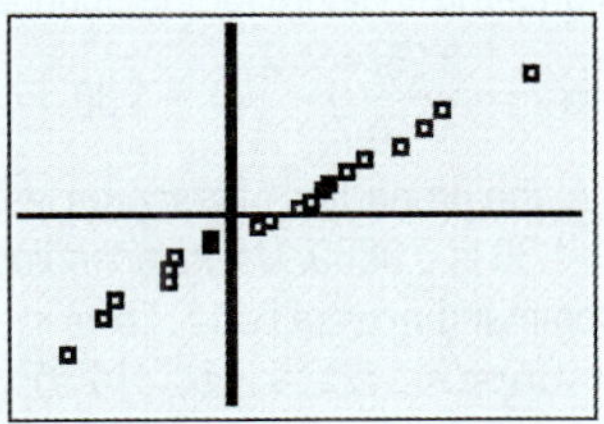

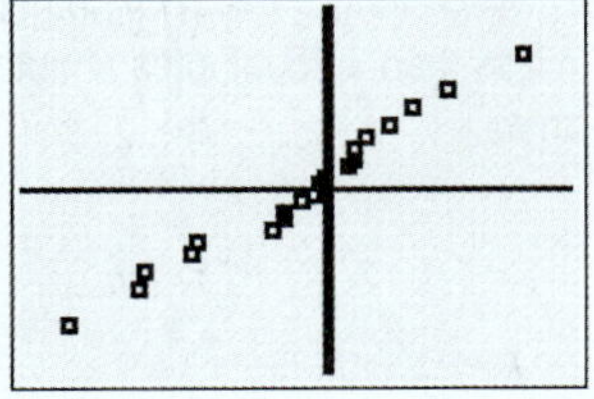

**ALTERNATE ACTIVITY**

*How linear should a Normal probability plot actually be?*

To answer this question, generate random samples from a Normal distribution and make Normal probability plots on a computer or graphing calculator. For example, on a TI-83/84:

- Press the `MATH` button and scroll over to the `PRB` menu. Choose the `randNorm` option and enter `randNorm(0,1,20)` → `L1` to generate a sample of 20 from the standard Normal distribution, and store them in list 1. The `STO>` command can be found just above the ON button.
- Make a Normal probability plot for these data to see what kinds of departures from Normality you can expect to happen just by chance.
- Repeat this process many times.

Here are some screen shots to illustrate what this should look like. As you can see, none of the plots were perfectly linear, even though the sample came from a Normal population.

## DATA EXPLORATION The vending machine problem

Have you ever purchased a hot drink from a vending machine? The intended sequence of events runs something like this. You insert your money into the machine and select your preferred beverage. A cup falls out of the machine, landing upright. Liquid pours out until the cup is nearly full. You reach in, grab the piping-hot cup, and drink happily.

Sometimes, things go wrong. The machine might swipe your money. Or the cup might fall over. More frequently, everything goes smoothly until the liquid begins to flow. It might stop flowing when the cup is only half full. Or the liquid might keep coming until your cup overflows. Neither of these results leaves you satisfied.

The vending machine company wants to keep customers happy. So they have decided to hire you as a statistical consultant. They provide you with the following summary of important facts about the vending machine:

- Cups will hold 8 ounces.
- The amount of liquid dispensed varies according to a Normal distribution centered at the mean $\mu$ that is set in the machine.
- $\sigma = 0.2$ ounces.

If a cup contains too much liquid, a customer may get burned from a spill. This could result in an expensive lawsuit for the company. On the other hand, customers may be irritated if they get a cup with too little liquid from the machine. Given these issues, what mean setting for the machine would you recommend? Write a brief report to the vending machine company president that explains your answer.

## Do You Sudoku?

In the chapter-opening Case Study (page 83), one of the authors played an online game of sudoku. At the end of his game, the graph on the next page was displayed. The density curve shown was constructed from a histogram of times from 4,000,000 games played in one week at this Web site. You will now use what you have learned in this chapter to analyze how well the author did.

1. State and interpret the percentile for the author's time of 3 minutes and 19 seconds. (Remember that smaller times indicate better performance.)
2. Explain why you cannot find the $z$-score corresponding to the author's time.
3. Suppose the author's time to finish the puzzle had been 5 minutes and 6 seconds instead.
   (a) Would his percentile be greater than 50%, equal to 50%, or less than 50%? Justify your answer.
   (b) Would his $z$-score be positive, negative, or zero? Explain.

### Data Exploration

Here are some notes for the Data Exploration:

- There isn't a single "correct" answer. However, a "good" answer should address both customer satisfaction and company profits.
- Consider having groups share their reports with classmates. This will give your students a chance to reinforce statistical concepts by explaining them to others.

### Answers to Case Closed

**1.** The author's time was at the 19th percentile. Only 19% of players had a faster time than he did.

**2.** We cannot find the $z$-score because we do not know the mean or the standard deviation. Also, we cannot convert the percentile to a $z$-score because the distribution of times isn't approximately Normal.

**3. (a)** If the author's time was 5 minutes and 6 seconds, his time would be equal to the average time. Because the distribution of times is skewed to the right, the mean time is greater than the median time. Thus, more than half of the times will be less than 5 minutes and 6 seconds, placing his time greater than the 50th percentile.
**(b)** Because his time is exactly equal to the average, his $z$-score would be 0.

**4.** *Step 1: State the distribution and the values of interest.* For the author, the amount of time to finish an easy puzzle follows a Normal distribution with mean 4.2 minutes and standard deviation 0.7 minutes. We want to find the percent of games that he finishes in less than 3.25 minutes (see graph below).
*Step 2: Perform calculations—show your work.* The standardized score for the boundary value is $z = \frac{3.25 - 4.2}{0.7} = -1.36$. From Table A, the proportion of $z$-scores less than $-1.36$ is 0.0869. *Using technology:* The command `normalcdf(lower:-1000,upper:3.25,μ:4.2,σ:0.7)` gives an area of 0.0874.
*Step 3: Answer the question.* The author will finish in less than 3.25 minutes about 9% of the time.

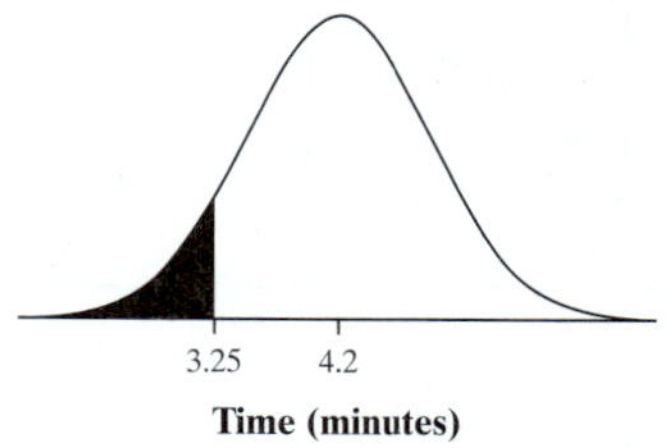


**5.** The author's time (3:19 = 3.317 minutes) has a standardized score of $z = \frac{3.317 - 4.2}{0.7} = -1.26$. His wife's time (3 minutes) has a standardized score of $z = \frac{3 - 3.8}{0.9} = -0.89$. Because it is better to have a below-average time, the author's performance was better, relatively speaking, even though his wife's time was faster overall.

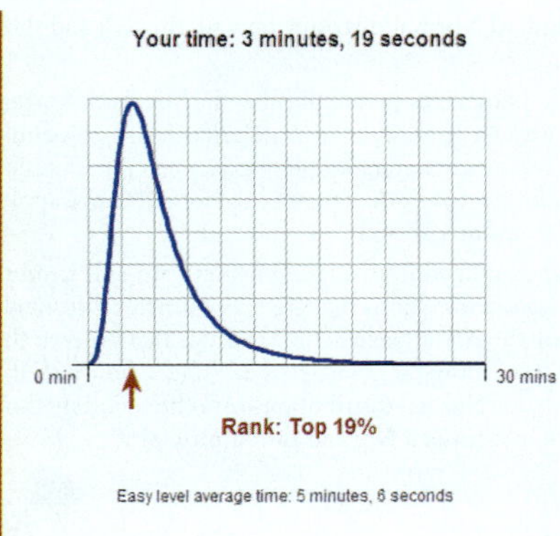


4. From long experience, the author's times to finish an easy sudoku puzzle at this Web site follow a Normal distribution with mean 4.2 minutes and standard deviation 0.7 minutes. In what percent of the games that he plays does the author finish an easy puzzle in less than 3 minutes and 15 seconds? Show your work. (*Note*: 3 minutes and 15 seconds is not the same as 3.15 seconds!)
5. The author's wife also enjoys playing sudoku online. Her times to finish an easy puzzle at this Web site follow a Normal distribution with mean 3.8 minutes and standard deviation 0.9 minutes. In her most recent game, she finished in 3 minutes. Whose performance is better, relatively speaking: the author's 3 minutes and 19 seconds or his wife's 3 minutes? Justify your answer.

## Section 2.2 Summary

- We can describe the overall pattern of a distribution by a **density curve**. A density curve always remains on or above the horizontal axis and has total area 1 underneath it. An area under a density curve gives the proportion of observations that fall in an interval of values.
- A density curve is an idealized description of the overall pattern of a distribution that smooths out the irregularities in the actual data. We write the **mean of a density curve** as $\mu$ and the **standard deviation of a density curve** as $\sigma$ to distinguish them from the mean $\bar{x}$ and the standard deviation $s_x$ of the actual data.
- The mean and the median of a density curve can be located by eye. The mean $\mu$ is the balance point of the curve. The median divides the area under the curve in half. The standard deviation $\sigma$ cannot be located by eye on most density curves.
- The mean and median are equal for symmetric density curves. The mean of a skewed curve is located farther toward the long tail than the median is.
- The **Normal distributions** are described by a special family of bell-shaped, symmetric density curves, called **Normal curves**. The mean $\mu$ and standard deviation $\sigma$ completely specify a Normal distribution $N(\mu,\sigma)$. The mean is the center of the curve, and $\sigma$ is the distance from $\mu$ to the change-of-curvature points on either side.
- All Normal distributions obey the **68–95–99.7 rule,** which describes what percent of observations lie within one, two, and three standard deviations of the mean.
- All Normal distributions are the same when measurements are standardized. If $x$ follows a Normal distribution with mean $\mu$ and standard deviation $\sigma$, we can standardize using

$$z = \frac{x - \mu}{\sigma}$$

**TRM Section 2.2 Quizzes**

There are three different quizzes available for this section in the Teacher's Resource Materials. Versions A and B are roughly parallel. Find them by clicking on the link in the TE-book, opening the TRCD, or logging onto the book's companion Web site www.whfreeman.com/tps5e (teacher log-in required).

## Teaching Tip

The solutions provided in the *Annotated Teacher's Edition* have been shortened to fit in the margin. Complete solutions are available in the printed *Solutions Manual*, or in the Teacher's e-Book or TRCD.

**TRM Full Solutions to Section 2.2 Exercises**

## Answers to Section 2.2 Exercises

**2.33** Sketches will vary, but here is one example:

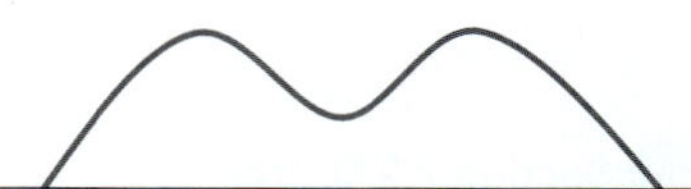

**2.34** Sketches will vary, but here is one example:

**2.35 (a)** It is on or above the horizontal axis everywhere, and the area beneath the curve is $\frac{1}{3} \times 3 = 1$. **(b)** $\frac{1}{3} \times 1 = \frac{1}{3}$. **(c)** Because $1.1 - 0.8 = 0.3$, the proportion is $\frac{1}{3} \times 0.3 = 0.1$.

**2.36 (a)** It is on or above the horizontal axis everywhere, and the area beneath the curve is $\frac{1}{10} \times 10 = 1$. **(b)** Because $10 - 8 = 2$, the percent is $\frac{1}{10} \times 2 = 0.2 = 20\%$. **(c)** Because $5.3 - 2.5 = 2.8$, the percent is $\frac{1}{10} \times 2.8 = 0.28 = 28\%$.

The variable $z$ has the **standard Normal distribution** with mean 0 and standard deviation 1.

- **Table A** at the back of the book gives percentiles for the standard Normal curve. By standardizing, we can use Table A to determine the percentile for a given $z$-score or the $z$-score corresponding to a given percentile in any Normal distribution. You can use your calculator or the *Normal Curve* applet to perform Normal calculations quickly.
- To perform certain inference procedures in later chapters, we will need to know that the data come from populations that are approximately Normally distributed. To assess Normality for a given set of data, we first observe the shape of a dotplot, stemplot, or histogram. Then we can check how well the data fit the 68–95–99.7 rule for Normal distributions. Another good method for assessing Normality is to construct a **Normal probability plot**.

### 2.2 TECHNOLOGY CORNERS

TI-Nspire instructions in Appendix B; HP Prime instructions on the book's Web site.



## Section 2.2 Exercises

33. **Density curves** Sketch a density curve that might describe a distribution that is symmetric but has two peaks.

34. **Density curves** Sketch a density curve that might describe a distribution that has a single peak and is skewed to the left.

*Exercises 35 to 38 involve a special type of density curve—one that takes constant height (looks like a horizontal line) over some interval of values. This density curve describes a variable whose values are distributed evenly (uniformly) over some interval of values. We say that such a variable has a* ***uniform distribution***.

35. **Biking accidents** Accidents on a level, 3-mile bike path occur uniformly along the length of the path. The figure below displays the density curve that describes the uniform distribution of accidents.

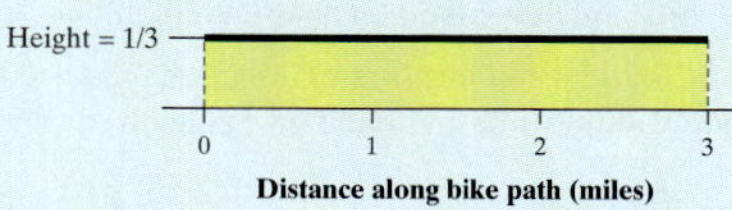


(a) Explain why this curve satisfies the two requirements for a density curve.

(b) The proportion of accidents that occur in the first mile of the path is the area under the density curve between 0 miles and 1 mile. What is this area?

(c) Sue's property adjoins the bike path between the 0.8 mile mark and the 1.1 mile mark. What proportion of accidents happen in front of Sue's property? Explain.

36. **Where's the bus?** Sally takes the same bus to work every morning. The amount of time (in minutes) that she has to wait for the bus to arrive is described by the uniform distribution below.

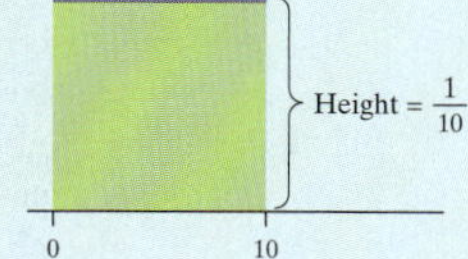


(a) Explain why this curve satisfies the two requirements for a density curve.

(b) On what percent of days does Sally have to wait more than 8 minutes for the bus?

(c) On what percent of days does Sally wait between 2.5 and 5.3 minutes for the bus?

**37. Biking accidents** What is the mean $\mu$ of the density curve pictured in Exercise 35? (That is, where would the curve balance?) What is the median? (That is, where is the point with area 0.5 on either side?)

**38. Where's the bus?** What is the mean $\mu$ of the density curve pictured in Exercise 36? What is the median?

**39. Mean and median** The figure below displays two density curves, each with three points marked. At which of these points on each curve do the mean and the median fall?

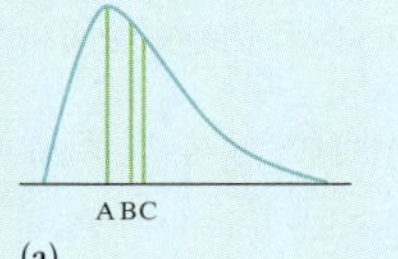


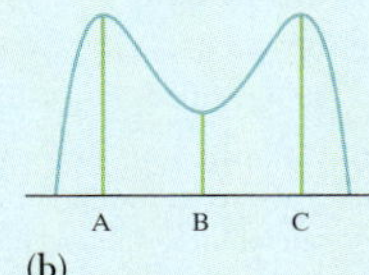


**40. Mean and median** The figure below displays two density curves, each with three points marked. At which of these points on each curve do the mean and the median fall?

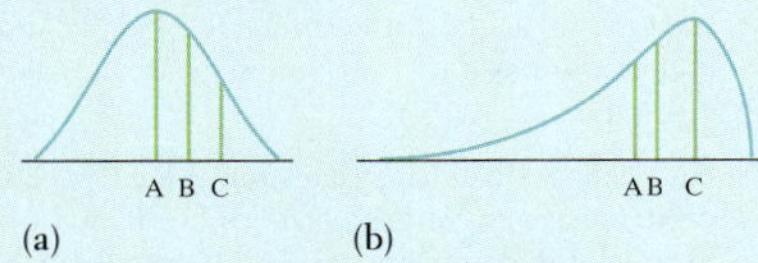


**41. Men's heights** The distribution of heights of adult American men is approximately Normal with mean 69 inches and standard deviation 2.5 inches. Draw an accurate sketch of the distribution of men's heights. Be sure to label the mean, as well as the points 1, 2, and 3 standard deviations away from the mean on the horizontal axis.

**42. Potato chips** The distribution of weights of 9-ounce bags of a particular brand of potato chips is approximately Normal with mean $\mu = 9.12$ ounces and standard deviation $\sigma = 0.05$ ounce. Draw an accurate sketch of the distribution of potato chip bag weights. Be sure to label the mean, as well as the points 1, 2, and 3 standard deviations away from the mean on the horizontal axis.

**43. Men's heights** Refer to Exercise 41. Use the 68–95–99.7 rule to answer the following questions. Show your work!

pg 111

(a) Between what heights do the middle 95% of men fall?

(b) What percent of men are taller than 74 inches?

(c) What percent of men are between 64 and 66.5 inches tall?

(d) A height of 71.5 inches corresponds to what percentile of adult male American heights?

**44. Potato chips** Refer to Exercise 42. Use the 68–95–99.7 rule to answer the following questions. Show your work!

(a) Between what weights do the middle 68% of bags fall?

(b) What percent of bags weigh less than 9.02 ounces?

(c) What percent of 9-ounce bags of this brand of potato chips weigh between 8.97 and 9.17 ounces?

(d) A bag that weighs 9.07 ounces is at what percentile in this distribution?

**45. Estimating SD** The figure below shows two Normal curves, both with mean 0. Approximately what is the standard deviation of each of these curves?

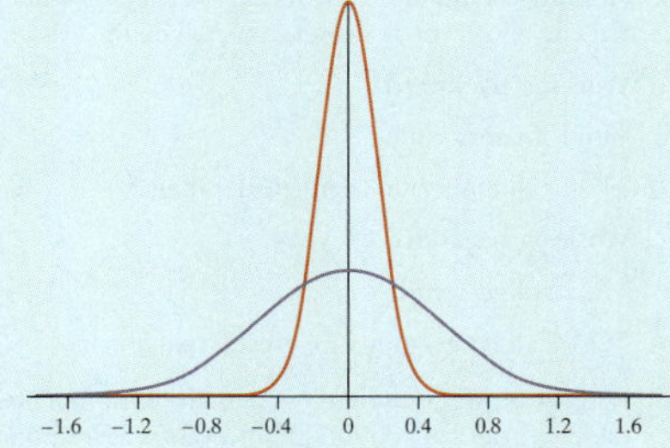


**46. A Normal curve** Estimate the mean and standard deviation of the Normal density curve in the figure below.

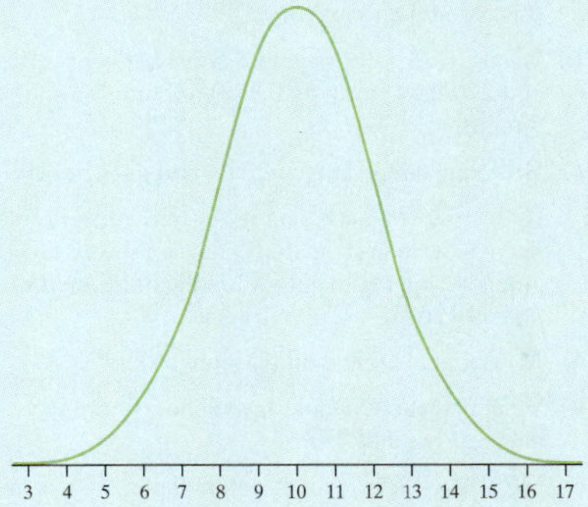


For Exercises 47 to 50, use Table A to find the proportion of observations from the standard Normal distribution that satisfies each of the following statements. In each case, sketch a standard Normal curve and shade the area under the curve that is the answer to the question.

**47. Table A practice**

(a) $z < 2.85$
(b) $z > 2.85$
(c) $z > -1.66$
(d) $-1.66 < z < 2.85$

**2.37** Both are 1.5.

**2.38** Both are 5.

**2.39** **(a)** Mean is C, median is B. **(b)** Mean is B, median is B.

**2.40** **(a)** Mean is A, median is A. **(b)** Mean is A, median is B.

**2.41** The graph is shown below.

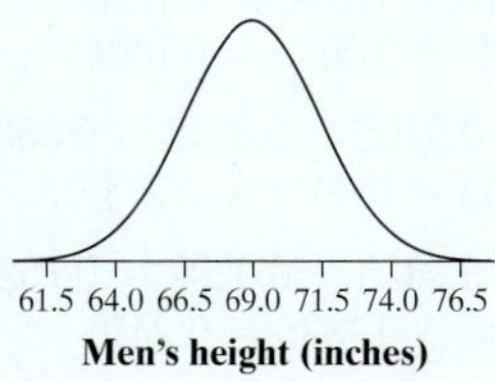


**2.42** The graph is shown below.

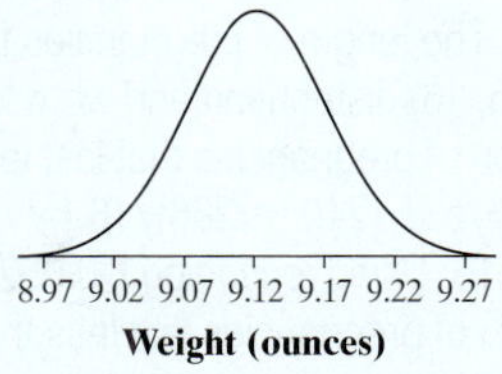


**2.43** **(a)** Between $69 - 2(2.5) = 64$ and $69 + 2(2.5) = 74$ inches. **(b)** About $\frac{100\% - 95\%}{2} = 2.5\%$. **(c)** About $\frac{100\% - 68\%}{2} = 16\%$ of men are shorter than 66.5 inches and $\frac{100\% - 95\%}{2} = 2.5\%$ are shorter than 64 inches, so approximately $16\% - 2.5\% = 13.5\%$ of men have heights between 64 inches and 66.5 inches. **(d)** Because $\frac{100\% - 68\%}{2} = 16\%$ of the area is to the right of 71.5, 71.5 is at the 84th percentile.

**2.44** **(a)** Between $9.12 - 0.05 = 9.07$ ounces and $9.12 + 0.05 = 9.17$ ounces. **(b)** About $\frac{100 - 95}{2} = 2.5\%$. **(c)** About $\frac{100\% - 68\%}{2} = 16\%$ of bags have weights greater than 9.17 ounces and $\frac{100\% - 99.7\%}{2} = 0.15\%$ of bags have weight less than 8.97 ounces, so approximately $100\% - 16\% - 0.15\% = 83.85\%$ of bags have weights between 8.97 and 9.17 ounces. **(d)** About $(100 - 68)/2 = 16\%$ of the bags weigh less than 9.07 ounces, so 9.07 is at the 16th percentile.

**2.45** Taller curve: standard deviation $\approx 0.2$. Shorter curve: standard deviation $\approx 0.5$.

**2.46** Mean $\approx 10$ and standard deviation $\approx 2$.

**2.47** **(a)** 0.9978 **(b)** $1 - 0.9978 = 0.0022$ **(c)** $1 - 0.0485 = 0.9515$ **(d)** $0.9978 - 0.0485 = 0.9493$

48. **Table A practice**

(a) $z < -2.46$ (c) $0.89 < z < 2.46$

(b) $z > 2.46$ (d) $-2.95 < z < -1.27$

pg 115 49. **More Table A practice**

(a) $z$ is between $-1.33$ and $1.65$

(b) $z$ is between $0.50$ and $1.79$

50. **More Table A practice**

(a) $z$ is between $-2.05$ and $0.78$

(b) $z$ is between $-1.11$ and $-0.32$

For Exercises 51 and 52, use Table A to find the value $z$ from the standard Normal distribution that satisfies each of the following conditions. In each case, sketch a standard Normal curve with your value of $z$ marked on the axis.

51. **Working backward**

(a) The 10th percentile.

(b) 34% of all observations are greater than $z$.

52. **Working backward**

(a) The 63rd percentile.

(b) 75% of all observations are greater than $z$.

53. **Length of pregnancies** The length of human pregnancies from conception to birth varies according to a distribution that is approximately Normal with mean 266 days and standard deviation 16 days.

pg 118 (a) At what percentile is a pregnancy that lasts 240 days (that's about 8 months)?

pg 119 (b) What percent of pregnancies last between 240 and 270 days (roughly between 8 months and 9 months)?

pg 120 (c) How long do the longest 20% of pregnancies last?

54. **IQ test scores** Scores on the Wechsler Adult Intelligence Scale (a standard IQ test) for the 20 to 34 age group are approximately Normally distributed with $\mu = 110$ and $\sigma = 25$.

(a) At what percentile is an IQ score of 150?

(b) What percent of people aged 20 to 34 have IQs between 125 and 150?

(c) MENSA is an elite organization that admits as members people who score in the top 2% on IQ tests. What score on the Wechsler Adult Intelligence Scale would an individual aged 20 to 34 have to earn to qualify for MENSA membership?

55. **Put a lid on it!** At some fast-food restaurants, customers who want a lid for their drinks get them from a large stack left near straws, napkins, and condiments. The lids are made with a small amount of flexibility so they can be stretched across the mouth of the cup and then snugly secured. When lids are too small or too large, customers can get very frustrated, especially if they end up spilling their drinks. At one particular restaurant, large drink cups require lids with a "diameter" of between 3.95 and 4.05 inches. The restaurant's lid supplier claims that the diameter of their large lids follows a Normal distribution with mean 3.98 inches and standard deviation 0.02 inches. Assume that the supplier's claim is true.

(a) What percent of large lids are too small to fit? Show your method.

(b) What percent of large lids are too big to fit? Show your method.

(c) Compare your answers to parts (a) and (b). Does it make sense for the lid manufacturer to try to make one of these values larger than the other? Why or why not?

56. **I think I can!** An important measure of the performance of a locomotive is its "adhesion," which is the locomotive's pulling force as a multiple of its weight. The adhesion of one 4400-horsepower diesel locomotive varies in actual use according to a Normal distribution with mean $\mu = 0.37$ and standard deviation $\sigma = 0.04$.

(a) For a certain small train's daily route, the locomotive needs to have an adhesion of at least 0.30 for the train to arrive at its destination on time. On what proportion of days will this happen? Show your method.

(b) An adhesion greater than 0.50 for the locomotive will result in a problem because the train will arrive too early at a switch point along the route. On what proportion of days will this happen? Show your method.

(c) Compare your answers to parts (a) and (b). Does it make sense to try to make one of these values larger than the other? Why or why not?

57. **Put a lid on it!** Refer to Exercise 55. The supplier is considering two changes to reduce the percent of its large-cup lids that are too small to 1%. One strategy is to adjust the mean diameter of its lids. Another option is to alter the production process, thereby decreasing the standard deviation of the lid diameters.

(a) If the standard deviation remains at $\sigma = 0.02$ inches, at what value should the supplier set the mean diameter of its large-cup lids so that only 1% are too small to fit? Show your method.

(b) If the mean diameter stays at $\mu = 3.98$ inches, what value of the standard deviation will result in only 1% of lids that are too small to fit? Show your method.

**2.48** **(a)** 0.0069 **(b)** $1 - 0.9931 = 0.0069$ **(c)** $0.9931 - 0.8133 = 0.1798$ **(d)** $0.1020 - 0.0016 = 0.1004$

**2.49** **(a)** $0.9505 - 0.0918 = 0.8587$ **(b)** $0.9633 - 0.6915 = 0.2718$

**2.50** **(a)** $0.7823 - 0.0202 = 0.7621$ **(b)** $0.3745 - 0.1335 = 0.2410$

**2.51** **(a)** $z = -1.28$ **(b)** $z = 0.41$

**2.52** **(a)** $z = 0.33$ **(b)** $z = -0.67$

**2.53** **(a)** The length of pregnancies follows an $N(266, 16)$ distribution and we want the proportion of pregnancies that last less than 240 days. $z = (240 - 266)/16 = -1.63 \rightarrow 0.0516$. *Using technology:* 0.0521. About 5% of pregnancies last less than 240 days, so 240 days is at the 5th percentile of pregnancy lengths. **(b)** We want the proportion of pregnancies that last between 240 and 270 days. $z = (240 - 266)/16 = -1.63$ and $z = (270 - 266)/16 = 0.25 \rightarrow 0.5987 - 0.0516 = 0.5471$. *Using technology:* 0.5466. About 55% of pregnancies last between 240 and 270 days. **(c)** We are looking for the boundary value $x$ that has an area of 0.20 to the right and 0.80 to the left. Solving $0.84 = (x - 266)/16$ gives $x = 279.44$. *Using technology:* 279.47. The longest 20% of pregnancies last longer than 279.47 days.

**2.54** **(a)** For the 20 to 34 age group, IQ scores follow an $N(110, 25)$ distribution and we want the proportion of people who have scores less than 150. $z = (150 - 110)/25 = 1.6 \rightarrow 0.9452$. *Using technology:* 0.9452. About 95% of people in this age group have IQ scores less than 150, so a score of 150 is at the 95th percentile. **(b)** We want the percent of people in this age group with IQ scores between 125 and 150. $z = (125 - 110)/25 = 0.6$ and $z = (150 - 110)/25 = 1.6 \rightarrow 0.9452 - 0.7257 = 0.2195$. *Using technology:* 0.2195. About 22% of 20- to 34-year-olds have IQ scores between 125 and 150. **(c)** Solving $2.05 = (x - 110)/25$ gives $x = 161.25$. *Using technology:* 161.34. Scores greater than 161.34 qualify for MENSA.

**2.55** **(a)** For large lids, the diameter follows an $N(3.98, 0.02)$ distribution and we want to find the percent of lids that have diameters less than 3.95. $z = (3.95 - 3.98)/0.02 = -1.5 \rightarrow 0.0668$. *Using technology:* 0.0668. About 7% of the large lids are too small to fit. **(b)** We want to find the percent of lids that have diameters greater than 4.05. $z = (4.05 - 3.98)/0.02 = 3.5 \rightarrow \approx 0$. *Using technology:* 0.0002. Approximately 0% of the large lids are too big to fit. **(c)** Make a larger proportion of lids too small. If lids are too small, customers will just try another lid. But if lids are too large, the customer may not notice and then spill the drink.

**2.56** **(a)** The adhesion of a locomotive follows an $N(0.37, 0.04)$ distribution and we want to find the proportion of days that the adhesion will be at least 0.30. $z = (0.3 - 0.37)/0.04 = -1.75 \rightarrow 1 - 0.0401 = 0.9599$. *Using technology:* 0.9599. The proportion of days that the locomotive will have an adhesion of at least 0.30 is 0.9599. **(b)** We want to find the proportion of days that the adhesion will be greater than 0.50. $z = (0.5 - 0.37)/0.04 = 3.25 \rightarrow 1 - 0.9994 = 0.0006$. *Using technology:* 0.0006. The proportion of days that the locomotive will have an adhesion of greater than 0.50 is 0.0006. **(c)** Make the value found in part (a) larger. We want the train to arrive at its destination on time, but not to arrive at the switch point early.

**2.57** **(a)** For large lids, the diameter follows an $N(\mu, 0.02)$ distribution and we want to find the value of $\mu$ that will result in only 1% of lids that are too small to fit. Solving $-2.33 = (3.95 - \mu)/0.02$ gives $\mu = 4.00$. *Using technology:* $z = -2.326$ and $\mu = 4.00$. The manufacturer should set the mean diameter to approximately $\mu = 4.00$ to ensure that only 1% of lids are too small. **(b)** We want to find the value of $\sigma$ that will result in only 1% of lids that are too small to fit. Solving $-2.33 = (3.95 - 3.98)/\sigma$ gives $\sigma = 0.013$. *Using technology:* $z = -2.326$ and $\sigma = 0.013$. A standard deviation of at most 0.013 will result in only 1% of lids that are too small to fit. **(c)** Reduce the standard deviation. This will reduce the number of lids that are too small and the number of lids that are too big. If we make the mean a little larger as in part (a) , we will reduce the number

(c) Which of the two options in parts (a) and (b) do you think is preferable? Justify your answer. (Be sure to consider the effect of these changes on the percent of lids that are too large to fit.)

58. **I think I can!** Refer to Exercise 56. The locomotive's manufacturer is considering two changes that could reduce the percent of times that the train arrives late. One option is to increase the mean adhesion of the locomotive. The other possibility is to decrease the variability in adhesion from trip to trip, that is, to reduce the standard deviation.

(a) If the standard deviation remains at $\sigma = 0.04$, to what value must the manufacturer change the mean adhesion of the locomotive to reduce its proportion of late arrivals to only 2% of days? Show your work.

(b) If the mean adhesion stays at $\mu = 0.37$, how much must the standard deviation be decreased to ensure that the train will arrive late only 2% of the time? Show your work.

(c) Which of the two options in parts (a) and (b) do you think is preferable? Justify your answer. (Be sure to consider the effect of these changes on the percent of days that the train arrives early to the switch point.)

59. **Deciles** The deciles of any distribution are the values at the 10th, 20th, . . . , 90th percentiles. The first and last deciles are the 10th and the 90th percentiles, respectively.

(a) What are the first and last deciles of the standard Normal distribution?

(b) The heights of young women are approximately Normal with mean 64.5 inches and standard deviation 2.5 inches. What are the first and last deciles of this distribution? Show your work.

60. **Outliers** The percent of the observations that are classified as outliers by the 1.5 × *IQR* rule is the same in any Normal distribution. What is this percent? Show your method clearly.

61. **Flight times** An airline flies the same route at the same time each day. The flight time varies according to a Normal distribution with unknown mean and standard deviation. On 15% of days, the flight takes more than an hour. On 3% of days, the flight lasts 75 minutes or more. Use this information to determine the mean and standard deviation of the flight time distribution.

62. **Brush your teeth** The amount of time Ricardo spends brushing his teeth follows a Normal distribution with unknown mean and standard deviation. Ricardo spends less than one minute brushing his teeth about 40% of the time. He spends more than two minutes brushing his teeth 2% of the time. Use this information to determine the mean and standard deviation of this distribution.

63. **Sharks** Here are the lengths in feet of 44 great white sharks:[11]

 pg 124

| 18.7 | 12.3 | 18.6 | 16.4 | 15.7 | 18.3 | 14.6 | 15.8 | 14.9 | 17.6 | 12.1 |
|---|---|---|---|---|---|---|---|---|---|---|
| 16.4 | 16.7 | 17.8 | 16.2 | 12.6 | 17.8 | 13.8 | 12.2 | 15.2 | 14.7 | 12.4 |
| 13.2 | 15.8 | 14.3 | 16.6 | 9.4 | 18.2 | 13.2 | 13.6 | 15.3 | 16.1 | 13.5 |
| 19.1 | 16.2 | 22.8 | 16.8 | 13.6 | 13.2 | 15.7 | 19.7 | 18.7 | 13.2 | 16.8 |

(a) Enter these data into your calculator and make a histogram. Include a sketch of the graph on your paper. Then calculate one-variable statistics. Describe the shape, center, and spread of the distribution of shark lengths.

(b) Calculate the percent of observations that fall within 1, 2, and 3 standard deviations of the mean. How do these results compare with the 68–95–99.7 rule?

(c) Use your calculator to construct a Normal probability plot. Include a sketch of the graph on your paper. Interpret this plot.

(d) Having inspected the data from several different perspectives, do you think these data are approximately Normal? Write a brief summary of your assessment that combines your findings from parts (a) through (c).

64. **Density of the earth** In 1798, the English scientist Henry Cavendish measured the density of the earth several times by careful work with a torsion balance. The variable recorded was the density of the earth as a multiple of the density of water. Here are Cavendish's 29 measurements:[12]

| 5.50 | 5.61 | 4.88 | 5.07 | 5.26 | 5.55 | 5.36 | 5.29 | 5.58 | 5.65 |
|---|---|---|---|---|---|---|---|---|---|
| 5.57 | 5.53 | 5.62 | 5.29 | 5.44 | 5.34 | 5.79 | 5.10 | 5.27 | 5.39 |
| 5.42 | 5.47 | 5.63 | 5.34 | 5.46 | 5.30 | 5.75 | 5.68 | 5.85 | |

(a) Enter these data into your calculator and make a histogram. Include a sketch of the graph on your paper. Then calculate one-variable statistics. Describe the shape, center, and spread of the distribution of density measurements.

(b) Calculate the percent of observations that fall within 1, 2, and 3 standard deviations of the mean. How do these results compare with the 68–95–99.7 rule?

(c) Use your calculator to construct a Normal probability plot. Include a sketch of the graph on your paper. Interpret this plot.

(d) Having inspected the data from several different perspectives, do you think these data are approximately Normal? Write a brief summary of your assessment that combines your findings from parts (a) through (c).

of lids that are too small, but we will increase the number of lids that are too big.

**2.58** **(a)** The adhesion of a locomotive follows an $N(\mu, 0.04)$ distribution and we want to find the value of $\mu$ that will result in only 2% of days with late arrivals. Solving $-2.05 = (0.30 - \mu)/0.04$ gives $\mu = 0.382$. *Using technology:* $z = -2.054$ and $\mu = 0.382$. The manufacturer should change the mean adhesion to approximately $\mu = 0.382$ to ensure that the train arrives late only 2% of days. **(b)** We want to find the value of $\sigma$ that will result in only 2% of days with late arrivals. Solving $-2.05 = (0.30 - 0.37)/\sigma$ gives $\sigma = 0.034$. *Using technology:* $z = -2.054$ and $\sigma = 0.034$. The standard deviation should be decreased to at most 0.034 to ensure that the train arrives late only 2% of days. **(c)** Reduce the standard deviation. This will reduce the proportion of late days and the proportion of early days. If we make the mean larger, we will reduce the proportion of late days, but increase the proportion of early days.

**2.59** **(a)** $z = -1.28$ and $z = 1.28$. **(b)** Solving $-1.28 = (x - 64.5)/2.5$ gives $x = 61.3$ inches and solving $1.28 = (x - 64.5)/2.5$ gives $x = 67.7$ inches.

**2.60** The value $z = -0.67$ has an area of about 0.25 to the left and the value $z = 0.67$ has an area of about 0.75 to the left. Thus, $Q_1 = -0.67$, $Q_3 = 0.67$, and $IQR = 0.67 - (-0.67) = 1.34$. Outliers are values less than $z = -0.67 - 1.5(1.34) = -2.68$ and greater than $z = 0.67 + 1.5(1.34) = 2.68$. Finally, the area to the left of $-2.68$ is 0.0037 and to the right of 2.68 is $1 - 0.9963 = 0.0037$. Thus, the percent of values that are outliers is $0.0037 + 0.0037 = 0.0074 = 0.74\%$.

**2.61** Solving $1.04 = (60 - \mu)/\sigma$ and $1.88 = (75 - \mu)/\sigma$ gives $\mu = 41.43$ minutes and $\sigma = 17.86$ minutes.

**2.62** Solving $-0.25 = (1 - \mu)/\sigma$ and $2.05 = (2 - \mu)/\sigma$ gives $\mu = 1.1087$ minutes and $\sigma = 0.4348$ minutes.

**2.63** **(a)** A histogram is given below. The distribution of shark lengths is roughly symmetric and somewhat bell-shaped with a mean of 15.586 feet and a standard deviation of 2.55 feet.

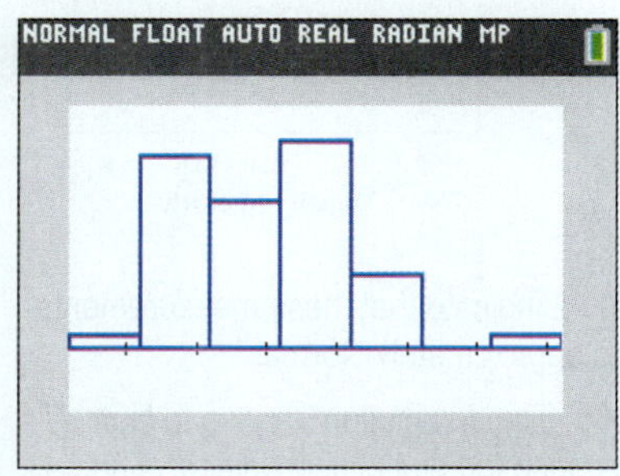


**(b)** 30/44 = 68.2%, 42/44 = 95.5%, and 44/44 = 100%. These are very close to the 68–95–99.7 rule. **(c)** A Normal probability plot is given below. Except for one small shark and one large shark, the plot is fairly linear, indicating that the distribution of shark lengths is approximately Normal.

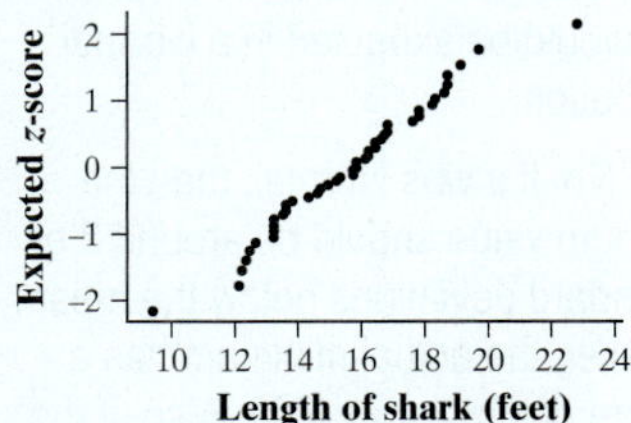


**(d)** All indicate that shark lengths are approximately Normal.

**2.64** **(a)** A histogram is given below. The measurements of the earth's density are roughly symmetric, single-peaked, and somewhat bell-shaped with a mean of 5.45 and a standard deviation of 0.2209.

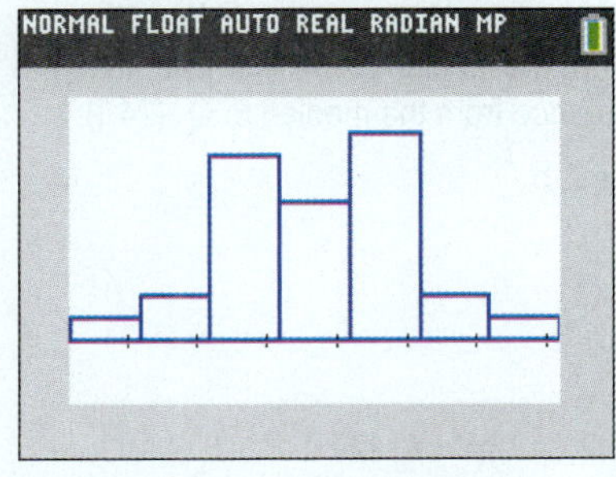


**(b)** 22/29 = 75.86%, 28/29 = 96.55%, and 29/29 = 100%. The densities follow the 68–95–99.7 rule reasonably well.

(c) A Normal probability plot (shown below) is roughly linear, indicating that the densities are approximately Normal.

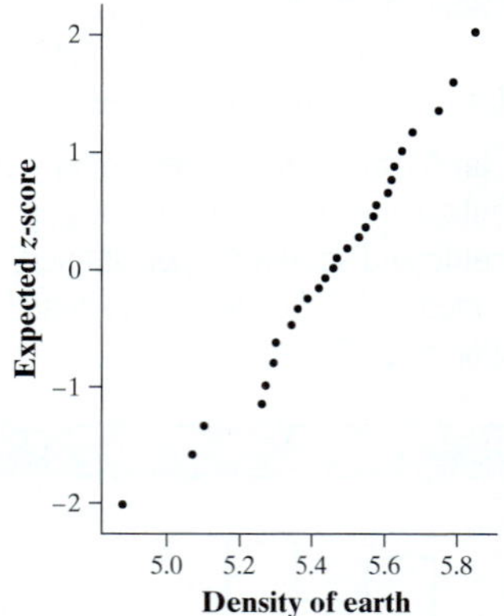


(d) All indicate that these measurements are approximately Normal.

**2.65** The distribution is close to Normal because the plot is nearly linear. There is a small "wiggle" between 120 and 130 with several values a little larger than would be expected in a Normal distribution. Also, the smallest value and the two largest values are a little farther from the mean than would be expected in a Normal distribution.

**2.66** The sharp curve in the plot suggests that the data are right-skewed. The values on the left are much closer to the mean than would be expected in a Normal distribution.

**2.67** No. If it was Normal, then the minimum value should be around 2 or 3 standard deviations below the mean. However, the actual minimum has a z-score of just $z = -1.09$. Also, if the distribution was Normal, the minimum and maximum should be about the same distance from the mean. However, the maximum is much farther from the mean (20,209) than the minimum (8741).

**2.68** The distribution is skewed to the right. In a Normal distribution, $Q_1$ and $Q_3$ should be about the same distance from the median. However, the distance from $Q_1$ to the median (14.9) is smaller than the distance from the median to $Q_3$ (24.1).

**2.69** b



65. **Runners' heart rates** The figure below is a Normal probability plot of the heart rates of 200 male runners after six minutes of exercise on a treadmill.[13] The distribution is close to Normal. How can you see this? Describe the nature of the small deviations from Normality that are visible in the plot.

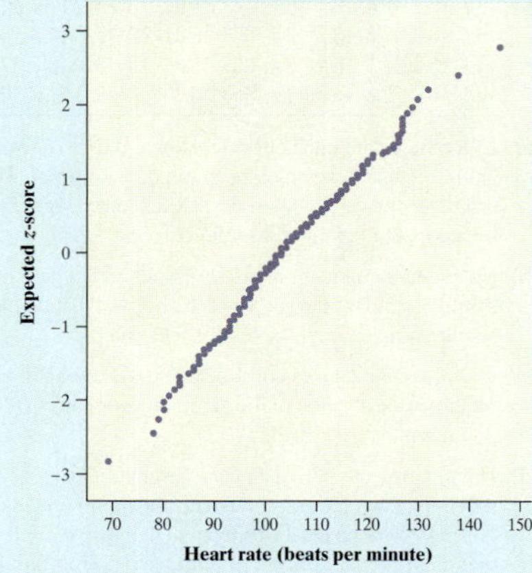


66. **Carbon dioxide emissions** The figure below is a Normal probability plot of the emissions of carbon dioxide per person in 48 countries.[14] In what ways is this distribution non-Normal?

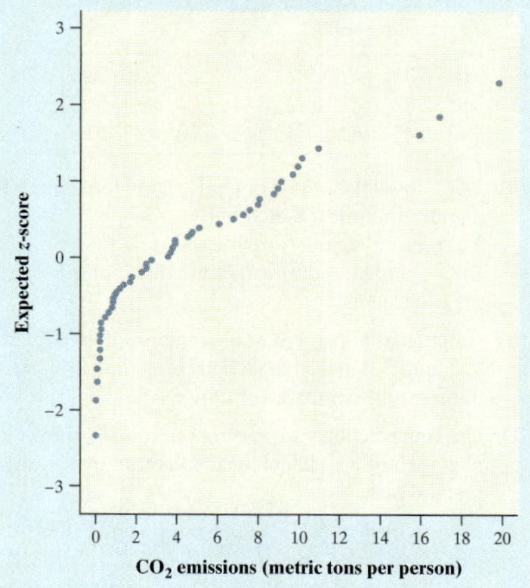


67. **Is Michigan Normal?** We collected data on the tuition charged by colleges and universities in Michigan. Here are some numerical summaries for the data:

| Mean | Std. Dev. | Min | Max |
|---|---|---|---|
| 10614 | 8049 | 1873 | 30823 |

Based on the relationship between the mean, standard deviation, minimum, and maximum, is it reasonable to believe that the distribution of Michigan tuitions is approximately Normal? Explain.

68. **Weights aren't Normal** The heights of people of the same gender and similar ages follow Normal distributions reasonably closely. Weights, on the other hand, are not Normally distributed. The weights of women aged 20 to 29 have mean 141.7 pounds and median 133.2 pounds. The first and third quartiles are 118.3 pounds and 157.3 pounds. What can you say about the shape of the weight distribution? Why?

*Multiple choice: Select the best answer for Exercises 69 to 74.*

69. Two measures of center are marked on the density curve shown. Which of the following is correct?

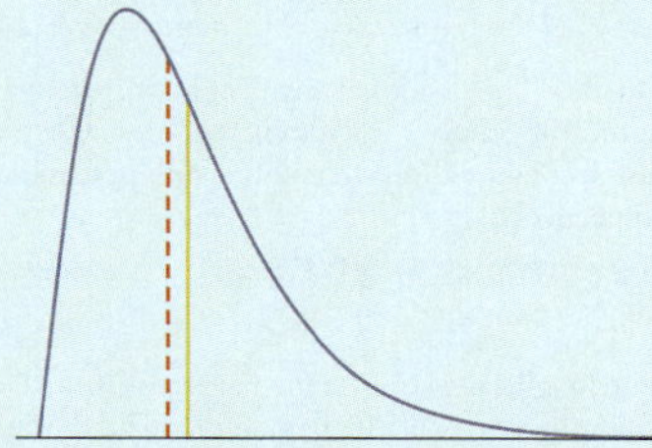

(a) The median is at the yellow line and the mean is at the red line.

(b) The median is at the red line and the mean is at the yellow line.

(c) The mode is at the red line and the median is at the yellow line.

(d) The mode is at the yellow line and the median is at the red line.

(e) The mode is at the red line and the mean is at the yellow line.

*Exercises 70 to 72 refer to the following setting.* The weights of laboratory cockroaches follow a Normal distribution with mean 80 grams and standard deviation 2 grams. The following figure is the Normal curve for this distribution of weights.

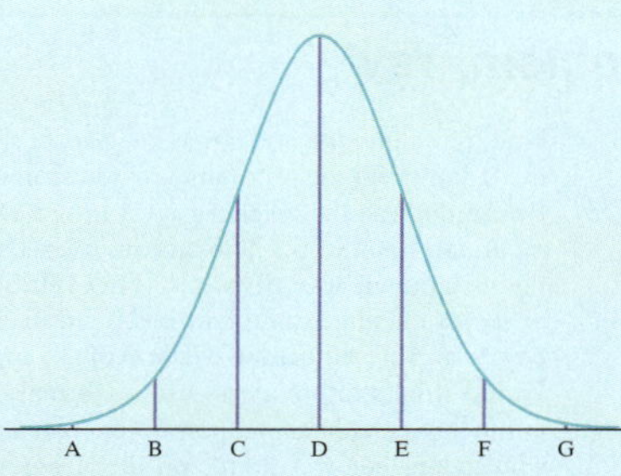


70. Point C on this Normal curve corresponds to
(a) 84 grams. (c) 78 grams. (e) 74 grams.
(b) 82 grams. (d) 76 grams.

71. About what percent of the cockroaches have weights between 76 and 84 grams?
(a) 99.7% (c) 68% (e) 34%
(b) 95% (d) 47.5%

72. About what proportion of the cockroaches will have weights greater than 83 grams?
(a) 0.0228 (c) 0.1587 (e) 0.0772
(b) 0.0668 (d) 0.9332

73. A different species of cockroach has weights that follow a Normal distribution with a mean of 50 grams. After measuring the weights of many of these cockroaches, a lab assistant reports that 14% of the cockroaches weigh more than 55 grams. Based on this report, what is the approximate standard deviation of weights for this species of cockroaches?
(a) 4.6 (d) 14.0
(b) 5.0 (e) Cannot determine without more information.
(c) 6.2

74. The following Normal probability plot shows the distribution of points scored for the 551 players in the 2011–2012 NBA season.

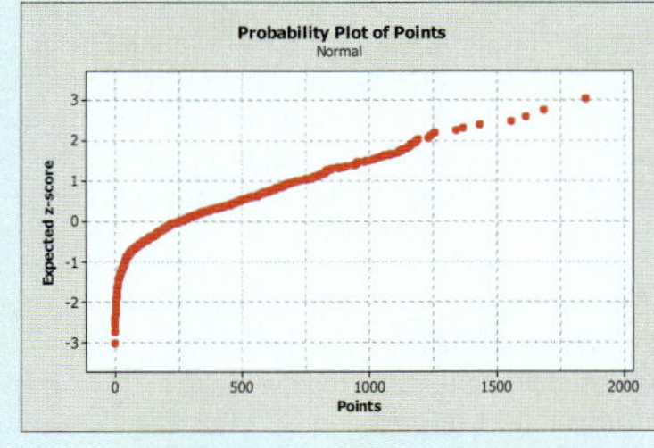


If the distribution of points was displayed in a histogram, what would be the best description of the histogram's shape?
(a) Approximately Normal
(b) Symmetric but not approximately Normal
(c) Skewed left
(d) Skewed right
(e) Cannot be determined

75. **Gas it up!** (1.3) Interested in a sporty car? Worried that it might use too much gas? The Environmental Protection Agency lists most such vehicles in its "two-seater" or "minicompact" categories. The figure shows boxplots for both city and highway gas mileages for our two groups of cars. Write a few sentences comparing these distributions.

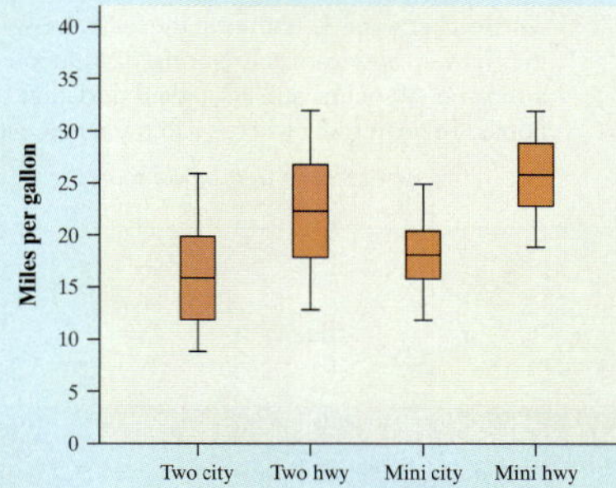


76. **Python eggs** (1.1) How is the hatching of water python eggs influenced by the temperature of the snake's nest? Researchers assigned newly laid eggs to one of three temperatures: hot, neutral, or cold. Hot duplicates the extra warmth provided by the mother python, and cold duplicates the absence of the mother. Here are the data on the number of eggs and the number that hatched:[15]

| | Cold | Neutral | Hot |
|---|---|---|---|
| Number of eggs | 27 | 56 | 104 |
| Number hatched | 16 | 38 | 75 |

(a) Make a two-way table of temperature by outcome (hatched or not).
(b) Calculate the percent of eggs in each group that hatched. The researchers believed that eggs would be less likely to hatch in cold water. Do the data support that belief?

**2.70** c

**2.71** b

**2.72** b

**2.73** a

**2.74** d

**2.75** For both kinds of cars we see that the highway mileage is greater than the city mileage. The two-seater cars have a more variable distribution, both on the highway and in the city. Also the mileage values are slightly lower for the two-seater cars than for the minicompact cars, both on the highway and in the city, with a greater difference on the highway. All four distributions are roughly symmetric.

**2.76** **(a)** The two-way table is shown below. **(b)** Cold: 16/27 = 59.26%, neutral: 38/56 = 67.86%, hot: 75/104 = 72.12%. The data support the belief that the eggs are less likely to hatch in cold water.

| | Cold | Neutral | Hot | Total |
|---|---|---|---|---|
| Hatched | 16 | 38 | 75 | **129** |
| Not hatched | 11 | 18 | 29 | **58** |
| **Total** | **27** | **56** | **104** | **187** |

## FRAPPY! Free Response AP® Problem, Yay!

The following problem is modeled after actual AP® Statistics exam free response questions. Your task is to generate a complete, concise response in 15 minutes.

*Directions: Show all your work. Indicate clearly the methods you use, because you will be scored on the correctness of your methods as well as on the accuracy and completeness of your results and explanations.*

The distribution of scores on a recent test closely followed a Normal distribution with a mean of 22 points and a standard deviation of 4 points.

(a) What proportion of the students scored at least 25 points on this test?

(b) What is the 31st percentile of the distribution of test scores?

(c) The teacher wants to transform the test scores so that they have an approximately Normal distribution with a mean of 80 points and a standard deviation of 10 points. To do this, she will use a formula in the form:

*new score = a + b (old score)*

Find the values of $a$ and $b$ that the teacher should use to transform the distribution of test scores.

(d) Before the test, the teacher gave a review assignment for homework. The maximum score on the assignment was 10 points. The distribution of scores on this assignment had a mean of 9.2 points and a standard deviation of 2.1 points. Would it be appropriate to use a Normal distribution to calculate the proportion of students who scored below 7 points on this assignment? Explain.

After you finish, you can view two example solutions on the book's Web site (www.whfreeman.com/tps5e). Determine whether you think each solution is "complete," "substantial," "developing," or "minimal." If the solution is not complete, what improvements would you suggest to the student who wrote it? Finally, your teacher will provide you with a scoring rubric. Score your response and note what, if anything, you would do differently to improve your own score.

# Chapter Review

### Section 2.1: Describing Location in a Distribution

In this section, you learned two different ways to describe the location of individuals in a distribution, percentiles and standardized scores ($z$-scores). Percentiles describe the location of an individual by measuring what percent of the observations in the distribution have a value less than the individual's value. A cumulative relative frequency graph is a handy tool for identifying percentiles in a distribution. You can use it to estimate the percentile for a particular value of a variable or estimate the value of the variable at a particular percentile.

Standardized scores ($z$-scores) describe the location of an individual in a distribution by measuring how many standard deviations the individual is above or below the mean. To find the standardized score for a particular observation, transform the value by subtracting the mean and dividing the difference by the standard deviation. Besides describing the location of an individual in a distribution, you can also use $z$-scores to compare observations from different distributions—standardizing the values puts them on a standard scale.

You also learned to describe the effects on the shape, center, and spread of a distribution when transforming data from one scale to another. Adding a positive constant to (or subtracting it from) each value in a data set changes the measures of location but not the shape or spread of the distribution. Multiplying or dividing each value in a data set by a positive constant changes the measures of location and measures of spread but not the shape of the distribution.

### TRM FRAPPY!

We present a model solution here. Please consult the Teacher's Resource Materials for sample student responses, a scoring rubric, and a black line master of the original question with space for students to write their responses.

**(a)** *Step 1: State the distribution and the values of interest.* The distribution of scores follows a Normal distribution with mean 22 and standard deviation 4. We want to find the proportion of scores that are at least 25 (see graph below).
*Step 2: Perform calculations—show your work.* The standardized score for the boundary value is $z = \frac{25 - 22}{4} = 0.75$.

From Table A, the proportion of $z$-scores above 0.75 is $1 - 0.7734 = 0.2266$.
*Using technology:* The command `normalcdf(lower:25,upper:1000,μ:22,σ:4)` gives an area of 0.2266.
*Step 3: Answer the question.* The proportion of students who scored at least 25 points is about 0.23.

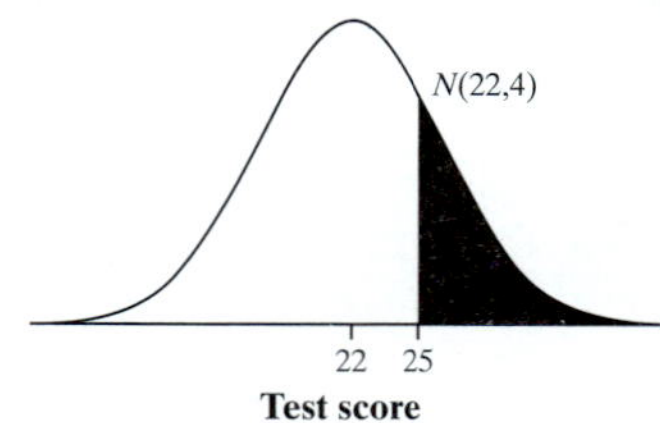


**(b)** *Step 1: State the distribution and the values of interest.* The distribution of scores follows a Normal distribution with mean 22 and standard deviation 4. The 31st percentile is the boundary value $x$ with 31% of the distribution to its left (see graph below).
*Step 2: Perform calculations—show your work.* Look in the body of Table A for the value closest to 0.31. A $z$-score of $-0.50$ gives the closest value (0.3085).

Solving $-0.50 = \frac{x - 22}{4}$ gives $x = 20$.

*Using technology:* The command `invNorm(area:0.31,μ:22,σ:4)` gives a value of 20.02.
*Step 3: Answer the question.* The 31st percentile is about 20 points.

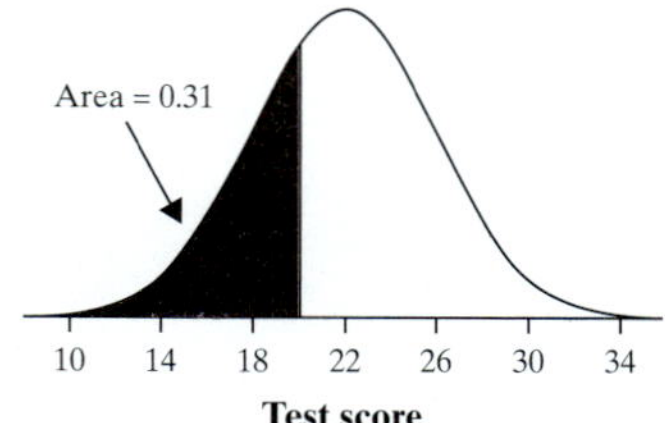


**(c)** Because adding a constant does not affect measures of spread:

$$\text{new SD} = b\,(\text{old SD}) \rightarrow 10 = b\,(4) \rightarrow b = 2.5$$

Because adding a constant and multiplying by a constant affect measures of center:

$$\text{new mean} = a + b\,(\text{old mean})$$
$$80 = a + 2.5\,(22) \rightarrow a = 25$$

The linear transformation should be old score = 25 + 2.5 (old score). **(d)** If the distribution was Normal, scores should go at least 2 standard deviations above the mean. However, in this distribution, the highest possible score is only $z = (10 - 9.2)/2.1 = 0.38$ standard deviations above the mean. Therefore, it would be inappropriate to use a Normal distribution to do calculations.

### Section 2.2: Density Curves and Normal Distributions

In this section, you learned how density curves are used to model distributions of data. An area under a density curve gives the proportion of observations that fall in a specified interval of values. The total area under a density curve is 1, or 100%.

The most commonly used density curve is called a Normal curve. The Normal curve is symmetric, single-peaked, and bell-shaped with mean $\mu$ and standard deviation $\sigma$. For any distribution of data that is approximately Normal in shape, about 68% of the observations will be within 1 standard deviation of the mean, about 95% of the observations will be within 2 standard deviations of the mean, and about 99.7% of the observations will be within 3 standard deviations of the mean. Conveniently, this relationship is called the 68–95–99.7 rule.

When observations do not fall exactly 1, 2, or 3 standard deviations from the mean, you learned how to use Table A (or technology) to identify the proportion of values in any specified interval under a Normal curve. You also learned how to use Table A (or technology) to determine the value of an individual that falls at a specified percentile in a Normal distribution. On the AP® exam, it is extremely important that you clearly communicate your methods when answering questions that involve the Normal distribution. You must specify the shape (Normal), center ($\mu$), and spread ($\sigma$) of the distribution; identify the region under the Normal curve that you are working with; and correctly calculate the answer with work shown. Shading a Normal curve with the mean, standard deviation, and boundaries clearly identified is a great start.

Finally, you learned how to determine whether a distribution of data is approximately Normal using graphs (dotplots, stemplots, histograms) and the 68–95–99.7 rule. You also learned that a Normal probability plot is a great way to determine whether the shape of a distribution is approximately Normal. The more linear the Normal probability plot, the more Normal the distribution of the data.

## What Did You Learn?

| Learning Objective | Section | Related Example on Page(s) | Relevant Chapter Review Exercise(s) |
|---|---|---|---|
| Find and interpret the percentile of an individual value within a distribution of data. | 2.1 | 86 | R2.1 |
| Estimate percentiles and individual values using a cumulative relative frequency graph. | 2.1 | 87, 88 | R2.2 |
| Find and interpret the standardized score (*z*-score) of an individual value within a distribution of data. | 2.1 | 90, 91 | R2.1 |
| Describe the effect of adding, subtracting, multiplying by, or dividing by a constant on the shape, center, and spread of a distribution of data. | 2.1 | 93, 94, 95 | R2.3 |
| Estimate the relative locations of the median and mean on a density curve. | 2.2 | Discussion on 106–107 | R2.4 |
| Use the 68–95–99.7 rule to estimate areas (proportions of values) in a Normal distribution. | 2.2 | 111 | R2.5 |
| Use Table A or technology to find (i) the proportion of *z*-values in a specified interval, or (ii) a *z*-score from a percentile in the standard Normal distribution. | 2.2 | 114, 115 Discussion on 116 | R2.6 |
| Use Table A or technology to find (i) the proportion of values in a specified interval, or (ii) the value that corresponds to a given percentile in any Normal distribution. | 2.2 | 118, 119, 120 | R2.7, R2.8, R2.9 |
| Determine whether a distribution of data is approximately Normal from graphical and numerical evidence. | 2.2 | 122, 123, 124 | R2.10, R2.11 |



### TRM Chapter 2 Tests

There are four different tests available for this section in the Teacher's Resource Materials. Versions A and B are roughly parallel and version D is cumulative. Find them by clicking on the link in the TE-book, opening the TRCD, or logging onto the book's companion Web site www.whfreeman.com/tps5e (teacher log-in required).

### TRM Learning Objectives Grid

# Chapter 2 Chapter Review Exercises

*These exercises are designed to help you review the important ideas and methods of the chapter.*

**R2.1 Is Paul tall?** According to the National Center for Health Statistics, the distribution of heights for 15-year-old males has a mean of 170 centimeters (cm) and a standard deviation of 7.5 cm. Paul is 15 years old and 179 cm tall.

(a) Find the $z$-score corresponding to Paul's height. Explain what this value means.

(b) Paul's height puts him at the 85th percentile among 15-year-old males. Explain what this means to someone who knows no statistics.

**R2.2 Computer use** Mrs. Causey asked her students how much time they had spent using a computer during the previous week. The following figure shows a cumulative relative frequency graph of her students' responses.

(a) At what percentile does a student who used her computer for 7 hours last week fall?

(b) Estimate the interquartile range ($IQR$) from the graph. Show your work.

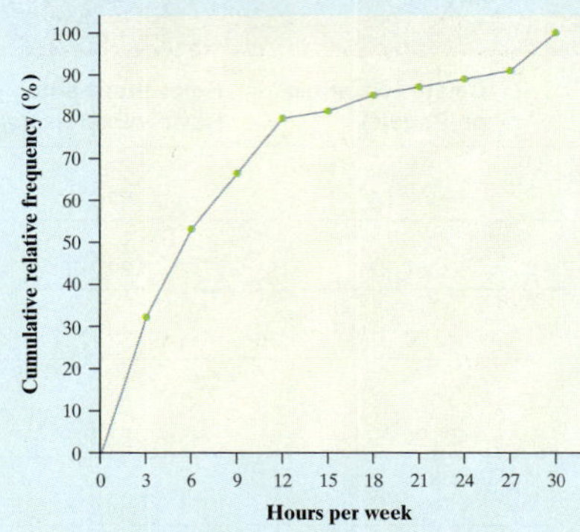


**R2.3 Aussie, Aussie, Aussie** A group of Australian students were asked to estimate the width of their classroom in feet. Use the dotplot and summary statistics below to answer the following questions.

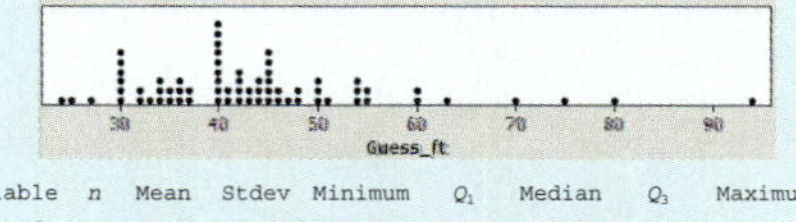


| Variable | n | Mean | Stdev | Minimum | $Q_1$ | Median | $Q_3$ | Maximum |
|---|---|---|---|---|---|---|---|---|
| Guess ft | 66 | 43.70 | 12.50 | 24.00 | 35.50 | 42.00 | 48.00 | 94.00 |

(a) Suppose we converted each student's guess from feet to meters (3.28 ft = 1 m). How would the shape of the distribution be affected? Find the mean, median, standard deviation, and $IQR$ for the transformed data.

(b) The actual width of the room was 42.6 feet. Suppose we calculated the error in each student's guess as follows: guess − 42.6. Find the mean and standard deviation of the errors. Justify your answers.

**R2.4 What the mean means** The figure below is a density curve. Trace the curve onto your paper.

(a) Mark the approximate location of the median. Explain your choice of location.

(b) Mark the approximate location of the mean. Explain your choice of location.

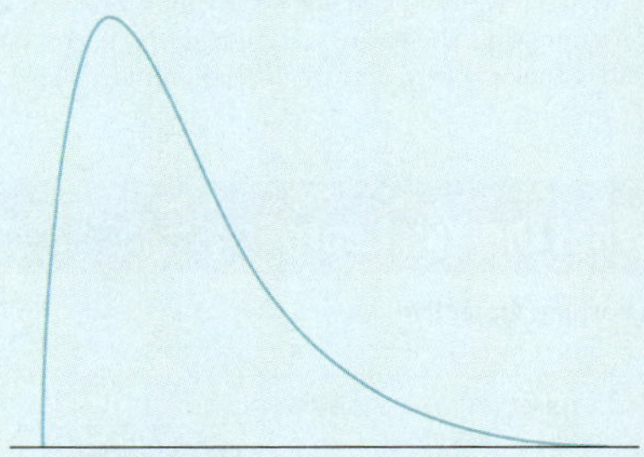

**R2.5 Horse pregnancies** Bigger animals tend to carry their young longer before birth. The length of horse pregnancies from conception to birth varies according to a roughly Normal distribution with mean 336 days and standard deviation 3 days. Use the 68–95–99.7 rule to answer the following questions.

(a) Almost all (99.7%) horse pregnancies fall in what interval of lengths?

(b) What percent of horse pregnancies are longer than 339 days? Show your work.

**R2.6 Standard Normal distribution** Use Table A (or technology) to find each of the following for a standard Normal distribution. In each case, sketch a standard Normal curve and shade the area of interest.

(a) The proportion of observations with $-2.25 < z < 1.77$

(b) The number $z$ such that 35% of all observations are greater than $z$

**TRM Full Solutions to Chapter 2 Review Exercises**

## Answers to Chapter 2 Review Exercises

**R2.1 (a)** $z = 1.20$. Paul's height is 1.20 standard deviations above the average male height for his age. **(b)** 85% of boys Paul's age are shorter than Paul.

**R2.2 (a)** 58th percentile. **(b)** $IQR = 11 - 2.5 = 8.5$ hours per week.

**R2.3 (a)** The shape of the distribution would not change. Mean $= 43.7/3.28 = 13.32$ meters, median $= 42/3.28 = 12.80$ meters, standard deviation $= 12.5/3.28 = 3.81$ meters, $IQR = 12.5/3.28 = 3.81$ meters. **(b)** Mean $= 43.7 - 42.6 = 1.1$ feet; standard deviation $= 12.5$ feet, because subtracting a constant from each observation does not change the spread.

**R2.4 (a)** The median (line A in the graph below) should be slightly to the right of the main peak with half of the area to the left and half to the right. **(b)** The mean (line B in the graph below) should be slightly to the right of the line for the median at the balancing point.

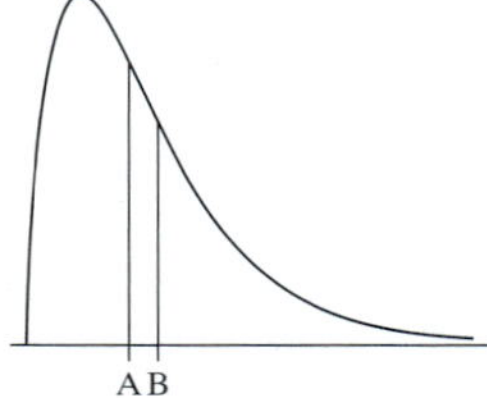


**R2.5 (a)** Between $336 - 3(3) = 327$ days and $336 + 3(3) = 345$ days. **(b)** About $(100\% - 68\%)/2 = 16\%$.

**R2.6 (a)** $0.9616 - 0.0122 = 0.9494$

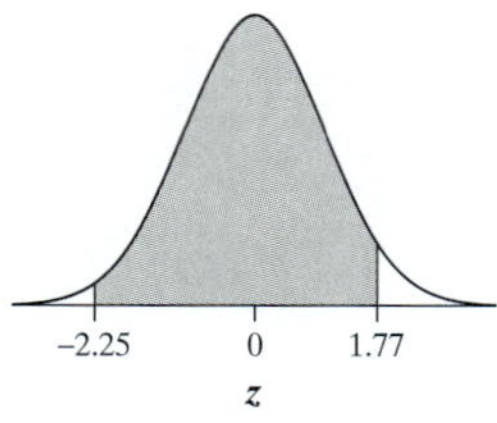


**(b)** If 35% of all values are greater than a particular $z$-value, then 65% are lower. A $z$-score of 0.39 gives the value closest to 0.65 (0.6517). *Using technology:* `invNorm(area:0.65,μ:0,σ:1)` gives $z = 0.385$.

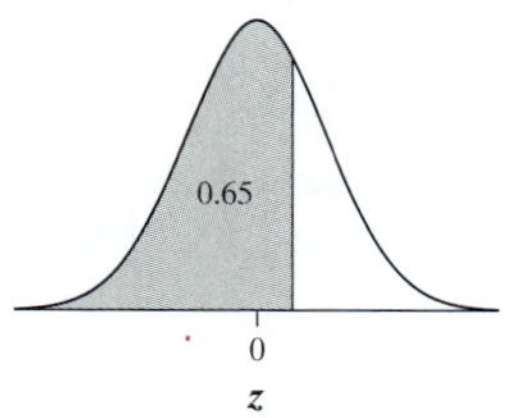


**R2.7 (a)** Birth weights follow an $N(3668, 511)$ distribution and we want to find the percent of babies with weights less than 2500 grams (see graph below). $z = (2500 - 3668)/511 = -2.29$. From Table A, the proportion of $z$-scores below $-2.29$ is 0.0110. *Using technology:* `normalcdf(lower:-1000,upper:2500,μ:3668,σ:511)` $= 0.0111$. About 1% of babies will be identified as low birth weight.

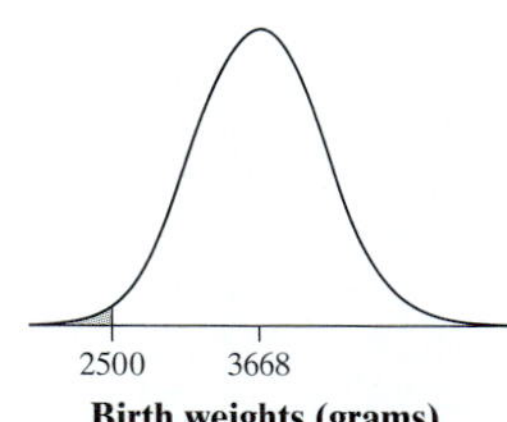


**(b)** Birth weights follow an $N(3668, 511)$ distribution. The first quartile is the boundary value with 25% of the area to its left. The third quartile is the boundary value with 75% of the area to its left (see graph below). A $z$-score of $-0.67$ gives the value closest to 0.25 (0.2514). Solving $-0.67 = \dfrac{x - 3668}{511}$ gives $Q_1 = 3325.63$. A $z$-score of 0.67 gives the value closest to 0.75 (0.7486). Solving $0.67 = \dfrac{x - 3668}{511}$ gives $Q_3 = 4010.37$. *Using technology:* `invNorm(area:0.25,μ:3668,σ:511)` gives $Q_1 = 3323.34$, and `invNorm(area:0.75,μ:3668,σ:511)` gives $Q_3 = 4012.66$. The quartiles are $Q_1 = 3323.34$ grams and $Q_3 = 4012.66$ grams.

**R2.7 Low-birth-weight babies** Researchers in Norway analyzed data on the birth weights of 400,000 newborns over a 6-year period. The distribution of birth weights is Normal with a mean of 3668 grams and a standard deviation of 511 grams.[16] Babies that weigh less than 2500 grams at birth are classified as "low birth weight."

(a) What percent of babies will be identified as low birth weight? Show your work.

(b) Find the quartiles of the birth weight distribution. Show your work.

**R2.8 Ketchup** A fast-food restaurant has just installed a new automatic ketchup dispenser for use in preparing its burgers. The amount of ketchup dispensed by the machine follows a Normal distribution with mean 1.05 ounces and standard deviation 0.08 ounce.

(a) If the restaurant's goal is to put between 1 and 1.2 ounces of ketchup on each burger, what percent of the time will this happen? Show your work.

(b) Suppose that the manager adjusts the machine's settings so that the mean amount of ketchup dispensed is 1.1 ounces. How much does the machine's standard deviation have to be reduced to ensure that at least 99% of the restaurant's burgers have between 1 and 1.2 ounces of ketchup on them?

**R2.9 Grading managers** Many companies "grade on a bell curve" to compare the performance of their managers and professional workers. This forces the use of some low performance ratings, so that not all workers are listed as "above average." Ford Motor Company's "performance management process" for a time assigned 10% A grades, 80% B grades, and 10% C grades to the company's 18,000 managers. Suppose that Ford's performance scores really are Normally distributed. This year, managers with scores less than 25 received C's, and those with scores above 475 received A's. What are the mean and standard deviation of the scores? Show your work.

**R2.10 Fruit fly thorax lengths** Here are the lengths in millimeters of the thorax for 49 male fruit flies:[17]

| 0.64 | 0.64 | 0.64 | 0.68 | 0.68 | 0.68 | 0.72 | 0.72 | 0.72 | 0.72 | 0.74 | 0.76 | 0.76 |
|---|---|---|---|---|---|---|---|---|---|---|---|---|
| 0.76 | 0.76 | 0.76 | 0.76 | 0.76 | 0.76 | 0.78 | 0.80 | 0.80 | 0.80 | 0.80 | 0.80 | 0.82 |
| 0.82 | 0.84 | 0.84 | 0.84 | 0.84 | 0.84 | 0.84 | 0.84 | 0.84 | 0.84 | 0.84 | 0.88 | 0.88 |
| 0.88 | 0.88 | 0.88 | 0.88 | 0.88 | 0.88 | 0.92 | 0.92 | 0.92 | 0.94 | | | |

Are these data approximately Normally distributed? Give appropriate graphical and numerical evidence to support your answer.

**R2.11 Assessing Normality** A Normal probability plot of a set of data is shown here. Would you say that these measurements are approximately Normally distributed? Why or why not?

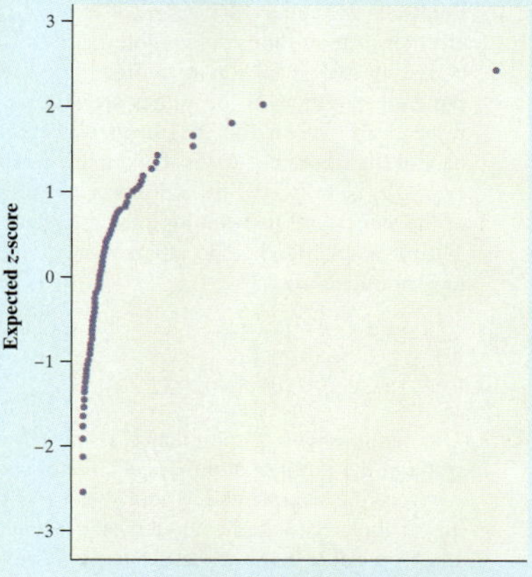


## Chapter 2 AP® Statistics Practice Test

**Section I: Multiple Choice** *Select the best answer for each question.*

**T2.1** Many professional schools require applicants to take a standardized test. Suppose that 1000 students take such a test. Several weeks after the test, Pete receives his score report: he got a 63, which placed him at the 73rd percentile. This means that

(a) Pete's score was below the median.

(b) Pete did worse than about 63% of the test takers.

(c) Pete did worse than about 73% of the test takers.

(d) Pete did better than about 63% of the test takers.

(e) Pete did better than about 73% of the test takers.

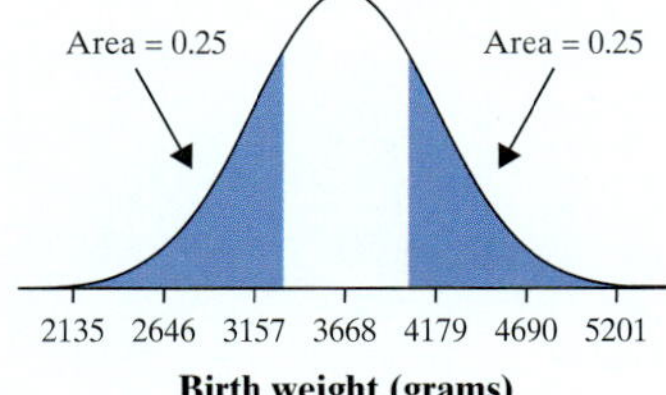


**R2.8 (a)** The amount of ketchup dispensed follows an $N(1.05, 0.08)$ distribution and we want to find the percent of times that the amount of ketchup dispensed will be between 1 and 1.2 ounces (see the following graph). $z = (1.2 - 1.05)/0.08 = 1.88$ and $z = (1 - 1.05)/0.08 = -0.63$. From Table A, the proportion of z-scores between $-0.63$ and 1.88 is $0.9699 - 0.2643 = 0.7056$. *Using technology:* `normalcdf(lower:1,upper:1.2,μ:1.05,σ:0.08)` = 0.7036. About 70% of the time the dispenser will put between 1 and 1.2 ounces of ketchup on a burger.

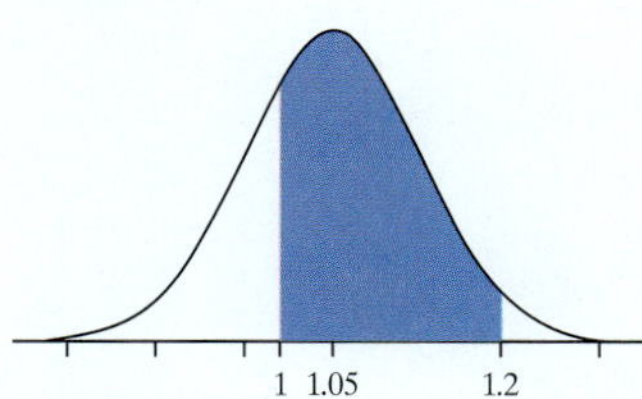


**(b)** The amount of ketchup dispensed follows an $N(1.1, \sigma)$ distribution and we want to find the value of $\sigma$ that will result in at least 99% of burgers getting between 1 and 1.2 ounces of ketchup (see graph above). Because the mean of 1.1 is in the middle of the interval from 1 to 1.2, we are looking for the middle 99% of the distribution. This leaves 0.5% in each tail. A z-score of $-2.58$ gives the value closest to 0.005(0.0049). Solving $-2.58 = (1 - 1.1)/\sigma$ gives $\sigma = 0.039$. *Using technology:* `invNorm(area:0.005,μ:0,σ:1)` gives $z = -2.576$. Solving $-2.576 = (1 - 1.1)/\sigma$ gives $\sigma = 0.039$. A standard deviation of at most 0.039 ounces will result in at least 99% of burgers getting between 1 and 1.2 ounces of ketchup.

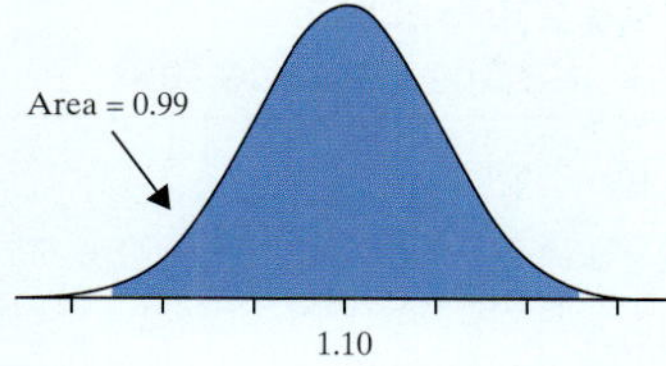


**R2.9** If the distribution is Normal, the 10th and 90th percentiles must be equal distances above and below the mean. Thus, the mean is $(25 + 475)/2 = 250$ points. The 10th percentile in a standard Normal distribution is $z = -1.28$. Solving $-1.28 = (25 - 250)/\sigma$, we get $\sigma = 175.8$. *Using technology*: `invNorm(area:0.10,μ:0,σ:1)` gives $z = -1.282$, so $-1.282 = (25 - 250)/\sigma$ and $\sigma = 175.5$.

**R2.10** A histogram and Normal probability plot are given below. The histogram is roughly symmetric but not very bell-shaped. The Normal probability plot, however, is roughly linear. For these data, the $\bar{x} = 0.8004$ and the $s_x = 0.0782$. Although the percentage within 1 standard deviation of the mean (55.1%) is less than expected (68%), the percentage within 2 (93.9%) and 3 standard deviations (100%) match the 68–95–99.7 rule quite well. It is reasonable to say that these data are approximately Normally distributed.

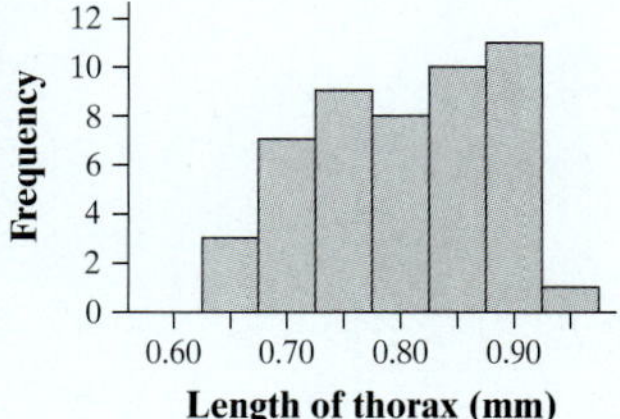


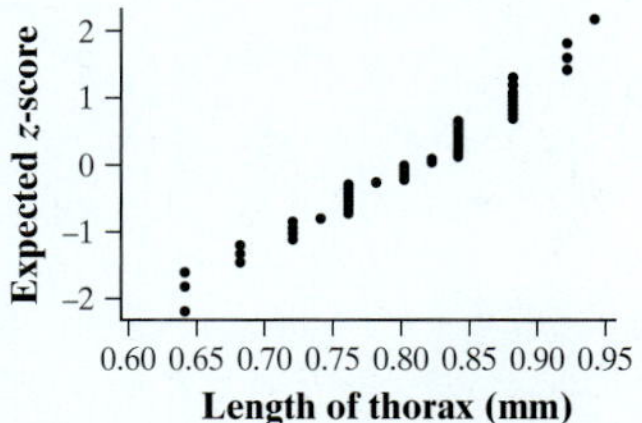


**R2.11** The steep, nearly vertical portion at the bottom and the clear bend to the right indicate that the distribution of the data is right-skewed with several outliers and not approximately Normally distributed.

**TRM Full Solutions to Chapter 2 AP® Statistics Practice Test**

Explanations for the multiple-choice questions in the Chapter AP® Practice Tests can be found in the *Solutions Manual* or the Teacher's e-Book.

## Answers to Chapter 2 AP® Statistics Practice Test

**T2.1** e

**T2.2** d

**T2.3** b

**T2.4** b

**T2.5** a

**T2.6** e

**T2.7** c

**T2.8** e

**T2.2** For the Normal distribution shown, the standard deviation is closest to

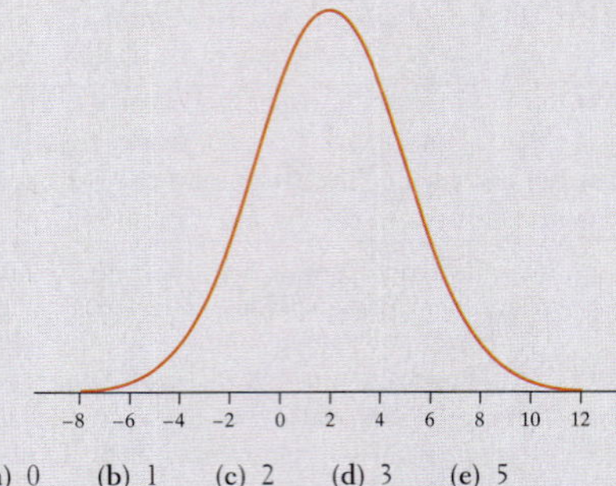


(a) 0 (b) 1 (c) 2 (d) 3 (e) 5

**T2.3** Rainwater was collected in water collectors at 30 different sites near an industrial complex, and the amount of acidity (pH level) was measured. The mean and standard deviation of the values are 4.60 and 1.10, respectively. When the pH meter was recalibrated back at the laboratory, it was found to be in error. The error can be corrected by adding 0.1 pH units to all of the values and then multiplying the result by 1.2. The mean and standard deviation of the corrected pH measurements are

(a) 5.64, 1.44
(b) 5.64, 1.32
(c) 5.40, 1.44
(d) 5.40, 1.32
(e) 5.64, 1.20

**T2.4** The figure shows a cumulative relative frequency graph of the number of ounces of alcohol consumed per week in a sample of 150 adults who report drinking alcohol occasionally. About what percent of these adults consume between 4 and 8 ounces per week?

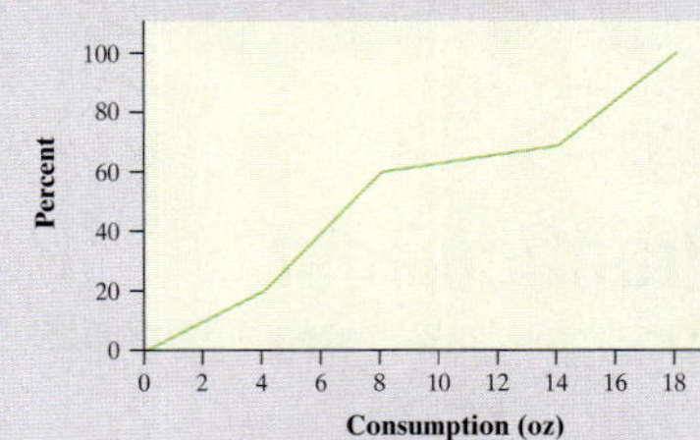


(a) 20% (b) 40% (c) 50% (d) 60% (e) 80%

**T2.5** The average yearly snowfall in Chillyville is Normally distributed with a mean of 55 inches. If the snowfall in Chillyville exceeds 60 inches in 15% of the years, what is the standard deviation?

(a) 4.83 inches
(b) 5.18 inches
(c) 6.04 inches
(d) 8.93 inches
(e) The standard deviation cannot be computed from the given information.

**T2.6** The figure shown is the density curve of a distribution. Seven values are marked on the density curve. Which of the following statements is true?

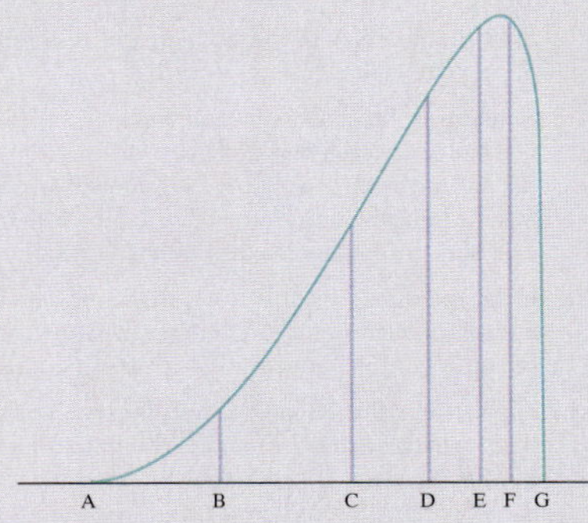


(a) The mean of the distribution is E.
(b) The area between B and F is 0.50.
(c) The median of the distribution is C.
(d) The 3rd quartile of the distribution is D.
(e) The area between A and G is 1.

**T2.7** If the heights of a population of men follow a Normal distribution, and 99.7% have heights between 5′0″ and 7′0″, what is your estimate of the standard deviation of the heights in this population?

(a) 1″ (b) 3″ (c) 4″ (d) 6″ (e) 12″

**T2.8** Which of the following is *not* correct about a standard Normal distribution?

(a) The proportion of scores that satisfy $0 < z < 1.5$ is 0.4332.
(b) The proportion of scores that satisfy $z < -1.0$ is 0.1587.
(c) The proportion of scores that satisfy $z > 2.0$ is 0.0228.
(d) The proportion of scores that satisfy $z < 1.5$ is 0.9332.
(e) The proportion of scores that satisfy $z > -3.0$ is 0.9938.

*Questions T2.9 and T2.10 refer to the following setting.* Until the scale was changed in 1995, SAT scores were based on a scale set many years ago. For Math scores, the mean under the old scale in the 1990s was 470 and the standard deviation was 110. In 2009, the mean was 515 and the standard deviation was 116.

**T2.9** What is the standardized score ($z$-score) for a student who scored 500 on the old SAT scale?

(a) −30 (b) −0.27 (c) −0.13 (d) 0.13 (e) 0.27

**T2.10** Gina took the SAT in 1994 and scored 500. Her cousin Colleen took the SAT in 2013 and scored 530. Who did better on the exam, and how can you tell?

(a) Colleen—she scored 30 points higher than Gina.

(b) Colleen—her standardized score is higher than Gina's.

(c) Gina—her standardized score is higher than Colleen's.

(d) Gina—the standard deviation was bigger in 2013.

(e) The two cousins did equally well—their $z$-scores are the same.

**Section II: Free Response** *Show all your work. Indicate clearly the methods you use, because you will be graded on the correctness of your methods as well as on the accuracy and completeness of your results and explanations.*

**T2.11** As part of the President's Challenge, students can attempt to earn the Presidential Physical Fitness Award or the National Physical Fitness Award by meeting qualifying standards in five events: curl-ups, shuttle run, sit and reach, one-mile run, and pull-ups. The qualifying standards are based on the 1985 School Population Fitness Survey. For the Presidential Award, the standard for each event is the 85th percentile of the results for a specific age group and gender among students who participated in the 1985 survey. For the National Award, the standard is the 50th percentile. To win either award, a student must meet the qualifying standard for all five events.

Jane, who is 9 years old, did 40 curl-ups in one minute. Matt, who is 12 years old, also did 40 curl-ups in one minute. The qualifying standard for the Presidential Award is 39 curl-ups for Jane and 50 curl-ups for Matt. For the National Award, the standards are 30 and 40, respectively.

(a) Compare Jane's and Matt's performances using percentiles. Explain in language simple enough for someone who knows little statistics to understand.

(b) Who has the higher standardized score ($z$-score), Jane or Matt? Justify your answer.

**T2.12** The army reports that the distribution of head circumference among male soldiers is approximately Normal with mean 22.8 inches and standard deviation 1.1 inches.

(a) A male soldier whose head circumference is 23.9 inches would be at what percentile? Show your method clearly.

(b) The army's helmet supplier regularly stocks helmets that fit male soldiers with head circumferences between 20 and 26 inches. Anyone with a head circumference outside that interval requires a customized helmet order. What percent of male soldiers require custom helmets?

(c) Find the interquartile range for the distribution of head circumference among male soldiers.

**T2.13** A study recorded the amount of oil recovered from the 64 wells in an oil field. Here are descriptive statistics for that set of data from Minitab.

**Descriptive Statistics: Oilprod**

| Variable | n | Mean | Median | StDev | Min | Max | $Q_1$ | $Q_3$ |
|---|---|---|---|---|---|---|---|---|
| Oilprod | 64 | 48.25 | 37.80 | 40.24 | 2.00 | 204.90 | 21.40 | 60.75 |

Does the amount of oil recovered from all wells in this field seem to follow a Normal distribution? Give appropriate statistical evidence to support your answer.

**T2.9** e

**T2.10** c

**T2.11 (a)** Jane's performance was better. Because her performance (40) exceeded the standard for the Presidential award (39), she performed above the 85th percentile. Matt's performance (40) met the standard for the National award (40), meaning he performed at the 50th percentile. **(b)** Because Jane's score has a higher percentile than Matt's score, she is farther to the right in her distribution than Matt is in his. Therefore, Jane's standardized score will likely be greater than Matt's.

**T2.12 (a)** For male soldiers, head circumference follows an $N(22.8, 1.1)$ distribution and we want to find the percent of soldiers with head circumference less than 23.9 inches (see graph below). $z = (23.9 - 22.8)/1.1 = 1$. From Table A, the proportion of $z$-scores below 1 is 0.8413. *Using technology*: `normalcdf(lower:-1000,upper:23.9,µ:22.8,σ:1.1)` = 0.8413. About 84% of soldiers have head circumferences less than 23.9 inches. Thus, 23.9 inches is at the 84th percentile.

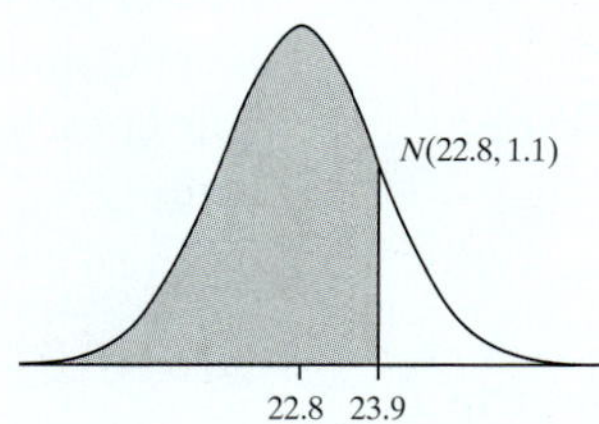


**(b)** For male soldiers, head circumference follows an $N(22.8, 1.1)$ distribution and we want to find the percent of soldiers with head circumferences less than 20 inches or greater than 26 inches (see graph below). $z = (20 - 22.8)/1.1 = -2.55$ and $z = (26 - 22.8/1.1 = 2.91$. From Table A, the proportion of $z$-scores below $z = -2.55$ is 0.0054 and the proportion of $z$-scores above 2.91 is $1 - 0.9982 = 0.0018$, for a total of $0.0054 + 0.0018 = 0.0072$. *Using technology:* 1 − `normalcdf(lower:20,upper:26,µ:22.8,σ:1.1)` = 1 − 0.9927 = 0.0073. A little less than 1% of soldiers have head circumferences less than 20 inches or greater than 26 inches and require custom helmets.

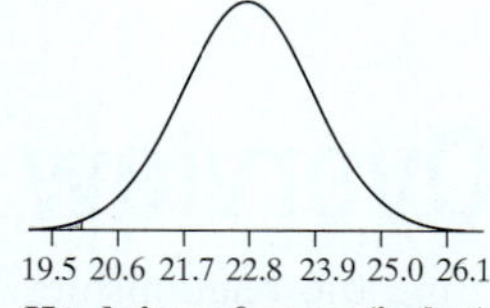


**(c)** For male soldiers, head circumference follows an $N(22.8, 1.1)$ distribution. The first quartile is the boundary value with 25% of the area to its left. The third quartile is the boundary value with 75% of the area to its left (see graph below). A $z$-score of −0.67 gives the value closest to 0.25 (0.2514). Solving $-0.67 = (x - 22.8)/1.1$ gives $Q_1 = 22.063$. A $z$-score of 0.67 gives the value closest to 0.75 (0.7486). Solving $0.67 = (x - 22.8)/1.1$ gives $Q_3 = 23.537$. *Using technology:* `invNorm(area:0.25,µ:22.8,σ:1.1)` gives $Q_1 = 22.058$, and `invNorm(area:0.75,µ:22.8,σ:1.1)` gives $Q_3 = 23.542$. Thus, $IQR = 23.542 - 22.058 = 1.484$ inches.

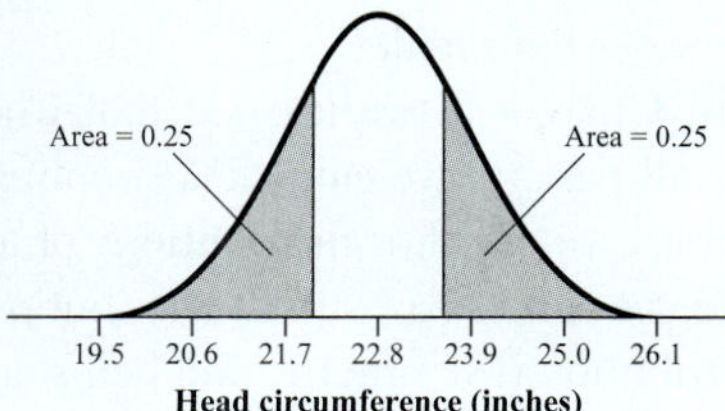


**T2.13** No. First, there is a large difference between the mean and the median. In a Normal distribution the mean and median are the same, but in this distribution the mean is 48.25 and the median is 37.80. Second, the distance between the minimum and the median is 35.80 but the distance between the median and the maximum is 167.10. In a Normal distribution, these distances should be about the same. Both of these facts suggest that the distribution is skewed to the right.

Chapter

3

# Describing Relationships

## PD Overview

Investigating the relationships between variables is central to what we do in statistics. In Section 1.1, we explored relationships between two categorical variables, such as gender and opinion about future wealth among young adults. We used two-way tables to summarize the data and bar graphs to visually display the relationship between these variables. We also introduced the idea of association: two variables have an association if knowing the value of one variable makes it easier to predict the value of the other variable. For young adults, we learned that males tended to be more optimistic about their future wealth than females.

In Sections 1.2 and 1.3, we explored the relationship between quantitative and categorical variables, such as the number of text messages sent and gender. Although we didn't explicitly use the terms "relationship" or "association"in these sections, we used summary statistics (like the mean and median) and graphs (like dotplots and boxplots) to compare the two genders. We learned that females typically sent and received more text messages than males.

In Chapter 2, we learned that when data follow a regular overall pattern, we can use a simplified model to describe the pattern we see. For distributions of a single quantitative variable, we can use a Normal distribution to model the data if the distribution is symmetric and bell-shaped.

In this chapter, we consider relationships between two quantitative variables, such as the amount of time between eruptions of Old Faithful and the duration of the previous eruption. We will use a scatterplot to display the relationship between two variables, correlation to measure the strength and direction of a linear association, and a least-squares regression line to model a linear relationship.

### Relationship to the AP® Statistics Course Description

Chapter 3 covers the following content as described in the Topic Outline from the AP® Statistics Course Description. Topics marked with a superscript are covered in other chapters as indicated.

I: Exploring Data: Describing patterns and departures from patterns

D. Exploring bivariate data

1. Analyzing patterns in scatterplots
2. Correlation and linearity
3. Least-squares regression line
4. Residual plots, outliers, and influential points
5. Transformations to achieve linearity: logarithmic and power transformations[ch12]

Note that we do not cover transformations to achieve linearity, the fifth topic in part D, until Chapter 12. We have chosen to delay this topic for several reasons. Experience has demonstrated that it is often an obstacle, especially for students with weaker backgrounds in mathematics. Students often obsess over the details of transformations and miss the big picture of what this chapter is about. Further, understanding this particular topic is not essential for any of the subsequent material. Finally, this topic has not been emphasized on the AP® exam in recent years.

Another topic that might seem to be missing is time plots. Although time plots can be useful for investigating the changes in a single quantitative variable over time, time plots are not included on the Topic Outline for the AP® Statistics exam, so we have decided to omit them from the printed text. For teachers who want to teach time plots, we have placed this content in the Teacher's Resource Materials.

## The Big Ideas

One of the challenges in teaching the AP® Statistics course is keeping students focused on the big picture and not just the details of each section. We outline the big ideas for the chapter here.

### Chapter 3 Introduction

Students should understand that knowing the distribution of a single variable often is not particularly useful. As the Old Faithful example illustrates, knowing the distribution of interval times isn't very helpful if you are trying to predict when the next eruption will occur. Instead, we look for relationships among variables to help explain patterns—and to make predictions.

The Activity in this section shows how we can use the relationship between height and hand size to help us solve a puzzling crime—who stole the cookies from the cookie jar?

## Section 3.1 Scatterplots and Correlation

In this section, we revisit the principles of data analysis introduced in Chapter 1:

- Plot the data, then add numerical summaries.
- Look for overall patterns and departures from those patterns.
- When there is a regular overall pattern, use a simplified model to describe it.

Students will often lose sight of these principles as they get more involved with the specific content of the chapter. It is a good idea to remind students of the big picture by occasionally showing them a scatterplot and asking, "What do you see?"

In Chapter 1, we learned four important characteristics to address when describing the distribution of a single quantitative variable: shape, center, spread, and outliers. When describing the association between two variables in a scatterplot, students also should address four important characteristics:

- Direction of the association
- Outliers—observations that fall outside the overall pattern
- Form of the association
- Strength of the association

You might help students remember these by using an acronym such as DOFS.

Students often find that describing the strength of an association in a scatterplot is challenging. The distinction between a moderately strong and a moderately weak association is much harder to recognize than the distinction between a positive and a negative association, for example. Fortunately, correlation is a numerical summary that can provide a more specific measure of the strength of a linear association. Instead of going straight to the formula, which can be difficult for some students to understand, we have chosen to focus on the properties of correlation. Once your students understand what correlation is all about, consider looking more carefully at the formula. Knowing how correlation is calculated can give insight into its properties. However, understanding the formula and calculating correlation by hand are not required for success on the AP® exam and will take valuable time away from other important topics.

Just as outliers can greatly influence the mean and standard deviation of a distribution, outliers can also influence the correlation. The Topic Outline for the AP® Statistics exam lists both outliers and influential points as topics that students need to know. Outliers are any points that deviate from the overall pattern of the data. Whether they are influential depends on where they are and what you are calculating. Although the AP® exam may not ask students to make this precise distinction, students should be able to address how a specific value will affect the correlation or any other measure. Doing the Activity on page 152 using the *Correlation and Regression* applet nicely illustrates the effects of influential points. APPLET

## Section 3.2 Least-Squares Regression

Once we determine that the data fall into a regular, overall pattern, we seek a simplified model to describe the pattern. In this section, we use least-squares regression lines as models for relationships between variables that have a linear association. We will learn how to create models for nonlinear relationships in Chapter 12.

As was the case with correlation, we de-emphasize formulas and hand calculations and focus instead on understanding and interpreting numerical summaries such as the slope, standard deviation of the residuals, and $r^2$. When data are provided, students should use technology to compute the least-squares regression line and summary statistics. Hand computation is time-consuming, tedious, and not required on the AP® exam.

As you teach this section, emphasize the difference between the actual data and the model used to describe the data. When we substitute a value for $x$ in the equation of a least-squares line, the result is a *predicted* value of $y$ (denoted $\hat{y}$), not an actual value of $y$. Likewise, if the slope of a least-squares regression line is 3, we should say, "For each 1 unit increase in $x$, the linear model *predicts* that the value of $y$ will increase by 3." Neglecting to include the word "predicts" suggests a deterministic relationship between $x$ and $y$ that lets us predict $y$ exactly for a given value of $x$. To emphasize that the model provides only predicted values, least-squares regression lines are always expressed in terms of $\hat{y}$ instead of $y$.

The difference between the observed value of $y$ and the predicted value of $y$ is called a *residual* (residual $= y - \hat{y}$). Residuals are the key to understanding most of the material in this section. To find the equation of the least-squares regression line, we find the line that minimizes the sum of the squared residuals. To see if a linear model is appropriate, we make a residual plot. To assess how well our line models the data, we calculate the standard deviation of the residuals $s$ to estimate the size of a typical prediction error. We also calculate $r^2$, which measures the decrease in the sum of squared residuals when using the least-squares regression line to make predictions. In many cases, closely analyzing the residuals can give us insights about the data, such as why individuals with certain characteristics regularly exceed our predictions.

Like correlation in Section 3.1, students should understand the effect that outliers and influential points have on the equation of the least-squares regression line, the standard deviation of the residuals $s$, and the value of $r^2$.

Students are also expected to use computer output to obtain the equation of the least-squares regression line and other numerical summaries. Ideally, students will have the opportunity to use a statistical software program to generate their own computer output. However, if you do not have the time or resources to have students use computer software for themselves, we have included examples of computer output in the narrative, examples, and exercises. There are many other examples in the Teacher's Resource Materials and released AP® Exam questions.

Emphasize that we cannot conclude that a cause-and-effect relationship exists between two variables just because there is an association between them. Although changes in one variable may accompany changes in the other, we cannot be certain of the cause unless we account for other variables that may have an effect on the association.

Users of the first three editions of *TPS* should note that we have omitted the distinction between "confounding" and "common response" in this section. We found that focusing on this distinction caused students to miss the big picture. We will address the issue of confounding more thoroughly in Chapter 4 when we learn about how to establish cause-and-effect relationships using experiments. We have also eliminated use of the term "lurking variable" in this chapter and throughout the book as the term doesn't have a commonly accepted meaning in the statistics world.

# Chapter 3: Resources

## TRM Additional *TPS* Teacher's Resource Materials

The following resources can be found by clicking on the links in the Teacher's e-Book (TE-book), opening the Teacher's Resource CD (TRCD), or logging onto the book's companion Web site www.whfreeman.com/tps5e (teacher log-in required).

- Materials for Introductory Activity: CSI Stats (page 142)
- Printable Chapter 3 Learning Objectives Grid (page 143)
- Additional Content: Time plots (page 149)
- Examples of Computer Output (page 181)
- Alternate Activity: Regression to the mean in sports (page 184)
- Chapter Project and Rubric (page 192)
- FRAPPY! materials: Question, Model Solution, Scoring Rubric, Sample Responses with commentary (page 200)
- Quizzes: 3 versions for each section; versions A and B are roughly parallel
  - Section 3.1 (page 158)
  - Section 3.2 (page 192)
- Tests: 4 versions; versions A and B are roughly parallel, version D is cumulative (page 200)

## Free-Response Questions from Previous AP® Exams

Questions can be found on the AP® Central Web site: http://apcentral.collegeboard.com/apc/members/exam/exam_questions/8357.html.

Students should be able to answer all the free-response questions listed below with material learned in this chapter. Questions that contain content from this chapter but also require content from later chapters are listed in the last chapter required to complete the entire question. This list will be updated after each AP® exam and will be posted on the book's Web site: www.whfreeman.com/tps5e.

Questions marked with an asterisk are from exams with released multiple-choice questions. You may want to save these questions until the end of the year so you can give your students a complete released exam for practice.

| Year | # | Content |
|---|---|---|
| 2013 | 6 | • Comparing distributions<br>• Describing trends in a scatterplot (time plot)<br>• Moving averages |
| 2012 | 1* | • Describing a scatterplot with nonlinear association<br>• Influential points<br>• Determining which points meet a consumer's criterion |
| 2007B | 4 | • Graphing a least-squares regression line<br>• Calculating a residual<br>• Influential points |
| 2005 | 3 | • Assessing linearity with residual plots<br>• Understanding and interpreting slope<br>• Interpreting $r^2$<br>• Extrapolation |
| 2003B | 1 | • Influential points |
| 2002 | 4* | • Using regression output to state the equation of a least-squares regression line<br>• Finding and interpreting the correlation from computer output<br>• Clusters and influential points |
| 2002B | 1 | • Making a scatterplot<br>• Interpreting correlation<br>• Assessing linearity<br>• Interpreting $r^2$ |
| 2000 | 1 | • Describing scatterplots |
| 1999 | 1 | • Using a residual plot to assess linearity<br>• Identifying slope and $y$ intercept from computer output<br>• Interpreting slope and $y$ intercept<br>• Making a prediction using a least-squares regression line<br>• Using residuals to estimate actual values |
| 1998 | 2 | • Making a histogram of one variable from a scatterplot<br>• Describing a histogram<br>• Describing a scatterplot |
| 1998 | 4 | • Using regression output to state the equation of a least-squares regression line<br>• Analyzing patterns in a residual plot (*Note*: The residual plot uses predicted values on the horizontal axis instead of the values of the explanatory variable.) |

## Applets 

- The *Correlation and Regression* applet allows students to add, delete, and drag points to investigate how outliers can influence the correlation and least-squares regression line. It also allows students to draw their own line and compare it with the actual least-squares regression line. www.whfreeman.com/tps5e (click on Statistical Applets and choose Correlation and Regression)
- The *Two-Variable Statistical Calculator* applet allows students to enter their own data or use data sets from the textbook. It will calculate the correlation and least-squares regression line, display the scatterplot, and display the residual plot. www.whfreeman.com/tps5e (click on Statistical Applets and choose Two-Variable Statistical Calculator).
- The *Guess the Correlation* applet asks students to guess the value of the correlation and keeps track of their results. www.rossmanchance.com/applets (choose Guess the Correlation).
- The *Least-Squares Regression* applet lets students fit a line to data and compare it with the least-squares regression line. They can also show the residuals and squared residuals for either line. Students can also add, delete, and drag points around. www.rossmanchance.com/applets (choose Least-Squares Regression).
- The *Descriptive Statistics and Graphs: Two Quantitative Variables* applet allows you to make scatterplots and calculate summary statistics. See www.lock5stat.com/statkey (look on the left-hand side for these applets; press "Edit Data" to enter your own data).

## TI-Nspire Activities

These Activities are designed to illustrate important statistical concepts using the TI-Nspire. The Activities can be found at www.whfreeman.com/tps5e.

- *Exploring Scatterplots*: Plotting data and looking at how correlation is affected by outliers and influential points.
- *Exploring Linear Regression, part I*: Using a movable line to explore the meaning of least squares.
- *Exploring Linear Regression, part II*: Investigating how outliers can affect the least-squares regression line.
- *Exploring Linear Regression, part III*: Using the Lists and Spreadsheet feature to calculate $r^2$ and plot the residuals to determine the quality of the linear model.

# Chapter 3: Pacing Guide, Learning Objectives, and Suggested Assignments

This pacing guide is based on a schedule with 110, 50-minute sessions before the AP® exam. The 110 days include 4 days for midterms and 10 days for AP® exam review. If you have a different number of sessions before the AP® exam, you can modify the pacing guide to suit your needs. If you have additional time, consider incorporating quizzes, released AP® free-response questions, or additional activities. See the Resources section above for suggestions.

The suggested homework assignments list odd-numbered exercises, whenever possible, so students can check their answers against the back-of-the-book answers. If you would rather students not have access to the answers while doing homework, adding 1 to the exercise numbers usually will do the trick, because the homework exercises typically are paired. For example, Exercises 1 and 2 will generally cover the same topics, but in different contexts. You may also choose to include the Recycle and Review questions at the end of each section, which review topics from previous sections or chapters. These questions are denoted with a symbol.

# Chapter 3

| Day | Topics | Learning Objectives: Students will be able to . . . | Suggested Assignment |
|---|---|---|---|
| 1 | Chapter 3 Introduction, 3.1 Explanatory and Response Variables, Displaying Relationships: Scatterplots, Describing Scatterplots | • Identify explanatory and response variables in situations where one variable helps to explain or influences the other.<br>• Make a scatterplot to display the relationship between two quantitative variables.<br>• Describe the direction, form, and strength of a relationship displayed in a scatterplot and recognize outliers in a scatterplot. | 1, 5, 7, 11, 13 |
| 2 | 3.1 Measuring Linear Association: Correlation, Facts about Correlation | • Interpret the correlation.<br>• Understand the basic properties of correlation, including how the correlation is influenced by outliers.<br>• Use technology to calculate correlation.<br>• Explain why association does not imply causation. | 14–18, 21 |
| 3 | 3.2 Least-Squares Regression, Interpreting a Regression Line, Prediction, Residuals | • Interpret the slope and *y* intercept of a least-squares regression line.<br>• Use the least-squares regression line to predict *y* for a given *x*. Explain the dangers of extrapolation.<br>• Calculate and interpret residuals. | 27–32, 35, 37, 39, 41, 45 |
| 4 | 3.2 Calculating the Equation of the Least-Squares Regression Line, Determining Whether a Linear Model Is Appropriate: Residual Plots | • Explain the concept of least squares.<br>• Determine the equation of a least-squares regression line using technology.<br>• Construct and interpret residual plots to assess if a linear model is appropriate. | 43, 47, 49, 51 |
| 5 | 3.2 How Well the Line Fits the Data: The Role of $s$ and $r^2$ in Regression | • Interpret the standard deviation of the residuals and $r^2$, and use these values to assess how well the least-squares regression line models the relationship between two variables. | 48, 50, 55, 58 |
| 6 | 3.2 Interpreting Computer Regression Output, Regression to the Mean, Correlation and Regression Wisdom | • Determine the equation of a least-squares regression line using computer output.<br>• Describe how the slope, *y* intercept, standard deviation of the residuals, and $r^2$ are influenced by outliers.<br>• Find the slope and *y* intercept of the least-squares regression line from the means and standard deviations of *x* and *y* and their correlation. | 59, 61, 63, 65, 69, 71–78 |
| 7 | Chapter 3 Review/FRAPPY! | | Chapter Review Exercises |
| 8 | Chapter 3 Test | | |

**PD Chapter 3 Overview**

Watch the video overview of Chapter 3 (for teachers) by clicking on the link in the TE-book, opening the TRCD, or logging onto the book's companion Web site www.whfreeman.com/tps5e (teacher log-in required).

# Chapter 3

# Describing Relationships

## case study

## How Faithful Is Old Faithful?

The Starnes family visited Yellowstone National Park in hopes of seeing the Old Faithful geyser erupt. They had only about four hours to spend in the park. When they pulled into the parking lot near Old Faithful, a large crowd of people was headed back to their cars from the geyser. Old Faithful had just finished erupting. How long would the Starnes family have to wait until the next eruption?

Let's look at some data. Figure 3.1 shows a histogram of times (in minutes) between consecutive eruptions of Old Faithful in the month before the Starnes family's visit. The shortest interval was 47 minutes, and the longest was 113 minutes. That's a lot of variability! The distribution has two clear peaks—one at about 60 minutes and the other at about 90 minutes.

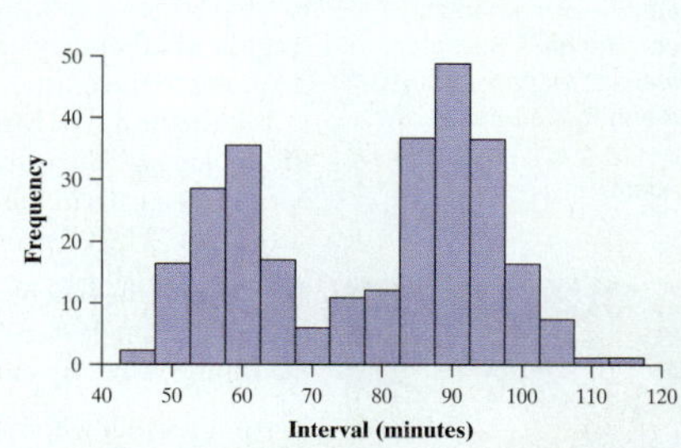


**FIGURE 3.1** Histogram of the interval (in minutes) between eruptions of the Old Faithful geyser in the month prior to the Starnes family's visit.

If the Starnes family hopes for a 60-minute gap between eruptions, but the actual interval is closer to 90 minutes, the kids will get impatient. If they plan for a 90-minute interval and go somewhere else in the park, they won't get back in time to see the next eruption if the gap is only about 60 minutes. What should the Starnes family do?

Later in the chapter, you'll answer this question. For now, keep this in mind: to understand one variable (like eruption interval), you often have to look at how it is related to other variables.

141

### Teaching Tip

Emphasize the importance of studying relationships between variables. Knowing the distribution of a single variable often doesn't tell us everything we want to know. In this Case Study, knowing just the distribution of interval times isn't enough to make a good prediction about how long the Starnes family will have to wait for the next eruption.

### WEB SITE

See www.geyserstudy.org for more information about Old Faithful and other geysers in Yellowstone National Park.

# Introduction

Investigating relationships between variables is central to what we do in statistics. When we understand the relationship between two variables, we can use the value of one variable to help us make predictions about the other variable. In Section 1.1, we explored relationships between categorical variables, such as the gender of a young person and his or her opinion about future income. The association between these two variables suggests that males are generally more optimistic about their future income than females.

In this chapter, we investigate relationships between two quantitative variables. Does knowing the number of points a football team scores per game tell us anything about how many wins it will have? What can we learn about the price of a used car from the number of miles it has been driven? Are there any variables that might help the Starnes family predict how long it will be until the next eruption of Old Faithful?

**Activity Overview**

You can watch an overview of this Activity (for teachers) by clicking on the link in the TE-book, opening the TRCD, or logging onto the book's companion Web site www.whfreeman.com/tps5e (teacher log-in required).

**TRM CSI Stats**

If you have time, consider modifying this activity to make it more personal. Use a handprint from one of the math teachers at your school and get the heights of each department member. If you have a treasured item in your room, such as a poster of your favorite NFL team, consider having that be the stolen item instead of cookies. If you are short on time, we have provided a handprint and roster in the Teacher's Resource Materials. Find these by clicking on the link in the TE-book, opening the TRCD, or logging onto the book's companion Web site www.whfreeman.com/tps5e (teacher log-in required).

**ALTERNATE ACTIVITY**

Collect data from students to see if there is a relationship between the amount of sleep they had the previous evening and the amount of time spent doing homework. Can you estimate the amount of time spent doing homework if you know how long they slept? What about the number of hours watching TV, playing video games, or going online?

## ACTIVITY CSI Stats: The case of the missing cookies

**MATERIALS:** Meterstick, handprint, and math department roster (from *Teacher's Resource Materials*) for each group of three to four students; one sheet of graph paper per student

Mrs. Hagen keeps a large jar full of cookies on her desk for her students. Over the past few days, a few cookies have disappeared. The only people with access to Mrs. Hagen's desk are the other math teachers at her school. She asks her colleagues whether they have been making withdrawals from the cookie jar. No one confesses to the crime.

But the next day, Mrs. Hagen catches a break—she finds a clear handprint on the cookie jar. The careless culprit has left behind crucial evidence! At this point, Mrs. Hagen calls in the CSI Stats team (your class) to help her identify the prime suspect in "The Case of the Missing Cookies."

1. Measure the height and hand span of each member of your group to the nearest centimeter (cm). (Hand span is the maximum distance from the tip of the thumb to the tip of the pinkie finger on a person's fully stretched-out hand.)

2. Your teacher will make a data table on the board with two columns, labeled as follows:

| Hand span (cm) | Height (cm) |
|---|---|

Send a representative to record the data for each member of your group in the table.

3. Copy the data table onto your graph paper very near the left margin of the page. Next, you will make a graph of these data. Begin by constructing a set of coordinate axes. Allow plenty of space on the page for your graph. Label the horizontal axis "Hand span (cm)" and the vertical axis "Height (cm)."

4. Since neither hand span nor height can be close to 0 cm, we want to start our horizontal and vertical scales at larger numbers. Scale the horizontal axis in 0.5-cm increments starting with 15 cm. Scale the vertical axis in 5-cm

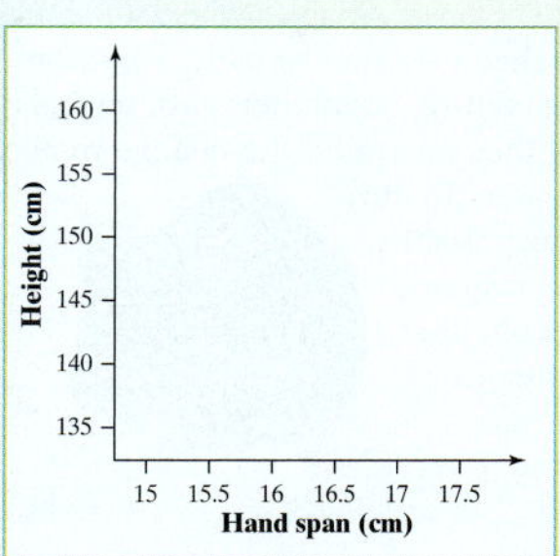


increments starting with 135 cm. Refer to the sketch in the margin for comparison.

5. Plot each point from your class data table as accurately as you can on the graph. Compare your graph with those of your group members.

6. As a group, discuss what the graph tells you about the relationship between hand span and height. Summarize your observations in a sentence or two.

7. Ask your teacher for a copy of the handprint found at the scene and the math department roster. Which math teacher does your group believe is the "prime suspect"? Justify your answer with appropriate statistical evidence.

# 3.1 Scatterplots and Correlation

**WHAT YOU WILL LEARN** By the end of the section, you should be able to:

- Identify explanatory and response variables in situations where one variable helps to explain or influences the other.
- Make a scatterplot to display the relationship between two quantitative variables.
- Describe the direction, form, and strength of a relationship displayed in a scatterplot and identify outliers in a scatterplot.
- Interpret the correlation.
- Understand the basic properties of correlation, including how the correlation is influenced by outliers.
- Use technology to calculate correlation.
- Explain why association does not imply causation.

Most statistical studies examine data on more than one variable. Fortunately, analysis of several-variable data builds on the tools we used to examine individual variables. The principles that guide our work also remain the same:

- Plot the data, then add numerical summaries.
- Look for overall patterns and departures from those patterns.
- When there's a regular overall pattern, use a simplified model to describe it.

## Explanatory and Response Variables

We think that car weight helps explain accident deaths and that smoking influences life expectancy. In these relationships, the two variables play different roles. Accident death rate and life expectancy are the **response variables** of interest. Car weight and number of cigarettes smoked are the **explanatory variables**.

**DEFINITION: Response variable, explanatory variable**

A **response variable** measures an outcome of a study. An **explanatory variable** may help explain or predict changes in a response variable.

### PD Section 3.1 Overview

Watch the video overview of Section 3.1 (for teachers) by clicking on the link in the TE-book, opening the TRCD, or logging onto the book's companion Web site www.whfreeman.com/tps5e (teacher log-in required).

### TRM Learning Objectives Grid

Remind students about the importance of the Learning Objectives. We include them at the beginning of each section so students know what they are expected to learn. At the end of the chapter, the Learning Objectives are included in a grid, along with references to examples and chapter review exercises. A printable version of this grid is included in the Teacher's Resource Materials.

### TRM Section 3.1 Alternate Examples

You can find the Alternate Examples for this section in Microsoft Word format by clicking on the link in the TE-book, opening the TRCD, or logging onto the book's companion Web site www.whfreeman.com/tps5e (teacher log-in required).

### Teaching Tip

Remind students how we used these principles in previous chapters:

- Make a graph such as a dotplot or histogram.
- Add numerical summaries such as the mean and standard deviation or the median and interquartile range.
- Look for outliers and explore their effects on the numerical summaries.
- Check for Normality—if the shape of the data is approximately Normal, use a Normal distribution as a model.

### Teaching Tip

You can also say that the explanatory variable "accounts for" changes in the response variable.

## Teaching Tip

When we talk about designing experiments in Chapter 4, we will use the word *factor* synonymously with *explanatory variable*. Either of these terms is better than "independent variable," because the word "independent" has another meaning in statistics.

## Teaching Tip

Remind students that the "play" icon at the top of this example (and many others) indicates that they can watch a video presentation of the example by clicking on the links in the e-Book or finding the videos at www.whfreeman.com/tps5e. Also, remind them that the icon in the lower-right corner indicates there is a link between this example and Exercise 1. If students have trouble with the exercise, they can look back to this example for help.

### ALTERNATE EXAMPLE

***Weight and height***

Tim wants to know if there is a relationship between height and weight. Kelly wants to know if she can predict a student's weight from his or her height. Information about height is easier to obtain than information about weight!

PROBLEM: For each student, identify the explanatory and response variables, if possible.

SOLUTION: Tim is just interested in the relationship between the two variables, so there is no clear explanatory or response variable. Kelly is treating a student's height as the explanatory variable and the student's weight as the response variable.

You will often see explanatory variables called *independent variables* and response variables called *dependent variables*. Because the words "independent" and "dependent" have other meanings in statistics, we won't use them here.

It is easiest to identify explanatory and response variables when we actually specify values of one variable to see how it affects another variable. For instance, to study the effect of alcohol on body temperature, researchers gave several different amounts of alcohol to mice. Then they measured the change in each mouse's body temperature 15 minutes later. In this case, amount of alcohol is the explanatory variable, and change in body temperature is the response variable. When we don't specify the values of either variable but just observe both variables, there may or may not be explanatory and response variables. Whether there are depends on how you plan to use the data.

## EXAMPLE Linking SAT Math and Critical Reading Scores

***Explanatory or response?***

Julie asks, "Can I predict a state's mean SAT Math score if I know its mean SAT Critical Reading score?" Jim wants to know how the mean SAT Math and Critical Reading scores this year in the 50 states are related to each other.

PROBLEM: For each student, identify the explanatory variable and the response variable if possible.

SOLUTION: Julie is treating the mean SAT Critical Reading score as the explanatory variable and the mean SAT Math score as the response variable. Jim is simply interested in exploring the relationship between the two variables. For him, there is no clear explanatory or response variable.

**For Practice** *Try Exercise* **1**

In many studies, the goal is to show that changes in one or more explanatory variables actually *cause* changes in a response variable. However, other explanatory-response relationships don't involve direct causation. In the alcohol and mice study, alcohol actually *causes* a change in body temperature. But there is no cause-and-effect relationship between SAT Math and Critical Reading scores. Because the scores are closely related, we can still use a state's mean SAT Critical Reading score to predict its mean Math score. We will learn how to make such predictions in Section 3.2.

### CHECK YOUR UNDERSTANDING

Identify the explanatory and response variables in each setting.

1. How does drinking beer affect the level of alcohol in people's blood? The legal limit for driving in all states is 0.08%. In a study, adult volunteers drank different numbers of cans of beer. Thirty minutes later, a police officer measured their blood alcohol levels.

2. The National Student Loan Survey provides data on the amount of debt for recent college graduates, their current income, and how stressed they feel about college debt. A sociologist looks at the data with the goal of using amount of debt and income to explain the stress caused by college debt.

## Teaching Tip

In Chapter 4, we will spend much more time talking about when it is appropriate to draw a cause-and-effect conclusion. For now, it's best to emphasize that we should never assume there is a cause-and-effect relationship just because an association exists between two variables.

### ANSWERS TO CYU

**1.** Explanatory: number of cans of beer. Response: blood alcohol level.

**2.** Explanatory: amount of debt and income. Response: stress caused by college debt.

## Displaying Relationships: Scatterplots

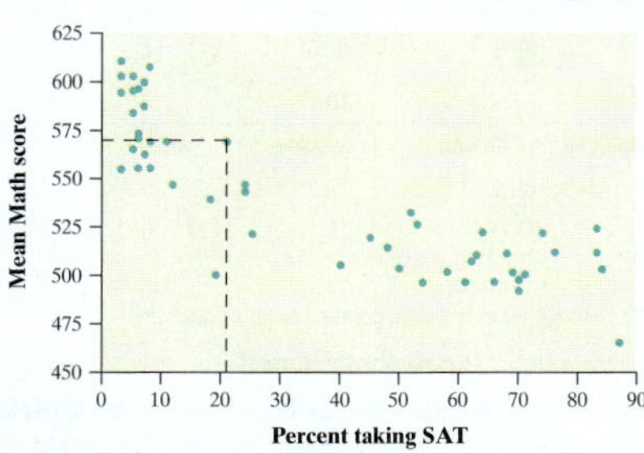


**FIGURE 3.2** Scatterplot of the mean SAT Math score in each state against the percent of that state's high school graduates who took the SAT. The dotted lines intersect at the point (21, 570), the values for Colorado.

The most useful graph for displaying the relationship between two quantitative variables is a **scatterplot.** Figure 3.2 shows a scatterplot of the percent of high school graduates in each state who took the SAT and the state's mean SAT Math score in a recent year. We think that "percent taking" will help explain "mean score." So "percent taking" is the explanatory variable and "mean score" is the response variable. We want to see how mean score changes when percent taking changes, so we put percent taking (the explanatory variable) on the horizontal axis. Each point represents a single state. In Colorado, for example, 21% took the SAT, and their mean SAT Math score was 570. Find 21 on the $x$ (horizontal) axis and 570 on the $y$ (vertical) axis. Colorado appears as the point (21, 570).

**DEFINITION: Scatterplot**

A **scatterplot** shows the relationship between two quantitative variables measured on the same individuals. The values of one variable appear on the horizontal axis, and the values of the other variable appear on the vertical axis. Each individual in the data appears as a point in the graph.

Here's a helpful way to remember: the e**X**planatory variable goes on the $x$ axis.

Always plot the explanatory variable, if there is one, on the horizontal axis (the $x$ axis) of a scatterplot. As a reminder, we usually call the explanatory variable $x$ and the response variable $y$. If there is no explanatory-response distinction, either variable can go on the horizontal axis.

We used computer software to produce Figure 3.2. For some problems, you'll be expected to make scatterplots by hand. Here's how to do it.

**HOW TO MAKE A SCATTERPLOT**

1. Decide which variable should go on each axis.
2. Label and scale your axes.
3. Plot individual data values.

The following example illustrates the process of constructing a scatterplot.

**EXAMPLE** **SEC Football**

*Making a scatterplot*

At the end of the 2011 college football season, the University of Alabama defeated Louisiana State University for the national championship. Interestingly, both of these teams were from the Southeastern Conference (SEC). Here are the average number of points scored per game and number of wins for each of the twelve teams in the SEC that season.[1]

### Teaching Tip

Point out that scatterplots are the only choice for displaying the relationship between *two* quantitative variables. For a single quantitative variable, there are many choices for displaying its distribution, including dotplots, histograms, boxplots, and stemplots.

### Teaching Tip

Tell students to pick "nice" values to mark on each axis, making sure to cover the values that each variable takes in the data set. Remind students that the axes usually do not start at 0 and are often not on the same scale. However, the scale on each axis must be consistent. Finally, remind them to clearly label the variable name on each axis. For a humorous look at the issue of labeling axes, show your students the cartoon at http://xkcd.com/833.

### AP® Exam Common Error

Many students lose credit because they forget to label their graphs. Including the proper labels is more important than graphing each point in precisely the right place.

## ALTERNATE EXAMPLE

### *Track and field day!*

Each member of a small statistics class ran a 40-yard sprint and then did a long jump (with a running start). The table below shows the sprint time (in seconds) and the long-jump distance (in inches):

| Sprint time (s) | Long-jump distance (in.) |
|---|---|
| 5.41 | 171 |
| 5.05 | 184 |
| 7.01 | 90 |
| 7.17 | 65 |
| 6.73 | 78 |
| 5.68 | 130 |
| 5.78 | 173 |
| 6.31 | 143 |
| 6.44 | 92 |
| 6.50 | 139 |
| 6.80 | 120 |
| 7.25 | 110 |

PROBLEM: Make a scatterplot of the relationship between sprint time and long-jump distance.

SOLUTION: Because the long-jump distance is somewhat dependent on how fast a student can run, we'll use sprint time as the explanatory variable and long-jump distance as the response variable. The sprint times vary from 5.05 to 7.25 seconds, so we will use a horizontal scale from 5 to 7.5, with tick marks every 0.25 seconds. The long-jump distances vary from 65 to 184, so we will use a vertical scale from 60 to 200 inches, with tick marks every 10 inches. The completed scatterplot is shown below.

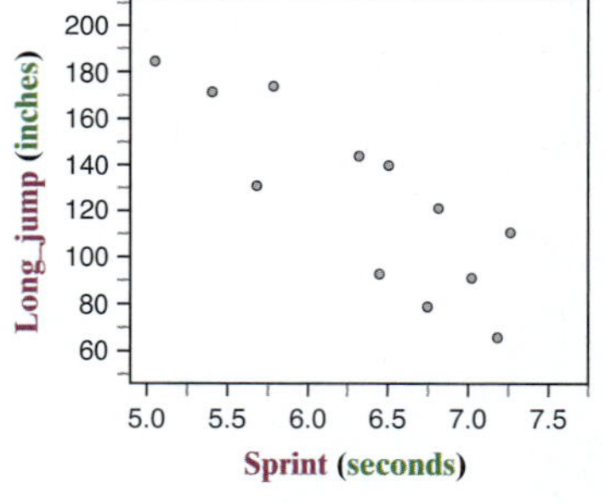


| Team | Alabama | Arkansas | Auburn | Florida | Georgia | Kentucky |
|---|---|---|---|---|---|---|
| Points per game | 34.8 | 36.8 | 25.7 | 25.5 | 32.0 | 15.8 |
| Wins | 12 | 11 | 8 | 7 | 10 | 5 |
| **Team** | **Louisiana State** | **Mississippi** | **Mississippi State** | **South Carolina** | **Tennessee** | **Vanderbilt** |
| Points per game | 35.7 | 16.1 | 25.3 | 30.1 | 20.3 | 26.7 |
| Wins | 13 | 2 | 7 | 11 | 5 | 6 |

PROBLEM: Make a scatterplot of the relationship between points per game and wins.

SOLUTION: We follow the steps described earlier to make the scatterplot.

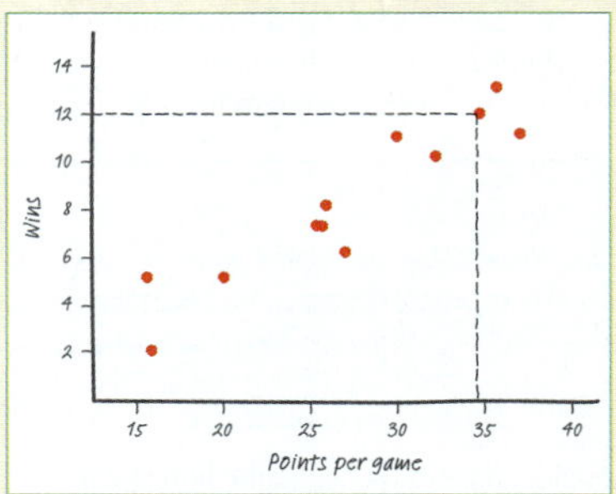


FIGURE 3.3 Completed scatterplot of points per game and wins for the teams in the SEC. The dotted lines intersect at the point (34.8, 12), the values for Alabama.

1. *Decide which variable should go on which axis.* The number of wins a football team has depends on the number of points they score. So we'll use points per game as the explanatory variable ($x$ axis) and wins as the response variable ($y$ axis).
2. *Label and scale your axes.* We labeled the $x$ axis "Points per game" and the $y$ axis "Wins." Because the teams' points per game vary from 15.8 to 36.8, we chose a horizontal scale starting at 15 points, with tick marks every 5 points. The teams' wins vary from 2 to 13, so we chose a vertical scale starting at 0 with tick marks every 2 wins.
3. *Plot individual data values.* The first team in the table, Alabama, scored 34.8 points per game and had 12 wins. We plot this point directly above 34.8 on the horizontal axis and to the right of 12 on the vertical axis, as shown in Figure 3.3. For the second team in the list, Arkansas, we add the point (36.8, 11) to the graph. By adding the points for the remaining ten teams, we get the completed scatterplot in Figure 3.3.

**For Practice** *Try Exercise* **5**

## Describing Scatterplots

To describe a scatterplot, follow the basic strategy of data analysis from Chapters 1 and 2: look for patterns and important departures from those patterns. Let's take a closer look at the scatterplot from Figure 3.2. What do we see?

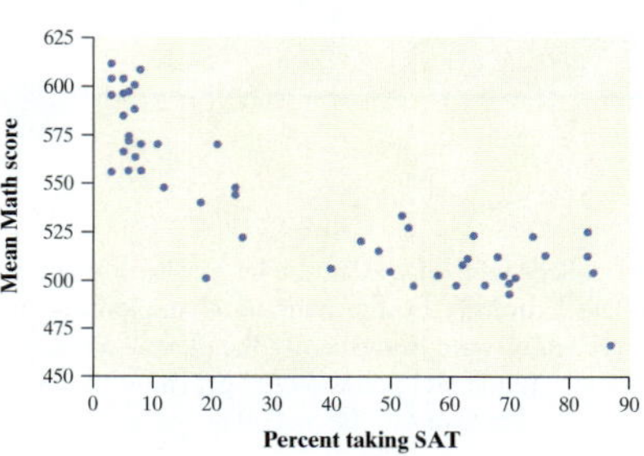


- The graph shows a clear **direction:** the overall pattern moves from upper left to lower right. That is, states in which higher percents of high school graduates take the SAT tend to have lower mean SAT Math scores. We call this a *negative association* between the two variables.
- The **form** of the relationship is slightly curved. More important, most states fall into one of two distinct *clusters.* In about half of the states, 25% or fewer graduates took the SAT. In the other half, more than 40% took the SAT.
- The **strength** of a relationship in a scatterplot is determined by how closely the points follow a clear form. The overall relationship in Figure 3.2 is moderately strong: states with similar percents taking the SAT tend to have roughly similar mean SAT Math scores.

## Teaching Tip

When discussing the direction of the association in the SAT example, emphasize that we are only looking at the overall tendency, not a deterministic relationship. The negative association won't hold true for every pair of points. For example, Missouri may have a higher percent taking the SAT than Arkansas *and* have a higher mean math score as well, even though states with a higher percent of students taking the SAT typically have lower mean math scores. The margin note on page 148 also addresses this.

- Two states stand out in the scatterplot: West Virginia at (19, 501) and Maine at (87, 466). These points can be described as **outliers** because they fall outside the overall pattern.

**What explains the clusters?** There are two widely used college entrance exams, the SAT and the American College Testing (ACT) exam. Each state usually favors one or the other. The ACT states cluster at the left of Figure 3.2 and the SAT states at the right. In ACT states, most students who take the SAT are applying to a selective college that prefers SAT scores. This select group of students has a higher mean score than the much larger group of students who take the SAT in SAT states.

### HOW TO EXAMINE A SCATTERPLOT

As in any graph of data, look for the *overall pattern* and for striking *departures* from that pattern.

- You can describe the overall pattern of a scatterplot by the **direction, form,** and **strength** of the relationship.
- An important kind of departure is an **outlier,** an individual value that falls outside the overall pattern of the relationship.

Let's practice examining scatterplots using the SEC football data from the previous example.

## EXAMPLE SEC Football

### *Describing a scatterplot*

In the last example, we constructed the scatterplot shown below that displays the average number of points scored per game and the number of wins for college football teams in the Southeastern Conference.

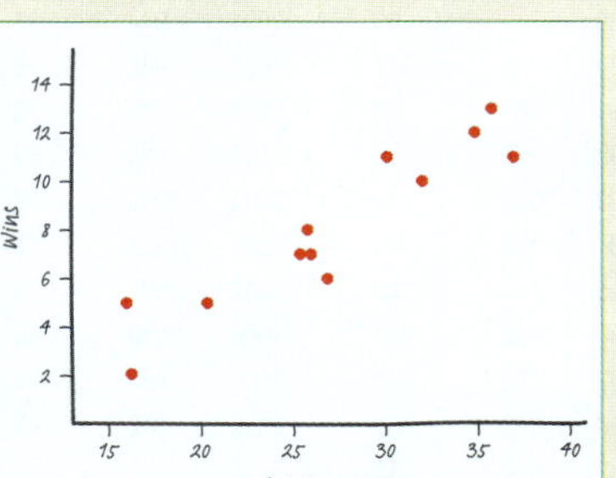


PROBLEM: Describe what the scatterplot reveals about the relationship between points per game and wins.

SOLUTION: *Direction:* In general, it appears that teams that score more points per game have more wins and teams that score fewer points per game have fewer wins. We say that there is a *positive association* between points per game and wins.

*Form:* There seems to be a linear pattern in the graph (that is, the overall pattern follows a straight line).

*Strength:* Because the points do not vary much from the linear pattern, the relationship is fairly strong. There do not appear to be any values that depart from the linear pattern, so there are no *outliers.*

**For Practice** Try Exercise  7

### Teaching Tip

When trying to identify outliers in a scatterplot, remind students to look for points that fall outside the overall pattern of the association. Unlike what happens with distributions of a single variable, there isn't a specific rule for identifying outliers on a scatterplot.

### AP® Exam Common Error

On the AP® exam, if students are asked to describe a scatterplot, they are expected to discuss the direction, form, and strength of the association *in context* as well as identify any possible outliers. To help students remember this, consider an acronym such as DOFS (direction, outliers, form, strength), similar to the SOCS (shape, outliers, center, spread) acronym from Chapter 1. Once we learn about $\hat{y}$, you can say something like "A gentleman always DOFS his *y*-hat when meeting a lady." You might have to explain what it means to doff a hat!

### ALTERNATE EXAMPLE

### *Track and field day!*

PROBLEM: Describe what the scatterplot in the previous example reveals about the relationship between sprint time and long-jump distances for students in the statistics class.

SOLUTION: *Direction:* Students who took longer to run 40 yards tended to have shorter long-jump distances, so there is a negative association between sprint times and long-jump distances.
*Form:* There seems to be a linear pattern in the graph.
*Strength:* Because the points don't vary too much from the linear pattern, the association is strong.
*Outliers:* There don't seem to be any outliers in this scatterplot.

### Teaching Tip

Students have limited experience describing the strength of an association and may have trouble deciding if an association is moderately strong or moderately weak. Tell them not to worry: we will learn a numerical way to measure strength in just a few pages!

Even when there is a clear association between two variables in a scatterplot, the direction of the relationship only describes the overall trend—not the relationship for each pair of points. For example, even though teams that score more points per game generally have more wins, Georgia and South Carolina are exceptions to the overall pattern. Georgia scored *more* points per game than South Carolina (32 versus 30.1) but had *fewer* wins (10 versus 11).

So far, we've seen relationships with two different directions. The number of wins generally increases as the points scored per game increases (**positive association**). The mean SAT score generally goes down as the percent of graduates taking the test increases (**negative association**). Let's give a careful definition for these terms.

**DEFINITION: Positive association, negative association**

Two variables have a **positive association** when above-average values of one tend to accompany above-average values of the other and when below-average values also tend to occur together.

Two variables have a **negative association** when above-average values of one tend to accompany below-average values of the other.

Of course, not all relationships have a clear direction that we can describe as a positive association or a negative association. Exercise 9 involves a relationship that doesn't have a single direction. This next example, however, illustrates a strong positive association with a simple and important form.

## EXAMPLE The Endangered Manatee

### *Pulling it all together*

Manatees are large, gentle, slow-moving creatures found along the coast of Florida. Many manatees are injured or killed by boats. The table below contains data on the number of boats registered in Florida (in thousands) and the number of manatees killed by boats for the years 1977 to 2010.[2]

**Florida boat registrations (thousands) and manatees killed by boats**

| YEAR | BOATS | MANATEES | YEAR | BOATS | MANATEES | YEAR | BOATS | MANATEES |
|---|---|---|---|---|---|---|---|---|
| 1977 | 447 | 13 | 1989 | 711 | 50 | 2001 | 944 | 81 |
| 1978 | 460 | 21 | 1990 | 719 | 47 | 2002 | 962 | 95 |
| 1979 | 481 | 24 | 1991 | 681 | 53 | 2003 | 978 | 73 |
| 1980 | 498 | 16 | 1992 | 679 | 38 | 2004 | 983 | 69 |
| 1981 | 513 | 24 | 1993 | 678 | 35 | 2005 | 1010 | 79 |
| 1982 | 512 | 20 | 1994 | 696 | 49 | 2006 | 1024 | 92 |
| 1983 | 526 | 15 | 1995 | 713 | 42 | 2007 | 1027 | 73 |
| 1984 | 559 | 34 | 1996 | 732 | 60 | 2008 | 1010 | 90 |
| 1985 | 585 | 33 | 1997 | 755 | 54 | 2009 | 982 | 97 |
| 1986 | 614 | 33 | 1998 | 809 | 66 | 2010 | 942 | 83 |
| 1987 | 645 | 39 | 1999 | 830 | 82 | | | |
| 1988 | 675 | 43 | 2000 | 880 | 78 | | | |

**ALTERNATE EXAMPLE**

### *Salads at McDonald's*

McDonald's restaurants offer a variety of salads. The table below lists 10 different salads, along with the amount of sodium (in milligrams) and the amount of fat (in grams).

| Salad | Sodium | Fat |
|---|---|---|
| Southwest Salad with Grilled Chicken | 650 | 8 |
| Southwest Salad with Crispy Chicken | 820 | 21 |
| Southwest Salad without Chicken | 150 | 4.5 |
| Bacon Ranch Salad with Grilled Chicken | 700 | 9 |
| Bacon Ranch Salad with Crispy Chicken | 870 | 22 |
| Bacon Ranch Salad without Chicken | 300 | 7 |
| Caesar Salad with Grilled Chicken | 580 | 5 |
| Caesar Salad with Crispy Chicken | 740 | 18 |
| Caesar Salad without Chicken | 180 | 4 |
| Side Salad | 10 | 0 |

*Source:* http://nutrition.mcdonalds.com/getnutrition/nutritionfacts.pdf.

PROBLEM: Make a scatterplot to display the relationship between the amount of sodium and the amount of fat in salads from McDonald's. Describe what you see.

SOLUTION: Here is a scatterplot of the data.

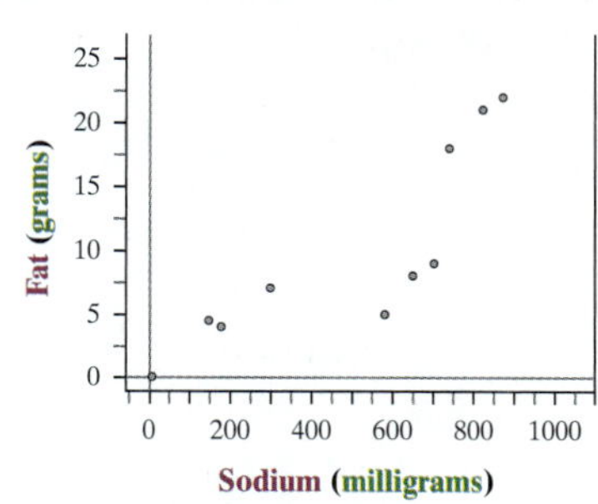


There is a positive association between sodium and fat—salads with more sodium tend to have more fat. The overall association is nonlinear, as the pattern does not follow a straight line. However, the association is fairly strong as the points do not deviate much from the nonlinear form. Finally, there are three distinct clusters of points, formed by salads with no chicken (lower left), salads with grilled chicken (lower right), and salads with crispy chicken (upper right). Within each cluster there is a positive, linear association between sodium and fat.

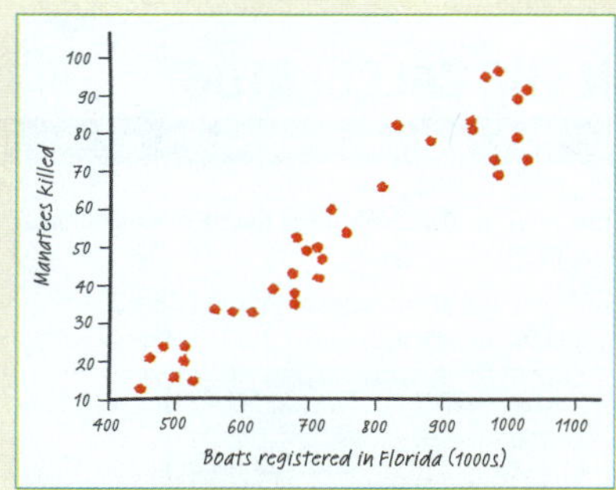


**FIGURE 3.4** Scatterplot of the number of Florida manatees killed by boats from 1977 to 2010 against the number of boats registered in Florida that year.

**PROBLEM:** Make a scatterplot to show the relationship between the number of manatees killed and the number of registered boats. Describe what you see.

**SOLUTION:** For the scatterplot, we'll use "boats registered" as the explanatory variable and "manatees killed" as the response variable. Figure 3.4 is our completed scatterplot. There is a positive association—more boats registered goes with more manatees killed. The form of the relationship is linear. That is, the overall pattern follows a straight line from lower left to upper right. The relationship is strong because the points don't deviate greatly from a line, except for the 4 years that have a high number of boats registered, but fewer deaths than expected based on the linear pattern.

**For Practice** *Try Exercise*  13

## CHECK YOUR UNDERSTANDING

In the chapter-opening Case Study (page 141), the Starnes family arrived at Old Faithful after it had erupted. They wondered how long it would be until the next eruption. Here is a scatterplot that plots the interval between consecutive eruptions of Old Faithful against the duration of the previous eruption, for the month prior to their visit.

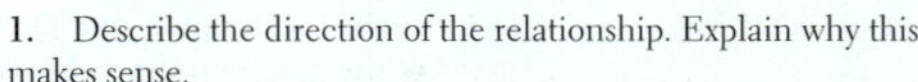

1. Describe the direction of the relationship. Explain why this makes sense.
2. What form does the relationship take? Why are there two clusters of points?
3. How strong is the relationship? Justify your answer.
4. Are there any outliers?
5. What information does the Starnes family need to predict when the next eruption will occur?

**TRM Extra Content on Time Plots**

Additional material about time plots can be found by clicking on the link in the TE-book, opening the TRCD, or logging onto the book's companion Web site www.whfreeman.com/tps5e (teacher log-in required).

**ANSWERS TO CYU**

**1.** Positive. The longer the duration of the eruption, the longer we should expect to wait between eruptions. Long eruptions use more energy and it will take longer to build up the energy needed to erupt again.

**2.** Roughly linear with two clusters. The clusters indicate that, in general, there are two types of eruptions—shorter eruptions that last around 2 minutes and longer eruptions that last around 4.5 minutes.

**3.** Fairly strong. The points don't deviate much from the linear form.

**4.** There are a few possible outliers around the clusters. However, there aren't many and potential outliers are not very distant from the main clusters of points.

**5.** How long the previous eruption was.

## Teaching Tip

Remind students that they can watch a video presentation of this technology corner (for the TI-83/84, TI-89, or TI-Nspire) at www.whfreeman.com/tps5e.

## Teaching Tip

It is often useful to investigate relationships for two different groups or populations, such as boys and girls, on one scatterplot. To do this, enter the boys' data in L1 and L2, and set up Plot1 to use the squares as marks. Then enter the girls' data in L3 and L4, and set up Plot2 to use the plus signs as marks. Or, if you are using a graphing calculator that has color, you can use different colors for each gender. Doing Zoom Stat will plot them both on the same scatterplot using different marks. See Exercises 11 and 12 for some examples.

## Teaching Tip

It might help to compare the strength of an association between two quantitative variables and the spread of the distribution of a single quantitative variable. For example, two distributions can have the same center and shape but vastly different spreads. Likewise, two associations can have the same direction and form but very different strengths. Also, both strength and spread are fairly difficult to describe without a numerical summary.

## Teaching Tip

Point out that the correlation measures direction (D) and strength (S), but only for linear relationships. The correlation alone cannot tell us the form (F) of the association or if there are any outliers (O).

## Teaching Tip

Some people refer to *r* as Pearson's correlation coefficient in honor of one of the fathers of statistics, Karl Pearson. Like many other things, it was not named after its creator (another instance of Stigler's law of eponymy: "No scientific discovery is named after its original discoverer"). It was developed by Francis Galton, a cousin of Charles Darwin, who also developed the ideas of linear regression that we will study in Section 3.2. The connection between correlation and regression is probably the reason why the symbol for correlation is *r* rather than *c*.

### 7. TECHNOLOGY CORNER — SCATTERPLOTS ON THE CALCULATOR

TI-Nspire instructions in Appendix B; HP Prime instructions on the book's Web site.

Making scatterplots with technology is much easier than constructing them by hand. We'll use the SEC football data from page 146 to show how to construct a scatterplot on a TI-83/84 or TI-89.

- Enter the data values into your lists. Put the points per game in L1/list1 and the number of wins in L2/list2.
- Define a scatterplot in the statistics plot menu (press F2 on the TI-89). Specify the settings shown below.

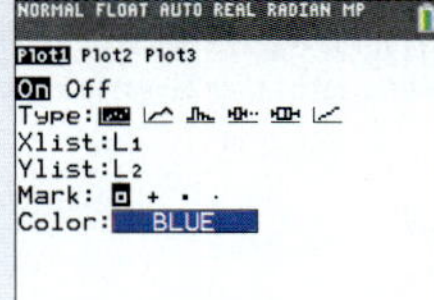


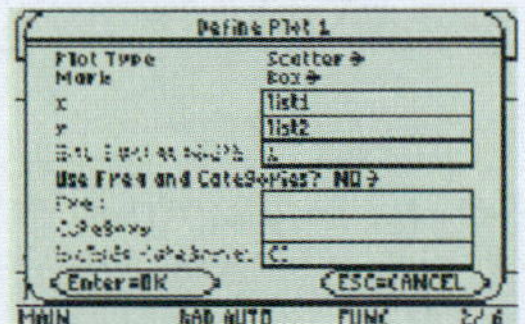


- Use ZoomStat (ZoomData on the TI-89) to obtain a graph. The calculator will set the window dimensions automatically by looking at the values in L1/list1 and L2/list2.

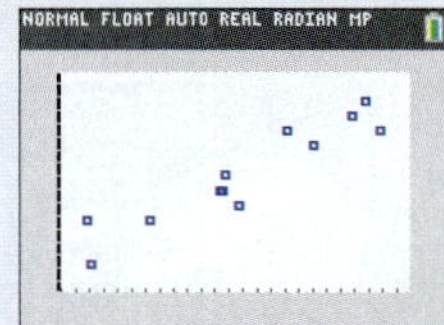


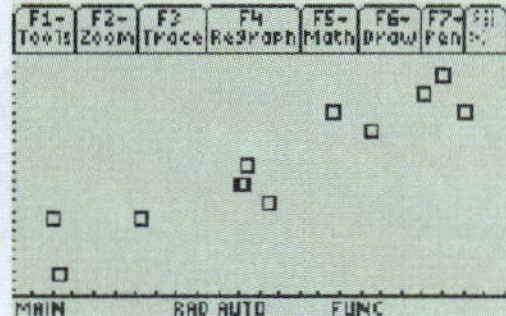


Notice that there are no scales on the axes and that the axes are not labeled. If you copy a scatterplot from your calculator onto your paper, make sure that you scale and label the axes.

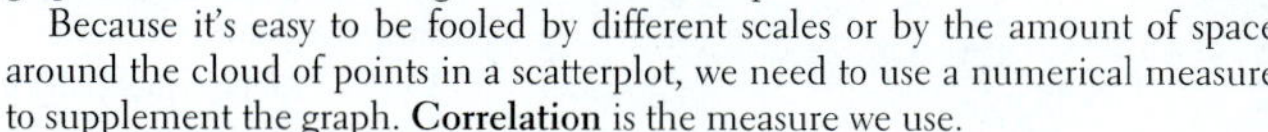
**AP® EXAM TIP** If you are asked to make a scatterplot on a free-response question, be sure to label and scale both axes. *Don't* just copy an unlabeled calculator graph directly onto your paper.

## Measuring Linear Association: Correlation

A scatterplot displays the direction, form, and strength of the relationship between two quantitative variables. Linear relationships are particularly important because a straight line is a simple pattern that is quite common. A linear relationship is strong if the points lie close to a straight line and weak if they are widely scattered about a line. Unfortunately, *our eyes are not good judges of how strong a linear relationship is.* The two scatterplots in Figure 3.5 (on the facing page) show the same data, but the graph on the right is drawn smaller in a large field. The right-hand graph seems to show a stronger linear relationship.

Because it's easy to be fooled by different scales or by the amount of space around the cloud of points in a scatterplot, we need to use a numerical measure to supplement the graph. **Correlation** is the measure we use.

Some people refer to *r* as the "correlation coefficient."

**DEFINITION: Correlation *r***

The **correlation *r*** measures the direction and strength of the linear relationship between two quantitative variables.

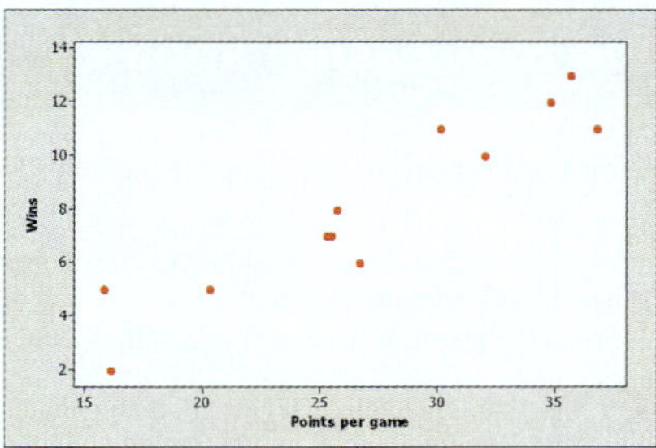


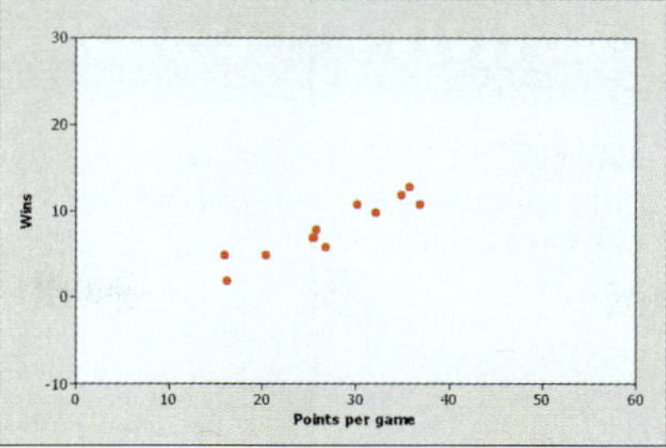


**FIGURE 3.5** Two Minitab scatterplots of the same data. The straight-line pattern in the graph on the right appears stronger because of the surrounding space.

How good are you at estimating the correlation by eye from a scatterplot? To find out, try an online applet. Just search for "guess the correlation applets."

*The correlation r is always a number between −1 and 1.* Correlation indicates the direction of a linear relationship by its sign: $r > 0$ for a positive association and $r < 0$ for a negative association. Values of $r$ near 0 indicate a very weak linear relationship. The strength of the linear relationship increases as $r$ moves away from 0 toward either −1 or 1. The extreme values $r = -1$ and $r = 1$ occur *only* in the case of a perfect linear relationship, when the points lie exactly along a straight line.

Figure 3.6 shows scatterplots that correspond to various values of $r$. To make the meaning of $r$ clearer, the standard deviations of both variables in these plots are equal, and the horizontal and vertical scales are the same. The correlation describes the direction and strength of the linear relationship in each graph.

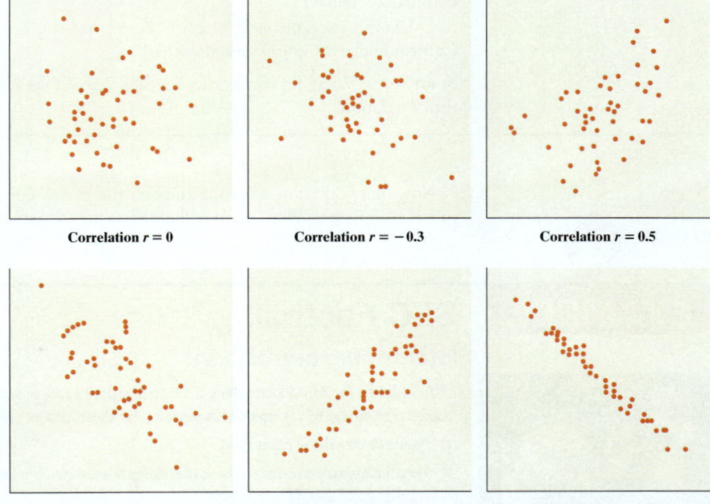


**FIGURE 3.6** How correlation measures the strength of a linear relationship. Patterns closer to a straight line have correlations closer to 1 or −1.

The following Activity lets you explore some important properties of the correlation.

## Teaching Tip

To estimate the correlation from a scatterplot, have students imagine drawing an oval around the points. The rounder the oval, the closer the correlation will be to 0. The longer and skinnier the oval, the closer the correlation will be to ±1. It helps that rounder ovals look like 0s and skinnier ovals look like tilted 1s!

## Teaching Tip

We delayed introducing the formula for correlation so students can get a feel for what correlation measures and how it works. In addition to doing the Activity using the applet at the book's Web site, we recommend that you spend some time conducting a "Guess the Correlation Contest." By the end of the contest, students will have a much better understanding of correlation.

- The *Guess the Correlation* applet at www.rossmanchance.com/applets asks students to guess the value of the correlation for randomly drawn scatterplots. We have tried the contest in two different ways, and they are both quite fun. In Version 1, present each student with a different scatterplot and have the student guess the correlation. Keep track of how much each student is off. Then give every student a second try, in reverse order so that the student who went first in the first round will go last in the second round. Give the top three students some extra credit. In Version 2, have a bracket-style tournament (think March Madness). For each head-to-head match-up, present one scatterplot. Have each student write down their guesses on notecards and reveal them at the same time. The student who is the closest survives until the next round.
- Another possible "correlation contest": Using the *Correlation and Regression* applet at the book's Web site, have students take turns trying to draw a scatterplot with a random correlation. To get a random value of $r$, use the command `2rand-1` on the TI graphing calculator (`rand` is found in the `MATH PRB` menu). Make sure to hide the actual correlation shown on the applet!

## Activity Overview

You can watch an overview of this Activity (for teachers) by clicking on the link in the TE-book, or looking on the TRCD or the book's companion Web site www.whfreeman.com/tps5e (teacher log-in required).

## Teaching Tip

It is very important that you do this Activity (or the Alternate Activity) with your students. Students are expected to understand the limitations of correlation on the AP® exam. In Step 1 of the activity, students should learn that the correlation does not tell us anything about the *form* of the association. In Step 2 of the activity, students should learn that a single point can have a large effect on the correlation.

## Teaching Tip

Here are two additional scatterplots that illustrate how a strong correlation doesn't imply that the association is linear. In the first scatterplot, the association is clearly nonlinear but the correlation is close to 1. In the second scatterplot, the association is clearly linear but the correlation is not close to 1 at all.

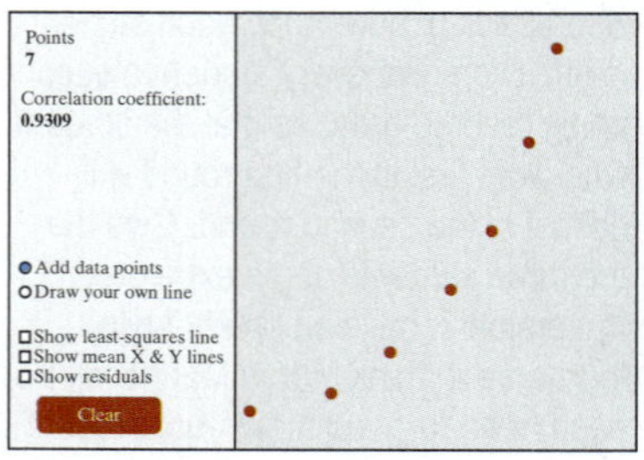


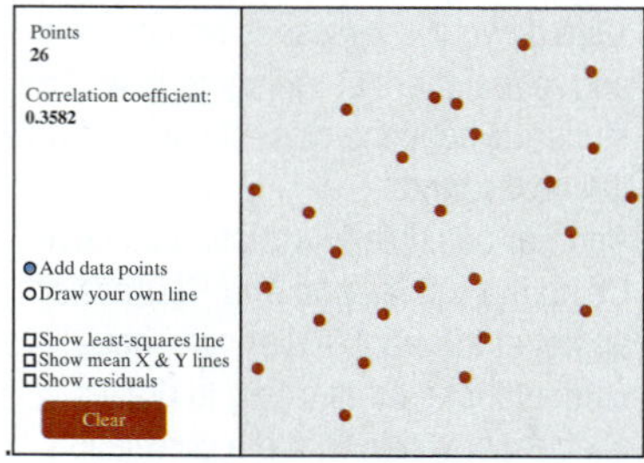


## Teaching Tip

Another way to demonstrate the influence of a single point on the correlation is to use the "oval method" described in a Teaching Tip on the previous page. If including the point makes the oval longer and skinnier, then that point makes the correlation get closer to ±1. If including the point makes the oval become more circular, that point makes the correlation get closer to 0.

## ACTIVITY | *Correlation and Regression applet*

**MATERIALS:**
Computer with Internet connection

Go to the book's Web site, www.whfreeman.com/tps5e, and launch the *Correlation and Regression* applet.

1. You are going to use the *Correlation and Regression* applet to make several scatterplots with 10 points that have correlation close to 0.7.

(a) Start by putting two points on the graph. What's the value of the correlation? Why does this make sense?

(b) Make a lower-left to upper-right pattern of 10 points with correlation about $r = 0.7$. (You can drag points up or down to adjust $r$ after you have 10 points.)

(c) Make another scatterplot: this one should have 9 points in a vertical stack at the left of the plot. Add 1 point far to the right and move it until the correlation is close to 0.7.

(d) Make a third scatterplot: make this one with 10 points in a curved pattern that starts at the lower left, rises to the right, then falls again at the far right. Adjust the points up or down until you have a very smooth curve with correlation close to 0.7.

*Summarize:* If you know that the correlation between two variables is $r = 0.7$, what can you say about the form of the relationship?

2. Click on the scatterplot to create a group of 10 points in the lower-left corner of the scatterplot with a strong straight-line pattern (correlation about 0.9).

(a) Add 1 point at the upper right that is in line with the first 10. How does the correlation change?

(b) Drag this last point straight down. How small can you make the correlation? Can you make the correlation negative?

*Summarize:* What did you learn from Step 2 about the effect of a single point on the correlation?

Now that you know what information the correlation provides—and doesn't provide—let's look at an example that shows how to interpret it.

## EXAMPLE SEC Football

### *Interpreting correlation*

PROBLEM: Our earlier scatterplot of the average points per game and number of wins for college football teams in the SEC is repeated at top right. For these data, $r = 0.936$.

(a) Interpret the value of $r$ in context.

(b) The point highlighted in red on the scatterplot is Mississippi. What effect does Mississippi have on the correlation? Justify your answer.

## ALTERNATE ACTIVITY

If you don't have a projector for your computer, here is an Activity you can do with your students using graphing calculators that explores outliers and influential points. The table shows the weight (in pounds) and cost (in dollars) of a sample of 11 stand mixers (from *Consumer Reports*, November 2005):

| Weight (lb) | Price ($) |
|---|---|
| 23 | 180 |
| 28 | 250 |
| 19 | 300 |
| 17 | 150 |
| 25 | 300 |
| 26 | 370 |
| 21 | 400 |
| 32 | 350 |
| 16 | 200 |
| 17 | 150 |
| 8 | 30 (Walmart mixer) |

1. Sketch a scatterplot of the data and describe what you see. (There is a moderately strong, positive linear association between weight and price.)
2. Calculate the correlation. ($r = 0.74$)

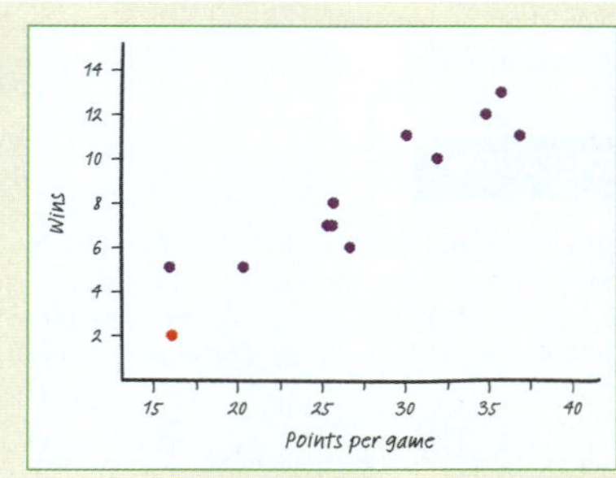


SOLUTION:

(a) The correlation of 0.936 confirms what we see in the scatterplot: there is a strong, positive linear relationship between points per game and wins in the SEC.

(b) Mississippi makes the correlation closer to 1 (stronger). If Mississippi were not included, the remaining points wouldn't be as tightly clustered in a linear pattern.

**For Practice** *Try Exercise* **21**

**AP® EXAM TIP** If you're asked to interpret a correlation, start by looking at a scatterplot of the data. Then be sure to address direction, form, strength, and outliers (sound familiar?) and put your answer in context.

## CHECK YOUR UNDERSTANDING

The scatterplots below show four sets of real data: (a) repeats the manatee plot in Figure 3.4 (page 149); (b) shows the number of named tropical storms and the number predicted before the start of hurricane season each year between 1984 and 2007 by William Gray of Colorado State University; (c) plots the healing rate in micrometers (millionths of a meter) per hour for the two front limbs of several newts in an experiment; and (d) shows stock market performance in consecutive years over a 56-year period. For each graph, estimate the correlation $r$. Then interpret the value of $r$ in context.

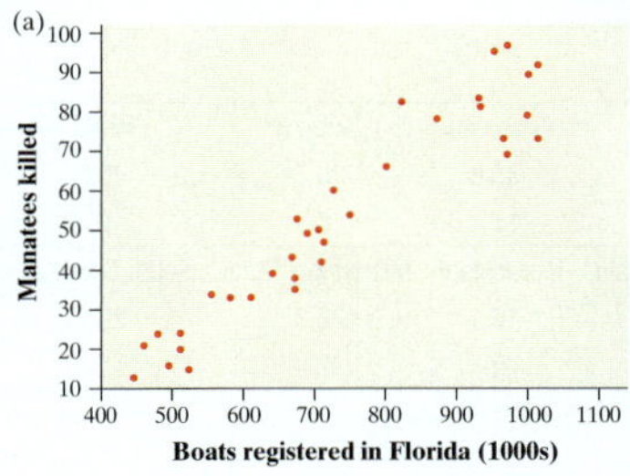


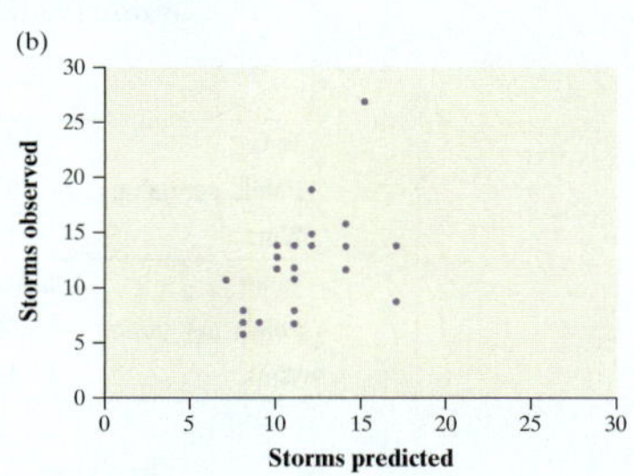


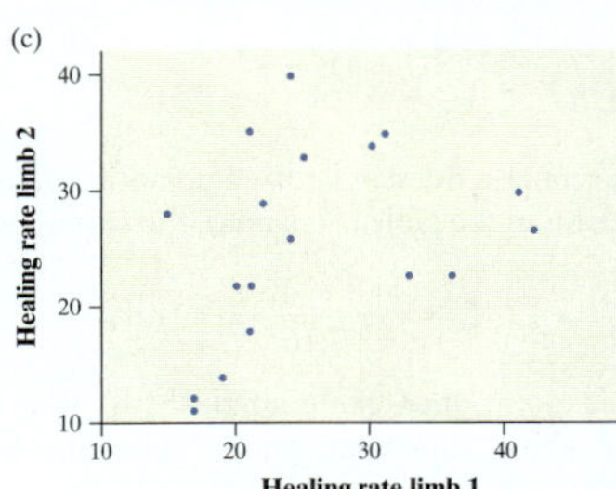


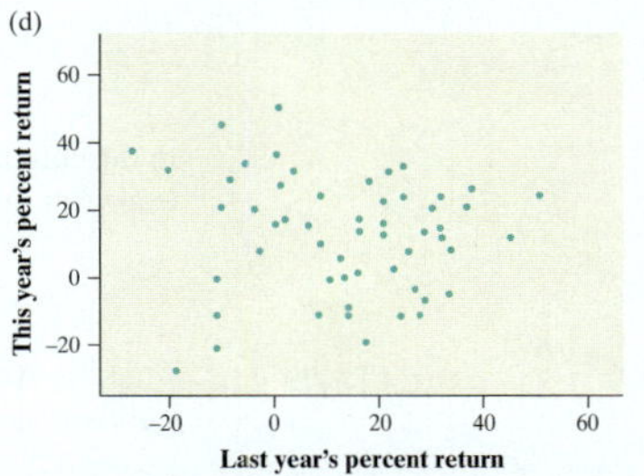


## ALTERNATE EXAMPLE

### *Back to the track!*

Here is a scatterplot of the sprint time and long-jump distance data from earlier. The correlation is $r = -0.84$.

PROBLEM:

**(a)** Interpret the value of $r$ in context.

**(b)** What effect does the point highlighted in red have on the correlation? Explain.

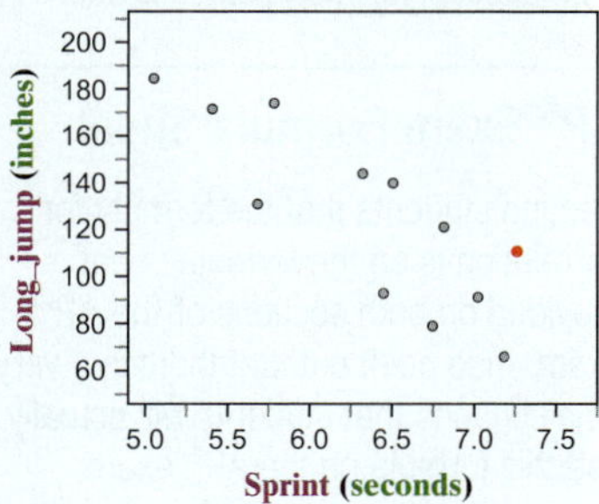


SOLUTION:

**(a)** The correlation of $r = -0.84$ confirms what we see in the scatterplot—there is a strong negative association between sprint time and long-jump distance.

**(b)** Because the highlighted point is a little higher than expected based on the pattern of the rest of the data, it makes the correlation closer to 0. The correlation without this point included is $r = -0.87$.

## ALTERNATE ACTIVITY *(continued)*

**3.** The last mixer in the table is from Walmart. What happens to the correlation when you remove this point? ($r = 0.56$)

**4.** What happens to the correlation if the Walmart mixer weighs 25 pounds instead of 8 pounds? Add the point (25,30) and recalculate the correlation. ($r = 0.34$)

**5.** Suppose that a new titanium mixer was introduced that weighed 8 pounds, but the cost was \$500. Remove the point (25,30) and add the point (8,500) and recalculate the correlation. ($r = -0.08$)

**6.** Summarize what you have learned about the effect of a single point on the correlation.

## ANSWERS TO CYU

**(a)** $r \approx 0.9$. This indicates that there is a strong, positive linear relationship between the number of boats registered in Florida and the number of manatees killed.

**(b)** $r \approx 0.5$. This indicates that there is a moderate, positive linear relationship between the number of named storms predicted and the actual number of named storms.

**(c)** $r \approx 0.3$. This indicates that there is a weak, positive linear relationship between the healing rate of the two front limbs of the newts.

**(d)** $r \approx -0.1$. This indicates that there is a weak, negative linear relationship between last year's percent return and this year's percent return in the stock market.

## AP® Exam Common Error

When asking students to interpret correlation, be careful when addressing form. While a strong positive linear relationship will have a correlation near 1, a correlation near 1 doesn't mean that the relationship is linear. It is easy to create curved data with a correlation close to 1. For example, graphing the function $y = x^2$ for the integers 1 to 10 gives a correlation of $r = 0.97$, even though the relationship between $x$ and $y$ is clearly not linear!

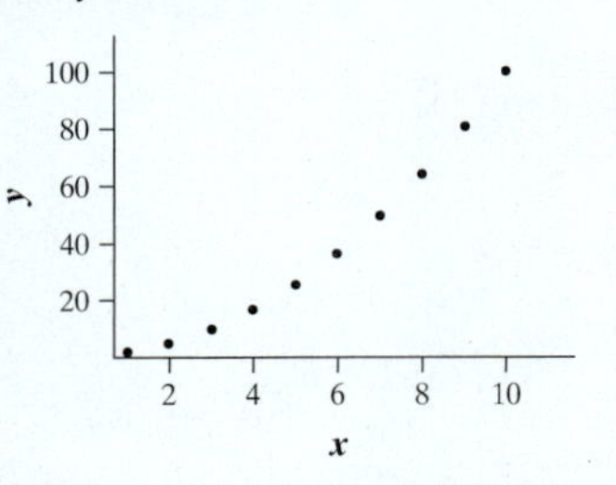

## Teaching Tip

Although understanding how the formula works illuminates some of the properties of the correlation (e.g., changing units won't affect the correlation, reversing the variables won't affect the correlation), you can move through it rather quickly if you are pressed for time. If the students never compute a correlation by hand, they can still be successful young statisticians!

## AP® Exam Formula Sheet

Remind students that the formula for correlation is on the formula sheet provided on both sections of the AP® exam. Also point out that there is a very small chance that students will actually use this formula on the AP® exam.

## TECHNOLOGY TIP

To learn how to calculate the correlation on your graphing calculator, see the Technology Corner on page 171. Unfortunately, the calculator will not let you calculate the correlation without calculating the least-squares regression line, so we will not present this method until we are ready to calculate the least-squares regression line in Section 3.2.



**Calculating Correlation** Now that you have some idea of how to interpret the correlation, let's look at how it's calculated.

### HOW TO CALCULATE THE CORRELATION $r$

Suppose that we have data on variables $x$ and $y$ for $n$ individuals. The values for the first individual are $x_1$ and $y_1$, the values for the second individual are $x_2$ and $y_2$, and so on. The means and standard deviations of the two variables are $\bar{x}$ and $s_x$ for the $x$-values, and $\bar{y}$ and $s_y$ for the $y$-values. The correlation $r$ between $x$ and $y$ is

$$r = \frac{1}{n-1}\left[\left(\frac{x_1-\bar{x}}{s_x}\right)\left(\frac{y_1-\bar{y}}{s_y}\right) + \left(\frac{x_2-\bar{x}}{s_x}\right)\left(\frac{y_2-\bar{y}}{s_y}\right) + \cdots + \left(\frac{x_n-\bar{x}}{s_x}\right)\left(\frac{y_n-\bar{y}}{s_y}\right)\right]$$

or, more compactly,

$$r = \frac{1}{n-1}\sum\left(\frac{x_i-\bar{x}}{s_x}\right)\left(\frac{y_i-\bar{y}}{s_y}\right)$$

The formula for the correlation $r$ is a bit complex. It helps us see what correlation is, but in practice, you should use your calculator or software to find $r$. Exercises 19 and 20 ask you to calculate a correlation step-by-step from the definition to solidify its meaning.

The formula for $r$ begins by standardizing the observations. Let's use the familiar SEC football data to perform the required calculations. The table below shows the values of points per game $x$ and number of wins $y$ for the SEC college football teams. For these data, $\bar{x} = 27.07$ and $s_x = 7.16$.

| Team | Alabama | Arkansas | Auburn | Florida | Georgia | Kentucky |
|---|---|---|---|---|---|---|
| Points per game | 34.8 | 36.8 | 25.7 | 25.5 | 32.0 | 15.8 |
| Wins | 12 | 11 | 8 | 7 | 10 | 5 |
| **Team** | **Louisiana State** | **Mississippi** | **Mississippi State** | **South Carolina** | **Tennessee** | **Vanderbilt** |
| Points per game | 35.7 | 16.1 | 25.3 | 30.1 | 20.3 | 26.7 |
| Wins | 13 | 2 | 7 | 11 | 5 | 6 |

The value

$$\frac{x_i-\bar{x}}{s_x}$$

in the correlation formula is the standardized points per game ($z$-score) of the $i$th team. For the first team in the table (Alabama), the corresponding $z$-score is

$$z_x = \frac{34.8-27.07}{7.16} = 1.08$$

That is, Alabama's points per game total (34.8) is a little more than 1 standard deviation above the mean points per game for the SEC teams.

Some people like to write the correlation formula as

$$r = \frac{1}{n-1}\sum z_x z_y$$

to emphasize the product of standardized scores in the calculation.

Standardized values have no units—in this example, they are no longer measured in points.

To standardize the number of wins, we use $\bar{y} = 8.08$ and $s_y = 3.34$. For Alabama, $z_y = \frac{12 - 8.08}{3.34} = 1.17$. Alabama's number of wins (12) is 1.17 standard deviations above the mean number of wins for SEC teams. When we multiply this team's two $z$-scores, we get a product of 1.2636. The correlation $r$ is an "average" of the products of the standardized scores for all the teams. Just as in the case of the standard deviation $s_x$, the average here divides by one fewer than the number of individuals. Finishing the calculation reveals that $r = 0.936$ for the SEC teams.

THINK ABOUT IT

**What does correlation measure?** The Fathom screen shots below provide more detail. At the left is a scatterplot of the SEC football data with two lines added—a vertical line at the group's mean points per game and a horizontal line at the mean number of wins of the group. Most of the points fall in the upper-right or lower-left "quadrants" of the graph. That is, teams with above-average points per game tend to have above-average numbers of wins, and teams with below-average points per game tend to have numbers of wins that are below average. This confirms the positive association between the variables.

Below on the right is a scatterplot of the standardized scores. To get this graph, we transformed both the $x$- and the $y$-values by subtracting their mean and dividing by their standard deviation. As we saw in Chapter 2, standardizing a data set converts the mean to 0 and the standard deviation to 1. That's why the vertical and horizontal lines in the right-hand graph are both at 0.

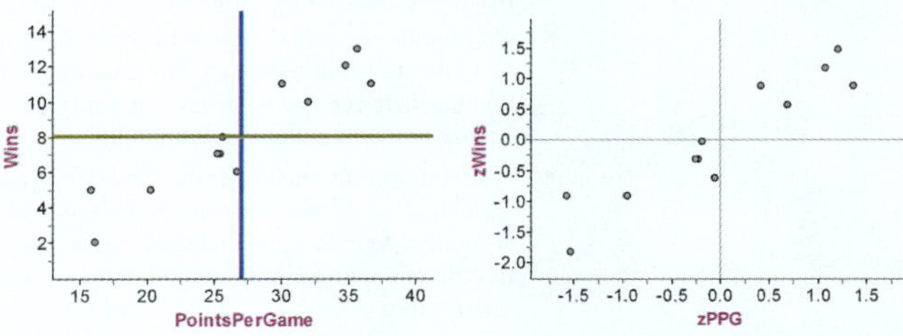


Notice that all the products of the standardized values will be positive—not surprising, considering the strong positive association between the variables. What if there was a negative association between two variables? Most of the points would be in the upper-left and lower-right "quadrants" and their $z$-score products would be negative, resulting in a negative correlation.

## Facts about Correlation

How correlation behaves is more important than the details of the formula. Here's what you need to know in order to interpret correlation correctly.

## Teaching Tip

For Fact 1, the commutative property guarantees that reversing the two variables results in the same correlation: $z_x z_y = z_y z_x$. You might also mention that if there is a strong correlation between height and weight, a strong correlation will also exist between weight and height.

## Teaching Tip

For Fact 2, remind students that changing the units doesn't change the way the scatterplot looks; it only changes the numbers on the scales.

## Teaching Tip

Make sure to emphasize the first bullet in the list of cautions. Humor is a good way to make the point clear.

- Did you know that eating chocolate makes winning a Nobel Prize more likely? The correlation between per capita chocolate consumption and the number of Nobel laureates per 10 million people for 23 selected countries is $r = 0.791$ (data from http://www.nejm.org/doi/full/10.1056/NEJMon1211064).
- Did you know that AP® Calculus is causing global warming? As the number of AP® Calculus exams has grown over the years, so has the average global temperature!
- Did you know that eating ice cream causes people to drown? In months with higher ice cream sales, there are also higher numbers of drowning deaths!
- Show a classic correlation comic strip at http://xkcd.com/552.

## Teaching Tip

Be careful with the word "correlation." Correlation is a numerical measure of the direction and strength of the relationship between two quantitative variables. If you are describing the relationship between two categorical variables (as in Section 1.1) or the relationship between a categorical and quantitative variable, use the word "association" instead.

"He says we've ruined his positive correlation between height and weight."

1. *Correlation makes no distinction between explanatory and response variables.* It makes no difference which variable you call $x$ and which you call $y$ in calculating the correlation. Can you see why from the formula?

$$r = \frac{1}{n-1}\sum\left(\frac{x_i - \bar{x}}{s_x}\right)\left(\frac{y_i - \bar{y}}{s_y}\right)$$

2. Because $r$ uses the standardized values of the observations, *$r$ does not change when we change the units of measurement of $x$, $y$, or both.* Measuring height in centimeters rather than inches and weight in kilograms rather than pounds does not change the correlation between height and weight.

3. *The correlation $r$ itself has no unit of measurement.* It is just a number.

Describing the relationship between two variables is more complex than describing the distribution of one variable. Here are some cautions to keep in mind when you use correlation.

- *Correlation does not imply causation.* Even when a scatterplot shows a strong linear relationship between two variables, we can't conclude that changes in one variable cause changes in the other. For example, looking at data from the last 10 years, there is a strong positive relationship between the number of high school students who own a cell phone and the number of students who pass the AP® Statistics exam. Does this mean that buying a cell phone will help you pass the AP® exam? Not likely. Instead, the correlation is positive because both of these variables are increasing over time.
- *Correlation requires that both variables be quantitative,* so that it makes sense to do the arithmetic indicated by the formula for $r$. We cannot calculate a correlation between the incomes of a group of people and what city they live in because city is a categorical variable.
- Correlation only measures the strength of a linear relationship between two variables. *Correlation does not describe curved relationships between variables, no matter how strong the relationship is.* A correlation of 0 doesn't guarantee that there's *no* relationship between two variables, just that there's no *linear* relationship.
- *A value of $r$ close to 1 or $-1$ does not guarantee a linear relationship between two variables.* A scatterplot with a clear curved form can have a correlation

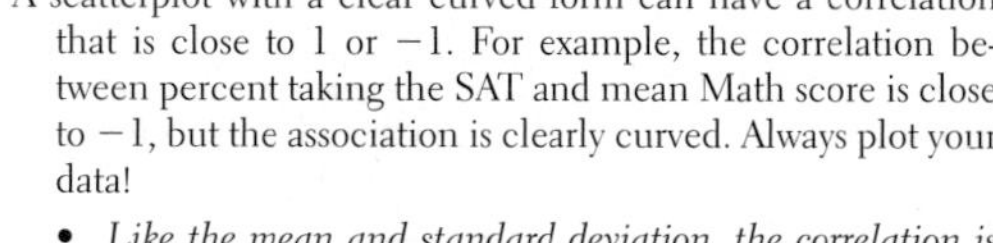

that is close to 1 or $-1$. For example, the correlation between percent taking the SAT and mean Math score is close to $-1$, but the association is clearly curved. Always plot your data!

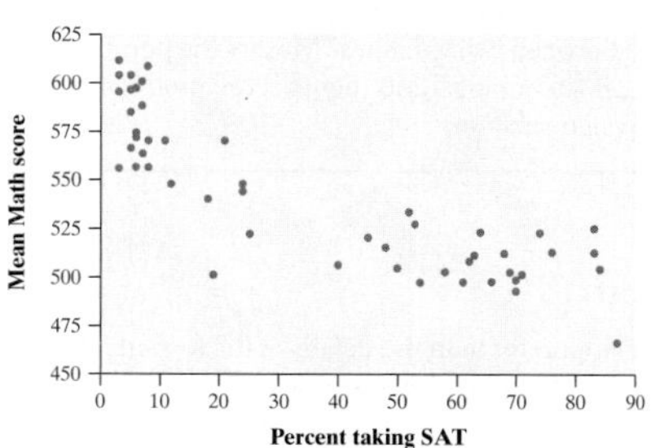


- *Like the mean and standard deviation, the correlation is not resistant: $r$ is strongly affected by a few outlying observations.* Use $r$ with caution when outliers appear in the scatterplot.
- *Correlation is not a complete summary of two-variable data,* even when the relationship between the variables is linear. You should give the means and standard deviations of both $x$ and $y$ along with the correlation.

## Teaching Tip

When you are discussing the third bullet, remind your students about the 68–95–99.7 rule. The 68–95–99.7 rule is handy when a distribution is approximately Normal, but it shouldn't be used otherwise. Similarly, the correlation $r$ is a handy way to measure strength of a linear association, but it shouldn't be used otherwise.

Of course, even giving means, standard deviations, and the correlation for "state SAT Math scores" and "percent taking" will not point out the clusters in Figure 3.2. Numerical summaries complement plots of data, but they do not replace them.

## EXAMPLE Scoring Figure Skaters

### *Why correlation doesn't tell the whole story*

Until a scandal at the 2002 Olympics brought change, figure skating was scored by judges on a scale from 0.0 to 6.0. The scores were often controversial. We have the scores awarded by two judges, Pierre and Elena, for many skaters. How well do they agree? We calculate that the correlation between their scores is $r = 0.9$. But the mean of Pierre's scores is 0.8 point lower than Elena's mean.

These facts don't contradict each other. They simply give different kinds of information. The mean scores show that Pierre awards lower scores than Elena. But because Pierre gives *every* skater a score about 0.8 point lower than Elena does, the correlation remains high. Adding the same number to all values of either $x$ or $y$ does not change the correlation. If both judges score the same skaters, the competition is scored consistently because Pierre and Elena agree on which performances are better than others. The high $r$ shows their agreement. But if Pierre scores some skaters and Elena others, we should add 0.8 point to Pierre's scores to arrive at a fair comparison.

## DATA EXPLORATION The SAT essay: Is longer better?

Following the debut of the new SAT Writing test in March 2005, Dr. Les Perelman from the Massachusetts Institute of Technology stirred controversy by reporting, "It appeared to me that regardless of what a student wrote, the longer the essay, the higher the score." He went on to say, "I have never found a quantifiable predictor in 25 years of grading that was anywhere as strong as this one. If you just graded them based on length without ever reading them, you'd be right over 90 percent of the time."[3] The table below shows the data that Dr. Perelman used to draw his conclusions.[4]

**Length of essay and score for a sample of SAT essays**

| | | | | | | | | | | | |
|---|---|---|---|---|---|---|---|---|---|---|---|
| Words: | 460 | 422 | 402 | 365 | 357 | 278 | 236 | 201 | 168 | 156 | 133 |
| Score: | 6 | 6 | 5 | 5 | 6 | 5 | 4 | 4 | 4 | 3 | 2 |
| Words: | 114 | 108 | 100 | 403 | 401 | 388 | 320 | 258 | 236 | 189 | 128 |
| Score: | 2 | 1 | 1 | 5 | 6 | 6 | 5 | 4 | 4 | 3 | 2 |
| Words: | 67 | 697 | 387 | 355 | 337 | 325 | 272 | 150 | 135 | | |
| Score: | 1 | 6 | 6 | 5 | 5 | 4 | 4 | 2 | 3 | | |

Does this mean that if students write a lot, they are guaranteed high scores? Carry out your own analysis of the data. How would you respond to each of Dr. Perelman's claims?

## ALTERNATE EXAMPLE

### *Scoring essays*

Peter and Elaine are both English teachers at a high school. Students think that Elaine is a harder grader, so Elaine and Peter decide to grade the same 10 essays and see how their scores compare. The correlation was $r = 0.98$, but Elaine's scores were about 5 points lower, on average. This can be seen in the scatterplot below, which includes the line $y = x$. If both teachers gave an essay the same score, the point for that essay would be on the $y = x$ line. All the points are below the $y = x$ line, however, which indicates that Elaine scored each essay lower than Peter did.

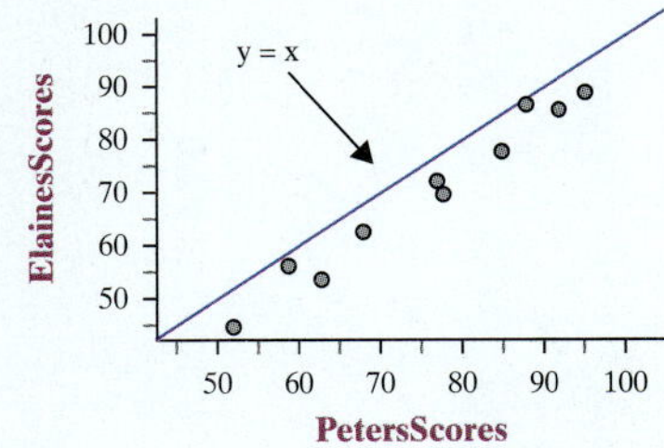


### Answers to Data Exploration

Here is a scatterplot of the data, using $x$ = number of words and $y$ = score.

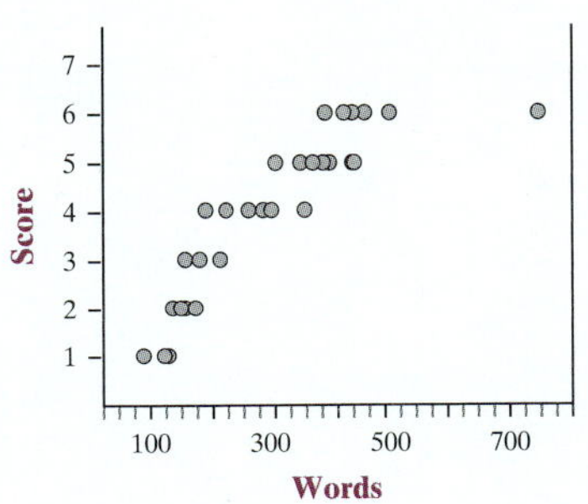


There is a strong, positive nonlinear association between the number of words in an essay and the score it received. This means that longer essays tend to get higher scores. However, if students write a long response, they are *not* guaranteed a good score. For example, repeating the same word over and over 700 times would be as long as any of the essays in the sample but would certainly receive the lowest possible score. The correlation for this association is $r = 0.881$. If Perelman made his 90% claim by rounding the correlation coefficient from 0.88 to 0.90, then he made a serious mistake in his interpretation of the correlation!

**TRM Section 3.1 Quizzes**

There are three different quizzes available for this section in the Teacher's Resource Materials. Versions A and B are roughly parallel. Find them by clicking on the link in the TE-book, opening the TRCD, or logging onto the book's companion Web site www.whfreeman.com/tps5e (teacher log-in required).

# Section 3.1 Summary

- A **scatterplot** displays the relationship between two quantitative variables measured on the same individuals. Mark values of one variable on the horizontal axis ($x$ axis) and values of the other variable on the vertical axis ($y$ axis). Plot each individual's data as a point on the graph.
- If we think that a variable $x$ may help explain, predict, or even cause changes in another variable $y$, we call $x$ an **explanatory variable** and $y$ a **response variable.** Always plot the explanatory variable, if there is one, on the $x$ axis of a scatterplot. Plot the response variable on the $y$ axis.
- In examining a scatterplot, look for an overall pattern showing the **direction, form,** and **strength** of the relationship and then look for **outliers** or other departures from this pattern.
- **Direction:** If the relationship has a clear direction, we speak of either **positive association** (above-average values of the two variables tend to occur together) or **negative association** (above-average values of one variable tend to occur with below-average values of the other variable).
- **Form:** Linear relationships, where the points show a straight-line pattern, are an important form of relationship between two variables. Curved relationships and clusters are other forms to watch for.
- **Strength:** The strength of a relationship is determined by how close the points in the scatterplot lie to a simple form such as a line.
- The **correlation** $r$ measures the strength and direction of the linear association between two quantitative variables $x$ and $y$. Although you can calculate a correlation for any scatterplot, $r$ measures strength for only straight-line relationships.
- Correlation indicates the direction of a linear relationship by its sign: $r > 0$ for a positive association and $r < 0$ for a negative association. Correlation always satisfies $-1 \leq r \leq 1$ and indicates the strength of a linear relationship by how close it is to $-1$ or 1. Perfect correlation, $r = \pm 1$, occurs only when the points on a scatterplot lie exactly on a straight line.
- Remember these important facts about $r$: Correlation does not imply causation. Correlation ignores the distinction between explanatory and response variables. The value of $r$ is not affected by changes in the unit of measurement of either variable. Correlation is not resistant, so outliers can greatly change the value of $r$.

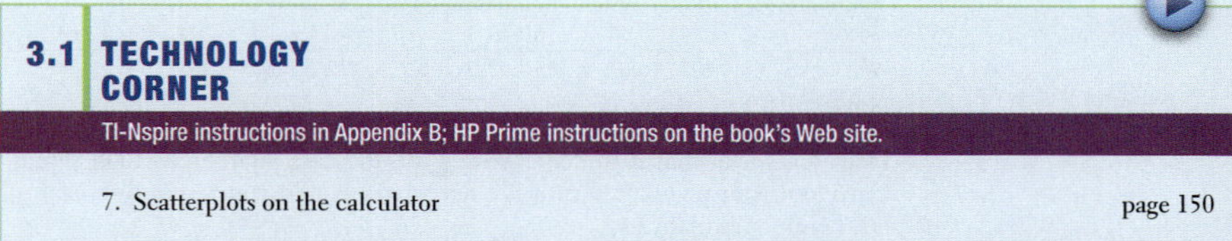

# Section 3.1 Exercises

1. **Coral reefs** How sensitive to changes in water temperature are coral reefs? To find out, measure the growth of corals in aquariums where the water temperature is controlled at different levels. Growth is measured by weighing the coral before and after the experiment. What are the explanatory and response variables? Are they categorical or quantitative?

pg 144

2. **Treating breast cancer** Early on, the most common treatment for breast cancer was removal of the breast. It is now usual to remove only the tumor and nearby lymph nodes, followed by radiation. The change in policy was due to a large medical experiment that compared the two treatments. Some breast cancer patients, chosen at random, were given one or the other treatment. The patients were closely followed to see how long they lived following surgery. What are the explanatory and response variables? Are they categorical or quantitative?

3. **IQ and grades** Do students with higher IQ test scores tend to do better in school? The figure below shows a scatterplot of IQ and school grade point average (GPA) for all 78 seventh-grade students in a rural midwestern school. (GPA was recorded on a 12-point scale with A+ = 12, A = 11, A− = 10, B+ = 9, . . . , D− = 1, and F = 0.)[5]

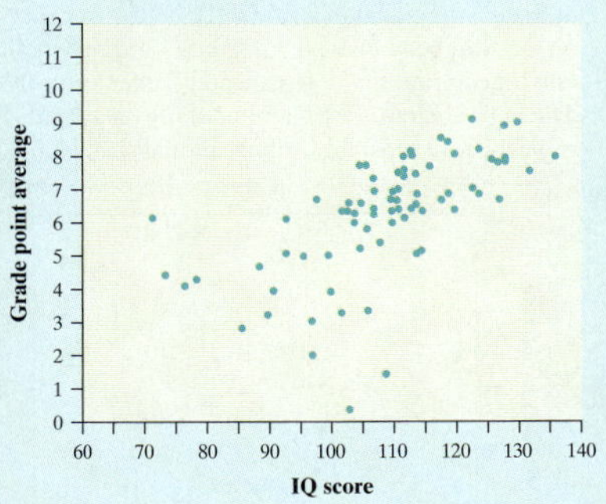


(a) Does the plot show a positive or negative association between the variables? Why does this make sense?

(b) What is the form of the relationship? Is it very strong? Explain your answers.

(c) At the bottom of the plot are several points that we might call outliers. One student in particular has a very low GPA despite an average IQ score. What are the approximate IQ and GPA for this student?

4. **How much gas?** Joan is concerned about the amount of energy she uses to heat her home. The graph below plots the mean number of cubic feet of gas per day that Joan used each month against the average temperature that month (in degrees Fahrenheit) for one heating season.

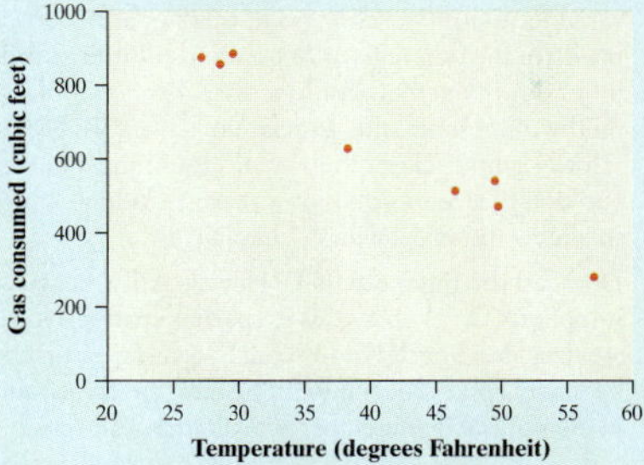


(a) Does the plot show a positive or negative association between the variables? Why does this make sense?

(b) What is the form of the relationship? Is it very strong? Explain your answers.

(c) Explain what the point at the bottom right of the plot represents.

5. **Heavy backpacks** Ninth-grade students at the Webb Schools go on a backpacking trip each fall. Students are divided into hiking groups of size 8 by selecting names from a hat. Before leaving, students and their backpacks are weighed. The data here are from one hiking group in a recent year. Make a scatterplot by hand that shows how backpack weight relates to body weight.

pg 145

| Body weight (lb): | 120 | 187 | 109 | 103 | 131 | 165 | 158 | 116 |
|---|---|---|---|---|---|---|---|---|
| Backpack weight (lb): | 26 | 30 | 26 | 24 | 29 | 35 | 31 | 28 |

6. **Bird colonies** One of nature's patterns connects the percent of adult birds in a colony that return from the previous year and the number of new adults that join the colony. Here are data for 13 colonies of sparrowhawks:[6]

| Percent return: | 74 | 66 | 81 | 52 | 73 | 62 | 52 | 45 | 62 | 46 | 60 | 46 | 38 |
|---|---|---|---|---|---|---|---|---|---|---|---|---|---|
| New adults: | 5 | 6 | 8 | 11 | 12 | 15 | 16 | 17 | 18 | 18 | 19 | 20 | 20 |

Make a scatterplot by hand that shows how the number of new adults relates to the percent of returning birds.

**TRM Full Solutions to Section 3.1 Exercises**

## Answers to Section 3.1 Exercises

**3.1** Explanatory: water temperature (quantitative). Response: weight change (quantitative).

**3.2** Explanatory: type of treatment (categorical). Response: survival time (quantitative).

**3.3 (a)** Positive. Students with higher IQs tend to have higher GPAs and vice versa, because both IQ and GPA are related to mental ability. **(b)** Roughly linear, because a line through the scatterplot of points would provide a good summary. Moderately strong, because most of the points would be close to the line. **(c)** IQ $\approx$ 103 and GPA $\approx$ 0.4.

**3.4 (a)** Negative. Months with high temperatures tend to have low gas consumption and vice versa, because as the temperature gets warmer, less gas would be needed for heating. **(b)** Roughly linear, because a line through the scatterplot of points would provide a good summary. Strong, because the points would be close to the line. **(c)** A month when an average temperature $\approx$ 58 degrees and gas usage $\approx$ 260 cubic feet.

**3.5** A scatterplot is shown below.

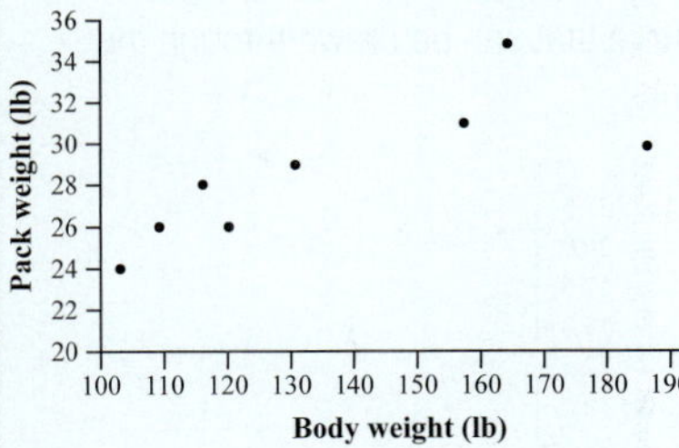


**3.6** A scatterplot is shown below.

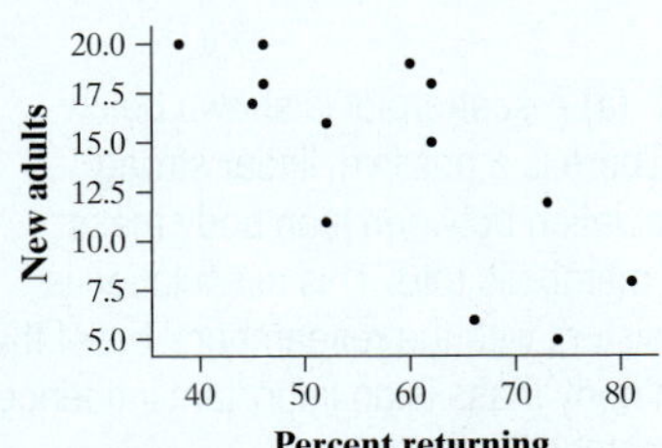

pg 147 7. **Heavy backpacks** Refer to your graph from Exercise 5.

(a) Describe the relationship between body weight and backpack weight for this group of hikers.

(b) One of the hikers is a possible outlier. Identify the body weight and backpack weight for this hiker. How does this hiker affect the form of the association?

8. **Bird colonies** Refer to your graph from Exercise 6.

(a) Describe the relationship between number of new sparrowhawks in a colony and percent of returning adults.

(b) For short-lived birds, the association between these variables is positive: changes in weather and food supply drive the populations of new and returning birds up or down together. For long-lived territorial birds, on the other hand, the association is negative because returning birds claim their territories in the colony and don't leave room for new recruits. Which type of species is the sparrowhawk? Explain.

9. **Does fast driving waste fuel?** How does the fuel consumption of a car change as its speed increases? Here are data for a British Ford Escort. Speed is measured in kilometers per hour, and fuel consumption is measured in liters of gasoline used per 100 kilometers traveled.[7]

| Speed (km/h) | Fuel used (liters/100 km) | Speed (km/h) | Fuel used (liters/100 km) |
|---|---|---|---|
| 10 | 21.00 | 90 | 7.57 |
| 20 | 13.00 | 100 | 8.27 |
| 30 | 10.00 | 110 | 9.03 |
| 40 | 8.00 | 120 | 9.87 |
| 50 | 7.00 | 130 | 10.79 |
| 60 | 5.90 | 140 | 11.77 |
| 70 | 6.30 | 150 | 12.83 |
| 80 | 6.95 | | |

(a) Use your calculator to help sketch a scatterplot.

(b) Describe the form of the relationship. Why is it not linear? Explain why the form of the relationship makes sense.

(c) It does not make sense to describe the variables as either positively associated or negatively associated. Why?

(d) Is the relationship reasonably strong or quite weak? Explain your answer.

10. **Do heavier people burn more energy?** Metabolic rate, the rate at which the body consumes energy, is important in studies of weight gain, dieting, and exercise. We have data on the lean body mass and resting metabolic rate for 12 women who are subjects in a study of dieting. Lean body mass, given in kilograms, is a person's weight leaving out all fat. Metabolic rate is measured in calories burned per 24 hours. The researchers believe that lean body mass is an important influence on metabolic rate.

| Mass: | 36.1 | 54.6 | 48.5 | 42.0 | 50.6 | 42.0 | 40.3 | 33.1 | 42.4 | 34.5 | 51.1 | 41.2 |
|---|---|---|---|---|---|---|---|---|---|---|---|---|
| Rate: | 995 | 1425 | 1396 | 1418 | 1502 | 1256 | 1189 | 913 | 1124 | 1052 | 1347 | 1204 |

(a) Use your calculator to help sketch a scatterplot to examine the researchers' belief.

(b) Describe the direction, form, and strength of the relationship.

11. **Southern education** For a long time, the South has lagged behind the rest of the United States in the performance of its schools. Efforts to improve education have reduced the gap. We wonder if the South stands out in our study of state average SAT Math scores. The figure below enhances the scatterplot in Figure 3.2 (page 145) by plotting 12 southern states in red.

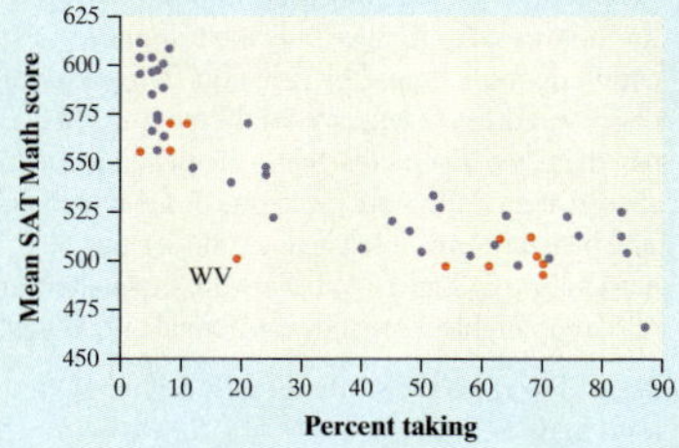


(a) What does the graph suggest about the southern states?

(b) The point for West Virginia is labeled in the graph. Explain how this state is an outlier.

12. **Do heavier people burn more energy?** The study of dieting described in Exercise 10 collected data on the lean body mass (in kilograms) and metabolic rate (in calories) for 12 female and 7 male subjects. The figure below is a scatterplot of the data for all 19 subjects, with separate symbols for males and females.

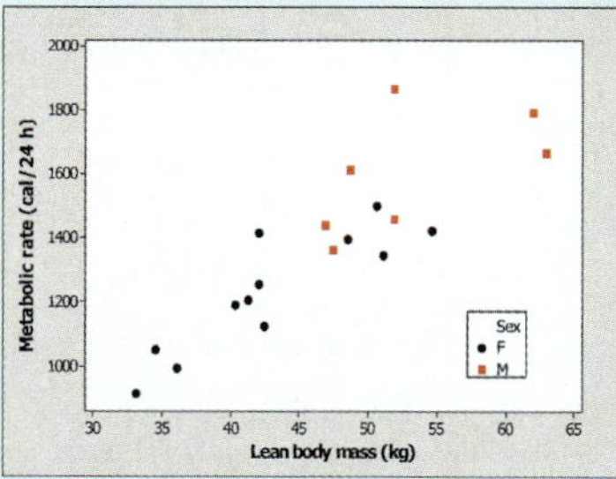


Does the same overall pattern hold for both women and men? What difference between the sexes do you see from the graph?

13. **Merlins breeding** The percent of an animal species in the wild that survives to breed again is often lower following a successful breeding season. A study of

pg 148

**3.7 (a)** There is a positive association between backpack weight and body weight. For students under 140 pounds, there seems to be a linear pattern in the graph. However, for students above 140 pounds, the association begins to curve. Because the points vary somewhat from the linear pattern, the relationship is only moderately strong. **(b)** The hiker with body weight 187 pounds and pack weight 30 pounds. This hiker makes the form appear to be nonlinear for weights above 140 pounds. Without this hiker, the association would look very linear for all body weights.

**3.8 (a)** There is a negative, linear, moderately strong association between percent returning and the number of new adults. **(b)** Because this association is negative, we conclude that the sparrowhawk is a long-lived territorial species.

**3.9 (a)** A scatterplot is shown below. **(b)** The relationship is curved. Large amounts of fuel were used for low and high values of speed and smaller amounts of fuel were used for moderate speeds. This makes sense because the best fuel efficiency is obtained by driving at moderate speeds. **(c)** Both directions are present in the scatterplot. The association is negative for lower speeds and positive for higher speeds. **(d)** The relationship is very strong, with little deviation from a curve that can be drawn through the points.

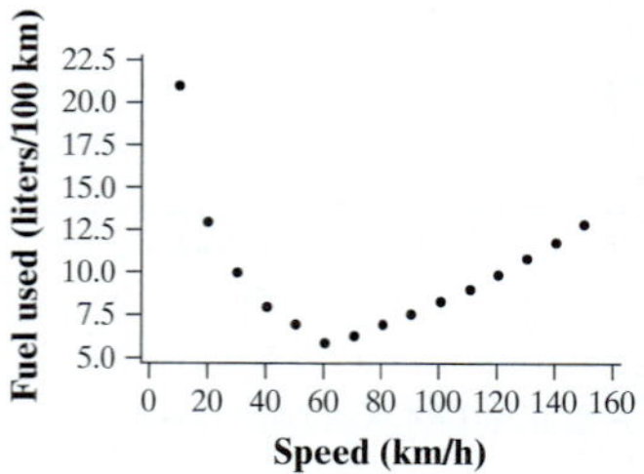


**3.10 (a)** A scatterplot is shown below. **(b)** There is a positive, linear strong association between lean body mass and metabolic rate. This association is consistent with the researchers' belief that lean body mass is an important influence on metabolic rate.

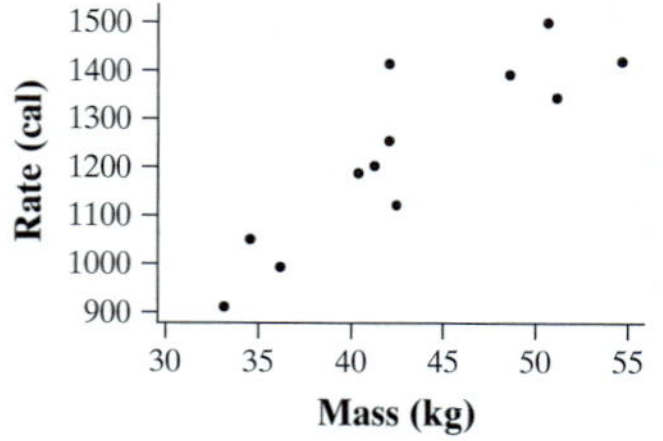


**3.11 (a)** Most of the southern states fall in the same pattern as the rest of the states. However, southern states typically have lower mean SAT math scores than other states with a similar percent of students taking the SAT. **(b)** West Virginia has a much lower mean SAT Math score than the other states that have a similar percent of students taking the exam.

**3.12** For both men and women, there is a relatively strong, positive linear association between lean body mass and metabolic rate. However, the relationship is not as strong for men. Also, the men tend to have higher lean body masses and higher metabolic rates.

merlins (small falcons) in northern Sweden observed the number of breeding pairs in an isolated area and the percent of males (banded for identification) that returned the next breeding season. Here are data for seven years:[8]

| Breeding pairs: | 28 | 29 | 29 | 29 | 30 | 32 | 33 |
|---|---|---|---|---|---|---|---|
| Percent return: | 82 | 83 | 70 | 61 | 69 | 58 | 43 |

Make a scatterplot to display the relationship between breeding pairs and percent return. Describe what you see.

14. **Does social rejection hurt?** We often describe our emotional reaction to social rejection as "pain." Does social rejection cause activity in areas of the brain that are known to be activated by physical pain? If it does, we really do experience social and physical pain in similar ways. Psychologists first included and then deliberately excluded individuals from a social activity while they measured changes in brain activity. After each activity, the subjects filled out questionnaires that assessed how excluded they felt. The table below shows data for 13 subjects.[9] "Social distress" is measured by each subject's questionnaire score after exclusion relative to the score after inclusion. (So values greater than 1 show the degree of distress caused by exclusion.) "Brain activity" is the change in activity in a region of the brain that is activated by physical pain. (So positive values show more pain.)

| Subject | Social distress | Brain activity |
|---|---|---|
| 1 | 1.26 | −0.055 |
| 2 | 1.85 | −0.040 |
| 3 | 1.10 | −0.026 |
| 4 | 2.50 | −0.017 |
| 5 | 2.17 | −0.017 |
| 6 | 2.67 | 0.017 |
| 7 | 2.01 | 0.021 |
| 8 | 2.18 | 0.025 |
| 9 | 2.58 | 0.027 |
| 10 | 2.75 | 0.033 |
| 11 | 2.75 | 0.064 |
| 12 | 3.33 | 0.077 |
| 13 | 3.65 | 0.124 |

Make a scatterplot to display the relationship between social distress and brain activity. Describe what you see.

15. **Matching correlations** Match each of the following scatterplots to the $r$ below that best describes it. (Some $r$'s will be left over.)

$r = -0.9 \quad r = -0.7 \quad r = -0.3 \quad r = 0$

$r = 0.3 \quad r = 0.7 \quad r = 0.9$

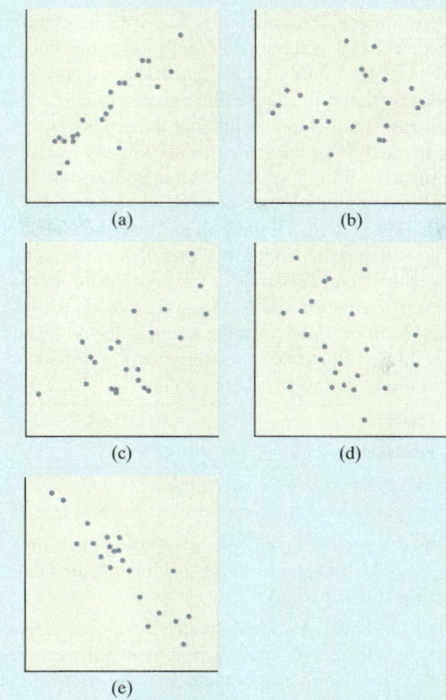


16. **Rank the correlations** Consider each of the following relationships: the heights of fathers and the heights of their adult sons, the heights of husbands and the heights of their wives, and the heights of women at age 4 and their heights at age 18. Rank the correlations between these pairs of variables from largest to smallest. Explain your reasoning.

17. **Correlation blunders** Each of the following statements contains an error. Explain what's wrong in each case.

(a) "There is a high correlation between the gender of American workers and their income."

(b) "We found a high correlation ($r = 1.09$) between students' ratings of faculty teaching and ratings made by other faculty members."

(c) "The correlation between planting rate and yield of corn was found to be $r = 0.23$ bushel."

18. **Teaching and research** A college newspaper interviews a psychologist about student ratings of the teaching of faculty members. The psychologist says, "The evidence indicates that the correlation between the research productivity and teaching rating of faculty members is close to zero." The paper reports this as "Professor McDaniel said that good researchers tend to be poor teachers, and vice versa." Explain why the paper's report is wrong. Write a statement in plain language (don't use the word "correlation") to explain the psychologist's meaning.

**3.13** A scatterplot is shown below. There is a negative, linear, moderately strong relationship between the percent returning and the number of breeding pairs.

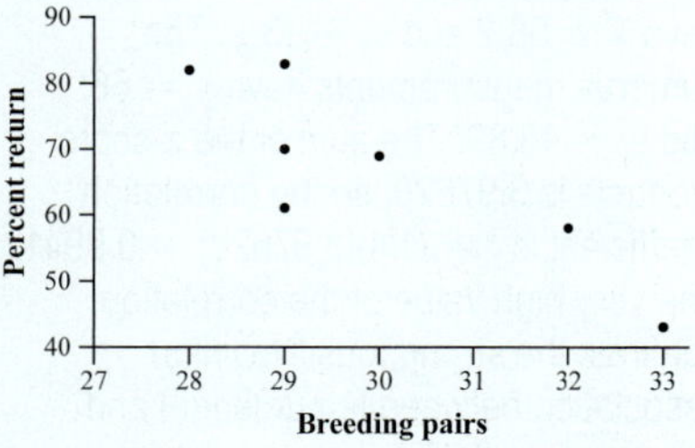


**3.14** A scatterplot is shown below. There is a positive, linear, strong relationship between brain activity and social distress score.

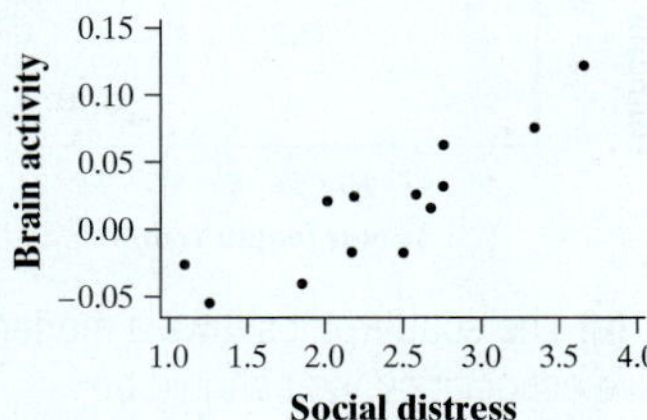


**3.15** **(a)** $r = 0.9$ **(b)** $r = 0$ **(c)** $r = 0.7$ **(d)** $r = -0.3$ **(e)** $r = -0.9$

**3.16** Largest: height of women at age 4 and their height as women at age 18 because taller children usually become taller adults and shorter children usually become shorter adults. Next highest: heights of fathers and their adult sons. Because they share genes, we expect a fairly strong association. Smallest: heights of husbands and their wives because some tall men may prefer to marry tall women, but this isn't always the case.

**3.17** **(a)** Gender is a categorical variable and correlation $r$ is for two quantitative variables. **(b)** The largest possible value of the correlation is $r = 1$. **(c)** The correlation $r$ has no units.

**3.18** The article suggests a negative association, but a correlation of $r \approx 0$ means that there is no linear association between research productivity and teaching rating. In other words, knowledge of a professor's research productivity will not help you predict her teaching rating.

**3.19 (a)** The scatterplot below shows a strong, positive linear relationship between the two measurements. It appears that all five specimens come from the same species. **(b)** The femur measurements have $\bar{x} = 58.2$ and $s_x = 13.2$. The humerus measurements have $\bar{y} = 66$ and $s_y = 15.89$. The sum of the z-score products is 3.97620, so the correlation coefficient is $r = (1/4)(3.97620) = 0.9941$. The very high value of the correlation confirms the strong, positive linear association between femur length and humerus length in the scatterplot from part (a).

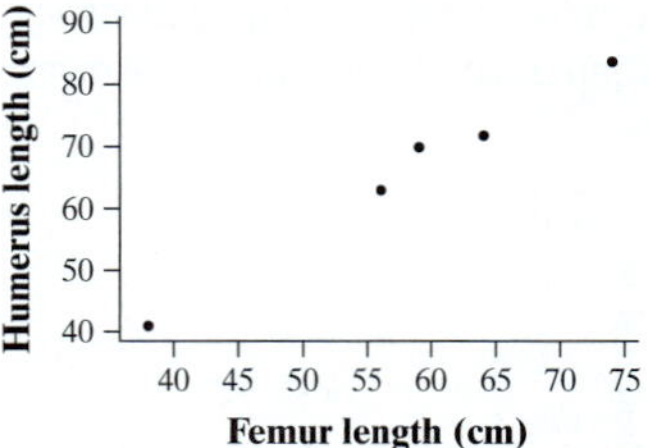


**3.20 (a)** The scatterplot shows a moderate positive association, so $r$ should be positive, but not close to 1. **(b)** For the women, $\bar{x} = 66$ and $s_x = 2.098$. For the men, $\bar{y} = 69$ and $s_y = 2.53$. The sum of the z-score products is 2.82596, so the correlation coefficient is $r = (1/5)(2.82596) = 0.5652$. Because the correlation is positive, it provides some evidence that taller women tend to date taller men.

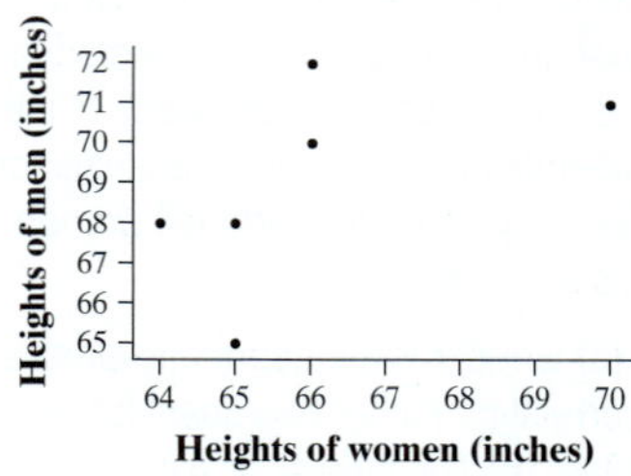




19. **Dem bones** Archaeopteryx is an extinct beast having feathers like a bird but teeth and a long bony tail like a reptile. Only six fossil specimens are known. Because these specimens differ greatly in size, some scientists think they are different species rather than individuals from the same species. We will examine some data. If the specimens belong to the same species and differ in size because some are younger than others, there should be a positive linear relationship between the lengths of a pair of bones from all individuals. An outlier from this relationship would suggest a different species. Here are data on the lengths in centimeters of the femur (a leg bone) and the humerus (a bone in the upper arm) for the five specimens that preserve both bones:[10]

| Femur ($x$): | 38 | 56 | 59 | 64 | 74 |
|---|---|---|---|---|---|
| Humerus ($y$): | 41 | 63 | 70 | 72 | 84 |

(a) Make a scatterplot. Do you think that all five specimens come from the same species? Explain.

(b) Find the correlation $r$ step by step, using the formula on page 154. Explain how your value for $r$ matches your graph in part (a).

20. **Data on dating** A student wonders if tall women tend to date taller men than do short women. She measures herself, her dormitory roommate, and the women in the adjoining rooms. Then she measures the next man each woman dates. Here are the data (heights in inches):

| Women ($x$): | 66 | 64 | 66 | 65 | 70 | 65 |
|---|---|---|---|---|---|---|
| Men ($y$): | 72 | 68 | 70 | 68 | 71 | 65 |

(a) Make a scatterplot of these data. Based on the scatterplot, do you expect the correlation to be positive or negative? Near ±1 or not?

(b) Find the correlation $r$ step by step, using the formula on page 154. Do the data show that taller women tend to date taller men?

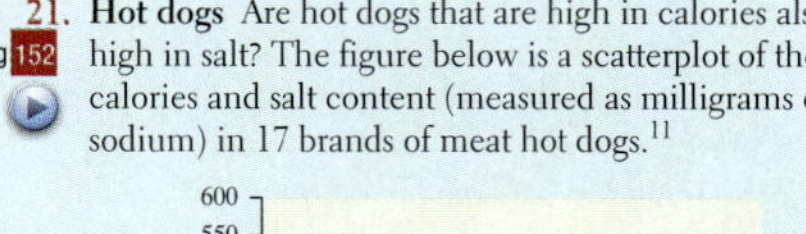

21. **Hot dogs** Are hot dogs that are high in calories also high in salt? The figure below is a scatterplot of the calories and salt content (measured as milligrams of sodium) in 17 brands of meat hot dogs.[11]

pg 152

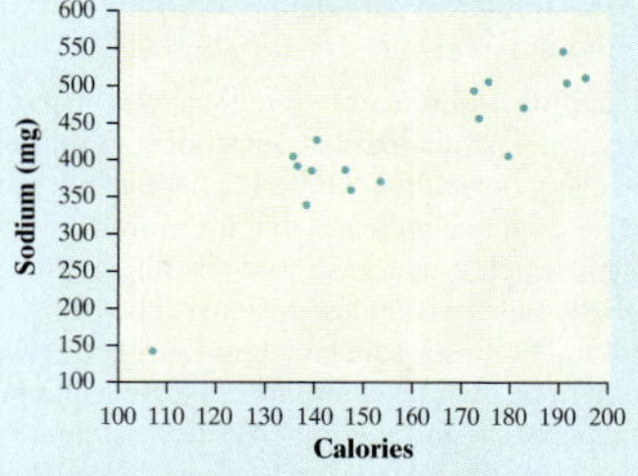


(a) The correlation for these data is $r = 0.87$. Explain what this value means.

(b) What effect does the hot dog brand with the lowest calorie content have on the correlation? Justify your answer.

22. **All brawn?** The figure below plots the average brain weight in grams versus average body weight in kilograms for 96 species of mammals.[12] There are many small mammals whose points overlap at the lower left.

(a) The correlation between body weight and brain weight is $r = 0.86$. Explain what this value means.

(b) What effect does the elephant have on the correlation? Justify your answer.

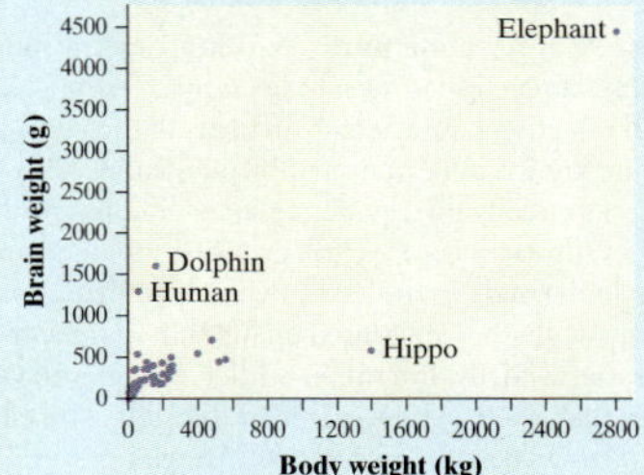


23. **Dem bones** Refer to Exercise 19.

(a) How would $r$ change if the bones had been measured in millimeters instead of centimeters? (There are 10 millimeters in a centimeter.)

(b) If the $x$ and $y$ variables are reversed, how would the correlation change? Explain.

24. **Data on dating** Refer to Exercise 20.

(a) How would $r$ change if all the men were 6 inches shorter than the heights given in the table? Does the correlation tell us if women tend to date men taller than themselves?

(b) If heights were measured in centimeters rather than inches, how would the correlation change? (There are 2.54 centimeters in an inch.)

25. **Strong association but no correlation** The gas mileage of an automobile first increases and then decreases as the speed increases. Suppose that this relationship is very regular, as shown by the following data on speed (miles per hour) and mileage (miles per gallon).

| Speed: | 20 | 30 | 40 | 50 | 60 |
|---|---|---|---|---|---|
| Mileage: | 24 | 28 | 30 | 28 | 24 |

(a) Make a scatterplot to show the relationship between speed and mileage.

(b) Calculate the correlation for these data by hand or using technology.

(c) Explain why the correlation has the value found in part (b) even though there is a strong relationship between speed and mileage.

**3.21 (a)** There is a strong, positive linear association between sodium and calories. **(b)** It increases the correlation. It falls in the linear pattern of the rest of the data and observations with unusually small or unusually large values of $x$ have a big influence on the correlation.

**3.22 (a)** There is a strong, positive linear association between body weight and brain weight of mammals. **(b)** It increases the correlation. It falls in the linear pattern of the rest of the data and observations with unusually small or unusually large values of $x$ have a big influence on the correlation.

**3.23 (a)** The correlation would not change, because correlation is not affected by a change of units for either variable. **(b)** The correlation would not change, because it does not distinguish between explanatory and response variables.

**3.24 (a)** The correlation would not change. The $y$-values would all shift down 6 inches, but the strength of the relationship won't change. The correlation tells us that there is a weak to moderate association between $x$ and $y$, not whether the $y$-values are larger than the $x$-values. **(b)** The correlation would not change, because correlation is not affected by a change of units for either variable.

**3.25 (a)** A scatterplot is shown below. **(b)** $r = 0$. **(c)** The correlation measures the strength of a *linear* association, but this plot shows a nonlinear relationship between speed and mileage.

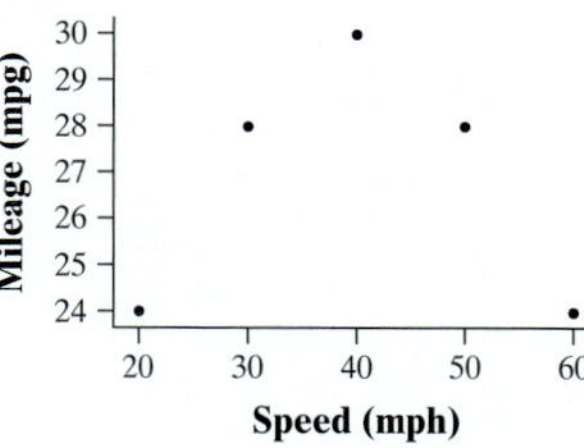

26. **What affects correlation?** Here are some hypothetical data:

| *x*: | 1 | 2 | 3 | 4 | 10 | 10 |
|---|---|---|---|---|---|---|
| *y*: | 1 | 3 | 3 | 5 | 1 | 11 |

(a) Make a scatterplot to show the relationship between $x$ and $y$.

(b) Calculate the correlation for these data by hand or using technology.

(c) What is responsible for reducing the correlation to the value in part (b) despite a strong straight-line relationship between $x$ and $y$ in most of the observations?

*Multiple choice: Select the best answer for Exercises 27 to 32.*

27. You have data for many years on the average price of a barrel of oil and the average retail price of a gallon of unleaded regular gasoline. If you want to see how well the price of oil predicts the price of gas, then you should make a scatterplot with ______ as the explanatory variable.

(a) the price of oil (c) the year (e) time
(b) the price of gas (d) either oil price or gas price

28. In a scatterplot of the average price of a barrel of oil and the average retail price of a gallon of gas, you expect to see

(a) very little association.
(b) a weak negative association.
(c) a strong negative association.
(d) a weak positive association.
(e) a strong positive association.

29. The following graph plots the gas mileage (miles per gallon) of various cars from the same model year versus the weight of these cars in thousands of pounds. The points marked with red dots correspond to cars made in Japan. From this plot, we may conclude that

(a) there is a positive association between weight and gas mileage for Japanese cars.
(b) the correlation between weight and gas mileage for all the cars is close to 1.
(c) there is little difference between Japanese cars and cars made in other countries.
(d) Japanese cars tend to be lighter in weight than other cars.
(e) Japanese cars tend to get worse gas mileage than other cars.

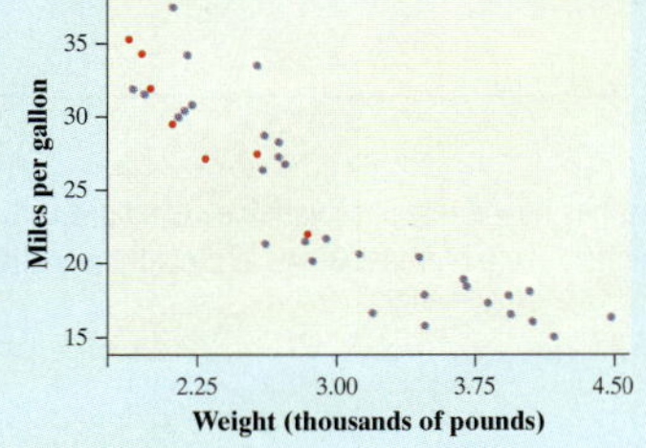


30. If women always married men who were 2 years older than themselves, what would the correlation between the ages of husband and wife be?

(a) 2
(b) 1
(c) 0.5
(d) 0
(e) Can't tell without seeing the data

31. The figure below is a scatterplot of reading test scores against IQ test scores for 14 fifth-grade children. There is one low outlier in the plot. What effect does this low outlier have on the correlation?

(a) It makes the correlation closer to 1.
(b) It makes the correlation closer to 0 but still positive.
(c) It makes the correlation equal to 0.
(d) It makes the correlation negative.
(e) It has no effect on the correlation.

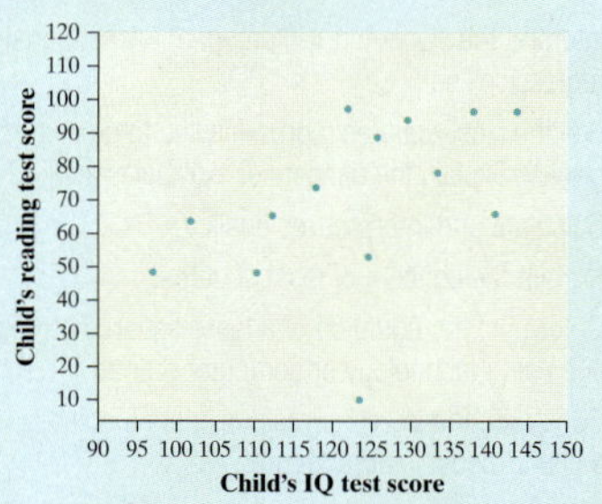


32. If we leave out the low outlier, the correlation for the remaining 13 points in the preceding figure is closest to

(a) −0.95. (c) 0. (e) 0.95.
(b) −0.5. (d) 0.5.

33. **Big diamonds** (1.2, 1.3) Here are the weights (in milligrams) of 58 diamonds from a nodule carried up to the earth's surface in surrounding rock. These data represent a population of diamonds formed in a single event deep in the earth.[13]

| 13.8 | 3.7 | 33.8 | 11.8 | 27.0 | 18.9 | 19.3 | 20.8 | 25.4 | 23.1 | 7.8 |
|---|---|---|---|---|---|---|---|---|---|---|
| 10.9 | 9.0 | 9.0 | 14.4 | 6.5 | 7.3 | 5.6 | 18.5 | 1.1 | 11.2 | 7.0 |
| 7.6 | 9.0 | 9.5 | 7.7 | 7.6 | 3.2 | 6.5 | 5.4 | 7.2 | 7.8 | 3.5 |
| 5.4 | 5.1 | 5.3 | 3.8 | 2.1 | 2.1 | 4.7 | 3.7 | 3.8 | 4.9 | 2.4 |
| 1.4 | 0.1 | 4.7 | 1.5 | 2.0 | 0.1 | 0.1 | 1.6 | 3.5 | 3.7 | 2.6 |
| 4.0 | 2.3 | 4.5 | | | | | | | | |

Make a graph that shows the distribution of weights of these diamonds. Describe what you see. Give appropriate numerical measures of center and spread.

**3.26 (a)** The scatterplot is shown below. **(b)** $r = 0.481$ **(c)** The outlier at (10, 1) is responsible for reducing the correlation because it falls far out of the linear pattern of the rest of the data.

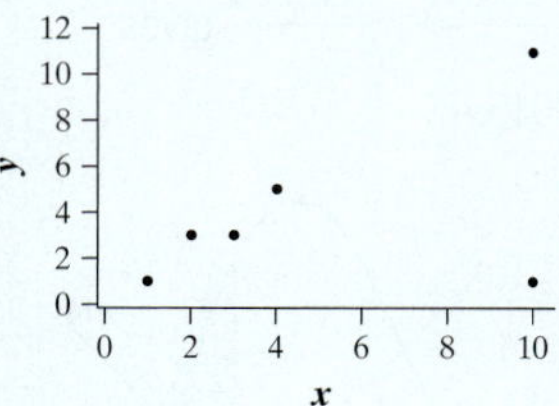


**3.27** a

**3.28** e

**3.29** d

**3.30** b

**3.31** b

**3.32** d

**3.33** A histogram is shown below. The distribution is right-skewed, with several outliers, so we should use the median (5.4 mg) and *IQR* (5.5 mg) to describe the center and spread.

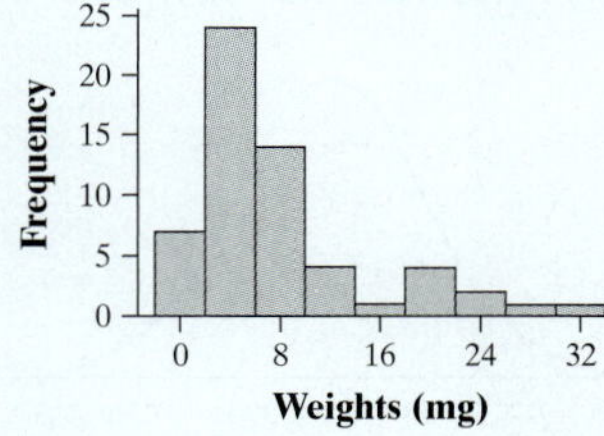

**34. College debt** (2.2) A report published by the Federal Reserve Bank of New York in 2012 reported the results of a nationwide study of college student debt. Researchers found that the average student loan balance per borrower is \$23,300. They also reported that about one-quarter of borrowers owe more than \$28,000.[14]

(a) Assuming that the distribution of student loan balances is approximately Normal, estimate the standard deviation of the distribution of student loan balances.

(b) Assuming that the distribution of student loan balances is approximately Normal, use your answer to part (a) to estimate the proportion of borrowers who owe more than \$54,000.

(c) In fact, the report states that about 10% of borrowers owe more than \$54,000. What does this fact indicate about the shape of the distribution of student loan balances?

(d) The report also states that the median student loan balance is \$12,800. Does this fact support your conclusion in part (c)? Explain.

# 3.2 Least-Squares Regression

**WHAT YOU WILL LEARN** By the end of the section, you should be able to:

- Interpret the slope and *y* intercept of a least-squares regression line.
- Use the least-squares regression line to predict *y* for a given *x*. Explain the dangers of extrapolation.
- Calculate and interpret residuals.
- Explain the concept of least squares.
- Determine the equation of a least-squares regression line using technology or computer output.
- Construct and interpret residual plots to assess whether a linear model is appropriate.
- Interpret the standard deviation of the residuals and $r^2$ and use these values to assess how well the least-squares regression line models the relationship between two variables.
- Describe how the slope, *y* intercept, standard deviation of the residuals, and $r^2$ are influenced by outliers.
- Find the slope and *y* intercept of the least-squares regression line from the means and standard deviations of *x* and *y* and their correlation.

Linear (straight-line) relationships between two quantitative variables are fairly common and easy to understand. In the previous section, we found linear relationships in settings as varied as sparrowhawk colonies, natural-gas consumption, and Florida manatee deaths. Correlation measures the direction and strength of these relationships. When a scatterplot shows a linear relationship, we'd like to summarize the overall pattern by drawing a line on the scatterplot. A **regression line** summarizes the relationship between two variables, but only in a specific setting: when one of the variables helps explain or predict the other. Regression, unlike correlation, requires that we have an explanatory variable and a response variable.

**DEFINITION: Regression line**

A **regression line** is a line that describes how a response variable *y* changes as an explanatory variable *x* changes. We often use a regression line to predict the value of *y* for a given value of *x*.

**3.34 (a)** Because 25% of borrowers owe more than \$28,000 (see graph below), we need to find the value of *z* that has 75% of the area to its left. Using Table A, this value is $z = 0.67$. Solving $0.67 = \frac{28{,}000 - 23{,}300}{\sigma}$ gives $\sigma = \$7015$.

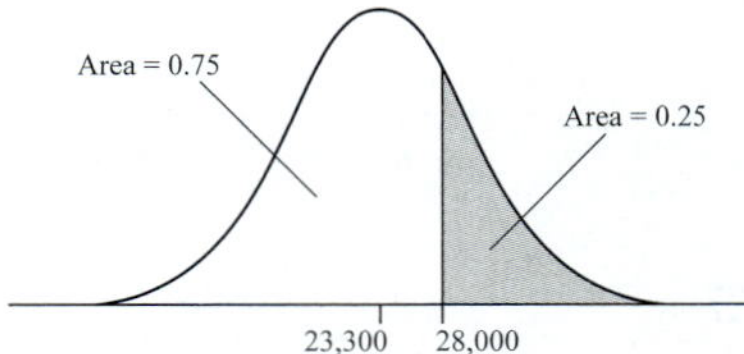


**(b)** For borrowers, loan balances follow a $N(23{,}300, 7015)$ distribution and we want to find the proportion of borrowers who owe more than \$54,000 (see graph below). $z = \frac{54{,}000 - 23{,}300}{7015} = 4.38$. From Table A, the proportion of z-scores above 4.38 is approximately 0. *Using technology:* `normalcdf(lower:54000,upper:100000,μ:23300,σ:7015)` ≈ 0. Approximately 0% of borrowers owe more than \$54,000.

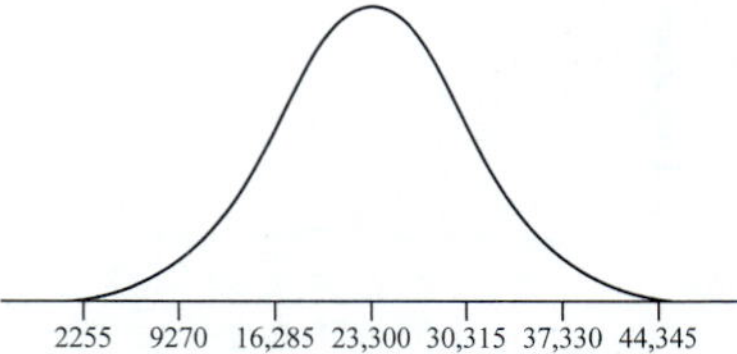


**(c)** If the distribution of loan balances is approximately Normal, then we would expect almost no one to have a balance that large. Because 10% of borrowers owe more than \$54,000, we can conclude that the distribution of loan balances isn't Normal and is right-skewed.

**(d)** Yes. Because the mean (\$23,300) is so much larger than the median (\$12,800).

**PD Section 3.2 Overview**

Watch the video overview of Section 3.2 (for teachers) by clicking on the link in the TE-book, opening the TRCD, or logging onto the book's companion Web site www.whfreeman.com/tps5e (teacher log-in required).

**TRM Section 3.2 Alternate Examples**

You can find the Alternate Examples for this section in Microsoft Word format by clicking on the link in the TE-book, opening the TRCD, or logging onto the book's companion Web site www.whfreeman.com/tps5e (teacher log-in required).

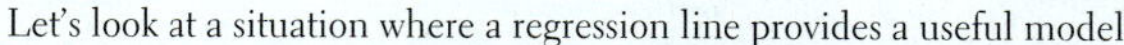

Let's look at a situation where a regression line provides a useful model.

## EXAMPLE How Much Is That Truck Worth?

### *Regression lines as models*

Everyone knows that cars and trucks lose value the more they are driven. Can we predict the price of a used Ford F-150 SuperCrew 4 × 4 if we know how many miles it has on the odometer? A random sample of 16 used Ford F-150 SuperCrew 4 × 4s was selected from among those listed for sale at autotrader.com. The number of miles driven and price (in dollars) were recorded for each of the trucks.[15] Here are the data:

| Miles driven | 70,583 | 129,484 | 29,932 | 29,953 | 24,495 | 75,678 | 8359 | 4447 |
|---|---|---|---|---|---|---|---|---|
| Price (in dollars) | 21,994 | 9500 | 29,875 | 41,995 | 41,995 | 28,986 | 31,891 | 37,991 |
| Miles driven | 34,077 | 58,023 | 44,447 | 68,474 | 144,162 | 140,776 | 29,397 | 131,385 |
| Price (in dollars) | 34,995 | 29,988 | 22,896 | 33,961 | 16,883 | 20,897 | 27,495 | 13,997 |

Figure 3.7 is a scatterplot of these data. The plot shows a moderately strong, negative linear association between miles driven and price with no outliers. The correlation is $r = -0.815$. The line on the plot is a regression line for predicting price from miles driven.

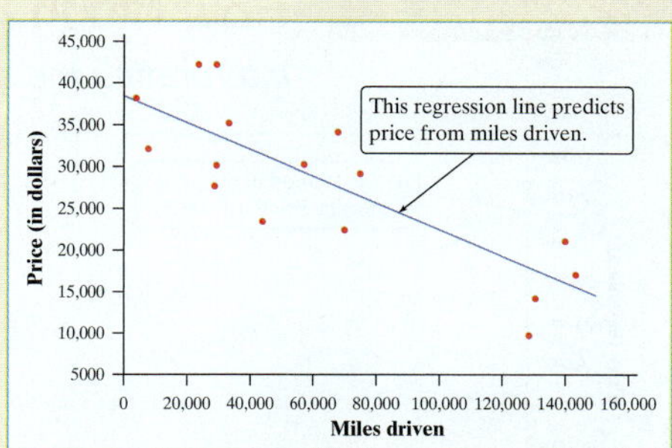


**FIGURE 3.7** Scatterplot showing the price and miles driven of used Ford F-150s, with a regression line added.

## Interpreting a Regression Line

A regression line is a *model* for the data, much like the density curves of Chapter 2. The equation of a regression line gives a compact mathematical description of what this model tells us about the relationship between the response variable $y$ and the explanatory variable $x$.

### ALTERNATE EXAMPLE

#### *Tapping on cans*

Don't you hate it when you open a can of soda and some of the contents spray out of the can? Two AP® Statistics students, Kerry and Danielle, wanted to investigate if tapping on a can of soda would reduce the amount of soda expelled after the can has been shaken. For their experiment, they vigorously shook 40 cans of soda and randomly assigned each can to be tapped for 0 seconds, 4 seconds, 8 seconds, or 12 seconds. Then, after opening the can and cleaning up the mess, the students measured the amount of soda left in each can (in milliliters). Here are the data and a scatterplot. The scatterplot shows a fairly strong, positive linear association between the amount of tapping time and the amount remaining in the can. The line on the plot is a regression line for predicting the amount remaining from the amount of tapping time.

| Amount of soda remaining (ml) | | | |
|---|---|---|---|
| 0 s | 4 s | 8 s | 12 s |
| 245 | 260 | 267 | 275 |
| 255 | 250 | 271 | 280 |
| 250 | 250 | 268 | 275 |
| 250 | 250 | 270 | 280 |
| 250 | 260 | 276 | 285 |
| 245 | 265 | 255 | 290 |
| 248 | 267 | 270 | 284 |
| 250 | 260 | 270 | 278 |
| 251 | 261 | 275 | 279 |
| 249 | 259 | 275 | 280 |

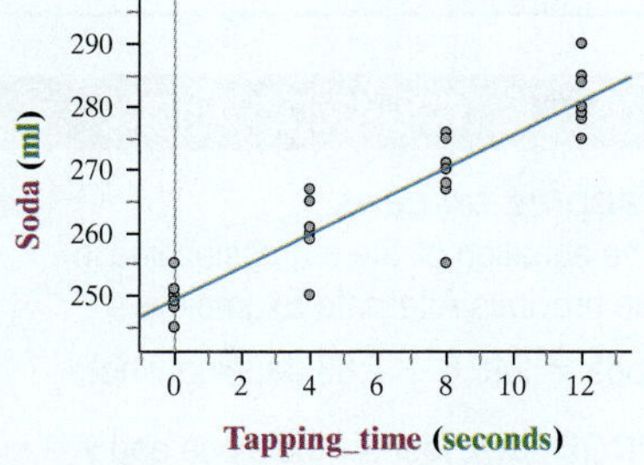

**DEFINITION: Regression line, predicted value, slope, *y* intercept**

Suppose that $y$ is a response variable (plotted on the vertical axis) and $x$ is an explanatory variable (plotted on the horizontal axis). A **regression line** relating $y$ to $x$ has an equation of the form

$$\hat{y} = a + bx$$

In this equation,

- $\hat{y}$ (read "y hat") is the **predicted value** of the response variable $y$ for a given value of the explanatory variable $x$.
- $b$ is the **slope**, the amount by which $y$ is predicted to change when $x$ increases by one unit.
- $a$ is the ***y* intercept**, the predicted value of $y$ when $x = 0$.

Although you are probably accustomed to the form $y = mx + b$ for the equation of a line from algebra, statisticians have adopted a different form for the equation of a regression line. Some use $\hat{y} = b_0 + b_1x$. We prefer $\hat{y} = a + bx$ for two reasons: (1) it's simpler and (2) your calculator uses this form. Don't get so caught up in the symbols that you lose sight of what they mean! The coefficient of $x$ is always the slope, no matter what symbol is used.

Many calculators and software programs will give you the equation of a regression line from keyed-in data. Understanding and using the line are more important than the details of where the equation comes from.

## EXAMPLE How Much Is That Truck Worth?

*Interpreting the slope and y intercept*

The equation of the regression line shown in Figure 3.7 is

$$\widehat{\text{price}} = 38{,}257 - 0.1629 \text{ (miles driven)}$$

The $y$ intercept of the regression line is $a = 38{,}257$.

Price (in dollars)

Miles driven

**PROBLEM:** Identify the slope and $y$ intercept of the regression line. Interpret each value in context.

**SOLUTION:** The slope $b = -0.1629$ tells us that the price of a used Ford F-150 is *predicted* to go down by 0.1629 dollars (16.29 cents) for each additional mile that the truck has been driven. The $y$ intercept $a = 38{,}257$ is the predicted price of a Ford F-150 that has been driven 0 miles.

**For Practice** *Try Exercise* 39(a) and (b)

The slope of a regression line is an important numerical description of the relationship between the two variables. Although we need the value of the $y$ intercept to draw the line, it is statistically meaningful only when the explanatory variable can actually take values close to zero, as in this setting.

### AP® Exam Common Error

When writing the equation of a regression line, students often forget to include the hat on the $y$. Make sure that students understand the difference between the actual value of $y$ (denoted $y$) and the predicted value of $y$ (denoted $\hat{y}$). This also applies to defining the variables used in a regression equation. In the Ford truck example, $y$ = actual price and $\hat{y}$ = predicted price.

### Teaching Tip

When writing the equation of a regression line, statisticians prefer to put the $y$ intercept (constant term) first because it makes things nicer when using multiple explanatory variables. For example, with two explanatory variables $x_1$ and $x_2$, it is better to write the regression equation as $\hat{y} = a + b_1x_1 + b_2x_2$ rather than $\hat{y} = b_1x_1 + a + b_2x_2$. *Note*: Multiple regression is not part of the AP® Statistics course.

### AP® Exam Formula Sheet

On the formula sheet provided on both sections of the AP® exam, the least-squares regression line is written $\hat{y} = b_0 + b_1x$. Likewise, formulas for the slope ($b_1$) and $y$ intercept ($b_0$) of the least-squares regression line also use this notation. The coefficient of $x$ is always the slope, no matter what notation is used.

### ALTERNATE EXAMPLE

***Tapping on cans***

The equation of the regression line in the previous Alternate Example is

$$\widehat{\text{soda}} = 248.6 + 2.63 \text{ (tapping time)}.$$

**PROBLEM:** Identify the slope and $y$ intercept of the regression line. Interpret each value in context.

**SOLUTION:** The slope $b = 2.63$ tells us that the amount of soda remaining in the can is predicted to increase by 2.63 ml for each additional second that the can is tapped. The $y$ intercept $a = 248.6$ is the predicted amount of soda remaining in the can if it isn't tapped at all (0 seconds).

### AP® Exam Common Error

Many students lose credit on the AP® exam for not stating that the slope is the *predicted* (or estimated or expected or average) change in the $y$ variable for each increase of 1 unit in the $x$ variable. Students will have to understand the difference between the actual data and the equation we are using to model the data. For example, a response that says, "The price will go down by 0.1629 dollars for each additional mile driven" mistakenly implies a deterministic relationship where all the data values will be on the regression line.

### Teaching Tip

Alert students that the $y$ axis is not always on the left edge of a scatterplot. In the scatterplot for the track and field day example on page 146, the $y$ intercept ($a = 415$) isn't visible on the scatterplot because the $x$-values only go down to $x = 5$.

**Does a small slope mean that there's no relationship?** For the miles driven and price regression line, the slope $b = -0.1629$ is a small number. This does *not* mean that change in miles driven has little effect on price. The size of the slope depends on the units in which we measure the two variables. In this setting, the slope is the predicted change in price (in dollars) when the distance driven increases by 1 mile. There are 100 cents in a dollar. If we measured price in cents instead of dollars, the slope would be 100 times larger, $b = 16.29$. *You can't say how strong a relationship is by looking at the size of the slope of the regression line.*

## Prediction

We can use a regression line to predict the response $\hat{y}$ for a specific value of the explanatory variable $x$. Here's how we do it.

### EXAMPLE How Much Is That Truck Worth?

*Predicting with a regression line*

For the Ford F-150 data, the equation of the regression line is

$$\widehat{\text{price}} = 38{,}257 - 0.1629\,(\text{miles driven})$$

If a used Ford F-150 has 100,000 miles driven, substitute $x = 100{,}000$ in the equation. The predicted price is

$$\widehat{\text{price}} = 38{,}257 - 0.1629(100{,}000) = 21{,}967 \text{ dollars}$$

This prediction is illustrated in Figure 3.8.

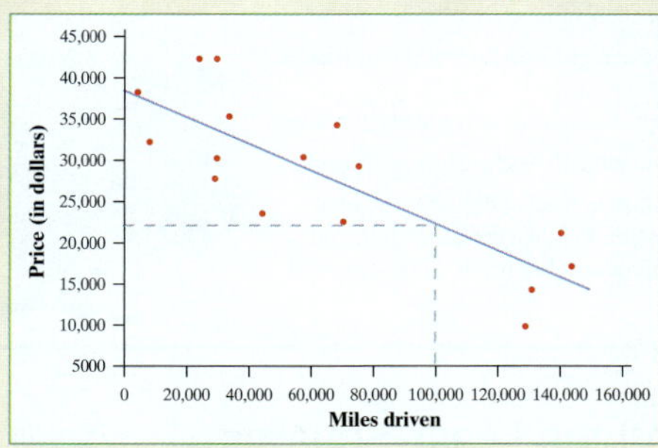


**FIGURE 3.8** Using the regression line to predict price for a Ford F-150 with 100,000 miles driven.

The accuracy of predictions from a regression line depends on how much the data scatter about the line. In this case, prices for trucks with similar mileage show a spread of about \$10,000. The regression line summarizes the pattern but gives only roughly accurate predictions.

Can we predict the price of a Ford F-150 with 300,000 miles driven? We can certainly substitute 300,000 into the equation of the line. The prediction is

$$\widehat{\text{price}} = 38{,}257 - 0.1629(300{,}000) = -10{,}613 \text{ dollars}$$

That is, we predict that we would need to pay someone else \$10,613 just to take the truck off our hands!

**ALTERNATE EXAMPLE**

***Tapping on cans***

For the soda example, the equation of the regression line is soda $= 248.6 + 2.63$ (tapping time). If we shook a can in the same way as the students did in their project and tapped on it for 10 seconds, the predicted amount of soda remaining would be $\widehat{\text{soda}} = 248.6 + 2.63(10) = 274.9$ ml. This is illustrated in the figure below.

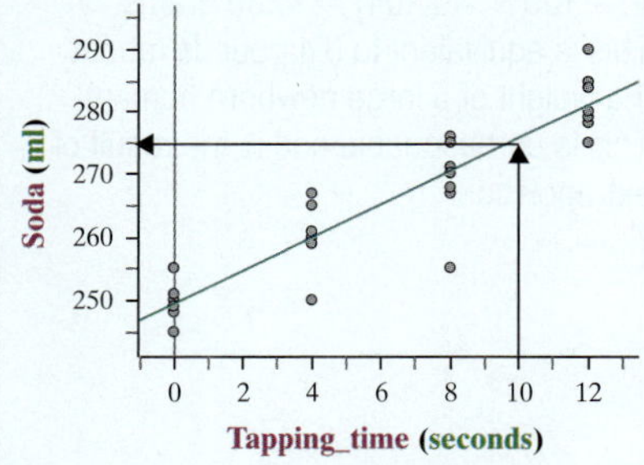


**ALTERNATE EXAMPLE**

***Extrapolation and tapping on cans***

Should we predict how much soda will be left after 60 seconds of tapping? No! We have data only for cans that were tapped between 0 and 12 seconds. We don't know if the linear pattern will continue beyond these values. In fact, if we did make a prediction for 60 seconds of tapping, we would get 406.4 ml, over 50 ml more than the can originally contained (355 ml)!

A negative price doesn't make much sense in this context. Look again at Figure 3.8. A truck with 300,000 miles driven is far outside the set of $x$ values for our data. We can't say whether the relationship between miles driven and price remains linear at such extreme values. Predicting price for a truck with 300,000 miles driven is an **extrapolation** of the relationship beyond what the data show.

Often, using the regression line to make a prediction for $x = 0$ is an extrapolation. That's why the $y$ intercept isn't always statistically meaningful.

**DEFINITION: Extrapolation**

**Extrapolation** is the use of a regression line for prediction far outside the interval of values of the explanatory variable $x$ used to obtain the line. Such predictions are often not accurate.

Few relationships are linear for all values of the explanatory variable. *Don't make predictions using values of x that are much larger or much smaller than those that actually appear in your data.*

## CHECK YOUR UNDERSTANDING

Some data were collected on the weight of a male white laboratory rat for the first 25 weeks after its birth. A scatterplot of the weight (in grams) and time since birth (in weeks) shows a fairly strong, positive linear relationship. The linear regression equation $\widehat{\text{weight}} = 100 + 40(\text{time})$ models the data fairly well.

1. What is the slope of the regression line? Explain what it means in context.
2. What's the $y$ intercept? Explain what it means in context.
3. Predict the rat's weight after 16 weeks. Show your work.
4. Should you use this line to predict the rat's weight at age 2 years? Use the equation to make the prediction and think about the reasonableness of the result. (There are 454 grams in a pound.)

**ANSWERS TO CYU**

**1.** 40. For each additional week, we predict that a rat will gain 40 grams of weight.

**2.** 100. The predicted weight for a newborn rat is 100 grams.

**3.** $\hat{y} = 100 + 40(16) = 740$ grams.

**4.** 2 years = 104 weeks, so $\hat{y} = 100 + 40(104) = 4260$ grams. This is equivalent to 9.4 pounds (about the weight of a large newborn human). This is unreasonable and is the result of extrapolation.

## Residuals and the Least-Squares Regression Line

In most cases, no line will pass exactly through all the points in a scatterplot. Because we use the line to predict $y$ from $x$, the prediction errors we make are errors in $y$, the vertical direction in the scatterplot. *A good regression line makes the vertical deviations of the points from the line as small as possible.*

Figure 3.9 shows a scatterplot of the Ford F-150 data with a regression line added. The prediction errors are

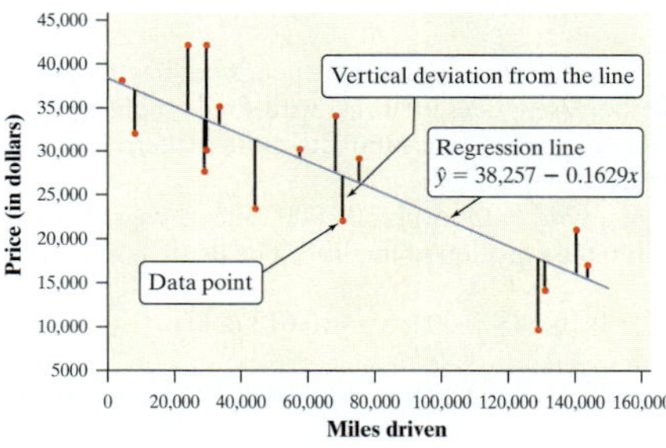


**FIGURE 3.9** Scatterplot of the Ford F-150 data with a regression line added. A good regression line should make the prediction errors (shown as bold vertical segments) as small as possible.

marked as bold segments in the graph. These vertical deviations represent "leftover" variation in the response variable after fitting the regression line. For that reason, they are called **residuals.**

**DEFINITION: Residual**

A **residual** is the difference between an observed value of the response variable and the value predicted by the regression line. That is,

$$\text{residual} = \text{observed } y - \text{predicted } y$$
$$= y - \hat{y}$$

The following example shows you how to calculate and interpret a residual.

## EXAMPLE How Much Is That Truck Worth?

### *Finding a residual*

PROBLEM: Find and interpret the residual for the Ford F-150 that had 70,583 miles driven and a price of $21,994.

SOLUTION: The regression line predicts a price of

$$\widehat{\text{price}} = 38,257 - 0.1629(70,583) = 26,759 \text{ dollars}$$

for this truck, but its actual price was $21,994. This truck's residual is

$$\text{residual} = \text{observed } y - \text{predicted } y$$
$$= y - \hat{y} = 21,994 - 26,759 = -4765 \text{ dollars}$$

That is, the actual price of this truck is $4765 lower than expected, based on its mileage. The actual price might be lower than predicted as a result of other factors. For example, the truck may have been in an accident or may need a new paint job.

**For Practice** *Try Exercise*  45

The line shown in Figure 3.9 makes the residuals for the 16 trucks "as small as possible." But what does that mean? Maybe this line minimizes the *sum* of the residuals. Actually, if we add up the prediction errors for all 16 trucks, the positive and negative residuals cancel out. That's the same issue we faced when we tried to measure deviation around the mean in Chapter 1. We'll solve the current problem in much the same way: by squaring the residuals. The regression line we want is the one that minimizes the sum of the squared residuals. That's what the line shown in Figure 3.9 does for the Ford F-150 data, which is why we call it the **least-squares regression line**.

**DEFINITION: Least-squares regression line**

The **least-squares regression line** of $y$ on $x$ is the line that makes the sum of the squared residuals as small as possible.

## Teaching Tip

Emphasize the connection between residuals and deviations from the mean. When calculating the standard deviation in Chapter 1, we used the quantity $x_i - \bar{x}$, which was the difference between the actual value of $x$ and the mean (expected) value of $x$. Likewise, in Chapter 2, when we calculated standardized scores, we found the difference between the actual value and the mean (expected) value. Looking ahead, we will see this same formulation again when we calculate the chi-square test statistic. An easy way to remember the order of subtraction is to think **AP** = **A**ctual − **P**redicted.

## ALTERNATE EXAMPLE

### *Tapping on cans*

PROBLEM: Find and interpret the residual for the can that was tapped for 4 seconds and had 260 ml of soda remaining.

SOLUTION: The regression line predicts a volume of $\widehat{\text{soda}} = 248.6 + 2.63(4) = 259.12$ ml. The actual amount of soda was 260 ml, so the residual $= y - \hat{y} = 260 - 259.12 = 0.88$ ml. This can had 0.88 ml more soda remaining than expected, based on how long it was tapped.

## Teaching Tip

When you have students calculate residuals, always ask them to interpret the value of the residual in context. Make sure they address both the distance from what we predicted and the direction of the error. For example, a residual of −15 means the actual value was 15 *below* the predicted value. You may also want to give students an example where the residual is known and they have to work backward to find the actual value of $y$. For example, what was the actual amount of soda remaining if a can was tapped for 8 seconds and the residual was −2.64? Because the predicted value for $x = 8$ seconds is $\widehat{\text{soda}} = 248.6 + 2.63(8) = 269.64$ ml, we can solve $y - 269.64 = -2.64$ and get an actual value of $y = 267$ ml.

## TECHNOLOGY TIP

If you have Fathom, you can re-create Figure 3.10b. Once you have a scatterplot with the least-squares regression line showing, right-click on the graph and choose "Show Squares." You might want to start, however, by creating the scatterplot and choosing "Add Movable Line" and then "Show Squares." Moving the line around will change the size of the squares and the sum of the squared residuals (shown below the equation of the line). As you make the line better, the squares will get smaller and it should be clear to students that we want to minimize the sum of the squared residuals. Once you have the best line you can get by hand, choose "Least-Squares Line" to see how close you are to the "least-squares" line.

The Core Math Tools from NCTM include a nice Data Analysis applet that will create scatterplots, draw residuals, and show squared residuals like those in Figure 3.10b. See www.nctm.org/standards/content.aspx?id=32702.

**Activity Overview**

You can watch an overview of this Activity (for teachers) by clicking on the link in the TE-book, or looking on the TRCD or the book's companion Web site www.whfreeman.com/tps5e (teacher log-in required).

## Teaching Tip

Make sure to do this Activity with your students (or modify the Alternate Activity on page 152 to investigate how outliers and influential points affect the least-squares regression line). On the AP® exam, students will be expected to know how outliers affect the equation of the least-squares regression line.

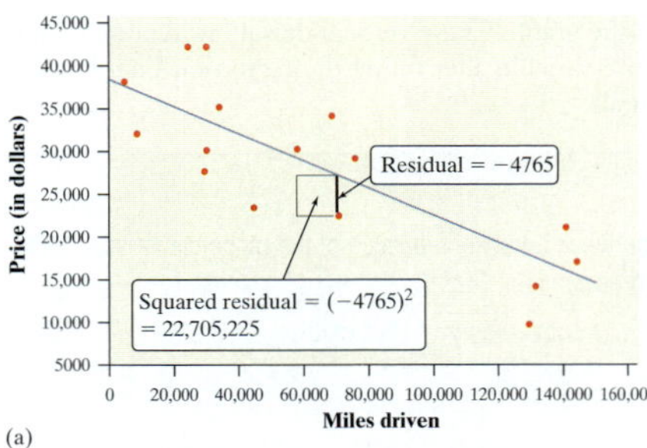


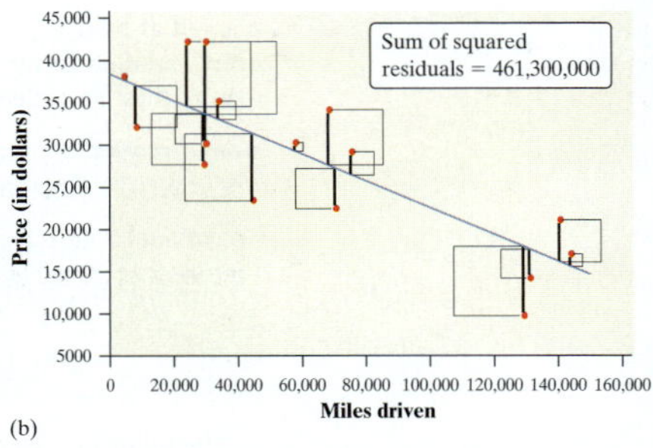


**FIGURE 3.10** The least-squares idea: make the errors in predicting *y* as small as possible by minimizing the sum of the squares of the residuals.

Figure 3.10 gives a geometric interpretation of the least-squares idea for the truck data. Figure 3.10(a) shows the "squared" residual for the truck with 70,583 miles driven and a price of \$21,994. The area of this square is $(-4765)(-4765) = 22{,}705{,}225$. Figure 3.10(b) shows the squared residuals for all the trucks. The sum of squared residuals is 461,300,000. No other regression line would give a smaller sum of squared residuals.

## ACTIVITY | Investigating properties of the least-squares regression line

**MATERIALS:**
Computer with Internet connection

In this Activity, you will use the *Correlation and Regression* applet at the book's Web site, www.whfreeman.com/tps5e, to explore some properties of the least-squares regression line.

1. Click on the scatterplot to create a group of 15 to 20 points from lower left to upper right with a clear positive straight-line pattern (correlation around 0.7).

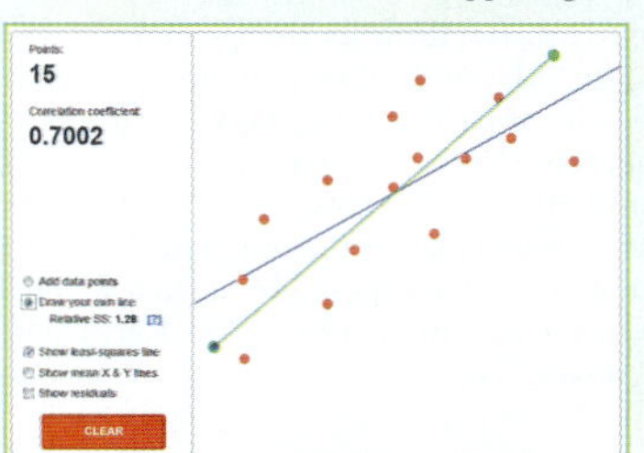


2. Click the "Draw your own line" button to select starting and ending points for your own line on the plot. Use the mouse to adjust the starting and ending points until you have a line that models the association well.
3. Click the "Show least-squares line" button. How do the two lines compare? One way to measure this is to compare the "Relative SS," the ratio of the sum of squared residuals from your line and the least-squares regression line. If the two lines are exactly the same, the relative sum of squares will be 1. Otherwise, the relative sum of squares will be larger than 1.
4. Press the "CLEAR" button and create another scatterplot as in Step 1. Then click on "Show least-squares line" and "Show mean X & Y lines." What do you notice? Move or add points, one at a time, in your scatterplot to see if this result continues to hold true.

## Teaching Tip

In Step 4, students should realize that the least-squares regression line always goes through the point $(\bar{x}, \bar{y})$. This fact is helpful when trying to anticipate how an outlier will affect the equation of the least-squares regression line. In Step 5, students should realize that an outlier that has an *x*-value close to $\bar{x}$ will not change the slope very much, but could move the intercept higher or lower depending on if the outlier is above or below the line. However, if an outlier has an *x*-value far from $\bar{x}$, it can greatly influence the slope. Here are two "analogies" that might help students visualize this fact:

- Imagine that the line is a metal rod and each point is attached to the line with a vertical spring. Points far from the line will want to "pull" the line toward them, so an outlier that is above the line and to the right of $\bar{x}$ will increase the slope.
- Imagine that you are floating on the ceiling and looking down on the door to your classroom. Then imagine that the top edge of the door is the least-squares regression line and the point $(\bar{x}, \bar{y})$ is the hinge where it is attached. It is easy to open the door if you push on it far from the hinge. However, if you push on the door close to the hinge, it is very difficult to move. Likewise, an outlier far from $(\bar{x}, \bar{y})$ can easily move the line, while a point closer to $(\bar{x}, \bar{y})$ won't be able to move the line as much.

5. Now click the "Show residuals" button. How does an outlier affect the slope and $y$ intercept of the least-squares regression line? Move or add points, one at a time, to investigate. Does it depend on whether the outlier has an $x$-value close to the center of the plot or toward the far edges of the plot?

Your calculator or statistical software will give the equation of the least-squares line from data that you enter. Then you can concentrate on understanding and using the regression line.

## 8. TECHNOLOGY CORNER — LEAST-SQUARES REGRESSION LINES ON THE CALCULATOR

TI-Nspire instructions in Appendix B; HP Prime instructions on the book's Web site.

Let's use the Ford F-150 data to show how to find the equation of the least-squares regression line on the TI-83/84 and TI-89. Here are the data again:

| Miles driven | 70,583 | 129,484 | 29,932 | 29,953 | 24,495 | 75,678 | 8359 | 4447 |
|---|---|---|---|---|---|---|---|---|
| **Price (in dollars)** | 21,994 | 9500 | 29,875 | 41,995 | 41,995 | 28,986 | 31,891 | 37,991 |
| **Miles driven** | 34,077 | 58,023 | 44,447 | 68,474 | 144,162 | 140,776 | 29,397 | 131,385 |
| **Price (in dollars)** | 34,995 | 29,988 | 22,896 | 33,961 | 16,883 | 20,897 | 27,495 | 13,997 |

1. Enter the miles driven data into L1/list1 and the price data into L2/list2. Then make a scatterplot. Refer to the Technology Corner on page 150.
2. To determine the least-squares regression line:

**TI-83/84**

- Press [STAT]; choose CALC and then `LinReg(a+bx)`. **OS 2.55 or later:** In the dialog box, enter the following: `Xlist:L1`, `Ylist:L2`, `FreqList (leave blank)`, `Store RegEQ:Y1`, and choose `Calculate`. **Older OS:** Finish the command to read `LinReg(a+bx)L1,L2,Y1` and press [ENTER]. (Y1 is found under VARS/Y-VARS/Function.)

**TI-89**

- In the Statistics/List Editor, press [F4] (CALC); choose `Regressions` and then `LinReg(a+bx)`.
- Enter list1 for the Xlist, list2 for the Ylist; choose to store the RegEqn to y1(x); and press [ENTER].

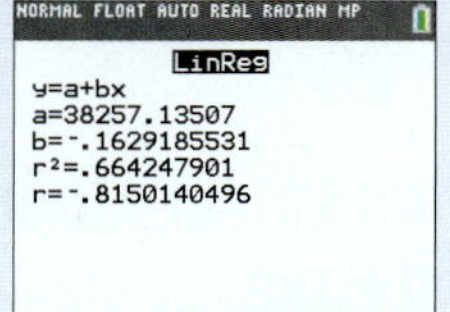


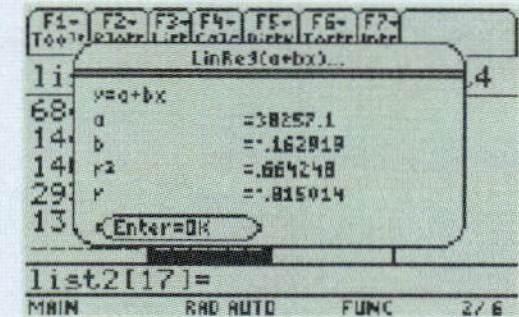


*Note:* If you do not want to store the equation to Y1, then leave the StoreRegEq prompt blank (OS 2.55 or later) or use the following command (older OS): `LinReg(a+bx) L1,L2`.

### TECHNOLOGY TIP

On older operating systems for the TI-83/84, the default lists for `LinReg(a+bx)` are `L1` and `L2`, so if you do not enter any lists, the calculator will automatically use `L1` and `L2`. Discourage your students from using this shortcut, because there will be times when they may not use `L1` and `L2` and forget to enter the correct lists.

3. Graph the regression line. Turn off all other equations in the Y= screen and use ZoomStat/ZoomData to add the least-squares line to the scatterplot.

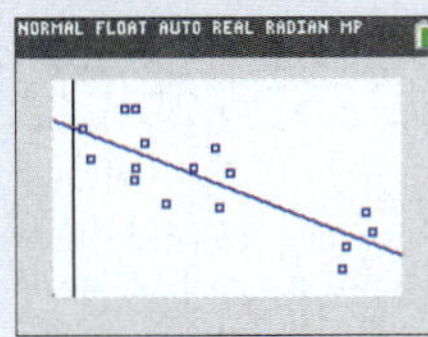

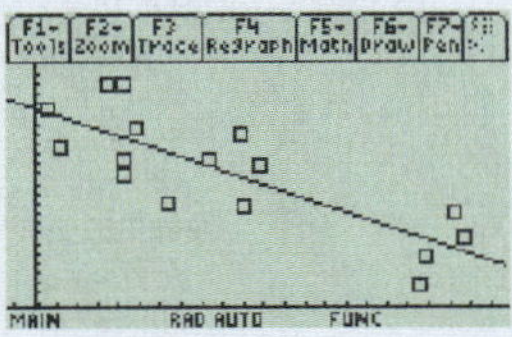

4. Save these lists for later use. On the home screen, use the STO▸ key to help execute the command `L1→MILES:L2→PRICE` (`list1→MILES:list2→PRICE` on the TI-89).

*Note:* If $r^2$ and $r$ do not appear on the TI-83/84 screen, do this one-time series of keystrokes: **OS 2.55 or later:** Press MODE and set STAT DIAGNOSTICS to ON. **Older OS:** Press 2nd 0 (CATALOG), scroll down to `DiagnosticOn`, and press ENTER. Press ENTER again to execute the command. The screen should say "Done." Then redo Step 2 to calculate the least-squares line. The $r^2$ and $r$ values should now appear.

**AP® EXAM TIP** When displaying the equation of a least-squares regression line, the calculator will report the slope and intercept with much more precision than we need. However, there is no firm rule for how many decimal places to show for answers on the AP® exam. Our advice: Decide how much to round based on the context of the problem you are working on.

## CHECK YOUR UNDERSTANDING

It's time to practice your calculator regression skills. Using the familiar SEC football data in the table below, repeat the steps in the previous Technology Corner. You should get $\hat{y} = -3.7506 + 0.4372x$ as the equation of the regression line.

| Team | Alabama | Arkansas | Auburn | Florida | Georgia | Kentucky |
|---|---|---|---|---|---|---|
| Points per game | 34.8 | 36.8 | 25.7 | 25.5 | 32.0 | 15.8 |
| Wins | 12 | 11 | 8 | 7 | 10 | 5 |

| Team | Louisiana State | Mississippi | Mississippi State | South Carolina | Tennessee | Vanderbilt |
|---|---|---|---|---|---|---|
| Points per game | 35.7 | 16.1 | 25.3 | 30.1 | 20.3 | 26.7 |
| Wins | 13 | 2 | 7 | 11 | 5 | 6 |

## Determining Whether a Linear Model Is Appropriate: Residual Plots

One of the first principles of data analysis is to look for an overall pattern and for striking departures from the pattern. A regression line describes the overall pattern of a linear relationship between an explanatory variable and a response variable. We see departures from this pattern by looking at the residuals.

### Teaching Tip

Make sure to point out the title of this subsection. The purpose of a residual plot is to assess if the form of the model we are using matches the form of the association. If a residual plot indicates that a linear model is *appropriate,* we can ask the follow-up question found in the next subsection: how *well* does the line fit the data?

# EXAMPLE How Much Is That Truck Worth?

## *Examining residuals*

Let's return to the Ford F-150 data about the number of miles driven and price for a random sample of 16 used trucks. In general, trucks with more miles driven have lower prices. In the Technology Corner, we confirmed that the equation of the least-squares regression line for these data is $\widehat{\text{price}} = 38{,}257 - 0.1629$ (miles driven). The calculator screen shot in the margin shows a scatterplot of the data with the least-squares line added.

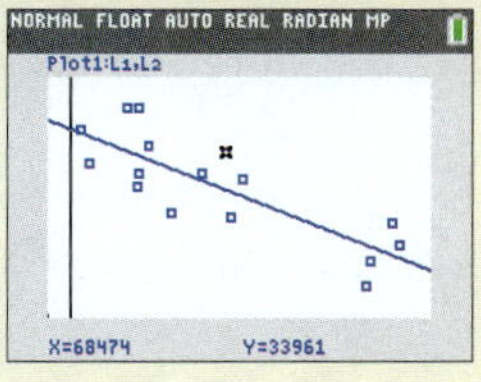


One truck had 68,474 miles driven and a price of \$33,961. This truck is marked on the scatterplot with an X. Because the point is above the line on the scatterplot, we know that its actual price is higher than the predicted price. To find out exactly how much higher, we calculate the residual for this truck. The predicted price for a Ford F-150 with 68,474 miles driven is

$$\hat{y} = 38{,}257 - 0.1629(68{,}474) = \$27{,}103$$

The residual for this truck is therefore

$$\text{residual} = \text{observed } y - \text{predicted } y = y - \hat{y} = 33{,}961 - 27{,}103 = \$6858$$

This truck costs \$6858 more than expected, based on its mileage.

The 16 points used in calculating the equation of the least-squares regression line produce 16 residuals. Rounded to the nearest dollar, they are

| | | | | | | | |
|---|---|---|---|---|---|---|---|
| −4765 | −7664 | −3506 | 8617 | 7728 | 3057 | −5004 | 458 |
| 2289 | 1183 | −8121 | 6858 | 2110 | 5572 | −5973 | −2857 |

Most graphing calculators and statistical software will calculate and store residuals for you.

Although residuals can be calculated from any model that is fitted to the data, the residuals from the least-squares line have a special property: *the mean of the least-squares residuals is always zero.* You can check that the sum of the residuals in the above example is −\$18. The sum is not exactly 0 because of rounding errors.

You can see the residuals in the scatterplot of Figure 3.11(a) on the next page by looking at the vertical deviations of the points from the line. The **residual plot** in Figure 3.11(b) makes it easier to study the residuals by plotting them against the explanatory variable, miles driven. Because the mean of the residuals is always zero, the horizontal line at zero in Figure 3.11(b) helps orient us. This "residual = 0" line corresponds to the regression line in Figure 3.11(a).

Some software packages prefer to plot the residuals against the predicted values $\hat{y}$ instead of against the values of the explanatory variable. The basic shape of the two plots is the same because $\hat{y}$ is linearly related to $x$.

**DEFINITION: Residual plot**

A **residual plot** is a scatterplot of the residuals against the explanatory variable. Residual plots help us assess whether a linear model is appropriate.

### ALTERNATE EXAMPLE

#### *Tapping on cans*

Here is a scatterplot showing the tapping time and amount of soda remaining for the 40 cans. The least-squares regression line, $\widehat{\text{soda}} = 248.6 + 2.63$ (tapping time), is shown on the scatterplot. The point in red is for the can that was tapped for 8 seconds and had 255 ml remaining after it was opened. The predicted amount remaining is $\widehat{\text{soda}} = 248.6 + 2.63(8) = 269.64$ ml. The residual is therefore $y - \hat{y} = 255 - 269.64 = -14.64$ ml. The amount of soda remaining in this can is 14.64 ml less than expected, based on the tapping time.

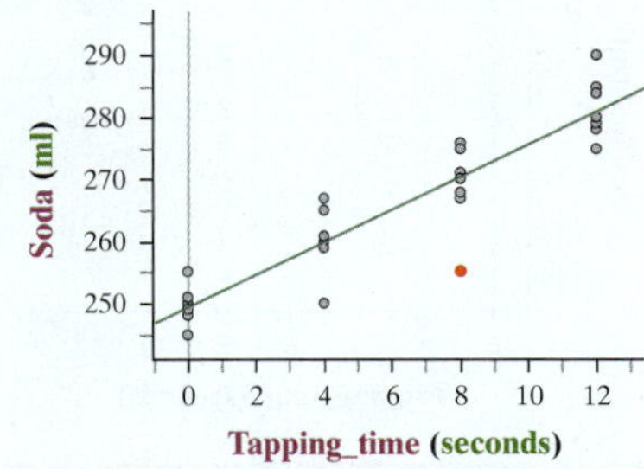


Here are all 40 residuals, in the same order as the original data:

| Residuals | | | |
|---|---|---|---|
| **0 s** | **4 s** | **8 s** | **12 s** |
| −3.6 | 0.88 | −2.64 | −5.16 |
| 6.4 | −9.12 | 1.36 | −0.16 |
| 1.4 | −9.12 | −1.64 | −5.16 |
| 1.4 | −9.12 | 0.36 | −0.16 |
| 1.4 | 0.88 | 6.36 | 4.84 |
| −3.6 | 5.88 | −14.64 | 9.84 |
| −0.6 | 7.88 | 0.36 | 3.84 |
| 1.4 | 0.88 | 0.36 | −2.16 |
| 2.4 | 1.88 | 5.36 | −1.16 |
| 0.4 | −0.12 | 5.36 | −0.16 |

### Teaching Tip

Software packages put the predicted values (sometimes called "fitted values") on the horizontal axis because they are built to do multiple regression. In multiple regression, many different explanatory variables are combined into one model to predict a response variable. Instead of picking one of the explanatory variables to put on the horizontal axis, software selects the predicted values, which are a function of all the explanatory variables.

ALTERNATE EXAMPLE

## *Tapping on cans*

Here are the scatterplot and residual plot for the can tapping data.

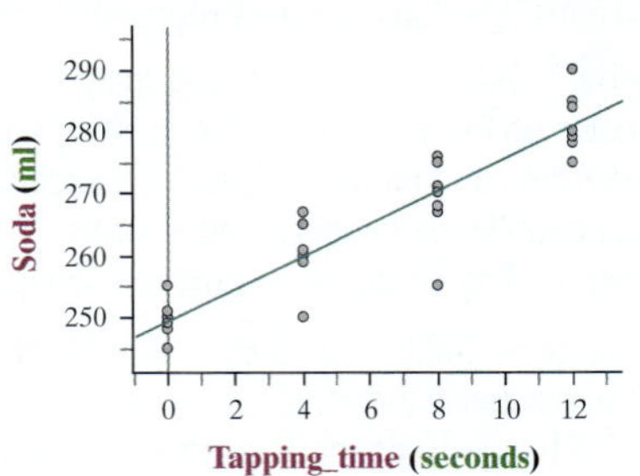


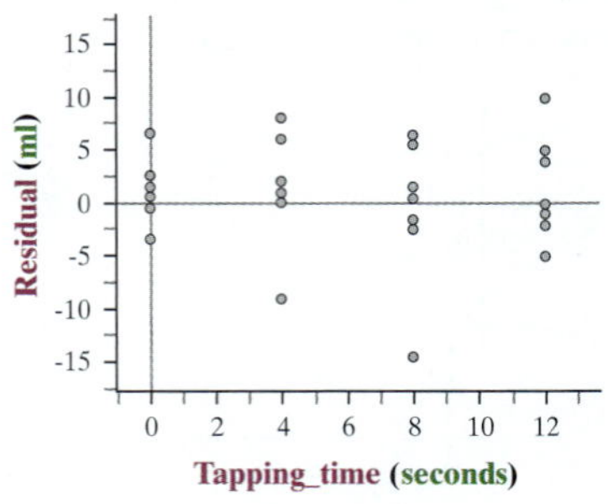


## ANSWERS TO CYU

1. $y - \hat{y} = 31{,}891 - 36{,}895 = -\$5004$.

2. The actual price of this truck is $5004 less than predicted based on the number of miles it has been driven.

3. The truck with 44,447 miles and a price of $22,896. This truck has a residual of −$8120, which means that the line over predicted the price by $8120. No other truck had a residual that was farther below 0 than this one.

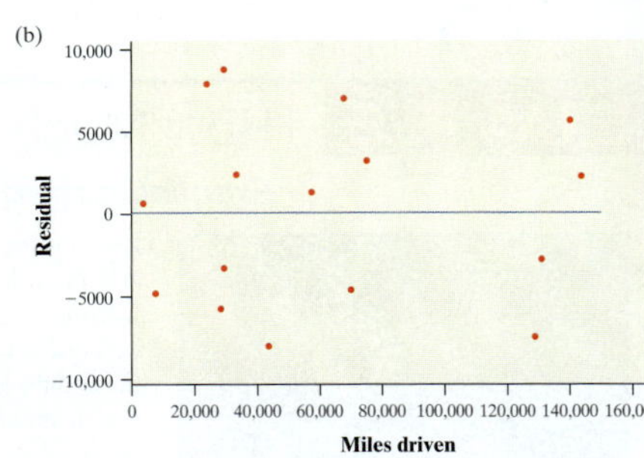


**FIGURE 3.11** (a) Scatterplot of price versus miles driven, with the least-squares line. (b) Residual plot for the regression line displayed in Figure 3.11(a). The line at $y = 0$ marks the sum (and mean) of the residuals.

## CHECK YOUR UNDERSTANDING

Refer to the Ford F-150 miles driven and price data.

1. Find the residual for the truck that had 8359 miles driven and a price of $31,891. Show your work.
2. Interpret the value of this truck's residual in context.
3. For which truck did the regression line overpredict price by the most? Justify your answer.

**Examining residual plots** A residual plot in effect turns the regression line horizontal. It magnifies the deviations of the points from the line, making it easier to see unusual observations and patterns. Because it is easier to see an unusual pattern in a residual plot than a scatterplot of the original data, we often use residual plots to determine if the model we are using is appropriate.

Figure 3.12(a) shows a nonlinear association between two variables and the least-squares regression line for these data. Figure 3.12(b) shows the residual plot for these data.

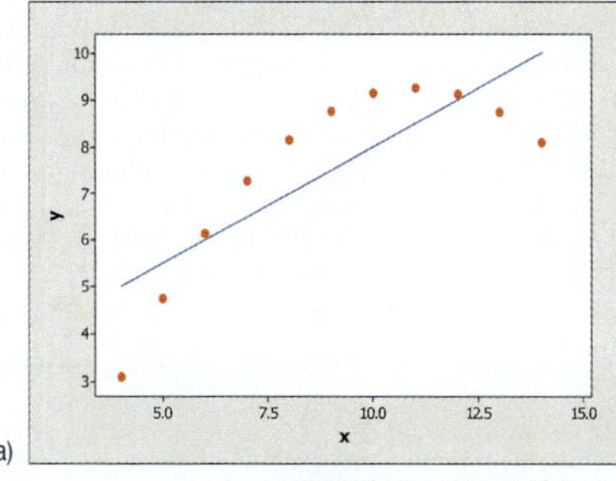


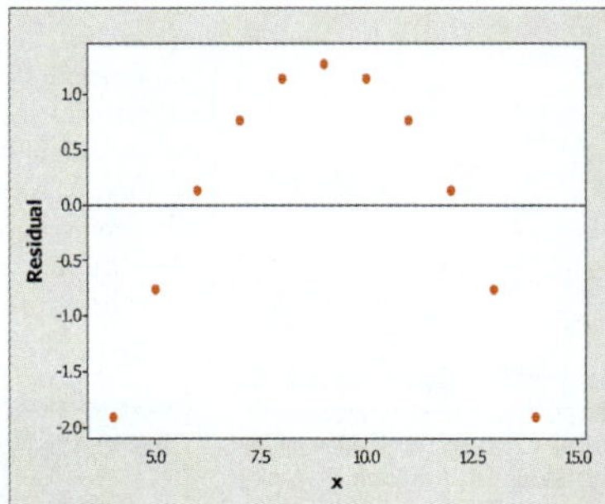


**FIGURE 3.12** (a) A straight line is not a good model for these data. (b) The residual plot has a curved pattern.

## AP® Exam Common Error

When asked if a linear model is appropriate, students will sometimes incorrectly use the correlation to justify linearity. However, a strong correlation doesn't mean an association is linear. An association can be clearly nonlinear and still have a correlation close to ±1. Only a residual plot can adequately address whether a line is an appropriate model for the data by showing the pattern of deviations from the line.

For example, graphing the function $y = x^2$ for the integers 1 to 10 yields a correlation of $r = 0.97$, but the residual plot shows an obvious pattern.

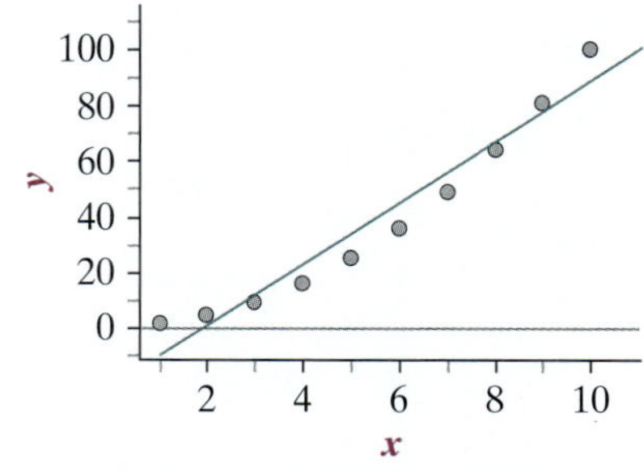


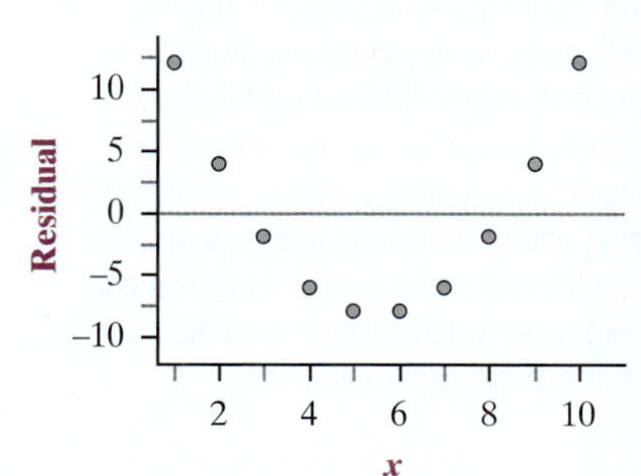

Because the form of our model (linear) is not the same as the form of the association (curved), there is an obvious leftover pattern in the residual plot. *When an obvious curved pattern exists in a residual plot, the model we are using is not appropriate.* We'll look at how to deal with curved relationships in Chapter 12.

When we use a line to model a linear association, there will be no leftover patterns in the residual plot, only random scatter. Figure 3.13 shows the residual plot for the Ford F-150 data. Because there is only random scatter in the residual plot, we know the linear model we used is appropriate.

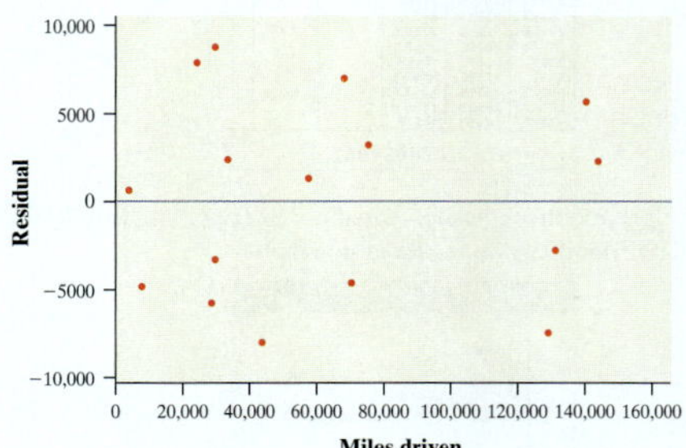


**FIGURE 3.13** The random scatter of points indicates that the regression line has the same form as the association, so the line is an appropriate model.

**Why do we look for patterns in residual plots?** The word *residual* comes from the Latin word *residuum*, meaning "left over." When we calculate a residual, we are calculating what is left over after subtracting the predicted value from the observed value:

$$\text{residual} = \text{observed } y - \text{predicted } y$$

Likewise, when we look at the form of a residual plot, we are looking at the form that is left over after subtracting the form of the model from the form of the association:

$$\text{form of residual plot} = \text{form of association} - \text{form of model}$$

When there is a leftover form in the residual plot, the form of the association and form of the model are not the same. However, if the form of the association and form of the model are the *same*, the residual plot should have no form, other than random scatter.

## RESIDUAL PLOTS ON THE CALCULATOR

TI-Nspire instructions in Appendix B; HP Prime instructions on the book's Web site.

Let's continue the analysis of the Ford F-150 miles driven and price data from the previous Technology Corner (page 171). You should have already made a scatterplot, calculated the equation of the least-squares regression line, and graphed the line on your plot. Now, we want to calculate residuals and make a residual plot. Fortunately, your calculator has already done most of the work. Each time the calculator computes a regression line, it also computes the residuals and stores them in a list named RESID. Make sure to calculate the equation of the regression line *before* using the RESID list!

### Teaching Tip

If the residual plot shows increasing spread but otherwise looks randomly scattered, a line is the right model to use. For example, see the residual plot in Exercise 54 on page 195. When residual plots have this pattern, the precision of predictions will vary depending on the value of *x*. Also, the presence of increasing spread will cause problems in Chapter 12 when we perform hypothesis tests and construct confidence intervals for the slope of a least-squares regression line.

### Teaching Tip

To help students understand the last paragraph in the Think About It feature, ask them what they get when they subtract 3 from 3 and subtract 10 from 10. In both cases, the result is 0—nothing. This is always the case when subtracting two things that are the same. Likewise, when the model and the association are the same, the residual plot should show no form except random scatter.

### TECHNOLOGY TIP

In previous editions of *TPS*, we didn't reveal the existence of the RESID list until Chapter 12. However, to put the emphasis on interpreting residual plots rather than constructing them, we decided to encourage students to use the RESID list whenever possible.

### TECHNOLOGY TIP

Emphasize the last sentence on this page—students must calculate the equation of the least-squares regression line before using the RESID list. Otherwise, the residuals in the RESID list will be from whatever line they calculated most recently.

| TI-83/84 | TI-89 |
|---|---|

1. Display the residuals in L3(list3).

- **TI-83/84:** With L3 highlighted, press [2nd] [STAT] (LIST) and select the RESID list.
- **TI-89:** With list3 highlighted, press [2nd] [-] (VAR-LINK), arrow down to STATVARS, and select the RESID list.

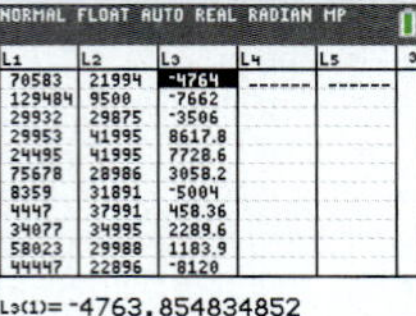

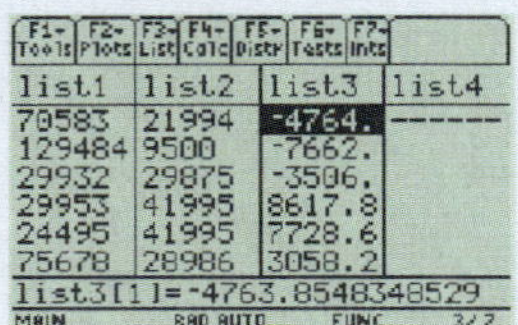

2. Turn off Plot1 and the regression equation. Specify Plot2 with L1/list1 as the *x* variable and L3/list3 as the *y* variable. Use ZoomStat (ZoomData) to see the residual plot.

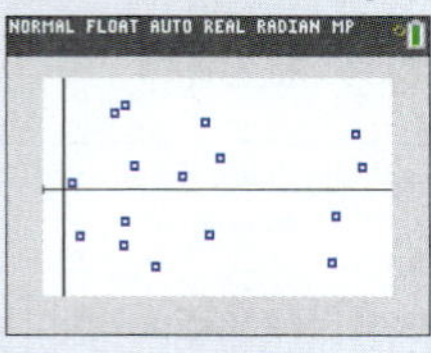

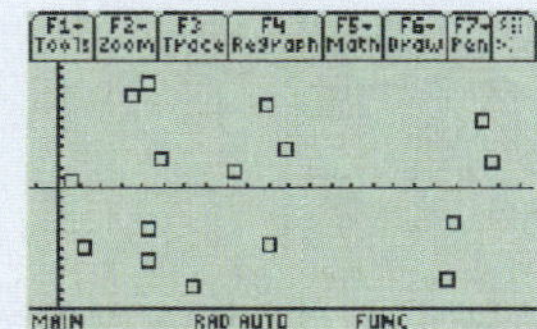

The *x* axis in the residual plot serves as a reference line: points above this line correspond to positive residuals and points below the line correspond to negative residuals.

*Note:* If you don't want to see the residuals in L3/list3, you can make a residual plot in one step by using the RESID list as the *y* variable in the scatterplot.

## CHECK YOUR UNDERSTANDING

In Exercises 5 and 7, we asked you to make and describe a scatterplot for the hiker data shown in the table below. Here is a residual plot for the least-squares regression of pack weight on body weight for the 8 hikers.

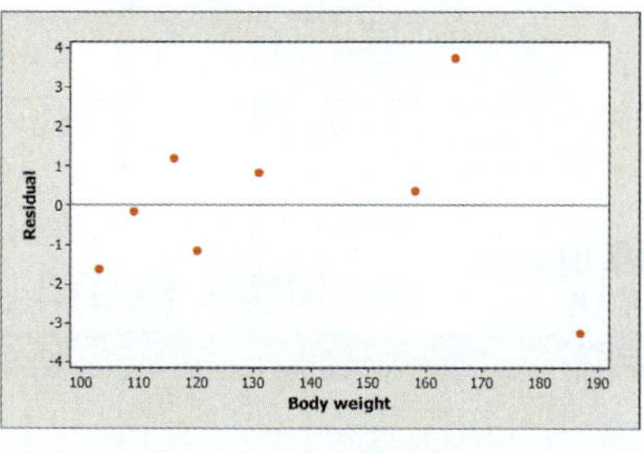


| Body weight (lb): | 120 | 187 | 109 | 103 | 131 | 165 | 158 | 116 |
|---|---|---|---|---|---|---|---|---|
| Backpack weight (lb): | 26 | 30 | 26 | 24 | 29 | 35 | 31 | 28 |

1. One of the hikers had a residual of nearly 4 pounds. Interpret this value.
2. Based on the residual plot, is a linear model appropriate for these data?

### ANSWERS TO CYU

1. The backpack for this hiker was almost 4 pounds heavier than expected based on the weight of the hiker.

2. Because there appears to be a negative-positive-negative pattern in the residual plot, a linear model is not appropriate for these data.

## How Well the Line Fits the Data: The Role of $s$ and $r^2$ in Regression

A residual plot is a graphical tool for determining if a least-squares regression line is an appropriate model for a relationship between two variables. Once we determine that a least-squares regression line is appropriate, it makes sense to ask a follow-up question: How well does the line work? That is, if we use the least-squares regression line to make predictions, how good will these predictions be?

**The Standard Deviation of the Residuals** We already know that a residual measures how far an observed $y$-value is from its corresponding predicted value $\hat{y}$. In an earlier example, we calculated the residual for the Ford F-150 with 68,474 miles driven and price \$33,961. The residual was \$6858, meaning that the actual price was \$6858 higher than we predicted.

To assess how well the line fits *all* the data, we need to consider the residuals for each of the 16 trucks, not just one. Using these residuals, we can estimate the "typical" prediction error when using the least-squares regression line. To do this, we calculate the **standard deviation of the residuals.**

$$s = \sqrt{\frac{\sum \text{residuals}^2}{n-2}}$$

For the Ford F-150 data, the sum of squared residuals is 461,300,000. So, the standard deviation of the residuals is

$$s = \sqrt{\frac{461{,}300{,}000}{14}} = 5740 \text{ dollars}$$

Did you recognize the number 461,300,000? We first encountered this number on page 170 when illustrating that the least-squares regression line minimized the sum of squared residuals. We'll see it again shortly.

When we use the least-squares regression line to predict the price of a Ford F-150 using the number of miles it has been driven, our predictions will typically be off by about \$5740. Looking at the residual plot, this seems like a reasonable value. Although some of the residuals are close to 0, others are close to \$10,000 or −\$10,000.

**DEFINITION: Standard deviation of the residuals ($s$)**

If we use a least-squares line to predict the values of a response variable $y$ from an explanatory variable $x$, the **standard deviation of the residuals ($s$)** is given by

$$s = \sqrt{\frac{\sum \text{residuals}^2}{n-2}} = \sqrt{\frac{\sum (y_i - \hat{y})^2}{n-2}}$$

This value gives the approximate size of a "typical" prediction error (residual).

**Does the formula for $s$ look slightly familiar?** It should. In Chapter 1, we defined the standard deviation of a set of quantitative data as

$$s_x = \sqrt{\frac{\sum (x_i - \bar{x})^2}{n-1}}$$

### Teaching Tip

Again, make sure to point out the title of this subsection. The standard deviation of the residuals and $r^2$ do *not* reveal if a linear model is appropriate. Instead, once a residual plot reveals that a linear model is appropriate, the values of $s$ and $r^2$ measure how good our predictions will be using the linear model.

### Teaching Tip

When introducing the idea of the standard deviation of the residuals, remind students about the standard deviation we learned earlier. In Chapter 1, the standard deviation was the typical distance of the actual values from the mean. In this chapter, the standard deviation is the typical distance of the actual values from the least-squares regression line. In general, a standard deviation is the typical distance that the actual values are from their expected values.

**ALTERNATE EXAMPLE**

*Tapping on cans*

For the can tapping data, the standard deviation of the residuals is

$$s = \sqrt{\frac{951.536}{40-2}} = 5.00 \text{ ml.}$$ When we use the least-squares regression line to predict the amount of soda remaining using the amount of tapping time, our predictions will typically be off by about 5 ml.

### Teaching Tip

Remind students that the standard deviation of the residuals is measured in the units of the response variable, while the correlation measures scatter on a standardized scale from −1 to 1. This makes the standard deviation a different, but more contextualized, way to measure how much scatter there is around the least-squares regression line.

### TECHNOLOGY TIP

To have your graphing calculator compute the standard deviation of the residuals, you can run the `LinRegTTest` by pressing `STAT`, choosing the `TESTS` menu, and scrolling to `LinRegTTest`. We recommend that you not show this to your students until Chapter 12.

## Teaching Tip

If your students are curious about where the $n - 2$ comes from, here is one way to explain it: if you had only two points, you could calculate a least-squares regression line but would have no way to estimate the amount of variability from the line. Only when you have three or more points can you begin to estimate variability from the line. This was also the case with $n - 1$ in Chapter 1. If we have one observation, we can estimate the mean but have no way of estimating the variability from the mean. Only when you have two or more observations can you begin estimating variability.

**ALTERNATE EXAMPLE**

### *Tapping on cans*

Suppose that we wandered in during the can tapping experiment and found a partially full can. Without measuring the contents, how could we predict how much soda is left in the can? We don't know how long it was tapped, so our best guess would be the mean amount remaining in all the cans: $\bar{y} = 264.45$ ml. The first scatterplot shows the squared prediction errors when using the mean amount $\bar{y}$ as our prediction. When using $\bar{y}$ as our predicted value, the sum of the squared prediction errors is 6506.

We could make much better predictions if we knew the tapping time. How much better? The second scatterplot shows the squared prediction errors when using the least-squares regression line. The sum of the squared residuals when using the least-squares regression line is 951.3 (the same quantity we used to calculate the standard deviation of the residuals, other than a small difference due to rounding error).

This means that only $\dfrac{951.3}{6506} = 14.6\%$ of the variation in amount of soda remaining is *un*accounted for by the least-squares regression line. The remaining variation is due to other factors, such as how vigorously the can was shaken. Therefore, $1 - \dfrac{951.3}{6506} = 85.4\%$ of the variability in amount of soda remaining is accounted for by the linear model relating amount of soda remaining to tapping time.



We interpreted the resulting value as the "typical" distance of the data points from the mean. In the case of two-variable data, we're interested in the typical (vertical) distance of the data points from the regression line. We find this value in much the same way: by adding up the squared deviations, then averaging (again in a funny way), and taking the square root to get back to the original units of measurement. Why do we divide by $n - 2$ this time instead of $n - 1$? You'll have to wait until Chapter 12 to find out.

**The Coefficient of Determination** There is another numerical quantity that tells us how well the least-squares line predicts values of the response variable $y$. It is $r^2$, the coefficient of determination. Some computer packages call it "R-sq." You may have noticed this value in some of the calculator and computer regression output that we showed earlier. Although it's true that $r^2$ is equal to the square of $r$, there is much more to this story.

**EXAMPLE**

## How Much Is That Truck Worth?

### *How can we predict y if we don't know x?*

Suppose that we randomly selected an additional used Ford F-150 that was on sale. What should we predict for its price? Figure 3.14 shows a scatterplot of the truck data that we have studied throughout this section, including the least-squares regression line. Another horizontal line has been added at the mean $y$-value, $\bar{y} = \$27{,}834$. If we don't know the number of miles driven for the additional truck, we can't use the regression line to make a prediction. What should we do? Our best strategy is to use the mean price of the other 16 trucks as our prediction.

**FIGURE 3.14** Scatterplot and least-squares regression line for the Ford F-150 data with a horizontal line added at the mean price, \$27,834.

Figure 3.15(a) on the facing page shows the prediction errors if we use the average price $\bar{y}$ as our prediction for the original group of 16 trucks. We can see that the sum of the squared residuals for this line is $\sum(y_i - \bar{y})^2 = 1{,}374{,}000{,}000$. This quantity measures the total variation in the $y$-values from their mean. This is also the same quantity we use to calculate the standard deviation of the prices, $s_y$.

If we learn the number of miles driven on the additional truck, then we could use the least-squares line to predict its price. How much better does the regression line do at predicting prices than simply using the average price $\bar{y}$ of all 16 trucks? Figure 3.15(b) reminds us that the sum of squared residuals for the least-squares line is $\sum\text{residuals}^2 = 461{,}300{,}000$. This is the same quantity we used to calculate the standard deviation of the residuals.

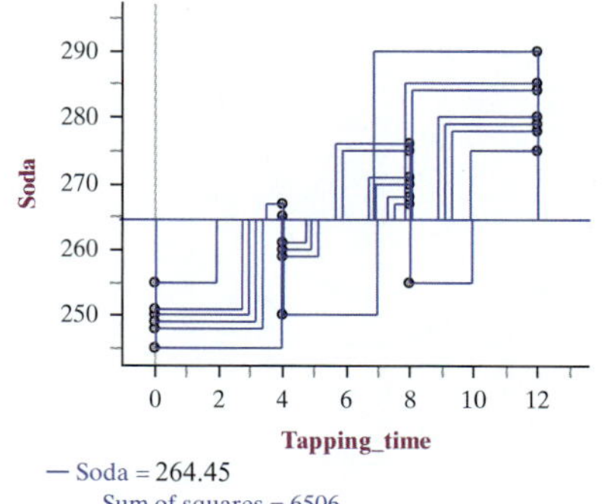


— Soda = 248.6 + 2.63 Tapping time; $r^2 = 0.85$
Sum of squares = 951.3

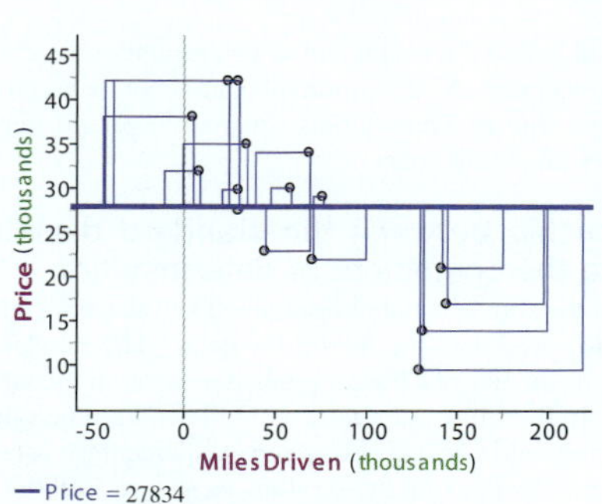


(a)

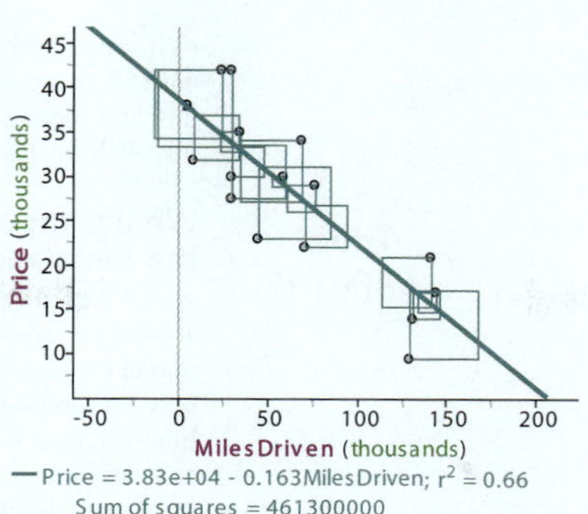


(b)

**FIGURE 3.15** (a) The sum of squared residuals is 1,374,000,000 if we use the mean price as our prediction for all 16 trucks. (b) The sum of squares from the least-squares regression line is 461,300,000.

The ratio of these two quantities tells us what proportion of the total variation in $y$ still remains after using the regression line to predict the values of the response variable. In this case,

$$\frac{461{,}300{,}000}{1{,}374{,}000{,}000} = 0.336$$

This means that 33.6% of the variation in price is *unaccounted for* by the least-squares regression line using $x$ = miles driven. This unaccounted-for variation is likely due to other factors, including the age of the truck or its condition. Taking this one step further, the proportion of the total variation in $y$ that *is accounted for* by the regression line is

$$1 - 0.336 = 0.664$$

We interpret this by saying that "*66.4% of the variation in price is accounted for by the linear model relating price to miles driven.*"

**DEFINITION: The coefficient of determination: $r^2$**

The **coefficient of determination** $r^2$ is the fraction of the variation in the values of $y$ that is accounted for by the least-squares regression line of $y$ on $x$. We can calculate $r^2$ using the following formula:

$$r^2 = 1 - \frac{\sum \text{residuals}^2}{\sum (y_i - \bar{y})^2}$$

If all the points fall directly on the least-squares line, the sum of squared residuals is 0 and $r^2 = 1$. Then all the variation in $y$ is accounted for by the linear relationship with $x$. Because the least-squares line yields the smallest possible sum of squared prediction errors, the sum of squared residuals can never be more than the sum of squared deviations from the mean of $y$. In the worst-case scenario, the least-squares line does no better at predicting $y$ than $y = \bar{y}$ does. Then the two sums of squares are the same and $r^2 = 0$.

## Teaching Tip

Students often struggle with the interpretation of $r^2$. It might help to start by saying something like, "The sum of squared residuals has been reduced by 66.4% when using the linear model relating asking price to number of miles." This interpretation follows nicely from the explanation on this page. Then move to the more traditional interpretation: "_____% of the variation in ($y$ variable) is accounted for by the linear model relating ($y$ variable) to ($x$ variable)." In this interpretation, the "variation in $y$" is measured by the sum of squared residuals.

## Teaching Tip

If you have data that fall on a perfectly horizontal line, your calculator will say that $r$ and $r^2$ are undefined. The formula for $r$ uses the standard deviation of $y$ in the denominator (zero in the case of a horizontal line). However, when we use $r^2$ to measure how much better our predictions will be using the least-squares regression line compared to simply using the mean of the response variable for our predictions, it is reasonable to say that $r^2 = 0$ because using a least-squares regression line will be no better than using the mean. In practice, however, it would be very unusual to have a response variable with no variability.

It seems fairly remarkable that the coefficient of determination is actually the correlation squared. This fact provides an important connection between correlation and regression. When you see a correlation, square it to get a better feel for how well the least-squares line fits the data.

**What's the relationship between the standard deviation of the residuals $s$ and the coefficient of determination $r^2$?** They are both calculated from the sum of squared residuals. They also both attempt to answer the question, "How well does the line fit the data?" The standard deviation of the residuals reports the size of a typical prediction error, in the same units as the response variable. In the truck example, $s = 5740$ *dollars*. The value of $r^2$, however, does not have units and is usually expressed as a percentage between 0% and 100%, such as $r^2 = 66.4\%$. Because these values assess how well the line fits the data in different ways, we recommend you follow the example of most statistical software and report them both.

Let's revisit the SEC football data to practice what we have learned.

## EXAMPLE SEC Football

### *Residual plots, s, and $r^2$*

In Section 3.1, we looked at the relationship between the average number of points scored per game $x$ and the number of wins $y$ for the 12 college football teams in the Southeastern Conference. A scatterplot with the least-squares regression line and a residual plot are shown. The equation of the least-squares regression line is $\hat{y} = -3.75 + 0.437x$. Also, $s = 1.24$ and $r^2 = 0.88$.

PROBLEM:

(a) Calculate and interpret the residual for South Carolina, which scored 30.1 points per game and had 11 wins.

(b) Is a linear model appropriate for these data? Explain.

(c) Interpret the value of $s$.

(d) Interpret the value of $r^2$.

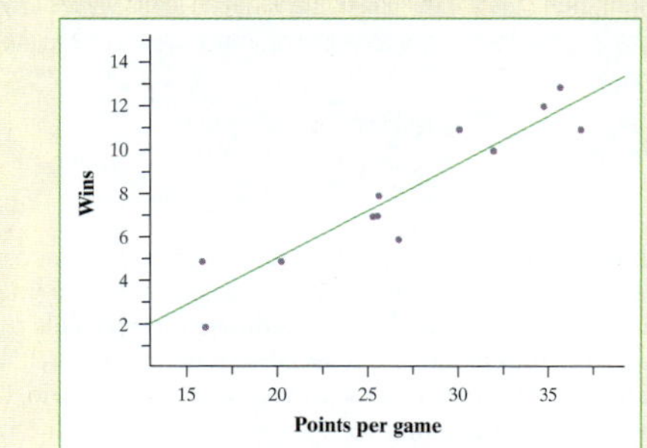


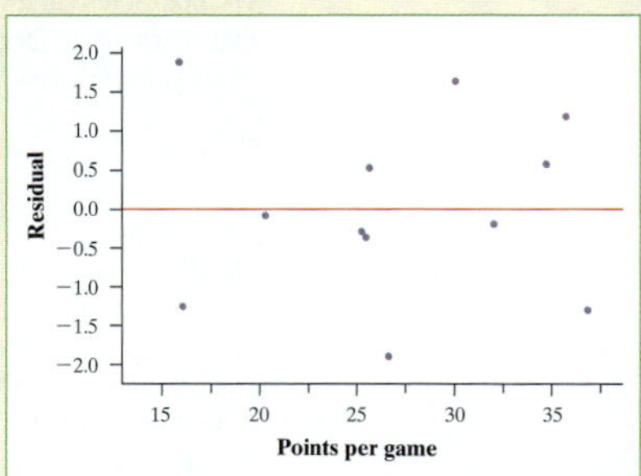


## Teaching Tip

Remind students that if they are given the value of $r^2$ and are asked to find the correlation, they have to consider the direction of the association. That is, $r = \pm\sqrt{r^2}$, where the + or − is determined by the direction of the association.

## ALTERNATE EXAMPLE

### *Back to the track!*

In Section 3.1, we looked at the relationship between the 40-yard sprint time (in seconds) and the long-jump distance (in inches) for a small statistics class with 12 students. A scatterplot with the least-squares regression line ($\hat{y} = 414.79 - 45.74x$) and a residual plot are shown below. Also, $s = 22.38$ and $r^2 = 0.702$.

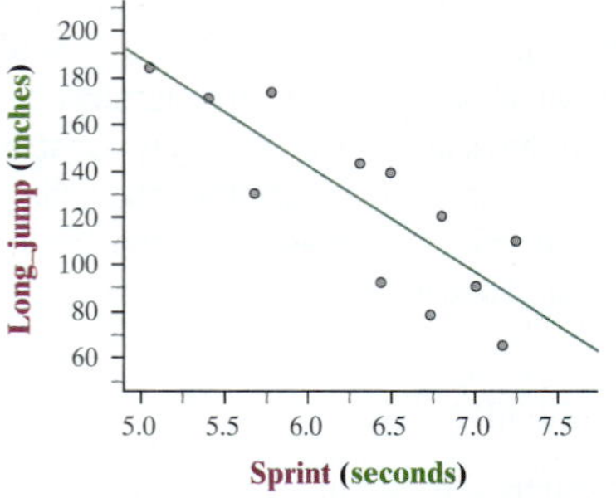


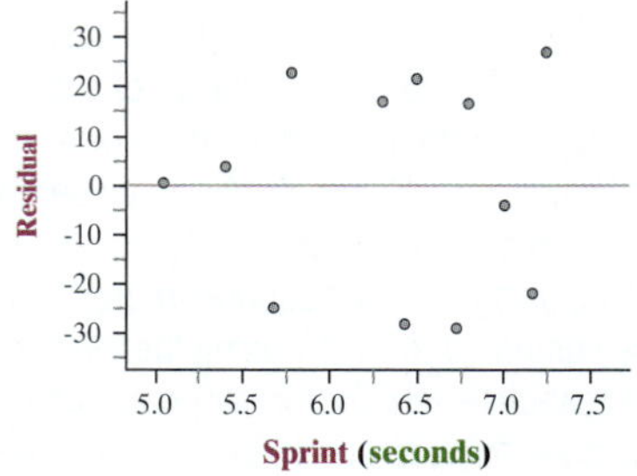


PROBLEM:

**(a)** Calculate and interpret the residual for Christian, who had a sprint time of 7.25 seconds and a long jump of 110 inches.

**(b)** Is a linear model appropriate for these data? Explain.

**(c)** Interpret the value of $s$.

**(d)** Interpret the value of $r^2$.

SOLUTION:

**(a)** The predicted length of the long jump for Christian is $\hat{y} = 414.79 - 45.74(7.25) = 83.175$ inches. Thus, the residual is $y - \hat{y} = 110 - 83.175 = 26.825$ inches. Christian's long jump was about 27 inches longer than expected, based on his sprint time.

**(b)** Because there is no obvious pattern left over in the residual plot, the linear model is appropriate.

**(c)** When using the least-squares regression line with $x$ = sprint time to predict $y$ = long-jump distance, we will typically be off by about 22.38 inches.

**(d)** 70.2% of the variation in long-jump distance is accounted for by the linear model relating long-jump distance to sprint time.

**AP® EXAM TIP** Students often have a hard time interpreting the value of $r^2$ on AP® exam questions. They frequently leave out key words in the definition. Our advice: Treat this as a fill-in-the-blank exercise. Write "____% of the variation in [response variable name] is accounted for by the linear model relating [response variable name] to [explanatory variable name]."

SOLUTION:

(a) The predicted amount of wins for South Carolina is

$$\hat{y} = -3.75 + 0.437(30.1) = 9.40 \text{ wins}$$

The residual for South Carolina is

$$\text{residual} = y - \hat{y} = 11 - 9.40 = 1.60 \text{ wins}$$

South Carolina won 1.60 more games than expected, based on the number of points they scored per game.

(b) Because there is no obvious pattern left over in the residual plot, the linear model is appropriate.

(c) When using the least-squares regression line with $x$ = points per game to predict $y$ = the number of wins, we will typically be off by about 1.24 wins.

(d) About 88% of the variation in wins is accounted for by the linear model relating wins to points per game.

**For Practice** *Try Exercise* **55**

## Interpreting Computer Regression Output

Figure 3.16 displays the basic regression output for the Ford F-150 data from two statistical software packages: Minitab and JMP. Other software produces very similar output. Each output records the slope and $y$ intercept of the least-squares line. The software also provides information that we don't yet need (or understand!), although we will use much of it later. Be sure that you can locate the slope, the $y$ intercept, and the values of $s$ and $r^2$ on both computer outputs. *Once you understand the statistical ideas, you can read and work with almost any software output.*

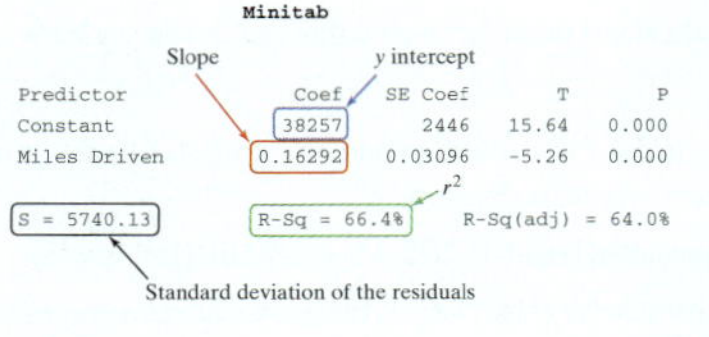
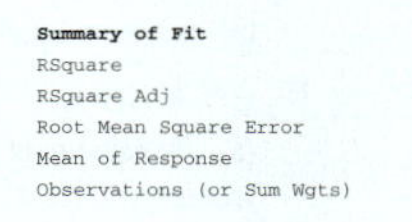
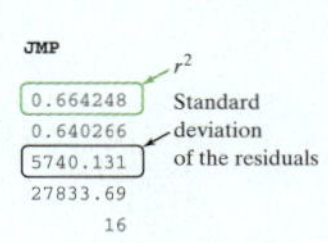

**Minitab**

```
Predictor          Coef     SE Coef       T        P
Constant          38257        2446   15.64    0.000
Miles Driven    0.16292     0.03096   -5.26    0.000

S = 5740.13     R-Sq = 66.4%     R-Sq(adj) = 64.0%
```



**JMP**

```
Summary of Fit
RSquare                       0.664248
RSquare Adj                   0.640266
Root Mean Square Error        5740.131
Mean of Response              27833.69
Observations (or Sum Wgts)          16

Parameter Estimates
Term            Estimate     Std Error   t Ratio   Prob>|t|
Intercept       38257.135    2445.813    15.64     <.0001
Miles Driven    -0.162919    0.030956    -5.26     0.0001
```



**FIGURE 3.16** Least-squares regression results for the Ford F-150 data from two statistical software packages. Other software produces similar output.

**EXAMPLE**

## Using Feet to Predict Height

*Interpreting regression output*

A random sample of 15 high school students was selected from the U.S. CensusAtSchool database. The foot length (in centimeters) and height (in centimeters) of each student in the sample were recorded. Least-squares

**TRM Computer Output for Alternate Examples**

See the Teacher's Resource Materials for examples of computer output using data from the SEC football, track and field day, and can tapping examples.

**TECHNOLOGY TIP**

In the JMP output, the standard deviation of the residuals is called "root mean square error." This is because it is calculated by finding the residuals (errors), squaring them, finding the average (sort of), and taking the square root. Also, your students may ask what the rest of the computer output means. Tell them that we will learn about most of it in Chapter 12. The only exception is R-Squared (adjusted), which is a measure that is used in multiple regression and isn't covered in AP® Statistics.

regression was performed on the data. A scatterplot with the regression line added, a residual plot, and some computer output from the regression are shown below.

| Predictor | Coef | SE Coef | T | P |
|---|---|---|---|---|
| Constant | 103.41 | 19.50 | 5.30 | 0.000 |
| Foot length | 2.7469 | 0.7833 | 3.51 | 0.004 |
| S = 7.95126 | R-Sq = 48.6% | R-Sq(adj) = 44.7% | | |

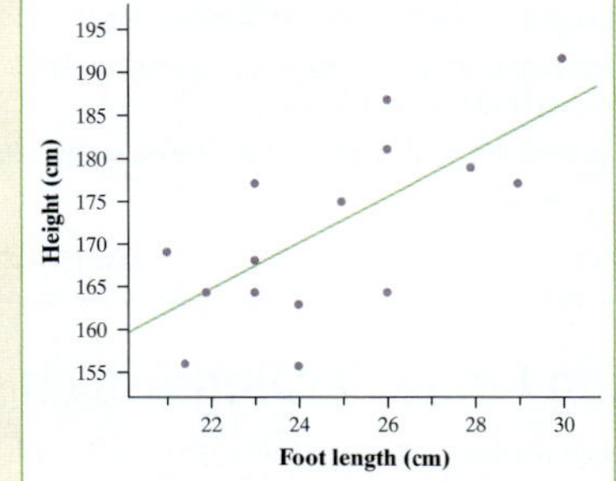


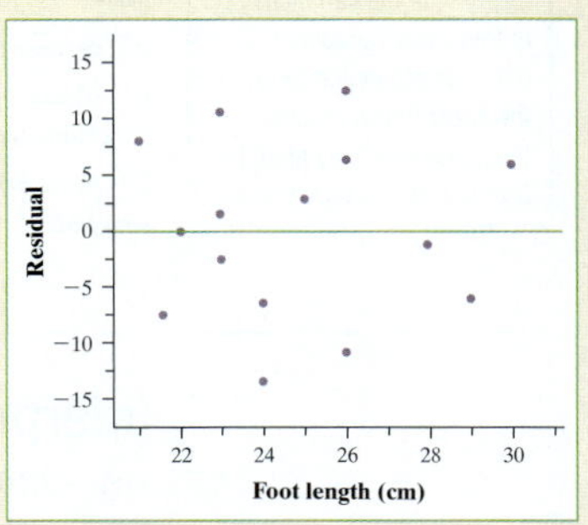


PROBLEM:

(a) What is the equation of the least-squares regression line that describes the relationship between foot length and height? Define any variables that you use.

(b) Interpret the slope of the regression line in context.

(c) Find the correlation.

(d) Is a line an appropriate model to use for these data? Explain how you know.

SOLUTION:

(a) The equation is $\hat{y} = 103.41 + 2.7469x$, where $\hat{y}$ = predicted height (in centimeters) and $x$ is foot length (in centimeters). We could also write

$$\text{predicted height} = 103.41 + 2.7469\,(\text{foot length})$$

(b) For each additional centimeter of foot length, the least-squares regression line predicts an increase of 2.7469 cm in height.

(c) To find the correlation, we take the square root of $r^2$: $r = \pm\sqrt{0.486} = \pm 0.697$. Because the scatterplot shows a positive association, $r = 0.697$.

(d) Because the scatterplot shows a linear association and the residual plot has no obvious leftover patterns, a line is an appropriate model to use for these data.

**For Practice** *Try Exercise*  59

## Regression to the Mean

Using technology is often the most convenient way to find the equation of a least-squares regression line. It is also possible to calculate the equation of the least-squares regression line using only the means and standard deviations of the two

### ALTERNATE EXAMPLE

***Does seat location affect grades?***

Many people believe that students learn better if they sit closer to the front of the classroom. Does sitting closer *cause* higher achievement, or do better students simply choose to sit in the front? To investigate, an AP® Statistics teacher randomly assigned students to seat locations in his classroom for a particular chapter. At the end of the chapter, he recorded the row number (Row 1 is closest to the front) and test score for each student. Least-squares regression was performed on the data. A scatterplot with the regression line added, a residual plot, and some computer output from the regression are shown below.

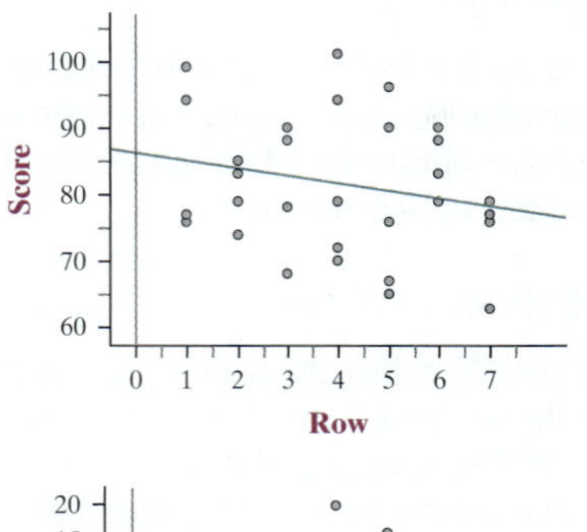


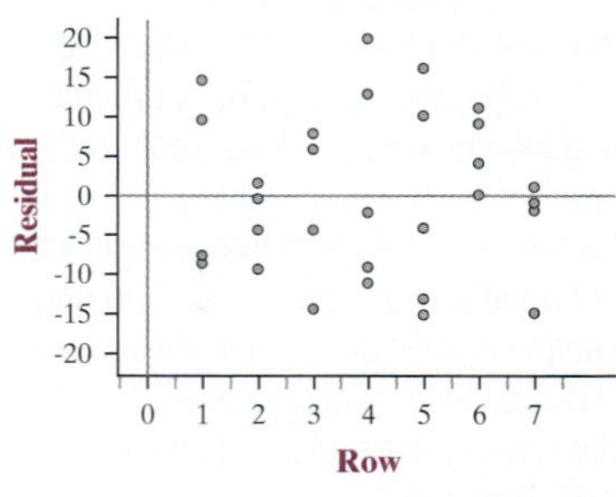


| Predictor | Coef | SECoef | T | P |
|---|---|---|---|---|
| Constant | 85.706 | 4.239 | 20.22 | 0.000 |
| Row | -1.1171 | 0.9472 | -1.18 | 0.248 |
| S = 10.0673 | R-Sq = 4.7% | R-Sq(adj) = 1.3% | | |

PROBLEM:

**(a)** What is the equation of the least-squares regression line that describes the relationship between row number and test score? Define any variables that you use.
**(b)** Interpret the slope of the regression line in context.
**(c)** Find the correlation.
**(d)** Is a line an appropriate model to use for these data? Explain how you know.

SOLUTION:

**(a)** The equation is $\hat{y} = 85.706 - 1.1171x$, where $\hat{y}$ = predicted test score and $x$ = row number.
**(b)** For each additional row from the front of the classroom, the least-squares regression line predicts a decrease of 1.1171 points on the student's test score.
**(c)** To find the correlation, we take the square root of $r^2$ and note that the association is negative. Thus, $r = -\sqrt{0.047} = -0.217$.
**(d)** Because there is no obvious leftover pattern in the residual plot, a line is an appropriate model for these data.

variables and their correlation. Exploring this method will highlight an important relationship between the correlation and the slope of a least-squares regression line—and reveal why we include the word "regression" in the expression "least-squares regression line."

**HOW TO CALCULATE THE LEAST-SQUARES REGRESSION LINE**

We have data on an explanatory variable $x$ and a response variable $y$ for $n$ individuals. From the data, calculate the means $\bar{x}$ and $\bar{y}$ and the standard deviations $s_x$ and $s_y$ of the two variables and their correlation $r$. The least-squares regression line is the line $\hat{y} = a + bx$ with **slope**

$$b = r\frac{s_y}{s_x}$$

and **$y$ intercept**

$$a = \bar{y} - b\bar{x}$$

**AP® EXAM TIP** The formula sheet for the AP® exam uses different notation for these equations: $b_1 = r\frac{s_y}{s_x}$ and $b_0 = \bar{y} - b_1\bar{x}$. That's because the least-squares line is written as $\hat{y} = b_0 + b_1x$. We prefer our simpler versions without the subscripts!

The formula for the $y$ intercept comes from the fact that the least-squares regression line always passes through the point $(\bar{x}, \bar{y})$. You discovered this in Step 4 of the Activity on page 170. Substituting $(\bar{x}, \bar{y})$ into the equation $\hat{y} = a + bx$ produces the equation $\bar{y} = a + b\bar{x}$. Solving this equation for $a$ gives the equation shown in the definition box, $a = \bar{y} - b\bar{x}$.

To see how these formulas work in practice, let's look at an example.

**EXAMPLE**

## Using Feet to Predict Height

### *Calculating the least-squares regression line*

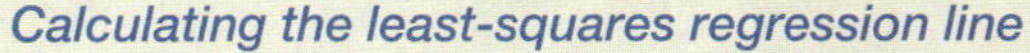

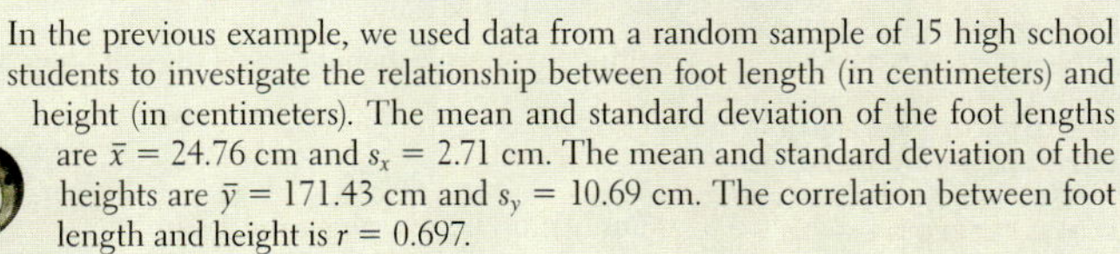

In the previous example, we used data from a random sample of 15 high school students to investigate the relationship between foot length (in centimeters) and height (in centimeters). The mean and standard deviation of the foot lengths are $\bar{x} = 24.76$ cm and $s_x = 2.71$ cm. The mean and standard deviation of the heights are $\bar{y} = 171.43$ cm and $s_y = 10.69$ cm. The correlation between foot length and height is $r = 0.697$.

PROBLEM: Find the equation of the least-squares regression line for predicting height from foot length. Show your work.

SOLUTION: The least-squares regression line of height $y$ on foot length $x$ has slope

$$b = r\frac{s_y}{s_x} = 0.697\frac{10.69}{2.71} = 2.75$$

The least-squares regression line has $y$ intercept

$$a = \bar{y} - b\bar{x} = 171.43 - 2.75(24.76) = 103.34$$

So, the equation of the least-squares regression line is $\hat{y} = 103.34 + 2.75x$.

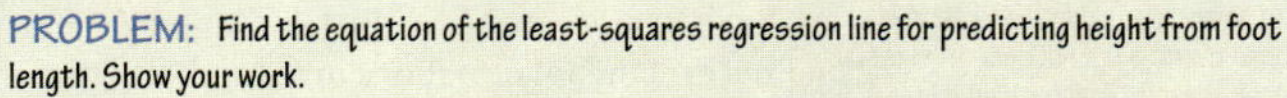

**For Practice** *Try Exercise* 61(a)

## AP® Exam Formula Sheet

Remind students that these two formulas are on the formula sheet provided on both parts of the AP® exam. Another formula to calculate the slope by hand is included on the formula sheet. However, it is very unlikely that students will need to use this additional formula.

## Teaching Tip

Remind students that the slope is always the coefficient of $x$, regardless of notation.

**ALTERNATE EXAMPLE**

### *Does seat location affect grades?*

In the previous Alternate Example, we investigated the relationship between test scores and seat location. The mean and standard deviation of the row numbers are $\bar{x} = 4.033$ and $s_x = 1.974$. The mean and standard deviation of the test scores are $\bar{y} = 81.2$ and $s_y = 10.135$. The correlation between row number and test score is $r = -0.218$. (Note that this value is slightly different than the previous example because of rounding in the computer output.)

PROBLEM: Find the equation of the least-squares regression line for predicting test score from row number. Show your work.

SOLUTION: The slope of the least-squares regression line is

$$b = r\frac{s_y}{s_x} = -0.218\frac{10.135}{1.974} = -1.119.$$

The $y$ intercept is $a = \bar{y} - b\bar{x}$ $= 81.2 - (-1.119)(4.033) = 85.71.$

So, the equation of the least-squares regression line is $\hat{y} = 85.71 - 1.119x$.

## Teaching Tip

Remind students that because of rounding errors, the slope and $y$ intercept may be a little different from what technology provides.

## Teaching Tip

There are several other ways to illustrate regression to the mean.

- Read the article at http://www.nytimes.com/2013/07/22/business/in-climbing-income-ladder-location-matters.html?hp&_r=1& and use its interactive feature to compare the income percentiles for parents and their children. For example, a child in Tucson, Arizona, whose parents' income is at the 10th percentile is predicted to have an income in the 36th percentile (still below the mean, but closer to the mean than the parents). Likewise, a child whose parents' income is at the 90th percentile is predicted to have an income in the 58th percentile (still above the mean, but closer to the mean than the parents).
- Make a scatterplot of your students' grades on the first two tests and include the line $y = x$. Typically, students who do well on the first exam don't do quite as well on the second. Likewise, students who do very poorly on the first exam tend to do better the second time around. Make sure to try this before you do it in front of your students, just in case.
- Assign and discuss Exercises 65 and 66. In both exercises, above-average values of $x$ lead to not-quite-as-far-above-average predicted values for $y$. Likewise, below-average values of $x$ lead to not-quite-as-far-below-average predicted values for $y$.

## TRM Regression to the Mean in Sports

The Alternate Activity *Regression to the Mean in Sports*, found in the Teacher's Resource Materials, is another nice way to illustrate regression to the mean. Find it by clicking on the link in the TE-book, opening the TRCD, or logging onto the book's companion Web site www.whfreeman.com/tps5e (teacher log-in required).

## Teaching Tip

Remind students that the correlation stays the same no matter what units are used for the $x$ and $y$ variables. The slope does not share this property, because it is measured in the units of the $x$ and $y$ variables. This can be seen in the formula for the slope: $b = r\frac{s_y}{s_x}$. The correlation $r$ has no units, but standard deviations are measured in the units of $y$ and $x$, so the slope is a rescaled version of the correlation.

There is a close connection between the correlation and the slope of the least-squares regression line. The slope is

$$b = r\frac{s_y}{s_x} = \frac{r \cdot s_y}{s_x}$$

This equation says that along the regression line, a change of 1 standard deviation in $x$ corresponds to a change of $r$ standard deviations in $y$. When the variables are perfectly correlated ($r = 1$ or $r = -1$), the change in the predicted response $\hat{y}$ is the same (in standard deviation units) as the change in $x$. For example, if $r = 1$ and $x$ is 2 standard deviations above its mean, then the corresponding value of $\hat{y}$ will be 2 standard deviations above the mean of $y$.

However, if the variables are not perfectly correlated ($-1 < r < 1$), the change in $\hat{y}$ is *less than* the change in $x$, when measured in standard deviation units. To illustrate this property, let's return to the foot length and height data from the previous example.

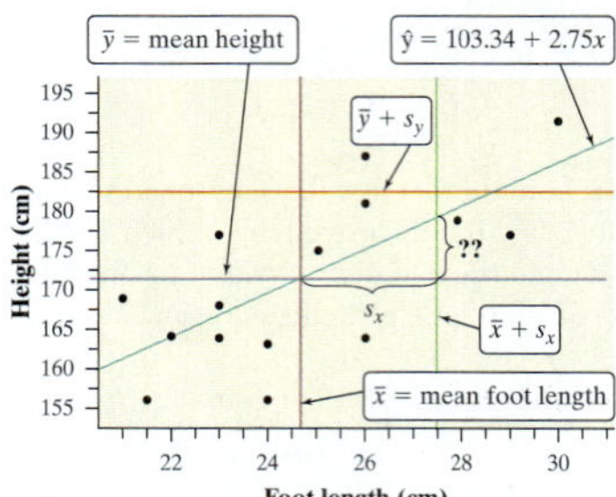


The figure at left shows the regression line $\hat{y} = 103.34 + 2.75x$. We have added four more lines to the graph: a vertical line at the mean foot length $\bar{x}$, a vertical line at $\bar{x} + s_x$ (1 standard deviation above the mean foot length), a horizontal line at the mean height $\bar{y}$, and a horizontal line at $\bar{y} + s_y$ (1 standard deviation above the mean height).

When a student's foot length is 1 standard deviation above the mean foot length $\bar{x}$, the predicted height $\hat{y}$ is above the mean height $\bar{y}$, but not an entire standard deviation above the mean. How far above the mean is the value of $\hat{y}$?

From the graph, we can see that

$$b = \text{slope} = \frac{\text{change in } y}{\text{change in } x} = \frac{??}{s_x}$$

From earlier, we know that

$$b = \frac{r \cdot s_y}{s_x}$$

Setting these two equations equal to each other, we have

$$\frac{??}{s_x} = \frac{r \cdot s_y}{s_x}$$

Thus, $\hat{y}$ must be $r \cdot s_y$ above the mean $\bar{y}$.

Sir Francis Galton (1822–1911) looked at data on the heights of children versus the heights of their parents. He found that taller-than-average parents tended to have children who were taller than average but not quite as tall as their parents. Likewise, shorter-than-average parents tended to have children who were shorter than average but not quite as short as their parents. Galton called this fact "regression to the mean" and used the symbol $r$ because of the correlation's important relationship to regression.

In other words, for an increase of 1 standard deviation in the value of the explanatory variable $x$, the least-squares regression line predicts an increase of *only* $r$ standard deviations in the response variable $y$. *When the correlation isn't $r = 1$ or $-1$, the predicted value of $y$ is closer to its mean $\bar{y}$ than the value of $x$ is to its mean $\bar{x}$. This is called regression to the mean, because the values of $y$ "regress" to their mean.*

**What happens if we standardize both variables?** Standardizing a variable converts its mean to 0 and its standard deviation to 1. Doing this to both $x$ and $y$ will transform the point $(\bar{x}, \bar{y})$ to (0, 0). So the least-squares line for the standardized values will pass through (0, 0). What about the slope of this line? From the formula, it's $b = rs_y/s_x$. Because we standardized, $s_x = s_y = 1$. That

### ALTERNATE EXAMPLE

#### *Netbook battery life*

Netbooks are a hybrid of a laptop computer and a tablet. They are smaller and have better battery life than a traditional laptop. They also have a separate keyboard, unlike most tablets. *Consumer Reports* did a study of 22 netbooks in their February 2010 issue. Among the variables measured were battery life (hours), weight (pounds), and cost. The data appear in the table:

| Battery life (hr) | Weight (lb) | Cost ($) |
|---|---|---|
| 6.00 | 2.8 | 370 |
| 7.75 | 2.9 | 350 |
| 7.25 | 2.8 | 330 |
| 5.50 | 2.4 | 370 |
| 8.25 | 2.9 | 360 |
| 9.50 | 2.9 | 400 |
| 7.75 | 2.9 | 340 |
| 7.75 | 2.7 | 340 |
| 8.00 | 2.8 | 350 |
| 7.00 | 2.8 | 350 |
| 6.50 | 3.2 | 360 |
| 6.25 | 2.7 | 310 |
| 5.25 | 2.9 | 330 |
| 5.00 | 2.7 | 320 |

means $b = r$. In other words, the slope is equal to the correlation. The Fathom screen shot confirms these results. It shows that $r^2 = 0.49$, so $r = \sqrt{0.49} = 0.7$, approximately the same value as the slope of 0.697.

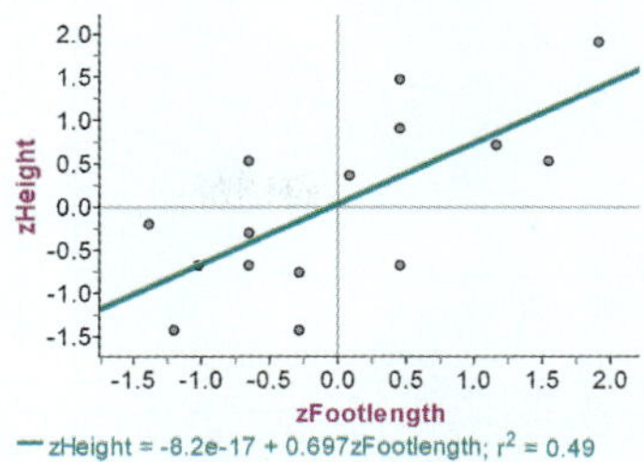


## Putting It All Together: Correlation and Regression

In Chapter 1, we introduced a four-step process for organizing a statistics problem. Here is another example of the four-step process in action.

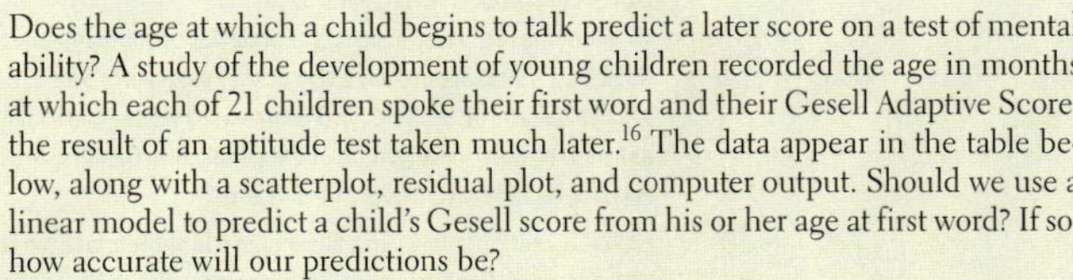

### EXAMPLE Gesell Scores

STEP 4

*Putting it all together*

Does the age at which a child begins to talk predict a later score on a test of mental ability? A study of the development of young children recorded the age in months at which each of 21 children spoke their first word and their Gesell Adaptive Score, the result of an aptitude test taken much later.[16] The data appear in the table below, along with a scatterplot, residual plot, and computer output. Should we use a linear model to predict a child's Gesell score from his or her age at first word? If so, how accurate will our predictions be?

| Age (months) at first word and Gesell score | | | | | | | | |
|---|---|---|---|---|---|---|---|---|
| CHILD | AGE | SCORE | CHILD | AGE | SCORE | CHILD | AGE | SCORE |
| 1 | 15 | 95 | 8 | 11 | 100 | 15 | 11 | 102 |
| 2 | 26 | 71 | 9 | 8 | 104 | 16 | 10 | 100 |
| 3 | 10 | 83 | 10 | 20 | 94 | 17 | 12 | 105 |
| 4 | 9 | 91 | 11 | 7 | 113 | 18 | 42 | 57 |
| 5 | 15 | 102 | 12 | 9 | 96 | 19 | 17 | 121 |
| 6 | 20 | 87 | 13 | 10 | 83 | 20 | 11 | 86 |
| 7 | 18 | 93 | 14 | 11 | 84 | 21 | 10 | 100 |

**ALTERNATE EXAMPLE *(continued)***

| | | |
|---|---|---|
| 3.75 | 2.6 | 380 |
| 4.50 | 2.8 | 335 |
| 2.75 | 2.4 | 350 |
| 2.75 | 2.4 | 300 |
| 2.50 | 2.6 | 280 |
| 2.50 | 2.5 | 500 |
| 2.00 | 2.5 | 290 |
| 2.75 | 2.5 | 500 |

Should we use a linear model to predict battery life of a new netbook based on its weight? If so, how accurate will our predictions be?

STATE: Is a linear model appropriate for these data? If so, how well does the least-squares regression line fit the data?

PLAN: To determine whether a linear model is appropriate, we will look at the scatterplot and residual plot to see if the association is linear or nonlinear. Then, if a linear model is appropriate, we will use the standard deviation of the residuals and $r^2$ to measure how well the least-squares line fits the data.

Here is some computer output from a regression analysis, including a scatterplot and residual plot:

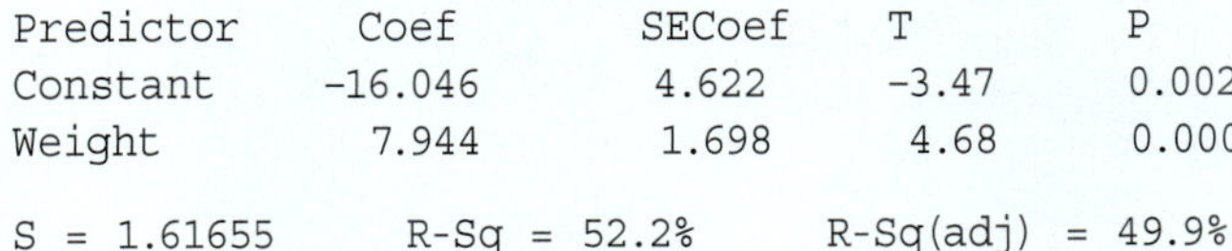

```
Predictor    Coef      SECoef    T       P
Constant   -16.046     4.622   -3.47   0.002
Weight       7.944     1.698    4.68   0.000

S = 1.61655    R-Sq = 52.2%    R-Sq(adj) = 49.9%
```

**ALTERNATE EXAMPLE *(continued)***

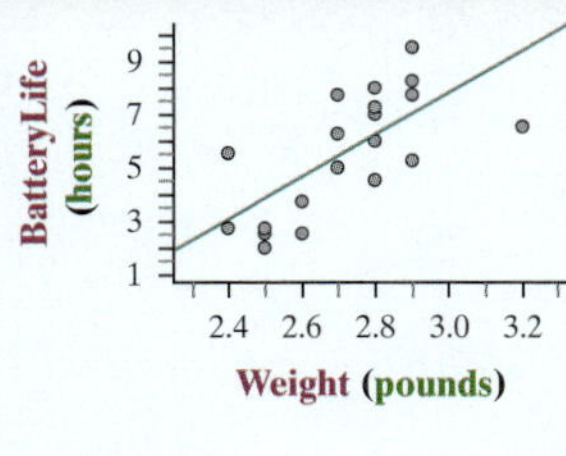


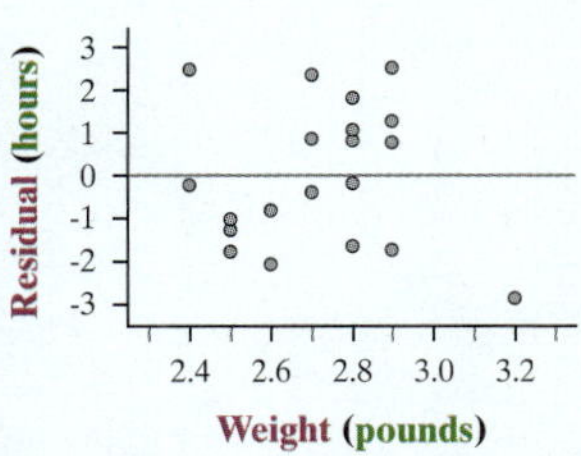


DO: The scatterplot shows a fairly strong, positive linear association between the weight of a netbook and its battery life. There is one possible outlier at 3.2 pounds and 6.50 hours. This netbook was much heavier than the others, but its battery life was only slightly above average. The residual plot does not show any obvious leftover patterns, so the linear model is appropriate for these data.

From the computer output, the equation of the least-squares regression line is $\widehat{\text{battery}} = -16.05 + 7.94\,(\text{weight})$. The standard deviation of the residuals is $s = 1.62$. This means that our predictions will be typically off by about 1.62 hours when we use the linear model to predict battery life from weight. Finally, 52% of the variation in battery life is accounted for by the linear model relating battery life to the weight of the netbook.

CONCLUDE: Although a linear model is appropriate for these data, our predictions will only be moderately accurate. The predictions will typically be off by over 1.6 hours and nearly half of the variation in battery life is unaccounted for by the linear model.

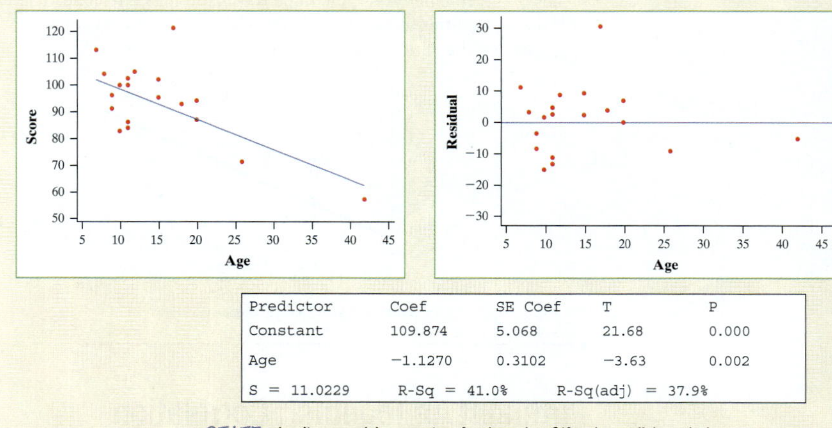


| Predictor | Coef | SE Coef | T | P |
|---|---|---|---|---|
| Constant | 109.874 | 5.068 | 21.68 | 0.000 |
| Age | −1.1270 | 0.3102 | −3.63 | 0.002 |
| S = 11.0229 | R-Sq = 41.0% | R-Sq(adj) = 37.9% | | |

STATE: Is a linear model appropriate for these data? If so, how well does the least-squares regression line fit the data?

PLAN: To determine whether a linear model is appropriate, we will look at the scatterplot and residual plot to see if the association is linear or nonlinear. Then, if a linear model is appropriate, we will use the standard deviation of the residuals and $r^2$ to measure how well the least-squares line fits the data.

DO: The scatterplot shows a moderately strong, negative linear association between age at first word and Gesell score. There are a couple of outliers in the scatterplot. Child 19 has a very high Gesell score for his or her age at first word. Also, child 18 didn't speak his or her first word until much later than the other children in the study and has a much lower Gesell score. The residual plot does not have any obvious patterns, confirming what we saw in the scatterplot—a linear model is appropriate for these data.

From the computer output, the equation of the least-squares regression line is $\hat{y} = 109.874 - 1.1270x$. The standard deviation of the residuals is $s = 11.0229$. This means that our predictions will typically be off by 11.0229 points when we use the linear model to predict Gesell scores from age at first word. Finally, 41% of the variation in Gesell score is accounted for by the linear model relating Gesell score to age at first word.

CONCLUDE: Although a linear model is appropriate for these data, our predictions might not be very accurate. Our typical prediction error is about 11 points, and more than half of the variation in Gesell score is still unaccounted for. Furthermore, we should be hesitant to use this model to make predictions until we understand the effect of the two outliers on the regression results.

**For Practice** *Try Exercise* 67

## Correlation and Regression Wisdom

Correlation and regression are powerful tools for describing the relationship between two variables. When you use these tools, you should be aware of their limitations.

1. *The distinction between explanatory and response variables is important in regression.* Least-squares regression makes the distances of the data points from the line small only in the *y* direction. If we reverse the roles of the two variables, we get a different least-squares regression line. This isn't true for correlation: switching *x* and *y* doesn't affect the value of *r*.

## EXAMPLE Predicting Price, Predicting Miles Driven

### *Two different regression lines*

Figure 3.17(a) repeats the scatterplot of the Ford F-150 data with the least-squares regression line for predicting price from miles driven. We might also use the data on these 16 trucks to predict the number of miles driven from the price of the truck. Now the roles of the variables are reversed: price is the explanatory variable and miles driven is the response variable. Figure 3.17(b) shows a scatterplot of these data with the least-squares regression line for predicting miles driven from price. The two regression lines are very different. The standard deviations of the residuals are different as well. In (a), the standard deviation is $s = 5740$ *dollars*, but in (b) the standard deviation is $s = 28{,}716$ *miles*. However, no matter which variable we put on the *x* axis, the value of $r^2$ is 66.4% and the correlation is $r = -0.815$.

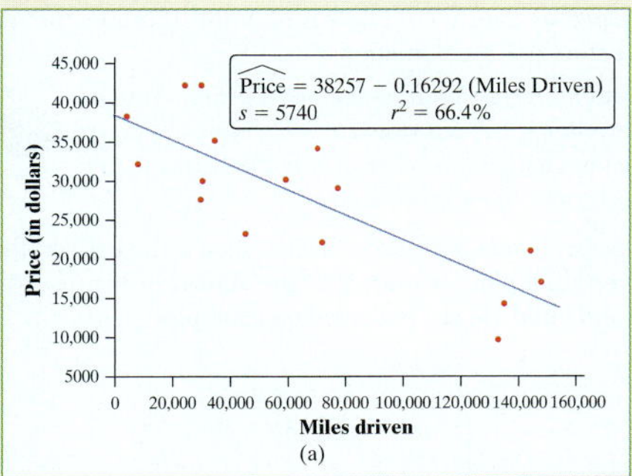


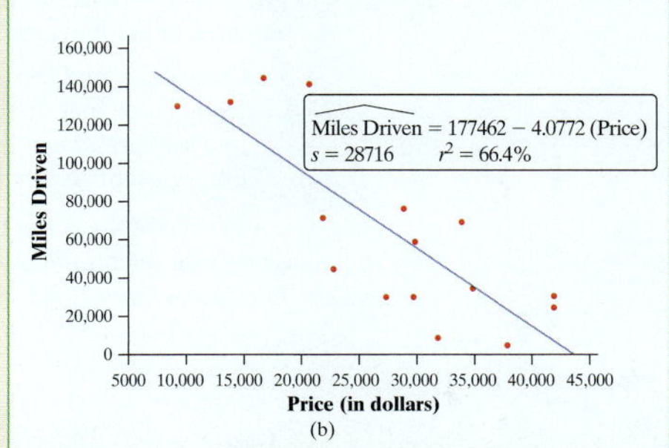


**FIGURE 3.17** (a) Scatterplot with least-squares regression line for predicting price from miles driven. (b) Scatterplot with least-squares regression line for predicting miles driven from price.

2. *Correlation and regression lines describe only linear relationships.* You can calculate the correlation and the least-squares line for any relationship between two quantitative variables, but the results are useful only if the scatterplot shows a linear pattern. *Always plot your data!*

The following four scatterplots show very different associations. Which do you think has the highest correlation?

## ALTERNATE EXAMPLE

### *Back to the track!*

Here is a scatterplot with the least-squares regression line for predicting the long-jump distance from sprint time and a scatterplot with the least-squares regression line for predicting sprint time from long-jump distance.

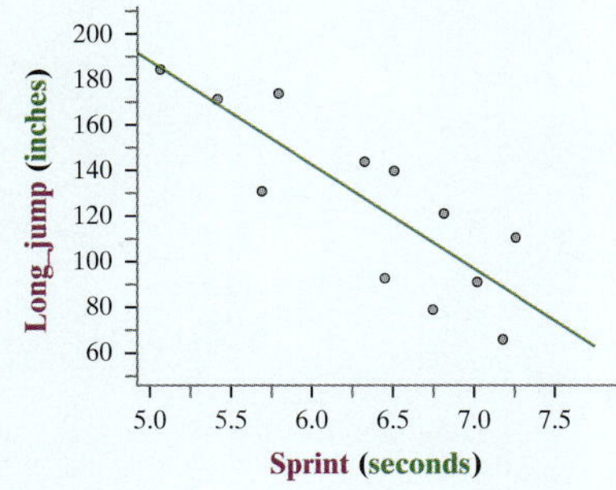


$\hat{y} = 414.79 - 45.74x$, $s = 22.38$, $r^2 = 0.702$.

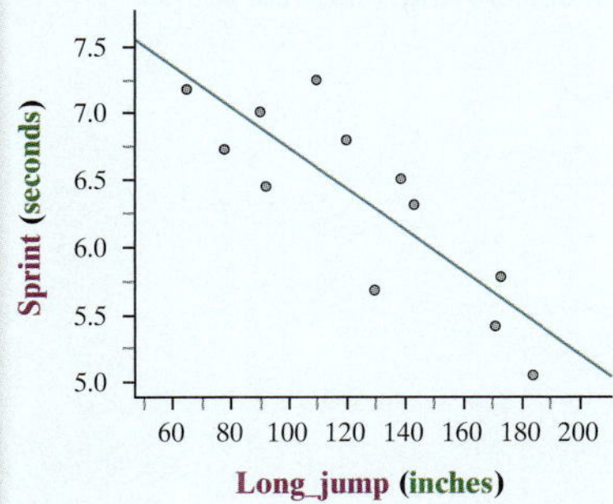


$\hat{y} = 8.257 - 0.0154x$, $s = 0.4099$, $r^2 = 0.702$.

## Teaching Tip

To illustrate that reversing the variables will change the equation of the least-squares regression line, remind students of the formula $b = r\frac{s_y}{s_x}$. Unless the standard deviations are the same, reversing the variables will change the slope.

## Teaching Tip

In the example on this page, ask students to compare the summary statistics for each of the scatterplots. They should notice that the $r^2$ values are the same because reversing the variables doesn't change the correlation. The values of *s* are different, however, because they are measured in the units of the response variable.

## Teaching Tip

Remind students that correlation measures direction and strength, but not form. Knowing that the correlation is $r = 0.816$ doesn't tell us anything about the form of the association, as illustrated in the scatterplots on this page.

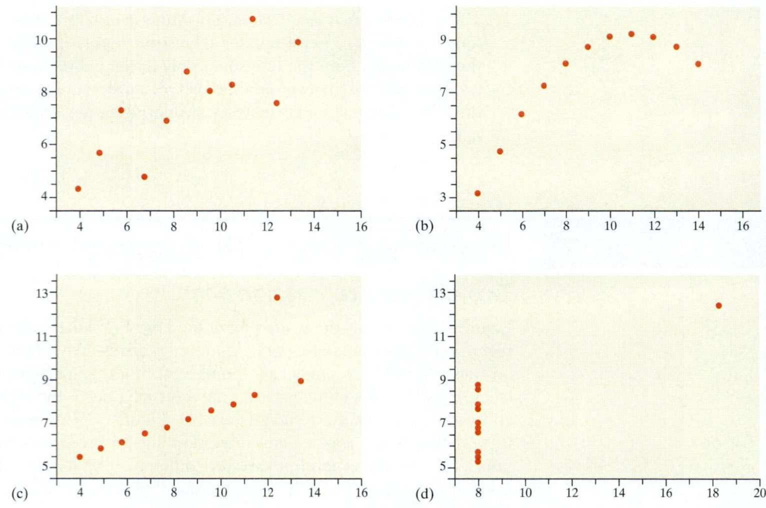

Answer: All four have the same correlation, $r = 0.816$. Furthermore, the least-squares regression line for each association is exactly the same, $\hat{y} = 3 + 0.5x$. These four data sets, developed by statistician Frank Anscombe, illustrate the importance of graphing data before doing calculations.[17]

3. *Correlation and least-squares regression lines are not resistant.* You already know that the correlation $r$ is not resistant. One unusual point in a scatterplot can greatly change the value of $r$. Is the least-squares line resistant? Not surprisingly, the answer is no.

## Teaching Tip

Make sure to look back at the Teaching Tips on pages 152 and 170 for helpful hints about outliers and influential points.

Let's revisit the age at first word and Gesell score data to shed some light on this issue. The scatterplot and residual plot for these data are shown in Figure 3.18. The two outliers, child 18 and child 19, are indicated on each plot.

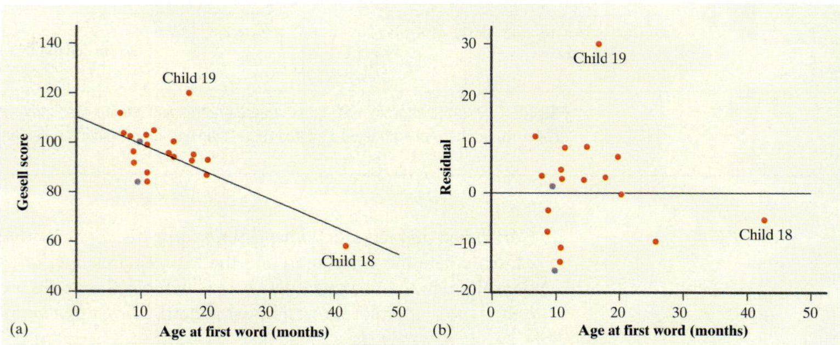


**FIGURE 3.18** (a) Scatterplot of Gesell Adaptive Scores versus the age at first word for 21 children. The line is the least-squares regression line for predicting Gesell score from age at first word. (b) Residual plot for the regression. Child 18 and Child 19 are outliers. Each blue point in the graphs stands for two individuals.

Child 19 has a very large residual because this point lies far from the regression line. However, Child 18 has a fairly small residual. That's because Child 18's point is close to the line. How do these two outliers affect the regression?

Figure 3.19 shows the results of removing each of these points on the correlation and the regression line. The graph adds two more regression lines, one calculated after leaving out Child 18 and the other after leaving out Child 19. You can see that removing the point for Child 18 moves the line quite a bit. (In fact, the equation of the new least-squares line is $\hat{y} = 105.630 - 0.779x$.) Because of Child 18's extreme position on the age scale, this point has a strong *influence* on the position of the regression line. However, removing Child 19 has little effect on the regression line.

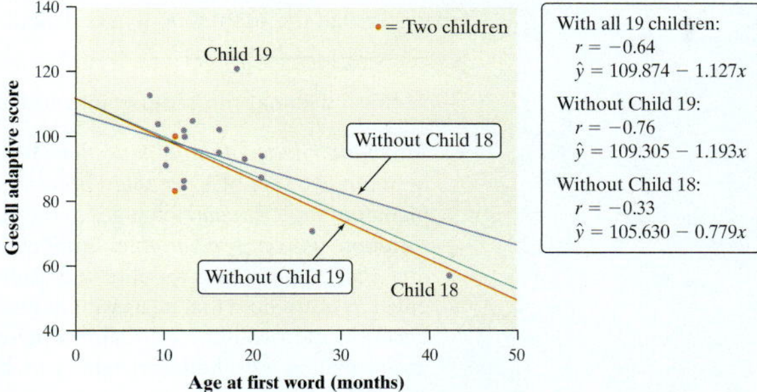


**FIGURE 3.19** Three least-squares regression lines of Gesell score on age at first word. The green line is calculated from all the data. The dark blue line is calculated leaving out Child 18. Child 18 is an influential observation because leaving out this point moves the regression line quite a bit. The red line is calculated leaving out only Child 19.

Least-squares lines make the sum of the squares of the vertical distances to the points as small as possible. A point that is extreme in the $x$ direction with no other points near it pulls the line toward itself. We call such points **influential.**

**DEFINITION: Outliers and influential observations in regression**

An **outlier** is an observation that lies outside the overall pattern of the other observations. Points that are outliers in the $y$ direction but not the $x$ direction of a scatterplot have large residuals. Other outliers may not have large residuals.

An observation is **influential** for a statistical calculation if removing it would markedly change the result of the calculation. Points that are outliers in the $x$ direction of a scatterplot are often influential for the least-squares regression line.

We did not need the distinction between outliers and influential observations in Chapter 1. A single large salary that pulls up the mean salary $\bar{x}$ for a group of workers is an outlier because it lies far above the other salaries. It is also influential, because the mean changes when it is removed. In the regression setting, however, not all outliers are influential. The least-squares line is most likely to be heavily influenced by observations that are outliers in the $x$ direction. The scatterplot will alert you to such observations. Influential points often have small residuals, because they pull the regression line toward themselves. If you look at just a residual plot, you may miss influential points.

The best way to verify that a point is influential is to find the regression line both with and without the unusual point, as in Figure 3.19. If the line moves more than a small amount when the point is deleted, the point is influential.

## Teaching Tip

Remind students that we are deleting individuals only to investigate their influence on the least-squares regression line and other summary measures. In practice, it would be wrong to delete inconvenient observations without a good justification.

## ALTERNATE EXAMPLE

### *More netbooks*

Here is a scatterplot and least-squares regression line for using cost to predict battery life. The equation is $\widehat{\text{battery}} = 6.6 - 0.00923\text{ (cost)}$ with $s = 2.33$ hours and $r^2 = 0.5\%$. According to the least-squares regression line, batteries last longer in cheaper netbooks, which seems counterintuitive. Looking at the scatterplot, we see two obvious influential points in the lower right. When these points are removed, the association becomes positive and the least-squares regression line is $\widehat{\text{battery}} = -8.54 + 0.0421\text{ (cost)}$ with $s = 1.83$ hours and $r^2 = 34\%$. Here is a scatterplot with both lines added.

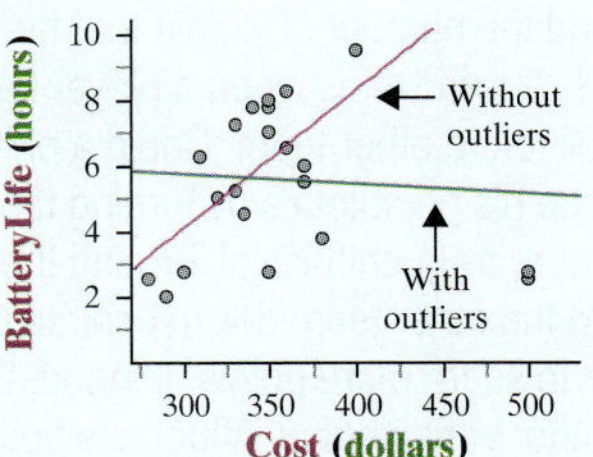


## Teaching Tip

Remind students that there is no rule for determining outliers or influential points in scatterplots. Students should be able to explain what effect any point has on the correlation, slope, $y$ intercept, standard deviation of the residuals, and $r^2$.

## Teaching Tip

Remind students that it is never a good idea to conclude that two variables have a cause-and-effect relationship based only on an association in a scatterplot. See the Teaching Tips on page 156 for some additional hints about this caution. We will learn how to establish cause-and-effect relationships in Chapter 4.

**ALTERNATE EXAMPLE**

### *Does committing more turnovers lead to more points?*

In the National Basketball Association, there is a strong *positive* association between the number of turnovers a player has and the number of points that he scores. A turnover is when a player loses the ball to the other team. Could a player increase his point totals by turning the ball over more frequently? No! Turning the ball over to the other team doesn't *cause* a player to score more points. Instead, there is another variable that influences both turnovers and points: playing time. Players who are on the court more often tend to score more points and have more turnovers than players who don't get much playing time.

THINK ABOUT IT

**How much difference can one point make?** The strong influence of Child 18 makes the original regression of Gesell score on age at first word misleading. The original data have $r^2 = 0.41$. That is, the least-squares line relating age at which a child begins to talk with Gesell score explains 41% of the variation on this later test of mental ability. This relationship is strong enough to be interesting to parents. If we leave out Child 18, $r^2$ drops to only 11%. The apparent strength of the association was largely due to a single influential observation.

What should the child development researcher do? She must decide whether Child 18 is so slow to speak that this individual should not be allowed to influence the analysis. If she excludes Child 18, much of the evidence for a connection between the age at which a child begins to talk and later ability score vanishes. If she keeps Child 18, she needs data on other children who were also slow to begin talking, so that the analysis no longer depends so heavily on just one child.

We finish with our most important caution about correlation and regression.

4. *Association does not imply causation.* When we study the relationship between two variables, we often hope to show that changes in the explanatory variable *cause* changes in the response variable. A *strong association between two variables is not enough to draw conclusions about cause and effect.* Sometimes an observed association really does reflect cause and effect. A household that heats with natural gas uses more gas in colder months because cold weather requires burning more gas to stay warm. In other cases, an association is explained by other variables, and the conclusion that $x$ causes $y$ is not valid.

**EXAMPLE**

## Does Having More Cars Make You Live Longer?

### *Association, not causation*

A serious study once found that people with two cars live longer than people who own only one car.[18] Owning three cars is even better, and so on. There is a substantial positive association between number of cars $x$ and length of life $y$.

The basic meaning of causation is that by changing $x$, we can bring about a change in $y$. Could we lengthen our lives by buying more cars? No. The study used number of cars as a quick indicator of wealth. Well-off people tend to have more cars. They also tend to live longer, probably because they are better educated, take better care of themselves, and get better medical care. The cars have nothing to do with it. There is no cause-and-effect link between number of cars and length of life.

Associations such as those in the previous example are sometimes called "nonsense associations." The association is real. What is nonsense is the conclusion that changing one of the variables causes changes in the other. Another variable—such as personal wealth in this example—that influences both $x$ and $y$ can create a strong association even though there is no direct connection between $x$ and $y$.

Remember: It only makes sense to talk about the *correlation* between two *quantitative* variables. If one or both variables are categorical, you should refer to the *association* between the two variables. To be safe, you can use the more general term "association" when describing the relationship between any two variables.

**ASSOCIATION DOES NOT IMPLY CAUSATION**

An association between an explanatory variable $x$ and a response variable $y$, even if it is very strong, is not by itself good evidence that changes in $x$ actually cause changes in $y$.

Here is a chance to use the skills you have gained to address the question posed at the beginning of the chapter.

case closed

# How Faithful Is Old Faithful?

In the chapter-opening Case Study (page 141), the Starnes family had just missed seeing Old Faithful erupt. They wondered how long it would be until the next eruption. The scatterplot below shows data on the duration (in minutes) and the interval of time until the next eruption (also in minutes) for each Old Faithful eruption in the month before their visit.

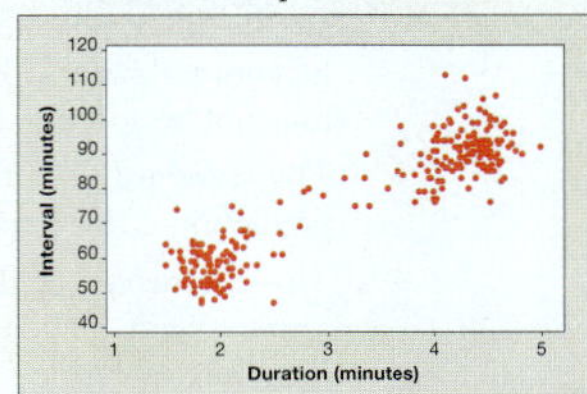


1. Describe the nature of the relationship between interval and duration.

Here is some computer output from a least-squares regression analysis on these data.

**Regression Analysis: Interval versus Duration**

| Predictor | Coef | SE Coef | T | P |
|---|---|---|---|---|
| Constant | 33.347 | 1.201 | 27.76 | 0.000 |
| Duration | 13.2854 | 0.3404 | 39.03 | 0.000 |

S = 6.49336 R-Sq = 85.4% R-Sq(adj) = 85.3%

2. Is a linear model appropriate? Justify your answer.
3. Give the equation of the least-squares regression line. Be sure to define any variables you use.
4. Park rangers indicated that the eruption of Old Faithful that just finished lasted 3.9 minutes. How long do you predict the Starnes family will have to wait for the next eruption? Show how you arrived at your answer.
5. The actual time that the Starnes family has to wait is probably not exactly equal to your prediction in Question 4. Based on the computer output, about how far off do you expect the prediction to be? Explain.

## Answers to Case Closed

1. The scatterplot shows a strong, positive linear association between the duration of the previous eruption and the time until the next eruption. There are two distinct clusters of points, one around $x = 2$ minutes and the other around $x = 4.5$ minutes.

2. Because there are no obvious leftover patterns in the residual plot, a linear model is appropriate for these data.

3. $\hat{y} = 33.347 + 13.2854x$, where $x$ = duration (in minutes) and $y$ = interval (in minutes).

4. $\hat{y} = 33.347 + 13.2854(3.9) =$ 85.16 minutes.

5. Because $s = 6.49336$, predictions using a linear model relating $y$ = interval to $x$ = duration will typically be off by about 6.5 minutes.

**TRM Section 3.2 Quizzes**

There are three different quizzes available for this section in the Teacher's Resource Materials. Versions A and B are roughly parallel. Find them by clicking on the link in the TE-book, opening the TRCD, or logging onto the book's companion Web site www.whfreeman.com/tps5e (teacher log-in required).

**TRM Student Project and Rubric**

If you would like to have your students do a project at the end of this chapter, see the Teacher's Resource Materials for a project description and rubric.



## Section 3.2 Summary

- A **regression line** is a straight line that describes how a response variable $y$ changes as an explanatory variable $x$ changes. You can use a regression line to **predict** the value of $y$ for any value of $x$ by substituting this $x$ into the equation of the line.
- The **slope** $b$ of a regression line $\hat{y} = a + bx$ is the rate at which the predicted response $\hat{y}$ changes along the line as the explanatory variable $x$ changes. Specifically, $b$ is the *predicted* change in $y$ when $x$ increases by 1 unit.
- The **y intercept** $a$ of a regression line $\hat{y} = a + bx$ is the predicted response $\hat{y}$ when the explanatory variable $x$ equals 0. This prediction is of no statistical use unless $x$ can actually take values near 0.
- Avoid **extrapolation,** the use of a regression line for prediction using values of the explanatory variable outside the range of the data from which the line was calculated.
- The most common method of fitting a line to a scatterplot is least squares. The **least-squares regression line** is the straight line $\hat{y} = a + bx$ that minimizes the sum of the squares of the vertical distances of the observed points from the line.
- You can examine the fit of a regression line by studying the **residuals,** which are the differences between the observed and predicted values of $y$. Be on the lookout for patterns in the **residual plot**, which indicate that a linear model may not be appropriate.
- The **standard deviation of the residuals** $s$ measures the typical size of the prediction errors (residuals) when using the regression line.
- The **coefficient of determination** $r^2$ is the fraction of the variation in the response variable that is accounted for by least-squares regression on the explanatory variable.
- The least-squares regression line of $y$ on $x$ is the line with slope $b = r(s_y/s_x)$ and intercept $a = \bar{y} - b\bar{x}$. This line always passes through the point $(\bar{x}, \bar{y})$.
- Correlation and regression must be interpreted with caution. Plot the data to be sure that the relationship is roughly linear and to detect **outliers.** Also look for **influential observations,** individual points that substantially change the correlation or the regression line. Outliers in $x$ are often influential for the regression line.
- Most of all, be careful not to conclude that there is a cause-and-effect relationship between two variables just because they are strongly associated.

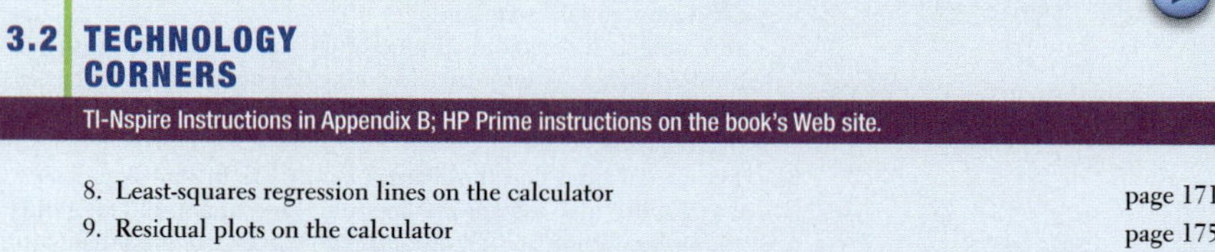

# Section 3.2 Exercises

35. **What's my line?** You use the same bar of soap to shower each morning. The bar weighs 80 grams when it is new. Its weight goes down by 6 grams per day on average. What is the equation of the regression line for predicting weight from days of use?

36. **What's my line?** An eccentric professor believes that a child with IQ 100 should have a reading test score of 50 and predicts that reading score should increase by 1 point for every additional point of IQ. What is the equation of the professor's regression line for predicting reading score from IQ?

37. **Gas mileage** We expect a car's highway gas mileage to be related to its city gas mileage. Data for all 1198 vehicles in the government's recent *Fuel Economy Guide* give the regression line: predicted highway mpg = 4.62 + 1.109 (city mpg).
    (a) What's the slope of this line? Interpret this value in context.
    (b) What's the $y$ intercept? Explain why the value of the intercept is not statistically meaningful.
    (c) Find the predicted highway mileage for a car that gets 16 miles per gallon in the city.

38. **IQ and reading scores** Data on the IQ test scores and reading test scores for a group of fifth-grade children give the following regression line: predicted reading score = −33.4 + 0.882(IQ score).
    (a) What's the slope of this line? Interpret this value in context.
    (b) What's the $y$ intercept? Explain why the value of the intercept is not statistically meaningful.
    (c) Find the predicted reading score for a child with an IQ score of 90.

39. **Acid rain** Researchers studying acid rain measured the acidity of precipitation in a Colorado wilderness area for 150 consecutive weeks. Acidity is measured by pH. Lower pH values show higher acidity. The researchers observed a linear pattern over time. They reported that the regression line $\widehat{pH} = 5.43 - 0.0053(\text{weeks})$ fit the data well.[19]

pg 166

    (a) Identify the slope of the line and explain what it means in this setting.
    (b) Identify the $y$ intercept of the line and explain what it means in this setting.
    (c) According to the regression line, what was the pH at the end of this study?

40. **How much gas?** In Exercise 4 (page 159), we examined the relationship between the average monthly temperature and the amount of natural gas consumed in Joan's midwestern home. The figure below shows the original scatterplot with the least-squares line added. The equation of the least-squares line is $\hat{y} = 1425 - 19.87x$.

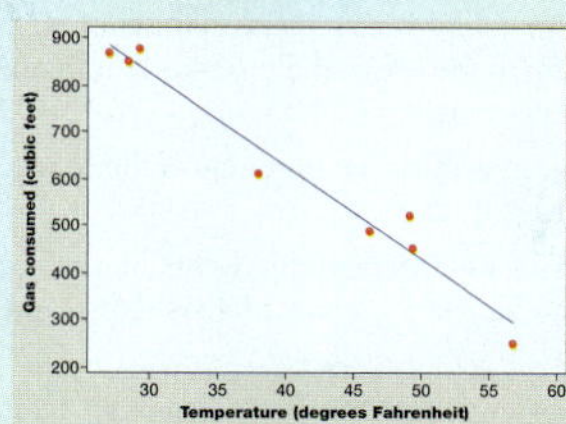


    (a) Identify the slope of the line and explain what it means in this setting.
    (b) Identify the $y$ intercept of the line. Explain why it's risky to use this value as a prediction.
    (c) Use the regression line to predict the amount of natural gas Joan will use in a month with an average temperature of 30°F.

41. **Acid rain** Refer to Exercise 39. Would it be appropriate to use the regression line to predict pH after 1000 months? Justify your answer.

42. **How much gas?** Refer to Exercise 40. Would it be appropriate to use the regression line to predict Joan's natural-gas consumption in a future month with an average temperature of 65°F? Justify your answer.

43. **Least-squares idea** The table below gives a small set of data. Which of the following two lines fits the data better: $\hat{y} = 1 - x$ or $\hat{y} = 3 - 2x$? Use the least-squares criterion to justify your answer. (*Note:* Neither of these two lines is the least-squares regression line for these data.)

| $x$: | −1 | 1 | 1 | 3 | 5 |
|---|---|---|---|---|---|
| $y$: | 2 | 0 | 1 | −1 | −5 |

44. **Least-squares idea** In Exercise 40, the line drawn on the scatterplot is the least-squares regression line. Explain the meaning of the phrase "least-squares" to Joan, who knows very little about statistics.

45. **Acid rain** In the acid rain study of Exercise 39, the actual pH measurement for Week 50 was 5.08. Find and interpret the residual for this week.

pg 169

46. **How much gas?** Refer to Exercise 40. During March, the average temperature was 46.4°F and Joan used 490 cubic feet of gas per day. Find and interpret the residual for this month.

**TRM Full Solutions to Section 3.2 Exercises**

## Answers to Section 3.2 Exercises

**3.35** predicted weight = 80 − 6 (days)

**3.36** predicted reading score = −50 + IQ

**3.37** **(a)** 1.109. For each 1-mpg increase in city mileage, the predicted highway mileage will increase by 1.109 mpg. **(b)** 4.62 mpg. This would represent the highway mileage for a car that gets 0 mpg in the city, which is impossible. **(c)** 22.36 mpg.

**3.38** **(a)** 0.882. For each 1-point increase in IQ, the predicted reading score will increase by 0.882. **(b)** −33.4. This represents the predicted reading score for a child with an IQ of 0, which is impossible. **(c)** 45.98.

**3.39** **(a)** −0.0053. For each additional week in the study, the predicted pH decreased by 0.0053 units. **(b)** 5.43. The predicted pH level at the beginning of the study (weeks = 0) is 5.43. **(c)** 4.635.

**3.40** **(a)** −19.87. For each additional 1-degree increase in the average monthly temperature, the predicted amount of gas consumed in Joan's home decreases by 19.87 cubic feet. **(b)** 1425. When the average monthly temperature is 0°F, the predicted gas consumption for Joan's home is 1425 cubic feet. **(c)** 828.9 cubic feet.

**3.41** No. 1000 months is well outside the observed time period and we can't be sure that the linear relationship continues after 150 weeks.

**3.42** No. 65°F is outside the interval of temperatures used to create the model and we can't be sure that the linear relationship continues above 57°F.

**3.43** The line $\hat{y} = 1 - x$ is a much better fit. The sum of squared residuals for this line is only 3, while the sum of squared residuals for $\hat{y} = 3 - 2x$ is 18.

**3.44** The least-squares line is the line that makes the sum of the squared prediction errors as small as possible.

**3.45** residual = 5.08 − 5.165 = −0.085. The actual pH value for that week was 0.085 less than predicted.

**3.46** residual = 490 − 503.032 = −13.032. The actual amount of gas consumed was 13.032 cubic feet less than predicted in the month of March.

47. **Bird colonies** Exercise 6 (page 159) examined the relationship between the number of new birds $y$ and percent of returning birds $x$ for 13 sparrowhawk colonies. Here are the data once again.

| Percent return: | 74 | 66 | 81 | 52 | 73 | 62 | 52 | 45 | 62 | 46 | 60 | 46 | 38 |
|---|---|---|---|---|---|---|---|---|---|---|---|---|---|
| New adults: | 5 | 6 | 8 | 11 | 12 | 15 | 16 | 17 | 18 | 18 | 19 | 20 | 20 |

(a) Use your calculator to help make a scatterplot.

(b) Use your calculator's regression function to find the equation of the least-squares regression line. Add this line to your scatterplot from (a).

(c) Explain in words what the slope of the regression line tells us.

(d) Calculate and interpret the residual for the colony that had 52% of the sparrowhawks return and 11 new adults.

48. **Do heavier people burn more energy?** Exercise 10 (page 160) presented data on the lean body mass and resting metabolic rate for 12 women who were subjects in a study of dieting. Lean body mass, given in kilograms, is a person's weight leaving out all fat. Metabolic rate, in calories burned per 24 hours, is the rate at which the body consumes energy. Here are the data again.

| Mass: | 36.1 | 54.6 | 48.5 | 42.0 | 50.6 | 42.0 | 40.3 | 33.1 | 42.4 | 34.5 | 51.1 | 41.2 |
|---|---|---|---|---|---|---|---|---|---|---|---|---|
| Rate: | 995 | 1425 | 1396 | 1418 | 1502 | 1256 | 1189 | 913 | 1124 | 1052 | 1347 | 1204 |

(a) Use your calculator to help make a scatterplot.

(b) Use your calculator's regression function to find the equation of the least-squares regression line. Add this line to your scatterplot from part (a).

(c) Explain in words what the slope of the regression line tells us.

(d) Calculate and interpret the residual for the woman who had a lean body mass of 50.6 kg and a metabolic rate of 1502.

49. **Bird colonies** Refer to Exercise 47.

(a) Use your calculator to make a residual plot. Describe what this graph tells you about the appropriateness of using a linear model.

(b) Which point has the largest residual? Explain what this residual means in context.

50. **Do heavier people burn more energy?** Refer to Exercise 48.

(a) Use your calculator to make a residual plot. Describe what this graph tells you about the appropriateness of using a linear model.

(b) Which point has the largest residual? Explain what the value of that residual means in context.

51. **Nahya infant weights** A study of nutrition in developing countries collected data from the Egyptian village of Nahya. Here are the mean weights (in kilograms) for 170 infants in Nahya who were weighed each month during their first year of life:

| Age (months): | 1 | 2 | 3 | 4 | 5 | 6 | 7 | 8 | 9 | 10 | 11 | 12 |
|---|---|---|---|---|---|---|---|---|---|---|---|---|
| Weight (kg): | 4.3 | 5.1 | 5.7 | 6.3 | 6.8 | 7.1 | 7.2 | 7.2 | 7.2 | 7.2 | 7.5 | 7.8 |

A hasty user of statistics enters the data into software and computes the least-squares line without plotting the data. The result is $\widehat{\text{weight}} = 4.88 + 0.267\,(\text{age})$. A residual plot is shown below. Would it be appropriate to use this regression line to predict $y$ from $x$? Justify your answer.

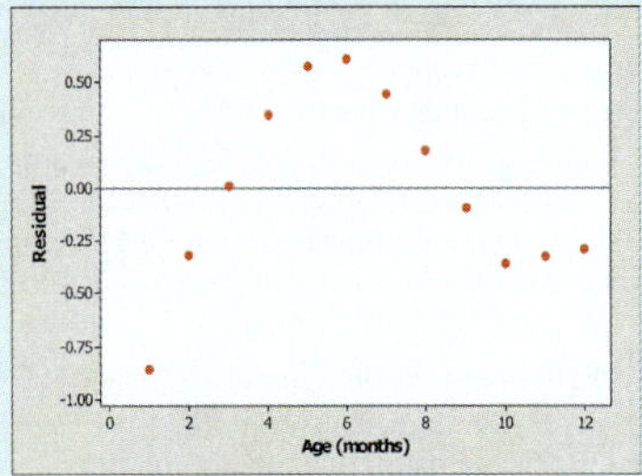


52. **Driving speed and fuel consumption** Exercise 9 (page 160) gives data on the fuel consumption $y$ of a car at various speeds $x$. Fuel consumption is measured in liters of gasoline per 100 kilometers driven and speed is measured in kilometers per hour. A statistical software package gives the least-squares regression line and the residual plot shown below. The regression line is $\hat{y} = 11.058 - 0.01466x$. Would it be appropriate to use the regression line to predict $y$ from $x$? Justify your answer.

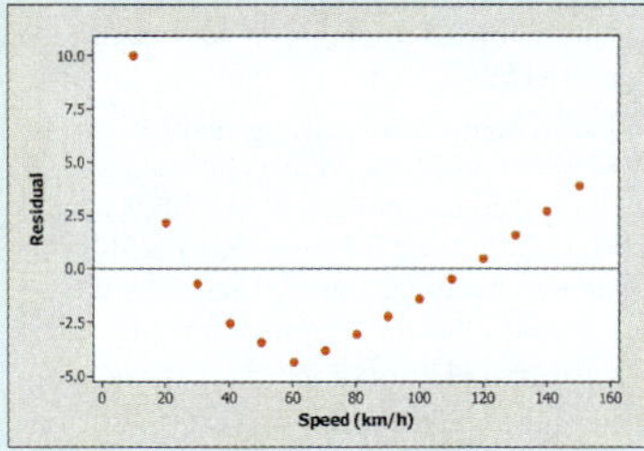


53. **Oil and residuals** The Trans-Alaska Oil Pipeline is a tube that is formed from 1/2-inch-thick steel and that carries oil across 800 miles of sensitive arctic and subarctic terrain. The pipe segments and the welds that join them were carefully examined before installation. How accurate are field measurements of the depth of small defects? The figure below compares the results of measurements on 100 defects made in the field with

**3.47** **(a)** The scatterplot (with regression line) is shown below. **(b)** $\hat{y} = 31.9 - 0.304x$. **(c)** For each increase of 1 in the percent of returning birds, the predicted number of new adult birds will decrease by 0.304. **(d)** residual $= 11 - 16.092 = -5.092$. In this colony, there were 5.092 fewer new adults than expected based on the percent of returning birds.

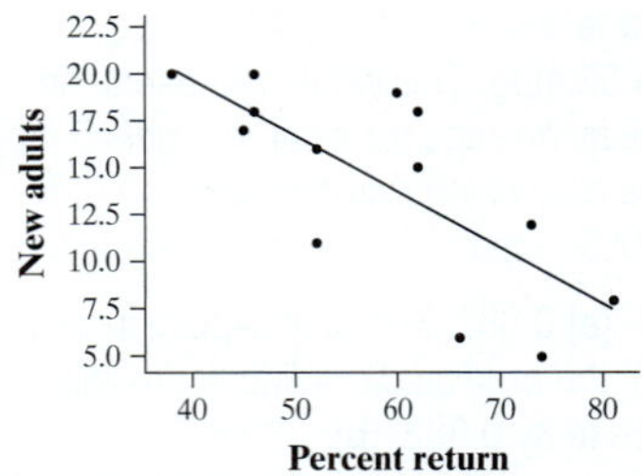


**3.48** **(a)** The scatterplot (with regression line) is shown below. **(b)** $\hat{y} = 201.2 + 24.026x$. **(c)** For each additional kilogram of body mass, the predicted metabolic rate increases by about 24 cal/day. **(d)** residual $= 1502 - 1416.9 = 85.1$. This woman's actual metabolic rate was 85.1 cal/day higher than predicted based on her lean body mass.

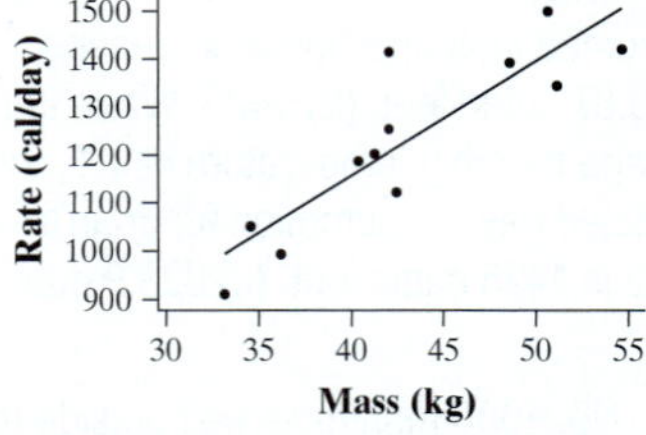


**3.49** **(a)** Because there is no obvious leftover pattern in the residual plot, a line is an appropriate model to use for these data. **(b)** The point with the largest residual (66% returning) has a residual of about $-6$. This means that the colony with 66% returning birds has about 6 fewer new adults than predicted based on the percent returning.

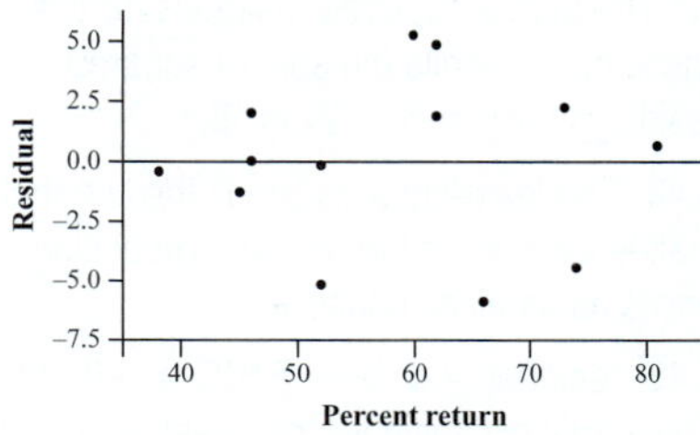


**3.50** **(a)** Because there is no obvious leftover pattern in the residual plot, a line is an appropriate model to use for these data. **(b)** The point with the largest residual (mass = 42) has a residual of about 200. This means that the person with a lean body mass of 42 kg has a metabolic rate that is about 200 cal/day higher than predicted based on the person's lean body mass.

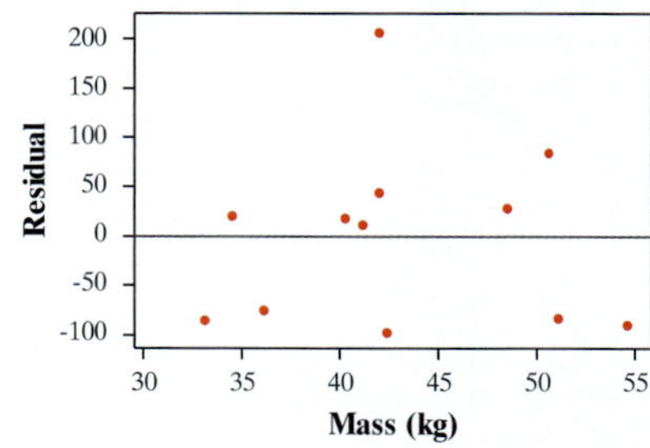


**3.51** No. Because there is an obvious negative-positive-negative pattern in the residual plot, a linear model is not appropriate for these data.

**3.52** No. Because there is an obvious positive-negative-positive pattern in the residual plot, a linear model is not appropriate for these data.

measurements of the same defects made in the laboratory.[20] The line $y = x$ is drawn on the scatterplot.

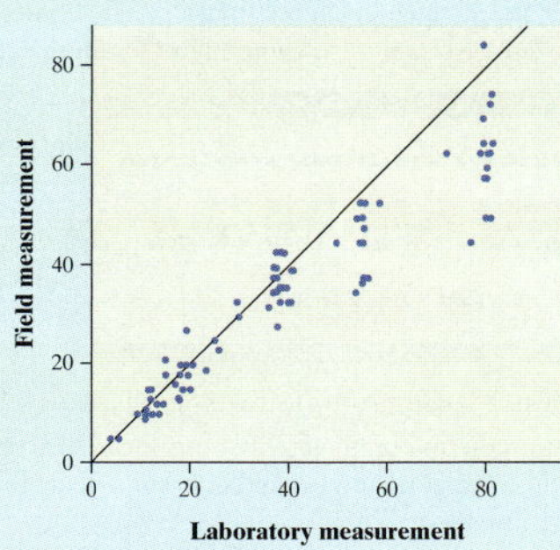


(a) Describe the overall pattern you see in the scatterplot, as well as any deviations from that pattern.

(b) If field and laboratory measurements all agree, then the points should fall on the $y = x$ line drawn on the plot, except for small variations in the measurements. Is this the case? Explain.

(c) The line drawn on the scatterplot ($y = x$) is *not* the least-squares regression line. How would the slope and $y$ intercept of the least-squares line compare? Justify your answer.

**54. Oil and residuals** Refer to Exercise 53. The following figure shows a residual plot for the least-squares regression line. Discuss what the residual plot tells you about the appropriateness of using a linear model.

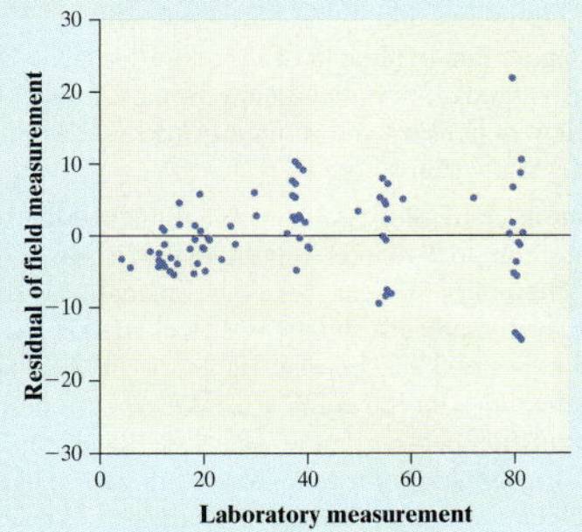


**55. Olympic figure skating** pg 180 For many people, the women's figure skating competition is the highlight of the Olympic Winter Games. Scores in the short program $x$ and scores in the free skate $y$ were recorded for each of the 24 skaters who competed in both rounds during the 2010 Winter Olympics in Vancouver, Canada.[21] A regression analysis was performed using these data. The scatterplot and residual plot follow. The equation of the least-squares regression line is $\hat{y} = -16.2 + 2.07x$. Also, $s = 10.2$ and $r^2 = 0.736$.

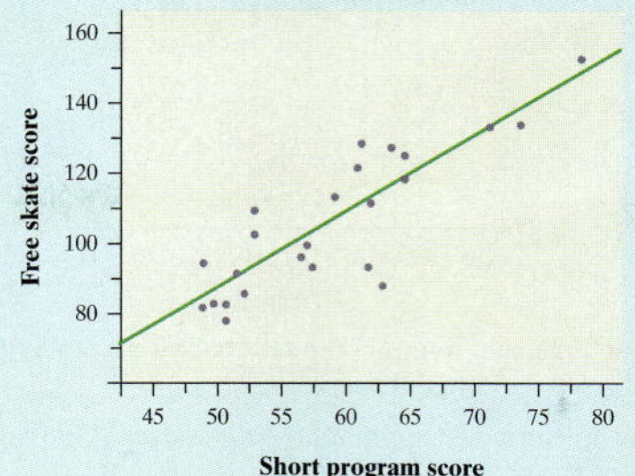


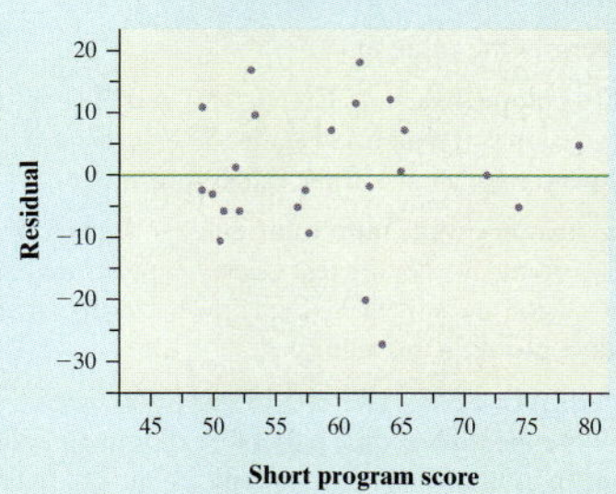


(a) Calculate and interpret the residual for the gold medal winner, Yu-Na Kim, who scored 78.50 in the short program and 150.06 in the free skate.

(b) Is a linear model appropriate for these data? Explain.

(c) Interpret the value of $s$.

(d) Interpret the value of $r^2$.

**56. Age and height** A random sample of 195 students was selected from the United Kingdom using the CensusAtSchool data selector. The age (in years) $x$ and height (in centimeters) $y$ was recorded for each of the students. A regression analysis was performed using these data. The scatterplot and residual plot are shown below. The equation of the least-squares regression line is $\hat{y} = 106.1 + 4.21x$. Also, $s = 8.61$ and $r^2 = 0.274$.

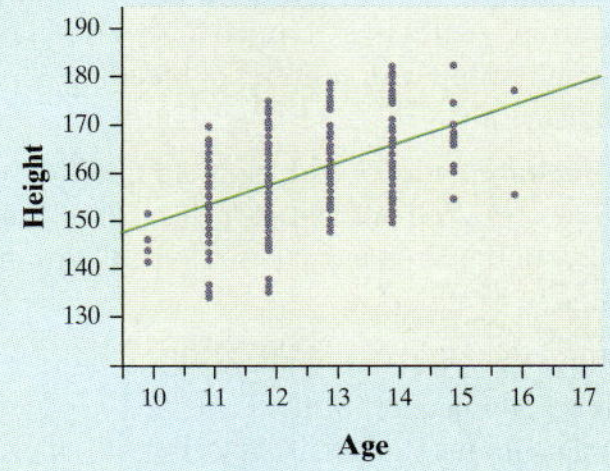


**3.53 (a)** There is a positive, linear association between the two variables. There is more variation in the field measurements for larger laboratory measurements. **(b)** No. The points for the larger depths fall systematically below the line $y = x$, showing that the field measurements are too small compared to the laboratory measurements. **(c)** The slope would be closer to 0 and the $y$ intercept would be larger.

**3.54** Because there is no obvious curved pattern leftover in the residual plot, a linear model is appropriate for these data. However, predictions using the line will be less accurate for larger measurements than for smaller measurements.

**3.55 (a)** residual $= 150.06 - 146.295 = 3.765$. Yu-Na Kim's free skate score was 3.765 points higher than predicted based on her short program score. **(b)** Because there is no leftover pattern in the residual plot, a linear model is appropriate for these data. **(c)** When using the least-squares regression line with $x =$ short program score to predict $y =$ free skate score, we will typically be off by about 10.2 points. **(d)** About 73.6% of the variation in free skate scores is accounted for by the linear model relating free skate scores to short program scores.

(a) Calculate and interpret the residual for the student who was 141 cm tall at age 10.

(b) Is a linear model appropriate for these data? Explain.

(c) Interpret the value of $s$.

(d) Interpret the value of $r^2$.

57. **Bird colonies** Refer to Exercises 47 and 49. For the regression you performed earlier, $r^2 = 0.56$ and $s = 3.67$. Explain what each of these values means in this setting.

58. **Do heavier people burn more energy?** Refer to Exercises 48 and 50. For the regression you performed earlier, $r^2 = 0.768$ and $s = 95.08$. Explain what each of these values means in this setting.

59. **Merlins breeding** Exercise 13 (page 160) gives data on the number of breeding pairs of merlins in an isolated area in each of seven years and the percent of males who returned the next year. The data show that the percent returning is lower after successful breeding seasons and that the relationship is roughly linear. The figure below shows Minitab regression output for these data.

pg 181

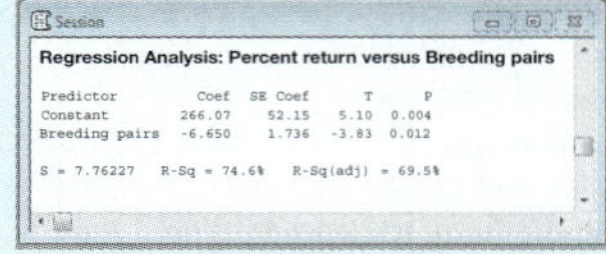
Regression Analysis: Percent return versus Breeding pairs

| Predictor | Coef | SE Coef | T | P |
|---|---|---|---|---|
| Constant | 266.07 | 52.15 | 5.10 | 0.004 |
| Breeding pairs | -6.650 | 1.736 | -3.83 | 0.012 |

S = 7.76227 R-Sq = 74.6% R-Sq(adj) = 69.5%

(a) What is the equation of the least-squares regression line for predicting the percent of males that return from the number of breeding pairs? Use the equation to predict the percent of returning males after a season with 30 breeding pairs.

(b) What percent of the year-to-year variation in percent of returning males is accounted for by the straight-line relationship with number of breeding pairs the previous year?

(c) Use the information in the figure to find the correlation $r$ between percent of males that return and number of breeding pairs. How do you know whether the sign of $r$ is + or −?

(d) Interpret the value of $s$ in this setting.

60. **Does social rejection hurt?** Exercise 14 (page 161) gives data from a study that shows that social exclusion causes "real pain." That is, activity in an area of the brain that responds to physical pain goes up as distress from social exclusion goes up. A scatterplot shows a moderately strong, linear relationship. The figure below shows Minitab regression output for these data.

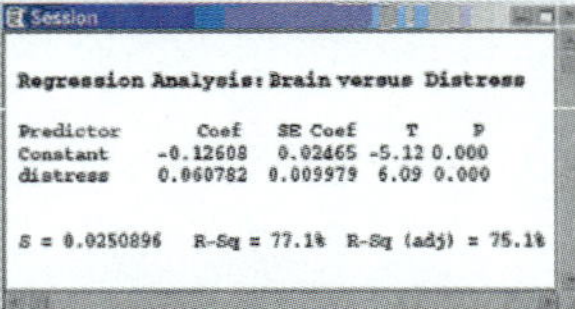
Regression Analysis: Brain versus Distress

| Predictor | Coef | SE Coef | T | P |
|---|---|---|---|---|
| Constant | -0.12608 | 0.02465 | -5.12 | 0.000 |
| distress | 0.060782 | 0.009979 | 6.09 | 0.000 |

S = 0.0250896 R-Sq = 77.1% R-Sq (adj) = 75.1%

(a) What is the equation of the least-squares regression line for predicting brain activity from social distress score? Use the equation to predict brain activity for social distress score 2.0.

(b) What percent of the variation in brain activity among these subjects is accounted for by the straight-line relationship with social distress score?

(c) Use the information in the figure to find the correlation $r$ between social distress score and brain activity. How do you know whether the sign of $r$ is + or −?

(d) Interpret the value of $s$ in this setting.

61. **Husbands and wives** The mean height of married American women in their early twenties is 64.5 inches and the standard deviation is 2.5 inches. The mean height of married men the same age is 68.5 inches, with standard deviation 2.7 inches. The correlation between the heights of husbands and wives is about $r = 0.5$.

pg 183

(a) Find the equation of the least-squares regression line for predicting a husband's height from his wife's height for married couples in their early 20s. Show your work.

(b) Suppose that the height of a randomly selected wife was 1 standard deviation below average. Predict the height of her husband *without* using the least-squares line. Show your work.

62. **The stock market** Some people think that the behavior of the stock market in January predicts its behavior for the rest of the year. Take the explanatory variable $x$ to be the percent change in a stock market index in January and the response variable $y$ to be the change in the index for the entire year. We expect a positive correlation between $x$ and $y$ because the change during January contributes to the full year's change. Calculation from data for an 18-year period gives

$$\bar{x} = 1.75\% \quad s_x = 5.36\% \quad \bar{y} = 9.07\%$$
$$s_y = 15.35\% \quad r = 0.596$$

(a) Find the equation of the least-squares line for predicting full-year change from January change. Show your work.

(b) Suppose that the percent change in a particular January was 2 standard deviations above average. Predict the percent change for the entire year, *without* using the least-squares line. Show your work.

**3.56** **(a)** residual = 141 − 148.2 = −7.2. This student's height was 7.2 cm less than predicted based on the student's age. **(b)** Because there is no leftover pattern in the residual plot, a linear model is appropriate for these data. **(c)** When using the least-squares regression line with $x$ = age to predict $y$ = height, we will typically be off by about 8.61 cm. **(d)** About 27.4% of the variation in height is accounted for by the linear model relating height to age.

**3.57** $r^2$: About 56% of the variation in the number of new adults is accounted for by the linear model relating number of new adults to the percent returning. $s$: When using the least-squares regression line with $x$ = percent returning to predict $y$ = number of new adults, we will typically be off by 3.67 adults.

**3.58** $r^2$: About 76.8% of the variation in the metabolic rate is accounted for by the linear model relating metabolic rate to lean body mass. $s$: When using the least-squares regression line with $x$ = lean body mass to predict $y$ = metabolic rate, we will typically be off by 95.08 cal/day.

**3.59** **(a)** $\hat{y} = 266.07 - 6.650x$, where $y$ = percent of males that return the next year and $x$ = number of breeding pairs. When $x = 30$, $\hat{y} = 66.57$. **(b)** R-Sq = 74.6% **(c)** $r = -\sqrt{0.746} = -0.864$. The sign is negative because the slope is negative. **(d)** When using the least-squares regression line with $x$ = number of breeding pairs to predict $y$ = percent returning, we will typically be off by 7.76%.

**3.60** **(a)** $\hat{y} = -0.126 + 0.0608x$, where $y$ = brain activity and $x$ = social distress score. When $x = 2$, $\hat{y} = -0.0044$. **(b)** R-sq = 77.1% **(c)** $r = +\sqrt{r^2} = 0.88$. The sign is positive because the slope is positive. **(d)** When using the least-squares regression line with $x$ = social distress score to predict $y$ = brain activity, we will typically be off by 0.0251.

**3.61** **(a)** $\hat{y} = 33.67 + 0.54x$ **(b)** If the value of $x$ is 1 standard deviation below $\bar{x}$, the predicted value of $y$ will be $r$ standard deviations of $y$ below $\bar{y}$. So, the predicted value for the husband is 68.5 − 0.5(2.7) = 67.15 inches.

**3.62** **(a)** $\hat{y} = 6.083 + 1.707x$ **(b)** If the value of $x$ is 2 standard deviations above $\bar{x}$, the predicted value of $y$ will be $2r$ standard deviations of $y$ above $\bar{y}$. So, the predicted value for the percent change for the entire year is 9.07 + 2(0.596)(15.35) = 27.4%.

**3.63** **(a)** $r^2 = 0.25$. About 25% of the variation in husbands' heights is accounted for by the linear model relating husband's height to wife's height. **(b)** When using the least-squares regression line with $x$ = wife's height to predict $y$ = husband's height, we will typically be off by 1.2 inches.

**3.64** **(a)** $r^2 = 0.3552$. About 35.52% of the variation in the percent change for the entire year is accounted for by the linear model relating the percent change for the entire year to the percent change in January. **(b)** When using the least-squares regression line with $x$ = percent change in January to predict $y$ = percent change for the entire year, we will typically be off by 8.3%.

**3.65** **(a)** $\hat{y} = x$, where $y$ = final and $x$ = midterm. **(b)** If $x = 50$, $\hat{y} = 67.1$. If $x = 100$, $\hat{y} = 87.6$. **(c)** The student who did poorly on the midterm (50) is predicted to do better on the final (closer to the mean), while the student who did very well on the midterm (100) is predicted to do worse on the final (closer to the mean).

**3.66** **(a)** $\hat{y} = x$ **(b)** If $x = 0.200$, $\hat{y} = 0.267$. If $x = 0.400$, $\hat{y} = 0.289$. **(c)** The player who hit poorly in the first month (0.200) is predicted to hit better the rest of the season (closer to the mean), while the player who hit very well in the first month (0.400) is predicted to do worse (closer to the mean).

63. **Husbands and wives** Refer to Exercise 61.

(a) Find $r^2$ and interpret this value in context.

(b) For these data, $s = 1.2$. Interpret this value.

64. **The stock market** Refer to Exercise 62.

(a) Find $r^2$ and interpret this value in context.

(b) For these data, $s = 8.3$. Interpret this value.

65. **Will I bomb the final?** We expect that students who do well on the midterm exam in a course will usually also do well on the final exam. Gary Smith of Pomona College looked at the exam scores of all 346 students who took his statistics class over a 10-year period.[22] Assume that both the midterm and final exam were scored out of 100 points.

(a) State the equation of the least-squares regression line if each student scored the same on the midterm and the final.

(b) The actual least-squares line for predicting final-exam score $y$ from midterm-exam score $x$ was $\hat{y} = 46.6 + 0.41x$. Predict the score of a student who scored 50 on the midterm and a student who scored 100 on the midterm.

(c) Explain how your answers to part (b) illustrate regression to the mean.

66. **It's still early** We expect that a baseball player who has a high batting average in the first month of the season will also have a high batting average the rest of the season. Using 66 Major League Baseball players from the 2010 season,[23] a least-squares regression line was calculated to predict rest-of-season batting average $y$ from first-month batting average $x$. *Note:* A player's batting average is the proportion of times at bat that he gets a hit. A batting average over 0.300 is considered very good in Major League Baseball.

(a) State the equation of the least-squares regression line if each player had the same batting average the rest of the season as he did in the first month of the season.

(b) The actual equation of the least-squares regression line is $\hat{y} = 0.245 + 0.109x$. Predict the rest-of-season batting average for a player who had a 0.200 batting average the first month of the season and for a player who had a 0.400 batting average the first month of the season.

(c) Explain how your answers to part (b) illustrate regression to the mean.

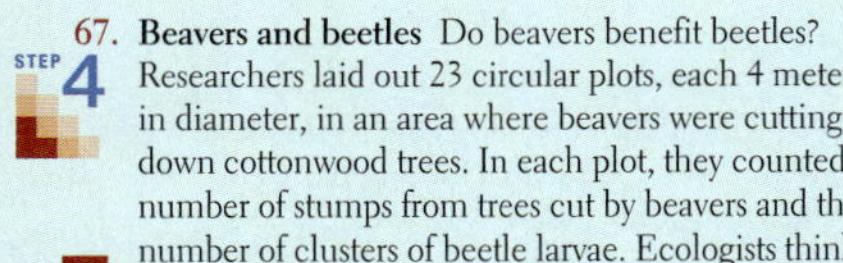

67. **Beavers and beetles** Do beavers benefit beetles? Researchers laid out 23 circular plots, each 4 meters in diameter, in an area where beavers were cutting down cottonwood trees. In each plot, they counted the number of stumps from trees cut by beavers and the number of clusters of beetle larvae. Ecologists think that the new sprouts from stumps are more tender than other cottonwood growth, so that beetles prefer them. If so, more stumps should produce more beetle larvae. Here are the data:[24]

pg 185

| Stumps: | 2 | 2 | 1 | 3 | 3 | 4 | 3 | 1 | 2 | 5 | 1 | 3 |
|---|---|---|---|---|---|---|---|---|---|---|---|---|
| Beetle larvae: | 10 | 30 | 12 | 24 | 36 | 40 | 43 | 11 | 27 | 56 | 18 | 40 |
| Stumps: | 2 | 1 | 2 | 2 | 1 | 1 | 4 | 1 | 2 | 1 | 4 | |
| Beetle larvae: | 25 | 8 | 21 | 14 | 16 | 6 | 54 | 9 | 13 | 14 | 50 | |

Can we use a linear model to predict the number of beetle larvae from the number of stumps? If so, how accurate will our predictions be? Follow the four-step process.

68. **Fat and calories** The number of calories in a food item depends on many factors, including the amount of fat in the item. The data below show the amount of fat (in grams) and the number of calories in 7 beef sandwiches at McDonalds.[25]

| Sandwich | Fat | Calories |
|---|---|---|
| Big Mac® | 29 | 550 |
| Quarter Pounder® with Cheese | 26 | 520 |
| Double Quarter Pounder® with Cheese | 42 | 750 |
| Hamburger | 9 | 250 |
| Cheeseburger | 12 | 300 |
| Double Cheeseburger | 23 | 440 |
| McDouble | 19 | 390 |

Can we use a linear model to predict the number of calories from the amount of fat? If so, how accurate will our predictions be? Follow the four-step process.

69. **Managing diabetes** People with diabetes measure their fasting plasma glucose (FPG; measured in units of milligrams per milliliter) after fasting for at least 8 hours. Another measurement, made at regular medical checkups, is called HbA. This is roughly the percent of red blood cells that have a glucose molecule attached. It measures average exposure to glucose over a period of several months. The table below gives data on both HbA and FPG for 18 diabetics five months after they had completed a diabetes education class.[27]

| Subject | HbA (%) | FPG (mg/mL) | Subject | HbA (%) | FPG (mg/mL) |
|---|---|---|---|---|---|
| 1 | 6.1 | 141 | 10 | 8.7 | 172 |
| 2 | 6.3 | 158 | 11 | 9.4 | 200 |
| 3 | 6.4 | 112 | 12 | 10.4 | 271 |
| 4 | 6.8 | 153 | 13 | 10.6 | 103 |
| 5 | 7.0 | 134 | 14 | 10.7 | 172 |
| 6 | 7.1 | 95 | 15 | 10.7 | 359 |
| 7 | 7.5 | 96 | 16 | 11.2 | 145 |
| 8 | 7.7 | 78 | 17 | 13.7 | 147 |
| 9 | 7.9 | 148 | 18 | 19.3 | 255 |

**3.67** STATE: Is a linear model appropriate for these data? If so, how well does the least-squares regression line fit the data? PLAN: We will look at the scatterplot and residual plot to see if the association is linear or nonlinear. Then if a linear model is appropriate, we will use $s$ and $r^2$ to measure how well the line fits the data. DO: The scatterplot shows a moderately strong, positive linear association between the number of stumps and the number of clusters of beetle larvae. The residual plot doesn't show any obvious leftover pattern, confirming that a linear model is appropriate.

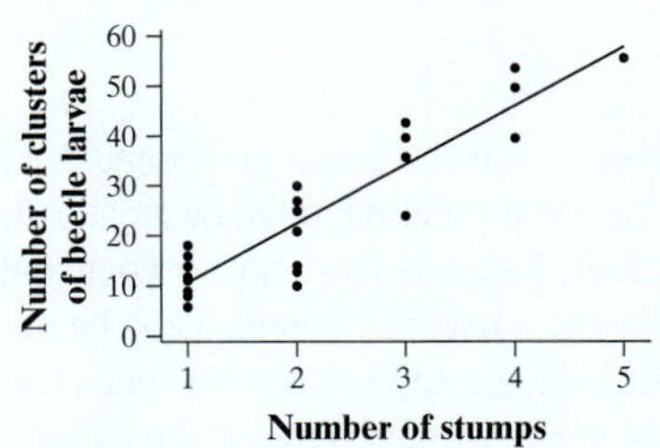


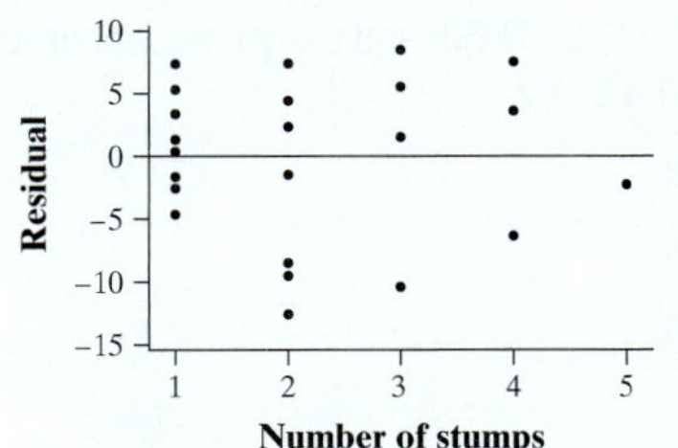


$\hat{y} = -1.29 + 11.89x$, where $y$ = number of clusters of beetle larvae and $x$ = number of stumps. $s = 6.42$, meaning that our predictions will typically be off by about 6.42 clusters when we use the line to predict the number of clusters of beetle larvae from the number of stumps. Finally, $r^2 = 0.839$, meaning 83.9% of the variation in the number of clusters of beetle larvae is accounted for by the linear model relating number of clusters of beetle larvae to the number of stumps. CONCLUDE: The linear model relating number of clusters of beetle larvae to the number of stumps is appropriate and fits the data well, accounting for more than 80% of the variation in number of clusters of beetle larvae.

**3.68** STATE: Is a linear model appropriate for these data? If so, how well does the least-squares regression line fit the data? PLAN: We will look at the scatterplot and residual plot to see if the association is linear or nonlinear. Then if a linear model is appropriate, we will use $s$ and $r^2$ to measure how well the line fits the data. DO: The scatterplot below shows a strong, positive linear association between fat and calories. The residual plot doesn't show any obvious leftover pattern, confirming that a linear model is appropriate.

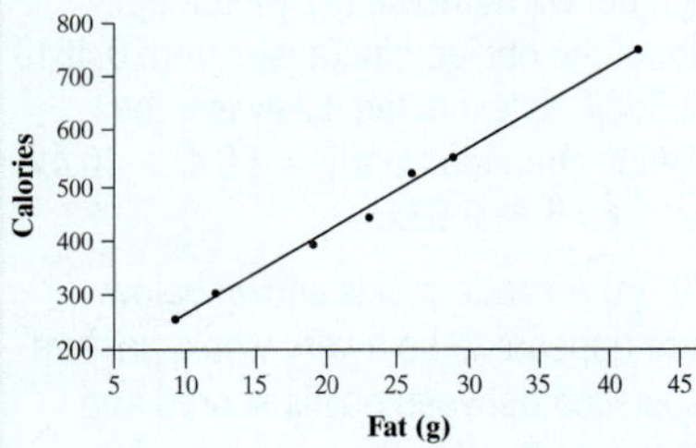


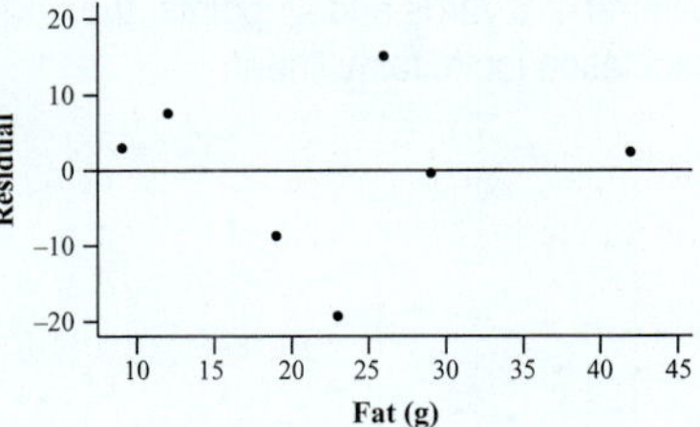


$\hat{y} = 110.44 + 15.1682x$, where $y$ = calories and $x$ = fat. $s = 12.25$, meaning that our predictions will typically be off by about 12.25 calories when we use the line to predict number of calories from grams of fat. Finally, $r^2 = 0.996$, meaning 99.6% of the variation in the number of calories is accounted for by the linear model relating number of calories to grams of fat. CONCLUDE: The linear model relating number of calories to the grams of fat is appropriate and fits the data very well, accounting for nearly all of the variation in number of calories.

**3.69 (a)** A scatterplot is shown below. There is a moderate, positive linear association between HbA and FBG. There are possible outliers to the far right (Subject 18) and near the top of the plot (Subject 15).

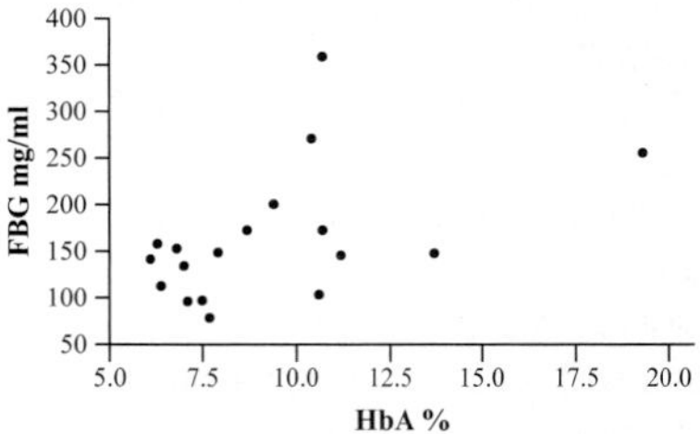


**(b)** Because the point is in the positive, linear pattern formed by most of the data values, it makes $r$ closer to 1. Also, because the point is likely to be below the least-squares regression line, it will "pull down" the line on the right side, making the slope closer to 0. Without the outlier, $r$ decreases from 0.4819 to 0.3837, as expected. Likewise, the equation changes from $\hat{y} = 66.4 + 10.4x$ to $\hat{y} = 52.3 + 12.1x$. **(c)** The point makes $r$ closer to 0 because it is out of the linear pattern formed by most of the data values. Because this point's $x$ coordinate is very close to $\bar{x}$ but the $y$ coordinate is far above $\bar{y}$, it won't influence the slope very much but will increase the $y$ intercept. Without the outlier, $r$ increases from 0.4819 to 0.5684, as expected. Likewise, the equation changes from $\hat{y} = 66.4 + 10.4x$ to $\hat{y} = 69.5 + 8.92x$.

**3.70 (a)** A scatterplot is shown below. There appears to be a very weak, positive association between points scored and rushing yards. With the exception of the outlier at 116 yards and 41 points, the association looks fairly linear.

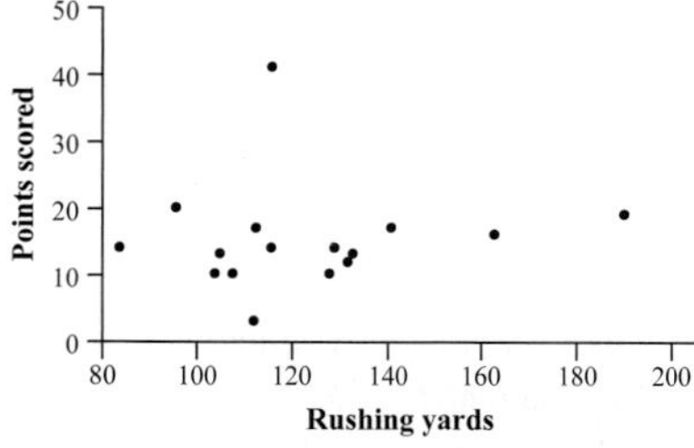


**(b)** Because this point is in the positive, linear pattern formed by most of the data values, it will make $r$ closer to 1. Also, because the point is likely to be above the least-squares regression line, it will "pull up" the line on the right side, making the slope a little steeper. Without the point, $r$ drops from 0.10 to 0.02. Likewise, the equation changes from $\hat{y} = 11.4 + 0.031x$ to $\hat{y} = 14 + 0.008x$.

**(c)** The point makes $r$ closer to 0 because it is out of the linear pattern formed by most of the data values. Because this point's $x$ coordinate is very close to $\bar{x}$ but the $y$ coordinate is far above $\bar{y}$, it won't influence the slope very much but will increase the $y$ intercept. Without the outlier, $r$ increases from 0.10 to 0.32, the slope changes from 0.031 to 0.050, and the $y$ intercept drops from 11.4 to 7.2.

**3.71** a

**3.72** a

**3.73** c



(a) Make a scatterplot with HbA as the explanatory variable. Describe what you see.

(b) Subject 18 is an outlier in the $x$ direction. What effect do you think this subject has on the correlation? What effect do you think this subject has on the equation of the least-squares regression line? Calculate the correlation and equation of the least-squares regression line with and without this subject to confirm your answer.

(c) Subject 15 is an outlier in the $y$ direction. What effect do you think this subject has on the correlation? What effect do you think this subject has on the equation of the least-squares regression line? Calculate the correlation and equation of the least-squares regression line with and without this subject to confirm your answer.

**70. Rushing for points** What is the relationship between rushing yards and points scored in the 2011 National Football League? The table below gives the number of rushing yards and the number of points scored for each of the 16 games played by the 2011 Jacksonville Jaguars.[26]

| Game | Rushing yards | Points scored |
|---|---|---|
| 1 | 163 | 16 |
| 2 | 112 | 3 |
| 3 | 128 | 10 |
| 4 | 104 | 10 |
| 5 | 96 | 20 |
| 6 | 133 | 13 |
| 7 | 132 | 12 |
| 8 | 84 | 14 |
| 9 | 141 | 17 |
| 10 | 108 | 10 |
| 11 | 105 | 13 |
| 12 | 129 | 14 |
| 13 | 116 | 41 |
| 14 | 116 | 14 |
| 15 | 113 | 17 |
| 16 | 190 | 19 |

(a) Make a scatterplot with rushing yards as the explanatory variable. Describe what you see.

(b) The number of rushing yards in Game 16 is an outlier in the $x$ direction. What effect do you think this game has on the correlation? On the equation of the least-squares regression line? Calculate the correlation and equation of the least-squares regression line with and without this game to confirm your answers.

(c) The number of points scored in Game 13 is an outlier in the $y$ direction. What effect do you think this game has on the correlation? On the equation of the least-squares regression line? Calculate the correlation and equation of the least-squares regression line with and without this game to confirm your answers.

*Multiple choice: Select the best answer for Exercises 71 to 78.*

71. Which of the following is *not* a characteristic of the least-squares regression line?

(a) The slope of the least-squares regression line is always between −1 and 1.

(b) The least-squares regression line always goes through the point $(\bar{x}, \bar{y})$.

(c) The least-squares regression line minimizes the sum of squared residuals.

(d) The slope of the least-squares regression line will always have the same sign as the correlation.

(e) The least-squares regression line is not resistant to outliers.

72. Each year, students in an elementary school take a standardized math test at the end of the school year. For a class of fourth-graders, the average score was 55.1 with a standard deviation of 12.3. In the third grade, these same students had an average score of 61.7 with a standard deviation of 14.0. The correlation between the two sets of scores is $r = 0.95$. Calculate the equation of the least-squares regression line for predicting a fourth-grade score from a third-grade score.

(a) $\hat{y} = 3.60 + 0.835x$

(b) $\hat{y} = 15.69 + 0.835x$

(c) $\hat{y} = 2.19 + 1.08x$

(d) $\hat{y} = -11.54 + 1.08x$

(e) Cannot be calculated without the data.

73. Using data from the 2009 LPGA tour, a regression analysis was performed using $x$ = average driving distance and $y$ = scoring average. Using the output from the regression analysis shown below, determine the equation of the least-squares regression line.

```
Predictor          Coef       SE Coef      T       P
Constant           87.974     2.391      36.78   0.000
Driving Distance  -0.060934   0.009536   -6.39   0.000

S = 1.01216   R-Sq = 22.1%   R-Sq(adj) = 21.6%
```

(a) $\hat{y} = 87.947 + 2.391x$

(b) $\hat{y} = 87.947 + 1.01216x$

(c) $\hat{y} = 87.947 - 0.060934x$

(d) $\hat{y} = -0.060934 + 1.01216x$

(e) $\hat{y} = -0.060934 + 87.947x$

*Exercises 74 to 78 refer to the following setting.* Measurements on young children in Mumbai, India, found this least-squares line for predicting height $y$ from arm span $x$:[28]

$$\hat{y} = 6.4 + 0.93x$$

Measurements are in centimeters (cm).

74. By looking at the equation of the least-squares regression line, you can see that the correlation between height and arm span is

(a) greater than zero.

(b) less than zero.

(c) 0.93.
(d) 6.4.
(e) Can't tell without seeing the data.

75. In addition to the regression line, the report on the Mumbai measurements says that $r^2 = 0.95$. This suggests that
(a) although arm span and height are correlated, arm span does not predict height very accurately.
(b) height increases by $\sqrt{0.95} = 0.97$ cm for each additional centimeter of arm span.
(c) 95% of the relationship between height and arm span is accounted for by the regression line.
(d) 95% of the variation in height is accounted for by the regression line.
(e) 95% of the height measurements are accounted for by the regression line.

76. One child in the Mumbai study had height 59 cm and arm span 60 cm. This child's residual is
(a) −3.2 cm.
(b) −2.2 cm.
(c) −1.3 cm.
(d) 3.2 cm.
(e) 62.2 cm.

77. Suppose that a tall child with arm span 120 cm and height 118 cm was added to the sample used in this study. What effect will adding this child have on the correlation and the slope of the least-squares regression line?
(a) Correlation will increase, slope will increase.
(b) Correlation will increase, slope will stay the same.
(c) Correlation will increase, slope will decrease.
(d) Correlation will stay the same, slope will stay the same.
(e) Correlation will stay the same, slope will increase.

78. Suppose that the measurements of arm span and height were converted from centimeters to meters by dividing each measurement by 100. How will this conversion affect the values of $r^2$ and $s$?
(a) $r^2$ will increase, $s$ will increase.
(b) $r^2$ will increase, $s$ will stay the same.
(c) $r^2$ will increase, $s$ will decrease.
(d) $r^2$ will stay the same, $s$ will stay the same.
(e) $r^2$ will stay the same, $s$ will decrease.

*Exercises 79 and 80 refer to the following setting.*

In its recent *Fuel Economy Guide*, the Environmental Protection Agency gives data on 1152 vehicles. There are a number of outliers, mainly vehicles with very poor gas mileage. If we ignore the outliers, however, the combined city and highway gas mileage of the other 1120 or so vehicles is approximately Normal with mean 18.7 miles per gallon (mpg) and standard deviation 4.3 mpg.

79. **In my Chevrolet (2.2)** The Chevrolet Malibu with a four-cylinder engine has a combined gas mileage of 25 mpg. What percent of all vehicles have worse gas mileage than the Malibu?

80. **The top 10% (2.2)** How high must a vehicle's gas mileage be in order to fall in the top 10% of all vehicles? (The distribution omits a few high outliers, mainly hybrid gas-electric vehicles.)

81. **Marijuana and traffic accidents (1.1)** Researchers in New Zealand interviewed 907 drivers at age 21. They had data on traffic accidents and they asked the drivers about marijuana use. Here are data on the numbers of accidents caused by these drivers at age 19, broken down by marijuana use at the same age:[29]

| | Marijuana use per year | | | |
|---|---|---|---|---|
| | Never | 1–10 times | 11–50 times | 51 + times |
| Drivers | 452 | 229 | 70 | 156 |
| Accidents caused | 59 | 36 | 15 | 50 |

(a) Make a graph that displays the accident rate for each class. Is there evidence of an association between marijuana use and traffic accidents?
(b) Explain why we can't conclude that marijuana use *causes* accidents.

## FRAPPY! Free Response AP® Problem, Yay!

The following problem is modeled after actual AP® Statistics exam free response questions. Your task is to generate a complete, concise response in 15 minutes.

*Directions: Show all your work. Indicate clearly the methods you use, because you will be scored on the correctness of your methods as well as on the accuracy and completeness of your results and explanations.*

Two statistics students went to a flower shop and randomly selected 12 carnations. When they got home, the students prepared 12 identical vases with exactly the same amount of water in each vase. They put one tablespoon of sugar in 3 vases, two tablespoons of sugar in 3 vases, and three tablespoons of sugar in 3 vases. In the remaining 3 vases, they put no sugar. After the vases were prepared, the students randomly assigned 1 carnation to each vase

**3.80** For these vehicles, the combined mileage follows a $N(18.7, 4.3)$ distribution. The boundary value that separates the top 10% from the rest of the distribution has an area of 0.90 to its left (see graph below). A $z$-score of 1.28 gives the value closest to 0.90 (0.8997). Solving $1.28 = \frac{x - 18.7}{4.3}$ gives $x = 24.2$. *Using technology:* `invNorm(area:0.9,μ:18.7,σ:4.3)` = 24.2. The top 10% of all vehicles get at least 24.2 mpg.

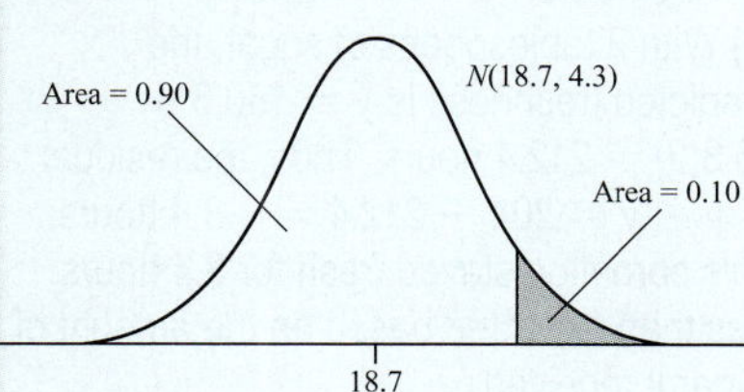


**3.81** **(a)** A bar graph is given below. The people who use marijuana more are more likely to have caused accidents. **(b)** Association does not imply causation. For example, it could be that drivers who use marijuana more often are more willing to take risks than other drivers and that the willingness to take risks is what is causing the higher accident rate.

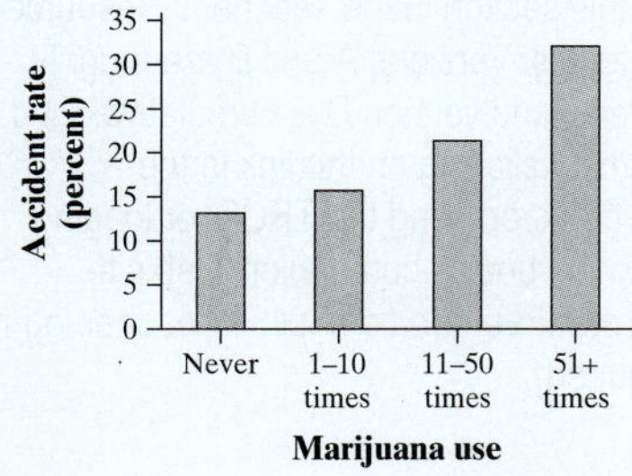


**3.74** a

**3.75** d

**3.76** a

**3.77** b

**3.78** e

**3.79** For these vehicles, the combined mileage follows a $N$ (18.7, 4.3) distribution and we want to find the percent of cars with lower mileage than 25 (see graph). $z = \frac{25 - 18.7}{4.3} = 1.47$. From Table A, the proportion of $z$-scores below 1.47 is 0.9292. *Using technology:* `normalcdf(lower:-1000,upper:25,μ:18.7,σ:4.3)` = 0.9286. About 93% percent of vehicles get worse combined mileage than the Chevrolet Malibu.

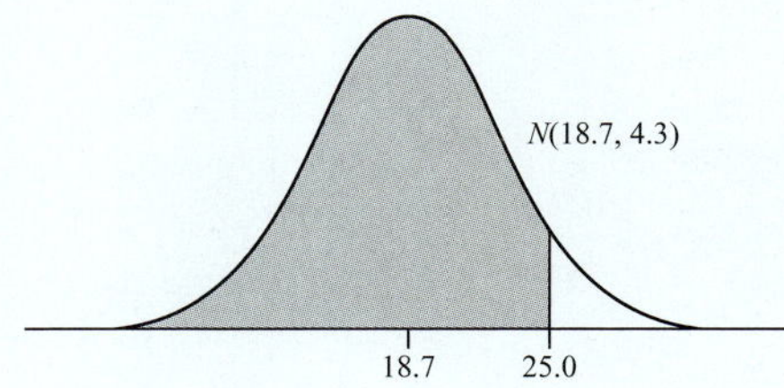

**TRM FRAPPY!**

We present a model solution here. Please consult the Teacher's Resource Materials for sample student responses, a scoring rubric, and a black line master of the original question with space for students to write their responses.

**(a)** There is a fairly strong, positive, linear relationship between sugar and freshness.

**(b)** For each increase of 1 tablespoon of sugar, the predicted freshness increases by 15.8 hours.

**(c)** With 2 tablespoons of sugar, the predicted freshness is $\hat{y} = 180.8 + 15.8(2) = 212.4$ hours. Thus, the residual $= y - \hat{y} = 204 - 212.4 = -8.4$ hours. This carnation stayed fresh for 8.4 hours *less* than expected based on the amount of sugar it received.

**(d)** Because other flowers will probably stay fresh for a longer or shorter period than carnations, this will add an additional source of unaccounted-for variability in freshness. Because there will be less variation in freshness that is accounted for, the value of $r^2$ will decrease.

**TRM Chapter 3 Tests**

There are four different tests available for this section in the Teacher's Resource Materials. Versions A and B are roughly parallel and version D is cumulative. Find them by clicking on the link in the TE-book, opening the TRCD, or logging onto the book's companion Web site www.whfreeman.com/tps5e (teacher log-in required).

and observed how many hours each flower continued to look fresh. A scatterplot of the data is shown below.

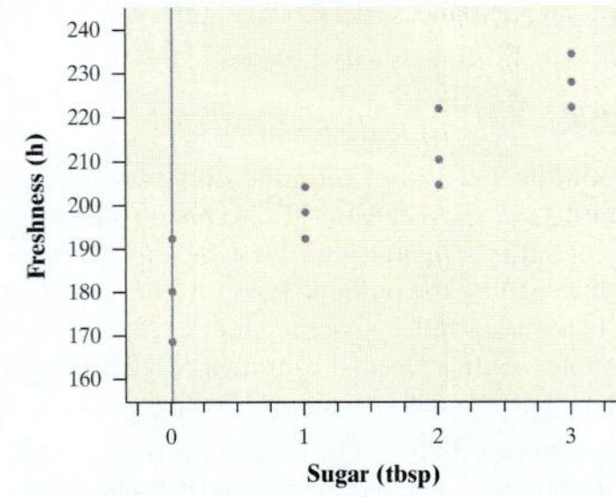


(a) Briefly describe the association shown in the scatterplot.

(b) The equation of the least-squares regression line for these data is $\hat{y} = 180.8 + 15.8x$. Interpret the slope of the line in the context of the study.

(c) Calculate and interpret the residual for the flower that had 2 tablespoons of sugar and looked fresh for 204 hours.

(d) Suppose that another group of students conducted a similar experiment using 12 flowers, but included different varieties in addition to carnations. Would you expect the value of $r^2$ for the second group's data to be greater than, less than, or about the same as the value of $r^2$ for the first group's data? Explain.

After you finish, you can view two example solutions on the book's Web site (www.whfreeman.com/tps5e). Determine whether you think each solution is "complete," "substantial," "developing," or "minimal." If the solution is not complete, what improvements would you suggest to the student who wrote it? Finally, your teacher will provide you with a scoring rubric. Score your response and note what, if anything, you would do differently to improve your own score.

# Chapter Review

### Section 3.1: Scatterplots and Correlation

In this section, you learned how to explore the relationship between two quantitative variables. As with distributions of a single variable, the first step is always to make a graph. A scatterplot is the appropriate type of graph to investigate associations between two quantitative variables. To describe a scatterplot, be sure to discuss four characteristics: direction, form, strength, and outliers. The direction of an association might be positive, negative, or neither. The form of an association can be linear or nonlinear. An association is strong if it closely follows a specific form. Finally, outliers are any points that clearly fall outside the pattern of the rest of the data.

The correlation $r$ is a numerical summary that describes the direction and strength of a linear association. When $r > 0$, the association is positive, and when $r < 0$, the association is negative. The correlation will always take values between $-1$ and 1, with $r = -1$ and $r = 1$ indicating a perfectly linear relationship. Strong linear associations have correlations near 1 or $-1$, while weak linear relationships have correlations near 0. However, it isn't possible to determine the form of an association from only the correlation. Strong nonlinear relationships can have a correlation close to 1 or a correlation close to 0, depending on the association. You also learned that outliers can greatly affect the value of the correlation and that correlation does not imply causation. That is, we can't assume that changes in one variable cause changes in the other variable, just because they have a correlation close to 1 or –1.

### Section 3.2: Least-Squares Regression

In this section, you learned how to use least-squares regression lines as models for relationships between variables that have a linear association. It is important to understand the difference between the actual data and the model used to describe the data. For example, when you are interpreting the slope of a least-squares regression line, describe the *predicted* change in the $y$ variable. To emphasize that the model only provides predicted values, least-squares regression lines are always expressed in terms of $\hat{y}$ instead of $y$.

The difference between the observed value of $y$ and the predicted value of $y$ is called a residual. Residuals are the key to understanding almost everything in this section. To find the equation of the least-squares regression line, find the line that minimizes the sum of the squared residuals. To see if a linear model is appropriate, make a residual plot. If there is no leftover pattern in the residual plot, you know the model is appropriate. To assess how well a line fits the data, calculate the standard deviation of the residuals $s$ to estimate the size of a typical prediction error. You can also calculate $r^2$, which measures the fraction of the variation in the $y$ variable that is accounted for by its linear relationship with the $x$ variable.

You also learned how to obtain the equation of a least-squares regression line from computer output and from summary statistics (the means and standard deviations of two variables and their correlation). As with the correlation, the equation of the least-squares regression line and the values of $s$ and $r^2$ can be greatly influenced by outliers, so be sure to plot the data and note any unusual values before making any calculations.

## What Did You Learn?

| Learning Objective | Section | Related Example on Page(s) | Relevant Chapter Review Exercise(s) |
|---|---|---|---|
| Identify explanatory and response variables in situations where one variable helps to explain or influences the other. | 3.1 | 144 | R3.4 |
| Make a scatterplot to display the relationship between two quantitative variables. | 3.1 | 145, 148 | R3.4 |
| Describe the direction, form, and strength of a relationship displayed in a scatterplot and recognize outliers in a scatterplot. | 3.1 | 147, 148 | R3.1 |
| Interpret the correlation. | 3.1 | 152 | R3.3, R3.4 |
| Understand the basic properties of correlation, including how the correlation is influenced by outliers. | 3.1 | 152, 156, 157 | R3.1, R3.2 |
| Use technology to calculate correlation. | 3.1 | Activity on 152, 171 | R3.4 |
| Explain why association does not imply causation. | 3.1 | Discussion on 156, 190 | R3.6 |
| Interpret the slope and $y$ intercept of a least-squares regression line. | 3.2 | 166 | R3.2, R3.4 |
| Use the least-squares regression line to predict $y$ for a given $x$. Explain the dangers of extrapolation. | 3.2 | 167, Discussion on 168 (for extrapolation) | R3.2, R3.4, R3.5 |
| Calculate and interpret residuals. | 3.2 | 169 | R3.3, R3.4 |
| Explain the concept of least squares. | 3.2 | Discussion on 169 | R3.5 |
| Determine the equation of a least-squares regression line using technology or computer output. | 3.2 | Technology Corner on 171, 181 | R3.3, R3.4 |
| Construct and interpret residual plots to assess whether a linear model is appropriate. | 3.2 | Discussion on 175, 180 | R3.3, R3.4 |
| Interpret the standard deviation of the residuals and $r^2$ and use these values to assess how well the least-squares regression line models the relationship between two variables. | 3.2 | 180 | R3.3, R3.5 |
| Describe how the slope, $y$ intercept, standard deviation of the residuals, and $r^2$ are influenced by outliers. | 3.2 | Discussion on 188 | R3.1 |
| Find the slope and $y$ intercept of the least-squares regression line from the means and standard deviations of $x$ and $y$ and their correlation. | 3.2 | 183 | R3.5 |



TRM Learning Objectives Grid

**TRM** **Full Solutions to Chapter 3 Review Exercises**

## Answers to Chapter 3 Review Exercises

**R3.1** **(a)** There is a moderate, positive linear association between gestation and life span. Without the outliers at the top and in the upper right, the association appears moderately strong, positive, and curved. **(b)** It makes $r$ closer to 0 because it decreases the strength of what would otherwise be a moderately strong positive association. Because this point is close to $\bar{x}$ but far above $\bar{y}$, it won't affect the slope much but will increase the $y$ intercept. Because it has such a large residual, it increases $s$. **(c)** Because it is in the positive, linear pattern formed by most of the data values, it will make $r$ closer to 1. Also, because the point is likely to be above the least-squares regression line, it will "pull up" the line on the right side, making the slope larger and the intercept smaller. Because this point is likely to have a small residual, it decreases $s$.

**R3.2** **(a)** 0.0138. For each increase of 1 meter in dive depth, the predicted duration increases by 0.0138 minutes. **(b)** The $y$ intercept suggests that a dive of 0 depth would last an average of 2.69 minutes; this obviously does not make any sense. **(c)** 5.45 minutes. **(d)** If the variables are reversed, the correlation will remain the same. However, the slope and $y$ intercept will be different.

**R3.3** **(a)** $\hat{y} = 3704 + 12{,}188x$, where $y$ represents the mileage of the cars and $x$ represents the age. **(b)** residual $= 65{,}000 - 76{,}832 = -11{,}832$. This teacher has driven 11,832 fewer miles than predicted based on the age of the car. **(c)** $r = +\sqrt{0.837} = 0.915$. This shows that there is a strong, positive linear association between the age of cars and their mileage. **(d)** Yes, because there is no leftover pattern in the residual plot. **(e)** $s = 20{,}870.5$: when using the least-squares regression line with $x$ = car's age to predict $y$ = number of miles it has been driven, we will typically be off by about 20,870.5 miles. $r^2 = 83.7\%$: about 83.7% of the variability in mileage is accounted for by the linear model relating mileage to age.

# Chapter 3 Chapter Review Exercises

*These exercises are designed to help you review the important ideas and methods of the chapter.*

**R3.1** **Born to be old?** Is there a relationship between the gestational period (time from conception to birth) of an animal and its average life span? The figure shows a scatterplot of the gestational period and average life span for 43 species of animals.[30]

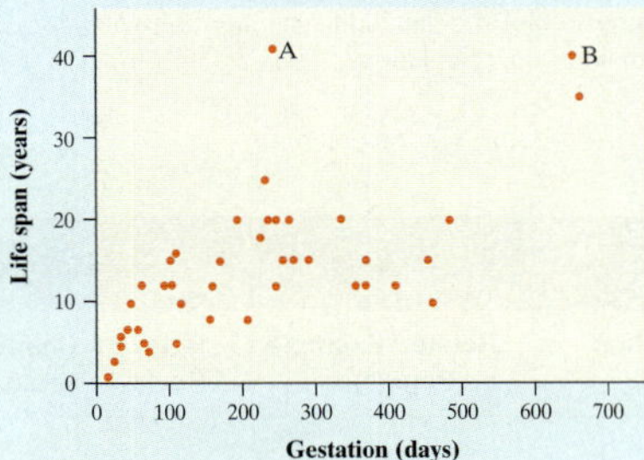


(a) Describe the association shown in the scatterplot.

(b) Point A is the hippopotamus. What effect does this point have on the correlation, the equation of the least-squares regression line, and the standard deviation of the residuals?

(c) Point B is the Asian elephant. What effect does this point have on the correlation, the equation of the least-squares regression line, and the standard deviation of the residuals?

**R3.2** **Penguins diving** A study of king penguins looked for a relationship between how deep the penguins dive to seek food and how long they stay under water.[31] For all but the shallowest dives, there is a linear relationship that is different for different penguins. The study gives a scatterplot for one penguin titled "The Relation of Dive Duration ($y$) to Depth ($x$)." Duration $y$ is measured in minutes and depth $x$ is in meters. The report then says, "The regression equation for this bird is: $\hat{y} = 2.69 + 0.0138x$."

(a) What is the slope of the regression line? Interpret this value.

(b) Does the $y$ intercept of the regression line make any sense? If so, interpret it. If not, explain why not.

(c) According to the regression line, how long does a typical dive to a depth of 200 meters last?

(d) Suppose that the researchers reversed the variables, using $x$ = dive duration and $y$ = depth. What effect will this have on the correlation? On the equation of the least-squares regression line?

**R3.3** **Stats teachers' cars** A random sample of AP® Statistics teachers was asked to report the age (in years) and mileage of their primary vehicles. A scatterplot of the data, a least-squares regression printout, and a residual plot are provided below.

```
Predictor    Coef     SE Coef    T       P
Constant     3704     8268       0.45    0.662
Age          12188    1492       8.17    0.000

S = 20870.5   R-Sq = 83.7%   R-Sq(adj) = 82.4%
```

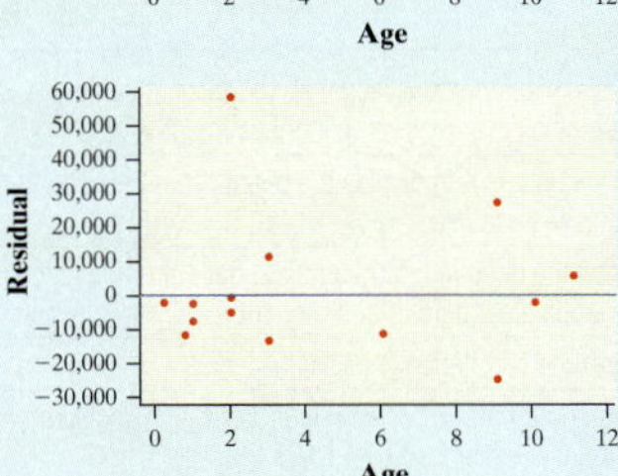


(a) Give the equation of the least-squares regression line for these data. Identify any variables you use.

(b) One teacher reported that her 6-year-old car had 65,000 miles on it. Find and interpret its residual.

(c) What's the correlation between car age and mileage? Interpret this value in context.

(d) Is a linear model appropriate for these data? Explain how you know.

(e) Interpret the values of $s$ and $r^2$.

**R3.4** **Late bloomers?** Japanese cherry trees tend to blossom early when spring weather is warm and later when spring weather is cool. Here are some data on the average March temperature (in °C) and the day in April when the first cherry blossom appeared over a 24-year period:[32]

| Temperature (°C): | 4.0 | 5.4 | 3.2 | 2.6 | 4.2 | 4.7 | 4.9 | 4.0 | 4.9 | 3.8 | 4.0 | 5.1 |
|---|---|---|---|---|---|---|---|---|---|---|---|---|
| Days in April to first bloom: | 14 | 8 | 11 | 19 | 14 | 14 | 14 | 21 | 9 | 14 | 13 | 11 |
| Temperature (°C): | 4.3 | 1.5 | 3.7 | 3.8 | 4.5 | 4.1 | 6.1 | 6.2 | 5.1 | 5.0 | 4.6 | 4.0 |
| Days in April to first bloom: | 13 | 28 | 17 | 19 | 10 | 17 | 3 | 3 | 11 | 6 | 9 | 11 |

(a) Make a well-labeled scatterplot that's suitable for predicting when the cherry trees will bloom from the temperature. Which variable did you choose as the explanatory variable? Explain.

(b) Use technology to calculate the correlation and the equation of the least-squares regression line. Interpret the correlation, slope, and $y$ intercept of the line in this setting.

(c) Suppose that the average March temperature this year was 8.2°C. Would you be willing to use the equation in part (b) to predict the date of first bloom? Explain.

(d) Calculate and interpret the residual for the year when the average March temperature was 4.5°C. Show your work.

(e) Use technology to help construct a residual plot. Describe what you see.

**R3.5 What's my grade?** In Professor Friedman's economics course, the correlation between the students' total scores prior to the final examination and their final-examination scores is $r = 0.6$. The pre-exam totals for all students in the course have mean 280 and standard deviation 30. The final-exam scores have mean 75 and standard deviation 8. Professor Friedman has lost Julie's final exam but knows that her total before the exam was 300. He decides to predict her final-exam score from her pre-exam total.

(a) Find the equation for the appropriate least-squares regression line for Professor Friedman's prediction.

(b) Use the least-squares regression line to predict Julie's final-exam score.

(c) Explain the meaning of the phrase "least squares" in the context of this question.

(d) Julie doesn't think this method accurately predicts how well she did on the final exam. Determine $r^2$. Use this result to argue that her actual score could have been much higher (or much lower) than the predicted value.

**R3.6 Calculating achievement** The principal of a high school read a study that reported a high correlation between the number of calculators owned by high school students and their math achievement. Based on this study, he decides to buy each student at his school two calculators, hoping to improve their math achievement. Explain the flaw in the principal's reasoning.

# Chapter 3 AP® Statistics Practice Test

**Section I: Multiple Choice** *Select the best answer for each question.*

**T3.1** A school guidance counselor examines the number of extracurricular activities that students do and their grade point average. The guidance counselor says, "The evidence indicates that the correlation between the number of extracurricular activities a student participates in and his or her grade point average is close to zero." A correct interpretation of this statement would be that

(a) active students tend to be students with poor grades, and vice versa.

(b) students with good grades tend to be students who are not involved in many extracurricular activities, and vice versa.

(c) students involved in many extracurricular activities are just as likely to get good grades as bad grades; the same is true for students involved in few extracurricular activities.

(d) there is no linear relationship between number of activities and grade point average for students at this school.

(e) involvement in many extracurricular activities and good grades go hand in hand.

**T3.2** The British government conducts regular surveys of household spending. The average weekly household spending (in pounds) on tobacco products and alcoholic beverages for each of 11 regions in Great Britain was recorded. A scatterplot of spending on alcohol versus spending on tobacco is shown below. Which of the following statements is true?

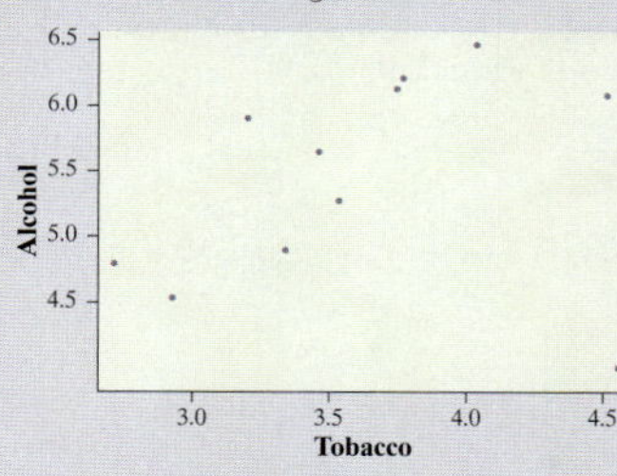


(a) The observation (4.5, 6.0) is an outlier.

(b) There is clear evidence of a negative association between spending on alcohol and tobacco.

(c) The equation of the least-squares line for this plot would be approximately $\hat{y} = 10 - 2x$.

(d) The correlation for these data is $r = 0.99$.

(e) The observation in the lower-right corner of the plot is influential for the least-squares line.

**R3.4 (a)** The scatterplot is shown below. Average March temperature, because changes in March temperature probably have an effect on the date of first bloom.

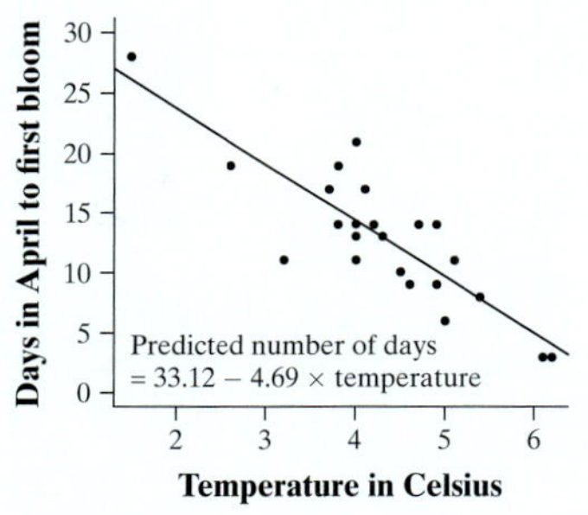


**(b)** $r = -0.85$ and $\hat{y} = 33.12 - 4.69x$, where $y$ represents the number of days and $x$ represents the temperature. *r*: there is a strong, negative linear association between the average March temperature and the days in April until first bloom. *slope*: for every 1-degree increase in average March temperature, the predicted number of days in April until first bloom decreases by 4.69. *y intercept*: if the average March temperature was 0°C, the predicted number of days in April to first bloom is 33.12 (May 3). **(c)** No, $x = 8.2$ is well beyond the values of $x$ we have in the data set. **(d)** residual $= 10 - 12.015 = -2.015$. In this year, the actual date of first bloom occurred about 2 days earlier than predicted based on the average March temperature. **(e)** There is no leftover pattern in the residual plot shown below, indicating that a linear model is appropriate.

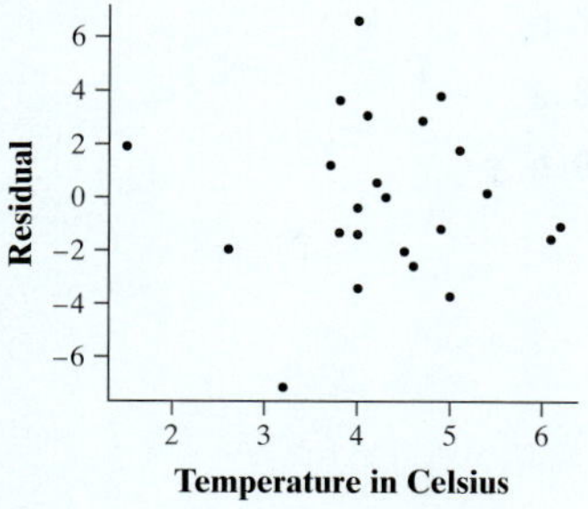


**R3.5 (a)** $\hat{y} = 30.2 + 0.16x$, where $y$ = final exam score and $x$ = total score before the final examination. **(b)** 78.2. **(c)** Of all the lines that the professor could use to summarize the relationship between final exam score and total points before the final exam, the least-squares regression line is the one that has the smallest sum of squared residuals. **(d)** Because $r^2 = 0.36$, only 36% of the variability in the final exam scores is accounted for by the linear model relating final exam scores to total score before the final exam. More than half (64%) of the variation in final exam scores is not accounted for, so Julie has reason to question this estimate.

**R3.6** Even though there is a high correlation between number of calculators and math achievement, we shouldn't conclude that increasing the number of calculators will *cause* an increase in math achievement. It is possible that students who are more serious about school have better math achievement and also have more calculators.

**TRM Full Solutions to Chapter 3 AP® Statistics Practice Test**

## Answers to Chapter 3 AP® Statistics Practice Test

**T3.1** d

**T3.2** e

T3.3 c

T3.4 a

T3.5 a

T3.6 c

T3.7 b

T3.8 e

T3.9 b



**T3.3** The fraction of the variation in the values of $y$ that is explained by the least-squares regression of $y$ on $x$ is

(a) the correlation.
(b) the slope of the least-squares regression line.
(c) the square of the correlation coefficient.
(d) the intercept of the least-squares regression line.
(e) the residual.

**T3.4** An AP® Statistics student designs an experiment to see whether today's high school students are becoming too calculator-dependent. She prepares two quizzes, both of which contain 40 questions that are best done using paper-and-pencil methods. A random sample of 30 students participates in the experiment. Each student takes both quizzes—one with a calculator and one without—in a random order. To analyze the data, the student constructs a scatterplot that displays the number of correct answers with and without a calculator for each of the 30 students. A least-squares regression yields the equation

$$\widehat{\text{Calculator}} = -1.2 + 0.865(\text{Pencil}) \quad r = 0.79$$

Which of the following statements is/are true?

I. If the student had used Calculator as the explanatory variable, the correlation would remain the same.
II. If the student had used Calculator as the explanatory variable, the slope of the least-squares line would remain the same.
III. The standard deviation of the number of correct answers on the paper-and-pencil quizzes was larger than the standard deviation on the calculator quizzes.

(a) I only (b) II only (c) III only (d) I and III only (e) I, II, and III

*Questions T3.5 and T3.6 refer to the following setting.* Scientists examined the activity level of 7 fish at different temperatures. Fish activity was rated on a scale of 0 (no activity) to 100 (maximal activity). The temperature was measured in degrees Celsius. A computer regression printout and a residual plot are given below. Notice that the horizontal axis on the residual plot is labeled "Fitted value."

| Predictor | Coef | SE Coef | T | P |
|---|---|---|---|---|
| Constant | 148.62 | 10.71 | 13.88 | 0.000 |
| Temperature | -3.2167 | 0.4533 | -7.10 | 0.001 |

S = 4.78505 R-Sq = 91.0% R-Sq(adj) = 89.2%

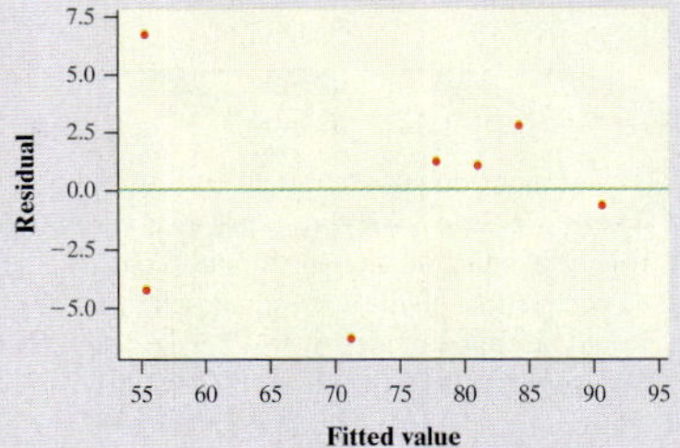


**T3.5** What was the activity level rating for the fish at a temperature of 20°C?

(a) 87 (b) 84 (c) 81 (d) 66 (e) 3

**T3.6** Which of the following gives a correct interpretation of $s$ in this setting?

(a) For every 1°C increase in temperature, fish activity is predicted to increase by 4.785 units.
(b) The typical distance of the temperature readings from their mean is about 4.785°C.
(c) The typical distance of the activity level ratings from the least-squares line is about 4.785 units.
(d) The typical distance of the activity level readings from their mean is about 4.785.
(e) At a temperature of 0°C, this model predicts an activity level of 4.785.

**T3.7** Which of the following statements is *not* true of the correlation $r$ between the lengths in inches and weights in pounds of a sample of brook trout?

(a) $r$ must take a value between $-1$ and 1.
(b) $r$ is measured in inches.
(c) If longer trout tend to also be heavier, then $r > 0$.
(d) $r$ would not change if we measured the lengths of the trout in centimeters instead of inches.
(e) $r$ would not change if we measured the weights of the trout in kilograms instead of pounds.

**T3.8** When we standardize the values of a variable, the distribution of standardized values has mean 0 and standard deviation 1. Suppose we measure two variables X and Y on each of several subjects. We standardize both variables and then compute the least-squares regression line. Suppose the slope of the least-squares regression line is $-0.44$. We may conclude that

(a) the intercept will also be $-0.44$.
(b) the intercept will be 1.0.
(c) the correlation will be $1/-0.44$.
(d) the correlation will be 1.0.
(e) the correlation will also be $-0.44$.

**T3.9** There is a linear relationship between the number of chirps made by the striped ground cricket and the air temperature. A least-squares fit of some data collected by a biologist gives the model $\hat{y} = 25.2 + 3.3x$, where $x$ is the number of chirps per minute and $\hat{y}$ is the estimated temperature in degrees Fahrenheit. What is the predicted increase in temperature for an increase of 5 chirps per minute?

(a) 3.3°F (b) 16.5°F (c) 25.2°F (d) 28.5°F (e) 41.7°F

**T3.10** A data set included the number of people per television set and the number of people per physician for 40 countries. The Fathom screen shot below displays a scatterplot of the data with the least-squares regression line added. In Ethiopia, there were 503 people per TV and 36,660 people per doctor. What effect would removing this point have on the regression line?

(a) Slope would increase; $y$ intercept would increase.
(b) Slope would increase; $y$ intercept would decrease.
(c) Slope would decrease; $y$ intercept would increase.
(d) Slope would decrease; $y$ intercept would decrease.
(e) Slope and $y$ intercept would stay the same.

**Section II: Free Response** *Show all your work. Indicate clearly the methods you use, because you will be graded on the correctness of your methods as well as on the accuracy and completeness of your results and explanations.*

**T3.11** Sarah's parents are concerned that she seems short for her age. Their doctor has the following record of Sarah's height:

| **Age (months):** | 36 | 48 | 51 | 54 | 57 | 60 |
|---|---|---|---|---|---|---|
| **Height (cm):** | 86 | 90 | 91 | 93 | 94 | 95 |

(a) Make a scatterplot of these data.
(b) Using your calculator, find the equation of the least-squares regression line of height on age.
(c) Use your regression line to predict Sarah's height at age 40 years (480 months). Convert your prediction to inches (2.54 cm = 1 inch).
(d) The prediction is impossibly large. Explain why this happened.

**T3.12** Drilling down beneath a lake in Alaska yields chemical evidence of past changes in climate. Biological silicon, left by the skeletons of single-celled creatures called diatoms, is a measure of the abundance of life in the lake. A rather complex variable based on the ratio of certain isotopes relative to ocean water gives an indirect measure of moisture, mostly from snow. As we drill down, we look further into the past. Here is a scatterplot of data from 2300 to 12,000 years ago:

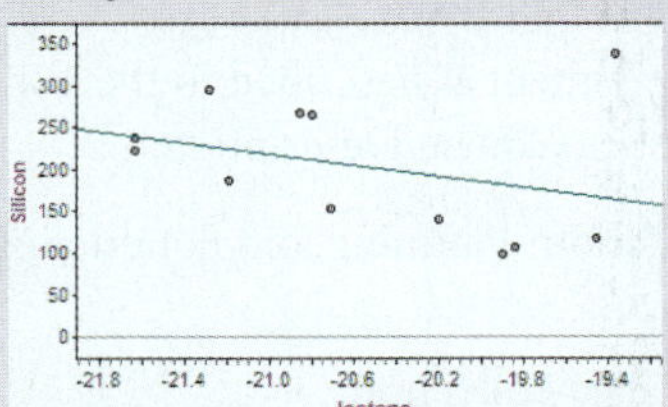


(a) Identify the unusual point in the scatterplot. Explain what's unusual about this point.
(b) If this point was removed, describe the effect on
i. the correlation.
ii. the slope and $y$ intercept of the least-squares line.
iii. the standard deviation of the residuals.

**T3.13** Long-term records from the Serengeti National Park in Tanzania show interesting ecological relationships. When wildebeest are more abundant, they graze the grass more heavily, so there are fewer fires and more trees grow. Lions feed more successfully when there are more trees, so the lion population increases. Researchers collected data on one part of this cycle, wildebeest abundance (in thousands of animals) and the percent of the grass area burned in the same year. The results of a least-squares regression on the data are shown here.[33]

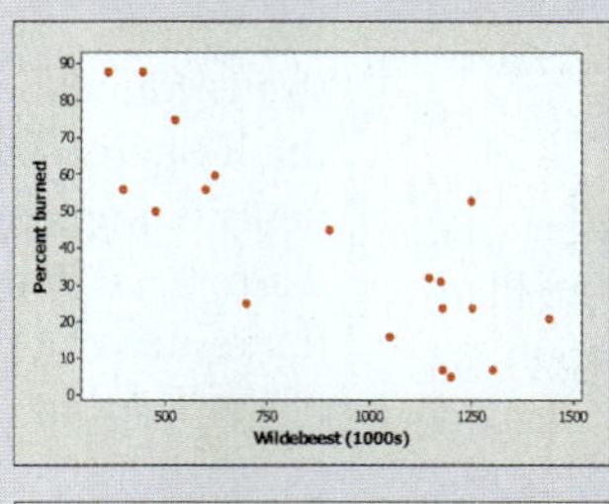


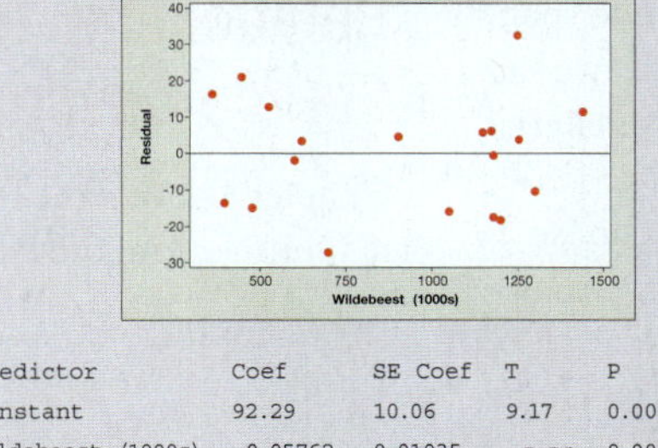


```
Predictor            Coef       SE Coef     T       P
Constant             92.29      10.06       9.17    0.000
Wildebeest (1000s)   -0.05762   0.01035    -5.56    0.000

S = 15.9880    R-Sq = 64.6%    R-Sq(adj) = 62.5%
```

(a) Give the equation of the least-squares regression line. Be sure to define any variables you use.
(b) Explain what the slope of the regression line means in this setting.
(c) Find the correlation. Interpret this value in context.
(d) Is a linear model appropriate for describing the relationship between wildebeest abundance and percent of grass area burned? Support your answer with appropriate evidence.

**T3.10** c

**T3.11** **(a)** A scatterplot with regression line is shown below. **(b)** $\hat{y} = 71.95 + 0.3833x$, where $y$ = height and $x$ = age. **(c)** 255.934 cm, or 100.76 inches. **(d)** This was an extrapolation. Our data were based only on the first 5 years of life and the linear trend will not continue forever.

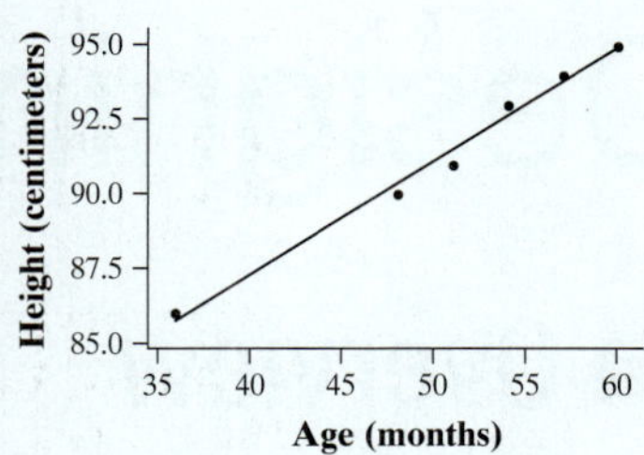


**T3.12** **(a)** The point in the upper-right-hand corner has a very high silicon value for its isotope value. **(b)** **(i)** $r$ would get closer to $-1$ because it does not follow the linear pattern of the other points. **(ii)** Because this point is "pulling up" the line on the right side of the plot, removing it will make the slope steeper (more negative) and the $y$ intercept smaller (note that the $y$ axis is to the *right* of the points in the scatterplot). **(iii)** Because this point has a large residual, removing it will make $s$ a little smaller.

**T3.13** **(a)** $\hat{y} = 92.29 - 0.05762x$, where $y$ is the percent of the grass burned and the variable $x$ is the number of wildebeest. **(b)** For every increase of 1000 wildebeest, the predicted percent of grassy area burned decreases by about 0.058. **(c)** $r = -\sqrt{0.646} = -0.804$. There is a strong, negative linear association between the percent of grass burned and the number of wildebeest. **(d)** Yes, because there is no obvious leftover pattern in the residual plot.

Chapter

4

# Designing Studies

## PD Overview

In this chapter, we cover every topic in Section II of the AP® Statistics Topic Outline, Sampling and Experimentation: Planning and Conducting a Study. The topics in this section are often new to students and include lots of vocabulary and difficult concepts. Between 10 and 15% of the questions on the AP® exam come from this content area, so make sure to spend adequate time here. As you can see in the pacing guide, we suggest spending 12 of the 110 class sessions covering this material, 3 more sessions than any other individual chapter.

Many teachers, including two of the authors, choose to teach Chapter 4 at the beginning of the school year. Because studies need to be designed before there are data to analyze, teaching this chapter first fits the order described in the *Guidelines for Assessment and Instruction in Statistics Education* (www.amstat.org/education/gaise/; see both the PreK–12 Report, especially Level C, and the College Report), endorsed by the American Statistical Association. This document suggests that "statistical problem solving is an investigative process that involves four components: Formulate Questions, Collect Data, Analyze Data, and Interpret Results." A second benefit of this approach is that students understand immediately that AP® Statistics is very different from any other "math" class they may have encountered in the past. Right away, students are confronted with real-world applications that show the relevance of what they are going to learn in the course. They will also quickly realize that there is much more reading and writing than in other math courses.

It is important that you move beyond written examples and exercises and have your students do some surveys and experiments.We have included several suggested Activities in the student text as well as here, in the *Annotated Teacher's Edition*, to get you started. Our students find it much easier to remember complicated concepts like stratified random sampling and blocking if they have taken part in studies that use these strategies. Taking time to do these Activities is another reason why we suggest spending a longer time in this chapter than others.

Two final suggestions: First, read the module called "Planning and Conducting a Study," available at AP® Central (under Classroom Resources: Sampling and Experimentation). This is an outstanding resource written by first-rate statisticians. It goes into more depth than students need, but it helps to build a strong foundation for teachers. Second, have your students do the Response Bias Activity on page 271. It is a great way for students to implement the principles of a good experiment while learning an important lesson about how easy it is to manipulate the results of a survey. There is a sample rubric for this Activity in the Teacher's Resource Materials.

### Relationship to the AP® Statistics Course Description

Chapter 4 covers the following content as described in the Topic Outline from the AP® Statistics Course Description:

II. Sampling and experimentation: Planning and conducting a study

A. Overview of methods of data collection
1. Census
2. Sample survey
3. Experiment
4. Observational study

B. Planning and conducting studies
1. Characteristics of a well-designed and well-conducted survey
2. Population, samples, and random selection
3. Sources of bias in sampling and surveys
4. Sampling methods, including simple random sampling, stratified random sampling, and cluster sampling

C. Planning and conducting experiments
1. Characteristics of a well-designed and well-conducted experiment
2. Treatments, control groups, experimental units, random assignments, and replication
3. Sources of bias and confounding, including placebo effect and blinding
4. Completely randomized design
5. Randomized block design, including matched pairs design

D. Generalizability of results and types of conclusions that can be drawn from observational studies, experiments, and surveys

# The Big Ideas

One of the challenges in teaching the AP® Statistics course is keeping students focused on the big picture and not just the details of each section. We outline the big ideas for the chapter here.

## Chapter 4 Introduction

The introduction gives several brief examples of the types of studies we will learn about in this chapter, including sample surveys, observational studies, and experiments. It also includes an Activity to introduce the basic ideas of an experiment.

## Section 4.1 Sampling and Surveys

In this section, we begin by describing the difference between a population and a sample. This is an extremely important distinction that will come up over and over again, especially in Chapters 7–12. In almost all cases, we take a sample from a population rather than conduct a census when we want to know something about a population. Unfortunately, there are many things that can go wrong along the way. Choosing people because they are convenient or letting people choose whether or not to be included in the sample are definite problems, because these groups are not likely to be representative of any larger population. These methods of obtaining a sample are called biased because they will provide estimates that are consistently too big or consistently too small, depending on the context.

To avoid these biased methods of obtaining a sample, the members of the sample should be chosen at random. One way to do this is with a simple random sample, which is equivalent to putting the name of each member of the population on a slip of paper, mixing all the slips in a hat, and choosing the sample by selecting names from the hat one at a time. The Federalist Papers Activity on page 218 is a great way to illustrate the benefits of using a simple random sample.

Next, we introduce two other sampling techniques: stratified random sampling and cluster sampling. Stratified random samples are a little more complicated to obtain and analyze than simple random samples but can produce estimates that have greater precision. Cluster samples can be easier to obtain than simple random samples, but if the clusters are too homogeneous, there is a chance that certain groups in the population might be over- or underrepresented just by chance. Students often get these two sampling techniques confused as they both start by dividing the population into smaller subgroups. However, the strategy for forming the groups and selecting the sample is quite different.

We also briefly introduce inference for sampling—using information from a sample to draw conclusions about a population. At this point, we want students to understand the overall idea of inference and begin developing the concept of sampling variability. That is, we want students to realize that different random samples from the same population will produce different estimates of a population characteristic, just by chance. We will expand on these ideas in much more depth beginning in Chapter 7.

We conclude this section with a discussion of other things that can go wrong in sample surveys—and how to avoid these problems. Undercoverage results when the sampling method systematically over- or underrepresents one part of the population. Nonresponse occurs when answers cannot be obtained from individuals chosen to be in the sample. Undercoverage and nonresponse both lead to bias—consistently underestimating or overestimating the value we want to know. Even when the sample is chosen well and everyone in the sample responds, the members of the sample don't always respond honestly. Bias can be introduced by the wording of the question, characteristics of the interviewer, or other factors.

## Section 4.2 Experiments

In this section, we begin by contrasting observational studies and experiments. Experiments deliberately impose a treatment to see whether there is a cause-and-effect relationship between two variables. Observational studies, on the other hand, look at relationships between two variables but cannot show cause and effect because of potential confounding. Two variables are confounded when it is impossible to determine which of them is causing a change in the response variable.

Next, we introduce the basic vocabulary of experiments and give an example of a poorly designed experiment about SAT prep courses. To fix the problems in this experiment, we address four principles of experimental design: comparison, random assignment, control, and replication.

Most well-designed experiments include at least two treatments, so that different values of the explanatory variable can be compared. To create groups of experimental units that are roughly equivalent at the beginning of the study, we randomly assign the units to the treatments. This random assignment should balance out the effects of other variables so that the only systematic difference between the groups is the treatment imposed. Once the treatments are assigned to experimental units, it is important that groups are treated in exactly the same ner, other than the treatments they were assigned. This is the principle of control: making sure all other variables stay the same for all units during the experiment. Without control, we can introduce confounding and additional variability, making it much harder to draw a cause-and-effect conclusion. Finally, the principle of replication says that we should use enough experimental units in each treatment group so that a difference in the response can be attributed to the treatments, not the chance variation in the random assignment.

The best way to reinforce these principles is by designing and conducting an experiment in class. We provide several suggestions, including an experiment investigating response bias (at the end of Section 4.3) and the Alternate Examples about caffeine and pulse rates.

To avoid a poorly designed experiment, the placebo effect should be taken into account. Whenever possible, both the subjects and the person measuring the response should not know who is receiving which treatment. The Physicians' Health Study on page 243 is a perfect example of a randomized, placebo-controlled, double-blind experiment.

As with sampling in Section 4.1, we briefly introduce inference for experiments. The results of an experiment are considered statistically significant if the difference in the response is too large to be accounted for by the random assignment of experimental units to treatments. The Distracted Driving Activity illustrates this nicely. The Alternate Activity about fish oil is similar, except that the response variable is quantitative rather than categorical.

Finally, we introduce an additional principle that can enhance the design of some experiments: blocking. Blocking in experiments is similar to stratifying in sampling. To form blocks, group experimental units that are similar with respect to variables that are most strongly associated with the response. Then randomly assign the units within each block to each treatment. When done correctly, an experiment with blocking will produce a more precise estimate than a completely randomized design. Blocks that have only two experimental units are called *matched pairs*.

### Section 4.3 Using Studies Wisely

In this section, we emphasize that the types of conclusions we can draw depend on how the data are produced. When samples are selected at random, we can make inferences about the population from which the sample was drawn. When treatments are applied to groups formed at random, we can conclude cause and effect.

Drawing a cause-and-effect conclusion is often difficult because it is impossible or unethical to impose certain treatments, such as smoking cigarettes. We briefly discuss how to justify a cause-and-effect conclusion in these circumstances and discuss some ethical issues related to collecting data. In a nutshell, studies should be approved by an institutional review board, subjects should give informed consent, and individual data must be kept confidential.

# Chapter 4: Resources

## TRM Additional *TPS* Teacher's Resource Materials

The following resources can be found by clicking on the links in the new Teacher's e-Book (TE-book), opening the Teacher's Resource CD (TRCD), or logging onto the book's companion Web site www.whfreeman.com/tps5e (teacher log-in required).

- Printable Chapter 4 Learning Objective Grid (page 209)
- Materials for The Federalist Papers Activity (page 218)
- Gallup Poll Worksheet (page 226)
- Alternate Activity: Reducing cholesterol: Completely randomized design versus matched pairs design (page 255)
- Template for Experiments (page 258)
- Rubric for Response Bias Project (page 271)
- FRAPPY! materials: Question, Model Solution, Scoring Rubric, Sample Responses with commentary (page 275)
- Quizzes: 3 versions for each section; versions A and B are roughly parallel
    - Section 4.1 (page 228)
    - Section 4.2 (page 258)
    - Section 4.3 (page 273)
- Tests: 4 versions; versions A and B are roughly parallel, version D is cumulative (page 276)

## Free-Response Questions from Previous AP® Exams

Questions can be found on the AP® Central Web site: http://apcentral.collegeboard.com/apc/members/exam/exam_questions/8357.html.

Students should be able to answer all the free-response questions listed on the following page with material learned in this chapter. Questions that contain content from this chapter but also require content from later chapters are listed in the last chapter required to complete the entire question. This list will be updated after each AP® exam and will be posted on the book's Web site: http://www.whfreeman.com/tps5e/.

Questions marked with an asterisk are from exams with released multiple-choice questions. You may want to save these questions until the end of the year so you can give your students a complete released exam for practice.

| Year | # | Content |
|---|---|---|
| 2013 | 2 | • Convenience sampling and bias<br>• Selecting an SRS with a random number generator<br>• Stratified sampling |
| 2011 | 3 | • Cluster sampling<br>• Stratified sampling |
| 2011B | 2 | • Observational study versus experiment<br>• Scope of inference<br>• Purpose of random assignment |
| 2010 | 1 | • Treatments<br>• Experimental units<br>• Response variable<br>• Scatterplots and linearity |
| 2010B | 2 | • Simple random sampling<br>• Stratified random sampling |
| 2009 | 3 | • Random assignment<br>• Nonrandom assignment |
| 2008 | 2 | • Nonresponse bias |
| 2007 | 2 | • Control groups<br>• Random assignment<br>• Blocking |
| 2007B | 3 | • Blocking<br>• Randomization |
| 2006 | 5 | • Treatments<br>• Randomization<br>• Sources of variability<br>• Generalizability |
| 2006B | 5 | • Response variable<br>• Treatments<br>• Experimental units<br>• Randomization<br>• Replication<br>• Confounding |
| 2005 | 1 | • Comparing distributions (stemplots)<br>• Generalizability of results<br>• Sampling variability |
| 2004 | 2 | • Blocking<br>• Random assignment within blocks |
| 2004B | 2 | • Selection and response bias |
| 2003 | 4 | • Random assignment<br>• Control groups<br>• Generalizability |
| 2002 | 2 | • Matched pairs experiment<br>• Double-blind |
| 2002B | 3 | • Designing experiment<br>• Blocking |
| 2001 | 4 | • Blocking<br>• Purpose of randomization |
| 2000 | 5 | • Designing experiment<br>• Blocking<br>• Double-blind |
| 1999 | 3 | • Experiment versus observational study<br>• Confounding<br>• Cause and effect |
| 1997 | 2 | • Designing experiment<br>• Blocking |

## Applets

- The *Simple Random Sample* applet allows students to randomly select a simple random sample from a population with labels from 1 to N. www.whfreeman.com/tps5e (click on Statistical applets and choose Simple Random Sample).
- Use the *Test for a Difference in Proportions* and *Test for a Difference in Means* applets at www.lock5stat.com/statkey to perform the randomization-based inference procedures used in the Distracted Driving Activity and the Fish Oil Alternate Activity.

## TI-Nspire Activities

These Activities are designed to illustrate important statistical concepts using the TI-Nspire. The Activities can be found at www.whfreeman.com/tps5e.

- *Random Circles*: In this Activity, students compare the results of using a judgment sample and a random sample to estimate the average area of the circles in a population.
- *Distracted Drivers*: In this Activity, students use the TI-Nspire to re-randomize the results of an experiment to investigate whether the results are statistically significant.

## Other Resources

- *Making Sense of Statistical Studies* by Roxy Peck and Daren Starnes, published by the American Statistical Association. This module consists of 15 hands-on investigations that provide students with valuable experience in designing and analyzing statistical studies.
- The *Planning and Conducting a Study* Module at AP® Central is an excellent resource that introduces teachers to the important concepts of data collection. It also goes just beyond the AP® curriculum and illustrates how to analyze the different types of designs encountered in AP® Statistics: http://apcentral.collegeboard.com/apc/public/repository/AP_Statistics_Module_Planning_and_Conducting_a_Study.pdf.
- Newspapers: The health sections of major newspapers often have stories reporting the results of different studies: http://www.nytimes.com/pages/health/index.html, http://www.latimes.com/news/health/, and http://www.usatoday.com/news/health.
- Correlation or Causation? A wonderful collection of links to articles that blur the lines between association and causation: http://jonathan.mueller.faculty.noctrl.edu/100/correlation_or_causation.htm.

# Chapter 4: Pacing Guide, Learning Objectives, and Suggested Assignments

This pacing guide is based on a schedule with 110, 50-minute sessions before the AP® exam. The 110 days include 4 days for midterms and 10 days for AP® exam review. If you have a different number of sessions before the AP® exam, you can modify the pacing guide to suit your needs. If you have additional time, consider incorporating quizzes, released AP® free-response questions, or additional activities. See the Resources section above for suggestions.

The suggested homework assignments list odd-numbered exercises, whenever possible, so students can check their answers against the back-of-the-book answers. If you would rather students not have access to the answers while doing homework, adding 1 to the exercise numbers usually will do the trick, because the homework exercises typically are paired. For example, Exercises 1 and 2 will generally cover the same topics, but in different contexts. You may also choose to include the Recycle and Review questions at the end of each section, which review topics from previous sections or chapters. These questions are denoted with a symbol.

# Chapter 4

| Day | Topics | Learning Objectives: Students will be able to. . . | Suggested Assignment |
|---|---|---|---|
| 1 | 4.1 Introduction, The Idea of a Sample Survey, How to Sample Badly, How to Sample Well: Simple Random Sampling | • Identify the population and sample in a statistical study.<br>• Identify voluntary response samples and convenience samples. Explain how these sampling methods can lead to bias.<br>• Describe how to obtain a random sample using slips of paper, technology, or a table of random digits. | 1, 3, 5, 7, 9, 11 |
| 2 | 4.1 Other Random Sampling Methods | • Distinguish a simple random sample from a stratified random sample or cluster sample. Give the advantages and disadvantages of each sampling method. | 13, 17, 19, 21, 23, 25 |
| 3 | 4.1 Inference for Sampling, Sample Surveys: What Can Go Wrong? | • Explain how undercoverage, nonresponse, question wording, and other aspects of a sample survey can lead to bias. | 27, 29, 31, 33, 35 |
| 4 | 4.2 Observational Study versus Experiment, The Language of Experiments | • Distinguish between an observational study and an experiment.<br>• Explain the concept of confounding and how it limits the ability to make cause-and-effect conclusions.<br>• Identify the experimental units, explanatory and response variables, and treatments. | 37–42, 45, 47, 49, 51, 53, 55 |
| 5 | 4.2 How to Experiment Badly, How to Experiment Well, Completely Randomized Designs | • Explain the purpose of comparison, random assignment, control, and replication in an experiment.<br>• Describe a completely randomized design for an experiment, including how to randomly assign treatments using slips of paper, technology, or a table of random digits. | 57, 59, 61, 63, 65 |
| 6 | 4.2 Experiments: What Can Go Wrong? Inference for Experiments | • Describe the placebo effect and the purpose of blinding in an experiment.<br>• Interpret the meaning of statistically significant in the context of an experiment. | 67, 69, 71, 73 |
| 7 | 4.2 Blocking | • Explain the purpose of blocking in an experiment. Describe a randomized block design or a matched pairs design for an experiment. | 75, 77, 79, 81, 85 |
| 8 | 4.2 Class Experiments | | 83, 87–94 |
| 9 | 4.3 Scope of Inference, The Challenges of Establishing Causation | • Describe the scope of inference that is appropriate in a statistical study. | 97–104 |
| 10 | 4.3 Data Ethics (optional topic) | • Evaluate whether a statistical study has been carried out in an ethical manner. | Chapter 4 Review Exercises |
| 11 | Chapter 4 Review/FRAPPY! | | Chapter 4 AP® Statistics Practice Test |
| 12 | Chapter 4 Test | | Cumulative AP® Practice Test 1 |

**PD** **Chapter 4 Overview**

Watch the video overview of Chapter 4 (for teachers) by clicking on the link in the TE-book, opening the TRCD, or logging onto the book's companion Web site www.whfreeman.com/tps5e (teacher log-in required).

# Chapter 4



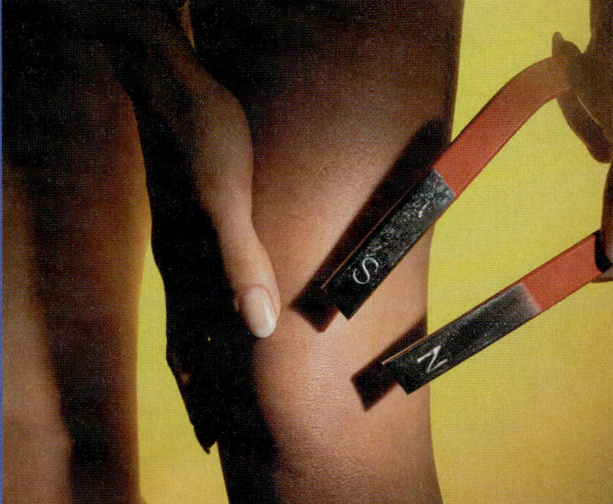

# Designing Studies

## case study

### Can Magnets Help Reduce Pain?

Early research showed that magnetic fields affected living tissue in humans. Some doctors have begun to use magnets to treat patients with chronic pain. Scientists wondered whether this type of therapy really worked. They designed a study to find out.

Fifty patients with chronic pain were recruited for the study. A doctor identified a painful site on each patient and asked him or her to rate the pain on a scale from 0 (mild pain) to 10 (severe pain). Then, the doctor selected a sealed envelope containing a magnet at random from a box with a mixture of active and inactive magnets. That way, neither the doctor nor the patient knew which type of magnet was being used. The chosen magnet was applied to the site of the pain for 45 minutes. After "treatment," each patient was again asked to rate the level of pain from 0 to 10.

In all, 29 patients were given active magnets and 21 patients received inactive magnets. Scientists decided to focus on the improvement in patients' pain ratings. Here they are, grouped by the type of magnet used:[1]

| | |
|---|---|
| **Active:** | 10 6 1 10 6 8 5 5 6 8 7 8 7 6 4 4 7 10 6 10 6 5 5 1 0 0 0 0 1 |
| **Inactive:** | 4 3 5 2 1 4 1 0 0 1 0 0 0 0 0 0 1 0 0 1 |

What do the data tell us about whether the active magnets helped reduce pain? By the end of the chapter, you'll be ready to interpret the results of this study.

207

## Introduction

You can hardly go a day without hearing the results of a statistical study. Here are some examples:

- The National Highway Traffic Safety Administration (NHTSA) reports that seat belt use in passenger vehicles increased from 84% in 2011 to 86% in 2012.[2]

- According to a recent survey, U.S. teens aged 13 to 18 spend an average of 26.8 hours per week online. Although 59% of the teens said that posting personal information or photos online is unsafe, 62% said they had posted photos of themselves.[3]
- A recent study suggests that lack of sleep increases the risk of catching a cold.[4]
- For their final project, two AP® Statistics students showed that listening to music while studying decreased subjects' performance on a memory task.[5]

Can we trust these results? As you'll learn in this chapter, that depends on how the data were produced. Let's take a closer look at where the data came from in each of these studies.

Each year, the NHTSA conducts an *observational study* of seat belt use in vehicles. The NHTSA sends trained observers to record the actual behavior of people in vehicles at randomly selected locations across the country. The idea of an observational study is simple: you can learn a lot just by watching. Or by asking a few questions, as in the survey of teens' online habits. Harris Interactive conducted this survey using a "representative sample" of 655 U.S. 13- to 18-year-olds. Both of these studies use information from a *sample* to draw conclusions about some larger *population*. Section 4.1 examines the issues involved in sampling and surveys.

In the sleep and catching a cold study, 153 volunteers took part. They answered questions about their sleep habits over a two-week period. Then, researchers gave them a virus and waited to see who developed a cold. This was a complicated observational study. Compare this with the *experiment* performed by the AP® Statistics students. They recruited 30 students and divided them into two groups of 15 by drawing names from a hat. Students in one group tried to memorize a list of words while listening to music. Students in the other group tried to memorize the same list of words while sitting in silence. Section 4.2 focuses on designing experiments.

The goal of many statistical studies is to show that changes in one variable *cause* changes in another variable. In Section 4.3, we'll look at why establishing causation is so difficult, especially in observational studies. We'll also consider some of the ethical issues involved in planning and conducting a study.

Here's an Activity that gives you a preview of what lies ahead.

### Teaching Tip

It is extremely helpful to supplement the examples in the text with current examples from the news. Your local newspaper probably includes summaries of statistical studies every day. You can also look at the Web sites of larger newspapers such as the *New York Times*, *Los Angeles Times*, or *USA Today* to find good articles. Members of the AP® Statistics Teacher Community often post links to interesting articles to use in class. Visit https://apcommunity.collegeboard.org to browse or contribute.

### ALTERNATE ACTIVITY

There are many ways to modify this Activity:

- Do students learn better when they aren't concentrating on anything else? Have one student stand on one leg and the other student stand on both legs. Read numbers to both students and see who can remember more numbers after 30 seconds.
- Do students learn better when the font size is bigger? Have one student read numbers in size 12 font and the other student read numbers in size 36 font.
- Do students learn better when the numbers are in color? Have one student read numbers printed in black ink on white paper and the other student read numbers printed in different colors on white paper.

You can also do either of the following activities from Chapter 1's Teacher's Resource Materials:

TRM Do You Know Your Geography?

TRM The Memory Game

### ALTERNATE ACTIVITY

Consider using The Federalist Papers Activity on page 218 *before* you begin Section 4.1. This Activity is a great way to motivate the need for random sampling and will probably work best if you haven't already introduced the idea of random sampling. There are black line masters available for this Activity in the Teacher's Resource Materials. Find them by clicking on the link in the TE-book, opening the TRCD, or logging onto the book's companion Web site www.whfreeman.com/tps5e (teacher log-in required).

TRM The Federalist Papers

## ACTIVITY | See no evil, hear no evil?

**MATERIALS:**
Two index cards, each with 10 distinct numbers from 00 to 99 written on it (prepared by your teacher); clock, watch, or stopwatch to measure 30 seconds; and a coin for each pair of students

Confucius said, "I hear and I forget. I see and I remember. I do and I understand." Do people really remember what they see better than what they hear?[6] In this Activity, you will perform an experiment to try to find out.

1. Divide the class into pairs of students by drawing names from a hat.
2. Your teacher will give each pair two index cards with 10 distinct numbers from 00 to 99 on them. *Do not look at the numbers* until it is time for you to do the experiment.
3. Flip a coin to decide which of you is Student 1 and which is Student 2. Shuffle the index cards and deal one face down to each partner.
4. Student 1 will be the first to attempt a memory task while Student 2 keeps time.

*Directions:* Study the numbers on the index card for 30 seconds. Then turn the card over. Recite the alphabet aloud (A, B, C, and so on). Then tell your partner what you think the numbers on the card are. *You may not say more than 10 numbers!* Student 2 will record how many numbers you recalled correctly.

5. Now it's Student 2's turn to do a memory task while Student 1 records the data.

*Directions:* Your partner will read the numbers on your index card aloud three times slowly. Next, you will recite the alphabet aloud (A, B, C, and so on) and then tell your partner what you think the numbers on the card are. *You may not say more than 10 numbers!* Student 1 will record how many numbers you recalled correctly.

6. Your teacher will scale and label axes on the board for parallel dotplots of the results. Plot how many numbers you remembered correctly on the appropriate graph.
7. Did students in your class remember numbers better when they saw them or when they heard them? Give appropriate evidence to support your answer.
8. Based on the results of this experiment, can we conclude that people in general remember better when they see than when they hear? Why or why not?

# 4.1 Sampling and Surveys

**WHAT YOU WILL LEARN** By the end of the section, you should be able to:

- Identify the population and sample in a statistical study.
- Identify voluntary response samples and convenience samples. Explain how these sampling methods can lead to bias.
- Describe how to obtain a random sample using slips of paper, technology, or a table of random digits.
- Distinguish a simple random sample from a stratified random sample or cluster sample. Give the advantages and disadvantages of each sampling method.
- Explain how undercoverage, nonresponse, question wording, and other aspects of a sample survey can lead to bias.

### Activity Overview

You can watch an overview of this Activity (for teachers) by clicking on the link in the TE-book, opening the TRCD, or logging onto the book's companion Web site www.whfreeman.com/tps5e (teacher log-in required).

### Teaching Tip

For the Activity, use 10 *unique* numbers from 00 to 99 on each card. To generate random numbers from 00 to 99, use pairs of digits from Table D in the back of the book or use `randInt(0,99)` on your calculator.

### PD Section 4.1 Overview

Watch the video overview of Section 4.1 (for teachers) by clicking on the link in the TE-book, opening the TRCD, or logging onto the book's companion Web site www.whfreeman.com/tps5e (teacher log-in required).

### TRM Learning Objectives Grid

Remind students about the importance of the Learning Objectives. We include them at the beginning of each section so students know what they are expected to learn. At the end of the chapter, the Learning Objectives are included in a grid, along with references to examples and chapter review exercises. A printable version of this grid is included in the Teacher's Resource Materials.

### TRM Section 4.1 Alternate Examples

You can find the Alternate Examples for this section in Microsoft Word format by clicking on the link in the TE-book, opening the TRCD, or logging onto the book's companion Web site www.whfreeman.com/tps5e (teacher log-in required).

Suppose we want to find out what percent of young drivers in the United States text while driving. To answer the question, we will survey 16- to 20-year-olds who live in the United States and drive. Ideally, we would ask them all (take a **census**). But contacting every driver in this age group wouldn't be practical: it would take too much time and cost too much money. Instead, we put the question to a **sample** chosen to represent the entire **population** of young drivers.

> **DEFINITION: Population, census, and sample**
>
> The **population** in a statistical study is the entire group of individuals we want information about. A **census** collects data from every individual in the population.
>
> A **sample** is a subset of individuals in the population from which we actually collect data.

The distinction between population and sample is basic to statistics. To make sense of any sample result, you must know what population the sample represents. Here's an example that illustrates this distinction and also introduces some major uses of sampling.

## EXAMPLE Sampling Hardwood and Humans

### *Populations and samples*

PROBLEM: Identify the population and the sample in each of the following settings.

(a) A furniture maker buys hardwood in large batches. The supplier is supposed to dry the wood before shipping (wood that isn't dry won't hold its size and shape). The furniture maker chooses five pieces of wood from each batch and tests their moisture content. If any piece exceeds 12% moisture content, the entire batch is sent back.

(b) Each week, the Gallup Poll questions a sample of about 1500 adult U.S. residents to determine national opinion on a wide variety of issues.

SOLUTION:

(a) The population is all the pieces of hardwood in a batch. The sample is the five pieces of wood that are selected from that batch and tested for moisture content.

(b) Gallup's population is all adult U.S. residents. Their sample is the 1500 adults who actually respond to the survey questions.

**For Practice** *Try Exercise* **1**

## The Idea of a Sample Survey

We often draw conclusions about a whole population on the basis of a sample. Have you ever tasted a sample of ice cream and ordered a cone if the sample tastes good? Because ice cream is fairly uniform, the single taste represents the whole. Choosing a representative sample from a large and varied population (like all young U.S. drivers) is not so easy. The first step in planning a **sample survey** is to say exactly *what population* we want to describe. The second step is to say exactly *what we want to measure*, that is, to give exact definitions of our variables.

### Teaching Tip

Emphasize that the population is the group of individuals we want to know about (sometimes called the population of interest), not necessarily the population we can make inferences about (sometimes called the population of inference). For example, if there were some individuals who were not eligible to be selected for the sample, then we should not make inferences about them, even if we want to.

### Teaching Tip

Remind students that the "play" icon at the top of this example (and many others) indicates that they can watch a video presentation of the example. Find the videos at www.whfreeman.com/tps5e. Also, remind them that the icon in the lower-right corner indicates that there is a link between this example and Exercise 1. If students are having trouble with the exercise, they can look back to this example for help.

### ALTERNATE EXAMPLE

#### *Sampling students and soda*

PROBLEM: Identify the population and sample in each of the following settings.

**(a)** The student government at a high school surveys 100 students at the school to get their opinions about a change to the bell schedule.

**(b)** The quality control manager at a bottling company selects a sample of 10 cans from the production line every hour to see whether the volume of the soda is within acceptable limits.

SOLUTION:

**(a)** Population is all students at the school; sample is the 100 students surveyed.

**(b)** Population is all cans produced that hour; sample is the 10 cans inspected.

We reserve the term "sample survey" for studies that use an organized plan to choose a sample that represents some specific population, like the pieces of hardwood and the U.S. adults in the previous example. By our definition, the population in a sample survey can consist of people, animals, or things. Some people use the terms "survey" or "sample survey" to refer only to studies in which people are asked one or more questions, like the Gallup Poll of the last example. We'll avoid this more restrictive terminology.

EXAMPLE

## How Does the Current Population Survey Work?

### *A sample survey*

One of the most important government sample surveys in the United States is the monthly Current Population Survey (CPS). The CPS contacts about 60,000 households each month. It produces the monthly unemployment rate and lots of other economic and social information. To measure unemployment, we must first specify the population we want to describe. The CPS defines its population as all U.S. residents (legal or not) 16 years of age and over who are civilians and are not in an institution such as a prison. The unemployment rate announced in the news refers to this specific population.

What does it mean to be "unemployed"? Someone who is not looking for work—for example, a full-time student—should not be called unemployed just because she is not working for pay. If you are chosen for the CPS sample, the interviewer first asks whether you are available to work and whether you actually looked for work in the past four weeks. If not, you are neither employed nor unemployed—you are not in the labor force.

If you are in the labor force, the interviewer goes on to ask about employment. If you did any work for pay or in your own business during the week of the survey, you are employed. If you worked at least 15 hours in a family business without pay, you are employed. You are also employed if you have a job but didn't work because of vacation, being on strike, or other good reason. An unemployment rate of 9.7% means that 9.7% of the sample was unemployed, using the exact CPS definitions of both "labor force" and "unemployed."

The final step in planning a sample survey is to decide how to choose a sample from the population. Let's take a closer look at some good and not-so-good sampling methods.

## How to Sample Badly

The sampling method that yields a convenience sample is called *convenience sampling*. Other sampling methods are named in similarly obvious ways!

Suppose we want to know how long students at a large high school spent doing homework last week. We might go to the school library and ask the first 30 students we see about their homework time. The sample we get is known as a **convenience sample.**

ALTERNATE EXAMPLE

### *How is consumer confidence measured?*

Each month, the Conference Board takes a representative sample of 5000 households and asks them five questions about their appraisal of current business and employment conditions and about their expectations regarding business conditions, employment conditions, and total family income 6 months from the time of the survey. For each question, there are three possible responses: positive, negative, and neutral. These are combined into an index that is calculated relative to the value of 100 in 1985. In July 2013 the index stood at 80.3, which was weak with respect to historical levels but much better than in the recent past. In May 2010, the index stood at 63.3!

**DEFINITION: Convenience sample**

Choosing individuals from the population who are easy to reach results in a **convenience sample**.

*Convenience sampling often produces unrepresentative data.* Consider our sample of 30 students from the school library. It's unlikely that this convenience sample accurately represents the homework habits of all students at the high school. In fact, if we were to repeat this sampling process again and again, we would almost always overestimate the average homework time in the population. Why? Because students who hang out in the library tend to be more studious. This is **bias:** using a method that favors some outcomes over others.

**DEFINITION: Bias**

The design of a statistical study shows **bias** if it would consistently underestimate or consistently overestimate the value you want to know.

**AP® EXAM TIP** If you're asked to describe how the design of a study leads to bias, you're expected to do two things: (1) identify a problem with the design, and (2) explain how this problem would lead to an underestimate or overestimate. Suppose you were asked, "Explain how using your statistics class as a sample to estimate the proportion of all high school students who own a graphing calculator could result in bias." You might respond, "This is a convenience sample. It would probably include a much higher proportion of students with a graphing calculator than in the population at large because a graphing calculator is required for the statistics class. So this method would probably lead to an overestimate of the actual population proportion."

*Bias is not just bad luck in one sample.* It's the result of a bad study design that will consistently miss the truth about the population in the same way. Convenience samples are almost guaranteed to show bias. So are **voluntary response samples**.

Voluntary response samples are also known as *self-selected samples.*

**DEFINITION: Voluntary response sample**

A **voluntary response sample** consists of people who choose themselves by responding to a general invitation.

The Internet brings voluntary response samples to the computer nearest you. Visit www.misterpoll.com to become part of the sample in any of dozens of online polls. As the site says, "None of these polls are 'scientific,' but do represent the collective opinion of everyone who participates." Unfortunately, such polls don't tell you anything about the views of the population.

Call-in, text-in, write-in, and many Internet polls rely on voluntary response samples. *People who choose to participate in such surveys are usually not representative of some larger population of interest.* Voluntary response samples attract people who feel strongly about an issue, and who often share the same opinion. That leads to bias.

## Teaching Tip

To say that a sampling method is biased, it has to *consistently* underestimate or *consistently* overestimate the value we want to know in repeated sampling. If a sampling method is *unbiased*, it will still produce estimates that differ from the value we want to know simply by chance. However, these estimates will be too small about half the time and too large about half the time.

## Teaching Tip

Emphasize the AP® Exam Tip on this page. For example, suppose that the parent of a football player is concerned about proposed budget cuts to the athletic program at the school, so she asks spectators at a game if they agree with the proposed cuts. This method of obtaining a sample is biased because it systematically favors a "No" outcome. That is, if the parent were to conduct a similar survey over and over again, each time the result will probably overestimate the proportion of people who would respond "No." See Question 2a on the 2013 AP® exam for a similar example.

## Teaching Tip

Mention that another problem with voluntary response samples is that people can respond more than once. Television shows such as *American Idol* and *Dancing with the Stars* are examples of polls that use voluntary response samples.

## Teaching Tip

In addition to www.misterpoll.com, visit the Web sites of your local newspaper and television stations. These often conduct polls that your students can participate in and see the results for the following day. Changing the privacy settings on your Internet browser to high may allow you to vote more than once.

## EXAMPLE Illegal Immigration

### *Online polls*

Former CNN commentator Lou Dobbs doesn't like illegal immigration. One of his shows was largely devoted to attacking a proposal to offer driver's licenses to illegal immigrants. During the show, Mr. Dobbs invited his viewers to go to loudobbs.com to vote on the question "Would you be more or less likely to vote for a presidential candidate who supports giving driver's licenses to illegal aliens? The result: 97% of the 7350 people who voted by the end of the show said, "Less likely."

**PROBLEM:** What type of sample did Mr. Dobbs use in his poll? Explain how this sampling method could lead to bias in the poll results.

**SOLUTION:** Mr. Dobbs used a voluntary response sample: people chose to go online and respond. Those who voted were viewers of Mr. Dobbs's program, which means that they are likely to support his views. The 97% poll result is probably an extreme overestimate of the percent of people in the population who would be less likely to support a presidential candidate with this position.

**For Practice** *Try Exercise* 9

### CHECK YOUR UNDERSTANDING

For each of the following situations, identify the sampling method used. Then explain how the sampling method could lead to bias.

1. A farmer brings a juice company several crates of oranges each week. A company inspector looks at 10 oranges from the top of each crate before deciding whether to buy all the oranges.
2. The ABC program *Nightline* once asked whether the United Nations should continue to have its headquarters in the United States. Viewers were invited to call one telephone number to respond "Yes" and another for "No." There was a charge for calling either number. More than 186,000 callers responded, and 67% said "No."

## How to Sample Well: Simple Random Sampling

In convenience sampling, the researcher chooses easy-to-reach members of the population. In voluntary response sampling, people decide whether to join the sample. Both sampling methods suffer from bias due to personal choice. The best way to avoid this problem is to let chance choose the sample. That's the idea of **random sampling**.

### ALTERNATE EXAMPLE

***More on illegal immigration***

In May 2010 the Los Angeles City Council voted to ban most travel and contracts with the state of Arizona to protest Arizona's new immigration enforcement law. The *Los Angeles Times* conducted an online poll that asked whether the City Council was right to pass a boycott of Arizona. The results showed that 96% of the 41,068 people in the sample said "No."

**PROBLEM:** What type of sample did the *Times* use in this poll? Explain how this sampling method could lead to bias in the poll results.

**SOLUTION:** This voluntary response sample represents only the opinions of those who participated in the poll. The members of the sample probably feel much more strongly about this issue than those who chose not to participate. The 96% value is likely to be greater than the true percentage of Los Angeles residents who would respond "No."

### ANSWERS TO CYU

1. Convenience sampling. This could lead the inspector to overestimate the quality of the oranges if the farmer puts the best oranges on top.
2. Voluntary response sampling. In this case, those who are happy that the UN has its headquarters in the United States already have what they want and so are less likely to respond. The proportion who answered "No" in the sample is likely to be higher than the true proportion in the United States who would answer "No."

**DEFINITION: Random Sampling**

**Random sampling** involves using a chance process to determine which members of a population are included in the sample.

In everyday life, some people use the word "random" to mean haphazard, as in "that's so random." In statistics, random means "due to chance." Don't say that a sample was chosen at random if a chance process wasn't used to select the individuals.

The easiest way to choose a random sample of $n$ people is to write their names on identical slips of paper, put the slips in a hat, mix them well, and pull out slips one at a time until you have $n$ of them. An alternative would be to give each member of the population a distinct number and to use the "hat method" with these numbers instead of people's names. Note that this version would work just as well if the population consisted of animals or things. The resulting sample is called a **simple random sample,** or SRS for short.

**DEFINITION: Simple Random Sample (SRS)**

A **simple random sample** (SRS) of size $n$ is chosen in such a way that every group of $n$ individuals in the population has an equal chance to be selected as the sample.

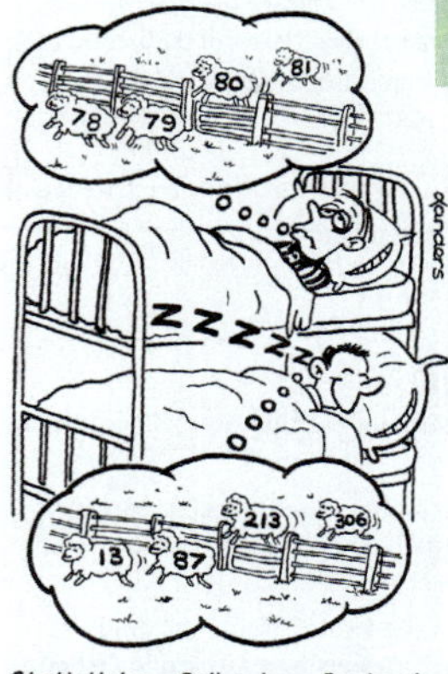


*Statisticians fall asleep faster by taking a random sample of sheep.*

An SRS gives every possible sample of the desired size an equal chance to be chosen. It also gives each member of the population an equal chance to be included in the sample. Picture drawing 20 slips (the sample) from a hat containing 200 identical slips (the population). Any 20 slips have the same chance as any other 20 to be chosen. Also, each slip has a 1-in-10 chance (20/200) of being selected.

Some other random sampling methods give each member of the population, but not each sample, an equal chance. We'll look at some of these later.

**How to Choose a Simple Random Sample** The hat method won't work well if the population is large. Imagine trying to take a simple random sample of 1000 U.S. adults! In practice, most people use random numbers generated by technology to choose samples.

## EXAMPLE Teens on the Internet

### *Choosing an SRS with technology*

The principal at Canyon del Oro High School in Arizona wants student input about limiting access to certain Internet sites on the school's computers. He asks the AP® Statistics teacher, Mr. Tabor, to select a "representative sample" of 10 students. Mr. Tabor decides to take an SRS from the 1750 students enrolled this year.

## Teaching Tip

To help illustrate the definition of an SRS, describe how to select a sample of size 4 from your class. To do this, you could put each student's name on a slip of paper, mix the slips up in a hat, and select 4 of them. Try this a few times and comment on the number of girls and boys in each sample. Then suggest that we might want to force the sample to have 2 girls and 2 boys. To do this, separate the names of the boys and girls into two hats and select 2 names from each hat. This is not an SRS because samples of 4 that do not contain 2 girls and 2 boys have no chance of being selected. In an SRS, each possible sample of 4 has the same chance of occurring.

**ALTERNATE EXAMPLE**

### *Going flat*

Environmental researchers want to know what proportion of drivers have cars with underinflated tires. They number the cars in the parking lot of a local mall from 1 to 583 and use the random integer generator at www.random.org/integer-sets to select 50 different integers from 1 to 583. The cars corresponding to the 50 numbers listed below will be inspected.

```
400 449 554 446 533 154 470 354 68 259 351 242 62 442 279 84 552 404 578 186 308 178 228 340 65
292 238 21 436 194 50 350 41 566 131 235 296 78 330 479 572 364 19 187 405 285 220 561 173 488
```

He gets an alphabetical roster from the registrar's office, and numbers the students from 1 to 1750. Then Mr. Tabor uses the random number generator at www.randomizer.org to choose 10 distinct numbers between 1 and 1750:

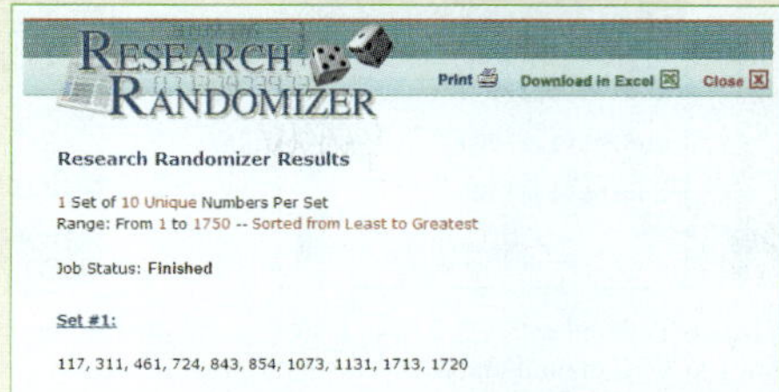


The 10 students on the roster that correspond to the chosen numbers will be on the principal's committee.

This example highlights the steps in **choosing a simple random sample with technology**.

It is standard practice to use *n* for the sample size and *N* for the population size.

**CHOOSING AN SRS WITH TECHNOLOGY**

**Step 1: Label.** Give each individual in the population a distinct numerical label from 1 to *N*.

**Step 2: Randomize.** Use a random number generator to obtain *n* different integers from 1 to *N*.

You can also use a graphing calculator to choose an SRS.

## 10. TECHNOLOGY CORNER — CHOOSING AN SRS

TI-Nspire instructions in Appendix B; HP Prime instructions on the book's Web site.

Let's use a graphing calculator to select an SRS of 10 students from the Canyon del Oro High School roster.

1. Check that your calculator's random number generator is working properly.

**TI-83/84**

- Press MATH, then select PRB and randInt(. Complete the command randInt(1,1750) and press ENTER.

**TI-89**

- Press CATALOG, then F3 (Flash Apps) and choose randInt(. Complete the command TIStat.randInt(1,1750) and press ENTER.

Compare your results with those of your classmates. If several students got the same number, you'll need to seed your calculator's random integer generator with different numbers before you proceed. Directions for doing this are given in the *Annotated Teacher's Edition*.

### AP® Exam Common Error

When describing how to select a sample using a random integer generator, many students forget to address what to do with repeated integers. Students must explicitly state that repeated integers should be ignored or say that they will generate random integers until they get *n* different numbers in the specified range.

### TECHNOLOGY TIP

Remind students that they can watch a video presentation of this technology corner (for the TI-83/84, TI-89, or TI-Nspire) at www.whfreeman.com/tps5e.

### TECHNOLOGY TIP

To seed the random number generator on the TI-83/84, have students enter a unique number on their home screen, such as their phone number or student ID number. Then have them press the STO→ button (just above the ON button). Finally, have them press the MATH button, scroll to the PROB menu, choose the first option, rand, and press ENTER, as shown.

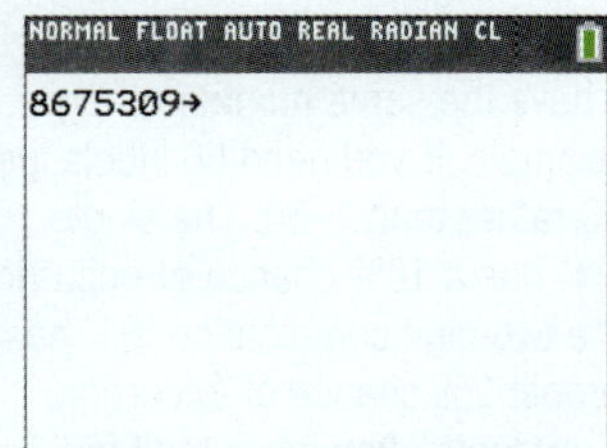


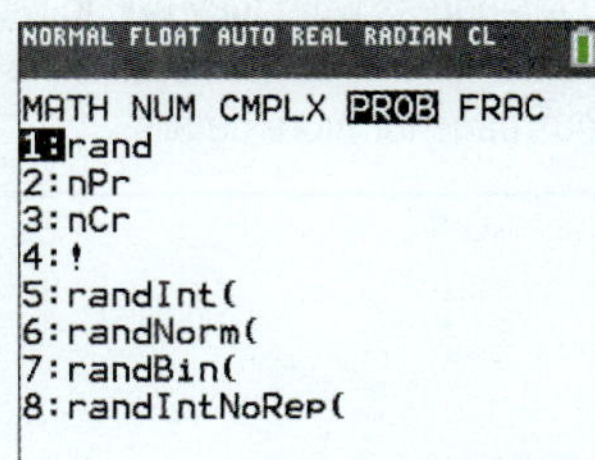


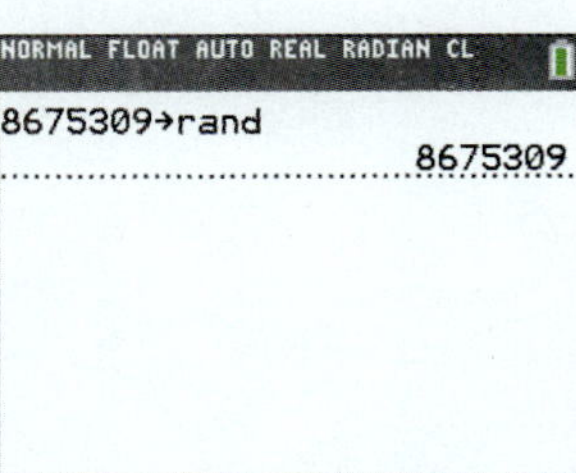


### TECHNOLOGY TIP

To seed the random number generator on the TI-89, from the home screen, press CATALOG and scroll down to RandSeed. Press ENTER to select the command. Then enter a unique number and press ENTER, as shown.

### TECHNOLOGY TIP

It is also possible to generate sets of random numbers at one time. For example, the command `randInt(1,28,4)` will generate four random integers from 1 to 28. However, because each number is independently generated, you may not get four *unique* numbers with this command.

### AP® Exam Tip

Occasionally, students are asked to use random digits on the AP® exam, so make sure to have your students use a table of random digits from time to time.

### AP® Exam Common Error

When working with a table of random digits, it is very important that each label have the same number of digits. For example, if you need 50 labels, use 01–50 rather than 1–50. The single digit "1" has a 10% chance of occurring but the two-digit combination "01" has the proper 2% chance of occurring. Also, students often lose credit for failing to address what they will do with a repeated label. See the note on the previous page for more details.

2. Randomly generate 10 distinct numbers from 1 to 1750.
   Do `randInt(1,1750)` again. Keep pressing ENTER until you have chosen 10 different labels.

```
NORMAL FLOAT AUTO REAL RADIAN CL
randInt(1,1750)
                139
randInt(1,1750)
               1126
randInt(1,1750)
                920
randInt(1,1750)
               1089
```

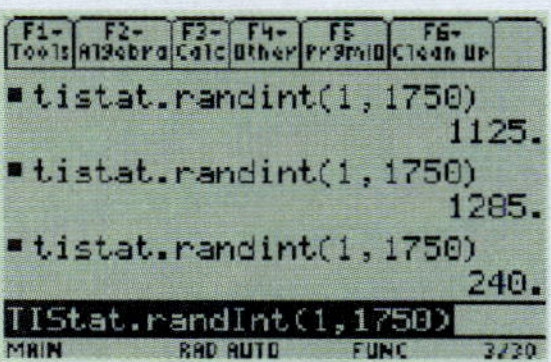


*Note:* If you have a TI-83/84 with OS 2.55 or later, you can use the command `RandIntNoRep(1,1750)` to sort the numbers from 1 to 1750 in random order. The first 10 numbers listed give the labels of the chosen students.

If you don't have technology handy, you can use a table of random digits to choose an SRS. We have provided a table of random digits at the back of the book (Table D). Here is an excerpt.

**Table D Random digits**

| LINE | | | | | | | | |
|---|---|---|---|---|---|---|---|---|
| 101 | 19223 | 95034 | 05756 | 28713 | 96409 | 12531 | 42544 | 82853 |
| 102 | 73676 | 47150 | 99400 | 01927 | 27754 | 42648 | 82425 | 36290 |
| 103 | 45467 | 71709 | 77558 | 00095 | 32863 | 29485 | 82226 | 90056 |

You can think of this table as the result of someone putting the digits 0 to 9 in a hat, mixing, drawing one, replacing it, mixing again, drawing another, and so on. The digits have been arranged in groups of five within numbered rows to make the table easier to read. The groups and rows have no special meaning—Table D is just a long list of randomly chosen digits. As with technology, there are two steps in using Table D to choose a random sample.

#### HOW TO CHOOSE AN SRS USING TABLE D

**Step 1: Label.** Give each member of the population a numerical label with the same number of digits. Use as few digits as possible.

**Step 2: Randomize.** Read consecutive groups of digits of the appropriate length from left to right across a line in Table D. Ignore any group of digits that wasn't used as a label or that duplicates a label already in the sample. Stop when you have chosen $n$ different labels.

Your sample contains the individuals whose labels you find.

Always use the shortest labels that will cover your population. For instance, you can label up to 100 individuals with two digits: 01, 02, . . . , 99, 00. As standard practice, we recommend that you begin with label 1 (or 01 or 001 or 0001, as needed). Reading groups of digits from the table gives all individuals the same chance to be chosen because all labels of the same length have the same chance

to be found in the table. For example, any pair of digits in the table is equally likely to be any of the 100 possible labels 01, 02, . . . , 99, 00. Here's an example that shows how this process works.

## EXAMPLE Spring Break!

### Choosing an SRS with Table D

The school newspaper is planning an article on family-friendly places to stay over spring break at a nearby beach town. The editors intend to call 4 randomly chosen hotels to ask about their amenities for families with children. They have an alphabetized list of all 28 hotels in the town.

PROBLEM: Use Table D at line 130 to choose an SRS of 4 hotels for the editors to call.

SOLUTION: We'll use the two-step process for selecting an SRS using Table D.

Step 1: Label. Two digits are needed to label the 28 hotels. We have added labels 01 to 28 to the alphabetized list of hotels below.

| | | | |
|---|---|---|---|
| 01 Aloha Kai | 08 Captiva | 15 Palm Tree | 22 Sea Shell |
| 02 Anchor Down | 09 Casa del Mar | 16 Radisson | 23 Silver Beach |
| 03 Banana Bay | 10 Coconuts | 17 Ramada | 24 Sunset Beach |
| 04 Banyan Tree | 11 Diplomat | 18 Sandpiper | 25 Tradewinds |
| 05 Beach Castle | 12 Holiday Inn | 19 Sea Castle | 26 Tropical Breeze |
| 06 Best Western | 13 Lime Tree | 20 Sea Club | 27 Tropical Shores |
| 07 Cabana | 14 Outrigger | 21 Sea Grape | 28 Veranda |

Step 2: Randomize. To use Table D, start at the left-hand side of line 130 and read two-digit groups. Skip any groups that aren't between 01 and 28, as well as any repeated groups. Continue until you have chosen four hotels. Here is the beginning of line 130:

69051 64817 87174 09517 84534 06489 87201 97245

The first 10 two-digit groups are

| 69 | 05 | 16 | 48 | 17 | 87 | 17 | 40 | 95 | 17 |
|---|---|---|---|---|---|---|---|---|---|
| Skip Too big | ✓ | ✓ | Skip Too big | ✓ | Skip Too big | Skip Repeat | Skip Too big | Skip Too big | Skip Repeat |

We skip 5 of these 10 groups because they are too high (over 28) and 2 because they are repeats (both 17s). The hotels labeled 05, 16, and 17 go into the sample. We need one more hotel to complete the sample. Continuing along line 130:

| 84 | 53 | 40 | 64 | 89 | 87 | 20 |
|---|---|---|---|---|---|---|
| Skip Too big | Skip Too big | Skip Too big | Skip Too big | Skip Too big | Skip Too big | ✓ |

Our SRS of 4 hotels for the editors to contact is 05 Beach Castle, 16 Radisson, 17 Ramada, and 20 Sea Club.

For Practice *Try Exercise* 11

### Teaching Tip

Some students might wonder why we decided to use line 130 in Table D as opposed to some other line. Tell them that it doesn't really matter which line you start on, as long as you don't "cheat" and pick a line based on the random digits it contains.

### ALTERNATE EXAMPLE

#### *Mall hours*

The management company of a local mall with 21 stores plans to survey 3 of them to determine the hours they would like to stay open during the holiday season.

PROBLEM: Use Table D at line 101 to select an SRS of 3 stores.

SOLUTION: Step 1: Label. Two digits are needed to label the 21 stores. The list of stores and their labels are shown below.

| | |
|---|---|
| 01 Aeropostale | 12 Just Sports |
| 02 All American Burger | 13 Mrs. Fields |
| 03 Arby's | 14 Nike Factory Store |
| 04 Barnes & Noble | 15 Old Navy |
| 05 Carter's for Kids | 16 Pac Sun |
| 06 Destination Tan | 17 Panda Express |
| 07 Famous Footwear | 18 Payless Shoes |
| 08 Forever 21 | 19 Star Jewelers |
| 09 GameStop | 20 Vitamin World |
| 10 Gymboree | 21 Zales Diamond Store |
| 11 Haggar | |

Step 2: Randomize. Using line 101, start on the left and read two-digit numbers until 3 distinct numbers from 01 to 21 are selected. Use the stores that correspond to these numbers in the sample. Using line 101, here are the selected stores: 19 (Star Jewelers), 22 (skip), 39 (skip), 50 (skip), 34 (skip), 05 (Carter's for Kids), 75 (skip), 62 (skip), 87 (skip), 13 (Mrs. Fields).

We can trust results from an SRS, as well as from other types of random samples that we will meet later, because the use of impersonal chance avoids bias. The following activity shows why random sampling is so important.

## ACTIVITY | Who Wrote the Federalist Papers?

The Federalist Papers are a series of 85 essays supporting the ratification of the U.S. Constitution. At the time they were published, the identity of the authors was a secret known to just a few people. Over time, however, the authors were identified as Alexander Hamilton, James Madison, and John Jay. The authorship of 73 of the essays is fairly certain, leaving 12 in dispute. However, thanks in some part to statistical analysis,[7] most scholars now believe that the 12 disputed essays were written by Madison alone or in collaboration with Hamilton.[8]

There are several ways to use statistics to help determine the authorship of a disputed text. One method is to estimate the average word length in a disputed text and compare it to the average word lengths of works where the authorship is not in dispute.

The following passage is the opening paragraph of Federalist Paper #51,[9] one of the disputed essays. The theme of this essay is the separation of powers between the three branches of government.

> To what expedient, then, shall we finally resort, for maintaining in practice the necessary partition of power among the several departments, as laid down in the Constitution? The only answer that can be given is, that as all these exterior provisions are found to be inadequate, the defect must be supplied, by so contriving the interior structure of the government as that its several constituent parts may, by their mutual relations, be the means of keeping each other in their proper places. Without presuming to undertake a full development of this important idea, I will hazard a few general observations, which may perhaps place it in a clearer light, and enable us to form a more correct judgment of the principles and structure of the government planned by the convention.

1. Choose 5 words from this passage. Count the number of letters in each of the words you selected, and find the average word length.
2. Your teacher will draw and label a horizontal axis for a class dotplot. Plot the average word length you obtained in Step 1 on the graph.
3. Use a table of random digits or a random number generator to select a simple random sample of 5 words from the 130 words in the opening passage. Count the number of letters in each of the words you selected, and find the average word length.
4. Your teacher will draw and label another horizontal axis with the same scale for a comparative class dotplot. Plot the average word length you obtained in Step 3 on the graph.
5. How do the dotplots compare? Can you think of any reasons why they might be different? Discuss with your classmates.

### Teaching Tip

Emphasize the importance of *random* sampling. Not only does it help us make generalizations about a population without bias, but it is also the foundation for many of the inference procedures we will learn about in Chapters 8–12.

 **Activity Overview**

You can watch an overview of this Activity (for teachers) by clicking on the link in the TE-book, opening the TRCD, or logging onto the book's companion Web site www.whfreeman.com/tps5e (teacher log-in required).

**TRM The Federalist Papers**

There are black line masters available for this Activity in the Teacher's Resource Materials. Find them by clicking on the link in the TE-book, opening the TRCD, or logging onto the book's companion Web site www.whfreeman.com/tps5e (teacher log-in required).

### Teaching Tip

For a more modern example of using statistics to help identify authorship, see the following article about how J. K. Rowling was identified as the author of *The Cuckoo's Calling*: http://phenomena.nationalgeographic.com/2013/07/19/how-forensic-linguistics-outed-j-k-rowling-not-to-mention-james-madison-barack-obama-and-the-rest-of-us/.

### Teaching Tip

When the authors have done this Activity in their classes, their students almost always overestimate the true average word length ($\mu = 4.9$ letters). Can you say bias? Here are dotplots showing the estimated averages when students chose the sample and when we let random chance choose the sample.

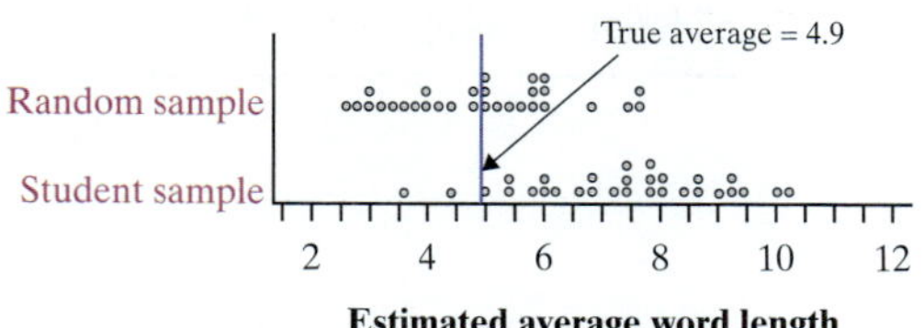

## Other Random Sampling Methods

The basic idea of sampling is straightforward: take an SRS from the population and use your sample results to gain information about the population. Unfortunately, it's usually difficult to get an SRS from the population of interest. Imagine trying to get a simple random sample of all the batteries produced in one day at a factory. Or an SRS of all U.S. high school students. In either case, it would be difficult to obtain an accurate list of the population from which to draw the sample. It would also be very time-consuming to collect data from each individual that's randomly selected. Sometimes, there are also statistical advantages to using more complex sampling methods.

One of the most common alternatives to an SRS involves sampling groups (**strata**) of similar individuals within the population separately. Then these separate "subsamples" are combined to form one **stratified random sample**.

Stratum is singular. Strata are plural.

**DEFINITION: Stratified random sample and strata**

To get a **stratified random sample**, start by classifying the population into groups of similar individuals, called **strata**. Then choose a separate SRS in each stratum and combine these SRSs to form the sample.

Choose the strata based on facts known before the sample is taken. For example, in a study of sleep habits on school nights, the population of students in a large high school might be divided into freshman, sophomore, junior, and senior strata. In a preelection poll, a population of election districts might be divided into urban, suburban, and rural strata. Stratified random sampling works best when the individuals within each stratum are similar with respect to what is being measured and when there are large differences between strata. The following Activity makes this point clear.

### ACTIVITY | Sampling sunflowers

**MATERIALS:**
Calculator for each student

A British farmer grows sunflowers for making sunflower oil. Her field is arranged in a grid pattern, with 10 rows and 10 columns as shown in the figure on the next page. Irrigation ditches run along the top and bottom of the field. The farmer would like to estimate the number of healthy plants in the field so she can project how much money she'll make from selling them. It would take too much time to count the plants in all 100 squares, so she'll accept an estimate based on a sample of 10 squares.

1. Use Table D or technology to take a simple random sample of 10 grid squares. Record the location (for example, B6) of each square you select.
2. This time, you'll take a stratified random sample using the *rows* as strata. Use Table D or technology to randomly select one square from each (horizontal) row. Record the location of each square—for example, Row 1: G, Row 2: B, and so on.

**Activity Overview**

You can watch an overview of this Activity (for teachers) by clicking on the link in the TE-book, opening the TRCD, or logging onto the book's companion Web site www.whfreeman.com/tps5e (teacher log-in required).

**ALTERNATE ACTIVITY**

*Rolling down the river*

For a similar, but more in-depth Activity illustrating the benefits of stratified random sampling and how to choose strata, see the Rolling Down the River Activity at http://courses.ncssm.edu/math/Stat_inst01/PDFS/river.pdf.

## Teaching Tip

After completing Step 3 of the Sampling Sunflowers Activity, ask students which method they believe will give the best estimate. Make sure they give a justification! In short, students should recognize that the irrigation ditches will be a source of variability in the number of sunflowers. Stratifying by row will give the best estimates because it will take a sample from each different distance from the ditches. In the other two methods, it is possible that squares selected will all be close to the ditches or far from the ditches. The dotplots on the bottom of the page confirm this—stratifying by row gives the most precise estimates.

## Teaching Tip

Make sure that students can explain how to choose a variable for stratification *and* why stratified random samples are sometimes better than simple random samples. To do this, introduce a context and ask students what variables would be good to stratify by and which variables would *not* be good. The best variables to use for stratification are those that would most accurately predict the response. For example, suppose that Congress was proposing a massive change to the Social Security program. Age would be a good variable to use when forming strata because older people probably have very different opinions about Social Security than younger people do. Using age to form strata will produce a more precise estimate of the entire population's opinion about the proposed change. Likewise, income would be a good variable to stratify by as wealthy people probably have very different opinions about Social Security than poor people. However, stratifying by gender would not be very useful as men and women aren't likely to differ much in their opinions about Social Security.

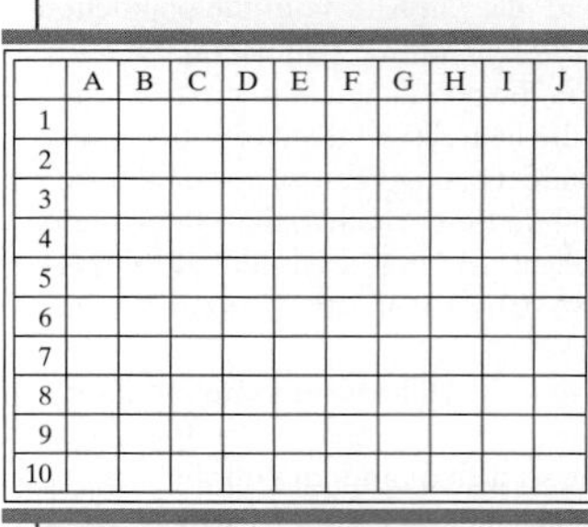

| | A | B | C | D | E | F | G | H | I | J |
|---|---|---|---|---|---|---|---|---|---|---|
| 1 | | | | | | | | | | |
| 2 | | | | | | | | | | |
| 3 | | | | | | | | | | |
| 4 | | | | | | | | | | |
| 5 | | | | | | | | | | |
| 6 | | | | | | | | | | |
| 7 | | | | | | | | | | |
| 8 | | | | | | | | | | |
| 9 | | | | | | | | | | |
| 10 | | | | | | | | | | |

3. Now, take a stratified random sample using the *columns* as strata. Use Table D or technology to randomly select one square from each (vertical) column. Record the location of each square—for example, Column A: 4, Column B: 1, and so on.

4. The table on page N/DS-5 in the back of the book gives the actual number of sunflowers in each grid square. Use the information provided to calculate your estimate of the mean number of sunflowers per square for each of your samples in Steps 1, 2, and 3.

5. Make comparative dotplots showing the mean number of sunflowers obtained using the three different sampling methods for all members of the class. Describe any similarities and differences you see.

6. Your teacher will provide you with the mean number of sunflowers in the population of all 100 grid squares in the field. How did the three sampling methods do?

The dotplots below show the mean number of healthy plants in 100 samples using each of the three sampling methods in the Activity: simple random sampling, stratified random sampling with rows of the field as strata, and stratified random sampling with columns of the field as strata. Notice that all three distributions are centered at about 102.5, the true mean number of healthy plants in all squares of the field. That makes sense because random sampling yields accurate estimates of unknown population values.

One other detail stands out in the graphs. There is much less variability in the estimates using stratified random sampling with the rows as strata. The table on page N/DS-5 shows the actual number of healthy sunflowers in each grid square. Notice that the squares within each row contain a similar number of healthy plants but there are big differences between rows. *When we can choose strata that are "similar within but different between," stratified random samples give more precise estimates than simple random samples of the same size.*

Why didn't using the columns as strata reduce the variability of the estimates in a similar way? Because the squares within each column have very different numbers of healthy plants.

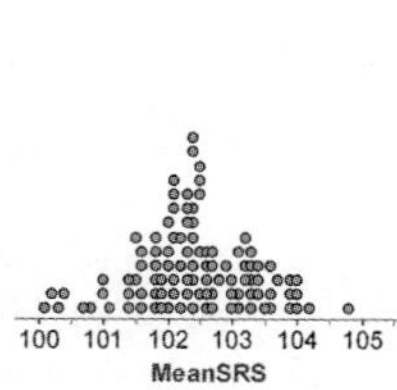


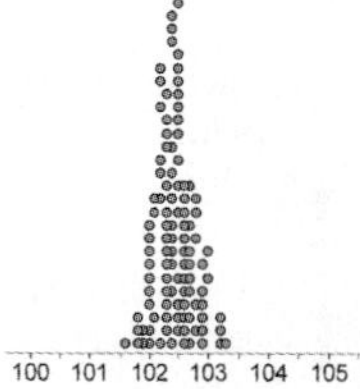


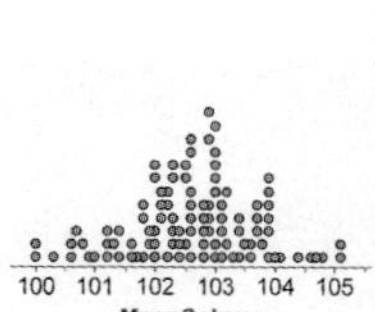


Both simple random sampling and stratified random sampling are hard to use when populations are large and spread out over a wide area. In that situation, we'd prefer a method that selects groups (**clusters**) of individuals that are "near" one another. That's the idea of a **cluster sample.**

**DEFINITION: Cluster sample and clusters**

To get a **cluster sample**, start by classifying the population into groups of individuals that are located near each other, called **clusters**. Then choose an SRS of the clusters. All individuals in the chosen clusters are included in the sample.

In a cluster sample, some people take an SRS from each cluster rather than including all members of the cluster.

Cluster samples are often used for practical reasons, like saving time and money. Cluster sampling works best when the clusters look just like the population but on a smaller scale. Imagine a large high school that assigns its students to homerooms alphabetically by last name. The school administration is considering a new schedule and would like student input. Administrators decide to survey 200 randomly selected students. It would be difficult to track down an SRS of 200 students, so the administration opts for a cluster sample of homerooms. The principal (who knows some statistics) takes a simple random sample of 8 homerooms and gives the survey to all 25 students in each homeroom.

Cluster samples don't offer the statistical advantage of better information about the population that stratified random samples do. That's because clusters are often chosen for ease so they may have as much variability as the population itself.

*Be sure you understand the difference between strata and clusters.* We want each stratum to contain similar individuals and for there to be large differences between strata. For a cluster sample, we'd *like* each cluster to look just like the population, but on a smaller scale. Here's an example that compares the random sampling methods we have discussed so far.

Remember: strata are ideally "similar within, but different between," while clusters are ideally "different within, but similar between."

## EXAMPLE Sampling at a School Assembly

### *Strata or clusters?*

The student council wants to conduct a survey during the first five minutes of an all-school assembly in the auditorium about use of the school library. They would like to announce the results of the survey at the end of the assembly. The student council president asks your statistics class to help carry out the survey.

PROBLEM: There are 800 students present at the assembly. A map of the auditorium is shown on the next page. Note that students are seated by grade level and that the seats are numbered from 1 to 800.

### Teaching Tip

Emphasize that efficiency is the primary benefit of cluster sampling. However, try to avoid the word "convenient" when describing this aspect of cluster sampling as students might associate it (incorrectly) with a convenience sample. Cluster sampling is perfectly valid if the clusters are selected at random.

### Teaching Tip

Ask students why it is better to use homeroom classes as clusters than to use math classes as clusters. Although both of these methods form clusters of units that are located close together, it is better if *each* cluster has the same characteristics as the population. Because math classes are formed by grouping students of the same ability, these clusters will be more homogeneous than the general population. Also, because only a few clusters will be selected, there is a good chance that certain types of classes (e.g., honors or remedial) will be over- or underrepresented.

### Teaching Tip

To illustrate the difference between stratified and cluster sampling, tell your students that you have a part-time job as a "salad quality inspector." Ask them which type of sampling would be better at a salad bar. Hopefully, they will say stratified because it would be best to take a small sample of each type of item to get an overall opinion. It would certainly not be a good idea to take a cluster sample and only inspect an entire jar of cherry tomatoes, an entire jar of shredded carrots, and an entire bowl of ranch dressing. However, in many fast-food restaurants, salads are prepared ahead of time and stored in plastic containers. In this case, a cluster sample of a few salads would be more efficient because each premade salad already includes all the different ingredients, such as lettuce, carrots, and cucumber. Sampling a little bit from each container would be very tedious and could possibly result in a sample made up entirely of lettuce.

### Teaching Tip

Here is another way to distinguish stratified sampling from cluster sampling: in a stratified sample, we divide the population into strata and take "some from all"; whereas in a cluster sample, we divide the population into clusters and take "all from some."

STAGE

| 9th grade: Seats 601–800 | 10th grade: Seats 401–600 |
|---|---|
| 11th grade: Seats 201–400 | 12th grade: Seats 1–200 |

Describe how you would use your calculator to select 80 students to complete the survey with each of the following:

(a) Simple random sample

(b) Stratified random sample

(c) Cluster sample

**SOLUTION:**

(a) To take an SRS, we need to choose 80 of the seat numbers at random. Use randInt(1,800) on your calculator until 80 different seats are selected. Then give the survey to the students in those seats.

(b) The students in the assembly are seated by grade level. Because students' library use might be similar within grade levels but different across grade levels, we'll use the grade level seating areas as our strata. Within each grade's seating area, we'll select 20 seats at random. For the 9th grade, use randInt(601,800) to select 20 different seats. Use randInt(401,600) to pick 20 different sophomore seats, randInt(201,400) to get 20 different junior seats, and randInt(1,200) to choose 20 different senior seats. Give the survey to the students in the selected seats.

(c) With the way students are seated, each column of seats from the stage to the back of the auditorium could be used as a cluster. Note that each cluster contains students from all four grade levels, so each should represent the population well. Because there are 20 clusters, each with 40 seats, we need to choose 2 clusters at random to get 80 students for the survey. Use randInt(1,20) to select two clusters, and then give the survey to all 40 students in each column of seats.

Note that cluster sampling is much more efficient than finding 80 seats scattered about the auditorium, as required by both of the other sampling methods.

**For Practice** *Try Exercise* **21**

Most large-scale sample surveys use *multistage samples* that combine two or more sampling methods. For example, the U.S. Census Bureau carries out a monthly Current Population Survey (CPS) of about 60,000 households. Researchers start by choosing a stratified random sample of neighborhoods in 756 of the 2007 geographical areas in the United States. Then they divide each neighborhood into clusters of four nearby households and select a cluster sample to interview.

Analyzing data from sampling methods more complex than an SRS takes us beyond basic statistics. But the SRS is the building block of more elaborate methods, and the principles of analysis remain much the same for these other methods.

## ALTERNATE EXAMPLE

### *A good read*

A school librarian wants to know the average number of pages in all the books in the library. The library has 20,000 books, arranged by type (fiction, biography, history, and so on) in shelves that hold about 50 books each. You want to select a random sample of 500 books.

**PROBLEM:**

**(a)** Explain how to select a simple random sample of 500 books.

**(b)** Explain how to select a stratified random sample of 500 books. Justify your choice of strata. Why might the librarian want to choose a stratified random sample?

**(c)** Explain how to select a cluster sample of 500 books. Justify your choice of clusters. Why might the librarian want to choose a cluster sample?

**SOLUTION:**

**(a)** Number the books from 1 to 20,000. Use the command `randInt(1,20000)` repeatedly until 500 different numbers between 1 and 20,000 are selected. Use the books that correspond to these numbers.

**(b)** Stratify by type because different types of books might be longer (or shorter) than other types. This will provide a more precise estimate of the average page length than a simple random sample. To select the sample, take an appropriately sized SRS of each type of book and combine the books selected from each type to form the sample. For example, if there are 1000 biographies (5% of 20,000), select an SRS of 25 biographies (5% of 500) using the method described in part (a).

**(c)** Clusters are formed by grouping books that are located near each other, making it easier for the librarian. We can use each shelf of 50 books as a cluster and randomly select 10 shelves to obtain the 500 books for our sample. Number the shelves from 1 to 400 and choose an SRS of 10 shelves using the method described in part (a). Then use all the books on the 10 selected shelves as the sample.

## Teaching Tip

Ask your students why it would *not* be a good idea to use horizontal rows as clusters in the school assembly example. Because only 4 clusters (rows) will be selected, it is quite possible that one of the grades will be completely left out of the sample. Also, ask why it wouldn't be a good idea to just use the kids in the aisle seats. Because the students in the aisle seats might be the last ones to the assembly, you could overrepresent the opinions of the less responsible students by using only these students.

## Teaching Tip

Although it is not in the AP® Statistics Topic Outline, you might mention another sampling design: a systematic sample. To get a systematic sample of 80 students, you could select every 10th student entering the auditorium. To make it a systematic *random* sample, randomly choose a number from 1 to 10 to be the first member of the sample and then survey every 10th student after that. For example, if Student 9 was selected, then survey student number 9, 19, 29, and so on.

## Teaching Tip

The inference methods we use in Chapters 8–12 assume that the data were collected using an SRS. To take advantage of the increased precision that results from stratified random sampling, we need to use a more complicated formula to estimate the standard deviation of our statistic than the ones we introduce in this book. These formulas can be found in a textbook about sampling, but you can see an example by looking at Questions 6(c) and 6(d) on the 2012 AP® exam.

## CHECK YOUR UNDERSTANDING

The manager of a sports arena wants to learn more about the financial status of the people who are attending an NBA basketball game. He would like to give a survey to a representative sample of the more than 20,000 fans in attendance. Ticket prices for the game vary a great deal: seats near the court cost over $100 each, while seats in the top rows of the arena cost $25 each. The arena is divided into 30 numbered sections, from 101 to 130. Each section has rows of seats labeled with letters from A (nearest the court) to ZZ (top row of the arena).

1. Explain why it might be difficult to give the survey to an SRS of 200 fans.
2. Which would be a better way to take a stratified random sample of fans: using the lettered rows or the numbered sections as strata? Explain.
3. Which would be a better way to take a cluster sample of fans: using the lettered rows or the numbered sections as clusters? Explain.

## Inference for Sampling

The purpose of a sample is to give us information about a larger population. The process of drawing conclusions about a population on the basis of sample data is called **inference** because we *infer* information about the population from what we *know* about the sample.

Inference from convenience samples or voluntary response samples would be misleading because these methods of choosing a sample are biased. We are almost certain that the sample does *not* fairly represent the population. *The first reason to rely on random sampling is to avoid bias in choosing a sample.*

Still, it is unlikely that results from a random sample are exactly the same as for the entire population. Sample results, like the unemployment rate obtained from the monthly Current Population Survey, are only estimates of the truth about the population. If we select two samples at random from the same population, we will almost certainly choose different individuals. So the sample results will differ somewhat, just by chance. Properly designed samples avoid systematic bias. But their results are rarely exactly correct, and we expect them to vary from sample to sample.

### EXAMPLE Going to class

#### How much do sample results vary?

Suppose that 70% of the students in a large university attended all their classes last week. Imagine taking a simple random sample of 100 students and recording the proportion of students in the sample who went to every class last week. Would the sample proportion be *exactly* 0.70? Probably not. Would the sample proportion be *close* to 0.70? That depends on what we mean by "close." The following graph shows the results of taking 500 SRSs, each of size 100, and recording the proportion of students who attended all their classes in each sample.

What do we see? The graph is centered at about 0.70, the population proportion. All of the sample proportions fall between 0.55 and 0.85. So we shouldn't be surprised if the difference between the sample proportion and the population proportion is as large as 0.15. The graph also has a very distinctive "bell shape."

### ANSWERS TO CYU

**1.** You would have to identify 200 different seats, go to those seats in the arena, and find the people who are sitting there, which would take too much time.

**2.** It is best to create strata where the people within a stratum are very similar to each other but different than the people in other strata. In this case, it would be better to take the lettered rows as the strata because each lettered row is the same distance from the court and so would contain only seats with the same (or nearly the same) ticket price.

**3.** It is best if the people in each cluster reflect the variability found in the population. In this case, it would be better to take the numbered sections as the clusters because they include all the different seat prices.

### ALTERNATE EXAMPLE

#### How much do sample results vary?

At a large high school, the number of pencils that students have in their possession has the distribution shown in the histogram below:

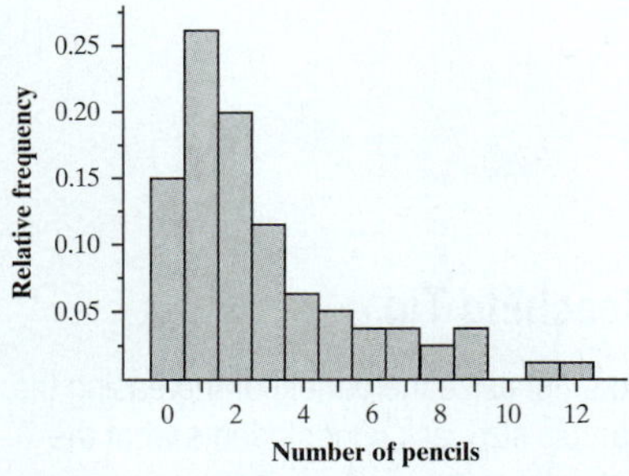


The mean of this distribution is 2.8 pencils. Suppose that each student in your class selects a random sample of 10 students and calculates the average number of pencils for the members of the sample. Will the sample averages always equal 2.8? If not, how much do we expect the sample mean to vary from 2.8?

Here are the sample means for 30 samples of size 10. You can see that the sample means vary quite a bit, from 1.3 pencils to 4.9 pencils.

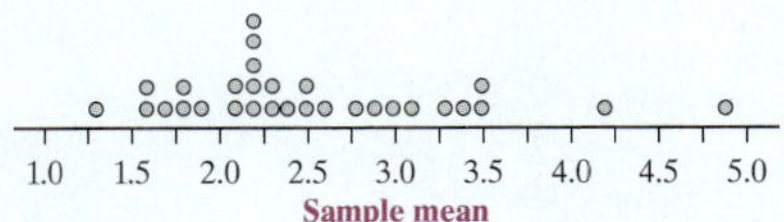


### ALTERNATE ACTIVITY

If you have time, you can do the following Activity to illustrate the idea of sampling variability.

*Materials*: At least 200 colored chips with exactly 60% of a particular color (say, red); large bag or other container that students can't see through.

In this Activity, you will explore how the results of repeated random samples vary in relation to the population truth.

**1.** Your teacher will prepare a large population of colored chips with 60% of a particular color.

**2.** One at a time, each student in the class will take an SRS of 20 chips and record the proportion of chips obtained that are red. The chips should be returned to the container before the next student chooses a sample.

**3.** Make a class dotplot of the sample proportions for this color of chip. Where is the graph centered? How much do the sample proportions vary around the center? Is there a clear shape?

Here is an example of what your class dotplot might look like:

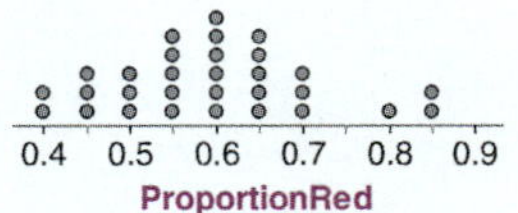

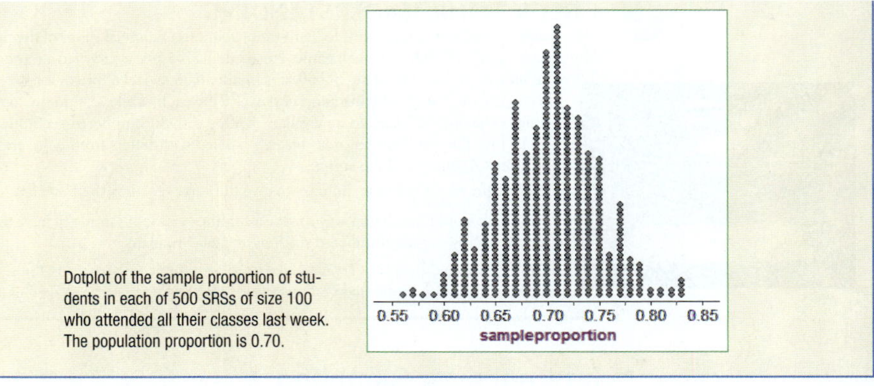


Dotplot of the sample proportion of students in each of 500 SRSs of size 100 who attended all their classes last week. The population proportion is 0.70.

Why can we trust random samples? As the previous example illustrates, the results of random sampling don't change haphazardly from sample to sample. Because we deliberately use chance, the results obey the laws of probability that govern chance behavior. These laws allow us to say how likely it is that sample results are close to the truth about the population. The second reason to use random sampling is that the laws of probability allow trustworthy inference about the population. Results from random samples come with a "margin of error" that sets bounds on the size of the likely error. We will discuss the details of inference for sampling later.

One point is worth making now: *larger random samples give better information about the population than smaller samples.* For instance, let's look at what happens if we increase the sample size in the example from 100 to 400 students. The dotplot below shows the results of taking 500 SRSs, each of size 400, and recording the proportion of students who attended all their classes in each sample. This graph is also centered at about 0.70. But now all the sample proportions fall between 0.63 and 0.77. So the difference between the sample proportion and the population proportion is at most 0.07. When using SRSs of size 100, this difference could be as much as 0.15. The moral of the story: by taking a very large random sample, you can be confident that the sample result is very close to the truth about the population.

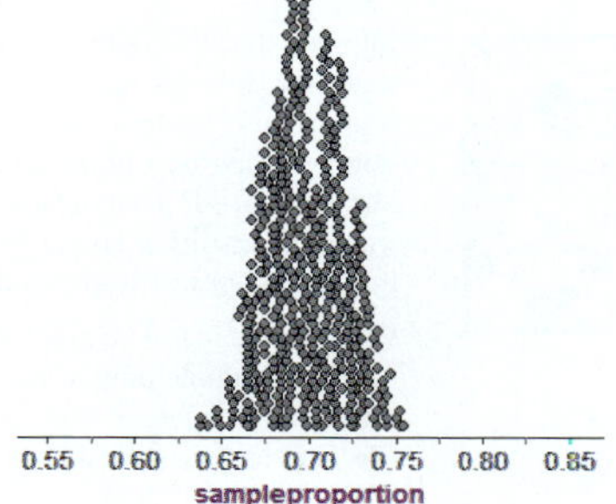


Dotplot of the sample proportion of students in each of 500 SRSs of size 400 who attended all their classes last week. The population proportion is 0.70.

## Teaching Tip

Point out that the dotplot is centered at 0.70. This means that there is no bias—the sample proportion isn't consistently less than 0.70 or consistently greater than 0.70. Using an unbiased sampling method doesn't guarantee perfect estimates, however!

## Teaching Tip

Mention to students that the word "error" in the phrase "margin of error" does not mean that a mistake has been made. The margin of error compensates for the variability that results from taking a random sample from a population. It does not account for a mistake made during the data collection process.

## Teaching Tip

To emphasize the benefit of increasing the sample size, ask your students what the dotplot would look like if they could survey *every* student at the university. Every dot would fall exactly at 0.70—there would be no variation at all.

The Current Population Survey contacts about 60,000 households, so we'd expect its estimate of the national unemployment rate to be within about 0.1% of the actual population value. Opinion polls that contact 1000 or 1500 people give less precise results—we expect the sample result to be within about 3% of the actual population percent with a given opinion. Of course, only samples chosen by chance carry this guarantee. Lou Dobbs's online sample tells us little about overall American public opinion even though 7350 people clicked a response.

## Sample Surveys: What Can Go Wrong?

The use of bad sampling methods (convenience or voluntary response) often leads to bias. Researchers can avoid bad methods by using random sampling to choose their samples. Other problems in conducting sample surveys are more difficult to avoid.

The list of individuals from which a sample will be drawn is called the *sampling frame.*

Sampling is often done using a list of individuals in the population. Such lists are seldom accurate or complete. The result is **undercoverage.**

**DEFINITION: Undercoverage**

**Undercoverage** occurs when some members of the population cannot be chosen in a sample.

Most samples suffer from some degree of undercoverage. A sample survey of households, for example, will miss not only homeless people but also prison inmates and students in dormitories. An opinion poll conducted by calling landline telephone numbers will miss households that have only cell phones as well as households without a phone. The results of national sample surveys therefore have some bias due to undercoverage if the people not covered differ from the rest of the population.

Well-designed sample surveys avoid bias in the sampling process. The real problems start *after* the sample is chosen.

One of the most serious sources of bias in sample surveys is **nonresponse**.

**DEFINITION: Nonresponse**

**Nonresponse** occurs when an individual chosen for the sample can't be contacted or refuses to participate.

Nonresponse to surveys often exceeds 50%, even with careful planning and several follow-up calls. If the people who respond differ from those who don't, in a way that is related to the response, bias results.

caution

*Some students misuse the term "voluntary response" to explain why certain individuals don't respond in a sample survey.* Their idea is that participation in the survey is optional (voluntary), so anyone can refuse to take part. What the students are describing is nonresponse. Think about it this way: nonresponse can occur only after a sample has been selected. In a voluntary response sample, every individual has opted to take part, so there won't be any nonresponse.

### Teaching Tip

To illustrate how undercoverage could lead to bias, ask students to imagine what would happen if they did a survey about increasing funding for Medicare by calling randomly selected landlines. Because many younger people only use cellphones, this method will likely overrepresent older people. Our estimate of the proportion of people who favor more funding will likely be too high.

### Teaching Tip

When discussing any type of bias, ask students to speculate on the direction of the bias. For example, if a survey about unemployment was conducted over the phone, nonresponse will lead to bias. It is likely that the estimated proportion of unemployed will be too high because people who are unemployed are more likely to be at home and available for the survey.

### AP® Exam Common Error

On the AP® exam, many students lose credit when describing what can go wrong in sample surveys because they use incorrect terminology. While it is important for students to understand and use the vocabulary of statistics correctly, they are rarely required to use specific vocabulary in their responses on the AP® exam. To be safe, tell your students to not worry about naming a specific problem or source of bias. Instead, have them clearly describe the problem and its consequences in the context of the question.

## EXAMPLE The ACS, GSS, and Opinion Polls

### *How bad is nonresponse?*

The Census Bureau's American Community Survey (ACS) has the lowest nonresponse rate of any poll we know: only about 1% of the households in the sample refuse to respond. The overall nonresponse rate, including "never at home" and other causes, is just 2.5%.[10] This monthly survey of about 250,000 households replaces the "long form" that in the past was sent to some households in the every-ten-years national census. Participation in the ACS is mandatory, and the Census Bureau follows up by telephone and then in person if a household doesn't return the mail questionnaire.

The University of Chicago's General Social Survey (GSS) is the nation's most important social science survey (see Figure 4.1). The GSS contacts its sample in person, and it is run by a university. Despite these advantages, its most recent survey had a 30% rate of nonresponse.

**FIGURE 4.1** The home page of the General Social Survey at the University of Chicago's National Opinion Research Center (http://www3.norc.org/GSS+Website/). The GSS has tracked opinions about a wide variety of issues since 1972.

What about opinion polls by news media and opinion-polling firms? We don't know their rates of nonresponse because they won't say. That's a bad sign. The Pew Research Center for the People and the Press imitated a careful random digit dialing survey and published the results: over 5 days, the survey reached 76% of the households in its chosen sample, but "because of busy schedules, skepticism and outright refusals, interviews were completed in just 38% of households that were reached." Combining households that could not be contacted with those who did not complete the interview gave a nonresponse rate of 73%.[11]

Another type of nonsampling problem occurs when people give inaccurate answers to survey questions. People may lie about their age, income, or drug use. They may misremember how many hours they spent on the Internet last week. Or they might make up an answer to a question that they don't understand.

**ALTERNATE EXAMPLE**

### *Women, love, and nonresponse*

In 1988 Shere Hite published a book entitled *Women and Love: A Cultural Revolution in Progress*, in which she claimed that 70% of women who had been married 5 years or more were having extramarital affairs. Hite's survey was sent to 100,000 women, and her conclusions were based on the 4500 women who responded. What was the nonresponse rate for this study? How might this nonresponse have affected the results of the survey?

*Possible answers:* The nonresponse rate is (100,000 − 4500)/(100,000), or 95.5%. If the women who responded to the survey differ from those who did not respond, in a way that is associated with having affairs, the estimated proportion of 0.70 will not be accurate.

**ALTERNATE EXAMPLE**

### *Presidential polls and nonresponse*

Here is a link to an interesting article that discusses the effects of nonresponse in presidential polls: www.huffingtonpost.com/2012/06/17/gallup-poll-race-barack-obama_n_1589937.html. Here is another article discussing the mistakes Gallup made in their 2012 presidential polls: www.washingtonpost.com/blogs/the-fix/wp/2013/06/04/gallup-explains-what-went-wrong-in-2012/.

### TRM How Polls Are Conducted

Consider having your students read the article "How Polls Are Conducted" at the Gallup Web site and answer the questions provided in the Teacher's Resource Materials. Do an Internet search for "How Gallup Polls Are Conducted" to find a pdf of the article or go to http://www.gallup.com/poll/File/125927/How%20Are%20Polls%20Conducted%20FINAL.pdf. Special thanks to Keyna McJilton, Downington (West) Senior High School, Downington, PA, who initially developed the questions and allowed us to use them.

### Teaching Tip

There is a great comic about response bias at http://blogs.courant.com/bob_englehart/2008/02/february-19-2008.html.

The gender, race, age, ethnicity, or behavior of the interviewer can also affect people's responses. A systematic pattern of inaccurate answers in a survey leads to **response bias**.

The **wording of questions** is the most important influence on the answers given to a sample survey. Confusing or leading questions can introduce strong bias. Changes in wording can greatly affect a survey's outcome.

## EXAMPLE How Do Americans Feel about Illegal Immigrants?

### *Question wording matters*

"Should illegal immigrants be prosecuted and deported for being in the U.S. illegally, or shouldn't they?" Asked this question in an opinion poll, 69% favored deportation. But when the very same sample was asked whether illegal immigrants who have worked in the United States for two years "should be given a chance to keep their jobs and eventually apply for legal status," 62% said that they should. Different questions give quite different impressions of attitudes toward illegal immigrants.

Even the order in which questions are asked matters. *Don't trust the results of a sample survey until you have read the exact questions asked.*

**Does the order matter?** Ask a sample of college students these two questions:

"How happy are you with your life in general?" (Answers on a scale of 1 to 5)
"How many dates did you have last month?"

There is almost no association between responses to the two questions when asked in this order. It appears that dating has little to do with happiness. Reverse the order of the questions, however, and a much stronger association appears: college students who say they had more dates tend to give higher ratings of happiness about life. Asking a question that brings dating to mind makes dating success a big factor in happiness.

## Teaching Tip

On page 271, there is an Activity about response bias. It asks students to design an experiment to estimate the effect of response bias in a survey. One of the authors has used this Activity as his first-semester project for many years and students love it. Look at the Activity description and corresponding Teaching Tips for some interesting examples of response bias.

## ALTERNATE EXAMPLE

### *How do Americans feel about health-care reform?*

During the middle of the drive to pass health-care reform legislation in July 2010, several different polling organizations attempted to measure the sentiments of Americans. A *New York Times*/CBS poll found that 66% supported "a government-administered health insurance plan—something like the Medicare coverage that people 65 and older get." However, a Fox News poll found that only 44% supported "a government-run health insurance plan."

## ALTERNATE EXAMPLE

### *Do Americans doubt the holocaust occurred?*

A badly worded question seemed to indicate that nearly one-third of all Americans had some doubt that the Holocaust occurred. Here was the question that the Roper polling organization asked in 1992: "Does it seem possible or does it seem impossible to you that the Nazi extermination of the Jews never happened?" Only 65% said it was impossible that it never happened. But, the wording of the question included a double-negative, making it difficult to understand. When the question was revised, the percentage who were certain that the Holocaust occurred dramatically increased. See http://www.nytimes.com/1994/07/08/us/poll-on-doubt-of-holocaust-is-corrected.html for more details.

## ANSWERS TO CYU

**1.** **(a)** Undercoverage. **(b)** Nonresponse. **(c)** Undercoverage.

**2.** By making it sound as if they are not a problem in the landfill, this question will result in fewer people suggesting that we should ban disposable diapers. The proportion who would say "Yes" to this survey question is likely to be smaller than the proportion who would say "Yes" to a more fairly worded question.

### TRM Section 4.1 Quizzes

There are three different quizzes available for this section in the *Teacher's Resource Materials*. Versions A and B are roughly parallel. Find them by clicking on the link in the TE-book, opening the TRCD, or logging onto the book's companion Web site www.whfreeman.com/tps5e (teacher log-in required).

## CHECK YOUR UNDERSTANDING

1. Each of the following is a possible source of bias in a sample survey. Name the type of bias that could result.

(a) The sample is chosen at random from a telephone directory.

(b) Some people cannot be contacted in five calls.

(c) Interviewers choose people walking by on the sidewalk to interview.

2. A survey paid for by makers of disposable diapers found that 84% of the sample opposed banning disposable diapers. Here is the actual question:

> It is estimated that disposable diapers account for less than 2% of the trash in today's landfills. In contrast, beverage containers, third-class mail, and yard wastes are estimated to account for about 21% of the trash in landfills. Given this, in your opinion, would it be fair to ban disposable diapers?[12]

Explain how the wording of the question could result in bias. Be sure to specify the direction of the bias.

## Section 4.1 Summary

- A **census** collects data from every individual in the **population**.
- A **sample survey** selects a **sample** from the population of all individuals about which we desire information. The goal of a sample survey is **inference:** we draw conclusions about the population based on data from the sample. It is important to specify exactly what population you are interested in and what variables you will measure.
- **Convenience samples** choose individuals who are easiest to reach. In **voluntary response samples**, individuals choose to join the sample in response to an open invitation. Both of these sampling methods usually lead to **bias**: they consistently underestimate or consistently overestimate the value you want to know.
- **Random sampling** uses chance to select a sample.
- The basic random sampling method is a **simple random sample (SRS).** An SRS gives every possible sample of a given size the same chance to be chosen. Choose an SRS by labeling the members of the population and using slips of paper, random digits, or technology to select the sample.
- To choose a **stratified random sample**, divide the population into **strata**, groups of individuals that are similar in some way that might affect their responses. Then choose a separate SRS from each stratum and combine these SRSs to form the sample. When strata are "similar within but different between," stratified random samples tend to give more precise estimates of unknown population values than simple random samples.
- To choose a **cluster sample,** divide the population into groups of individuals that are located near each other, called **clusters.** Randomly select some of these clusters. All the individuals in the chosen clusters are included in the sample. Ideally, clusters are "different within but similar between." Cluster

sampling saves time and money by collecting data from entire groups of individuals that are close together.

- Random sampling helps avoid bias in choosing a sample. Bias can still occur in the sampling process due to **undercoverage**, which happens when some members of the population cannot be chosen.
- The most serious errors in sample surveys, however, are ones that occur after the sample is chosen. The single biggest problem is **nonresponse:** when people can't be contacted or refuse to answer. Incorrect answers by respondents can lead to **response bias**. Finally, the **wording of questions** has a big influence on the answers.

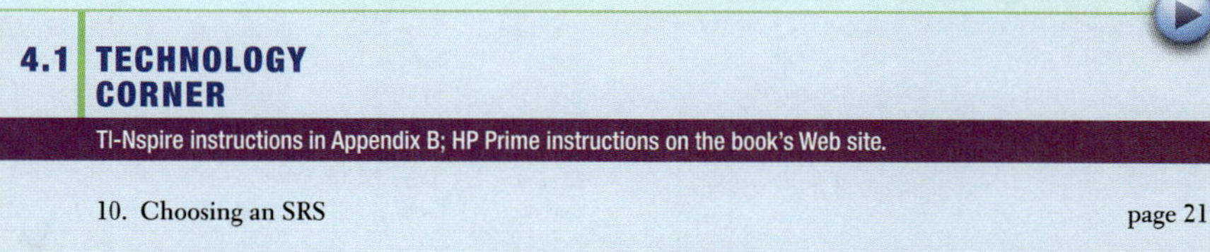




## Section 4.1 Exercises

pg 210

1. **Students as customers** A high school's student newspaper plans to survey local businesses about the importance of students as customers. From an alphabetical list of all local businesses, the newspaper staff chooses 150 businesses at random. Of these, 73 return the questionnaire mailed by the staff. Identify the population and the sample.
2. **Student archaeologists** An archaeological dig turns up large numbers of pottery shards, broken stone tools, and other artifacts. Students working on the project classify each artifact and assign it a number. The counts in different categories are important for understanding the site, so the project director chooses 2% of the artifacts at random and checks the students' work. Identify the population and the sample.
3. **Sampling stuffed envelopes** A large retailer prepares its customers' monthly credit card bills using an automatic machine that folds the bills, stuffs them into envelopes, and seals the envelopes for mailing. Are the envelopes completely sealed? Inspectors choose 40 envelopes at random from the 1000 stuffed each hour for visual inspection. Identify the population and the sample.
4. **Customer satisfaction** A department store mails a customer satisfaction survey to people who make credit card purchases at the store. This month, 45,000 people made credit card purchases. Surveys are mailed to 1000 of these people, chosen at random, and 137 people return the survey form. Identify the population and the sample.
5. **Call the shots** An advertisement for an upcoming TV show asked: "Should handgun control be tougher? You call the shots in a special call-in poll tonight. If yes, call 1-900-720-6181. If no, call 1-900-720-6182. Charge is 50 cents for the first minute." Over 90% of people who called in said "Yes." Explain why this opinion poll is almost certainly biased.
6. **Explain it to the congresswoman** You are on the staff of a member of Congress who is considering a bill that would provide government-sponsored insurance for nursing-home care. You report that 1128 letters have been received on the issue, of which 871 oppose the legislation. "I'm surprised that most of my constituents oppose the bill. I thought it would be quite popular," says the congresswoman. Are you convinced that a majority of the voters oppose the bill? How would you explain the statistical issue to the congresswoman?
7. **Instant opinion** A recent online poll posed the question "Should female athletes be paid the same as men for the work they do?" In all, 13,147 (44%) said "Yes," 15,182 (50%) said "No," and the

**TRM Full Solutions to Section 4.1 Exercises**

### Answers to Section 4.1 Exercises

**4.1** Population: all local businesses. Sample: the 73 businesses that return the questionnaire.

**4.2** Population: all the artifacts discovered at the dig. Sample: those artifacts that are chosen for inspection.

**4.3** Population: the 1000 envelopes stuffed during a given hour. Sample: the 40 randomly selected envelopes.

**4.4** Population: all 45,000 people who made credit card purchases. Sample: the 137 people who returned the survey form.

**4.5** This is a voluntary response sample. In this case, it appears that people who strongly support gun control volunteered more often, causing the proportion in the sample to be greater than the proportion in the population.

**4.6** This is a voluntary response sample—the proportion of letters opposed to the insurance is probably not a fair representation of the attitudes of her constituents. Only those who have very strong opinions will write in. It is likely that the true proportion of constituents who oppose the bill is less than 871/1128.

remaining 1448 said "Don't know." In spite of the large sample size for this survey, we can't trust the result. Why not?

8. **Sampling at the mall** You have probably seen the mall interviewer, approaching people passing by with clipboard in hand. Explain why even a large sample of mall shoppers would not provide a trustworthy estimate of the current unemployment rate.

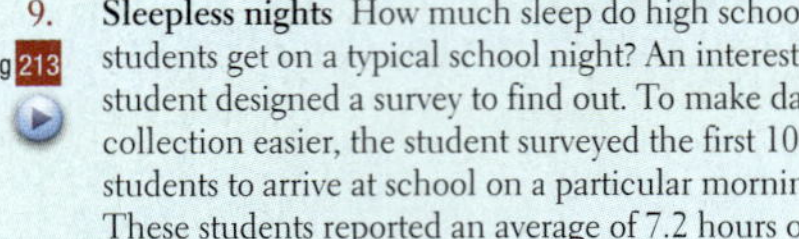

9. **Sleepless nights** How much sleep do high school students get on a typical school night? An interested student designed a survey to find out. To make data collection easier, the student surveyed the first 100 students to arrive at school on a particular morning. These students reported an average of 7.2 hours of sleep on the previous night.
   (a) What type of sample did the student obtain?
   (b) Explain why this sampling method is biased. Is 7.2 hours probably higher or lower than the true average amount of sleep last night for all students at the school? Why?

pg 213

10. **Online polls** In June 2008, *Parade* magazine posed the following question: "Should drivers be banned from using all cell phones?" Readers were encouraged to vote online at parade.com. The July 13, 2008, issue of *Parade* reported the results: 2407 (85%) said "Yes" and 410 (15%) said "No."
   (a) What type of sample did the *Parade* survey obtain?
   (b) Explain why this sampling method is biased. Is 85% probably higher or lower than the true percent of all adults who believe that cell phone use while driving should be banned? Why?

11. **Do you trust the Internet?** You want to ask a sample of high school students the question "How much do you trust information about health that you find on the Internet—a great deal, somewhat, not much, or not at all?" You try out this and other questions on a pilot group of 5 students chosen from your class. The class members are listed below.
   (a) Explain how you would use a line of Table D to choose an SRS of 5 students from the following list. Explain your method clearly enough for a classmate to obtain your results.
   (b) Use line 107 to select the sample. Show how you use each of the digits.

pg 217

| Anderson | Deng | Glaus | Nguyen | Samuels |
|---|---|---|---|---|
| Arroyo | De Ramos | Helling | Palmiero | Shen |
| Batista | Drasin | Husain | Percival | Tse |
| Bell | Eckstein | Johnson | Prince | Velasco |
| Burke | Fernandez | Kim | Puri | Wallace |
| Cabrera | Fullmer | Molina | Richards | Washburn |
| Calloway | Gandhi | Morgan | Rider | Zabidi |
| Delluci | Garcia | Murphy | Rodriguez | Zhao |

12. **Apartment living** You are planning a report on apartment living in a college town. You decide to select three apartment complexes at random for in-depth interviews with residents.
   (a) Explain how you would use a line of Table D to choose an SRS of 3 complexes from the list below. Explain your method clearly enough for a classmate to obtain your results.
   (b) Use line 117 to select the sample. Show how you use each of the digits.

| Ashley Oaks | Chauncey Village | Franklin Park | Richfield |
|---|---|---|---|
| Bay Pointe | Country Squire | Georgetown | Sagamore Ridge |
| Beau Jardin | Country View | Greenacres | Salem Courthouse |
| Bluffs | Country Villa | Lahr House | Village Manor |
| Brandon Place | Crestview | Mayfair Village | Waterford Court |
| Briarwood | Del-Lynn | Nobb Hill | Williamsburg |
| Brownstone | Fairington | Pemberly Courts | |
| Burberry | Fairway Knolls | Peppermill | |
| Cambridge | Fowler | Pheasant Run | |

13. **Sampling the forest** To gather data on a 1200-acre pine forest in Louisiana, the U.S. Forest Service laid a grid of 1410 equally spaced circular plots over a map of the forest. A ground survey visited a sample of 10% of these plots.[13]
   (a) Explain how you would use your calculator or Table D to choose an SRS of 141 plots. Your description should be clear enough for a classmate to carry out your plan.
   (b) Use your method from (a) to choose the first 3 plots.

14. **Sampling gravestones** The local genealogical society in Coles County, Illinois, has compiled records on all 55,914 gravestones in cemeteries in the county for the years 1825 to 1985. Historians plan to use these records to learn about African Americans in Coles County's history. They first choose an SRS of 395 records to check their accuracy by visiting the actual gravestones.[14]
   (a) Explain how you would use your calculator or Table D to choose the SRS. Your description should be clear enough for a classmate to carry out your plan.
   (b) Use your method from (a) to choose the first 3 gravestones.

15. **Random digits** Which of the following statements are true of a table of random digits, and which are false? Briefly explain your answers.
   (a) There are exactly four 0s in each row of 40 digits.
   (b) Each pair of digits has chance 1/100 of being 00.
   (c) The digits 0000 can never appear as a group, because this pattern is not random.

16. **Random digits** In using Table D repeatedly to choose random samples, you should not always begin at the same place, such as line 101. Why not?

**4.7** This is a voluntary response sample and overrepresents the opinions of those who feel most strongly about the issue being surveyed.

**4.8** This is a convenience sample. The sample is likely to overestimate the unemployment rate because people without jobs have more time to be at the mall than those who are employed.

**4.9** **(a)** A convenience sample. **(b)** The first 100 students to arrive at school likely had to wake up earlier than other students, so 7.2 hours is probably less than the true average.

**4.10** **(a)** A voluntary response sample. **(b)** The reported value of 85% is probably higher than the true percent because readers who have been involved in an accident caused by cell phone use are more likely to respond to the poll and say "Yes."

**4.11** **(a)** Number the 40 students from 01 to 40. Pick a starting point on the random number table. Record two-digit numbers, skipping numbers that aren't between 01 and 40 and any repeated numbers, until you have 5 unique numbers between 01 and 40. Use the 5 students corresponding to these numbers. **(b)** Using line 107, skip the numbers not in bold: 82 73 95 78 90 **20** 80 74 75 **11** 81 67 65 53 00 94 **38** **31** 48 93 60 94 **07**. Select Johnson (20), Drasin (11), Washburn (38), Rider (31), and Calloway (07).

**4.12** **(a)** Number the 33 complexes from 01 to 33. Pick a starting point on the random number table. Record two-digit numbers, skipping any that aren't between 01 and 33 and any repeated numbers, until you have 3 unique numbers between 01 and 33. Use the 3 complexes corresponding to these numbers. **(b)** Using line 117, skip the numbers not in bold: 38 **16** 79 85 **32** 62 **18**. Select Fairington (16), Waterford Court (32), and Fowler (18).

**4.13** **(a)** *Using a calculator*: Number the plots from 1 to 1410. Use the command `randInt(1,1410)` to select 141 different integers from 1 to 1410 and use the corresponding 141 plots. **(b)** Answers will vary.

**4.14** **(a)** *Using a calculator*: Number the gravestones from 1 to 55,914. Use the command `randInt(1,55914)` to select 395 different integers from 1 to 55,914 and use the corresponding 395 plots. **(b)** Answers will vary.

**4.15** **(a)** False—although on average there will be four 0s in every set of 40 digits, the number of 0s can be less than 4 or greater than 4 by chance. **(b)** True—there are 100 pairs of digits 00 through 99, and all are equally likely. **(c)** False—0000 is just as likely as any other string of four digits.

**4.16** The results would not be random, because you would use the same sample in every case.

**17. iPhones** Suppose 1000 iPhones are produced at a factory today. Management would like to ensure that the phones' display screens meet their quality standards before shipping them to retail stores. Since it takes about 10 minutes to inspect an individual phone's display screen, managers decide to inspect a sample of 20 phones from the day's production.

(a) Explain why it would be difficult for managers to inspect an SRS of 20 iPhones that are produced today.

(b) An eager employee suggests that it would be easy to inspect the last 20 iPhones that were produced today. Why isn't this a good idea?

(c) Another employee recommends a different sampling method: Randomly choose one of the first 50 iPhones produced. Inspect that phone and every fiftieth iPhone produced afterward. (This method is known as **systematic random sampling**.) Explain carefully why this sampling method is *not* an SRS.

**18. Dead trees** On the west side of Rocky Mountain National Park, many mature pine trees are dying due to infestation by pine beetles. Scientists would like to use sampling to estimate the proportion of all pine trees in the area that have been infected.

(a) Explain why it wouldn't be practical for scientists to obtain an SRS in this setting.

(b) A possible alternative would be to use every pine tree along the park's main road as a sample. Why is this sampling method biased?

(c) Suppose that a more complicated random sampling plan is carried out, and that 35% of the pine trees in the sample are infested by the pine beetle. Can scientists conclude that exactly 35% of *all* the pine trees on the west side of the park are infested? Why or why not?

**19. Who goes to the convention?** A club has 30 student members and 10 faculty members. The students are

| | | | | |
|---|---|---|---|---|
| Abel | Fisher | Huber | Miranda | Reinmann |
| Carson | Ghosh | Jimenez | Moskowitz | Santos |
| Chen | Griswold | Jones | Neyman | Shaw |
| David | Hein | Kim | O'Brien | Thompson |
| Deming | Hernandez | Klotz | Pearl | Utts |
| Elashoff | Holland | Liu | Potter | Varga |

The faculty members are

| | | | | |
|---|---|---|---|---|
| Andrews | Fernandez | Kim | Moore | West |
| Besicovitch | Gupta | Lightman | Phillips | Yang |

The club can send 4 students and 2 faculty members to a convention. It decides to choose those who will go by random selection. Describe a method for using Table D to select a stratified random sample of 4 students and 2 faculty. Then use line 123 to select the sample.

**20. Sampling by accountants** Accountants often use stratified samples during audits to verify a company's records of such things as accounts receivable. The stratification is based on the dollar amount of the item and often includes 100% sampling of the largest items. One company reports 5000 accounts receivable. Of these, 100 are in amounts over \$50,000; 500 are in amounts between \$1000 and \$50,000; and the remaining 4400 are in amounts under \$1000. Using these groups as strata, you decide to verify all the largest accounts and to sample 5% of the midsize accounts and 1% of the small accounts. Describe a method for using Table D to select a stratified random sample of the midsize and small accounts. Then use line 115 to select only the first 3 accounts from each of these strata.

**21. Go Blue!** pg 221 Michigan Stadium, also known as "The Big House," seats over 100,000 fans for a football game. The University of Michigan athletic department plans to conduct a survey about concessions that are sold during games. Tickets are most expensive for seats on the sidelines. The cheapest seats are in the end zones (where one of the authors sat as a student). A map of the stadium is shown.

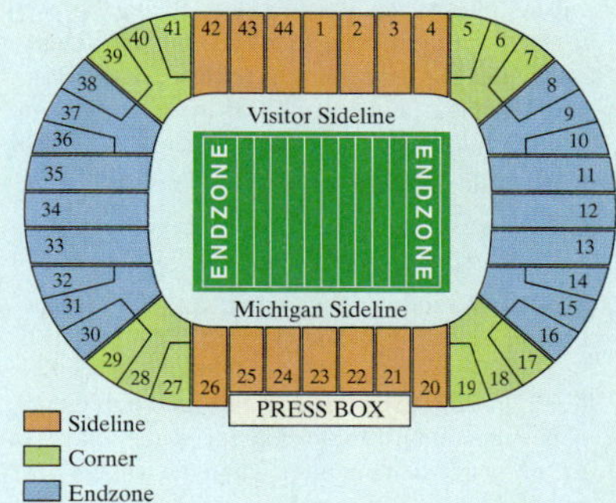


(a) The athletic department is considering a stratified random sample. What would you recommend as the strata? Why?

(b) Explain why a cluster sample might be easier to obtain. What would you recommend for the clusters? Why?

**22. How was your stay?** A hotel has 30 floors with 40 rooms per floor. The rooms on one side of the hotel face the water, while rooms on the other side face a golf course. There is an extra charge for the rooms with a water view. The hotel manager wants to survey 120 guests who stayed at the hotel during a convention about their overall satisfaction with the property.

**4.17 (a)** It might be difficult to locate the 20 phones from among the 1000 produced that day. **(b)** The quality of the phones produced may change during the day, so that the last phones manufactured are not representative of the day's production. **(c)** Because each sample of 20 phones does not have the same probability of being selected. In an SRS, it is possible for 2 consecutive phones to be selected in a sample, but this is not possible with a systematic random sample.

**4.18 (a)** Each tree would need to be identified and numbered, which would take too much time. **(b)** This convenience sampling method is biased because trees along the main road are more likely to be damaged by cars and people, and may be more susceptible to infestation. **(c)** The percentage is unlikely to be exactly 35% because of sampling variability; however, it should be close to 35%.

**4.19** Assign numbers 01 to 30 to the students. Pick a starting point on the random digit table. Record two-digit numbers, skipping any that aren't between 01 and 30 and any repeated numbers, until you have 4 unique numbers between 01 and 30. Use the corresponding four students. Then assign numbers 0 to 9 to the faculty members. Continuing on the table, record one-digit numbers, skipping any repeated numbers, until you have 2 unique numbers between 0 and 9. Use the corresponding faculty members. Starting on line 123 gives 08-Ghosh, 15-Jones, 07-Fisher, and 27-Shaw for the students and 1-Besicovitch and 0-Andrews for the faculty.

**4.20** Label the 500 midsize accounts with numbers from 001 to 500. Pick a starting point on the random digit table. Record three-digit numbers, skipping any that aren't between 001 and 500 and any repeated numbers, until you have 25 unique numbers between 001 and 500. Verify the corresponding 25 accounts. Then label the 4400 small accounts with numbers from 0001 to 4400. Continuing on the random digit table, record four-digit numbers, skipping any that aren't between 0001 and 4400 and any repeated numbers, until you have 44 unique numbers between 0001 and 4400. Verify the corresponding 44 accounts. Starting at line 115, the first three midsize accounts are 417, 494, and 322. The first three small accounts are 2470, 1893, and 3259.

**4.21 (a)** Use the three types of seats as the strata because people who can afford more expensive tickets probably have different opinions about the concessions than people who can only afford the cheaper tickets. **(b)** A stratified random sample will include seats from all over the stadium, which would make it very time-consuming to obtain. A cluster sample of numbered sections would be easier to obtain, because the people selected for the sample would be sitting close together.

(a) Explain why choosing a stratified random sample might be preferable to an SRS in this case. What would you use as strata?

(b) Why might a cluster sample be a simpler option? What would you use as clusters?

23. **Is it an SRS?** A corporation employs 2000 male and 500 female engineers. A stratified random sample of 200 male and 50 female engineers gives each engineer 1 chance in 10 to be chosen. This sample design gives every individual in the population the same chance to be chosen for the sample. Is it an SRS? Explain your answer.

24. **Attitudes toward alcohol** At a party there are 30 students over age 21 and 20 students under age 21. You choose at random 3 of those over 21 and separately choose at random 2 of those under 21 to interview about attitudes toward alcohol. You have given every student at the party the same chance to be interviewed. Why is your sample not an SRS?

25. **High-speed Internet** Laying fiber-optic cable is expensive. Cable companies want to make sure that if they extend their lines out to less dense suburban or rural areas, there will be sufficient demand and the work will be cost-effective. They decide to conduct a survey to determine the proportion of households in a rural subdivision that would buy the service. They select a simple random sample of 5 blocks in the subdivision and survey each family that lives on one of those blocks.

(a) What is the name for this kind of sampling method?

(b) Give a possible reason why the cable company chose this method.

26. **Timber!** A lumber company wants to estimate the proportion of trees in a large forest that are ready to be cut down. They use an aerial map to divide the forest into 200 equal-sized rectangles. Then they choose a random sample of 20 rectangles and examine every tree that's in one of those rectangles.

(a) What is the name for this kind of sampling method?

(b) Give a possible reason why the lumber company chose this method.

27. **Tweet, tweet!** What proportion of students at your school use Twitter? To find out, you survey a simple random sample of students from the school roster.

(a) Will your sample result be exactly the same as the true population proportion? Explain.

(b) Which would be more likely to get your sample result closer to the true population value: an SRS of 50 students or an SRS of 100 students? Explain.

28. **Far from home?** A researcher wants to estimate the average distance that students at a large community college live from campus. To find out, she surveys a simple random sample of students from the registrar's database.

(a) Will the researcher's sample result be exactly the same as the true population mean? Explain.

(b) Which would be more likely to get the researcher's sample result closer to the true population value: an SRS of 100 students or an SRS of 50 students? Explain.

29. **Baseball tickets** Suppose you want to know the average amount of money spent by the fans attending opening day for the Cleveland Indians baseball season. You get permission from the team's management to conduct a survey at the stadium, but they will not allow you to bother the fans in the club seating or box seats (the most expensive seating). Using a computer, you randomly select 500 seats from the rest of the stadium. During the game, you ask the fans in those seats how much they spent that day.

Give a reason why this survey might yield a biased result. Explain the likely direction of the bias.

30. **Rise and shine** How long before school starts do students get out of bed, on average? Administrators survey a random sample of students on each school bus one morning.

Give a reason why this survey might yield a biased result. Explain the likely direction of the bias.

31. **Nonresponse** A survey of drivers began by randomly sampling all listed residential telephone numbers in the United States. Of 45,956 calls to these numbers, 5029 were completed. The goal of the survey was to estimate how far people drive, on average, per day.[15]

(a) What was the rate of nonresponse for this sample?

(b) Explain how nonresponse can lead to bias in this survey. Be sure to give the direction of the bias.

32. **Ring-no-answer** A common form of nonresponse in telephone surveys is "ring-no-answer." That is, a call is made to an active number but no one answers. The Italian National Statistical Institute looked at nonresponse to a government survey of households in Italy during the periods January 1 to Easter and July 1 to August 31. All calls were made between 7 and 10 P.M., but 21.4% gave "ring-no-answer" in one period versus 41.5% "ring-no-answer" in the other period.[16] Which period do you think had the higher rate of no answers? Why? Explain why a high rate of nonresponse makes sample results less reliable.

33. **Running red lights** The sample described in Exercise 31 produced a list of 5024 licensed drivers. The investigators then chose an SRS of 880 of these drivers to answer questions about their driving habits. One question asked was: "Recalling the last ten traffic lights you drove through, how many of them were red when you entered the intersections?" Of the 880 respondents, 171 admitted that at least one light had been red. A practical problem with this survey is that

**4.22 (a)** Because satisfaction with the property is likely to vary depending on the location of the room, we should stratify by floor and view. Using a stratified random sample would assure the manager that he got opinions from each type of room and provide a more precise estimate of customer satisfaction. **(b)** We could use floors as clusters. This would be a simpler option because the manager would only need to survey guests on three floors instead of having to survey guests all over the hotel.

**4.23** No. In an SRS, each possible sample of 250 engineers is equally likely to be selected, including samples that aren't exactly 200 males and 50 females.

**4.24** In an SRS, each possible sample of 5 students is equally likely to be selected, including samples that aren't exactly 3 students over 21 and 2 students under 21.

**4.25 (a)** Cluster sampling. **(b)** To save time and money. In an SRS, the company would have to visit individual homes all over the rural subdivision instead of only 5 locations.

**4.26 (a)** Cluster sampling. **(b)** To save time and money. In an SRS, the company would have to inspect trees all over the entire forest instead of only 20 locations.

**4.27 (a)** It is unlikely, because different random samples will include different students and produce different estimates of the proportion of students who use Twitter. **(b)** An SRS of 100 students. Larger random samples give us better information about the population than smaller random samples.

**4.28 (a)** It is unlikely, because different random samples will include different students and produce different estimates of the average distance students live from the college. **(b)** An SRS of 100 students. Larger random samples give us better information about the population than smaller random samples.

**4.29** Because you are sampling only from the lower-priced ticket holders, this will likely produce an estimate that is too small, as fans in the club seats and box seats probably spend more money at the game than fans in cheaper seats.

**4.30** Because administrators are only sampling from students who take the bus and not including other students, this will likely produce an estimate that is too large, as students who take the bus probably need to wake up earlier so they don't miss the bus.

**4.31 (a)** 89.1%. **(b)** Because the people who have long commutes are less likely to be at home and be included in the sample, this will likely produce an estimate that is too small.

**4.32** The higher no-answer rate was probably the second period—when families are likely to be vacationing or spending time outdoors. A high rate of nonresponse makes sample results less reliable because the people away from their phones might answer differently than the people who respond.

people may not give truthful answers. What is the likely direction of the bias? Explain.

34. **Seat belt use** A study in El Paso, Texas, looked at seat belt use by drivers. Drivers were observed at randomly chosen convenience stores. After they left their cars, they were invited to answer questions that included questions about seat belt use. In all, 75% said they always used seat belts, yet only 61.5% were wearing seat belts when they pulled into the store parking lots.[17] Explain the reason for the bias observed in responses to the survey. Do you expect bias in the same direction in most surveys about seat belt use?

35. **Wording bias** Comment on each of the following as a potential sample survey question. Is the question clear? Is it slanted toward a desired response?
(a) "Some cell phone users have developed brain cancer. Should all cell phones come with a warning label explaining the danger of using cell phones?"
(b) "Do you agree that a national system of health insurance should be favored because it would provide health insurance for everyone and would reduce administrative costs?"
(c) "In view of escalating environmental degradation and incipient resource depletion, would you favor economic incentives for recycling of resource–intensive consumer goods?"

36. **Checking for bias** Comment on each of the following as a potential sample survey question. Is the question clear? Is it slanted toward a desired response?
(a) Which of the following best represents your opinion on gun control?
1. The government should confiscate our guns.
2. We have the right to keep and bear arms.
(b) A freeze in nuclear weapons should be favored because it would begin a much-needed process to stop everyone in the world from building nuclear weapons now and reduce the possibility of nuclear war in the future. Do you agree or disagree?

*Multiple choice: Select the best answer for Exercises 37 to 42.*

37. The Web portal AOL places opinion poll questions next to many of its news stories. Simply click your response to join the sample. One of the questions in January 2008 was "Do you plan to diet this year?" More than 30,000 people responded, with 68% saying "Yes." You can conclude that
(a) about 68% of Americans planned to diet in 2008.
(b) the poll used a convenience sample, so the results tell us little about the population of all adults.
(c) the poll uses voluntary response, so the results tell us little about the population of all adults.
(d) the sample is too small to draw any conclusion.
(e) None of these.

38. To gather information about the validity of a new standardized test for tenth-grade students in a particular state, a random sample of 15 high schools was selected from the state. The new test was administered to every 10th-grade student in the selected high schools. What kind of sample is this?
(a) A simple random sample
(b) A stratified random sample
(c) A cluster sample
(d) A systematic random sample
(e) A voluntary response sample

39. Your statistics class has 30 students. You want to call an SRS of 5 students from your class to ask where they use a computer for the online quizzes. You label the students 01, 02, ..., 30. You enter the table of random digits at this line:

14459 26056 31424 80371 65103 62253 22490 61181

Your SRS contains the students labeled
(a) 14, 45, 92, 60, 56.
(b) 14, 31, 03, 10, 22.
(c) 14, 03, 10, 22, 22.
(d) 14, 03, 10, 22, 06.
(e) 14, 03, 10, 22, 11.

40. Suppose that 35% of the registered voters in a state are registered as Republicans, 40% as Democrats, and 25% as Independents. A newspaper wants to select a sample of 1000 registered voters to predict the outcome of the next election. If they randomly select 350 Republicans, randomly select 400 Democrats, and randomly select 250 Independents, did this sampling procedure result in a simple random sample of registered voters from this district?
(a) Yes, because each registered voter had the same chance of being chosen.
(b) Yes, because random chance was involved.
(c) No, because not all registered voters had the same chance of being chosen.
(d) No, because there were a different number of registered voters selected from each party.
(e) No, because not all possible groups of 1000 registered voters had the same chance of being chosen.

41. A local news agency conducted a survey about unemployment by randomly dialing phone numbers until they had gathered responses from 1000 adults in their state. In the survey, 19% of those who responded said they were not currently employed. In reality, only 6% of the adults in the state were not currently employed

**4.33** We would not expect very many people to claim they have run red lights when they haven't, but some people will deny running red lights when they have. Thus, we expect that the sample proportion underestimates the true proportion of drivers who have run a red light.

**4.34** People likely claim to wear their seat belts because they know they should, even if they don't. Such overestimations are likely in most surveys about seat belt use (and similar topics).

**4.35** **(a)** The wording is clear, but the question is slanted in favor of warning labels because of the first sentence stating that some cell phone users have developed brain cancer. **(b)** The question is clear, but it is slanted in favor of national health insurance by asserting it would reduce administrative costs and not providing any counter-arguments. **(c)** The wording is too technical for many people to understand. For those who do understand the question, it is slanted because it suggests reasons why one should support recycling.

**4.36** **(a)** The question is clear, but the two options presented are too extreme; no middle position on gun control is allowed. Also, the language used in Option 2 is from the Constitution and people might avoid Option 1 because they don't like the idea of government confiscating personal property. **(b)** The question isn't clear. The phrasing of this question will tend to make people respond in favor of a nuclear freeze because only one side of the issue is presented.

**4.37** c

**4.38** c

**4.39** d

**4.40** e

**4.41** d

**4.42** c

**4.43** **(a)** For each additional day, the predicted sleep debt increases by about 3.17 hours. **(b)** The predicted sleep debt for a 5-day school week is $2.23 + 3.17(5) = 18.08$ hours. This is about 3 hours more than the researcher claimed for a 5-day week, so the students have reason to be skeptical of the research study's reported results.

**4.44** **(a)** 95% of the other bandwidth measurements will be less than this amount. **(b)** 98th percentile, because it would suggest a higher usage of bandwidth by the company.

at the time of the survey. Which of the following best explains the difference in the two percentages?

(a) The difference is due to sampling variability. We shouldn't expect the results of a random sample to match the truth about the population every time.

(b) The difference is due to response bias. Adults who are employed are likely to lie and say that they are unemployed.

(c) The difference is due to undercoverage bias. The survey included only adults and did not include teenagers who are eligible to work.

(d) The difference is due to nonresponse bias. Adults who are employed are less likely to be available for the sample than adults who are unemployed.

(e) The difference is due to voluntary response. Adults are able to volunteer as a member of the sample.

42. A simple random sample of 1200 adult Americans is selected, and each person is asked the following question: "In light of the huge national deficit, should the government at this time spend additional money to establish a national system of health insurance?" Only 39% of those responding answered "Yes." This survey

(a) is reasonably accurate since it used a large simple random sample.

(b) needs to be larger since only about 24 people were drawn from each state.

(c) probably understates the percent of people who favor a system of national health insurance.

(d) is very inaccurate but neither understates nor overstates the percent of people who favor a system of national health insurance. Because simple random sampling was used, it is unbiased.

(e) probably overstates the percent of people who favor a system of national health insurance.

43. **Sleep debt** (3.2) A researcher reported that the typical teenager needs 9.3 hours of sleep per night but gets only 6.3 hours.[18] By the end of a 5-day school week, a teenager would accumulate about 15 hours of "sleep debt." Students in a high school statistics class were skeptical, so they gathered data on the amount of sleep debt (in hours) accumulated over time (in days) by a random sample of 25 high school students. The resulting least-squares regression equation for their data is $\widehat{\text{Sleep debt}} = 2.23 + 3.17(\text{days})$.

(a) Interpret the slope of the regression line in context.

(b) Are the students' results consistent with the researcher's report? Explain.

44. **Internet charges** (2.1) Some Internet service providers (ISPs) charge companies based on how much bandwidth they use in a month. One method that ISPs use for calculating bandwidth is to find the 95th percentile of a company's usage based on samples of hundreds of 5-minute intervals during a month.

(a) Explain what "95th percentile" means in this setting.

(b) Which would cost a company more: the 95th percentile method or a similar approach using the 98th percentile? Justify your answer.

## 4.2 Experiments

**WHAT YOU WILL LEARN** By the end of the section, you should be able to:

- Distinguish between an observational study and an experiment.
- Explain the concept of confounding and how it limits the ability to make cause-and-effect conclusions.
- Identify the experimental units, explanatory and response variables, and treatments in an experiment.
- Explain the purpose of comparison, random assignment, control, and replication in an experiment.
- Describe a completely randomized design for an experiment, including how to randomly assign treatments using slips of paper, technology, or a table of random digits.
- Describe the placebo effect and the purpose of blinding in an experiment.
- Interpret the meaning of statistically significant in the context of an experiment.
- Explain the purpose of blocking in an experiment. Describe a randomized block design or a matched pairs design for an experiment.

**PD Section 4.2 Overview**

Watch the video overview of Section 4.2 (for teachers) by clicking on the link in the TE-book, opening the TRCD, or logging onto the book's companion Web site www.whfreeman.com/tps5e (teacher log-in required).

**TRM Section 4.2 Alternate Examples**

You can find the Alternate Examples for this section in Microsoft Word format by clicking on the link in the TE-book, opening the TRCD, or logging onto the book's companion Web site www.whfreeman.com/tps5e (teacher log-in required).

A sample survey aims to gather information about a population without disturbing the population in the process. Sample surveys are one kind of **observational study.** Other observational studies watch the behavior of animals in the wild or the interactions between teacher and students in the classroom. This section is about statistical designs for **experiments,** a very different way to produce data.

## Observational Study versus Experiment

In contrast to observational studies, experiments don't just observe individuals or ask them questions. They actively impose some *treatment* to measure the response. Experiments can answer questions like "Does aspirin reduce the chance of a heart attack?" and "Can yoga help dogs live longer?"

**DEFINITION: Observational study and experiment**

An **observational study** observes individuals and measures variables of interest but does not attempt to influence the responses.

An **experiment** deliberately imposes some treatment on individuals to measure their responses.

The goal of an observational study can be to describe some group or situation, to compare groups, or to examine relationships between variables. The purpose of an experiment is to determine whether the treatment causes a change in the response. An observational study, even one based on a random sample, is a poor way to gauge the effect that changes in one variable have on another variable. To see the response to a change, we must actually impose the change. *When our goal is to understand cause and effect, experiments are the only source of fully convincing data.* For this reason, the distinction between observational study and experiment is one of the most important in statistics.

### EXAMPLE Does Taking Hormones Reduce Heart Attack Risk after Menopause?

*Observation versus experiment*

Should women take hormones such as estrogen after menopause, when natural production of these hormones ends? In 1992, several major medical organizations said "Yes." Women who took hormones seemed to reduce their risk of a heart attack by 35% to 50%. The risks of taking hormones appeared small compared with the benefits.

The evidence in favor of hormone replacement came from a number of *observational studies* that compared women who were taking hormones with others who were not. But the women who chose to take hormones were richer and better educated and saw doctors more often than women who didn't take hormones. Because the women who took hormones did many other things to better maintain their health, it isn't surprising that they had fewer heart attacks.

### Teaching Tip

During the 2013 AP® Reading, Professor Jessica Utts of the University of California–Irvine gave a talk entitled "What do Future Senators, Scientists, Social Workers, and Sales Clerks Need to Learn from *Your* Statistics Class?" In the talk, she listed the top10 most important ideas for students to learn in a statistics class. At #1, she listed the distinction between observational studies and experiments and the concept of confounding. If students remember nothing else from your course, they will be much better citizens if they understand this distinction. It helps if you can read articles from the newspaper throughout the year. Ask students to identify the type of study being described and what type of inference can be made from the study.

### Teaching Tip

Consider using some of the investigations in *Making Sense of Statistical Studies* by Roxy Peck and Daren Starnes to help your students understand the important differences between observational studies and experiments. These investigations start with a research question on some topic of interest. Students are then led through a series of questions that help them examine the study design, analyze data, and interpret results. Later investigations ask students to design, carry out, and analyze results from their own studies. See http://www.amstat.org/education/msss/index.cfm for more information about this resource.

### Teaching Tip

When discussing the menopause study (or any other observational study), it is important to focus on the big picture: there is more than one difference between the group of women who took hormones and the group of women who didn't take hormones. Any of these differences could be the cause of the reduction in heart attack rate.

### ALTERNATE EXAMPLE

*Soy good for you?*

The November 2009 issue of *Nutrition Action* discusses what the current research tells us about the supposed benefits of soy. For a long time, scientists have believed that the soy foods in Asian diets explain the lower rates of breast cancer, prostate cancer, osteoporosis, and heart disease in places like China and Japan. However, when experiments were conducted, soy either had no effect or a very small effect on the health of the participants. For example, several different studies randomly assigned elderly women to either soy or placebo, and none of the studies showed that soy was more beneficial for preventing osteoporosis. So what explains the lower rates of osteoporosis in Asian cultures? We still don't know. It could be due to genetics, other dietary factors, or any other difference between Asian cultures and non-Asian cultures.

### ALTERNATE EXAMPLE

*The buzz about coffee*

The article "Coffee Buzz: Study Finds Java Drinkers Live Longer" from the *Arizona Daily Star* discusses a very large study of coffee drinkers. It suggests that coffee drinkers live a little longer than non-drinkers, whether they drink regular or decaf. Previous studies had indicated that drinking coffee might increase the risk of heart disease, but these studies didn't take into account that coffee drinkers were also more likely to smoke, drink more alcohol, eat more red meat, and exercise less than non-coffee-drinkers. The new study is still an observational study, however, so we can't be sure that drinking coffee is the cause of longer life—it could be something else associated with drinking coffee that is the cause. From http://azstarnet.com/news/science/health-med-fit/coffee-buzz-study-finds-java-drinkers-live-longer/article_c8de3c15-67da-5c57-8efd-a120d86ca967.html.

To get convincing data on the link between hormone replacement and heart attacks, we should do an *experiment*. Experiments don't let women decide what to do. They assign women to either hormone replacement pills or to placebo pills that look and taste the same as the hormone pills. The assignment is done by a coin toss, so that all kinds of women are equally likely to get either treatment. By 2002, several experiments with women of different ages agreed that hormone replacement does *not* reduce the risk of heart attacks. The National Institutes of Health, after reviewing the evidence, concluded that the first studies were wrong. Taking hormones after menopause quickly fell out of favor.[19]

From Chapter 3: A **response variable** measures an outcome of a study. An **explanatory variable** may help explain or predict changes in a response variable.

For each of these studies, the *explanatory variable* was whether or not a woman took hormones, and the *response variable* was whether or not the woman had a heart attack. Researchers wanted to argue that changes in the explanatory variable (hormone status) actually caused changes in the response variable (heart attack status). In the early observational studies, however, the effect of taking hormones was mixed up with the characteristics of women who chose to take them. These other variables make it hard to see the true relationship between the explanatory and response variables.

Let's consider two other variables from the observational studies of hormone replacement: number of doctor visits per year and age. The women who chose to take hormones visited their doctors more often than the women who didn't take hormones. Did the women in the hormone group have fewer heart attacks because they got better health care or because they took hormones? We can't be sure. A situation like this, in which the effects of two variables on a response variable cannot be separated from each other, is called **confounding**.

Some people call a variable that results in confounding, like the number of doctor visits per year in this case, a *confounding variable*.

What about age? Older women are at greater risk of having a heart attack than younger women. If the women who took hormones were generally younger than those who didn't, we'd have more confounding. That wasn't the case, however. There was no link between age and group membership (hormones or not) in the observational studies. If there is no difference between the groups with respect to the other variable, there can be no confounding.

**DEFINITION: Confounding**

**Confounding** occurs when two variables are associated in such a way that their effects on a response variable cannot be distinguished from each other.

**AP® EXAM TIP** If you are asked to identify a possible confounding variable in a given setting, you are expected to explain how the variable you choose (1) is associated with the explanatory variable and (2) affects the response variable.

Caution !

*Observational studies of the effect of an explanatory variable on a response variable often fail because of confounding between the explanatory variable and one or more other variables.* Well-designed experiments take steps to prevent confounding. The later hormone therapy experiments avoided confounding by letting chance decide who took hormones and who didn't. That way, women who took better care of themselves were split about evenly between the two groups. So were older women and younger women. When these experiments found no reduction in heart attack risk for women taking hormones, researchers began to doubt the results of the earlier observational studies. The moral of the story is simple: *beware the influence of other variables!*

## Teaching Tip

Your students might enjoy the tongue-in-cheek study at the link below. In short, the authors question the evidence that parachutes prevent fatal falls because the existing studies are observational. There are no experiments that assign some people to use parachutes and others not. From http://www.ncbi.nlm.nih.gov/pmc/articles/PMC300808/.

## Teaching Tip

In many other textbooks, including previous editions of *The Practice of Statistics*, the term "lurking variable" is introduced in the discussion about confounding. As we have taught the course, however, we discovered that the term doesn't help student understanding. In fact, having an additional label for certain variables tended to make students more confused. We still think it is vital that students consider other variables that might affect the response variable; we simply don't give these variables a special name.

## Teaching Tip

Another way to define confounding is to say that "two variables are confounded if it is impossible to determine which of the two variables is causing a change in the response variable."

## Teaching Tip

To verify that students understand the AP® Exam Tip on this page, ask them to evaluate this statement in the context of the menopause example: "Diet is a confounding variable because people with bad diets are more likely to have a heart attack." This statement would not receive full credit on the AP® exam because it doesn't explain how diet is also associated with the choice to take hormones. In other words, talking about diet doesn't fully explain why women who choose to take hormones are less likely to have a heart attack. To get full credit, the answer would need to also address why a better diet might be more common in the group of women who chose to take hormones.

## AP® Exam Common Error

Many students lose credit on the AP® exam for using statistical vocabulary without additional explanation. Instead of relying on vocabulary, students should explain the concept in the context of the problem. For example, just saying that "wealth is a confounding variable" will probably not earn credit if students were asked to explain the concept of confounding in this study. Students need to explain not only that wealthier people tend to go to the doctor more often and to be healthier overall, but also that wealthier people are more likely to get hormone replacement. Thus, it could actually be the overall healthiness of wealthy women that caused the reduction in heart attacks, not any hormones they might have taken. In fact, any variable that is more common in the group of women taking replacement hormones could be the cause of the reduction in heart attacks.

## CHECK YOUR UNDERSTANDING

1. Does reducing screen brightness increase battery life in laptop computers? To find out, researchers obtained 30 new laptops of the same brand. They chose 15 of the computers at random and adjusted their screens to the brightest setting. The other 15 laptop screens were left at the default setting—moderate brightness. Researchers then measured how long each machine's battery lasted. Was this an observational study or an experiment? Justify your answer.

*Questions 2 to 4 refer to the following setting.* Does eating dinner with their families improve students' academic performance? According to an ABC News article, "Teenagers who eat with their families at least five times a week are more likely to get better grades in school."[20] This finding was based on a sample survey conducted by researchers at Columbia University.

2. Was this an observational study or an experiment? Justify your answer.
3. What are the explanatory and response variables?
4. Explain clearly why such a study cannot establish a cause-and-effect relationship. Suggest a variable that may be confounded with whether families eat dinner together.

## The Language of Experiments

An experiment is a statistical study in which we actually do something (a **treatment**) to people, animals, or objects (the **experimental units**) to observe the response. Here is the basic vocabulary of experiments.

**DEFINITION: Treatment, experimental units, subjects**

A specific condition applied to the individuals in an experiment is called a **treatment**. If an experiment has several explanatory variables, a treatment is a combination of specific values of these variables.

The **experimental units** are the smallest collection of individuals to which treatments are applied. When the units are human beings, they often are called **subjects**.

The best way to learn the language of experiments is to practice using it.

**EXAMPLE** When Will I Ever Use This Stuff?

### *Vocabulary of experiments*

Researchers at the University of North Carolina were concerned about the increasing dropout rate in the state's high schools, especially for low-income students. Surveys of recent dropouts revealed that many of these students had started to lose interest during middle school. They said they saw little connection between what they were studying in school and their future plans. To change

### ANSWERS TO CYU

**1.** Experiment, because a treatment (brightness of screen) was imposed on the laptops.

**2.** Observational study, because students were not assigned to eat a particular number of meals with their family per week.

**3.** Explanatory: number of meals per week eaten with their family. Response: GPA.

**4.** There are probably other variables that influence the response variable. For example, students who have part-time jobs may not be able to eat many meals with their families and may not have much time to study, leading to lower grades.

### ALTERNATE EXAMPLE

### *A louse-y situation*

A study published in the *New England Journal of Medicine* (March 11, 2010) compared two medicines to treat head lice: an oral medication called ivermectin and a topical lotion containing malathion. Researchers studied 812 people in 376 households in seven areas around the world. Of the 185 households randomly assigned to ivermectin, 171 were free from head lice after 2 weeks compared with only 151 of the 191 households randomly assigned to malathion.

PROBLEM: Identify the experimental units, explanatory and response variables, and the treatments in this experiment.

SOLUTION: The experimental units are the 376 households, not the 812 people, because the treatments were assigned to entire households, not separately to individuals within the household. The explanatory variable is type of medication, and the response variable is whether the household was lice-free. The treatments were ivermectin and malathion.

this perception, researchers developed a program called CareerStart. The big idea of the program is that teachers show students how the topics they learn get used in specific careers.

To test the effectiveness of CareerStart, the researchers recruited 14 middle schools in Forsyth County to participate in an experiment. Seven of the schools, determined at random, used CareerStart along with the district's standard curriculum. The other seven schools just followed the standard curriculum. Researchers followed both groups of students for several years, collecting data on students' attendance, behavior, standardized test scores, level of engagement in school, and whether or not the students graduated from high school.

*Results:* Students at schools that used CareerStart generally had better attendance and fewer discipline problems, earned higher test scores, reported greater engagement in their classes, and were more likely to graduate.[21]

**PROBLEM:** Identify the experimental units, explanatory and response variables, and the treatments in the CareerStart experiment.

**SOLUTION:** The experimental units are 14 middle schools in Forsyth County, NC. The explanatory variable is whether the school used the CareerStart program with its students. Several response variables were measured, including test scores, attendance, behavior, student engagement, and graduation rates. This experiment compares two treatments: (1) the standard middle school curriculum and (2) the standard curriculum plus CareerStart.

**For Practice** *Try Exercise* **51**

Note that the experimental units in the CareerStart example are the schools, not individual students. Experimental units are the smallest collection of individuals to which treatments are applied. The curricular treatments were administered to entire schools, so those are the experimental units.

The previous example illustrates the big advantage of experiments over observational studies: *experiments can give good evidence for causation.* In an experiment, we study the effects of the specific treatments we are interested in, while trying to control for the effects of other variables. For instance, the students in all 14 schools followed the standard curriculum. To ensure that the two groups of schools were as similar as possible before the treatments were administered, researchers let chance decide which 7 schools would use CareerStart. The only systematic difference between the schools was the educational treatment. When students from the CareerStart schools did much better, researchers were able to conclude that the program made the difference.

Sometimes, the explanatory variables in an experiment are called **factors**. Many experiments study the joint effects of several factors. In such an experiment, each treatment is formed by combining a specific value (often called a **level**) of each of the factors. Here's an example of a multifactor experiment.

### AP® Exam Common Error

Students often have trouble identifying the experimental units in an experiment. In the Example on this page, the experimental units are the schools, not the students. In the Alternate Example on page 237, students might mistakenly identify the individual people or the lice as the experimental units. However, the decision about which treatment to apply was made for each household, not for each individual person or louse.

### Teaching Tip

When there is only one factor (explanatory variable) in an experiment, the levels of the factor are equivalent to the treatments. However, when there is more than one factor, the combinations of each level of each factor form the treatments.

## EXAMPLE TV Advertising

### *Experiments with multiple explanatory variables*

What are the effects of repeated exposure to an advertising message? The answer may depend on both the length of the ad and on how often it is repeated. An experiment investigated this question using 120 undergraduate students who volunteered to participate. All subjects viewed a 40-minute television program that included ads for a digital camera. Some subjects saw a 30-second commercial; others, a 90-second version. The same commercial was shown either 1, 3, or 5 times during the program. After viewing, all the subjects answered questions about their recall of the ad, their attitude toward the camera, and their intention to purchase it.[22]

Factor B Repetitions

| Factor A Length | 1 time | 3 times | 5 times |
|---|---|---|---|
| 30 seconds | 1 | 2 | 3 |
| 90 seconds | 4 | 5 | 6 |

Subjects assigned to Treatment 3 see a 30-second ad five times during the program.

**FIGURE 4.2** The six treatments in the TV ad experiment. Combinations of values of the two explanatory variables (factors) form six treatments.

PROBLEM: For the advertising study, identify the experimental units or subjects, explanatory and response variables, and the treatments.

SOLUTION:

The subjects are the 120 undergraduate students. This experiment has 2 explanatory variables (factors): length of the commercial and number of repetitions. The response variables include measures of subjects' recall of the ad, their attitudes about the digital camera, and whether they intend to purchase it.

There are 2 different lengths of commercial (30 and 90 seconds) and three different numbers of repetitions (1, 3, and 5). The 6 combinations consisting of one level of each factor form the 6 treatments shown in Figure 4.2: (1) 30 seconds, 1 time; (2) 30 seconds, 3 times; (3) 30 seconds, 5 times; (4) 90 seconds, 1 time; (5) 90 seconds, 3 times; (6) 90 seconds, 5 times.

**For Practice** *Try Exercise* 55

This example shows how experiments allow us to study the combined effect of several factors. The interaction of several factors can produce effects that could not be predicted from looking at the effect of each factor alone. Perhaps longer commercials increase interest in a product, and more commercials also increase interest. But if we both make a commercial longer and show it more often, viewers get annoyed and their interest in the product drops. The two-factor experiment in the TV advertising example will help us find out.

## How to Experiment Badly

Experiments are the preferred method for examining the effect of one variable on another. By imposing the specific treatment of interest and controlling other influences, we can pin down cause and effect. Good designs are essential for effective experiments, just as they are for sampling. To see why, let's start with an example of a bad experimental design.

### ALTERNATE EXAMPLE

#### *Growing tomatoes*

Does adding fertilizer affect the productivity of tomato plants? How about the amount of water given to the plants? To answer these questions, a gardener plants 24 similar tomato plants in identical pots in his greenhouse. He will add fertilizer to the soil in half of the pots. Also, he will water 8 of the plants with 0.5 gallon of water per day, 8 of the plants with 1 gallon of water per day, and the remaining 8 plants with 1.5 gallons of water per day. At the end of 3 months, he will record the total weight of tomatoes produced on each plant.

PROBLEM: Identify the experimental units or subjects, explanatory and response variables, and the treatments.

SOLUTION: The experimental units are the tomato plants. The two explanatory variables (factors) are whether fertilizer is applied and amount of water. The response variable is the weight of tomatoes produced. There are 6 treatments: (1) fertilizer, 0.5 gallon; (2) fertilizer, 1 gallon; (3) fertilizer, 1.5 gallons; (4) no fertilizer, 0.5 gallon; (5) no fertilizer, 1 gallon; (6) no fertilizer, 1.5 gallons.

### ALTERNATE EXAMPLE

#### *How to buy happiness*

Watch the short TED talk "How to Buy Happiness" by Michael Norton at http://www.ted.com/talks/michael_norton_how_to_buy_happiness.html. He describes a neat multifactor experiment to determine if money can buy happiness. After he describes the treatments, pause the video and ask students how many factors the experiment involves. Many students will say "4" because they are confusing factors and treatments. In fact, there are two factors: amount of money and how it is spent.

### Teaching Tip

To determine the number of treatments, multiply the number of levels for each factor in the experiment. In the TV advertising example, the factor length had 2 levels and the factor frequency had 3 levels. Thus, there are $2 \times 3 = 6$ treatments.

ALTERNATE EXAMPLE

***Does caffeine affect pulse rate?***

Many students regularly consume caffeine to help them stay alert. So it seems plausible that taking caffeine might increase an individual's pulse rate. Is this true? One way to investigate this claim is to ask volunteers to measure their pulse rates, drink some cola with caffeine, measure their pulse rates again after 10 minutes, and calculate the increase in pulse rate. Unfortunately, even if the pulse rate of every student went up, we couldn't attribute the increase to caffeine. Perhaps the excitement of being in an experiment made their pulse rates increase. Perhaps it was the sugar in the cola and not the caffeine. Perhaps their teacher told them a funny joke during the 10-minute waiting period and made everyone laugh! In other words, there are many other variables that are potentially confounded with taking caffeine.

## EXAMPLE Are Online SAT Prep Courses Effective?

### *A bad experiment*

A high school regularly offers a review course to prepare students for the SAT. This year, budget cuts will allow the school to offer only an online version of the course. Suppose the group of students who take the online course earn an average increase of 45 points in their math scores from a pre-test to the actual SAT test. Can we conclude that the online course is effective?

This experiment has a very simple design. A group of subjects (the students) were exposed to a treatment (the online course), and the outcome (increase in math scores) was observed. Here is the design:

Students → Online course → increase in math scores

A closer look showed that many of the students in the online review course were taking advanced math classes in school. Maybe the students in the online course improved their math scores because of what they were learning in their school math classes, not because of the online course. This confounding prevents us from concluding that the online course is effective.

Many laboratory experiments use a design like the one in the online SAT course example:

Experimental units → Treatment → Measure response

In the lab environment, simple designs often work well. Field experiments and experiments with animals or people deal with more varied conditions. *Outside the lab, badly designed experiments often yield worthless results because of confounding.*

## How to Experiment Well

The remedy for the confounding in the SAT prep course example is to do a *comparative experiment* in which some students are taught the SAT course in the classroom and other, similar students take the course online. Most well-designed experiments compare two or more treatments.

**Comparison** alone isn't enough to produce results we can trust. If the treatments are given to groups that differ greatly when the experiment begins, *bias* will result. For example, if we allow students to select online or classroom instruction, more self-motivated students are likely to sign up for the online course. Allowing

### Teaching Tip

Emphasize that a good experiment needs comparison to help avoid confounding. In the caffeine experiment, it would be a worthwhile idea to have some students drink cola with caffeine while others drink cola without caffeine.

personal choice will bias our results in the same way that volunteers bias the results of online opinion polls. The solution to the problem of bias in sampling is random selection. In experiments, the solution is **random assignment.**

**DEFINITION: Random assignment**

In an experiment, **random assignment** means that experimental units are assigned to treatments using a chance process.

Let's look at how random assignment can be used to improve the design of the SAT prep course experiment.

## EXAMPLE SAT Prep: Online versus Classroom

### *How random assignment works*

This year, the high school has enough budget money to compare the online SAT course with the classroom SAT course. Fifty students have agreed to participate in an experiment comparing the two instructional methods.

PROBLEM: Describe how you would randomly assign 25 students to each of the two methods:

(a) Using 50 identical slips of paper

(b) Using technology

(c) Using Table D

SOLUTION:

(a) The simplest way would be to use the "hat method." Write each subject's name on one of the slips. Put all the slips in a hat and mix them thoroughly. Draw them out one at a time until you have 25 slips. These 25 students will take the online course. The remaining 25 students will take the classroom course. Alternatively, you could write "online" on 25 of the slips and "classroom" on the other 25 slips. Then put the slips in a hat and mix them well. Have students come up one by one and (without looking) pick a slip from the hat. This guarantees 25 students per group, with the treatments assigned by chance.

(b) Give numbers 1, 2, 3, . . . , 49, 50 to the subjects in alphabetical order by last name. Then use your calculator's randInt command or a computer's random number generator to produce numbers between 1 and 50. Ignore any repeated numbers. The first 25 different numbers chosen select the students for the online course. The remaining 25 subjects will take the classroom course.

(c) Give labels 01, 02, 03, . . . , 49, 50 to the subjects in alphabetical order by last name. Go to a line of Table D and read two-digit groups moving from left to right. The first 25 distinct labels between 01 and 50 identify the 25 students that are assigned to the online course. The remaining 25 students will take the classroom course. Ignore repeated labels and groups of digits from 51 to 00.

**For Practice** *Try Exercise* 59

### ALTERNATE EXAMPLE

### *Does caffeine affect pulse rates?*

Suppose you have a class of 30 students who volunteer to be subjects in the caffeine experiment described earlier.

PROBLEM: Explain how you would randomly assign 15 students to each of the two treatments:

**(a)** Using 30 identical slips of paper

**(b)** Using technology

**(c)** Using Table D

SOLUTION:

**(a)** Using 30 identical slips of paper, write A on 15 pieces of paper and B on the other 15. Mix them thoroughly in a hat and have each student select one slip of paper. Then ask each student who received an A to drink the cola with caffeine and each student who received a B to drink the cola without caffeine.

**(b)** Number the students from 1 to 30. Use `randInt(1,30)` to select 15 *different* numbers from 1 to 30. These students will drink the cola with caffeine and the remaining 15 will drink the cola without caffeine.

**(c)** Number the students from 01 to 30. Use a line from Table D and read two-digit numbers moving from left to right. The first 15 *different* numbers from 01 to 30 will identify the students who will drink the cola with caffeine. The remaining 15 students will drink the cola without caffeine.

### Teaching Tip

Emphasize that the goal of random assignment is to create groups of experimental units that are roughly equivalent at the beginning of the study. Random assignment accomplishes this by balancing the effects of other variables among the treatment groups. In the menopause example, the initial observational studies couldn't draw a cause-and-effect conclusion because there were many differences between the group of women who chose to take hormones and the group of women who didn't. When treatments are randomly assigned in an experiment, the only systematic difference between the two groups should be the treatment assigned.

### AP® Exam Common Error

Many students lose credit on the AP® exam for failing to adequately describe how they assign the treatments to experimental units in an experiment. Most important, the method the students use must be *random*. In addition, the method must be described in *sufficient detail* so that two knowledgeable users of statistics could follow the student's description and carry out the method in exactly the same way. For example, saying "Assign students to the two groups using random digits" isn't sufficiently detailed because there are many ways to use random digits. If a student chooses to use random digits, he or she must use labels of the same length (e.g., 01–30, not 1–30). Students must also address how they will deal with repeated numbers that come up when using a random digit table or random number generator. For example, they can say "ignoring repeats" or state that they will generate 25 "different" numbers from 1 to 50.

## Teaching Tip

Don't say (or let your students say) that random assignment *eliminates* the effects of other variables. It doesn't. It simply balances their effects among the treatment groups. For example, in the SAT study, the math course that a student is taking will still affect his or her improvement in SAT score. However, the number of students taking advanced math classes in the two groups should be about the same after the random assignment. Likewise, the number of students taking on-level math classes should be about the same in both groups, as will the number of boys, the number of people with after-school jobs, and so on.

## Teaching Tip

Many students think that "control" refers only to a control group. While good experiments will always include comparison, an experimenter should do whatever is possible to control other variables by making them the same for all treatment groups. After all, there are still plenty of things that can go wrong even when there are randomly assigned treatment groups. See the Think About It feature on the next page for more details.

## Teaching Tip

To illustrate the benefits of replication, ask your students to imagine using 6 subjects in the SAT experiment, 2 of which are in an advanced math class. There is a 50% chance that these 2 subjects will end up in the same treatment group. However, if we had 60 students in the experiment and 20 of them were in an advanced math class, there is practically no chance that they would all end up in the same treatment group. The more replication, the more balanced the treatments groups will be after the random assignment.

Random assignment should distribute the students taking advanced math classes in roughly equal numbers to each group. It should also balance out the number of students with lots of extracurricular activities and those with part-time jobs in the classroom and online SAT prep courses. Random assignment helps ensure that the effects of other variables (such as current math course or amount of available study time) are spread evenly among the two groups.

Although random assignment should create two groups of students that are roughly equivalent to begin with, we still have to ensure that the only consistent difference between the groups during the experiment is the type of SAT prep they receive. We can **control** for the effects of some variables by keeping them the same for both groups. For instance, we should give all students the same pretest and actual SAT test at the same times on the same days. The length, timing, content, and instructor of the SAT prep classes should also be the same.

Because the two groups are alike except for the treatments, any difference in their average math score improvements must be due either to the treatments themselves or to the random assignment. We can't say that any difference between the average SAT scores of students enrolled online and in the classroom must be caused by a difference in the effectiveness of the two types of instruction. There would be *some* difference, even if both groups received the same instruction, because of variation among students in background and study habits. Chance assigns students to one group or the other, which results in a chance difference between the groups.

In statistics, replication means "use enough subjects." In other fields, the term "replication" has a different meaning. When one experiment is conducted and then the same or a similar experiment is independently conducted in a different location by different investigators, this is known as replication. That is, replication means repeatability.

We would not trust an experiment with just one student in each group. The results would depend too much on which group got lucky and received the stronger student. If we assign many subjects to each group, however, the effects of chance will balance out, and there will be little difference in the average responses in the two groups unless the treatments themselves cause a difference. This is the idea of **replication:** use enough experimental units to distinguish a difference in the effects of the treatments from chance variation due to the random assignment.

### PRINCIPLES OF EXPERIMENTAL DESIGN

The basic principles for designing experiments are as follows:

1. **Comparison.** Use a design that compares two or more treatments.
2. **Random assignment.** Use chance to assign experimental units to treatments. Doing so helps create roughly equivalent groups of experimental units by balancing the effects of other variables among the treatment groups.
3. **Control.** Keep other variables that might affect the response the same for all groups.
4. **Replication.** Use enough experimental units in each group so that any differences in the effects of the treatments can be distinguished from chance differences between the groups.

**Why is control important in an experiment?** For two reasons. Suppose we used two different instructors in the SAT experiment. If Mrs. McDonald taught the online group and Mr. Tyson taught the classroom group, then course type will be confounded with instructor. We won't know if the difference in average improvement for the two groups was due to the difference in instructor or the difference in course type. So one reason we need to control other variables is to prevent confounding.

The second reason we should control other variables is to reduce variability in the response variable. Suppose that we allow students in both groups to choose how many class sessions to attend. Their choices will increase the variation in the response variable (improvement) for both groups. Some students will attend fewer sessions and experience smaller improvements than they would have otherwise. Other students will attend as many sessions as possible and experience bigger improvements than they might have otherwise. This increase in variation will make it harder to see if one treatment is really more effective.

The dotplots on the left show the results of an experiment in which the number of class sessions was the same for all participating students. From these graphs, it seems clear that the online course is more effective than the classroom course. The dotplots on the right show the results of an experiment in which the students were permitted to choose the number of class sessions they attended. Notice that the centers of the distributions haven't changed, but the distributions are much more variable. The increased overlap in the graphs makes the evidence supporting the online course less convincing.

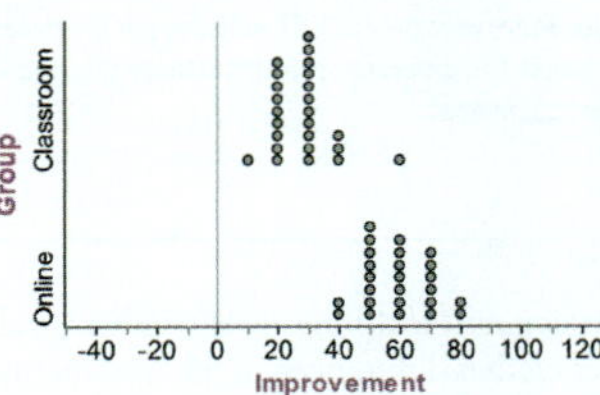


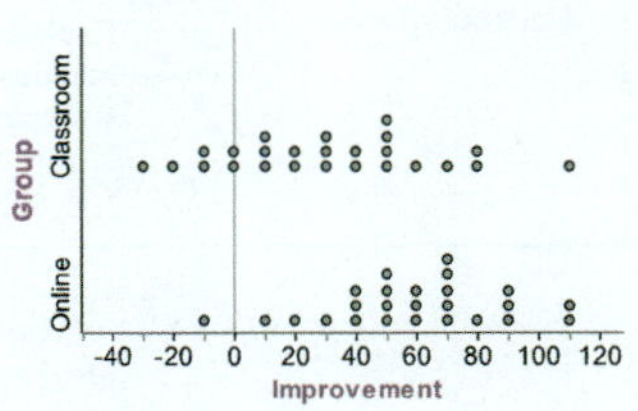

Let's see how these principles were used in designing a famous medical experiment.

## EXAMPLE The Physicians' Health Study

### *A well-designed experiment*

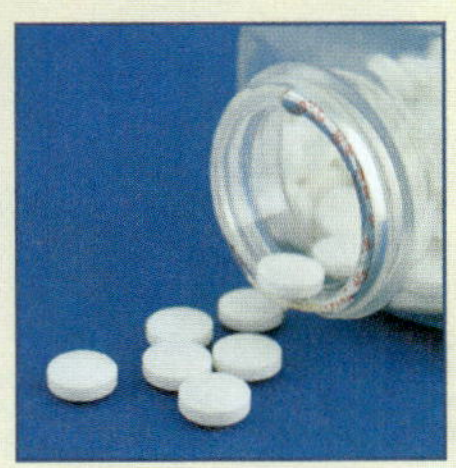

Does regularly taking aspirin help protect people against heart attacks? The Physicians' Health Study was a medical experiment that helped answer this question. In fact, the Physicians' Health Study looked at the effects of two drugs: aspirin and beta-carotene. Researchers wondered whether beta-carotene would help prevent some forms of cancer. The subjects in this experiment were 21,996 male physicians. There were two explanatory variables (factors), each having two levels: aspirin (yes or no) and beta-carotene (yes or no). Combinations of the levels of these factors form the four treatments shown in Figure 4.3 on the next page. One-fourth of the subjects were assigned at random to each of these treatments.

## Teaching Tip

Many students think that confounding is the only possible problem that other variables can cause. In this case, however, "number of class sessions" is not confounded with course type because students in both types of courses were allowed to choose the number of sessions they attended. This means that there is no systematic difference between the two groups—and no confounding.

## Teaching Tip

Although we don't formally introduce the concept of power until Chapter 9, you can talk about it informally here. If one treatment is more effective than the other, controlling other variables makes it more likely to find convincing evidence that it is more effective. Leaving other variables uncontrolled makes is harder to detect a difference in the effectiveness of the treatments, making a Type II error more likely.

## ALTERNATE EXAMPLE

### Multitasking

Researchers in Canada performed an experiment with university students to examine the effects of in-class laptop use on student learning. All participants in the study were asked to attend a university-style lecture and take notes with their laptops. Half of the participants were assigned to complete other non-lecture-related online tasks during the lecture. These tasks were meant to imitate typical student Web browsing during classes. The remaining students simply took notes with their laptops. To assign the treatments, the researchers printed 40 papers with instructions (20 with multitasking and 20 without), shuffled them, and handed them out at random to students in the classroom. At the end of the lecture, all participants took a comprehension test to measure how much they learned from it. The results: students who were assigned to multitask did significantly worse (11%) than students who were not assigned to multitask. From http://www.sciencedirect.com/science/article/pii/S0360131512002254.

PROBLEM: Explain how each of the four principles of experimental design was used in the multitasking study.

SOLUTION: *Comparison*: The researchers compared students who were multitasking to other students who were not multitasking.
*Random assignment*: It was determined at random which students (and which seat locations) received the instructions to multitask and which students (and seat locations) received the regular instructions.
*Control*: The experiment used undergraduate students from the same university in Canada. All participants listened to the same lecture and took the same comprehension test at the end of the lecture.
*Replication*: There were 20 students in each treatment group. Although this is not a very large number of subjects, the difference in performance was big enough to determine that the difference was due to multitasking and not the chance variation in random assignment.

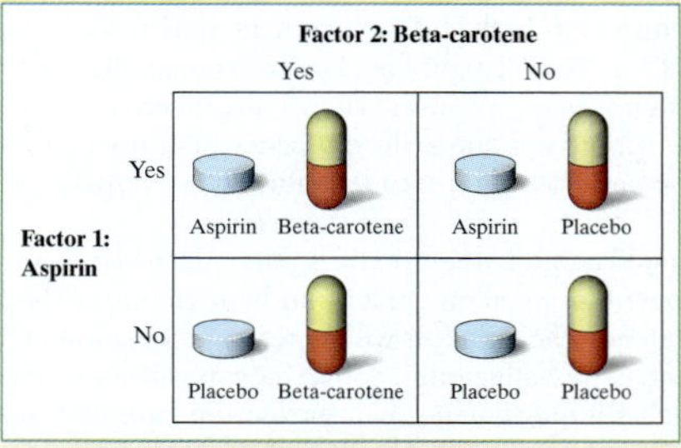


**FIGURE 4.3** The treatments in the Physicians' Health Study.

On odd-numbered days, the subjects took either a tablet that contained aspirin or a dummy pill that looked and tasted like the aspirin but had no active ingredient (a **placebo**). On even-numbered days, they took either a capsule containing beta-carotene or a placebo. There were several response variables—the study looked for heart attacks, several kinds of cancer, and other medical outcomes. After several years, 239 of the placebo group but only 139 of the aspirin group had suffered heart attacks. This difference is large enough to give good evidence that taking aspirin does reduce heart attacks.[23] It did not appear, however, that beta-carotene had any effect on preventing cancer.

Why did researchers decide to do the Physicians' Health Study (PHS)? The interesting history that led to this experiment is detailed at the PHS Web site. You can also find out about the Physicians' Health Study II, which ended in December 2007.

PROBLEM: Explain how each of the four principles of experimental design was used in the Physicians' Health Study.

SOLUTION:

Comparison: Researchers used a design that compared both of the active treatments to a placebo.

Random assignment: Was used to determine which subjects received each of the four treatment combinations. This helped ensure that the treatment groups were roughly equivalent to begin with.

Control: The experiment used subjects of the same gender and occupation. All subjects followed the same schedule of pill taking.

Replication: There were over 5000 subjects per treatment group. This large number of subjects helped ensure that the difference in heart attacks was due to the aspirin and not to chance variation in the random assignment.

**For Practice** *Try Exercise* 63

The Physicians' Health Study shows how well-designed experiments can yield good evidence that differences in the treatments cause the differences we observe in the response.

## Completely Randomized Designs

The diagram in Figure 4.4 presents the details of the SAT prep experiment: random assignment, the sizes of the groups and which treatment they receive, and the response variable. There are, as we will see later, statistical reasons for using treatment groups that are about equal in size. This type of design is called a **completely randomized design**.

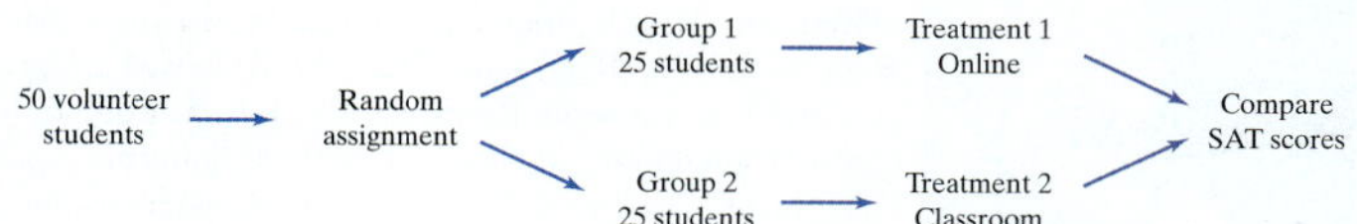


**FIGURE 4.4** Outline of a completely randomized design to compare online and classroom instruction.

### Teaching Tip

After reading the example about the Physicians' Health Study, ask your students to identify who the results of this experiment apply to. Because the study only used male physicians, the results of the study apply only to male physicians like the ones in the study. Applying the results to females or males of other professions might be reasonable, but it isn't justified by the design of the study. We will learn more about what inferences we can make from different types of studies in Section 4.3.

### Teaching Tip

Having equal group sizes results in slightly more power when we perform significance tests. That is, if one treatment is really more effective than the other, we will be more likely to find convincing evidence of this when the group sizes are the same.

**DEFINITION: Completely randomized design**

In a **completely randomized design,** the experimental units are assigned to the treatments completely by chance.

Notice that the definition of a completely randomized design does not require that each treatment be assigned to an equal number of experimental units. It does specify that the assignment of treatments must occur completely at random.

THINK ABOUT IT

**Does using chance to assign treatments in an experiment guarantee a completely randomized design?** Actually, no. Let's return to the SAT prep course experiment. Another way to randomly assign the 50 students to the two treatments is by tossing a coin. Have each student come forward and toss a coin. If it's heads, then the student will take the course online. If it's tails, then the student will take the classroom course.

As long as all 50 students toss a coin, this is still a completely randomized design. Of course, the two experimental groups are unlikely to contain exactly 25 students each due to the chance variation in coin tosses.

The problem comes if we try to force the two groups to have equal sizes. Suppose we let the coin tossing continue until one of the groups has 25 students and then place the remaining students in the other group. This is no longer a completely randomized design, because the last few students aren't being assigned to one of the treatments by chance. In fact, these students will all end up in the same group, which could lead to bias if these individuals share some characteristic that would systematically affect the response variable. For example, if the students came to toss the coin last because they're lazier than the other students who volunteered, then the SAT prep class that they're in will seem less effective than it really is.

Completely randomized designs can compare any number of treatments. Here is an experiment that compares three treatments.

## EXAMPLE Conserving Energy

### *A completely randomized design*

Many utility companies have introduced programs to encourage energy conservation among their customers. An electric company considers placing small digital displays in households to show current electricity use and what the cost would be if this use continued for a month. Will the displays reduce electricity use? One cheaper approach is to give customers a chart and information about monitoring their electricity use from their outside meter. Would this method work almost as well? The company decides to conduct an experiment to compare these two approaches (display, chart) with a group of customers who receive information about energy consumption but no help in monitoring electricity use.

### Teaching Tip

A completely randomized design in an experiment is somewhat analogous to an SRS when selecting a sample. In an SRS, there are no restrictions on who can be included in the sample as is the case with a stratified random sample or a cluster sample. Likewise, in a completely randomized design, there are no restrictions on who can be assigned to each treatment. There are other designs that we will discuss shortly that place restrictions on who can be assigned to each group.

### ALTERNATE EXAMPLE

**_Television and young children_**

Psychologists want to know how different types of television shows impact young children. They recruit 60 four-year-olds and have them watch 9 minutes of a fast-paced children's program (defined as scene changes every 10–15 seconds), watch 9 minutes of a slow-paced children's program (defined as scene changes every 30–45 seconds), or draw for 9 minutes. After the 9 minutes, each child will complete several tasks, including tests for mental ability and impulse control.

PROBLEM: Describe a completely randomized design for this experiment. Write a few sentences describing how you would implement your design.

SOLUTION: Here is an outline of the design:

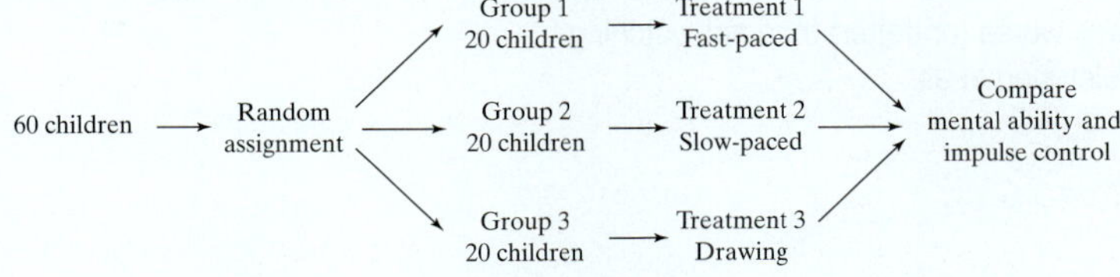


To implement the design, use 60 equally sized slips of paper. Label 20 of the slips "1," 20 of the slips "2," and 20 of the slips "3." Mix the numbers in a hat and have each child draw a number without looking. The number that each child chooses will be the group to which he or she is assigned. At the end of the study, compare the mental ability and impulse control for the three treatment groups.

### AP® Exam Common Error

On several different AP® free-response questions, students have lost credit for using random assignment methods like the one described in the Think About It feature on this page. Encourage your students to use the "hat" method whenever possible. It is easy to describe and helps students avoid making the mistakes that are possible when using random digits or coin flips. Just make sure that the slips of paper are the same size and that they are well mixed!

PROBLEM: Describe a completely randomized design involving 60 single-family residences in the same city whose owners are willing to participate in such an experiment. Write a few sentences explaining how you would implement your design.

SOLUTION: Figure 4.5 outlines the design. We'll randomly assign 20 houses to each of three treatments: digital display, chart plus information, and information only. Our response variable is the total amount of electricity used in a year.

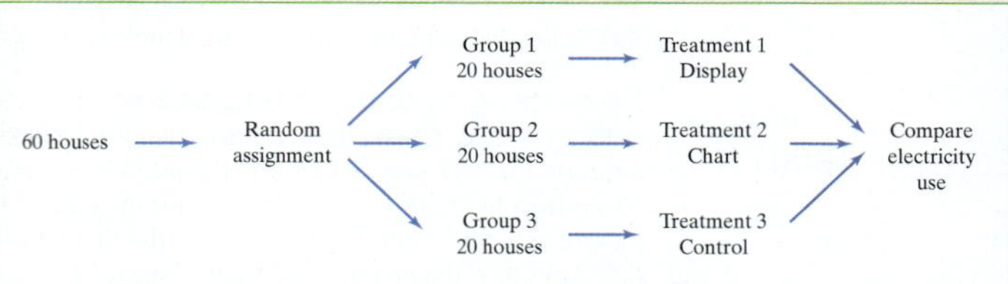


**FIGURE 4.5** Outline of a completely randomized design to compare three energy-saving programs.

To implement the design, start by labeling each house with a distinct number from 1 to 60. Write the labels on 60 identical slips of paper, put them in a hat, and mix them well. Draw out 20 slips. The corresponding homes will be given digital displays showing current electricity use. Now draw out 20 more slips. Those homes will use a chart. The remaining 20 houses will be given information about energy consumption but no way to monitor their usage. At the end of the year, compare how much electricity was used by the homes in the three groups.

**AP® EXAM TIP** If you are asked to describe the design of an experiment on the AP® exam, you won't get full credit for providing only a diagram like Figure 4.5. You are expected to describe how the treatments are assigned to the experimental units and to clearly state what will be measured or compared. Some students prefer to start with a diagram and then add a few sentences. Others choose to skip the diagram and put their entire response in narrative form.

**For Practice** *Try Exercise* 65

Why did we include a **control group** of 20 houses in the energy conservation experiment? *The main purpose of a control group is to provide a baseline for comparing the effects of the other treatments.* Without such a comparison group, we wouldn't be able to tell whether homes with digital displays or charts used less electricity than homes without such aids.

**Was a control group really necessary?** You might be thinking that the change in electricity use from last year to this year in the houses with displays and charts would tell us whether these treatments helped. Unfortunately, it's not that simple. Suppose last year's temperatures were more extreme than this year's. Then many households might show a decrease in electricity use, but we couldn't be sure whether this change was due to the weather or to the treatments. (Can you say confounding?!)

## Teaching Tip

If you did the Alternate Example on television and young children, ask your students why the researchers had 20 of the children draw for 9 minutes when the study was about different types of television viewing. In this case, the children who were drawing served as a control group that could be used for comparison. If there was no control group, the researchers could compare fast- and slow-paced television shows, but they would not be able to determine if either of these types of shows was worse (or better) than not watching television at all.

Many experiments (like the one in the previous example) include a control group that receives an inactive treatment. However, a control group can also be given an active treatment. Suppose we want to compare the effectiveness of a newly developed drug for treating a serious disease with a drug that's already known to work. In that case, the experimental units that receive the existing drug form the control group.

Some experimental designs don't include a control group. That's appropriate if researchers simply want to compare the effects of several treatments, rather than determining whether any of them works better than an inactive treatment. For instance, a state's highway department wants to see which of three brands of paint will last longest when marking lane lines on the freeway. Putting no paint on the highway is clearly not a good option!

### CHECK YOUR UNDERSTANDING

Music students often don't evaluate their own performances accurately. Can small-group discussions help? The subjects were 29 students preparing for the end-of-semester performance that is an important part of their grade. The 15 students in one group each videotaped a practice performance, evaluated it themselves, and then discussed the tape with a small group of other students. The remaining 14 students watched and evaluated their tapes alone. At the end of the semester, the discussion-group students evaluated their final performance more accurately.[24]

1. Describe a completely randomized design for this experiment. Write a few sentences describing how you would implement your design.
2. What is the purpose of the control group in this experiment?

## Experiments: What Can Go Wrong?

The logic of a randomized comparative experiment depends on our ability to treat all the subjects the same in every way except for the actual treatments being compared. Good experiments, therefore, require careful attention to details to ensure that all subjects really are treated identically.

If some subjects in a medical experiment take a pill each day and a control group takes no pill, the subjects are not treated identically. Many medical experiments are therefore "placebo-controlled," like the Physicians' Health Study. On odd-numbered days, all the subjects took an aspirin or a placebo. On even-numbered days, all of them took either a beta-carotene pill or a placebo.

Many patients respond favorably to any treatment, even a placebo, perhaps because they trust the doctor. The response to a dummy treatment is called the **placebo effect**. If some subjects in the Physicians' Health Study did not take any pills, the effect of aspirin or beta-carotene would be confounded with the placebo effect, the effect of simply taking pills.

EXAMPLE

### Curing Baldness and Soothing Pain

*Do placebos work?*

Want to help balding men keep their hair? Give them a placebo. One study found that 42% of balding men maintained or increased the amount of hair on their heads when they took a placebo. In another study, researchers zapped the wrists

### Teaching Tip

An inactive treatment is often called a *placebo*. We will introduce this term shortly, but it is OK to use it here. Many of your students will already know about placebos from other classes, such as AP® Psychology.

### ANSWERS TO CYU

**1.** Randomly assign the 29 students to two treatments: evaluating the performance in small groups or evaluating the performance alone. The response variable will be the accuracy of their final performance evaluations. To implement this design, use 29 equally sized slips of paper. Label 15 of them "small group" and 14 of them "alone." Then shuffle the papers and hand them out at random to the 29 students to assign them to a treatment.

**2.** The purpose of the control group is to provide a baseline for comparison. Without a group to compare to, it is impossible to determine if the small group treatment is more effective.

### Teaching Tip

A good illustration of the placebo effect is a parent kissing a child's "boo-boo" when the child gets injured. Even though the kiss has no "active treatment," it makes the child feel better!

ALTERNATE EXAMPLE

***A more expensive placebo?***

In a study reported by the *New York Times* on March 5, 2008 ("More Expensive Placebos Bring More Relief"), researchers discovered that placebos have a stronger effect when they are perceived to be more expensive. The study had volunteers rate the pain of an electric shock before and after taking a new medication. However, half of the subjects were told the medication cost $2.50 per dose, while the other half were told the medication cost $0.10 per dose. In reality, both medications were placebos, and both had a strong effect. Of the "cheap" placebo users, 61% experienced pain relief, while 85% of the "expensive" placebo users experienced pain relief. The researchers suggested that people are accustomed to paying more for better medications, which may account for the difference in response. As with any placebo, it's all about the expectations of the subjects.

ALTERNATE EXAMPLE

### *Antidepressants and the placebo effect*

The CBS News program *60 Minutes* aired an excellent segment about the placebo effect called "Treating Depression: Is There a Placebo Effect?" Watch the 20-minute video at http://www.cbsnews.com/video/watch/?id=7399362n.

ALTERNATE EXAMPLE

### *Prescribing placebos*

The article at the link below describes how the placebo effect works, even when subjects know they are receiving a placebo: www.medpagetoday.com/PainManagement/PainManagement/24056.

### Teaching Tip

Blinding the subjects and those who interact with them and measure the results is another form of control. In this case, we are keeping the expectations the same for all treatment groups.

### Teaching Tip

The caffeine experiment can be conducted in a double-blind manner. To make sure the subjects are blind, the two colas must look and taste exactly the same. To be double-blind, the people measuring the pulse rates and interacting with the subjects should also be unaware of which subjects are receiving which treatments. To make this happen, have the subjects measure their own pulse as they are already blind. Also, ask another teacher to prepare the colas with labels A and B and then leave the room before anyone else gets there. Only after the experiment is complete will this teacher come back to reveal which treatment is which.

### AP® Exam Common Error

Many students believe that an experiment cannot be double-blind because "someone has to know who gets which treatment." Although there is always someone who knows which subjects are receiving which treatment, as long as this person doesn't interact with the subjects or measure the response, the experiment can still be double-blind.

of 24 test subjects with a painful jolt of electricity. Then they rubbed a cream with no active medicine on subjects' wrists and told them the cream should help soothe the pain. When researchers shocked them again, 8 subjects said they experienced significantly less pain.[25]

When the ailment is vague and psychological, like depression, some experts think that the placebo effect accounts for about three-quarters of the effect of the most widely used drugs.[26] Others disagree. In any case, "placebos work" is a good place to start when you think about planning medical experiments.

The strength of the placebo effect is a strong argument for randomized comparative experiments. In the baldness study, 42% of the placebo group kept or increased their hair, but 86% of the men getting a new drug to fight baldness did so. The drug beats the placebo, so it has something besides the placebo effect going for it. Of course, the placebo effect is still part of the reason this and other treatments work.

Because the placebo effect is so strong, it would be foolish to tell subjects in a medical experiment whether they are receiving a new drug or a placebo. Knowing that they are getting "just a placebo" might weaken the placebo effect and bias the experiment in favor of the other treatments.

It is also foolish to tell doctors and other medical personnel what treatment each subject received. If they know that a subject is getting "just a placebo," they may expect less than if they know the subject is receiving a promising experimental drug. Doctors' expectations change how they interact with patients and even the way they diagnose a patient's condition. Whenever possible, experiments with human subjects should be **double-blind**.

Reports in medical journals regularly begin with words like these, from a study of a flu vaccine given as a nose spray: "This study was a randomized, double-blind, placebo-controlled trial. Participants were enrolled from 13 sites across the continental United States between mid-September and mid-November."[27] Doctors are supposed to know what this means. Now you know, too.

**DEFINITION: Double-blind**

In a **double-blind** experiment, neither the subjects nor those who interact with them and measure the response variable know which treatment a subject received.

The idea of a double-blind design is simple. Until the experiment ends and the results are in, only the study's statistician knows for sure which treatment a subject is receiving. However, some experiments cannot be carried out in a double-blind manner. If researchers are comparing the effects of exercise and dieting on weight loss, then subjects will know which treatment they are receiving. Such an experiment can still be **single-blind** if the individuals who are interacting with the subjects and measuring the response variable don't know who is dieting and who is exercising. In other single-blind experiments, the subjects are unaware of which treatment they are receiving, but the people interacting with them and measuring the response variable do know.

## CHECK YOUR UNDERSTANDING

In an interesting experiment, researchers examined the effect of ultrasound on birth weight. Pregnant women participating in the study were randomly assigned to one of two groups. The first group of women received an ultrasound; the second group did not. When the subjects' babies were born, their birth weights were recorded. The women who received the ultrasounds had heavier babies.[28]

1. Did the experimental design take the placebo effect into account? Why is this important?
2. Was the experiment double-blind? Why is this important?
3. Based on your answers to Questions 1 and 2, describe an improved design for this experiment.

## Inference for Experiments

In an experiment, researchers usually hope to see a difference in the responses so large that it is unlikely to happen just because of chance variation. We can use the laws of probability, which describe chance behavior, to learn whether the treatment effects are larger than we would expect to see if only chance were operating. If they are, we call them **statistically significant**.

**DEFINITION: Statistically significant**

An observed effect so large that it would rarely occur by chance is called **statistically significant.**

If we observe statistically significant differences among the groups in a randomized comparative experiment, we have good evidence that the treatments caused these differences. You will often see the phrase "statistically significant" in published research reports in many fields. The great advantage of randomized comparative experiments is that they can produce data that give good evidence for a cause-and-effect relationship between the explanatory and response variables. We know that in general a strong association does not imply causation. A statistically significant association in data from a well-designed experiment *does* imply causation.

## ACTIVITY | Distracted driving

MATERIALS: Set of 48 index cards or standard deck of playing cards for each pair of students

Is talking on a cell phone while driving more distracting than talking to a passenger? David Strayer and his colleagues at the University of Utah designed an experiment to help answer this question. They used 48 undergraduate students as subjects. The researchers randomly assigned half of the subjects to drive in a simulator while talking on a cell phone, and the other half to drive in the simulator

## ✓ ANSWERS TO CYU

**1.** No. Perhaps seeing the image of their unborn child encouraged the mothers who had an ultrasound to eat a better diet, resulting in healthier babies.

**2.** No. While the people weighing the babies at birth may not have known whether that particular mother underwent an ultrasound or not, the mothers knew. This might have affected the outcome because the mothers knew whether they had received the treatment or not.

**3.** Treat all mothers as if they had an ultrasound, but for some mothers the ultrasound machine wouldn't be turned on. To prevent mothers from figuring out the machine was turned off, the ultrasound screen would have to be turned away from all the mothers.

### Teaching Tip

We will learn much more about statistical significance in Chapters 9–12. For now, emphasize the connection between statistical significance and the variability that arises from randomly assigning the treatments to experimental units. Because no random assignment will create perfectly equivalent treatment groups at the beginning of an experiment, there will likely be differences in the response variable just by chance. Results are called statistically significant only if the difference in the response is bigger than what would be expected due to chance variation in the random assignment.

### AP® Exam Common Error

Although a statistically significant association in data from a well-designed experiment does imply causation, many students incorrectly respond that a statistically significant association in data from an observational study implies causation as well. This isn't true—the design of the study is what determines if a cause-and-effect conclusion is possible. We will address this in more detail in Section 4.3.

### Activity Overview

You can watch an overview of this Activity (for teachers) by clicking on the link in the TE-book, opening the TRCD, or logging onto the book's companion Web site www.whfreeman.com/tps5e (teacher log-in required).

### Teaching Tip

This Activity (and the Alternate Activity) use simulation to determine if the results of an experiment are statistically significant. This is exactly what students are expected to do in the Common Core State Standards (IC.B.5). After performing the activities by hand, you can use the *Randomization Test* applets at www.lock5stat.com/statkey to automate the simulation. Choose the *Test for a Difference in Proportions* to do the Distracted Driving Activity and the *Test for Difference in Means* for the Fish Oil Alternate Activity.

## Teaching Tip

After looking at the results of the experiment displayed in the two-way table, make sure students can identify the "evidence" that cell phones are more distracting. They should be able to point out that only half of the cell phone users (12 of 24) stopped at the rest area, while nearly all of the subjects who were talking to a passenger (21 of the 24) stopped at the rest area. To determine if this evidence is *convincing*, we must rule out the possibility that the difference was due only to the chance involved in the random assignment. To make this determination, we conduct a simulation to see what outcomes could happen by chance alone.

### ALTERNATE ACTIVITY

*Fish oil*

To see if fish oil can help reduce blood pressure, 14 males with high blood pressure were recruited and randomly assigned to one of two treatments. The first treatment was a 4-week diet that included fish oil, and the second was a 4-week diet that included regular oil. At the end of the 4 weeks, each volunteer's blood pressure was measured again and the reduction in diastolic blood pressure was recorded. The results of this study are shown below. Note that a negative value means that the subject's blood pressure *increased*. [*Source: New England Journal of Medicine*, 320 (1989), pp. 1037–1043; cited in Fred Ramsey and Daniel Schafer, *The Statistical Sleuth*, Pacific Grove, CA: Duxbury Press, 2002, p. 23.]

| | | | | | | | | |
|---|---|---|---|---|---|---|---|---|
| Fish oil: | 8 | 12 | 10 | 14 | 2 | 0 | 0 | $\bar{x}_F = 6.57$ |
| Regular oil: | −6 | 0 | 1 | 2 | −3 | −4 | 2 | $\bar{x}_R = -1.14$ |
| | | | | | | | | $\bar{x}_F - \bar{x}_R = 7.71$ |

Is there convincing evidence that fish oil is better than regular oil in reducing diastolic blood pressure? In other words, is the difference in mean reduction statistically significant? To find out, let's see what would happen if we randomly assign the 14 people in the experiment to the two groups many times, *assuming the two treatments have the same effect on blood pressure*.

**1.** Write each person's reduction in blood pressure on a separate but equally sized slip of paper.

**2.** Shuffle the papers and deal two piles of 7 each. The first pile will represent the fish oil group and the second pile will represent the regular oil group. Calculate the mean decrease for each group and then find the simulated difference in means ($\bar{x}_F - \bar{x}_R$).

while talking to a passenger. One response variable was whether or not the driver stopped at a rest area that was specified by researchers before the simulation started. The table below shows the results:[29]

| Stopped at rest area? | Distraction: Cell phone | Distraction: Passenger |
|---|---|---|
| Yes | 12 | 21 |
| No | 12 | 3 |

Are these results statistically significant? To find out, let's see what would happen just by chance if we randomly reassign the 48 people in this experiment to the two groups many times, *assuming the treatment received doesn't affect whether a driver stops at the rest area.*

1. We need 48 cards to represent the drivers in this study. In the original experiment, 33 drivers stopped at the rest area and 15 didn't. Because we're assuming that the treatment received won't change whether each driver stops at the rest area, we use 33 cards to represent drivers who stop and 15 cards to represent those who don't.
   - *Using index cards:* Write "Yes" on 33 cards and "No" on 15 cards.
   - *Using playing cards:* Remove jokers and other specialty cards from the deck, as well as the ace of spades and any three of the 2s. All cards with denominations 2 through 10 represent drivers who stop. All jacks, queens, kings, and aces represent drivers who don't stop.

Why did we only ask you to count the number of drivers who didn't stop at the rest area in the cell-phone group? Suppose you get 10 in one trial of the simulation. That means the other 24 − 10 drivers in the cell-phone group did stop at the rest area. Also, there must be 15 − 10 = 5 drivers in the passenger group who failed to stop, leaving 24 − 5 = 19 drivers in the group who did stop. Recording the one number tells you all the others.

2. Shuffle and deal two piles of 24 cards each—the first pile represents the cell phone group and the second pile represents the passenger group. The shuffling reflects our assumption that the outcome for each subject is not affected by the treatment. Record the number of drivers who failed to stop at the rest area in the cell-phone group.
3. Your teacher will draw and label axes for a class dotplot. Plot the result you got in Step 2 on the graph.
4. Repeat Steps 2 and 3 if needed to get a total of at least 40 repetitions of the simulation for your class.
5. In the original experiment, 12 of the 24 drivers using cell phones didn't stop at the rest area. Based on the class's simulation results, how surprising would it be to get a result this large or larger simply due to the chance involved in the random assignment? Is the result statistically significant?
6. What conclusion would you draw about whether talking on a cell phone is more distracting than talking to a passenger?

Here is an example of what the class dotplot in the Activity might look like after 100 trials. In the 100 trials, only once did 12 or more people fail to stop when using a cell phone. Because a result of 12 or more is unlikely to happen by chance alone, the results of this study should be considered statistically significant.

**3.** Repeat this process many times and make a class dotplot of the simulated differences. How often did you get a difference of 7.71 or more? Here are the results of 100 trials of this simulation. In these 100 trials, a difference of 7.71 or greater didn't occur a single time, suggesting that a difference of 7.71 is statistically significant.

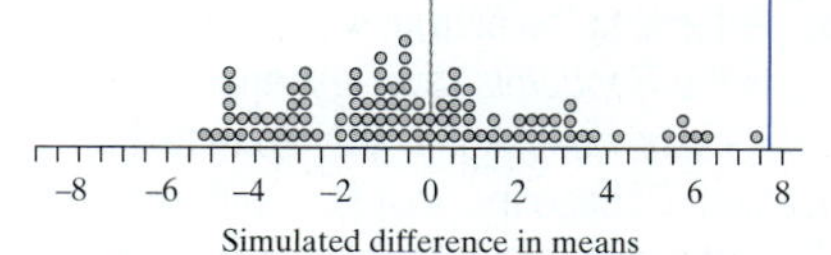


## Teaching Tip

If you choose to conduct the caffeine experiment (or some other experiment) in class, do a simulation to determine whether the results are statistically significant. If the response variable in your study is categorical, follow the distracted driver example. If the response variable in your study is quantitative, follow the fish oil example.

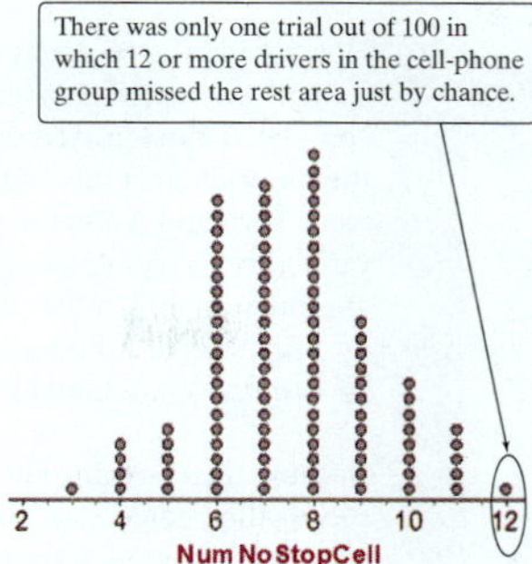

**Can an "unlucky" random assignment lead to confounding?**
Let's return to the distracted-driver Activity. Some people are more forgetful than others. Suppose that the random assignment happens to put most of the forgetful subjects in one group. If more drivers in that group fail to stop at the rest area, we don't know if it's because of the treatment they received (cell phone or passenger) or their forgetfulness. Is this confounding?

You might be surprised that the answer is "No!" Although people's memory is a variable that might affect whether or not they stop at the rest area (the response variable), the design of the experiment takes care of this by randomly assigning subjects to the two treatment groups. The "unlucky" random assignments are taken into account in determining statistical significance. In an experiment, confounding occurs when the design doesn't account for existing differences in the experimental units that might systematically affect their response to the treatments.

## Blocking

Completely randomized designs are the simplest statistical designs for experiments. They illustrate clearly the principles of comparison, random assignment, control, and replication. But just as with sampling, there are times when the simplest method doesn't yield the most precise results. When a population consists of groups of individuals that are "similar within but different between," a stratified random sample gives a better estimate than a simple random sample. This same logic applies in experiments.

**EXAMPLE**

### A Smarter Design?

*The idea of blocking*

Suppose that a mobile phone company is considering two different keyboard designs (A and B) for its new smart phone. The company decides to perform an experiment to compare the two keyboards using a group of 10 volunteers. The response variable is typing speed, measured in words per minute.

### Teaching Tip

Some students might ask, "How unlikely does a result need to be for it to be considered statistically significant?" This is a great question! Tell them that it depends on several factors, including the consequences of making a wrong decision. We will discuss this issue in much more depth in Chapter 9. If the students really want an answer, you can say that 5% is a commonly used boundary.

**ALTERNATE EXAMPLE**

### *More SAT prep*

In the previous examples about online versus classroom SAT preparation courses, we acknowledged that the student's current math class might be a source of variability in their improvement. In our completely randomized design, we hoped that the random assignment would distribute the Precalculus students roughly equally between the online and classroom SAT courses. Likewise, we hope that the random assignment would roughly balance the Algebra 2 students, the Geometry students, and so on. Still, there might be lots of variability in the results for each course if the students in the Precalculus class experience greater improvements than students in a Geometry class, regardless of which SAT preparation course they took. This might make it hard to detect a difference in the effectiveness of the two courses.

To account for this potential variability, we can separate the students into groups (blocks) based on their current math class and randomly divide the members of each block into the online and classroom SAT courses. This randomized block design will help us account for the variation in improvement due to one's current math class.

How should the company deal with the fact that four of the volunteers already use a smart phone, whereas the remaining six volunteers do not? They could use a completely randomized design and hope that the random assignment distributes the smart-phone users and non-smart-phone users about evenly between the group using keyboard A and the group using keyboard B. Even so, there might be a lot of variability in typing speed in both groups because some members of each group are much more familiar with smart phones than others. This additional variability might make it difficult to detect a difference in the effectiveness of the two keyboards. What should the researchers do?

Because the company knows that experience with smart phones will affect typing speed, they could start by separating the volunteers into two groups—one with experienced smart-phone users and one with inexperienced smart-phone users. Each of these groups of similar subjects is known as a **block**. Within each block, the company could then randomly assign half of the subjects to use keyboard A and the other half to use keyboard B. To control other variables, each subject should be given the same passage to type while in a quiet room with no distractions. This **randomized block design** helps account for the variation in typing speed that is due to experience with smart phones.

## AP® Exam Common Error

Many students confuse blocks and treatment groups. Blocks are *not* formed at random. Instead, they are formed by grouping experimental units that are similar in some way that is expected to systematically affect the response to the treatments. This means that each block should be very different from other blocks. However, a treatment group is formed at random, with the goal that the treatment groups be as similar as possible to each other, other than the treatment.

Figure 4.6 outlines the randomized block design for the smart-phone experiment. The subjects are first separated into blocks based on their experience with smart phones. Then the two treatments are randomly assigned within each block.

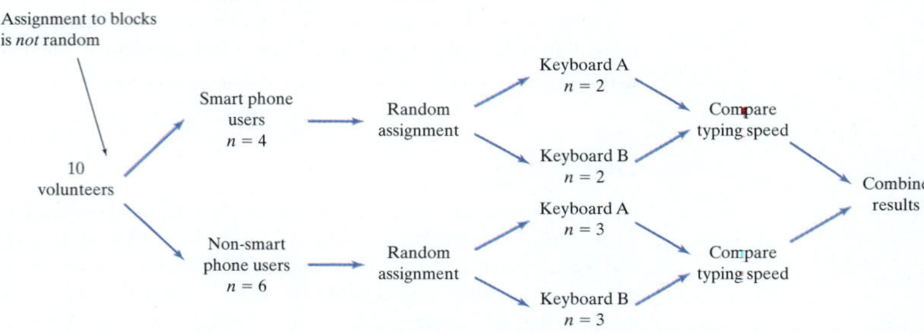


**Figure 4.6** Outline of a randomized block design for the smart-phone experiment. The blocks consist of volunteers who have used smart phones and volunteers who have not used smart phones. The treatments are keyboard A and keyboard B.

**DEFINITION: Block and randomized block design**

A **block** is a group of experimental units that are known before the experiment to be similar in some way that is expected to affect the response to the treatments.

In a **randomized block design**, the random assignment of experimental units to treatments is carried out separately within each block.

## Teaching Tip

Notice that the last stage in the diagram for an experiment with blocking is "combine results" not "compare blocks." The discussion on the following page illustrates how we combine the results after accounting for the difference in the blocks.

## Teaching Tip

Make sure students are able to explain why they chose a particular variable to use when forming blocks. The best blocking variables are those with the strongest association with the response variable. That is, the best variables to use are those that are the best predictors of the response variable. Of course, students need to give their explanation in the context of the question! For example, "I will block by smart-phone experience because this variable is a good predictor of typing speed on a smart phone."

Using a randomized block design allows us to account for the variation in the response that is due to the blocking variable. This makes it easier to determine if one treatment is really more effective than the other.

To see how blocking helps, let's look at the results of an experiment using 10 volunteers, 4 who already use a smart phone and 6 who do not. In the block of 4 smart-phone users, 2 will be randomly assigned to use keyboard A and the other 2 will be assigned to use keyboard B. Likewise, in the block of 6 non-smart-phone users, 3 will be randomly assigned to use keyboard A and the other 3 will be assigned to use keyboard B. Each of the 10 volunteers will type the same passage and the typing speed will be recorded.

Here are the results:

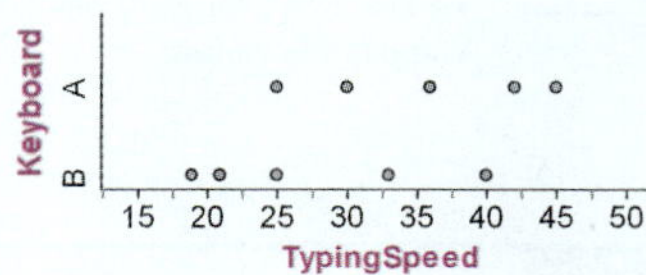


There is some evidence that keyboard A results in higher typing speeds, but the evidence isn't that convincing. Enough overlap occurs in the two distributions that the differences might simply be due to the chance variation in the random assignment.

If we compare the results for the two keyboards within each block, however, a different story emerges. Among the 4 smart-phone users (indicated by the blue squares), keyboard A was the clear winner. Likewise, among the 6 non-smart-phone users (indicated by the gray dots), keyboard A was also the clear winner.

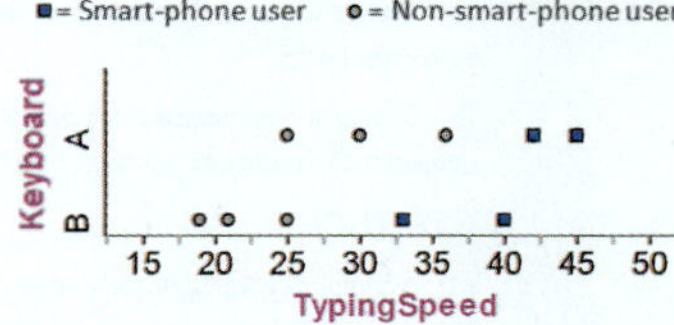


The overlap in the first set of dotplots was due almost entirely to the variation in smart-phone experience—smart-phone users were generally faster than non-smart-phone users, regardless of which keyboard they used. In fact, the average typing speed for the smart-phone users was 40, while the average typing speed for non-smart-phone users was only 26, a difference of 14 words per minute. To account for the variation created by the difference in smart-phone experience, let's subtract 14 from each of the typing speeds in the block of smart-phone users to "even the playing field."

Here are the results:

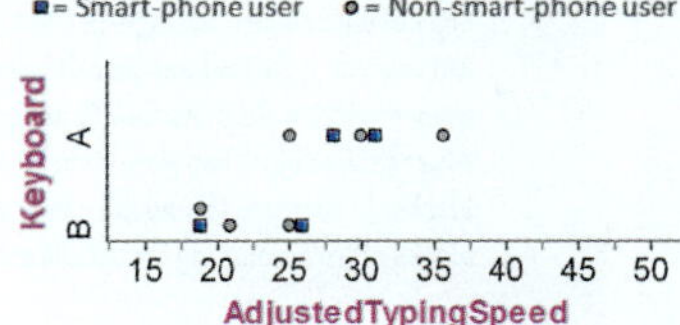


## Teaching Tip

Students are not expected to analyze the results of a blocked experiment on the AP® exam, except when the blocks are matched pairs (more on these coming very soon). We did the informal analysis on this page to illustrate how blocking helps to account for unwanted variability in an experiment.

### ALTERNATE EXAMPLE

#### *More SAT prep*

Here are the results from a randomized block experiment comparing online and classroom SAT preparation courses, using current math class as a blocking variable. The response variable is improvement in SAT score. Of the 20 subjects, 10 were in Precalculus, 6 were in Algebra 2, and 4 were in Geometry.

Here are the results without accounting for current math class:

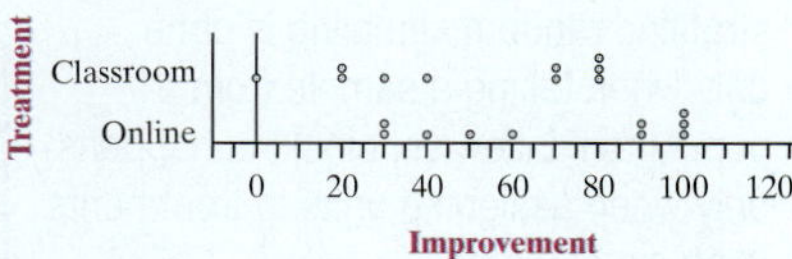


There is quite a bit of overlap, meaning that the difference in mean improvement for students in the online course and students in the classroom course is unlikely to be statistically significant.

Here are the same dotplots, this time with the current math class identified for each student:

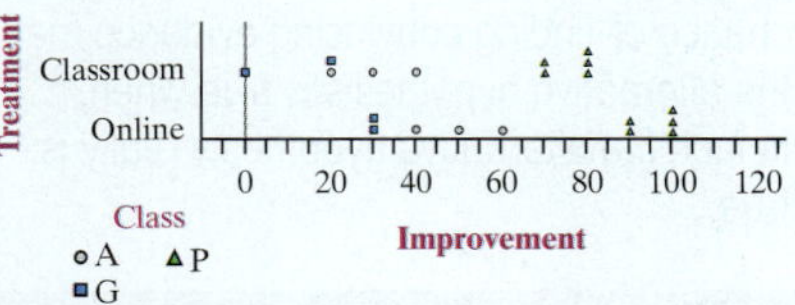


Looking at each class separately, it appears that the online course was clearly better within each block. The overlap in the original dotplot was mostly due to the fact that Precalculus students improved much more than Algebra 2 students and Geometry students. The mean improvement for Precalculus students was 86, while the mean improvement for Algebra 2 students was 40 and for Geometry students was 20. To level the playing field, let's subtract 66 points from each student in Precalculus and subtract 20 from each student in Algebra 2. Here are the adjusted results, on the same scale:

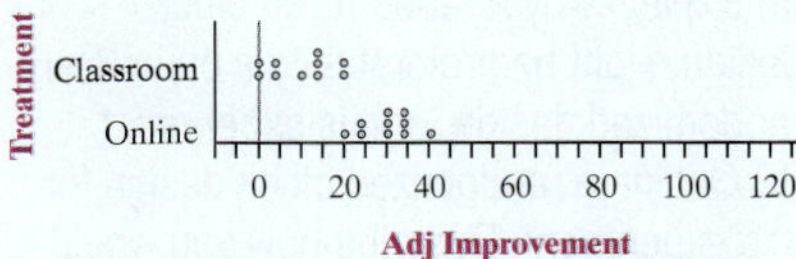


Now that the variability due to current math class has been accounted for, it is much more obvious that the online course is better than the classroom course.

## AP® Exam Common Error

Confusing stratified random sampling with a randomized block design is another way that students mix up the language of sample surveys and experiments. These two ideas are similar in some ways, because they both involve forming groups of similar subjects before random selection or random assignment. Both methods also help to account for the variability created by other variables. However, stratified random sampling is done only when taking a sample from a population. Likewise, blocking happens only when assigning units to treatments in an experiment.

## Teaching Tip

When we say that it is easier to find convincing evidence that one treatment is more effective than another, we are saying that we have more statistical power. As we will learn in Chapter 9, a significance test is more powerful when it has a greater chance of finding convincing evidence that the alternative hypothesis is true when, in fact, the alternative hypothesis really is true.

**AP® EXAM TIP** Don't mix the language of experiments and the language of sample surveys or other observational studies. You will lose credit for saying things like "use a randomized block design to select the sample for this survey" or "this experiment suffers from nonresponse since some subjects dropped out during the study."

Because we accounted for the variation due to the difference in smart-phone experience, the variation in each of the distributions has been reduced. There is now almost no overlap between the two distributions, meaning that the evidence in favor of keyboard A is much more convincing. *When blocks are formed wisely, it is easier to find convincing evidence that one treatment is more effective than another.*

The idea of blocking is an important additional principle of experimental design. A wise experimenter will form blocks based on the most important unavoidable sources of variation (other variables) among the experimental units. Randomization will then average out the effects of the remaining other variables and allow an unbiased comparison of the treatments. The moral of the story is: *control what you can, block on what you can't control, and randomize to create comparable groups.*

## EXAMPLE Men, Women, and Advertising

### *Blocking in an experiment*

Women and men respond differently to advertising. Researchers would like to design an experiment to compare the effectiveness of three advertisements for the same product.

**PROBLEM:**

(a) Explain why a randomized block design might be preferable to a completely randomized design for this experiment.

(b) Outline a randomized block design using 300 volunteers (180 men and 120 women) as subjects. Describe how you would carry out the random assignment required by your design.

**SOLUTION:**

(a) A completely randomized design considers all subjects, both men and women, as a single pool. The random assignment would send subjects to three treatment groups without regard to their gender. This ignores the differences between men and women, which would probably result in a great deal of variability in responses to the advertising in all three groups. For example, if an ad appealed much more to men, you would get a wide range of reactions to that ad from the two genders. That would make it harder to determine whether one ad was more effective.

A randomized block design would consider women and men separately. In this case, the random assignment would occur separately in each block. Blocking will account for the variability in responses to advertising due to gender. This will allow researchers to look separately at the reactions of men and women, as well as to more effectively assess the overall response to the ads.

(b) Figure 4.7 outlines the randomized block design. We randomly assign the 120 women into three groups of 40, one for each of the advertising treatments. Write the women's names on 120 identical slips of paper, place the slips in a hat, and mix them well. Pull out 40 slips to determine which women will view Ad 1. Pull out another 40 slips to determine which women will view Ad 2. The remaining 40 women will view Ad 3. Randomly assign the 180 men into three groups of 60 using a similar process. After each subject has viewed the assigned ad, compare reactions to the three ads within the gender blocks. To compare the overall effectiveness of the three ads, combine the results from the two blocks after accounting for the difference in response for the men and women.

### ALTERNATE EXAMPLE

### *Microwave popcorn*

A popcorn lover wants to know if it is better to use the "popcorn button" on her microwave oven or use the amount of time recommended on the bag of popcorn. To measure how well each method works, she will count the number of unpopped kernels remaining after popping. To obtain the experimental units, she goes to the store and buys 10 bags each of 4 different varieties of microwave popcorn (butter, cheese, natural, and kettle corn), for a total of 40 bags.

**PROBLEM:**

**(a)** Explain why a randomized block design might be preferable to a completely randomized design for this experiment.

**(b)** Outline a randomized block design for this experiment. Describe how you would carry out the random assignment required by your design.

**SOLUTION:**

**(a)** A completely randomized design ignores the difference between the four types of popcorn, which will probably result in a great deal of variability in the number of unpopped kernels for both treatments. For example, if there are many more unpopped kernels in bags of cheese popcorn and kettle corn than the other two varieties, it will be harder to tell if there is a difference in the methods of popping. A randomized block design considers each variety of popcorn separately, which allows us to account for the variability in the number of unpopped kernels created by the difference in varieties.

**(b)** We will randomly assign 5 bags of each variety to each of the two treatments. To perform the random assignment, place all 10 bags of a particular variety in a large grocery bag. Shake the bag and then randomly select 5 bags to be popped using the "popcorn button." The remaining 5 bags will be popped using the instructions on the bags. Repeat this process for the remaining 3 varieties. After popping each of the 40 bags in random order, count the number of unpopped kernels in each bag and compare the results within each variety. Then combine the results from the 4 varieties after accounting for the difference in average response for each variety.

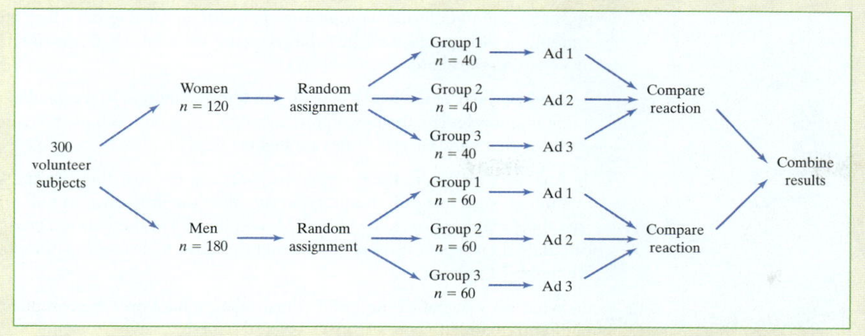


**FIGURE 4.7** Randomized block design for comparing responses to three advertisements. The blocks consist of male and female subjects.

**For Practice** *Try Exercise* 77

**Matched Pairs Design:** A common type of randomized block design for comparing two treatments is a **matched pairs design**. The idea is to create blocks by matching pairs of similar experimental units. Then we can use chance to decide which member of a pair gets the first treatment. The other subject in that pair receives the other treatment. That is, the random assignment of subjects to treatments is done within each matched pair. Just as with other forms of blocking, matching helps account for the variation among the experimental units.

Sometimes each "pair" in a matched pairs design consists of just one experimental unit that gets both treatments one after the other. In that case, each experimental unit serves as its own control. The *order* of the treatments can influence the response, so we randomize the order for each experimental unit.

## ACTIVITY | Get your heart beating

MATERIALS:
Clock or stopwatch

Are standing pulse rates generally higher than sitting pulse rates? In this Activity, you will perform two experiments to try to answer this question.

1. *Completely randomized design* For the first experiment, your teacher will randomly assign half of the students in your class to stand and the other half to sit. Once the two treatment groups have been formed, students should stand or sit as required. Then they should measure their pulses for one minute. Have the subjects in each group record their data on the board.

2. *Matched pairs design* In a matched pairs design, each student should receive both treatments in a random order. Because you already sat or stood in Step 1, you just need to do the opposite now. As before, everyone should measure their

### Teaching Tip

A matched pairs design is a special case of a randomized block design that uses blocks of size 2.

### TRM Reducing Cholesterol

Consider using the Reducing Cholesterol: Completely Randomized Design versus Matched Pairs Design Alternate Activity found in the Teacher's Resource Materials. This Activity has students simulate an experiment to compare two cholesterol drugs using two different designs. The Activity clearly shows how using a paired design can greatly improve an experiment.

### Activity Overview

You can watch an overview of this Activity (for teachers) by clicking on the link in the TE-book, opening the TRCD, or logging onto the book's companion Web site www.whfreeman.com/tps5e (teacher log-in required).

pulses for one minute after completing the treatment (that is, once they are standing or sitting). Have all the subjects record their data (both measurements) in a chart on the board.

3. Analyze the data for the completely randomized design. Make parallel dotplots and calculate the mean pulse rate for each group. Is there convincing evidence that standing pulse rates are higher? Explain.

4. Analyze the data for the matched pairs design. Because the data are paired by student, your first step should be to calculate the difference in pulse rate (standing – sitting) for each subject. Make a dotplot of these differences and calculate their mean. Is there convincing evidence that standing pulse rates are higher? Explain.

5. What advantage does the matched pairs design have over the completely randomized design?

An AP® Statistics class with 24 students performed the "Get Your Heart Beating" Activity. We'll analyze the results of their experiment in the following example.

## EXAMPLE Standing and Sitting Pulse Rate

### *Design determines analysis*

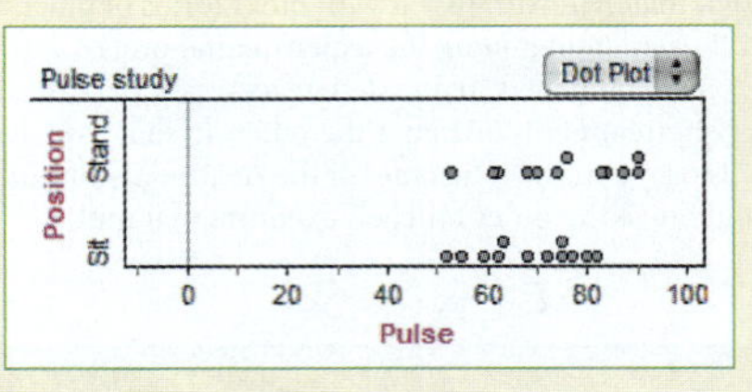


A Fathom dotplot of the pulse rates for their completely randomized design is shown. The mean pulse rate for the standing group is 74.83; the mean for the sitting group is 68.33. So the average pulse rate is 6.5 beats per minute higher in the standing group. However, the variability in pulse rates for the two groups creates a lot of overlap in the dotplots. These data don't provide convincing evidence that standing pulse rates tend to be higher.

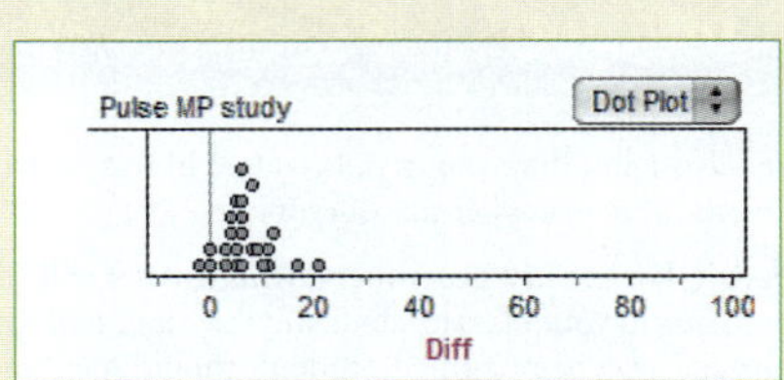


What about the class's matched pairs experiment? The Fathom dotplot shows their data on the difference in pulse rates (standing – sitting). For these 24 students, the mean difference was 6.8 beats per minute. In addition, 21 of the 24 students recorded a positive difference (meaning the standing pulse rate was higher). These data provide convincing evidence that people's standing pulse rates tend to be higher than their sitting pulse rates.

Let's take one more look at the two Fathom dotplots in the example. Notice that we used the same scale for both graphs. This is to help you visually compare the amount of variability in the response variable for each of the two experimental designs. Blocking by subject in the matched pairs design greatly reduced the

ALTERNATE EXAMPLE

### *Cholesterol drugs*

Here are Fathom dotplots for the two different designs in the Reducing Cholesterol Alternate Activity. The dotplots from the completely randomized design show that the new drug has a slightly greater center, but there is lots of overlap in the distributions. Based on the results of the completely randomized experiment, there is *not* convincing evidence that the new drug is better than the current drug. However, in the design where pairs were formed by initial cholesterol level, the distribution was much less variable and each difference was greater than 0. This means that in each pair, the subject with the new drug did better. This *is* convincing evidence that the new drug is better. The lesson for students is that careful planning during the design of an experiment can greatly increase the chances of getting statistically significant results if one treatment is really better than the other.

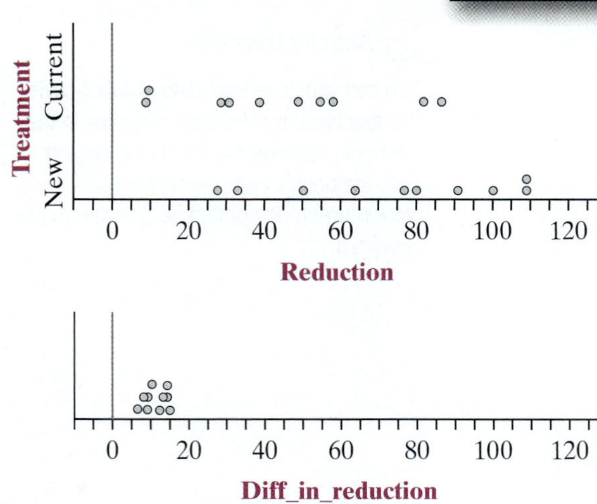

variability in the response variable. That made it easier to detect the fact that standing causes an increase in pulse rate. With the large amount of variability in the completely randomized design, we were unable to draw such a conclusion.

Another important lesson to take away from the example is this: *the design of the study determines the appropriate method of analysis*. For the completely randomized design, it makes sense to compare pulse rates for the two groups with parallel dotplots and means. In the matched pairs design, each student is a block. We compare the effects of the treatments within each block by examining the differences in standing and sitting pulse rates for each student. Then we combine the results from each block (student) and examine the distribution of differences.

The following Data Exploration asks you to apply what you have learned about analyzing data from an experiment.

## DATA EXPLORATION Nitrogen in tires–a lot of hot air?

Most automobile tires are inflated with compressed air, which consists of about 78% nitrogen. Aircraft tires are filled with pure nitrogen, which is safer than air in case of fire. Could filling automobile tires with nitrogen improve safety, performance, or both?

Consumers Union designed a study to test whether nitrogen-filled tires would maintain pressure better than air-filled tires. They obtained two tires from each of several brands and then filled one tire in each pair with air and one with nitrogen. All tires were inflated to a pressure of 30 pounds per square inch and then placed outside for a year. At the end of the year, Consumers Union measured the pressure in each tire. The amount of pressure lost (in pounds per square inch) during the year for the air-filled and nitrogen-filled tires of each brand is shown in the table below.[30]

| Brand | Air | Nitrogen | Brand | Air | Nitrogen |
|---|---|---|---|---|---|
| BF Goodrich Traction T/A HR | 7.6 | 7.2 | Pirelli P6 Four Seasons | 4.4 | 4.2 |
| Bridgestone HP50 (Sears) | 3.8 | 2.5 | Sumitomo HTR H4 | 1.4 | 2.1 |
| Bridgestone Potenza G009 | 3.7 | 1.6 | Yokohama Avid H4S | 4.3 | 3.0 |
| Bridgestone Potenza RE950 | 4.7 | 1.5 | BF Goodrich Traction T/A V | 5.5 | 3.4 |
| Bridgestone Potenza EL400 | 2.1 | 1.0 | Bridgestone Potenza RE950 | 4.1 | 2.8 |
| Continental Premier Contact H | 4.9 | 3.1 | Continental ContiExtreme Contact | 5.0 | 3.4 |
| Cooper Lifeliner Touring SLE | 5.2 | 3.5 | Continental ContiProContact | 4.8 | 3.3 |
| Dayton Daytona HR | 3.4 | 3.2 | Cooper Lifeliner Touring SLE | 3.2 | 2.5 |
| Falken Ziex ZE-512 | 4.1 | 3.3 | General Exclaim UHP | 6.8 | 2.7 |
| Fuzion Hrl | 2.7 | 2.2 | Hankook Ventus V4 H105 | 3.1 | 1.4 |
| General Exclaim | 3.1 | 3.4 | Michelin Energy MXV4 Plus | 2.5 | 1.5 |
| Goodyear Assurance Tripletred | 3.8 | 3.2 | Michelin Pilot Exalto A/S | 6.6 | 2.2 |
| Hankook Optimo H418 | 3.0 | 0.9 | Michelin Pilot HX MXM4 | 2.2 | 2.0 |
| Kumho Solus KH16 | 6.2 | 3.4 | Pirelli P6 Four Seasons | 2.5 | 2.7 |
| Michelin Energy MXV4 Plus | 2.0 | 1.8 | Sumitomo HTR+ | 4.4 | 3.7 |
| Michelin Pilot XGT H4 | 1.1 | 0.7 | | | |

Does filling tires with nitrogen instead of compressed air reduce pressure loss? Give appropriate graphical and numerical evidence to support your answer.

### Teaching Tip

Ask your students to identify the blocks in the Nitrogen in Tires Data Exploration. In this case, the blocks are formed by brand, and two tires of each brand form a block. Then one tire in each block is randomly assigned to nitrogen, and the other tire in the block is assigned to air.

### Answers to Data Exploration

Because these data are paired by brand, we should look at the difference in pressure loss for each brand (air−nitrogen). Below are a dotplot and boxplot of these differences.

Overall, 28 of the 31 brands had positive differences, meaning that almost all the brands lost more pressure when the tires were filled with air than when they were filled with nitrogen. The typical difference was about 1 pound per square inch (psi), with differences as small as −0.7 psi and as large as 4.4 psi. The distribution was roughly symmetric, with two brands having unusually large differences (General Exclaim UHP and Michelin Pilot Exalto A/S).

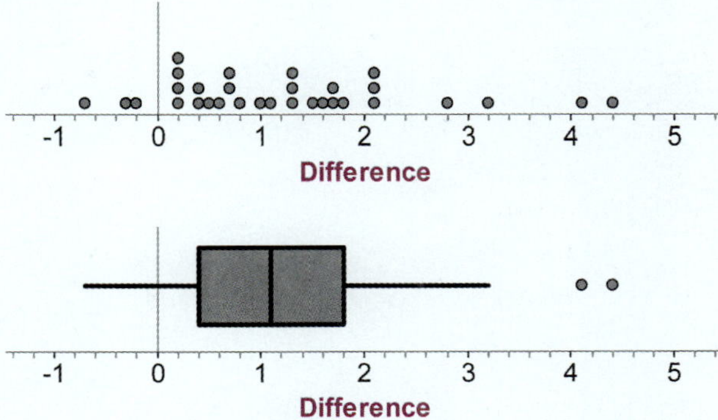

### TRM Template for Experiments

There is a handy template for experiments in the Teacher's Resource Materials. Find it by clicking on the link in the new TE-book, opening the TRCD, or logging onto the book's companion Web site www.whfreeman.com/tps5e (teacher log-in required).

### TRM Section 4.2 Quizzes

There are three different quizzes available for this section in the Teacher's Resource Materials. Versions A and B are roughly parallel.



## Section 4.2 Summary

- We can produce data intended to answer specific questions by **observational studies** or **experiments**. An observational study gathers data on individuals as they are. Experiments actively do something to people, animals, or objects in order to measure their response.
- Statistical studies often try to show that changing one variable (the **explanatory variable**) causes changes in another variable (the **response variable**). Variables are **confounded** when their effects on a response variable can't be distinguished from each other. Observational studies and uncontrolled experiments often fail to show that changes in an explanatory variable cause changes in a response variable because the explanatory variable is confounded with other variables.
- In an experiment, we impose one or more **treatments** on a group of **experimental units** (sometimes called **subjects** if they are human). Each treatment is a combination of values of the explanatory variables (also called **factors**).
- The basic principles of experimental design are as follows:
  1. *Comparison:* Use a design that compares two or more treatments.
  2. *Random assignment:* Use chance to assign experimental units to treatments. This helps create roughly equivalent groups before treatments are imposed.
  3. *Control:* Keep as many other variables as possible the same for all groups. Control helps avoid confounding and reduces the variation in responses, making it easier to decide whether a treatment is effective.
  4. *Replication:* Impose each treatment on enough experimental units so that the effects of the treatments can be distinguished from chance differences between the groups.
- In a **completely randomized design**, all of the experimental units are assigned to the treatments completely by chance.
- Some experiments give a **placebo** (fake treatment) to a control group. That helps prevent confounding due to the **placebo effect**, in which some patients get better because they expect the treatment to work.
- Many behavioral and medical experiments are **double-blind**. That is, neither the subjects nor those interacting with them and measuring their responses know who is receiving which treatment. If one party knows and the other doesn't, then the experiment is **single-blind**.
- When an observed difference in responses between the groups in an experiment is too large to be explained by chance variation in the random assignment, we say that the result is **statistically significant**.
- A **randomized block design** forms groups (**blocks**) of experimental units that are similar with respect to a variable that is expected to affect the response. Treatments are assigned at random within each block. Responses are then compared within each block and combined with the responses of other blocks after accounting for the differences between the blocks. When blocks are chosen wisely, it is easier to determine if one treatment is more effective than another.

- A **matched pairs design** is a common form of blocking for comparing just two treatments. In some matched pairs designs, each subject receives both treatments in a random order. In others, two very similar subjects are paired, and the two treatments are randomly assigned within each pair.

## Section 4.2 Exercises

45. **Learning biology with computers** An educator wants to compare the effectiveness of computer software for teaching biology with that of a textbook presentation. She gives a biology pretest to each of a group of high school juniors, then randomly divides them into two groups. One group uses the computer, and the other studies the text. At the end of the year, she tests all the students again and compares the increase in biology test scores in the two groups. Is this an observational study or an experiment? Justify your answer.

46. **Cell phones and brain cancer** One study of cell phones and the risk of brain cancer looked at a group of 469 people who have brain cancer. The investigators matched each cancer patient with a person of the same age, gender, and race who did not have brain cancer, then asked about the use of cell phones. Result: "Our data suggest that the use of handheld cellular phones is not associated with risk of brain cancer."[31] Is this an observational study or an experiment? Justify your answer.

47. **Chocolate and happy babies** A University of Helsinki (Finland) study wanted to determine if chocolate consumption during pregnancy had an effect on infant temperament at age 6 months. Researchers began by asking 305 healthy pregnant women to report their chocolate consumption. Six months after birth, the researchers asked mothers to rate their infants' temperament, including smiling, laughter, and fear. The babies born to women who had been eating chocolate daily during pregnancy were found to be more active and "positively reactive"—a measure that the investigators said encompasses traits like smiling and laughter.[32]
(a) Was this an observational study or an experiment? Justify your answer.
(b) What are the explanatory and response variables?
(c) Does this study show that eating chocolate regularly during pregnancy helps produce infants with good temperament? Explain.

48. **Child care and aggression** A study of child care enrolled 1364 infants and followed them through their sixth year in school. Later, the researchers published an article in which they stated that "the more time children spent in child care from birth to age four-and-a-half, the more adults tended to rate them, both at age four-and-a-half and at kindergarten, as less likely to get along with others, as more assertive, as disobedient, and as aggressive."[33]
(a) Is this an observational study or an experiment? Justify your answer.
(b) What are the explanatory and response variables?
(c) Does this study show that child care causes children to be more aggressive? Explain.

49. **Effects of class size** Do smaller classes in elementary school really benefit students in areas such as scores on standardized tests, staying in school, and going on to college? We might do an observational study that compares students who happened to be in smaller and larger classes in their early school years. Identify a variable that may lead to confounding with the effects of small classes. Explain how confounding might occur.

50. **Effects of binge drinking** A common definition of "binge drinking" is 5 or more drinks at one sitting for men and 4 or more for women. An observational study finds that students who binge drink have lower average GPA than those who don't. Identify a variable that may be confounded with the effects of binge drinking. Explain how confounding might occur.

*For the experiments described in Exercises 51 to 56, identify the experimental units, the explanatory and response variables, and the treatments.*

51. pg 237  **Growing in the shade** Ability to grow in shade may help pines found in the dry forests of Arizona to resist drought. How well do these pines grow in shade? Investigators planted pine seedlings in a greenhouse in either full light, light reduced to 25% of normal by shade cloth, or light reduced to 5% of normal. At the end of the study, they dried the young trees and weighed them.

52. **Internet telephone calls** You can use Voice over Internet Protocol (VoIP) to make long-distance calls over the Internet. One of the most popular VoIP services is Skype. How will the appearance of ads during calls affect the use of this service? Researchers design an experiment to find out. They recruit 300 people who

**TRM Full Solutions to Section 4.2 Exercises**

### Answers to Section 4.2 Exercises

**4.45** Experiment, because students were randomly assigned to the different teaching methods.

**4.46** Observational study, because the researchers did not assign people to either use or not use cell phones.

**4.47** **(a)** Observational study, because mothers weren't assigned to eat different amounts of chocolate. **(b)** Explanatory: the mother's chocolate consumption. Response: the baby's temperament. **(c)** No, this study is an observational study so we cannot draw a cause-and-effect conclusion. It is possible that women who eat chocolate daily have less stressful lives and the lack of stress helps their babies develop better temperaments.

**4.48** **(a)** Observational study, because children weren't assigned to different amounts of child care. **(b)** Explanatory: amount of time in child care from birth to age $4\frac{1}{2}$. Response: adult ratings of their behavior. **(c)** No, this study is an observational study so we cannot make a cause-and-effect conclusion. It is possible that children who spend more time in child care have less time with their parents and get less instruction about proper behavior.

**4.49** Type of school. For example, private schools tend to have smaller class sizes and students that come from families with higher socioeconomic status. If these students do better in the future, we wouldn't know if the better performance was due to smaller class sizes or higher socioeconomic status.

**4.50** Level of academic motivation. For example, students with low academic motivation will spend less time studying, leaving time to go binge drinking. We can't know if the lower GPAs were caused by the binge drinking or by low academic motivation.

**4.51** Experimental units: pine seedlings. Explanatory variable: light intensity. Response variable: dry weight at the end of the study. Treatments: full light, 25% light, and 5% light.

have not used Skype before to participate. Some people get the current version of Skype with no ads. Others see ads whenever they make calls. The researchers are interested in frequency and length of phone calls.

53. **Improving response rate** How can we reduce the rate of refusals in telephone surveys? Most people who answer at all listen to the interviewer's introductory remarks and then decide whether to continue. One study made telephone calls to randomly selected households to ask opinions about the next election. In some calls, the interviewer gave her name; in others, she identified the university she was representing; and in still others, she identified both herself and the university. For each type of call, the interviewer either did or did not offer to send a copy of the final survey results to the person interviewed. Do these differences in the introduction affect whether the interview is completed?

54. **Eat well and exercise** Most American adolescents don't eat well and don't exercise enough. Can middle schools increase physical activity among their students? Can they persuade students to eat better? Investigators designed a "physical activity intervention" to increase activity in physical education classes and during leisure periods throughout the school day. They also designed a "nutrition intervention" that improved school lunches and offered ideas for healthy home-packed lunches. Each participating school was randomly assigned to one of the interventions, both interventions, or no intervention. The investigators observed physical activity and lunchtime consumption of fat.

pg 239

55. **Fabric science** A maker of fabric for clothing is setting up a new line to "finish" the raw fabric. The line will use either metal rollers or natural-bristle rollers to raise the surface of the fabric; a dyeing-cycle time of either 30 or 40 minutes; and a temperature of either 150° or 175°C. An experiment will compare all combinations of these choices. Three specimens of fabric will be subjected to each treatment and scored for quality.

56. **Exercise and heart rate** A student project measured the increase in the heart rates of fellow students when they stepped up and down for 3 minutes to the beat of a metronome. The step was either 5.75 or 11.5 inches high and the metronome beat was 14, 21, or 28 steps per minute. Thirty students took part in the experiment. Five of them stepped at each combination of height and speed.

57. **Cocoa and blood flow** A study conducted by Norman Hollenberg, professor of medicine at Brigham and Women's Hospital and Harvard Medical School, involved 27 healthy people aged 18 to 72. Each subject consumed a cocoa beverage containing 900 milligrams of flavonols (a class of flavonoids) daily for 5 days. Using a finger cuff, blood flow was measured on the first and fifth days of the study. After 5 days, researchers measured what they called "significant improvement" in blood flow and the function of the cells that line the blood vessels.[34] What flaw in the design of this experiment makes it impossible to say whether the cocoa really caused the improved blood flow? Explain.

58. **Reducing unemployment** Will cash bonuses speed the return to work of unemployed people? A state department of labor notes that last year 68% of people who filed claims for unemployment insurance found a new job within 15 weeks. As an experiment, this year the state offers $500 to people filing unemployment claims if they find a job within 15 weeks. The percent who do so increases to 77%. What flaw in the design of this experiment makes it impossible to say whether the bonus really caused the increase? Explain.

pg 241

59. **Layoffs and "survivor guilt"** Workers who survive a layoff of other employees at their location may suffer from "survivor guilt." A study of survivor guilt and its effects used as subjects 120 students who were offered an opportunity to earn extra course credit by doing proofreading. Each subject worked in the same cubicle as another student, who was an accomplice of the experimenters. At a break midway through the work, one of three things happened:

   *Treatment 1:* The accomplice was told to leave; it was explained that this was because she performed poorly.

   *Treatment 2:* It was explained that unforeseen circumstances meant there was only enough work for one person. By "chance," the accomplice was chosen to be laid off.

   *Treatment 3:* Both students continued to work after the break.

   The subjects' work performance after the break was compared with performance before the break.[35] Describe how you would randomly assign the subjects to the treatments

   (a) using slips of paper.
   (b) using technology.
   (c) using Table D.

60. **Effects of TV advertising** Figure 4.2 (page 239) displays the 6 treatments for a two-factor experiment on TV advertising. Suppose we have 150 students who are willing to serve as subjects. Describe how you would randomly assign the subjects to the treatments

   (a) using slips of paper.
   (b) using technology.
   (c) using Table D.

61. **Stronger players** A football coach hears that a new exercise program will increase upper-body strength better than lifting weights. He is eager to test this new program in the off-season with the players on his high school team. The coach decides to let his players choose which of the two treatments they will undergo for 3 weeks—exercise or weight lifting. He will use the

**4.52** Experimental units: 300 people who haven't used Skype before. Explanatory variable: whether ads are present. Response variables: frequency and length of phone calls. Treatments: no ads shown during phone calls and ads shown during phone calls.

**4.53** Experimental units: the individuals who were called. Explanatory variables: (1) information provided by interviewer; (2) whether caller offered survey results. Response variable: whether or not the call was completed. Treatments: (1) name/no offer; (2) university/no offer; (3) name and university/no offer; (4) name/offer; (5) university/offer; (6) name and university/offer.

**4.54** Experimental units: middle schools. Explanatory variables: whether a physical activity program was offered and whether a nutrition program was offered. Response variables: physical activity and lunchtime consumption of fat. Treatments: (1) activity intervention only; (2) nutrition intervention only; (3) both interventions; (4) neither intervention.

**4.55** Experimental units: 24 fabric specimens. Explanatory variables: (1) roller type; (2) dyeing cycle time; (3) temperature. Response variable: a quality score. Treatments: (1) metal, 30 min, 150°; (2) natural, 30 min, 150°; (3) metal, 40 min, 150°; (4) natural, 40 min, 150°; (5) metal, 30 min, 175°; (6) natural, 30 min, 175°; (7) metal, 40 min, 175°; (8) natural, 40 min, 175°.

**4.56** Experimental units: 30 students. Explanatory variables: (1) step height; (2) metronome pace. Response variable: increase in heart rate. Treatments: (1) 5.75", 14 steps/min; (2) 5.75", 21 steps/min; (3) 5.75", 28 steps/min; (4) 11.5", 14 steps/min; (5) 11.5", 21 steps/min; (6) 11.5", 28 steps/min.

**4.57** There was no control group. We don't know if the improvement was due to the placebo effect or if the flavonols actually affected the blood flow.

**4.58** There was no control group this year. The increase could have been due to the bonus or to other changes during the year, including the state of the economy and the costs of hiring new workers.

**4.59** **(a)** Write all names on slips of paper, put them in a container, and mix thoroughly. Pull out 40 slips of paper and assign these subjects to Treatment 1. Then pull out 40 more slips of paper and assign these subjects to Treatment 2. The remaining 40 subjects are assigned to Treatment 3. **(b)** Assign the students numbers from 1 to 120. Using the command `RandInt(1,120)` on the calculator, assign the students corresponding to the first 40 unique numbers chosen to Treatment 1, the students corresponding to the next 40 unique numbers chosen to Treatment 2, and the remaining 40 students to Treatment 3. **(c)** Assign the students numbers from 001 to 120. Pick a spot on Table D and read off the first 40 unique numbers between 001 and 120. The students corresponding to these numbers are assigned to Treatment 1. The students corresponding to the next 40 unique numbers between 001 and 120 are assigned to Treatment 2. The remaining 40 students are assigned to Treatment 3.

**4.60** **(a)** Write all names on slips of paper, put them in a container, and mix thoroughly. Pull out 25 slips of paper and assign these subjects to treatment 1 (30 sec, 1 rep). Then pull out 25 more slips of paper and assign these subjects to treatment 2 (30 sec, 3 reps). Continue in the same manner until all 150 subjects have been assigned to one of the 6 treatments. **(b)** Assign the students numbers from 1 to 150. Using the command `RandInt(1,150)` on the calculator, assign the students corresponding to the first 25 unique numbers chosen to Treatment 1, the students corresponding to the next 25 unique numbers chosen to Treatment 2, and so on. **(c)** Assign the students numbers from 001 to 150. Pick a spot on Table D and read off the first 25 unique numbers between 001 and 150. The students corresponding to these numbers are assigned to Treatment 1. The students corresponding to the next 25 unique numbers from 001 to 150 are assigned to Treatment 2, and so on.

number of push-ups a player can do at the end of the experiment as the response variable. Which principle of experimental design does the coach's plan violate? Explain how this violation could lead to confounding.

62. **Killing weeds** A biologist would like to determine which of two brands of weed killer, X or Y, is less likely to harm the plants in a garden at the university. Before spraying near the plants, the biologist decides to conduct an experiment using 24 individual plants. Which of the following two plans for randomly assigning the treatments should the biologist use? Why?

*Plan A:* Choose the 12 healthiest-looking plants. Then flip a coin. If it lands heads, apply Brand X weed killer to these plants and Brand Y weed killer to the remaining 12 plants. If it lands tails, do the opposite.

*Plan B:* Choose 12 of the 24 plants at random. Apply Brand X weed killer to those 12 plants and Brand Y weed killer to the remaining 12 plants.

63. **Do diets work?** Dr. Linda Stern and her colleagues recruited 132 obese adults at the Philadelphia Veterans Affairs Medical Center in Pennsylvania. Half the participants were randomly assigned to a low-carbohydrate diet and the other half to a low-fat diet. Researchers measured each participant's change in weight and cholesterol level after six months and again after one year. Explain how each of the four principles of experimental design was used in this study.

pg 243

64. **The effects of day care** Does day care help low-income children stay in school and hold good jobs later in life? The Carolina Abecedarian Project (the name suggests the ABCs) has followed a group of 111 children since 1972. Back then, these individuals were all healthy but low-income black infants in Chapel Hill, North Carolina. All the infants received nutritional supplements and help from social workers. Half were also assigned at random to an intensive preschool program.[36] Explain how each of the four principles of experimental design was used in this study.

65. **Headache relief** Doctors identify "chronic tension-type headaches" as headaches that occur almost daily for at least six months. Can antidepressant medications or stress management training reduce the number and severity of these headaches? Are both together more effective than either alone? Researchers want to compare four treatments: antidepressant alone, placebo alone, antidepressant plus stress management, and placebo plus stress management. Describe a completely randomized design involving 36 headache sufferers who are willing to participate in this experiment. Write a few sentences describing how you would implement your design.

pg 245

66. **More rain for California?** The changing climate will probably bring more rain to California, but we don't know whether the additional rain will come during the winter wet season or extend into the long dry season in spring and summer. Kenwyn Suttle of the University of California at Berkeley and his coworkers wanted to compare the effects of three treatments: added water equal to 20% of annual rainfall either during January to March (winter) or during April to June (spring), and no added water (control). Eighteen plots of open grassland, each with area 70 square meters, were available for this study. One response variable was total plant biomass, in grams per square meter, produced in a plot over a year.[37]

Describe a completely randomized design for this experiment. Write a few sentences describing how you would implement your design.

67. **Treating prostate disease** A large study used records from Canada's national health care system to compare the effectiveness of two ways to treat prostate disease. The two treatments are traditional surgery and a new method that does not require surgery. The records described many patients whose doctors had chosen each method. The study found that patients treated by the new method were significantly more likely to die within 8 years.[38]

(a) Further study of the data showed that this conclusion was wrong. The extra deaths among patients who got the new method could be explained by other variables. What other variables might be confounded with a doctor's choice of surgical or nonsurgical treatment?

(b) You have 300 prostate patients who are willing to serve as subjects in an experiment to compare the two methods. Describe a completely randomized design for this experiment. Write a few sentences describing how you would implement your design.

68. **Getting teachers to come to school** Elementary schools in rural India are usually small, with a single teacher. The teachers often fail to show up for work. Here is an idea for improving attendance: give the teacher a digital camera with a tamperproof time and date stamp and ask a student to take a photo of the teacher and class at the beginning and end of the day. Offer the teacher better pay for good attendance, verified by the photos. Will this work? Researchers obtained permission to use 120 rural schools in Rajasthan for an experiment to find out.[39]

(a) Explain why it would not be a good idea to offer better pay for good attendance to the teachers in all 120 schools and then to compare this year's attendance with last year's.

(b) Describe a completely randomized design for an experiment involving these 120 schools. Write a few sentences describing how you would implement your design.

69. **Do placebos really work?** Researchers in Japan conducted an experiment on 13 individuals who were extremely allergic to poison ivy. On one arm,

**4.61** Random assignment. If players are allowed to choose which treatment they receive, perhaps the more motivated players will choose the new method. If they improve more by the end of the study, the coach can't be sure if it was the exercise program or player motivation that caused the improvement.

**4.62** Plan B. If the biologist uses Plan A, the brand that is assigned to the healthier plants might appear safer. However, we won't know if this is due to the brand or to the fact that these plants were healthier to begin with. With Plan B, the two groups of plants should be roughly equivalent at the beginning of the experiment so that a difference in response can be attributed to the type of weed killer.

**4.63** *Comparison*: Researchers used a design that compared a low-carbohydrate diet with a low-fat diet. *Random assignment*: Subjects were randomly assigned to one of the two diets. *Control*: The experiment used subjects who were all obese at the beginning of the study and who all lived in the same area. *Replication*: There were 66 subjects in each treatment group.

**4.64** *Comparison*: Researchers used a design that compared children who were assigned to an intensive preschool program to children who were not enrolled in an intensive preschool program. *Random assignment*: Subjects were randomly assigned to be enrolled in the intensive preschool program or not. *Control*: All subjects were healthy, low-income, and black. Also, all subjects received nutritional supplements and help from social workers. *Replication*: There were over 50 subjects in each treatment group.

**4.65** Write the names of the patients on 36 identical slips of paper, put them in a hat, and mix them well. Draw out 9 slips. The corresponding patients will receive the antidepressant. Draw out 9 more slips. Those patients will receive the antidepressant plus stress management. The patients corresponding to the next 9 slips drawn will receive the placebo, and the remaining 9 patients will receive the placebo plus stress management. At the end of the experiment, record the number and severity of chronic tension-type headaches for each of the 36 subjects and compare the results for the 4 groups.

**4.66** Assign the plots the labels 01 through 18. Write the labels on 18 identical slips of paper, put them in a hat, and mix them well. Draw out 6 slips. The corresponding plots will be in Group 1 and receive added water in winter. Draw out 6 more slips. These plots will be in Group 2 and receive added water in spring. The remaining 6 plots will be in Group 3 and receive no extra water. At the end of the experiment, record the total plant biomass for each plot and compare the results for the 3 groups.

**4.67** **(a)** Other variables include expense and condition of the patient. For example, if a patient is in very poor health, a doctor might choose not to recommend surgery because of the added complications. Then we won't know if a higher death rate is due to the treatment or the initial health of the subjects. **(b)** Write the names of all 300 patients on identical slips of paper, put them in a hat, and mix them well. Draw out 150 slips and assign the corresponding subjects to receive surgery. The remaining 150 subjects receive the new method. At the end of the study, count how many patients survived in each group.

**4.68** **(a)** There may be other variables that have changed over time. For example, if the weather is better this year, we wouldn't know if the better attendance is due to the financial incentive or better weather. **(b)** Write the names of all 120 schools on identical slips of paper, put them in a hat, and mix them well. Draw out 60 slips and assign the corresponding schools to offer better pay to the teachers who have good attendance. The remaining schools do not get to offer the financial incentive. At the end of the study period, compare the attendance of the teachers.

each subject was rubbed with a poison ivy leaf and told the leaf was harmless. On the other arm, each subject was rubbed with a harmless leaf and told it was poison ivy. All the subjects developed a rash on the arm where the harmless leaf was rubbed. Of the 13 subjects, 11 did not have any reaction to the real poison ivy leaf.[40] Explain how the results of this study support the idea of a placebo effect.

70. **Pain relief study** Fizz Laboratories, a pharmaceutical company, has developed a new drug for relieving chronic pain. Sixty patients suffering from arthritis and needing pain relief are available. Each patient will be treated and asked an hour later, "About what percent of pain relief did you experience?"

(a) Why should Fizz not simply give the new drug to 30 patients and no treatment to the other 30 patients, and then record the patients' responses?

(b) Should the patients be told whether they are getting the new drug or a placebo? How would this knowledge probably affect their reactions?

71. **Meditation for anxiety** An experiment that claimed to show that meditation lowers anxiety proceeded as follows. The experimenter interviewed the subjects and rated their level of anxiety. Then the subjects were randomly assigned to two groups. The experimenter taught one group how to meditate and they meditated daily for a month. The other group was simply told to relax more. At the end of the month, the experimenter interviewed all the subjects again and rated their anxiety level. The meditation group now had less anxiety. Psychologists said that the results were suspect because the ratings were not blind. Explain what this means and how lack of blindness could bias the reported results.

72. **Testosterone for older men** As men age, their testosterone levels gradually decrease. This may cause a reduction in lean body mass, an increase in fat, and other undesirable changes. Do testosterone supplements reverse some of these effects? A study in the Netherlands assigned 237 men aged 60 to 80 with low or low-normal testosterone levels to either a testosterone supplement or a placebo. The report in the *Journal of the American Medical Association* described the study as a "double-blind, randomized, placebo-controlled trial."[41] Explain each of these terms to someone who knows nothing about statistics.

73. **Do diets work?** Refer to Exercise 63. Subjects in the low-carb diet group lost significantly more weight than subjects in the low-fat diet group during the first six months. At the end of a year, however, the average weight loss for subjects in the two groups was not significantly different.[42]

(a) Why did researchers randomly assign the subjects to the diet treatments?

(b) Explain to someone who knows little statistics what "lost significantly more weight" means.

(c) The subjects in the low-carb diet group lost an average of 5.1 kg in a year. The subjects in the low-fat diet group lost an average of 3.1 kg. Explain how this information could be consistent with the fact that weight loss in the two groups was not significantly different.

74. **Acupuncture and pregnancy** A study sought to determine whether the ancient Chinese art of acupuncture could help infertile women become pregnant.[43] One hundred sixty healthy women undergoing assisted reproductive therapy were recruited for the study. Half of the subjects were randomly assigned to receive acupuncture treatment 25 minutes before embryo transfer and again 25 minutes after the transfer. The remaining 80 subjects were instructed to lie still for 25 minutes after the embryo transfer. *Results:* In the acupuncture group, 34 women became pregnant. In the control group, 21 women became pregnant.

(a) Why did researchers randomly assign the subjects to the two treatments?

(b) The difference in the percent of women who became pregnant in the two groups is statistically significant. Explain what this means to someone who knows little statistics.

(c) Explain why the design of the study prevents us from concluding that acupuncture caused the difference in pregnancy rates.

75. **Doctors and nurses** Nurse-practitioners are nurses with advanced qualifications who often act much like primary-care physicians. Are they as effective as doctors at treating patients with chronic conditions? An experiment was conducted with 1316 patients who had been diagnosed with asthma, diabetes, or high blood pressure. Within each condition, patients were randomly assigned to either a doctor or a nurse-practitioner. The response variables included measures of the patients' health and of their satisfaction with their medical care after 6 months.[44]

(a) Which are the blocks in this experiment: the different diagnoses (asthma, and so on) or the type of care (nurse or doctor)? Why?

(b) Explain why a randomized block design is preferable to a completely randomized design here.

76. **Comparing cancer treatments** The progress of a type of cancer differs in women and men. Researchers want to design an experiment to compare three therapies for this cancer. They recruit 500 male and 300 female patients who are willing to serve as subjects.

(a) Which are the blocks in this experiment: the cancer therapies or the two sexes? Why?

(b) What are the advantages of a randomized block design over a completely randomized design using these 800 subjects?

**4.69** The subjects developed rashes on the arm exposed to the placebo (a harmless leaf) simply because they thought they were being exposed to a poison ivy leaf. Likewise, most of the subjects didn't develop rashes on the arm that was exposed to poison ivy because they didn't believe they were being exposed to the real thing.

**4.70** **(a)** The company won't know if pain relief was caused by the drug or by the expectation of pain relief (the placebo effect). **(b)** The subjects should not know what drug they are getting. For example, a patient told that she is receiving a placebo will probably not experience any pain relief, while a patient told that she is receiving a real treatment might experience pain relief due to the drug and the placebo effect. We won't know if any difference in response between the groups was due to the drug or to expectations of the subjects.

**4.71** Because the experimenter knew which subjects had learned the meditation techniques, he is not blind. If the experimenter believed that meditation was beneficial, he may subconsciously rate subjects in the meditation group as being less anxious.

**4.72** "Double-blind" means that the treatment assigned to a subject (testosterone or placebo) was unknown to both the subject and those responsible for assessing the effectiveness of that treatment. "Randomized" means that patients were randomly assigned to a treatment. "Placebo-controlled" means that some of the subjects were given placebos so researchers could separate the effects of testosterone from the expectations of subjects receiving a treatment.

**4.73** **(a)** To make sure that the two groups were as similar as possible before the treatments were administered. **(b)** The difference in weight loss was larger than would be expected due to the chance variation created by the random assignment to treatments. **(c)** Even though the low-carb dieters lost 2 kg more over the year than the low-fat group, a difference of 2 kg could be due just to chance variation created by the random assignment.

**4.74** **(a)** To make sure that the two groups were as similar as possible before the treatments were administered. **(b)** The difference in pregnancy rates was greater than would be expected due to the chance variation created by the random assignment to treatments. **(c)** Because the women were aware of which treatment they received, we don't know if their expectations or the treatment was the cause of the increase in pregnancy rates.

**4.75** **(a)** The different diagnoses, because the treatments were randomly assigned to patients within each diagnosis. **(b)** Using a randomized block design allows us to account for the variability in response due to differences in diagnosis by initially comparing the results within each block. In a completely randomized design, this variability will be unaccounted for, making it harder to determine if there is a difference in health and satisfaction due to the difference between doctors and nurse-practitioners.

**4.76** **(a)** The sexes, because researchers will randomly assign all three therapies to each gender. **(b)** Using a randomized block design allows us to account for the variability in response due to differences in gender by initially comparing the results of the three therapies within each gender. In a completely randomized design, the variability due to differences in gender will be unaccounted for, making it harder to determine if there is a difference in the response to the three therapies.

(c) Suppose the researchers had 800 male and no female subjects available for the study. What advantage would this offer? What disadvantage?

77. **In the cornfield** An agriculture researcher wants to compare the yield of 5 corn varieties: A, B, C, D, and E. The field in which the experiment will be carried out increases in fertility from north to south. The researcher therefore divides the field into 25 plots of equal size, arranged in 5 east–west rows of 5 plots each, as shown in the diagram. pg 254

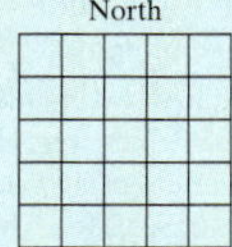


(a) Explain why a randomized block design would be better than a completely randomized design in this setting.

(b) Should the researcher use the rows or the columns of the field as blocks? Justify your answer.

(c) Use technology or Table D to carry out the random assignment required by your design. Explain your method clearly.

78. **Comparing weight-loss treatments** Twenty overweight females have agreed to participate in a study of the effectiveness of four weight-loss treatments: A, B, C, and D. The researcher first calculates how overweight each subject is by comparing the subject's actual weight with her "ideal" weight. The subjects and their excess weights in pounds are as follows:

| | | | | | | | |
|---|---|---|---|---|---|---|---|
| Birnbaum | 35 | Hernandez | 25 | Moses | 25 | Smith | 29 |
| Brown | 34 | Jackson | 33 | Nevesky | 39 | Stall | 33 |
| Brunk | 30 | Kendall | 28 | Obrach | 30 | Tran | 35 |
| Cruz | 34 | Loren | 32 | Rodriguez | 30 | Wilansky | 42 |
| Deng | 24 | Mann | 28 | Santiago | 27 | Williams | 22 |

The response variable is the weight lost after 8 weeks of treatment. Previous studies have shown that the effects of a diet may vary based on a subject's initial weight.

(a) Explain why a randomized block design would be better than a completely randomized design in this setting.

(b) Should researchers form blocks of size 4 based on subjects' last names in alphabetical order or by how overweight the subjects are? Explain.

(c) Use technology or Table D to carry out the random assignment required by your design. Explain your method clearly.

79. **Aw, rats!** A nutrition experimenter intends to compare the weight gain of newly weaned male rats fed Diet A with that of rats fed Diet B. To do this, she will feed each diet to 10 rats. She has available 10 rats from one litter and 10 rats from a second litter. Rats in the first litter appear to be slightly healthier.

(a) If the 10 rats from Litter 1 were fed Diet A, the effects of genetics and diet would be confounded, and the experiment would be biased. Explain this statement carefully.

(b) Describe a better design for this experiment.

80. **Technology for teaching statistics** The Brigham Young University (BYU) statistics department is performing experiments to compare teaching methods. Response variables include students' final-exam scores and a measure of their attitude toward statistics. One study compares two levels of technology for large lectures: standard (overhead projectors and chalk) and multimedia. There are eight lecture sections of a basic statistics course at BYU, each with about 200 students. There are four instructors, each of whom teaches two sections.[45] Suppose the sections and lecturers are as follows:

| Section | Lecturer | Section | Lecturer |
|---|---|---|---|
| 1 | Hilton | 5 | Tolley |
| 2 | Christensen | 6 | Hilton |
| 3 | Hadfield | 7 | Tolley |
| 4 | Hadfield | 8 | Christensen |

(a) Suppose we randomly assign two lecturers to use standard technology in their sections and the other two lecturers to use multimedia technology. Explain how this could lead to confounding.

(b) Describe a better design for this experiment.

81. **Look, Ma, no hands!** Does talking on a hands-free cell phone distract drivers? Researchers recruit 40 student subjects for an experiment to investigate this question. They have a driving simulator equipped with a hands-free phone for use in the study. Each subject will complete two sessions in the simulator: one while talking on the hands-free phone and the other while just driving. The order of the two sessions for each subject will be determined at random. The route, driving conditions, and traffic flow will be the same in both sessions.

(a) What type of design did the researchers use in their study?

(b) Explain why the researchers chose this design instead of a completely randomized design.

(c) Why is it important to randomly assign the order of the treatments?

(d) Explain how and why researchers controlled for other variables in this experiment.

82. **Chocolate gets my heart pumping** Cardiologists at Athens Medical School in Greece wanted to test whether chocolate affected blood flow in the blood vessels. The researchers recruited 17 healthy young volunteers, who were each given a 3.5-ounce bar of dark chocolate, either bittersweet or fake chocolate. On another day, the volunteers received the other treatment. The order in which subjects received the bittersweet and fake chocolate was determined at random. The subjects had no chocolate outside the study, and investigators didn't know whether a subject had eaten the real or the fake chocolate. An ultrasound was taken of each

**4.76 (c)** There would be no variability in responses caused by differences in gender. However, we would only be able to make conclusions about how the treatments work for males.

**4.77 (a)** A randomized block design would help us account for the variability in yield that is due to the differences in fertility in the field, making it easier to determine if one variety is better than the others. **(b)** The rows. There should be a stronger association between row number and yield than column number and yield. **(c)** Let the digits 1 to 5 correspond to the five corn varieties A to E. Begin with line 111 on the random digit table, and assign the letters to the top row from left to right, ignoring numbers 0 and 6–9 and repeated numbers. Use a different line (111, 112, 113, 114, and 115) for each row. Top row (left to right): ADECB, second row: ECDAB, third row: BEDCA, fourth row: DEACB, bottom row: ADCBE.

**4.78 (a)** A randomized block design would help us account for the variability in weight loss that is due to the differences in initial weight, making it easier to determine if one diet plan is better than the others. **(b)** How overweight the subjects are. There should be a stronger association between amount overweight and future weight loss than last name and future weight loss. **(c)** Ordered by increasing weight, the five blocks are (1) Williams-22, Deng-24, Hernandez-25, and Moses-25; (2) Santiago-27, Kendall-28, Mann-28, and Smith-29; (3) Brunk-30, Obrach-30, Rodriguez-30, and Loren-32; (4) Jackson-33, Stall-33, Brown-34, and Cruz-34; (5) Birnbaum-35, Tran-35, Nevesky-39, and Wilansky-42. For each block, number the subjects from 1 to 4. At the beginning of line 101 of Table D, look for a digit from 1 to 4. Assign the corresponding person to Treatment A. Continue reading digits from the same line until you get to a different number from 1 to 4. Assign this person to treatment B, and so on. Here are the results, using lines 101 to 105 of Table D: (1) Williams-A, Deng-B, Hernandez-C, and Moses-D; (2) Santiago-C, Kendall-D, Mann-A, and Smith-B; (3) Brunk-B, Obrach-D, Rodriguez-C, and Loren-A; (4) Jackson-B, Stall-A, Brown-C, and Cruz-D; (5) Birnbaum-C, Tran-A, Nevesky-D, and Wilansky-B.

**4.79 (a)** If all rats from Litter 1 were fed diet A and if these rats gained more weight, we would not know if this was because of the diet or because of genetics and initial health. **(b)** Use a randomized block design with the litters as blocks. For each of the litters, randomly assign half of the rats to receive Diet A and the other half to receive Diet B. This will allow researchers to account for the differences in weight gain caused by the differences in genetics and initial health.

**4.80 (a)** If we find a difference in response between the two treatment groups, we will not know if the difference is due to the technology or to the instructor. **(b)** For each instructor, randomly assign one of the sections to be taught with multimedia and one section to be taught with standard methods. This will allow researchers to account for the differences in final exam scores and attitudes toward statistics caused by the differences in instructor.

**4.81 (a)** Matched pairs design. **(b)** In a completely randomized design, the differences between the students will add variability to the response, making it harder to detect if there is a difference caused by the treatments. In a matched pairs design, each student is compared with himself (or herself), so the differences between students are accounted for. **(c)** If all the students used the hands-free phone during the first session and performed worse, we wouldn't know if the better performance during the second session is due to the lack of phone or learning from their mistakes the first time. By randomizing the order, some students will use the hands-free phone during the first session and others during the second session. **(d)** The simulator, route, driving conditions, and traffic flow were all kept the same for both sessions, preventing these variables from adding variability to the response variable.

volunteer's upper arm to see the functioning of the cells in the walls of the main artery. The researchers found that blood vessel function was improved when the subjects ate bittersweet chocolate, and that there were no such changes when they ate the placebo (fake chocolate).[46]

(a) What type of design did the researchers use in their study?

(b) Explain why the researchers chose this design instead of a completely randomized design.

(c) Why is it important to randomly assign the order of the treatments for the subjects?

(d) Explain how and why researchers controlled for other variables in this experiment.

83. **Room temperature and dexterity** An expert on worker performance is interested in the effect of room temperature on the performance of tasks requiring manual dexterity. She chooses temperatures of 70°F and 90°F as treatments. The response variable is the number of correct insertions, during a 30-minute period, in a peg-and-hole apparatus that requires the use of both hands simultaneously. Each subject is trained on the apparatus and then asked to make as many insertions as possible in 30 minutes of continuous effort.

(a) Describe a completely randomized design to compare dexterity at 70° and 90° using 20 volunteer subjects.

(b) Because individuals differ greatly in dexterity, the wide variation in individual scores may hide the systematic effect of temperature unless there are many subjects in each group. Describe in detail the design of a matched pairs experiment in which each subject serves as his or her own control.

84. **Carbon dioxide and tree growth** The concentration of carbon dioxide ($CO_2$) in the atmosphere is increasing rapidly due to our use of fossil fuels. Because plants use $CO_2$ to fuel photosynthesis, more $CO_2$ may cause trees and other plants to grow faster. An elaborate apparatus allows researchers to pipe extra $CO_2$ to a 30-meter circle of forest. We want to compare the growth in base area of trees in treated and untreated areas to see if extra $CO_2$ does in fact increase growth. We can afford to treat three circular areas.[47]

(a) Describe the design of a completely randomized experiment using six well-separated 30-meter circular areas in a pine forest. Sketch the circles and carry out the randomization your design calls for.

(b) Areas within the forest may differ in soil fertility. Describe a matched pairs design using three pairs of circles that will account for the extra variation due to different fertility. Sketch the circles and carry out the randomization your design calls for.

85. **Got deodorant?** A group of students wants to perform an experiment to determine whether Brand A or Brand B deodorant lasts longer. One group member suggests the following design: Recruit 40 student volunteers—20 male and 20 female. Separate by gender, because male and female bodies might respond differently to deodorant. Give all the males Brand A deodorant and all the females Brand B. Have each student rate how well the deodorant is still working at the end of the school day on a 0 to 10 scale. Then compare ratings for the two treatments.

(a) Identify any flaws you see in the proposed design for this experiment.

(b) Describe how you would design the experiment. Explain how your design addresses each of the problems you identified in part (a).

86. **Close shave** Which of two brands (X or Y) of electric razor shaves closer? Researchers want to design and carry out an experiment to answer this question using 50 adult male volunteers. Here's one idea: Have all 50 subjects shave the left sides of their faces with the Brand X razor and shave the right sides of their faces with the Brand Y razor. Then have each man decide which razor gave the closer shave and compile the results.

(a) Identify any flaws you see in the proposed design for this experiment.

(b) Describe how you would design the experiment. Explain how your design addresses each of the problems you identified in part (a).

*Multiple choice: Select the best answer for Exercises 87 to 94.*

87. Can changing diet reduce high blood pressure? Vegetarian diets and low-salt diets are both promising. Men with high blood pressure are assigned at random to four diets: (1) normal diet with unrestricted salt; (2) vegetarian with unrestricted salt; (3) normal with restricted salt; and (4) vegetarian with restricted salt. This experiment has

(a) one factor, the type of diet.

(b) two factors, high blood pressure and type of diet.

(c) two factors, normal/vegetarian diet and unrestricted/restricted salt.

(d) three factors, men, high blood pressure, and type of diet.

(e) four factors, the four diets being compared.

88. In the experiment of the previous exercise, the subjects were randomly assigned to the different treatments. What is the most important reason for this random assignment?

(a) Random assignment eliminates the effects of other variables such as stress and body weight.

(b) Random assignment is a good way to create groups of subjects that are roughly equivalent at the beginning of the experiment.

(c) Random assignment makes it possible to make a conclusion about all men.

(d) Random assignment reduces the amount of variation in blood pressure.

**4.82 (a)** Matched pairs design. **(b)** In a completely randomized design, the differences between the volunteers will add variability to the response, making it harder to detect if there is a difference caused by the treatments. In a matched pairs design, each volunteer is compared with himself (or herself), so the differences between volunteers are accounted for. **(c)** If the weather is warmer on one of the days, we won't know if the difference in blood vessel function is due to the difference in treatment or the difference in weather. By randomizing the order, some volunteers will have the bittersweet chocolate on the warmer day and some will have it on the colder day. **(d)** The method of measurement and the total amount of chocolate were kept the same for all subjects, preventing these variables from adding variability to the response variable.

**4.83 (a)** Randomly assign the 20 subjects into two groups of 10. Write the name of each subject on a note card, shuffle the cards, and select 10 to be assigned to the 70° environment. The remaining 10 subjects will be assigned to the 90° environment. Then the number of correct insertions will be recorded for each subject and the two groups will be compared. **(b)** All subjects will perform the task twice, once in each temperature condition. Randomly choose the order by flipping a coin. Heads: 70°, then 90°. Tails: 90°, then 70°. For each subject, compare the number of correct insertions in each environment.

**4.84 (a)** Number the circles from 1 to 6 and randomly assign three of the circles to be treated with additional $CO_2$ and the other three circles to be left untreated. At the end of the study, compare tree growth in the treated and untreated areas. Line 104 of Table D was used to select 3 circular areas for the treatment. The first 4 digits are 5 ($CO_2$) 2 ($CO_2$) 7 (skip) 1 ($CO_2$). Leave areas 3, 4, and 6 untreated.

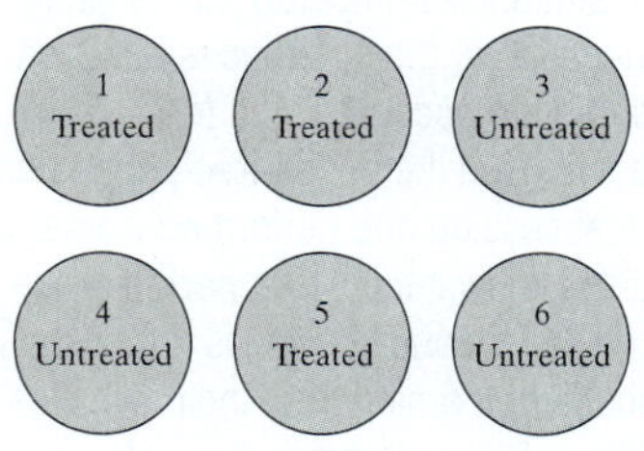


**(b)** Create pairs of circles that are similar in fertility but different than the circles in other pairs. For each pair, randomly assign one of the two areas to receive additional $CO_2$ and the other to be left untreated. Compare tree growth for the treated and untreated area in each pair. A coin was flipped for each pair. Heads: top treated, tails: bottom treated.

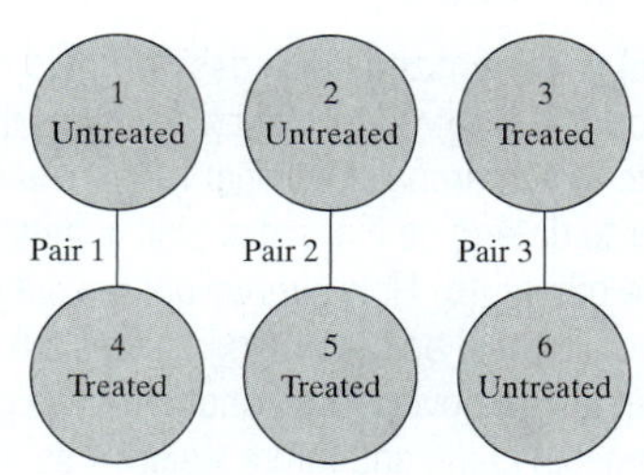


**4.85 (a)** If the students find a difference between the two groups, they will not know if the difference is due to gender or the deodorant. **(b)** Each student should have one armpit randomly assigned to receive Deodorant A and the other Deodorant B. Because each gender uses both deodorants, there is no longer any confounding between gender and deodorant.

**4.86 (a)** If Brand Y appears better, we won't know if the difference is due to the brand of razor or because most men can shave better on the right side (because most are right-handed). **(b)** Have each man shave one half of his face with Brand X and the other half with Brand Y, but randomly assign which half of each man's face receives which brand of razor. Because each brand is used on each half of the face, there is no longer any confounding between brand of razor and side of face.

**4.87** c

**4.88** b

(e) Random assignment prevents the placebo effect from ruining the results of the study.

**89.** To investigate whether standing up while studying affects performance in an algebra class, a teacher assigns half of the 30 students in his class to stand up while studying and assigns the other half to not stand up while studying. To determine who receives which treatment, the teacher identifies the two students who did best on the last exam and randomly assigns one to stand and one to not stand. The teacher does the same for the next two highest-scoring students and continues in this manner until each student is assigned a treatment. Which of the following best describes this plan?

(a) This is an observational study.
(b) This is an experiment with blocking.
(c) This is a completely randomized experiment.
(d) This is a stratified random sample.
(e) This is a cluster sample.

**90.** A gardener wants to try different combinations of fertilizer (none, 1 cup, 2 cups) and mulch (none, wood chips, pine needles, plastic) to determine which combination produces the highest yield for a variety of green beans. He has 60 green-bean plants to use in the experiment. If he wants an equal number of plants to be assigned to each treatment, how many plants will be assigned to each treatment?

(a) 1 (b) 3 (c) 4 (d) 5 (e) 12

**91.** Corn variety 1 yielded 140 bushels per acre last year at a research farm. This year, corn variety 2, planted in the same location, yielded only 110 bushels per acre. Based on these results, is it reasonable to conclude that corn variety 1 is more productive than corn variety 2?

(a) Yes, because 140 bushels per acre is greater than 110 bushels per acre.
(b) Yes, because the study was done at a research farm.
(c) No, because there may be other differences between the two years besides the corn variety.
(d) No, because there was no use of a placebo in the experiment.
(e) No, because the experiment wasn't double-blind.

**92.** A report in a medical journal notes that the risk of developing Alzheimer's disease among subjects who regularly opted to take the drug ibuprofen was about half the risk among those who did not. Is this good evidence that ibuprofen is effective in preventing Alzheimer's disease?

(a) Yes, because the study was a randomized, comparative experiment.
(b) No, because the effect of ibuprofen is confounded with the placebo effect.
(c) Yes, because the results were published in a reputable professional journal.
(d) No, because this is an observational study. An experiment would be needed to confirm (or not confirm) the observed effect.
(e) Yes, because a 50% reduction can't happen just by chance.

**93.** A farmer is conducting an experiment to determine which variety of apple tree, Fuji or Gala, will produce more fruit in his orchard. The orchard is divided into 20 equally sized square plots. He has 10 trees of each variety and randomly assigns each tree to a separate plot in the orchard. What are the experimental unit(s) in this study?

(a) The trees (c) The apples (e) The orchard
(b) The plots (d) The farmer

**94.** Two essential features of all statistically designed experiments are

(a) compare several treatments; use the double-blind method.
(b) compare several treatments; use chance to assign subjects to treatments.
(c) always have a placebo group; use the double-blind method.
(d) use a block design; use chance to assign subjects to treatments.
(e) use enough subjects; always have a control group.

**95. Seed weights** (2.2) Biological measurements on the same species often follow a Normal distribution quite closely. The weights of seeds of a variety of winged bean are approximately Normal with mean 525 milligrams (mg) and standard deviation 110 mg.

(a) What percent of seeds weigh more than 500 mg? Show your method.
(b) If we discard the lightest 10% of these seeds, what is the smallest weight among the remaining seeds? Show your method.

**96. Twins** (1.3, 3.1) A researcher studied a group of identical twins who had been separated and adopted at birth. In each case, one twin (Twin A) was adopted by a low-income family and the other (Twin B) by a high-income family. Both twins were given an IQ test as adults. Here are their scores:[48]

| | | | | | | | | | | | | |
|---|---|---|---|---|---|---|---|---|---|---|---|---|
| **Twin A:** | 120 | 99 | 99 | 94 | 111 | 97 | 99 | 94 | 104 | 114 | 113 | 100 |
| **Twin B:** | 128 | 104 | 108 | 100 | 116 | 105 | 100 | 100 | 103 | 124 | 114 | 112 |

(a) How well does one twin's IQ predict the other's? Give appropriate evidence to support your answer.
(b) Do identical twins living in low-income homes tend to have lower IQs later in life than their twins who live in high-income homes? Give appropriate evidence to support your answer.

**4.89** b

**4.90** d

**4.91** c

**4.92** d

**4.93** b

**4.94** b

**4.95 (a)** For these seeds, the weights follow a $N(525, 110)$ distribution and we want the proportion of seeds that weigh more than 500 mg (see graph). $z = \frac{500 - 525}{110} = -0.23$. From Table A, the proportion of $z$-scores greater than $-0.23$ is 0.5910. *Using technology*: `normalcdf(lower:500,upper:10000,μ:525,σ:110)` = 0.5899. About 59% of seeds will weigh more than 500 mg.

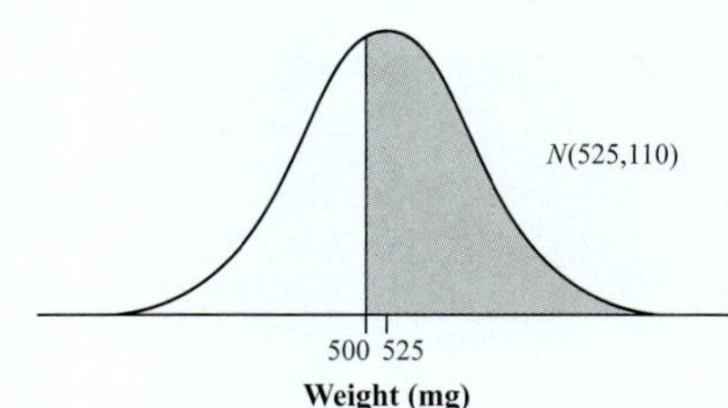


**(b)** For these seeds, the weights follow a $N(525, 110)$ distribution and we are looking for the boundary value $x$ that has an area of 0.10 to the left (see graph below). A $z$-score of $-1.28$ gives the closest value to 0.10 (0.1003). Solving $-1.28 = \frac{x - 525}{110}$ gives $x = 384.2$. *Using technology*: `invNorm(area:0.10,μ:525,σ:110)` = 384.0. The smallest weight among the remaining seeds should be about 384 mg.

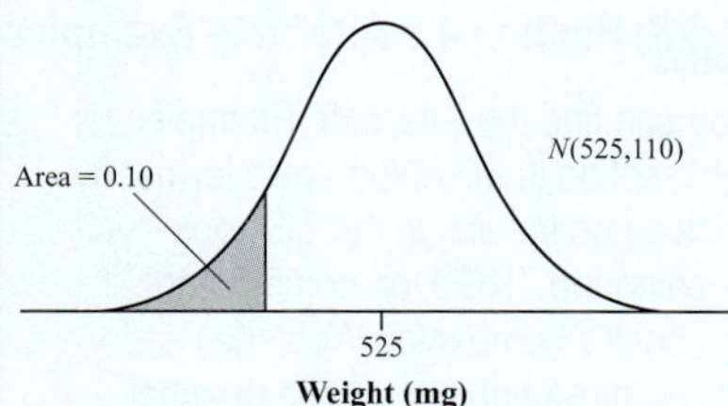


**4.96 (a)** Fairly well, because there is a reasonably strong linear relationship between the IQ's of the twins with a correlation of $r = 0.91$ (see graph).

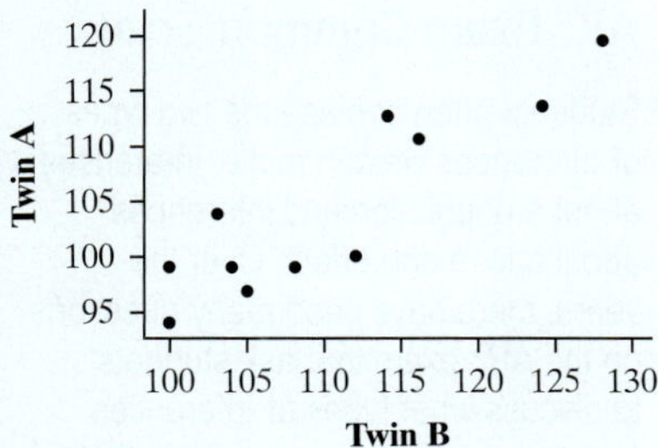


**(b)** Here is a dotplot of the difference in IQ (Twin B − Twin A). Because all but one of the differences are positive, Twin B (the one living in the higher-income homes) tends to have a higher IQ than Twin A.

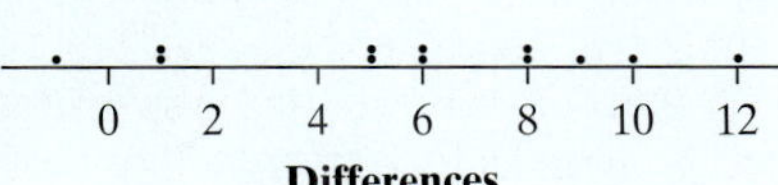

**PD Section 4.3 Overview**

Watch the video overview of Section 4.3 (for teachers) by clicking on the link in the TE-book, opening the TRCD, or logging onto the book's companion Web site www.whfreeman.com/tps5e (teacher log-in required).

**TRM Section 4.3 Alternate Examples**

You can find the Alternate Examples for this section in Microsoft Word format by clicking on the link in the TE-book, opening the TRCD, or logging onto the book's companion Web site www.whfreeman.com/tps5e (teacher log-in required).

**AP® Exam Common Error**

Students often confuse the two types of inferences we can make: inferences about a population and inferences about cause and effect. Over the years, there have been many questions on the AP® exam that ask students to discuss what types of inferences (conclusions) are appropriate based on the design of the study. Make sure students understand how the role of randomization in a study helps them decide which type of inference, if any, is appropriate.

# 4.3 Using Studies Wisely

**WHAT YOU WILL LEARN** By the end of the section, you should be able to:

- Describe the scope of inference that is appropriate in a statistical study.
- Evaluate whether a statistical study has been carried out in an ethical manner.*

Researchers who conduct statistical studies often want to draw conclusions (make inferences) that go beyond the data they produce. Here are two examples.

- The U.S. Census Bureau carries out a monthly Current Population Survey of about 60,000 households. Their goal is to use data from these randomly selected households to estimate the percent of unemployed individuals in the population.
- Scientists performed an experiment that randomly assigned 21 volunteer subjects to one of two treatments: sleep deprivation for one night or unrestricted sleep. The experimenters hoped to show that sleep deprivation causes a decrease in performance two days later.[49]

What type of inference can be made from a particular study? The answer depends on the design of the study.

## Scope of Inference

In the Census Bureau's sample survey, the individuals who responded were chosen at random from the population of interest. Random sampling avoids bias and produces trustworthy estimates of the truth about the population. The Census Bureau should be safe making an *inference about the population* based on the results of the sample.

In the sleep deprivation experiment, subjects were randomly assigned to the sleep deprivation and unrestricted sleep treatments. Random assignment helps ensure that the two groups of subjects are as alike as possible before the treatments are imposed. If the unrestricted sleep group performs much better than the sleep deprivation group, and the difference is too large to be explained by chance variation in the random assignment, it must be due to the treatments. In that case, the scientists could safely conclude that sleep deprivation caused the decrease in performance. That is, they can make an *inference about cause and effect*. However, because the experiment used volunteer subjects, this limits scientists' ability to generalize their findings to some larger population of individuals.

Let's recap what we've learned about the scope of inference in a statistical study. Random selection of individuals allows inference about the population. Random assignment of individuals to groups permits inference about cause and effect. The following chart summarizes the possibilities.[50]

*This is an important topic, but it is not required for the AP® Statistics exam.

Both random sampling and random assignment introduce chance variation into a statistical study. When performing inference, statisticians use the laws of probability to describe this chance variation. You'll learn how this works later in the book.

| | | Were individuals randomly assigned to groups? | |
|---|---|---|---|
| | | **Yes** | **No** |
| **Were individuals randomly selected?** | **Yes** | Inference about the population: YES<br>Inference about cause and effect: YES | Inference about the population: YES<br>Inference about cause and effect: NO |
| | **No** | Inference about the population: NO<br>Inference about cause and effect: YES | Inference about the population: NO<br>Inference about cause and effect: NO |

Well-designed experiments randomly assign individuals to treatment groups. However, most experiments don't select experimental units at random from the larger population. That limits such experiments to inference about cause and effect. Observational studies don't randomly assign individuals to groups, which rules out inference about cause and effect. An observational study that uses random sampling can make an inference about the population. The following example illustrates all four cases from the table above in a single setting.

## EXAMPLE Vitamin C and Canker Sores

### *Determining scope of inference*

A small-town dentist wants to know if a daily dose of 500 milligrams (mg) of vitamin C will result in fewer canker sores in the mouth than taking no vitamin C.[51]

The dentist is considering the following four study designs:

*Design 1:* Get all dental patients in town with appointments in the next two weeks to take part in a study. Give each patient a survey with two questions: (1) Do you take at least 500 mg of vitamin C each day? (2) Do you frequently have canker sores? Based on patients' answers to Question 1, divide them into two groups: those who take at least 500 mg of vitamin C daily and those who don't.

*Design 2:* Get all dental patients in town with appointments in the next two weeks to take part in a study. Randomly assign half of them to take 500 mg of vitamin C each day and the other half to abstain from taking vitamin C for three months.

*Design 3:* Select a random sample of dental patients in town and get them to take part in a study. Divide the patients into two groups as in Design 1.

*Design 4:* Select a random sample of dental patients in town and get them to take part in a study. Randomly assign half of them to take 500 mg of vitamin C each day and the other half to abstain from taking vitamin C for three months.

For whichever design the dentist chooses, suppose she compares the proportion of patients in each group who complain of canker sores. Also suppose that she finds a statistically significant difference, with a smaller proportion of those taking vitamin C having canker sores.

**PROBLEM:** What can the dentist conclude for each design?

**SOLUTION:**

*Design 1:* Because the patients were not randomly selected, the dentist cannot infer that this result holds for a larger population of dental patients. This was an observational study because no treatments were deliberately imposed on the patients. With no random assignment to the two

### ALTERNATE EXAMPLE

### *Silence is golden?*

Many students insist that they study better when listening to music. A teacher doubts this claim and suspects that listening to music actually hurts academic performance. Here are four possible study designs to address this question at your school. In each case, the response variable will be the students' GPA at the end of the semester.

**1.** Get all the students in your AP® Statistics class to participate in a study. Ask them whether or not they study with music on and divide them into two groups based on their answer to this question.

**2.** Select a random sample of students from your school to participate in a study. Then divide them into two groups as in Design 1.

**3.** Get all the students in your AP® Statistics class to participate in a study. Randomly assign half of the students to listen to music while studying for the entire semester and have the remaining half abstain from listening to music while studying.

**4.** Select a random sample of students from your school to participate in a study. Randomly assign half of the students to listen to music while studying for the entire semester and have the remaining half abstain from listening to music while studying.

### ALTERNATE EXAMPLE *(continued)*

For each design, suppose that the mean GPA for students who listen to music while studying was significantly lower than the mean GPA of students who didn't listen to music while studying.

**PROBLEM:** What can we conclude for each design?

**SOLUTION:**

**1.** With no random selection, the results of the study should be applied only to the AP® Statistics students in the study. With no random assignment, we should not conclude anything about cause and effect. All we can conclude is that the students in the AP® Statistics class who listen to music while studying have lower GPAs than those who do not listen to music while studying. We don't know why and we can't apply these results to any larger group of students.

**2.** With random selection, the results of the study can be applied to the entire population—in this case, all the students at this school. With no random assignment, however, we should not conclude anything about cause and effect. All we can conclude is that students at this school who listen to music while studying have lower GPAs than those who do not listen to music while studying. We don't know why their GPAs are lower, however.

**3.** With no random selection, the results of the study should be applied only to the AP® Statistics students in the study. With random assignment, however, we can conclude that there is a cause-and-effect relationship between listening to music while studying and GPA, but only for the AP® Statistics students who took part in the study.

**4.** With random selection, the results of the study can be applied to the entire population—in this case, all the students at this school. With random assignment, we can conclude that there is a cause-and-effect relationship between listening to music while studying and GPA for all the students at the school.

groups, no inference about cause and effect can be made. The dentist just knows that for these patients, those who took vitamin C had fewer canker sores than those who didn't.

*Design 2:* As in Design 1, the dentist can't make any inference about this result holding for a larger population of dental patients. However, the treatments were randomly assigned to the subjects. Assuming proper control in the experiment, she can conclude that taking vitamin C reduced the chance of getting canker sores in her subjects.

*Design 3:* Because the patients were randomly selected from the population of dental patients in the town, the dentist can generalize the results of this study to the population. Because this was an observational study, no inference about cause and effect can be made. The dentist would conclude that for the population of dental patients in this town, those taking vitamin C have fewer canker sores than those who don't. She can't say whether the vitamin C causes this reduction or some other confounding variable.

*Design 4:* As in Design 3, the random sampling allows the dentist to generalize the results of this study to the population of dental patients in the town. As in Design 2, the random assignment would allow her to conclude (assuming proper control in the experiment) that taking vitamin C reduced the chance of getting canker sores. So the dentist would conclude that for the population of dental patients in this town, those taking vitamin C will tend to have fewer canker sores than those who don't due to the vitamin C.

**For Practice** *Try Exercise* 101

## The Challenges of Establishing Causation

A well-designed experiment tells us that changes in the explanatory variable cause changes in the response variable. More precisely, it tells us that this happened for specific individuals in the specific environment of this specific experiment. A serious threat is that the treatments, the subjects, or the environment of our experiment may not be realistic. *Lack of realism* can limit our ability to apply the conclusions of an experiment to the settings of greatest interest.

### EXAMPLE Do Center Brake Lights Reduce Rear-End Crashes?

*Lack of realism*

Do those high center brake lights, required on all cars sold in the United States since 1986, really reduce rear-end collisions? Randomized comparative experiments with fleets of rental and business cars, done before the lights were required, showed that the third brake light reduced rear-end collisions by as much as 50%. But requiring the third light in all cars led to only a 5% drop.

What happened? Most cars did not have the extra brake light when the experiments were carried out, so it caught the eye of following drivers. Now that almost all cars have the third light, they no longer capture attention.

In some cases, it isn't practical or even ethical to do an experiment. Consider these important questions:

**ALTERNATE EXAMPLE**

***Do calories matter?***

Does including the number of calories on a menu encourage diners to make healthier choices? Two AP® Statistics students designed a study to find out. They randomly assigned students to look at one of two different menus and indicate what they would order if they dined at the restaurant. One of the menus listed the number of calories for each item and the other menu did not. Result: no significant difference in the average number of calories ordered. The two students speculated that there wasn't a difference because the students weren't actually at the restaurant and didn't get to eat the food they "ordered." In other words, the experiment suffered from a lack of realism.

- Does texting while driving increase the risk of having an accident?
- Does going to church regularly help people live longer?
- Does smoking cause lung cancer?

To answer these cause-and-effect questions, we just need to perform a randomized comparative experiment. Unfortunately, we can't randomly assign people to text while driving or to attend church or to smoke cigarettes. The best data we have about these and many other cause-and-effect questions come from observational studies.

It is sometimes possible to build a strong case for causation in the absence of experiments. The evidence that smoking causes lung cancer is about as strong as nonexperimental evidence can be.

## EXAMPLE Does Smoking Cause Lung Cancer?

### *Living with observational studies*

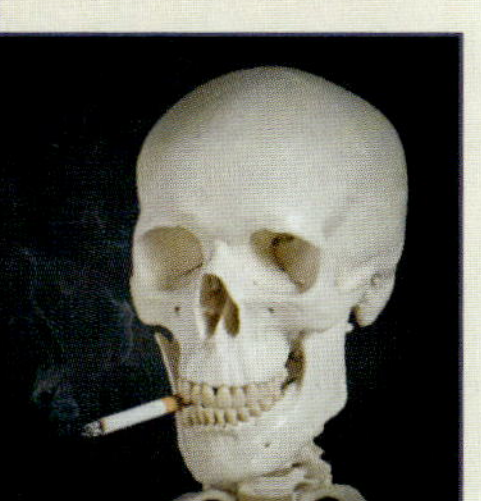

Doctors had long observed that most lung cancer patients were smokers. Comparison of smokers and similar nonsmokers showed a very strong association between smoking and death from lung cancer. Could the association be due to some other variable? Is there some genetic factor that makes people both more likely to get addicted to nicotine and to develop lung cancer? If so, then smoking and lung cancer would be strongly associated even if smoking had no direct effect on the lungs. Or maybe confounding is to blame. It might be that smokers live unhealthy lives in other ways (diet, alcohol, lack of exercise) and that some other habit confounded with smoking is a cause of lung cancer. How were these objections overcome?

What are the criteria for establishing causation when we can't do an experiment?

- *The association is strong.* The association between smoking and lung cancer is very strong.
- *The association is consistent.* Many studies of different kinds of people in many countries link smoking to lung cancer. That reduces the chance that some other variable specific to one group or one study explains the association.
- *Larger values of the explanatory variable are associated with stronger responses.* People who smoke more cigarettes per day or who smoke over a longer period get lung cancer more often. People who stop smoking reduce their risk.
- *The alleged cause precedes the effect in time.* Lung cancer develops after years of smoking. The number of men dying of lung cancer rose as smoking became more common, with a lag of about 30 years. Lung cancer kills more men than any other form of cancer. Lung cancer was rare among women until women began to smoke. Lung cancer in women rose along with smoking, again with a lag of about 30 years, and has now passed breast cancer as the leading cause of cancer death among women.
- *The alleged cause is plausible.* Experiments with animals show that tars from cigarette smoke do cause cancer.

Medical authorities do not hesitate to say that smoking causes lung cancer. The U.S. Surgeon General states that cigarette smoking is "the largest avoidable

### ALTERNATE EXAMPLE

### *Do tanning beds cause skin cancer?*

Doctors have noticed that people who frequently visit tanning salons are at a much greater risk for skin cancer. But are the tanning beds the cause? Could the association be due to another variable, such as sun exposure? After all, people who like to tan indoors also probably like to tan outdoors. An experiment would help to settle this question, but forcing people to use tanning beds to see if they get skin cancer is unethical.

## Teaching Tip

The section on data ethics is optional, because it is not covered on the AP® exam. However, it is a very important part of statistics. We recommend that you spend some time discussing the topics in this section.

cause of death and disability in the United States."[53] The evidence for causation is overwhelming—but it is not as strong as the evidence provided by well-designed experiments. Conducting an experiment in which some subjects were forced to smoke and others were not allowed to would be unethical. In cases like this, observational studies are our best source of reliable information.

## Data Ethics*

Medical professionals are taught to follow the basic principle "First, do no harm." Shouldn't those who carry out statistical studies follow the same principle? Most reasonable people think so. But this may not always be as simple as it sounds. Decide whether you think each of the following studies is ethical or unethical:

- A promising new drug has been developed for treating cancer in humans. Before giving the drug to human subjects, researchers want to administer the drug to animals to see if there are any potentially serious side effects.
- Are companies discriminating against some individuals in the hiring process? To find out, researchers prepare several equivalent résumés for fictitious job applicants, with the only difference being the gender of the applicant. They send the fake résumés to companies advertising positions and keep track of the number of males and females who are contacted for interviews.
- Will people try to stop someone from driving drunk? A television news program hires an actor to play a drunk driver and uses a hidden camera to record the behavior of individuals who encounter the driver.

The most complex issues of data ethics arise when we collect data from people. The ethical difficulties are more severe for experiments that impose some treatment on people than for sample surveys that simply gather information. Trials of new medical treatments, for example, can do harm as well as good to their subjects. Here are some basic standards of data ethics that must be obeyed by all studies that gather data from human subjects, both observational studies and experiments.

**BASIC DATA ETHICS**

All planned studies must be reviewed in advance by an **institutional review board** charged with protecting the safety and well-being of the subjects.

All individuals who are subjects in a study must give their **informed consent** before data are collected.

All individual data must be kept **confidential**. Only statistical summaries for groups of subjects may be made public.

The law requires that studies carried out or funded by the federal government obey these principles.[54] But neither the law nor the consensus of experts is completely clear about the details of their application.

**Institutional review boards** The purpose of an institutional review board is not to decide whether a proposed study will produce valuable information or whether it is statistically sound. The board's purpose is, in the words of one uni-

*This is an important topic, but it is not required for the AP® Statistics exam.

versity's board, "to protect the rights and welfare of human subjects (including patients) recruited to participate in research activities." The board reviews the plan of the study and can require changes. It reviews the consent form to be sure that subjects are informed about the nature of the study and about any potential risks. Once research begins, the board monitors its progress at least once a year.

**Informed consent** Both words in the phrase "informed consent" are important, and both can be controversial. Subjects must be *informed* in advance about the nature of a study and any risk of harm it may bring. In the case of a sample survey, physical harm is not possible. But a survey on sensitive issues could result in emotional harm. The participants should be told what kinds of questions the survey will ask and about how much of their time it will take. Experimenters must tell subjects the nature and purpose of the study and outline possible risks. Subjects must then *consent* in writing.

**Confidentiality** Ethical problems do not disappear once a study has been cleared by the review board, has obtained consent from its participants, and has actually collected data about them. It is important to protect individuals' privacy by keeping all data about them **confidential.** The report of an opinion poll may say what percent of the 1200 respondents believed that legal immigration should be reduced. It may not report what *you* said about this or any other issue.

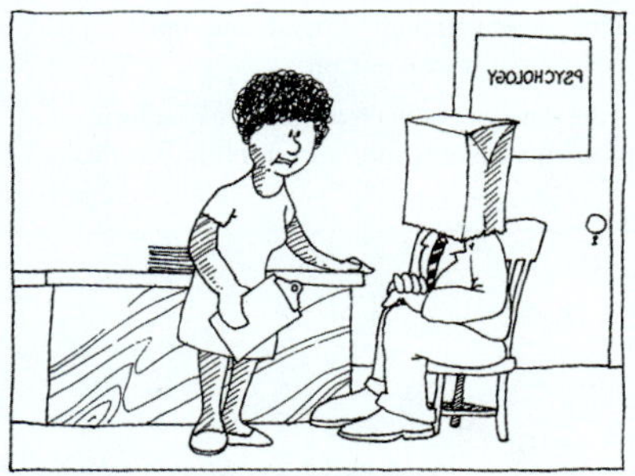

"I realize the participants in this study are to be anonymous, but you're going to have to expose your eyes."

Confidentiality is not the same as **anonymity.** Anonymity means that individuals are anonymous—their names are not known even to the director of the study. Anonymity is rare in statistical studies. Even where anonymity is possible (mainly in surveys conducted by mail), it prevents any follow-up to improve nonresponse or inform individuals of results.

Any breach of confidentiality is a serious violation of data ethics. The best practice is to separate the identity of the study's participants from the rest of the data at once. A clever computer search of several data bases might be able, by combining information, to identify you and learn a great deal about you even if your name and other identification have been removed from the data available for search. Privacy and confidentiality of data are hot issues among statisticians in the computer age.

## ACTIVITY | Response bias

In this Activity, your team will design and conduct an experiment to investigate the effects of response bias in surveys.[52] You may choose the topic for your surveys, but you must design your experiment so that it can answer at least one of the following questions:

- Can the wording of a question create response bias?
- Do the characteristics of the interviewer create response bias?
- Does anonymity change the responses to sensitive questions?
- Does manipulating the answer choices change the response?

**TRM Response Bias Activity**

The Response Bias Activity works well as a major project. One of the authors has his students work in pairs and present their results with a poster that is hung in the classroom. When other students see the poster, they always ask, "Can we do cool projects like that?" The Teacher's Resource Materials include a sample rubric that you can use or modify for this project.

 **Activity Overview**

You can watch an overview of this Activity (for teachers) by clicking on the link in the TE-book, opening the TRCD, or logging onto the book's companion Web site www.whfreeman.com/tps5e (teacher log-in required).

1. Write a proposal describing the design of your experiment. Be sure to include
   (a) your chosen topic and which of the above questions you'll try to answer.
   (b) a detailed description of how you will obtain your subjects (minimum of 50). Your plan must be practical!
   (c) what your questions will be and how they will be asked.
   (d) a clear explanation of how you will implement your design.
   (e) precautions you will take to collect data ethically.

   Here are two examples of successful student experiments:

   *"Make-Up," by Caryn S. and Trisha T.* (all questions asked to males)
   i. "Do you find females who wear makeup attractive?" (questioner wearing makeup: 75% answered yes)
   ii. "Do you find females who wear makeup attractive?" (questioner not wearing makeup: 30% answered yes)

   *"Cartoons" by Sean W. and Brian H.*
   i. "Do you watch cartoons?" (90% answered yes)
   ii. "Do you *still* watch cartoons?" (60% answered yes)

2. Once your teacher has approved your design, carry out the experiment. Record your data in a table.
3. Analyze your data. What conclusion do you draw? Provide appropriate graphical and numerical evidence to support your answer.
4. Prepare a report that includes the data you collected, your analysis from Step 3, and a discussion of any problems you encountered and how you dealt with them.

## Can Magnets Help Reduce Pain?

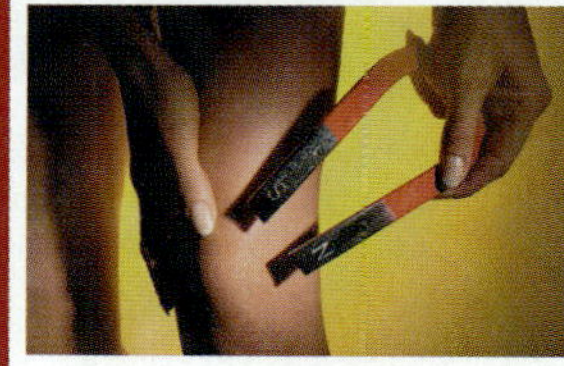

Re-read the chapter-opening Case Study on page 207. Then use what you have learned in this chapter to help answer the following questions.

1. Why is the magnet study an experiment, not an observational study?
2. What type of design was used in this experiment? Identify the experimental units, the treatments, and the response variable.
3. There were two distinct purposes for having the doctors select a sealed envelope at random from the box. Describe them both.

### Teaching Tip

Here are some additional examples of successful response bias projects:

- *Weight. Wait, What?* Females were asked how much they weigh anonymously and in person. The average weight was higher for the anonymous responses.
- *How Do Eye Look?* Females were asked whether they prefer blue eyes or green eyes by either an interviewer with blue eyes or an interviewer with green eyes. When asked by the blue-eyed interviewer, more females preferred blue eyes than when asked by the green-eyed interviewer.
- *Does the Order Matter?* Students were asked a series of two questions, in random order. The questions were "How hungry are you on a scale from 1 to 10?" and "How long has it been since your last meal?" When students were asked about their last meal first, their average hunger rating was higher.
- *Aid for the Poor?* People were asked if there should be more funding to help the poor, but half of the people were shown a picture of a homeless mother and child when asked the question. When the photo was shown, more people supported increased funding for the poor.

### Answers to Case Closed

**1.** This was an experiment because researchers imposed a treatment (active or inactive magnets) on the subjects.

**2.** This was a completely randomized design because the treatments were assigned completely at random. In other words, there were no restrictions on who could end up in which group. The experimental units were 50 patients with chronic pain. The treatments were the active magnets and the inactive magnets. The response variable is the improvement in pain ratings.

**3.** One purpose of selecting the sealed envelopes at random is so the subjects could be randomly assigned to the treatments. The second reason sealed envelopes were used was so the doctors and patients did not know who received which treatment. In other words, it made the experiment double-blind.

**4.** The distribution of improvements for the group with active magnets is roughly symmetric, while the distribution for the group with inactive magnets is strongly skewed to the right. The center of the distribution for the active group is much higher than the center for the inactive group. Also, the distribution for the active group is much more variable than the distribution for the inactive group.

**5.** Because this was an experiment and the difference of 4.14 is larger than what could be expected due to the chance variation in random assignment, we can conclude that magnets are effective for reducing pain for patients like the ones in the experiment.

4. The dotplot shows the improvement in pain ratings for both groups. Write a few sentences comparing the two distributions.

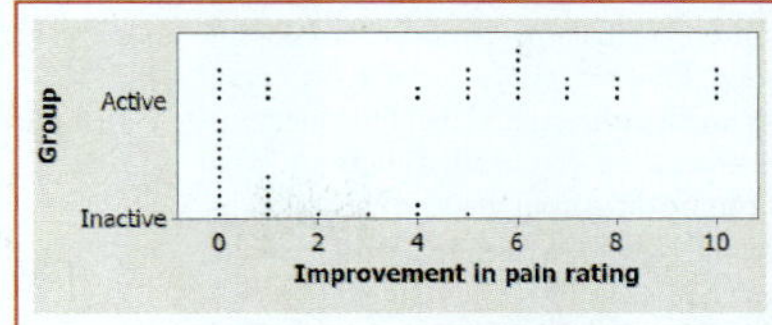


5. The mean difference in pain ratings was 5.24 for the active-magnet group and 1.10 for the inactive-magnet group. This difference is statistically significant. What conclusion should we draw?

## Section 4.3 Summary

- Most statistical studies aim to make inferences that go beyond the data actually produced. **Inference about a population** requires that the individuals taking part in a study be randomly selected from the population. A well-designed experiment that randomly assigns experimental units to treatments allows **inference about cause and effect**.
- **Lack of realism** in an experiment can prevent us from generalizing its results.
- In the absence of an experiment, good evidence of causation requires a strong association that appears consistently in many studies, a clear explanation for the alleged causal link, and careful examination of other variables.
- Studies involving humans must be screened in advance by an **institutional review board**. All participants must give their **informed consent** before taking part. Any information about the individuals in the study must be kept **confidential**.

## Section 4.3 Exercises

97. **Random sampling versus random assignment** Explain the difference between the types of inference that can be made as a result of random sampling and random assignment.

98. **Observation versus experimentation** Explain the difference between the types of inference than can usually be made from an observational study and an experiment.

99. **Foster care versus orphanages** Do abandoned children placed in foster homes do better than similar children placed in an institution? The Bucharest Early Intervention Project found that the answer is a clear "Yes." The subjects were 136 young children abandoned at birth and living in orphanages in Bucharest, Romania. Half of the children, chosen at random, were placed in foster homes. The other half remained in the orphanages.[55] (Foster care was not easily available in Romania at the time and so was paid for by the study.) What conclusion can we draw from this study? Explain.

100. **Frozen batteries** Will storing batteries in a freezer make them last longer? To find out, a company that

### TRM Section 4.3 Quizzes

There are three different quizzes available for this section in the Teacher's Resource Materials. Versions A and B are roughly parallel. Find them by clicking on the link in the new TE-book, opening the TRCD, or logging onto the book's companion Web site www.whfreeman.com/tps5e (teacher log-in required).

### TRM Full Solutions to Section 4.3 Exercises

### Answers to Section 4.3 Exercises

**4.97** If the study involves random sampling, we can make inferences about the population from which we sampled. If the study involves random assignment, we can make inferences about cause and effect.

**4.98** In a well-designed experiment, we can make inferences about cause and effect, but not in an observational study. In either case, we can make inferences about a larger population if the individuals were randomly selected from that population.

**4.99** Because this study involved random assignment to the treatments, we can infer that the difference between foster care or institutional care caused the difference in response.

**4.100** Because this study involved random assignment to the treatments, we can infer that being stored in the freezer caused the increase in average charge. Also, because the batteries were randomly chosen from the warehouse, we can generalize this result to the entire population of batteries in the warehouse.

**4.101** Because this study did not involve random assignment to a treatment, we cannot infer cause and effect. Also, because the individuals were not randomly chosen, we cannot generalize to a larger population.

**4.102** Because this study involved a random sample of adults, we can make an inference that adults who attend religious services regularly have a lower risk of dying. However, because subjects were not randomly assigned to attend religious services or not, we cannot infer cause and effect.

**4.103** As daytime running lights become more common, they may be less effective at catching the attention of other drivers. Also, a driving simulator might not be very realistic.

**4.104** No. The disappointment associated with losing games during one evening is not likely to be equivalent to the disappointment felt after months of hard work.

**4.105** Answers will vary.

**4.106** Answers will vary.

**4.107** Answers will vary.

**4.108** Answers will vary.



produces batteries takes a random sample of 100 AA batteries from its warehouse. The company statistician randomly assigns 50 batteries to be stored in the freezer and the other 50 to be stored at room temperature for 3 years. At the end of that time period, each battery's charge is tested. *Result*: Batteries stored in the freezer had a higher average charge, and the difference between the groups was statistically significant. What conclusion can we draw from this study? Explain.

101. **Who talks more—women or men?** According to Louann Brizendine, author of *The Female Brain*, women say nearly three times as many words per day as men. Skeptical researchers devised a study to test this claim. They used electronic devices to record the talking patterns of 396 university students who volunteered to participate in the study. The device was programmed to record 30 seconds of sound every 12.5 minutes without the carrier's knowledge. According to a published report of the study in *Scientific American*, "Men showed a slightly wider variability in words uttered.... But in the end, the sexes came out just about even in the daily averages: women at 16,215 words and men at 15,669."[56] This difference was not statistically significant. What conclusion can we draw from this study? Explain.

pg 267

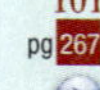

102. **Attend church, live longer?** One of the better studies of the effect of regular attendance at religious services gathered data from a random sample of 3617 adults. The researchers then measured lots of variables, not just the explanatory variable (religious activities) and the response variable (length of life). A news article said: "Churchgoers were more likely to be nonsmokers, physically active, and at their right weight. But even after health behaviors were taken into account, those not attending religious services regularly still were about 25% more likely to have died."[57] What conclusion can we draw from this study? Explain.

103. **Daytime running lights** Canada and the European Union require that cars be equipped with "daytime running lights," headlights that automatically come on at a low level when the car is started. Many manufacturers are now equipping cars sold in the United States with running lights. Will running lights reduce accidents by making cars more visible? An experiment conducted in a driving simulator suggests that the answer may be "Yes." What concerns would you have about generalizing the results of such an experiment?

104. **Studying frustration** A psychologist wants to study the effects of failure and frustration on the relationships among members of a work team. She forms a team of students, brings them to the psychology lab, and has them play a game that requires teamwork. The game is rigged so that they lose regularly. The psychologist observes the students through a one-way window and notes the changes in their behavior during an evening of game playing. Can the psychologist generalize the results of her study to a team of employees that spends months developing a new product that never works right and is finally abandoned by their company? Explain.

105.* **Minimal risk?** You have been invited to serve on a college's institutional review board. You must decide whether several research proposals qualify for lighter review because they involve only minimal risk to subjects. Federal regulations say that "minimal risk" means the risks are no greater than "those ordinarily encountered in daily life or during the performance of routine physical or psychological examinations or tests." That's vague. Which of these do you think qualifies as "minimal risk"?

(a) Draw a drop of blood by pricking a finger to measure blood sugar.

(b) Draw blood from the arm for a full set of blood tests.

(c) Insert a tube that remains in the arm, so that blood can be drawn regularly.

106.* **Who reviews?** Government regulations require that institutional review boards consist of at least five people, including at least one scientist, one nonscientist, and one person from outside the institution. Most boards are larger, but many contain just one outsider.

(a) Why should review boards contain people who are not scientists?

(b) Do you think that one outside member is enough? How would you choose that member? (For example, would you prefer a medical doctor? A member of the clergy? An activist for patients' rights?)

107.* **No consent needed?** In which of the circumstances below would you allow collecting personal information without the subjects' consent?

(a) A government agency takes a random sample of income tax returns to obtain information on the average income of people in different occupations. Only the incomes and occupations are recorded from the returns, not the names.

(b) A social psychologist attends public meetings of a religious group to study the behavior patterns of members.

(c) A social psychologist pretends to be converted to membership in a religious group and attends private meetings to study the behavior patterns of members.

108.* **Surveys of youth** A survey asked teenagers whether they had ever consumed an alcoholic beverage. Those who said "Yes" were then asked, "How old were you when you first consumed an alcoholic beverage?" Should consent of parents be required to ask minors about alcohol, drugs, and other such issues, or is consent of the minors themselves enough? Give reasons for your opinion.

109.* **Anonymous? Confidential?** One of the most important nongovernment surveys in the United States is the National Opinion Research Center's General Social

*Exercises 105 to 112: This is an important topic, but it is not required for the AP® Statistics exam.

Survey. The GSS regularly monitors public opinion on a wide variety of political and social issues. Interviews are conducted in person in the subject's home. Are a subject's responses to GSS questions anonymous, confidential, or both? Explain your answer.

110. *Anonymous? Confidential? Texas A&M, like many universities, offers screening for HIV, the virus that causes AIDS. Students may choose either anonymous or confidential screening. An announcement says, "Persons who sign up for screening will be assigned a number so that they do not have to give their name." They can learn the results of the test by telephone, still without giving their name. Does this describe the *anonymous* or the *confidential* screening? Why?

111. *The Willowbrook hepatitis studies In the 1960s, children entering the Willowbrook State School, an institution for the intellectually disabled on Staten Island in New York, were deliberately infected with hepatitis. The researchers argued that almost all children in the institution quickly became infected anyway. The studies showed for the first time that two strains of hepatitis existed. This finding contributed to the development of effective vaccines. Despite these valuable results, the Willowbrook studies are now considered an example of unethical research. Explain why, according to current ethical standards, useful results are not enough to allow a study.

112. *Unequal benefits Researchers on aging proposed to investigate the effect of supplemental health services on the quality of life of older people. Eligible patients on the rolls of a large medical clinic were to be randomly assigned to treatment and control groups. The treatment group would be offered hearing aids, dentures, transportation, and other services not available without charge to the control group. The review board felt that providing these services to some but not other persons in the same institution raised ethical questions. Do you agree?

113. **Animal testing** (1.1) "It is right to use animals for medical testing if it might save human lives." The General Social Survey asked 1152 adults to react to this statement. Here is the two-way table of their responses:

| | Male | Female |
|---|---|---|
| Strongly agree | 76 | 59 |
| Agree | 270 | 247 |
| Neither agree nor disagree | 87 | 139 |
| Disagree | 61 | 123 |
| Strongly disagree | 22 | 68 |

How do the distributions of opinion differ between men and women? Give appropriate graphical and numerical evidence to support your answer.

114. **Initial public offerings** (1.3) The business magazine *Forbes* reports that 4567 companies sold their first stock to the public between 1990 and 2000. The *mean* change in the stock price of these companies since the first stock was issued was +111%. The *median* change was −31%.[58] Explain how this could happen.

*Exercises 105 to 112: This is an important topic, but it is not required for the AP® Statistics exam.

## FRAPPY! Free Response AP® Problem, Yay!

The following problem is modeled after actual AP® Statistics exam free response questions. Your task is to generate a complete, concise response in 15 minutes.

*Directions: Show all your work. Indicate clearly the methods you use, because you will be scored on the correctness of your methods as well as on the accuracy and completeness of your results and explanations.*

In a recent study, 166 adults from the St. Louis area were recruited and randomly assigned to receive one of two treatments for a sinus infection. Half of the subjects received an antibiotic (amoxicillin) and the other half received a placebo.[59]

(a) Describe how the researchers could have assigned treatments to subjects if they wanted to use a completely randomized design.

(b) All the subjects in the experiment had moderate, severe, or very severe symptoms at the beginning of the study. Describe one statistical benefit and one statistical drawback for using subjects with moderate, severe, or very severe symptoms instead of just using subjects with very severe symptoms.

(c) At different stages during the next month, all subjects took the sino-nasal outcome test. After 10 days, the difference in average test scores was *not* statistically significant. In this context, explain what it means for the difference to be not statistically significant.

(d) One possible way that researchers could have improved the study is to use a randomized block design. Explain how the researchers could have incorporated blocking in their design.

After you finish, you can view two example solutions on the book's Web site (www.whfreeman.com/tps5e). Determine whether you think each solution is "complete," "substantial," "developing," or "minimal." If the solution is not complete, what improvements would you suggest to the student who wrote it? Finally, your teacher will provide you with a scoring rubric. Score your response and note what, if anything, you would do differently to improve your own score.

**4.109** Confidential. The person taking the survey knows who is answering the questions, but will not share the results of individuals with anyone else.

**4.110** Anonymous. The patient never gives their name, but rather is just assigned a number.

**4.111** The subjects were not able to give informed consent. They did not know what was happening to them and they were not old enough to understand the ramifications.

**4.112** Answers will vary.

**4.113** The conditional distributions for males and females are displayed in the table and graph. Men are more likely to view animal testing as justified if it might save human lives: over two-thirds of men agree or strongly agree with this statement, compared to slightly less than half of the women. The percentages who disagree or strongly disagree tell a similar story: 16% of men versus 30% of women.

| Response | Male | Female |
|---|---|---|
| Strongly agree | 14.7% | 9.3% |
| Agree | 52.3% | 38.8% |
| Neither | 16.9% | 21.9% |
| Disagree | 11.8% | 19.3% |
| Strongly disagree | 4.3% | 10.7% |

**4.114** The median is resistant to outliers, while the mean is not. Because there is no upper limit for an increase in stock price, it is likely that there were some very high outliers that had a big effect on the mean.

**TRM FRAPPY!**

We present a model solution here. Please consult the Teacher's Resource Materials for sample student responses, a scoring rubric, and a black line master of the original question with space for students to write their responses.

**(a)** Label 83 note cards with A and 83 note cards with B. Shuffle the cards well and hand one card to each subject at random. Subjects with A cards will receive the antibiotic and subjects with B cards will receive the placebo.

**(b)** Benefit: We can make conclusions about a wider variety of subjects. If we only used subjects with very severe symptoms, we couldn't make conclusions about subjects with less severe symptoms. Drawback: Because these subjects varied in their initial condition, the responses to the treatments will vary more than if all the subjects had the same initial condition. This additional variability will make it harder to determine if the antibiotic is really better.

**(c)** If the difference is not statistically significant, then the difference wasn't large enough to rule out random chance as a plausible explanation. That is, the difference could be due to the random assignment and not to the effects of the treatments.

**(d)** To incorporate blocking, form blocks based on the initial conditions of the patients. That is, put all the patients with very severe symptoms into one block and so on. Then within each block, randomly assign the subjects to treatments as in part (a). Blocking by initial severity will help us account for the additional variability in test scores caused by the differences in severity.

**TRM Chapter 4 Tests**

There are four different tests available for this section in the Teacher's Resource Materials. Versions A and B are roughly parallel and version D is cumulative. Find them by clicking on the link in the TE-book, opening the TRCD, or logging onto the book's companion Web site www.whfreeman.com/tps5e (teacher log-in required).

# Chapter Review

### Section 4.1: Sampling and Surveys

In this section, you learned that a population is the group of all individuals that we want information about. A sample is the subset of the population that we use to gather this information. The goal of most sample surveys is to use sample information to draw conclusions about the population. Choosing people for a sample because they are located nearby or letting people choose whether or not to be in the sample are poor ways to choose a sample. Because convenience samples and voluntary response samples will produce estimates that are consistently too large or consistently too small, these methods of choosing a sample are biased.

To avoid bias in the way the sample is formed, the members of the sample should be chosen at random. One way to do this is with a simple random sample (SRS), which is equivalent to pulling well-mixed slips of paper from a hat. It is often more convenient to select an SRS using technology or a table of random digits.

Two other random sampling methods are stratified sampling and cluster sampling. To obtain a stratified random sample, divide the population into groups (strata) of similar individuals, take an SRS from each stratum, and combine the chosen individuals to form the sample. Stratified random samples can produce estimates with much greater precision than simple random samples. To obtain a cluster sample, divide the population into groups (clusters) of individuals that are in similar locations, randomly select clusters, and use every individual in the chosen clusters. Cluster samples are easier to obtain than simple random samples or stratified random samples, but they may not produce very precise estimates.

Finally, you learned about other issues in sample surveys that can lead to bias: undercoverage occurs when the sampling method systematically excludes one part of the population. Nonresponse describes when answers cannot be obtained from some people that were chosen to be in the sample. Bias can also result when some people in the sample don't give accurate responses due to question wording, interviewer characteristics, or other factors.

### Section 4.2: Experiments

In this section, you learned about the difference between observational studies and experiments. Experiments deliberately impose a treatment to see if there is a cause-and-effect relationship between two variables. Observational studies look at relationships between two variables, but cannot show cause and effect because other variables may be confounded with the explanatory variable. Variables are confounded when it is impossible to determine which of the variables is causing a change in the response variable.

A common type of experiment uses a completely randomized design. In this type of design, the experimental units are divided into groups, one group for each of the treatments. To determine which experimental units are in which group, we use random assignment. With random assignment, the effects of variables (other than the explanatory variable) are roughly balanced out between the groups. Replication means giving each treatment to as many experimental units as possible. This makes it easier to see the effects of the treatments because the effects of other variables are more likely to be balanced among the treatment groups.

During an experiment, it is important that other variables be controlled (kept the same) for each experimental unit. Doing so helps avoid confounding and removes a possible source of variation in the response variable. Also, beware of the placebo effect—the tendency for people to improve because they expect to, not because of the treatment they are receiving. One way to make sure that all experimental units have the same expectations is to make them blind—unaware of which treatment they are receiving. When the person measuring the response variable is also blind, the experiment is called double-blind.

The results of an experiment are statistically significant if the difference in the response is too large to be accounted for by the random assignment of experimental units to treatments. To make it more likely to obtain statistically significant results, experiments can incorporate blocking. Blocking in experiments is similar to stratifying in sampling. To form blocks, group together experimental units that are similar with respect to a variable that is associated with the response. Then randomly assign the treatments within each block. A design that uses blocks with two experimental units is called a matched pairs design. Blocking helps us estimate the effects of the treatments more precisely because we can account for the variability introduced by the variables used to form the blocks.

### Section 4.3: Using Studies Wisely

In this section, you learned that the different types of conclusions we can draw depend on how the data are produced. When samples are selected at random, we can make inferences about the population from which the sample was drawn. When treatments are applied to groups formed at random, we can conclude cause and effect.

Making a cause-and-effect conclusion is often difficult because it is impossible or unethical to perform certain types of experiments. Good data ethics requires that studies should be approved by an institutional review board, subjects should give informed consent, and individual data must be kept confidential.

## What Did You Learn?

| Learning Objective | Section | Related Example on Page(s) | Relevant Chapter Review Exercise(s) |
|---|---|---|---|
| Identify the population and sample in a statistical study. | 4.1 | 210 | R4.1 |
| Identify voluntary response samples and convenience samples. Explain how these sampling methods can lead to bias. | 4.1 | 213 | R4.2 |
| Describe how to obtain a random sample using slips of paper, technology, or a table of random digits. | 4.1 | 214, 217 | R4.3 |
| Distinguish a simple random sample from a stratified random sample or cluster sample. Give the advantages and disadvantages of each sampling method. | 4.1 | 221 | R4.4 |
| Explain how undercoverage, nonresponse, question wording, and other aspects of a sample survey can lead to bias. | 4.1 | 226, 227 | R4.5 |
| Distinguish between an observational study and an experiment. | 4.2 | 235 | R4.6 |
| Explain the concept of confounding and how it limits the ability to make cause-and-effect conclusions. | 4.2 | 235 | R4.6 |
| Identify the experimental units, explanatory and response variables, and treatments in an experiment. | 4.2 | 237, 239 | R4.7 |
| Explain the purpose of comparison, random assignment, control, and replication in an experiment. | 4.2 | 243 | R4.9 |
| Describe a completely randomized design for an experiment, including how to randomly assign treatments using slips of paper, technology, or a table of random digits. | 4.2 | 246 | R4.7, R4.10 |
| Describe the placebo effect and the purpose of blinding in an experiment. | 4.2 | 247 | R4.9 |
| Interpret the meaning of statistically significant in the context of an experiment. | 4.2 | 249 (Activity) | R4.9 |
| Explain the purpose of blocking in an experiment. Describe a randomized block design or a matched pairs design for an experiment. | 4.2 | 251, 254 | R4.7, R4.10 |
| Describe the scope of inference that is appropriate in a statistical study. | 4.3 | 267 | R4.8 |
| *Evaluate whether a statistical study has been carried out in an ethical manner. | 4.3 | Discussion on 270 | R4.11 |

*This is an important topic, but it is not required for the AP® Statistics exam.



TRM **Learning Objectives Grid**

TRM **Full Solutions to Chapter 4 Review Exercises**

## Answers to Chapter 4 Review Exercises

**R4.1** **(a)** Population: all Ontario residents. Sample: the 61,239 people interviewed. **(b)** Because different samples will produce different estimates, it is unlikely that the percentages in the entire population would be exactly the same as the percentages in the sample. However, they should be fairly close.

**R4.2** **(a)** Announce in daily bulletin that there is a survey concerning student parking available in the main office for students who want to respond. Because those who feel strongly are more likely to respond, their opinions will be over-represented. **(b)** Interview a group of students as they come in from the parking lot. People who already can park on campus might have different opinions about the parking situation than those who cannot.

**R4.3** **(a)** Number the players from 01 to 25 in alphabetical order. Move from left to right, reading pairs of digits until you find three different pairs between 01 and 25, and select the corresponding players. **(b)** 17 (Musselman), 09 (Fuhrmann), and 23 (Smith).

**R4.4** Stratified, because it is likely that the opinions of professors will vary based on which type of institution they teach at. Then a stratified random sample will provide a more precise estimate than the other methods. Furthermore, the other methods might miss faculty from one particular type of institution.

**R4.5** **(a)** People may not remember how many movies they watched in a movie theater in the past year. So, shorten the amount of time that they ask about, perhaps 3 or 6 months. **(b)** This will underrepresent younger adults who use only cell phones. If younger adults go to movies more often than older adults, the estimated mean will be too small. **(c)** Because the frequent moviegoers will not be at home to respond, the estimated mean will be too small.

**R4.6** **(a)** Different anesthetics were not randomly assigned to the subjects. **(b)** Type of surgery. If Anesthesia C is used more often with a type of surgery that has a higher death rate, we wouldn't know if the death rate was higher because of the anesthesia or the type of surgery.

# Chapter 4 Chapter Review Exercises

*These exercises are designed to help you review the important ideas and methods of the chapter.*

**R4.1 Ontario Health Survey** The Ministry of Health in the province of Ontario, Canada, wants to know whether the national health care system is achieving its goals in the province. Much information about health care comes from patient records, but that source doesn't allow us to compare people who use health services with those who don't. So the Ministry of Health conducted the Ontario Health Survey, which interviewed a random sample of 61,239 people who live in the province of Ontario.[60]

(a) What is the population for this sample survey? What is the sample?

(b) The survey found that 76% of males and 86% of females in the sample had visited a general practitioner at least once in the past year. If a census were conducted, do you think that the percentages would be the same as in the sample? Explain.

**R4.2 Bad sampling** A large high school wants to gather student opinion about parking for students on campus. It isn't practical to contact all students.

(a) Give an example of a way to choose a voluntary response sample of students. Explain how this method could lead to bias.

(b) Give an example of a way to choose a convenience sample of students. Explain how this method could lead to bias.

**R4.3 Drug testing** A baseball team regularly conducts random drug tests on its players. The 25 members of the team are listed below.

| | | | | |
|---|---|---|---|---|
| Agarwal | Chen | Healy | Moser | Roberts |
| Andrews | Frank | Hixson | Musselman | Shen |
| Baer | Fuest | Lee | Pavnica | Smith |
| Berger | Fuhrmann | Lynch | Petrucelli | Sundheim |
| Brockman | Garcia | Milhalko | Reda | Wilson |

(a) Explain how you would use the line of random digits below to select an SRS of 3 team members for a random drug test.

(b) Use your method from part (a) to choose the SRS using the digits below. Show your work.

17521 78009 46239 84569 03316

**R4.4 Polling the faculty** A researcher wants to study the attitudes of college faculty members about the work habits of entering freshmen. These attitudes appear to differ depending on the type of college. The American Association of University Professors classifies colleges as follows:

Class I: Offer doctorate degrees and award at least 15 per year.

Class IIA: Award degrees above the bachelor's but are not in Class I.

Class IIB: Award no degrees beyond the bachelor's.

Class III: Two-year colleges.

The researcher would like to survey about 200 faculty members. Would you recommend a simple random sample, stratified random sample, or cluster sample? Justify your answer.

**R4.5 Been to the movies?** An opinion poll calls 2000 randomly chosen residential telephone numbers, then asks to speak with an adult member of the household. The interviewer asks, "How many movies have you watched in a movie theater in the past 12 months?" In all, 1131 people responded. The researchers used the responses to estimate the mean number of movies adults have watched in a movie theater in the past 12 months.

(a) Describe a potential source of bias related to the wording of the question. Suggest a change that would help fix this problem.

(b) Describe how using only residential phone numbers might lead to bias and how this will affect the estimate.

(c) Describe how nonresponse might lead to bias and how this will affect the estimate.

**R4.6 Are anesthetics safe?** The National Halothane Study was a major investigation of the safety of anesthetics used in surgery. Records of over 850,000 operations performed in 34 major hospitals showed the following death rates for four common anesthetics:[61]

| Anesthetic: | A | B | C | D |
|---|---|---|---|---|
| Death rate: | 1.7% | 1.7% | 3.4% | 1.9% |

There seems to be a clear association between the anesthetic used and the death rate of patients. Anesthetic C appears to be more dangerous.

(a) Explain why we call the National Halothane Study an observational study rather than an experiment, even though it compared the results of using different anesthetics in actual surgery.

(b) When the study looked at other variables that are related to a doctor's choice of anesthetic, it found that Anesthetic C was not causing extra deaths. Explain the concept of confounding in this context and identify a variable that might be confounded with the doctor's choice of anesthetic.

**R4.7 Ugly fries** Few people want to eat discolored french fries. Potatoes are kept refrigerated before being cut for french fries to prevent spoiling and preserve flavor. But immediate processing of cold potatoes causes discoloring due to complex chemical reactions. The potatoes must therefore be brought to room temperature before processing. Researchers want to design an experiment in which tasters will rate the color and flavor of french fries prepared from several groups of potatoes. The potatoes will be freshly picked or stored for a month at room temperature or stored for a month refrigerated. They will then be sliced and cooked either immediately or after an hour at room temperature.

(a) Identify the experimental units, the explanatory and response variables, and the treatments.

(b) The researchers plan to use a completely randomized design. Describe how they should assign treatments to the experimental units if there are 300 potatoes available for the experiment.

(c) The researchers decided to do a follow-up experiment using sweet potatoes as well as regular potatoes. Describe how they should change the design of the experiment to account for the addition of sweet potatoes.

**R4.8 Don't catch a cold!** A recent study of 1000 students at the University of Michigan investigated how to prevent catching the common cold. The students were randomly assigned to three different cold prevention methods for 6 weeks. Some wore masks, some wore masks and used hand sanitizer, and others took no precautions. The two groups who used masks reported 10–50% fewer cold symptoms than those who did not wear a mask.[62]

(a) Does this study allow for inference about a population? Explain.

(b) Does this study allow for inference about cause and effect? Explain.

**R4.9 An herb for depression?** Does the herb Saint-John's-wort relieve major depression? Here is an excerpt from the report of a study of this issue: "Design: Randomized, Double-Blind, Placebo-Controlled Clinical Trial."[63] The study concluded that the difference in effectiveness of Saint-John's-wort and a placebo was not statistically significant.

(a) How did the design of this experiment account for the placebo effect?

(b) Explain the purpose of the random assignment.

(c) Why is a double-blind design a good idea in this setting?

(d) Explain what "not statistically significant" means in this context.

**R4.10 How long did I work?** A psychologist wants to know if the difficulty of a task influences our estimate of how long we spend working at it. She designs two sets of mazes that subjects can work through on a computer. One set has easy mazes and the other has difficult mazes. Subjects work until told to stop (after 6 minutes, but subjects do not know this). They are then asked to estimate how long they worked. The psychologist has 30 students available to serve as subjects.

(a) Describe an experiment using a completely randomized design to learn the effect of difficulty on estimated time.

(b) Describe a matched pairs experimental design using the same 30 subjects.

(c) Which design would be more likely to detect a difference in the effects of the treatments? Explain.

**R4.11* Deceiving subjects** Students sign up to be subjects in a psychology experiment. When they arrive, they are told that interviews are running late and are taken to a waiting room. The experimenters then stage a theft of a valuable object left in the waiting room. Some subjects are alone with the thief, and others are in pairs—these are the treatments being compared. Will the subject report the theft?

(a) The students had agreed to take part in an unspecified study, and the true nature of the experiment is explained to them afterward. Does this meet the requirement of informed consent? Explain.

(b) What two other ethical principles should be followed in this study?

*This is an important topic, but it is not required for the AP® Statistics exam.

# Chapter 4 AP® Statistics Practice Test

**Section I: Multiple Choice** *Select the best answer for each question.*

**T4.1** When we take a census, we attempt to collect data from

(a) a stratified random sample.
(b) every individual chosen in a simple random sample.
(c) every individual in the population.
(d) a voluntary response sample.
(e) a convenience sample.

**T4.2** You want to take a simple random sample (SRS) of 50 of the 816 students who live in a dormitory on

**R4.7 (a)** Units: potatoes. Explanatory: storage method and time from slicing until cooking. Response: ratings of color and flavor. Treatments: (1) fresh/immediately, (2) fresh/after an hour, (3) room temperature/immediately, (4) room temperature/after an hour, (5) refrigerator/immediately, (6) refrigerator/after an hour. **(b)** Using 300 identical slips of paper, write "1" on 50 of them, "2" on 50 of them, and so on. Put the papers in a hat and mix well. Then select a potato and randomly select a slip from the hat to determine which treatment that potato will receive. Repeat this process for the remaining 299 potatoes, making sure not to replace the slips of paper into the hat. **(c)** Use a randomized block design with regular potatoes in one block and sweet potatoes in the other block. Randomly assign the 6 treatments within each block as in part (b).

**R4.8 (a)** No. The 1000 students were not randomly selected from any larger population. **(b)** Yes. The students were randomly assigned to the three treatments.

**R4.9 (a)** By giving some patients a treatment that should have no effect at all, but appears like Saint-John's-wort, the researchers can account for the expectations of patients (the placebo effect) by comparing the results for the two groups. **(b)** To create two groups of subjects that are roughly equivalent at the beginning of the experiment. **(c)** The subjects should not know which treatment they are getting so that the researchers can account for the placebo effect. The researchers should be unaware of which subjects received which treatment so they cannot influence how the results are measured. **(d)** The difference in improvement between the two groups wasn't large enough to rule out the chance variation caused by the random assignment to treatments.

**R4.10 (a)** Randomly assign 15 students to easy mazes and the other 15 to hard mazes. Use 30 identical slips of paper and write the name of each subject on a slip. Mix the slips in a hat, select 15 of them at random, and assign these subjects to hard mazes. The remaining 15 will be assigned to easy mazes. After the experiment, compare the time estimates of the two groups. **(b)** Each student does the activity twice, once with each type of maze. Randomly determine which set of mazes is used first by flipping a coin for each subject. Heads: easy, then hard. Tails: hard, then easy. After the experiment, compare each student's easy maze and hard maze time estimate. **(c)** The matched pairs design would be more likely to detect a difference because it accounts for the variability between subjects.

**R4.11 (a)** This does not meet the requirements of informed consent because the subjects did not know the nature of the experiment before they agreed to participate. **(b)** All individual data should be kept confidential and the experiment should go before an institutional review board before being implemented.

## Teaching Tip

Brief explanations for the AP® Statistics Practice Test multiple-choice exercises are provided in the Solutions Manual or Teacher's E-book.

**TRM Full Solutions to Chapter 4 AP® Statistics Practice Test**

## Answers to Chapter 4 AP® Statistics Practice Test

**T4.1** c

T4.2 e

T4.3 d

T4.4 c

T4.5 b

T4.6 b

T4.7 d

T4.8 d



campus. You label the students 001 to 816 in alphabetical order. In the table of random digits, you read the entries

95592 94007 69769 33547 72450 16632 81194 14873

The first three students in your sample have labels

(a) 955, 929, 400.
(b) 400, 769, 769.
(c) 559, 294, 007.
(d) 929, 400, 769.
(e) 400, 769, 335.

**T4.3** A study of treatments for angina (pain due to low blood supply to the heart) compared bypass surgery, angioplasty, and use of drugs. The study looked at the medical records of thousands of angina patients whose doctors had chosen one of these treatments. It found that the average survival time of patients given drugs was the highest. What do you conclude?

(a) This study proves that drugs prolong life and should be the treatment of choice.
(b) We can conclude that drugs prolong life because the study was a comparative experiment.
(c) We can't conclude that drugs prolong life because the patients were volunteers.
(d) We can't conclude that drugs prolong life because this was an observational study.
(e) We can't conclude that drugs prolong life because no placebo was used.

**T4.4** Tonya wanted to estimate the average amount of time that students at her school spend on Facebook each day. She gets an alphabetical roster of students in the school from the registrar's office and numbers the students from 1 to 1137. Then Tonya uses a random number generator to pick 30 distinct labels from 1 to 1137. She surveys those 30 students about their Facebook use. Tonya's sample is a simple random sample because

(a) it was selected using a chance process.
(b) it gave every individual the same chance to be selected.
(c) it gave every possible sample of the same size an equal chance to be selected.
(d) it doesn't involve strata or clusters.
(e) it is guaranteed to be representative of the population.

**T4.5** Consider an experiment to investigate the effectiveness of different insecticides in controlling pests and their impact on the productivity of tomato plants. What is the best reason for randomly assigning treatment levels (spraying or not spraying) to the experimental units (farms)?

(a) Random assignment allows researchers to generalize conclusions about the effectiveness of the insecticides to all farms.
(b) Random assignment will tend to average out all other uncontrolled factors such as soil fertility so that they are not confounded with the treatment effects.
(c) Random assignment eliminates the effects of other variables, like soil fertility.
(d) Random assignment eliminates chance variation in the responses.
(e) Random assignment helps avoid bias due to the placebo effect.

**T4.6** The most important advantage of experiments over observational studies is that

(a) experiments are usually easier to carry out.
(b) experiments can give better evidence of causation.
(c) confounding cannot happen in experiments.
(d) an observational study cannot have a response variable.
(e) observational studies cannot use random samples.

**T4.7** A TV station wishes to obtain information on the TV viewing habits in its market area. The market area contains one city of population 170,000, another city of 70,000, and four towns of about 5000 inhabitants each. The station suspects that the viewing habits may be different in larger and smaller cities and in the rural areas. Which of the following sampling designs would give the type of information that the station requires?

(a) A cluster sample using the cities and towns as clusters
(b) A convenience sample from the market area
(c) A simple random sample from the market area
(d) A stratified sample from the cities and towns in the market area
(e) An online poll that invites all people from the cities and towns in the market area to participate

**T4.8** *Bias* in a sampling method is

(a) any difference between the sample result and the truth about the population.
(b) the difference between the sample result and the truth about the population due to using chance to select a sample.
(c) any difference between the sample result and the truth about the population due to practical difficulties such as contacting the subjects selected.
(d) any difference between the sample result and the truth about the population that tends to occur in the same direction whenever you use this sampling method.
(e) racism or sexism on the part of those who take the sample.

**T4.9** You wonder if TV ads are more effective when they are longer or repeated more often or both. So you design an experiment. You prepare 30-second and 60-second ads for a camera. Your subjects all watch the same TV program, but you assign them at random to four groups. One group sees the 30-second ad once during the program; another sees it three times; the third group sees the 60-second ad once; and the last group sees the 60-second ad three times. You ask all subjects how likely they are to buy the camera.

(a) This is a randomized block design, but not a matched pairs design.
(b) This is a matched pairs design.
(c) This is a completely randomized design with one explanatory variable (factor).
(d) This is a completely randomized design with two explanatory variables (factors).
(e) This is a completely randomized design with four explanatory variables (factors).

T4.10 A researcher wishes to compare the effects of two fertilizers on the yield of soybeans. She has 20 plots of land available for the experiment, and she decides to use a matched pairs design with 10 pairs of plots. To carry out the random assignment for this design, the researcher should

(a) use a table of random numbers to divide the 20 plots into 10 pairs and then, for each pair, flip a coin to assign the fertilizers to the 2 plots.
(b) subjectively divide the 20 plots into 10 pairs (making the plots within a pair as similar as possible) and then, for each pair, flip a coin to assign the fertilizers to the 2 plots.
(c) use a table of random numbers to divide the 20 plots into 10 pairs and then use the table of random numbers a second time to decide on the fertilizer to be applied to each member of the pair.
(d) flip a coin to divide the 20 plots into 10 pairs and then, for each pair, use a table of random numbers to assign the fertilizers to the 2 plots.
(e) use a table of random numbers to assign the two fertilizers to the 20 plots and then use the table of random numbers a second time to place the plots into 10 pairs.

T4.11 You want to know the opinions of American high school teachers on the issue of establishing a national proficiency test as a prerequisite for graduation from high school. You obtain a list of all high school teachers belonging to the National Education Association (the country's largest teachers' union) and mail a survey to a random sample of 2500 teachers. In all, 1347 of the teachers return the survey. Of those who responded, 32% say that they favor some kind of national proficiency test. Which of the following statements about this situation is true?

(a) Because random sampling was used, we can feel confident that the percent of all American high school teachers who would say they favor a national proficiency test is close to 32%.
(b) We cannot trust these results, because the survey was mailed. Only survey results from face-to-face interviews are considered valid.
(c) Because over half of those who were mailed the survey actually responded, we can feel fairly confident that the actual percent of all American high school teachers who would say they favor a national proficiency test is close to 32%.
(d) The results of this survey may be affected by nonresponse bias.
(e) The results of this survey cannot be trusted due to voluntary response bias.

**Section II: Free Response** *Show all your work. Indicate clearly the methods you use, because you will be graded on the correctness of your methods as well as on the accuracy and completeness of your results and explanations.*

T4.12 Elephants sometimes damage trees in Africa. It turns out that elephants dislike bees. They recognize beehives in areas where they are common and avoid them. Can this be used to keep elephants away from trees? Will elephant damage be less in trees with hives? Will even empty hives keep elephants away? Researchers want to design an experiment to answer these questions using 72 acacia trees.[64]

(a) Identify the experimental units, treatments, and the response variable.
(b) Describe how the researchers could carry out a completely randomized design for this experiment. Include a description of how the treatments should be assigned.

T4.13 A *New York Times* article on public opinion about steroid use in baseball discussed the results of a sample survey. The survey found that 34% of adults think that at least half of Major League Baseball (MLB) players "use steroids to enhance their athletic performance." Another 36% thought that about a quarter of MLB players use steroids; 8% had no opinion. Here is part of the *Times*'s statement on "How the Poll Was Conducted":

> The latest New York Times/CBS News Poll is based on telephone interviews conducted March 15 through March 18 with 1,067 adults throughout the United States.... The sample of telephone numbers called was randomly selected by a computer from a list of more than 42,000 active residential exchanges across the country. The exchanges were chosen to ensure that each region of the country was represented in proportion to its population. In each exchange, random digits were added to form a complete telephone number, thus permitting access to listed and unlisted numbers. In each household, one adult was designated by a random procedure to be the respondent for the survey.[65]

(a) Explain why the sampling method used in this survey was *not* a simple random sample.

**T4.9** d

**T4.10** b

**T4.11** d

**T4.12** **(a)** Experimental units: acacia trees. Treatments: placing either active beehives, empty beehives, or nothing in the trees. Response: damage to the trees caused by elephants. **(b)** Assign the trees numbers from 01 to 72 and use a random number table to pick 24 different two-digit numbers in this range. Those trees will get the active beehives. The trees corresponding to the next 24 different two-digit numbers from 01 to 72 will get the empty beehives, and the remaining 24 trees will remain empty. Compare the damage caused by elephants to the three groups of trees.

**T4.13** **(a)** Not all possible samples of size 1067 were possible. For example, using their method they could not have had a sample which included respondents who were all from the east coast.

**T4.13 (b)** If the household members who typically answer the phone have a different opinion than those who don't typically answer the phone, their opinions will be overrepresented. **(c)** If people without phones or with cell phones only have different opinions than the group of people with residential lines, these opinions will be underrepresented.

**T4.14 (a)** Each of the 11 individuals will be a block in this matched pairs design, with the order of treatments randomly assigned. This was to help account for the variability in tapping speed caused by the differences in subjects. **(b)** If all the subjects got caffeine the second time, the researchers wouldn't know if the increase was due to the caffeine or due to practice with the task. **(c)** Yes. Neither the subjects nor the people who come into contact with them during the experiment (including those who record the number of taps) need to know the order in which the caffeine or placebo was administered.

**TRM Full Solutions to Cumulative AP® Practice Test 1**

## Answers to Cumulative AP® Practice Test 1

**AP1.1** d

**AP1.2** e

**AP1.3** b

**AP1.4** c

**AP1.5** a



(b) Why was one adult chosen at random in each household to respond to the survey?

(c) Explain how undercoverage could lead to bias in this sample survey.

T4.14 Many people start their day with a jolt of caffeine from coffee or a soft drink. Most experts agree that people who take in large amounts of caffeine each day may suffer from physical withdrawal symptoms if they stop ingesting their usual amounts of caffeine. Researchers recruited 11 volunteers who were caffeine dependent and who were willing to take part in a caffeine withdrawal experiment. The experiment was conducted on two 2-day periods that occurred one week apart. During one of the 2-day periods, each subject was given a capsule containing the amount of caffeine normally ingested by that subject in one day. During the other study period, the subjects were given placebos. The order in which each subject received the two types of capsules was randomized. The subjects' diets were restricted during each of the study periods. At the end of each 2-day study period, subjects were evaluated using a tapping task in which they were instructed to press a button 200 times as fast as they could.[66]

(a) How and why was blocking used in the design of this experiment?

(b) Why did researchers randomize the order in which subjects received the two treatments?

(c) Could this experiment have been carried out in a double-blind manner? Explain.

# Cumulative AP® Practice Test 1

**Section I: Multiple Choice** *Choose the best answer for Questions AP1.1 to AP1.14.*

AP1.1 You look at real estate ads for houses in Sarasota, Florida. Many houses range from $200,000 to $400,000 in price. The few houses on the water, however, have prices up to $15 million. Which of the following statements best describes the distribution of home prices in Sarasota?

(a) The distribution is most likely skewed to the left, and the mean is greater than the median.

(b) The distribution is most likely skewed to the left, and the mean is less than the median.

(c) The distribution is roughly symmetric with a few high outliers, and the mean is approximately equal to the median.

(d) The distribution is most likely skewed to the right, and the mean is greater than the median.

(e) The distribution is most likely skewed to the right, and the mean is less than the median.

AP1.2 A child is 40 inches tall, which places her at the 90th percentile of all children of similar age. The heights for children of this age form an approximately Normal distribution with a mean of 38 inches. Based on this information, what is the standard deviation of the heights of all children of this age?

(a) 0.20 inches (b) 0.31 inches (c) 0.65 inches (d) 1.21 inches (e) 1.56 inches

AP1.3 A large set of test scores has mean 60 and standard deviation 18. If each score is doubled, and then 5 is subtracted from the result, the mean and standard deviation of the new scores are

(a) mean 115; std. dev. 31.
(b) mean 115; std. dev. 36.
(c) mean 120; std. dev 6.
(d) mean 120; std. dev. 31.
(e) mean 120; std. dev. 36.

AP1.4 For a certain experiment, the available experimental units are eight rats, of which four are female (F1, F2, F3, F4) and four are male (M1, M2, M3, M4). There are to be four treatment groups, A, B, C, and D. If a randomized block design is used, with the experimental units blocked by gender, which of the following assignments of treatments is impossible?

(a) A → (F1, M1), B → (F2, M2), C → (F3, M3), D → (F4, M4)

(b) A → (F1, M2), B → (F2, M3), C → (F3, M4), D → (F4, M1)

(c) A → (F1, M2), B → (F3, F2), C → (F4, M1), D → (M3, M4)

(d) A → (F4, M1), B → (F2, M3), C → (F3, M2), D → (F1, M4)

(e) A → (F4, M1), B → (F1, M4), C → (F3, M2), D → (F2, M3)

AP1.5 For a biology project, you measure the weight in grams (g) and the tail length in millimeters (mm) of a group of mice. The equation of the least-squares line for predicting tail length from weight is

$$\text{predicted tail length} = 20 + 3 \times \text{weight}$$

Which of the following is *not* correct?

(a) The slope is 3, which indicates that a mouse's weight should increase by about 3 grams for each additional millimeter of tail length.

(b) The predicted tail length of a mouse that weighs 38 grams is 134 millimeters.

(c) By looking at the equation of the least-squares line, you can see that the correlation between weight and tail length is positive.

(d) If you had measured the tail length in centimeters instead of millimeters, the slope of the regression line would have been 3/10 = 0.3.

(e) One mouse weighed 29 grams and had a tail length of 100 millimeters. The residual for this mouse is −7.

**AP1.6** The figure below shows a Normal density curve. Which of the following gives the best estimates for the mean and standard deviation of this Normal distribution?

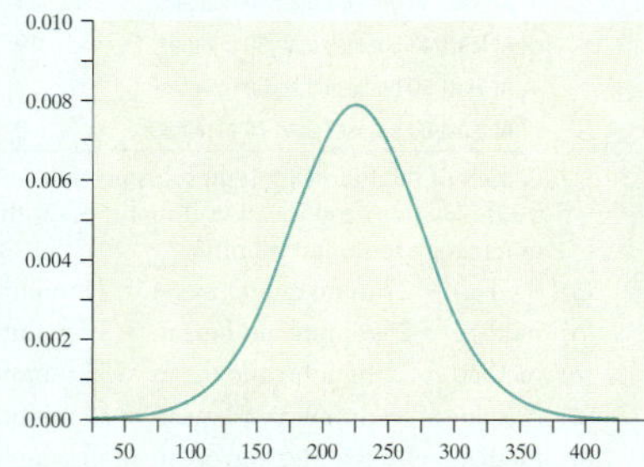


(a) $\mu = 200, \sigma = 50$
(b) $\mu = 200, \sigma = 25$
(c) $\mu = 225, \sigma = 50$
(d) $\mu = 225, \sigma = 25$
(e) $\mu = 225, \sigma = 275$

**AP1.7** The owner of a chain of supermarkets notices that there is a positive correlation between the sales of beer and the sales of ice cream over the course of the previous year. During seasons when sales of beer were above average, sales of ice cream also tended to be above average. Likewise, during seasons when sales of beer were below average, sales of ice cream also tended to be below average. Which of the following would be a valid conclusion from these facts?

(a) Sales records must be in error. There should be no association between beer and ice cream sales.

(b) Evidently, for a significant proportion of customers of these supermarkets, drinking beer causes a desire for ice cream or eating ice cream causes a thirst for beer.

(c) A scatterplot of monthly ice cream sales versus monthly beer sales would show that a straight line describes the pattern in the plot, but it would have to be a horizontal line.

(d) There is a clear negative association between beer sales and ice cream sales.

(e) The positive correlation is most likely a result of the variable temperature; that is, as temperatures increase, so do both beer sales and ice cream sales.

**AP1.8** Here are the IQ scores of 10 randomly chosen fifth-grade students:

145 139 126 122 125 130 96 110 118 118

Which of the following statements about this data set is *not* true?

(a) The student with an IQ of 96 is considered an outlier by the $1.5 \times IQR$ rule.

(b) The five-number summary of the 10 IQ scores is 96, 118, 123.5, 130, 145.

(c) If the value 96 were removed from the data set, the mean of the remaining 9 IQ scores would be greater than the mean of all 10 IQ scores.

(d) If the value 96 were removed from the data set, the standard deviation of the remaining 9 IQ scores would be less than the standard deviation of all 10 IQ scores.

(e) If the value 96 were removed from the data set, the *IQR* of the remaining 9 IQ scores would be less than the *IQR* of all 10 IQ scores.

**AP1.9** Before he goes to bed each night, Mr. Kleen pours dishwasher powder into his dishwasher and turns it on. Each morning, Mrs. Kleen weighs the box of dishwasher powder. From an examination of the data, she concludes that Mr. Kleen dispenses a rather consistent amount of powder each night. Which of the following statements is true?

I. There is a high positive correlation between the number of days that have passed since the box of dishwasher powder was opened and the amount of powder left in the box.

II. A scatterplot with days since purchase as the explanatory variable and amount of dishwasher powder used as the response variable would display a strong positive association.

III. The correlation between the amount of powder left in the box and the amount of powder used should be −1.

(a) I only
(b) II only
(c) III only
(d) II and III only
(e) I, II, and III

**AP1.10** The General Social Survey (GSS), conducted by the National Opinion Research Center at the University of Chicago, is a major source of data on social attitudes in the United States. Once each year, 1500 adults are interviewed in their homes all across the country. The subjects are asked their opinions about sex and marriage, attitudes toward women, welfare, foreign policy, and many other issues. The GSS begins by selecting a sample of counties from the 3000 counties in the country. The counties are divided into urban, rural, and suburban; a separate sample is chosen at random from each group. This is a

(a) simple random sample.
(b) systematic random sample.
(c) cluster sample.
(d) stratified random sample.
(e) voluntary response sample.

**AP1.6** c
**AP1.7** e
**AP1.8** e
**AP1.9** d
**AP1.10** d

AP1.11 d

AP1.12 b

AP1.13 b

AP1.14 a



**AP1.11** You are planning an experiment to determine the effect of the brand of gasoline and the weight of a car on gas mileage measured in miles per gallon. You will use a single test car, adding weights so that its total weight is 3000, 3500, or 4000 pounds. The car will drive on a test track at each weight using each of Amoco, Marathon, and Speedway gasoline. Which is the best way to organize the study?

(a) Start with 3000 pounds and Amoco and run the car on the test track. Then do 3500 and 4000 pounds. Change to Marathon and go through the three weights in order. Then change to Speedway and do the three weights in order once more.

(b) Start with 3000 pounds and Amoco and run the car on the test track. Then change to Marathon and then to Speedway without changing the weight. Then add weights to get 3500 pounds and go through the three gasolines in the same order. Then change to 4000 pounds and do the three gasolines in order again.

(c) Choose a gasoline at random, and run the car with this gasoline at 3000, 3500, and 4000 pounds in order. Choose one of the two remaining gasolines at random and again run the car at 3000, then 3500, then 4000 pounds. Do the same with the last gasoline.

(d) There are nine combinations of weight and gasoline. Run the car several times using each of these combinations. Make all these runs in random order.

(e) Randomly select an amount of weight and a brand of gasoline, and run the car on the test track. Repeat this process a total of 30 times.

**AP1.12** A linear regression was performed using the five following data points: A(2, 22), B(10, 4), C(6, 14), D(14, 2), E(18, −4). The residual for which of the five points has the largest absolute value?

(a) A (b) B (c) C (d) D (e) E

**AP1.13** The frequency table below summarizes the times in the last month that patients at the emergency room of a small-city hospital waited to receive medical attention.

| Waiting time | Frequency |
|---|---|
| Less than 10 minutes | 5 |
| At least 10 but less than 20 minutes | 24 |
| At least 20 but less than 30 minutes | 45 |
| At least 30 but less than 40 minutes | 38 |
| At least 40 but less than 50 minutes | 19 |
| At least 50 but less than 60 minutes | 7 |
| At least 60 but less than 70 minutes | 2 |

Which of the following represents possible values for the median and mean waiting times for the emergency room last month?

(a) median = 27 minutes and mean = 24 minutes

(b) median = 28 minutes and mean = 30 minutes

(c) median = 31 minutes and mean = 35 minutes

(d) median = 35 minutes and mean = 39 minutes

(e) median = 45 minutes and mean = 46 minutes

**AP1.14** Boxplots of two data sets are shown.

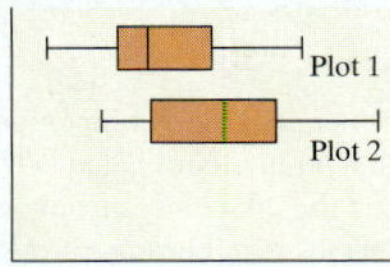


Based on the boxplots, which statement below is true?

(a) The range of both plots is about the same.

(b) The means of both plots are approximately equal.

(c) Plot 2 contains more data points than Plot 1.

(d) The medians are approximately equal.

(e) Plot 1 is more symmetric than Plot 2.

**Section II: Free Response** Show all your work. Indicate clearly the methods you use, because you will be graded on the correctness of your methods as well as on the accuracy and completeness of your results and explanations.

**AP1.15** The manufacturer of exercise machines for fitness centers has designed two new elliptical machines that are meant to increase cardiovascular fitness. The two machines are being tested on 30 volunteers at a fitness center near the company's headquarters. The volunteers are randomly assigned to one of the machines and use it daily for two months. A measure of cardiovascular fitness is administered at the start of the experiment and again at the end. The following table contains the differences in the two scores (After – Before) for the two machines. Note that higher scores indicate larger gains in fitness.

| Machine A | | Machine B |
|---|---|---|
| | 0 | 2 |
| 5 4 | 1 | 0 |
| 8 7 6 3 2 0 | 2 | 1 5 9 |
| 9 7 4 1 1 | 3 | 2 4 8 9 |
| 6 1 | 4 | 2 5 7 |
| | 5 | 3 5 9 |

(a) Write a few sentences comparing the distributions of cardiovascular fitness gains from the two elliptical machines.

(b) Which machine should be chosen if the company wants to advertise it as achieving the highest overall gain in cardiovascular fitness? Explain your reasoning.

(c) Which machine should be chosen if the company wants to advertise it as achieving the most consistent gain in cardiovascular fitness? Explain your reasoning.

(d) Give one reason why the advertising claims of the company (the scope of inference) for this experiment would be limited. Explain how the company could broaden that scope of inference.

**AP1.16** Those who advocate for monetary incentives in a work environment claim that this type of incentive has the greatest appeal because it allows the winners to do what they want with their winnings. Those in favor of tangible incentives argue that money lacks the emotional appeal of, say, a weekend for two at a romantic country inn or elegant hotel or a weeklong trip to Europe.

A few years ago a national tire company, in an effort to improve sales of a new line of tires, decided to test which method—offering cash incentives or offering non-cash prizes such as vacations—was more successful in increasing sales. The company had 60 retail sales districts of various sizes across the country and data on the previous sales volume for each district.

(a) Describe a completely randomized design using the 60 retail sales districts that would help answer this question.

(b) Explain how you would use the table of random digits below to do the randomization that your design requires. Then use your method to make the first three assignments. Show your work clearly.

07511 88915 41267 16853 84569 79367 32337 03316
81486 69487 60513 09297 00412 71238 27649 39950

(c) One of the company's officers suggested that it would be better to use a matched pairs design instead of a completely randomized design. Explain how you would change your design to accomplish this.

**AP1.17** In retail stores, there is a lot of competition for shelf space. There are national brands for most products, and many stores carry their own line of in-house brands, too. Since shelf space is not infinite, the question is how many linear feet to allocate to each product and which shelf (top, bottom, or somewhere in the middle) to put it on. The middle shelf is the most popular and lucrative, because many shoppers, if undecided, will simply pick the product that is at eye level.

A local store that sells many upscale goods is trying to determine how much shelf space to allocate to its own brand of men's personal-grooming products. The middle shelf space is randomly varied between 3 and 6 linear feet over the next 12 weeks, and weekly sales revenue (in dollars) from the store's brand of personal-grooming products for men is recorded. Below is some computer output from the study, along with a scatterplot.

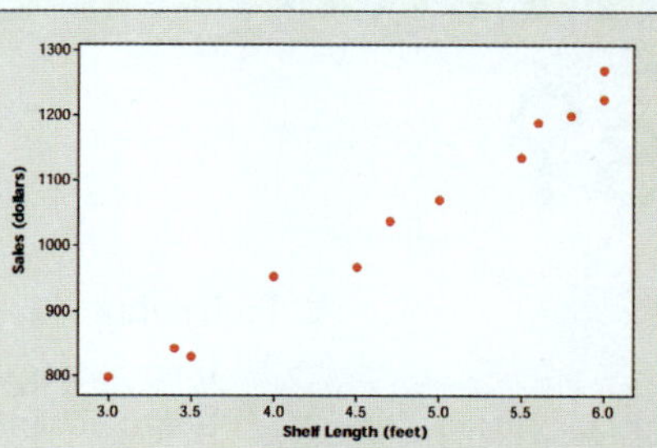


```
Predictor       Coef    SE Coef      T       P
Constant      317.94     31.32   10.15   0.000
Shelf length  152.680     6.445   23.69   0.000
S = 22.9212  R-Sq = 98.2%  R-Sq(adj) = 98.1%
```

(a) Describe the relationship between shelf length and sales.

(b) Write the equation of the least-squares regression line. Be sure to define any variables you use.

(c) If the store manager were to decide to allocate 5 linear feet of shelf space to the store's brand of men's grooming products, what is the best estimate of the weekly sales revenue?

(d) Interpret the value of $s$.

(e) Identify and interpret the coefficient of determination.

(f) The store manager questions the intercept of the regression line: "Am I supposed to believe that this analysis tells me that I can sell these products with no shelf space?" How do you answer her?

**AP1.15** **(a)** The distribution of gains for subjects using Machine A is roughly symmetric, while the distribution of gains for subjects using Machine B is skewed to the left. The center of the distribution is greater for Machine B than for Machine A. The distribution for Machine B is more variable than the distribution for Machine A. **(b)** B. The typical gain using Machine B is greater than the typical gain using Machine A. **(c)** A. The spread for Machine A is less than the spread for Machine B. **(d)** Volunteers from one fitness center were used and these volunteers may be different in some way from the general population of those who are interested in cardiovascular fitness. To broaden their scope of inference, they should randomly select people from the population they would like to draw an inference about.

**AP1.16** **(a)** Number the 60 retail sales districts with a two-digit number from 01 to 60. Using a table of random digits, read two-digit numbers until 30 unique numbers from 01 to 60 have been selected. The corresponding 30 districts are assigned to the monetary incentives group and the remaining 30 to the tangible incentives group. After a specified period of time, record the change in sales for each district and compare the two groups. **(b)** The districts labeled 07, 51, and 18 are the first three to be assigned to the monetary incentives group. **(c)** Pair the two districts with the largest sales, the next two largest, down to the two smallest districts. For each pair, pick one of the districts and flip a coin. If the flip is "heads," this district is assigned to the monetary incentives group. If it is "tails," this district is assigned to the tangible incentives group. The other district in the pair is assigned to the other group. After a specified period of time, record the change in sales for each district and compare within each pair.

**AP1.17** **(a)** There is a very strong, positive, linear association between sales and shelf length. **(b)** $\hat{y} = 317.94 + 152.68x$, where $y$ = weekly sales (in dollars) and $x$ = shelf length (in feet). **(c)** \$1081 **(d)** When using the least-squares regression line with $x$ = shelf space to predict $y$ = sales, we will typically be off by about $s$ = \$23. **(e)** $r^2 = 0.982$. About 98.2% of the variation in weekly sales revenue can be accounted for by the linear model relating sales to shelf length. **(f)** It would be inappropriate to interpret the intercept, because the data represent sales based on shelf lengths of 3 to 6 feet and 0 feet falls substantially outside that domain.

Chapter

5

# Probability: What Are the Chances?

## PD Overview

This chapter begins a three-chapter section about probability. It also begins the transition from the descriptive methods we learned about in Chapters 1–3 to the inferential methods we will learn about in Chapters 8–12.

Some of the concepts in this chapter will be familiar to students, but there will be new and very important ideas as well. In Section 5.1, we introduce simulation as a very useful way to estimate probabilities. In Section 5.2, we introduce the idea of a probability model and some basic probability rules. In addition, we revisit two-way tables as a good way to investigate the relationship between two events. In Section 5.3, we introduce two concepts that we will be using throughout the rest of the book: conditional probability and independence.

You may notice that many traditional probability topics such as counting rules, permutations, and combinations are not included in this chapter (or any other chapter). The reasons are twofold: (1) these topics are not on the AP® Statistics Topic Outline, and (2) this is a course about statistics, so we spend time addressing only the probability topics that will help us do statistics.

### Relationship to the AP® Statistics Course Description

Chapter 5 covers the following content as described in the Topic Outline from the AP® Statistics Course Description. Topics marked with a superscript are covered in other chapters as indicated.

I. Exploring data: Describing patterns and departures from patterns

- E. Exploring categorical data
  1. Frequency tables and bar charts[ch1]
  2. Marginal and joint frequencies for two-way tables
  3. Conditional relative frequencies and association
  4. Comparing distributions using bar charts[ch1]

III. Anticipating patterns: Exploring random phenomena using probability and simulation

- A. Probability
  1. Interpreting probability, including long-run relative frequency interpretation
  2. "Law of Large Numbers" concept
  3. Addition rule, multiplication rule, conditional probability, and independence
  4. Discrete random variables and their probability distributions, including binomial and geometric[ch6]
  5. Simulation of random behavior and probability distributions
  6. Mean (expected value) and standard deviation of a random variable and linear transformation of a random variable[ch6]

## The Big Ideas

One of the challenges in teaching the AP® Statistics course is keeping students focused on the big picture and not just the details of each section. We outline the big ideas for the chapter here.

### Chapter 5 Introduction

Chance is all around us, whether we realize it or not. We introduce this chapter with several quick examples of how chance shows up in our daily lives. The mathematics of chance is called *probability*, and probability is the basis for the confidence intervals and significance tests we will learn about in Chapters 8–12. This makes probability a vital bridge between designing studies and drawing conclusions from them.

The Activity in this section highlights several important ideas, including simulation as a powerful tool to estimate probabilities. It also continues to foreshadow the logic of inference, which we will encounter in later chapters.

### Section 5.1 Randomness, Probability, and Simulation

In this section, we introduce the idea of probability, dispel some myths about randomness, and show how simulations can be used to estimate probabilities and make decisions.

The probability of an outcome in a chance process describes the proportion of times the outcome would occur in a very long series of repetitions. For example, when we say that the probability of getting heads when flipping a coin is 0.5, we do not mean that exactly 1 out of every 2 flips will result in heads. Instead, a probability of 0.5 means that if we were to flip the coin millions of times, then the proportion of heads should be very close to 0.5. The fact that the proportion of heads will get very close to the true probability of heads after millions of flips is called the *law of large numbers*.

Many students (and adults!) have misconceptions about randomness, including a belief in short-term regularity. That is, streaks of the same outcome are often viewed with suspicion even though they occur quite often just by chance. Also, many people mistakenly believe in the "law of averages," which suggests that outcomes will "even out" in a short number of trials. For example, upon seeing three heads in a row, many people think that the next flip is more likely to be tails because there should be an equal number of heads and tails.

Simulations are powerful tools that we can use to imitate chance processes and estimate probabilities. Simulation is especially useful when calculating a probability theoretically is difficult. It is also a very versatile tool that we will continue to use throughout the book to illustrate many different statistical concepts. To perform a simulation, we will use the four-step process we introduced in Chapter 1 and revisited in Chapter 3. We state the question of interest, plan how to imitate a chance process, do many repetitions, and make a conclusion based on the results.

## Section 5.2 Probability Rules

In this section, we introduce probability models. Just as some distributions of data can be described by a Normal model and some relationships between quantitative variables can be described by a linear model, chance behavior can be described by a probability model. Probability models have two parts: a list of possible outcomes and the probability of each outcome.

We also introduce some basic rules of probability, including that probabilities must be between 0 and 1 and that the probabilities of all the outcomes in the sample space must add to 1. Other basic rules include the complement rule and the addition rule for mutually exclusive events.

A helpful way to organize information involving two events is a two-way table. We first saw two-way tables in Section 1.1 when we learned about relationships between categorical variables. Working with two-way tables leads to the general addition rule for two events, which works whether or not the events are mutually exclusive. We also present another useful way to organize information involving two events: a Venn diagram.

## Section 5.3 Conditional Probability and Independence

In this section, we introduce two very important ideas: conditional probability and independence. These ideas will appear in every remaining chapter of the book, so make sure your students develop a solid understanding of them. A conditional probability describes the probability that an event occurs given that another event has already occurred. If knowing that event *A* occurs doesn't change the probability that event *B* occurs, then events *A* and *B* are independent.

There is a connection between conditional probability and the conditional distributions we learned about in Chapter 1. In Section 1.1, we looked at the distribution of one categorical variable for a specific value of another categorical variable. In this chapter, we look at the probability of a specific event occurring given that another event has already occurred.

Two-way tables and tree diagrams are two very useful ways to organize the information provided in a conditional probability problem. Two-way tables are best when the problem describes the number or proportion of cases with certain characteristics. Tree diagrams are best when the problem provides conditional probabilities of different events or describes events that are sequential.

In this section, we also introduce three formulas: the general multiplication rule, the multiplication rule for independent events, and the conditional probability formula. As always, it is very important for students to recognize which formula to use in a particular situation and to be able to justify their choice.

Finally, an important type of conditional probability question involves going "backward" in a tree diagram. For example, we may know the probability that a person who has a disease should receive a positive result when given a test for the disease, but what is the probability that a person who receives a positive result actually has the disease? These two probabilities are not the same.

# Chapter 5: Resources

## TRM Additional *TPS* Teacher's Resource Materials

The following resources can be found by clicking on the links in the new Teacher's e-Book (TE-book), opening the Teacher's Resource CD (TRCD), or logging onto the book's companion Web site www.whfreeman.com/tps5e (teacher log-in required).

- Printable Chapter 5 Learning Objective Grid (page 289)
- Additional Simulation Questions (page 299)
  - Feel the Power
  - The Duck Hunters
  - Airline Overbooking
- FRAPPY! materials: Question, Model Solution, Scoring Rubric, Sample Responses with commentary (page 338)
- Quizzes: 3 versions for each section; versions A and B are roughly parallel
  - Section 5.1 (page 299)
  - Section 5.2 (page 314)
  - Section 5.3 (page 333)
- Tests: 4 versions; versions A and B are roughly parallel, version D is cumulative (page 338)

## Free-Response Questions from Previous AP® Exams

Questions can be found on the AP® Central Web site: http://apcentral.collegeboard.com/apc/members/exam/exam_questions/8357.html.

Students should be able to answer all the free-response questions listed below with material learned in this chapter. Questions that contain content from this chapter but also require content from later chapters are listed in the last chapter required to complete the entire question. This list will be updated after each AP® exam and will be posted on the book's Web site: www.whfreeman.com/tps5e

Questions marked with an asterisk are from exams with released multiple-choice questions. You may want to save these questions until the end of the year so you can give your students a complete released exam for practice.

| Year | # | Content |
|---|---|---|
| 2011 | 2 | • Conditional probability from a two-way table<br>• Independence of two events<br>• Segmented bar charts and independence |
| 2009B | 2 | • Conditional probability<br>• Multiplication rule |
| 2003B | 2 | • Two-way tables<br>• Conditional probability<br>• Independence |
| 2001 | 3 | • Simulation |
| 1997 | 3* | • Conditional probability |

## Applets

- The *Probability* applet simulates the flip of a coin and keeps track of the proportion of heads. It also allows you to change the probability of heads. www.whfreeman.com/tps5e
- The *Random Babies* applet simulates delivering four babies at random to four houses and recording the number of correct deliveries. The applet displays a histogram of the results and shows how the probability of each outcome changes as more trials are run. www.rossmanchance.com/applets

## TI-Nspire Activities

These Activities are designed to illustrate important statistical concepts using the TI-Nspire. The Activities can be found at www.whfreeman.com/tps5e.

- *Short-Run and Long-Run Behavior*: In this Activity, students will use simulation to investigate how the proportion of heads varies in a small number of tosses and in a large number of tosses.

Chapter

5

# Chapter 5: Pacing Guide, Learning Objectives, and Suggested Assignments

This pacing guide is based on a schedule with 110, 50-minute sessions before the AP® exam. The 110 days include 4 days for midterms and 10 days for AP® exam review. If you have a different number of sessions before the AP® exam, you can modify the pacing guide to suit your needs. If you have additional time, consider incorporating quizzes, released AP® free-response questions, or additional activities. See the Resources section above for suggestions.

The suggested homework assignments list odd-numbered exercises, whenever possible, so students can check their answers against the back-of-the-book answers. If you would rather students not have access to the answers while doing homework, adding 1 to the exercise numbers usually will do the trick, because the homework exercises typically are paired. For example, Exercises 1 and 2 will generally cover the same topics, but in different contexts. You may also choose to include the Recycle and Review questions at the end of each section, which review topics from previous sections or chapters. These questions are denoted with a symbol.

| Day | Topics | Learning Objectives: Students will be able to. . . | Suggested Assignment |
|---|---|---|---|
| 1 | 5.1 The Idea of Probability, Myths about Randomness | • Interpret probability as a long-run relative frequency. | 1, 3, 7, 9, 11 |
| 2 | 5.1 Simulation | • Use simulation to model chance behavior. | 15, 17, 19, 23, 25 |
| 3 | 5.2 Probability Models, Basic Rules of Probability | • Determine a probability model for a chance process.<br>• Use basic probability rules, including the complement rule and the addition rule for mutually exclusive events. | 27, 31, 32, 39, 41, 43, 45, 47 |
| 4 | 5.2 Two-Way Tables, Probability, and the General Addition Rule, Venn Diagrams and Probability | • Use a two-way table or Venn diagram to model a chance process and calculate probabilities involving two events.<br>• Use the general addition rule to calculate probabilities. | 29, 33–36, 49, 51, 53, 55 |
| 5 | 5.3 What Is Conditional Probability?, The General Multiplication Rule and Tree Diagrams | • Calculate and interpret conditional probabilities.<br>• Use the general multiplication rule to calculate probabilities.<br>• Use tree diagrams to model a chance process and calculate probabilities involving two or more events. | 57–60, 63, 65, 67, 71, 73, 77, 79 |
| 6 | 5.3 Conditional Probability and Independence: A Special Multiplication Rule | • Determine whether two events are independent.<br>• When appropriate, use the multiplication rule for independent events to compute probabilities. | 81, 83, 85, 89, 91, 93, 95, 97–99 |
| 7 | Chapter 5 Review/FRAPPY! | | Chapter 5 Review Exercises |
| 8 | Chapter 5 Test | | |

**PD Chapter 5 Overview**

Watch the video overview of Chapter 5 (for teachers) by clicking on the link in the TE-book, opening the TRCD, or logging onto the book's companion Web site www.whfreeman.com/tps5e (teacher log-in required).

# Chapter 5

# Probability: What Are the Chances?

case study

## Calculated Risks

Many high schools now have drug-testing programs for athletes. The main goal of these programs is to reduce the use of banned substances by students who play sports. It is not practical to test every athlete for drug use regularly. Instead, school administrators give drug tests to randomly selected student athletes at unannounced times during the school year. Students who test positive face serious consequences, including letters to their parents, required counseling, and suspension from athletic participation.

Drug tests aren't perfect. Sometimes the tests say that athletes took a banned substance when they did not. This is known as a *false positive*. Other times, drug tests say that athletes are "clean" when they did take a banned substance. This is called a *false negative*.

Suppose that 16% of the high school athletes in a large school district have taken a banned substance. The drug test used by this district has a false positive rate of 5% and a false negative rate of 10%. If a randomly chosen athlete tests positive, what's the chance that the student actually took a banned substance?

By the end of the chapter, you will be ready to answer questions that involve drug and disease testing, false positives, and false negatives.



### Teaching Tip

Make sure to read this Case Study with your students and discuss what it means to have a false positive and a false negative. This will help your students better understand Type I and Type II errors when we get to Chapter 9.

### TECHNOLOGY TIP

There is a great application called Prob Sim for the TI-83+ and TI-84. You can use it to toss coins, roll dice, pick marbles, spin a spinner, draw cards, or generate random numbers, with or without replacement. If it isn't already loaded onto the students' calculators, they can link it from another calculator or download it at http://education.ti.com/educationportal/sites/US/productDetail/us_prob_sim_83_84.html.

### Activity Overview

You can watch an overview of this Activity (for teachers) by clicking on the link in the TE-book, opening the TRCD, or logging onto the book's companion Web site www.whfreeman.com/tps5e (teacher log-in required).

### Teaching Tip

It is a good idea to introduce your students to the logic of significance testing early and often. Although The "1 in 6 Wins" Game Activity doesn't mention anything about hypotheses, test statistics, or *P*-values, it is fundamentally a significance test. The null hypothesis is that the 1-in-6 claim is correct, while the alternative is that more than 1 in 6 win. The "test statistic" is 3, the observed number of wins in 7 attempts. The *P*-value is the probability that a group of 7 friends will have 3 or more winners, just by chance, assuming that the company's 1-in-6 claim is correct. Here is a dotplot showing the results of 40 trials of this simulation. According to these results, getting 3 winners isn't very likely, but isn't unusual enough to conclude that the company is lying.

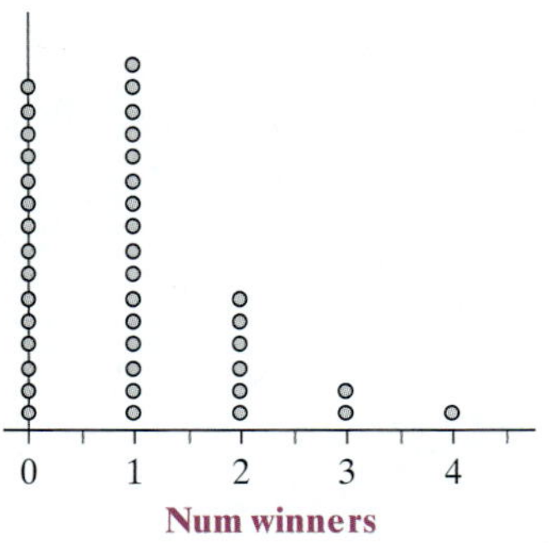




# Introduction

Chance is all around us. You and your friend play rock-paper-scissors to determine who gets the last slice of pizza. A coin toss decides which team gets to receive the ball first in a football game. Many adults regularly play the lottery, hoping to win a big jackpot with a few lucky numbers. Others head to casinos or racetracks, hoping that some combination of luck and skill will pay off. People young and old play games of chance involving cards or dice or spinners. The traits that children inherit—gender, hair and eye color, blood type, handedness, dimples, whether they can roll their tongues—are determined by the chance involved in which genes get passed along by their parents.

A roll of a die, a simple random sample, and even genetic inheritance represent chance behavior that we can understand and work with. We can roll the die again and again and again. The outcomes are governed by chance, but in many repetitions a pattern emerges. We use mathematics to understand the regular patterns of chance behavior when we repeat the same chance process again and again.

The mathematics of chance is called *probability*. Probability is the topic of this chapter. Here is an Activity that gives you some idea of what lies ahead.

## ACTIVITY | The "1 in 6 wins" game

**MATERIALS:** One six-sided die for each student

As a special promotion for its 20-ounce bottles of soda, a soft drink company printed a message on the inside of each bottle cap. Some of the caps said, "Please try again!" while others said, "You're a winner!" The company advertised the promotion with the slogan "1 in 6 wins a prize." The prize is a free 20-ounce bottle of soda, which comes out of the store owner's profits.

Seven friends each buy one 20-ounce bottle at a local convenience store. The store clerk is surprised when three of them win a prize. The store owner is concerned about losing money from giving away too many free sodas. She wonders if this group of friends is just lucky or if the company's 1-in-6 claim is inaccurate. In this Activity, you and your classmates will perform a *simulation* to help answer this question.

For now, let's assume that the company is telling the truth, and that every 20-ounce bottle of soda it fills has a 1-in-6 chance of getting a cap that says, "You're a winner!" We can model the status of an individual bottle with a six-sided die: let 1 through 5 represent "Please try again!" and 6 represent "You're a winner!"

1. Roll your die seven times to imitate the process of the seven friends buying their sodas. How many of them won a prize?

2. Your teacher will draw and label axes for a class dotplot. Plot the number of prize winners you got in Step 1 on the graph.

### Teaching Tip

The probability of getting 3 or more winners can be found theoretically using the binomial distribution (it is approximately 0.096). When students learn about the binomial distribution in the next chapter, it would be a good idea to revisit this problem to illustrate the differences between using probability theory and simulation to estimate a probability.

3. Repeat Steps 1 and 2 if needed to get a total of at least 40 repetitions of the simulation for your class.

4. Discuss the results with your classmates. What percent of the time did the friends come away with three or more prizes, just by chance? Does it seem plausible that the company is telling the truth, but that the seven friends just got lucky? Explain.

As the Activity shows, simulation is a powerful method for modeling chance behavior. Section 5.1 begins by examining the idea of probability and then illustrates how simulation can be used to estimate probabilities. In Sections 5.2 and 5.3, we develop the basic rules of probability. Along the way, we introduce some helpful tools for displaying possible outcomes from a chance process: two-way tables, Venn diagrams, and tree diagrams.

Probability calculations are the basis for inference. When we produce data by random sampling or randomized comparative experiments, the laws of probability answer the question "What would happen if we repeated the random sampling or random assignment process many times?" Many of the examples, exercises, and activities in this chapter focus on the connection between probability and inference.

# 5.1 Randomness, Probability, and Simulation

**WHAT YOU WILL LEARN** By the end of the section, you should be able to:

- Interpret probability as a long-run relative frequency.
- Use simulation to model chance behavior.

Toss a coin 10 times. How likely are you to get a run of 3 or more consecutive heads or tails? An airline knows that a certain percent of customers who purchased tickets will not show up for a flight. If the airline overbooks a particular flight, what are the chances that they'll have enough seats for the passengers who show up? A couple plans to have children until they have at least one child of each gender. How many children should they expect to have? To answer these questions, you need a better understanding of how chance behavior operates.

## The Idea of Probability

In football, a coin toss helps determine which team gets the ball first. Why do the rules of football require a coin toss? Because tossing a coin seems a "fair" way to decide. That's one reason

**ALTERNATE ACTIVITY**

*Whose book is this?*

Suppose that four friends get together to study at Leigh's house for their next test in AP® Statistics. When they go for a snack in the kitchen, Leigh's 3-year-old brother makes a tower using their textbooks. Unfortunately, none of the students wrote their name in their book, so when they leave, each student takes one of the books at random. When the students return the books at the end of the year and the clerk scans their barcodes, the students are surprised to learn that none of the four had their own book. How likely is it that none of the four students ended up with the correct book?

**1.** On four equally sized slips of paper, write "Student 1," "Student 2," "Student 3," and "Student 4." Likewise, on four equally sized slips of paper, write "Book 1," "Book 2," and so on.

**2.** Place the four papers with the student numbers on your desk. Then shuffle the papers with book numbers and randomly place one paper on each "student." If the book number matches the student number, this represents a student choosing his own book from the tower of textbooks.

**3.** Count the number of students who get the correct book and share your results with the class.

**4.** Repeat Steps 2 and 3 again if needed to get a total of at least 40 repetitions.

**5.** How likely is it for none of the students to end up with their own book?

### Teaching Tip

Here is a dotplot showing the results of 40 trials of the "Whose Book Is This?" Alternate Activity. In 19 of the 40 trials (47.5%), none of the students ended up with their own book.

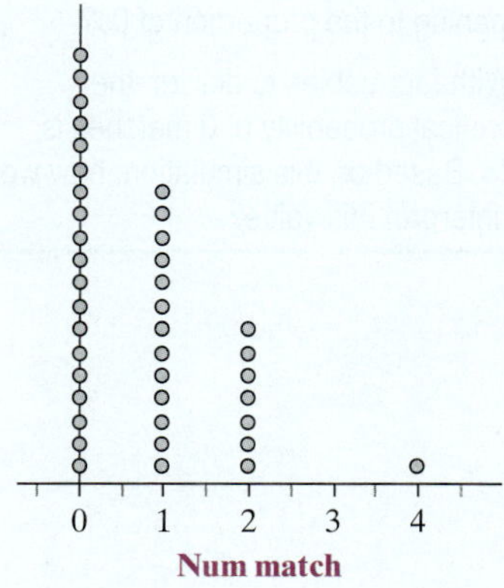


### PD Section 5.1 Overview

Watch the video overview of Section 5.1 (for teachers) by clicking on the link in the TE-book), opening the TRCD, or logging onto the book's companion Web site www.whfreeman.com/tps5e (teacher log-in required).

### TRM Learning Objectives Grid

Remind students about the importance of the Learning Objectives. We include them at the beginning of each section so students know what they are expected to learn. At the end of the chapter, the Learning Objectives are included in a grid, along with references to examples and chapter review exercises. A printable version of this grid is included in the Teacher's Resource Materials.

### TRM Section 5.1 Alternate Examples

You can find the Alternate Examples for this section in Microsoft Word format by clicking on the link in the TE-book, opening the TRCD, or logging onto the book's companion Web site www.whfreeman.com/tps5e (teacher log-in required).

### Activity Overview

You can watch an overview of this Activity (for teachers) by clicking on the link in the TE-book, opening the TRCD, or logging onto the book's companion Web site www.whfreeman.com/tps5e (teacher log-in required).

### Teaching Tip

Before you use this Activity, it is a good idea to use a real coin to do 10–15 flips. Graph the results by hand so students will understand what the applet is doing.

**ALTERNATE ACTIVITY**

*Random babies*

The "Whose Book Is This?" Activity can be explored using the *Random Babies* applet at www.rossmanchance.com/applets. This applet simulates a stork randomly delivering four babies to four different houses and counting the number of correct deliveries. Both of these problems involve a concept called "derangements." If you're curious, do an Internet search for "derangements" to learn more. One interesting fact: the probability of no matches is approximately the same no matter how many babies are being delivered—and approaches $1/e$ as the number of babies increases.

1. Press "Randomize" to have the stork deliver the babies. If there is a correct match, the sun will shine on the house; otherwise, there will be a storm cloud. The number of matches will be recorded in the histogram.
2. After animating several deliveries, change the "Number of trials" to 10 and press "Randomize." Click inside the bar above 0 to see a plot that records the proportion of 0s after each trial.
3. Keep pressing "Randomize" to add more trials to the plot. What seems to be happening to the proportion of 0s?
4. With four babies to deliver, the theoretical probability of 0 matches is 0.375. Based on this simulation, how would you interpret this value?



why statisticians recommend random samples and randomized experiments. They avoid bias by letting chance decide who gets selected or who receives which treatment.

A big fact emerges when we watch coin tosses or the results of random sampling and random assignment closely: *chance behavior is unpredictable in the short run but has a regular and predictable pattern in the long run.* This remarkable fact is the basis for the idea of probability.

## ACTIVITY | *Probability* applet

**MATERIALS:** Computer with Internet connection

If you toss a fair coin, what's the chance that it shows heads? It's 1/2, or 0.5, right? In this Activity, you'll use the *Probability* applet at www.whfreeman.com/tps5e to investigate what probability really means.

1. If you toss a fair coin 10 times, how many heads will you get? Before you answer, launch the *Probability* applet. Set the number of tosses at 10 and click "Toss." What proportion of the tosses were heads? Click "Reset" and toss the coin 10 more times. What proportion of heads did you get this time? Repeat this process several more times. What do you notice?
2. What if you toss the coin 100 times? Reset the applet and have it do 100 tosses. Is the proportion of heads exactly equal to 0.5? Close to 0.5?
3. Keep on tossing without hitting "Reset." What happens to the proportion of heads?
4. As a class, discuss what the following statement means: "If you toss a fair coin, the probability of heads is 0.5."
5. Predict what will happen if you change the probability of heads to 0.3 (an unfair coin). Then use the applet to test your prediction.
6. If you toss a coin, it can land heads or tails. If you "toss" a thumbtack, it can land with the point sticking up or with the point down. Does that mean that the probability of a tossed thumbtack landing point up is 0.5? How could you find out? Discuss with your classmates.

We might suspect that a coin has probability 0.5 of coming up heads just because the coin has two sides. But we can't be sure. In fact, spinning a penny on a flat surface, rather than tossing the coin, gives heads a probability of about 0.45 rather than 0.5.[1] What about thumbtacks? They also have two ways to land—point up or point down—but the chance that a tossed thumbtack lands point up isn't 0.5. How do we know that? From tossing a thumbtack over and over and over again. Probability describes what happens in very many trials, and we must actually observe many tosses of a coin or thumbtack to pin down a probability.

### Teaching Tip

Make sure to point out the pun about *pinning* down the thumbtack probability in the last line of this page. We're sure that the *sharp* students will get the *point*.

## EXAMPLE Tossing Coins

### *Short-run and long-run behavior*

When you toss a coin, there are only two possible outcomes, heads or tails. Figure 5.1(a) shows the results of tossing a coin 20 times. For each number of tosses from 1 to 20, we have plotted the proportion of those tosses that gave a head. You can see that the proportion of heads starts at 1 on the first toss, falls to 0.5 when the second toss gives a tail, then rises to 0.67, and then falls to 0.5, and 0.4 as we get two more tails. After that, the proportion of heads continues to fluctuate but never exceeds 0.5 again.

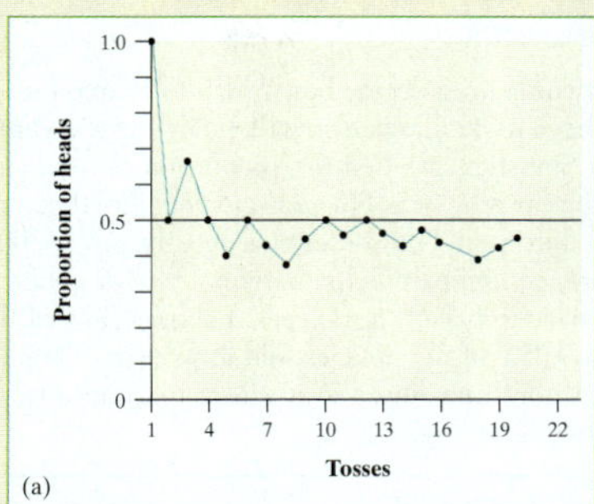


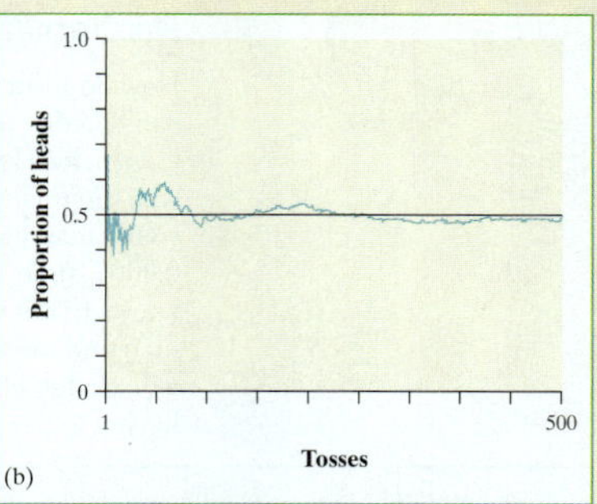


**FIGURE 5.1** (a) The proportion of heads in the first 20 tosses of a coin. (b) The proportion of heads in the first 500 tosses of a coin.

Suppose we keep tossing the coin until we have made 500 tosses. Figure 5.1(b) shows the results. The proportion of tosses that produce heads is quite variable at first. As we make more and more tosses, however, the proportion of heads gets close to 0.5 and stays there.

The fact that the proportion of heads in many tosses eventually closes in on 0.5 is guaranteed by the **law of large numbers**. This important result says that if we observe more and more repetitions of any chance process, the proportion of times that a specific outcome occurs approaches a single value. We call this value the **probability**. The previous example confirms that the probability of getting a head when we toss a fair coin is 0.5. Probability 0.5 means "occurs half the time in a very large number of trials."

**DEFINITION: Probability**

The **probability** of any outcome of a chance process is a number between 0 and 1 that describes the proportion of times the outcome would occur in a very long series of repetitions.

### ALTERNATE EXAMPLE

#### *Whose book is this?*

The graphs below show the short-run and long-run behavior of the proportion of trials in which there are no matches when four students choose a book at random. The blue line is the correct probability of 0.375. As you can see, in the first 20 trials, there is quite a bit of variability. However, after 500 trials, the proportion of times no match occurred is quite close to the actual value.

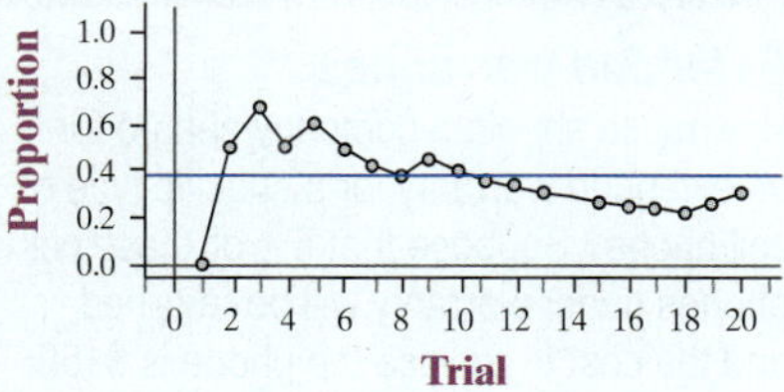


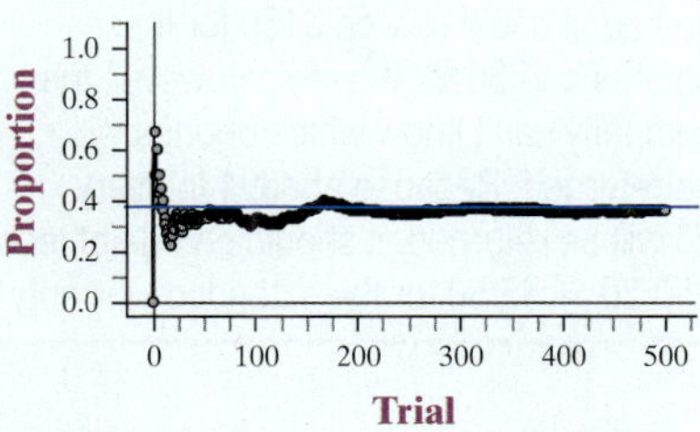


### Teaching Tip

Another way to interpret the probability of an outcome is its *predicted* long-run relative frequency. For example, if the probability of heads is 0.5, then we would predict that after a large number of coin flips, the proportion of heads will be about 0.5.

### AP® Exam Common Error

Students sometimes report probabilities as a number greater than 1. In many cases, this is because their calculator is expressing a very small number in scientific notation, such as 2.3844832E−8. Remind students to check that their answer makes sense. Probabilities must be between 0 and 1!

### ALTERNATE EXAMPLE

***Extended warranties***

How much should a company charge for an extended warranty for a specific type of cell phone? Suppose that 5% of these cell phones under warranty will be returned, and the cost to replace the phone is $150. If the company knew which phones would go bad, it could charge $150 for these phones and $0 for the rest. However, the company can't know which phones will be returned. Because about 1 in every 20 will be returned, it should charge at least 150/20 = $7.50 for the extended warranty.

### ANSWERS TO CYU

**1. (a)** If you asked a large sample of U.S. adults whether they usually eat breakfast, about 61% of them will answer yes. **(b)** The exact number of breakfast eaters will vary from sample to sample.

**2. (a)** 0. If an outcome can never occur, then it will occur in 0% of the trials. **(b)** 1. If an outcome will occur on every trial, then it will occur in 100% of the trials. **(c)** 0.01. An outcome that occurs in 1% of the trials is very unlikely, but will occur every once in a while. **(d)** 0.6. An outcome that occurs in 60% of the trials will happen more than half of the time.

Outcomes that never occur have probability 0. An outcome that happens on every repetition has probability 1. An outcome that happens half the time in a very long series of trials has probability 0.5. Of course, we can never observe a probability exactly. We could always continue tossing the coin, for example. The mathematics of probability is based on imagining what would happen in an indefinitely long series of trials.

Recall from Chapter 4 that "random" doesn't mean haphazard. In statistics, random means "by chance."

Probability gives us a language to describe the long-term regularity of random behavior. The outcome of a coin toss and the gender of the next baby born in a local hospital are both random. So is the result of a random sample or a random assignment. Even life insurance, for example, is based on the fact that deaths occur at random among many individuals.

## EXAMPLE Life Insurance

***Probability and risk***

How do insurance companies decide how much to charge for life insurance? We can't predict whether a particular person will die in the next year. But the National Center for Health Statistics says that the proportion of men aged 20 to 24 years who die in any one year is 0.0015. This is the *probability* that a randomly selected young man will die next year. For women that age, the probability of death is about 0.0005. If an insurance company sells many policies to people aged 20 to 24, it knows that it will have to pay off next year on about 0.15% of the policies sold to men and on about 0.05% of the policies sold to women. Therefore, the company will charge about three times more to insure a man because the probability of having to pay is three times higher.

We often encounter the unpredictable side of randomness in our everyday lives, but we rarely see enough repetitions of the same chance process to observe the long-run regularity that probability describes. Life insurance companies, casinos, and others who make important decisions based on probability rely on the long-run predictability of random behavior.

### CHECK YOUR UNDERSTANDING

1. According to the Book of Odds Web site www.bookofodds.com, the probability that a randomly selected U.S. adult usually eats breakfast is 0.61.

(a) Explain what probability 0.61 means in this setting.

(b) Why doesn't this probability say that if 100 U.S. adults are chosen at random, exactly 61 of them usually eat breakfast?

2. Probability is a measure of how likely an outcome is to occur. Match one of the probabilities that follow with each statement. Be prepared to defend your answer.

0 0.01 0.3 0.6 0.99 1

(a) This outcome is impossible. It can never occur.

(b) This outcome is certain. It will occur on every trial.

(c) This outcome is very unlikely, but it will occur once in a while in a long sequence of trials.

(d) This outcome will occur more often than not.

## Myths about Randomness

The idea of probability seems straightforward. It answers the question "What would happen if we did this many times?" In fact, both the behavior of random phenomena and the idea of probability are a bit subtle.

### ACTIVITY | Investigating randomness

1. Pretend that you are flipping a fair coin. Without actually flipping a coin, *imagine* the first toss. Write down the result you see in your mind, heads (H) or tails (T).
2. Imagine a second coin flip. Write down the result.
3. Keep doing this until you have recorded the results of 50 imaginary flips. Write your results in groups of 5 to make them easier to read, like this: HTHTH TTHHT, etc.
4. A *run* is a repetition of the same result. In the example in Step 3, there is a run of two tails followed by a run of two heads in the first 10 coin flips. Read through your 50 imagined coin flips and find the longest run.
5. Your teacher will draw and label a number line for a class dotplot. Plot the length of the longest run you got in Step 4 on the graph.
6. Use Table D, technology, or a coin to generate a similar list of 50 coin flips. Find the longest run that you have.
7. Your teacher will draw and label a number line with the same scale immediately above or below the one in Step 5. Plot the length of the longest run you got in Step 6 on the new dotplot.
8. Compare the distributions of longest run from imagined tosses and random tosses. What do you notice?

The idea of probability is that randomness is predictable in the long run. Unfortunately, our intuition about randomness tries to tell us that random phenomena should also be predictable in the short run. When they aren't, we look for some explanation other than chance variation.

### EXAMPLE Runs in Coin Tossing

#### *What looks random?*

Toss a coin six times and record heads (H) or tails (T) on each toss. Which of the following outcomes is more probable?

HTHTTH TTTHHH

Almost everyone says that HTHTTH is more probable, because TTTHHH does not "look random." In fact, both are equally likely. That heads and tails are equally

### Teaching Tip

Have students complete Steps 1–3 of this Activity *before* getting to this page in the book. Some students might change their responses based on the title of this subsection.

### Activity Overview

You can watch an overview of this Activity (for teachers) by clicking on the link in the TE-book, opening the TRCD, or logging onto the book's companion Web site www.whfreeman.com/tps5e (teacher log-in required).

### ALTERNATE ACTIVITY

#### *Streakiness*

Suppose that a basketball announcer suggests a certain player is streaky. That is, the announcer believes that if the player makes a shot, then the player is more likely to make his next shot. As evidence, the announcer points to a recent game where the player took 30 shots and had a streak of 7 made shots in a row. Is this convincing evidence of streakiness, or could it have occurred simply by chance? Assuming that this player makes 50% of his shots and the results of a shot don't depend on previous shots, how likely is it for the player to have a streak of 7 or more made shots in a row?

**1.** Using a coin, let heads represent a made shot and tails represent a missed shot.

**2.** Flip a coin 30 times, writing down the outcome after each flip. Record the length of the longest streak of made shots.

**3.** Combine your results with those of your classmates. In what proportion of the trials did the player have a streak of at least 7 in a row?

Here are the results of 50 trials of this simulation. The player had a streak of 7 or more made shots in 7 of the 50 simulated games (14%).

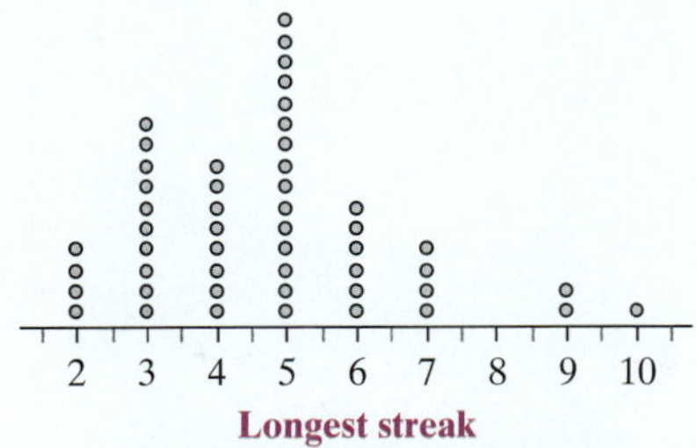


### ALTERNATE EXAMPLE

#### *Runs in die rolling*

Roll a die 12 times and record the result of each roll. Which of the following outcomes is more probable?

1 2 3 4 5 6 6 5 4 3 2 1 or 1 5 4 5 2 4 3 3 6 1 2 6

These outcomes are both equally (un)likely, even though the first set of rolls has a more noticeable pattern.

probable says only that about half of a very long sequence of tosses will be heads. It doesn't say that heads and tails must come close to alternating in the short run. The coin has no memory. It doesn't know what past outcomes were, and it can't try to create a balanced sequence.

The outcome TTTHHH in tossing six coins looks unusual because of the runs of 3 straight tails and 3 straight heads. Runs seem "not random" to our intuition but are quite common. Here's a more striking example than tossing coins.

## EXAMPLE That Shooter Seems "Hot"

### *Chance variation or skill?*

Is there such a thing as a "hot hand" in basketball? Belief that runs must result from something other than "just chance" influences behavior. If a basketball player makes several consecutive shots, both the fans and her teammates believe that she has a "hot hand" and is more likely to make the next shot. Several studies have shown that runs of baskets made or missed are no more frequent in basketball than would be expected if the result of each shot is unrelated to the outcomes of the player's previous shots. If a player makes half her shots in the long run, her made shots and misses behave just like tosses of a coin—and that means that runs of makes and misses are more common than our intuition expects.[2]

Free throws may be a different story. A recent study suggests that players who shoot two free throws are *slightly* more likely to make the second shot if they make the first one.[3]

Once, at a convention in Las Vegas, one of the authors roamed the gambling floors, watching money disappear into the drop boxes under the tables. You can see some interesting human behavior in a casino. When the shooter in the dice game called craps rolls several winners in a row, some gamblers think she has a "hot hand" and bet that she will keep on winning. Others say that "the law of averages" means that she must now lose so that wins and losses will balance out. Believers in the law of averages think that if you toss a coin six times and get TTTTTT, the next toss must be more likely to give a head. It's true that in the long run heads will appear half the time. What is a myth is that future outcomes must make up for an imbalance like six straight tails.

Don't confuse the law of large numbers, which describes the big idea of probability, with the "law of averages" described here.

Coins and dice have no memories. A coin doesn't know that the first six outcomes were tails, and it can't try to get a head on the next toss to even things out. Of course, things do even out *in the long run*. That's the law of large numbers in action. After 10,000 tosses, the results of the first six tosses don't matter. They are overwhelmed by the results of the next 9994 tosses.

### ALTERNATE EXAMPLE

#### *Joe DiMaggio's hitting streak*

There was an interesting discussion of the "hot hand" in an article in the *New York Times* (March 30, 2008) written by Samuel Arbesman and Steven Strogatz (http://www.nytimes.com/2008/03/30/opinion/30strogatz.html). In the article, the authors claim that one of the most remarkable streaks in baseball history, Joe DiMaggio's 56-consecutive-game hitting streak, was actually not very remarkable at all. Obviously, it is *extremely* unlikely for any particular individual to have a hitting streak this long. But, when considering all the players and all the seasons in baseball history, we should expect some very unusual performances every now and then. To investigate, they simulated the performances of every baseball player in every season a total of 10,000 times. In each of those 10,000 simulated "histories of baseball," they recorded the longest hitting streak. In about 42% of the trials of the simulation, someone had a hitting streak of at least 56 games in a row, with the longest being an amazing 109 games in a row!

## EXAMPLE Aren't We Due for a Boy?

### *Don't fall for the "law of averages"*

"So the law of averages doesn't guarantee me a girl after seven straight boys, but can't I at least get a group discount on the delivery fee?"

Belief in this phony "law of averages" can lead to serious consequences. A few years ago, an advice columnist published a letter from a distraught mother of eight girls. She and her husband had planned to limit their family to four children, but they wanted to have at least one boy. When the first four children were all girls, they tried again—and again and again. After seven straight girls, even her doctor had assured her that "the law of averages was in our favor 100 to 1." Unfortunately for this couple, having children is like tossing coins. Eight girls in a row is highly unlikely, but once seven girls have been born, it is not at all unlikely that the next child will be a girl—and it was.

## Simulation

The imitation of chance behavior, based on a model that accurately reflects the situation, is called a **simulation.** You already have some experience with simulations. In Chapter 1's "Hiring Discrimination—It Just Won't Fly!" Activity (page 5), you drew beads or slips of paper to imitate a random lottery to choose which pilots would become captains. Chapter 4's "Distracted Driving" Activity (page 249) asked you to shuffle and deal piles of cards to mimic the random assignment of subjects to treatments. The "1 in 6 wins" game that opened this chapter had you roll a die several times to simulate buying 20-ounce sodas and looking under the cap. These simulations involved different chance "devices"—beads, slips of paper, cards, or dice. But the same basic strategy was followed in all three simulations. We can summarize this strategy using our familiar four-step process: State, Plan, Do, Conclude.

### PERFORMING A SIMULATION

**State:** Ask a question of interest about some chance process.

**Plan:** Describe how to use a chance device to imitate one repetition of the process. Tell what you will record at the end of each repetition.

**Do:** Perform many repetitions of the simulation.

**Conclude:** Use the results of your simulation to answer the question of interest.

The following table summarizes this four-step process for each of our previous simulations.

### ALTERNATE EXAMPLE

***Red is due!***

In casinos, there is often a large display next to every roulette table showing the outcomes of the previous spins of the wheel. The results of previous spins reveal nothing about the results of future spins, so why do the casinos pay for these displays? The casinos know that many people will be more willing to place a bet if they observe a pattern in the previous outcomes. And as long as people are placing bets, the casino is making money.

### Teaching Tip

Remind students that Statistics Problems Demand Consistency!

## Teaching Tip

Here are the four steps for the "Whose Book Is This?" Alternate Activity:
STATE: When four students mix up their AP® Stats books and distribute them at random, what is the probability that none of the students end up with their own book?
PLAN: On four equally sized slips of paper, write "Student 1," "Student 2," "Student 3," and "Student 4." Likewise, on four equally sized slips of paper, write "Book 1," "Book 2," and so on. Place the four papers with the student numbers on your desk. Then shuffle the papers with book numbers and randomly place one paper on each "student." Count the number of students who get the matching book.
DO: Do this process many times, recording the number of students who get the correct book in each trial.
CONCLUDE: Out of 40 total repetitions, there were 19 times when none of the students ended up with their own book. So the *estimated* probability is 19/40 = 47.5%.

## Teaching Tip

When students write the conclusion to a simulation question, emphasize that the answer is always an approximation. For example, instead of saying that the probability *is* 19/40, it is better to say that the probability is *approximately* 19/40.

## Teaching Tip

Remind students that the "play" icon at the top of this example (and many others) indicates that they can watch a video presentation of the example. Find the videos at www.whfreeman.com/tps5e or in the e-Book. Also, remind them that the icon in the lower-right corner indicates that there is a link between this example and Exercise 19. If students are having trouble with the exercise, they can look back to this example for help.

| | Hiring Discrimination | Distracted Driving | "1 In 6 Wins" Game |
|---|---|---|---|
| **State** | How likely is it that a fair lottery would result in 5 or more female pilots being selected from an initial pool of 15 male and 10 female pilots? | Is it plausible that just by the luck of the random assignment of 48 subjects into two groups of 24 each, 12 of the 15 subjects who were going to miss the rest area anyway ended up in the cell-phone group, while only 3 of the 15 were assigned to the passenger conversation group? | What's the probability that 3 or more of 7 people who buy a 20-ounce bottle of soda win a prize if each bottle has a 1/6 chance of saying, "You're a winner!"? |
| **Plan** | Prepare a bag with 25 beads (15 of one color and 10 of another) or 25 identical slips of paper (15 labeled "M" and 10 labeled "F"). Mix well. Then without looking, remove 8 beads/slips from the bag. Record the number of female pilots selected. | • *Using index cards:* Write "Stop" on 33 cards and "Don't" on 15 cards. • *Using playing cards:* Remove jokers, specialty cards, the ace of spades, and three 2s from the deck. J, Q, K, A = miss rest area 2 through 10 = stop at rest area Shuffle well and deal two piles of 24 cards—the cell-phone group and the passenger group. Record the number of drivers who miss the rest area in the cell-phone group. | Use a six-sided die to determine the outcome for each person's bottle of soda. 6 = wins a prize 1 to 5 = no prize Roll the die seven times, once for each person. Record the number of people who win a prize. |
| **Do** | Have each student in class do this several times. | Have each pair of students repeat the process several times. | Have each student perform several repetitions. |
| **Conclude** | In 100 repetitions, 18 yielded 5 or more female pilots. So about 18% of the time, a fair lottery would choose at least 5 female pilots to become captains. It seems plausible that the company carried out a fair lottery. | In 300 repetitions of the random assignment, there were only 2 times when 12 or more drivers who missed the rest area ended up in the cell-phone group. That's less than 1% of the time! It doesn't seem plausible that the random assignment is the explanation for the difference between the groups. | Out of 125 total repetitions of the simulation, there were 15 times when three or more of the seven people won a prize. So our estimate of the probability is 15/125, or about 12%. It seems plausible that the company is telling the truth. |

So far, we have used physical devices for our simulations. Using random numbers from Table D or technology is another option, as the following examples illustrate.

## Golden Ticket Parking Lottery

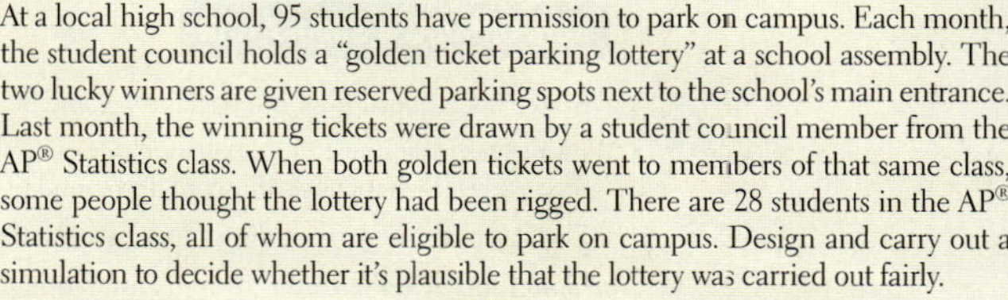
At a local high school, 95 students have permission to park on campus. Each month, the student council holds a "golden ticket parking lottery" at a school assembly. The two lucky winners are given reserved parking spots next to the school's main entrance. Last month, the winning tickets were drawn by a student council member from the AP® Statistics class. When both golden tickets went to members of that same class, some people thought the lottery had been rigged. There are 28 students in the AP® Statistics class, all of whom are eligible to park on campus. Design and carry out a simulation to decide whether it's plausible that the lottery was carried out fairly.

STATE: What's the probability that a fair lottery would result in two winners from the AP® Statistics class?

PLAN: We'll use Table D to simulate choosing the golden ticket lottery winners. Because there are 95 eligible students in the lottery, we'll label the students in the AP® Statistics class from 01 to 28,

## AP® Exam Common Error

When working with a random digit table, students often make mistakes or do not clearly communicate the method they are using. Remind students that every label needs to be the same length. For example, if you are labeling 1500 items, the labels should be 0001–1500, not 1–1500. Also, if students are not using all the labels of a certain length, they should explain that the remaining labels will be ignored. Finally, if the process involves sampling without replacement, then students should clearly acknowledge they will ignore repeated labels or look for a certain number of different labels.

and the remaining students from 29 to 95. Numbers from 96 to 00 will be skipped. Moving left to right across the row, we'll look at pairs of digits until we come across two *different* labels from 01 to 95. The two students with these labels will win the prime parking spaces. We will record the number of winners from the AP® Statistics class.

DO: Let's perform many repetitions of our simulation. We'll use Table D starting at line 139. The digits from that row are shown below. We have drawn vertical bars to separate the pairs of digits. Underneath each pair, we have marked a √ if the chosen student is in the AP® Statistics class, × if the student is not in the class, and "skip" if the number isn't between 01 and 95 or if that student was already selected. (Note that if the consecutive "70" labels had been in the same repetition, we would have skipped the second one.) We also recorded the number of students in the AP® Statistics class for each repetition.

| Rep 1 | Rep 2 | Rep 3 | Rep 4 | Rep 5 | Rep 6 | Rep 7 | Rep 8 | Rep 9 | Rep 10 |
|---|---|---|---|---|---|---|---|---|---|
| 55\|58 | 89\|94 | 04\|70 | 70\|84 | 10\|98\|43 | 56\|35 | 69\|34 | 48\|39 | 45\|17 | 19\|12 |
| × × | × × | ✓ × | × × | ✓ skip × | × × | × × | × × | × ✓ | ✓ ✓ |
| 0 | 0 | 1 | 0 | 1 | 0 | 0 | 0 | 1 | 2 |

In the first 10 repetitions, there was one time when the two winners were both from the AP® Statistics class. But 10 isn't many repetitions of the simulation. Continuing where we left off,

| Rep 11 | Rep 12 | Rep 13 | Rep 14 | Rep 15 | Rep 16 | Rep 17 | Rep 18 | Rep 19 | Rep 20 |
|---|---|---|---|---|---|---|---|---|---|
| 97\|51\|32 | 58\|13 | 04\|84 | 51\|44 | 72\|32 | 18\|19 | 40\|00\|36 | 00\|24\|28 | 96\|76\|73 | 59\|64 |
| skip × × | × ✓ | ✓ × | × × | × × | ✓ ✓ | × skip × | skip ✓ ✓ | skip × × | × × |
| 0 | 1 | 1 | 0 | 0 | 2 | 0 | 2 | 0 | 0 |

So after 20 repetitions, there have been 3 times when both winners were in the AP® Statistics class. If we keep going for 30 more repetitions (to bring our total to 50), we find 28 more "No" and 2 more "Yes" results. All totaled, that's 5 "Yes" and 45 "No" results.

CONCLUDE: In our simulation of a fair lottery, both winners came from the AP® Statistics class in 10% of the repetitions. So about 1 in every 10 times the student council holds the golden ticket lottery, this will happen just by chance. It seems plausible that the lottery was conducted fairly.

**For Practice** *Try Exercise* 

**AP® EXAM TIP** On the AP® exam, you may be asked to describe how you will perform a simulation using rows of random digits. If so, provide a clear enough description of your simulation process for the reader to get the same results you did from *only* your written explanation.

In the previous example, we could have saved a little time by using `randInt(1,95)` repeatedly instead of Table D (so we wouldn't have to worry about numbers 96 to 00). We'll take this alternate approach in the next example.

EXAMPLE

## NASCAR Cards and Cereal Boxes

### *Simulations with technology*

In an attempt to increase sales, a breakfast cereal company decides to offer a NASCAR promotion. Each box of cereal will contain a collectible card featuring one of these NASCAR drivers: Jeff Gordon, Dale Earnhardt, Jr., Tony Stewart, Danica Patrick, or Jimmie Johnson. The company says that each of the 5 cards is equally likely to appear in any box of cereal. A NASCAR fan decides to keep buying boxes of the cereal until she has all 5 drivers' cards. She is surprised when it takes her 23 boxes to get the full set of cards. Should she be surprised? Design and carry out a simulation to help answer this question.

**ALTERNATE EXAMPLE**

### *Stratified sampling*

Suppose I want to choose a simple random sample of size 6 from a group of 60 seniors and 30 juniors. To do this, I write each person's name on an equally sized piece of paper and mix the papers in a large grocery bag. Just as I am about to select the first name, a thoughtful student suggests that I should stratify by class. I agree, and we decide it would be appropriate to select 4 seniors and 2 juniors. However, because I have already mixed up the names, I don't want to have to separate them all again. Instead, I will select names one at a time from the bag until I get 4 seniors and 2 juniors. Design and carry out a simulation using Table D to estimate the probability that you must draw 10 or more names to get 4 seniors and 2 juniors.

STATE: What is the probability that it takes 10 or more selections to get 4 seniors and 2 juniors?

PLAN: Using pairs of digits from Table D, we'll label the 60 seniors 01–60 and the 30 juniors 61–90. Numbers 00 and 91–99 will be skipped. Moving left to right across a row, we'll look at pairs of digits until we have 4 *different* labels from 01–60 and 2 *different* labels from 61–90. Then we will count how many *different* labels from 01–90 we looked at.

DO: Here is an example of one repetition, using line 101 from Table D:

19 (senior) 22 (senior) 39 (senior)
50 (senior) 34 (senior) 05 (senior)
75 (junior) 62 (junior)

In this trial, it took 8 selections to get at least 4 seniors and at least 2 juniors. Here are the results of 50 trials:

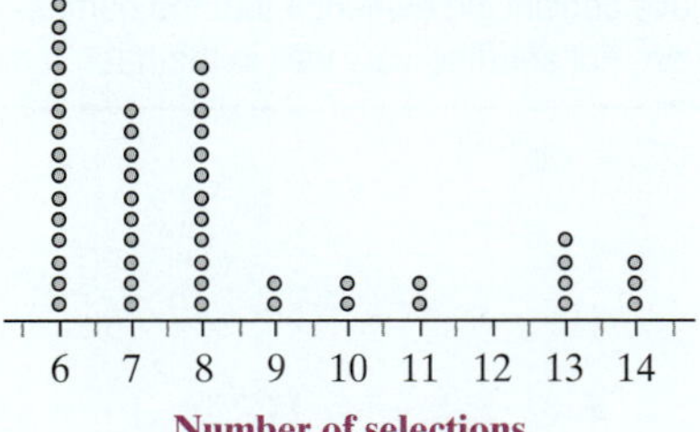


CONCLUDE: In the simulation, 11 of the 50 trials required 10 or more selections to get 4 seniors and 2 juniors, so the probability that it takes 10 or more selections is *approximately* 0.22.

**STATE:** What is the probability that it will take 23 or more boxes to get a full set of 5 NASCAR collectible cards?

**PLAN:** We need five numbers to represent the five possible cards. Let's let 1 = Jeff Gordon; 2 = Dale Earnhardt, Jr.; 3 = Tony Stewart; 4 = Danica Patrick; and 5 = Jimmie Johnson. We'll use randInt(1,5) to simulate buying one box of cereal and looking at which card is inside. Because we want a full set of cards, we'll keep pressing Enter until we get all five of the labels from 1 to 5. We'll record the number of boxes that we had to open.

**DO:** It's time to perform many repetitions of the simulation. Here are our first few results:

*Rep 1:* 3 5 2 1 5 2 3 5 4 9 boxes
*Rep 2:* 5 1 2 5 1 4 1 4 1 2 2 2 4 4 5 3 16 boxes
*Rep 3:* 5 5 5 2 4 1 2 1 5 3 10 boxes
*Rep 4:* 4 3 5 3 5 1 1 1 5 3 1 5 4 5 2 15 boxes
*Rep 5:* 3 3 2 2 1 2 4 3 3 4 2 2 3 3 3 2 3 3 4 2 2 5 22 boxes

The Fathom dotplot shows the number of boxes we had to buy in 50 repetitions of the simulation.

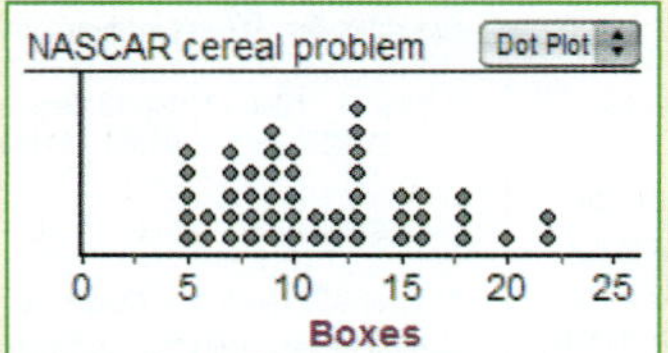


**CONCLUDE:** We never had to buy more than 22 boxes to get the full set of NASCAR drivers' cards in 50 repetitions of our simulation. So our estimate of the probability that it takes 23 or more boxes to get a full set is roughly 0. The NASCAR fan should be surprised by how many boxes she had to buy.

**For Practice** *Try Exercise* 

In the golden ticket lottery example, we ignored repeated numbers from 01 to 95 within a given repetition. That's because the chance process involved sampling students *without* replacement. In the NASCAR example, we allowed repeated numbers from 1 to 5 in a given repetition. That's because we are selecting a small number of cards from a very large population of cards in thousands of cereal boxes. So the probability of getting, say, a Danica Patrick card in the next box of cereal is still very close to 1/5 even if we have already selected a Danica Patrick card.

**What *don't* these simulations tell us?** For the golden ticket parking lottery, we concluded that it's plausible the drawing was done fairly. Does that mean the lottery *was* conducted fairly? Not necessarily. All we did was estimate that the probability of getting two winners from the AP® Statistics class was about 10% *if* the drawing was fair. So the result isn't unlikely enough to convince us that the lottery was rigged. What about the cereal box simulation? It took our NASCAR fan 23 boxes to complete the set of 5 cards. Does that mean the company didn't

## ALTERNATE EXAMPLE

### *Picking teams*

At their annual picnic, 18 students in the mathematics/statistics department at a university decide to play a softball game. Twelve of the 18 students are math majors and 6 are stats majors. To divide into two teams of 9, one of the professors put all the players' names into a hat and drew out 9 players to form one team, with the remaining 9 players forming the other team. The players were surprised when one team was made up entirely of math majors. Is it possible that the names weren't adequately mixed in the hat, or could this have happened by chance? Design and carry out a simulation to help answer this question.

**STATE:** What is the probability that there will be one team with all math majors when randomly assigning 12 math majors and 6 stats majors to two teams of 9?

**PLAN:** Number the math majors 1–12 and the stats majors 13–18. Use technology to pick 9 unique integers between 1 and 18 for one team. Record whether the 9 numbers on either team are all between 1 and 12 (all math majors).

**DO:** Here is an example of one trial:

**Team A:** M M M S M M M M M (8 math majors)
**Team B:** M S M M S S S S M (4 math majors)

In this trial, neither team was made up entirely of math majors. In 29 additional trials, there was only one trial where an entire team was made up of math majors.

**CONCLUDE:** Only 1 trial in 30 resulted in a team with all math majors, so the probability of getting an entire team of math majors is approximately 0.033. Because getting a team of all math majors is very unlikely to happen by chance alone, we have convincing evidence that the names were not shuffled very well in the hat.

## AP® Exam Common Error

When students make conclusions, they often lose credit for suggesting that a claim is *definitely* true or saying that the evidence *proves* that a claim is incorrect. If the conclusion is based on a probability, students should always acknowledge that they can't be certain that their conclusion is correct. A better response would be to say that there is convincing evidence (or there isn't convincing evidence) to support a particular claim.

tell the truth about how the cards were distributed? Not necessarily. Our simulation says that it's very unlikely for someone to have to buy 23 boxes to get a full set *if* each card is equally likely to appear in a box of cereal. The evidence suggests that the company's statement is incorrect. It's still possible, however, that the NASCAR fan was just very unlucky.

## CHECK YOUR UNDERSTANDING

1. Refer to the golden ticket parking lottery example. At the following month's school assembly, the two lucky winners were once again members of the AP® Statistics class. This raised suspicions about how the lottery was being conducted. How would you modify the simulation in the example to estimate the probability of getting two winners from the AP® Statistics class in back-to-back months just by chance?

2. Refer to the NASCAR and breakfast cereal example. What if the cereal company decided to make it harder to get some drivers' cards than others? For instance, suppose the chance that each card appears in a box of the cereal is Jeff Gordon, 10%; Dale Earnhardt, Jr., 30%; Tony Stewart, 20%; Danica Patrick, 25%; and Jimmie Johnson, 15%. How would you modify the simulation in the example to estimate the chance that a fan would have to buy 23 or more boxes to get the full set?

## Section 5.1 Summary

- A chance process has outcomes that we cannot predict but that nonetheless have a regular distribution in very many repetitions. The **law of large numbers** says that the proportion of times that a particular outcome occurs in many repetitions will approach a single number. This long-run relative frequency of a chance outcome is its **probability.** A probability is a number between 0 (never occurs) and 1 (always occurs).
- Probabilities describe only what happens in the long run. Short runs of random phenomena like tossing coins or shooting a basketball often don't look random to us because they do not show the regularity that emerges only in very many repetitions.
- A **simulation** is an imitation of chance behavior, most often carried out with random numbers. To perform a simulation, follow the four-step process:

STATE: Ask a question of interest about some chance process.

PLAN: Describe how to use a chance device to imitate one repetition of the process. Tell what you will record at the end of each repetition.

DO: Perform many repetitions of the simulation.

CONCLUDE: Use the results of your simulation to answer the question of interest.

### TRM Additional Simulation Problems

See the Teacher's Resource Materials for some additional simulation problems you can use with your class. Find them by clicking on the link in the new TE-book, opening the TRCD, or logging onto the book's companion Web site www.whfreeman.com/tps5e (teacher log-in required).

### ANSWERS TO CYU

**1.** Assign the members of the AP® Statistics class the numbers 01–28 and the rest of the students the numbers 29–95. Ignore the numbers 96–99 and 00. In Table D, read off 4 two-digit numbers, making sure that the second number is different than the first and that the fourth number is different than the third. Record whether all four numbers are between 01 and 28 or not.

**2.** Assign the numbers 1–10 to Jeff Gordon, 11–40 to Dale Earnhardt, Jr., 41–60 to Tony Stewart, 61–85 to Danica Patrick, and 86–100 to Jimmie Johnson. Then proceed as in the example.

### TRM Section 5.1 Quizzes

There are three different quizzes available for this section in the Teacher's Resource Materials. Versions A and B are roughly parallel. Find them by clicking on the link in the TE-book, opening the TRCD, or logging onto the book's companion Web site www.whfreeman.com/tps5e (teacher log-in required).

TRM Full Solutions to Section 5.1 Exercises

## Answers to Section 5.1 Exercises

**5.1** **(a)** If we use a polygraph machine on many, many people who are all telling the truth, the machine will say about 8% of the people are lying. **(b)** Answers will vary. A false positive would mean that a person telling the truth is found to be lying. A false negative would mean that a person lying is found to be telling the truth.

**5.2** **(a)** If we test many, many women with breast cancer, about 10% of the time the test will indicate that the woman does not have breast cancer. **(b)** A false negative, because a woman with breast cancer will not get potentially life-saving treatment. A false positive would result in temporary stress until a more thorough examination is performed.

**5.3** **(a)** If we look at many families like this, approximately 25% of them will have a first-born child who develops cystic fibrosis. **(b)** No. The number of children with cystic fibrosis could be smaller or larger than 1 by random chance.

**5.4** **(a)** If we look at many, many hands of poker in which you hold a pair, you will make four of a kind about 8.8% of the time. **(b)** No. The number of hands that get four of a kind could be smaller or larger than 88 by random chance.

**5.5** **(a)** Answers will vary. **(b)** Spin the coin many more times.

**5.6** **(a)** Answers will vary. **(b)** Knock down the coin many more times.

**5.7** In the short run, there was quite a bit of variability in the percentage of made free throws. However, this percentage became less variable and approached 0.30 as the number of shots increased.

**5.8** In the short run, there was lots of variability in the proportion of heads. In the long run, this proportion became less variable and settled down around 0.50 for both sets of 5000 tosses.

# Section 5.1 Exercises

1. **Liar, liar!** Sometimes police use a lie detector (also known as a polygraph) to help determine whether a suspect is telling the truth. A lie detector test isn't foolproof—sometimes it suggests that a person is lying when he or she is actually telling the truth (a "false positive"). Other times, the test says that the suspect is being truthful when the person is actually lying (a "false negative"). For one brand of polygraph machine, the probability of a false positive is 0.08.
   (a) Interpret this probability as a long-run relative frequency.
   (b) Which is a more serious error in this case: a false positive or a false negative? Justify your answer.
2. **Mammograms** Many women choose to have annual mammograms to screen for breast cancer after age 40. A mammogram isn't foolproof. Sometimes the test suggests that a woman has breast cancer when she really doesn't (a "false positive"). Other times the test says that a woman doesn't have breast cancer when she actually does (a "false negative"). Suppose the false negative rate for a mammogram is 0.10.
   (a) Interpret this probability as a long-run relative frequency.
   (b) Which is a more serious error in this case: a false positive or a false negative? Justify your answer.
3. **Genetics** Suppose a married man and woman both carry a gene for cystic fibrosis but don't have the disease themselves. According to the laws of genetics, the probability that their first child will develop cystic fibrosis is 0.25.
   (a) Explain what this probability means.
   (b) If the couple has 4 children, is one of them guaranteed to get cystic fibrosis? Explain.
4. **Texas hold 'em** In the popular Texas hold 'em variety of poker, players make their best five-card poker hand by combining the two cards they are dealt with three of five cards available to all players. You read in a book on poker that if you hold a pair (two cards of the same rank) in your hand, the probability of getting four of a kind is 88/1000.
   (a) Explain what this probability means.
   (b) If you play 1000 such hands, will you get four of a kind in exactly 88 of them? Explain.
5. **Spinning a quarter** With your forefinger, hold a new quarter (with a state featured on the reverse) upright, on its edge, on a hard surface. Then flick it with your other forefinger so that it spins for some time before it falls and comes to rest. Spin the coin a total of 25 times, and record the results.
   (a) What's your estimate for the probability of heads? Why?
   (b) Explain how you could get an even better estimate.
6. **Nickels falling over** You may feel it's obvious that the probability of a head in tossing a coin is about 1/2 because the coin has two faces. Such opinions are not always correct. Stand a nickel on edge on a hard, flat surface. Pound the surface with your hand so that the nickel falls over. Do this 25 times, and record the results.
   (a) What's your estimate for the probability that the coin falls heads up? Why?
   (b) Explain how you could get an even better estimate.
7. **Free throws** The figure below shows the results of a virtual basketball player shooting several free throws. Explain what this graph says about chance behavior in the short run and long run.

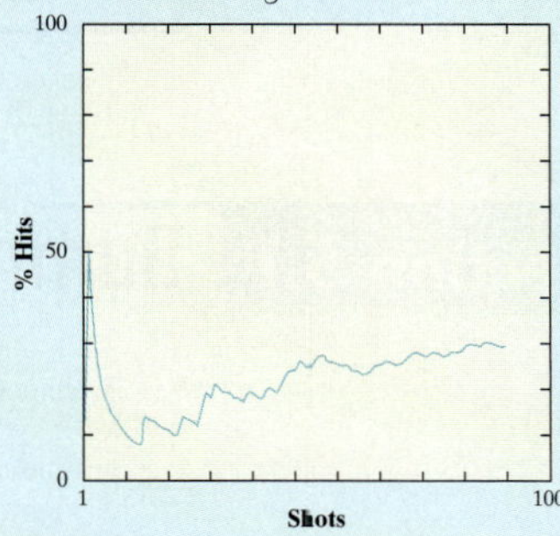


8. **Keep on tossing** The figure below shows the results of two different sets of 5000 coin tosses. Explain what this graph says about chance behavior in the short run and the long run.

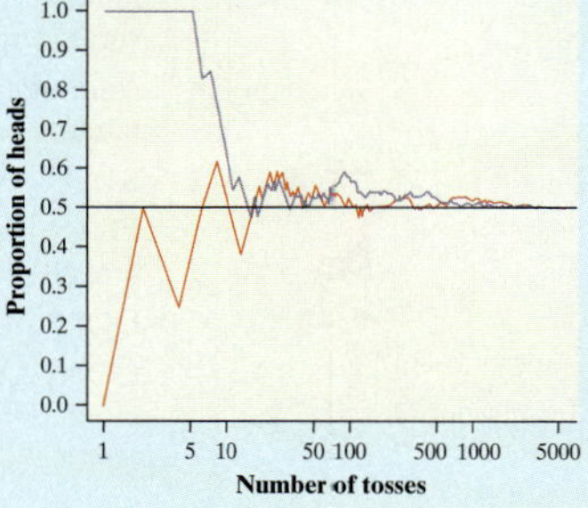

9. **Due for a hit** A very good professional baseball player gets a hit about 35% of the time over an entire season. After the player failed to hit safely in six straight at-bats, a TV commentator said, "He is due for a hit by the law of averages." Is that right? Why?

10. **Cold weather coming** A TV weather man, predicting a colder-than-normal winter, said, "First, in looking at the past few winters, there has been a lack of really cold weather. Even though we are not supposed to use the law of averages, we are due." Do you think that "due by the law of averages" makes sense in talking about the weather? Why or why not?

11. **Playing "Pick 4"** The Pick 4 games in many state lotteries announce a four-digit winning number each day. You can think of the winning number as a four-digit group from a table of random digits. You win (or share) the jackpot if your choice matches the winning number. The winnings are divided among all players who matched the winning number. That suggests a way to get an edge.

(a) The winning number might be, for example, either 2873 or 9999. Explain why these two outcomes have exactly the same probability.

(b) If you asked many people whether 2873 or 9999 is more likely to be the randomly chosen winning number, most would favor one of them. Use the information in this section to say which one and to explain why. How might this affect the four-digit number you would choose?

12. **An unenlightened gambler**

(a) A gambler knows that red and black are equally likely to occur on each spin of a roulette wheel. He observes five consecutive reds occur and bets heavily on black at the next spin. Asked why, he explains that black is "due by the law of averages." Explain to the gambler what is wrong with this reasoning.

(b) After hearing you explain why red and black are still equally likely after five reds on the roulette wheel, the gambler moves to a poker game. He is dealt five straight red cards. He remembers what you said and assumes that the next card dealt in the same hand is equally likely to be red or black. Is the gambler right or wrong, and why?

13. **Free throws** A basketball player has probability 0.75 of making a free throw. Explain how you would use each chance device to simulate one free throw by the player.

(a) A standard deck of playing cards

(b) Table D of random digits

(c) A calculator or computer's random integer generator

14. **Stoplight** On her drive to work every day, Ilana passes through an intersection with a traffic light. The light has probability 1/3 of being green when she gets to the intersection. Explain how you would use each chance device to simulate whether the light is red or green on a given day.

(a) A six-sided die

(b) Table D of random digits

(c) A calculator or computer's random integer generator

15. **Simulation blunders** Explain what's wrong with each of the following simulation designs.

(a) A roulette wheel has 38 colored slots—18 red, 18 black, and 2 green. To simulate one spin of the wheel, let numbers 00 to 18 represent red, 19 to 37 represent black, and 38 to 40 represent green.

(b) About 10% of U.S. adults are left-handed. To simulate randomly selecting one adult at a time until you find a left-hander, use two digits. Let 00 to 09 represent being left-handed and 10 to 99 represent being right-handed. Move across a row in Table D, two digits at a time, skipping any numbers that have already appeared, until you find a number between 00 and 09. Record the number of people selected.

16. **Simulation blunders** Explain what's wrong with each of the following simulation designs.

(a) According to the Centers for Disease Control and Prevention, about 36% of U.S. adults were obese in 2012. To simulate choosing 8 adults at random and seeing how many are obese, we could use two digits. Let 00 to 35 represent obese and 36 to 99 represent not obese. Move across a row in Table D, two digits at a time, until you find 8 distinct numbers (no repeats). Record the number of obese people selected.

(b) Assume that the probability of a newborn being a boy is 0.5. To simulate choosing a random sample of 9 babies who were born at a local hospital today and observing their gender, use one digit. Use `randInt(0,9)` on your calculator to determine how many babies in the sample are male.

17. **Is this valid?** Determine whether each of the following simulation designs is valid. Justify your answer.

(a) According to a recent poll, 75% of American adults regularly recycle. To simulate choosing a random sample of 100 U.S. adults and seeing how many of them recycle, roll a 4-sided die 100 times. A result of 1, 2, or 3 means the person recycles; a 4 means that the person doesn't recycle.

(b) An archer hits the center of the target with 60% of her shots. To simulate having her shoot 10 times, use a coin. Flip the coin once for each of the 10 shots. If it lands heads, then she hits the center of the target. If the coin lands tails, she doesn't.

**5.9** No, he is incorrectly applying the law of large numbers to a small number of at-bats.

**5.10** No, the weather in one year might be related to the weather in previous years. Also, he is incorrectly applying the law of large numbers to a small number of years.

**5.11** **(a)** There are 10,000 four-digit numbers (0000, 0001, . . . , 2873, . . . , 9999), and each is equally likely. **(b)** 2873. To many, 2873 "looks" more random than 9999—we don't "expect" to get the same number four times in a row. It would be best to choose a number that others would avoid so you don't have to split the pot with many other people.

**5.12** **(a)** The wheel is not affected by its past outcomes—it has no memory. So on any one spin, black and red remain equally likely. **(b)** Wrong. Removing 5 red cards changes the composition of the remaining deck, so the probability of being dealt another red card decreases.

**5.13** **(a)** Let diamonds, spades, and clubs represent making a free throw and hearts represent missing. Deal one card from the deck. **(b)** Let 00–74 represent making the free throw and 75–99 represent missing. Read a two-digit number from Table D. **(c)** Let 1–3 represent the player making the free throw and 4 represent a miss. Generate a random integer from 1–4.

**5.14** **(a)** Let 1 and 2 represent green and 3–6 represent red. Roll the die once. **(b)** Let the numbers 1–3 represent green and 4–9 represent red. Ignore 0. Look up one number in the table. **(c)** Let 1 represent green and 2 and 3 represent red. Generate a random integer from 1–3.

**5.15** **(a)** There are 19 (not 18) numbers from 00 to 18, 19 (not 18) numbers from 19 to 37, and 3 (not 2) numbers from 38 to 40. **(b)** Repeats should not be skipped. For example, if the first number selected was 08, then the probability of selecting a left-hander on the next selection would be 9% (instead of 10%).

**5.16** **(a)** Repeats should not be skipped. For example, if the first number selected was 18, then the probability of selecting an obese adult on the next selection would be 35/99 = 35.35% (instead of 36%). **(b)** This gives the numbers 0 through 9 an equal chance of occurring. However, it is more likely to have 4, 5, or 6 boys than 0 or 9 boys.

**5.17** **(a)** Valid. The chance of rolling a 1, 2, or 3 is 75% on a 4-sided die and the 100 rolls represent the 100 randomly selected U.S. adults. **(b)** Not valid. The probability of heads is 50% rather than 60%. This method will underestimate the number of times she hits the center of the target.

**5.18 (a)** Not valid. You are not putting the card back in the deck after dealing. If the first card was red, then there is only a 25/51 (49%) probability that the next person selected will be addicted to texting. **(b)** Valid. There is a 95% probability of getting a number between 00 and 94 and the five pairs of digits represent the five serve attempts.

**5.19 (a)** What is the probability that, in a random selection of 10 passengers, none from first class are chosen? **(b)** Number the first-class passengers 01–12 and the other passengers 13–76. Look up two-digit numbers in Table D until you have 10 unique numbers from 01 to 76. Count the number between 01 and 12. **(c)** 71 48 70 ~~99 84~~ 29 07 ~~71 48~~ 63 61 68 34 ~~70~~ 52. There is one person selected who is in first class. **(d)** It seems plausible that the actual selection was random, because 15/100 is not very small.

**5.20 (a)** What is the probability that, in selecting 7 tiles from 100, all 7 are vowels? **(b)** Let the numbers 01–42 represent the vowels, 43–98 represent the consonants, and 99 and 00 represent the blank tiles. Look up two-digit numbers in Table D until you have 7 unique numbers. Record whether all 7 numbers are between 01 and 42 or not. **(c)** The numbers read in pairs are: 00 69 40 59 77 19 66. There are two vowels, so the whole sample is not just vowels. **(d)** Because 2/1000 is a very small probability, it is very unlikely to get a sample like this by chance.

**5.21 (a)** Use a random integer generator to select 30 numbers from 1 to 365. Record whether or not there were any repeats in the sample. **(b)** Answers will vary. **(c)** Answers will vary.

**5.22 (a)** Answers will vary. **(b)** Marilyn—you are twice as likely to win if you switch doors.

18. **Is this valid?** Determine whether each of the following simulation designs is valid. Justify your answer.

(a) According to a recent survey, 50% of people aged 13 and older in the United States are addicted to texting. To simulate choosing a random sample of 20 people in this population and seeing how many of them are addicted to texting, use a deck of cards. Shuffle the deck well, and then draw one card at a time. A red card means that person is addicted to texting; a black card means he isn't. Continue until you have drawn 20 cards (without replacement) for the sample.

(b) A tennis player gets 95% of his serves in play during practice (that is, the ball doesn't go out of bounds). To simulate the player hitting 5 serves, look at 5 pairs of digits going across a row in Table D. If the number is between 00 and 94, the serve is in; numbers between 95 and 99 indicate that the serve is out.

19. pg 296 **Airport security** The Transportation Security Administration (TSA) is responsible for airport safety. On some flights, TSA officers randomly select passengers for an extra security check prior to boarding. One such flight had 76 passengers—12 in first class and 64 in coach class. Some passengers were surprised when none of the 10 passengers chosen for screening were seated in first class. We can use a simulation to see if this result is likely to happen by chance.

(a) State the question of interest using the language of probability.

(b) How would you use random digits to imitate one repetition of the process? What variable would you measure?

(c) Use the line of random digits below to perform one repetition. Copy these digits onto your paper. Mark directly on or above them to show how you determined the outcomes of the chance process.

71487 09984 29077 14863 61683 47052 62224 51025

(d) In 100 repetitions of the simulation, there were 15 times when none of the 10 passengers chosen was seated in first class. What conclusion would you draw?

20. **Scrabble** In the game of Scrabble, each player begins by drawing 7 tiles from a bag containing 100 tiles. There are 42 vowels, 56 consonants, and 2 blank tiles in the bag. Cait chooses her 7 tiles and is surprised to discover that all of them are vowels. We can use a simulation to see if this result is likely to happen by chance.

(a) State the question of interest using the language of probability.

(b) How would you use random digits to imitate one repetition of the process? What variable would you measure?

(c) Use the line of random digits below to perform one repetition. Copy these digits onto your paper. Mark directly on or above them to show how you determined the outcomes of the chance process.

00694 05977 19664 65441 20903 62371 22725 53340

(d) In 1000 repetitions of the simulation, there were 2 times when all 7 tiles were vowels. What conclusion would you draw?

21. **The birthday problem** What's the probability that in a randomly selected group of 30 unrelated people, at least two have the same birthday? Let's make two assumptions to simplify the problem. First, we'll ignore the possibility of a February 29 birthday. Second, we assume that a randomly chosen person is equally likely to be born on each of the remaining 365 days of the year.

(a) How would you use random digits to imitate one repetition of the process? What variable would you measure?

(b) Use technology to perform 5 repetitions. Record the outcome of each repetition.

(c) Would you be surprised to learn that the theoretical probability is 0.71? Why or why not?

22. **Monty Hall problem** In *Parade* magazine, a reader posed the following question to Marilyn vos Savant and the "Ask Marilyn" column:

> Suppose you're on a game show, and you're given the choice of three doors. Behind one door is a car, behind the others, goats. You pick a door, say #1, and the host, who knows what's behind the doors, opens another door, say #3, which has a goat. He says to you, "Do you want to pick door #2?" Is it to your advantage to switch your choice of doors?[4]

The game show in question was *Let's Make a Deal* and the host was Monty Hall. Here's the first part of Marilyn's response: "Yes; you should switch. The first door has a 1/3 chance of winning, but the second door has a 2/3 chance." Thousands of readers wrote to Marilyn to disagree with her answer. But she held her ground.

(a) Use an online *Let's Make a Deal* applet to perform at least 50 repetitions of the simulation. Record whether you stay or switch (try to do each about half the time) and the outcome of each repetition.

(b) Do you agree with Marilyn or her readers? Explain.

23. **Recycling** Do most teens recycle? To find out, an AP® Statistics class asked an SRS of 100 students at their school whether they regularly recycle. In the sample, 55 students said that they recycle. Is this convincing evidence that more than half of the students at the school would say they regularly recycle? The Fathom dotplot below shows the results of

taking 200 SRSs of 100 students from a population in which the true proportion who recycle is 0.50.

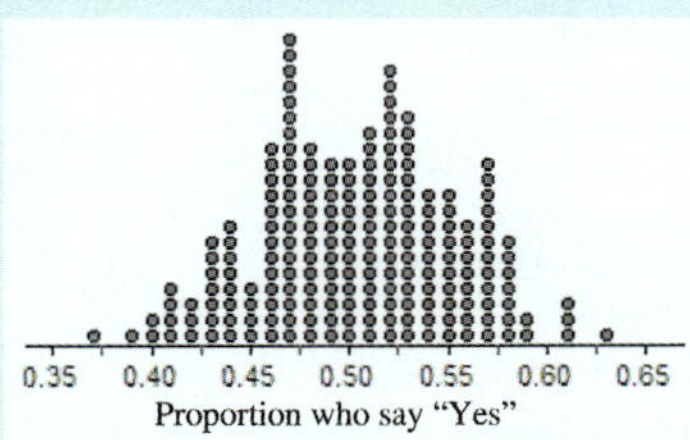


(a) Explain why the sample result does not give convincing evidence that more than half of the school's students recycle.

(b) Suppose instead that 63 students in the class's sample had said "Yes." Explain why this result would give strong evidence that a majority of the school's students recycle.

**24. Brushing teeth, wasting water?** A recent study reported that fewer than half of young adults turn off the water while brushing their teeth. Is the same true for teenagers? To find out, a group of statistics students asked an SRS of 60 students at their school if they usually brush with the water off. In the sample, 27 students said "No." The Fathom dotplot below shows the results of taking 200 SRSs of 60 students from a population in which the true proportion who brush with the water off is 0.50.

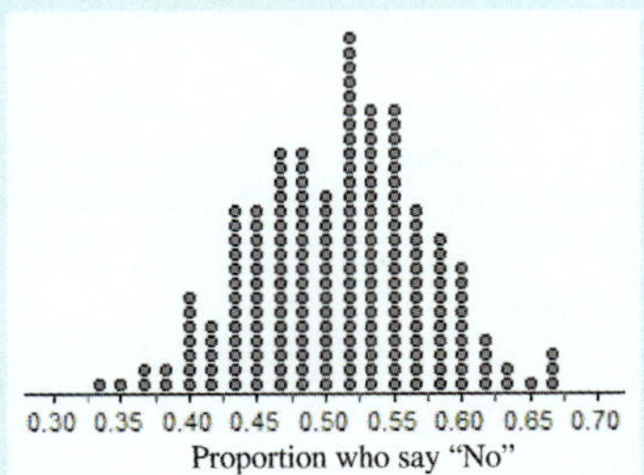


(a) Explain why the sample result does not give convincing evidence that fewer than half of the school's students brush their teeth with the water off.

(b) Suppose instead that 18 students in the class's sample had said "No." Explain why this result would give strong evidence that fewer than 50% of the school's students brush their teeth with the water off.

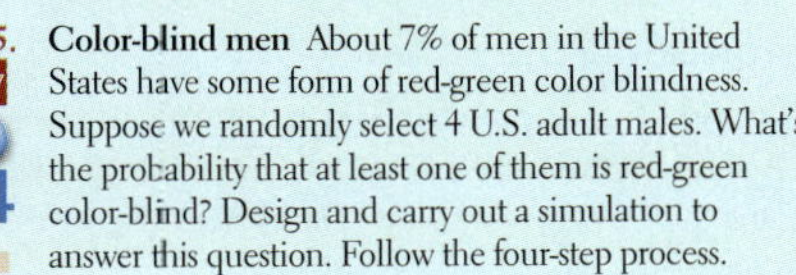

**25. Color-blind men** About 7% of men in the United States have some form of red-green color blindness. Suppose we randomly select 4 U.S. adult males. What's the probability that at least one of them is red-green color-blind? Design and carry out a simulation to answer this question. Follow the four-step process.

pg 297

**26. Lotto** In the United Kingdom's Lotto game, a player picks six numbers from 1 to 49 for each ticket. Rosemary bought one ticket for herself. She had the lottery computer randomly select the six numbers. When the six winning numbers were drawn, Rosemary was surprised to find that none of these numbers appeared on the Lotto ticket she had bought. Should she be? Design and carry out a simulation to answer this question. Follow the four-step process.

**27. Color-blind men** Refer to Exercise 25. Suppose we randomly select one U.S. adult male at a time until we find one who is red-green color-blind. Should we be surprised if it takes us 20 or more men? Design and carry out a simulation to answer this question. Follow the four-step process.

**28. Scrabble** Refer to Exercise 20. About 3% of the time, the first player in Scrabble can "bingo" by playing all 7 tiles on the first turn. Should we be surprised if it takes 30 or more games for this to happen? Design and carry out a simulation to answer this question. Follow the four-step process.

**29. Random assignment** Researchers recruited 20 volunteers—8 men and 12 women—to take part in an experiment. They randomly assigned the subjects into two groups of 10 people each. To their surprise, 6 of the 8 men were randomly assigned to the same treatment. Should they be surprised? Design and carry out a simulation to estimate the probability that the random assignment puts 6 or more men in the same group. Follow the four-step process.

**30. Taking the train** According to New Jersey Transit, the 8:00 A.M. weekday train from Princeton to New York City has a 90% chance of arriving on time. To test this claim, an auditor chooses 6 weekdays at random during a month to ride this train. The train arrives late on 2 of those days. Does the auditor have convincing evidence that the company's claim isn't true? Design and carry out a simulation to estimate the probability that a train with a 90% chance of arriving on time each day would be late on 2 or more of 6 days. Follow the four-step process.

*Multiple choice: Select the best answer for Exercises 31 to 36.*

31. You read in a book about bridge that the probability that each of the four players is dealt exactly one ace is about 0.11. This means that

(a) in every 100 bridge deals, each player has one ace exactly 11 times.

(b) in 1 million bridge deals, the number of deals on which each player has one ace will be exactly 110,000.

**5.23 (a)** Obtaining a sample percentage of 55% or higher is not particularly unusual (probability ≈ 43/200) when 50% of all students recycle. **(b)** Obtaining a sample percentage of at least 63% is very unlikely (probability ≈ 1/200) when 50% of all students recycle.

**5.24 (a)** Obtaining a sample percentage of 27/60 = 45% or lower is not particularly unusual (probability ≈ 44/200) when 50% of all people brush with the water off. **(b)** Obtaining a sample percentage of at most 18/60 = 30% is very unlikely (probability ≈ 0) when 50% of all people brush with the water off.

**5.25** STATE: What is the probability that, in a sample of 4 randomly selected U.S. adult males, at least one of them is red-green colorblind? PLAN: Let 00–06 denote a colorblind man and 07–99 denote a non-colorblind man. Read 4 two-digit numbers from Table D for each sample and record whether or not the sample had at least one red-green colorblind man in it. DO: In our 50 samples, 15 had at least one colorblind man in them. CONCLUDE: The probability that a sample of 4 men would have at least one colorblind man is approximately 15/50 = 0.30.

**5.26** STATE: What is the probability that none of the winning numbers will appear on a Lotto ticket with six numbers randomly selected? PLAN: Let 1–6 represent the winning numbers. Then using a random integer generator, pick 6 unique numbers from 1–49. Record how many numbers are 1–6. DO: In our 50 samples, 27 had no matches. CONCLUDE: Not surprised. The probability of getting no matches on a lottery ticket is large (approximately 27/50 = 0.54).

**5.27** STATE: What is the probability that it takes 20 or more selections in order to find one man who is red-green colorblind? PLAN: Let 0–6 denote a colorblind man and 7–99 denote a non-colorblind man. Use technology to pick integers from 0 to 99 until we get a number between 0 and 6. Count how many numbers there are in the sample. DO: In 16 of our 50 samples, it took 20 or more selections to get one colorblind man. CONCLUDE: Not surprised. The probability of needing 20 or more selections to get one colorblind man is fairly large (approximately 16/50 = 0.32).

**5.28** STATE: What is the probability that it takes 30 or more games to have a player bingo on the first turn? PLAN: Let 0–2 denote a bingo and 3–99 denote a nonbingo. Use technology to pick integers from 0 to 99 until we get a number between 0 and 2. Count how many numbers there are in the sample. DO: In 21 of our 50 repetitions, it took 30 or more games to get a bingo on the first turn. CONCLUDE: Not surprised. The probability of needing 30 or more games to get a bingo on the first turn is fairly large (approximately 21/50 = 0.42).

**5.29** STATE: What is the probability that the random assignment will result in at least 6 men in the same group? PLAN: Number the men 1–8 and the women 9–20. Use technology to pick 10 unique integers between 1 and 20 for one group. Record if there are at least 6 numbers between 1 and 8 in either group. DO: In our 50 repetitions, 9 had one group with 6 or more men in it. CONCLUDE: Not surprised. The probability of getting 6 or more men in one group is fairly large (approximately 9/50 = 0.18).

**5.30** STATE: What is the probability that the train will be late on at least 2 out of 6 randomly selected days? PLAN: Let 1–9 represent on-time days and 0 represent late days. Use Table D to pick 6 digits and record whether there were at least two 0s among the 6 digits. DO: In 5 of our 50 repetitions, the train arrived late on 2 or more of the 6 days. CONCLUDE: The probability of having the train arrive late twice or more in 6 days is approximately 5/50 = 0.10. Because this is not a very small probability, we do not have convincing evidence that the company's claim is false.

**5.31** c

**5.32** a

**5.33** b

**5.34** c

**5.35** c

**5.36** b

**5.37** **(a)** Population: adult U.S. residents. Sample: the 353,564 adults who were interviewed. **(b)** The people who do not have a telephone were excluded. This would lead to an underestimate of the proportion in the population who experienced stress a lot of the day yesterday if the people without phones are poorer and consequently experience more stress.

**5.38** **(a)** Both distributions are skewed to the right. The median time to leave was greater for drivers with someone waiting. There is more variability for the drivers with someone waiting as well. Finally, there were two high outliers for those with no one waiting. **(b)** Not necessarily. The researchers did not randomly assign the treatments of either having a person waiting or not to the drivers of the cars leaving the lot.



(c) in a very large number of bridge deals, the percent of deals on which each player has one ace will be very close to 11%.

(d) in a very large number of bridge deals, the average number of aces in a hand will be very close to 0.11.

(e) If each player gets an ace in only 2 of the first 50 deals, then each player should get an ace in more than 11% of the next 50 deals.

**32.** If I toss a fair coin five times and the outcomes are TTTTT, then the probability that tails appears on the next toss is

(a) 0.5. (c) greater than 0.5. (e) 1.

(b) less than 0.5. (d) 0.

*Exercises 33 to 35 refer to the following setting.* A basketball player claims to make 47% of her shots from the field. We want to simulate the player taking sets of 10 shots, assuming that her claim is true.

**33.** To simulate the number of makes in 10 shot attempts, you would perform the simulation as follows:

(a) Use 10 random one-digit numbers, where 0–4 are a make and 5–9 are a miss.

(b) Use 10 random two-digit numbers, where 00–46 are a make and 47–99 are a miss.

(c) Use 10 random two-digit numbers, where 00–47 are a make and 48–99 are a miss.

(d) Use 47 random one-digit numbers, where 0 is a make and 1–9 are a miss.

(e) Use 47 random two-digit numbers, where 00–46 are a make and 47–99 are a miss.

**34.** Twenty-five repetitions of the simulation were performed. The simulated number of makes in each set of 10 shots was recorded on the dotplot below. What is the approximate probability that a 47% shooter makes 5 or more shots in 10 attempts?

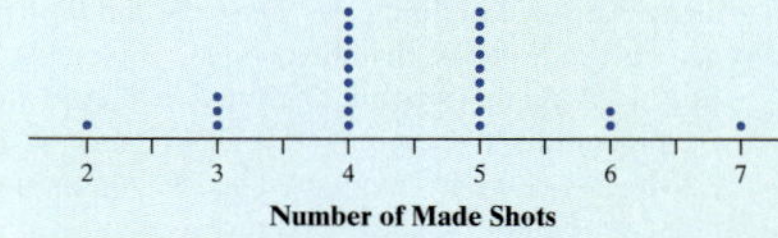


(a) 5/10 (b) 3/10 (c) 12/25 (d) 3/25 (e) 47/100

**35.** Suppose this player attempts 10 shots in a game and only makes 3 of them. Does this provide convincing evidence that she is less than a 47% shooter?

(a) Yes, because 3/10 (30%) is less than 47%.

(b) Yes, because she never made 47% of her shots in the simulation.

(c) No, because it is plausible that she would make 3 or fewer shots by chance alone.

(d) No, because the simulation was only repeated 25 times.

(e) No, because the distribution is approximately symmetric.

**36.** Ten percent of U.S. households contain 5 or more people. You want to simulate choosing a household at random and recording "Yes" if it contains 5 or more people. Which of these are correct assignments of digits for this simulation?

(a) Odd = Yes; Even = No

(b) 0 = Yes; 1–9 = No

(c) 0–5 = Yes; 6–9 = No

(d) 0–4 = Yes; 5–9 = No

(e) None of these

**37.** **Are you feeling stressed?** (4.1) A Gallup Poll asked whether people experienced stress "a lot of the day yesterday." About 41 percent said they did. Gallup's report said, "Results are based on telephone interviews conducted ... Jan. 1–Dec. 31, 2012, with a random sample of 353,564 adults aged 18 and older."[5]

(a) Identify the population and the sample.

(b) Explain how undercoverage could lead to bias in this survey.

**38.** **Waiting to park** (1.3, 4.2) Do drivers take longer to leave their parking spaces when someone is waiting? Researchers hung out in a parking lot and collected some data. The graphs and numerical summaries below display information about how long it took drivers to exit their spaces.

(a) Write a few sentences comparing these distributions.

(b) Can we conclude that having someone waiting causes drivers to leave their spaces more slowly? Why or why not?

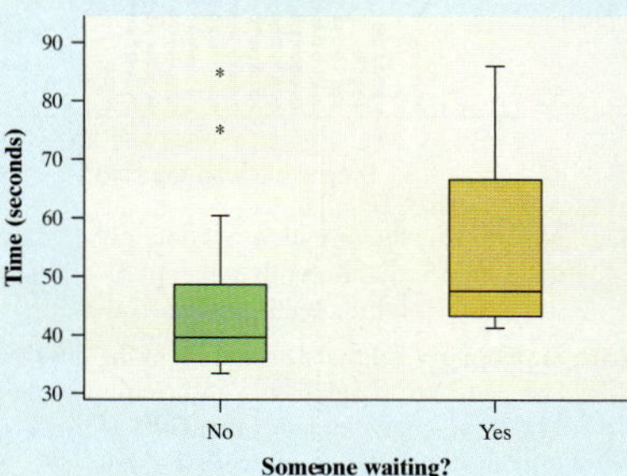


Descriptive Statistics: Time

| Waiting | $n$ | Mean | StDev | Minimum | $Q_1$ | Median | $Q_3$ | Maximum |
|---|---|---|---|---|---|---|---|---|
| No | 20 | 44.42 | 14.10 | 33.76 | 35.61 | 39.56 | 48.48 | 84.92 |
| Yes | 20 | 54.11 | 14.39 | 41.61 | 43.41 | 47.14 | 66.44 | 85.97 |

# 5.2 Probability Rules

**WHAT YOU WILL LEARN** **By the end of the section, you will be able to:**

- Describe a probability model for a chance process.
- Use basic probability rules, including the complement rule and the addition rule for mutually exclusive events.
- Use a two-way table or Venn diagram to model a chance process and calculate probabilities involving two events.
- Use the general addition rule to calculate probabilities.

The idea of probability [illegible] he fact that chance behavior is predictable in the long run. In Sectio[illegible] imulation to imitate chance behavior. Do we always need to [illegible] ss many times to determine the probability of a partic[illegible] e answer is no.

[illegible] could be used to model some [illegible] ar relationships between two [illegible] line. Now we're ready to

[illegible] in once. When we toss [illegible] now? We are willing [illegible] that each of these [illegible] two parts:

[illegible] asic vocabu-

[illegible] set of all possible outcomes.

[illegible] ne chance process that consists of two parts: a [illegible] ility for each outcome.

A sample space $S$ can be very simple or very complex. When we toss a coin once, there are only two possible outcomes, heads and tails. We can write the sample space using set notation as $S = \{H, T\}$. When Gallup draws a random sample of 1523 adults and asks a survey question, the sample space contains all possible sets of responses from 1523 of the 235 million adults in the country. This $S$ is extremely large. Each member of $S$ lists the answers from one possible sample, which explains the term *sample space*.

Let's look at how to set up a probability model in a familiar setting—rolling a pair of dice.

**PD Section 5.2 Overview**

Watch the video overview of Section 5.2 (for teachers) by clicking on the link in the TE-book, opening the TRCD, or logging onto the book's companion Web site www.whfreeman.com/tps5e (teacher log-in required).

**TRM Section 5.2 Alternate Examples**

You can find the Alternate Examples for this section in Microsoft Word format by clicking on the link in the TE-book, opening the TRCD, or logging onto the book's companion Web site www.whfreeman.com/tps5e (teacher log-in required).

## Teaching Tip

There are $\binom{235{,}000{,}000}{1523} = 2.1 \times 10^{8561}$ possible samples of size 1523 when choosing from 235 million adults, according to www.wolframalpha.com.

ALTERNATE EXAMPLE

### *Flipping coins*

Imagine flipping a fair coin three times.

PROBLEM: Give a probability model for this chance process.

SOLUTION: There are eight possible outcomes when we flip the coin three times:

HHH HHT HTH HTT TTT TTH THT THH

Because the coin is fair, each of these eight outcomes will be equally likely and have probability 1/8.

## Teaching Tip

Here are some possible events and their probabilities involving three flips of a coin that parallel the discussion on pages 306–307. Define event *A* as getting two or more heads. Then $P(A) = 4/8$ because there are four outcomes (HHH, HHT, HTH, THH) that make up this event. Define event *B* as getting no heads. Then $P(B) = 1/8$ because only one outcome (TTT) has no heads. Define event *C* as getting at least one head. Then $P(C) = 7/8$ because all the outcomes except TTT have at least one head. Notice that $P(B) + P(C) = 1$. What's the probability of getting no heads or two or more heads, that is, $P(A \text{ or } C)$? Because these events have no outcomes in common, we can add their probabilities: $P(\text{no heads or two or more heads}) = P(\text{no heads}) + P(\text{two or more heads}) = 1/8 + 4/8 = 5/8$.

EXAMPLE

## Roll the Dice

### *Building a probability model*

Many board games involve rolling dice. Imagine rolling two fair, six-sided dice—one that's red and one that's green.

PROBLEM: *Give a probability model for this chance process.*

SOLUTION: *There are 36 possible outcomes when we roll two dice and record the number of spots showing on the up-faces. Figure 5.2 displays these outcomes. They make up the sample space S. If the dice are perfectly balanced, all 36 outcomes will be equally likely. That is, each of the 36 outcomes will come up on one-thirty-sixth of all rolls in the long run. So each outcome has probability 1/36.*

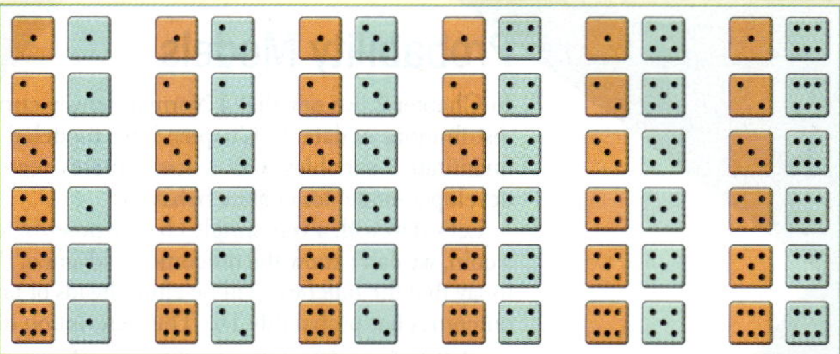

**FIGURE 5.2** The 36 possible outcomes in rolling two dice. If the dice are carefully made, all of these outcomes have the same probability.

**For Practice** *Try Exercise* **39**

A probability model does more than just assign a probability to each outcome. It allows us to find the probability of any collection of outcomes, which we call an **event.**

**DEFINITION: Event**

An **event** is any collection of outcomes from some chance process. That is, an event is a subset of the sample space. Events are usually designated by capital letters, like *A*, *B*, *C*, and so on.

If A is any event, we write its probability as P(A). In the dice-rolling example, suppose we define event A as "sum is 5." What's P(A), the probability that event A occurs? There are four outcomes in event A:

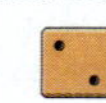

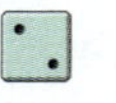

Because each of these outcomes has probability 1/36, $P(A) = 4/36$. Now consider event *B*: sum is not 5. To find $P(B)$, we could list all the outcomes that make up event $B$, but that would take a while. Fortunately, there's an easier way. Of the 36 equally likely outcomes in Figure 5.2, event A (sum is 5) occurs in 4 of them. So event A does *not* occur in 32 of these outcomes. Then $P(B) = P(\text{sum isn't } 5) = P(\text{not } A) = 32/36$. Notice that $P(A) + P(B) = 1$.

Let's consider one more event, which we'll call C: sum is 6. The outcomes in event C are

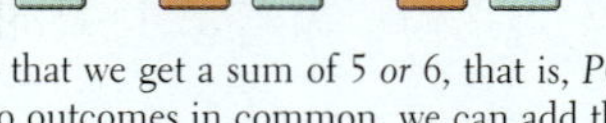

So $P(C) = 5/36$. What's the probability that we get a sum of 5 *or* 6, that is, $P(A \text{ or } C)$? Because these two events have no outcomes in common, we can add the probabilities of the individual events:

$$P(\text{sum is 5 or sum is 6}) = P(\text{sum is 5}) + P(\text{sum is 6}) = 4/36 + 5/36 = 9/36$$

In other words, $P(A \text{ or } C) = P(A) + P(C)$.

## Basic Rules of Probability

Our dice-rolling scenario revealed some basic rules that any probability model must obey:

- *The probability of any event is a number between 0 and 1.* The probability of an event is the long-run proportion of repetitions on which that event occurs. Any proportion is a number between 0 and 1, so any probability is also a number between 0 and 1. An event with probability 0 never occurs, and an event with probability 1 occurs on every trial. An event with probability 0.5 occurs in half the trials in the long run.
- *All possible outcomes together must have probabilities that add up to 1.* Because some outcome must occur on every trial, the sum of the probabilities for all possible outcomes must be exactly 1.
- *If all outcomes in the sample space are equally likely, the probability that event A occurs can be found using the formula*

$$P(A) = \frac{\text{number of outcomes corresponding to event A}}{\text{total number of outcomes in sample space}}$$

- *The probability that an event does not occur is 1 minus the probability that the event does occur.* If an event occurs in (say) 70% of all trials, it fails to occur in the other 30%. The probability that an event occurs and the probability that it does not occur always add to 100%, or 1. (This explains why $P(\text{sum isn't } 5) = 1 - P(\text{sum is } 5)$ in the dice-rolling example.) We refer to the event "not A" as the **complement** of A and denote it by $A^C$.
- *If two events have no outcomes in common, the probability that one or the other occurs is the sum of their individual probabilities.* If one event occurs in 40% of all trials, a different event occurs in 25% of all trials, and the two can never occur together, then one or the other occurs on 65% of all trials because 40% + 25% = 65%. When two events have no outcomes in common, we refer to them as **mutually exclusive** or **disjoint**.

**DEFINITION: Mutually exclusive (disjoint)**

Two events $A$ and $B$ are **mutually exclusive (disjoint)** if they have no outcomes in common and so can never occur together—that is, if $P(A \text{ and } B) = 0$.

### Teaching Tip

If students express a probability as a fraction, it's OK if they don't reduce it. In fact, it's often easier for students (and teachers) to keep the denominator the same for the probability of each outcome. This makes it easier to compare the probabilities and verify that they add up to 1.

### Teaching Tip

Emphasize that this formula for the probability of event $A$ works only if all the outcomes in the sample space are equally likely. For example, there are two possible outcomes regarding tomorrow's weather: rain or no rain. However, the probability of rain isn't necessarily 1/2 because these two outcomes may not be equally likely.

### Teaching Tip

Ask students if they have used the word "complement" in any other classes. Hopefully, someone will remember that two complementary angles have a sum of 90 degrees. In probability, the probabilities of two complementary events have a sum of 1.

### AP® Exam Common Error

When using the complement rule, some students use incorrect notation such as $0.25^C = 0.75$ or $P(A)^C = 0.75$. Remind students that events have complements, not probabilities. Here is a correct statement: If $P(A) = 0.25$, then $P(A^C) = 1 - 0.25 = 0.75$. Notice that the "$C$" appears in the superscript of event $A$, not its probability.

We can summarize the basic probability rules more concisely in symbolic form.

**BASIC PROBABILITY RULES**

- For any event A, $0 \leq P(A) \leq 1$.
- If S is the sample space in a probability model, $P(S) = 1$.
- In the case of *equally likely* outcomes,

$$P(A) = \frac{\text{number of outcomes corresponding to event A}}{\text{total number of outcomes in sample space}}$$

- **Complement rule:** $P(A^C) = 1 - P(A)$.
- **Addition rule for mutually exclusive events:** If A and B are mutually exclusive, $P(A \text{ or } B) = P(A) + P(B)$.

The earlier dice-rolling example involved equally likely outcomes. Here's an example that illustrates use of the basic probability rules when the outcomes of a chance process are not equally likely.

**ALTERNATE EXAMPLE**

### AP® Statistics scores

Randomly select a student who took the 2013 AP® Statistics exam and record the student's score. Here is the probability model:

| Score | Probability |
|---|---|
| 1 | 0.235 |
| 2 | 0.188 |
| 3 | 0.249 |
| 4 | 0.202 |
| 5 | 0.126 |

PROBLEM:
**(a)** Show that this is a legitimate probability model.
**(b)** Find the probability that the chosen student scored 3 or better.

SOLUTION:
**(a)** All the probabilities are between 0 and 1 and the sum of the probabilities is 1, so this is a legitimate probability model.
**(b)** There are two ways to find this probability: By the addition rule, $P(3 \text{ or better}) = 0.249 + 0.202 + 0.126 = 0.577$. By the complement rule and addition rule, $P(3 \text{ or better}) = 1 - P(2 \text{ or less}) = 1 - (0.235 + 0.188) = 1 - 0.423 = 0.577$.

**EXAMPLE**

## Distance Learning

### *Applying probability rules*

Distance-learning courses are rapidly gaining popularity among college students. Randomly select an undergraduate student who is taking a distance-learning course for credit, and record the student's age. Here is the probability model:[6]

| Age group (yr): | 18 to 23 | 24 to 29 | 30 to 39 | 40 or over |
|---|---|---|---|---|
| Probability: | 0.57 | 0.17 | 0.14 | 0.12 |

PROBLEM:
(a) Show that this is a legitimate probability model.
(b) Find the probability that the chosen student is not in the traditional college age group (18 to 23 years).

SOLUTION:
(a) The probability of each outcome is a number between 0 and 1, and the probabilities of all the possible outcomes add to 1, so this is a legitimate probability model.
(b) There are two ways to find this probability. By the complement rule,

$$P(\text{not 18 to 23 years}) = 1 - P(\text{18 to 23 years}) = 1 - 0.57 = 0.43$$

That is, if 57% of distance learners are 18 to 23 years old, then the remaining 43% are not in this age group.

Using the addition rule for mutually exclusive events,

$$P(\text{not 18 to 23 years}) = P(\text{24 to 29 years}) + P(\text{30 to 39 years}) + P(\text{40 years or over})$$
$$= 0.17 + 0.14 + 0.12 = 0.43$$

There is a 43% chance that the chosen student is not in the traditional college age group.

**For Practice** *Try Exercise* **45**

### CHECK YOUR UNDERSTANDING

Choose an American adult at random. Define two events:

A = the person has a cholesterol level of 240 milligrams per deciliter of blood (mg/dl) or above (high cholesterol)
B = the person has a cholesterol level of 200 to 239 mg/dl (borderline high cholesterol)
According to the American Heart Association, $P(A) = 0.16$ and $P(B) = 0.29$.

1. Explain why events A and B are mutually exclusive.
2. Say in plain language what the event "A or B" is. What is $P(A \text{ or } B)$?
3. If C is the event that the person chosen has normal cholesterol (below 200 mg/dl), what's $P(C)$?

## Two-Way Tables, Probability, and the General Addition Rule

When we're trying to find probabilities involving two events, a two-way table can display the sample space in a way that makes probability calculations easier.

## EXAMPLE Who Has Pierced Ears?

### *Two-way tables and probability*

Students in a college statistics class wanted to find out how common it is for young adults to have their ears pierced. They recorded data on two variables—gender and whether the student had a pierced ear—for all 178 people in the class. The two-way table below displays the data.

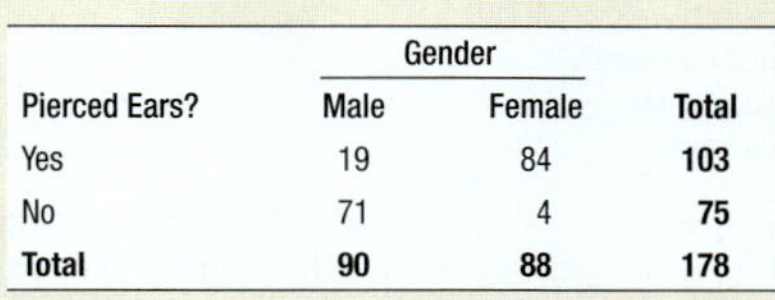

| | Gender | | |
|---|---|---|---|
| Pierced Ears? | Male | Female | Total |
| Yes | 19 | 84 | **103** |
| No | 71 | 4 | **75** |
| **Total** | **90** | **88** | **178** |

PROBLEM: Suppose we choose a student from the class at random. Find the probability that the student

(a) has pierced ears.
(b) is male and has pierced ears.
(c) is male or has pierced ears.

SOLUTION: We'll define events $A$: is male and $B$: has pierced ears.

(a) Because each of the 178 students in the class is equally likely to be chosen, and there are 103 students with pierced ears, $P(\text{pierced ears}) = P(B) = 103/178$.

(b) We want to find $P(\text{male } and \text{ pierced ears})$, that is, $P(A \text{ and } B)$. Looking at the intersection of the "Male" column and "Yes" row, we see that there are 19 males with pierced ears. So $P(\text{male and pierced ears}) = P(A \text{ and } B) = 19/178$.

(c) This time, we're interested in $P(\text{male } or \text{ pierced ears})$, that is, $P(A \text{ or } B)$. (Note the mathematical use of the word *or* here—the person could be a male or have pierced ears or both.) From the two-way table,

### ✓ ANSWERS TO CYU

**1.** A person cannot have a cholesterol level of both 240 or above and between 200 and 239 at the same time.

**2.** A person has either a cholesterol level of 240 or above or they have a cholesterol level between 200 and 239. $P(A \text{ or } B) = 0.16 + 0.29 = 0.45$.

**3.** $P(C) = 1 - 0.45 = 0.55$

### Teaching Tip

As in Chapter 1, when there is an explanatory variable, we put the values of this variable across the top of a two-way table and put the values of the response variable down the left side. This matches how we display the relationship between two quantitative variables on a scatterplot: explanatory variable on the horizontal axis and response variable on the vertical axis.

### Teaching Tip

To help students understand the data summarized in the table, consider using 178 notecards to represent the students in the class. For example, on 19 cards write "Male" on one side and "Yes" on the other side. Then, on 84 cards write "Female" on one side and "Yes" on the other side, and so on. These cards will prove useful over the next two sections.

### ALTERNATE EXAMPLE

### *Who owns a home?*

What is the relationship between educational achievement and home ownership? A random sample of 500 U.S. adults was selected. Each member of the sample was identified as a high school graduate (or not) and as a homeowner (or not). The two-way table displays the data.

| | High school graduate | Not a high school graduate | Total |
|---|---|---|---|
| Homeowner | 221 | 119 | **340** |
| Not a homeowner | 89 | 71 | **160** |
| **Total** | **310** | **190** | **500** |

PROBLEM: Suppose we choose a member of the sample at random. Find the probability that the member
**(a)** is a high school graduate.
**(b)** is a high school graduate and owns a home.
**(c)** is a high school graduate or owns a home.

SOLUTION: We will define event $G$ as being a high school graduate and event $H$ as being a homeowner.

**(a)** Because 310 of the 500 members of the sample graduated from high school, $P(G) = 310/500$.
**(b)** Because 221 of the 500 members of the sample graduated from high school and own a home, $P(G \text{ and } H) = 221/500$.
**(c)** Because there are $221 + 89 + 119 = 429$ people who graduated from high school or own a home, $P(G \text{ or } H)$ is 429/500. Note that it is inappropriate to compute $P(G) + P(H)$ to find this probability because the events $G$ and $H$ are not mutually exclusive—221 people are both high school graduates and homeowners. If you did add these probabilities, the result would be 650/500, which is clearly wrong because it results in a probability that is greater than 1.

we see that there are 90 males in the class, so $P(A) = 90/178$. Can we just add $P(A)$ to $P(B)$ to get the correct answer? No! These two events are *not* mutually exclusive, because there are 19 males with pierced ears. (If we did add the two probabilities, we'd get $90/178 + 103/178 = 193/178$, which is clearly wrong, because the probability is bigger than 1!) From the two-way table, we see that there are $19 + 71 + 84 = 174$ students who are male *or* have pierced ears. So $P(A \text{ or } B) = 174/178$.

**For Practice** *Try Exercise* **49**

When we found the probability of getting a male with pierced ears in the example, we could have described this as either $P(A \text{ and } B)$ or $P(B \text{ and } A)$. Why? Because "$A$ and $B$" describes the same event as "$B$ and $A$." Likewise, $P(A \text{ or } B)$ is the same as $P(B \text{ or } A)$. *Don't get so caught up in the notation that you lose sight of what's really happening!*

The previous example revealed two important facts about finding the probability P(A or B) when the two events are not mutually exclusive. First, the use of the word "or" in probability questions is different from that in everyday life. If someone says, "I'll either watch a movie or go to the football game," that usually means they'll do one thing or the other, but not both. In statistics, "A or B" could mean one or the other or both. Second, we can't use the addition rule for mutually exclusive events unless two events have no outcomes in common.

If events A and B are *not* mutually exclusive, they can occur together. The probability that one or the other occurs is then *less* than the sum of their probabilities. As Figure 5.3 illustrates, outcomes common to both are counted twice when we add probabilities.

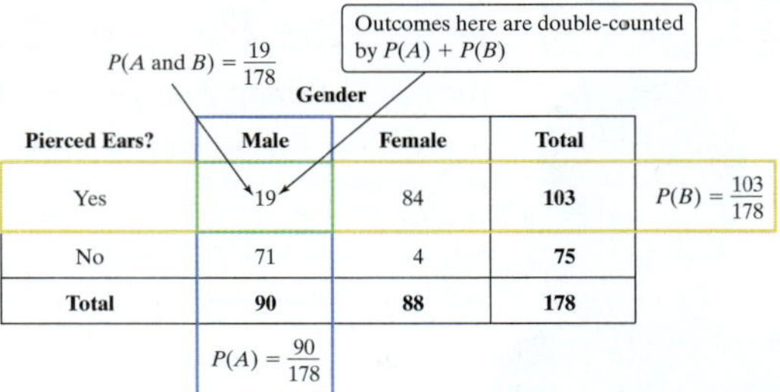


| Pierced Ears? | Male | Female | Total |
|---|---|---|---|
| Yes | 19 | 84 | 103 |
| No | 71 | 4 | 75 |
| Total | 90 | 88 | 178 |

**FIGURE 5.3** Two-way table showing events $A$ and $B$ from the pierced-ears example. These events are *not* mutually exclusive, so we can't find $P(A \text{ or } B)$ by just adding the probabilities of the two events.

We can fix the double-counting problem illustrated in the two-way table by subtracting the probability P(A and B) from the sum. That is,

$$P(A \text{ or } B) = P(A) + P(B) - P(A \text{ and } B)$$

This result is known as the **general addition rule.** Let's check that it works for the pierced-ears example:

$$\begin{aligned} P(A \text{ or } B) &= P(A) + P(B) - P(A \text{ and } B) \\ &= 90/178 + 103/178 - 19/178 \\ &= 174/178 \end{aligned}$$

This matches our earlier result.

**GENERAL ADDITION RULE FOR TWO EVENTS**

If A and B are any two events resulting from some chance process, then

$$P(A \text{ or } B) = P(A) + P(B) - P(A \text{ and } B)$$

## Teaching Tip

Using the notecards described on page 309, show how the 19 males with pierced ears are double-counted by attempting to separate the cards into a pile for "Yes" and a pile for "Male." The 19 cards for the males with pierced ears can't be put in both piles, but they are included in the totals on the table.

## Teaching Tip

Here are some examples of the two different uses of the word "or." In a restaurant, when you are asked if you want "soup or salad," the waiter wants you to choose one or the other, but not both. However, when you order coffee and are asked if you want "cream or sugar," it's OK to ask for one or the other or both.

**ALTERNATE EXAMPLE**

***Who owns a home?***

The general addition rule works for the homeowner example as well:

$P(G \text{ or } H) = P(G) + P(H) - P(G \text{ and } H)$
$= 310/500 + 340/500 - 221/500$
$= 429/500.$

This matches our previous result.

## AP® Exam Formula Sheet

The general addition rule is included on the formula sheet that the students are provided on both sections of the AP® exam. However, the formula on the AP® exam uses the union (∪) and intersection (∩) symbols instead of the words "or" and "and":

$$P(A \cup B) = P(A) + P(B) - P(A \cap B)$$

We will introduce these symbols shortly when discussing Venn diagrams.

What happens if we use the general addition rule for two mutually exclusive events A and B? In that case, P(A and B) = 0, and the formula reduces to P(A or B) = P(A) + P(B). In other words, the addition rule for mutually exclusive events is just a special case of the general addition rule.

You might be wondering whether there is also a rule for finding P(A and B). There is, but it's not quite as intuitive. Stay tuned for that later.

## CHECK YOUR UNDERSTANDING

Sometimes it's easier to designate events with letters that relate to the context, like *F* for "face card" and *H* for "heart."

A standard deck of playing cards (with jokers removed) consists of 52 cards in four suits—clubs, diamonds, hearts, and spades. Each suit has 13 cards, with denominations ace, 2, 3, 4, 5, 6, 7, 8, 9, 10, jack, queen, and king. The jack, queen, and king are referred to as "face cards." Imagine that we shuffle the deck thoroughly and deal one card. Let's define events *F*: getting a face card and *H*: getting a heart.

1. Make a two-way table that displays the sample space.
2. Find P(*F* and *H*).
3. Explain why P(*F* or *H*) ≠ P(*F*) + P(*H*). Then use the general addition rule to find P(*F* or *H*).

## Venn Diagrams and Probability

We have already seen that two-way tables can be used to illustrate the sample space of a chance process involving two events. So can **Venn diagrams**. Because Venn diagrams have uses in other branches of mathematics, some standard vocabulary and notation have been developed.

- We introduced the *complement* of an event earlier. In Figure 5.4(a), the complement $A^C$ contains exactly the outcomes that are not in A.
- The events A and B in Figure 5.4(b) are *mutually exclusive* (*disjoint*) because they do not overlap; that is, they have no outcomes in common.

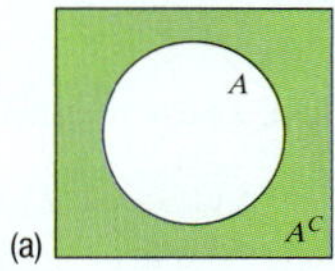

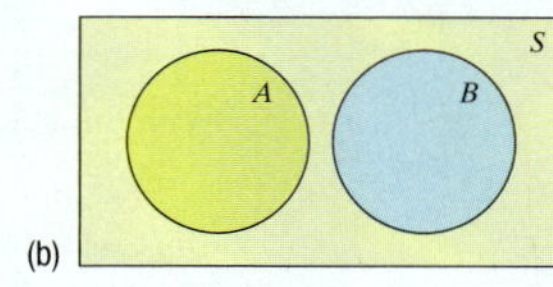


FIGURE 5.4 Venn diagrams showing: (a) event *A* and its complement $A^C$ and (b) mutually exclusive (disjoint) events *A* and *B*.

- Figure 5.5(a) shows the event "*A and B*." You can see why this event is also called the **intersection** of A and B. The corresponding notation is $A \cap B$.
- The event "A or B" is shown in Figure 5.5(b). This event is also known as the **union** of A and B. The corresponding notation is $A \cup B$.

Here's a way to keep the symbols straight: ∪ for **u**nion; ∩ for **in**tersection.

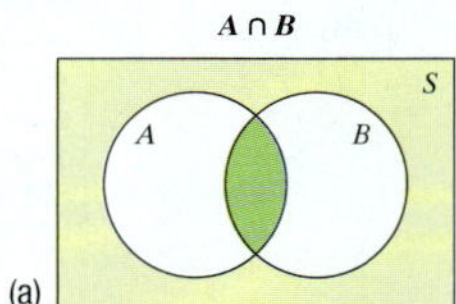

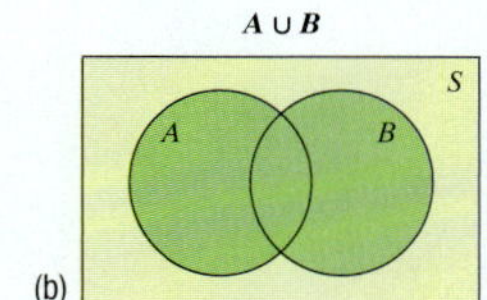


FIGURE 5.5 Venn diagrams showing (a) the intersection and (b) the union of events *A* and *B*.

### ANSWERS TO CYU

**1.**

| | Face Card | Non-face Card | Total |
|---|---|---|---|
| Heart | 3 | 10 | **13** |
| Non-heart | 9 | 30 | **39** |
| **Total** | **12** | **40** | **52** |

**2.** *P*(*F* and *H*) = 3/52 = 0.058.

**3.** The face cards that are hearts will be double-counted because *F* and *H* are not mutually exclusive.

$$P(F \text{ or } H) = \frac{12}{52} + \frac{13}{52} - \frac{3}{52} = \frac{22}{52}$$
$$= 0.423$$

### Teaching Tip

For these Venn diagrams, the sample space *S* is represented by the rectangular box.

### Teaching Tip

Ask your students what mutually exclusive "looks like" on a two-way table. Here is an example of a survey of 30 high school students. The events "is a freshman" and "is taking AP® Statistics" are mutually exclusive because there are 0 students who are in both events.

| | Freshman | Not a freshman | Total |
|---|---|---|---|
| In AP® Statistics | 0 | 3 | **7** |
| Not in AP® Statistics | 8 | 19 | **27** |
| **Total** | **8** | **22** | **30** |

## EXAMPLE Who Has Pierced Ears?

### *Understanding Venn diagrams*

In the preceding example, we looked at data from a survey on gender and ear piercings for a large group of college students. The chance process came from selecting a student in the class at random. Our events of interest were A: is male and B: has pierced ears. Here is the two-way table that summarizes the sample space:

| | Gender | |
|---|---|---|
| **Pierced Ears?** | **Male** | **Female** |
| Yes | 19 | 84 |
| No | 71 | 4 |

How would we construct a Venn diagram that displays the information in the two-way table?

There are four distinct regions in the Venn diagram shown in Figure 5.6. These regions correspond to the four cells in the two-way table. We can describe this correspondence in tabular form as follows:

| Region in Venn diagram | In words | In symbols | Count |
|---|---|---|---|
| In the intersection of two circles | Male and pierced ears | $A \cap B$ | 19 |
| Inside circle *A*, outside circle *B* | Male and no pierced ears | $A \cap B^C$ | 71 |
| Inside circle *B*, outside circle *A* | Female and pierced ears | $A^C \cap B$ | 84 |
| Outside both circles | Female and no pierced ears | $A^C \cap B^C$ | 4 |

We have added the appropriate counts of students to the four regions in Figure 5.6.

**FIGURE 5.6** Completed Venn diagram for the large group of college students. The circles represent the two events *A*: is male and *B*: has pierced ears.

With this new notation, we can rewrite the general addition rule in symbols as

$$P(A \cup B) = P(A) + P(B) - P(A \cap B)$$

This Venn diagram shows why the formula works in the pierced-ears example.

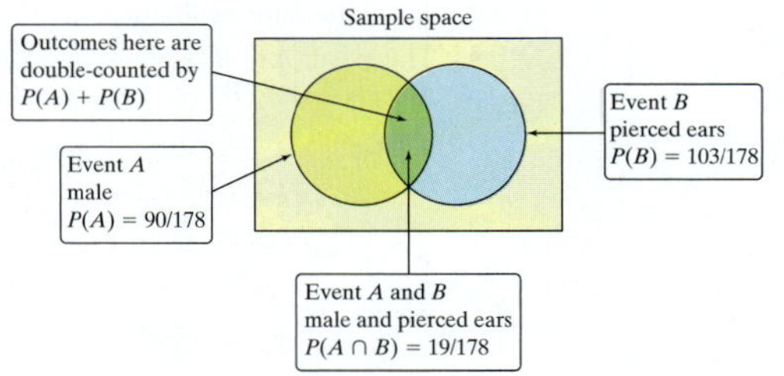


The following example ties all this together.

### Teaching Tip

Using the cards from page 309, separate them into 4 groups (male/yes, male/no, female/yes, female/no) and physically place them into the appropriate places on a Venn diagram.

### ALTERNATE EXAMPLE

#### *Who owns a home?*

Here is the two-way table summarizing the relationship between educational status and home ownership from the preceding Alternate Example. Our events of interest were *G*: is a high school graduate and *H*: is a homeowner.

| | High school graduate | Not a high school graduate | Total |
|---|---|---|---|
| Homeowner | 221 | 119 | **340** |
| Not a homeowner | 89 | 71 | **160** |
| **Total** | **310** | **190** | **500** |

The four distinct regions in the Venn diagram at right correspond to the four cells in the two-way table as follows:

| Region in Venn diagram | In words | In symbols | Count |
|---|---|---|---|
| In the intersection of two circles | Graduate and owns home | $G \cap H$ | 221 |
| Inside circle *G*, outside circle *H* | Graduate and doesn't own home | $G \cap H^C$ | 89 |
| Inside circle *H*, outside circle *G* | Not a graduate but owns home | $G^C \cap H$ | 119 |
| Outside both circles | Not a graduate and doesn't own home | $G^C \cap H^C$ | 71 |

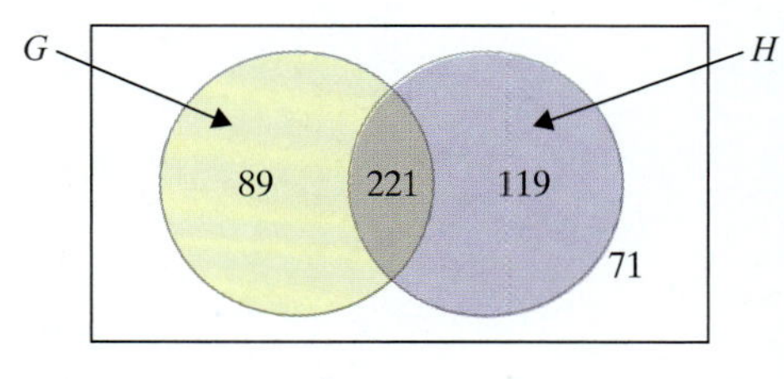

# EXAMPLE Who Reads the Paper?

## *Venn diagrams, two-way tables, and probability*

In an apartment complex, 40% of residents read *USA Today*. Only 25% read the *New York Times*. Five percent of residents read both papers. Suppose we select a resident of the apartment complex at random and record which of the two papers the person reads.

PROBLEM:

(a) Make a two-way table that displays the sample space of this chance process.

(b) Construct a Venn diagram to represent the outcomes of this chance process.

(c) Find the probability that the person reads at least one of the two papers.

(d) Find the probability that the person doesn't read either paper.

SOLUTION: We'll define events $A$: reads *USA Today* and $B$: reads *New York Times*. From the problem statement, we know that $P(A) = 0.40$, $P(B) = 0.25$, and $P(A \cap B) = 0.05$.

(a) We can enter the value 0.40 as the total for the "Yes" column, 0.25 as the total for the "Yes" row, 0.05 in the "Yes, Yes" cell, and 1 as the grand total in the two-way table shown here. This gives us enough information to fill in the empty cells of the table, starting with the missing row total for "No" $(1 - 0.25 = 0.75)$ and the missing column total for "No" $(1 - 0.40 = 0.60)$. In a similar way, we can determine the missing number in the "Yes" row $(0.25 - 0.05 = 0.20)$ and the "Yes" column $(0.40 - 0.05 = 0.35)$. That leaves 0.40 for the "No, No" cell.

| | Reads *USA Today*? | | |
|---|---|---|---|
| Reads *New York Times* | Yes | No | **Total** |
| Yes | 0.05 | 0.20 | **0.25** |
| No | 0.35 | 0.40 | **0.75** |
| **Total** | **0.40** | **0.60** | **1.00** |

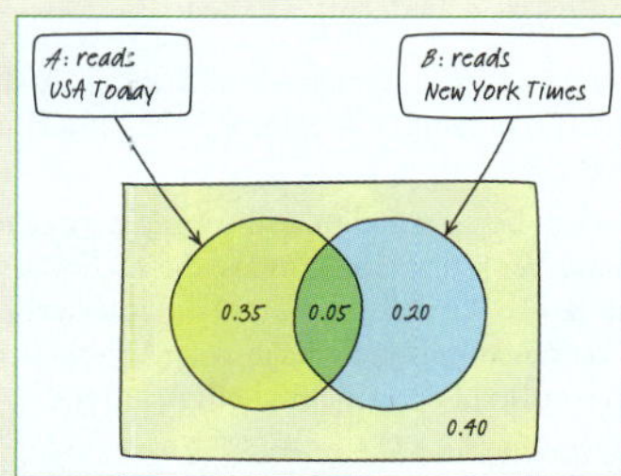


**FIGURE 5.7** Venn diagram showing the residents of an apartment complex who $A$: read *USA Today* and $B$: read the *New York Times*.

(b) Figure 5.7 shows the Venn diagram that corresponds to the completed two-way table from (a).

(c) If the randomly selected person reads *at least one* of the two papers, then he or she reads *USA Today*, the *New York Times*, or both papers. But that's the same as the event $A \cup B$. From the two-way table, the Venn diagram, or the general addition rule, we have

$$P(A \cup B) = P(A) + P(B) - P(A \cap B)$$
$$= 0.40 + 0.25 - 0.05 = 0.60$$

So there's a 60% chance that the randomly selected resident reads at least one of the two papers.

(d) From the two-way table or Venn diagram, $P(\text{reads neither paper}) = P(A^C \cap B^C) = 0.40$.

**For Practice** *Try Exercise*  55

**AP® EXAM TIP** Many probability problems involve simple computations that you can do on your calculator. It may be tempting to just write down your final answer without showing the supporting work. Don't do it! A "naked answer," even if it's correct, will usually earn you no credit on a free-response question.

## ALTERNATE EXAMPLE

### *Phone usage*

According to the National Center for Health Statistics (http://www.cdc.gov/nchs/data/nhis/earlyrelease/wireless201306.pdf), in December 2012, 60% of U.S. households had a traditional landline telephone, 89% of households had cell phones, and 51% had both. Suppose we randomly selected a household in December 2012.

PROBLEM:

**(a)** Make a two-way table that displays the sample space of this chance process.
**(b)** Construct a Venn diagram to represent the outcomes of this chance process.
**(c)** Find the probability that the household has at least one of the two types of phones.
**(d)** Find the probability that the household has a cell phone only.

SOLUTION: We will define events $L$: has a landline and $C$: has a cell phone.

**(a)**

| | Cell phone | No cell phone | Total |
|---|---|---|---|
| Landline | 0.51 | 0.09 | **0.60** |
| No landline | 0.38 | 0.02 | **0.40** |
| **Total** | **0.89** | **0.11** | **1.00** |

**(b)**

*L: Landline* *C: Cell phone*

0.09 0.51 0.38 0.02

**(c)** To find the probability that the household has at least one of the two types of phones, we need to find the probability that the household has a landline, a cell phone, or both.

$$P(L \cup C) = P(L) + P(C) - P(L \cap C)$$
$$= 0.60 + 0.89 - 0.51 = 0.98$$

There is a 0.98 probability that the household has at least one of the two types of phones.
**(d)** $P(\text{cell phone only}) = P(L^C \cap C) = 0.38$. There is a 0.38 probability that the household is cell-phone only.

## Teaching Tip

Emphasize the AP® Exam Tip on this page and make sure to grade your own exams this way to train your students.

In the previous example, the event "reads neither paper" is the complement of the event "reads at least one of the papers." To solve part (d) of the problem, we could have used our answer from (c) and the complement rule:

$$P(\text{reads neither paper}) = 1 - P(\text{reads at least one paper}) = 1 - 0.60 = 0.40$$

As you'll see in Section 5.3, the fact that "none" is the opposite of "at least one" comes in handy for a variety of probability questions.

**TRM Section 5.2 Quizzes**

There are three different quizzes available for this section in the Teacher's Resource Materials. Versions A and B are roughly parallel. Find them by clicking on the link in the TE-book, opening the TRCD, or logging onto the book's companion Web site www.whfreeman.com/tps5e (teacher log-in required).

## Section 5.2 Summary

- A **probability model** describes chance behavior by listing the possible outcomes in the **sample space S** and giving the probability that each outcome occurs.
- An **event** is a subset of the possible outcomes in the sample space. To find the probability that an event A happens, we can rely on some basic probability rules:
  - For any event A, $0 \le P(A) \le 1$.
  - $P(S) = 1$, where $S$ = the sample space
  - If all outcomes in the sample space are equally likely,

    $$P(A) = \frac{\text{number of outcomes corresponding to event A}}{\text{total number of outcomes in sample space}}$$

  - *Complement rule:* $P(A^C) = 1 - P(A)$, where $A^C$ is the **complement** of event A; that is, the event that A does not happen.
  - *Addition rule for mutually exclusive events:* Events A and B are **mutually exclusive (disjoint)** if they have no outcomes in common. If A and B are disjoint, $P(A \text{ or } B) = P(A) + P(B)$.
- A **two-way table** or a **Venn diagram** can be used to display the sample space for a chance process. Two-way tables and Venn diagrams can also be used to find probabilities involving events A and B, like the **union** $(A \cup B)$ and **intersection** $(A \cap B)$. The event $A \cup B$ ("A or B") consists of all outcomes in event A, event B, or both. The event $A \cap B$ ("A and B") consists of outcomes in both A and B.
- The **general addition rule** can be used to find $P(A \text{ or } B)$:

  $$P(A \text{ or } B) = P(A \cup B) = P(A) + P(B) - P(A \cap B)$$

**TRM Full Solutions to Section 5.2 Exercises**

## Section 5.2 Exercises

39. pg 306  **Role-playing games** Computer games in which the players take the roles of characters are very popular. They go back to earlier tabletop games such as Dungeons & Dragons. These games use many different types of dice. A four-sided die has faces with 1, 2, 3, and 4 spots.

(a) List the sample space for rolling the die twice (spots showing on first and second rolls).

(b) What is the assignment of probabilities to outcomes in this sample space? Assume that the die is perfectly balanced.

### Answers to Section 5.2 Exercises

**5.39 (a)** (1, 1), (1, 2), (1, 3), (1, 4), (2, 1), (2, 2), (2, 3), (2, 4), (3, 1), (3, 2), (3, 3), (3, 4), (4, 1), (4, 2), (4, 3), (4, 4).

**(b)** Each outcome has probability $\frac{1}{16}$.

**40. Tossing coins** Imagine tossing a fair coin 3 times.

(a) What is the sample space for this chance process?

(b) What is the assignment of probabilities to outcomes in this sample space?

**41. Role-playing games** Refer to Exercise 39. Define event *A*: sum is 5. Find *P*(*A*).

**42. Tossing coins** Refer to Exercise 40. Define event *B*: get more heads than tails. Find *P*(*B*).

**43. Probability models?** In each of the following situations, state whether or not the given assignment of probabilities to individual outcomes is legitimate, that is, satisfies the rules of probability. If not, give specific reasons for your answer.

(a) Roll a 6-sided die and record the count of spots on the up-face: $P(1) = 0$, $P(2) = 1/6$, $P(3) = 1/3$, $P(4) = 1/3$, $P(5) = 1/6$, $P(6) = 0$.

(b) Choose a college student at random and record gender and enrollment status: *P*(female full-time) = 0.56, *P*(male full-time) = 0.44, *P*(female part-time) = 0.24, *P*(male part-time) = 0.17.

(c) Deal a card from a shuffled deck: *P*(clubs) = 12/52, *P*(diamonds) = 12/52, *P*(hearts) = 12/52, *P*(spades) = 16/52.

**44. Rolling a die** The following figure displays several possible probability models for rolling a die. Some of the models are not *legitimate*. That is, they do not obey the rules. Which are legitimate and which are not? In the case of the illegitimate models, explain what is wrong.

| | Probability | | | |
|---|---|---|---|---|
| Outcome | Model 1 | Model 2 | Model 3 | Model 4 |
| ⚀ | 1/7 | 1/3 | 1/3 | 1 |
| ⚁ | 1/7 | 1/6 | 1/6 | 1 |
| ⚂ | 1/7 | 1/6 | 1/6 | 2 |
| ⚃ | 1/7 | 0 | 1/6 | 1 |
| ⚄ | 1/7 | 1/6 | 1/6 | 1 |
| ⚅ | 1/7 | 1/6 | 1/6 | 2 |

**45. Blood types** All human blood can be typed as one of O, A, B, or AB, but the distribution of the types varies a bit with race. Here is the distribution of the blood type of a randomly chosen black American:

pg 308

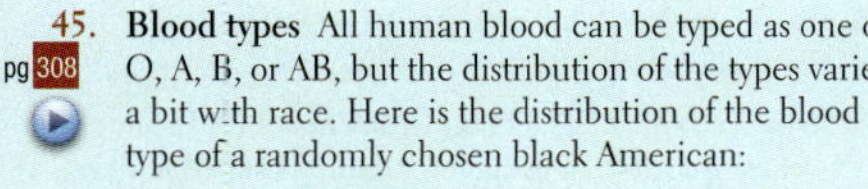

| Blood type: | O | A | B | AB |
|---|---|---|---|---|
| Probability: | 0.49 | 0.27 | 0.20 | ? |

(a) What is the probability of type AB blood? Why?

(b) What is the probability that the person chosen does not have type AB blood?

(c) Maria has type B blood. She can safely receive blood transfusions from people with blood types O and B. What is the probability that a randomly chosen black American can donate blood to Maria?

**46. Languages in Canada** Canada has two official languages, English and French. Choose a Canadian at random and ask, "What is your mother tongue?" Here is the distribution of responses, combining many separate languages from the broad Asia/Pacific region:[7]

| Language: | English | French | Asian/Pacific | Other |
|---|---|---|---|---|
| Probability: | 0.63 | 0.22 | 0.06 | ? |

(a) What probability should replace "?" in the distribution? Why?

(b) What is the probability that a Canadian's mother tongue is not English?

(c) What is the probability that a Canadian's mother tongue is a language other than English or French?

**47. Education among young adults** Choose a young adult (aged 25 to 29) at random. The probability is 0.13 that the person chosen did not complete high school, 0.29 that the person has a high school diploma but no further education, and 0.30 that the person has at least a bachelor's degree.

(a) What must be the probability that a randomly chosen young adult has some education beyond high school but does not have a bachelor's degree? Why?

(b) What is the probability that a randomly chosen young adult has at least a high school education? Which rule of probability did you use to find the answer?

**48. Preparing for the GMAT** A company that offers courses to prepare students for the Graduate Management Admission Test (GMAT) has the following information about its customers: 20% are currently undergraduate students in business; 15% are undergraduate students in other fields of study; 60% are college graduates who are currently employed; and 5% are college graduates who are not employed. Choose a customer at random.

(a) What's the probability that the customer is currently an undergraduate? Which rule of probability did you use to find the answer?

(b) What's the probability that the customer is not an undergraduate business student? Which rule of probability did you use to find the answer?

**49. Who eats breakfast?** Students in an urban school were curious about how many children regularly eat breakfast. They conducted a survey, asking, "Do you

pg 309

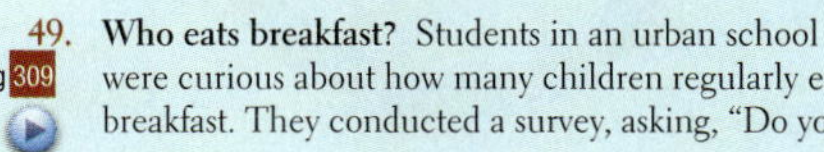

**5.40 (a)** HHH, HHT, HTH, THH, HTT, THT, TTH, TTT. **(b)** Each outcome has probability $\frac{1}{8}$.

**5.41** $P(A) = \frac{1}{16} + \frac{1}{16} + \frac{1}{16} + \frac{1}{16} = \frac{4}{16} = 0.25$

**5.42** $P(B) = \frac{1}{8} + \frac{1}{8} + \frac{1}{8} + \frac{1}{8} = \frac{4}{8} = 0.50$

**5.43 (a)** Legitimate **(b)** Not legitimate: the total is more than 1. **(c)** Legitimate

**5.44** Model 1: not legitimate because the probabilities have sum $6/7 \neq 1$. Model 2: legitimate. Model 3: not legitimate because the probabilities have sum $7/6 \neq 1$. Model 4: not legitimate because probabilities cannot be greater than 1 and the sum of the probabilities is more than 1.

**5.45 (a)** *P*(type AB) = 1 - 0.96 = 0.04 **(b)** *P*(not type AB) = 1 − *P*(type AB) = 1 − 0.04 = 0.96 **(c)** *P*(type O or B) = 0.49 + 0.20 = 0.69

**5.46 (a)** *P*(other) = 1 − 0.91 = 0.09 **(b)** *P*(not English) = 1 − 0.63 = 0.37 **(c)** *P*(neither English nor French) = 1 − 0.63 − 0.22 = 0.15

**5.47 (a)** 1 − 0.13 − 0.29 − 0.30 = 0.28 **(b)** Using the complement rule, 1 − 0.13 = 0.87.

**5.48 (a)** Using the addition rule for mutually exclusive events, 0.20 + 0.15 = 0.35. **(b)** Using the complement rule, 1 − 0.20 = 0.80.

**5.49 (a)** $P(\text{Female}) = \frac{275}{595} = 0.462$

**(b)** $P(\text{Eats breakfast regularly}) = \frac{300}{595} = 0.504$

**(c)** $P(\text{Female and breakfast}) = \frac{110}{595} = 0.185$

**(d)** $P(\text{Female or breakfast}) = \frac{275}{595} + \frac{300}{595} - \frac{110}{595} = \frac{465}{595} = 0.782$

**5.50 (a)** $P(\text{Democrat}) = \frac{(47 + 13)}{100} = \frac{60}{100} = 0.6$ **(b)** $P(\text{Female}) = \frac{(13 + 4)}{100} = \frac{17}{100} = 0.17$

**(c)** $P(\text{Female and Democrat}) = \frac{13}{100} = 0.13$

**(d)** $P(\text{Female or Democrat}) = \frac{60}{100} + \frac{17}{100} - \frac{13}{100} = \frac{64}{100} = 0.64$

**5.51 (a)**

| | B | Not B | Total |
|---|---|---|---|
| E | 10 | 10 | **20** |
| Not E | 8 | 10 | **18** |
| **Total** | **18** | **20** | **38** |

**(b)** $P(B) = \frac{18}{38} = 0.474$; $P(E) = \frac{20}{38} = 0.526$. **(c)** The ball lands in a spot that is black and even. $P(B \text{ and } E) = \frac{10}{38} = 0.263$. **(d)** If we add the probabilities of $B$ and $E$, the spots that are black and even will be double-counted. $P(B \text{ or } E) = \frac{18}{38} + \frac{20}{38} - \frac{10}{38} = \frac{28}{38} = 0.737$.

**5.52 (a)**

| | J | Not J | Total |
|---|---|---|---|
| R | 2 | 24 | **26** |
| Not R | 2 | 24 | **26** |
| **Total** | **4** | **48** | **52** |

**(b)** $P(J) = \frac{4}{52} = 0.077$; $P(R) = \frac{26}{52} = 0.50$. **(c)** We deal a card that is red and a jack. $P(J \text{ and } R) = \frac{2}{52} = 0.038$. **(d)** If we add the probabilities of $J$ and $R$, the cards that are jacks and red will be double-counted. $P(J \text{ or } R) = \frac{4}{52} + \frac{26}{52} - \frac{2}{52} = \frac{28}{52} = 0.538$.



eat breakfast on a regular basis?" All 595 students in the school responded to the survey. The resulting data are shown in the two-way table below.[8]

| | Male | Female | Total |
|---|---|---|---|
| Eats breakfast regularly | 190 | 110 | **300** |
| Doesn't eat breakfast regularly | 130 | 165 | **295** |
| **Total** | **320** | **275** | **595** |

If we select a student from the school at random, what is the probability that the student is

(a) a female?

(b) someone who eats breakfast regularly?

(c) a female and eats breakfast regularly?

(d) a female or eats breakfast regularly?

50. **Sampling senators** The two-way table below describes the members of the U.S Senate in a recent year.

| | Male | Female |
|---|---|---|
| Democrats | 47 | 13 |
| Republicans | 36 | 4 |

If we select a U.S. senator at random, what's the probability that the senator is

(a) a Democrat?

(b) a female?

(c) a female and a Democrat?

(d) a female or a Democrat?

51. **Roulette** An American roulette wheel has 38 slots with numbers 1 through 36, 0, and 00, as shown in the figure. Of the numbered slots, 18 are red, 18 are black, and 2—the 0 and 00—are green. When the wheel is spun, a metal ball is dropped onto the middle of the wheel. If the wheel is balanced, the ball is equally likely to settle in any of the numbered slots. Imagine spinning a fair wheel once. Define events $B$: ball lands in a black slot, and $E$: ball lands in an even-numbered slot. (Treat 0 and 00 as even numbers.)

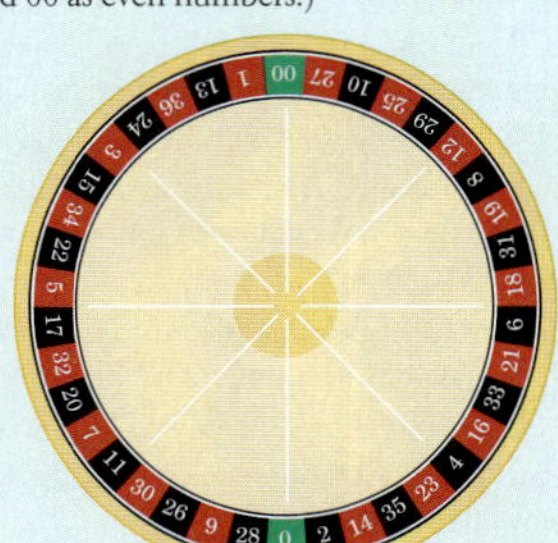

(a) Make a two-way table that displays the sample space in terms of events $B$ and $E$.

(b) Find $P(B)$ and $P(E)$.

(c) Describe the event "$B$ and $E$" in words. Then find $P(B \text{ and } E)$.

(d) Explain why $P(B \text{ or } E) \neq P(B) + P(E)$. Then use the general addition rule to compute $P(B \text{ or } E)$.

52. **Playing cards** Shuffle a standard deck of playing cards and deal one card. Define events $J$: getting a jack, and $R$: getting a red card.

(a) Construct a two-way table that describes the sample space in terms of events $J$ and $R$.

(b) Find $P(J)$ and $P(R)$.

(c) Describe the event "$J$ and $R$" in words. Then find $P(J \text{ and } R)$.

(d) Explain why $P(J \text{ or } R) \neq P(J) + P(R)$. Then use the general addition rule to compute $P(J \text{ or } R)$.

53. **Who eats breakfast?** Refer to Exercise 49.

(a) Construct a Venn diagram that models the chance process using events $B$: eats breakfast regularly, and $M$: is male.

(b) Find $P(B \cup M)$. Interpret this value in context.

(c) Find $P(B^C \cap M^C)$. Interpret this value in context.

54. **Sampling senators** Refer to Exercise 50.

(a) Construct a Venn diagram that models the chance process using events $R$: is a Republican, and $F$: is female.

(b) Find $P(R \cup F)$. Interpret this value in context.

(c) Find $P(R^C \cap F^C)$. Interpret this value in context.

55. **Facebook versus YouTube** A recent survey suggests that 85% of college students have posted a profile on Facebook, 73% use YouTube regularly, and 66% do both. Suppose we select a college student at random. pg 313

(a) Make a two-way table for this chance process.

(b) Construct a Venn diagram to represent this setting.

(c) Consider the event that the randomly selected college student has posted a profile on Facebook or uses YouTube regularly. Write this event in symbolic form based on your Venn diagram in part (b).

(d) Find the probability of the event described in part (c). Explain your method.

56. **Mac or PC?** A recent census at a major university revealed that 40% of its students mainly used Macintosh computers (Macs). The rest mainly used PCs. At the time of the census, 67% of the school's students were undergraduates. The rest were graduate students.

**5.53 (a)**

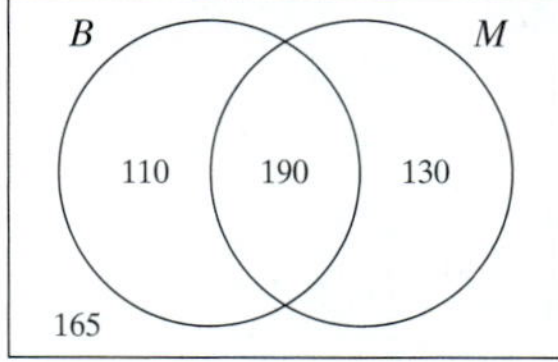


**(b)** $P(B \cup M) = \frac{430}{595} = 0.723$. There is a 0.723 probability that we select a person who is a breakfast eater, a male, or both.

**(c)** $P(B^c \cap M^c) = \frac{165}{595} = 0.277$. There is a 0.277 probability that we select a female who is not a breakfast eater.

**5.54 (a)**

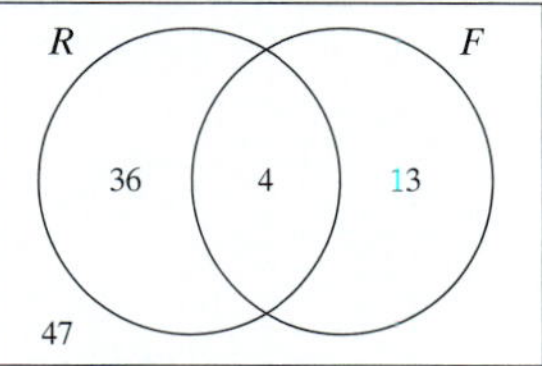


**(b)** $P(R \cup F) = \frac{53}{100} = 0.53$. There is a 0.53 probability that we select a senator who is a Republican, a female, or both.

**(c)** $P(R^c \cap F^c) = \frac{47}{100} = 0.47$. There is a 0.47 probability that we select a senator who is a male Democrat.

In the census, 23% of respondents were graduate students who said that they used PCs as their main computers. Suppose we select a student at random from among those who were part of the census.

(a) Make a two-way table for this chance process.

(b) Construct a Venn diagram to represent this setting.

(c) Consider the event that the randomly selected student is a graduate student and uses a Mac. Write this event in symbolic form based on your Venn diagram in part (b).

(d) Find the probability of the event described in part (c). Explain your method.

***Multiple choice: Select the best answer for Exercises 57 to 60.***

**57.** In government data, a household consists of all occupants of a dwelling unit. Choose an American household at random and count the number of people it contains. Here is the assignment of probabilities for the outcome:

| **Number of persons:** | 1 | 2 | 3 | 4 | 5 | 6 | 7+ |
|---|---|---|---|---|---|---|---|
| **Probability:** | 0.25 | 0.32 | ??? | ??? | 0.07 | 0.03 | 0.01 |

The probability of finding 3 people in a household is the same as the probability of finding 4 people. These probabilities are marked ??? in the table of the distribution. The probability that a household contains 3 people must be

(a) 0.68. (b) 0.32. (c) 0.16. (d) 0.08.

(e) between 0 and 1, and we can say no more.

**58.** In a sample of 275 students, 20 say they are vegetarians. Of the vegetarians, 9 eat both fish and eggs, 3 eat eggs but not fish, and 7 eat neither. Choose one of the vegetarians at random. What is the probability that the chosen student eats fish or eggs?

(a) 9/20 (c) 22/20 (e) 22/275

(b) 13/20 (d) 9/275

*Exercises 59 and 60 refer to the following setting.* The casino game craps is based on rolling two dice. Here is the assignment of probabilities to the sum of the numbers on the up-faces when two dice are rolled:

| **Outcome:** | 2 | 3 | 4 | 5 | 6 | 7 | 8 | 9 | 10 | 11 | 12 |
|---|---|---|---|---|---|---|---|---|---|---|---|
| **Probability:** | 1/36 | 2/36 | 3/36 | 4/36 | 5/36 | 6/36 | 5/36 | 4/36 | 3/36 | 2/36 | 1/36 |

**59.** The most common bet in craps is the "pass line." A pass line bettor wins immediately if either a 7 or an 11 comes up on the first roll. This is called a *natural*. What is the probability of a natural?

(a) 2/36 (c) 8/36 (e) 20/36

(b) 6/36 (d) 12/36

**60.** If a player rolls a 2, 3, or 12, it is called *craps*. What is the probability of getting craps or an even sum on one roll of the dice?

(a) 4/36 (c) 20/36 (e) 32/36

(b) 18/36 (d) 22/36

**61. Crawl before you walk** (3.2) At what age do babies learn to crawl? Does it take longer to learn in the winter, when babies are often bundled in clothes that restrict their movement? Perhaps there might even be an association between babies' crawling age and the average temperature during the month they first try to crawl (around six months after birth). Data were collected from parents who brought their babies to the University of Denver Infant Study Center to participate in one of a number of studies. Parents reported the birth month and the age at which their child was first able to creep or crawl a distance of 4 feet within one minute. Information was obtained on 414 infants (208 boys and 206 girls). Crawling age is given in weeks, and average temperature (in °F) is given for the month that is six months after the birth month.[9]

| Birth month | Average crawling age | Average temperature |
|---|---|---|
| January | 29.84 | 66 |
| February | 30.52 | 73 |
| March | 29.70 | 72 |
| April | 31.84 | 63 |
| May | 28.58 | 52 |
| June | 31.44 | 39 |
| July | 33.64 | 33 |
| August | 32.82 | 30 |
| September | 33.83 | 33 |
| October | 33.35 | 37 |
| November | 33.38 | 48 |
| December | 32.32 | 57 |

Analyze the relationship between average crawling age and average temperature. What do you conclude about when babies learn to crawl?

**62. Treating low bone density** (4.2) Fractures of the spine are common and serious among women with advanced osteoporosis (low mineral density in the bones). Can taking strontium ranelate help? A large medical trial assigned 1649 women to take either strontium ranelate or a placebo each day. All of the subjects had osteoporosis and had had at least one fracture. All were taking calcium supplements and receiving standard medical care. The response variables were measurements of bone density and counts of new fractures over three years. The subjects were treated at 10 medical centers in 10 different countries.[10] Outline an appropriate design for this experiment. Explain why this is the proper design.

**5.57** c

**5.58** b

**5.59** c

**5.60** c

**5.61** The scatterplot for the average crawling age and average temperature is given below.

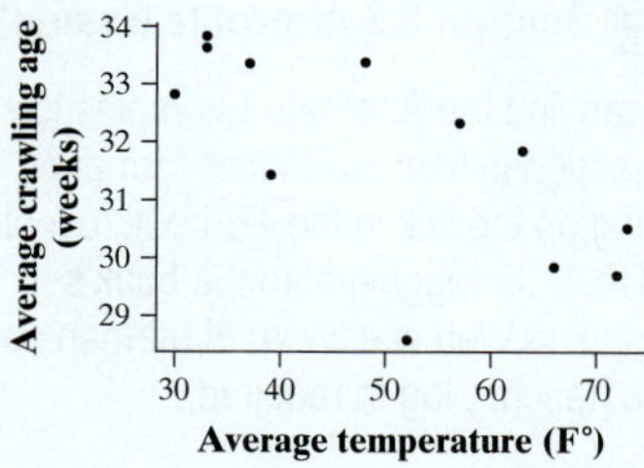


In this scatterplot, there appears to be a moderately strong, negative linear relationship between average temperature and average crawling age. The equation for the least-squares regression line is $\widehat{\text{age}} = 35.7 - 0.077\ \text{temp}$. We predict that babies will walk 0.077 weeks earlier for every degree warmer it gets.

**5.62** Use a randomized block design because using patients from different countries might increase the variation in the response variables. Use the 10 medical centers as blocks and randomly assign half of the women in each center to receive ranelate and the other half to receive the placebo. At each center, write each woman's name on a slip of paper, mix the papers in a hat, and choose half to get the ranelate. The remaining women at each center receive the placebo. At the end of the study, measure the bone density and counts of new fractures.

**5.55 (a)**

| | F | Not F | Total |
|---|---|---|---|
| Y | 0.66 | 0.07 | **0.73** |
| Not Y | 0.19 | 0.08 | **0.27** |
| **Total** | **0.85** | **0.15** | **1** |

**(b)**

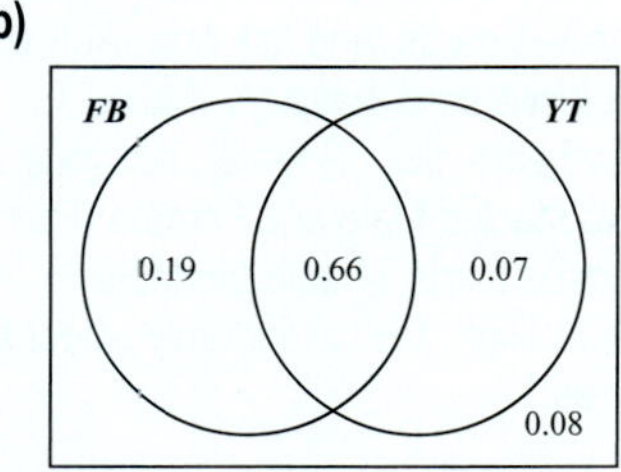


**(c)** F ∪ Y

**(d)** $P(F \cup Y) = 0.85 + 0.73 - 0.66 = 0.92$.

**5.56 (a)**

| | MAC | PC | Total |
|---|---|---|---|
| Undergrad | 0.30 | 0.37 | **0.67** |
| Graduate | 0.10 | 0.23 | **0.33** |
| **Total** | **0.40** | **0.60** | **1** |

**(b)**

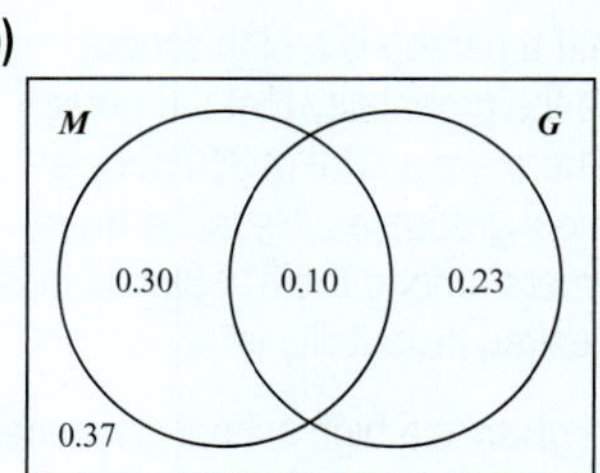


**(c)** (M ∩ G)

**(d)** Looking at the intersection, $P(M \cap G) = 0.10$

**PD Section 5.3 Overview**

Watch the video overview of Section 5.3 (for teachers) by clicking on the link in the TE-book, opening the TRCD, or logging onto the book's companion Web site www.whfreeman.com/tps5e (teacher log-in required).

**TRM Section 5.3 Alternate Examples**

You can find the Alternate Examples for this section in Microsoft Word format by clicking on the link in the TE-book, opening the TRCD, or logging onto the book's companion Web site www.whfreeman.com/tps5e (teacher log-in required).

### Teaching Tip

When introducing the idea of conditional probability, display the entire two-way table and then physically cover up the rows or columns that are not part of the "given." This will help students focus on only the row or column that represents the event that we know has occurred. See the top of the next page for a similar technique.

### Teaching Tip

Emphasize that order matters when asking conditional probability questions, just as it does with other mathematical operations such as subtraction or division.

# 5.3 Conditional Probability and Independence

**WHAT YOU WILL LEARN** By the end of the section, you will be able to:

- Calculate and interpret conditional probabilities.
- Use the general multiplication rule to calculate probabilities.
- Use tree diagrams to model a chance process and calculate probabilities involving two or more events.
- Determine if two events are independent.
- When appropriate, use the multiplication rule for independent events to compute probabilities.

The probability we assign to an event can change if we know that some other event has occurred. This idea is the key to many applications of probability.

## What Is Conditional Probability?

Let's return to the setting of the pierced-ears example (page 309). Earlier, we used the two-way table below to find probabilities involving events $A$: is male and $B$: has pierced ears for a randomly selected student. Here is a summary of our previous results:

$P(A) = P(\text{male}) = 90/178$  $P(A \cap B) = P(\text{male and pierced ears}) = 19/178$

$P(B) = P(\text{pierced ears}) = 103/178$  $P(A \cup B) = P(\text{male or pierced ears}) = 174/178$

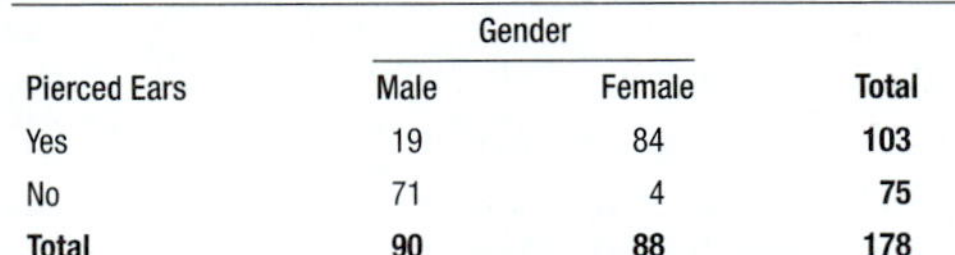

| | Gender | | |
|---|---|---|---|
| Pierced Ears | Male | Female | Total |
| Yes | 19 | 84 | **103** |
| No | 71 | 4 | **75** |
| **Total** | **90** | **88** | **178** |

Now let's turn our attention to some other interesting probability questions.

## EXAMPLE Who Has Pierced Ears?

*The idea of conditional probability*

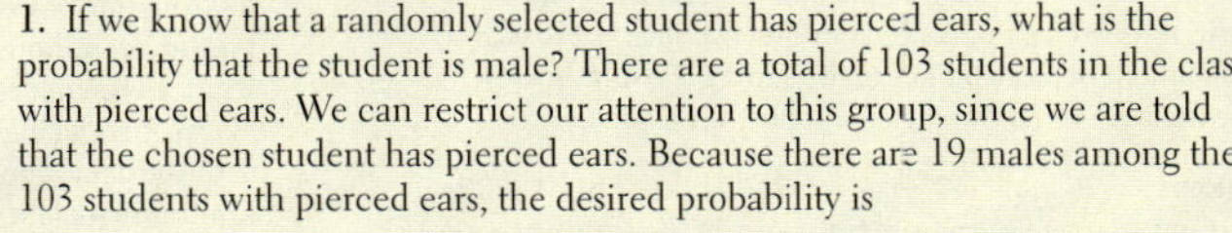

1. If we know that a randomly selected student has pierced ears, what is the probability that the student is male? There are a total of 103 students in the class with pierced ears. We can restrict our attention to this group, since we are told that the chosen student has pierced ears. Because there are 19 males among the 103 students with pierced ears, the desired probability is

$$P(\text{is male } \textit{given} \text{ has pierced ears}) = 19/103, \text{ or about } 18.4\%$$

2. If we know that a randomly selected student is male, what's the probability that the student has pierced ears? This time, our attention is focused on the males in the class. Because 19 of the 90 males in the class have pierced ears,

$$P(\text{has pierced ears } \textit{given} \text{ is male}) = 19/90, \text{ or about } 21.1\%$$

These two questions sound alike, but they're asking about two very different things.

**ALTERNATE EXAMPLE**

***Who owns a home?***

Here is the two-way table from the homeowner Alternate Example.

| | High school graduate | Not a high school graduate | Total |
|---|---|---|---|
| Homeowner | 221 | 119 | **340** |
| Not a homeowner | 89 | 71 | **160** |
| **Total** | **310** | **190** | **500** |

**1.** If we know that a person owns a home, what is the probability that the person is a high school graduate? There are a total of 340 people in the sample who own a home. Because there are 221 high school graduates among the 340 homeowners, the desired probability is

$P(\text{is a high school graduate } \textit{given} \text{ owns a home}) = 221/340$, or 65%

**2.** If we know that a person is a high school graduate, what is the probability that the person owns a home? There are a total of 310 people who are high school graduates. Because there are 221 homeowners among the 310 high school graduates, the desired probability is

$P(\text{owns a home } \textit{given} \text{ is a high school graduate}) = 221/310$, or about 71%

### Teaching Tip

To illustrate the idea of conditional probability, use the cards from page 309. To illustrate $P$(is male *given* has pierced ears), separate the cards into two stacks: Yes and No. Then, set aside the No cards, shuffle the Yes cards, and ask how likely it is to randomly select a card that says "Male." To illustrate $P$(pierced ears *given* is male), separate the cards into two stacks: Male and Female. Then, set aside the Female cards, shuffle the Male cards, and ask how likely it is to randomly select a card that says "Yes."

A probability like "the probability that a randomly selected student is male *given that* the student has pierced ears" is known as a **conditional probability**. The name comes from the fact that we are trying to find the probability that one event will happen under the *condition* that some other event is already known to have occurred. We often use the phrase "given that" to signal the condition. There's even a special notation to indicate a conditional probability. In the example above, we would write $P(\text{is male} \mid \text{has pierced ears})$, where the | means "given that" or "under the condition that." Because we already defined the events A: is male and B: has pierced ears, we could write the conditional probability as $P(A \mid B) = 19/103$.

**DEFINITION: Conditional probability**

The probability that one event happens given that another event is already known to have happened is called a **conditional probability**. Suppose we know that event *A* has happened. Then the probability that event *B* happens given that event *A* has happened is denoted by $P(B \mid A)$.

Let's look more closely at how conditional probabilities are calculated. From the two-way table below, we see that

$$P(\text{male} \mid \text{pierced ears}) = \frac{\text{Number of students who are male and have pierced ears}}{\text{Number of students with pierced ears}} = \frac{19}{103}$$

| | Gender | | | |
|---|---|---|---|---|
| Pierced Ears? | Male | Female | Total | |
| Yes | 19 | 84 | **103** | $P(\text{male} \mid \text{pierced ears}) = 19/103$ |
| No | 71 | 4 | **75** | |
| **Total** | **90** | **88** | **178** | |

What if we focus on probabilities instead of numbers of students? Notice that

$$\frac{P(\text{male and pierced ears})}{P(\text{pierced ears})} = \frac{\frac{19}{178}}{\frac{103}{178}} = \frac{19}{103} = P(\text{male} \mid \text{pierced ears})$$

In symbols, $P(A \mid B) = \frac{P(A \cap B)}{P(B)}$. This observation leads to a general formula for computing a conditional probability.

**CALCULATING CONDITIONAL PROBABILITIES**

To find the conditional probability $P(A \mid B)$, use the formula

$$P(A \mid B) = \frac{P(A \cap B)}{P(B)}$$

The conditional probability $P(B \mid A)$ is given by

$$P(B \mid A) = \frac{P(B \cap A)}{P(A)}$$

## Teaching Tip

Using conditional probability notation, we can restate the previous answers:

$$P(\text{graduate} \mid \text{owns a home}) = P(G \mid H) = 221/340$$

$$P(\text{owns a home} \mid \text{graduate}) = P(H \mid G) = 221/310$$

## AP® Exam Formula Sheet

The conditional probability formula is included on the formula sheet that is provided on the AP® exam. The conditional probability formula on the formula sheet matches the first formula in the definition box:

$$P(A \mid B) = \frac{P(A \cap B)}{P(B)}$$

## Teaching Tip

Emphasize how these two versions of the conditional probability formula are equivalent. The numerator is the same and they are both in the form

$$\frac{P(\text{both events occur})}{P(\text{given event occurs})}$$

Here's an example that illustrates the use of this formula in a familiar setting.

## EXAMPLE Who Reads the Paper?

### *Conditional probability formula*

On page 313, we classified the residents of a large apartment complex based on the events A: reads *USA Today*, and B: reads the *New York Times*. The completed Venn diagram is reproduced here.

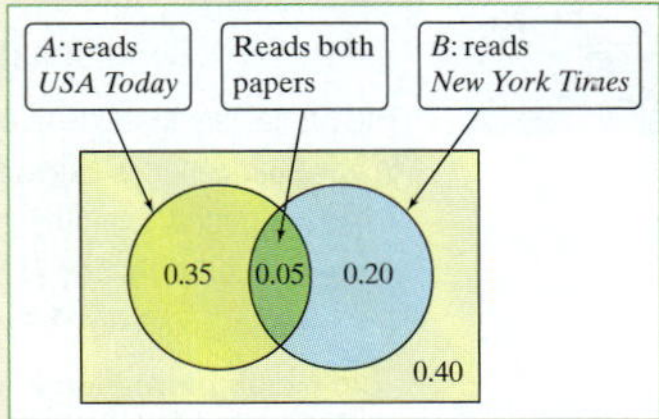


**PROBLEM:** What's the probability that a randomly selected resident who reads *USA Today* also reads the *New York Times*?

**SOLUTION:** Because we're given that the randomly chosen resident reads *USA Today*, we want to find *P*(reads *New York Times* | reads *USA Today*), or $P(B \mid A)$. By the conditional probability formula,

$$P(B \mid A) = \frac{P(B \cap A)}{P(A)}$$

Because $P(B \cap A) = 0.05$ and $P(A) = 0.40$, we have

$$P(B \mid A) = \frac{0.05}{0.40} = 0.125$$

There's a 12.5% chance that a randomly selected resident who reads *USA Today* also reads the *New York Times*.

**For Practice** *Try Exercise* **71**

**AP® EXAM TIP** You can write statements like $P(B \mid A)$ if events *A* and *B* are defined clearly, or you can use a verbal equivalent, such as *P*(reads *New York Times* | reads *USA Today*). Use the approach that makes the most sense to you.

**Is there a connection between conditional probability and the conditional distributions of Chapter 1?** Of course! For the college statistics class that we discussed earlier, Figure 5.8 shows the conditional distribution of ear-piercing status for each gender. Above, we found that *P*(pierced ears | male) = 19/90, or about 21%. Note that *P*(pierced ears | female) = 84/88, or about 95%. You can see these values displayed in the "Yes" bars of Figure 5.8.

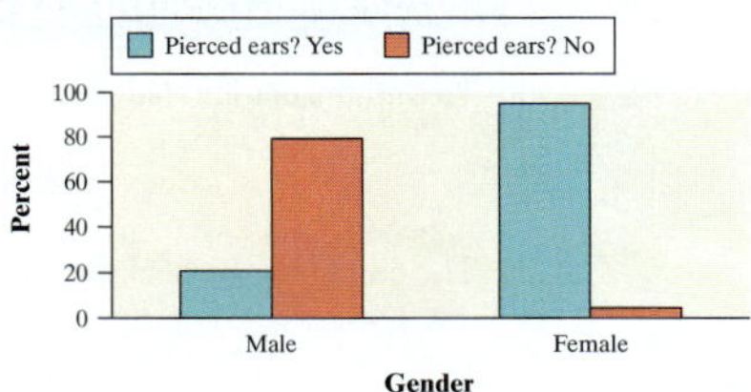


**FIGURE 5.8** Conditional distribution of ear-piercing status for each gender in a large college statistics class.

## ALTERNATE EXAMPLE

### *Phone usage*

In an Alternate Example in Section 5.2, we classified U.S. households according to the types of phones they used.

| | Cell phone | No cell phone | Total |
|---|---|---|---|
| Landline | 0.51 | 0.09 | **0.60** |
| No landline | 0.38 | 0.02 | **0.40** |
| **Total** | **0.89** | **0.11** | **1.00** |

**PROBLEM:** What is the probability that a randomly selected household with a landline also has a cell phone?

**SOLUTION:** We want to find P(cell phone | landline). Using the conditional probability formula,

$$P(\text{cell phone} \mid \text{landline}) = \frac{P(\text{cell phone and landline})}{P(\text{landline})} = \frac{0.51}{0.60} = 0.85$$

There is a 0.85 probability that a landline user also has a cell phone.

## AP® Exam Common Error

Many students lose credit on probability questions because they do not show their work. When using the conditional probability formula, students should show the numerator and the denominator of the probability, not just the decimal answer.

## Teaching Tip

Notice that we are conditioning on the explanatory variable in the Think About It (gender) and in the following Alternate Example (education level).

## ALTERNATE EXAMPLE

### *Who owns a home?*

Here is a segmented bar graph displaying the conditional distribution of home ownership for high school graduates and non-graduates.

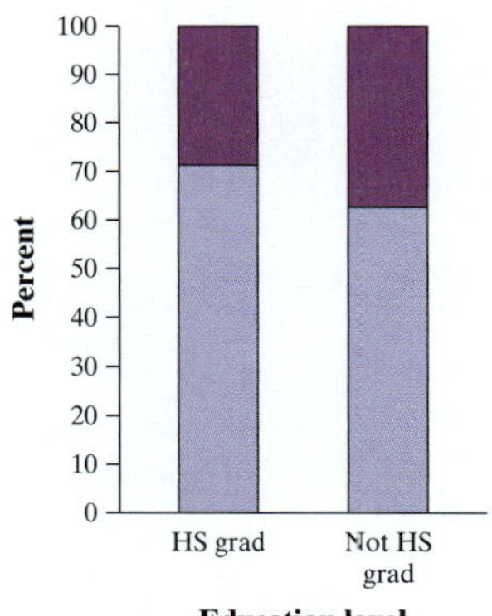

## CHECK YOUR UNDERSTANDING

Students at the University of New Harmony received 10,000 course grades last semester. The two-way table below breaks down these grades by which school of the university taught the course. The schools are Liberal Arts, Engineering and Physical Sciences, and Health and Human Services.

| | Grade Level | | |
|---|---|---|---|
| **School** | **A** | **B** | **Below B** |
| Liberal Arts | 2,142 | 1,890 | 2,268 |
| Engineering and Physical Sciences | 368 | 432 | 800 |
| Health and Human Services | 882 | 630 | 588 |

(This table is based closely on grade distributions at an actual university, simplified a bit for clarity.)[11]

College grades tend to be lower in engineering and the physical sciences (EPS) than in liberal arts and social sciences (which includes Health and Human Services). Choose a University of New Harmony course grade at random. Consider the two events $E$: the grade comes from an EPS course, and $L$: the grade is lower than a B.

1. Find $P(L)$. Interpret this probability in context.
2. Find $P(E \mid L)$ and $P(L \mid E)$. Which of these conditional probabilities tells you whether this college's EPS students tend to earn lower grades than students in liberal arts and social sciences? Explain.

## The General Multiplication Rule and Tree Diagrams

Suppose that A and B are two events resulting from the same chance process. We can find the probability P(A or B) with the general addition rule:

$$P(\text{A or B}) = P(A \cup B) = P(A) + P(B) - P(A \cap B)$$

How do we find the probability that both events happen, P(A and B)?

Start with the conditional probability formula

$$P(B \mid A) = \frac{P(B \cap A)}{P(A)}$$

The numerator, $P(B \cap A)$, is the probability we want because P(B and A) is the same as P(A and B). Multiply both sides of the above equation by P(A) to get

$$P(A) \cdot P(B \mid A) = P(B \cap A) = P(A \cap B) = P(\text{A and B})$$

This formula is known as the **general multiplication rule**.

**GENERAL MULTIPLICATION RULE**

The probability that events A and B both occur can be found using the **general multiplication rule**

$$P(\text{A and B}) = P(A \cap B) = P(A) \cdot P(B \mid A)$$

where $P(B \mid A)$ is the conditional probability that event B occurs given that event A has already occurred.

### ANSWERS TO CYU

**1.** $P(L) = \frac{3656}{10{,}000} = 0.3656$. There is a 0.3656 probability of selecting a course grade that is lower than a B.

**2.** $P(E \mid L) = \frac{800}{3656} = 0.219$.

$P(L \mid E) = \frac{800}{1600} = 0.50$. $P(L \mid E)$ gives the probability of getting a lower grade given that the student is studying engineering or physical science. Because this probability (0.50) is greater than $P(L) = 0.3656$, we can conclude that grades are lower in engineering and physical sciences.

### Teaching Tip

Unlike the addition rule for mutually exclusive events and the general addition rule, we introduce the general multiplication rule before the special multiplication rule for independent events. We do this because the general multiplication rule flows very nicely from the conditional probability formula.

### AP® Exam Formula Sheet

The general multiplication rule is *not* on the formula sheets provided on the AP® exam. However, the conditional probability formula that we introduced earlier *is* on the formula sheet. Fortunately, the formula for the general multiplication rule is actually the same as the conditional probability formula, just rearranged.

In words, this rule says that for both of two events to occur, first one must occur, and then given that the first event has occurred, the second must occur. This is just common sense expressed in the language of probability, as the following example illustrates.

## EXAMPLE Teens with Online Profiles

### *Using the general multiplication rule*

The Pew Internet and American Life Project find that 93% of teenagers (ages 12 to 17) use the Internet, and that 55% of online teens have posted a profile on a social-networking site.[12]

PROBLEM: Find the probability that a randomly selected teen uses the Internet and has posted a profile. Show your work.

SOLUTION: We know that $P(\text{online}) = 0.93$ and $P(\text{profile} \mid \text{online}) = 0.55$. Use the general multiplication rule:

$$P(\text{online and have profile}) = P(\text{online}) \cdot P(\text{profile} \mid \text{online}) = (0.93)(0.55) = 0.5115$$

There is about a 51% chance that a randomly selected teen uses the Internet and has posted a profile on a social-networking site.

**For Practice** *Try Exercise* **73**

The general multiplication rule is especially useful when a chance process involves a sequence of outcomes. In such cases, we can use a **tree diagram** to display the sample space.

## EXAMPLE Serve It Up!

### *Tree diagrams and the general multiplication rule*

Tennis great Roger Federer made 63% of his first serves in the 2011 season. When Federer made his first serve, he won 78% of the points. When Federer missed his first serve and had to serve again, he won only 57% of the points.[13] Suppose we randomly choose a point on which Federer served.

Figure 5.9 on the facing page shows a tree diagram for this chance process. There are only two possible outcomes on Federer's first serve, a make or a miss. The first set of branches in the tree diagram displays these outcomes with their probabilities. The second set of branches shows the two possible results of the point for Federer—win or lose—and the chance of each result based on the outcome of the first serve. Note that the probabilities on the second set of branches are *conditional probabilities*, like $P(\text{win point} \mid \text{make first serve}) = 0.78$.

## ALTERNATE EXAMPLE

### *Playing in the NCAA*

About 55% of high school students participate in a school athletic team at some level, and about 5% of these athletes go on to play on a college team in the NCAA (http://www.washingtonpost.com/wp-dyn/content/article/2009/09/23/AR2009092301947.html, http://www.collegesportsscholarships.com/percentage-high-school-athletes-ncaa-college.htm).

PROBLEM: What percent of high school students play a sport in high school *and* go on to play a sport in the NCAA?

SOLUTION: We know $P(\text{high school sport}) = 0.55$ and $P(\text{NCAA sport} \mid \text{high school sport}) = 0.05$, so $P(\text{high school sport and NCAA sport}) = P(\text{high school sport}) \times P(\text{NCAA sport} \mid \text{high school sport}) = (0.55)(0.05) = 0.0275$. Almost 3% of high school students will play a sport in high school and in the NCAA.

## ALTERNATE EXAMPLE

### *Late for school*

Shannon hits the snooze bar on her alarm clock on 60% of school days. If she doesn't hit the snooze bar, there is a 0.90 probability that she makes it to class on time. However, if she hits the snooze bar, there is only a 0.70 probability that she makes it to class on time. On a randomly chosen day, what is the probability that Shannon is late for class?

The figure below shows the tree diagram for this chance process.

0.60 → Hits snooze bar
- 0.70 → On time: $P$(hits snooze bar and is on time) $= 0.60 \times 0.70 = 0.42$
- 0.30 → Late: $P$(hits snooze bar and is late) $= 0.60 \times 0.30 = 0.18$

0.40 → Doesn't hit snooze bar
- 0.90 → On time: $P$(doesn't hit snooze bar and is on time) $= 0.40 \times 0.90 = 0.36$
- 0.10 → Late: $P$(doesn't hit snooze bar and is late) $= 0.40 \times 0.10 = 0.04$

The probability that she hits the snooze bar and is late is $(0.60)(0.30) = 0.18$. The probability that she doesn't hit the snooze bar and is late is $(0.40)(0.10) = 0.04$. Because these outcomes are mutually exclusive, $P(\text{late}) = 0.18 + 0.04 = 0.22$. Shannon is late on about 22% of school days.

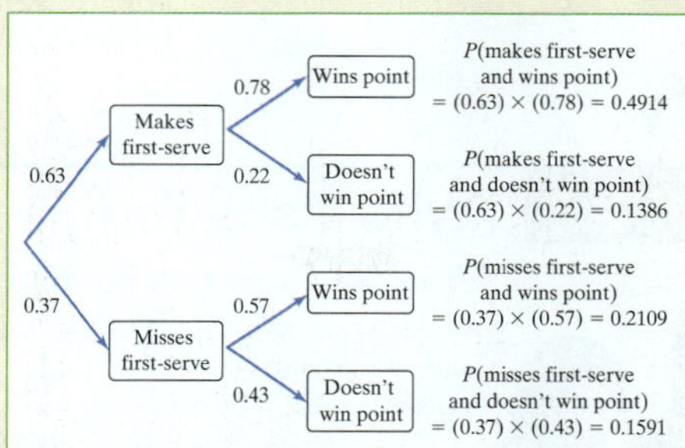


**FIGURE 5.9** A tree diagram displaying the sample space of randomly choosing a point on which Roger Federer served.

What's the probability that Federer makes the first serve and wins the point? From the tree diagram, Federer makes 63% of his first serves. Of this 63%, he wins the point 78% of the time. Because 78% of 63% = (0.63)(0.78) = 0.4914, Federer makes his first serve and wins the point about 49.14% of the time.

The previous calculation amounts to multiplying probabilities along the branches of the tree diagram. Why does this work? The general multiplication rule provides the answer:

$$P(\text{make first serve and win point}) = P(\text{make first serve}) \cdot P(\text{win point} \mid \text{make first serve})$$
$$= (0.63)(0.78) = 0.4914$$

When Federer is serving, what's the probability that he wins the point? From the tree diagram, there are two ways Federer can win the point. He can make the first serve and win the point, or he can miss the first serve and win the point. Because these outcomes are mutually exclusive,

$$P(\text{win point}) = P(\text{makes first serve and wins point}) + P(\text{misses first serve and wins point})$$
$$= 0.4914 + 0.2109 = 0.7023$$

Federer wins about 70% of the points when he is serving.

Some interesting conditional probability questions involve "going in reverse" on a tree diagram. Here's one related to the previous example. Suppose you are watching a recording of one of Federer's matches from 2011 and he is serving in the current game. You get distracted before seeing his first serve but look up in time to see Federer win the point. How likely is it that he missed his first serve?

To find this probability, we start with the result of the point, which is displayed on the second set of branches in Figure 5.9, and ask about the outcome of the serve, which is shown on the first set of branches. We can use the information from the tree diagram and the conditional probability to do the required calculation:

Some people use a result known as Bayes's theorem to solve probability questions that require "going backward" in a tree diagram, like the one in this example. To be honest, Bayes's theorem is just a complicated formula for computing conditional probabilities. For that reason, we won't introduce it.

$$P(\text{missed first serve} \mid \text{wins point}) = \frac{P(\text{missed first serve and wins point})}{P(\text{wins point})}$$
$$= \frac{0.2109}{0.4914 + 0.2109} = \frac{0.2109}{0.7023} = 0.3003$$

Given that Federer won the point, there is about a 30% chance that he missed his first serve.

**ALTERNATE EXAMPLE**

***Late for school***

Suppose that Shannon is late for school. What is the probability that she hit the snooze bar that morning? We want to find

$$P(\text{hits snooze bar} \mid \text{late for school})$$
$$= \frac{P(\text{hits snooze bar and late for school})}{P(\text{late for school})}$$
$$= \frac{0.18}{0.18 + 0.04} = \frac{0.18}{0.22} = 0.818$$

When Shannon is late, there is a 0.818 probability that she hit the snooze bar that morning.

Here is another example where we need to reverse the conditioning.

## EXAMPLE Who Visits YouTube?

### *Tree diagrams and conditional probability*

Video-sharing sites, led by YouTube, are popular destinations on the Internet. Let's look only at adult Internet users, aged 18 and over. About 27% of adult Internet users are 18 to 29 years old, another 45% are 30 to 49 years old, and the remaining 28% are 50 and over. The Pew Internet and American Life Project finds that 70% of Internet users aged 18 to 29 have visited a video-sharing site, along with 51% of those aged 30 to 49 and 26% of those 50 or older. Do most Internet users visit YouTube and similar sites?

**PROBLEM:** Suppose we select an adult Internet user at random.

(a) Draw a tree diagram to represent this situation.

(b) Find the probability that this person has visited a video-sharing site. Show your work.

(c) Given that this person has visited a video-sharing site, find the probability that he or she is aged 18 to 29. Show your work.

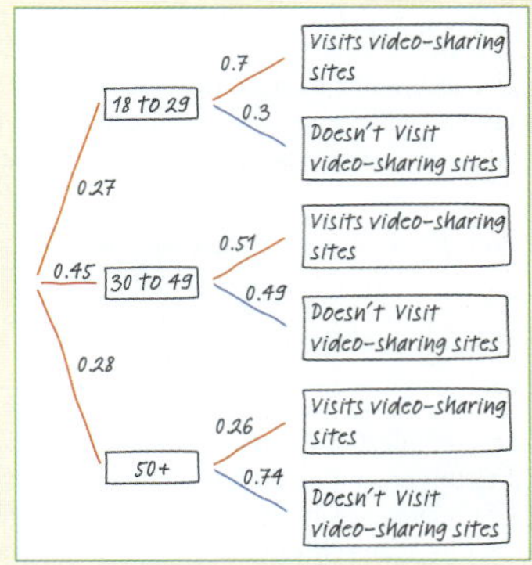


**FIGURE 5.10** Tree diagram for use of the Internet and video-sharing sites such as YouTube. The three disjoint paths to the outcome that an adult Internet user visits video-sharing sites are colored red.

**SOLUTION:**

(a) The tree diagram in Figure 5.10 organizes the given information.

(b) There are three disjoint paths to "visits video-sharing sites," one for each of the three age groups. These paths are colored red in Figure 5.10. Because the three paths are disjoint, the probability that an adult Internet user visits video-sharing sites is the sum of their probabilities:

$$P(\text{visits video-sharing sites}) = (0.27)(0.70) + (0.45)(0.51) + (0.28)(0.26)$$
$$= 0.1890 + 0.2295 + 0.0728 = 0.4913$$

About 49% of all adult Internet users have visited a video-sharing site.

(c) Use the tree diagram and the definition of conditional probability:

$$P(\text{18 to 29} \mid \text{visits video-sharing site}) = \frac{P(\text{18 to 29 and visits video-sharing site})}{P(\text{visits video-sharing site})}$$
$$= \frac{0.1890}{0.4913} = 0.3847$$

Given that the person visits video-sharing sites, there is about a 38.5% chance that he or she is aged 18 to 29.

**For Practice** *Try Exercise* 77

One of the most important applications of tree diagrams and conditional probability is in the area of drug and disease testing.

### ALTERNATE EXAMPLE

### *Media usage and good grades*

The Kaiser Family Foundation recently released a study about the influence of media in the lives of young people aged 8–18 (www.kff.org/entmedia/mh012010pkg.cfm). In the study, 17% of the youth were classified as light media users, 62% were classified as moderate media users, and 21% were classified as heavy media users. Of the light users who responded, 74% described their grades as good (As and Bs), while only 68% of the moderate users and 52% of the heavy users described their grades as good. Suppose that we selected one young person at random.

**PROBLEM:**

**(a)** Draw a tree diagram to represent this situation.

**(b)** Find the probability that this person describes his or her grades as good.

**(c)** Given that this person describes his or her grades as good, what is the probability that he or she is a heavy user of media?

**SOLUTION:**

**(a)** Here is a tree diagram for this situation.

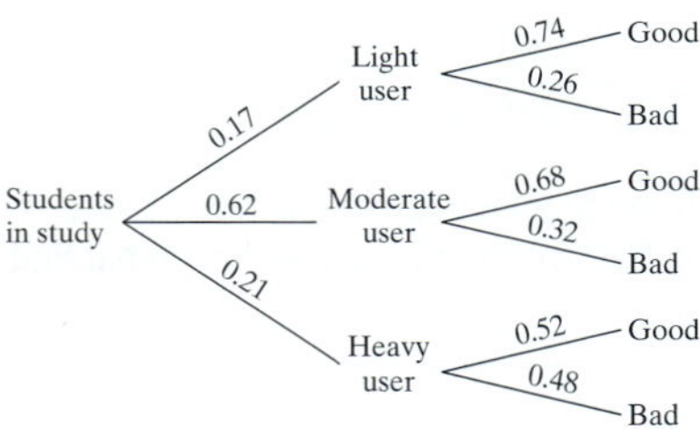


**(b)** There are three groups of students who say they get good grades: those who are light users and get good grades, those who are moderate users and get good grades, and those who are heavy users and get good grades. Because these groups are mutually exclusive, we can add the probabilities of being in one of these three groups. $P(\text{good grades}) = (0.17)(0.74) + (0.62)(0.68) + (0.21)(0.52) = 0.1258 + 0.4216 + 0.1092 = 0.6566$. About 66% of the young people would describe their grades as good.

**(c)** Using the tree diagram,

$$P(\text{heavy user} \mid \text{good grades}) = \frac{P(\text{heavy user } \textit{and} \text{ good grades})}{P(\text{good grades})}$$
$$= \frac{0.1092}{0.1258 + 0.4216 + 0.1092} = \frac{0.1092}{0.6566} = 0.166$$

About 17% of young people who say they have good grades are heavy users of media.

# EXAMPLE Mammograms

## Conditional probability in real life

Many women choose to have annual mammograms to screen for breast cancer after age 40. A mammogram isn't foolproof. Sometimes, the test suggests that a woman has breast cancer when she really doesn't (a "false positive"). Other times, the test says that a woman doesn't have breast cancer when she actually does (a "false negative").

Suppose that we know the following information about breast cancer and mammograms in a particular region:

- One percent of the women aged 40 or over in this region have breast cancer.
- For women who have breast cancer, the probability of a negative mammogram is 0.03.
- For women who don't have breast cancer, the probability of a positive mammogram is 0.06.

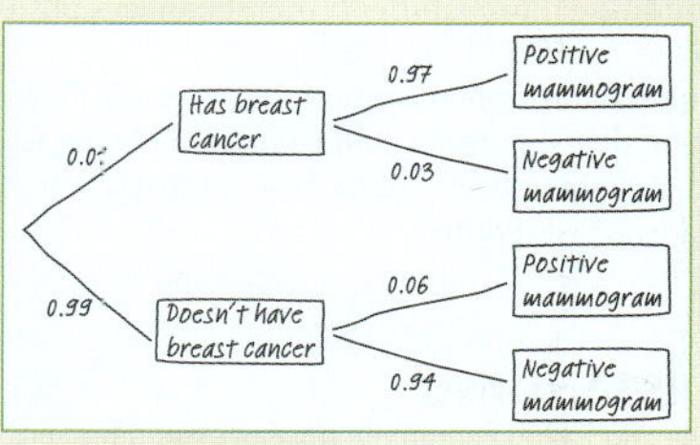


**FIGURE 5.11** Tree diagram showing whether or not a woman has breast cancer and the likelihood of her receiving a positive or a negative test result from a mammogram.

PROBLEM: A randomly selected woman aged 40 or over from this region tests positive for breast cancer in a mammogram. Find the probability that she actually has breast cancer. Show your work.

SOLUTION: The tree diagram in Figure 5.11 summarizes the situation. Because 1% of women in this region have breast cancer, 99% don't. Of those women who do have breast cancer, 3% would test negative on a mammogram. The remaining 97% would (correctly) test positive. Among the women who don't have breast cancer, 6% would test positive on a mammogram. The remaining 94% would (correctly) test negative.

We want to find $P$(breast cancer | positive mammogram). By the conditional probability formula,

$$P(\text{breast cancer} \mid \text{positive mammogram}) = \frac{P(\text{breast cancer and positive mammogram})}{P(\text{positive mammogram})}$$

- To find $P$(breast cancer and positive mammogram), we use the general multiplication rule along with the information displayed in the tree diagram:

$$P(\text{breast cancer and positive mammogram}) = P(\text{breast cancer}) \cdot P(\text{positive mammogram} \mid \text{breast cancer}) = (0.01)(0.97) = 0.0097$$

- To find $P$(positive mammogram), we need to calculate the probability that a randomly selected woman aged 40 or over from this region gets a positive mammogram. There are two ways this can happen: (1) if the woman has breast cancer and the test result is positive, and (2) if the woman doesn't have cancer, but the mammogram gives a false positive. From the tree diagram, the desired probability is

$$P(\text{positive mammogram}) = (0.01)(0.97) + (0.99)(0.06) = 0.0691$$

Using these two results, we can find the conditional probability:

$$P(\text{breast cancer} \mid \text{positive mammogram}) = \frac{P(\text{breast cancer and positive mammogram})}{P(\text{positive mammogram})} = \frac{0.0097}{0.0691} = 0.14$$

Given that a randomly selected woman from the region has a positive mammogram, there is only about a 14% chance that she actually has breast cancer!

For Practice Try Exercise 81

## ALTERNATE EXAMPLE

### False positives and drug testing

Many employers require prospective employees to take a drug test. A positive result on this test indicates that the prospective employee uses illegal drugs. However, not all people who test positive actually use drugs. Suppose that 4% of prospective employees use drugs, the false positive rate is 5%, and the false negative rate is 10% (http://www.cbsnews.com/stories/2010/06/01/health/webmd/main6537635.shtml).

PROBLEM: A randomly selected prospective employee tests positive for drugs. What is the probability that he actually took drugs?

SOLUTION: The tree diagram below summarizes this situation.

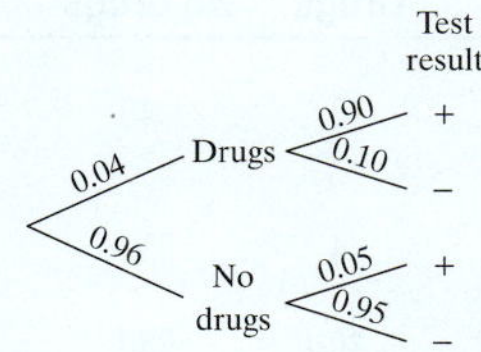


We want to find

$$P(\text{took drugs} \mid \text{positive test}) = \frac{P(\text{took drugs and tested positive})}{P(\text{tested positive})}$$

Two groups tested positive: those who take drugs and test positive (probability $= (0.04)(0.90) = 0.036$) and those who don't take drugs and test positive (probability $= (0.96)(0.05) = 0.048$), so the probability of testing positive is $0.036 + 0.048 = 0.084$. Thus,

$$P(\text{took drugs} \mid \text{positive test}) = \frac{0.036}{0.036 + 0.048} = \frac{0.036}{0.084} = 0.429$$

That is, 42.9% of the prospective employees who test positive actually took drugs.

Are you surprised by the final result of the example? Most people are. Sometimes a two-way table that includes counts is more convincing. To make calculations simple, we'll suppose that there were exactly 10,000 women aged 40 or over in this region, and that exactly 100 have breast cancer (that's 1% of the women).

How many of those 100 would have a positive mammogram? It would be 97% of 100, or 97 of them. That leaves 3 who would test negative. How many of the 9900 women who don't have breast cancer would get a positive mammogram? Six percent of them, or $(0.06)(9900) = 594$ women. The remaining $9900 - 594 = 9306$ would test negative. In total, $97 + 594 = 691$ women would have positive mammograms and $3 + 9306 = 9309$ women would have negative mammograms. This information is summarized in the two-way table below:

| | | Has breast cancer? | | |
|---|---|---|---|---|
| | | Yes | No | Total |
| Mammogram result | Positive | 97 | 594 | **691** |
| | Negative | 3 | 9306 | **9309** |
| | **Total** | **100** | **9900** | **10,000** |

Given that a randomly selected woman has a positive mammogram, the two-way table shows that the conditional probability $P(\text{breast cancer} \mid \text{positive mammogram}) = 97/691 = 0.14$.

This example illustrates an important fact when considering proposals for widespread testing for serious diseases or illegal drug use: if the condition being tested is uncommon in the population, many positives will be false positives. The best remedy is to retest any individual who tests positive.

## CHECK YOUR UNDERSTANDING

A computer company makes desktop and laptop computers at factories in three states—California, Texas, and New York. The California factory produces 40% of the company's computers, the Texas factory makes 25%, and the remaining 35% are manufactured in New York. Of the computers made in California, 75% are laptops. Of those made in Texas and New York, 70% and 50%, respectively, are laptops. All computers are first shipped to a distribution center in Missouri before being sent out to stores. Suppose we select a computer at random from the distribution center.[14]

1. Construct a tree diagram to represent this situation.
2. Find the probability that the computer is a laptop. Show your work.
3. Given that a laptop is selected, what is the probability that it was made in California?

## Conditional Probability and Independence

Suppose you toss a fair coin twice. Define events $A$: first toss is a head, and $B$: second toss is a head. We know that $P(A) = 1/2$ and $P(B) = 1/2$. What's $P(B \mid A)$? It's the conditional probability that the second toss is a head given that the first toss was a head. The coin has no memory, so $P(B \mid A) = 1/2$. In this case, $P(B \mid A) = P(B)$. Knowing that the first toss was a head does not change the probability that the second toss is a head.

### Teaching Tip

Ask your students to think about the relationship between false positive and false negative rates. To make the false positive rate go down, the threshold that the test uses must be reduced. This will result in more women with breast cancer testing negative, increasing the false negative rate. The relationship between false positive rates and false negative rates is very similar to the relationship between the probability of a Type I error and the probability of a Type II error, which we will discuss in Chapter 9. Discussing the relationship between false positive rates and false negative rates now will definitely pay off later.

### Teaching Tip

Here is a two-way table summarizing the previous Alternate Example, assuming there were 1000 prospective employees tested.

| | Drugs | No drugs | Total |
|---|---|---|---|
| Tested positive | 36 | 48 | **84** |
| Tested negative | 4 | 912 | **916** |
| **Total** | **40** | **960** | **1000** |

From the table, $P(\text{took drugs} \mid \text{tested positive}) = 36/84 = 0.429$.

### Teaching Tip

Most conditional probability questions can be solved using a tree diagram or a two-way table. However, if the problem provides conditional probabilities or describes a sequence of events, it is often easier to use a tree diagram. If the problem provides counts or proportions of people in different categories, a two-way table is often easier to use.

### ANSWERS TO CYU

**1.**

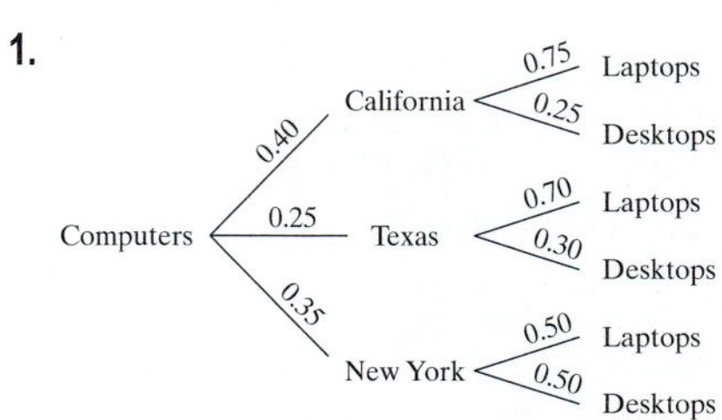


**2.** $P(\text{laptop}) = 0.30 + 0.175 + 0.175 = 0.65$

**3.** $P(\text{made in CA} \mid \text{laptop}) = \dfrac{0.30}{0.65} = 0.462$

### ALTERNATE EXAMPLE

#### *Who owns a home?*

The events of interest in this scenario were $G$: is a high school graduate and $H$: owns a home. We already learned that $P(H) = 340/500 = 68\%$ and that $P(H \mid G) = 221/310 = 71.3\%$. That is, we know that a randomly selected member of the sample has a 68% probability of owning a home. However, if we know that the randomly selected member is a high school graduate, the probability of owning a home increases to 71.3%. Because knowing the outcome of one event changes the probability of the other event, these two events are *not* independent.

Let's contrast the coin-toss scenario with our earlier pierced-ears example. In that case, the chance process involved randomly selecting a student from a college statistics class. The events of interest were A: is male, and B: has pierced ears. We already found that $P(A) = 90/178$, $P(B) = 103/178$, and $P(B \mid A) = 19/90$. If we know that the chosen student is male, the probability that he has pierced ears is $19/90 = 0.211$. This conditional probability is very different from the unconditional probability $P(B) = 103/178 = 0.579$ that a randomly selected student from the class has pierced ears.

| | Gender | | |
|---|---|---|---|
| **Pierced Ears** | **Male** | **Female** | **Total** |
| Yes | 19 | 84 | **103** |
| No | 71 | 4 | **75** |
| **Total** | **90** | **88** | **178** |

To recap, $P(B \mid A) = P(B)$ for events A and B in the coin-toss setting. For the pierced-ears scenario, however, $P(B \mid A) \neq P(B)$. When knowledge that one event has happened does not change the likelihood that another event will happen, we say that the two events are **independent**.

**DEFINITION: Independent events**

Two events $A$ and $B$ are **independent** if the occurrence of one event does not change the probability that the other event will happen. In other words, events $A$ and $B$ are independent if $P(A \mid B) = P(A)$ and $P(B \mid A) = P(B)$.

Determining whether two events related to the same chance process are independent requires us to compute probabilities. Here's an example that shows what we mean.

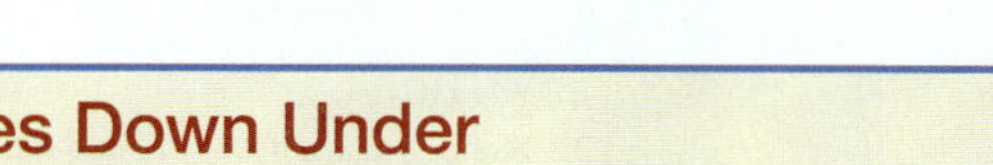

## EXAMPLE Lefties Down Under

### *Checking for independence*

Is there a relationship between gender and handedness? To find out, we used CensusAtSchool's Random Data Selector to choose an SRS of 100 Australian high school students who completed a survey. The two-way table displays data on the gender and dominant hand of each student.

| | Gender | | |
|---|---|---|---|
| **Dominant Hand** | **Male** | **Female** | **Total** |
| Right | 39 | 51 | **90** |
| Left | 7 | 3 | **10** |
| **Total** | **46** | **54** | **100** |

PROBLEM: Are the events "male" and "left-handed" independent? Justify your answer.

SOLUTION: To check whether the two events are independent, we want to find out if knowing that one event has happened changes the probability that the other event occurs. Suppose we are told that the chosen student is male. From the two-way table, $P(\text{left-handed} \mid \text{male}) = 7/46 = 0.152$. The unconditional probability $P(\text{left-handed}) = 10/100 = 0.10$. These two probabilities are close, but they're not equal. So the events "male" and "left-handed" are not independent. Knowing that the student is male increases the probability that the student is left-handed.

For Practice Try Exercise 85

## Teaching Tip

To help students understand the meaning of independence, sketch a scatterplot that displays no association. When there is no association between the variables, knowing the value of the $x$ variable won't help you predict the value of the $y$ variable. When two events are independent, knowing that one event occurred won't help you predict if the other event will occur. We will be talking about independence in every remaining chapter of the book, so really emphasize what it means. The sooner students are comfortable with the idea of independence, the better.

## Teaching Tip

Also, two events $A$ and $B$ are independent if $P(A \mid B) = P(A \mid B^c)$. In the context of the pierced ears example, we could note that $P(\text{pierced ears} \mid \text{male}) \neq P(\text{pierced ears} \mid \text{female})$ to show events $A$ and $B$ are not independent. Illustrate with the cards from page 309. Divide the cards into two stacks: Male and Female. Hold up each pile and ask whether the probability of selecting a "Yes" card is the same for both stacks.

## Teaching Tip

Notice the last sentence in the example. Although it isn't essential to include this contextual conclusion, we believe it helps students understand the two probabilities that they are comparing.

## ALTERNATE EXAMPLE

### *Finger length*

Is there a relationship between gender and relative finger length? To find out, we used the random sampler at the United States CensusAtSchool Web site (www.amstat.org/censusatschool) to randomly select 452 U.S. high school students who completed a survey. The two-way table shows the gender of each student and which finger was longer on their left hand (index finger or ring finger).

| | **Female** | **Male** | **Total** |
|---|---|---|---|
| Index finger | 78 | 45 | **123** |
| Ring finger | 82 | 152 | **234** |
| Same length | 52 | 43 | **95** |
| **Total** | **212** | **240** | **452** |

PROBLEM: Are the events "female" and "has a longer ring finger" independent? Justify your answer.

SOLUTION: To check whether two events are independent, we need to check whether knowing a student's gender changes the probability that the student has a longer ring finger. If a student is female, then the probability she has a longer ring finger is $P(\text{ring finger} \mid \text{female}) = 82/212 = 0.387$. However, the unconditional probability of having a longer ring finger is $P(\text{ring finger}) = 234/452 = 0.518$. These two probabilities are not equal, so the events "female" and "has a longer ring finger" are not independent. Knowing that a student is female makes it less likely that her ring finger is longer than her index finger.

You might have thought, "Surely there's no connection between gender and handedness. The events 'male' and 'left-handed' are bound to be independent." As the example shows, you can't use your intuition to check whether events are independent. To be sure, you have to calculate some probabilities.

In the preceding example, we could have also compared P(male | left-handed) with P(male). Of the 10 left-handed students in the sample, 7 were male. So P(male | left-handed) = 7/10 = 0.70. We can see from the two-way table that P(male) = 46/100 = 0.46. Once again, the two probabilities are not equal. Knowing that a person is left-handed makes it more likely that the person is male.

**Is there a connection between independence of events and association between two variables?** In the previous example, we found that the events "male" and "left-handed" were *not* independent. Does that mean there actually *is* a relationship between the variables gender and handedness in the larger population? Maybe or maybe not. If there is *no* association between the variables, it would be surprising to choose a random sample of 100 students for which P(left-handed | male) was *exactly* equal to P(left-handed). But these two probabilities should be close to equal if there's no association between the variables. How close is close? You'll have to wait a few chapters to find out.

### Teaching Tip

Remind your students about the concept of sampling variability when discussing the Think About It on this page. If you were to take another random sample, the table would probably look a little different, as would the probabilities we calculated. It is certainly possible that there is no association between gender and handedness (or gender and relative finger length) in the larger population and that we got these small differences in probabilities due to sampling variability. In Chapter 11, we will use a chi-square test for independence to address this issue more completely.

## CHECK YOUR UNDERSTANDING

For each chance process below, determine whether the events are independent. Justify your answer.

1. Shuffle a standard deck of cards, and turn over the top card. Put it back in the deck, shuffle again, and turn over the top card. Define events A: first card is a heart, and B: second card is a heart.

2. Shuffle a standard deck of cards, and turn over the top two cards, one at a time. Define events A: first card is a heart, and B: second card is a heart.

3. The 28 students in Mr. Tabor's AP® Statistics class completed a brief survey. One of the questions asked whether each student was right- or left-handed. The two-way table summarizes the class data. Choose a student from the class at random. The events of interest are "female" and "right-handed."

| | Gender | |
|---|---|---|
| **Handedness** | **Female** | **Male** |
| Left | 3 | 1 |
| Right | 18 | 6 |

### ANSWERS TO CYU

1. Independent. Because we are replacing the cards, knowing what the first card was will not help us predict what the second card will be.

2. Not independent. Once we know the suit of the first card, then the probability of getting a heart on the second card will change depending on what the first card was.

3. Independent. *P*(right-handed) = 24/28 = 6/7 is the same as *P*(right-handed | female) = 18/21 = 6/7.

## Independence: A Special Multiplication Rule

What happens to the general multiplication rule in the special case when events A and B are independent? In that case, $P(B \mid A) = P(B)$. We can simplify the general multiplication rule as follows:

$$\begin{aligned} P(A \cap B) &= P(A) \cdot P(B \mid A) \\ &= P(A) \cdot P(B) \end{aligned}$$

This result is known as the **multiplication rule for independent events.**

**DEFINITION: Multiplication rule for independent events**

If *A* and *B* are independent events, then the probability that *A* and *B* both occur is

$$P(A \cap B) = P(A) \cdot P(B)$$

### Teaching Tip

Remind students how the general addition rule simplifies to *P*(*A* or *B*) = *P*(*A*) + *P*(*B*) when the events are mutually exclusive.

Note that this rule *only* applies to independent events. Let's look at an example that uses the multiplication rule for independent events to analyze an important historical event.

## EXAMPLE The *Challenger* Disaster

### *Independence and the multiplication rule*

On January 28, 1986, Space Shuttle *Challenger* exploded on takeoff. All seven crew members were killed. Following the disaster, scientists and statisticians helped analyze what went wrong. They determined that the failure of O-ring joints in the shuttle's booster rockets was to blame. Under the cold conditions that day, experts estimated that the probability that an individual O-ring joint would function properly was 0.977. But there were six of these O-ring joints, and all six had to function properly for the shuttle to launch safely.

PROBLEM: Assuming that O-ring joints succeed or fail independently, find the probability that the shuttle would launch safely under similar conditions.

SOLUTION: For the shuttle to launch safely, all six O-ring joints need to function properly. The chance that this happens is given by

$$P(\text{joint 1 OK } and \text{ joint 2 OK } and \text{ joint 3 OK } and \text{ joint 4 OK } and \text{ joint 5 OK } and \text{ joint 6 OK})$$

By the multiplication rule for independent events, this probability is

$$P(\text{joint 1 OK}) \cdot P(\text{joint 2 OK}) \cdot P(\text{joint 3 OK}) \cdot P(\text{joint 4 OK}) \cdot P(\text{joint 5 OK}) \cdot P(\text{joint 6 OK})$$
$$= (0.977)(0.977)(0.977)(0.977)(0.977)(0.977) = 0.87$$

There's an 87% chance that the shuttle would launch safely under similar conditions (and a 13% chance that it wouldn't).

*Note:* As a result of the statistical analysis following the *Challenger* disaster, NASA made important safety changes to the design of the shuttle's booster rockets.

**For Practice** *Try Exercise*  89

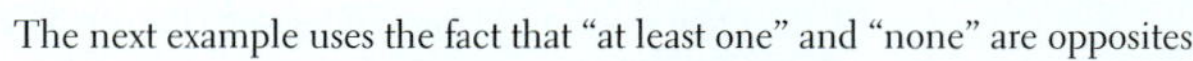

The next example uses the fact that "at least one" and "none" are opposites.

## EXAMPLE Rapid HIV Testing

### *Finding the probability of "at least one"*

Many people who come to clinics to be tested for HIV, the virus that causes AIDS, don't come back to learn the test results. Clinics now use "rapid HIV tests" that give a result while the client waits. In a clinic in Malawi, for example,

**ALTERNATE EXAMPLE**

### *Perfect games*

In baseball, a perfect game is when a pitcher doesn't allow any hitters to reach base in all nine innings. Historically, pitchers throw a perfect inning—an inning where no hitters reach base—about 40% of the time (http://www.baseballprospectus.com/article.php?articleid=11110). So, to throw a perfect game, a pitcher needs to have nine perfect innings in a row.

PROBLEM: What is the probability that a pitcher throws nine perfect innings in a row, assuming the pitcher's performance in an inning is independent of his performance in other innings?

SOLUTION: The probability of having nine perfect innings in a row is

$$P(\text{inning 1 perfect and inning 2 perfect and} \ldots \text{and inning 9 perfect})$$

$$= P(\text{inning 1 perfect}) \cdot P(\text{inning 2 perfect}) \cdot \ldots \cdot P(\text{inning 9 perfect})$$

$$= (0.4)(0.4) \ldots (0.4) = (0.4)^9 = 0.00026$$

With 30 teams playing 162 games per season, this means we would expect to see about (30)(162)(0.00026) = 1.3 perfect games per season. However, in baseball history, perfect games are much rarer, occurring once every 4 or 5 seasons, on average. This discrepancy suggests that our assumption of independence may not have been a good one.

## Teaching Tip

In the example on this page, make sure to emphasize that the false positive rate is a *conditional* probability. That is, $P(\text{positive test} \mid \text{no HIV antibodies}) = 0.004$. Thus, this probability applies only to people with no HIV antibodies.

## ALTERNATE EXAMPLE

### *First trimester screen*

The First Trimester Screen is a noninvasive test given during the first trimester of pregnancy to determine if there are specific chromosomal abnormalities in the fetus. According to a study published in the *New England Journal of Medicine* in November 2005 (http://www.americanpregnancy.org/prenataltesting/firstscreen.html), approximately 5% of normal pregnancies will receive a positive result.

PROBLEM: If 100 women with normal pregnancies are tested with the First Trimester Screen, what is the probability that at least 1 woman will receive a positive result?

SOLUTION: It is reasonable to assume that the test results for different women are independent. To find the probability of at least one false positive, we can use the complement rule and the probability that none of the women will receive a positive test result:

$$P(\text{at least one positive}) = 1 - P(\text{no positive results})$$

For women with normal pregnancies, the probability that a single test is negative is $1 - 0.05 = 0.95$. The probability that all 100 women will get negative results is $(0.95)(0.95)\cdots(0.95) = (0.95)^{100} = 0.0059$. Thus, $P(\text{at least one positive}) = 1 - 0.0059 = 0.9941$. There is over a 99% probability that at least 1 of the 100 women with normal pregnancies will receive a false positive on the First Trimester Screen.

## Teaching Tip

Remind your students that they should always think about the conditions necessary to use a particular formula or method. This will be a very important part of the inference procedures that we will learn about in Chapters 8–12.

use of rapid tests increased the percent of clients who learned their test results from 69% to 99.7%.

The trade-off for fast results is that rapid tests are less accurate than slower laboratory tests. Applied to people who have no HIV antibodies, one rapid test has probability about 0.004 of producing a false positive (that is, of falsely indicating that antibodies are present).[15]

PROBLEM: If a clinic tests 200 randomly selected people who are free of HIV antibodies, what is the chance that at least one false positive will occur?

SOLUTION: It is reasonable to assume that the test results for different individuals are independent. We have 200 independent events, each with probability 0.004. "At least one" combines many possible outcomes. It will be easier to use the fact that

$$P(\text{at least one positive}) = 1 - P(\text{no positives})$$

We'll find $P(\text{no positives})$ first. The probability of a negative result for any one person is $1 - 0.004 = 0.996$. To find the probability that all 200 people tested have negative results, use the multiplication rule for independent events:

$$\begin{aligned} P(\text{no positives}) &= P(\text{all 200 negative}) \\ &= (0.996)(0.996)\ldots(0.996) \\ &= 0.996^{200} = 0.4486 \end{aligned}$$

The probability we want is therefore

$$P(\text{at least one positive}) = 1 - 0.4486 = 0.5514$$

There is more than a 50% chance that at least 1 of the 200 people will test positive for HIV, even though no one has the virus.

**For Practice** *Try Exercise* **91**

The multiplication rule $P(A \text{ and } B) = P(A) \cdot P(B)$ holds if $A$ and $B$ are *independent* but not otherwise. The addition rule $P(A \text{ or } B) = P(A) + P(B)$ holds if $A$ and $B$ are *mutually exclusive* but not otherwise. Resist the temptation to use these simple rules when the conditions that justify them are not present.

## EXAMPLE Sudden Infant Death

### *Condemned by independence*

Assuming independence when it isn't true can lead to disaster. Several mothers in England were convicted of murder simply because two of their children had died in their cribs with no visible cause. An "expert witness" for the prosecution said that the probability of an unexplained crib death in a nonsmoking middle-class family is 1/8500. He then multiplied 1/8500 by 1/8500 to claim that there is only a 1-in-72-million chance that two children in the same family could have died naturally. This is nonsense: it assumes that crib deaths are independent, and data suggest that they are not. Some common genetic or environmental cause, not murder, probably explains the deaths.

THINK ABOUT IT

**Is there a connection between mutually exclusive and independent?** Let's start with a new chance process. Choose a U.S. adult at random. Define event $A$: the person is male, and event $B$: the person is pregnant. It's fairly clear that these two events are mutually exclusive (can't happen together)! What about independence? If you know that event $A$ has occurred, does this change the probability that event $B$ happens? Of course! If we know the person is male, then the chance that the person is pregnant is 0. Because $P(B \mid A) \neq P(B)$, the two events are not independent. *Two mutually exclusive events can never be independent*, because if one event happens, the other event is guaranteed not to happen.

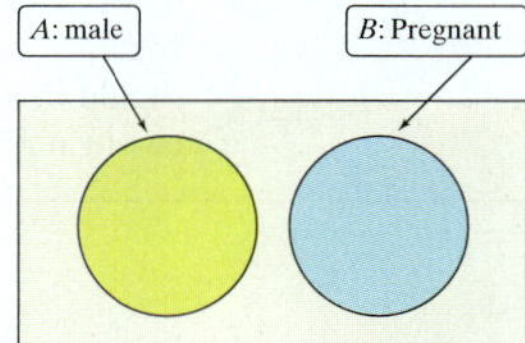

### CHECK YOUR UNDERSTANDING

1. During World War II, the British found that the probability that a bomber is lost through enemy action on a mission over occupied Europe was 0.05. Assuming that missions are independent, find the probability that a bomber returned safely from 20 missions.

2. Government data show that 8% of adults are full-time college students and that 30% of adults are age 55 or older. Because $(0.08)(0.30) = 0.024$, can we conclude that about 2.4% of adults are college students 55 or older? Why or why not?

### ANSWERS TO CYU

**1.** $P$ (returned safely) = 0.95. So, $P$(safe return on all 20 missions) = $0.95^{20} = 0.3585$

**2.** No. Being a college student and being 55 or older are not independent events.

### ALTERNATE EXAMPLE

#### *Weather conditions*

Hacienda Heights and La Puente are two neighboring suburbs in the Los Angeles area. According to the local newspaper, there is a 50% chance of rain tomorrow in Hacienda Heights and a 50% chance of rain in La Puente. Does this mean that there is a $(0.5)(0.5) = 0.25$ probability that it will rain in both cities tomorrow? No. It is not appropriate to multiply the two probabilities, because the events aren't independent. If it is raining in one of these locations, there is a very high probability that it is raining in the other location. However, suppose that there was also a 50% chance of rain in New York tomorrow. To find the probability that it will rain in Hacienda Heights and in New York, it *would* be appropriate to multiply the probabilities, because it is reasonable to believe that knowledge of rain in Hacienda Heights won't help us predict rain in New York.

### Teaching Tip

Many students confuse when two events are independent and when two events are mutually exclusive. Here are some examples to illustrate the difference.

- Select one card from a standard deck and define the events $A$: the card is red and $B$: the card is a club. Because there are no red clubs, $A$ and $B$ are mutually exclusive. However, because $P(A) = 0.5$ and $P(A \mid B) = 0$, these events are not independent.
- Select one card from a standard deck and define the events $A$: the card is red and $B$: the card is a 7. Because there are two red 7s, $A$ and $B$ are not mutually exclusive. However, because $P(A) = 0.5$ and $P(A \mid B) = 0.5$, these events are independent.
- Select one card from a standard deck and define the events $A$: the card is red and $B$: the card is a heart. Because all the hearts are red, $A$ and $B$ are not mutually exclusive. Also, because $P(A) = 0.5$ and $P(A \mid B) = 1$, these events are not independent.

Overall, because of its prominence in later chapters, it is much more important that students understand the idea of independence than the idea of mutual exclusivity.

## Calculated Risks

The chapter-opening Case Study on page 287 described drug-testing programs for high school athletes. Suppose that 16% of the high school athletes in a large school district have taken a banned substance. The drug test used by this district has a false positive rate of 5% and a false negative rate of 10%. Use what you have learned in this chapter to help answer the following questions about the district's drug-testing program. Show your method clearly.

1. What's the probability that a randomly chosen athlete tests positive for banned substances?
2. If two athletes are randomly selected, what's the probability that at least one of them tests positive?
3. If a randomly chosen athlete tests positive, what's the probability that the student did not take a banned substance? Based on your answer, do you think that an athlete who tests positive should be suspended from athletic competition for a year? Why or why not?
4. If a randomly chosen athlete tests negative, what's the probability that the student took a banned substance? Explain why it makes sense for the drug-testing process to be designed so that this probability is less than the one you found in Question 3.
5. The district decides to immediately retest any athlete who tests positive. Assume that the results of an athlete's two tests are independent. Find the probability that a student who gets a positive result on both tests actually took a banned substance. Based on your answer, do you think that an athlete who tests positive twice should be suspended from athletic competition for a year? Why or why not?

## Section 5.3 Summary

- If one event has happened, the chance that another event will happen is a **conditional probability**. The notation $P(B \mid A)$ represents the probability that event $B$ occurs given that event $A$ has occurred.
- You can calculate conditional probabilities with the **conditional probability formula**

$$P(A \mid B) = \frac{P(A \cap B)}{P(B)}$$

### Answers to Case Closed

**1.** Here is a tree diagram that summarizes the information provided.

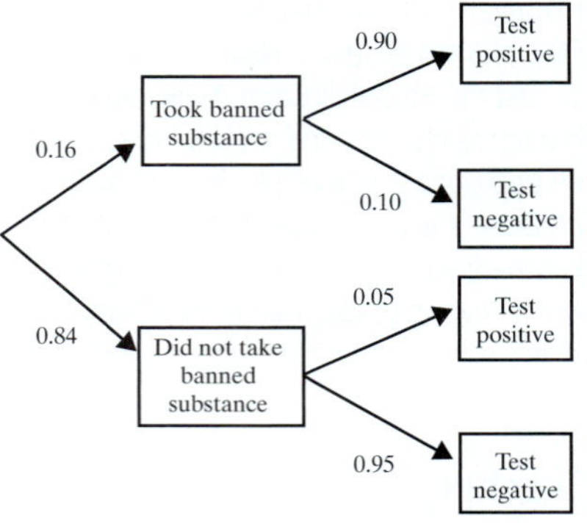


There are two groups who test positive: those who take banned substances and test positive and those who do not take banned substances and test positive. $P(\text{test positive}) = (0.16)(0.90) + (0.84)(0.05) = 0.144 + 0.042 = 0.186$. There is a 0.186 probability that a randomly chosen athlete tests positive.

**2.** $P(\text{at least 1 tests positive}) = 1 - P(\text{neither tests positive}) = 1 - (1 - 0.186)^2 = 1 - (0.814)^2 = 1 - 0.663 = 0.337$. There is a 0.337 probability that at least one of them tests positive.

**3.** $P(\text{did not take banned substance} \mid \text{test positive})$

$$= \frac{P(\text{did not take a banned substance and tests positive})}{P(\text{tests positive})}$$

$$= \frac{0.042}{0.186} = 0.226$$

Given that an athlete tests positive, there is a 0.226 probability that he or she did not take banned substances. Because this a relatively high probability, suspending everyone who tests positive will unfairly penalize a sizable number of clean athletes.

**4.** $P(\text{took a banned substance} \mid \text{test negative})$

$$= \frac{P\,(\text{took a banned substance and tests negative})}{P\,(\text{tests negative})}$$

$$= \frac{(0.16)(0.10)}{(0.16)(0.10) + (0.84)(0.95)} = \frac{0.016}{0.814} = 0.020$$

Given that an athlete tests negative, there is a 0.020 probability that he or she actually took a banned substance. We want this probability to be smaller because we want to minimize the number of athletes who take banned substances that escape detection.

**5.** $P(\text{took a banned substance} \mid \text{two positive tests})$

$$= \frac{P(\text{took a banned substance and tests positive twice})}{P(\text{tests positive twice})}$$

$$= \frac{(0.16)(0.90)(0.90)}{(0.16)(0.90)(0.90) + (0.84)(0.05)(0.05)} = \frac{0.1296}{0.1317} = 0.984$$

It would be more reasonable to suspend an athlete who tests positive twice, as nearly all (98.4%) of these athletes are guilty.

- The **general multiplication rule** states that the probability of events A and B occurring together is

$$P(A \text{ and } B) = P(A \cap B) = P(A) \cdot P(B \mid A)$$

- When chance behavior involves a sequence of outcomes, a **tree diagram** can be used to describe the sample space. Tree diagrams can also help in finding the probability that two or more events occur together. We simply multiply along the branches that correspond to the outcomes of interest.
- When knowing that one event has occurred does not change the probability that another event happens, we say that the two events are **independent**. For independent events A and B, $P(A \mid B) = P(A)$ and $P(B \mid A) = P(B)$. If two events A and B are mutually exclusive (disjoint), they cannot be independent.
- In the special case of *independent* events, the multiplication rule becomes

$$P(A \text{ and } B) = P(A \cap B) = P(A) \cdot P(B)$$

## Section 5.3 Exercises

**63. Get rich** A survey of 4826 randomly selected young adults (aged 19 to 25) asked, "What do you think are the chances you will have much more than a middle-class income at age 30?" The two-way table shows the responses.[16] Choose a survey respondent at random.

| | Gender | | |
|---|---|---|---|
| **Opinion** | **Female** | **Male** | **Total** |
| Almost no chance | 96 | 98 | **194** |
| Some chance but probably not | 426 | 286 | **712** |
| A 50-50 chance | 696 | 720 | **1416** |
| A good chance | 663 | 758 | **1421** |
| Almost certain | 486 | 597 | **1083** |
| **Total** | **2367** | **2459** | **4826** |

(a) Given that the person selected is male, what's the probability that he answered "almost certain"?

(b) If the person selected said "some chance but probably not," what's the probability that the person is female?

**64. A *Titanic* disaster** In 1912 the luxury liner *Titanic*, on its first voyage across the Atlantic, struck an iceberg and sank. Some passengers got off the ship in lifeboats, but many died. The two-way table gives information about adult passengers who lived and who died, by class of travel. Suppose we choose an adult passenger at random.

| | Survival Status | |
|---|---|---|
| **Class of Travel** | **Survived** | **Died** |
| First class | 197 | 122 |
| Second class | 94 | 167 |
| Third class | 151 | 476 |

(a) Given that the person selected was in first class, what's the probability that he or she survived?

(b) If the person selected survived, what's the probability that he or she was a third-class passenger?

**65. Sampling senators** The two-way table describes the members of the U.S. Senate in a recent year. Suppose we select a senator at random. Consider events D: is a democrat, and F: is female.

| | Male | Female |
|---|---|---|
| Democrats | 47 | 13 |
| Republicans | 36 | 4 |

(a) Find $P(D \mid F)$. Explain what this value means.

(b) Find $P(F \mid D)$. Explain what this value means.

**66. Who eats breakfast?** The following two-way table describes the 595 students who responded to a school survey about eating breakfast. Suppose we select a student at random. Consider events B: eats breakfast regularly, and M: is male.

### TRM Section 5.3 Quizzes

There are three different quizzes available for this section in the Teacher's Resource Materials. Versions A and B are roughly parallel. Find them by clicking on the link in the TE-book, opening the TRCD, or logging onto the book's companion Web site www.whfreeman.com/tps5e (teacher log-in required).

### TRM Full Solutions to Section 5.3 Exercises

### Answers to Section 5.3 Exercises

**5.63 (a)**

$$P(\text{almost certain} \mid M) = \frac{597}{2459} = 0.2428$$

**(b)** $P(F \mid \text{Some chance}) = \frac{426}{712} = 0.5983$

**5.64 (a)**

$$P(\text{survived} \mid \text{first class}) = \frac{197}{319} = 0.6176$$

**(b)**

$$P(\text{third class} \mid \text{survived}) = \frac{151}{442} = 0.3416$$

**5.65 (a)** $P(D \mid F) = \frac{13}{17} = 0.7647$. Given that a senator is female, there is a 0.7647 probability that she is a Democrat.

**(b)** $P(F \mid D) = \frac{13}{60} = 0.2167$. Given that a senator is a Democrat, there is a 0.2167 probability that she is a female.

| | Male | Female | Total |
|---|---|---|---|
| Eats breakfast regularly | 190 | 110 | **300** |
| Doesn't eat breakfast regularly | 130 | 165 | **295** |
| **Total** | **320** | **275** | **595** |

(a) Find $P(B \mid M)$. Explain what this value means.

(b) Find $P(M \mid B)$. Explain what this value means.

67. **Foreign-language study** Choose a student in grades 9 to 12 at random and ask if he or she is studying a language other than English. Here is the distribution of results:

| **Language:** | Spanish | French | German | All others | None |
|---|---|---|---|---|---|
| **Probability:** | 0.26 | 0.09 | 0.03 | 0.03 | 0.59 |

(a) What's the probability that the student is studying a language other than English?

(b) What is the conditional probability that a student is studying Spanish given that he or she is studying some language other than English?

68. **Income tax returns** Here is the distribution of the adjusted gross income (in thousands of dollars) reported on individual federal income tax returns in a recent year:

| **Income:** | <25 | 25–49 | 50–99 | 100–499 | ≥500 |
|---|---|---|---|---|---|
| **Probability:** | 0.431 | 0.248 | 0.215 | 0.100 | 0.006 |

(a) What is the probability that a randomly chosen return shows an adjusted gross income of \$50,000 or more?

(b) Given that a return shows an income of at least \$50,000, what is the conditional probability that the income is at least \$100,000?

69. **Tall people and basketball players** Select an adult at random. Define events $T$: person is over 6 feet tall, and $B$: person is a professional basketball player. Rank the following probabilities from smallest to largest. Justify your answer.

$P(T) \quad P(B) \quad P(T \mid B) \quad P(B \mid T)$

70. **Teachers and college degrees** Select an adult at random. Define events $A$: person has earned a college degree, and $T$: person's career is teaching. Rank the following probabilities from smallest to largest. Justify your answer.

$P(A) \quad P(T) \quad P(A \mid T) \quad P(T \mid A)$

71. pg 320  **Facebook versus YouTube** A recent survey suggests that 85% of college students have posted a profile on Facebook, 73% use YouTube regularly, and 66% do both. Suppose we select a college student at random and learn that the student has a profile on Facebook. Find the probability that the student uses YouTube regularly. Show your work.

72. **Mac or PC?** A recent census at a major university revealed that 40% of its students mainly used Macintosh computers (Macs). The rest mainly used PCs. At the time of the census, 67% of the school's students were undergraduates. The rest were graduate students. In the census, 23% of the respondents were graduate students who said that they used PCs as their primary computers. Suppose we select a student at random from among those who were part of the census and learn that the student mainly uses a PC. Find the probability that this person is a graduate student. Show your work.

73. pg 322 **Free downloads?** Illegal music downloading has become a big problem: 29% of Internet users download music files, and 67% of downloaders say they don't care if the music is copyrighted.[17] What percent of Internet users download music and don't care if it's copyrighted? Write the information given in terms of probabilities, and use the general multiplication rule.

74. **At the gym** Suppose that 10% of adults belong to health clubs, and 40% of these health club members go to the club at least twice a week. What percent of all adults go to a health club at least twice a week? Write the information given in terms of probabilities, and use the general multiplication rule.

75. **Box of chocolates** According to Forrest Gump, "Life is like a box of chocolates. You never know what you're gonna get." Suppose a candy maker offers a special "Gump box" with 20 chocolate candies that look the same. In fact, 14 of the candies have soft centers and 6 have hard centers. Choose 2 of the candies from a Gump box at random.

(a) Draw a tree diagram that shows the sample space of this chance process.

(b) Find the probability that one of the chocolates has a soft center and the other one doesn't.

76. **Inspecting switches** A shipment contains 10,000 switches. Of these, 1000 are bad. An inspector draws 2 switches at random, one after the other.

(a) Draw a tree diagram that shows the sample space of this chance process.

(b) Find the probability that both switches are defective.

77. pg 324 **Fill 'er up!** In a recent month, 88% of automobile drivers filled their vehicles with regular gasoline, 2% purchased midgrade gas, and 10% bought premium gas.[18] Of those who bought regular gas, 28% paid with a credit card; of customers who bought midgrade and premium gas, 34% and 42%, respectively, paid with a credit card. Suppose we select a customer at random.

**5.66 (a)** $P(B \mid M) = \dfrac{190}{320} = 0.5938$. Given that a student is male, there is a 0.5938 probability that he eats breakfast regularly. **(b)** $P(M \mid B) = \dfrac{190}{300} = 0.6333$. Given that a student is a regular breakfast eater, there is a 0.6333 probability that he is a male.

**5.67 (a)** $P(\text{not English}) = 1 - 0.59 = 0.41$.
**(b)** $P(\text{Spanish} \mid \text{other than English}) = \dfrac{0.26}{0.41} = 0.6341$

**5.68 (a)** $P(\$50{,}000 \text{ or more}) = 0.321$.
**(b)** $P(\text{at least } \$100{,}000 \mid \text{at least } \$50{,}000) = \dfrac{0.106}{0.321} = 0.3302$.

**5.69** $P(B) < P(B \mid T) < P(T) < P(T \mid B)$. There are very few pro basketball players, so $P(B)$ should be smallest. If you are a pro basketball player, it is quite likely that you are tall, so $P(T \mid B)$ should be largest. Finally, it's much more likely to be over 6 feet tall than it is to be a pro basketball player if you're over 6 feet tall.

**5.70** $P(T) < P(T \mid A) < P(A) < P(A \mid T)$. Nearly everyone whose career is teaching has a college degree, so $P(A \mid T)$ will be largest. A larger percentage of the population will have college degrees than will be teachers, so $P(T) < P(A)$. Finally, the proportion of teachers among the college-educated is smaller than the proportion of college-educated among all people, so $P(T \mid A) < P(A)$.

**5.71** $P(YT \mid FB) = \dfrac{0.66}{0.85} = 0.7765$

**5.72** $P(G \mid PC) = \dfrac{0.23}{0.60} = 0.383$

**5.73** $P(\text{download music}) = 0.29$, $P(\text{don't care} \mid \text{download music}) = 0.67$. $P(\text{download music} \cap \text{don't care}) = (0.29)(0.67) = 0.1943 = 19.43\%$.

**5.74** $P(\text{belong to health club}) = 0.10$, $P(\text{go twice/week} \mid \text{belong}) = 0.40$. $P(\text{belong to health club} \cap \text{go twice a week}) = (0.40)(0.10) = 0.04 = 4\%$.

**5.75 (a)** A tree diagram is below.

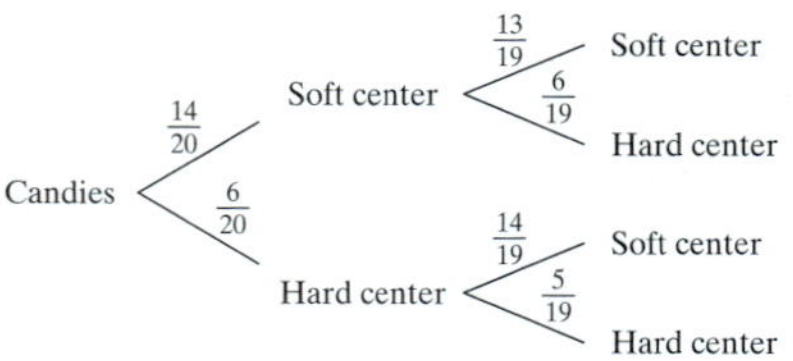


**(b)** $P(\text{one soft} \cap \text{one hard}) = \left(\dfrac{14}{20}\right)\left(\dfrac{6}{19}\right) + \left(\dfrac{6}{20}\right)\left(\dfrac{14}{19}\right) = \dfrac{168}{380} = 0.4421$

**5.76 (a)** A tree diagram is below.

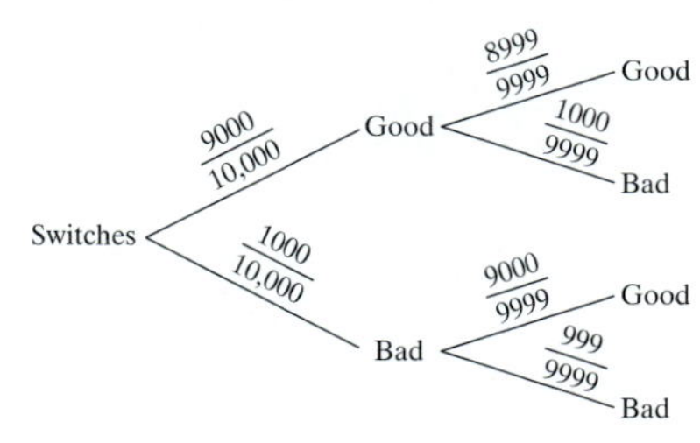


**(b)** $P(\text{defective} \cap \text{defective}) = \left(\dfrac{1000}{10{,}000}\right)\left(\dfrac{999}{9999}\right) = 0.00999$

(a) Draw a tree diagram to represent this situation.

(b) Find the probability that the customer paid with a credit card. Show your work.

(c) Given that the customer paid with a credit card, find the probability that she bought premium gas. Show your work.

78. **Urban voters** The voters in a large city are 40% white, 40% black, and 20% Hispanic. (Hispanics may be of any race in official statistics, but here we are speaking of political blocks.) A mayoral candidate anticipates attracting 30% of the white vote, 90% of the black vote, and 50% of the Hispanic vote. Suppose we select a voter at random.

(a) Draw a tree diagram to represent this situation.

(b) Find the probability that this voter votes for the mayoral candidate. Show your work.

(c) Given that the chosen voter plans to vote for the candidate, find the probability that the voter is black. Show your work.

79. **Lactose intolerance** Lactose intolerance causes difficulty in digesting dairy products that contain lactose (milk sugar). It is particularly common among people of African and Asian ancestry. In the United States (ignoring other groups and people who consider themselves to belong to more than one race), 82% of the population is white, 14% is black, and 4% is Asian. Moreover, 15% of whites, 70% of blacks, and 90% of Asians are lactose intolerant.[19] Suppose we select a U.S. person at random.

(a) What is the probability that the person is lactose intolerant? Show your work.

(b) Given that the person is lactose intolerant, find the probability that he or she is Asian. Show your work.

80. **Fundraising by telephone** Tree diagrams can organize problems having more than two stages. The figure at top right shows probabilities for a charity calling potential donors by telephone.[20] Each person called is either a recent donor, a past donor, or a new prospect. At the next stage, the person called either does or does not pledge to contribute, with conditional probabilities that depend on the donor class to which the person belongs. Finally, those who make a pledge either do or don't actually make a contribution. Suppose we randomly select a person who is called by the charity.

(a) What is the probability that the person contributed to the charity? Show your work.

(b) Given that the person contributed, find the probability that he or she is a recent donor. Show your work.

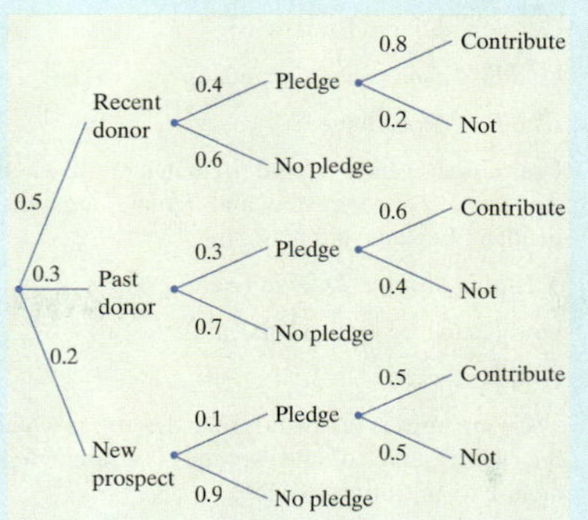


81. **HIV testing** Enzyme immunoassay (EIA) tests are used to screen blood specimens for the presence of antibodies to HIV, the virus that causes AIDS. Antibodies indicate the presence of the virus. The test is quite accurate but is not always correct. Here are approximate probabilities of positive and negative EIA outcomes when the blood tested does and does not actually contain antibodies to HIV:[21]

pg 325

| | Test Result | |
|---|---|---|
| **Truth** | + | − |
| Antibodies present | 0.9985 | 0.0015 |
| Antibodies absent | 0.006 | 0.994 |

Suppose that 1% of a large population carries antibodies to HIV in their blood. We choose a person from this population at random. Given that the EIA test is positive, find the probability that the person has the antibody. Show your work.

82. **Testing the test** Are false positives too common in some medical tests? Researchers conducted an experiment involving 250 patients with a medical condition and 750 other patients who did not have the medical condition. The medical technicians who were reading the test results were unaware that they were subjects in an experiment.

(a) Technicians correctly identified 240 of the 250 patients with the condition. They also identified 50 of the healthy patients as having the condition. What were the false positive and false negative rates for the test?

(b) Given that a patient got a positive test result, what is the probability that the patient actually had the medical condition? Show your work.

**5.79 (a)** $P(\text{lactose intolerant}) = (0.82)(0.15) + (0.14)(0.70) + (0.04)(0.90) = 0.257$

**(b)** $P(\text{Asian} \mid \text{lactose intolerant}) = \dfrac{0.036}{0.257} = 0.1401$

**5.80 (a)** $P(\text{contributed}) = (0.5)(0.4)(0.8) + (0.3)(0.3)(0.6) + (0.2)(0.1)(0.5) = 0.224$

**(b)** $P(\text{recent donor} \mid \text{contribute}) = \dfrac{0.16}{0.224} = 0.7143$

**5.81** $P(\text{antibody} \mid \text{positive}) = \dfrac{P(\text{antibody} \cap \text{positive})}{P(\text{positive})} = \dfrac{(0.01)(0.9985)}{(0.01)(0.9985) + (0.99)(0.006)} = 0.6270$

**5.82 (a)** False-positive rate $= \dfrac{50}{750} = 0.0667$. False-negative rate $= \dfrac{10}{250} = 0.04$.

**(b)** $P(\text{have condition} \mid \text{positive test result}) = 240/290 = 0.8276$

**5.77 (a)** A tree diagram is below.

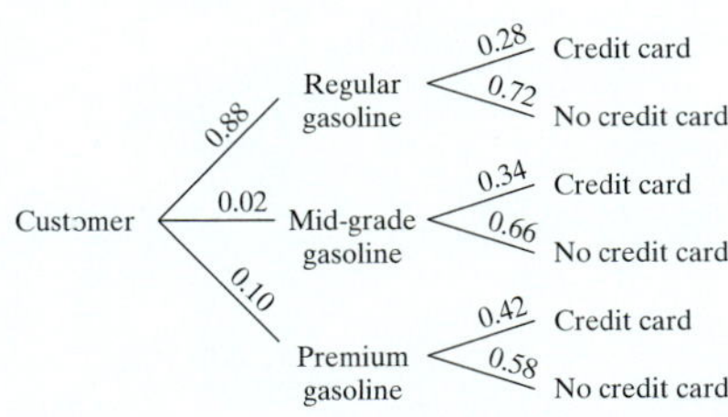


**(b)** $P(\text{credit card}) = (0.88)(0.28) + (0.02)(0.34) + (0.10)(0.42) = 0.2952$

**(c)** $P(\text{premium gasoline} \mid \text{credit card}) = \dfrac{0.0420}{0.2952} = 0.142$

**5.78 (a)** A tree diagram is below.

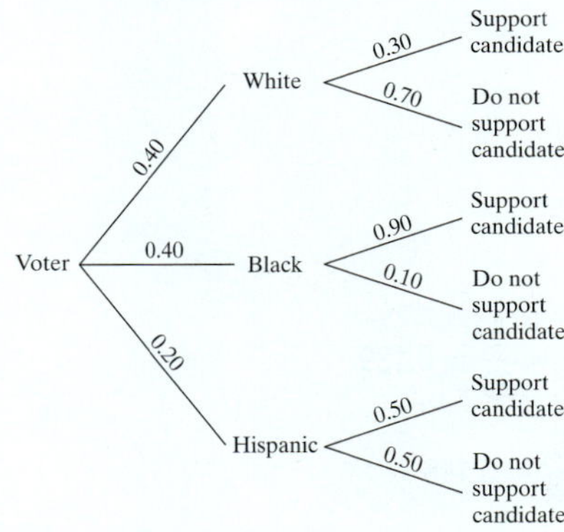


**(b)** $P(\text{support}) = (0.40)(0.30) + (0.40)(0.90) + (0.20)(0.50) = 0.58$

**(c)** $P(\text{black} \mid \text{support}) = \dfrac{0.36}{0.58} = 0.6207$

83. **Get rich** Refer to Exercise 63.

(a) Find $P$("a good chance" | female).

(b) Find $P$("a good chance").

(c) Use your answers to (a) and (b) to determine whether the events "a good chance" and "female" are independent. Explain your reasoning.

84. **A *Titanic* disaster** Refer to Exercise 64.

(a) Find $P$(survived | second class).

(b) Find $P$(survived).

(c) Use your answers to (a) and (b) to determine whether the events "survived" and "second class" are independent. Explain your reasoning.

pg 327 85. **Sampling senators** Refer to Exercise 65. Are events $D$ and $F$ independent? Justify your answer.

86. **Who eats breakfast?** Refer to Exercise 66. Are events $B$ and $M$ independent? Justify your answer.

87. **Rolling dice** Suppose you roll two fair, six-sided dice—one red and one green. Are the events "sum is 7" and "green die shows a 4" independent? Justify your answer.

88. **Rolling dice** Suppose you roll two fair, six-sided dice—one red and one green. Are the events "sum is 8" and "green die shows a 4" independent? Justify your answer.

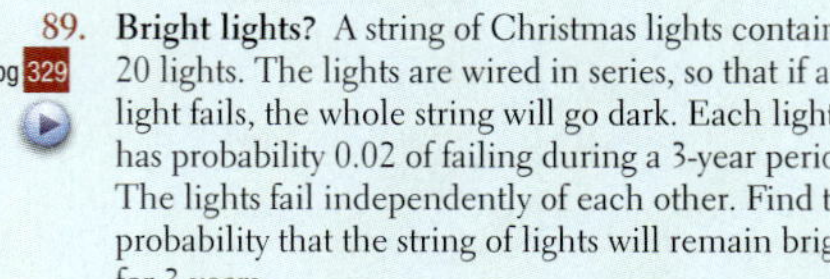

pg 329 89. **Bright lights?** A string of Christmas lights contains 20 lights. The lights are wired in series, so that if any light fails, the whole string will go dark. Each light has probability 0.02 of failing during a 3-year period. The lights fail independently of each other. Find the probability that the string of lights will remain bright for 3 years.

90. **Common names** The Census Bureau says that the 10 most common names in the United States are (in order) Smith, Johnson, Williams, Brown, Jones, Miller, Davis, Garcia, Rodriguez, and Wilson. These names account for 9.6% of all U.S. residents. Out of curiosity, you look at the authors of the textbooks for your current courses. There are 9 authors in all. Would you be surprised if none of the names of these authors were among the 10 most common? (Assume that authors' names are independent and follow the same probability distribution as the names of all residents.)

pg 329 91. **Universal blood donors** People with type O-negative blood are universal donors. That is, any patient can receive a transfusion of O-negative blood. Only 7.2% of the American population have O-negative blood. If we choose 10 Americans at random who gave blood, what is the probability that at least 1 of them is a universal donor?

92. **Lost Internet sites** Internet sites often vanish or move, so that references to them can't be followed. In fact, 13% of Internet sites referenced in major scientific journals are lost within two years after publication.[22] If we randomly select seven Internet references, from scientific journals, what is the probability that at least one of them doesn't work two years later?

93. **Late shows** Some TV shows begin after their scheduled times when earlier programs run late. According to a network's records, about 3% of its shows start late. To find the probability that three consecutive shows on this network start on time, can we multiply (0.97)(0.97)(0.97)? Why or why not?

94. **Late flights** An airline reports that 85% of its flights arrive on time. To find the probability that its next four flights into LaGuardia Airport all arrive on time, can we multiply (0.85)(0.85)(0.85)(0.85)? Why or why not?

95. **The geometric distributions** You are tossing a pair of fair, six-sided dice in a board game. Tosses are independent. You land in a danger zone that requires you to roll doubles (both faces showing the same number of spots) before you are allowed to play again. How long will you wait to play again?

(a) What is the probability of rolling doubles on a single toss of the dice? (If you need review, the possible outcomes appear in Figure 5.2 (page 306). All 36 outcomes are equally likely.)

(b) What is the probability that you do not roll doubles on the first toss, but you do on the second toss?

(c) What is the probability that the first two tosses are not doubles and the third toss is doubles? This is the probability that the first doubles occurs on the third toss.

(d) Now you see the pattern. What is the probability that the first doubles occurs on the fourth toss? On the fifth toss? Give the general result: what is the probability that the first doubles occurs on the $k$th toss?

96. **The probability of a flush** A poker player holds a flush when all 5 cards in the hand belong to the same suit. We will find the probability of a flush when 5 cards are dealt. Remember that a deck contains 52 cards, 13 of each suit, and that when the deck is well shuffled, each card dealt is equally likely to be any of those that remain in the deck.

**5.83 (a)** $\frac{663}{2367} = 0.2801$ **(b)** $\frac{1421}{4826} = 0.2944$ **(c)** The events are not independent because the probabilities in parts (a) and (b) are not the same.

**5.84 (a)** $\frac{94}{261} = 0.3602$ **(b)** $\frac{442}{1207} = 0.3662$ **(c)** The events are not independent because the two probabilities in parts (a) and (b) are not the same.

**5.85** Not independent. From Exercise 5.65, we saw that $P(D \mid F) = 0.7647$, which is not the same as $P(D) = \frac{60}{100} = 0.60$.

**5.86** Not independent. From Exercise 5.66, we saw that $P(B \mid M) = 0.5938$, which is not the same as $P(B) = \frac{300}{595} = 0.504$.

**5.87** Independent. $P$(sum of 7 | green is 4) = 1 / 6 = 0.1667, which equals $P$(sum of 7) = 6 / 36 = 0.1667

**5.88** Not independent. $P$(sum of 8 | green is 4) = 1 / 6 = 0.1667 is not equal to $P$(sum of 8) = 5 / 36 = 0.1389.

**5.89** $P$(all remain bright) = $(0.98)^{20}$ = 0.6676

**5.90** $P$(none common) = $(0.904)^9$ = 0.4032. Because this probability is fairly large, we should not be surprised.

**5.91** $P$(at least one universal donor) = $1 - (0.928)^{10} = 0.5263$

**5.92** $P$(at least one doesn't work) = $1 - (0.87)^7 = 0.6227$

**5.93** No, because the events are not independent. If one show starts late, we can predict that the next show will start late as well.

**5.94** No, because the events are not independent. If one flight is late, we can predict that others are likely to be late as well.

**5.95 (a)** $P(\text{doubles}) = \frac{6}{36} = \frac{1}{6} = 0.167$

**(b)** $P(\text{no doubles first} \cap \text{doubles second}) = \frac{5}{6}\left(\frac{1}{6}\right) = \frac{5}{36} = 0.139$

**(c)** $P(\text{first doubles on third roll}) = \frac{5}{6}\left(\frac{5}{6}\right)\left(\frac{1}{6}\right) = \frac{25}{216} = 0.116$

**(d)** 4th: $\left(\frac{5}{6}\right)^3\left(\frac{1}{6}\right)$. 5th: $\left(\frac{5}{6}\right)^4\left(\frac{1}{6}\right)$. The probability that the first doubles are rolled on the $k$th roll is $\left(\frac{5}{6}\right)^{k-1}\left(\frac{1}{6}\right)$.

(a) We will concentrate on spades. What is the probability that the first card dealt is a spade? What is the conditional probability that the second card is a spade given that the first is a spade?

(b) Continue to count the remaining cards to find the conditional probabilities of a spade on the third, the fourth, and the fifth card given in each case that all previous cards are spades.

(c) The probability of being dealt 5 spades is the product of the five probabilities you have found. Why? What is this probability?

(d) The probability of being dealt 5 hearts or 5 diamonds or 5 clubs is the same as the probability of being dealt 5 spades. What is the probability of being dealt a flush?

*Multiple choice: Select the best answer for Exercises 97 to 99.*

97. An athlete suspected of using steroids is given two tests that operate independently of each other. Test A has probability 0.9 of being positive if steroids have been used. Test B has probability 0.8 of being positive if steroids have been used. What is the probability that neither test is positive if steroids have been used?

(a) 0.72 (b) 0.38 (c) 0.02 (d) 0.28 (e) 0.08

98. In an effort to find the source of an outbreak of food poisoning at a conference, a team of medical detectives carried out a study. They examined all 50 people who had food poisoning and a random sample of 200 people attending the conference who didn't get food poisoning. The detectives found that 40% of the people with food poisoning went to a cocktail party on the second night of the conference, while only 10% of the people in the random sample attended the same party. Which of the following statements is appropriate for describing the 40% of people who went to the party? (Let $F$ = got food poisoning and $A$ = attended party.)

(a) $P(F \mid A) = 0.40$
(b) $P(A \mid F^C) = 0.40$
(c) $P(F \mid A^C) = 0.40$
(d) $P(A^C \mid F) = 0.40$
(e) $P(A \mid F) = 0.40$

99. Suppose a loaded die has the following probability model:

| Outcome: | 1 | 2 | 3 | 4 | 5 | 6 |
|---|---|---|---|---|---|---|
| Probability: | 0.3 | 0.1 | 0.1 | 0.1 | 0.1 | 0.3 |

If this die is thrown and the top face shows an odd number, what is the probability that the die shows a 1?

(a) 0.10 (b) 0.17 (c) 0.30 (d) 0.50 (e) 0.60

*Exercises 100 and 101 refer to the following setting.* Your body mass index (BMI) is your weight in kilograms divided by the square of your height in meters. Online BMI calculators allow you to enter weight in pounds and height in inches. High BMI is a common but controversial indicator of being overweight or obese. A study by the National Center for Health Statistics found that the BMI of American young women (ages 20 to 29) is approximately Normal with mean 26.8 and standard deviation 7.4.[23]

100. **BMI** (2.2) People with BMI less than 18.5 are often classed as "underweight." What percent of young women are underweight by this criterion? Sketch and shade the area of interest under a Normal curve.

101. **BMI** (5.2) Suppose we select two American young women in this age group at random. Find the probability that at least one of them is classified as underweight. Show your work.

102. **Life at work** (1.1) The University of Chicago's General Social Survey asked a representative sample of adults this question: "Which of the following statements best describes how your daily work is organized? (1) I am free to decide how my daily work is organized. (2) I can decide how my daily work is organized, within certain limits. (3) I am not free to decide how my daily work is organized." Here is a two-way table of the responses for three levels of education:[24]

| | Highest Degree Completed | | |
|---|---|---|---|
| Response | Less than High School | High School | Bachelor's |
| 1 | 31 | 161 | 81 |
| 2 | 49 | 269 | 85 |
| 3 | 47 | 112 | 14 |

Do these data suggest that there is an association between level of education and freedom to organize one's work in the adult population? Give appropriate evidence to support your answer.

**5.96 (a)** $P(\text{1st card } \spadesuit) = \frac{13}{52} = 0.25$,
$P(\text{2nd card } \spadesuit \mid \spadesuit \text{ picked}) = \frac{12}{51} = 0.2353$.
**(b)** $P(\text{3rd card } \spadesuit \mid 2\ \spadesuit\text{s picked}) = \frac{11}{50} = 0.22$,
$P(\text{4th card } \spadesuit \mid 3\ \spadesuit\text{s picked}) = \frac{10}{49} = 0.2041$,
$P(\text{5th card } \spadesuit \mid 4\ \spadesuit\text{s picked}) = \frac{9}{48} = 0.1875$.
**(c)** Using the general multiplication rule,
$P(\text{5 spades}) = \left(\frac{13}{52}\right)\left(\frac{12}{51}\right)\left(\frac{11}{50}\right)\left(\frac{10}{49}\right)\left(\frac{9}{48}\right) = 0.0004952$.
**(d)** $4(0.0004952) = 0.001981$

**5.97** c

**5.98** e

**5.99** e

**5.100** BMI for American young women follows a $N(26.8, 7.4)$ distribution and we want to find the percent of women with BMI less than 18.5. $z = \frac{18.5 - 26.8}{7.4} = -1.12$. From Table A, the proportion of z-scores below −1.12 is 0.1314. *Using technology*: `normalcdf(lower:−1000, upper:18.5,μ:26.8,σ:7.4)` = 0.1310. About 13% of American young women are classified as underweight.

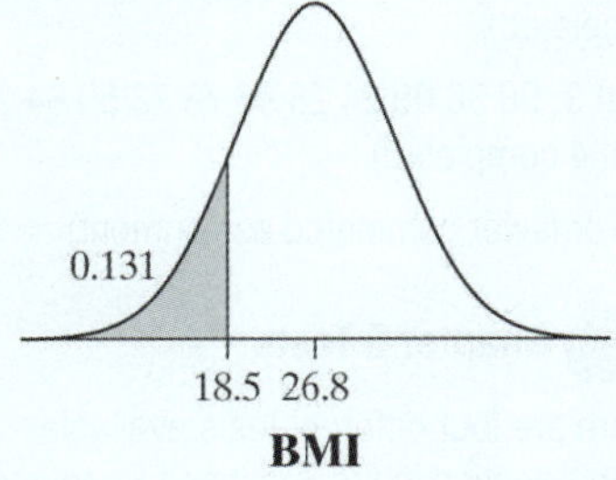


**5.101** $P(\text{at least one is underweight}) = 1 - (1 - 0.131)^2 = 0.2448$

**5.102** There appears to be an association in the population of adults. Generally, more education means more freedom. For example, 45% of those with bachelor's degrees felt free to organize their work, compared to 29.7% of those who finished high school and 24.4% of those who did not finish high school.

**TRM FRAPPY!**

We present a model solution here. Please consult the Teacher's Resource Materials for sample student responses, a scoring rubric, and a black line master of the original question with space for students to write their responses.

**(a)** Here is a two-way table that organizes the given information:

| | Male | Female | Total |
|---|---|---|---|
| Completed assignment | 11 | 19 | **30** |
| Did not complete assignment | 6 | 4 | **10** |
| **Total** | **17** | **23** | **40** |

*P*(complete) = 30/40 = 0.75. There is a 0.75 probability that the student completed the assignment.

**(b)** No. *P*(complete | female) = 19/23 = 0.826 is not equal to *P*(complete) = 30/40 = 0.75. Knowing that the student is a female increases the probability that the assignment was completed.

**(c)** Let 01–30 = completed assignment and 31–40 = didn't complete assignment. Skip 00, 41–99. Going left to right on the random digit table, choose 4 *different* two-digit numbers from 01–40. Count the number of students among the four chosen who completed the assignment (i.e., count how many of the numbers are 01–30). Repeat this process many times and divide the number of trials where 2 or fewer completed the assignment by the total number of trials.

**(d) Trial 1:** 12 ~~97~~ ~~51~~ 32 ~~58~~ 13 04 (3 of 4 completed)

**Trial 2:** ~~84~~ ~~51~~ ~~44~~ ~~72~~ 32 18 19 40 (2 of 4 completed)

**Trial 3:** ~~00~~ 36 ~~00~~ 24 28 ~~96~~ ~~76~~ ~~73~~ ~~59~~ ~~64~~ 23 (3 of 4 completed)

*P*(2 or fewer completed assignment) ≈ 1/3

**TRM Chapter 5 Tests**

There are four different tests available for this section in the Teacher's Resource Materials. Versions A and B are roughly parallel and version D is cumulative. Find them by clicking on the link in the TE-book, opening the TRCD, or logging onto the book's companion Web site www.whfreeman.com/tps5e (teacher log-in required).

## FRAPPY! Free Response AP® Problem, Yay!

**The following problem is modeled after actual AP® Statistics exam free response questions. Your task is to generate a complete, concise response in 15 minutes.**

*Directions: Show all your work. Indicate clearly the methods you use, because you will be scored on the correctness of your methods as well as on the accuracy and completeness of your results and explanations.*

A statistics teacher has 40 students in his class, 23 females and 17 males. At the beginning of class on a Monday, the teacher planned to spend time reviewing an assignment due that day. Unknown to the teacher, only 19 of the females and 11 of the males had completed the assignment. The teacher plans to randomly select students to do problems from the assignment on the whiteboard.

(a) What is the probability that a randomly selected student has completed the assignment?

(b) Are the events "selecting a female" and "selecting a student who completed the assignment" independent? Justify your answer.

Suppose that the teacher randomly selects 4 students to do a problem on the whiteboard and only 2 of the students had completed the assignment.

(c) Describe how to use a table of random digits to estimate the probability that 2 or fewer of the 4 randomly selected students completed the assignment.

(d) Complete three repetitions of your simulation using the random digits below and use the results to estimate the probability described in part (c).

12975 13258 13048 45144 72321 81940 00360 02428
96767 35964 23822 96012 94951 65194 50842 55372
37609 59057 66967 83401 60705 02384 90597 93600

**After you finish, you can view two example solutions on the book's Web site (www.whfreeman.com/tps5e) Determine whether you think each solution is "complete," "substantial," "developing," or "minimal." If the solution is not complete, what improvements would you suggest to the student who wrote it? Finally, your teacher will provide you with a scoring rubric Score your response and note what, if anything, you would do differently to improve your own score.**

# Chapter Review

### Section 5.1: Randomness, Probability, and Simulation

In this section, you learned about the law of large numbers and the idea of probability. The law of large numbers says that when you repeat a chance process many, many times, the relative frequency of an outcome will approach a single number. This single number is called the probability of the outcome—how often we expect the outcome to occur in a very large number of repetitions of the chance process. Make sure to remember the "large" part of the law of large numbers. Although clear patterns emerge in a large number of repetitions, we shouldn't expect such regularity in a small number of repetitions.

Simulations are powerful tools that we can use to imitate chance processes and estimate probabilities. To perform a simulation, use the familiar four-step process: state the question of interest, plan how to use a chance device to imitate a process, do many repetitions, and make a conclusion based on the results. If you are using random digits to perform your simulation, be sure to consider whether or not digits can be repeated within each trial.

### Section 5.2: Probability Rules

In this section, you learned that chance behavior can be described by a probability model. Probability models have two parts, a list of possible outcomes (the sample space) and a probability for each outcome. The probability of each outcome in a probability model must be between 0 and 1, and the probabilities of all the outcomes in the sample space must add to 1.

An event is a subset of the possible outcomes of a chance process. The complement rule says that the probability that an event occurs is 1 minus the probability that the event doesn't occur. In symbols, the complement rule says that $P(E) = 1 - P(E^C)$. Given two events A and B from some chance process, use the general addition rule to find the probability that event A or event B occurs:

$$P(A \text{ or } B) = P(A \cup B) = P(A) + P(B) - P(A \cap B)$$

If the events A and B have no outcomes in common, use the addition rule for mutually exclusive events: $P(A \cup B) = P(A) + P(B)$.

Finally, you learned how to use two-way tables and Venn diagrams to display the sample space for a chance process involving two events. Using a two-way table or a Venn diagram is a helpful way to organize information and calculate probabilities involving the union (A or B) and the intersection (A and B) of two events.

### Section 5.3: Conditional Probability and Independence

In this section, you learned that a conditional probability describes the probability of an event occurring given that another event is known to have already occurred. To calculate the probability that event A occurs given that event B has occurred, use the formula

$$P(A \mid B) = \frac{P(A \cap B)}{P(B)} = \frac{P(A \text{ and } B)}{P(B)}$$

Two-way tables and tree diagrams are useful ways to organize the information provided in a conditional probability problem. Two-way tables are best when the problem describes the number or proportion of cases with certain characteristics. Tree diagrams are best when the problem provides the conditional probabilities of different events or describes a sequence of events.

Use the general multiplication rule for calculating the probability that event A and event B both occur:

$$P(A \text{ and } B) = P(A \cap B) = P(A) \cdot P(B \mid A)$$

If knowing that event B occurs doesn't change the probability that event A occurs, then events A and B are independent. That is, events A and B are independent if $P(A \mid B) = P(A)$. If events A and B are independent, use the multiplication rule for independent events to find the probability that events A and B both occur: $P(A \cap B) = P(A) \cdot P(B)$.

## What Did You Learn?

| Learning Objective | Section | Related Example on Page(s) | Relevant Chapter Review Exercise(s) |
|---|---|---|---|
| Interpret probability as a long-run relative frequency. | 5.1 | 291, 292, 293, 294, 295 | R5.1 |
| Use simulation to model chance behavior. | 5.1 | 296, 297 | R5.2 |
| Determine a probability model for a chance process. | 5.2 | 306 | R5.3, R5.10 |
| Use basic probability rules, including the complement rule and the addition rule for mutually exclusive events. | 5.2 | 308 | R5.4, R5.10 |
| Use a two-way table or Venn diagram to model a chance process and calculate probabilities involving two events. | 5.2 | 309, 312, 313 | R5.4, R5.5, R5.7, R5.8 |
| Use the general addition rule to calculate probabilities. | 5.2 | 313 | R5.4, R5.5 |
| Calculate and interpret conditional probabilities. | 5.3 | 318, 320 | R5.6, R5.8 |
| Use the general multiplication rule to calculate probabilities. | 5.3 | 322 | R5.6 |
| Use tree diagrams to model a chance process and calculate probabilities involving two or more events. | 5.3 | 322, 324, 325 | R5.6 |
| Determine whether two events are independent. | 5.3 | 327 | R5.7, R5.8 |
| When appropriate, use the multiplication rule for independent events to compute probabilities. | 5.3 | 329, 331 | R5.9, R5.10 |



TRM Learning Objectives Grid

## TRM Full Solutions to Chapter 5 Review Exercises

## Answers to Chapter 5 Review Exercises

**R5.1** When the weather conditions are like those seen today, it has rained on the following day about 30% of the time.

**R5.2** **(a)** Let the numbers 00–14 represent not wearing a seat belt and 15–99 represent wearing a seat belt. Read 10 sets of two-digit numbers. For each set of 10 two-digit numbers, record whether there are two consecutive numbers between 00–14 or not. **(b)** The first sample is 29 **07** 71 48 63 61 68 34 70 52 (not two consecutive). The second sample is 62 22 45 **10** 25 95 **05** 29 **09 08** (two consecutive). The third sample is 73 59 27 51 86 87 **13** 69 57 61 (not two consecutive).

**R5.3** **(a)**

| Difference | Probability |
|---|---|
| 1 | $\frac{18}{36}$ |
| 5 | $\frac{6}{36}$ |
| −3 | $\frac{12}{36}$ |

**(b)** Die A. $P(A > B) = P(\text{positive difference}) = \frac{18}{36} + \frac{6}{36} = \frac{24}{36}$.

**R5.4** **(a)** Legitimate. All probabilities are between 0 and 1 and they add up to 1. **(b)** $P(\text{Hispanic}) = 0.001 + 0.006 + 0.139 + 0.003 = 0.149$. **(c)** $P(\text{not a non-Hispanic white}) = 1 - 0.674 = 0.326$. **(d)** People who are white and Hispanic will be double-counted. $P(\text{white or Hispanic}) = 0.813 + 0.149 - 0.139 = 0.823$.

**R5.5** **(a)**

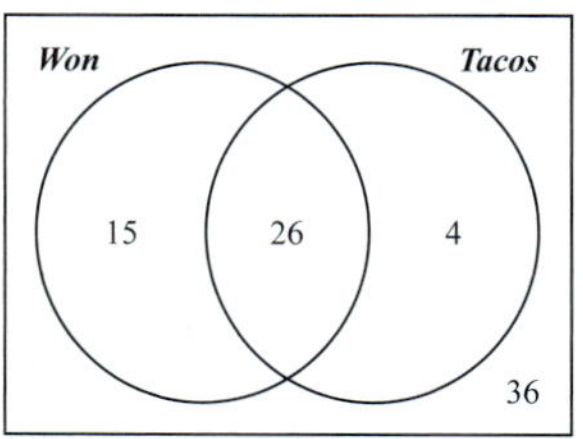


**(b)** $P(\text{lost and no tacos}) = 36/81 = 0.444$ **(c)** $P(\text{won or tacos}) = \frac{41}{81} + \frac{30}{81} - \frac{26}{81} = \frac{45}{81} = 0.556$



# Chapter 5 Chapter Review Exercises

*These exercises are designed to help you review the important ideas and methods of the chapter.*

**R5.1 Rainy days** The TV weatherman says, "There's a 30% chance of rain tomorrow." Explain what this statement means.

**R5.2 Click it or else** From police records, it has been determined that 15% of drivers stopped for routine license checks are not wearing seat belts. If a police officer stops 10 vehicles, how likely is it that two consecutive drivers won't be wearing their seat belts?

(a) Describe the design of a simulation to estimate this probability. Explain clearly how you will use the partial table of random digits below to carry out your simulation.

(b) Carry out three repetitions of the simulation. Copy the random digits below onto your paper. Then mark on or directly above the table to show your results.

| | | | | | |
|---|---|---|---|---|---|
| 29077 | 14863 | 61683 | 47052 | 62224 | 51025 |
| 95052 | 90908 | 73592 | 75186 | 87136 | 95761 |
| 27102 | 56027 | 55892 | 33063 | 41842 | 81868 |
| 43367 | 49497 | 72719 | 96758 | 27611 | 91596 |

**R5.3 Weird dice** Nonstandard dice can produce interesting distributions of outcomes. Suppose you have two balanced, six-sided dice. Die A has faces with 2, 2, 2, 2, 6, and 6 spots. Die B has faces with 1, 1, 1, 5, 5, and 5 spots. Imagine that you roll both dice at the same time.

(a) Find a probability model for the difference (Die A − Die B) in the total number of spots on the up-faces.

(b) Which die is more likely to roll a higher number? Justify your answer.

**R5.4 Race and ethnicity** The Census Bureau allows each person to choose from a long list of races. That is, in the eyes of the Census Bureau, you belong to whatever race you say you belong to. Hispanic (also called Latino) is a separate category. Hispanics may be of any race. If we choose a resident of the United States at random, the Census Bureau gives these probabilities:[25]

| | Hispanic | Not Hispanic |
|---|---|---|
| Asian | 0.001 | 0.044 |
| Black | 0.006 | 0.124 |
| White | 0.139 | 0.674 |
| Other | 0.003 | 0.009 |

(a) Verify that this is a legitimate assignment of probabilities.

(b) What is the probability that a randomly chosen American is Hispanic?

(c) Non-Hispanic whites are the historical majority in the United States. What is the probability that a randomly chosen American is *not* a member of this group?

(d) Explain why $P(\text{white or Hispanic}) \neq P(\text{white}) + P(\text{Hispanic})$. Then find $P(\text{white or Hispanic})$.

**R5.5** In 2012, fans at Arizona Diamondbacks home games would win 3 free tacos from Taco Bell if the Diamondbacks scored 6 or more runs. In the 2012 season, the Diamondbacks won 41 of their 81 home games and gave away free tacos in 30 of their 81 home games. In 26 of the games, the Diamondbacks won and gave away free tacos. Choose a Diamondbacks home game at random.

(a) Make a Venn diagram to model this chance process.

(b) What is the probability that the Diamondbacks lost and did not give away free tacos?

(c) What is the probability that the Diamondbacks won the game or fans got free tacos?

**R5.6 Steroids** A company has developed a drug test to detect steroid use by athletes. The test is accurate 95% of the time when an athlete has taken steroids. It is 97% accurate when an athlete hasn't taken steroids. Suppose that the drug test will be used in a population of athletes in which 10% have actually taken steroids. Let's choose an athlete at random and administer the drug test.

(a) Make a tree diagram showing the sample space of this chance process.

(b) What's the probability that the randomly selected athlete tests positive? Show your work.

(c) Suppose that the chosen athlete tests positive. What's the probability that he or she actually used steroids? Show your work.

**R5.7 Mike's pizza** You work at Mike's pizza shop. You have the following information about the 7 pizzas in the oven: 3 of the 7 have thick crust and 2 of the 3 thick crust pizzas have mushrooms. Of the remaining 4 pizzas, 2 have mushrooms. Choose a pizza at random from the oven.

(a) Make a two-way table to model this chance process.

(b) Are the events "getting a thick-crust pizza" and "getting a pizza with mushrooms" independent? Explain.

(c) You add an eighth pizza to the oven. This pizza has thick crust with only cheese. Now are the events "getting a thick-crust pizza" and "getting a pizza with mushrooms" independent? Explain.

**R5.8 Deer and pine seedlings** As suburban gardeners know, deer will eat almost anything green. In a study of pine seedlings at an environmental center in Ohio, researchers noted how deer damage varied with how much of the seedling was covered by thorny undergrowth:[26]

| | Deer Damage | |
|---|---|---|
| Thorny Cover | Yes | No |
| None | 60 | 151 |
| <1/3 | 76 | 158 |
| 1/3 to 2/3 | 44 | 177 |
| >2/3 | 29 | 176 |

(a) What is the probability that a randomly selected seedling was damaged by deer?

(b) What are the conditional probabilities that a randomly selected seedling was damaged, given each level of cover?

(c) Does knowing about the amount of thorny cover on a seedling change the probability of deer damage? Justify your answer.

**R5.9 A random walk on Wall Street?** The "random walk" theory of stock prices holds that price movements in disjoint time periods are independent of each other. Suppose that we record only whether the price is up or down each year, and that the probability that our portfolio rises in price in any one year is 0.65. (This probability is approximately correct for a portfolio containing equal dollar amounts of all common stocks listed on the New York Stock Exchange.)

(a) What is the probability that our portfolio goes up for three consecutive years?

(b) What is the probability that the portfolio's value moves in the same direction (either up or down) for three consecutive years?

**R5.10 Blood types** Each of us has an ABO blood type, which describes whether two characteristics called A and B are present. Every human being has two blood type alleles (gene forms), one inherited from our mother and one from our father. Each of these alleles can be A, B, or O. The two that we inherit determine our blood type. The table shows what our blood type is for each combination of two alleles. We inherit each of a parent's two alleles with probability 0.5. We inherit independently from our mother and father.

| Alleles inherited | Blood type |
|---|---|
| A and A | A |
| A and B | AB |
| A and O | A |
| B and B | B |
| B and O | B |
| O and O | O |

(a) Hannah and Jacob both have alleles A and B. Diagram the sample space that shows the alleles that their next child could receive. Then give the possible blood types that this child could have, along with the probability for each blood type.

(b) Jennifer has alleles A and O. Jose has alleles A and B. They have two children. What is the probability that at least one of the two children has blood type B? Show your method.

**R5.8 (a)** $P(\text{damage}) = \frac{209}{871} = 0.24$

**(b)** $P(\text{damage} \mid \text{no cover}) = \frac{60}{211} = 0.2844$, $P\left(\text{damage} \mid < \frac{1}{3}\right) = \frac{76}{234} = 0.3248$, $P\left(\text{damage} \mid \frac{1}{3} \text{ to } \frac{2}{3}\right) = \frac{44}{221} = 0.1991$, and $P\left(\text{damage} \mid > \frac{2}{3}\right) = \frac{29}{205} = 0.1415$.

**(c)** Yes. It appears that deer do much more damage when there is no cover or less than 1/3 cover than when there is more cover.

**R5.9 (a)** $P(\text{up three consecutive years}) = (0.65)^3 = 0.274625$ **(b)** $P(\text{same direction for 3 years}) = (0.65)^3 + (0.35)^3 = 0.3175$

**R5.10 (a)** (A, A), (A, B), (B, A), (B, B)

| Blood type: | A | AB | B |
|---|---|---|---|
| Probability: | 0.25 | 0.5 | 0.25 |

**(b)** (A, A), (A, B), (O, A), (O, B)

| Blood type: | A | AB | B |
|---|---|---|---|
| Probability: | 0.5 | 0.25 | 0.25 |

$P(\text{at least 1 type B}) = 1 - P(\text{neither are type B}) = 1 - (0.75)^2 = 0.4375$.

**R5.6 (a)**

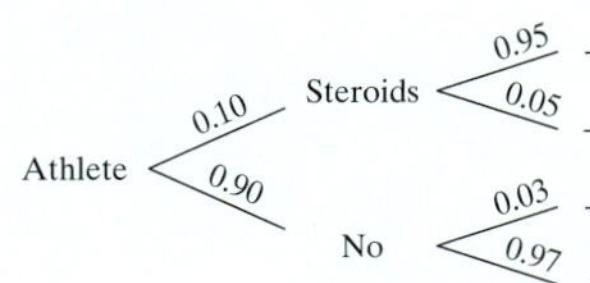


**(b)** $P(+) = (0.10)(0.95) + (0.90)(0.03) = 0.122$

**(c)** $P(\text{steroids} \mid +) = \frac{0.095}{0.122} = 0.7787$

**R5.7 (a)**

| | Thick | Thin | Total |
|---|---|---|---|
| Mushrooms | 2 | 2 | **4** |
| No mushrooms | 1 | 2 | **3** |
| **Total** | **3** | **4** | **7** |

**(b)** Not independent:

$P(\text{mushrooms}) = \frac{4}{7} = 0.571$ does not equal

$P(\text{mushrooms} \mid \text{thick crust}) = \frac{2}{3} = 0.667$

**(c)** Independent:

$P(\text{mushrooms}) = \frac{4}{8} = \frac{1}{2} = 0.50$ is equal to

$P(\text{mushrooms} \mid \text{thick crust}) = \frac{2}{4} = \frac{1}{2} = 0.50$

**TRM Full Solutions to Chapter 5 AP® Statistics Practice Test**

## Answers to Chapter 5 AP® Statistics Practice Test

**T5.1** e
**T5.2** d
**T5.3** c
**T5.4** b
**T5.5** b
**T5.6** c
**T5.7** e
**T5.8** e



# Chapter 5 AP® Statistics Practice Test

**Section I: Multiple Choice** *Select the best answer for each question.*

**T5.1** Dr. Stats plans to toss a fair coin 10,000 times in the hope that it will lead him to a deeper understanding of the laws of probability. Which of the following statements is true?

(a) It is unlikely that Dr. Stats will get more than 5000 heads.
(b) Whenever Dr. Stats gets a string of 15 tails in a row, it becomes more likely that the next toss will be a head.
(c) The fraction of tosses resulting in heads should be exactly 1/2.
(d) The chance that the 100th toss will be a head depends somewhat on the results of the first 99 tosses.
(e) It is likely that Dr. Stats will get about 50% heads.

**T5.2** China has 1.2 billion people. Marketers want to know which international brands they have heard of. A large study showed that 62% of all Chinese adults have heard of Coca-Cola. You want to simulate choosing a Chinese at random and asking if he or she has heard of Coca-Cola. One correct way to assign random digits to simulate the answer is:

(a) One digit simulates one person's answer; odd means "Yes" and even means "No."
(b) One digit simulates one person's answer; 0 to 6 mean "Yes" and 7 to 9 mean "No."
(c) One digit simulates the result; 0 to 9 tells how many in the sample said "Yes."
(d) Two digits simulate one person's answer; 00 to 61 mean "Yes" and 62 to 99 mean "No."
(e) Two digits simulate one person's answer; 00 to 62 mean "Yes" and 63 to 99 mean "No."

**T5.3** Choose an American household at random and record the number of vehicles they own. Here is the probability model if we ignore the few households that own more than 5 cars:

| Number of cars: | 0 | 1 | 2 | 3 | 4 | 5 |
|---|---|---|---|---|---|---|
| Probability: | 0.09 | 0.36 | 0.35 | 0.13 | 0.05 | 0.02 |

A housing company builds houses with two-car garages. What percent of households have more cars than the garage can hold?

(a) 7% (b) 13% (c) 20% (d) 45% (e) 55%

**T5.4** Computer voice recognition software is getting better. Some companies claim that their software correctly recognizes 98% of all words spoken by a trained user. To simulate recognizing a single word when the probability of being correct is 0.98, let two digits simulate one word; 00 to 97 mean "correct." The program recognizes words (or not) independently. To simulate the program's performance on 10 words, use these random digits:

60970 70024 17868 29843 61790 90656 87964

The number of words recognized correctly out of the 10 is

(a) 10 (b) 9 (c) 8 (d) 7 (e) 6

*Questions T5.5 to T5.7 refer to the following setting. One thousand students at a city high school were classified according to both GPA and whether or not they consistently skipped classes. The two-way table below summarizes the data. Suppose that we choose a student from the school at random.*

| | GPA | | |
|---|---|---|---|
| **Skipped Classes** | **<2.0** | **2.0–3.0** | **>3.0** |
| Many | 80 | 25 | 5 |
| Few | 175 | 450 | 265 |

**T5.5** What is the probability that a student has a GPA under 2.0?

(a) 0.227 (b) 0.255 (c) 0.450 (d) 0.475 (e) 0.506

**T5.6** What is the probability that a student has a GPA under 2.0 or has skipped many classes?

(a) 0.080 (b) 0.281 (c) 0.285 (d) 0.365 (e) 0.727

**T5.7** What is the probability that a student has a GPA under 2.0 given that he or she has skipped many classes?

(a) 0.080 (b) 0.281 (c) 0.285 (d) 0.314 (e) 0.727

**T5.8** For events A and B related to the same chance process, which of the following statements is true?

(a) If A and B are mutually exclusive, then they must be independent.
(b) If A and B are independent, then they must be mutually exclusive.
(c) If A and B are not mutually exclusive, then they must be independent.
(d) If A and B are not independent, then they must be mutually exclusive.
(e) If A and B are independent, then they cannot be mutually exclusive.

**T5.9** Choose an American adult at random. The probability that you choose a woman is 0.52. The probability that the person you choose has never married is 0.25. The probability that you choose a woman who has never married is 0.11. The probability that the person you choose is either a woman or has never been married (or both) is therefore about

(a) 0.77. (b) 0.66. (c) 0.44. (d) 0.38. (e) 0.13.

**T5.10** A deck of playing cards has 52 cards, of which 12 are face cards. If you shuffle the deck well and turn over the top 3 cards, one after the other, what's the probability that all 3 are face cards?

(a) 0.001 (b) 0.005 (c) 0.010 (d) 0.012 (e) 0.02

**Section II: Free Response** *Show all your work. Indicate clearly the methods you use, because you will be graded on the correctness of your methods as well as on the accuracy and completeness of your results and explanations.*

**T5.11** Your teacher has invented a "fair" dice game to play. Here's how it works. Your teacher will roll one fair *eight-sided* die, and you will roll a fair *six-sided* die. Each player rolls once, and the winner is the person with the higher number. In case of a tie, neither player wins. The table shows the sample space of this chance process.

| | Teacher Rolls | | | | | | | |
|---|---|---|---|---|---|---|---|---|
| **You Roll** | 1 | 2 | 3 | 4 | 5 | 6 | 7 | 8 |
| 1 | | | | | | | | |
| 2 | | | | | | | | |
| 3 | | | | | | | | |
| 4 | | | | | | | | |
| 5 | | | | | | | | |
| 6 | | | | | | | | |

(a) Let A be the event "your teacher wins." Find $P(A)$.

(b) Let $B$ be the event "you get a 3 on your first roll." Find $P(A \cup B)$.

(c) Are events A and $B$ independent? Justify your answer.

**T5.12** Three machines—A, B, and C—are used to produce a large quantity of identical parts at a factory. Machine A produces 60% of the parts, while Machines B and C produce 30% and 10% of the parts, respectively. Historical records indicate that 10% of the parts produced by Machine A are defective, compared with 30% for Machine B and 40% for Machine C.

(a) Draw a tree diagram to represent this chance process.

(b) If we choose a part produced by one of these three machines, what's the probability that it's defective? Show your work.

(c) If a part is inspected and found to be defective, which machine is most likely to have produced it? Give appropriate evidence to support your answer.

**T5.13** Researchers are interested in the relationship between cigarette smoking and lung cancer. Suppose an adult male is randomly selected from a particular population. The following table shows the probabilities of some events related to this chance process:

| Event | Probability |
|---|---|
| Smokes | 0.25 |
| Smokes and gets cancer | 0.08 |
| Does not smoke and does not get cancer | 0.71 |

(a) Find the probability that the individual gets cancer given that he is a smoker. Show your work.

(b) Find the probability that the individual smokes or gets cancer. Show your work.

(c) Two adult males are selected at random. Find the probability that at least one of the two gets cancer. Show your work.

**T5.14** Based on previous records, 17% of the vehicles passing through a tollbooth have out-of-state plates. A bored tollbooth worker decides to pass the time by counting how many vehicles pass through until he sees two with out-of-state plates.[27]

(a) Describe the design of a simulation to estimate the average number of vehicles it takes to find two with out-of-state plates. Explain clearly how you will use the partial table of random digits below to carry out your simulation.

(b) Perform three repetitions of the simulation you described in part (a). Copy the random digits below onto your paper. Then mark on or directly above the table to show your results.

```
41050 92031 06449 05059 59884 31880
53115 84469 94868 57967 05811 84514
84177 06757 17613 15582 51506 81435
75011 13006 63395 55041 15866 06589
```

**T5.9** b

**T5.10** c

**T5.11 (a)** Here is a completed table, with T indicating the teacher wins and Y indicating that you win. $P(\text{teacher wins}) = \frac{27}{48} = 0.5625$.

| | 1 | 2 | 3 | 4 | 5 | 6 | 7 | 8 |
|---|---|---|---|---|---|---|---|---|
| **1** | — | T | T | T | T | T | T | T |
| **2** | Y | — | T | T | T | T | T | T |
| **3** | Y | Y | — | T | T | T | T | T |
| **4** | Y | Y | Y | — | T | T | T | T |
| **5** | Y | Y | Y | Y | — | T | T | T |
| **6** | Y | Y | Y | Y | Y | — | T | T |

**(b)** $P(A \cup B) = P(A) + P(B) - P(A \cap B)$

$$= \frac{27}{48} + \frac{8}{48} - \frac{5}{48} = \frac{30}{48}$$

**(c)** Not independent. $P(A) = \frac{27}{48} = 0.5625$ does not equal $P(A \mid B) = \frac{5}{8} = 0.625$.

**T5.12 (a)**

Part → A (0.60): Defective (0.10), OK (0.90)
Part → B (0.30): Defective (0.30), OK (0.70)
Part → C (0.10): Defective (0.40), OK (0.60)

**(b)** $P(\text{defective}) = (0.60)(0.10) + (0.30)(0.30) + (0.10)(0.40) = 0.19$

**(c)** Machine B

$$P(\text{A} \mid \text{defective}) = \frac{0.06}{0.19} = 0.3158$$

$$P(\text{B} \mid \text{defective}) = \frac{0.09}{0.19} = 0.4737$$

$$P(\text{C} \mid \text{defective}) = \frac{0.04}{0.19} = 0.2105$$

**T5.13 (a)** Here is a two-way table that summarizes this information:

| | Smokes | Does not smoke | Total |
|---|---|---|---|
| Cancer | 0.08 | 0.04 | **0.12** |
| No cancer | 0.17 | 0.71 | **0.88** |
| **Total** | **0.25** | **0.75** | **1.00** |

$$P(\text{gets cancer} \mid \text{smoker}) = \frac{0.08}{0.25} = 0.32$$

**(b)** $P(\text{smokes} \cup \text{gets cancer}) = 0.25 + 0.12 - 0.08 = 0.29$

**(c)** $P(\text{cancer}) = 0.12$, so $P(\text{at least one gets cancer}) = 1 - P(\text{neither gets cancer}) = 1 - 0.88^2 = 0.2256$

**T5.14 (a)** Let 00–16 represent out-of-state and 17–99 represent in-state. Read two-digit numbers until you have found two numbers between 00 and 16. Record how many two-digit numbers you had to read.

**(b)** The first sample is 41 **05 09** (it took three cars). The second sample is 20 31 **06** 44 90 50 59 59 88 43 18 80 53 **11** (it took 14 cars). The third sample is 58 44 69 94 86 85 79 67 **05** 81 18 45 **14** (it took 13 cars).

Chapter

6

# Random Variables

## PD Overview

In this chapter, we continue our study of probability with a focus on random variables. A random variable is a numerical outcome of a chance process, such as the height of a randomly selected student, the number of heads in three flips of a coin, or the difference between the diameter of a randomly chosen cup and the diameter of a randomly chosen lid.

Many students have a tough time with this chapter, possibly because of the large number of rules and formulas that are introduced. We suggest that you allow your students to use the formula sheet that is provided on the AP® exam during your classroom quizzes and tests. First, it allows students to spend less time memorizing and more time learning how and when to use the formulas. Second, because students will be allowed to use the formula sheets on the real AP® exam, it is helpful if students get practice using them.

In addition to learning how to calculate probabilities of events involving random variables, we will also focus on identifying the shape, center, and spread of the probability distribution of a random variable. Being able to anticipate the shape, center, and spread of a distribution of a random variable or combination of random variables will be extremely important as we proceed into Chapter 7 and beyond. To estimate a sample proportion with 95% confidence, we need to know how this random variable (sample proportion) will behave in repeated sampling. Likewise, to investigate whether a difference in sample means is statistically significant, we need to know how the difference in these random variables (sample means) will behave when the true means are really the same.

*Note*: Consider starting the Penny for Your Thoughts Activity from Chapter 7 during this chapter. Make sure to read the Teaching Tips that go along with the Activity for suggestions.

### Relationship to the AP® Statistics Course Description

Chapter 6 covers the following content as described in the Topic Outline from the AP® Statistics Course Description. Topics marked with a superscript are covered in other chapters as indicated.

III. Anticipating patterns: Exploring random phenomena using probability and simulation

A. Probability

1. Interpreting probability, including long-run relative frequency interpretation[ch5]
2. "Law of Large Numbers" concept[ch5]
3. Addition rule, multiplication rule, conditional probability, and independence[ch5]
4. Discrete random variables and their probability distributions, including binomial and geometric
5. Simulation of random behavior and probability distributions[ch5]
6. Mean (expected value) and standard deviation of a random variable, and linear transformation of a random variable

B. Combining independent random variables

1. Notion of independence versus dependence
2. Mean and standard deviation for sums and differences of independent random variables

## The Big Ideas

One of the challenges in teaching the AP® Statistics course is keeping students focused on the big picture and not just the details of each section. We outline the big ideas for the chapter here.

### Chapter 6 Introduction

This chapter begins by introducing the idea of a random variable, a numerical outcome of a chance process. The Bottled Water versus Tap Water Activity has students identify which of three cups contains bottled water. This Activity is a nice way to introduce random variables and a great way to review how to conduct a simulation. It also foreshadows one of the basic ideas of statistical inference—comparing the results obtained in a study with what might happen by chance alone.

### Section 6.1 Discrete and Continuous Random Variables

In this section, we introduce probability distributions for random variables. These distributions give the possible values of a random variable and their probabilities. This is essentially the same definition for "distribution" we used in Chapter 1, where we said that a distribution tells us what values a variable takes and how often it takes these values.

Random variables are divided into two categories, discrete and continuous. Discrete random variables take on a fixed set of values, whereas continuous random variables take on all values in an interval of numbers. Most discrete random variables result from counting something, such as the number of pencils that students have in their backpacks. Continuous random variables typically result from measuring something, such as the lengths of pencils or weights of backpacks.

As in Chapter 1, we are often interested in the shape, center, and spread of a probability distribution. The shape can be identified by graphing a histogram of the probability distribution, with the height of each bar representing the probability of that outcome. The center of the distribution is usually identified by the mean (expected value) of the random variable. The expected value is the average value of the random variable if the chance process could be repeated many, many times. The spread of a probability distribution is usually identified by the standard deviation of the random variable, which describes how much the values of the variable typically differ from the mean value.

Continuous probability distributions, such as the Normal distribution, describe the distribution of continuous random variables. A density curve is used to graph a continuous probability distribution rather than a histogram, and probabilities are determined by finding the area under the density curve and above the values of interest.

## Section 6.2 Transforming and Combining Random Variables

In this section, we introduce rules for transforming and combining random variables. To do this, we include several examples from a fun context (Pete's Jeep Tours) to discover how a linear transformation affects the shape, center, and spread of the distribution of a random variable. We then continue with the same context to discover how we can predict the shape, center, and spread of a combination of two or more random variables. We also provide a series of alternate examples that use a different context (El Dorado Community College) to introduce the same rules.

Linear transformations have the same effect on the probability distribution of a random variable as they have on a distribution of data. In Chapter 2, we learned that linear transformations do not change the shape, as long as we don't multiply by a negative number. However, linear transformations can change the center, spread, or both depending on the type of transformation. Multiplying each value by a constant $b$ multiplies measures of center by $b$ and most measures of spread by $|b|$. Adding a constant $a$ to each value adds $a$ to measures of center but doesn't change measures of spread.

Combining two or more random variables is a new and very important topic. Many of the methods we use in this book involve differences in random variables, such as the difference in two means ("Is there a significant difference in the mean salaries for men and women in a certain industry?") or a difference in two proportions ("Does a new drug have a higher survival rate than the currently used drug?"). To analyze questions like these, students need to be able to anticipate what the distribution of a combination of random variables will look like.

We conclude this section by looking at several examples where Normal random variables are combined and the resulting distribution is used to calculate a probability.

Because students can feel overwhelmed by the quantity of formulas presented in this section, it is important that you help distinguish linear transformations (taking the values of *one* random variable and multiplying them by a constant and/or adding a constant) and combinations of random variables (adding or subtracting *two or more* random variables).

## Section 6.3 Binomial and Geometric Random Variables

In this section, we introduce two common types of discrete random variables: binomial random variables and geometric random variables. Make sure to pay special attention to binomial random variables, because questions about binomial random variables frequently show up on the AP® exam. Also, the binomial distribution is closely related to the sampling distribution of a sample proportion, which we will learn more about in Chapter 7.

Binomial random variables count the number of successes in a fixed number of trials, while geometric random variables count the number of trials needed to achieve one success. Otherwise, the binomial and geometric settings have the same conditions: there must be two possible outcomes for each trial (success or failure), the trials must be independent, and the probability of success must stay the same throughout all trials.

A very common application of the binomial distribution in statistics is counting the number of times a particular outcome occurs in a random sample from some population (e.g., the number of defective flash drives in a sample of size 10 from a shipment of 10,000). In cases like this, the sampling is almost always done without replacement. This means that the trials are no longer independent. However, if the sample is a small fraction of the population, the lack of independence has a very small effect on the probabilities we calculate.

Also included in this section is an optional topic: Normal approximations to binomial distributions. Although this topic is a nice connection between discrete and continuous probability distributions, it is not essential for the AP® exam and can be skipped if you are short on time. If you do choose to skip this topic, many of the same ideas will be presented in Chapter 7 when we cover the sampling distribution of a sample proportion.

# Chapter 6: Resources

## TRM Additional *TPS* Teacher's Resource Materials

The following resources can be found by clicking on the links in the new Teacher's e-Book (TE-book), opening the Teacher's Resource CD (TRCD), or logging onto the book's companion Web site www.whfreeman.com/tps5e (teacher log-in required).

- Printable Chapter 3 Learning Objective Grid (page 347)
- The Casino Lab (page 378)
- A Fair Coin? (page 400)
- Airline Overbooking, Part II (page 400)
- FRAPPY! materials: Question, Model Solution, Scoring Rubric, Sample Responses with commentary (page 414)
- Quizzes: 3 versions for each section; versions A and B are roughly parallel
  - Section 6.1 (page 358)
  - Section 6.2 (page 382)
  - Section 6.3 (page 409)
- Tests: 4 versions; versions A and B are roughly parallel, version D is cumulative (page 415)

## Free-Response Questions from Previous AP® Exams

Questions can be found on the AP® Central Web site: http://apcentral.collegeboard.com/apc/members/exam/exam_questions/8357.html.

Students should be able to answer all the free-response questions listed below with material learned in this chapter. Questions that contain content from this chapter but also require content from later chapters are listed in the last chapter required to complete the entire question. This list will be updated after each AP® exam and will be posted on the book's Web site: www.whfreeman.com/tps5e.

Questions marked with an asterisk are from exams with released multiple-choice questions. You may want to save these questions until the end of the year, so you can give your students a complete released exam for practice.

| Year | # | Content |
|---|---|---|
| 2013 | 3 | • Normal probability calculation<br>• Mean and standard deviation of a sum of random variables |
| 2012 | 2* | • Discrete probability distributions<br>• Expected value of a discrete random variable<br>• Application of expected value<br>• Normal probability calculation |
| 2011B | 3 | • Geometric probability<br>• Binomial probability<br>• Cumulative binomial probability |
| 2010 | 4 | • Mean and standard deviation of a binomial distribution<br>• Binomial calculations<br>• Stratified sampling |
| 2010B | 3 | • Binomial distribution<br>• Expected value<br>• Binomial calculations |
| 2008 | 3 | • Expected value<br>• Basic probability rules |
| 2008B | 5 | • Combining normal random variables<br>• Normal calculations |
| 2006B | 3 | • Normal calculations<br>• Binomial calculations<br>• Inverse normal calculations |
| 2005 | 2 | • Expected value<br>• Median of a discrete random variable<br>• Relationship of mean and median |
| 2005B | 2 | • Mean and standard deviation of a discrete random variable<br>• Combining independent random variables<br>• Linear transformations of a random variable |

# Chapter 6

| Year | # | Content |
|---|---|---|
| 2004 | 3 | • Binomial conditions<br>• Multiplication rule<br>• Interpreting probability<br>• Generalizability |
| 2004 | 4 | • Conditional probability<br>• Expected value |
| 2003 | 3 | • Normal calculations<br>• Binomial calculations |
| 2002 | 3* | • Normal calculations<br>• Combining independent random variables |
| 2002B | 2 | • Addition rule<br>• Expected value<br>• Conditional probability |
| 2001 | 2 | • Expected value |
| 1999 | 4 | • Normal calculations<br>• Binomial calculations<br>• Outlier rules |
| 1999 | 5 | • Sample space<br>• Expected value |
| 1998 | 6 | • Normal calculations<br>• Simulation<br>• Expected value |

## Applets

- The *Law of Large Numbers* applet allows students to roll dice repeatedly and keep track of the mean of their rolls. After many repetitions, the mean of their rolls should approach the true mean, or expected value. www.whfreeman.com/tps5e (click on Statistical Applets and choose Law of Large Numbers).
- The *Normal Curve* applet allows students to find the area under a Normal curve. Students can choose the mean, standard deviation, and boundaries. www.whfreeman.com/tps5e (click on Statistical Applets and choose Normal Curve).
- The *Z-Table Normal Distribution* applet allows students to find the area under a Normal curve and to work backward to find percentiles. Students can enter the boundaries, mean, and standard deviation. http://davidmlane.com/hyperstat/z_table.html.
- The *Normal* applet allows students to find the area under a Normal curve and to work backwards to find percentiles. Students can enter the boundaries, area, mean, and standard deviation. www.lock5stat.com/statkey (under Theoretical Distributions choose Normal).
- The *Normal Approximation to Binomial* applet allows students to try different combinations of $n$ and $p$ to see when it would be reasonable to use a Normal curve to model a binomial probability distribution. www.whfreeman.com/tps5e (click on Statistical Applets and choose Normal Approximation to Binomial).

## TI-Nspire Activities

These Activities are designed to illustrate important statistical concepts using the TI-Nspire. The Activities can be found at-www.whfreeman.com/tps5e.

- *Exploring Random Variables*: Students investigate the shape, center, and spread of the probability distribution of a random variable.
- *Transforming and Combining Random Variables*: Students explore what happens to the mean and standard deviation when random variables are transformed or combined.
- *Binomial Distributions*: Students explore different values of $n$ and $p$ and their effect on the histogram of the binomial probability distribution, leading to the Normal approximation.

# Chapter 6: Pacing Guide, Learning Objectives, and Suggested Assignments

This pacing guide is based on a schedule with 110, 50-minute sessions before the AP® exam. The 110 days include 4 days for midterms and 10 days for AP® exam review. If you have a different number of sessions before the AP® exam, you can modify the pacing guide to suit your needs. If you have additional time, consider incorporating quizzes, released AP® free-response questions, or additional activities. See the Resources section above for suggestions.

The suggested homework assignments list odd-numbered exercises, whenever possible, so students can check their answers against the back-of-the-book answers. If you would rather students not have access to the answers while doing homework, adding 1 to the exercise numbers usually will do the trick, because the homework exercises typically are paired. For example, Exercises 1 and 2 will generally cover the same topics, but in different contexts. You may also choose to include the Recycle and Review questions at the end of each section, which review topics from previous sections or chapters. These questions are denoted with a symbol.

| Day | Topics | Learning Objectives: Students will be able to . . . | Suggested Assignment |
|---|---|---|---|
| 1 | Chapter 6 Introduction, 6.1 Discrete Random Variables, Mean (Expected Value) of a Discrete Random Variable | • Compute probabilities using the probability distribution of a discrete random variable.<br>• Calculate and interpret the mean (expected value) of a discrete random variable. | 1, 3, 5, 7, 9, 11, 13 |
| 2 | 6.1 Standard Deviation (and Variance) of a Discrete Random Variable, Continuous Random Variables | • Calculate and interpret the standard deviation of a discrete random variable.<br>• Compute probabilities using the probability distribution of a continuous random variable. | 14, 15, 17, 18, 21, 23, 25 |
| 3 | 6.2 Linear Transformations | • Describe the effects of transforming a random variable by adding or subtracting a constant and multiplying or dividing by a constant. | 27–30, 35, 37, 39–41, 43, 45 |
| 4 | 6.2 Combining Random Variables, Combining Normal Random Variables | • Find the mean and standard deviation of the sum or difference of independent random variables.<br>• Find probabilities involving the sum or difference of independent Normal random variables. | 47, 49, 51, 53, 55, 57–59, 61 |
| 5 | 6.3 Binomial Settings and Binomial Random Variables, Binomial Probabilities | • Determine whether the conditions for using a binomial random variable are met.<br>• Compute and interpret probabilities involving binomial distributions. | 63, 65, 66, 69, 71, 73, 75, 77 |
| 6 | 6.3 Mean and Standard Deviation of a Binomial Distribution, Binomial Distributions in Statistical Sampling | • Calculate the mean and standard deviation of a binomial random variable. Interpret these values in context. | 79, 81, 83, 85, 87, 89 |
| 7 | 6.3 Geometric Random Variables | • Find probabilities involving geometric random variables. | 93, 95, 97, 99, 101–104 |
| 8 | Chapter 6 Review/FRAPPY! | | Chapter 6 Review Exercises |
| 9 | Chapter 6 Test | | |

Chapter

NOTES

**PD** **Chapter 6 Overview**

Watch the video overview of Chapter 6 (for teachers) by clicking on the link in the TE-book, opening the TRCD, or logging onto the book's companion Web site www.whfreeman.com/tps5e (teacher log-in required).

# Chapter 6

# Random Variables

case study

## A Jury of Your Peers?

Are accused criminals in the United States entitled to a "jury of their peers"? Sort of. The Sixth Amendment to the U.S. Constitution begins, "In all criminal prosecutions, the accused shall enjoy the right to a speedy and public trial, by an impartial jury of the State and district wherein the crime shall have been committed...." There is no mention of a "jury of your peers" in the Constitution or any of its amendments. However, an 1879 U.S. Supreme Court decision said that a jury should be chosen from a group "composed of the peers or equals [of the accused]; that is, of his neighbors, fellows, associates, persons having the same legal status in society as he holds."[1]

To meet the Sixth Amendment requirement of impartiality, most courts start by randomly selecting a large jury pool from the citizens who live in the court's jurisdiction. The jurors for a given trial are then chosen from the jury pool in a process known as voir dire. Each prospective juror answers a set of questions posed by the judge and the lawyers for both the prosecution and the defense. Depending on their answers, prospective jurors are excluded or seated on the jury.

In one case that made it all the way to the Supreme Court, a defense lawyer in Michigan challenged the process of selecting the jury pool in the trial of his accused client. Here are the facts:[2]

- About 7.28% of the citizens in the court's jurisdiction were black.
- The jury pool had between 60 and 100 members, only 3 of whom were black.

Is it plausible that a jury pool with so few black citizens could be chosen just by chance?

By the end of this chapter, you will be ready to analyze the results of the jury selection process and to render your own verdict about the defense attorney's challenge.

## Activity Overview

You can watch an overview of this Activity (for teachers) by clicking on the link in the TE-book, opening the TRCD, or logging onto the book's companion Web site www.whfreeman.com/tps5e (teacher log-in required).

## Teaching Tip

For the Bottled Water versus Tap Water Activity, set up three stations. At Station 1, fill the cups labeled B with bottled water and the others with tap water. At Station 2, fill the cups labeled A with bottled water and the others with tap water. At Station 3, fill the cups labeled C with bottled water and the others with tap water. Prepare an index card for each student, with roughly equal numbers labeled "Station 1," "Station 2," and "Station 3." Shuffle the cards and hand them out at random when students arrive to class.

For the simulation in Step 8, use the `randInt` feature on your calculator to generate $n$ values ($n$ = number of students who participated) from 1 to 3, letting 1 represent a correct guess and 2 and 3 represent an incorrect guess. Count the number of 1s and record this value. Repeat this process many times to determine how many correct answers would indicate that the students aren't just guessing.

For example, in Mr. Bullard's class (see the paragraph following the Activity), we would randomly generate 21 integers from 1 to 3 and count the number of 1s. Here is a dotplot showing 100 trials of the simulation for his class. Not once in 100 trials were there 13 or more correct guesses. This gives convincing evidence that the students in Mr. Bullard's class did better than random guessing.

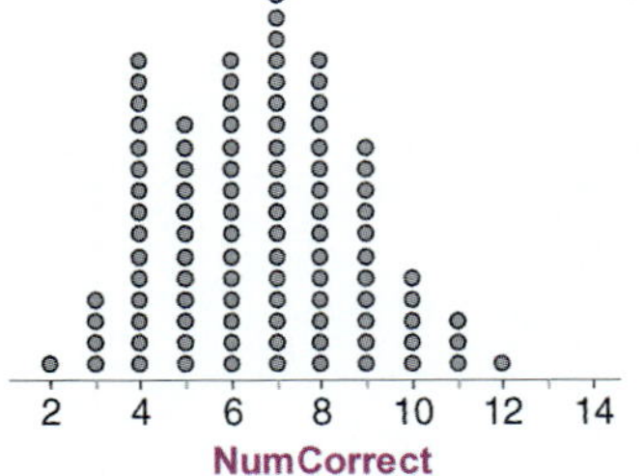


# Introduction

Do you drink bottled water or tap water? According to a recent report in *U.S. Mayor Newspaper*, about 75% of people drink bottled water regularly. Some people do so because they believe bottled water is safer than tap water. (There's little evidence to support this belief.) Others say they prefer the taste of bottled water. Can people really tell the difference?

## ACTIVITY Bottled Water versus Tap Water

**MATERIALS:**
3 small paper cups per student; enough tap water for 2 cups per student and enough bottled water for 1 cup per student; 1 six-sided die and 1 index card per student

The ABC News program *20/20* set up a blind taste test in which people were asked to rate four different brands of bottled water and New York City tap water without knowing which they were drinking. Can you guess the result? Tap water came out the clear winner in terms of taste.

This Activity will give you and your classmates a chance to discover whether or not you can taste the difference between bottled water and tap water.

1. Before class begins, your teacher will prepare numbered stations with cups of water. You will be given an index card with a station number on it.
2. Go to the corresponding station. Pick up three cups (labeled A, B, and C) and take them back to your seat.
3. Your task is to determine which one of the three cups contains the bottled water. Drink all the water in Cup A first, then the water in Cup B, and finally the water in Cup C. Write down the letter of the cup that you think held the bottled water. Do not discuss your results with any of your classmates yet!
4. While you taste, your teacher will make a chart on the board like this one:

| Station number | Bottled water cup? | Truth |
|---|---|---|

5. When you are told to do so, go to the board and record your station number and the letter of the cup you identified as containing bottled water.
6. Your teacher will now reveal the truth about the cups of drinking water. How many students in the class identified the bottled water correctly? What percent of the class is this?
7. Let's assume that no one in your class can distinguish tap water from bottled water. In that case, students would just be guessing which cup of water tastes different. If so, what's the probability that an individual student would guess correctly?
8. How many correct identifications would you need to see to be convinced that the students in your class aren't just guessing? With your classmates, design and carry out a simulation to answer this question. What do you conclude about your class's ability to distinguish tap water from bottled water?

When Mr. Bullard's class did the preceding Activity, 13 out of 21 students made correct identifications. If we assume that the students in his class can't tell tap water from bottled water, then each one is basically guessing, with a 1/3 chance of being correct. So we'd expect about one-third of his 21 students, that is, about 7 students, to guess correctly. How likely is it that 13 or more of his 21 students would guess correctly? To answer this question without a simulation, we need a different kind of probability model from the ones we saw in Chapter 5.

## ALTERNATE ACTIVITY

There are many ways to modify the Bottled Water versus Tap Water Activity to suit the interests of your students. For example, using an identical setup you could investigate whether students can correctly tell the difference between name-brand potato chips and store-brand potato chips. You could also modify the Activity to see whether students prefer one type over the other rather than just investigating whether they can tell the difference. Or you can do both, first asking students which they prefer and then asking which they think is the name brand. See the Alternate Example on page 399 for another possibility.

Section 6.1 introduces the concept of a *random variable*, a numerical outcome of some chance process (like the 13 students who guessed correctly in Mr. Bullard's class). Each random variable has a *probability distribution* that gives us information about the likelihood that a specific event happens (like 13 or more correct guesses out of 21) and about what's expected to happen if the chance behavior is repeated many times. Section 6.2 examines the effect of transforming and combining random variables on the shape, center, and spread of their probability distributions. In Section 6.3, we'll look at two random variables with probability distributions that are used enough to have their own names—*binomial* and *geometric*.

# 6.1 Discrete and Continuous Random Variables

**WHAT YOU WILL LEARN** By the end of the section, you should be able to:

- Compute probabilities using the probability distribution of a discrete random variable.
- Calculate and interpret the mean (expected value) of a discrete random variable.
- Calculate and interpret the standard deviation of a discrete random variable.
- Compute probabilities using the probability distribution of certain continuous random variables.

A probability model describes the possible outcomes of a chance process and the likelihood that those outcomes will occur. For example, suppose we toss a fair coin 3 times. The sample space for this chance process is

HHH HHT HTH THH HTT THT TTH TTT

Because there are 8 equally likely outcomes, the probability is 1/8 for each possible outcome. Define the variable $X$ = the number of heads obtained. The value of $X$ will vary from one set of tosses to another but will always be one of the numbers 0, 1, 2, or 3. How likely is $X$ to take each of those values? It will be easier to answer this question if we group the possible outcomes by the number of heads obtained:

$X = 0$: TTT
$X = 1$: HTT THT TTH
$X = 2$: HHT HTH THH
$X = 3$: HHH

We can summarize the **probability distribution** of $X$ as follows:

| Value: | 0 | 1 | 2 | 3 |
|---|---|---|---|---|
| Probability: | 1/8 | 3/8 | 3/8 | 1/8 |

Figure 6.1 on next page shows the probability distribution of $X$ in graphical form. Notice the symmetric shape.

We can use the probability distribution to answer questions about the variable $X$.

## PD Section 6.1 Overview

Watch the video overview of Section 6.1 (for teachers) by clicking on the link in the TE-book, opening the TRCD, or logging onto the book's companion Web site www.whfreeman.com/tps5e (teacher log-in required).

## TRM Learning Objectives Grid

Remind students about the importance of the Learning Objectives. We include them at the beginning of each section, so students know what they are expected to learn. At the end of the chapter, the Learning Objectives are included in a grid, along with references to examples and chapter review exercises. A printable version of this grid is included in the Teacher's Resource Materials.

## Teaching Tip

Remind your students that we could estimate this probability distribution using the simulation methods from Section 5.1. Looking ahead, we will also learn how to create this probability distribution using the binomial distributions in Section 6.3.

## TRM Section 6.1 Alternate Examples

You can find the Alternate Examples for this section in Microsoft Word format by clicking on the link in the TE-book, opening the TRCD, or logging onto the book's companion Web site www.whfreeman.com/tps5e (teacher log-in required).

## Teaching Tip

Ask your students to interpret $P(X \geq 1) = 7/8$. This probability means that if we were to *repeatedly* flip 3 coins and record the number of heads, we will get 1 or more heads about 87.5% of the time.

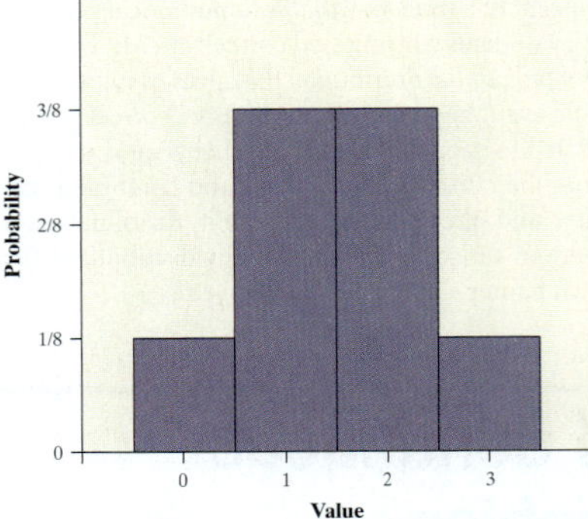


FIGURE 6.1 Histogram of the probability distribution for $X$ = number of heads in three tosses of a fair coin.

What's the probability that we get at least one head in three tosses of the coin? In symbols, we want to find $P(X \geq 1)$. We could add probabilities to get the answer:

$$P(X \geq 1) = P(X = 1) + P(X = 2) + P(X = 3)$$
$$= 1/8 + 3/8 + 3/8 = 7/8$$

Or we could use the complement rule from Chapter 5:

$$P(X \geq 1) = 1 - P(X < 1) = 1 - P(X = 0)$$
$$= 1 - 1/8 = 7/8$$

A numerical variable that describes the outcomes of a chance process (like X in the coin-tossing scenario) is called a **random variable.** The probability model for a random variable is its probability distribution.

**DEFINITION: Random variable and probability distribution**

A **random variable** takes numerical values that describe the outcomes of some chance process. The **probability distribution** of a random variable gives its possible values and their probabilities.

## Teaching Tip

Emphasize that this use of the word "distribution" is the same as it has been since Chapter 1. A distribution describes the possible values a variable can take and how often it takes those values.

There are two main types of random variables, corresponding to two types of probability distributions: *discrete* and *continuous*.

# Discrete Random Variables

We have learned several rules of probability but only one way of assigning probabilities to events: assign a probability to every individual outcome, then add these probabilities to find the probability of any event. This idea works well if we can find a way to list all possible outcomes. We will call random variables having probability assigned in this way **discrete random variables.**[3] The probability distribution for a discrete random variable must have outcome probabilities that are between 0 and 1 and that add up to 1.

**DISCRETE RANDOM VARIABLES AND THEIR PROBABILITY DISTRIBUTIONS**

A **discrete random variable** X takes a fixed set of possible values with gaps between. The probability distribution of a discrete random variable X lists the values $x_i$ and their probabilities $p_i$:

| **Value:** | $x_1$ | $x_2$ | $x_3$ | ... |
|---|---|---|---|---|
| **Probability:** | $p_1$ | $p_2$ | $p_3$ | ... |

The probabilities $p_i$ must satisfy two requirements:

1. Every probability $p_i$ is a number between 0 and 1.
2. The sum of the probabilities is 1: $p_1 + p_2 + p_3 + \cdots = 1$.

To find the probability of any event, add the probabilities $p_i$ of the particular values $x_i$ that make up the event.

## Teaching Tip

Even though a discrete random variable takes a fixed set of possible values, the set of possible values can be infinite. For example, if $X$ = number of rolls of a fair die needed to get a 6, there is no upper limit for the values of $X$. To illustrate the idea of "gaps between" the possible values, draw two number lines, one representing possible shoe sizes and one representing possible foot lengths. On the shoe size number line, draw solid dots at each half size (e.g., 6, 6.5, 7, 7.5, and so on) to represent the possible values. On the foot length number line, shade an entire interval (e.g., 6 to 12 inches) to represent the possible values. Shoe size is discrete because there are gaps between shoe sizes—there are no sizes between 6 and 6.5, for example. Foot length is continuous because there are no gaps between values—there are many possible foot lengths between 6 and 6.5 inches.

The variable $X$ in the coin-tossing example is a discrete random variable. We can list the possible values of $X$ as 0, 1, 2, 3. Note that there are gaps between these values on a number line. The corresponding probabilities are all between 0 and 1, and their sum is $1/8 + 3/8 + 3/8 + 1/8 = 1$.

Here's an example of a discrete random variable that involves something a bit more serious than tossing coins.

## EXAMPLE Apgar Scores: Babies' Health at Birth

### *Discrete random variables*

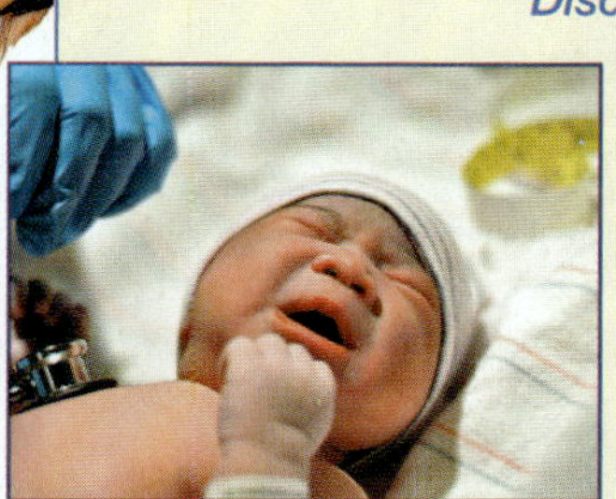

In 1952, Dr. Virginia Apgar suggested five criteria for measuring a baby's health at birth: skin color, heart rate, muscle tone, breathing, and response when stimulated. She developed a 0-1-2 scale to rate a newborn on each of the five criteria. A baby's Apgar score is the sum of the ratings on each of the five scales, which gives a whole-number value from 0 to 10. Apgar scores are still used today to evaluate the health of newborns.

What Apgar scores are typical? To find out, researchers recorded the Apgar scores of over 2 million newborn babies in a single year.[4] Imagine selecting one of these newborns at random. (That's our chance process.) Define the random variable $X$ = Apgar score of a randomly selected baby one minute after birth. The table below gives the probability distribution for $X$.

| Value: | 0 | 1 | 2 | 3 | 4 | 5 | 6 | 7 | 8 | 9 | 10 |
|---|---|---|---|---|---|---|---|---|---|---|---|
| Probability: | 0.001 | 0.006 | 0.007 | 0.008 | 0.012 | 0.020 | 0.038 | 0.099 | 0.319 | 0.437 | 0.053 |

**PROBLEM:**

(a) Show that the probability distribution for $X$ is legitimate.

(b) Make a histogram of the probability distribution. Describe what you see.

(c) Doctors decided that Apgar scores of 7 or higher indicate a healthy baby. What's the probability that a randomly selected baby is healthy?

**SOLUTION:**

(a) The probabilities are all between 0 and 1, and they add up to 1. So this is a legitimate probability distribution.

(b) Figure 6.2 shows a histogram of the probability distribution of $X$. *Shape:* The graph is skewed to the left and single-peaked. A randomly selected newborn will most likely have an Apgar score on the high end of the scale, which means that the baby was fairly healthy at birth. *Center:* From the probability distribution, we see that the median is 8. *Spread:* Apgar scores vary from 0 to 10. But most newborns receive scores between 4 and 10.

(c) The probability of choosing a healthy baby is $P(X \geq 7)$. We can calculate this probability as follows:

$$P(X \geq 7) = P(X = 7) + P(X = 8) + P(X = 9) + P(X = 10)$$
$$= 0.099 + 0.319 + 0.437 + 0.053 = 0.908$$

That is, we'd have about a 91% chance of randomly choosing a healthy baby.

**FIGURE 6.2** Histogram showing the probability distribution of the random variable $X$ = Apgar score of a randomly selected newborn one minute after birth.

**For Practice** *Try Exercise*  5

### Teaching Tip

Remind students that the "play" icon at the top of this example (and many others) indicates that they can watch a video presentation of the example. Find the videos at www.whfreeman.com/tps5e. Also, remind them that the icon in the lower-right corner indicates that a link exists between this example and Exercise 5. If students have trouble with the exercise, they can look back to this example for help.

### ALTERNATE EXAMPLE

#### *How many languages?*

Imagine selecting a U.S. high school student at random. Define the random variable $X$ = number of languages spoken by the randomly selected student. The table below gives the probability distribution of $X$, based on a sample of students from the U.S. Census at School database:

| Languages: | 1 | 2 | 3 | 4 | 5 |
|---|---|---|---|---|---|
| Probability: | 0.630 | 0.295 | 0.065 | 0.008 | 0.002 |

**PROBLEM:**

**(a)** Show that the probability distribution for $X$ is legitimate.
**(b)** Make a histogram of the probability distribution. Describe what you see.
**(c)** What is the probability that a randomly selected student speaks at least three languages?

**SOLUTION:**

**(a)** All the probabilities are between 0 and 1 and they add to 1, so this is a legitimate probability distribution.
**(b)** *Shape*: The histogram is strongly skewed to the right. *Center*: The median is 1, but the mean will be slightly higher due to the skewness. *Spread*: The number of languages varies from 1 to 5, but nearly all of the students speak just one or two languages.

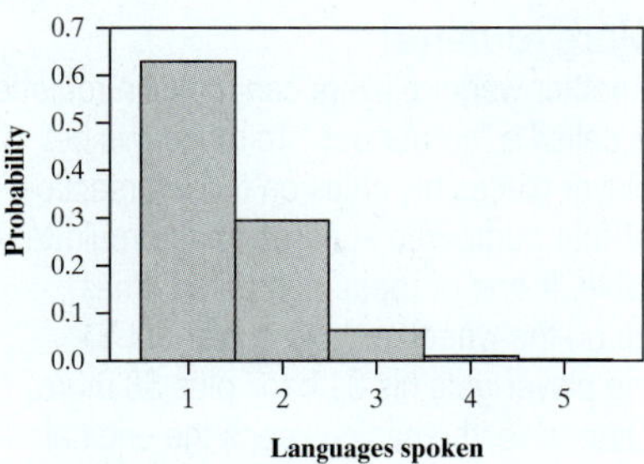


**(c)** $P(X \geq 3) = P(X = 3) + P(X = 4) + P(X = 5) = 0.065 + 0.008 + 0.002 = 0.075$. There is a 0.075 probability of randomly selecting a student who speaks three or more languages.

Although this procedure was later named for Dr. Apgar, the acronym APGAR also represents the five scales: Appearance, Pulse, Grimace, Activity, and Respiration.

*Note that the probability of randomly selecting a newborn whose Apgar score is greater than or equal to 7 is not the same as the probability that the baby's Apgar score is strictly greater than 7.* The latter probability is

$$P(X > 7) = P(X = 8) + P(X = 9) + P(X = 10)$$
$$= 0.319 + 0.437 + 0.053 = 0.809$$

The outcome $X = 7$ is included in "greater than or equal to" and is not included in "greater than." Be sure to confirm the values of interest when dealing with discrete random variables.

## CHECK YOUR UNDERSTANDING

North Carolina State University posts the grade distributions for its courses online.[5] Students in Statistics 101 in a recent semester received 26% A's, 42% B's, 20% C's, 10% D's, and 2% F's. Choose a Statistics 101 student at random. The student's grade on a four-point scale (with A = 4) is a discrete random variable X with this probability distribution:

| Value of X: | 0 | 1 | 2 | 3 | 4 |
|---|---|---|---|---|---|
| Probability: | 0.02 | 0.10 | 0.20 | 0.42 | 0.26 |

1. Say in words what the meaning of $P(X \geq 3)$ is. What is this probability?
2. Write the event "the student got a grade worse than C" in terms of values of the random variable X. What is the probability of this event?
3. Sketch a graph of the probability distribution. Describe what you see.

## Mean (Expected Value) of a Discrete Random Variable

When we analyzed distributions of quantitative data in Chapter 1, we made it a point to discuss their shape, center, and spread. We'll follow the same strategy with probability distributions of random variables. You can use what you learned earlier to describe the shape of a probability distribution histogram. We've already seen examples of symmetric (number of heads in three coin tosses) and left-skewed (Apgar score of a randomly chosen baby) probability distributions. What about center and spread?

The mean $\bar{x}$ of a set of observations is their average. The **mean $\mu_X$ of a discrete random variable** X is also an average of the possible values of X but with an important change to take into account the fact that not all outcomes may be equally likely. A simple example will show what we need to do.

### EXAMPLE Winning (and Losing) at Roulette

***Finding the mean of a discrete random variable***

On an American roulette wheel, there are 38 slots numbered 1 through 36, plus 0 and 00. Half of the slots from 1 to 36 are red; the other half are black. Both the 0 and 00 slots are green. Suppose that a player places a simple $1 bet on red. If the ball lands in a red slot, the player gets the original dollar back, plus an additional

### Teaching Tip

Emphasize the difference between $P(X \geq 7)$ and $P(X > 7)$ in this example. When working with discrete random variables, these probabilities will be different. However, when working with continuous random variables, these probabilities will be the same. Students often have trouble remembering when the "equal to" makes a difference.

### ANSWERS TO CYU

**1.** $P(X \geq 3)$ is the probability that the student got either an A or a B. $P(X \geq 3) = 0.68$.

**2.** $P(X < 2) = 0.02 + 0.10 = 0.12$

**3.** The histogram below is skewed to the left. Higher grades are more likely, but there are a few lower grades. The median grade is a B ($X = 3$) and the grades vary from A ($X = 4$) to F ($X = 0$).

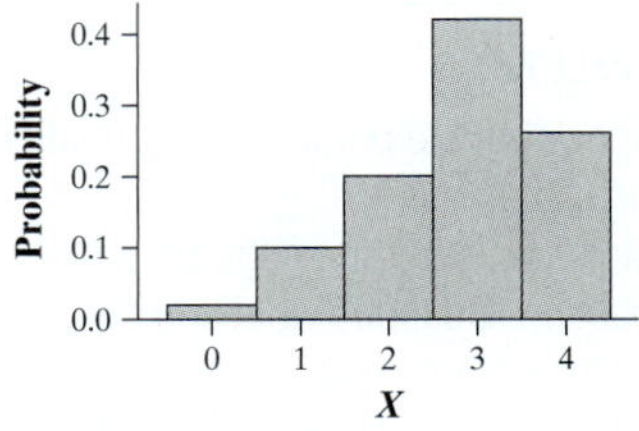


### Teaching Tip

This is a good time to remind your students about SOCS (shape, outliers, center, spread). Although outliers were a big focus when we studied data distributions, we spend less time addressing them in the study of probability distributions of random variables because outliers are less common.

### ALTERNATE EXAMPLE

***More roulette***

Another wager players can make in roulette is called a "corner bet." To make this bet, a player places his chips on the intersection of four numbered squares on the roulette table. If one of these numbers comes up on the wheel and the player bet $1, the player gets his $1 back plus $8 more. Otherwise, the casino keeps the original $1 bet. If $X$ = net gain from a single $1 corner bet, the possible outcomes are $X = -1$ or $X = 8$. Here is the probability distribution of $X$:

| Value: | −$1 | $8 |
|---|---|---|
| Probability: | 34/38 | 4/38 |

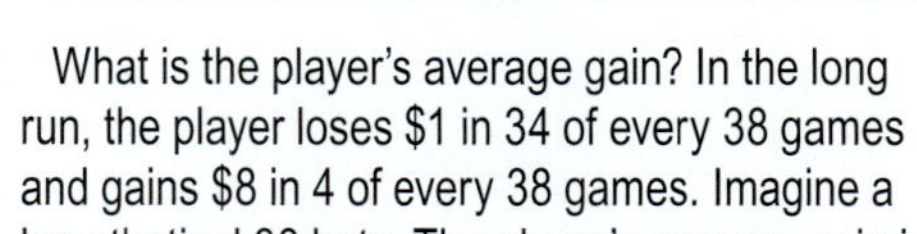

What is the player's average gain? In the long run, the player loses $1 in 34 of every 38 games and gains $8 in 4 of every 38 games. Imagine a hypothetical 38 bets. The player's average gain is

$$\mu_X = \frac{-1 + -1 + \cdots - 1 + 8 + 8 + 8 + 8}{38}$$
$$= \frac{34(-1) + 4(8)}{38}$$
$$= \frac{34(-1)}{38} + \frac{4(8)}{38}$$
$$= (-1)\left(\frac{34}{38}\right) + (8)\left(\frac{4}{38}\right)$$
$$= -\$0.05$$

If a player were to make $1 corner bets many, many times, the average gain would be about −$0.05 per bet. In other words, the casino keeps about 5 cents of every dollar bet in roulette, on average.

*Note*: All of the bets in roulette have the same expected value! However, the bets have different standard deviations, with the red/black bet having the smallest standard deviation and the "straight-up" bet on a single number having the largest standard deviation.

dollar for winning the bet. If the ball lands in a different-colored slot, the player loses the dollar bet to the casino.

Let's define the random variable $X$ = net gain from a single \$1 bet on red. The possible values of $X$ are −\$1 and \$1. (The player either gains a dollar or loses a dollar.) What are the corresponding probabilities? The chance that the ball lands in a red slot is 18/38. The chance that the ball lands in a different-colored slot is 20/38. Here is the probability distribution of $X$:

| Value: | −\$1 | \$1 |
|---|---|---|
| Probability: | 20/38 | 18/38 |

What is the player's average gain? The ordinary average of the two possible outcomes −\$1 and \$1 is \$0. But \$0 isn't the average winnings because the player is less likely to win \$1 than to lose \$1. In the long run, the player gains a dollar 18 times in every 38 games played and loses a dollar on the remaining 20 of 38 bets. The player's long-run average gain for this simple bet is

$$\mu_X = (-\$1)\left(\frac{20}{38}\right) + (\$1)\left(\frac{18}{38}\right) = -\$0.05$$

You see that the player loses (and the casino gains) an average of five cents per \$1 bet in many, many plays of the game.

If someone played several games of roulette, we would call the mean amount the person gained $\bar{x}$. The mean in the previous example is a different quantity—it is the long-run average gain we'd expect if someone played roulette a very large number of times. For this reason, the mean of a random variable is often referred to as its **expected value.** Just as probabilities describe the proportion of times that an outcome occurs in many repetitions of a chance process, the mean of a discrete random variable describes the long-run average outcome.

There are two ways of denoting the mean of a random variable $X$. We can use the notation $\mu_X$, or we can write $E(X)$, as in the "expected value of $X$." In the roulette example, $\mu_X = E(X) = -\$0.05$.

The mean of any discrete random variable is found just as in the roulette example. It is an average of the possible outcomes, but a weighted average in which each outcome is weighted by its probability. Here (finally!) is the definition.

**DEFINITION: Mean (expected value) of a discrete random variable**

Suppose that $X$ is a discrete random variable with probability distribution

| Value: | $x_1$ | $x_2$ | $x_3$ | ... |
|---|---|---|---|---|
| Probability: | $p_1$ | $p_2$ | $p_3$ | ... |

To find the **mean (expected value)** of $X$, multiply each possible value by its probability, then add all the products:

$$\mu_X = E(X) = x_1p_1 + x_2p_2 + x_3p_3 + \ldots$$
$$= \sum x_ip_i$$

## Teaching Tip

In this example, students might ask about using the values $X = 0$ and $X = 2$ instead of $X = -1$ and $X = 1$. In this case, $X$ = winnings, not net gain. The mean value of $X$ in this case is $\mu_X = \$0.95$. After subtracting the dollar you bet, the result is the same. The casino still keeps about 5 cents of every dollar bet in the long run.

## Teaching Tip

Consider showing your students the "long version" of calculating an expected value shown in the Alternate Example. This will help them see that the expected-value formula isn't so mysterious.

## APPLET

The *Law of Large Numbers* applet allows students to roll dice repeatedly and keep track of the mean of their rolls. After many repetitions, the mean of their rolls should approach the true mean, or expected value. www.whfreeman.com/tps5e (click on Statistical Applets and choose Law of Large Numbers).

## AP® Exam Formula Sheet

Let your students know that the formula for the mean (expected value) of a discrete random variable is included on the formula sheet that is given to students on both sections of the AP® exam.

## Teaching Tip

When reading this example, students may ask why we use an uppercase $X$ in some places and lowercase $x$ in other places. It is a convention in statistics to use an uppercase letter to represent a random variable (e.g., $X$ = the Apgar score of a randomly selected baby) and lowercase letters to represent values of that variable (e.g., $x = 8$ or $x = 9$). Although we have made an effort to follow this convention, students will not be expected to make this distinction on the AP® exam.

### ALTERNATE EXAMPLE

*How many languages?*

In an earlier Alternate Example, we defined the random variable $X$ to be the number of languages spoken by a randomly selected U.S. high school student. The table below gives the probability distribution of $X$:

| Languages: | 1 | 2 | 3 | 4 | 5 |
|---|---|---|---|---|---|
| Probability: | 0.630 | 0.295 | 0.065 | 0.008 | 0.002 |

PROBLEM: Compute the mean of the random variable $X$ and interpret this value in context.

SOLUTION:

$$\mu_X = E(X) = (1)(0.630) + (2)(0.295) + (3)(0.065) + (4)(0.008) + (5)(0.002) = 1.457$$

If we were to randomly select many, many U.S. high school students, the average number of languages spoken would be about 1.457.

Let's put the definition to use in calculating the mean of a familiar random variable.

## EXAMPLE Apgar Scores: What's Typical?

*Mean and expected value as an average*

In our earlier example, we defined the random variable X to be the Apgar score of a randomly selected baby. The table below gives the probability distribution for X once again.

| Value $x_i$: | 0 | 1 | 2 | 3 | 4 | 5 | 6 | 7 | 8 | 9 | 10 |
|---|---|---|---|---|---|---|---|---|---|---|---|
| Probability $p_i$: | 0.001 | 0.006 | 0.007 | 0.008 | 0.012 | 0.020 | 0.038 | 0.099 | 0.319 | 0.437 | 0.053 |

PROBLEM: Compute the mean of the random variable $X$. Interpret this value in context.

SOLUTION: From the probability distribution for $X$, we see that 1 in every 1000 babies would have an Apgar score of 0; 6 in every 1000 babies would have an Apgar score of 1; and so on. So the mean (expected value) of $X$ is

$$\mu_X = E(X) = \sum x_i p_i$$
$$= (0)(0.001) + (1)(0.006) + (2)(0.007) + \cdots + (10)(0.053) = 8.128$$

The mean Apgar score of a randomly selected newborn is 8.128. This is the average Apgar score of many, many randomly chosen babies.

For Practice *Try Exercise* 9

**AP® EXAM TIP** If the mean of a random variable has a non-integer value, but you report it as an integer, your answer will not get full credit.

Notice that the mean Apgar score, 8.128, is not a possible value of the random variable X. It's also not an integer. If you think of the mean as a long-run average over many repetitions, these facts shouldn't bother you.

## Standard Deviation (and Variance) of a Discrete Random Variable

With the mean as our measure of center for a discrete random variable, it shouldn't surprise you that we'll use the standard deviation as our measure of spread. In Chapter 1, we first defined the sample variance $s_x^2$ as the "typical" squared deviation from the mean and then took the square root of the variance to get the sample standard deviation $s_x$. The definition of the **variance of a random variable** $\sigma_X^2$ is similar to the definition of the variance for a set of quantitative data. That is,

### AP® Exam Common Error

Emphasize the AP® Exam Tip on this page. Many students incorrectly believe that the *expected* value of a random variable must be equal to one of the *possible* values of the variable. This is not the case. In the roulette examples, the expected value was −$0.05, even though this was not a possible gain from a single $1 bet.

Recall that the formula for the sample variance is

$$s_x^2 = \frac{\sum(x_i - \bar{x})^2}{n-1}$$

the variance is an "average" of the squared deviation $(x_i - \mu_X)^2$ of the values of the variable X from its mean $\mu_X$. As with the mean, the average we use is a weighted average. Each outcome is weighted by its probability to take account of outcomes that are not equally likely. To get the **standard deviation of a random variable,** we take the square root of the variance. Here are the details.

**DEFINITION: Variance and standard deviation of a discrete random variable**

Suppose that $X$ is a discrete random variable with probability distribution

| Value: | $x_1$ | $x_2$ | $x_3$ | ... |
|---|---|---|---|---|
| Probability: | $p_1$ | $p_2$ | $p_3$ | ... |

and that $\mu_X$ is the mean of $X$. The **variance** of $X$ is

$$\text{Var}(X) = \sigma_X^2 = (x_1 - \mu_X)^2 p_1 + (x_2 - \mu_X)^2 p_2 + (x_3 - \mu_X)^2 p_3 + \ldots$$
$$= \sum(x_i - \mu_X)^2 p_i$$

The **standard deviation** of $X$, $\sigma_X$, is the square root of the variance.

$$\sigma_X = \sqrt{\sum(x_i - \mu_X)^2 p_i}$$

The standard deviation of a random variable X is a measure of how much the values of the variable typically vary from the mean $\mu_X$. Let's compute the variance and standard deviation of a familiar discrete random variable.

EXAMPLE

## Apgar Scores: How Variable Are They?

### *Calculating measures of spread*

In the last example, we calculated the mean Apgar score of a randomly chosen newborn to be $\mu_X = 8.128$. The table below gives the probability distribution for X one more time.

| Value $x_i$: | 0 | 1 | 2 | 3 | 4 | 5 | 6 | 7 | 8 | 9 | 10 |
|---|---|---|---|---|---|---|---|---|---|---|---|
| Probability $p_i$: | 0.001 | 0.006 | 0.007 | 0.008 | 0.012 | 0.020 | 0.038 | 0.099 | 0.319 | 0.437 | 0.053 |

PROBLEM: Compute and interpret the standard deviation of the random variable $X$.

SOLUTION: The formula for the variance of $X$ is $\sigma_X^2 = \sum(x_i - \mu_X)^2 p_i$. Plugging in values gives

$$\sigma_X^2 = (0 - 8.128)^2(0.001) + (1 - 8.128)^2(0.006)$$
$$+ (2 - 8.128)^2(0.007) + \cdots + (10 - 8.128)^2(0.053)$$
$$\sigma_X^2 = 2.066$$

The standard deviation of $X$ is $\sigma_X = \sqrt{2.066} = 1.437$. A randomly selected baby's Apgar score will typically differ from the mean (8.128) by about 1.4 units.

**For Practice** *Try Exercise* 15

### AP® Exam Formula Sheet

Let your students know that the formula for the variance of a discrete random variable is included on the formula sheet that is given to students on both sections of the AP® exam. However, the formula for the standard deviation of a discrete random variable is not provided on the formula sheet. This shouldn't be a problem if students remember that the standard deviation is always the square root of the variance.

### ALTERNATE EXAMPLE

***How many languages?***

In the last Alternate Example, we calculated the mean number of languages spoken for a randomly selected U.S. high school student to be $\mu_X = 1.457$. The table below gives the probability distribution for $X$ one more time:

| Languages: | 1 | 2 | 3 | 4 | 5 |
|---|---|---|---|---|---|
| Probability: | 0.630 | 0.295 | 0.065 | 0.008 | 0.002 |

PROBLEM: Compute and interpret the standard deviation of the random variable $X$.

SOLUTION:

$$\sigma_X^2 = (1 - 1.457)^2(0.630) + (2 - 1.457)^2(0.295) + \cdots = 0.450$$

The standard deviation of $X$ is $\sigma_X = \sqrt{0.450} = 0.671$. The number of languages spoken by a randomly selected U.S. high school student typically varies by about 0.671 languages from the mean (1.457).

### Teaching Tip

It might help students understand the formula better if they calculate the variance the "long way." In the Apgar example, imagine that we have a representative set of 1000 newborn babies. We would expect 1 baby to have a score of 0, 6 babies to have a score of 1, and so on. Using the method from Chapter 1, we get

Notice that we divided by 1000 ($n$) rather than 999 ($n - 1$) in the calculation. Because we have no sampling variability in this context, we do not need to divide by $n - 1$ to compensate for having to estimate the mean of the distribution.

$$Var(X) = \frac{(0 - 8.128)^2 + (1 - 8.128)^2 + (1 - 8.128)^2 + \cdots}{1000}$$
$$= \frac{1(0 - 8.128)^2 + 6(1 - 8.128)^2 + \cdots}{1000}$$
$$= (0 - 8.128)^2\left(\frac{1}{1000}\right) + (1 - 8.128)^2\left(\frac{6}{1000}\right) + \cdots = 2.066$$

### TECHNOLOGY TIP

Remind students that they can watch a video presentation of this technology corner (for the TI-83/84, TI-89, or TI-Nspire) at www.whfreeman.com/tps5e.

### TECHNOLOGY TIP

Remind your students to change the "Freq" entry back to "1" after graphing a probability distribution. Otherwise, the next time they try to graph a distribution of data, the calculator will be looking for frequencies (or relative frequencies) in L2.

### TECHNOLOGY TIP

Conveniently, the calculator doesn't report a value for $s_x$ (sample standard deviation). Because it thinks $n = 1$, the denominator of the sample standard deviation is equal to $n - 1 = 1 - 1 = 0$, resulting in an undefined value. Using the sample standard deviation would be inappropriate here in any case because we are not working with values obtained in a sample from some population.

### AP® Exam Common Error

When students are asked to calculate the mean (expected value) or standard deviation of a random variable, they often lose credit for not showing adequate work—or for showing no work at all. Because not all AP® readers are familiar with every brand of calculator, students need to go beyond just reporting the commands they used on their calculator. In this case, showing the first couple of terms with an ellipsis (. . .) for the calculation of the mean or standard deviation is sufficient.

You can use your calculator to graph the probability distribution of a discrete random variable and to calculate measures of center and spread, as the following Technology Corner illustrates.

## 11. TECHNOLOGY CORNER — ANALYZING RANDOM VARIABLES ON THE CALCULATOR

TI-Nspire instructions in Appendix B; HP Prime instructions on the book's Web site.

Let's explore what the calculator can do using the random variable X = Apgar score of a randomly selected newborn.

**TI-83/84** | **TI-89**

1. Start by entering the values of the random variable in L1/list1 and the corresponding probabilities in L2/list2.

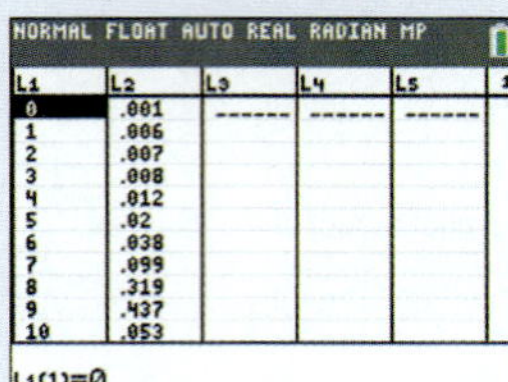

2. To graph a histogram of the probability distribution:
   - Set up a statistics plot with Xlist: L1/list1 and Freq: L2/list2.
   - Adjust your window settings as follows: Xmin = −1, Xmax = 11, Xscl = 1,Ymin = −0.1, Ymax = 0.5, Yscl = 0.1.
   - Press GRAPH (◆ F3 on the TI-89).

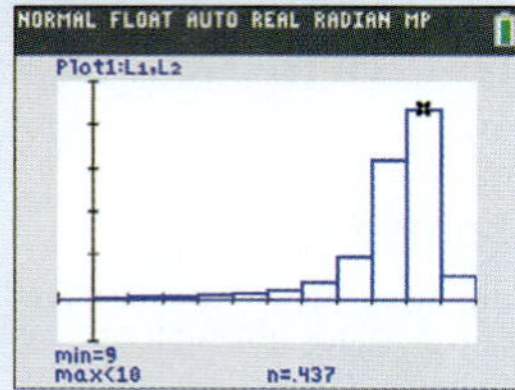

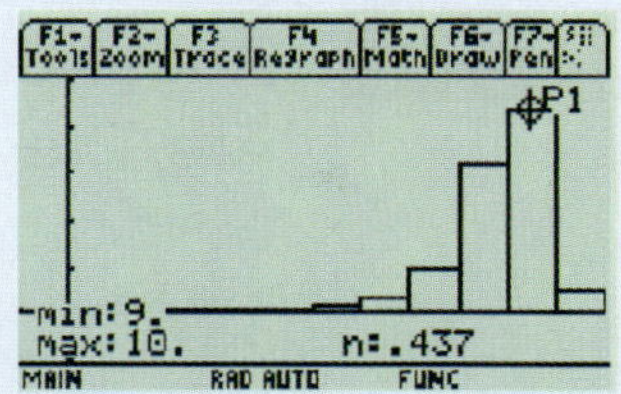

3. To calculate the mean and standard deviation of the random variable, use one-variable statistics with the values in L1/list1 and the probabilities (relative frequencies) in L2/list2.

- **OS 2.55 or later:** In the dialog box, specify List: L1 and FreqList: L2. Then choose Calculate. **Older OS:** Execute the command `1-Var Stats L1,L2`.
- In the Statistics/List Editor, press F4 (Calc) and choose `1-Var Stats...` Use the inputs List: list1 and Freq: list2.

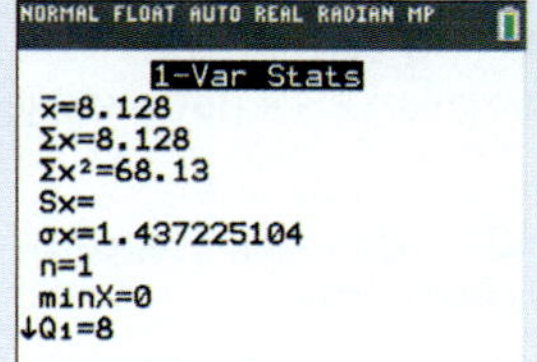


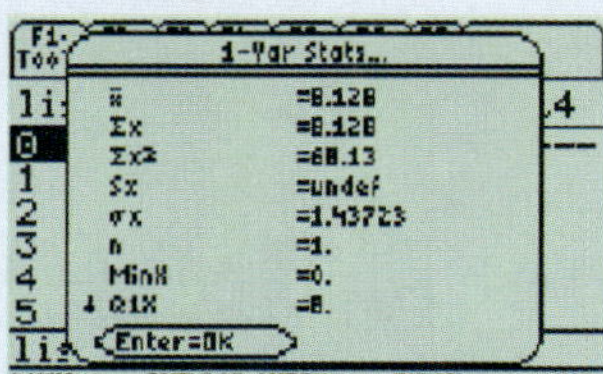

## CHECK YOUR UNDERSTANDING

A large auto dealership keeps track of sales made during each hour of the day. Let X = the number of cars sold during the first hour of business on a randomly selected Friday. Based on previous records, the probability distribution of X is as follows:

| Cars sold: | 0 | 1 | 2 | 3 |
|---|---|---|---|---|
| Probability: | 0.3 | 0.4 | 0.2 | 0.1 |

1. Compute and interpret the mean of X.
2. Compute and interpret the standard deviation of X.

## Continuous Random Variables

When we use the table of random digits to select a digit between 0 and 9, the result is a discrete random variable (call it X). The probability model assigns probability 1/10 to each of the 10 possible values of X.

The calculator command `rand` will generate a random number from 0 to 1. Can you figure out how to modify the command to find a random number between, say, 1 and 3?

Suppose we want to choose a number at random between 0 and 1, allowing *any* number between 0 and 1 as the outcome (like 0.84522 or 0.1111119). Calculator and computer random number generators will do this. The sample space of this chance process is an entire interval of numbers:

$$S = \text{all numbers between 0 and 1}$$

Call the outcome of the random number generator Y for short. How can we find probabilities of events like $P(0.3 \le Y \le 0.7)$? As in the case of selecting a random digit, we would like all possible outcomes to be equally likely. But we cannot assign probabilities to each individual value of Y and then add them, because there are infinitely many possible values.

In situations like this, we use a different way of assigning probabilities directly to events—as *areas under a density curve*. Recall from Chapter 2 that any density curve has area exactly 1 underneath it, corresponding to total probability 1.

### EXAMPLE Random Numbers

#### *Density curves and probability distributions*

The random number generator will spread its output uniformly across the entire interval from 0 to 1 as we allow it to generate a long sequence of random numbers. The results of many trials are represented by the density curve of a *uniform distribution*. This density curve appears in purple in Figure 6.3 on the next page. It has height 1 over the interval from 0 to 1. The area under the density curve is 1, and the probability of any event is the area under the density curve and above the event in question.

### ANSWERS TO CYU

1. $\mu_X = 1.1$. If many, many Fridays are randomly selected, the average number of cars sold will be about 1.1.

2. $\sigma_X = \sqrt{0.89} = 0.943$. The number of cars sold on a randomly selected Friday will typically vary from the mean (1.1) by about 0.943 cars.

### Teaching Tip

Ask students if they can answer the question in the margin note on this page. To generate a random number from 1 to 3, use the command `2rand+1`. Doubling a random number from 0 to 1 will produce a random number from 0 to 2. Adding 1 to this number will produce a random number from 1 to 3.

### ALTERNATE EXAMPLE

*Random Normal*

Many calculators and statistical software packages can generate random numbers from a Normal distribution. For example, on the TI-84, pressing `randNorm(0,1)` over and over will generate a distribution that can be represented by the standard Normal curve. If $Z$ = the result when `randNorm(0,1)` is pressed, the probability that $Z$ is between $-1$ and 1 is about 0.68. This is shown in the following figure, where the area under the curve between $-1$ and 1 is about 0.68.

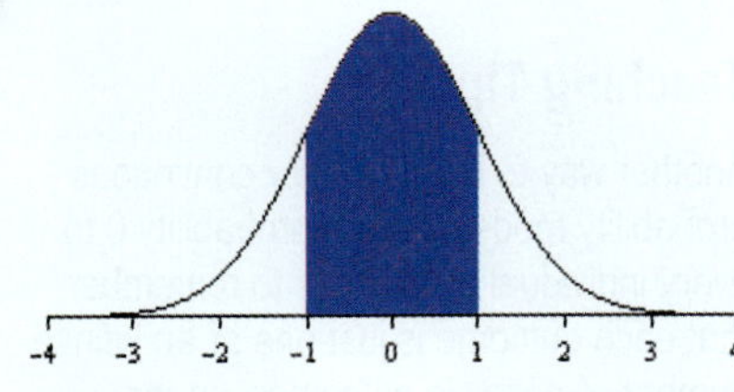

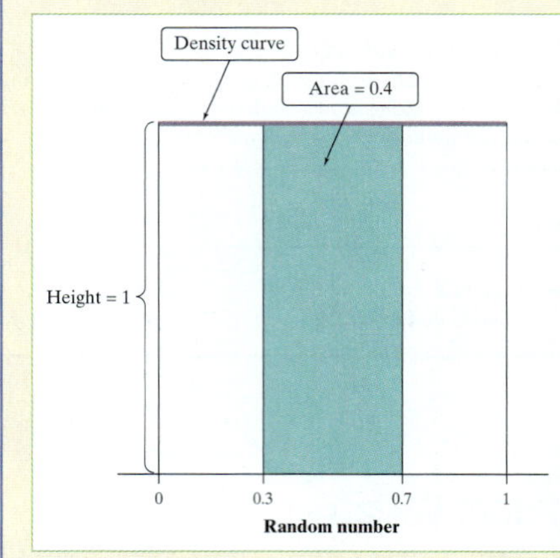


As Figure 6.3 shows, the probability that the random number generator produces a number Y between 0.3 and 0.7 is

$$P(0.3 \le Y \le 0.7) = 0.4$$

That's because the area of the shaded rectangle is

$$\text{length} \times \text{width} = 0.4 \times 1 = 0.4$$

**FIGURE 6.3** Assigning probabilities for generating a random number between 0 and 1. The probability of any interval of numbers is the area above the interval and under the density curve. The shaded area represents $P(0.3 \le Y \le 0.7)$

In many cases, discrete random variables arise from counting something—for instance, the number of siblings that a randomly selected student has. Continuous random variables often arise from measuring something—for instance, the height or time to run a mile for a randomly selected student.

Figure 6.3 shows the probability distribution of the random variable Y = random number between 0 and 1. We call Y a **continuous random variable** because its values are not isolated numbers but rather an entire interval of numbers.

**DEFINITION: Continuous random variable**

A **continuous random variable** $X$ takes all values in an interval of numbers. The probability distribution of $X$ is described by a density curve. The probability of any event is the area under the density curve and above the values of $X$ that make up the event.

The probability distribution for a continuous random variable assigns probabilities to intervals of outcomes rather than to individual outcomes. In fact, *all continuous probability models assign probability 0 to every individual outcome.* Only intervals of values have positive probability. To see that this is true, consider a specific outcome from the random number generator of the previous example, such as $P(Y = 0.7)$. The probability of this event is the area under the density curve that's above the point 0.70000 ... on the horizontal axis. But this vertical line segment has no width, so the area is 0. For that reason,

$$P(0.3 \le Y \le 0.7) = P(0.3 \le Y < 0.7) = P(0.3 < Y < 0.7) = 0.4$$

We can use any density curve to assign probabilities. The density curves that are most familiar to us are the Normal curves of Chapter 2. We learned how to find areas in any Normal distribution on page 118. Normal distributions can be probability distributions as well as models for data. The following example shows the connection between the two.

## Teaching Tip

Remind students about the shoe size versus foot length example from the Teaching Tip on page 348 to illustrate the difference between discrete and continuous random variables. Other examples include the amount of time it takes to run the 110-meter hurdles (continuous) versus the number of hurdles cleanly jumped (discrete) and a student's age (continuous) versus the number of birthdays he or she has had (discrete).

## Teaching Tip

Some students may argue that variables like height are discrete rather than continuous, because they are typically reported as values rounded to the nearest inch or half-inch. However, we still consider height a continuous random variable because a student's actual height (not rounded) could be any value in an interval of numbers (e.g., 65.283499843 inches).

## Teaching Tip

Another way to illustrate why continuous probability models assign probability 0 to every individual outcome is to remember that each outcome is just one of an infinite number of possible outcomes, so the probability of any particular outcome is $\frac{1}{\infty}$.

## EXAMPLE Young Women's Heights

### *Normal probability distributions*

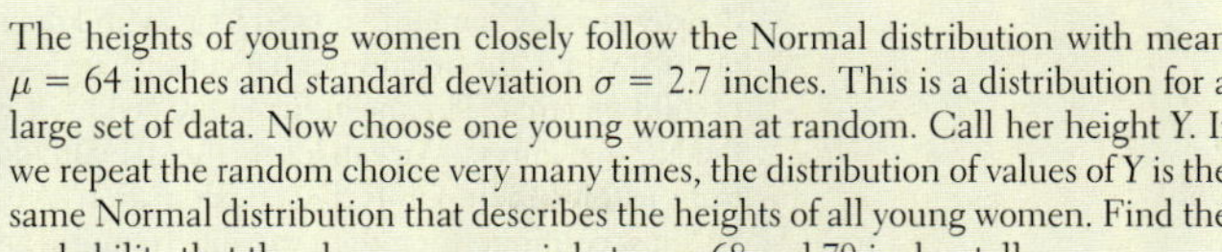

The heights of young women closely follow the Normal distribution with mean $\mu = 64$ inches and standard deviation $\sigma = 2.7$ inches. This is a distribution for a large set of data. Now choose one young woman at random. Call her height Y. If we repeat the random choice very many times, the distribution of values of Y is the same Normal distribution that describes the heights of all young women. Find the probability that the chosen woman is between 68 and 70 inches tall.

PROBLEM: What's the probability that a randomly chosen young woman has height between 68 and 70 inches?

SOLUTION:

Step 1: State the distribution and the values of interest. The height Y of a randomly chosen young woman has the $N(64, 2.7)$ distribution. We want to find $P(68 \le Y \le 70)$. Figure 6.4 shows the distribution with the area of interest shaded and the mean, standard deviation, and boundary values labeled.

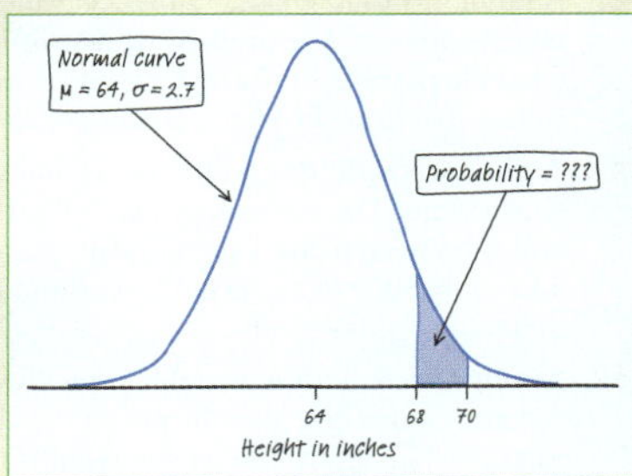


**FIGURE 6.4** The probability that a randomly chosen young woman has height between 68 and 70 inches as an area under a Normal curve.

Step 2: Perform calculations—show your work! The standardized scores for the two boundary values are

$$z = \frac{68 - 64}{2.7} = 1.48 \quad \text{and} \quad z = \frac{70 - 64}{2.7} = 2.22$$

The random variable Z follows a standard Normal distribution, and the desired probability is $P(1.48 \le Z \le 2.22)$. From Table A, we find that $P(Z \le 2.22) = 0.9868$ and $P(Z \le 1.48) = 0.9306$. So we have

$$P(1.48 \le Z \le 2.22) = P(Z \le 2.22) - P(Z \le 1.48) = 0.9868 - 0.9306 = 0.0562$$

*Using technology:* The command `normalcdf(lower:68, upper:70, μ:64, σ:2.7)` gives an area of 0.0561.

Step 3: Answer the question. There's about a 5.6% chance that a randomly chosen young woman has a height between 68 and 70 inches.

**For Practice** *Try Exercise* **23**

**AP® EXAM TIP** When showing your work on a free response question, you must include more than a calculator command. Writing `normalcdf(68,70,64,2.7)` will *not* earn you full credit for a Normal calculation. At a minimum, you must indicate what each of those calculator inputs represents. Better yet, sketch and label a Normal curve to show what you're finding.

### Teaching Tip

Remind your students about the three-step process for computing probabilities from a Normal distribution. We will be using the same three-step process when we compute binomial probabilities in Section 6.3.

### Teaching Tip

When reading this example with your students, mention that $P(68 \le X \le 70)$ is really the same as $P(68 < X < 70)$, because there is no area directly above $x = 68$ or $x = 70$.

### ALTERNATE EXAMPLE

#### *Weights of 3-year-old females*

The weights of 3-year-old females closely follow a Normal distribution with a mean of $\mu = 30.7$ pounds and a standard deviation of 3.6 pounds. Randomly choose one 3-year-old female and call her weight X.

PROBLEM: What is the probability that a randomly selected 3-year-old female weighs at least 30 pounds?

SOLUTION:

Step 1: State the distribution and the values of interest. The weight X of a randomly chosen 3-year-old female has the $N(30.7, 3.6)$ distribution. We want to find $P(X \ge 30)$ as shown in the figure below.

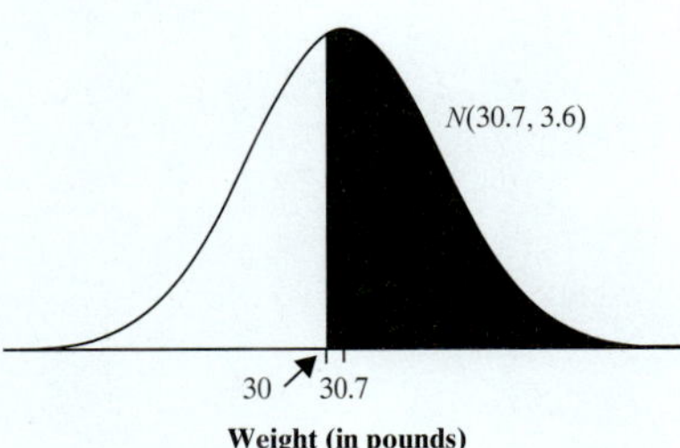


Step 2: Perform calculations–show your work! The standardized score for the boundary value is $z = \frac{30 - 30.7}{3.6} = -0.19$. From Table A, $P(Z < -0.19) = 0.4247$, so $P(Z \ge -0.19) = 1 - 0.4247 = 0.5753$. *Using technology*: The command `normalcdf(lower:30,upper:1000,μ:30.7,σ:3.6)` gives an area of 0.5771.

Step 3: Answer the question. There is about a 58% chance that the randomly selected 3-year-old female will weigh at least 30 pounds.

## Teaching Tip

See the Notes and Data Sources section at the end of the book for more details on the "advanced mathematics" mentioned at the conclusion of this paragraph.

### TRM Section 6.1 Quizzes

There are three different quizzes available for this section in the Teacher's Resource Materials. Versions A and B are roughly parallel. Find them by clicking on the link in the TE-book, opening the TRCD, or logging onto the book's companion Web site www.whfreeman.com/tps5e (teacher log-in required).

The calculation in the preceding example is the same as those we did in Chapter 2. Only the language of probability is new.

What about the mean and standard deviation for continuous random variables? The probability distribution of a continuous random variable $X$ is described by a density curve. Chapter 2 showed how to find the mean of the distribution: it is the point at which the area under the density curve would balance if it were made out of solid material. The mean lies at the center of symmetric density curves such as the Normal curves. We can locate the standard deviation of a Normal distribution from its inflection points. Exact calculation of the mean and standard deviation for most continuous random variables requires advanced mathematics.[6]

## Section 6.1 Summary

- A **random variable** takes numerical values determined by the outcome of a chance process. The **probability distribution** of a random variable $X$ tells us what the possible values of $X$ are and how probabilities are assigned to those values. There are two types of random variables: *discrete* and *continuous*.
- A **discrete random variable** has a fixed set of possible values with gaps between them. The probability distribution assigns each of these values a probability between 0 and 1 such that the sum of all the probabilities is exactly 1. The probability of any event is the sum of the probabilities of all the values that make up the event.
- A **continuous random variable** takes all values in some interval of numbers. A density curve describes the probability distribution of a continuous random variable. The probability of any event is the area under the curve above the values that make up the event.
- The **mean of a random variable** $\mu_X$ is the balance point of the probability distribution histogram or density curve. Because the mean is the long-run average value of the variable after many repetitions of the chance process, it is also known as the **expected value** of the random variable, $E(X)$.
- If $X$ is a discrete random variable, the mean is the average of the values of $X$, each weighted by its probability:

$$\mu_X = E(X) = \sum x_i p_i = x_1 p_1 + x_2 p_2 + x_3 p_3 + \dots$$

- The **variance of a random variable** $\sigma_X^2$ is the "average" squared deviation of the values of the variable from their mean. The **standard deviation** $\sigma_X$ is the square root of the variance. The standard deviation measures the typical distance of the values in the distribution from the mean.
- For a discrete random variable $X$, the variance is

$$\sigma_X^2 = \Sigma(x_i - \mu_X)^2 p_i = (x_1 - \mu_X)^2 p_1 + (x_2 - \mu_X)^2 p_2 + (x_3 - \mu_X)^2 p_3 + \cdots$$

and the standard deviation is

$$\sigma_X = \sqrt{\Sigma(x_i - \mu_X)^2 p_i}$$

**6.1 TECHNOLOGY CORNER**

TI-Nspire instructions in Appendix B; HP Prime instructions on the book's Web site.



# Section 6.1 Exercises

1. **Toss 4 times** Suppose you toss a fair coin 4 times. Let X = the number of heads you get.
   (a) Find the probability distribution of X.
   (b) Make a histogram of the probability distribution. Describe what you see.
   (c) Find $P(X \le 3)$ and interpret the result.

2. **Pair-a-dice** Suppose you roll a pair of fair, six-sided dice. Let $T$ = the sum of the spots showing on the up-faces.
   (a) Find the probability distribution of $T$.
   (b) Make a histogram of the probability distribution. Describe what you see.
   (c) Find $P(T \ge 5)$ and interpret the result.

3. **Spell-checking** Spell-checking software catches "nonword errors," which result in a string of letters that is not a word, as when "the" is typed as "teh." When undergraduates are asked to write a 250-word essay (without spell-checking), the number X of nonword errors has the following distribution:

| Value: | 0 | 1 | 2 | 3 | 4 |
|---|---|---|---|---|---|
| Probability: | 0.1 | 0.2 | 0.3 | 0.3 | 0.1 |

   (a) Write the event "at least one nonword error" in terms of X. What is the probability of this event?
   (b) Describe the event $X \le 2$ in words. What is its probability? What is the probability that $X < 2$?

4. **Kids and toys** In an experiment on the behavior of young children, each subject is placed in an area with five toys. Past experiments have shown that the probability distribution of the number X of toys played with by a randomly selected subject is as follows:

| Number of toys $x_i$: | 0 | 1 | 2 | 3 | 4 | 5 |
|---|---|---|---|---|---|---|
| Probability $p_i$: | 0.03 | 0.16 | 0.30 | 0.23 | 0.17 | 0.11 |

   (a) Write the event "plays with at most two toys" in terms of X. What is the probability of this event?
   (b) Describe the event $X > 3$ in words. What is its probability? What is the probability that $X \ge 3$?

5. pg 349  **Benford's law** Faked numbers in tax returns, invoices, or expense account claims often display patterns that aren't present in legitimate records. Some patterns, like too many round numbers, are obvious and easily avoided by a clever crook. Others are more subtle. It is a striking fact that the first digits of numbers in legitimate records often follow a model known as Benford's law.[7] Call the first digit of a randomly chosen record X for short. Benford's law gives this probability model for X (note that a first digit can't be 0):

| First digit: | 1 | 2 | 3 | 4 | 5 | 6 | 7 | 8 | 9 |
|---|---|---|---|---|---|---|---|---|---|
| Probability: | 0.301 | 0.176 | 0.125 | 0.097 | 0.079 | 0.067 | 0.058 | 0.051 | 0.046 |

   (a) Show that this is a legitimate probability distribution.
   (b) Make a histogram of the probability distribution. Describe what you see.
   (c) Describe the event $X \ge 6$ in words. What is $P(X \ge 6)$?
   (d) Express the event "first digit is at most 5" in terms of X. What is the probability of this event?

6. **Working out** Choose a person aged 19 to 25 years at random and ask, "In the past seven days, how many times did you go to an exercise or fitness center or work out?" Call the response Y for short. Based on a large sample survey, here is a probability model for the answer you will get:[8]

| Days: | 0 | 1 | 2 | 3 | 4 | 5 | 6 | 7 |
|---|---|---|---|---|---|---|---|---|
| Probability: | 0.68 | 0.05 | 0.07 | 0.08 | 0.05 | 0.04 | 0.01 | 0.02 |

   (a) Show that this is a legitimate probability distribution.
   (b) Make a histogram of the probability distribution. Describe what you see.

**TRM Full Solutions to Section 6.1 Exercises**

## Answers to Section 6.1 Exercises

**6.1 (a)**

| Value: | 0 | 1 | 2 | 3 | 4 |
|---|---|---|---|---|---|
| Probability: | 1/16 | 4/16 | 6/16 | 4/16 | 1/16 |

**(b)** The histogram below shows that this distribution is symmetric with a center at 2.

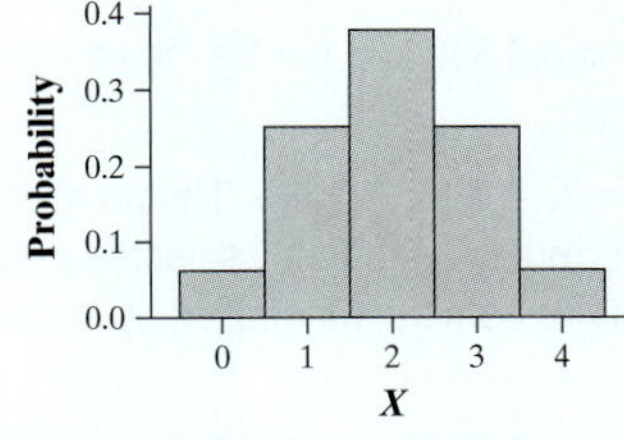


**(c)** $P(X \le 3) = 15/16 = 0.9375$. There is a 0.9375 probability that you will get three or fewer heads in 4 tosses of a fair coin.

**6.2 (a)**

| Value: | 2 | 3 | 4 | 5 | 6 | 7 | 8 | 9 | 10 | 11 | 12 |
|---|---|---|---|---|---|---|---|---|---|---|---|
| Probability: | $\frac{1}{36}$ | $\frac{2}{36}$ | $\frac{3}{36}$ | $\frac{4}{36}$ | $\frac{5}{36}$ | $\frac{6}{36}$ | $\frac{5}{36}$ | $\frac{4}{36}$ | $\frac{3}{36}$ | $\frac{2}{36}$ | $\frac{1}{36}$ |

**(b)** The histogram below shows that the distribution is symmetric with a center at 7.

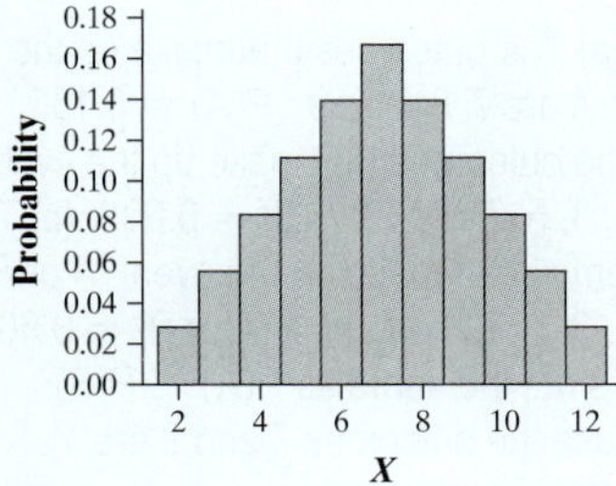


**(c)** $P(T \ge 5) = \frac{30}{36} = 0.833$. There is a 0.833 probability that you will get a sum of 5 or more when you roll a pair of 6-sided dice.

**6.3 (a)** $P(X \ge 1) = 0.9$ **(b)** The event $X \le 2$ is "at most two nonword errors." $P(X \le 2) = 0.6$. $P(X < 2) = 0.3$.

**6.4 (a)** $P(X \le 2) = 0.49$ **(b)** The event $X > 3$ is "the child plays with more than three toys." $P(X > 3) = 0.28$. $P(X \ge 3) = 0.51$.

**6.5 (a)** All of the probabilities are between 0 and 1 and they sum to 1. **(b)** The histogram below is unimodal and skewed to the right.

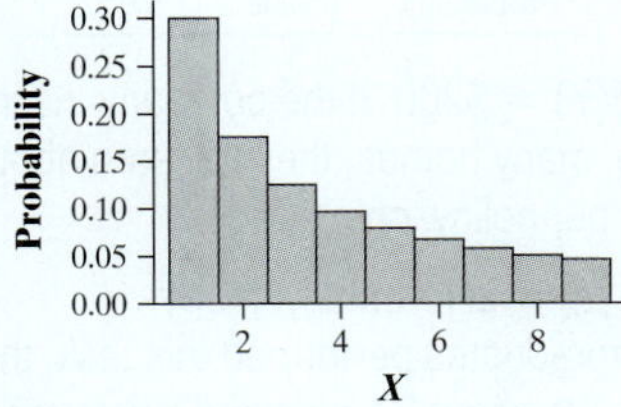


**(c)** The event $X \ge 6$ is the event that "the first digit in a randomly chosen record is a 6 or higher." $P(X \ge 6) = 0.222$.
**(d)** $P(X \le 5) = 0.778$

**6.6 (a)** All of the probabilities are between 0 and 1 and they sum to 1. **(b)** The histogram below shows a strongly right skewed distribution with a peak at 0 days.

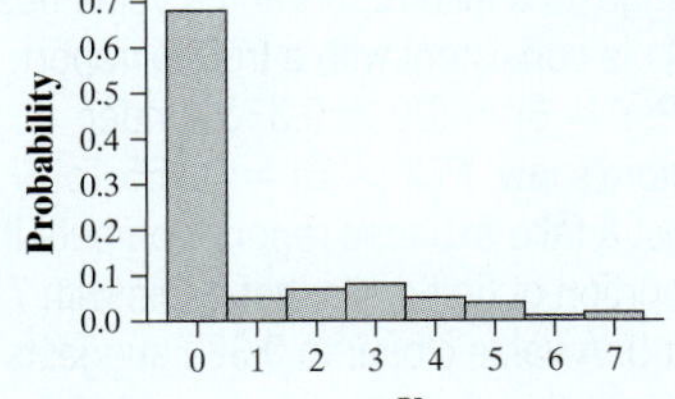

(c) Describe the event $Y < 7$ in words. What is $P(Y < 7)$?

(d) Express the event "worked out at least once" in terms of Y. What is the probability of this event?

7. **Benford's law** Refer to Exercise 5. The first digit of a randomly chosen expense account claim follows Benford's law. Consider the events A = first digit is 7 or greater and B = first digit is odd.

(a) What outcomes make up the event A? What is $P(A)$?

(b) What outcomes make up the event B? What is $P(B)$?

(c) What outcomes make up the event "A or B"? What is $P(A \text{ or } B)$? Why is this probability not equal to $P(A) + P(B)$?

8. **Working out** Refer to Exercise 6. Consider the events A = works out at least once and B = works out less than 5 times per week.

(a) What outcomes make up the event A? What is $P(A)$?

(b) What outcomes make up the event B? What is $P(B)$?

(c) What outcomes make up the event "A and B"? What is $P(A \text{ and } B)$? Why is this probability not equal to $P(A) \cdot P(B)$?

9. **Keno** Keno is a favorite game in casinos, and similar games are popular with the states that operate lotteries. Balls numbered 1 to 80 are tumbled in a machine as the bets are placed, then 20 of the balls are chosen at random. Players select numbers by marking a card. The simplest of the many wagers available is "Mark 1 Number." Your payoff is \$3 on a \$1 bet if the number you select is one of those chosen. Because 20 of 80 numbers are chosen, your probability of winning is 20/80, or 0.25. Let X = the net amount you gain on a single play of the game. pg 352

(a) Make a table that shows the probability distribution of X.

(b) Compute the expected value of X. Explain what this result means for the player.

10. **Fire insurance** Suppose a homeowner spends \$300 for a home insurance policy that will pay out \$200,000 if the home is destroyed by fire. Let Y = the profit made by the company on a single policy. From previous data, the probability that a home in this area will be destroyed by fire is 0.0002.

(a) Make a table that shows the probability distribution of Y.

(b) Compute the expected value of Y. Explain what this result means for the insurance company.

11. **Spell-checking** Refer to Exercise 3. Calculate the mean of the random variable X and interpret this result in context.

12. **Kids and toys** Refer to Exercise 4. Calculate the mean of the random variable X and interpret this result in context.

13. **Benford's law and fraud** A not-so-clever employee decided to fake his monthly expense report. He believed that the first digits of his expense amounts should be equally likely to be any of the numbers from 1 to 9. In that case, the first digit Y of a randomly selected expense amount would have the probability distribution shown in the histogram.

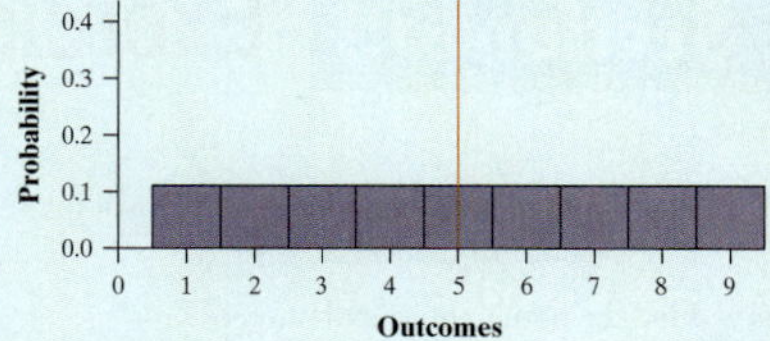


(a) Explain why the mean of the random variable Y is located at the solid red line in the figure.

(b) The first digits of randomly selected expense amounts actually follow Benford's law (Exercise 5). According to Benford's law, what's the expected value of the first digit? Explain how this information could be used to detect a fake expense report.

(c) What's $P(Y > 6)$ in the above distribution? According to Benford's law, what proportion of first digits in the employee's expense amounts should be greater than 6? How could this information be used to detect a fake expense report?

14. **Life insurance** A life insurance company sells a term insurance policy to a 21-year-old male that pays \$100,000 if the insured dies within the next 5 years. The probability that a randomly chosen male will die each year can be found in mortality tables. The company collects a premium of \$250 each year as payment for the insurance. The amount Y that the company earns on this policy is \$250 per year, less the \$100,000 that it must pay if the insured dies. Here is a partially completed table that shows information about risk of mortality and the values of Y = profit earned by the company:

| Age at death: | 21 | 22 | 23 | 24 | 25 | 26 or more |
|---|---|---|---|---|---|---|
| Profit: | −\$99,750 | −\$99,500 | −\$99,250 | −\$99,000 | −\$98,750 | \$1250 |
| Probability: | 0.00183 | 0.00186 | 0.00189 | 0.00191 | 0.00193 | |

(a) Explain why the company suffers a loss of \$98,750 on such a policy if a client dies at age 25.

(b) Find the missing probability. Show your work.

(c) Calculate the mean $\mu_Y$. Interpret this value in context.

**(c)** The event $Y < 7$ is the event that the person "did not work out all 7 days." $P(Y < 7) = 0.98$. **(d)** $P(Y \geq 1) = 0.32$

**6.7 (a)** The outcomes that make up the event *A* are 7, 8, and 9. $P(A) = 0.155$. **(b)** The outcomes that make up the event *B* are 1, 3, 5, 7 and 9. $P(B) = 0.609$. **(c)** The outcomes that make up the event "*A* or *B*" are 1, 3, 5, 7, 8 and 9. $P(A \text{ or } B) = 0.660$. This is not the same as $P(A) + P(B)$ because the outcomes 7 and 9 are included in both events.

**6.8 (a)** The outcomes that make up the event *A* are 1, 2, 3, 4, 5, 6, 7. $P(A) = 0.32$. **(b)** The outcomes that make up the event *B* are 0, 1, 2, 3, 4. $P(B) = 0.93$. **(c)** The event "*A* and *B*" has the outcomes 1, 2, 3, 4. So $P(A \text{ and } B) = 0.25$. This is not the same as $P(A) \cdot P(B) = 0.2976$ because *A* and *B* are not independent.

**6.9 (a)**

| X: | −\$1 | \$2 |
|---|---|---|
| Probability: | 0.75 | 0.25 |

**(b)** $E(X) = -\$0.25$. If the player makes many \$1 bets, he will lose about \$0.25 per \$1 bet, on average.

**6.10 (a)**

| Y: | \$300 | −199,700 |
|---|---|---|
| Probability: | 0.9998 | 0.0002 |

**(b)** $E(Y) = \$260$. If the company insures many, many homes, they will earn about \$260 per policy, on average.

**6.11** $\mu_X = 2.1$. If many, many undergraduates performed this task, they will make about 2.1 nonword errors, on average.

**6.12** $\mu_X = 2.68$. If many, many kids participated in this experiment, they would play with about 2.68 toys, on average.

**6.13 (a)** This distribution is symmetric and 5 is located at the center. **(b)** According to Benford's law, $E(X) = 3.441$. To detect a fake expense report, compute the sample mean of the first digits. A value closer to 5 suggests a fake report and a value near 3.441 is consistent with a truthful report. **(c)** $P(Y > 6) = 3/9 = 0.333$. Under Benford's law, $P(X > 6) = 0.155$. To detect a fake expense report, compute the proportion of first digits that begin with 7, 8, or 9. A value closer to 0.333 suggests a fake report and a value closer to 0.155 is consistent with a truthful report.

**6.14 (a)** The company has collected 5 payments of \$250 each (for a total of \$1250) and has to pay out \$100,000. **(b)** 0.99058 **(c)** $\mu_Y = \$303.35$. If the company insures many, many people, they will make about \$303.35 per life insurance policy, on average.

**6.15** $\sigma_X = \sqrt{1.29} = 1.1358$. The number of nonword errors in a randomly selected essay will typically differ from the mean (2.1) by about 1.14 words.

**6.16** $\sigma_X = \sqrt{1.7176} = 1.3106$. The number of toys played with by a child will typically differ from the mean (2.68) by about 1.31 toys.

**6.17 (a)** $\sigma_Y = \sqrt{6.667} = 2.58$

**(b)** $\sigma_X = \sqrt{6.0605}$ This would not be the best way to tell the difference between a fake and a real expense report because the standard deviations are similar.

**6.18 (a)** Even though the company will lose a large amount of money on a small number of policyholders who die, it will gain a small amount from many thousands of 21-year-old men. Because they insure so many men, the insurance company can expect to make close to \$303.35 per insurance policy, on average.

**(b)** $\sigma_Y = \sqrt{94{,}236{,}826.64} = \$9{,}707.57$

**6.19 (a)** See the histograms on the next page. The distribution of the number of rooms is roughly symmetric for owners and skewed to the right for renters. Renter-occupied units tend to have fewer rooms than owner-occupied units. There is more variability in the number of rooms for owner-occupied units.

15. **Spell-checking** Refer to Exercise 3. Calculate and interpret the standard deviation of the random variable X. Show your work.
pg 353

16. **Kids and toys** Refer to Exercise 4. Calculate and interpret the standard deviation of the random variable X. Show your work.

17. **Benford's law and fraud** Refer to Exercise 13. It might also be possible to detect an employee's fake expense records by looking at the variability in the first digits of those expense amounts.

(a) Calculate the standard deviation $\sigma_Y$. This gives us an idea of how much variation we'd expect in the employee's expense records if he assumed that first digits from 1 to 9 were equally likely.

(b) Now calculate the standard deviation of first digits that follow Benford's law (Exercise 5). Would using standard deviations be a good way to detect fraud? Explain.

18. **Life insurance**

(a) It would be quite risky for you to insure the life of a 21-year-old friend under the terms of Exercise 14. There is a high probability that your friend would live and you would gain $1250 in premiums. But if he were to die, you would lose almost $100,000. Explain carefully why selling insurance is not risky for an insurance company that insures many thousands of 21-year-old men.

(b) The risk of an investment is often measured by the standard deviation of the return on the investment. The more variable the return is, the riskier the investment. We can measure the great risk of insuring a single person's life in Exercise 14 by computing the standard deviation of the income Y that the insurer will receive. Find $\sigma_Y$ using the distribution and mean found in Exercise 14.

19. **Housing in San Jose** How do rented housing units differ from units occupied by their owners? Here are the distributions of the number of rooms for owner-occupied units and renter-occupied units in San Jose, California:[9]

| | Number of Rooms | | | | | | | | | |
|---|---|---|---|---|---|---|---|---|---|---|
| | 1 | 2 | 3 | 4 | 5 | 6 | 7 | 8 | 9 | 10 |
| Owned | 0.003 | 0.002 | 0.023 | 0.104 | 0.210 | 0.224 | 0.197 | 0.149 | 0.053 | 0.035 |
| Rented | 0.008 | 0.027 | 0.287 | 0.363 | 0.164 | 0.093 | 0.039 | 0.013 | 0.003 | 0.003 |

Let X = the number of rooms in a randomly selected owner-occupied unit and Y = the number of rooms in a randomly chosen renter-occupied unit.

(a) Make histograms suitable for comparing the probability distributions of X and Y. Describe any differences that you observe.

(b) Find the mean number of rooms for both types of housing unit. Explain why this difference makes sense.

(c) Find and interpret the standard deviations of both X and Y.

20. **Size of American households** In government data, a household consists of all occupants of a dwelling unit, while a family consists of two or more persons who live together and are related by blood or marriage. So all families form households, but some households are not families. Here are the distributions of household size and family size in the United States:

| | Number of Persons | | | | | | |
|---|---|---|---|---|---|---|---|
| | 1 | 2 | 3 | 4 | 5 | 6 | 7 |
| Household probability | 0.25 | 0.32 | 0.17 | 0.15 | 0.07 | 0.03 | 0.01 |
| Family probability | 0 | 0.42 | 0.23 | 0.21 | 0.09 | 0.03 | 0.02 |

Let X = the number of people in a randomly selected U.S. household and Y = the number of people in a randomly chosen U.S. family.

(a) Make histograms suitable for comparing the probability distributions of X and Y. Describe any differences that you observe.

(b) Find the mean for each random variable. Explain why this difference makes sense.

(c) Find and interpret the standard deviations of both X and Y.

21. **Random numbers** Let X be a number between 0 and 1 produced by a random number generator. Assuming that the random variable X has a uniform distribution, find the following probabilities:

(a) $P(X > 0.49)$

(b) $P(X \geq 0.49)$

(c) $P(0.19 \leq X < 0.37 \text{ or } 0.84 < X \leq 1.27)$

22. **Random numbers** Let Y be a number between 0 and 1 produced by a random number generator. Assuming that the random variable Y has a uniform distribution, find the following probabilities:

(a) $P(Y \leq 0.4)$

(b) $P(Y < 0.4)$

(c) $P(0.1 < Y \leq 0.15 \text{ or } 0.77 \leq Y < 0.88)$

23. **Running a mile** A study of 12,000 able-bodied male students at the University of Illinois found that their times for the mile run were approximately Normal with mean 7.11 minutes and standard deviation 0.74 minute.[10] Choose a student at random from this group and call his time for the mile Y. Find $P(Y < 6)$ and interpret the result.
pg 357

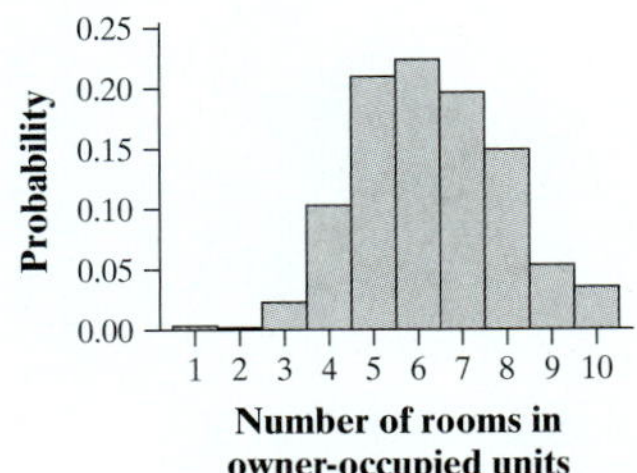


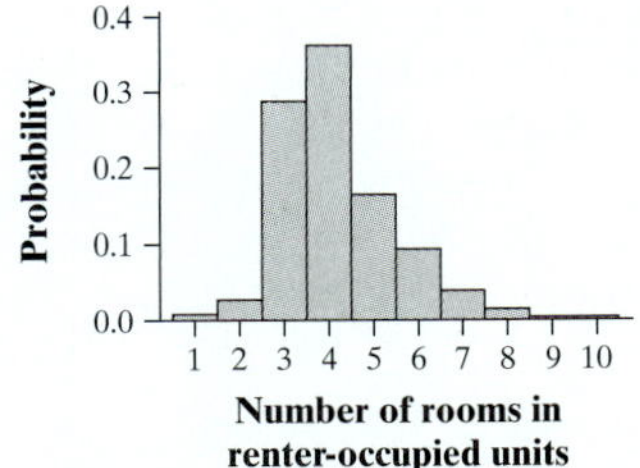


**(b)** Owner: $\mu_X = 6.284$ rooms. Renter: $\mu_Y = 4.187$ rooms. Single people and younger people are more likely to rent and need less space than people with families.

**(c)** $\sigma_X = \sqrt{2.68934} = 1.6399$. The number of rooms in a randomly selected owner-occupied unit will typically differ from the mean (6.284) by about 1.6399 rooms. $\sigma_Y = \sqrt{1.71003} = 1.3077$. The number of rooms in a randomly selected renter-occupied unit will typically differ from the mean (4.187) by about 1.3077 rooms.

**6.20 (a)** Both distributions are skewed to the right. The center for the "family" distribution is greater than the center for the "household" distribution, but the spread of the "family" distribution is less than the spread of the "household" distribution. Also, the event $X = 1$ has a much higher probability in the "household" distribution.

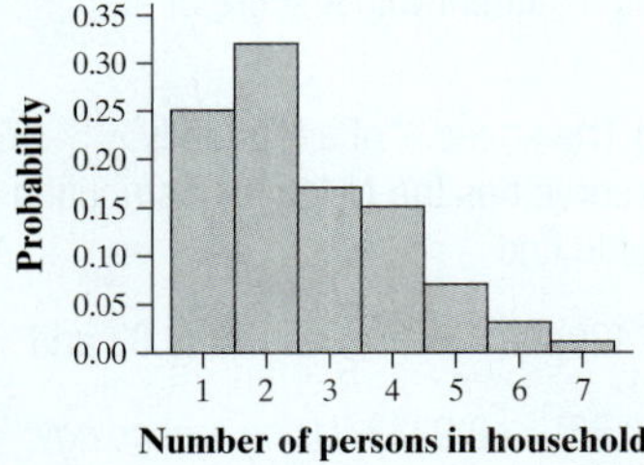


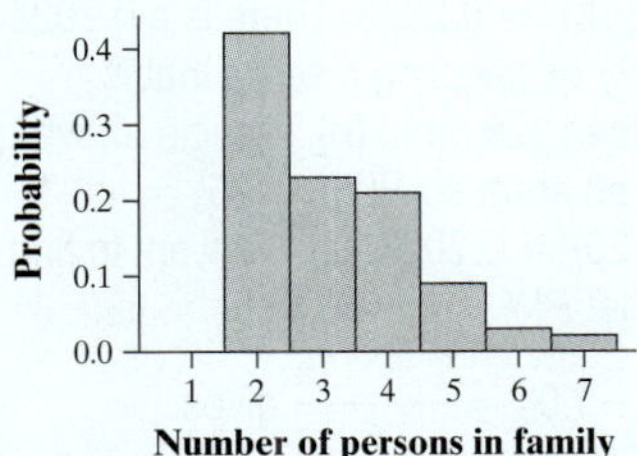


**(b)** Household: $\mu_X = 2.6$ people. Family: $\mu_Y = 3.14$ people. The family mean is higher because there are no values equal to 1 in the family distribution, but a value of 1 is very common in the household distribution.

**(c)** $\sigma_X = \sqrt{2.02} = 1.421$. The number of people in a randomly selected household will typically differ from the mean (2.6) by about 1.421 people. $\sigma_Y = \sqrt{1.5604} = 1.249$. The number of people in a randomly selected family will typically differ from the mean (3.14) by about 1.249 people.

**6.21 (a)** $P(X > 0.49) = 0.51$ **(b)** $P(X \geq 0.49) = 0.51$ **(c)** $P(0.19 \leq X < 0.37 \text{ or } 0.84 < X \leq 1.27) = 0.18 + 0.16 = 0.34$

**6.22 (a)** $P(Y \leq 0.4) = 0.4$ **(b)** $P(Y < 0.4) = 0.4$ **(c)** $P(0.1 < Y \leq 0.15 \text{ or } 0.77 \leq Y < 0.88) = 0.05 + 0.11 = 0.16$

**6.23** The time Y of a randomly chosen student has the $N(7.11, 0.74)$ distribution. We want to find $P(Y < 6)$.

$$z = \frac{6 - 7.11}{0.74} = -1.50 \text{ and}$$

$P(Z < -1.50) = 0.0668$. *Using technology*: `normalcdf(lower:−1000,upper:6,μ:7.11,σ:0.74)` $= 0.0668$. There is about a 7% chance that this student will run the mile in under 6 minutes.

**6.24** The score $X$ of a randomly chosen student has the $N(6.8, 1.6)$ distribution. We want to find $P(X \geq 9)$. $z = \frac{9 - 6.8}{1.6} = 1.38$ and $P(Z \geq 1.38) = 0.0838$. *Using technology*: `normalcdf(lower:9,upper:1000,μ:6.8,σ:1.6)` $= 0.0846$. There is a 0.0846 probability of selecting a student with a score of at least 9.

**6.25 (a)** The speed $Y$ of a randomly chosen serve has the $N(115, 6)$ distribution. We want to find $P(Y > 120)$. $z = \frac{120 - 115}{6} = 0.83$ and $P(Z > 0.83) = 0.2033$. *Using technology*: `normalcdf(lower:120,upper:1000,μ:115,σ:6)` $= 0.2023$. There is a 0.2023 probability of selecting a serve that is greater than 120 mph. **(b)** The line above 120 has no area, so $P(Y \geq 120) = P(Y > 120) = 0.2023$. **(c)** We want to find $c$ such that $P(Y \leq c) = 0.15$. Solving $-1.04 = \frac{c - 115}{6}$ gives $c = 108.76$. *Using technology*: `invNorm(area:0.15,μ:115,σ:6)` $=$ 108.78. 15% of Nadal's serves will be less than or equal to 108.78 mph.

**6.26 (a)** The length $X$ of a randomly chosen woman's pregnancy has the $N(266, 16)$ distribution. We want to find $P(X \geq 240)$. $z = \frac{240 - 266}{16} = -1.63$ and $P(Z \geq -1.63) = 0.9484$. *Using technology*: `normalcdf(lower:240,upper:1000,μ:266,σ:16)` $= 0.9479$. There is a 0.9479 probability of selecting a woman whose pregnancy lasts at least 240 days. **(b)** The line above 240 has no area, so $P(X \geq 240) = P(X > 240) = 0.9479$. **(c)** We want to find $c$ such that $P(X \geq c) = 0.20$. Solving $0.84 = \frac{c - 266}{16}$ gives $c = 279.44$. *Using technology*: `invNorm(area:0.80,μ:266,σ:16)` $=$ 279.47. 20% of pregnancies last at least 279.47 days.

**6.27** b

**6.28** b

**6.29** c

**6.30** c

24. **ITBS scores** The Normal distribution with mean $\mu = 6.8$ and standard deviation $\sigma = 1.6$ is a good description of the Iowa Test of Basic Skills (ITBS) vocabulary scores of seventh-grade students in Gary, Indiana. Call the score of a randomly chosen student X for short. Find $P(X \geq 9)$ and interpret the result.

25. **Ace!** Professional tennis player Rafael Nadal hits the ball extremely hard. His first-serve speeds follow a Normal distribution with mean 115 miles per hour (mph) and standard deviation 6 mph. Choose one of Nadal's first serves at random. Let $Y$ = its speed, measured in miles per hour.
   (a) Find $P(Y > 120)$ and interpret the result.
   (b) What is $P(Y \geq 120)$? Explain.
   (c) Find the value of $c$ such that $P(Y \leq c) = 0.15$. Show your work.

26. **Pregnancy length** The length of human pregnancies from conception to birth follows a Normal distribution with mean 266 days and standard deviation 16 days. Choose a pregnant woman at random. Let $X$ = the length of her pregnancy.
   (a) Find $P(X \geq 240)$ and interpret the result.
   (b) What is $P(X > 240)$? Explain.
   (c) Find the value of $c$ such that $P(X \geq c) = 0.20$. Show your work.

***Multiple choice: Select the best answer for Exercises 27 to 30.***

*Exercises 27 to 29 refer to the following setting.* Choose an American household at random and let the random variable $X$ be the number of cars (including SUVs and light trucks) they own. Here is the probability model if we ignore the few households that own more than 5 cars:

| Number of cars X: | 0 | 1 | 2 | 3 | 4 | 5 |
|---|---|---|---|---|---|---|
| Probability: | 0.09 | 0.36 | 0.35 | 0.13 | 0.05 | 0.02 |

27. What's the expected number of cars in a randomly selected American household?
   (a) 1.00 (b) 1.75 (c) 1.84 (d) 2.00 (e) 2.50

28. The standard deviation of $X$ is $\sigma_X = 1.08$. If many households were selected at random, which of the following would be the best interpretation of the value 1.08?
   (a) The mean number of cars would be about 1.08.
   (b) The number of cars would typically be about 1.08 from the mean.
   (c) The number of cars would be at most 1.08 from the mean.
   (d) The number of cars would be within 1.08 from the mean about 68% of the time.
   (e) The mean number of cars would be about 1.08 from the expected value.

29. About what percentage of households have a number of cars within 2 standard deviations of the mean?
   (a) 68% (b) 71% (c) 93% (d) 95% (e) 98%

30. A deck of cards contains 52 cards, of which 4 are aces. You are offered the following wager: Draw one card at random from the deck. You win $10 if the card drawn is an ace. Otherwise, you lose $1. If you make this wager very many times, what will be the mean amount you win?
   (a) About −$1, because you will lose most of the time.
   (b) About $9, because you win $10 but lose only $1.
   (c) About −$0.15; that is, on average you lose about 15 cents.
   (d) About $0.77; that is, on average you win about 77 cents.
   (e) About $0, because the random draw gives you a fair bet.

*Exercises 31 to 34 refer to the following setting.* Many chess masters and chess advocates believe that chess play develops general intelligence, analytical skill, and the ability to concentrate. According to such beliefs, improved reading skills should result from study to improve chess-playing skills. To investigate this belief, researchers conducted a study. All of the subjects in the study participated in a comprehensive chess program, and their reading performances were measured before and after the program. The graphs and numerical summaries below provide information on the subjects' pretest scores, posttest scores, and the difference (post – pre) between these two scores.

Descriptive Statistics: Pretest, Posttest, Post – pre

| Variable | N | Mean | Median | StDev | Min | Max | $Q_1$ | $Q_3$ |
|---|---|---|---|---|---|---|---|---|
| Pretest | 53 | 57.70 | 58.00 | 17.84 | 23.00 | 99.00 | 44.50 | 70.50 |
| Posttest | 53 | 63.08 | 64.00 | 18.70 | 28.00 | 99.00 | 48.00 | 76.00 |
| Post-pre | 53 | 5.38 | 3.00 | 13.02 | -19.00 | 42.00 | -3.50 | 14.00 |

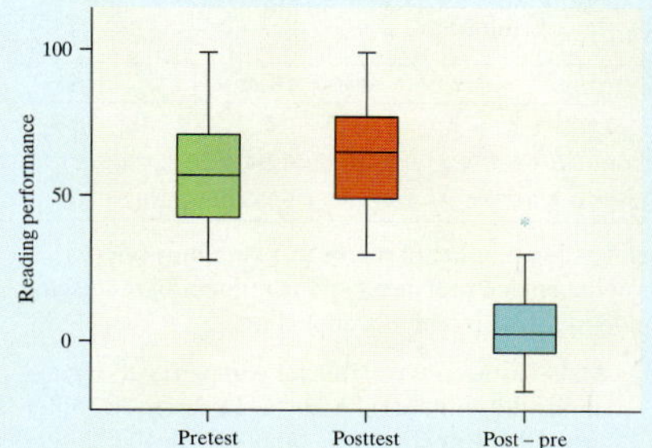

**31. Better readers?** (1.3) Did students have higher reading scores after participating in the chess program? Give appropriate statistical evidence to support your answer.

**32. Chess and reading** (4.3) If the study found a statistically significant improvement in reading scores, could you conclude that playing chess causes an increase in reading skills? Justify your answer.

*Some graphical and numerical information about the relationship between pretest and posttest scores is provided below.*

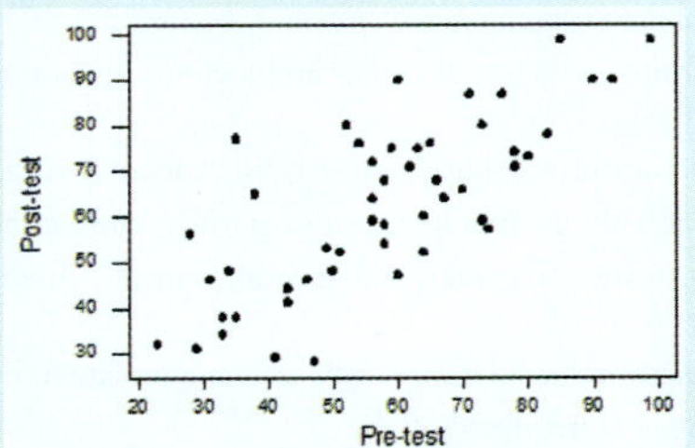


Residual
30
20
10
0
-10
-20
-30
30 40 50 60 70 80 90 100
Fitted Value

Regression Analysis: Posttest versus Pretest

| Predictor | Coef | SE Coef | T | P |
|---|---|---|---|---|
| Constant | 17.897 | 5.889 | 3.04 | 0.004 |
| Pretest | 0.78301 | 0.09758 | 8.02 | 0.000 |
| S = 12.55 | R-Sq = 55.8% | | R-Sq(adj) = 54.9% | |

**33. Predicting posttest scores** (3.2) What is the equation of the linear regression model relating posttest and pretest scores? Define any variables used.

**34. How well does it fit?** (3.2) Discuss what $s$, $r^2$, and the residual plot tell you about this linear regression model.

**6.31** Yes. The mean difference (post − pre) was 5.38 and the median difference was 3. This means that at least half of the students improved their reading scores.

**6.32** No. We do not have a control group that did not participate in the chess program for comparison. It may be that children of this age improve their reading scores for other reasons (e.g., regular school) and that the chess program had nothing to do with their improvement.

**6.33** predicted posttest = 17.897 + 0.78301(pretest).

**6.34** This linear model is appropriate because the residual plot does not show any leftover patterns. 55.8% of the variation in posttest score is accounted for by this linear model relating the posttest score to pretest score. Finally, when using this model to predict posttest scores from pretest scores, the actual values will typically be about 12.55 from the predicted values.

# 6.2 Transforming and Combining Random Variables

**WHAT YOU WILL LEARN** By the end of the section, you should be able to:

- Describe the effects of transforming a random variable by adding or subtracting a constant and multiplying or dividing by a constant.
- Find the mean and standard deviation of the sum or difference of independent random variables.
- Find probabilities involving the sum or difference of independent Normal random variables.

In Section 6.1, we looked at several examples of random variables and their probability distributions. We also saw that the mean $\mu_X$ and standard deviation $\sigma_X$ give us important information about a random variable. For instance, for $X$ = the amount gained on a single \$1 bet on red in a game of roulette, we already showed that $\mu_X = -\$0.05$. You can verify that the standard deviation is $\sigma_X = \$1.00$. That is, a player can expect to lose an average of 5 cents per \$1 bet if he plays many games. But if he plays only a few games, his actual gain could be much better or worse than this expected value.

**PD Section 6.2 Overview**

Watch the video overview of Section 6.2 (for teachers) by clicking on the link in the TE-book, opening the TRCD, or logging onto the book's companion Web site www.whfreeman.com/tps5e (teacher log-in required).

**TRM Section 6.2 Alternate Examples**

You can find the Alternate Examples for this section in Microsoft Word format by clicking on the link in the TE-book, opening the TRCD, or logging onto the book's companion Web site www.whfreeman.com/tps5e (teacher log-in required).

## Teaching Tip

Ask your students: If $X$ = sum of the outcomes of two rolls of a die and $Y$ = twice the outcome of one roll of a die, do $X$ and $Y$ have the same distribution? No. The possible values of $X$ are 2, 3, 4, 5, 6, 7, 8, 9, 10, 11, and 12, but the possible values for $Y$ are only 2, 4, 6, 8, 10, and 12.

## Teaching Tip

Remind students that the effect of multiplying by a constant is different for the variance. If we multiply each value in distribution by a positive constant $b$, then most measures of spread (range, *IQR*, standard deviation) are also multiplied by $b$. However, the variance is multiplied by $b^2$. Refer to the Think About It on page 366 for more details.

## Teaching Tip

If you are short on time and your students understood the effects of a linear transformation of data in Section 2.1, you can move through the first half of this section very quickly. Make sure to look at the examples, however, as we use the same context for examples later in the section when discussing combinations of random variables. Even if you don't speed up, make sure to emphasize that the rules are no different than in Section 2.1.

Would the player be better off playing one game of roulette with a \$2 bet on red or playing two games and betting \$1 on red each time? To find out, we need to compare the probability distributions of the random variables $Y$ = gain from a \$2 bet and $T$ = total gain from two \$1 bets. Which random variable (if either) has the higher expected gain in the long run? Which has the larger variability? By the end of this section, you'll be able to answer questions like these.

## Linear Transformations

In Chapter 2, we studied the effects of transformations on the shape, center, and spread of a distribution of data. Recall what we discovered:

1. *Adding (or subtracting) a constant:* Adding the same positive number $a$ to (subtracting $a$ from) each observation:
   - Adds $a$ to (subtracts $a$ from) measures of center and location (mean, median, quartiles, percentiles).
   - Does not change shape or measures of spread (range, *IQR*, standard deviation).
2. Multiplying (or dividing) each observation by the same positive number $b$:
   - Multiplies (divides) measures of center and location (mean, median, quartiles, percentiles) by $b$.
   - Multiplies (divides) measures of spread (range, *IQR*, standard deviation) by $b$.
   - Does not change the shape of the distribution.

How are the probability distributions of random variables affected by similar transformations to the values of the variable? For reasons that will be clear later, we'll start by considering multiplication (or division) by a constant.

**Effect of multiplying or dividing by a constant** Let's start with a simple example of a discrete random variable. Pete's Jeep Tours offers a popular half-day trip in a tourist area. There must be at least 2 passengers for the trip to run, and the vehicle will hold up to 6 passengers. The number of passengers $X$ on a randomly selected day has the following probability distribution.

| No. of passengers $x_i$: | 2 | 3 | 4 | 5 | 6 |
|---|---|---|---|---|---|
| Probability $p_i$: | 0.15 | 0.25 | 0.35 | 0.20 | 0.05 |

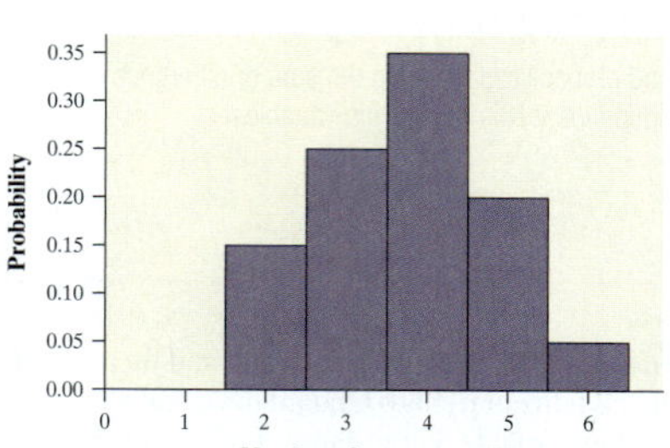


**FIGURE 6.5** The probability distribution of the random variable $X$ = the number of passengers on Pete's trip on a randomly chosen day.

Figure 6.5 shows a histogram of the probability distribution. Using what we learned in Section 6.1, the mean of $X$ is

$$\mu_X = \Sigma x_i p_i = (2)(0.15) + (3)(0.25) + (4)(0.35) + (5)(0.20) + (6)(0.05) = 3.75$$

That is, Pete expects an average of 3.75 passengers per trip. The variance of $X$ is given by

$$\sigma_X^2 = \Sigma(x_i - \mu_X)^2 p_i = (2 - 3.75)^2(0.15) + (3 - 3.75)^2(0.25) + \cdots + (6 - 3.75)^2(0.05) = 1.1875$$

So the standard deviation of $X$ is

$$\sigma_X = \sqrt{1.1875} = 1.0897$$

On a randomly selected day, the number of people on a trip typically differs from the mean by about 1.09 passengers.

### ALTERNATE EXAMPLE

### *El Dorado Community College*

El Dorado Community College considers a student to be full-time if he or she is taking between 12 and 18 units. The number of units $X$ that a randomly selected El Dorado Community College full-time student is taking in the fall semester has the following distribution:

| Number of units: | 12 | 13 | 14 | 15 | 16 | 17 | 18 |
|---|---|---|---|---|---|---|---|
| Probability: | 0.25 | 0.10 | 0.05 | 0.30 | 0.10 | 0.05 | 0.15 |

Here is a histogram of the probability distribution along with the mean and standard deviation:

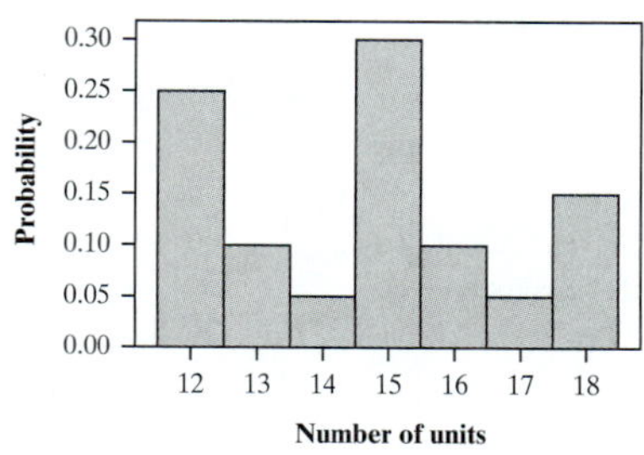


$$\mu_X = 12(0.25) + \cdots + 18(0.15) = 14.65$$

$$\sigma_X^2 = (12 - 14.65)^2(0.25) + \cdots + (18 - 14.65)^2(0.15) = 4.23$$

$$\sigma_X = \sqrt{4.23} = 2.06$$

At El Dorado Community College, the tuition for full-time students is \$50 per unit. That is, if $T$ = tuition charge for a randomly selected full-time student, $T = 50X$. Here is the probability distribution for $T$ and a histogram of the probability distribution:

| Tuition charge: | \$600 | \$650 | \$700 | \$750 | \$800 | \$850 | \$900 |
|---|---|---|---|---|---|---|---|
| Probability: | 0.25 | 0.10 | 0.05 | 0.30 | 0.10 | 0.05 | 0.15 |

## EXAMPLE Pete's Jeep Tours

### *Multiplying a random variable by a constant*

Pete charges \$150 per passenger. Let C = the total amount of money that Pete collects on a randomly selected trip. Because the amount of money Pete collects is just \$150 times the number of passengers, we can write $C = 150X$. From the probability distribution of X, we can see that the chance of having two people ($X = 2$) on the trip is 0.15. In that case, $C = (150)(2) = 300$. So one possible value of C is \$300, and its corresponding probability is 0.15. If $X = 3$, then $C = (150)(3) = 450$, and the corresponding probability is 0.25. Thus, the probability distribution of C is

| Total collected $c_i$: | 300 | 450 | 600 | 750 | 900 |
|---|---|---|---|---|---|
| Probability $p_i$: | 0.15 | 0.25 | 0.35 | 0.20 | 0.05 |

Figure 6.6 is a histogram of this probability distribution.

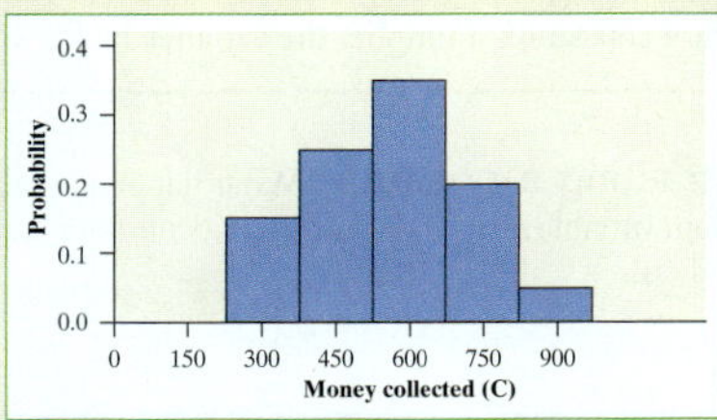


**FIGURE 6.6** The probability distribution of the random variable $C$ = the amount of money Pete collects from his trip on a randomly chosen day.

The mean of C is $\mu_C = \Sigma c_i p_i = (300)(0.15) + (450)(0.25) + \cdots + (900)(0.05) = 562.50$.

On average, Pete will collect a total of \$562.50 from the half-day trip. The variance of C is

$$\sigma_C^2 = \Sigma(c_i - \mu_C)^2 p_i$$
$$= (300 - 562.50)^2(0.15) + (450 - 562.50)^2(0.25)$$
$$+ \cdots + (900 - 562.50)^2(0.05) = 26{,}718.75$$

So the standard deviation of C is $\sigma_C = \sqrt{26{,}718.75} = \$163.46$.

In the previous example, the random variable C was obtained by multiplying the values of our earlier random variable X by 150. To understand the effect of multiplying by a constant, let's compare the probability distributions of the random variables X and C.

**Shape:** The two probability distributions have the same shape.
**Center:** The mean of X is $\mu_X = 3.75$. The mean of C is $\mu_C = 562.50$, which is (150)(3.75). That is, $\mu_C = 150\mu_X$.
**Spread:** The standard deviation of X is $\sigma_X = 1.0897$. The standard deviation of C is $\sigma_C = 163.46$, which is (150)(1.0897). That is, $\sigma_C = 150\sigma_X$.

Let's summarize what we've learned so far about transforming a random variable.

**ALTERNATE EXAMPLE (*continued*)**

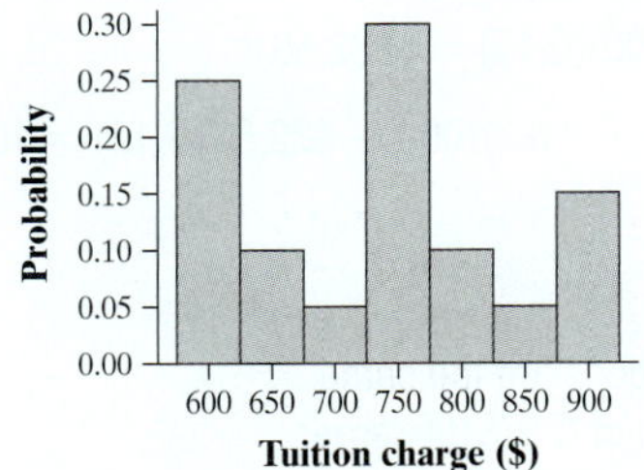


The mean and standard deviation of this distribution are

$$\mu_T = 600(0.25) + \cdots + (900)(0.15) = \$732.50$$
$$\sigma_T^2 = (600 - 732.5)^2(0.25) + \cdots + (900 - 732.5)^2(0.15) = 10{,}568$$
$$\sigma_T = \sqrt{10{,}568} = \$103$$

Notice that

- The shapes of both distributions are the same.
- The mean of the distribution of $T$ is 50 times bigger than the mean of $X$: $732.50 = 50(14.65)$. That is, $\mu_T = 50\mu_X$.
- The standard deviation of $T$ is 50 times bigger than the standard deviation of $X$: $103 = 50(2.06)$. That is, $\sigma_T = 50\sigma_X$.

## Teaching Tip

Emphasize that these results are exactly the same as we discovered in Chapter 2.



### EFFECT ON A RANDOM VARIABLE OF MULTIPLYING (OR DIVIDING) BY A CONSTANT

Multiplying (or dividing) each value of a random variable by a positive number $b$:

- Multiplies (divides) measures of center and location (mean, median, quartiles, percentiles) by $b$.
- Multiplies (divides) measures of spread (range, *IQR*, standard deviation) by $b$.
- Does not change the shape of the distribution.

As with data, if we multiply a random variable by a negative constant $b$, our common measures of spread are multiplied by $|b|$.

**How does multiplying by a constant affect the variance?** For Pete's Jeep Tours, the variance of the number of passengers on a randomly selected trip is $\sigma_X^2 = 1.1875$. The variance of the total amount of money that Pete collects from such a trip is $\sigma_C^2 = 26{,}718.75$. That's $(22{,}500)(1.1875)$. So $\sigma_C^2 = 22{,}500\sigma_X^2$. Where did 22,500 come from? It's just $(150)^2$. In other words, $\sigma_C^2 = (150)^2\sigma_X^2$. Multiplying a random variable by a constant $b$ multiplies the variance by $b^2$.

**Effect of adding or subtracting a constant** What happens to the probability distribution of a random variable if we add or subtract a constant? Let's return to Pete's Jeep Tours to find out.

## EXAMPLE Pete's Jeep Tours

### *Effect of adding or subtracting a constant*

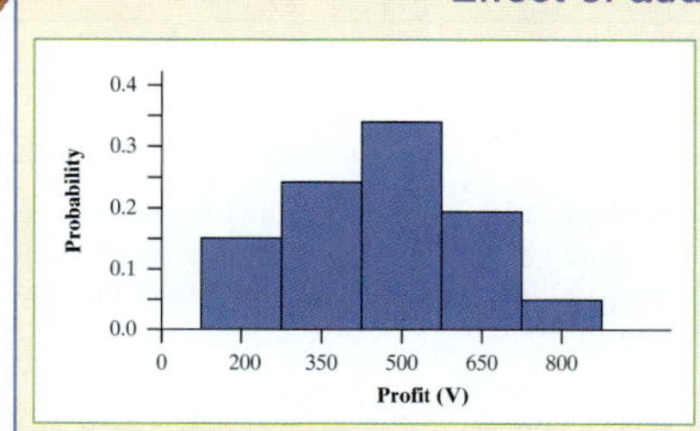


**FIGURE 6.7** The probability distribution of the random variable $V$ = the profit that Pete makes from his trip on a randomly chosen day.

It costs Pete \$100 to buy permits, gas, and a ferry pass for each half-day trip. The amount of profit $V$ that Pete makes from the trip is the total amount of money $C$ that he collects from passengers minus \$100. That is, $V = C - 100$. If Pete has only two passengers on the trip ($X = 2$), then $C = 2(150) = 300$ and $V = 200$. From the probability distribution of $C$, the chance that this happens is 0.15. So the smallest possible value of $V$ is \$200; its corresponding probability is 0.15. If $X = 3$, then $C = 450$ and $V = 350$, and the corresponding probability is 0.25. The probability distribution of $V$ is

| Profit $v_i$: | 200 | 350 | 500 | 650 | 800 |
|---|---|---|---|---|---|
| Probability $p_i$: | 0.15 | 0.25 | 0.35 | 0.20 | 0.05 |

Figure 6.7 shows a histogram of this probability distribution.

## ALTERNATE EXAMPLE

### *El Dorado Community College*

In addition to tuition charges, each full-time student at El Dorado Community College is assessed student fees of \$100 per semester. If $C$ = overall cost for a randomly selected full-time student, $C = 100 + T$.

Here is the probability distribution for $C$ and a histogram of the probability distribution:

| Overall cost: | \$700 | \$750 | \$800 | \$850 | \$900 | \$950 | \$1000 |
|---|---|---|---|---|---|---|---|
| Probability: | 0.25 | 0.10 | 0.05 | 0.30 | 0.10 | 0.05 | 0.15 |

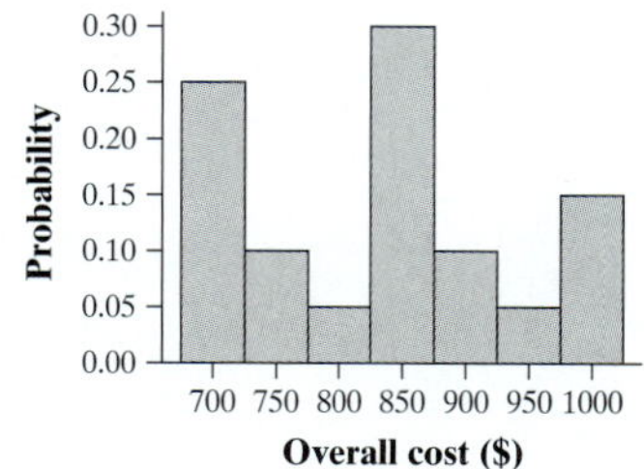


The mean and standard deviation of this distribution are

$$\mu_C = 700(0.25) + \cdots + (1000)(0.15) = \$832.50$$

$$\sigma_C^2 = (700 - 832.5)^2(0.25) + \cdots + (1000 - 832.5)^2(0.15) = 10{,}568$$

$$\sigma_C = \sqrt{10{,}568} = \$103$$

Notice that

- The shapes of both distributions are the same.
- The mean of the distribution of $C$ is \$100 larger than the mean of $T$: $832.5 = 100 + 732.50$. That is, $\mu_C = 100 + \mu_T$.
- The standard deviation of $C$ is the *same* as the standard deviation of $T$. That is, $\sigma_C = \sigma_T$.

The mean of $V$ is $\mu_V = \Sigma v_i p_i = (200)(0.15) + (350)(0.25) + \cdots + (800)(0.05) = 462.50$. On average, Pete will make a profit of \$462.50 from the trip. The variance of $V$ is

$$\sigma_V^2 = \Sigma(v_i - \mu_V)^2 p_i$$
$$= (200 - 462.50)^2(0.15) + (350 - 462.50)^2(0.25)$$
$$+ \cdots + (800 - 462.50)^2(0.05) = 26{,}718.75$$

So the standard deviation of $V$ is

$$\sigma_V = \sqrt{26{,}718.75} = \$163.46$$

It's fairly clear from the previous example that subtracting 100 from the values of the random variable $C$ just shifts the probability distribution to the left by 100. This transformation decreases the mean by 100 (from \$562.50 to \$462.50) but doesn't change the standard deviation (\$163.46) or the shape. These results can be generalized for any random variable.

**EFFECT ON A RANDOM VARIABLE OF ADDING (OR SUBTRACTING) A CONSTANT**

Adding the same positive number $a$ to (subtracting $a$ from) each value of a random variable:

- Adds $a$ to (subtracts $a$ from) measures of center and location (mean, median, quartiles, percentiles).
- Does not change shape or measures of spread (range, *IQR*, standard deviation).

## CHECK YOUR UNDERSTANDING

A large auto dealership keeps track of sales made during each hour of the day. Let $X$ = the number of cars sold during the first hour of business on a randomly selected Friday. Based on previous records, the probability distribution of $X$ is as follows:

| Cars sold: | 0 | 1 | 2 | 3 |
|---|---|---|---|---|
| Probability: | 0.3 | 0.4 | 0.2 | 0.1 |

The random variable $X$ has mean $\mu_X = 1.1$ and standard deviation $\sigma_X = 0.943$.

1. Suppose the dealership's manager receives a \$500 bonus from the company for each car sold. Let $Y$ = the bonus received from car sales during the first hour on a randomly selected Friday. Find the mean and standard deviation of $Y$.
2. To encourage customers to buy cars on Friday mornings, the manager spends \$75 to provide coffee and doughnuts. The manager's net profit $T$ on a randomly selected Friday is the bonus earned minus this \$75. Find the mean and standard deviation of $T$.

**Putting it all together: Adding/subtracting and multiplying/dividing** What happens if we transform a random variable by both adding or subtracting a constant and multiplying or dividing by a constant? Let's consider

### Teaching Tip

Emphasize that these results are exactly the same as we discovered in Chapter 2.

### ANSWERS TO CYU

1. $Y = 500X$. $\mu_Y = 500(1.1) = \$550$. $\sigma_Y = 500(0.943) = \$471.50$.
2. $T = Y - 75$. $\mu_T = 550 - 75 = \$475$. $\sigma_T = \$471.50$.

## AP® Exam Common Error

When students are asked about the effect of a linear transformation on summary statistics, some students forget that adding (subtracting) a constant to (from) every value in the distribution has no effect on measures of spread, including the standard deviation, variance, range, and interquartile range.

## Teaching Tip

Emphasize that these rules don't just apply to means and standard deviations by asking questions about other summary statistics such as the median and *IQR*.

Pete's Jeep Tours again. We could have gone directly from the number of passengers $X$ on a randomly selected jeep tour to Pete's profit $V$ with the equation $V = 150X - 100$ or, equivalently, $V = -100 + 150X$. This **linear transformation** of the random variable $X$ includes both of the transformations that we performed earlier: (1) multiplying by 150 and (2) subtracting 100. (In general, a linear transformation can be written in the form $Y = a + bX$, where $a$ and $b$ are constants.) The net effect of this sequence of transformations is as follows:

For the linear transformation $V = -100 + 150X$, it would *not* be correct to apply the transformations in the reverse order: subtract 100 and then multiply by 150. Doing so would yield the same standard deviation but a different (wrong) mean. Just follow the order of operations from algebra.

**Shape:** Neither transformation changes the shape of the probability distribution.
**Center:** The mean of $X$ is multiplied by 150 and then decreased by 100; that is, $\mu_V = 150\mu_X - 100 = -100 + 150\mu_X$.
**Spread:** The standard deviation of $X$ is multiplied by 150 and is unchanged by the subtraction: $\sigma_V = 150\sigma_X$.

This logic generalizes to any linear transformation.

### EFFECTS OF A LINEAR TRANSFORMATION ON A RANDOM VARIABLE

If $Y = a + bX$ is a linear transformation of the random variable $X$, then

- the probability distribution of $Y$ has the same shape as the probability distribution of $X$ if $b > 0$.
- $\mu_Y = a + b\mu_X$.
- $\sigma_Y = |b|\sigma_X$ (because $b$ could be a negative number).

Can you see why this is called a "linear" transformation? The equation describing the sequence of transformations has the form $Y = a + bX$, which you should recognize as a linear equation.

The bottom two rules in the summary box don't just apply to means and standard deviations. Linear transformations have similar effects on other measures of center or location (median, quartiles, percentiles) and spread (range, *IQR*). *Whether we're dealing with data or random variables, the effects of a linear transformation are the same.* Note that these results apply to both discrete and continuous random variables.

## EXAMPLE The Baby and the Bathwater

### *Linear transformations*

PROBLEM: One brand of bathtub comes with a dial to set the water temperature. When the "babysafe" setting is selected and the tub is filled, the temperature $X$ of the water follows a Normal distribution with a mean of 34°C and a standard deviation of 2°C.

(a) Define the random variable $Y$ to be the water temperature in degrees Fahrenheit (recall that $F = \frac{9}{5}C + 32$) when the dial is set on "babysafe." Find the mean and standard deviation of $Y$.

(b) According to Babies R Us, the temperature of a baby's bathwater should be between 90°F and 100°F. Find the probability that the water temperature on a randomly selected day when the "babysafe" setting is used meets the Babies R Us recommendation. Show your work.

## ALTERNATE EXAMPLE

### *Scaling a test*

PROBLEM: In a large introductory statistics class, the distribution of raw scores on a test $X$ follows a Normal distribution with a mean of 17.2 and a standard deviation of 3.8. The professor decides to scale the scores by multiplying the raw scores by 4 and adding 10.

**(a)** Define the variable $Y$ to be the scaled score of a randomly selected student from this class. Find the mean and standard deviation of $Y$.

**(b)** What is the probability that a randomly selected student has a scaled test score of at least 90?

SOLUTION:

**(a)** Because $Y = 10 + 4X$, $\mu_Y = 10 + 4\mu_X = 10 + 4(17.2) = 78.8$ and $\sigma_Y = 4\sigma_X = 4(3.8) = 15.2$

**(b)** **Step 1: State the distribution and the values of interest.** Because linear transformations do not change the shape, $Y$ follows a $N(78.8, 15.2)$ distribution. We want to find $P(Y \geq 90)$ as shown below.

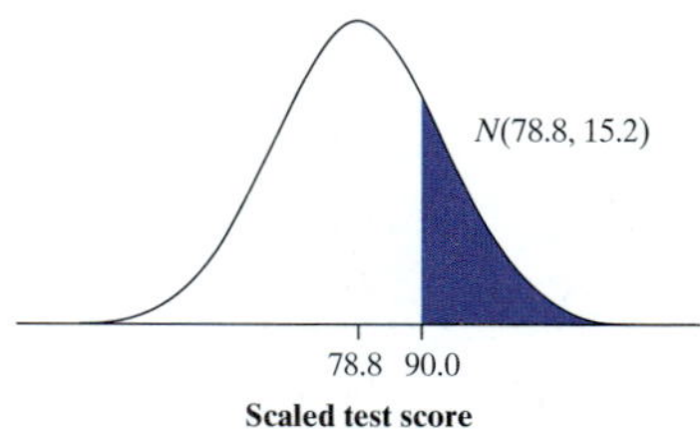


**Step 2: Perform calculations—show your work!** The standardized score for a scaled score of 90 is $z = \frac{90 - 78.8}{15.2} = 0.74$.

According to Table A, $P(Z < 0.74) = 0.7704$. Thus, $P(Y \geq 90) = 1 - 0.7704 = 0.2296$. *Using technology:* The command `normalcdf(lower:90,upper:1000,μ:78.8,σ:15.2)` gives an area of 0.2306.

**Step 3: Answer the question.** There is about a 23% chance that a randomly selected student has a scaled test score of at least 90.

SOLUTION:

(a) According to the formula for converting Celsius to Fahrenheit, $Y = \frac{9}{5}X + 32$. We could also write this in the form $Y = 32 + \frac{9}{5}X$. The mean of $Y$ is

$$\mu_Y = 32 + \frac{9}{5}\mu_X = 32 + \frac{9}{5}(34) = 93.2°\text{F}$$

The standard deviation of $Y$ is

$$\sigma_Y = \frac{9}{5}\sigma_X = \frac{9}{5}(2) = 3.6°\text{F}$$

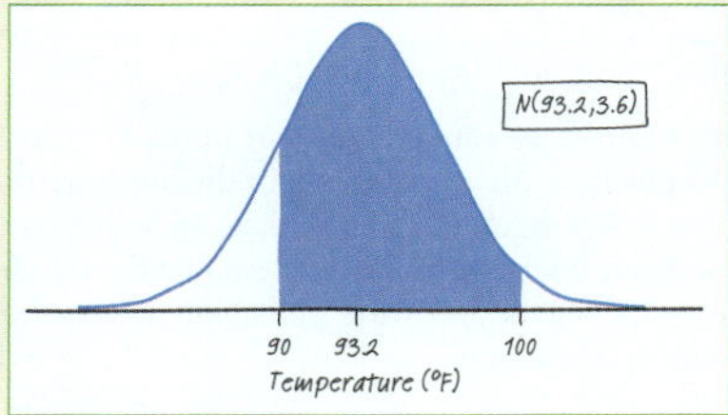


**FIGURE 6.8** The Normal probability distribution of the random variable $Y$ = the temperature (in °F) of the bathwater when the dial is set on "babysafe." The shaded area is the probability that the water temperature is between 90°F and 100°F.

(b) **Step 1: State the distribution and the values of interest.** The linear transformation doesn't change the shape of the probability distribution, so the random variable $Y$ is Normally distributed with a mean of 93.2 and a standard deviation of 3.6. We want to find 380$P(90 \leq Y \leq 100)$. The shaded area in Figure 6.8 shows the desired probability.

**Step 2: Perform calculations—show your work!** To find this area, we can standardize the boundary values and use Table A:

$$z = \frac{90 - 93.2}{3.6} = -0.89 \quad \text{and} \quad z = \frac{100 - 93.2}{3.6} = 1.89$$

Then $P(-0.89 \leq Z \leq 1.89) = 0.9706 - 0.1867 = 0.7839$.

*Using technology:* The command `normalcdf(lower:90, upper:100, μ:93.2, σ:3.6)` gives an area of 0.7835.

**Step 3: Answer the question.** There's about a 78% chance that the water temperature meets the recommendation on a randomly selected day.

**For Practice** *Try Exercise*  45

## Combining Random Variables

So far, we have looked at settings that involved a single random variable. Many interesting statistics problems require us to combine two or more random variables.

### EXAMPLE Pete's Jeeps and Erin's Adventures

*When one random variable isn't enough*

Earlier, we examined the probability distribution for the random variable $X$ = the number of passengers on a randomly selected half-day trip with Pete's Jeep Tours. Here's a brief recap:

| No. of passengers $x_i$: | 2 | 3 | 4 | 5 | 6 |
|---|---|---|---|---|---|
| Probability $p_i$: | 0.15 | 0.25 | 0.35 | 0.20 | 0.05 |

Mean: $\mu_X = 3.75$ Standard deviation: $\sigma_X = 1.0897$

**ALTERNATE EXAMPLE**

***El Dorado Community College***

El Dorado Community College also has a campus downtown, specializing in just a few fields of study. Full-time students at the downtown campus take only 3-unit classes. Let $Y$ = number of units taken in the fall semester by a randomly selected full-time student at the downtown campus. Here is the probability distribution of $Y$ and a probability histogram:

| Number of units: | 12 | 15 | 18 |
|---|---|---|---|
| Probability: | 0.3 | 0.4 | 0.3 |

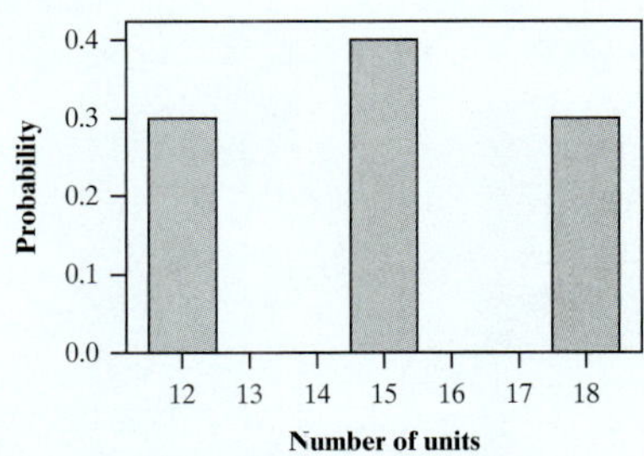


The mean of this distribution is $\mu_Y = 15$ units, the variance is $\sigma_Y^2 = 5.40$ units$^2$, and the standard deviation is $\sigma_Y = 2.3$ units.

If you were to randomly select one full-time student from the main campus and one full-time student from the downtown campus and add their number of units, the expected value of the sum ($S = X + Y$) would be $\mu_S = \mu_X + \mu_Y = 14.65 + 15 = 29.65$.

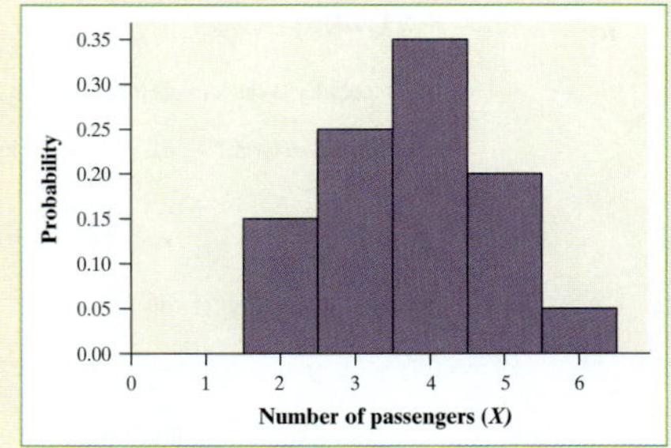


Pete's sister Erin, who lives near a tourist area in another part of the country, is impressed by the success of Pete's business. She decides to join the business, running tours on the same days as Pete in her slightly smaller vehicle, under the name Erin's Adventures. After a year of steady bookings, Erin discovers that the number of passengers Y on her half-day tours has the following probability distribution. Figure 6.9 displays this distribution as a histogram.

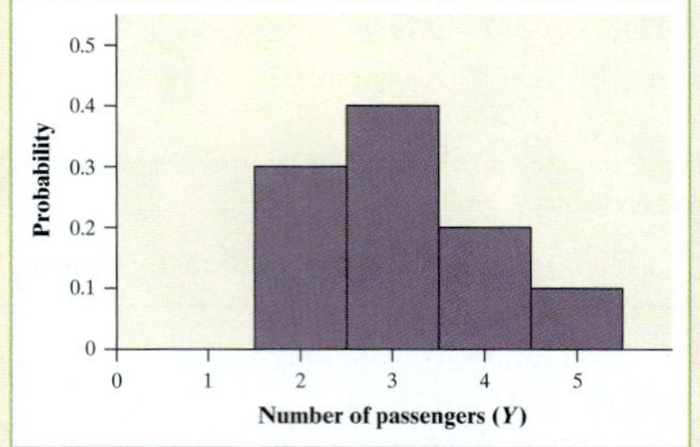


FIGURE 6.9 The probability distribution of the random variable $Y$ = the number of passengers on Erin's trip on a randomly chosen day.

| No. of passengers $y_i$: | 2 | 3 | 4 | 5 |
|---|---|---|---|---|
| Probability $p_i$: | 0.3 | 0.4 | 0.2 | 0.1 |

Mean: $\mu_Y = 3.10$ Standard deviation: $\sigma_Y = 0.943$

How many total passengers $T$ will Pete and Erin have on their tours on a randomly selected day? To answer this question, we need to know about the distribution of the random variable $T = X + Y$.

How many more or fewer passengers $D$ will Pete have than Erin on a randomly selected day? To answer this question, we need to know about the distribution of the random variable $D = X - Y$.

As the example suggests, we want to investigate what happens when we add or subtract random variables.

**Sums of random variables** How many total passengers $T$ can Pete and Erin expect to have on their tours on a randomly selected day? Because Pete averages $\mu_X = 3.75$ passengers per trip and Erin averages $\mu_Y = 3.10$ passengers per trip, they will average a total of $\mu_T = 3.75 + 3.10 = 6.85$ passengers per day. We can generalize this result for any two random variables as follows: if $T = X + Y$, then $\mu_T = \mu_X + \mu_Y$. In other words, the expected value (mean) of the sum of two random variables is equal to the sum of their expected values (means).

**MEAN OF THE SUM OF RANDOM VARIABLES**

For any two random variables X and Y, if $T = X + Y$, then the expected value of $T$ is

$$E(T) = \mu_T = \mu_X + \mu_Y$$

In general, the mean of the sum of several random variables is the sum of their means.

How much variability is there in the total number of passengers who go on Pete's and Erin's tours on a randomly chosen day? Let's think about the possible values of $T = X + Y$. The number of passengers X on Pete's tour is between 2 and 6, and the number of passengers Y on Erin's tour is between 2 and 5. So the total number of passengers $T$ is between 4 and 11. Thus, the range of $T$ is $11 - 4 = 7$. How is this value related to the ranges of X and Y? The range of X is 4 and the range of Y is 3, so

$$\text{range of } T = \text{range of } X + \text{range of } Y$$

That is, there's more variability in the values of $T$ than in the values of X or Y alone. This makes sense, because the variation in X and the variation in Y both contribute to the variation in $T$.

What about the standard deviation $\sigma_T$? If we had the probability distribution of the random variable $T$, then we could calculate $\sigma_T$. Let's try to construct this probability distribution starting with the smallest possible value, $T = 4$. The only way to get a total of 4 passengers is if Pete has $X = 2$ passengers and Erin has $Y = 2$ passengers. We know that $P(X = 2) = 0.15$ and that $P(Y = 2) = 0.3$. If the two events $X = 2$ and $Y = 2$ are independent, then we can multiply these two probabilities. Otherwise, we're stuck. In fact, we can't calculate the probability for any value of $T$ unless X and Y are **independent random variables.**

**DEFINITION: Independent random variables**

If knowing whether any event involving *X* alone has occurred tells us nothing about the occurrence of any event involving *Y* alone, and vice versa, then *X* and *Y* are **independent random variables**.

Probability models often assume independence when the random variables describe outcomes that appear unrelated to each other. You should always ask whether the assumption of independence seems reasonable. For instance, it's reasonable to treat the random variables X = number of passengers on Pete's trip and Y = number of passengers on Erin's trip on a randomly chosen day as independent, because the siblings operate their trips in different parts of the country. Now we can calculate the probability distribution of the total number of passengers that day.

## Teaching Tip

This formula can also be written as $\mu_{X+Y} = \mu_X + \mu_Y$. Unfortunately, it is not on the formula sheets provided on the AP® exam.

## Teaching Tip

Emphasize that more variables mean more variability.

## Teaching Tip

Remind students what we learned about two independent events in Chapter 5 —knowing that one event occurred doesn't change the probability that the other event occurs. You can also remind students how we described "no association" in a scatterplot—knowing the value of one variable doesn't help us predict the value of the other variable. Many of the formulas and methods we learn assume that the random variables involved are independent, so make sure students really understand the concept of independence.

## Teaching Tip

If students want to see all possible combinations of $X$ and $Y$ for Pete's Jeep Tours and Erin's Adventures, see the table below. It lists each combination, along with its sum and associated probability.

| $x_i$ | $p_i$ | $y_i$ | $p_i$ | $t_i = x_i + y_i$ | $p_i$ |
|---|---|---|---|---|---|
| 2 | 0.15 | 2 | 0.3 | 4 | 0.045 |
| 2 | 0.15 | 3 | 0.4 | 5 | 0.060 |
| 2 | 0.15 | 4 | 0.2 | 6 | 0.030 |
| 2 | 0.15 | 5 | 0.1 | 7 | 0.015 |
| 3 | 0.25 | 2 | 0.3 | 5 | 0.075 |
| 3 | 0.25 | 3 | 0.4 | 6 | 0.100 |
| 3 | 0.25 | 4 | 0.2 | 7 | 0.050 |
| 3 | 0.25 | 5 | 0.1 | 8 | 0.025 |
| 4 | 0.35 | 2 | 0.3 | 6 | 0.105 |
| 4 | 0.35 | 3 | 0.4 | 7 | 0.140 |
| 4 | 0.35 | 4 | 0.2 | 8 | 0.070 |
| 4 | 0.35 | 5 | 0.1 | 9 | 0.035 |
| 5 | 0.20 | 2 | 0.3 | 7 | 0.060 |
| 5 | 0.20 | 3 | 0.4 | 8 | 0.080 |
| 5 | 0.20 | 4 | 0.2 | 9 | 0.040 |
| 5 | 0.20 | 5 | 0.1 | 10 | 0.020 |
| 6 | 0.05 | 2 | 0.3 | 8 | 0.015 |
| 6 | 0.05 | 3 | 0.4 | 9 | 0.020 |
| 6 | 0.05 | 4 | 0.2 | 10 | 0.010 |
| 6 | 0.05 | 5 | 0.1 | 11 | 0.005 |

## EXAMPLE Pete's Jeep Tours and Erin's Adventures

### *Sum of two random variables*

Let $T = X + Y$, as before. Because X and Y are independent random variables, $P(T = 4) = P(X = 2 \text{ and } Y = 2) = P(X = 2) \times P(Y = 2) = (0.15)(0.3) = 0.045$. There are two ways to get a total of $T = 5$ passengers on a randomly selected day: $X = 3, Y = 2$ or $X = 2, Y = 3$. So $P(T = 5) = P(X = 2 \text{ and } Y = 3) + P(X = 3 \text{ and } Y = 2) = (0.15)(0.4) + (0.25)(0.3) = 0.06 + 0.075 = 0.135$.

We can construct the probability distribution by listing all combinations of X and Y that yield each possible value of T and adding the corresponding probabilities. Here is the result.

| Value $t_i$: | 4 | 5 | 6 | 7 | 8 | 9 | 10 | 11 |
|---|---|---|---|---|---|---|---|---|
| Probability $p_i$: | 0.045 | 0.135 | 0.235 | 0.265 | 0.190 | 0.095 | 0.030 | 0.005 |

You can check that the probabilities add to 1. A histogram of the probability distribution is shown in Figure 6.10.



**FIGURE 6.10** The probability distribution of the random variable $T$ = the total number of passengers on Pete's and Erin's trips on a randomly chosen day.

The mean of $T$ is $\mu_T = \sum t_i p_i = (4)(0.045) + (5)(0.135) + \cdots + (11)(0.005) = 6.85$. Recall that $\mu_X = 3.75$ and $\mu_Y = 3.10$. Our calculation confirms that

$$\mu_T = \mu_X + \mu_Y = 3.75 + 3.10 = 6.85$$

What about the variance of $T$? It's

$$\begin{aligned}\sigma_T^2 &= \sum(t_i - \mu_T)^2 p_i \\ &= (4 - 6.85)^2(0.045) + (5 - 6.85)^2(0.135) \\ &\quad + \cdots + (11 - 6.85)^2(0.005) = 2.0775\end{aligned}$$

Recalling that $\sigma_X^2 = 1.1875$ and $\sigma_Y^2 = 0.89$, we see that $1.1875 + 0.89 = 2.0775$. That is,

$$\sigma_T^2 = \sigma_X^2 + \sigma_Y^2$$

To find the standard deviation of $T$, take the square root of the variance

$$\sigma_T = \sqrt{2.0775} = 1.441$$

## ALTERNATE EXAMPLE

### *El Dorado Community College*

Let $S = X + Y$ as before. It is reasonable to assume that $X$ and $Y$ are independent, because each student was selected at random. Thus, $P(S = 24) = P(X = 12 \text{ and } Y = 12) = P(X = 12) \times P(Y = 12) = (0.25)(0.3) = 0.075$. That is, there is a 0.075 probability that the sum is 24 units.

We can construct the probability distribution of $S$ by listing all the combinations of $X$ and $Y$, the sum $S$ of each combination, and the probability of each combination. Here is the probability distribution of $S$:

| Value: | 24 | 25 | 26 | 27 | 28 | 29 | 30 | 31 | 32 | 33 | 34 | 35 | 36 |
|---|---|---|---|---|---|---|---|---|---|---|---|---|---|
| Probability: | 0.075 | 0.03 | 0.015 | 0.19 | 0.07 | 0.035 | 0.24 | 0.07 | 0.035 | 0.15 | 0.03 | 0.015 | 0.045 |

$\mu_S = 24(0.075) + 25(0.03) + \cdots + 36(0.045) = 29.65$ units

$\sigma_S^2 = (24 - 29.65)^2(0.075) + (25 - 29.65)^2(0.03) + \cdots + (36 - 29.65)^2(0.045) = 9.63$

$\sigma_S = \sqrt{9.63} = 3.10$ units

Notice that $\mu_S = \mu_X + \mu_Y$ (29.65 = 14.65 + 15) and that $\sigma_S^2 = \sigma_X^2 + \sigma_Y^2$ (9.63 = 4.23 + 5.40).

As the preceding example illustrates, when we add two *independent* random variables, their variances add. *Standard deviations do not add.* For Pete's and Erin's passenger totals,

$$\sigma_X + \sigma_Y = 1.0897 + 0.943 = 2.0327$$

This is very different from $\sigma_T = 1.441$.

**VARIANCE OF THE SUM OF INDEPENDENT RANDOM VARIABLES**

For any two *independent* random variables X and Y, if $T = X + Y$, then the variance of $T$ is

$$\sigma_T^2 = \sigma_X^2 + \sigma_Y^2$$

In general, the variance of the sum of several *independent* random variables is the sum of their variances.

You might be wondering whether there's a formula for computing the variance of the sum of two random variables that are *not* independent. There is, but it's beyond the scope of this course. *Just remember that you can add variances only if the two random variables are independent and that you can never add standard deviations.*

## EXAMPLE SAT Scores

### *The role of independence*

A college uses SAT scores as one criterion for admission. Experience has shown that the distribution of SAT scores among its entire population of applicants is such that

| | | |
|---|---|---|
| SAT Math score X: | $\mu_X = 519$ | $\sigma_X = 115$ |
| SAT Critical Reading score Y: | $\mu_Y = 507$ | $\sigma_Y = 111$ |

PROBLEM: What are the mean and standard deviation of the total score $X + Y$ for a randomly selected applicant to this college?

SOLUTION: The mean total score is

$$\mu_{X+Y} = \mu_X + \mu_Y = 519 + 507 = 1026$$

The variance and standard deviation of the total *cannot be computed* from the information given. SAT Math and Critical Reading scores are not independent, because students who score high on one exam tend to score high on the other also.

**For Practice** *Try Exercise* 47

The next example involves two independent random variables and some transformations.

### AP® Exam Common Error

On the AP® exam, many students lose credit because they add standard deviations instead of adding variances when combining random variables.

### Teaching Tip

This formula can also be written as $\sigma_{X+Y}^2 = \sigma_X^2 + \sigma_Y^2$. Unfortunately, it is not on the formula sheets provided on the AP® exam.

### Teaching Tip

Point out that the formula on this page looks like another formula students have learned in math: the Pythagorean theorem. And just as the Pythagorean theorem applies only to right triangles, this formula for the variance of a sum of random variables applies only to *independent* random variables.

If your students seem interested in the connection with the Pythagorean theorem, remind them that to solve any triangle, they can use the law of cosines: $a^2 = b^2 + c^2 - 2bc(\cos A)$. The variance formula for the *sum* of *any* two random variables is $\sigma_{X+Y}^2 = \sigma_X^2 + \sigma_Y^2 + 2\sigma_X\sigma_Y\rho$, where $\rho$ = the correlation between the two variables. If the variables are independent, then correlation = 0 and the last term of the formula becomes 0. This simplifies to the rule for independent random variables that we present in the book. *Note*: The AP® exam only requires that students know how to work with combinations of independent random variables.

### Teaching Tip

The formula on this page generalizes to the sum of more than two independent random variables as well. So, if Pete and Erin had another sibling in the business, the variance of the total number of passengers would just be the sum of all three individual variances (assuming independence, of course!).

### ALTERNATE EXAMPLE

#### *Tuition, fees, and books*

Let $B$ = the amount spent on books in the fall semester for a randomly selected full-time student at El Dorado Community College. Suppose that $\mu_B = 153$ and $\sigma_B = 32$. Recall from earlier that $C$ = overall cost for tuition and fees for a randomly selected full-time student at El Dorado Community College and that $\mu_C = \$832.50$ and $\sigma_C = \$103$.

PROBLEM: Find the mean and standard deviation of the cost of tuition, fees, and books $(C + B)$ for a randomly selected full-time student at El Dorado Community College.

SOLUTION: The mean is $\mu_{C+B} = \mu_C + \mu_B = 832.50 + 153 = \$985.50$. The standard deviation cannot be calculated because the cost for tuition and fees and the cost for books are not independent. Students who take more units will typically have to buy more books.

## Teaching Tip

In this example, we want to know the mean and standard deviation of the random variable $W$, where $W = 150X + 175Y$. These computations involve two linear transformations ($150 \times X$, $175 \times Y$) and a combination of the resulting random variables ($150X + 175Y$). Make sure that students recognize the difference between a linear transformation of one random variable and a combination of two random variables. Also, the "order of operations" is important. We always address the multiplying (linear transformations) before the adding (combining random variables).

## ALTERNATE EXAMPLE

### El Dorado Community College

Earlier we defined $X$ = the number of units for a randomly selected full-time student at the main campus and $Y$ = the number of units for a randomly selected full-time student at the downtown campus. Also, $\mu_X = 14.65$, $\sigma_X = 2.06$ and $\mu_Y = 15$, $\sigma_Y = 2.3$.

At the main campus, full-time students pay \$50 per unit. At the downtown campus, full-time students pay \$55 per unit.

PROBLEM: Calculate the mean and standard deviation of the total amount of tuition for a randomly selected full-time student at the main campus and for a randomly selected full-time student at the downtown campus.

SOLUTION: From earlier, $T$ = amount collected from a randomly selected full-time student at the main campus with $\mu_T = \$732.50$ and $\sigma_T = \$103$. Let $U$ = amount collected from a randomly selected full-time student at the downtown campus. Then $\mu_U = 55\mu_Y = 55(15) = \$825$ and $\sigma_U = 55\sigma_Y = 55(2.3) = \$126.50$. Finally, $\mu_{T+U} = \mu_T + \mu_U = 732.50 + 825 = \$1557.50$. Because the students are randomly selected, $T$ and $U$ are independent random variables. Thus, $\sigma^2_{T+U} = \sigma^2_T + \sigma^2_U = 10{,}568 + 16{,}002 = 26{,}570$ and $\sigma_{T+U} = \sqrt{26{,}570} = \$163$.

## EXAMPLE Pete's and Erin's Tours

### Rules for adding random variables

Earlier, we defined X = the number of passengers that Pete has and Y = the number of passengers that Erin has on a randomly selected day. Recall that

$$\mu_X = 3.75,\ \sigma_X = 1.0897 \qquad \mu_Y = 3.10,\ \sigma_Y = 0.943$$

Pete charges \$150 per passenger and Erin charges \$175 per passenger.

PROBLEM: Calculate the mean and the standard deviation of the total amount that Pete and Erin collect on a randomly chosen day.

SOLUTION: Let $W$ = the total amount collected. Then $W = 150X + 175Y$. If we let $C = 150X$ and $G = 175Y$, then we can write $W$ as the sum of two random variables: $W = C + G$. We can use what we learned earlier about the effect of multiplying by a constant to find the mean and standard deviation of $C$ and $G$. For $C = 150X$,

$$\mu_C = 150\mu_X = 150(3.75) = \$562.50 \text{ and } \sigma_C = 150(1.0897) = \$163.46$$

For $G = 175Y$,

$$\mu_G = 175\mu_Y = 175(3.10) = \$542.50 \text{ and } \sigma_G = 175(0.943) = \$165.03$$

We know that the mean of the sum of two random variables equals the sum of their means:

$$\mu_W = \mu_C + \mu_G = 562.50 + 542.50 = 1105$$

On average, Pete and Erin expect to collect a total of \$1105 per day.

Because the number of passengers $X$ and $Y$ are *independent* random variables, so are the amounts of money collected $C$ and $G$. Therefore, the variance of $W$ is the sum of the variances of $C$ and $G$.

$$\sigma^2_W = \sigma^2_C + \sigma^2_G = (163.46)^2 + (165.03)^2 = 53{,}954.07$$

To get the standard deviation, we take the square root of the variance:

$$\sigma_W = \sqrt{53{,}954.07} = 232.28$$

The standard deviation of the total amount they collect is \$232.28.

**For Practice** *Try Exercise* **51**

We can extend our rules for adding random variables to situations involving repeated observations from the same chance process. For instance, suppose a gambler plays two games of roulette, each time placing a \$1 bet on either red or black. What can we say about his total gain (or loss) from playing two games? Earlier, we showed that if X = the amount gained on a single \$1 bet on red or black, then $\mu_X = -\$0.05$ and $\sigma_X = \$1.00$. Because we're interested in the player's total gain over two games, we'll define $X_1$ as the amount he gains from the first game and $X_2$ as the amount he gains from the second game. Then, his total gain $T = X_1 + X_2$. Both $X_1$ and $X_2$ have the

## Teaching Tip

We use subscripts on the $X$'s because the value of $X$ isn't always the same from game to game. In other words, $X$ is a variable! This is different than how we use "variables" in algebra. In the algebra equation $2x + x = 9$, $x$ is a "variable," but there is only one value of $x$ that makes the equation true. For more on this topic, see the Think About It on the next page and read the article "Random Variables vs. Algebraic Variables" on AP® Central.

same probability distribution as X and, therefore, the same mean (−\$0.05) and standard deviation (\$1.00). The player's expected gain in two games is

$$\mu_T = \mu_{X_1} + \mu_{X_2} = (-\$0.05) + (-\$0.05) = -\$0.10$$

Because knowing the result of one game tells the player nothing about the result of the other game, $X_1$ and $X_2$ are independent random variables. As a result,

$$\sigma_T^2 = \sigma_{X_1}^2 + \sigma_{X_2}^2 = (1.00)^2 + (1.00)^2 = 2.00$$

and the standard deviation of the player's total gain is

$$\sigma_T = \sqrt{2.00} = \$1.41$$

**$X_1 + X_2$ is not the same as $2X$** At the beginning of the section, we asked whether a roulette player would be better off placing two separate \$1 bets on red or a single \$2 bet on red. The player's total gain from two \$1 bets on red is $T = X_1 + X_2$. This sum of random variables has mean $\mu_T = -\$0.10$ and standard deviation $\sigma_T = \$1.41$. Now think about what happens if the gambler places a \$2 bet on red in a single game of roulette. Because the random variable X represents a player's gain from a \$1 bet, the random variable $Y = 2X$ represents his gain from a \$2 bet.

What's the player's expected gain from a single \$2 bet on red? It's

$$\mu_Y = 2\mu_X = 2(-\$0.05) = -\$0.10$$

That's the same as his expected gain from playing two games of roulette with a \$1 bet each time. But the standard deviation of the player's gain from a single \$2 bet is

$$\sigma_Y = 2\sigma_X = 2(\$1.00) = \$2.00$$

Compare this result to $\sigma_T = \$1.41$. There's more variability in the gain from a single \$2 bet than in the total gain from two \$1 bets.

Let's take this one step further. Would it be better for the player to place a single \$100 bet on red or to play 100 games and bet \$1 each time on red? For the single \$100 bet, the mean and standard deviation of the amount gained would be

$$\text{mean} = 100\mu_X = 100(-\$0.05) = -\$5.00$$
$$\text{standard deviation} = 100\sigma_X = 100(\$1.00) = \$100.00$$

For 100 games with a \$1 bet, the mean and standard deviation of the amount gained would be

$$\begin{aligned}\text{mean} &= \mu_{X_1} + \mu_{X_2} + \cdots + \mu_{X_{100}}\\ &= (-\$0.05) + (-\$0.05) + \cdots + (-\$0.05) = -\$5.00\\ \text{variance} &= \sigma_{X_1}^2 + \sigma_{X_2}^2 + \cdots + \sigma_{X_{100}}^2 = (1)^2 + (1)^2 + \cdots + (1)^2 = 100\\ \text{standard deviation} &= \sqrt{100} = \$10.00\end{aligned}$$

The player has a much better chance of winning (or losing) big with a single \$100 bet than with 100 separate \$1 bets. Of course, the casino accepts thousands of bets each day, so it can count on being fairly close to its expected return of 5 cents per dollar bet.

## Teaching Tip

Ask students to think of *T* from the casino's perspective. After two bets, their expected gain is \$0.10 with a standard deviation of \$1.41. Here, the standard deviation is much bigger than the expected gain, meaning the outcome for the casino is quite uncertain.

After 100 bets, their expected gain is 0.05 + 0.05 + ... = 100(0.05) = \$5.00 with a standard deviation of $\sigma_T = \sqrt{1^2 + 1^2 + \cdots} = \sqrt{100} = \$10$. After 10,000 bets, their expected gain is \$500 with a standard deviation of \$100. Now, the outcome for the casino is much more certain. To lose money, the value of *T* would need to be more than 5 standard deviations below the mean! In other words, as long as lots of bets are being made, the casinos will make money.

## ANSWERS TO CYU

1. $\mu_T = 1.1 + 0.7 = 1.8$. Over many Fridays, this dealership sells or leases about 1.8 cars in the first hour of business, on average.

2. $\sigma_T^2 = (0.943)^2 + (0.64)^2 = 1.2988$, so $\sigma_T = \sqrt{1.2988} = 1.14$.

3. $\mu_B = 500(1.1) + 300(0.7) = \$760$. $\sigma_B^2 = (500)^2(0.943)^2 + (300)^2(0.64)^2 = 259{,}176.25$, so $\sigma_B = \sqrt{259{,}176.25} = \$509.09$.

## CHECK YOUR UNDERSTANDING

A large auto dealership keeps track of sales and lease agreements made during each hour of the day. Let X = the number of cars sold and Y = the number of cars leased during the first hour of business on a randomly selected Friday. Based on previous records, the probability distributions of X and Y are as follows:

| Cars sold $x_i$: | 0 | 1 | 2 | 3 |
|---|---|---|---|---|
| Probability $p_i$: | 0.3 | 0.4 | 0.2 | 0.1 |

Mean: $\mu_X = 1.1$ Standard deviation: $\sigma_X = 0.943$

| Cars leased $y_i$: | 0 | 1 | 2 |
|---|---|---|---|
| Probability $p_i$: | 0.4 | 0.5 | 0.1 |

Mean: $\mu_Y = 0.7$ Standard deviation: $\sigma_Y = 0.64$

Define $T = X + Y$. Assume that X and Y are independent.

1. Find and interpret $\mu_T$.
2. Compute $\sigma_T$. Show your work.
3. The dealership's manager receives a \$500 bonus for each car sold and a \$300 bonus for each car leased. Find the mean and standard deviation of the manager's total bonus $B$. Show your work.

**Differences of random variables** Now that we've examined sums of random variables, it's time to investigate the difference of two random variables. Let's start by looking at the difference in the number of passengers that Pete and Erin have on their tours on a randomly selected day, $D = X - Y$. Because Pete averages $\mu_X = 3.75$ passengers per trip and Erin averages $\mu_Y = 3.10$ passengers per trip, the average difference is $\mu_D = 3.75 - 3.10 = 0.65$ passengers. That is, Pete averages 0.65 more passengers per day than Erin does. We can generalize this result for any two random variables as follows: if $D = X - Y$, then $\mu_D = \mu_X - \mu_Y$. In other words, the mean (expected value) of the difference of two random variables is equal to the difference of their means (expected values).

### MEAN OF THE DIFFERENCE OF RANDOM VARIABLES

For any two random variables X and Y, if $D = X - Y$, then the mean of $D$ is

$$\mu_D = E(D) = \mu_X - \mu_Y$$

## Teaching Tip

This formula can also be written as $\mu_{X-Y} = \mu_X - \mu_Y$. Unfortunately, it is not on the formula sheets provided on the AP® exam.

*The order of subtraction is important.* If we had defined $D = Y - X$, then $\mu_D = \mu_Y - \mu_X = 3.10 - 3.75 = -0.65$. In other words, Erin averages 0.65 fewer passengers than Pete does on a randomly chosen day.

Earlier, we saw that the variance of the sum of two independent random variables is the sum of their variances. Can you guess what the variance of the *difference* of two independent random variables will be? (If you were thinking something like "the difference of their variances," think again!) Let's return to the jeep tours scenario. On a randomly selected day, the number of passengers X on Pete's tour is between 2 and 6, and the number of passengers Y on Erin's tour is between

2 and 5. So the difference in the number of passengers $D = X - Y$ is between $-3$ and 4. Thus, the range of $D$ is $4 - (-3) = 7$. How is this value related to the ranges of $X$ and $Y$? The range of $X$ is 4 and the range of $Y$ is 3, so

$$\text{range of } D = \text{range of } X + \text{range of } Y$$

As with sums of random variables, there's more variability in the values of the difference $D$ than in the values of $X$ or $Y$ alone. This should make sense, because the variation in $X$ and the variation in $Y$ both contribute to the variation in $D$.

If you follow the process we used earlier with the random variable $T = X + Y$, you can build the probability distribution of $D = X - Y$. Here it is.

| **Value $d_i$:** | −3 | −2 | −1 | 0 | 1 | 2 | 3 | 4 |
|---|---|---|---|---|---|---|---|---|
| **Probability $p_i$:** | 0.015 | 0.055 | 0.145 | 0.235 | 0.260 | 0.195 | 0.080 | 0.015 |

You can use the probability distribution to confirm that:

1. $\mu_D = \mu_X - \mu_Y = 3.75 - 3.10 = 0.65$
2. $\sigma_D^2 = \sigma_X^2 + \sigma_Y^2 = 1.1875 + 0.89 = 2.0775$
3. $\sigma_D = \sqrt{2.0775} = 1.441$

Result 2 shows that, just like with addition, when we subtract two independent random variables, variances add.

**VARIANCE OF THE DIFFERENCE OF RANDOM VARIABLES**

For any two *independent* random variables $X$ and $Y$, if $D = X - Y$, then the variance of $D$ is

$$\sigma_D^2 = \sigma_X^2 + \sigma_Y^2$$

Let's put our new rules for subtracting random variables to use in a familiar setting.

**EXAMPLE**

## Pete's Jeep Tours and Erin's Adventures

### *Difference of random variables*

We have defined several random variables related to Pete's and Erin's tour businesses. For a randomly selected day,

$C$ = amount of money that Pete collects   $G$ = amount of money that Erin collects

Here are the means and standard deviations of these random variables:

$$\mu_C = 562.50 \quad \mu_G = 542.50$$
$$\sigma_C = 163.46 \quad \sigma_G = 165.03$$

### Teaching Tip

Remind your students that more variables mean more variability, whether you are adding or *subtracting* the random variables.

### Teaching Tip

To create the distribution of $D$ shown on this page, you could create a table like the one in the Teaching Tip on page 372. Just make sure to subtract the values of $X$ and $Y$ rather than adding them.

### Teaching Tip

This formula can also be written as $\sigma_{X-Y}^2 = \sigma_X^2 + \sigma_Y^2$. Unfortunately, it is not on the formula sheets provided on the AP® exam.

### Teaching Tip

Emphasize that this formula for the variance of a difference of random variables applies to *independent* random variables only. The general formula for the variance of the difference of *any* two random variables is $\sigma_{X+Y}^2 = \sigma_X^2 + \sigma_Y^2 - 2\sigma_X\sigma_Y\rho$, where $\rho$ = the correlation between the two variables. If the variables are independent, then correlation = 0 and the last term of the formula becomes 0. This simplifies to the rule for independent random variables that we present in the book. *Note*: The AP® exam only requires that students know how to work with combinations of independent random variables.

PROBLEM: Calculate the mean and the standard deviation of the difference $D = C - G$ in the amounts that Pete and Erin collect on a randomly chosen day. Interpret each value in context.

SOLUTION: We know that the mean of the difference of two random variables is the difference of their means. That is,

$$\mu_D = \mu_C - \mu_G = 562.50 - 542.50 = 20.00$$

On average, Pete collects \$20 more per day than Erin does. Some days the difference will be more than \$20, other days it will be less, but the average difference after lots of days will be about \$20.

Because the number of passengers $X$ and $Y$ are *independent* random variables, so are the amounts of money collected $C$ and $G$. Therefore, the variance of $D$ is the sum of the variances of $C$ and $G$:

$$\sigma_D^2 = \sigma_C^2 + \sigma_G^2 = (163.46)^2 + (165.03)^2 = 53{,}954.07$$
$$\sigma_D = \sqrt{53{,}954.07} = 232.28$$

The value $\sigma_D = \$232.28$ in the example should look familiar. It's the same value we got earlier when we calculated the standard deviation of the total amount that Pete and Erin collect on a randomly chosen day, $\sigma_T = \$232.28$.

The standard deviation of the difference in the amounts collected by Pete and Erin is \$232.28. Even though the *average* difference in the amounts collected is \$20, the difference on individual days will typically vary from the mean by about \$232.

**For Practice** *Try Exercise* **55**

## CHECK YOUR UNDERSTANDING

A large auto dealership keeps track of sales and lease agreements made during each hour of the day. Let $X$ = the number of cars sold and $Y$ = the number of cars leased during the first hour of business on a randomly selected Friday. Based on previous records, the probability distributions of $X$ and $Y$ are as follows:

| Cars sold $x_i$: | 0 | 1 | 2 | 3 |
|---|---|---|---|---|
| Probability $p_i$: | 0.3 | 0.4 | 0.2 | 0.1 |

Mean: $\mu_X = 1.1$ Standard deviation: $\sigma_X = 0.943$

| Cars leased $y_i$: | 0 | 1 | 2 |
|---|---|---|---|
| Probability $p_i$: | 0.4 | 0.5 | 0.1 |

Mean: $\mu_Y = 0.7$ Standard deviation: $\sigma_Y = 0.64$

Define $D = X - Y$. Assume that $X$ and $Y$ are independent.

1. Find and interpret $\mu_D$.
2. Compute $\sigma_D$. Show your work.
3. The dealership's manager receives a \$500 bonus for each car sold and a \$300 bonus for each car leased. Find the mean and standard deviation of the difference in the manager's bonus for cars sold and leased. Show your work.

## Combining Normal Random Variables

So far, we have concentrated on finding rules for means and variances of random variables. If a random variable is Normally distributed, we can use its mean

### ALTERNATE EXAMPLE

*El Dorado Community College*

Earlier we defined $T$ = amount collected from a randomly selected full-time student at the main campus and $U$ = amount collected from a randomly selected full-time student at the downtown campus. The means and standard deviations are

$\mu_T = \$732.50$ $\sigma_T = \$103$
$\mu_U = \$825$ $\sigma_U = \$126.50$

PROBLEM: Suppose we randomly select one full-time student from each of the two campuses. What are the mean and standard deviation of the difference in tuition charges, $T - U$? Interpret each of these values.

SOLUTION: $\mu_{T-U} = \mu_T - \mu_U = 732.50 - 825 = -\$92.50$. This means that, on average, full-time students at the main campus pay \$92.50 less in tuition than full-time students at the downtown campus.

Because the students are randomly selected, $T$ and $U$ are independent. Thus, $\sigma_{T-U}^2 = \sigma_T^2 + \sigma_U^2 = 10{,}568 + 16{,}002 = 26{,}570$ and $\sigma_{T-U} = \sqrt{26{,}570} = \$163$. Although the average difference in tuition for the two campuses is $-\$92.50$, the difference in tuition for a randomly selected full-time student from each college will typically vary from the average difference by about \$163. Notice that the standard deviation is the same for the sum of tuition costs and the difference of tuition costs.

**TRM Casino Lab**

At this point in the chapter, you might consider doing the Casino Lab, found in the Teacher's Resource Materials. It includes activities exploring craps, roulette, blackjack, the Monty Hall problem, and how to make your own casino game. Find it by clicking on the link in the TE-book, opening the TRCD, or logging onto the book's companion Web site www.whfreeman.com/tps5e (teacher log-in required).

### ANSWERS TO CYU

1. $\mu_D = 1.1 - 0.7 = 0.4$. Over many Fridays, this dealership sells about 0.4 cars more than it leases during the first hour of business, on average.

2. $\sigma_D^2 = (0.943)^2 + (0.64)^2 = 1.2998$, so $\sigma_D = \sqrt{1.2998} = 1.14$.

3. $\mu_B = 500(1.1) - 300(0.7) = \$340$. $\sigma_B^2 = (500)^2(0.943)^2 + (300)^2(0.64)^2 = 259{,}176.25$, so $\sigma_B = \sqrt{259{,}176.25} = \$509.09$.

and standard deviation to compute probabilities. The earlier example on young women's heights (page 357) shows the method. What if we combine two Normal random variables?

**EXAMPLE**

## A Computer Simulation

### *Sums and differences of Normal random variables*

We used Fathom software to simulate taking independent SRSs of 1000 observations from each of two Normally distributed random variables, X and Y. Figure 6.11(a) shows the results. The random variable X is N(3, 0.9) and the random variable Y is N(1, 1.2). What do we know about the sum and difference of these two random variables? The histograms in Figure 6.11(b) came from adding and subtracting the values of X and Y for the 1000 randomly generated observations from each distribution.

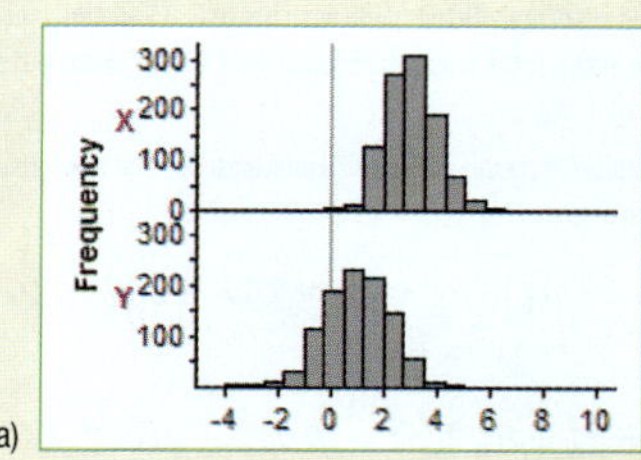

(a)

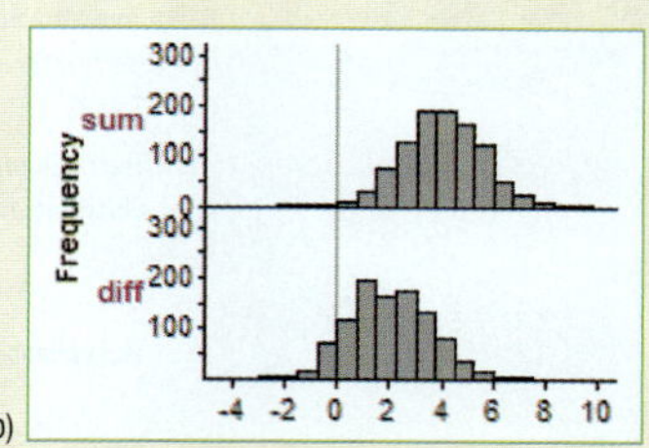

(b)

**FIGURE 6.11** (a) Histograms showing the results of randomly selecting 1000 values from two different Normal random variables *X* and *Y*. (b) Histograms of the sum and difference of the 1000 randomly selected values of *X* and *Y*.

Let's summarize what we see:

| | Sum X + Y | Difference X − Y |
|---|---|---|
| **Shape:** | Looks approximately Normal | Looks approximately Normal |
| **Center:** | About 4, which makes sense because $\mu_{X+Y} = \mu_X + \mu_Y = 3+1 = 4$ | About 2, which makes sense because $\mu_{X-Y} = \mu_X - \mu_Y = 3 - 1 = 2$ |

**Spread:** The spreads of the two distributions are about the same. That makes sense because

$$\sigma^2_{X+Y} = \sigma^2_{X-Y} = \sigma^2_X + \sigma^2_Y$$

As the previous example illustrates, *any sum or difference of independent Normal random variables is also Normally distributed.* The mean and standard deviation of the resulting Normal distribution can be found using the appropriate rules for means and variances.

**ALTERNATE EXAMPLE**

### *A graphing calculator simulation*

You can imitate the Computer Simulation example on a graphing calculator. Start by having each student generate a value for *X* with the command `randNorm(3,0.9)` and a value for *Y* with the command `randNorm(1,1.2)`. Then have them compute $X + Y$ and $X - Y$ and share their values with the class. Make a dotplot for $X + Y$ and a dotplot for $X - Y$ (on the same scale) to help students understand what is being displayed in the graphs on this page.

Now, increase the number of sums and differences using the lists in the calculator. In the heading of L1, enter `randNorm(3,0.9,500)` to randomly generate 500 values from a Normal distribution with a mean of 3 and a standard deviation of 0.9. Likewise, enter `randNorm(1,1.2,500)` in the heading of L2 to randomly generate 500 values from a Normal distribution with a mean of 1 and a standard deviation of 1.2. In the heading of L3, enter `L1+L2`, and in the heading of L4, enter `L1-L2`. Look at histograms of the values in L1, L2, L3, and L4 and you should see the same things as in the Computer Simulation example.

### Teaching Tip

Point out that the standard deviations form a Pythagorean triple: $0.9^2 + 1.2^2 = 1.5^2$. This will also be true in the Speed Dating Alternate Example on page 381.

### Teaching Tip

Using the simulation approach described in the example (and alternate example) on this page can make problems involving sums and differences of random variables less abstract. For example, in the Give Me Some Sugar example on the next page, have each student randomly generate the weights of four sugar packets and find the sum. Then create a dotplot of the sums and find the proportions that are between 8.5 and 9 grams. To generate the weights of four packets, use the command `randNorm(2.17,0.08,4)`. After showing the simulation approach, get a more precise answer using the methods from this section.

## EXAMPLE Give Me Some Sugar!

### Sums of Normal random variables

Mr. Starnes likes sugar in his hot tea. From experience, he needs between 8.5 and 9 grams of sugar in a cup of tea for the drink to taste right. While making his tea one morning, Mr. Starnes adds four randomly selected packets of sugar. Suppose the amount of sugar in these packets follows a Normal distribution with mean 2.17 grams and standard deviation 0.08 grams.

**PROBLEM:** What's the probability that Mr. Starnes's tea tastes right?

**SOLUTION:**

**Step 1: State the distribution and the values of interest.** Let $X$ = the amount of sugar in a randomly selected packet. Then $X_1$ = amount of sugar in Packet 1, $X_2$ = amount of sugar in Packet 2, $X_3$ = amount of sugar in Packet 3, and $X_4$ = amount of sugar in Packet 4. Each of these random variables has a Normal distribution with mean 2.17 grams and standard deviation 0.08 grams. We're interested in the total amount of sugar that Mr. Starnes puts in his tea, which is given by $T = X_1 + X_2 + X_3 + X_4$.

The random variable $T$ is a sum of four independent Normal random variables. So $T$ follows a Normal distribution with mean

$$\mu_T = \mu_{X_1} + \mu_{X_2} + \mu_{X_3} + \mu_{X_4} = 2.17 + 2.17 + 2.17 + 2.17 = 8.68 \text{ grams}$$

and variance

$$\sigma_T^2 = \sigma_{X_1}^2 + \sigma_{X_2}^2 + \sigma_{X_3}^2 + \sigma_{X_4}^2 = (0.08)^2 + (0.08)^2 + (0.08)^2 + (0.08)^2 = 0.0256$$

The standard deviation of $T$ is

$$\sigma_T = \sqrt{0.0256} = 0.16 \text{ grams}$$

We want to find the probability that the total amount of sugar in Mr. Starnes's tea is between 8.5 and 9 grams. Figure 6.12 shows this probability as the area under a Normal curve.

N(8.68,0.16)

8.50 8.68 9.00

Total amount of sugar (grams)

**FIGURE 6.12** Normal distribution of the total amount of sugar in Mr. Starnes's tea.

**Step 2: Perform calculations—show your work!** To find this area, we can standardize the boundary values and use Table A:

$$z = \frac{8.5 - 8.68}{0.16} = -1.13 \quad \text{and} \quad z = \frac{9 - 8.68}{0.16} = 2.00$$

Then $P(-1.13 \le Z \le 2.00) = 0.9772 - 0.1292 = 0.8480$.

*Using Technology:* The command `normalcdf(lower:8.5, upper:9, μ:8.68, σ:0.16)` gives an area of 0.8470.

**Step 3: Answer the question.** There's about an 85% chance that Mr. Starnes's tea will taste right.

**For Practice** *Try Exercise* 61

Here's an example that involves subtracting two Normal random variables.

## ALTERNATE EXAMPLE

### Apples

Suppose that a certain variety of apples have weights that follow a Normal distribution with a mean of 9 ounces and a standard deviation of 1.5 ounces.

**PROBLEM:** If bags of apples are filled by randomly selecting 12 apples, what is the probability that the sum of the weights of the 12 apples is less than 100 ounces?

**SOLUTION:**

**Step 1: State the distribution and the values of interest.** Let $X$ = weight of a randomly selected apple. Then $X_1$ = weight of first randomly selected apple, and so on. We are interested in the total weight $T = X_1 + X_2 + \ldots + X_{12}$. Because $T$ is a sum of 12 independent Normal random variables, $T$ follows a Normal distribution with mean $\mu_T = \mu_{X_1} + \mu_{X_2} + \cdots + \mu_{X_{12}} = 9 + 9 + \cdots + 9 = 108$ ounces and variance $\sigma_T^2 = \sigma_{X_1}^2 + \sigma_{X_2}^2 + \cdots + \sigma_{X_{12}}^2 = 1.5^2 + 1.5^2 + \cdots + 1.5^2 = 27$. The standard deviation is $\sigma_T = \sqrt{27} = 5.2$ ounces. We want to find $P(T < 100)$.

**Step 2: Perform calculations—show your work!** The standardized score for the boundary value is $z = \frac{100 - 108}{5.2} = -1.54$. From Table A, $P(Z < -1.54) = 0.0618$. *Using technology*: The command `normalcdf(lower:−1000,upper:100,μ:108,σ:5.2)` gives an area of 0.0620.

**Step 3: Answer the question.** There is about a 6.2% chance that the 12 randomly selected apples will have a total weight of less than 100 ounces.

### Teaching Tip

In Chapter 7, we will use the sampling distribution of the sample mean to solve problems like those on this page. For the Sugar example, we would want to know the probability that the *average* amount of sugar in each packet is between 8.5/4 = 2.125 and 9/4 = 2.25 grams.

### Teaching Tip

To see how well your students understand the rules for combining random variables, have them try Question 3 on the 2013 AP® exam. It is challenging because it gives students the mean and standard deviation of a combination of random variables and asks students to work backward to find the mean and standard deviation of one of the variables.

## EXAMPLE Put a Lid on It!

### *Differences of Normal random variables*

The diameter $C$ of a randomly selected large drink cup at a fast-food restaurant follows a Normal distribution with a mean of 3.96 inches and a standard deviation of 0.01 inches. The diameter $L$ of a randomly selected large lid at this restaurant follows a Normal distribution with mean 3.98 inches and standard deviation 0.02 inches. For a lid to fit on a cup, the value of $L$ has to be bigger than the value of $C$, but not by more than 0.06 inches.

**PROBLEM:** What's the probability that a randomly selected large lid will fit on a randomly chosen large drink cup?

**SOLUTION:**

**Step 1: State the distribution and the values of interest.** We'll define the random variable $D = L - C$ to represent the difference between the lid's diameter and the cup's diameter. The random variable $D$ is the difference of two independent Normal random variables. So $D$ follows a Normal distribution with mean

$$\mu_D = \mu_L - \mu_C = 3.98 - 3.96 = 0.02$$

and variance

$$\sigma_D^2 = \sigma_L^2 + \sigma_C^2 = (0.02)^2 + (0.01)^2 = 0.0005$$

The standard deviation of $D$ is

$$\sigma_D = \sqrt{0.0005} = 0.0224$$

We want to find the probability that the difference $D$ is between 0 and 0.06 inches. Figure 6.13 shows this probability as the area under a Normal curve.

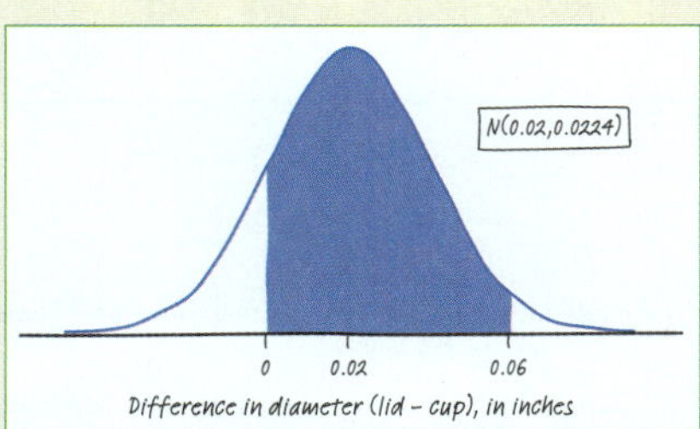


**FIGURE 6.13** Normal distribution of the difference in lid diameter and cup diameter at a fast-food restaurant.

**Step 2: Perform calculations—show your work!** To find this area, we can standardize the boundary values and use Table A:

$$z = \frac{0 - 0.02}{0.0224} = -0.89 \quad \text{and} \quad z = \frac{0.06 - 0.02}{0.0224} = 1.79$$

Then $P(-0.89 \leq Z \leq 1.79) = 0.9633 - 0.1867 = 0.7766$.

*Using Technology:* The command `normalcdf(lower:0, upper:0.06, μ:0.02, σ:0.0224)` gives an area of 0.7770.

**Step 3: Answer the question.** There's about a 78% chance that a randomly selected large lid will fit on a randomly chosen large drink cup at this fast-food restaurant. Roughly 22% of the time, the lid won't fit. This seems like an unreasonably high chance of getting a lid that doesn't fit. Maybe the restaurant should find a new supplier!

**For Practice** *Try Exercise* 63

## ALTERNATE EXAMPLE

### *Speed dating*

To save time and money, many single people have decided to try speed dating. At a speed dating event, women sit in a circle and men spend about 10 minutes getting to know a woman before moving on to the next one. Suppose that the height $M$ of male speed daters follows a Normal distribution with a mean of 69.5 inches and a standard deviation of 4 inches, and that the height $F$ of female speed daters follows a Normal distribution with a mean of 65 inches and a standard deviation of 3 inches.

**PROBLEM:** What is the probability that a randomly selected male speed dater is taller than the randomly selected female speed dater with whom he is paired?

**SOLUTION:**

**Step 1: State the distribution and the values of interest.** We'll define the random variable $D = M - F$ to represent the difference between the male's height and the female's height. Because $D$ is the difference of two independent Normal random variables, $D$ follows a Normal distribution with mean $\mu_D = \mu_M - \mu_F = 69.5 - 65 = 4.5$ inches and variance $\sigma_D^2 = \sigma_M^2 + \sigma_F^2 = 4^2 + 3^2 = 25$. Thus, the standard deviation is $\sigma_D = \sqrt{25} = 5$ inches. Our goal is to find $P(M > F)$, or $P(D > 0)$.

**Step 2: Perform calculations—show your work!** The standardized score for the boundary value is $z = \frac{0 - 4.5}{5} = -0.90$. From Table A, $P(Z > -0.90) = 1 - 0.1841 = 0.8159$. *Using technology:* The command `normalcdf(lower:0, upper:1000,μ:4.5,σ:5)` gives an area of 0.8159.

**Step 3: Answer the question.** There is about an 82% chance that a randomly selected male speed dater will be taller than the randomly selected female speed dater with whom he is paired.

**TRM Section 6.2 Quizzes**

There are three different quizzes available for this section in the Teacher's Resource Materials. Versions A and B are roughly parallel. Find them by clicking on the link in the TE-book, opening the TRCD, or logging onto the book's companion Web site www.whfreeman.com/tps5e (teacher log-in required).

**TRM Full Solutions to Section 6.2 Exercises**

## Answers to Section 6.2 Exercises

**6.35** $\mu_Y = 2.54(1.2) = 3.048$ cm and $\sigma_Y = 2.54(0.25) = 0.635$ cm.

**6.36** $\mu_J = 12(5.8) = 69.6$ inches and $\sigma_J = 12(0.24) = 2.88$ inches.



## Section 6.2 Summary

- Adding a positive constant $a$ to (subtracting $a$ from) a random variable increases (decreases) the mean of the random variable by $a$ but does not affect its standard deviation or the shape of its probability distribution.
- Multiplying (dividing) a random variable by a positive constant $b$ multiplies (divides) the mean of the random variable by $b$ and the standard deviation by $b$ but does not change the shape of its probability distribution.
- A **linear transformation** of a random variable involves adding or subtracting a constant $a$, multiplying or dividing by a constant $b$, or both. We can write a linear transformation of the random variable $X$ in the form $Y = a + bX$. The shape, center, and spread of the probability distribution of $Y$ are as follows:

  **Shape:** Same as the probability distribution of $X$ if $b > 0$.

  **Center:** $\mu_Y = a + b\mu_X$

  **Spread:** $\sigma_Y = |b|\sigma_X$
- If $X$ and $Y$ are *any* two random variables,

  $\mu_{X+Y} = \mu_X + \mu_Y$: The mean of the sum of two random variables is the sum of their means.

  $\mu_{X-Y} = \mu_X - \mu_Y$: The mean of the difference of two random variables is the difference of their means.
- If $X$ and $Y$ are **independent random variables**, then knowing the value of one variable tells you nothing about the value of the other. In that case, variances add:

  $\sigma^2_{X+Y} = \sigma^2_X + \sigma^2_Y$: The variance of the sum of two independent random variables is the sum of their variances.

  $\sigma^2_{X-Y} = \sigma^2_X + \sigma^2_Y$: The variance of the difference of two independent random variables is the sum of their variances.
- The sum or difference of independent Normal random variables follows a Normal distribution.

## Section 6.2 Exercises

35. **Crickets** The length in inches of a cricket chosen at random from a field is a random variable $X$ with mean 1.2 inches and standard deviation 0.25 inches. Find the mean and standard deviation of the length $Y$ of a randomly chosen cricket from the field in centimeters. There are 2.54 centimeters in an inch.

36. **Men's heights** A report of the National Center for Health Statistics says that the height of a 20-year-old man chosen at random is a random variable $H$ with mean 5.8 feet and standard deviation 0.24 feet. Find the mean and standard deviation of the height $J$ of a randomly selected 20-year-old man in inches. There are 12 inches in a foot.

37. **Get on the boat!** A small ferry runs every half hour from one side of a large river to the other. The number of cars $X$ on a randomly chosen ferry trip has the probability distribution shown below. You can check that $\mu_X = 3.87$ and $\sigma_X = 1.29$.

| Cars: | 0 | 1 | 2 | 3 | 4 | 5 |
|---|---|---|---|---|---|---|
| Probability: | 0.02 | 0.05 | 0.08 | 0.16 | 0.27 | 0.42 |

(a) The cost for the ferry trip is \$5. Make a graph of the probability distribution for the random variable $M$ = money collected on a randomly selected ferry trip. Describe its shape.

(b) Find and interpret $\mu_M$.

(c) Find and interpret $\sigma_M$.

38. **Skee Ball** Ana is a dedicated Skee Ball player (see photo) who always rolls for the 50-point slot. The probability distribution of Ana's score X on a single roll of the ball is shown below. You can check that $\mu_X = 23.8$ and $\sigma_X = 12.63$.

| Score: | 10 | 20 | 30 | 40 | 50 |
|---|---|---|---|---|---|
| Probability: | 0.32 | 0.27 | 0.19 | 0.15 | 0.07 |

(a) A player receives one ticket from the game for every 10 points scored. Make a graph of the probability distribution for the random variable $T$ = number of tickets Ana gets on a randomly selected throw. Describe its shape.

(b) Find and interpret $\mu_T$.

(c) Find and interpret $\sigma_T$.

*Exercises 39 and 40 refer to the following setting.* Ms. Hall gave her class a 10-question multiple-choice quiz. Let X = the number of questions that a randomly selected student in the class answered correctly. The computer output below gives information about the probability distribution of X. To determine each student's grade on the quiz (out of 100), Ms. Hall will multiply his or her number of correct answers by 5 and then add 50. Let G = the grade of a randomly chosen student in the class.

| N | Mean | Median | StDev | Min | Max | $Q_1$ | $Q_3$ |
|---|---|---|---|---|---|---|---|
| 30 | 7.6 | 8.5 | 1.32 | 4 | 10 | 8 | 9 |

39. **Easy quiz**

(a) Find the mean of G. Show your method.

(b) Find the standard deviation of G. Show your method.

(c) How do the variance of X and the variance of G compare? Justify your answer.

40. **Easy quiz**

(a) Find the median of G. Show your method.

(b) Find the *IQR* of G. Show your method.

(c) What shape would the probability distribution of G have? Justify your answer.

41. **Get on the boat!** Refer to Exercise 37. The ferry company's expenses are \$20 per trip. Define the random variable Y to be the amount of profit (money collected minus expenses) made by the ferry company on a randomly selected trip. That is, Y = M − 20.

(a) Find and interpret the mean of Y.

(b) Find and interpret the standard deviation of Y.

42. **The Tri-State Pick 3** Most states and Canadian provinces have government-sponsored lotteries. Here is a simple lottery wager, from the Tri-State Pick 3 game that New Hampshire shares with Maine and Vermont. You choose a number with 3 digits from 0 to 9; the state chooses a three-digit winning number at random and pays you \$500 if your number is chosen. Because there are 1000 numbers with three digits, you have probability 1/1000 of winning. Taking X to be the amount your ticket pays you, the probability distribution of X is

| Payoff: | \$0 | \$500 |
|---|---|---|
| Probability: | 0.999 | 0.001 |

(a) Show that the mean and standard deviation of X are $\mu_X = \$0.50$ and $\sigma_X = \$15.80$.

(b) If you buy a Pick 3 ticket, your winnings are W = X − 1, because it costs \$1 to play. Find the mean and standard deviation of W. Interpret each of these values in context.

43. **Get on the boat!** Based on the analysis in Exercise 41, the ferry company decides to increase the cost of a trip to \$6. We can calculate the company's profit Y on a randomly selected trip from the number of cars X. Find the mean and standard deviation of Y. Show your work.

44. **Making a profit** Rotter Partners is planning a major investment. From experience, the amount of profit X (in millions of dollars) on a randomly selected investment of this type is uncertain, but an estimate gives the following probability distribution:

| Profit: | 1 | 1.5 | 2 | 4 | 10 |
|---|---|---|---|---|---|
| Probability: | 0.1 | 0.2 | 0.4 | 0.2 | 0.1 |

Based on this estimate, $\mu_X = 3$ and $\sigma_X = 2.52$. Rotter Partners owes its lender a fee of \$200,000 plus 10% of the profits X. So the firm actually retains Y = 0.9X − 0.2 from the investment. Find the mean and standard deviation of Y. Show your work.

**6.39 (a)** $\mu_G = 5(7.6) + 50 = 88$ **(b)** $\sigma_G = 5(1.32) = 6.6$ **(c)** $\sigma_G^2 = (5\sigma_X)^2 = 25\sigma_X^2$. The variance of $G$ is 25 times the variance of $X$.

**6.40 (a)** $\text{median}_G = 5(8.5) + 50 = 92.5$ **(b)** $IQR_G = 5(9 - 8) = 5$ **(c)** The shape will be the same as the distribution of $X$, because multiplying by a constant and adding a constant don't change the shape.

**6.41 (a)** $\mu_Y = -\$0.65$. If many ferry trips were selected at random, the ferry would lose about \$0.65 per trip, on average. **(b)** $\sigma_Y = \$6.45$. The amount of profit on randomly selected ferry trips will typically vary by about \$6.45 from the mean (−\$0.65).

**6.42 (a)** $\mu_X = 0(0.999) + 500(0.001) = \$0.50$. $\sigma_X^2 = (0 - 0.50)^2(0.999) + (500 - 0.50)^2(0.001) = 249.75$, so $\sigma_X = \sqrt{249.75} = \$15.80$. **(b)** $\mu_W = -\$0.50$. If you play this game many times, you will lose \$0.50 per game, on average. $\sigma_W = \$15.80$. The amount you win per play will typically vary by about \$15.80 from the mean (−\$0.50).

**6.43** $\mu_Y = 6(3.87) - 20 = \$3.22$. $\sigma_Y = 6(1.29) = \$7.74$.

**6.44** $\mu_Y = 0.9(3) - 0.2 = \$2.5$ million and $\sigma_Y = 0.9(2.52) = \$2.268$ million.

**6.37 (a)** The distribution shown below is skewed to the left. Most of the time, the ferry makes \$20 or \$25.

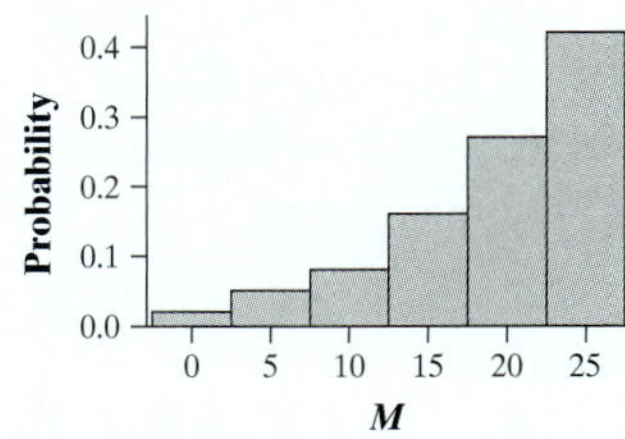


**(b)** $\mu_M = \$19.35$. If many ferry trips were selected at random, the ferry would collect about \$19.35 per trip, on average. **(c)** $\sigma_M = \$6.45$. The amounts collected on randomly selected ferry trips will typically vary by about \$6.45 from the mean (\$19.35).

**6.38 (a)** The distribution shown below is skewed to the right. Ana is more likely to get 1 or 2 tickets than 4 or 5.

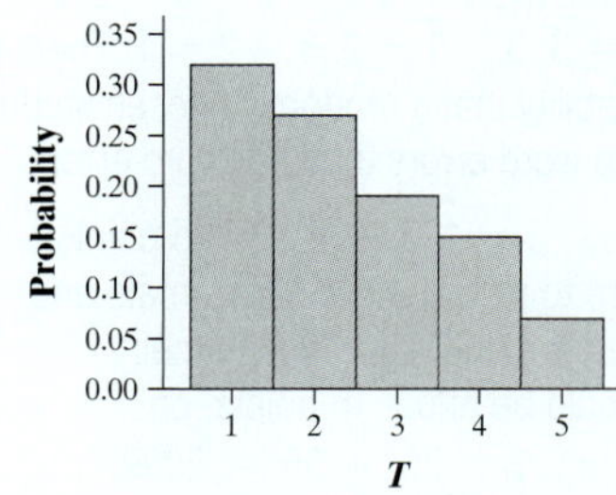


**(b)** $\mu_T = 2.38$ tickets. After many rolls, Ana will receive about 2.38 tickets per roll, on average. **(c)** $\sigma_T = 1.263$ tickets. The number of tickets that Ana wins will typically vary by about 1.263 from the mean (2.38).

**45. Too cool at the cabin?** During the winter months, the temperatures at the Starneses' Colorado cabin can stay well below freezing (32°F or 0°C) for weeks at a time. To prevent the pipes from freezing, Mrs. Starnes sets the thermostat at 50°F. She also buys a digital thermometer that records the indoor temperature each night at midnight. Unfortunately, the thermometer is programmed to measure the temperature in degrees Celsius. Based on several years' worth of data, the temperature $T$ in the cabin at midnight on a randomly selected night follows a Normal distribution with mean 8.5°C and standard deviation 2.25°C. pg 368

(a) Let $Y$ = the temperature in the cabin at midnight on a randomly selected night in degrees Fahrenheit (recall that $F = (9/5)C + 32$). Find the mean and standard deviation of $Y$.

(b) Find the probability that the midnight temperature in the cabin is below 40°F. Show your work.

**46. Cereal** A company's single-serving cereal boxes advertise 9.63 ounces of cereal. In fact, the amount of cereal $X$ in a randomly selected box follows a Normal distribution with a mean of 9.70 ounces and a standard deviation of 0.03 ounces.

(a) Let $Y$ = the excess amount of cereal beyond what's advertised in a randomly selected box, measured in grams (1 ounce = 28.35 grams). Find the mean and standard deviation of $Y$.

(b) Find the probability of getting at least 3 grams more cereal than advertised. Show your work.

**47. His and her earnings** Researchers randomly select a married couple in which both spouses are employed. Let $X$ be the income of the husband and $Y$ be the income of the wife. Suppose that you know the means $\mu_X$ and $\mu_Y$ and the variances $\sigma_X^2$ and $\sigma_Y^2$ of both variables. pg 373

(a) Is it reasonable to take the mean of the total income $X + Y$ to be $\mu_X + \mu_Y$? Explain your answer.

(b) Is it reasonable to take the variance of the total income to be $\sigma_X^2 + \sigma_Y^2$? Explain your answer.

**48. Rainy days** Imagine that we randomly select a day from the past 10 years. Let $X$ be the recorded rainfall on this date at the airport in Orlando, Florida, and $Y$ be the recorded rainfall on this date at Disney World just outside Orlando. Suppose that you know the means $\mu_X$ and $\mu_Y$ and the variances $\sigma_X^2$ and $\sigma_Y^2$ of both variables.

(a) Is it reasonable to take the mean of the total rainfall $X + Y$ to be $\mu_X + \mu_Y$? Explain your answer.

(b) Is it reasonable to take the variance of the total rainfall to be $\sigma_X^2 + \sigma_Y^2$? Explain your answer.

**49. Get on the boat!** Refer to Exercise 41. Find the expected value and standard deviation of the total amount of profit made on two randomly selected days. Show your work.

**50. The Tri-State Pick 3** Refer to Exercise 42. Suppose you buy one Pick 3 ticket on each of two consecutive days. Find the expected value and standard deviation of your total winnings. Show your work.

**51. Essay errors** Typographical and spelling errors can be either "nonword errors" or "word errors." A nonword error is not a real word, as when "the" is typed as "teh." A word error is a real word, but not the right word, as when "lose" is typed as "loose." When students are asked to write a 250-word essay (without spell-checking), the number of nonword errors $X$ in a randomly selected essay has the following probability distribution: pg 374

| Value: | 0 | 1 | 2 | 3 | 4 |
|---|---|---|---|---|---|
| Probability: | 0.1 | 0.2 | 0.3 | 0.3 | 0.1 |

$\mu_X = 2.1 \quad \sigma_X = 1.136$

The number of word errors $Y$ has this probability distribution:

| Value: | 0 | 1 | 2 | 3 |
|---|---|---|---|---|
| Probability: | 0.4 | 0.3 | 0.2 | 0.1 |

$\mu_Y = 1.0 \quad \sigma_Y = 1.0$

Assume that $X$ and $Y$ are independent.

An English professor deducts 3 points from a student's essay score for each nonword error and 2 points for each word error. Find the mean and standard deviation of the total score deductions for a randomly selected essay. Show your work.

**52. The Tri-State Pick 3** Refer to Exercise 42. You and a friend decide to play Pick 3, but with two different strategies. Your friend buys a \$1 Pick 3 ticket on each of five consecutive days. You bet \$5 on a single number on your Pick 3 ticket. Find the mean and standard deviation of the total winnings for you and your friend. Show your work.

**53. Essay errors** Refer to Exercise 51.

(a) Find the mean and standard deviation of the difference $Y - X$ in the number of errors made by a randomly selected student. Interpret each value in context.

(b) From the information given, can you find the probability that a randomly selected student makes more word errors than nonword errors? If so, find this probability. If not, explain why not.

**54. Study habits** The Survey of Study Habits and Attitudes (SSHA) is a psychological test that measures academic motivation and study habits. The distribution of SSHA scores among the women at a college has mean 120 and standard deviation 28, and the distribution of scores among male students has mean

**6.45 (a)** $\mu_Y = 47.3$°F. $\sigma_Y = 4.05$°F. **(b)** $Y$ has the $N(47.3, 4.05)$ distribution. We want to find $P(Y < 40)$. $z = \dfrac{40 - 47.3}{4.05} = -1.80$ and $P(Z < -1.80) = 0.0359$. *Using technology*: `normalcdf(lower:−1000, upper:40,μ:47.3,σ:4.05)` = 0.0357. There is a 0.0357 probability that the midnight temperature in the cabin is below 40°F.

**6.46 (a)** $\mu_Y = 28.35(9.70) - 273.01 = 1.985$ grams. $\sigma_Y = 28.35(0.03) = 0.8505$ grams. **(b)** $Y$ has the $N(1.985, 0.8505)$ distribution. We want to find $P(Y \geq 3)$. $z = \dfrac{3 - 1.985}{0.8505} = 1.19$ and $P(Z \geq 1.19) = 0.1170$. *Using technology*: `normalcdf(lower:3,upper:1000,μ:1.985, σ:0.8505)` = 0.1164. There is a 0.1164 probability of getting at least 3 grams more cereal than advertised.

**6.47 (a)** Yes. The mean of a sum is always equal to the sum of the means. **(b)** No, because it is not reasonable to assume that $X$ and $Y$ are independent.

**6.48 (a)** Yes. The mean of a sum is always equal to the sum of the means. **(b)** No, because it is not reasonable to assume that $X$ and $Y$ are independent.

**6.49** $\mu_{Y_1+Y_2} = (-0.65) + (-0.65) = -\$1.30$. $\sigma^2_{Y_1+Y_2} = 6.45^2 + 6.45^2 = 83.205$, so $\sigma_{Y_1+Y_2} = \sqrt{83.205} = \$9.12$.

**6.50** $\mu_{W_1+W_2} = (-0.50) + (-0.50) = -\$1.00$. $\sigma^2_{W_1+W_2} = 15.80^2 + 15.80^2 = 499.28$, so $\sigma_{W_1+W_2} = \sqrt{499.28} = \$22.34$.

**6.51** $\mu_{3X} = 3(2.1) = 6.3$ and $\sigma_{3X} = 3(1.136) = 3.408$. $\mu_{2Y} = 2(1.0) = 2.0$ and $\sigma_{2Y} = 2(1.0) = 2.0$. Thus, $\mu_{3X+2Y} = 6.3 + 2.0 = 8.3$ and $\sigma^2_{3X+2Y} = 3.408^2 + 2.0^2 = 15.6145$, so $\sigma_{3X+2Y} = \sqrt{15.6145} = 3.95$.

**6.52** $T = W_1 + W_2 + W_3 + W_4 + W_5 + 5W_6$. The mean and standard deviation of $5W_6$ are $\mu_{5W_6} = 5(-0.50) = -2.50$ and $\sigma_{5W_6} = 5(15.80) = 79$. Thus, $\mu_T = -0.50 + -0.50 + -0.50 + -0.50 + -0.50 + -2.50 = -\$5.00$ and $\sigma_T^2 = 15.80^2 + 15.80^2 + 15.80^2 + 15.80^2 + 15.80^2 + 79^2 = 7489.2$ so $\sigma_T = \sqrt{7489.2} = \$86.54$.

**6.53 (a)** $\mu_{Y-X} = 1.0 - 2.1 = -1.1$. If you were to select many essays, there would be about 1.1 fewer word errors than nonword errors, on average. $\sigma^2_{Y-X} = (1.0)^2 + (1.136)^2 = 2.2905$, so $\sigma_{Y-X} = \sqrt{2.2905} = 1.51$. The difference in the number of errors will typically vary by about 1.51 from the mean (−1.1). **(b)** The outcomes that make up this event are $1 - 0 = 1$, $2 - 0 = 2$, $2 - 1 = 1$, $3 - 0 = 3$, $3 - 1 = 2$, $3 - 2 = 1$. There is a 0.15 probability that a randomly chosen student will have more word errors than nonword errors.

**6.54 (a)** $\mu_{F-M} = 120 - 105 = 15$ points. If you were to repeatedly select a male and female student, the difference in their scores would be about 15 points, on average. $\sigma^2_{F-M} = 28^2 + 35^2 = 2009$, so $\sigma_{F-M} = \sqrt{2009} = 44.8219$ points. The difference in scores will typically vary by about 44.8219 from the mean (15). **(b)** No. We cannot assume that the distributions are Normal without additional information.

**6.55** The difference in score deductions for a randomly selected essay is $3X - 2Y$. $\mu_{3X} = 3(2.1) = 6.3$ and $\sigma_{3X} = 3(1.136) = 3.408$. $\mu_{2Y} = 2(1.0) = 2.0$ and $\sigma_{2Y} = 2(1.0) = 2.0$. Thus, $\mu_{3X-2Y} = 6.3 - 2.0 = 4.3$ and $\sigma^2_{3X-2Y} = 3.408^2 + 2.0^2 = 15.6145$, so $\sigma_{3X-2Y} = \sqrt{15.6145} = 3.95$.

**6.56** The difference in winnings for you and your friend is $D = 5W_6 - (W_1 + W_2 + W_3 + W_4 + W_5)$. $\mu_{5W_6} = 5(-0.50) = -2.50$ and $\sigma_{5W_6} = 5(15.80) = 79$. $\mu_D = -2.50 - (-0.50 + -0.50 + -0.50 + -0.50 + -0.50) = 0$. $\sigma_D^2 = 79^2 + 15.80^2 + 15.80^2 + 15.80^2 + 15.80^2 + 15.80^2 = 7489.2$ so $\sigma_D = \sqrt{7489.2} = 86.54$.

105 and standard deviation 35. You select a single male student and a single female student at random and give them the SSHA test.

(a) Find the mean and standard deviation of the difference (female minus male) between their scores. Interpret each value in context.

(b) From the information given, can you find the probability that the woman chosen scores higher than the man? If so, find this probability. If not, explain why you cannot.

55. **Essay scores** Refer to Exercise 51. Find the mean and standard deviation of the difference in score deductions (nonword – word) for a randomly selected essay. Show your work. pg 377

56. **The Tri-State Pick 3** Refer to Exercise 52. Find the mean and standard deviation of the difference (you – your friend) in winnings. Show your work.

*Exercises 57 and 58 refer to the following setting.* In Exercises 14 and 18 of Section 6.1, we examined the probability distribution of the random variable $X$ = the amount a life insurance company earns on a randomly chosen 5-year term life policy. Calculations reveal that $\mu_X = \$303.35$ and $\sigma_X = \$9707.57$.

57. **Life insurance** The risk of insuring one person's life is reduced if we insure many people. Suppose that we insure two 21-year-old males, and that their ages at death are independent. If $X_1$ and $X_2$ are the insurer's income from the two insurance policies, the insurer's average income $W$ on the two policies is

$$W = \frac{X_1 + X_2}{2}$$

Find the mean and standard deviation of $W$. (You see that the mean income is the same as for a single policy but the standard deviation is less.)

58. **Life insurance** If four 21-year-old men are insured, the insurer's average income is

$$V = \frac{X_1 + X_2 + X_3 + X_4}{4}$$

where $X_i$ is the income from insuring one man. Assuming that the amount of income earned on individual policies is independent, find the mean and standard deviation of $V$. (If you compare with the results of Exercise 57, you should see that averaging over more insured individuals reduces risk.)

59. **Time and motion** A time-and-motion study measures the time required for an assembly-line worker to perform a repetitive task. The data show that the time required to bring a part from a bin to its position on an automobile chassis varies from car to car according to a Normal distribution with mean 11 seconds and standard deviation 2 seconds. The time required to attach the part to the chassis follows a Normal distribution with mean 20 seconds and standard deviation 4 seconds. The study finds that the times required for the two steps are independent. A part that takes a long time to position, for example, does not take more or less time to attach than other parts.

(a) What is the distribution of the time required for the entire operation of positioning and attaching a randomly selected part?

(b) Management's goal is for the entire process to take less than 30 seconds. Find the probability that this goal will be met for a randomly selected part.

60. **Electronic circuit** The design of an electronic circuit for a toaster calls for a 100-ohm resistor and a 250-ohm resistor connected in series so that their resistances add. The components used are not perfectly uniform, so that the actual resistances vary independently according to Normal distributions. The resistance of 100-ohm resistors has mean 100 ohms and standard deviation 2.5 ohms, while that of 250-ohm resistors has mean 250 ohms and standard deviation 2.8 ohms.

(a) What is the distribution of the total resistance of the two components in series for a randomly selected toaster?

(b) Find the probability that the total resistance for a randomly selected toaster lies between 345 and 355 ohms.

61. **Swim team** Hanover High School has the best women's swimming team in the region. The 400-meter freestyle relay team is undefeated this year. In the 400-meter freestyle relay, each swimmer swims 100 meters. The times, in seconds, for the four swimmers this season are approximately Normally distributed with means and standard deviations as shown. Assume that the swimmer's individual times are independent. Find the probability that the total team time in the 400-meter freestyle relay for a randomly selected race is less than 220 seconds. pg 380

| Swimmer | Mean | Std. dev. |
|---|---|---|
| Wendy | 55.2 | 2.8 |
| Jill | 58.0 | 3.0 |
| Carmen | 56.3 | 2.6 |
| Latrice | 54.7 | 2.7 |

62. **Toothpaste** Ken is traveling for his business. He has a new 0.85-ounce tube of toothpaste that's supposed to last him the whole trip. The amount of toothpaste Ken squeezes out of the tube each time he brushes varies according to a Normal distribution with mean 0.13 ounces and standard deviation 0.02 ounces. If Ken brushes his teeth six times on a randomly selected trip, what's the probability that he'll use all the toothpaste in the tube?

**6.57** $\mu_{X_1+X_2} = 303.35 + 303.35 = \$606.70$ and $\sigma^2_{X_1+X_2} = 9707.57^2 + 9707.57^2 = 188{,}473{,}830.6$, so $\sigma_{X_1+X_2} = \sqrt{188{,}473{,}830.6} = \$13{,}728.58$. $W = \frac{1}{2}(X_1 + X_2)$, so $\mu_W = \frac{1}{2}(606.70) = \$303.35$ and $\sigma_W = \frac{1}{2}(13{,}728.58) = \$6864.29$.

**6.58** $\mu_{X_1+X_2+X_3+X_4} = 303.35 + 303.35 + 303.35 + 303.35 = \$1213.40$ and $\sigma^2_{X_1+X_2+X_3+X_4} = 9707.57^2 + 9707.57^2 + 9707.57^2 + 9707.57^2 = 376{,}947{,}661.2$, so $\sigma_{X_1+X_2+X_3+X_4} = \sqrt{376{,}947{,}661.2} = \$19{,}415.14$. $V = \frac{1}{4}(X_1 + X_2 + X_3 + X_4)$, so $\mu_V = \frac{1}{4}(1213.40) = \$303.35$ and $\sigma_V = \frac{1}{4}(19{,}415.14) = \$4853.785$.

**6.59** **(a)** Normal with mean $= 11 + 20 = 31$ seconds and standard deviation $= \sqrt{2^2 + 4^2} = 4.4721$ seconds.

**(b)** We want to find the probability that the total time is less than 30 seconds. $z = \frac{30 - 31}{4.4721} = -0.22$ and $P(Z < -0.22) = 0.4129$. *Using technology*: `normalcdf(lower:−1000,upper:30,μ:31,σ:4.4721)` $= 0.4115$. There is a 0.4115 probability of completing the process in less than 30 seconds for a randomly selected part.

**6.60** **(a)** Normal with mean $100 + 250 = 350$ ohms and standard deviation $= \sqrt{2.5^2 + 2.8^2} = 3.7537$ ohms. **(b)** We want to find the probability that the total resistance is between 345 and 355 ohms. $z = \frac{345 - 350}{3.7537} = -1.33$ and $z = \frac{355 - 350}{3.7537} = 1.33$ and $P(-1.33 < Z < 1.33) = 0.8164$. *Using technology*: `normalcdf(lower:345,upper:355,μ:350,σ:3.7537)` $= 0.8171$. There is a 0.8171 probability that the total resistance is between 345 and 355 for a randomly selected toaster.

**6.61** Let $T$ = the total team swim time. $\mu_T = 55.2 + 58.0 + 56.3 + 54.7 = 224.2$ seconds and $\sigma_T^2 = (2.8)^2 + (3.0)^2 + (2.6)^2 + (2.7)^2 = 30.89$, so $\sigma_T = \sqrt{30.89} = 5.56$ seconds. Thus, $T$ has the $N(224.2, 5.56)$ distribution. We want to find $P(T < 220)$. $z = \frac{220 - 224.2}{5.56} = -0.76$ and $P(Z < -0.76) = 0.2236$. *Using technology*: `normalcdf(lower:−1000,upper:220,μ:224.2,σ:5.56)`$= 0.2250$. There is a 0.2250 probability that the total team time is less than 220 seconds in a randomly selected race.

**6.62** Let $T$ be the total amount of toothpaste. $\mu_T = 0.13 + 0.13 + 0.13 + 0.13 + 0.13 + 0.13 = 0.78$ ounces and $\sigma_T^2 = 0.02^2 + 0.02^2 + 0.02^2 + 0.02^2 + 0.02^2 + 0.02^2 = 0.0024$, so $\sigma_T = \sqrt{0.0024} = 0.049$ ounces. Thus, $T$ has the $N(0.78, 0.049)$ distribution. We want to find $P(T > 0.85)$. $z = \frac{0.85 - 0.78}{0.049} = 1.43$ and $P(Z > 1.43) = 0.0764$. *Using technology*: `normalcdf(lower:0.85,upper:1000,μ:0.78,σ:0.049)` $= 0.0766$. There is a 0.0766 probability that Ken will use the entire tube of toothpaste on a randomly selected trip.

63. **Auto emissions** The amount of nitrogen oxides (NOX) present in the exhaust of a particular type of car varies from car to car according to a Normal distribution with mean 1.4 grams per mile (g/mi) and standard deviation 0.3 g/mi. Two randomly selected cars of this type are tested. One has 1.1 g/mi of NOX; the other has 1.9 g/mi. The test station attendant finds this difference in emissions between two similar cars surprising. If the NOX levels for two randomly chosen cars of this type are independent, find the probability that the difference is at least as large as the value the attendant observed.

pg 381

64. **Loser buys the pizza** Leona and Fred are friendly competitors in high school. Both are about to take the ACT college entrance examination. They agree that if one of them scores 5 or more points better than the other, the loser will buy the winner a pizza. Suppose that in fact Fred and Leona have equal ability, so that each score varies Normally with mean 24 and standard deviation 2. (The variation is due to luck in guessing and the accident of the specific questions being familiar to the student.) The two scores are independent. What is the probability that the scores differ by 5 or more points in either direction?

***Multiple choice: Select the best answer for Exercises 65 and 66, which refer to the following setting.*** The number of calories in a 1-ounce serving of a certain breakfast cereal is a random variable with mean 110 and standard deviation 10. The number of calories in a cup of whole milk is a random variable with mean 140 and standard deviation 12. For breakfast, you eat 1 ounce of the cereal with 1/2 cup of whole milk. Let $T$ be the random variable that represents the total number of calories in this breakfast.

65. The mean of $T$ is

(a) 110. (b) 140. (c) 180. (d) 195. (e) 250.

66. The standard deviation of $T$ is

(a) 22. (b) 16. (c) 15.62. (d) 11.66. (e) 4.

67. **Statistics for investing** (3.1) Joe's retirement plan invests in stocks through an "index fund" that follows the behavior of the stock market as a whole, as measured by the Standard & Poor's (S&P) 500 stock index. Joe wants to buy a mutual fund that does not track the index closely. He reads that monthly returns from Fidelity Technology Fund have correlation $r = 0.77$ with the S&P 500 index and that Fidelity Real Estate Fund has correlation $r = 0.37$ with the index.

(a) Which of these funds has the closer relationship to returns from the stock market as a whole? How do you know?

(b) Does the information given tell Joe anything about which fund has had higher returns?

68. **Buying stock** (5.3, 6.1) You purchase a hot stock for \$1000. The stock either gains 30% or loses 25% each day, each with probability 0.5. Its returns on consecutive days are independent of each other. You plan to sell the stock after two days.

(a) What are the possible values of the stock after two days, and what is the probability for each value? What is the probability that the stock is worth more after two days than the \$1000 you paid for it?

(b) What is the mean value of the stock after two days? (*Comment:* You see that these two criteria give different answers to the question "Should I invest?")

# 6.3 Binomial and Geometric Random Variables

**WHAT YOU WILL LEARN** **By the end of the section, you should be able to:**

- Determine whether the conditions for using a binomial random variable are met.
- Compute and interpret probabilities involving binomial distributions.
- Calculate the mean and standard deviation of a binomial random variable. Interpret these values in context.
- Find probabilities involving geometric random variables.
- When appropriate, use the Normal approximation to the binomial distribution to calculate probabilities.*

*This topic is not required for the AP® Statistics exam. Some teachers prefer to discuss this topic when presenting the sampling distribution of $\hat{p}$ (Chapter 7).

**6.63** Let $D = X_1 - X_2$ = the difference in NOX levels. $\mu_D = 1.4 - 1.4 = 0$ and $\sigma^2_{X_1-X_2} = \sigma^2_{X_1} + \sigma^2_{X_2} = 0.3^2 + 0.3^2 = 0.18$, so $\sigma_{X_1-X_2} = \sqrt{0.18} = 0.4243$. Thus, $D$ has the $N(0, 0.4243)$ distribution. We want to find $P(D > 0.8 \text{ or } D < -0.8) = P(D > 0.8) + P(D < -0.8)$. $z = \frac{0.8 - 0}{0.4243} = 1.89$ and $z = \frac{-0.8 - 0}{0.4243} = -1.89$ and $P(Z < -1.89 \text{ or } Z > 1.89) = 0.0588$. *Using technology:* `1 − normalcdf(lower:−0.8,upper:0.8,μ:0,σ:0.4243)` = 0.0594. There is a 0.0594 probability that the difference is at least as large as the attendant observed.

**6.64** Let $D$ = the difference in ACT scores. $\mu_D = 24 - 24 = 0$ and $\sigma^2_{X_1-X_2} = 2^2 + 2^2 = 8$, so $\sigma_{X_1-X_2} = \sqrt{8} = 2.83$. Thus, $D$ has the $N(0, 2.83)$ distribution. We want to find $P(D \geq 5 \text{ or } D \leq -5) = P(D \geq 5) + P(D \leq -5)$. $z = \frac{5 - 0}{2.83} = 1.77$ and $z = \frac{-5 - 0}{2.83} = -1.77$ and $P(Z \leq -1.77 \text{ or } Z \geq 1.77) = 0.0768$. *Using technology:* `1 − normalcdf(lower:−5,upper:5,μ:0,σ:2.83)` = 0.0773. There is a 0.0773 probability that the scores differ by 5 or more points in either direction.

**6.65** c

**6.66** d

**6.67 (a)** Fidelity Technology Fund, because its correlation is larger. **(b)** No, the correlation doesn't tell us anything about the values of the variables, only about the strength of the linear relationship between them.

**6.68 (a)**

| Amount: | \$562.50 | \$975 | \$1690 |
|---|---|---|---|
| Probability: | 0.25 | 0.50 | 0.25 |

$P(\text{amount} > 1000) = 0.25$

**(b)** $\mu = 562.50(0.25) + 975(0.5) + 1690(0.25) = \$1,050.63$

**PD Section 6.3 Overview**

Watch the video overview of Section 6.3 (for teachers) by clicking on the link in the TE-book, opening the TRCD, or logging onto the book's companion Web site www.whfreeman.com/tps5e (teacher log-in required).

When the same chance process is repeated several times, we are often interested in whether a particular outcome does or doesn't happen on each repetition. Here are some examples:

- To test whether someone has extrasensory perception (ESP), choose one of four cards at random—a star, wave, cross, or circle. Ask the person to identify the card without seeing it. Do this a total of 50 times and see how many cards the person identifies correctly. *Chance process:* choose a card at random. *Outcome of interest:* person identifies card correctly. *Random variable:* $X$ = number of correct identifications.
- A shipping company claims that 90% of its shipments arrive on time. To test this claim, take a random sample of 100 shipments made by the company last month and see how many arrived on time. *Chance process:* Randomly select a shipment and check when it arrived. *Outcome of interest:* arrived on time. *Random variable:* $Y$ = number of on-time shipments.
- In the game of Pass the Pigs, a player rolls a pair of pig-shaped dice. On each roll, the player earns points according to how the pigs land. If the player gets a "pig out," in which the two pigs land on opposite sides, she loses all points earned in that round and must pass the pigs to the next player. A player can choose to stop rolling at any point during her turn and to keep the points that she has earned before passing the pigs. *Chance process:* roll the pig dice. *Outcome of interest:* pig out. *Random variable:* $T$ = number of rolls until the player pigs out.

Some random variables, like $X$ and $Y$ in the first two examples above, count the number of times the outcome of interest occurs in a fixed number of repetitions. They are called *binomial random variables.* Other random variables, like $T$ in the Pass the Pigs setting, count the number of repetitions of the chance process it takes for the outcome of interest to occur. They are known as *geometric random variables.* These two special types of discrete random variables are the focus of this section.

## Binomial Settings and Binomial Random Variables

What do the following scenarios have in common?

- Toss a coin 5 times. Count the number of heads.
- Spin a roulette wheel 8 times. Record how many times the ball lands in a red slot.
- Take a random sample of 100 babies born in U.S. hospitals today. Count the number of females.

In each case, we're performing repeated *trials* of the same chance process. The number of trials is fixed in advance. Also, knowing the outcome of one trial tells us nothing about the outcome of any other trial. That is, the trials are *independent.* We're interested in the number of times that a specific event (we'll call it a "success") occurs. Our chances of getting a "success" are the same on each trial. When these conditions are met, we have a **binomial setting.**

### TRM Section 6.3 Alternate Examples

You can find the Alternate Examples for this section in Microsoft Word format by clicking on the link in the TE-book, opening the TRCD, or logging onto the book's companion Web site www.whfreeman.com/tps5e (teacher log-in required).

### AP® Exam Common Error

On many questions involving binomial settings, students do not recognize that using a binomial distribution is appropriate. In fact, free-response questions about the binomial distribution are often among the lowest-scoring questions on the exam. Make sure to spend plenty of time learning how to identify a binomial distribution. You can also suggest that students should check to see if it is a binomial setting whenever they aren't sure how to answer a probability question.

### Teaching Tip

When describing the independent condition, try to avoid using statements like "one trial doesn't affect another." When students hear this, they might think that the independence condition is violated only when there is a cause-and-effect relationship between the trials. Instead, use the wording that we do here and elsewhere: knowing the outcome of one trial tells us nothing about the outcomes of other trials.

## Teaching Tip

In the BINS acronym, the S can also stand for Same probability of success.

## Teaching Tip

When there is only 1 trial (i.e., $n = 1$), the trial is often called a Bernoulli trial. However, it is not important to make this distinction with your students.

The boldface letters in the box give you a helpful way to remember the conditions for a binomial setting: just check the BINS!

**DEFINITION: Binomial setting**

A **binomial setting** arises when we perform several independent trials of the same chance process and record the number of times that a particular outcome occurs. The four conditions for a binomial setting are

- **B**inary? The possible outcomes of each trial can be classified as "success" or "failure."
- **I**ndependent? Trials must be independent; that is, knowing the result of one trial must not tell us anything about the result of any other trial.
- **N**umber? The number of trials $n$ of the chance process must be fixed in advance.
- **S**uccess? There is the same probability $p$ of success on each trial.

When checking the Binary condition, note that there can be more than two possible outcomes per trial—a roulette wheel has numbered slots of *three* colors: red, black, and green. If we define "success" as having the ball land in a red slot, then "failure" occurs when the ball lands in a black or a green slot.

Think of tossing a coin $n$ times as an example of the binomial setting. Each toss gives either heads or tails. Knowing the outcome of one toss doesn't change the probability of a head on any other toss, so the tosses are independent. If we call heads a success, then $p$ is the probability of a head and remains the same as long as we toss the same coin. For tossing a fair coin, $p$ is 0.5. The number of heads we count is a **binomial random variable** X. The probability distribution of X is called a **binomial distribution.**

## Teaching Tip

Emphasize that the binomial random variable is the number (count) of successes in a fixed number of trials. This will help students recognize when a binomial distribution is appropriate.

**DEFINITION: Binomial random variable and binomial distribution**

The count $X$ of successes in a binomial setting is a **binomial random variable**. The probability distribution of $X$ is a **binomial distribution** with parameters $n$ and $p$, where $n$ is the number of trials of the chance process and $p$ is the probability of a success on any one trial. The possible values of $X$ are the whole numbers from 0 to $n$.

Later in the section, we'll learn how to assign probabilities to outcomes and how to find the mean and standard deviation of a binomial random variable. For now, it's important to be able to distinguish situations in which a binomial distribution does and doesn't apply.

## EXAMPLE From Blood Types to Aces

### *Binomial settings and random variables*

PROBLEM: Here are three scenarios involving chance behavior. In each case, determine whether or not the given random variable has a binomial distribution. Justify your answer.

(a) *Genetics says that children receive genes from each of their parents independently. Each child of a particular set of parents has probability 0.25 of having type O blood. Suppose these parents have 5 children. Let $X$ = the number of children with type O blood.*

## ALTERNATE EXAMPLE

### *Dice, cars, and hoops*

PROBLEM: Determine whether the random variables below have a binomial distribution. Justify your answer.

**(a)** Roll a fair die 10 times and let $X$ = the number of 6s.

**(b)** Shoot a basketball 20 times from various distances on the court. Let $Y$ = number of shots made.

**(c)** Observe the next 100 cars that go by and let $C$ = color.

SOLUTION:

**(a)** *Binary?* Yes; success = 6, failure = not a 6. *Independent?* Yes; knowing the outcomes of past rolls tells you nothing about the outcomes of future rolls. *Number?* Yes; there are $n = 10$ trials. *Success?* Yes; the probability of success is always $p = 1/6$. This is a binomial setting. The number of 6s $X$ is a binomial random variable with $n = 10$ and $p = 1/6$.

**(b)** *Binary?* Yes; success = make the shot, failure = miss the shot. *Independent?* Yes; evidence suggests that it is reasonable to assume that knowing the outcome of a shot tells us nothing about the outcome of other shots. *Number?* Yes; there are $n = 20$ trials. *Success?* No; the probability of success changes because the shots are taken from various distances. Because the probability of success is not constant, $Y$ is not a binomial random variable.

**(c)** *Binary?* No. There are more than two possible colors. *Independent?* Yes, knowing the color of one car tells you nothing about the color of other cars. *Number?* Yes; there are $n = 100$ trials. *Success?* A success hasn't been defined, so we cannot determine if the probability of success is always the same. Because there are more than two possible outcomes, $C$ is not a binomial random variable.

(b) Shuffle a deck of cards. Turn over the first 10 cards, one at a time. Let $Y$ = the number of aces you observe.

(c) Shuffle a deck of cards. Turn over the top card. Put the card back in the deck, and shuffle again. Repeat this process until you get an ace. Let $W$ = the number of cards required.

**SOLUTION:**

(a) To see if this is a binomial setting, we'll check the BINS:

- Binary? "Success" = has type O blood. "Failure" = doesn't have type O blood.
- Independent? Knowing one child's blood type tells you nothing about another's because children inherit genes determining blood type independently from each of their parents.
- Number? There are $n = 5$ trials of this chance process.
- Success? The probability of a success is $p = 0.25$ on each trial.

This is a binomial setting. Because $X$ counts the number of successes, it is a binomial random variable with parameters $n = 5$ and $p = 0.25$.

(b) Let's check the BINS:

- Binary? "Success" = get an ace. "Failure" = don't get an ace.
- Independent? No. If the first card you turn over is an ace, then the next card is less likely to be an ace because you're not replacing the top card in the deck. Similarly, if the first card isn't an ace, the second card is more likely to be an ace.
- Number? There are $n = 10$ trials of this chance process.
- Success? The probability that any particular card in a shuffled deck is an ace is $p = 4/52$.

Because the trials are not independent, this is not a binomial setting.

The Independent condition involves *conditional* probabilities. $P$(2nd card ace | 1st card ace) = 3/51 ≠ $P$(2nd card ace) = 4/52, so the trials are not independent. The Success condition is about *unconditional* probabilities. $P$($k$th card in a shuffled deck is an ace) = 4/52, so this condition is met.

(c) Let's check the BINS:

- Binary? "Success" = get an ace. "Failure" = don't get an ace.
- Independent? Because you are replacing the card in the deck and shuffling each time, the result of one trial does not tell you anything about the outcome of any other trial.
- Number? The number of trials is not set in advance. You could get an ace on the first card you turn over, or it may take many cards to get an ace.
- Success? The probability of getting an ace is $p = 4/52$ on each trial.

Because there is no fixed number of trials, this is not a binomial setting.

**For Practice** *Try Exercises*  69 *and*  71

Part (c) of the example raises an important point about binomial random variables. Besides checking the BINS, make sure that you're being asked to count the number of successes in a certain number of trials. In part (c), you're asked to count the number of *trials* until you get a success. That can't be a binomial random variable. (As you'll see later, W is a *geometric* random variable.)

## CHECK YOUR UNDERSTANDING

For each of the following situations, determine whether the given random variable has a binomial distribution or not. Justify your answer.

1. Shuffle a deck of cards. Turn over the top card. Put the card back in the deck, and shuffle again. Repeat this process 10 times. Let X = the number of aces you observe.

### Teaching Tip

Make sure to read the margin note in this example. Many students have trouble distinguishing violations of the independence condition with violations of the same probability of success condition. When sampling is done without replacement, the independent condition is violated.

### ✓ ANSWERS TO CYU

**1.** Binomial. *Binary?* "Success" = get an ace. "Failure" = don't get an ace. *Independent?* Because you are replacing the card in the deck and shuffling each time, the result of one trial does not tell you anything about the outcome of any other trial. *Number?* $n = 10$. *Success?* The probability of success is $p = 4/52$ for each trial.

2. Choose students at random from your class. Let $Y$ = the number who are over 6 feet tall.

3. Roll a fair die 100 times. Sometime during the 100 rolls, one corner of the die chips off. Let $W$ = the number of 5s you roll.

## ✓ ANSWERS TO CYU *(continued)*

**2.** Not binomial. *Binary?* "Success" = over 6 feet. "Failure" = not over 6 feet. *Independent?* Because we are selecting without replacement from a small number of students, the observations are not independent. *Number?* $n = 3$. *Success?* The probability of success will not change from trial to trial.

**3.** Not binomial. *Binary?* "Success" = roll a 5. "Failure" = don't roll a 5. *Independent?* Because you are rolling a die, the outcome of any one trial does not tell you anything about the outcome of any other trial. *Number?* $n = 100$. *Success?* No. The probability of success changes when the corner of the die is chipped off.

## ALTERNATE EXAMPLE

### *Rolling doubles*

In many games involving dice, rolling doubles is desirable. Rolling doubles means that the outcomes of two dice are the same, such as 1 and 1 or 5 and 5. The probability of rolling doubles when rolling two dice is $6/36 = 1/6$. If $X$ = the number of doubles in 4 rolls of two dice, then $X$ is binomial with $n = 4$ and $p = 1/6$.

What is $P(X = 0)$? That is, what is the probability that all 4 rolls are *not* doubles? Because the probability of not getting doubles on a particular roll is $1 - 1/6 = 5/6$, $P(X = 0) = P(\text{FFFF}) = (5/6)(5/6)(5/6)(5/6) = (5/6)^4 = 0.482$. (*Note*: F represents a failure and S represents a success.)

What is $P(X = 1)$? There are four different ways to roll doubles 1 time in 4 tries. The doubles could occur on the first try (SFFF), the second try (FSFF), the third try (FFSF), or the fourth try (FFFS).

$$P(\text{SFFF}) = (1/6)(5/6)(5/6)(5/6) = (1/6)(5/6)^3$$

$$P(\text{FSFF}) = (5/6)(1/6)(5/6)(5/6) = (1/6)(5/6)^3$$

$$P(\text{FFSF}) = (5/6)(5/6)(1/6)(5/6) = (1/6)(5/6)^3$$

$$P(\text{FFFS}) = (5/6)(5/6)(5/6)(1/6) = (1/6)(5/6)^3$$

Thus, the probability of rolling doubles 1 time in 4 attempts is $P(X = 1) = 4(1/6)(5/6)^3 = 0.386$.

## Binomial Probabilities

In a binomial setting, we can define a random variable (say, $X$) as the number of successes in $n$ independent trials. What's the probability distribution of $X$? Let's see if an example can help shed some light on this question.

## EXAMPLE Inheriting Blood Type

### *Calculating binomial probabilities*

Each child of a particular set of parents has probability 0.25 of having type O blood. Genetics says that children receive genes from each of their parents independently. If these parents have 5 children, the count $X$ of children with type O blood is a binomial random variable with $n = 5$ trials and probability $p = 0.25$ of success on each trial. In this setting, a child with type O blood is a "success" (S) and a child with another blood type is a "failure" (F).

What's $P(X = 0)$? That is, what's the probability that *none* of the 5 children has type O blood? It's the chance that all 5 children *don't* have type O blood. The probability that any one of this couple's children doesn't have type O blood is $1 - 0.25 = 0.75$ (complement rule). By the multiplication rule for independent events (Chapter 5),

$$P(X = 0) = P(\text{FFFFF}) = (0.75)(0.75)(0.75)(0.75)(0.75) = (0.75)^5 = 0.2373$$

How about $P(X = 1)$? There are several different ways in which exactly 1 of the 5 children could have type O blood. For instance, the first child born might have type O blood, while the remaining 4 children don't have type O blood. The probability that this happens is

$$P(\text{SFFFF}) = (0.25)(0.75)(0.75)(0.75)(0.75) = (0.25)(0.75)^4$$

Alternatively, Child 2 could be the one that has type O blood. The corresponding probability is

$$P(\text{FSFFF}) = (0.75)(0.25)(0.75)(0.75)(0.75) = (0.25)(0.75)^4$$

There are three more possibilities to consider:

$$P(\text{FFSFF}) = (0.75)(0.75)(0.25)(0.75)(0.75) = (0.25)(0.75)^4$$

$$P(\text{FFFSF}) = (0.75)(0.75)(0.75)(0.25)(0.75) = (0.25)(0.75)^4$$

$$P(\text{FFFFS}) = (0.75)(0.75)(0.75)(0.75)(0.25) = (0.25)(0.75)^4$$

In all, there are five different ways in which exactly 1 child would have type O blood, each with the same probability of occurring. As a result,

$$\begin{aligned} P(X = 1) &= P(\text{exactly 1 child with type O blood}) \\ &= P(SFFFF) + P(FSFFF) + P(FFSFF) + P(FFFSF) + P(FFFFS) \\ &= 5(0.25)(0.75)^4 = 0.39551 \end{aligned}$$

There's about a 40% chance that exactly 1 of the couple's 5 children will have type O blood.

Let's continue with the scenario from the previous example. What if we wanted to find $P(X = 2)$, the probability that exactly 2 of the couple's children have type O blood? Because the method doesn't depend on the specific setting, we use "S" for success and "F" for failure for short.

Do the work in two stages, as shown in the example.

- Find the probability that a specific 2 of the 5 tries—say, the first and the third—give successes. This is the outcome SFSFF. Because tries are independent, the multiplication rule for independent events applies. The probability we want is

$$\begin{aligned} P(SFSFF) &= P(S)P(F)P(S)P(F)P(F) \\ &= (0.25)(0.75)(0.25)(0.75)(0.75) \\ &= (0.25)^2(0.75)^3 \end{aligned}$$

- Observe that *any one arrangement* of 2 S's and 3 F's has this same probability. This is true because we multiply together 0.25 twice and 0.75 three times whenever we have 2 S's and 3 F's. The probability that $X = 2$ is the probability of getting 2 S's and 3 F's in any arrangement whatsoever. Here are all the possible arrangements:

| | | | | |
|---|---|---|---|---|
| SSFFF | SFSFF | SFFSF | SFFFS | FSSFF |
| FSFSF | FSFFS | FFSSF | FFSFS | FFFSS |

There are 10 of them, all with the same probability. The overall probability of 2 successes is therefore

$$P(X = 2) = 10(0.25)^2(0.75)^3 = 0.26367$$

The pattern of this calculation works for any binomial probability. That is,

$$\begin{aligned} P(X = k) &= P(\text{exactly } k \text{ successes in } n \text{ trials}) \\ &= \text{number of arrangements} \cdot p^k(1 - p)^{n-k} \end{aligned}$$

To use this formula, we must count the number of arrangements of $k$ successes in $n$ observations. This number is called the **binomial coefficient**. We use the following fact to do the counting without actually listing all the arrangements.

## Teaching Tip

Students may already be familiar with binomial coefficients if they have studied how to expand binomials such as $(a + b)^{12}$. Another way to identify the binomial coefficients is to use Pascal's triangle:

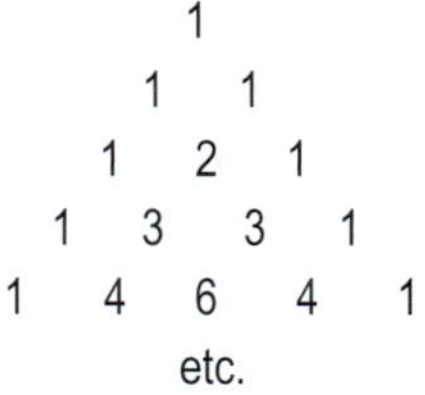


If $n = 4$, look in the row where the second value is 4. The coefficient for $P(X = 0)$ is 1, the coefficient for $P(X = 1)$ is 4, and so on.

**DEFINITION: Binomial coefficient**

The number of ways of arranging $k$ successes among $n$ observations is given by the **binomial coefficient**

$$\binom{n}{k} = \frac{n!}{k!(n-k)!}$$

for $k = 0, 1, 2, \ldots, n$ where

$$n! = n(n-1)(n-2) \cdot \ldots \cdot (3)(2)(1)$$

and $0! = 1$.

The formula for binomial coefficients uses the **factorial** notation. For any positive whole number $n$, its factorial $n!$ is

$$n! = n(n-1)(n-2) \cdot \ldots \cdot (3)(2)(1)$$

We also define $0! = 1$.

The larger of the two factorials in the denominator of a binomial coefficient will cancel much of the $n!$ in the numerator. For example, the binomial coefficient we need to find the probability that exactly 2 of the couple's 5 children inherit type O blood is

$$\binom{5}{2} = \frac{5!}{2!3!} = \frac{(5)(4)(\cancel{3})(\cancel{2})(\cancel{1})}{(2)(1)(\cancel{3})(\cancel{2})(\cancel{1})} = \frac{(5)(4)}{(2)(1)} = 10$$

Some people prefer the notation ${}_5C_2$ instead of $\binom{5}{2}$ for the binomial coefficient.

*The binomial coefficient is* $\binom{5}{2}$ *not related to the fraction* $\frac{5}{2}$. A helpful way to remember its meaning is to read it as "5 choose 2." Binomial coefficients have many uses, but we are interested in them only as an aid to finding binomial probabilities. If you need to compute a binomial coefficient, use your calculator.

## TECHNOLOGY TIP

Remind students that they can watch a video presentation of this technology corner (for the TI-83/84, TI-89, or TI-Nspire) at www.whfreeman.com/tps5e.

## Teaching Tip

Students aren't expected to calculate binomial coefficients by hand on the AP® exam. This is definitely a job for technology!

## 12. TECHNOLOGY CORNER — BINOMIAL COEFFICIENTS ON THE CALCULATOR

TI-Nspire instructions in Appendix B; HP Prime instructions on the book's Web site.

To calculate a binomial coefficient like $\binom{5}{2}$ on the TI-83/84 or TI-89, proceed as follows:

**TI-83/84**

- Type 5, press MATH, arrow over to PRB, choose nCr, and press ENTER. Then type 2 and press ENTER again to execute the command 5 nCr 2.

```
NORMAL FLOAT AUTO REAL RADIAN CL
5 nCr 2
                                10
```

**TI-89**

- From the home screen, press 2nd 5 (MATH), choose Probability, nCr(, and press ENTER. Complete the command nCr(5,2) and press ENTER.

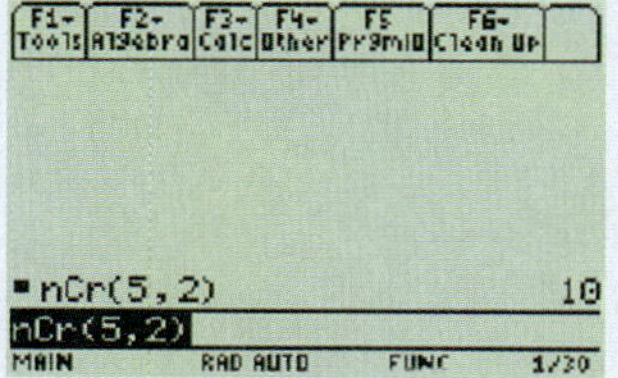

The binomial coefficient $\binom{n}{k}$ counts the number of different ways in which $k$ successes can be arranged among $n$ trials. The binomial probability $P(X = k)$ is this count multiplied by the probability of any one specific arrangement of the $k$ successes. Here is the result we seek.

**BINOMIAL PROBABILITY FORMULA**

If X has the binomial distribution with $n$ trials and probability $p$ of success on each trial, the possible values of X are 0, 1, 2, . . . , $n$. If $k$ is any one of these values,

$$P(X = k) = \binom{n}{k} p^k (1 - p)^{n-k}$$

With our formula in hand, we can now calculate any binomial probability.

## EXAMPLE Inheriting Blood Type

### *Using the binomial probability formula*

PROBLEM: Each child of a particular set of parents has probability 0.25 of having type O blood. Suppose the parents have 5 children.

(a) Find the probability that exactly 3 of the children have type O blood.

(b) Should the parents be surprised if more than 3 of their children have type O blood? Justify your answer.

SOLUTION: Let $X$ = the number of children with type O blood. We know that $X$ has a binomial distribution with $n = 5$ and $p = 0.25$.

(a) We want to find $P(X = 3)$.

$$P(X = 3) = \binom{5}{3}(0.25)^3(0.75)^2 = 10(0.25)^3(0.75)^2 = 0.08789$$

There is about a 9% chance that exactly 3 of the 5 children have type O blood.

(b) To answer this question, we need to find $P(X > 3)$.

$$P(X > 3) = P(X = 4) + P(X = 5) = \binom{5}{4}(0.25)^4(0.75)^1 + \binom{5}{5}(0.25)^5(0.75)^0$$
$$= 5(0.25)^4(0.75)^1 + 1(0.25)^5(0.75)^0 = 0.01465 + 0.00098 = 0.01563$$

Because there's only about a 1.5% chance of having more than 3 children with type O blood, the parents should definitely be surprised if this happens.

**For Practice** *Try Exercises*  75 *and* 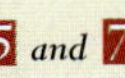 77

We could also use the calculator's `binompdf` and `binomcdf` commands to perform the calculations in the previous example. The following Technology Corner shows how to do it.

**AP® Exam Formula Sheet**

Let your students know that the formula for binomial probability is included on the formula sheet that is given to students on both sections of the AP® exam. However, the formula for the binomial coefficient is *not* provided on the formula sheets.

**ALTERNATE EXAMPLE**

### *Rolling doubles*

PROBLEM: When rolling two dice, the probability of rolling doubles is 1/6. Suppose that a game player rolls the dice 4 times, hoping to roll doubles.

**(a)** Find the probability that the player gets doubles twice in 4 attempts.

**(b)** Should the player be surprised if he gets doubles more than twice in 4 attempts? Justify your answer.

SOLUTION: Let $X$ = number of doubles. $X$ has a binomial distribution with $n = 4$ and $p = 1/6$.

**(a)** $P(X = 2) = \binom{4}{2}\left(\frac{1}{6}\right)^2\left(\frac{5}{6}\right)^2 = 6\left(\frac{1}{6}\right)^2\left(\frac{5}{6}\right)^2 = 0.116$. There is an 11.6% chance of getting exactly two doubles in 4 attempts.

**(b)** $P(X > 2) = P(X = 3) + P(X = 4) = \binom{4}{3}\left(\frac{1}{6}\right)^3\left(\frac{5}{6}\right)^1 + \binom{4}{4}\left(\frac{1}{6}\right)^4\left(\frac{5}{6}\right)^0 = 0.015 + 0.001 = 0.016$.

Because there is only a 1.6% chance of getting more than two doubles in 4 rolls, the player should be surprised if this happens.

### TECHNOLOGY TIP

Make sure students understand the difference between binompdf and binomcdf. The "c" in cdf stands for "cumulative," which should help students remember that cdf finds $P(X \leq k)$. In other words, it finds the sum $P(X = 0) + P(X = 1) + \cdots + P(X = k)$.

### TECHNOLOGY TIP

For students with older operating systems on the TI-83+ and TI-84, the Catalog Help application helps students remember the syntax for the `binompdf` and `binomcdf` functions. Newer calculators come equipped with this application, but you will have to link it to students with older calculators. Once the application is turned on, students can press the + button instead of the ENTER button after highlighting the `binompdf` or `binomcdf` function to be reminded in what order to enter the values. This also works for the Normal and geometric distribution options in the distribution menu.

### TECHNOLOGY TIP

If you enter only the values of $n$ and $p$ in the `binompdf` command, the calculator will display the probabilities for all possible outcomes of the binomial random variable, from $P(X = 0)$ to $P(X = n)$. Likewise, if you enter only the values of $n$ and $p$ in the `binomcdf` command, the calculator will display the cumulative probabilities for all possible outcomes of the binomial random variable, from $P(X \leq 0)$ to $P(X \leq n)$.

### TECHNOLOGY TIP

When using the `binomcdf` command on the TI-89, students can choose both the lower bound and the upper bound. To find $P(X > 3)$ in the example, enter `NumTrials,n:5,Prob Success,p:0.25,lower value:4,uppervalue:5.`

## 13. TECHNOLOGY CORNER BINOMIAL PROBABILITY ON THE CALCULATOR

**TI-Nspire instructions in Appendix B; HP Prime instructions on the book's Web site.**

There are two handy commands on the TI-83/84 and TI-89 for finding binomial probabilities: `binompdf` and `binomcdf`. The inputs for both commands are the number of trials $n$, the success probability $p$, and the values of interest for the binomial random variable $X$.

`binompdf(n,p,k)` computes $P(X = k)$
`binomcdf(n,p,k)` computes $P(X \leq k)$

Let's use these commands to confirm our answers in the previous example.

(a) Find the probability that exactly 3 of the children have type O blood.

**TI-83/84**

- Press [2nd] [VARS] (DISTR) and choose `binompdf(`.
- **OS 2.55 or later:** In the dialog box, enter these values: `trials:5, p:0.25, x value:3`, choose Paste, and then press [ENTER]. **Older OS:** Complete the command `binompdf(5,0.25,3)` and press [ENTER].

**TI-89**

- In the Stats/List Editor, Press [F5] `(Distr)` and choose `Binomial Pdf`.
- In the dialog box, enter these values: `Num Trials, n:5, Prob Success,p:0.25, X value:3`, and then choose [ENTER].

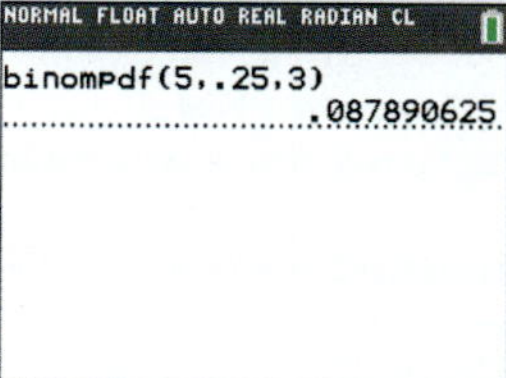


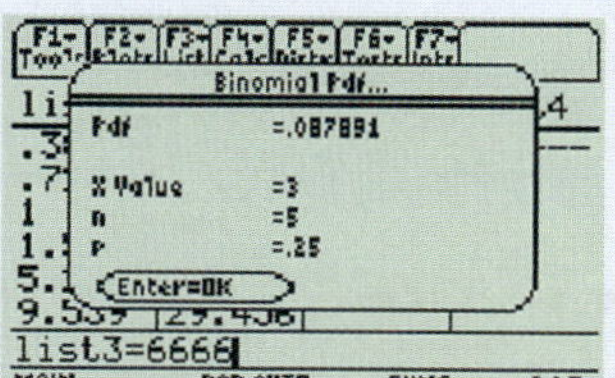


These results agree with our previous answer using the binomial probability formula: 0.08789.

(b) Should the parents be surprised if more than 3 of their children have type O blood?
To find $P(X > 3)$, use the complement rule:

$P(X > 3) = 1 - P(X \leq 3) = 1 -$ `binomcdf(5,0.25,3)`

- Press [2nd] [VARS] (DISTR) and choose `binomcdf(`.
- **OS 2.55 or later:** In the dialog box, enter these values: `trials:5, p:0.25, x value:3`, choose Paste, and then press [ENTER]. Subtract this result from 1 to get the answer. **Older OS:** Complete the command `binomcdf(5,0.25,3)` and press [ENTER]. Subtract this result from 1 to get the answer.

- In the Stats/List Editor, Press [F5] `(Distr)` and choose `Binomial Cdf...`.
- In the dialog box, enter these values: `Num Trials, n:5, Prob Success,p:0.25, lower value:0, upper value:3` and then choose [ENTER]. Subtract this result from 1 to get the answer.

We could also have done the calculation for part (b) as $P(X > 3) = P(X = 4) + P(X = 5)$ = `binompdf(5,0.25,4) + binompdf(5,0.25,5)` = 0.01465 + 0.00098 = 0.01563.

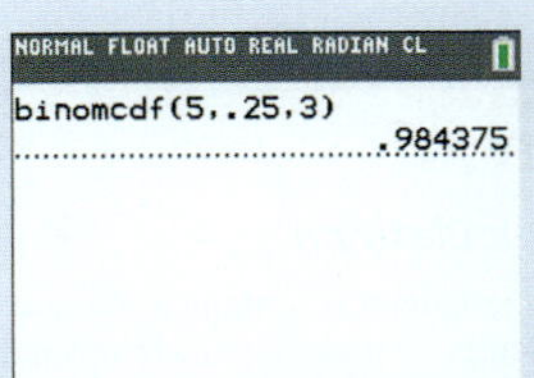


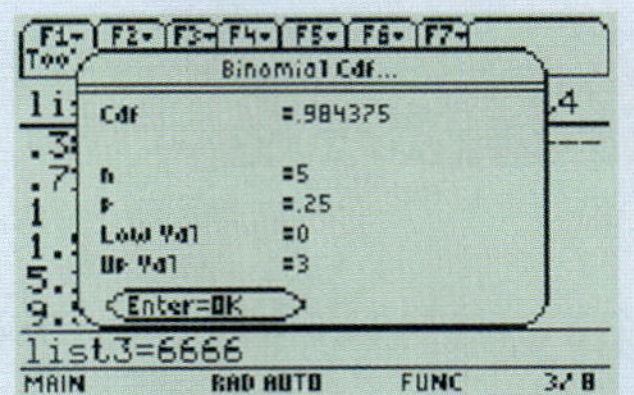


Now we subtract from 1 to get the desired answer: 1 − 0.984375 = 0.015625. This result agrees with our previous answer using the binomial probability formula: 0.01563.

**AP® EXAM TIP** Don't rely on "calculator speak" when showing your work on free-response questions. Writing `binompdf(5,0.25,3)` = 0.08789 will *not* earn you full credit for a binomial probability calculation. At the very least, you must indicate what each of those calculator inputs represents. For example, "I used binompdf(trials:5,p:0.25,x value:3)."

Note the use of the complement rule to find $P(X > 3)$ in the Technology Corner: $P(X > 3) = 1 - P(X \leq 3)$. This is necessary because the calculator's `binomcdf(n,p,k)` command only computes the probability of getting $k$ *or fewer* successes in $n$ trials. Students often have trouble identifying the correct third input for the `binomcdf` command when a question asks them to find the probability of getting less than, more than, or at least so many successes.

Here's a helpful tip to avoid making such a mistake: write out the possible values of the variable, circle the ones you want to find the probability of, and cross out the rest. In the previous example, X can take values from 0 to 5 and we want to find $P(X > 3)$:

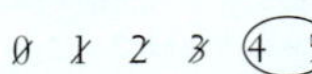


Crossing out the values from 0 to 3 shows why the correct calculation is $1 - P(X \leq 3)$.

Take another look at the solution in the blood-type example. The structure is much like the one we used when doing Normal calculations. Here is a revised summary box that describes the process.

### HOW TO FIND BINOMIAL PROBABILITIES

**Step 1: State the distribution and the values of interest.** Specify a binomial distribution with the number of trials $n$, success probability $p$, and the values of the variable clearly identified.

**Step 2: Perform calculations—show your work!** Do one of the following: (i) Use the binomial probability formula to find the desired probability; or (ii) use the `binompdf` or `binomcdf` command and label each of the inputs.

**Step 3: Answer the question.**

Here's an example that shows the method at work.

## AP® Exam Common Error

Emphasize the AP® exam tip on this page. Many students lose credit for not clearly identifying the distribution and the values of interest. Having students follow the three steps listed in the summary box on this page is the best way to train students to avoid this mistake.

ALTERNATE EXAMPLE

### *The last kiss*

Do people have a preference for the last thing they taste? Researchers at the University of Michigan designed a study to find out. The researchers gave 22 students five different Hershey's Kisses (milk chocolate, dark chocolate, crème, caramel, and almond) in random order and asked the student to rate each one. Participants were not told how many Kisses they would be tasting. However, when the 5th and final Kiss was presented, participants were told that it would be their last one. Of the 22 students, 14 of them gave the final Kiss the highest rating. http://www.sitemaker.umich.edu/eob/files/obrienellsworth2012.pdf

PROBLEM: Assume that the participants in the study don't have a special preference for the last thing they taste. That is, assume that the probability a person would prefer the last Kiss tasted is $p = 0.20$.

**(a)** What is the probability that exactly 5 of the 22 participants would prefer the last Kiss they tried?

**(b)** What is the probability that 14 or more of the 22 participants would prefer the last Kiss they tried?

SOLUTION:

**(a)** Step 1: State the distribution and the values of interest. Let $X$ = the number of participants who prefer the last Kiss they taste. $X$ has a binomial distribution with $n = 22$ and $p = 0.20$. We want to find $P(X = 5)$.

Step 2: Perform calculations—show your work!

$P(X = 5) = \binom{22}{5}(0.20)^5(0.80)^{17} =$ 0.1898. *Using technology*: The command `binompdf(trials:22,p:0.20, x value:5)` gives 0.1898.

Step 3: Answer the question. There is about a 19% chance that exactly 5 participants would choose the last Kiss, assuming that they have no special preference for the last thing they taste.

**(b)** Step 1: State the distribution and the values of interest. Let $X$ = the number of participants who prefer the last Kiss they taste. $X$ has a binomial distribution with $n = 22$ and $p = 0.20$. We want to find $P(X \geq 14)$.

Step 2: Perform calculations—show your work! $P(X \geq 14) = 1 - P(X \leq 13)$. *Using technology*: The command `1 − binomcdf(trials:22,p:0.20, x value:13)` gives 0.00001.

Step 3: Answer the question. There is about a 0.001% chance that 14 or more participants would choose the last Kiss, assuming that they have no special preference for the last thing they taste. Because this probability is so small, there is convincing evidence that the participants have a preference for the last thing they taste. It is almost impossible to get 14 or more just by chance.

## EXAMPLE Free Lunch?

### *Binomial calculations*

A local fast-food restaurant is running a "Draw a three, get it free" lunch promotion. After each customer orders, a touch-screen display shows the message "Press here to win a free lunch." A computer program then simulates one card being drawn from a standard deck. If the chosen card is a 3, the customer's order is free. Otherwise, the customer must pay the bill.

PROBLEM:

(a) All 12 players on a school's basketball team place individual orders at the restaurant. What is the probability that exactly 2 of them win a free lunch?

(b) If 250 customers place lunch orders on the first day of the promotion, what's the probability that fewer than 10 win a free lunch?

SOLUTION:

(a) Step 1: State the distribution and the values of interest. Let $X$ = the number of players who win a free lunch. There are 12 independent trials of the chance process, each with success probability 4/52 (because there are 4 threes in a standard deck of 52 cards). So $X$ has a binomial distribution with $n = 12$ and $p = 4/52$. We want to find $P(X = 2)$.

Step 2: Perform calculations—show your work! The binomial probability formula gives

$$P(X = 2) = \binom{12}{2}\left(\frac{4}{52}\right)^2\left(\frac{48}{52}\right)^{10} = 0.1754$$

*Using technology:* The command `binompdf(trials:12,p:4/52,x value:2)` also gives 0.1754.

Step 3: Answer the question. There is about a 17.5% probability that exactly 2 players will win a free lunch.

(b) Step 1: State the distribution and the values of interest. Let $Y$ = the number of customers who win a free lunch. There are 250 independent trials of the chance process, each with success probability 4/52. So $Y$ has a binomial distribution with $n = 250$ and $p = 4/52$. We want to find $P(Y < 10)$.

Step 2: Perform calculations—show your work! The values of $Y$ that interest us are

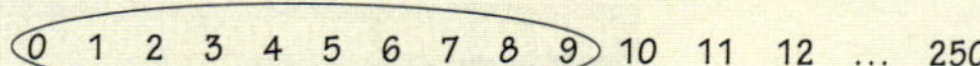

0 1 2 3 4 5 6 7 8 9 10 11 12 ... 250

To use the binomial formula, we would have to add up the probabilities for $Y = 0, Y = 1, \ldots, Y = 9$. That's too much work! The better option is to use technology:

$P(Y < 10) = P(Y \leq 9) =$ `binomcdf(trials:250,p:4/52,xvalue:9)` $= 0.00613$

Step 3: Answer the question. There is almost no chance that fewer than 10 customers will win a free lunch. If this actually happened, the customers should be suspicious about the restaurant's claim.

**For Practice** *Try Exercise* 79

## CHECK YOUR UNDERSTANDING

To introduce her class to binomial distributions, Mrs. Desai gives a 10-item, multiple-choice quiz. The catch is, students must simply guess an answer (A through E) for each question. Mrs. Desai uses her computer's random number generator to produce the answer key, so that each possible answer has an equal chance to be chosen. Patti is one of the students in this class. Let X = the number of Patti's correct guesses.

1. Show that X is a binomial random variable.
2. Find $P(X = 3)$. Explain what this result means.
3. To get a passing score on the quiz, a student must guess correctly at least 6 times. Would you be surprised if Patti earned a passing score? Compute an appropriate probability to support your answer.

## Mean and Standard Deviation of a Binomial Distribution

What does the probability distribution of a binomial random variable look like? The table below shows the possible values and corresponding probabilities for X = the number of children with type O blood. This is a binomial random variable with $n = 5$ and $p = 0.25$. Figure 6.14 shows a histogram of the probability distribution.

| Value $x_i$: | 0 | 1 | 2 | 3 | 4 | 5 |
|---|---|---|---|---|---|---|
| Probability $p_i$: | 0.23730 | 0.39551 | 0.26367 | 0.08789 | 0.01465 | 0.00098 |

Let's describe what we see.

*Shape:* The probability distribution of X is skewed to the right. Because the chance that any one of the couple's children inherits type O blood is 0.25, it's quite likely that 0, 1, or 2 of the children will have type O blood. Larger values of X are much less likely.

*Center:* The median number of children with type O blood is 1 because that's where the 50th percentile of the distribution falls. How about the mean? It's

$$\mu_X = \Sigma x_i p_i = (0)(0.23730) + (1)(0.39551) + \cdots + (5)(0.00098) = 1.25$$

**FIGURE 6.14** Histogram showing the probability distribution of the binomial random variable $X$ = number of children with type O blood in a family with 5 children.

So the expected number of children with type O blood in families like this one with 5 children is 1.25.

*Spread:* The variance of X is

$$\sigma_X^2 = \Sigma(x_i - \mu_X)^2 p_i = (0 - 1.25)^2(0.23730) + (1 - 1.25)^2(0.39551) + \cdots + (5 - 1.25)^2(0.00098) = 0.9375$$

So the standard deviation of X is

$$\sigma_X = \sqrt{0.9375} = 0.968$$

The number of children with type O blood will typically differ from the mean by about 0.968 in families like this one with 5 children.

### ANSWERS TO CYU

1. *Binary?* "Success" = question answered correctly. "Failure" = question not answered correctly. *Independent?* The computer randomly assigned correct answers to the questions, so knowing the result of one trial (question) should not tell you anything about the result on any other trial. *Number?* $n = 10$. *Success?* The probability of success is $p = 0.20$ for each trial.

2. $P(X = 3) = \binom{10}{3}(0.2)^3(0.8)^7 = 0.2013$. There is a 20% chance that Patti will answer exactly 3 questions correctly.

3. $P(X \geq 6) = 1 - P(X \leq 5) = 1 - 0.9936 = 0.0064$. There is only a 0.0064 probability that a student would get 6 or more correct, so we would be quite surprised if Patti was able to pass.

### ALTERNATE EXAMPLE

#### *Rolling doubles*

If $X$ = number of doubles in 4 attempts, then $X$ follows a binomial distribution with $n = 4$ and $p = 1/6$. Here is the probability distribution and a histogram of the probability distribution.

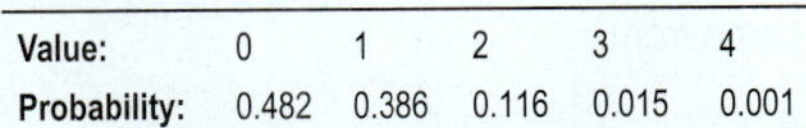

| Value: | 0 | 1 | 2 | 3 | 4 |
|---|---|---|---|---|---|
| Probability: | 0.482 | 0.386 | 0.116 | 0.015 | 0.001 |

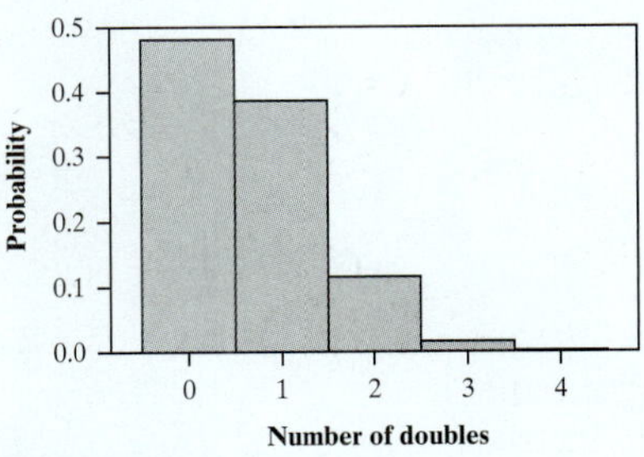


The mean of this distribution is $\mu_X = 0(0.482) + 1(0.386) + \cdots + 4(0.001) = 4(1/6) = 0.667$. If many students were to attempt to roll doubles 4 times, the average number of doubles would be about 0.667.

The variance of this distribution is $\sigma_X^2 = (0 - 0.667)^2(0.482) + (1 - 0.667)^2(0.386) + \cdots + (4 - 0.667)^2(0.001) = 4(1/6)(5/6) = 0.556$, so the standard deviation is $\sigma_X = \sqrt{0.556} = 0.746$. If many students were to attempt to roll doubles 4 times, the number of doubles would typically vary by about 0.746 from the mean (0.667).

Did you think about why the mean is $\mu_X = 1.25$? Because each child has a 0.25 chance of inheriting type O blood, we'd expect one-fourth of the 5 children to have this blood type. In other words, $\mu_X = 5(0.25) = 1.25$. This method can be used to find the mean of any binomial random variable with parameters $n$ and $p$:

$$\mu_X = np$$

There are fairly simple formulas for the variance and standard deviation, too, but they aren't as easy to explain:

$$\sigma_X^2 = np(1-p) \quad \text{and} \quad \sigma_X = \sqrt{np(1-p)}$$

For our family with 5 children,

$$\sigma_X^2 = 5(0.25)(0.75) = 0.9375 \quad \text{and} \quad \sigma_X = \sqrt{0.9375} = 0.968$$

**MEAN AND STANDARD DEVIATION OF A BINOMIAL RANDOM VARIABLE**

If a count $X$ of successes has the binomial distribution with number of trials $n$ and probability of success $p$, the **mean** and **standard deviation** of $X$ are

$$\mu_X = np$$
$$\sigma_X = \sqrt{np(1-p)}$$

*Remember that these formulas work only for binomial distributions. They can't be used for other distributions.*

THINK ABOUT IT

**Where do the binomial mean and variance formulas come from?** We can derive the formulas for the mean and variance of a binomial distribution using what we learned about combining random variables in Section 6.2. Let's start with the random variable $B$ that's described by the following probability distribution:

| Value $b_i$: | 0 | 1 |
|---|---|---|
| Probability $p_i$: | $1-p$ | $p$ |

You can think of $B$ as representing the result of a single trial of some chance process. If a success occurs (probability $p$), then $B = 1$. If a failure occurs, then $B = 0$. Notice that the mean of $B$ is

$$\mu_B = \sum b_i p_i = (0)(1-p) + (1)(p) = p$$

and that the variance of $B$ is

$$\begin{aligned}\sigma_B^2 = \sum (b_i - \mu_B)^2 p_i &= (0-p)^2(1-p) + (1-p)^2 p \\ &= p(1-p)[p + (1-p)] \\ &= p(1-p)\end{aligned}$$

Now consider the random variable $X = B_1 + B_2 + \cdots + B_n$. We can think of $X$ as counting the number of successes in $n$ independent trials of this chance process, with each trial having success probability $p$. In other words, $X$ is a bino-

## AP® Exam Formula Sheet

Let your students know that the formulas for the mean and standard deviation of a binomial distribution are included on the formula sheet that is given to students on both sections of the AP® exam.

## Teaching Tip

We used $B$ to represent the random variable in this Think About It because a single trial in a binomial setting is called a Bernoulli trial. Students are not expected to derive these formulas on the AP® exam, but it is nice to show them where they came from.

mial random variable with parameters $n$ and $p$. By the rules from Section 6.2, the mean of X is

$$\mu_X = \mu_{B_1} + \mu_{B_2} + \cdots + \mu_{B_n} = p + p + \cdots + p = np$$

and the variance of X is

$$\sigma_X^2 = \sigma_{B_1}^2 + \sigma_{B_2}^2 + \cdots + \sigma_{B_n}^2$$
$$= p(1-p) + p(1-p) + \cdots + p(1-p)$$
$$= np(1-p)$$

The standard deviation of X is therefore

$$\sigma_X = \sqrt{np(1-p)}$$

## EXAMPLE Bottled Water versus Tap Water

### *Binomial distribution in action*

Mr. Bullard's AP® Statistics class did the Activity on page 346. There were 21 students in the class. If we assume that the students in his class could *not* tell tap water from bottled water, then each one was basically guessing, with a 1/3 chance of being correct. Let X = the number of students who correctly identified the cup containing the different type of water.

PROBLEM:

(a) Explain why $X$ is a binomial random variable.

(b) Find the mean and standard deviation of $X$. Interpret each value in context.

(c) Of the 21 students in the class, 13 made correct identifications. Are you convinced that Mr. Bullard's students could tell bottled water from tap water? Justify your answer.

SOLUTION:

(a) Assuming that students were just guessing, the Activity consisted of 21 repetitions of this chance process. Let's check the BINS:

- Binary? On each trial, "success" = correct identification; "failure" = incorrect identification.
- Independent? One student's result should tell us nothing about any other student's result.
- Number? There were $n = 21$ trials.
- Success? Each student had a $p = 1/3$ chance of guessing correctly.

Because $X$ is counting the number of successes in this binomial setting, it is a binomial random variable.

(b) The mean of $X$ is

$$\mu_X = np = 21(1/3) = 7$$

If the Activity were repeated many times with groups of 21 students who were just guessing, the average number of students who guess correctly would be about 7. The standard deviation of $X$ is

$$\sigma_X = \sqrt{np(1-p)} = \sqrt{21(1/3)(2/3)} = 2.16$$

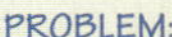

## ALTERNATE EXAMPLE

### *Tastes as good as the real thing?*

The makers of a diet cola claim that its taste is indistinguishable from the taste of the full-calorie version of the same cola. To investigate, an AP® Statistics student named Emily prepared small samples of each type of soda in identical cups. Then she had volunteers taste each cola in random order and try to identify which was the diet cola and which was the regular cola. Overall, 23 of the 30 subjects made the correct identification.

If we assume that the volunteers really couldn't tell the difference, then each one was guessing with a 1/2 chance of being correct. Let $X$ = the number of volunteers who correctly identify the colas.

PROBLEM:

**(a)** Explain why $X$ is a binomial random variable.

**(b)** Find the mean and the standard deviation of $X$. Interpret each value in context.

**(c)** Of the 30 volunteers, 23 made correct identifications. Does this give convincing evidence that the volunteers can taste the difference between the diet and regular colas?

SOLUTION:

**(a)** The chance process is each volunteer guessing which sample is the diet cola. *Binary?* Yes; guesses are either correct or incorrect. *Independent?* Yes; the results of one volunteer's guess tell us nothing about the results of other volunteers' guesses. *Number?* Yes; there are 30 trials. *Success?* Yes; the probability of guessing correctly is always 50%. Because $X$ is counting the number of successful guesses, $X$ is a binomial random variable with $n = 30$ and $p = 0.50$.

**(b)** The mean of $X$ is $\mu_X = np = 30(0.5) = 15$, and the standard deviation of $X$ is $\sigma_X = \sqrt{np(1-p)} = \sqrt{30(0.5)(1-0.5)} = 2.74$. If this experiment were repeated many times and the volunteers were randomly guessing, the average number of correct guesses would be about 15. Also, the number of correct guesses would typically vary by about 2.74 from the mean (15).

**(c)** $P(X \geq 23) = 1 - P(X \leq 22) =$ 1 − `binomcdf(trials:30,p:0.5, x value:22)` = 1 − 0.9974 = 0.0026. There is a very small chance that there would be 23 or more correct guesses if the volunteers couldn't tell the difference in the colas. Therefore, we have convincing evidence that the volunteers can taste the difference.

If the Activity were repeated many times with groups of 21 students who were just guessing, the number of correct identifications would typically differ from 7 by about 2.16.

(c) The class's result corresponds to $X = 13$, a value that's nearly 3 standard deviations above the mean. How likely is it that 13 or more of Mr. Bullard's students would guess correctly? It's

$$P(X \geq 13) = 1 - P(X \leq 12)$$

Using the calculator's `binomcdf(trials:21,p:1/3,xvalue:12)` command:

$$P(X \geq 13) = 1 - 0.9932 = 0.0068$$

The students had less than a 1% chance of getting so many right if they were all just guessing. This is strong evidence that some of the students in the class could tell bottled water from tap water.

**For Practice** *Try Exercise* 85

Figure 6.15 shows the probability distribution for the number of correct guesses in Mr. Bullard's class if no one can tell bottled water from tap water. As you can see from the graph, the chance of 13 or more guessing correctly is quite small.

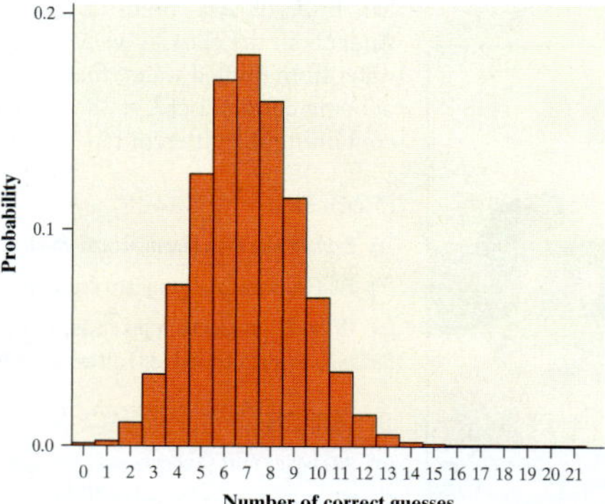


Although the histogram is slightly right-skewed (there's a long tail that extends out to $X = 21$), it looks like a Normal density curve might fit the bulk of the distribution fairly well.

**FIGURE 6.15** Histogram of the probability distribution for the binomial random variable $X$ = number of correct guesses in Mr. Bullard's class.

## CHECK YOUR UNDERSTANDING

Refer to the previous Check Your Understanding (page 397) about Mrs. Desai's special multiple-choice quiz on binomial distributions. We defined $X$ = the number of Patti's correct guesses.

1. Find $\mu_X$. Interpret this value in context.
2. Find $\sigma_X$. Interpret this value in context.
3. What's the probability that the number of Patti's correct guesses is more than 2 standard deviations above the mean? Show your method.

**TRM Airline Overbooking, Part II**

**TRM A Fair Coin?**

There are two additional resources that you can use when teaching the binomial distribution. The Airline Overbooking problem is a continuation of the Airline Overbooking problem from Chapter 5, but you can use this one even if you didn't do the first one. The A Fair Coin? problem reviews binomial probabilities and foreshadows the ideas of Type I and Type II errors. Find them by clicking on the link in the TE-book, opening the TRCD, or logging onto the book's companion Web site www.whfreeman.com/tps5e (teacher log-in required).

### ANSWERS TO CYU

1. $\mu_X = 10(0.20) = 2$. If many students took the quiz, we would expect students to get about 2 answers correct, on average.
2. $\sigma_X = \sqrt{10(0.20)(0.80)} = 1.265$. If many students took the quiz, we would expect individual students' scores to typically vary from the mean of 2 correct answers by about 1.265 correct answers.
3. $P(X > 2 + 2(1.265)) = P(X > 4.53) = 1 - P(X \leq 4) = 1 - 0.9672 = 0.0328$

## Binomial Distributions in Statistical Sampling

The binomial distributions are important in statistics when we wish to make inferences about the proportion $p$ of successes in a population. Here is an example involving a familiar product.

EXAMPLE

### Bad Flash Drives

*Binomial distributions and sampling*

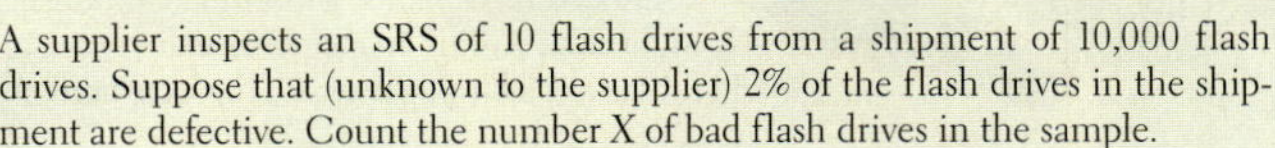

A supplier inspects an SRS of 10 flash drives from a shipment of 10,000 flash drives. Suppose that (unknown to the supplier) 2% of the flash drives in the shipment are defective. Count the number X of bad flash drives in the sample.

*This is not quite a binomial setting.* Removing 1 flash drive changes the proportion of bad flash drives remaining in the shipment. The conditional probability that the second flash drive chosen is bad changes when we know whether the first is good or bad. But removing 1 flash drive from a shipment of 10,000 changes the makeup of the remaining 9999 flash drives very little. The distribution of X is very close to the binomial distribution with $n = 10$ and $p = 0.02$. To illustrate this, let's compute the probability that none of the 10 flash drives is defective. Using the binomial distribution, it's

$$P(X = 0) = \binom{10}{0}(0.02)^0(0.98)^{10} = 0.8171$$

The actual probability of getting no defective flash drives is

$$P(\text{no defectives}) = \frac{9800}{10{,}000} \times \frac{9799}{9999} \times \frac{9798}{9998} \times \cdots \times \frac{9791}{9991} = 0.8170$$

Those two probabilities are quite close!

Almost all real-world sampling, such as taking an SRS from a population of interest, is done without replacement. As the previous example illustrates, sampling without replacement leads to a violation of the independence condition.

The flash drives example shows how we can use binomial distributions in the statistical setting of selecting an SRS. When the population is much larger than the sample, a count of successes in an SRS of size $n$ has approximately the binomial distribution with $n$ equal to the sample size and $p$ equal to the proportion of successes in the population. What counts as "much larger"? In practice, the binomial distribution gives a good approximation as long as we don't sample more than 10% of the population. We refer to this as the **10% condition**.

**10% CONDITION**

When taking an SRS of size $n$ from a population of size N, we can use a binomial distribution to model the count of successes in the sample as long as $n \leq \frac{1}{10}N$.

### Teaching Tip

Because all real-world sampling is done without replacement, we will always need to check the 10% condition when working with sample data in Chapters 6–12.

### Teaching Tip

Some students mistakenly believe that the 10% condition implies that we want *small* samples. This is almost never the case. If we have a sample that is larger than 10% of the population, it just means that we shouldn't use the binomial distribution. We would need to use the hypergeometric distribution instead (which isn't covered on the AP® exam).

ALTERNATE EXAMPLE

### *NASCAR cards and cereal boxes*

In the NASCAR Cards and Cereal Boxes example from Section 5.1, we read about a cereal company that put 1 of 5 different cards into each box of cereal. Each card featured a different driver: Jeff Gordon, Dale Earnhardt, Jr., Tony Stewart, Danica Patrick, or Jimmie Johnson. Suppose that the company printed 20,000 of each card, so there were 100,000 total boxes of cereal with a card inside. If a person bought 6 boxes at random, what is the probability of getting no Danica Patrick cards?

Let $X$ be the number of Danica Patrick cards obtained from 6 different boxes of cereal. Because we are sampling without replacement, the trials are not independent. The distribution of $X$ is not quite binomial—but it is close. If we assume $X$ is binomial with $n = 6$ and $p = 0.2$, then

$$P(X = 0) = \binom{6}{0}(0.2)^0(0.8)^6 = 0.262144$$

The actual probability, using the general multiplication rule, is

$$P(\text{no Danica Patrick cards}) = \left(\frac{80{,}000}{100{,}000}\right)\left(\frac{79{,}999}{99{,}999}\right)\left(\frac{79{,}998}{99{,}998}\right)\left(\frac{79{,}997}{99{,}997}\right)\left(\frac{79{,}996}{99{,}996}\right)\left(\frac{79{,}995}{99{,}995}\right) = 0.262134$$

These two probabilities are quite close!

ALTERNATE EXAMPLE

### *Dead batteries*

Almost everyone has one—a drawer that holds miscellaneous batteries of all sizes. Suppose that your drawer contains 8 AAA batteries but only 6 of them are good. You need to choose 4 for your graphing calculator. If you randomly select 4 batteries, what is the probability that all 4 of them will work?

PROBLEM: Explain why the answer isn't

$$P(X = 4) = \binom{4}{4}(0.75)^4(0.25)^0 = 0.3164.$$

(The actual probability is 0.2143.)

SOLUTION: Because we are sampling without replacement, the selections of batteries aren't independent. We can ignore this problem if the sample we are selecting is less than 10% of the population. However, in this case we are sampling 50% of the population (4/8), so it is not reasonable to ignore the lack of independence and use the binomial distribution. This explains why the binomial probability is so different from the actual probability.

## Teaching Tip

The Normal approximation to the binomial distribution is not required for the AP® exam. If you are having trouble keeping up the pace, this would be a good topic to skip. Also, if your students are having trouble with binomial distributions, adding an alternate approach might make things more difficult. However, using a continuous probability distribution to approximate a discrete probability distribution is fairly common, so the connection between continuous and discrete distributions is nice to discuss if you have the time. If you do choose to skip this topic, make sure you don't skip the geometric distribution, which follows on page 404.

## Teaching Tip

If you choose to teach the Normal approximation to the binomial distribution, students might ask why they need to use a Normal approximation when the binomial distribution is more exact. This is a good question! We introduce it here because we will use the Normal approximation in later chapters when we perform inference about the proportion of successes in a population.

Here's an example that shows why it's important to check the 10% condition before calculating a binomial probability. You might recognize the setting from the first activity in the book (page 5).

EXAMPLE

## Hiring Discrimination—It Just Won't Fly!

### *Sampling without replacement*

An airline has just finished training 25 first officers—15 male and 10 female—to become captains. Unfortunately, only eight captain positions are available right now. Airline managers decide to use a lottery to determine which pilots will fill the available positions. Of the 8 captains chosen, 5 are female and 3 are male.

PROBLEM: Explain why the probability that 5 female pilots are chosen in a fair lottery is *not*

$$P(X = 5) = \binom{8}{5}(0.40)^5(0.60)^3 = 0.124$$

(The correct probability is 0.106.)

SOLUTION: The managers are sampling without replacement when they do the lottery. There's a 0.40 chance that the first pilot selected for a captain position is female. Once that person is chosen, the probability that the next pilot selected will be female is no longer 0.40. The binomial formula assumes that the conditional probability of success stays constant at 0.40 throughout the eight trials of this chance process. This calculation will be approximately correct if the success probability doesn't change too much—as long as we don't sample more than 10% of the population. In this case, managers are sampling 8 out of 25 pilots—almost 1/3 of the population. That explains why the binomial probability is off by about 17% (0.018/0.106) from the correct answer.

**For Practice** *Try Exercise*  87

**The Normal approximation to binomial distributions*** As $n$ gets larger, something interesting happens to the shape of the binomial distribution. Figure 6.16 shows histograms of binomial distributions for different values of $n$ and $p$. As the number of observations $n$ becomes larger, the binomial distribution gets close to a Normal distribution. You can investigate the relationship between $n$ and $p$ yourself using the *Normal Approximation to Binomial Distributions* applet at the book's Web site.

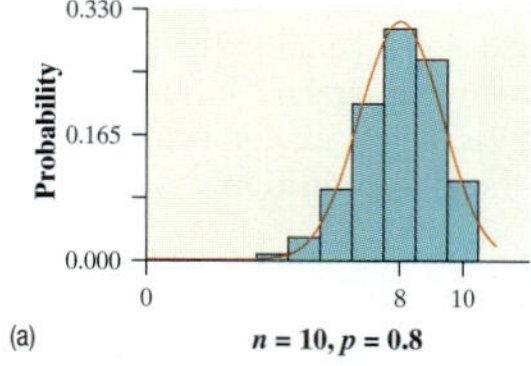


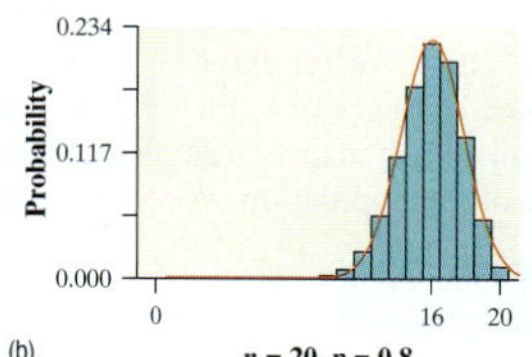


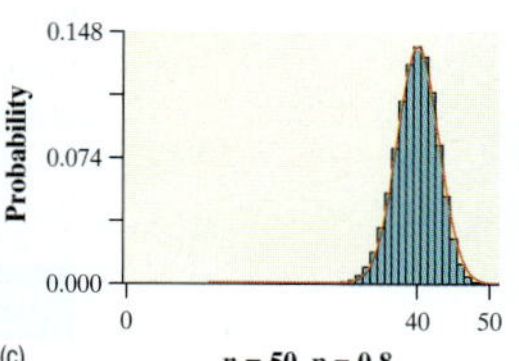


**FIGURE 6.16** Histograms of binomial distributions with (a) $n = 10$ and $p = 0.8$, (b) $n = 20$ and $p = 0.8$, and (c) $n = 50$ and $p = 0.8$. As $n$ increases, the shape of the probability distribution gets more and more Normal.

*This topic is not required for the AP® Statistics exam. Some teachers prefer to discuss this topic when presenting the sampling distribution of $\hat{p}$ (Chapter 7).

## APPLET

If you teach the Normal approximation, make sure to begin with the applet described on this page. It dynamically shows how the shape of the binomial distribution changes for different values of $n$ and $p$. Make sure to show combinations that illustrate that the distribution looks quite Normal when both $np$ and $n(1 - p)$ are at least 10 (the Large Counts condition discussed on the next page) but not nearly as Normal when this condition isn't met.

## Teaching Tip

In many older statistics books, the continuity correction is taught alongside the Normal approximation to the binomial. However, this is not a topic on the AP® exam, and we suggest that you avoid the unnecessary complications with your students.

When $n$ is large, we can use Normal probability calculations to approximate binomial probabilities. Here are the facts.

**NORMAL APPROXIMATION FOR BINOMIAL DISTRIBUTIONS: THE LARGE COUNTS CONDITION**

Suppose that a count $X$ of successes has the binomial distribution with $n$ trials and success probability $p$. When $n$ is large, the distribution of $X$ is approximately Normal with

$$\text{mean: } \mu_X = np \quad \text{and} \quad \text{standard deviation: } \sigma_X = \sqrt{np(1-p)}$$

As a rule of thumb, we will use the Normal approximation when $n$ is so large that

$$np \geq 10 \qquad \text{and} \qquad n(1-p) \geq 10$$

That is, the expected number of successes and failures are both at least 10. We refer to this as the **Large Counts condition**.

The Normal approximation is easy to remember because it says to act as if $X$ is Normal with exactly the same mean and standard deviation as the binomial. The accuracy of the Normal approximation improves as the sample size $n$ increases. It is most accurate for any fixed $n$ when $p$ is close to 1/2 and least accurate when $p$ is near 0 or 1. This is why the rule of thumb in the box depends on $p$ as well as $n$.

## EXAMPLE Attitudes toward Shopping

### *Normal approximation to a binomial*

Sample surveys show that fewer people enjoy shopping than in the past. A survey asked a nationwide random sample of 2500 adults if they agreed or disagreed that "I like buying new clothes, but shopping is often frustrating and time-consuming."[11] The population that the poll wants to draw conclusions about is all U.S. residents aged 18 and over.

PROBLEM: Suppose that exactly 60% of all adult U.S. residents would say "Agree" if asked the same question. Let $X$ = the number in the sample who agree.

(a) Show that $X$ is approximately a binomial random variable.

(b) Check the conditions for using a Normal approximation in this setting.

(c) Use a Normal distribution to estimate the probability that 1520 or more of the sample agree.

SOLUTION:

(a) Let's check the BINS.

- Binary? Success = agree that shopping is frustrating, failure = don't agree.
- Independent? The trials are not independent: the conditional probability of a success changes due to the sampling without replacement. But the *10% condition* is met because 2500 people is much less than 10% of all U.S. adult residents.
- Number? There are $n$ = 2500 trials of this chance process.
- Success? There is the same probability of selecting an adult who agrees on each trial: $p = 0.6$.

**ALTERNATE EXAMPLE**

### *Teens and debit cards*

In a survey of 506 teenagers aged 14 to 18, subjects were asked a variety of questions about personal finance (http://www.nclnet.org/personal-finance/66-teens-and-money/120-ncl-survey-teens-and-financial-education).One question asked teens if they had a debit card.

PROBLEM: Suppose that exactly 10% of teens aged 14 to 18 have debit cards. Let $X$ = the number of teens in a random sample of size 506 who have a debit card.

**(a)** Show that the distribution of $X$ is approximately binomial.

**(b)** Check the conditions for using a Normal approximation in this setting.

**(c)** Use a Normal distribution to estimate the probability that 40 or fewer teens in the sample have debit cards.

SOLUTION:

**(a)** *Binary?* Yes; teens either have a debit card or they don't. *Independent?* No; because we are sampling without replacement. However, because the sample size ($n$ = 506) is much less than 10% of the population size (there are millions of teens aged 14 to 18), the responses will be very close to independent. *Number?* Yes; there is a fixed sample size, $n$ = 506. *Success?* Yes; the unconditional probability of selecting a teen with a debit card is 10%.

**(b)** We need to check the Large Counts condition. Because $np = 506(0.10) = 50.6$ and $n(1-p) = 506(0.90) = 455.4$ are both at least 10, we should be safe using the Normal approximation.

**(c)** Step 1: State the distribution and the values of interest. $\mu_X = np = 506(0.10) = 50.6$ and $\sigma_X = \sqrt{np(1-p)} = \sqrt{506(0.1)(0.9)} = 6.75$. Thus, $X$ is approximately Normally distributed with mean 50.6 and standard deviation 6.75. We want to find $P(X \leq 40)$.

Step 2: Perform calculations—show your work! Standardizing the boundary value gives $z = \frac{40 - 50.6}{6.75} = -1.57$. Using Table A, $P(Z \leq -1.57) = 0.0582$. *Using technology*: The command `normalcdf(lower:−1000,upper:40,μ:50.6,σ:6.75)` gives an area of 0.0582. (*Note*: The probability using the binomial distribution is 0.064.)

Step 3: Answer the question. There is about a 6% chance that 40 or fewer teens in a sample of size 506 will have a debit card.

So the number in our sample who agree that shopping is frustrating is a random variable $X$ having roughly the binomial distribution with $n = 2500$ and $p = 0.6$.

(b) We need to check the Large Counts condition. Because $np = 2500(0.6) = 1500$ and $n(1 - p) = 2500(0.4) = 1000$ are both at least 10, we should be safe using the Normal approximation.

(c) **Step 1: State the distribution and the values of interest.** Act as though the count $X$ has the Normal distribution with the same mean and standard deviation as the binomial distribution:

$$\mu = np = 2500(0.6) = 1500 \quad \text{and} \quad \sigma = \sqrt{np(1-p)} = \sqrt{2500(0.6)(0.4)} = 24.49$$

We want to find $P(X \geq 1520)$. Figure 6.17 shows this probability as the area under a Normal curve.

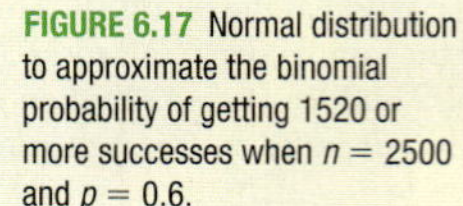


**FIGURE 6.17** Normal distribution to approximate the binomial probability of getting 1520 or more successes when $n = 2500$ and $p = 0.6$.

**Step 2: Perform calculations—show your work!** Standardizing the boundary value gives

$$z = \frac{1520 - 1500}{24.49} = 0.82$$

Using Table A, the probability we want is

$$P(Z \geq 0.82) = 1 - 0.7939 = 0.2061$$

*Using technology:* The command `normalcdf(lower:1520, upper:10000, μ:1500, σ:24.49)` gives an area of 0.2071.

**Step 3: Answer the question.** There is about a 21% chance of getting a sample in which 1520 or more agree with the statement.

**For Practice** *Try Exercise* **91**

We can also find the probability that 1520 or more of the sample agree that shopping is often frustrating and time-consuming using the command `1-binomcdf(2500,0.6,1519)`, which yields 0.2131. The Normal approximation, 0.2061, misses the more accurate binomial probability by about 0.007.

## Geometric Random Variables

In a binomial setting, the number of trials $n$ is fixed in advance, and the binomial random variable X counts the number of successes. The possible values of $X$ are 0, 1, 2, . . . , $n$. In other situations, the goal is to repeat a chance process *until a success occurs:*

- Roll a pair of dice until you get doubles.
- In basketball, attempt a three-point shot until you make one.
- Keep placing a $1 bet on the number 15 in roulette until you win.

These are all examples of a **geometric setting.** Although the number of trials isn't fixed in advance, the trials are independent and the probability of success remains constant.

**DEFINITION: Geometric setting**

A **geometric setting** arises when we perform independent trials of the same chance process and record the number of trials it takes to get one success. On each trial, the probability $p$ of success must be the same.

### Teaching Tip

Emphasize the difference between binomial and geometric random variables. Binomial random variables count the number of successes in a fixed number of trials. Geometric random variables count the number of trials it takes to get one success. This is why a geometric distribution is often referred to as a "wait time" distribution, because we are waiting for a success to occur.

Here's an Activity your class can try that involves a geometric setting.

## ACTIVITY | Is This Your Lucky Day?

**MATERIALS:**
Calculator or computer random number generator to select student names and days of the week

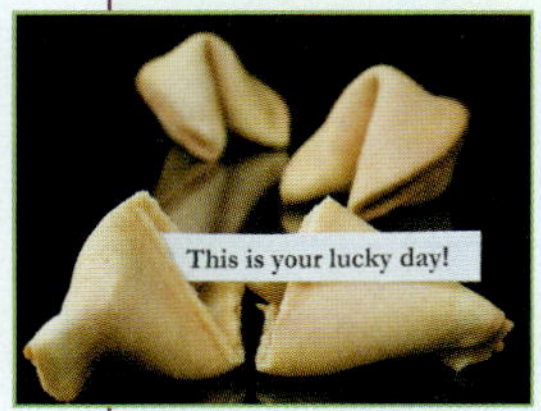


Your teacher is planning to give you 10 problems for homework. As an alternative, you can agree to play the Lucky Day Game. Here's how it works. A student will be selected at random from your class and asked to pick a day of the week (for instance, Thursday). Then your teacher will use technology to randomly choose a day of the week as the "lucky day." If the student picks the correct day, the class will have only one homework problem. If the student picks the wrong day, your teacher will select another student from the class at random. The chosen student will pick a day of the week and your teacher will use technology to choose a "lucky day." If this student gets it right, the class will have two homework problems. The game continues until a student correctly guesses the lucky day. Your teacher will assign a number of homework problems that is equal to the total number of guesses made by members of your class. Are you ready to play the Lucky Day Game?

1. Decide as a class about whether to "gamble" on the number of homework problems you will receive. You have 30 seconds.
2. Play the Lucky Day Game and see what happens!

In a geometric setting, if we define the random variable $Y$ to be the number of trials needed to get the first success, then $Y$ is called a **geometric random variable.** The probability distribution of $Y$ is a **geometric distribution.**

**DEFINITION: Geometric random variable and geometric distribution**

The number of trials $Y$ that it takes to get a success in a geometric setting is a **geometric random variable**. The probability distribution of $Y$ is a **geometric distribution** with parameter $p$, the probability of a success on any trial. The possible values of $Y$ are 1, 2, 3, . . . .

As with binomial random variables, it's important to be able to distinguish situations in which a geometric distribution does and doesn't apply.

## EXAMPLE The Lucky Day Game

### *Geometric settings and random variables*

The random variable of interest in this game is $Y$ = the number of picks it takes to correctly match the lucky day. Each pick is one trial of the chance process. Knowing the result of one student's pick tells us nothing about the result of any other pick. On each trial, the probability of a correct pick is 1/7.

This is a geometric setting. Because $Y$ counts the number of trials to get the first success, it is a geometric random variable with parameter $p = 1/7$.

**Activity Overview**

You can watch an overview of this Activity (for teachers) by clicking on the link in the TE-book, opening the TRCD, or logging onto the book's companion Web site www.whfreeman.com/tps5e (teacher log-in required).

### Teaching Tip

The value of a geometric random variable includes the trial on which the success occurs. For example, if the first success occurs on the 5th trial, the value of the geometric random variable is $Y = 5$. For this reason, it is helpful to define the variable as the "number of trials it takes to get a success" rather than the "number of trials until you get a success." Because of the somewhat ambiguous word "until," some students might not include the success trial.

**ALTERNATE EXAMPLE**

### *Monopoly*

In the board game Monopoly, one way to get out of jail is to roll doubles. Suppose that this was the only way a player could get out of jail. The random variable of interest in this example is $Y$ = number of attempts it takes to roll doubles one time. Each attempt is one trial of the chance process. Knowing the outcome of previous rolls does not tell us anything about future rolls. On each roll, the probability of success is 1/6. This is a geometric setting. Because $Y$ counts the number of attempts it takes to get doubles, it is a geometric random variable with parameter $p = 1/6$.

What is the probability that the first student picks correctly and wins the Lucky Day Game? It's $P(Y = 1) = 1/7$. That's also the class's chance of getting only one homework problem. For the class to have two homework problems, the first student selected must pick an incorrect day of the week and the second student must pick the lucky day correctly. The probability that this happens is $P(Y = 2) = (6/7)(1/7) = 0.1224$. Likewise, $P(Y = 3) = (6/7)(6/7)(1/7) = 0.1050$. In general, the probability that the first correct pick comes on the $k$th trial is $P(Y = k) = (6/7)^{k-1}(1/7)$. Let's summarize what we've learned about calculating a **geometric probability.**

**GEOMETRIC PROBABILITY FORMULA**

If $Y$ has the geometric distribution with probability $p$ of success on each trial, the possible values of $Y$ are 1, 2, 3, . . . . If $k$ is any one of these values,

$$P(Y = k) = (1 - p)^{k-1}p$$

With our formula in hand, we can now compute any geometric probability.

## EXAMPLE The Lucky Day Game

### *Calculating geometric probabilities*

PROBLEM: Let the random variable $Y$ be defined as in the previous example.

(a) Find the probability that the class receives exactly 10 homework problems as a result of playing the Lucky Day Game.

(b) Find $P(Y < 10)$ and interpret this value in context.

SOLUTION: $Y$ = the number of attempts it takes to get a correct pick = the number of homework problems.

(a) $P(Y = 10) = (6/7)^9(1/7) = 0.0357$.

(b) $P(Y < 10) = P(Y = 1) + P(Y = 2) + P(Y = 3) + \ldots + P(Y = 9) = 1/7 + (6/7)(1/7) + (6/7)^2(1/7) + \ldots + (6/7)^8(1/7) = 0.7503$. There's about a 75% chance that the class will get less homework by playing the Lucky Day Game.

For Practice *Try Exercise*  97

As you probably guessed, we used the calculator's `geometpdf` and `geometcdf` commands for the computations in the previous example. The following Technology Corner shows you how we did it.

## 14. TECHNOLOGY CORNER GEOMETRIC PROBABILITY ON THE CALCULATOR

TI-Nspire instructions in Appendix B; HP Prime instructions on the book's Web site.

There are two handy commands on the TI-83/84 and TI-89 for finding geometric probabilities: `geometpdf` and `geometcdf`. The inputs for both commands are the success probability $p$ and the value(s) of interest for the geometric random variable $Y$.

`geometpdf(p,k)` computes $P(Y = k)$

`geometcdf(p,k)` computes $P(Y \leq k)$

## Teaching Tip

Ask students why this type of variable is called a *geometric* random variable. Hopefully, students will make the connection between geometric sequences and geometric probability distributions. In a geometric sequence, each term of the sequence is determined by multiplying the previous term by a constant value called the common ratio. In a geometric probability distribution, the probability of each value of the random variable can be calculated by multiplying the probability of the previous value by $(1 - p)$. Furthermore, the sum of the probabilities of all the possible values of a geometric random variable form an infinite geometric series. Because the sum of an infinite geometric series is (first term)/(1 − common ratio), we can see that the sum of all the probabilities in a geometric probability distribution is $\frac{p}{1-(1-p)} = \frac{p}{p} = 1$. This is good news, because the sum of the probabilities in a legitimate probability distribution must equal 1.

## AP® Exam Formula Sheet

The formula for geometric probability is *not* on the formula sheet provided to students on the AP® exam. However, geometric probabilities can be found using the complement rule and the multiplication rule for independent events.

## ALTERNATE EXAMPLE

### *Monopoly*

PROBLEM: Let the random variable $Y$ be defined as in the previous Alternate Example.

**(a)** Find the probability that it takes 3 turns to roll doubles.

**(b)** Find the probability that it takes more than 3 turns to roll doubles, and interpret this value in context.

SOLUTION:

**(a)** $P(Y = 3) = (5/6)^2(1/6) = 0.116$

**(b)** Because there are an infinite number of possible values of $Y$ greater than 3, we will use the complement rule.

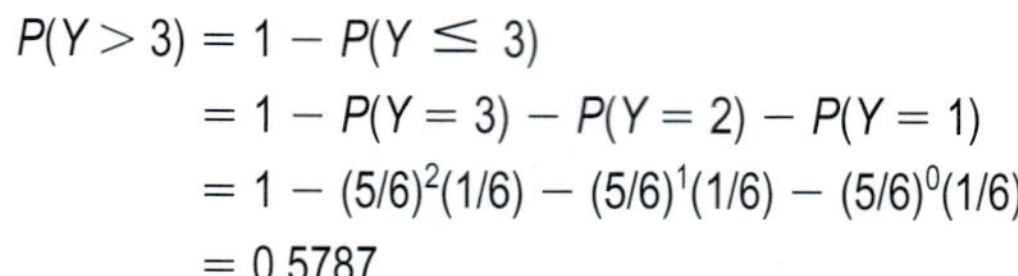

$$\begin{aligned} P(Y > 3) &= 1 - P(Y \leq 3) \\ &= 1 - P(Y = 3) - P(Y = 2) - P(Y = 1) \\ &= 1 - (5/6)^2(1/6) - (5/6)^1(1/6) - (5/6)^0(1/6) \\ &= 0.5787 \end{aligned}$$

If a player tried to get out of jail many, many times by trying to roll doubles, about 58% of the time it would take more than 3 attempts.

Let's use these commands to confirm our answers in the previous example.

(a) Find the probability that the class receives exactly 10 homework problems as a result of playing the Lucky Day Game.

**TI-83/84**

- Press [2nd] [VARS] (DISTR) and choose `geometpdf(`.
- **OS 2.55 or later:** In the dialog box, enter these values: `p:1/7`, `xvalue:10`, choose Paste, and then press [ENTER]. **Older OS:** Complete the command `geometpdf(1/7,10)` and press [ENTER].

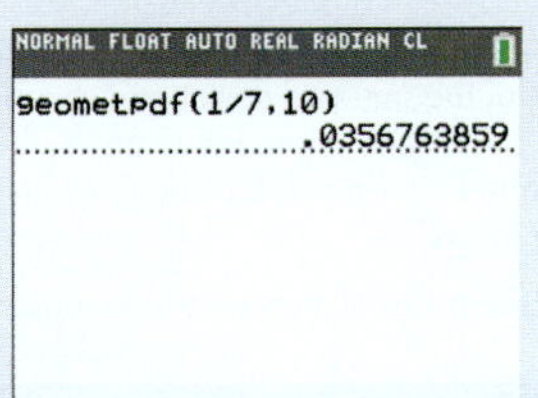


**TI-89**

- In the Stats/List Editor, Press [F5] (Distr) and choose `Geometric Pdf...`.
- In the dialog box, enter these values: `Prob Success, p:1/7`, `X value:10`, and then choose [ENTER].

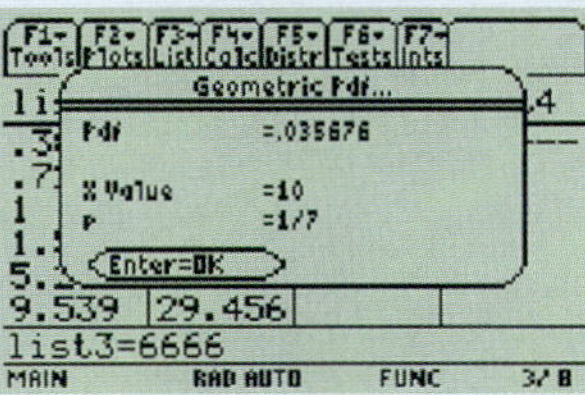


These results agree with our previous answer using the geometric probability formula: 0.0357.

(b) Find $P(Y < 10)$ and interpret this value in context. To find $P(Y < 10)$, use the `geometcdf` command:

$$P(Y < 10) = P(Y \leq 9) = \texttt{geometcdf(1/7,9)}$$

**TI-83/84**

- Press [2nd] [VARS] (DISTR) and choose `geometcdf(`.
- **OS 2.55 or later:** In the dialog box, enter these values: `p:1/7`, `x value:9`, choose `Paste`, and then press [ENTER]. **Older OS:** Complete the command `geometcdf(1/7,9)` and press [ENTER].

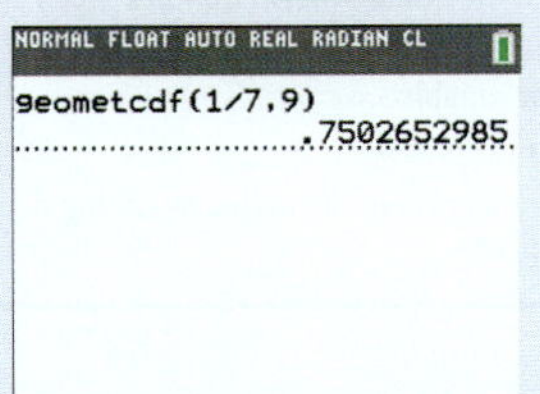


**TI-89**

- In the Stats/List Editor, Press [F5] (Distr) and choose `Geometric Cdf....`
- In the dialog box, enter these values: `Prob Success, p:1/7`, `Lower value:0`, `Upper value:9`, and then choose [ENTER].

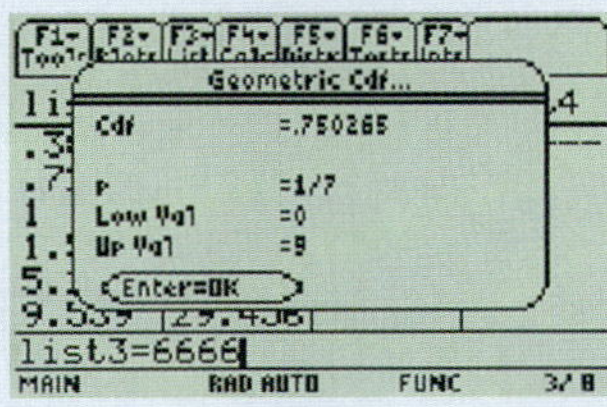


These results agree with our previous answer using the geometric probability formula: 0.7503.

The table below shows part of the probability distribution of $Y$. We can't show the entire distribution, because the number of trials it takes to get the first success could be a very large number.

| Value $y_i$: | 1 | 2 | 3 | 4 | 5 | 6 | 7 | 8 | 9 | ... |
|---|---|---|---|---|---|---|---|---|---|---|
| Probability $p_i$: | 0.143 | 0.122 | 0.105 | 0.090 | 0.077 | 0.066 | 0.057 | 0.049 | 0.042 | |

Figure 6.18 is a histogram of the probability distribution for values of $Y$ from 1 to 26. Let's describe what we see.

## TECHNOLOGY TIP

If your students are feeling overwhelmed with calculator commands, you can skip over the `geometpdf` and `geometcdf` commands and have your students just use the complement rule and multiplication rule for independent events. If you do teach these commands, remind students of the difference between pdf and cdf (the c stands for cumulative).

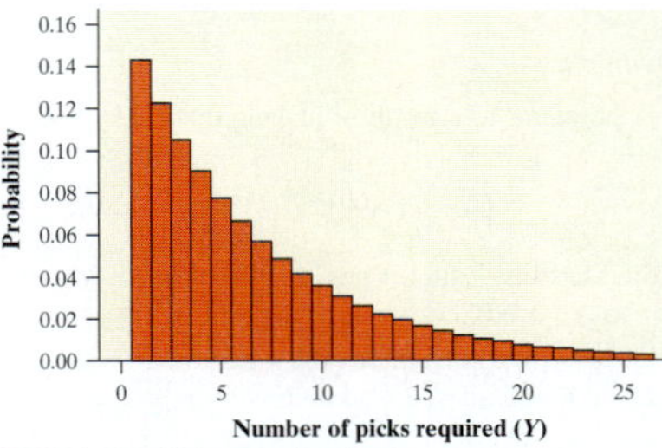


**FIGURE 6.18** Histogram showing the probability distribution of the geometric random variable $Y$ = number of trials needed for students to pick correctly in the Lucky Day Game.

*Shape:* The heavily right-skewed shape is characteristic of any geometric distribution. That's because the most likely value of a geometric random variable is 1. The probability of each successive value decreases by a factor of $(1 - p)$.

*Center:* The mean of $Y$ is $\mu_Y = 7$. (Due to the infinite number of possible values of $Y$, the calculation of the mean is beyond the scope of this text.) If the class played the Lucky Day Game many times, they would receive an average of 7 homework problems. It's no coincidence that $p = 1/7$ and $\mu_Y = 7$. With probability of success 1/7 on each trial, we'd expect it to take an average of 7 trials to get the first success.

*Spread:* The standard deviation of $Y$ is $\sigma_Y = 6.48$. (Due to the infinite number of possible values of $Y$, the calculation of the standard deviation is beyond the scope of this text.) If the class played the Lucky Day game many times, the number of homework problems they receive would typically differ from 7 by about 6.5 problems. That could mean a lot of homework!

We can generalize the result for the mean of a geometric random variable.

### MEAN (EXPECTED VALUE) OF A GEOMETRIC RANDOM VARIABLE

If $Y$ is a geometric random variable with probability of success $p$ on each trial, then its **mean** (expected value) is $\mu_Y = E(Y) = \frac{1}{p}$. That is, the expected number of trials required to get the first success is $1/p$.

### CHECK YOUR UNDERSTANDING

Suppose you roll a pair of fair, six-sided dice until you get doubles. Let $T$ = the number of rolls it takes.

1. Show that $T$ is a geometric random variable.
2. Find $P(T = 3)$. Interpret this result in context.
3. In the game of Monopoly, a player can get out of jail free by rolling doubles within 3 turns. Find the probability that this happens.

## A Jury of Your Peers?

In the chapter-opening Case Study on page 345, a defense attorney challenged the jury-pool selection process in his accused client's trial. Here are the facts:

- About 7.28% of the citizens in the court's jurisdiction were black.
- The jury pool had between 60 and 100 members, 3 of whom were black.

Use what you have learned in this chapter to help answer the following questions.

### Teaching Tip

All geometric distributions have this shape, with 1 being the most likely value. Binomial distributions, on the other hand, can be skewed to the right, skewed to the left, or roughly symmetric. Also, the mode of a binomial distribution will be near $np$, the expected value.

### AP® Exam Formula Sheet

The formula for the mean of a geometric distribution is *not* on the formula sheet provided to students on the AP® exam.

### ANSWERS TO CYU

1. Die rolls are independent, the probability of getting doubles is the same on each roll (1/6), and we are repeating the chance process until we get a success (doubles).

2. $P(T = 3) = \left(\frac{5}{6}\right)^2\left(\frac{1}{6}\right) = 0.1157$.

There is a 0.1157 probability that you will get the first set of doubles on the third roll of the dice.

3. $P(T \leq 3) = \frac{1}{6} + \left(\frac{5}{6}\right)\left(\frac{1}{6}\right) + \left(\frac{5}{6}\right)^2\left(\frac{1}{6}\right) = 0.4213$

For now, assume that the court carried out a proper random-selection process to obtain a jury pool with 100 members.

1. Let X = the number of black citizens in the jury pool. What distribution does the random variable X have? Justify your answer.
2. Find the mean and standard deviation of X. Interpret these values in context.
3. If a jury pool has 3 or fewer blacks, should we be suspicious that the court did not carry out the random selection process correctly? Compute $P(X \leq 3)$ and use this result to support your answer.

What if the jury pool had 60 members? Assume once again that the court carried out a proper random-selection process. Let Y = the number of black citizens in the jury pool.

4. Without doing any calculations, decide if $P(Y \leq 3)$ is greater than, equal to, or less than $P(X \leq 3)$. Justify your answer.
5. Using the logic of Question 4, explain why you do not have to consider jury pools with 61, 62, . . . , 99 members to render a verdict about whether or not the jury-selection process was carried out properly. What is your verdict?

## Section 6.3 Summary

- A **binomial setting** consists of $n$ independent trials of the same chance process, each resulting in a success or a failure, with probability of success $p$ on each trial. Remember to check the BINS! The count X of successes is a **binomial random variable.** Its probability distribution is a **binomial distribution.**
- The **binomial coefficient**

$$\binom{n}{k} = \frac{n!}{k!(n-k)!}$$

counts the number of ways $k$ successes can be arranged among $n$ trials. The **factorial** of $n$ is

$$n! = n(n-1)(n-2) \cdot \ldots \cdot (3)(2)(1)$$

for positive whole numbers $n$, and $0! = 1$.
- If X has the binomial distribution with parameters $n$ and $p$, the possible values of X are the whole numbers 0, 1, 2, . . . , $n$. The **binomial probability** of observing $k$ successes in $n$ trials is

$$P(X = k) = \binom{n}{k} p^k (1-p)^{n-k}$$

- Binomial probabilities are best found using technology.

### TRM Section 6.3 Quizzes

There are three different quizzes available for this section in the Teacher's Resource Materials. Versions A and B are roughly parallel. Find them by clicking on the link in the TE-book, opening the TRCD, or logging onto the book's companion Web site www.whfreeman.com/tps5e (teacher log-in required).

### Answers to Case Closed

**1.** *X* has a binomial distribution with $n = 100$ and $p = 0.0728$. Checking the BINS: *Binary?* "Success" = black and "Failure" = not black. *Independent?* The sample of size $n = 100$ is less than 10% of all people in the jurisdiction, so we can assume that the outcomes of trials are independent. *Number?* There is a fixed number of trials, $n = 100$. *Success?* The probability of success is constant, $p = 0.0728$.

**2.** $\mu_X = np = 100(0.0728) = 7.28$. If we were to select many jury pools of 100 people, on average, there would be about 7.28 black citizens per pool. $\sigma_X = \sqrt{np(1-p)} = \sqrt{100(0.0728)(0.9272)} = 2.60$. If we were to select many jury pools of 100 people, the number of black citizens would typically vary by about 2.60 from the mean (7.28).

**3.** **Step 1: State the distribution and the values of interest.** *X* has a binomial distribution with $n = 100$ and $p = 0.0728$. We want to find $P(X \leq 3)$. **Step 2: Perform calculations—show your work!** $P(X \leq 3) = P(X = 0) + P(X = 1) + P(X = 2) + P(X = 3) =$

$$\binom{100}{0}(0.0728)^0(0.9272)^{100} + \cdots + \binom{100}{3}(0.0728)^3(0.9272)^{97} = 0.0614.$$

*Using technology*: The command `binomcdf(trials:100,p:0.0728, x value:3)` gives 0.0614. **Step 3: Answer the question.** There is about a 6% chance that 3 or fewer black citizens would be selected. Because this probability isn't very small, it is plausible that the court carried out the random selection process correctly. About 6 jury pools in every 100 would have 3 or fewer black citizens just by chance.

**4.** $P(Y \leq 3)$ is greater than $P(X \leq 3)$. Getting a small number of black citizens (i.e., 3 or fewer) is more likely when the jury pool itself is smaller. Also, 3 black citizens is closer to the expected number of black citizens when $n = 60$ ($np = 4.368$) than when $n = 100$ ($np = 7.28$).

**5.** Because the probability of selecting 3 or fewer blacks gets bigger as $n$ get smaller and because there wasn't convincing evidence of improper selection when $n = 100$, we don't need to consider any sample sizes smaller than $n = 100$. So, with a jury pool of between 60 and 100 members, we do not have convincing evidence that the random selection process was done improperly.

- The **mean** and **standard deviation** of a binomial random variable X are

$$\mu_X = np \qquad \sigma_X = \sqrt{np(1-p)}$$

- The binomial distribution with $n$ trials and probability $p$ of success gives a good approximation to the count of successes in an SRS of size $n$ from a large population containing proportion $p$ of successes. This is true as long as the sample size $n$ is no more than 10% of the population size N (the **10% condition**).
- The **Normal approximation** to the binomial distribution* says that if X is a count of successes having the binomial distribution with parameters $n$ and $p$, then when $n$ is large, X is approximately Normally distributed with mean $np$ and standard deviation $\sqrt{np(1-p)}$. We will use this approximation when $np \geq 10$ and $n(1-p) \geq 10$ (the **Large Counts condition**).
- A **geometric setting** consists of repeated trials of the same chance process in which the probability $p$ of success is the same on each trial, and the goal is to count the number of trials it takes to get one success. If Y = the number of trials required to obtain the first success, then Y is a **geometric random variable.** Its probability distribution is called a **geometric distribution.**
- If Y has the geometric distribution with probability of success $p$, the possible values of Y are the positive integers 1, 2, 3, . . . . The **geometric probability** that Y takes any value is

$$P(Y = k) = (1-p)^{k-1}p$$

- The **mean** (expected value) of a geometric random variable Y is $\mu_Y = 1/p$.

*This topic is not required for the AP® Statistics exam. Some teachers prefer to discuss this topic when presenting the sampling distribution of $\hat{p}$ (Chapter 7).

## 6.3 TECHNOLOGY CORNERS

TI-Nspire Instructions in Appendix B; HP Prime instructions on the book's Web site.



# Section 6.3 Exercises

*In Exercises 69 to 72, determine whether the given random variable has a binomial distribution. Justify your answer.*

69. **Sowing seeds** Seed Depot advertises that its new flower seeds have an 85% chance of germinating (growing). Suppose that the company's claim is true. Judy gets a packet with 20 randomly selected new flower seeds from Seed Depot and plants them in her garden. Let X = the number of seeds that germinate.

pg 388

70. **Long or short?** Put the names of all the students in your class in a hat. Mix them up, and draw four names without looking. Let Y = the number whose last names have more than six letters.

**TRM Full Solutions to Section 6.3 Exercises**

## Answers to Section 6.3 Exercises

**6.69** Binomial. *Binary?* "Success" = seed germinates and "Failure" = seed does not germinate. *Independent?* Yes, because the seeds were randomly selected, knowing the outcome of one seed shouldn't tell us anything about the outcomes of other seeds. *Number?* $n$ = 20 seeds. *Success?* $p$ = 0.85.

**6.70** Not binomial. *Binary?* "Success" = name has more than 6 letters and "Failure" = name has 6 letters or less. *Independent?* Because we are selecting without replacement from a small number of students, the observations are not independent. *Number?* $n$ = 4. *Success?* The probability that a randomly chosen student's name has more than 6 letters is constant.

**6.71** Not binomial. *Binary?* "Success" = person is left-handed and "Failure" = person is right-handed. *Independent?* Because students are selected randomly, their handedness is independent. *Number?* There is not a fixed number of trials for this chance process because you continue until you find a left-handed student. *Success?* $p$ = 0.10.

**6.72** Binomial. *Binary?* "Success" = train is late and "Failure" = train is on time. *Independent?* Because the days were randomly selected, the arrival times are independent. *Number?* $n$ = 6. *Success?* $p$ = 0.10.

**6.73 (a)** Binomial. *Binary?* "Success" = reaching a live person and "Failure" = any other outcome. *Independent?* Knowing whether or not one call was completed tells us nothing about the outcome on any other call. *Number?* $n$ = 15. *Success?* $p$ = 0.2. **(b)** This is not a binomial setting because there are not a fixed number of attempts. The binary, independent, and success conditions are satisfied, however, as in part (a).

**6.74 (a)** Not binomial. *Binary?* "Success" = making the field goal and "Failure" = missing the field goal. *Independent?* It is reasonable to think that the outcomes of field goal attempts are independent. *Number?* $n$ = 20. *Success?* The probability of success is likely to change from one attempt to another. **(b)** Binomial. *Binary?* "Success" = making a free throw and "Failure" = missing a free throw. *Independent?* It is reasonable to believe that each shot is independent of the others. *Number?* $n$ = 150. *Success?* $p$ = 0.8.

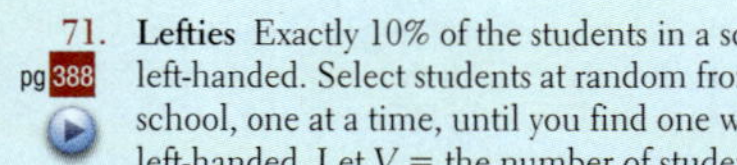

71. **Lefties** Exactly 10% of the students in a school are left-handed. Select students at random from the school, one at a time, until you find one who is left-handed. Let $V$ = the number of students chosen.

72. **Taking the train** According to New Jersey Transit, the 8:00 A.M. weekday train from Princeton to New York City has a 90% chance of arriving on time on a randomly selected day. Suppose this claim is true. Choose 6 days at random. Let $W$ = the number of days on which the train arrives late.

73. **Binomial setting?** A binomial distribution will be approximately correct as a model for one of these two settings and not for the other. Explain why by briefly discussing both settings.

(a) When an opinion poll calls residential telephone numbers at random, only 20% of the calls reach a person. You watch the random digit dialing machine make 15 calls. $X$ is the number that reach a person.

(b) When an opinion poll calls residential telephone numbers at random, only 20% of the calls reach a live person. You watch the random digit dialing machine make calls. $Y$ is the number of calls needed to reach a live person.

74. **Binomial setting?** A binomial distribution will be approximately correct as a model for one of these two sports settings and not for the other. Explain why by briefly discussing both settings.

(a) A National Football League kicker has made 80% of his field goal attempts in the past. This season he attempts 20 field goals. The attempts differ widely in distance, angle, wind, and so on.

(b) A National Basketball Association player has made 80% of his free-throw attempts in the past. This season he takes 150 free throws. Basketball free throws are always attempted from 15 feet away with no interference from other players.

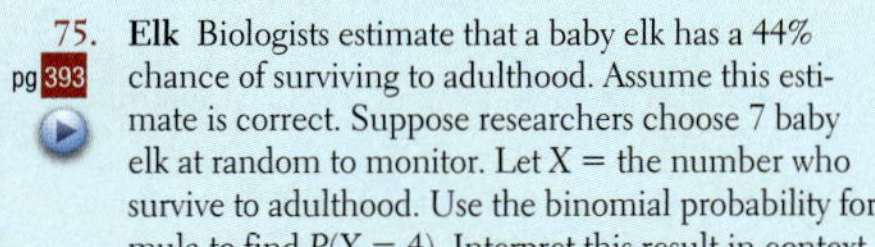

75. **Elk** Biologists estimate that a baby elk has a 44% chance of surviving to adulthood. Assume this estimate is correct. Suppose researchers choose 7 baby elk at random to monitor. Let $X$ = the number who survive to adulthood. Use the binomial probability formula to find $P(X = 4)$. Interpret this result in context.

76. **Rhubarb** Suppose you purchase a bundle of 10 bare-root rhubarb plants. The sales clerk tells you that 5% of these plants will die before producing any rhubarb. Assume that the bundle is a random sample of plants and that the sales clerk's statement is accurate. Let $Y$ = the number of plants that die before producing any rhubarb. Use the binomial probability formula to find $P(Y = 1)$. Interpret this result in context.

77. **Elk** Refer to Exercise 75. How surprising would it be for more than 4 elk in the sample to survive to adulthood? Calculate an appropriate probability to support your answer.

78. **Rhubarb** Refer to Exercise 76. Would you be surprised if 3 or more of the plants in the bundle die before producing any rhubarb? Calculate an appropriate probability to support your answer.

79. **Sowing seeds** Refer to Exercise 69.

(a) Find the probability that exactly 17 seeds germinate. Show your work.

(b) If only 12 seeds actually germinate, should Judy be suspicious that the company's claim is not true? Compute $P(X \le 12)$ and use this result to support your answer.

80. **Taking the train** Refer to Exercise 72.

(a) Find the probability that the train arrives late on exactly 2 days. Show your work.

(b) Would you be surprised if the train arrived late on 2 or more days? Compute $P(W \ge 2)$ and use this result to support your answer.

81. **Random digit dialing** When an opinion poll calls a residential telephone number at random, there is only a 20% chance that the call reaches a live person. You watch the random digit dialing machine make 15 calls. Let $X$ = the number of calls that reach a live person.

(a) Find and interpret $\mu_X$.

(b) Find and interpret $\sigma_X$.

82. **Lie detectors** A federal report finds that lie detector tests given to truthful persons have probability about 0.2 of suggesting that the person is deceptive.[12] A company asks 12 job applicants about thefts from previous employers, using a lie detector to assess their truthfulness. Suppose that all 12 answer truthfully. Let $X$ = the number of people who the lie detector says are being deceptive.

(a) Find and interpret $\mu_X$.

(b) Find and interpret $\sigma_X$.

83. **Random digit dialing** Refer to Exercise 81. Let $Y$ = the number of calls that *don't* reach a live person.

(a) Find the mean of $Y$. How is it related to the mean of $X$? Explain why this makes sense.

(b) Find the standard deviation of $Y$. How is it related to the standard deviation of $X$? Explain why this makes sense.

**6.80 (a)** $P(W = 2) = \binom{6}{2}(0.10)^2(0.90)^4 = 0.0984$

**(b)** $P(W \ge 2) = \binom{6}{2}(0.10)^2(0.90)^4 + \cdots = 0.1143$. Because this probability is not very small, it would not be surprising if the train arrived late 2 or more times.

**6.81 (a)** $\mu_X = 15(0.20) = 3$. If we watched the machine make many sets of 15 calls, we would expect about 3 calls to reach a live person, on average.

**(b)** $\sigma_X = \sqrt{15(0.20)(0.80)} = 1.55$. If we watched the machine make many sets of 15 calls, we would expect the number of calls that reach a live person to typically vary by about 1.55 from the mean (3).

**6.82 (a)** $\mu_X = 12(0.20) = 2.4$. If we tested many groups of 12 truthful applicants, we would expect about 2.4 people to be declared deceptive, on average.

**(b)** $\sigma_X = \sqrt{12(0.20)(0.80)} = 1.39$. If we tested many groups of 12 truthful applicants, we would expect the number of people declared deceptive to typically vary by about 1.39 from the mean (2.4).

**6.83 (a)** $\mu_Y = 15(0.80) = 12$. Notice that $\mu_X = 3$ and $12 + 3 = 15$ (the total number of calls).

**(b)** $\sigma_Y = \sqrt{15(0.80)(0.20)} = 1.55$. This is the same value as $\sigma_X$, because $Y = 15 - X$ and adding a constant to a random variable doesn't change the spread.

**6.75** $P(X = 4) = \binom{7}{4}(0.44)^4(0.56)^3 = 0.2304$. There is a 0.2304 probability that exactly 4 of the 7 elk survive to adulthood.

**6.76** $P(Y = 1) = \binom{10}{1}(0.05)(0.95)^9 = 0.3151$. There is a 0.3151 probability that exactly 1 of the 10 rhubarb plants will die before producing any rhubarb.

**6.77** $P(X > 4) = \binom{7}{5}(0.44)^5(0.56)^2 + \cdots = 0.1402$

Because this probability isn't very small, it is not surprising for more than 4 elk to survive to adulthood.

**6.78** $P(Y \ge 3) = \binom{10}{3}(0.05)^3(0.95)^7 + \cdots = 0.0115$. Because this is a small probability, it would be surprising if 3 of more of the plants die before producing any rhubarb.

**6.79 (a)** $P(X = 17) = \binom{20}{17}(0.85)^{17}(0.15)^3 = 0.2428$

**(b)** $P(X \le 12) = \binom{20}{0}(0.85)^0(0.15)^{20} + \cdots + \binom{20}{12}(0.85)^{12}(0.15)^8 = 0.0059$. Because this is such a low probability, Judy should be suspicious.

**6.84** **(a)** $P(Y \geq 10) = \binom{12}{10}(0.8)^{10}(0.2)^2 + \cdots = 0.5583$. This is the same as $P(X \leq 2)$ because finding 2 or fewer lying is the same as finding 10 or more who are telling the truth.
**(b)** $\mu_Y = 12(0.8) = 9.6$. Notice that $\mu_X = 2.4$ and $2.4 + 9.6 = 12$ (the total number of people).
$\sigma_Y = \sigma_X = \sqrt{12(0.8)(0.2)} = 1.39$. This is not surprising because adding a constant to a random variable doesn't change the spread.

**6.85** **(a)** *Binary?* "Success" = win a prize and "Failure" = don't win a prize. *Independent?* Knowing whether one bottle wins or not should not tell us anything about the caps on other bottles. *Number?* $n = 7$. *Success?* $p = 1/6$.
**(b)** $\mu_X = 1.167$. If we were to buy many sets of 7 bottles, we would get 1.167 winners per set, on average. $\sigma_X = 0.986$. If we were to buy many sets of 7 bottles, the number of winning bottles would typically differ from the mean (1.167) by 0.986. **(c)** $P(X \geq 3) = 1 - P(X \leq 2) = 0.0958$. Because 0.0958 isn't a very small probability, the clerk shouldn't be surprised. It is plausible to get 3 or more winners in a sample of 7 bottles by chance alone.

**6.86** **(a)** *Binary?* "Success" = operates for an hour without failure and "Failure" = does not operate for an hour without failure. *Independent?* Whether one engine operates for an hour or not should not tell us whether any other engine does or not. *Number?* $n = 350$. *Success?* $p = 0.999$.
**(b)** $\mu_X = 349.65$. If we were to test many sets of 350 engines, we would expect that about 349.65 of them would operate for an hour without failure, on average. $\sigma_X = 0.591$. If we were to test many sets of 350 engines, we would expect the number that operate for an hour without failure to typically vary by about 0.591 from the mean (349.65). **(c)** $P(X \leq 348) = 1 - 0.2468 - 0.7046 = 0.0486$. Because this is a small probability, there is convincing evidence that the engines are less reliable than they are supposed to be.

**6.87** No. Because we are sampling without replacement and the sample size (10) is more than 10% of the population size (76), we should not treat the observations as independent.

**6.88** Yes. We can use the binomial distribution in this case because the sample size (7) is less than 10% of the population size (100).

84. **Lie detectors** Refer to Exercise 82. Let Y = the number of people who the lie detector says are telling the truth.
   (a) Find $P(Y \geq 10)$. How is this related to $P(X \leq 2)$? Explain.
   (b) Calculate $\mu_Y$ and $\sigma_Y$. How do they compare with $\mu_X$ and $\sigma_X$? Explain why this makes sense.

85. **1 in 6 wins** As a special promotion for its 20-ounce bottles of soda, a soft drink company printed a message on the inside of each cap. Some of the caps said, "Please try again," while others said, "You're a winner!" The company advertised the promotion with the slogan "1 in 6 wins a prize." Suppose the company is telling the truth and that every 20-ounce bottle of soda it fills has a 1-in-6 chance of being a winner. Seven friends each buy one 20-ounce bottle of the soda at a local convenience store. Let X = the number who win a prize. pg 399
   (a) Explain why X is a binomial random variable.
   (b) Find the mean and standard deviation of X. Interpret each value in context.
   (c) The store clerk is surprised when three of the friends win a prize. Is this group of friends just lucky, or is the company's 1-in-6 claim inaccurate? Compute $P(X \geq 3)$ and use the result to justify your answer.

86. **Aircraft engines** Engineers define reliability as the probability that an item will perform its function under specific conditions for a specific period of time. A certain model of aircraft engine is designed so that each engine has probability 0.999 of performing properly for an hour of flight. Company engineers test an SRS of 350 engines of this model. Let X = the number that operate for an hour without failure.
   (a) Explain why X is a binomial random variable.
   (b) Find the mean and standard deviation of X. Interpret each value in context.
   (c) Two engines failed the test. Are you convinced that this model of engine is less reliable than it's supposed to be? Compute $P(X \leq 348)$ and use the result to justify your answer.

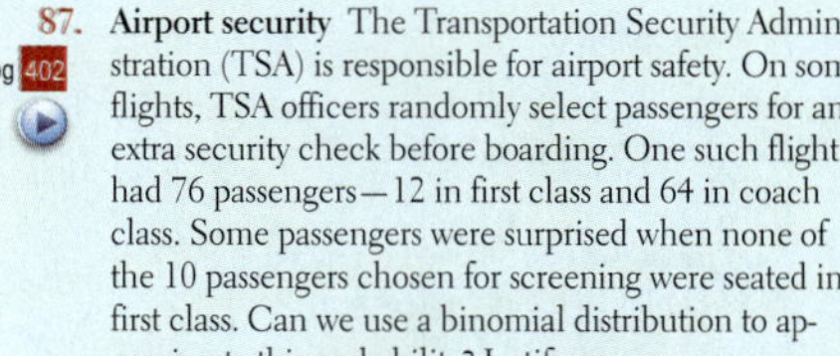

87. **Airport security** The Transportation Security Administration (TSA) is responsible for airport safety. On some flights, TSA officers randomly select passengers for an extra security check before boarding. One such flight had 76 passengers—12 in first class and 64 in coach class. Some passengers were surprised when none of the 10 passengers chosen for screening were seated in first class. Can we use a binomial distribution to approximate this probability? Justify your answer. pg 402

88. **Scrabble** In the game of Scrabble, each player begins by drawing 7 tiles from a bag containing 100 tiles. There are 42 vowels, 56 consonants, and 2 blank tiles in the bag. Cait chooses her 7 tiles and is surprised to discover that all of them are vowels. Can we use a binomial distribution to approximate this probability? Justify your answer.

89. **10% condition** To use a binomial distribution to approximate the count of successes in an SRS, why do we require that the sample size $n$ be no more than 10% of the population size $N$?

90. *__Large Counts condition__ To use a Normal distribution to approximate binomial probabilities, why do we require that both $np$ and $n(1 - p)$ be at least 10?

91. *__On the Web__ What kinds of Web sites do males aged 18 to 34 visit most often? Half of male Internet users in this age group visit an auction site such as eBay at least once a month.[13] A study of Internet use interviews a random sample of 500 men aged 18 to 34. Let X = the number in the sample who visit an auction site at least once a month. pg 403
   (a) Show that X is approximately a binomial random variable.
   (b) Check the conditions for using a Normal approximation in this setting.
   (c) Use a Normal distribution to estimate the probability that at least 235 of the men in the sample visit an online auction site at least once a month.

92. *__Checking for survey errors__ One way of checking the effect of undercoverage, nonresponse, and other sources of error in a sample survey is to compare the sample with known facts about the population. About 12% of American adults identify themselves as black. Suppose we take an SRS of 1500 American adults and let X be the number of blacks in the sample.
   (a) Show that X is approximately a binomial random variable.
   (b) Check the conditions for using a Normal approximation in this setting.
   (c) Use a Normal distribution to estimate the probability that the sample will contain between 165 and 195 blacks.

93. **Using Benford's law** According to Benford's law (Exercise 5, page 359), the probability that the first digit of the amount on a randomly chosen invoice is a 1 or a 2 is 0.477. Suppose you examine an SRS of 90 invoices from a vendor and find 29 that have first digits 1 or 2. Do you suspect that the invoice

*This topic is not required for the AP® Statistics exam. Some teachers prefer to discuss this topic when presenting the sampling distribution of $\hat{p}$ (Chapter 7).

**6.89** If the sample is a small fraction of the population (less than 10%), the make-up of the population doesn't change enough to make the lack of independent trials an issue.

**6.90** When $np \geq 10$ and $n(1 - p) \geq 10$, then we know that the combination of $n$ and $p$ is such that the shape of the binomial distribution is close enough to a Normal distribution.

**6.91** **(a)** *Binary?* "Success" = visit an auction site at least once a month and "Failure" = don't visit an auction site at least once a month. *Independent?* We are sampling without replacement, but the sample size (500) is far less than 10% of all males aged 18 to 34. *Number?* $n = 500$. *Success?* $p = 0.50$. **(b)** $np = 250$ and $n(1 - p) = 250$ are both at least 10.
**(c)** $\mu_X = 250$ and $\sigma_X = 11.18$. Thus, $X$ has approximately the $N(250, 11.18)$ distribution. We want to find $P(X \geq 235)$.
$z = \frac{235 - 250}{11.18} = -1.34$ and $P(Z \geq -1.34) = 0.9099$. *Using technology*: `normalcdf(lower:235,upper:1000,μ:250,σ:11.18)` = 0.9102. There is a 0.9102 probability that at least 235 of the men in the sample visit an online auction site.

**6.92** **(a)** *Binary?* "Success" = person identifies themself as black and "Failure" = person does not identify themself as black. *Independent?* We are sampling without replacement, but the sample size (1500) is far less than 10% of all American adults. *Number?* $n = 1500$. *Success?* $p = 0.12$.

amounts are not genuine? Compute an appropriate probability to support your answer.

**94. A .300 hitter** In baseball, a 0.300 hitter gets a hit in 30% of times at bat. When a baseball player hits 0.300, fans tend to be impressed. Typical Major Leaguers bat about 500 times a season and hit about 0.260. A hitter's successive tries seem to be independent. Could a typical Major Leaguer hit 0.300 just by chance? Compute an appropriate probability to support your answer.

**95. Geometric or not?** Determine whether each of the following scenarios describes a geometric setting. If so, define an appropriate geometric random variable.

(a) A popular brand of cereal puts a card with 1 of 5 famous NASCAR drivers in each box. There is a 1/5 chance that any particular driver's card ends up in any box of cereal. Buy boxes of the cereal until you have all 5 drivers' cards.

(b) In a game of 4-Spot Keno, Lola picks 4 numbers from 1 to 80. The casino randomly selects 20 winning numbers from 1 to 80. Lola wins money if she picks 2 or more of the winning numbers. The probability that this happens is 0.259. Lola decides to keep playing games of 4-Spot Keno until she wins some money.

**96. Geometric or not?** Determine whether each of the following scenarios describes a geometric setting. If so, define an appropriate geometric random variable.

(a) Shuffle a standard deck of playing cards well. Then turn over one card at a time from the top of the deck until you get an ace.

(b) Lawrence likes to shoot a bow and arrow in his free time. On any shot, he has about a 10% chance of hitting the bull's-eye. As a challenge one day, Lawrence decides to keep shooting until he gets a bull's-eye.

**97. 1-in-6 wins** pg 406 Alan decides to use a different strategy for the 1-in-6 wins game of Exercise 85. He keeps buying one 20-ounce bottle of the soda at a time until he gets a winner.

(a) Find the probability that he buys exactly 5 bottles. Show your work.

(b) Find the probability that he buys no more than 8 bottles. Show your work.

**98. Cranky mower** To start her old lawn mower, Rita has to pull a cord and hope for some luck. On any particular pull, the mower has a 20% chance of starting.

(a) Find the probability that it takes her exactly 3 pulls to start the mower. Show your work.

(b) Find the probability that it takes her more than 10 pulls to start the mower. Show your work.

**99. Using Benford's law** According to Benford's law (Exercise 5, page 359), the probability that the first digit of the amount of a randomly chosen invoice is an 8 or a 9 is 0.097. Suppose you examine randomly selected invoices from a vendor until you find one whose amount begins with an 8 or a 9.

(a) How many invoices do you expect to examine until you get one that begins with an 8 or 9? Justify your answer.

(b) In fact, you don't get an amount starting with an 8 or 9 until the 40th invoice. Do you suspect that the invoice amounts are not genuine? Compute an appropriate probability to support your answer.

**100. Roulette** Marti decides to keep placing a \$1 bet on number 15 in consecutive spins of a roulette wheel until she wins. On any spin, there's a 1-in-38 chance that the ball will land in the 15 slot.

(a) How many spins do you expect it to take until Marti wins? Justify your answer.

(b) Would you be surprised if Marti won in 3 or fewer spins? Compute an appropriate probability to support your answer.

*Multiple choice: Select the best answer for Exercises 101 to 105.*

**101.** Joe reads that 1 out of 4 eggs contains salmonella bacteria. So he never uses more than 3 eggs in cooking. If eggs do or don't contain salmonella independently of each other, the number of contaminated eggs when Joe uses 3 chosen at random has the following distribution:

(a) binomial; $n = 4$ and $p = 1/4$

(b) binomial; $n = 3$ and $p = 1/4$

(c) binomial; $n = 3$ and $p = 1/3$

(d) geometric; $p = 1/4$

(e) geometric; $p = 1/3$

*Exercises 102 and 103 refer to the following setting.* A fast-food restaurant runs a promotion in which certain food items come with game pieces. According to the restaurant, 1 in 4 game pieces is a winner.

**102.** If Jeff gets 4 game pieces, what is the probability that he wins exactly 1 prize?

(a) 0.25

(b) 1.00

(c) $\binom{4}{1}(0.25)^1(0.75)^3$

(d) $\binom{4}{1}(0.25)^3(0.75)^1$

(e) $(0.75)^3(0.25)^1$

**(b)** $np = 180$ and $n(1 - p) = 1320$ are both at least 10.

**(c)** $\mu_X = np = 1500(0.12) = 180$ and $\sigma_X = \sqrt{1500(0.12)(0.88)} = 12.5857$. Thus, $X$ has approximately the $N(180, 12.5857)$ distribution. We want to find $P(165 \le X \le 195)$. $z = \frac{165 - 180}{12.5857} = -1.19$ and $z = \frac{195 - 180}{12.5857} = 1.19$ and $P(Z \le 1.19) - P(Z \le -1.19) = 0.7660$. *Using technology:* `normalcdf(lower:165,upper:195,μ:180,σ:12.5857)` = 0.7667. There is a 0.7667 probability that the sample will contain between 165 and 195 blacks.

**6.93** Let $X$ be the number of 1s and 2s. Then $X$ has a binomial distribution with $n = 90$ and $p = 0.477$ in the absence of fraud. $P(X \le 29) = 0.0021$. Because the probability of getting 29 or fewer invoices that begin with the digits 1 or 2 is quite small, we have reason to be suspicious that the invoice amounts are not genuine.

**6.94** Let $X$ be the number of hits out of 500 times at bat. Then $X$ has a binomial distribution with $n = 500$ and $p = 0.260$. $P(X \ge 150) = 0.0246$. Because this probability is fairly small, it is unlikely for a 0.260 hitter to hit 0.300 just by chance. However, if there are a large number of 0.260 hitters, we would expect about 2% of them to hit 0.300 or higher just by chance.

**6.95 (a)** Not geometric. We can't classify the possible outcomes on each trial (card) as "success" or "failure" and we are not selecting cards until we get a *single* success. **(b)** Games of 4-Spot Keno are independent, the probability of winning is the same in each game ($p = 0.259$), and Lola is repeating a chance process until she gets a success. $X$ = number of games needed to win once is a geometric random variable with $p = 0.259$.

**6.96 (a)** Not geometric. Because we are selecting cards without replacement, the trials are not independent. **(b)** Assuming his shots are independent, the probability of success is the same for each shot ($p = 0.10$) and he continues to shoot until he gets a bull's-eye. $X$ = number of shots needed to get a bull's-eye is a geometric random variable with $p = 0.10$.

**6.97 (a)** Let $X$ = the number of bottles Alan purchases to find one winner. $P(X = 5) = (5/6)^4(1/6) = 0.0804$. **(b)** $P(X \le 8) = (1/6) + \cdots + (5/6)^7(1/6) = 0.7674$.

**6.98 (a)** Let $X$ = the number of pulls that Rita needs to start the lawn mower. $P(X = 3) = (0.8)^2(0.2) = 0.128$. **(b)** $P(X > 10) = 1 - P(X \le 10) = 1 - (0.20) - \cdots - (0.8)^9(0.20) = 0.1074$.

**6.99 (a)** $\mu_X = \frac{1}{0.097} = 10.31$ **(b)** $P(X \ge 40) = 1 - P(X \le 39) = 0.0187$. Because the probability of not getting an 8 or 9 before the 40th invoice is small, we may begin to worry that the invoice amounts are fraudulent.

**6.100 (a)** $\mu_X = \frac{1}{\frac{1}{38}} = 38$. We would expect it to take about 38 games for the ball to land in the 15 slot. **(b)** $P(X \le 3) = 0.0769$. Because the probability is not that small, I wouldn't be surprised if Marti won in 3 or fewer spins.

**6.101** b

**6.102** c

**6.103** d

**6.104** d

**6.105** c

**6.106** This is an experiment because a treatment was deliberately imposed on the students, randomly assigned at the time they visited the site. Explanatory: type of login box, response: the student's action.

**6.107 (a)**

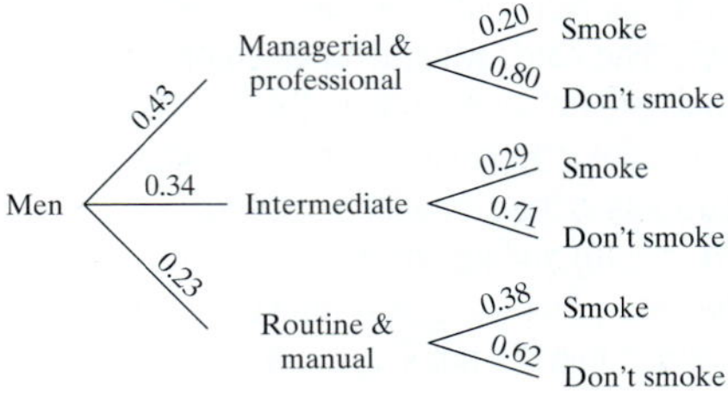


$P(\text{smoke}) = 0.43(0.20) + 0.34(0.29) + 0.23(0.38) = 0.272 = 27.2\%$

**(b)** $P(\text{routine and manual} \mid \text{smoke}) = \dfrac{(0.23)(0.38)}{0.272} = 0.321 = 32.1\%$



**103.** If Jeff keeps playing until he wins a prize, what is the probability that he has to play the game exactly 5 times?

(a) $(0.25)^5$

(b) $(0.75)^4$

(c) $(0.75)^5$

(d) $(0.75)^4(0.25)$

(e) $\binom{5}{1}(0.75)^4(0.25)$

**104.** Each entry in a table of random digits like Table D has probability 0.1 of being a 0, and the digits are independent of one another. If many lines of 40 random digits are selected, the mean and standard deviation of the number of 0s will be approximately

(a) mean = 0.1, standard deviation = 0.05.

(b) mean = 0.1, standard deviation = 0.1.

(c) mean = 4, standard deviation = 0.05.

(d) mean = 4, standard deviation = 1.90.

(e) mean = 4, standard deviation = 3.60.

**105.** *In which of the following situations would it be appropriate to use a Normal distribution to approximate probabilities for a binomial distribution with the given values of $n$ and $p$?

(a) $n = 10, p = 0.5$

(b) $n = 40, p = 0.88$

(c) $n = 100, p = 0.2$

(d) $n = 100, p = 0.99$

(e) $n = 1000, p = 0.003$

*This topic is not required for the AP® Statistics exam. Some teachers prefer to discuss this topic when presenting the sampling distribution of $\hat{p}$ (Chapter 7).

**106. Spoofing** (4.2) To collect information such as passwords, online criminals use "spoofing" to direct Internet users to fraudulent Web sites. In one study of Internet fraud, students were warned about spoofing and then asked to log in to their university account starting from the university's home page. In some cases, the login link led to the genuine dialog box. In others, the box looked genuine but in fact was linked to a different site that recorded the ID and password the student entered. The box that appeared for each student was determined at random. An alert student could detect the fraud by looking at the true Internet address displayed in the browser status bar, but most just entered their ID and password. Is this study an experiment? Why? What are the explanatory and response variables?

**107. Smoking and social class** (5.3) As the dangers of smoking have become more widely known, clear class differences in smoking have emerged. British government statistics classify adult men by occupation as "managerial and professional" (43% of the population), "intermediate" (34%), or "routine and manual" (23%). A survey finds that 20% of men in managerial and professional occupations smoke, 29% of the intermediate group smoke, and 38% in routine and manual occupations smoke.[14]

(a) Use a tree diagram to find the percent of all adult British men who smoke.

(b) Find the percent of male smokers who have routine and manual occupations.

## FRAPPY! Free Response AP® Problem, Yay!

**The following problem is modeled after actual AP® Statistics exam free response questions. Your task is to generate a complete, concise response in 15 minutes.**

*Directions: Show all your work. Indicate clearly the methods you use, because you will be scored on the correctness of your methods as well as on the accuracy and completeness of your results and explanations.*

Buckley Farms produces homemade potato chips that it sells in bags labeled "16 ounces." The total weight of each bag follows an approximately Normal distribution with a mean of 16.15 ounces and a standard deviation of 0.12 ounces.

(a) If you randomly selected 1 bag of these chips, what is the probability that the total weight is less than 16 ounces?

(b) If you randomly selected 10 bags of these chips, what is the probability that exactly 2 of the bags will have a total weight less than 16 ounces?

(c) Buckley Farms ships its chips in boxes that contain 6 bags. The empty boxes have a mean weight of 10 ounces and a standard deviation of 0.05 ounces. Calculate the mean and standard deviation of the total weight of a box containing 6 bags of chips.

(d) Buckley Farms decides to increase the mean weight of each bag of chips so that only 5% of the bags have weights that are less than 16 ounces. Assuming that the standard deviation remains 0.12 ounces, what mean weight should Buckley Farms use?

**After you finish, you can view two example solutions on the book's Web site (www.whfreeman.com/tps5e). Determine whether you think each solution is "complete," "substantial," "developing," or "minimal." If the solution is not complete, what improvements would you suggest to the student who wrote it? Finally, your teacher will provide you with a scoring rubric. Score your response and note what, if anything, you would do differently to improve your own score.**

**TRM FRAPPY!**

We present a model solution here. Please consult the Teacher's Resource Materials for sample student responses, a scoring rubric, and a black line master of the original question with space for students to write their responses.

**(a) Step 1: State the distribution and values of interest.** Let $X$ = the total weight of a bag of chips. $X$ follows a Normal distribution with mean 16.15 ounces and standard deviation 0.12 ounces. We want to find $P(X < 16)$ as shown below.

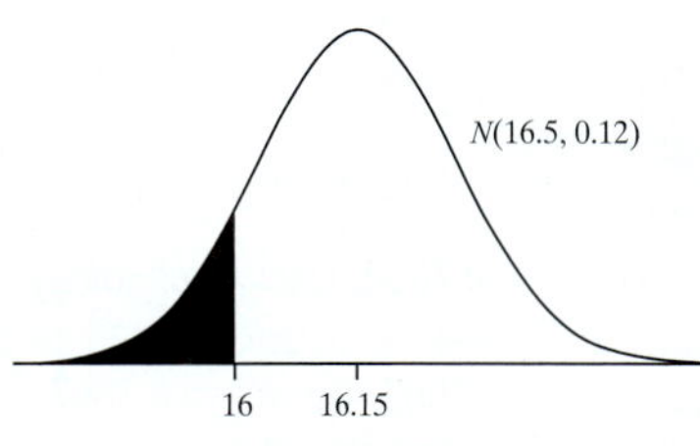

# Chapter Review

**Section 6.1: Discrete and Continuous Random Variables**

A random variable assigns numerical values to the outcomes of a chance process. The probability distribution of a random variable describes its possible values and their probabilities. There are two types of random variables: discrete and continuous. Discrete random variables take on a fixed set of values with gaps in between. Continuous random variables take on all values in an interval of numbers.

As in Chapter 1, we are often interested in the shape, center, and spread of a probability distribution. The shape of a discrete probability distribution can be identified by graphing a probability histogram, with the height of each bar representing the probability of a single value. The center is usually identified by the mean (expected value) of the random variable. The expected value is the average value of the random variable if the chance process is repeated many times. The spread of a probability distribution is usually identified by the standard deviation, which describes how much the values of a random variable typically differ from the mean value, in many repetitions of the chance process.

Continuous probability distributions, such as the Normal distribution, describe the distribution of continuous random variables. A density curve is used to display a continuous probability distribution. Probabilities for continuous random variables are determined by finding the area under the density curve and above the values of interest.

**Section 6.2: Transforming and Combining Random Variables**

In this section, you learned how linear transformations of a random variable affect the shape, center, and spread of its probability distribution. As you learned in Chapter 2, a linear transformation does not change the shape (unless you multiply by a negative number) but can change the center and spread depending on the type of transformation. Multiplying (or dividing) each value of the random variable by a positive constant $b$ multiplies (divides) the mean and standard deviation by $b$. Adding a constant $a$ to (subtracting $a$ from) each value of the random variable adds $a$ to (subtracts $a$ from) the mean but doesn't change the standard deviation.

You also learned how to calculate the mean and standard deviation for a combination of two or more random variables. If you are adding two random variables, X and Y, the mean and standard deviation of X + Y are

$$\mu_{X+Y} = \mu_X + \mu_Y \text{ and } \sigma_{X+Y} = \sqrt{\sigma_X^2 + \sigma_Y^2}$$

Likewise, if you are subtracting two random variables, X and Y, the mean and standard deviation of X − Y are

$$\mu_{X-Y} = \mu_X - \mu_Y \text{ and } \sigma_{X-Y} = \sqrt{\sigma_X^2 + \sigma_Y^2}$$

The formulas for the standard deviation of X + Y and X − Y are only correct if X and Y are independent, that is, if knowing the value of one variable doesn't provide any additional information about the other variable. Also, if X and Y are both Normally distributed, then X + Y and X − Y are both Normally distributed as well.

To determine which formulas to use for a particular problem, it is important to be able to distinguish linear transformations and combinations of random variables. Linear transformations take the values of *one* random variable and add, subtract, multiply, or divide them by a constant. Combinations of random variables take *two or more* random variables and add or subtract them. When a problem involves both linear transformations and a combination of random variables, remember to do the linear transformations first.

**Section 6.3: Binomial and Geometric Random Variables**

In this section, you learned about two common types of discrete random variables, binomial random variables and geometric random variables. Binomial random variables count the number of successes in a fixed number of trials ($n$), whereas geometric random variables count the number of trials needed to get one success. Otherwise, the binomial and geometric settings have the same conditions: there must be two possible outcomes for each trial (success or failure), the trials must be independent, and the probability of success $p$ must stay the same throughout all trials.

To calculate probabilities for a binomial distribution with $n$ trials and probability of success $p$ on each trial, use technology or the binomial probability formula

$$P(X = k) = \binom{n}{k} p^k (1-p)^{n-k}$$

The mean and standard deviation of a binomial random variable X are

$$\mu_X = np \text{ and } \sigma_X = \sqrt{np(1-p)}$$

The shape of a binomial distribution depends on both the number of trials $n$ and the probability of success $p$. When the number of trials is large enough that both $np$ and $n(1-p)$ are at least 10, the distribution of the binomial random variable X has an approximately Normal distribution. Be sure to check the large counts condition before using a Normal approximation to a binomial distribution.

A common application of the binomial distribution is when we count the number of times a particular outcome occurs in a random sample from some population. Because sampling is almost always done without replacement, the independence condition is violated. However, if the sample size is a small fraction of the population size (less than 10%), the lack of independence isn't a concern. Be sure to check the 10% condition when sampling is done without replacement before using a binomial distribution.

Finally, to calculate probabilities for a geometric distribution with probability of success $p$ on each trial, use technology or the geometric probability formula

$$P(Y = k) = (1-p)^{k-1} p$$



**TRM Chapter 6 Tests**

There are four different tests available for this section in the Teacher's Resource Materials. Versions A and B are roughly parallel and version D is cumulative. Find them by clicking on the link in the TE-book, opening the TRCD, or logging onto the book's companion Web site www.whfreeman.com/tps5e (teacher log-in required).

**Step 2: Perform calculations. Show your work.** The standardized score for the boundary value is $z = \frac{16 - 16.15}{0.12} = -1.25$. From Table A, $P(Z < -1.25) = 0.1057$. *Using technology*: The command `normalcdf(lower:−1000,upper:16,μ:16.15,σ:0.12)` gives an area of 0.1057.

**Step 3: Answer the question.** There is about an 11% probability of randomly selecting a bag that weighs less than 16 ounces.

**(b) Step 1: State the distribution and the values of interest.** Let $Y$ = the number of bags that weigh less than 16 ounces. $Y$ is binomial with $n = 10$ and $p = 0.1057$. We want to find $P(Y = 2)$.

**Step 2: Perform calculations—show your work!** $P(Y = 2) = \binom{10}{2}(0.1057)^2(0.8943)^8 = 0.2057$. *Using technology*: The command `binompdf(trials:10,p:0.1057,x value:2)` gives 0.2057.

**Step 3: Answer the question.** There is about a 21% probability that exactly 2 out of 10 randomly selected bags weigh less than 16 ounces.

**(c)** Let $B$ = weight of box and $C$ = combined weight of box and 6 bags of chips. Then $C = B + X_1 + X_2 + X_3 + X_4 + X_5 + X_6$. The mean is $\mu_C = 10 + 16.15 + 16.15 + 16.15 + 16.15 + 16.15 + 16.15 = 106.9$ ounces. $\sigma_C^2 = 0.05^2 + 0.12^2 + 0.12^2 + 0.12^2 + 0.12^2 + 0.12^2 + 0.12^2 = 0.1364$, so the standard deviation is $\sigma_C = \sqrt{0.1364} = 0.369$ ounces.

**(d) Step 1: State the distribution and values of interest.** Let $X$ = the total weight of a bag of chips. $X$ follows a Normal distribution with mean M and standard deviation 0.12 ounces. We want to find the value of M such that $P(X < 16) = 0.05$. See the graph at the end of Step 3.

**Step 2: Perform calculations.** Show your work. Look in the body of Table A for a value closest to 0.05. A z-score of −1.64 gives the closest value (0.0505). Solving $-1.64 = \frac{16 - M}{0.12}$ gives M = 16.1968 ounces. *Using technology*: The command `invNorm(area:0.05,μ:0,σ:1)` gives a value of −1.645. Solving $-1.645 = \frac{16 - M}{0.12}$ gives M = 16.1974 ounces.

**Step 3: Answer the question.** Buckley Farms should use a mean weight of about 16.2 ounces to make sure less than 5% of bags are underweight.

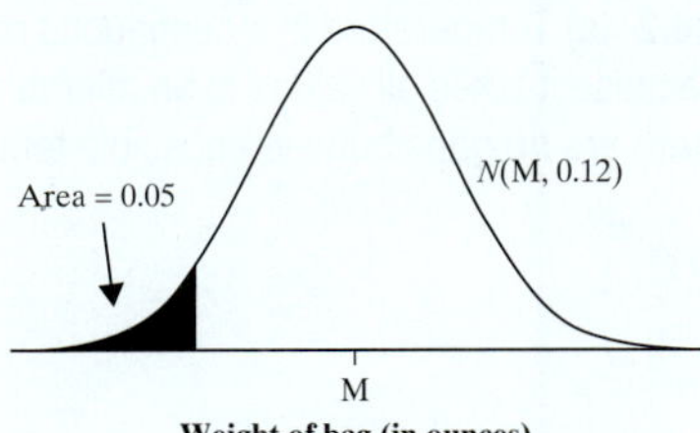

TRM Learning Objectives Grid

## What Did You Learn?

| Learning Objective | Section | Related Example on Page(s) | Relevant Chapter Review Exercise(s) |
|---|---|---|---|
| Compute probabilities using the probability distribution of a discrete random variable. | 6.1 | 349 | R6.1 |
| Calculate and interpret the mean (expected value) of a discrete random variable. | 6.1 | 350, 352 | R6.1, R6.3 |
| Calculate and interpret the standard deviation of a discrete random variable. | 6.1 | 353 | R6.1, R6.3 |
| Compute probabilities using the probability distribution of certain continuous random variables. | 6.1 | 355, 357 | R6.4 |
| Describe the effects of transforming a random variable by adding or subtracting a constant and multiplying or dividing by a constant. | 6.2 | 365, 366, 368 | R6.2, R6.3 |
| Find the mean and standard deviation of the sum or difference of independent random variables. | 6.2 | 372, 373, 374, 377 | R6.3, R6.4 |
| Find probabilities involving the sum or difference of independent Normal random variables. | 6.2 | 380, 381 | R6.4 |
| Determine whether the conditions for using a binomial random variable are met. | 6.3 | 388 | R6.5 |
| Compute and interpret probabilities involving binomial distributions. | 6.3 | 390, 393, 396 | R6.6 |
| Calculate the mean and standard deviation of a binomial random variable. Interpret these values in context. | 6.3 | 399 | R6.5 |
| Find probabilities involving geometric random variables. | 6.3 | 406 | R6.7 |
| *When appropriate, use the Normal approximation to the binomial distribution to calculate probabilities. | 6.3 | 403 | R6.8 |

*This topic is not required for the AP® Statistics exam.

# Chapter 6 Chapter Review Exercises

*These exercises are designed to help you review the important ideas and methods of the chapter.*

**R6.1 Knees** Patients receiving artificial knees often experience pain after surgery. The pain is measured on a subjective scale with possible values of 1 (low) to 5 (high). Let X be the pain score for a randomly selected patient. The following table gives part of the probability distribution for X.

| **Value:** | 1 | 2 | 3 | 4 | 5 |
|---|---|---|---|---|---|
| **Probability:** | 0.1 | 0.2 | 0.3 | 0.3 | ?? |

(a) Find $P(X = 5)$.

(b) Is pain score a discrete or continuous random variable? Explain.

(c) Find $P(X \leq 2)$. Is this the same as $P(X < 2)$? Explain.

(d) Compute the expected pain score and the standard deviation of the pain scores.

**R6.2 A glass act** In a process for manufacturing glassware, glass stems are sealed by heating them in a flame. Let X be the temperature (in degrees Celsius) for a randomly chosen glass. The mean and standard deviation of X are $\mu_X = 550°C$ and $\sigma_X = 5.7°C$.

416

TRM Full Solutions to Chapter 6 Review Exercises

## Answers to Chapter 6 Review Exercises

**R6.1 (a)** $P(X = 5) = 1 - 0.1 - 0.2 - 0.3 - 0.3 = 0.1$ **(b)** Discrete, because it takes a fixed set of values with gaps in between. **(c)** $P(X \leq 2) = 0.3$. $P(X < 2) = 0.1$. These are not the same because the outcome $X = 2$ is included in the first calculation but not the second. **(d)** $\mu_X = 1(0.1) + \cdots + 5(0.1) = 3.1$. $\sigma_X^2 = (1 - 3.1)^2(0.1) + \cdots + (5 - 3.1)^2(0.1) = 1.29$, so $\sigma_X = \sqrt{1.29} = 1.136$.

**R6.2 (a)** Temperature is a continuous random variable because it takes all values in an interval of numbers—there are no gaps between possible temperatures. **(b)** $P(X < 540) = P(X \leq 540)$ because $X$ is a continuous random variable. In this case, $P(X = 540) = 0$ because the line segment above $X = 540$ has no area. **(c)** Mean $= 550 - 550 = 0°C$. The standard deviation stays the same, 5.7°C, because subtracting a constant does not change the variability. **(d)** In degrees Fahrenheit, the mean is $\mu_Y = \frac{9}{5}(550) + 32 = 1022°F$ and the standard deviation is $\sigma_Y = \left(\frac{9}{5}\right)(5.7) = 10.26°F$.

**R6.3 (a)** If you were to play many games of 4-Spot Keno, you would get a payout of about \$0.70 per game, on average. If you were to play many games of 4-Spot Keno, the payout amounts would typically vary by about \$6.58 from the mean (\$0.70). **(b)** Let $Y$ be the amount

(a) Is temperature a discrete or continuous random variable? Explain.

(b) How is $P(X < 540)$ related to $P(X \le 540)$? Explain.

(c) The target temperature is 550°C. What are the mean and standard deviation of the number of degrees off target, $D = X - 550$?

(d) A manager asks for results in degrees Fahrenheit. The conversion of X into degrees Fahrenheit is given by $Y = \frac{9}{5}X + 32$. What are the mean $\mu_Y$ and the standard deviation $\sigma_Y$ of the temperature of the flame in the Fahrenheit scale?

**R6.3 Keno** In a game of 4-Spot Keno, the player picks 4 numbers from 1 to 80. The casino randomly selects 20 winning numbers from 1 to 80. The table below shows the possible outcomes of the game and their probabilities, along with the amount of money (Payout) that the player wins for a \$1 bet. If X = the payout for a single \$1 bet, you can check that $\mu_X = \$0.70$ and $\sigma_X = \$6.58$.

| **Matches:** | 0 | 1 | 2 | 3 | 4 |
|---|---|---|---|---|---|
| **Payout $x_i$:** | \$0 | \$0 | \$1 | \$3 | \$120 |
| **Probability $p_i$:** | 0.308 | 0.433 | 0.213 | 0.043 | 0.003 |

(a) Interpret the values of $\mu_X$ and $\sigma_X$ in context.

(b) Jerry places a single \$5 bet on 4-Spot Keno. Find the expected value and the standard deviation of his winnings.

(c) Marla plays five games of 4-Spot Keno, betting \$1 each time. Find the expected value and the standard deviation of her total winnings.

(d) Based on your answers to (b) and (c), which player would the casino prefer? Justify your answer.

**R6.4 Applying torque** A machine fastens plastic screw-on caps onto containers of motor oil. If the machine applies more torque than the cap can withstand, the cap will break. Both the torque applied and the strength of the caps vary. The capping-machine torque T follows a Normal distribution with mean 7 inch-pounds and standard deviation 0.9 inch-pounds. The cap strength C (the torque that would break the cap) follows a Normal distribution with mean 10 inch-pounds and standard deviation 1.2 inch-pounds.

(a) What is the probability that a randomly selected cap has a strength greater than 11 inch-pounds?

(b) Explain why it is reasonable to assume that the cap strength and the torque applied by the machine are independent.

(c) Let the random variable $D = C - T$. Find its mean and standard deviation.

(d) What is the probability that a randomly selected cap will break while being fastened by the machine? Show your work.

*Exercises R6.5 and R6.6 refer to the following setting.* According to Mars, Incorporated, 20% of its plain M&M'S candies are orange. Assume that the company's claim is true. Suppose that you reach into a large bag of plain M&M'S (without looking) and pull out 8 candies. Let X = the number of orange candies you get.

**R6.5 Orange M&M'S**

(a) Explain why it is reasonable to use the binomial distribution for probability calculations involving X.

(b) Find and interpret the expected value of X.

(c) Find and interpret the standard deviation of X.

**R6.6 Orange M&M'S**

(a) Would you be surprised if none of the candies were orange? Compute an appropriate probability to support your answer.

(b) How surprising would it be to get 5 or more orange candies? Compute an appropriate probability to support your answer.

**R6.7** ***Sushi Roulette*** In the Japanese game show *Sushi Roulette*, the contestant spins a large wheel that's divided into 12 equal sections. Nine of the sections have a sushi roll, and three have a "wasabi bomb." When the wheel stops, the contestant must eat whatever food is on that section. To win the game, the contestant must eat one wasabi bomb. Find the probability that it takes 3 or fewer spins for the contestant to get a wasabi bomb. Show your method clearly.

**R6.8* Is this coin balanced?** While he was a prisoner of war during World War II, John Kerrich tossed a coin 10,000 times. He got 5067 heads. If the coin is perfectly balanced, the probability of a head is 0.5.

(a) Find the mean and the standard deviation of the number of heads in 10,000 tosses, assuming the coin is perfectly balanced.

(b) Explain why the Normal approximation is appropriate for calculating probabilities involving the number of heads in 10,000 tosses.

(c) Is there reason to think that Kerrich's coin was not balanced? To answer this question, use a Normal distribution to estimate the probability that tossing a balanced coin 10,000 times would give a count of heads at least this far from 5000 (that is, at least 5067 heads or at most 4933 heads).

*This topic is not required for the AP® Statistics exam. Some teachers prefer to discuss this topic when presenting the sampling distribution of $\hat{p}$ (Chapter 7).

of Jerry's payout. $\mu_Y = 5(0.70) = \$3.50$ and $\sigma_Y = 5(6.58) = \$32.90$. **(c)** Let *W* be the amount of Marla's payout. $\mu_W = 0.70 + 0.70 + 0.70 + 0.70 + 0.70 = \$3.50$ and $\sigma_W^2 = 6.58^2 + 6.58^2 + 6.58^2 + 6.58^2 + 6.58^2 = 216.482$, so $\sigma_W = \sqrt{216.482} = \$14.71$. **(d)** Even though their expected values are the same, the casino would probably prefer Marla because there is less variability in her strategy and her winnings are more predictable.

**R6.4 (a)** *C* follows an *N*(10, 1.2) distribution and we want to find $P(C > 11)$. $z = \frac{11 - 10}{1.2} = 0.83$ and $P(Z > 0.83) = 0.2033$. *Using technology*: `normalcdf(lower:11,upper:1000,μ:10,σ:1.2)` = 0.2023. There is a 0.2023 probability that a randomly selected cap has a strength greater than 11 inch-pounds. **(b)** The machine that makes the caps and the machine that applies the torque are not the same. **(c)** $C - T$ is Normal with mean $10 - 7 = 3$ inch-pounds and standard deviation $\sqrt{0.9^2 + 1.2^2} = 1.5$ inch-pounds. **(d)** We want to find $P(C - T < 0)$. $z = \frac{0 - 3}{1.5} = -2$ and $P(Z < -2) = 0.0228$. *Using technology*: `normalcdf(lower:-1000,upper:0,μ:3,σ:1.5)` = 0.0228. There is a 0.0228 probability that a randomly selected cap will break when being fastened by the machine.

**R6.5 (a)** *Binary?* "Success" = orange and "Failure" = not orange. *Independent?* The sample of size $n = 8$ is less than 10% of the large bag, so we can assume the outcomes of trials are independent. *Number?* $n = 8$. *Success?* $p = 0.20$. **(b)** $\mu_X = 8(0.2) = 1.6$. If we were to select many samples of size 8, we would expect to get about 1.6 orange M&M'S, on average. **(c)** $\sigma_X = \sqrt{8(0.2)(0.8)} = 1.13$. If we were to select many samples of size 8, the number of orange M&M'S would typically vary by about 1.13 from the mean (1.6).

**R6.6**

**(a)** $P(X = 0) = \binom{8}{0}(0.2)^0(0.8)^8 = 0.1678$. Because the probability is not that small, it would not be surprising to get no orange M&M'S in a sample of size 8.

**(b)** $P(X \ge 5) = \binom{8}{5}(0.20)^5(0.80)^3 + \cdots = 0.0104$. Because the probability is small, it would be surprising to find 5 or more orange M&M'S in a sample of size 8.

**R6.7** Let *Y* be the number of spins to get a "wasabi bomb." *Y* is a geometric random variable with $p = \frac{3}{12} = 0.25$. $P(Y \le 3) = (0.75)^2(0.25) + (0.75)(0.25) + 0.25 = 0.5781$.

**R6.8 (a)** Let *X* be the number of heads in 10,000 tosses. $\mu_X = 10{,}000(0.5) = 5000$ and $\sigma_X = \sqrt{10{,}000(0.5)(0.5)} = 50$. **(b)** $np = 10{,}000(0.5) = 5000$ and $n(1 - p) = 10{,}000(0.5) = 5000$ are both at least 10. **(c)** We want to find $P(X \le 4933$ or $X \ge 5067)$. $z = \frac{4933 - 5000}{50} = -1.34$ and $z = \frac{5067 - 5000}{50} = 1.34$ and $P(Z \le -1.34) + P(Z \ge 1.34) = 0.1802$. *Using technology*: `1 - normalcdf(lower:4933,upper:5067,μ:5000,σ:50)` = 0.1802. Because this probability isn't small, we don't have convincing evidence that Kerrich's coin was unbalanced—a difference this far from 5000 could be due to chance alone.

TRM Full Solutions to Chapter 6 AP® Statistics Practice Test

## Answers to Chapter 6 AP® Statistics Practice Test

**T6.1** b

**T6.2** d

**T6.3** d

**T6.4** e

**T6.5** d

**T6.6** b

**T6.7** c

**T6.8** b



# Chapter 6 AP® Statistics Practice Test

**Section I: Multiple Choice** *Select the best answer for each question.*

*Questions T6.1 to T6.3 refer to the following setting. A psychologist studied the number of puzzles that subjects were able to solve in a five-minute period while listening to soothing music. Let X be the number of puzzles completed successfully by a randomly chosen subject. The psychologist found that X had the following probability distribution:*

| Value: | 1 | 2 | 3 | 4 |
|---|---|---|---|---|
| Probability: | 0.2 | 0.4 | 0.3 | 0.1 |

**T6.1** What is the probability that a randomly chosen subject completes more than the expected number of puzzles in the five-minute period while listening to soothing music?

(a) 0.1
(b) 0.4
(c) 0.8
(d) 1
(e) Cannot be determined

**T6.2** The standard deviation of X is 0.9. Which of the following is the best interpretation of this value?

(a) About 90% of subjects solved 3 or fewer puzzles.
(b) About 68% of subjects solved between 0.9 puzzles less and 0.9 puzzles more than the mean.
(c) The typical subject solved an average of 0.9 puzzles.
(d) The number of puzzles solved by subjects typically differed from the mean by about 0.9 puzzles.
(e) The number of puzzles solved by subjects typically differed from one another by about 0.9 puzzles.

**T6.3** Let $D$ be the difference in the number of puzzles solved by two randomly selected subjects in a five-minute period. What is the standard deviation of $D$?

(a) 0 (b) 0.81 (c) 0.9 (d) 1.27 (e) 1.8

**T6.4** Suppose a student is randomly selected from your school. Which of the following pairs of random variables are most likely independent?

(a) X = student's height; Y = student's weight
(b) X = student's IQ; Y = student's GPA
(c) X = student's PSAT Math score; Y = student's PSAT Verbal score
(d) X = average amount of homework the student does per night; Y = student's GPA
(e) X = average amount of homework the student does per night; Y = student's height

**T6.5** A certain vending machine offers 20-ounce bottles of soda for $1.50. The number of bottles X bought from the machine on any day is a random variable with mean 50 and standard deviation 15. Let the random variable Y equal the total revenue from this machine on a given day. Assume that the machine works properly and that no sodas are stolen from the machine. What are the mean and standard deviation of Y?

(a) $\mu_Y = \$1.50, \sigma_Y = \$22.50$
(b) $\mu_Y = \$1.50, \sigma_Y = \$33.75$
(c) $\mu_Y = \$75, \sigma_Y = \$18.37$
(d) $\mu_Y = \$75, \sigma_Y = \$22.50$
(e) $\mu_Y = \$75, \sigma_Y = \$33.75$

**T6.6** The weight of tomatoes chosen at random from a bin at the farmer's market follows a Normal distribution with mean $\mu = 10$ ounces and standard deviation $\sigma = 1$ ounce. Suppose we pick four tomatoes at random from the bin and find their total weight $T$. The random variable $T$ is

(a) Normal, with mean 10 ounces and standard deviation 1 ounce.
(b) Normal, with mean 40 ounces and standard deviation 2 ounces.
(c) Normal, with mean 40 ounces and standard deviation 4 ounces.
(d) binomial, with mean 40 ounces and standard deviation 2 ounces.
(e) binomial, with mean 40 ounces and standard deviation 4 ounces.

**T6.7** Which of the following random variables is geometric?

(a) The number of times I have to roll a die to get two 6s.
(b) The number of cards I deal from a well-shuffled deck of 52 cards until I get a heart.
(c) The number of digits I read in a randomly selected row of the random digits table until I find a 7.
(d) The number of 7s in a row of 40 random digits.
(e) The number of 6s I get if I roll a die 10 times.

**T6.8** Seventeen people have been exposed to a particular disease. Each one independently has a 40% chance of contracting the disease. A hospital has the capacity to handle 10 cases of the disease. What is the probability that the hospital's capacity will be exceeded?

(a) 0.011
(b) 0.035
(c) 0.092
(d) 0.965
(e) 0.989

**T6.9** The figure shows the probability distribution of a discrete random variable X. Note that the cursor is on the histogram bar representing a value of 6. Which of the following best describes this random variable?

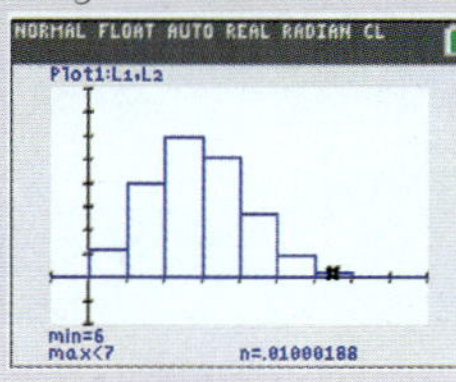


(a) Binomial with $n = 8$, $p = 0.1$
(b) Binomial with $n = 8$, $p = 0.3$
(c) Binomial with $n = 8$, $p = 0.8$
(d) Geometric with $p = 0.1$
(e) Geometric with $p = 0.2$

**T6.10** A test for extrasensory perception (ESP) involves asking a person to tell which of 5 shapes—a circle, star, triangle, diamond, or heart—appears on a hidden computer screen. On each trial, the computer is equally likely to select any of the 5 shapes. Suppose researchers are testing a person who does not have ESP and so is just guessing on each trial. What is the probability that the person guesses the first 4 shapes incorrectly but gets the fifth correct?

(a) 1/5
(b) $\left(\frac{4}{5}\right)^4$
(c) $\left(\frac{4}{5}\right)^4 \cdot \left(\frac{1}{5}\right)$
(d) $\binom{5}{1} \cdot \left(\frac{4}{5}\right)^4 \cdot \left(\frac{1}{5}\right)$
(e) 4/5

**Section II: Free Response** *Show all your work. Indicate clearly the methods you use, because you will be graded on the correctness of your methods as well as on the accuracy and completeness of your results and explanations.*

**T6.11** Let Y denote the number of broken eggs in a randomly selected carton of one dozen "store brand" eggs at a local supermarket. Suppose that the probability distribution of Y is as follows.

| Value $y_i$: | 0 | 1 | 2 | 3 | 4 |
|---|---|---|---|---|---|
| Probability $p_i$: | 0.78 | 0.11 | 0.07 | 0.03 | 0.01 |

(a) What is the probability that at least 10 eggs in a randomly selected carton are *unbroken*?
(b) Calculate and interpret $\mu_Y$.
(c) Calculate and interpret $\sigma_Y$. Show your work.
(d) A quality control inspector at the store keeps looking at randomly selected cartons of eggs until he finds one with at least 2 broken eggs. Find the probability that this happens in one of the first three cartons he inspects.

**T6.12** *Ladies Home Journal* magazine reported that 66% of all dog owners greet their dog before greeting their spouse or children when they return home at the end of the workday. Assume that this claim is true. Suppose 12 dog owners are selected at random. Let X = the number of owners who greet their dogs first.

(a) Explain why it is reasonable to use the binomial distribution for probability calculations involving X.
(b) Only 4 of the owners in the random sample greeted their dogs first. Does this give convincing evidence against the *Ladies Home Journal* claim? Calculate an appropriate probability to support your answer.

**T6.13** Ed and Adelaide attend the same high school, but are in different math classes. The time E that it takes Ed to do his math homework follows a Normal distribution with mean 25 minutes and standard deviation 5 minutes. Adelaide's math homework time A follows a Normal distribution with mean 50 minutes and standard deviation 10 minutes. Assume that E and A are independent random variables.

(a) Randomly select one math assignment of Ed's and one math assignment of Adelaide's. Let the random variable D be the difference in the amount of time each student spent on their assignments: $D = A - E$. Find the mean and the standard deviation of D. Show your work.
(b) Find the probability that Ed spent longer on his assignment than Adelaide did on hers. Show your work.

**T6.14** According to the Census Bureau, 13% of American adults (aged 18 and over) are Hispanic. An opinion poll plans to contact an SRS of 1200 adults.

(a) What is the mean number of Hispanics in such samples? What is the standard deviation?
(b) Should we be suspicious if the sample selected for the opinion poll contains 15% Hispanic people? Compute an appropriate probability to support your answer.

**T6.9** b

**T6.10** c

**T6.11** **(a)** $P(Y \le 2) = 0.96$ **(b)** $\mu_Y = 0(0.78) + \cdots = 0.38$. If we were to randomly select many cartons of eggs, we would expect about 0.38 to be broken, on average. **(c)** $\sigma_Y^2 = (0 - 0.38)^2(0.78) + \cdots = 0.6756$ Thus, $\sigma_Y = \sqrt{0.6756} = 0.8219$. If we were to randomly select many cartons of eggs, the number of broken eggs would typically vary by about 0.8219 from the mean (0.38). **(d)** Let $X$ stand for the number of cartons inspected to find one carton with at least 2 broken eggs. $X$ is a geometric random variable with $p = 0.11$. $P(X \le 3) = (0.11) + (0.89)(0.11) + (0.89)^2(0.11) = 0.2950$.

**T6.12** **(a)** *Binary?* "Success" = dog first and "Failure" = not dog first. *Independent?* We are sampling without replacement, but 12 is less than 10% of all dog owners. *Number?* $n = 12$. *Success?* $p = 0.66$.

**(b)** $P(X \le 4) = \binom{12}{0}(0.66)^0(0.34)^{12} + \cdots + \binom{12}{4}(0.66)^4(0.34)^8 = 0.0213$.

Because this probability is small, it is unlikely that only 4 or fewer owners greet their dogs first by chance alone. This gives convincing evidence that the claim by the *Ladies Home Journal* is incorrect.

**T6.13** **(a)** $\mu_D = 50 - 25 = 25$ minutes, $\sigma_D^2 = 100 + 25 = 125$, and $\sigma_D = \sqrt{125} = 11.18$ minutes. **(b)** $D$ follows an $N(25, 11.18)$ distribution and we want to find $P(D < 0)$. $z = \frac{0 - 25}{11.18} = -2.24$ and $P(Z < -2.24) = 0.0125$. *Using technology*: `normalcdf(lower:−1000,upper:0,μ:25,σ:11.18)` = 0.0127. There is a 0.0127 probability that Ed spent longer on his assignment than Adelaide did on hers.

**T6.14** **(a)** Let $X$ stand for the number of Hispanics in the sample. $\mu_X = 1200(0.13) = 156$ and $\sigma_X = \sqrt{1200(0.13)(0.87)} = 11.6499$. **(b)** 15% of 1200 is 180, so we want to find $P(X \ge 180) = \binom{1200}{180}(0.13)^{180}(0.87)^{1020} + \cdots = 0.0235$. Because this probability is small, it is unlikely that 180 or more Hispanics will be in the sample just by chance. This gives us reason to be suspicious about the sampling process.

Chapter

7

# Sampling Distributions

## PD Overview

Sampling distributions are the culmination of the first three major content areas of the course—exploring data, planning studies, and probability—and the final piece needed for the formal study of statistical inference. Sampling distributions are also typically very difficult for students to understand. To help your students through this difficult topic, we have included an extra share of Activities and applets that nicely illustrate the important concepts. Make sure to do as many of these as possible.

We spend quite a bit of time in the Introduction and in Section 8.1 developing the idea of a sampling distribution. In short, a sampling distribution describes the possible values of a statistic and how often it takes those values. A statistic is any quantity that can be calculated from a sample, such as a sample mean $\bar{x}$ or a sample proportion $\hat{p}$.

Starting in this chapter and continuing in Chapters 8, 9, and 10, we present methods for proportions before we present methods for means. This conscious choice to be consistent in all four chapters was based on our experiences with students. Because we use the Normal distribution when doing inference procedures for proportions, students can focus on the logic of inference without having to learn about the $t$ distributions we use when performing inference for means.

Finally, we highly recommend that you download and read the Special Focus materials on Sampling Distributions available on AP® Central http://apcentral.collegeboard.com/apc/public/repository/Statistics_Sampling_Distributions_SF.pdf. These materials include an overview of sampling distributions, along with some great activities.

### Relationship to the AP® Statistics Course Description

Chapter 7 covers the following content as described in the Topic Outline from the AP® Statistics Course Description. Topics marked with a superscript are covered in other chapters as indicated.

III. Anticipating patterns: Exploring random phenomena using probability and simulation

D. Sampling distributions

1. Sampling distribution of a sample proportion
2. Sampling distribution of a sample mean
3. Central Limit Theorem
4. Sampling distribution of a difference between two independent sample proportions[ch10]
5. Sampling distribution of a difference between two independent sample means[ch10]
6. Simulation of sampling distributions
7. $t$-distribution[ch9]
8. Chi-square distribution[ch11]

## The Big Ideas

One of the challenges in teaching the AP® Statistics course is keeping students focused on the big picture and not just the details of each section. We outline the big ideas for the chapter here.

### Chapter 7 Introduction

Two of the best Activities of the year show up in this chapter, so make sure to take the time to do them with your students. The German Tank Problem Activity is a wonderful way to show how statisticians used creativity—and sampling distributions—to provide valuable information to the Allies during World War II. Likewise, the Penny for Your Thoughts Activity is an excellent way for students to discover properties of the sampling distribution of a sample proportion and of the sampling distribution of a sample mean.

### Section 7.1 What Is a Sampling Distribution?

In this section, we introduce the "big ideas" of sampling distributions. The first big idea is the difference between a statistic and a parameter. A parameter is a number that describes some characteristic of a population. A statistic estimates the value of a parameter using a sample from the population. Although it may seem simple to students when introduced, the concept is a crucial distinction that students often fail to make when they perform statistical inference.

The second big idea is that statistics vary. For example, the mean weight in a sample of high school students is a variable that will change from sample to sample. This means that statistics have distributions, whereas parameters do not.

The third big idea is the difference between the distribution of the population, the distribution of the sample, and the sampling distribution of a sample statistic. The figure on page 428 and the Activity on page 453 illustrate this distinction nicely.

The fourth big idea is how to describe a sampling distribution. To adequately describe a sampling distribution, we need to address its shape, center, and spread. If the mean of the sampling distribution is the same as the value of the parameter being estimated, then the statistic is called an unbiased estimator. Ideally, the spread of a sampling distribution will be very small, meaning that the statistic is very precise. Larger sample sizes result in sampling distributions with smaller spreads.

## Section 7.2 Sample Proportions

In this section, we focus on the sampling distribution of a sample proportion. Using The Candy Machine Activity or the Penny for Your Thoughts . . . on Proportions Alternate Activity is a great way to start this section. They are well worth the time it takes to do them in class.

A sample proportion is simply the number of successes in the sample divided by the sample size. And because the number of successes in a sample often follows a binomial distribution, the sampling distribution of a sample proportion should already be somewhat familiar to students. In fact, the sampling distribution of a proportion is just a rescaled (by a factor of $1/n$) version of a binomial distribution.

This makes it easy to calculate the mean and standard deviation of the sampling distribution if we know the sample size $n$ and the proportion of successes in the population $p$. Also, if the sample size is large enough and the value of $p$ isn't too close to 0 or 1, the sampling distribution of $\hat{p}$ is approximately Normal, allowing us to do probability calculations.

## Section 7.3 Sample Means

In this section, we focus on the sampling distribution of a sample mean. To introduce and reinforce the important lessons of this section, we highly recommend that you do the Activities on pages 450 and 453.

The sampling distribution of a sample mean behaves in much the same way as the sampling distribution of a sample proportion. Both the sample mean and sample proportion are unbiased estimators, so their sampling distributions are always centered at the value of the population parameter they are trying to estimate. Also, as the sample size increases, both sampling distributions become less variable and more Normal. The very useful fact that the distribution of $\bar{x}$ becomes more Normal as the sample size increases is called the *central limit theorem*. However, in the special case of a Normally distributed population, the sampling distribution of a sample mean will always be Normal for any sample size.

# Chapter 7: Resources

## TRM Additional *TPS* Teacher's Resource Materials

The following resources can be found by clicking on the links in the new Teacher's e-Book (TE-book), opening the Teacher's Resource CD (TRCD), or logging onto the book's companion Web site www.whfreeman.com/tps5e (teacher log-in required).

- Printable Chapter 7 Learning Objective Grid (page 424)
- Sampling Distribution of a Sample Proportion (page 441)
- Sampling Distribution of a Sample Mean (page 450)
- FRAPPY! materials: Question, Model Solution, Scoring Rubric, Sample Responses with commentary (page 464)
- Quizzes: 3 versions for each section; versions A and B are roughly parallel
  - Section 7.1 (page 436)
  - Section 7.2 (page 446)
  - Section 7.3 (page 461)
- Tests: 4 versions; versions A and B are roughly parallel, version D is cumulative (page 466)

## Free-Response Questions from Previous AP® Exams

Questions can be found on the AP® Central Web site: http://apcentral.collegeboard.com/apc/members/exam/exam_questions/8357.html.

Students should be able to answer all the free-response questions listed below with material learned in this chapter. Questions that contain content from this chapter but also require content from later chapters are listed in the last chapter required to complete the entire question. This list will be updated after each AP® exam and will be posted on the book's Web site: http://www.whfreeman.com/tps5e.

Questions marked with an asterisk are from exams with released multiple-choice questions. You may want to save these questions until the end of the year so you can give your students a complete released exam for practice.

| Year | # | Content |
|---|---|---|
| 2012 | 6* | • Selecting an SRS<br>• Standard error of the mean for a simple random sample<br>• Standard error of the mean for a stratified random sample<br>• How stratified random sampling reduces variability |
| 2010 | 2 | • Sampling distribution of the sample mean<br>• Probability calculation for a total |
| 2009 | 2 | • Inverse Normal calculation<br>• Binomial probability calculation<br>• Probability calculation for the sample mean |
| 2008B | 2 | • Properties of estimators: bias and variability |
| 2007 | 3* | • Sampling distribution of the sample mean<br>• Probability calculation for the sample mean<br>• Central Limit Theorem |
| 2007B | 2 | • Addition rule<br>• Binomial probability calculation<br>• Sampling distribution of the sample mean |
| 2006 | 3 | • Normal probability calculation<br>• Binomial probability calculation<br>• Probability calculation for a sample mean |
| 2004B | 3 | • Normal probability calculation and interpretation<br>• Probability calculation and interpretation for a sample mean |
| 1998 | 1 | • Sampling distribution of the sample mean<br>• Effect of sample size on shape of sampling distribution |

## Applets

- The *Reese's Pieces* applet at www.rossmanchance.com/applets allows students to investigate the sampling distribution of a sample proportion of orange candies. Students can change the sample size and proportion of successes to see how these factors affect the sampling distribution of the sample proportion.
- The *Normal Approximation to Binomial Distributions* applet at www.whfreeman.com/tps5e allows students to use sliders to discover that the Normal approximation gets better as the sample size increases and the proportion of successes gets closer to 0.5 and farther from 0 and 1.
- The *Sampling Distribution* applet at http://onlinestatbook.com/stat_sim/sampling_dist/index.html allows students to take samples of different sizes from populations of different shapes to explore how these factors affect the sampling distribution of the sample mean.
- The *Sampling Pennies* applet at www.rossmanchance.com/applets allows students to investigate how changing the sample size affects the sampling distribution of the sample mean when sampling from a skewed population.

## TI-Nspire Activities

These Activities are designed to illustrate important statistical concepts using the TI-Nspire and can be found at www.whfreeman.com/tps5e

- *Exploring Sampling Distributions*: Students explore the distribution of sample means for different values of *n*, focusing on the center and spread of the sampling distribution.

Chapter

7

# Chapter 7: Pacing Guide, Learning Objectives, and Suggested Assignments

This pacing guide is based on a schedule with 110, 50-minute sessions before the AP® exam. The 110 days include 4 days for midterms and 10 days for AP® exam review. If you have a different number of sessions before the AP® exam, you can modify the pacing guide to suit your needs. If you have additional time, consider incorporating quizzes, released AP® free-response questions, or additional activities. See the Resources section above for suggestions.

The suggested homework assignments list odd-numbered exercises, whenever possible, so students can check their answers against the back-of-the-book answers. If you would rather students not have access to the answers while doing homework, adding 1 to the exercise numbers usually will do the trick, because the homework exercises typically are paired. For example, Exercises 1 and 2 will generally cover the same topics, but in different contexts. You may also choose to include the Recycle and Review questions at the end of each section, which review topics from previous sections or chapters. These questions are denoted with a symbol.

| Day | Topics | **Learning Objectives:** Students will be able to. . . | Suggested Assignment |
|---|---|---|---|
| 1 | Introduction: German Tank Problem, 7.1 Parameters and Statistics | • Distinguish between a parameter and a statistic. | 1, 3, 5 |
| 2 | 7.1 Sampling Variability, Describing Sampling Distributions | • Distinguish among the distribution of a population, the distribution of a sample, and the sampling distribution of a statistic.<br>• Use the sampling distribution of a statistic to evaluate a claim about a parameter.<br>• Determine whether or not a statistic is an unbiased estimator of a population parameter.<br>• Describe the relationship between sample size and the variability of a statistic. | 7, 9, 11, 13, 15, 17, 19 |
| 3 | 7.2 The Sampling Distribution of $\hat{p}$, Using the Normal Approximation for $\hat{p}$. | • Find the mean and standard deviation of the sampling distribution of a sample proportion $\hat{p}$. Check the 10% condition before calculating $\sigma_{\hat{p}}$.<br>• Determine if the sampling distribution of $\hat{p}$ is approximately Normal.<br>• If appropriate, use a Normal distribution to calculate probabilities involving $\hat{p}$. | 21–24, 27, 29, 33, 35, 37, 39 |
| 4 | 7.3 The Sampling Distribution of $\bar{x}$: Mean and Standard Deviation, Sampling from a Normal Population | • Find the mean and standard deviation of the sampling distribution of a sample mean $\bar{x}$. Check the 10% condition before calculating $\sigma_{\bar{x}}$.<br>• If appropriate, use a Normal distribution to calculate probabilities involving $\bar{x}$. | 43–46, 49, 51, 53, 55 |
| 5 | 7.3 The Central Limit Theorem | • Explain how the shape of the sampling distribution of $\bar{x}$ is affected by the shape of the population distribution and the sample size.<br>• If appropriate, use a Normal distribution to calculate probabilities involving $\bar{x}$. | 57, 59, 61, 63, 65–68 |
| 6 | Chapter 7 Review/FRAPPY! | | Chapter 7 Review Exercises |
| 7 | Chapter 7 Test | | Cumulative AP® Practice Test 2 |

# Chapter 7

**PD Chapter Overview**

Watch the video overview of Chapter 7 (for teachers) by clicking on the link in the TE-book, opening the TRCD, or logging onto the book's companion Web site www.whfreeman.com/tps5e (teacher log-in required).

### Teaching Tip

We highly recommend that you download and read the Special Focus materials on Sampling Distributions available on AP® Central. These materials include an overview of sampling distributions, along with some great activities including the German Tank Activity that we do on page 422. The materials on AP® Central also include instructions for using Fathom to simulate the German Tank Problem. http://apcentral.collegeboard.com/apc/public/repository/Statistics_Sampling_Distributions_SF.pdf.

# Sampling Distributions

## case study

## Building Better Batteries

Everyone wants to have the latest technological gadget. That's why iPods, digital cameras, smartphones, Game Boys, and the Wii have sold millions of units. These devices require lots of power and can drain batteries quickly. Battery manufacturers are constantly searching for ways to build longer-lasting batteries.

A particular manufacturer produces AA batteries that are designed to last an average of 17 hours with a standard deviation of 0.8 hours. Quality control inspectors select a random sample of 50 batteries during each hour of production, and they then drain them under conditions that mimic normal use. Here are the lifetimes (in hours) of the batteries from one such sample:

| | | | | | | | | | | | | |
|---|---|---|---|---|---|---|---|---|---|---|---|---|
| 16.73 | 15.60 | 16.31 | 17.57 | 16.14 | 17.28 | 16.67 | 17.28 | 17.27 | 17.50 | 15.46 | 16.50 | 16.19 |
| 15.59 | 17.54 | 16.46 | 15.63 | 16.82 | 17.16 | 16.62 | 16.71 | 16.69 | 17.98 | 16.36 | 17.80 | 16.61 |
| 15.99 | 15.64 | 17.20 | 17.24 | 16.68 | 16.55 | 17.48 | 15.58 | 17.61 | 15.98 | 16.99 | 16.93 | 16.01 |
| 17.54 | 17.41 | 16.91 | 16.60 | 16.78 | 15.75 | 17.31 | 16.50 | 16.72 | 17.55 | 16.46 | | |

Do these data suggest that the production process is working properly? Is it safe for plant managers to send out all the batteries produced in this hour for sale? In this chapter, you will develop the tools you need to help answer questions like this.



### Teaching Tip

Consider doing the Penny for Your Thoughts . . . on Proportions Alternate Activity on page 441 and the Penny for Your Thoughts Activity on page 450 simultaneously before beginning this chapter. Here are some additional suggestions for these Activities:

- Start collecting pennies early in the year. Consider sending an email to your fellow teachers, asking them to contribute pennies they may have accumulated.
- Use easel-size grid paper to create the dotplots for each Activity. Hang them on the wall so you can refer to them when you discuss the concepts of this chapter. These graphs will be extremely valuable when introducing challenging concepts such as $\mu_{\bar{x}}$ and $\sigma_{\hat{p}}$.
- Rather than marking their values on the dotplot with dots, have students use $\hat{p}$'s and $\bar{x}$'s instead. This makes it easier for students to understand that we are studying the distribution of $\hat{p}$ and $\bar{x}$.
- Instead of spending time during class to complete the Activities, consider starting them in Chapter 5 or Chapter 6 and do a little bit each day. For example, during the first week, students take a sample of size 1 each day and record the year of the penny. Each day during the second week, students take a sample of size 5 and record the average year and proportion from the 2000s. The next week, students take samples of size 25. This will save time later and allow students to gradually learn the important lessons: how increasing the sample size affects the shape, center, and spread of a sampling distribution. Also, it prevents a student who happens to be absent on the day of the Activity from missing the lessons to be learned.

# Introduction

The battery manufacturer in the Case Study could find the true mean lifetime $\mu$ of all the batteries produced in an hour. Quality control inspectors would simply measure the lifetime of each battery (by draining it) and then calculate the average. With this method, the company would know the truth about the population mean $\mu$, but it would have no batteries left to sell!

Instead of taking a census, the manufacturer collects data from a random sample of 50 batteries produced that hour. The company's goal is to use the sample mean lifetime $\bar{x}$ to estimate the unknown population mean $\mu$. This is an example of *statistical inference*: we use information from a sample to draw conclusions about a larger population.

To make such an inference, we need to know how close the sample mean $\bar{x}$ is likely to be to the population mean $\mu$. After all, different random samples of 50 batteries from the same hour of production would yield different values of $\bar{x}$. How can we describe this *sampling distribution* of possible $\bar{x}$-values? We can think of $\bar{x}$ as a random variable because it takes numerical values that describe the outcomes of the random sampling process. As a result, we can examine its probability distribution using what we learned in Chapter 6.

This same reasoning applies to other types of inference settings. Here are a few examples.

- Each month, the Current Population Survey (CPS) interviews a random sample of individuals in about 60,000 U.S. households. The CPS uses the proportion of unemployed people in the sample $\hat{p}$ to estimate the national unemployment rate $p$.
- Tom is cooking a large turkey breast for a holiday meal. He wants to be sure that the turkey is safe to eat, which requires a minimum internal temperature of 165°F. Tom uses a thermometer to measure the temperature of the turkey meat at four randomly chosen points. If the minimum reading in the sample is 170°F, can Tom safely serve the turkey?
- How much do gasoline prices vary in a large city? To find out, a reporter records the price per gallon of regular unleaded gasoline at a random sample of 10 gas stations in the city on the same day. The range (maximum – minimum) of the prices in the sample is 25 cents. What can the reporter say about the range of gas prices at all the city's stations?

The following Activity gives you a chance to estimate an unknown population value based on data from a random sample.

## ACTIVITY The German Tank Problem

**MATERIALS:**
Tags or pieces of cardstock numbered 1 to *N*; small brown paper bag; index card and prelabeled graph grid for each team; prizes for the winners

During World War II, the Allies captured several German tanks. Each tank had a serial number on it. Allied commanders wanted to know how many tanks the Germans had so that they could allocate their forces appropriately. They sent the serial numbers of the captured tanks to a group of mathematicians in Washington, D.C., and asked for an estimate of the total number of German tanks N. In this Activity, you and your teammates will play the role of the mathematicians.

### Teaching Tip

Whenever we use information from a sample to estimate a characteristic of a population, our estimates will almost always vary from the truth, just by chance. As you discuss these three examples, point out that we could ask similar questions about any quantity calculated from a sample.

 **Activity Overview**

You can watch an overview of this Activity (for teachers) by clicking on the link in the TE-book, opening the TRCD, or logging onto the book's companion Web site www.whfreeman.com/tps5e (teacher log-in required).

### Teaching Tip

Students can do an Internet search and quickly discover "the answers" to the German Tank Problem, so you might want to change the context. The Alternate Activity on the next page is essentially the same as the German Tank Problem, but in a different context and with a slightly different approach.

### Teaching Tip

In the German Tank Problem Activity, you can use whatever value of $N$ you prefer. However, we suggest that you use a value of $N$ that is greater than 100. In the example on page 435, we use $N = 342$.

### Teaching Tip

See page 435 for the results of a simulation comparing five different statistics: maximum + minimum, $\bar{x} + 2s_x$, $2\bar{x}$, $2 \times$ median, $\left(\frac{n+1}{n}\right)$ maximum. Other reasonable statistics include $Q_1 + Q_3$, $2IQR$, $\bar{x}$ + *median*, and *IQR* + *median*. One year, a group won the competition by guessing the maximum value. Although the sample maximum will almost always be too small, it is usually close to the true maximum and very easy to calculate!

1. Your teacher will create tags numbered 1, 2, 3, . . . , N to represent the German tanks and place them in a bag. The class will be divided into teams of three or four students.

2. The teacher will mix the tags well and ask four students to draw one tag each from the bag. Each selected tag represents the serial number of a captured German tank. All four numbers should be written on the board for everyone to see. Return the tags to the bag.

3. Each team will have 15 minutes to come up with a statistical formula for estimating the total number of tanks N in the bag. You should have time to try several ideas. When you are satisfied with your method, calculate your estimate of N. Write your team members' names, your formula, and your estimate on the index card provided.

4. When time is called, each team must give its index card to your teacher. The teacher will make a chart on the board showing the formulas and estimates. Each team will have one minute to explain why it chose the formula it did.

5. The teacher will reveal the actual number of tanks. Which team came closest to the correct answer?

6. What if the Allies had captured four other German tanks? Which team's formula would consistently produce the best estimate? Students should help choose nine more simple random samples of four tanks from the bag. After each sample is taken, the four serial numbers chosen should be written on the board, and the tags should be returned to the bag and mixed thoroughly.

7. Each team should use its formula to estimate the total number of tanks N for each of the nine new samples. The team should then make a dotplot of its 10 estimates.

8. Compare the teams' dotplots. As a class, decide which team used the best method for estimating the number of tanks.

More recently, people used the mathematicians' method from the German tank problem to estimate the number of iPhones produced.

Sampling distributions are the key to inference when data are produced by random sampling. Because the results of random samples include an element of chance, we can't guarantee that our inferences are correct. What we can guarantee is that our methods usually give correct answers. The reasoning of statistical inference rests on asking, "How often would this method give a correct answer if I used it very many times?" If our data come from random sampling, the laws of probability answer the question "What would happen if we did this many times?"

Section 7.1 presents the basic ideas of sampling distributions. The most common applications of statistical inference involve proportions and means. Section 7.2 focuses on sampling distributions of sample proportions. Section 7.3 investigates sampling distributions of sample means.

## Teaching Tip

When introducing the idea of a sampling distribution, emphasize that every statistic has a sampling distribution. That is, anything that can be calculated from a sample has its own distribution that can be approximated with simulation.

## ALTERNATE ACTIVITY

### *The textbook problem*

In this Activity, your job is to develop a method to estimate the total number of Algebra 2 books at your school based on a random sample of books and the assumption that the books are numbered sequentially starting with 1. For example, suppose a random sample of $n = 7$ books gives the following book numbers: 10, 38, 59, 61, 74, 90, 94. One possible method to estimate the total number of books ($N$) would be to take the median value and double it. Using these data, the estimate for the total number of books is $\hat{N} = \text{median} \cdot 2 = 61 \cdot 2 = 122$.

You and your teammates will create three other methods to estimate the total number of books. You may use any combination of the summary statistics we already know (for example, mean, median, minimum, maximum, quartiles, *IQR*, standard deviation) or invent your own. The goal is to find a relatively *simple* method that reliably predicts the total number of books.

To determine which of your three methods gives the best predictions, perform a simulation to generate multiple estimates using each method. For the purposes of the simulation, assume that there are 100 books total ($N = 100$) and that you will be taking samples of size 7 ($n = 7$). For each trial of your simulation, generate 7 random numbers from 1 to 100 and compute the estimates using each of your three methods. Repeat this process many times and display the distribution of estimates for each method on parallel dotplots. Based on the results of your simulation, decide which method is best.

At the end of class, we will have a competition to see which team has the best method. During each round of the competition, I will secretly choose $N$ and give you a sample of $n$ numbers from 1 to $N$. Each team will compute an estimate of $N$ using their preferred method. The team that comes the closest gets 3 points, the second closest gets 2 points, and the third closest gets 1 point. After several rounds, the team with the most points will receive a prize.

**PD Section 7.1 Overview**

Watch the video overview of Section 7.1 (for teachers) by clicking on the link in the TE-book, opening the TRCD, or logging onto the book's companion Web site www.whfreeman.com/tps5e (teacher log-in required).

**TRM Learning Objectives Grid**

Remind students about the importance of the Learning Objectives. We include them at the beginning of each section so students know what they are expected to learn. At the end of the chapter, the Learning Objectives are included in a grid, along with references to examples and chapter review exercises. A printable version of this grid is included in the Teacher's Resource Materials.

### AP® Exam Common Error

On the AP® exam, many students lose credit because they fail to make the distinction between parameters and statistics. For example, in a significance test, stating hypotheses in terms of statistics instead of parameters will always lose credit. Likewise, when defining a parameter, it is important that the student use phrases like "population mean" or "true mean" rather than simply saying "mean," which can refer to the mean of a population or the mean of a sample.

# 7.1 What Is a Sampling Distribution?

**WHAT YOU WILL LEARN** By the end of the section, you should be able to:

- Distinguish between a parameter and a statistic.
- Use the sampling distribution of a statistic to evaluate a claim about a parameter.
- Distinguish among the distribution of a population, the distribution of a sample, and the sampling distribution of a statistic.
- Determine whether or not a statistic is an unbiased estimator of a population parameter.
- Describe the relationship between sample size and the variability of a statistic.

What is the average income of American households? Each March, the government's Current Population Survey (CPS) asks detailed questions about income. The random sample of about 60,000 households contacted in March 2012 had a mean "total money income" of $69,677 in 2011.[1] (The median income was lower, of course, at $50,054.) That $69,677 describes the sample, but we use it to estimate the mean income of all households.

## Parameters and Statistics

As we begin to use sample data to draw conclusions about a wider population, we must be clear about whether a number describes a sample or a population. For the sample of households contacted by the CPS, the mean income was $\bar{x}$ = $69,677. The number $69,677 is a **statistic** because it describes this one CPS sample. The population that the poll wants to draw conclusions about is all 121 million U.S. households. In this case, the **parameter** of interest is the mean income $\mu$ of all these households. We don't know the value of this parameter.

**DEFINITION: Parameter, statistic**

A **parameter** is a number that describes some characteristic of the population.

A **statistic** is a number that describes some characteristic of a sample.

The value of a parameter is usually not known because we cannot examine the entire population. The value of a statistic can be computed directly from the sample data. We often use a statistic to estimate an unknown parameter.

Remember **s** and **p**: statistics come from samples, and parameters come from populations. As long as we were doing data analysis, the distinction between population and sample rarely came up. Now, however, it is essential. The notation we use should reflect this distinction. For instance, we write $\mu$ (the Greek letter mu) for the population mean and $\bar{x}$ for the sample mean. We use $p$ to represent a population proportion. The sample proportion $\hat{p}$ is used to estimate the unknown parameter $p$.

It is common practice to use Greek letters for parameters and Roman letters for statistics. In that case, the population proportion would be $\pi$ (pi, the Greek letter for "p") and the sample proportion would be $p$. We'll stick with the notation that's used on the AP® exam, however.

**TRM Section 7.1 Alternate Examples**

You can find the Alternate Examples for this section in Microsoft Word format by clicking on the link in the TE-book, opening the TRCD, or logging onto the book's companion Web site www.whfreeman.com/tps5e (teacher log-in required).

## EXAMPLE From Ghosts to Cold Cabins

### Parameters and statistics

PROBLEM: Identify the population, the parameter, the sample, and the statistic in each of the following settings.

(a) The Gallup Poll asked a random sample of 515 U.S. adults whether or not they believe in ghosts. Of the respondents, 160 said "Yes."[2]

(b) During the winter months, the temperatures outside the Starneses' cabin in Colorado can stay well below freezing (32°F, or 0°C) for weeks at a time. To prevent the pipes from freezing, Mrs. Starnes sets the thermostat at 50°F. She wants to know how low the temperature actually gets in the cabin. A digital thermometer records the indoor temperature at 20 randomly chosen times during a given day. The minimum reading is 38°F.

SOLUTION:

(a) The population is all U.S. adults, and the parameter of interest is $p$, the proportion of all U.S. adults who believe in ghosts. The sample is the 515 people who were interviewed in this Gallup Poll. The statistic is $\hat{p} = \frac{160}{515} = 0.31$, the proportion of the sample who say they believe in ghosts.

(b) The population is all times during the day in question; the parameter of interest is the true minimum temperature in the cabin that day. The sample consists of the 20 temperature readings at randomly selected times. The statistic is the sample minimum, 38°F.

**For Practice** *Try Exercise* **1**

## CHECK YOUR UNDERSTANDING

*Each boldface number in Questions 1 and 2 is the value of either a **parameter** or a **statistic**. In each case, state which it is and use appropriate notation to describe the number.*

1. On Tuesday, the bottles of Arizona Iced Tea filled in a plant were supposed to contain an average of **20** ounces of iced tea. Quality control inspectors sampled 50 bottles at random from the day's production. These bottles contained an average of **19.6** ounces of iced tea.

2. On a New York–to–Denver flight, **8%** of the 125 passengers were selected for random security screening before boarding. According to the Transportation Security Administration, **10%** of passengers at this airport are chosen for random screening.

## Sampling Variability

How can $\bar{x}$, based on a sample of only a few thousand of the 121 million American households, be an accurate estimate of $\mu$? After all, a second random sample taken at the same time would choose different households and likely produce a different value of $\bar{x}$. This basic fact is called **sampling variability:** the value of a statistic varies in repeated random sampling.

### Teaching Tip

Remind students that the "play" icon at the top of this example (and many others) indicates that they can watch a video presentation of the example. Find the videos at www.whfreeman.com/tps5e. Also, remind them that the icon in the lower-right corner indicates that there is a link between this example and Exercise 1. If students are having trouble with the exercise, they can look back to this example for help.

### ALTERNATE EXAMPLE

#### *Heights and cell phones*

PROBLEM:
Identify the population, the parameter, the sample, and the statistic in each of the following settings:
**(a)** A pediatrician wants to know the 75th percentile for the distribution of heights of 10-year-old boys, so she takes a sample of 50 patients and calculates that the 75th percentile in the sample is 56 inches.
**(b)** A Pew Research Center Poll asked 1102 12- to 17-year-olds in the United States if they have a cell phone. Of the respondents, 71% said Yes. http://www.pewinternet.org/Reports/2009/14--Teens-and-Mobile-Phones-Data-Memo.aspx.

SOLUTION:
**(a)** The population is all 10-year-old boys; the parameter of interest is the 75th percentile for all 10-year-old boys. The sample is the 50 10-year-old boys included in the sample; the statistic is 56 inches, the 75th percentile of the heights in the sample.
**(b)** The population is all 12- to 17-year-olds in the United States; the parameter is $p$, the proportion of all 12- to 17-year-olds with cell phones. The sample is the 1102 12- to 17-year-olds in the sample; the statistic is the sample proportion with a cell phone, $\hat{p} = 0.71$.

### ANSWERS TO CYU

**1.** Parameter: $\mu = 20$ ounces. Statistic: $\bar{x} = 19.6$ ounces.

**2.** Parameter: $\hat{p} = 0.10$, or 10% of passengers. Statistic: $\hat{p} = 0.08$, or 8% of the sample of passengers.

To make sense of sampling variability, we ask, "What would happen if we took many samples?" Here's how to answer that question:

- Take a large number of samples from the same population.
- Calculate the statistic (like the sample mean $\bar{x}$ or sample proportion $\hat{p}$) for each sample.
- Make a graph of the values of the statistic.
- Examine the distribution displayed in the graph for shape, center, and spread, as well as outliers or other unusual features.

The following Activity gives you a chance to see sampling variability in action.

## ACTIVITY | Reaching for Chips

**MATERIALS:**
200 colored chips, including 100 of the same color; large bag or other container

Before class, your teacher will prepare a population of 200 colored chips, with 100 having the same color (say, red). The parameter is the actual proportion $p$ of red chips in the population: $p = 0.50$. In this Activity, you will investigate sampling variability by taking repeated random samples of size 20 from the population.

1. After your teacher has mixed the chips thoroughly, each student in the class should take a sample of 20 chips and note the sample proportion $\hat{p}$ of red chips. When finished, the student should return all the chips to the bag, stir them up, and pass the bag to the next student.

*Note:* If your class has fewer than 25 students, have some students take two samples.

2. Each student should record the $\hat{p}$-value in a chart on the board and plot this value on a class dotplot. Label the graph scale from 0.10 to 0.90 with tick marks spaced 0.05 units apart.

3. Describe what you see: shape, center, spread, and any outliers or other unusual features.

When Mr. Caldwell's class did the "Reaching for Chips" Activity, his 35 students produced the graph shown in Figure 7.1. Here's what the class said about its distribution of $\hat{p}$-values.

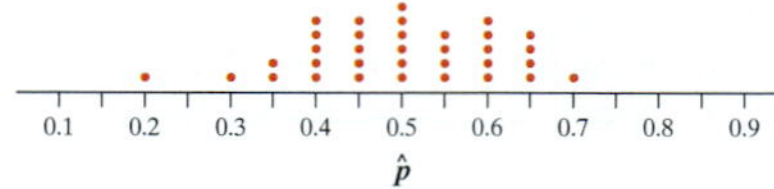


**FIGURE 7.1** Dotplot of sample proportions obtained by the 35 students in Mr. Caldwell's class.

*Shape:* The graph is roughly symmetric with a single peak at 0.5.

*Center:* The mean of our sample proportions is 0.499. This is the balance point of the distribution.

*Spread:* The standard deviation of our sample proportions is 0.112. The values of $\hat{p}$ are typically about 0.112 away from the mean.

*Outliers:* There are no obvious outliers or other unusual features.

Of course, the class only took 35 different simple random samples of 20 chips. There are many, many possible SRSs of size 20 from a population of size 200 (about $1.6 \cdot 10^{27}$, actually). If we took every one of those possible samples, calculated the value of $\hat{p}$ for each, and graphed all those $\hat{p}$-values, then we'd have a **sampling distribution**.

### Teaching Tip

Emphasize the four steps listed here whenever you create a sampling distribution. Students often get lost in the details of different sampling distributions and lose sight of the big picture.

### Activity Overview

You can watch an overview of this Activity (for teachers) by clicking on the link in the TE-book, opening the TRCD, or logging onto the book's companion Web site www.whfreeman.com/tps5e (teacher log-in required).

### Teaching Tip

In Step 2 of this Activity, when students record their sample proportions on the dotplot, have them use $\hat{p}$'s instead of dots.

**ALTERNATE ACTIVITY**

### *Choosing cards*

*MATERIALS*: A deck of cards with aces and face cards removed so that only the cards 2 through 10 remain.

**1.** Have each student shuffle the deck, randomly select 5 cards, and note the median value of the cards. For example, if the selected cards were 2, 2, 4, 5, and 9, the median would be 4.

**2.** Have each student record the value of the sample median on a class dotplot going from 2 to 10.

**3.** Describe what you see: shape, center, spread, and any unusual values.

Here is an example from a class of 25 students:

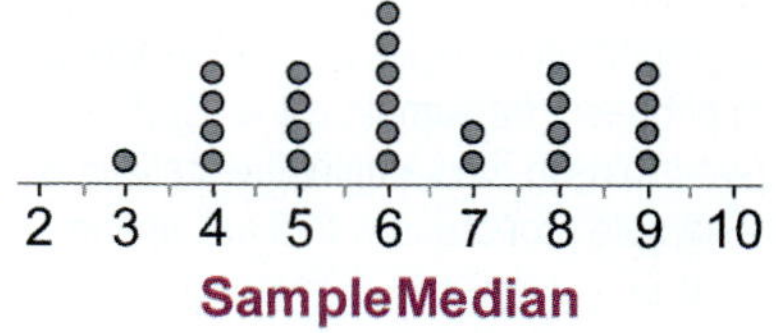


***Shape***: The graph is roughly symmetric with a single peak at 6.

***Center***: The mean of the distribution is 6.28, which is the balance point of this distribution.

***Spread***: The standard deviation of the distribution is 1.84, indicating that the values of the sample median are typically about 1.84 away from the mean (6.28).

***Outliers***: There don't appear to be any outliers.

**DEFINITION: Sampling distribution**

The **sampling distribution** of a statistic is the distribution of values taken by the statistic in all possible samples of the same size from the same population.

It's usually too difficult to take all possible samples of size $n$ to obtain the sampling distribution of a statistic. Instead, we can use simulation to imitate the process of taking many, many samples and create an approximate sampling distribution.

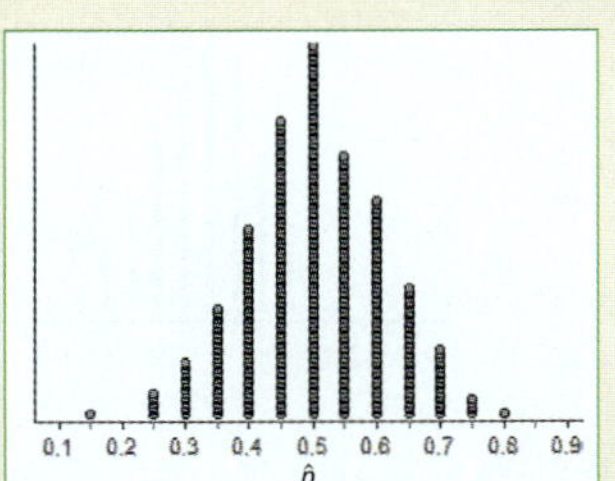

## EXAMPLE Reaching for Chips

### *Simulating a sampling distribution*

We used Fathom software to simulate choosing 500 SRSs of size $n = 20$ from a population of 200 chips, 100 red and 100 blue. Figure 7.2 is a dotplot of the values of $\hat{p}$, the sample proportion of red chips, from these 500 samples.

**FIGURE 7.2** Dotplot of the sample proportion $\hat{p}$ of red chips in 500 simulated SRSs, created by Fathom software.

PROBLEM:

(a) There is one dot on the graph at 0.15. Explain what this value represents.

(b) Describe the distribution. Are there any obvious outliers?

(c) Would it be surprising to get a sample proportion of 0.85 or higher in an SRS of size 20 when $p = 0.5$? Justify your answer.

(d) Suppose your teacher prepares a bag with 200 chips and claims that half of them are red. A classmate takes an SRS of 20 chips; 17 of them are red. What would you conclude about your teacher's claim? Explain.

SOLUTION:

(a) In one SRS of 20 chips, there were 3 red chips. So $\hat{p} = 3/20 = 0.15$ for this sample.

(b) *Shape:* Symmetric, unimodal, and somewhat bell-shaped. *Center:* Around 0.5. *Spread:* The values of $\hat{p}$ fall mostly between 0.25 and 0.75. *Outliers:* One sample with $\hat{p} = 0.15$ stands out.

(c) It is very unlikely to obtain an SRS of 20 chips in which $\hat{p} = 0.85$ from a population in which $p = 0.5$. A value of $\hat{p}$ this large or larger never occurred in 500 simulated samples.

(d) This student's result gives strong evidence against the teacher's claim. As noted in part (c), it is very unlikely to get a sample proportion of 0.85 or higher when $p = 0.5$.

**For Practice** *Try Exercise* 9

Strictly speaking, the sampling distribution is the ideal pattern that would emerge if we looked at *all* possible samples of size 20 from our population of chips. A distribution obtained from simulating a smaller number of random samples, like the 500 values of $\hat{p}$ in Figure 7.2, is only an approximation to the sampling distribution. One of the uses of probability theory in statistics is to obtain sampling distributions without simulation. We'll get to the theory later. The interpretation of a sampling distribution is the same, however, whether we obtain it by simulation or by the mathematics of probability.

### Teaching Tip

Remind students about the definition of a distribution we introduced in Chapter 1: a list of the possible values a variable can take and how often it takes those values. In this case, the variable is a statistic calculated from a sample.

## ALTERNATE EXAMPLE

### *Choosing cards*

We used Fathom to simulate choosing 100 SRSs of size 5 from the deck of cards described in the Alternate Activity on the previous page. The graph below shows the distribution of the sample median for these 100 samples.

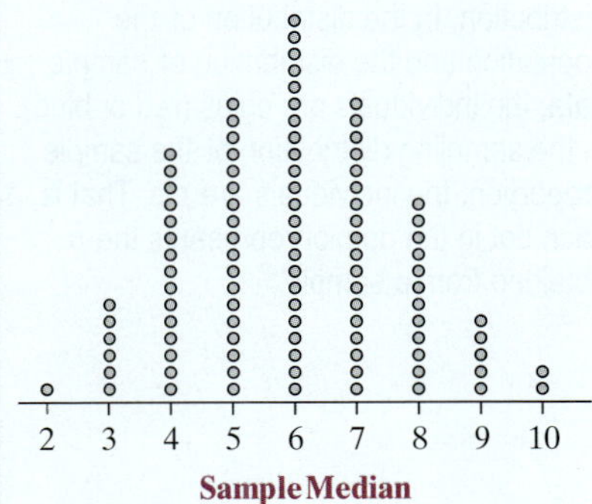


PROBLEM:

**(a)** There is one dot on the graph at 2. Explain what this value represents.

**(b)** Describe the distribution. Are there any obvious outliers?

**(c)** Would it be surprising to get a sample median of 4 or smaller in an SRS of size 5 from this deck? Justify your answer.

**(d)** Suppose that another student prepared a different deck of cards and claimed that it was exactly the same as the one used in the Activity. When you take an SRS of size 5, the median is 4. Does this provide convincing evidence that the student's deck is different? Explain.

SOLUTION:

**(a)** In one SRS of size 5, the median of the sample was 2.

**(b)** *Shape*: The graph is roughly symmetric with a single peak at 6. *Center*: The mean of the distribution is about 6. *Spread*: The values fall mostly between 4 and 8, although there are values as low as 2 and as high as 10. *Outliers*: There don't seem to be any outliers.

**(c)** No. Getting a sample median of 4 or smaller occurs fairly often just by chance when taking random samples of size 5 from a deck of cards with the aces and face cards removed. A median of 4 or smaller occurred in 22 of the 100 trials of the simulation.

**(d)** Getting a sample median of 4 does *not* provide convincing evidence that the student's deck is different. Because getting a median of 4 or smaller is somewhat likely to happen by chance alone, it is plausible that the deck is the same.

## Teaching Tip

Make sure your students completely understand the figure on this page. Students often have trouble making the distinction between the distribution of the population, the distribution of sample data, and the distribution of the sample statistic. One way to help make this distinction is to ask who the individuals are in each distribution. In the distribution of the population and the distribution of sample data, the individuals are chips (red or blue). In the sampling distribution of the sample proportion, the individuals are $\hat{p}$'s. That is, each dot in the dotplot represents the $\hat{p}$ obtained from a sample.

Figure 7.3 illustrates the process of choosing many random samples of 20 chips and finding the sample proportion of red chips $\hat{p}$ for each one. Follow the flow of the figure from the population at the left, to choosing an SRS and finding the $\hat{p}$ for this sample, to collecting together the $\hat{p}$'s from many samples. The first sample has $\hat{p} = 0.40$. The second sample is a different group of chips, with $\hat{p} = 0.55$, and so on. The dotplot at the right of the figure shows the distribution of the values of $\hat{p}$ from 500 separate SRSs of size 20. This dotplot displays the approximate sampling distribution of the statistic $\hat{p}$.

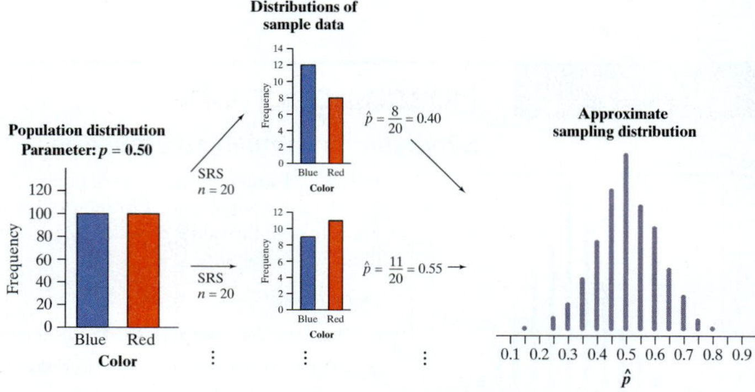


**FIGURE 7.3** The idea of a sampling distribution: take many samples from the same population, collect the $\hat{p}$'s from all the samples, and display the distribution of the $\hat{p}$'s. The dotplot shows the results of 500 samples.

As Figure 7.3 shows, there are three distinct distributions involved when we sample repeatedly and measure a variable of interest. The **population distribution** gives the values of the variable for all individuals in the population. In this case, the individuals are the 200 chips and the variable we're recording is color. Our parameter of interest is the proportion of red chips in the population, $p = 0.50$. Each random sample that we take consists of 20 chips.

The **distribution of sample data** shows the values of the variable "color" for the individuals in the sample. For each sample, we record a value for the statistic $\hat{p}$, the sample proportion of red chips. Finally, we collect the values of $\hat{p}$ from all possible samples of the same size and display them in the *sampling distribution*.

*Be careful: The population distribution and the distribution of sample data describe* individuals. *A sampling distribution describes how a* statistic *varies in many samples from the population.*

**AP® EXAM TIP** Terminology matters. Don't say "sample distribution" when you mean sampling distribution. You will lose credit on free response questions for misusing statistical terms.

## CHECK YOUR UNDERSTANDING

Mars, Incorporated, says that the mix of colors in its M&M'S® Milk Chocolate Candies is 24% blue, 20% orange, 16% green, 14% yellow, 13% red, and 13% brown. Assume that the company's claim is true. We want to examine the proportion of orange M&M'S in repeated random samples of 50 candies.

1. Graph the population distribution. Identify the individuals, the variable, and the parameter of interest.

## AP® Exam Common Error

On the AP® exam, a common error is to write an ambiguous statement such as "the variability decreases when the sample size increases." This statement may be true for the distribution of a sample statistic, but it almost certainly isn't true for the distribution of the sample. Whenever students are describing a distribution, they should always say "the distribution of ____" to avoid ambiguity.

## ANSWERS TO CYU

**1.** Individuals: M&M'S® Milk Chocolate Candies; variable: color; and parameter of interest: proportion of orange M&M'S. The graph below shows the population distribution.

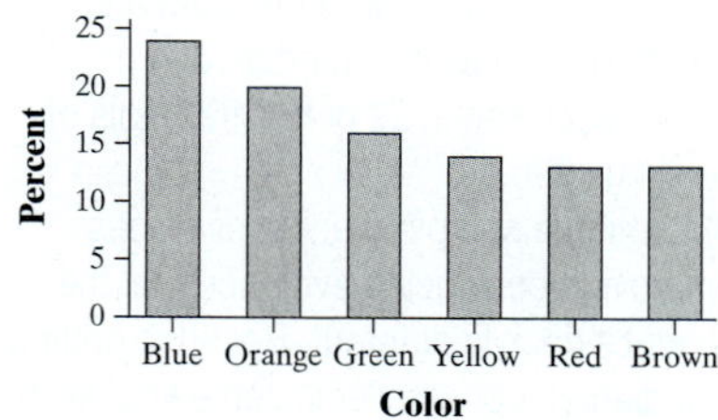

2. Imagine taking an SRS of 50 M&M'S. Make a graph showing a possible distribution of the sample data. Give the value of the appropriate statistic for this sample.

3. Which of the graphs that follow could be the approximate sampling distribution of the statistic? Explain your choice.

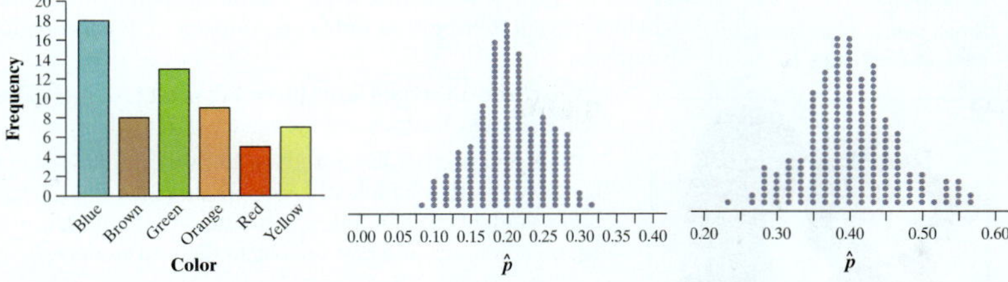


## Describing Sampling Distributions

The fact that statistics from random samples have definite sampling distributions allows us to answer the question "How trustworthy is a statistic as an estimate of a parameter?" To get a complete answer, we consider the shape, center, and spread of the sampling distribution. For reasons that will be clear later, we'll save shape for last.

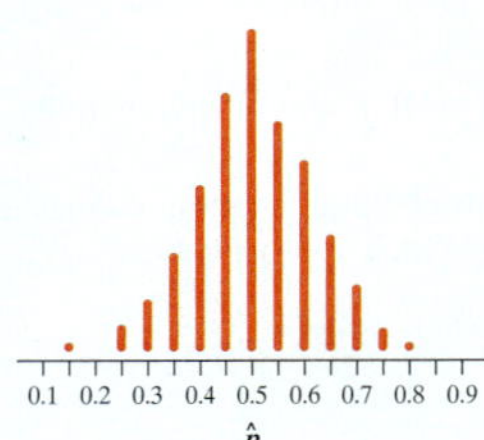


**Center: Biased and unbiased estimators** Let's return to the familiar chips example. How well does the sample proportion of red chips estimate the population proportion of red chips, $p = 0.5$? The dotplot in the margin shows the approximate sampling distribution of $\hat{p}$ once again. We noted earlier that the center of this distribution is very close to 0.5, the parameter value. In fact, if we took all possible samples of 20 chips from the population, calculated $\hat{p}$ for each sample, and then found the mean of all those $\hat{p}$-values, we'd get *exactly* 0.5. For this reason, we say that $\hat{p}$ is an **unbiased estimator** of $p$.

**DEFINITION: Unbiased estimator**

A statistic used to estimate a parameter is an **unbiased estimator** if the mean of its sampling distribution is equal to the value of the parameter being estimated.

If we take many samples, the value of an unbiased estimator will sometimes exceed the value of the parameter and sometimes be less. However, because the sampling distribution of the statistic is centered at the true value, we will not consistently overestimate or underestimate the parameter. This is consistent with our definition of bias from Chapter 4.

We will confirm in Section 7.2 that the sample proportion $\hat{p}$ is an unbiased estimator of the population proportion $p$. This is a very helpful result if we're dealing with a categorical variable (like color). With quantitative variables, we might be interested in estimating the population mean, median, minimum, maximum, $Q_1$, $Q_3$, variance, standard deviation, $IQR$, or range. Which (if any) of these are unbiased estimators? The following Activity should shed some light on this question.

### ANSWERS TO CYU (continued)

2. The graph below shows a possible distribution of sample data. For this sample, there are 11 orange M&M'S, so $\hat{p} = \frac{11}{50} = 0.22$.

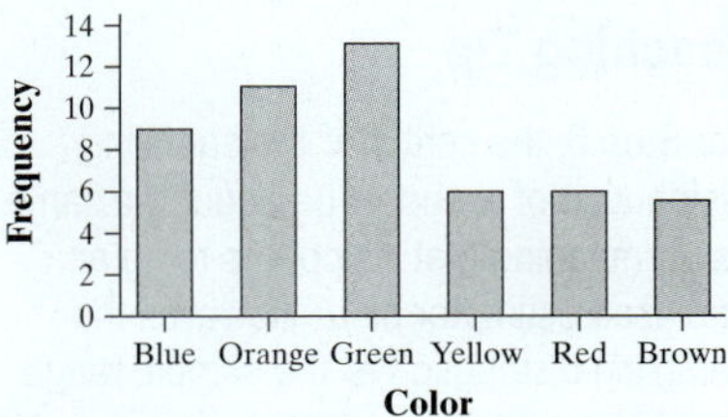


3. The middle graph is the approximate sampling distribution of $\hat{p}$ because the center of the distribution should be at approximately 0.20. The first graph shows the distribution of the colors for one sample and the third graph is centered at 0.40 rather than 0.20.

### Teaching Tip

Tell your students that "unbiased" doesn't mean perfect. An unbiased estimator will almost always provide an estimate that is *not* equal to the value of the population parameter. It is called "unbiased" because the estimates won't consistently be too high or consistently be too low. In other words, the center of the statistic's sampling distribution will be in the right place.

### Teaching Tip

Remind students that when we talk about biased and unbiased estimators, we are assuming that the sampling *process* we are using has no bias. For example, when we say that $\hat{p}$ is an unbiased estimator of $p$, we are assuming the value of $\hat{p}$ came from an SRS, not a convenience sample or voluntary response sample. Using a convenience sample of AP® Statistics students to estimate the proportion of all students at a high school with a graphing calculator will consistently produce estimates that are too high. Likewise, we are also assuming that there are no problems with response bias or nonresponse. For example, even though $\bar{x}$ is an unbiased estimator of $\mu$, asking people in person how much they weigh will consistently produce estimates that are too small.

## ACTIVITY | Sampling heights

**MATERIALS:**
Small piece of cardstock for each student; bag

In this Activity, you will use a population of quantitative data to investigate whether a given statistic is an unbiased estimator of its corresponding population parameter.

1. Each student should write his or her height (in inches) neatly on a small piece of cardstock and then place it in the bag.
2. After your teacher has mixed the cards thoroughly, each student in the class should take a sample of four cards and record the heights of the four chosen students. When finished, the student should return the cards to the bag, mix them up, and pass the bag to the next student.

   *Note:* If your class has fewer than 25 students, have some students take two samples.
3. For your SRS of four students from the class, calculate the sample mean $\bar{x}$ and the sample range (maximum − minimum) of the heights. Then go to the board and record the heights of the four students in your sample, the sample mean $\bar{x}$, and the sample range in a chart like the one below.

| Height (in.) | Sample mean ($\bar{x}$) | Sample range (max − min) |
|---|---|---|
| 62, 75, 68, 63 | 67 | 75 − 62 = 13 |

4. Plot the values of your sample mean and sample range on the two class dotplots drawn by your teacher.
5. Once everyone has finished, find the population mean $\mu$ and the population range.
6. Based on your approximate sampling distributions of $\bar{x}$ and the sample range, which statistic appears to be an unbiased estimator? Which appears to be a *biased estimator*?

When Mrs. Washington's class did the "Sampling Heights" Activity, they produced the graphs shown in Figure 7.4. Her students concluded that the sample mean $\bar{x}$ is probably an unbiased estimator of the population mean $\mu$. Their reason: the center of the approximate sampling distribution of $\bar{x}$, 65.67 inches, is close to the population mean of 66.07 inches. On the other hand, Mrs. Washington's

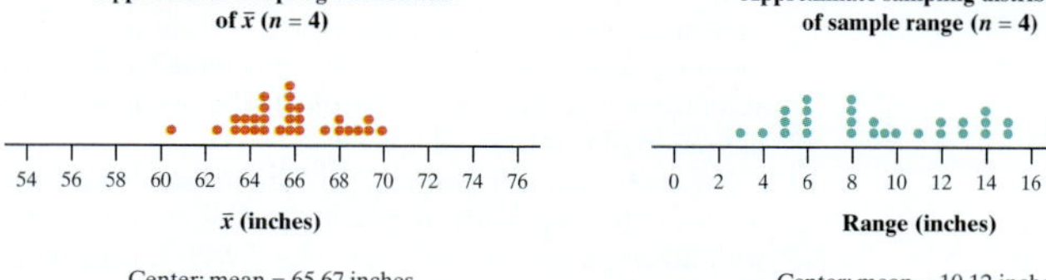


**FIGURE 7.4** Results from Mrs. Washington's class. The sample mean appears to be an unbiased estimator. The sample range appears to be a biased estimator.

### Activity Overview

You can watch an overview of this Activity (for teachers) by clicking on the link in the TE-book, opening the TRCD, or logging onto the book's companion Web site www.whfreeman.com/tps5e (teacher log-in required).

### Teaching Tip

For Step 6, the center of the sampling distribution of $\bar{x}$ should be about the same as $\mu$, meaning that $\bar{x}$ appears to be an unbiased estimator of $\mu$. However, the sampling distribution of the sample range should be centered at a value that is much smaller than the population range, making the sample range a biased estimator of the population range.

**ALTERNATE ACTIVITY**

### *Sampling sleep*

1. Each student should write his or her name and the number of hours of sleep he or she got last night (e.g., 7 hours, 8.5 hours) on a small piece of cardstock and place it in the bag provided by the teacher.
2. Mix the cards thoroughly and have each student take a sample of size 4, record the hours of sleep for the 4 selected students, and return the cards to the bag. Then the next student should mix up the cards again and repeat the same process.
3. Have each student calculate the sample *IQR* of sleep hours and the sample maximum of sleep hours, and plot the values of these statistics on dotplots made by the teacher.
4. Calculate the true values of the *IQR* and maximum. Based on these values and the approximate sampling distributions, do either of these statistics appear to be unbiased estimators?

Here is an example. The population of sleep hours is shown below. The true *IQR* is 1.5 hours and the true maximum is 10 hours.

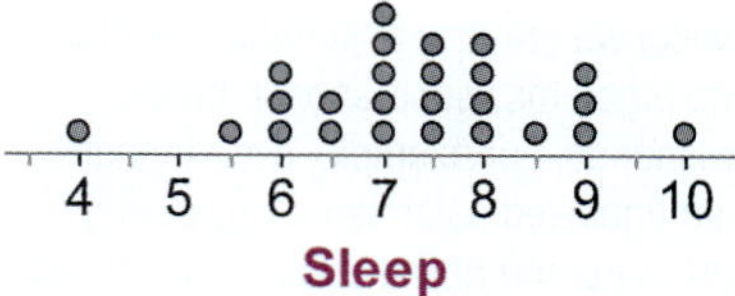


Here are the approximate sampling distributions of the sample *IQR* and sample maximum, using 25 SRSs of size 4, with the mean of each distribution indicated by a vertical line.

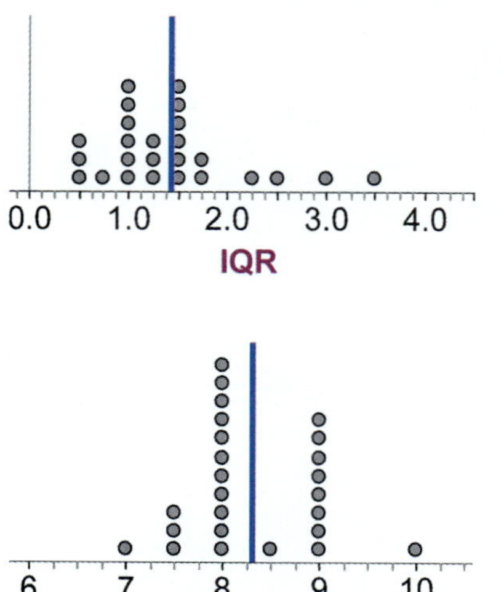


Because the approximate sampling distribution of the sample *IQR* is centered near the true *IQR* of 1.5, it appears that the sample *IQR* might be an unbiased estimator of the true *IQR*. However, because the center of the approximate sampling distribution of the sample maximum is much lower than the true maximum value of 10, the sample maximum is a biased estimator of the true maximum. The sample maximum is consistently lower than the true maximum.

*Note:* After using 10,000 SRSs of size 4, the center of the approximate sampling distribution of the *IQR* was 1.70, meaning that the sample *IQR* is probably a biased estimator as well, but the bias is relatively small. Interestingly, the median of the sampling distribution of the sample *IQR* was 1.50, the same as the true value of the *IQR*.

students decided that the sample range is a **biased estimator** of the population range. Why? Because the center of the sampling distribution for this statistic was 10.12 inches, much less than the corresponding parameter value of 21 inches.

To confirm the class's conclusions, we used Fathom software to simulate taking 250 SRSs of $n = 4$ students. For each sample, we plotted the mean height $\bar{x}$ and the range of the heights. Figure 7.5 shows the approximate sampling distributions for these two statistics. It looks like the class was right: $\bar{x}$ is an unbiased estimator of $\mu$, but the sample range is clearly a biased estimator. The range of the sample heights tends to be much lower, on average, than the population range.

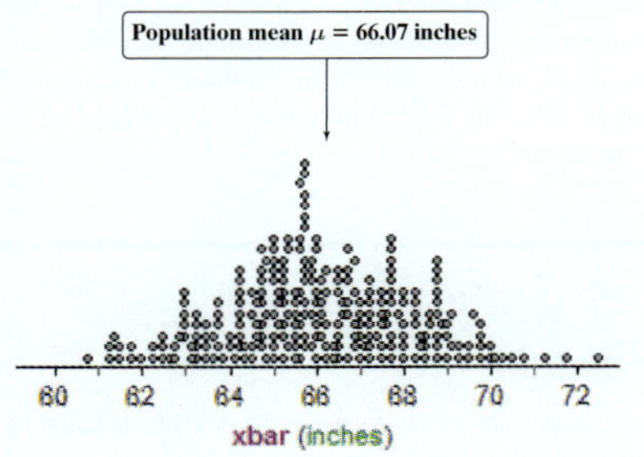


Center: mean = 66.15 inches

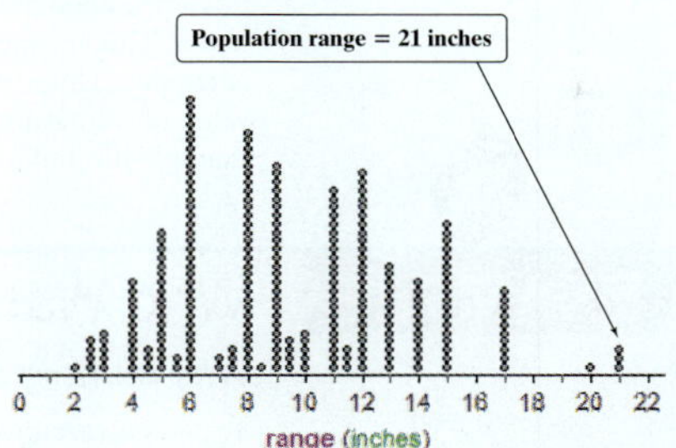


Center: mean = 10.21 inches

**FIGURE 7.5** Results from a Fathom simulation of 250 SRSs of size $n = 4$ from the students in Mrs. Washington's class. The sample mean is an unbiased estimator. The sample range is a biased estimator.

**Why do we divide by $n - 1$ when calculating the sample variance?** In Chapter 1, we introduced the sample variance $s_x^2$ as a measure of spread for a set of quantitative data. The idea of $s_x^2$ is simple: it's a number that describes the "average" squared deviation of the values in the sample from their mean $\bar{x}$. It probably surprised you when we computed this average by dividing by $n - 1$ instead of $n$. Now we're ready to tell you why we defined $s_x^2 = \frac{1}{n-1}\Sigma(x_i - \bar{x})^2$.

In an inference setting involving a quantitative variable, we might be interested in estimating the variance $\sigma^2$ of the population distribution. The most logical choice for our estimator is the sample variance $s_x^2$. We used Fathom software to take 500 SRSs of size $n = 4$ from the population distribution of heights in Mrs. Washington's class. Note that the population variance is $\sigma^2 = 22.19$. For each sample, we recorded the value of two statistics:

$$\text{var} = s_x^2 = \frac{1}{n-1}\Sigma(x_i - \bar{x})^2 \text{ (the sample variance)}$$

$$\text{varn} = \frac{1}{n}\Sigma(x_i - \bar{x})^2$$

Figure 7.6 shows the approximate sampling distributions of these two statistics. We used histograms to show the overall pattern more clearly. The vertical lines mark the means of these two distributions.

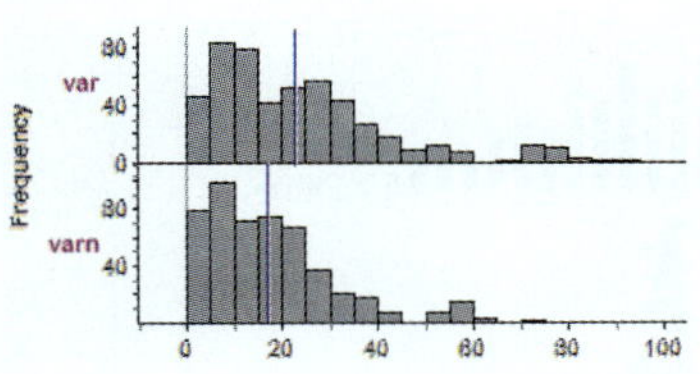


Value of variance from sample data

**FIGURE 7.6** Results from a Fathom simulation of 500 SRSs of size $n = 4$ from the population distribution of heights in Mrs. Washington's class. The sample variance $s_x^2$ (labeled "var" in the figure) is an unbiased estimator. The "varn" statistic (dividing by $n$ instead of $n - 1$) is a biased estimator.

## Teaching Tip

Even though the sample variance is an unbiased estimator of the population variance when we use $n - 1$ in the denominator, the sample standard deviation with a denominator of $n - 1$ is still a biased estimator of the true standard deviation. However, the bias is smaller when using $n - 1$ than when using only $n$ in the denominator.

We used Fathom software to take 5000 SRSs of size 4 from the sleep data in the earlier Alternate Activity. For each sample, we calculated the sample standard deviation using $n - 1$ in the denominator ($s$) and the standard deviation using $n$ in the denominator ($sn$). The blue line shows the population standard deviation $\sigma = 1.298$. The green lines show the means of each approximate sampling distribution. As you can see, when using $n - 1$ in the denominator, the center of the approximate sampling distribution is closer to the true standard deviation.

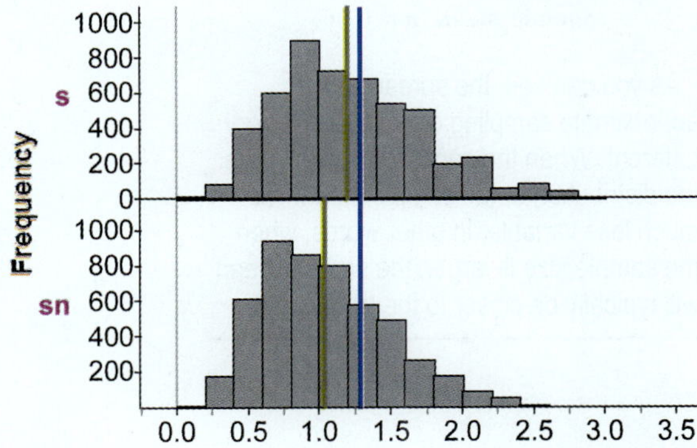

We can see that "varn" is a *biased* estimator of the population variance. The mean of its sampling distribution (marked with a blue line segment) is clearly less than the value of the population parameter, 22.19. However, the statistic "var" (otherwise known as the sample variance $s_x^2$) is an unbiased estimator. Its values are centered at 22.19. That's why we divide by $n - 1$ and not $n$ when calculating the sample variance: to get an unbiased estimator of the population variance.

**Spread: Low variability is better!** To get a trustworthy estimate of an unknown population parameter, start by using a statistic that's an unbiased estimator. This ensures that you won't consistently overestimate or underestimate the parameter. Unfortunately, using an unbiased estimator doesn't guarantee that the value of your statistic will be close to the actual parameter value. The following example illustrates what we mean.

## EXAMPLE Who Watches *Survivor*?

### *Why sample size matters*

Television executives and companies who advertise on TV are interested in how many viewers watch particular shows. According to Nielsen ratings, *Survivor* was one of the most-watched television shows in the United States during every week that it aired. Suppose that the true proportion of U.S. adults who have watched *Survivor* is $p = 0.37$.

The top dotplot in Figure 7.7 shows the results of drawing 400 SRSs of size $n = 100$ from a population with $p = 0.37$. We see that a sample of 100 people often gave a $\hat{p}$ quite far from the population parameter. That is why a Gallup Poll asked not 100, but 1000 people whether they had watched *Survivor*. Let's repeat our simulation, this time taking 400 SRSs of size $n = 1000$ from a population with proportion $p = 0.37$ who have watched *Survivor*. The bottom dotplot in Figure 7.7 displays the distribution of the 400 values of $\hat{p}$ from these new samples. Both graphs are drawn on the same horizontal scale to make comparison easy.

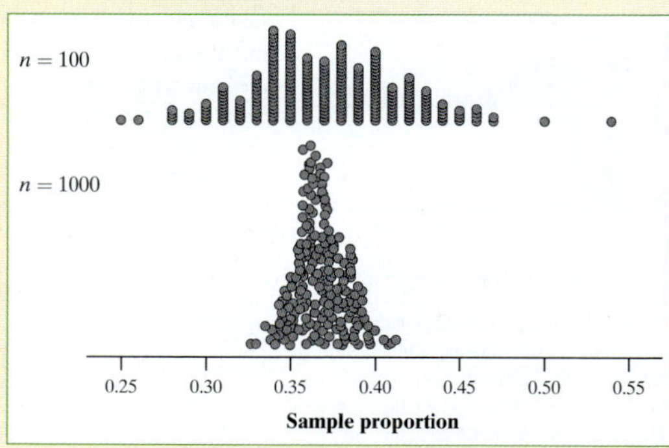


**FIGURE 7.7** The approximate sampling distribution of the sample proportion $\hat{p}$ from SRSs of size $n = 100$ and $n = 1000$ drawn from a population with proportion $p = 0.37$ who have watched *Survivor*. Both dotplots show the results of 400 SRSs.

**ALTERNATE EXAMPLE**

### *Sampling heights*

Suppose that the heights of adult males are approximately Normally distributed with a mean of 70 inches and a standard deviation of 3 inches. To see why sample size matters, we took 200 SRSs of size 100 and calculated the sample mean height and then took 200 SRSs of size 1500 and calculated the sample mean height. Here are the results, graphed on the same scale for easy comparisons:

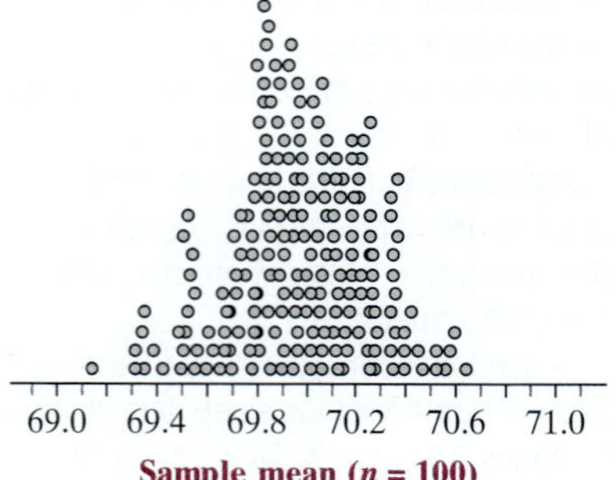

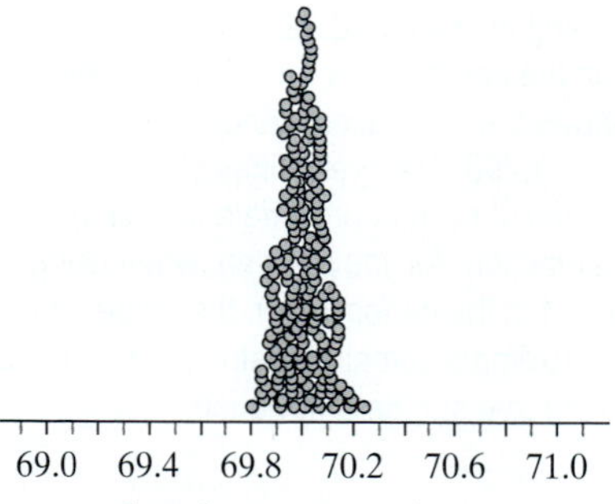


As you can see, the spread of the approximate sampling distributions is very different. When the sample size was larger, the distribution of the sample mean was much less variable. In other words, when the sample size is larger, the sample mean will typically be closer to the true mean.

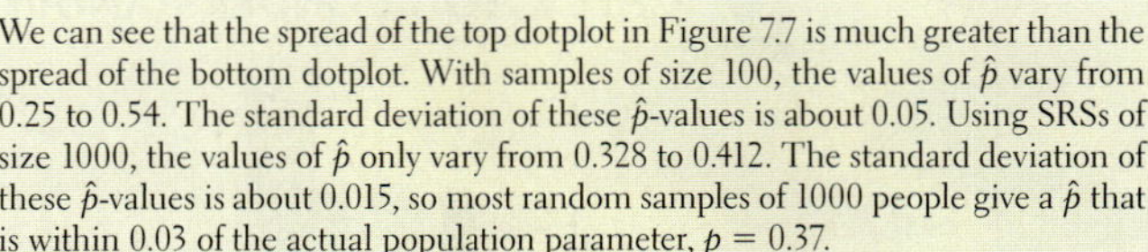

We can see that the spread of the top dotplot in Figure 7.7 is much greater than the spread of the bottom dotplot. With samples of size 100, the values of $\hat{p}$ vary from 0.25 to 0.54. The standard deviation of these $\hat{p}$-values is about 0.05. Using SRSs of size 1000, the values of $\hat{p}$ only vary from 0.328 to 0.412. The standard deviation of these $\hat{p}$-values is about 0.015, so most random samples of 1000 people give a $\hat{p}$ that is within 0.03 of the actual population parameter, $p = 0.37$.

The sample proportion $\hat{p}$ from a random sample of any size is an unbiased estimator of the parameter $p$. As we can see from the previous example, though, larger random samples have a clear advantage. They are much more likely to produce an estimate close to the true value of the parameter. Said another way, larger random samples give us more *precise* estimates than smaller random samples. That's because a large random sample gives us more information about the underlying population than a smaller sample does.

*Taking a larger sample doesn't fix bias. Remember that even a very large voluntary response sample or convenience sample is worthless because of bias.*

There are general rules for describing how the spread of the sampling distribution of a statistic decreases as the sample size increases. In Sections 7.2 and 7.3, we'll reveal these rules for the sampling distributions of $\hat{p}$ and $\bar{x}$. One important and surprising fact is that the variability of a statistic in repeated sampling does *not* depend very much on the size of the population.

### VARIABILITY OF A STATISTIC

The **variability of a statistic** is described by the spread of its sampling distribution. This spread is determined mainly by the size of the random sample. Larger samples give smaller spreads. The spread of the sampling distribution does not depend much on the size of the population, as long as the population is at least 10 times larger than the sample.

Why does the size of the population have little influence on the behavior of statistics from random samples? Imagine sampling harvested corn by thrusting a scoop into a large sack of corn kernels. The scoop doesn't know whether it is surrounded by a bag of corn or by an entire truckload. As long as the corn is well mixed (so that the scoop selects a random sample), the variability of the result depends only on the size of the scoop.

The fact that the variability of a statistic is controlled by the size of the sample has important consequences for designing samples. Suppose a researcher wants to estimate the proportion of all U.S. adults who use Twitter regularly. A random sample of 1000 or 1500 people will give a fairly precise estimate of the parameter because the sample size is large. Now consider another researcher who wants to estimate the proportion of all Ohio State University students who use Twitter regularly. It can take just as large an SRS to estimate the proportion of Ohio State University students who use Twitter regularly as to estimate with equal precision the proportion of all U.S. adults who use Twitter regularly. We can't expect to need a smaller SRS at Ohio State just because there are about 60,000 Ohio State students and about 235 million adults in the United States.

## Teaching Tip

Remind your students that taking a larger sample will reduce the variability of a statistic, but it won't eliminate bias.

## Teaching Tip

The population size *does* matter when the sample size is greater than 10% of the population size. To see why, consider what happens when we take a census. Because we include every member of the population in the "sample," the statistic will have no variability at all—it will always be equal to the value of the parameter. There are methods we can use to account for cases when the sample is a large fraction of the population, but these are beyond the scope of this book (and the AP® Statistics exam). Consequently, we will stick to situations where the sample is a small fraction of the population.

## CHECK YOUR UNDERSTANDING

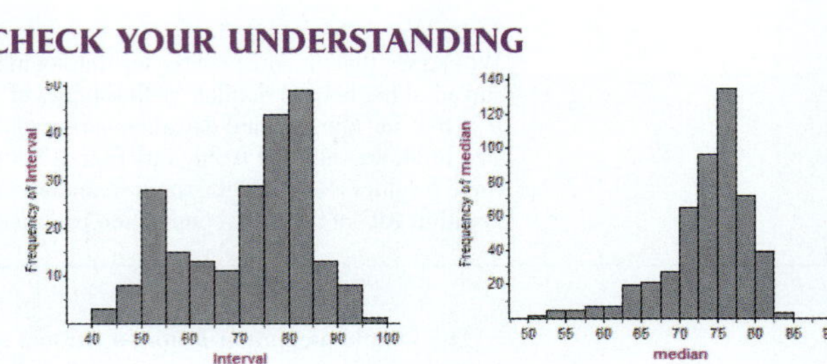


The histogram above left shows the intervals (in minutes) between eruptions of Old Faithful geyser for all 222 recorded eruptions during a particular month. For this population, the median is 75 minutes. We used Fathom software to take 500 SRSs of size 10 from the population. The 500 values of the sample median are displayed in the histogram above right. The mean of these 500 values is 73.5.

1. Is the sample median an unbiased estimator of the population median? Justify your answer.
2. Suppose we had taken samples of size 20 instead of size 10. Would the spread of the sampling distribution be larger, smaller, or about the same? Justify your answer.
3. Describe the shape of the sampling distribution.

### ✓ ANSWERS TO CYU

1. No. The mean of the approximate sampling distribution of the sample median (73.5) is not equal to the median of the population (75).
2. Smaller. Larger samples provide more precise estimates because larger samples include more information about the population distribution.
3. Skewed to the left and unimodal.

**Bias, variability, and shape** We can think of the true value of the population parameter as the bull's-eye on a target and of the sample statistic as an arrow fired at the target. Both bias and variability describe what happens when we take many shots at the target. *Bias* means that our aim is off and we consistently miss the bull's-eye in the same direction. Our sample values do not center on the population value. *High variability* means that repeated shots are widely scattered on the target. Repeated samples do not give very similar results. Figure 7.8 shows this target illustration of the two types of error.

Notice that low variability (shots are close together) can accompany high bias (shots are consistently away from the bull's-eye in one direction). And low or no bias (shots center on the bull's-eye) can accompany high variability (shots are widely scattered). Ideally, we'd like our estimates to be *accurate* (unbiased) and *precise* (have low variability). See Figure 7.8(d).

The following example attempts to tie these ideas together in a familiar setting.

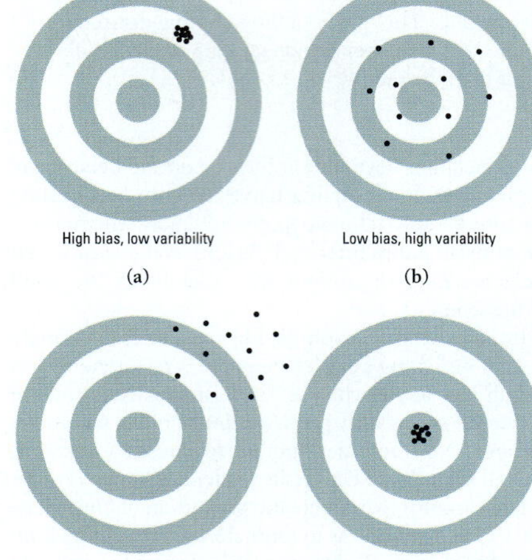


**FIGURE 7.8** Bias and variability. (a) High bias, low variability. (b) Low bias, high variability. (c) High bias, high variability. (d) The ideal: no bias, low variability.

### Teaching Tip

The heading of this section, "Bias, variability, and shape," corresponds to our familiar characteristics of a distribution: center, spread, and shape.

### Teaching Tip

Emphasize the difference between accurate (centered in the right place) and precise (having low variability). Students often say "accurate" when they mean "precise." For example, it isn't correct to say, "Our estimate will be more accurate if we increase the sample size." A correct statement is "Our estimate will be more *precise* if we increase the sample size."

## EXAMPLE The German Tank Problem

### Evaluating estimators: Shape, center, spread

Refer to the Activity on page 422. Mrs. Friedman's student teams came up with four different methods for estimating the number of tanks in the bag: (1) "maxmin" = maximum + minimum, (2) "meanpl2sd" = $\bar{x} + 2s_x$, (3) "twicemean" = $2\bar{x}$, and (4) "twomedian" = 2(median). She added one more method, called "partition." Figure 7.9 shows the results of taking 250 SRSs of 4 tanks and recording the value of the five statistics for each sample. The vertical line marks the actual value of the population parameter N: there were 342 tanks in the bag.

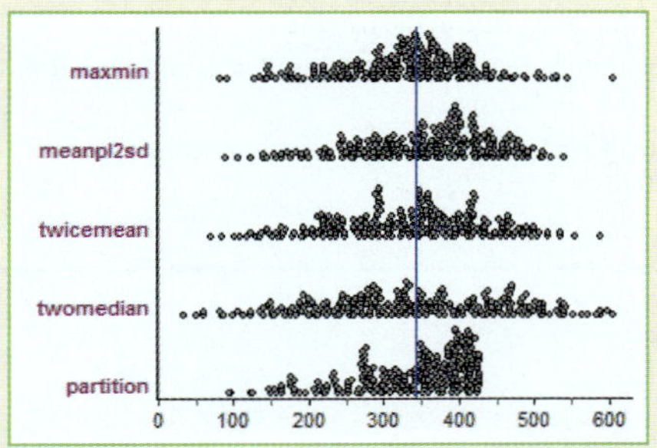


**FIGURE 7.9** Results from a Fathom simulation of 250 SRSs of 4 tanks. The approximate sampling distributions of five different statistics are shown.

PROBLEM: Use the information in Figure 7.9 to help answer these questions.

(a) Which of the four statistics proposed by the student teams is the best estimator? Justify your answer.

(b) Why was the partition method, which uses the statistic (5/4) · maximum, recommended by the mathematicians in Washington, D.C.?

SOLUTION:

(a) Meanpl2sd is a biased estimator: the center of its sampling distribution is too high. This statistic produces consistent overestimates of the number of tanks. The other three statistics proposed by the students appear to be unbiased estimators. All three sampling distributions have roughly symmetric shapes, so these statistics are about equally likely to underestimate or overestimate the number of tanks. Because maxmin has the smallest variability among the three, it would generally produce estimates that are closer to the actual number of tanks. Among the students' proposed statistics, maxmin would be the best estimator.

(b) The partition method uses a statistic (5/4 · maximum) that is an unbiased estimator and that has much less variability than any of the student teams' statistics. Its sampling distribution is left-skewed, so the mean of the distribution is less than its median. Because more than half of the dots in the graph are to the right of the mean, the statistic is more likely to overestimate than underestimate the number of tanks. The mathematicians believed that it would be better to err on the side of caution and give the military commanders an estimate that is slightly too high.

**For Practice** *Try Exercise* **19**

The lesson about center and spread is clear: given a choice of statistics to estimate an unknown parameter, choose one with no or low bias and minimum variability. Shape is a more complicated issue. We have seen sampling distributions that are left-skewed, right-skewed, roughly symmetric, and even approximately Normal. The same statistic can have sampling distributions with different shapes depending on the population distribution and the sample size. Our advice: be sure to consider the shape of the sampling distribution before doing inference.

### ALTERNATE EXAMPLE

***More tanks***

Here are four additional methods for estimating the total number of tanks. The partition method that was chosen by the mathematicians is also included for comparison.

**(1)** Partition = (5/4)maximum

**(2)** Max = maximum

**(3)** MeanMedian = mean + median

**(4)** SumQuartiles = $Q_1 + Q_3$

**(5)** TwiceIQR = $2IQR$

The graph below shows the approximate sampling distribution for each of these statistics when taking 250 samples of size 4 from a population of 342 tanks.

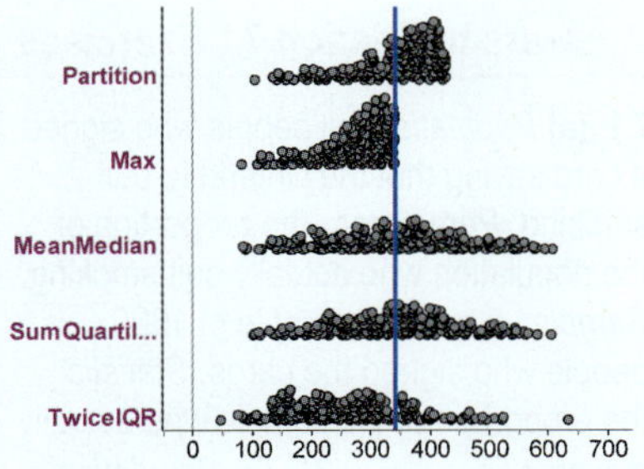


PROBLEM:

**(a)** Which of these statistics appear to be biased estimators? Explain.

**(b)** Of the unbiased estimators, which is best? Explain.

**(c)** Explain why a biased estimator might be preferred over an unbiased estimator.

SOLUTION:

**(a)** The statistics Max and TwiceIQR appear to be biased estimators because they are consistently too low. That is, the centers of their sampling distributions appear to be below the correct value of 342.

**(b)** Of the three unbiased statistics, Partition is best since it has the least variability.

**(c)** Even though Max is a biased estimator, it often produces estimates very close to the truth. MeanMedian, although unbiased, is quite variable and not close to the true value as often. For example, in 120 of the 250 SRSs, Max produced an estimate within 50 of the true value. However, MeanMedian was this close in only 79 of the 250 SRSs.

**TRM Section 7.1 Quizzes**

There are three different quizzes available for this section in the Teacher's Resource Materials. Versions A and B are roughly parallel. Find them by clicking on the link in the TE-book, opening the TRCD, or logging onto the book's companion Web site www.whfreeman.com/tps5e (teacher log-in required).

### Teaching Tip

Full solutions are available in the Solutions Manual. The solutions included here have been shortened to fit in the margin.

**TRM Full Solutions to Section 7.1 Exercises**

### Answers to Section 7.1 Exercises

**7.1** **(a)** *Population*: all people who signed a card saying that they intend to quit smoking. *Parameter*: the proportion of the population who actually quit smoking. *Sample*: a random sample of 1000 people who signed the cards. *Statistic*: the proportion of the sample who actually quit smoking; $\hat{p} = 0.21$. **(b)** *Population*: all the turkey meat. *Parameter*: minimum temperature in all of the turkey meat. *Sample*: four randomly chosen locations in the turkey. *Statistic*: minimum temperature in the sample of four locations; sample minimum = 170°F.

**7.2** **(a)** *Population*: individuals in U.S. households. *Parameter*: proportion of the U.S. population who were unemployed. *Sample*: a random sample of individuals from 60,000 U.S. households. *Statistic*: the proportion of the sample who were unemployed; $\hat{p} = 0.079$. **(b)** *Population*: all gasoline stations in a large city. *Parameter*: range of gas prices at all gasoline stations in the city. *Sample*: a random sample of 10 gas stations in the city. *Statistic*: the range of prices in the sample; sample range = 25 cents.

**7.3** $\mu = 2.5003$ is a parameter and $\bar{x} = 2.5009$ is a statistic.

**7.4** $p = 0.41$ is a parameter and $\hat{p} = 0.33$ is a statistic.

**7.5** $\hat{p} = 0.48$ is a statistic and $p = 0.52$ is a parameter.

**7.6** $\bar{x} = 64.5$ inches is a statistic and $\mu = 63$ inches is a parameter.

## Section 7.1 Summary

- A **parameter** is a number that describes a population. To estimate an unknown parameter, use a **statistic** calculated from a sample.
- The **population distribution** of a variable describes the values of the variable for all individuals in a population. The **sampling distribution** of a statistic describes the values of the statistic in all possible samples of the same size from the same population. Don't confuse the sampling distribution with a **distribution of sample data**, which gives the values of the variable for all individuals in a particular sample.
- A statistic can be an **unbiased estimator** or a **biased estimator** of a parameter. A statistic is a biased estimator if the center (mean) of its sampling distribution is not equal to the true value of the parameter.
- The **variability** of a statistic is described by the spread of its sampling distribution. Larger samples give smaller spread.
- When trying to estimate a parameter, choose a statistic with low or no bias and minimum variability.

## Section 7.1 Exercises

*For Exercises 1 and 2, identify the population, the parameter, the sample, and the statistic in each setting.*

1. **Healthy living**

pg 425

(a) A random sample of 1000 people who signed a card saying they intended to quit smoking were contacted 9 months later. It turned out that 210 (21%) of the sampled individuals had not smoked over the past 6 months.

(b) Tom is cooking a large turkey breast for a holiday meal. He wants to be sure that the turkey is safe to eat, which requires a minimum internal temperature of 165°F. Tom uses a thermometer to measure the temperature of the turkey meat at four randomly chosen points. The minimum reading in the sample is 170°F.

2. **The economy**

(a) Each month, the Current Population Survey interviews a random sample of individuals in about 60,000 U.S. households. One of their goals is to estimate the national unemployment rate. In October 2012, 7.9% of those interviewed were unemployed.

(b) How much do gasoline prices vary in a large city? To find out, a reporter records the price per gallon of regular unleaded gasoline at a random sample of 10 gas stations in the city on the same day. The range (maximum − minimum) of the prices in the sample is 25 cents.

*For each boldface number in Exercises 3 to 6, (1) state whether it is a parameter or a statistic and (2) use appropriate notation to describe each number; for example, $p = 0.65$.*

3. **Get your bearings** A large container is full of ball bearings with mean diameter **2.5003** centimeters (cm). This is within the specifications for acceptance of the container by the purchaser. By chance, an inspector chooses 100 bearings from the container that have mean diameter **2.5009** cm. Because this is outside the specified limits, the container is mistakenly rejected.

4. **Voters** Voter registration records show that **41%** of voters in a state are registered as Democrats. To test a random digit dialing device, you use it to call 250 randomly chosen residential telephones in the state. Of the registered voters contacted, **33%** are registered Democrats.

5. **Unlisted numbers** A telemarketing firm in a large city uses a device that dials residential telephone numbers in that city at random. Of the first 100 numbers dialed, **48%** are unlisted. This is not surprising because **52%** of all residential phones in the city are unlisted.

6. **How tall?** A random sample of female college students has a mean height of **64.5** inches, which is greater than the **63**-inch mean height of all adult American women.

*Exercises 7 and 8 refer to the small population {2, 6, 8, 10, 10, 12} with mean $\mu = 8$ and range 10.*

7. **Sampling distribution**
   (a) List all 15 possible SRSs of size $n = 2$ from the population. Find the value of $\bar{x}$ for each sample.
   (b) Make a graph of the sampling distribution of $\bar{x}$. Describe what you see.

8. **Sampling distribution**
   (a) List all 15 possible SRSs of size $n = 2$ from the population. Find the value of the range for each sample.
   (b) Make a graph of the sampling distribution of the sample range. Describe what you see.

9. **Doing homework** A school newspaper article claims that 60% of the students at a large high school did all their assigned homework last week. Some skeptical AP® Statistics students want to investigate whether this claim is true, so they choose an SRS of 100 students from the school to interview. What values of the sample proportion $\hat{p}$ would be consistent with the claim that the population proportion of students who completed all their homework is $p = 0.60$? To find out, we used Fathom software to simulate choosing 250 SRSs of size $n = 100$ students from a population in which $p = 0.60$. The figure below is a dotplot of the sample proportion $\hat{p}$ of students who did all their homework.

pg 427 

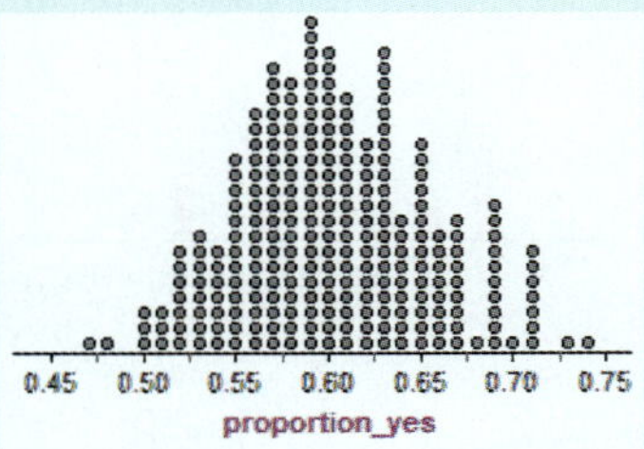


   (a) There is one dot on the graph at 0.73. Explain what this value represents.
   (b) Describe the distribution. Are there any obvious outliers?
   (c) Would it be surprising to get a sample proportion of 0.45 or lower in an SRS of size 100 when $p = 0.6$? Justify your answer.
   (d) Suppose that 45 of the 100 students in the actual sample say that they did all their homework last week. What would you conclude about the newspaper article's claim? Explain.

10. **Tall girls** According to the National Center for Health Statistics, the distribution of heights for 16-year-old females is modeled well by a Normal density curve with mean $\mu = 64$ inches and standard deviation $\sigma = 2.5$ inches. To see if this distribution applies at their high school, an AP® Statistics class takes an SRS of 20 of the 300 16-year-old females at the school and measures their heights. What values of the sample mean $\bar{x}$ would be consistent with the population distribution being $N(64, 2.5)$? To find out, we used Fathom software to simulate choosing 250 SRSs of size $n = 20$ students from a population that is $N(64, 2.5)$. The figure below is a dotplot of the sample mean height $\bar{x}$ of the students in each sample.

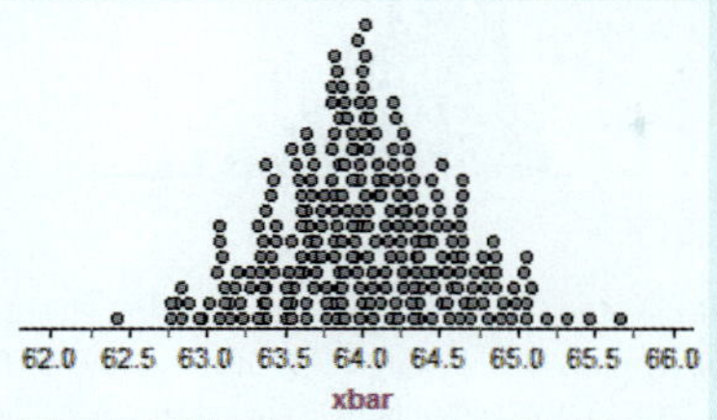


   (a) There is one dot on the graph at 62.4. Explain what this value represents.
   (b) Describe the distribution. Are there any obvious outliers?
   (c) Would it be surprising to get a sample mean of 64.7 or more in an SRS of size 20 when $\mu = 64$? Justify your answer.
   (d) Suppose that the average height of the 20 girls in the class's actual sample is $\bar{x} = 64.7$. What would you conclude about the population mean height $\mu$ for the 16-year-old females at the school? Explain.

11. **Doing homework** Refer to Exercise 9.
   (a) Make a bar graph of the population distribution given that the newspaper's claim is correct.
   (b) Sketch a possible graph of the distribution of sample data for the SRS of size 100 taken by the AP® Statistics students.

12. **Tall girls** Refer to Exercise 10.
   (a) Make a graph of the population distribution.
   (b) Sketch a possible dotplot of the distribution of sample data for the SRS of size 20 taken by the AP® Statistics class.

*Exercises 13 and 14 refer to the following setting.* During the winter months, outside temperatures at the Starneses' cabin in Colorado can stay well below freezing (32°F, or 0°C) for weeks at a time. To prevent the pipes from freezing, Mrs. Starnes sets the thermostat at 50°F. The manufacturer claims that the thermostat allows variation in home temperature that follows a Normal distribution with $\sigma = 3$°F. To test this claim, Mrs. Starnes programs her digital thermometer to take an SRS of $n = 10$ readings during a 24-hour period. Suppose the thermostat is

**7.7 (a)** 2 and 6 ($\bar{x} = 4$), 2 and 8 (5), 2 and 10 (6), 2 and 10 (6), 2 and 12 (7), 6 and 8 (7), 6 and 10 (8), 6 and 10 (8), 6 and 12 (9), 8 and 10 (9), 8 and 10 (9), 8 and 12 (10), 10 and 10 (10), 10 and 12 (11), 10 and 12 (11). **(b)** The sampling distribution of $\bar{x}$ is skewed to the left and unimodal. The mean of the sampling distribution is 8, which is equal to the mean of the population. The values of $\bar{x}$ vary from 4 to 11.

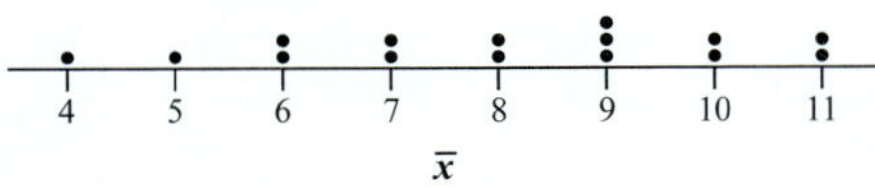


**7.8 (a)** 2 and 6 (range = 4), 2 and 8 (6), 2 and 10 (8), 2 and 10 (8), 2 and 12 (10), 6 and 8 (2), 6 and 10 (4), 6 and 10 (4), 6 and 12 (6), 8 and 10 (2), 8 and 10 (2), 8 and 12 (4), 10 and 10 (0), 10 and 12 (2), 10 and 12 (2). **(b)** The sampling distribution of the sample range is skewed to the right and unimodal. The mean of the sampling distribution is 4.27, which is much less than the range of the population (10). The values of the sample range vary from 0 to 10.

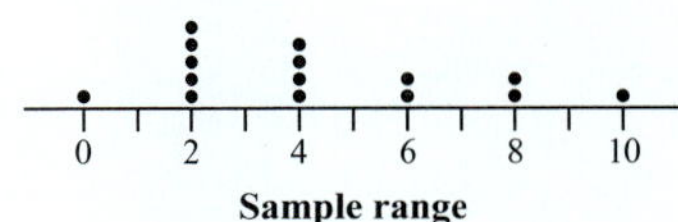


**7.9 (a)** In one simulated SRS of 100 students, there were 73 students who did all of their assigned homework. **(b)** The distribution is reasonably symmetric and bell-shaped. It is centered at about 0.60. Values vary from about 0.47 to 0.74. There don't appear to be any outliers. **(c)** Yes, because there were no values of $\hat{p}$ less than or equal to 0.45 in the simulation. **(d)** Because it would be very surprising to get a sample proportion of 0.45 or less in an SRS of size 100 when $p = 0.60$, we should be skeptical of the newspaper's claim.

**7.10 (a)** In one simulated SRS of 20 students, the average height was $\bar{x} = 62.4$ inches. **(b)** The distribution is reasonably symmetric and bell-shaped. It is centered at about 64. Values vary from about 62.4 to 65.7. There do not seem to be any outliers. **(c)** No, because about 10% of the values of $\bar{x}$ were 64.7 or greater. **(d)** Because it isn't surprising to get a sample mean of 64.7 or greater in an SRS of size 20 from a Normal population with $\mu = 64$, we do not have convincing evidence that the population mean height at the school is different than $\mu = 64$.

**7.11 (a)** A graph of the population distribution is shown below.

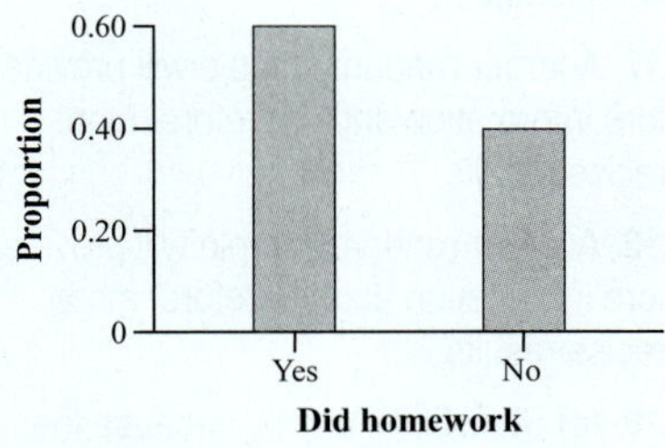


**(b)** Answers will vary. An example bar graph is given.

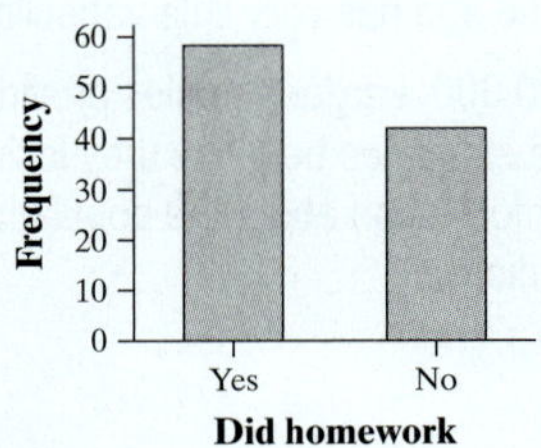


**7.12 (a)** A graph of the population distribution is shown below.

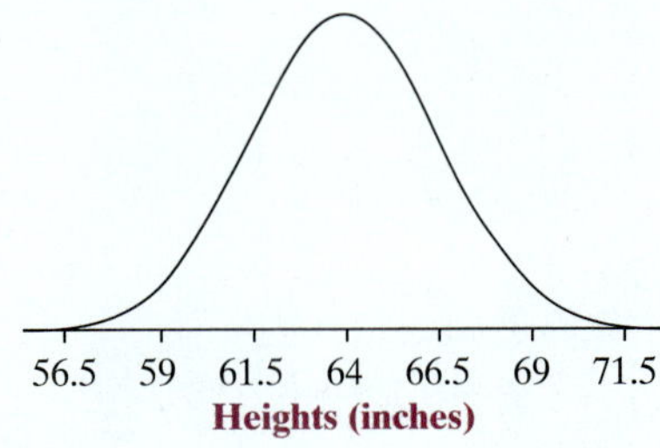


**(b)** Answers will vary. An example dotplot is given below.

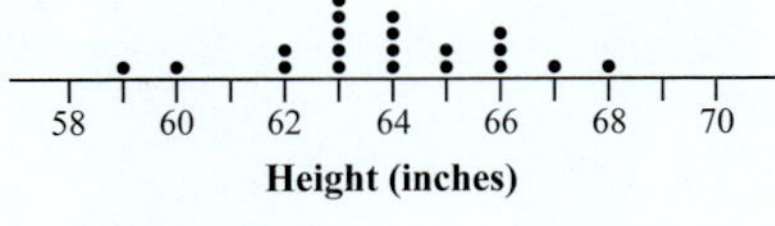

**7.13 (a)** Skewed to the right with a center at 9(°F)$^2$. The values vary from about 2 to 27.5(°F)$^2$. **(b)** A sample variance of 25(°F)$^2$ provides convincing evidence that the manufacturer's claim is false and that the thermostat actually has more variability than claimed because a value this large was rare in the simulation.

**7.14 (a)** Slightly left-skewed and centered at 45.5°F. The values vary from about 39 to 50°F. **(b)** A sample minimum of 40°F provides convincing evidence that the manufacturer's claim is false because a value this small was rare in the simulation.

**7.15** If we chose many SRSs and calculated the sample mean $\bar{x}$ for each sample, we will not consistently underestimate $\mu$ or consistently overestimate $\mu$.

**7.16** If we chose many random samples and calculated the sample proportion $\hat{p}$ for each sample, we will not consistently underestimate $p$ or consistently overestimate $p$.

**7.17** A larger random sample will provide more information and, therefore, more precise results.

**7.18** A larger random sample will provide more information and, therefore, more precise results.

**7.19 (a)** Statistics ii and iii, because the means of their sampling distributions appear to be equal to the population parameter. **(b)** Statistic ii, because it is unbiased and has very little variability.

**7.20** 20,000. Larger samples provide more precise estimates because they include more information about the population distribution.



working properly and that the actual temperatures in the cabin vary according to a Normal distribution with mean $\mu = 50°F$ and standard deviation $\sigma = 3°F$.

13. **Cold cabin?** The Fathom screen shot below shows the results of taking 500 SRSs of 10 temperature readings from a population distribution that is N(50, 3) and recording the sample variance $s_x^2$ each time.

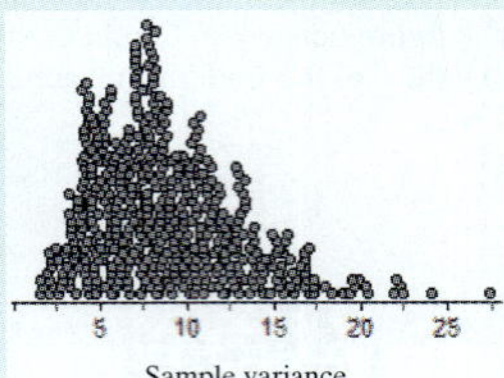


(a) Describe the approximate sampling distribution.

(b) Suppose that the variance from an actual sample is $s_x^2 = 25$. What would you conclude about the thermostat manufacturer's claim? Explain.

14. **Cold cabin?** The Fathom screen shot below shows the results of taking 500 SRSs of 10 temperature readings from a population distribution that is N(50, 3) and recording the sample minimum each time.

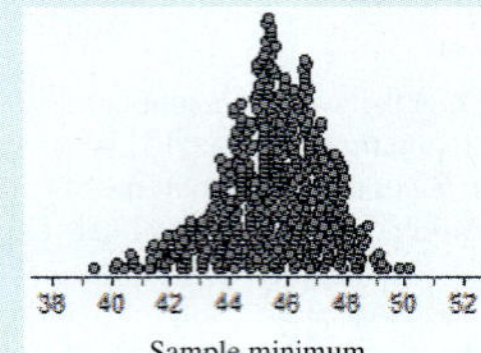


(a) Describe the approximate sampling distribution.

(b) Suppose that the minimum of an actual sample is 40°F. What would you conclude about the thermostat manufacturer's claim? Explain.

15. **A sample of teens** A study of the health of teenagers plans to measure the blood cholesterol levels of an SRS of 13- to 16-year-olds. The researchers will report the mean $\bar{x}$ from their sample as an estimate of the mean cholesterol level $\mu$ in this population. Explain to someone who knows little about statistics what it means to say that $\bar{x}$ is an unbiased estimator of $\mu$.

16. **Predict the election** A polling organization plans to ask a random sample of likely voters who they plan to vote for in an upcoming election. The researchers will report the sample proportion $\hat{p}$ that favors the incumbent as an estimate of the population proportion $p$ that favors the incumbent. Explain to someone who knows little about statistics what it means to say that $\hat{p}$ is an unbiased estimator of $p$.

17. **A sample of teens** Refer to Exercise 15. The sample mean $\bar{x}$ is an unbiased estimator of the population mean $\mu$ no matter what size SRS the study chooses. Explain to someone who knows nothing about statistics why a large random sample will give more trustworthy results than a small random sample.

18. **Predict the election** Refer to Exercise 16. The sample proportion $\hat{p}$ is an unbiased estimator of the population proportion $p$ no matter what size random sample the polling organization chooses. Explain to someone who knows nothing about statistics why a large random sample will give more trustworthy results than a small random sample.

19. **Bias and variability** The figure below shows histograms of four sampling distributions of different statistics intended to estimate the same parameter.

pg 435

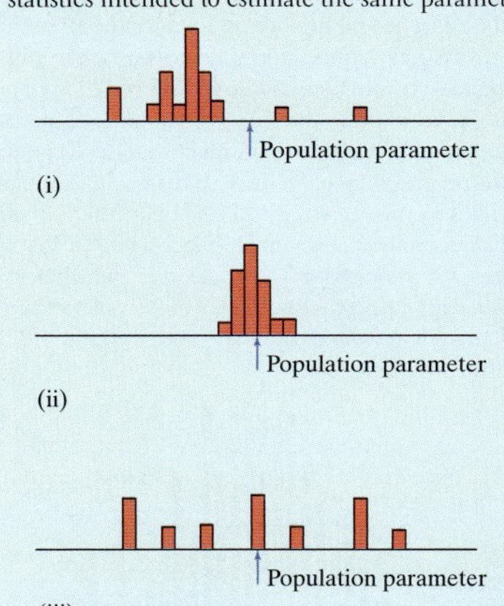


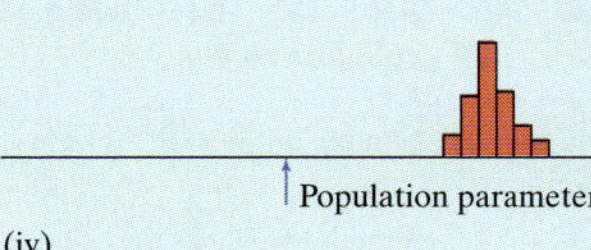


(a) Which statistics are unbiased estimators? Justify your answer.

(b) Which statistic does the best job of estimating the parameter? Explain.

20. **IRS audits** The Internal Revenue Service plans to examine an SRS of individual federal income tax returns. The parameter of interest is the proportion of all returns claiming itemized deductions. Which would be better for estimating this parameter: an SRS of 20,000 returns or an SRS of 2000 returns? Justify your answer.

*Multiple choice: Select the best answer for Exercises 21 to 24.*

21. At a particular college, 78% of all students are receiving some kind of financial aid. The school newspaper selects a random sample of 100 students and 72% of the respondents say they are receiving some sort of financial aid. Which of the following is true?

(a) 78% is a population and 72% is a sample.
(b) 72% is a population and 78% is a sample.
(c) 78% is a parameter and 72% is a statistic.
(d) 72% is a parameter and 78% is a statistic.
(e) 78% is a parameter and 100 is a statistic.

22. A statistic is an unbiased estimator of a parameter when

(a) the statistic is calculated from a random sample.
(b) in a single sample, the value of the statistic is equal to the value of the parameter.
(c) in many samples, the values of the statistic are very close to the value of the parameter.
(d) in many samples, the values of the statistic are centered at the value of the parameter.
(e) in many samples, the distribution of the statistic has a shape that is approximately Normal.

23. In a residential neighborhood, the median value of a house is $200,000. For which of the following sample sizes is the sample median most likely to be above $250,000?

(a) $n = 10$
(b) $n = 50$
(c) $n = 100$
(d) $n = 1000$
(e) Impossible to determine without more information.

24. Increasing the sample size of an opinion poll will reduce the

(a) bias of the estimates made from the data collected in the poll.
(b) variability of the estimates made from the data collected in the poll.
(c) effect of nonresponse on the poll.
(d) variability of opinions in the sample.
(e) variability of opinions in the population.

25. **Dem bones** (2.2) Osteoporosis is a condition in which the bones become brittle due to loss of minerals. To diagnose osteoporosis, an elaborate apparatus measures bone mineral density (BMD). BMD is usually reported in standardized form. The standardization is based on a population of healthy young adults. The World Health Organization (WHO) criterion for osteoporosis is a BMD score that is 2.5 standard deviations below the mean for young adults. BMD measurements in a population of people similar in age and gender roughly follow a Normal distribution.

(a) What percent of healthy young adults have osteoporosis by the WHO criterion?
(b) Women aged 70 to 79 are, of course, not young adults. The mean BMD in this age group is about −2 on the standard scale for young adults. Suppose that the standard deviation is the same as for young adults. What percent of this older population has osteoporosis?

26. **Squirrels and their food supply** (3.2) Animal species produce more offspring when their supply of food goes up. Some animals appear able to anticipate unusual food abundance. Red squirrels eat seeds from pinecones, a food source that sometimes has very large crops. Researchers collected data on an index of the abundance of pinecones and the average number of offspring per female over 16 years.[3] Computer output from a least-squares regression on these data and a residual plot are shown below.

```
Predictor     Coef     SE Coef     T      P
Constant      1.4146   0.2517     5.62   0.000
Cone index    0.4399   0.1016     4.33   0.001
S = 0.600309   R-Sq = 57.2%   R-Sq(adj) = 54.2%
```

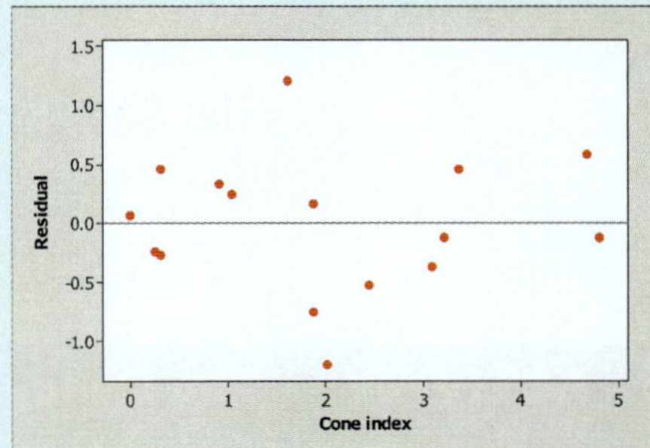


(a) Give the equation for the least-squares regression line. Define any variables you use.
(b) Is a linear model appropriate for these data? Explain.
(c) Interpret the values of $r^2$ and $s$ in context.

**7.21** c

**7.22** d

**7.23** a

**7.24** b

**7.25 (a)** We are looking for the percentage of values that are 2.5 standard deviations or farther below the mean in a Normal distribution. In other words, we are looking for $P(Z \leq -2.5)$. Using Table A, $P(Z \leq -2.5) = 0.0062$. *Using technology*: `normalcdf(lower:-1000,upper:-2.50,μ:0,σ:1)` = 0.0062. Less than 1% of healthy young adults have osteoporosis. **(b)** Let $X$ be the BMD for women aged 70 to 79 on the standard scale. Then $X$ follows an $N(-2, 1)$ distribution and we want to find $P(X \leq -2.5)$. $z = \frac{-2.5 - (-2)}{1} = -0.5$ and $P(Z \leq -0.5) = 0.3085$. *Using technology*: `normalcdf(lower:-1000,upper:-2.5,μ:-2,σ:1)` = 0.3085. About 31% of women aged 70–79 have osteoporosis.

**7.26 (a)** $\hat{y} = 1.4146 + 0.4399x$, where $\hat{y}$ is the predicted average number of offspring per female and $x$ is the index of the abundance of pine cones. **(b)** Yes, because there is no leftover pattern in the residual plot. **(c)** $r^2$: 57.2% of the variability in the average number of offspring can be accounted for by the linear model relating average number of offspring to abundance of pine cones. $s$: When using this model to predict average number of offspring from abundance of pine cones, we will typically be off by about 0.60 offspring.

**PD Section 7.2 Overview**

Watch the video overview of Section 7.2 (for teachers) by clicking on the link in the TE-book, opening the TRCD, or logging onto the book's companion Web site www.whfreeman.com/tps5e (teacher log-in required).

**TRM Section 7.2 Alternate Examples**

You can find the Alternate Examples for this section in Microsoft Word format by clicking on the link in the TE-book, opening the TRCD, or logging onto the book's companion Web site www.whfreeman.com/tps5e (teacher log-in required).

# 7.2 Sample Proportions

**WHAT YOU WILL LEARN** By the end of the section, you should be able to:

- Find the mean and standard deviation of the sampling distribution of a sample proportion $\hat{p}$. Check the 10% condition before calculating $\sigma_{\hat{p}}$.
- Determine if the sampling distribution of $\hat{p}$ is approximately Normal.
- If appropriate, use a Normal distribution to calculate probabilities involving $\hat{p}$.

What proportion of U.S. teens know that 1492 was the year in which Columbus "discovered" America? A Gallup Poll found that 210 out of a random sample of 501 American teens aged 13 to 17 knew this historically important date.[4] The sample proportion

$$\hat{p} = \frac{210}{501} = 0.42$$

is the statistic that we use to gain information about the unknown population proportion $p$. Because another random sample of 501 teens would likely result in a different estimate, we can only say that "about" 42% of U.S. teenagers know that Columbus discovered America in 1492. In this section, we'll use sampling distributions to clarify what "about" means.

## The Sampling Distribution of $\hat{p}$

How good is the statistic $\hat{p}$ as an estimate of the parameter $p$? To find out, we ask, "What would happen if we took many samples?" The **sampling distribution of $\hat{p}$** answers this question. How do we determine the shape, center, and spread of the sampling distribution of $\hat{p}$? Let's start with a simulation.

**Activity Overview**

You can watch an overview of this Activity (for teachers) by clicking on the link in the TE-book, opening the TRCD, or logging onto the book's companion Web site www.whfreeman.com/tps5e (teacher log-in required).

### ACTIVITY The Candy Machine

**MATERIALS:** Computer with Internet access—one for the class or one per pair of students

Imagine a very large candy machine filled with orange, brown, and yellow candies. When you insert money, the machine dispenses a sample of candies. In this Activity, you will use an applet to investigate the sample-to-sample variability in the proportion of orange candies dispensed by the machine.

### Teaching Tip

When you use The Candy Machine Activity, make sure to emphasize the distinction between the distribution of the population (the machine full of candy), the distribution of the sample (the three bins below the candy machine), and the distribution of the sample proportion of orange candies (shown with the dotplot).

## Sampling Reese's Pieces

1. Launch the *Reese's Pieces*® applet at www.rossmanchance.com. Change the population proportion of orange candies to $p = 0.45$ (the applet calls this value $\pi$ instead of $p$).

2. Click on the "Draw Samples" button. An animated simple random sample of $n = 25$ candies should be dispensed. Figure 7.10 shows the results of one such sample. Was your sample proportion of orange candies close to the actual population proportion, $p = 0.45$? Look at the value of $\hat{p}$ in the applet window.

3. Click "Draw Samples" 9 more times, so that you have a total of 10 sample results. Look at the dotplot of your $\hat{p}$-values. What is the mean of your 10 sample proportions? What is their standard deviation?

4. To take many more samples quickly, enter 390 in the "number of samples" box. Click on the Animate box to turn the animation off. Then click "Draw Samples." You have now taken a total of 400 samples of 25 candies from the machine. Describe the shape, center, and spread of the approximate sampling distribution of $\hat{p}$ shown in the dotplot.

5. How would the sampling distribution of the sample proportion $\hat{p}$ change if the machine dispensed $n = 50$ candies each time instead of 25? "Reset" the applet. Take 400 samples of 50 candies. Describe the shape, center, and spread of the approximate sampling distribution.

6. How would the sampling distribution of $\hat{p}$ change if the proportion of orange candies in the machine was $p = 0.15$ instead of $p = 0.45$? Does your answer depend on whether $n = 25$ or $n = 50$? Use the applet to investigate these questions. Then write a brief summary of what you learned.

7. For what combinations of $n$ and $p$ is the sampling distribution of $\hat{p}$ approximately Normal? Use the applet to investigate.

**FIGURE 7.10** The result of taking one SRS of 25 candies from a large candy machine in which 45% of the candies are orange.

Figure 7.11 shows one set of possible results from Step 4 of "The Candy Machine" Activity. Let's describe what we see.

*Shape:* Roughly symmetric and somewhat bell-shaped. It looks as though a Normal curve would approximate this distribution fairly well.

*Center:* The mean of the 400 sample proportions is 0.449. This is quite close to the actual population proportion, $p = 0.45$.

*Spread:* The standard deviation of the 400 values of $\hat{p}$ from these samples is 0.105.

The dotplot in Figure 7.11 is the approximate sampling distribution of $\hat{p}$. If we took all possible SRSs of $n = 25$ candies from the machine and graphed the value of $\hat{p}$ for each sample, then we'd have the sampling distribution of $\hat{p}$. We can get an idea of its shape, center, and spread from Figure 7.11.

**FIGURE 7.11** The result of taking 400 SRSs of 25 candies from a large candy machine in which 45% of the candies are orange. The dotplot shows the approximate sampling distribution of $\hat{p}$.

### Teaching Tip

Some books use $\pi$ to represent the population proportion instead of $p$. This is consistent with the use of Greek letters for population parameters ($\mu$ for mean, $\sigma$ for standard deviation, and so on). However, we use $p$ for the population proportion to match the notation used on the formula sheets provided on the AP® exam.

### ALTERNATE ACTIVITY

***Penny for your thoughts . . . on proportions***

1. Have each student take a sample of size 5 from a large collection of pennies.
2. Calculate $\hat{p}$ = the proportion of pennies in the sample that were minted in the 2000s.
3. Have each student record their value on a class dotplot using a $\hat{p}$ symbol instead of a dot. For the dotplot, use increments of 0.05 starting at 0 and ending at 1.
4. Have students take a sample of size 20 and repeat Steps 2 and 3.
5. Ask students to describe the changes in the distribution of $\hat{p}$ as the sample size increases.

### TRM Sampling Distribution of a Sample Proportion

This assignment is built around the *Reese's Pieces* applet found at www.rossmanchance.com/applets and previews many of the examples in the section. It helps students explore the shape, center, and spread of the sampling distribution of $\hat{p}$ for different sample sizes and population proportions. You can assign it to students at the beginning of the section to preview the examples in the section or assign it after using the examples as a way to reinforce what was covered in class. Find it by clicking on the link in the TE-book, opening the TRCD, or logging onto the book's companion Web site www.whfreeman.com/tps5e (teacher log-in required).

# EXAMPLE Sampling Candies

## *Effect of n and p on shape, center, and spread*

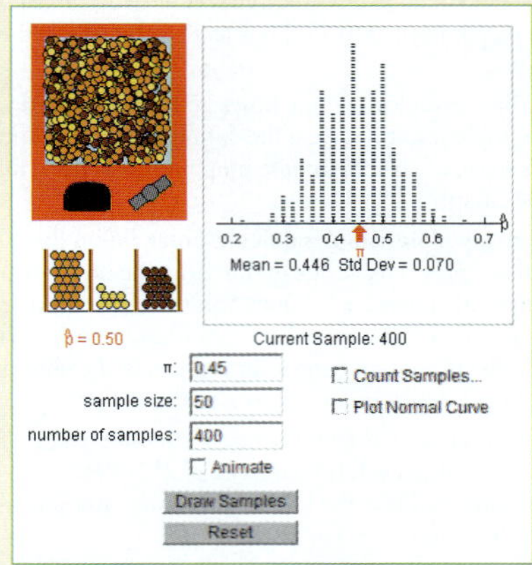


**FIGURE 7.12** The approximate sampling distribution of $\hat{p}$ for 400 SRSs of 50 candies from a population in which $p = 0.45$ of the candies are orange.

In a similar way, we can explore the sampling distribution of $\hat{p}$ when $n = 50$ (Step 5 of the Activity). As Figure 7.12 shows, the dotplot is once again roughly symmetric and somewhat bell-shaped. This graph is also centered at about 0.45. With samples of size 50, however, there is less spread in the values of $\hat{p}$. The standard deviation in Figure 7.12 is 0.070. For the samples of size 25 in Figure 7.11, it is 0.105. To repeat what we said earlier, larger samples give the sampling distribution a smaller spread.

What if the actual proportion of orange candies in the machine were $p = 0.15$? Figure 7.13(a) shows the approximate sampling distribution of $\hat{p}$ when $n = 25$. Notice that the dotplot is slightly right-skewed. The graph is centered close to the population parameter, $p = 0.15$. As for the spread, it's similar to the standard deviation in Figure 7.12, where $n = 50$ and $p = 0.45$. If we increase the sample size to $n = 50$, the sampling distribution of $\hat{p}$ should show less variability. The standard deviation in Figure 7.13(b) confirms this. Note that we can't just visually compare the graphs because the horizontal scales are different. The dotplot is more symmetrical than the graph in Figure 7.13(a) and is once again centered at a value that is close to $p = 0.15$.

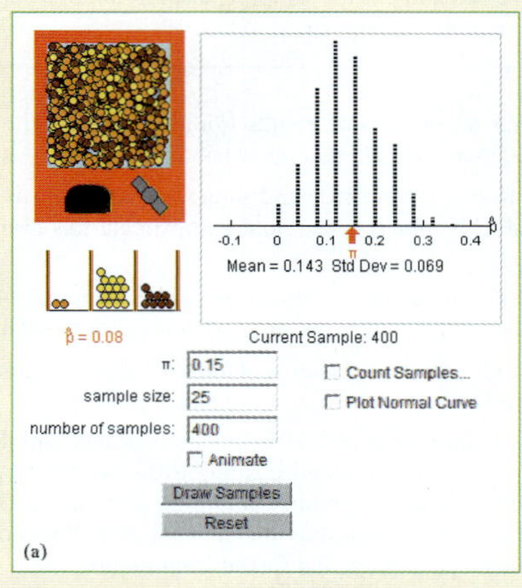


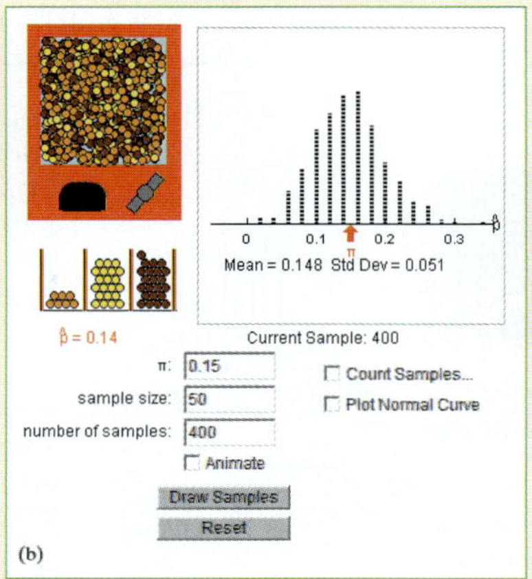


**FIGURE 7.13** The result of taking 400 SRSs of (a) size $n = 25$ and (b) size $n = 50$ candies from a large candy machine in which 15% of the candies are orange. The dotplots show the approximate sampling distribution of $\hat{p}$ in each case.

## ALTERNATE EXAMPLE

### *Penny for your thoughts . . . on proportions*

Here are the results of 500 SRSs of size 5 and 500 SRSs of size 20 when sampling from a population of 2341 pennies where the true proportion of pennies minted in the 2000s is $p = 0.293$.

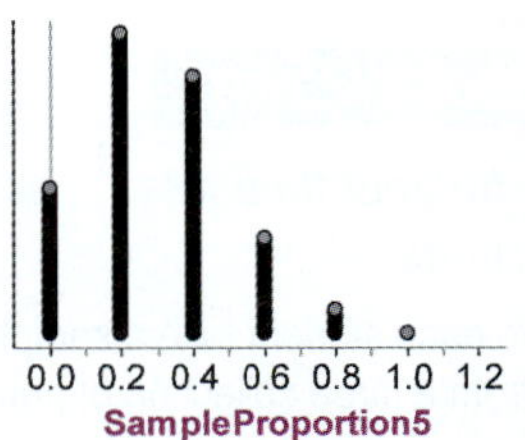


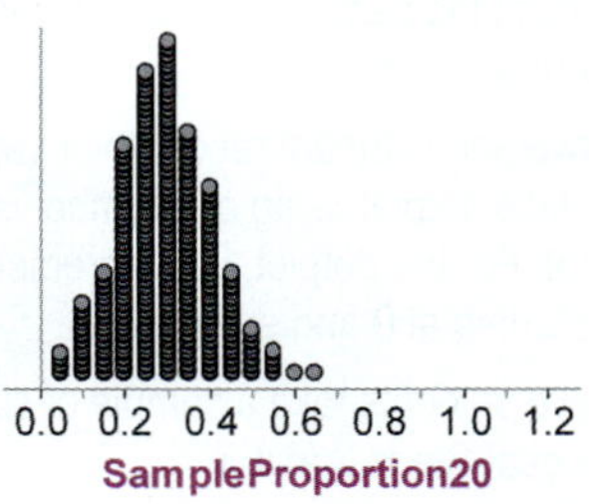


Notice that both distributions have a mean of about 0.293, the true proportion of pennies minted in the 2000s in the population. The spread, however, is smaller for samples of size 20. Finally, the shape is more symmetric and less skewed to the right with the larger sample size.

Now, suppose we wanted to estimate the proportion of pennies minted before 1976. In the same population, the true proportion of pennies minted before 1976 is $p = 0.092$. Here are the results of 500 SRSs of size 5 and 500 SRSs of size 20:

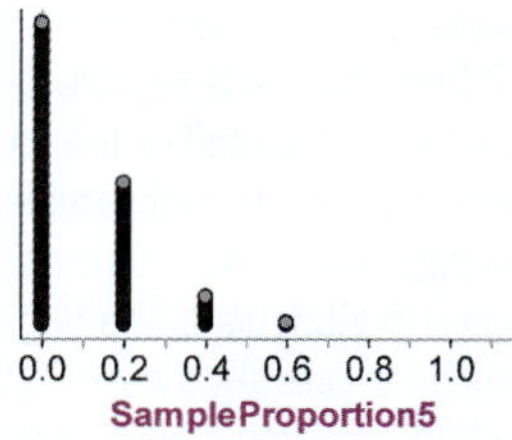


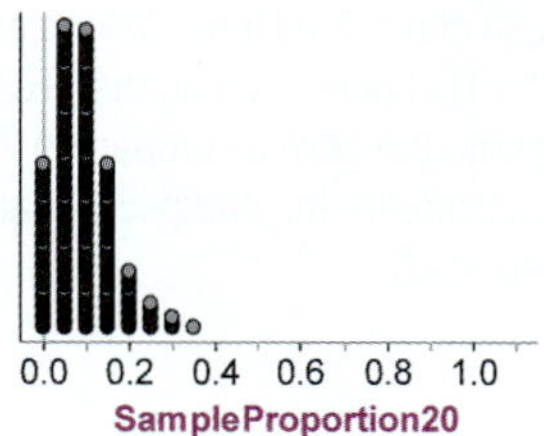


Notice that the means of the distributions are about the same and approximately equal to the true proportion, $p = 0.092$. Also, the spread of the distribution is smaller for the larger sample size. Finally, the shape of the distribution is more symmetric and less skewed to the right for the larger sample size.

For both sample sizes, the distributions were more skewed when $p = 0.092$ than when $p = 0.293$. In general, the closer $p$ is to 0 or 1, the more skewed the sampling distribution of $\hat{p}$ will be for samples of a given size.

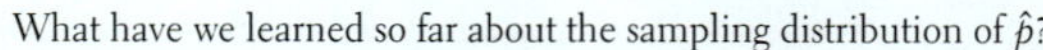

What have we learned so far about the sampling distribution of $\hat{p}$?

*Shape:* In some cases, the sampling distribution of $\hat{p}$ can be approximated by a Normal curve. This seems to depend on both the sample size $n$ and the population proportion $p$.

*Center:* The mean of the distribution is $\mu_{\hat{p}} = p$. This makes sense because the sample proportion $\hat{p}$ is an *unbiased estimator* of $p$.

*Spread:* For a specific value of $p$, the standard deviation $\sigma_{\hat{p}}$ gets smaller as $n$ gets larger. The value of $\sigma_{\hat{p}}$ depends on both $n$ and $p$.

To sort out the details of shape and spread, we need to make an important connection between the sample proportion $\hat{p}$ and the number of "successes" $X$ in the sample.

In the candy machine example, we started by taking repeated SRSs of $n = 25$ candies from a population with proportion $p = 0.45$ of orange candies. For any such sample, we can think of each candy that comes out of the machine as a trial of this chance process. A "success" occurs when we get an orange candy. Let $X =$ the number of orange candies obtained. As long as the number of candies in the machine is very large, $X$ will have close to a binomial distribution with $n = 25$ and $p = 0.45$. (Refer to the 10% condition on page 401.) The sample proportion of successes is closely related to $X$:

$$\hat{p} = \frac{\text{count of successes in sample}}{\text{size of sample}} = \frac{X}{n}$$

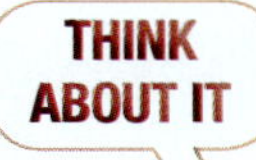


**How is the sampling distribution of $\hat{p}$ related to the binomial count $X$?** From Chapter 6, we know that the mean and standard deviation of a binomial random variable $X$ are

$$\mu_X = np \quad \text{and} \quad \sigma_X = \sqrt{np(1-p)}$$

Because $\hat{p} = X/n = (1/n)X$, we're just multiplying the random variable $X$ by a constant $(1/n)$ to get the random variable $\hat{p}$. Recall from Chapter 6 that multiplying by a constant multiplies both the mean and the standard deviation of the new random variable by that constant. We have

$$\mu_{\hat{p}} = \frac{1}{n}(np) = p \quad \text{(confirming that } \hat{p} \text{ is an unbiased estimator of } p\text{)}$$

$$\sigma_{\hat{p}} = \frac{1}{n}\sqrt{np(1-p)} = \sqrt{\frac{np(1-p)}{n^2}} = \sqrt{\frac{p(1-p)}{n}}$$

(as sample size increases, spread decreases)

That takes care of center and spread. What about shape? Multiplying a random variable by a constant doesn't change the shape of the probability distribution. So the sampling distribution of $\hat{p}$ will have the same shape as the distribution of the binomial random variable $X$.

If you studied the optional material in Chapter 6 about the Normal approximation to a binomial distribution, then you already know the punch line. Whenever $np$ and $n(1-p)$ are at least 10, a Normal distribution can be used to approximate the sampling distribution of $\hat{p}$.

## Teaching Tip

Students are sometimes confused by the symbols $\mu_{\hat{p}}$ and $\sigma_{\hat{p}}$. These are simply the mean and standard deviation of the sampling distribution of the sample proportion. To illustrate, show your students one of the graphs you made in the Penny for Your Thoughts . . . on Proportions Alternate Activity or The Candy Machine Activity, and remind them that each of the values plotted on the graph is a $\hat{p}$. Then tell them that the way we denote the mean of the $\hat{p}$ distribution is $\mu_{\hat{p}}$ and the way we denote the standard deviation of the $\hat{p}$ distribution is $\sigma_{\hat{p}}$.

## APPLET

If you didn't cover the Normal approximation to the binomial distribution in Chapter 6, you might want to briefly explore the *Normal Approximation to Binomial Distributions* applet at www.whfreeman.com/tps5e. Using the sliders helps students to see that the Normal approximation gets better as the sample size increases and as the proportion of successes gets closer to 0.5 and farther from 0 and 1.

## AP® Exam Formula Sheet

Let your students know that the formulas for the mean and standard deviation of the sampling distribution of a sample proportion are included on the formula sheet that is given to students on both sections of the AP® exam. They are listed in the section with the formulas for the binomial distribution, as they are very closely related!

## Teaching Tip

We will be referring to the "10% condition" and the "Large Counts condition" quite often in the remainder of the book, so it is a good idea to begin using these titles right away.

## Teaching Tip

Remind students that each dot in Figure 7.14 represents the value of $\hat{p}$ from a particular SRS of size $n$ from the population.

Here's a summary of the important facts about the sampling distribution of $\hat{p}$.

### SAMPLING DISTRIBUTION OF A SAMPLE PROPORTION

Choose an SRS of size $n$ from a population of size N with proportion $p$ of successes. Let $\hat{p}$ be the sample proportion of successes. Then:

- The **mean** of the sampling distribution of $\hat{p}$ is $\mu_{\hat{p}} = p$.
- The **standard deviation** of the sampling distribution of $\hat{p}$ is

$$\sigma_{\hat{p}} = \sqrt{\frac{p(1-p)}{n}}$$

  as long as the *10% condition* is satisfied: $n \leq \frac{1}{10}N$.
- As $n$ increases, the sampling distribution of $\hat{p}$ becomes **approximately Normal**. Before you perform Normal calculations, check that the *Large Counts condition* is satisfied: $np \geq 10$ and $n(1-p) \geq 10$.

Figure 7.14 displays the facts in a form that helps you recall the big idea of a sampling distribution. The mean of the sampling distribution of $\hat{p}$ is the true value of the population proportion $p$. The standard deviation of $\hat{p}$ gets smaller as the sample size $n$ increases. In fact, because the sample size $n$ is under the square root sign, we'd have to take a sample four times as large to cut the standard deviation in half.

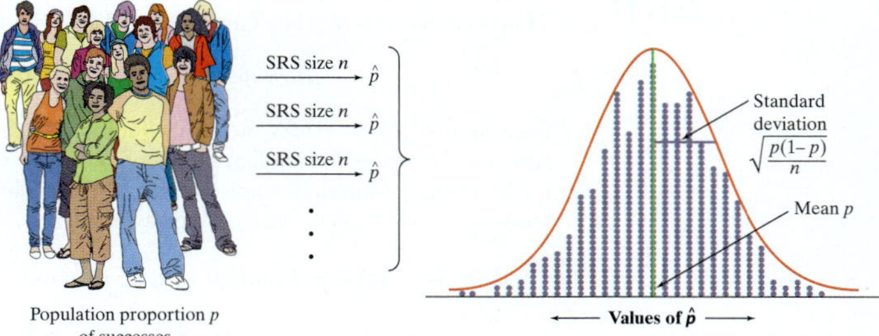


**FIGURE 7.14** Select a large SRS from a population in which proportion $p$ are successes. The sampling distribution of the proportion $\hat{p}$ of successes in the sample is approximately Normal. The mean is $p$ and the standard deviation is $\sqrt{p(1-p)/n}$.

The two conditions in the preceding box are very important. (1) *Large Counts condition:* If we assume that the sampling distribution of $\hat{p}$ is approximately Normal when it isn't, any calculations we make using a Normal distribution will be flawed. (2) *10% condition:* When we're sampling without replacement from a (finite) population, the observations are not independent, because knowing the outcome of one trial helps us predict the outcome of future trials. But the standard deviation formula assumes that the observations are independent. If we sample too large a fraction of the population, our calculated value of $\sigma_{\hat{p}}$ will be inaccurate.

Because larger random samples give better information, it sometimes makes sense to sample more than 10% of a population. In such a case, there's a more accurate formula for calculating the standard deviation $\sigma_{\hat{p}}$. It uses something called a *finite population correction* (FPC). We'll avoid situations that require the FPC in this text.

## CHECK YOUR UNDERSTANDING

About 75% of young adult Internet users (ages 18 to 29) watch online videos. Suppose that a sample survey contacts an SRS of 1000 young adult Internet users and calculates the proportion $\hat{p}$ in this sample who watch online videos.

1. What is the mean of the sampling distribution of $\hat{p}$? Explain.
2. Find the standard deviation of the sampling distribution of $\hat{p}$. Check that the 10% condition is met.
3. Is the sampling distribution of $\hat{p}$ approximately Normal? Check that the Large Counts condition is met.
4. If the sample size were 9000 rather than 1000, how would this change the sampling distribution of $\hat{p}$?

## Using the Normal Approximation for $\hat{p}$

Inference about a population proportion $p$ is based on the sampling distribution of $\hat{p}$. When the sample size is large enough for $np$ and $n(1-p)$ to both be at least 10 (the *Large Counts condition*), the sampling distribution of $\hat{p}$ is approximately Normal. In that case, we can use a Normal distribution to calculate the probability of obtaining an SRS in which $\hat{p}$ lies in a specified interval of values. Here is an example.

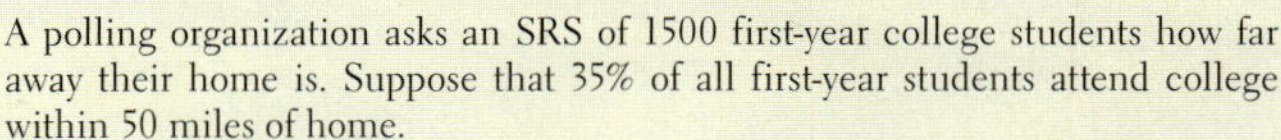

**EXAMPLE**

## Going to College

*Normal calculations involving $\hat{p}$*

A polling organization asks an SRS of 1500 first-year college students how far away their home is. Suppose that 35% of all first-year students attend college within 50 miles of home.

**PROBLEM:** Find the probability that the random sample of 1500 students will give a result within 2 percentage points of this true value. Show your work.

**SOLUTION:**

**Step 1: State the distribution and the values of interest.** We want to find the probability that $\hat{p}$ falls between 0.33 and 0.37 (within 2 percentage points, or 0.02, of 0.35). In symbols, that's $P(0.33 \leq \hat{p} \leq 0.37)$. We have an SRS of size $n = 1500$ drawn from a population in which the proportion $p = 0.35$ attend college within 50 miles of home. What do we know about the sampling distribution of $\hat{p}$?

- Its mean is $\mu_{\hat{p}} = p = 0.35$.

### ANSWERS TO CYU

1. $\mu_{\hat{p}} = p = 0.75$

2. The standard deviation of the sampling distribution of $\hat{p}$ is $\sigma_{\hat{p}} = \sqrt{\frac{p(1-p)}{n}} = \sqrt{\frac{0.75(0.25)}{1000}} = 0.0137$. There are more than 10(1000) = 10,000 young adult Internet users, so the 10% condition has been met.

3. Yes. Both $np = 1000(0.75) = 750$ and $n(1-p) = 1000(0.25) = 250$ are at least 10.

4. The sampling distribution would still be approximately Normal with mean 0.75. However, the standard deviation would be smaller by a factor of 3: $\sigma_{\hat{p}} = \sqrt{\frac{p(1-p)}{n}} = \sqrt{\frac{0.75(0.25)}{9000}} = 0.0046$

### Teaching Tip

The finite population correction (FPC) factor is $\sqrt{1-\frac{n}{N}}$, where $n$ = the sample size and $N$ = the population size. So the standard deviation of the sampling distribution of a sample proportion *when sampling without replacement* is actually $\sigma_{\hat{p}} = \sqrt{\frac{p(1-p)}{n}}\sqrt{1-\frac{n}{N}}$. Here are a few observations you might mention to your students:

1. The formula for the standard deviation presented in the text is actually conservative, meaning that the value provided by the formula in the text will always be bigger than the actual standard deviation of the sampling distribution of $\hat{p}$.

2. When the sample is small compared with the population, the FPC factor is close to 1, so it doesn't change the value of $\sigma_{\hat{p}}$ very much. For example, if the sample is exactly 10% of the population, the value of the FPC factor is $\sqrt{1-\frac{1}{10}} = 0.95$.

3. If a census is conducted, then there is *no sampling variability,* so the value of $\sigma_{\hat{p}}$ should be 0. However, if the FPC factor isn't used, $\sigma_{\hat{p}}$ will be greater than 0. For example, suppose a census was taken in a high school of 1000 students and the proportion of students with Internet access at home was $p = 0.793$. Incorrectly using the formula in the student text in this setting gives a standard deviation of $\sigma_{\hat{p}} = \sqrt{\frac{0.793(1-0.793)}{1000}} = 0.0128$ when the "sample" proportion actually does not vary at all.

4. Although interesting, the FPC is not covered on the AP® exam.

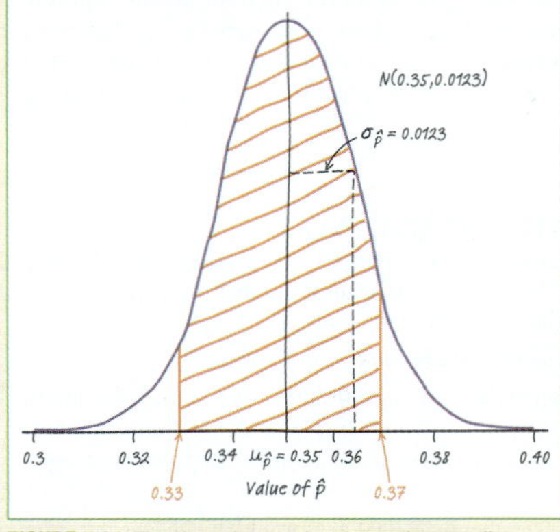


**FIGURE 7.15** The Normal approximation to the sampling distribution of $\hat{p}$.

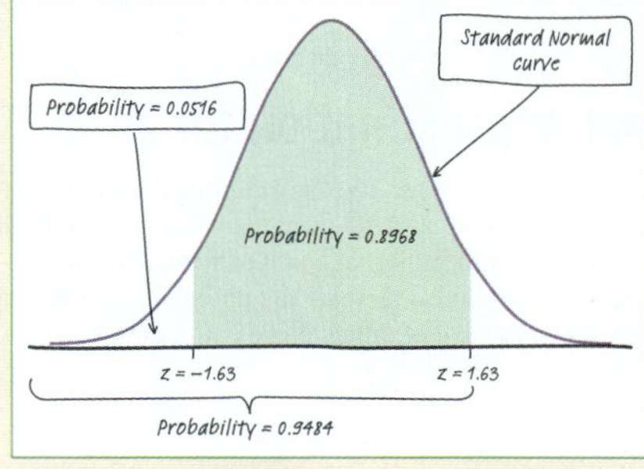


**FIGURE 7.16** Probabilities as areas under the standard Normal curve.

- What about the standard deviation? We need to check the 10% condition. To use the standard deviation formula we derived, the population must contain at least 10(1500) = 15,000 people. There are over 1.7 million first-year college students, so

$$\sigma_{\hat{p}} = \sqrt{\frac{p(1-p)}{n}} = \sqrt{\frac{(0.35)(0.65)}{1500}} = 0.0123$$

- Can we use a Normal distribution to approximate the sampling distribution of $\hat{p}$? Check the Large Counts condition: $np = 1500(0.35) = 525$ and $n(1-p) = 1500(0.65) = 975$. Both are much larger than 10, so the Normal approximation will be quite accurate.

Figure 7.15 shows the Normal distribution that we'll use with the area of interest shaded and the mean, standard deviation, and boundary values labeled.

**Step 2: Perform calculations—show your work!** The standardized scores for the two boundary values are

$$z = \frac{0.33 - 0.35}{0.0123} = -1.63 \text{ and } z = \frac{0.37 - 0.35}{0.0123} = 1.63$$

Figure 7.16 shows the area under the standard Normal curve corresponding to these standardized values. Using Table A, the desired probability is

$$P(0.33 \le \hat{p} \le 0.37) = P(-1.63 \le Z \le 1.63)$$
$$= 0.9484 - 0.0516 = 0.8968$$

*Using technology:* The command `normalcdf(lower:0.33, upper:0.37, μ:0.35, σ:0.0123)` gives an area of 0.8961.

**Step 3: Answer the question.** About 90% of all SRSs of size 1500 will give a result within 2 percentage points of the truth about the population.

**For Practice** *Try Exercise* 39

## Section 7.2 Summary

- When we want information about the population proportion $p$ of successes, we often take an SRS and use the sample proportion $\hat{p}$ to estimate the unknown parameter $p$. The **sampling distribution** of $\hat{p}$ describes how the sample proportion varies in all possible samples from the population.
- The **mean** of the sampling distribution of $\hat{p}$ is equal to the population proportion $p$. That is, $\hat{p}$ is an unbiased estimator of $p$.

**ALTERNATE EXAMPLE**

### Planning for college

The superintendent of a large school district wants to know what proportion of middle school students in her district are planning to attend a four-year college or university. Suppose that 80% of all middle school students in her district are planning to attend a four-year college or university.

**PROBLEM:**

What is the probability that an SRS of size 125 will give a result within 7 percentage points of the true value?

**SOLUTION:**

**Step 1: State the distribution and the values of interest.** We want to find the probability that the percentage of middle school students who plan to attend a four-year college or university falls between 73% and 87%. Converting to proportions, we want to find $P(0.73 \le \hat{p} \le 0.87)$. The mean of the sampling distribution of $\hat{p}$ is $\mu_{\hat{p}} = p = 0.80$. Because the school district is large, we can assume that there are more than 10(125) = 1250 middle school students, so $\sigma_{\hat{p}} = \sqrt{\frac{0.80(0.20)}{125}} = 0.036$. We can consider the distribution of $\hat{p}$ to be approximately Normal because $np = 125(0.80) = 100 \ge 10$ and $n(1-p) = 125(0.20) = 25 \ge 10$.

**Step 2: Perform calculations—show your work.** The standardized scores for the boundary values are

$$z = \frac{0.73 - 0.80}{0.036} = -1.94 \text{ and}$$

$$z = \frac{0.87 - 0.80}{0.036} = 1.94.$$

$P(0.73 \le \hat{p} \le 0.87) = 0.9476$. *Using technology*: The command `normalcdf(lower:0.73,upper:0.87,μ:0.80,σ:0.036)` gives an area of 0.9482.

**Step 3: Answer the question.** About 95% of all SRSs of size 125 will give a sample proportion within 7 percentage points of the true proportion of middle school students who are planning to attend a four-year college or university.

### Teaching Tip

You might mention the idea of a confidence interval at the end of this example. Simply say that because 90% of all SRSs will give a sample proportion within 2 percentage points of the truth, then in 90% of the SRSs, the true proportion will be within 2 percentage points of the sample proportion. This is a crucial connection, because we almost always know the *sample* proportion but *not* the *true* proportion.

### TRM Section 7.2 Quizzes

There are three different quizzes available for this section in the Teacher's Resource Materials. Versions A and B are roughly parallel. Find them by clicking on the link in the TE-book, opening the TRCD, or logging onto the book's companion Web site www.whfreeman.com/tps5e (teacher log-in required).

- The **standard deviation** of the sampling distribution of $\hat{p}$ is $\sqrt{p(1-p)/n}$ for an SRS of size $n$. This formula can be used if the population is at least 10 times as large as the sample (the *10% condition*). The standard deviation of $\hat{p}$ gets smaller as the sample size $n$ gets larger. Because of the square root, a sample four times larger is needed to cut the standard deviation in half.
- When the sample size $n$ is large, the sampling distribution of $\hat{p}$ is close to a Normal distribution with mean $p$ and standard deviation $\sqrt{p(1-p)/n}$. In practice, use this **Normal approximation** when both $np \geq 10$ and $n(1-p) \geq 10$ (the *Large Counts condition*).

## Section 7.2 Exercises

27. **The candy machine** Suppose a large candy machine has 45% orange candies. Use Figures 7.11 and 7.12 (pages 441 and 442) to help answer the following questions.

(a) Would you be surprised if a sample of 25 candies from the machine contained 8 orange candies (that's 32% orange)? How about 5 orange candies (20% orange)? Explain.

(b) Which is more surprising: getting a sample of 25 candies in which 32% are orange or getting a sample of 50 candies in which 32% are orange? Explain.

28. **The candy machine** Suppose a large candy machine has 15% orange candies. Use Figure 7.13 (page 442) to help answer the following questions.

(a) Would you be surprised if a sample of 25 candies from the machine contained 8 orange candies (that's 32% orange)? How about 5 orange candies (20% orange)? Explain.

(b) Which is more surprising: getting a sample of 25 candies in which 32% are orange or getting a sample of 50 candies in which 32% are orange? Explain.

29. **The candy machine** Suppose a large candy machine has 45% orange candies. Imagine taking an SRS of 25 candies from the machine and observing the sample proportion $\hat{p}$ of orange candies.

(a) What is the mean of the sampling distribution of $\hat{p}$? Why?

(b) Find the standard deviation of the sampling distribution of $\hat{p}$. Check to see if the 10% condition is met.

(c) Is the sampling distribution of $\hat{p}$ approximately Normal? Check to see if the Large Counts condition is met.

(d) If the sample size were 100 rather than 25, how would this change the sampling distribution of $\hat{p}$?

30. **The candy machine** Suppose a large candy machine has 15% orange candies. Imagine taking an SRS of 25 candies from the machine and observing the sample proportion $\hat{p}$ of orange candies.

(a) What is the mean of the sampling distribution of $\hat{p}$? Why?

(b) Find the standard deviation of the sampling distribution of $\hat{p}$. Check to see if the 10% condition is met.

(c) Is the sampling distribution of $\hat{p}$ approximately Normal? Check to see if the Large Counts condition is met.

(d) If the sample size were 225 rather than 25, how would this change the sampling distribution of $\hat{p}$?

31. **Airport security** The Transportation Security Administration (TSA) is responsible for airport safety. On some flights, TSA officers randomly select passengers for an extra security check before boarding. One such flight had 76 passengers—12 in first class and 64 in coach class. TSA officers selected an SRS of 10 passengers for screening. Let $\hat{p}$ be the proportion of first-class passengers in the sample.

(a) Is the 10% condition met in this case? Justify your answer.

(b) Is the Large Counts condition met in this case? Justify your answer.

**TRM Full Solutions to Section 7.2 Exercises**

### Answers to Section 7.2 Exercises

**7.27** **(a)** We would not be surprised to find 8 (32%) orange candies because values this small happened fairly often in the simulation. However, there were few samples in which there were 5 (20%) or fewer orange candies. Thus, getting 5 orange candies would be surprising. **(b)** A sample of 50, because we expect to be closer to $p = 0.45$ in larger samples.

**7.28** **(a)** We would be surprised to find 32% orange candies, because very few of the samples of size 25 had 32% or more orange candies. However, 20% is very near the center of the distribution and not unusual at all. **(b)** A sample of 50, because we expect to be closer to $p = 0.15$ in larger samples.

**7.29** **(a)** $\mu_{\hat{p}} = p = 0.45$ **(b)** $\sigma_{\hat{p}} = \sqrt{\frac{p(1-p)}{n}} = \sqrt{\frac{0.45(0.55)}{25}} = 0.0995$. The 10% condition is met because there are more than $10(25) = 250$ candies in the large machine. **(c)** Yes, because $np = 25(0.45) = 11.25$ and $n(1-p) = 25(0.55) = 13.75$ are both at least 10. **(d)** The sampling distribution would still be approximately Normal with a mean of $\mu_{\hat{p}} = 0.45$. However, the standard deviation decreases to $\sigma_{\hat{p}} = \sqrt{\frac{p(1-p)}{n}} = \sqrt{\frac{0.45(0.55)}{100}} = 0.0497$.

**7.30** **(a)** $\mu_{\hat{p}} = p = 0.15$ **(b)** $\sigma_{\hat{p}} = \sqrt{\frac{p(1-p)}{n}} = \sqrt{\frac{0.15(0.85)}{25}} = 0.0714$. The 10% condition is met because there are more than $10(25) = 250$ candies in the large machine. **(c)** No, because $np = 25(0.15) = 3.75$ is less than 10. **(d)** The sampling distribution would now be approximately Normal because $np = 225(0.15) = 33.75$ and $n(1-p) = 225(0.85) = 191.25$ are both at least 10. The mean is still $\mu_{\hat{p}} = 0.15$, but the standard deviation decreases to $\sigma_{\hat{p}} = \sqrt{\frac{p(1-p)}{n}} = \sqrt{\frac{0.15(0.85)}{225}} = 0.0238$.

**7.31** **(a)** No, because more than 10% of the population ($10/76 = 13\%$) was selected. **(b)** No, because the sample size was only $n = 10$. Neither $np$ nor $n(1-p)$ will be at least 10.

**7.32 (a)** Yes, because less than 10% of the population (7/100 = 7%) was selected. **(b)** No, because the sample size was only $n = 7$. Neither $np$ nor $n(1 - p)$ will be at least 10.

**7.33** The Large Counts condition is not met because $np = 15(0.3) = 4.5 < 10$.

**7.34** The 10% condition is not met because more than 10% of the population (50/316 = 15.8%) was selected.

**7.35 (a)** $\mu_{\hat{p}} = p = 0.70$ **(b)** $\sigma_{\hat{p}} = \sqrt{\dfrac{p(1-p)}{n}} = \sqrt{\dfrac{0.7(0.3)}{1012}} = 0.0144$. The 10% condition is met because the sample of size 1012 is less than 10% of the population of all U.S. adults. **(c)** Yes, because $np = 1012(0.70) = 708.4$ and $n(1 - p) = 1012(0.30) = 303.6$ are both at least 10. **(d)** We want to find $P(\hat{p} \le 0.67)$. $z = \dfrac{0.67 - 0.70}{0.0144} = -2.08$ and $P(Z \le -2.08) = 0.0188$. *Using technology*: `normalcdf(lower:-1000, upper:0.67,μ:0.70,σ:0.0144)` = 0.0186. There is a 0.0186 probability of obtaining a sample in which 67% or fewer say they drink the milk. Because this is a small probability, there is convincing evidence against the claim.

**7.36 (a)** $\mu_{\hat{p}} = p = 0.4$ **(b)** $\sigma_{\hat{p}} = \sqrt{\dfrac{0.4(0.6)}{1785}} = 0.0116$. The 10% condition is met because the sample of size 1785 is less than 10% of the population of adults. **(c)** Yes, because $np = 1785(0.4) = 714$ and $n(1 - p) = 1785(0.6) = 1071$ are both at least 10. **(d)** We want to find $P(\hat{p} \ge 0.44)$. $z = \dfrac{0.44 - 0.40}{0.0116} = 3.45$ and $P(Z \ge 3.45) = 0.0003$. *Using technology*: `normalcdf(lower:0.44,upper: 1000,μ:0.40,σ:0.0116)` = 0.0003. There is a 0.0003 probability of obtaining a sample in which 44% or more say they attended church last week. Because this is a small probability, there is convincing evidence against the claim.

**7.37** 4048, because using $4n$ for the sample size halves the standard deviation ($\sqrt{4n} = 2\sqrt{n}$).

**7.38** 16,065, because using $9n$ for the sample size reduces the standard deviation of the sampling distribution to one-third of the previous value ($\sqrt{9n} = 3\sqrt{n}$).



32. **Scrabble** In the game of Scrabble, each player begins by drawing 7 tiles from a bag containing 100 tiles. There are 42 vowels, 56 consonants, and 2 blank tiles in the bag. Cait chooses an SRS of 7 tiles. Let $\hat{p}$ be the proportion of vowels in her sample.

    (a) Is the 10% condition met in this case? Justify your answer.

    (b) Is the Large Counts condition met in this case? Justify your answer.

*In Exercises 33 and 34, explain why you cannot use the methods of this section to find the desired probability.*

33. **Hispanic workers** A factory employs 3000 unionized workers, of whom 30% are Hispanic. The 15-member union executive committee contains 3 Hispanics. What would be the probability of 3 or fewer Hispanics if the executive committee were chosen at random from all the workers?

34. **Studious athletes** A university is concerned about the academic standing of its intercollegiate athletes. A study committee chooses an SRS of 50 of the 316 athletes to interview in detail. Suppose that 40% of the athletes have been told by coaches to neglect their studies on at least one occasion. What is the probability that at least 15 in the sample are among this group?

35. **Do you drink the cereal milk?** A *USA Today* Poll asked a random sample of 1012 U.S. adults what they do with the milk in the bowl after they have eaten the cereal. Let $\hat{p}$ be the proportion of people in the sample who drink the cereal milk. A spokesman for the dairy industry claims that 70% of all U.S. adults drink the cereal milk. Suppose this claim is true.

    (a) What is the mean of the sampling distribution of $\hat{p}$? Why?

    (b) Find the standard deviation of the sampling distribution of $\hat{p}$. Check to see if the 10% condition is met.

    (c) Is the sampling distribution of $\hat{p}$ approximately Normal? Check to see if the Large Counts condition is met.

    (d) Of the poll respondents, 67% said that they drink the cereal milk. Find the probability of obtaining a sample of 1012 adults in which 67% or fewer say they drink the cereal milk if the milk industry spokesman's claim is true. Does this poll give convincing evidence against the claim? Explain.

36. **Do you go to church?** The Gallup Poll asked a random sample of 1785 adults whether they attended church during the past week. Let $\hat{p}$ be the proportion of people in the sample who attended church. A newspaper report claims that 40% of all U.S. adults went to church last week. Suppose this claim is true.

    (a) What is the mean of the sampling distribution of $\hat{p}$? Why?

    (b) Find the standard deviation of the sampling distribution of $\hat{p}$. Check to see if the 10% condition is met.

    (c) Is the sampling distribution of $\hat{p}$ approximately Normal? Check to see if the Large Counts condition is met.

    (d) Of the poll respondents, 44% said they did attend church last week. Find the probability of obtaining a sample of 1785 adults in which 44% or more say they attended church last week if the newspaper report's claim is true. Does this poll give convincing evidence against the claim? Explain.

37. **Do you drink the cereal milk?** What sample size would be required to reduce the standard deviation of the sampling distribution to one-half the value you found in Exercise 35(b)? Justify your answer.

38. **Do you go to church?** What sample size would be required to reduce the standard deviation of the sampling distribution to one-third the value you found in Exercise 36(b)? Justify your answer.

39. pg 445 **Students on diets** A sample survey interviews an SRS of 267 college women. Suppose that 70% of college women have been on a diet within the past 12 months. What is the probability that 75% or more of the women in the sample have been on a diet? Show your work.

40. **Who owns a Harley?** Harley-Davidson motorcycles make up 14% of all the motorcycles registered in the United States. You plan to interview an SRS of 500 motorcycle owners. How likely is your sample to contain 20% or more who own Harleys? Show your work.

41. **On-time shipping** A mail-order company advertises that it ships 90% of its orders within three working days. You select an SRS of 100 of the 5000 orders received in the past week for an audit. The audit reveals that 86 of these orders were shipped on time.

    (a) If the company really ships 90% of its orders on time, what is the probability that the proportion in an SRS of 100 orders is 0.86 or less? Show your work.

    (b) A critic says, "Aha! You claim 90%, but in your sample the on-time percentage is lower than that.

**7.39** $\mu_{\hat{p}} = 0.70$. Because 267 is less than 10% of the population of college women, $\sigma_{\hat{p}} = \sqrt{\dfrac{0.7(0.3)}{267}} = 0.0280$. Because $np = 267(0.7) = 186.9$ and $n(1 - p) = 267(0.3) = 80.1$ are both at least 10, the sampling distribution of $\hat{p}$ can be approximated by a Normal distribution. We want to find $P(\hat{p} \ge 0.75)$. $z = \dfrac{0.75 - 0.7}{0.0280} = 1.79$ and $P(Z \ge 1.79) = 0.0367$. *Using technology*: `normalcdf(lower:0.75, upper:1000,μ:0.7,σ:0.0280)` = 0.0371. There is a 0.0371 probability that 75% or more of the women in the sample have been on a diet within the last 12 months.

**7.40** $\mu_{\hat{p}} = 0.14$. Because 500 is less than 10% of the population of registered motorcyles, $\sigma_{\hat{p}} = \sqrt{\dfrac{0.14(0.86)}{500}} = 0.0155$. Because $np = 500(0.14) = 70$ and $n(1 - p) = 500(0.86) = 430$ are both at least 10, the sampling distribution of $\hat{p}$ can be approximated by a Normal distribution. We want to find $P(\hat{p} \ge 0.20)$. $z = \dfrac{0.20 - 0.14}{0.0155} = 3.87$and $P(Z \ge 3.87) \approx 0$. *Using technology*: `normalcdf(lower:0.20,upper:1000, μ:0.14,σ:0.0155)` = 0.0001. There is a 0.0001 probability that 20% or more of the motorcycles in the sample are Harleys.

So the 90% claim is wrong." Explain in simple language why your probability calculation in (a) shows that the result of the sample does not refute the 90% claim.

42. **Underage drinking** The Harvard College Alcohol Study finds that 67% of college students support efforts to "crack down on underage drinking." Does this result hold at a large local college? To find out, college administrators survey an SRS of 100 students and find that 62 support a crackdown on underage drinking.

(a) Suppose that the proportion of all students attending this college who support a crackdown is 67%, the same as the national proportion. What is the probability that the proportion in an SRS of 100 students is 0.62 or less? Show your work.

(b) A writer in the college's student paper says that "support for a crackdown is lower at our school than nationally." Write a short letter to the editor explaining why the survey does not support this conclusion.

***Multiple choice: Select the best answer for Exercises 43 to 46.*** *Exercises 43 to 45 refer to the following setting.* The magazine *Sports Illustrated* asked a random sample of 750 Division I college athletes, "Do you believe performance-enhancing drugs are a problem in college sports?" Suppose that 30% of all Division I athletes think that these drugs are a problem. Let $\hat{p}$ be the sample proportion who say that these drugs are a problem.

43. Which of the following are the mean and standard deviation of the sampling distribution of the sample proportion $\hat{p}$?

(a) Mean = 0.30, SD = 0.017

(b) Mean = 0.30, SD = 0.55

(c) Mean = 0.30, SD = 0.0003

(d) Mean = 225, SD = 12.5

(e) Mean = 225, SD = 157.5

44. Decreasing the sample size from 750 to 375 would multiply the standard deviation by

(a) 2. (b) $\sqrt{2}$. (c) 1/2. (d) $1/\sqrt{2}$. (e) none of these.

45. The sampling distribution of $\hat{p}$ is approximately Normal because

(a) there are at least 7500 Division I college athletes.

(b) $np = 225$ and $n(1-p) = 525$ are both at least 10.

(c) a random sample was chosen.

(d) the athletes' responses are quantitative.

(e) the sampling distribution of $\hat{p}$ always has this shape.

46. In a congressional district, 55% of the registered voters are Democrats. Which of the following is equivalent to the probability of getting less than 50% Democrats in a random sample of size 100?

(a) $P\left(Z < \frac{0.50 - 0.55}{100}\right)$

(b) $P\left(Z < \frac{0.50 - 0.55}{\sqrt{\frac{0.55(0.45)}{100}}}\right)$

(c) $P\left(Z < \frac{0.55 - 0.50}{\sqrt{\frac{0.55(0.45)}{100}}}\right)$

(d) $P\left(Z < \frac{0.50 - 0.55}{\sqrt{100(0.55)(0.45)}}\right)$

(e) $P\left(Z < \frac{0.55 - 0.50}{\sqrt{100(0.55)(0.45)}}\right)$

47. **Sharing music online** (5.2) A sample survey reports that 29% of Internet users download music files online, 21% share music files from their computers, and 12% both download and share music.[5] Make a Venn diagram that displays this information. What percent of Internet users neither download nor share music files?

48. **California's endangered animals** (4.1) The California Department of Fish and Game publishes a list of the state's endangered animals. The reptiles on the list are given below.

| | |
|---|---|
| Desert tortoise | Southern rubber boa |
| Olive Ridley sea turtle | Loggerhead sea turtle |
| Island night lizard | Barefoot banded gecko |
| Flat-tailed horned lizard | Coachella Valley fringe-toed lizard |
| Green sea turtle | Blunt-nosed leopard lizard |
| Leatherback sea turtle | Giant garter snake |
| Alameda whip snake | San Francisco garter snake |

(a) Describe how you would use Table D at line 111 to choose an SRS of 3 of these reptiles to study.

(b) Use your method from part (a) to select your sample. Identify the reptiles you chose.

**(b)** Because the probability isn't very small, it is plausible that the 67% claim is correct and that the lower than expected percentage is due to chance alone.

**7.43** a

**7.44** b

**7.45** b

**7.46** b

**7.47** The Venn diagram is shown below.

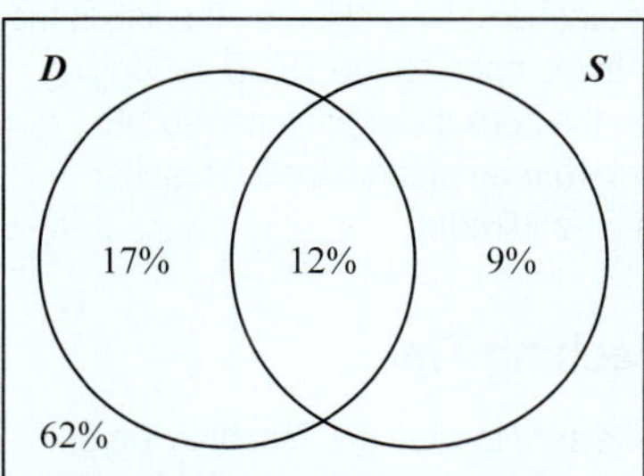


62% neither download nor share music files.

**7.48** **(a)** Assign numbers 01–14 to the animals (01 to the desert tortoise, and so on). Starting at line 111 in Table D, read pairs of numbers until you get three different numbers between 01 and 14. These numbers represent the animals chosen. **(b)** The animals chosen are 12, 04, and 11, which represent the blunt-nosed leopard lizard, the flat-tailed horned lizard, and the Coachella Valley fringe-toed lizard.

**7.41** **(a)** $\mu_{\hat{p}} = 0.90$. Because 100 is less than 10% of the population of orders, $\sigma_{\hat{p}} = \sqrt{\frac{0.90(0.10)}{100}} = 0.03$. Because $np = 100(0.90) = 90$ and $n(1-p) = 100(0.10) = 10$ are both at least 10, the sampling distribution of $\hat{p}$ can be approximated by a Normal distribution. We want to find $P(\hat{p} \le 0.86)$. $z = \frac{0.86 - 0.90}{0.03} = -1.33$ and $P(Z \le -1.33) = 0.0918$. *Using technology*: `normalcdf(lower:-1000,upper:0.86,μ:0.90,σ:0.03)` $= 0.0912$. There is a 0.0912 probability that 86% or fewer of orders in an SRS of 100 were shipped within 3 working days. **(b)** Because the probability isn't very small, it is plausible that the 90% claim is correct and that the lower than expected percentage is due to chance alone.

**7.42** **(a)** $\mu_{\hat{p}} = 0.67$. Because 100 is less than 10% of the population of students at a large college, $\sigma_{\hat{p}} = \sqrt{\frac{0.67(0.33)}{100}} = 0.0470$. Because $np = 100(0.67) = 67$ and $n(1-p) = 100(0.33) = 33$ are both at least 10, the sampling distribution of $\hat{p}$ can be approximated by a Normal distribution. We want to find $P(\hat{p} \le 0.62)$. $z = \frac{0.62 - 0.67}{0.0470} = -1.06$ and $P(Z \le -1.06) = 0.1446$. *Using technology*: `normalcdf(lower:-1000,upper:0.62,μ:0.67,σ:0.0470)` $= 0.1437$. There is a 0.1437 probability that 62% or fewer students in an SRS of 100 support a crackdown on underage drinking.

**PD** **Section 7.3 Overview**

Watch the video overview of Section 7.3 (for teachers) by clicking on the link in the TE-book, opening the TRCD, or logging onto the book's companion Web site www.whfreeman.com/tps5e (teacher log-in required).

**Activity Overview**

You can watch an overview of this Activity (for teachers) by clicking on the link in the TE-book, opening the TRCD, or logging onto the book's companion Web site www.whfreeman.com/tps5e (teacher log-in required).

### Teaching Tip

Make sure to read the Teaching Tip on page 421 for suggestions regarding the Penny for Your Thoughts Activity. Here are some additional suggestions for this specific part of the Activity:

- Make sure to use the same scale for all three distributions: the population distribution and the two sampling distributions.
- If your population of pennies is very large, then prepare the graph of the population ahead of time or just have the students take a large sample and call the graph the "approximate" population. Don't skip making a graph of the population, however. It is very important that students see the shape, center, and spread of the population distribution so they can compare them with the shape, center, and spread of the sampling distributions of $\bar{x}$ for different sample sizes.

**TRM** **Sampling Distribution of a Sample Mean**

This assignment is built around the *Sampling Distribution Simulation* applet at the Rice Virtual Lab in Statistics. It helps students explore the shape, center, and spread of the sampling distribution of $\bar{x}$ for different sample sizes and population shapes. You can assign it to students at the beginning of the section to preview the examples in the section or assign it after using the examples as a way to reinforce what was covered in class. Find it by clicking on the link in the TE-book, opening the TRCD, or logging onto the book's companion Web site www.whfreeman.com/tps5e (teacher log-in required).



## 7.3 Sample Means

**WHAT YOU WILL LEARN** By the end of the section, you should be able to:

- Find the mean and standard deviation of the sampling distribution of a sample mean $\bar{x}$. Check the 10% condition before calculating $\sigma_{\bar{x}}$.
- Explain how the shape of the sampling distribution of $\bar{x}$ is affected by the shape of the population distribution and the sample size.
- If appropriate, use a Normal distribution to calculate probabilities involving $\bar{x}$.

Sample proportions arise most often when we are interested in categorical variables. We then ask questions like "What proportion of U.S. adults have watched *Survivor*?" or "What percent of the adult population attended church last week?" But when we record quantitative variables—household income, lifetime of car brake pads, blood pressure—we are interested in other statistics, such as the median or mean or standard deviation of the variable. The sample mean $\bar{x}$ is the most common statistic computed from quantitative data. This section describes the sampling distribution of the sample mean. The following Activity and the subsequent example give you a sense of what lies ahead.

### ACTIVITY | Penny for Your Thoughts

**MATERIALS:**
Large container with several hundred pennies

Your teacher will assemble a large population of pennies of various ages.[6] In this Activity, your class will investigate the sampling distribution of the mean year $\bar{x}$ in a sample of pennies for SRSs of several different sizes. Then, you will compare these distributions of the mean year with the population distribution.

1. Your teacher will provide a dotplot of the population distribution of penny years.
2. Have each member of the class take an SRS of 5 pennies from the population and record the year on each penny. Be sure to replace these coins in the container before the next student takes a sample. If your class has fewer than 25 students, have each person take two samples.
3. Calculate the mean year $\bar{x}$ of the 5 pennies in your sample.
4. Make a class dotplot of the sample mean years for SRSs of size 5 using the same scale as you did for the population distribution. Use $\bar{x}$'s instead of dots when making the graph.
5. Repeat the process in Steps 2 to 4 for samples of size 25. Use the same scale for your dotplot and place it beside the graph for samples of size 5.
6. Compare the population distribution with the two approximate sampling distributions of $\bar{x}$. What do you notice about shape, center, and spread as the sample size increases?

**TRM** **Section 7.3 Alternate Examples**

You can find the Alternate Examples for this section in Microsoft Word format by clicking on the link in the TE-book, opening the TRCD, or logging onto the book's companion Web site www.whfreeman.com/tps5e (teacher log-in required).

# EXAMPLE Making Money

### *A first look at the sampling distribution of $\bar{x}$*

Figure 7.17(a) is a histogram of the earnings of a population of 61,742 households that had earned income greater than zero in a recent year.[7]

As we expect, the distribution of earned incomes is strongly skewed to the right and very spread out. The right tail of the distribution is even longer than the histogram shows because there are too few high incomes for their bars to be visible on this scale. We cut off the earnings scale at $400,000 to save space. The mean earnings for these 61,742 households was $\mu$ = $69,750.

Take an SRS of 100 households. The mean earnings in this sample is $\bar{x}$ = $66,807. That's less than the mean of the population. Take another SRS of size 100. The mean for this sample is $\bar{x}$ = $70,820. That's higher than the mean of the population. What would happen if we did this many times? Figure 7.17(b) is a histogram of the mean earnings for 500 samples, each of size $n = 100$. The scales in Figures 7.17(a) and 7.17(b) are the same, for easy comparison. Although the distribution of individual earnings is skewed and very spread out, the distribution of sample means is roughly symmetric and much less spread out. Both distributions are centered at $\mu$ = $69,750.

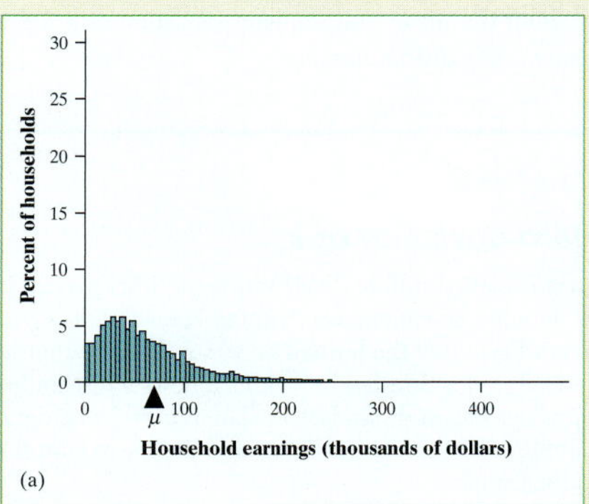


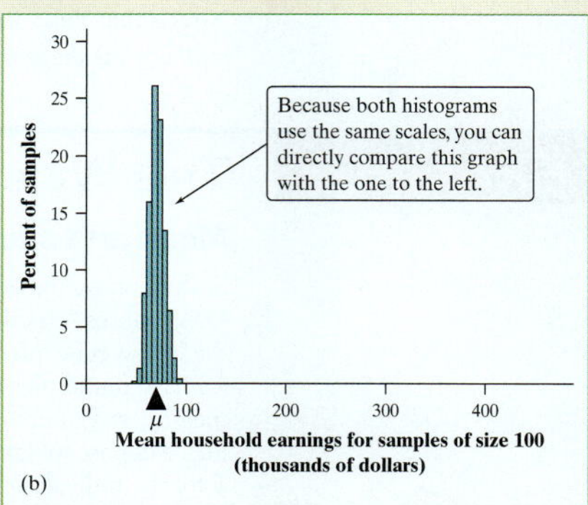


**FIGURE 7.17** (a) The distribution of earned income in a population of 61,472 households. (b) The distribution of the mean earnings $\bar{x}$ for 500 SRSs of $n = 100$ households from this population.

This example illustrates an important fact that we will make precise in this section: averages are less variable than individual observations.

## The Sampling Distribution of $\bar{x}$: Mean and Standard Deviation

Figure 7.17 suggests that when we choose many SRSs from a population, the sampling distribution of the sample mean is centered at the population mean $\mu$ and is less spread out than the population distribution. Here are the facts.

**ALTERNATE EXAMPLE**

### *Stealing bases*

The histogram below shows the distribution of stolen bases for the 1341 Major League Baseball players who had at least one plate appearance in a recent season. The right tail of the histogram actually extends to 70, but we cut off the scale at 20 to be able to focus on the majority of the observations.

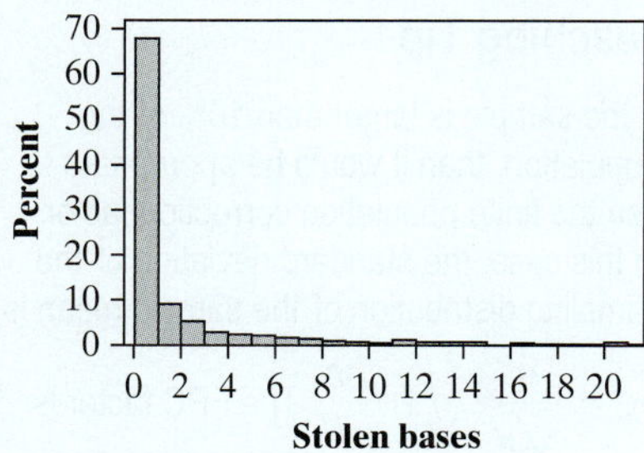


Here is a histogram showing the distribution of the sample mean number of stolen bases for 100 SRSs of size $n = 45$. It is graphed on the same scale to make it easier to see the difference in variability.

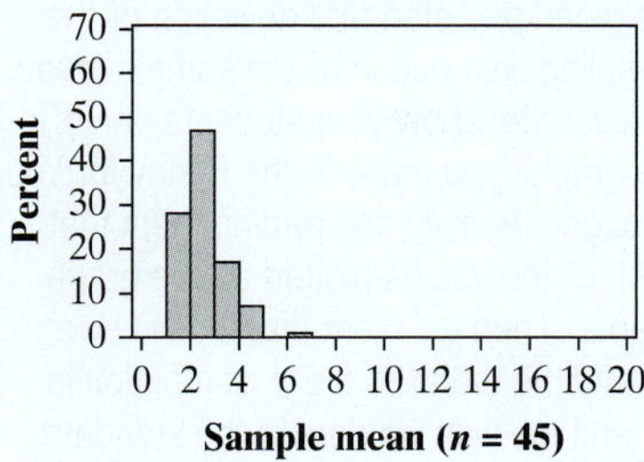


In addition to being much less spread out, the distribution of $\bar{x}$ is also more symmetric (less skewed) than the population distribution. However, the center of each distribution is about the same.

### AP® Exam Formula Sheet

Let your students know that the formulas for the mean and standard deviation of the sampling distribution of a sample mean are included on the formula sheet that is given to students on both sections of the AP® exam.

### Teaching Tip

If the sample is larger than 10% of the population, then it would be appropriate to use the finite population correction factor. In this case, the standard deviation of the sampling distribution of the sample mean is $\sigma_{\bar{x}} = \frac{\sigma}{\sqrt{n}} \cdot \sqrt{1 - \frac{n}{N}}$. The FPC factor is not covered on the AP® exam, however.

### Teaching Tip

Students are sometimes confused by the symbols $\mu_{\bar{x}}$ and $\sigma_{\bar{x}}$. These are simply the mean and standard deviation of the sampling distribution of the sample mean. To illustrate, show your students one of the graphs you made in the Penny for Your Thoughts Activity and remind them that each of the values plotted on the graph is an $\bar{x}$. Then tell them that the way we denote the mean of the $\bar{x}$ distribution is $\mu_{\bar{x}}$ and the way we denote the standard deviation of the $\bar{x}$ distribution is $\sigma_{\bar{x}}$.

### AP® Exam Common Error

Many students lose credit for using inappropriate symbols or formulas on the AP® exam. The best way to avoid losing credit is to know exactly what each of the symbols in the AP® Exam Tip mean. When in doubt, however, leave them out. Students will still get credit for showing work if they start with numbers substituted into a formula and don't risk using an incorrect symbol.

**MEAN AND STANDARD DEVIATION OF THE SAMPLING DISTRIBUTION OF $\bar{x}$**

Suppose that $\bar{x}$ is the mean of an SRS of size $n$ drawn from a large population with mean $\mu$ and standard deviation $\sigma$. Then:

- The **mean** of the sampling distribution of $\bar{x}$ is $\mu_{\bar{x}} = \mu$.
- The **standard deviation** of the sampling distribution of $\bar{x}$ is

$$\sigma_{\bar{x}} = \frac{\sigma}{\sqrt{n}}$$

as long as the *10% condition* is satisfied: $n \leq \frac{1}{10}N$.

**AP® EXAM TIP** Notation matters. The symbols $\hat{p}$, $\bar{x}$, $p$, $\mu$, $\sigma$, $\mu_{\hat{p}}$, $\sigma_{\hat{p}}$, $\mu_{\bar{x}}$, and $\sigma_{\bar{x}}$ all have specific and different meanings. Either use notation correctly—or don't use it at all. You can expect to lose credit if you use incorrect notation.

The behavior of $\bar{x}$ in repeated samples is much like that of the sample proportion $\hat{p}$:

- The sample mean $\bar{x}$ is an *unbiased estimator* of the population mean $\mu$.
- The values of $\bar{x}$ are less spread out for larger samples. Their standard deviation decreases at the rate $\sqrt{n}$, so you must take a sample four times as large to cut the standard deviation of the distribution of $\bar{x}$ in half.
- You should use the formula $\sigma/\sqrt{n}$ for the standard deviation of $\bar{x}$ only when the population is at least 10 times as large as the sample (the *10% condition*).

Notice that these facts about the mean and standard deviation of $\bar{x}$ are true *no matter what shape the population distribution has.*

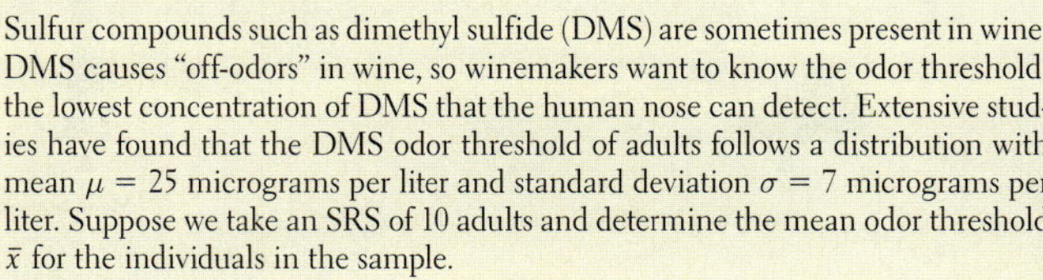

## EXAMPLE This Wine Stinks

### *Mean and standard deviation of $\bar{x}$*

Sulfur compounds such as dimethyl sulfide (DMS) are sometimes present in wine. DMS causes "off-odors" in wine, so winemakers want to know the odor threshold, the lowest concentration of DMS that the human nose can detect. Extensive studies have found that the DMS odor threshold of adults follows a distribution with mean $\mu = 25$ micrograms per liter and standard deviation $\sigma = 7$ micrograms per liter. Suppose we take an SRS of 10 adults and determine the mean odor threshold $\bar{x}$ for the individuals in the sample.

PROBLEM:

(a) What is the mean of the sampling distribution of $\bar{x}$? Explain.

(b) What is the standard deviation of the sampling distribution of $\bar{x}$? Check that the 10% condition is met.

SOLUTION:

(a) Because $\bar{x}$ is an unbiased estimator of $\mu$, $\mu_{\bar{x}} = \mu = 25$ micrograms per liter.

(b) The standard deviation is $\sigma_{\bar{x}} = \frac{\sigma}{\sqrt{n}} = \frac{7}{\sqrt{10}} = 2.214$ because there are at least $10(10) = 100$ adults in the population.

**For Practice** *Try Exercise*  49

### ALTERNATE EXAMPLE

#### *Movies*

Suppose that the number of movies viewed in the last year by high school students has an average of 19.3 with a standard deviation of 15.8. Suppose we take an SRS of 100 high school students and calculate the mean number of movies viewed by the members of the sample.

PROBLEM:

**(a)** What is the mean of the sampling distribution of $\bar{x}$? Explain.

**(b)** What is the standard deviation of the sampling distribution of $\bar{x}$? Check that the 10% condition is met.

SOLUTION:

**(a)** Because $\bar{x}$ is an unbiased estimator of $\mu$, $\mu_{\bar{x}} = \mu = 19.3$ movies.

**(b)** $\sigma_{\bar{x}} = \frac{\sigma}{\sqrt{n}} = \frac{15.8}{\sqrt{100}} = 1.58$ movies. The 10% condition is met because there are more than $10(100) = 1000$ high school students.

**Can we confirm the formulas for the mean and standard deviation of $\bar{x}$?** Choose an SRS of size $n$ from a population, and measure a variable $X$ on each individual in the sample. Call the individual measurements $X_1, X_2, \ldots, X_n$. If the population is large relative to the sample, we can think of these $X_i$'s as independent random variables, each with mean $\mu$ and standard deviation $\sigma$. Because

$$\bar{x} = \frac{1}{n}(X_1 + X_2 + \cdots + X_n)$$

we can use the rules for random variables from Chapter 6 to find the mean and standard deviation of $\bar{x}$. If we let $T = X_1 + X_2 + \cdots + X_n$, then $\bar{x} = \frac{1}{n}T$.

Using the addition rules for means and variances, we get

$$\mu_T = \mu_{X_1} + \mu_{X_2} + \cdots + \mu_{X_n} = \mu + \mu + \cdots + \mu = n\mu$$
$$\sigma_T^2 = \sigma_{X_1}^2 + \sigma_{X_2}^2 + \cdots + \sigma_{X_n}^2 = \sigma^2 + \sigma^2 + \cdots + \sigma^2 = n\sigma^2$$
$$\Rightarrow \sigma_T = \sqrt{n\sigma^2} = \sigma\sqrt{n}$$

Because $\bar{x}$ is just a constant multiple of the random variable $T$,

$$\mu_{\bar{x}} = \frac{1}{n}\mu_T = \frac{1}{n}(n\mu) = \mu$$

$$\sigma_{\bar{x}} = \frac{1}{n}\sigma_T = \frac{1}{n}(\sigma\sqrt{n}) = \sigma\sqrt{\frac{n}{n^2}} = \sigma\sqrt{\frac{1}{n}} = \sigma\frac{1}{\sqrt{n}} = \frac{\sigma}{\sqrt{n}}$$

## Sampling from a Normal Population

We have described the mean and standard deviation of the sampling distribution of a sample mean $\bar{x}$ but not its shape. That's because the shape of the distribution of $\bar{x}$ depends on the shape of the population distribution. In one important case, there is a simple relationship between the two distributions. The following Activity shows what we mean.

### ACTIVITY | Exploring the Sampling Distribution of $\bar{x}$ for a Normal Population

**MATERIALS:** Computer with Internet access—one for the class or one per pair of students

Professor David Lane of Rice University has developed a wonderful applet for investigating the sampling distribution of $\bar{x}$. It's dynamic, and it's fun to play with. In this Activity, you'll use Professor Lane's applet to explore the shape of the sampling distribution when the population is Normally distributed.

1. Search for "online statbook sampling distributions applet" and go to the Web site. When the BEGIN button appears on the left side of the screen, click on it. You will then see a yellow page entitled "Sampling Distributions" like the one in the following figure.

### Teaching Tip

Students are expected to recognize that problems involving totals can also be solved with means and vice versa. For example, to find the probability that the total weight of 4 men is greater than 800 pounds, you can find the probability that the mean weight of the 4 men is greater than 200 pounds. However, students are not expected to derive the formulas for the mean and standard deviation of the sampling distribution of the sample mean as shown in this Think About It.

### Activity Overview

You can watch an overview of this Activity (for teachers) by clicking on the link in the TE-book, opening the TRCD, or logging onto the book's companion Web site www.whfreeman.com/tps5e (teacher log-in required).

### Teaching Tip

When you do this Activity, make sure to emphasize the distinctions between the first three graphs shown on the following page. The graph on top shows the distribution of the population, the second graph shows the distribution of the sample data, and the third graph shows the approximate sampling distribution of the sample mean.

ALTERNATE ACTIVITY

### Sampling by hand from an approximately Normal population

There are advantages to starting with physical simulations before moving to the calculator/computer, which beginning students often regard as a "magic box." Here's a by-hand simulation Activity. Prepare a population of 100 identical index cards. Write numbers on the cards as follows:

| Write each of these numbers . . . | . . . on this many cards |
|---|---|
| 50 | 10 |
| 49, 51 | 9 |
| 48, 52 | 9 |
| 47, 53 | 8 |
| 46, 54 | 6 |
| 45, 55 | 5 |
| 44, 56 | 3 |
| 43, 57 | 2 |
| 42, 58 | 1 |
| 41, 59 | 1 |
| 40, 60 | 1 |

The 100 cards form a population. The distribution of values (the numbers on the cards) in this population is roughly Normal with mean $\mu = 50$ and standard deviation $\sigma = 4$. Make a dotplot of the 100 population values and find their mean (it is exactly $\mu = 50$ because of the symmetry). This is a population distribution, and its mean is a parameter. Now put the cards in a box and have a student take a random sample of size 4 by drawing 4 cards blindly and recording the numbers on them. Calculate the mean $\bar{x}$ of the 4 observations. This is a statistic. Return these cards to the box and shuffle the cards in the box thoroughly. Draw another random sample of size 4, record the numbers, and find $\bar{x}$. Repeat this as many times as is convenient, preferably about 100 times. Make a dotplot of the $\bar{x}$ -values and find their mean and standard deviation. This is an approximation of the sampling distribution of $\bar{x}$.

If you have time, repeat this simulation for samples of different sizes as well. Students should see that the center of the $\bar{x}$ distribution is always around 50, that the shape is always approximately Normal, and that the spread of the $\bar{x}$ distribution decreases as the sample size increases.

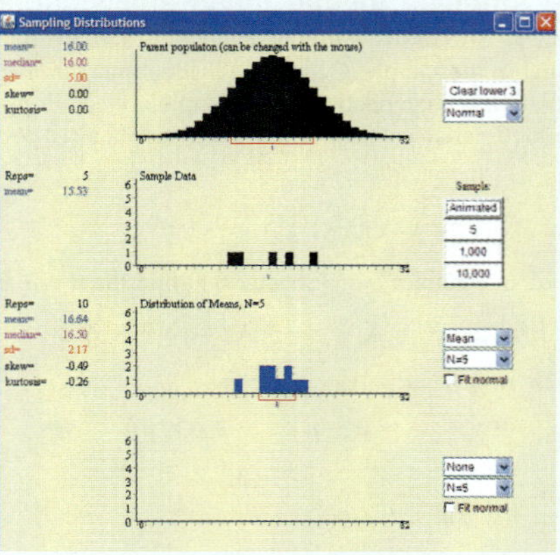


2. There are choices for the population distribution: Normal, uniform, skewed, and custom. The default is Normal. Click the "Animated" button. What happens? Click the button several more times. What do the black boxes represent? What is the blue square that drops down onto the plot below? What does the red horizontal band under the population histogram tell us?

Look at the left panel. Important numbers are displayed there. Did you notice that the colors of the numbers match up with the objects to the right? As you make things happen, the numbers change accordingly, like an automatic scorekeeper.

3. Click on "Clear lower 3" to start clean. Then click on the "1,000" button under "Sample:" repeatedly until you have simulated taking 10,000 SRSs of size $n = 5$ from the population (look for "Reps = 10000" on the left panel in black letters). Answer these questions:

- Does the approximate sampling distribution (blue bars) have a recognizable shape? Click the box next to "Fit normal."
- Compare the mean of the approximate sampling distribution with the mean of the population.
- How is the standard deviation of the approximate sampling distribution related to the standard deviation of the population?

4. Click "Clear lower 3." Use the drop-down menus to set up the bottom graph to display the mean for samples of size $n = 20$. Then sample 10,000 times. How do the two distributions of $\bar{x}$ compare: shape, center, and spread?

5. What have you learned about the shape of the sampling distribution of $\bar{x}$ when the population has a Normal shape?

As the previous Activity demonstrates, if the population distribution is Normal, then so is the sampling distribution of $\bar{x}$. *This is true no matter what the sample size is.*

**SAMPLING DISTRIBUTION OF A SAMPLE MEAN FROM A NORMAL POPULATION**

Suppose that a population is Normally distributed with mean $\mu$ and standard deviation $\sigma$. Then the sampling distribution of $\bar{x}$ has the Normal distribution with mean $\mu$ and standard deviation (provided the 10% condition is met) $\sigma/\sqrt{n}$.

We already knew the mean and standard deviation of the sampling distribution. All we have added is the Normal shape. Now we have enough information to calculate probabilities involving $\bar{x}$ when the population distribution is Normal.

## EXAMPLE Young Women's Heights

### *Finding probabilities involving the sample mean*

PROBLEM: The height of young women follows a Normal distribution with mean $\mu = 64.5$ inches and standard deviation $\sigma = 2.5$ inches.

(a) Find the probability that a randomly selected young woman is taller than 66.5 inches. Show your work.

(b) Find the probability that the mean height of an SRS of 10 young women exceeds 66.5 inches. Show your work.

SOLUTION:

(a) **Step 1: State the distribution and the values of interest.** Let $X$ be the height of a randomly selected young woman. The random variable $X$ follows a Normal distribution with $\mu = 64.5$ inches and $\sigma = 2.5$ inches. We want to find $P(X > 66.5)$. Figure 7.18 shows the distribution (purple curve) with the area of interest shaded and the mean, standard deviation, and boundary value labeled.

**Step 2: Perform calculations—show your work!** The standardized score for the boundary value is $z = \frac{66.5 - 64.5}{2.5} = 0.80$. *Using Table A*, $P(X > 66.5) = P(Z > 0.80) = 1 - 0.7881 = 0.2119$.

*Using technology:* The command `normalcdf(lower:66.5, upper:10000, μ:64.5, σ:2.5)` gives an area of 0.2119.

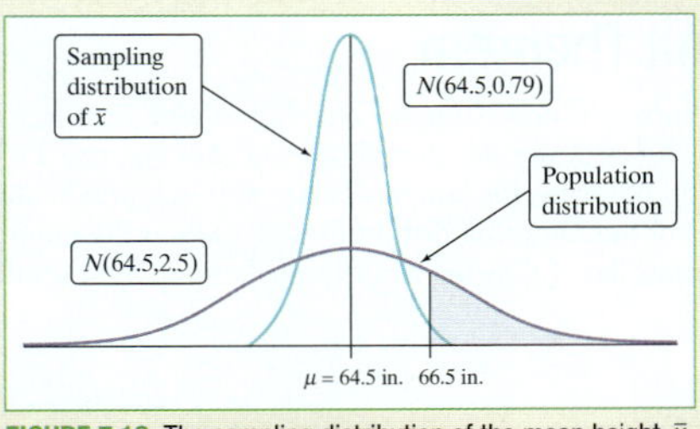


FIGURE 7.18 The sampling distribution of the mean height $\bar{x}$ for SRSs of 10 young women compared with the population distribution of young women's heights.

**Step 3: Answer the question.** The probability of choosing a young woman at random whose height exceeds 66.5 inches is about 0.21.

(b) **Step 1: State the distribution and the values of interest.** For an SRS of 10 young women, the sampling distribution of their sample mean height $\bar{x}$ will have mean $\mu_{\bar{x}} = \mu = 64.5$ inches. The 10% condition is met because there are at least 10(10) = 100 young women in the population. So the standard deviation is $\sigma_{\bar{x}} = \frac{\sigma}{\sqrt{n}} = \frac{2.5}{\sqrt{10}} = 0.79$. Because the population distribution is Normal, the values of $\bar{x}$ will follow an $N(64.5, 0.79)$ distribution. We want to find $P(\bar{x} > 66.5)$ inches. Figure 7.18 shows the distribution (blue curve) with the area of interest shaded and the mean, standard deviation, and boundary value labeled.

Figure 7.18 compares the population distribution and the sampling distribution of $\bar{x}$. It also shows the areas corresponding to the probabilities that we computed. You can see that it is much less likely for the average height of 10 randomly selected young women to exceed 66.5 inches than it is for the height of one randomly selected young woman to exceed 66.5 inches.

**Step 2: Perform calculations—show your work!** The standardized score for the boundary value is

$$z = \frac{66.5 - 64.5}{0.79} = 2.53.$$

*Using Table A.* $P(\bar{x} > 66.5) = P(Z > 2.53) = 1 - 0.9943 = 0.0057$.

*Using technology:* The command `normalcdf(lower:66.5, upper:10000, μ:64.5, σ:0.79)` gives an area of 0.0057.

**Step 3: Answer the question.** It is very unlikely (less than a 1% chance) that we would choose an SRS of 10 young women whose average height exceeds 66.5 inches.

**For Practice** *Try Exercise* 55

### AP® Exam Common Error

Students often forget to divide by the square root of the sample size when finding probabilities involving the sample mean. When reading the example on this page, make sure to discuss the distinction between parts (a) and (b) with this in mind. In part (a), we are asked to find a probability involving $X$, the height of *one* woman. In part (b), we are asked to find a probability involving $\bar{x}$, the *average* height of a sample of women.

### ALTERNATE EXAMPLE

#### *Buy me some peanuts and sample means*

PROBLEM: At the P. Nutty Peanut Company, dry-roasted, shelled peanuts are placed in jars by a machine. The distribution of weights in the jars is approximately Normal, with a mean of 16.1 ounces and a standard deviation of 0.15 ounces.

**(a)** Without doing any calculations, explain which outcome is more likely: randomly selecting a single jar and finding that the contents weigh less than 16 ounces, or randomly selecting 10 jars and finding that the average contents weigh less than 16 ounces.

**(b)** Find the probability that a randomly selected jar contains less than 16 ounces of peanuts.

**(c)** Find the probability that 10 randomly selected jars contain less than 16 ounces of peanuts, on average.

SOLUTION:

**(a)** Because averages are less variable than individual measurements, you would expect the sample mean of 10 jars to be closer, on average, to the true mean of 16.1 ounces. Thus, it is more likely that a single jar would weigh less than 16 ounces than for the average weight of 10 jars to be less than 16 ounces.

**(b)** **Step 1: State the distribution and values of interest.** Let $X$ = weight of the contents of a randomly selected jar of peanuts. $X$ has roughly the $N(16.1, 0.15)$ distribution and we want to find $P(X < 16)$.

**Step 2: Perform calculations—show your work!** The standardized score for the boundary value is $z = \frac{16 - 16.1}{0.15} = -0.67$ and $P(Z < -0.67) = 0.2514$. *Using technology*: The command `normalcdf(lower:-1000,upper:16,μ:16.1,σ:0.15)` gives an area of 0.2525.

**Step 3: Answer the question.** There is about a 25% chance that a single jar will contain less than 16 ounces of peanuts.

**(c)** **Step 1: State the distribution and values of interest.** Let $\bar{x}$ = average weight of the contents of 10 randomly selected jars of peanuts. $\mu_{\bar{x}} = \mu = 16.1$ and because 10 is less than 10% of all jars of peanuts produced, $\sigma_{\bar{x}} = \frac{\sigma}{\sqrt{n}} = \frac{0.15}{\sqrt{10}} = 0.047$. Thus, $\bar{x}$ has roughly the $N(16.1, 0.047)$ distribution and we want to find $P(\bar{x} < 16)$.

**Step 2: Perform calculations—show your work!** The standardized score for the boundary value is $z = \frac{16 - 16.1}{0.047} = -2.13$ and $P(Z < -2.13) = 0.0166$. *Using technology*: The command `normalcdf(lower:-1000,upper:16,μ:16.1,σ:0.047)` gives an area of 0.0167.

**Step 3: Answer the question.** There is about a 2% chance that a random sample of 10 jars will contain less than 16 ounces of peanuts, on average. This confirms our answer to part (a).

## Teaching Tip

Mention to your students that doctors may measure a patient's blood pressure more than once and use the average. This is to reduce the variability in the doctor's estimate of the patient's true blood pressure.

### ANSWERS TO CYU

1. $X$ = length of pregnancy follows an $N(266, 16)$ distribution and we want to find $P(X > 270)$. $z = \frac{270 - 266}{16} = 0.25$ and $P(Z > 0.25) = 0.4013$. *Using technology*: normalcdf(lower:270,upper:1000, $\mu$:266,$\sigma$:16) = 0.4013. There is a 0.4013 probability of selecting a woman whose pregnancy lasts for more than 270 days.

2. $\mu_{\bar{x}} = \mu = 266$ days

3. The sample of size 6 is less than 10% of all pregnant women, so $\sigma_{\bar{x}} = \frac{\sigma}{\sqrt{n}} = \frac{16}{\sqrt{6}} = 6.532$ days.

4. $\bar{x}$ follows an $N(266, 6.532)$ distribution and we want to find $P(\bar{x} > 270)$. $z = \frac{270 - 266}{6.532} = 0.61$ and $P(Z > 0.61) = 0.2709$. *Using technology*: normalcdf (lower:270,upper:1000,$\mu$:266, $\sigma$:6.532) = 0.2701. There is a 0.2701 probability of selecting a sample of 6 women whose mean pregnancy length exceeds 270 days.

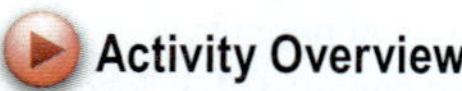

### Activity Overview

You can watch an overview of this Activity (for teachers) by clicking on the link in the TE-book, opening the TRCD, or logging onto the book's companion Web site www.whfreeman.com/tps5e (teacher log-in required).

## Teaching Tip

When you are doing this Activity with your students, ask them to describe clearly what each of the three graphs represents. The graph on top shows the distribution of the population, the second graph shows the distribution of the sample data, and the third graph shows the approximate sampling distribution of the sample mean.

The fact that averages of several observations are less variable than individual observations is important in many settings. For example, it is common practice to repeat a measurement several times and report the average of the results. Think of the results of $n$ repeated measurements as an SRS from the population of outcomes we would get if we repeated the measurement forever. The average of the $n$ results (the sample mean $\bar{x}$) is less variable than a single measurement.

### CHECK YOUR UNDERSTANDING

The length of human pregnancies from conception to birth varies according to a distribution that is approximately Normal with mean 266 days and standard deviation 16 days.

1. Find the probability that a randomly chosen pregnant woman has a pregnancy that lasts for more than 270 days. Show your work.

Suppose we choose an SRS of 6 pregnant women. Let $\bar{x}$ = the mean pregnancy length for the sample.

2. What is the mean of the sampling distribution of $\bar{x}$? Explain.
3. Compute the standard deviation of the sampling distribution of $\bar{x}$. Check that the 10% condition is met.
4. Find the probability that the mean pregnancy length for the women in the sample exceeds 270 days. Show your work.

## The Central Limit Theorem

Most population distributions are not Normal. The household incomes in Figure 7.17(a) on page 451, for example, are strongly skewed. Yet Figure 7.17(b) suggests that the distribution of means for samples of size 100 is approximately Normal. What is the shape of the sampling distribution of $\bar{x}$ when the population distribution isn't Normal? The following Activity sheds some light on this question.

### ACTIVITY Exploring the Sampling Distribution of $\bar{x}$ for a Non-Normal Population

**MATERIALS:** Computer with Internet access—one for the class or one per pair of students

Let's use the sampling distributions applet from the previous Activity (page 453) to investigate what happens when we start with a non-Normal population distribution.

1. Go to the Web site and launch the applet. Select "Skewed" population. Set the bottom two graphs to display the mean—one for samples of size 2 and the other for samples of size 5. Click the Animated button a few times to be sure you see what's happening. Then "Clear lower 3" and take 10,000 SRSs. Describe what you see.
2. Change the sample sizes to $n$ = 10 and $n$ =16 and repeat Step 1. What do you notice?

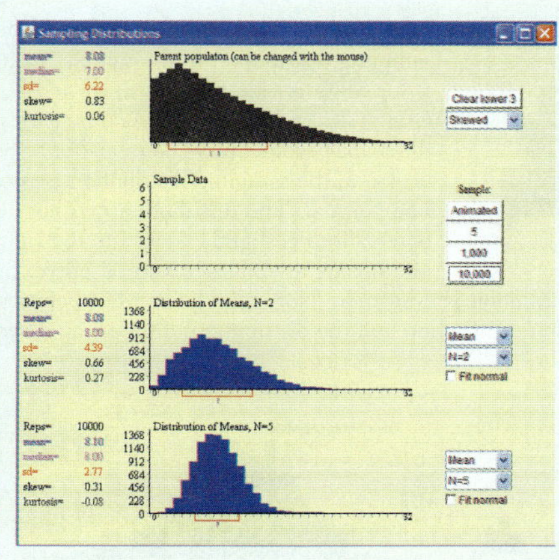

3. Now change the sample sizes to $n = 20$ and $n = 25$ and take 10,000 more samples. Did this confirm what you saw in Step 2?

4. Clear the page, and select "Custom" distribution. Click on a point on the population graph to insert a bar of that height. Or click on a point on the horizontal axis, and drag up to define a bar. Make a distribution that looks as strange as you can. (*Note:* You can shorten a bar or get rid of it completely by clicking on the top of the bar and dragging down to the axis.) Then repeat Steps 1 to 3 for your custom distribution. Cool, huh?

5. Summarize what you learned about the shape of the sampling distribution of $\bar{x}$.

It is a remarkable fact that as the sample size increases, the sampling distribution of $\bar{x}$ changes shape: it looks less like that of the population and more like a Normal distribution. When the sample size is large enough, the sampling distribution of $\bar{x}$ is very close to Normal. This is true no matter what shape the population distribution has, as long as the population has a finite standard deviation $\sigma$. This famous fact of probability theory is called the **central limit theorem** (sometimes abbreviated as CLT).

**DEFINITION: Central limit theorem (CLT)**

Draw an SRS of size $n$ from any population with mean $\mu$ and finite standard deviation $\sigma$. The **central limit theorem (CLT)** says that when $n$ is large, the sampling distribution of the sample mean $\bar{x}$ is approximately Normal.

How large a sample size $n$ is needed for the sampling distribution of $\bar{x}$ to be close to Normal depends on the population distribution. More observations are required if the shape of the population distribution is far from Normal. In that case, the sampling distribution of $\bar{x}$ will also be very non-Normal if the sample size is small. *Be sure you understand what the CLT does—and doesn't—say.*

## A Strange Population Distribution

### *The CLT in action*

We used the sampling distribution applet to create a population distribution with a very strange shape. See the graph at the top of the next page.

**ALTERNATE ACTIVITY**

***Sampling by hand from a skewed population***

Follow the directions for the Alternate Activity on page 454, but modify the population so it is skewed. For example, use 100 index cards with the following numbers:

| Write each of these numbers . . . | . . . on this many cards |
|---|---|
| 50 | 30 |
| 51 | 20 |
| 52 | 15 |
| 53 | 10 |
| 54 | 8 |
| 55 | 7 |
| 56 | 4 |
| 57 | 3 |
| 58 | 1 |
| 59 | 1 |
| 60 | 1 |

Make sure to use several different sample sizes, so students clearly see the change in shape as well as the change in spread.

**APPLET**

The *Sampling Pennies* applet at www.rossmanchance.com/applets allows students to investigate how changing the sample size affects the sampling distribution of the sample mean when sampling from a skewed population.

**AP® Exam Common Error**

Students often refer to the central limit theorem when describing how the variability of a sampling distribution decreases as the sample size increases. While this is true, it isn't what the central limit theorem says. The CLT is only about the *shape* of the sampling distribution of the sample mean, not the spread.

**Teaching Tip**

Some distributions do not have a finite standard deviation, which is why this requirement is mentioned. For example, both the mean and variance of the Cauchy distribution are undefined. If $U$ and $V$ are both standard Normal random variables, then the ratio $U/V$ has a standard Cauchy distribution, which happens to coincide with a $t$ distribution with 1 degree of freedom. http://en.wikipedia.org/wiki/Cauchy_distribution.

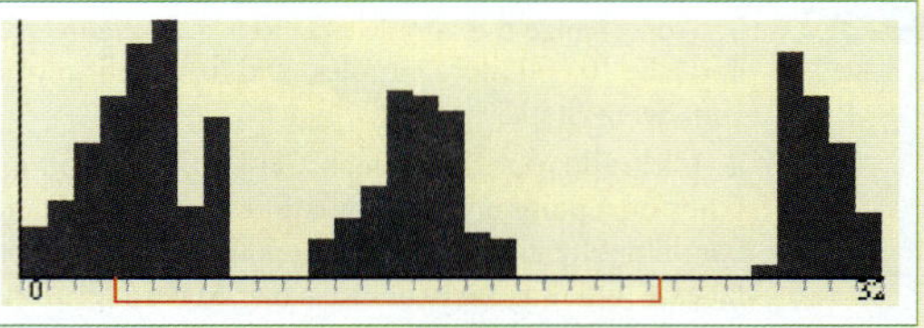


Figure 7.19 below shows the approximate sampling distribution of the sample mean $\bar{x}$ for SRSs of size (a) $n = 2$, (b) $n = 5$, (c) $n = 10$, and (d) $n = 25$. As $n$ increases, the shape becomes more Normal. For SRSs of size 2, the sampling distribution is very non-Normal. The distribution of $\bar{x}$ for 10 observations is slightly skewed to the right but already resembles a Normal curve. By $n = 25$, the sampling distribution is even more Normal. The contrast between the shapes of the population distribution and the distribution of the mean when $n = 10$ or 25 is striking.

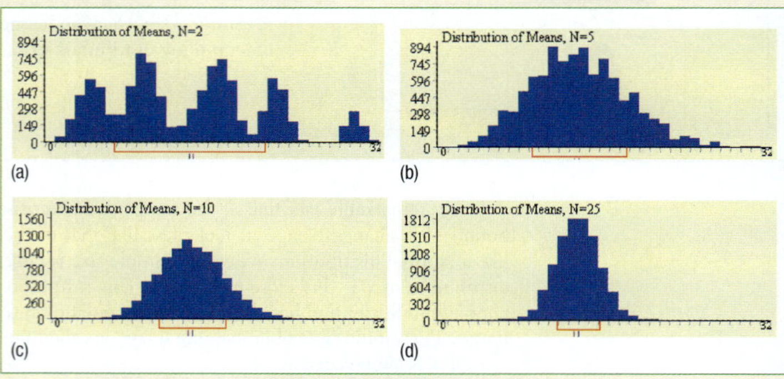


**FIGURE 7.19** The central limit theorem in action: the distribution of sample means $\bar{x}$ from a strongly non-Normal population becomes more Normal as the sample size increases. (a) The distribution of $\bar{x}$ for samples of size 2. (b) The distribution of $\bar{x}$ for samples of size 5. (c) The distribution of $\bar{x}$ for samples of size 10. (d) The distribution of $\bar{x}$ for samples of size 25.

As the previous example illustrates, even when the population distribution is very non-Normal, the sampling distribution of $\bar{x}$ often looks approximately Normal with sample sizes as small as $n = 25$. To be safe, we'll require that $n$ be at least 30 to invoke the CLT. With that issue settled, we can now state the *Normal/Large Sample condition* for sample means.

**NORMAL/LARGE SAMPLE CONDITION FOR SAMPLE MEANS**

- If the population distribution is Normal, then so is the sampling distribution of $\bar{x}$. This is true no matter what the sample size $n$ is.
- If the population distribution is not Normal, the central limit theorem tells us that the sampling distribution of $\bar{x}$ will be approximately Normal in most cases if $n \geq 30$.

The central limit theorem allows us to use Normal probability calculations to answer questions about sample means from many observations even when the population distribution is not Normal.

## Teaching Tip

We will be referring to the "Normal/Large Sample condition" quite often in the remainder of the book, so it is a good idea to begin using this title right away.

**ALTERNATE EXAMPLE**

### *Another strange population*

Here is another population distribution with a strange shape:

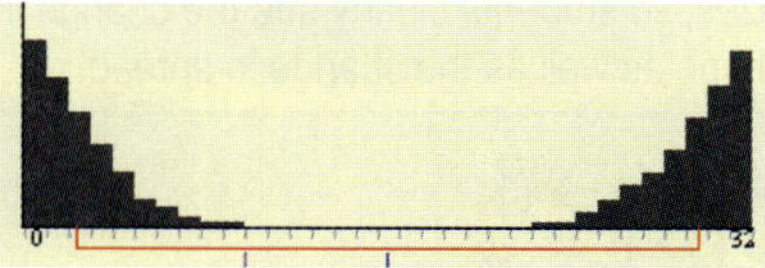


What do you think the sampling distribution of $\bar{x}$ will look like for samples of size 2? What about samples of size 5? Size 25?

Here are the results of 10,000 SRSs of each size. The first graph has three peaks, since there are only 4 basic outcomes for a sample: two small values, which gives a small mean; two large values, which gives a large mean; or one of each, which gives a mean in the middle. Because there are two ways to get one of each, the middle pile is roughly twice as big. By the time $n = 25$, the $\bar{x}$ distribution is roughly Normal!

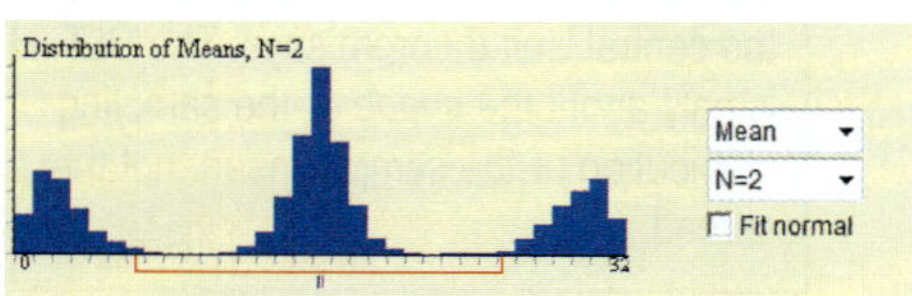


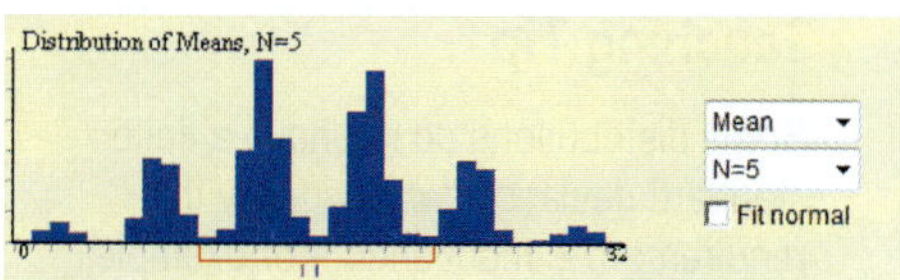


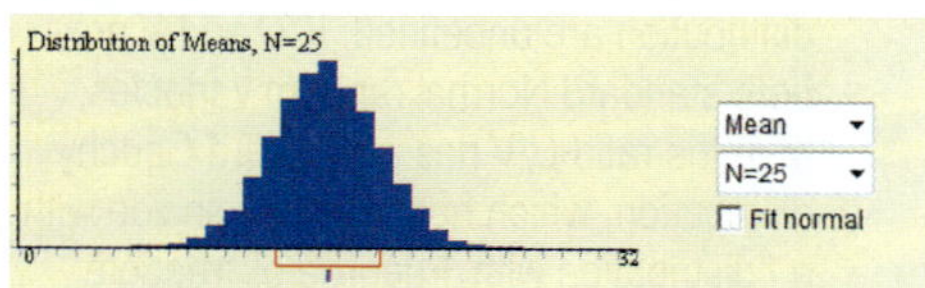

# EXAMPLE Servicing Air Conditioners

## Calculations using the CLT

Your company has a contract to perform preventive maintenance on thousands of air-conditioning units in a large city. Based on service records from the past year, the time (in hours) that a technician requires to complete the work follows a strongly right-skewed distribution with $\mu = 1$ hour and $\sigma = 1$ hour. In the coming week, your company will service an SRS of 70 air-conditioning units in the city. You plan to budget an average of 1.1 hours per unit for a technician to complete the work. Will this be enough?

**PROBLEM:** What is the probability that the average maintenance time $\bar{x}$ for 70 units exceeds 1.1 hours? Show your work.

**SOLUTION:**

**Step 1: State the distribution and the values of interest.** The sampling distribution of the sample mean time $\bar{x}$ spent working on 70 units has

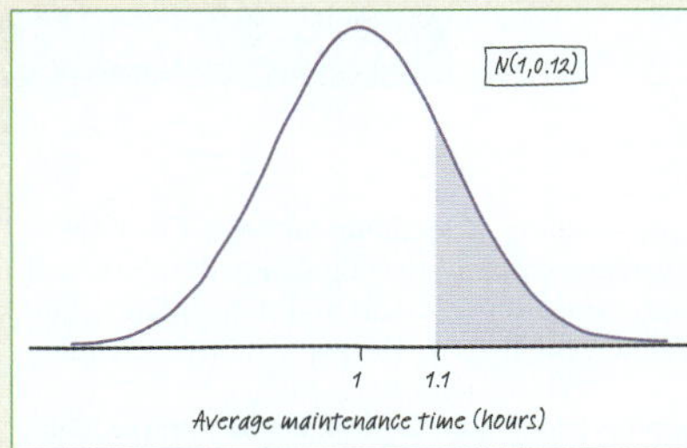


**FIGURE 7.20** The Normal approximation from the central limit theorem for the average time needed to maintain an air conditioner.

- mean $\mu_{\bar{x}} = \mu = 1$ hour
- standard deviation $\sigma_{\bar{x}} = \frac{\sigma}{\sqrt{70}} = \frac{1}{\sqrt{70}} = 0.12$ because the 10% condition is met (there are more than 10(70) = 700 air-conditioning units in the population)
- an approximately Normal shape because the Normal/Large Sample condition is met: $n = 70 \geq 30$

The distribution of $\bar{x}$ is therefore approximately $N(1, 0.12)$. We want to find $P(\bar{x} > 1.1)$. Figure 7.20 shows the Normal curve with the area of interest shaded and the mean, standard deviation, and boundary value labeled.

**Step 2: Perform calculations—show your work!** The standardized score for the boundary value is

$$z = \frac{1.1 - 1}{0.12} = 0.83$$

Using Table A, $P(\bar{x} > 1.1) = P(Z > 0.83) = 1 - 0.7967 = 0.2033$.

*Using technology:* The command `normalcdf(lower:1.1, upper:10000, μ:1, σ:0.12)` gives an area of 0.2023.

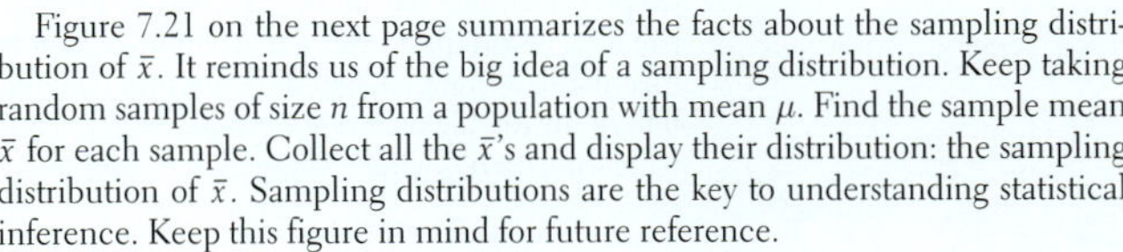

**Step 3: Answer the question.** If you budget 1.1 hours per unit, there is about a 20% chance that the technicians will not complete the work within the budgeted time. You will have to decide if this risk is worth taking or if you should schedule more time for the work.

**For Practice** *Try Exercise* 63

Figure 7.21 on the next page summarizes the facts about the sampling distribution of $\bar{x}$. It reminds us of the big idea of a sampling distribution. Keep taking random samples of size $n$ from a population with mean $\mu$. Find the sample mean $\bar{x}$ for each sample. Collect all the $\bar{x}$'s and display their distribution: the sampling distribution of $\bar{x}$. Sampling distributions are the key to understanding statistical inference. Keep this figure in mind for future reference.

## ALTERNATE EXAMPLE

### *Mean texts*

Suppose that the number of texts sent during a typical day by students at a large high school follows a right-skewed distribution with a mean of 45 and a standard deviation of 35.

**PROBLEM:** How likely is it that a random sample of 50 students will have sent more than a total of 2500 texts in the last day?

**SOLUTION:**

**Step 1: State the distribution and values of interest.** Let $\bar{x}$ = average number of texts for 50 randomly selected students. We want to find $P(\bar{x} > 2500/50) = P(\bar{x} > 50)$. $\mu_{\bar{x}} = \mu = 45$ and because 50 is less than 10% of all students at this school, $\sigma_{\bar{x}} = \frac{\sigma}{\sqrt{n}} = \frac{35}{\sqrt{50}} = 4.95$. Thus, $\bar{x}$ has approximately the $N(45, 4.95)$ distribution.

**Step 2: Perform calculations—show your work!** The standardized score for the boundary value is $z = \frac{50 - 45}{4.95} = 1.01$ and $P(Z > 1.01) = 0.1562$. *Using technology*: The command `normalcdf(lower:50,upper:1000,μ:45,σ:4.95)` gives an area of 0.1562.

**Step 3: Answer the question.** There is about a 16% chance that a random sample of 50 students will have sent over 2500 texts in the last day.

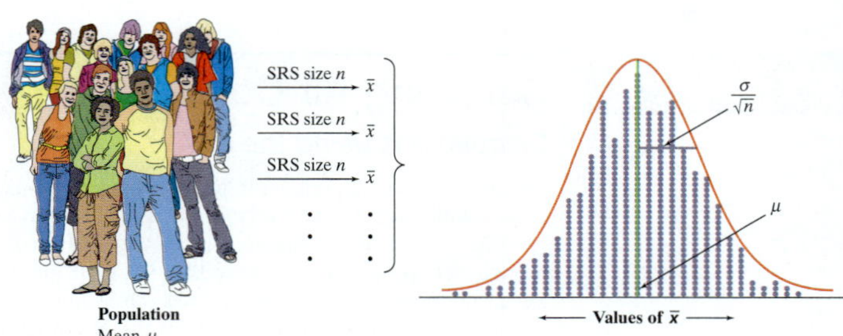


**FIGURE 7.21** The sampling distribution of a sample mean $\bar{x}$ has mean $\mu$ and standard deviation $\sigma/\sqrt{n}$. It has a Normal shape if the population distribution is Normal. If the population distribution isn't Normal, the sampling distribution of $\bar{x}$ is approximately Normal if the sample size is large enough.

case closed

## Building Better Batteries

Refer to the chapter-opening Case Study on page 421. Assuming the process is working properly, the population distribution of battery lifetimes has mean $\mu = 17$ hours and standard deviation $\sigma = 0.8$. We don't know the shape of the population distribution.

1. Make an appropriate graph to display the sample data. Describe what you see.
2. Assume that the battery production process is working properly. Describe the shape, center, and spread of the sampling distribution of $\bar{x}$ for random samples of 50 batteries. Justify your answers.

For the random sample of 50 batteries, the average lifetime was $\bar{x} = 16.718$ hours.

3. Find the probability of obtaining a random sample of 50 batteries with a mean lifetime of 16.718 hours or less if the production process is working properly. Show your work. Based on your answer, do you believe that the process is working properly? Why or why not?

The plant manager also wants to know what proportion $p$ of all the batteries produced that day lasted less than 16.5 hours, which he has declared "unsuitable." From past experience, about 27% of batteries made at the plant are unsuitable. If the manager does not find convincing evidence that the proportion of unsuitable batteries $p$ produced that day is greater than 0.27, the whole batch of batteries will be shipped to customers.

### Teaching Tip

In the figure on this page, remind students that each dot represents the value of $\bar{x}$ from a particular SRS of size $n$ from the population.

### Answers to Case Closed

**1.** A histogram is shown below. The distribution is roughly symmetric with a mean of 16.718 hours and a standard deviation of 0.664 hours. There don't appear to be any outliers.

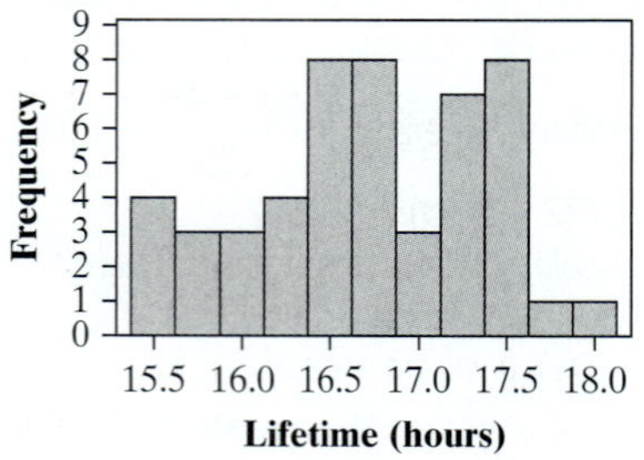


**2.** The shape of the sampling distribution of the sample mean is approximately Normal because the sample size is large ($n = 50 \geq 30$). The mean of the sampling distribution is $\mu_{\bar{x}} = \mu = 17$. Because $n = 50$ is less than 10% of all batteries produced, $\sigma_{\bar{x}} = \frac{\sigma}{\sqrt{n}} = \frac{0.8}{\sqrt{50}} = 0.113$.

**3.** *Step 1: State the distribution and values of interest.* We want to find $P(\bar{x} \leq 16.718)$ using the $N(17, 0.113)$ distribution.

*Step 2: Perform calculations—show your work!* The standardized score for the boundary value is

$$z = \frac{16.718 - 17}{0.113} = -2.50$$ and

$P(Z \leq -2.50) = 0.0062$. *Using technology*: The command `normalcdf(lower:-1000,upper:16.718,μ:17,σ:0.113)` gives an area of 0.0063.

*Step 3: Answer the question.* There is less than a 1% chance that a random sample of 50 batteries would have a mean lifetime of at most 16.718 hours if the process is working correctly. Because this probability is so small, it isn't plausible that the process is working correctly.

**4.** The shape of the sampling distribution of the sample proportion is approximately Normal because $np = 50(0.27) = 13.5 \geq 10$ and $n(1 - p) = 50(0.73) = 36.5 \geq 10$. The mean of the sampling distribution is $\mu_{\hat{p}} = p = 0.27$. Because $n = 50$ is less than 10% of all batteries produced,

$$\sigma_{\hat{p}} = \sqrt{\frac{p(1-p)}{n}} = \sqrt{\frac{0.27(0.73)}{50}} = 0.0628.$$

**5.** *Step 1: State the distribution and values of interest.* We want to find $P(\hat{p} \geq 0.32)$ using the $N(0.27, 0.0628)$ distribution.

*Step 2: Perform calculations—show your work!* The standardized score for the boundary value is $z = \frac{0.32 - 0.27}{0.0628} = 0.80$ and $P(Z \geq 0.80) = 0.2119$. *Using technology*: The command `normalcdf(lower:0.32,upper:1000,μ:0.27,σ:0.0628)` gives an area of 0.2130.

*Step 3: Answer the question.* There is about a 21% chance that in a random sample of 50 batteries, 32% or more would last less than 16.50 hours if the process is working correctly. Because this probability isn't very small, there isn't convincing evidence that the proportion of unsuitable batteries is greater than 0.27. It is plausible that the process is working correctly and we got a sample proportion of 0.32 by chance alone. Thus, the entire batch should be shipped to customers.

4. Assume that the actual proportion of unsuitable batteries produced that day is $p = 0.27$. Describe the shape, center, and spread of the sampling distribution of $\hat{p}$ for random samples of 50 batteries. Justify your answers.

For the random sample of 50 batteries, the sample proportion with lifetimes less than 16.5 hours was $\hat{p} = 0.32$.

5. Find the probability of obtaining a random sample of 50 batteries in which 32% or more of the batteries are unsuitable if $p = 0.27$. Show your work. Based on your answer, should the entire batch of batteries be shipped to customers? Why or why not?

## Section 7.3 Summary

- When we want information about the population mean $\mu$ for some variable, we often take an SRS and use the sample mean $\bar{x}$ to estimate the unknown parameter $\mu$. The **sampling distribution of** $\bar{x}$ describes how the statistic $\bar{x}$ varies in *all* possible samples of the same size from the population.
- The **mean** of the sampling distribution is $\mu$, so $\bar{x}$ is an unbiased estimator of $\mu$.
- The **standard deviation** of the sampling distribution of $\bar{x}$ is $\sigma/\sqrt{n}$ for an SRS of size $n$ if the population has standard deviation $\sigma$. That is, averages are less variable than individual observations. This formula can be used if the population is at least 10 times as large as the sample (*10% condition*).
- Choose an SRS of size $n$ from a population with mean $\mu$ and standard deviation $\sigma$. If the population distribution is Normal, then so is the sampling distribution of the sample mean $\bar{x}$. If the population distribution is not Normal, the **central limit theorem (CLT)** states that when $n$ is large, the sampling distribution of $\bar{x}$ is approximately Normal.
- We can use a Normal distribution to calculate approximate probabilities for events involving $\bar{x}$ whenever the *Normal/Large Sample condition* is met:
  - If the population distribution is Normal, so is the sampling distribution of $\bar{x}$.
  - If $n \geq 30$, the CLT tells us that the sampling distribution of $\bar{x}$ will be approximately Normal in most cases.

## Section 7.3 Exercises

49. pg 452 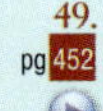 **Songs on an iPod** David's iPod has about 10,000 songs. The distribution of the play times for these songs is heavily skewed to the right with a mean of 225 seconds and a standard deviation of 60 seconds. Suppose we choose an SRS of 10 songs from this population and calculate the mean play time $\bar{x}$ of these songs. What are the mean and the standard deviation of the sampling distribution of $\bar{x}$? Explain.

**TRM Section 7.3 Quizzes**

There are three different quizzes available for this section in the Teacher's Resource Materials. Versions A and B are roughly parallel. Find them by clicking on the link in the TE-book, opening the TRCD, or logging onto the book's companion Web site www.whfreeman.com/tps5e (teacher log-in required).

**TRM Full Solutions to Section 7.3 Exercises**

### Answers to Section 7.3 Exercises

**7.49** $\mu_{\bar{x}} = \mu = 0.225$ seconds. Because the sample size (10) is less than 10% of the population of songs on David's iPod, $\sigma_{\bar{x}} = \frac{\sigma}{\sqrt{n}} = \frac{60}{\sqrt{10}} = 18.974$ seconds.

50. **Making auto parts** A grinding machine in an auto parts plant prepares axles with a target diameter $\mu = 40.125$ millimeters (mm). The machine has some variability, so the standard deviation of the diameters is $\sigma = 0.002$ mm. The machine operator inspects a random sample of 4 axles each hour for quality control purposes and records the sample mean diameter $\bar{x}$. Assuming that the process is working properly, what are the mean and standard deviation of the sampling distribution of $\bar{x}$? Explain.

51. **Songs on an iPod** Refer to Exercise 49. How many songs would you need to sample if you wanted the standard deviation of the sampling distribution of $\bar{x}$ to be 30 seconds? Justify your answer.

52. **Making auto parts** Refer to Exercise 50. How many axles would you need to sample if you wanted the standard deviation of the sampling distribution of $\bar{x}$ to be 0.0005 mm? Justify your answer.

53. **Larger sample** Suppose that the blood cholesterol level of all men aged 20 to 34 follows the Normal distribution with mean $\mu = 188$ milligrams per deciliter (mg/dl) and standard deviation $\sigma = 41$ mg/dl.

(a) Choose an SRS of 100 men from this population. Describe the sampling distribution of $\bar{x}$.

(b) Find the probability that $\bar{x}$ estimates $\mu$ within $\pm 3$ mg/dl. (This is the probability that $\bar{x}$ takes a value between 185 and 191 mg/dl.) Show your work.

(c) Choose an SRS of 1000 men from this population. Now what is the probability that $\bar{x}$ falls within $\pm 3$ mg/dl of $\mu$? Show your work. In what sense is the larger sample "better"?

54. **Dead battery?** A car company has found that the lifetime of its batteries varies from car to car according to a Normal distribution with mean $\mu = 48$ months and standard deviation $\sigma = 8.2$ months. The company installs a new brand of battery on an SRS of 8 cars.

(a) If the new brand has the same lifetime distribution as the previous type of battery, describe the sampling distribution of the mean lifetime $\bar{x}$.

(b) The average life of the batteries on these 8 cars turns out to be $\bar{x} = 42.2$ months. Find the probability that the sample mean lifetime is 42.2 months or less if the lifetime distribution is unchanged. What conclusion would you draw?

55. **Bottling cola** A bottling company uses a filling machine to fill plastic bottles with cola. The bottles are supposed to contain 300 milliliters (ml). In fact, the contents vary according to a Normal distribution with mean $\mu = 298$ ml and standard deviation $\sigma = 3$ ml.

pg 455

(a) What is the probability that a randomly selected bottle contains less than 295 ml? Show your work.

(b) What is the probability that the mean contents of six randomly selected bottles are less than 295 ml? Show your work.

56. **Cereal** A company's cereal boxes advertise 9.65 ounces of cereal. In fact, the amount of cereal in a randomly selected box follows a Normal distribution with mean $\mu = 9.70$ ounces and standard deviation $\sigma = 0.03$ ounces.

(a) What is the probability that a randomly selected box of the cereal contains less than 9.65 ounces of cereal? Show your work.

(b) Now take an SRS of 5 boxes. What is the probability that the mean amount of cereal $\bar{x}$ in these boxes is 9.65 ounces or less? Show your work.

57. **What does the CLT say?** Asked what the central limit theorem says, a student replies, "As you take larger and larger samples from a population, the histogram of the sample values looks more and more Normal." Is the student right? Explain your answer.

58. **What does the CLT say?** Asked what the central limit theorem says, a student replies, "As you take larger and larger samples from a population, the spread of the sampling distribution of the sample mean decreases." Is the student right? Explain your answer.

59. **Songs on an iPod** Refer to Exercise 49.

(a) Explain why you cannot safely calculate the probability that the mean play time $\bar{x}$ is more than 4 minutes (240 seconds) for an SRS of 10 songs.

(b) Suppose we take an SRS of 36 songs instead. Explain how the central limit theorem allows us to find the probability that the mean play time is more than 240 seconds. Then calculate this probability. Show your work.

60. **Lightning strikes** The number of lightning strikes on a square kilometer of open ground in a year has mean 6 and standard deviation 2.4. The National Lightning Detection Network (NLDN) uses automatic sensors to watch for lightning in a random sample of 10 one-square-kilometer plots of land.

(a) What are the mean and standard deviation of the sampling distribution of $\bar{x}$, the sample mean number of strikes per square kilometer?

(b) Explain why you cannot safely calculate the probability that $\bar{x} < 5$ based on a sample of size 10.

(c) Suppose the NLDN takes a random sample of $n = 50$ square kilometers instead. Explain how the central limit theorem allows us to find the probability that the mean number of lightning strikes per square kilometer is less than 5. Then calculate this probability. Show your work.

**7.50** $\mu_{\bar{x}} = 40.125$ mm. Assuming the sample size (4) is less than 10% of all axles produced this hour, $\sigma_{\bar{x}} = 0.002 / \sqrt{4} = 0.001$ mm.

**7.51** $30 = 60 / \sqrt{n} \rightarrow n = 4$

**7.52** $0.0005 = 0.002 / \sqrt{n} \rightarrow n = 16$

**7.53 (a)** Normal with $\mu_{\bar{x}} = 188$ mg/dl. Because the sample size (100) is less than 10% of all men aged 20 to 34, $\sigma_{\bar{x}} = 41 / \sqrt{100} = 4.1$ mg/dl.
**(b)** We want to find $P(185 \leq \bar{x} \leq 191)$. $P(-0.73 \leq Z \leq 0.73) = 0.5346$. *Using technology*: `normalcdf(lower:185, upper:191,μ:188,σ:4.1)` = 0.5357. There is a 0.5357 probability that $\bar{x}$ estimates $\mu$ within $\pm 3$ mg/dl.
**(c)** $\sigma_{\bar{x}} = 41 / \sqrt{1000} = 1.30$mg/dl. So $\bar{x}$ follows an $N(188, 1.30)$ distribution and we want to find $P(185 \leq \bar{x} \leq 191)$. $P(-2.31 \leq Z \leq 2.31) = 0.9792$. *Using technology*: `normalcdf(lower:185, upper:191,μ:188,σ:1.30)` = 0.9790. There is a 0.9790 probability that $\bar{x}$ estimates $\mu$ within $\pm 3$ mg/dl. The larger sample is better because it is more likely to produce a sample mean within 3 mg/dl of the population mean.

**7.54 (a)** Normal with $\mu_{\bar{x}} = 48$ months. Because the sample size (8) is less than 10% of all batteries, $\sigma_{\bar{x}} = 8.2 / \sqrt{8} = 2.899$ months. **(b)** We want to find $P(\bar{x} \leq 42.2)$. $P(Z \leq -2.00) = 0.0228$. *Using technology*: `normalcdf(lower:-1000,upper: 42.2,μ:48,σ:2.899)` = 0.0227. There is a 0.0227 probability that the sample mean lifetime is 42.2 months or less if the lifetime distribution is unchanged. Because this probability is very small, there is convincing evidence that the lifetime distribution has changed.

**7.55 (a)** Let $X$ = amount of cola in a randomly selected bottle. $X$ follows the $N(298, 3)$ distribution and we want to find $P(X < 295)$. $P(Z < -1) = 0.1587$. *Using technology*: `normalcdf(lower:-1000, upper:295,μ:298,σ:3)` = 0.1587. There is a 0.1587 probability that a randomly selected bottle contains less than 295 ml. **(b)** $\mu_{\bar{x}} = 298$ ml. Because 6 is less than 10% of all bottles produced, $\sigma_{\bar{x}} = 3 / \sqrt{6} = 1.2247$ ml. We want to find $P(\bar{x} < 295)$ using the $N(298, 1.2247)$ distribution. $P(Z < -2.45) = 0.0071$. *Using technology*: `normalcdf(lower: -1000,upper:295,μ:298,σ:1.2247)` = 0.0072. There is a 0.0072 probability that the mean contents of 6 randomly selected bottles is less than 295 ml.

**7.56 (a)** Let $X$ = amount of cereal in a randomly selected box. $X$ follows the $N(9.70, 0.03)$ distribution and we want to find $P(X < 9.65)$. $P(Z < -1.67) = 0.0475$. *Using technology*: `normalcdf(lower:-1000, upper:9.65,μ:9.70,σ:0.03)` = 0.0478. There is a 0.0478 probability that a randomly selected box contains less than 9.65 ounces.
**(b)** $\mu_{\bar{x}} = 9.70$ ounces. Because 5 is less than 10% of all boxes produced, $\sigma_{\bar{x}} = 0.03 / \sqrt{5} = 0.0134$ ounces. We want to find $P(\bar{x} < 9.65)$ using the $N(9.70, 0.0134)$ distribution. $P(Z < -3.73) \approx 0$. *Using technology*: `normalcdf(lower:-1000, upper:9.65,μ:9.70,σ:0.0134)` = 0.0001. There is a 0.0001 probability that the mean contents of 5 randomly selected boxes is less than 9.65 ounces.

**7.57** No. The histogram of the sample values will look like the population distribution. The CLT says that the histogram of the sampling distribution of the *sample mean* will look more and more Normal as the sample size increases.

**7.58** Although this is a correct statement about the standard deviation of the sampling distribution of $\bar{x}$, the CLT addresses the *shape* of the sampling distribution, not the spread.

**7.59 (a)** Because the distribution of the play times of the population of songs is heavily skewed to the right and $n = 10 < 30$. **(b)** Because $n = 36 \geq 30$, the CLT applies. $\mu_{\bar{x}} = 225$ seconds. Because 36 is less than 10% of all songs on David's iPod, $\sigma_{\bar{x}} = 60 / \sqrt{36} = 10$ seconds. We want to find $P(\bar{x} > 240)$ using the $N(225, 10)$ distribution. $P(Z > 1.50) = 0.0668$. *Using technology*: `normalcdf(lower:240,upper:1000, μ:225,σ:10)` = 0.0668. There is a 0.0668 probability that the mean play time is more than 240 seconds.

**61. Airline passengers get heavier** In response to the increasing weight of airline passengers, the Federal Aviation Administration (FAA) told airlines to assume that passengers average 190 pounds in the summer, including clothes and carry-on baggage. But passengers vary, and the FAA did not specify a standard deviation. A reasonable standard deviation is 35 pounds. Weights are not Normally distributed, especially when the population includes both men and women, but they are not very non-Normal. A commuter plane carries 30 passengers.

(a) Explain why you cannot calculate the probability that a randomly selected passenger weighs more than 200 pounds.

(b) Find the probability that the total weight of 30 randomly selected passengers exceeds 6000 pounds. Show your work. (*Hint:* To apply the central limit theorem, restate the problem in terms of the mean weight.)

**62. How many people in a car?** A study of rush-hour traffic in San Francisco counts the number of people in each car entering a freeway at a suburban interchange. Suppose that this count has mean 1.5 and standard deviation 0.75 in the population of all cars that enter at this interchange during rush hours.

(a) Could the exact distribution of the count be Normal? Why or why not?

(b) Traffic engineers estimate that the capacity of the interchange is 700 cars per hour. Find the probability that 700 randomly selected cars at this freeway entrance will carry more than 1075 people. Show your work. (*Hint:* Restate this event in terms of the mean number of people $\bar{x}$ per car.)

**63. More on insurance** pg 459 An insurance company claims that in the entire population of homeowners, the mean annual loss from fire is $\mu = \$250$ and the standard deviation of the loss is $\sigma = \$1000$. The distribution of losses is strongly right-skewed: many policies have \$0 loss, but a few have large losses. An auditor examines a random sample of 10,000 of the company's policies. If the company's claim is correct, what's the probability that the average loss from fire in the sample is no greater than \$275? Show your work.

**64. Bad carpet** The number of flaws per square yard in a type of carpet material varies with mean 1.6 flaws per square yard and standard deviation 1.2 flaws per square yard. The population distribution cannot be Normal, because a count takes only whole-number values. An inspector studies a random sample of 200 square yards of the material, records the number of flaws found in each square yard, and calculates $\bar{x}$, the mean number of flaws per square yard inspected. Find the probability that the mean number of flaws exceeds 1.8 per square yard. Show your work.

*Multiple choice: Select the best answer for Exercises 65 to 68.*

65. Scores on the mathematics part of the SAT exam in a recent year were roughly Normal with mean 515 and standard deviation 114. You choose an SRS of 100 students and average their SAT Math scores. Suppose that you do this many, many times. Which of the following are the mean and standard deviation of the sampling distribution of $\bar{x}$?

(a) Mean = 515, SD = 114

(b) Mean = 515, SD = $114/\sqrt{100}$

(c) Mean = 515/100, SD = 114/100

(d) Mean = 515/100, SD = $114/\sqrt{100}$

(e) Cannot be determined without knowing the 100 scores.

66. Why is it important to check the 10% condition before calculating probabilities involving $\bar{x}$?

(a) To reduce the variability of the sampling distribution of $\bar{x}$.

(b) To ensure that the distribution of $\bar{x}$ is approximately Normal.

(c) To ensure that we can generalize the results to a larger population.

(d) To ensure that $\bar{x}$ will be an unbiased estimator of $\mu$.

(e) To ensure that the observations in the sample are close to independent.

67. A newborn baby has extremely low birth weight (ELBW) if it weighs less than 1000 grams. A study of the health of such children in later years examined a random sample of 219 children. Their mean weight at birth was $\bar{x} = 810$ grams. This sample mean is an *unbiased estimator* of the mean weight $\mu$ in the population of all ELBW babies, which means that

(a) in all possible samples of size 219 from this population, the mean of the values of $\bar{x}$ will equal 810.

(b) in all possible samples of size 219 from this population, the mean of the values of $\bar{x}$ will equal $\mu$.

(c) as we take larger and larger samples from this population, $\bar{x}$ will get closer and closer to $\mu$.

(d) in all possible samples of size 219 from this population, the values of $\bar{x}$ will have a distribution that is close to Normal.

(e) the person measuring the children's weights does so without any error.

**7.60** **(a)** $\mu_{\bar{x}} = 6$ strikes/km$^2$ and because 10 is less than 10% of all one-square-kilometer plots of land, $\sigma_{\bar{x}} = 2.4/\sqrt{10} = 0.7589$ strikes/km$^2$. **(b)** Because we do not know the shape of the distribution of the number of lightning strikes and $n = 10 < 30$. **(c)** Because $n = 50 \geq 30$, the CLT applies. $\mu_{\bar{x}} = 6$ strikes/km$^2$. Because 50 is less than 10% of all 1 km$^2$ plots, $\sigma_{\bar{x}} = 2.4/\sqrt{50} = 0.3394$ strikes/km$^2$. We want to find $P(\bar{x} < 5)$ using the $N(6, 0.3394)$ distribution. $P(Z < -2.95) = 0.0016$. *Using technology*: `normalcdf(lower:-1000, upper:5,μ:6,σ:0.3394)` = 0.0016. There is a 0.0016 probability that the mean number of lightning strikes per kilometer is less than 5.

**7.61** **(a)** We do not know the shape of the distribution of passenger weights. **(b)** We want to find $P(\bar{x} > 6000/30) = P(\bar{x} > 200)$. Because the sample size is large ($n = 30 \geq 30$), the distribution of $\bar{x}$ is approximately Normal with $\mu_{\bar{x}} = 190$ pounds. Because $n = 30$ is less than 10% of all possible passengers, $\sigma_{\bar{x}} = 35/\sqrt{30} = 6.3901$ pounds. $P(Z > 1.56) = 0.0594$. *Using technology*: `normalcdf(lower:200,upper:1000,μ:190, σ:6.3901)` = 0.0588. There is a 0.0588 probability that the mean weight exceeds 200 pounds.

**7.62** **(a)** No. A count only takes on whole-number values, so it cannot be Normally distributed. **(b)** We want to find $P(\bar{x} > 1075/700) = P(\bar{x} > 1.5357)$. Because the sample size is large ($n = 700 \geq 30$), the distribution of $\bar{x}$ is approximately Normal with $\mu_{\bar{x}} = 1.5$ people. Because 700 is less than 10% of all cars entering this interchange, $\sigma_{\bar{x}} = 0.75/\sqrt{700} = 0.0283$ people. $P(Z > 1.26) = 0.1038$. *Using technology*: `normalcdf(lower:1.5357,upper: 1000,μ:1.5,σ:0.0283)` = 0.1036. There is a 0.1036 probability that the mean number of people exceeds 1.5357.

**7.63** Because the sample size is large ($n = 10{,}000 \geq 30$), the sampling distribution of $\bar{x}$ is approximately Normal. $\mu_{\bar{x}} = \$250$. Assuming 10,000 is less than 10% of all homeowners with fire insurance, $\sigma_{\bar{x}} = 1000/\sqrt{10{,}000} = \$10$. We want to find $P(\bar{x} \leq 275)$ using the $N(250, 10)$ distribution. $P(Z \leq 2.50) = 0.9938$. *Using technology*: `normalcdf(lower:-1000, upper:275,μ:250,σ:10)` = 0.9938. There is a 0.9938 probability that the mean annual loss from a sample of 10,000 policies is no greater than \$275.

**7.64** Because the sample size is large ($n = 200 \geq 30$), the sampling distribution of $\bar{x}$ is approximately Normal with $\mu_{\bar{x}} = 1.6$. Because 200 is less than 10% of all square yards of this carpet, $\sigma_{\bar{x}} = 1.2/\sqrt{200} = 0.085$. We want to find $P(\bar{x} > 1.8)$ using the $N(1.6, 0.085)$ distribution. $P(Z > 2.35) = 0.0094$. *Using technology*: `normalcdf(lower: 1.8,upper:1000,μ:1.6,σ:0.085)` = 0.0093. There is a 0.0093 probability of finding that the mean number of flaws is greater than 1.8 in a random sample of 200 square yards of carpet.

**7.65** b

**7.66** e

**7.67** b

68. The number of hours a lightbulb burns before failing varies from bulb to bulb. The population distribution of burnout times is strongly skewed to the right. The central limit theorem says that

(a) as we look at more and more bulbs, their average burnout time gets ever closer to the mean $\mu$ for all bulbs of this type.

(b) the average burnout time of a large number of bulbs has a sampling distribution with the same shape (strongly skewed) as the population distribution.

(c) the average burnout time of a large number of bulbs has a sampling distribution with similar shape but not as extreme (skewed, but not as strongly) as the population distribution.

(d) the average burnout time of a large number of bulbs has a sampling distribution that is close to Normal.

(e) the average burnout time of a large number of bulbs has a sampling distribution that is exactly Normal.

*Exercises 69 to 72 refer to the following setting.* In the language of government statistics, you are "in the labor force" if you are available for work and either working or actively seeking work. The unemployment rate is the proportion of the labor force (not of the entire population) who are unemployed. Here are data from the Current Population Survey for the civilian population aged 25 years and over in a recent year. The table entries are counts in thousands of people.

| Highest education | Total population | In labor force | Employed |
|---|---|---|---|
| Didn't finish high school | 27,669 | 12,470 | 11,408 |
| High school but no college | 59,860 | 37,834 | 35,857 |
| Less than bachelor's degree | 47,556 | 34,439 | 32,977 |
| College graduate | 51,582 | 40,390 | 39,293 |

69. **Unemployment** (1.1) Find the unemployment rate for people with each level of education. How does the unemployment rate change with education?

70. **Unemployment** (5.1) What is the probability that a randomly chosen person 25 years of age or older is in the labor force? Show your work.

71. **Unemployment** (5.3) If you know that a randomly chosen person 25 years of age or older is a college graduate, what is the probability that he or she is in the labor force? Show your work.

72. **Unemployment** (5.3) Are the events "in the labor force" and "college graduate" independent? Justify your answer.

## FRAPPY! Free Response AP® Problem, Yay!

The following problem is modeled after actual AP® Statistics exam free response questions. Your task is to generate a complete, concise response in 15 minutes.

*Directions: Show all your work. Indicate clearly the methods you use, because you will be scored on the correctness of your methods as well as on the accuracy and completeness of your results and explanations.*

The principal of a large high school is concerned about the number of absences for students at his school. To investigate, he prints a list showing the number of absences during the last month for each of the 2500 students at the school. For this population of students, the distribution of absences last month is skewed to the right with a mean of $\mu = 1.1$ and a standard deviation of $\sigma = 1.4$.

Suppose that a random sample of 50 students is selected from the list printed by the principal and the sample mean number of absences is calculated.

(a) What is the shape of the sampling distribution of the sample mean? Explain.

(b) What are the mean and standard deviation of the sampling distribution of the sample mean?

(c) What is the probability that the mean number of absences in a random sample of 50 students is less than 1?

(d) Because the population distribution is skewed, the principal is considering using the median number of absences last month instead of the mean number of absences to summarize the distribution. Describe how the principal could use a simulation to estimate the standard deviation of the sampling distribution of the sample median for samples of size 50.

After you finish, you can view two example solutions on the book's Web site (www.whfreeman.com/tps5e). Determine whether you think each solution is "complete," "substantial," "developing," or "minimal." If the solution is not complete, what improvements would you suggest to the student who wrote it? Finally, your teacher will provide you with a scoring rubric. Score your response and note what, if anything, you would do differently to improve your own score.

**7.68** d

**7.69** Didn't finish high school: $\frac{1062}{12,470} = 0.0852$; high school but no college: $\frac{1977}{37,834} = 0.0523$; less than a bachelor's degree: $\frac{1462}{34,439} = 0.0425$; college graduate: $\frac{1097}{40,390} = 0.0272$. The unemployment rate decreases with additional education.

**7.70** $\frac{125,133}{186,667} = 0.6704$

**7.71** $P(\text{in labor force} \mid \text{college graduate}) = \frac{40,390}{51,582} = 0.7830$

**7.72** The events are not independent, because the probability of being in the labor force (0.6704) does not equal the probability of being in the labor force given that the person is a college graduate (0.7830). College graduates are more likely to be in the labor force than other people.

**TRM FRAPPY!**

We present a model solution here. Please consult the Teacher's Resource Materials for sample student responses, a scoring rubric, and a black line master of the original question with space for students to write their responses.

**(a)** The sampling distribution of the sample mean will be approximately Normal, because the sample size is large ($50 \geq 30$).

**(b)** $\mu_{\bar{x}} = \mu = 1.1$ and because the sample size is less than 10% of all students ($50 < 10\%$ of 2500), $\sigma_{\bar{x}} = \frac{\sigma}{\sqrt{n}} = \frac{1.4}{\sqrt{50}} = 0.198$.

**(c)** **Step 1: State the distribution and values of interest.** $\bar{x}$ has roughly the $N(1.1, 0.198)$ distribution and we want to find $P(\bar{x} < 1)$.

**Step 2: Perform calculations—show your work!** The standardized score for the boundary value is $z = \frac{1 - 1.1}{0.198} = -0.51$ and $P(Z < -0.51) = 0.3050$.

*Using technology*: The command `normalcdf(lower:-1000,upper:1,μ:1.1,σ:0.198)` gives an area of 0.3068.

**Step 3: Answer the question.** There is about a 31% chance that the mean number of absences in a random sample of 50 students is less than 1.

**(d)** Take many, many random samples of size 50. For each sample, calculate the sample median. Finally, calculate the standard deviation of the distribution of sample medians.

# Chapter Review

### Section 7.1: What Is a Sampling Distribution?

In this section, you learned the "big ideas" of sampling distributions. The first big idea is the difference between a statistic and a parameter. A parameter is a number that describes some characteristic of a population. A statistic estimates the value of a parameter using a sample from the population. Making the distinction between a statistic and a parameter will be crucial throughout the rest of the course.

The second big idea is that statistics vary. For example, the mean weight in a sample of high school students is a variable that will change from sample to sample. This means that statistics have distributions, but parameters do not. The distribution of a statistic in all possible samples of the same size is called the sampling distribution of the statistic.

The third big idea is the distinction between the distribution of the population, the distribution of the sample, and the sampling distribution of a sample statistic. Reviewing the illustration on page 428 will help you understand the difference between these three distributions. When you are writing your answers, be sure to indicate which distribution you are referring to. Don't make ambiguous statements like "the distribution will become less variable."

The fourth big idea is how to describe a sampling distribution. To adequately describe a sampling distribution, you need to address shape, center, and spread. If the center (mean) of the sampling distribution is the same as the value of the parameter being estimated, then the statistic is called an unbiased estimator. An estimator is unbiased if it doesn't consistently under- or overestimate the parameter in many samples. Ideally, the spread of a sampling distribution will be very small, meaning that the statistic provides precise estimates of the parameter. Larger sample sizes result in sampling distributions with smaller spreads.

### Section 7.2: Sample Proportions

In this section, you learned about the shape, center, and spread of the sampling distribution of a sample proportion. When the Large Counts condition ($np \geq 10$ and $n(1 - p) \geq 10$) is met, the sampling distribution of $\hat{p}$ will be approximately Normal. The mean of the sampling distribution of $\hat{p}$ is $\mu_{\hat{p}} = p$, the population proportion. As a result, the sample proportion $\hat{p}$ is an unbiased estimator of the population proportion $p$. When the 10% condition $\left(n \leq \frac{1}{10}N\right)$ is met, the standard deviation of the sampling distribution of the sample proportion is $\sigma_{\hat{p}} = \sqrt{\frac{p(1-p)}{n}}$. This formula tells us that the variability of the distribution of $\hat{p}$ is smaller when the sample size is larger.

### Section 7.3: Sample Means

In this section, you learned about the shape, center, and spread of the sampling distribution of a sample mean. When the population is Normal, the sampling distribution of $\bar{x}$ will also be Normal for any sample size. When the population is not Normal and the sample size is small, the sampling distribution of $\bar{x}$ will resemble the population shape. However, the central limit theorem says that the sampling distribution of $\bar{x}$ will become approximately Normal for larger sample sizes (typically when $n \geq 30$), no matter what the population shape. When you are using a Normal distribution to calculate probabilities involving the sampling distribution of $\bar{x}$, make sure that the Normal/Large Sample condition is met.

The mean of the sampling distribution of $\bar{x}$ is $\mu_{\bar{x}} = \mu$, the population mean. As a result, the sample mean $\bar{x}$ is an unbiased estimator of the population mean $\mu$. When the 10% condition $\left(n \leq \frac{1}{10}N\right)$ is met, the standard deviation of the sampling distribution of the sample mean is $\sigma_{\bar{x}} = \frac{\sigma}{\sqrt{n}}$. This formula tells us that the variability of the distribution of $\bar{x}$ is smaller when the sample size is larger.

Finally, when you are using a Normal distribution to calculate probabilities involving the sampling distribution of $\hat{p}$ or $\bar{x}$, make sure that you (1) state the distribution and values of interest, (2) perform calculations—show your work, and (3) answer the question.

**TRM Chapter 7 Tests**

There are four different tests available for this section in the Teacher's Resource Materials. Versions A and B are roughly parallel and version D is cumulative. Find them by clicking on the link in the TE-book, opening the TRCD, or logging onto the book's companion Web site www.whfreeman.com/tps5e (teacher log-in required).

**TRM Learning Objectives Grid**

## What Did You Learn?

| Learning Objective | Section | Related Example on Page(s) | Relevant Chapter Review Exercise(s) |
|---|---|---|---|
| Distinguish between a parameter and a statistic. | 7.1 | 425 | R7.1 |
| Use the sampling distribution of a statistic to evaluate a claim about a parameter. | 7.1 | 427 | R7.5, R7.7 |
| Distinguish among the distribution of a population, the distribution of a sample, and the sampling distribution of a statistic. | 7.1 | Discussion on 428 | R7.2 |
| Determine whether or not a statistic is an unbiased estimator of a population parameter. | 7.1 | Discussion on 430–431; 435 | R7.3 |
| Describe the relationship between sample size and the variability of a statistic. | 7.1 | 432 | R7.3 |
| Find the mean and standard deviation of the sampling distribution of a sample proportion $\hat{p}$. Check the 10% condition before calculating $\sigma_{\hat{p}}$. | 7.2 | 445 | R7.4 |
| Determine if the sampling distribution of $\hat{p}$ is approximately Normal. | 7.2 | 445 | R7.4 |
| If appropriate, use a Normal distribution to calculate probabilities involving $\hat{p}$. | 7.2 | 445 | R7.4, R7.5 |
| Find the mean and standard deviation of the sampling distribution of a sample mean $\bar{x}$. Check the 10% condition before calculating $\sigma_{\bar{x}}$. | 7.3 | 452 | R7.6 |
| Explain how the shape of the sampling distribution of $\bar{x}$ is affected by the shape of the population distribution and the sample size. | 7.3 | 457 | R7.6, R7.7 |
| If appropriate, use a Normal distribution to calculate probabilities involving $\bar{x}$. | 7.3 | 455, 459 | R7.6, R7.7 |

# Chapter 7 Chapter Review Exercises

*These exercises are designed to help you review the important ideas and methods of the chapter.*

**R7.1 Bad eggs** Sale of eggs that are contaminated with salmonella can cause food poisoning in consumers. A large egg producer takes an SRS of 200 eggs from all the eggs shipped in one day. The laboratory reports that 9 of these eggs had salmonella contamination. Unknown to the producer, 3% of all eggs shipped had salmonella. Identify the population, the parameter, the sample, and the statistic.

*Exercises R7.2 and R7.3 refer to the following setting.* Researchers in Norway analyzed data on the birth weights of 400,000 newborns over a 6-year period. The distribution of birth weights is approximately Normal



**TRM Full Solutions to Chapter 7 Review Exercises**

## Answers to Chapter 7 Review Exercises

**R7.1** The population is the set of all eggs shipped in one day. The sample consists of the 200 eggs examined. The parameter is the proportion $p = 0.03$ of eggs shipped that day that had salmonella. The statistic is the proportion $\hat{p} = \dfrac{9}{200} = 0.045$ of eggs in the sample that had salmonella.

**R7.2 (a)** A sketch of the population distribution is given below.

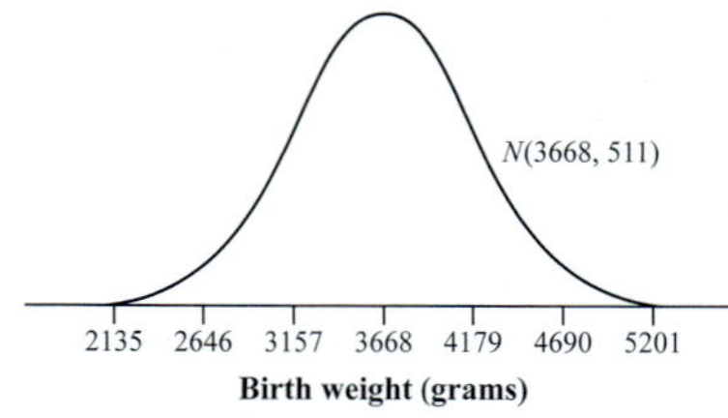


**(b)** Answers will vary. An example dotplot is given.

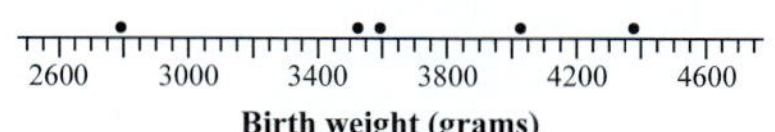


**(c)** The dot at 2750 represents one SRS of size 5 from this population, where the sample range was 2750 grams.

**R7.3 (a)** No, because sample range is always less than the actual range (3417). If it were unbiased, the distribution would be centered at 3417. **(b)** Take larger samples.

**R7.4 (a)** $\mu_{\hat{p}} = p = 0.15$ **(b)** Because the sample size of $n = 1540$ is less than 10% of the population of all adults, $\sigma_{\hat{p}} = \sqrt{\dfrac{0.15(0.85)}{1540}} = 0.0091$. **(c)** Yes, because $np = 1540(0.15) = 231$ and $n(1 - p) = 1540(0.85) = 1309$ are both at least 10. **(d)** We want to find $P(0.13 \le \hat{p} \le 0.17)$. $z = \dfrac{0.13 - 0.15}{0.0091} = -2.20$ and $z = \dfrac{0.17 - 0.15}{0.0091} = 2.20$. The desired probability is $P(-2.20 \le Z \le 2.20) = 0.9722$. *Using technology*: `normalcdf(lower:0.13, upper:0.17,μ:0.15,σ:0.0091)` = 0.9720. There is a 0.9720 probability of obtaining a sample in which between 13% and 17% are joggers.

**R7.5 (a)** $\mu_{\hat{p}} = p = 0.30$. Because 100 is less than 10% of the population of travelers, $\sigma_{\hat{p}} = \sqrt{\dfrac{0.30(0.70)}{100}} = 0.0458$.

with a mean of 3668 grams and a standard deviation of 511 grams.[8] In this population, the range (maximum – minimum) of birth weights is 3417 grams. We used Fathom software to take 500 SRSs of size $n = 5$ and calculate the range (maximum – minimum) for each sample. The dotplot below shows the results.

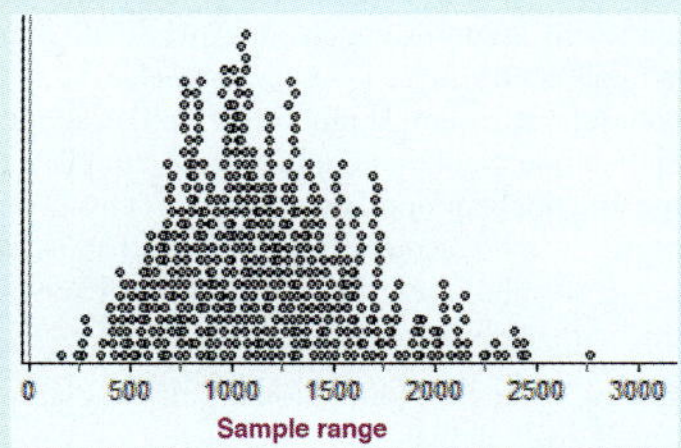


**R7.2 Birth weights**

(a) Sketch a graph that displays the distribution of birth weights for this population.

(b) Sketch a possible graph of the distribution of birth weights for an SRS of size 5.

(c) In the graph above, there is a dot at approximately 2750. Explain what this value represents.

**R7.3 Birth weights**

(a) Is the sample range an unbiased estimator of the population range? Give evidence from the graph above to support your answer.

(b) Explain how we could decrease the variability of the sampling distribution of the sample range.

**R7.4. Do you jog?** The Gallup Poll once asked a random sample of 1540 adults, "Do you happen to jog?" Suppose that the true proportion of all adults who jog is $p = 0.15$.

(a) What is the mean of the sampling distribution of $\hat{p}$? Justify your answer.

(b) Find the standard deviation of the sampling distribution of $\hat{p}$. Check that the 10% condition is met.

(c) Is the sampling distribution of $\hat{p}$ approximately Normal? Justify your answer.

(d) Find the probability that between 13% and 17% of a random sample of 1540 adults are joggers.

**R7.5 Bag check** Thousands of travelers pass through the airport in Guadalajara, Mexico, each day. Before leaving the airport, each passenger must pass through the Customs inspection area. Customs agents want to be sure that passengers do not bring illegal items into the country. But they do not have time to search every traveler's luggage. Instead, they require each person to press a button. Either a red or a green bulb lights up. If the red light shows, the passenger will be searched by Customs agents. A green light means "go ahead." Customs agents claim that the proportion of all travelers who will be stopped (red light) is 0.30, because the light has probability 0.30 of showing red on any push of the button. To test this claim, a concerned citizen watches a random sample of 100 travelers push the button. Only 20 get a red light.

(a) Assume that the Customs agents' claim is true. Find the probability that the proportion of travelers who get a red light is as small as or smaller than the result in this sample. Show your work.

(b) Based on your results in (a), do you believe the Customs agents' claim? Explain.

**R7.6 IQ tests** The Wechsler Adult Intelligence Scale (WAIS) is a common "IQ test" for adults. The distribution of WAIS scores for persons over 16 years of age is approximately Normal with mean 100 and standard deviation 15.

(a) What is the probability that a randomly chosen individual has a WAIS score of 105 or higher? Show your work.

(b) Find the mean and standard deviation of the sampling distribution of the average WAIS score $\bar{x}$ for an SRS of 60 people.

(c) What is the probability that the average WAIS score of an SRS of 60 people is 105 or higher? Show your work.

(d) Would your answers to any of parts (a), (b), or (c) be affected if the distribution of WAIS scores in the adult population were distinctly non-Normal? Explain.

**R7.7 Detecting gypsy moths** The gypsy moth is a serious threat to oak and aspen trees. A state agriculture department places traps throughout the state to detect the moths. Each month, an SRS of 50 traps is inspected, the number of moths in each trap is recorded, and the mean number of moths is calculated. Based on years of data, the distribution of moth counts is discrete and strongly skewed, with a mean of 0.5 and a standard deviation of 0.7.

(a) Explain why it is reasonable to use a Normal distribution to approximate the sampling distribution of $\bar{x}$ for SRSs of size 50.

(b) Estimate the probability that the mean number of moths in a sample of size 50 is greater than or equal to 0.6.

(c) In a recent month, the mean number of moths in an SRS of size 50 was 0.6. Based on this result, should the state agricultural department be worried that the moth population is getting larger in their state? Explain.

Because $np = 100(0.30) = 30$ and $n(1 - p) = 100(0.70) = 70$ are both at least 10, the sampling distribution of $\hat{p}$ can be approximated by a Normal distribution. We want to find $P(\hat{p} \leq 0.20)$. $z = \frac{0.20 - 0.30}{0.0458} = -2.18$ and $P(Z \leq -2.18) = 0.0146$. *Using technology*: `normalcdf(lower:-1000,upper:0.20,` `μ:0.30,σ:0.0458)` $= 0.0145$. There is a 0.0145 probability that 20% or fewer of the travelers get a red light. **(b)** Because this is a small probability, there is convincing evidence against the agents' claim—it isn't plausible to get a sample proportion of travelers with a red light this small by chance alone.

**R7.6 (a)** $X$ = WAIS score for a randomly selected individual follows an $N(100, 15)$ distribution and we want to find $P(X \geq 105)$. $z = \frac{105 - 100}{15} = 0.33$ and $P(Z \geq 0.33) = 0.3707$. *Using technology*: `normalcdf(lower:105,upper:1000,` `μ:100,σ:15)` $= 0.3694$. There is a 0.3694 probability of selecting an individual with a WAIS score of at least 105. **(b)** $\mu_{\bar{x}} = \mu = 100$. Because the sample of size 60 is less than 10% of all adults, $\sigma_{\bar{x}} = \frac{\sigma}{\sqrt{n}} = \frac{15}{\sqrt{60}} = 1.9365$. **(c)** $\bar{x}$ follows an $N(100, 1.9365)$ distribution and we want to find $P(\bar{x} \geq 105)$. $z = \frac{105 - 100}{1.9365} = 2.58$ and $P(Z \geq 2.58) = 0.0049$. *Using technology*: `normalcdf(lower:105,upper:1000,` `μ:100,σ:1.9365)` $= 0.0049$. There is a 0.0049 probability of selecting a sample of 60 adults whose mean WAIS score is at least 105. **(d)** The answer to part (a) could be quite different depending on the shape of the population distribution. The answer to part (b) would be the same because the mean and standard deviation do not depend on the shape of the population distribution. Because of the large sample size ($60 \geq 30$), the answer for part (c) would still be fairly reliable due to the central limit theorem.

**R7.7 (a)** Because $n = 50 \geq 30$. **(b)** $\mu_{\bar{x}} = \mu = 0.5$. Because 50 is less than 10% of all traps, the standard deviation is $\sigma_{\bar{x}} = \frac{\sigma}{\sqrt{n}} = \frac{0.7}{\sqrt{50}} = 0.0990$. Thus, $\bar{x}$ follows an $N(0.5, 0.0990)$ distribution and we want to find $P(\bar{x} \geq 0.6)$. $z = \frac{0.6 - 0.5}{0.0990} = 1.01$ and $P(Z \geq 1.01) = 0.1562$. *Using technology*: `normalcdf` `(lower:0.6,upper:1000,μ:0.5,` `σ:0.0990)` $= 0.1562$. There is a 0.1562 probability that the mean number of moths is greater than or equal to 0.6. **(c)** No. Because this probability is not small, it is plausible that the sample mean number of moths is this high by chance alone.

## Teaching Tip

Explanations for the multiple-choice questions in the AP® Statistics Practice Test can be found in the Solutions Manual.

**TRM Full Solutions to Chapter 7 AP® Statistics Practice Test**

### Answers to Chapter 7 AP® Statistics Practice Test

**T7.1** c
**T7.2** c
**T7.3** c
**T7.4** a
**T7.5** b
**T7.6** b
**T7.7** b

# Chapter 7 AP® Statistics Practice Test

**Section I: Multiple Choice** *Select the best answer for each question.*

**T7.1** A study of voting chose 663 registered voters at random shortly after an election. Of these, **72%** said they had voted in the election. Election records show that only **56%** of registered voters voted in the election. Which of the following statements is true about the boldface numbers?

(a) 72% is a sample; 56% is a population.
(b) 72% and 56% are both statistics.
(c) 72% is a statistic and 56% is a parameter.
(d) 72% is a parameter and 56% is a statistic.
(e) 72% and 56% are both parameters.

**T7.2** The Gallup Poll has decided to increase the size of its random sample of voters from about 1500 people to about 4000 people right before an election. The poll is designed to estimate the proportion of voters who favor a new law banning smoking in public buildings. The effect of this increase is to

(a) reduce the bias of the estimate.
(b) increase the bias of the estimate.
(c) reduce the variability of the estimate.
(d) increase the variability of the estimate.
(e) reduce the bias and variability of the estimate.

**T7.3** Suppose we select an SRS of size $n = 100$ from a large population having proportion $p$ of successes. Let $\hat{p}$ be the proportion of successes in the sample. For which value of $p$ would it be safe to use the Normal approximation to the sampling distribution of $\hat{p}$?

(a) 0.01 (b) 0.09 (c) 0.85 (d) 0.975 (e) 0.999

**T7.4** The central limit theorem is important in statistics because it allows us to use the Normal distribution to find probabilities involving the sample mean

(a) if the sample size is reasonably large (for any population).
(b) if the population is Normally distributed and the sample size is reasonably large.
(c) if the population is Normally distributed (for any sample size).
(d) if the population is Normally distributed and the population standard deviation is known (for any sample size).
(e) if the population size is reasonably large (whether the population distribution is known or not).

**T7.5** The number of undergraduates at Johns Hopkins University is approximately 2000, while the number at Ohio State University is approximately 60,000. At both schools, a simple random sample of about 3% of the undergraduates is taken. Each sample is used to estimate the proportion $p$ of all students at that university who own an iPod. Suppose that, in fact, $p = 0.80$ at both schools. Which of the following is the best conclusion?

(a) The estimate from Johns Hopkins has less sampling variability than that from Ohio State.
(b) The estimate from Johns Hopkins has more sampling variability than that from Ohio State.
(c) The two estimates have about the same amount of sampling variability.
(d) It is impossible to make any statement about the sampling variability of the two estimates because the students surveyed were different.
(e) None of the above.

**T7.6** A researcher initially plans to take an SRS of size $n$ from a population that has mean 80 and standard deviation 20. If he were to double his sample size (to $2n$), the standard deviation of the sampling distribution of the sample mean would be multiplied by

(a) $\sqrt{2}$. (b) $1/\sqrt{2}$. (c) 2. (d) 1/2. (e) $1/\sqrt{2n}$.

**T7.7** The student newspaper at a large university asks an SRS of 250 undergraduates, "Do you favor eliminating the carnival from the term-end celebration?" All in all, 150 of the 250 are in favor. Suppose that (unknown to you) 55% of all undergraduates favor eliminating the carnival. If you took a very large number of SRSs of size $n = 250$ from this population, the sampling distribution of the sample proportion $\hat{p}$ would be

(a) exactly Normal with mean 0.55 and standard deviation 0.03.
(b) approximately Normal with mean 0.55 and standard deviation 0.03.
(c) exactly Normal with mean 0.60 and standard deviation 0.03.
(d) approximately Normal with mean 0.60 and standard deviation 0.03.
(e) heavily skewed with mean 0.55 and standard deviation 0.03.

T7.8 Which of the following statements about the sampling distribution of the sample mean is *incorrect?*

(a) The standard deviation of the sampling distribution will decrease as the sample size increases.
(b) The standard deviation of the sampling distribution is a measure of the variability of the sample mean among repeated samples.
(c) The sample mean is an unbiased estimator of the population mean.
(d) The sampling distribution shows how the sample mean will vary in repeated samples.
(e) The sampling distribution shows how the sample was distributed around the sample mean.

T7.9 A machine is designed to fill 16-ounce bottles of shampoo. When the machine is working properly, the amount poured into the bottles follows a Normal distribution with mean 16.05 ounces and standard deviation 0.1 ounce. Assume that the machine is working properly. If four bottles are randomly selected and the number of ounces in each bottle is measured, then there is about a 95% chance that the sample mean will fall in which of the following intervals?

(a) 16.05 to 16.15 ounces
(b) 16.00 to 16.10 ounces
(c) 15.95 to 16.15 ounces
(d) 15.90 to 16.20 ounces
(e) 15.85 to 16.25 ounces

T7.10 Suppose that you are a student aide in the library and agree to be paid according to the "random pay" system. Each week, the librarian flips a coin. If the coin comes up heads, your pay for the week is \$80. If it comes up tails, your pay for the week is \$40. You work for the library for 100 weeks. Suppose we choose an SRS of 2 weeks and calculate your average earnings $\bar{x}$. The shape of the sampling distribution of $\bar{x}$ will be

(a) Normal.
(b) approximately Normal.
(c) right-skewed.
(d) left-skewed.
(e) symmetric but not Normal.

**Section II: Free Response** *Show all your work. Indicate clearly the methods you use, because you will be graded on the correctness of your methods as well as on the accuracy and completeness of your results and explanations.*

T7.11 Below are histograms of the values taken by three sample statistics in several hundred samples from the same population. The true value of the population parameter is marked with an arrow on each histogram.

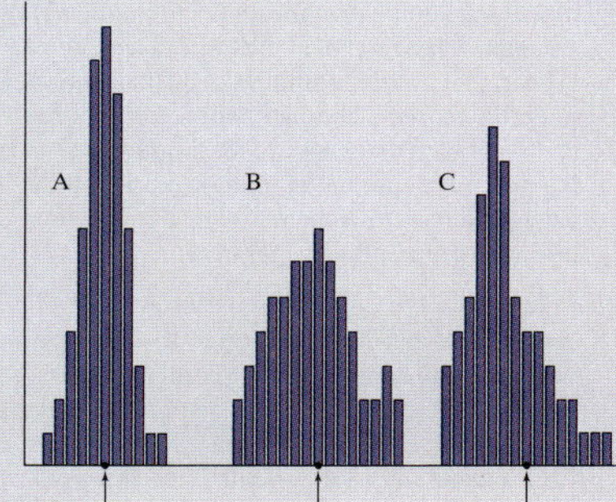


Which statistic would provide the best estimate of the parameter? Justify your answer.

T7.12 The amount that households pay service providers for access to the Internet varies quite a bit, but the mean monthly fee is \$38 and the standard deviation is \$10. The distribution is not Normal: many households pay a base rate for low-speed access, but some pay much more for faster connections. A sample survey asks an SRS of 500 households with Internet access how much they pay. Let $\bar{x}$ be the mean amount paid.

(a) Explain why you can't determine the probability that the amount a randomly selected household pays for access to the Internet exceeds \$39.
(b) What are the mean and standard deviation of the sampling distribution of $\bar{x}$?
(c) What is the shape of the sampling distribution of $\bar{x}$? Justify your answer.
(d) Find the probability that the average fee paid by the sample of households exceeds \$39. Show your work.

T7.13 According to government data, 22% of American children under the age of six live in households with incomes less than the official poverty level. A study of learning in early childhood chooses an SRS of 300 children. Find the probability that more than 20% of the sample are from poverty-level households. Be sure to check that you can use the Normal approximation.

**T7.8** e

**T7.9** c

**T7.10** e

**T7.11** A. Both A and B appear to be unbiased and A has less variability than B.

**T7.12** **(a)** We do not know the shape of the population distribution of monthly fees. **(b)** $\mu_{\bar{x}} = \mu = \$38$. Because the sample of size 500 is less than 10% of all households with Internet access,

$$\sigma_{\bar{x}} = \frac{\sigma}{\sqrt{n}} = \frac{10}{\sqrt{500}} = \$0.4472.$$

**(c)** Because the sample size is large ($n = 500 \geq 30$), the distribution of $\bar{x}$ will be approximately Normal. **(d)** We want to find $P(\bar{x} > 39)$. $z = \frac{39 - 38}{0.4472} = 2.24$ and $P(Z > 2.24) = 0.0125$. *Using technology*: `normalcdf(lower:39,upper:1000,μ:38,σ:0.4472)` $= 0.0127$. There is a 0.0127 probability that the mean monthly fee exceeds \$39.

**T7.13** $\mu_{\hat{p}} = p = 0.22$. Because 300 is less than 10% of children under the age of 6, $\sigma_{\hat{p}} = \sqrt{\frac{0.22(0.78)}{300}} = 0.0239$. Because $np = 300(0.22) = 66$ and $n(1 - p) = 300(0.78) = 234$ are both at least 10, the sampling distribution of $\hat{p}$ can be approximated by a Normal distribution. We want to find $P(\hat{p} > 0.20)$.

$$z = \frac{0.20 - 0.22}{0.0239} = -0.84$$ and $P(Z > -0.84) = 0.7995$. *Using technology*: `normalcdf(lower:0.20,upper:1000,μ:0.22,σ:0.0239)` $=$ 0.7987. There is a 0.7987 probability that more than 20% of the sample are from poverty-level households.

**TRM Full Solutions to Cumulative AP® Statistics Practice Test 2**

## Answers to Cumulative AP® Statistics Practice Test 2

AP2.1 a

AP2.2 d

AP2.3 e

AP2.4 b

AP2.5 c

AP2.6 e

AP2.7 c



# Cumulative AP® Practice Test 2

**Section I: Multiple Choice** *Choose the best answer for each question.*

**AP2.1** The five-number summary for a data set is given by min = 5, $Q_1$ = 18, median = 20, $Q_3$ = 40, max = 75. If you wanted to construct a boxplot for the data set (that is, one that would show outliers, if any existed), what would be the maximum possible length of the right-side "whisker"?

(a) 33 (b) 35 (c) 45 (d) 53 (e) 55

**AP2.2** The probability distribution for the number of heads in four tosses of a coin is given by

| Number of heads: | 0 | 1 | 2 | 3 | 4 |
|---|---|---|---|---|---|
| Probability: | 0.0625 | 0.2500 | 0.3750 | 0.2500 | 0.0625 |

The probability of getting at least one *tail* in four tosses of a coin is

(a) 0.2500.
(b) 0.3125.
(c) 0.6875.
(d) 0.9375.
(e) 0.0625.

**AP2.3** In a certain large population of adults, the distribution of IQ scores is strongly left-skewed with a mean of 122 and a standard deviation of 5. Suppose 200 adults are randomly selected from this population for a market research study. The distribution of the sample mean of IQ scores is

(a) left-skewed with mean 122 and standard deviation 0.35.
(b) exactly Normal with mean 122 and standard deviation 5.
(c) exactly Normal with mean 122 and standard deviation 0.35.
(d) approximately Normal with mean 122 and standard deviation 5.
(e) approximately Normal with mean 122 and standard deviation 0.35.

**AP2.4** A 10-question multiple-choice exam offers 5 choices for each question. Jason just guesses the answers, so he has probability 1/5 of getting any one answer correct. You want to perform a simulation to determine the number of correct answers that Jason gets. One correct way to use a table of random digits to do this is the following:

(a) One digit from the random digit table simulates one answer, with 5 = right and all other digits = wrong. Ten digits from the table simulate 10 answers.
(b) One digit from the random digit table simulates one answer, with 0 or 1 = right and all other digits = wrong. Ten digits from the table simulate 10 answers.
(c) One digit from the random digit table simulates one answer, with odd = right and even = wrong. Ten digits from the table simulate 10 answers.
(d) One digit from the random digit table simulates one answer, with 0 or 1 = right and all other digits = wrong, ignoring repeats. Ten digits from the table simulate 10 answers.
(e) Two digits from the random digit table simulate one answer, with 00 to 20 = right and 21 to 99 = wrong. Ten pairs of digits from the table simulate 10 answers.

**AP2.5** Suppose we roll a fair die four times. The probability that a 6 occurs on exactly one of the rolls is

(a) $4\left(\frac{1}{6}\right)^3\left(\frac{5}{6}\right)^1$
(b) $\left(\frac{1}{6}\right)^3\left(\frac{5}{6}\right)^1$
(c) $4\left(\frac{1}{6}\right)^1\left(\frac{5}{6}\right)^3$
(d) $\left(\frac{1}{6}\right)^1\left(\frac{5}{6}\right)^3$
(e) $6\left(\frac{1}{6}\right)^1\left(\frac{5}{6}\right)^3$

**AP2.6** You want to take an SRS of 50 of the 816 students who live in a dormitory on a college campus. You label the students 001 to 816 in alphabetical order. In the table of random digits, you read the entries

95592 94007 69769 33547 72450 16632 81194 14873

The first three students in your sample have labels

(a) 955, 929, 400.
(b) 400, 769, 769.
(c) 559, 294, 007.
(d) 929, 400, 769.
(e) 400, 769, 335.

**AP2.7** The number of unbroken charcoal briquets in a 20-pound bag filled at the factory follows a Normal distribution with a mean of 450 briquets and a standard deviation of 20 briquets. The company expects that a certain number of the bags will be underfilled, so the company will replace for free the 5% of bags that have too few briquets. What is the minimum number of unbroken briquets the bag would have to contain for the company to avoid having to replace the bag for free?

(a) 404 (b) 411 (c) 418 (d) 425 (e) 448

**AP2.8** You work for an advertising agency that is preparing a new television commercial to appeal to women. You have been asked to design an experiment to compare the effectiveness of three versions of the commercial. Each subject will be shown one of the three versions and then asked about her attitude toward the product. You think there may be large differences between women who are employed and those who are not. Because of these differences, you should use

(a) a block design, but not a matched pairs design.
(b) a completely randomized design.
(c) a matched pairs design.

(d) a simple random sample.
(e) a stratified random sample.

**AP2.9** Suppose that you have torn a tendon and are facing surgery to repair it. The orthopedic surgeon explains the risks to you. Infection occurs in 3% of such operations, the repair fails in 14%, and both infection and failure occur together 1% of the time. What is the probability that the operation is successful for someone who has an operation that is free from infection?

(a) 0.8342 (c) 0.8600 (e) 0.9900
(b) 0.8400 (d) 0.8660

**AP2.10** Social scientists are interested in the association between high school graduation rate (HSGR, measured as a percent) and the percent of U.S. families living in poverty (POV). Data were collected from all 50 states and the District of Columbia, and a regression analysis was conducted.

The resulting least-squares regression line is given by $\widehat{POV} = 59.2 - 0.620(HSGR)$ with $r^2 = 0.802$. Based on the information, which of the following is the best interpretation for the slope of the least-squares regression line?

(a) For each 1% increase in the graduation rate, the percent of families living in poverty is predicted to decrease by approximately 0.896.
(b) For each 1% increase in the graduation rate, the percent of families living in poverty is predicted to decrease by approximately 0.802.
(c) For each 1% increase in the graduation rate, the percent of families living in poverty is predicted to decrease by approximately 0.620.
(d) For each 1% increase in the percent of families living in poverty, the graduation rate is predicted to increase by approximately 0.802.
(e) For each 1% increase in the percent of families living in poverty, the graduation rate is predicted to decrease by approximately 0.620.

Here is a dotplot of the adult literacy rates in 177 countries in a recent year, according to the United Nations. For example, the lowest literacy rate was 23.6%, in the African country of Burkina Faso. Mali had the next lowest literacy rate at 24.0%. Use the graph to answer Questions AP2.11 to AP2.13.

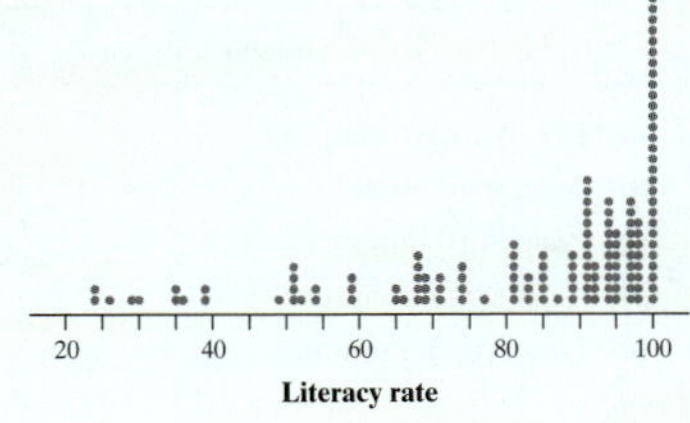


**AP2.11** The overall shape of this distribution is
(a) clearly skewed to the right.
(b) clearly skewed to the left.
(c) roughly symmetric.
(d) uniform.
(e) There is no clear shape.

**AP2.12** The mean of this distribution (*don't* try to find it) will be
(a) very close to the median.
(b) greater than the median.
(c) less than the median.
(d) You can't say, because distribution isn't symmetric.
(e) You can't say, because the distribution isn't Normal.

**AP2.13** Based on the shape of this distribution, what measures of center and spread would be most appropriate to report?
(a) The mean and standard deviation
(b) The mean and the interquartile range
(c) The median and the standard deviation
(d) The median and the interquartile range
(e) The mean and the range

**AP2.14** The correlation between the age and height of children under the age of 12 is found to be $r = 0.60$. Suppose we use the age $x$ of a child to predict the height $y$ of the child. What can we conclude?
(a) The height is generally 60% of a child's weight.
(b) About 60% of the time, age will accurately predict height.
(c) Thirty-six percent of the variation in height is accounted for by the linear model relating height to age.
(d) For every 1 year older a child is, the regression line predicts an increase of 0.6 feet in height.
(e) Thirty-six percent of the time, the least-squares regression line accurately predicts height from age.

**AP2.15** An agronomist wants to test three different types of fertilizer (A, B, and C) on the yield of a new variety of wheat. The yield will be measured in bushels per acre. Six 1-acre plots of land were randomly assigned to each of the three fertilizers. The treatment, experimental unit, and response variable are, respectively,
(a) a specific fertilizer, bushels per acre, a plot of land.
(b) a plot of land, bushels per acre, a specific fertilizer.
(c) random assignment, a plot of land, wheat yield.
(d) a specific fertilizer, a plot of land, wheat yield.
(e) a specific fertilizer, the agronomist, wheat yield.

**AP2.16** According to the U.S. Census, the proportion of adults in a certain county who owned their own home was 0.71. An SRS of 100 adults in a certain

**AP2.8** a
**AP2.9** d
**AP2.10** c
**AP2.11** b
**AP2.12** c
**AP2.13** d
**AP2.14** c
**AP2.15** d

AP2.16 c
AP2.17 e
AP2.18 a
AP2.19 c
AP2.20 b
AP2.21 a



section of the county found that 65 owned their home. Which one of the following represents the approximate probability of obtaining a sample of 100 adults in which fewer than 65 own their home, assuming that this section of the county has the same overall proportion of adults who own their home as does the entire county?

(a) $\binom{100}{65}(0.71)^{65}(0.29)^{35}$

(b) $\binom{100}{65}(0.29)^{65}(0.71)^{35}$

(c) $P\left(Z < \frac{0.65-0.71}{\sqrt{\frac{(0.71)(0.29)}{100}}}\right)$

(d) $P\left(Z < \frac{0.65-0.71}{\sqrt{\frac{(0.65)(0.35)}{100}}}\right)$

(e) $P\left(Z < \frac{0.65-0.71}{\frac{(0.71)(0.29)}{\sqrt{100}}}\right)$

**AP2.17** Which one of the following would be a correct interpretation if you have a *z*-score of +2.0 on an exam?

(a) It means that you missed two questions on the exam.
(b) It means that you got twice as many questions correct as the average student.
(c) It means that your grade was 2 points higher than the mean grade on this exam.
(d) It means that your grade was in the upper 2% of all grades on this exam.
(e) It means that your grade is 2 standard deviations above the mean for this exam.

**AP2.18** Records from a random sample of dairy farms yielded the information below on the number of male and female calves born at various times of the day.

| | Day | Evening | Night | Total |
|---|---|---|---|---|
| Males | 129 | 15 | 117 | **261** |
| Females | 118 | 18 | 116 | **252** |
| **Total** | **247** | **33** | **233** | **513** |

What is the probability that a randomly selected calf was born in the night or was a female?

(a) $\frac{369}{513}$ (b) $\frac{485}{513}$ (c) $\frac{116}{513}$ (d) $\frac{116}{252}$ (e) $\frac{116}{233}$

**AP2.19** When people order books from a popular online source, they are shipped in standard-sized boxes. Suppose that the mean weight of the boxes is 1.5 pounds with a standard deviation of 0.3 pounds, the mean weight of the packing material is 0.5 pounds with a standard deviation of 0.1 pounds, and the mean weight of the books shipped is 12 pounds with a standard deviation of 3 pounds. Assuming that the weights are independent, what is the standard deviation of the total weight of the boxes that are shipped from this source?

(a) 1.84 (b) 2.60 (c) 3.02 (d) 3.40 (e) 9.10

**AP2.20** A grocery chain runs a prize game by giving each customer a ticket that may win a prize when the box is scratched off. Printed on the ticket is a dollar value ($500, $100, $25) or the statement "This ticket is not a winner." Monetary prizes can be redeemed for groceries at the store. Here is the probability distribution of the amount won on a randomly selected ticket:

| **Amount won:** | $500 | $100 | $25 | $0 |
|---|---|---|---|---|
| **Probability:** | 0.01 | 0.05 | 0.20 | 0.74 |

Which of the following are the mean and standard deviation, respectively, of the winnings?

(a) $15.00, $2900.00
(b) $15.00, $53.85
(c) $15.00, $26.93
(d) $156.25, $53.85
(e) $156.25, $26.93

**AP2.21** A large company is interested in improving the efficiency of its customer service and decides to examine the length of the business phone calls made to clients by its sales staff. A cumulative relative frequency graph is shown below from data collected over the past year. According to the graph, the shortest 80% of calls will take how long to complete?

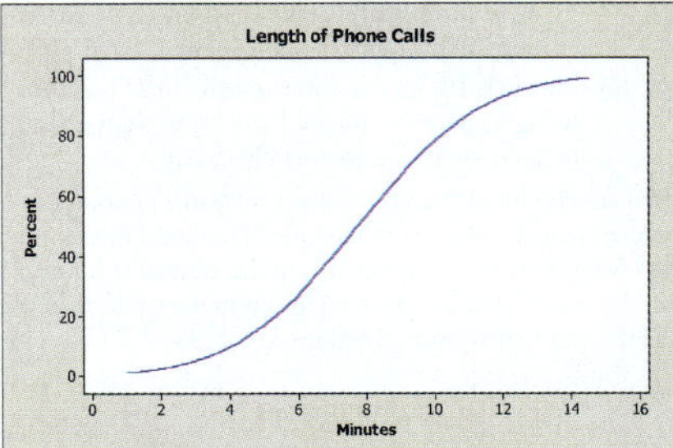


(a) Less than 10 minutes
(b) At least 10 minutes
(c) Exactly 10 minutes
(d) At least 5.5 minutes
(e) Less than 5.5 minutes

**Section II: Free Response** *Show all your work. Indicate clearly the methods you use, because you will be graded on the correctness of your methods as well as on the accuracy and completeness of your results and explanations.*

**AP2.22** A health worker is interested in determining if omega-3 fish oil can help reduce cholesterol in adults. She obtains permission to examine the health records of 200 people in a large medical clinic and classifies them according to whether or not they take omega-3 fish oil. She also obtains their latest cholesterol readings and finds that the mean cholesterol reading for those who are taking omega-3 fish oil is 18 points lower than the mean for the group not taking omega-3 fish oil.

(a) Is this an observational study or an experiment? Justify your answer.

(b) Explain the concept of confounding in the context of this study and give one example of a variable that could be confounded with whether or not people take omega-3 fish oil.

(c) Researchers find that the 18-point difference in the mean cholesterol readings of the two groups is statistically significant. Can they conclude that omega-3 fish oil is the cause? Why or why not?

**AP2.23** There are four major blood types in humans: O, A, B, and AB. In a study conducted using blood specimens from the Blood Bank of Hawaii, individuals were classified according to blood type and ethnic group. The ethnic groups were Hawaiian, Hawaiian-White, Hawaiian-Chinese, and White. Suppose that a blood bank specimen is selected at random.

| | Ethnic Group | | | | |
|---|---|---|---|---|---|
| Blood type | Hawaiians | Hawaiian-White | Hawaiian-Chinese | White | Total |
| O | 1903 | 4469 | 2206 | 53,759 | **62,337** |
| A | 2490 | 4671 | 2368 | 50,008 | **59,537** |
| B | 178 | 606 | 568 | 16,252 | **17,604** |
| AB | 99 | 236 | 243 | 5001 | **5579** |
| **Total** | **4670** | **9982** | **5385** | **125,020** | **145,057** |

(a) Find the probability that the specimen contains type O blood or comes from the Hawaiian-Chinese ethnic group. Show your work.

(b) What is the probability that the specimen contains type AB blood, given that it comes from the Hawaiian ethnic group? Show your work.

(c) Are the events "type B blood" and "Hawaiian ethnic group" independent? Give appropriate statistical evidence to support your answer.

(d) Now suppose that two blood bank specimens are selected at random. Find the probability that at least one of the specimens contains type A blood from the White ethnic group.

**AP2.24** Every 17 years, swarms of cicadas emerge from the ground in the eastern United States, live for about six weeks, and then die. (There are several different "broods," so we experience cicada eruptions more often than every 17 years.) There are so many cicadas that their dead bodies can serve as fertilizer and increase plant growth. In a study, a researcher added 10 dead cicadas under 39 randomly selected plants in a natural plot of American bellflowers on the forest floor, leaving other plants undisturbed. One of the response variables measured was the size of seeds produced by the plants. Here are the boxplots and summary statistics of seed mass (in milligrams) for 39 cicada plants and 33 undisturbed (control) plants:

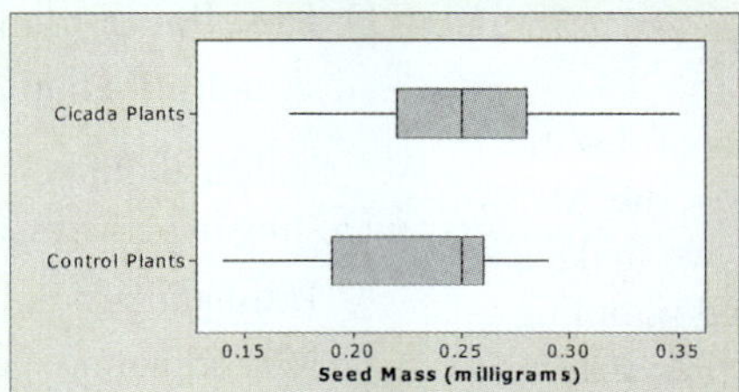


| Variable: | $n$ | Minimum | $Q_1$ | Median | $Q_3$ | Maximum |
|---|---|---|---|---|---|---|
| Cicada plants: | 39 | 0.17 | 0.22 | 0.25 | 0.28 | 0.35 |
| Control plants: | 33 | 0.14 | 0.19 | 0.25 | 0.26 | 0.29 |

(a) Write a few sentences comparing the distributions of seed mass for the two groups of plants.

(b) Based on the graphical displays, which distribution has the larger mean? Justify your answer.

(c) Explain the purpose of the random assignment in this study.

(d) Name one benefit and one drawback of only using American bellflowers in the study.

**AP2.25** In a city library, the mean number of pages in a novel is 525 with a standard deviation of 200. Approximately 30% of the novels have fewer than 400 pages. Suppose that you randomly select 50 novels from the library.

(a) What is the probability that the total number of pages is fewer than 25,000? Show your work.

(b) What is the probability that at least 20 of the novels have fewer than 400 pages? Show your work.

**AP2.22** **(a)** Observational study, because no treatments were imposed on the subjects. **(b)** Two variables are confounded when their effects on the cholesterol level cannot be distinguished from one another. For example, people who take omega-3 fish oil might also exercise more. Researchers would not know whether it was the omega-3 fish oil or the exercise that was the real explanation for lower cholesterol. **(c)** No. Even though the difference was statistically significant, this wasn't an experiment and taking fish oil is possibly confounded with exercise.

**AP2.23** **(a)** $P$(type O or Hawaiian-Chinese) = 65,516/145,057 = 0.452 **(b)** $P$(type AB | Hawaiian) = 99/4670 = 0.021 **(c)** $P$(Hawaiian) = 4670/145,057 = 0.032; $P$(Hawaiian | type B) = 178/17,604 = 0.010. Because these probabilities are not equal, the two events are not independent. **(d)** $P$(type A and white) = 50,008/145,057 = 0.345. $P$(at least one type A and white) = 1 − $P$(neither are type A and white) = $1 - (1 - 0.345)^2 = 0.571$.

**AP2.24** **(a)** The distribution of seed mass for the cicada plants is roughly symmetric, while the distribution for the control plants is skewed to the left. The median seed mass is the same for both groups. The cicada plants had a bigger range in seed mass, but the control plants had a bigger *IQR*. Neither group had any outliers. **(b)** The cicada plants. The distribution of seed mass for the cicada plants is roughly symmetric, which suggests that the mean should be about the same as the median. However, the distribution of seed mass for the control plants is skewed to the left, which will pull the mean of this distribution below its median toward the lower values. Because the medians of both distributions are equal, the mean for the cicada plants is greater than the mean for the control plants. **(c)** The purpose of the random assignment is to create two groups of plants that are roughly equivalent at the beginning of the experiment. **(d)** Benefit: controlling a source of variability. Different types of flowers will have different seed masses, making the response more variable if other types of plants were used. Drawback: we can't make inferences about the effect of cicadas on other types of plants, because other plants might respond differently to cicadas.

**AP2.25** **(a)** We want to find $P(\bar{x} < 25{,}000/50) = P(\bar{x} < 500)$. Because the sample size is large ($n = 50 \geq 30$), the distribution of $\bar{x}$ is approximately Normal with $\mu_{\bar{x}} = \mu = 525$ pages. Because $n = 50$ is less than 10% of all novels in the library, $\sigma_{\bar{x}} = \frac{\sigma}{\sqrt{n}} = \frac{200}{\sqrt{50}} = 28.28$ pages. $z = \frac{500 - 525}{28.28} = -0.88$ and $P(Z < -0.88) = 0.1894$. *Using technology*: normalcdf(lower:-1000, upper:500,μ:525,σ:28.28) = 0.1883. There is a 0.1883 probability that the total number of pages in 50 novels is fewer than 25,000. **(b)** Let $X$ be the number of novels that have fewer than 400 pages. $X$ is a binomial random variable with $n = 50$ and $p = 0.30$. We want to find $P(X \geq 20)$. *Using technology*: $P(X \geq 20) = 1 - P(X \leq 19)$ = 1 − binomcdf(trials:50,p:0.30,x value:19) = 0.0848. There is a 0.0848 probability of selecting at least 20 novels that have fewer than 400 pages. *Note*: Using the Normal approximation, $P(X \geq 20) = 0.0614$.

Chapter

# 8

# Estimating with Confidence

## PD Overview

Chapter 8 marks the beginning of our formal study of statistical inference, the fourth and final content area in the AP® Statistics Topic Outline. Starting in this chapter, we make a major transition from Chapter 7. When we studied sampling distributions in Chapter 7, we pretended to know the value of the population parameter and then asked questions about the distribution of a statistic. For example, "Suppose that 35% of all first-year students actually attend college within 50 miles of home. What is the probability that the random sample of 1500 students will give a result within 2 percentage points of this true value?" In Chapter 8, we no longer pretend to know the truth. Instead, we use information from a random sample or randomized experiment to create an estimate for the parameter.

In this chapter, we introduce the idea of a confidence interval—an interval of plausible values for a parameter. We also provide the specific details for constructing confidence intervals for a population proportion and confidence intervals for a population mean. In later chapters, we will introduce confidence intervals for a difference in means, confidence intervals for a difference in proportions, and confidence intervals for the slope of a least-squares regression line. Fortunately, many of the big ideas of this chapter apply to all confidence intervals. This should make it easier to get through Chapters 9–12, assuming students have a solid understanding of the material in this chapter.

An excellent resource for you to learn more about inference is available for free at AP® Central. On the AP® Statistics homepage, look for the "Special Focus Materials" about inference. These materials were originally developed for use in College Board workshops and were written by people very familiar with the AP® Statistics course.

### Relationship to the AP® Statistics Course Description

Chapter 8 covers the following content as described in the Topic Outline from the AP® Statistics Course Description. Topics marked with a superscript are covered in other chapters as indicated.

III. Anticipating patterns: Exploring random phenomena using probability and simulation

D. Sampling distributions

7. *t*-distribution

IV. Statistical inference: Estimating population parameters and testing hypotheses

A. Estimation (point estimators and confidence intervals)

1. Estimating population parameters and margins of error
2. Properties of point estimators, including unbiasedness and variability
3. Logic of confidence intervals, meaning of confidence level and confidence intervals, and properties of confidence intervals
4. Large-sample confidence interval for a proportion
5. Large-sample confidence interval for a difference between two proportions[ch10]
6. Confidence interval for a mean
7. Confidence interval for a difference between two means (paired[ch9] and unpaired[ch10])
8. Confidence interval for the slope of a least-squares regression line[ch12]

## The Big Ideas

One of the challenges in teaching the AP® Statistics course is keeping students focused on the big picture and not just the details of each section. We outline the big ideas for the chapter here.

### Chapter 8 Introduction

We strongly recommend that you start this chapter with the Activity on page 476. It is a great way to get students thinking about estimating with confidence—how to use information from a sample to estimate the value of a population parameter. We also refer back to this Activity throughout Section 8.1, so it will definitely help if your students are familiar with it.

### Section 8.1 Confidence Intervals: The Basics

In this section, we start by introducing the idea of a point estimate—a single best guess for the value of a population parameter. Because of sampling variability, however, it is very unlikely that our point estimate will be correct. To increase our chances of getting a correct estimate, we prefer to use an interval of plausible values to estimate the value of the parameter rather than a single point estimate.

To create this interval of plausible values, we rely on our understanding of sampling distributions from Chapter 7. When the distribution of a statistic (such as $\bar{x}$ or $\hat{p}$) is approximately Normal, we know that in about 95% of possible samples, the value of the statistic will be within 2 standard deviations of the population parameter (such as $\mu$ or $p$). This implies that in about 95% of possible samples, the population parameter ($\mu$ or $p$) will be within 2 standard deviations of the statistic ($\bar{x}$ or $\hat{p}$).

In this section, we have chosen to focus on the big picture and avoid details about the calculations of different types of confidence intervals. For example, we believe that using 2 as the critical value in this section allows students to focus on most important concepts, such as interpreting a confidence interval and interpreting a confidence level. We always interpret a confidence interval as an interval of plausible values for a parameter. Likewise, we always interpret a confidence level as the long-run capture rate of the method we use to compute the interval. Make sure to do the Activity on page 482 to reinforce this idea.

The basic structure of a confidence interval is: point estimate ± margin of error. The size of the margin of error is determined primarily by two factors. Increasing the sample size decreases the margin of error, making our intervals narrower. Increasing the confidence level, however, increases the margin of error, making our intervals wider. Finally, the margin of error only accounts for sampling variability—it does not account for other problems in the data collection process that can lead to bias, such as undercoverage or nonresponse.

## Section 8.2 Estimating a Population Proportion

In this section, we focus exclusively on confidence intervals for population proportions. We begin with an Activity that encourages students to think about how to create a confidence interval for the proportion of red beads in a container.

Next, we discuss what conditions students need to check when calculating a confidence interval for a population proportion. Students need to verify that the data come from a well-designed random sample or randomized experiment (the Random condition), that the sample is a small fraction of the population when sampling without replacement (the 10% condition), and that the sample size is large enough so the distribution of $\hat{p}$ is approximately Normal (the Large Counts condition).

Although we used a 95% confidence level and a critical value of 2 in Section 8.1, these are not our only choices. In this section, we introduce other common confidence levels and show how to calculate the critical values for them.

The four-step process is ideally suited for confidence interval questions, and we ask students to use this process each time they calculate and interpret a confidence interval. Students should *state* the parameter they are estimating and at what confidence level, *plan* their work by naming the type of interval they will use and checking the appropriate conditions, *do* the calculations, and make a *conclusion* in the context of the problem.

Finally, we consider how to choose an appropriate sample size when planning a study. The necessary sample size is based on the confidence level, the proportion of successes, and the desired margin of error.

## Section 8.3 Estimating a Population Mean

In this section, we focus exclusively on confidence intervals for a population mean. Although the basic ideas remain the same, estimating a population mean comes with an extra complication. Because we typically don't know the population standard deviation $\sigma$ when estimating the population mean $\mu$, we have to use the sample standard deviation $s_x$ instead. As we show in the Calculator BINGO! Activity on page 509, using critical values from the standard Normal distribution when replacing $\sigma$ with $s_x$ makes the capture rate too small. To fix the problem, we get critical values from a $t$ distribution rather than the standard Normal distribution. The $t$ distributions are bell-shaped, symmetric, centered at 0, and slightly wider than the standard Normal distribution.

The form of the one-sample $t$ interval for $\mu$ is the same as that of the one-sample $z$ interval for $p$. The way we interpret confidence intervals and confidence levels remains the same as well, as do the Random and 10% conditions. However, the Normal/Large Sample condition that we introduced in Section 7.3 replaces the Large Counts condition that we used in Section 8.2.

Finally, we consider how to choose an appropriate sample size when planning a study. The necessary sample size is based on the confidence level, the standard deviation, and the desired margin of error.

# Chapter 8: Resources

## TRM Additional *TPS* Teacher's Resource Materials

The following resources can be found by clicking on the links in the new Teacher's e-Book (TE-book), opening the Teacher's Resource CD (TRCD), or logging onto the book's companion Web site www.whfreeman.com/tps5e (teacher log-in required).

- Printable Chapter 8 Learning Objectives Grid (page 477)
- Hershey's Kisses Alternate Activity (page 493)
- Additional Material: The Plus Four Interval for a Population Proportion (page 500)
- Good Books Alternate Activity (page 507)
- FRAPPY! materials: Question, Model Solution, Scoring Rubric, Sample Responses with commentary (page 530)
- Quizzes: 3 versions for each section; versions A and B are roughly parallel
  - Section 8.1 (page 488)
  - Section 8.2 (page 503)
  - Section 8.3 (page 526)
- Tests: 4 versions; versions A and B are roughly parallel, version D is cumulative (page 531)

## Free-Response Questions from Previous AP® Exams

Questions can be found on the AP® Central Web site: http://apcentral.collegeboard.com/apc/members/exam/exam_questions/8357.html.

Students should be able to answer all the free-response questions listed below with material learned in this chapter. Questions that contain content from this chapter but also require content from later chapters are listed in the last chapter required to complete the entire question. This list will be updated after each AP® exam and will be posted on the book's Web site: www.whfreeman.com/tps5e.

Questions marked with an asterisk are from exams with released multiple-choice questions. You may want to save these questions until the end of the year so you can give your students a complete released exam for practice.

| Year | # | Content |
|---|---|---|
| 2013 | 1 | • Interpreting stemplots<br>• One-sample *t* interval for a population mean |
| 2011 | 6 | • One-sample *z* interval for a proportion<br>• Tree diagrams<br>• Using information from a tree diagram to create a new confidence interval |
| 2011B | 5 | • One-sample *z* interval for a proportion<br>• Using a CI to assess a claim<br>• Determining sample size |
| 2010 | 3 | • Interpreting confidence level<br>• Using confidence intervals to make decisions<br>• Determining sample size (CI for a proportion) |
| 2010B | 4 | • One-sample *z* interval for a population proportion<br>• Effect of sampling without replacement |
| 2008B | 3 | • Determining sample size (CI for a mean)<br>• Practical constraints |
| 2005 | 5 | • Sources of bias in a survey<br>• Determining sample size (CI for a proportion)<br>• Stratified random sampling |
| 2003 | 6 | • Interpreting a graph<br>• One-sample *z* interval for a population proportion<br>• Using confidence intervals to make decisions |
| 2003B | 6 | • One-sample *z* interval for a population proportion<br>• Interpreting confidence level<br>• Determining sample sizes for different subgroups (CI for a proportion) |
| 2002 | 1* | • Precision of interval estimates<br>• Using confidence intervals to make decisions |
| 2002B | 4 | • One-sample *z* interval for a population proportion<br>• Interpreting confidence level<br>• Using confidence intervals to make decisions |
| 2000 | 2 | • Conditions for a one-sample t interval for a population mean |
| 2000B | 6 | • One-sample *z* interval for a population proportion<br>• Combining Normal random variables<br>• Independence<br>• Anticipating patterns in a scatterplot |

## Applets

- The *Confidence Intervals* applet at www.whfreeman.com/tps5e allows students to investigate the success rate of confidence intervals for population means using different confidence levels and sample sizes.
- The *Confidence Intervals for Proportions* applet at www.whfreeman.com/tps5e allows students to investigate the success rate of confidence intervals for population proportions using different population proportions, confidence levels, and sample sizes.
- The *Simulating Confidence Intervals for Population Parameter* applet at www.rossmanchance.com/applets allows students to investigate the success rate of confidence intervals for population proportions and population means under different circumstances.
- The *Simulating t-Intervals for Different Population Shapes* applet at www.rossmanchance.com/applets allows students to investigate the success rate of confidence intervals for population means for different sample sizes and populations of different shapes.

## TI-Nspire Activities

These Activities are designed to illustrate important statistical concepts using the TI-Nspire and can be found at www.whfreeman.com/tps5e.

- *Investigating Confidence Intervals*: Students generate random samples, calculate confidence intervals for $\mu$, and estimate how often their intervals capture the true mean.
- *What's the Difference between z and t?*: Students generate random samples and explore the differences between the $z$ statistic and the $t$ statistic.

# Chapter 8: Pacing Guide, Learning Objectives, and Suggested Assignments

This pacing guide is based on a schedule with 110, 50-minute sessions before the AP® exam. The 110 days include 4 days for midterms and 10 days for AP® exam review. If you have a different number of sessions before the AP® exam, you can modify the pacing guide to suit your needs. If you have additional time, consider incorporating quizzes, released AP® free-response questions, or additional activities. See the Resources section above for suggestions.

The suggested homework assignments list odd-numbered exercises, whenever possible, so students can check their answers against the back-of-the-book answers. If you would rather students not have access to the answers while doing homework, adding 1 to the exercise numbers usually will do the trick, because the homework exercises typically are paired. For example, Exercises 1 and 2 will generally cover the same topics, but in different contexts. You may also choose to include the Recycle and Review questions at the end of each section, which review topics from previous sections or chapters. These questions are denoted with a symbol.

| Day | Topics | Learning Objectives: Students will be able to. . . | Suggested Assignment |
|---|---|---|---|
| 1 | Chapter 8 Introduction; 8.1 The Idea of a Confidence Interval, Interpreting Confidence Intervals and Confidence Levels | • Interpret a confidence interval in context.<br>• Interpret a confidence level in context. | 1, 3, 5, 7, 9 |
| 2 | 8.1 Constructing a Confidence Interval; Using Confidence Intervals Wisely | • Determine the point estimate and margin of error from a confidence interval.<br>• Describe how the sample size and confidence level affect the length of a confidence interval.<br>• Explain how practical issues like nonresponse, undercoverage, and response bias can affect the interpretation of a confidence interval. | 10, 11, 13, 15, 17, 19 |
| 3 | 8.2 Conditions for Estimating $p$, Constructing a Confidence Interval for $p$, Putting It All Together: The Four-Step Process | • State and check the Random, 10%, and Large Counts conditions for constructing a confidence interval for a population proportion.<br>• Determine critical values for calculating a $C$% confidence interval for a population proportion using a table or technology.<br>• Construct and interpret a confidence interval for a population proportion. | 20–24, 31, 33, 35, 37 |
| 4 | 8.2 Choosing the Sample Size | • Determine the sample size required to obtain a $C$% confidence interval for a population proportion with a specified margin of error. | 39, 41, 43, 45, 47 |
| 5 | 8.3 The Problem of Unknown $\sigma$, When $\sigma$ Is Unknown: The $t$ Distributions, Conditions for Estimating $\mu$ | • Explain how the $t$ distributions are different from the standard Normal distribution, and why it is necessary to use a $t$ distribution when calculating a confidence interval for a population mean.<br>• Determine critical values for calculating a $C$% confidence interval for a population mean using a table or technology.<br>• State and check the Random, 10%, and Normal/Large Sample conditions for constructing a confidence interval for a population mean. | 49–52, 55, 57, 59 |
| 6 | 8.3 Constructing a Confidence Interval for $\mu$, Choosing a Sample Size | • Construct and interpret a confidence interval for a population mean.<br>• Determine the sample size required to obtain a $C$% confidence interval for a population mean with a specified margin of error. | 61, 65, 69, 71, 73, 75–78 |
| 7 | Chapter 8 Review/ FRAPPY! | | Chapter 8 Review Exercises |
| 8 | Chapter 8 Test | | |

Chapter
8

NOTES

**PD** **Chapter 8 Overview**

Watch the video overview of Chapter 8 (for teachers) by clicking on the link in the TE-book, opening the TRCD, or logging onto the book's companion Web site www.whfreeman.com/tps5e (teacher log-in required).

# Chapter 8

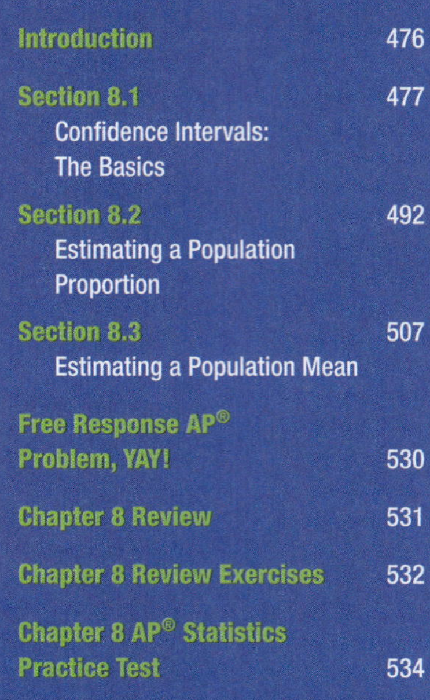

# Estimating with Confidence

## case study

### Need Help? Give Us a Call!

If your cable television goes out, you phone the cable company to get it fixed. Does a real person answer your call? These days, probably not. It is far more likely that you will get an automated response. You will probably be offered several options, such as: to order cable service, press 1; for questions about your bill, press 2; to add new channels, press 3; (and finally) to speak with a customer service agent, press 4. Customers will get frustrated if they have to wait too long before speaking to a live person. So companies try hard to minimize the time required to connect to a customer service representative.

A large bank decided to study the call response times in its customer service department. The bank's goal was to have a representative answer an incoming call in less than 30 seconds. Figure 8.1 is a histogram of the response times in a random sample of 241 calls to the bank's customer service center in a given month. What does the graph suggest about how well the bank is meeting its goal?

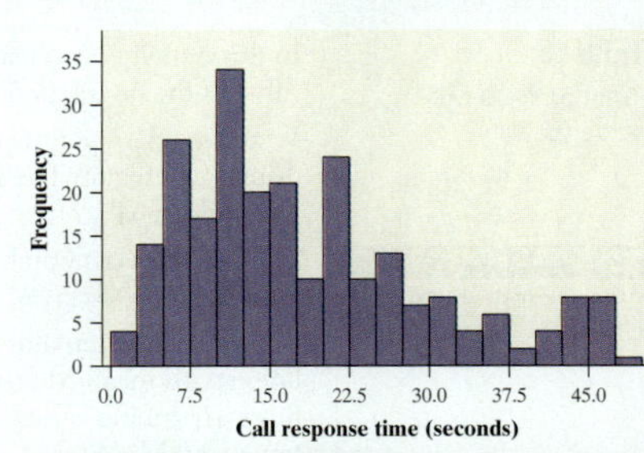


**FIGURE 8.1** Histogram showing the response time (in seconds) at a bank's customer service center for a random sample of 241 calls.

By the end of this chapter, you will be able to use the sample data to make inferences about the proportion $p$ of all calls to the customer service department that are answered within 30 seconds and the mean call response time $\mu$.

475

### Teaching Tip

Emphasize that we are making a big transition as we move from Chapter 7 to Chapter 8. In Chapter 7, we assumed that we knew the true value of a parameter and then asked questions about the distribution of the statistic used to estimate that parameter. For example, we asked questions like "Assuming $\mu = 10$, what is the probability that $\bar{x}$ is greater than 12?" In Chapter 8, we no longer pretend to know the true value of a parameter. We start with the more realistic situation where we know only the value of a sample statistic. Then we use the value of the statistic to estimate the value of the population parameter. For example, we ask questions like "If $\bar{x}$ is 123.8, what are the plausible values of $\mu$?"

**Activity Overview**

You can watch an overview of this Activity (for teachers) by clicking on the link in the TE-book, opening the TRCD, or logging onto the book's companion Web site www.whfreeman.com/tps5e (teacher log-in required).

### TECHNOLOGY TIP

To store a value for M before you begin the Activity, enter a number (e.g., 250), press the STO> button (just above the ON button), press the ALPHA key, and press the M key.

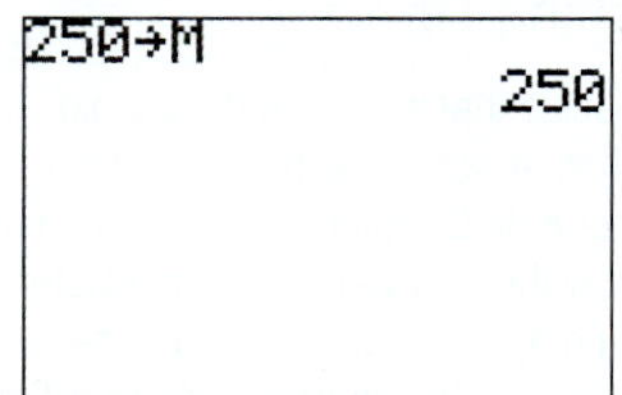


In step 2, you can find the mean( command by pressing the LIST button (2nd STAT) and arrowing over to the MATH menu. You can also find the command in CATALOG(2nd 0).

### Teaching Tip

In Step 3 of the Activity, students may come up with several different intervals of plausible values, depending on how many standard deviations they extend in each direction. This is OK for now, but you might ask students why they might prefer a longer interval compared to a shorter one. This will set the stage for discussing the trade-off of increased confidence and decreased precision. Also, make sure they are using the standard deviation of $\bar{x}$, $\sigma_{\bar{x}} = \frac{20}{\sqrt{16}} = 5$, not the standard deviation of $X$, $\sigma = 20$. Hopefully, they will come up with something in the form $\bar{x} \pm 2\frac{20}{\sqrt{16}} = \bar{x} \pm 10$.

### Teaching Tip

After completing the Mystery Mean Activity, students might want you to reveal the true value of M. Respond by saying, "In real life, we need to conduct a census to find out the true value of a parameter. And a census costs a lot of money. So, I will tell you the value of M, but it will cost $100." If they don't pay, don't reveal the value of M!

# Introduction

How long does a new model of laptop battery last? What proportion of college undergraduates have engaged in binge drinking? How much does the weight of a quarter-pound hamburger at a fast-food restaurant vary after cooking? These are the types of questions we would like to be able to answer.

It wouldn't be practical to determine the lifetime of *every* laptop battery, to ask *all* college undergraduates about their drinking habits, or to weigh *every* burger after cooking. Instead, we choose a sample of individuals (batteries, college students, burgers) to represent the population and collect data from those individuals. Our goal in each case is to use a sample statistic to estimate an unknown population parameter. From what we learned in Chapter 4, if we randomly select the sample, we should be able to generalize our results to the population of interest.

We cannot be certain that our conclusions are correct—a different sample would probably yield a different estimate. Statistical inference uses the language of probability to express the strength of our conclusions. Probability allows us to take chance variation due to random selection or random assignment into account. The following Activity gives you an idea of what lies ahead.

## ACTIVITY | The Mystery Mean

**MATERIALS:**
TI-83/84 or TI-89 with display capability

In this Activity, each team of three to four students will try to estimate the mystery value of the population mean $\mu$ that your teacher entered before class.[1]

1. Before class, your teacher will store a value of $\mu$ (represented by M) in the display calculator. The teacher will then clear the home screen so you can't see the value of M.

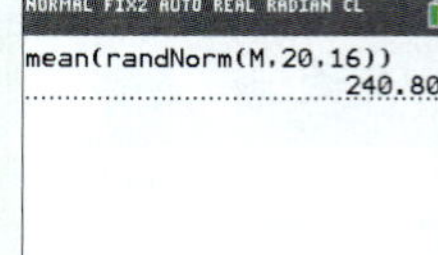


2. With the class watching, the teacher will execute the following command: mean(randNorm(M,20,16)).
This tells the calculator to choose an SRS of 16 observations from a Normal population with mean M and standard deviation 20 and then compute the mean $\bar{x}$ of those 16 sample values. Is the sample mean shown likely to be equal to the mystery mean M? Why or why not?
3. Now for the challenge! Your group must determine an interval of *reasonable* values for the population mean $\mu$. Use the result from Step 2 and what you learned about sampling distributions in the previous chapter.
4. Share your team's results with the class.

In this chapter and the next, we will meet the two most common types of formal statistical inference. Chapter 8 concerns *confidence intervals* for estimating the value of a parameter. Chapter 9 presents *significance tests*, which assess the evidence for a claim about a parameter. Both types of inference are based on the sampling distributions of statistics. That is, both report probabilities that state *what would happen if we used the inference method many times.*

### ALTERNATE ACTIVITY

***The mystery proportion***

This Activity closely parallels The Mystery Mean Activity in the Student Edition but asks students to come up with an interval of plausible values for a population proportion $p$. Before class, choose a value from 0 to 1 for the mystery value of $p$. It is best if the value is close to 0.5, so the Large Counts condition will be met. To do this, enter the value of $p$ (e.g., $p = 0.43$), press the STO> button (just above the ON button), press the ALPHA key, and press the P key (see the screen shot at right).

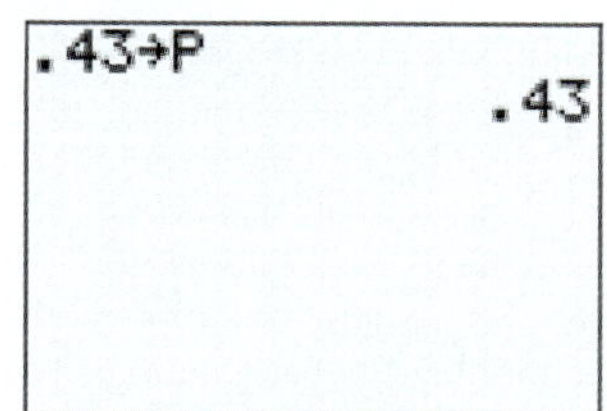


Then, with the class watching, enter the command randBin(100,P)/100. This will generate a value of $\hat{p}$ for a random sample of size 100 from a population where the true proportion of successes is $p$. In the following example, the sample proportion is $\hat{p} = 0.45$.

Section 8.1 examines the idea of a confidence interval. We start by presenting the reasoning of confidence intervals in a general way that applies to estimating any unknown parameter. In Section 8.2, we show how to estimate a population proportion. Section 8.3 focuses on confidence intervals for a population mean.

# 8.1 Confidence Intervals: The Basics

**WHAT YOU WILL LEARN** By the end of the section, you should be able to:

- Determine the point estimate and margin of error from a confidence interval.
- Interpret a confidence interval in context.
- Interpret a confidence level in context.
- Describe how the sample size and confidence level affect the length of a confidence interval.
- Explain how practical issues like nonresponse, undercoverage, and response bias can affect the interpretation of a confidence interval.

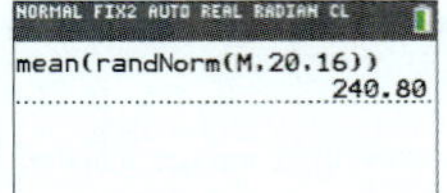


Mr. Schiel's class did the mystery mean Activity from the Introduction. The TI screen shot displays the information that the students received about the unknown population mean $\mu$. Here is a summary of what the class said about the calculator output:

- The population distribution is Normal and its standard deviation is $\sigma = 20$.
- A simple random sample of $n = 16$ observations was taken from this population.
- The sample mean is $\bar{x} = 240.80$.

If we had to give a single number to estimate the value of M that Mr. Schiel chose, what would it be? Such a value is known as a **point estimate.** How about 240.80? That makes sense, because the sample mean $\bar{x}$ is an unbiased estimator of the population mean $\mu$. We are using the statistic $\bar{x}$ as a **point estimator** of the parameter $\mu$.

**DEFINITION: Point estimator and point estimate**

A **point estimator** is a statistic that provides an estimate of a population parameter. The value of that statistic from a sample is called a **point estimate**.

As we saw in Chapter 7, the ideal point estimator will have no bias and low variability. Here's an example involving some of the more common point estimators.

**PD Section 8.1 Overview**

Watch the video overview of Section 8.1 (for teachers) by clicking on the link in the TE-book, opening the TRCD, or logging onto the book's companion Web site www.whfreeman.com/tps5e (teacher log-in required).

**TRM Learning Objectives Grid**

Remind students about the importance of the Learning Objectives. We include them at the beginning of each section so students know what they are expected to learn. At the end of the chapter, the Learning Objectives are included in a grid, along with references to examples and chapter review exercises. A printable version of this grid is included in the Teacher's Resource Materials.

**TRM Section 8.1 Alternate Examples**

You can find the Alternate Examples for this section in Microsoft Word format by clicking on the link in the TE-book, opening the TRCD, or logging onto the book's companion Web site www.whfreeman.com/tps5e (teacher log-in required).

## Teaching Tip

A single estimate of a population parameter is called a "point estimate" because it represents a single point on the number line. We will soon learn about "interval estimates" (i.e., confidence intervals) that cover an entire interval of points on the number line.

```
randBin(100,P)/1
00
                .45
```

Ask your students to come up with an interval of plausible values for $p$ based on the value of $\hat{p}$. Hopefully, they will come up with something in the form $\hat{p} \pm 2\sqrt{\frac{\hat{p}(1-\hat{p})}{100}}$.

## Teaching Tip

Remind students that the "play" icon at the top of this example (and many others) indicates that they can watch a video presentation of the example. Find the videos at www.whfreeman.com/tps5e. Also, remind them that the icon in the lower right corner indicates that there is a link between this example and Exercises 1 and 3. If students are having trouble with either exercise, they can look back to this example for help.

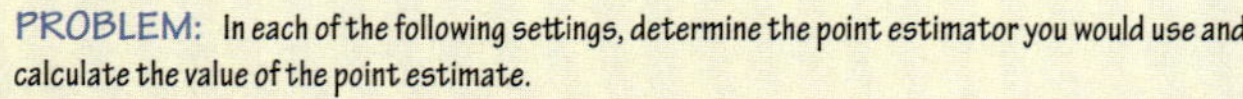

## EXAMPLE From Batteries to Smoking

### *Point estimators*

PROBLEM: In each of the following settings, determine the point estimator you would use and calculate the value of the point estimate.

(a) Quality control inspectors want to estimate the mean lifetime $\mu$ of the AA batteries produced in an hour at a factory. They select a random sample of 50 batteries during each hour of production and then drain them under conditions that mimic normal use. Here are the lifetimes (in hours) of the batteries from one such sample:

16.73 15.60 16.31 17.57 16.14 17.28 16.67 17.28 17.27 17.50 15.46 16.50 16.19
15.59 17.54 16.46 15.63 16.82 17.16 16.62 16.71 16.69 17.98 16.36 17.80 16.61
15.99 15.64 17.20 17.24 16.68 16.55 17.48 15.58 17.61 15.98 16.99 16.93 16.01
17.54 17.41 16.91 16.60 16.78 15.75 17.31 16.50 16.72 17.55 16.46

(b) What proportion $p$ of U.S. high school students smoke? The 2011 Youth Risk Behavioral Survey questioned a random sample of 15,425 students in grades 9 to 12. Of these, 2792 said they had smoked cigarettes at least one day in the past month.

(c) The quality control inspectors in part (a) want to investigate the variability in battery lifetimes by estimating the population variance $\sigma^2$.

SOLUTION:

(a) Use the sample mean $\bar{x}$ as a point estimator for the population mean $\mu$. For these data, our point estimate is $\bar{x} = 16.718$ hours.

(b) Use the sample proportion $\hat{p}$ as a point estimator for the population proportion $p$. For this survey, our point estimate is $\hat{p} = \frac{2792}{15{,}425} = 0.181$.

(c) Use the sample variance $s_x^2$ as a point estimator for the population variance $\sigma^2$. For the battery life data, our point estimate is $s_x^2 = 0.441$ hours$^2$.

For Practice Try Exercises  1 and  3

## ALTERNATE EXAMPLE

### *From golf balls to graphing calculators*

PROBLEM: In each of the following settings, determine the point estimator you would use and calculate the value of the point estimate.

**(a)** The makers of a new golf ball want to estimate the median distance the new balls will travel when hit by a mechanical driver. They select a random sample of 10 balls and measure the distance each ball travels after being hit by the mechanical driver. Here are the distances (in yards):

285 286 284 285 282 284 287 290 288 285

**(b)** The golf ball manufacturer would also like to investigate the variability of the distance traveled by the golf balls by estimating the interquartile range.

**(c)** The math department wants to know what proportion of its students own a graphing calculator, so they take a random sample of 100 students and find that 28 own a graphing calculator.

SOLUTION:

**(a)** Use the sample median as a point estimator for the population median. The sample median is 285 yards.

**(b)** Use the sample *IQR* as a point estimator for the population *IQR*. The sample *IQR* is 287 − 284 = 3 yards.

**(c)** Use the sample proportion $\hat{p}$ as a point estimator for the population proportion $p$. The sample proportion is $\hat{p} = 0.28$.

## The Idea of a Confidence Interval

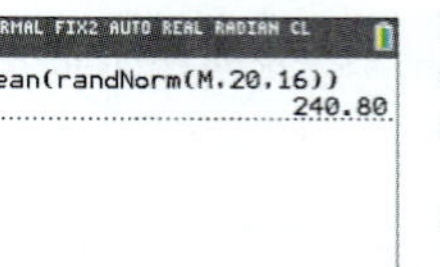


Is the value of the population mean $\mu$ that Mr. Schiel entered in his calculator exactly 240.80? Probably not. Because $\bar{x} = 240.80$, we guess that $\mu$ is "somewhere around 240.80." How close to 240.80 is $\mu$ likely to be?

To answer this question, we ask another: *How would the sample mean $\bar{x}$ vary if we took many SRSs of size 16 from this same population?*

The sampling distribution of $\bar{x}$ describes how the values of $\bar{x}$ vary in repeated samples. Recall the facts about this sampling distribution from Chapter 7:

*Shape:* Because the population distribution is Normal, so is the sampling distribution of $\bar{x}$. Thus, the *Normal/Large Sample condition* is met.
*Center:* The mean of the sampling distribution of $\bar{x}$ is the same as the unknown mean $\mu$ of the entire population. That is, $\mu_{\bar{x}} = \mu$.
*Spread:* The standard deviation of the sampling distribution of $\bar{x}$ for samples of size $n = 16$ is

$$\sigma_{\bar{x}} = \frac{\sigma}{\sqrt{n}} = \frac{20}{\sqrt{16}} = 5$$

## Teaching Tip

Here is a visual way to introduce the idea of a confidence interval:

1. Ask a student volunteer to stand up.
2. Ask if the distance from you to the student is the same as the distance from the student to you.
3. Have the student stand up-front in the classroom, representing $\mu$, the population mean. If possible, make a sign with the $\mu$ symbol for the student to wear.
4. Have the student extend his or her arms, representing 2 standard deviations on either side.
5. You will represent $\bar{x}$—the sample mean. If possible, make a sign with the $\bar{x}$ symbol for you to wear.
6. Ask, "If many samples were taken, how often would I be within the student's arm span?" Around 95%, assuming the distribution of $\bar{x}$ is approximately Normal. Demonstrate this by moving repeatedly back and forth within the student's arm span.
7. Say, "Unfortunately, we do not know the true value of $\mu$." Have the student put his or her arms down.
8. Say, "But, if I am within 2 standard deviations of $\mu$ in 95% of all samples, then $\mu$ will be within 2 standard deviations of me in 95% of all samples."
9. Extend your arms and move back and forth as before, making sure the student is within your arm span 95% of the time.
10. Say, "Thus, if I construct an interval of $\bar{x} \pm$ 2 standard deviations, about 95% of the possible samples will produce intervals that capture $\mu$."

because the *10% condition* is met—we are sampling from an infinite population in this case.

Figure 8.2 summarizes these facts.

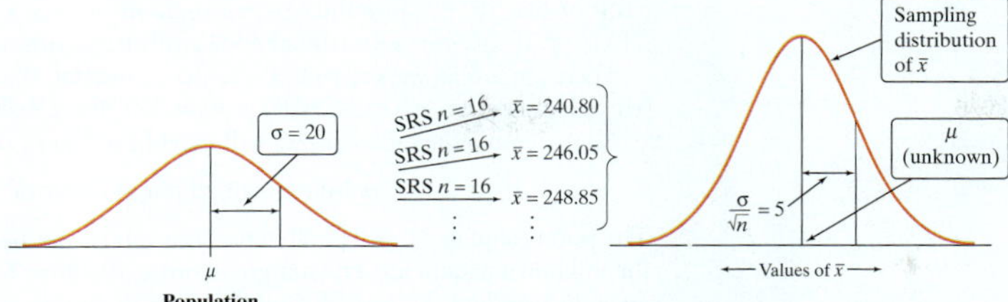


**FIGURE 8.2** The sampling distribution of the mean score $\bar{x}$ for SRSs of 16 observations from a Normally distributed population with unknown mean $\mu$ and standard deviation $\sigma = 20$.

The next example gives the reasoning of statistical estimation in a nutshell.

## EXAMPLE The Mystery Mean

### *Moving beyond a point estimate*

When Mr. Schiel's class discussed the results of the mystery mean Activity, students used the following logic to come up with an "interval estimate" for the unknown population mean $\mu$.

1. The sample mean $\bar{x} = 240.80$ is our point estimate for $\mu$. We don't expect $\bar{x}$ to be exactly equal to $\mu$, so we want to say how precise this estimate is.
2. In repeated samples, the values of $\bar{x}$ follow a Normal distribution with mean $\mu$ and standard deviation 5, as in Figure 8.2.
3. The 95 part of the 68–95–99.7 rule for Normal distributions says that $\bar{x}$ is within $2(5) = 10$ (that's 2 standard deviations) of the population mean $\mu$ in about 95% of all samples of size $n = 16$. See Figure 8.3.
4. Whenever $\bar{x}$ is within 10 points of $\mu$, $\mu$ is within 10 points of $\bar{x}$. This happens in about 95% of all possible samples. So the interval from $\bar{x} - 10$ to $\bar{x} + 10$ "captures" the population mean $\mu$ in about 95% of all samples of size 16.
5. If we estimate that $\mu$ lies somewhere in the interval from

$$\bar{x} - 10 = 240.80 - 10 = 230.80$$

to

$$\bar{x} + 10 = 240.80 + 10 = 250.80$$

we'd be calculating this interval using a method that captures the true $\mu$ in about 95% of all possible samples of this size.

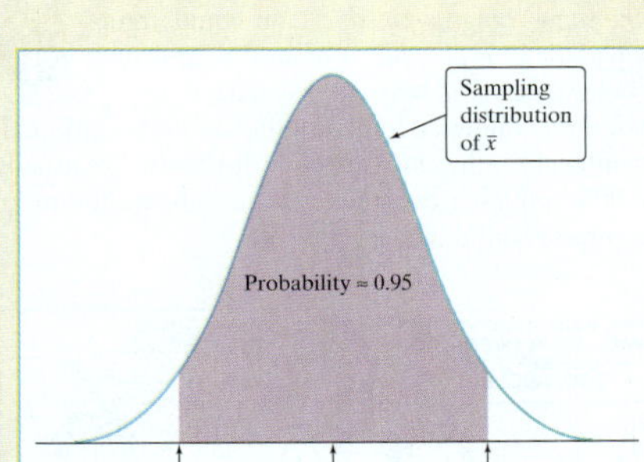


**FIGURE 8.3** In about 95% of all samples, $\bar{x}$ lies within $\pm 10$ of the unknown population mean $\mu$. So $\mu$ also lies within $\pm 10$ of $\bar{x}$ in those samples.

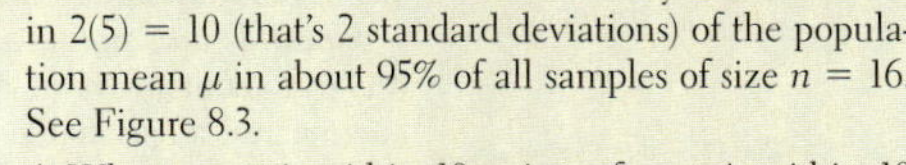

### Teaching Tip

For the distribution of $\bar{x}$ pictured on the right, have students mentally populate the area under the Normal curve with $\bar{x}$'s. This will remind them that the sampling distribution of $\bar{x}$ shows the possible values of $\bar{x}$ and how likely those values are.

### ALTERNATE EXAMPLE

#### *The mystery proportion*

In The Mystery Proportion Alternate Activity, students should have used the following logic to come up with an "interval estimate" for the unknown population proportion $p$:

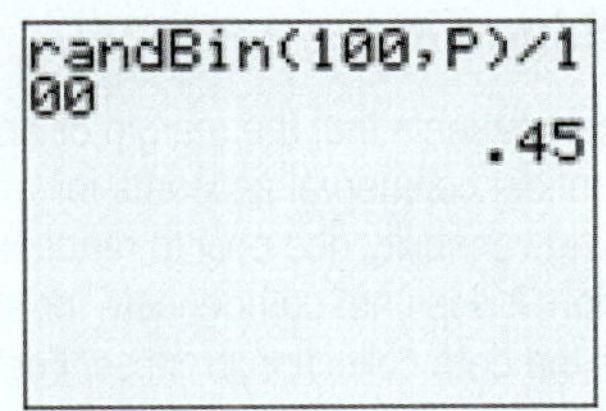


1. Use the sample proportion $\hat{p} = 0.45$ as the point estimate for $p$.
2. In repeated samples, the values of $\hat{p}$ follow an approximately Normal distribution (because the Large Counts condition is met) with mean $p$ and standard deviation $\sigma_{\hat{p}} = \sqrt{\frac{p(1-p)}{100}}$.
Because we don't actually know the value of $p$, we can use $\hat{p}$ to help estimate the standard deviation of the sampling distribution of the sample proportion:
$$\sigma_{\hat{p}} \approx \sqrt{\frac{0.45(1-0.45)}{100}} = 0.05.$$
3. In about 95% of samples of size 100, $\hat{p}$ will be within $2(0.05) = 0.10$ of $p$.
4. Thus, in about 95% of all possible samples of size 100, $p$ will be within 0.10 of $\hat{p}$. We estimate that $p$ lies somewhere between $0.45 - 0.10 = 0.35$ and $0.45 + 0.10 = 0.55$.
5. The interval from 0.35 to 0.55 is an approximate 95% confidence interval for $p$. In other words, it wouldn't be surprising if we found out that the true value of $p$ was any value from 0.35 to 0.55.

The big idea is that the sampling distribution of $\bar{x}$ tells us how close to $\mu$ the sample mean $\bar{x}$ is likely to be. Statistical estimation just turns that information around to say how close to $\bar{x}$ the unknown population mean $\mu$ is likely to be. In the mystery mean example, the value of $\mu$ is usually within $2(5) = 10$ of $\bar{x}$ for SRSs of size 16. Because the class's sample mean was $\bar{x} = 240.80$, the interval $240.80 \pm 10$ gives an approximate **95% confidence interval** for $\mu$.

There are several ways to write a confidence interval. We can give the interval for Mr. Schiel's mystery mean as $240.80 \pm 10$, as 230.80 to 250.80, or as (230.80, 250.80).

All the confidence intervals we will meet have a form similar to this:

$$\text{point estimate} \pm \text{margin of error}$$

The point estimate ($\bar{x} = 240.80$ in our example) is our best guess for the value of the unknown parameter. The **margin of error,** 10, shows how close we believe our guess is, based on the variability of the estimate in repeated SRSs of size 16. We say that our **confidence level** is about 95% because the interval $\bar{x} \pm 10$ catches the unknown parameter in about 95% of all possible samples.

A confidence interval is sometimes referred to as an *interval estimate*. This is consistent with our earlier use of the term *point estimate*.

**DEFINITION: Confidence interval, margin of error, confidence level**

A ***C*% confidence interval** gives an interval of plausible values for a parameter. The interval is calculated from the data and has the form

$$\text{point estimate} \pm \text{margin of error}$$

The difference between the point estimate and the true parameter value will be less than the **margin of error** in *C*% of all samples.

The **confidence level *C*** gives the overall success rate of the method for calculating the confidence interval. That is, in *C*% of all possible samples, the method would yield an interval that captures the true parameter value.

The interval from 230.80 to 250.80 gives the set of plausible values for Mr. Schiel's mystery mean $\mu$ with 95% confidence. We wouldn't be surprised if any of the values in this interval turned out to be the actual value of $\mu$.

*Plausible does not mean the same thing as possible.* You could argue that just about any value of a parameter is *possible*. A *plausible* value of a parameter is a reasonable or believable value based on the data.

There is a trade-off between the confidence level and the amount of information provided by a confidence interval, as the cartoon below illustrates. We usually choose a confidence level of 90% or higher because we want to be quite sure of our conclusions. The most common confidence level is 95%.

## Teaching Tip

When writing a confidence interval in the form (230.80, 250.80), we are not using formal interval notation, which is [230.80, 250.80]. Even though we can be very precise with our calculations, the endpoints of a confidence interval are always "fuzzy" in practice, due to small violations of our conditions (e.g., lack of independent observations due to sampling without replacement).

## Teaching Tip

Students sometimes have trouble distinguishing the margin of error from the standard error. The standard error describes how far the sample statistic typically is from the population parameter. The margin of error is a multiple of the standard error, calculated so that the statistic will be within the margin of error of the parameter in *C*% of samples.

## Teaching Tip

Tell your students that the margin of error in a confidence interval accounts for sampling variability due only to random selection; it does not compensate for any bias in the data collection process. For example, if the sample wasn't random or the questions were worded poorly, you shouldn't have much confidence at all!

## Teaching Tip

Emphasize the word "plausible" when describing the values in a confidence interval. "Plausible" means that if we were to do a census and actually know the value of the population parameter, we wouldn't be surprised if it was any one of the values in the interval. We would be surprised, however, if the true value of the parameter was not one of the values in the interval.

## Teaching Tip

Use the Garfield cartoon on this page to point out two facts about using an interval estimate:

1. Using an interval of values rather than a point estimate greatly increases the chances of being correct.

2. Using an interval that is really wide isn't very helpful (e.g., the weather report in the cartoon won't help you pick out clothes for the day). This is why statisticians typically prefer 95% confidence intervals: they are wide enough to be correct most of the time, but not wider than they need to be. For example, moving from 95% to 99.7% confidence increases our success rate by a little, but increases the width of the interval by 50%.

## Interpreting Confidence Intervals and Confidence Levels

Our 95% confidence interval for Mr. Schiel's mystery mean was (230.80, 250.80). How do we interpret this interval? We say, "We are 95% confident that the interval from 230.80 to 250.80 captures the mystery mean chosen by Mr. Schiel."

**INTERPRETING CONFIDENCE INTERVALS**

To interpret a C% confidence interval for an unknown parameter, say, "We are C% confident that the interval from ____ to ____ captures the [parameter in context]."

Here's an example that involves interpreting a confidence interval for a proportion.

**EXAMPLE**

### Who Will Win the Election?

*Interpreting a confidence interval*

Two weeks before a presidential election, a polling organization asked a random sample of registered voters the following question: "If the presidential election were held today, would you vote for candidate A or candidate B?" Based on this poll, the 95% confidence interval for the population proportion who favor candidate A is (0.48, 0.54).

**PROBLEM:**

(a) Interpret the confidence interval.

(b) What is the point estimate that was used to create the interval? What is the margin of error?

(c) Based on this poll, a political reporter claims that the majority of registered voters favor candidate A. Use the confidence interval to evaluate this claim.

**SOLUTION:**

(a) We are 95% confident that the interval from 0.48 to 0.54 captures the true proportion of all registered voters who favor candidate A in the election.

(b) A confidence interval has the form

point estimate ± margin of error

The point estimate is at the midpoint of the interval. Here, the point estimate is $\hat{p} = 0.51$. The margin of error gives the distance from the point estimate to either end of the interval. So the margin of error for this interval is 0.03.

(c) Any value from 0.48 to 0.54 is a plausible value for the population proportion $p$ that favors candidate A. Because there are plausible values of $p$ less than 0.50, the confidence interval does not give convincing evidence to support the reporter's claim that a majority (more than 50%) of registered voters favor candidate A.

**For Practice** *Try Exercise*  9

**ALTERNATE EXAMPLE**

### *Keep on walkin'*

A large company is concerned that many of its employees are in poor physical condition, which can result in decreased productivity. To determine how many steps each employee takes per day, on average, the company provides a pedometer to 50 randomly selected employees to use for one 24-hour period. After collecting the data, the company statistician reports a 95% confidence interval of 4547 steps to 8473 steps.

**PROBLEM:**

**(a)** Interpret the confidence interval.

**(b)** What is the point estimate that was used to create the interval? What is the margin of error?

**(c)** Recent guidelines suggest that people aim for 10,000 steps per day. Is there convincing evidence that the employees of this company are not meeting the guideline, on average? Explain.

**SOLUTION:**

**(a)** We are 95% confident that the interval from 4547 to 8473 captures the true mean number of steps taken per day for employees at this company.

**(b)** The point estimate is the midpoint of the interval: $\frac{4547 + 8473}{2} = 6510$.

The margin of error is the distance from the point estimate to each endpoint of the interval: $8473 - 6510 = 1963$.

**(c)** Because all the values in the interval are less than 10,000, there is convincing evidence that the employees of this company are not taking 10,000 steps per day, on average. That is, there is convincing evidence that the employees are not meeting the guideline.

### Teaching Tip

The way we interpret a confidence interval is slightly different than you will see in many other books. Our interpretation of Mr. Schiel's interval is "We are 95% confident that the interval from 230.80 to 250.80 captures the mystery mean chosen by Mr. Schiel." Other books might interpret the interval this way: "We are 95% confident that the mystery mean chosen by Mr. Schiel is between 230.80 and 250.80." Both interpretations are fine on the AP® exam, but we prefer our interpretation because it emphasizes that the interval captures the parameter rather than saying that the parameter falls into the interval. After all, the parameter doesn't move—it is a constant. The interval is what varies from sample to sample. We believe this slight change in wording will also help students interpret the confidence level correctly and avoid a very common error on the AP® exam.

The following Activity gives you a chance to explore the meaning of the confidence *level*.

## ACTIVITY | The *Confidence Intervals* Applet

**MATERIALS:**
Computer with Internet connection and display capability

The *Confidence Intervals* applet at the book's Web site will quickly generate many confidence intervals. In this Activity, you will use the applet to investigate the idea of a confidence level.

1. Go to www.whfreeman.com/tps5e and launch the applet. Use the default settings: confidence level 95% and sample size $n = 20$.
2. Click "Sample" to choose an SRS and display the resulting confidence interval. Did the interval capture the population mean $\mu$ (what the applet calls a "hit")? Do this a total of 10 times. How many of the intervals captured the population mean $\mu$? *Note*: So far, you have used the applet to take 10 SRSs, each of size $n = 20$. Be sure you understand the difference between sample size and the number of samples taken.
3. Reset the applet. Click "Sample 25" twice to choose 50 SRSs and display the confidence intervals based on those samples. How many captured the parameter $\mu$? Keep clicking "Sample 25" and observe the value of "Percent hit." What do you notice?
4. Repeat Step 3 using a 90% confidence level.
5. Repeat Step 3 using an 80% confidence level.
6. Summarize what you have learned about the relationship between confidence level and "Percent hit" after taking many samples.

We will investigate the effect of changing the sample size later.

As the Activity confirms, *the confidence level is the overall capture rate if the method is used many times.* Figure 8.4 illustrates the behavior of the confidence interval $\bar{x} \pm 10$ for Mr. Schiel's mystery mean. Starting with the population, imagine taking many SRSs of 16 observations. The first sample has $\bar{x} = 240.80$, the second has $\bar{x} = 246.05$, the third has $\bar{x} = 248.85$, and so on. The sample mean varies from sample to sample, but when we use the formula $\bar{x} \pm 10$ to get an interval based on each sample, about 95% *of these intervals capture the unknown population mean* $\mu$.

### Teaching Tip

It is essential that you do the Activity on this page with your students. The Activity makes it much easier for students to understand the meaning of 95% confidence—in other words, how to interpret the confidence level. Make sure to emphasize that the capture rate of the intervals ("Percent hit") should be the same as the confidence level in the long run.

### Activity Overview

You can watch an overview of this Activity (for teachers) by clicking on the link in the TE-book, opening the TRCD, or logging onto the book's companion Web site www.whfreeman.com/tps5e (teacher log-in required).

### Teaching Tip

In Step 2 of this Activity, ask your students to explain what is happening. Hopefully, they'll say something like, "The computer is selecting an SRS of size 20, computing the sample mean $\bar{x}$, and constructing a 95% confidence interval to estimate $\mu$. If the interval crosses the vertical line, the interval succeeded in capturing $\mu$."

### Teaching Tip

In Steps 4 and 5, ask students why the intervals decrease in length when the confidence level is decreased. Hopefully, they will say that the intervals don't need to be as long because the desired capture rate (confidence level) is smaller.

### ALTERNATE ACTIVITY

***Another confidence interval applet***

1. Go to www.rossmanchance.com/applets and launch the *Simulating Confidence Intervals for a Population Parameter* applet.
2. Keep the default settings of "Proportions" and "Wald." The Wald interval is another name for the standard confidence interval for a proportion that we learn in AP® Statistics. The true proportion is designated $\pi$ instead of $p$. Leave this as 0.5 and keep $n$ as 100.
3. Press "Sample" to take one sample of size 100, and use the sample proportion to compute a 95% confidence interval for the true proportion. Make sure your students understand what is going on in this step, including what the vertical line represents.
4. Keep pressing "Sample" and noting whether the interval captures the true proportion.
5. Now change the number of intervals to 100 and press "Sample." Point out that most of the intervals capture the true proportion, and a few of the intervals do not capture the true proportion. Keep pressing "Sample" and watch how the running total approaches a value close to 95%.

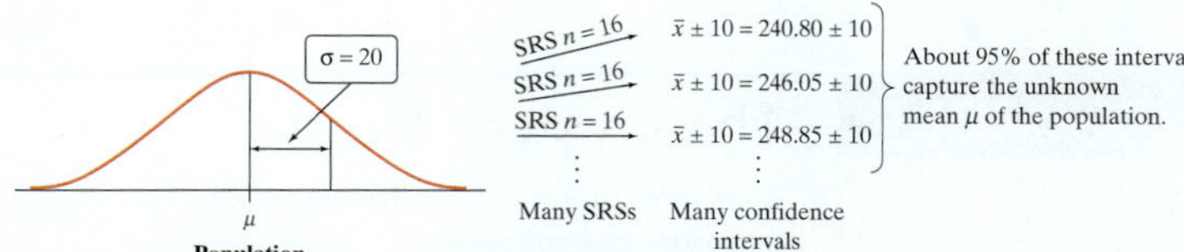


**FIGURE 8.4** To say that $\bar{x} \pm 10$ is a 95% confidence interval for the population mean $\mu$ is to say that, in repeated samples, about 95% of these intervals capture $\mu$.

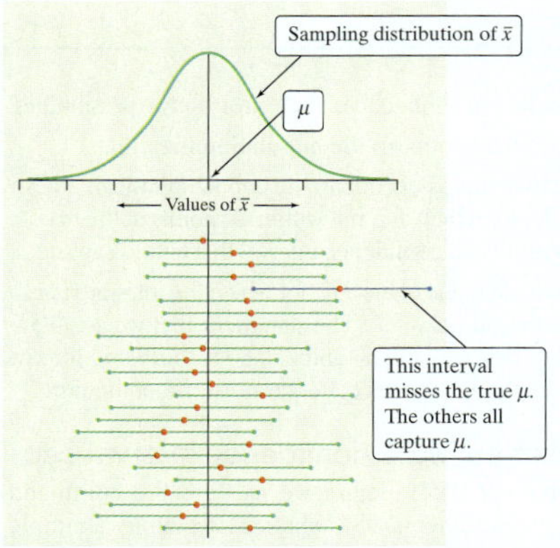


**FIGURE 8.5** Twenty-five samples of the same size from the same population gave these 95% confidence intervals. In the long run, about 95% of samples give an interval that captures the population mean $\mu$.

Figure 8.5 illustrates the idea of a confidence interval in a different form. It shows the result of drawing many SRSs from the same population and calculating a 95% confidence interval from each sample. The center of each interval is at $\bar{x}$ and therefore varies from sample to sample. The sampling distribution of $\bar{x}$ appears at the top of the figure to show the long-term pattern of this variation. The 95% confidence intervals from 25 SRSs appear below.

Here's what you should notice:

- The center $\bar{x}$ of each interval is marked by a dot.
- The distance from the dot to either endpoint of the interval is the margin of error.
- 24 of these 25 intervals (that's 96%) contain the true value of $\mu$. If we took many samples, about 95% of the resulting confidence intervals would capture $\mu$.

Figures 8.4 and 8.5 give us the insight we need to interpret a confidence level.

### INTERPRETING CONFIDENCE LEVELS

To say that we are 95% *confident* is shorthand for "If we take many samples of the same size from this population, about 95% of them will result in an interval that captures the actual parameter value."

The confidence level tells us how likely it is that the method we are using will produce an interval that captures the population parameter *if we use it many times*. However, in practice we tend to calculate only a single confidence interval for a given situation. *The confidence level does not tell us the chance that a particular confidence interval captures the population parameter.* Instead, the confidence interval gives us a set of plausible values for the parameter.

caution

## Teaching Tip

When asked to interpret the confidence level, tell your students to describe Figure 8.5. If many samples are taken and 95% confidence intervals are constructed based on these samples, then about 95% of the intervals will capture the true parameter being estimated. Also, remind your students that in real life we usually calculate just one interval. The figure describes what would happen if we were to take many, many samples.

## AP® Exam Common Error

Many students confuse interpreting confidence *levels* with interpreting confidence *intervals*. Every time students construct a confidence interval, they are expected to interpret the *interval* they calculated (see page 481). They are expected to interpret the confidence *level* only when they are specifically asked to do so. If students interpret the confidence *level* when they aren't asked to, they won't gain any credit and will lose credit if the interpretation is incorrect.

## Teaching Tip

A good interpretation of confidence *level* will not include the endpoints of a specific interval. In fact, we can interpret a confidence level *before* the data are obtained because our confidence is in the method, not a particular interval.

### ALTERNATE EXAMPLE

***The mystery proportion***

The 95% confidence *level* in the mystery proportion example tells us that in about 95% of samples of size 100 from the mystery population, the interval $\hat{p} \pm 0.10$ will contain the population proportion $p$.

### AP® Exam Common Error

When interpreting a confidence interval (or confidence level), many students incorrectly offer statements such as "There is a 95% *probability* that the true mean falls in the interval from 230.80 to 250.80." In AP® Statistics, this is an incorrect statement because we interpret probabilities as long-run relative frequencies. For there to be a 95% probability, the true mean would need to be in the interval 95% of the time and not in the interval 5% of the time. However, because the true mean is a constant and the calculated confidence interval is also constant, the true mean is either in the interval or it isn't. We don't recommend the following, but it is acceptable to use the phrase "95% probability" in reference to a *future* event such as: "If we were to take an SRS and calculate a 95% confidence interval, then there is a 95% probability that the interval we get will contain the true mean."

### ALTERNATE EXAMPLE

***Median household income***

According to the American Community Survey, a 95% confidence interval for the median household income in Texas during the years 2009–2011 is $58,929 ± $218. http://www.census.gov/hhes/www/income/data/statemedian/.

**PROBLEM:** Interpret the confidence interval and the confidence level.

**SOLUTION:** *Confidence interval*: We are 95% confident that the interval from $58,711 to $59,147 captures the true median income for households in Texas. *Confidence level*: If many samples of Texas households were selected, the resulting confidence intervals would capture the true median income for households in Texas for about 95% of those samples.

## EXAMPLE The Mystery Mean

***Interpreting a confidence level***

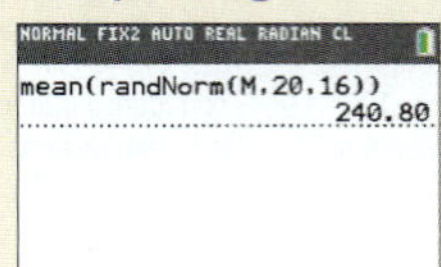


The confidence level in the mystery mean example—roughly 95%—tells us that if we take many SRSs of size 16 from Mr. Schiel's mystery population, the interval $\bar{x} \pm 10$ will capture the population mean $\mu$ for about 95% of those samples.

Be sure you understand the basis for our confidence. There are only two possibilities:

1. The interval from 230.80 to 250.80 contains the population mean $\mu$.
2. The interval from 230.80 to 250.80 does *not* contain the population mean $\mu$. Our SRS was one of the few samples for which $\bar{x}$ is not within 10 points of the true $\mu$. Only about 5% of all samples result in a confidence interval that fails to capture $\mu$.

We cannot know whether our sample is one of the 95% for which the interval $\bar{x} \pm 10$ catches $\mu$ or whether it is one of the unlucky 5%. The statement that we are "95% confident" that the unknown $\mu$ lies between 230.80 and 250.80 is shorthand for saying, "We got these numbers by a method that gives correct results 95% of the time."

**What's the probability that our 95% confidence interval captures the parameter?** It's *not* 95%! Before we execute the command `mean(randNorm(M,20,16))`, we have a 95% chance of getting a sample mean that's within $2\sigma_{\bar{x}}$ of the mystery $\mu$, which would lead to a confidence interval that captures $\mu$. Once we have chosen a random sample, the sample mean $\bar{x}$ either is or isn't within $2\sigma_{\bar{x}}$ of $\mu$. And the resulting confidence interval either does or doesn't contain $\mu$. *After* we construct a confidence interval, the probability that it captures the population parameter is either *1 (it does)* or *0 (it doesn't)*.

We interpret confidence intervals and confidence levels in much the same way whether we are estimating a population mean, proportion, or some other parameter.

## EXAMPLE Do You Use Twitter?

***Interpreting a confidence interval and a confidence level***

The Pew Internet and American Life Project asked a random sample of 2253 U.S. adults, "Do you ever . . . use Twitter or another service to share updates about yourself or to see updates about others?" Of the sample, 19% said "Yes." According to Pew, the resulting 95% confidence interval is (0.167, 0.213).[2]

PROBLEM: Interpret the confidence interval and the confidence level.

SOLUTION:

*Confidence interval:* We are 95% confident that the interval from 0.167 to 0.213 captures the true proportion $p$ of all U.S. adults who use Twitter or another service for updates.

*Confidence level:* If many samples of 2253 U.S. adults were taken, the resulting confidence intervals would capture the true proportion of all U.S. adults who use Twitter or another service for updates for about 95% of those samples.

**For Practice** *Try Exercise* **15**

**AP® EXAM TIP** On a given problem, you may be asked to interpret the confidence interval, the confidence level, or both. Be sure you understand the difference: the confidence interval gives a set of plausible values for the parameter and the confidence level describes the long-run capture rate of the method.

Confidence intervals are statements about *parameters*. In the previous example, it would be wrong to say, "We are 95% confident that the interval from 0.167 to 0.213 contains the *sample* proportion of U.S. adults who use Twitter." Why? Because we *know* that the *sample* proportion, $\hat{p} = 0.19$, is in the interval. Likewise, in the mystery mean example, it would be wrong to say that "95% of the values are between 230.80 and 250.80," whether we are referring to the sample or the population. All we can say is, "Based on Mr. Schiel's sample, we believe that the *population mean* is somewhere between 230.80 and 250.80."

When interpreting a confidence interval, make it clear that you are describing a parameter and not a statistic. And be sure to include context.

## CHECK YOUR UNDERSTANDING

How much does the fat content of Brand X hot dogs vary? To find out, researchers measured the fat content (in grams) of a random sample of 10 Brand X hot dogs. A 95% confidence interval for the population standard deviation $\sigma$ is 2.84 to 7.55.

1. Interpret the confidence interval.
2. Interpret the confidence level.
3. True or false: The interval from 2.84 to 7.55 has a 95% chance of containing the actual population standard deviation $\sigma$. Justify your answer.

## Constructing a Confidence Interval

Why settle for 95% confidence when estimating an unknown parameter? Do larger random samples yield "better" intervals? The *Confidence Intervals* applet might shed some light on these questions.

### ACTIVITY | The *Confidence Intervals* Applet

**MATERIALS:** Computer with Internet connection and display capability

In this Activity, you will use the applet to explore the relationship between the confidence level, the sample size, and the confidence interval.

1. Go to www.whfreeman.com/tps5e and launch the *Confidence Intervals* applet. Use the default settings: confidence level 95% and sample size $n = 20$. Click "Sample 25."

### AP® Exam Common Error

Many students lose credit for referring to the sample when interpreting a confidence interval. For example, the following interpretation is incorrect because it makes a conclusion about the members of the sample rather than the members of the population: "We are 95% confident that the interval from 0.167 to 0.213 captures the true proportion of U.S. adults who said that they use Twitter." Replacing the word "said" with "would say" makes the conclusion about the larger population and not just the people who were in the sample.

### ANSWERS TO CYU

1. We are 95% confident that the interval from 2.84 to 7.55 g captures the population standard deviation of the fat content of Brand X hot dogs.

2. If this sampling process were repeated many times, approximately 95% of the resulting confidence intervals would capture the population standard deviation of the fat content of Brand X hot dogs.

3. False. Once the interval is calculated, it either contains $\sigma$ or it does not contain $\sigma$.

### Activity Overview

You can watch an overview of this Activity (for teachers) by clicking on the link in the TE-book, opening the TRCD, or logging onto the book's companion Web site www.whfreeman.com/tps5e (teacher log-in required).

### ALTERNATE ACTIVITY

#### *Another confidence interval applet*

1. Go to www.rossmanchance.com/applets and launch the *Simulating Confidence Intervals for a Population Parameter* applet.

2. Using the default values of $\pi = 0.5$ and $n = 100$, set the number of intervals to 100 and the confidence level to 95%. Press "Sample" many times. What is the capture rate for these intervals?

3. Keeping the other values the same, change the confidence level to 50%. Press "Sample" over and over. What is the capture rate for these intervals? What happened to the length of the intervals?

4. Summarize what you have learned about the relationship between the confidence level and the length of a confidence interval for a fixed sample size.

5. Change the confidence level back to 95% and increase the sample size to 400. How does increasing the sample size affect the length of the confidence intervals? How does increasing the sample size affect the confidence level (capture rate)?

## AP® Exam Common Error

Make sure to discuss Step 7 of this Activity with your students. Increasing the sample size should *not* affect the confidence level. The intervals will be shorter (more precise), but the capture rate should still be around 95%. Consequently, it would be incorrect to say, "Increasing the sample size makes us more confident." A correct statement would be that "increasing the sample size makes our estimate more precise."

2. Change the confidence level to 99%. What happens to the length of the confidence intervals?
3. Now change the confidence level to 90%. What happens to the length of the confidence intervals?
4. Finally, change the confidence level to 80%. What happens to the length of the confidence intervals?
5. Summarize what you learned about the relationship between the confidence level and the length of the confidence intervals for a fixed sample size.
6. Reset the applet and change the confidence level to 95%. What happens to the length of the confidence intervals as you increase the sample size?
7. Does increasing the sample size increase the capture rate (percent hit)? Use the applet to investigate.

As the Activity illustrates, the price we pay for greater confidence is a wider interval. If we're satisfied with 80% confidence, then our interval of plausible values for the parameter will be much narrower than if we insist on 90%, 95%, or 99% confidence. But we'll also be much less confident in our estimate. Taking the idea of confidence to an extreme, what if we want to estimate with 100% confidence the proportion $p$ of all U.S. adults who use Twitter? That's easy: we're 100% confident that the interval from 0 to 1 captures the true population proportion!

The activity also shows that we can get a more precise estimate of a parameter by increasing the sample size. Larger samples yield narrower confidence intervals. This result holds for any confidence level.

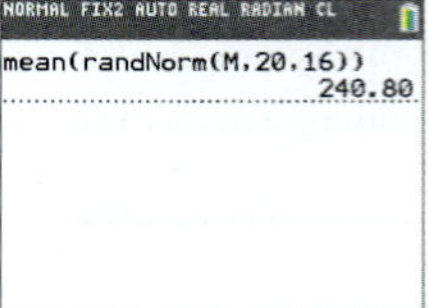


Let's look a bit more closely at the method we used earlier to calculate an approximate 95% confidence interval for Mr. Schiel's mystery mean. We started with

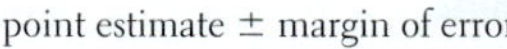

$$\text{point estimate} \pm \text{margin of error}$$

Our point estimate came from the sample statistic $\bar{x} = 240.80$. What about the margin of error? Because the population distribution is Normal, so is the sampling distribution of $\bar{x}$. About 95% of the values of $\bar{x}$ will lie within 2 standard deviations ($2\sigma_{\bar{x}}$) of the mystery mean $\mu$. See the figure below. We could rewrite our interval as

$$240.80 \pm 2 \cdot 5 = \bar{x} \pm 2 \cdot \sigma_{\bar{x}}$$

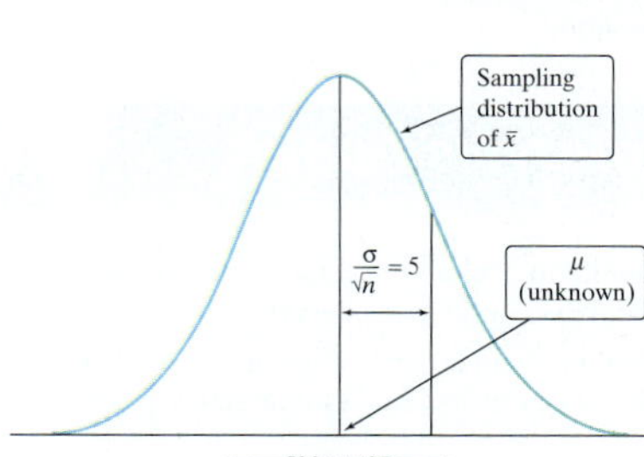


This leads to the more general formula for a confidence interval:

$$\text{statistic} \pm (\text{critical value}) \cdot (\text{standard deviation of statistic})$$

The **critical value** is a multiplier that makes the interval wide enough to have the stated capture rate. The critical value depends on both the confidence level C and the sampling distribution of the statistic.

## Teaching Tip

Point out that using a critical value of 1 instead of 2 would result in 68% confidence. Likewise, using a critical value of 3 would result in 99.7% confidence. Don't worry about calculating other critical values in this section, however. We will learn how to do this in Sections 8.2 and 8.3.

**CALCULATING A CONFIDENCE INTERVAL**

The confidence interval for estimating a population parameter has the form

statistic ± (critical value) · (standard deviation of statistic)

where the statistic we use is the point estimator for the parameter.

### AP® Exam Formula Sheet

This general formula for calculating a confidence interval is included on the formula sheet that is given to students on both sections of the AP® exam. Formulas are also provided for the standard deviations of the different statistics that we use in AP® Statistics.

The confidence interval for the mystery mean $\mu$ of Mr. Schiel's population illustrates several important properties that are shared by all confidence intervals in common use. The user chooses the confidence level, and the margin of error follows from this choice. We would like high confidence and also a small margin of error. High confidence says that our method almost always gives correct answers. A small margin of error says that we have pinned down the parameter quite precisely.

Our general formula for a confidence interval is

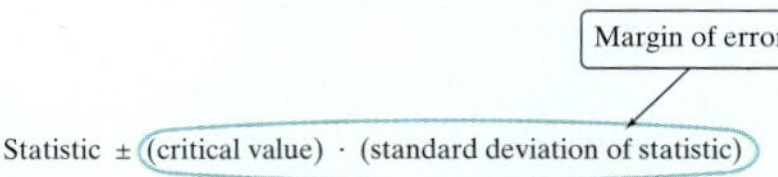


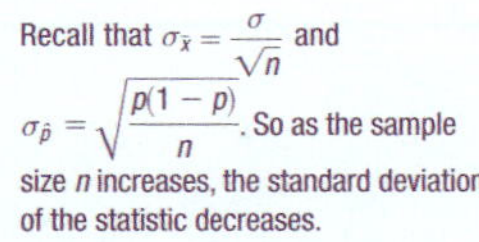

Recall that $\sigma_{\bar{x}} = \frac{\sigma}{\sqrt{n}}$ and $\sigma_{\hat{p}} = \sqrt{\frac{p(1-p)}{n}}$. So as the sample size $n$ increases, the standard deviation of the statistic decreases.

We can see that the margin of error depends on the critical value and the standard deviation of the statistic. The critical value is tied directly to the confidence level: greater confidence requires a larger critical value. The standard deviation of the statistic depends on the sample size $n$: larger samples give more precise estimates, which means less variability in the statistic.

So the margin of error gets smaller when:

- *The confidence level decreases.* There is a trade-off between the confidence level and the margin of error. To obtain a smaller margin of error from the same data, you must be willing to accept lower confidence. Earlier, we found that a 95% confidence interval for Mr. Schiel's mystery mean $\mu$ is 230.80 to 250.80. The 80% confidence interval for $\mu$ is 234.39 to 247.21. Figure 8.6 compares these two intervals.
- *The sample size n increases.* Increasing the sample size $n$ reduces the margin of error for any fixed confidence level.

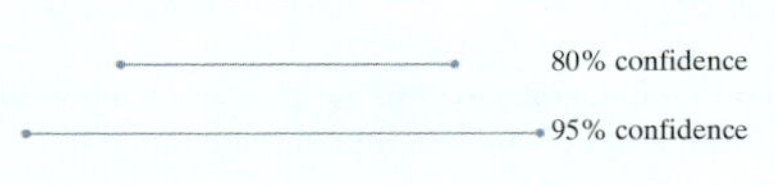


**FIGURE 8.6** 80% and 95% confidence intervals for Mr. Schiel's mystery mean. Higher confidence requires a longer interval.

### Teaching Tip

Ask your students whether there is a trade-off for increasing the sample size. The trade-off is that increasing the sample size is often very costly and time-consuming!

## Using Confidence Intervals Wisely

Our goal in this section has been to introduce you to the big ideas of confidence intervals without getting bogged down in details. You may have noticed that we only calculated intervals in a contrived setting: estimating an unknown population mean $\mu$ when we somehow knew the population standard deviation $\sigma$. In practice, when we don't know $\mu$, we don't know $\sigma$ either. We'll learn to construct confidence intervals for a population mean in this more realistic setting in Section 8.3. First, we will study confidence intervals for a population proportion $p$ in Section 8.2. Although it is possible to estimate other parameters, confidence intervals for means and proportions are the most common tools in everyday use.

### Teaching Tip

If we use the standard formula for a confidence interval based on an SRS when data were obtained from a well-designed stratified random sample, our intervals will be wider than they need to be. In other words, the intervals calculated in this way will have a capture rate that is higher than the stated confidence level. See #6 on the 2012 AP® exam for an example of how the formula for the standard error of the sample mean is calculated when a stratified random sample is used.

### TRM Section 8.1 Quizzes

There are three different quizzes available for this section in the Teacher's Resource Materials. Versions A and B are roughly parallel. Find them by clicking on the link in the TE-book, opening the TRCD, or logging onto the book's companion Web site www.whfreeman.com/tps5e (teacher log-in required).

Here are two important cautions to keep in mind when constructing and interpreting confidence intervals.

- *Our method of calculation assumes that the data come from an SRS of size n from the population of interest.* Other types of random samples (stratified or cluster, say) might be preferable to an SRS in a given setting, but they require more complex calculations than the ones we'll use.
- *The margin of error in a confidence interval covers only chance variation due to random sampling or random assignment.* The margin of error is obtained from the sampling distribution. It indicates how close our estimate is likely to be to the unknown parameter if we repeat the random sampling or random assignment process many times. Practical difficulties, such as undercoverage and nonresponse in a sample survey, can lead to additional errors that may be larger than this chance variation. Remember this unpleasant fact when reading the results of an opinion poll or other sample survey. The way in which a survey or experiment is conducted influences the trustworthiness of its results in ways that are not included in the announced margin of error.

## Section 8.1 Summary

- To estimate an unknown population parameter, start with a statistic that provides a reasonable guess. The chosen statistic is a **point estimator** for the parameter. The specific value of the point estimator that we use gives a **point estimate** for the parameter.
- A **C% confidence interval** uses sample data to estimate an unknown population parameter with an indication of how precise the estimate is and of how confident we are that the result is correct.
- A confidence interval gives an interval of plausible values for the parameter. The interval is computed from the data and has the form

$$\text{point estimate} \pm \text{margin of error}$$

  When calculating a confidence interval, it is common to use the form

$$\text{statistic} \pm (\text{critical value}) \cdot (\text{standard deviation of statistic})$$

- To interpret a C% confidence interval, say, "We are C% confident that the interval from ____ to ____ captures the [parameter in context]." Be sure that your interpretation describes a parameter and not a statistic.
- The **confidence level C** is the success rate of the method that produces the interval. If you use 95% confidence intervals often, in the long run about 95% of your intervals will contain the true parameter value. You don't know whether a 95% confidence interval calculated from a particular set of data actually captures the true parameter value.

- Other things being equal, the **margin of error** of a confidence interval gets smaller as
  - the confidence level C decreases.
  - the sample size $n$ increases.
- Remember that the margin of error for a confidence interval includes only chance variation, not other sources of error like nonresponse and undercoverage.

## Section 8.1 Exercises

*In Exercises 1 to 4, determine the point estimator you would use and calculate the value of the point estimate.*

1. **Got shoes?** How many pairs of shoes, on average, do female teens have? To find out, an AP® Statistics class conducted a survey. They selected an SRS of 20 female students from their school. Then they recorded the number of pairs of shoes that each student reported having. Here are the data:

pg 478 

| 50 | 26 | 26 | 31 | 57 | 19 | 24 | 22 | 23 | 38 |
|---|---|---|---|---|---|---|---|---|---|
| 13 | 50 | 13 | 34 | 23 | 30 | 49 | 13 | 15 | 51 |

2. **Got shoes?** The class in Exercise 1 wants to estimate the variability in the number of pairs of shoes that female students have by estimating the population variance $\sigma^2$.

3. **Going to the prom** Tonya wants to estimate what proportion of the seniors in her school plan to attend the prom. She interviews an SRS of 50 of the 750 seniors in her school and finds that 36 plan to go to the prom.

pg 478 

4. **Reporting cheating** What proportion of students are willing to report cheating by other students? A student project put this question to an SRS of 172 undergraduates at a large university: "You witness two students cheating on a quiz. Do you go to the professor?" Only 19 answered "Yes."[3]

5. **NAEP scores** Young people have a better chance of full-time employment and good wages if they are good with numbers. How strong are the quantitative skills of young Americans of working age? One source of data is the National Assessment of Educational Progress (NAEP) Young Adult Literacy Assessment Survey, which is based on a nationwide probability sample of households. The NAEP survey includes a short test of quantitative skills, covering mainly basic arithmetic and the ability to apply it to realistic problems. Scores on the test range from 0 to 500. For example, a person who scores 233 can add the amounts of two checks appearing on a bank deposit slip; someone scoring 325 can determine the price of a meal from a menu; a person scoring 375 can transform a price in cents per ounce into dollars per pound.[4]

   Suppose that you give the NAEP test to an SRS of 840 people from a large population in which the scores have mean 280 and standard deviation $\sigma = 60$. The mean $\bar{x}$ of the 840 scores will vary if you take repeated samples.

   (a) Describe the shape, center, and spread of the sampling distribution of $\bar{x}$.

   (b) Sketch the sampling distribution of $\bar{x}$. Mark its mean and the values 1, 2, and 3 standard deviations on either side of the mean.

   (c) According to the 68–95–99.7 rule, about 95% of all values of $\bar{x}$ lie within a distance $m$ of the mean of the sampling distribution. What is $m$? Shade the region *on the axis* of your sketch that is within $m$ of the mean.

   (d) Whenever $\bar{x}$ falls in the region you shaded, the population mean $\mu$ lies in the confidence interval $\bar{x} \pm m$. For what percent of all possible samples does the interval capture $\mu$?

6. **Auto emissions** Oxides of nitrogen (called NOX for short) emitted by cars and trucks are important contributors to air pollution. The amount of NOX emitted by a particular model varies from vehicle to vehicle. For one light-truck model, NOX emissions vary with mean $\mu = 1.8$ grams per mile and standard deviation $\sigma = 0.4$ gram per mile. You test an SRS of 50 of these trucks. The sample mean NOX level $\bar{x}$ will vary if you take repeated samples.

**TRM Full Solutions to Section 8.1 Exercises**

### Answers to Section 8.1 Exercises

**8.1** Sample mean, $\bar{x} = 30.35$

**8.2** Sample variance, $s_x^2 = 202.77$

**8.3** Sample proportion, $\hat{p} = \frac{36}{50} = 0.72$

**8.4** Sample proportion, $\hat{p} = \frac{19}{172} = 0.11$

**8.5** **(a)** Approximately Normal with mean $\mu_{\bar{x}} = 280$ and standard deviation $\sigma_{\bar{x}} = \frac{60}{\sqrt{840}} = 2.1$. **(b)** See graph below.

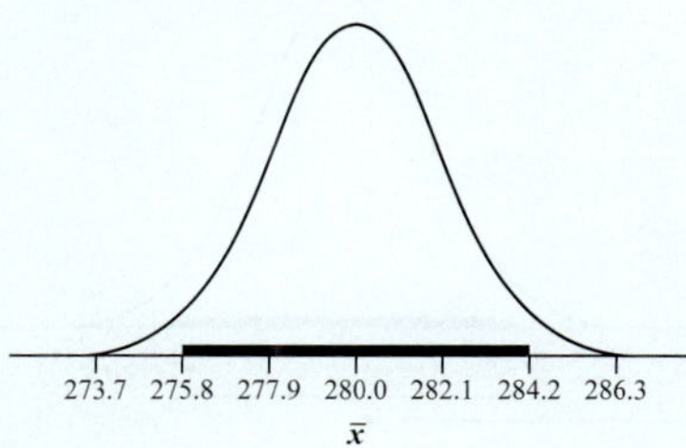


**(c)** About 95% of the $\bar{x}$ values will be within 2 standard deviations of the mean. Therefore, $m = 2(2.1) = 4.2$. **(d)** About 95%

**8.6** **(a)** Approximately Normal with mean $\mu_{\bar{x}} = 1.8$ and standard deviation $\sigma_{\bar{x}} = \dfrac{0.4}{\sqrt{50}} = 0.057$. **(b)** See graph below.

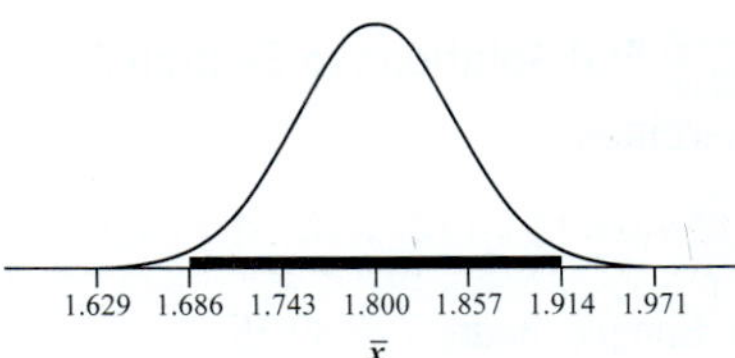


**(c)** About 95% of the $\bar{x}$ values will be within 2 standard deviations of the mean. Therefore, $m = 2(0.057) = 0.114$.
**(d)** About 95%

**8.7** The sketch is given below. The interval with the value of $\bar{x}$ in the shaded region will contain the population mean (280), while the interval with the value of $\bar{x}$ outside the shaded region will not contain the population mean (280).

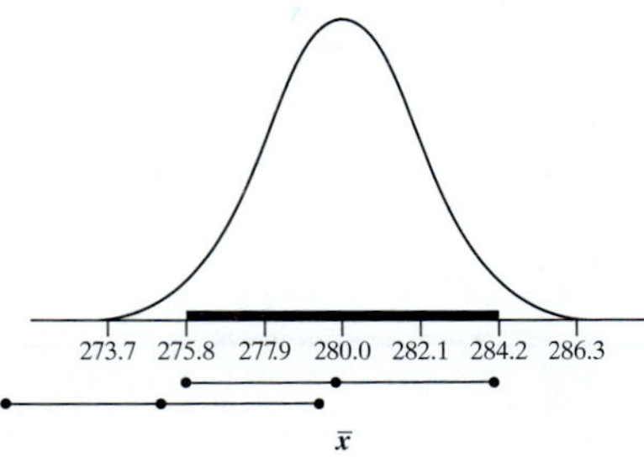


**8.8** The sketch is given below. The interval with the value of $\bar{x}$ in the shaded region will contain the population mean (1.8), while the interval with the value of $\bar{x}$ outside the shaded region will not contain the population mean (1.8).

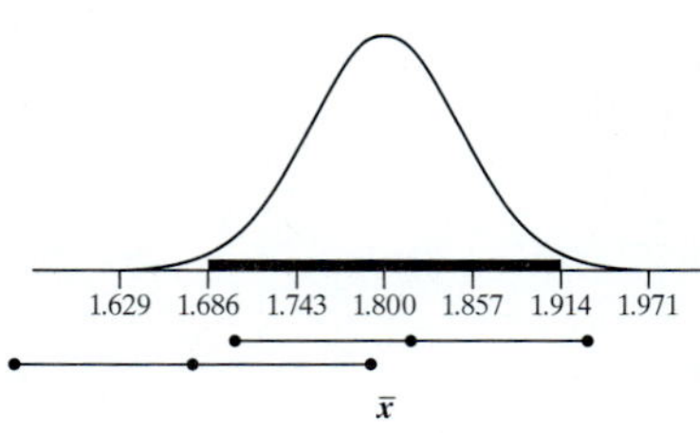


**8.9** **(a)** We are 95% confident that the interval from 0.63 to 0.69 captures the true proportion of those who favor an amendment to the Constitution that would permit organized prayer in public schools. **(b)** Point estimate $= \hat{p} = \dfrac{0.63 + 0.69}{2} = 0.66$ and margin of error $= 0.69 - 0.66 = 0.03$.
**(c)** Because the value $2/3 = 0.667$ (and values less than 2/3) are in the interval of plausible values, there is not convincing evidence that more than two-thirds of U.S. adults favor such an amendment.



(a) Describe the shape, center, and spread of the sampling distribution of $\bar{x}$.

(b) Sketch the sampling distribution of $\bar{x}$. Mark its mean and the values 1, 2, and 3 standard deviations on either side of the mean.

(c) According to the 68–95–99.7 rule, about 95% of all values of $\bar{x}$ lie within a distance $m$ of the mean of the sampling distribution. What is $m$? Shade the region *on the axis* of your sketch that is within $m$ of the mean.

(d) Whenever $\bar{x}$ falls in the region you shaded, the unknown population mean $\mu$ lies in the confidence interval $\bar{x} \pm m$. For what percent of all possible samples does the interval capture $\mu$?

7. **NAEP scores** Refer to Exercise 5. Below your sketch, choose one value of $\bar{x}$ inside the shaded region and draw its corresponding confidence interval. Do the same for one value of $\bar{x}$ outside the shaded region. What is the most important difference between these intervals? (Use Figure 8.5, on page 483, as a model for your drawing.)

8. **Auto emissions** Refer to Exercise 6. Below your sketch, choose one value of $\bar{x}$ inside the shaded region and draw its corresponding confidence interval. Do the same for one value of $\bar{x}$ outside the shaded region. What is the most important difference between these intervals? (Use Figure 8.5, on page 483, as a model for your drawing.)

pg 481

9. **Prayer in school** A *New York Times*/CBS News Poll asked a random sample of U.S. adults the question, "Do you favor an amendment to the Constitution that would permit organized prayer in public schools?" Based on this poll, the 95% confidence interval for the population proportion who favor such an amendment is (0.63, 0.69).

(a) Interpret the confidence interval.

(b) What is the point estimate that was used to create the interval? What is the margin of error?

(c) Based on this poll, a reporter claims that more than two-thirds of U.S. adults favor such an amendment. Use the confidence interval to evaluate this claim.

10. **Losing weight** A Gallup Poll asked a random sample of U.S. adults, "Would you like to lose weight?" Based on this poll, the 95% confidence interval for the population proportion who want to lose weight is (0.56, 0.62).[5]

(a) Interpret the confidence interval.

(b) What is the point estimate that was used to create the interval? What is the margin of error?

(c) Based on this poll, Gallup claims that more than half of U.S. adults want to lose weight. Use the confidence interval to evaluate this claim.

11. **How confident?** The figure below shows the result of taking 25 SRSs from a Normal population and constructing a confidence interval for each sample. Which confidence level—80%, 90%, 95%, or 99%—do you think was used? Explain.

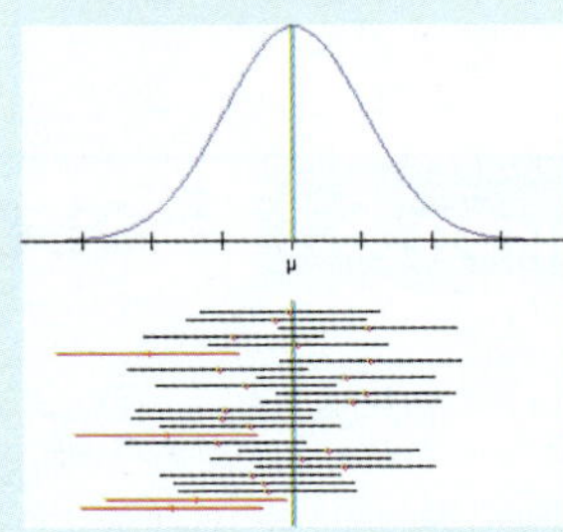

12. **How confident?** The figure below shows the result of taking 25 SRSs from a Normal population and constructing a confidence interval for each sample. Which confidence level—80%, 90%, 95%, or 99%—do you think was used? Explain.

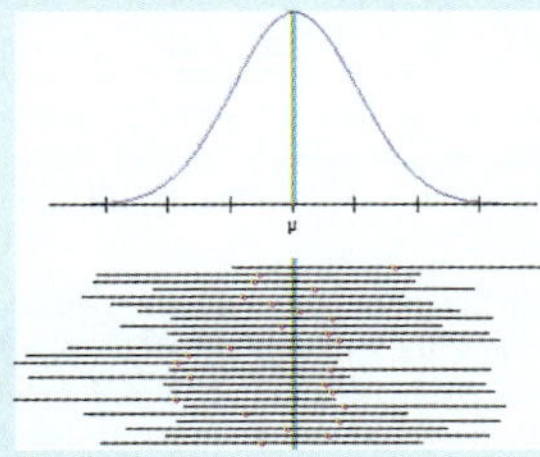

13. **Prayer in school** Refer to Exercise 9. The news article goes on to say: "The theoretical errors do not take into account · · · additional error resulting from the various practical difficulties in taking any survey of public opinion." List some of the "practical difficulties" that may cause errors which are not included in the ±3 percentage point margin of error.

14. **Losing weight** Refer to Exercise 10. As Gallup indicates, the 3 percentage point margin of error for this poll includes only sampling variability (what they call "sampling error"). What other potential sources of error (Gallup calls these "nonsampling errors") could affect the accuracy of the 95% confidence interval?

pg 484

15. **Shoes** The AP® Statistics class in Exercise 1 also asked an SRS of 20 boys at their school how many

**8.10** **(a)** We are 95% confident that the interval from 0.56 to 0.62 captures the true proportion of U.S. adults who would like to lose weight.

**(b)** Point estimate $= \hat{p} = \dfrac{0.56 + 0.62}{2} = 0.59$ and margin of error $= 0.62 - 0.59 = 0.03$.
**(c)** Because all of the plausible values in the interval are above 0.5, there is convincing evidence that more than half of U.S. adults want to lose weight.

**8.11** Because only 84% of the intervals actually contained the true parameter, these were probably 80% or 90% confidence intervals.

**8.12** Because 100% of the intervals contained the true mean, the confidence level was probably 99%.

**8.13** Answers will vary. One practical difficulty is response bias: people might answer "yes" because they think they should, even if they don't really support the amendment.

**8.14** Answers will vary. One practical difficulty is response bias: people might answer "yes" because they know that they should lose weight, even if they really don't want to.

pairs of shoes they have. A 95% confidence interval for the difference in the population means (girls – boys) is 10.9 to 26.5. Interpret the confidence interval and the confidence level.

16. **Lying online** Many teens have posted profiles on sites such as Facebook. A sample survey asked random samples of teens with online profiles if they included false information in their profiles. Of 170 younger teens (ages 12 to 14) polled, 117 said "Yes." Of 317 older teens (ages 15 to 17) polled, 152 said "Yes."[6] A 95% confidence interval for the difference in the population proportions (younger teens – older teens) is 0.120 to 0.297. Interpret the confidence interval and the confidence level.

17. **Shoes** Refer to Exercise 15. Does the confidence interval give convincing evidence of a difference in the population mean number of pairs of shoes for boys and girls at the school? Justify your answer.

18. **Lying online** Refer to Exercise 16. Does the confidence interval give convincing evidence of a difference in the population proportions of younger and older teens who include false information in their profiles? Justify your answer.

19. **Explaining confidence** A 95% confidence interval for the mean body mass index (BMI) of young American women is 26.8 ± 0.6. Discuss whether each of the following explanations is correct.

(a) We are confident that 95% of all young women have BMI between 26.2 and 27.4.

(b) We are 95% confident that future samples of young women will have mean BMI between 26.2 and 27.4.

(c) Any value from 26.2 to 27.4 is believable as the true mean BMI of young American women.

(d) If we take many samples, the population mean BMI will be between 26.2 and 27.4 in about 95% of those samples.

(e) The mean BMI of young American women cannot be 28.

20. **Explaining confidence** The admissions director from Big City University found that (107.8, 116.2) is a 95% confidence interval for the mean IQ score of all freshmen. Discuss whether each of the following explanations is correct.

(a) There is a 95% probability that the interval from 107.8 to 116.2 contains $\mu$.

(b) There is a 95% chance that the interval (107.8, 116.2) contains $\bar{x}$.

(c) This interval was constructed using a method that produces intervals that capture the true mean in 95% of all possible samples.

(d) If we take many samples, about 95% of them will contain the interval (107.8, 116.2).

(e) The probability that the interval (107.8, 116.2) captures $\mu$ is either 0 or 1, but we don't know which.

***Multiple choice: Select the best answer for Exercises 21 to 24.***

*Exercises 21 and 22 refer to the following setting.* A researcher plans to use a random sample of families to estimate the mean monthly family income for a large population.

21. The researcher is deciding between a 95% confidence level and a 99% confidence level. Compared to a 95% confidence interval, a 99% confidence interval will be

(a) narrower and would involve a larger risk of being incorrect.

(b) wider and would involve a smaller risk of being incorrect.

(c) narrower and would involve a smaller risk of being incorrect.

(d) wider and would involve a larger risk of being incorrect.

(e) wider and would have the same risk of being incorrect.

22. The researcher is deciding between a sample of size $n = 500$ and a sample of size $n = 1000$. Compared to using a sample size of $n = 500$, a 95% confidence interval based on a sample size of $n = 1000$ will be

(a) narrower and would involve a larger risk of being incorrect.

(b) wider and would involve a smaller risk of being incorrect.

(c) narrower and would involve a smaller risk of being incorrect.

(d) wider and would involve a larger risk of being incorrect.

(e) narrower and would have the same risk of being incorrect.

23. In a poll,

I. Some people refused to answer questions.

II. People without telephones could not be in the sample.

III. Some people never answered the phone in several calls.

Which of these possible sources of bias is included in the ±2% margin of error announced for the poll?

(a) I only (b) II only (c) III only (d) I, II, and III (e) None of these

24. You have measured the systolic blood pressure of an SRS of 25 company employees. A 95% confidence interval for the mean systolic blood pressure for the employees of this company is (122, 138). Which of the following statements is true?

**8.15** *Interval*: We are 95% confident that the interval from 10.9 to 26.5 captures the true difference (girls − boys) in the mean number of pairs of shoes owned by girls and boys. *Level*: If this sampling process were repeated many times, approximately 95% of the resulting confidence intervals would capture the true difference (girls − boys) in the mean number of pairs of shoes owned by girls and boys.

**8.16** *Interval*: We are 95% confident that the interval from 0.120 to 0.297 captures the true difference (younger − older) in the proportions of younger teens and older teens who include false information on their profiles. *Level*: If this sampling process were repeated many times, approximately 95% of the resulting confidence intervals would capture the true difference (younger − older) in the proportions of younger teens and older teens who include false information on their profiles.

**8.17** Yes. Because the interval does not include 0 as a plausible value, there is convincing evidence of a difference in the mean number of shoes for boys and girls.

**8.18** Yes. Because the interval does not include 0 as a plausible value, there is convincing evidence of a difference in the proportion of younger teens and older teens who include false information on their profiles.

**8.19** **(a)** Incorrect. The interval provides plausible values for the *mean* BMI of all women, not plausible values for individual BMI measurements. **(b)** Incorrect. We shouldn't use the results of one sample to predict the results for future samples. **(c)** Correct. A confidence interval provides an interval of plausible values for a parameter. **(d)** Incorrect. The population mean doesn't change and will either be a value between 26.2 and 27.4 100% of the time or 0% of the time. **(e)** Incorrect. We are 95% confident that the population mean is between 26.2 and 27.4, but that does not absolutely rule out any other possibility.

**8.20** **(a)** Incorrect. The population mean is always the same, so the probability that $\mu$ is in the interval is either 0 or 1. **(b)** Incorrect. The point estimate $\bar{x}$ will always be in the center of the confidence interval. **(c)** Correct. This is the meaning of 95% confidence. **(d)** Incorrect. It doesn't make sense to say that a sample contains an interval. **(e)** Correct. The value of $\mu$ is always the same, so it is either always in the interval or always not in the interval.

**8.21** b

**8.22** e

**8.23** e

**8.24** c

**8.25** **(a)** Observational study, because there was no treatment imposed on the pregnant women or the children. **(b)** No. We cannot make any conclusions about cause and effect because this was not an experiment.

**8.26** **(a)** The scatterplot (below) shows a weak, positive linear association between the two heights.

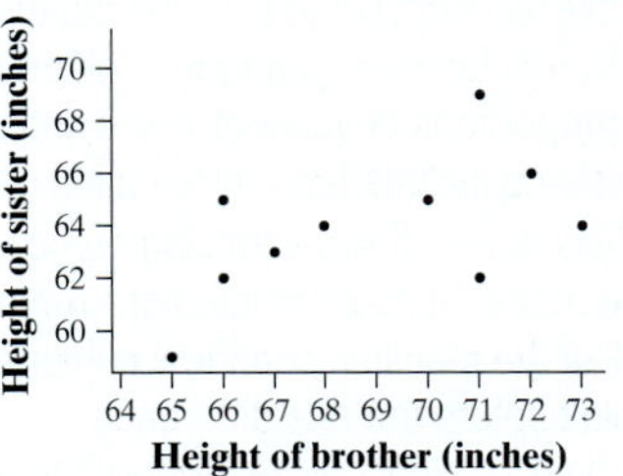


**(b)** $\hat{y} = 27.64 + 0.5270x$, where $y$ = sister's height and $x$ = brother's height. For each increase of 1 inch in the brother's height, the predicted height of the sister will increase by 0.5270 inches. **(c)** $\hat{y} = 27.64 + 0.5270(70) = 64.53$ inches **(d)** Because the association is fairly weak, I don't think that the predicted height will be very close to the actual height. However, because the association is linear, it should be about equally likely for you to get a prediction that is too high or a prediction that is too low.



(a) 95% of the sample of employees have a systolic blood pressure between 122 and 138.

(b) 95% of the population of employees have a systolic blood pressure between 122 and 138.

(c) If the procedure were repeated many times, 95% of the resulting confidence intervals would contain the population mean systolic blood pressure.

(d) If the procedure were repeated many times, 95% of the time the population mean systolic blood pressure would be between 122 and 138.

(e) If the procedure were repeated many times, 95% of the time the sample mean systolic blood pressure would be between 122 and 138.

25. **Power lines and cancer** (4.2, 4.3) Does living near power lines cause leukemia in children? The National Cancer Institute spent 5 years and $5 million gathering data on this question. The researchers compared 638 children who had leukemia with 620 who did not. They went into the homes and measured the magnetic fields in children's bedrooms, in other rooms, and at the front door. They recorded facts about power lines near the family home and also near the mother's residence when she was pregnant. Result: no connection between leukemia and exposure to magnetic fields of the kind produced by power lines was found.[7]

(a) Was this an observational study or an experiment? Justify your answer.

(b) Does this study show that living near power lines doesn't cause cancer? Explain.

26. **Sisters and brothers** (3.1, 3.2) How strongly do physical characteristics of sisters and brothers correlate? Here are data on the heights (in inches) of 11 adult pairs:[8]

| | | | | | | | | | | | |
|---|---|---|---|---|---|---|---|---|---|---|---|
| **Brother:** | 71 | 68 | 66 | 67 | 70 | 71 | 70 | 73 | 72 | 65 | 66 |
| **Sister:** | 69 | 64 | 65 | 63 | 65 | 62 | 65 | 64 | 66 | 59 | 62 |

(a) Construct a scatterplot using brother's height as the explanatory variable. Describe what you see.

(b) Use your calculator to compute the least-squares regression line for predicting sister's height from brother's height. Interpret the slope in context.

(c) Damien is 70 inches tall. Predict the height of his sister Tonya.

(d) Do you expect your prediction in (c) to be very accurate? Give appropriate evidence to support your answer.

# 8.2 Estimating a Population Proportion

**WHAT YOU WILL LEARN** **By the end of the section, you should be able to:**

- State and check the Random, 10%, and Large Counts conditions for constructing a confidence interval for a population proportion.
- Determine critical values for calculating a *C*% confidence interval for a population proportion using a table or technology.
- Construct and interpret a confidence interval for a population proportion.
- Determine the sample size required to obtain a *C*% confidence interval for a population proportion with a specified margin of error.

In Section 8.1, we saw that a confidence interval can be used to estimate an unknown population parameter. We are often interested in estimating the proportion $p$ of some outcome in the population. Here are some examples:

- What proportion of U.S. adults are unemployed right now?
- What proportion of high school students have cheated on a test?

**PD Section 8.2 Overview**

Watch the video overview of Section 8.2 (for teachers) by clicking on the link in the TE-book, opening the TRCD, or logging onto the book's companion Web site www.whfreeman.com/tps5e (teacher log-in required).

**TRM Section 8.2 Alternate Examples**

You can find the Alternate Examples for this section in Microsoft Word format by clicking on the link in the TE-book, opening the TRCD, or logging onto the book's companion Web site www.whfreeman.com/tps5e (teacher log-in required).

- What proportion of pine trees in a national park are infested with beetles?
- What proportion of college students pray daily?
- What proportion of a company's laptop batteries last as long as the company claims?

This section shows you how to construct and interpret a confidence interval for a population proportion. The following Activity gives you a taste of what lies ahead.

## ACTIVITY | The Beads

**MATERIALS:** Several thousand small plastic beads of at least two colors, container to hold all the beads, small cup for sampling, several small bowls

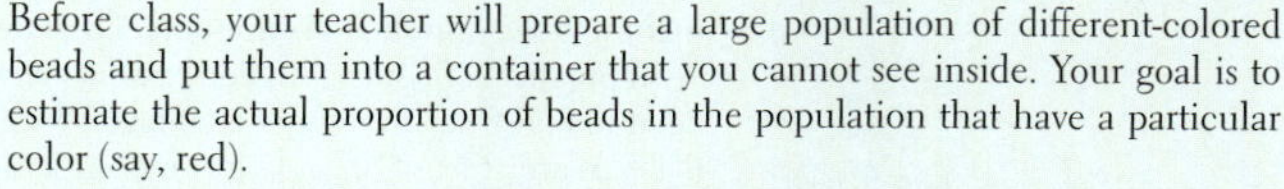

Before class, your teacher will prepare a large population of different-colored beads and put them into a container that you cannot see inside. Your goal is to estimate the actual proportion of beads in the population that have a particular color (say, red).

1. As a class, discuss how to use the cup provided to get a simple random sample of beads from the container. Think this through carefully, because you will get to take only one sample.

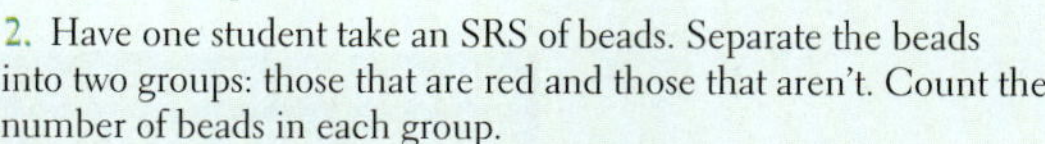

2. Have one student take an SRS of beads. Separate the beads into two groups: those that are red and those that aren't. Count the number of beads in each group.

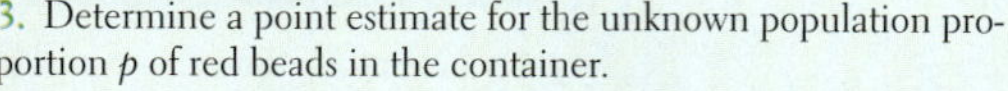

3. Determine a point estimate for the unknown population proportion $p$ of red beads in the container.

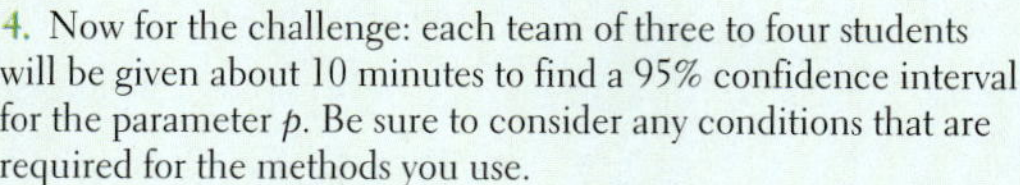

4. Now for the challenge: each team of three to four students will be given about 10 minutes to find a 95% confidence interval for the parameter $p$. Be sure to consider any conditions that are required for the methods you use.
5. Compare the results with those of the other teams in the class. Discuss any problems you encountered and how you dealt with them.

## Conditions for Estimating $p$

Before constructing a confidence interval for $p$, you should check some important conditions:

- **Random:** The data should come from a well-designed random sample or randomized experiment. Otherwise, there's no scope for inference to a population (sampling) or inference about cause and effect (experiment). If we can't draw conclusions beyond the data at hand, there's not much point in constructing a confidence interval!

  Another important reason for random selection or random assignment is to introduce chance into the data-production process. We can model chance behavior with a probability distribution, like the sampling distributions of Chapter 7. The probability distribution helps us calculate a confidence interval.

  - **10%:** The procedure for calculating confidence intervals assumes that individual observations are independent. Well-conducted studies that use random sampling or random assignment can help ensure independent

### Activity Overview

You can watch an overview of this Activity (for teachers) by clicking on the link in the TE-book, opening the TRCD, or logging onto the book's companion Web site www.whfreeman.com/tps5e (teacher log-in required).

### Teaching Tip

After you finish the Activity on this page, students will probably want to know the true proportion of red beads. Even if you know the answer, don't tell your students. After all, in real life we would know the true proportion only if we did a census.

### ALTERNATE ACTIVITY

***The pennies***

Here is another great use for the jar of pennies you collected for the Penny for Your Thoughts Activity in Chapter 7. The goal is to estimate what proportion of the pennies in your collection are more than 10 years old.

1. As a class, think about how to select an SRS of pennies from the jar.
2. Take an SRS of approximately 100 pennies. Separate the pennies into two piles: those that are more than 10 years old and those that aren't. Count the number in each pile.
3. Determine a point estimate ($\hat{p}$) for the unknown population proportion of pennies more than 10 years old.
4. See Step 4 in the Student Edition.
5. See Step 5 in the Student Edition.

### TRM Hershey's Kisses Alternate Activity

For an additional Activity, see the Hershey's Kisses Alternate Activity in the Teacher's Resource Materials. Find it by clicking on the link in the TE-book, opening the TRCD, or logging onto the book's companion Web site www.whfreeman.com/tps5e (teacher log-in required).

observations. However, our formula for the standard deviation of the sampling distribution of $\hat{p}$, $\sigma_{\hat{p}} = \sqrt{\frac{p(1-p)}{n}}$, acts as if we are sampling *with replacement* from a population. That's rarely the case. When sampling without replacement from a finite population, be sure to check that $n \le \frac{1}{10}N$. Sampling more than 10% of the population would give a more precise estimate of the parameter $p$ but would require us to use a different formula to calculate the standard deviation of the sampling distribution.

- **Large Counts:** The method that we use to construct a confidence interval for $p$ depends on the fact that the sampling distribution of $\hat{p}$ is approximately Normal. From what we learned in Chapter 7, we can use the Normal approximation to the sampling distribution of $\hat{p}$ as long as $np \ge 10$ and $n(1-p) \ge 10$.
  In practice, of course, we don't know the value of $p$. If we did, we wouldn't need to construct a confidence interval for it! So we cannot check if $np$ and $n(1-p)$ are at least 10. In large random samples, $\hat{p}$ will tend to be close to $p$. So we replace $p$ by $\hat{p}$ in checking the Large Counts condition: $n\hat{p} \ge 10$ and $n(1-\hat{p}) \ge 10$.

**CONDITIONS FOR CONSTRUCTING A CONFIDENCE INTERVAL ABOUT A PROPORTION**

- **Random:** The data come from a well-designed random sample or randomized experiment.
  - **10%:** When sampling without replacement, check that $n \le \frac{1}{10}N$.
- **Large Counts:** Both $n\hat{p}$ and $n(1-\hat{p})$ are at least 10.

When Mr. Vignolini's class did the beads Activity, they got 107 red beads and 144 white beads. Their point estimate for the unknown proportion $p$ of red beads in the population is

$$\hat{p} = \frac{107}{251} = 0.426$$

Let's see how the conditions play out for Mr. Vignolini's class.

**EXAMPLE**

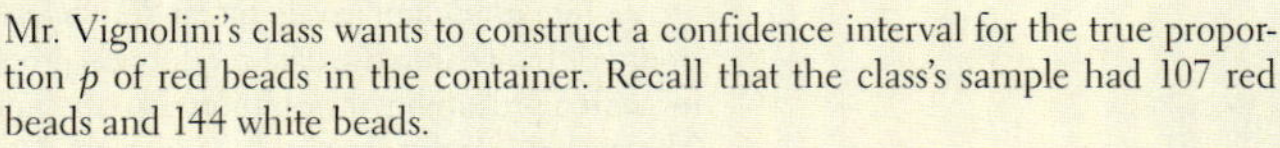

## The Beads

### *Checking conditions*

Mr. Vignolini's class wants to construct a confidence interval for the true proportion $p$ of red beads in the container. Recall that the class's sample had 107 red beads and 144 white beads.

PROBLEM: Check that the conditions for constructing a confidence interval for $p$ are met.

SOLUTION: There are three conditions to check:

## Teaching Tip

If we sample more than 10% of the population, the formula we use for the standard deviation of $\hat{p}$ includes the finite population correction factor:

$$\sigma_{\hat{p}} = \sqrt{\frac{p(1-p)}{n}} \cdot \sqrt{1 - \frac{n}{N}}$$

The finite population correction factor is not a topic on the AP® exam, however. For more details, see the discussion at the top of page 496.

## Teaching Tip

If the Large Counts condition is violated, the actual capture rate of our intervals will usually be less than the stated capture rate. See the next page for more details.

## Teaching Tip

It might help students to remember that the three conditions correspond to the three components of a confidence interval (statistic, critical value, standard deviation of the statistic). The Random condition helps ensure that the statistic is unbiased, the Large Counts condition ensures that we are using the correct critical value, and the 10% condition ensures that we are using the correct formula for the standard deviation of the statistic.

**ALTERNATE EXAMPLE**

### *The pennies*

Mrs. Quinn's class wants to construct a confidence interval for the proportion $p$ of pennies more than 10 years old in their collection. Their sample included 57 pennies that were more than 10 years old and 45 pennies that were at most 10 years old.

PROBLEM: Check that the conditions for constructing a confidence interval for $p$ are met.

SOLUTION:

- *Random*: The class took an SRS of 102 pennies from the collection.
  - *10%*: Because the class sampled without replacement, the number of pennies in the population must be at least 10(102) = 1020. Because there are more than 2000 pennies in Mrs. Quinn's collection, the 10% condition is met.

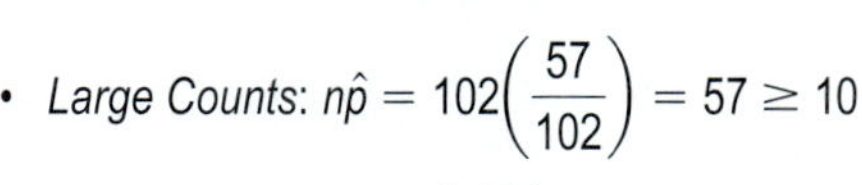

- *Large Counts*: $n\hat{p} = 102\left(\frac{57}{102}\right) = 57 \ge 10$ and $n(1-\hat{p}) = 102\left(\frac{45}{102}\right) = 45 \ge 10$. Both the number of successes and the number of failures are at least 10.

- **Random:** The class took an SRS of 251 beads from the container.
  - **10%:** Because the class sampled without replacement, they need to check that there are at least 10(251) = 2510 beads in the population. Mr. Vignolini reveals that there are 3000 beads in the container.
- **Large Counts:** To use a Normal approximation for the sampling distribution of $\hat{p}$, we need both $np$ and $n(1-p)$ to be at least 10. Because we don't know $p$, we check

$$n\hat{p} = 251\left(\frac{107}{251}\right) = 107 \geq 10 \text{ and } n(1-\hat{p}) = 251\left(1-\frac{107}{251}\right) = 251\left(\frac{144}{251}\right) = 144 \geq 10$$

That is, the counts of successes (red beads) and failures (non-red beads) are both at least 10.
All conditions are met, so it should be safe for the class to construct a confidence interval.

**For Practice** *Try Exercise* 

Notice that $n\hat{p}$ and $n(1-\hat{p})$ should be whole numbers. You don't really need to calculate these values since they are just the number of successes and failures in the sample. In the previous example, we could address the Large Counts condition simply by saying, "The numbers of successes (107) and failures (144) in the sample are both at least 10."

**What happens if one of the conditions is violated?** If the data come from a convenience sample or a poorly designed experiment, there's no point constructing a confidence interval for $p$. Violation of the Random condition severely limits our ability to make any inference beyond the data at hand.

The figure below shows a screen shot from the *Confidence Intervals for Proportions* applet at the book's Web site, www.whfreeman.com/tps5e. We set $n = 20$ and $p = 0.25$. The Large Counts condition is not met because $np = 20(0.25) = 5$. We used the applet to generate 1000 95% confidence intervals for $p$. Only 902 of those 1000 intervals contained $p = 0.25$, a capture rate of 90.2%. When the Large Counts condition is violated, the capture rate will be lower than the one advertised by the confidence level if the method is used many times.

## Teaching Tip

Use the applet described on this page (or the *Simulating Confidence Intervals for a Population Parameter* at www.rossmanchance.com/applets) to explore other combinations of $n$ and $p$. For example, if we take samples of size $n = 30$ from a population where $p = 0.01$ and calculate 95% confidence intervals, only about 25% of the intervals will actually capture the value of the population parameter!

## Teaching Tip

The correct formula for the standard deviation of $\hat{p}$ is given in the first Teaching Tip on page 494. Using the formula without the finite population correction factor produces a standard deviation (and margin of error) that is larger than it should be. For example, consider taking random samples of size $n = 500$ from a population of size $N = 1000$, where $p = 0.80$. If we calculate 95% confidence intervals without the finite population correction factor, about 99.5% of the intervals actually capture the value $p = 0.80$.

## ✓ ANSWERS TO CYU

1. *Random*: not met because this was a convenience sample. *10%*: met because the sample of 100 is less than 10% of the population at a large high school. *Large Counts*: met because 17 successes and 83 failures are both at least 10.

2. *Random*: met because the inspector chose an SRS of bags. *10%*: met because the sample of 25 is less than 10% of the thousands of bags filled in an hour. *Large Counts*: not met because there were only 3 successes, which is less than 10.

### AP® Exam Formula Sheet

The formula for the standard deviation of the sample proportion,

$\sigma_{\hat{p}} = \sqrt{\frac{p(1-p)}{n}}$, is included on the formula sheet that is given to students on both sections of the AP® exam. However, the formula for the standard error of the sample proportion,

$SE_{\hat{p}} = \sqrt{\frac{\hat{p}(1-\hat{p})}{n}}$, is not included on the formula sheet. Students are expected to know that they should use $\hat{p}$ in the formula when the value of $p$ is unknown.

## Teaching Tip

Because there isn't universal agreement on when to use the term *standard error*, it is very unlikely that students will be expected to make the distinction between the standard deviation of a statistic and the standard error of a statistic on the AP® exam.

That leaves just the 10% condition when sampling without replacement from a finite population. Large random samples give more precise estimates than small random samples. So randomly selecting more than 10% of a population should be a good thing! Unfortunately, the formula for the standard deviation of $\hat{p}$ that we developed in Chapter 7, $\sigma_{\hat{p}} = \sqrt{\frac{p(1-p)}{n}}$, is not correct when the 10% condition is violated. The formula gives a value that is too large. Confidence intervals based on this formula are longer than they need to be. If many 95% confidence intervals for a population proportion are constructed in this way, more than 95% of them will capture $p$. The actual capture rate is *greater* than the reported confidence level when the 10% condition is violated.

### CHECK YOUR UNDERSTANDING

In each of the following settings, check whether the conditions for calculating a confidence interval for the population proportion $p$ are met.

1. An AP® Statistics class at a large high school conducts a survey. They ask the first 100 students to arrive at school one morning whether or not they slept at least 8 hours the night before. Only 17 students say "Yes."

2. A quality control inspector takes a random sample of 25 bags of potato chips from the thousands of bags filled in an hour. Of the bags selected, 3 had too much salt.

## Constructing a Confidence Interval for $p$

When the conditions are met, the sampling distribution of $\hat{p}$ will be approximately Normal with mean $\mu_{\hat{p}} = p$ and standard deviation $\sigma_{\hat{p}} = \sqrt{\frac{p(1-p)}{n}}$. Figure 8.7 displays this distribution. Inference about a population proportion $p$ is based on the sampling distribution of $\hat{p}$.

We can use the general formula from Section 8.1 to construct a confidence interval for an unknown population proportion $p$:

statistic ± (critical value) · (standard deviation of statistic)

The sample proportion $\hat{p}$ is the statistic we use to estimate $p$. Doing so makes sense if the data came from a well-designed random sample or randomized experiment (the Random condition).

The standard deviation of the sampling distribution of $\hat{p}$ is

$$\sigma_{\hat{p}} = \sqrt{\frac{p(1-p)}{n}}$$

if the 10% condition is met. Because we don't know the value of $p$, we replace it with the sample proportion $\hat{p}$. The resulting quantity is called the **standard error (SE)** of the sample proportion $\hat{p}$.

$$SE_{\hat{p}} = \sqrt{\frac{\hat{p}(1-\hat{p})}{n}}$$

It describes how close the sample proportion $\hat{p}$ will typically be to the population proportion $p$ in repeated SRSs of size $n$.



**FIGURE 8.7** Suppose that a population has proportion $p$ of successes. When the conditions for inference are met, the sampling distribution of the proportion of successes $\hat{p}$ in a sample is approximately Normal with mean $p$ and standard deviation $\sqrt{\frac{p(1-p)}{n}}$.

Some books refer to $\sigma_{\hat{p}}$ as the "standard error" of $\hat{p}$ and to what we call the standard error as the "estimated standard error."

**DEFINITION: Standard error**

When the standard deviation of a statistic is estimated from data, the result is called the **standard error** of the statistic.

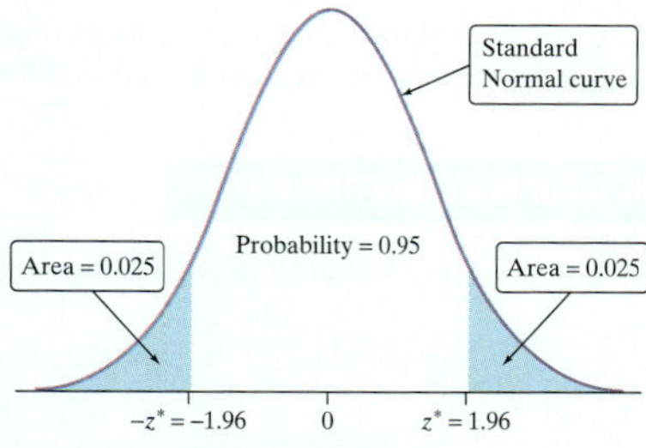


**FIGURE 8.8** Finding the critical value for a 95% confidence interval. The correct value is $z^* = 1.96$, which is more precise than the value of 2 we had been using for 95% confidence.

How do we get the critical value for our confidence interval? If the Large Counts condition is met, we can use a Normal curve. For the approximate 95% confidence intervals of Section 8.1, we used a critical value of 2 based on the 68–95–99.7 rule for Normal distributions. We can get a more precise critical value from Table A or a calculator. As Figure 8.8 shows, the central 95% of the standard Normal distribution is marked off by 2 points, $z^* = 1.96$ and $-z^* = -1.96$. We use the * to remind you that this is a critical value, not a standardized score that has been calculated from data.

To find a level C confidence interval, we need to catch the central C% under the standard Normal curve. Here's an example that shows how to get the **critical value** $z^*$ for a different confidence level.

## EXAMPLE 80% Confidence

### *Finding a critical value*

PROBLEM: Use Table A or technology to find the critical value $z^*$ for an 80% confidence interval. Assume that the Large Counts condition is met.

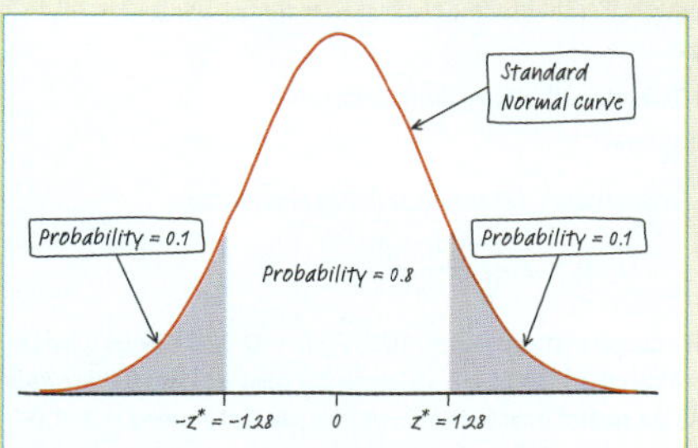


**FIGURE 8.9** Finding the critical value for an 80% confidence interval.

SOLUTION: For an 80% confidence level, we need to capture the central 80% of the standard Normal distribution. In capturing the central 80%, we leave out 20%, or 10% in each tail. So the desired critical value $z^*$ is the point with area 0.1 to its right under the standard Normal curve. Figure 8.9 shows the details in picture form.

Search the body of Table A to find the point $-z^*$ with area 0.1 to its left. The closest entry is $z = -1.28$. (See the excerpt from Table A below.) So the critical value we want is $z^* = 1.28$.

| z | .07 | .08 | .09 |
|---|---|---|---|
| −1.3 | .0853 | .0838 | .0823 |
| −1.2 | .1020 | .1003 | .0985 |
| −1.1 | .1210 | .1190 | .1170 |

*Using technology:* The command `invNorm(area:0.1,μ:0,σ:1)` gives $z = -1.28$. The critical value is $z^* = 1.28$, which matches what we got from Table A.

**For Practice** *Try Exercise* **31**

### Teaching Tip

We will use $z$ (without the asterisk) in Chapter 9 when we calculate the test statistic for a significance test. Make sure to use $z^*$ (with the asterisk) whenever you are calculating a critical value for a confidence interval so students aren't confused in the next chapter.

### TECHNOLOGY TIP

If your students are using a TI-83/84 with an older operating system, they can use the command `invNorm(0.1)` to find the critical value for an 80% confidence interval. The calculator will default to $\mu = 0$ and $\sigma = 1$ automatically, because these are the values for the standard Normal distribution.

### ALTERNATE EXAMPLE

#### *96% confidence*

PROBLEM: Use Table A or technology to find the critical value $z^*$ for a 96% confidence interval. Assume that the Large Counts condition is met.

SOLUTION: For a 96% confidence interval, we need to capture the middle 96% of the standard Normal distribution. This leaves out 2% in each tail, so we want to find the z-score with an area of 0.02 to its left. The closest entry is $z = -2.05$, so the critical value we want is $z^* = 2.05$.

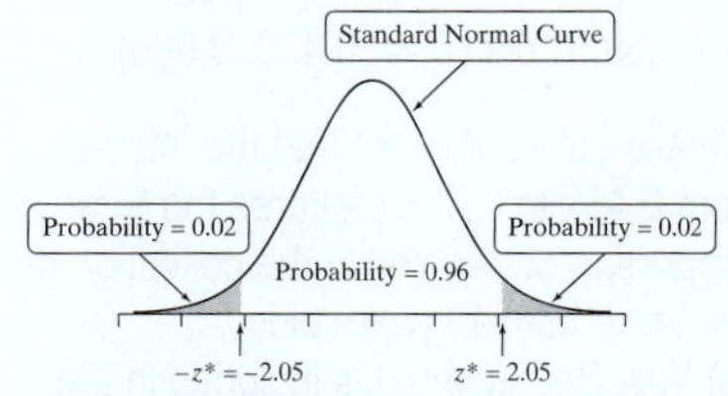


*Using technology*: The command `invNorm(area:0.02,μ:0,σ:1)` gives $z = -2.05$, so $z^* = 2.05$, which matches the value we got from Table A.

Technically, the correct formula for a confidence interval is statistic ± (critical value) · (standard *error* of statistic). We are following the convention used on the AP® Statistics exam formula sheet.

Once we find the critical value $z^*$, our confidence interval for the population proportion $p$ is

$$\text{statistic} \pm (\text{critical value}) \cdot (\text{standard deviation of statistic})$$

$$= \hat{p} \pm z^*\sqrt{\frac{\hat{p}(1-\hat{p})}{n}}$$

Notice that we replaced the standard deviation of $\hat{p}$ with the formula for its standard error. The resulting interval is sometimes called a **one-sample $z$ interval for a population proportion.**

**ONE-SAMPLE $z$ INTERVAL FOR A POPULATION PROPORTION**

When the conditions are met, a C% confidence interval for the unknown proportion $p$ is

$$\hat{p} \pm z^*\sqrt{\frac{\hat{p}(1-\hat{p})}{n}}$$

where $z^*$ is the critical value for the standard Normal curve with C% of its area between $-z^*$ and $z^*$.

Now we can get the desired confidence interval for Mr. Vignolini's class.

## EXAMPLE The Beads

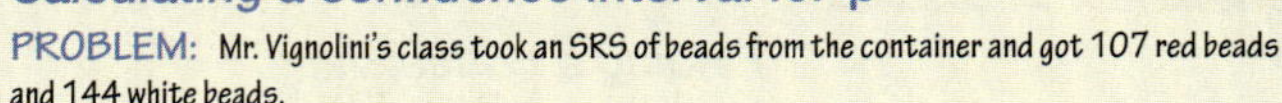

### *Calculating a confidence interval for p*

PROBLEM: Mr. Vignolini's class took an SRS of beads from the container and got 107 red beads and 144 white beads.

(a) Calculate and interpret a 90% confidence interval for $p$.

(b) Mr. Vignolini claims that exactly half of the beads in the container are red. Use your result from part (a) to comment on this claim.

SOLUTION: We checked conditions for calculating the interval earlier.

(a) Our confidence interval has the form

$$\text{statistic} \pm (\text{critical value}) \cdot (\text{standard deviation of statistic})$$

$$= \hat{p} \pm z^*\sqrt{\frac{\hat{p}(1-\hat{p})}{n}}$$

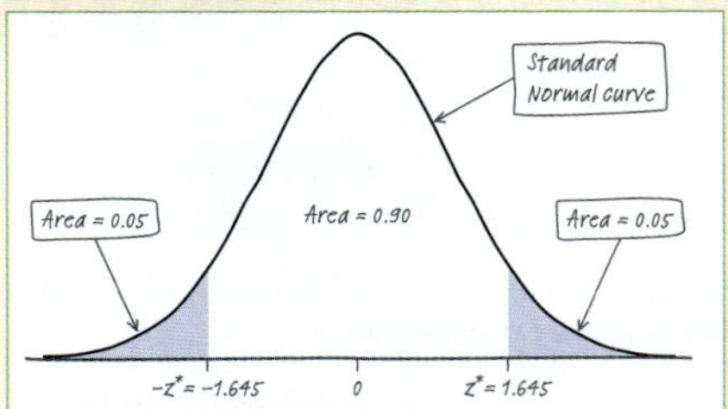


The sample statistic is $\hat{p} = 107/251 = 0.426$. Now let's find the critical value. From Table A, we look for the point with area 0.05 to its left. As the excerpt from Table A shows, this point is between $z = -1.64$ and $z = -1.65$. The calculator's `invNorm(area:0.05,μ:0,σ:1)` gives $z = -1.645$. So we use $z^* = 1.645$ as our critical value.

| z | .03 | .04 | .05 |
|---|---|---|---|
| −1.7 | .0418 | .0409 | .0401 |
| −1.6 | .0516 | .0505 | .0495 |
| −1.5 | .0630 | .0618 | .0606 |

**AP® Exam Formula Sheet**

The specific formula for a one-sample $z$ interval for a population proportion is *not* on the formula sheet that is given to students on the AP® exam. However, the formula sheet does include the general form for a confidence interval and the formula for the standard deviation of the statistic $\hat{p}$.

**ALTERNATE EXAMPLE**

### *The pennies*

PROBLEM: Mrs. Quinn's class took an SRS of 102 pennies and discovered that 57 of the pennies were more than 10 years old.

**(a)** Calculate and interpret a 99% confidence interval for $p$ = the true proportion of pennies from the collection that are more than 10 years old.

**(b)** Is it plausible that exactly 60% of all the pennies in the collection are more than 10 years old? Explain.

SOLUTION:

**(a)** The proportion of pennies more than 10 years old in the sample was $\hat{p} = 57/102 = 0.559$. The critical value for a 99% confidence interval can be found by looking for the point that has an area of 0.005 to the left. The calculator command `invNorm(area:0.005,μ:0,σ:1)` gives $z = -2.576$, so the appropriate critical value for 99% confidence is $z^* = 2.576$. The 99% confidence interval is

$$\hat{p} \pm z^*\sqrt{\frac{\hat{p}(1-\hat{p})}{n}}$$

$$= 0.559 \pm 2.576\sqrt{\frac{0.559(1-0.559)}{102}}$$

$$= 0.559 \pm 0.127 = (0.432, 0.686)$$

We are 99% confident that the interval from 0.432 to 0.686 captures the true proportion of pennies in the collection that are more than 10 years old.

**(b)** Yes. Because 0.6 is included in the confidence interval, it is plausible that 60% of all the pennies in the collection are more than 10 years old.

*The resulting 90% confidence interval is*

$$= 0.426 \pm 1.645\sqrt{\frac{(0.426)(1-0.426)}{251}}$$
$$= 0.426 \pm 0.051$$
$$= (0.375, 0.477)$$

*We are 90% confident that the interval from 0.375 to 0.477 captures the true proportion of red beads in Mr. Vignolini's container.*

*(b) The confidence interval in part (a) gives a set of plausible values for the population proportion of red beads. Because 0.5 is not contained in the interval, it is not a plausible value for p. We have good reason to doubt Mr. Vignolini's claim.*

**For Practice** *Try Exercise* 35

## CHECK YOUR UNDERSTANDING

Alcohol abuse has been described by college presidents as the number one problem on campus, and it is an important cause of death in young adults. How common is it? A survey of 10,904 randomly selected U.S. college students collected information on drinking behavior and alcohol-related problems.[9] The researchers defined "frequent binge drinking" as having five or more drinks in a row three or more times in the past two weeks. According to this definition, 2486 students were classified as frequent binge drinkers.

1. Identify the parameter of interest.
2. Check conditions for constructing a confidence interval for the parameter.
3. Find the critical value for a 99% confidence interval. Show your method. Then calculate the interval.
4. Interpret the interval in context.

## Putting It All Together: The Four-Step Process

Taken together, the examples about Mr. Vignolini's class and the beads Activity show you how to get a confidence interval for an unknown population proportion $p$. We can use the familiar four-step process whenever a problem asks us to construct and interpret a confidence interval.

**CONFIDENCE INTERVALS: A FOUR-STEP PROCESS**

STEP 4

**State:** What *parameter* do you want to estimate, and at what *confidence level*?

**Plan:** Identify the appropriate inference *method*. Check *conditions*.

**Do:** If the conditions are met, perform *calculations*.

**Conclude:** *Interpret* your interval in the context of the problem.

### ✓ ANSWERS TO CYU

**1.** $p$ = the true proportion of all U.S. college students who are classified as frequent binge drinkers.

**2.** *Random*: met because the statement says that the students were chosen randomly. *10%*: met because the sample of 10,904 is less than 10% of all U.S. college students. *Large Counts*: met because 2486 successes and 8418 failures are both at least 10.

**3.** $\frac{1-0.99}{2} = 0.005$ and the closest area in Table A is 0.0051 (or 0.0049), corresponding to a critical value of $z^* = 2.57$ (or 2.58). *Using technology*: `invNorm(area:0.005,μ:0,σ:1)` $= -2.576$, so $z^* = 2.576$.

$$0.228 \pm 2.576\sqrt{\frac{0.228(1-0.228)}{10{,}904}}$$
$$= 0.228 + 0.010 = (0.218, 0.238).$$

**4.** We are 99% confident that the interval from 0.218 to 0.238 captures the true proportion of all U.S. college students who are classified as frequent binge drinkers.

### Teaching Tip

It is essential that your students follow this four-step process whenever a problem asks them to calculate a confidence interval. These steps were carefully designed based on the scoring rubrics for confidence interval questions on the AP® exam. Remind your students that **S**tatistics **P**roblems **D**emand **C**onsistency! This would also be a good time to show your students the wealth of information on the inside back cover of their book. This Inference Summary repeats the four-step process for confidence intervals (and significance tests) at the top and then presents the details for each of the different inference procedures in a nicely organized grid.

### Teaching Tip

Point out that the Conclude step includes an interpretation of the confidence *interval*, not of the confidence *level*. Including an interpretation of the confidence level is risky—if it isn't correct, students will lose credit. If it is correct, it will be ignored by graders.

The next example illustrates the four-step process in action.

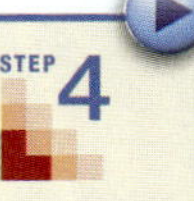

## EXAMPLE Teens Say Sex Can Wait

### Confidence interval for p

STEP 4

The Gallup Youth Survey asked a random sample of 439 U.S. teens aged 13 to 17 whether they thought young people should wait to have sex until marriage.[10] Of the sample, 246 said "Yes." Construct and interpret a 95% confidence interval for the proportion of all teens who would say "Yes" if asked this question.

STATE: We want to estimate the true proportion $p$ of all 13- to 17-year-olds in the United States who would say that young people should wait to have sex until they get married with 95% confidence.

PLAN: We should use a one-sample $z$ interval for $p$ if the conditions are met.

- Random: Gallup surveyed a random sample of U.S. teens.
  - 10%: Because Gallup is sampling without replacement, we need to check the 10% condition: there are at least $10(439) = 4390$ U.S. teens aged 13 to 17.
- Large Counts: We check the counts of "successes" and "failures":

$$n\hat{p} = 246 \geq 10 \quad \text{and} \quad n(1-\hat{p}) = 193 \geq 10$$

DO: The sample statistic is $\hat{p} = 246/439 = 0.56$. A 95% confidence interval for $p$ is given by

$$\hat{p} \pm z^*\sqrt{\frac{\hat{p}(1-\hat{p})}{n}} = 0.56 \pm 1.96\sqrt{\frac{(0.56)(0.44)}{439}}$$
$$= 0.56 \pm 0.046$$
$$= (0.514, 0.606)$$

CONCLUDE: We are 95% confident that the interval from 0.514 to 0.606 captures the true proportion of 13- to 17-year-olds in the United States who would say that teens should wait until marriage to have sex.

**AP® EXAM TIP** If a free-response question asks you to construct and interpret a confidence interval, you are expected to do the entire four-step process. That includes clearly defining the parameter, identifying the procedure, and checking conditions.

**For Practice** *Try Exercise* 39

Simulation studies have shown that a variation of our method for calculating a 95% confidence interval for $p$ can result in closer to a 95% capture rate in the long run, especially for small sample sizes. This simple adjustment, first suggested by Edwin Bidwell Wilson in 1927, is sometimes called the "plus four" estimate. Just pretend we have four additional observations, two of which are successes and two of which are failures. Then calculate the "plus four interval" using the plus four estimate in place of $\hat{p}$ in our usual formula.

*Remember that the margin of error in a confidence interval includes only sampling variability!* There are other sources of error that are not taken into account. As is the case with many surveys, we are forced to assume that the teens answered truthfully. If they didn't, then our estimate may be biased. Other problems like nonresponse and question wording can also affect the results of this or any other poll. *Lesson:* Sampling beads is much easier than sampling people!

Your calculator will handle the "Do" part of the four-step process, as the following Technology Corner illustrates.

**AP® EXAM TIP** You may use your calculator to compute a confidence interval on the AP® exam. But there's a risk involved. If you just give the calculator answer with no work, you'll get either full credit for the "Do" step (if the interval is correct) or no credit (if it's wrong). If you opt for the calculator-only method, be sure to name the procedure (e.g., one-proportion $z$ interval) and to give the interval (e.g., 0.514 to 0.607).

**ALTERNATE EXAMPLE**

***Spinning the globe***

In her first-grade social studies class, Jordan learned that 70% of Earth's surface was covered in water. She wondered if this was really true and asked her dad for help. To investigate, he tossed an inflatable globe to her 50 times, being careful to spin the globe each time. When she caught it, he recorded where her right index finger was pointing. In 50 tosses, her finger was pointing to water 33 times. Construct and interpret a 95% confidence interval for the proportion of Earth's surface that is covered in water.

STATE: We want to estimate $p$ = the true proportion of Earth's surface that is covered in water with 95% confidence.

PLAN: We should use a one-sample $z$ interval for $p$ if the conditions are met.

- *Random*: The 50 locations are a random sample of all possible locations on the globe.
  - *10%*: We do not need to check the 10% condition because the locations were not selected without replacement.
- *Large Counts*: $n\hat{p} = 33 \geq 10$ and $n(1-\hat{p}) = 17 \geq 10$.

DO: $\hat{p} \pm z^*\sqrt{\frac{\hat{p}(1-\hat{p})}{n}}$

$$= 0.66 \pm 1.96\sqrt{\frac{0.66(1-0.66)}{50}}$$
$$= 0.66 \pm 0.131 = (0.529, 0.791)$$

CONCLUDE: We are 95% confident that the interval from 0.529 to 0.791 captures the true proportion of Earth's surface that is covered in water. This is consistent with the claim that 70% of Earth's surface is covered in water, because 0.70 is one of the plausible values in the interval.

## Teaching Tip

After reading the conclusion in this example, make sure to emphasize the phrase "who *would say* teens should wait." It would be incorrect to make a conclusion about those "who *said* teens should wait" because this refers to the sample, not the population of all 13- to 17-year-olds.

**TRM The Plus Four Interval for a Population Proportion**

After reading the marginal note on this page, students may be wondering why we might need other variations of confidence intervals for a proportion. Studies have shown that when the usual formula for a confidence interval is used, the actual capture rate is often below the stated capture rate, especially when the sample size is small. For example, a "95% confidence interval" may capture the true proportion in only 92% of all possible samples. This discrepancy occurs because the distribution of $\hat{p}$ is discrete, not continuous. Because we use the standard Normal distribution, which is continuous, to calculate the critical values, these critical values are only approximations. If you or your students want to know more about the "plus four interval," there is additional content in the Teacher's Resource Materials. Just know that the "plus four interval" is not on the AP® exam.

## 15. TECHNOLOGY CORNER — CONFIDENCE INTERVAL FOR A POPULATION PROPORTION

TI-Nspire instructions in Appendix B; HP Prime instructions on the book's Web site.

The TI-83/84 and TI-89 can be used to construct a confidence interval for an unknown population proportion. We'll demonstrate using the previous example. Of $n = 439$ teens surveyed, $X = 246$ said they thought that young people should wait to have sex until after marriage. To construct a confidence interval:

**TI-83/84**

- Press STAT, then choose TESTS and `1-PropZInt`.

**TI-89**

- In the Statistics/List Editor, press 2nd F2 ([F7]) and choose `1-PropZInt`.

- When the 1-PropZInt screen appears, enter $x = 246$, $n = 439$, and confidence level 0.95.

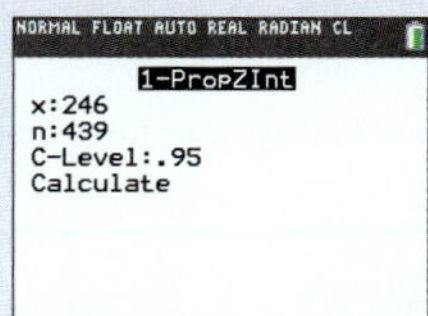


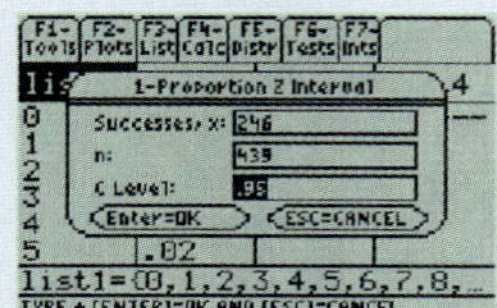


- Highlight "Calculate" and press ENTER. The 95% confidence interval for $p$ is reported, along with the sample proportion $\hat{p}$ and the sample size, as shown here.

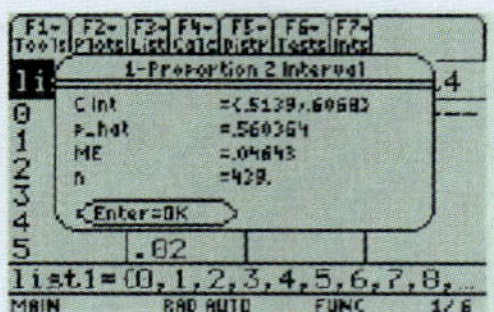


## Choosing the Sample Size

In planning a study, we may want to choose a sample size that allows us to estimate a population proportion within a given margin of error. National survey organizations like the Gallup Poll typically sample between 1000 and 1500 American adults, who are interviewed by telephone. Why do they choose such sample sizes?

The margin of error ($ME$) in the confidence interval for $p$ is

$$ME = z^* \sqrt{\frac{\hat{p}(1-\hat{p})}{n}}$$

Here, $z^*$ is the standard Normal critical value for the confidence level we want. Because the margin of error involves the sample proportion of successes $\hat{p}$, we have to guess the value of $\hat{p}$ when choosing $n$. Here are two ways to do this:

1. Use a guess for $\hat{p}$ based on a pilot study or on past experience with similar studies. You should do several calculations that cover the range of $\hat{p}$-values you might get.
2. Use $\hat{p} = 0.5$ as the guess. The margin of error $ME$ is largest when $\hat{p} = 0.5$, so this guess is conservative in the sense that if we get any other $\hat{p}$ when we do our study, we will get a margin of error smaller than planned.

### TECHNOLOGY TIP

Remind students that they can watch a video presentation of this technology corner (for the TI-83/84, TI-89, or TI-Nspire) at www.whfreeman.com/tps5e.

### TECHNOLOGY TIP

Remind students to look for the option that says "1-PropZInt," not the option that says "ZInterval." The latter option is for calculating a confidence interval for $\mu$ when $\sigma$ is known. We prefer the name "one-sample $z$ interval for $p$" to highlight that this procedure is based on data from one sample. We have also found that students will sometimes mistakenly choose the "2-PropZInt" because there are two proportions that can be computed from the sample: the proportion of successes and the proportion of failures. The 2-PropZInt option is for the two-sample $z$ interval for a difference in proportions that we will learn about in Chapter 10.

### TECHNOLOGY TIP

The value of "$x$" that students enter into the calculator must be an integer or else the calculator will return a domain error. Students often enter the value of $\hat{p}$ instead of the number of successes. Remind your students that $\hat{p} = x/n$ and that the calculator wants both $x$ and $n$.

### Teaching Tip

Even though it is acceptable to use the calculator's 1-PropZInt feature on the AP® exam, you can insist that students show additional work in your class. While the formulas are still relatively simple, we recommend this practice. It will help students understand the structure of confidence intervals and prepare them for multiple-choice questions that focus on the way confidence intervals are constructed.

### AP® Exam Common Error

Many students use the 1-PropZInt feature to correctly calculate the confidence interval and then try to "show their work" with an incorrect formula. This will result in a loss of credit because the two attempts are considered "parallel solutions" and students are scored on the worse response. If students want to include a formula in their response, they should make sure it produces the same results as the calculator. If the results aren't the same, students should choose one of the answers and cross the other out. Also, if students choose to include a formula, they should skip the symbolic formula and start with numbers substituted in. In some cases, students choose the right formula and use the correct values, but include an incorrect symbol (e.g., $p$ instead of $\hat{p}$) and lose credit.

### AP® Exam Formula Sheet

The specific formula for the margin of error on this page is not included on the formula sheet provided to students on the AP® exam. However, the general formula for a confidence interval and the formula for the standard deviation of the statistic $\hat{p}$ are included, so students should be able to put it together.

### Teaching Tip

Ask students why the margin of error is largest when $\hat{p} = 0.5$. The $ME$ is directly proportional to $\hat{p}(1-\hat{p})$, and $\hat{p}(1-\hat{p})$ is largest when $\hat{p} = 0.5$. You can investigate this by trying different values of $\hat{p}$ or by graphing the function $y = x(1-x) = x - x^2$ and finding the maximum.

### ALTERNATE EXAMPLE

#### Tattoos

Suppose that you want to estimate $p$ = the true proportion of students at your school who have a tattoo with 95% confidence and a margin of error of no more than 0.10.

PROBLEM: Determine how many students should be surveyed to estimate $p$ within 0.10 with 95% confidence.

SOLUTION: Because we don't have any previous knowledge about the proportion of students with a tattoo, we will use $\hat{p} = 0.5$ to estimate the sample size needed:

$$1.96\sqrt{\frac{0.5(1-0.5)}{n}} \leq 0.10$$

$$\rightarrow \left(\frac{1.96}{0.10}\right)^2 (0.5)(1-0.5) \leq n$$

$$\rightarrow n \geq 96.04$$

We need to survey at least 97 students to estimate the true proportion of students with a tattoo with 95% confidence and a margin of error of at most 0.10.

Once you have a guess for $\hat{p}$, the formula for the margin of error can be solved to give the sample size $n$ needed.

**SAMPLE SIZE FOR DESIRED MARGIN OF ERROR**

To determine the sample size $n$ that will yield a $C$% confidence interval for a population proportion $p$ with a maximum margin of error $ME$, solve the following inequality for $n$:

$$z^*\sqrt{\frac{\hat{p}(1-\hat{p})}{n}} \leq ME$$

where $\hat{p}$ is a guessed value for the sample proportion. The margin of error will always be less than or equal to $ME$ if you use $\hat{p} = 0.5$.

Here's an example that shows you how to determine the sample size.

## EXAMPLE Customer Satisfaction

### *Determining sample size*

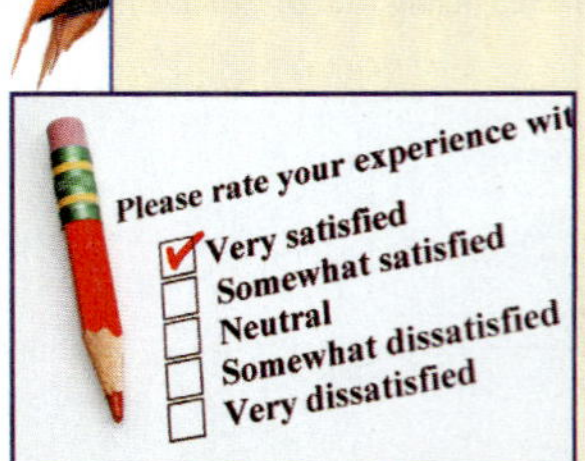


A company has received complaints about its customer service. The managers intend to hire a consultant to carry out a survey of customers. Before contacting the consultant, the company president wants some idea of the sample size that she will be required to pay for. One critical question is the degree of satisfaction with the company's customer service, measured on a 5-point scale. The president wants to estimate the proportion $p$ of customers who are satisfied (that is, who choose either "somewhat satisfied" or "very satisfied," the two highest levels on the 5-point scale). She decides that she wants the estimate to be within 3% (0.03) at a 95% confidence level. How large a sample is needed?

PROBLEM: Determine the sample size needed to estimate $p$ within 0.03 with 95% confidence.

SOLUTION: The critical value for 95% confidence is $z^* = 1.96$. We have no idea about the true proportion $p$ of satisfied customers, so we decide to use $\hat{p} = 0.5$ as our guess. Because the company president wants a margin of error of no more than 0.03, we need to solve the inequality

$$1.96\sqrt{\frac{0.5(1-0.5)}{n}} \leq 0.03$$

for $n$. Multiplying both sides by $\sqrt{n}$ and then dividing both sides by 0.03 yields

$$\frac{1.96}{0.03}\sqrt{0.5(1-0.5)} \leq \sqrt{n}$$

Squaring both sides gives

$$\left(\frac{1.96}{0.03}\right)^2 (0.5)(1-0.5) \leq n$$

$$1067.111 \leq n$$

We round up to 1068 respondents to ensure that the margin of error is no more than 3%.

For Practice Try Exercise 43

Why not round to the nearest whole number—in this case, 1067? Because a smaller sample size will result in a larger margin of error, possibly more than the desired 3% for the poll.

If you want a 2.5% margin of error rather than 3%, then

$$n \geq \left(\frac{1.96}{0.025}\right)^2 (0.5)(1-0.5) = 1536.64 \Rightarrow n = 1537$$

For a 2% margin of error, the sample size you need is

$$n \geq \left(\frac{1.96}{0.02}\right)^2 (0.5)(1-0.5) = 2401$$

As usual, smaller margins of error call for larger samples.

News reports frequently describe the results of surveys with sample sizes between 1000 and 1500 and a margin of error of about 3%. These surveys generally use sampling procedures more complicated than a simple random sample, so the calculation of confidence intervals is more involved than what we have studied in this section. The calculations of the previous example still give you a rough idea of how such surveys are planned.

## CHECK YOUR UNDERSTANDING

Refer to the previous example about the company's customer satisfaction survey.

1. In the company's prior-year survey, 80% of customers surveyed said they were "somewhat satisfied" or "very satisfied." Using this value as a guess for $\hat{p}$, find the sample size needed for a margin of error of 3% at a 95% confidence level.

2. What if the company president demands 99% confidence instead? Determine how this would affect your answer to Question 1.

## Section 8.2 Summary

- The conditions for constructing a confidence interval about a population proportion are
  - **Random:** The data were produced by a well-designed random sample or randomized experiment.
    - **10%:** When sampling without replacement, we check that the population is at least 10 times as large as the sample.
  - **Large Counts:** The sample is large enough that $n\hat{p}$ and $n(1-\hat{p})$, the counts of successes and failures in the sample, are both at least 10.
- Confidence intervals for a population proportion $p$ are based on the sampling distribution of the sample proportion $\hat{p}$. When the conditions for inference are met, the sampling distribution of $\hat{p}$ is approximately Normal with mean $p$ and standard deviation $\sqrt{p(1-p)/n}$.

### Teaching Tip

When solving for $n$ in problems like this, using the inequality $\leq$ or $\geq$ will help students understand why they need to always round up if the result is a decimal.

### ANSWERS TO CYU

1. Solving $1.96\sqrt{\frac{0.80(0.20)}{n}} \leq 0.03$ for $n$ gives $n \geq 682.95$. We should select a sample of at least 683 customers.

2. The required sample size will be larger because the critical value is larger for 99% confidence (2.576) versus 95% confidence (1.96). The company would need to select at least 1180 customers.

### TRM Section 8.2 Quizzes

There are three different quizzes available for this section in the Teacher's Resource Materials. Versions A and B are roughly parallel. Find them by clicking on the link in the TE-book, opening the TRCD, or logging onto the book's companion Web site www.whfreeman.com/tps5e (teacher log-in required).

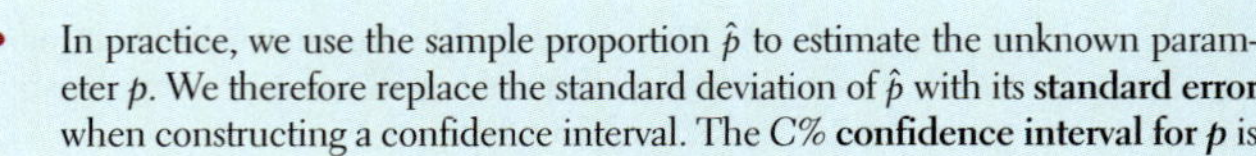

- In practice, we use the sample proportion $\hat{p}$ to estimate the unknown parameter $p$. We therefore replace the standard deviation of $\hat{p}$ with its **standard error** when constructing a confidence interval. The $C\%$ **confidence interval for $p$** is

$$\hat{p} \pm z^*\sqrt{\frac{\hat{p}(1-\hat{p})}{n}}$$

where $z^*$ is the standard Normal **critical value** with $C\%$ of its area between $-z^*$ and $z^*$.

- When constructing a confidence interval, follow the four-step process:

**STATE:** What *parameter* do you want to estimate, and at what *confidence level*?

**PLAN:** Identify the appropriate inference *method*. Check *conditions*.

**DO:** If the conditions are met, perform *calculations*.

**CONCLUDE:** *Interpret* your interval in the context of the problem.

- The sample size needed to obtain a confidence interval with approximate margin of error $ME$ for a population proportion involves solving

$$z^*\sqrt{\frac{\hat{p}(1-\hat{p})}{n}} \leq ME$$

for $n$, where $\hat{p}$ is a guessed value for the sample proportion, and $z^*$ is the critical value for the confidence level you want. Use $\hat{p} = 0.5$ if you don't have a good idea about the value of $\hat{p}$.

**8.2 TECHNOLOGY CORNER**

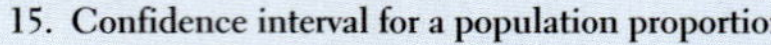

TI-Nspire Instructions in Appendix B; HP Prime instructions on the book's Web site.

15. Confidence interval for a population proportion page 501

## Section 8.2 Exercises

*For Exercises 27 to 30, check whether each of the conditions is met for calculating a confidence interval for the population proportion p.*

27. **Rating school food** Latoya wants to estimate what proportion of the seniors at her boarding high school like the cafeteria food. She interviews an SRS of 50 of the 175 seniors living in the dormitory. She finds that 14 think the cafeteria food is good.

pg 494

28. **High tuition costs** Glenn wonders what proportion of the students at his school believe that tuition is too high. He interviews an SRS of 50 of the 2400 students at his college. Thirty-eight of those interviewed think tuition is too high.

29. **AIDS and risk factors** In the National AIDS Behavioral Surveys sample of 2673 adult heterosexuals, 0.2% had both received a blood transfusion and had a sexual partner from a group at high risk of AIDS. We want to estimate the proportion $p$ in the population who share these two risk factors.

30. **Whelks and mussels** The small round holes you often see in sea shells were drilled by other sea creatures, who ate the former dwellers of the shells.

**TRM Full Solutions to Section 8.2 Exercises**

### Answers to Section 8.2 Exercises

**8.27** *Random*: met because Latoya selected an SRS of students. *10%*: not met because the sample size (50) is more than 10% of the population of seniors in the dormitory (175). *Large Counts*: met because $n\hat{p} = 14 \geq 10$ and $n(1-\hat{p}) = 36 \geq 10$.

**8.28** *Random*: met because the sample was an SRS. *10%*: met because the sample size (50) is less than 10% of the population of students at his college (2400). *Large Counts*: met because $n\hat{p} = 38 \geq 10$ and $n(1-\hat{p}) = 12 \geq 10$.

**8.29** *Random*: may not be met because we do not know if the people contacted were a random sample. *10%*: met because the sample size (2673) is less than 10% of the population of adult heterosexuals. *Large Counts*: not met because $n\hat{p} = 2673(0.002) \approx 5$ is not at least 10.

**8.30** *Random*: may not be met because we do not know if the whelk eggs were a random sample. *10%*: met because the sample size (98) is less than 10% of all whelk eggs. *Large Counts*: not met because $n\hat{p} = 9$, which is not at least 10.

**8.31** $\frac{1-0.98}{2} = 0.01$, and the closest area is 0.0099, corresponding to a critical value of $z^* = 2.33$. *Using technology*: `invNorm(area:0.01, μ:0,σ:1)` $= -2.326$, so $z^* = 2.326$.

**8.32** $\frac{1-0.93}{2} = 0.035$, and the closest area is 0.0351, corresponding to a critical value of $z^* = 1.81$. *Using technology*: `invNorm(area:0.035, μ:0,σ:1)` $= -1.812$, so $z^* = 1.812$.

**8.33** **(a)** *Population*: seniors at Tonya's high school. *Parameter*: true proportion of all seniors who plan to attend the prom. **(b)** *Random*: The sample is a simple random sample. *10%*: The sample size (50) is less than 10% of the population size (750). *Large Counts*: $n\hat{p} = 36 \geq 10$ and $n(1-\hat{p}) = 14 \geq 10$. **(c)** $0.72 \pm 1.645\sqrt{\frac{0.72(0.28)}{50}} = 0.72 \pm 0.10 = (0.62, 0.82)$. **(d)** We are 90% confident that the interval from 0.62 to 0.82 captures the true proportion of all seniors at Tonya's high school who plan to attend the prom.

**8.34** **(a)** *Population*: undergraduates at a large university. *Parameter*: proportion of all undergraduates who would be willing to report cheating. **(b)** *Random*: The sample was a simple random sample. *10%*: The sample size (172) is less than 10% of the population of all undergraduates at a large university. *Large Counts*: $n\hat{p} = 19 \geq 10$ and $n(1-\hat{p}) = 153 \geq 10$.

Whelks often drill into mussels, but this behavior appears to be more or less common in different locations. Researchers collected whelk eggs from the coast of Oregon, raised the whelks in the laboratory, then put each whelk in a container with some delicious mussels. Only 9 of 98 whelks drilled into a mussel.[11] The researchers want to estimate the proportion $p$ of Oregon whelks that will spontaneously drill into mussels.

31. **98% confidence** Find $z^*$ for a 98% confidence interval using Table A or your calculator. Show your method. pg 497

32. **93% confidence** Find $z^*$ for a 93% confidence interval using Table A or your calculator. Show your method.

33. **Going to the prom** Tonya wants to estimate what proportion of her school's seniors plan to attend the prom. She interviews an SRS of 50 of the 750 seniors in her school and finds that 36 plan to go to the prom.
   (a) Identify the population and parameter of interest.
   (b) Check conditions for constructing a confidence interval for the parameter.
   (c) Construct a 90% confidence interval for $p$. Show your method.
   (d) Interpret the interval in context.

34. **Reporting cheating** What proportion of students are willing to report cheating by other students? A student project put this question to an SRS of 172 undergraduates at a large university: "You witness two students cheating on a quiz. Do you go to the professor?" Only 19 answered "Yes."[12]
   (a) Identify the population and parameter of interest.
   (b) Check conditions for constructing a confidence interval for the parameter.
   (c) Construct a 99% confidence interval for $p$. Show your method.
   (d) Interpret the interval in context.

35. **Binge drinking** In a recent National Survey of Drug Use and Health, 2312 of 5914 randomly selected full-time U.S. college students were classified as binge drinkers.[13] pg 498
   STEP 4
   (a) Calculate and interpret a 99% confidence interval for the population proportion $p$ that are binge drinkers.
   (b) A newspaper article claims that 45% of full-time U.S. college students are binge drinkers. Use your result from part (a) to comment on this claim.

36. **Teens' texting** A Pew Internet and American Life Project survey found that 392 of 799 randomly selected teens reported texting with their friends every day.

   (a) Calculate and interpret a 95% confidence interval for the population proportion $p$ that would report texting with their friends every day.
   (b) Is it plausible that the true proportion of American teens who text with their friends every day is 0.45? Use your result from part (a) to support your answer.

37. **Binge drinking** Describe a possible source of error that is not included in the margin of error for the 99% confidence interval in Exercise 35.

38. **Teens' texting** Describe a possible source of error that is not included in the margin of error for the 95% confidence interval in Exercise 36.

39. **How common is SAT coaching?** A random sample of students who took the SAT college entrance examination twice found that 427 of the respondents had paid for coaching courses and that the remaining 2733 had not.[14] Construct and interpret a 99% confidence interval for the proportion of coaching among students who retake the SAT. pg 500
   STEP 4

40. **2010 begins** In January 2010 a Gallup Poll asked a random sample of adults, "In general, are you satisfied or dissatisfied with the way things are going in the United States at this time?" In all, 256 said that they were satisfied and the remaining 769 said they were not. Construct and interpret a 90% confidence interval for the proportion of adults who are satisfied with how things are going.
   STEP 4

41. **Equality for women?** Have efforts to promote equality for women gone far enough in the United States? A poll on this issue by the cable network MSNBC contacted 1019 adults. A newspaper article about the poll said, "Results have a margin of sampling error of plus or minus 3 percentage points."[15]
   (a) The news article said that 65% of men, but only 43% of women, think that efforts to promote equality have gone far enough. Explain why we do not have enough information to give confidence intervals for men and women separately.
   (b) Would a 95% confidence interval for women alone have a margin of error less than 0.03, about equal to 0.03, or greater than 0.03? Why? (You see that the news article's statement about the margin of error for poll results is a bit misleading.)

42. **A TV poll** A television news program conducts a call-in poll about a proposed city ban on handgun ownership. Of the 2372 calls, 1921 oppose the ban.

**8.37** Answers will vary. Response bias is one possibility.

**8.38** Answers will vary. Response bias is one possibility.

**8.39** *S*: $p =$ the true proportion of all students retaking the SAT who receive coaching. *P*: One-sample $z$ interval for $p$. *Random*: The students were selected randomly. *10%*: The sample size (3160) is less than 10% of the population of all students taking the SAT twice. *Large Counts*: $n\hat{p} = 427 \geq 10$ and $n(1-\hat{p}) = 2733 \geq 10$. *D*: (0.119, 0.151). *C*: We are 99% confident that the interval from 0.119 to 0.151 captures the true proportion of students retaking the SAT who receive coaching.

**8.40** *S*: $p =$ the true proportion of all U.S. adults who were satisfied with the way things were going in the United States in January 2010. *P*: One-sample $z$ interval for $p$. *Random*: The adults were selected randomly. *10%*: The sample size (1025) is less than 10% of the population of U.S. adults. *Large Counts*: $n\hat{p} = 256 \geq 10$ and $n(1-\hat{p}) = 769 \geq 10$. *D*: (0.23, 0.27). *C*: We are 90% confident that the interval from 0.23 to 0.27 captures the true proportion of all U.S. adults who were satisfied with the way things were going in the United States in January 2010.

**8.41** **(a)** We do not know the sample sizes for the men and for the women. **(b)** The margin of error for women alone would be greater than 0.03 because the sample size for women alone is smaller than 1019.

**(c)** $0.11 \pm 2.576\sqrt{\dfrac{0.11(0.89)}{172}} = 0.11 \pm 0.06 =$ (0.05, 0.17). **(d)** We are 99% confident that the interval from 0.05 to 0.17 captures the true proportion of all undergraduates at this university who would be willing to report cheating.

**8.35** **(a)** *S*: $p =$ the true proportion of all full-time U.S. college students who are binge drinkers. *P*: One-sample $z$ interval for $p$. *Random*: The students were selected randomly. *10%*: The sample size (5914) is less than 10% of the population of all college students. *Large Counts*: $n\hat{p} = 2312 \geq 10$ and $n(1-\hat{p}) = 3602 \geq 10$. *D*: (0.375, 0.407). *C*: We are 99% confident that the interval from 0.375 to 0.407 captures the true proportion of full-time U.S. college students who are binge drinkers. **(b)** Because the value 0.45 does not appear in our 99% confidence interval, it isn't plausible that 45% of full-time U.S. college students are binge drinkers.

**8.36** **(a)** *S*: $p =$ the population proportion of all teens who would report texting with their friends every day. *P*: One-sample $z$ interval for $p$. *Random*: The teens were selected randomly. *10%*: The sample size (799) is less than 10% of the population of all teens. *Large Counts*: $n\hat{p} = 392 \geq 10$ and $n(1-\hat{p}) = 407 \geq 10$. *D*: (0.456, 0.526). *C*: We are 95% confident that the interval from 0.456 to 0.526 captures the true proportion of teens who would report texting with their friends every day. **(b)** Because the value 0.45 does not appear in our 95% confidence interval, it isn't plausible that 45% of American teens would report texting with their friends every day.

**8.42** No, the data are not from a random sample, so the conditions for inference have not been met.

**8.43** **(a)** Solving $1.645\sqrt{\frac{0.75(0.25)}{n}} \leq 0.04$ gives $n \geq 318$. **(b)** Solving $1.645\sqrt{\frac{0.5(0.5)}{n}} \leq 0.04$ gives $n \geq 423$. In this case, the sample size needed is 105 people larger.

**8.44** **(a)** Solving $2.576\sqrt{\frac{0.44(0.56)}{n}} \leq 0.03$ gives $n \geq 1817$. **(b)** Solving $2.576\sqrt{\frac{0.5(0.5)}{n}} \leq 0.03$ gives $n \geq 1844$ adults. The conservative approach requires 27 more adults.

**8.45** Solving $1.96\sqrt{\frac{0.5(0.5)}{n}} \leq 0.03$ gives $n \geq 1068$.

**8.46** Solving $1.645\sqrt{\frac{0.5(0.5)}{n}} \leq 0.04$ gives $n \geq 423$.

**8.47** **(a)** Solving $0.03 = z^*\sqrt{\frac{0.64(0.36)}{1028}}$ gives $z^* = 2.00$. The confidence level is likely 95%, because 2.00 is very close to 1.96. **(b)** Teens are hard to reach and often unwilling to participate in surveys, so nonresponse is a major "practical difficulty" for this type of poll.

**8.48** **(a)** Solving $0.01 = z^*\sqrt{\frac{0.6341(0.3659)}{5594}}$ gives $z^* = 1.55$. $P(-1.55 \leq Z \leq 1.55) = 0.8788$, so the confidence level is likely 88%.
**(b)** We do not know if those who *did* respond can reliably represent those who did not.

**8.49** a



The station, following recommended practice, makes a confidence statement: "81% of the Channel 13 Pulse Poll sample opposed the ban. We can be 95% confident that the true proportion of citizens opposing a handgun ban is within 1.6% of the sample result." Is the station's conclusion justified? Explain.

pg 502

43. **Can you taste PTC?** PTC is a substance that has a strong bitter taste for some people and is tasteless for others. The ability to taste PTC is inherited. About 75% of Italians can taste PTC, for example. You want to estimate the proportion of Americans who have at least one Italian grandparent and who can taste PTC.
 (a) How large a sample must you test to estimate the proportion of PTC tasters within 0.04 with 90% confidence? Answer this question using the 75% estimate as the guessed value for $\hat{p}$.
 (b) Answer the question in part (a) again, but this time use the conservative guess $\hat{p} = 0.5$. By how much do the two sample sizes differ?

44. **School vouchers** A national opinion poll found that 44% of all American adults agree that parents should be given vouchers that are good for education at any public or private school of their choice. The result was based on a small sample.
 (a) How large an SRS is required to obtain a margin of error of 0.03 (that is, ±3%) in a 99% confidence interval? Answer this question using the previous poll's result as the guessed value for $\hat{p}$.
 (b) Answer the question in part (a) again, but this time use the conservative guess $\hat{p} = 0.5$. By how much do the two sample sizes differ?

45. **Election polling** Gloria Chavez and Ronald Flynn are the candidates for mayor in a large city. We want to estimate the proportion $p$ of all registered voters in the city who plan to vote for Chavez with 95% confidence and a margin of error no greater than 0.03. How large a random sample do we need? Show your work.

46. **Starting a nightclub** A college student organization wants to start a nightclub for students under the age of 21. To assess support for this proposal, they will select an SRS of students and ask each respondent if he or she would patronize this type of establishment. What sample size is required to obtain a 90% confidence interval with an approximate margin of error of 0.04? Show your work.

47. **Teens and their TV sets** According to a Gallup Poll report, 64% of teens aged 13 to 17 have TVs in their rooms. Here is part of the footnote to this report:

 *These results are based on telephone interviews with a randomly selected national sample of 1028 teenagers in the Gallup Poll Panel of households, aged 13 to 17. For results based on this sample, one can say . . . that the maximum error attributable to sampling and other random effects is ±3 percentage points. In addition to sampling error, question wording and practical difficulties in conducting surveys can introduce error or bias into the findings of public opinion polls.*[16]

 (a) We omitted the confidence level from the footnote. Use what you have learned to determine the confidence level, assuming that Gallup took an SRS.
 (b) Give an example of a "practical difficulty" that could lead to biased results for this survey.

48. **Gambling and the NCAA** Gambling is an issue of great concern to those involved in college athletics. Because of this concern, the National Collegiate Athletic Association (NCAA) surveyed randomly selected student athletes concerning their gambling-related behaviors.[17] Of the 5594 Division I male athletes in the survey, 3547 reported participation in some gambling behavior. This includes playing cards, betting on games of skill, buying lottery tickets, betting on sports, and similar activities. A report of this study cited a 1% margin of error.
 (a) The confidence level was not stated in the report. Use what you have learned to find the confidence level, assuming that the NCAA took an SRS.
 (b) The study was designed to protect the anonymity of the student athletes who responded. As a result, it was not possible to calculate the number of students who were asked to respond but did not. How does this fact affect the way that you interpret the results?

***Multiple choice:** Select the best answer for Exercises 49 to 52.*

49. A Gallup Poll found that only 28% of American adults expect to inherit money or valuable possessions from a relative. The poll's margin of error was ±3 percentage points at a 95% confidence level. This means that
 (a) the poll used a method that gets an answer within 3% of the truth about the population 95% of the time.
 (b) the percent of all adults who expect an inheritance is between 25% and 31%.
 (c) if Gallup takes another poll on this issue, the results of the second poll will lie between 25% and 31%.
 (d) there's a 95% chance that the percent of all adults who expect an inheritance is between 25% and 31%.
 (e) Gallup can be 95% confident that between 25% and 31% of the sample expect an inheritance.

50. Most people can roll their tongues, but many can't. The ability to roll the tongue is genetically determined. Suppose we are interested in determining what proportion of students can roll their tongues. We test a simple random sample of 400 students and find that 317 can roll their tongues. The margin of error for a 95% confidence interval for the true proportion of tongue rollers among students is closest to

(a) 0.0008. (c) 0.03. (e) 0.05.
(b) 0.02. (d) 0.04.

51. You want to design a study to estimate the proportion of students at your school who agree with the statement, "The student government is an effective organization for expressing the needs of students to the administration." You will use a 95% confidence interval, and you would like the margin of error to be 0.05 or less. The minimum sample size required is

(a) 22. (b) 271. (c) 385. (d) 769. (e) 1795.

52. A newspaper reporter asked an SRS of 100 residents in a large city for their opinion about the mayor's job performance. Using the results from the sample, the C% confidence interval for the proportion of all residents in the city who approve of the mayor's job performance is 0.565 to 0.695. What is the value of C?

(a) 82 (b) 86 (c) 90 (d) 95 (e) 99

*Exercises 53 and 54 refer to the following setting.* The following table displays the number of accidents at a factory during each hour of a 24-hour shift (1 = 1:00 A.M.).

| Hour | Number of accidents | Hour | Number of accidents |
|---|---|---|---|
| 1 | 5 | 13 | 21 |
| 2 | 8 | 14 | 12 |
| 3 | 17 | 15 | 10 |
| 4 | 31 | 16 | 1 |
| 5 | 24 | 17 | 0 |
| 6 | 18 | 18 | 1 |
| 7 | 12 | 19 | 3 |
| 8 | 7 | 20 | 21 |
| 9 | 1 | 21 | 23 |
| 10 | 0 | 22 | 18 |
| 11 | 2 | 23 | 11 |
| 12 | 14 | 24 | 2 |

53. **Accidents happen** (1.2, 3.1)

(a) Construct a plot that displays the distribution of the number of accidents effectively.

(b) Construct a plot that shows the relationship between the number of accidents and the time when they occurred.

(c) Describe something that the plot in part (a) tells you about the data that the plot in part (b) does not.

(d) Describe something that the plot in part (b) tells you about the data that the plot in part (a) does not.

54. **Accidents happen** (1.3) Plant managers are concerned that the number of accidents may be significantly higher during the midnight to 8:00 A.M. shift than during the 4:00 P.M. to midnight shift. What would you tell them? Give appropriate statistical evidence to support your conclusion.

# 8.3 Estimating a Population Mean

**WHAT YOU WILL LEARN** By the end of the section, you should be able to:

- State and check the Random, 10%, and Normal/Large Sample conditions for constructing a confidence interval for a population mean.
- Explain how the *t* distributions are different from the standard Normal distribution and why it is necessary to use a *t* distribution when calculating a confidence interval for a population mean.
- Determine critical values for calculating a *C*% confidence interval for a population mean using a table or technology.
- Construct and interpret a confidence interval for a population mean.
- Determine the sample size required to obtain a *C*% confidence interval for a population mean with a specified margin of error.

**8.50** d

**8.51** c

**8.52** a

**8.53 (a)** A histogram of the number of accidents per hour is given below.

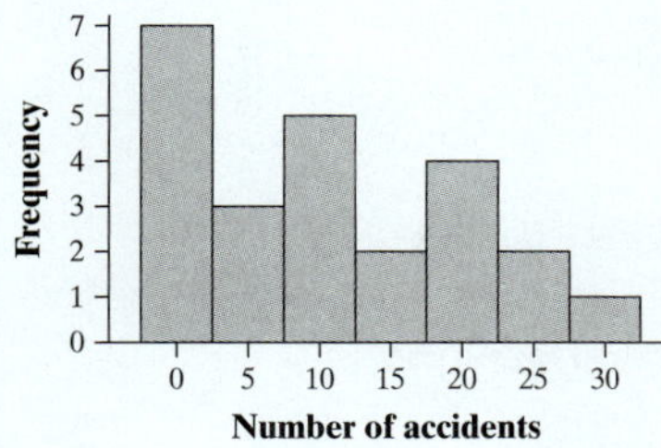


**(b)** A graph of the number of accidents is given below.

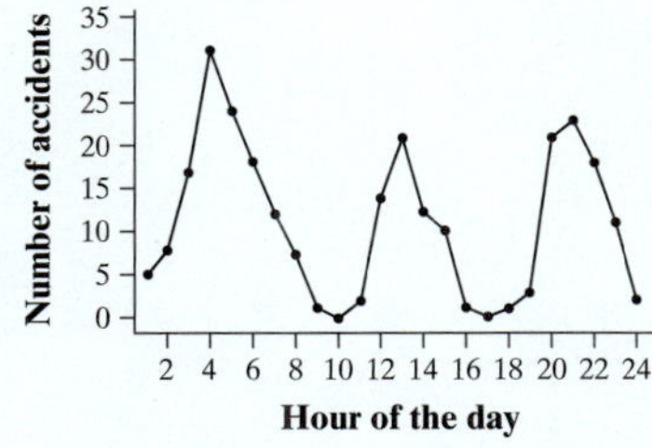


**(c)** The histogram in part (a) shows that the number of accidents has a distribution that is skewed to the right. **(d)** The graph in (b) shows that there is a cyclical nature to the number of accidents.

**8.54** The boxplots below show that both distributions are slightly skewed to the right. The median number of accidents is greater for the midnight to 8 A.M. shift (Shift 1) than for the 4 P.M. to midnight shift (Shift 3), confirming the manager's concern. The variability of the two distributions is similar, with Shift 1 having a slightly larger range and Shift 3 having a slightly larger *IQR*.

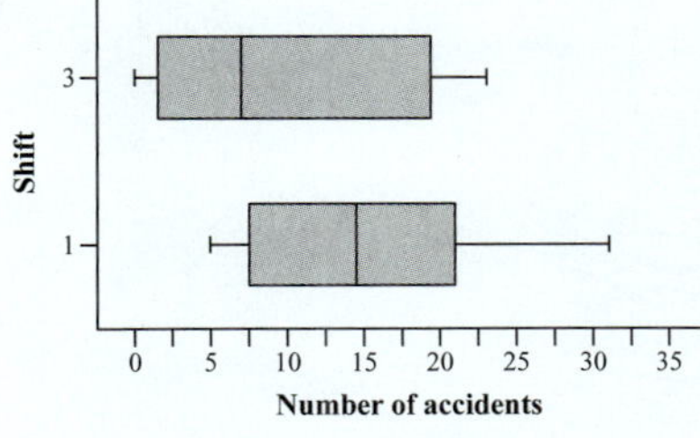


**PD Section 8.3 Overview**

Watch the video overview of Section 8.3 (for teachers) by clicking on the link in the TE-book, opening the TRCD, or logging onto the book's companion Web site www.whfreeman.com/tps5e (teacher log-in required).

**TRM "Good Books" Alternate Activity**

For a nice Activity to introduce this section, see the Good Books Activity in the Teacher's Resource Materials. Find it by clicking on the link in the TE-book, opening the TRCD, or logging onto the book's companion Web site www.whfreeman.com/tps5e (teacher log-in required).

**TRM Section 8.3 Alternate Examples**

You can find the Alternate Examples for this section in Microsoft Word format by clicking on the link in the TE-book, opening the TRCD, or logging onto the book's companion Web site www.whfreeman.com/tps5e (teacher log-in required).

Inference about a population proportion usually arises when we study *categorical* variables. We learned how to construct and interpret confidence intervals for a population proportion $p$ in Section 8.2. To estimate a population mean, we have to record values of a *quantitative* variable for a sample of individuals. It makes sense to try to estimate the mean amount of sleep that students at a large high school got last night but not their mean eye color! In this section, we'll examine confidence intervals for a population mean $\mu$.

## The Problem of Unknown $\sigma$

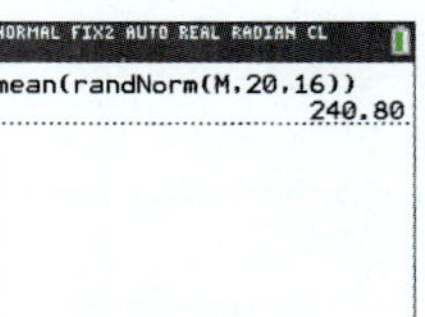


Mr. Schiel's class did the mystery mean Activity (page 476) and got a value of $\bar{x} = 240.80$ from an SRS of size 16, as shown.

Their task was to estimate the unknown population mean $\mu$. They knew that the population distribution was Normal and that its standard deviation was $\sigma = 20$. Their estimate was based on the sampling distribution of $\bar{x}$. Figure 8.10 shows this Normal sampling distribution once again.

**FIGURE 8.10** The Normal sampling distribution of $\bar{x}$ for the mystery mean Activity.

To calculate a 95% confidence interval for $\mu$, we use our familiar formula:

$$\text{statistic} \pm (\text{critical value}) \cdot (\text{standard deviation of statistic})$$

The critical value, $z^* = 1.96$, tells us how many standardized units we need to go out to catch the middle 95% of the sampling distribution. Our interval is

$$\bar{x} \pm z^* \cdot \frac{\sigma}{\sqrt{n}} = 240.80 \pm 1.96 \cdot \frac{20}{\sqrt{16}} = 240.80 \pm 9.80 = (231.00, 250.60)$$

We call such an interval a *one-sample z interval for a population mean.*

This method isn't very useful in practice, however. In most real-world settings, if we don't know the population mean $\mu$, then we don't know the population standard deviation $\sigma$ either.

How do we estimate $\mu$ when the population standard deviation $\sigma$ is unknown? Our best guess for the value of $\sigma$ is the sample standard deviation $s_x$. Maybe we could use the one-sample $z$ interval for a population mean with $s_x$ in place of $\sigma$:

$$\bar{x} \pm z^* \frac{s_x}{\sqrt{n}}$$

Let's try it.

### Teaching Tip

Emphasize that we rarely, if ever, use $z$ procedures to calculate a confidence interval for a population mean $\mu$. The only time $z$ procedures are valid is when the population standard deviation is known. We briefly introduce the one-sample $z$ interval for a population mean to provide a link between the $z$ intervals we have been using in Sections 8.1 and 8.2.

## ACTIVITY | Calculator BINGO!

**MATERIALS:**
TI-83/84 for each student

A farmer wants to estimate the mean weight (in grams) of all tomatoes grown on his farm. To do so, he will select a random sample of 4 tomatoes, calculate the mean weight (in grams), and use the sample mean $\bar{x}$ to create a 99% confidence interval for the population mean $\mu$. Suppose that the weights of tomatoes on his farm are approximately Normally distributed with a mean of 100 grams and a standard deviation of 40 grams.

1. Use your calculator to simulate taking an SRS of size 4 from this population and creating a one-sample $z$ interval for $\mu$: $\bar{x} \pm z^*\frac{\sigma}{\sqrt{n}} = \bar{x} \pm 2.576\frac{40}{\sqrt{4}}$. Enter the command shown below and press ENTER.

```
randNorm(100,40,4)→L1:ZInterval 40,mean(L1),4,99
```

Check to see whether the resulting interval captures $\mu = 100$. If it does not, shout "BINGO!"

To get the `randNorm` command, press [MATH] and arrow to PRB. The `ZInterval` command is in the Catalog. To get the `mean` and `stdDev` commands, press [2nd] [STAT] (LIST) and arrow to the MATH menu.

Keep pressing ENTER to generate more 99% confidence intervals. Check each interval to see whether it captures $\mu = 100$. If it does not, shout "BINGO!" If this method of constructing confidence intervals is working properly, about what percent of the time should you get a BINGO? Does the method seem to be working?

The method in Step 1 works well if we know the population standard deviation $\sigma$. That's rarely the case in real life. What happens if we use the sample standard deviation $s_x$ in place of $\sigma$ when calculating a confidence interval for the population mean?

2. Use your calculator to simulate taking an SRS of size 4 from this population and creating a "modified" one-sample $z$ interval for $\mu$: $\bar{x} \pm z^*\frac{s_x}{\sqrt{n}} = \bar{x} \pm 2.576\frac{s_x}{\sqrt{4}}$. Enter the command shown below and press ENTER.

```
randNorm(100,40,4)→L1:ZInterval stdDev(L1),mean(L1),4,99
```

Check to see whether the resulting interval captures $\mu = 100$. If it does not, shout "BINGO!"

Keep pressing ENTER to generate more 99% confidence intervals. Check each interval to see whether it captures $\mu = 100$. If it does not, shout "BINGO!" If this method of constructing confidence intervals is working properly, about what percent of the time should you get a BINGO? Does the method seem to be working?

The figure on the next page shows the results of using an applet from www.rossmanchance.com to repeatedly construct confidence intervals as described in Step 2 of the Activity. Of the 1000 intervals constructed, only 923 (that's 92.3%) captured the population mean. That's far below our desired 99% confidence level. What went wrong? The intervals that missed (those in red) came from samples with small standard deviations $s_x$ and from samples in which $\bar{x}$ was far from the population mean $\mu$. In those cases, multiplying $s_x/\sqrt{4}$ by $z^* = 2.576$ didn't produce long enough intervals to reach $\mu = 100$. We need to multiply by a larger critical value to achieve a 99% capture rate. But what critical value should we use?

### Activity Overview

You can watch an overview of this Activity (for teachers) by clicking on the link in the TE-book, opening the TRCD, or logging onto the book's companion Web site www.whfreeman.com/tps5e (teacher log-in required).

### Teaching Tip

Although it is a little more complicated than other Activities, we highly recommend that you do the Activity on this page. It should make it clear to students why we can't use $z$ critical values when we use the sample standard deviation to estimate the population standard deviation. If you are short on time or have a variety of different calculators in your classroom, you might opt for the related Alternate Activity described on the next page.

### TECHNOLOGY TIP

Joining commands with a colon tells the calculator to execute each command in order with the push of a single button. Make sure students understand what is happening in each of the commands before moving on!

### TECHNOLOGY TIP

Here are the commands for students with a TI-Nspire:

- Open a Lists and Spreadsheet page.
- Name the first column "samp."
- Add a Calculator page.
- Press `Menu`, choose `5:Probability`, choose `4:Random` and enter the command `Randnorm(100,40,4)→samp:`
- Press `Menu`, choose `6:Statistics`, choose `6:Confidence Intervals`, and choose `1:z interval`.
- Change the input method from data to stats and enter the following values in the dialogue boxes: $\sigma$`:40,`$\bar{x}$`:mean(samp),n:4,c-level:0.99`. Press `OK`. (*Note*: For Step 2, enter `stDevSamp(samp)` in the dialogue box for $\sigma$.)
- Keep pressing `Enter` to obtain more intervals.

ALTERNATE ACTIVITY

***Computer BINGO!***

This activity parallels the Calculator BINGO! Activity, but uses the *Simulating Confidence Intervals for a Population Parameter* applet at www.rossmanchance.com/applets.

**Step 1:** Use the applet to simulate taking an SRS of size 4 from this population and creating a one-sample $z$ interval for $\mu: \bar{x} \pm z^* \frac{\sigma}{\sqrt{n}} = \bar{x} \pm 2.576 \frac{40}{\sqrt{4}}$.

Choose "Means" and "z with sigma" from the drop-down menus. Enter $\mu = 100$, $\sigma = 40$, $n = 4$, Intervals = 100, and conf level = 99. Press Sample over and over. Was the overall capture rate close to 99%?

**Step 2:** Use the applet to simulate taking an SRS of size 4 from this population and creating a "modified" one-sample $z$ interval for $\mu: \bar{x} \pm z^* \frac{s_x}{\sqrt{n}} = \bar{x} \pm 2.576 \frac{s_x}{\sqrt{4}}$.

Choose "Means" and "z with sigma" from the drop-down menus. Enter $\mu = 100$, $\sigma = 40$, $n = 4$, Intervals = 100, and conf level = 99. Press Sample over and over. Was the overall capture rate close to 99%? What do you notice about the intervals that don't capture $\mu = 100$?

**Step 3** (*optional*): Use the applet to simulate taking an SRS of size 4 from this population and creating a one-sample $t$ interval for $\mu: \bar{x} \pm t^* \frac{s_x}{\sqrt{n}} = \bar{x} \pm t^* \frac{s_x}{\sqrt{4}}$.

Choose "Means" and "t" from the drop-down menus. Enter $\mu = 100$, $\sigma = 40$, $n = 4$, Intervals = 100, and conf level = 99. Press Sample over and over. Was the overall capture rate close to 99%? What do you notice about the lengths of the intervals?

**Simulating Confidence Intervals**

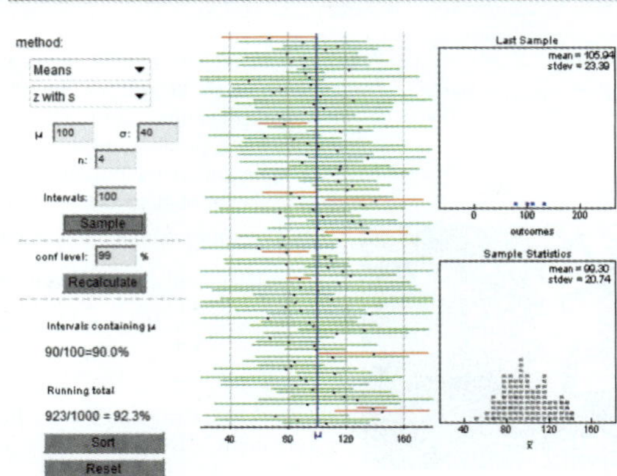


## When $\sigma$ Is Unknown: The $t$ Distributions

When the sampling distribution of $\bar{x}$ is close to Normal, we can find probabilities involving $\bar{x}$ by standardizing:

$$z = \frac{\bar{x} - \mu}{\sigma/\sqrt{n}}$$

Recall that a statistic is a number computed from sample data. We know that the sample mean $\bar{x}$ is a statistic. So is the standardized value $z = \frac{\bar{x} - \mu}{\sigma/\sqrt{n}}$. The sampling distribution of $z$ shows the values this statistic takes in all possible SRSs of size $n$ from the population.

Recall that the sampling distribution of $\bar{x}$ has mean $\mu$ and standard deviation $\sigma/\sqrt{n}$, as shown in Figure 8.11(a). What are the shape, center, and spread of the sampling distribution of the new statistic $z$?

From what we learned in Chapter 6, subtracting the constant $\mu$ from the values of the random variable $\bar{x}$ shifts the distribution left by $\mu$ units, making the mean 0. This transformation doesn't affect the shape or spread of the distribution. Dividing

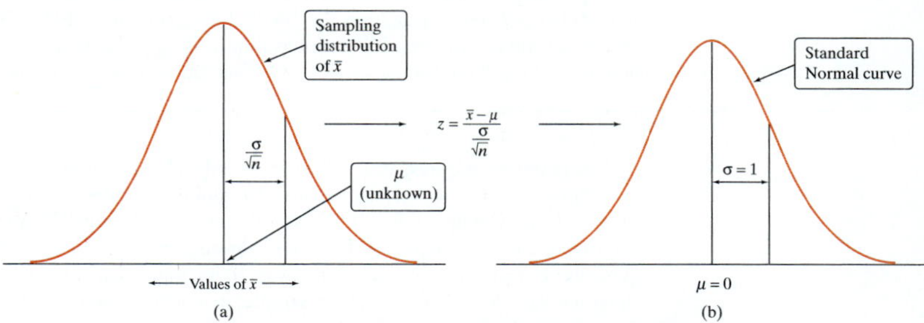


**FIGURE 8.11** (a) Sampling distribution of $\bar{x}$ when the Normal/Large Sample condition is met. (b) Standardized values of $\bar{x}$ lead to the statistic $z$, which follows the standard Normal distribution.

by the constant $\sigma/\sqrt{n}$ keeps the mean at 0, makes the standard deviation 1, and leaves the shape unchanged. As shown in Figure 8.11(b), $z$ has the standard Normal distribution N(0, 1). Therefore, we can use Table A or a calculator to find the related probability involving $z$. That's how we have gotten the critical values for our confidence intervals so far.

When we don't know $\sigma$, we estimate it using the sample standard deviation $s_x$. What happens now when we standardize?

$$?? = \frac{\bar{x} - \mu}{s_x/\sqrt{n}}$$

To find out, let's start with a Normal population having mean $\mu = 100$ and standard deviation $\sigma = 5$. We'll simulate choosing an SRS of size $n = 4$ and calculating the sample mean $\bar{x}$. Then we will standardize the result in two ways:

$$z = \frac{\bar{x} - 100}{5/\sqrt{4}} \quad \text{and} \quad ?? = \frac{\bar{x} - 100}{s_x/\sqrt{4}}$$

Figure 8.12 shows the results of taking 500 SRSs of size $n = 4$ and standardizing the value of the sample mean $\bar{x}$ in both ways. The values of $z$ follow a standard Normal distribution, as expected. The standardized values we get, using the sample standard deviation $s_x$ in place of the population standard deviation $\sigma$, show much greater spread. In fact, in a few samples, the statistic

$$?? = \frac{\bar{x} - \mu}{s_x/\sqrt{n}}$$

took values below −6 or above 6.

This statistic has a distribution that is new to us, called a **$t$ distribution.** It has a *different shape* than the standard Normal curve: still symmetric with a single peak at 0, but with much more area in the tails.

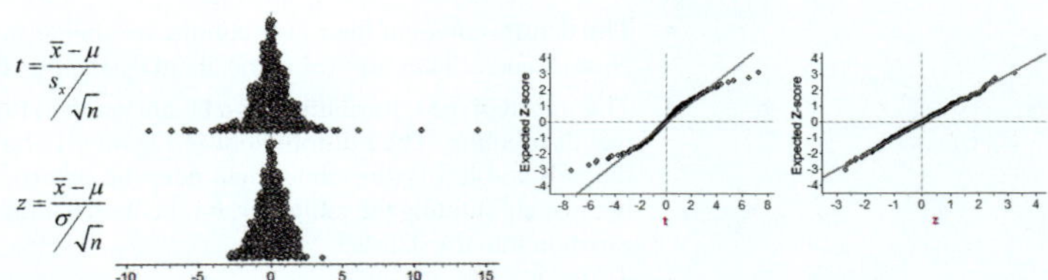

**FIGURE 8.12** Fathom simulation showing standardized values of the sample mean $\bar{x}$ in 500 SRSs. The statistic $z$ follows a standard Normal distribution. Replacing $\sigma$ with $s_x$ yields a statistic with much greater variability that doesn't follow the standard Normal curve. The Normal probability plot for the $t$ statistic shows the departure from Normality in the tails of the $t$ distribution.

The statistic $t$ has the same interpretation as any standardized statistic: it says how far $\bar{x}$ is from its mean $\mu$ in standard deviation units. There is a different $t$ distribution for each sample size. We specify a particular $t$ distribution by giving its **degrees of freedom (df).** When we perform inference about a population mean $\mu$ using a $t$ distribution, the appropriate degrees of freedom are found by subtracting 1 from the sample size $n$, making df $= n - 1$. We will write the $t$ distribution with $n - 1$ degrees of freedom as $t_{n-1}$ for short.

## Teaching Tip

In Chapter 6, we learned that transforming a variable by subtracting a constant and then dividing by a constant changes the center and spread of the distribution of that variable but not the shape. However, we are now dividing by a *variable* ($s_x$), which changes the shape in addition to the spread.

## Teaching Tip

Make sure your students understand what variable follows a $t$ distribution. When a sample is selected from a Normal population, the sample mean $\bar{x}$ follows a Normal distribution, but the standardized $t = \frac{\bar{x} - \mu}{s_x/\sqrt{n}}$ statistic follows a $t$ distribution. For example, a $t$ distribution with df = 3 would nicely model the dotplot showing the values of $t$ on this page.

## Teaching Tip

Students will probably ask you to explain what degrees of freedom are all about. Unfortunately, there is no simple answer. For now, simply explain that the shape and spread of the $t$ distributions depend on the degrees of freedom, which depend on the sample size (df $= n - 1$).

The $t$ distribution and the $t$ inference procedures were invented by William S. Gosset (1876–1937). Gosset worked for the Guinness brewery, and his goal in life was to make better beer. He used his new $t$ procedures to find the best varieties of barley and hops. Gosset's statistical work helped him become head brewer. Because Gosset published under the pen name "Student," you will often see the $t$ distribution called "Student's $t$" in his honor.

## THE $t$ DISTRIBUTIONS; DEGREES OF FREEDOM

Draw an SRS of size $n$ from a large population that has a Normal distribution with mean $\mu$ and standard deviation $\sigma$. The statistic

$$t = \frac{\bar{x} - \mu}{s_x/\sqrt{n}}$$

has the **$t$ distribution** with **degrees of freedom** df $= n - 1$. When the population distribution isn't Normal, this statistic will have approximately a $t_{n-1}$ distribution if the sample size is large enough.

Figure 8.13 compares the density curves of the standard Normal distribution and the $t$ distributions with 2 and 9 degrees of freedom. The figure illustrates these facts about the $t$ distributions:

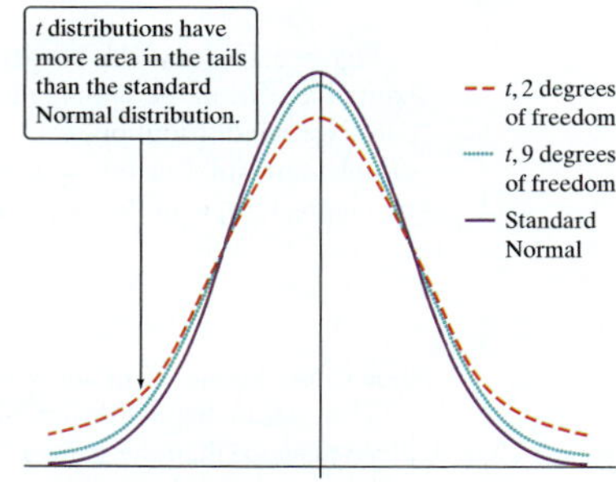


**FIGURE 8.13** Density curves for the $t$ distributions with 2 and 9 degrees of freedom and the standard Normal distribution. All are symmetric with center 0. The $t$ distributions are somewhat more spread out.

- The density curves of the $t$ distributions are similar in shape to the standard Normal curve. They are symmetric about 0, single-peaked, and bell-shaped.
- The spread of the $t$ distributions is a bit greater than that of the standard Normal distribution. The $t$ distributions in Figure 8.13 have more probability in the tails and less in the center than does the standard Normal. This is true because substituting the estimate $s_x$ for the fixed parameter $\sigma$ introduces more variation into the statistic.
- As the degrees of freedom increase, the $t$ density curve approaches the standard Normal density curve ever more closely. This happens because $s_x$ estimates $\sigma$ more accurately as the sample size increases. So using $s_x$ in place of $\sigma$ causes little extra variation when the sample is large.

Table B in the back of the book gives critical values $t^*$ for the $t$ distributions. Each row in the table contains critical values for the $t$ distribution whose degrees of freedom appear at the left of the row. For convenience, several of the more common confidence levels C are given at the bottom of the table. By looking down any column, you can check that the $t$ critical values approach the Normal critical values $z^*$ as the degrees of freedom increase.

When you use Table B to determine the correct value of $t^*$ for a given confidence interval, all you need to know are the confidence level C and the degrees

### Teaching Tip

Because the spread of the $t$ distributions is slightly larger than the spread of a standard Normal distribution, we need to go farther out in the $t$ distributions than $\pm 1.96$ to capture the central 95%. To determine exactly how far, we use Table B (the $t$ table) or technology. If we continued to use 1.96, the intervals would be too short and our capture rate would be smaller than 95%, as illustrated in the Calculator BINGO! Activity.

of freedom (df). Unfortunately, Table B does not include every possible sample size. *When the actual df does not appear in the table, use the greatest df available that is less than your desired df.* This guarantees a wider confidence interval than you need to justify a given confidence level. Better yet, use technology to find an accurate value of $t^*$ for any df.

## EXAMPLE Finding $t^*$ Critical Values

### *Using Table B*

PROBLEM: What critical value $t^*$ from Table B should be used in constructing a confidence interval for the population mean in each of the following settings?

(a) A 95% confidence interval based on an SRS of size $n = 12$.

(b) A 90% confidence interval from a random sample of 48 observations.

SOLUTION:

(a) In Table B, we consult the row corresponding to df $= 12 - 1 = 11$. We move across that row to the entry that is directly above 95% confidence level on the bottom of the chart. The desired critical value is $t^* = 2.201$.

(b) With 48 observations, we want to find the $t^*$ critical value for df $= 48 - 1 = 47$ and 90% confidence. There is no df $= 47$ row in Table B, so we use the more conservative df $= 40$. The corresponding critical value is $t^* = 1.684$.

The bottom row of Table B gives $z^*$ critical values and is labeled as df $= \infty$. That's because the $t$ distributions approach the standard Normal distribution as the degrees of freedom approach infinity.

| | Upper-tail probability $p$ | | | |
|---|---|---|---|---|
| df | .05 | .025 | .02 | .01 |
| 10 | 1.812 | 2.228 | 2.359 | 2.764 |
| 11 | 1.796 | 2.201 | 2.328 | 2.718 |
| 12 | 1.782 | 2.179 | 2.303 | 2.681 |
| $\infty$ | 1.645 | 1.960 | 2.054 | 2.326 |
| | 90% | 95% | 96% | 98% |
| | Confidence level $C$ | | | |

| | Upper-tail probability $p$ | | | |
|---|---|---|---|---|
| df | .10 | .05 | .025 | .02 |
| 30 | 1.310 | 1.697 | 2.042 | 2.147 |
| 40 | 1.303 | 1.684 | 2.021 | 2.123 |
| 50 | 1.299 | 1.676 | 2.009 | 2.109 |
| $\infty$ | 1.282 | 1.645 | 1.960 | 2.054 |
| | 80% | 90% | 95% | 96% |
| | Confidence level $C$ | | | |

For Practice *Try Exercise* 55

For part (a) of the example, the corresponding standard Normal critical value for 95% confidence is $z^* = 1.96$. We have to go out farther than 1.96 standard deviations to capture the central 95% of the $t$ distribution with 11 degrees of freedom.

Technology will quickly produce $t^*$ critical values for any sample size.

## 16. TECHNOLOGY CORNER INVERSE *t* ON THE CALCULATOR

TI-Nspire instructions in Appendix B; HP Prime instructions on the book's Web site.

Most newer TI-84 and TI-89 calculators allow you to find critical values $t^*$ using the inverse $t$ command. As with the calculator's inverse Normal command, you have to enter the *area to the left* of the desired critical value. Let's use the inverse $t$ command to find the critical values in parts (a) and (b) of the example.

### ALTERNATE EXAMPLE

#### *Finding $t^*$*

PROBLEM: What critical value $t^*$ should be used in constructing a confidence interval in the following settings?

**(a)** A 90% confidence interval based on an SRS of size 10

**(b)** A 99% confidence interval based on a random sample of 95 observations

SOLUTION:

**(a)** Using the row for df $= 10 - 1 = 9$ and the column that is above the 90% confidence level, the desired critical value is $t^* = 1.833$.

**(b)** With $n = 95$, df $= 95 - 1 = 94$. Because there is no row for df $= 94$, we use the more conservative df $= 80$. Using the column for 99% confidence, $t^* = 2.639$.

### Teaching Tip

If you did the Calculator BINGO! Activity, ask your students to look up the correct critical value for 99% confidence and df $= 4 - 1 = 3$. It is $t^* = 5.841$, more than double the $z^*$ critical value of $z^* = 2.576$. The critical value needs to be this large to make 99% of the confidence intervals capture the true mean.

### Teaching Tip

As mentioned in the margin note, the last line of critical values in Table B corresponds to the standard Normal distribution. On the AP® exam, the $t$ table provided to students also provides the critical values for df $= \infty$ in the last line.

### TECHNOLOGY TIP

If your students have a TI-83 or older TI-84, they may not have the `invT` command. Installing an upgraded operating system (OS) will usually do the trick. Visit education.ti.com or copy the newer OS directly from another calculator using the Link function.

**TI-83/84**

- Press [2nd][VARS] (DISTR) and choose `invT(`.
- For part (a), **OS 2.55 or later:** In the dialog box, enter these values: `area:.025, df:11`, choose `Paste`, and then press [ENTER]. **Older OS:** Complete the command `invT(.025,11)` and press [ENTER].
- For part (b), use the command `invT(.05,47)`.

**TI-89**

- In the Statistics/List Editor, press [F5], choose `Inverse` and `Inverse t....`
- For part (a), enter Area: .025 and Deg of Freedom, df: 11, and then press [ENTER].
- For part (b), use Area:.05 and df: 47.

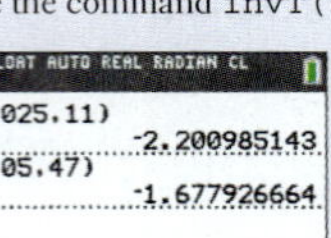

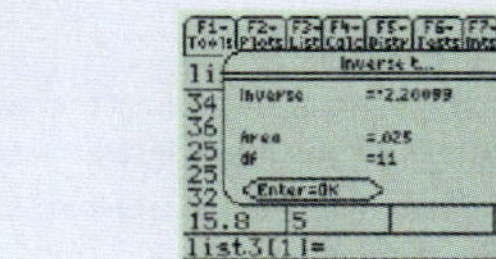

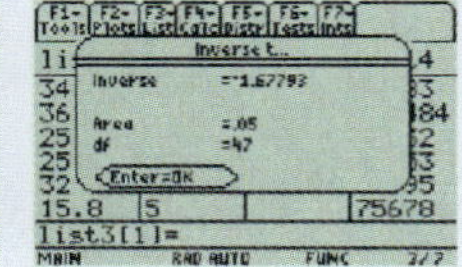

Note that the $t$ critical values are $t^* = 2.201$ and $t^* = 1.678$, respectively.

### CHECK YOUR UNDERSTANDING

Use Table B to find the critical value $t^*$ that you would use for a confidence interval for a population mean $\mu$ in each of the following settings. If possible, check your answer with technology.

1. A 96% confidence interval based on a random sample of 22 observations.
2. A 99% confidence interval from an SRS of 71 observations.

### ANSWERS TO CYU

**1.** df = 21, $t^* = 2.189$. *Using technology*: `invT(area:0.02,df = 21)` $= -2.189$, so $t^* = 2.189$.

**2.** df = 70, $t^* = 2.660$ (using df = 60). *Using technology*: `invT(area:0.005, df = 70)` $= -2.648$, so $t^* = 2.648$.

## Conditions for Estimating $\mu$

As with proportions, you should check some important conditions before constructing a confidence interval for a population mean. Two of the conditions should be familiar by now. The *Random condition* is crucial for doing inference. If the data don't come from a well-designed random sample or randomized experiment, you can't draw conclusions about a larger population or about cause and effect. When sampling without replacement, the *10% condition* ensures that our formula for the standard deviation of the statistic $\bar{x}$

$$\sigma_{\bar{x}} = \frac{\sigma}{\sqrt{n}}$$

is approximately correct.

The method we use to construct a confidence interval for $\mu$ depends on the fact that the sampling distribution of $\bar{x}$ is approximately Normal. From Chapter 7, we know that the sampling distribution of $\bar{x}$ is Normal if the population distribution is Normal. When the population distribution is not Normal, the central limit theorem tells us that the sampling distribution of $\bar{x}$ will be approximately Normal if the sample size is large enough ($n \geq 30$). Be sure to check this *Normal/Large Sample condition* before calculating a confidence interval.

**CONDITIONS FOR CONSTRUCTING A CONFIDENCE INTERVAL ABOUT A MEAN**

- **Random:** The data come from a well-designed random sample or randomized experiment.
  - **10%:** When sampling without replacement, check that $n \leq \frac{1}{10}N$.
- **Normal/Large Sample:** The population has a Normal distribution or the sample size is large ($n \geq 30$). If the population distribution has unknown shape and $n < 30$, use a graph of the sample data to assess the Normality of the population. Do not use $t$ procedures if the graph shows strong skewness or outliers.

### Teaching Tip

Make sure your students understand the purpose of each of these conditions, and the consequences of violating them. Like violating the Large Counts condition, when we violate the Normal/Large Sample condition, the actual capture rate of our intervals will be too small. See the discussion on pages 493–496 for more details.

Larger samples improve the accuracy of critical values from the $t$ distributions when the population is not Normal. This is true for two reasons:

1. The sampling distribution of $\bar{x}$ for large sample sizes is close to Normal.
2. As the sample size $n$ grows, the sample standard deviation $s_x$ will give a more accurate estimate of $\sigma$. This is important because we use $\frac{s_x}{\sqrt{n}}$ in place of $\frac{\sigma}{\sqrt{n}}$ when doing calculations.

The Normal/Large Sample condition is obviously met if we know that the population distribution is Normal or that the sample size is at least 30. What if we don't know the shape of the population distribution and $n < 30$? In that case, we have to graph the sample data. Our goal is to answer the question, "Is it reasonable to believe that these data came from a Normal population?"

How should graphs of data from small samples look if the population has a Normal distribution? The following Activity sheds some light on this question.

## ACTIVITY Sampling from a Normal Population

**MATERIALS:**
TI-83/84 or TI-89 for each student

Let's use the calculator to simulate choosing random samples of size $n = 20$ from a Normal distribution with $\mu = 100$ and $\sigma = 15$ and then to plot the data.

1. Choose an SRS of 20 observations from this Normal population.

TI-83/84: Press MATH, arrow to PRB and choose `randNorm(`. Complete the command `randNorm(100,15,20)→L1` and press ENTER.

TI-89: Press CATALOG F3 (Flash Apps), press alpha 2 (R) to jump to the r's, and choose `randNorm(`... Complete the command `tistat.randNorm(100,15,20)→list1` and press ENTER.

2. Make a histogram, a boxplot, and a Normal probability plot of the data in L1/list1. Do you see any obvious departures from Normality in the graphs of the sample data?

### Activity Overview

You can watch an overview of this Activity (for teachers) by clicking on the link in the TE-book, opening the TRCD, or logging onto the book's companion Web site www.whfreeman.com/tps5e (teacher log-in required).

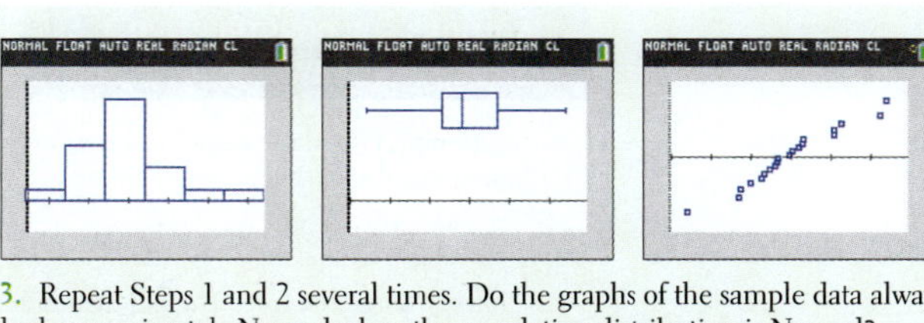

3. Repeat Steps 1 and 2 several times. Do the graphs of the sample data always look approximately Normal when the population distribution is Normal?

4. Compare the results with those of your classmates. How easy do you think it will be to use a graph of sample data to determine whether or not a population has a Normal distribution?

Did you expect that a random sample from a Normal population would yield a graph that looked Normal? Unfortunately, that's usually not the case. The figure below shows boxplots from three different SRSs of size 20 chosen in Step 3 of the Activity. The left-hand graph is skewed to the right. The right-hand graph shows three outliers in the sample. Only the middle graph looks symmetric and has no outliers.

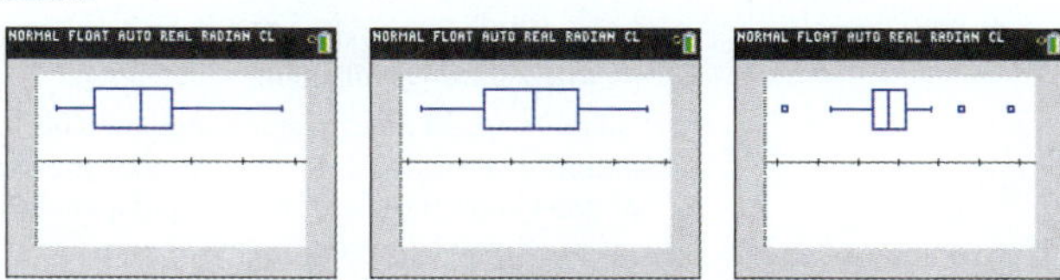

As the Activity shows, it is very difficult to use a graph of sample data to assess the Normality of a population distribution. If the graph has a skewed shape or if there are outliers present, it could be because the population distribution isn't Normal. Skewness or outliers could also occur naturally in a random sample from a Normal population. To be safe, you should *only use a t distribution for small samples with no outliers or strong skewness.*

caution

What constitutes strong skewness in a distribution? The following example gives you some idea.

### Teaching Tip

On the AP® exam, students will not be asked to construct a confidence interval when the conditions for inference are clearly violated. If a student thinks that there is a problem with one of the conditions when asked to construct a confidence interval, the student should state his or her concerns and proceed with calculating and interpreting the interval. In many cases, the student has misremembered the condition and would lose out on the remaining points if he or she didn't complete the problem.

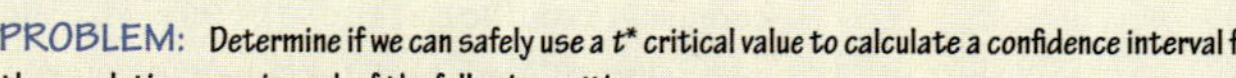

## EXAMPLE GPAs, Wood, and SATs

### *Can we use t?*

PROBLEM: Determine if we can safely use a $t^*$ critical value to calculate a confidence interval for the population mean in each of the following settings.

(a) To estimate the average GPA of students at your school, you randomly select 50 students from classes you take. Figure 8.14(a) is a histogram of their GPAs.

(b) How much force does it take to pull wood apart? Figure 8.14(b) shows a stemplot of the force (in pounds) required to pull apart a random sample of 20 pieces of Douglas fir.

(c) Suppose you want to estimate the mean SAT Math score at a large high school. Figure 8.14(c) is a boxplot of the SAT Math scores for a random sample of 20 students at the school.

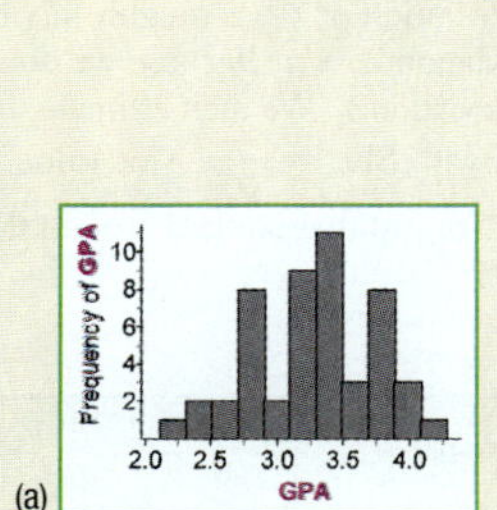


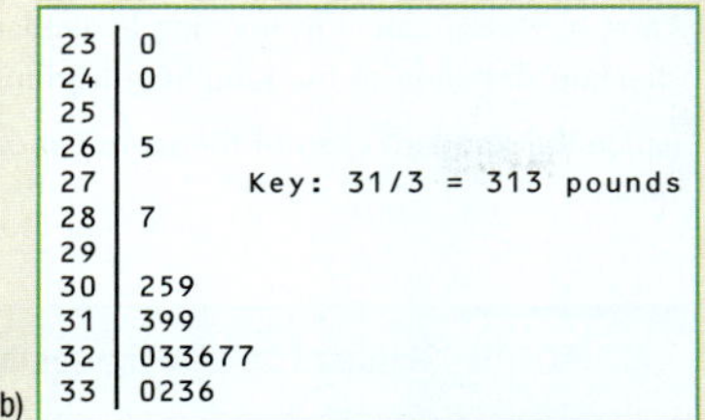


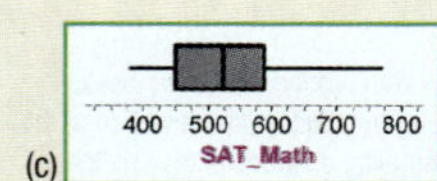


**FIGURE 8.14** Can we use a *t* distribution for these data? (a) GPAs of 50 randomly selected students in your classes at school. (b) Force required to pull apart a random sample of 20 pieces of Douglas fir. (c) SAT Math scores for a random sample of 20 students at a large high school.

SOLUTION:

(a) No. Although the histogram is roughly symmetric with no outliers, the random sample of 50 students was only from *your* classes and not from all students at your school. So we should not use these data to calculate a confidence interval for the mean GPA of all students at the school.

(b) No. The graph is strongly skewed to the left with possible low outliers. We cannot trust a critical value from a *t* distribution with df = 19 in this case.

(c) Yes. The distribution is only moderately skewed to the right and there are no outliers present.

**AP® EXAM TIP** If a question on the AP® exam asks you to calculate a confidence interval, all the conditions should be met. However, you are still required to state the conditions and show evidence that they are met.

**For Practice** *Try Exercise* 

**What's the difference between "strongly skewed" and "moderately skewed"?** Look at the stemplot in Figure 8.14(b) and the boxplot in Figure 8.14(c). Compare the distance from the maximum to the median and from the median to the minimum in both graphs. In Figure 8.14(b), maximum − median = 336 − 319.5 = 16.5 and median − minimum = 319.5 − 230 = 89.5. The half of the stemplot with smaller values is more than five times as long as the half of the stemplot with larger values. In Figure 8.14(c), maximum − median ≈ 775 − 525 = 250 and median − minimum ≈ 525 − 375 = 150. The right half of the boxplot is not quite twice as long as the left half.

There is no accepted rule of thumb for identifying strong skewness. For that reason, we have chosen the data sets in examples and exercises to avoid borderline cases. You should be able to tell easily if strong skewness is present in a graph of data from a small sample.

ALTERNATE EXAMPLE

## *Coffee, height, and homework*

PROBLEM: Determine whether we can safely use a $t^*$ critical value to calculate a confidence interval for the population mean in each of the following settings:

**(a)** We want to estimate the average time (in minutes) to order and receive a regular coffee at a local coffee shop. The times for five randomly selected visits to a local coffee shop are shown in the dotplot below.

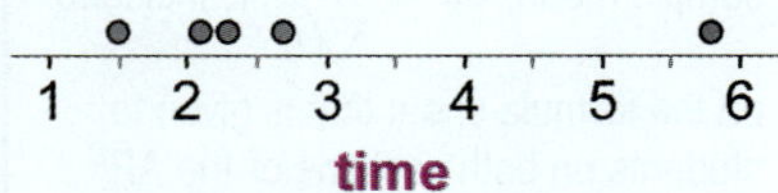


**(b)** We want to estimate the average height of high school students. The boxplot below shows the distribution of heights (in centimeters) for 100 randomly selected students.

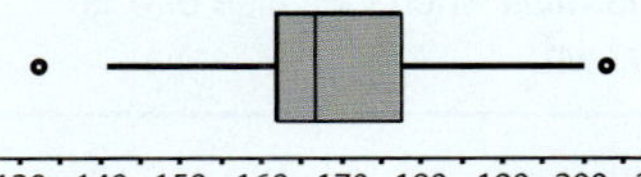


**(c)** A middle school counselor wants to estimate how many minutes, on average, students spend doing homework. A histogram of the amount of time doing homework the previous evening for a random sample of 20 students is shown below.

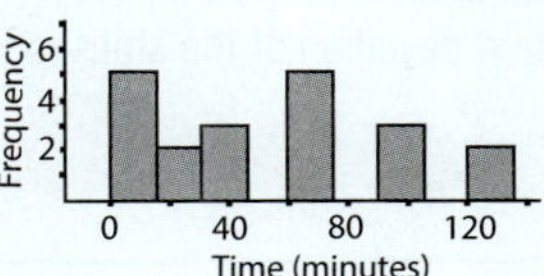


SOLUTION:

**(a)** No. Because the sample size is small and there is a possible outlier, we should not use a $t^*$ critical value to calculate the interval.

**(b)** Yes. Even though there are outliers, the sample size is large ($n = 100 \geq 30$).

**(c)** Yes. Although the sample size is small, there is no strong skewness or outliers.

## Constructing a Confidence Interval for $\mu$

When the conditions are met, the sampling distribution of $\bar{x}$ has roughly a Normal distribution with mean $\mu$ and standard deviation $\sigma/\sqrt{n}$. Because we don't know $\sigma$, we estimate it by the sample standard deviation $s_x$. We then estimate the standard deviation of the sampling distribution with $\text{SE}_{\bar{x}} = \frac{s_x}{\sqrt{n}}$. This value is called the **standard error of the sample mean** $\bar{x}$, or just the standard error of the mean.

As with proportions, some books refer to the standard deviation of the sampling distribution of $\bar{x}$ as the "standard error" and what we call the standard error of the mean as the "estimated standard error." The standard error of the mean is often abbreviated SEM.

**DEFINITION: Standard error of the sample mean**

The **standard error of the sample mean** $\bar{x}$ is $\text{SE}_{\bar{x}} = \frac{s_x}{\sqrt{n}}$, where $s_x$ is the sample standard deviation. It describes how far $\bar{x}$ will typically be from $\mu$ in repeated SRSs of size $n$.

To construct a confidence interval for $\mu$, replace the standard deviation $\sigma/\sqrt{n}$ of $\bar{x}$ by its standard error $s_x/\sqrt{n}$ in the formula for the one-sample $z$ interval for a population mean. Use critical values from the $t$ distribution with $n-1$ degrees of freedom in place of the $z$ critical values. That is,

$$\text{statistic} \pm (\text{critical value}) \cdot (\text{standard deviation of statistic})$$
$$= \bar{x} \pm t^* \frac{s_x}{\sqrt{n}}$$

This **one-sample $t$ interval for a population mean** is similar in both reasoning and computational detail to the one-sample $z$ interval for a population proportion of Section 8.2. So we will now pay more attention to questions about using these methods in practice.

**THE ONE-SAMPLE $t$ INTERVAL FOR A POPULATION MEAN**

When the conditions are met, a C% confidence interval for the unknown mean $\mu$ is

$$\bar{x} \pm t^* \frac{s_x}{\sqrt{n}}$$

where $t^*$ is the critical value for the $t_{n-1}$ distribution, with C% of the area between $-t^*$ and $t^*$.

The following example shows you how to construct a confidence interval for a population mean when $\sigma$ is unknown. By now, you should recognize the four-step process.

### AP® Exam Formula Sheet

The formula for the standard deviation of the sampling distribution of the sample mean, $\sigma_{\bar{x}} = \frac{\sigma}{\sqrt{n}}$, is included on the formula sheet that is given to students on both sections of the AP® exam. However, the formula for the *standard error* of the sample mean, $\text{SE}_{\bar{x}} = \frac{s_x}{\sqrt{n}}$, is not included on the formula sheet. Students are expected to know that it is OK to use $s_x$ in the formula when the value of $\sigma$ is unknown.

### AP® Exam Formula Sheet

The specific formula for a one-sample *t* interval for a population mean is not on the formula sheet that is given to students on the AP® exam. However, the formula sheet does include the general formula for a confidence interval and the formula for the standard deviation of the statistic $\bar{x}$, $\sigma_{\bar{x}} = \frac{\sigma}{\sqrt{n}}$.

### Teaching Tip

Emphasize to your students how the form of the one-sample $t$ interval for $\mu$ is very similar to the form of the one-sample $z$ interval for $p$.

## EXAMPLE Auto Pollution

### *A one-sample t interval for μ*

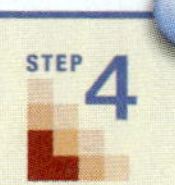

STEP 4

Environmentalists, government officials, and vehicle manufacturers are all interested in studying the auto exhaust emissions produced by motor vehicles. The major pollutants in auto exhaust from gasoline engines are hydrocarbons, carbon monoxide, and nitrogen oxides (NOX). Researchers collected data on the NOX levels (in grams/mile) for a random sample of 40 light-duty engines of the same type. The mean NOX reading was 1.2675 and the standard deviation was 0.3332.[18]

PROBLEM:

(a) Construct and interpret a 95% confidence interval for the mean amount of NOX emitted by light-duty engines of this type.

(b) The Environmental Protection Agency (EPA) sets a limit of 1.0 gram/mile for average NOX emissions. Are you convinced that this type of engine violates the EPA limit? Use your interval from (a) to support your answer.

SOLUTION:

(a) STATE: We want to estimate the true mean amount $\mu$ of NOX emitted by all light-duty engines of this type at a 95% confidence level.

PLAN: We should construct a one-sample $t$ interval for $\mu$ if the conditions are met.

- Random: The data come from a random sample of 40 light-duty engines of this type.
  - 10%: We are sampling without replacement, so we need to assume that there are at least $10(40) = 400$ light-duty engines of this type.
- Normal/Large Sample: We don't know whether the population distribution of NOX emissions is Normal. Because the sample size is large ($n = 40 \geq 30$), we should be safe using a $t$ distribution.

DO: The formula for the one-sample $t$ interval is

$$\bar{x} \pm t^* \frac{s_x}{\sqrt{n}}$$

From the information given, $\bar{x} = 1.2675$ g/mi and $s_x = 0.3332$ g/mi. To find the critical value $t^*$, we use the $t$ distribution with df $= 40 - 1 = 39$. Unfortunately, there is no row corresponding to 39 degrees of freedom in Table B. We can't pretend we have a larger sample size than we actually do, so we use the more conservative df $= 30$.

| | Upper-tail probability *p* | | |
|---|---|---|---|
| df | .05 | .025 | .02 |
| 29 | 1.699 | 2.045 | 2.150 |
| 30 | 1.697 | 2.042 | 2.147 |
| 40 | 1.684 | 2.021 | 2.123 |
| | 90% | 95% | 96% |
| | Confidence level C | | |

At a 95% confidence level, the critical value is $t^* = 2.042$. So the 95% confidence interval for $\mu$ is

$$\bar{x} \pm t^* \frac{s_x}{\sqrt{n}} = 1.2675 \pm 2.042 \frac{0.3332}{\sqrt{40}} = 1.2675 \pm 0.1076 = (1.1599, 1.3751)$$

*Using technology:* The command `invT(.025,39)` gives $t = -2.023$. Using the critical value $t^* = 2.023$ for the 95% confidence interval gives

$$\bar{x} \pm t^* \frac{s_x}{\sqrt{n}} = 1.2675 \pm 2.023 \frac{0.3332}{\sqrt{40}} = 1.2675 \pm 0.1066 = (1.1609, 1.3741)$$

This interval is slightly narrower than the one found using Table B.

CONCLUDE: We are 95% confident that the interval from 1.1609 to 1.3741 grams/mile captures the true mean level of nitrogen oxides emitted by this type of light-duty engine.

## ALTERNATE EXAMPLE

### *Milk's favorite cookie*

For their second-semester project in AP® Statistics, Ann and Tori wanted to estimate the average weight of an Oreo cookie to determine if the average weight was less than advertised. They selected a random sample of 36 cookies and found the weight of each cookie (in grams). The mean weight was $\bar{x} = 11.3921$ grams with a standard deviation of $s_x = 0.0817$ grams.

PROBLEM:

**(a)** Construct and interpret a 95% confidence interval for the mean weight of an Oreo cookie.

**(b)** On the packaging, the stated serving size is 3 cookies (34 grams). Does the interval in part (a) provide convincing evidence that the average weight of an Oreo cookie is less than advertised? Explain.

SOLUTION:

**(a)** STATE: We want to estimate $\mu$ = the true mean weight of an Oreo cookie at a 95% confidence level.

PLAN: We will construct a one-sample $t$ interval for $\mu$ if the conditions are met:

- *Random*: The cookies were randomly selected.
  - *10%*: $n = 36$ is less than 10% of all Oreo cookies.
- *Normal/Large Sample*: Because the sample size is large ($n = 36 \geq 30$), we should be safe using a $t$ distribution.

DO: Because there are $36 - 1 = 35$ degrees of freedom and we want 95% confidence, we will use the $t$ table and a conservative degrees of freedom of 30 to get a critical value of $t^* = 2.042$:

$$\bar{x} \pm t^* \frac{s_x}{\sqrt{n}} = 11.3921 \pm 2.042 \frac{0.0817}{\sqrt{36}}$$
$$= 11.3921 \pm 0.0278$$
$$= (11.3643, 11.4199)$$

*Using technology*: The command `invT(0.025,35)` gives $t = -2.030$. The resulting interval is

$$11.3921 \pm 2.030 \frac{0.0817}{\sqrt{36}} = 11.3921 \pm 0.0276 = (11.3645, 11.4197).$$

CONCLUDE: We are 95% confident that the interval from 11.3645 grams to 11.4197 grams captures the true mean weight of an Oreo cookie. **(b)** The stated serving size is 34 grams for 3 cookies, or 34/3 = 11.33 grams/cookie. Because all of the plausible values in the interval are greater than 11.33 grams, there is no evidence that the average weight of an Oreo cookie is less than advertised. In fact, there is convincing evidence that the average weight is greater than advertised!

## Teaching Tip

A nice way for students to remember to use $z$ for proportions and $t$ for means is the acronym "zap ta$\bar{x}$." It is never correct, even with a large sample size, to use a $z$ interval for $\mu$ when the population standard deviation is unknown.

### ALTERNATE EXAMPLE

### *Can you Spare a Square?*

As part of their final project in AP® Statistics, Christina and Rachel randomly selected 18 rolls of a generic brand of toilet paper to measure how well this brand could absorb water. To do this, they poured 1/4 cup of water onto a hard surface and counted how many squares of toilet paper it took to completely absorb the water. Here are the results from their 18 rolls:

| 29 | 20 | 25 | 29 | 21 | 24 | 27 | 25 | 24 |
|---|---|---|---|---|---|---|---|---|
| 29 | 24 | 27 | 28 | 21 | 25 | 26 | 22 | 23 |

Construct and interpret a 99% confidence interval for $\mu$ = the true mean number of squares of generic toilet paper needed to absorb 1/4 cup of water.

STATE: We want to estimate $\mu$ = the true mean number of squares of generic toilet paper needed to absorb 1/4 cup of water with 99% confidence.

PLAN: We will construct a one-sample $t$ interval if the conditions are met:

- *Random*: The students selected the rolls of generic toilet paper at random.
  - *10%*: It is safe to assume that there are at least 10(18) = 180 rolls of generic toilet paper.
- *Normal/Large Sample*: Because the sample size is small ($n$ = 18), we need to check whether it is reasonable to believe that the population has a Normal distribution. The dotplot below doesn't show any outliers or strong skewness, so it is reasonable to assume that the population distribution is approximately Normal.

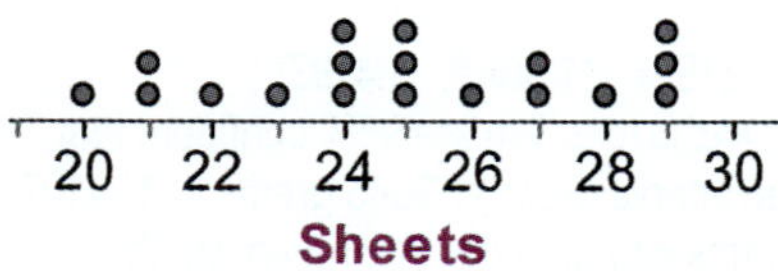


DO: The sample mean for these data is $\bar{x} = 24.94$ and the sample standard deviation is $s_x = 2.86$. Because there are $18 - 1 = 17$ degrees of freedom and we want 99% confidence, we will use a critical value of $t^* = 2.898$:

$$\bar{x} \pm t^* \frac{s_x}{\sqrt{n}} = 24.94 \pm 2.898 \frac{2.86}{\sqrt{18}}$$

$$= 24.94 \pm 1.95 = (22.99, 26.89)$$

CONCLUDE: We are 99% confident that the interval from 22.99 squares to 26.89 squares captures the true mean number of squares of generic toilet paper needed to absorb 1/4 cup of water.

(b) The confidence interval from (a) tells us that any value from 1.1609 to 1.3741 g/mi is a plausible value of the mean NOX level $\mu$ for this type of engine. Because the entire interval exceeds 1.0, it appears that this type of engine violates EPA limits.

**For Practice** *Try Exercise* 65

Now that we've calculated our first confidence interval for a population mean $\mu$, it's time to make a simple observation. Inference for *proportions* uses $z$; inference for *means* uses $t$. That's one reason why distinguishing categorical from quantitative variables is so important.

Here is another example, this time with a smaller sample size.

## EXAMPLE Video Screen Tension

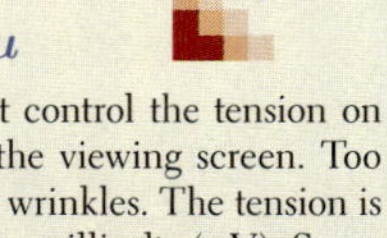


### *Constructing a confidence interval for $\mu$*

A manufacturer of high-resolution video terminals must control the tension on the mesh of fine wires that lies behind the surface of the viewing screen. Too much tension will tear the mesh, and too little will allow wrinkles. The tension is measured by an electrical device with output readings in millivolts (mV). Some variation is inherent in the production process. Here are the tension readings from a random sample of 20 screens from a single day's production:

| 269.5 | 297.0 | 269.6 | 283.3 | 304.8 | 280.4 | 233.5 | 257.4 | 317.5 | 327.4 |
|---|---|---|---|---|---|---|---|---|---|
| 264.7 | 307.7 | 310.0 | 343.3 | 328.1 | 342.6 | 338.8 | 340.1 | 374.6 | 336.1 |

Construct and interpret a 90% confidence interval for the mean tension $\mu$ of all the screens produced on this day.

STATE: We want to estimate the true mean tension $\mu$ of all the video terminals produced this day with 90% confidence.

PLAN: If the conditions are met, we should use a one-sample $t$ interval to estimate $\mu$.

- Random: We are told that the data come from a random sample of 20 screens produced that day.
  - 10%: Because we are sampling without replacement, we must assume that at least 10(20) = 200 video terminals were produced this day.
- Normal/Large Sample: Because the sample size is small ($n$ = 20), we must check whether it's reasonable to believe that the population distribution is Normal. So we examine the sample data. Figure 8.15 shows (a) a dotplot, (b) a boxplot, and (c) a Normal probability plot of the tension

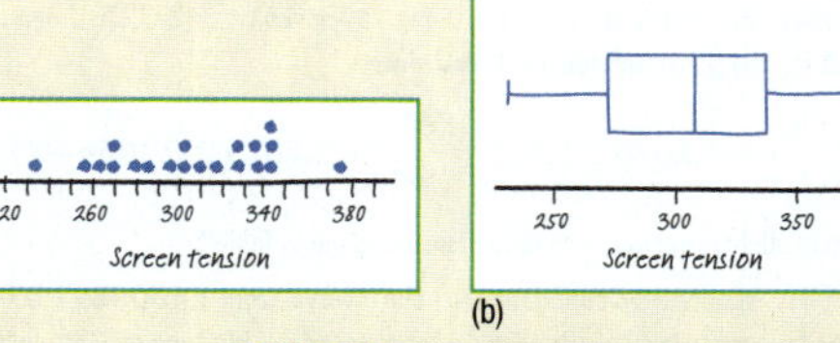


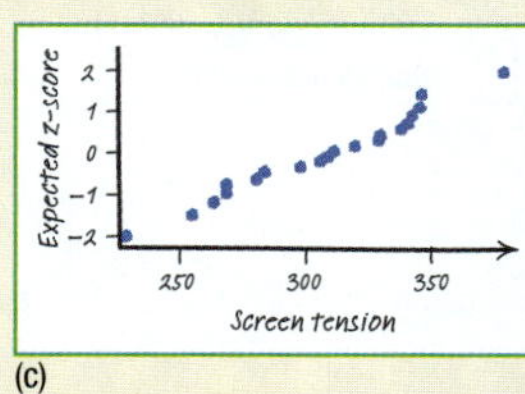


**FIGURE 8.15** (a) A dotplot, (b) boxplot, and (c) Normal probability plot of the video screen tension readings.

When the sample size is small ($n < 30$), as in this example, the Normal condition is about the shape of the *population* distribution. We look at a graph of the sample data to see if it's believable that the data came from a Normal population.

| df | .10 | .05 | .025 |
|---|---|---|---|
| | Upper-tail probability *p* | | |
| 18 | 1.330 | 1.734 | 2.101 |
| 19 | 1.328 | 1.729 | 2.093 |
| 20 | 1.325 | 1.725 | 2.086 |
| | 80% | 90% | 95% |
| | Confidence level *C* | | |

readings in the sample. Neither the dotplot nor the boxplot shows strong skewness or any outliers. The Normal probability plot looks roughly linear. These graphs give us no reason to doubt the Normality of the population.

DO: We used our calculator to find the mean and standard deviation of the tension readings for the 20 screens in the sample: $\bar{x} = 306.32$ mV and $s_x = 36.21$ mV. We use the $t$ distribution with df = 19 to find the critical value. For a 90% confidence level, the critical value is $t^* = 1.729$. So the 90% confidence interval for $\mu$ is

$$\bar{x} \pm t^* \frac{s_x}{\sqrt{n}} = 306.32 \pm 1.729 \frac{36.21}{\sqrt{20}} = 306.32 \pm 14.00 = (292.32, 320.32)$$

*Using technology:* The calculator's invT(.05,19) gives $t = -1.729$, which matches the $t^* = 1.729$ critical value we got from the table.

CONCLUDE: We are 90% confident that the interval from 292.32 to 320.32 mV captures the true mean tension in the entire batch of video terminals produced that day.

**For Practice** *Try Exercise* 69

**AP® EXAM TIP** It is not enough just to make a graph of the data on your calculator when assessing Normality. You must *sketch* the graph on your paper to receive credit. You don't have to draw multiple graphs—any appropriate graph will do.

As you probably guessed, your calculator will compute a one-sample $t$ interval for a population mean from sample data or summary statistics.

## 17. TECHNOLOGY CORNER — ONE-SAMPLE *t* INTERVALS FOR $\mu$ ON THE CALCULATOR

TI-Nspire instructions in Appendix B; HP Prime instructions on the book's Web site.

Confidence intervals for a population mean using $t$ distributions can be constructed on the TI-83/84 and TI-89, thus avoiding the use of Table B. Here is a brief summary of the techniques when you have the actual data values and when you have only numerical summaries.

1. Using summary statistics (see auto pollution example, page 519)

**TI-83/84**

From the home screen,

- Press STAT, arrow over to TESTS, and choose TInterval....
- On the TInterval screen, adjust your settings as shown and choose Calculate.

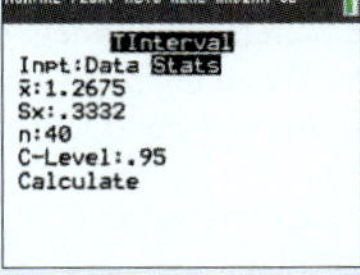


**TI-89**

From inside the Statistics/List Editor,

- Press 2nd F2 ([F7]) to go into the intervals (Ints) menu, then choose TInterval....
- Choose "Stats" as the Data Input Method.
- On the TInterval screen, adjust your settings as shown and press ENTER.

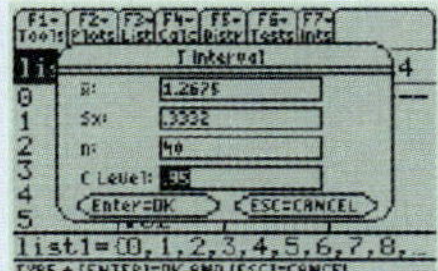


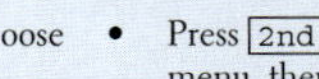

## AP® Exam Common Error

Many students incorrectly state the Normal/Large Sample condition by saying that the *sample* must have a Normal distribution. This isn't true (or even possible!). Instead, we use the sample data to make an inference about the shape of the population from which the sample came.

## TECHNOLOGY TIP

The TI-83/84 and TI-89 also list a ZInterval for calculating a confidence interval for a population mean. This procedure is appropriate only if the population standard deviation is known (which is very rare).

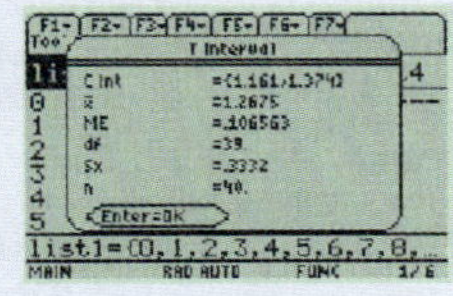

2. Using raw data (see video screen tension example, page 520)

Enter the 20 video screen tension readings data in L1/list1. Proceed to the TInterval screen as in Step 1, but choose Data as the input method. Then adjust your settings as shown and calculate the interval.

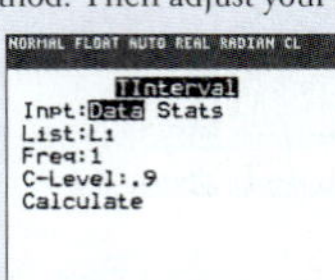


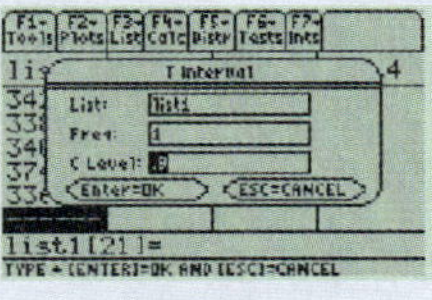

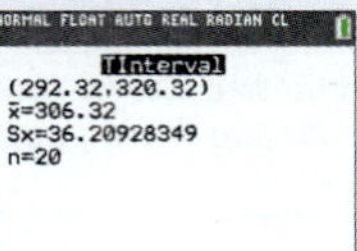


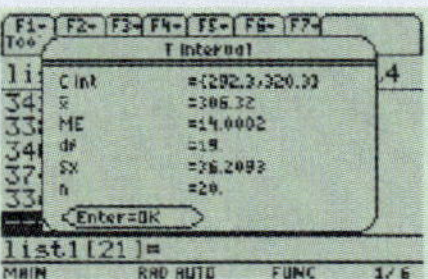

## Teaching Tip

Students are allowed to use the TInterval command on the AP® exam. However, make sure to review the AP® Exam Common Error discussed on page 501 to help your students avoid losing credit.

## ANSWERS TO CYU

*S*: We are trying to estimate $\mu$ = the true mean healing rate at a 95% confidence level. *P*: We should use a one-sample *t* interval for $\mu$ if the conditions are met. *Random*: The description says that the newts were randomly chosen. *10%*: The sample size (18) is less than 10% of the population of newts. *Normal/Large Sample*: The histogram below shows no strong skewness or outliers, so this condition is met.

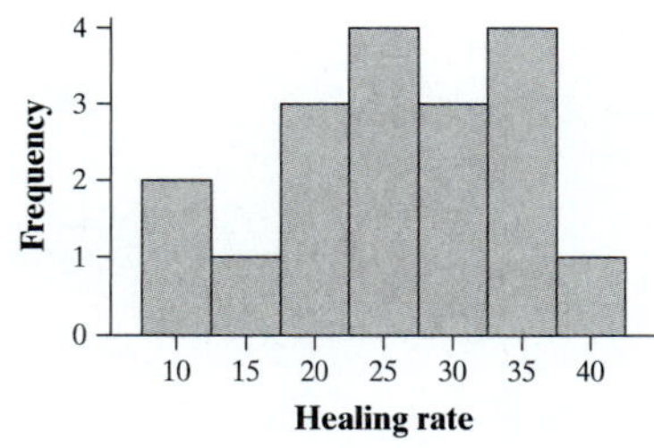


*D*: $25.67 \pm 2.110\left(\frac{8.32}{\sqrt{18}}\right) = 25.67 \pm 4.14 = (21.53, 29.81)$ *C*: We are 95% confident that the interval from 21.53 to 29.81 micrometers per hour captures the true mean healing rate for newts.

## CHECK YOUR UNDERSTANDING

Biologists studying the healing of skin wounds measured the rate at which new cells closed a cut made in the skin of an anesthetized newt. Here are data from a random sample of 18 newts, measured in micrometers (millionths of a meter) per hour:[19]

| 29 | 27 | 34 | 40 | 22 | 28 | 14 | 35 | 26 | 35 | 12 | 30 | 23 | 18 | 11 | 22 | 23 | 33 |
|---|---|---|---|---|---|---|---|---|---|---|---|---|---|---|---|---|---|

Calculate and interpret a 95% confidence interval for the mean healing rate $\mu$.

## Choosing the Sample Size

A wise user of statistics never plans data collection without planning the inference at the same time. You can arrange to have both high confidence and a small margin of error by taking enough observations. When the population standard deviation $\sigma$ is unknown and conditions are met, the $C$% confidence interval for $\mu$ is

$$\bar{x} \pm t^* \frac{s_x}{\sqrt{n}}$$

where $t^*$ is the critical value for confidence level C and degrees of freedom df $= n - 1$. The margin of error ($ME$) of the confidence interval is

$$ME = t^* \frac{s_x}{\sqrt{n}}$$

To determine the sample size for a desired margin of error, it makes sense to set the expression for $ME$ less than or equal to the specified value and solve the inequality for $n$. There are two problems with this approach:

1. We don't know the sample standard deviation $s_x$ because we haven't produced the data yet.
2. The critical value $t^*$ depends on the sample size $n$ that we choose.

The second problem is more serious. To get the correct value of $t^*$, we need to know the sample size. But that's what we're trying to find! There is no easy solution to this problem.

One alternative (the one we recommend!) is to come up with a reasonable estimate for the *population* standard deviation $\sigma$ from a similar study that was done in the past or from a small-scale pilot study. By pretending that $\sigma$ is known, we can use the one-sample $z$ interval for $\mu$:

$$\bar{x} \pm z^* \frac{\sigma}{\sqrt{n}}$$

Using the appropriate standard Normal critical value $z^*$ for confidence level C, we can solve

$$z^* \frac{\sigma}{\sqrt{n}} \leq ME$$

for $n$. Here is a summary of this strategy.

There are other methods of determining sample size that do not require us to use a known value of the population standard deviation $\sigma$. These methods are beyond the scope of this text. Our advice: consult with a statistician when planning your study!

**CHOOSING SAMPLE SIZE FOR A DESIRED MARGIN OF ERROR WHEN ESTIMATING $\mu$**

To determine the sample size $n$ that will yield a C% confidence interval for a population mean with a specified margin of error $ME$:

- Get a reasonable value for the population standard deviation $\sigma$ from an earlier or pilot study.
- Find the critical value $z^*$ from a standard Normal curve for confidence level C.
- Set the expression for the margin of error to be less than or equal to $ME$ and solve for $n$:

$$z^* \frac{\sigma}{\sqrt{n}} \leq ME$$

The procedure is best illustrated with an example.

## Teaching Tip

The method we present for estimating the sample size actually underestimates the sample size needed by using a $z^*$ critical value instead of a slightly larger $t^*$ critical value. However, when the anticipated sample size is large, the difference between the $z^*$ and $t^*$ critical values—and the difference in the sample size calculations—is very small.

ALTERNATE EXAMPLE

### Texting at school

A counselor wants to estimate the average number of text messages sent by students at his school during school hours. He wants to estimate $\mu$ at the 99% confidence level with a margin of error of at most 2 texts. A pilot study indicated that the number of texts sent during school hours has a standard deviation of about 9 texts.

**PROBLEM:** How many students need to be surveyed to estimate the mean number of texts sent during school hours with 99% confidence and a margin of error of at most 2 texts?

**SOLUTION:**

$$2.576\frac{9}{\sqrt{n}} \le 2 \rightarrow \left(2.576 \cdot \frac{9}{2}\right)^2 \le n \rightarrow 134.4 \le n$$

The counselor needs to survey at least 135 students.

## EXAMPLE How Many Monkeys?

### Determining sample size from margin of error

Researchers would like to estimate the mean cholesterol level $\mu$ of a particular variety of monkey that is often used in laboratory experiments. They would like their estimate to be within 1 milligram per deciliter (mg/dl) of the true value of $\mu$ at a 95% confidence level. A previous study involving this variety of monkey suggests that the standard deviation of cholesterol level is about 5 mg/dl.

**PROBLEM:** Obtaining monkeys for research is time-consuming, expensive, and controversial. What is the minimum number of monkeys the researchers will need to get a satisfactory estimate?

**SOLUTION:** For 95% confidence, $z^* = 1.96$. We will use $\sigma = 5$ as our best guess for the standard deviation of the monkeys' cholesterol level. Set the expression for the margin of error to be at most 1 and solve for $n$:

$$1.96\frac{5}{\sqrt{n}} \le 1$$

$$\frac{(1.96)(5)}{1} \le \sqrt{n}$$

$$96.04 \le n$$

Remember: always round up to the next whole number when finding $n$.

Because 96 monkeys would give a slightly larger margin of error than desired, the researchers would need 97 monkeys to estimate the cholesterol levels to their satisfaction. (On learning the cost of getting this many monkeys, the researchers might want to consider studying rats instead!)

**For Practice** *Try Exercise* 73

Taking observations costs time and money. The required sample size may be impossibly expensive. *Notice that it is the size of the sample that determines the margin of error. The size of the population does not influence the sample size we need.* This is true as long as the population is much larger than the sample.

### CHECK YOUR UNDERSTANDING

Administrators at your school want to estimate how much time students spend on homework, on average, during a typical week. They want to estimate $\mu$ at the 90% confidence level with a margin of error of at most 30 minutes. A pilot study indicated that the standard deviation of time spent on homework per week is about 154 minutes.

How many students need to be surveyed to meet the administrators' goal? Show your work.

### ANSWERS TO CYU

Using $\sigma = 154$ and $z^* = 1.645$ for 90% confidence, $30 \ge 1.645\frac{154}{\sqrt{n}}$. Thus, $n \ge \left(\frac{1.645(154)}{30}\right)^2 = 71.3$, so take a sample of 72 students.

## Need Help? Give Us a Call!

Refer to the chapter-opening Case Study on page 475. The bank manager wants to know whether or not the bank's customer service agents generally met the goal of answering incoming calls in less than 30 seconds. We can approach this question in two ways: by estimating the proportion $p$ of all calls that were answered within 30 seconds or by estimating the mean response time $\mu$.

Some graphs and numerical summaries of the data are provided below.

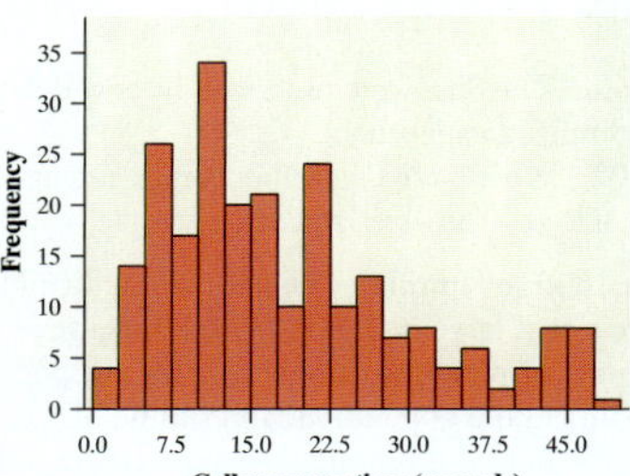


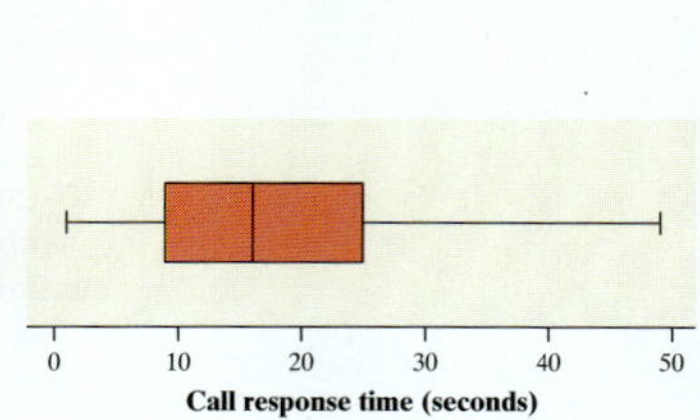


**Descriptive Statistics: Call response time (sec)**

| Variable | N | Mean | SE Mean | StDev | Minimum | Q1 | Median | Q3 | Maximum |
|---|---|---|---|---|---|---|---|---|---|
| Call response time (sec) | 241 | 18.353 | 0.758 | 11.761 | 1.000 | 9.000 | 16.000 | 25.000 | 49.000 |

1. Describe the distribution of call response times for the random sample of 241 calls.
2. About what proportion of the call response times in the sample were less than 30 seconds? Explain how you got your answer.
3. The bank's manager would like to estimate the true proportion $p$ of calls to the bank's customer service center that are answered in less than 30 seconds.
   (a) What conditions must be met to calculate a 95% confidence interval for $p$? Show that the conditions are met in this case.
   (b) Explain the meaning of 95% confidence in this setting.
   (c) A 95% confidence interval for $p$ is (0.783, 0.877). Give the margin of error and show how it was calculated.
   (d) Interpret the interval from part (c) in context.
4. Construct and interpret a 95% confidence interval for the true mean response time of calls to the bank's customer service center.
5. Is the customer service center meeting its goal of answering calls in less than 30 seconds? Give appropriate evidence to support your answer.

### Answers to Case Closed

**1**. The distribution of call response times is skewed to the right. It has a center of $\bar{x} = 18.353$ minutes with a standard deviation of 11.761 minutes. There don't appear to be any outliers.

**2**. About $200/241 = 0.830$ of the call wait times were less than 30 seconds. To find this answer, I estimated the heights of all the bars less than 30 seconds and added the frequencies.

**3**. **(a)** *Random*: This was a random sample of calls. *10%*: It is reasonable to assume that there are more than 10(241) calls to the customer service center. *Large Counts*: $n\hat{p} = 200 \geq 10$ and $n(1 - \hat{p}) = 41 \geq 10$.

**(b)** If many, many samples of 241 calls are selected, then about 95% of the resulting confidence intervals will contain the true proportion of calls to the bank's customer service center that are answered in less than 30 seconds.

**(c)** The margin of error is

$$1.96\sqrt{\frac{0.830(0.170)}{241}} = 0.047.$$

**(d)** We are 95% confident that the interval from 0.783 to 0.877 captures the true proportion of calls to the bank's customer service center that are answered in less than 30 seconds.

**4**. STATE: We want to estimate the true mean response time of calls to the bank's customer service center with 95% confidence. PLAN: If the conditions are met, we will use a one-sample $t$ interval for $\mu$. *Random*: This was a random sample of calls. *10%*: It is reasonable to assume that there are more than 10(241) calls to the customer service center. *Normal/Large Sample*: $n = 241 \geq 30$. DO: Using df = 100, $18.353 \pm 1.984\frac{11.761}{\sqrt{241}} = (16.850, 19.856)$. *Using technology*: With df = 240: (16.861, 19.845). CONCLUDE: We are 95% confident that the interval from 16.861 to 19.845 captures the true mean response time of calls to the bank's customer service center.

**5**. There is convincing evidence that the bank is answering the calls in less than 30 seconds, on average. All of the plausible values in the interval for the true mean response time were less than 30 seconds. However, not every call is answered in less than 30 seconds. According to the interval for the true proportion of calls answered in less than 30 seconds, between 78.3% and 87.7% of all calls are answered in less than 30 seconds. This means that between 12.3% and 21.7% are answered in 30 seconds or more.

**TRM Section 8.3 Quizzes**

There are three different quizzes available for this section in the Teacher's Resource Materials. Versions A and B are roughly parallel. Find them by clicking on the link in the TE-book, opening the TRCD, or logging onto the book's companion Web site www.whfreeman.com/tps5e (teacher log-in required).

# Section 8.3 Summary

- **Confidence intervals for the mean $\mu$** of a Normal population are based on the sample mean $\bar{x}$ of an SRS. If we somehow know $\sigma$, we use the $z$ critical value and the standard Normal distribution to help calculate confidence intervals.
- In practice, we usually don't know $\sigma$. Replace the standard deviation $\sigma/\sqrt{n}$ of the sampling distribution of $\bar{x}$ by the **standard error** $SE_{\bar{x}} = s_x/\sqrt{n}$ and use the $t$ distribution with $n - 1$ **degrees of freedom (df).**
- There is a ***t* distribution** for every positive degrees of freedom. All $t$ distributions are unimodal, symmetric, and centered at 0. The $t$ distributions approach the standard Normal distribution as the number of degrees of freedom increases.
- The conditions for constructing a confidence interval about a population mean are
  - **Random:** The data were produced by a well-designed random sample or randomized experiment.
    - **10%:** When sampling without replacement, check that the population is at least 10 times as large as the sample.
  - **Normal/Large Sample:** The population distribution is Normal or the sample size is large ($n \geq 30$). When the sample size is small ($n < 30$), examine a graph of the sample data for any possible departures from Normality in the population. You should be safe using a $t$ distribution as long as there is no strong skewness and no outliers are present.
- When conditions are met, a C% confidence interval for the mean $\mu$ is given by the **one-sample *t* interval**

$$\bar{x} \pm t^* \frac{s_x}{\sqrt{n}}$$

The critical value $t^*$ is chosen so that the $t$ curve with $n - 1$ degrees of freedom has C% of the area between $-t^*$ and $t^*$.

- Follow the four-step process—State, Plan, Do, Conclude—whenever you are asked to construct and interpret a confidence interval for a population mean Remember: inference for proportions uses $z$; inference for means uses $t$.
- The sample size needed to obtain a confidence interval with approximate margin of error $ME$ for a population mean involves solving

$$z^* \frac{\sigma}{\sqrt{n}} \leq ME$$

for $n$, where the standard deviation $\sigma$ is a reasonable value from a previous or pilot study, and $z^*$ is the critical value for the level of confidence we want.

**8.3 TECHNOLOGY CORNERS**

TI-Nspire Instructions in Appendix B; HP Prime instructions on the book's Web site.

# Section 8.3 Exercises

**55. Critical values** What critical value $t^*$ from Table B would you use for a confidence interval for the population mean in each of the following situations?

pg 513

(a) A 95% confidence interval based on $n = 10$ randomly selected observations

(b) A 99% confidence interval from an SRS of 20 observations

(c) A 90% confidence interval based on a random sample of 77 individuals

**56. Critical values** What critical value $t^*$ from Table B should be used for a confidence interval for the population mean in each of the following situations?

(a) A 90% confidence interval based on $n = 12$ randomly selected observations

(b) A 95% confidence interval from an SRS of 30 observations

(c) A 99% confidence interval based on a random sample of size 58

**57. Pulling wood apart** How heavy a load (pounds) is needed to pull apart pieces of Douglas fir 4 inches long and 1.5 inches square? A random sample of 20 similar pieces of Douglas fir from a large batch was selected for a science class. The Fathom boxplot below shows the class's data. Explain why it would not be wise to use a $t$ critical value to construct a confidence interval for the population mean $\mu$.

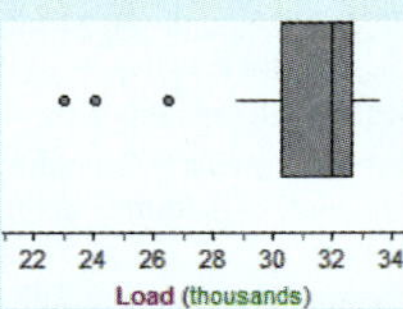


**58. Weeds among the corn** Velvetleaf is a particularly annoying weed in cornfields. It produces lots of seeds, and the seeds wait in the soil for years until conditions are right for sprouting. How many seeds do velvetleaf plants produce? The Fathom histogram below shows the counts from a random sample of 28 plants that came up in a cornfield when no herbicide was used.[20] Explain why it would not be wise to use a $t$ critical value to construct a confidence interval for the mean number of seeds $\mu$ produced by velvetleaf plants.

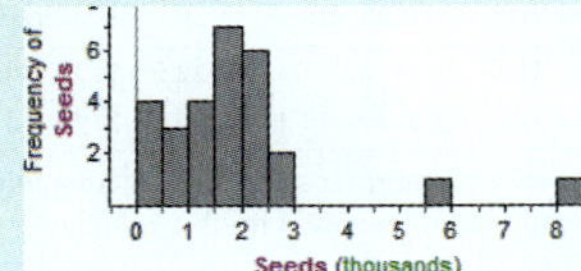


**59. Should we use $t$?** Determine whether we can safely use a $t^*$ critical value to calculate a confidence interval for the population mean in each of the following settings.

pg 516

(a) We collect data from a random sample of adult residents in a state. Our goal is to estimate the overall percent of adults in the state who are college graduates.

(b) The coach of a college men's basketball team records the resting heart rates of the 15 team members. We use these data to construct a confidence interval for the mean resting heart rate of all male students at this college.

(c) Do teens text more than they call? To find out, an AP® Statistics class at a large high school collected data on the number of text messages and calls sent or received by each of 25 randomly selected students. The Fathom boxplot below displays the difference (texts − calls) for each student.

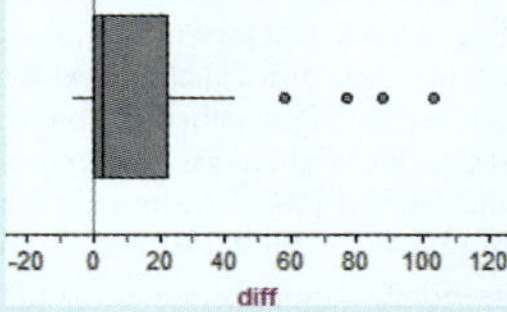


**60. Should we use $t$?** Determine whether we can safely use a $t^*$ critical value to calculate a confidence interval for the population mean in each of the following settings.

(a) We want to estimate the average age at which U.S. presidents have died. So we obtain a list of all U.S. presidents who have died and their ages at death.

(b) How much time do students spend on the Internet? We collect data from the 32 members of our AP® Statistics class and calculate the mean amount of time that each student spent on the Internet yesterday.

(c) Judy is interested in the reading level of a medical journal. She records the length of a random sample of 100 words. The Minitab histogram below displays the data.

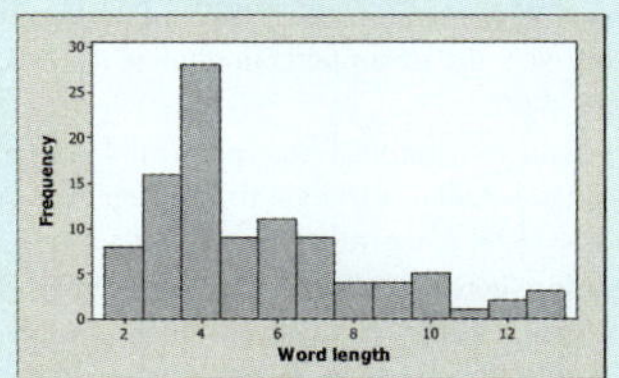


**61. Blood pressure** A medical study finds that $\bar{x} = 114.9$ and $s_x = 9.3$ for the seated systolic blood pressure of the 27 members of one treatment group. What is the standard error of the mean? Interpret this value in context.

**TRM Full Solutions to Section 8.3 Exercises**

## Answers to Section 8.3 Exercises

**8.55 (a)** $t^* = 2.262$ **(b)** $t^* = 2.861$ **(c)** $t^* = 1.671$ (*Using technology*: $t^* = 1.665$)

**8.56 (a)** $t^* = 1.796$ **(b)** $t^* = 2.045$ **(c)** $t^* = 2.678$ (*Using technology*: $t^* = 2.665$)

**8.57** Because the sample size is small ($n = 20 < 30$) and there are outliers in the data.

**8.58** Because the sample size is small ($n = 28 < 30$) and there are outliers in the data.

**8.59 (a)** No, because we are trying to estimate a population proportion, not a population mean. **(b)** No, because the 15 team members are not a random sample from the population. **(c)** No, because the sample size is small ($n = 25 < 30$) and there are outliers in the sample.

**8.60 (a)** No, because we have measurements on the entire population. **(b)** No, because members of the AP® Statistics class are not a random sample of all students. **(c)** Yes. *Random*: The words were selected at random. *10%*: The sample size (100) is less than 10% of the number of words in the entire medical journal. *Normal/Large Sample*: The sample size is large ($n = 100 \geq 30$).

**8.61** $SE_{\bar{x}} = \frac{9.3}{\sqrt{27}} = 1.7898$. If we take many samples of size 27, the sample mean blood pressure will typically vary by about 1.7898 from the population mean blood pressure.

62. **Travel time to work** A study of commuting times reports the travel times to work of a random sample of 20 employed adults in New York State. The mean is $\bar{x} = 31.25$ minutes, and the standard deviation is $s_x = 21.88$ minutes. What is the standard error of the mean? Interpret this value in context.

63. **Willows in Yellowstone** Writers in some fields summarize data by giving $\bar{x}$ and its standard error rather than $\bar{x}$ and $s_x$. Biologists studying willow plants in Yellowstone National Park reported their results in a table with columns labeled $\bar{x} \pm$ SE. The table entry for the heights of willow plants (in centimeters) in one region of the park was $61.55 \pm 19.03$.[21] The researchers measured a total of 23 plants.

(a) Find the sample standard deviation $s_x$ for these measurements. Show your work.

(b) A hasty reader believes that the interval given in the table is a 95% confidence interval for the mean height of willow plants in this region of the park. Find the actual confidence level for the given interval.

64. **Blink** When two lights close together blink alternately, we "see" one light moving back and forth if the time between blinks is short. What is the longest interval of time between blinks that preserves the illusion of motion? Ask subjects to turn a knob that slows the blinking until they "see" two lights rather than one light moving. A report gives the results in the form "mean plus or minus the standard error of the mean."[22] Data for 12 subjects are summarized as $251 \pm 45$ (in milliseconds).

(a) Find the sample standard deviation $s_x$ for these measurements. Show your work.

(b) A hasty reader believes that the interval given in the report is a 95% confidence interval for the population mean. Find the actual confidence level for the given interval.

65. **Bone loss by nursing mothers** Breast-feeding mothers secrete calcium into their milk. Some of the calcium may come from their bones, so mothers may lose bone mineral. Researchers measured the percent change in bone mineral content (BMC) of the spines of 47 randomly selected mothers during three months of breast-feeding.[23] The mean change in BMC was −3.587% and the standard deviation was 2.506%.

pg 519

STEP 4

(a) Construct and interpret a 99% confidence interval to estimate the mean percent change in BMC in the population.

(b) Based on your interval from part (a), do these data give good evidence that on the average nursing mothers lose bone mineral? Explain.

66. **Reading scores in Atlanta** The Trial Urban District Assessment (TUDA) is a government-sponsored study of student achievement in large urban school districts. TUDA gives a reading test scored from 0 to 500. A score of 243 is a "basic" reading level and a score of 281 is "proficient." Scores for a random sample of 1470 eighth-graders in Atlanta had $\bar{x} = 240$ with standard deviation 42.17.[24]

STEP 4

(a) Calculate and interpret a 99% confidence interval for the mean score of all Atlanta eighth-graders.

(b) Based on your interval from part (a), is there good evidence that the mean for all Atlanta eighth-graders is less than the basic level? Explain.

67. **Men and muscle** Ask young men to estimate their own degree of body muscle by choosing from a set of 100 photos. Then ask them to choose what they believe women prefer. The researchers know the actual degree of muscle, measured as kilograms per square meter of fat-free mass, for each of the photos. They can therefore measure the difference between what a subject thinks women prefer and the subject's own self-image. Call this difference the "muscle gap." Here are summary statistics for the muscle gap from a random sample of 200 American and European young men: $\bar{x} = 2.35$ and $s_x = 2.5$.[25]

STEP 4

(a) Calculate and interpret a 95% confidence interval for the mean size of the muscle gap for the population of American and European young men.

(b) A graph of the sample data is strongly skewed to the right. Explain why this information does not invalidate the interval you calculated in part (a).

68. **A big-toe problem** A bunion on the big toe is fairly uncommon in youth and often requires surgery. Doctors used X-rays to measure the angle (in degrees) of deformity on the big toe in a random sample of 37 patients under the age of 21 who came to a medical center for surgery to correct a bunion. The angle is a measure of the seriousness of the deformity. For these 37 patients, the mean angle of deformity was 24.76 degrees and the standard deviation was 6.34 degrees. A dotplot of the data revealed no outliers or strong skewness.[26]

STEP 4

(a) Construct and interpret a 90% confidence interval for the mean angle of deformity in the population of all such patients.

(b) Researchers omitted one patient with a deformity angle of 50 degrees from the analysis due to a measurement issue. What effect would including this outlier have on the confidence interval in part (a)? Justify your answer without doing any calculations.

69. **Give it some gas!** Computers in some vehicles calculate various quantities related to performance. One of these is fuel efficiency, or gas mileage, usually expressed as miles per gallon (mpg). For one vehicle equipped in this way, the miles per gallon were recorded each time the gas tank was filled and the computer was then reset.[27] Here are the mpg values for a random sample of 20 of these records:

pg 520

STEP 4

| | | | | | | | | | |
|---|---|---|---|---|---|---|---|---|---|
| 15.8 | 13.6 | 15.6 | 19.1 | 22.4 | 15.6 | 22.5 | 17.2 | 19.4 | 22.6 |
| 19.4 | 18.0 | 14.6 | 18.7 | 21.0 | 14.8 | 22.6 | 21.5 | 14.3 | 20.9 |

Construct and interpret a 95% confidence interval for the mean fuel efficiency $\mu$ for this vehicle.

**8.62** $SE_{\bar{x}} = 21.88/\sqrt{20} = 4.8925$. If we take many samples of size 20, the sample mean commute time will typically vary by about 4.8925 from the population mean commute time.

**8.63** **(a)** Because $19.03 = s_x/\sqrt{23}$, $s_x = 91.26$ cm. **(b)** They are using a critical value of $t^* = 1$. With df = 22, the area between $t = -1$ and $t = 1$ is approximately `tcdf(lower:-1, upper:1, df:22)` = 0.67. So, the confidence level is 67%.

**8.64** **(a)** Because, $45 = s_x/\sqrt{12}$, $s_x = 155.88$ milliseconds. **(b)** They are using a critical value of $t^* = 1$. With df = 11, the area between $t = -1$ and $t = 1$ is approximately `tcdf(lower:−1, upper:1, df:11)` = 0.66. So, the confidence level is 66%.

**8.65** **(a)** *S*: $\mu$ = the true mean percent change in BMC for breast-feeding mothers. *P*: One-sample *t* interval for $\mu$. *Random*: Random sample. *10%*: 47 is less than 10% of all breast-feeding mothers. *Normal/Large Sample*: $n = 47 \geq 30$. *D*: Using df = 40, (−4.575, −2.605). *Using technology*: (−4.569,−2.605) with df = 46. *C*: We are 99% confident that the interval from −4.569 to −2.605 captures the true mean percent change in BMC for breast-feeding mothers. **(b)** Because all of the plausible values in the interval are negative (indicating bone loss), the data give convincing evidence that breast-feeding mothers lose bone mineral, on average.

**8.66** **(a)** *S*: $\mu$ = the true mean score for all Atlanta eighth-graders. *P*: One-sample *t* interval for $\mu$. *Random*: Random sample. *10%*: 1470 is less than 10% of eighth-graders in Atlanta. *Normal/Large Sample*: $n = 1470 \geq 30$. *D*: Using df = 1000, (237.16, 242.84). *Using technology*: (237.16, 242.84) with df = 1469. *C*: We are 99% confident that the interval from 237.16 to 242.84 captures the true mean score for all Atlanta eighth-graders. **(b)** Because all of the plausible values in the interval are below 243, there is convincing evidence that the mean score for all Atlanta eighth-graders is below the basic level.

**8.67** **(a)** *S*: $\mu$ = the true mean size of the muscle gap for the population of American and European young men. *P*: One-sample *t* interval for $\mu$. *Random*: Random sample. *10%*: 200 is less than 10% of young men in America and Europe. *Normal/Large Sample*: $n = 200 \geq 30$. *D*: Using df = 100, (1.999, 2.701). *Using technology*: (2.001, 2.699) with df = 199. *C*: We are 95% confident that the interval from 2.001 to 2.699 captures the true mean size of the muscle gap for the population of American and European young men. **(b)** The large sample size ($n = 200 \geq 30$) allows us to use a *t* interval for $\mu$.

**8.68** **(a)** *S*: $\mu$ = the true mean angle of deformity in the population of all such patients. *P*: One-sample *t* interval for $\mu$. *Random*: Random sample. *10%*: The sample size (37) is less than 10% of the population of people under age 21 with a bunion on the big toe. *Normal/Large Sample*: $n = 37 \geq 30$. *D*: Using df = 30, (22.991, 26.529). *Using technology*: (23.00, 26.52) with df = 36. *C*: We are 90% confident that the interval from 23.00 to 26.52 degrees captures the true mean angle of deformity for the population of all such patients. **(b)** Adding back the outlier would increase the mean, making the center of the interval larger. It would also increase the standard deviation, making the interval wider.

**8.69** *S*: $\mu$ = the true mean fuel efficiency for this vehicle. *P*: One-sample *t* interval for $\mu$. *Random*: Random sample. *10%*: It is reasonable to assume that 20 is less than 10% of all records for this vehicle. *Normal/Large Sample*: No strong skewness or outliers.

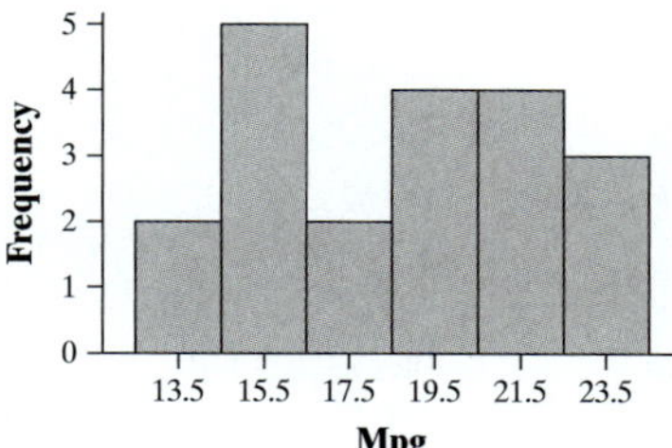

**70. Vitamin C content** Several years ago, the U.S. Agency for International Development provided 238,300 metric tons of corn-soy blend (CSB) for emergency relief in countries throughout the world. CSB is a highly nutritious, low-cost fortified food. As part of a study to evaluate appropriate vitamin C levels in this food, measurements were taken on samples of CSB produced in a factory.[28] The following data are the amounts of vitamin C, measured in milligrams per 100 grams (mg/100 g) of blend, for a random sample of size 8 from one production run:

| 26 | 31 | 23 | 22 | 11 | 22 | 14 | 31 |
|---|---|---|---|---|---|---|---|

Construct and interpret a 95% confidence interval for the mean amount of vitamin C $\mu$ in the CSB from this production run.

**71. Paired tires** Researchers were interested in comparing two methods for estimating tire wear. The first method used the amount of weight lost by a tire. The second method used the amount of wear in the grooves of the tire. A random sample of 16 tires was obtained. Both methods were used to estimate the total distance traveled by each tire. The table below provides the two estimates (in thousands of miles) for each tire.[29]

(a) Construct and interpret a 95% confidence interval for the mean difference $\mu$ in the estimates from these two methods in the population of tires.

(b) Does your interval in part (a) give convincing evidence of a difference in the two methods of estimating tire wear? Justify your answer.

| Tire | Weight | Groove | Diff. | Tire | Weight | Groove | Diff. |
|---|---|---|---|---|---|---|---|
| 1 | 45.9 | 35.7 | 10.2 | 9 | 30.4 | 23.1 | 7.3 |
| 2 | 41.9 | 39.2 | 2.7 | 10 | 27.3 | 23.7 | 3.6 |
| 3 | 37.5 | 31.1 | 6.4 | 11 | 20.4 | 20.9 | −0.5 |
| 4 | 33.4 | 28.1 | 5.3 | 12 | 24.5 | 16.1 | 8.4 |
| 5 | 31.0 | 24.0 | 7.0 | 13 | 20.9 | 19.9 | 1.0 |
| 6 | 30.5 | 28.7 | 1.8 | 14 | 18.9 | 15.2 | 3.7 |
| 7 | 30.9 | 25.9 | 5.0 | 15 | 13.7 | 11.5 | 2.2 |
| 8 | 31.9 | 23.3 | 8.6 | 16 | 11.4 | 11.2 | 0.2 |

**72. Water** Trace metals found in wells affect the taste of drinking water, and high concentrations can pose a health risk. Researchers measured the concentration of zinc (in milligrams/liter) near the top and the bottom of 10 randomly selected wells in a large region. The data are provided in the table below.[30]

(a) Construct and interpret a 95% confidence interval for the mean difference $\mu$ in the zinc concentrations from these two locations in the wells.

(b) Does your interval in part (a) give convincing evidence of a difference in zinc concentrations at the top and bottom of wells in the region? Justify your answer.

| Well: | 1 | 2 | 3 | 4 | 5 | 6 | 7 | 8 | 9 | 10 |
|---|---|---|---|---|---|---|---|---|---|---|
| Bottom: | 0.430 | 0.266 | 0.567 | 0.531 | 0.707 | 0.716 | 0.651 | 0.589 | 0.469 | 0.723 |
| Top: | 0.415 | 0.238 | 0.390 | 0.410 | 0.605 | 0.609 | 0.632 | 0.523 | 0.411 | 0.612 |
| Difference: | 0.015 | 0.028 | 0.177 | 0.121 | 0.102 | 0.107 | 0.019 | 0.066 | 0.058 | 0.111 |

**73. Estimating BMI** pg 524 The body mass index (BMI) of all American young women is believed to follow a Normal distribution with a standard deviation of about 7.5. How large a sample would be needed to estimate the mean BMI $\mu$ in this population to within ±1 with 99% confidence? Show your work.

**74. The SAT again** High school students who take the SAT Math exam a second time generally score higher than on their first try. Past data suggest that the score increase has a standard deviation of about 50 points. How large a sample of high school students would be needed to estimate the mean change in SAT score to within 2 points with 95% confidence? Show your work.

*Multiple choice: Select the best answer for Exercises 75 to 78.*

75. One reason for using a $t$ distribution instead of the standard Normal curve to find critical values when calculating a level C confidence interval for a population mean is that

(a) $z$ can be used only for large samples.

(b) $z$ requires that you know the population standard deviation $\sigma$.

(c) $z$ requires that you can regard your data as an SRS from the population.

(d) $z$ requires that the sample size is at most 10% of the population size.

(e) a $z$ critical value will lead to a wider interval than a $t$ critical value.

76. You have an SRS of 23 observations from a large population. The distribution of sample values is roughly symmetric with no outliers. What critical value would you use to obtain a 98% confidence interval for the mean of the population?

(a) 2.177 (b) 2.183 (c) 2.326 (d) 2.500 (e) 2.508

77. A quality control inspector will measure the salt content (in milligrams) in a random sample of bags of potato chips from an hour of production. Which of the following would result in the smallest margin of error in estimating the mean salt content $\mu$?

(a) 90% confidence; $n = 25$

(b) 90% confidence; $n = 50$

(c) 95% confidence; $n = 25$

(d) 95% confidence; $n = 50$

(e) $n = 100$ at any confidence level

78. Scientists collect data on the blood cholesterol levels (milligrams per deciliter of blood) of a random sample of 24 laboratory rats. A 95% confidence interval for the mean blood cholesterol level $\mu$ is 80.2 to 89.8. Which of the following would cause the most worry about the validity of this interval?

(a) There is a clear outlier in the data.

(b) A stemplot of the data shows a mild right skew.

*D*: Using df = 19, (17.022, 19.938). *C*: We are 95% confident that the interval from 17.022 to 19.938 captures the true mean fuel efficiency for this vehicle.

**8.70** *S*: $\mu$ = the true mean amount of vitamin C in the CSB from this production run. *P*: One-sample $t$ interval for $\mu$. *Random*: Random sample. *10%*: We assume the sample size (8) is less than 10% of the CSB from this run. *Normal/Large Sample*: No strong skewness or outliers.

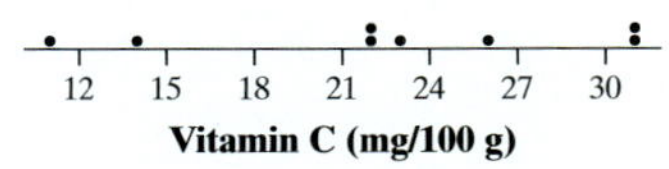


*D*: Using df = 7, (16.49, 28.51). *C*: We are 95% confident that the interval from 16.49 to 28.51 mg/100 g captures the true mean amount of vitamin C in this production run.

**8.71 (a)** *S*: $\mu$ = the true mean difference in the estimates from these two methods in the population of tires. *P*: One-sample $t$ interval for $\mu$. *Random*: Random sample. *10%*: The sample size (16) is less than 10% of all tires. *Normal/Large Sample*: No strong skewness or outliers.

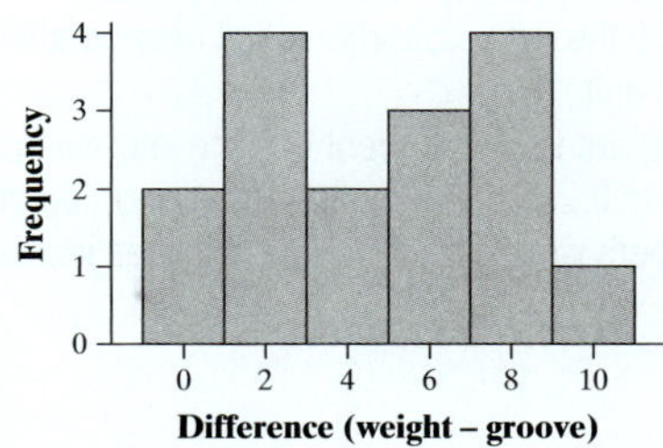


*D*: Using df = 15, (2.837, 6.275). *C*: We are 95% confident that the interval from 2.837 to 6.275 thousands of miles captures the true mean difference in the estimates from these two methods in the population of tires. **(b)** Because 0 is not included in the confidence interval, there is convincing evidence of a difference in the two methods of estimating tire wear.

**8.72 (a)** *S*: $\mu$ = the true mean difference in the amount of zinc in the top and the bottom of wells in this large region. *P*: One-sample $t$ interval for $\mu$. *Random*: Random sample. *10%*: The sample size (10) is less than 10% of the population of wells in this large region. *Normal/Large Sample*: No strong skewness or outliers.

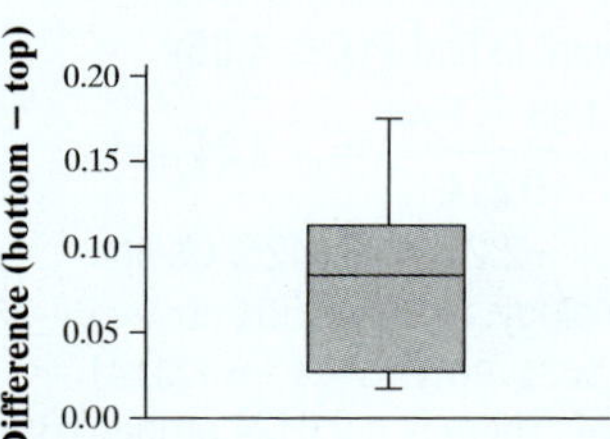


*D*: Using df = 9, (0.043, 0.1178). *C*: We are 95% confident that the interval from 0.043 to 0.1178 mg/l captures the true mean difference in the amount of zinc between the bottom and the top of the wells in this large region. **(b)** Because 0 is not included in the confidence interval, there is convincing evidence of a difference in zinc concentrations at the top and bottom of wells in this large region.

**8.73** Solving $2.576\frac{7.5}{\sqrt{n}} \le 1$ gives $n \ge 374$.

**8.74** Solving $1.96\frac{50}{\sqrt{n}} \le 2$ gives $n \ge 2401$.

**8.75** b

**8.76** e

**8.77** b

**8.78** a

**8.79** **(a)** Because the sum of the probabilities must be 1, $P(X = 7) = 0.57$. **(b)** $\mu_X = 5.44$. If we were to randomly select many young people, the average number of days they watched television in the past 7 days would be about 5.44. **(c)** Because the sample size is large ($n = 100 \geq 30$), we expect the mean number of days $\bar{x}$ for 100 randomly selected young people (aged 19 to 25) to be approximately Normally distributed with mean $\mu_{\bar{x}} = \mu = 5.44$. Because the sample size (100) is less than 10% of all young people aged 19 to 25, the standard deviation is $\sigma_{\bar{x}} = \frac{\sigma}{\sqrt{n}} = \frac{2.14}{\sqrt{100}} = 0.214$. We want to find $P(\bar{x} \leq 4.96)$. $z = \frac{4.96 - 5.44}{0.214} = -2.24$ and $P(Z \leq -2.24) = 0.0125$. *Using technology*: `normalcdf(lower:-1000, upper:4.96,μ:5.44,σ:0.214)` = 0.0124. There is a 0.0124 probability of getting a sample mean of 4.96 or smaller. Because this probability is small, a sample mean of 4.96 or smaller would be surprising.

**8.80** **(a)** There will be 8 treatment groups, with 25 people randomly assigned to each group. Researchers will compare the mean attractiveness rating given by individuals in the eight groups. **(b)** Label the subjects 001, 002, . . . , 200. Moving from left to right through the table of random digits, select three-digit numbers. Ignore labels 000 and 201 to 999 and any repeated labels. Once we have 25 subjects for the first treatment, we select 25 subjects for the second treatment, and so on, until all subjects have been assigned to a treatment group. Using the digits provided, the first three subjects selected are labeled 180, 020, and 193. **(c)** The range "40% to 70% off" slowly decreases in attractiveness to customers as the percent of food on sale increases. However, the precise "60% off" grows increasingly attractive to customers as the percent of food on sale increases.



(c) You do not know the population standard deviation $\sigma$.

(d) The population distribution is not exactly Normal.

(e) None of these are a problem when using a $t$ interval.

79. **Watching TV** (6.1, 7.3) Choose a young person (aged 19 to 25) at random and ask, "In the past seven days, how many days did you watch television?" Call the response X for short. Here is the probability distribution for X:[31]

| Days: | 0 | 1 | 2 | 3 | 4 | 5 | 6 | 7 |
|---|---|---|---|---|---|---|---|---|
| Probability: | 0.04 | 0.03 | 0.06 | 0.08 | 0.09 | 0.08 | 0.05 | ??? |

(a) What is the probability that X = 7? Justify your answer.

(b) Calculate the mean of the random variable X. Interpret this value in context.

(c) Suppose that you asked 100 randomly selected young people (aged 19 to 25) to respond to the question and found that the mean $\bar{x}$ of their responses was 4.96. Would this result surprise you? Justify your answer.

80. **Price cuts** (4.2) Stores advertise price reductions to attract customers. What type of price cut is most attractive? Experiments with more than one factor allow insight into interactions between the factors. A study of the attractiveness of advertised price discounts had two factors: percent of all foods on sale (25%, 50%, 75%, or 100%) and whether the discount was stated precisely (as in, for example, "60% off") or as a range (as in "40% to 70% off"). Subjects rated the attractiveness of the sale on a scale of 1 to 7.

(a) Describe a completely randomized design using 200 student subjects.

(b) Explain how you would use the partial table of random digits below to assign subjects to treatment groups. Then use your method to select the first 3 subjects for one of the treatment groups. Show your work clearly on your paper.

45740 41807 65561 33302 07051 93623 18132 09547
12975 13258 13048 45144 72321 81940 00360 02428

(c) The figure below shows the mean ratings for the eight treatments formed from the two factors.[32] Based on these results, write a careful description of how percent on sale and precise discount versus range of discounts influence the attractiveness of a sale.

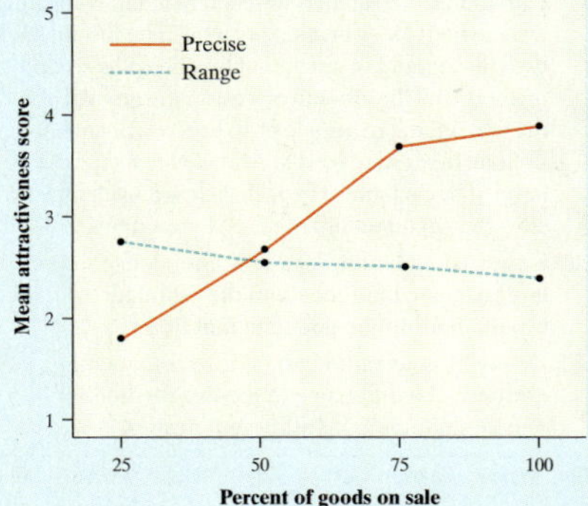


## FRAPPY! Free Response AP® Problem, Yay!

The following problem is modeled after actual AP® Statistics exam free response questions. Your task is to generate a complete, concise response in 15 minutes.

*Directions: Show all your work. Indicate clearly the methods you use, because you will be scored on the correctness of your methods as well as on the accuracy and completeness of your results and explanations.*

Members at a popular fitness club currently pay a $40 per month membership fee. The owner of the club wants to raise the fee to $50 but is concerned that some members will leave the gym if the fee increases. To investigate, the owner plans to survey a random sample of the club members and construct a 95% confidence interval for the proportion of all members who would quit if the fee was raised to $50.

(a) Explain the meaning of "95% confidence" in the context of the study.

(b) After the owner conducted the survey, he calculated the confidence interval to be $0.18 \pm 0.075$. Interpret this interval in the context of the study.

(c) According to the club's accountant, the fee increase will be worthwhile if fewer than 20% of the members quit. According to the interval from part (b), can the owner be confident that the fee increase will be worthwhile? Explain.

(d) One of the conditions for calculating the confidence interval in part (b) is that $n\hat{p} \geq 10$ and $n(1 - \hat{p}) \geq 10$. Explain why it is necessary to check this condition.

After you finish, you can view two example solutions on the book's Web site (www.whfreeman.com/tps5e). Determine whether you think each solution is "complete," "substantial," "developing," or "minimal." If the solution is not complete, what improvements would you suggest to the student who wrote it? Finally, your teacher will provide you with a scoring rubric. Score your response and note what, if anything, you would do differently to improve your own score.

### TRM FRAPPY!

We present a model solution here. Please consult the Teacher's Resource Materials for sample student responses, a scoring rubric, and a black line master of the original question with space for students to write their responses.

**(a)** If many samples of members were selected and many 95% confidence intervals were constructed, about 95% of the intervals would capture the true proportion of all members who would quit.

**(b)** We are 95% confident that the interval from 0.105 to 0.255 captures the true proportion of all members who would quit if the fee was increased.

**(c)** No, there are plausible values in the interval above 0.20. So, the owner cannot be confident that the true proportion is less than 0.20.

**(d)** We check this condition to make sure $\hat{p}$ has an approximately Normal distribution, which allows us to use a Normal distribution to calculate the $z^*$ critical value used in the interval.

# Chapter Review

**TRM Chapter 8 Tests**

There are four different tests available for this chapter in the Teacher's Resource Materials. Versions A and B are roughly parallel and version D is cumulative. Find them by clicking on the link in the TE-book, opening the TRCD, or logging onto the book's companion Web site www.whfreeman.com/tps5e (teacher log-in required).

### Section 8.1: Confidence Intervals: The Basics

In this section, you learned that a point estimate is the single best guess for the value of a population parameter. You also learned that a confidence interval provides an interval of plausible values for a parameter. To interpret a confidence interval, say, "We are *C*% confident that the interval from ___ to ___ captures the [parameter in context]," where C is the confidence level of the interval.

The confidence level C describes the percentage of confidence intervals that we expect to capture the value of the parameter. To interpret a *C*% confidence level, we say, "If we took many samples of the same size and used them to construct *C*% confidence intervals, about *C*% of those intervals would capture the [parameter in context]."

Confidence intervals are formed by including a margin of error on either side of the point estimate. The size of the margin of error is determined by several factors, including the confidence level C and the sample size $n$. Increasing the sample size $n$ makes the standard deviation of our estimate smaller, decreasing the margin of error. Increasing the confidence level C makes the margin of error larger, to ensure that the capture rate of the interval increases to *C*%. Remember that the margin of error only accounts for sampling variability—it does not account for any bias in the data collection process.

### Section 8.2: Estimating a Population Proportion

In this section, you learned how to construct and interpret confidence intervals for a population proportion. Several important conditions must be met for this type of confidence interval to be valid. First, the data used to calculate the interval must come from a well-designed random sample or randomized experiment (the Random condition). When the sample is taken without replacement from the population, the sample size should be no more than 10% of the population size (the 10% condition). Finally, the observed number of successes $n\hat{p}$ and observed number of failures $n(1 - \hat{p})$ must both be at least 10 (the Large Counts condition).

The formula for calculating a confidence interval for a population proportion is

$$\hat{p} \pm z^* \sqrt{\frac{\hat{p}(1 - \hat{p})}{n}}$$

where $\hat{p}$ is the sample proportion, $z^*$ is the critical value, and $n$ is the sample size. The value of $z^*$ is based on the confidence level C. To find $z^*$, use Table A or technology to determine the values of $z^*$ and $-z^*$ that capture the middle *C*% of the standard Normal distribution.

The four-step process (State, Plan, Do, Conclude) is perfectly suited for problems that ask you to construct and interpret a confidence interval. You should *state* the parameter you are estimating and at what confidence level, *plan* your work by naming the type of interval you will use and checking the appropriate conditions, *do* the calculations, and make a *conclusion* in the context of the problem. You can use technology for the Do step, but make sure that you identify the procedure you are using and type in the values correctly.

Finally, an important part of planning a study is determining the size of the sample to be selected. The necessary sample size is based on the confidence level, the proportion of successes, and the desired margin of error. To calculate the minimum sample size, solve the following inequality for $n$, where $\hat{p}$ is a guessed value for the sample proportion:

$$z^* \sqrt{\frac{\hat{p}(1 - \hat{p})}{n}} \leq ME$$

If you do not have an approximate value of $\hat{p}$ from a previous study or a pilot study, use $\hat{p} = 0.5$ to determine the sample size that will yield a value less than or equal to the desired margin of error.

### Section 8.3: Estimating a Population Mean

In this section, you learned how to construct and interpret confidence intervals for a population mean. Remember that you have to check conditions before doing calculations. The Random and 10% conditions are the same as those for proportions. There's one new condition for means: the population must be Normally distributed or the sample size must be at least 30 (the Normal/Large Sample condition). If the population shape is unknown and the sample size is less than 30, graph the sample data and check for strong skewness or outliers. If there is no strong skewness or outliers, it is reasonable to assume that the population distribution is approximately Normal.

The formula for calculating a confidence interval for a population mean is

$$\bar{x} \pm t^* \frac{s_x}{\sqrt{n}}$$

where $\bar{x}$ is the sample mean, $t^*$ is the critical value, $s_x$ is the sample standard deviation, and $n$ is the sample size. We use a $t$ critical value instead of a $z$ critical value when the population standard deviation is unknown—which is almost always the case. The value of $t^*$ is based on the confidence level C and the degrees of freedom ($df = n - 1$). To find $t^*$, use Table B or technology to determine the values of $t^*$ and $-t^*$ that capture the middle *C*% of the appropriate $t$ distribution. The $t$ distributions are bell-shaped, symmetric, and centered at 0. However, they are more variable and have a shape slightly different from that of the standard Normal distribution.

You also learned how to estimate the sample size when planning a study, as in Section 8.2. To calculate the minimum sample size, solve the following inequality for $n$, where $\sigma$ is a guessed value for the population standard deviation:

$$z^* \frac{\sigma}{\sqrt{n}} \leq ME$$

TRM Learning Objectives Grid

## What Did You Learn?

| Learning Objective | Section | Related Example on Page(s) | Relevant Chapter Review Exercise(s) |
|---|---|---|---|
| Determine the point estimate and margin of error from a confidence interval. | 8.1 | 481 | R8.2 |
| Interpret a confidence interval in context. | 8.1 | 481, 484 | R8.3, R8.4, R8.6, R8.7 |
| Interpret a confidence level in context. | 8.1 | 484 | R8.2 |
| Describe how the sample size and confidence level affect the length of a confidence interval. | 8.1 | Discussion on 487 | R8.9 |
| Explain how practical issues like nonresponse, undercoverage, and response bias can affect the interpretation of a confidence interval. | 8.1 | Discussion on 488 | R8.6 |
| State and check the Random, 10%, and Large Counts conditions for constructing a confidence interval for a population proportion. | 8.2 | 494 | R8.3 |
| Determine critical values for calculating a *C*% confidence interval for a population proportion using a table or technology. | 8.2 | 497 | R8.1 |
| Construct and interpret a confidence interval for a population proportion. | 8.2 | 498, 500 | R8.3, R8.6 |
| Determine the sample size required to obtain a *C*% confidence interval for a population proportion with a specified margin of error. | 8.2 | 502 | R8.5 |
| State and check the Random, 10%, and Normal/Large Sample conditions for constructing a confidence interval for a population mean. | 8.3 | 516 | R8.4 |
| Explain how the *t* distributions are different from the standard Normal distribution and why it is necessary to use a *t* distribution when calculating a confidence interval for a population mean. | 8.3 | Discussion on 511–512 | R8.10 |
| Determine critical values for calculating a *C*% confidence interval for a population mean using a table or technology. | 8.3 | 513 | R8.1 |
| Construct and interpret a confidence interval for a population mean. | 8.3 | 519–520 | R8.4, R8.7 |
| Determine the sample size required to obtain a *C*% confidence interval for a population mean with a specified margin of error. | 8.3 | 524 | R8.8 |

# Chapter 8 Chapter Review Exercises

*These exercises are designed to help you review the important ideas and methods of the chapter.*

**R8.1 It's critical** Find the appropriate critical value for constructing a confidence interval in each of the following settings.

(a) Estimating a population proportion $p$ at a 94% confidence level based on an SRS of size 125.

(b) Estimating a population mean $\mu$ at a 99% confidence level based on an SRS of size 58.

**R8.2 Batteries** A company that produces AA batteries tests the lifetime of a random sample of 30 batteries using a special device designed to imitate real-world use. Based on the testing, the company makes the following statement: "Our AA batteries last an average of 430 to 470 minutes, and our confidence in that interval is 95%."[33]

532

TRM **Full Solutions to Chapter 8 Review Exercises**

## Answers to Chapter Review Exercises

**R8.1 (a)** $\frac{1-0.94}{2} = 0.03$, and the closest area is 0.0301, corresponding to a critical value of $z^* = 1.88$. *Using technology*: `invNorm(area:0.03,μ:0,σ:1)` $= -1.881$, so $z^* = 1.881$. **(b)** Using Table B and 50 degrees of freedom, $t^* = 2.678$. *Using technology:* `invT(area:0.005, df = 57)` $= -2.665$, so $t^* = 2.665$.

**R8.2 (a)** $\bar{x} = \frac{430 + 470}{2} = 450$ minutes. Margin of error $= 470 - 450 = 20$ minutes. Because $n = 30$, df $= 29$ and $t^* = 2.045$.

Because $20 = 2.045\frac{s_x}{\sqrt{30}}$, standard error $=$

$\frac{s_x}{\sqrt{30}} = 9.780$ minutes and $s_x = 53.57$ minutes.

**(b)** The confidence interval provided gives an interval estimate for the *mean* lifetime of batteries produced by this company, not individual lifetimes. **(c)** No. A confidence interval provides a statement about an unknown population mean, not another sample mean. **(d)** If we were to take many samples of 30 batteries and compute 95% confidence intervals for the mean lifetime, about 95% of these intervals will capture the true mean lifetime of the batteries.

**R8.3 (a)** $p =$ the proportion of all adults aged 18 and older who would say that football is their favorite sport to watch on television. It may not equal 0.37 because the proportion who choose football will vary from sample to sample. **(b)** *Random*: The sample was random. *10%*: The sample size (1000) is less than 10% of all adults. *Large Counts*: $n\hat{p} = 370 \geq 10$ and $n(1-\hat{p}) = 630 \geq 10$. **(c)** $0.37 \pm 1.96\sqrt{\frac{0.37(0.63)}{1000}} = (0.3401, 0.3999)$. **(d)** We are 95% confident that the interval from 0.3401 to 0.3999 captures the true proportion of all adults who would say that football is their favorite sport to watch on television.

(a) Determine the point estimate, margin of error, standard error, and sample standard deviation.

(b) A reporter translates the statistical announcement into "plain English" as follows: "If you buy one of this company's AA batteries, there is a 95% chance that it will last between 430 and 470 minutes." Comment on this interpretation.

(c) Your friend, who has just started studying statistics, claims that if you select 40 more AA batteries at random from those manufactured by this company, there is a 95% probability that the mean lifetime will fall between 430 and 470 minutes. Do you agree? Explain.

(d) Give a statistically correct interpretation of the confidence level that could be published in a newspaper report.

**R8.3 We love football!** A recent Gallup Poll conducted telephone interviews with a random sample of adults aged 18 and older. Data were obtained for 1000 people. Of these, 37% said that football is their favorite sport to watch on television.

(a) Define the parameter $p$ in this setting. Explain to someone who knows nothing about statistics why we can't just say that 37% of all adults would say that football is their favorite sport to watch on television.

(b) Check the conditions for constructing a confidence interval for $p$.

(c) Construct a 95% confidence interval for $p$.

(d) Interpret the interval in context.

**R8.4 Smart kids** A school counselor wants to know how smart the students in her school are. She gets funding from the principal to give an IQ test to an SRS of 60 of the over 1000 students in the school. The mean IQ score was 114.98 and the standard deviation was 14.80.[34]

(a) Define the parameter $\mu$ in this setting.

(b) Check the conditions for constructing a confidence interval for $\mu$.

(c) Construct a 90% confidence interval for the mean IQ score of students at the school.

(d) Interpret your result from part (c) in context.

**R8.5 Do you go to church?** The Gallup Poll plans to ask a random sample of adults whether they attended a religious service in the last 7 days. How large a sample would be required to obtain a margin of error of at most 0.01 in a 99% confidence interval for the population proportion who would say that they attended a religious service? Show your work.

**R8.6 Running red lights** A random digit dialing telephone survey of 880 drivers asked, "Recalling the last ten traffic lights you drove through, how many of them were red when you entered the intersections?" Of the 880 respondents, 171 admitted that at least one light had been red.[35]

(a) Construct and interpret a 95% confidence interval for the population proportion.

(b) Nonresponse is a practical problem for this survey—only 21.6% of calls that reached a live person were completed. Another practical problem is that people may not give truthful answers. What is the likely direction of the bias: Do you think more or fewer than 171 of the 880 respondents really ran a red light? Why? Are these sources of bias included in the margin of error?

**R8.7 Engine parts** Here are measurements (in millimeters) of a critical dimension on an SRS of 16 of the more than 200 auto engine crankshafts produced in one day:

| | | | | | | | |
|---|---|---|---|---|---|---|---|
| 224.120 | 224.001 | 224.017 | 223.982 | 223.989 | 223.961 | 223.960 | 224.089 |
| 223.987 | 223.976 | 223.902 | 223.980 | 224.098 | 224.057 | 223.913 | 223.999 |

(a) Construct and interpret a 95% confidence interval for the process mean at the time these crankshafts were produced.

(b) The process mean is supposed to be $\mu = 224$ mm but can drift away from this target during production. Does your interval from part (a) suggest that the process mean has drifted? Explain.

**R8.8 Good wood?** A lab supply company sells pieces of Douglas fir 4 inches long and 1.5 inches square for force experiments in science classes. From experience, the strength of these pieces of wood follows a Normal distribution with standard deviation 3000 pounds. You want to estimate the mean load needed to pull apart these pieces of wood to within 1000 pounds with 95% confidence. How large a sample is needed? Show your work.

**R8.9 It's about *ME*** Explain how each of the following would affect the margin of error of a confidence interval, if all other things remained the same.

(a) Increasing the confidence level

(b) Quadrupling the sample size

**R8.10 *t* time** When constructing confidence intervals for a population mean, we almost always use critical values from a $t$ distribution rather than the standard Normal distribution.

(a) When is it necessary to use a $t$ critical value rather than a $z$ critical value when constructing a confidence interval for a population mean?

(b) Describe two ways that the $t$ distributions are different from the standard Normal distribution.

(c) Explain what happens to the $t$ distributions as the degrees of freedom increase.

**R8.4 (a)** $\mu$ = mean IQ score for the 1000 students in the school. **(b)** *Random*: The data are from an SRS. *10%*: The sample size (60) is less than 10% of the 1000 students at the school. *Normal/Large Sample*: $n = 60 \geq 30$. **(c)** Using df = 50,

$$114.98 \pm 1.676\left(\frac{14.8}{\sqrt{60}}\right) = (111.778, 118.182).$$

*Using technology*: (111.79, 118.17) with df = 59. **(d)** We are 90% confident that the interval from 111.79 to 118.17 captures the true mean IQ score for the 1000 students in the school.

**R8.5** Solving $2.576\sqrt{\frac{0.5(0.5)}{n}} \leq 0.01$ gives $n \geq 16{,}590$.

**R8.6 (a)** *S*: $p$ = the true proportion of all drivers who have run at least one red light in the last 10 intersections they have entered. *P*: One-sample $z$ interval for $p$. *Random*: The drivers were selected at random. *10%*: The sample size (880) is less than 10% of all drivers. *Large Counts*: $n\hat{p} = 171 \geq 10$ and $n(1-\hat{p}) = 709 \geq 10$. *D*: (0.168, 0.220). *C*: We are 95% confident that the interval from 0.168 to 0.220 captures the true proportion of all drivers who have run at least one red light in the last 10 intersections they have entered. **(b)** It is likely that more than 171 respondents have run red lights because some people may lie and say they haven't run a red light. The margin of error does not account for these sources of bias, only sampling variability.

**R8.7 (a)** *S*: $\mu$ = the true mean measurement of the critical dimension for the engine crankshafts produced in one day. *P*: One-sample $t$ interval for $\mu$. *Random*: The data come from an SRS. *10%*: The sample size (16) is less than 10% of all crankshafts produced in one day. *Normal/Large Sample*: The histogram shows no strong skewness or outliers.

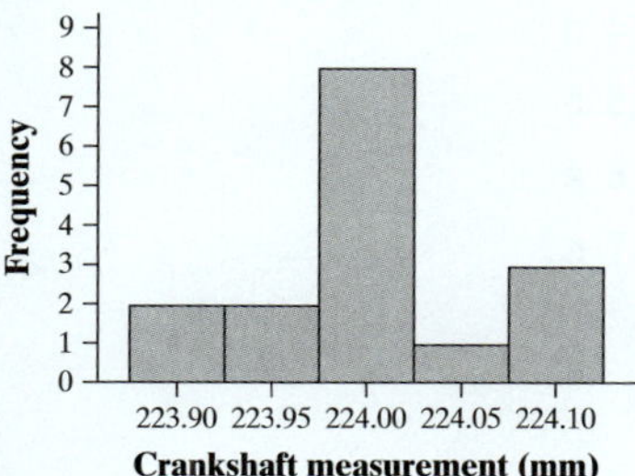


*D*: Using df = 15, (223.969, 224.035) and $t^* = 2.131$. *C*: We are 95% confident that the interval from 223.969 to 224.035 mm captures the true mean measurement of the critical dimension for engine crankshafts produced on this day. **(b)** Because 224 is a plausible value in this interval, we don't have convincing evidence that the process mean has drifted.

**R8.8** Solving $1.96\left(\frac{3000}{\sqrt{n}}\right) \leq 1000$ gives $n \geq 35$.

**R8.9 (a)** The margin of error must get larger to increase the capture rate of the intervals. **(b)** If we quadruple the sample size, the margin of error will decrease by a factor of 2.

**R8.10 (a)** When we use the sample standard deviation $s_x$ to estimate the population standard deviation $\sigma$. **(b)** The $t$ distributions are wider than the standard Normal distribution and they have a slightly different shape with more area in the tails. **(c)** As the degrees of freedom increase, the spread and shape of the $t$ distributions become more like the standard Normal distribution.

**TRM Full Solutions to Chapter 8 AP® Statistics Practice Test**

## Answers to Chapter 8 AP® Statistics Practice Test

**T8.1** a

**T8.2** d

**T8.3** c

**T8.4** d

**T8.5** b

**T8.6** a

**T8.7** c



# Chapter 8 AP® Statistics Practice Test

**Section I: Multiple Choice** *Select the best answer for each question.*

**T8.1** The Gallup Poll interviews 1600 people. Of these, 18% say that they jog regularly. The news report adds: "The poll had a margin of error of plus or minus three percentage points at a 95% confidence level." You can safely conclude that

(a) 95% of all Gallup Poll samples like this one give answers within ±3% of the true population value.

(b) the percent of the population who jog is certain to be between 15% and 21%.

(c) 95% of the population jog between 15% and 21% of the time.

(d) we can be 95% confident that the sample proportion is captured by the confidence interval.

(e) if Gallup took many samples, 95% of them would find that 18% of the people in the sample jog.

**T8.2** The weights (in pounds) of three adult males are 160, 215, and 195. The standard error of the mean of these three weights is

(a) 190. (b) 27.84. (c) 22.73. (d) 16.07. (e) 13.13.

**T8.3** In preparing to construct a one-sample *t* interval for a population mean, suppose we are not sure if the population distribution is Normal. In which of the following circumstances would we *not* be safe constructing the interval based on an SRS of size 24 from the population?

(a) A stemplot of the data is roughly bell-shaped.

(b) A histogram of the data shows slight skewness.

(c) A boxplot of the data has a large outlier.

(d) The sample standard deviation is large.

(e) A Normal probability plot of the data is fairly linear.

**T8.4** Many television viewers express doubts about the validity of certain commercials. In an attempt to answer their critics, Timex Group USA wishes to estimate the proportion of consumers who believe what is shown in Timex television commercials. Let $p$ represent the true proportion of consumers who believe what is shown in Timex television commercials. What is the smallest number of consumers that Timex can survey to guarantee a margin of error of 0.05 or less at a 99% confidence level?

(a) 550 (b) 600 (c) 650 (d) 700 (e) 750

**T8.5** You want to compute a 90% confidence interval for the mean of a population with unknown population standard deviation. The sample size is 30. The value of $t^*$ you would use for this interval is

(a) 1.645. (b) 1.699. (c) 1.697. (d) 1.96. (e) 2.045.

**T8.6** A radio talk show host with a large audience is interested in the proportion $p$ of adults in his listening area who think the drinking age should be lowered to eighteen. To find this out, he poses the following question to his listeners: "Do you think that the drinking age should be reduced to eighteen in light of the fact that eighteen-year-olds are eligible for military service?" He asks listeners to phone in and vote "Yes" if they agree the drinking age should be lowered and "No" if not. Of the 100 people who phoned in, 70 answered "Yes." Which of the following conditions for inference about a proportion using a confidence interval are violated?

I. The data are a random sample from the population of interest.

II. The population is at least 10 times as large as the sample.

III. $n$ is so large that both $n\hat{p}$ and $n(1-\hat{p})$ are at least 10.

(a) I only (b) II only (c) III only (d) I and II only (e) I, II, and III

**T8.7** A 90% confidence interval for the mean $\mu$ of a population is computed from a random sample and is found to be $9 \pm 3$. Which of the following *could* be the 95% confidence interval based on the same data?

(a) $9 \pm 2$ (b) $9 \pm 3$ (c) $9 \pm 4$ (d) $9 \pm 8$

(e) Without knowing the sample size, any of the above answers could be the 95% confidence interval.

**T8.8** Suppose we want a 90% confidence interval for the average amount spent on books by freshmen in their first year at a major university. The interval is to have a margin of error of \$2. Based on last year's book sales, we estimate that the standard deviation of the amount spent will be close to \$30. The number of observations required is closest to

(a) 25. (b) 30. (c) 608. (d) 609. (e) 865.

**T8.9** A telephone poll of an SRS of 1234 adults found that 62% are generally satisfied with their lives. The announced margin of error for the poll was 3%. Does the margin of error account for the fact that some adults do not have telephones?

(a) Yes. The margin of error includes all sources of error in the poll.
(b) Yes. Taking an SRS eliminates any possible bias in estimating the population proportion.
(c) Yes. The margin of error includes undercoverage but not nonresponse.
(d) No. The margin of error includes nonresponse but not undercoverage.
(e) No. The margin of error only includes sampling variability.

**T8.10** A Census Bureau report on the income of Americans says that with 90% confidence the median income of all U.S. households in a recent year was \$57,005 with a margin of error of ±\$742. This means that

(a) 90% of all households had incomes in the range \$57,005 ± \$742.
(b) we can be sure that the median income for all households in the country lies in the interval \$57,005 ± \$742.
(c) 90% of the households in the sample interviewed by the Census Bureau had incomes in the interval \$57,005 ± \$742.
(d) the Census Bureau got the result \$57,005 ± \$742 using a method that will capture the true median income 90% of the time when used repeatedly.
(e) 90% of all possible samples of this same size would result in a sample median that falls within \$742 of \$57,005.

**Section II: Free Response** *Show all your work. Indicate clearly the methods you use, because you will be graded on the correctness of your methods as well as on the accuracy and completeness of your results and explanations.*

**T8.11** The U.S. Forest Service is considering additional restrictions on the number of vehicles allowed to enter Yellowstone National Park. To assess public reaction, the service asks a random sample of 150 visitors if they favor the proposal. Of these, 89 say "Yes."

(a) Construct and interpret a 99% confidence interval for the proportion of all visitors to Yellowstone who favor the restrictions.
(b) Based on your work in part (a), can the U.S. Forest Service conclude that more than half of visitors to Yellowstone National Park favor the proposal? Justify your answer.

**T8.12** How many people live in South African households? To find out, we collected data from an SRS of 48 out of the over 700,000 South African students who took part in the CensusAtSchool survey project. The mean number of people living in a household was 6.208; the standard deviation was 2.576.

(a) Is the Normal/Large Sample condition met in this case? Justify your answer.
(b) Maurice claims that a 95% confidence interval for the population mean is $6.208 \pm 1.96\frac{2.576}{\sqrt{47}}$. Explain why this interval is wrong. Then give the correct interval.

**T8.13** A milk processor monitors the number of bacteria per milliliter in raw milk received at the factory. A random sample of 10 one-milliliter specimens of milk supplied by one producer gives the following data:

| 5370 | 4890 | 5100 | 4500 | 5260 | 5150 | 4900 | 4760 | 4700 | 4870 |
|---|---|---|---|---|---|---|---|---|---|

Construct and interpret a 90% confidence interval for the population mean $\mu$.

**T8.8** d

**T8.9** e

**T8.10** d

**T8.11** **(a)** *S*: $p$ = the true proportion of all visitors to Yellowstone who would say they favor the restrictions. *P*: One-sample $z$ interval for $p$. *Random*: The visitors were selected randomly. *10%*: The sample size (150) is less than 10% of all visitors to Yellowstone National Park. *Large Counts*: $n\hat{p} = 89 \geq 10$ and $n(1-\hat{p}) = 61 \geq 10$. *D*: (0.490, 0.696). *C*: We are 99% confident that the interval from 0.490 to 0.696 captures the true proportion of all visitors who would say that they favor the restrictions. **(b)** Because there are values smaller than 0.50 in the confidence interval, the U.S. Forest Service cannot conclude that more than half of visitors to Yellowstone National Park favor the proposal.

**T8.12** **(a)** Because the sample size is large ($n = 48 \geq 30$), the Normal/Large Sample condition is met. **(b)** Maurice's interval uses a $z$ critical value instead of a $t$ critical value. Also, Maurice used the wrong value in the square root—it should be $n = 48$. *Correct*: Using df = 40,

$$6.208 \pm 2.021\left(\frac{2.576}{\sqrt{48}}\right) = (5.457, 6.959).$$

*Using technology*: (5.46, 6.956) with df = 47.

**T8.13** *S*: $\mu$ = the true mean number of bacteria per milliliter in raw milk received at the factory. *P*: One-sample $t$ interval for $\mu$. *Random*: The data come from a random sample. *10%*: The sample size (10) is less than 10% of all 1-milliliter specimens that arrive at the factory. *Normal/Large Sample*: The dotplot shows that there is no strong skewness or outliers.

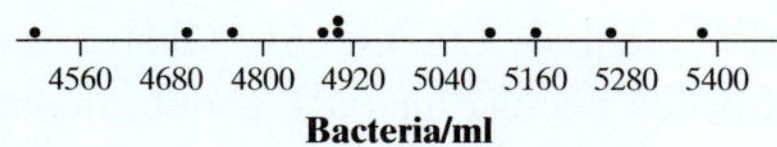


*D*: Using df = 9, (4794.37, 5105.63). *C*: We are 90% confident that the interval from 4794.37 to 5105.63 bacteria/ml captures the true mean number of bacteria in the milk received at this factory.

Chapter

9

# Testing a Claim

## PD Overview

In the last chapter, we used data from random samples to estimate population parameters with confidence intervals. In this chapter, we will use data from random samples and randomized experiments to test claims about a population parameter. This type of inference is called a significance test.

The logic of significance testing should already be somewhat familiar to students, especially if you did the Hiring Discrimination Activity from Chapter 1 and Distracted Driving Activity from Chapter 4 with your class. In both of these Activities, we used simulation to investigate whether the observed results gave convincing evidence in support of a claim or if they were simply the results of random chance. Likewise, many of the simulation questions in Section 5.1 asked students to use the logic of significance testing to determine whether an observed result was unusual enough to cast doubt on a claim, such as in the 1-in-6 Wins Game Activity.

Beginning in this chapter, students will learn how to make these decisions using a more formal, four-step process. We introduce the basics of significance testing in Section 9.1. In Section 9.2,we focus on one-sample tests for a population proportion.In Section 9.3, we focus on one-sample tests for a population mean. By the end of the course, students will have learned eight different significance tests. And while each test varies in the details, they all use the same logic and four-step process.

If you didn't already use this in Chapter 8, an excellent resource to learn more about inference is available for free at AP® Central. On the AP® Statistics homepage, look for the "Special Focus Materials" about inference. These materials were originally developed for use in College Board workshops and were written by people very familiar with the AP® Statistics course.

### Relationship to the AP® Statistics Course Description

Chapter 9 covers the following content as described in the Topic Outline from the AP® Statistics Course Description. Topics marked with a superscript are covered in other chapters as indicated.

IV. Statistical inference: Estimating population parameters and testing hypotheses

A. Estimation (point estimators and confidence intervals)

7. Confidence interval for a difference between two means (unpaired[ch10] and paired)

B. Tests of significance

1. Logic of significance testing, null and alternative hypotheses; *p*-values; one-and two-sided tests; concepts of Type I and Type II errors; concept of power
2. Large sample test for a proportion
3. Large sample test for a difference between two proportions[ch10]
4. Test for a mean
5. Test for a difference between two means (unpaired[ch10] and paired)
6. Chi-square test for goodness of fit, homogeneity of proportions, and independence (one- and two-way tables)[ch11]
7. Test for the slope of a least-squares regression line[ch12]

## The Big Ideas

One of the challenges in teaching the AP® Statistics course is keeping students focused on the big picture and not just the details of each section. We outline the big ideas for the chapter here.

### Chapter 9 Introduction

The purpose of the I'm a Great Free-Throw Shooter Activity in the Introduction is to get students thinking about how to test a claim. The basketball player in the applet claims to make 80% of his free throws, but we claim that he is exaggerating. To decide which of these claims is more reasonable, we collect data. Students should learn that it is easier to find convincing evidence that the player is exaggerating when we have lots of data and when the shooter's actual free-throw percentage is very different from his claimed percentage.

### Section 9.1 Significance Tests: The Basics

In this section, we introduce the basic ideas of significance testing. The first step in any significance test is stating the hypotheses that we want to test. The null hypothesis typically is a statement of "no difference" such as "the treatments are equally effective" or "the player's true free-throw percentage is no different than what he claims." The alternative hypothesis describes what we suspect is true and what we are looking for evidence to support: for example, "Treatment A is more effective than Treatment B" or "the player's true free-throw percentage is less than he claims."

At the beginning of each significance test, make sure your students understand that there are two ways to explain why the sample data disagree with the null hypothesis: (1) The null hypothesis is true, and the disagreement is due to chance—sampling variability when using data from a random sample or chance variation in the random assignment when using data from a randomized experiment; or (2) the null hypothesis is false, and the sample data are consistent with an alternative value of the parameter. In a significance test, we always start with the belief that the null hypothesis is true and choose Explanation 2 only when we are convinced that chance is not a plausible explanation for the difference between the sample statistic and the hypothesized parameter value.

A *P*-value measures how likely it is to get a value of the sample statistic at least as extreme as the observed sample statistic by chance alone, assuming the null hypothesis is true. If the *P*-value is large, then it is plausible that the difference between the sample statistic and hypothesized parameter value could have been due to chance alone. However, if the *P*-value is small, we can rule out chance as a plausible explanation for the difference and conclude that the hypothesized parameter value is incorrect. To determine if a *P*-value is "small," we often compare it to a predetermined significance level, such as $\alpha = 0.05$.

Because we are basing our conclusions on sample data, there is always a chance that our conclusions will be in error. We can make two types of errors in a significance test: a Type I error occurs if we find convincing evidence that the alternative hypothesis is true, when in reality it isn't true. A Type II error occurs if we fail to find convincing evidence that the alternative hypothesis is true, when in reality it is true.

## Section 9.2 Tests about a Population Proportion

In this section, we present the details for conducting a significance test for a population proportion. Not surprisingly, we suggest that students perform significance tests using the State–Plan–Do–Conclude four-step process.

In the "State" step, students are expected to state the hypotheses they are testing, state the significance level they are using, and define the parameter they are using in their hypotheses. In the "Plan" step, students are expected to name the procedure they plan to use (e.g., one-sample *z* test for a population proportion) and check the appropriate conditions to see if the procedure is appropriate. In the "Do" step, students should calculate the test statistic and *P*-value. The test statistic measures how far the sample statistic is from the hypothesized parameter value, in standardized units. Finally, in the "Conclude" step, students should use the *P*-value to make an appropriate decision about the hypotheses in the context of the problem.

We also introduce two-sided tests in this section. We perform a two-sided test when looking for convincing evidence that the true value of the parameter is different from the hypothesized value of the parameter, in either direction. Thus, the alternative hypothesis in a two-sided test will be in the form $H_a: p \neq$ the hypothesized value. When an alternative hypothesis is two-sided, there is a nice connection between confidence intervals and significance tests. Because confidence intervals provide a range of plausible values for a population parameter, we can use confidence intervals to reject or fail to reject a null hypothesis.

Finally, we revisit Type I and Type II errors. The probability that we avoid making a Type II error when an alternative value of the parameter is true is called the power of the test. Power is good—if the alternative hypothesis is true, we want to maximize the probability of coming to this conclusion. We can increase the power of a significance test by increasing the sample size, increasing the significance level, and minimizing sources of variability in the data collection process.

## Section 9.3 Tests about a Population Mean

In this section, we present the details for conducting a significance test for a population mean. We also introduce paired *t* tests to analyze data from studies involving paired data. As with significance tests for a population proportion, we expect students to use the State–Plan–Do–Conclude four-step process when performing significance tests for a population mean or mean difference.

There is actually not much new in this section. The four-step process is the same as in Section 9.2, and the conditions for conducting a one-sample *t* test for $\mu$ are the same as those used for constructing a one-sample *t* interval for $\mu$, discussed in Section 8.3. Make sure to point this out to your students to help them avoid getting overwhelmed.

As with any significance test, always make sure your students can provide the two explanations for why the value of the sample mean differs from the hypothesized population mean. The first explanation is that the hypothesized population mean is correct, and the sample mean differs because of sampling variability or chance variation due to the random assignment. The second explanation is that the sample mean differs because the hypothesized population mean isn't correct. We believe the first explanation until we have convincing evidence that the difference didn't occur by chance alone.

Finally, we introduce paired *t* tests to analyze the results of comparative experiments and observational studies that use paired designs. In an experiment that uses a matched pairs design, such as the caffeine dependence example on page 586, it is best to analyze the difference in each pair. Even though there are initially two sets of responses, we end up having a single set of differences. This allows us to use the one-sample *t* procedures of this section instead of the two independent samples procedure that we will learn about in Chapter 10.

# Chapter 9: Resources

## TRM Additional *TPS* Teacher's Resource Materials

The following resources can be found by clicking on the links in the new Teacher's e-Book (TE-book), opening the Teacher's Resource CD (TRCD), or logging onto the book's companion Web site www.whfreeman.com/tps5e (teacher log-in required).

- Printable Chapter 9 Learning Objective Grid (page 539)
- *Rolling Sixes* (page 566)
- FRAPPY! materials: Question, Model Solution, Scoring Rubric, Sample Responses with commentary (page 601)
- Quizzes: 3 versions for each section; versions A and B are roughly parallel
    - Section 9.1 (page 550)
    - Section 9.2 (page 569)
    - Section 9.3 (page 594)
- Tests: 4 versions; versions A and B are roughly parallel, version D is cumulative (page 602)

## Free-Response Questions from Previous AP® Exams

Questions can be found on the AP® Central Web site: http://apcentral.collegeboard.com/apc/members/exam/exam_questions/8357.html.

Students should be able to answer all the free-response questions listed below with material learned in this chapter. Questions that contain content from this chapter but also require content from later chapters are listed in the last chapter required to complete the entire question. This list will be updated after each AP® exam and will be posted on the book's Web site: www.whfreeman.com/tps5e.

Questions marked with an asterisk are from exams with released multiple-choice questions. You may want to save these questions until the end of the year so you can give your students a complete released exam for practice.

| Year | # | Content |
|---|---|---|
| 2012 | 5* | • Type II error and consequence<br>• Conclusion to a significance test for a single proportion<br>• Voluntary response bias |
| 2009 | 6 | • Stating hypotheses<br>• Relationship between mean and median<br>• Testing for skewness<br>• Creating a test statistic |
| 2009B | 4 | • Random assignment in blocks<br>• Increasing the power of a test |
| 2009B | 5 | • One-sample *t* test for a mean<br>• Using simulation to test a standard deviation |
| 2008B | 4 | • Experimental design<br>• Type I and II errors and consequences |
| 2008B | 6 | • Interpreting scatterplots<br>• Paired *t* test<br>• Creating a classification rule |
| 2007 | 4* | • Paired *t* test |
| 2006B | 4 | • Paired *t* test |
| 2006B | 6 | • Stating hypotheses<br>• Conditions for a one-sample *z* test for a proportion<br>• Binomial probability calculations<br>• Significance levels<br>• Calculating *P*-values and drawing conclusions<br>• Improving a study |
| 2005 | 4 | • One-sample *z* test for a proportion |
| 2005B | 4 | • Paired *t* interval<br>• Using a confidence interval to assess significance |
| 2005B | 6 | • One-sample *t* test for a mean<br>• Normal probability calculation<br>• Multiplication rule for independent events<br>• Using simulation to estimate a probability |

Chapter

# 9

| Year | # | Content |
|---|---|---|
| 2004 | 6 | • One-sample $t$ interval for a mean<br>• Relationship between confidence intervals and significance tests<br>• One-sided confidence intervals |
| 2003 | 1 | • Constructing boxplots<br>• Using boxplots to compare variability<br>• Stating hypotheses |
| 2003 | 2 | • Stating hypotheses<br>• Type I and II errors and consequences |
| 2001 | 5 | • Paired $t$ test |
| 1999 | 6 | • One-sample $t$ test for a mean<br>• Paired $t$ test<br>• Displaying relationships with scatterplots |
| 1998 | 5 | • One-sample $z$ test for a proportion<br>• Effect of nonresponse |
| 1997 | 5* | • Paired $t$ test |

## Applets

- *The Reasoning of a Statistical Test* applet at www.whfreeman.com/tps5e allows students to investigate the logic of hypothesis testing by simulating free-throw attempts for a player who claims to make 80% of his shots.
- The *Statistical Power* applet at www.whfreeman.com/tps5e allows students to investigate how the power of a significance test for a population mean is affected by changes in the sample size, significance level, and alternative parameter value.
- The *Power* applet at http://www.amstat.org/publications/jse/v11n3/java/Power allows students to investigate how the power of a significance test is affected by changes in the sample size, significance level, and alternative parameter value.
- The *Improved Batting Averages (Power)* applet at www.rossmanchance.com/applets estimates the power of a one-sample $z$ test for a population proportion using simulation. It allows students to investigate how the power of the test is affected by changes in the sample size, significance level, and alternative parameter value.
- The *WISE Statistical Power* applet at http://wise.cgu.edu/powermod/power_applet.asp allows students to dynamically investigate how the power of a significance test for a population mean is affected by changes in the sample size, significance level, and alternative parameter value.
- The *Errors in Hypothesis Testing* applet at www.intuitor.com/statistics/CurveApplet.html allows students to use sliders to dynamically investigate how the power of a significance test for a population mean is affected by changes in the sample size, significance level, and alternative parameter value.

## TI-Nspire Activities

These activities are designed to illustrate important statistical concepts using the TI-Nspire and can be found at the book's companion Web site www.whfreeman.com/tps5e.

- *Exploring Significance Tests*: Students are introduced to the logic of significance testing.
- *Exploring t Tests*: Students are introduced to $t$ tests for a population mean.

# Chapter 9: Pacing Guide, Learning Objectives, and Suggested Assignments

This pacing guide is based on a schedule with 110, 50-minute sessions before the AP® exam. The 110 days include 4 days for midterms and 10 days for AP® exam review. If you have a different number of sessions before the AP® exam, you can modify the pacing guide to suit your needs. If you have additional time, consider incorporating quizzes, released AP® free-response questions, or additional activities. See the Resources section above for suggestions.

The suggested homework assignments list odd-numbered exercises, whenever possible, so students can check their answers against the back-of-the-book answers. If you would rather students not have access to the answers while doing homework, adding 1 to the exercise numbers usually will do the trick, because the homework exercises typically are paired. For example, Exercises 1 and 2 will generally cover the same topics, but in different contexts. You may also choose to include the Recycle and Review questions at the end of each section, which review topics from previous sections or chapters. These questions are denoted with a symbol.

| Day | Topics | Learning Objectives: Students will be able to . . . | Suggested Assignment |
|---|---|---|---|
| 1 | 9.1 Stating Hypotheses, The Reasoning of Significance Tests, Interpreting *P*-values, Statistical Significance | • State the null and alternative hypotheses for a significance test about a population parameter.<br>• Interpret a *P*-value in context.<br>• Determine if the results of a study are statistically significant and draw an appropriate conclusion using a significance level. | 1, 3, 5, 7, 9, 11, 15 |
| 2 | 9.1 Type I and Type II Errors | • Interpret a Type I and a Type II error in context, and give a consequence of each. | 13, 17, 19, 21, 23 |
| 3 | 9.2 Carrying Out a Significance Test, The One-Sample *z* Test for a Proportion | • State and check the Random, 10%, and Large Counts conditions for performing a significance test about a population proportion.<br>• Perform a significance test about a population proportion. | 25–28, 31, 35, 39, 41 |
| 4 | 9.2 Two-Sided Tests, Why Confidence Intervals Give More Information, Type II Error and the Power of a Test | • Use a confidence interval to draw a conclusion for a two-sided test about a population parameter.<br>• Interpret the power of a test and describe what factors affect the power of a test.<br>• Describe the relationship among the probability of a Type I error (significance level), the probability of a Type II error, and the power of a test. | 43, 45, 47, 51, 53, 55, 57 |
| 5 | 9.3 Carrying Out a Significance Test for $\mu$, The One Sample *t* Test, Two-Sided Tests and Confidence Intervals | • State and check the Random, 10%, and Normal/Large Sample conditions for performing a significance test about a population mean.<br>• Perform a significance test about a population mean.<br>• Use a confidence interval to draw a conclusion for a two-sided test about a population parameter. | 59–62, 65, 69, 73, 77, 79 |
| 6 | 9.3 Inference for Means: Paired Data, Using Tests Wisely | • Perform a significance test about a mean difference using paired data. | 83, 85, 87, 89–91, 93, 95–102 |
| 7 | Chapter 9 Review/FRAPPY! | | Chapter 9 Review Exercises |
| 8 | Chapter 9 Test | | |

Chapter

NOTES

# Chapter 9

**PD Chapter 9 Overview**

Watch the video overview of Chapter 9 (for teachers) by clicking on the link in the TE-book, opening the TRCD, or logging onto the book's companion Web site www.whfreeman.com/tps5e (teacher log-in required).

# Testing a Claim

case study

## Do You Have a Fever?

Sometimes when you're sick, your forehead feels really warm. You might have a fever. How can you find out whether you do? By taking your temperature, of course. But what temperature should the thermometer show if you're healthy? Is this temperature the same for everyone?

Several years ago, researchers conducted a study to determine whether the "accepted" value for normal body temperature, 98.6°F, is accurate. They used an oral thermometer to measure the temperatures of a random sample of healthy men and women aged 18 to 40. As is often the case, the researchers did not provide their original data.

Allen Shoemaker, from Calvin College, produced a data set with the same properties as the original temperature readings. His data set consists of one oral temperature reading for each of 130 randomly chosen, healthy 18- to 40-year-olds.[1] A dotplot of Shoemaker's temperature data is shown below. We have added a vertical line at 98.6°F for reference.

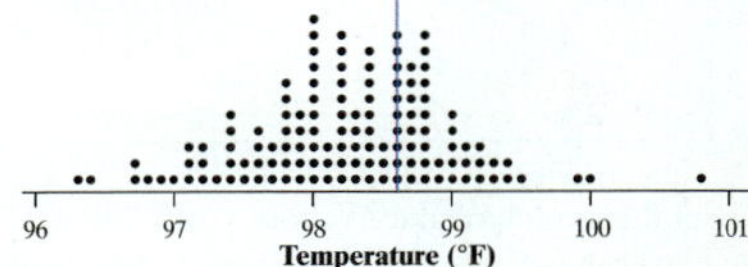


Exploratory data analysis revealed several interesting facts about this data set:

- The mean temperature was $\bar{x} = 98.25$°F.
- The standard deviation of the temperature readings was $s_x = 0.73$°F.
- 62.3% of the temperature readings were less than 98.6°F.

Based on the results of this study, can we conclude that "normal" body temperature in the population of healthy 18- to 40-year-olds is *not* 98.6°F? By the end of this chapter, you will have developed the necessary tools for answering this question.

537

### Teaching Tip

Point out that this question can be addressed by comparing the sample mean temperature to $\mu = 98.6$°F or the sample proportion of subjects with temperatures less than 98.6°F to $p = 0.5$.

# Introduction

Confidence intervals are one of the two most common types of statistical inference. Use a confidence interval when your goal is to estimate a population parameter. The second common type of inference, called *significance tests*, has a different goal: to assess the evidence provided by data about some claim concerning a parameter. Here is an Activity that illustrates the reasoning of statistical tests.

## ACTIVITY | I'm a Great Free-Throw Shooter!

**MATERIALS:**
Computer with Internet access and projection capability

A basketball player claims to make 80% of the free throws that he attempts. We think he might be exaggerating. To test this claim, we'll ask him to shoot some free throws—virtually—using *The Reasoning of a Statistical Test* applet at the book's Web site.

1. Go to www.whfreeman.com/tps5e and launch the applet.

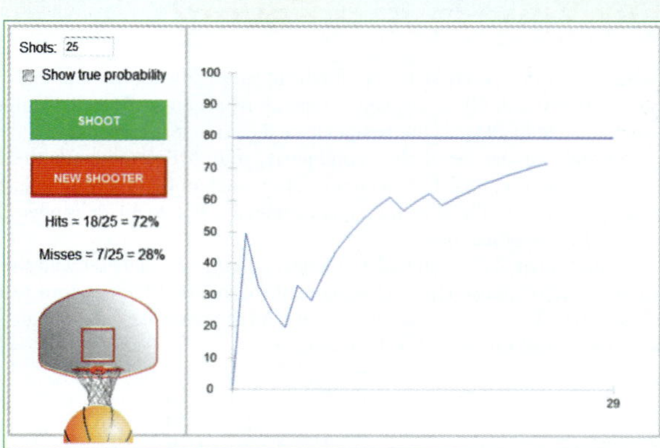


2. Set the applet to take 25 shots. Click "Shoot." How many of the 25 shots did the player make? Do you have enough data to decide whether the player's claim is valid?
3. Click "Shoot" again for 25 more shots. Keep doing this until you are convinced *either* that the player makes less than 80% of his shots *or* that the player's claim is true. How large a sample of shots did you need to make your decision?
4. Click "Show true probability" to reveal the truth. Was your conclusion correct?
5. If time permits, choose a new shooter and repeat Steps 2 through 4. Is it easier to tell that the player is exaggerating when his actual proportion of free throws made is closer to 0.8 or farther from 0.8?

### Activity Overview

You can watch an overview of this Activity (for teachers) by clicking on the link in the TE-book, opening the TRCD, or logging onto the book's companion Web site www.whfreeman.com/tps5e (teacher log-in required).

### Teaching Tip

This Activity nicely illustrates two important aspects of significance testing:

**1.** Increasing the sample size helps us be more certain that the difference between the observed proportion of successes and the hypothesized proportion of successes is not due to sampling variability.

**2.** It is easier to reject a hypothesized proportion of successes when there is a greater difference between the hypothesized proportion of successes and the actual proportion of successes. That is, the power of a test will be higher when the true value of a parameter is farther from the hypothesized value of a parameter.

### Teaching Tip

If you don't do the I'm a Great Free-Throw Shooter Activity on this page, make sure to introduce the context to your students. We will refer back to our free-throw shooter (and his 80% claim) throughout Section 9.1.

### ALTERNATE ACTIVITY

***Kissing the right way***

According to an article in the San Gabriel Valley Tribune (February 13, 2003), "Most people are kissing the 'right way.'" That is, according to a study, the majority of couples prefer to tilt their heads to the right when kissing. In the study, a researcher observed a random sample of 124 kissing couples and found that 83/124 ($\hat{p} = 0.669$) of the couples tilted to the right. Is this convincing evidence that couples really do prefer to kiss the right way? Or, it is possible that couples have no preference and the sample proportion was greater than $p = 0.5$ by chance alone?

If couples have no directional preference, then we can use a coin to simulate observing a kissing couple. Let heads represent a couple tilting to the right and tails represent a couple tilting to the left.

**1.** Flip the coin 124 times to represent observing 124 couples. Record how many times the coin landed on heads (the number of couples who tilted to the right).

**2.** Share your result with the class and create a class dotplot showing the simulated number of right-tilting couples.

**3.** Based on the results of the simulation, how likely is it to observe 83 or more couples (out of 124) tilting to the right, if they really have no directional preference?

**4.** Based on your answer to Question 3, is there convincing evidence that couples prefer to kiss the right way? Explain.

*Note*: Here is what your class dotplot might look like. Based on these results, it is very unlikely you will get 83 or more couples tilting to the right just by chance. Because chance is not a plausible explanation for the observed results, there is convincing evidence that couples do prefer to kiss the right way.

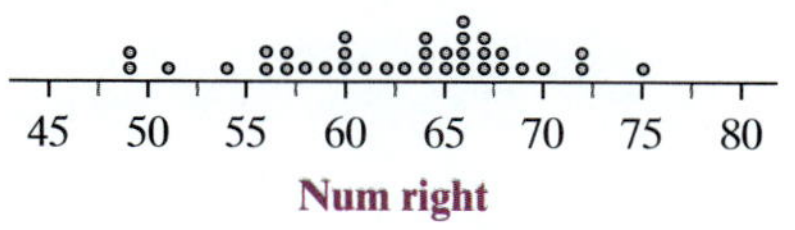

In the free-throw shooter Activity, the parameter of interest is the proportion $p$ of free throws that the player will make if he shoots forever. Our player claims that $p = 0.80$. To test his claim, we let the applet simulate 25 shots. If the player makes only 40% of his free throws (10 of 25 shots made), we have fairly strong evidence that he doesn't shoot 80%. But what if he makes 76% of his free throws (19 of 25 shots made)? This provides *some* evidence that his true long-term percent may be less than 80%, but it's not nearly as convincing as $\hat{p} = 0.40$. Statistical tests weigh the evidence *against* a claim like $p = 0.8$ and in favor of a counter-claim like $p < 0.80$.

Section 9.1 focuses on the underlying logic of statistical tests. Once the foundation is laid, we consider the implications of using these tests to make decisions—about everything from free-throw shooting to the effectiveness of a new drug. In Section 9.2, we present the details of performing a test about a population proportion. Section 9.3 shows how to test a claim about a population mean. Along the way, we examine the connection between confidence intervals and tests.

# 9.1 Significance Tests: The Basics

**WHAT YOU WILL LEARN** By the end of the section, you should be able to:

- State the null and alternative hypotheses for a significance test about a population parameter.
- Interpret a *P*-value in context.
- Determine whether the results of a study are statistically significant and make an appropriate conclusion using a significance level.
- Interpret a Type I and a Type II error in context and give a consequence of each.

A **significance test** is a formal procedure for using observed data to decide between two competing claims (also called *hypotheses*). The claims are often statements about a parameter, like the population proportion $p$ or the population mean $\mu$. Let's start by taking a closer look at how to state hypotheses.

## Stating Hypotheses

A significance test starts with a careful statement of the claims we want to compare. In our free-throw shooter example, the virtual player claims that his long-run proportion of made free throws is $p = 0.80$. This is the claim we seek evidence *against*. We call it the **null hypothesis**, abbreviated $H_0$. Usually, the null hypothesis is a statement of "no difference." For the free-throw shooter, no difference from what he claimed gives $H_0: p = 0.80$.

Remember: the null hypothesis is the dull hypothesis!

The claim we hope or suspect to be true instead of the null hypothesis is called the **alternative hypothesis**. We abbreviate the alternative hypothesis as $H_a$. In this case, we believe the player might be exaggerating, so our alternative hypothesis is $H_a: p < 0.80$.

### PD Section 9.1 Overview

Watch the video overview of Section 9.1 (for teachers) by clicking on the link in the TE-book, opening the TRCD, or logging onto the book's companion Web site www.whfreeman.com/tps5e (teacher log-in required).

### TRM Learning Objectives Grid

Remind students about the importance of the Learning Objectives. We include them at the beginning of each section so students know what they are expected to learn. At the end of the chapter, the Learning Objectives are included in a grid, along with references to examples and chapter review exercises. A printable version of this grid is included in the Teacher's Resource Materials.

### TRM Section 9.1 Alternate Examples

You can find the Alternate Examples for this section in Microsoft Word format by clicking on the link in the TE-book, opening the TRCD, or logging onto the book's companion Web site www.whfreeman.com/tps5e (teacher log-in required).

### Teaching Tip

Based on their experiences in science classes, students may be surprised to hear that we state two hypotheses. Students are probably accustomed to stating a single research hypothesis, which is the hypothesis that they hope to find evidence to support (what we call the alternative hypothesis). This also may initially lead to confusion about which hypothesis is the null and which is the alternative.

**DEFINITION: Null hypothesis $H_0$, alternative hypothesis $H_a$**

The claim we weigh evidence against in a statistical test is called the **null hypothesis ($H_0$)**. Often the null hypothesis is a statement of "no difference."

The claim about the population that we are trying to find evidence *for* is the **alternative hypothesis ($H_a$)**.

Some people insist that all three possibilities—greater than, less than, and equal to—be accounted for in the hypotheses. For the free-throw shooter example, since the alternative hypothesis is $H_a: p < 0.80$, they would write the null hypothesis as $H_0: p \geq 0.80$. In spite of the mathematical appeal of covering all three cases, we use only the value $p = 0.80$ in our calculations. So we'll stick with $H_0: p = 0.80$.

In the free-throw shooter example, our hypotheses are

$$H_0: p = 0.80$$
$$H_a: p < 0.80$$

where $p$ is the long-run proportion of made free throws. The alternative hypothesis is **one-sided** because we are interested only in whether the player is overstating his free-throw shooting ability. Because $H_a$ expresses the effect that we hope to find evidence *for*, it is sometimes easier to begin by stating $H_a$ and then set up $H_0$ as the statement that the hoped-for effect is not present.

Here is an example in which the alternative hypothesis is **two-sided**.

## EXAMPLE Juicy Pineapples

### *Stating hypotheses*

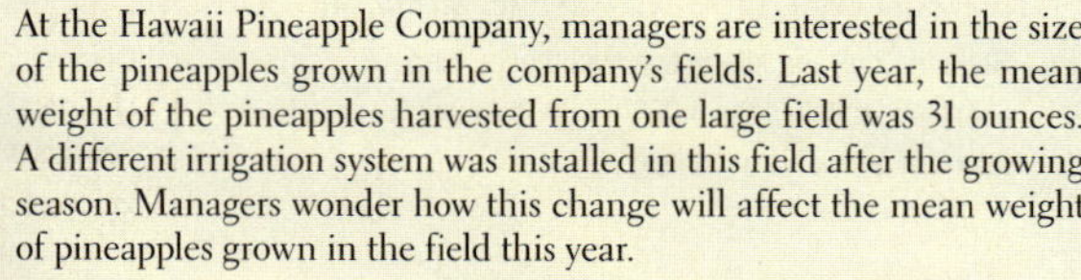

At the Hawaii Pineapple Company, managers are interested in the size of the pineapples grown in the company's fields. Last year, the mean weight of the pineapples harvested from one large field was 31 ounces. A different irrigation system was installed in this field after the growing season. Managers wonder how this change will affect the mean weight of pineapples grown in the field this year.

PROBLEM: State appropriate hypotheses for performing a significance test. Be sure to define the parameter of interest.

SOLUTION: The parameter of interest is the mean weight $\mu$ of all pineapples grown in the field this year. Because managers wonder whether the mean weight of this year's pineapples will differ from last year's mean weight of 31 ounces, the alternative hypothesis is two-sided; that is, either $\mu < 31$ or $\mu > 31$. For simplicity, we write this as $\mu \neq 31$. The null hypothesis says that there is no difference in the mean weight of the pineapples after the irrigation system was changed. That is,

$$H_0: \mu = 31$$
$$H_a: \mu \neq 31$$

**For Practice** *Try Exercise* **1**

*The hypotheses should express the hopes or suspicions we have* ***before*** *we see the data. It is cheating to look at the data first and then frame hypotheses to fit what the data show.* For example, the data for the pineapple study

## Teaching Tip

If you did the Kissing the Right Way Alternate Activity, the hypotheses are

$$H_0: p = 0.5$$
$$H_a: p > 0.5$$

where $p$ = the true proportion of couples who tilt their heads to the right when kissing.

## Teaching Tip

Remind students that the "play" icon at the top of this example (and many others) indicates that they can watch a video presentation of the example. Find the videos at www.whfreeman.com/tps5e. Also, remind them that the icon in the lower-right corner indicates that a link exists between this example and Exercise 1. If students are having trouble with the exercise, they can look back to this example for help.

## ALTERNATE EXAMPLE

### *A better golf club?*

Mike is an avid golfer who would like to improve his play. A friend suggests getting new clubs and lets Mike try out his 7-iron. Based on years of experience, Mike has established that the mean distance that balls travel when hit with his old 7-iron is $\mu$ = 175 yards with a standard deviation of $\sigma$ = 15 yards. He is hoping that this new club will make his shots with a 7-iron more consistent (less variable), so he goes to the driving range and hits 50 shots with the new 7-iron.

PROBLEM: State appropriate hypotheses for performing a significance test. Be sure to define the parameter of interest.

SOLUTION: Mike is interested in being more consistent, so the parameter of interest is the true standard deviation $\sigma$ of the distance he hits the ball when using the new 7-iron. Because Mike wants the distances to be less variable, he should test the following hypotheses:

$$H_0: \sigma = 15$$
$$H_a: \sigma < 15$$

## AP® Exam Common Error

On the AP® exam, many students incorrectly use the data to form the hypotheses rather than the wording of the question. Remind your students that in real life the hypotheses are formed *before* the data are collected. If we always waited to see the data first, there would never be a two-sided test—we would always make the alternative hypothesis go in the same direction as the sample statistic suggests.

showed that $\bar{x} = 31.935$ ounces for a random sample of 50 pineapples grown in the field this year. You should *not* change the alternative hypothesis to $H_a: \mu > 31$ after looking at the data. If you do not have a specific direction firmly in mind in advance, use a two-sided alternative hypothesis.

**DEFINITION: One-sided alternative hypothesis and two-sided alternative hypothesis**

The alternative hypothesis is **one-sided** if it states that a parameter is *larger than* the null hypothesis value or if it states that the parameter is *smaller than* the null value. It is **two-sided** if it states that the parameter is *different from* the null hypothesis value (it could be either larger or smaller).

It is common to refer to a significance test with a one-sided alternative hypothesis as a *one-sided test* or *one-tailed test* and to a test with a two-sided alternative hypothesis as a *two-sided test* or *two-tailed test*.

The null hypothesis has the form $H_0$: parameter = value. The alternative hypothesis has one of the forms $H_a$: parameter < value, $H_a$: parameter > value, or $H_a$: parameter ≠ value. To determine the correct form of $H_a$, read the problem carefully.

Hypotheses always refer to a *population*, not to a sample. Be sure to state $H_0$ and $H_a$ in terms of *population parameters*. It is *never* correct to write a hypothesis about a sample statistic, such as $\hat{p} = 0.64$ or $\bar{x} > 85$.

**CHECK YOUR UNDERSTANDING**

For each of the following settings, (a) describe the parameter of interest, and (b) state appropriate hypotheses for a significance test.

1. According to the Web site sleepdeprivation.com, 85% of teens are getting less than eight hours of sleep a night. Jannie wonders whether this result holds in her large high school. She asks an SRS of 100 students at the school how much sleep they get on a typical night. In all, 75 of the responders said less than 8 hours.
2. As part of its 2010 census marketing campaign, the U.S. Census Bureau advertised "10 questions, 10 minutes—that's all it takes." On the census form itself, we read, "The U.S. Census Bureau estimates that, for the average household, this form will take about 10 minutes to complete, including the time for reviewing the instructions and answers." We suspect that the actual time it takes to complete the form may be longer than advertised.

## The Reasoning of Significance Tests

A significance test is sometimes referred to as a *test of significance*, a *hypothesis test*, or a *test of hypotheses*.

Significance tests ask if sample data give convincing evidence against the null hypothesis and in favor of the alternative hypothesis. A test answers the question, "How likely is it to get a result like this just by chance when the null hypothesis is true?" The answer comes in the form of a probability.

Here is an activity that introduces the underlying logic of significance tests.

### Teaching Tip

To recognize that a problem is calling for a two-sided test, look for key phrases such as "has the true value *changed*?" or "is there convincing evidence that the true value is *different*?" Also, a question might ask the students to "test the claim that $\mu = 10$" without suggesting a direction for the alternative. In cases like this, use a two-sided test.

### AP® Exam Common Error

On the AP® exam, many students incorrectly use sample statistics such as $\bar{x}$ or $\hat{p}$ when stating their hypotheses. Remind your students that we already know the value of the sample statistic—it is the population parameter that we are trying to make an inference about.

### ANSWERS TO CYU

1. **(a)** $p$ = proportion of all students at Jannie's high school who get less than 8 hours of sleep at night. **(b)** $H_0: p = 0.85$ and $H_a: p \neq 0.85$.
2. **(a)** $\mu$ = true mean amount of time that it takes to complete the census form. **(b)** $H_0: \mu = 10$ and $H_a: \mu > 10$.

## Activity Overview

You can watch an overview of this Activity (for teachers) by clicking on the link in the TE-book, opening the TRCD, or logging onto the book's companion Web site www.whfreeman.com/tps5e (teacher log-in required).

## Teaching Tip

Ask your students if they can think of any other ways to calculate the probability that the shooter will make 32 or fewer shots in 50 attempts. Hopefully, they will recognize that this probability can be calculated with the binomial distribution, the Normal approximation to the binomial distribution, or the sampling distribution of $\hat{p}$.

**ALTERNATE ACTIVITY**

*Kissing the right way*

If you didn't do the Alternate Activity on page 538, it would work here as well.

**ALTERNATE EXAMPLE**

*Kissing the right way*

If you did the Kissing the Right Way Alternate Activity, here are the results of 500 trials of the simulation. Only once did we observe 83 or more couples that tilted to the right when kissing. That is, assuming that couples have no directional preference, there is about a 0.002 probability that we would observe 83 or more couples kissing the right way in a sample of 124 couples. Because this probability is so small, we have very convincing evidence that the majority of couples prefer to kiss the right way.

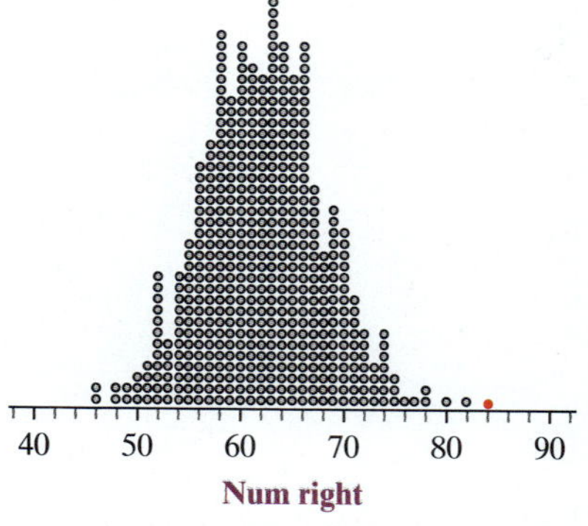


## ACTIVITY I'm a Great Free-Throw Shooter!

**MATERIALS:**
Copy of pie chart with 80% shaded and paper clip for each student

Our virtual basketball player in the previous Activity claimed to be an 80% free-throw shooter. Suppose that he shoots 50 free throws and makes 32 of them. His sample proportion of made shots is $\hat{p} = \frac{32}{50} = 0.64$. This result suggests that he may really make less than 80% of his free throws in the long run. But do we have convincing evidence that $p < 0.80$? In this activity, you and your classmates will perform a simulation to find out.

1. Make a spinner that gives the shooter an 80% chance of making a free throw. Using the pie chart provided by your teacher, label the 80% region "made shot" and the 20% region "missed shot." Straighten out one of the ends of a paper clip so that there is a loop on one side and a pointer on the other. On a flat surface, place a pencil through the loop, and put the tip of the pencil on the center of the pie chart. Then flick the paper clip and see where the pointed end lands.
2. Simulate a random sample of 50 shots. Flick the paper clip 50 times, and count the number of times that the pointed end lands in the "made shot" region.
3. Compute the sample proportion $\hat{p}$ of made shots in your simulation from Step 2. Plot this value on the class dotplot drawn by your teacher.
4. Repeat Steps 2 and 3 as needed to get at least 40 trials of the simulation for your class.
5. Based on the results of your simulation, how likely is it for an 80% shooter to make 64% or less when he shoots 50 free throws? Does the observed $\hat{p} = 0.64$ result give convincing evidence that the player is exaggerating?

Figure 9.1 shows what sample proportions are likely to occur by chance alone, assuming that $p = 0.80$.

Our reasoning in the Activity is based on asking what would happen if the player's claim ($p = 0.80$) were true and we observed many samples of 50 free throws. We used Fathom software to simulate 400 sets of 50 shots assuming that the player is really an 80% shooter. Figure 9.1 shows a dotplot of the results. Each dot on the graph represents the proportion of made shots in one set of 50 attempts. For example, if the player makes 43/50 shots in one trial, the dot would be placed at $\hat{p} = 0.86$.

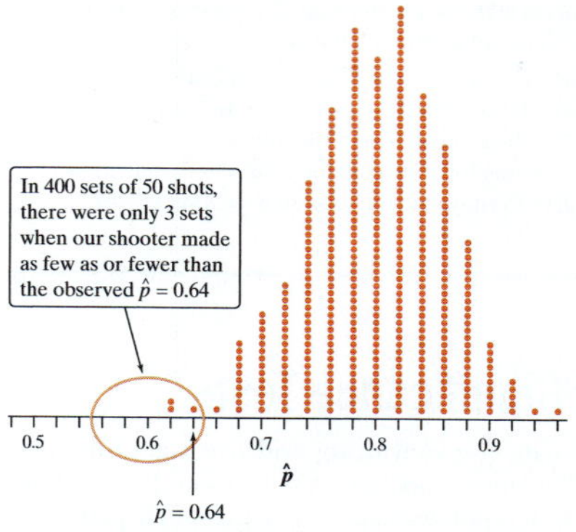


**FIGURE 9.1** Fathom dotplot of the sample proportion $\hat{p}$ of free throws made by an 80% shooter in 400 sets of 50 shots.

You can say how strong the evidence against the player's claim is by giving the probability that he would make as few as 32 out of 50 free throws if he really makes 80% in the long run. Based on the simulation, our estimate of this probability is $3/400 = 0.0075$. The observed statistic, $\hat{p} = 0.64$, is so unlikely if the actual parameter value is $p = 0.80$ that it gives convincing evidence that the player's claim is not true.

Be sure that you understand why this evidence is convincing. There are two possible explanations of the fact that our virtual player made only $\hat{p} = 32/50 = 0.64$ of his free throws:

1. The null hypothesis is correct ($p = 0.8$), and just by chance, a very unlikely outcome occurred.
2. The alternative hypothesis is correct—the population proportion is less than 0.8, so the sample result is not an unlikely outcome.

## Teaching Tip

For each significance test that you discuss in class, make sure your students can identify the two possible explanations for why the sample statistic differs from the hypothesized value of the population parameter. One explanation is that the hypothesized value of the population parameter is correct, and the sample statistic differs by chance alone. The other explanation is that the hypothesized value of the population parameter is incorrect, and the sample statistic is consistent with the true value of the population parameter.

In the Kissing example, the sample proportion $\hat{p} = 83/124$ was greater than the hypothesized proportion of $p = 0.5$. One explanation for the difference is that couples have no directional preference and we got a sample proportion of $\hat{p} = 83/124$ by chance alone. The other explanation is that couples really do prefer to kiss the right way. Because the simulation showed that getting a sample proportion of at least $\hat{p} = 83/124$ would be unlikely to happen by chance alone, it is more plausible that couples prefer to kiss the right way.

Explanation 1 might be correct—the result of our random sample of 50 shots *could* be due to chance alone. But the probability that such a result would occur by chance is so small (less than 1 in a 100) that we are quite confident that Explanation 2 is right.

Statistical tests use an elaborate vocabulary, but the basic idea is: *an outcome that would rarely happen if the null hypothesis were true is good evidence that the null hypothesis is not true.*

## Interpreting *P*-Values

The idea of stating a null hypothesis that we want to find evidence *against* seems odd at first. It may help to think of a criminal trial. The defendant is "innocent until proven guilty." That is, the null hypothesis is innocence and the prosecution must try to provide convincing evidence against this hypothesis and in favor of the alternative hypothesis: guilt. That's exactly how statistical tests work, although in statistics we deal with evidence provided by data and use a probability to say how strong the evidence is.

The null hypothesis $H_0$ states the claim that we are seeking evidence against. The probability that measures the strength of the evidence against $H_0$ and in favor of $H_a$ is called a **P-value**.

**DEFINITION: *P*-value**

The probability, computed assuming $H_0$ is true, that the statistic (such as $\hat{p}$ or $\bar{x}$) would take a value as extreme as or more extreme than the one actually observed, in the direction specified by $H_a$, is called the ***P*-value** of the test.

Small *P*-values are evidence against $H_0$ because they say that the observed result is unlikely to occur when $H_0$ is true. Large *P*-values fail to give convincing evidence against $H_0$ and in favor of $H_a$ because they say that the observed result is likely to occur by chance alone when $H_0$ is true.

We'll show you how to calculate *P*-values later. For now, let's focus on interpreting them.

### EXAMPLE I'm a Great Free-Throw Shooter!

*Interpreting a P-value*

The *P*-value is the probability of getting a sample result at least as extreme as the one we did if $H_0$ were true. Because the alternative hypothesis is $H_a: p < 0.80$, the sample results that count as "at least as extreme" are those with $\hat{p} \le 0.64$. In other words, the *P*-value is the conditional probability $P(\hat{p} \le 0.64 \mid p = 0.80)$. Earlier, we used a simulation to estimate this probability as $3/400 = 0.0075$. So *if $H_0$ is true and the player makes 80% of his free throws in the long run, there's about a 0.0075 probability that the player would make 32 or fewer of 50 shots by chance alone.* The small probability gives strong evidence against $H_0$ and in favor of the alternative $H_a: p < 0.80$ because it would be so unlikely for this result to occur just by chance if $H_0$ were true.

### Teaching Tip

The analogy with a criminal trial is a very useful way to discuss the logic of significance testing. For a defendant to be brought to trial, there must be some evidence that he is guilty. It is the job of the jury to decide if the evidence is *convincing*—that is, to decide that there are no other plausible explanations for how the crime was committed. We do a significance test when the sample statistic provides evidence that the alternative hypothesis is true. To determine if the evidence is convincing, we calculate a *P*-value. If the *P*-value is small, then there is convincing evidence in favor of the alternative hypothesis because we can rule out chance as a plausible explanation.

### ALTERNATE EXAMPLE

***Kissing the right way***

In the Kissing example on the previous page, we used simulation to estimate the probability that we would observe 83 *or more* couples tilting to the right when observing a sample of 124 couples with no directional preference. That is, we estimated the probability that we got a statistic as extreme *or more extreme* than the one observed ($\hat{p} = 83/124$) in the direction specified by the alternative hypothesis ($H_a : p > 0.5$), assuming that the null hypothesis is true ($H_0 : p = 0.5$). From the simulation, $P(\hat{p} \ge 83/124 \mid p = 0.5) \approx 0.002$. So *if couples really have no directional preference, there is about a 0.002 probability that we would observe 83 or more couples out of 124 kiss the right way.* Because it is unlikely for this result to occur if couples have no directional preference when kissing, we have convincing evidence that couples prefer to kiss the right way.

ALTERNATE EXAMPLE

### A better golf club?

When Mike was testing a new 7-iron, the hypotheses were

$$H_0: \sigma = 15$$
$$H_a: \sigma < 15$$

where $\sigma$ = the true standard deviation of the distances Mike hits golf balls using the new 7-iron. Based on 50 shots with the new 7-iron, the standard deviation was $s_x = 13.9$ yards. After checking that the conditions were met, Mike performed a significance test and obtained a *P*-value of 0.25.

PROBLEM:

**(a)** Explain what it would mean for the null hypothesis to be true in this setting.
**(b)** Interpret the *P*-value in this context.

SOLUTION:

**(a)** In this setting, $H_0: \sigma = 15$ means that the true standard deviation with the new 7-iron is 15 yards. That is, Mike is equally consistent with the new and old 7-iron.
**(b)** Assuming the true standard deviation is 15 yards, there is about a 0.25 probability that the sample standard deviation would be 13.9 yards or less by chance alone in a sample of 50 shots with the new 7-iron.

The alternative hypothesis sets the direction that counts as evidence against $H_0$. In the previous example, only values of $\hat{p}$ that are much less than 0.80 count as evidence against the null hypothesis because the alternative is one-sided on the low side. If the alternative is two-sided, both directions count.

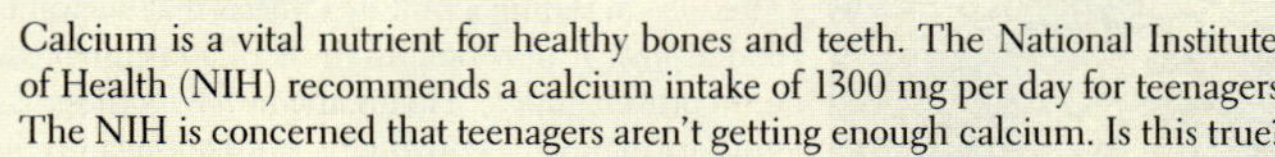

EXAMPLE

## Healthy Bones

### Interpreting a P-value

Calcium is a vital nutrient for healthy bones and teeth. The National Institutes of Health (NIH) recommends a calcium intake of 1300 mg per day for teenagers. The NIH is concerned that teenagers aren't getting enough calcium. Is this true?

Researchers want to perform a test of

$$H_0: \mu = 1300$$
$$H_a: \mu < 1300$$

where $\mu$ is the true mean daily calcium intake in the population of teenagers. They ask a random sample of 20 teens to record their food and drink consumption for 1 day. The researchers then compute the calcium intake for each student. Data analysis reveals that $\bar{x} = 1198$ mg and $s_x = 411$ mg. After checking that conditions were met, researchers performed a significance test and obtained a *P*-value of 0.1404.

PROBLEM:

(a) **Explain what it would mean for the null hypothesis to be true in this setting.**
(b) **Interpret the *P*-value in context.**

SOLUTION:

(a) In this setting, $H_0: \mu = 1300$ says that the mean daily calcium intake in the population of teenagers is 1300 mg. If $H_0$ is true, then teenagers are getting enough calcium, on average.
(b) Assuming that the mean daily calcium intake in the teen population is 1300 mg, there is a 0.1404 probability of getting a sample mean of 1198 mg or less just by chance in a random sample of 20 teens.

**For Practice** *Try Exercise*  11

## Statistical Significance

The final step in performing a significance test is to draw a conclusion about the competing claims you were testing. We will make one of two decisions based on the strength of the evidence against the null hypothesis (and in favor of the alternative hypothesis)—**reject $H_0$** or **fail to reject $H_0$**. If our sample result is too unlikely to have happened by chance assuming $H_0$ is true, then we'll reject $H_0$ and say that there is convincing evidence for $H_a$. Otherwise, we will fail to reject $H_0$ and say that there is *not* convincing evidence for $H_a$.

This wording may seem unusual at first, but it's consistent with what happens in a criminal trial. Once the jury has weighed the evidence against the null hypothesis of innocence, they return one of two verdicts: "guilty" (reject $H_0$) or "not guilty" (fail to reject $H_0$). A not-guilty verdict doesn't guarantee that the defendant is innocent, just that there's not convincing evidence of guilt. Likewise, a fail-to-reject $H_0$ decision in a significance test doesn't mean that $H_0$ is true. For that reason, *you should never "accept $H_0$" or use language implying that you believe $H_0$ is true.*

## EXAMPLE Free Throws and Healthy Bones

### *Drawing conclusions*

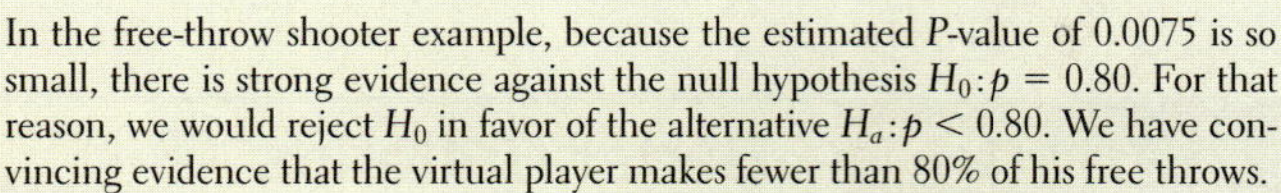

In the free-throw shooter example, because the estimated $P$-value of 0.0075 is so small, there is strong evidence against the null hypothesis $H_0: p = 0.80$. For that reason, we would reject $H_0$ in favor of the alternative $H_a: p < 0.80$. We have convincing evidence that the virtual player makes fewer than 80% of his free throws.

For the teen calcium study, however, the large $P$-value of 0.1404 gives weak evidence against $H_0: \mu = 1300$ and in favor of $H_a: \mu < 1300$. We therefore fail to reject $H_0$. Researchers do not have convincing evidence that teens are getting less than 1300 mg of calcium per day, on average.

In a nutshell, our conclusion in a significance test comes down to

$P$-value small → reject $H_0$ → convincing evidence for $H_a$ (in context)

$P$-value large → fail to reject $H_0$ → not convincing evidence for $H_a$ (in context)

There is no rule for how small a $P$-value is required to reject $H_0$—it's a matter of judgment and depends on the specific circumstances. But we can compare the $P$-value with a fixed value that we regard as decisive, called the **significance level**. We write it as $\alpha$, the Greek letter alpha.

If we choose $\alpha = 0.05$, we are requiring that the data give evidence against $H_0$ so strong that it would happen less than 5% of the time just by chance when $H_0$ is true. If we choose $\alpha = 0.01$, we are insisting on stronger evidence against the null hypothesis, a result that would occur less often than 1 in every 100 times by chance alone when $H_0$ is true.

In Chapter 4, we said that an observed result is "statistically significant" if it would rarely occur by chance alone. When our $P$-value is less than the chosen $\alpha$ in a significance test, we say that the result is **statistically significant at level $\alpha$.**

**DEFINITION: Statistically significant at level $\alpha$**

If the $P$-value is smaller than alpha, we say that the results of a study are **statistically significant at level $\alpha$.** In that case, we reject the null hypothesis $H_0$ and conclude that there is convincing evidence in favor of the alternative hypothesis $H_a$.

*"Significant" in the statistical sense does not necessarily mean "important."* It means simply "not likely to happen just by chance." The significance level $\alpha$ makes "not likely" more exact.

### AP® Exam Common Error

Instead of failing to reject a null hypothesis, many students "accept the null hypothesis" or use equivalent language such as "we can conclude that teens are getting enough calcium." Accepting the null hypothesis will always lose credit on the AP® exam. Remind students that when we do a significance test, we have *some* evidence that the null hypothesis is true. If the $P$-value is large, it simply means that the evidence we have is not convincing enough to rule out chance. It does *not* mean that there is convincing evidence that the null hypothesis is true.

### ALTERNATE EXAMPLE

#### *Kissing and golf clubs*

In the kissing example, the estimated $P$-value was 0.002. Because the $P$-value is small, we reject the null hypothesis that $p = 0.5$. We have convincing evidence that couples prefer to kiss the right way.

In the 7-iron example, the estimated $P$-value was 0.25. Because the $P$-value is large, we fail to reject the null hypothesis that $\sigma = 15$. We do not have convincing evidence that Mike is more consistent when using the new 7-iron.

### Teaching Tip

We strongly recommend that you require students to formulate their conclusions as shown on this page, just below the example. If you restrict them to these two formulations, your students will be much less likely to "accept the null hypothesis" or make other errors in their conclusions.

### Teaching Tip

Remind your students that the word "significant" has a specific meaning in statistics, so they should not use it in the more casual sense as a synonym for "important." When readers are scoring AP® exams, they will assume that a student who uses this word is using it in the statistical sense. Likewise, students should assume that when a question uses the word "significant," it means "statistically significant."

Significance at level 0.01 is often expressed by the statement "The results were significant ($P < 0.01$)." Here, $P$ stands for the $P$-value. The actual $P$-value is more informative than a statement of significance because it allows us to assess significance at any level we choose. For example, a result with $P = 0.03$ is significant at the $\alpha = 0.05$ level but is not significant at the $\alpha = 0.01$ level. When we use a fixed significance level to draw a conclusion in a statistical test,

$P$-value $< \alpha \rightarrow$ reject $H_0 \rightarrow$ convincing evidence for $H_a$ (in context)

$P$-value $\geq \alpha \rightarrow$ fail to reject $H_0 \rightarrow$ not convincing evidence for $H_a$ (in context)

## EXAMPLE Better Batteries

### *Statistical significance*

A company has developed a new deluxe AAA battery that is supposed to last longer than its regular AAA battery.[2] However, these new batteries are more expensive to produce, so the company would like to be convinced that they really do last longer. Based on years of experience, the company knows that its regular AAA batteries last for 30 hours of continuous use, on average. The company selects an SRS of 15 new batteries and uses them continuously until they are completely drained. The sample mean lifetime is $\bar{x} = 33.9$ hours. A significance test is performed using the hypotheses

$$H_0: \mu = 30 \text{ hours}$$
$$H_a: \mu > 30 \text{ hours}$$

where $\mu$ is the true mean lifetime of the new deluxe AAA batteries. The resulting $P$-value is 0.0729.

**AP® EXAM TIP** The conclusion to a significance test should always include three components: (1) an explicit comparison of the $P$-value to a stated significance level, (2) a decision about the null hypothesis: reject or fail to reject $H_0$, and (3) a statement in the context of the problem about whether or not there is convincing evidence for $H_a$.

PROBLEM: **What conclusion would you make for each of the following significance levels? Justify your answer.**

(a) $\alpha = 0.10$ (b) $\alpha = 0.05$

SOLUTION:

(a) **Because the $P$-value, 0.0729, is less than $\alpha = 0.10$, we reject $H_0$. We have convincing evidence that the company's deluxe AAA batteries last longer than 30 hours, on average.**

(b) **Because the $P$-value, 0.0729, is greater than $\alpha = 0.05$, we fail to reject $H_0$. We do not have convincing evidence that the company's deluxe AAA batteries last longer than 30 hours, on average.**

**For Practice** *Try Exercise*  13

In practice, the most commonly used significance level is $\alpha = 0.05$. This is mainly due to Sir Ronald A. Fisher, a famous statistician who worked on agricultural experiments in England during the early twentieth century. Fisher was the first to suggest deliberately using random assignment in an experiment. In a paper published in 1926, Fisher wrote that it is convenient to draw the line at

### Teaching Tip

There are two schools of thought about how to use a $P$-value to make a conclusion in a significance test. The "strength of evidence" school doesn't "reject" or "fail to reject" the null hypothesis but instead uses the $P$-value to make statements about how strongly the evidence supports the alternative hypothesis (e.g., "because the $P$-value is so small, we have very convincing evidence to support the alternative hypothesis"). The "fixed significance level" school uses a preset boundary (e.g., $\alpha = 0.05$) to decide whether the null hypothesis should be rejected in favor of the alternative hypothesis. To help students avoid making mistakes and losing credit on the AP® exam, we recommend that you use the fixed significance level approach. We will model this approach in the student text.

### Teaching Tip

In the solution to part (b) of the example on this page, point out to your students that failing to reject the null hypothesis does not mean that we believe the true mean really is 30 hours. Rather, it means that the evidence we had wasn't strong enough to rule out sampling variability as a plausible explanation for why the sample mean was greater than 30 hours.

### ALTERNATE EXAMPLE

#### *Tasty chips*

For his second semester project in AP® Statistics, Zenon decided to investigate whether students at his school prefer name-brand potato chips to generic potato chips. He randomly selected 50 students and had each student try both types of chips, in random order. Overall, 32 of the 50 ($\hat{p} = 0.64$) students preferred the name-brand chips. Zenon performed a significance test using the hypotheses

$$H_0: p = 0.5$$
$$H_a: p > 0.5$$

where $p$ = the true proportion of students at his school who prefer name-brand chips. The resulting $P$-value was 0.0239.

PROBLEM:
What conclusion would you make at each of the following significance levels?
**(a)** $\alpha = 0.05$
**(b)** $\alpha = 0.01$

SOLUTION:
**(a)** Because the $P$-value of 0.0239 is less than $\alpha = 0.05$, we reject $H_0$. There is convincing evidence that students at Zenon's school prefer name-brand chips.
**(b)** Because the $P$-value of 0.0239 is greater than $\alpha = 0.01$, we fail to reject $H_0$. There is not convincing evidence that students at Zenon's school prefer name-brand chips.

about the level at which we can say: "Either there is something in the treatment, or a coincidence has occurred such as does not occur more than once in twenty trials."[3]

Sometimes it may be preferable to choose $\alpha = 0.01$ or $\alpha = 0.10$, for reasons we will discuss shortly. *Warning:* if you are going to draw a conclusion based on a significance level $\alpha$, then $\alpha$ should be stated *before* the data are produced. Otherwise, a deceptive user of statistics might set an $\alpha$ level *after* the data have been analyzed in an attempt to manipulate the conclusion. This is just as inappropriate as choosing an alternative hypothesis to be one-sided in a particular direction *after* looking at the data.

THINK ABOUT IT

**How do you choose a significance level?** The purpose of a significance test is to give a clear statement of the strength of evidence provided by the data against the null hypothesis and in favor of the alternative hypothesis. The *P*-value does this. But how small a *P*-value is convincing evidence against the null hypothesis? This depends mainly on two circumstances:

- *How plausible is $H_0$?* If $H_0$ represents an assumption that the people you must convince have believed for years, strong evidence (a very small *P*-value) will be needed to persuade them.
- *What are the consequences of rejecting $H_0$?* If rejecting $H_0$ in favor of $H_a$ means making an expensive change of some kind, you need strong evidence that the change will be beneficial.

These criteria are a bit subjective. Different people will insist on different levels of significance. Giving the *P*-value allows each of us to decide individually if the evidence is sufficiently strong.

Users of statistics have often emphasized standard significance levels such as 10%, 5%, and 1%. The 5% level, $\alpha = 0.05$, is very common. For example, courts have tended to accept 5% as a standard in discrimination cases.[4]

Beginning users of statistical tests generally find it easier to compare a *P*-value to a significance level than to interpret the *P*-value correctly in context. For that reason, we will include stating a significance level as a required part of every significance test. We'll also ask you to explain what a *P*-value means in a variety of settings.

## Type I and Type II Errors

When we draw a conclusion from a significance test, we hope our conclusion will be correct. But sometimes it will be wrong. There are two types of mistakes we can make. We can reject the null hypothesis when it's actually true, known as a **Type I error**, or we can fail to reject $H_0$ when the alternative hypothesis is true, which is a **Type II error**.

**DEFINITION: Type I error and Type II error**

If we reject $H_0$ when $H_0$ is true, we have committed a **Type I error**.

If we fail to reject $H_0$ when $H_a$ is true, we have committed a **Type II error**.

### Teaching Tip

Roxy Peck suggested the Alternate Activity on this page as a way to get students thinking about when an outcome was too unlikely to be due to chance. To prepare for the Activity, you need to form a single deck consisting entirely of red cards from two identical decks of cards. If you can return the newly formed deck to a sealed box and have a student break the seal to open the box, students will be "hooked" from the outset. The best way to do this is to cut the plastic packaging on the bottom of the box and carefully tape it back together.

Invite students to take part in a challenge: draw one card from the shuffled deck. If the card is black, you will give them a cookie. If not, the card will be returned to the deck and the deck shuffled again for the next player. Allow several students to draw. At what point do students start to suspect that something is amiss? After 1 draw? Probably not. After 2 draws? Probably not. In most classes, by the time the 4th or 5th consecutive red card has been drawn, students are beginning to doubt that they suffer from bad luck. Because the probability of 4 consecutive red cards is 1/16 and the probability of 5 consecutive red cards is 1/32 if the deck is "fair," students naturally gravitate toward the popular 5% significance level.

### ALTERNATE ACTIVITY

***Pick a card!***

*Materials*: Deck of playing cards (provided by your teacher) and cookies or some other prize for the winners.

Your teacher will offer you the chance to compete for a reward. All you have to do is get lucky! The teacher will shuffle a deck of cards and announce the winning color (red or black). Any student who wants to take the challenge may choose one card (without looking!) from the deck. If the card has the winning color, that student wins a prize. The chosen card will be replaced in the deck, and the cards will be shuffled again before the next student plays. Continue playing until all prizes have been awarded or all interested students have played.

### Teaching Tip

Continuing with the analogy of a criminal trial, ask your students to explain how the jury's conclusion might be wrong. Hopefully, they will say that the jury could convict an innocent person (and make a Type I error) or let a guilty person go free (and make a Type II error). You might also ask students which error is worse. The judicial system is set up to make Type I errors less common by requiring strong evidence for a conviction. In other contexts, such as a disease test, a Type II error (false-negative) might be worse.

### AP® Exam Common Error

On the AP® exam, students often reverse Type I and Type II errors. To avoid this mistake, tell students that a Type II error occurs when they fail "II" reject the null hypothesis when they really should.

## Teaching Tip

When you are describing Type II errors, make sure to avoid language that suggests you are accepting the null hypothesis. For example, using the potato context from the example on this page, avoid saying, "We conclude that the true proportion of potatoes with blemishes is 0.08, when in reality the true proportion is greater than 0.08." A better response would be to say that "we don't find convincing evidence that the true proportion of potatoes with blemishes is greater than 0.08, when in reality the true proportion is greater than 0.08."

## Teaching Tip

Point out how the description of a Type I or Type II error follows naturally from the conclusion in a significance test. For example, our conclusion might be that "we have convincing evidence that the proportion of potatoes with blemishes is greater than $p = 0.08$." Then a Type I error would be *"finding convincing evidence that the proportion of potatoes with blemishes is greater than $p = 0.08$*, when the actual proportion is 0.08 or less." In essence, the description of a Type I or Type II error involves stating a conclusion we could make, and then stating that the opposite is actually true.

| | | Truth about the population | |
|---|---|---|---|
| | | $H_0$ true | $H_a$ true |
| **Conclusion based on sample** | Reject $H_0$ | Type I error | Correct conclusion |
| | Fail to reject $H_0$ | Correct conclusion | Type II error |

**FIGURE 9.2** The two types of errors in significance tests.

The possibilities are summarized in Figure 9.2. If $H_0$ is true, our conclusion is correct if we fail to reject $H_0$, but it is a Type I error if we reject $H_0$. If $H_a$ is true, our conclusion is correct if we reject $H_0$, but is a Type II error if we fail to reject $H_0$. Only one error is possible at a time.

It is important to be able to describe Type I and Type II errors in the context of a problem. Considering the consequences of each of these types of error is also important, as the following example shows.

## EXAMPLE Perfect Potatoes

### *Type I and Type II errors*

A potato chip producer and its main supplier agree that each shipment of potatoes must meet certain quality standards. If the producer determines that more than 8% of the potatoes in the shipment have "blemishes," the truck will be sent away to get another load of potatoes from the supplier. Otherwise, the entire truckload will be used to make potato chips. To make the decision, a supervisor will inspect a random sample of potatoes from the shipment. The producer will then perform a significance test using the hypotheses

$$H_0: p = 0.08$$
$$H_a: p > 0.08$$

where $p$ is the actual proportion of potatoes with blemishes in a given truckload.

PROBLEM: Describe a Type I and a Type II error in this setting, and explain the consequences of each.

SOLUTION: A Type I error occurs if we reject $H_0$ when $H_0$ is true. That would happen if the producer finds convincing evidence that the proportion of potatoes with blemishes is greater than 0.08 when the actual proportion is 0.08 (or less). Consequence: The potato-chip producer sends the truckload of acceptable potatoes away, which may result in lost revenue for the supplier. Furthermore, the producer will have to wait for another shipment of potatoes before producing the next batch of potato chips.

A Type II error occurs if we fail to reject $H_0$ when $H_a$ is true. That would happen if the producer does not find convincing evidence that more than 8% of the potatoes in the shipment have blemishes when that is actually the case. Consequence: The producer uses the truckload of potatoes to make potato chips. More chips will be made with blemished potatoes, which may upset customers.

Here's a helpful reminder to keep the two types of errors straight. "Fail to" goes with Type II.

For Practice Try Exercise 21

Which is more serious: a Type I error or a Type II error? That depends on the situation. For the potato-chip producer, a Type II error could result in upset customers, leading to decreased sales. A Type I error, turning away a shipment even though 8% or less of the potatoes have blemishes, may not have much impact if additional shipments of potatoes can be obtained fairly easily. However, the supplier won't be too happy with a Type I error.

## ALTERNATE EXAMPLE

### *Faster fast food?*

The manager of a fast-food restaurant wants to reduce the proportion of drive-through customers who have to wait longer than 2 minutes to receive their food after placing an order. Based on store records, the proportion of customers who had to wait longer than 2 minutes was $p = 0.63$. To reduce this proportion, the manager assigns an additional employee to assist with drive-through orders. During the next month, the manager will collect a random sample of drive-through times and test the following hypotheses:

$$H_0: p = 0.63$$
$$H_a: p < 0.63$$

where $p$ = the true proportion of drive-through customers who have to wait longer than 2 minutes to receive their food.

PROBLEM: Describe a Type I and a Type II error in this setting and explain the consequences of each.

SOLUTION: A Type I error would occur if the manager finds convincing evidence that the true proportion of drive-through customers who have to wait longer than 2 minutes has been reduced, when it really hasn't been reduced. A consequence is that the manager will have to pay unnecessarily for an additional employee. A Type II error would occur if the manager didn't find convincing evidence that the true proportion of customers who had to wait longer than 2 minutes had been reduced, when it really had been reduced. A consequence is that the restaurant wouldn't be serving customers more quickly when it could.

## Teaching Tip

Here's another way to remember which error is which. When the *P*-value is *small*, it is possible that we make a Type I error (I is *smaller* than II). When the *P*-value is *large*, it is possible that we make a Type II error (II is *larger* than I).

## Teaching Tip

When you are discussing Type I and Type II errors, make sure your students don't think these errors are caused by a mistake made by the person conducting the test. Although we can take steps to minimize their probabilities, Type I and Type II errors are ultimately caused by the variability in the data collection process.

## CHECK YOUR UNDERSTANDING

Refer to the "Better Batteries" example on page 546.

1. Describe a Type I error in this setting.
2. Describe a Type II error in this setting.
3. Which type of error is more serious in this case? Justify your answer.

**Error Probabilities** We can assess the performance of a significance test by looking at the probabilities of the two types of error. That's because statistical inference is based on asking, "What would happen if I repeated the data-production process many times?" We cannot (without inspecting the whole truckload) guarantee that good shipments of potatoes will never be sent away and bad shipments will never be used to make chips. But we can think about our chances of making each of these mistakes.

## EXAMPLE Perfect Potatoes

### *Type I error probability*

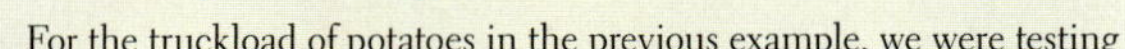

For the truckload of potatoes in the previous example, we were testing

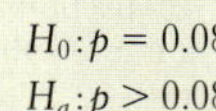

$$H_0: p = 0.08$$
$$H_a: p > 0.08$$

where $p$ is the proportion of all potatoes with blemishes in the shipment. Suppose that the potato-chip producer decides to carry out this test based on a random sample of 500 potatoes using a 5% significance level ($\alpha = 0.05$).

A Type I error is to reject $H_0$ when $H_0$ is actually true. If our sample results in a value of $\hat{p}$ that is much larger than 0.08, we will reject $H_0$. How large would $\hat{p}$ need to be? The 5% significance level tells us to count results that could happen less than 5% of the time by chance if $H_0$ is true as evidence that $H_0$ is false.

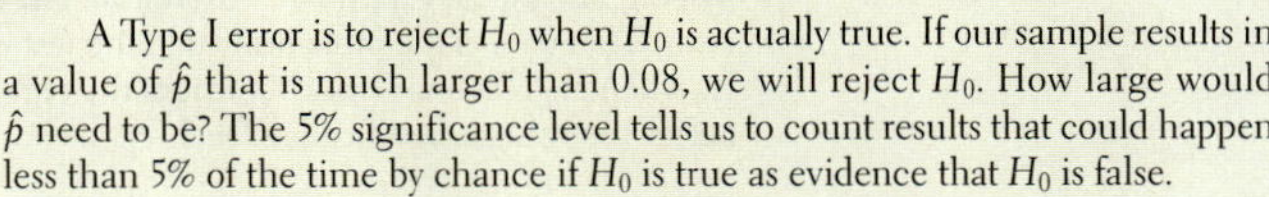

Assuming $H_0: p = 0.08$ is true, the sampling distribution of $\hat{p}$ will have

*Shape:* Approximately Normal because $np = 500(0.08) = 40$ and $n(1-p) = 500(0.92) = 460$ are both at least 10.

*Center:* $\mu_{\hat{p}} = p = 0.08$

*Spread:* $\sigma_{\hat{p}} = \sqrt{\frac{p(1-p)}{n}} = \sqrt{\frac{0.08(0.92)}{500}} = 0.01213$, assuming that there are at least $10(500) = 5000$ potatoes in the shipment.

Figure 9.3 on the next page shows the Normal curve that approximates this sampling distribution.

The shaded area in the right tail of Figure 9.3 is 5%. We used the calculator command `invNorm(area: .95, μ: .08, σ: .01213)` to get the boundary value $\hat{p} = 0.10$. Values of $\hat{p}$ to the right of the green line at $\hat{p} = 0.10$ will cause us to reject $H_0$ even though $H_0$ is true. This will happen in 5% of all possible samples. That is, the probability of making a Type I error is 0.05.

## ANSWERS TO CYU

**1.** Finding convincing evidence that the new batteries last longer than 30 hours, on average, when in reality their true mean lifetime is 30 hours.

**2.** Not finding convincing evidence that the new batteries last longer than 30 hours, on average, when in reality their true mean lifetime > 30 hours.

**3.** Answers will vary. A consequence of a Type I error would be that the company spends the extra money to produce these new batteries when they aren't any better than the older, cheaper type. A consequence of a Type II error would be that the company would not produce the new batteries, even though they were better.

## ALTERNATE EXAMPLE

### *Faster fast food?*

In the fast-food Alternate Example on page 548, we tested the following hypotheses:

$$H_0: p = 0.63$$
$$H_a: p < 0.63$$

where $p$ = the true proportion of drive-through customers who have to wait longer than 2 minutes to receive their food after placing an order. Suppose that the manager decided to carry out this test using a random sample of 250 orders and a significance level of $\alpha = 0.10$. What is the probability of making a Type I error?

To make a Type I error means that we reject $H_0$ when $H_0$ is actually true. In this case, a Type I error occurs when the true proportion of customers who have to wait at least 2 minutes remains $p = 0.63$, but we get a value of $\hat{p}$ small enough that the *P*-value is less than 0.10. When $H_0$ is true, this will happen 10% of the time just by chance. In other words, $P(\text{Type I error}) = \alpha$.

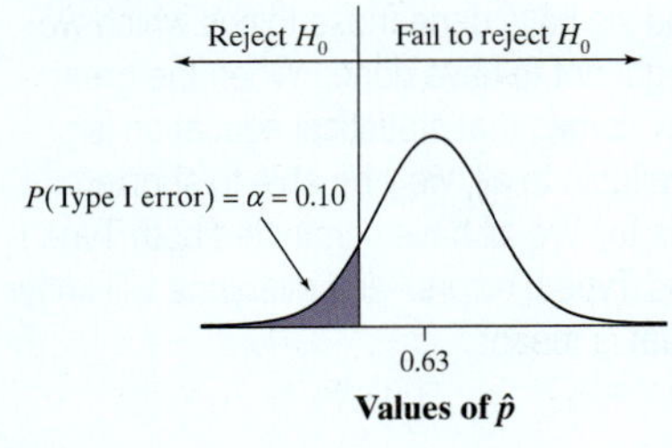

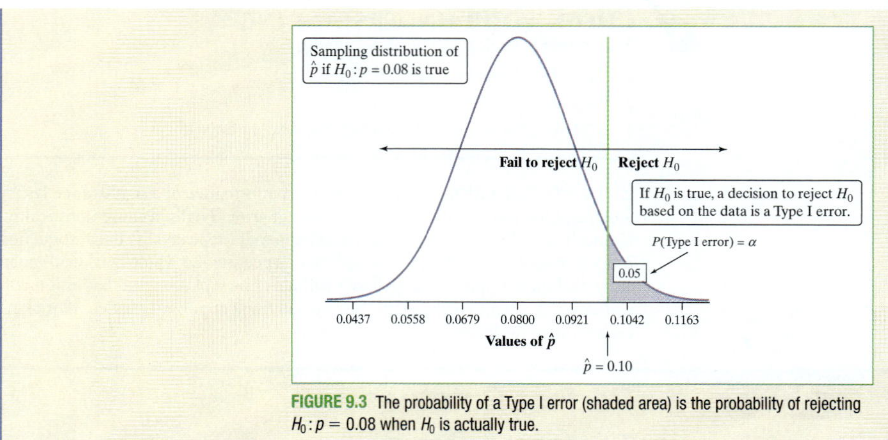


**FIGURE 9.3** The probability of a Type I error (shaded area) is the probability of rejecting $H_0: p = 0.08$ when $H_0$ is actually true.

The probability of a Type I error is the probability of rejecting $H_0$ when it is true. As the previous example showed, this is exactly the significance level of the test.

**SIGNIFICANCE AND TYPE I ERROR**

The significance level $\alpha$ of any fixed-level test is the probability of a Type I error. That is, $\alpha$ is the probability that the test will reject the null hypothesis $H_0$ when $H_0$ is actually true. Consider the consequences of a Type I error before choosing a significance level.

What about Type II errors? We'll discuss them at the end of Section 9.2, after you have learned how to carry out a significance test.

## Section 9.1 Summary

- **A significance test** assesses the evidence provided by data against a **null hypothesis** $H_0$ and in favor of an **alternative hypothesis** $H_a$.
- The hypotheses are usually stated in terms of population parameters. Often, $H_0$ is a statement of no change or no difference. The alternative hypothesis states what we hope or suspect is true.
- A **one-sided alternative** $H_a$ says that a parameter differs from the null hypothesis value in a specific direction. A **two-sided alternative** $H_a$ says that a parameter differs from the null value in either direction.

### Teaching Tip

Remind your students that the probability of a Type I error is a *conditional* probability: $P(\text{Type I error}) = P(\text{reject } H_0 \mid H_0 \text{ is true})$.

### Teaching Tip

If the conditions for inference are not satisfied, the actual Type I error rate might be higher or lower than the significance level. In Chapter 8, we learned the same lesson: when the conditions for inference aren't met, the actual capture rate of a confidence interval may vary from the stated confidence level.

### Teaching Tip

In *How to Tell the Liars from the Statisticians* (Marcel Dekker, New York, 1983, page 68), Robert Hooke writes the following: "Nothing can be farther from statistics then the Book of Common Prayer, wherein it is stated 'We have left undone those things which we ought to have done; and we have done those things which we ought not to have done.' When the great day comes that statistical education is available to all, we'll be able to shorten this to, 'We all have committed both Type I and Type II errors,' and everyone will know what is meant."

### TRM Section 9.1 Quizzes

There are three different quizzes available for this section in the Teacher's Resource Materials. Versions A and B are roughly parallel. Find them by clicking on the link in the TE-book, opening the TRCD, or logging onto the book's companion Web site www.whfreeman.com/tps5e (teacher log-in required).

- The reasoning of a significance test is as follows. Suppose that the null hypothesis is true. If we repeated our data production many times, would we often get data as inconsistent with $H_0$, in the direction specified by $H_a$, as the data we actually have? If the data are unlikely when $H_0$ is true, they provide evidence against $H_0$ and in favor of $H_a$.
- The **P-value** of a test is the probability, computed supposing $H_0$ to be true, that the statistic will take a value at least as extreme as the observed result in the direction specified by $H_a$.
- Small *P*-values indicate strong evidence against $H_0$. To calculate a *P*-value, we must know the sampling distribution of the test statistic when $H_0$ is true.
- If the *P*-value is smaller than a specified value $\alpha$ (called the **significance level**), the data are **statistically significant at level $\alpha$**. In that case, we can **reject** $H_0$ and say that we have convincing evidence for $H_a$. If the *P*-value is greater than or equal to $\alpha$, we **fail to reject** $H_0$ and say that we do *not* have convincing evidence for $H_a$.
- A **Type I error** occurs if we reject $H_0$ when it is in fact true. In other words, the data give convincing evidence for $H_a$ when the null hypothesis is correct. A **Type II error** occurs if we fail to reject $H_0$ when $H_a$ is true. In other words, the data don't give convincing evidence for $H_a$, even though the alternative hypothesis is correct.
- In a fixed level $\alpha$ significance test, the probability of a Type I error is the significance level $\alpha$.

## Section 9.1 Exercises

*In Exercises 1 to 6, each situation calls for a significance test. State the appropriate null hypothesis $H_0$ and alternative hypothesis $H_a$ in each case. Be sure to define your parameter each time.*

pg 540

1. **Attitudes** The Survey of Study Habits and Attitudes (SSHA) is a psychological test that measures students' attitudes toward school and study habits. Scores range from 0 to 200. The mean score for U.S. college students is about 115. A teacher suspects that older students have better attitudes toward school. She gives the SSHA to an SRS of 45 of the over 1000 students at her college who are at least 30 years of age.
2. **Anemia** Hemoglobin is a protein in red blood cells that carries oxygen from the lungs to body tissues. People with less than 12 grams of hemoglobin per deciliter of blood (g/dl) are anemic. A public health official in Jordan suspects that Jordanian children are at risk of anemia. He measures a random sample of 50 children.
3. **Lefties** Simon reads a newspaper report claiming that 12% of all adults in the United States are left-handed. He wonders if the proportion of lefties at his large community college is really 12%. Simon chooses an SRS of 100 students and records whether each student is right- or left-handed.
4. **Don't argue!** A Gallup Poll report revealed that 72% of teens said they seldom or never argue with their friends.[5] Yvonne wonders whether this result holds true in her large high school. So she surveys a random sample of 150 students at her school.
5. **Cold cabin?** During the winter months, the temperatures at the Colorado cabin owned by the Starnes family can stay well below freezing (32°F or 0°C) for weeks at a time. To prevent the pipes from freezing, Mrs. Starnes sets the thermostat at 50°F. The manufacturer claims that the thermostat allows variation in home temperature of $\sigma = 3$°F. Mrs. Starnes suspects that the manufacturer is overstating how well the thermostat works.
6. **Ski jump** When ski jumpers take off, the distance they fly varies considerably depending on their speed, skill, and wind conditions. Event organizers

**TRM Full Solutions to Section 9.1 Exercises**

### Answers to Section 9.1 Exercises

**9.1** $H_0: \mu = 115$; $H_a: \mu > 115$, where $\mu$ is the true mean score on the SSHA for all students at least 30 years of age at the teacher's college.

**9.2** $H_0: \mu = 12$; $H_a: \mu < 12$, where $\mu$ is the true mean amount of hemoglobin in Jordanian children.

**9.3** $H_0: p = 0.12$; $H_a: p \neq 0.12$, where $p$ is the true proportion of lefties at his large community college.

**9.4** $H_0: p = 0.72$; $H_a: p \neq 0.72$, where $p$ is the true proportion of teens in Yvonne's school who rarely or never argue with their friends.

**9.5** $H_0: \sigma = 3$; $H_a: \sigma > 3$, where $\sigma$ is the true standard deviation of the temperature in the cabin.

**9.6** $H_0: \sigma = 10$; $H_a: \sigma > 10$, where $\sigma$ is the true standard deviation of the distance jumped by the ski jumpers.

**9.7** The null hypothesis is always that there is "no difference" or "no change" and the alternative hypothesis is what we suspect is true. Correct: $H_0: p = 0.37$; $H_a: p > 0.37$.

**9.8** Hypotheses are always about population parameters. Also, we are only interested in whether the situation has improved. Correct: $H_0: p = 0.37$; $H_a: p > 0.37$.

**9.9** Hypotheses are always about population parameters. Correct: $H_0: \mu = 1000$ grams; $H_a: \mu < 1000$ grams.

**9.10** The values in both hypotheses must be the same. Also, the null hypothesis should have the equals sign. Correct: $H_0: \mu = 1000$ grams; $H_a: \mu < 1000$ grams.

**9.11** **(a)** The attitudes of older students do not differ from other students, on average. **(b)** Assuming the mean score on the SSHA for students at least 30 years of age at this school is really 115, there is a 0.0101 probability of getting a sample mean of at least 125.7 just by chance in an SRS of 45 older students.

**9.12** **(a)** Jordanian children are not at risk of anemia, on average. **(b)** Assuming the mean hemoglobin level of Jordanian children is really 12, there is a 0.0016 probability of getting a sample mean of 11.3 or smaller just by chance in a random sample of 50 children.

**9.13** $\alpha = 0.10$: Because the $P$-value of $0.2184 > \alpha = 0.10$, we fail to reject $H_0$. We do not have convincing evidence that the proportion of left-handed students at Simon's college is different from the national proportion. $\alpha = 0.05$: Because the $P$-value of $0.2184 > \alpha = 0.05$, we fail to reject $H_0$. We do not have convincing evidence that the proportion of left-handed students at Simon's college is different from the national proportion.

**9.14** $\alpha = 0.05$ : Because the $P$-value of $0.0291 < \alpha = 0.05$, we reject $H_0$. We have convincing evidence that the true proportion of teens in Yvonne's school who rarely or never argue with their friends is different from 0.72. $\alpha = 0.01$ : Because the $P$-value of $0.0291 > \alpha = 0.01$, we fail to reject $H_0$. We do not have convincing evidence that the true proportion of teens in Yvonne's school who rarely or never argue with their friends is different from 0.72.

**9.15** $\alpha = 0.05$: Because the $P$-value of $0.0101 < \alpha = 0.05$, we reject $H_0$. We have convincing evidence that the true mean score on the SSHA for all students at least 30 years of age at the teacher's college $> 115$. $\alpha = 0.01$: Because the $P$-value of 0.0101 $> \alpha = 0.01$, we fail to reject $H_0$. We do not have convincing evidence that the true mean score on the SSHA for all students at least 30 years of age at the teacher's college $> 115$.

**9.16** $\alpha = 0.05$ : Because the $P$-value of $0.0016 < \alpha = 0.05$, we reject $H_0$. We have convincing evidence that the true mean amount of hemoglobin in Jordanian children $< 12$ g/dl. $\alpha = 0.01$: Because the $P$-value of $0.0016 < \alpha = 0.01$, we reject $H_0$. We have convincing evidence that the true mean amount of hemoglobin in Jordanian children $< 12$ g/dl.

**9.17** Either $H_0$ is true or $H_0$ is false—it isn't true some of the time and not true at other times.

**9.18** The interpretation doesn't include a statement that the probability was calculated assuming that the null hypothesis is true. Also, the $P$-value is the probability that the sample proportion is *at least* as far away from 0.72 as the sample result, in either direction.

**9.19** The $P$-value should be compared with a significance level (such as $\alpha = 0.05$), not the hypothesized value of $p$. Also, the data never "prove" that a hypothesis is true, no matter how large or small the $P$-value.

**9.20** It is never correct to "accept the null hypothesis." If the $P$-value is large, the data do not provide convincing evidence that the alternative hypothesis is true. However, lacking evidence for the alternative hypothesis does not provide convincing evidence that the null hypothesis is true.



must position the landing area to allow for differences in the distances that the athletes fly. For a particular competition, the organizers estimate that the variation in distance flown by the athletes will be $\sigma = 10$ meters. An experienced jumper thinks that the organizers are underestimating the variation.

*In Exercises 7 to 10, explain what's wrong with the stated hypotheses. Then give correct hypotheses.*

7. **Better parking** A change is made that should improve student satisfaction with the parking situation at a local high school. Right now, 37% of students approve of the parking that's provided. The null hypothesis $H_0: p > 0.37$ is tested against the alternative $H_a: p = 0.37$.

8. **Better parking** A change is made that should improve student satisfaction with the parking situation at your school. Right now, 37% of students approve of the parking that's provided. The null hypothesis $H_0: \hat{p} = 0.37$ is tested against the alternative $H_a: \hat{p} \neq 0.37$.

9. **Birth weights** In planning a study of the birth weights of babies whose mothers did not see a doctor before delivery, a researcher states the hypotheses as

$$H_0: \bar{x} = 1000 \text{ grams}$$
$$H_a: \bar{x} < 1000 \text{ grams}$$

10. **Birth weights** In planning a study of the birth weights of babies whose mothers did not see a doctor before delivery, a researcher states the hypotheses as

$$H_0: \mu < 1000 \text{ grams}$$
$$H_a: \mu = 900 \text{ grams}$$

11. **Attitudes** In the study of older students' attitudes from Exercise 1, the sample mean SSHA score was 125.7 and the sample standard deviation was 29.8. A significance test yields a $P$-value of 0.0101.
(a) Explain what it would mean for the null hypothesis to be true in this setting.
(b) Interpret the $P$-value in context.

12. **Anemia** For the study of Jordanian children in Exercise 2, the sample mean hemoglobin level was 11.3 g/dl and the sample standard deviation was 1.6 g/dl. A significance test yields a $P$-value of 0.0016.
(a) Explain what it would mean for the null hypothesis to be true in this setting.
(b) Interpret the $P$-value in context.

13. **Lefties** Refer to Exercise 3. In Simon's SRS, 16 of the students were left-handed. A significance test yields a $P$-value of 0.2184. What conclusion would you make if $\alpha = 0.10$? If $\alpha = 0.05$? Justify your answers.

14. **Don't argue!** Refer to Exercise 4. For Yvonne's survey, 96 students in the sample said they rarely or never argue with friends. A significance test yields a $P$-value of 0.0291. What conclusion would you make if $\alpha = 0.05$? If $\alpha = 0.01$? Justify your answers.

15. **Attitudes** Refer to Exercise 11. What conclusion would you make if $\alpha = 0.05$? If $\alpha = 0.01$? Justify your answers.

16. **Anemia** Refer to Exercise 12. What conclusion would you make if $\alpha = 0.05$? If $\alpha = 0.01$? Justify your answers.

17. **Interpreting a P-value** When asked to explain the meaning of the $P$-value in Exercise 13, a student says, "This means there is about a 22% chance that the null hypothesis is true." Explain why the student's explanation is wrong.

18. **Interpreting a P-value** When asked to explain the meaning of the $P$-value in Exercise 14, a student says, "There is a 0.0291 probability of getting a sample proportion of $\hat{p} = 96/150 = 0.64$ by chance alone." Explain why the student's explanation is wrong.

19. **Drawing conclusions** A student performs a test of $H_0: p = 0.75$ versus $H_a: p > 0.75$ and gets a $P$-value of 0.99. The student writes: "Because the $P$-value is greater than 0.75, we reject $H_0$. The data prove that $H_a$ is true." Explain what is wrong with this conclusion.

20. **Drawing conclusions** A student performs a test of $H_0: p = 0.5$ versus $H_a: p \neq 0.5$ and gets a $P$-value of 0.63. The student writes: "Because the $P$-value is greater than $\alpha = 0.05$, we accept $H_0$. The data provide convincing evidence that the null hypothesis is true." Explain what is wrong with this conclusion.

*Exercises 21 and 22 refer to the following setting.* Slow response times by paramedics, firefighters, and policemen can have serious consequences for accident victims. In the case of life-threatening injuries, victims generally need medical attention within 8 minutes of the accident. Several cities have begun to monitor emergency response times. In one such city, the mean response time to all accidents involving life-threatening injuries last year was $\mu = 6.7$ minutes. Emergency personnel arrived within 8 minutes on 78% of all calls involving life-threatening injuries last year. The city manager shares this information and encourages these first responders to "do better." At the end of the year, the city manager selects an SRS of 400 calls involving life-threatening injuries and examines the response times.

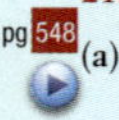

21. **Awful accidents**
(a) State hypotheses for a significance test to determine whether the average response time has decreased. Be sure to define the parameter of interest.

(b) Describe a Type I error and a Type II error in this setting, and explain the consequences of each.

(c) Which is more serious in this setting: a Type I error or a Type II error? Justify your answer.

22. **Awful accidents**

(a) State hypotheses for a significance test to determine whether first responders are arriving within 8 minutes of the call more often. Be sure to define the parameter of interest.

(b) Describe a Type I error and a Type II error in this setting and explain the consequences of each.

(c) Which is more serious in this setting: a Type I error or a Type II error? Justify your answer.

23. **Opening a restaurant** You are thinking about opening a restaurant and are searching for a good location. From research you have done, you know that the mean income of those living near the restaurant must be over \$85,000 to support the type of upscale restaurant you wish to open. You decide to take a simple random sample of 50 people living near one potential location. Based on the mean income of this sample, you will decide whether to open a restaurant there.[6]

(a) State appropriate null and alternative hypotheses. Be sure to define your parameter.

(b) Describe a Type I and a Type II error, and explain the consequences of each.

24. **Blood pressure screening** Your company markets a computerized device for detecting high blood pressure. The device measures an individual's blood pressure once per hour at a randomly selected time throughout a 12-hour period. Then it calculates the mean systolic (top number) pressure for the sample of measurements. Based on the sample results, the device determines whether there is convincing evidence that the individual's actual mean systolic pressure is greater than 130. If so, it recommends that the person seek medical attention.

(a) State appropriate null and alternative hypotheses in this setting. Be sure to define your parameter.

(b) Describe a Type I and a Type II error, and explain the consequences of each.

***Multiple choice: Select the best answer for Exercises 25 to 28.***

25. Experiments on learning in animals sometimes measure how long it takes mice to find their way through a maze. The mean time is 18 seconds for one particular maze. A researcher thinks that a loud noise will cause the mice to complete the maze faster. She measures how long each of 10 mice takes with a noise as stimulus. The appropriate hypotheses for the significance test are

(a) $H_0: \mu = 18$; $H_a: \mu \neq 18$.

(b) $H_0: \mu = 18$; $H_a: \mu > 18$.

(c) $H_0: \mu < 18$; $H_a: \mu = 18$.

(d) $H_0: \mu = 18$; $H_a: \mu < 18$.

(e) $H_0: \bar{x} = 18$; $H_a: \bar{x} < 18$.

*Exercises 26–28 refer to the following setting.* Members of the city council want to know if a majority of city residents supports a 1% increase in the sales tax to fund road repairs. To investigate, they survey a random sample of 300 city residents and use the results to test the following hypotheses:

$$H_0: p = 0.50$$
$$H_a: p > 0.50$$

where $p$ is the proportion of all city residents who support a 1% increase in the sales tax to fund road repairs.

26. A Type I error in the context of this study occurs if the city council

(a) finds convincing evidence that a majority of residents supports the tax increase, when in reality there isn't convincing evidence that a majority supports the increase.

(b) finds convincing evidence that a majority of residents supports the tax increase, when in reality at most 50% of city residents support the increase.

(c) finds convincing evidence that a majority of residents supports the tax increase, when in reality more than 50% of city residents do support the increase.

(d) does not find convincing evidence that a majority of residents supports the tax increase, when in reality more than 50% of city residents do support the increase.

(e) does not find convincing evidence that a majority of residents supports the tax increase, when in reality at most 50% of city residents do support the increase.

27. In the sample, $\hat{p} = 158/300 = 0.527$. The resulting *P*-value is 0.18. What is the correct interpretation of this *P*-value?

(a) Only 18% of the city residents support the tax increase.

(b) There is an 18% chance that the majority of residents supports the tax increase.

(c) Assuming that 50% of residents support the tax increase, there is an 18% probability that the sample proportion would be 0.527 or higher by chance alone.

(d) Assuming that more than 50% of residents support the tax increase, there is an 18% probability that the sample proportion would be 0.527 or higher by chance alone.

(e) Assuming that 50% of residents support the tax increase, there is an 18% chance that the null hypothesis is true by chance alone.

**9.21 (a)** $H_0: \mu = 6.7$; $H_a: \mu < 6.7$, where $\mu$ represents the mean response time for all accidents involving life-threatening injuries in the city.
**(b)** I: Finding convincing evidence that the mean response time has decreased when it really hasn't. A consequence is that the city may not investigate other ways to reduce the mean response time and more people could die. II: Not finding convincing evidence that the mean response time has decreased when it really has. A consequence is that the city spends time and money investigating other methods to reduce the mean response time when they aren't necessary.
**(c)** Type I, because people may end up dying as a result.

**9.22 (a)** $H_0: p = 0.78$; $H_a: p > 0.78$, where $p$ represents the proportion of all calls in which first responders arrived within 8 minutes. **(b)** I: Finding convincing evidence that the proportion of all calls in which first responders arrived within 8 minutes had increased when it really hasn't. A consequence is that the city may not investigate other ways to reduce the response time and more people could die. II: Not finding convincing evidence that the proportion of all calls in which first responders arrived within 8 minutes had increased when it really has. A consequence is that the city spends time and money investigating other methods to reduce the response time when they aren't necessary. **(c)** Type I, because people may end up dying as a result.

**9.23 (a)** $H_0: \mu = \$85{,}000$; $H_a: \mu > \$85{,}000$, where $\mu$ = the mean income of all residents near the restaurant.
**(b)** I: Finding convincing evidence that the mean income of all residents near the restaurant exceeds \$85,000 when in reality it does not. The consequence is that you will open your restaurant in a location where the residents will not be able to support it. II: Not finding convincing evidence that the mean income of all residents near the restaurant exceeds \$85,000 when in reality it does. The consequence of this error is that you will not open your restaurant in a location where the residents would have been able to support it and you lose potential income.

**9.24 (a)** $H_0: \mu = 130$; $H_a: \mu > 130$, where $\mu$ represents the true mean systolic blood pressure of the individual.
**(b)** I: Finding convincing evidence that the individual's true mean systolic blood pressure $> 130$ when in reality it isn't. A consequence is that the individual will seek medical attention when it isn't necessary. II: Not finding convincing evidence that the individual's true mean systolic blood pressure $> 130$ when in reality it is. A consequence is that the individual will not seek medical attention when it is needed, increasing the risk of a stroke or some other problem.

**9.25** d

**9.26** b

**9.27** c

9.28 e

**9.29** **(a)** $P(\text{woman}) = 0.4168$, so $(24{,}611) \times (0.4168) = 10{,}258$ degrees were awarded to women. **(b)** No. $P(\text{woman}) = 0.4168$, which is not equal to $P(\text{woman} \mid \text{bachelor's}) = 0.43$. **(c)** $P(\text{at least 1 of the 2 degrees earned by a woman}) = 1 - P(\text{neither degree is earned by a woman})$

$$= 1 - \left(\frac{14{,}353}{24{,}611}\right)\left(\frac{14{,}352}{24{,}610}\right) = 0.6599.$$

**9.30** It should say that other polls would have results within 3 percentage points *of the true proportion,* not "of the results of this survey." Also, it should say "about 19" rather than "at least 19" of the 20 surveys will be within 3 percentage points.

**PD** **Section 9.2 Overview**

Watch the video overview of Section 9.2 (for teachers) by clicking on the link in the TE-book, opening the TRCD, or logging onto the book's companion Web site www.whfreeman.com/tps5e (teacher log-in required).

**TRM** **Section 9.2 Alternate Examples**

You can find the Alternate Examples for this section in Microsoft Word format by clicking on the link in the TE-book, opening the TRCD, or logging onto the book's companion Web site www.whfreeman.com/tps5e (teacher log-in required).



28. Based on the *P*-value in Exercise 27, which of the following would be the most appropriate conclusion?
(a) Because the *P*-value is large, we reject $H_0$. We have convincing evidence that more than 50% of city residents support the tax increase.
(b) Because the *P*-value is large, we fail to reject $H_0$. We have convincing evidence that more than 50% of city residents support the tax increase.
(c) Because the *P*-value is large, we reject $H_0$. We have convincing evidence that at most 50% of city residents support the tax increase.
(d) Because the *P*-value is large, we fail to reject $H_0$. We have convincing evidence that at most 50% of city residents support the tax increase.
(e) Because the *P*-value is large, we fail to reject $H_0$. We do not have convincing evidence that more than 50% of city residents support the tax increase.

29. **Women in math** (5.3) Of the 24,611 degrees in mathematics given by U.S. colleges and universities in a recent year, 70% were bachelor's degrees, 24% were master's degrees, and the rest were doctorates. Moreover, women earned 43% of the bachelor's degrees, 41% of the master's degrees, and 29% of the doctorates.[7]
(a) How many of the mathematics degrees given in this year were earned by women? Justify your answer.
(b) Are the events "degree earned by a woman" and "degree was a bachelor's degree" independent? Justify your answer using appropriate probabilities.
(c) If you choose 2 of the 24,611 mathematics degrees at random, what is the probability that at least 1 of the 2 degrees was earned by a woman? Show your work.

30. **Explaining confidence** (8.2) Here is an explanation from a newspaper concerning one of its opinion polls. Explain what is wrong with the following statement.

*For a poll of 1,600 adults, the variation due to sampling error is no more than three percentage points either way. The error margin is said to be valid at the 95 percent confidence level. This means that, if the same questions were repeated in 20 polls, the results of at least 19 surveys would be within three percentage points of the results of this survey.*

# 9.2 Tests about a Population Proportion

**WHAT YOU WILL LEARN** By the end of the section, you should be able to:

- State and check the Random, 10%, and Large Counts conditions for performing a significance test about a population proportion.
- Perform a significance test about a population proportion.
- Interpret the power of a test and describe what factors affect the power of a test.
- Describe the relationship among the probability of a Type I error (significance level), the probability of a Type II error, and the power of a test.

Confidence intervals and significance tests are based on the sampling distributions of statistics. That is, both use probability to say what would happen if we used the inference method many times. Section 9.1 presented the reasoning of significance tests, including the idea of a *P*-value. In this section, we focus on the details of testing a claim about a population proportion.

## Carrying Out a Significance Test

In Section 9.1, we met a virtual basketball player who claimed to make 80% of his free throws. We thought that he might be exaggerating. In an SRS of 50 shots, the player made only 32. His sample proportion of made free throws was therefore

$$\hat{p} = \frac{32}{50} = 0.64$$

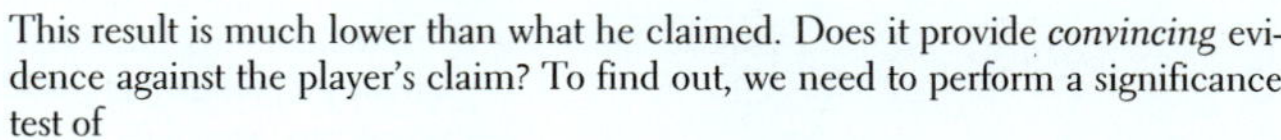

This result is much lower than what he claimed. Does it provide *convincing* evidence against the player's claim? To find out, we need to perform a significance test of

$$H_0: p = 0.80$$
$$H_a: p < 0.80$$

where $p$ = the actual proportion of free throws that the shooter makes in the long run.

**Conditions** In Chapter 8, we introduced conditions that should be met before we construct a confidence interval for an unknown population proportion: Random, 10% when sampling without replacement, and Large Counts. These same conditions must be verified before carrying out a significance test.

The Large Counts condition for proportions requires that both $np$ and $n(1 - p)$ be at least 10. Because we assume $H_0$ is true when performing a significance test, we use the parameter value specified by the null hypothesis (sometimes called $p_0$) when checking the Large Counts condition. In this case, the Large Counts condition says that the expected count of successes $np_0$ and failures $n(1 - p_0)$ are both at least 10.

**CONDITIONS FOR PERFORMING A SIGNIFICANCE TEST ABOUT A PROPORTION**

- **Random:** The data come from a well-designed random sample or randomized experiment.
  - **10%:** When sampling without replacement, check that $n \leq \frac{1}{10}N$.
- **Large Counts:** Both $np_0$ and $n(1 - p_0)$ are at least 10.

Here's an example that shows how to check the conditions.

**EXAMPLE**

## I'm a Great Free-Throw Shooter!

### *Checking conditions*

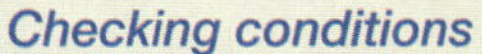

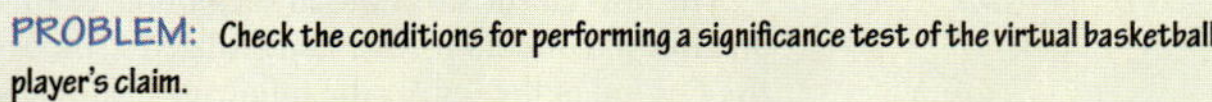

PROBLEM: Check the conditions for performing a significance test of the virtual basketball player's claim.

SOLUTION: The required conditions are

- *Random:* We can view this set of 50 computer-generated shots as a simple random sample from the population of all possible shots that the virtual shooter takes.
  - 10%: We're not sampling without replacement from a finite population (because the applet can keep on shooting), so we don't need to check the 10% condition. (Note that the outcomes of individual shots are independent because they are determined by the computer's random number generator.)
- *Large Counts:* Assuming $H_0$ is true, $p = 0.80$. Then $np_0 = (50)(0.80) = 40$ and $n(1 - p_0) = (50)(0.20) = 10$ are both at least 10, so this condition is met.

**For Practice** *Try Exercise* **31**

## Teaching Tip

Emphasize the difference between evidence and *convincing* evidence. When the basketball player made only 64% of his free throws, we have some evidence that his true success rate is less than 80%. To determine if this evidence is convincing, we must do a significance test to account for the possibility that he really is an 80% shooter and his poor performance was due to chance alone.

## Teaching Tip

Emphasize that the Random, 10%, and Large Counts conditions are essentially the same for significance tests and confidence intervals. For inference about a population proportion, the only difference occurs with the Large Counts condition. For confidence intervals, we check $n\hat{p} \geq 10$ and $n(1 - \hat{p}) \geq 10$ using $\hat{p}$ rather than $p$, because we don't know the value of $p$. For significance tests, we assume the hypothesized value $p_0$ is correct when doing calculations, so we check $np_0 \geq 10$ and $n(1 - p_0) \geq 10$ instead. This also means that all three conditions can be addressed during the planning stages—before the data are collected.

**ALTERNATE EXAMPLE**

### *Kissing the right way*

PROBLEM: Check the conditions for performing a significance test using the kissing data from the previous Alternate Examples.

SOLUTION:

The required conditions are:

- *Random*: The 124 kissing couples were randomly selected.
  - *10%*: 124 is less than 10% of all kissing couples in the population.
- *Large Counts*: Assuming $H_0$ is true, $p_0 = 0.5$. Thus, $np_0 = 124(0.5) = 62 \geq 10$ and $n(1 - p_0) = 124(0.5) = 62 \geq 10$.

## Teaching Tip

Tell your students that Figure 9.4 on this page is a model for the simulated sampling distribution of $\hat{p}$ in Figure 9.1 on page 542. That is, it shows the possible values of $\hat{p}$ and how often they will occur, assuming the shooter makes 80% of his shots in the long run. It's never wrong to use a simulation (with a large number of trials) to estimate a *P*-value, but it's more traditional and often easier to estimate a *P*-value using a probability distribution (e.g., Normal) that models the sampling distribution of the statistic.

## Teaching Tip

In the second bullet, it says that only values "in the direction specified by the alternative hypothesis" give evidence against the null hypothesis. Ask your students to think about why. If they have trouble, tell them that we wouldn't bother doing a significance test if the basketball shooter made more than 80% of his shots in the 50 attempts. If $\hat{p} > 0.80$, then there is *no* evidence that the alternative hypothesis is true. A significance test is designed to see whether the initial evidence in favor of the alternative is convincing or not. The Think About It feature on page 560 addresses this issue more completely.

### AP® Exam Formula Sheet

The formula for a standardized test statistic is included on the formula sheet that is given to students on both parts of the AP® exam. The formula sheet also lists the standard deviation of several statistics, including the ones we will use in this chapter, $\sigma_{\hat{p}}$ and $\sigma_{\bar{x}}$.



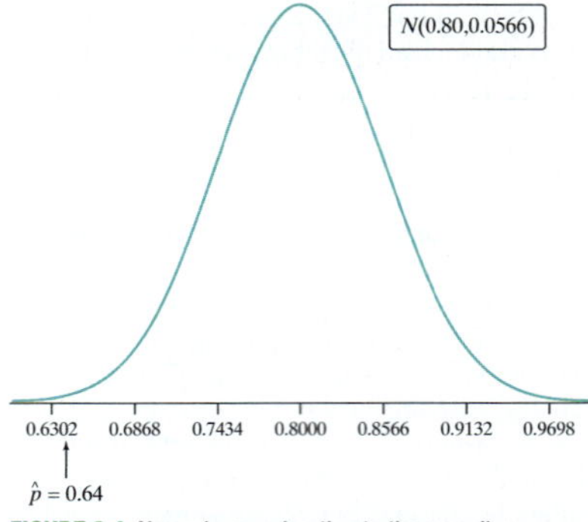


**FIGURE 9.4** Normal approximation to the sampling distribution of the proportion $\hat{p}$ of made shots in random samples of 50 free throws by an 80% shooter.

If the null hypothesis $H_0: p = 0.80$ is true, then the player's sample proportion $\hat{p}$ of made free throws in an SRS of 50 shots would vary according to an approximately Normal sampling distribution with mean $\mu_{\hat{p}} = p = 0.80$ and standard deviation

$$\sigma_{\hat{p}} = \sqrt{\frac{p(1-p)}{n}} = \sqrt{\frac{(0.80)(0.20)}{50}} = 0.0566.$$

Figure 9.4 displays this distribution. We have added the player's sample result, $\hat{p} = \frac{32}{50} = 0.64$.

**Calculations: Test Statistic and *P*-Value** A significance test uses sample data to measure the strength of evidence against $H_0$ and in favor of $H_a$. Here are some principles that apply to most tests:

- The test compares a statistic calculated from sample data with the value of the parameter stated by the null hypothesis.
- Values of the statistic far from the null parameter value in the direction specified by the alternative hypothesis give strong evidence against $H_0$.
- To assess *how far* the statistic is from the parameter, standardize the statistic. This standardized value is called the **test statistic**:

On the AP® exam formula sheet, this value is referred to as the "standardized test statistic."

$$\text{test statistic} = \frac{\text{statistic} - \text{parameter}}{\text{standard deviation of statistic}}$$

**DEFINITION: Test statistic**

A **test statistic** measures how far a sample statistic diverges from what we would expect if the null hypothesis $H_0$ were true, in standardized units. That is,

$$\text{test statistic} = \frac{\text{statistic} - \text{parameter}}{\text{standard deviation of statistic}}$$

The test statistic says how far the sample result is from the null parameter value, and in what direction, on a standardized scale. You can use the test statistic to find the *P*-value of the test, as the following example shows.

## EXAMPLE I'm a Great Free-Throw Shooter!

### *Computing the test statistic*

PROBLEM: In an SRS of 50 free throws, the virtual player made 32.

(a) Calculate the test statistic.

(b) Find the *P*-value using Table A or technology. Show this result as an area under a standard Normal curve.

## ALTERNATE EXAMPLE

### *Kissing the right way*

PROBLEM: In a random sample of 124 couples, 83 were observed tilting to the right when kissing.

**(a)** Calculate the test statistic.

**(b)** Find the *P*-value using Table A or technology. Show this result as an area under a standard Normal curve.

SOLUTION: **(a)** Because the hypothesized value is $p_0 = 0.5$ and $\hat{p} = 83/124 = 0.669$, the standardized test statistic is

$$z = \frac{0.669 - 0.5}{\sqrt{\frac{0.5(1-0.5)}{124}}} = 3.76$$

**(b)** Using Table A, $P(Z \geq 3.76) \approx 0$.

*Using technology*: The command `normalcdf(lower:3.76,upper:1000,μ:0,σ:1)` gives a *P*-value of 0.0001.

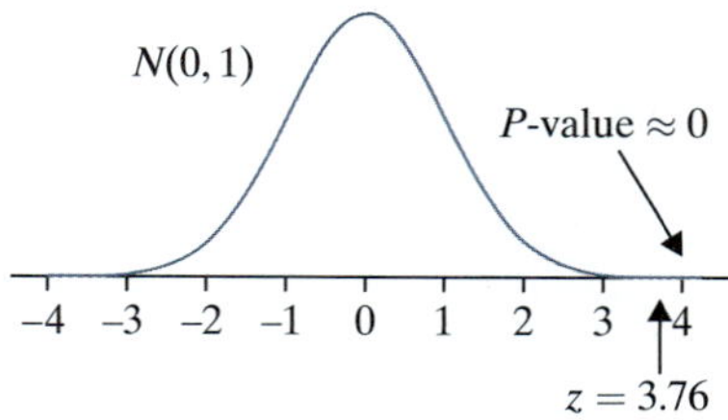


If couples really have no directional preference when kissing, there is a 0.0001 probability of observing at least 83 of 124 couples kissing the right way by chance alone.

SOLUTION:

(a) His sample proportion of made shots is $\hat{p} = 0.64$. Standardizing, we get

$$\text{test statistic} = \frac{\text{statistic} - \text{parameter}}{\text{standard deviation of statistic}}$$

$$z = \frac{0.64 - 0.80}{\sqrt{\frac{(0.80)(0.20)}{50}}} = \frac{0.64 - 0.80}{0.0566} = -2.83$$

(b) The shaded area under the curve in Figure 9.5(a) shows the *P*-value. Figure 9.5(b) shows the corresponding area on the standard Normal curve, which displays the distribution of the *z* test statistic. From Table A, we find that the *P*-value is $P(Z \leq -2.83) = 0.0023$.

*Using technology:* The command `normalcdf(lower: −10000, upper: −2.83, μ: 0, σ: 1)` also gives a *P*-value of 0.0023.

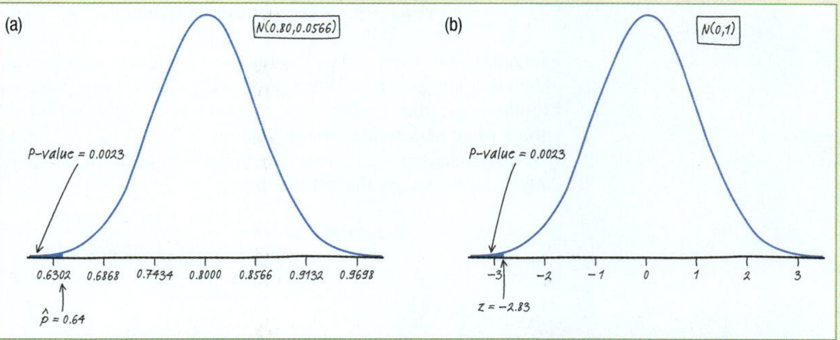


**FIGURE 9.5** The shaded area shows the *P*-value for the player's sample proportion of made shots (a) on the Normal approximation to the sampling distribution of $\hat{p}$ from Figure 9.4 and (b) on the standard Normal curve.

If $H_0$ is true and the player makes 80% of his free throws in the long run, there's only about a 0.0023 probability that he would make 32 or fewer of 50 shots by chance alone.

**For Practice** *Try Exercise* **35**

Earlier, we used simulation to estimate the *P*-value as 0.0075. As the example shows, the *P*-value is even smaller, 0.0023. So if $H_0$ is true, and the player makes 80% of his free throws in the long run, there's only about a 0.0023 probability that the player would make 32 or fewer of 50 shots by chance alone. This result confirms our earlier decision to reject $H_0$ and gives convincing evidence that the player is exaggerating.

## The One-Sample *z* Test for a Proportion

To perform a significance test, we state hypotheses, check conditions, calculate a test statistic and *P*-value, and draw a conclusion in the context of the problem. The four-step process is ideal for organizing our work.

### Teaching Tip

Some students might ask why we need to calculate a *z*-score (test statistic) when we can use the command `normalcdf(lower:-1000,upper:0.64,μ:0.80,σ:0.0566)` to get the *P*-value directly. This is a good question! The best answer is that including the value of the test statistic shows better communication and has been a long-established standard of statistical reporting. A more practical answer is that rubrics for the AP® exam typically require a test statistic in addition to the *P*-value.

### Teaching Tip

In the free-throw shooter example, both the simulated *P*-value (0.0075) and the *P*-value calculated using the Normal distribution (0.0023) are *estimates* of the *P*-value. In this case, the exact *P*-value can be calculated using the binomial distribution. On the TI-84, the command would be `binomcdf(trials:50,p:0.8, x value:32)` = 0.0063.

## Teaching Tip

Remind students that **S**tatistics **P**roblems **D**emand **C**onsistency! The four-step process we describe for significance tests matches up extremely well with the four-part rubrics for these types of questions on the AP® exam.

## Teaching Tip

When performing significance tests, we always do the calculations assuming that the null hypothesis is true. This is equivalent to assuming that the defendant is innocent until proven guilty in a criminal trial. It is also why we typically state the null hypothesis as an equality, not an inequality such as $H_0: p \geq 0.8$. In the latter case, assuming that the null hypothesis is true doesn't tell us what value of $p$ to use in the calculations.

### AP® Exam Formula Sheet

The formula for the test statistic in a one-sample $z$ test for a proportion is *not* included on the formula sheet that is given to students on the AP® exam. Rather, the formula sheet includes the general formula for a test statistic and lists the different standard deviations students might need, including $\sigma_{\hat{p}}$. Students are expected to figure out which standard deviation to use based on the context of the question.

### SIGNIFICANCE TESTS: A FOUR-STEP PROCESS

**State:** What *hypotheses* do you want to test, and at what *significance level*? Define any *parameters* you use.

**Plan:** Choose the appropriate inference *method*. Check *conditions*.

**Do:** If the conditions are met, perform *calculations*.

- Compute the **test statistic**.
- Find the ***P*-value**.

**Conclude:** Make a *decision* about the hypotheses in the context of the problem.

When the conditions are met—Random, 10%, and Large Counts—the sampling distribution of $\hat{p}$ is approximately Normal with

$$\text{mean } \mu_{\hat{p}} = p \quad \text{and} \quad \text{standard deviation } \sigma_{\hat{p}} = \sqrt{\frac{p(1-p)}{n}}$$

For confidence intervals, we substitute $\hat{p}$ for $p$ in the standard deviation formula to obtain the standard error. When performing a significance test, however, the null hypothesis specifies a value for $p$, which we call $p_0$. We assume that this value is correct when performing our calculations.

If we standardize the statistic $\hat{p}$ by subtracting its mean and dividing by its standard deviation, we get the test statistic:

$$\text{test statistic} = \frac{\text{statistic} - \text{parameter}}{\text{standard deviation of statistic}}$$

$$z = \frac{\hat{p} - p_0}{\sqrt{\frac{p_0(1-p_0)}{n}}}$$

This $z$ statistic has approximately the standard Normal distribution when $H_0$ is true and the conditions are met. *P*-values therefore come from the standard Normal distribution.

Here is a summary of the details for a **one-sample $z$ test for a proportion**.

The AP® Statistics course outline calls this test a *large-sample test for a proportion* because it is based on a Normal approximation to the sampling distribution of $\hat{p}$ that becomes more accurate as the sample size increases.

### ONE-SAMPLE $z$ TEST FOR A PROPORTION

Suppose the conditions are met. To test the hypothesis $H_0: p = p_0$, compute the $z$ statistic

$$z = \frac{\hat{p} - p_0}{\sqrt{\frac{p_0(1-p_0)}{n}}}$$

Find the *P*-value by calculating the probability of getting a $z$ statistic this large or larger in the direction specified by the alternative hypothesis $H_a$:

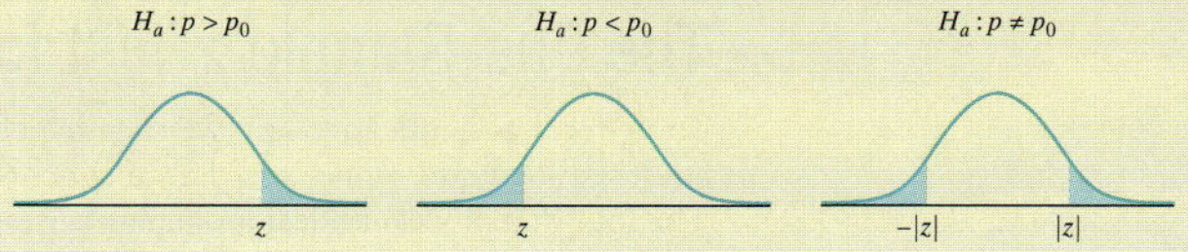

Here is an example of the test in action.

## EXAMPLE One Potato, Two Potato

STEP 4

### Performing a significance test about p

The potato-chip producer of Section 9.1 has just received a truckload of potatoes from its main supplier. Recall that if the producer finds convincing evidence that more than 8% of the potatoes in the shipment have blemishes, the truck will be sent away to get another load from the supplier. A supervisor selects a random sample of 500 potatoes from the truck. An inspection reveals that 47 of the potatoes have blemishes. Carry out a significance test at the $\alpha = 0.05$ significance level. What should the producer conclude?

STATE: We want to perform a test of

$$H_0: p = 0.08$$
$$H_a: p > 0.08$$

where $p$ is the actual proportion of potatoes in this shipment with blemishes. We'll use an $\alpha = 0.05$ significance level.

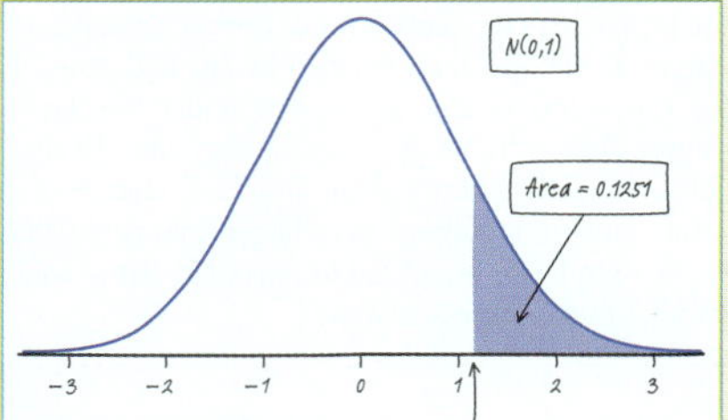


**FIGURE 9.6** The *P*-value for the one-sided test.

PLAN: If conditions are met, we should do a one-sample $z$ test for the population proportion $p$.

- *Random:* The supervisor took a random sample of 500 potatoes from the shipment.
  - *10%:* It seems reasonable to assume that there are at least $10(500) = 5000$ potatoes in the shipment.
- *Large Counts:* Assuming $H_0: p = 0.08$ is true, the expected counts of blemished and unblemished potatoes are $np_0 = 500(0.08) = 40$ and $n(1 - p_0) = 500(0.92) = 460$, respectively. Because both of these values are at least 10, we should be safe doing Normal calculations.

DO: The sample proportion of blemished potatoes is $\hat{p} = 47/500 = 0.094$

- Test statistic $z = \dfrac{\hat{p} - p_0}{\sqrt{\dfrac{p_0(1-p_0)}{n}}} = \dfrac{0.094 - 0.08}{\sqrt{\dfrac{0.08(0.92)}{500}}} = 1.15$
- *P*-value Figure 9.6 displays the *P*-value as an area under the standard Normal curve for this one-sided test. Table A gives the *P*-value as $P(Z \geq 1.15) = 1 - 0.8749 = 0.1251$.
- *Using technology:* The command `normalcdf(lower:1.15, upper:10000, μ:0, σ:1)` also gives a *P*-value of 0.1251.

CONCLUDE: Because our *P*-value, 0.1251, is greater than $\alpha = 0.05$, we fail to reject $H_0$. There is not convincing evidence that the shipment contains more than 8% blemished potatoes. As a result, the producer will use this truckload of potatoes to make potato chips.

**AP® EXAM TIP** When a significance test leads to a fail to reject $H_0$ decision, as in this example, be sure to interpret the results as "we don't have convincing evidence to conclude $H_a$." Saying anything that sounds like you believe $H_0$ is (or might be) true will lead to a loss of credit. And don't write text-message-type responses, like "FTR the $H_0$."

**For Practice** *Try Exercise* 39

## Teaching Tip

After reading the first paragraph of the example on this page, ask students to provide the two possible explanations for why the sample proportion of blemished potatoes ($\hat{p} = 47/500 = 0.094$) is greater than $p = 0.08$. The first explanation is that the overall proportion for the load is 0.08, and we got a sample proportion this high because of sampling variability. The second explanation is that the actual proportion of blemished potatoes *is* greater than 0.08. We always start with the belief that the first explanation is true until we have convincing evidence that it isn't.

## AP® Exam Common Error

Emphasize the AP® Exam Tip on this page! Many students lose credit on the AP® exam for making statements such as "We conclude that the true proportion of potatoes with blemishes is 0.08" or "We conclude that the shipment of potatoes is OK." Both of these statements suggest that there is evidence the null hypothesis is true. Instead, the results of the test only say that there isn't convincing evidence the proportion of blemished potatoes is greater than 0.08.

## ALTERNATE EXAMPLE

### Better to be last?

On TV shows like *American Idol*, contestants often wonder if there is an advantage to performing last. To investigate, researchers selected a random sample of 600 college students and showed each student the audition video of 12 different singers. For each student, the videos were shown in random order. So we would expect approximately 1/12 of the students to prefer the last singer they view, assuming the order doesn't matter. In this study, 59 of the 600 students preferred the last singer they viewed. Do these data provide convincing evidence at the 5% significance level that there is an advantage to going last?

STATE: We want to test the following hypotheses at the $\alpha = 0.05$ significance level:

$$H_0: p = 1/12$$
$$H_a: p > 1/12$$

where $p$ = the true proportion of students who prefer the last singer they see.

PLAN: If conditions are met, we will perform a one-sample $z$ test for $p$:

- *Random*: A random sample of students was selected, and the order in which the videos were viewed was randomized for each subject.
  - *10%*: It is reasonable to assume that there are more than $10(600) = 6000$ students.
- *Large Counts*: $np_0 = 600(1/12) = 50 \geq 10$, $n(1 - p_0) = 600(1 - 1/12) = 550 \geq 10$.

DO: The sample proportion of fans who preferred the last contestant is $\hat{p} = 59/600 = 0.098$.

- *Test statistic*
$$z = \frac{0.098 - 0.083}{\sqrt{\dfrac{0.083(1-0.083)}{600}}} = 1.33$$
- *P-value* $P(Z > 1.33) = 1 - 0.9082 = 0.0918$. *Using technology*: The command `normalcdf(lower:1.33, upper:1000,μ:0,σ:1)` gives a *P*-value of 0.0918.

CONCLUDE: Because the *P*-value of 0.0918 is greater than $\alpha = 0.05$, we fail to reject the null hypothesis. There is not convincing evidence that the true proportion of students who prefer the final singer is greater than $p = 1/12$. That is, there is not convincing evidence of an advantage to performing last.

The preceding example reminds us why significance tests are important. The sample proportion of blemished potatoes was $\hat{p} = 47/500 = 0.094$. This result gave evidence against $H_0$ in favor of $H_a$. To see whether such an outcome is unlikely to occur just by chance when $H_0$ is true, we had to carry out a significance test. The *P*-value told us that a sample proportion this large or larger would occur in about 12.5% of all random samples of 500 potatoes when $H_0$ is true. So we can't rule out sampling variability as a plausible explanation for getting a sample proportion of $\hat{p} = 0.094$.

**What happens when the data don't support $H_a$?** Suppose the supervisor had inspected a random sample of 500 potatoes from the shipment and found 33 with blemishes. This yields a sample proportion of $\hat{p} = 33/500 = 0.066$. Such a sample doesn't give any evidence to support the alternative hypothesis $H_a: p > 0.08$! There's no need to continue with the significance test. The conclusion is clear: we should fail to reject $H_0: p = 0.08$. This truckload of potatoes will be used by the potato-chip producer.

If you weren't paying attention, you might end up carrying out the test. Let's see what would happen. The corresponding test statistic is

$$z = \frac{\hat{p} - p_0}{\sqrt{\dfrac{p_0(1-p_0)}{n}}} = \frac{0.066 - 0.08}{\sqrt{\dfrac{0.08(0.92)}{500}}} = -1.15$$

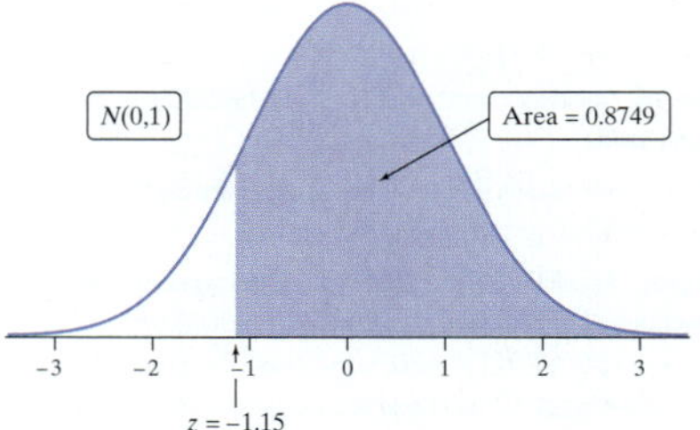


**FIGURE 9.7** The *P*-value for the one-sided test.

What's the *P*-value? It's the probability of getting a *z* statistic this large or larger in the direction specified by $H_a$, $P(Z \geq -1.15)$. Figure 9.7 shows this *P*-value as an area under the standard Normal curve. Using Table A or technology, the *P*-value is $1 - 0.1251 = 0.8749$. There's about an 87.5% chance of getting a sample proportion as large as or larger than $\hat{p} = 0.066$ if $p = 0.08$. As a result, we would fail to reject $H_0$. Same conclusion, but with lots of unnecessary work!

Always check to see whether the data give evidence against $H_0$ in the direction specified by $H_a$ before you do calculations.

## CHECK YOUR UNDERSTANDING

According to the National Campaign to Prevent Teen and Unplanned Pregnancy, 20% of teens aged 13 to 19 say that they have electronically sent or posted sexually suggestive images of themselves.[8] The counselor at a large high school worries that the actual figure might be higher at her school. To find out, she administers an anonymous survey to a random sample of 250 of the school's 2800 students. All 250 respond, and 63 admit to sending or posting sexual images. Carry out a significance test at the $\alpha = 0.05$ significance level. What conclusion should the counselor draw?

Your calculator will handle the "Do" part of the four-step process, as the following Technology Corner illustrates. However, be sure to read the AP® Exam Tip on the next page.

### Teaching Tip

Many older statistics books use a "rejection region" approach to decide what conclusion to make rather than calculating a *P*-value. To use this approach, a critical value is identified for the test statistic, such as $z^* = 1.645$ when $\alpha = 0.05$ and the test is one-sided. If the observed *z* test statistic is greater than $z^* = 1.645$, then the null hypothesis is rejected. We do not recommend this approach, because knowing the *P*-value gives us more information than just whether to reject or fail to reject the null hypothesis.

### Teaching Tip

Make sure to discuss the Think About It on this page with your students. Many students immediately start doing a significance test without thinking about whether it is necessary. For example, if fewer than 1/12 of the fans chose the last contestant in the *Better to be Last?* Alternate Example, we could stop right there because no evidence exists indicating an advantage to going last. We only worry about doing a significance test when there is *some* evidence in favor of the alternative hypothesis.

### ANSWERS TO CYU

S: $H_0: p = 0.20$ versus $H_a: p > 0.20$, where *p* is the true proportion of all teens at the school who would say they have electronically sent or posted sexually suggestive images of themselves.
P: One-sample *z* test for *p*. *Random*: Random sample. *10%*: The sample size (250) < 10% of the 2800 students. *Large Counts*: $250(0.2) = 50 \geq 10$ and $250(0.8) = 200 \geq 10$.
D: $z = \dfrac{0.252 - 0.20}{\sqrt{\dfrac{0.20(0.80)}{250}}} = 2.06$ and $P(Z \geq 2.06) = 0.0197$. C: Because the *P*-value of $0.0197 < \alpha = 0.05$, we reject $H_0$. We have convincing evidence that more than 20% of the teens in her school would say they have electronically sent or posted sexually suggestive images of themselves.

## 18. TECHNOLOGY CORNER — ONE-PROPORTION $z$ TEST ON THE CALCULATOR

TI-Nspire instructions in Appendix B; HP Prime instructions on the book's Web site.

The TI-83/84 and TI-89 can be used to test a claim about a population proportion. We'll demonstrate using the previous example. In a random sample of size $n = 500$, the supervisor found $X = 47$ potatoes with blemishes. To perform a significance test:

**TI-83/84**

- Press STAT, then choose TESTS and `1-PropZTest`.

**TI-89**

- In the Statistics/List Editor, press 2nd F1 ([F6]) and choose `1-PropZTest`.

- On the 1-PropZTest screen, enter the values shown: $p_0 = 0.08$, $x = 47$, and $n = 500$. Specify the alternative hypothesis as "prop $> p_0$." *Note:* $x$ is the number of successes and $n$ is the number of trials. Both must be whole numbers!

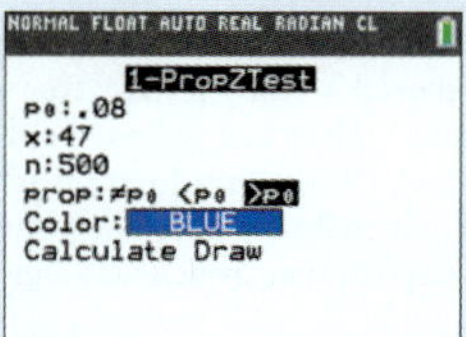


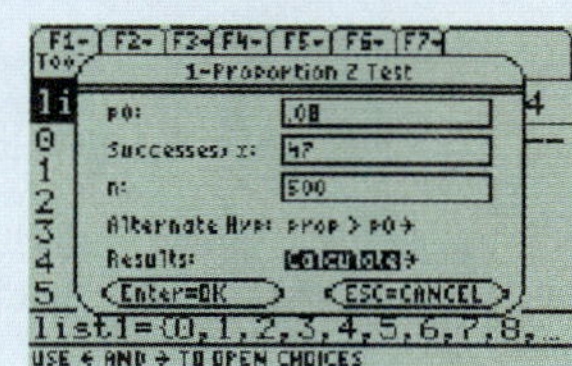


- If you select "Calculate" and press ENTER, you will see that the test statistic is $z = 1.15$ and the $P$-value is 0.1243.

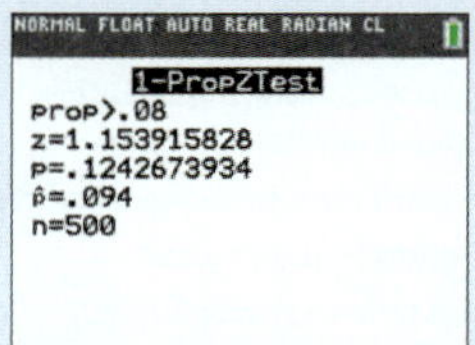


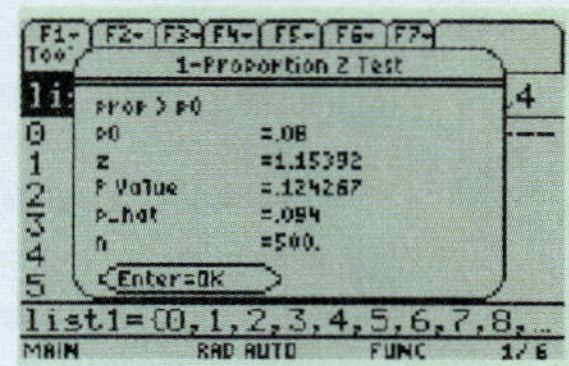


- If you select the "Draw" option, you will see the screen shown here. Compare these results with those in the example on page 559.

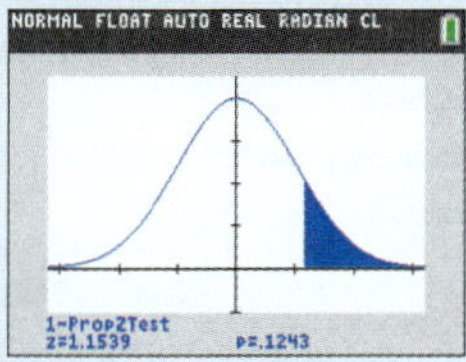

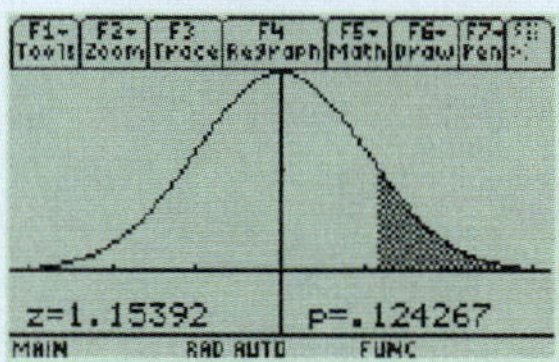


**AP® EXAM TIP** You can use your calculator to carry out the mechanics of a significance test on the AP® exam. But there's a risk involved. If you just give the calculator answer with no work, and one or more of your values is incorrect, you will probably get no credit for the "Do" step. If you opt for the calculator-only method, be sure to name the procedure (one-proportion $z$ test) and to report the test statistic ($z = 1.15$) and $P$-value (0.1243).

### TECHNOLOGY TIP

Remind students that they can watch a video presentation of this technology corner (for the TI-83/84, TI-89, or TI-Nspire) at www.whfreeman.com/tps5e.

### TECHNOLOGY TIP

The calculator abbreviates the one-sample $z$ test for a proportion as "1-PropZTest." Make sure your students don't accidentally choose the $Z$-Test option, which is for testing a population mean when the population standard deviation $\sigma$ is known.

### TECHNOLOGY TIP

Pay special attention to the note in the second bullet. Students often use the value of the sample proportion of successes $\hat{p}$ for $x$ instead of the number of successes.

### Teaching Tip

Even though it is acceptable to use the calculator's 1PropZTest feature on the AP® exam, you can insist that students show additional work in your class. While the formulas are still relatively simple, we recommend this practice. It will help students understand the structure of a standardized test statistic and prepare them for multiple-choice questions that focus on the way test statistics are calculated.

## Teaching Tip

Your students may wonder why we have to find the area in both tails when performing a significance test with a two-sided alternative hypothesis. Explain that when we use a two-sided alternative hypothesis, a sample statistic less than or greater than the hypothesized value of the parameter gives evidence against the null hypothesis. Thus, the *P*-value is the probability of getting a sample statistic at least as extreme (in either direction) as the observed statistic when the null hypothesis is true.

### TECHNOLOGY TIP

When students use the 1Prop*Z*Test feature and choose the two-sided alternative hypothesis, the calculator automatically finds the two-sided *P*-value. Many students incorrectly double the *P*-value provided by the calculator even when they chose the two-sided alternative.

### AP® Exam Common Error

Students often use a one-sided alternative hypothesis when $H_a$ should be two-sided. Remind students to use the wording of the question to determine the alternative hypothesis, not the data! In this case, Taeyeon wonders if the result will hold true at his school. Because there was no direction specified (e.g., Taeyeon suspects the proportion is greater at his school), he should use a two-sided alternative hypothesis.

### AP® Exam Common Error

Many students incorrectly refer to the sample when defining the parameter or stating their conclusion. Notice how we define *p* as the proportion who *would say* they have never smoked rather than the proportion *who said* they have never smoked. The proportion who said they have never smoked is the sample proportion, $\hat{p} = 90/150$.

## Two-Sided Tests

Both the free-throw shooter and blemished-potato examples involved one-sided tests. The *P*-value in a one-sided test is the area in one tail of a standard Normal distribution—the tail specified by $H_a$. In a two-sided test, the alternative hypothesis has the form $H_a: p \neq p_0$. The *P*-value in such a test is the probability of getting a sample proportion as far as or farther from $p_0$ *in either direction* than the observed value of $\hat{p}$. As a result, you have to find the area in both tails of a standard Normal distribution to get the *P*-value. The following example shows how this process works.

### EXAMPLE Nonsmokers

*A two-sided test*

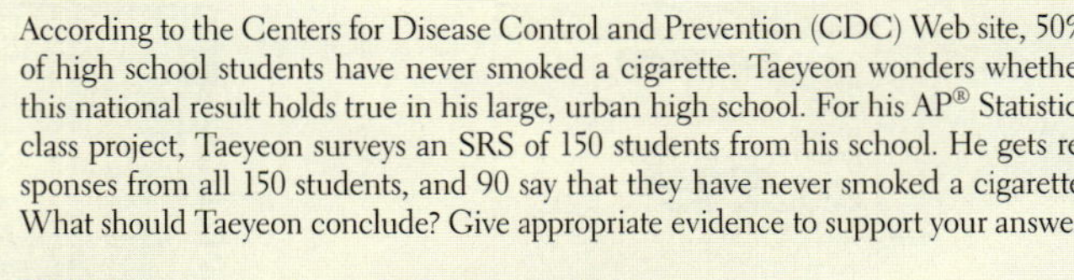

According to the Centers for Disease Control and Prevention (CDC) Web site, 50% of high school students have never smoked a cigarette. Taeyeon wonders whether this national result holds true in his large, urban high school. For his AP® Statistics class project, Taeyeon surveys an SRS of 150 students from his school. He gets responses from all 150 students, and 90 say that they have never smoked a cigarette. What should Taeyeon conclude? Give appropriate evidence to support your answer.

STATE: We want to perform a significance test using the hypotheses

$$H_0: p = 0.50$$
$$H_a: p \neq 0.50$$

where $p$ = the proportion of all students in Taeyeon's school who would say they have never smoked cigarettes. Because no significance level was stated, we will use $\alpha = 0.05$.

PLAN: If conditions are met, we'll do a one-sample $z$ test for the population proportion $p$.

- *Random:* Taeyeon surveyed an SRS of 150 students from his school.
  - *10%:* It seems reasonable to assume that there are at least 10(150) = 1500 students in a large high school.
- *Large Counts:* Assuming $H_0: p = 0.50$ is true, the expected counts of smokers and nonsmokers in the sample are $np_0 = 150(0.50) = 75$ and $n(1 - p_0) = 150(0.50) = 75$. Because both of these values are at least 10, we should be safe doing Normal calculations.

DO: The sample proportion is $\hat{p} = 90/150 = 0.60$.

- *Test statistic*

$$z = \frac{\hat{p} - p_0}{\sqrt{\frac{p_0(1-p_0)}{n}}} = \frac{0.60 - 0.50}{\sqrt{\frac{0.50(0.50)}{150}}} = 2.45$$

- *P-value* Figure 9.8 displays the *P*-value as an area under the standard Normal curve for this two-sided test. To compute this *P*-value, we find the area in one tail and double it. Table A gives $P(z \geq 2.45) = 0.0071$ (the right-tail area). So the desired *P*-value is $2(0.0071) = 0.0142$.

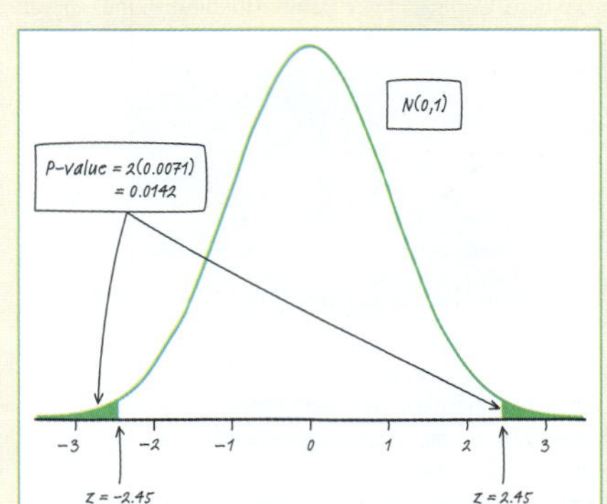


**FIGURE 9.8** The *P*-value for the two-sided test.

### ALTERNATE EXAMPLE

***Benford's law and fraud***

When the accounting firm AJL and Associates audits a company's financial records for fraud, it often uses a test based on Benford's law. Benford's law states that the distribution of first digits in many real-life sources of data is not uniform. In fact, when there is no fraud, about 30.1% of the numbers in financial records begin with the digit 1. If the proportion of first digits that are 1 is significantly different from 0.301 in a random sample of records, AJL and Associates does a much more thorough investigation of the company. Suppose that a random sample of 300 expenses from a company's financial records results in only 68 expenses that begin with the digit 1. Should AJL and Associates do a more thorough investigation of this company? Justify your answer.

STATE: We want to test the following hypotheses at the $\alpha = 0.05$ significance level:

$$H_0: p = 0.301$$
$$H_a: p \neq 0.301$$

where $p$ = the proportion of all expenses that begin with the digit 1 for this company.

PLAN: If conditions are met, we will perform a one-sample $z$ test for $p$.

- *Random*: A random sample of expenses was selected.
  - *10%*: It is reasonable to assume that there are more than 10(300) = 3000 expenses in this company's financial records.
- *Large Counts*: $np_0 = 300(0.301) = 90.3 \geq 10$, $n(1 - p_0) = 300(1 - 0.301) = 209.7 \geq 10$.

DO: The sample proportion of expenses that began with the digit 1 is $\hat{p} = 68/300 = 0.227$.

- *Test statistic*

$$z = \frac{0.227 - 0.301}{\sqrt{\frac{0.301(1-0.301)}{300}}} = -2.79$$

*Using technology:* The calculator's 1-PropZTest gives $z = 2.449$ and *P*-value $= 0.0143$.

CONCLUDE: Because our *P*-value, 0.0143, is less than $\alpha = 0.05$, we reject $H_0$. We have convincing evidence that the proportion of all students at Taeyeon's school who would say they have never smoked differs from the national result of 0.50.

**For Practice** *Try Exercise*  45

## CHECK YOUR UNDERSTANDING

According to the National Institute for Occupational Safety and Health, job stress poses a major threat to the health of workers. A news report claims that 75% of restaurant employees feel that work stress has a negative impact on their personal lives.[9] Managers of a large restaurant chain wonder whether this claim is valid for their employees. A random sample of 100 employees finds that 68 answer "Yes" when asked, "Does work stress have a negative impact on your personal life?" Is this good reason to think that the proportion of all employees in this chain who would say "Yes" differs from 0.75? Support your answer with a significance test.

## Why Confidence Intervals Give More Information

The result of a significance test begins with a decision to reject $H_0$ or fail to reject $H_0$. In Taeyeon's smoking study, for instance, the data led us to reject $H_0: p = 0.50$ because we found convincing evidence that the proportion of students at his school who would say they have never smoked cigarettes differs from the national value. We're left wondering what the actual proportion $p$ might be. A confidence interval might shed some light on this issue.

**EXAMPLE**

### Nonsmokers

*A confidence interval gives more info*

Taeyeon found that 90 of an SRS of 150 students said that they had never smoked a cigarette. We checked the conditions for performing the significance test earlier. Before we construct a confidence interval for the population proportion $p$, we should check that both $n\hat{p}$ and $n(1-\hat{p})$ are at least 10. Because the number of successes and the number of failures in the sample are 90 and 60, respectively, we can proceed with calculations.

Our 95% confidence interval is

$$\hat{p} \pm z^*\sqrt{\frac{\hat{p}(1-\hat{p})}{n}} = 0.60 \pm 1.96\sqrt{\frac{0.60(0.40)}{150}} = 0.60 \pm 0.078 = (0.522, 0.678)$$

We are 95% confident that the interval from 0.522 to 0.678 captures the true proportion of students at Taeyeon's high school who would say that they have never smoked a cigarette.

**ALTERNATE EXAMPLE**

### Benford's law and fraud

PROBLEM:

**(a)** Find and interpret a confidence interval for the proportion of all expenses that begin with the digit 1 for the company in the previous Alternate Example.

**(b)** Does your interval from part (a) lead to the same conclusion as the significance test? Explain.

SOLUTION:

**(a)** STATE: We want to estimate $p$ = the proportion of all expenses that begin with the digit 1 for this company at the 95% confidence level.

PLAN: We will use a one-sample $z$ interval for $p$ if the following conditions are satisfied:

- *Random*: A random sample of expenses was selected.
  - *10%*: It is reasonable to assume that there are more than 10(300) = 3000 expenses in this company's financial records.
- *Large Counts*: $n\hat{p} = 68 \geq 10$ and $n(1-\hat{p}) = 232 \geq 10$.

DO: $\hat{p} \pm z^*\sqrt{\frac{\hat{p}(1-\hat{p})}{n}}$

$= 0.227 \pm 1.96\sqrt{\frac{0.227(1-0.227)}{300}}$

$= 0.227 \pm 0.047 = (0.180, 0.274)$

CONCLUDE: We are 95% confident that the interval from 0.180 to 0.274 captures the proportion of all expenses at this company that begin with the digit 1.

**(b)** Yes. Because 0.301 is not in the interval, it is not a plausible value for the true proportion of expenses that begin with the digit 1. Thus, this company should be investigated for fraud.

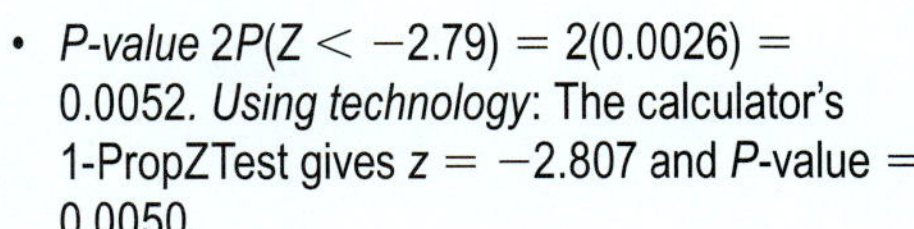

- *P-value* $2P(Z < -2.79) = 2(0.0026) = 0.0052$. *Using technology*: The calculator's 1-PropZTest gives $z = -2.807$ and *P*-value $= 0.0050$.

CONCLUDE: Because the *P*-value of 0.0050 is less than $\alpha = 0.05$, we reject the null hypothesis. There is convincing evidence that the proportion of all expenses that have a first digit of 1 for this company is not 0.301.Therefore, AJL and Associates should do a more thorough investigation of the company.

**ANSWERS TO CYU**

*S*: $H_0: p = 0.75$ versus $H_a: p \neq 0.75$, where $p$ is the true proportion of all restaurant employees at this chain who would say that work stress has a negative impact on their personal lives. *P*: One-sample $z$ test for $p$. *Random*: Random sample. *10%*: The sample size (100) < 10% of all employees. *Large Counts*: $100(0.75) = 75 \geq 10$ and $100(0.25) = 25 \geq 10$.

*D*: $z = \dfrac{0.68 - 0.75}{\sqrt{\dfrac{0.75(0.25)}{100}}} = -1.62$ and $2P(Z \leq -1.62) = 0.1052$. *C*: Because the *P*-value of $0.1052 > \alpha = 0.05$, we fail to reject $H_0$. We do not have convincing evidence that the true proportion of all restaurant employees at this large restaurant chain who would say that work stress has a negative impact on their personal lives is different from 0.75.

The confidence interval in this example is much more informative than the significance test we performed earlier. The interval gives the values of $p$ that are plausible based on the sample data. We would not be surprised if the true proportion of students at Taeyeon's school who would say they have never smoked cigarettes was as low as 0.522 or as high as 0.678. However, we would be surprised if the true proportion was 0.50 because this value is not contained in the confidence interval. Figure 9.9 gives computer output from Minitab software that includes both the results of the significance test and the confidence interval.

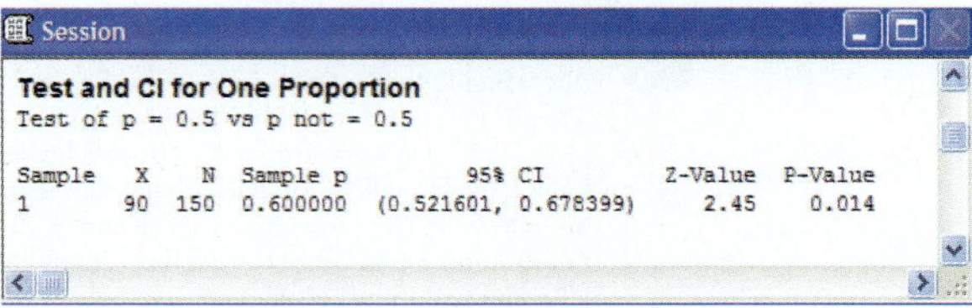
Session

**Test and CI for One Proportion**
Test of p = 0.5 vs p not = 0.5

| Sample | X | N | Sample p | 95% CI | Z-Value | P-Value |
|---|---|---|---|---|---|---|
| 1 | 90 | 150 | 0.600000 | (0.521601, 0.678399) | 2.45 | 0.014 |

**FIGURE 9.9** Minitab output for the two-sided significance test at $\alpha = 0.05$ and a 95% confidence interval for Taeyeon's smoking study.

There is a link between confidence intervals and two-sided tests. The 95% confidence interval (0.522, 0.678) gives an approximate set of $p_0$'s that would not be rejected by a two-sided test at the $\alpha = 0.05$ significance level. With proportions, the link isn't perfect because the standard error used for the confidence interval is based on the sample proportion $\hat{p}$, while the denominator of the test statistic is based on the value $p_0$ from the null hypothesis.

Test statistic: $z = \dfrac{\hat{p} - p_0}{\sqrt{\dfrac{p_0(1-p_0)}{n}}}$ Confidence interval: $\hat{p} \pm z^* \sqrt{\dfrac{\hat{p}(1-\hat{p})}{n}}$

The big idea is still worth considering: a two-sided test at significance level $\alpha$ and a $100(1-\alpha)\%$ confidence interval (a 95% confidence interval if $\alpha = 0.05$) give similar information about the population parameter.

## CHECK YOUR UNDERSTANDING

The figure below shows Minitab output from a significance test and confidence interval for the restaurant worker data in the previous Check Your Understanding (page 563). Explain how the confidence interval is consistent with, but gives more information than, the test.

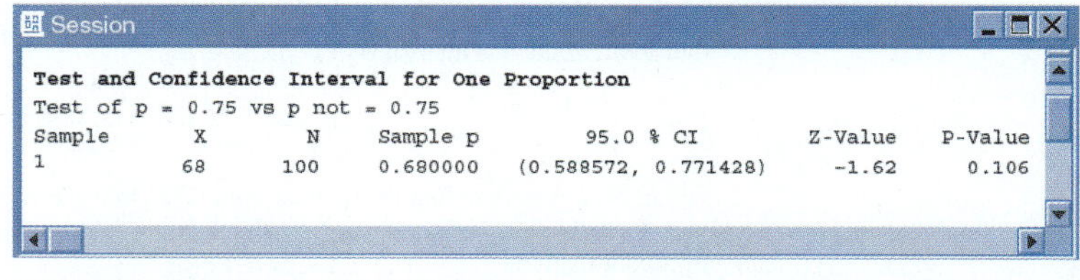
Session

**Test and Confidence Interval for One Proportion**
Test of p = 0.75 vs p not = 0.75

| Sample | X | N | Sample p | 95.0 % CI | Z-Value | P-Value |
|---|---|---|---|---|---|---|
| 1 | 68 | 100 | 0.680000 | (0.588572, 0.771428) | -1.62 | 0.106 |

### AP® Exam Common Error

On the AP® exam, it is acceptable for students to use a confidence interval rather than the test statistic and *P*-value to address a *two*-sided alternative hypothesis. However, if the alternative hypothesis is one-sided, students will lose credit for using a confidence interval approach unless they explicitly address the imperfect link between one-sided tests and confidence intervals (e.g., by adjusting the confidence level appropriately). Our recommendation is to stick with a significance test whenever the alternative hypothesis is one-sided.

### ANSWERS TO CYU

The confidence interval given in the output includes 0.75, which means that 0.75 is a plausible value for the population proportion we are seeking. So both the significance test (which didn't rule out 0.75 as the proportion) and the confidence interval give the same conclusion. The confidence interval, however, gives a range of plausible values for the population proportion instead of only making a decision about a single value.

## Type II Error and the Power of a Test

A significance test makes a Type II error when it fails to reject a null hypothesis $H_0$ that really is false. There are many values of the parameter that make the alternative hypothesis $H_a$ true, so we concentrate on one value. Consider the potato-chip example on page 559 that involved a test of $H_0: p = 0.08$ versus $H_a: p > 0.08$. If the true proportion of blemished potatoes in the shipment was $p = 0.09$, we made a Type II error by failing to reject $H_0$ based on the sample data. Of course, we also made a Type II error if $p = 0.11$ or $p = 0.15$.

The probability of making a Type II error depends on several factors, including the actual value of the parameter. In the potato-chip example, our test will be more likely to reject $H_0: p = 0.08$ in favor of $H_a: p > 0.08$ if the true proportion of blemished potatoes in the shipment is $p = 0.11$ than if it is $p = 0.09$. Why? because $p = 0.11$ is farther away from the null value than is $p = 0.09$. So we will be less likely to make a Type II error if 11% of potatoes in the shipment are blemished than if only 9% are blemished. A high probability of Type II error for a specific alternative parameter value means that the test is not sensitive enough to usually detect that alternative.

It is more common to report the probability that a significance test *does* reject $H_0$ when an alternative parameter value is true. This probability is called the **power** of the test against that specific alternative.

**DEFINITION: Power**

The **power** of a test against a specific alternative is the probability that the test will reject $H_0$ at a chosen significance level $\alpha$ when the specified alternative value of the parameter is true.

As the following example illustrates, Type II error and power are closely linked.

EXAMPLE

### Perfect Potatoes

*Type II error and power*

The potato-chip producer wonders whether the significance test of $H_0: p = 0.08$ versus $H_a: p > 0.08$ based on a random sample of 500 potatoes has enough power to detect a shipment with, say, 11% blemished potatoes. In this case, a particular Type II error is to fail to reject $H_0: p = 0.08$ when $p = 0.11$. Figure 9.10 on the next page shows two sampling distributions of $\hat{p}$, one when $p = 0.08$ and the other when $p = 0.11$.

Earlier, we decided to reject $H_0$ if our sample yielded a value of $\hat{p}$ to the right of the green line at $\hat{p} = 0.10$. That decision was based on using a significance level (Type I error probability) of $\alpha = 0.05$.

Now look at the sampling distribution for $p = 0.11$. The shaded area to the right of the green line represents the probability of correctly rejecting $H_0$ when $p = 0.11$. That is, the *power* of this test to detect $p = 0.11$ is about 0.76. In other words, the potato-chip producer has roughly a 3-in-4 chance of rejecting a truckload with 11% blemished potatoes based on a random sample of 500 potatoes from the shipment.

We would fail to reject $H_0$ if the sample proportion $\hat{p}$ falls to the left of the green line. The white area under the bottom Normal distribution shows the probability

### Teaching Tip

When introducing the concept of power, emphasize that power is the probability of avoiding a Type II error. You might also want to "preview" the factors that affect power in an informal way before reading the potato example.

- Ask your students if they are more likely to avoid making a Type II error if they inspect many potatoes or just a few. Having more data provides better information about the proportion of potatoes with blemishes and helps us avoid a Type II error.
- Ask your students if it is easier to reject the null hypothesis if $\alpha = 0.10$ or 0.01. When $\alpha$ is larger, it is easier to reject $H_0$ because the *P*-value doesn't need to be as large. This makes it easier to avoid committing a Type II error (failing to find convincing evidence that the proportion of potatoes with blemishes is greater than 0.08, when it really is).
- Ask your students if it will be easier to detect a bad shipment if the proportion of blemished potatoes is $p = 0.09$ or $p = 0.90$. If 90% of the potatoes have blemishes, it should be fairly easy to avoid incorrectly accepting the shipment! When the alternative value of the parameter is farther from the hypothesized value, it is easier to avoid a Type II error.

### Teaching Tip

Remind your students that the power of a test against a specific alternative is a *conditional* probability: power = *P*(reject $H_0$ | parameter = some alternative value). In other words, power is the probability that we find convincing evidence the alternative hypothesis is true, *given that* the alternative hypothesis really is true.

### Teaching Tip

Students will not be expected to calculate the probability of a Type II error or power on the AP® exam as we do in this example (and the corresponding Alternate Example on the next page). We do the calculations here to illustrate the concept of power and the factors that affect it. However, students could be asked to estimate the power of a test using the results of a simulation or to identify the power based on a diagram such as Figure 9.10 on the next page.

ALTERNATE EXAMPLE

### *Spinning heads?*

When a fair coin is flipped, we all know that the probability the coin lands on heads is 0.50. However, what if a coin is spun? According to the article "Euro Coin Accused of Unfair Flipping" in the *New Scientist* (January 4, 2002), two Polish math professors and their students spun a Belgian euro coin 250 times. It landed heads 140 times. One of the professors concluded that the coin was minted asymmetrically. A representative from the Belgian mint indicated the result was just chance.

PROBLEM:

**(a)** State the hypotheses we are interested in testing.
**(b)** Assuming the conditions for inference are met, what values of $\hat{p}$ would lead to a rejection of $H_0$ at the 5% significance level? Explain how you obtained your answer.
**(c)** Based on your answer to part (b) and the results of the study, what conclusion would you draw?
**(d)** Suppose that the actual probability of heads when spinning a Belgian euro coin is $p = 0.55$. What is the probability that a Type II error will be committed when using the results of 250 spins? That is, what is the probability that $\hat{p}$ does not fall in the rejection region from part (c) when $p$ really equals 0.55?
**(e)** Based on your answer to part (d), what is the probability that you find convincing evidence that $p \neq 0.5$ when spinning a coin 250 times, assuming that $p$ really equals 0.55?

SOLUTION:

**(a)** $H_0: p = 0.5$ versus $H_a: p \neq 0.5$, where $p$ = the long-run proportion of heads when spinning a Belgian euro coin.
**(b)** The sampling distribution of $\hat{p}$ is approximately Normal with $\mu_{\hat{p}} = 0.5$ and $\sigma_{\hat{p}} = \sqrt{\dfrac{0.5(1-0.5)}{250}} = 0.0316$. Because the alternative hypothesis is two-sided, any values in the outer 2.5% of either tail of the sampling distribution of $\hat{p}$ would lead to a rejection of $H_0$ when $\alpha = 0.05$, as shown in the diagram below.

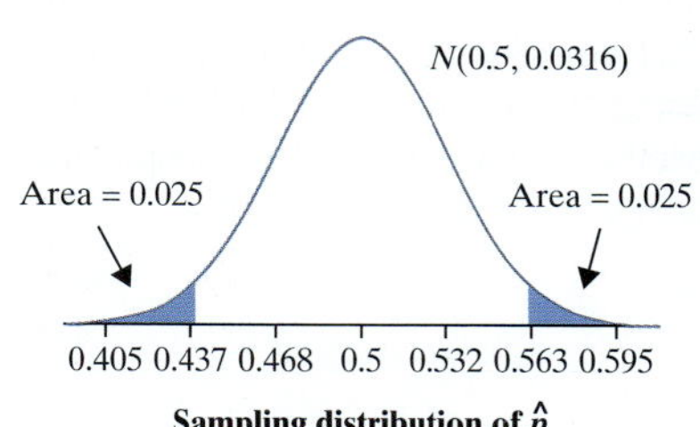


Using Table A, the boundary values are at $z = -1.96$ and $z = 1.96$. Solving $\dfrac{\hat{p}-0.5}{0.0316} < -1.96$ and $1.96 < \dfrac{\hat{p}-0.5}{0.0316}$ indicates we should reject $H_0$ if $\hat{p} < 0.438$ or $\hat{p} > 0.562$.
**(c)** Because $\hat{p} = 140/250 = 0.56$ is not in either rejection region, we fail to reject $H_0$. These data do not provide convincing evidence that the long-run proportion of heads is different than 0.5 when spinning this Belgian euro coin.
**(d)** If $p = 0.55$, then $\mu_{\hat{p}} = 0.55$ and $\sigma_{\hat{p}} = \sqrt{\dfrac{0.55(1-0.55)}{250}} = 0.0315$. Thus, $P(\text{Type II error}) = P(0.438 < \hat{p} < 0.562)$
$= P\left(\dfrac{0.438-0.55}{0.0315} < Z < \dfrac{0.562-0.55}{0.0315}\right)$
$= P(-3.56 < Z < 0.38) = 0.6480 - 0.0002 = 0.6478$. Assuming $p = 0.55$, there is a 0.6478 probability that we do not find convincing evidence that the long-run proportion of heads differs from 0.5.
**(e)** If $p = 0.55$, there is a $1 - 0.6478 = 0.3522$ probability that we find convincing evidence that the long-run proportion of heads differs from 0.50.

of failing to reject $H_0$ when $H_0$ is false. This is the probability of a Type II error. The potato-chip producer has about a 1-in-4 chance of failing to send away a shipment with 11% blemished potatoes.

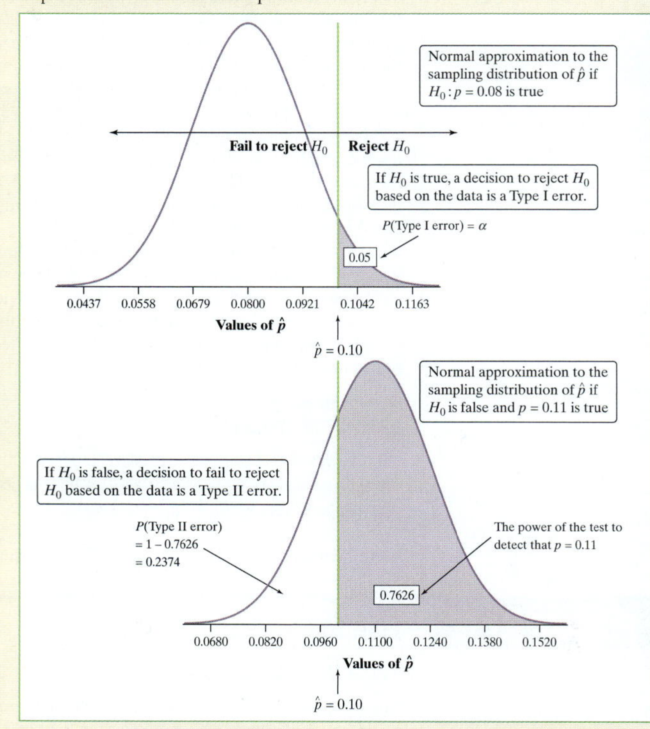


**FIGURE 9.10** In the bottom graph, the power of the test (shaded area) is the probability that it correctly rejects $H_0: p = 0.08$ when the truth is $p = 0.11$. In this case, power = 0.7626. The probability of making a Type II error (white area) is $1 - 0.7626 = 0.2374$.

After reading the example, you might be wondering whether 0.76 is a high power or a low power. That depends on how certain the potato-chip producer wants to be to detect a shipment with 11% blemished potatoes. The power of a test against a specific alternative value of the parameter (like $p = 0.11$) is a number between 0 and 1. A power close to 0 means the test has almost no chance of correctly detecting that $H_0$ is false. A power near 1 means the test is very likely to reject $H_0$ in favor of $H_a$ when $H_0$ is false.

The significance level of a test is the probability of reaching the *wrong* conclusion when the null hypothesis is true. The power of a test to detect a specific alternative is the probability of reaching the *right* conclusion when that alternative is true. We can just as easily describe the test by giving the probability of making a Type II error (sometimes called $\beta$).

### TRM Rolling Sixes

In this problem, students use a binomial distribution to calculate a *P*-value when the Large Counts condition isn't satisfied for a one-sample *z* test. After making their conclusion, students are then asked to

**POWER AND TYPE II ERROR**

The power of a test against any alternative is 1 minus the probability of a Type II error for that alternative; that is, power $= 1 - \beta$.

Calculating a Type II error probability or power by hand is possible but unpleasant. It's better to let technology do the work for you.

## ACTIVITY | What Affects the Power of a Test?

**MATERIALS:**
Computer with Internet access and projection capability

A virtual basketball player claims to make 80% of his free throws. Suppose that the player is exaggerating—he really makes less than 80% in the long run. We have the computer player shoot 50 shots and record the sample proportion $\hat{p}$ of made free throws. We then use the sample result to perform a test of

$$H_0: p = 0.80$$
$$H_a: p < 0.80$$

at the $\alpha = 0.05$ significance level. How can we increase the power of the test to detect that the player is exaggerating? In this Activity, we will use an applet to investigate.

1. Go to www.amstat.org/publications/jse/v11n3/java/Power and select the *Proportions* applet at the bottom of the page.
2. Adjust the applet settings as follows: choose "One Sample" for the Test, "Less Than" for the Alternative, enter 0.8 for p1, 50 for n1, and 0.05 for $\alpha$. The Null distribution should appear in the applet window.

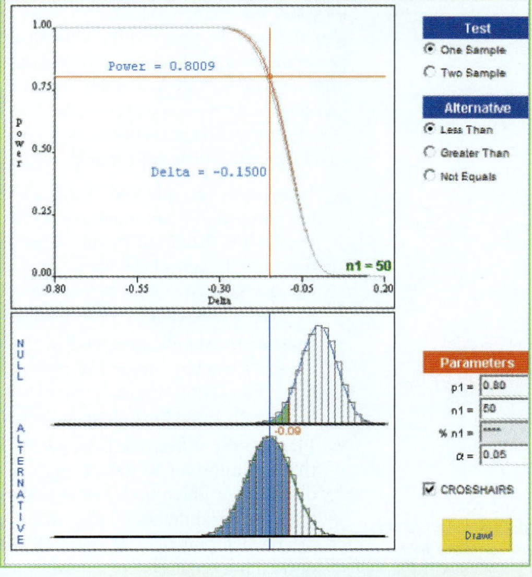


### Activity Overview

You can watch an overview of this Activity (for teachers) by clicking on the link in the TE-book, opening the TRCD, or logging onto the book's companion Web site www.whfreeman.com/tps5e (teacher log-in required).

**ALTERNATE ACTIVITY**

### *Spinning heads*

In the previous Alternate Example, we estimated the power to detect that the long-run proportion of heads is different than $p = 0.5$ when spinning a Belgian euro. If the true long-run proportion is $p = 0.55$, we calculated that the power of the test is approximately 0.35. How can we increase the power of the test? Let's use an applet to investigate.

**1.** Go to http://www.amstat.org/publications/jse/v11n3/java/Power/ and launch the *Proportions* applet at the bottom of the page.

**2.** Leave the test as One Sample and the alternative hypothesis as Not Equals. Enter 0.50 for p1, 250 for n1, and 0.05 for $\alpha$. Press "Draw!" and move the crosshairs to Delta = 0.0500. This makes the alternative value of the parameter = 0.50 + 0.05 = 0.55. The power of the test is 0.3518, matching the value we calculated in the previous Alternate Example.

**3.** How does increasing the sample size affect power? Change n1 to 500 and press "Draw!" How does the power change? What happens to the sampling distributions shown at the bottom when you increase the sample size?

**4.** How does changing the significance level affect power? Change n1 back to 250 and change $\alpha$ to 0.01. Press "Draw!" Now, change $\alpha$ to 0.10 and press "Draw!" How does changing $\alpha$ affect power?

**5.** How does the difference between the alternative value of the parameter and the hypothesized value affect power? Reset $\alpha$ to 0.05 and increase Delta to 0.10 using the crosshairs. This means that the true long-run proportion of heads is 0.50 + 0.10 = 0.60. How does this affect power? Now, move the crosshairs so Delta = 0.01. How does this affect power?

**6.** Summarize what you have learned about the factors that affect the power of a test.

determine which values of the sample statistic would lead to a rejection of $H_0$ and to use this information to calculate the probability of a Type II error. Although using the binomial distribution to calculate a *P*-value or the probability of a Type II error is not required on the AP® exam, having students go through this problem should help them solidify their understanding of the logic of significance tests.

### Teaching Tip

To help illustrate the relationships among power, the probability of a Type I error, and the probability of a Type II error, you may want to draw a new version of Figure 9.2 from page 548, placing "power" in the upper-right-hand cell. Then explain that there are only two possible states of reality: $H_0$ is true or $H_a$ is true. If $H_a$ is true, then the only possible outcomes are listed in the right-hand column. Either we make a correct decision to reject $H_0$ (probability of this outcome = power) or we don't make a correct decision (probability of this outcome = $\beta$). Thus, power $= 1 - \beta$. Notice that $\alpha$ and $\beta$ are *not* complementary, even though they are inversely related.

### Teaching Tip

In the screen shot, the value −0.09 under the first Normal curve is the distance from $p = 0.80$ that $\hat{p}$ needs to be to reject the null hypothesis. In other words, a $\hat{p}$ value less than 0.80 − 0.09 = 0.71 will lead to a rejection of $H_0$ at $\alpha = 0.05$.

3. Let's assume for now that the virtual player really makes 65% of his free throws ($p = 0.65$). Drag your mouse to the left in the applet screen and watch as an Alternative distribution appears. Keep dragging until the Delta value in the top panel shows −0.1500. This sets the alternative parameter value to be 0.15 less than the null parameter value of 0.80. Click your mouse to set the Alternative distribution. The Power of the test to detect $p = 0.65$ is shown in the top panel: 0.8009.

4. *Sample size* Change the sample size from $n = 50$ shots to $n = 100$ shots. What happens in the bottom panel of the applet? Does the power increase or decrease? Explain why this makes sense.

5. *Significance level* Reset the sample size to $n = 50$.

(a) Change the significance level to $\alpha = 0.01$. What happens in the bottom panel of the applet? Does the power increase or decrease?

(b) Make a guess about what will happen if you change the significance level to $\alpha = 0.10$. Use the applet to test your conjecture.

(c) Explain what the results in parts (a) and (b) tell you about the relationship between Type I error probability and Type II error probability.

6. *Difference between null parameter value and alternative parameter value* Reset the sample size to $n = 50$ and the significance level to $\alpha = 0.05$. Will we be more likely to detect that the player is really a 65% shooter or that he is really a 70% shooter? Use your mouse to adjust the location of the Alternative distribution. How does the power change? Explain why this makes sense.

## Teaching Tip

Make sure students understand the relationship between the probability of a Type I error ($\alpha$), the probability of a Type II error ($\beta$), and power ($1 - \beta$). Understanding these relationships will help students choose a significance level when they are performing a significance test. If it is more important to avoid a Type I error, choose a small value of $\alpha$, such as $\alpha = 0.01$. If it is more important to avoid a Type II error, choose a larger value of $\alpha$, such as alpha = 0.10.

Step 5 of the Activity reveals an important link between Type I and Type II error probabilities. Because $P(\text{Type I error}) = \alpha$, increasing the significance level increases the chance of making a Type I error. As the applet shows, this change also increases the power of the test. Because $P(\text{Type II error}) = 1 - \text{Power}$, higher power means a smaller chance of making a Type II error. So increasing the Type I error probability $\alpha$ decreases the Type II error probability $\beta$. By the same logic, decreasing the chance of a Type I error results in a higher chance of a Type II error.

Let's summarize what the Activity reveals about how to increase the power of a significance test to detect when $H_0$ is false and $H_a$ is true.

- **Increase the sample size.** As Step 4 of the Activity confirms, we get better information about the virtual player's free-throw shooting from a random sample of 100 shots than from a random sample of 50 shots. Increasing the sample size decreases the spread of both the Null and Alternative distributions. This change decreases the amount of overlap between the two distributions, making it easier to detect a difference between the null and alternative parameter values.
- **Increase the significance level $\alpha$.** Using a larger value of $\alpha$ increases the area of the green and blue "reject $H_0$" regions in both the Null and Alternative distributions. This change makes it more likely to get a sample proportion that leads us to correctly reject the null hypothesis when the shooter is exaggerating.
- **Increase the difference between the null and alternative parameter values that is important to detect.** Step 6 of the Activity shows that it is easier to detect large differences between the null and alternative parameter values than smaller differences. The size of difference that is important to detect is usually determined by experts in the field, so the statistician usually gets little or no input on this factor.

Many researchers who design statistical studies refer to the difference that's important to detect as the *effect size.*

## Teaching Tip

Increasing the power of a test is generally a good thing. However, there are very real trade-offs. Ask your students for the drawback to increasing the sample size (more time and money) and increasing the significance level (you are more likely to commit a Type I error).

In addition to these three factors, we can also gain power by making wise choices when collecting data. For example, using blocking in an experiment or stratified random sampling can greatly increase the power of a test in some circumstances. Our best advice for maximizing the power of a test is to choose as high an $\alpha$ level (Type I error probability) as you are willing to risk and as large a sample size as you can afford.

### CHECK YOUR UNDERSTANDING

Refer to the Perfect Potatoes example on page 565.

1. Which is more serious for the potato-chip producer in this setting: a Type I error or a Type II error? Based on your answer, would you choose a significance level of $\alpha = 0.01$, 0.05, or 0.10?

2. Tell if each of the following would increase or decrease the power of the test. Justify your answers.

(a) Change the significance level to $\alpha = 0.10$.

(b) Take a random sample of 250 potatoes instead of 500 potatoes.

(c) Insist on being able to detect that $p = 0.10$ instead of $p = 0.11$.

## Section 9.2 Summary

- The conditions for performing a significance test of $H_0: p = p_0$ are:
  - **Random:** The data were produced by a well-designed random sample or randomized experiment.
    - **10%:** When sampling without replacement, check that the population is at least 10 times as large as the sample.
  - **Large Counts:** The sample is large enough to satisfy $np_0 \geq 10$ and $n(1 - p_0) \geq 10$ (that is, the expected counts of successes and failures are both at least 10).
- The **one-sample *z* test for a population proportion** is based on the **test statistic**

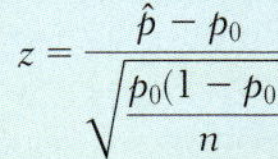

$$z = \frac{\hat{p} - p_0}{\sqrt{\frac{p_0(1 - p_0)}{n}}}$$

with *P*-values calculated from the standard Normal distribution.

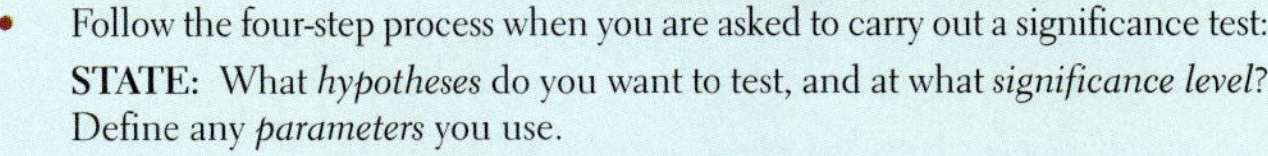

- Follow the four-step process when you are asked to carry out a significance test:

STATE: What *hypotheses* do you want to test, and at what *significance level*? Define any *parameters* you use.

PLAN: Choose the appropriate inference *method*. Check *conditions*.

DO: If the conditions are met, perform *calculations*.

- Compute the **test statistic**.
- Find the ***P*-value**.

CONCLUDE: Make a *decision* about the hypotheses in the context of the problem.

### Teaching Tip

Wisely using blocking in an experiment or stratified random sampling in an observational study or survey helps account for variation in the response variable, reducing the variability in the sampling distribution of the statistic. This has the same effect as increasing the sample size—and it's a lot cheaper!

### ANSWERS TO CYU

**1.** A Type II error. If a Type I error occurred, they would reject a good shipment of potatoes and have to wait to receive a new delivery. However, if a Type II error occurred, they would accept a bad batch and make potato chips with blemishes. This might upset consumers and decrease sales. To minimize the probability of a Type II error, choose a large significance level such as $\alpha = 0.10$.

**2. (a)** Increase. Increasing $\alpha$ to 0.10 makes it easier to reject the null hypothesis, which increases power. **(b)** Decrease. Decreasing the sample size means we don't have as much information to use when making the decision, which makes it less likely to correctly reject $H_0$. **(c)** Decrease. It is harder to detect a difference of 0.02 (0.10 − 0.08) than a difference of 0.03 (0.11 − 0.08).

### TRM Section 9.2 Quizzes

There are three different quizzes available for this section in the Teacher's Resource Materials. Versions A and B are roughly parallel. Find them by clicking on the link in the TE-book, opening the TRCD, or logging onto the book's companion Web site www.whfreeman.com/tps5e (teacher log-in required).

- Confidence intervals provide additional information that significance tests do not—namely, a set of plausible values for the true population parameter $p$. A two-sided test of $H_0: p = p_0$ at significance level $\alpha$ gives roughly the same conclusion as a $100(1 - \alpha)\%$ confidence interval.
- The **power** of a significance test against a specific alternative is the probability that the test will reject $H_0$ when the alternative is true. Power measures the ability of the test to detect an alternative value of the parameter. For a specific alternative, $P(\text{Type II error}) = 1 - \text{power}$.
- There is an important link between the probabilities of Type I and Type II error in a significance test: as one increases, the other decreases.
- We can increase the power of a significance test by increasing the sample size, increasing the significance level, or increasing the difference that is important to detect between the null and alternative parameter values.

## 9.2 TECHNOLOGY CORNER

TI-Nspire Instructions in Appendix B; HP Prime instructions on the book's Web site.



## Section 9.2 Exercises

*In Exercises 31 and 32, check that the conditions for carrying out a one-sample z test for the population proportion p are met.*

**31. Home computers** pg 555 Jason reads a report that says 80% of U.S. high school students have a computer at home. He believes the proportion is smaller than 0.80 at his large rural high school. Jason chooses an SRS of 60 students and records whether they have a computer at home.

**32. Walking to school** A recent report claimed that 13% of students typically walk to school.[10] DeAnna thinks that the proportion is higher than 0.13 at her large elementary school, so she surveys a random sample of 100 students to find out.

*In Exercises 33 and 34, explain why we aren't safe carrying out a one-sample z test for the population proportion p.*

**33. No test** You toss a coin 10 times to perform a test of $H_0: p = 0.5$ that the coin is balanced against $H_a: p \neq 0.5$.

**34. No test** A college president says, "99% of the alumni support my firing of Coach Boggs." You contact an SRS of 200 of the college's 15,000 living alumni to perform a test of $H_0: p = 0.99$ versus $H_a: p < 0.99$.

**35. Home computers** pg 556 Refer to Exercise 31. In Jason's SRS, 41 of the students had a computer at home.

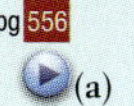

(a) Calculate the test statistic.

(b) Find the $P$-value using Table A or technology. Show this result as an area under a standard Normal curve.

**36. Walking to school** Refer to Exercise 32. For DeAnna's survey, 17 students in the sample said they typically walk to school.

(a) Calculate the test statistic.

(b) Find the $P$-value using Table A or technology. Show this result as an area under a standard Normal curve.

**37. Significance tests** A test of $H_0: p = 0.5$ versus $H_a: p > 0.5$ has test statistic $z = 2.19$.

(a) What conclusion would you draw at the 5% significance level? At the 1% level?

(b) If the alternative hypothesis were $H_a: p \neq 0.5$, what conclusion would you draw at the 5% significance level? At the 1% level?

### TRM Full Solutions to Section 9.2 Exercises

### Answers to Section 9.2 Exercises

**9.31** *Random*: Random sample. *10%*: The sample size (60) < 10% of all students. *Large Counts*: $60(0.80) = 48 \geq 10$ and $60(0.20) = 12 \geq 10$.

**9.32** *Random*: Random sample. *10%*: The sample size (100) < 10% of the students. *Large Counts*: $100(0.13) = 13 \geq 10$ and $100(0.87) = 87 \geq 10$.

**9.33** $np_0 = 10(0.5) = 5$ and $n(1 - p_0) = 10(0.5) = 5$ are both less than 10.

**9.34** $n(1 - p_0) = 200(0.01) = 2 < 10$

**9.35 (a)** $z = \dfrac{0.683 - 0.80}{\sqrt{\dfrac{0.80(0.20)}{60}}} = -2.27$

**(b)** $P(Z \leq -2.27) = 0.0116$. *Using technology*: normalcdf(lower: -1000,upper:-2.27,μ:0,σ:1) = 0.0116. The graph is given below.

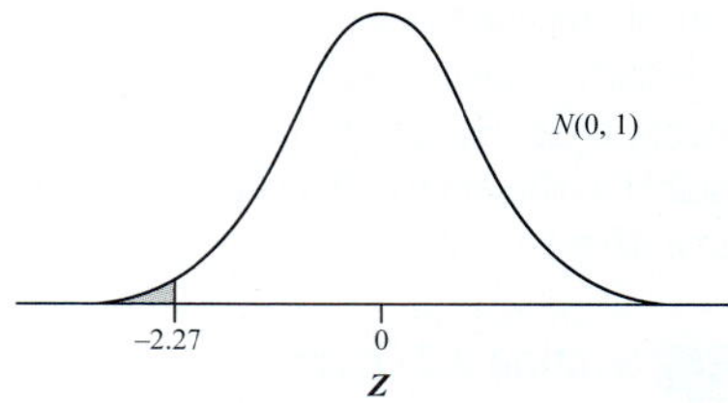


**9.36 (a)** $z = \dfrac{0.17 - 0.13}{\sqrt{\dfrac{0.13(0.87)}{100}}} = 1.19$

**(b)** $P$-value is $P(Z \geq 1.19) = 0.1170$. *Using technology*: normalcdf(lower:1.19, upper:1000,μ:0,σ:1)= 0.1170. The graph is given below.

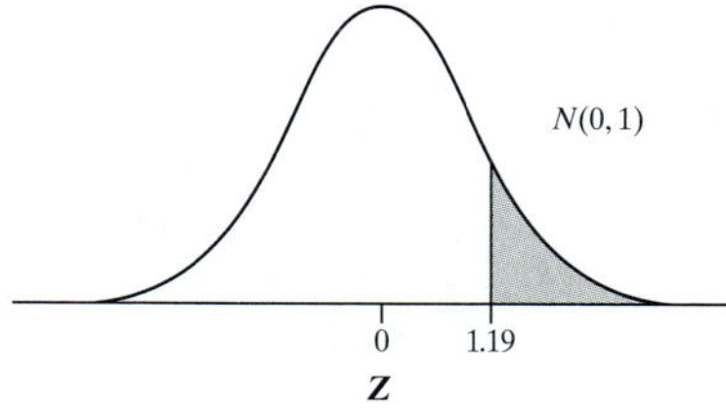


**9.37 (a)** $P$-value = 0.0143. *5%:* Because the $P$-value of $0.0143 < \alpha = 0.05$, we reject $H_0$. There is convincing evidence that $p > 0.5$. *1%*: Because the $P$-value of $0.0143 > \alpha = 0.01$, we fail to reject $H_0$. There is not convincing evidence that $p > 0.5$. **(b)** $P$-value = 0.0286. Because this $P$-value is still less than $\alpha = 0.05$ and greater than $\alpha = 0.01$, we would again reject $H_0$ at the 5% significance level and fail to reject $H_0$ at the 1% significance level.

38. **Significance tests** A test of $H_0: p = 0.65$ against $H_a: p < 0.65$ has test statistic $z = -1.78$.

(a) What conclusion would you draw at the 5% significance level? At the 1% level?

(b) If the alternative hypothesis were $H_a: p \neq 0.65$, what conclusion would you draw at the 5% significance level? At the 1% level?

39. **Better parking** A local high school makes a change that should improve student satisfaction with the parking situation. Before the change, 37% of the school's students approved of the parking that was provided. After the change, the principal surveys an SRS of 200 of the over 2500 students at the school. In all, 83 students say that they approve of the new parking arrangement. The principal cites this as evidence that the change was effective. Perform a test of the principal's claim at the $\alpha = 0.05$ significance level.

pg 559

40. **Side effects** A drug manufacturer claims that less than 10% of patients who take its new drug for treating Alzheimer's disease will experience nausea. To test this claim, researchers conduct an experiment. They give the new drug to a random sample of 300 out of 5000 Alzheimer's patients whose families have given informed consent for the patients to participate in the study. In all, 25 of the subjects experience nausea. Use these data to perform a test of the drug manufacturer's claim at the $\alpha = 0.05$ significance level.

41. **Are boys more likely?** We hear that newborn babies are more likely to be boys than girls. Is this true? A random sample of 25,468 firstborn children included 13,173 boys.[11]

(a) Do these data give convincing evidence that firstborn children are more likely to be boys than girls?

(b) To what population can the results of this study be generalized: all children or all firstborn children? Justify your answer.

42. **Fresh coffee** People of taste are supposed to prefer fresh-brewed coffee to the instant variety. On the other hand, perhaps many coffee drinkers just want their caffeine fix. A skeptic claims that only half of all coffee drinkers prefer fresh-brewed coffee. To test this claim, we ask a random sample of 50 coffee drinkers in a small city to take part in a study. Each person tastes two unmarked cups—one containing instant coffee and one containing fresh-brewed coffee—and says which he or she prefers. We find that 36 of the 50 choose the fresh coffee.

(a) Do these results give convincing evidence that coffee drinkers favor fresh-brewed over instant coffee?

(b) We presented the two cups to each coffee drinker in a random order, so that some people tasted the fresh coffee first, while others drank the instant coffee first. Why do you think we did this?

43. **Bullies in middle school** A University of Illinois study on aggressive behavior surveyed a random sample of 558 middle school students. When asked to describe their behavior in the last 30 days, 445 students said their behavior included physical aggression, social ridicule, teasing, name-calling, and issuing threats. This behavior was not defined as bullying in the questionnaire.[12] Is this evidence that more than three-quarters of middle school students engage in bullying behavior? To find out, Maurice decides to perform a significance test. Unfortunately, he made a few errors along the way. Your job is to spot the mistakes and correct them.

$$H_0: p = 0.75$$
$$H_a: \hat{p} > 0.797$$

where $p$ = the true mean proportion of middle school students who engaged in bullying.

- A random sample of 558 middle school students was surveyed.
- $558(0.797) = 444.73$ is at least 10.

$$z = \frac{0.75 - 0.797}{\sqrt{\frac{0.797(0.203)}{445}}} = -2.46; P\text{-value} = 2(0.0069) = 0.0138$$

The probability that the null hypothesis is true is only 0.0138, so we reject $H_0$. This proves that more than three-quarters of the school engaged in bullying behavior.

44. **Is this coin fair?** The French naturalist Count Buffon (1707–1788) tossed a coin 4040 times. He got 2048 heads. That's a bit more than one-half. Is this evidence that Count Buffon's coin was not balanced? To find out, Luisa decides to perform a significance test. Unfortunately, she made a few errors along the way. Your job is to spot the mistakes and correct them.

$$H_0: \mu > 0.5$$
$$H_a: \bar{x} = 0.5$$

- **10%:** $4040(0.5) = 2020$ and $4040(1 - 0.5) = 2020$ are both at least 10.
- **Large Counts:** There are at least 40,400 coins in the world.

$$t = \frac{0.5 - 0.507}{\sqrt{\frac{0.5(0.5)}{4040}}} = -0.89; P\text{-value} = 1 - 0.1867 = 0.8133$$

Reject $H_0$ because the $P$-value is so large and conclude that the coin is fair.

**9.38 (a)** $P$-value = 0.0375. *5%:* Because the $P$-value of $0.0375 < \alpha = 0.05$, we reject $H_0$. There is convincing evidence that $p < 0.65$. *1%:* Because the $P$-value of $0.0375 > \alpha = 0.01$, we fail to reject $H_0$. There is not convincing evidence that $p < 0.65$. **(b)** $P$-value = 0.0750. Because this $P$-value > both $\alpha = 0.05$ and $\alpha = 0.01$, we would fail to reject $H_0$ at both the 5% and 1% significance levels.

**9.39** *S*: $H_0: p = 0.37$ vs. $H_a: p > 0.37$, where $p$ = true proportion of all students who are satisfied with the parking situation after the change. *P*: One-sample $z$ test for $p$. *Random*: Random sample. *10%*: The sample size (200) < 10% of the population of size 2500. *Large Counts*: $200(0.37) = 74 \geq 10$ and $200(0.63) = 126 \geq 10$. *D*: $z = 1.32$, $P$-value = 0.0934. *C*: Because the $P$-value of $0.0934 > \alpha = 0.05$, we fail to reject $H_0$. We do not have convincing evidence that the true proportion of all students who are satisfied with the parking situation after the change > 0.37.

**9.40** *S*: $H_0: p = 0.10$ vs. $H_a: p < 0.10$, where $p$ is the true proportion of the 5000 Alzheimer's patients who would experience nausea while taking the new drug. *P*: One-sample $z$ test for $p$. *Random*: Random sample. *10%*: The sample size (300) < 10% of the population size (5000). *Large Counts*: $300(0.10) = 30 \geq 10$ and $300(0.90) = 270 \geq 10$. *D*: $z = -0.96$, $P$-value = 0.1685. *C*: Because the $P$-value of $0.1685 > \alpha = 0.05$, we fail to reject $H_0$. We do not have convincing evidence that the true proportion of these 5000 Alzheimer's patients who would experience nausea while taking the new drug < 10%.

**9.41 (a)** *S*: $H_0: p = 0.50$ vs. $H_a: p > 0.50$, where $p$ is the true proportion of boys among first-born children. *P*: One-sample $z$ test for $p$. *Random*: Random sample. *10%*: The sample size (25,468) < 10% of all first-borns. *Large Counts:* $25{,}468(0.50) = 12{,}734 \geq 10$ and $25{,}468(0.50) = 12{,}734 \geq 10$. *D*: $z = 5.49$, $P$-value ≈ 0. *C*: Because the $P$-value of approximately $0 < \alpha = 0.05$, we reject $H_0$. There is convincing evidence that first-born children are more likely to be boys. **(b)** First-born children, because that is the group we sampled from.

**9.42 (a)** *S*: $H_0: p = 0.50$ vs. $H_a: p > 0.50$, where $p$ is the true proportion of coffee drinkers in this small city who prefer fresh-brewed. *P*: One-sample $z$ test for $p$. *Random*: Random sample. *10%*: The sample size (50) < 10% of the population of coffee drinkers. *Large Counts*: $50(0.50) = 25 \geq 10$ and $50(0.50) = 25 \geq 10$. *D*: $z = 3.11$, $P$-value = 0.0009. *C*: Because the $P$-value of $0.0009 < 0.05$, we reject $H_0$. There is convincing evidence that the proportion of coffee drinkers in this small city who prefer fresh-brewed coffee > 0.5. **(b)** If people tend to prefer the most recent sample tasted and we always gave the fresh-brewed coffee second, we wouldn't know if the fresh-brewed was preferred because of its taste or because of the order.

**9.43** Here are the corrections: $H_a: p > 0.75$; $p$ = the true proportion of middle school students who engage in bullying behavior; *10%*: the sample size (558) < 10% of the population of middle school students; $np_0 = 558(0.75) = 418.5 \geq 10$ and $n(1 - p_0) = 558(0.25) = 139.5 \geq 10$;
$$z = \frac{0.7975 - 0.75}{\sqrt{\frac{(0.75)(0.25)}{558}}} = 2.59;$$
$P$-value = 0.0048. Because the $P$-value of $0.0048 < \alpha = 0.05$, we reject $H_0$. We have convincing evidence that more than three-quarters of middle school students engage in bullying behavior.

**9.44** Here are the corrections: $H_0: p = 0.5$ and $H_a: p \neq 0.5$; reverse the 10% and Large Counts labels;
$$z = \frac{0.507 - 0.5}{\sqrt{\frac{0.5(0.5)}{4040}}} = 0.89;$$
$P$-value = 0.3734. Because the $P$-value of $0.3734 > \alpha = 0.05$, we fail to reject $H_0$. There is not convincing evidence that Buffon's coin is not balanced.

45. **Teen drivers** A state's Division of Motor Vehicles (DMV) claims that 60% of teens pass their driving test on the first attempt. An investigative reporter examines an SRS of the DMV records for 125 teens; 86 of them passed the test on their first try. Is there convincing evidence at the $\alpha$ = 0.05 significance level that the DMV's claim is incorrect? pg 562

46. **We want to be rich** In a recent year, 73% of first-year college students responding to a national survey identified "being very well-off financially" as an important personal goal. A state university finds that 132 of an SRS of 200 of its first-year students say that this goal is important. Is there convincing evidence at the $\alpha$ = 0.05 significance level that the proportion of all first-year students at this university who think being very well-off is important differs from the national value, 73%?

47. **Teen drivers** Refer to Exercise 45.
    (a) Construct and interpret a 95% confidence interval for the proportion of all teens in the state who passed their driving test on the first attempt.
    (b) Explain what the interval in part (a) tells you about the DMV's claim.

48. **We want to be rich** Refer to Exercise 46.
    (a) Construct and interpret a 95% confidence interval for the true proportion $p$ of all first-year students at the university who would identify being well-off as an important personal goal.
    (b) Explain what the interval in part (a) tells you about whether the national value holds at this university.

49. **Do you Tweet?** In early 2012, the Pew Internet and American Life Project asked a random sample of U.S. adults, "Do you ever . . . use Twitter or another service to share updates about yourself or to see updates about others?" According to Pew, the resulting 95% confidence interval is (0.123, 0.177).[13] Does this interval provide convincing evidence that the actual proportion of U.S. adults who would say they use Twitter differs from 0.16? Justify your answer.

50. **Losing weight** A Gallup Poll found that 59% of the people in its sample said "Yes" when asked, "Would you like to lose weight?" Gallup announced: "For results based on the total sample of national adults, one can say with 95% confidence that the margin of (sampling) error is ±3 percentage points."[14] Does this interval provide convincing evidence that the actual proportion of U.S. adults who would say they want to lose weight differs from 0.55? Justify your answer.

51. **Teens and sex** The Gallup Youth Survey asked a random sample of U.S. teens aged 13 to 17 whether they thought that young people should wait to have sex until marriage.[15] The Minitab output below shows the results of a significance test and a 95% confidence interval based on the survey data.

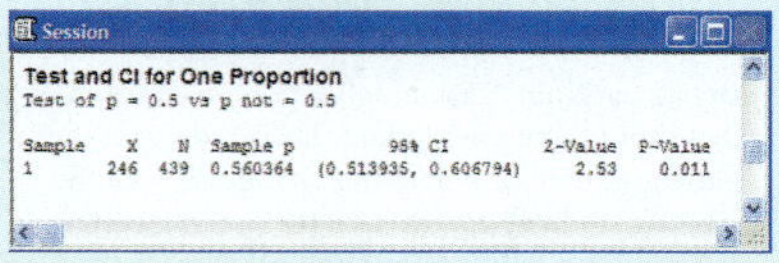
Session

**Test and CI for One Proportion**

Test of p = 0.5 vs p not = 0.5

| Sample | X | N | Sample p | 95% CI | Z-Value | P-Value |
|---|---|---|---|---|---|---|
| 1 | 246 | 439 | 0.560364 | (0.513935, 0.606794) | 2.53 | 0.011 |

(a) Define the parameter of interest.
(b) Check that the conditions for performing the significance test are met in this case.
(c) Interpret the $P$-value in context.
(d) Do these data give convincing evidence that the actual population proportion differs from 0.5? Justify your answer with appropriate evidence.

52. **Reporting cheating** What proportion of students are willing to report cheating by other students? A student project put this question to an SRS of 172 undergraduates at a large university: "You witness two students cheating on a quiz. Do you go to the professor?" The Minitab output below shows the results of a significance test and a 95% confidence interval based on the survey data.[16]

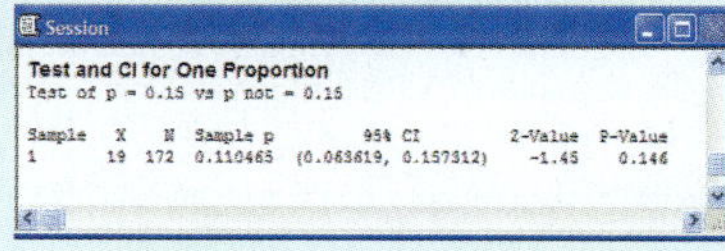
Session

**Test and CI for One Proportion**

Test of p = 0.15 vs p not = 0.15

| Sample | X | N | Sample p | 95% CI | Z-Value | P-Value |
|---|---|---|---|---|---|---|
| 1 | 19 | 172 | 0.110465 | (0.063619, 0.157312) | -1.45 | 0.146 |

(a) Define the parameter of interest.
(b) Check that the conditions for performing the significance test are met in this case.
(c) Interpret the $P$-value in context.
(d) Do these data give convincing evidence that the actual population proportion differs from 0.15? Justify your answer with appropriate evidence.

53. **Better parking** Refer to Exercise 39.
    (a) Describe a Type I error and a Type II error in this setting, and explain the consequences of each.
    (b) The test has a power of 0.75 to detect that $p = 0.45$. Explain what this means.
    (c) Identify two ways to increase the power in part (b).

**9.45** *S*: $H_0: p = 0.60$ vs. $H_a: p \neq 0.60$, where $p$ is the true proportion of teens who pass their driving test on the first attempt. *P*: One-sample $z$ test for $p$. *Random*: Random sample. *10%*: The sample size (125) < 10% of all teens. *Large Counts:* 125(0.60) = 75 ≥ 10 and 125(0.40) = 50 ≥ 10. *D*: $z = 2.01$, *P*-value = 0.0444. *C*: Because our *P*-value of 0.0444 < $\alpha$ = 0.05, we reject $H_0$. There is convincing evidence that the true proportion of teens who pass the driving test on their first attempt is different from 0.60.

**9.46** *S*: $H_0: p = 0.73$ vs. $H_a: p \neq 0.73$, where $p$ is the true proportion of first-year students at this university who think being very well-off financially is an important personal goal. *P*: One-sample $z$ test for $p$. *Random*: Random sample. *10%*: The sample size (200) < 10% of first-year students. *Large Counts*: 200(0.73) = 146 ≥ 10 and 200(0.27) = 54 ≥ 10. *D*: $z = -2.23$, *P*-value = 0.0258. *C*: Because our *P*-value of 0.0258 < $\alpha$ = 0.05, we reject $H_0$. We have convincing evidence that the proportion of first-year students at this university who think that being very well-off financially is an important goal is different from 0.73.

**9.47** **(a)** *D*: (0.607, 0.769). *C*: We are 95% confident that the interval from 0.607 to 0.769 captures the true proportion of teens who pass the driving test on their first attempt. **(b)** Because 0.60 is not in the interval, we have convincing evidence that the true proportion of teens who pass the driving test on their first attempt is different from 0.60.

**9.48** **(a)** *D*: (0.594, 0.726). *C*: We are 95% confident that the interval from 0.594 to 0.726 captures the true proportion of first-year students at this university who identify being very well-off financially as an important personal goal. **(b)** Because 0.73 is not in the interval, we have convincing evidence that the proportion of first-year students at this university who think that being very well-off financially is an important goal is different from 0.73.

**9.49** No. Because the value 0.16 is included in the interval, we do not have convincing evidence that the true proportion of U.S. adults who would say they use Twitter differs from 0.16.

**9.50** Yes. Because 0.55 is not included in the interval (0.56, 0.62), we have convincing evidence that the true proportion of U.S. adults who would say that they want to lose weight differs from 0.55.

**9.51** **(a)** $p$ = the true proportion of U.S. teens aged 13 to 17 who think that young people should wait to have sex until marriage. **(b)** *Random*: Random sample. *10%*: The sample size (439) < 10% of the population of all U.S. teens. *Large Counts*: 439(0.5) = 219.5 ≥ 10 and 439(0.5) = 219.5 ≥ 10. **(c)** Assuming that the true proportion of U.S. teens aged 13 to 17 who think that young people should wait to have sex until marriage is 0.50, there is a 0.011 probability of getting a sample proportion that is at least as different from 0.5 as the proportion in the sample. **(d)** Yes. Because the *P*-value of 0.011 < $\alpha$ = 0.05, we reject $H_0$. There is convincing evidence that the true proportion of U.S. teens aged 13 to 17 who think that young people should wait to have sex until marriage differs from 0.5.

**9.52** **(a)** $p$ = the true proportion of undergraduates at this large university who would be willing to report cheating by other students. **(b)** *Random*: Random sample. *10%*: The sample size (172) < 10% of all undergraduates. *Large Counts*: 172(0.15) = 25.8 ≥ 10 and 172(0.85) = 146.2 ≥ 10. **(c)** Assuming that the true proportion of undergraduates at this large university who would report cheating is 0.15, there is a 0.146 probability of getting a sample proportion that is at least as different from 0.15 as the proportion in the sample. **(d)** No. Because the *P*-value of 0.146 > $\alpha$ = 0.05, we fail to reject $H_0$. We do not have convincing evidence that the true proportion of undergraduates at this large university who would report cheating differs from 0.15.

**9.53** **(a)** I: Finding convincing evidence that more than 37% of students were satisfied with the new parking arrangement, when in reality only 37% were satisfied. *Consequence*: The

54. **Side effects** Refer to Exercise 40.

(a) Describe a Type I error and a Type II error in this setting, and explain the consequences of each.

(b) The test has a power of 0.54 to detect that $p = 0.07$. Explain what this means.

(c) Identify two ways to increase the power in part (b).

55. **Error probabilities** You read that a statistical test at significance level $\alpha = 0.05$ has power 0.78. What are the probabilities of Type I and Type II errors for this test?

56. **Error probabilities** You read that a statistical test at the $\alpha = 0.01$ level has probability 0.14 of making a Type II error when a specific alternative is true. What is the power of the test against this alternative?

57. **Power** A drug manufacturer claims that fewer than 10% of patients who take its new drug for treating Alzheimer's disease will experience nausea. To test this claim, a significance test is carried out of

$$H_0: p = 0.10$$
$$H_a: p < 0.10$$

You learn that the power of this test at the 5% significance level against the alternative $p = 0.08$ is 0.29.

(a) Explain in simple language what "power = 0.29" means in this setting.

(b) You could get higher power against the same alternative with the same $\alpha$ by changing the number of measurements you make. Should you make more measurements or fewer to increase power? Explain.

(c) If you decide to use $\alpha = 0.01$ in place of $\alpha = 0.05$, with no other changes in the test, will the power increase or decrease? Justify your answer.

(d) If you shift your interest to the alternative $p = 0.07$ with no other changes, will the power increase or decrease? Justify your answer.

58. **What is power?** You manufacture and sell a liquid product whose electrical conductivity is supposed to be 5. You plan to make six measurements of the conductivity of each lot of product. If the product meets specifications, the mean of many measurements will be 5. You will therefore test

$$H_0: \mu = 5$$
$$H_a: \mu \neq 5$$

If the true conductivity is 5.1, the liquid is not suitable for its intended use. You learn that the power of your test at the 5% significance level against the alternative $\mu = 5.1$ is 0.23.

(a) Explain in simple language what "power = 0.23" means in this setting.

(b) You could get higher power against the same alternative with the same $\alpha$ by changing the number of measurements you make. Should you make more measurements or fewer to increase power?

(c) If you decide to use $\alpha = 0.10$ in place of $\alpha = 0.05$, with no other changes in the test, will the power increase or decrease? Justify your answer.

(d) If you shift your interest to the alternative $\mu = 5.2$, with no other changes, will the power increase or decrease? Justify your answer.

*Multiple choice: Select the best answer for Exercises 59 to 62.*

59. After once again losing a football game to the archrival, a college's alumni association conducted a survey to see if alumni were in favor of firing the coach. An SRS of 100 alumni from the population of all living alumni was taken, and 64 of the alumni in the sample were in favor of firing the coach. Suppose you wish to see if a majority of living alumni are in favor of firing the coach. The appropriate test statistic is

(a) $z = \dfrac{0.64 - 0.5}{\sqrt{\dfrac{0.64(0.36)}{100}}}$

(b) $t = \dfrac{0.64 - 0.5}{\sqrt{\dfrac{0.64(0.36)}{100}}}$

(c) $z = \dfrac{0.64 - 0.5}{\sqrt{\dfrac{0.5(0.5)}{100}}}$

(d) $z = \dfrac{0.64 - 0.5}{\sqrt{\dfrac{0.64(0.36)}{64}}}$

(e) $z = \dfrac{0.5 - 0.64}{\sqrt{\dfrac{0.5(0.5)}{100}}}$

60. Which of the following is *not* a condition for performing a significance test about a population proportion $p$?

(a) The data should come from a random sample or randomized experiment.

(b) Both $np_0$ and $n(1 - p_0)$ should be at least 10.

(c) If you are sampling without replacement from a finite population, then you should sample no more than 10% of the population.

(d) The population distribution should be approximately Normal, unless the sample size is large.

(e) All of the above are conditions for performing a significance test about a population proportion.

principal believes that students are satisfied and takes no further action. II: Failing to find convincing evidence that more than 37% are satisfied with the new parking arrangement, when in reality more than 37% are satisfied. *Consequence*: The principal takes further action on parking when none is needed. **(b)** If the true proportion of students who are satisfied with the new arrangement is really 0.45, there is a 0.75 probability that the survey provides convincing evidence the true proportion > 0.37. **(c)** Increase the sample size or significance level.

**9.54** **(a)** I: Finding convincing evidence that less than 10% of patients would suffer nausea after using the drug, when in reality at least 10% would. *Consequence*: The new drug is used on more Alzheimer's patients, leading to excessive nausea. II: Failing to find convincing evidence that less than 10% would suffer from nausea, when in reality less than 10% would. *Consequence*: The new drug is used on fewer Alzheimer's patients, even though it rarely causes nausea. **(b)** If the true proportion of Alzheimer's patients who would experience nausea is really 0.07, there is a 0.54 probability that the results of the study would provide convincing evidence the true proportion < 0.10. **(c)** Increase the sample size or significance level.

**9.55** $P(\text{Type I}) = \alpha = 0.05$ and $P(\text{Type II}) = 0.22$

**9.56** Power $= 1 - P(\text{Type II error}) = 0.86$

**9.57** **(a)** If the true proportion of Alzheimer's patients who would experience nausea is really 0.08, there is a 0.29 probability that the results of the study would provide convincing evidence the true proportion < 0.10. **(b)** Increase the number of measurements taken ($n$) to get more information. **(c)** Decrease. If $\alpha$ is smaller, it becomes harder to reject the null hypothesis. This makes it harder to correctly reject $H_0$. **(d)** Increase. Because 0.07 is further from the null hypothesis value of 0.10, it will be easier to detect a difference between the null value and actual value.

**9.58** **(a)** If the true mean is really 5.1, there is a 0.23 probability that the sample will provide convincing evidence the true mean is different from 5. **(b)** Increase the number of measurements taken ($n$) to get more information. **(c)** Increase. If $\alpha$ is greater, it becomes easier to reject the null hypothesis. This makes it easier to correctly reject $H_0$. **(d)** Increase. Because 5.2 is further from the null hypothesis value of 5.0, it will be easier to detect a difference between the null value and the actual value.

**9.59** c

**9.60** d

**9.61** b

**9.62** a

**9.63** **(a)** $X - Y$ has a Normal distribution with $\mu_{X-Y} = -0.2$ and standard deviation $\sigma_{X-Y} = \sqrt{(0.1)^2 + (0.05)^2} = 0.112$. To fit in a case, $X - Y$ must take on a negative number. **(b)** We want to find $P(X - Y < 0)$ using the $N(-0.2, 0.112)$ distribution. $z = \dfrac{0 - (-0.2)}{0.112} = 1.79$ and $P(Z < 1.79) = 0.9633$. *Using technology*: 0.9629. There is a 0.9629 probability that a randomly selected CD will fit in a randomly selected case. **(c)** $P(\text{all fit}) = (0.9629)^{100} = 0.0228$. There is a 0.0228 probability that all 100 CDs will fit in their cases.

**9.64** **(a)** Randomly assign the people to the three treatments, with 3355 people receiving each treatment ($500 bonus for worker, $500 bonus for employer, no bonus). After the experiment, compare the mean time to return to work in the three groups or the proportion of each group who find a job within 11 weeks.
**(b)** Label the people from 00001 to 10065 in alphabetical order and then take five digits at a time from the table, ignoring repeated numbers and the numbers 10066 to 99999 and 00000. The first 3 people chosen are 06565, 00795, and 08727.
**(c)** The purpose of a control group is to give the researchers a baseline for comparison. This will allow them to see if a financial incentive is better than no incentive at all.

**PD Section 9.3 Overview**

Watch the video overview of Section 9.3 (for teachers) by clicking on the link in the TE-book, opening the TRCD, or logging onto the book's companion Web site www.whfreeman.com/tps5e (teacher log-in required).

**TRM Section 9.3 Alternate Examples**

You can find the Alternate Examples for this section in Microsoft Word format by clicking on the link in the TE-book, opening the TRCD, or logging onto the book's companion Web site www.whfreeman.com/tps5e (teacher log-in required).

### Teaching Tip

In a Chapter 8 Teaching Tip, we recommended that students use the phrase "zap ta$\bar{x}$" to remember that they should use $z$ for proportions and $t$ for means ($\bar{x}$'s). The same rules apply in Chapter 9 as well.

**61.** The $z$ statistic for a test of $H_0: p = 0.4$ versus $H_a: p \neq 0.4$ is $z = 2.43$. This test is

(a) not significant at either $\alpha = 0.05$ or $\alpha = 0.01$.

(b) significant at $\alpha = 0.05$ but not at $\alpha = 0.01$.

(c) significant at $\alpha = 0.01$ but not at $\alpha = 0.05$.

(d) significant at both $\alpha = 0.05$ and $\alpha = 0.01$.

(e) inconclusive because we don't know the value of $\hat{p}$.

**62.** Which of the following 95% confidence intervals would lead us to reject $H_0: p = 0.30$ in favor of $H_a: p \neq 0.30$ at the 5% significance level?

(a) (0.19, 0.27) (b) (0.24, 0.30) (c) (0.27, 0.31) (d) (0.29, 0.38) (e) None of these

**63. Packaging CDs** (6.2, 5.3) A manufacturer of compact discs (CDs) wants to be sure that their CDs will fit inside the plastic cases they have bought for packaging. Both the CDs and the cases are circular. According to the supplier, the plastic cases vary Normally with mean diameter $\mu = 4.2$ inches and standard deviation $\sigma = 0.05$ inches. The CD manufacturer decides to produce CDs with mean diameter $\mu = 4$ inches. Their diameters follow a Normal distribution with $\sigma = 0.1$ inches.

(a) Let $X$ = the diameter of a randomly selected CD and $Y$ = the diameter of a randomly selected case. Describe the shape, center, and spread of the distribution of the random variable $X - Y$. What is the importance of this random variable to the CD manufacturer?

(b) Compute the probability that a randomly selected CD will fit inside a randomly selected case.

(c) The production process actually runs in batches of 100 CDs. If each of these CDs is paired with a randomly chosen plastic case, find the probability that all the CDs fit in their cases.

**64. Cash to find work?** (4.2) Will cash bonuses speed the return to work of unemployed people? The Illinois Department of Employment Security designed an experiment to find out. The subjects were 10,065 people aged 20 to 54 who were filing claims for unemployment insurance. Some were offered $500 if they found a job within 11 weeks and held it for at least 4 months. Others could tell potential employers that the state would pay the employer $500 for hiring them. A control group got neither kind of bonus.[17]

(a) Describe a completely randomized design for this experiment.

(b) How will you label the subjects for random assignment? Use Table D at line 127 to choose the first 3 subjects for the first treatment.

(c) Explain the purpose of a control group in this setting.

# 9.3 Tests about a Population Mean

**WHAT YOU WILL LEARN** By the end of the section, you should be able to:

- State and check the Random, 10%, and Normal/Large Sample conditions for performing a significance test about a population mean.
- Perform a significance test about a population mean.
- Use a confidence interval to draw a conclusion for a two-sided test about a population parameter.
- Perform a significance test about a mean difference using paired data.

Confidence intervals and significance tests for a population proportion $p$ are based on $z$-values from the standard Normal distribution. Inference about a population mean $\mu$ uses a $t$ distribution with $n - 1$ degrees of freedom, except in the rare case when the population standard deviation $\sigma$ is known. We learned how to construct confidence intervals for a population mean in Section 8.3. Now we'll examine the details of testing a claim about a population mean $\mu$.

## Carrying Out a Significance Test for $\mu$

In an earlier example, a company claimed to have developed a new AAA battery that lasts longer than its regular AAA batteries. Based on years of experience, the company knows that its regular AAA batteries last for 30 hours of continuous use, on average. An SRS of 15 new batteries lasted an average of 33.9 hours with a standard deviation of 9.8 hours. Do these data give convincing evidence that the new batteries last longer on average? To find out, we perform a test of

$$H_0: \mu = 30 \text{ hours}$$
$$H_a: \mu > 30 \text{ hours}$$

where $\mu$ is the true mean lifetime of the new deluxe AAA batteries.

**Conditions** In Chapter 8, we introduced conditions that should be met before we construct a confidence interval for a population mean: Random, 10% when sampling without replacement, and Normal/Large Sample. These same three conditions must be verified before performing a significance test about a population mean.

As in the previous chapter, the Normal/Large Sample condition for means is

population distribution is Normal or sample size is large ($n \geq 30$)

We often don't know whether the population distribution is Normal. But if the sample size is large ($n \geq 30$), we can safely carry out a significance test. If the sample size is small, we should examine the sample data for any obvious departures from Normality, such as strong skewness and outliers.

**CONDITIONS FOR PERFORMING A SIGNIFICANCE TEST ABOUT A MEAN**

- **Random:** The data come from a well-designed random sample or randomized experiment.
  - **10%:** When sampling without replacement, check that $n \leq \frac{1}{10}N$.
- **Normal/Large Sample:** The population has a Normal distribution or the sample size is large ($n \geq 30$). If the population distribution has unknown shape and $n < 30$, use a graph of the sample data to assess the Normality of the population. Do not use $t$ procedures if the graph shows strong skewness or outliers.

Here's an example that shows how to check the conditions.

**EXAMPLE** 

### Better Batteries

*Checking conditions*

Figure 9.11 on the next page shows a dotplot, boxplot, and Normal probability plot of the battery lifetimes for an SRS of 15 batteries.

**ALTERNATE EXAMPLE**

### *Less music?*

A classic rock radio station claims to play an average of 50 minutes of music every hour. However, it seems that every time you turn to this station, there is a commercial playing. To investigate the station's claim, you randomly select 12 different hours during the next week and record what the radio station plays in each of the 12 hours. Here are the number of minutes of music played during each of these hours:

| 44 | 49 | 45 | 51 | 49 | 53 |
|---|---|---|---|---|---|
| 49 | 44 | 47 | 50 | 46 | 48 |

PROBLEM: Check the conditions for carrying out a significance test of the station's claim that it plays an average of 50 minutes of music per hour.

SOLUTION:

- *Random*: A random sample of hours was selected.
  - *10%*: There are more than 10(12) = 120 hours of music played during the week.
- *Normal/Large Sample*: We don't know if the population distribution of music times is approximately Normal. Because we have a small sample size ($12 < 30$), we will graph the data and look for any departures from Normality.

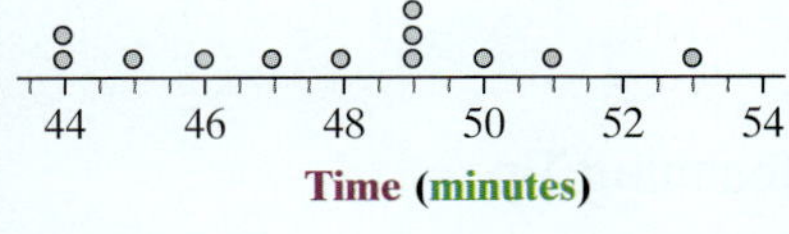


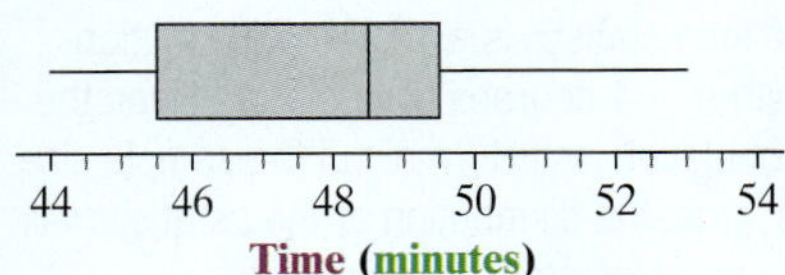


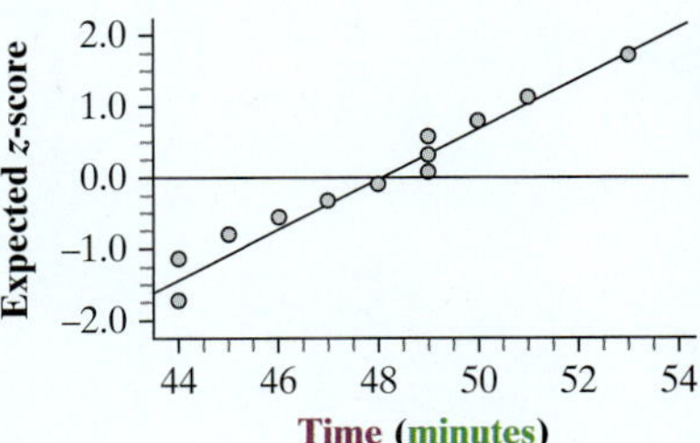


The graphs do not show much skewness and there are no outliers, so it is reasonable to use $t$ procedures for these data.

### Teaching Tip

Emphasize that the Random, 10%, and Normal/Large Sample conditions are exactly the same for significance tests and confidence intervals for a population mean $\mu$.

### Teaching Tip

In this Example (and the corresponding Alternate Example), we provide three different graphs to assess Normality. On the AP® exam, however, students only need to use a single graph. Also, they must include a sketch of the graph in their response—it isn't enough to simply claim that there are no outliers or strong skewness.

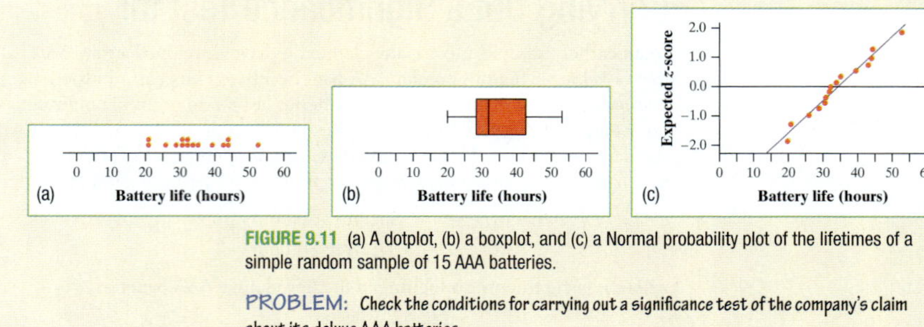


**FIGURE 9.11** (a) A dotplot, (b) a boxplot, and (c) a Normal probability plot of the lifetimes of a simple random sample of 15 AAA batteries.

PROBLEM: Check the conditions for carrying out a significance test of the company's claim about its deluxe AAA batteries.

SOLUTION:

- *Random:* The company tested a simple random sample of 15 new AAA batteries.
  - 10%: Because the batteries are being sampled without replacement, we need to check that there are at least $10(15) = 150$ new AAA batteries. This seems reasonable to believe.
- *Normal/Large Sample:* We don't know if the population distribution of battery lifetimes for the company's new AAA batteries is Normal. With such a small sample size ($n = 15$), we need to graph the data to look for any departures from Normality. The dotplot and boxplot show slight right-skewness but no outliers. The Normal probability plot is fairly linear. Because none of the graphs shows any strong skewness or outliers, we should be safe performing a test about the population mean lifetime $\mu$.

**For Practice** *Try Exercise* 65

There is a small number of real-world situations in which we might know the population standard deviation $\sigma$. When this is the case, the test statistic

$$z = \frac{\bar{x} - \mu_0}{\sigma/\sqrt{n}}$$

will follow a standard Normal distribution if the Normal/Large Sample condition is met. If so, then we can calculate *P*-values using Table A or technology. The TI-83/84 and TI-89's *Z*-Test option in the TESTS menu is designed for this special situation.

**Calculations: Test Statistic and *P*-Value** When performing a significance test, we do calculations assuming that the null hypothesis $H_0$ is true. The test statistic measures how far the sample result diverges from the parameter value specified by $H_0$, in standardized units. As before,

$$\text{test statistic} = \frac{\text{statistic} - \text{parameter}}{\text{standard deviation of statistic}}$$

For a test of $H_0: \mu = \mu_0$, our statistic is the sample mean $\bar{x}$. Its standard deviation is

$$\sigma_{\bar{x}} = \frac{\sigma}{\sqrt{n}}$$

In an ideal world, our test statistic would be

$$z = \frac{\bar{x} - \mu_0}{\sigma/\sqrt{n}}$$

Because the population standard deviation $\sigma$ is usually unknown, we use the sample standard deviation $s_x$ in its place. The resulting test statistic has the standard error of $\bar{x}$ in the denominator

$$t = \frac{\bar{x} - \mu_0}{s_x/\sqrt{n}}$$

As we saw earlier, when the Normal/Large Sample condition is met, this statistic has approximately a $t$ distribution with $n - 1$ degrees of freedom.

## Teaching Tip

Remind students that the basic form of the standardized test statistic is always the same, regardless of which parameter you are testing.

## Teaching Tip

If the population is Normal, the distribution of the $t$ statistic is exactly a $t$ distribution with $n - 1$ degrees of freedom. When the population isn't Normal but the sample size is large, the distribution of the $t$ statistic will be approximately a $t$ distribution.

## AP® Exam Formula Sheet

The formula for the test statistic in a one-sample $t$ test for a mean is not included on the formula sheet that is given to students on the AP® exam. Rather, the formula sheet includes the general formula for a test statistic and lists the different standard deviations students might need, including $\sigma_{\bar{x}}$. Students are expected to figure out which standard deviation to use based on the context of the question.

In Section 8.3, we used Table B to find critical values from the $t$ distributions when constructing confidence intervals about an unknown population mean $\mu$. Once we have calculated the test statistic, we can use Table B to find the $P$-value for a significance test about $\mu$. The following example shows how this works.

## EXAMPLE Better Batteries

### *Computing the test statistic and P-value*

The battery company wants to test $H_0: \mu = 30$ versus $H_a: \mu > 30$ based on an SRS of 15 new AAA batteries with mean lifetime $\bar{x} = 33.9$ hours and standard deviation $s_x = 9.8$ hours. The test statistic is

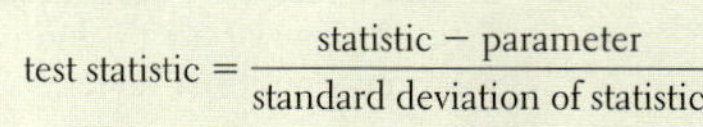

$$\text{test statistic} = \frac{\text{statistic} - \text{parameter}}{\text{standard deviation of statistic}}$$

$$t = \frac{\bar{x} - \mu_0}{s_x/\sqrt{n}} = \frac{33.9 - 30}{9.8/\sqrt{15}} = 1.54$$

The $P$-value is the probability of getting a result this large or larger in the direction indicated by $H_a$, that is, $P(t \geq 1.54)$. Figure 9.12 shows this probability as an area under the $t$ distribution curve with df = 15 − 1 = 14. We can find this $P$-value using Table B.

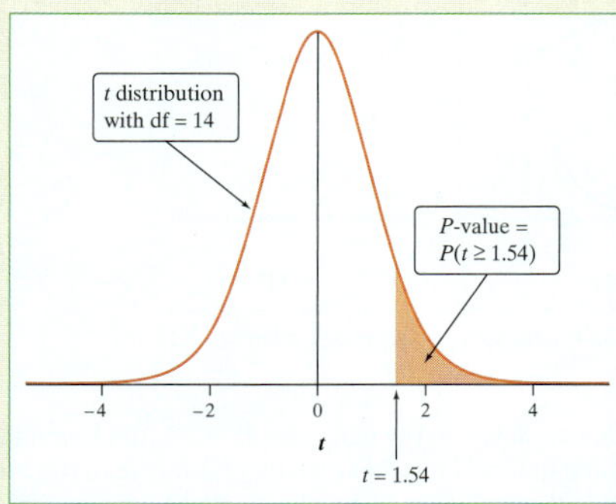


**FIGURE 9.12** The $P$-value for a one-sided test with $t = 1.54$.

Go to the df = 14 row. The $t$ statistic falls between the values 1.345 and 1.761. If you look at the top of the corresponding columns in Table B, you'll find that the "Upper-tail probability $p$" is between 0.10 and 0.05. (See the excerpt from Table B at right.) Because we are looking for $P(t \geq 1.54)$, this is the probability we seek. That is, the $P$-value for this test is between 0.05 and 0.10.

| | Upper-tail probability $p$ | | |
|---|---|---|---|
| **df** | **.10** | **.05** | **.025** |
| 13 | 1.350 | 1.771 | 2.160 |
| 14 | 1.345 | 1.761 | 2.145 |
| 15 | 1.341 | 1.753 | 2.131 |
| | 80% | 90% | 95% |
| | **Confidence level *C*** | | |

As you can see, Table B gives an interval of possible $P$-values for a significance test. We can still draw a conclusion from the test in much the same way as if we had a single probability. In the case of the new AAA batteries, for instance, we would fail to reject $H_0: \mu = 30$ because the $P$-value exceeds our default $\alpha = 0.05$ significance level. We don't have convincing evidence that the company's new AAA batteries last longer than 30 hours, on average.

Table B has other limitations for finding $P$-values. It includes probabilities only for $t$ distributions with degrees of freedom from 1 to 30 and then skips to df = 40, 50, 60, 80, 100, and 1000. (The bottom row gives probabilities for df = ∞, which corresponds to the standard Normal distribution.) Also, Table B shows probabilities only for positive values of $t$. To find a $P$-value for a negative value of $t$, we use the symmetry of the $t$ distributions. The next example shows how we deal with both of these issues.

### AP® Exam Common Error

When calculating the value of a test statistic on the AP® exam, many students use incorrect symbols in the formula. Even if correct numbers are substituted in a later step, using an incorrect formula such as $t = \frac{\bar{x} - \mu}{\sigma/\sqrt{n}}$ will result in a loss of credit. To avoid this error, we suggest that students start with the numbers substituted into the equation. For example, begin with $t = \frac{33.9 - 30}{9.8/\sqrt{15}}$.

### ALTERNATE EXAMPLE

#### *Less music?*

In the Alternate Example on page 575, we wanted to test $H_0: \mu = 50$ versus $H_a: \mu < 50$ using a sample of 12 hours of music.

PROBLEM: Compute the test statistic and $P$-value for these data.

SOLUTION: The sample mean for these data is $\bar{x} = 47.9$ with a standard deviation of $s_x = 2.81$. Thus, the test statistic is $t = \frac{47.9 - 50}{2.81/\sqrt{12}} = -2.59$. Using Table B and a $t$ distribution with 12 − 1 = 11 degrees of freedom, the $P$-value is between 0.01 and 0.02.

ALTERNATE EXAMPLE

### *More practice with Table B*

PROBLEM:

**(a)** Find the *P*-value for a test of $H_0: \mu = 10$ versus $H_a: \mu > 10$ that uses a sample of size 75 and has a test statistic of $t = 2.33$.

**(b)** Find the *P*-value for a test of $H_0: \mu = 300$ versus $H_a: \mu \neq 300$ that uses a sample of size 10 and has a test statistic of $t = -0.51$.

SOLUTION:

**(a)** Table B does not include a line for df $= 75 - 1 = 74$, so we will be conservative and use df $= 60$. Because $t = 2.33$ is between 2.099 and 2.390, the *P*-value is between 0.01 and 0.02.

**(b)** Using the line for df $= 10 - 1 = 9$, we find that the value 0.51 is smaller than any *t*-value provided in that line. This means that the area under the *t* curve to the right of 0.51 is greater than 0.25. Because this is a two-sided test, the *P*-value must be at least 0.50.

EXAMPLE

## Two-Sided Tests, Negative *t*-Values, and More

### *Using Table B wisely*

What if you were performing a test of $H_0: \mu = 5$ versus $H_a: \mu \neq 5$ based on a sample size of $n = 37$ and obtained $t = -3.17$? Because this is a two-sided test, you are interested in the probability of getting a value of *t* less than or equal to $-3.17$ or greater than or equal to 3.17. Figure 9.13 shows the desired *P*-value as an area under the *t* distribution curve with 36 degrees of freedom. Notice that $P(t \leq -3.17) = P(t \geq 3.17)$ due to the symmetric shape of the density curve. Table B shows only positive *t*-values, so we will focus on $t = 3.17$.

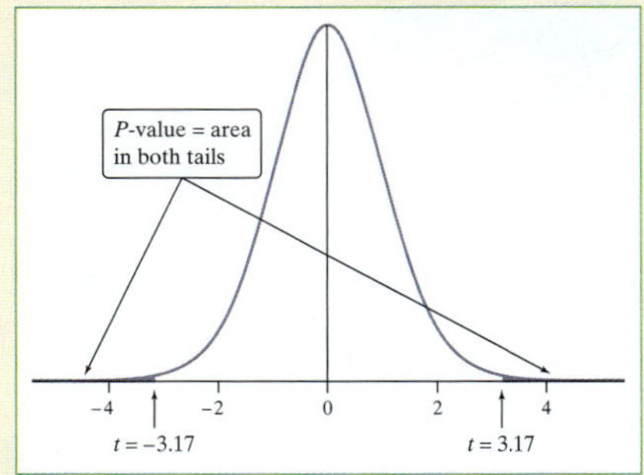


**FIGURE 9.13** The *P*-value for a two-sided test with $t = -3.17$.

| | Upper-tail probability *p* | | |
|---|---|---|---|
| df | .005 | .0025 | .001 |
| 29 | 2.756 | 3.038 | 3.396 |
| 30 | 2.750 | 3.030 | 3.385 |
| 40 | 2.704 | 2.971 | 3.307 |
| | 99% | 99.5% | 99.8% |
| | Confidence level *C* | | |

Because df $= 37 - 1 = 36$ is not available on the table, use df $= 30$. You might be tempted to use df $= 40$, but doing so would result in a smaller *P*-value than you are entitled to with df $= 36$. (In other words, you'd be cheating!) Move across the df $= 30$ row, and notice that $t = 3.17$ falls between 3.030 and 3.385. The corresponding "Upper-tail probability *p*" is between 0.001 and 0.0025. (See the excerpt from Table B.) For this two-sided test, the corresponding *P*-value would be between $2(0.001) = 0.002$ and $2(0.0025) = 0.005$.

caution

One point from the example deserves repeating: *if the df you need isn't provided in Table B, use the next lower df that is available.* It's no fair "rounding up" to a larger df. This is like pretending that your sample size is larger than it really is. Doing so would give you a smaller *P*-value than is true and would make you more likely to incorrectly reject $H_0$ when it's true (make a Type I error).

Given the limitations of Table B, our advice is to use technology to find *P*-values when carrying out a significance test about a population mean.

**19. TECHNOLOGY CORNER**

## COMPUTING *P*-VALUES FROM *t* DISTRIBUTIONS ON THE CALCULATOR

TI-Nspire instructions in Appendix B; HP Prime instructions on the book's Web site.

You can use the `tcdf` command on the TI-83/84 and TI-89 to calculate areas under a $t$ distribution curve. The syntax is `tcdf(lower bound,upper bound,df)`.

Let's use the `tcdf` command to compute the *P*-values from the last two examples.

*Better batteries:* To find $P(t \geq 1.54)$,

**TI-83/84**

- Press 2nd VARS (DISTR) and choose `tcdf(`.

**OS 2.55 or later:** In the dialog box, enter these values: `lower:1.54, upper: 10000, df:14`, choose `Paste`, and then press ENTER. **Older OS:** Complete the command `tcdf(1.54,10000,14)` and press ENTER.

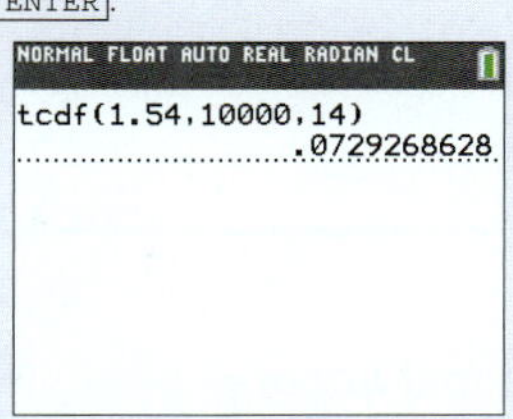


**TI-89**

- In the Stats/List Editor, press F5 (Distr) and choose `t Cdf....`
- In the dialog box, enter these values: `Lower value:1.54, Upper value:10000, Deg of Freedom, df:14`, and then choose ENTER.

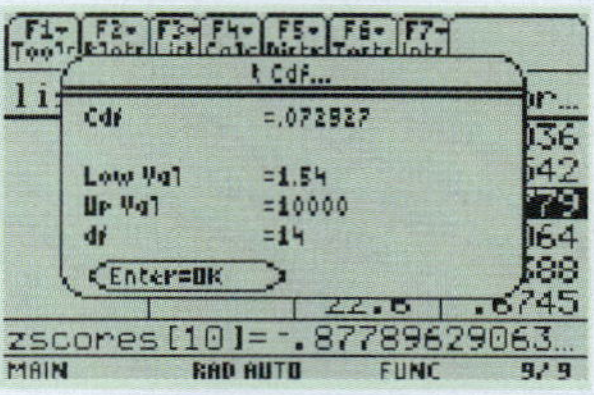


*Two-sided test:* To find the *P*-value for the two-sided test with df = 36 and $t = -3.17$, do `tcdf(-10000,-3.17,36)` and multiply the result by 2.

### CHECK YOUR UNDERSTANDING

The makers of Aspro brand aspirin want to be sure that their tablets contain the right amount of active ingredient (acetylsalicylic acid). So they inspect a random sample of 36 tablets from a batch in production. When the production process is working properly, Aspro tablets have an average of $\mu = 320$ milligrams (mg) of active ingredient. The amount of active ingredient in the 36 selected tablets has mean 319 mg and standard deviation 3 mg.

1. State appropriate hypotheses for a significance test in this setting.
2. Check that the conditions are met for carrying out the test.
3. Calculate the test statistic. Show your work.
4. Use Table B to find the *P*-value. Then use technology to get a more accurate result. What conclusion would you draw?

## The One-Sample *t* Test

When the conditions are met, we can test a claim about a population mean $\mu$ using a **one-sample *t* test for a mean**. Here are the details.

### TECHNOLOGY TIP

Remind students that they can watch a video presentation of this technology corner (for the TI-83/84, TI-89, or TI-Nspire) at www.whfreeman.com/tps5e.

### TECHNOLOGY TIP

Your students may wonder why we use the value 10000 or −10000 when using the `tcdf` command. Technically, we would like to use ∞ or −∞, but the TI-84 calculator doesn't have an ∞ button. Instead, we use a value that is extremely large. Choosing a value 10,000 standard deviations above the mean seems plenty large to us!

### TECHNOLOGY TIP

If your students are using a TI-84 with OS 2.55 or later and aren't given dialog boxes when they use the `tcdf` command, have them turn on the Stat Wizards by pressing Mode and scrolling down to the second page. If students are using a TI-83 Plus or TI-84 with an older operating system, they can run an app called *Catalog Help* (ctlghelp) that will remind them what values they need to enter for certain commands. Press the APPS button to see if it is already loaded. If it is, press Enter and read the directions. If it isn't, download the app from education.ti.com or copy it by linking from another calculator.

### ANSWERS TO CYU

1. $H_0: \mu = 320$ vs. $H_a: \mu \neq 320$, where $\mu$ = the true mean amount of active ingredient (in milligrams) contained in Aspro tablets from this batch of production.
2. *Random*: Random sample. *10%*: The sample of size $36 < 10\%$ of the population of all tablets in this batch. *Normal/Large Sample*: $n = 36 \geq 30$.
3. $t = \dfrac{319 - 320}{3/\sqrt{36}} = -2$.
4. For this test, df = 35. Using Table B and df = 30, the tail area is between 0.025 and 0.05. Thus, the *P*-value for the two-sided test is between 0.05 and 0.10. *Using technology*: `2tcdf(lower:-1000,upper:-2,df:35)` = 2(0.0267) = 0.0534. Because the *P*-value of $0.0534 > \alpha = 0.05$, we fail to reject $H_0$. There is not convincing evidence that the true mean amount of the active ingredient in Aspro tablets from this batch of production differs from 320 mg.

## ONE-SAMPLE t TEST FOR A MEAN

Suppose the conditions are met. To test the hypothesis $H_0: \mu = \mu_0$, compute the one-sample $t$ statistic

$$t = \frac{\bar{x} - \mu_0}{s_x/\sqrt{n}}$$

Find the P-value by calculating the probability of getting a $t$ statistic this large or larger in the direction specified by the alternative hypothesis $H_a$ in a $t$ distribution with df $= n - 1$:

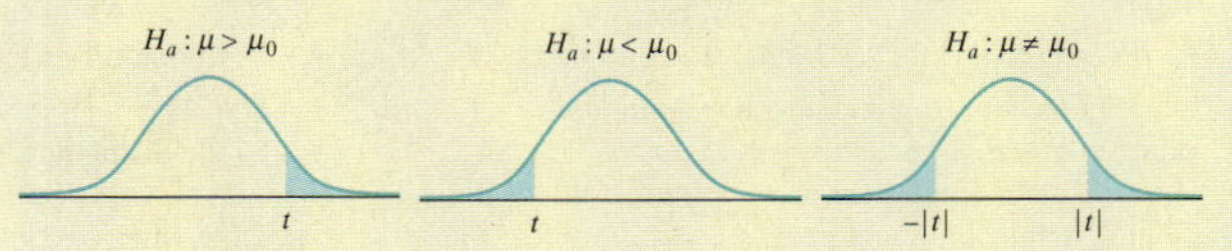


Now we are ready to test a claim about an unknown population mean. Once again, we follow the four-step process.

## EXAMPLE Healthy Streams

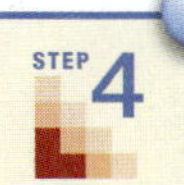

### Performing a significance test about $\mu$

The level of dissolved oxygen (DO) in a stream or river is an important indicator of the water's ability to support aquatic life. A researcher measures the DO level at 15 randomly chosen locations along a stream. Here are the results in milligrams per liter (mg/l):

| | | | | | | | |
|---|---|---|---|---|---|---|---|
| 4.53 | 5.04 | 3.29 | 5.23 | 4.13 | 5.50 | 4.83 | 4.40 |
| 5.42 | 6.38 | 4.01 | 4.66 | 2.87 | 5.73 | 5.55 | |

A dissolved oxygen level below 5 mg/l puts aquatic life at risk.

PROBLEM:

(a) Do we have convincing evidence at the $\alpha = 0.05$ significance level that aquatic life in this stream is at risk?

(b) Given your conclusion in part (a), which kind of mistake—a Type I error or a Type II error—could you have made? Explain what this mistake would mean in context.

SOLUTION:

(a) We will follow the four-step process.

STATE: We want to test a claim about the true mean dissolved oxygen level $\mu$ in this stream at the $\alpha = 0.05$ level. Our hypotheses are

$$H_0: \mu = 5$$
$$H_a: \mu < 5$$

## ALTERNATE EXAMPLE

### Short subs

Abby and Raquel like to eat sub sandwiches. However, they noticed that the lengths of the "6-inch sub" sandwiches they get at their favorite restaurant seemed shorter than the advertised length. To investigate, they randomly selected 24 different times during the next month and ordered a "6-inch" sub. Here are the actual lengths of each of the 24 sandwiches (in inches):

| | | | | | | | |
|---|---|---|---|---|---|---|---|
| 4.50 | 4.75 | 4.75 | 5.00 | 5.00 | 5.00 | 5.50 | 5.50 |
| 5.50 | 5.50 | 5.50 | 5.50 | 5.75 | 5.75 | 5.75 | 6.00 |
| 6.00 | 6.00 | 6.00 | 6.00 | 6.50 | 6.75 | 6.75 | 7.00 |

PROBLEM:

**(a)** Do these data provide convincing evidence at the $\alpha = 0.10$ level that the sandwiches at this restaurant are shorter than advertised, on average?

**(b)** Given your conclusion in part (a), which kind of mistake—a Type I or a Type II error—could you have made? Explain what this mistake would mean in context.

SOLUTION:

**(a)** STATE: We want to test the following hypotheses at the $\alpha = 0.10$ significance level:

$$H_0: \mu = 6$$
$$H_a: \mu < 6$$

where $\mu$ = the true mean length of "6-inch" subs from this restaurant.

PLAN: If conditions are met, we will perform a one-sample $t$ test for $\mu$.

- *Random*: Times were randomly selected to order a sub, so this should be a random sample of 6-inch subs from the restaurant.
  - *10%*: This restaurant makes more than 10(24) = 240 subs in a month.
- *Normal/Large Sample*: We don't know if the population distribution of lengths is approximately Normal and we don't have a large sample size, so we will graph the data and look for any departures from Normality.

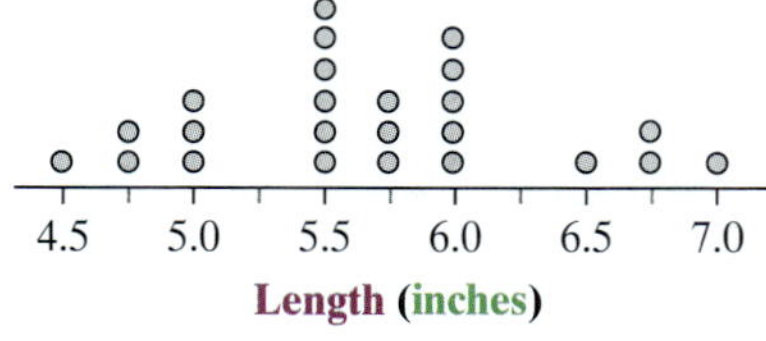


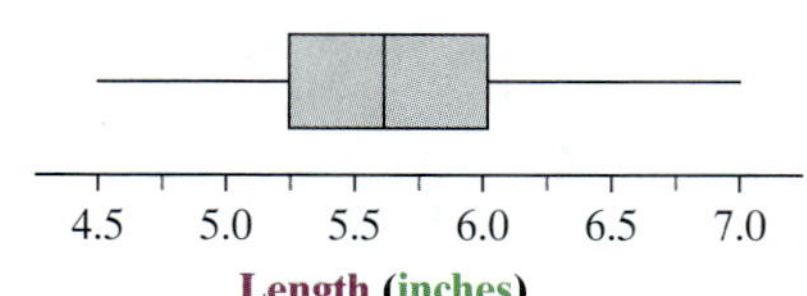


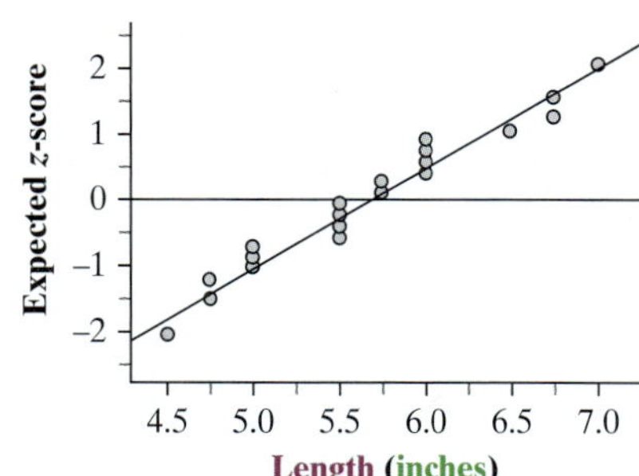


The graphs do not show much skewness and there are no outliers, so it is reasonable to use $t$ procedures for these data.

DO: The sample mean length is $\bar{x} = 5.68$ inches with a standard deviation of $s_{\bar{x}} = 0.66$ inches.

**AP® EXAM TIP** It is not enough just to make a graph of the data on your calculator when assessing Normality. You must *sketch* the graph on your paper to receive credit. You don't have to draw multiple graphs—any appropriate graph will do.

PLAN: If the conditions are met, we should do a one-sample $t$ test for $\mu$.

- *Random:* The researcher measured the DO level at 15 randomly chosen locations.
  - 10%: There is an infinite number of possible locations along the stream, so it isn't necessary to check the 10% condition.
- *Normal/Large Sample:* We don't know whether the population distribution of DO levels at all points along the stream is Normal. With such a small sample size ($n = 15$), we need to graph the data to see if it's safe to use $t$ procedures. Figure 9.14 shows our hand sketches of a calculator histogram, boxplot, and Normal probability plot for these data. The histogram looks roughly symmetric; the boxplot shows no outliers; and the Normal probability plot is fairly linear. With no outliers or strong skewness, the $t$ procedures should be pretty accurate.

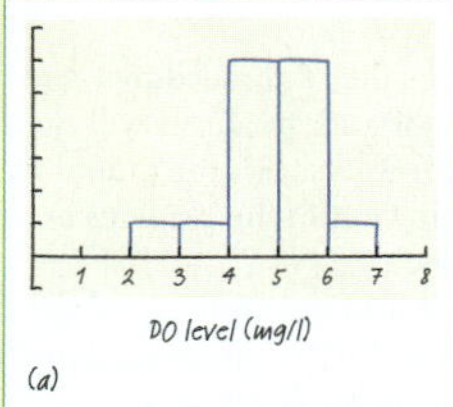


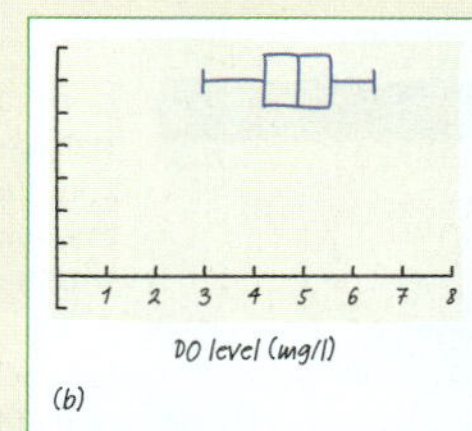


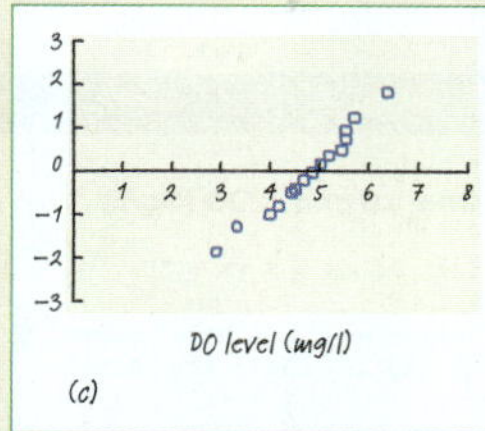


**FIGURE 9.14** Sketches of (a) a histogram, (b) a boxplot, and (c) a Normal probability plot for the dissolved oxygen (DO) readings in the sample, in mg/l.

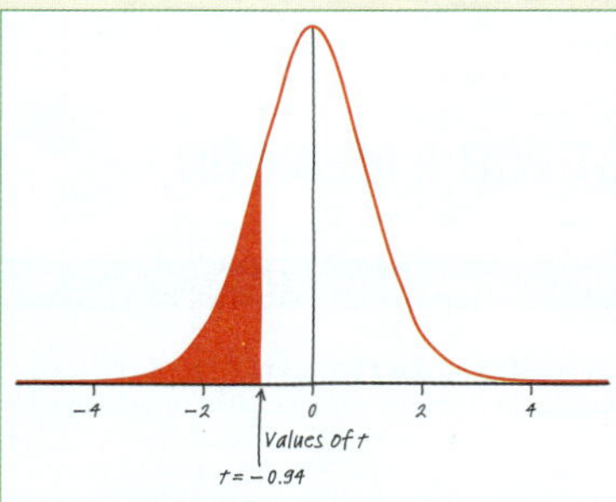


**FIGURE 9.15** The *P*-value for a one-sided test when $t = -0.94$.

DO: We entered the data into our calculator and did 1-Var Stats (see screen shot). The sample mean is $\bar{x} = 4.771$ and the sample standard deviation is $s_x = 0.9396$.

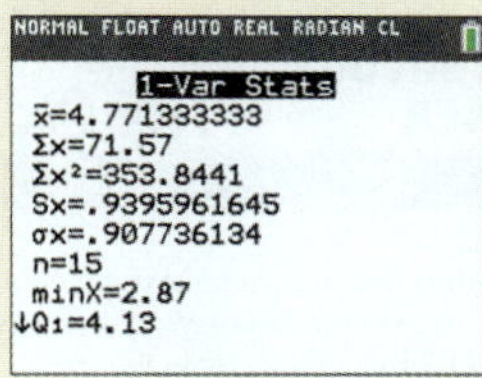


- *Test statistic*

$$t = \frac{\bar{x} - \mu_0}{s_x/\sqrt{n}} = \frac{4.771 - 5}{0.9396/\sqrt{15}} = -0.94$$

- *P-value* The *P*-value is the area to the left of $t = -0.94$ under the $t$ distribution curve with degrees of freedom df $= 15 - 1 = 14$. Figure 9.15 shows this probability.

*Using the table:* Table B shows only areas in the upper tail of the distribution. Because the $t$ distributions are symmetric, $P(t \le -0.94) = P(t \ge 0.94)$. Search the df = 14 row of Table B for entries that bracket $t = 0.94$ (see the excerpt at right). Because the observed $t$ lies between 0.868 and 1.076, the *P*-value lies between 0.15 and 0.20.

| | Upper-tail probability $p$ | | |
|---|---|---|---|
| df | .25 | .20 | .15 |
| 13 | .694 | .870 | 1.079 |
| 14 | .692 | .868 | 1.076 |
| 15 | .691 | .866 | 1.074 |
| | 50% | 60% | 70% |
| | Confidence level $C$ | | |

- *Test statistic* $t = \dfrac{5.68 - 6}{0.66/\sqrt{24}} = -2.38$
- *P-value* Using the $t$ distribution with $24 - 1 = 23$ degrees of freedom, the *P*-value is between 0.01 and 0.02. Using technology: *P*-value = `tcdf(lower:-10000,upper:-2.38,df:23)` = 0.0130.

CONCLUDE: Because the *P*-value of 0.0130 is less than $\alpha = 0.10$, we reject the null hypothesis. There is convincing evidence that the true mean length of "6-inch" subs at this restaurant is less than 6 inches.

**(b)** Because we rejected the null hypothesis, it is possible that we made a Type I error. In other words, it is possible that we found convincing evidence that the mean length was less than 6 inches when in reality the mean length is 6 inches.

## Teaching Tip

In the "Conclude" step of the example on this page, remind students that it isn't correct to conclude that "the mean dissolved oxygen level *is* 5 mg/l."

*Using technology:* We can find the exact *P*-value using a calculator: `tcdf(lower:-100, upper:-0.94, df:14)` = 0.1816.

CONCLUDE: Because the *P*-value, 0.1816, is greater than our $\alpha = 0.05$ significance level, we fail to reject $H_0$. We don't have convincing evidence that the mean DO level in the stream is less than 5 mg/l.

(b) Because we decided not to reject $H_0$ in part (a), we could have made a Type II error (failing to reject $H_0$ when $H_0$ is false). If we did, then the mean dissolved oxygen level $\mu$ in the stream is actually less than 5 mg/l, but we didn't find convincing evidence of that with our significance test.

**For Practice** *Try Exercise* 

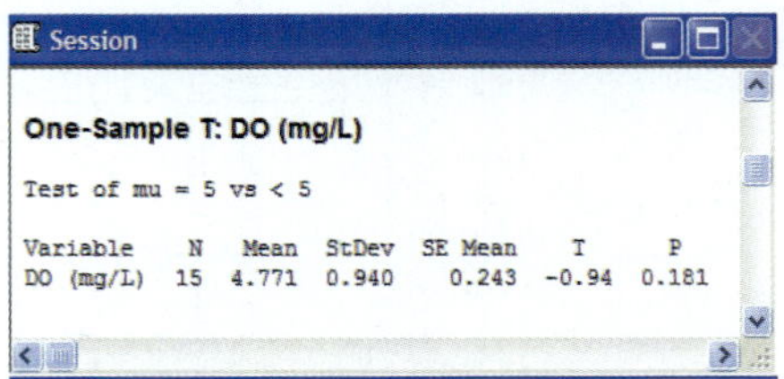


**FIGURE 9.16** Minitab output for the one-sample *t* test from the dissolved oxygen example.

Because the *t* procedures are so common, all statistical software packages will do the calculations for you. Figure 9.16 shows the output from Minitab for the one-sample *t* test in the previous example. Note that the results match!

You can also use your calculator to carry out a one-sample *t* test. But be sure to read the AP® exam tip at the end of the Technology Corner.

## TECHNOLOGY TIP

Remind students that they can watch a video presentation of this technology corner (for the TI-83/84, TI-89, or TI-Nspire) at www.whfreeman.com/tps5e.

## 20. TECHNOLOGY CORNER ONE-SAMPLE *t* TEST FOR A MEAN ON THE CALCULATOR

TI-Nspire instructions in Appendix B; HP Prime instructions on the book's Web site.

You can perform a one-sample *t* test using either raw data or summary statistics on the TI-83/84 or TI-89. Let's use the calculator to carry out the test of $H_0$: $\mu = 5$ versus $H_a$: $\mu < 5$ from the dissolved oxygen example. Start by entering the sample data in L1/list1. Then, to do the test:

**TI-83/84**

- Press [STAT], choose TESTS and `T-Test`.
- Adjust your settings as shown.

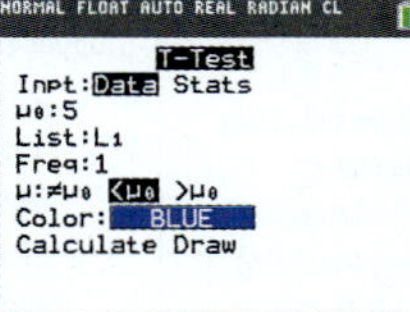


**TI-89**

- Press [2nd] [F1] ([F6]) and choose `T-Test`.
- Adjust your settings as shown.

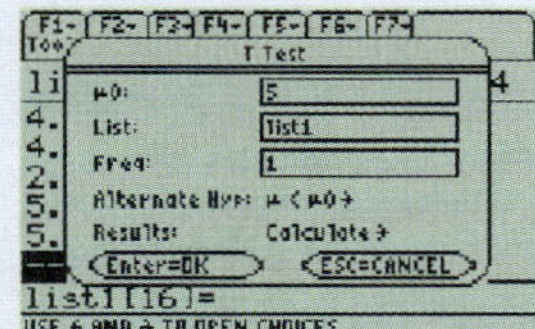

If you select "Calculate," the following screen appears:

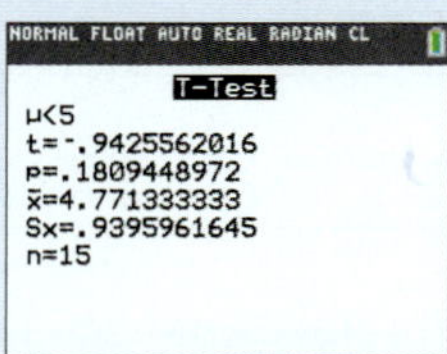


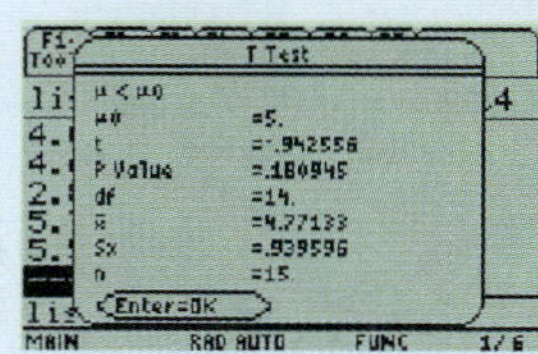


The test statistic is $t = -0.94$ and the *P*-value is 0.1809.

If you specify "Draw," you see a $t$ distribution curve (df = 14) with the lower tail shaded.

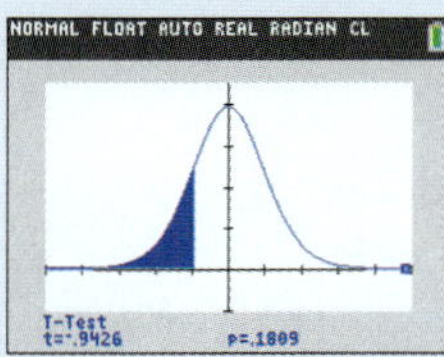


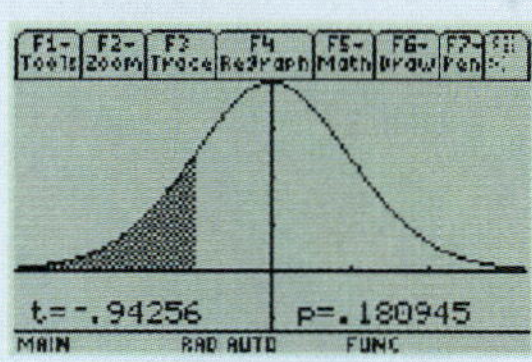


*Note:* If you are given summary statistics instead of the original data, you would select the option "Stats" instead of "Data" in the first screen.

**AP® EXAM TIP** Remember: if you just give calculator results with no work, and one or more values are wrong, you probably won't get any credit for the "Do" step. If you opt for the calculator-only method, name the procedure ($t$ test) and report the test statistic ($t = -0.94$), degrees of freedom (df = 14), and *P*-value (0.1809).

## CHECK YOUR UNDERSTANDING

A college professor suspects that students at his school are getting less than 8 hours of sleep a night, on average. To test his belief, the professor asks a random sample of 28 students, "How much sleep did you get last night?" Here are the data (in hours):

9 6 8 6 8 8 6 6.5 6 7 9 4 3 4 5 6 11 6 3 6 6 10 7 8 4.5 9 7 7

Do these data provide convincing evidence at the $\alpha = 0.05$ significance level in support of the professor's suspicion?

## Two-Sided Tests and Confidence Intervals

Now let's look at an example involving a two-sided test.

### EXAMPLE Juicy Pineapples

STEP 4

*A two-sided test*

At the Hawaii Pineapple Company, managers are interested in the sizes of the pineapples grown in the company's fields. Last year, the mean weight of the pineapples harvested from one large field was 31 ounces. A different irrigation system was installed in this field after the growing season. Managers wonder how this change will affect the mean weight of future pineapples grown in the field. To

### TECHNOLOGY TIP

Make sure to highlight the last sentence of this Technology Corner. In many cases, students will be given summary statistics instead of the actual data.

### Teaching Tip

Even though it is acceptable to use the calculator's T-Test feature on the AP® exam, you can insist that students show additional work in your class. While the formulas are still relatively simple, we recommend this practice. It will help students understand the structure of a standardized test statistic and prepare them for multiple-choice questions that focus on the way test statistics are calculated.

### ANSWERS TO CYU

*S*: $H_0: \mu = 8$ vs. $H_a: \mu < 8$, where $\mu$ is the true mean amount of sleep that students at the professor's school get each night. *P*: One-sample $t$ test for $\mu$. *Random*: Random sample. *10%*: The sample size (28) < 10% of the population of students. *Normal/Large Sample*: The histogram below indicates that there is not much skewness and no outliers.

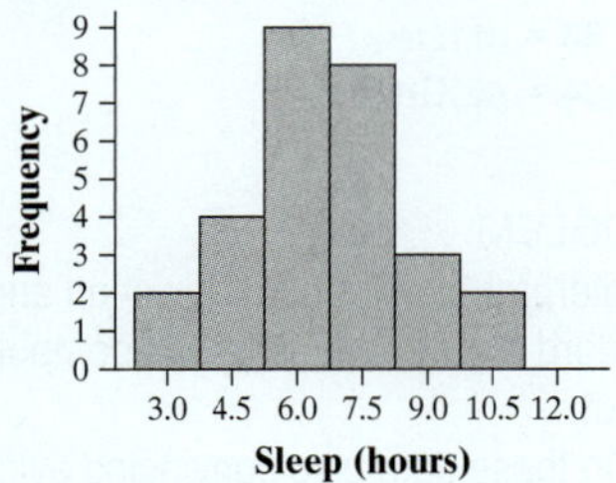


*D*: $\bar{x} = 6.643$ and $s_x = 1.981$. $t = -3.625$ and the *P*-value is between 0.0005 and 0.001. *Using technology*: *P*-value = 0.0006. *C*: Because our *P*-value of $0.0006 < \alpha = 0.05$, we reject $H_0$. There is convincing evidence that students at this university get less than 8 hours of sleep, on average.

find out, they select and weigh a random sample of 50 pineapples from this year's crop. The Minitab output below summarizes the data.

Descriptive Statistics: Weight (oz)

| Variable | N | Mean | SE Mean | StDev | Minimum | Q1 | Median | Q3 | Maximum |
|---|---|---|---|---|---|---|---|---|---|
| Weight (oz) | 50 | 31.935 | 0.339 | 2.394 | 26.491 | 29.990 | 31.739 | 34.115 | 35.547 |

PROBLEM:

(a) Do these data give convincing evidence that the mean weight of pineapples produced in the field has changed this year?

(b) Can we conclude that the different irrigation system caused a change in the mean weight of pineapples produced? Explain your answer.

SOLUTION:

(a) STATE: We want to perform a test of

$$H_0: \mu = 31$$
$$H_a: \mu \neq 31$$

where $\mu$ = the mean weight (in ounces) of all pineapples grown in the field this year. Because no significance level is given, we'll use $\alpha = 0.05$.

PLAN: If the conditions are met, we should conduct a one-sample *t* test for $\mu$.

- *Random:* The data came from a random sample of 50 pineapples from this year's crop.
  - *10%:* There need to be at least $10(50) = 500$ pineapples in the field because managers are sampling without replacement. We would expect many more than 500 pineapples in a "large field."
- *Normal/Large Sample:* We don't know whether the population distribution of pineapple weights this year is Normally distributed. But $n = 50 \geq 30$, so the large sample size makes it OK to use *t* procedures.

DO: From the Minitab output, $\bar{x} = 31.935$ ounces and $s_x = 2.394$ ounces.

- *Test statistic*

$$t = \frac{\bar{x} - \mu_0}{s_x/\sqrt{n}} = \frac{31.935 - 31}{2.394/\sqrt{50}} = 2.762$$

- *P-value* Figure 9.17 displays the *P*-value for this two-sided test as an area under the *t* distribution curve with $50 - 1 = 49$ degrees of freedom.

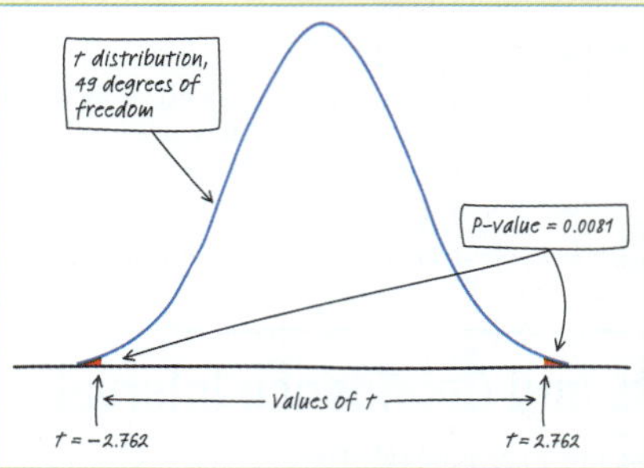


FIGURE 9.17 The *P*-value for the two-sided test with $t = 2.762$.

*Using the table:* Table B doesn't have an entry for df = 49, so we have to use the more conservative df = 40. As the excerpt below shows, the upper-tail probability is between 0.0025 and 0.005. So the desired *P*-value for this two-sided test is between $2(0.0025) = 0.005$ and $2(0.005) = 0.01$.

| | Upper-tail probability *p* | | |
|---|---|---|---|
| df | .005 | .0025 | .001 |
| 30 | 2.750 | 3.030 | 3.385 |
| 40 | 2.704 | 2.971 | 3.307 |
| 50 | 2.678 | 2.937 | 3.261 |
| | 99% | 99.5% | 99.8% |
| | Confidence level *C* | | |

*Using technology:* The calculator's T-Test command gives $t = 2.762$ and *P*-value 0.0081 using df = 49.

CONCLUDE: Because the *P*-value, 0.0081, is less than $\alpha = 0.05$, we reject $H_0$. We have convincing evidence that the mean weight of the pineapples grown this year is not 31 ounces.

(b) No. This was not a comparative experiment, so we cannot infer causation. It is possible that other things besides the irrigation system changed from last year's growing season. Maybe the weather was different this year, and that's why the pineapples have a different mean weight than last year.

For Practice *Try Exercise* 77

ALTERNATE EXAMPLE

## Don't break the ice

In the children's game Don't Break the Ice, small plastic ice cubes are squeezed into a square frame. Each child takes turns tapping out a cube of "ice" with a plastic hammer, hoping that the remaining cubes don't collapse. For the game to work correctly, the cubes must be big enough so that they hold each other in place in the plastic frame, but not so big that they are too difficult to tap out. The machine that produces the plastic cubes is designed to make cubes that are 29.5 millimeters (mm) wide, but the actual width varies a little. To ensure that the machine is working well, a supervisor inspects a random sample of 50 cubes every hour and measures their width. The Fathom output below summarizes the data from a sample taken during 1 hour:

Collection 1

| | |
|---|---|
| Width | 29.4846 mm |
| | 50 |
| | 0.0900274 mm |
| | 0.0127318 mm |

S1 = mean ( )
S2 = count ( )
S3 = stdDev ( )
S4 = stdError ( )

PROBLEM:

**(a)** Interpret the standard deviation and the standard error provided by the computer output.

**(b)** Do these data give convincing evidence that the mean width of cubes produced this hour is different from 29.5 mm?

SOLUTION:

**(a)** Standard deviation: The widths of the cubes are typically about 0.090 mm from the mean width (29.4846). Standard error: In random samples of size 50, the sample mean will typically be about 0.013 mm from the true mean.

**(b)** STATE: We want to test the following hypotheses at the $\alpha = 0.05$ significance level:

$$H_0: \mu = 29.5$$
$$H_a: \mu \neq 29.5$$

where $\mu$ = the true mean width of plastic ice cubes.

PLAN: If conditions are met, we will perform a one-sample *t* test for $\mu$.

- *Random*: A random sample of plastic ice cubes was selected.
  - *10%*: It is reasonable to assume that there are more than $10(50) = 500$ cubes produced by this machine each hour.
- *Normal/Large Sample*: The sample size is large ($n = 50 \geq 30$).

DO:

- *Test statistic* $t = \dfrac{29.4846 - 29.5}{0.0900274/\sqrt{50}} = -1.21$
- *P-value* Using the *t* distribution with 40 degrees of freedom, the *P*-value is between $2(0.10) = 0.20$ and $2(0.15) = 0.30$. *Using technology*: With df = 49, the calculator's T-Test gives *P*-value = 0.2322.

CONCLUDE: Because the *P*-value of 0.2322 is greater than $\alpha = 0.05$, we fail to reject the null hypothesis. There is not convincing evidence that the true width of the plastic ice cubes produced this hour is different from 29.5 mm.

## TECHNOLOGY TIP

When computer output provides the "standard error of the mean," we can use this value in the denominator of the *t* statistic instead of $s_x/\sqrt{n}$. Using the Minitab output on page 584, show your students that $2.394/\sqrt{50} = 0.339$, so $t = \dfrac{31.935 - 31}{0.339}$.

The significance test in the previous example gives convincing evidence that the mean weight $\mu$ of the pineapples grown in the field this year differs from last year's 31 ounces. Unfortunately, the test doesn't give us an idea of what the actual value of $\mu$ is. For that, we need a confidence interval.

**EXAMPLE**

## Juicy Pineapples

### *Confidence intervals give more information*

Minitab output for a significance test and confidence interval based on the pineapple data is shown below. The test statistic and *P*-value match what we got earlier (up to rounding).

**One-Sample T: Weight (oz)**

```
Test of mu = 31 vs not = 31
Variable      N   Mean   StDev  SE Mean        95% CI           T      P
Weight (oz)  50  31.935  2.394   0.339   (31.255, 32.616)  2.76  0.008
```

The 95% confidence interval for the mean weight of all the pineapples grown in the field this year is 31.255 to 32.616 ounces. We are 95% confident that this interval captures the true mean weight $\mu$ of this year's pineapple crop.

As with proportions, there is a link between a two-sided test at significance level $\alpha$ and a $100(1 - \alpha)\%$ confidence interval for a population mean $\mu$. For the pineapples, the two-sided test at $\alpha = 0.05$ rejects $H_0: \mu = 31$ in favor of $H_a: \mu \neq 31$. The corresponding 95% confidence interval does not include 31 as a plausible value of the parameter $\mu$. In other words, the test and interval lead to the same conclusion about $H_0$. But the confidence interval provides much more information: a set of plausible values for the population mean.

The connection between two-sided tests and confidence intervals is even stronger for means than it was for proportions. That's because both inference methods for means use the standard error of $\bar{x}$ in the calculations.

$$\text{test statistic: } t = \frac{\bar{x} - \mu_0}{s_x/\sqrt{n}} \qquad \text{Confidence interval: } \bar{x} \pm t^* \frac{s_x}{\sqrt{n}}$$

When the two-sided significance test at level $\alpha$ rejects $H_0: \mu = \mu_0$, the $100(1-\alpha)\%$ confidence interval for $\mu$ will not contain the hypothesized value $\mu_0$. And when the test fails to reject the null hypothesis, the confidence interval will contain $\mu_0$.

**THINK ABOUT IT**

**Is there a connection between one-sided tests and confidence intervals for a population mean?** As you might expect, the answer is yes. But the link is more complicated. Consider a one-sided test of $H_0: \mu = 10$ versus $H_a: \mu > 10$ based on an SRS of 30 observations. With df $= 30 - 1 = 29$, Table B says that the test will reject $H_0$ at $\alpha = 0.05$ if the test statistic $t$ is greater than 1.699. For this to happen, the sample mean $\bar{x}$ would have to exceed $\mu_0 = 10$ by more than 1.699 standardized units.

Table B also shows that $t^* = 1.699$ is the critical value for a 90% confidence interval. That is, a 90% confidence interval will extend 1.699 standardized units

**ALTERNATE EXAMPLE**

### *Don't break the ice*

Here is Fathom output for a 95% confidence interval for the true mean width of plastic ice cubes produced this hour:

```
Estimate of Collection 1                              Estimate Mean
Attribute (numeric): Width
Interval estimate for population mean of Width

Count:              50
Mean:               29.4846 mm
Std dev:            0.0900274 mm
Std error:          0.0127318 mm
Confidence level:   95.0 %
Estimate:           29.4846 mm +/- 0.0255855 mm
Range:              29.459 mm to 29.5102 mm
```

PROBLEM:

**(a)** Interpret the confidence interval. Would you make the same conclusion with the confidence interval as you did with the significance test in the previous Alternate Example?

**(b)** Interpret the confidence level.

SOLUTION:

**(a)** We are 95% confident that the interval from 29.4590 to 29.5102 mm captures the true mean width of plastic ice cubes produced this hour. Because the interval includes 29.5 as a plausible value for the true mean width, we do not have convincing evidence that the true mean is different from 29.5 mm. This is the same conclusion we made in the significance test earlier.

**(b)** If we were to take many random samples of 50 plastic ice cubes and make 95% confidence intervals for the true mean width of the cubes, then about 95% of the intervals we construct will include the true mean width.

## Teaching Tip

The Investigative Task (Question 6) from the 2004 AP® exam explores the link between one-sided tests and confidence intervals and introduces students to one-sided confidence intervals. Although one-sided confidence intervals are not on the AP® Statistics Topic Outline, discussing this question is a nice way to explore the relationship between confidence intervals and significance tests. Remember that the Investigative Task expects students to apply their knowledge in new contexts or nonroutine ways, including in areas not on the Topic Outline.

on either side of the sample mean $\bar{x}$. If $\bar{x}$ exceeds 10 by more than 1.699 standardized units, the resulting interval will not include 10. And the one-sided test will reject $H_0: \mu = 10$. There's the link: our one-sided test at $\alpha = 0.05$ gives the same conclusion about $H_0$ as a 90% confidence interval for $\mu$.

## CHECK YOUR UNDERSTANDING

The health director of a large company is concerned about the effects of stress on the company's middle-aged male employees. According to the National Center for Health Statistics, the mean systolic blood pressure for males 35 to 44 years of age is 128. The health director examines the medical records of a random sample of 72 male employees in this age group. The Minitab output displays the results of a significance test and a confidence interval.

```
Session

One-Sample T

Test of mu = 128 vs not = 128

 N    Mean  StDev  SE Mean          95% CI          T      P
72  129.93  14.90     1.76  (126.43, 133.43)  1.10  0.275
```

1. Do the results of the significance test give convincing evidence that the mean blood pressure for all the company's middle-aged male employees differs from the national average? Justify your answer.

2. Interpret the 95% confidence interval in context. Explain how the confidence interval leads to the same conclusion as in Question 1.

## Inference for Means: Paired Data

Study designs that involve making two observations on the same individual, or one observation on each of two similar individuals, yield **paired data**. When paired data result from measuring the same quantitative variable twice, we can make comparisons by analyzing the differences in each pair. If the conditions for inference are met, we can use one-sample $t$ procedures to perform inference about the mean difference $\mu_d$. (These methods are sometimes called **paired $t$ procedures**.) An example should help illustrate what we mean.

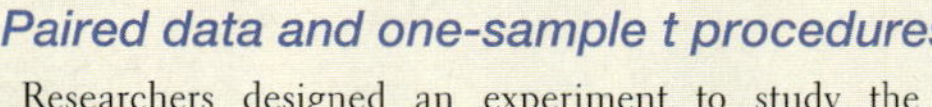

### EXAMPLE Is Caffeine Dependence Real?

*Paired data and one-sample t procedures*

Researchers designed an experiment to study the effects of caffeine withdrawal. They recruited 11 volunteers who were diagnosed as being caffeine dependent to serve as subjects. Each subject was barred from coffee, colas, and other substances with caffeine for the duration of the experiment. During one 2-day period, subjects took capsules containing their normal caffeine intake. During another 2-day period, they took placebo capsules. The order in which subjects took caffeine and the placebo was randomized. At the end of each 2-day period, a test for depression was given to all 11 subjects. Researchers wanted to know whether being deprived of caffeine would lead to an increase in depression.[18]

### ANSWERS TO CYU

1. *S*: $H_0: \mu = 128$ vs. $H_a: \mu \neq 128$, where $\mu$ is the true mean systolic blood pressure for the company's middle-aged male employees. *P*: One-sample $t$ test for $\mu$. *Random*: Random sample. *10%*: The sample size (72) < 10% of the population of middle-aged male employees. *Normal/Large Sample*: $n = 72 \geq 30$. *D*: $t = 1.10$, $P$-value $= 0.275$. *C*: Because our $P$-value of $0.275 > \alpha = 0.05$, we fail to reject $H_0$. There is not convincing evidence that the mean systolic blood pressure for this company's middle-aged male employees differs from the national average of 128.

2. We are 95% confident that the interval from 126.43 to 133.43 captures the true mean systolic blood pressure for the company's middle-aged male employees. The value of 128 is in this interval and is therefore a plausible mean systolic blood pressure for the males 35 to 44 years of age.

### Teaching Tip

Paired data can be produced in observational studies and in experiments. In an experiment, a matched pairs design is a special case of a randomized block design where each block has two units. In experiments or observational studies, using paired data can increase power by accounting for a source of variability in the response variable. In the example beginning on this page, using a matched pairs design helps account for the variability introduced by using different subjects in the experiment.

### ALTERNATE EXAMPLE

*Is the express lane faster?*

For their second-semester project in AP® Statistics, Libby and Kathryn decided to investigate which line was faster in the supermarket: the express lane or a regular lane. To collect their data, they randomly selected 15 times during a week, went to the same store, and bought the same item. One of the students used the express lane and the other used the closest regular lane. To decide which lane each of them would use, they flipped a coin. They entered the lanes at the same time, paid with the same method, and recorded the time in seconds it took them to complete the transaction.

| Time in Regular Lane (Seconds) | Time in Express Lane (Seconds) |
|---|---|
| 342 | 337 |
| 472 | 226 |
| 456 | 502 |
| 529 | 408 |
| 181 | 151 |
| 339 | 284 |
| 229 | 150 |
| 263 | 357 |
| 332 | 349 |
| 352 | 257 |
| 341 | 321 |
| 397 | 383 |
| 694 | 565 |
| 324 | 363 |
| 127 | 85 |

PROBLEM: Carry out a test to see if there is convincing evidence that the express lane is faster, on average.

SOLUTION: Because these data are paired, we will consider the differences in time (regular − express). Here are the 15 differences, with a positive difference indicating that the express lane was faster (took less time):

| | | | | | | | |
|---|---|---|---|---|---|---|---|
| 5 | 246 | −46 | 121 | 30 | 55 | 79 | −94 |
| −17 | 95 | 20 | 14 | 129 | −39 | 42 | |

The table below contains data on the subjects' scores on the depression test. Higher scores show more symptoms of depression. For each subject, we calculated the difference in test scores following each of the two treatments (placebo − caffeine). We chose this order of subtraction to get mostly positive values.

**Results of a caffeine-deprivation study**

| Subject | Depression (caffeine) | Depression (placebo) | Difference (placebo − caffeine) |
|---|---|---|---|
| 1 | 5 | 16 | 11 |
| 2 | 5 | 23 | 18 |
| 3 | 4 | 5 | 1 |
| 4 | 3 | 7 | 4 |
| 5 | 8 | 14 | 6 |
| 6 | 5 | 24 | 19 |
| 7 | 0 | 6 | 6 |
| 8 | 0 | 3 | 3 |
| 9 | 2 | 15 | 13 |
| 10 | 11 | 12 | 1 |
| 11 | 1 | 0 | −1 |

PROBLEM:

(a) Why did researchers randomly assign the order in which subjects received placebo and caffeine?

(b) Carry out a test to investigate the researchers' question.

SOLUTION:

(a) Researchers want to be able to conclude that any statistically significant change in depression score is due to the treatments themselves and not to some other variable. One obvious concern is the order of the treatments. Suppose that caffeine were given to all the subjects during the first 2-day period. What if the weather were nicer on these 2 days than during the second 2-day period when all subjects were given a placebo? Researchers wouldn't be able to tell if a large increase in the mean depression score is due to the difference in weather or due to the treatments. Random assignment of the caffeine and placebo to the two time periods in the experiment should help ensure that no other variable (like the weather) is systematically affecting subjects' responses.

(b) We'll follow the four-step process.

STATE: If caffeine deprivation has no effect on depression, then we would expect the actual mean difference in depression scores to be 0. We therefore want to test the hypotheses

$$H_0: \mu_d = 0$$
$$H_a: \mu_d > 0$$

where $\mu_d$ is the true mean difference (placebo − caffeine) in depression score for subjects like these. Because no significance level is given, we'll use $\alpha = 0.05$.

It is uncommon for the subjects in an experiment to be randomly selected from some larger population. In fact, most experiments use recruited volunteers as subjects. When there is no sampling, we don't need to check the 10% condition.

PLAN: If the conditions are met, we should conduct a paired *t* test for $\mu_d$.

- *Random:* Researchers randomly assigned the treatments—placebo then caffeine, caffeine then placebo—to the subjects.
  - 10%: We aren't sampling, so it isn't necessary to check the 10% condition.

STATE: We want to test the following hypotheses at the $\alpha = 0.05$ significance level:

$$H_0: \mu_d = 0$$
$$H_a: \mu_d > 0$$

where $\mu_d$ = the true mean difference (regular − express) in time required to purchase an item at the supermarket.

PLAN: If conditions are met, we will perform a paired *t* test for $\mu_d$.

- *Random*: A random sample of times to make the purchases was selected.
  - *10%*: 15 is less than 10% of all possible visiting times at the store.
- *Normal/Large Sample*: We don't know if the population distribution of differences is approximately Normal and we don't have a large sample size, so we will graph the differences and look for any departures from Normality.

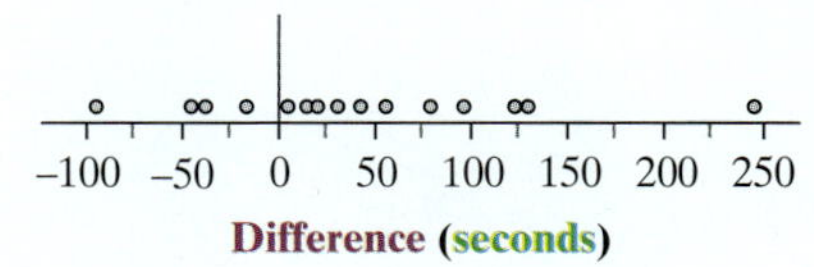


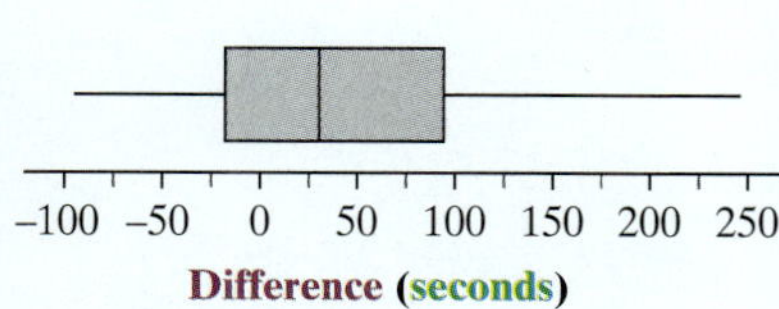


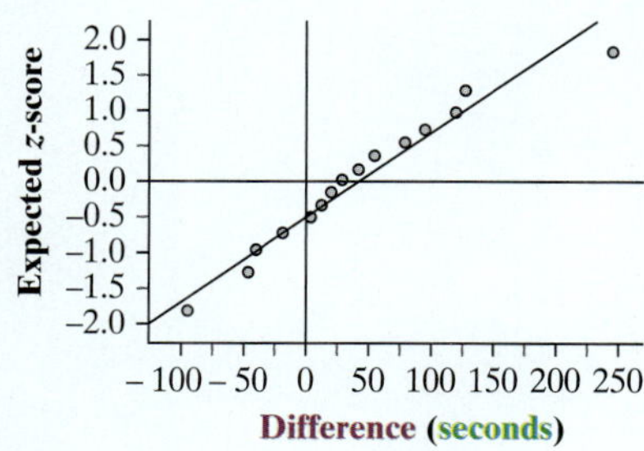


The graphs do not show much skewness and there are no outliers, so it is reasonable to use *t* procedures for these data.

DO: The sample mean difference is $\bar{x} = 42.7$ seconds with a standard deviation of $s_x = 84.0$ seconds.

- *Test statistic* $t = \dfrac{42.7 - 0}{84.0/\sqrt{15}} = 1.97$
- *P-value* Using the *t* distribution with $15 - 1 = 14$ degrees of freedom, the *P*-value is between 0.025 and 0.05. *Using technology*: The calculator's T-Test gives *P*-value = 0.035.

CONCLUDE: Because the *P*-value of 0.035 is less than $\alpha = 0.05$, we reject the null hypothesis. There is convincing evidence that the express lane is faster than the regular lane, on average.

**AP® Exam Common Error**

In problems involving paired data, students often graph the two original lists of data instead of making a single graph of the differences. If the problem involves paired data, then students should use the differences for each step of the four-step process.

- *Normal/Large Sample:* We don't know whether the actual distribution of difference in depression scores (placebo − caffeine) for subjects like these is Normal. With such a small sample size ($n = 11$), we need to graph the data to see if it's safe to use *t* procedures. Figure 9.18 shows hand sketches of a calculator histogram, boxplot, and Normal probability plot for these data. The histogram has an irregular shape with so few values; the boxplot shows some right skewness but no outliers; and the Normal probability plot is slightly curved, indicating mild skewness. With no outliers or strong skewness, the *t* procedures should be fairly accurate.

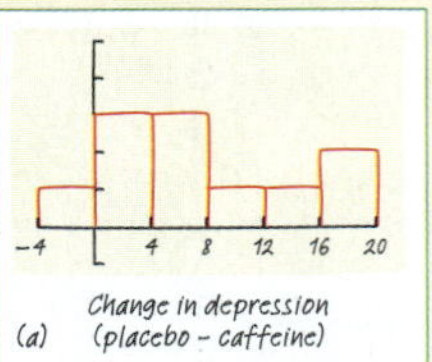


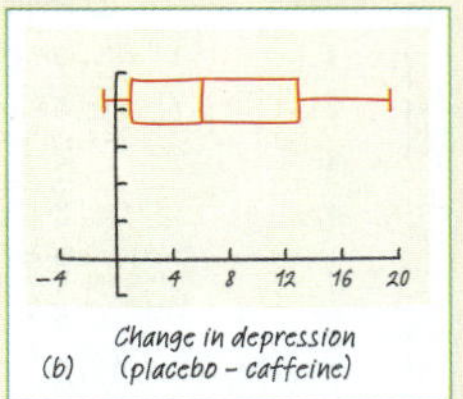


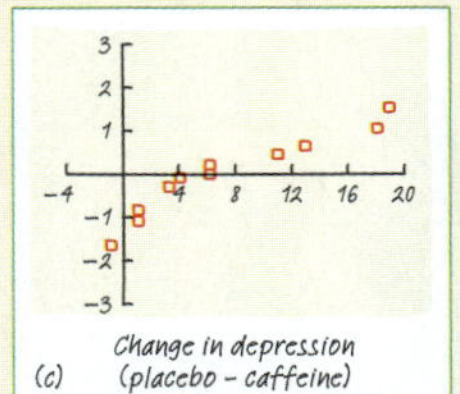


**FIGURE 9.18** Sketches of (a) a histogram, (b) a boxplot, and (c) a Normal probability plot of the change in depression scores (placebo − caffeine) for the 11 subjects in the caffeine experiment.

DO: We entered the differences in list1 and then used the calculator's *t*test command with the "Draw" option.

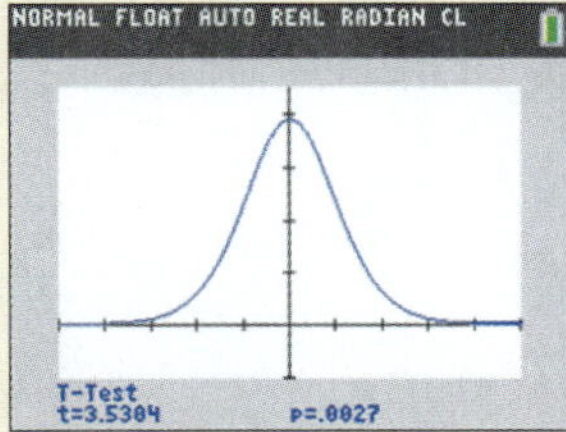


- *Test statistic* $t = 3.53$
- *P-value* 0.0027, which is the area to the right of $t = 3.53$ on the *t* distribution curve with $df = 11 - 1 = 10$.

*Note:* The calculator doesn't report the degrees of freedom, but you should.

Just by looking at the data, it appears that the true mean change in depression score $\mu_d$ is greater than 0. However, it's possible that there has been no change and we got a result this much larger than $\mu_d = 0$ by the luck of the random assignment. The significance test tells us whether this explanation is plausible.

CONCLUDE: Because the *P*-value of 0.0027 is less than $\alpha = 0.05$, we reject $H_0: \mu_d = 0$. We have convincing evidence that the true mean difference (placebo − caffeine) in depression score is positive for subjects like these.

**For Practice** *Try Exercise* 85

A few follow-up comments about this example are in order.

1. We could have calculated the test statistic in the example using the formula

$$t = \frac{\bar{x}_d - \mu_0}{s_d/\sqrt{n}} = \frac{7.364 - 0}{6.918/\sqrt{11}} = 3.53$$

   and obtained the *P*-value using Table B or technology. Check with your teacher on whether the calculator-only method is acceptable. *Be sure to report the degrees of freedom with any t procedure, even if technology doesn't.*

2. The subjects in this experiment were *not* chosen at random from the population of caffeine-dependent individuals. As a result, we can't generalize our findings to *all* caffeine-dependent people—only to people like the ones who took part in this experiment.

## AP® Exam Common Error

Students often mix up the role of random selection and the role of random assignment in a statistical study. From Section 4.3, random selection from a population allows us to make an inference about the population based on the sample data. Random assignment of subjects to treatments allows us to make cause-and-effect conclusions.

3. Because researchers randomly assigned the treatments, they can make an inference about cause and effect. The data from this experiment provide convincing evidence that depriving caffeine-dependent subjects like these of caffeine causes an average increase in depression scores.

Until now, we have only used one-sample $t$ procedures in settings involving random sampling. The paired data in the caffeine example came from a matched pairs experiment, in which each subject received both treatments in a random order. A coin toss or some other chance process was used to carry out the random assignment. Why is it legitimate to use a $t$ distribution to perform inference about the parameter $\mu$ in a randomized experiment? The answer to that question will have to wait until the next chapter.

## CHECK YOUR UNDERSTANDING

Refer to the Data Exploration from Chapter 4 on page 257. Do the data give convincing evidence at the $\alpha = 0.05$ significance level that filling tires with nitrogen instead of air decreases pressure loss?

## Using Tests Wisely

Significance tests are widely used in reporting the results of research in many fields. New drugs require significant evidence of effectiveness and safety. Courts ask about statistical significance in hearing discrimination cases. Marketers want to know whether a new ad campaign significantly outperforms the old one, and medical researchers want to know whether a new therapy performs significantly better. In all these uses, statistical significance is valued because it points to an effect that is unlikely to occur simply by chance.

Carrying out a significance test is often quite simple, especially if you use technology. Using tests wisely is not so simple. Here are some points to keep in mind when using or interpreting significance tests.

**Determining Sample Size** How large a sample should researchers take when they plan to carry out a significance test? The answer depends on three factors: significance level, effect size, and the desired power of the test. Here are the questions that researchers must answer to decide how many observations they need:

1. *Significance level.* How much risk of a Type I error—rejecting the null hypothesis when $H_0$ is actually true—are we willing to accept? If a Type I error has serious consequences, we might opt for $\alpha = 0.01$. Otherwise, we should choose $\alpha = 0.05$ or $\alpha = 0.10$. Recall that using a higher significance level would decrease the Type II error probability and increase the power.

### ANSWERS TO CYU

*S*: $H_0: \mu_d = 0$ vs. $H_a: \mu_d > 0$, where $\mu_d$ is the true mean difference (air − nitrogen) in pressure lost. *P*: Paired $t$ test for $\mu_d$. *Random*: Treatments were assigned at random to each pair of tires. *Normal/Large Sample*: $n = 31 \geq 30$. *D*: $\bar{x} = 1.252$ and $s_x = 1.202$. $t = 5.80$ and *P*-value $\approx 0$. *C*: Because the *P*-value of approximately $0 < \alpha = 0.05$, we reject $H_0$. We have convincing evidence that the true mean difference in pressure lost (air − nitrogen) $> 0$. In other words, we have convincing evidence that tires lose less pressure when filled with nitrogen than when filled with air, on average.

### Teaching Tip

Determining the sample size is a very important part of the planning process. Many studies fail because of a lack of power—something that a statistician might have predicted before the study began. Lesson: always consult a statistician before investing time and money in a research project.

2. *Effect size.* How large a difference between the null parameter value and the actual parameter value is important for us to detect?
3. *Power.* What chance do we want our study to have to detect a difference of the size we think is important?

Let's illustrate typical answers to these questions using an example.

## EXAMPLE Developing Stronger Bones

### *Planning a study*

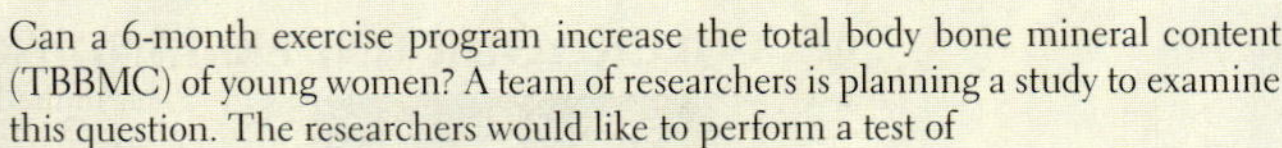

Can a 6-month exercise program increase the total body bone mineral content (TBBMC) of young women? A team of researchers is planning a study to examine this question. The researchers would like to perform a test of

$$H_0: \mu = 0$$
$$H_a: \mu > 0$$

where $\mu$ is the true mean percent change in TBBMC due to the exercise program. To decide how many subjects they should include in their study, researchers begin by answering the three questions above.

1. *Significance level.* The researchers decide that $\alpha = 0.05$ gives enough protection against declaring that the exercise program increases bone mineral content when it really doesn't (a Type I error).
2. *Effect size.* A mean increase in TBBMC of 1% would be considered important to detect.
3. *Power.* The researchers want probability at least 0.9 that a test at the chosen significance level will reject the null hypothesis $H_0: \mu = 0$ when the truth is $\mu = 1$.

The following Activity gives you a chance to investigate the sample size needed to achieve a power of 0.9 in the bone mineral content study.

## ACTIVITY Investigating Power

**MATERIALS:** Computer with Internet connection and display capability

In this Activity, you will use the *Statistical Power* applet at the book's Web site to determine the sample size needed for the exercise study of the previous example. Based on the results of a previous study, researchers are willing to assume that $\sigma = 2$ for the percent change in TBBMC over the 6-month period. We'll start by seeing whether or not 25 subjects are enough.

## ALTERNATE EXAMPLE

### *Computer-based learning*

The superintendent of a large school district wants to buy a computer-based math program for his district. Because the program is very expensive, he will randomly select some classes to pilot-test the program. The response variable will be the average increase in grade-level equivalent for the students in the class. For example, if a student went from 3.1 (third grade, 1 month) to 4.3 (fourth grade, 3 months) by the end of the year, the student increased 1.2 grade levels during the year. The hypotheses that the superintendent will test are: $H_0: \mu = 1$ vs. $H_a: \mu > 1$, where $\mu$ = the mean increase in grade-level equivalent for classes in the district. To determine the number of classes he should use, the superintendent considers the following:

- *Significance level*: The superintendent wants to avoid a Type I error (and spending money on a program that doesn't help), so he chooses $\alpha = 0.01$.
- *Effect size*: A mean increase of one month (1.1) would be important to detect.
- *Power*: The superintendent wants a probability of at least 0.80 that a test will detect an increase of 1 month in grade-level equivalent.

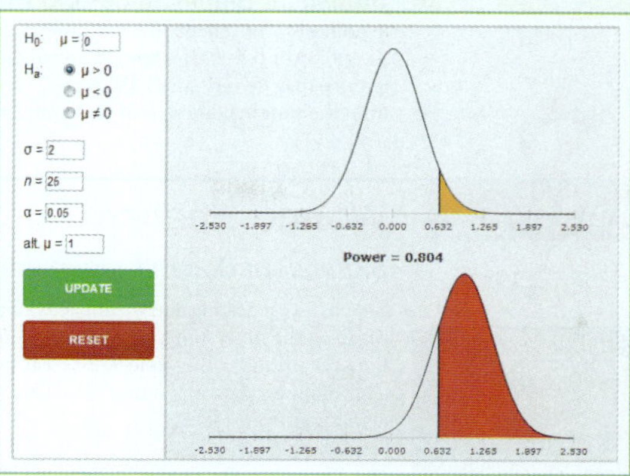


1. Go to www.whfreeman.com/tps5e and launch the *Statistical Power* applet. Enter the values: $H_0: \mu = 0$, $H_a: \mu > 0$, $\sigma = 2$, $n = 25$, $\alpha = 0.05$, and alternate $\mu = 1$. Then click "Update." What is the power? As a class, discuss what this number means in simple terms.
2. Change the significance level to 0.01. What effect does this have on the power of the test to detect $\mu = 1$? Why?
3. The researchers decide that they are willing to risk a 5% chance of making a Type I error. Change the significance level back to $\alpha = 0.05$. Now increase the sample size to 30. What happens to the power? Why?
4. Keep increasing the sample size until the power is at least 0.90. What minimum sample size should the researchers use for their study?
5. Would the researchers need a smaller or a larger sample size to detect a mean increase of 1.5% in TBBMC? A 0.85% increase? Use the applet to investigate.
6. Summarize what you have learned about how significance level, effect size, and power influence the sample size needed for a significance test.

Here is a summary of influences on "How large a sample do I need?" from the Activity.

- If you insist on a smaller significance level (such as 1% rather than 5%), you have to take a larger sample. A smaller significance level requires stronger evidence to reject the null hypothesis.
- If you insist on higher power (such as 0.99 rather than 0.90), you will need a larger sample. Higher power gives a better chance of detecting a difference when it really exists.
- At any significance level and desired power, detecting a small difference between the null and alternative parameter values requires a larger sample than detecting a large difference.

## ALTERNATE ACTIVITY

### *Computer-based learning*

Let's use an applet to calculate the sample size needed for the computer-based learning study in the previous Alternate Example.

**1.** Go to www.whfreeman.com/tps5e and launch the *Statistical Power* applet. Enter the values: $H_0$: $\mu = 1$, $H_a$: $\mu > 1$, $\sigma = 0.2$, $n = 10$, $\alpha = 0.01$, and alternate $\mu = 1.1$. Then click "Update." What is the power?

**2.** Increase the sample size ($n$) until the power is at least 0.80. What is the smallest sample size that the superintendent can use if he wants the power to be at least 0.80?

**3.** If the superintendent cannot afford to use a sample this large but insists that the power be at least 0.80, what other options does he have? Be specific. Are there any trade-offs to these options?

ALTERNATE EXAMPLE

### *Improving SAT scores*

A national chain of SAT-preparation schools wants to know if using a smartphone app in addition to their regular program will help increase student scores more than using just the regular program. On average, the students in their regular program increase their scores by 128 points during the 3-month class. To investigate using the smartphone app, they have 5000 students use the app along with the regular program and measure their improvement. Then they will test the following hypotheses: $H_0$: $\mu = 128$ vs. $H_a$: $\mu > 128$, where $\mu$ is the true mean improvement in the SAT score for students at this school.

After 3 months, the average improvement was $\bar{x} = 130$ with a standard deviation of $s_x = 65$. The test statistic is $t = 2.18$ with a *P*-value of 0.0148. The increase was statistically significant, but not practically important. An increase of only 2 points on the SAT isn't a very big deal.

## Teaching Tip

The CBS News program *60 Minutes* aired an excellent segment about the placebo effect called "Treating Depression: Is There a Placebo Effect?" The 20-minute video addresses several of the issues we discuss in this section, including the difference between statistical significance and practical (clinical) importance and the use of multiple tests. Find the video at http://www.cbsnews.com/video/watch/?id=7399362n.

**Statistical Significance and Practical Importance** When a null hypothesis ("no effect" or "no difference") can be rejected at the usual levels ($\alpha = 0.05$ or $\alpha = 0.01$), there is convincing evidence of a difference. But that difference may be very small. When large samples are available, even tiny deviations from the null hypothesis will be significant.

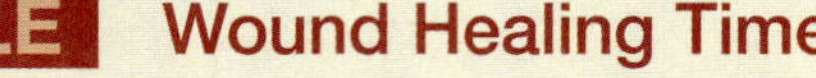

EXAMPLE

## Wound Healing Time

### *Significant doesn't mean important*

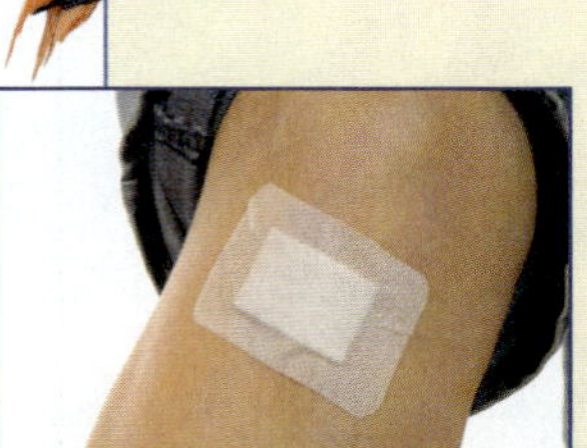

Suppose we're testing a new antibacterial cream, "Formulation NS," on a small cut made on the inner forearm. We know from previous research that with no medication, the mean healing time (defined as the time for the scab to fall off) is 7.6 days. The claim we want to test here is that Formulation NS speeds healing. We will use a 5% significance level.

*Procedure:* We cut a random sample of 250 college students and apply Formulation NS to the wounds. The mean healing time for these subjects is $\bar{x} = 7.5$ days and the standard deviation is $s_x = 0.9$ days.

*Discussion:* We want to test a claim about the mean healing time $\mu$ in the population of college students whose cuts are treated with Formulation NS. Our hypotheses are

$$H_0: \mu = 7.6 \text{ days}$$
$$H_a: \mu < 7.6 \text{ days}$$

An examination of the data reveals no outliers or strong skewness, so the conditions for performing a one-sample *t* test are met. We carry out the test and find that $t = -1.76$ and *P*-value $= 0.04$ with df $= 249$. Because 0.04 is less than $\alpha = 0.05$, we reject $H_0$. We have convincing evidence that Formulation NS reduces the average healing time. However, this result is not practically important. Having your scab fall off one-tenth a day sooner is no big deal.

Remember the wise saying: *Statistical significance is not the same thing as practical importance.* The remedy for attaching too much importance to statistical significance is to pay attention to the actual data as well as to the *P*-value. Plot your data and examine them carefully. Are there outliers or other departures from a consistent pattern? A few outlying observations can produce highly significant results if you blindly apply common significance tests. Outliers can also destroy the significance of otherwise-convincing data.

*The foolish user of statistics who feeds the data to a calculator or computer without exploratory analysis will often be embarrassed.* Is the difference you are seeking visible in your plots? If not, ask yourself whether the difference is large enough to be practically important. Give a confidence interval for the parameter in which you are interested. A confidence interval gives a set of plausible values for the parameter rather than simply asking if the observed result is too surprising to occur by chance alone when $H_0$ is true. Confidence intervals are not used as often as they should be, whereas significance tests are perhaps overused.

**Beware of Multiple Analyses** Statistical significance ought to mean that you have found a difference that you were looking for. The reasoning behind statistical significance works well if you decide what difference you are seeking, design a study to search for it, and use a significance test to weigh the evidence you get. In other settings, significance may have little meaning.

## EXAMPLE Cell Phones and Brain Cancer

***Don't search for significance!***

Might the radiation from cell phones be harmful to users? Many studies have found little or no connection between using cell phones and various illnesses. Here is part of a news account of one study:

> A hospital study that compared brain cancer patients and a similar group without brain cancer found no statistically significant difference between brain cancer rates for the two groups. But when 20 distinct types of brain cancer were considered separately, a significant difference in brain cancer rates was found for one rare type. Puzzlingly, however, this risk appeared to decrease rather than increase with greater mobile phone use.[19]

Think for a moment. Suppose that the 20 null hypotheses for these 20 significance tests are all true. Then each test has a 5% chance of being significant at the 5% level. That's what $\alpha = 0.05$ means: results this extreme occur only 5% of the time just by chance when the null hypothesis is true. We expect about 1 of 20 tests to give a significant result just by chance. Running one test and reaching the $\alpha = 0.05$ level is reasonably good evidence that you have found something; running 20 tests and reaching that level only once is not.

For more on the pitfalls of multiple analyses, do an Internet search for the XKCD comic about jelly beans causing acne.

Searching data for patterns is certainly legitimate. Exploratory data analysis is an important part of statistics. But the reasoning of formal inference does not apply when your search for a striking effect in the data is successful. The remedy is clear. Once you have a hypothesis, design a study to search specifically for the effect you now think is there. If the result of this study is statistically significant, you have real evidence.

## case closed

## Do You Have a Fever?

At the beginning of the chapter, we described a study investigating whether "normal" human body temperature is really 98.6°F. Here is a summary of the details we provided in the chapter-opening Case Study (page 537).

### ALTERNATE EXAMPLE

***More cell phones and brain cancer***

Suppose that 20 significance tests were conducted and in each case the null hypothesis was true. What is the probability that we avoid a Type I error in all 20 tests? If we are using a 5% significance level, each individual test has a 0.95 probability of avoiding a Type I error.

Assuming that the results of the tests are independent, the probability of avoiding Type I errors in each of the tests is $(0.95)(0.95)\cdots(0.95) = (0.95)^{20} = 0.36$. This means that there is a 64% chance we will make at least one Type I error in these 20 tests. So finding 1 significant result in 20 is definitely not a surprise.

To avoid this problem, we can adjust the significance level for each individual test so that the cumulative probability of avoiding a Type I error is 0.95. Using the same logic as in the last calculation, we should solve the following for $\alpha$:

$$(1-\alpha)(1-\alpha)\cdots = (1-\alpha)^{20} = 0.95$$

$$\alpha = 1 - \sqrt[20]{0.95} = 0.00256$$

Thus, if we plan to do 20 tests, use a significance level of $\alpha = 0.00256$ to get the probability of at least one Type I error to be 0.05.

### Teaching Tip

As of this printing, the web address for the XKCD comic is xkcd.com/882. Look it up!

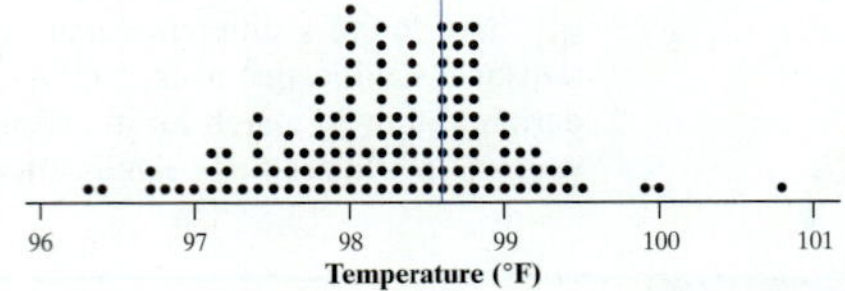


- The mean temperature was $\bar{x} = 98.25$°F.
- The standard deviation of the temperature readings was $s_x = 0.73$°F.
- 62.3% of the temperature readings were less than 98.6°F.

1. If "normal" body temperature really is 98.6°F, we would expect that about half of all healthy 18- to 40-year-olds will have a body temperature less than 98.6°F. Do the data from this study provide convincing evidence at the $\alpha = 0.05$ level that this is not the case?
2. The test in Question 1 has power 0.66 to detect that the actual population proportion is 0.60. Describe two changes that could be made to increase the power of the test.

Do the data provide convincing evidence that average normal body temperature is *not* 98.6°F? The computer output below shows the results of a one-sample $t$ test and a 95% confidence interval for the population mean $\mu$.

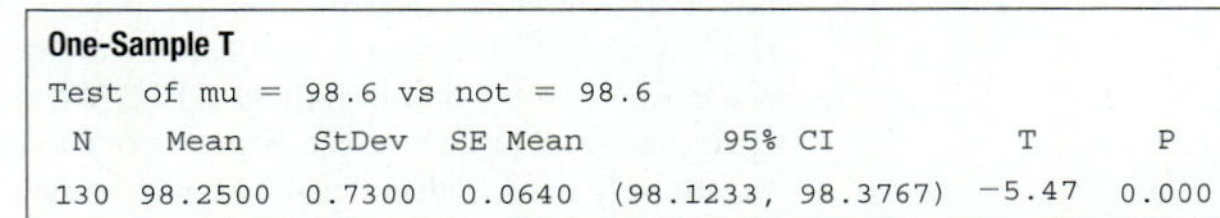

**One-Sample T**

Test of mu = 98.6 vs not = 98.6

| N | Mean | StDev | SE Mean | 95% CI | T | P |
|---|---|---|---|---|---|---|
| 130 | 98.2500 | 0.7300 | 0.0640 | (98.1233, 98.3767) | -5.47 | 0.000 |

3. What conditions must be satisfied for a one-sample $t$ test to give valid results? Show that these conditions are met in this setting.
4. Explain how the $P$-value and the confidence interval lead to the same conclusion for the significance test.
5. Based on the conclusion in Question 4, which type of error could have been made: a Type I error or a Type II error? Justify your answer.

## Section 9.3 Summary

- The conditions for performing a significance test of $H_0$: $\mu = \mu_0$ are:
  - **Random:** The data were produced by a well-designed random sample or randomized experiment.
    - **10%:** When sampling without replacement, check that the population is at least 10 times as large as the sample.

### Answers to Case Closed

**1.** *S*: $H_0: p = 0.50$ vs. $H_a: p > 0.50$, where $p$ is the true proportion of all healthy 18- to 40-year-olds who have a body temperature less than 98.6°F. *P*: One-sample $z$ test for $p$. *Random*: Random sample. *10%*: The sample size (130) < 10% of all healthy 18- to 40-year-olds. *Large Counts*: $130(0.5) = 65 \geq 10$ and $130(0.5) = 65 \geq 10$.
*D*: $z = \dfrac{0.623 - 0.50}{\sqrt{\dfrac{0.50(0.50)}{130}}} = 2.80$ and $P(Z \geq 2.80) = 0.0026$. *C*: Because the $P$-value of $0.0026 < \alpha = 0.05$, we reject $H_0$. We have convincing evidence that more than 50% of healthy 18- to 40-year-olds have a body temperature less than 98.6°F.

**2.** Increase the sample size ($n$) or increase the significance level ($\alpha$).

**3.** *Random*: A random sample of healthy 18- to 40-year-olds was selected. *10%*: 130 is less than 10% of all healthy 18- to 40-year-olds. *Normal/Large Sample*: $n = 130 \geq 30$.

**4.** Because the $P$-value of 0.000 is less than $\alpha = 0.05$, we reject $H_0$: $\mu = 98.6$. We have convincing evidence that the true mean body temperature of healthy 18- to 40-year-olds is not 98.6°F. Also, because the 95% confidence interval does not include 98.6°F as a plausible value, we have convincing evidence that the true mean body temperature of healthy 18- to 40-year-olds is not 98.6°F.

**5.** Because we found convincing evidence that the average body temperature of healthy 18- to 40-year-olds is not 98.6°F, it is possible we made a Type I error. That is, it is possible we found convincing evidence that the average body temperature of healthy 18- to 40-year-olds is not 98.6°F, when in reality it is 98.6°F.

### TRM Section 9.3 Quizzes

There are three different quizzes available for this section in the Teacher's Resource Materials. Versions A and B are roughly parallel. Find them by clicking on the link in the TE-book, opening the TRCD, or logging onto the book's companion Web site www.whfreeman.com/tps5e (teacher log-in required).

- **Normal/Large Sample:** The population distribution is Normal *or* the sample size is large ($n \geq 30$). When the sample size is small ($n < 30$), examine a graph of the sample data for any possible departures from Normality in the population. You should be safe using a *t* distribution as long as there is no strong skewness and no outliers are present.

- The **one-sample *t* test for a mean** uses the test statistic

$$t = \frac{\bar{x} - \mu_0}{s_x/\sqrt{n}}$$

with *P*-values calculated from the *t* distribution with $n - 1$ degrees of freedom.

- Confidence intervals provide additional information that significance tests do not—namely, a set of plausible values for the parameter $\mu$. A two-sided test of $H_0: \mu = \mu_0$ at significance level $\alpha$ gives the same conclusion as a $100(1 - \alpha)\%$ confidence interval for $\mu$.
- Analyze **paired data** by first taking the difference within each pair to produce a single sample. Then use one-sample *t* procedures.
- There are three factors that influence the sample size required for a statistical test: significance level, effect size, and the desired power of the test.
- Very small differences can be highly significant (small *P*-value) when a test is based on a large sample. A statistically significant difference need not be practically important.
- Many tests run at once will probably produce some significant results by chance alone, even if all the null hypotheses are true.

## 9.3 TECHNOLOGY CORNERS

TI-Nspire Instructions in Appendix B; HP Prime instructions on the book's Web site.



# Section 9.3 Exercises

**65. Attitudes** The Survey of Study Habits and Attitudes (SSHA) is a psychological test that measures students' attitudes toward school and study habits. Scores range from 0 to 200. Higher scores indicate more positive attitudes. The mean score for U.S. college students is about 115. A teacher suspects that older students have better attitudes toward school. She gives the SSHA to an SRS of 45 of the over 1000 students at her college who are at least 30 years of age. Check the conditions for carrying out a significance test of the teacher's suspicion.

pg 575

**66. Anemia** Hemoglobin is a protein in red blood cells that carries oxygen from the lungs to body tissues. People with fewer than 12 grams of hemoglobin per deciliter of blood (g/dl) are anemic. A public health official in Jordan suspects that Jordanian children are at risk of anemia. He measures a random sample of

**TRM Full Solutions to Section 9.3 Exercises**

## Answers to Section 9.3 Exercises

**9.65** *Random*: Random sample. *10%*: The sample size (45) < 10% of the population size of 1000. *Normal/Large Sample*: $n = 45 \geq 30$.

**9.66** *Random*: Random sample. *10%*: The sample size (50) < 10% of the population of children in Jordan. *Normal/Large Sample*: $n = 50 \geq 30$.

50 children. Check the conditions for carrying out a significance test of the official's suspicion.

67. **Ancient air** The composition of the earth's atmosphere may have changed over time. To try to discover the nature of the atmosphere long ago, we can examine the gas in bubbles inside ancient amber. Amber is tree resin that has hardened and been trapped in rocks. The gas in bubbles within amber should be a sample of the atmosphere at the time the amber was formed. Measurements on 9 specimens of amber from the late Cretaceous era (75 to 95 million years ago) give these percents of nitrogen:[20]

63.4 65.0 64.4 63.3 54.8 64.5 60.8 49.1 51.0

Explain why we should not carry out a one-sample $t$ test in this setting.

68. **Paying high prices?** A retailer entered into an exclusive agreement with a supplier who guaranteed to provide all products at competitive prices. The retailer eventually began to purchase supplies from other vendors who offered better prices. The original supplier filed a lawsuit claiming violation of the agreement. In defense, the retailer had an audit performed on a random sample of 25 invoices. For each audited invoice, all purchases made from other suppliers were examined and compared with those offered by the original supplier. The percent of purchases on each invoice for which an alternative supplier offered a lower price than the original supplier was recorded.[21] For example, a data value of 38 means that the price would be lower with a different supplier for 38% of the items on the invoice. A histogram and some computer output for these data are shown below. Explain why we should not carry out a one-sample $t$ test in this setting.

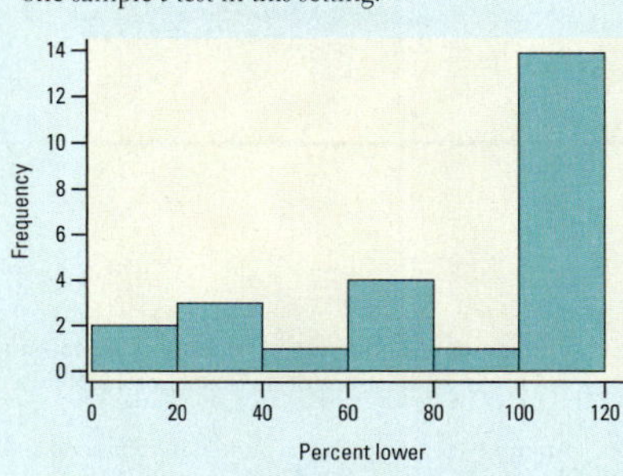


**Summary statistics**

| Column | n | Mean | Std. Dev. | Std. Err. | Median | Min | Max | Q1 | Q3 |
|---|---|---|---|---|---|---|---|---|---|
| pctlower | 25 | 77.76 | 32.6768 | 6.5353603 | 100 | 0 | 100 | 68 | 100 |

69. **Attitudes** In the study of older students' attitudes from Exercise 65, the sample mean SSHA score was 125.7 and the sample standard deviation was 29.8.

(a) Calculate the test statistic.

(b) Find the $P$-value using Table B. Then obtain a more precise $P$-value from your calculator.

70. **Anemia** For the study of Jordanian children in Exercise 66, the sample mean hemoglobin level was 11.3 mg/dl and the sample standard deviation was 1.6 mg/dl.

(a) Calculate the test statistic.

(b) Find the $P$-value using Table B. Then obtain a more precise $P$-value from your calculator.

71. **One-sided test** Suppose you carry out a significance test of $H_0: \mu = 5$ versus $H_a: \mu < 5$ based on a sample of size $n = 20$ and obtain $t = -1.81$.

(a) Find the $P$-value for this test using Table B or technology. What conclusion would you draw at the 5% significance level? At the 1% significance level?

(b) Redo part (a) using an alternative hypothesis of $H_a: \mu \neq 5$.

72. **Two-sided test** The one-sample $t$ statistic from a sample of $n = 25$ observations for the two-sided test of

$$H_0: \mu = 64$$
$$H_a: \mu \neq 64$$

has the value $t = -1.12$.

(a) Find the $P$-value for this test using Table B or technology. What conclusion would you draw at the 5% significance level? At the 1% significance level?

(b) Redo part (a) using an alternative hypothesis of $H_a: \mu < 64$.

73. pg 580 **Construction zones** Every road has one at some point—construction zones that have much lower speed limits. To see if drivers obey these lower speed limits, a police officer uses a radar gun to measure the speed (in miles per hours, or mph) of a random sample of 10 drivers in a 25 mph construction zone. Here are the data:

27 33 32 21 30 30 29 25 27 34

(a) Is there convincing evidence that the average speed of drivers in this construction zone is greater than the posted speed limit?

(b) Given your conclusion in part (a), which kind of mistake—a Type I error or a Type II error—could you have made? Explain what this mistake would mean in context.

74. **Heat through the glass** How well materials conduct heat matters when designing houses, for example. Conductivity is measured in terms of watts of heat

**9.67** The Random condition may not be met, because we don't know if this is a random sample of the atmosphere in the Cretaceous era. Also, the Normal/Large Sample condition is not met. The sample size < 30 and the histogram shows that the data are skewed to the left.

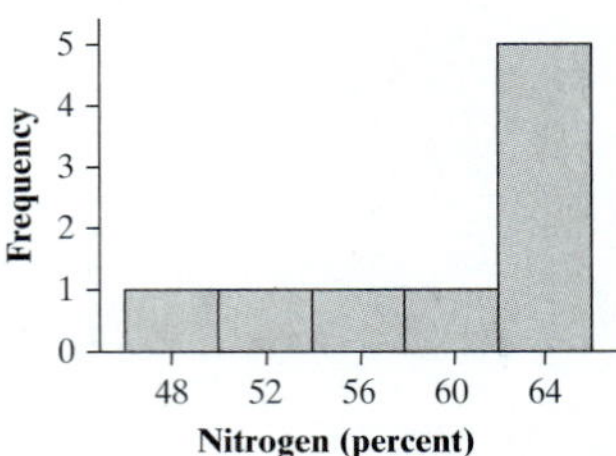


**9.68** The Normal/Large Sample condition is not met. The sample size < 30 and the histogram shows that the data are strongly skewed to the left.

**9.69 (a)** $t = \dfrac{125.7 - 115}{29.8/\sqrt{45}} = 2.409$

**(b)** For this test, df = 44. Using Table B and df = 40, 0.01 < $P$-value < 0.02. *Using technology: P*-value = `tcdf(lower:2.409,upper:1000, df:44)` = 0.0101.

**9.70 (a)** $t = \dfrac{11.3 - 12}{1.6/\sqrt{50}} = -3.094$

**(b)** For this test, df = 49. Using Table B and df = 40, 0.001 < $P$-value < 0.0025. *Using technology: P*-value = `tcdf(lower:-1000,upper:-3.094,df=49)` = 0.0016.

**9.71 (a)** Using Table B and df = 19, 0.025 < $P$-value < 0.05. *Using technology: P*-value =0.043. *5%*: Because the $P$-value of 0.043 < $\alpha = 0.05$, we reject $H_0$. There is convincing evidence that $\mu < 5$. *1%*: Because the $P$-value of 0.043 > $\alpha = 0.01$, we fail to reject $H_0$. There is not convincing evidence that $\mu < 5$.
**(b)** *Using technology: P*-value = 0.086. *5%:* Because the $P$-value of 0.086 > $\alpha = 0.05$, we fail to reject $H_0$. There is not convincing evidence that $\mu \neq 5$. *1%*: Same conclusion as part (a).

**9.72 (a)** Using Table B and df = 24, 2(0.10) < $P$-value < 2(0.15) or 0.20 < $P$-value < 0.30. *Using technology: P*-value = 0.2738. *5%*: Because the $P$-value of 0.2738 > $\alpha = 0.05$, we fail to reject $H_0$. There is not convincing evidence that $\mu \neq 64$. *1%:* Same as 5%.
**(b)** *Using technology: P*-value = 0.1369. Same conclusions as part (a).

**9.73 (a)** *S*: $H_0$: $\mu = 25$ vs. $H_a$: $\mu > 25$, where $\mu$ is the true mean speed of all drivers in a construction zone. *P*: One-sample $t$ test for $\mu$. *Random*: Random sample. *10%*: The sample size (10) < 10% of all drivers. *Normal/Large Sample*: No strong skewness or outliers in the sample.

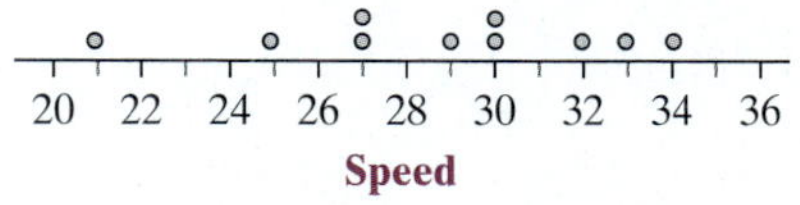


*D*: $\bar{x} = 28.8$ and $s_x = 3.94$. $t = 3.05$, df = 9, and the $P$-value is between 0.005 and 0.01 (0.0069). *C*: Because the $P$-value of 0.0069 < $\alpha = 0.05$, we reject $H_0$. We have convincing evidence that the true mean speed of all drivers in the construction zone > 25 mph. **(b)** Because we rejected $H_0$, it is possible we made a Type I error—finding convincing evidence that the true mean speed > 25 mph when it really isn't.

**9.74 (a)** *S*: $H_0$: $\mu = 1$ vs. $H_a$: $\mu > 1$, where $\mu$ is the true mean conductivity for this type of glass. *P*: One-sample $t$ test for $\mu$. *Random*: Random sample. *10%*: The sample size (11) < 10% of all pieces. *Normal/Large Counts*: The dotplot below indicates that there is not much skewness and no outliers.

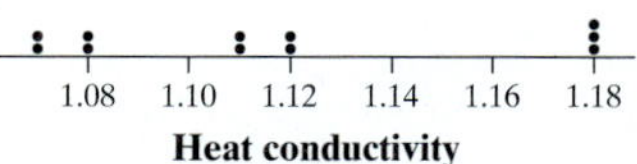


*D*: $\bar{x} = 1.1182$ and $s_x = 0.0438$. $t = 8.95$, df = 10, and $P$-value ≈ 0. *C*: Because the $P$-value of approximately 0 < $\alpha = 0.05$, we reject $H_0$. There is convincing evidence that this glass has a true mean conductivity greater than 1. **(b)** Because we rejected $H_0$, it is possible we made a Type I error—finding convincing evidence that the true mean conductivity > 1 when it really isn't.

power transmitted per square meter of surface per degree Celsius of temperature difference on the two sides of the material. In these units, glass has conductivity about 1. The National Institute of Standards and Technology provides exact data on properties of materials. Here are measurements of the heat conductivity of 11 randomly selected pieces of a particular type of glass:[22]

1.11 1.07 1.11 1.07 1.12 1.08 1.08 1.18 1.18 1.18 1.12

(a) Is there convincing evidence that the mean conductivity of this type of glass is greater than 1?

(b) Given your conclusion in part (a), which kind of mistake—a Type I error or a Type II error—could you have made? Explain what this mistake would mean in context.

75. **Healthy bones** The recommended daily allowance (RDA) of calcium for women between the ages of 18 and 24 years is 1200 milligrams (mg). Researchers who were involved in a large-scale study of women's bone health suspected that their participants had significantly lower calcium intakes than the RDA. To test this suspicion, the researchers measured the daily calcium intake of a random sample of 36 women from the study who fell in the desired age range. The Minitab output below displays the results of a significance test.

**One-Sample T: Calcium intake (mg)**

Test of mu = 1200 vs < 1200

| Variable | N | Mean | StDev | SE Mean | T | P |
|---|---|---|---|---|---|---|
| Calcium | 36 | 856.2 | 306.7 | 51.1 | −6.73 | 0.000 |

(a) Do these data give convincing evidence to support the researchers' suspicion? Justify your answer.

(b) Interpret the $P$-value in context.

76. **Taking stock** An investor with a stock portfolio worth several hundred thousand dollars sued his broker due to the low returns he got from the portfolio at a time when the stock market did well overall. The investor's lawyer wants to compare the broker's performance against the market as a whole. He collects data on the broker's returns for a random sample of 36 weeks. Over the 10-year period that the broker has managed portfolios, stocks in the Standard & Poor's 500 index gained an average of 0.95% per week. The Minitab output below displays the results of a significance test.

**One-Sample T: Return (percent)**

Test of mu = 0.95 vs < 0.95

| Variable | N | Mean | StDev | SE Mean | T | P |
|---|---|---|---|---|---|---|
| Return (percent) | 36 | −1.441 | 4.810 | 0.802 | −2.98 | 0.003 |

(a) Do these data give convincing evidence to support the lawyer's case? Justify your answer.

(b) Interpret the $P$-value in context.

77. **Pressing pills** pg 583  A drug manufacturer forms tablets by compressing a granular material that contains the active ingredient and various fillers. The hardness of a sample from each batch of tablets produced is measured to control the compression process. The target value for the hardness is $\mu = 11.5$. The hardness data for a random sample of 20 tablets are

| | | | | |
|---|---|---|---|---|
| 11.627 | 11.613 | 11.493 | 11.602 | 11.360 |
| 11.374 | 11.592 | 11.458 | 11.552 | 11.463 |
| 11.383 | 11.715 | 11.485 | 11.509 | 11.429 |
| 11.477 | 11.570 | 11.623 | 11.472 | 11.531 |

Is there convincing evidence at the 5% level that the mean hardness of the tablets differs from the target value?

78. **Filling cola bottles** Bottles of a popular cola are supposed to contain 300 milliliters (ml) of cola. There is some variation from bottle to bottle because the filling machinery is not perfectly precise. An inspector measures the contents of six randomly selected bottles from a single day's production. The results are

299.4 297.7 301.0 298.9 300.2 297.0

Do these data provide convincing evidence that the mean amount of cola in all the bottles filled that day differs from the target value of 300 ml?

79. **Pressing pills** Refer to Exercise 77. Construct and interpret a 95% confidence interval for the population mean $\mu$. What additional information does the confidence interval provide?

80. **Filling cola bottles** Refer to Exercise 78. Construct and interpret a 95% confidence interval for the population mean $\mu$. What additional information does the confidence interval provide?

81. **Fast connection?** How long does it take for a chunk of information to travel from one server to another and back on the Internet? According to the site internettrafficreport.com, a typical response time is 200 milliseconds (about one-fifth of a second). Researchers collected data on response times of a random sample of 14 servers in Europe. A graph of the data reveals no strong skewness or outliers. The following figure displays Minitab output for a one-sample $t$ interval for the population mean. Is there convincing evidence at the 5% significance level that the site's claim is incorrect? Justify your answer.

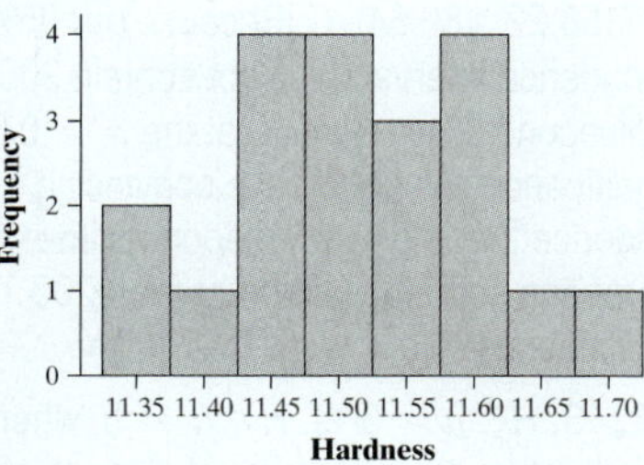


*D*: $\bar{x} = 11.5164$ and $s_x = 0.0950$. $t = 0.77$, df = 19, and the $P$-value is between 0.40 and 0.50 (0.4494). *C*: Because our $P$-value of $0.4494 > \alpha = 0.05$, we fail to reject $H_0$. We do not have convincing evidence that the true mean hardness of these tablets is different from 11.5.

**9.75 (a)** *S*: $H_0: \mu = 1200$ vs. $H_a: \mu < 1200$, where $\mu$ is the true mean daily calcium intake of women 18 to 24 years of age. *P*: One-sample *t* test for $\mu$. *Random*: Random sample. *10%*: The sample size (36) < 10% of all women aged 18 to 24. *Normal/Large Sample*: $n = 36 \geq 30$. *D*: $t = -6.73$ and $P$-value = 0.000. *C*: Because the $P$-value of approximately $0 < \alpha = 0.05$, we reject $H_0$. There is convincing evidence that women aged 18 to 24 are getting less than 1200 mg of calcium daily, on average. **(b)** Assuming that women aged 18 to 24 get 1200 mg of calcium per day, on average, there is about a 0 probability that we would observe a sample mean ≤ 856.2 mg by chance alone.

**9.76 (a)** *S*: $H_0: \mu = 0.95$ vs. $H_a: \mu < 0.95$, where $\mu$ is the true mean weekly return on the broker's portfolios. *P*: One-sample *t* test for $\mu$. *Random*: Random sample. *10%*: The sample size (36) < 10% of the total number of weeks in the 10 years (520). *Normal/Large Sample*: $n = 36 \geq 30$. *D*: $t = -2.98$ and $P$-value = 0.003. *C*: Because our $P$-value of $0.003 < \alpha = 0.05$, we reject $H_0$. We have convincing evidence that the portfolios managed by this broker had a smaller true mean return than Standard and Poor's mean return. **(b)** Assuming that the broker's true mean weekly return is 0.95%, there is a 0.003 probability of getting a sample mean weekly return of −1.441% or less by chance alone.

**9.77** *S*: $H_0: \mu = 11.5$ vs. $H_a: \mu \neq 11.5$, where $\mu$ is the true mean hardness of the tablets. *P*: One-sample *t* test for $\mu$. *Random*: Random sample. *10%*: The sample size (20) < 10% of all tablets in the batch. *Normal/Large Sample*: No strong skewness or outliers.

**9.78** *S*: $H_0: \mu = 300$ ml vs. $H_a: \mu \neq 300$ ml, where $\mu$ is the true mean amount of cola in the bottles. *P*: One-sample *t* test for $\mu$. *Random*: Random sample. *10%*: The sample size (6) < 10% of all bottles produced this day. *Normal/Large Sample*: No strong skewness or outliers.

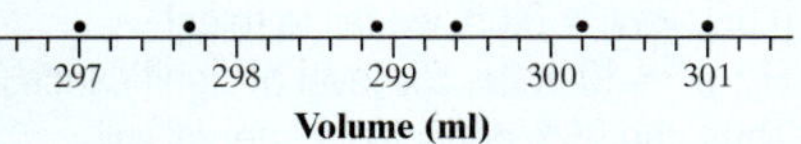


*D*: $\bar{x} = 299.033$ and $s_x = 1.503$. $t = -1.58$, df = 5, and the $P$-value is between 0.10 and 0.20 (0.1760). *C*: Because our $P$-value of $0.1760 > \alpha = 0.05$, we fail to reject $H_0$. We do not have convincing evidence that the true mean amount of cola in all the bottles differs from the target value of 300 ml.

**9.79** *D*: With df = 19, (11.472, 11.561). *C*: We are 95% confident that the interval from 11.472 to 11.561 captures the true mean hardness measurement for this type of pill. The confidence interval gives 11.5 as a plausible value for the true mean hardness $\mu$, but it gives other plausible values as well.

**9.80** *D*: With df = 5, (297.455, 300.611). *C*: We are 95% confident that the interval from 297.455 to 300.611 ml captures the true mean amount of cola in the bottles. The confidence interval gives 300 ml as a plausible value for the true mean amount of cola $\mu$, but it gives other plausible values as well.

**9.81** *S*: $H_0: \mu = 200$ vs. $H_a: \mu \neq 200$, where $\mu$ is the true mean response time of European servers. *P*: One-sample *t* interval to help us perform a two-sided test for $\mu$. *Random*: The servers were selected randomly. *10%*: The sample size (14) < 10% of all servers in Europe. *Normal/Large Sample*: The sample size is small, but a graph of the data reveals no strong skewness or outliers.

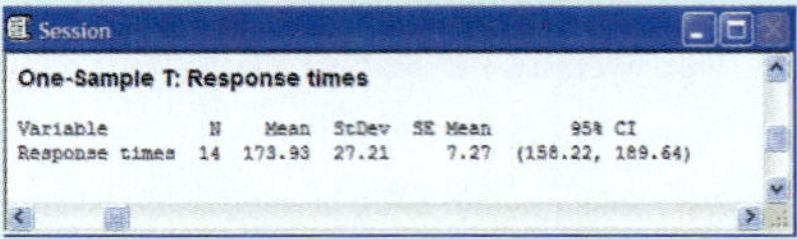

Session

One-Sample T: Response times

| Variable | N | Mean | StDev | SE Mean | 95% CI |
|---|---|---|---|---|---|
| Response times | 14 | 173.93 | 27.21 | 7.27 | (158.22, 189.64) |

82. **Water!** A blogger claims that U.S. adults drink an average of five 8-ounce glasses of water per day. Skeptical researchers ask a random sample of 24 U.S. adults about their daily water intake. A graph of the data shows a roughly symmetric shape with no outliers. The figure below displays Minitab output for a one-sample *t* interval for the population mean. Is there convincing evidence at the 10% significance level that the blogger's claim is incorrect? Justify your answer.

Session

One-Sample T: Water intake (oz)

| Variable | N | Mean | StDev | SE Mean | 90% CI |
|---|---|---|---|---|---|
| Water intake (oz) | 24 | 4.204 | 1.173 | 0.240 | (3.794, 4.615) |

83. **Tests and CIs** The *P*-value for a two-sided test of the null hypothesis $H_0: \mu = 10$ is 0.06.

(a) Does the 95% confidence interval for $\mu$ include 10? Why or why not?

(b) Does the 90% confidence interval for $\mu$ include 10? Why or why not?

84. **Tests and CIs** The *P*-value for a two-sided test of the null hypothesis $H_0: \mu = 15$ is 0.03.

(a) Does the 99% confidence interval for $\mu$ include 15? Why or why not?

(b) Does the 95% confidence interval for $\mu$ include 15? Why or why not?

85. pg 586 **Right versus left** The design of controls and instruments affects how easily people can use them. A student project investigated this effect by asking 25 right-handed students to turn a knob (with their right hands) that moved an indicator. There were two identical instruments, one with a right-hand thread (the knob turns clockwise) and the other with a left-hand thread (the knob must be turned counterclockwise). Each of the 25 students used both instruments in a random order. The following table gives the times in seconds each subject took to move the indicator a fixed distance.[23] Note that smaller times are better.

| Subject | Right thread | Left thread |
|---|---|---|
| 1 | 113 | 137 |
| 2 | 105 | 105 |
| 3 | 130 | 133 |
| 4 | 101 | 108 |
| 5 | 138 | 115 |
| 6 | 118 | 170 |
| 7 | 87 | 103 |
| 8 | 116 | 145 |
| 9 | 75 | 78 |
| 10 | 96 | 107 |
| 11 | 122 | 84 |
| 12 | 103 | 148 |
| 13 | 116 | 147 |
| 14 | 107 | 87 |
| 15 | 118 | 166 |
| 16 | 103 | 146 |
| 17 | 111 | 123 |
| 18 | 104 | 135 |
| 19 | 111 | 112 |
| 20 | 89 | 93 |
| 21 | 78 | 76 |
| 22 | 100 | 116 |
| 23 | 89 | 78 |
| 24 | 85 | 101 |
| 25 | 88 | 123 |

(a) Explain why it was important to randomly assign the order in which each subject used the two knobs.

(b) The project designers hoped to show that right-handed people find right-hand threads easier to use, on average. Carry out a test at the 5% significance level to investigate this claim.

86. **Floral scents and learning** We hear that listening to Mozart improves students' performance on tests. Maybe pleasant odors have a similar effect. To test this idea, 21 subjects worked two different but roughly equivalent paper-and-pencil mazes while wearing a mask. The mask was either unscented or carried a floral scent. Each subject used both masks, in a random order. The table below gives the subjects' times (in seconds) with both masks.[24] Note that smaller times are better.

| Subject | Unscented | Scented |
|---|---|---|
| 1 | 30.60 | 37.97 |
| 2 | 48.43 | 51.57 |
| 3 | 60.77 | 56.67 |
| 4 | 36.07 | 40.47 |

*D*: (158.22, 189.64). *C*: Because our 95% confidence interval does not contain 200 milliseconds, we reject $H_0$ at the $\alpha = 0.05$ significance level. We have convincing evidence that the mean response time of European servers is different from 200 milliseconds.

**9.82** *S*: $H_0: \mu = 5$ vs. $H_a: \mu \neq 5$, where $\mu$ is the true mean number of glasses of water that a U.S. adult drinks per day. *P*: One-sample *t* interval to help us perform a two-sided test for $\mu$. *Random*: The adults were selected randomly. *10%*: The sample size (24) < 10% of the population of U.S. adults. *Normal/Large Sample*: The sample size is small, but a graph of the data reveals a roughly symmetric shape with no outliers. *D*: (3.794, 4.615). *C*: Because our 90% confidence interval does not contain 5 glasses of water, we reject $H_0$. We have convincing evidence that the true mean number of glasses of water that U.S. adults drink in a day is different from 5.

**9.83 (a)** Yes. Because the *P*-value of $0.06 > \alpha = 0.05$, we fail to reject $H_0$: $\mu = 10$ at the 5% level of significance. Thus, the 95% confidence interval will include 10. **(b)** No. Because the *P*-value of $0.06 < \alpha = 0.10$, we reject $H_0: \mu = 10$ at the 10% level of significance. Thus, the 90% confidence interval would not include 10 as a plausible value.

**9.84 (a)** Yes. Because the *P*-value of $0.03 > \alpha = 0.01$, we fail to reject $H_0$: $\mu = 15$ at the 1% level of significance. Thus, the 99% confidence interval will include 15. **(b)** No. Because the *P*-value of $0.03 < \alpha = 0.05$, we reject $H_0: \mu = 15$ at the 5% level of significance. Thus, the 95% confidence interval would not include 15 as a plausible value.

**9.85 (a)** If all the subjects used the right thread first and they were tired when they used the left thread, then we wouldn't know if the difference in times was because of tiredness or the direction of the thread. **(b)** *S*: $H_0: \mu_d = 0$ vs. $H_a: \mu_d > 0$, where $\mu_d$ is the true mean difference (left − right) in the time (in seconds) it takes to turn the knob with the left-hand thread and the right-hand thread. *P*: Paired *t* test for $\mu_d$. *Random*: The order of treatments was determined at random. *Normal/Large Sample*: No strong skewness or outliers.

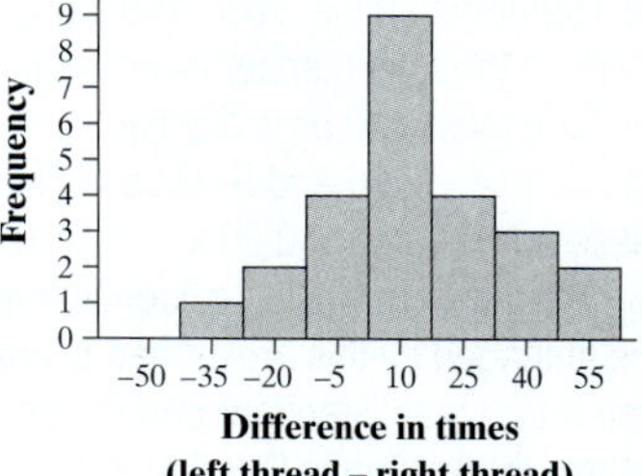


*D*: $\bar{x} = 13.32$ and $s_x = 22.94$. $t = 2.903$, df = 24, and the *P*-value is between 0.0025 and 0.005 (0.0039). *C*: Because the *P*-value of $0.0039 < \alpha = 0.05$, we reject $H_0$. We have convincing evidence that the true mean difference (left − right) in time it takes to turn the knob > 0.

**9.86 (a)** If all the subjects used the scented mask first and they were tired when they used the unscented mask, then we wouldn't know if the difference in times was because of tiredness or the mask they were using. **(b)** *S*: $H_0: \mu_d = 0$ seconds vs. $H_a: \mu_d < 0$ seconds, where $\mu_d$ is the true mean difference (scented − unscented) in time it takes to complete the mazes with the scented and unscented masks. *P*: Paired *t* test for $\mu_d$. *Random*: The order of treatments was determined at random. *Normal/Large Sample*: No strong skewness or outliers.

| Subject | Unscented | Scented |
|---|---|---|
| 5 | 68.47 | 49.00 |
| 6 | 32.43 | 43.23 |
| 7 | 43.70 | 44.57 |
| 8 | 37.10 | 28.40 |
| 9 | 31.17 | 28.23 |
| 10 | 51.23 | 68.47 |
| 11 | 65.40 | 51.10 |
| 12 | 58.93 | 83.50 |
| 13 | 54.47 | 38.30 |
| 14 | 43.53 | 51.37 |
| 15 | 37.93 | 29.33 |
| 16 | 43.50 | 54.27 |
| 17 | 87.70 | 62.73 |
| 18 | 53.53 | 58.00 |
| 19 | 64.30 | 52.40 |
| 20 | 47.37 | 53.63 |
| 21 | 53.67 | 47.00 |

(a) Explain why it was important to randomly assign the order in which each subject used the two masks.

(b) Do these data provide convincing evidence that the floral scent improved performance, on average?

87. **Growing tomatoes** Researchers suspect that Variety A tomato plants have a higher average yield than Variety B tomato plants. To find out, researchers randomly select 10 Variety A and 10 Variety B tomato plants. Then the researchers divide in half each of 10 small plots of land in different locations. For each plot, a coin toss determines which half of the plot gets a Variety A plant; a Variety B plant goes in the other half. After harvest, they compare the yield in pounds for the plants at each location. The 10 differences in yield (Variety A − Variety B) are recorded. A graph of the differences looks roughly symmetric and single-peaked with no outliers. A paired $t$ test on the differences yields $t = 1.295$ and $P$-value $= 0.1138$.

(a) State appropriate hypotheses for the paired $t$ test. Be sure to define your parameter.

(b) What are the degrees of freedom for the paired $t$ test?

(c) Interpret the $P$-value in context. What conclusion should the researchers draw?

(d) Describe a Type I error and a Type II error in this setting. Which mistake could researchers have made based on your answer to part (c)?

88. **Music and memory** Does listening to music while studying hinder students' learning? Two AP® Statistics students designed an experiment to find out. They selected a random sample of 30 students from their medium-sized high school to participate. Each subject was given 10 minutes to memorize two different lists of 20 words, once while listening to music and once in silence. The order of the two word lists was determined at random; so was the order of the treatments. The difference in the number of words recalled (music − silence) was recorded for each subject. A paired $t$ test on the differences yielded $t = -3.01$ and $P$-value $= 0.0027$.

(a) State appropriate hypotheses for the paired $t$ test. Be sure to define your parameter.

(b) What are the degrees of freedom for the paired $t$ test?

(c) Interpret the $P$-value in context. What conclusion should the students draw?

(d) Describe a Type I error and a Type II error in this setting. Which mistake could students have made based on your answer to part (c)?

89. **The power of tomatoes** Refer to Exercise 87. Explain two ways that the researchers could have increased the power of the test to detect $\mu = 0.5$.

90. **Music and memory** Refer to Exercise 88. Which of the following changes would give the test a higher power to detect $\mu = -1$: using $\alpha = 0.01$ or $\alpha = 0.10$? Explain.

91. **Significance and sample size** A study with 5000 subjects reported a result that was statistically significant at the 5% level. Explain why this result might not be particularly large or important.

92. **Sampling shoppers** A marketing consultant observes 50 consecutive shoppers at a supermarket, recording how much each shopper spends in the store. Explain why it would not be wise to use these data to carry out a significance test about the mean amount spent by all shoppers at this supermarket.

93. **Do you have ESP?** A researcher looking for evidence of extrasensory perception (ESP) tests 500 subjects. Four of these subjects do significantly better ($P < 0.01$) than random guessing.

(a) Is it proper to conclude that these four people have ESP? Explain your answer.

(b) What should the researcher now do to test whether any of these four subjects have ESP?

94. **Ages of presidents** Joe is writing a report on the backgrounds of American presidents. He looks up the ages of all the presidents when they entered office. Because Joe took a statistics course, he uses these numbers to perform a significance test about the mean age of all U.S. presidents. Explain why this makes no sense.

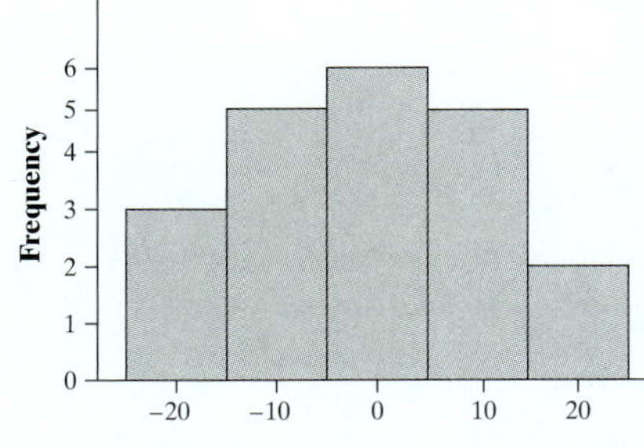


*D*: $\bar{x} = -0.96$ and $s_x = 12.55$. $t = -0.351$, df $= 20$, and $P$-value $> 0.25$ (0.3646). *C*: Because the $P$-value of $0.3646 > \alpha = 0.05$, we fail to reject $H_0$. We do not have convincing evidence that the true mean difference in time (scented − unscented) $< 0$.

**9.87** **(a)** $H_0: \mu_d = 0$ vs. $H_a: \mu_d > 0$, where $\mu_d$ is the true mean difference in tomato yield (A − B). **(b)** df $= 9$ **(c)** *Interpretation*: Assuming that the average yield for both varieties is the same, there is a 0.1138 probability of getting a mean difference as large or larger than the one observed in this experiment. *Conclusion*: Because the $P$-value of $0.1138 > \alpha = 0.05$, we fail to reject $H_0$. We do not have convincing evidence that the true mean difference in tomato yield (A − B) $> 0$.
**(d)** I: Finding convincing evidence that Variety A tomato plants have a greater mean yield, when in reality there is no difference. II: Not finding convincing evidence that Variety A tomato plants have a higher mean yield, when in reality Variety A does have a greater mean yield. They might have made a Type II error.

**9.88** **(a)** $H_0: \mu_d = 0$ vs. $H_a: \mu_d < 0$, where $\mu_d$ is the true mean difference in the number of words recalled (music − silence). **(b)** df $= 29$ **(c)** *Interpretation*: Assuming that the average number of words recalled is the same when listening to music and in silence, there is a 0.0027 probability of getting a mean difference as small or even smaller than the mean difference observed in our experiment. *Conclusion*: Because the $P$-value of $0.0027 < \alpha = 0.05$, we reject $H_0$. We have convincing evidence that the true mean difference in the number of words recalled $< 0$. **(d)** I: Finding convincing evidence that the average number of words recalled when listening to music $<$ the average number of words recalled in silence, when in reality there is no difference. II: Not finding convincing evidence that the average number of words recalled when listening to music $<$ the average number of words recalled in silence, when in reality the average number of words recalled is smaller when listening to music. They might have made a Type I error.

**9.89** Increase the significance level $\alpha$ or increase the sample size $n$.

**9.90** $\alpha = 0.10$. A higher significance level also makes it easier to reject $H_0$, leading to more correct rejections of $H_0$ when it is false.

**9.91** When the sample size is very large, rejecting the null hypothesis is very likely, even if the actual parameter is only slightly different from the hypothesized value.

**9.92** This sample wasn't randomly selected—it was a convenience sample.

**9.93** **(a)** No, in a sample of size $n = 500$, we expect to see about $(500)(0.01) = 5$ people who do better than random guessing, with a significance level of 0.01. **(b)** The researcher should repeat the procedure on these four to see if they again perform well.

**9.94** We have information about all presidents—the whole population of interest—so Joe can calculate the exact value of $\mu$.

**9.95** b
**9.96** a
**9.97** d
**9.98** c
**9.99** c
**9.100** a
**9.101** a

*Multiple choice: Select the best answer for Exercises 95 to 102.*

95. The reason we use $t$ procedures instead of $z$ procedures when carrying out a test about a population mean is that

(a) $z$ requires that the sample size be large.

(b) $z$ requires that you know the population standard deviation $\sigma$.

(c) $z$ requires that the data come from a random sample or randomized experiment.

(d) $z$ requires that the population distribution be perfectly Normal.

(e) $z$ can only be used for proportions.

96. You are testing $H_0: \mu = 10$ against $H_a: \mu < 10$ based on an SRS of 20 observations from a Normal population. The $t$ statistic is $t = -2.25$. The $P$-value

(a) falls between 0.01 and 0.02.

(b) falls between 0.02 and 0.04.

(c) falls between 0.04 and 0.05.

(d) falls between 0.05 and 0.25.

(e) is greater than 0.25.

97. You are testing $H_0: \mu = 10$ against $H_a: \mu \neq 10$ based on an SRS of 15 observations from a Normal population. What values of the $t$ statistic are statistically significant at the $\alpha = 0.005$ level?

(a) $t > 3.326$

(b) $t > 3.286$

(c) $t > 2.977$

(d) $t < -3.326$ or $t > 3.326$

(e) $t < -3.286$ or $t > 3.286$

98. After checking that conditions are met, you perform a significance test of $H_0: \mu = 1$ versus $H_a: \mu \neq 1$. You obtain a $P$-value of 0.022. Which of the following must be true?

(a) A 95% confidence interval for $\mu$ will include the value 1.

(b) A 95% confidence interval for $\mu$ will include the value 0.

(c) A 99% confidence interval for $\mu$ will include the value 1.

(d) A 99% confidence interval for $\mu$ will include the value 0.

(e) None of these is necessarily true.

99. Does Friday the 13th have an effect on people's behavior? Researchers collected data on the number of shoppers at a sample of 45 nearby grocery stores on Friday the 6th and Friday the 13th in the same month. The dotplot and computer output below summarize the data on the difference in the number of shoppers at each store on these two days (subtracting in the order 6th minus 13th).[25]

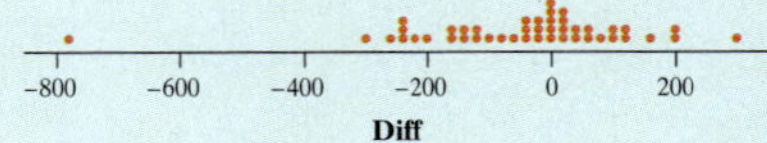


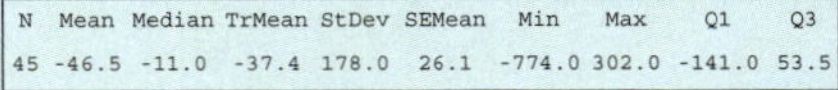

| N | Mean | Median | TrMean | StDev | SEMean | Min | Max | Q1 | Q3 |
|---|---|---|---|---|---|---|---|---|---|
| 45 | -46.5 | -11.0 | -37.4 | 178.0 | 26.1 | -774.0 | 302.0 | -141.0 | 53.5 |

Researchers would like to carry out a test of $H_0: \mu_d = 0$ versus $H_a: \mu_d \neq 0$, where $\mu_d$ is the true mean difference in the number of grocery shoppers on these two days. Which of the following conditions for performing a paired $t$ test are clearly satisfied?

I. Random II. 10% III. Normal/Large Sample

(a) I only

(b) II only

(c) III only

(d) I and II only

(e) I, II, and III

100. The most important condition for sound conclusions from statistical inference is that

(a) the data come from a well-designed random sample or randomized experiment.

(b) the population distribution be exactly Normal.

(c) the data contain no outliers.

(d) the sample size be no more than 10% of the population size.

(e) the sample size be at least 30.

101. Vigorous exercise helps people live several years longer (on average). Whether mild activities like slow walking extend life is not clear. Suppose that the added life expectancy from regular slow walking is just 2 months. A statistical test is more likely to find a significant increase in mean life expectancy if

(a) it is based on a very large random sample and a 5% significance level is used.

(b) it is based on a very large random sample and a 1% significance level is used.

(c) it is based on a very small random sample and a 5% significance level is used.

(d) it is based on a very small random sample and a 1% significance level is used.

(e) the size of the sample doesn't have any effect on the significance of the test.

**102.** A researcher plans to conduct a significance test at the $\alpha = 0.01$ significance level. She designs her study to have a power of 0.90 at a particular alternative value of the parameter of interest. The probability that the researcher will commit a Type II error for the particular alternative value of the parameter at which she computed the power is

(a) 0.01. (b) 0.10. (c) 0.89. (d) 0.90. (e) 0.99.

**103. Is your food safe?** (8.1) "Do you feel confident or not confident that the food available at most grocery stores is safe to eat?" When a Gallup Poll asked this question, 87% of the sample said they were confident.[26] Gallup announced the poll's margin of error for 95% confidence as ±3 percentage points. Which of the following sources of error are included in this margin of error? Explain.

(a) Gallup dialed landline telephone numbers at random and so missed all people without landline phones, including people whose only phone is a cell phone.

(b) Some people whose numbers were chosen never answered the phone in several calls or answered but refused to participate in the poll.

(c) There is chance variation in the random selection of telephone numbers.

**104. Spinning for apples** (6.3 or 7.3) In the "Ask Marilyn" column of *Parade* magazine, a reader posed this question: "Say that a slot machine has five wheels, and each wheel has five symbols: an apple, a grape, a peach, a pear, and a plum. I pull the lever five times. What are the chances that I'll get at least one apple?" Suppose that the wheels spin independently and that the five symbols are equally likely to appear on each wheel in a given spin.

(a) Find the probability that the slot player gets at least one apple in one pull of the lever. Show your method clearly.

(b) Now answer the reader's question. Show your method clearly.

## FRAPPY! Free Response AP® Problem, Yay!

The following problem is modeled after actual AP® Statistics exam free response questions. Your task is to generate a complete, concise response in 15 minutes.

*Directions: Show all your work. Indicate clearly the methods you use, because you will be scored on the correctness of your methods as well as on the accuracy and completeness of your results and explanations.*

Anne reads that the average price of regular gas in her state is \$4.06 per gallon. To see if the average price of gas is different in her city, she selects 10 gas stations at random and records the price per gallon for regular gas at each station. The data, along with the sample mean and standard deviation, are listed in the table below.

| Station | Price |
|---|---|
| 1 | 4.13 |
| 2 | 4.01 |
| 3 | 4.09 |
| 4 | 4.05 |
| 5 | 3.97 |
| 6 | 3.99 |
| 7 | 4.05 |
| 8 | 3.98 |
| 9 | 4.09 |
| 10 | 4.02 |
| **Mean** | **4.038** |
| **SD** | **0.0533** |

Do the data provide convincing evidence that the average price of gas in Anne's city is different from \$4.06 per gallon?

After you finish, you can view two example solutions on the book's Web site (www.whfreeman.com/tps5e). Determine whether you think each solution is "complete," "substantial," "developing," or "minimal." If the solution is not complete, what improvements would you suggest to the student who wrote it? Finally, your teacher will provide you with a scoring rubric. Score your response and note what, if anything, you would do differently to improve your own score.

**9.102** b

**9.103** **(a)** Not included. The margin of error does not account for undercoverage. **(b)** Not included. The margin of error does not account for nonresponse. **(c)** Included. The margin of error is calculated to account for sampling variability.

**9.104** **(a)** $P$(at least one apple) $=$ $1 - P$(no apples) $= 1 - (0.80)^5 = 0.67232$. **(b)** $P$(no apples on any of 5 pulls) $=$ $(0.32768)^5 = 0.00378$ and the probability of at least one apple in 5 pulls of the lever is $1 - 0.00378 = 0.99622$.

**TRM FRAPPY!**

We present a model solution here. Please consult the Teacher's Resource Materials for sample student responses, a scoring rubric, and a black line master of the original question with space for students to write their responses.

STATE: We want to test the following hypotheses at the $\alpha = 0.05$ significance level:

$$H_0: \mu = 4.06$$
$$H_a: \mu \neq 4.06$$

where $\mu$ = the true mean price per gallon of regular gas in Anne's city (in dollars).

PLAN: If conditions are met, we will perform a one-sample $t$ test for $\mu$.

- *Random*: A random sample of gas stations was selected.
  - *10%*: It is reasonable to assume that there are more than $10(10) = 100$ gas stations in her city.
- *Normal/Large Sample*: The sample size is small, so we will graph the sample data and look for departures from Normality. The graphs below show no outliers or strong skewness, so it is reasonable to use $t$ procedures.

3.96 4.00 4.04 4.08 4.12
Price (dollars)

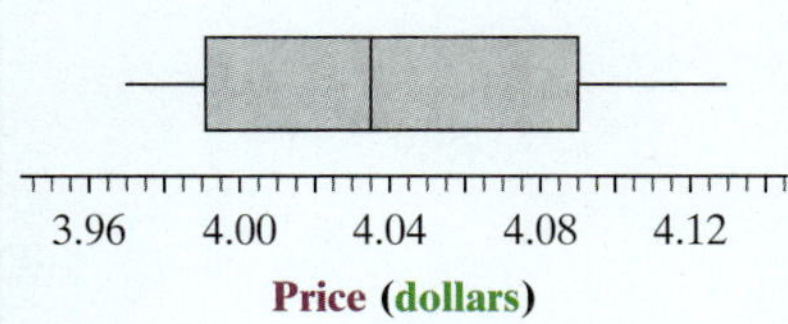


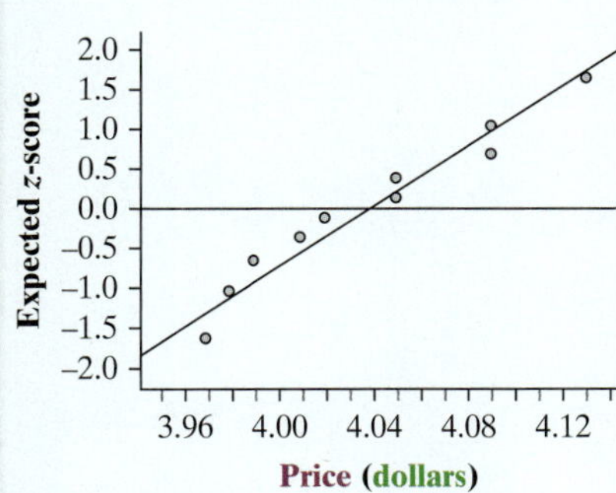


DO:

- *Test statistic*
$$t = \frac{4.038 - 4.06}{0.0533/\sqrt{10}} = -1.31$$
- *P-value* Using the $t$ distribution with 9 degrees of freedom, the $P$-value is between $2(0.10) = 0.20$ and $2(0.15) = 0.30$. *Using technology*: With df $= 9$, the calculator's T-Test gives $P$-value $= 0.2242$.

CONCLUDE: Because the $P$-value of 0.2242 is greater than $\alpha = 0.05$, we fail to reject the null hypothesis. There is not convincing evidence that the true mean price per gallon of regular gas in Anne's city (in dollars) is different from \$4.06.

**TRM Chapter 9 Tests**

There are four different tests available for this chapter in the Teacher's Resource Materials. Versions A and B are roughly parallel and version D is cumulative. Find them by clicking on the link in the TE-book, opening the TRCD, or logging onto the book's companion Web site www.whfreeman.com/tps5e (teacher log-in required).

# Chapter Review

### Section 9.1: Significance Tests: The Basics

In this section, you learned the basic ideas of significance testing. Start by stating the hypotheses that you want to test. The null hypothesis ($H_0$) is typically a statement of "no difference" and the alternative hypothesis ($H_a$) describes what we suspect is true. Remember that hypotheses are always about parameters, not statistics.

When sample data provide support for the alternative hypothesis, there are two possible explanations: (1) the null hypothesis is true, and data supporting the alternative hypothesis occurred just by chance, or (2) the alternative hypothesis is true, and the data are consistent with an alternative value of the parameter. In a significance test, always start with the belief that the null hypothesis is true. If you can rule out chance as a plausible explanation for the observed data, there is convincing evidence that the alternative hypothesis is true.

The *P*-value in a significance test measures how likely it is to get results at least as extreme as the observed results by chance alone, assuming the null hypothesis is true. To determine if the *P*-value is small enough to reject $H_0$, compare it to a predetermined significance level such as $\alpha = 0.05$. If *P*-value $< \alpha$, reject $H_0$—there is convincing evidence that the alternative hypothesis is true. However, if *P*-value $\geq \alpha$, fail to reject $H_0$—there is not convincing evidence that the alternative hypothesis is true.

Because conclusions are based on sample data, there is a possibility that the conclusion will be incorrect. You can make two types of errors in a significance test: a Type I error occurs if you find convincing evidence for the alternative hypothesis when, in reality, the null hypothesis is true. A Type II error occurs when you don't find convincing evidence that the alternative hypothesis is true when, in reality, the alternative hypothesis is true. The probability of making a Type I error is equal to the significance level ($\alpha$) of the test.

### Section 9.2: Tests about a Population Proportion

In this section, you learned the details of conducting a significance test for a population proportion $p$. Whenever you are asked if there is convincing evidence for a claim about a population proportion, you are expected to respond using the familiar four-step process.

**STATE:** Give the hypotheses you are testing in terms of $p$, state the significance level, and define the parameter $p$.

**PLAN:** Name the procedure you plan to use (one-sample $z$ test for a population proportion) and check the appropriate conditions (Random, 10%, Large Counts) to see if the procedure is appropriate.

- Random: The data come from a well-designed random sample or randomized experiment.
  - **10%:** The sample size should be no larger than 10% of the population when sampling without replacement.
- Large Counts: Both $np_0$ and $n(1-p_0)$ must be at least 10, where $p_0$ is the value of $p$ in the null hypothesis.

**DO:** Calculate the test statistic and *P*-value. The test statistic $z$ measures how far away the sample statistic is from the hypothesized parameter value in standardized units:

$$z = \frac{\hat{p} - p_0}{\sqrt{\frac{p_0(1 - p_0)}{n}}}$$

To calculate the *P*-value, use Table A or technology.

**CONCLUDE:** Use the *P*-value to make an appropriate conclusion about the hypotheses in context.

Perform a two-sided test when looking for convincing evidence that the true value of the parameter is *different* from the hypothesized value, in either direction. The *P*-value for a two-sided test is calculated by finding the probability of getting a sample statistic at least as extreme as the observed statistic, in either direction, assuming the null hypothesis is true.

You can also use a confidence interval to make a conclusion for a two-sided test. If the null parameter value is one of the plausible values in the interval, there isn't convincing evidence that the alternative hypothesis is true. However, if the null parameter value is not one of the plausible values in the interval, there is convincing evidence that the alternative hypothesis is true. Besides helping you draw a conclusion, the interval tells you which alternative parameter values are plausible.

The probability that you avoid making a Type II error when an alternative value of the parameter is true is called the power of the test. Power is good—if the alternative hypothesis is true, we want to maximize the probability of finding convincing evidence that it is true. We can increase the power of a significance test by increasing the sample size or by increasing the significance level. The power of a test will also be greater when the alternative value of the parameter is farther away from the null hypothesis value.

### Section 9.3: Tests about a Population Mean

In this section, you learned the details of conducting a significance test for a population mean. Although some of the details are different, the reasoning and structure of the tests in this section are the same as in Section 9.2. In fact, the "State" and "Conclude" steps are exactly the same, other than the switch from proportions to means.

**PLAN:** Name the procedure you are using (one-sample $t$ test for a population mean), and check the conditions (Random, 10%, and Normal/Large Sample). The Random and 10% conditions are the same as in Section 9.2. The Normal/Large Sample condition states that the population distribution must be Normal or the sample size must be large ($n \geq 30$). If the sample is small and the population shape is unknown, graph the sample data to make sure there is no strong skewness or outliers.

**DO:** Calculate the test statistic and $P$-value. The test statistic $t$ measures how far away the sample statistic is from the hypothesized parameter value in standardized units:

$$t = \frac{\bar{x} - \mu}{s_x/\sqrt{n}}$$

To calculate the $P$-value, determine the degrees of freedom ($df = n - 1$) and use Table B or technology.

Use a paired $t$ test to analyze the results of comparative experiments and observational studies that produce paired data. Start by calculating the difference for each pair and use the set of differences to check the Normal/Large Sample condition and to calculate the test statistic and $P$-value.

Remember to use significance tests wisely. When planning a study, use a large enough sample size so the test will have adequate power. Also, remember that statistically significant results aren't always "practically" important. Finally, be aware that the probability of making at least one Type I error goes up dramatically when conducting multiple tests.

## What Did You Learn?

| Learning Objective | Section | Related Example on Page(s) | Relevant Chapter Review Exercise(s) |
|---|---|---|---|
| State the null and alternative hypotheses for a significance test about a population parameter. | 9.1 | 540 | R9.1 |
| Interpret a $P$-value in context. | 9.1 | 543, 544 | R9.5 |
| Determine if the results of a study are statistically significant and draw an appropriate conclusion using a significance level. | 9.1 | 546 | R9.5 |
| Interpret a Type I and a Type II error in context, and give a consequence of each. | 9.1 | 548 | R9.3, R9.4 |
| State and check the Random, 10%, and Large Counts conditions for performing a significance test about a population proportion. | 9.2 | 555 | R9.4 |
| Perform a significance test about a population proportion. | 9.2 | 559, 562 | R9.4 |
| Interpret the power of a test and describe what factors affect the power of a test. | 9.2 | 565, Discussion on 568 | R9.3 |
| Describe the relationship among the probability of a Type I error (significance level), the probability of a Type II error, and the power of a test. | 9.2 | 565 | R9.3 |
| State and check the Random, 10%, and Normal/Large Sample conditions for performing a significance test about a population mean. | 9.3 | 575 | R9.2, R9.6, R9.7 |
| Perform a significance test about a population mean. | 9.3 | 580, 583 | R9.6 |
| Use a confidence interval to draw a conclusion for a two-sided test about a population parameter. | 9.2, 9.3 | 563, 585 | R9.5, R9.6 |
| Perform a significance test about a mean difference using paired data. | 9.3 | 586 | R9.7 |



TRM Learning Objectives Grid

# Chapter 9 Chapter Review Exercises

*These exercises are designed to help you review the important ideas and methods of the chapter.*

**R9.1 Stating hypotheses** State the appropriate null and alternative hypotheses in each of the following settings. Be sure to define the parameter.

(a) The average height of 18-year-old American women is 64.2 inches. You wonder whether the mean height of this year's female graduates from a large local high school (over 3000 students) differs from the national average. You measure an SRS of 48 female graduates and find that $\bar{x} = 63.1$ inches.

(b) Mr. Starnes believes that less than 75% of the students at his school completed their math homework last night. The math teachers inspect the homework assignments from a random sample of students at the school to help Mr. Starnes test his claim.

**R9.2 Fonts and reading ease** Does the use of fancy type fonts slow down the reading of text on a computer screen? Adults can read four paragraphs of text in the common Times New Roman font in an average time of 22 seconds. Researchers asked a random sample of 24 adults to read this text in the ornate font named Gigi. Here are their times, in seconds:

| 23.2 | 21.2 | 28.9 | 27.7 | 29.1 | 27.3 | 16.1 | 22.6 | 25.6 | 34.2 | 23.9 | 26.8 |
|---|---|---|---|---|---|---|---|---|---|---|---|
| 20.5 | 34.3 | 21.4 | 32.6 | 26.2 | 34.1 | 31.5 | 24.6 | 23.0 | 28.6 | 24.4 | 28.1 |

State and check the conditions for performing a significance test using these data.

**R9.3 Strong chairs?** A company that manufactures classroom chairs for high school students claims that the mean breaking strength of the chairs that they make is 300 pounds. One of the chairs collapsed beneath a 220-pound student last week. You wonder whether the manufacturer is exaggerating the breaking strength of the chairs.

(a) State appropriate null and alternative hypotheses in this setting. Be sure to define your parameter.

(b) Describe a Type I error and a Type II error in this setting, and give the consequences of each.

(c) Would you recommend a significance level of 0.01, 0.05, or 0.10 for this test? Justify your choice.

(d) The power of the test to detect $\mu = 294$ using $\alpha = 0.05$ is 0.71. Interpret this value in context.

(e) Explain two ways that you could increase the power of the test from (d).

**R9.4 Flu vaccine** A drug company has developed a new vaccine for preventing the flu. The company claims that fewer than 5% of adults who use its vaccine will get the flu. To test the claim, researchers give the vaccine to a random sample of 1000 adults. Of these, 43 get the flu.

(a) Do these data provide convincing evidence to support the company's claim?

(b) Which kind of mistake—a Type I error or a Type II error—could you have made in (a)? Explain.

(c) From the company's point of view, would a Type I error or Type II error be more serious? Why?

**R9.5 Roulette** An American roulette wheel has 18 red slots among its 38 slots. To test if a particular roulette wheel is fair, you spin the wheel 50 times and the ball lands in a red slot 31 times. The resulting *P*-value is 0.0384.

(a) Interpret the *P*-value in context.

(b) Are the results statistically significant at the $\alpha = 0.05$ level? Explain. What conclusion would you make?

(c) The casino manager uses your data to produce a 99% confidence interval for $p$ and gets (0.44, 0.80). He says that this interval provides convincing evidence that the wheel is fair. How do you respond?

**R9.6 Radon detectors** Radon is a colorless, odorless gas that is naturally released by rocks and soils and may concentrate in tightly closed houses. Because radon is slightly radioactive, there is some concern that it may be a health hazard. Radon detectors are sold to homeowners worried about this risk, but the detectors may be inaccurate. University researchers placed a random sample of 11 detectors in a chamber where they were exposed to 105 picocuries per liter of radon over 3 days. A graph of the radon readings from the 11 detectors shows no strong skewness or outliers. The mean reading is 104.82 and the standard deviation of the readings is 9.54.

(a) Is there convincing evidence at the 10% level that the mean reading differs from the true value 105?

(b) A 90% confidence interval for the true mean reading is (99.61, 110.03). Is this interval consistent with your conclusion from part (a)? Explain.

**R9.7 Better barley** Does drying barley seeds in a kiln increase the yield of barley? A famous experiment by William S. Gosset (who discovered the $t$ distributions) investigated this question. Eleven pairs of adjacent plots were marked out in a large field. For each pair, regular barley seeds were planted in one plot and kiln-dried seeds were planted in the other. The following table displays the data on yield (lb/acre).[27]

## TRM Full Solutions to Chapter 9 Review Exercises

## Answers to Chapter 9 Review Exercises

**R9.1 (a)** $H_0: \mu = 64.2$; $H_a: \mu \neq 64.2$, where $\mu$ = the true mean height of this year's female graduates from the local high school. **(b)** $H_0: p = 0.75$; $H_a: p < 0.75$, where $p$ = the true proportion of all students at Mr. Starnes' school who completed their math homework last night.

**R9.2** *Random*: Random sample. *10%*: The sample size (24) < 10% of the population of adults. *Normal/Large Sample*: The histogram below shows that the distribution is roughly symmetric with no outliers.

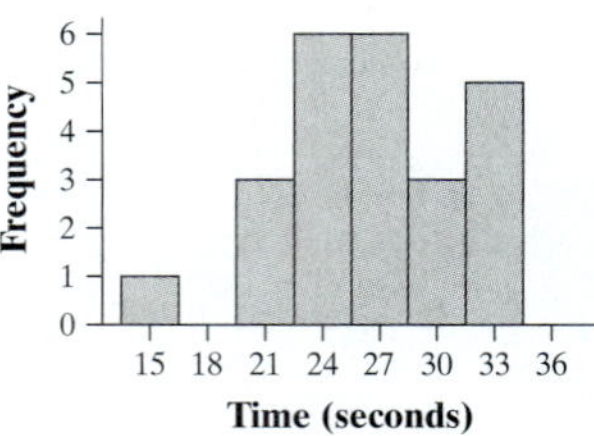


**R9.3 (a)** $H_0: \mu = 300$ vs. $H_a: \mu < 300$, where $\mu$ = the true mean breaking strength of these chairs. **(b)** I: Finding convincing evidence that the mean breaking strength < 300 pounds, when in reality it is 300 pounds or higher. *Consequence*: Falsely accusing the company of lying. II: Not finding convincing evidence that the mean breaking strength < 300 pounds, when in reality it is < 300 pounds. *Consequence*: Allowing the company to continue to sell chairs that have a breaking strength less than advertised. **(c)** Because a Type II error is more serious, increase the probability of a Type I error by using $\alpha = 0.10$. **(d)** If the true mean breaking strength is 294 pounds, there is a 0.71 probability that we will find convincing evidence the true mean breaking strength is < 300 pounds. **(e)** Increase the sample size or increase the significance level.

**R9.4 (a)** *S*: $H_0: p = 0.05$ vs. $H_a: p < 0.05$, where $p$ is the true proportion of adults who will get the flu after using the vaccine. *P*: One-sample $z$ test for $p$. *Random*: Random sample. *10%*: The sample size (1000) < 10% of the population of adults. *Large Counts:* $1000(0.05) = 50 \geq 10$ and $1000(0.95) = 950 \geq 10$. *D*: $z = -1.02$ and *P*-value = 0.1539. *C*: Because the *P*-value of $0.1539 > \alpha = 0.05$, we fail to reject $H_0$. We do not have convincing evidence that fewer than 5% of adults who receive this vaccine will get the flu. **(b)** Because we failed to reject the null hypothesis, we could have made a Type II error—not finding convincing evidence that the true proportion of adults who get the flu after using this vaccine < 0.05, when in reality the true proportion < 0.05. **(c)** Answers will vary.

**R9.5 (a)** Assuming that the roulette wheel is fair, there is a 0.0384 probability that we would get a sample proportion of reds at least this different from the expected proportion of reds (18/38) by chance alone. **(b)** Because the *P*-value of $0.0384 < \alpha = 0.05$, the results are statistically significant at the $\alpha = 0.05$ level. This means that we reject $H_0$ and have convincing evidence that the true proportion of reds is different than $p = 18/38$. **(c)** Because $18/38 = 0.474$ is one of the plausible values in the interval, this interval does not provide convincing evidence that the wheel is unfair. It does not, however, prove that the wheel is fair as there are many other plausible values in the interval that are not equal to 18/38. Also, the conclusion here is inconsistent with the conclusion in part (b) because the manager used a 99% confidence interval, which is equivalent to a test using $\alpha = 0.01$.

**R9.6 (a)** *S*: $H_0: \mu = 105$ vs. $H_a: \mu \neq 105$, where $\mu$ is the true mean reading from radon detectors. *P*: One-sample $t$ test for $\mu$. *Random*: Random sample. *10%*: The sample size (11) < 10% of all radon detectors. *Normal/Large Sample*: A graph of the data shows no strong skewness or outliers. *D*: $t = -0.06$, df = 10, and *P*-value > 0.50 (0.9513). *C*: Because the *P*-value of $0.9513 > \alpha = 0.10$, we fail to reject $H_0$.

| Plot | Regular | Kiln |
|---|---|---|
| 1 | 1903 | 2009 |
| 2 | 1935 | 1915 |
| 3 | 1910 | 2011 |
| 4 | 2496 | 2463 |
| 5 | 2108 | 2180 |
| 6 | 1961 | 1925 |
| 7 | 2060 | 2122 |
| 8 | 1444 | 1482 |
| 9 | 1612 | 1542 |
| 10 | 1316 | 1443 |
| 11 | 1511 | 1535 |

(a) How can the Random condition be satisfied in this study?

(b) Assuming that the Random condition has been met, do these data provide convincing evidence that drying barley seeds in a kiln increases the yield of barley, on average? Justify your answer.

# Chapter 9 AP® Statistics Practice Test

**Section I: Multiple Choice** *Select the best answer for each question.*

**T9.1** An opinion poll asks a random sample of adults whether they favor banning ownership of handguns by private citizens. A commentator believes that more than half of all adults favor such a ban. The null and alternative hypotheses you would use to test this claim are

(a) $H_0: \hat{p} = 0.5; H_a: \hat{p} > 0.5$
(b) $H_0: p = 0.5; H_a: p > 0.5$
(c) $H_0: p = 0.5; H_a: p < 0.5$
(d) $H_0: p = 0.5; H_a: p \neq 0.5$
(e) $H_0: p > 0.5; H_a: p = 0.5$

**T9.2** You are thinking of conducting a one-sample $t$ test about a population mean $\mu$ using a 0.05 significance level. Which of the following statements is correct?

(a) You should not carry out the test if the sample does not have a Normal distribution.
(b) You can safely carry out the test if there are no outliers, regardless of the sample size.
(c) You can carry out the test if a graph of the data shows no strong skewness, regardless of the sample size.
(d) You can carry out the test only if the population standard deviation is known.
(e) You can safely carry out the test if your sample size is at least 30.

**T9.3** To determine the reliability of experts who interpret lie detector tests in criminal investigations, a random sample of 280 such cases was studied. The results were

| | Suspect's True Status | |
|---|---|---|
| Examiner's Decision | Innocent | Guilty |
| "Innocent" | 131 | 15 |
| "Guilty" | 9 | 125 |

If the hypotheses are $H_0$: suspect is innocent versus $H_a$: suspect is guilty, then we could estimate the probability that experts who interpret lie detector tests will make a Type II error as

(a) 15/280.
(b) 9/280.
(c) 15/140.
(d) 9/140.
(e) 15/146.

**T9.4** A significance test allows you to reject a null hypothesis $H_0$ in favor of an alternative $H_a$ at the 5% significance level. What can you say about significance at the 1% level?

(a) $H_0$ can be rejected at the 1% significance level.
(b) There is insufficient evidence to reject $H_0$ at the 1% significance level.
(c) There is sufficient evidence to accept $H_0$ at the 1% significance level.

We do not have convincing evidence that the true mean reading from the radon detectors is different than 105. **(b)** Yes. Because 105 is in the interval from 99.61 to 110.03, both the confidence interval and the significance test agree that 105 is a plausible value for the true mean reading from the radon detectors.

**R9.7 (a)** The random condition can be satisfied by randomly allocating which plot got the regular barley seeds and which one got the kiln-dried seeds within each pair of adjacent plots. **(b)** *S*: $H_0: \mu_d = 0$ vs. $H_a: \mu_d < 0$, where $\mu_d$ is the true mean difference (regular − kiln) in yield between regular barley seeds and kiln-dried barley seeds. *P*: Paired $t$ test for $\mu_d$. *Random:* Assumed. *Normal/Large Sample*: The histogram below shows no strong skewness or outliers.

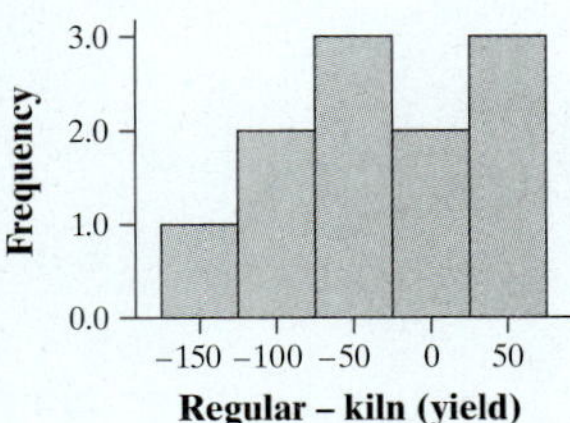


*D*: $\bar{x} = -33.7$ and $s_x = 66.2$. $t = -1.690$, df = 10, and the *P*-value is between 0.05 and 0.10 (0.0609). *C*: Because the *P*-value of $0.0609 > \alpha = 0.05$, we fail to reject $H_0$. We do not have convincing evidence that the true mean difference (regular − kiln) in yield < 0.

**TRM Full Solutions to Chapter 9 AP® Statistics Practice Test**

## Answers to Chapter 9 AP® Statistics Practice Test

**T9.1** b
**T9.2** e
**T9.3** c
**T9.4** e

T9.5 b

T9.6 c

T9.7 e

T9.8 d

T9.9 a

T9.10 c



(d) $H_a$ can be rejected at the 1% significance level.

(e) The answer can't be determined from the information given.

**T9.5** A random sample of 100 likely voters in a small city produced 59 voters in favor of Candidate A. The observed value of the test statistic for testing the null hypothesis $H_0: p = 0.5$ versus the alternative hypothesis $H_a: p > 0.5$ is

(a) $z = \dfrac{0.59 - 0.5}{\sqrt{\dfrac{0.59(0.41)}{100}}}$

(b) $z = \dfrac{0.59 - 0.5}{\sqrt{\dfrac{0.5(0.5)}{100}}}$

(c) $z = \dfrac{0.5 - 0.59}{\sqrt{\dfrac{0.59(0.41)}{100}}}$

(d) $z = \dfrac{0.5 - 0.59}{\sqrt{\dfrac{0.5(0.5)}{100}}}$

(e) $t = \dfrac{0.59 - 0.5}{\sqrt{\dfrac{0.5(0.5)}{100}}}$

**T9.6** A researcher claims to have found a drug that causes people to grow taller. The coach of the basketball team at Brandon University has expressed interest but demands evidence. Over 1000 Brandon students volunteer to participate in an experiment to test this new drug. Fifty of the volunteers are randomly selected, their heights are measured, and they are given the drug. Two weeks later, their heights are measured again. The power of the test to detect an average increase in height of 1 inch could be increased by

(a) using only volunteers from the basketball team in the experiment.

(b) using $\alpha = 0.01$ instead of $\alpha = 0.05$.

(c) using $\alpha = 0.05$ instead of $\alpha = 0.01$.

(d) giving the drug to 25 randomly selected students instead of 50.

(e) using a two-sided test instead of a one-sided test.

**T9.7** A 95% confidence interval for a population mean $\mu$ is calculated to be (1.7, 3.5). Assume that the conditions for performing inference are met. What conclusion can we draw for a test of $H_0: \mu = 2$ versus $H_a: \mu \neq 2$ at the $\alpha = 0.05$ level based on the confidence interval?

(a) None. We cannot carry out the test without the original data.

(b) None. We cannot draw a conclusion at the $\alpha =$ 0.05 level because this test corresponds to the 97.5% confidence interval.

(c) None. Confidence intervals and significance tests are unrelated procedures.

(d) We would reject $H_0$ at level $\alpha = 0.05$.

(e) We would fail to reject $H_0$ at level $\alpha = 0.05$.

**T9.8** In a test of $H_0: p = 0.4$ against $H_a: p \neq 0.4$, a random sample of size 100 yields a test statistic of $z = 1.28$. The P-value of the test is approximately equal to

(a) 0.90. (b) 0.40. (c) 0.05. (d) 0.20. (e) 0.10.

**T9.9** An SRS of 100 postal employees found that the average time these employees had worked at the postal service was 7 years with standard deviation 2 years. Do these data provide convincing evidence that the mean time of employment $\mu$ for the population of postal employees has changed from the value of 7.5 that was true 20 years ago? To determine this, we test the hypotheses $H_0: \mu = 7.5$ versus $H_a: \mu \neq 7.5$ using a one-sample $t$ test. What conclusion should we draw at the 5% significance level?

(a) There is convincing evidence that the mean time working with the postal service has changed.

(b) There is not convincing evidence that the mean time working with the postal service has changed.

(c) There is convincing evidence that the mean time working with the postal service is still 7.5 years.

(d) There is convincing evidence that the mean time working with the postal service is now 7 years.

(e) We cannot draw a conclusion at the 5% significance level. The sample size is too small.

**T9.10** Are TV commercials louder than their surrounding programs? To find out, researchers collected data on 50 randomly selected commercials in a given week. With the television's volume at a fixed setting, they measured the maximum loudness of each commercial and the maximum loudness in the first 30 seconds of regular programming that followed. Assuming conditions for inference are met, the most appropriate method for answering the question of interest is

(a) a one-proportion $z$ test.

(b) a one-proportion $z$ interval.

(c) a paired $t$ test.

(d) a paired $t$ interval.

(e) None of these.

**Section II: Free Response** *Show all your work. Indicate clearly the methods you use, because you will be graded on the correctness of your methods as well as on the accuracy and completeness of your results and explanations.*

**T9.11** A software company is trying to decide whether to produce an upgrade of one of its programs. Customers would have to pay \$100 for the upgrade. For the upgrade to be profitable, the company needs to sell it to more than 20% of their customers. You contact a random sample of 60 customers and find that 16 would be willing to pay \$100 for the upgrade.

(a) Do the sample data give good evidence that more than 20% of the company's customers are willing to purchase the upgrade? Carry out an appropriate test at the $\alpha = 0.05$ significance level.

(b) Which would be a more serious mistake in this setting—a Type I error or a Type II error? Justify your answer.

(c) Describe two ways to increase the power of the test in part (a).

**T9.12** "I can't get through my day without coffee" is a common statement from many students. Assumed benefits include keeping students awake during lectures and making them more alert for exams and tests. Students in a statistics class designed an experiment to measure memory retention with and without drinking a cup of coffee one hour before a test. This experiment took place on two different days in the same week (Monday and Wednesday). Ten students were used. Each student received no coffee or one cup of coffee one hour before the test on a particular day. The test consisted of a series of words flashed on a screen, after which the student had to write down as many of the words as possible. On the other day, each student received a different amount of coffee (none or one cup).

(a) One of the researchers suggested that all the subjects in the experiment drink no coffee before Monday's test and one cup of coffee before Wednesday's test. Explain to the researcher why this is a bad idea *and* suggest a better method of deciding when each subject receives the two treatments.

(b) The data from the experiment are provided in the table below. Set up and carry out an appropriate test to determine whether there is convincing evidence that drinking coffee improves memory.

| Student | No cup | One cup |
|---|---|---|
| 1 | 24 | 25 |
| 2 | 30 | 31 |
| 3 | 22 | 23 |
| 4 | 24 | 24 |
| 5 | 26 | 27 |
| 6 | 23 | 25 |
| 7 | 26 | 28 |
| 8 | 20 | 20 |
| 9 | 27 | 27 |
| 10 | 28 | 30 |

**T9.13** A government report says that the average amount of money spent per U.S. household per week on food is about \$158. A random sample of 50 households in a small city is selected, and their weekly spending on food is recorded. The sample data have a mean of \$165 and a standard deviation of \$20. Is there convincing evidence that the mean weekly spending on food in this city differs from the national figure of \$158?

**T9.11 (a)** *S*: $H_0: p = 0.20$ vs. $H_a: p > 0.20$, where *p* is the true proportion of customers who would pay \$100 for the upgrade. *P*: One-sample *z* test for *p*. *Random*: Random sample. *10%*: The sample size (60) $<$ 10% of this company's customers. *Large Counts:* $60(0.20) = 12 \geq 10$ and $60(0.8) = 48 \geq 10$. *D*: $z = 1.29$, *P*-value $= 0.0984$. *C*: Because the *P*-value of $0.0984 > \alpha = 0.05$, we fail to reject $H_0$. We do not have convincing evidence that more than 20% of customers would pay \$100 for the upgrade.
**(b)** I: Finding convincing evidence that more than 20% of customers would pay for the upgrade, when in reality they would not. II: Not finding convincing evidence that more than 20% of customers would pay for the upgrade, when in reality more than 20% would. For the company, a Type I error is worse because they would go ahead with the upgrade and lose money.
**(c)** Increase the sample size or increase the significance level.

**T9.12 (a)** Students may improve from Monday to Wednesday just because they have already done the task once. Then we wouldn't know if the experience with the test or the caffeine is the cause of the difference in scores. A better way to run the experiment would be to randomly assign half the students to get 1 cup of coffee on Monday and the other half to get no coffee on Monday. Then have each student receive the opposite treatment on Wednesday. **(b)** *S*: $H_0: \mu_d = 0$ vs. $H_a: \mu_d < 0$, where $\mu_d$ is the true mean difference (no coffee − coffee) in the number of words recalled without coffee and with coffee. *P*: Paired *t* test for $\mu_d$. *Random*: The treatments were assigned at random. *Normal/Large Sample*: The histogram below shows a symmetric distribution with no outliers.

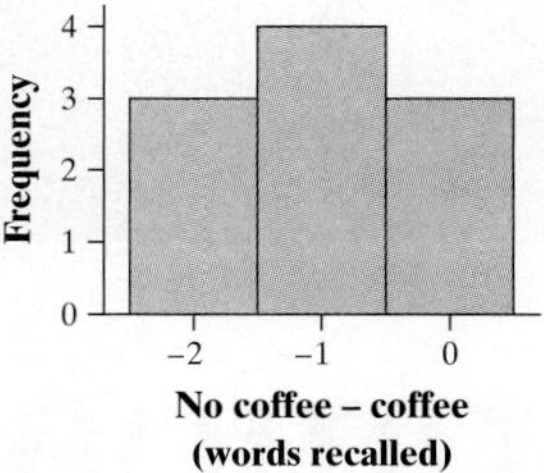


*D*: $\bar{x} = -1$ and $s_x = 0.816$. $t = -3.873$, df = 9, and the *P*-value is between 0.001 and 0.0025 (0.0019). *C*: Because the *P*-value of $0.0019 < \alpha = 0.05$, we reject $H_0$. We have convincing evidence that the mean difference (no coffee − coffee) in word recall $< 0$.

**T9.13** *S*: $H_0: \mu = \$158$ vs. $H_a: \mu \neq \$158$, where $\mu$ is the true mean amount spent on food by households in this city. *P*: One-sample *t* test for $\mu$. *Random*: Random sample. *10%*: The sample size (50) $<$ 10% of households in this small city. *Normal/Large Sample*: $n = 50 \geq 30$. *D*: $t = 2.47$; using df = 40, the *P*-value is between 0.01 and 0.02 (using df = 49, *P*-value = 0.0168). *C*: Because the *P*-value of $0.0168 < \alpha = 0.05$, we reject $H_0$. We have convincing evidence that the true mean amount spent on food per household in this city is different from the national average of \$158.

Chapter

10

# Comparing Two Populations or Groups

## PD Overview

Almost all statistical studies are comparative. In this chapter, we learn how to compare the proportion of successes for two populations or treatments using both confidence intervals and significance tests. We also learn how to use confidence intervals and significance tests to compare means for two populations or treatments. As in Chapters 7 through 9, we start first with proportions (categorical data) and then move to means (quantitative data). Within each type of data, we start with confidence intervals and then move to significance tests.

*Every* time students are asked to construct a confidence interval or perform a significance test, they should use the four-step process. Rubrics on previous AP® exams have demanded all four steps, even if the question doesn't specifically ask for them. To help prepare students for these expectations, we no longer prompt the students to do all four steps in the exercises.

Make sure to do the Is Yawning Contagious? and Does Polyester Decay? activities in this chapter. They are excellent for introducing the two-sample contexts of the chapter and reinforcing the logic of significance testing. The examples and exercises in this chapter also provide great opportunities to review important aspects of designing studies, such as the difference in the types of conclusions we can make from observational studies and experiments.

Finally, take advantage of articles and stories in the news. Almost every day there are summaries of comparative experiments and observational studies that use the inferential methods we learn about in this chapter. Even if the data aren't provided in the article, you can ask your students to consider how the study was designed and to determine what inference procedure would be most appropriate to analyze the results.

### Relationship to the AP® Statistics Course Description

Chapter 10 covers the following content as described in the Topic Outline from the AP® Statistics Course Description. Topics marked with a superscript are covered in other chapters as indicated.

III. Anticipating patterns: Exploring random phenomena using probability and simulation

- D. Sampling distributions
  4. Sampling distribution of a difference between two independent sample proportions
  5. Sampling distribution of a difference between two independent sample means

IV. Statistical inference: Estimating population parameters and testing hypotheses

- A. Estimation (point estimators and confidence intervals)
  5. Large sample confidence interval for a difference between two proportions
  7. Confidence interval for a difference between two means (unpaired and paired[ch9])
- B. Tests of significance
  3. Large sample test for a difference between two proportions
  7. Test for a difference between two means (unpaired and paired[ch9])

## The Big Ideas

One of the challenges in teaching the AP® Statistics course is keeping students focused on the big picture and not just the details of each section. We outline the big ideas for the chapter here.

### Chapter 10 Introduction

The introduction highlights the types of studies that we analyze in this chapter. We will learn how to compare two proportions, such as the proportion of subjects who would have a heart attack under each of two treatments. We will also learn how to compare two means, like the mean amount of time males and females spend studying. In both cases, we will learn how to analyze the results of these types of studies using confidence intervals and significance tests. The introduction also includes an excellent activity designed to introduce inference for a difference in proportions.

## Section 10.1 Comparing Two Proportions

In Section 10.1, we focus on comparing two proportions. We begin by exploring the sampling distribution of a difference between two proportions. Understanding how the difference between two sample proportions varies due to the chance involved in random selection (or random assignment) is essential for drawing conclusions based on an observed difference in proportions in a study. We can reliably model the distribution of $\hat{p}_1 - \hat{p}_2$ (and perform inference about $p_1 - p_2$) when the Random, 10%, and Large Counts conditions are met.

A confidence interval for a difference between two proportions provides an interval of plausible values for the true difference in proportions. The logic of confidence intervals, including how to construct and interpret a confidence interval, what conditions to check, and how to interpret a confidence level, is the same as it was in Chapter 8 when confidence intervals were first introduced. Reminding your students of these similarities will help them avoid becoming overwhelmed with the new procedures we learn in this chapter.

Likewise, a significance test for a difference between two proportions uses the same logic as the significance tests we learned about in Chapter 9. We start by assuming that the null hypothesis is true and asking how likely it would be to get results at least as unusual as the results observed in a study by chance alone. If it is plausible that a difference in proportions could be the result of sampling variability or chance variation due to the random assignment, we do not have convincing evidence that the alternative hypothesis is true. However, if the difference is too big to attribute to chance, there is convincing evidence that the alternative hypothesis is true.

At the end of this section, we introduce inference for completely randomized experiments. Fortunately, the methods used for analyzing the difference in the proportions from two independent random samples work very well for analyzing the difference in the proportions from two groups in a completely randomized experiment.

## Section 10.2 Comparing Two Means

Almost everything we do in Section 10.2 is parallel to what we did in Section 10.1, except that we are comparing two means instead of two proportions. We start with an activity designed to introduce inference for a difference in means and reinforce the logic of significance testing. Next, we develop the sampling distribution of the difference between two means—the foundation for confidence intervals and significance tests for a difference in means. We follow with the details of confidence intervals and significance tests for $\mu_1 - \mu_2$.

However, there are some additional details to be concerned with when doing inference for means. The use of the $t$ statistic requires that we know the proper number of degrees of freedom to use, and the option of pooling creates confusion for some students. We give students clear guidance on these issues. We also remind students that many comparative experiments use paired data and should not be analyzed with a two-sample $t$ test. In such cases, the paired $t$ test that we learned about in Chapter 9 would be more appropriate.

# Chapter 10: Resources

## TRM Additional *TPS* Teacher's Resource Materials

The following resources can be found by clicking on the links in the new Teacher's e-Book (TE-book), opening the Teacher's Resource CD (TRCD), or logging onto the book's companion Web site www.whfreeman.com/tps5e (teacher log-in required).

- Printable Chapter 10 Learning Objective Grid (page 612)
- Comparing Pulse Rates: This Activity highlights why we need to use significance tests when comparing two means. (page 634)
- Paper Airplanes Experiment: This Activity has students test two different models of paper airplane to see if one model flies significantly farther than the other. (page 647)
- Exercising to Lose Weight: This problem reviews important aspects of experimental design and asks students to use the results of a real experiment to determine if one exercise method is significantly better. (page 647)
- FRAPPY! materials: Question, Model Solution, Scoring Rubric, Sample Responses with commentary (page 662)
- Quizzes: 3 versions for each section; versions A and B are roughly parallel
  - Section 10.1 (page 628)
  - Section 10.2 (page 652)
- Tests: 4 versions; versions A and B are roughly parallel, version D is cumulative (page 662)

## Free-Response Questions from Previous AP® Exams

Questions can be found on the AP® Central Web site: apcentral.collegeboard.com/apc/members/exam/exam_questions/8357.html.

Students should be able to answer all the free-response questions listed below with material learned in this chapter. Questions that contain content from this chapter but also require content from later chapters are listed in the last chapter required to complete the entire question. This list will be updated after each AP® exam and will be posted on the book's Web site: www.whfreeman.com/tps5e.

Questions marked with an asterisk are from exams with released multiple-choice questions. You may want to save these questions until the end of the year so you can give your students a complete released exam for practice.

| Year | # | Content |
|---|---|---|
| 2013 | 5 | • Scope of inference<br>• Conditions for a two-sample *z* test for a difference in proportions<br>• Logic of inference, simulation of sampling distribution |
| 2012 | 3* | • Comparing histograms<br>• Conditions for two-sample *t* procedures |
| 2012 | 4* | • Two-sample *z* test for a difference in proportions |
| 2011 | 4 | • Two-sample *t* test for the difference between two means |
| 2010 | 5 | • Two-sample *t* test for the difference between two means |
| 2009 | 4 | • Two-sample *t* interval for the difference between two means<br>• Using a confidence interval to test hypotheses |
| 2009 | 5 | • Interpreting a *P*-value for a two-sample *z* test for the difference between two proportions<br>• Using a *P*-value to make a conclusion<br>• Type I and Type II errors and consequence |
| 2009B | 3 | • Two-sample *z* test for the difference between two proportions |
| 2009B | 6 | • Double-blind experiments<br>• Two-sample *z* interval for the difference between two proportions<br>• Relative risk |
| 2008 | 4 | • Constructing and interpreting scatterplots<br>• Standard error of the average of two proportions |
| 2008B | 1 | • Constructing and comparing dotplots<br>• Logic of hypothesis tests |
| 2007 | 1* | • Interpreting standard deviation<br>• Comparing center<br>• Using a confidence interval to test hypotheses |

| Year | # | Content |
|---|---|---|
| 2007 | 5* | • Experiment versus observational study<br>• Stating hypotheses<br>• Two-sample *z* test for the difference between two proportions (conditions only)<br>• Interpreting a *P*-value and making a conclusion |
| 2007B | 5 | • Two-sample *t* test for the difference between two means |
| 2006 | 4 | • Two-sample *t* interval for the difference between two means<br>• Using a confidence interval to test hypotheses |
| 2006B | 2 | • Two-sample *z* interval for the difference between two proportions<br>• Using a confidence interval to test hypotheses |
| 2005 | 6 | • Two-sample *t* interval for the difference between two means<br>• Constructing and interpreting an interaction plot |
| 2005B | 3 | • Completely randomized design versus matched pairs design<br>• Two-sample *t* test versus paired *t* test |
| 2004B | 4 | • Two-sample *t* interval for the difference between two means<br>• Two-sample versus paired *t* interval |
| 2004B | 5 | • Boxplots<br>• One-sample *t* interval for a mean (conditions only)<br>• Two-sample *t* test for the difference between two means (conditions only) |
| 2004B | 6 | • Two-sample *z* test for the difference between two proportions<br>• Estimating a total<br>• Random condition |
| 2003B | 3 | • Experiment versus observational study<br>• Two-sample *z* test for the difference between two proportions (identification and hypotheses only) |
| 2003B | 4 | • Random assignment<br>• Control groups<br>• Choosing a correct inference procedure<br>• Sources of variability |
| 2002 | 5* | • Stating hypotheses<br>• Two-sample *t* test for the difference between two means |
| 2002 | 6* | • One-sample *z* interval for a proportion<br>• Interpreting confidence level<br>• Two-sample *z* test for the difference between two proportions<br>• Pooling |
| 2000 | 4 | • Two-sample *t* test for the difference between two means<br>• Inference about cause and effect |
| 1997 | 4* | • Two-sample *z* test for the difference between two proportions |

Chapter

# 10

## Applets 

- The *Yawning Study* applet at www.rossmanchance.com/applets allows students to automate the re-randomization process for the Is Yawning Contagious? Activity.
- The *Test for a Difference in Means* and *Test for a Difference in Proportions* applets at www.lock5stat.com/statkey automate the re-randomization process we use in the Yawning and Polyester Activities.

## TI-Nspire Activities

These Activities are designed to illustrate important statistical concepts using the TI-Nspire. The Activities can be found at www.whfreeman.com/tps5e.

- *Comparing Two Means*: Uses the TI-Nspire as a statistical tool for an exploration of jumping distances between different sizes of origami frogs. This Activity also explores the distinction between a two-sample $t$ test and a paired $t$ test.

# Chapter 10: Pacing Guide, Learning Objectives, and Suggested Assignments

This pacing guide is based on a schedule with 110, 50-minute sessions before the AP® exam. The 110 days include 4 days for midterms and 10 days for AP® exam review. If you have a different number of sessions before the AP® exam, you can modify the pacing guide to suit your needs. If you have additional time, consider incorporating quizzes, released AP® free-response questions, or additional activities. See the Resources section above for suggestions.

The suggested homework assignments list odd-numbered exercises, whenever possible, so students can check their answers against the back-of-the-book answers. If you would rather students not have access to the answers while doing homework, adding 1 to the exercise numbers usually will do the trick, because the homework exercises typically are paired. For example, Exercises 1 and 2 will generally cover the same topics, but in different contexts. You may also choose to include the Recycle and Review questions at the end of each section, which review topics from previous sections or chapters. These questions are denoted with a symbol.

| Day | Topics | **Learning Objectives:** Students will be able to . . . | Suggested Assignment |
|---|---|---|---|
| 1 | Is Yawning Contagious? Activity, 10.1 The Sampling Distribution of a Difference between Two Proportions | • Describe the shape, center, and spread of the sampling distribution of $\hat{p}_1 - \hat{p}_2$ | 1, 3 |
| 2 | 10.1 Confidence Intervals for $p_1 - p_2$ | • Determine whether the conditions are met for doing inference about $p_1 - p_2$.<br>• Construct and interpret a confidence interval to compare two proportions. | 5, 7, 9, 11 |
| 3 | 10.1 Significance Tests for $p_1 - p_2$, Inference for Experiments | • Perform a significance test to compare two proportions. | 13, 15, 17, 21, 23 |
| 4 | 10.2 Does Polyester Decay? Activity, The Sampling Distribution of a Difference between Two Means | • Describe the shape, center, and spread of the sampling distribution of $\bar{x}_1 - \bar{x}_2$<br>• Determine whether the conditions are met for doing inference about $\mu_1 - \mu_2$. | 31, 33, 35, 51 |
| 5 | 10.2 The Two-Sample $t$ Statistic, Confidence Intervals for $\mu_1 - \mu_2$ | • Construct and interpret a confidence interval to compare two means. | 25–28, 37, 39 |
| 6 | 10.2 Significance Tests for $\mu_1 - \mu_2$, Using Two-Sample $t$ Procedures Wisely | • Perform a significance test to compare two means.<br>• Determine when it is appropriate to use two-sample $t$ procedures versus paired $t$ procedures. | 41, 43, 45, 47, 53, 57–60 |
| 7 | Chapter 10 Review/FRAPPY! | | Chapter 10 Review Exercises |
| 8 | Chapter 10 Test | | Cumulative AP® Practice Test 3 |

**PD** **Chapter 10 Overview**

Watch the video overview of Chapter 10 (for teachers) by clicking on the link in the TE-book, opening the TRCD, or logging onto the book's companion Web site www.whfreeman.com/tps5e (teacher log-in required).

# Chapter 10

# Comparing Two Populations or Groups

## case study

## Fast-Food Frenzy!

More than $70 billion is spent each year in the drive-thru lanes of America's fast-food restaurants. Having quick, accurate, and friendly service at a drive-thru window translates directly into revenue for the restaurant. According to Jack Greenberg, former CEO of McDonald's, sales increase 1% for every six seconds saved at the drive-thru. So industry executives, stockholders, and analysts closely follow the ratings of fast-food drive-thru lanes that appear annually in *QSR*, a publication that reports on the quick-service restaurant industry.

The 2012 *QSR* magazine drive-thru study involved visits to a random sample of restaurants in the 20 largest fast-food chains in all 50 states. During each visit, the researcher ordered a modified main item (for example, a hamburger with no pickles), a side item, and a drink. If any item was not received as ordered, or if the restaurant failed to give the correct change or supply a straw and a napkin, then the order was considered "inaccurate." Service time, which is the time from when the car stopped at the speaker to when the entire order was received, was measured each visit. Researchers also recorded whether or not each restaurant had an order-confirmation board in its drive-thru.[1]

Here are some results from the 2012 *QSR* study:

- For restaurants with order-confirmation boards, 1169 of 1327 visits (88.1%) resulted in accurate orders. For restaurants with no order-confirmation board, 655 of 726 visits (90.2%) resulted in accurate orders.
- McDonald's average service time for 362 drive-thru visits was 188.83 seconds with a standard deviation of 17.38 seconds. Burger King's service time for 318 drive-thru visits had a mean of 201.33 seconds and a standard deviation of 18.85 seconds.

Was there a significant difference in accuracy at restaurants with and without order-confirmation boards? How much better was the average service time at McDonald's than at Burger King restaurants in 2012? By the end of the chapter, you should have acquired the tools to help answer questions like these.



### Teaching Tip

One of the biggest challenges students face on the AP® exam is picking the correct inference procedure. Use the fast-food example on this page to highlight the difference between analyzing categorical data (proportion of accurate orders) and analyzing quantitative data (mean service time).

**Activity Overview**

You can watch an overview of this Activity (for teachers) by clicking on the link in the TE-book, opening the TRCD, or logging onto the book's companion Web site www.whfreeman.com/tps5e (teacher log-in required).

## Teaching Tip

Here are some tips for the Activity on this page:

- Emphasize that the logic of significance testing is the same in this chapter as it was in Chapter 9, including the initial assumption that the null hypothesis is true. When we randomly reassign the 50 people into two groups, we are assuming that the treatment received doesn't affect whether or not a person yawns. That is, if a person who yawned in one group is reassigned to the other treatment, he or she would also yawn.
- A sample dotplot is provided just below the Activity on page 611. Assuming the yawn seed has no effect, a difference of at least 0.044 is quite likely to happen by chance alone.
- When we estimate a *P*-value using a simulation like the one in this activity, it is called a "randomization test." There have been several questions on the AP® exam in recent years that use this approach, including 2013 #5, 2010 #6, 2009 #6, and 2009B #5.
- The *exact P*-value can be obtained by calculating how often the difference in proportions is at least 0.044 in *all* possible reassignments. The procedure is called Fisher's exact test and it gives a *P*-value of 0.5128. *Note*: Fisher's exact test is not included on the AP® Statistics Topic Outline.

# Introduction

Which of two popular drugs—Lipitor or Pravachol—helps lower "bad cholesterol" more? Researchers designed an experiment, called the PROVE-IT Study, to find out. They used about 4000 people with heart disease as subjects. These individuals were randomly assigned to one of two treatment groups: Lipitor or Pravachol. At the end of the study, researchers compared the proportion of subjects in each group who died, had a heart attack, or suffered other serious consequences within two years. For those using Pravachol, the proportion was 0.263; for those using Lipitor, it was 0.224.[2] Could such a difference have occurred purely by the chance involved in the random assignment? This is a question about *comparing two proportions*.

Who studies more in college—men or women? Researchers asked separate random samples of 30 males and 30 females at a large university how many minutes they studied on a typical weeknight. The females reported studying an average of 165.17 minutes; the male average was 117.17 minutes. How large is the difference in the corresponding population means? This is a question about *comparing two means*.

Comparing two proportions or means based on random sampling or a randomized experiment is one of the most common situations encountered in statistical practice. In the PROVE-IT experiment, the goal of inference is to determine whether the treatments (Lipitor and Pravachol) *caused* the observed difference in the proportion of subjects who experienced serious consequences in the two groups. For the college studying survey, the goal of inference is to draw a conclusion about the actual mean study times for *all* women and *all* men at the university.

The following Activity gives you a taste of what lies ahead in this chapter.

## ACTIVITY Is Yawning Contagious?

**MATERIALS:** Set of 50 index cards or standard deck of playing cards for each pair of students

According to the popular TV show *Mythbusters*, the answer is "Yes." The *Mythbusters* team conducted an experiment involving 50 subjects. Each subject was placed in a booth for an extended period of time and monitored by hidden camera. Thirty-four subjects were given a "yawn seed" by one of the experimenters; that is, the experimenter yawned in the subject's presence before leaving the room. The remaining 16 subjects were given no yawn seed.

What happened in the experiment? The table below shows the results:[3]

| | Yawn Seed? | | |
|---|---|---|---|
| **Subject Yawned?** | **Yes** | **No** | **Total** |
| Yes | 10 | 4 | **14** |
| No | 24 | 12 | **36** |
| **Total** | **34** | **16** | **50** |

Ten of the 34 subjects (29.4%) in the yawn-seed group yawned, compared to 4 of the 16 subjects (25.0%) in the no-yawn-seed group. The difference in the proportions who yawned for the

**APPLET**

- The *Yawning Study* applet at www.rossmanchance.com/applets allows students to automate the re-randomization process for this Activity. Students can do one repetition at a time or have the applet do multiple repetitions and plot the results. To match the Activity, make sure to choose "Difference in Proportion of Face Cards/Yawners."
- The *Test for a Difference in Proportions* applet at www.lock5stat.com/statkey can also automate the re-randomization process we use in the Yawning Activity (or the Response Bias Alternate Activity on the next page). Click Edit Data to enter the Group 1 count (10), Group 1 sample size (34), Group 2 count (4), and Group 2 sample size (16).

two groups is 10/34 − 4/16 = 0.044. Adam Savage and Jamie Hyneman, the co-hosts of *MythBusters*, used this difference as evidence that yawning is contagious. But is the evidence *convincing?*

In this Activity, your class will investigate whether the results of the experiment were really statistically significant. Let's see what would happen just by chance if we randomly reassign the 50 people in this experiment to the two groups (yawn seed and no yawn seed) many times, *assuming the treatment received doesn't affect whether or not a person yawns.*

1. We need 50 cards to represent the subjects in this study. In the *MythBusters* experiment, 14 people yawned and 36 didn't. Because we're assuming that the treatment received won't change whether each subject yawns, we use 14 cards to represent people who yawn and 36 cards to represent those who don't.
   - *Using index cards:* Write "Yes" on 14 cards and "No" on 36 cards.
   - *Using playing cards:* Remove the ace of spades and ace of clubs from the deck. All jacks, queens, kings, and aces represent subjects who yawn. All remaining cards represent subjects who don't yawn.
2. Shuffle and deal two piles of cards—one with 34 cards and one with 16 cards. The first pile represents the yawn-seed group and the second pile represents the no-yawn-seed group. The shuffling reflects our assumption that the outcome for each subject is not affected by the treatment.

   Calculate the difference in the proportions who yawned for the two groups (yawn seed – no yawn seed). For example, if you get 9 yawners in the yawn-seed group and 5 yawners in the no-yawn-seed group, the resulting difference in proportions is

$$\frac{9}{34} - \frac{5}{16} = -0.048$$

   A negative difference would mean that a smaller proportion of people in the yawn-seed group yawned during the experiment than in the no-yawn-seed group.
3. Your teacher will draw and label axes for a class dotplot. Plot the result you got in Step 2 on the graph.
4. Repeat Steps 2 and 3 if needed to get a total of at least 40 repetitions of the simulation for your class.
5. Based on the class's simulation results, how surprising would it be to get a difference in proportions of 0.044 (what the *Mythbusters* got in their experiment) or larger simply due to the chance involved in the random assignment?
6. What conclusion would you draw about whether yawning is contagious? Explain.

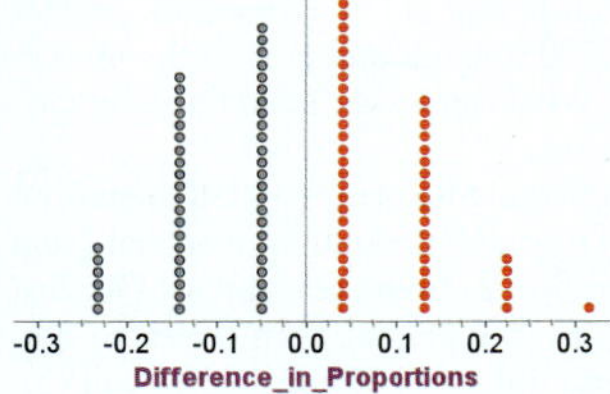


Here is an example of what the class dotplot in the Activity might look like after 100 trials. In this simulation, 50 of the 100 trials (in red) produced a difference in proportions of at least 0.044, so the approximate *P*-value is 0.50. It is very likely that a difference this big could occur just due to the chance variation in random assignment! This result is not statistically significant and does not provide convincing evidence that yawning is contagious.

ALTERNATE ACTIVITY

## *Drinking age and response bias*

In Chapter 4, we learned how the wording of a question can create bias. Two AP® Statistics students, Kala and Liz, decided to investigate this issue by asking high school students two different versions of a question about underage drinking. Here are the questions they asked:

**1.** Each year, approximately 5000 people under the age of 21 die as a result of underage alcohol consumption. Should minors (under age 21) be allowed to consume alcohol?

**2.** You are legally considered an adult when you turn 18, gaining voting rights and other privileges. Should minors (under age 21) be allowed to consume alcohol?

Because of the wording, Kala and Liz speculated that a higher proportion of students would answer "Yes" to the second version of the question. Using 44 subjects from their school, Kala and Liz randomly assigned 21 students to the first question and 23 students to the second question. Here are the results:

| | Question 1 | Question 2 | Total |
|---|---|---|---|
| Yes | 6 | 16 | 22 |
| No | 15 | 7 | 22 |
| Total | 21 | 23 | 44 |

The difference in the proportions of students who responded "Yes" (Question 2 − Question 1) was 16/23 − 6/21 = 0.696 − 0.286 = 0.410. Does this difference provide convincing evidence that the wording of the questions has an effect on the response, or could the difference be due to the chance variation in random assignment?

**1.** Using 44 index cards or equally sized slips of paper, write "Yes" on 22 and "No" on 22. Shuffle the cards and divide them at random into a group of 21 (to represent those who were asked Question 1) and a group of 23 (to represent those who were asked Question 2).

**2.** Calculate the difference in proportion of "Yes" responses for the two groups (Question 2 − Question 1) and record this difference on a dotplot.

**3.** Repeat Steps 1 and 2 many times.

**4.** How often did a difference of 0.410 or greater occur due only to the chance variation of the random assignment? What conclusion would you make about the effect of the wording of these questions?

Here is an example of what your class dotplot might look like after 100 trials. Only 2 trials out of 100 resulted in a difference of 0.410 or greater, meaning the *P*-value is approximately 0.02. Because the *P*-value is small and this was a randomized comparative experiment, we have convincing evidence that the wording of the questions caused a higher proportion of students to answer "Yes" to the second question.

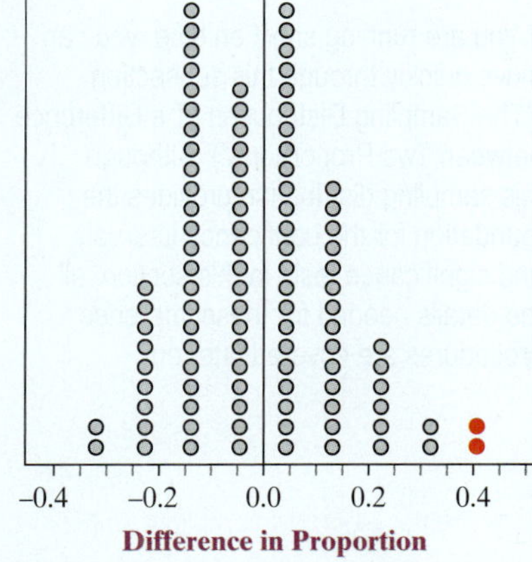

**PD Section 10.1 Overview**

Watch the video overview of Section 10.1 (for teachers) by clicking on the link in the TE-book, opening the TRCD, or logging onto the book's companion Web site www.whfreeman.com/tps5e (teacher log-in required).

**TRM Learning Objectives Grid**

Remind students about the importance of the Learning Objectives. We include them at the beginning of each section so students know what they are expected to learn. At the end of the chapter, the Learning Objectives are included in a grid, along with references to examples and chapter review exercises. A printable version of this grid is included in the Teacher's Resource Materials.

**TRM Section 10.1 Alternate Examples**

You can find the Alternate Examples for this section in Microsoft Word format by clicking on the link in the TE-book, opening the TRCD, or logging onto the book's companion Web site www.whfreeman.com/tps5e (teacher log-in required).

### Teaching Tip

After teaching the section "The Sampling Distribution of a Difference between Two Proportions," some teachers prefer to reverse our order and teach significance tests before confidence intervals. Significance tests follow naturally from the Is Yawning Contagious? Activity and confidence intervals are a good follow-up analysis as they provide more information than a significance test. This section, including the exercises, was written so teachers could take either approach.

### Teaching Tip

If you are running short on time, you can move quickly through this subsection ("The Sampling Distribution of a Difference between Two Proportions"). Although this sampling distribution provides the foundation for the confidence intervals and significance tests in this section, all the details needed for these inference procedures are covered later on.

# 10.1 Comparing Two Proportions

**WHAT YOU WILL LEARN** By the end of the section, you should be able to:

- Describe the shape, center, and spread of the sampling distribution of $\hat{p}_1 - \hat{p}_2$.
- Determine whether the conditions are met for doing inference about $p_1 - p_2$.
- Construct and interpret a confidence interval to compare two proportions.
- Perform a significance test to compare two proportions.

Suppose we want to compare the proportions of individuals with a certain characteristic in Population 1 and Population 2. Let's call these parameters of interest $p_1$ and $p_2$. The ideal strategy is to take a separate random sample from each population and to compare the sample proportions $\hat{p}_1$ and $\hat{p}_2$ with that characteristic.

What if we want to compare the effectiveness of Treatment 1 and Treatment 2 in a completely randomized experiment? This time, the parameters $p_1$ and $p_2$ that we want to compare are the true proportions of successful outcomes for each treatment. We use the proportions of successes in the two treatment groups, $\hat{p}_1$ and $\hat{p}_2$, to make the comparison.

Here's a table that summarizes these two situations:

| Population or treatment | Parameter | Statistic | Sample size |
|---|---|---|---|
| 1 | $p_1$ | $\hat{p}_1$ | $n_1$ |
| 2 | $p_2$ | $\hat{p}_2$ | $n_2$ |

We compare the populations or treatments by doing inference about the difference $p_1 - p_2$ between the parameters. The statistic that estimates this difference is the difference between the two sample proportions, $\hat{p}_1 - \hat{p}_2$. To use $\hat{p}_1 - \hat{p}_2$ for inference, we must know its sampling distribution.

## The Sampling Distribution of a Difference between Two Proportions

To explore the sampling distribution of $\hat{p}_1 - \hat{p}_2$, let's start with two populations having a known proportion of successes. Suppose that there are two large high schools, each with over 2000 students, in a certain town. At School 1, 70% of students did their homework last night. Only 50% of the students at School 2 did their homework last night. The counselor at School 1 takes an SRS of 100 students and records the proportion $\hat{p}_1$ that did the homework. School 2's counselor takes an SRS of 200 students and records the proportion $\hat{p}_2$ that did the homework. What can we say about the difference $\hat{p}_1 - \hat{p}_2$ in the sample proportions?

We used Fathom software to take an SRS of $n_1 = 100$ students from School 1 and a separate SRS of $n_2 = 200$ students from School 2 and to plot the values of $\hat{p}_1$, $\hat{p}_2$, and $\hat{p}_1 - \hat{p}_2$ from each sample. Our first set of simulated samples gave $\hat{p}_1 = 0.68$ and $\hat{p}_2 = 0.505$, so dots were placed above each of those values in Figure 10.1(a) and (b).

### Teaching Tip

Remind your students that the sampling distribution of a statistic tells us the possible values of the statistic and how likely they are to occur. Also, remind your students that we are temporarily pretending to know the true proportion of successes in each population so we can investigate the sampling distribution of the difference in sample proportions. We will return to the real world (where we don't know the values of the parameters) in a couple of pages.

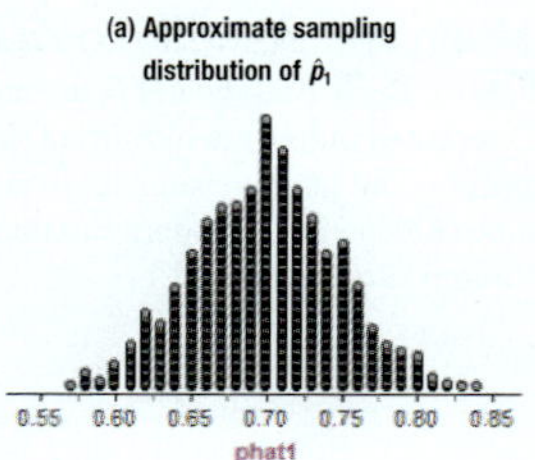


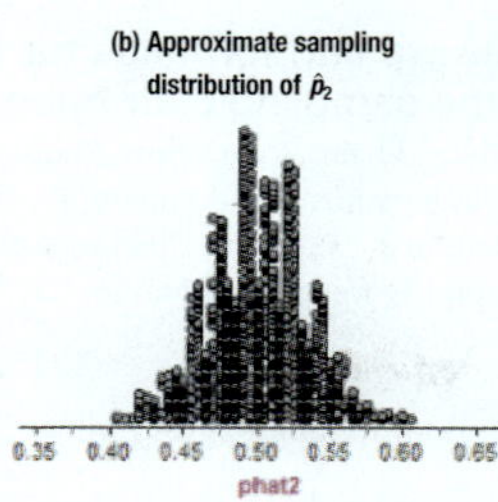


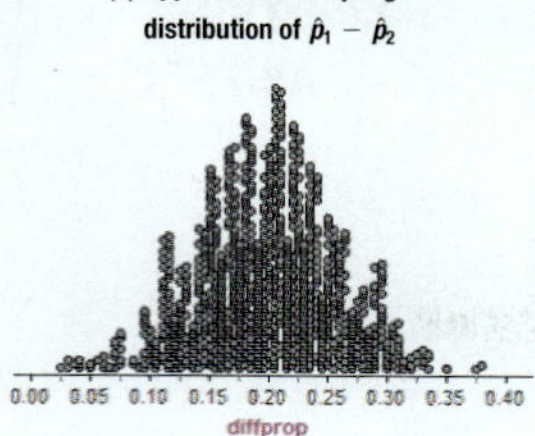


**FIGURE 10.1** Simulated sampling distributions of (a) the sample proportion $\hat{p}_1$ of successes in 1000 SRSs of size $n_1 = 100$ from a population with $p_1 = 0.70$, (b) the sample proportion $\hat{p}_2$ of successes in 1000 SRSs of size $n_2 = 200$ from a population with $p_2 = 0.50$, and (c) the difference in sample proportions $\hat{p}_1 - \hat{p}_2$ for each of the 1000 repetitions.

The difference in the sample proportions for this first set of samples is $\hat{p}_1 - \hat{p}_2 = 0.68 - 0.505 = 0.175$. A dot for this value appears in Figure 10.1(c). The three dotplots in Figure 10.1 show the results of repeating this process 1000 times. These are the approximate sampling distributions of $\hat{p}_1$, $\hat{p}_2$, and $\hat{p}_1 - \hat{p}_2$.

In Chapter 7, we saw that the sampling distribution of a sample proportion $\hat{p}$ has the following properties:

*Shape:* Approximately Normal if $np \geq 10$ and $n(1-p) \geq 10$

*Center:* $\mu_{\hat{p}} = p$

*Spread:* $\sigma_{\hat{p}} = \sqrt{\dfrac{p(1-p)}{n}}$ if $n \leq \dfrac{1}{10}N$

For the sampling distributions of $\hat{p}_1$ and $\hat{p}_2$ in this case:

| | Sampling distribution of $\hat{p}_1$ | Sampling distribution of $\hat{p}_2$ |
|---|---|---|
| **Shape** | Approximately Normal; $n_1p_1 = 100(0.70) = 70 \geq 10$ and $n_1(1-p_1) = 100(0.30) = 30 \geq 10$ | Approximately Normal; $n_2p_2 = 200(0.50) = 100 \geq 10$ and $n_2(1-p_2) = 200(0.50) = 100 \geq 10$ |
| **Center** | $\mu_{\hat{p}_1} = p_1 = 0.70$ | $\mu_{\hat{p}_2} = p_2 = 0.50$ |
| **Spread** | $\sigma_{\hat{p}_1} = \sqrt{\dfrac{p_1(1-p_1)}{n_1}} = \sqrt{\dfrac{0.7(0.3)}{100}} = 0.0458$ because School 1 has a population of over 10(100) = 1000 students. | $\sigma_{\hat{p}_2} = \sqrt{\dfrac{p_2(1-p_2)}{n_2}} = \sqrt{\dfrac{0.5(0.5)}{200}} = 0.0354$ because School 2 has a population of over 10(200) = 2000 students. |

The approximate sampling distributions in Figures 10.1(a) and (b) give similar results.

What about the sampling distribution of $\hat{p}_1 - \hat{p}_2$? Figure 10.1(c) suggests that it has an approximately Normal shape, is centered at about 0.198, and has standard deviation about 0.0572. The shape makes sense because we are combining two independent random variables, $\hat{p}_1$ and $\hat{p}_2$, that have approximately Normal distributions. How about the center? The true proportion of students who did last night's homework at School 1 is $p_1 = 0.70$ and at School 2 is $p_2 = 0.50$. We expect the difference $\hat{p}_1 - \hat{p}_2$ to center on the actual difference in the population proportions, $p_1 - p_2 = 0.70 - 0.50 = 0.20$. The spread, however, is a bit more complicated.

## Teaching Tip

Before showing the Fathom dotplots on this page, make sure your students understand how these graphs are created. One way to do this would be to create two large populations of beads to represent the two schools, one with 70% of a certain color (e.g., red) and one with 50% of a certain color. Then take a random sample of size 100 from the first population and calculate the proportion of red beads. Do the same thing with a random sample of size 200 from the second population. Finally, subtract the two proportions. The three dotplots on this page show the results of doing this many times.

## Teaching Tip

Ask your students to interpret the standard deviation for each of these distributions. For example, if $\sigma_{\hat{p}_1 - \hat{p}_2} \approx 0.0572$, then the difference in the *sample* proportions of students who did their homework at the two schools will typically be about 0.0572 from the *true* difference in proportions.

### Teaching Tip

Students won't be asked to derive the formulas for the mean and standard deviation of a difference in sample proportions as we do here. However, reading through this Think About It is a great way to review the rules for combining random variables that we learned in Chapter 6 and the properties of the sampling distribution of a sample proportion that we learned in Chapter 7.

**How can we find formulas for the mean and standard deviation of the sampling distribution of $\hat{p}_1 - \hat{p}_2$?** Both $\hat{p}_1$ and $\hat{p}_2$ are random variables. That is, their values would vary in repeated independent SRSs of size $n_1$ and $n_2$. Independent random samples yield independent random variables $\hat{p}_1$ and $\hat{p}_2$. The statistic $\hat{p}_1 - \hat{p}_2$ is the difference of these two independent random variables.

In Chapter 6, we learned that for any two random variables X and Y,

$$\mu_{X-Y} = \mu_X - \mu_Y$$

For the random variables $\hat{p}_1$ and $\hat{p}_2$, we have

$$\mu_{\hat{p}_1-\hat{p}_2} = \mu_{\hat{p}_1} - \mu_{\hat{p}_2} = p_1 - p_2$$

In the school homework survey,

$$\mu_{\hat{p}_1-\hat{p}_2} = p_1 - p_2 = 0.70 - 0.50 = 0.20$$

We also learned in Chapter 6 that for *independent* random variables X and Y,

$$\sigma^2_{X-Y} = \sigma^2_X + \sigma^2_Y$$

For the random variables $\hat{p}_1$ and $\hat{p}_2$, we have

$$\sigma^2_{\hat{p}_1-\hat{p}_2} = \sigma^2_{\hat{p}_1} + \sigma^2_{\hat{p}_2} = \left(\sqrt{\frac{p_1(1-p_1)}{n_1}}\right)^2 + \left(\sqrt{\frac{p_2(1-p_2)}{n_2}}\right)^2 = \frac{p_1(1-p_1)}{n_1} + \frac{p_2(1-p_2)}{n_2}$$

So $\sigma_{\hat{p}_1-\hat{p}_2} = \sqrt{\frac{p_1(1-p_1)}{n_1} + \frac{p_2(1-p_2)}{n_2}}$.

In the school homework survey,

$$\sigma_{\hat{p}_1-\hat{p}_2} = \sqrt{\frac{p_1(1-p_1)}{n_1} + \frac{p_2(1-p_2)}{n_2}} = \sqrt{\frac{0.7(0.3)}{100} + \frac{0.5(0.5)}{200}} = 0.058$$

This is similar to the result from the Fathom simulation.

Here are the facts we need.

### Teaching Tip

As always, to know the shape, center, and spread of a sampling distribution, certain conditions must be met. Fortunately, these conditions should sound very familiar by now! See Question #5b from the 2013 AP® exam for a problem that asks about the conditions necessary to use a Normal approximation to the sampling distribution of $\hat{p}_1 - \hat{p}_2$.

**THE SAMPLING DISTRIBUTION OF $\hat{p}_1 - \hat{p}_2$**

Choose an SRS of size $n_1$ from Population 1 with proportion of successes $p_1$ and an independent SRS of size $n_2$ from Population 2 with proportion of successes $p_2$.

- **Shape:** When $n_1p_1$, $n_1(1 - p_1)$, $n_2p_2$, and $n_2(1 - p_2)$ are all at least 10, the sampling distribution of $\hat{p}_1 - \hat{p}_2$ is approximately Normal.
- **Center:** The mean of the sampling distribution of $\hat{p}_1 - \hat{p}_2$ is $p_1 - p_2$.
- **Spread:** The standard deviation of the sampling distribution of $\hat{p}_1 - \hat{p}_2$ is

$$\sqrt{\frac{p_1(1-p_1)}{n_1} + \frac{p_2(1-p_2)}{n_2}}$$

as long as each sample is no more than 10% of its population.

When conditions are met, the sampling distribution of $\hat{p}_1 - \hat{p}_2$ will be approximately Normal with mean $\mu_{\hat{p}_1-\hat{p}_2} = p_1 - p_2$ and standard deviation $\sigma_{\hat{p}_1-\hat{p}_2} = \sqrt{\frac{p_1(1-p_1)}{n_1} + \frac{p_2(1-p_2)}{n_2}}$. Figure 10.2 displays this distribution.

The formula for the standard deviation of the sampling distribution involves the unknown parameters $p_1$ and $p_2$. Just as in Chapters 8 and 9, we must replace these by estimates to do inference. And just as before, we do this a bit differently for confidence intervals and for tests. We'll get to inference shortly. For now, let's focus on the sampling distribution of $\hat{p}_1 - \hat{p}_2$.

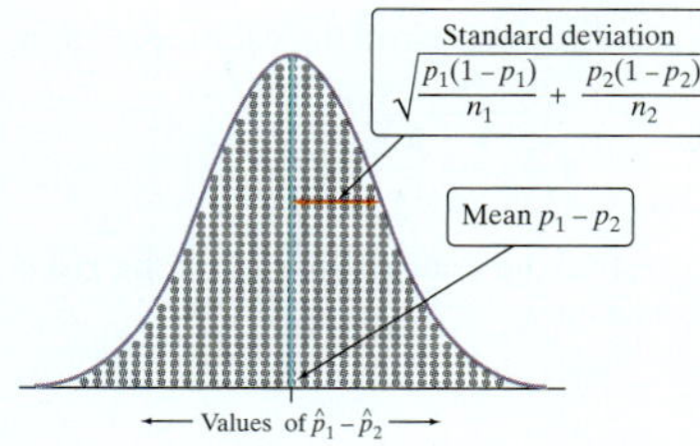
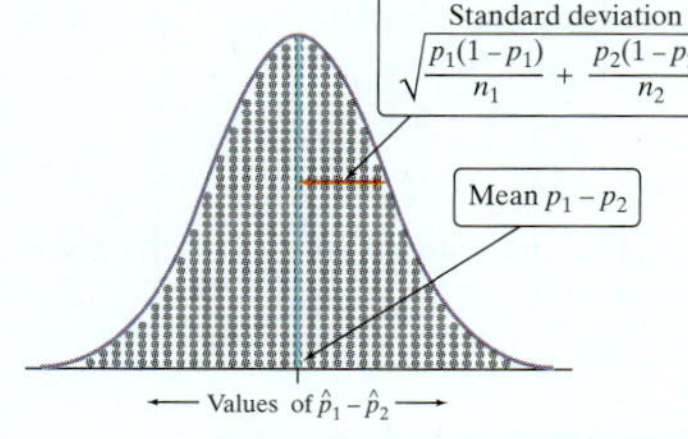


**FIGURE 10.2** Select independent SRSs from two populations having proportions of successes $p_1$ and $p_2$. The proportions of successes in the two samples are $\hat{p}_1$ and $\hat{p}_2$. When the samples are large, the sampling distribution of the difference $\hat{p}_1 - \hat{p}_2$ is approximately Normal.

## EXAMPLE Yummy Goldfish!

### *Describing the sampling distribution of $\hat{p}_1 - \hat{p}_2$*

Your teacher brings two bags of colored goldfish crackers to class. Bag 1 has 25% red crackers and Bag 2 has 35% red crackers. Each bag contains more than 1000 crackers. Using a paper cup, your teacher takes an SRS of 50 crackers from Bag 1 and a separate SRS of 40 crackers from Bag 2. Let $\hat{p}_1 - \hat{p}_2$ be the difference in the sample proportions of red crackers.

PROBLEM:

(a) What is the shape of the sampling distribution of $\hat{p}_1 - \hat{p}_2$? Why?

(b) Find the mean of the sampling distribution. Show your work.

(c) Find the standard deviation of the sampling distribution. Show your work.

SOLUTION:

(a) Because $n_1p_1 = 50(0.25) = 12.5$, $n_1(1 - p_1) = 50(0.75) = 37.5$, $n_2p_2 = 40(0.35) = 14$, and $n_2(1 - p_2) = 40(0.65) = 26$ are all at least 10, the sampling distribution of $\hat{p}_1 - \hat{p}_2$ is approximately Normal.

(b) The mean is $\mu_{\hat{p}_1-\hat{p}_2} = p_1 - p_2 = 0.25 - 0.35 = -0.10$.

(c) Because there are at least $10(50) = 500$ crackers in Bag 1 and $10(40) = 400$ crackers in Bag 2, the standard deviation is

$$\sigma_{\hat{p}_1-\hat{p}_2} = \sqrt{\frac{p_1(1-p_1)}{n_1} + \frac{p_2(1-p_2)}{n_2}} = \sqrt{\frac{0.25(0.75)}{50} + \frac{0.35(0.65)}{40}} = 0.0971$$

**For Practice** *Try Exercise* **1**

## Teaching Tip

Remind your students that Figure 10.2 on this page is a theoretical model for dotplots such as Figure 10.1(c). It might help to remind your students that each dot in Figure 10.2 represents one possible observation of $\hat{p}_1 - \hat{p}_2$.

## Teaching Tip

Remind students that the "play" icon at the top of this example (and many others) indicates that they can watch a video presentation of the example. Find the videos at www.whfreeman.com/tps5e. Also, remind them that the icon in the lower right corner indicates a link between this example and Exercise 1. If students have trouble with the exercise, they can look back to this example for help.

## ALTERNATE EXAMPLE

### *American-made cars*

Nathan and Kyle both work for the Department of Motor Vehicles (DMV), but they live in different states. In Nathan's state, 80% of the registered cars are made by American manufacturers. In Kyle's state, only 60% of the registered cars are made by American manufacturers. Nathan selects a random sample of 100 cars in his state and Kyle selects a random sample of 70 cars in his state. Let $\hat{p}_N - \hat{p}_K$ be the difference in the sample proportion of cars made by American manufacturers.

PROBLEM:

**(a)** What is the shape of the sampling distribution of $\hat{p}_N - \hat{p}_K$? Why?

**(b)** Find the mean of the sampling distribution. Show your work.

**(c)** Find the standard deviation of the sampling distribution. Show your work.

SOLUTION:

**(a)** Because $100(0.80) = 80$, $100(1 - 0.80) = 20$, $70(0.60) = 42$, and $70(1 - 0.60) = 28$ are all $\geq 10$, the sampling distribution of $\hat{p}_N - \hat{p}_K$ is approximately Normal.

**(b)** The mean is $\mu_{\hat{p}_N-\hat{p}_K} = 0.80 - 0.60 = 0.20$.

**(c)** Because there are at least $10(100) = 1000$ cars in Nathan's state and at least $10(70) = 700$ cars in Kyle's state, the standard deviation is $\sigma_{\hat{p}_N-\hat{p}_K}$

$$= \sqrt{\frac{0.80(1-0.80)}{100} + \frac{0.60(1-0.60)}{70}} = 0.0709.$$

## Confidence Intervals for $p_1 - p_2$

When data come from two independent random samples or two groups in a randomized experiment (the Random condition), the statistic $\hat{p}_1 - \hat{p}_2$ is our best guess for the value of $p_1 - p_2$. We can use our familiar formula to calculate a confidence interval for $p_1 - p_2$:

statistic ± (critical value) · (standard deviation of statistic)

When the 10% condition is met, the standard deviation of the statistic $\hat{p}_1 - \hat{p}_2$ is

$$\sigma_{\hat{p}_1-\hat{p}_2} = \sqrt{\frac{p_1(1-p_1)}{n_1} + \frac{p_2(1-p_2)}{n_2}}$$

If the Large Counts condition is met, we find the critical value $z^*$ for the given confidence level from the standard Normal curve.

**CONDITIONS FOR CONSTRUCTING A CONFIDENCE INTERVAL ABOUT A DIFFERENCE IN PROPORTIONS**

- **Random:** The data come from two independent random samples or from two groups in a randomized experiment.
  - **10%:** When sampling without replacement, check that $n_1 \leq \frac{1}{10}N_1$ and $n_2 \leq \frac{1}{10}N_2$.
- **Large Counts:** The counts of "successes" and "failures" in each sample or group—$n_1\hat{p}_1$, $n_1(1-\hat{p}_1)$, $n_2\hat{p}_2$, $n_2(1-\hat{p}_2)$—are all at least 10.

Because we don't know the values of the parameters $p_1$ and $p_2$, we replace them in the standard deviation formula with the sample proportions. The result is the **standard error** (also called the *estimated standard deviation*) of the statistic $\hat{p}_1 - \hat{p}_2$:

$$SE_{\hat{p}_1-\hat{p}_2} = \sqrt{\frac{\hat{p}_1(1-\hat{p}_1)}{n_1} + \frac{\hat{p}_2(1-\hat{p}_2)}{n_2}}$$

This value tells us how far the difference in sample proportions will typically be from the difference in population proportions if we repeat the random sampling or random assignment many times.

When the conditions are met, our confidence interval for $p_1 - p_2$ is therefore

statistic ± (critical value) · (standard deviation of statistic)

$$(\hat{p}_1 - \hat{p}_2) \pm z^*\sqrt{\frac{\hat{p}_1(1-\hat{p}_1)}{n_1} + \frac{\hat{p}_2(1-\hat{p}_2)}{n_2}}$$

This is often called a **two-sample $z$ interval for a difference between two proportions**.

### Teaching Tip

Point out that the conditions for doing inference for a difference in proportions are very similar to the conditions for doing inference for a single proportion. The main difference is that we need to check the conditions for *both* samples. The Random condition now requires that the two samples are independent. The Random condition can also be met by using two groups in a randomized experiment. We will address inference for experiments in detail beginning on page 625.

### Teaching Tip

Remind your students that the standard error of a statistic uses sample data to estimate the standard deviation of the statistic. The formula for the standard error of the statistic $\hat{p}_1 - \hat{p}_2$ is *not* included on the AP® exam formula sheet, but it is very closely related to the formula for the standard deviation of the statistic $\hat{p}_1 - \hat{p}_2$, which is on the formula sheet.

**TWO-SAMPLE $z$ INTERVAL FOR A DIFFERENCE BETWEEN TWO PROPORTIONS**

When the conditions are met, an approximate C% confidence interval for $\hat{p}_1 - \hat{p}_2$ is

$$(\hat{p}_1 - \hat{p}_2) \pm z^* \sqrt{\frac{\hat{p}_1(1-\hat{p}_1)}{n_1} + \frac{\hat{p}_2(1-\hat{p}_2)}{n_2}}$$

where $z^*$ is the critical value for the standard Normal curve with C% of its area between $-z^*$ and $z^*$.

The following example shows how to construct and interpret a confidence interval for a difference in proportions. As usual with inference problems, we follow the four-step process. Because you are expected to include these four steps whenever you construct a confidence interval or perform a significance test, *we will limit our use of the four-step icon to examples from this point forward.*

EXAMPLE

## Teens and Adults on Social Networking Sites

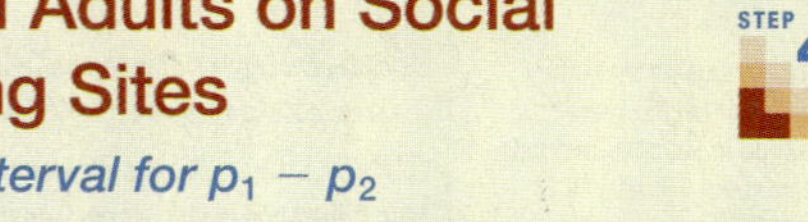

STEP 4

*Confidence interval for $p_1 - p_2$*

As part of the Pew Internet and American Life Project, researchers conducted two surveys in 2012. The first survey asked a random sample of 799 U.S. teens about their use of social media and the Internet. A second survey posed similar questions to a random sample of 2253 U.S. adults. In these two studies, 80% of teens and 69% of adults used social-networking sites.

PROBLEM:

(a) Calculate the standard error of the sampling distribution of the difference in the sample proportions (teens − adults). What information does this value provide?

(b) Construct and interpret a 95% confidence interval for the difference between the proportion of all U.S. teens and adults who use social-networking sites.

SOLUTION:

(a) The sample proportions of teens and adults who use social-networking sites are $\hat{p}_1 = 0.80$ and $\hat{p}_2 = 0.69$, respectively. The standard error of the sampling distribution of $\hat{p}_1 - \hat{p}_2$ is

$$SE_{\hat{p}_1 - \hat{p}_2} = \sqrt{\frac{\hat{p}_1(1-\hat{p}_1)}{n_1} + \frac{\hat{p}_2(1-\hat{p}_2)}{n_2}} = \sqrt{\frac{0.80(0.20)}{799} + \frac{0.69(0.31)}{2253}} = 0.0172$$

If we were to take many random samples of 799 teens and 2253 adults, the difference in the sample proportions of teens and adults who use social-networking sites will typically be 0.0172 from the true difference in proportions of all teens and adults who use social-networking sites.

(b) STATE: Our parameters of interest are $p_1$ = the proportion of all U.S. teens who use social-networking sites and $p_2$ = the proportion of all U.S. adults who use social-networking sites. We want to estimate the difference $p_1 - p_2$ at a 95% confidence level.

### AP® Exam Formula Sheet

The formula for the confidence interval for a difference of proportions is *not* included on the AP® exam formula sheet. However, the general formula for a confidence interval is included on the formula sheet that is provided with both sections of the AP® exam.

### Teaching Tip

Make sure to point out the last sentence before the example. Because students will not be prompted to follow the four-step process on the AP® exam, we no longer will remind them to perform all four steps in the exercises.

### Teaching Tip

As you read this example with your students, ask them why it would not be appropriate to conclude that the difference in the proportions of *all* U.S. teens and *all* adults is *exactly* 0.11. Because we are using random samples to estimate the proportions of teens and adults who use social networking sites, the proportions we are using are just estimates of the true proportions. We expect the sample proportions to vary somewhat from the true proportions, which is why we must include a margin of error when we estimate the difference.

### Teaching Tip

In this example, notice that the parameters clearly refer to the population of *all* U.S. teens and *all* U.S. adults. Referring to the sample (e.g., the proportion of teens who *said* they use social networking sites) when defining the parameters or stating a conclusion will result in lost credit.

## ALTERNATE EXAMPLE

### Presidential approval

Many news organizations conduct polls asking adults in the United States if they approve of the job the president is doing. How did President Obama's approval rating change from October 2012 to October 2013? According to a Gallup poll of 1500 randomly selected U.S. adults on October 2–4, 2012, 52% approved of Obama's job performance. A Gallup poll of 1500 randomly selected U.S. adults on October 5–7, 2013, showed that 46% approved of Obama's job performance.

PROBLEM:

**(a)** Calculate the standard error of the sampling distribution of the difference in the sample proportions (2013 − 2012). Interpret this value.
**(b)** Use the results of these polls to construct and interpret a 90% confidence interval for the change in Obama's approval rating among all U.S. adults from October 2012 to October 2013.
**(c)** Based on your interval, is there convincing evidence that Obama's job approval rating has changed?

SOLUTION:

**(a)** $SE_{\hat{p}_{13}-\hat{p}_{12}} = \sqrt{\frac{0.46(1-0.46)}{1500} + \frac{0.52(1-0.52)}{1500}} = 0.0182$. If we were to take many random samples of 1500 U.S. adults in October 2012 and October 2013, the difference in the sample proportions who approve of President Obama will typically be 0.0182 from the true difference in proportions.

**(b)** STATE: We want to estimate $p_{13} - p_{12}$ at the 90% confidence level, where $p_{13}$ = the true proportion of all U.S. adults who approved of President Obama's job performance in October 2013 and $p_{12}$ = the true proportion of all U.S. adults who approved of President Obama's job performance in October 2012.

PLAN: We should use a two-sample $z$ interval for $p_{13} - p_{12}$ if the conditions are met.

- *Random*: The data came from independent random samples.
  - *10%*: There were more than 10(1500) = 15,000 U.S. adults in both October 2012 and October 2013.
- *Large Counts*: $n_{13}\hat{p}_{13} = 1500(0.46) = 690$, $n_{13}(1-\hat{p}_{13}) = 1500(0.54) = 810$, $n_{12}\hat{p}_{12} = 1500(0.52) = 780$, $n_{12}(1-\hat{p}_{12}) = 1500(0.48) = 720$ are all at least 10.

DO: $(0.46-0.52) \pm 1.645 \times \sqrt{\frac{0.46(1-0.46)}{1500} + \frac{0.52(1-0.52)}{1500}} = -0.06 \pm 0.030 = (-0.090, -0.030)$.

CONCLUDE: We are 90% confident that the interval from −0.090 to −0.030 captures the true change in the proportion of U.S. adults who approve of President Obama's job performance from October 2012 to October 2013.

**(c)** Because 0 is not included in the interval, we have convincing evidence that President Obama's approval rating has changed from October 2012 to October 2013.

PLAN: We should use a two-sample $z$ interval for $p_1 - p_2$ if the conditions are met.

- *Random:* The data come from independent random samples of 799 U.S. teens and 2253 U.S. adults.
  - *10%:* The researchers are sampling without replacement, so we must check the 10% condition: there are at least 10(799) = 7990 U.S. teens and at least 10(2253) = 22,530 U.S. adults.
- *Large Counts:* We check the counts of "successes" and "failures":

$n_1\hat{p}_1 = 799(0.80) = 639.2 \rightarrow 639$ $\quad n_1(1-\hat{p}_1) = 799(1-0.80) = 159.8 \rightarrow 160$

$n_2\hat{p}_2 = 2253(0.69) = 1554.57 \rightarrow 1555$ $\quad n_2(1-\hat{p}_2) = 2253(1-0.69) = 698.43 \rightarrow 698$

Note that the observed counts have to be whole numbers! Because all four values are at least 10, this condition is met.

DO: We know that $n_1 = 799$, $\hat{p}_1 = 0.80$, $n_2 = 2253$, and $\hat{p}_2 = 0.69$. For a 95% confidence level, the critical value is $z^* = 1.96$. So our 95% confidence interval for $p_1 - p_2$ is

$$(\hat{p}_1 - \hat{p}_2) \pm z^*\sqrt{\frac{\hat{p}_1(1-\hat{p}_1)}{n_1} + \frac{\hat{p}_2(1-\hat{p}_2)}{n_2}} = (0.80 - 0.69) \pm 1.96\sqrt{\frac{0.80(0.20)}{799} + \frac{0.69(0.31)}{2253}}$$
$$= 0.11 \pm 1.96(0.0172)$$
$$= 0.11 \pm 0.034$$
$$= (0.076, 0.144)$$

*Using technology:* Refer to the Technology Corner that follows the example. The calculator's `2-PropZInt` gives (0.07588, 0.14324).

CONCLUDE: We are 95% confident that the interval from 0.07588 to 0.14324 captures the true difference in the proportion of all U.S. teens and adults who use social-networking sites.

This interval suggests that more teens than adults in the United States engage in social networking by between about 7.6 and 14.3 percentage points.

**For Practice** *Try Exercise* **9**

The researchers in the previous example selected independent random samples from the two populations they wanted to compare. In practice, it's common to take one random sample that includes individuals from both populations of interest and then to separate the chosen individuals into two groups. The two-sample $z$ procedures for comparing proportions are still valid in such situations, provided that the two groups can be viewed as independent samples from their respective populations of interest.

You can use technology to perform the calculations in the "Do" step. Remember that this comes with potential benefits and risks on the AP® exam.

## 21. TECHNOLOGY CORNER — CONFIDENCE INTERVAL FOR A DIFFERENCE IN PROPORTIONS

TI-Nspire instructions in Appendix B; HP Prime instructions on the book's Web site.

The TI-83/84 and TI-89 can be used to construct a confidence interval for $p_1 - p_2$. We'll demonstrate using the previous example. Of $n_1 = 799$ teens surveyed, $X = 639$ said they used social-networking sites. Of $n_2 = 2253$ adults surveyed, $X = 1555$ said they engaged in social networking. To construct a confidence interval:

### Teaching Tip

In this chapter, we do not make a distinction between taking two independent random samples and taking one random sample and separating it into two groups. However, in Chapter 11 this will become an important distinction when we study chi-square tests.

### TECHNOLOGY TIP

Remind students that they can watch a video presentation of this technology corner (for the TI-83/84, TI-89, or TI-Nspire) at www.whfreeman.com/tps5e.

**TI-83/84**

- Press STAT, then choose TESTS and 2-PropZInt.

**TI-89**

- In the Statistics/List Editor, press 2nd F2 ([F7]) and choose 2-PropZInt.

- When the 2-PropZInt screen appears, enter the values shown.

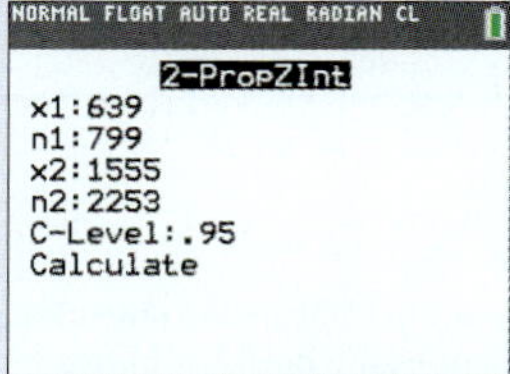

- Highlight "Calculate" and press ENTER.

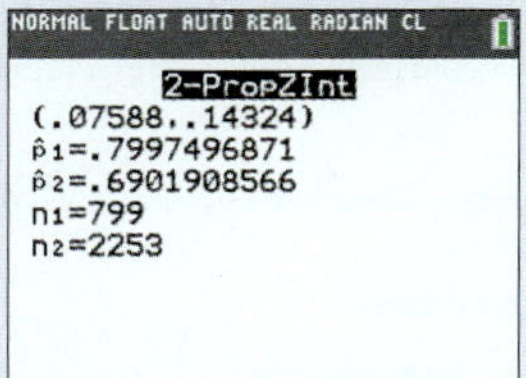


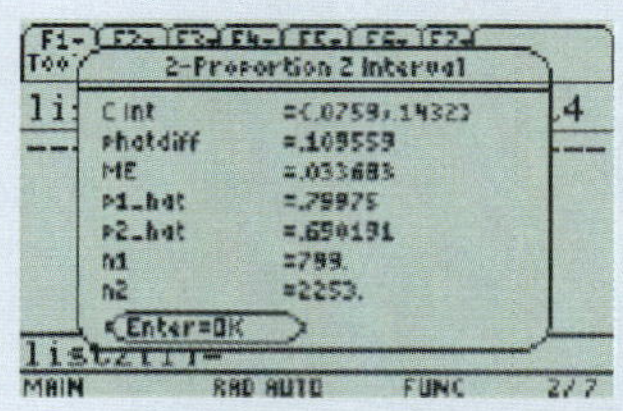


**AP® EXAM TIP** The formula for the two-sample $z$ interval for $p_1 - p_2$ often leads to calculation errors by students. As a result, we recommend using the calculator's 2-PropZInt feature to compute the confidence interval on the AP® exam. Be sure to name the procedure (two-proportion $z$ interval) and to give the interval (0.076, 0.143) as part of the "Do" step.

## CHECK YOUR UNDERSTANDING

Are teens or adults more likely to go online daily? The Pew Internet and American Life Project asked a random sample of 799 teens and a separate random sample of 2253 adults how often they use the Internet. In these two surveys, 63% of teens and 68% of adults said that they go online every day. Construct and interpret a 90% confidence interval for $p_1 - p_2$.

## Significance Tests for $p_1 - p_2$

An observed difference between two sample proportions can reflect an actual difference in the parameters, or it may just be due to chance variation in random sampling or random assignment. Significance tests help us decide which explanation makes more sense.

The null hypothesis has the general form

$$H_0: p_1 - p_2 = \text{hypothesized value}$$

### TECHNOLOGY TIP

Warn your students not to choose the wrong procedure on their calculators. The 2-SampZInt is for comparing means when the population standard deviations are known. What we want is the 2-PropZInt.

### TECHNOLOGY TIP

The values of x1 and x2 are the numbers of successes, not the proportions of successes. Also, they must be integers!

### AP® Exam Common Error

After using their calculator's 2-PropZInt feature to correctly construct the interval, some students try to show additional work by stating a formula or showing numbers substituted into a formula. This often results in a loss of credit because students use the wrong formula or include a wrong symbol in the correct formula (e.g., using $p$ instead of $\hat{p}$ in the standard error). For two-sample procedures, we recommend that students use their calculator to do the calculations, finish the exam, and then double-check their calculations if there is time. If they show additional work, it must be consistent with the interval from the calculator!

### ANSWERS TO CYU

*S*: $p_1$ = true proportion of teens who go online every day and $p_2$ = true proportion of adults who go online every day.
*P*: Two-sample $z$ interval for $p_1 - p_2$. *Random*: Independent random samples. *10%*: $n_1 = 799 < 10\%$ of teens and $n_2 = 2253 < 10\%$ of adults. *Large Counts*: 503, 296, 1532, and 721 are all $\geq 10$. *D*:

$$(0.63 - 0.68) \pm 1.645\sqrt{\frac{0.63(0.37)}{799} + \frac{0.68(0.32)}{2253}} = (-0.0824, -0.0176).$$

*C*: We are 90% confident that the interval from −0.0824 to −0.0176 captures the true difference in the proportion of U.S. adults and teens who go online every day.

We'll restrict ourselves to situations in which the hypothesized difference is 0. Then the null hypothesis says that there is no difference between the two parameters:

$$H_0: p_1 - p_2 = 0 \text{ or, alternatively, } H_0: p_1 = p_2$$

The alternative hypothesis says what kind of difference we expect.

## EXAMPLE Hungry Children

### *Stating hypotheses*

Researchers designed a survey to compare the proportions of children who come to school without eating breakfast in two low-income elementary schools. An SRS of 80 students from School 1 found that 19 had not eaten breakfast. At School 2, an SRS of 150 students included 26 who had not had breakfast. More than 1500 students attend each school. Do these data give convincing evidence of a difference in the population proportions?

PROBLEM: State appropriate hypotheses for a significance test to answer this question. Define any parameters you use.

SOLUTION: We should carry out a test of

$$H_0: p_1 - p_2 = 0$$
$$H_a: p_1 - p_2 \neq 0$$

where $p_1$ = the true proportion of students at School 1 who did not eat breakfast and $p_2$ = the true proportion of students at School 2 who did not eat breakfast.

**For Practice** *Try Exercise* **13**

The conditions for performing a significance test about $p_1 - p_2$ are the same as for constructing a confidence interval.

**CONDITIONS FOR PERFORMING A SIGNIFICANCE TEST ABOUT A DIFFERENCE IN PROPORTIONS**

- **Random:** The data come from two independent random samples or from two groups in a randomized experiment.
  - **10%:** When sampling without replacement, check that $n_1 \leq \frac{1}{10}N_1$ and $n_2 \leq \frac{1}{10}N_2$.
- **Large Counts:** The counts of "successes" and "failures" in each sample or group—$n_1\hat{p}_1$, $n_1(1-\hat{p}_1)$, $n_2\hat{p}_2$, $n_2(1-\hat{p}_2)$—are all at least 10.

### Teaching Tip

There are occasions when the hypothesized difference is not 0. For example, suppose that a pharmaceutical company will decide to market a new drug if its success rate is more than 0.05 higher than the currently used drug. In this case, the null hypothesis would be $H_0: p_{new} - p_{current} = 0.05$ and the alternative hypothesis would be $H_a: p_{new} - p_{current} > 0.05$.

### Teaching Tip

Ask your students to give two explanations for why the sample proportions of students who didn't eat breakfast weren't the same at the two schools. One explanation is that the true proportions at the two schools are the same and the difference was due to sampling variability. The other explanation is that the true proportions are actually different at the two schools.

### ALTERNATE EXAMPLE

#### *Hearing loss*

Are teenagers going deaf? In a study of 3000 randomly selected teenagers in 1988–1994, 15% showed some hearing loss. In a similar study of 1800 teenagers in 2005–2006, 19.5% showed some hearing loss. Do these data give convincing evidence that the proportion of all teens with hearing loss has increased? (These data were reported in the *Arizona Daily Star,* August 18, 2010.)

PROBLEM:
State the hypotheses we are interested in testing. Define any parameters you use.

SOLUTION:
$H_0: p_1 - p_2 = 0$ and $H_a: p_1 - p_2 < 0$, where $p_1$ = the proportion of all teenagers with hearing loss in 1988–1994 and $p_2$ = the proportion of all teenagers with hearing loss in 2005–2006.

### Teaching Tip

One of the nice features of the re-randomization approach that we used in the Is Yawning Contagious? Activity is that there are fewer conditions that need to be met. Specifically, there is no Large Counts condition because we never use a Normal distribution to do calculations, unlike our procedure in a two-sample z test for a difference between two proportions. This is especially important for the yawning problem because the number of successes in the "no seed" group is 4, which is too small to satisfy the Large Counts condition.

If the conditions are met, we can proceed with calculations. To do a test, standardize $\hat{p}_1 - \hat{p}_2$ to get a $z$ statistic:

$$\text{test statistic} = \frac{\text{statistic} - \text{parameter}}{\text{standard deviation of statistic}}$$

$$z = \frac{(\hat{p}_1 - \hat{p}_2) - 0}{\text{standard deviation of statistic}}$$

If $H_0: p_1 = p_2$ is true, the two parameters are the same. We call their common value $p$. But now we need a way to estimate $p$, so it makes sense to combine the data from the two samples as if they came from one larger sample. This **pooled** (or **combined**) **sample proportion** is

$$\hat{p}_C = \frac{\text{count of successes in both samples combined}}{\text{count of individuals in both samples combined}} = \frac{X_1 + X_2}{n_1 + n_2}$$

In other words, $\hat{p}_C$ gives the overall proportion of successes in the combined samples.

Let's look at how to calculate $\hat{p}_C$ in the hungry children example. The two-way table below summarizes the survey data. We have combined the independent SRSs from the two schools in the right-hand Total column.

| | School | | |
|---|---|---|---|
| Breakfast? | 1 | 2 | **Total** |
| No | 19 | 26 | **45** |
| Yes | 61 | 124 | **185** |
| **Total** | **80** | **150** | **230** |

Because researchers want to compare the proportions of students at School 1 and School 2 who have not eaten breakfast, we treat the individuals in the "No" row as successes. It is easy to see from the table that the overall proportion of successes in the combined samples is $\hat{p}_C = \frac{45}{230} = 0.1957$. We can also get this result using the formula above:

$$\hat{p}_C = \frac{X_1 + X_2}{n_1 + n_2} = \frac{19 + 26}{80 + 150} = \frac{45}{230} = 0.1957$$

Recall that the standard deviation of $\hat{p}_1 - \hat{p}_2$ is

$$\sigma_{\hat{p}_1 - \hat{p}_2} = \sqrt{\frac{p_1(1-p_1)}{n_1} + \frac{p_2(1-p_2)}{n_2}}$$

Use $\hat{p}_C$ in place of both $p_1$ and $p_2$ in this expression for the denominator of the test statistic:

$$z = \frac{(\hat{p}_1 - \hat{p}_2) - 0}{\sqrt{\frac{\hat{p}_C(1-\hat{p}_C)}{n_1} + \frac{\hat{p}_C(1-\hat{p}_C)}{n_2}}}$$

We can use a little algebra to rewrite the denominator of the test statistic:

$$\sqrt{\frac{\hat{p}_C(1-\hat{p}_C)}{n_1} + \frac{\hat{p}_C(1-\hat{p}_C)}{n_2}} = \sqrt{\hat{p}_C(1-\hat{p}_C)\left(\frac{1}{n_1} + \frac{1}{n_2}\right)} = \sqrt{\hat{p}_C(1-\hat{p}_C)}\sqrt{\frac{1}{n_1} + \frac{1}{n_2}}$$

The final formula looks like the one given on the AP® exam formula sheet.

When the Large Counts condition is met, this will yield a $z$ statistic that has approximately the standard Normal distribution when $H_0$ is true. Here are the details for the **two-sample $z$ test for the difference between two proportions**.

## Teaching Tip

If students ask why it makes sense to combine the data from the two samples, remind them that we can get a more precise estimate of $p$ when we use more data. Using the two-way table approach shown below makes the formula for the combined proportion less mysterious.

## AP® Exam Formula Sheet

Remind your students that the general formula for a test statistic is included on the formula sheet that is given to students on both sections of the AP® exam. Students are also given the formula for the standard deviation of the statistic $\hat{p}_1 - \hat{p}_2$ using the pooled proportion, $\sqrt{p(1-p)}\sqrt{\frac{1}{n_1} + \frac{1}{n_2}}$. We prefer to write the formula as $\sqrt{\frac{\hat{p}_C(1-\hat{p}_C)}{n_1} + \frac{\hat{p}_C(1-\hat{p}_C)}{n_2}}$ because it is very similar to the standard error formula we use to construct a confidence interval for a difference in proportions.

Some people prefer to use $\hat{p}_C$ to check the Large Counts condition. If the expected counts $n_1\hat{p}_C$, $n_1(1-\hat{p}_C)$, $n_2\hat{p}_C$, and $n_2(1-\hat{p}_C)$ are all at least 10, the sampling distribution of $\hat{p}_1 - \hat{p}_2$ is approximately Normal. Checking the observed counts of successes and failures is more conservative, as the expected counts will always be at least 10 if the observed counts are at least 10.

### TWO-SAMPLE $z$ TEST FOR THE DIFFERENCE BETWEEN TWO PROPORTIONS

Suppose the conditions are met. To test the hypothesis $H_0: p_1 - p_2 = 0$, first find the pooled proportion $\hat{p}_C$ of successes in both samples combined. Then compute the $z$ statistic

$$z = \frac{(\hat{p}_1 - \hat{p}_2) - 0}{\sqrt{\frac{\hat{p}_C(1-\hat{p}_C)}{n_1} + \frac{\hat{p}_C(1-\hat{p}_C)}{n_2}}}$$

Find the $P$-value by calculating the probability of getting a $z$ statistic this large or larger in the direction specified by the alternative hypothesis $H_a$:

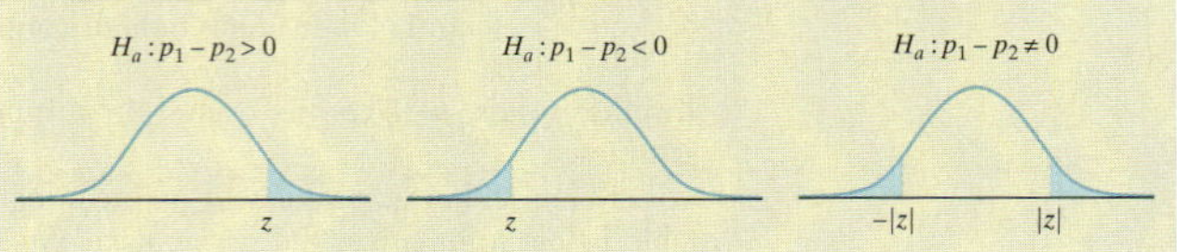


Now we can finish the test we started earlier.

## EXAMPLE Hungry Children

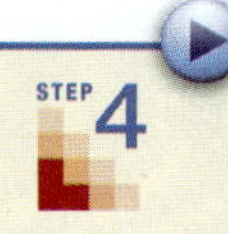


### *Significance test for $p_1 - p_2$*

Researchers designed a survey to compare the proportions of children who come to school without eating breakfast in two low-income elementary schools. An SRS of 80 students from School 1 found that 19 had not eaten breakfast. At School 2, an SRS of 150 students included 26 who had not had breakfast. More than 1500 students attend each school. Do these data give convincing evidence at the $\alpha = 0.05$ level of a difference in the population proportions?

STATE: Our hypotheses are

$$H_0: p_1 - p_2 = 0$$
$$H_a: p_1 - p_2 \neq 0$$

where $p_1$ = the true proportion of students at School 1 who did not eat breakfast and $p_2$ = the true proportion of students at School 2 who did not eat breakfast.

PLAN: If conditions are met, we should perform a two-sample $z$ test for $p_1 - p_2$.

- *Random:* The data were produced using two independent random samples—80 students from School 1 and 150 students from School 2.
  - *10%:* The researchers are sampling without replacement, so we check the 10% condition: there are at least 10(80) = 800 students at School 1 and at least 10(150) = 1500 students at School 2.
- *Large Counts:* We check the counts of "successes" and "failures":

$$n_1\hat{p}_1 = 19,\ n_1(1-\hat{p}_1) = 61,\ n_2\hat{p}_2 = 26,\ n_2(1-\hat{p}_2) = 124$$

All four values are at least 10, so this condition is met.

## Teaching Tip

Emphasize that the conditions for doing a significance test for the difference between two proportions are the same as the conditions for constructing a confidence interval for the difference between two proportions. However, confidence intervals and significance tests use different formulas for the standard error of the difference in proportions.

## ALTERNATE EXAMPLE

### *Hearing loss*

Are teenagers going deaf? In a study of 3000 randomly selected teenagers in 1988–1994, 15% showed some hearing loss. In a similar study of 1800 teenagers in 2005–2006, 19.5% showed some hearing loss. (These data are reported in the *Arizona Daily Star*, August 18, 2010.)

PROBLEM:

**(a)** Do these data give convincing evidence that the proportion of all teens with hearing loss has increased?

**(b)** Between the two studies, Apple introduced the iPod. If the results of the test are statistically significant, can we blame iPods for the increased hearing loss in teenagers?

SOLUTION:

**(a)** STATE: We will test $H_0: p_1 - p_2 = 0$ versus $H_a: p_1 - p_2 < 0$ at the $\alpha = 0.05$ significance level, where $p_1$ = the proportion of all teenagers with hearing loss in 1988–1994 and $p_2$ = the proportion of all teenagers with hearing loss in 2005–2006.

PLAN: We should use a two-sample $z$ test for $p_1 - p_2$ if the conditions are satisfied.

- *Random*: The data came from two independent random samples.
  - *10%*: There were more than 10(3000) = 30,000 teenagers in 1988–1994 and 10(1800) = 18,000 teenagers in 2005–2006.
- *Large Counts*: $n_1\hat{p}_1 = 450$, $n_1(1-\hat{p}_1) = 2550$, $n_2\hat{p}_2 = 351$, and $n_2(1-\hat{p}_2) = 1449$ are all at least 10.

DO: $\hat{p}_C = \frac{351 + 450}{1800 + 3000} = 0.167$,

$$z = \frac{(0.15 - 0.195) - 0}{\sqrt{\frac{0.167(1-0.167)}{3000} + \frac{0.167(1-0.167)}{1800}}} = -4.05,\ P\text{-value} \approx 0.$$

CONCLUDE: Because the $P$-value of approximately 0 is less than $\alpha = 0.05$, we reject $H_0$. We have convincing evidence that the proportion of all teens with hearing loss has increased from 1988–1994 to 2005–2006.

**(b)** No. Because we didn't do an experiment where we randomly assigned some teens to listen to iPods and other teens to avoid listening to iPods, we cannot conclude that iPods are the cause. It is possible that teens who listen to iPods also like to listen to music in their cars, and perhaps their car stereos are causing the hearing loss.

| Breakfast? | School 1 | School 2 | Total |
|---|---|---|---|
| No | 19 | 26 | 45 |
| Yes | 61 | 124 | 185 |
| Total | 80 | 150 | 230 |

DO: We know that $n_1 = 80$, $\hat{p}_1 = \frac{19}{80} = 0.2375$, $n_2 = 150$, and $\hat{p}_2 = \frac{26}{150} = 0.1733$. Our point estimate for the difference in population proportions is $\hat{p}_1 - \hat{p}_2 = 0.2375 - 0.1733 = 0.0642$. The pooled proportion of students who didn't eat breakfast in the two samples is

$$\hat{p}_C = \frac{19 + 26}{80 + 150} = \frac{45}{230} = 0.1957$$

See the two-way table in the margin for confirmation.

- *Test statistic*

$$z = \frac{(\hat{p}_1 - \hat{p}_2) - 0}{\sqrt{\frac{\hat{p}_C(1-\hat{p}_C)}{n_1} + \frac{\hat{p}_C(1-\hat{p}_C)}{n_2}}} = \frac{0.0642 - 0}{\sqrt{\frac{0.1957(1-0.1957)}{80} + \frac{0.1957(1-0.1957)}{150}}} = 1.17$$

- *P-value* Figure 10.3 displays the *P*-value as an area under the standard Normal curve for this two-tailed test. Using Table A or `normalcdf`, the desired *P*-value is $2P(Z \geq 1.17) = 2(1 - 0.8790) = 0.2420$.

*Using technology:* Refer to the Technology Corner that follows the example. The calculator's `2-PropZTest` gives $z = 1.1683$ and *P*-value $= 0.2427$.

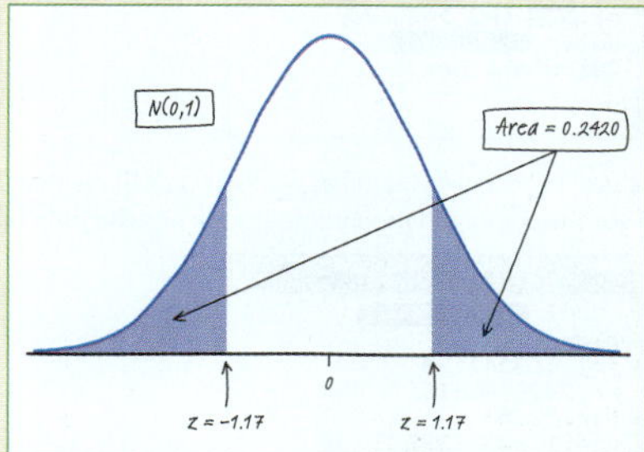


**FIGURE 10.3** The *P*-value for the two-sided test.

CONCLUDE: Because our *P*-value, 0.2427, is greater than $\alpha = 0.05$, we fail to reject $H_0$. There is not convincing evidence that the true proportions of students at the two schools who didn't eat breakfast are different.

**For Practice** *Try Exercise*  15

Exactly what does the *P*-value in the previous example tell us? If we repeated the random sampling process many times, we'd get a difference in sample proportions as large as or larger than 0.0642 in either direction about 24% of the time when $H_0$: $p_1 - p_2 = 0$ is true. With such a high probability of getting a result like this just by chance when the null hypothesis is true, we don't have enough evidence to reject $H_0$.

We can get additional information about the difference between the population proportions at School 1 and School 2 with a confidence interval. The TI-84's `2-PropZInt` gives the 95% confidence interval for $p_1 - p_2$ as $(-0.047, 0.175)$. That is, we are 95% confident that the difference in the true proportions of students who ate breakfast at the two schools is between 4.7 percentage points lower at School 1 and 17.5 percentage points higher at School 1. This is consistent with our "fail to reject $H_0$" conclusion in the example because 0 is included in the interval of plausible values for $p_1 - p_2$.

The two-sample *z* test and two-sample *z* interval for the difference between two proportions don't always give consistent results. That's because the "standard deviation of the statistic" used in calculating the test statistic is

$$\sqrt{\frac{\hat{p}_C(1-\hat{p}_C)}{n_1} + \frac{\hat{p}_C(1-\hat{p}_C)}{n_2}}$$

but for the confidence interval, it's

$$\sqrt{\frac{\hat{p}_1(1-\hat{p}_1)}{n_1} + \frac{\hat{p}_2(1-\hat{p}_2)}{n_2}}$$

## AP® Exam Common Error

When students fail to reject the null hypothesis, they often state that the null hypothesis is true (i.e., they accept the null hypothesis). In this example, it would be incorrect to conclude that the proportion of students who didn't eat breakfast is the *same* at the two schools. We don't have evidence that this is true; we just don't have convincing evidence that it isn't. To emphasize that 0 is just one of many plausible values for the true difference in proportions, point out the confidence interval just below the example on this page. Any of the values from $-0.047$ to 0.175 could be the true difference in proportions—there is nothing extra-special about 0.

## Teaching Tip

Emphasize that there is a difference between saying that a proportion is 20% higher and saying that the proportion is 20 percentage points higher. For example, if $p_1 = 0.4$ and $p_2 = 0.6$, then $p_2$ is *50%* higher than $p_1$, but only *20 percentage points* higher than $p_1$.

### TECHNOLOGY TIP

Warn your students not to choose the wrong procedure on their calculators. The 2-SampZTest is for comparing means when the population standard deviations are known. What we want is the 2-PropZTest.

### TECHNOLOGY TIP

The values of x1 and x2 are the numbers of successes, not the proportions of successes. Also, they must be integers!

## 22. TECHNOLOGY CORNER SIGNIFICANCE TEST FOR A DIFFERENCE IN PROPORTIONS

TI-Nspire instructions in Appendix B; HP Prime instructions on the book's Web site.

The TI-83/84 and TI-89 can be used to perform significance tests for comparing two proportions. Here, we use the data from the hungry children example. To perform a test of $H_0: p_1 - p_2 = 0$ versus $H_a: p_1 - p_2 \neq 0$:

**TI-83/84**

- Press [STAT], then choose TESTS and `2-PropZTest`.

**TI-89**

- In the Statistics/List Editor, press [2nd] [F1] ([F6]) and choose `2-PropZTest`.

- When the `2-PropZTest` screen appears, enter the values $x_1 = 19$, $n_1 = 80$, $x_2 = 26$, $n_2 = 150$. Specify the alternative hypothesis $p_1 \neq p_2$, as shown.

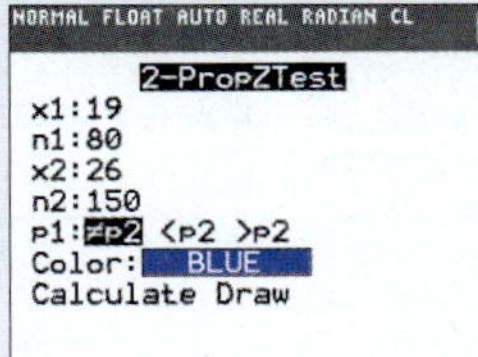


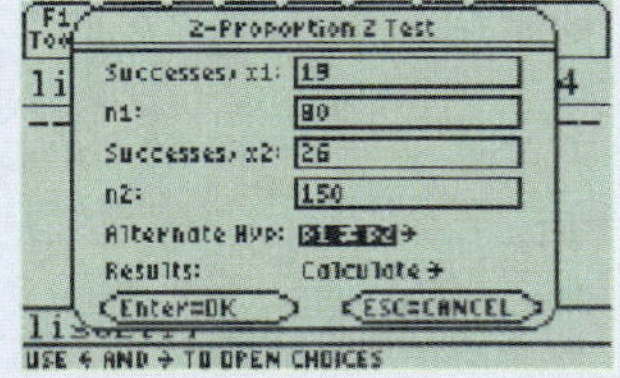


- If you select "Calculate" and press [ENTER], you will see that the test statistic is $z = 1.168$ and the $P$-value is 0.2427. Do you see the combined proportion of students who didn't eat breakfast? It's the value labeled $\hat{p}$, 0.1957.

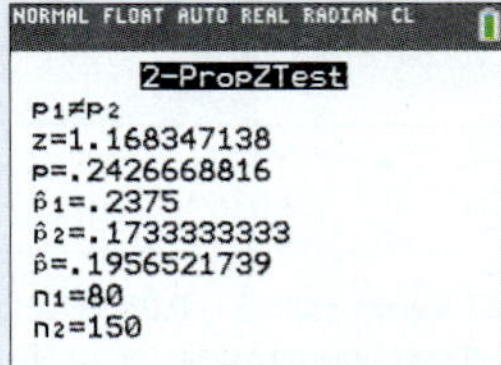


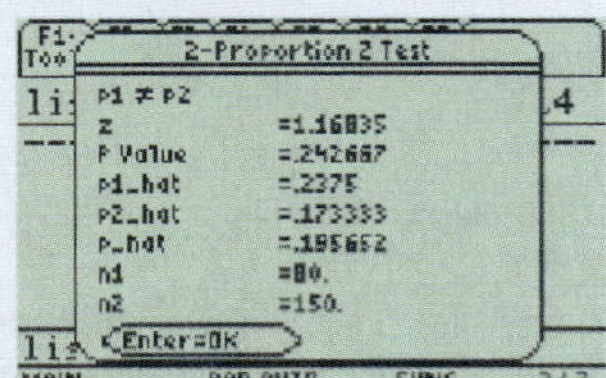


- If you select the "Draw" option, you will see the screen shown here.

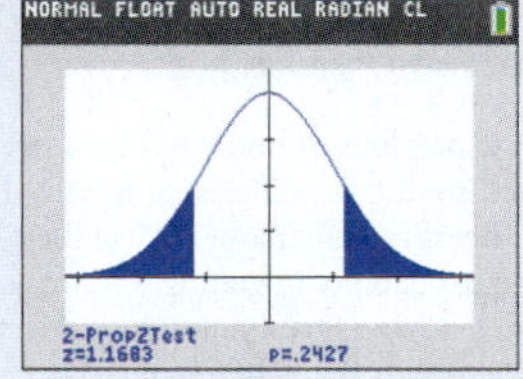


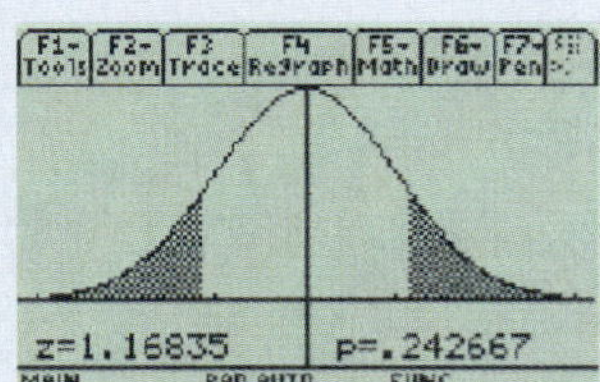


**AP® EXAM TIP** The formula for the two-sample $z$ statistic for a test about $p_1 - p_2$ often leads to calculation errors by students. As a result, we recommend using the calculator's `2-PropZTest` feature to perform calculations on the AP® exam. Be sure to name the procedure (two-proportion $z$ test) and to report the test statistic ($z = 1.17$) and $P$-value (0.2427) as part of the "Do" step.

### AP® Exam Common Error

After using their calculator's 2-PropZTest feature to correctly calculate the test statistic and $P$-value, some students try to show additional work by stating a formula or showing numbers substituted into a formula. This often results in a loss of credit because students use the wrong formula or include a wrong symbol in the correct formula (e.g., using $p_1$ or $\hat{p}_1$ instead of $\hat{p}_C$ in the standard error). For two-sample procedures, we recommend that students use their calculator to do the calculations, finish the exam, and then double-check their calculations if there is time. If they show additional work, it must be consistent with the values from the calculator!

## Inference for Experiments

Most of the examples in this section have involved doing inference about $p_1 - p_2$ using data that were produced by random sampling. In such cases, the parameters $p_1$ and $p_2$ are the true proportions of successes in the corresponding populations. However, many important statistical results come from randomized comparative experiments. Defining the parameters in experimental settings is more challenging.

The "Is Yawning Contagious?" Activity on page 610 describes an experiment that used 50 volunteer adults as subjects. Researchers randomly assigned 34 subjects to get a yawn seed and 16 subjects to get no yawn seed. Then researchers compared the proportions of people in the two groups who yawned. The parameters in this setting are:

$p_1$ = the true proportion of people like these who would yawn when given a yawn seed

$p_2$ = the true proportion of people like these who would yawn when no yawn seed is given

Most experiments on people use recruited volunteers as subjects. When subjects are not randomly selected, researchers cannot generalize the results of an experiment to some larger populations of interest. But researchers can draw cause-and-effect conclusions that apply to people like those who took part in the experiment. This same logic applies to experiments on animals or things. Also note that unless the experimental units are randomly selected, we don't need to check the 10% condition when performing inference about an experiment.

Here is an example that involves comparing two proportions.

### EXAMPLE Cholesterol and Heart Attacks

*Significance test in an experiment*

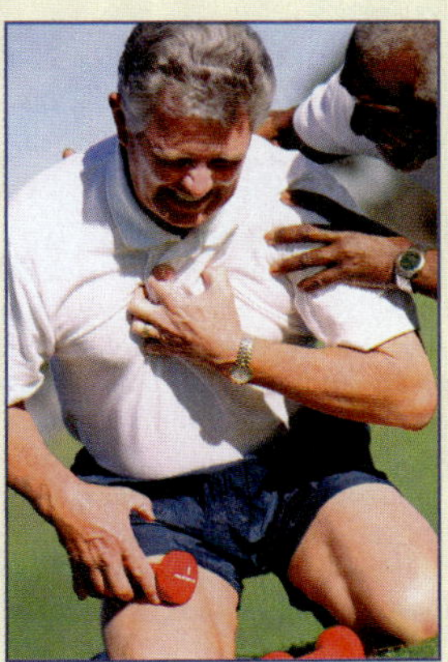

High levels of cholesterol in the blood are associated with higher risk of heart attacks. Will using a drug to lower blood cholesterol reduce heart attacks? The Helsinki Heart Study recruited middle-aged men with high cholesterol but no history of other serious medical problems to investigate this question. The volunteer subjects were assigned at random to one of two treatments: 2051 men took the drug gemfibrozil to reduce their cholesterol levels, and a control group of 2030 men took a placebo. During the next five years, 56 men in the gemfibrozil group and 84 men in the placebo group had heart attacks. Is this difference statistically significant at the $\alpha = 0.01$ level?

STATE: We hope to show that gemfibrozil reduces heart attacks, so we have a one-sided alternative:

$H_0: p_1 - p_2 = 0$  
$H_a: p_1 - p_2 < 0$

or, equivalently,

$H_0: p_1 = p_2$  
$H_a: p_1 < p_2$

where $p_1$ is the actual heart attack rate for middle-aged men like the ones in this study who take gemfibrozil, and $p_2$ is the actual heart attack rate for middle-aged men like the ones in this study who take only a placebo. We'll use $\alpha = 0.01$.

### AP® Exam Common Error

Many students don't use correct language when defining parameters, especially for problems about experiments. In the yawning Activity, it would be incorrect to define $p_1$ asthe proportion of subjects who *yawned* when given a yawn seed. The proportion who *yawned* is $\hat{p}_1$, not $p_1$. Likewise, in the Cholesterol and Heart Attacks example, it is incorrect to define $p_1$ as the heart attack rate for men who *took* gemfibrozil.

### ALTERNATE EXAMPLE

**Cash for quitters**

In an effort to reduce health care costs, General Motors sponsored a study to help employees stop smoking. In the study, half of the subjects were randomly assigned to receive up to $750 for quitting smoking for a year, while the other half were simply encouraged to use traditional methods to stop smoking. None of the 878 volunteers knew that there was a financial incentive when they first signed up. At the end of one year, 15% of those in the financial rewards group had quit smoking, while only 5% in the traditional group had quit smoking. Do the results of this study give convincing evidence that a financial incentive helps people quit smoking? (These data are reported in the *Arizona Daily Star*, February 11, 2009.)

STATE: We will test $H_0: p_F - p_T = 0$ versus $H_a: p_F - p_T > 0$ at the $\alpha = 0.05$ significance level, where $p_F$ = the true quitting rate for employees like these who get a financial incentive to quit smoking and $p_T$ = the true quitting rate for employees like these who are encouraged to use traditional methods to quit smoking.

PLAN: We should use a two-sample $z$ test for $p_F - p_T$ if the conditions are met.

- *Random*: The data come from two groups in a randomized experiment.
  - *10%*: There was no sampling without replacement, so this condition doesn't need to be checked.
- *Large Counts*: $n_F\hat{p}_F = 439(0.15) = 66$, $n_F(1 - \hat{p}_F) = 439(0.85) = 373$, $n_T\hat{p}_T = 439(0.05) = 22$, $n_T(1 - \hat{p}_T) = 439(0.95) = 417$ are all at least 10.

DO: $\hat{p}_C = \dfrac{66 + 22}{439 + 439} = 0.100$,

$$z = \frac{(0.15 - 0.05) - 0}{\sqrt{\dfrac{0.1(1 - 0.1)}{439} + \dfrac{0.1(1 - 0.1)}{439}}}$$

$= 4.94$, $P$-value $\approx 0$.

CONCLUDE: Because the $P$-value of approximately 0 is less than $\alpha = 0.05$, we reject $H_0$. We have convincing evidence that employees like these have a greater quitting rate when given a financial incentive.

## Teaching Tip

Even though the difference in proportions is relatively small ($0.0414 - 0.0273 = 0.0141$), the difference is statistically significant. In the media, the results of a study like this are often reported using a ratio rather than a proportion. For example, a news report about this study might report that taking gemfibrozil reduces heart attack rates by 34% (0.0141/0.0414). Although relative risk is not a topic on the AP® exam, there was an interesting investigative task (Question 6) on the 2009 overseas exam (Form B) that explored this topic.



Note that we did not need to check the 10% condition because the subjects in the experiment were not sampled without replacement from some larger population.

PLAN: If conditions are met, we will do a two-sample $z$ test for $p_1 - p_2$.

- *Random:* The data come from two groups in a randomized experiment.
  - 10%: Don't need to check because there was no sampling.
- *Large Counts:* The number of successes (heart attacks!) and failures in the two groups are 56, 1995, 84, and 1946. These are all at least 10, so this condition is met.

DO: The proportions of men who had heart attacks in each group are

$$\hat{p}_1 = \frac{56}{2051} = 0.0273 \text{ (gemfibrozil group) and } \hat{p}_2 = \frac{84}{2030} = 0.0414 \text{ (placebo group)}$$

The pooled proportion of heart attacks for the two groups is

$$\hat{p}_C = \frac{\text{count of heart attacks in both samples combined}}{\text{count of subjects in both samples combined}} = \frac{56 + 84}{2051 + 2030} = \frac{140}{4081} = 0.0343$$

See the two-way table in the margin.

| Heart attack? | Drug taken: Gemfibrozil | Placebo | Total |
|---|---|---|---|
| Yes | 56 | 84 | 140 |
| No | 1995 | 1946 | 3941 |
| Total | 2051 | 2030 | 4081 |

We'll use the calculator's 2-PropZTest to perform calculations.

- *Test statistic* $z = -2.47$
- *P-value* This is the area under the standard Normal curve to the left of $z = -2.47$, shown in Figure 10.4.

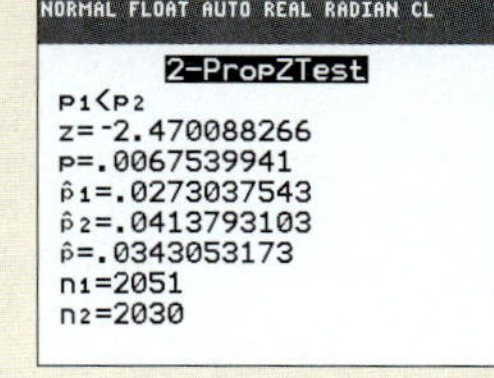


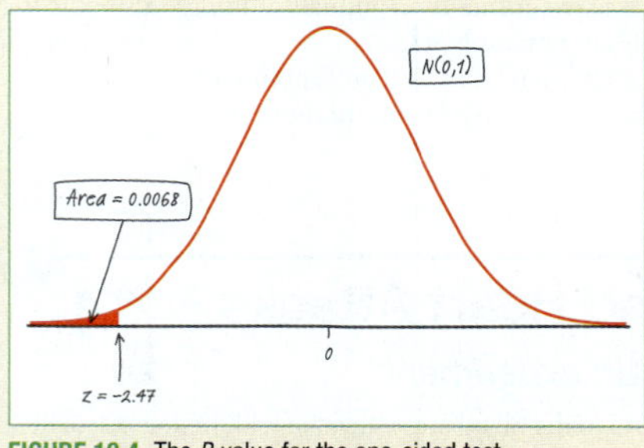


**FIGURE 10.4** The *P*-value for the one-sided test.

CONCLUDE: Because the *P*-value, 0.0068, is less than 0.01, we can reject $H_0$. The results are statistically significant at the $\alpha = 0.01$ level. There is convincing evidence of a lower heart attack rate for middle-aged men like these who take gemfibrozil than for those who take only a placebo.

**For Practice** *Try Exercise* **21**

We chose $\alpha = 0.01$ in the example to reduce the chance of making a Type I error—finding convincing evidence that gemfibrozil reduces heart attack risk when it actually doesn't. This error could have serious consequences if an ineffective drug was given to lots of middle-aged men with high cholesterol!

The random assignment in the Helsinki Heart Study allowed researchers to draw a cause-and-effect conclusion. They could say that gemfibrozil reduces the rate of heart attacks for middle-aged men like those who took part in the experiment. Because the subjects were not randomly selected from a larger population, researchers could not generalize the findings of this study any further. No conclusions could be drawn about the effectiveness of gemfibrozil at preventing heart attacks for all middle-aged men, for older men, or for women.

## Teaching Tip

It's a good idea to ask students about the types of conclusions they can make after each inference procedure they conduct. If the data come from independent random samples, then conclusions can be made about the populations from which the samples were selected. If the data come from two groups in a randomized experiment, then conclusions can be made about cause and effect. Questions about the scope of inference are very common on the AP® exam.

**Why do the inference methods for random sampling work for randomized experiments?** Confidence intervals and tests for $p_1 - p_2$ are based on the sampling distribution of $\hat{p}_1 - \hat{p}_2$. But in experiments, we aren't sampling at random from any larger populations. We can think about what would happen if the random assignment were repeated many times under the assumption that $H_0: p_1 - p_2 = 0$ is true. That is, we assume that the specific treatment received doesn't affect an individual subject's response.

Let's see what would happen just by chance if we randomly reassign the 4081 subjects in the Helsinki Heart Study to the two groups many times, assuming the drug received *doesn't affect* whether or not each individual has a heart attack. We used Fathom software to redo the random assignment 500 times. The approximate **randomization distribution** of $\hat{p}_1 - \hat{p}_2$ is shown in Figure 10.5. It has an approximately Normal shape with mean 0 and standard deviation 0.0058. These are roughly the same as the shape, center, and spread of the sampling distribution of $\hat{p}_1 - \hat{p}_2$ that we used to perform calculations in the previous example because

$$\sqrt{\frac{\hat{p}_C(1-\hat{p}_C)}{n_1} + \frac{\hat{p}_C(1-\hat{p}_C)}{n_2}} = \sqrt{\frac{0.0343(1-0.0343)}{2051} + \frac{0.0343(1-0.0343)}{2030}} = 0.0057$$

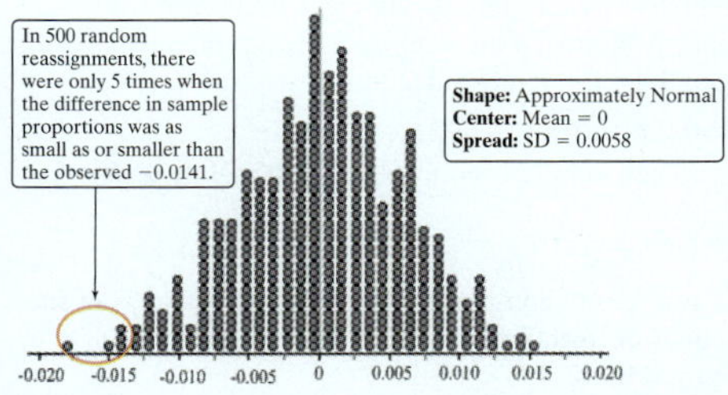


**FIGURE 10.5** Fathom simulation showing the approximate randomization distribution of $\hat{p}_1 - \hat{p}_2$ from 500 random reassignments of subjects to treatment groups in the Helsinki Heart Study.

In the Helsinki Heart Study, the difference in the proportions of subjects who had a heart attack in the gemfibrozil and placebo groups was $0.0273 - 0.0414 = -0.0141$. How likely is it that a difference this large or larger would happen just by chance when $H_0$ is true? Figure 10.5 provides a rough answer: 5 of the 500 random reassignments yielded a difference in proportions less than or equal to $-0.0141$. That is, our estimate of the *P*-value is 0.01. This is quite close to the 0.0068 *P*-value that we calculated in the previous example.

Figure 10.6 shows the value of the *z* test statistic for each of the 500 re-randomizations, calculated using our familiar formula

$$z = \frac{(\hat{p}_1 - \hat{p}_2) - 0}{\sqrt{\frac{\hat{p}_C(1-\hat{p}_C)}{n_1} + \frac{\hat{p}_C(1-\hat{p}_C)}{n_2}}}$$

The standard Normal density curve is shown in blue. We can see that the *z* test statistic has approximately the standard Normal distribution in this case.

Whenever the conditions are met, the randomization distribution of $\hat{p}_1 - \hat{p}_2$ looks much like its sampling distribution. We are therefore safe using two-sample *z* procedures for comparing two proportions in a randomized experiment.

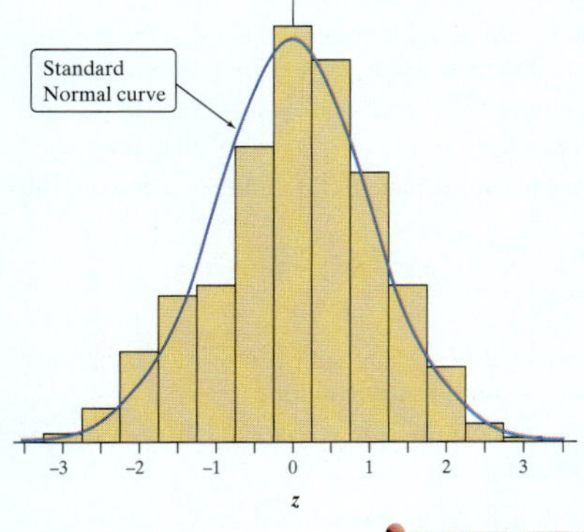


**FIGURE 10.6** The distribution of the *z* test statistic for the 500 random reassignments in Figure 10.5.

## Teaching Tip

Both of these *P*-values are estimates of the exact *P*-value, which can be found using Fisher's exact test. Using the applet at http://www.langsrud.com/fisher.htm, the exact *P*-value is 0.0084. Given enough trials, the estimate of the *P*-value using Fathom will approach the exact value.

## ANSWERS TO CYU

*S*: $H_0: p_1 - p_2 = 0$ vs. $H_a: p_1 - p_2 > 0$, where $p_1$ is the true proportion of children like those in the study who do not attend preschool that use social services later and $p_2$ is the true proportion of children like those in the study who attend preschool that use social services later. *P*: Two-sample *z* test for $p_1 - p_2$. *Random*: Two groups in a randomized experiment. *Large Counts*: 49, 12, 38, 24 are all $\geq 10$. *D*:

$$z = \frac{(0.8033 - 0.6129) - 0}{\sqrt{\frac{0.7073(0.2927)}{61} + \frac{0.7073(0.2927)}{62}}}$$

$= 2.32$ and *P*-value $= 0.0102$. *C*: Because the *P*-value of $0.0102 < \alpha = 0.05$, we reject $H_0$. There is convincing evidence that the true proportion of children like those in the study who do not attend preschool that use social services later > the true proportion of children like those in the study who attend preschool that use social services later.

### Teaching Tip

In the Answers to CYU above, there was no need to check the 10% condition. This condition only applies when a sample is selected without replacement from a population. In this experiment, the subjects weren't selected at random.

### TRM Section 10.1 Quizzes

There are three different quizzes available for this section in the Teacher's Resource Materials. Versions A and B are roughly parallel. Find them by clicking on the link in the TE-book, opening the TRCD, or logging onto the book's companion Web site www.whfreeman.com/tps5e (teacher log-in required)

## CHECK YOUR UNDERSTANDING

To study the long-term effects of preschool programs for poor children, researchers designed an experiment. They recruited 123 children who had never attended preschool from low-income families in Michigan. Researchers randomly assigned 62 of the children to attend preschool (paid for by the study budget) and the other 61 to serve as a control group who would not go to preschool. One response variable of interest was the need for social services as adults. Over a 10-year period, 38 children in the preschool group and 49 in the control group have needed social services.[4]

Does this study provide convincing evidence that preschool reduces the later need for social services? Justify your answer.

## Section 10.1 Summary

- Choose independent SRSs of size $n_1$ from Population 1 with proportion of successes $p_1$ and of size $n_2$ from Population 2 with proportion of successes $p_2$. The sampling distribution of $\hat{p}_1 - \hat{p}_2$ has the following properties:
  - **Shape** Approximately Normal if the samples are large enough that $n_1p_1$, $n_1(1 - p_1)$, $n_2p_2$, and $n_2(1 - p_2)$ are all at least 10.
  - **Center** The mean is $p_1 - p_2$.
  - **Spread** As long as each sample is no more than 10% of its population, the standard deviation is $\sqrt{\frac{p_1(1-p_1)}{n_1} + \frac{p_2(1-p_2)}{n_2}}$.
- Confidence intervals and tests to compare the proportions $p_1$ and $p_2$ of successes for two populations or treatments are based on the difference $\hat{p}_1 - \hat{p}_2$ between the sample proportions.
- Before estimating or testing a claim about $p_1 - p_2$, check that these conditions are met:
  - **Random:** The data come from two independent random samples or from two groups in a randomized experiment.
    - **10%:** When sampling without replacement, check that the two populations are at least 10 times as large as the corresponding samples.
  - **Large Counts:** The counts of "successes" and "failures" in each sample or group—$n_1\hat{p}_1$, $n_1(1 - \hat{p}_1)$, $n_2\hat{p}_2$, and $n_2(1 - \hat{p}_2)$—are all at least 10.
- When conditions are met, an approximate C% confidence interval for $p_1 - p_2$ is

$$(\hat{p}_1 - \hat{p}_2) \pm z^*\sqrt{\frac{\hat{p}_1(1-\hat{p}_1)}{n_1} + \frac{\hat{p}_2(1-\hat{p}_2)}{n_2}}$$

  where $z^*$ is the standard Normal critical value with C% of its area between $-z^*$ and $z^*$. This is called a **two-sample *z* interval for $p_1 - p_2$**.
- Significance tests of $H_0: p_1 - p_2 = 0$ use the **pooled (combined) sample proportion** in the standard error formula:

$$\hat{p}_C = \frac{\text{count of successes in both samples combined}}{\text{count of individuals in both samples combined}} = \frac{X_1 + X_2}{n_1 + n_2}$$

When conditions are met, the **two-sample z test for $p_1 - p_2$** uses the test statistic

$$z = \frac{(\hat{p}_1 - \hat{p}_2) - 0}{\sqrt{\frac{\hat{p}_C(1-\hat{p}_C)}{n_1} + \frac{\hat{p}_C(1-\hat{p}_C)}{n_2}}}$$

with *P*-values calculated from the standard Normal distribution.

- Inference about the difference $p_1 - p_2$ in the effectiveness of two treatments in a completely randomized experiment is based on the **randomization distribution** of $\hat{p}_1 - \hat{p}_2$. When conditions are met, our usual inference procedures based on the sampling distribution of $\hat{p}_1 - \hat{p}_2$ will be approximately correct.
- Be sure to follow the four-step process whenever you construct a confidence interval or perform a significance test for comparing two proportions.

### 10.1 TECHNOLOGY CORNERS

TI-Nspire Instructions in Appendix B; HP Prime instructions on the book's Web site.



## Section 10.1 Exercises

*Remember: We are no longer reminding you to use the four-step process in exercises that require you to perform inference.*

1. **Goldfish** Refer to the example on page 615. Suppose that your teacher decides to take SRSs of 100 crackers from both bags instead.
   pg 615
   (a) What is the shape of the sampling distribution of $\hat{p}_1 - \hat{p}_2$? Why?
   (b) Find the mean of the sampling distribution. Show your work.
   (c) Find the standard deviation of the sampling distribution. Show your work.

2. **Homework** Refer to page 612. Suppose that both school counselors decide to take SRSs of 150 students instead.
   (a) What is the shape of the sampling distribution of $\hat{p}_1 - \hat{p}_2$? Why?
   (b) Find the mean of the sampling distribution. Show your work.
   (c) Find the standard deviation of the sampling distribution. Show your work.

3. **I want red!** A candy maker offers Child and Adult bags of jelly beans with different color mixes. The company claims that the Child mix has 30% red jelly beans, while the Adult mix contains 15% red jelly beans. Assume that the candy maker's claim is true. Suppose we take a random sample of 50 jelly beans from the Child mix and a separate random sample of 100 jelly beans from the Adult mix. Let $\hat{p}_C$ and $\hat{p}_A$ be the sample proportions of red jelly beans from the Child and Adult mixes, respectively.
   (a) What is the shape of the sampling distribution of $\hat{p}_C - \hat{p}_A$? Why?
   (b) Find the mean of the sampling distribution. Show your work.
   (c) Find the standard deviation of the sampling distribution. Show your work.

4. **Literacy** A researcher reports that 80% of high school graduates, but only 40% of high school dropouts, would pass a basic literacy test.[5] Assume that the researcher's claim is true. Suppose we give

**TRM Full Solutions to Section 10.1 Exercises**

### Answers to Section 10.1 Exercises

**10.1** **(a)** Approximately Normal, because 100(0.25) = 25, 100(0.75) = 75, 100(0.35) = 35, and 100(0.65) = 65 are all at least 10. **(b)** $\mu_{\hat{p}_1-\hat{p}_2} =$ 0.25 − 0.35 = −0.10 **(c)** Because $n_1 = 100 <$ 10% of the first bag and $n_2 = 100 <$ 10% of the second bag, $\sigma_{\hat{p}_1-\hat{p}_2} = \sqrt{\frac{0.25(0.75)}{100} + \frac{0.35(0.65)}{100}} = 0.0644$.

**10.2** **(a)** Approximately Normal, because 150(0.70) = 105, 150(0.30) = 45, 150(0.50) = 75, and 150(0.50) = 75 are all at least 10. **(b)** $\mu_{\hat{p}_1-\hat{p}_2} =$ 0.70 − 0.50 = 0.20 **(c)** Because $n_1 = 150 <$ 10% of the students at school 1 and $n_2 = 150 <$ 10% of the students at school 2, $\sigma_{\hat{p}_1-\hat{p}_2} = \sqrt{\frac{0.70(0.30)}{150} + \frac{0.50(0.50)}{150}} = 0.0554$.

**10.3** **(a)** Approximately Normal, because 15, 35, 15, and 85 are all at least 10. **(b)** $\mu_{\hat{p}_C-\hat{p}_A} = 0.30 - 0.15 = 0.15$ **(c)** $\sigma_{\hat{p}_C-\hat{p}_A} = \sqrt{\frac{0.3(0.7)}{50} + \frac{0.15(0.85)}{100}} = 0.0740$

a basic literacy test to a random sample of 60 high school graduates and a separate random sample of 75 high school dropouts. Let $\hat{p}_G$ and $\hat{p}_D$ be the sample proportions of graduates and dropouts, respectively, who pass the test.

(a) What is the shape of the sampling distribution of $\hat{p}_G - \hat{p}_D$? Why?

(b) Find the mean of the sampling distribution. Show your work.

(c) Find the standard deviation of the sampling distribution. Show your work.

*Explain why the conditions for constructing a two-sample z interval for $p_1 - p_2$ are not met in the settings of Exercises 5 through 8.*

5. **Don't drink the water!** The movie *A Civil Action* (Touchstone Pictures, 1998) tells the story of a major legal battle that took place in the small town of Woburn, Massachusetts. A town well that supplied water to eastern Woburn residents was contaminated by industrial chemicals. During the period that residents drank water from this well, 16 of the 414 babies born had birth defects. On the west side of Woburn, 3 of the 228 babies born during the same time period had birth defects.

6. **In-line skaters** A study of injuries to in-line skaters used data from the National Electronic Injury Surveillance System, which collects data from a random sample of hospital emergency rooms. The researchers interviewed 161 people who came to emergency rooms with injuries from in-line skating. Wrist injuries (mostly fractures) were the most common.[6] The interviews found that 53 people were wearing wrist guards and 6 of these had wrist injuries. Of the 108 who did not wear wrist guards, 45 had wrist injuries.

7. **Shrubs and fire** Fire is a serious threat to shrubs in dry climates. Some shrubs can resprout from their roots after their tops are destroyed. One study of resprouting took place in a dry area of Mexico.[7] The investigators randomly assigned shrubs to treatment and control groups. They clipped the tops of all the shrubs. They then applied a propane torch to the stumps of the treatment group to simulate a fire. All 12 of the shrubs in the treatment group resprouted. Only 8 of the 12 shrubs in the control group resprouted.

8. **Broken crackers** We don't like to find broken crackers when we open the package. How can makers reduce breaking? One idea is to microwave the crackers for 30 seconds right after baking them. Breaks start as hairline cracks called "checking." Randomly assign 65 newly baked crackers to the microwave and another 65 to a control group that is not microwaved. After one day, none of the microwave group and 16 of the control group show checking.[8]

9. pg 617  **Who tweets?** Do younger people use Twitter more often than older people? In a random sample of 316 adult Internet users aged 18 to 29, 26% used Twitter. In a separate random sample of 532 adult Internet users aged 30 to 49, 14% used Twitter.[9]

(a) Calculate the standard error of the sampling distribution of the difference in the sample proportions (younger adults − older adults). What information does this value provide?

(b) Construct and interpret a 90% confidence interval for the difference between the true proportions of adult Internet users in these age groups who use Twitter.

10. **Listening to rap** Is rap music more popular among young blacks than among young whites? A sample survey compared 634 randomly chosen blacks aged 15 to 25 with 567 randomly selected whites in the same age group. It found that 368 of the blacks and 130 of the whites listened to rap music every day.[10]

(a) Calculate the standard error of the sampling distribution of the difference in the sample proportions (blacks − whites). What information does this value provide?

(b) Construct and interpret a 95% confidence interval for the difference between the proportions of black and white young people who listen to rap every day.

11. **Young adults living at home** A surprising number of young adults (ages 19 to 25) still live in their parents' homes. A random sample by the National Institutes of Health included 2253 men and 2629 women in this age group.[11] The survey found that 986 of the men and 923 of the women lived with their parents.

(a) Construct and interpret a 99% confidence interval for the difference in the true proportions of men and women aged 19 to 25 who live in their parents' homes.

(b) Does your interval from part (a) give convincing evidence of a difference between the population proportions? Explain.

12. **Fear of crime** The elderly fear crime more than younger people, even though they are less likely to be victims of crime. One study recruited separate random samples of 56 black women and 63 black men over the age of 65 from Atlantic City, New Jersey. Of the women, 27 said they "felt vulnerable" to crime; 46 of the men said this.[12]

**10.4 (a)** Approximately Normal, because 48, 12, 30, and 45 are all at least 10.
**(b)** $\mu_{\hat{p}_G-\hat{p}_D} = 0.80 - 0.40 = 0.40$
**(c)** $\sigma_{\hat{p}_G-\hat{p}_D}$

$$= \sqrt{\frac{0.80(0.20)}{60} + \frac{0.40(0.60)}{75}} = 0.0766$$

**10.5** No random samples or randomized experiment. Also, $3 < 10$.

**10.6** $6 < 10$

**10.7** 0, 8, and 4 are all $< 10$.

**10.8** $0 < 10$

**10.9 (a)** $SE_{\hat{p}_1-\hat{p}_2} = 0.0289$. In repeated samples like these, the difference in the sample proportions of young and older adults who use Twitter will typically be 0.0289 from the true difference.
**(b)** *S*: $p_1$ = true proportion of young adults who use Twitter and $p_2$ = true proportion of older adults .... *P*: Two-sample *z* interval for $p_1 - p_2$. *Random:* Two independent random samples. *10%:* 316 < 10% of all young adults and 532 < 10% of all older adults. *Large Counts:* 82, 234, 74, 458 ≥ 10. *D:* (0.072, 0.168). *C:* We are 90% confident that the interval from 0.072 to 0.168 captures the true difference in the proportions of young adults and older adults who use Twitter.

**10.10 (a)** $SE_{\hat{p}_1-\hat{p}_2} = 0.0264$. In repeated samples like these, the difference in the sample proportions of young blacks and whites who listen to rap every day will typically be 0.0264 from the true difference.
**(b)** *S*: $p_1$ = true proportion of blacks who listen to rap music every day and $p_2$ = true proportion of young whites ....
*P*: Two-sample *z* interval for $p_1 - p_2$.
*Random:* Independent random samples.
*10%*: 634 < 10% of all young blacks and 567 < 10% of all young whites. *Large Counts:* 368, 266, 130, 437 ≥ 10.
*D*: (0.299, 0.403). *C*: We are 95% confident that the interval from 0.299 to 0.403 captures the true difference in the proportions of young blacks and young whites who listen to rap music every day.

**10.11 (a)** *S*: $p_1$ = true proportion of young men who live in their parents' home and $p_2$ = true proportion of young women ....
*P*: Two-sample *z* interval for $p_1 - p_2$.
*Random*: Reasonable to consider these independent random samples. *10%*: 2253 < 10% of the population of young men and 2629 < 10% of the population of young women. *Large Counts*: 986, 1267, 923, 1706 ≥ 10. *D*: (0.051, 0.123). *C*: We are 99% confident that the interval from 0.051 to 0.123 captures the true difference in the proportions of young men and young women who live in their parents' home.
**(b)** Because the interval does not contain 0, there is convincing evidence that the true proportion of young people who live in their parents' home is different for men and women.

**10.12 (a)** *S*: $p_1$ = true proportion of older black men in Atlantic City who feel vulnerable to crime and $p_2$ = true proportion of older black women .... *P*: Two-sample *z* interval for $p_1 - p_2$.
*Random*: Independent random samples. *10%*: 63 < 10% of all older black men from Atlantic City and 56 < 10% of all older black women from Atlantic City. *Large Counts*: 46, 17, 27, and 29 ≥ 10. *D*: (0.105, 0.391). *C*: We are 90% confident that the interval from 0.105 to 0.391 captures the true difference in the proportions of older black men and women who feel vulnerable to crime.
**(b)** Because the interval does not contain 0, there is convincing evidence that the true proportion of older blacks who feel vulnerable to crime is different for men and women.

**10.13** $H_0: p_1 - p_2 = 0$ vs. $H_a: p_1 - p_2 \neq 0$, where $p_1$ is the true proportion of all teens who would say that they own an iPod or MP3 player and $p_2$ is the true proportion of all young adults ....

**10.14** $H_0: p_1 - p_2 = 0$ vs. $H_a: p_1 - p_2 \neq 0$, where $p_1$ is the true proportion of freshmen in Illinois who have used anabolic steroids and $p_2$ is the true proportion of seniors ....

**10.15** *P*: Two-sample *z* test for $p_1 - p_2$.
*Random*: Independent random samples.
*10%*: 800 < 10% of all teens and 400 < 10% of all young adults. *Large Counts*: 632, 168, 268, and 132 ≥ 10. *D*: $z = 4.53$ and *P*-value ≈ 0.

(a) Construct and interpret a 90% confidence interval for the difference in the true proportions of black women and black men in Atlantic City who would say they felt vulnerable to crime.

(b) Does your interval from part (a) give convincing evidence of a difference between the population proportions? Explain.

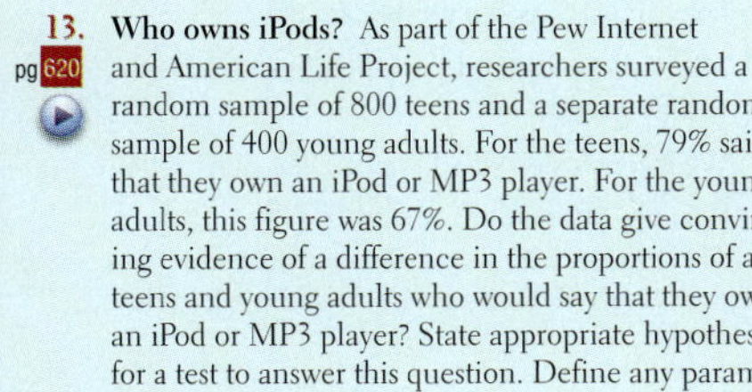

**13. Who owns iPods?** As part of the Pew Internet and American Life Project, researchers surveyed a random sample of 800 teens and a separate random sample of 400 young adults. For the teens, 79% said that they own an iPod or MP3 player. For the young adults, this figure was 67%. Do the data give convincing evidence of a difference in the proportions of all teens and young adults who would say that they own an iPod or MP3 player? State appropriate hypotheses for a test to answer this question. Define any parameters you use.

**14. Steroids in high school** A study by the National Athletic Trainers Association surveyed random samples of 1679 high school freshmen and 1366 high school seniors in Illinois. Results showed that 34 of the freshmen and 24 of the seniors had used anabolic steroids. Steroids, which are dangerous, are sometimes used in an attempt to improve athletic performance.[13] Do the data give convincing evidence of a difference in the proportion of all Illinois high school freshmen and seniors who have used anabolic steroids? State appropriate hypotheses for a test to answer this question. Define any parameters you use.

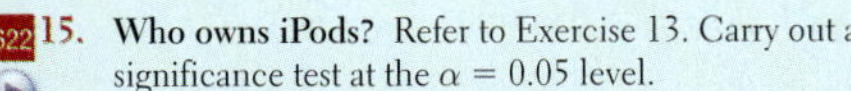

**15. Who owns iPods?** Refer to Exercise 13. Carry out a significance test at the $\alpha = 0.05$ level.

**16. Steroids in high school** Refer to Exercise 14. Carry out a significance test at the $\alpha = 0.05$ level.

**17. Who owns iPods?** Refer to Exercise 13. Construct and interpret a 95% confidence interval for the difference between the population proportions. Explain how the confidence interval is consistent with the results of the test in Exercise 15.

**18. Steroids in high school** Refer to Exercise 14. Construct and interpret a 95% confidence interval for the difference between the population proportions. Explain how the confidence interval is consistent with the results of the test in Exercise 16.

**19. Children make choices** Many new products introduced into the market are targeted toward children. The choice behavior of children with regard to new products is of particular interest to companies that design marketing strategies for these products. As part of one study, randomly selected children in different age groups were compared on their ability to sort new products into the correct product category (milk or juice).[14] Here are some of the data:

| Age group | N | Number who sorted correctly |
|---|---|---|
| 4- to 5-year-olds | 50 | 10 |
| 6- to 7-year-olds | 53 | 28 |

Did a significantly higher proportion of the 6- to 7-year-olds than the 4- to 5-year-olds sort correctly? Give appropriate evidence to justify your answer.

**20. Marriage and status** "Would you marry a person from a lower social class than your own?" Researchers asked this question of a random sample of 385 black, never-married college students. Of the 149 men in the sample, 91 said "Yes." Among the 236 women, 117 said "Yes."[15] Did a significantly higher proportion of the men than the women who were surveyed say "Yes"? Give appropriate evidence to justify your answer.

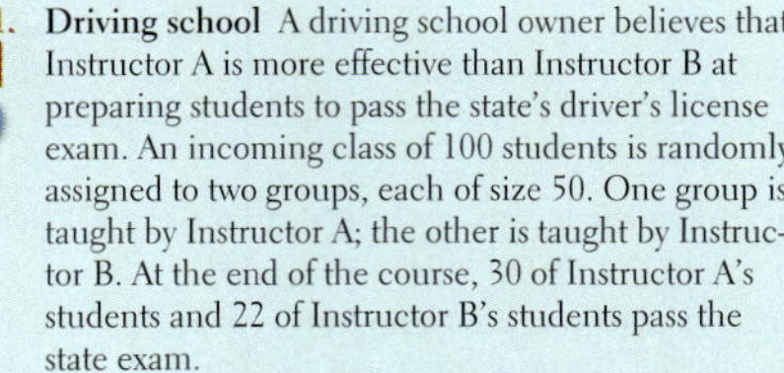

**21. Driving school** A driving school owner believes that Instructor A is more effective than Instructor B at preparing students to pass the state's driver's license exam. An incoming class of 100 students is randomly assigned to two groups, each of size 50. One group is taught by Instructor A; the other is taught by Instructor B. At the end of the course, 30 of Instructor A's students and 22 of Instructor B's students pass the state exam.

(a) Do these results give convincing evidence at the $\alpha = 0.05$ level that Instructor A is more effective?

(b) Describe a Type I and a Type II error in this setting. Which error could you have made in part (a)?

**22. Preventing strokes** Aspirin prevents blood from clotting and so helps prevent strokes. The Second European Stroke Prevention Study asked whether adding another anticlotting drug, named dipyridamole, would be more effective for patients who had already had a stroke. Here are the data on strokes during the two years of the study:[16]

| | Number of patients | Number of strokes |
|---|---|---|
| Aspirin alone | 1649 | 206 |
| Aspirin + dipyridamole | 1650 | 157 |

The study was a randomized comparative experiment.

(a) Is there convincing evidence at the $\alpha = 0.05$ level that adding dipyridamole helps reduce the risk of stroke?

(b) Describe a Type I and a Type II error in this setting. Which is more serious? Explain.

*C*: Because the *P*-value of approximately $0 < \alpha = 0.05$, we reject $H_0$. There is convincing evidence that the true proportion of people who would say that they own an iPod or MP3 player is different for teens and young adults.

**10.16** *P*: Two-sample *z* test for $p_1 - p_2$. *Random*: Independent random samples. *10%*: 1679 < 10% of freshmen in Illinois and 1366 < 10% of seniors in Illinois. *Large Counts*: 34, 1645, 24, 1342 ≥ 10. *D*: $z = 0.54$ and *P*-value = 0.5892. *C*: Because the *P*-value of $0.5892 > \alpha = 0.05$, we fail to reject $H_0$. We do not have convincing evidence that there is a difference in the true proportion of all Illinois freshmen and seniors who have used anabolic steroids.

**10.17** *D*: (0.066, 0.174). *C*: We are 95% confident that the interval from 0.066 to 0.174 captures the true difference in the proportion of teens and young adults who own iPods or MP3 players. Consistent because 0 is not included in the interval.

**10.18** *D*: (−0.007, 0.0124). *C*: We are 95% confident that the interval from −0.007 to 0.0124 captures the true difference in the proportion of all Illinois freshmen and seniors who have used anabolic steroids. Consistent because 0 is included in the interval.

**10.19** *S*: $H_0: p_1 - p_2 = 0$ vs. $H_a: p_1 - p_2 > 0$, where $p_1$ is the true proportion of 6- to 7-year-olds who would sort correctly and $p_2$ is the true proportion of 4- to 5-year-olds .... *P*: Two-sample *z* test for $p_1 - p_2$. *Random*: Independent random samples. *10%*: 53 < 10% of all 6- to 7-year olds and 50 < 10% of all 4- to 5-year-olds. *Large Counts*: 28, 25, 10, 40 ≥ 10. *D*: $z = 3.45$ and *P*-value = 0.0003. *C*: Because the *P*-value of $0.0003 < \alpha = 0.05$, we reject $H_0$. We have convincing evidence that the true proportion who would sort correctly is greater for 6- to 7-year-olds than for 4- to 5-year-olds.

**10.20** *S*: $H_0: p_1 - p_2 = 0$ vs. $H_a: p_1 - p_2 > 0$, where $p_1$ is the true proportion of male, black, never-married college students who would say "Yes" and $p_2$ is the true proportion of female, black, never-married college students .... *P*: Two-sample *z* test for $p_1 - p_2$. *Random*: Reasonable to consider these two independent random samples. *10%*: 149 < 10% of all male, black, never-married college students and 236 < 10% of all female, black, never-married college students. *Large Counts*: 91, 58, 117, 119 ≥ 10. *D*: $z = 2.21$ and *P*-value = 0.0136. *C*: Because the *P*-value of $0.0136 < \alpha = 0.05$, we reject $H_0$. We have convincing evidence that the true proportion of black, never-married college students who would say "Yes" is greater for males than for females.

**10.21 (a)** *S*: $H_0: p_A - p_B = 0$ vs. $H_a: p_A - p_B > 0$, where $p_A$ is the true proportion of students like these who would pass the driver's license exam when taught by Instructor A and $p_B$ is the true proportion of students like these who would pass the driver's license exam when taught by Instructor B. *P*: Two-sample *z* test for $p_A - p_B$. *Random*: Two groups in a randomized experiment. *Large Counts*: 30, 20, 22, 28 ≥ 10. *D*: $z = 1.60$ and *P*-value = 0.0547. *C*: Because the *P*-value of $0.0547 > \alpha = 0.05$, we fail to reject $H_0$. There is not convincing evidence that the true proportion of students like these who would pass is greater for A than for B. **(b)** I: Finding convincing evidence that A is more effective than B, when in reality the instructors are equally effective. II: Not finding convincing evidence that A is better, when in reality A is more effective. Type II error.

**10.22 (a)** *S*: $H_0: p_1 - p_2 = 0$ vs. $H_a: p_1 - p_2 > 0$, where $p_1$ is the true proportion of patients like those in the study who would have a stroke when taking aspirin alone and $p_2$ is the true proportion of patients like those in the study who would have a stroke when taking both drugs. *P*: Two-sample *z* test for $p_1 - p_2$. *Random*: Two groups in a randomized experiment. *Large Counts*: 206, 1443, 157, 1493 ≥ 10. *D*: $z = 2.73$ and *P*-value = 0.0031. *C*: Because the *P*-value of $0.0031 < \alpha = 0.05$, we reject $H_0$.

We have convincing evidence that the true proportion of patients like those in the study who would have a stroke is greater for those taking aspirin alone than for those taking both drugs. **(b)** I: Finding convincing evidence that adding dipyridamole helps prevent strokes, when in reality adding it doesn't help. II: Not finding convincing evidence that adding dipyridamole helps prevent strokes, when in reality adding it does help. A Type II error because we would be missing out on a drug that could be preventing people from having a stroke.

**10.23** **(a)** Two-sample z test for $p_1 - p_2$. *Random*: Two groups in a randomized experiment. *Large Counts*: 44, 44, 21, 60 are all $\geq 10$. **(b)** If there is no difference in the true pregnancy rates of women who are being prayed for and those who are not, there is a 0.0007 probability of getting a difference in pregnancy rates as large or larger than the one observed in the experiment by chance alone. **(c)** Because the *P*-value of $0.0007 < \alpha = 0.05$, we reject $H_0$. There is convincing evidence that the pregnancy rates among women like these who are prayed for is higher than the pregnancy rates for those who are not prayed for. **(d)** Knowing they were being prayed for might have affected their behavior in some way that would have affected whether they became pregnant or not. Then we wouldn't know if it was the prayer or the other behaviors that caused the higher pregnancy rate.

**10.24** **(a)** Two-sample z test for $p_1 - p_2$. *Random*: Two groups in a randomized experiment. *Large Counts*: 34, 46, 21, 59 are all $\geq 10$. **(b)** If there is no difference in the pregnancy rates of women who receive acupuncture and those who don't, there is a 0.0152 probability of getting a difference in pregnancy rates as large as or larger than the one observed in the experiment by chance alone. **(c)** Because the *P*-value of $0.0152 < \alpha = 0.05$, we reject $H_0$. We have convincing evidence that the pregnancy rate among women like these who receive acupuncture is higher than the pregnancy rate for those who do not receive acupuncture. **(d)** Knowing whether or not they received acupuncture might have affected the women's behavior in some way that would have affected whether they became pregnant or not. Then we wouldn't know if it was the acupuncture or the other behaviors that caused the higher pregnancy rate.

**10.25** a

*Exercises 23 and 24 involve the following setting.* Some women would like to have children but cannot do so for medical reasons. One option for these women is a procedure called in vitro fertilization (IVF), which involves injecting a fertilized egg into the woman's uterus.

23. **Prayer and pregnancy** Two hundred women who were about to undergo IVF served as subjects in an experiment. Each subject was randomly assigned to either a treatment group or a control group. Women in the treatment group were intentionally prayed for by several people (called *intercessors*) who did not know them, a process known as intercessory prayer. The praying continued for three weeks following IVF. The intercessors did not pray for the women in the control group. Here are the results: 44 of the 88 women in the treatment group got pregnant, compared to 21 out of 81 in the control group.[17]

    Is the pregnancy rate significantly higher for women who received intercessory prayer? To find out, researchers perform a test of $H_0: p_1 = p_2$ versus $H_a: p_1 > p_2$, where $p_1$ and $p_2$ are the actual pregnancy rates for women like those in the study who do and don't receive intercessory prayer, respectively.

    (a) Name the appropriate test and check that the conditions for carrying out this test are met.

    (b) The appropriate test from part (a) yields a *P*-value of 0.0007. Interpret this *P*-value in context.

    (c) What conclusion should researchers draw at the $\alpha = 0.05$ significance level? Explain.

    (d) The women in the study did not know whether they were being prayed for. Explain why this is important.

24. **Acupuncture and pregnancy** A study reported in the medical journal *Fertility and Sterility* sought to determine whether the ancient Chinese art of acupuncture could help infertile women become pregnant.[18] One hundred sixty healthy women who planned to have IVF were recruited for the study. Half of the subjects (80) were randomly assigned to receive acupuncture 25 minutes before embryo transfer and again 25 minutes after the transfer. The remaining 80 women were assigned to a control group and instructed to lie still for 25 minutes after the embryo transfer. Results are shown in the table below.

| | Acupuncture group | Control group |
|---|---|---|
| Pregnant | 34 | 21 |
| Not pregnant | 46 | 59 |
| Total | 80 | 80 |

    Is the pregnancy rate significantly higher for women who received acupuncture? To find out, researchers perform a test of $H_0: p_1 = p_2$ versus $H_a: p_1 > p_2$, where $p_1$ and $p_2$ are the actual pregnancy rates for women like those in the study who do and don't receive acupuncture, respectively.

    (a) Name the appropriate test and check that the conditions for carrying out this test are met.

    (b) The appropriate test from part (a) yields a *P*-value of 0.0152. Interpret this *P*-value in context.

    (c) What conclusion should researchers draw at the $\alpha = 0.05$ significance level? Explain.

    (d) The women in the study knew whether or not they received acupuncture. Explain why this is important.

***Multiple choice: Select the best answer for Exercises 25 to 28.***

*Exercises 25 to 27 refer to the following setting.* A sample survey interviews SRSs of 500 female college students and 550 male college students. Researchers want to determine whether there is a difference in the proportion of male and female college students who worked for pay last summer. In all, 410 of the females and 484 of the males say they worked for pay last summer.

25. Take $p_M$ and $p_F$ to be the proportions of all college males and females who worked last summer. The hypotheses to be tested are

    (a) $H_0: p_M - p_F = 0$ versus $H_a: p_M - p_F \neq 0$.

    (b) $H_0: p_M - p_F = 0$ versus $H_a: p_M - p_F > 0$.

    (c) $H_0: p_M - p_F = 0$ versus $H_a: p_M - p_F < 0$.

    (d) $H_0: p_M - p_F > 0$ versus $H_a: p_M - p_F = 0$.

    (e) $H_0: p_M - p_F \neq 0$ versus $H_a: p_M - p_F = 0$.

26. The researchers report that the results were statistically significant at the 1% level. Which of the following is the most appropriate conclusion?

    (a) Because the *P*-value is less than 1%, fail to reject $H_0$. There is not convincing evidence that the proportion of male college students in the study who worked for pay last summer is different from the proportion of female college students in the study who worked for pay last summer.

    (b) Because the *P*-value is less than 1%, fail to reject $H_0$. There is not convincing evidence that the proportion of all male college students who worked for pay last summer is different from the proportion of all female college students who worked for pay last summer.

(c) Because the P-value is less than 1%, reject $H_0$. There is convincing evidence that the proportion of all male college students who worked for pay last summer is the same as the proportion of all female college students who worked for pay last summer.

(d) Because the P-value is less than 1%, reject $H_0$. There is convincing evidence that the proportion of all male college students in the study who worked for pay last summer is different from the proportion of all female college students in the study who worked for pay last summer.

(e) Because the P-value is less than 1%, reject $H_0$. There is convincing evidence that the proportion of all male college students who worked for pay last summer is different from the proportion of all female college students who worked for pay last summer.

**27.** Which of the following is the correct margin of error for a 99% confidence interval for the difference in the proportion of male and female college students who worked for pay last summer?

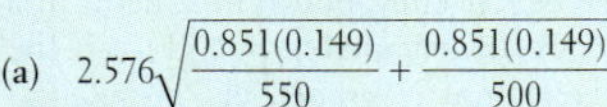

(a) $2.576\sqrt{\dfrac{0.851(0.149)}{550}+\dfrac{0.851(0.149)}{500}}$

(b) $2.576\sqrt{\dfrac{0.851(0.149)}{1050}}$

(c) $2.576\sqrt{\dfrac{0.880(0.120)}{550}+\dfrac{0.820(0.180)}{500}}$

(d) $1.960\sqrt{\dfrac{0.851(0.149)}{550}+\dfrac{0.851(0.149)}{500}}$

(e) $1.960\sqrt{\dfrac{0.880(0.120)}{550}+\dfrac{0.820(0.180)}{500}}$

**28.** In an experiment to learn whether Substance M can help restore memory, the brains of 20 rats were treated to damage their memories. First, the rats were trained to run a maze. After a day, 10 rats (determined at random) were given M and 7 of them succeeded in the maze. Only 2 of the 10 control rats were successful. The two-sample $z$ test for "no difference" against "a significantly higher proportion of the M group succeeds"

(a) gives $z = 2.25$, $P < 0.02$.

(b) gives $z = 2.60$, $P < 0.005$.

(c) gives $z = 2.25$, $P < 0.04$ but not $< 0.02$.

(d) should not be used because the Random condition is violated.

(e) should not be used because the Large Counts condition is violated.

*Exercises 29 and 30 refer to the following setting.* Thirty randomly selected seniors at Council High School were asked to report the age (in years) and mileage of their main vehicles. Here is a scatterplot of the data:

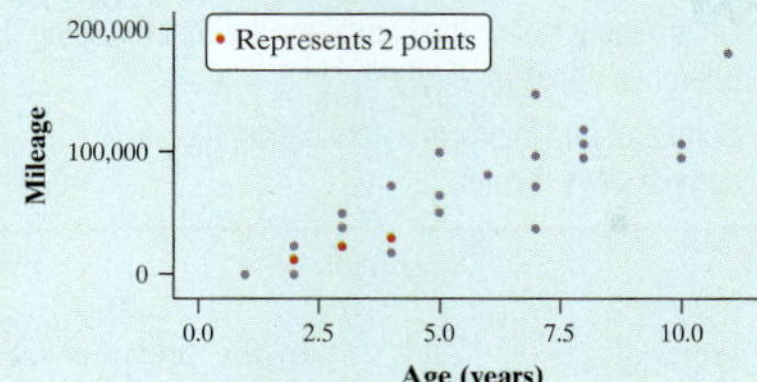


We used Minitab to perform a least-squares regression analysis for these data. Part of the computer output from this regression is shown below.

```
Predictor    Coef     Stdev    t-ratio       P
Constant    -13832    8773     -1.58       0.126
Age          14954    1546      9.67       0.000
s = 22723   R-sq = 77.0%     R-sq(adj) = 76.1%
```

**29. Drive my car** (3.2)

(a) What is the equation of the least-squares regression line? Be sure to define any symbols you use.

(b) Interpret the slope of the least-squares line in the context of this problem.

(c) One student reported that her 10-year-old car had 110,000 miles on it. Find and interpret the residual for this data value. Show your work.

**30. Drive my car** (3.2, 4.3)

(a) Explain what the value of $r^2$ tells you about how well the least-squares line fits the data.

(b) The mean age of the students' cars in the sample was $\bar{x} = 8$ years. Find the mean mileage of the cars in the sample. Show your work.

(c) Interpret the value of $s$ in the context of this setting.

(d) Would it be reasonable to use the least-squares line to predict a car's mileage from its age for a Council High School teacher? Justify your answer.

**10.26** e

**10.27** c

**10.28** e

**10.29 (a)** $\hat{y} = -13{,}832 + 14{,}954x$, where $\hat{y}$ = the predicted mileage and $x$ = the age in years of the cars. **(b)** For each year older the car is, the predicted mileage will increase by 14,954 miles. **(c)** Residual = −25,708. The student's car had 25,708 fewer miles than expected, based on its age.

**10.30 (a)** 77% of the variation in mileage is accounted for by the linear model relating mileage to age. **(b)** The least-squares regression line always goes through the point $(\bar{x}, \bar{y})$, so $\bar{y} = -13{,}832 + 14{,}954(8) = 105{,}800$ miles. **(c)** When using the least-squares regression line with $x$ = age to predict $y$ = mileage, we will typically be off by about 22,723 miles. **(d)** No, the least-squares line is based on a sample of cars owned and driven by students, not teachers.

**PD** **Section 10.2 Overview**

Watch the video overview of Section 10.2 (for teachers) by clicking on the link in the TE-book, opening the TRCD, or logging onto the book's companion Web site www.whfreeman.com/tps5e (teacher log-in required).

**TRM** **Section 10.2 Alternate Examples**

You can find the Alternate Examples for this section in Microsoft Word format by clicking on the link in the TE-book, opening the TRCD, or logging onto the book's companion Web site www.whfreeman.com/tps5e (teacher log-in required).

## Teaching Tip

Make sure to do the Activity on this page (or the Alternate Activity on the next page). It is a great way to reinforce the logic of inference and introduce significance tests for a difference in two means.

**TRM** **Comparing Pulse Rates**

The Comparing Pulse Rates Activity is another way to get students thinking about how random chance can affect the difference between two means. Find it by clicking on the link in the TE-book, opening the TRCD, or logging onto the book's companion Web site www.whfreeman.com/tps5e (teacher log-in required).

 **Activity Overview**

You can watch an overview of this Activity (for teachers) by clicking on the link in the TE-book, opening the TRCD, or logging onto the book's companion Web site www.whfreeman.com/tps5e (teacher log-in required).

## Teaching Tip

When you do the Activity on this page, remind your students that we always begin a significance test with the assumption that the null hypothesis is true. It is possible that the burial time has no effect on breaking strength and polyester appears to decay over time because the stronger strips happened to be assigned to the 2-week group. To investigate how likely this is, we assume that each individual strip would have the same breaking strength if it was assigned to the other group and see how often we get a random assignment that results in a difference at least as extreme as the one in the actual study.

# 10.2 Comparing Two Means

**WHAT YOU WILL LEARN** By the end of the section, you should be able to:

- Describe the shape, center, and spread of the sampling distribution of $\bar{x}_1 - \bar{x}_2$.
- Determine whether the conditions are met for doing inference about $\mu_1 - \mu_2$.
- Construct and interpret a confidence interval to compare two means.
- Perform a significance test to compare two means.
- Determine when it is appropriate to use two-sample *t* procedures versus paired *t* procedures.

In the previous section, we developed methods for comparing two proportions. What if we want to compare the mean of some quantitative variable for the individuals in Population 1 and Population 2? Our parameters of interest are the population means $\mu_1$ and $\mu_2$. Once again, the best approach is to take separate random samples from each population and to compare the sample means $\bar{x}_1$ and $\bar{x}_2$.

Suppose we want to compare the average effectiveness of two treatments in a completely randomized experiment. In this case, the parameters $\mu_1$ and $\mu_2$ are the true mean responses for Treatment 1 and Treatment 2, respectively. We use the mean response in the two groups, $\bar{x}_1$ and $\bar{x}_2$, to make the comparison. Here's a table that summarizes these two situations:

| Population or treatment | Parameter | Statistic | Sample size |
|---|---|---|---|
| 1 | $\mu_1$ | $\bar{x}_1$ | $n_1$ |
| 2 | $\mu_2$ | $\bar{x}_2$ | $n_2$ |

We compare the populations or treatments by doing inference about the difference $\mu_1 - \mu_2$ between the parameters. The statistic that estimates this difference is the difference between the two sample means, $\bar{x}_1 - \bar{x}_2$. To use $\bar{x}_1 - \bar{x}_2$ for inference, we must know its sampling distribution. Here is an Activity that gives you a preview of what lies ahead.

## ACTIVITY Does Polyester Decay?

**MATERIALS:** 10 small pieces of card stock (or index cards) per pair of students

How quickly do synthetic fabrics such as polyester decay in landfills? A researcher buried polyester strips in the soil for different lengths of time, then dug up the strips and measured the force required to break them. Breaking strength is easy to measure and is a good indicator of decay. Lower strength means the fabric has decayed.

The researcher buried 10 strips of polyester fabric in well-drained soil in the summer. The strips were randomly assigned to two groups: 5 of them were buried for 2 weeks and the other 5 were buried for 16 weeks. Here are the breaking strengths in pounds:[19]

| Group 1 (2 weeks): | 118 | 126 | 126 | 120 | 129 |
|---|---|---|---|---|---|
| Group 2 (16 weeks): | 124 | 98 | 110 | 140 | 110 |

Do the data give convincing evidence that polyester decays more in 16 weeks than in 2 weeks?

1. The Fathom dotplot displays the data from the experiment. Discuss what this graph shows with your classmates.

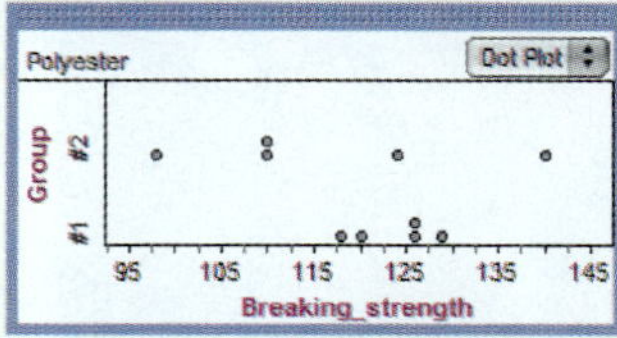


For Group 1, the mean breaking strength was $\bar{x}_1 = 123.8$ pounds. For Group 2, the mean breaking strength was $\bar{x}_2 = 116.4$ pounds. The observed difference in average breaking strength for the two groups is $\bar{x}_1 - \bar{x}_2 = 123.8 - 116.4 = 7.4$ pounds. Is it plausible that this difference is due to the chance involved in the random assignment and not to the treatments themselves? To find out, your class will perform a simulation.

Suppose that the length of time in the ground has no effect on the breaking strength of the polyester specimens. Then each specimen would have the same breaking strength regardless of whether it was assigned to Group 1 or Group 2. In that case, we could examine the results of repeated random assignments of the specimens to the two groups.

2. Write each of the 10 breaking-strength measurements on a separate card. Mix the cards well and deal them face down into two piles of 5 cards each. Be sure to decide which pile is Group 1 and which is Group 2 *before* you look at the cards. Calculate the difference in the mean breaking strength (Group 1 − Group 2). Record this value.
3. Your teacher will draw and label axes for a class dotplot. Plot the result you got in Step 2 on the graph.
4. Repeat Steps 2 and 3 if needed to get a total of at least 40 repetitions of the simulation for your class.
5. Based on the class's simulation results, how surprising would it be to get a difference in means of 7.4 or larger simply due to the chance involved in the random assignment?
6. What conclusion would you draw about whether polyester decays more when left in the ground for longer periods of time? Explain.

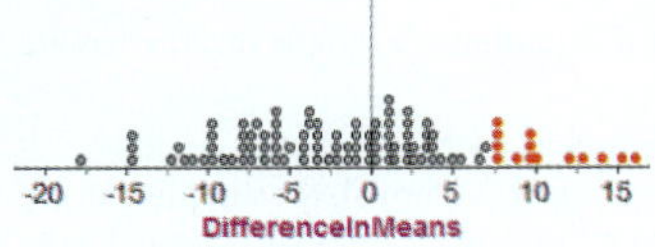


In this simulation, 14 of the 100 trials (in red) produced a difference in means of at least 7.4 pounds, so the approximate *P*-value is 0.14. It is likely that a difference this big could have happened just due to the chance variation in random assignment. The observed difference is not statistically significant and does not provide convincing evidence that polyester decays more in 16 weeks than in 2 weeks.

## ALTERNATE ACTIVITY

### *Caffeine and pulse rates*

An AP® Statistics class with 20 students performed the caffeine and pulse rates experiment described in an Alternate Example in Section 4.2. Half of the students were given regular Pepsi, and the other half were given caffeine-free Pepsi. Each student recorded his or her initial pulse rates, drank the soda, and measured his or her final pulse rate. The increase in pulse rates for each student is shown below (a negative value indicates a decrease in pulse rate).

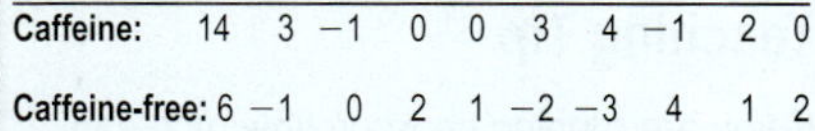

| **Caffeine:** | 14 | 3 | −1 | 0 | 0 | 3 | 4 | −1 | 2 | 0 |
|---|---|---|---|---|---|---|---|---|---|---|
| **Caffeine-free:** | 6 | −1 | 0 | 2 | 1 | −2 | −3 | 4 | 1 | 2 |

Here is a graph of the two distributions:

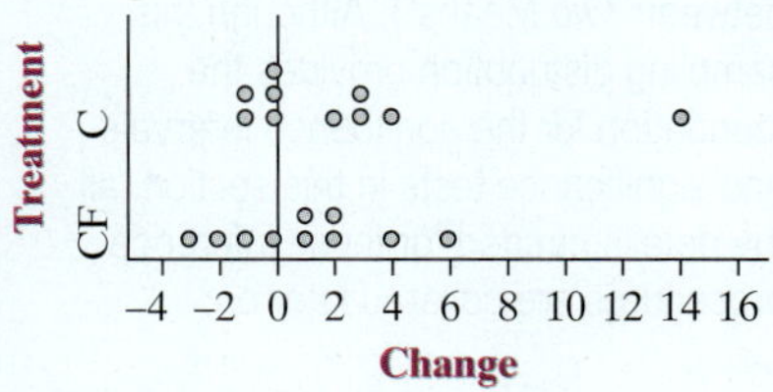


The mean change for the caffeine group was $\bar{x}_C = 2.4$ and the mean change for the caffeine-free group was $\bar{x}_{CF} = 1.0$. Does the difference $\bar{x}_C - \bar{x}_{CF} = 2.4 - 1.0 = 1.4$ provide convincing evidence that caffeine increases pulse rates?

Write each of the 20 increases on a separate index card. Shuffle the cards and randomly divide them into two piles of 10. Find the mean of each pile and subtract the mean of the second pile from the mean of the first pile. Record this value and repeat the process many times. How often did you get a difference of at least 1.4? What conclusion would you draw about the effect of caffeine on pulse rates?

Here is an example of what your class's dotplot might look like after 100 trials.

In this simulation, the approximate *P*-value was 0.27 because 27 of the 100 trials (in red) produced a difference in means of at least 1.4. It is very likely that a difference this big could have resulted just from the chance variation in random assignment. This difference is not statistically significant and does not provide convincing evidence that caffeine increases pulse rates.

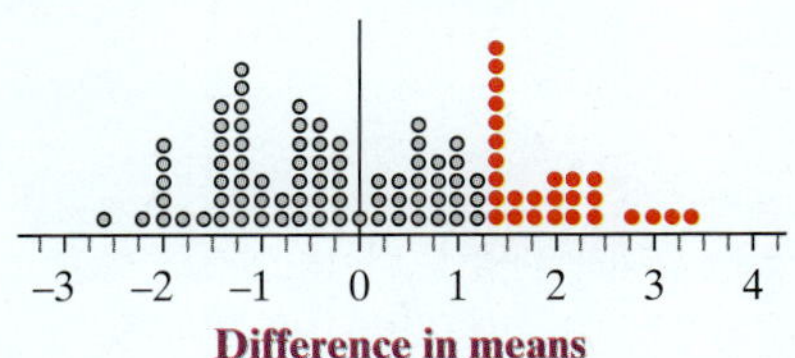

**APPLET**

The *Test for a Difference in Means* applet at www.lock5stat.com/statkey automates the re-randomization process we use in the Does Polyester Decay? activity. To enter the data, press the Edit Data button and use the following format:

label, value
two, 118
two, 126
two, 126
two, 120
two, 129
sixteen, 124
sixteen, 98
sixteen, 110
sixteen, 140
sixteen, 110

## Teaching Tip

After reading that there isn't convincing evidence that polyester decays more in 16 weeks than in 2 weeks, ask your students which type of error could have been committed (Type II). Then ask them what could be done to reduce the probability of a Type II error (increase the sample size for both groups).

## Teaching Tip

After teaching the section "The Sampling Distribution of a Difference between Two Means," some teachers prefer to reverse the order of the text here and teach significance tests before confidence intervals. Significance tests follow naturally from the Does Polyester Decay? activity and confidence intervals are a good follow-up analysis as they provide more information than a significance test. This section, including the exercises, was written so teachers could take either approach.

## Teaching Tip

If you are running short on time, you can move quickly through this subsection ("The Sampling Distribution of a Difference between Two Means"). Although this sampling distribution provides the foundation for the confidence intervals and significance tests in this section, all the details needed for these inference procedures are covered later on.

## Teaching Tip

To make sure students understand what each of the dotplots on this page represents, simulate taking a random sample of 12 girls on a graphing calculator using the command `randNorm(56.4,2.7,12)`. Store these values in List1 and use `1-Var Stats` to find the mean. Likewise, simulate a random sample of 8 boys using the command `randNorm(55.7,3.8,8)`. Store these values in List2 and use `1-Var Stats` to find the mean. Then subtract the two means to get a value of $\bar{x}_F - \bar{x}_M$. Doing this over and over and making dotplots for each of these values will result in graphs like those shown in Figure 10.7.

## Teaching Tip

Ask your students to interpret the standard deviation for each of these distributions. For example, if $\sigma_{\bar{x}_F - \bar{x}_M} \approx 1.56$ inches, then the difference in the *sample* mean heights of females and males will be about 1.56 inches from the *true* difference in means, on average.

# The Sampling Distribution of a Difference between Two Means

To explore the sampling distribution of $\bar{x}_1 - \bar{x}_2$, let's start with two Normally distributed populations having known means and standard deviations. Based on information from the U.S. National Health and Nutrition Examination Survey (NHANES), the heights of 10-year-old girls follow a Normal distribution with mean $\mu_F = 56.4$ inches and standard deviation $\sigma_F = 2.7$ inches. The heights of 10-year-old boys follow a Normal distribution with mean $\mu_M = 55.7$ inches and standard deviation $\sigma_M = 3.8$ inches.[20]

Suppose we take independent SRSs of 12 girls and 8 boys of this age and measure their heights. What can we say about the difference $\bar{x}_F - \bar{x}_M$ in the average heights of the sample of girls and the sample of boys?

We used Fathom software to take an SRS of 12 ten-year-old girls and 8 ten-year-old boys and to plot the values of $\bar{x}_F$, $\bar{x}_M$, and $\bar{x}_F - \bar{x}_M$ for each sample. Our first set of simulated samples gave $\bar{x}_F = 56.09$ inches and $\bar{x}_M = 54.68$ inches, so dots were placed above each of those values in Figure 10.7(a) and (b). The difference in the sample means is $\bar{x}_F - \bar{x}_M = 56.09 - 54.68 = 1.41$ inches. A dot for this value appears in Figure 10.7(c). The three dotplots in Figure 10.7 show the results of repeating this process 1000 times. These are the approximate sampling distributions of $\bar{x}_F$, $\bar{x}_M$, and $\bar{x}_F - \bar{x}_M$.

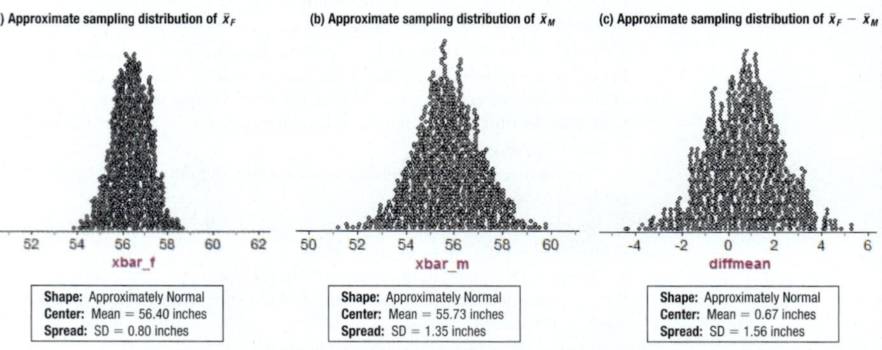


**FIGURE 10.7** Simulated sampling distributions of (a) the sample mean height $\bar{x}_F$ in 1000 SRSs of size $n_F = 12$ from the population of 10-year-old girls, (b) the sample mean height $\bar{x}_M$ in 1000 SRSs of size $n_M = 8$ from the population of 10-year-old boys, and (c) the difference in sample means $\bar{x}_F - \bar{x}_M$ for each of the 1000 repetitions.

In Chapter 7, we saw that the sampling distribution of a sample mean $\bar{x}$ has the following properties:

*Shape:* (1) If the population distribution is Normal, then so is the sampling distribution of $\bar{x}$; (2) if the population distribution isn't Normal, the sampling distribution of $\bar{x}$ will be approximately Normal if the sample size is large enough (say, $n \geq 30$) by the central limit theorem (CLT).

*Center:* $\mu_{\bar{x}} = \mu$

*Spread:* $\sigma_{\bar{x}} = \frac{\sigma}{\sqrt{n}}$ if the sample is no more than 10% of the population

For the sampling distributions of $\bar{x}_F$ and $\bar{x}_M$ in this case:

| | Sampling distribution of $\bar{x}_F$ | Sampling distribution of $\bar{x}_M$ |
|---|---|---|
| **Shape** | Normal, because the population distribution is Normal | Normal, because the population distribution is Normal |
| **Center** | $\mu_{\bar{x}_F} = \mu_F = 56.4$ inches | $\mu_{\bar{x}_M} = \mu_M = 55.7$ inches |
| **Spread** | $\sigma_{\bar{x}_F} = \frac{\sigma_F}{\sqrt{n_F}} = \frac{2.7}{\sqrt{12}} = 0.78$ inches because there are way more than 10(12) = 120 ten-year-old girls in the United States. | $\sigma_{\bar{x}_M} = \frac{\sigma_M}{\sqrt{n_M}} = \frac{3.8}{\sqrt{8}} = 1.34$ inches because there are way more than 10(8) = 80 ten-year-old boys in the United States. |

The approximate sampling distributions in Figures 10.7(a) and (b) give similar results.

What about the sampling distribution of $\bar{x}_F - \bar{x}_M$? Figure 10.7(c) suggests that it has a roughly Normal shape, is centered at about 0.67 inches, and has standard deviation about 1.56 inches. The shape makes sense because we are combining two independent Normal random variables, $\bar{x}_F$ and $\bar{x}_M$. How about the center? The actual mean height of 10-year-old girls is $\mu_F = 56.4$ inches. For 10-year-old boys, the actual mean height is $\mu_M = 55.7$ inches. We'd expect the difference $\bar{x}_F - \bar{x}_M$ to center on the actual difference in the population means, $\mu_F - \mu_M = 56.4 - 55.7 = 0.7$ inches. The spread, however, is a bit more complicated.

**THINK ABOUT IT**

**How can we find formulas for the mean and standard deviation of the sampling distribution of $\bar{x}_1 - \bar{x}_2$?** Both $\bar{x}_1$ and $\bar{x}_2$ are random variables. That is, their values would vary in repeated independent SRSs of size $n_1$ and $n_2$. Independent random samples yield independent random variables $\bar{x}_1$ and $\bar{x}_2$. The statistic $\bar{x}_1 - \bar{x}_2$ is the difference of these two independent random variables.

In Chapter 6, we learned that for any two random variables X and Y,

$$\mu_{X-Y} = \mu_X - \mu_Y$$

For the random variables $\bar{x}_1$ and $\bar{x}_2$, we have

$$\mu_{\bar{x}_1-\bar{x}_2} = \mu_{\bar{x}_1} - \mu_{\bar{x}_2} = \mu_1 - \mu_2$$

In the observational study of the heights of 10-year-olds,

$$\mu_{\bar{x}_F-\bar{x}_M} = \mu_F - \mu_M = 56.4 - 55.7 = 0.70 \text{ inches}$$

We also learned in Chapter 6 that for *independent* random variables X and Y,

$$\sigma^2_{X-Y} = \sigma^2_X + \sigma^2_Y$$

For the random variables $\bar{x}_1$ and $\bar{x}_2$, we have

$$\sigma^2_{\bar{x}_1-\bar{x}_2} = \sigma^2_{\bar{x}_1} + \sigma^2_{\bar{x}_2} = \left(\frac{\sigma_1}{\sqrt{n_1}}\right)^2 + \left(\frac{\sigma_2}{\sqrt{n_2}}\right)^2 = \frac{\sigma_1^2}{n_1} + \frac{\sigma_2^2}{n_2}$$

So $\sigma_{\bar{x}_1-\bar{x}_2} = \sqrt{\frac{\sigma_1^2}{n_1} + \frac{\sigma_2^2}{n_2}}$.

## Teaching Tip

Students won't be asked to derive the formulas for the mean and standard deviation of a difference in sample means as we do here. However, reading through this Think About It is a great way to review the rules for combining random variables that we learned in Chapter 6 and the properties of the sampling distribution of a sample mean that we learned in Chapter 7.

In the observational study of the heights of 10-year-olds,

$$\sigma_{\bar{x}_F - \bar{x}_M} = \sqrt{\frac{\sigma_F^2}{n_F} + \frac{\sigma_M^2}{n_M}} = \sqrt{\frac{2.7^2}{12} + \frac{3.8^2}{8}} = 1.55$$

This is similar to the result from the Fathom simulation.

Here are the facts we need.

### THE SAMPLING DISTRIBUTION OF $\bar{x}_1 - \bar{x}_2$

Choose an SRS of size $n_1$ from Population 1 with mean $\mu_1$ and standard deviation $\sigma_1$ and an independent SRS of size $n_2$ from Population 2 with mean $\mu_2$ and standard deviation $\sigma_2$.

- **Shape:** When the population distributions are Normal, the sampling distribution of $\bar{x}_1 - \bar{x}_2$ is Normal. In other cases, the sampling distribution of $\bar{x}_1 - \bar{x}_2$ will be approximately Normal if the sample sizes are large enough ($n_1 \geq 30$ and $n_2 \geq 30$).
- **Center:** The mean of the sampling distribution of $\bar{x}_1 - \bar{x}_2$ is $\mu_1 - \mu_2$.
- **Spread:** The standard deviation of the sampling distribution of $\bar{x}_1 - \bar{x}_2$ is

$$\sqrt{\frac{\sigma_1^2}{n_1} + \frac{\sigma_2^2}{n_2}}$$

as long as each sample is no more than 10% of its population.

## Teaching Tip

Remind your students that Figure 10.8 is a theoretical model for dotplots such as Figure 10.7(c). It might also help to remind your students that each dot in Figure 10.2 represents one possible observation of $\bar{x}_1 - \bar{x}_2$.

**FIGURE 10.8** Select independent SRSs from two populations having means $\mu_1$ and $\mu_2$ and standard deviations $\sigma_1$ and $\sigma_2$. The two sample means are $\bar{x}_1$ and $\bar{x}_2$. If the population distributions are both Normal, the sampling distribution of the difference $\bar{x}_1 - \bar{x}_2$ is Normal. The sampling distribution will be approximately Normal in other cases if both samples are large enough ($n_1 \geq 30$ and $n_2 \geq 30$).

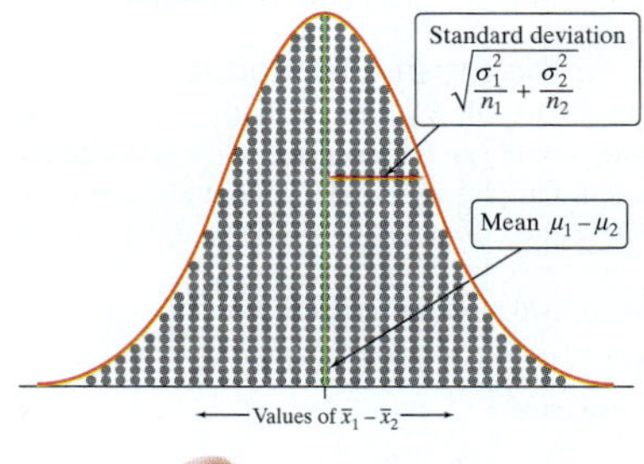


When conditions are met, the sampling distribution of $\bar{x}_1 - \bar{x}_2$ will be approximately Normal with mean $\mu_{\bar{x}_1 - \bar{x}_2} = \mu_1 - \mu_2$ and standard deviation $\sigma_{\bar{x}_1 - \bar{x}_2} = \sqrt{\frac{\sigma_1^2}{n_1} + \frac{\sigma_2^2}{n_2}}$. Figure 10.8 displays this distribution.

The formula for the standard deviation of the sampling distribution involves the parameters $\sigma_1$ and $\sigma_2$, which are usually unknown. Just as in Chapters 8 and 9, we must replace these by estimates to do inference. We'll get to confidence intervals and significance tests shortly. For now, let's focus on the sampling distribution of $\bar{x}_1 - \bar{x}_2$.

## EXAMPLE Medium or Large Drink?

### *Describing the sampling distribution of $\bar{x}_1 - \bar{x}_2$*

A fast-food restaurant uses an automated filling machine to pour its soft drinks. The machine has different settings for small, medium, and large drink cups. According to the machine's manufacturer, when the large setting is chosen, the amount of liquid $L$ dispensed by the machine follows a Normal distribution with mean 27 ounces and

## ALTERNATE EXAMPLE

### *Potato chips*

A potato chip manufacturer buys potatoes from two different suppliers, Riderwood Farms and Camberley, Inc. The weights of potatoes from Riderwood Farms are approximately Normally distributed with a mean of 175 grams and a standard deviation of 25 grams. The weights of potatoes from Camberley are approximately Normally distributed with a mean of 180 grams and a standard deviation of 30 grams. When shipments arrive at the factory, inspectors randomly select a sample of 20 potatoes from each shipment and weigh them. Let $\bar{x}_C - \bar{x}_R$ be the difference in the sample mean weight of potatoes from the two suppliers.

PROBLEM:

**(a)** What is the shape of the sampling distribution of $\bar{x}_C - \bar{x}_R$? Why?

**(b)** Find the mean of the sampling distribution. Show your work.

**(c)** Find the standard deviation of the sampling distribution. Show your work.

SOLUTION:

**(a)** The shape of the sampling distribution of $\bar{x}_C - \bar{x}_R$ is approximately Normal because both population distributions are approximately Normal.

**(b)** The mean of the sampling distribution is $\mu_{\bar{x}_C - \bar{x}_R} = 180 - 175 = 5$ grams.

**(c)** Because 20 potatoes < 10% of each shipment, the standard deviation is

$$\sigma_{\bar{x}_C - \bar{x}_R} = \sqrt{\frac{25^2}{20} + \frac{30^2}{20}} = 8.73 \text{ grams.}$$

standard deviation 0.8 ounces. When the medium setting is chosen, the amount of liquid $M$ dispensed follows a Normal distribution with mean 17 ounces and standard deviation 0.5 ounces. To test the manufacturer's claim, the restaurant manager measures the amount of liquid in each of 20 cups filled with the large setting and 25 cups filled with the medium setting. Let $\bar{x}_L - \bar{x}_M$ be the difference in the sample mean amount of liquid under the two settings.

**PROBLEM:**

(a) What is the shape of the sampling distribution of $\bar{x}_L - \bar{x}_M$? Why?

(b) Find the mean of the sampling distribution. Show your work.

(c) Find the standard deviation of the sampling distribution. Show your work.

**SOLUTION:**

(a) The sampling distribution of $\bar{x}_L - \bar{x}_M$ is Normal because both population distributions are Normal.

(b) The mean is $\mu_{\bar{x}_L - \bar{x}_M} = \mu_L - \mu_M = 27 - 17 = 10$ ounces.

(c) The standard deviation is $\sigma_{\bar{x}_L - \bar{x}_M} = \sqrt{\frac{\sigma_L^2}{n_L} + \frac{\sigma_M^2}{n_M}} = \sqrt{\frac{(0.80)^2}{20} + \frac{(0.50)^2}{25}} = 0.205$ ounces.

Note that we do not need to check the 10% condition because we are not sampling without replacement from a finite population.

**For Practice** *Try Exercise*  31

## The Two-Sample *t* Statistic

When data come from two independent random samples or two groups in a randomized experiment (the Random condition), the statistic $\bar{x}_1 - \bar{x}_2$ is our best guess for the value of $\mu_1 - \mu_2$. If the 10% condition is met, the standard deviation of the sampling distribution of $\bar{x}_1 - \bar{x}_2$ is

$$\sigma_{\bar{x}_1 - \bar{x}_2} = \sqrt{\frac{\sigma_1^2}{n_1} + \frac{\sigma_2^2}{n_2}}$$

If the Normal/Large Sample condition is met, we can standardize the observed difference $\bar{x}_1 - \bar{x}_2$ to obtain a $z$ statistic that is modeled well by a standard Normal distribution:

$$z = \frac{(\bar{x}_1 - \bar{x}_2) - (\mu_1 - \mu_2)}{\sqrt{\frac{\sigma_1^2}{n_1} + \frac{\sigma_2^2}{n_2}}}$$

In the unlikely event that both population standard deviations are known, this *two-sample z statistic* is the basis for inference about $\mu_1 - \mu_2$.

### Teaching Tip

On the AP® exam, it is very unlikely that students will be asked to do inference for a difference of means when the population standard deviations are known. Other than briefly mentioning the two-sample $z$ statistic and when it might be used, we don't recommend spending any additional time on this topic.

### AP® Exam Formula Sheet

The formula for the standard deviation of the sampling distribution of $\bar{x}_1 - \bar{x}_2$, $\sigma_{\bar{x}_1-\bar{x}_2} = \sqrt{\frac{\sigma_1^2}{n_1} + \frac{\sigma_2^2}{n_2}}$, is included on the formula sheet that is provided on both sections of the AP® exam. However, the formula for the standard error, $SE_{\bar{x}_1-\bar{x}_2} = \sqrt{\frac{s_1^2}{n_1} + \frac{s_2^2}{n_2}}$, is *not* included on the formula sheet.

### Teaching Tip

Our two-sample *t* statistic doesn't quite follow a *t* distribution, even when the conditions are met. However, when an additional condition is met (equal population standard deviations), the *pooled* version of the two-sample *t* statistic does follow a *t* distribution with df $= n_1 - 1 + n_2 - 1$. We don't recommend using the pooled version for two reasons: (1) The equal population standard deviation condition is rarely met. (2) The approximation using our (unpooled) two-sample *t* statistic is quite accurate. For more about pooling when doing inference about a difference in means, see page 650.

### Teaching Tip

Point out that the conditions for doing inference for a difference in means are very similar to the conditions for doing inference for a single mean. The main difference is that we need to check the conditions for *both* samples. Also, the Random condition requires that the two samples are independent or that the data come from two groups in a randomized experiment.

### Teaching Tip

Although we will illustrate both options calculating the degrees of freedom, we recommend that you use technology whenever possible. This will result in narrower confidence intervals, smaller *P*-values, and consequently more power (at no extra cost). *Note*: Students are not expected to calculate the degrees of freedom using this formula by hand, nor is the formula provided on the AP® exam.

Suppose now that the population standard deviations $\sigma_1$ and $\sigma_2$ are not known. We estimate them by the standard deviations $s_1$ and $s_2$ from our two samples. The result is the **standard error** (also called the *estimated standard deviation*) of $\bar{x}_1 - \bar{x}_2$:

$$SE_{\bar{x}_1-\bar{x}_2} = \sqrt{\frac{s_1^2}{n_1} + \frac{s_2^2}{n_2}}$$

Now when we standardize the point estimate $\bar{x}_1 - \bar{x}_2$, the result is the **two-sample *t* statistic:**

$$t = \frac{(\bar{x}_1 - \bar{x}_2) - (\mu_1 - \mu_2)}{\sqrt{\frac{s_1^2}{n_1} + \frac{s_2^2}{n_2}}}$$

The statistic *t* has the same interpretation as any *z* or *t* statistic: it says how far $\bar{x}_1 - \bar{x}_2$ is from its mean in standard deviation units. When the Normal/Large Sample condition is met, the two-sample *t* statistic has approximately a *t* distribution. It does not have exactly a *t* distribution even if the populations are both exactly Normal. In practice, however, the approximation is very accurate.

#### CONDITIONS FOR PERFORMING INFERENCE ABOUT $\mu_1 - \mu_2$

- **Random:** The data come from two independent random samples or from two groups in a randomized experiment.
  - **10%:** When sampling without replacement, check that $n_1 \le \frac{1}{10}N_1$ and $n_2 \le \frac{1}{10}N_2$.
- **Normal/Large Sample:** Both population distributions (or the true distributions of responses to the two treatments) are Normal or both sample sizes are large ($n_1 \ge 30$ and $n_2 \ge 30$). If either population (treatment) distribution has unknown shape and the corresponding sample size is less than 30, use a graph of the sample data to assess the Normality of the population (treatment) distribution. Do not use two-sample *t* procedures if the graph shows strong skewness or outliers.

There are two practical options for using the two-sample *t* procedures when the conditions are met. The two options are exactly the same except for the degrees of freedom used for *t* critical values and *P*-values.

**Option 1 (Technology):** Use the *t* distribution with degrees of freedom calculated from the data by the formula below. Note that the df given by this formula is usually not a whole number.

$$\text{df} = \frac{\left(\frac{s_1^2}{n_1} + \frac{s_2^2}{n_2}\right)^2}{\frac{1}{n_1 - 1}\left(\frac{s_1^2}{n_1}\right)^2 + \frac{1}{n_2 - 1}\left(\frac{s_2^2}{n_2}\right)^2}$$

You can thank statisticians B. L. Welch and F. E. Satterthwaite for discovering this fairly remarkable formula.

**Option 2 (Conservative):** Use the *t* distribution with degrees of freedom equal to the *smaller* of $n_1 - 1$ and $n_2 - 1$. With this option, the resulting confidence interval has a margin of error *as large as or larger than* is needed for the desired confidence level. The significance test using this option gives a *P*-value *equal to or*

*greater than* the true *P*-value. As the sample sizes increase, confidence levels and *P*-values from Option 2 become more accurate.[21]

## Confidence Intervals for $\mu_1 - \mu_2$

If the Random, 10%, and Normal/Large Sample conditions are met, we can use our standard formula to construct a confidence interval for $\mu_1 - \mu_2$:

$$\text{statistic} \pm (\text{critical value}) \cdot (\text{standard deviation of statistic})$$

$$= (\bar{x}_1 - \bar{x}_2) \pm t^*\sqrt{\frac{s_1^2}{n_1} + \frac{s_2^2}{n_2}}$$

We can use either technology or the conservative approach with Table B to find the critical value $t^*$ for the given confidence level. This method is called a **two-sample *t* interval for a difference between two means.**

**TWO-SAMPLE *t* INTERVAL FOR A DIFFERENCE BETWEEN TWO MEANS**

When the conditions are met, an approximate *C*% confidence interval for $\mu_1 - \mu_2$ is

$$(\bar{x}_1 - \bar{x}_2) \pm t^*\sqrt{\frac{s_1^2}{n_1} + \frac{s_2^2}{n_2}}$$

Here, $t^*$ is the critical value with *C*% of its area between $-t^*$ and $t^*$ for the *t* distribution with degrees of freedom using either Option 1 (technology) or Option 2 (the smaller of $n_1 - 1$ and $n_2 - 1$).

**AP® Exam Formula Sheet**

Although the specific formula for a confidence interval for the difference between two means is *not* included on the AP® exam formula sheet, students are given the general formula for a confidence interval and the formula for the standard deviation of the sampling distribution of $\bar{x}_1 - \bar{x}_2$.

The following example shows how to construct and interpret a confidence interval for a difference in means. As usual with inference problems, we follow the four-step process.

EXAMPLE

### Big Trees, Small Trees, Short Trees, Tall Trees

STEP 4

*Confidence interval for $\mu_1 - \mu_2$*

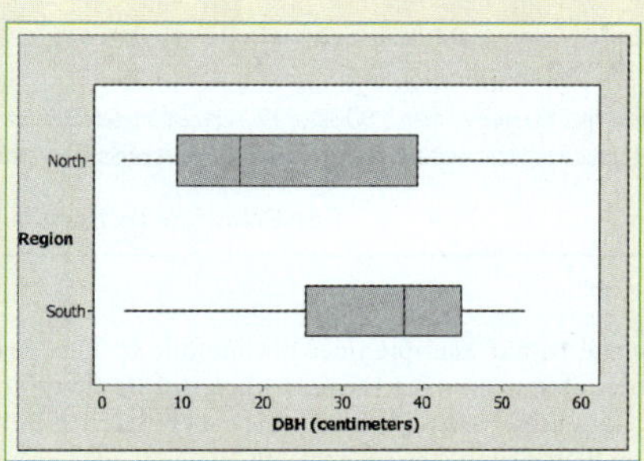


The Wade Tract Preserve in Georgia is an old-growth forest of longleaf pines that has survived in a relatively undisturbed state for hundreds of years. One question of interest to foresters who study the area is "How do the sizes of longleaf pine trees in the northern and southern halves of the forest compare?" To find out, researchers took random samples of 30 trees from each half and measured the diameter at breast height (DBH) in centimeters.[22] Here are comparative boxplots of the data and summary statistics from Minitab.

**Descriptive Statistics: North, South**

| Variable | N | Mean | StDev |
|---|---|---|---|
| North | 30 | 23.70 | 17.50 |
| South | 30 | 34.53 | 14.26 |

ALTERNATE EXAMPLE

## Chocolate chips

Ashtyn and Olivia wanted to know if generic chocolate chip cookies have as many chocolate chips as name-brand chocolate chip cookies, on average. To investigate, they randomly selected 10 bags of Chips Ahoy cookies and 10 bags of Great Value cookies and randomly selected 1 cookie from each bag. Then they carefully broke apart each cookie and counted the number of chocolate chips in each. Here are their results:

| | | | | | | | | | | |
|---|---|---|---|---|---|---|---|---|---|---|
| **Chips Ahoy:** | 17 | 19 | 21 | 16 | 17 | 18 | 20 | 21 | 17 | 18 |
| **Great Value:** | 22 | 20 | 14 | 17 | 21 | 22 | 15 | 19 | 26 | 18 |

PROBLEM:

**(a)** Construct and interpret a 99% confidence interval for the difference in the mean number of chocolate chips in Chips Ahoy and Great Value cookies.

**(b)** Does your interval provide convincing evidence that there is a difference in the mean number of chocolate chips?

SOLUTION:

**(a)** STATE: We want to estimate $\mu_{CA} - \mu_{GV}$ at the 99% confidence level, where $\mu_{CA}$ = the true mean number of chocolate chips in Chips Ahoy cookies and $\mu_{GV}$ = the true mean number of chocolate chips in Great Value cookies.

PLAN: If the conditions are met, we will calculate a two-sample *t* interval for $\mu_{CA} - \mu_{GV}$.

- *Random*: The data come from independent random samples.
  - *10%*: There are more than $10(10) = 100$ Chips Ahoy cookies and more than $10(10) = 100$ Great Value cookies.
- *Normal/Large Sample*: Because there is no obvious skewness or outliers in the graphs below, it is safe to use *t* procedures.

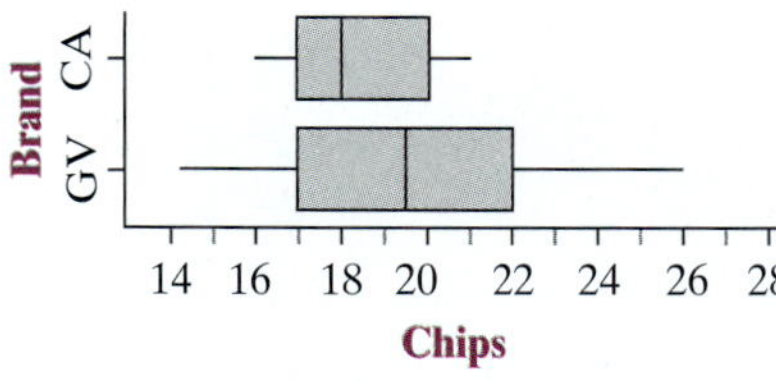


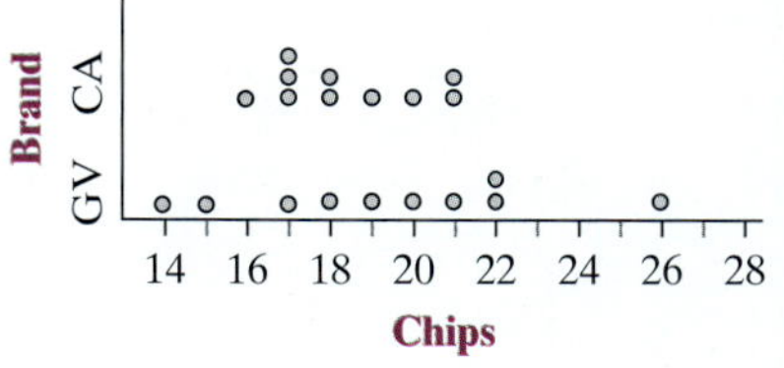


PROBLEM:

(a) Based on the graph and numerical summaries, write a few sentences comparing the sizes of longleaf pine trees in the two halves of the forest.

(b) Construct and interpret a 90% confidence interval for the difference in the mean DBH of longleaf pines in the northern and southern halves of the Wade Tract Preserve.

SOLUTION:

(a) The distribution of DBH measurements in the northern sample is skewed to the right, while the distribution of DBH measurements in the southern sample is skewed to the left. It appears that trees in the southern half of the forest have larger diameters. The mean and median DBH for the southern sample are both much larger than the corresponding measures of center for the northern sample. Furthermore, the boxplots show that more than 75% of the southern trees have diameters that are above the northern sample's median. There is more variability in the diameters of the northern longleaf pines, as we can see from the larger range, *IQR*, and standard deviation for this sample. No outliers are present in either sample.

(b) STATE: Our parameters of interest are $\mu_1$ = the true mean DBH of all trees in the southern half of the forest and $\mu_2$ = the true mean DBH of all trees in the northern half of the forest. We want to estimate the difference $\mu_1 - \mu_2$ at a 90% confidence level.

PLAN: If conditions are met, we'll construct a two-sample *t* interval for $\mu_1 - \mu_2$.

- *Random:* The data came from independent random samples of 30 trees each from the northern and southern halves of the forest.
  - *10%:* Because sampling without replacement was used, there have to be at least $10(30) = 300$ trees in each half of the forest. This is fairly safe to assume.
- *Normal/Large Sample:* The boxplots give us reason to believe that the population distributions of DBH measurements may not be Normal. However, because both sample sizes are at least 30, we are safe using two-sample *t* procedures.

DO: From the Minitab output, $\bar{x}_1 = 34.53$, $s_1 = 14.26$, $n_1 = 30$, $\bar{x}_2 = 23.70$, $s_2 = 17.50$, and $n_2 = 30$. We'll use the conservative df = the smaller of $n_1 - 1$ and $n_2 - 1$, which is 29. For a 90% confidence level the critical value from Table B is $t^* = 1.699$. So a 90% confidence interval for $\mu_1 - \mu_2$ is

| Upper-tail probability *p* | | | |
|---|---|---|---|
| df | .10 | .05 | .025 |
| 28 | 1.313 | 1.701 | 2.048 |
| 29 | 1.311 | 1.699 | 2.045 |
| 30 | 1.310 | 1.697 | 2.042 |
| | 80% | 90% | 95% |
| | Confidence level *C* | | |

$$(\bar{x}_1 - \bar{x}_1) \pm t^*\sqrt{\frac{s_1^2}{n_1} + \frac{s_2^2}{n_2}} = (34.53 - 23.70) \pm 1.699\sqrt{\frac{14.26^2}{30} + \frac{17.50^2}{30}}$$
$$= 10.83 \pm 7.00 = (3.83, 17.83)$$

*Using technology:* Refer to the Technology Corner that follows the example. The calculator's `2-SampTInt` gives (3.9362, 17.724) using df = 55.728.

CONCLUDE: We are 90% confident that the interval from 3.9362 to 17.724 centimeters captures the difference in the actual mean DBH of the southern trees and the actual mean DBH of the northern trees.

**For Practice** *Try Exercise* 37

The 90% confidence interval in the example does not include 0. This gives convincing evidence that the difference in the mean diameter of northern and southern trees in the Wade Tract Preserve isn't 0. However, the confidence interval provides more information than a simple reject or fail to reject $H_0$ conclusion. It gives a set of plausible values for $\mu_1 - \mu_2$. The interval suggests that the mean diameter of the southern trees is between 3.83 and 17.83 cm larger than the mean diameter of the northern trees.

DO: For these data, $\bar{x}_{CA} = 18.4$, $s_{CA} = 1.78$, $\bar{x}_{GV} = 19.4$, $s_{GV} = 3.60$. Using df = 10 − 1, the critical value for 99% confidence is $t^* = 3.250$. Thus, the confidence interval is

$$(18.4 - 19.4) \pm 3.250\sqrt{\frac{1.78^2}{10} + \frac{3.60^2}{10}} =$$

$-1 \pm 4.13 = (-5.13, 3.13)$. *Using technology:* Using df = 13.145, (−4.81, 2.81).

CONCLUDE: We are 99% confident that the interval from −4.81 to 2.81 captures the true difference in the mean number of chocolate chips in Chips Ahoy and Great Value cookies.

**(b)** Because the interval includes 0, there is not convincing evidence that there is a difference in the mean number of chocolate chips in Chips Ahoy and Great Value chocolate chip cookies.

### AP® Exam Common Error

Many students lose credit for not including the word "mean" or "average" in their interpretations of a confidence interval for a difference in means (or the conclusion to a significance test). Remember, the interval is about the difference in means, not the difference in variables.

We chose the parameters in the DBH example so that $\bar{x}_1 - \bar{x}_2$ would be positive. What if we had defined $\mu_1$ as the mean DBH of the northern trees and $\mu_2$ as the mean DBH of the southern trees? The 90% confidence interval for $\mu_1 - \mu_2$ would be

$$(23.70 - 34.53) \pm 1.699\sqrt{\frac{17.50^2}{30} + \frac{14.26^2}{30}} = -10.83 \pm 7.00 = (-17.83, -3.83)$$

This interval suggests that the mean diameter of the northern trees is between 3.83 and 17.83 cm smaller than the mean diameter of the southern trees. Changing the order of subtraction doesn't change the result.

As with other inference procedures, you can use technology to perform the calculations in the "Do" step. Remember that technology comes with potential benefits and risks on the AP® exam.

## 23. TECHNOLOGY CORNER — TWO-SAMPLE *t* INTERVALS ON THE CALCULATOR

TI-Nspire instructions in Appendix B; HP Prime instructions on the book's Web site.

You can use the two-sample *t* interval command on the TI-83/84 or TI-89 to construct a confidence interval for the difference between two means. We'll show you the steps using the summary statistics from the pine trees example.

**TI-83/84**

- Press STAT, then choose TESTS and 2-SampTInt....

**TI-89**

- Press 2nd F2 ([F7]) Ints and choose 2-SampTInt....

- Choose Stats as the input method and enter the summary statistics as shown.

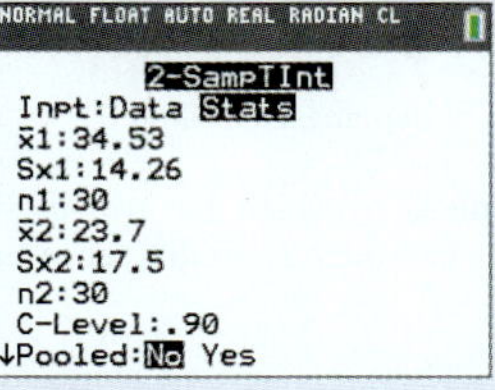


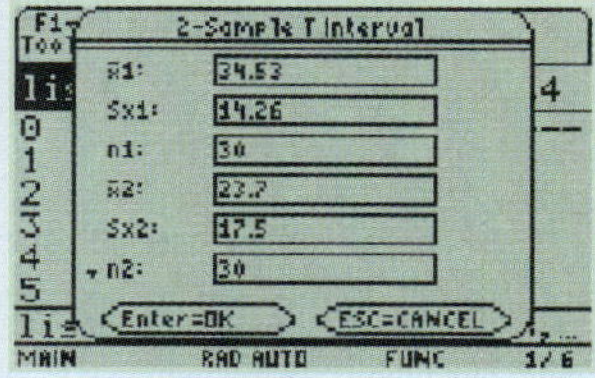


- Enter the confidence level: C-level: .90. For Pooled: choose "No." We'll discuss pooling later.
- Highlight Calculate and press ENTER.

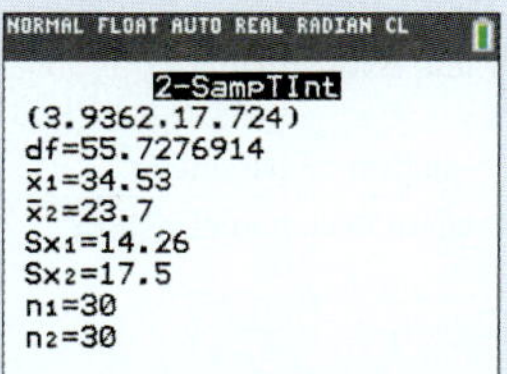


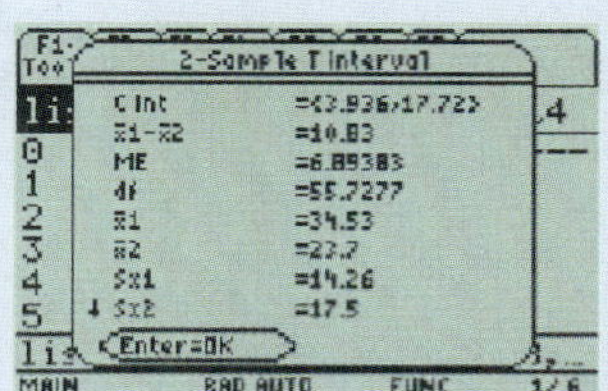


**AP® EXAM TIP** The formula for the two-sample *t* interval for $\mu_1 - \mu_2$ often leads to calculation errors by students. As a result, we recommend using the calculator's 2-SampTInt feature to compute the confidence interval on the AP® exam. Be sure to name the procedure (two-sample *t* interval) and to give the interval (3.9362, 17.724) and df (55.728) as part of the "Do" step.

### Teaching Tip

Remind your students that when they check their answers, it is OK if the endpoints of their interval have the opposite signs as the endpoints for the interval in the back of the book. Either subtraction order is fine, as long as students clearly identify the order they are using.

### Teaching Tip

We strongly recommend that students use technology for two-sample *t* procedures to ease the computations and take advantage of the larger number of degrees of freedom.

### AP® Exam Common Error

Many students use their calculators to construct a two-sample *t* interval and also write out the equation using a $t^*$ critical value calculated with the conservative degrees of freedom. This is dangerous because these methods will not produce the same answer. If you want students to write out the formula with the numbers substituted in, have them leave $t^*$ in the formula instead of using the conservative critical value. Then report the calculator's interval and degrees of freedom.

### TECHNOLOGY TIP

For two-sample *t* procedures, we always recommend saying "No" to pooling. We'll discuss why on page 650, but for now tell your students to "just say No!"

## Teaching Tip

If you use the conservative degrees of freedom, the confidence interval will be wider than necessary. This means that the actual capture rate of intervals using the conservative df will be greater than the stated confidence level.

### ANSWERS TO CYU

*S*: $\mu_1$ = the true mean price of wheat in July and $\mu_2$ = the true mean price of wheat in September. *P*: Two-sample *t* interval for $\mu_1 - \mu_2$. *Random*: Independent random samples. *10%*: $n_1 = 90 < 10\%$ of all wheat producers in July and $n_2 = 45 < 10\%$ of all wheat producers in September. *Normal/Large Sample*: $n_1 = 90 \geq 30$ and $n_2 = 45 \geq 30$. *D*: Using df = 40, (−0.759, −0.561). Using df = 100.45, (−0.756, −0.564). *C*: We are 99% confident that the interval from −0.756 to −0.564 captures the true difference in mean wheat prices in July and September.

## Teaching Tip

See Exercises 49 and 50 for examples of settings where the hypothesized difference is something different than 0.

## Teaching Tip

The Random, 10%, and Normal/Large Sample conditions are exactly the same for significance tests for a difference in means as they were for a confidence interval for a difference in means. Yay!

### AP® Exam Formula Sheet

Although the formula for the two-sample *t* statistic is not given on the formula sheet that is provided on the AP® exam, the general formula for a test statistic and the formula for the standard deviation of $\bar{x}_1 - \bar{x}_2$ are included. Students will need to remember to calculate the standard error using the sample standard deviations ($s_1$ and $s_2$) instead of the population standard deviations ($\sigma_1$ and $\sigma_2$), however.

The calculator's 90% confidence interval for $\mu_1 - \mu_2$ is (3.936, 17.724). This interval is narrower than the one we found by hand earlier: (3.83, 17.83). Why the difference? We used the conservative df = 29, but the calculator used df = 55.73. With more degrees of freedom, the calculator's critical value is smaller than our $t^* = 1.699$, which results in a smaller margin of error and a narrower interval.

### CHECK YOUR UNDERSTANDING

The U.S. Department of Agriculture (USDA) conducted a survey to estimate the average price of wheat in July and in September of the same year. Independent random samples of wheat producers were selected for each of the two months. Here are summary statistics on the reported price of wheat from the selected producers, in dollars per bushel:[23]

| Month | $n$ | $\bar{x}$ | $s_x$ |
|---|---|---|---|
| July | 90 | \$2.95 | \$0.22 |
| September | 45 | \$3.61 | \$0.19 |

Construct and interpret a 99% confidence interval for the difference in the true mean wheat price in July and in September.

## Significance Tests for $\mu_1 - \mu_2$

An observed difference between two sample means can reflect an actual difference in the parameters $\mu_1$ and $\mu_2$, or it may just be due to chance variation in random sampling or random assignment. Significance tests help us decide which explanation makes more sense. The null hypothesis has the general form

$$H_0: \mu_1 - \mu_2 = \text{hypothesized value}$$

We're often interested in situations in which the hypothesized difference is 0. Then the null hypothesis says that there is no difference between the two parameters:

$$H_0: \mu_1 - \mu_2 = 0 \quad \text{or, alternatively,} \quad H_0: \mu_1 = \mu_2$$

The alternative hypothesis says what kind of difference we expect.

If the Random, 10%, and Normal/Large Sample conditions are met, we can proceed with calculations. To do a test, standardize $\bar{x}_1 - \bar{x}_2$ to get a two-sample *t* statistic:

$$\text{test statistic} = \frac{\text{statistic} - \text{parameter}}{\text{standard deviation of statistic}}$$

$$t = \frac{(\bar{x}_1 - \bar{x}_2) - (\mu_1 - \mu_2)}{\sqrt{\frac{s_1^2}{n_1} + \frac{s_2^2}{n_2}}}$$

To find the *P*-value, use the *t* distribution with degrees of freedom given by Option 1 (technology) or Option 2 (df = smaller of $n_1 - 1$ and $n_2 - 1$). Here are the details for the **two-sample *t* test for the difference between two means.**

### TWO-SAMPLE *t* TEST FOR THE DIFFERENCE BETWEEN TWO MEANS

Suppose the conditions are met. To test the hypothesis $H_0: \mu_1 - \mu_2 =$ hypothesized value, compute the two-sample *t* statistic

$$t = \frac{(\bar{x}_1 - \bar{x}_2) - (\mu_1 - \mu_2)}{\sqrt{\frac{s_1^2}{n_1} + \frac{s_2^2}{n_2}}}$$

Find the *P*-value by calculating the probability of getting a *t* statistic this large or larger in the direction specified by the alternative hypothesis $H_a$. Use the *t* distribution with degrees of freedom approximated by technology or the smaller of $n_1 - 1$ and $n_2 - 1$.

$H_a: \mu_1 - \mu_2 >$ hypothesized value $H_a: \mu_1 - \mu_2 <$ hypothesized value $H_a: \mu_1 - \mu_2 \neq$ hypothesized value

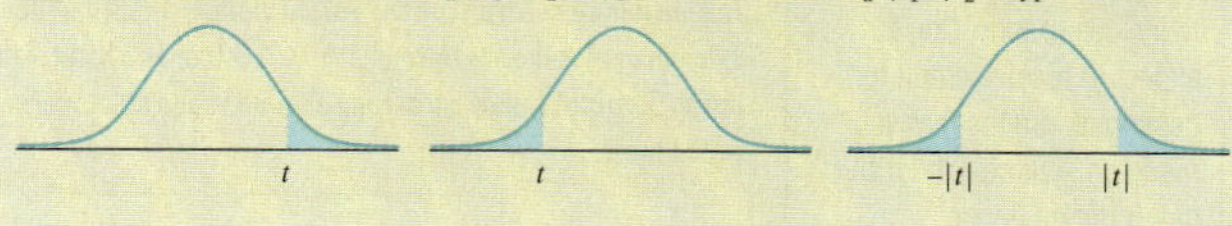


Here's an example that shows how to perform a two-sample *t* test for a randomized experiment.

## EXAMPLE Calcium and Blood Pressure

### *Comparing two means*

Does increasing the amount of calcium in our diet reduce blood pressure? Examination of a large sample of people revealed a relationship between calcium intake and blood pressure. The relationship was strongest for black men. Such observational studies do not establish causation. Researchers therefore designed a randomized comparative experiment.

The subjects were 21 healthy black men who volunteered to take part in the experiment. They were randomly assigned to two groups: 10 of the men received a calcium supplement for 12 weeks, while the control group of 11 men received a placebo pill that looked identical. The experiment was double-blind. The response variable is the decrease in systolic (top number) blood pressure for a subject after 12 weeks, in millimeters of mercury. An increase appears as a negative number.[24] Here are the data:

| Group 1 (calcium): | 7 | −4 | 18 | 17 | −3 | −5 | 1 | 10 | 11 | −2 | |
|---|---|---|---|---|---|---|---|---|---|---|---|
| Group 2 (placebo): | −1 | 12 | −1 | −3 | 3 | −5 | 5 | 2 | −11 | −1 | −3 |

PROBLEM:

(a) **Do the data provide convincing evidence that a calcium supplement reduces blood pressure more than a placebo? Carry out an appropriate test to support your answer.**

(b) **Interpret the *P*-value you got in part (a) in the context of this experiment.**

### Teaching Tip

When planning a study that will result in a two-sample *t* test for a difference in means, it is helpful to keep the two sample sizes about the same. This makes the *P*-value calculations more accurate and decreases the standard error of the difference in means, making the power of the test higher (assuming the standard deviations for both populations are roughly the same). For example, suppose you have 60 subjects for an experiment and the standard deviation of the response variable in both groups is 10. Trying different values for $n_1$ in the formula $\sqrt{\frac{10^2}{n_1} + \frac{10^2}{60 - n_1}}$ shows that the standard error will be smallest when the sample sizes are the same.

### Teaching Tip

Here are some things to do with your students when reading the example on this page:

- Based only on the boxplots shown in the solution on the next page, have your students speculate whether the difference in means will be statistically significant. Because the distributions overlap quite a bit and the group sizes are relatively small, it is unlikely that the means are significantly different. In other words, it is plausible that the difference in these two distributions could be due only to chance variation in the random assignment.
- Discuss why the Random condition is different for experiments and observational studies.
- Ask your students if a cause-and-effect conclusion can be made if the difference in means turns out to be statistically significant.
- Remind your students not to use the past tense when defining the parameters in an experiment. In this case, the mean decrease for men who *took* calcium is $\bar{x}_1$ not $\mu_1$.

SOLUTION:

(a) STATE: We want to perform a test of

$$H_0: \mu_1 - \mu_2 = 0 \quad \text{or, equivalently,} \quad H_0: \mu_1 = \mu_2$$
$$H_a: \mu_1 - \mu_2 > 0 \qquad\qquad\qquad\qquad H_a: \mu_1 > \mu_2$$

where $\mu_1$ is the true mean decrease in systolic blood pressure for healthy black men like the ones in this study who take a calcium supplement and $\mu_2$ is the true mean decrease in systolic blood pressure for healthy black men like the ones in this study who take a placebo. No significance level was specified, so we'll use $\alpha = 0.05$.

PLAN: If conditions are met, we will carry out a two sample *t* test for $\mu_1 - \mu_2$.

- *Random:* The 21 subjects were randomly assigned to the two treatments.
    - 10%: Don't need to check because there was no sampling.
- *Normal/Large Sample:* With such small sample sizes, we need to graph the data to see if it's reasonable to believe that the actual distributions of differences in blood pressure when taking calcium or placebo are Normal. Figure 10.9 shows hand sketches of calculator boxplots for these data. The graphs show no strong skewness and no outliers. So we are safe using two-sample *t* procedures.

**AP® EXAM TIP** When checking the Normal condition on an AP® exam question involving inference about means, be sure to include graphs. Don't expect to receive credit for describing graphs that you made on your calculator but didn't put on paper.

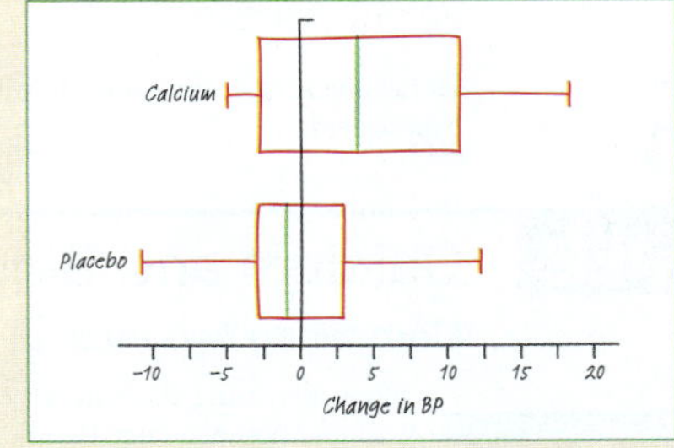


**FIGURE 10.9** Sketches of boxplots of the changes in blood pressure for the two groups of subjects in the calcium and blood pressure experiment.

DO: From the data, we calculated summary statistics:

| Group | Treatment | $n$ | $\bar{x}$ | $s_x$ |
|---|---|---|---|---|
| 1 | Calcium | 10 | 5.000 | 8.743 |
| 2 | Placebo | 11 | −0.273 | 5.901 |

- *Test statistic*

$$t = \frac{(\bar{x}_1 - \bar{x}_2) - (\mu_1 - \mu_2)}{\sqrt{\frac{s_1^2}{n_1} + \frac{s_2^2}{n_2}}} = \frac{[5.000 - (-0.273)] - 0}{\sqrt{\frac{8.743^2}{10} + \frac{5.901^2}{11}}} = \frac{5.273}{3.2878} = 1.604$$

- *P-value* By the conservative method, the smaller of $n_1 - 1$ and $n_2 - 1$ gives df = 9. Because $H_a$ counts only positive values of *t* as evidence against $H_0$, the *P*-value is the area to the right of $t = 1.604$ under the *t* distribution curve with df = 9. Figure 10.10 illustrates this *P*-value. Table B shows that the *P*-value lies between 0.05 and 0.10.

| | Upper-tail probability *p* | | |
|---|---|---|---|
| df | .10 | .05 | .025 |
| 8 | 1.397 | 1.860 | 2.306 |
| 9 | 1.383 | 1.833 | 2.262 |
| 10 | 1.372 | 1.812 | 2.228 |

*Using technology:* Refer to the Technology Corner that follows the example. The calculator's `2-SampTTest` gives $t = 1.60$ and *P*-value = 0.0644 using df = 15.59.

## ALTERNATE EXAMPLE

### *Leaking helium*

After buying many helium balloons only to see them deflate within a couple of days, Erin and Jenna decided to test whether helium-filled balloons deflate more quickly than air-filled balloons. To find out, they bought 60 balloons and randomly divided them into two piles of 30, filling the balloons in the first pile with helium and the balloons in the second pile with air. Then they measured the circumference of each balloon immediately after it was filled and again 3 days later. The average decrease in circumference of the helium-filled balloons was 26.5 cm with a standard deviation of 1.92 cm. The average decrease of the air-filled balloons was 2.1 cm with a standard deviation of 2.79 cm.

PROBLEM:

**(a)** Do these data provide convincing evidence that helium-filled balloons deflate more quickly than air-filled balloons?
**(b)** Interpret the *P*-value you got in part (a) in the context of this study.

SOLUTION:

**(a)** STATE: We want to perform a test of $H_0: \mu_H - \mu_A = 0$ versus $H_a: \mu_H - \mu_A > 0$ at the 5% level of significance, where $\mu_H$ = the mean decrease in circumference of helium-filled balloons after 3 days and $\mu_A$ = the mean decrease in circumference of air-filled balloons after 3 days.

PLAN: If the conditions are met, we will conduct a two-sample *t* test for $\mu_H - \mu_A$.

- *Random*: The data came from two groups in a randomized experiment.
    - *10%*: Not necessary because there was no sampling without replacement.
- *Normal/Large Sample*: $n_H = 30 \geq 30$ and $n_A = 30 \geq 30$.

DO: Test statistic: $t = \dfrac{(26.5 - 2.1) - 0}{\sqrt{\dfrac{1.92^2}{30} + \dfrac{2.79^2}{30}}}$ = 39.46. *P*-value: With df = 30 − 1 = 29, *P*-value ≈ 0. *Using technology:* With df = 51.4, *P*-value ≈ 0.

CONCLUDE: Because the *P*-value of approximately 0 is less than $\alpha = 0.05$, we reject $H_0$. There is convincing evidence that helium-filled balloons deflate more quickly than air-filled balloons. **(b)** Assuming that the mean decrease in circumference is the same for helium-filled and air-filled balloons, there is approximately 0 probability of getting a difference of 24.4 cm or more by chance alone.

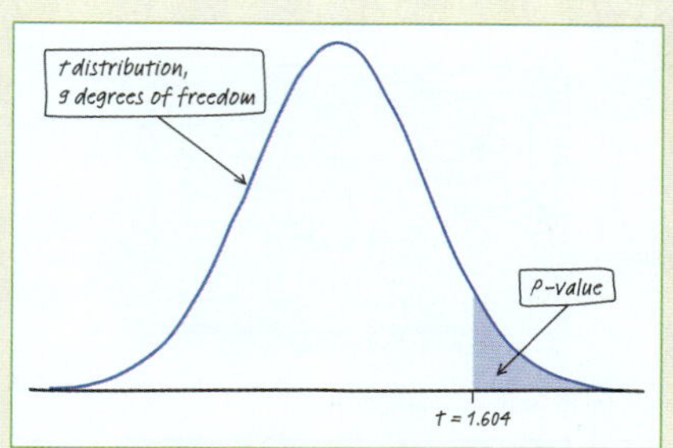


**FIGURE 10.10** The *P*-value for the one-sided test using the conservative method, which leads to the *t* distribution with 9 degrees of freedom.

CONCLUDE: Because the *P*-value is greater than $\alpha = 0.05$, we fail to reject $H_0$. The experiment does not provide convincing evidence that the true mean decrease in systolic blood pressure is higher for men like these who take calcium than for men like these who take a placebo.

(b) Assuming $H_0: \mu_1 - \mu_2 = 0$ is true, there is a 0.0644 probability of getting a difference in mean blood pressure reduction for the two groups (calcium − placebo) of 5.273 or greater just by the chance involved in the random assignment.

**For Practice** *Try Exercise* **41**

When a significance test leads to a fail to reject $H_0$ decision, as in the previous example, be sure to interpret the results as "We don't have convincing evidence to conclude $H_a$." Saying anything that sounds like you believe $H_0$ is (or might be) true is incorrect.

**THINK ABOUT IT**

**Why didn't researchers find a significant difference in the calcium and blood pressure experiment?** The difference in mean systolic blood pressures for the two groups was 5.273 millimeters of mercury. This seems like a fairly large difference. With the small group sizes, however, this difference wasn't large enough to reject $H_0: \mu_1 - \mu_2 = 0$ in favor of the one-sided alternative. We suspect that larger groups might show a similar difference in mean blood pressure reduction, which would indicate that calcium has a significant effect. If so, then the researchers in this experiment made a Type II error—failing to reject a false $H_0$. In fact, later analysis of data from an experiment with more subjects resulted in a *P*-value of 0.008. *Sample size strongly affects the power of a test.* It is easier to detect an actual difference in the effectiveness of two treatments if both are applied to large numbers of subjects.

## 24. TECHNOLOGY CORNER — TWO-SAMPLE *t* TESTS ON THE CALCULATOR

TI-Nspire instructions in Appendix B; HP Prime instructions on the book's Web site.

Technology gives smaller *P*-values for two-sample *t* tests than the conservative method. That's because calculators and software use the more complicated formula on page 640 to obtain a larger number of degrees of freedom.

- Enter the Group 1 (calcium) data in L1/list1 and the Group 2 (placebo) data in L2/list2.
- To perform the significance test, go to STAT/TESTS (Tests menu in the Statistics/List Editor on the TI-89) and choose `2-SampTTest`.
- In the 2-SampTTest screen, specify "Data" and adjust your other settings as shown.

## Teaching Tip

Here are two nice activities that you can do with your students. Find them by clicking on the link in the TE-book, opening the TRCD, or logging onto the book's companion Web site www.whfreeman.com/tps5e (teacher log-in required).

### TRM Paper Airplanes Experiment

This Activity has students test two different models of paper airplane to see if one model flies significantly farther than the other. If you are planning to teach Chapter 13 (ANOVA) after the AP® exam, consider modifying the Activity to include three or more airplane designs. Although students will only be able to analyze the difference between two means in Chapter 10, this activity can be a nice preview of what's to come. For now, if you choose to include more than two designs, have students do a separate test for each pair of airplane designs. Find templates for different models at www.funpaperairplanes.com.

### TRM Exercising to Lose Weight

When the results of studies are reported in academic journals, the authors often use specialized notation to describe their results. This problem provides students with some details about a weight loss experiment and asks them to discuss aspects of the experimental design and confirm the results of the study.

## Teaching Tip

We strongly recommend that students use technology for two-sample *t* procedures to ease the computations and take advantage of the larger number of degrees of freedom.

TI-83/84 TI-89

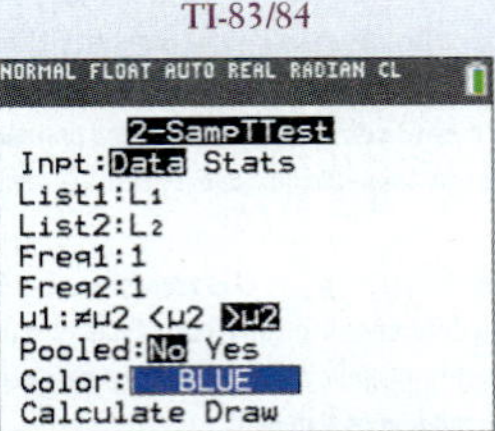


- Highlight "Calculate" and press ENTER. (The Pooled option will be discussed shortly.)

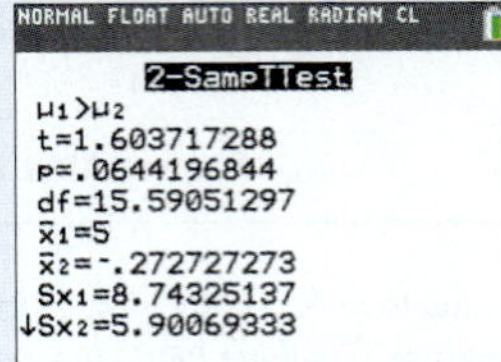


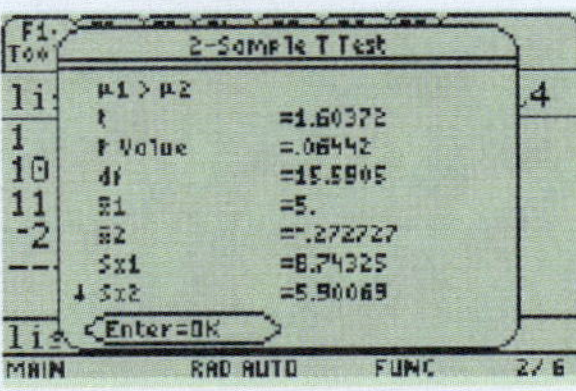


If you select "Draw" instead of "Calculate," the appropriate $t$ distribution will be displayed, showing the test statistic and the shaded area corresponding to the $P$-value.

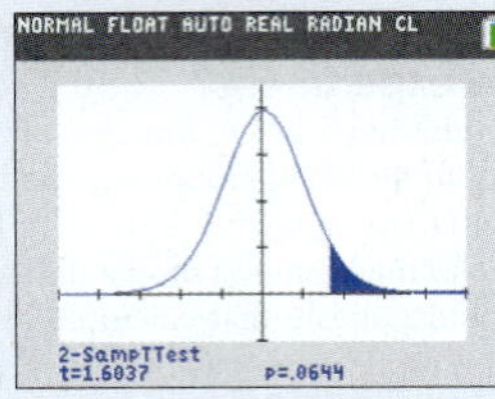


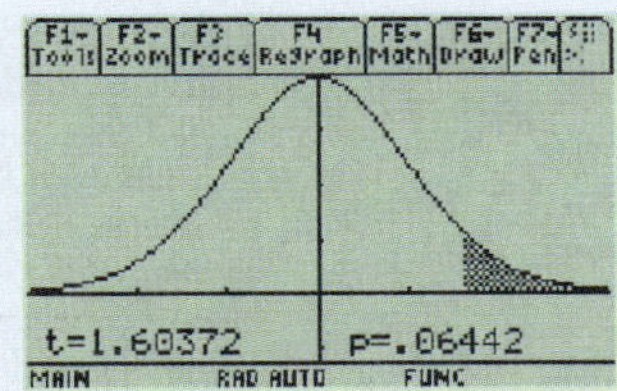


**AP® EXAM TIP** The formula for the two-sample $t$ statistic for a test about $\mu_1 - \mu_2$ often leads to calculation errors by students. As a result, we recommend using the calculator's `2-SampTTest` feature to perform calculations on the AP® exam. Be sure to name the procedure (two-sample $t$ test) and to report the test statistic ($t = 1.60$), $P$-value (0.0644), and df (15.59) as part of the "Do" step.

**Inference for Experiments** Confidence intervals and tests for $\mu_1 - \mu_2$ are based on the sampling distribution of $\bar{x}_1 - \bar{x}_2$. But in experiments, we aren't sampling at random from any larger populations. We can think about what would happen if the random assignment were repeated many times under the assumption that $H_0\colon \mu_1 - \mu_2 = 0$ is true. That is, we assume that the specific treatment received doesn't affect an individual subject's response.

Let's see what would happen just by chance if we randomly reassign the 21 subjects in the calcium and blood pressure experiment to the two groups many times, assuming the drug received *doesn't affect* each individual's change in systolic blood pressure. We used Fathom software to redo the random assignment 1000 times. The approximate *randomization distribution* of $\bar{x}_1 - \bar{x}_2$ is shown in Figure 10.11. It has an approximately Normal shape with mean 0 (no difference) and standard deviation 3.42.

## TECHNOLOGY TIP

For two-sample $t$ procedures, we always recommend saying "No" to pooling. We'll discuss why on page 650, but for now tell your students to "just say No!"

## AP® Exam Common Error

Make sure students report the degrees of freedom that they use to calculate the $P$-value. This will help the grader understand which method, conservative or technology, the student is using.

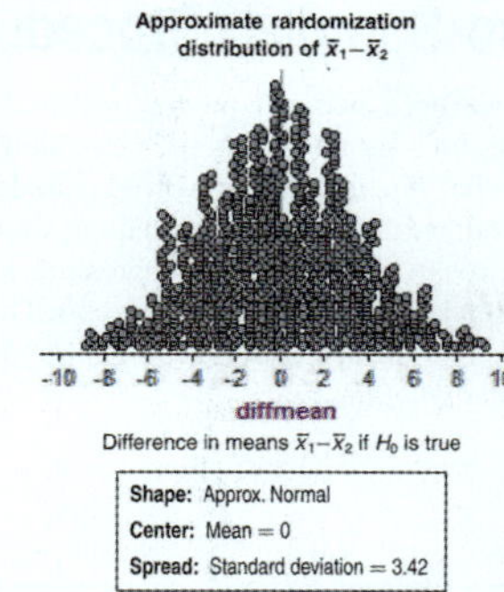


**FIGURE 10.11** Fathom simulation showing the approximate randomization distribution of $\bar{x}_1 - \bar{x}_2$ from 1000 random reassignments of subjects to treatment groups in the calcium and blood pressure experiment.

In the actual experiment, the difference in the mean change in blood pressure in the calcium and placebo groups was $5.000 - (-0.273) = 5.273$. How likely is it that a difference this large or larger would happen just by chance when $H_0$ is true? Figure 10.11 provides a rough answer: 61 of the 1000 random reassignments yielded a difference in means greater than or equal to 5.273. That is, our estimate of the *P*-value is 0.061. This is quite close to the 0.0644 *P*-value that we obtained in the Technology Corner.

If Figure 10.11 displayed the results of *all* possible random reassignments of subjects to treatment groups, it would be the actual randomization distribution of $\bar{x}_1 - \bar{x}_2$. The *P*-value obtained from this distribution would be *exactly correct*. Using the two-sample *t* test to calculate the *P*-value gives only approximately correct results.

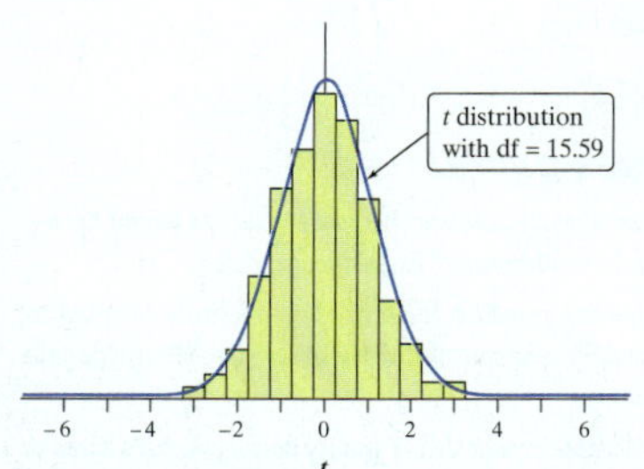

**FIGURE 10.12** The distribution of the two-sample *t* test statistic for the 1000 random reassignments in Figure 10.11.

Figure 10.12 shows the value of the two-sample *t* test statistic for each of the 1000 re-randomizations, calculated using our familiar formula

$$t = \frac{(\bar{x}_1 - \bar{x}_2) - (\mu_1 - \mu_2)}{\sqrt{\frac{s_1^2}{n_1} + \frac{s_2^2}{n_2}}}$$

The density curve for the *t* distribution with df = 15.59 is shown in blue. We can see that the test statistic follows the *t* distribution quite closely in this case.

Whenever the conditions are met, the randomization distribution of $\bar{x}_1 - \bar{x}_2$ looks much like its sampling distribution. We are therefore safe using two-sample *t* procedures for comparing two means in a randomized experiment.

## CHECK YOUR UNDERSTANDING

How quickly do synthetic fabrics such as polyester decay in landfills? A researcher buried polyester strips in the soil for different lengths of time, then dug up the strips and measured the force required to break them. Breaking strength is easy to measure and is a good indicator of decay. Lower strength means the fabric has decayed.

For one part of the study, the researcher buried 10 strips of polyester fabric in well-drained soil in the summer. The strips were randomly assigned to two groups: 5 of them were buried for 2 weeks and the other 5 were buried for 16 weeks. Here are the breaking strengths in pounds:[25]

| | | | | | |
|---|---|---|---|---|---|
| Group 1 (2 weeks): | 118 | 126 | 126 | 120 | 129 |
| Group 2 (16 weeks): | 124 | 98 | 110 | 140 | 110 |

Do the data give convincing evidence that polyester decays more in 16 weeks than in 2 weeks?

### Teaching Tip

One of the nice features of the re-randomization approach shown here and in the Does Polyester Decay? Activity is that there are fewer conditions that need to be met. Specifically, there is no Normal/Large Sample condition, because we never need to assume that the "populations" follow a Normal distribution as we do in a two-sample *t* test. If the Normal/Large Sample condition happens to be met, then both approaches (re-randomization, two-sample *t*) will give very similar results.

### Teaching Tip

To get a better estimate of the *P*-value, have Fathom do many more than 1000 re-randomizations. For example, after 10,000 re-randomizations, the *P*-value is approximately 0.0557, and after 100,000 re-randomizations, the *P*-value is approximately 0.0572.

### ANSWERS TO CYU

*S*: $H_0: \mu_1 - \mu_2 = 0$ vs. $H_a: \mu_1 - \mu_2 > 0$, where $\mu_1$ is the true mean breaking strength for polyester fabric buried for 2 weeks and $\mu_2$ is the true mean breaking strength for polyester fabric buried for 16 weeks. *P*: Two-sample *t* test. *Random*: Two groups in a randomized experiment. *Normal/Large Sample*: The dotplots below show no strong skewness or outliers in either group.

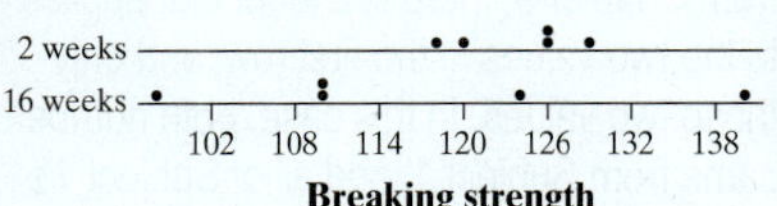


*D*: $\bar{x}_1 = 123.8$, $s_1 = 4.60$, $\bar{x}_2 = 116.4$, $s_2 = 16.09$. $t = 0.989$. Using df = 4, the *P*-value is between 0.15 and 0.20. Using df = 4.65, *P*-value = 0.1857. *C*: Because the *P*-value of $0.1857 > \alpha = 0.05$, we fail to reject $H_0$. We do not have convincing evidence that the true mean breaking strength of polyester fabric that is buried for 2 weeks > the true mean breaking strength for polyester fabric that is buried for 16 weeks.

## Using Two-Sample $t$ Procedures Wisely

In Chapter 9, we used paired $t$ procedures to compare the mean change in depression scores for a group of caffeine-dependent individuals when taking caffeine and a placebo. The inference involved paired data because the same 11 subjects received both treatments. In this chapter, we used two-sample $t$ procedures to compare the mean change in blood pressure for a group of healthy black men when taking calcium and a placebo. This time, the inference involved two distinct groups of subjects. *The proper method of analysis depends on the design of the study.*

### EXAMPLE Comparing Tires and Comparing Workers

*Independent samples versus paired data*

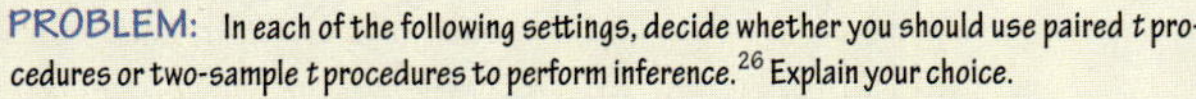

PROBLEM: In each of the following settings, decide whether you should use paired $t$ procedures or two-sample $t$ procedures to perform inference.[26] Explain your choice.

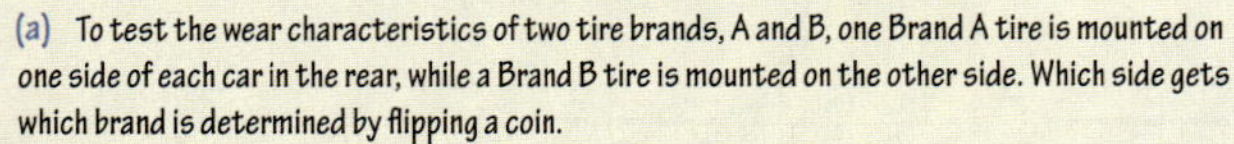

(a) To test the wear characteristics of two tire brands, A and B, one Brand A tire is mounted on one side of each car in the rear, while a Brand B tire is mounted on the other side. Which side gets which brand is determined by flipping a coin.

(b) Can listening to music while working increase productivity? Twenty factory workers agree to take part in a study to investigate this question. Researchers randomly assign 10 workers to do a repetitive task while listening to music and the other 10 workers to do the task in silence.

SOLUTION:

(a) Paired $t$ procedures. This is a matched pairs experiment, with the two treatments (Brand A and Brand B) being randomly assigned to the rear pair of wheels on each car.

(b) Two-sample $t$ procedures. The data are being produced using two distinct groups of workers in a randomized experiment.

**For Practice** *Try Exercise* 55

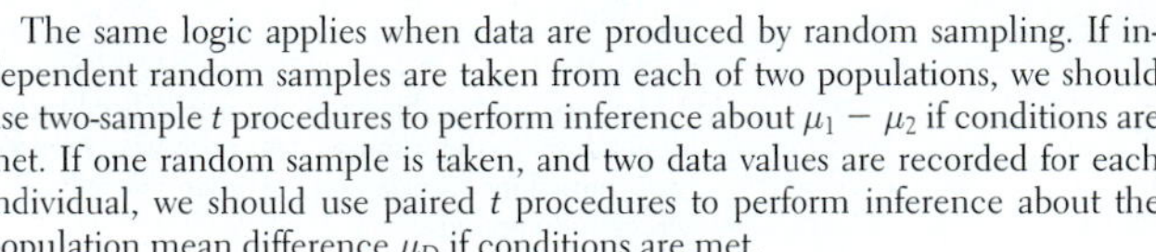

The same logic applies when data are produced by random sampling. If independent random samples are taken from each of two populations, we should use two-sample $t$ procedures to perform inference about $\mu_1 - \mu_2$ if conditions are met. If one random sample is taken, and two data values are recorded for each individual, we should use paired $t$ procedures to perform inference about the population mean difference $\mu_D$ if conditions are met.

**The Pooled Two-Sample $t$ Procedures (Don't Use Them!)** Most software offers a choice of two-sample $t$ statistics. One is often labeled "unequal" variances; the other, "equal" variances. The "unequal" variance procedure uses our two-sample $t$ statistic. *This test is valid whether or not the population variances are equal.*

### AP® Exam Common Error

Many students have trouble deciding when it is appropriate to use a two-sample $t$ test and when it is appropriate to use a paired $t$ test. The key distinction is how the data were produced. In an experiment, if groups were formed using a completely randomized design, then a two-sample $t$ test is the right choice. However, if subjects were paired and then split at random into the two treatment groups, or if each subject received both treatments, then a paired $t$ test is appropriate. Paired data can also arise when the data are produced by random sampling. For example, if wells are selected at random and the quality of water is measured at the top and the bottom of each well, we should treat these measurements as paired data.

### Teaching Tip

Here is another way to tell the difference between settings that call for a paired $t$ test and settings that call for a two-sample $t$ test. List the data in two columns and ask: "Is there a label that would apply to the two values in each row, and only the two values in that row?"

For example, in the polyester example, there is no label that would apply to the two values in the first row and only those two values:

| Label? | 2 weeks | 16 weeks |
|---|---|---|
| ? | 118 | 124 |
| ? | 126 | 98 |
| ⋮ | ⋮ | ⋮ |

However, in the caffeine deprivation study from Chapter 9, there is a label that applies to the two values in the first row, and only those two values. In this case, both numbers came from Subject 1, and all of Subject 1's numbers appear in the first row:

| Label? | Depression (caffeine) | Depression (placebo) |
|---|---|---|
| Subject 1 | 5 | 16 |
| Subject 2 | 5 | 23 |
| ⋮ | ⋮ | ⋮ |

### ALTERNATE EXAMPLE

*Testing with distractions*

Suppose you are designing an experiment to determine if students perform better on tests when there are no distractions, such as a teacher talking on the phone. You have access to two classrooms and 30 volunteers who are willing to participate in your experiment.

PROBLEM:

**(a)** Design an experiment so that a two-sample $t$ test would be the appropriate inference procedure.

**(b)** Design an experiment so that a paired $t$ test would be the appropriate inference procedure.

**(c)** Which experimental design is better? Explain.

SOLUTION:

**(a)** On 15 index cards write "A" and on 15 index cards write "B." Shuffle the cards and hand them out at random to the 30 volunteers. All 30 subjects will take the same reading comprehension test. Subjects who receive A cards will go to a classroom with no distractions, and subjects who receive B cards will go to a classroom where the proctor is talking on the phone during the test. At the end of the experiment, compare the mean score for subjects in Room A with the mean score for subjects in Room B.

**(b)** Using the same procedure as in part (a), divide the subjects into two rooms and give them the same reading comprehension test. One room will be distraction-free, and in the other room a proctor will be talking on the phone. After a short break, give all 30 subjects a similar reading comprehension test but have the teacher talking on the phone in the opposite room. At the end of the experiment, calculate the difference in the two reading comprehension scores for each subject and compare the mean difference with 0.

**(c)** The experimental design in part (b) is better because it accounts for an important source of variability—the reading comprehension skills of the individual subjects.

The other choice is a special version of the two-sample $t$ statistic that assumes that the two populations have the same variance. This procedure combines (the statistical term is *pools*) the two sample variances to estimate the common population variance. The resulting statistic is called the *pooled two-sample t statistic.*

The pooled $t$ statistic has exactly the $t$ distribution with $n_1 + n_2 - 2$ degrees of freedom *if* the two population variances really are equal and the population distributions are exactly Normal. This method offers more degrees of freedom than Option 1 (technology), which leads to narrower confidence intervals and smaller $P$-values. The pooled $t$ procedures were in common use before software made it easy to use Option 1 for our two-sample $t$ statistic.

In the real world, distributions are not exactly Normal, and population variances are not exactly equal. In practice, the Option 1 two-sample $t$ procedures are almost always more accurate than the pooled procedures. Our advice: *Never use the pooled t procedures if you have software that will carry out Option 1.* caution

Remember, we always use the pooled sample proportion $\hat{p}_C$ when performing a significance test for comparing two proportions. But we don't recommend pooling when comparing two means.

## Fast-Food Frenzy!

Let's return to the chapter-opening Case Study (page 609) about drive-thru service at fast-food restaurants. Here, once again, are some results from the 2012 QSR study.

- For restaurants with order-confirmation boards, 1169 of 1327 visits (88.1%) resulted in accurate orders. For restaurants with no order-confirmation board, 655 of 726 visits (90.2%) resulted in accurate orders.
- McDonald's average service time for 362 drive-thru visits was 188.83 seconds with a standard deviation of 17.38 seconds. Burger King's service time for 318 drive-thru visits had a mean of 201.33 seconds and a standard deviation of 18.85 seconds.

You are now ready to use what you have learned about comparing population parameters to perform inference about accuracy and average service time in the drive-thru lane.

1. Is there a significant difference in order accuracy between restaurants with and without order-confirmation boards? Carry out an appropriate test at the $\alpha = 0.05$ level to help answer this question.

A 95% confidence interval for the difference in the population proportions of accurate orders at restaurants with and without order-confirmation boards is (−0.049, 0.00649).

2. Interpret the meaning of "95% confident" in the context of this study.

### Teaching Tip

If your students really want to know how pooling works, here is the explanation.

Suppose we have two Normally distributed populations with the same variance $\sigma^2$. If we take independent random samples from the two populations, both sample variances $s_1^2$ and $s_2^2$ estimate $\sigma^2$. The best way to combine (pool) these two estimates is to average them with weights equal to their degrees of freedom. This gives more weight to the information from the larger sample, which is reasonable. The resulting estimator of $\sigma^2$ is $s_p^2 = \dfrac{(n_1-1)s_1^2 + (n_2-1)s_2^2}{n_1+n_2-2}$.

This is called the pooled estimator of $\sigma^2$ because it combines the information in both samples.

To perform inference about $\mu_1 - \mu_2$ in this setting, we start by simplifying our formula for the standard deviation of the statistic:

$$\sigma_{\bar{x}_1-\bar{x}_2} = \sqrt{\frac{\sigma_1^2}{n_1}+\frac{\sigma_2^2}{n_2}} = \sqrt{\frac{\sigma^2}{n_1}+\frac{\sigma^2}{n_2}} = \sqrt{\sigma^2\left(\frac{1}{n_1}+\frac{1}{n_2}\right)} = \sigma\sqrt{\frac{1}{n_1}+\frac{1}{n_2}}.$$

We then substitute the pooled estimate $s_p$ for $\sigma$. The resulting pooled two-sample $t$ statistic is

$$t = \frac{(\bar{x}_1-\bar{x}_2)-(\mu_1-\mu_2)}{s_p\sqrt{\dfrac{1}{n_1}+\dfrac{1}{n_2}}}$$

This statistic has a $t$ distribution with $(n_1-1)+(n_2-1) = n_1+n_2-2$ degrees of freedom.

### AP® Exam Formula Sheet

The formula for the pooled estimator $s_p$ is included on the formula sheet given to students on the AP® exam. We recommend that you point out this formula to your students and have them cross it out on their copy of the formula sheet. If you ignore it, students may mistakenly think that it is a formula they need to understand.

### Teaching Tip

After emphasizing that we do not recommend pooling for two-sample $t$ procedures, many students will ask why we pool for a two-sample $z$ test for a difference in proportions. The difference is in the standard errors.

In a two-sample $z$ test for a difference in proportions, we assume that the two population proportions are the same when doing the calculations. Because the formula for the standard error includes these proportions, we are required to assume that they are the same when calculating the standard error. We pool the sample proportions to get a better estimate of the true proportion for this calculation.

In a two-sample $t$ test for a difference in means, we assume that the two population means are the same when doing calculations. However, the formula for the standard error does not include these means, so assuming that the null hypothesis is true has no effect on the standard error calculations.

3. Explain how the confidence interval is consistent with your conclusion from Question 1.

Now turn your attention to the speed-of-service data.

4. Construct and interpret a 99% confidence interval for the difference in the mean service times at McDonald's and Burger King drive-thrus.

## Section 10.2 Summary

- Choose independent SRSs of size $n_1$ from Population 1 and size $n_2$ from Population 2. The sampling distribution of $\bar{x}_1 - \bar{x}_2$ has the following properties:
  - **Shape:** Normal if both population distributions are Normal; approximately Normal otherwise if both samples are large enough ($n_1 \geq 30$ and $n_2 \geq 30$) by the central limit theorem.
  - **Center:** Its mean is $\mu_1 - \mu_2$.
  - **Spread:** As long as each sample is no more than 10% of its population, its standard deviation is $\sqrt{\frac{\sigma_1^2}{n_1} + \frac{\sigma_2^2}{n_2}}$.
- Confidence intervals and tests for the difference between the means of two populations or the mean responses to two treatments $\mu_1$ and $\mu_2$ are based on the difference $\bar{x}_1 - \bar{x}_2$ between the sample means.
- Because we almost never know the population standard deviations in practice, we use the **two-sample *t* statistic**

$$t = \frac{(\bar{x}_1 - \bar{x}_2) - (\mu_1 - \mu_2)}{\sqrt{\frac{s_1^2}{n_1} + \frac{s_2^2}{n_2}}}$$

This statistic has approximately a *t* distribution. There are two options for using a *t* distribution to approximate the distribution of the two-sample *t* statistic:
  - **Option 1 (Technology)** Use the *t* distribution with degrees of freedom calculated from the data by a somewhat messy formula. The degrees of freedom probably won't be a whole number.
  - **Option 2 (Conservative)** Use the *t* distribution with degrees of freedom equal to the *smaller* of $n_1 - 1$ and $n_2 - 1$. This method gives wider confidence intervals and larger *P*-values than Option 1.
- Before estimating or testing a claim about $\mu_1 - \mu_2$, check that these conditions are met:
  - **Random:** The data are produced by independent random samples of size $n_1$ from Population 1 and of size $n_2$ from Population 2 or by two groups of size $n_1$ and $n_2$ in a randomized experiment.

### Answers to Case Closed

1. STATE: We will test $H_0: p_1 - p_2 = 0$ versus $H_a: p_1 - p_2 \neq 0$ at the $\alpha = 0.05$ significance level, where $p_1$ = the true proportion of accurate orders at restaurants with order-confirmation boards and $p_2$ = the true proportion of accurate orders at restaurants without order-confirmation boards. PLAN: We should use a two-sample *z* test for $p_1 - p_2$ if the conditions are satisfied.
- *Random*: The data came from two independent random samples.
  - *10%*: There were more than 10(1327) = 13,270 visits to restaurants with order-confirmation boards and more than 10(726) = 7260 visits to restaurants without order-confirmation boards.
- *Large Counts*: $n_1\hat{p}_1 = 1169$, $n_1(1 - \hat{p}_1) = 158$, $n_2\hat{p}_2 = 655$, and $n_2(1 - \hat{p}_2) = 71$ are all at least 10.

DO: $\hat{p}_C = \frac{1169 + 655}{1327 + 726} = 0.888$, $z = \frac{(0.881 - 0.902) - 0}{\sqrt{\frac{0.888(1 - 0.888)}{1327} + \frac{0.888(1 - 0.888)}{726}}} = -1.44$, *P*-value = 0.1498. *Using technology*: $z = -1.46$, *P*-value = 0.1433. CONCLUDE: Because the *P*-value of 0.1422 is greater than $\alpha = 0.05$, we fail to reject $H_0$. We do not have convincing evidence that the true proportion of accurate orders is different at restaurants with order-confirmation boards and restaurants without order-confirmation boards.

2. If we were to take many samples of size 1327 from restaurants with order-confirmation boards and many samples of size 726 from restaurants without order-confirmation boards, and construct 95% confidence intervals for the difference in the proportion of accurate orders, about 95% of the confidence intervals will capture the true difference in the proportion of accurate orders.

3. Because the interval includes 0 as a plausible value for the true difference in proportions, we do not have convincing evidence that there is a difference in the true proportion of accurate orders at restaurants with order-confirmation boards and restaurants without order-confirmation boards.

4. STATE: We want to estimate $\mu_M - \mu_B$ at the 99% confidence level, where $\mu_M$ = the true mean service time at McDonald's and $\mu_B$ = the true mean service time at Burger King. PLAN: If the conditions are met, we will calculate a two-sample *t* interval for $\mu_M - \mu_B$.
- *Random*: The data come from independent random samples.
  - *10%*: There are more than 10(362) = 3620 visits to the drive-thru at McDonald's and more than 10(318) = 3180 visits to the drive-thru at Burger King.
- *Normal/Large Sample*: $n_M = 362 \geq 30$ and $n_B = 318 \geq 30$.

DO: Using df = 100,

$$(188.83 - 201.33) \pm 2.626\sqrt{\frac{17.38^2}{362} + \frac{18.85^2}{318}}$$

= (−16.17, −8.83). *With technology:* Using df = 649.3, (−16.11, −8.89).

CONCLUDE: We are 99% confident that the interval from −16.11 seconds to −8.89 seconds captures the true difference in the mean response time at McDonald's and Burger King.

### TRM Section 10.2 Quizzes

There are three different quizzes available for this section in the Teacher's Resource Materials. Versions A and B are roughly parallel. Find them by clicking on the link in the TE-book, opening the TRCD, or logging onto the book's companion Web site www.whfreeman.com/tps5e (teacher log-in required).

- ○ **10%:** When sampling without replacement, check that the two populations are at least 10 times as large as the corresponding samples.
- **Normal/Large Sample:** Both population distributions (or the true distributions of responses to the two treatments) are Normal or both sample sizes are large ($n_1 \geq 30$ and $n_2 \geq 30$). If either population (treatment) distribution has unknown shape and the corresponding sample size is less than 30, use a graph of the sample data to assess the Normality of the population (treatment) distribution. Do not use two-sample $t$ procedures if the graph shows strong skewness or outliers.

- An approximate C% confidence interval for $\mu_1 - \mu_2$ is

$$(\bar{x}_1 - \bar{x}_2) \pm t^* \sqrt{\frac{s_1^2}{n_1} + \frac{s_2^2}{n_2}}$$

where $t^*$ is the critical value with C% of its area between $-t^*$ and $t^*$ for the $t$ distribution with degrees of freedom from either Option 1 (technology) or Option 2 (the smaller of $n_1 - 1$ and $n_2 - 1$). This is called a **two-sample $t$ interval for $\mu_1 - \mu_2$**.

- To test $H_0: \mu_1 - \mu_2 =$ hypothesized value, use a **two-sample $t$ test for $\mu_1 - \mu_2$**. The test statistic is

$$t = \frac{(\bar{x}_1 - \bar{x}_2) - (\mu_1 - \mu_2)}{\sqrt{\frac{s_1^2}{n_1} + \frac{s_2^2}{n_2}}}$$

$P$-values are calculated using the $t$ distribution with degrees of freedom from either Option 1 (technology) or Option 2 (the smaller of $n_1 - 1$ and $n_2 - 1$).

- Inference about the difference $\mu_1 - \mu_2$ in the effectiveness of two treatments in a completely randomized experiment is based on the **randomization distribution** of $\bar{x}_1 - \bar{x}_2$. When the conditions are met, our usual inference procedures based on the sampling distribution of $\bar{x}_1 - \bar{x}_2$ will be approximately correct.
- Don't use two-sample $t$ procedures to compare means for paired data.

- Be sure to follow the four-step process whenever you construct a confidence interval or perform a significance test for comparing two means.

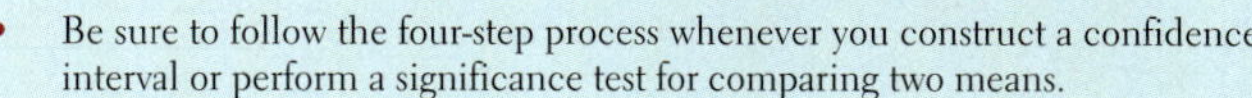

## 10.2 TECHNOLOGY CORNERS

TI-Nspire Instructions in Appendix B; HP Prime instructions on the book's Web site.



**APPLET**

Exercises 61 and 62 ask students to choose the correct inference procedure for different settings. You can use the excellent applet at http://www.ltcconline.net/greenl/java/Statistics/catStatProb/categorizingStatProblems12.html to practice choosing the correct inference procedure. Uncheck the procedures we don't know yet (prediction intervals, the three chi-square tests, and 1-way ANOVA), press Submit, and have fun!

TRM **Full Solutions to Section 10.2 Exercises**

## Answers to Section 10.2 Exercises

**10.31 (a)** Because the distributions of *M* and *B* are Normal, the distribution of $\bar{x}_M - \bar{x}_B$ is also Normal.
**(b)** $\mu_{\bar{x}_M - \bar{x}_B} = 188 - 170 = 18$ mg/dl
**(c)** Because $25 < 10\%$ of all 20- to 34-year-old males and $36 < 10\%$ of all 14-year-old boys, $\sigma_{\bar{x}_M - \bar{x}_B}$
$= \sqrt{\frac{(41)^2}{25} + \frac{(30)^2}{36}} = 9.60$ mg/dl.

**10.32 (a)** Because the distributions of *M* and *W* are Normal, the distribution of $\bar{x}_M - \bar{x}_W$ is also Normal.
**(b)** $\mu_{\bar{x}_M - \bar{x}_W} = 69.3 - 64.5 = 4.8$ inches.
**(c)** Because $16 < 10\%$ of all young men and $9 < 10\%$ of all young women, $\sigma_{\bar{x}_M - \bar{x}_W}$
$= \sqrt{\frac{(2.8)^2}{16} + \frac{(2.5)^2}{9}} = 1.09$ inches.

**10.33** *Random*: Two independent random samples. *10%*: $20 < 10\%$ of all males at the school and $20 < 10\%$ of all females at the school. *Normal/Large Sample*: Not met because there are fewer than 30 observations in each group and the stemplot for males shows several outliers.

**10.34** *Random*: Two independent random samples. *10%*: $50 < 10\%$ of students in the United Kingdom and $50 < 10\%$ of students in South Africa. *Normal/Large Sample*: Met because $50 \geq 30$ and $50 \geq 30$, even though there is an outlier in the South African distribution.

# Section 10.2 Exercises

*Remember: We are no longer reminding you to use the four-step process in exercises that require you to perform inference.*

**31. Cholesterol** The level of cholesterol in the blood for all men aged 20 to 34 follows a Normal distribution with mean 188 milligrams per deciliter (mg/dl) and standard deviation 41 mg/dl. For 14-year-old boys, blood cholesterol levels follow a Normal distribution with mean 170 mg/dl and standard deviation 30 mg/dl. Suppose we select independent SRSs of 25 men aged 20 to 34 and 36 boys aged 14 and calculate the sample mean cholesterol levels $\bar{x}_M$ and $\bar{x}_B$.

pg 638

(a) What is the shape of the sampling distribution of $\bar{x}_M - \bar{x}_B$? Why?

(b) Find the mean of the sampling distribution. Show your work.

(c) Find the standard deviation of the sampling distribution. Show your work.

**32. How tall?** The heights of young men follow a Normal distribution with mean 69.3 inches and standard deviation 2.8 inches. The heights of young women follow a Normal distribution with mean 64.5 inches and standard deviation 2.5 inches. Suppose we select independent SRSs of 16 young men and 9 young women and calculate the sample mean heights $\bar{x}_M$ and $\bar{x}_W$.

(a) What is the shape of the sampling distribution of $\bar{x}_M - \bar{x}_W$? Why?

(b) Find the mean of the sampling distribution. Show your work.

(c) Find the standard deviation of the sampling distribution. Show your work.

*In Exercises 33 to 36, determine whether or not the conditions for using two-sample t procedures are met.*

**33. Shoes** How many pairs of shoes do teenagers have? To find out, a group of AP® Statistics students conducted a survey. They selected a random sample of 20 female students and a separate random sample of 20 male students from their school. Then they recorded the number of pairs of shoes that each respondent reported having. The back-to-back stemplot displays the data.

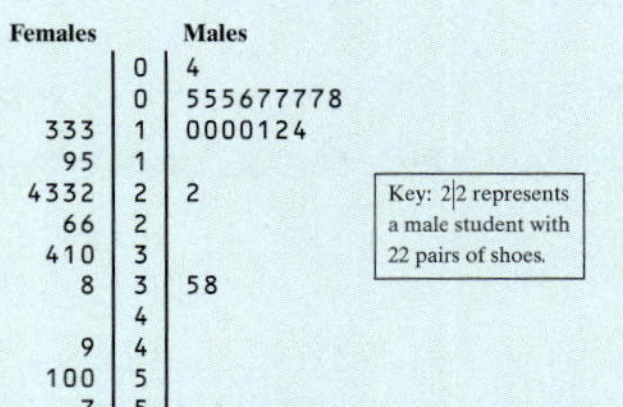


**34. Household size** How do the numbers of people living in households in the United Kingdom (U.K.) and South Africa compare? To help answer this question, we used CensusAtSchool's random data selector to choose independent samples of 50 students from each country. Here is a Fathom dotplot of the household sizes reported by the students in the survey.

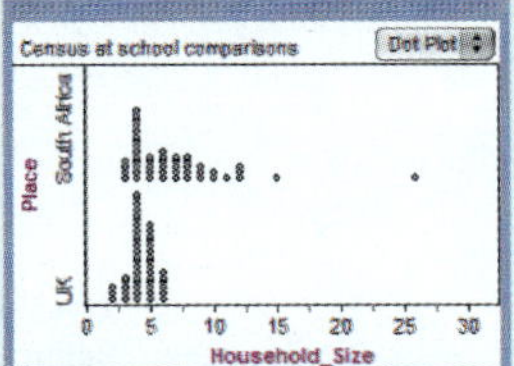


**35. Literacy rates** Do males have higher average literacy rates than females in Islamic countries? The table below shows the percent of men and women who were literate in the major Islamic nations at the time of this writing.[27] (We omitted countries with populations of less than 3 million.)

| Country | Male (%) | Female (%) |
|---|---|---|
| Afghanistan | 43 | 13 |
| Algeria | 80 | 60 |
| Azerbaijan | 99.9 | 99.7 |
| Bangladesh | 61 | 52 |
| Egypt | 80 | 64 |
| Indonesia | 94 | 86.8 |
| Iran | 84 | 70 |
| Iraq | 86 | 71 |
| Jordan | 96 | 89 |
| Kazakhstan | 100 | 99 |
| Kyrgyzstan | 99.3 | 98.1 |
| Lebanon | 93 | 82 |
| Libya | 96 | 83 |
| Malaysia | 92 | 85 |
| Morocco | 69 | 44 |
| Pakistan | 68.6 | 30.3 |
| Saudi Arabia | 90 | 81 |

| Country | Male (%) | Female (%) |
|---|---|---|
| Syria | 86 | 74 |
| Tajikistan | 100 | 100 |
| Tunisia | 83 | 65 |
| Turkey | 98 | 90 |
| Turkmenistan | 99.3 | 98.3 |
| Uzbekistan | 100 | 99 |
| Yemen | 81 | 47 |

36. **Long words** Mary was interested in comparing the mean word length in articles from a medical journal and an airline's in-flight magazine. She counted the number of letters in the first 400 words of an article in the medical journal and in the first 100 words of an article in the airline magazine. Mary then used Minitab statistical software to produce the histograms shown. Note that J is for journal and M is for magazine.

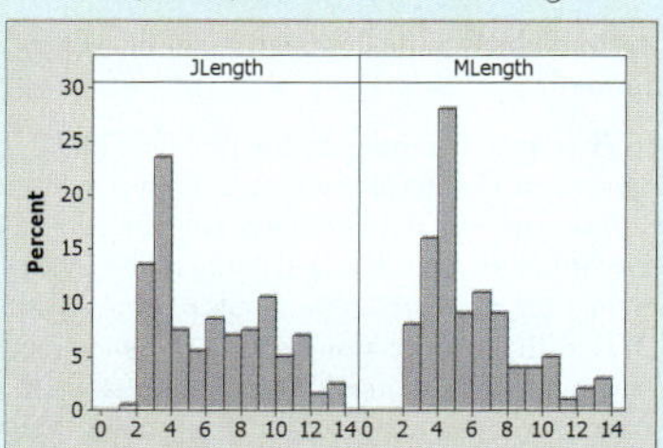


37. **Is red wine better than white wine?** Observational studies suggest that moderate use of alcohol by adults reduces heart attacks and that red wine may have special benefits. One reason may be that red wine contains polyphenols, substances that do good things to cholesterol in the blood and so may reduce the risk of heart attacks. In an experiment, healthy men were assigned at random to drink half a bottle of either red or white wine each day for two weeks. The level of polyphenols in their blood was measured before and after the two-week period. Here are the percent changes in level for the subjects in both groups:[28]

pg 641

| Red wine: | 3.5 | 8.1 | 7.4 | 4.0 | 0.7 | 4.9 | 8.4 | 7.0 | 5.5 |
|---|---|---|---|---|---|---|---|---|---|
| White wine: | 3.1 | 0.5 | −3.8 | 4.1 | −0.6 | 2.7 | 1.9 | −5.9 | 0.1 |

(a) A Fathom dotplot of the data is shown below. Write a few sentences comparing the distributions.

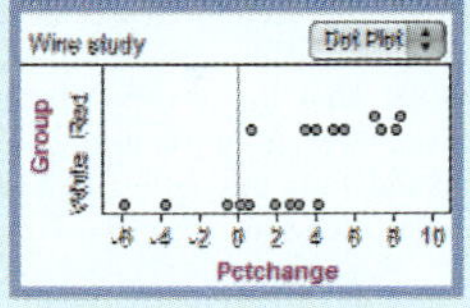


(b) Construct and interpret a 90% confidence interval for the difference in mean percent change in polyphenol levels for the red wine and white wine treatments.

(c) Does the interval in part (b) suggest that red wine is more effective than white wine? Explain.

38. **Tropical flowers** Different varieties of the tropical flower *Heliconia* are fertilized by different species of hummingbirds.

Researchers believe that over time, the lengths of the flowers and the forms of the hummingbirds' beaks have evolved to match each other. Here are data on the lengths in millimeters for random samples of two color varieties of the same species of flower on the island of Dominica:[29]

| *H. caribaea* red | | | | | | | |
|---|---|---|---|---|---|---|---|
| 41.90 | 42.01 | 41.93 | 43.09 | 41.17 | 41.69 | 39.78 | 40.57 |
| 39.63 | 42.18 | 40.66 | 37.87 | 39.16 | 37.40 | 38.20 | 38.07 |
| 38.10 | 37.97 | 38.79 | 38.23 | 38.87 | 37.78 | 38.01 | |

| *H. caribaea* yellow | | | | | | | |
|---|---|---|---|---|---|---|---|
| 36.78 | 37.02 | 36.52 | 36.11 | 36.03 | 35.45 | 38.13 | 37.10 |
| 35.17 | 36.82 | 36.66 | 35.68 | 36.03 | 34.57 | 34.63 | |

(a) A Fathom dotplot of the data is shown below. Write a few sentences comparing the distributions.

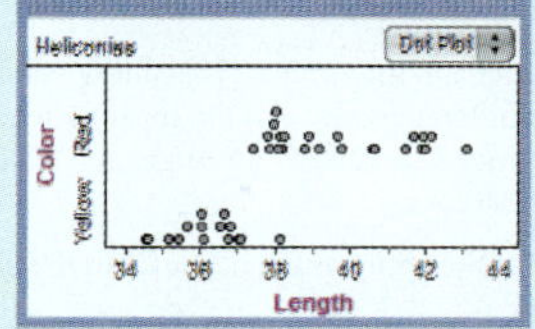


(b) Construct and interpret a 95% confidence interval for the difference in the mean lengths of these two varieties of flowers.

(c) Does the interval support the researchers' belief that the two flower varieties have different average lengths? Explain.

**10.35** *Random*: Not met because these data are not from two *independent* random samples. Knowing the literacy percent for females in a country helps us predict the literacy percent for males in that country. *10%*: Not met because 24 is more than 10% of Islamic countries. *Normal/Large Sample*: Not met because the sample sizes are both small and both distributions are skewed to the left and have an outlier (see the boxplots below).

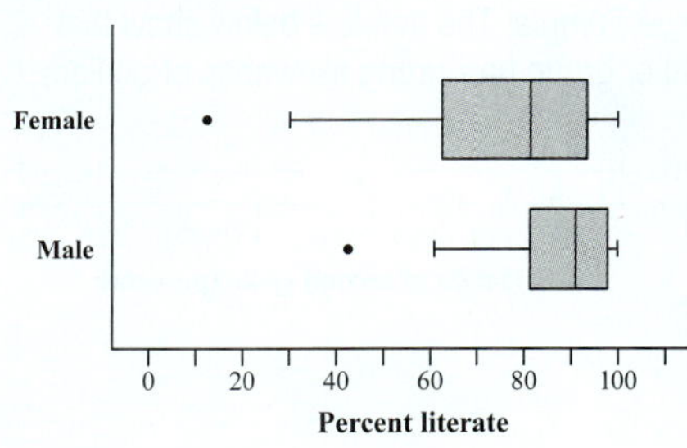


**10.36** *Random*: Not met because the words chosen from each article were the first words in the article, not a random sample of words. *10%*: 400 < 10% of the words in a medical journal and 100 < 10% of the words in an airline magazine. *Normal/Large Sample*: 100 ≥ 30 and 400 ≥ 30.

**10.37 (a)** The distributions of percent change are both slightly skewed to the left. People drinking red wine generally have more polyphenols in their blood, on average. The distribution of percent change for the white wine drinkers is a little bit more variable. **(b)** *S*: $\mu_1$ = the true mean change in polyphenol level in the blood of people like those in the study who drink red wine and $\mu_2$ = the true mean polyphenol level in the blood of people like those in the study who drink white wine. *P*: Two-sample *t* interval for $\mu_1 - \mu_2$. *Random*: Two groups in a randomized experiment. *Normal/Large Sample*: The dotplots given in the problem do not show strong skewness or outliers. *D*: $\bar{x}_1 = 5.5$, $s_1 = 2.517$, $\bar{x}_2 = 0.23$, $s_2 = 3.292$. Using df = 8, (2.701, 7.839). Using df = 14.97, (2.845, 7.689). *C*: We are 90% confident that the interval from 2.845 to 7.689 captures the true difference in mean change in polyphenol level for men like these who drink red wine and men like these who drink white wine. **(c)** Because all of the plausible values in the interval are positive, this interval supports the researcher's belief that red wine is more effective than white wine.

**10.38 (a)** The distribution of length for the red flowers is skewed to the right, while the distribution of length for the yellow flowers is roughly symmetric. The center of the distribution of length is much greater for the red flowers. The lengths of the red flowers are also more variable than those for the yellow flowers. **(b)** *S*: $\mu_1$ = the true mean length of red flowers and $\mu_2$ = the true mean length of yellow flowers. *P*: Two-sample *t* interval for $\mu_1 - \mu_2$. *Random*: Two independent random samples. *10%*: $n_1 = 23 <$ 10% of all red flowers and $n_2 = 15 <$ 10% of all yellow flowers. *Normal/Large Sample*: The dotplots do not indicate strong skewness or outliers. *D*: $\bar{x}_1 = 39.698, s_1 = 1.786$, $\bar{x}_2 = 36.18$, and $s_2 = 0.975$. Using df = 14, (2.554, 4.482). Using df = 35.16, (2.606, 4.431). *C*: We are 95% confident that the interval from 2.606 to 4.431 captures the true difference in mean length of red flowers and yellow flowers. **(c)** Because 0 is not in this interval, this interval does support the researchers' belief that the two varieties have different lengths.

39. **Paying for college** College financial aid offices expect students to use summer earnings to help pay for college. But how large are these earnings? One large university studied this question by asking a random sample of 1296 students who had summer jobs how much they earned. The financial aid office separated the responses into two groups based on gender. Here are the data in summary form:[30]

| Group | $n$ | $\bar{x}$ | $s_x$ |
|---|---|---|---|
| Males | 675 | $1884.52 | $1368.37 |
| Females | 621 | $1360.39 | $1037.46 |

(a) How can you tell from the summary statistics that the distribution of earnings in each group is strongly skewed to the right? The use of two-sample *t* procedures is still justified. Why?

(b) Construct and interpret a 90% confidence interval for the difference between the mean summer earnings of male and female students at this university.

(c) Interpret the 90% confidence level in the context of this study.

40. **Happy customers** As the Hispanic population in the United States has grown, businesses have tried to understand what Hispanics like. One study interviewed a random sample of customers leaving a bank. Customers were classified as Hispanic if they preferred to be interviewed in Spanish or as Anglo if they preferred English. Each customer rated the importance of several aspects of bank service on a 10-point scale.[31] Here are summary results for the importance of "reliability" (the accuracy of account records and so on):

| Group | $n$ | $\bar{x}$ | $s_x$ |
|---|---|---|---|
| Anglo | 92 | 6.37 | 0.60 |
| Hispanic | 86 | 5.91 | 0.93 |

(a) The distribution of reliability ratings in each group is not Normal. The use of two-sample *t* procedures is still justified. Why?

(b) Construct and interpret a 95% confidence interval for the difference between the mean ratings of the importance of reliability for Anglo and Hispanic bank customers.

(c) Interpret the 95% confidence level in the context of this study.

41. pg 645  **Baby birds** Do birds learn to time their breeding? Blue titmice eat caterpillars. The birds would like lots of caterpillars around when they have young to feed, but they must breed much earlier. Do the birds learn from one year's experience when to time their breeding next year? Researchers randomly assigned 7 pairs of birds to have the natural caterpillar supply supplemented while feeding their young and another 6 pairs to serve as a control group relying on natural food supply. The next year, they measured how many days after the caterpillar peak the birds produced their nestlings.[32] The investigators expected the control group to adjust their breeding date the next year, whereas the well-fed supplemented group had no reason to change. Here are the data (days after caterpillar peak):

| | | | | | | | |
|---|---|---|---|---|---|---|---|
| Control: | 4.6 | 2.3 | 7.7 | 6.0 | 4.6 | −1.2 | |
| Supplemented: | 15.5 | 11.3 | 5.4 | 16.5 | 11.3 | 11.4 | 7.7 |

(a) Do the data provide convincing evidence to confirm the researchers' belief?

(b) Interpret the *P*-value from part (a) in the context of this study.

42. **DDT in rats** Poisoning by the pesticide DDT causes convulsions in humans and other mammals. Researchers seek to understand how the convulsions are caused. In a randomized comparative experiment, they compared 6 white rats poisoned with DDT with a control group of 6 unpoisoned rats. Electrical measurements of nerve activity are the main clue to the nature of DDT poisoning. When a nerve is stimulated, its electrical response shows a sharp spike followed by a much smaller second spike. The researchers measured the height of the second spike as a percent of the first spike when a nerve in the rat's leg was stimulated.[33] For the poisoned rats, the results were

12.207 16.869 25.050 22.429 8.456 20.589

The control group data were

11.074 9.686 12.064 9.351 8.182 6.642

(a) Do these data provide convincing evidence that DDT affects the mean relative height of the second spike's electrical response?

(b) Interpret the *P*-value from part (a) in the context of this study.

43. **Who talks more—men or women?** Researchers equipped random samples of 56 male and 56 female students from a large university with a small device that secretly records sound for a random 30 seconds during each 12.5-minute period over two days. Then they counted the number of words spoken by each subject during each recording period and, from this, estimated how many words per day each subject speaks. The female

**10.39 (a)** Earnings amounts cannot be negative, yet the standard deviation is almost as large as the distance between the mean and 0. However, the sample sizes are both very large ($675 \geq 30$ and $621 \geq 30$). **(b)** *S*: $\mu_1$ = the true mean summer earnings of male students and $\mu_2$ = the true mean summer earnings of female students. *P*: Two-sample *t* interval for $\mu_1 - \mu_2$. *Random*: Reasonable to consider these independent random samples. *10%*: $n_1 = 675 <$ 10% of male students at a large university and $n_2 = 621 <$ 10% of female students at a large university. *Normal/Large Sample*: $n_1 = 675 \geq 30$ and $n_2 = 621 \geq 30$. *D*: Using df = 100, (412.68, 635.58). Using df = 1249.21, (413.62, 634.64). *C*: We are 90% confident that the interval from $413.62 to $634.64 captures the true difference in mean summer earnings of male students and female students at this large university. **(c)** If we took many random samples of 675 males and 621 females from this university and each time constructed a 90% confidence interval in the same way, about 90% of the resulting intervals would capture the true difference in mean earnings for males and females.

**10.40 (a)** Both sample sizes are large ($92 \geq 30$ and $86 \geq 30$). **(b)** *S*: $\mu_1$ = the true mean reliability rating of Anglo customers and $\mu_2$ = the true mean reliability rating of Hispanic customers. *P*: Two-sample *t* interval for $\mu_1 - \mu_2$. *Random*: Reasonable to consider these independent random samples. *10%*: $n_1 = 92 <$ 10% of Anglo customers and $n_2 = 86 <$ 10% of Hispanic customers. *Normal/Large Sample*: $n_1 = 92 \geq 30$ and $n_2 = 86 \geq 30$. *D*: Using df = 80, (0.225, 0.695). Using df = 143.69: (0.226, 0.694). *C*: We are 95% confident that the interval from 0.226 to 0.694 captures the true difference in mean reliability rating for Anglos and Hispanics. **(c)** If we took many random samples of 92 Anglos and 86 Hispanics and each time constructed a 95% confidence interval in this same way, about 95% of the resulting intervals would capture the true difference in mean reliability rating for Anglos and Hispanics.

**10.41 (a)** *S*: $H_0: \mu_1 - \mu_2 = 0$ vs. $H_a: \mu_1 - \mu_2 < 0$, where $\mu_1$ is the true mean time to breeding for the birds relying on natural food supply and $\mu_2$ is the true mean time to breeding for birds with food-supplementation. *P*: Two-sample *t* test. *Random*: Two groups in a randomized experiment. *Normal/Large Sample*: Neither distribution displays strong skewness or outliers.

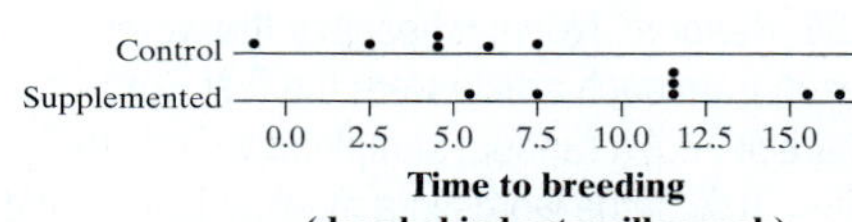


*D*: $\bar{x}_1 = 4.0$, $s_1 = 3.11$, $\bar{x}_2 = 11.3$, $s_2 = 3.93$. $t = -3.74$. Using df = 5, the *P*-value is between 0.005 and 0.01. Using df = 10.95, *P*-value = 0.0016. *C*: Because the *P*-value of $0.0016 < \alpha = 0.05$, we reject $H_0$. We have convincing evidence that the true mean time to breeding is less for birds relying on natural food supply than for birds given food supplements. **(b)** Assuming that the true mean time to breeding is the same for birds relying on natural food supply and birds given food supplements, there is a 0.0016 probability that we would observe a difference in sample means of −7.3 or smaller by chance alone.

**10.42 (a)** *S*: $H_0: \mu_1 - \mu_2 = 0$ vs. $H_a: \mu_1 - \mu_2 \neq 0$, where $\mu_1$ is the true mean relative height of the second spike in rats given DDT and $\mu_2$ is the true mean relative height of the second spike in rats not given DDT. *P*: Two-sample *t* test. *Random*: These data come from two groups in a randomized experiment. *Normal/Large Sample*: The dotplots below show that neither group has strong skewness or outliers.

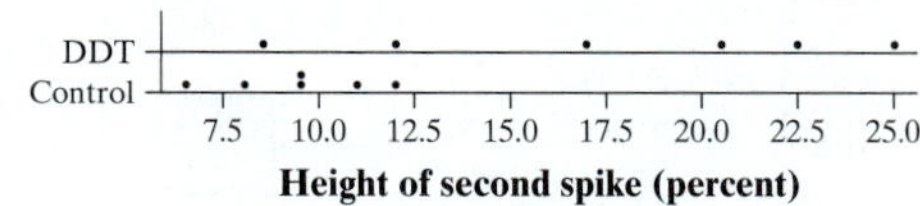

estimates had a mean of 16,177 words per day with a standard deviation of 7520 words per day. For the male estimates, the mean was 16,569 and the standard deviation was 9108. Do these data provide convincing evidence of a difference in the average number of words spoken in a day by male and female students at this university?

44. **Competitive rowers** What aspects of rowing technique distinguish between novice and skilled competitive rowers? Researchers compared two randomly selected groups of female competitive rowers: a group of skilled rowers and a group of novices. The researchers measured many mechanical aspects of rowing style as the subjects rowed on a Stanford Rowing Ergometer. One important variable is the angular velocity of the knee, which describes the rate at which the knee joint opens as the legs push the body back on the sliding seat. The data show no outliers or strong skewness. Here is the SAS computer output:[34]

TTEST PROCEDURE

Variable: KNEE

| GROUP | N | Mean | Std Dev | Std Error |
|---|---|---|---|---|
| SKILLED | 10 | 4.182 | 0.479 | 0.151 |
| NOVICE | 8 | 3.010 | 0.959 | 0.339 |

The researchers believed that the knee velocity would be higher for skilled rowers. Do the data provide convincing evidence to support this belief?

45. **Teaching reading** An educator believes that new reading activities in the classroom will help elementary school pupils improve their reading ability. She recruits 44 third-grade students and randomly assigns them into two groups. One group of 21 students does these new activities for an 8-week period. A control group of 23 third-graders follows the same curriculum without the activities. At the end of the 8 weeks, all students are given the Degree of Reading Power (DRP) test, which measures the aspects of reading ability that the treatment is designed to improve. Comparative boxplots and summary statistics for the data from Fathom are shown below.[35]

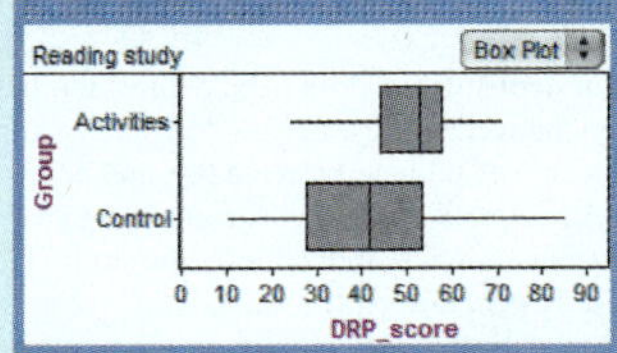


Reading study

| | Group | |
|---|---|---|
| | Activities | Control |
| DRP_score | 51.4762 | 41.5217 |
| | 21 | 23 |
| | 11.0074 | 17.1487 |

S1 = mean ( )
S2 = count ( )
S3 = stdDev ( )

(a) Based on the graph and numerical summaries, write a few sentences comparing the DRP scores for the two groups.

(b) Is the mean DRP score significantly higher for the students who did the reading activities? Give appropriate evidence to justify your answer.

(c) Can we conclude that the new reading activities caused an increase in the mean DRP score? Explain.

46. **Does breast-feeding weaken bones?** Breast-feeding mothers secrete calcium into their milk. Some of the calcium may come from their bones, so mothers may lose bone mineral. Researchers compared a random sample of 47 breast-feeding women with a random sample of 22 women of similar age who were neither pregnant nor lactating. They measured the percent change in the bone mineral content (BMC) of the women's spines over three months. Comparative boxplots and summary statistics for the data from Fathom are shown below.[36]

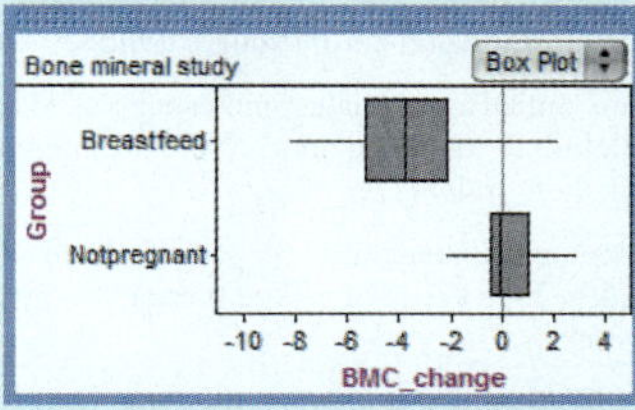


Bone mineral study

| | Group | |
|---|---|---|
| | Breastfeed | Notpregnant |
| BMC_change | -3.58723 | 0.309091 |
| | 47 | 22 |
| | 2.50561 | 1.29832 |

S1 = mean ( )
S2 = count ( )
S3 = stdDev ( )

(a) Based on the graph and numerical summaries, write a few sentences comparing the percent changes in BMC for the two groups.

*D*: $\bar{x}_1 = 17.6$, $s_1 = 6.34$, $\bar{x}_2 = 9.50$, $s_2 = 1.95$. $t = 2.99$. Using df = 5, the *P*-value is between 0.02 and 0.04. Using df = 5.9, *P*-value = 0.0246. *C*: Because the *P*-value of $0.0246 < \alpha = 0.05$, we reject $H_0$. We have convincing evidence that the true mean relative height of the second spike is different for rats poisoned with DDT and control rats. **(b)** Assuming that the true mean relative height of the second spike is the same for rats with DDT and rats without DDT, there is a 0.0246 probability of observing a difference between the sample means of 8.10 or greater by chance alone.

**10.43** *S*: $H_0: \mu_1 - \mu_2 = 0$ vs. $H_a: \mu_1 - \mu_2 \neq 0$, where $\mu_1$ is the true mean number of words spoken per day by female students and $\mu_2$ is the true mean number of words spoken per day by male students. *P*: Two-sample *t* test. *Random*: Independent random samples. *10%*: $n_1 = 56 <$ 10% of females at a large university and $n_2 = 56 <$ 10% of males at a large university. *Normal/Large Sample*: $n_1 = 56 \geq 30$ and $n_2 = 56 \geq 30$. *D*: $t = -0.248$. Using df = 50, *P*-value > 0.50. Using df = 106.20, *P*-value = 0.8043. *C*: Because the *P*-value of $0.8043 > \alpha = 0.05$, we fail to reject $H_0$. We do not have convincing evidence that the true mean number of words spoken per day by female students is different than the true mean number of words spoken per day by male students at this university.

**10.44** *S*: $H_0: \mu_1 - \mu_2 = 0$ vs. $H_a: \mu_1 - \mu_2 > 0$, where $\mu_1$ is the true mean knee velocity for skilled rowers and $\mu_2$ is the true mean knee velocity for novice rowers. *P*: Two-sample *t* test. *Random*: Independent random samples. *10%*: $n_1 = 10 <$ 10% of skilled rowers and $n_2 = 8 <$ 10% of novice rowers. *Normal/Large Sample*: No outliers or strong skewness. *D*: $t = 3.16$. Using df = 7, the *P*-value is between 0.005 and 0.01. Using df = 9.77, *P*-value = 0.0053. *C*: Because the *P*-value of $0.0053 < \alpha = 0.05$, we reject $H_0$. We have convincing evidence that the mean knee velocity is greater for skilled rowers than for novice rowers.

**10.45 (a)** The distribution for the activities group is slightly skewed to the left, while the distribution for the control group is slightly skewed to the right. The center of the activities group is higher than the center of the control group. The scores in the activities group are less variable than the scores in the control group. **(b)** *S*: $H_0: \mu_1 - \mu_2 = 0$ vs. $H_a: \mu_1 - \mu_2 > 0$, where $\mu_1$ is the true mean DRP score for third-grade students like those in the experiment who do the activities and $\mu_2$ is the true mean DRP score for third-grade students like those in the experiment who don't do the activities. *P*: Two-sample *t* test. *Random*: Two groups in a randomized experiment. *Normal/Large Sample*: No strong skewness or outliers in either boxplot. *D*: $t = 2.311$. Using df = 20, the *P*-value is between 0.01 and 0.02. Using df = 37.86, *P*-value = 0.0132. *C*: Because the *P*-value of $0.0132 < \alpha = 0.05$, we reject $H_0$. We have convincing evidence that the true mean DRP score for third-grade students like those in the experiment who do the activities > the true mean DRP score for third-grade students like those in the experiment who don't do the activities. **(c)** Because this was a randomized controlled experiment, we can conclude that the activities caused the increase in the mean DRP score.

**10.46 (a)** Both distributions are slightly skewed to the right. The center of the BMC change distribution is much smaller for breast-feeding mothers than for women who are not pregnant or lactating. The BMC changes are much more variable for breast-feeding mothers than for women who are not pregnant or lactating.

(b) Is the mean change in BMC significantly lower for the mothers who are breast-feeding? Give appropriate evidence to justify your answer.

(c) Can we conclude that breast-feeding causes a mother's bones to weaken? Why or why not?

47. **Who talks more—men or women?** Refer to Exercise 43. Construct and interpret a 95% confidence interval for the difference in mean number of words spoken in a day. Explain how this interval provides more information than the significance test in Exercise 43.

48. **DDT in rats** Refer to Exercise 42. Construct and interpret a 95% confidence interval for the difference in mean relative height of the second spike's electrical response. Explain how this interval provides more information than the significance test in Exercise 42.

49. **A better drug?** In a pilot study, a company's new cholesterol-reducing drug outperforms the currently available drug. If the data provide convincing evidence that the mean cholesterol reduction with the new drug is more than 10 milligrams per deciliter of blood (mg/dl) greater than with the current drug, the company will begin the expensive process of mass-producing the new drug. For the 14 subjects who were assigned at random to the current drug, the mean cholesterol reduction was 54.1 mg/dl with a standard deviation of 11.93 mg/dl. For the 15 subjects who were randomly assigned to the new drug, the mean cholesterol reduction was 68.7 mg/dl with a standard deviation of 13.3 mg/dl. Graphs of the data reveal no outliers or strong skewness.

(a) Carry out an appropriate significance test. What conclusion would you draw? (Note that the null hypothesis is *not* $H_0$: $\mu_1 - \mu_2 = 0$.)

(b) Based on your conclusion in part (a), could you have made a Type I error or a Type II error? Justify your answer.

50. **Down the toilet** A company that makes hotel toilets claims that its new pressure-assisted toilet reduces the average amount of water used by more than 0.5 gallon per flush when compared to its current model. To test this claim, the company randomly selects 30 toilets of each type and measures the amount of water that is used when each toilet is flushed once. For the current-model toilets, the mean amount of water used is 1.64 gal with a standard deviation of 0.29 gal. For the new toilets, the mean amount of water used is 1.09 gal with a standard deviation of 0.18 gal.

(a) Carry out an appropriate significance test. What conclusion would you draw? (Note that the null hypothesis is *not* $H_0$: $\mu_1 - \mu_2 = 0$.)

(b) Based on your conclusion in part (a), could you have made a Type I error or a Type II error? Justify your answer.

51. **Rewards and creativity** Dr. Teresa Amabile conducted a study involving 47 college students who were randomly assigned to two treatment groups. The 23 students in one group were given a list of statements about external reasons (E) for writing, such as public recognition, making money, or pleasing their parents. The 24 students in the other group were given a list of statements about internal reasons (I) for writing, such as expressing yourself and enjoying playing with words. Both groups were then instructed to write a poem about laughter. Each student's poem was rated separately by 12 different poets using a creativity scale.[37] The 12 poets' ratings of each student's poem were averaged to obtain an overall creativity score.

We used Fathom software to randomly reassign the 47 subjects to the two groups 1000 times, assuming the treatment received doesn't affect each individual's average creativity rating. The dotplot shows the approximate randomization distribution of $\bar{x}_I - \bar{x}_E$.

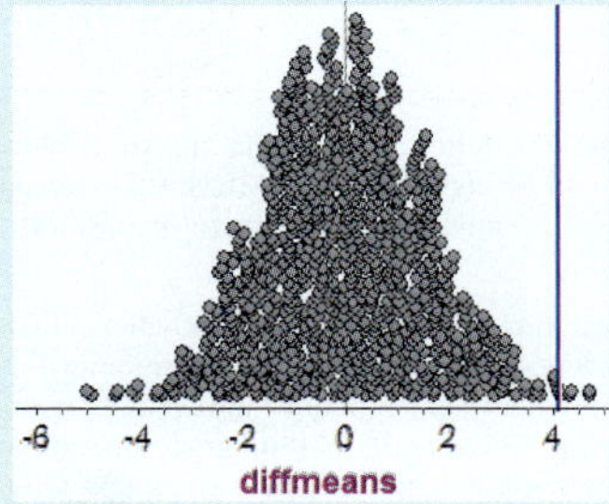


(a) Why did researchers randomly assign the subjects to the two treatment groups?

(b) In the actual experiment, $\bar{x}_I - \bar{x}_E = 4.15$. This value is marked with a blue line in the figure. What conclusion would you draw? Justify your answer with appropriate evidence.

(c) Based on your conclusion in part (b), could you have made a Type I error or a Type II error? Justify your answer.

52. **Sleep deprivation** Does sleep deprivation linger for more than a day? Researchers designed a study using 21 volunteer subjects between the ages of 18 and 25. All 21 participants took a computer-based visual discrimination test at the start of the study. Then the subjects were randomly assigned into two groups. The 11 subjects in one group, D, were deprived of sleep for an entire night in a laboratory setting. The

**10.46 (b)** *S*: $H_0: \mu_1 - \mu_2 = 0$ vs. $H_a: \mu_1 - \mu_2 < 0$, where $\mu_1$ is the true mean percent change in mineral content for breast-feeding women and $\mu_2$ is the true mean percent change in mineral content for women who were neither pregnant nor lactating. *P*: Two-sample *t* test. *Random*: Independent random samples. *10%*: $n_1 = 47 < 10\%$ of breast-feeding mothers and $n_2 = 22 <$ 10% of women who are not pregnant or lactating. *Normal/Large Sample*: Neither boxplot displays strong skewness or outliers. *D*: $t = -8.499$. Using df = 21 or df = 66.20, *P*-value $\approx$ 0. *C*: Because the *P*-value of approximately $0 < \alpha = 0.05$, we reject $H_0$. We have convincing evidence that breast-feeding women experience a greater mean percent bone mineral loss than women who are neither pregnant nor lactating. **(c)** Because this was not a randomized controlled experiment, we cannot conclude that breast-feeding is the cause of the bone mineral loss.

**10.47** *D*: Using df = 50, (−3563, 2779). Using df = 106.2, (−3521, 2737). *C*: We are 95% confident that the interval from −3521 to 2737 captures the true difference in mean number of words spoken per day by female students and mean number of words spoken per day by male students. This interval allows us to determine if 0 is a plausible value for the difference in means and also provides other plausible values for the difference in mean words spoken per day.

**10.48** *D*: Using df = 5, (1.14, 15.06). Using df = 5.94, (1.46, 14.74). *C*: We are 95% confident that the interval from 1.46 to 14.74 captures the true difference in mean relative height of the second spike in rats given DDT and mean relative height of the second spike in rats not given DDT. This interval allows us to determine if 0 is a plausible value for the difference in means and also provides other plausible values for the difference in mean relative height of the second spike.

**10.49 (a)** *S*: $H_0: \mu_1 - \mu_2 = 10$ vs. $H_a: \mu_1 - \mu_2 > 10$, where $\mu_1$ is the true mean cholesterol reduction for people like those in the study when using the new drug and $\mu_2$ is the true mean cholesterol reduction for people like those in the study when using the current drug. *P*: Two-sample *t* test. *Random*: Two groups in a randomized experiment. *Normal/Large Sample*: No strong skewness or outliers. *D*: $t = 0.982$. Using df = 13, the *P*-value is between 0.15 and 0.20. Using df = 26.96, *P*-value = 0.1675. *C*: Because the *P*-value of $0.1675 > \alpha = 0.05$, we fail to reject $H_0$. We do not have convincing evidence that the true mean cholesterol reduction is more than 10 mg/dl greater for the new drug than for the current drug. **(b)** Type II error. It is possible that the difference in mean cholesterol reduction is more than 10 mg/dl greater for the new drug than the current drug, but we didn't find convincing evidence that it was.

**10.50 (a)** *S*: $H_0: \mu_1 - \mu_2 = 0.5$ versus $H_a: \mu_1 - \mu_2 > 0.5$, where $\mu_1$ is the true mean amount of water used in the current model toilets and $\mu_2$ is the true mean amount of water used in the new model toilets. *P*: Two-sample *t* test. *Random*: Independent random samples. *10%*: $n_1 = 30 < 10\%$ of current model toilets and $n_2 = 30 < 10\%$ of new model toilets. *Normal/Large Sample*: $n_1 = 30 \geq 30$ and $n_2 = 30 \geq 30$. *D*: $t = 0.802$. Using df = 29, the *P*-value is between 0.20 and 0.25. Using df = 48.46, *P*-value = 0.2131. *C*: Because the *P*-value of $0.2131 > \alpha = 0.05$, we fail to reject $H_0$. We do not have convincing evidence that the new model toilet reduces the amount of water used by more than 0.50 gallons per flush than the current model toilet, on average. **(b)** Type II error. That is, it is possible that the new model toilet does reduce the amount of water used by more than 0.50 gallons, on average, but we did not find convincing evidence of this fact.

**10.51 (a)** The researchers randomly assigned the subjects to create two groups that were roughly equivalent at the beginning of the experiment. **(b)** Only about 5 out of the 1000 differences were $\geq 4.15$, *P*-value $\approx$ 0.005. Because the *P*-value of $0.005 < \alpha = 0.05$, we have convincing evidence that the true mean rating for students

10 subjects in the other group, A, were allowed unrestricted sleep for the night. Both groups were allowed as much sleep as they wanted for the next two nights. On Day 4, all the subjects took the same visual discrimination test on the computer. Researchers recorded the improvement in time (measured in milliseconds) from Day 1 to Day 4 on the test for each subject.[38]

We used Fathom software to randomly reassign the 21 subjects to the two groups 1000 times, assuming the treatment received doesn't affect each individual's time improvement on the test. The dotplot shows the approximate randomization distribution of $\bar{x}_A - \bar{x}_D$.

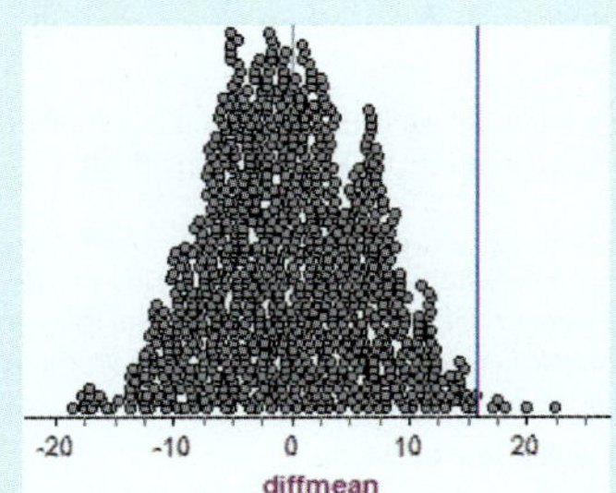


(a) Explain why the researchers didn't let the subjects choose whether to be in the sleep deprivation group or the unrestricted sleep group.

(b) In the actual experiment, $\bar{x}_A - \bar{x}_D = 15.92$. This value is marked with a blue line in the figure. What conclusion would you draw? Justify your answer with appropriate evidence.

(c) Based on your conclusion in part (b), could you have made a Type I error or a Type II error? Justify your answer.

**53. Paired or unpaired?** In each of the following settings, decide whether you should use paired $t$ procedures or two-sample $t$ procedures to perform inference. Explain your choice.[39]

pg 650

(a) To test the wear characteristics of two tire brands, A and B, each brand of tire is randomly assigned to 50 cars of the same make and model.

(b) To test the effect of background music on productivity, factory workers are observed. For one month, each subject works without music. For another month, the subject works while listening to music on an MP3 player. The month in which each subject listens to music is determined by a coin toss.

(c) A study was designed to compare the effectiveness of two weight-reducing diets. Fifty obese women who volunteered to participate were randomly assigned into two equal-sized groups. One group used Diet A and the other used Diet B. The weight of each woman was measured before the assigned diet and again after 10 weeks on the diet.

**54. Paired or unpaired?** In each of the following settings, decide whether you should use paired $t$ procedures or two-sample $t$ procedures to perform inference. Explain your choice.[40]

(a) To compare the average weight gain of pigs fed two different rations, nine pairs of pigs were used. The pigs in each pair were littermates. A coin toss was used to decide which pig in each pair got Ration A and which got Ration B.

(b) Separate random samples of male and female college professors are taken. We wish to compare the average salaries of male and female teachers.

(c) To test the effects of a new fertilizer, 100 plots are treated with the new fertilizer, and 100 plots are treated with another fertilizer. A computer's random number generator is used to determine which plots get which fertilizer.

*Exercises 55 and 56 refer to the following setting.* Coaching companies claim that their courses can raise the SAT scores of high school students. Of course, students who retake the SAT without paying for coaching generally raise their scores. A random sample of students who took the SAT twice found 427 who were coached and 2733 who were uncoached.[41] Starting with their Verbal scores on the first and second tries, we have these summary statistics:

| | Try 1 | | | Try 2 | | Gain | |
|---|---|---|---|---|---|---|---|
| | $n$ | $\bar{x}$ | $s_x$ | $\bar{x}$ | $s_x$ | $\bar{x}$ | $s_x$ |
| Coached | 427 | 500 | 92 | 529 | 97 | 29 | 59 |
| Uncoached | 2733 | 506 | 101 | 527 | 101 | 21 | 52 |

**55. Coaching and SAT scores** Let's first ask if students who are coached increased their scores significantly.

(a) You could use the information on the Coached line to carry out either a two-sample $t$ test comparing Try 1 with Try 2 for coached students or a paired $t$ test using Gain. Which is the correct test? Why?

(b) Carry out the proper test. What do you conclude?

**56. Coaching and SAT scores** What we really want to know is whether coached students improve more than uncoached students, and whether any advantage is large enough to be worth paying for. Use the information above to answer these questions:

(a) How much more do coached students gain on the average? Construct and interpret a 99% confidence interval.

(b) Does the interval in part (a) give convincing evidence that coached students gain more, on average, than uncoached students? Explain.

(c) Based on your work, what is your opinion: do you think coaching courses are worth paying for?

**10.55 (a)** Paired, because we have two scores for each student. **(b)** *S*: $H_0: \mu_d = 0$ vs. $H_a: \mu_d > 0$, where $\mu_d$ is the true mean increase in SAT verbal scores of students who were coached. *P*: Paired *t* test for $\mu_d$. *Random*: Random sample. *10%*: $n_d = 427 < 10\%$ of students who are coached. *Normal/Large Sample*: $427 \geq 30$. *D*: $t = 10.16$. Using df $= 426$, *P*-value $\approx 0$. *C*: Because the *P*-value of approximately $0 < \alpha = 0.05$, we reject $H_0$. There is convincing evidence that students who are coached increase their scores on the SAT verbal test, on average.

**10.56 (a)** *S*: $\mu_1$ is the true mean increase in SAT verbal scores for coached students and $\mu_2$ is the true mean increase in SAT verbal scores for students who were not coached. *P*: Two-sample *t* interval for $\mu_1 - \mu_2$. *Random*: Independent random samples. *10%*: $n_1 = 427 < 10\%$ of all coached students and $n_2 = 2733 < 10\%$ of all uncoached students. *Normal/Large Sample*: $n_1 = 427 \geq 30$ and $n_2 = 2733 \geq 30$. *D*: Using df $= 100$, (0.06, 15.94). Using df $= 534.45$, (0.18, 15.82). *C*: We are 99% confident that the interval from 0.18 to 15.82 captures the true difference in mean increase in scores on the SAT verbal exam for students who are coached and students who are not coached.
**(b)** Because the interval does not include 0, there is convincing evidence to conclude that students who receive coaching have a higher mean increase in scores on the SAT verbal exam than those who do not.
**(c)** Answers will vary.

like these who are provided with internal reasons is higher than the true mean rating for students like these who are provided with external reasons. **(c)** Because we found convincing evidence that the mean is higher for students with internal reasons when it is possible that there is no difference in the means, we could have made a Type I error.

**10.52 (a)** If people were allowed to choose which group they wanted to be in, it is likely that all those who choose one particular treatment (sleep deprivation, for example) might be systematically different from those who choose to be in the other treatment group. **(b)** Only about 5 out of the 1000 differences were $\geq 15.92$, *P*-value $\approx 0.005$. Because the *P*-value of $0.005 < \alpha = 0.05$, we have convincing evidence that the mean increase in score is higher for subjects like these who are allowed to sleep than for subjects like these who are sleep deprived. **(c)** Because we found convincing evidence that the mean is higher for subjects with unrestricted sleep when it is possible that there is no difference in the means, we could have made a Type I error.

**10.53 (a)** Two-sample. Two distinct groups of cars in a randomized experiment. **(b)** Paired. Both treatments are applied to each subject.
**(c)** Two-sample. Two distinct groups of women.

**10.54 (a)** Paired. One pig in each pair received one treatment and the other pig in the pair received the other treatment. **(b)** Two-sample. The samples of men and women are independent.
**(c)** Two-sample. Two distinct groups of plots in a randomized experiment.

**10.57** a

**10.58** d

**10.59** b

**10.60** d

**10.61** **(a)** One-sample $z$ interval for a proportion. **(b)** Paired $t$ test for the mean difference. **(c)** Two-sample $z$ interval for the difference in proportions.
**(d)** Two-sample $t$ test for a difference in means.

*Multiple choice: Select the best answer for Exercises 57 to 60.*

57. There are two common methods for measuring the concentration of a pollutant in fish tissue. Do the two methods differ, on average? You apply both methods to each fish in a random sample of 18 carp and use

(a) the paired $t$ test for $\mu_d$.

(b) the one-sample $z$ test for $p$.

(c) the two-sample $t$ test for $\mu_1 - \mu_2$.

(d) the two-sample $z$ test for $p_1 - p_2$.

(e) none of these.

*Exercises 58 to 60 refer to the following setting.* A study of road rage asked random samples of 596 men and 523 women about their behavior while driving. Based on their answers, each person was assigned a road rage score on a scale of 0 to 20. The participants were chosen by random digit dialing of phone numbers. The researchers performed a test of the following hypotheses: $H_0: \mu_M = \mu_F$ versus $H_a: \mu_M \neq \mu_F$.

58. Which of the following describes a Type II error in the context of this study?

(a) Finding convincing evidence that the true means are different for males and females, when in reality the true means are the same

(b) Finding convincing evidence that the true means are different for males and females, when in reality the true means are different

(c) Not finding convincing evidence that the true means are different for males and females, when in reality the true means are the same

(d) Not finding convincing evidence that the true means are different for males and females, when in reality the true means are different

(e) Not finding convincing evidence that the true means are different for males and females, when in reality there is convincing evidence that the true means are different

59. The $P$-value for the stated hypotheses is 0.002. Interpret this value in the context of this study.

(a) Assuming that the true mean road rage score is the same for males and females, there is a 0.002 probability of getting a difference in sample means.

(b) Assuming that the true mean road rage score is the same for males and females, there is a 0.002 probability of getting an observed difference at least as extreme as the observed difference.

(c) Assuming that the true mean road rage score is different for males and females, there is a 0.002 probability of getting an observed difference at least as extreme as the observed difference.

(d) Assuming that the true mean road rage score is the same for males and females, there is a 0.002 probability that the null hypothesis is true.

(e) Assuming that the true mean road rage score is the same for males and females, there is a 0.002 probability that the alternative hypothesis is true.

60. Based on the $P$-value in Exercise 59, which of the following must be true?

(a) A 90% confidence interval for $\mu_M - \mu_F$ will contain 0.

(b) A 95% confidence interval for $\mu_M - \mu_F$ will contain 0.

(c) A 99% confidence interval for $\mu_M - \mu_F$ will contain 0.

(d) A 99.9% confidence interval for $\mu_M - \mu_F$ will contain 0.

(e) It is impossible to determine whether any of these statements is true based only on the $P$-value.

*In each part of Exercises 61 and 62, state which inference procedure from Chapter 8, 9, or 10 you would use. Be specific. For example, you might say, "Two-sample z test for the difference between two proportions." You do not need to carry out any procedures.*

61. **Which inference method?**

(a) Drowning in bathtubs is a major cause of death in children less than 5 years old. A random sample of parents was asked many questions related to bathtub safety. Overall, 85% of the sample said they used baby bathtubs for infants. Estimate the percent of all parents of young children who use baby bathtubs.

(b) How seriously do people view speeding in comparison with other annoying behaviors? A large random sample of adults was asked to rate a number of behaviors on a scale of 1 (no problem at all) to 5 (very severe problem). Do speeding drivers get a higher average rating than noisy neighbors?

(c) You have data from interviews with a random sample of students who failed to graduate from a particular college in 7 years and also from a random sample of students who entered at the same time and did graduate. You will use these data to compare the percents of students from rural backgrounds among dropouts and graduates.

(d) Do experienced computer game players earn higher scores when they play with someone present to cheer them on or when they play alone? Fifty teenagers with experience playing a particular computer game have volunteered for a study. We randomly assign 25 of them to play the game alone and the other 25 to play the game with a supporter present. Each player's score is recorded.

62. **Which inference method?**

(a) How do young adults look back on adolescent romance? Investigators interviewed 40 couples in their midtwenties. The female and male partners were interviewed separately. Each was asked about his or her current relationship and also about a romantic relationship that lasted at least two months when they were aged 15 or 16. One response variable was a measure on a numerical scale of how much the attractiveness of the adolescent partner mattered. You want to find out how much men and women differ on this measure.

(b) Are more than 75% of Toyota owners generally satisfied with their vehicles? Let's design a study to find out. We'll select a random sample of 400 Toyota owners. Then we'll ask each individual in the sample: "Would you say that you are generally satisfied with your Toyota vehicle?"

(c) Are male college students more likely to binge drink than female college students? The Harvard School of Public Health surveys random samples of male and female undergraduates at four-year colleges and universities about whether they have engaged in binge drinking.

(d) A bank wants to know which of two incentive plans will most increase the use of its credit cards and by how much. It offers each incentive to a group of current credit card customers, determined at random, and compares the amount charged during the following six months.

63. **Quality control** (2.2, 5.3, 6.3) Many manufacturing companies use statistical techniques to ensure that the products they make meet standards. One common way to do this is to take a random sample of products at regular intervals throughout the production shift. Assuming that the process is working properly, the mean measurement $\bar{x}$ from a random sample varies according to a Normal distribution with mean $\mu_{\bar{x}}$ and standard deviation $\sigma_{\bar{x}}$. For each question that follows, assume that the process is working properly.

(a) What's the probability that at least one of the next two sample means will fall more than $2\sigma_{\bar{x}}$ from the target mean $\mu_{\bar{x}}$? Show your work.

(b) What's the probability that the first sample mean that is greater than $\mu_{\bar{x}} + 2\sigma_{\bar{x}}$ is the one from the fourth sample taken?

Plant managers are trying to develop a criterion for determining when the process is not working properly. One idea they have is to look at the 5 most recent sample means. If at least 4 of the 5 fall outside the interval $(\mu_{\bar{x}} - \sigma_{\bar{x}}, \mu_{\bar{x}} + \sigma_{\bar{x}})$, they will conclude that the process isn't working.

(c) Find the probability that at least 4 of the 5 most recent sample means fall outside the interval, assuming the process is working properly. Is this a reasonable criterion? Explain.

64. **Information online** (8.2, 10.1) A random digit dialing sample of 2092 adults found that 1318 used the Internet.[42] Of the users, 1041 said that they expect businesses to have Web sites that give product information; 294 of the 774 nonusers said this.

(a) Construct and interpret a 95% confidence interval for the proportion of all adults who use the Internet.

(b) Construct and interpret a 95% confidence interval to compare the proportions of users and nonusers who expect businesses to have Web sites that give product information.

65. **Coaching and SAT scores: Critique** (4.1, 4.3) The data in Exercises 55 and 56 came from a random sample of students who took the SAT twice. The response rate was 63%, which is fairly good for nongovernment surveys.

(a) Explain how nonresponse could lead to bias in this study.

(b) We can't be sure that coaching actually *caused* the coached students to gain more than the uncoached students. Explain briefly but clearly why this is so.

**10.62 (a)** Paired *t* interval for the mean difference. **(b)** One-sample *z* test for a proportion. **(c)** Two-sample *z* test for a difference in proportions. **(d)** Two-sample *t* interval for a difference in means.

**10.63 (a)** *P*(at least one mean outside interval) = 1 − *P*(neither mean outside interval) = $1 - (0.95)^2 = 1 - 0.9025 = 0.0975$. **(b)** Let $X$ = the number of samples that must be taken to observe one falling above $\mu_{\bar{x}} + 2\sigma_{\bar{x}}$. Then $X$ is a geometric random variable with $p = 0.025$. $P(X = 4) = (1 - 0.025)^3(0.025) = 0.0232$. **(c)** Let $X$ = the number of sample means out of 5 that fall outside this interval. $X$ is a binomial random variable with $n = 5$ and $p = 1 - 0.68 = 0.32$. We want $P(X \geq 4) = 1 - P(X \leq 3) =$ `1 - binomcdf(trials: 5, p: 0.32, x value: 3)` $= 1 - 0.961 = 0.039$. This is a reasonable criterion because when the process is under control, we would only get a "false alarm" about 4% of the time.

**10.64 (a)** *S*: $p$ = the true proportion of all adults who use the Internet. *P*: One-sample *z* interval. *Random*: Random sample. *10%*: $n = 2092 <$ 10% of all adults. *Large Counts*: 1318 and 774 are both ≥ 10. *D*: (0.609, 0.651). *C*: We are 95% confident that the interval from 0.609 to 0.651 captures the true proportion of adults who use the Internet. **(b)** *S*: $p_1$ = true proportion of adult Internet users who expect businesses to have Web sites and $p_2$ = true proportion of adult non-Internet users who expect businesses to have Web sites. *P*: Two-sample *z* interval. *Random*: Reasonable to consider these independent random samples. *10%*: $n_1 = 1318 <$ 10% of adult Internet users and $n_2 = 774 <$ 10% of adult Internet nonusers. *Large Counts*: 1041, 277, 294, 480 are all ≥ 10. *D*: (0.369, 0.451). *C*: We are 95% confident that the interval from 0.369 to 0.451 captures the true difference in the proportion of Internet users and non-Internet users who expect businesses to have a Web site.

**10.65 (a)** Perhaps the people who responded are prouder of their improvements and are more willing to share. This could lead to an overestimate of the true mean improvement. **(b)** This was an observational study, not an experiment. The students (or their parents) chose whether or not to be coached; students who choose coaching might have other motivating factors that help them do better the second time.

### TRM FRAPPY!

We present a model solution here. Please consult the Teacher's Resource Materials for sample student responses, a scoring rubric, and a black line master of the original question with space for students to write their responses.

STATE: We want to perform a test of $H_0: \mu_N - \mu_S = 0$ versus $H_a: \mu_N - \mu_S > 0$ at the $\alpha = 0.05$ level of significance, where $\mu_N$ = the true mean percentage of popped kernels for name-brand microwave popcorn and $\mu_S$ = the true mean percentage of popped kernels for store-brand microwave popcorn.

PLAN: If the conditions are met, we will conduct a two-sample *t* test for $\mu_N - \mu_S$.

- *Random*: The data come from independent random samples.
  - *10%*: There are more than 10(10) = 100 bags of name-brand microwave popcorn and more than 10(10) = 100 bags of store-brand microwave popcorn.
- *Normal/Large Sample*: There are no outliers or strong skewness in either sample, so it is reasonable to use *t* procedures.

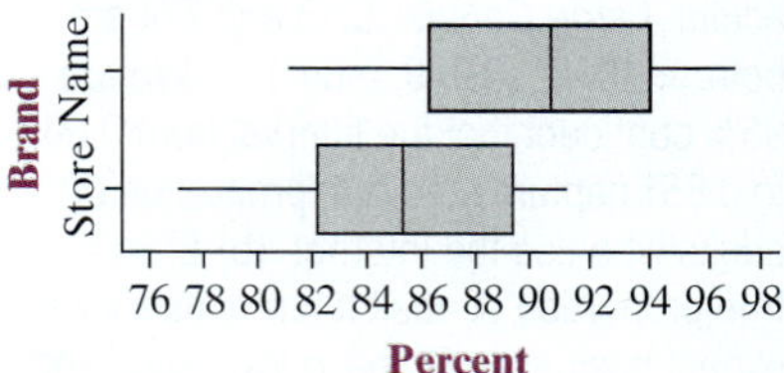


DO: For these data, $\bar{x}_N = 89.9$, $s_N = 5.13$, $\bar{x}_S = 84.5$, and $s_S = 4.81$.

Test statistic: $t = \dfrac{(89.9 - 84.5) - 0}{\sqrt{\dfrac{5.13^2}{10} + \dfrac{4.81^2}{10}}} = 2.43$. *P*-value: With df = 9, the *P*-value is between 0.01 and 0.02. *Using technology:* With df = 17.93, *P*-value = 0.013.

CONCLUDE: Because the *P*-value of 0.013 is less than $\alpha = 0.05$, we reject $H_0$. We have convincing evidence that the true mean percentage of popped kernels is greater for name-brand microwave popcorn than for store-brand microwave popcorn.

## FRAPPY! Free Response AP® Problem, Yay!

**The following problem is modeled after actual AP® Statistics exam free response questions. Your task is to generate a complete, concise response in 15 minutes.**

*Directions: Show all your work. Indicate clearly the methods you use, because you will be scored on the correctness of your methods as well as on the accuracy and completeness of your results and explanations.*

Will using name-brand microwave popcorn result in a greater percentage of popped kernels than using store-brand microwave popcorn? To find out, Briana and Maggie randomly selected 10 bags of name-brand microwave popcorn and 10 bags of store-brand microwave popcorn. The chosen bags were arranged in a random order. Then each bag was popped for 3.5 minutes, and the percentage of popped kernels was calculated. The results are displayed in the following table.

| Name-brand | Store-brand | Name-brand | Store-brand |
|---|---|---|---|
| 95 | 91 | 90 | 78 |
| 88 | 89 | 97 | 84 |
| 84 | 82 | 93 | 86 |
| 94 | 82 | 91 | 86 |
| 81 | 77 | 86 | 90 |

Do the data provide convincing evidence that using name-brand microwave popcorn will result in a greater mean percentage of popped kernels?

**After you finish, you can view two example solutions on the book's Web site (www.whfreeman.com/tps5e). Determine whether you think each solution is "complete," "substantial," "developing," or "minimal." If the solution is not complete, what improvements would you suggest to the student who wrote it? Finally, your teacher will provide you with a scoring rubric. Score your response and note what, if anything, you would do differently to improve your own score.**

# Chapter Review

### Section 10.1: Comparing Two Proportions

In this section, you learned how to construct confidence intervals and perform significance tests for a difference between two proportions. Inference for a difference in proportions is based on the sampling distribution of $\hat{p}_1 - \hat{p}_2$. When the conditions are met, the sampling distribution of $\hat{p}_1 - \hat{p}_2$ is approximately Normal with a mean of $\mu_{\hat{p}_1-\hat{p}_2} = p_1 - p_2$ and a standard deviation of $\sigma_{\hat{p}_1-\hat{p}_2} = \sqrt{\dfrac{p_1(1-p_1)}{n_1} + \dfrac{p_2(1-p_2)}{n_2}}$.

The conditions for inference about a difference in proportions are the same for confidence intervals and significance tests. The Random condition says that the data must be from two independent random samples or two groups in a randomized experiment. The 10% condition says that each sample size should be less than 10% of the corresponding population size when sampling without replacement. The Large Counts condition says that the number of successes and number of failures from each sample/group should be at least 10. That is, $n_1\hat{p}_1$, $n_1(1-\hat{p}_1)$, $n_2\hat{p}_2$, $n_2(1-\hat{p}_2)$ are all ≥10.

A confidence interval for a difference between two proportions provides an interval of plausible values for the true difference in proportions. The formula is

$$(\hat{p}_1 - \hat{p}_2) \pm z^*\sqrt{\frac{\hat{p}_1(1-\hat{p}_1)}{n_1} + \frac{\hat{p}_2(1-\hat{p}_2)}{n_2}}$$

The logic of confidence intervals, including how to interpret the confidence interval and the confidence level, is the same as it was in Chapter 8, when you first learned about confidence intervals.

Likewise, a significance test for a difference between two proportions uses the same logic as the significance tests you learned about in Chapter 9. We start by assuming the null hypothesis is true and asking how likely it would be to get results at least as unusual as the results observed in a study by chance alone. If it is plausible that a difference in proportions could be the result of sampling variability or the chance variation due to random assignment, we do not have convincing evidence that the alternative hypothesis is true. However, if the difference is too big to attribute to chance, there is convincing evidence to believe that the alternative hypothesis is true. For a test of $H_0: p_1 - p_2 = 0$, the test statistic is

$$z = \frac{(\hat{p}_1 - \hat{p}_2) - 0}{\sqrt{\dfrac{\hat{p}_C(1-\hat{p}_C)}{n_1} + \dfrac{\hat{p}_C(1-\hat{p}_C)}{n_2}}}$$

### TRM Chapter 10 Tests

There are four different tests available for this chapter in the Teacher's Resource Materials. Versions A and B are roughly parallel and version D is cumulative. Find them by clicking on the link in the TE-book, opening the TRCD, or logging onto the book's companion Web site www.whfreeman.com/tps5e (teacher log-in required).

where $\hat{p}_C$ is the combined (overall) proportion of successes: $\hat{p}_C = \frac{X_1 + X_2}{n_1 + n_2}$.

Finally, you learned that the inference techniques used for analyzing a difference in proportions from two independent random samples work very well for analyzing a difference in proportions from two groups in a completely randomized experiment.

**Section 10.2: Comparing Two Means**

In this section, you learned how to construct confidence intervals and perform significance tests for a difference in two means. Inference for a difference in means is based on the sampling distribution of $\bar{x}_1 - \bar{x}_2$. When the conditions are met, the sampling distribution of $\bar{x}_1 - \bar{x}_2$ is approximately Normal with a mean of $\mu_{\bar{x}_1-\bar{x}_2} = \mu_1 - \mu_2$ and a standard deviation of $\sigma_{\bar{x}_1-\bar{x}_2} = \sqrt{\frac{\sigma_1^2}{n_1} + \frac{\sigma_2^2}{n_2}}$.

The conditions for inference about a difference in means are the same for confidence intervals and significance tests. The Random condition says that the data must be from two independent random samples or two groups in a randomized experiment. The 10% condition says that each sample size should be less than 10% of the corresponding population size when sampling without replacement. The Normal/Large Sample condition says that the two populations are Normal or that the two sample/group sizes are large ($n_1 \geq 30$, $n_2 \geq 30$). If the sample/group sizes are small and the population shapes are unknown, graph both sets of data to make sure there is no strong skewness or outliers.

As in Chapters 8 and 9, inference techniques for means are based on the $t$ distributions. There are two options for calculating the number of degrees of freedom to use. The first option is to use technology to calculate the degrees of freedom. The second option is to use the smaller of $n_1 - 1$ and $n_2 - 1$. The technology option is preferred because it produces a larger number of degrees of freedom, resulting in narrower confidence intervals and smaller *P*-values. If you are using technology, always choose the *un*pooled option.

A confidence interval for a difference between two means provides an interval of plausible values for the true difference in means. The formula is

$$(\bar{x}_1 - \bar{x}_2) \pm t^* \sqrt{\frac{s_1^2}{n_1} + \frac{s_1^2}{n_2}}$$

Use a significance test to decide between two competing hypotheses about a difference in true means. The test statistic is

$$t = \frac{(\bar{x}_1 - \bar{x}_2) - (\mu_1 - \mu_2)}{\sqrt{\frac{s_1^2}{n_1} + \frac{s_2^2}{n_2}}}$$

where $\mu_1 - \mu_2$ is the difference specified by the null hypothesis.

When constructing confidence intervals or performing significance tests for a difference in means, make sure that the data are not paired. If the data are paired, use the paired $t$ procedures from Chapter 9.

## What Did You Learn?

| Learning Objective | Section | Related Example on Page(s) | Relevant Chapter Review Exercise(s) |
|---|---|---|---|
| Describe the shape, center, and spread of the sampling distribution of $\hat{p}_1 - \hat{p}_2$. | 10.1 | 615 | R10.2 |
| Determine whether the conditions are met for doing inference about $p_1 - p_2$. | 10.1 | 617 | R10.5, R10.6 |
| Construct and interpret a confidence interval to compare two proportions. | 10.1 | 617 | R10.2 |
| Perform a significance test to compare two proportions. | 10.1 | 622, 625 | R10.5 |
| Describe the shape, center, and spread of the sampling distribution of $\bar{x}_1 - \bar{x}_2$. | 10.2 | 638 | R10.3 |
| Determine whether the conditions are met for doing inference for $\mu_1 - \mu_2$. | 10.2 | 641 | R10.3, R10.4, R10.6 |
| Construct and interpret a confidence interval to compare two means. | 10.2 | 641 | R10.4 |
| Perform a significance test to compare two means. | 10.2 | 645 | R10.7 |
| Determine when it is appropriate to use two-sample *t* procedures versus paired *t* procedures. | 10.2 | 650 | R10.1, R10.7 |



**TRM** Learning Objectives Grid

# Chapter 10 Chapter Review Exercises

*These exercises are designed to help you review the important ideas and methods of the chapter.*

**R10.1 Which procedure?** For each of the following settings, say which inference procedure from Chapter 8, 9, or 10 you would use. Be specific. For example, you might say, "Two-sample $z$ test for the difference between two proportions." You do not need to carry out any procedures.[43]

(a) Do people smoke less when cigarettes cost more? A random sample of 500 smokers was selected. The number of cigarettes each person smoked per day was recorded over a one-month period before a 30% cigarette tax was imposed and again for one month after the tax was imposed.

(b) How much greater is the percent of senior citizens who attend a play at least once per year than the percent of people in their twenties who do so? Random samples of 100 senior citizens and 100 people in their twenties were surveyed.

(c) You have data on rainwater collected at 16 locations in the Adirondack Mountains of New York State. One measurement is the acidity of the water, measured by pH on a scale of 0 to 14 (the pH of distilled water is 7.0). Estimate the average acidity of rainwater in the Adirondacks.

(d) Consumers Union wants to see which of two brands of calculator is easier to use. They recruit 100 volunteers and randomly assign them to two equal-sized groups. The people in one group use Calculator A and those in the other group use Calculator B. Researchers record the time required for each volunteer to carry out the same series of routine calculations (such as figuring discounts and sales tax, totaling a bill) on the assigned calculator.

**R10.2 Seat belt use** The proportion of drivers who use seat belts depends on things like age (young people are more likely to go unbelted) and gender (women are more likely to use belts). It also depends on local law. In New York City, police can stop a driver who is not belted. In Boston at the time of the study, police could cite a driver for not wearing a seat belt only if the driver had been stopped for some other violation. Here are data from observing random samples of female Hispanic drivers in these two cities:[44]

| City | Drivers | Belted |
|---|---|---|
| New York | 220 | 183 |
| Boston | 117 | 68 |

(a) Calculate the standard error of the sampling distribution of the difference in the proportions of female Hispanic drivers in the two cities who wear seat belts. What information does this value provide?

(b) Construct and interpret a 95% confidence interval for the difference in the proportions of female Hispanic drivers in the two cities who wear seat belts.

**R10.3 Expensive ads** Consumers who think a product's advertising is expensive often also think the product must be of high quality. Can other information undermine this effect? To find out, marketing researchers did an experiment. The subjects were 90 women from the clerical and administrative staff of a large organization. All subjects read an ad that described a fictional line of food products called "Five Chefs." The ad also described the major TV commercials that would soon be shown, an unusual expense for this type of product. The 45 women who were randomly assigned to the control group read nothing else. The 45 in the "undermine group" also read a news story headlined "No Link between Advertising Spending and New Product Quality." All the subjects then rated the quality of Five Chefs products on a 7-point scale. The study report said, "The mean quality ratings were significantly lower in the undermine treatment ($\bar{x}_A = 4.56$) than in the control treatment ($\bar{x}_C = 5.05$; $t = 2.64$, $P < 0.01$)."[45]

(a) The 90 women who participated in the study were not randomly selected from a population. Explain why the Random condition is still satisfied.

(b) The distribution of individual responses is not Normal, because there is only a 7-point scale. What is the shape of the sampling distribution of $\bar{x}_C - \bar{x}_A$? Explain.

(c) Interpret the $P$-value in context.

**R10.4 Men versus women** The National Assessment of Educational Progress (NAEP) Young Adult Literacy Assessment Survey interviewed a random sample of 1917 people 21 to 25 years old. The sample contained 840 men and 1077 women.[46] The mean and standard deviation of scores on the NAEP's test of quantitative skills were $\bar{x}_1 = 272.40$ and $s_1 = 59.2$ for the men in the sample. For the women, the results were $\bar{x}_2 = 274.73$ and $s_2 = 57.5$.

(a) Construct and interpret a 90% confidence interval for the difference in mean score for male and female young adults.

(b) Based only on the interval from part (a), is there convincing evidence of a difference in mean score for male and female young adults?

**TRM Full Solutions to Chapter 10 Review Exercises**

## Answers to Chapter 10 Review Exercises

**R10.1 (a)** Paired $t$ test for the mean difference. **(b)** Two-sample $z$ interval for the difference in proportions. **(c)** One-sample $t$ interval for the mean. **(d)** Two-sample $t$ interval for the difference between two means.

**R10.2 (a)** $SE_{\hat{p}_1-\hat{p}_2} = \sqrt{\frac{0.832(1-0.832)}{220} + \frac{0.581(1-0.581)}{117}} = 0.0521$. If we were to take many random samples of 220 Hispanic female drivers in New York and 117 Hispanic female drivers in Boston, the difference in the sample proportions who wear seat belts will typically be 0.0521 from the true difference in proportions of all Hispanic female drivers in New York and Boston who wear seat belts. **(b)** *S*: $p_1$ = proportion of all Hispanic female drivers in New York who wear seat belts and $p_2$ = proportion of all Hispanic female drivers in Boston who wear seat belts. *P*: Two-sample $z$ interval for $p_1 - p_2$. *Random*: Independent random samples. *10%*: $n_1 = 220 <$ 10% of all Hispanic female drivers in New York and $n_2 = 117 <$ 10% of all Hispanic female drivers in Boston. *Large Counts*: 183, 37, 68, 49 are all $\geq 10$. *D*: (0.149, 0.353). *C*: We are 95% confident that the interval from 0.149 to 0.353 captures the true difference in the proportions of Hispanic women drivers in New York and Boston who wear their seat belts.

**R10.3 (a)** The women in the study were randomly assigned to one of the two treatments. **(b)** Because both groups are large ($n_C = 45 \geq 30$ and $n_A = 45 \geq 30$), the sampling distribution of $\bar{x}_C - \bar{x}_A$ should be approximately Normal. **(c)** Assuming there is no difference in the true mean ratings of the product for women like these who read or don't read the news story, there is $<$ a 0.01 probability of observing a difference as large as or larger than 0.49 by chance alone.

**R10.4 (a)** *S*: $\mu_1$ = the true mean NAEP quantitative skills test score for young men and $\mu_2$ = the true mean NAEP quantitative skills test score for young women. *P*: Two-sample $t$ interval for $\mu_1 - \mu_2$. *Random*: Reasonable to consider these independent random samples. *10%*: $n_1 = 840 <$ 10% of all young men and $n_2 = 1077 <$ 10% of all young women. *Normal/Large Sample*: $n_1 = 840 \geq 30$ and $n_2 = 1077 \geq 30$. *D*: Using df = 100, (−6.80, 2.14). Using df = 1777.52, (−6.76, 2.10). *C*: We are 90% confident that the interval from −6.76 to 2.10 captures the true difference in the mean NAEP quantitative skills test score for young men and the mean NAEP quantitative skills test score for young women. **(b)** Because 0 is in the interval, we do not have convincing evidence of a difference in the mean score for male and female young adults.

**R10.5 Treating AIDS** The drug AZT was the first drug that seemed effective in delaying the onset of AIDS. Evidence for AZT's effectiveness came from a large randomized comparative experiment. The subjects were 870 volunteers who were infected with HIV, the virus that causes AIDS, but did not yet have AIDS. The study assigned 435 of the subjects at random to take 500 milligrams of AZT each day and another 435 to take a placebo. At the end of the study, 38 of the placebo subjects and 17 of the AZT subjects had developed AIDS.

(a) Do the data provide convincing evidence at the $\alpha = 0.05$ level that taking AZT lowers the proportion of infected people who will develop AIDS in a given period of time?

(b) Describe a Type I error and a Type II error in this setting and give a consequence of each error. Based on your conclusion in part (a), which error could have been made in this study?

**R10.6 Conditions** Explain why it is not safe to use the methods of this chapter to perform inference in each of the following settings.

(a) Lyme disease is spread in the northeastern United States by infected ticks. The ticks are infected mainly by feeding on mice, so more mice result in more infected ticks. The mouse population in turn rises and falls with the abundance of acorns, their favored food. Experimenters studied two similar forest areas in a year when the acorn crop failed. They added hundreds of thousands of acorns to one area to imitate an abundant acorn crop, while leaving the other area untouched. The next spring, 54 of the 72 mice trapped in the first area were in breeding condition, versus 10 of the 17 mice trapped in the second area.[47]

(b) Who texts more—males or females? For their final project, a group of AP® Statistics students investigated their belief that females text more than males. They asked a random sample of 31 students—15 males and 16 females—from their school to record the number of text messages sent and received over a 2-day period. Boxplots of their data are shown below.

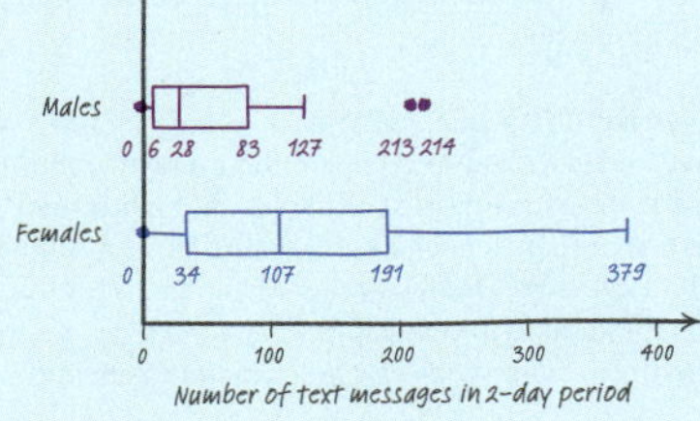


**R10.7 Each day I am getting better in math** A "subliminal" message is below our threshold of awareness but may nonetheless influence us. Can subliminal messages help students learn math? A group of 18 students who had failed the mathematics part of the City University of New York Skills Assessment Test agreed to participate in a study to find out. All received a daily subliminal message, flashed on a screen too rapidly to be consciously read. The treatment group of 10 students (assigned at random) was exposed to "Each day I am getting better in math." The control group of 8 students was exposed to a neutral message, "People are walking on the street." All 18 students participated in a summer program designed to raise their math skills, and all took the assessment test again at the end of the program. The table below gives data on the subjects' scores before and after the program.[48]

| Treatment Group | | | Control Group | | |
|---|---|---|---|---|---|
| Pretest | Posttest | Difference | Pretest | Posttest | Difference |
| 18 | 24 | 6 | 18 | 29 | 11 |
| 18 | 25 | 7 | 24 | 29 | 5 |
| 21 | 33 | 12 | 20 | 24 | 4 |
| 18 | 29 | 11 | 18 | 26 | 8 |
| 18 | 33 | 15 | 24 | 38 | 14 |
| 20 | 36 | 16 | 22 | 27 | 5 |
| 23 | 34 | 11 | 15 | 22 | 7 |
| 23 | 36 | 13 | 19 | 31 | 12 |
| 21 | 34 | 13 | | | |
| 17 | 27 | 10 | | | |

(a) Explain why a two-sample $t$ test is more appropriate than a paired $t$ test for analyzing these data.

(b) The Fathom boxplots below display the differences in pretest and posttest scores for the students in the control (Cdiff) and treatment (Tdiff) groups. Write a few sentences comparing the performance of these two groups.

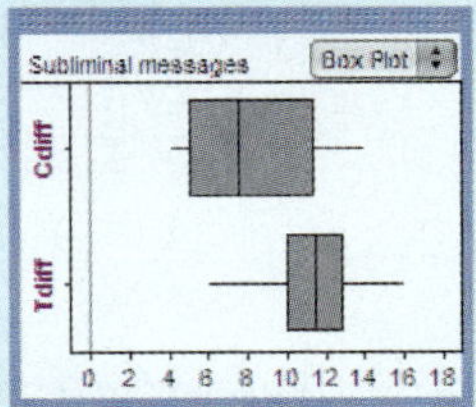


(c) Do the data provide convincing evidence that subliminal messages help students learn math?

(d) Can we generalize these results to the population of all students who failed the mathematics part of the City University of New York Skills Assessment Test? Why or why not?

**R10.5 (a)** *S*: $H_0: p_1 - p_2 = 0$ vs. $H_a: p_1 - p_2 < 0$, where $p_1$ is the true proportion of patients like these who take AZT and develop AIDS and $p_2$ is the true proportion of patients like these who take a placebo and develop AIDS. *P*: Two-sample *z* test for $p_1 - p_2$. *Random*: Two groups in a randomized experiment. *Large Counts*: 17, 418, 38, 397 are all $\geq 10$. *D*: $z = -2.91$, *P*-value $= 0.0018$. *C*: Because the *P*-value of $0.0018 < \alpha = 0.05$, we reject $H_0$. We have convincing evidence that taking AZT lowers the proportion of patients like these who develop AIDS compared to a placebo. **(b)** I: Finding convincing evidence that AZT lowers the risk of developing AIDS, when in reality it does not. Consequence: Patients will pay for a drug that doesn't help. II: Not finding convincing evidence that AZT lowers the risk of developing AIDS, when in reality it does. Consequence: Patients won't take the drug when it could actually delay the onset of AIDS. It is possible that we made a Type I error.

**R10.6 (a)** The Large Counts condition is not met because there are only 7 failures in the control area. **(b)** The Normal/Large Sample condition is not met because both sample sizes are small and there are outliers in the male distribution.

**R10.7 (a)** Even though each subject has two scores (before and after), the two groups of students are independent. **(b)** The distribution for the control group is slightly skewed to the right, while the distribution for the treatment group is roughly symmetric. The center for the treatment group > the center for the control group. The differences in the control group are more variable than the differences in the treatment group. **(c)** *S*: $H_0: \mu_1 - \mu_2 = 0$ vs. $H_a: \mu_1 - \mu_2 > 0$, where $\mu_1 =$ the true mean difference in test scores for students like these who get the treatment message and $\mu_2 =$ the true mean difference in test scores for students like these who get the neutral message. *P*: Two-sample *t* test for $\mu_1 - \mu_2$. *Random*: Two groups in a randomized experiment. *Normal/Large Sample*: Neither boxplot showed strong skewness or any outliers. *D*: Using the differences, $\bar{x}_1 = 11.4$, $s_1 = 3.169$, $\bar{x}_2 = 8.25, s_2 = 3.69$. $t = 1.91$. Using df $= 7$, the *P*-value is between 0.025 and 0.05. Using df $= 13.92$, *P*-value $= 0.0382$. *C*: Because the *P*-value of $0.0382 < \alpha = 0.05$, we reject $H_0$. There is convincing evidence that the true mean difference in test scores for students like these who get the treatment message > the true mean difference in test scores for students like these who get the neutral message. **(d)** We cannot generalize to all students who failed the test because our sample was not a random sample of all students who failed the test.

**TRM** Full Solutions to Chapter 10 AP® Statistics Practice Test

## Answers to Chapter 10 AP® Statistics Practice Test

T10.1 e

T10.2 b

T10.3 a

T10.4 a

# Chapter 10 AP® Statistics Practice Test

**Section I: Multiple Choice** *Select the best answer for each question.*

**T10.1** A study of road rage asked separate random samples of 596 men and 523 women about their behavior while driving. Based on their answers, each respondent was assigned a road rage score on a scale of 0 to 20. Are the conditions for performing a two-sample $t$ test satisfied?

(a) Maybe; we have independent random samples, but we need to look at the data to check Normality.

(b) No; road rage scores in a range between 0 and 20 can't be Normal.

(c) No; we don't know the population standard deviations.

(d) Yes; the large sample sizes guarantee that the corresponding population distributions will be Normal.

(e) Yes; we have two independent random samples and large sample sizes.

**T10.2** Thirty-five people from a random sample of 125 workers from Company A admitted to using sick leave when they weren't really ill. Seventeen employees from a random sample of 68 workers from Company B admitted that they had used sick leave when they weren't ill. A 95% confidence interval for the difference in the proportions of workers at the two companies who would admit to using sick leave when they weren't ill is

(a) $0.03 \pm \sqrt{\frac{(0.28)(0.72)}{125} + \frac{(0.25)(0.75)}{68}}$

(b) $0.03 \pm 1.96\sqrt{\frac{(0.28)(0.72)}{125} + \frac{(0.25)(0.75)}{68}}$

(c) $0.03 \pm 1.645\sqrt{\frac{(0.28)(0.72)}{125} + \frac{(0.25)(0.75)}{68}}$

(d) $0.03 \pm 1.96\sqrt{\frac{(0.269)(0.731)}{125} + \frac{(0.269)(0.731)}{68}}$

(e) $0.03 \pm 1.645\sqrt{\frac{(0.269)(0.731)}{125} + \frac{(0.269)(0.731)}{68}}$

**T10.3** The power takeoff driveline on tractors used in agriculture is a potentially serious hazard to operators of farm equipment. The driveline is covered by a shield in new tractors, but for a variety of reasons, the shield is often missing on older tractors. Two types of shields are the bolt-on and the flip-up. It was believed that the bolt-on shield was perceived as a nuisance by the operators and deliberately removed, but the flip-up shield is easily lifted for inspection and maintenance and may be left in place. In a study initiated by the U.S. National Safety Council, random samples of older tractors with both types of shields were taken to see what proportion of shields were removed. Of 183 tractors designed to have bolt-on shields, 35 had been removed. Of the 136 tractors with flip-up shields, 15 were removed. We wish to perform a test of $H_0: p_b = p_f$ versus $H_a: p_b > p_f$, where $p_b$ and $p_f$ are the proportions of all tractors with the bolt-on and flip-up shields removed, respectively. Which of the following is not a condition for performing the significance test?

(a) Both populations are Normally distributed.

(b) The data come from two independent samples.

(c) Both samples were chosen at random.

(d) The counts of successes and failures are large enough to use Normal calculations.

(e) Both populations are at least 10 times the corresponding sample sizes.

**T10.4** A quiz question gives random samples of $n = 10$ observations from each of two Normally distributed populations. Tom uses a table of $t$ distribution critical values and 9 degrees of freedom to calculate a 95% confidence interval for the difference in the two population means. Janelle uses her calculator's two-sample $t$ interval with 16.87 degrees of freedom to compute the 95% confidence interval. Assume that both students calculate the intervals correctly. Which of the following is true?

(a) Tom's confidence interval is wider.

(b) Janelle's confidence interval is wider.

(c) Both confidence intervals are the same.

(d) There is insufficient information to determine which confidence interval is wider.

(e) Janelle made a mistake; degrees of freedom has to be a whole number.

*Exercises T10.5 and T10.6 refer to the following setting.* A researcher wished to compare the average amount of time spent in extracurricular activities by high school students in a suburban school district with that in a school district of a large city. The researcher obtained an SRS of 60 high school students in a large suburban school district and found the mean time spent in extracurricular activities per week to be 6 hours with a standard deviation of 3 hours. The researcher also obtained an independent SRS of 40 high school students in a large city school district and found the mean time spent in extracurricular activities per week to be 5 hours with a standard deviation of 2 hours. Suppose that the researcher decides to carry out a significance test of $H_0: \mu_{suburban} = \mu_{city}$ versus a two-sided alternative.

**T10.5** The correct test statistic is

(a) $z = \dfrac{(6-5)-0}{\sqrt{\dfrac{3}{60}+\dfrac{2}{40}}}$

(b) $z = \dfrac{(6-5)-0}{\sqrt{\dfrac{3^2}{60}+\dfrac{2^2}{40}}}$

(c) $t = \dfrac{(6-5)-0}{\dfrac{3}{\sqrt{60}}+\dfrac{2}{\sqrt{40}}}$

(d) $t = \dfrac{(6-5)-0}{\sqrt{\dfrac{3}{60}+\dfrac{2}{40}}}$

(e) $t = \dfrac{(6-5)-0}{\sqrt{\dfrac{3^2}{60}+\dfrac{2^2}{40}}}$

**T10.6** The *P*-value for the test is 0.048. A correct conclusion is to

(a) fail to reject $H_0$ at the $\alpha = 0.05$ level. There is convincing evidence of a difference in the average time spent on extracurricular activities by students in the suburban and city school districts.

(b) fail to reject $H_0$ at the $\alpha = 0.05$ level. There is not convincing evidence of a difference in the average time spent on extracurricular activities by students in the suburban and city school districts.

(c) fail to reject $H_0$ at the $\alpha = 0.05$ level. There is convincing evidence that the average time spent on extracurricular activities by students in the suburban and city school districts is the same.

(d) reject $H_0$ at the $\alpha = 0.05$ level. There is not convincing evidence of a difference in the average time spent on extracurricular activities by students in the suburban and city school districts.

(e) reject $H_0$ at the $\alpha = 0.05$ level. There is convincing evidence of a difference in the average time spent on extracurricular activities by students in the suburban and city school districts.

**T10.7** At a baseball game, 42 of 65 randomly selected people own an iPod. At a rock concert occurring at the same time across town, 34 of 52 randomly selected people own an iPod. A researcher wants to test the claim that the proportion of iPod owners at the two venues is different. A 90% confidence interval for the difference in population proportions (game − concert) is (−0.154, 0.138). Which of the following gives the correct outcome of the researcher's test of the claim?

(a) Because the confidence interval includes 0, the researcher can conclude that the proportion of iPod owners at the two venues is the same.

(b) Because the center of the interval is −0.008, the researcher can conclude that a higher proportion of people at the rock concert own iPods than at the baseball game.

(c) Because the confidence interval includes 0, the researcher cannot conclude that the proportion of iPod owners at the two venues is different.

(d) Because the confidence interval includes more negative than positive values, the researcher can conclude that a higher proportion of people at the rock concert own iPods than at the baseball game.

(e) The researcher cannot draw a conclusion about a claim without performing a significance test.

**T10.8** An SRS of size 100 is taken from Population A with proportion 0.8 of successes. An independent SRS of size 400 is taken from Population B with proportion 0.5 of successes. The sampling distribution for the difference (Population A − Population B) in sample proportions has what mean and standard deviation?

(a) mean = 0.3; standard deviation = 1.3

(b) mean = 0.3; standard deviation = 0.40

(c) mean = 0.3; standard deviation = 0.047

(d) mean = 0.3; standard deviation = 0.0022

(e) mean = 0.3; standard deviation = 0.0002

**T10.9** How much more effective is exercise and drug treatment than drug treatment alone at reducing the rate of heart attacks among men aged 65 and older? To find out, researchers perform a completely randomized experiment involving 1000 healthy males in this age group. Half of the subjects are assigned to receive drug treatment only, while the other half are assigned to exercise regularly and to receive drug treatment. The most appropriate inference method for answering the original research question is

(a) one-sample $z$ test for a proportion.

(b) two-sample $z$ interval for $p_1 - p_2$.

(c) two-sample $z$ test for $p_1 - p_2$.

(d) two-sample $t$ interval for $\mu_1 - \mu_2$.

(e) two-sample $t$ test for $\mu_1 - \mu_2$.

**T10.5** e

**T10.6** e

**T10.7** c

**T10.8** c

**T10.9** b

**T10.10** Researchers are interested in evaluating the effect of a natural product on reducing blood pressure. This will be done by comparing the mean reduction in blood pressure of a treatment (natural product) group and a placebo group using a two-sample $t$ test. The researchers would like to be able to detect whether the natural product reduces blood pressure by at least 7 points more, on average, than the placebo. If groups of size 50 are used in the experiment, a two-sample $t$ test using $\alpha = 0.01$ will have a power of 80% to detect a 7-point difference in mean blood pressure reduction. If the researchers want to be able to detect a 5-point difference instead, then the power of the test

(a) would be less than 80%.

(b) would be greater than 80%.

(c) would still be 80%.

(d) could be either less than or greater than 80%, depending on whether the natural product is effective.

(e) would vary depending on the standard deviation of the data.

**Section II: Free Response** *Show all your work. Indicate clearly the methods you use, because you will be graded on the correctness of your methods as well as on the accuracy and completeness of your results and explanations.*

**T10.11** Researchers wondered whether maintaining a patient's body temperature close to normal by heating the patient during surgery would affect wound infection rates. Patients were assigned at random to two groups: the normothermic group (patients' core temperatures were maintained at near normal, 36.5°C, with heating blankets) and the hypothermic group (patients' core temperatures were allowed to decrease to about 34.5°C). If keeping patients warm during surgery alters the chance of infection, patients in the two groups should have hospital stays of very different lengths. Here are summary statistics on hospital stay (in number of days) for the two groups:

| Group | $n$ | $\bar{x}$ | $s_x$ |
|---|---|---|---|
| Normothermic | 104 | 12.1 | 4.4 |
| Hypothermic | 96 | 14.7 | 6.5 |

(a) Construct and interpret a 95% confidence interval for the difference in the true mean length of hospital stay for normothermic and hypothermic patients.

(b) Does your interval in part (a) suggest that keeping patients warm during surgery affects the average length of patients' hospital stays? Justify your answer.

(c) Interpret the meaning of "95% confidence" in the context of this study.

**T10.12** A random sample of 100 of a certain popular car model last year found that 20 had a certain minor defect in the brakes. The car company made an adjustment in the production process to try to reduce the proportion of cars with the brake problem. A random sample of 350 of this year's model found that 50 had the minor brake defect.

(a) Was the company's adjustment successful? Carry out an appropriate test to support your answer.

(b) Describe a Type I error and a Type II error in this setting, and give a possible consequence of each.

**T10.13** Pat wants to compare the cost of one- and two-bedroom apartments in the area of her college campus. She collects data for a random sample of 10 advertisements of each type. The table below shows the rents (in dollars per month) for the selected apartments.

| | | | | | | | | | | |
|---|---|---|---|---|---|---|---|---|---|---|
| **1 bedroom:** | 500 | 650 | 600 | 505 | 450 | 550 | 515 | 495 | 650 | 395 |
| **2 bedroom:** | 595 | 500 | 580 | 650 | 675 | 675 | 750 | 500 | 495 | 670 |

Pat wonders if two-bedroom apartments rent for significantly more, on average, than one-bedroom apartments.

(a) State an appropriate pair of hypotheses for a significance test. Be sure to define any parameters you use.

(b) Name the appropriate test and show that the conditions for carrying out this test are met.

(c) The appropriate test from part (b) yields a $P$-value of 0.029. Interpret this $P$-value in context.

(d) What conclusion should Pat draw at the $\alpha = 0.05$ significance level? Explain.

**T10.10** a

**T10.11** **(a)** *S*: $\mu_1$ = the true mean hospital stay for patients like these who get heating blankets during surgery and $\mu_2$ = the true mean hospital stay for patients like these who have their core temperatures reduced during surgery. *P*: Two-sample *t* interval for $\mu_1 - \mu_2$. *Random*: Two groups in a randomized experiment. *Normal/Large Sample*: $n_1 = 104 \geq 30$ and $n_2 = 96 \geq 30$. *D*: Using df = 80, $(-4.17, -1.03)$. Using df = 165.12, $(-4.16, -1.04)$. *C*: We are 95% confident that the interval from $-4.16$ to $-1.04$ captures the true difference in mean length of hospital stay for patients like these who get heating blankets during surgery and those who have their core temperatures reduced during surgery. **(b)** Yes. Because 0 is not in the interval, we have convincing evidence that the true mean hospital stay for patients like these who get heating blankets during surgery is different than the true mean hospital stay for patients like these who have their core temperatures reduced during surgery. **(c)** If we were to repeat this experiment many times and calculate 95% confidence intervals for the difference in means each time, about 95% of the intervals would capture the true difference in mean hospital stay for patients like these who get heating blankets during surgery and mean hospital stay for patients like these who have their core temperatures reduced during surgery.

**T10.12** **(a)** *S*: $H_0: p_1 - p_2 = 0$ vs. $H_a: p_1 - p_2 > 0$, where $p_1$ is the true proportion of cars that have the brake defect in last year's model and $p_2$ is the true proportion of cars that have the brake defect in this year's model. *P*: Two-sample *z* test for $p_1 - p_2$. *Random*: Independent random samples. *10%*: $n_1 = 100 < 10\%$ of last year's model and $n_2 = 350 < 10\%$ of this year's model. *Large Counts*: 20, 80, 50, 300 are all $\geq 10$. *D*: $z = 1.39$, $P$-value = 0.0822. *C*: Because the $P$-value of 0.0822 $> \alpha = 0.05$, we fail to reject $H_0$. We do not have convincing evidence that the true proportion of brake defects is smaller in this year's model compared to last year's model. **(b)** I: Finding convincing evidence that there is a smaller proportion of brake defects in this year's car model, when in reality there is not. This might result in more accidents because people think that their brakes are safe. II: Not finding convincing evidence that there is a smaller proportion of brake defects in this year's model, when in reality there is. This might result in reduced sales of this year's model.

**T10.13** **(a)** $H_0: \mu_1 - \mu_2 = 0$ vs. $H_a: \mu_1 - \mu_2 < 0$, where $\mu_1$ = the true mean rent for one-bedroom apartments in the area of her college campus and $\mu_2$ = the true mean rent for two-bedroom apartments in the area of her college campus. **(b)** Two-sample *t* test for $\mu_1 - \mu_2$. *Random*: Independent random samples. *10%*: $n_1 = 10 < 10\%$ of all one-bedroom apartments in this area and $n_2 = 10 < 10\%$ of all two-bedroom apartments in this area. *Normal/Large Sample*: The dotplots below show no strong skewness or outliers in either distribution.

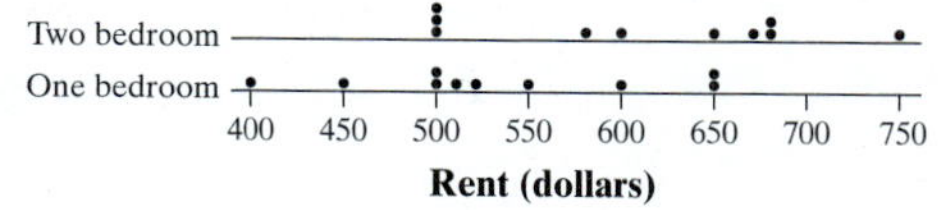


**(c)** Assuming the true mean rent of the two types of apartments is really the same, there is a 0.029 probability of getting an observed difference in mean rents as large as or larger than the one in this study. **(d)** Because the $P$-value of 0.029 $< \alpha = 0.05$, Pat should reject $H_0$. She has convincing evidence that the true mean rent of two-bedroom apartments > the true mean rent of one-bedroom apartments in the area of her college campus.

# Cumulative AP® Practice Test 3

**Section I: Multiple Choice** *Choose the best answer.*

**AP3.1** Suppose the probability that a softball player gets a hit in any single at-bat is 0.300. Assuming that her chance of getting a hit on a particular time at bat is independent of her other times at bat, what is the probability that she will not get a hit until her fourth time at bat in a game?

(a) $\binom{4}{3}(0.3)^1(0.7)^3$ (d) $(0.3)^3(0.7)^1$

(b) $\binom{4}{3}(0.3)^3(0.7)^1$ (e) $(0.3)^1(0.7)^3$

(c) $\binom{4}{1}(0.3)^3(0.7)^1$

**AP3.2** The probability that Color Me Dandy wins a horse race at Batavia Downs given good track conditions is 0.60. The probability of good track conditions on any given day is 0.85. What is the probability that Color Me Dandy wins or the track conditions are good?

(a) 0.94 (b) 0.51 (c) 0.49 (d) 0.06

(e) The answer cannot be determined from the given information.

**AP3.3** *Sports Illustrated* planned to ask a random sample of Division I college athletes, "Do you believe performance-enhancing drugs are a problem in college sports?" How many athletes must be interviewed to estimate the proportion concerned about use of drugs within ±2% with 90% confidence?

(a) 17 (c) 1680 (e) 2401
(b) 21 (d) 1702

**AP3.4** The distribution of grade point averages for a certain college is approximately Normal with a mean of 2.5 and a standard deviation of 0.6. Within which of the following intervals would we expect to find approximately 81.5% of all GPAs for students at this college?

(a) (0.7, 3.1) (c) (1.9, 3.7) (e) (0.7, 4.3)
(b) (1.3, 3.7) (d) (1.9, 4.3)

**AP3.5** Which of the following will increase the power of a significance test?

(a) Increase the Type II error probability.
(b) Decrease the sample size.
(c) Reject the null hypothesis only if the *P*-value is smaller than the level of significance.
(d) Increase the significance level $\alpha$.
(e) Select a value for the alternative hypothesis closer to the value of the null hypothesis.

**AP3.6** You can find some interesting polls online. Anyone can become part of the sample just by clicking on a response. One such poll asked, "Do you prefer watching first-run movies at a movie theater, or waiting until they are available to watch at home or on a digital device?" In all, 8896 people responded, with only 12% (1118 people) saying they preferred theaters. You can conclude that

(a) American adults strongly prefer watching movies at home or on their digital devices.
(b) the high nonresponse rate prevents us from drawing a conclusion.
(c) the sample is too small to draw any conclusion.
(d) the poll uses voluntary response, so the results tell us little about all American adults.
(e) American adults strongly prefer seeing movies at a movie theater.

**AP3.7** A certain candy has different wrappers for various holidays. During Holiday 1, the candy wrappers are 30% silver, 30% red, and 40% pink. During Holiday 2, the wrappers are 50% silver and 50% blue. Forty pieces of candy are randomly selected from the Holiday 1 distribution, and 40 pieces are randomly selected from the Holiday 2 distribution. What are the expected value and standard deviation of the total number of silver wrappers?

(a) 32, 18.4 (c) 32, 4.29 (e) 80, 4.29
(b) 32, 6.06 (d) 80, 18.4

**AP3.8** A beef rancher randomly sampled 42 cattle from her large herd to obtain a 95% confidence interval to estimate the mean weight of the cows in the herd. The interval obtained was (1010, 1321). If the rancher had used a 98% confidence interval instead, the interval would have been

(a) wider and would have less precision than the original estimate.
(b) wider and would have more precision than the original estimate.
(c) wider and would have the same precision as the original estimate.
(d) narrower and would have less precision than the original estimate.
(e) narrower and would have more precision than the original estimate.

**TRM Full Solutions to Cumulative AP® Practice Test 3**

## Answers to Cumulative AP® Practice Test 3

**AP3.1** e
**AP3.2** e
**AP3.3** d
**AP3.4** c
**AP3.5** d
**AP3.6** d
**AP3.7** c
**AP3.8** a

AP3.9 d

AP3.10 c

AP3.11 b

AP3.12 c

AP3.13 c

AP3.14 d



**AP3.9** School A has 400 students and School B has 2700 students. A local newspaper wants to compare the distributions of SAT scores for the two schools. Which of the following would be the most useful for making this comparison?

(a) Back-to-back stemplots for A and B
(b) A scatterplot of A versus B
(c) Dotplots for A and B drawn on the same scale
(d) Two relative frequency histograms of A and B drawn on the same scale
(e) Two bar graphs for A and B drawn on the same scale

**AP3.10** Let X represent the outcome when a fair six-sided die is rolled. For this random variable, $\mu_X = 3.5$ and $\sigma_X = 1.71$. If the die is rolled 100 times, what is the approximate probability that the total score is at least 375?

(a) 0.0000 (c) 0.0721 (e) 0.9279
(b) 0.0017 (d) 0.4420

**AP3.11** An agricultural station is testing the yields for six different varieties of seed corn. The station has four large fields available, which are located in four distinctly different parts of the county. The agricultural researchers consider the climatic and soil conditions in the four parts of the county as being unequal but are reasonably confident that the conditions within each field are fairly similar throughout. The researchers divide each field into six sections and then randomly assign one variety of corn seed to each section in that field. This procedure is done for each field. At the end of the growing season, the corn will be harvested, and the yield, measured in tons per acre, will be compared. Which one of the following statements about the design is correct?

(a) This is an observational study because the researchers are watching the corn grow.
(b) This a randomized block design with fields as blocks and seed types as treatments.
(c) This is a randomized block design with seed types as blocks and fields as treatments.
(d) This is a completely randomized design because the six seed types were randomly assigned to the four fields.
(e) This is a completely randomized design with 24 treatments—6 seed types and 4 fields.

**AP3.12** The correlation between the heights of fathers and the heights of their (grownup) sons is $r = 0.52$, both measured in inches. If fathers' heights were measured in feet instead, the correlation between heights of fathers and heights of sons would be

(a) much smaller than 0.52.
(b) slightly smaller than 0.52.
(c) unchanged; equal to 0.52.
(d) slightly larger than 0.52.
(e) much larger than 0.52.

**AP3.13** A random sample of 200 New York State voters included 88 Republicans, while a random sample of 300 California voters produced 141 Republicans. Which of the following represents the 95% confidence interval that should be used to estimate the true difference in the proportions of Republicans in New York State and California?

(a) $(0.44 - 0.47) \pm 1.96\frac{(0.44)(0.56) + (0.47)(0.53)}{\sqrt{200 + 300}}$

(b) $(0.44 - 0.47) \pm 1.96\frac{(0.44)(0.56)}{\sqrt{200}} + \frac{(0.47)(0.53)}{\sqrt{300}}$

(c) $(0.44 - 0.47) \pm 1.96\sqrt{\frac{(0.44)(0.56)}{200} + \frac{(0.47)(0.53)}{300}}$

(d) $(0.44 - 0.47) \pm 1.96\sqrt{\frac{(0.44)(0.56) + (0.47)(0.53)}{200 + 300}}$

(e) $(0.44 - 0.47) \pm 1.96\sqrt{\frac{(0.45)(0.55)}{200} + \frac{(0.45)(0.55)}{300}}$

**AP3.14** Which of the following is *not* a property of a binomial setting?

(a) Outcomes of different trials are independent.
(b) The chance process consists of a fixed number of trials, $n$.
(c) The probability of success is the same for each trial.
(d) Trials are repeated until a success occurs.
(e) Each trial can result in either a success or a failure.

**AP3.15** Mrs. Woods and Mrs. Bryan are avid vegetable gardeners. They use different fertilizers, and each claims that hers is the best fertilizer to use when growing tomatoes. Both agree to do a study using the weight of their tomatoes as the response variable. They had each planted the same varieties of tomatoes on the same day and fertilized the plants on the same schedule throughout the growing season. At harvest time, they each randomly select 15 tomatoes from their respective gardens and weigh them. After performing a two-sample $t$ test on the difference in mean weights of tomatoes, they get $t = 5.24$ and $P = 0.0008$. Can the gardener with the larger mean claim that her fertilizer caused her tomatoes to be heavier?

(a) Yes, because a different fertilizer was used on each garden.

(b) Yes, because random samples were taken from each garden.

(c) Yes, because the *P*-value is so small.

(d) No, because the soil conditions in the two gardens is a potential confounding variable.

(e) No, because there was no replication.

**AP3.16** The Environmental Protection Agency is charged with monitoring industrial emissions that pollute the atmosphere and water. So long as emission levels stay within specified guidelines, the EPA does not take action against the polluter. If the polluter is in violation of the regulations, the offender can be fined, forced to clean up the problem, or possibly closed. Suppose that for a particular industry the acceptable emission level has been set at no more than 5 parts per million (5 ppm). The null and alternative hypotheses are $H_0: \mu = 5$ versus $H_a: \mu > 5$. Which of the following describes a Type II error?

(a) The EPA fails to find convincing evidence that emissions exceed acceptable limits when, in fact, they are within acceptable limits.

(b) The EPA finds convincing evidence that emissions exceed acceptable limits when, in fact, they are within acceptable limits.

(c) The EPA finds convincing evidence that emissions exceed acceptable limits when, in fact, they do exceed acceptable limits.

(d) The EPA takes more samples to ensure that they make the correct decision.

(e) The EPA fails to find convincing evidence that emissions exceed acceptable limits when, in fact, they do exceed acceptable limits.

**AP3.17** Which of the following is *false*?

(a) A measure of center alone does not completely describe the characteristics of a set of data. Some measure of spread is also needed.

(b) If the original measurements are in inches, converting them to centimeters will not change the mean or standard deviation.

(c) One of the disadvantages of a histogram is that it doesn't show each data value.

(d) Between the range and the interquartile range, the *IQR* is a better measure of spread if there are outliers.

(e) If a distribution is skewed, the median and interquartile range should be reported rather than the mean and standard deviation.

**AP3.18** A 96% confidence interval for the proportion of the labor force that is unemployed in a certain city is (0.07, 0.10). Which of the following statements about this interval is true?

(a) The probability is 0.96 that between 7% and 10% of the labor force is unemployed.

(b) About 96% of the intervals constructed by this method will contain the true proportion of unemployed in the city.

(c) In repeated samples of the same size, there is a 96% chance that the sample proportion will fall between 0.07 and 0.10.

(d) The true rate of unemployment lies within this interval 96% of the time.

(e) Between 7% and 10% of the labor force is unemployed 96% of the time.

**AP3.19** A large toy company introduces a lot of new toys to its product line each year. The company wants to predict the demand as measured by *y*, first-year sales (in millions of dollars) using *x*, awareness of the product (as measured by the percent of customers who had heard of the product by the end of the second month after its introduction). A random sample of 65 new products was taken, and a correlation of 0.96 was computed. Which of the following is a correct interpretation of this value?

(a) Ninety-six percent of the time, the least-squares regression line accurately predicts first-year sales.

(b) About 92% of the time, the percent of people who have heard of the product by the end of the second month will correctly predict first-year sales.

(c) About 92% of first-year sales can be accounted for by the percent of people who have heard of the product by the end of the second month.

(d) For each increase of 1% in awareness of the new product, the predicted sales will go up by 0.96 million dollars.

(e) About 92% of the variation in first-year sales can be accounted for by the least-squares regression line with percent of people who have heard of the product by the end of the second month as the explanatory variable.

**AP3.20** Final grades for a class are approximately Normally distributed with a mean of 76 and a standard deviation of 8. A professor says that the top 10% of the class will receive an A, the next 20% a B, the next 40% a C, the next 20% a D, and the bottom 10% an F. What is the approximate maximum grade a student could attain and still receive an F for the course?

(a) 70 (b) 69.27 (c) 65.75 (d) 62.84 (e) 57

**AP3.15** d
**AP3.16** e
**AP3.17** b
**AP3.18** b
**AP3.19** e
**AP3.20** c

AP3.21 a
AP3.22 d
AP3.23 b
AP3.24 e
AP3.25 a



AP3.21 National Park rangers keep data on the bears that inhabit their park. Below is a histogram of the weights of 143 bears measured in a recent year.

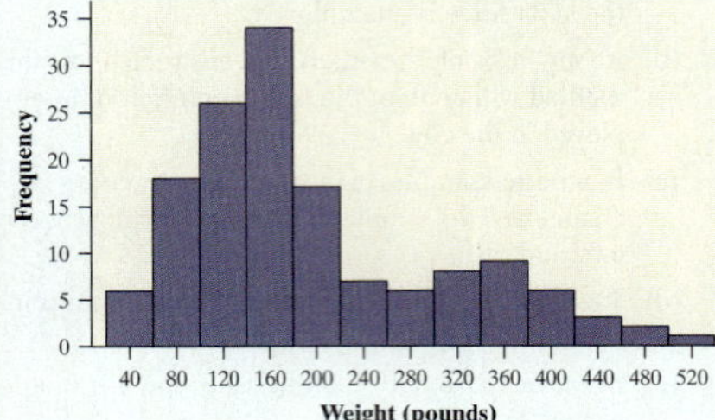


Which statement below is correct?

(a) The median will lie in the interval (140, 180), and the mean will lie in the interval (180, 220).
(b) The median will lie in the interval (140, 180), and the mean will lie in the interval (260, 300).
(c) The median will lie in the interval (100, 140), and the mean will lie in the interval (180, 220).
(d) The mean will lie in the interval (140, 180), and the median will lie in the interval (260, 300).
(e) The mean will lie in the interval (100, 140), and the median will lie in the interval (180, 220).

AP3.22 A random sample of size $n$ will be selected from a population, and the proportion of those in the sample who have a Facebook page will be calculated. How would the margin of error for a 95% confidence interval be affected if the sample size were increased from 50 to 200?

(a) It remains the same.
(b) It is multiplied by 2.
(c) It is multiplied by 4.
(d) It is divided by 2.
(e) It is divided by 4.

AP3.23 A scatterplot and a least-squares regression line are shown in the figure below. What effect does point P have on the slope of the regression line and the correlation?

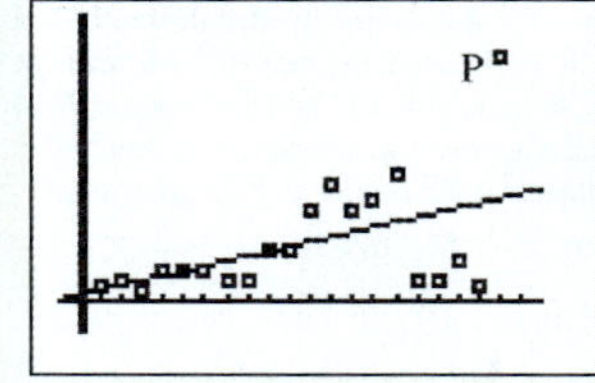


(a) Point P increases the slope and increases the correlation.
(b) Point P increases the slope and decreases the correlation.
(c) Point P decreases the slope and decreases the correlation.
(d) Point P decreases the slope and increases the correlation.
(e) No conclusion can be drawn because the other coordinates are unknown.

AP3.24 The following dotplots show the average high temperatures (in degrees Fahrenheit) for a sample of tourist cities from around the world. Both the January and July average high temperatures are shown. What is one statement that can be made with certainty from an analysis of the graphical display?

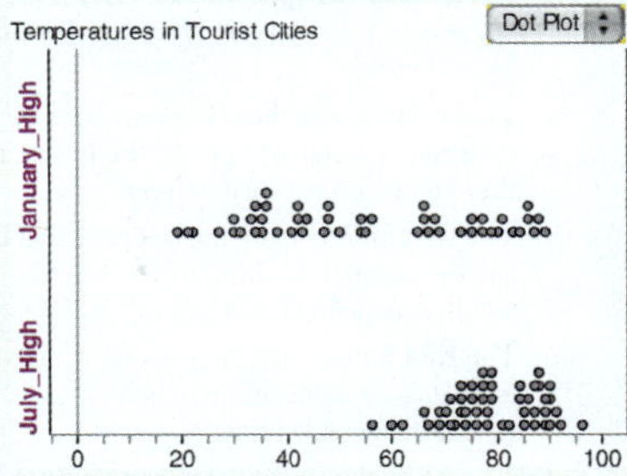


(a) Every city has a larger average high temperature in July than in January.
(b) The distribution of temperatures in July is skewed right, while the distribution of temperatures in January is skewed left.
(c) The median average high temperature for January is higher than the median average high temperature for July.
(d) There appear to be outliers in the average high temperatures for January and July.
(e) There is more variability in average high temperatures in January than in July.

AP3.25 Suppose the null and alternative hypotheses for a significance test are defined as

$$H_0: \mu = 40$$
$$H_a: \mu < 40$$

Which of the following specific values for $H_a$ will give the highest power?

(a) $\mu = 38$
(b) $\mu = 39$
(c) $\mu = 40$
(d) $\mu = 41$
(e) $\mu = 42$

**AP3.26** A large university is considering the establishment of a schoolwide recycling program. To gauge interest in the program by means of a questionnaire, the university takes separate random samples of undergraduate students, graduate students, faculty, and staff. This is an example of what type of sampling design?

(a) Simple random sample
(b) Stratified random sample
(c) Convenience sample
(d) Cluster sample
(e) Randomized block design

**AP3.27** Suppose the true proportion of people who use public transportation to get to work in the Washington, D.C., area is 0.45. In a simple random sample of 250 people who work in Washington, about how far do you expect the sample proportion to be from the true proportion?

(a) 0.4975 (b) 0.2475 (c) 0.0315 (d) 0.0009 (e) 0

***Questions 28 and 29 refer to the following setting.*** According to sleep researchers, if you are between the ages of 12 and 18 years old, you need 9 hours of sleep to be fully functional. A simple random sample of 28 students was chosen from a large high school, and these students were asked how much sleep they got the previous night. The mean of the responses was 7.9 hours, with a standard deviation of 2.1 hours.

**AP3.28** If we are interested in whether students at this high school are getting too little sleep, which of the following represents the appropriate null and alternative hypotheses?

(a) $H_0: \mu = 7.9$ and $H_a: \mu < 7.9$
(b) $H_0: \mu = 7.9$ and $H_a: \mu \neq 7.9$
(c) $H_0: \mu = 9$ and $H_a: \mu \neq 9$
(d) $H_0: \mu = 9$ and $H_a: \mu < 9$
(e) $H_0: \mu \leq 9$ and $H_a: \mu \geq 9$

**AP3.29** Which of the following is the test statistic for the hypothesis test?

(a) $t = \dfrac{7.9 - 9}{\frac{2.1}{\sqrt{28}}}$ (b) $t = \dfrac{9 - 7.9}{\frac{2.1}{\sqrt{28}}}$ (c) $t = \dfrac{7.9 - 9}{\sqrt{\frac{2.1}{28}}}$

(d) $t = \dfrac{7.9 - 9}{\frac{2.1}{\sqrt{27}}}$ (e) $t = \dfrac{9 - 7.9}{\frac{2.1}{\sqrt{27}}}$

**AP3.30** Shortly before the 2012 presidential election, a survey was taken by the school newspaper at a very large state university. Randomly selected students were asked, "Whom do you plan to vote for in the upcoming presidential election?" Here is a two-way table of the responses by political persuasion for 1850 students:

| Candidate of choice | Political Persuasion: Democrat | Republican | Independent | Total |
|---|---|---|---|---|
| Obama | 925 | 78 | 26 | **1029** |
| Romney | 78 | 598 | 19 | **695** |
| Other | 2 | 8 | 11 | **21** |
| Undecided | 32 | 28 | 45 | **105** |
| **Total** | **1037** | **712** | **101** | **1850** |

Which of the following statements about these data is true?

(a) The percent of Republicans among the respondents is 41%.
(b) The marginal distribution of the variable choice of candidate is given by Obama: 55.6%; Romney: 37.6%; Other: 1.1%; Undecided: 5.7%.
(c) About 11.2% of Democrats reported that they planned to vote for Romney.
(d) About 44.6% of those who are undecided are Independents.
(e) The conditional distribution of political persuasion among those for whom Romney is the candidate of choice is Democrat: 7.5%; Republican: 84.0%; Independent: 18.8%

**Section II: Free Response** *Show all your work. Indicate clearly the methods you use, because you will be graded on the correctness of your methods as well as on the accuracy and completeness of your results and explanations.*

**AP3.31** A researcher wants to determine whether or not a five-week crash diet is effective over a long period of time. A random sample of 15 dieters is selected. Each person's weight is recorded before starting the diet and one year after it is concluded. Based on the data shown at right (weight in pounds), can we conclude that the diet has a long-term effect, that is, that dieters manage to not regain the weight they lose? Include appropriate statistical evidence to justify your answer.

| | 1 | 2 | 3 | 4 | 5 | 6 | 7 | 8 |
|---|---|---|---|---|---|---|---|---|
| Before | 158 | 185 | 176 | 172 | 164 | 234 | 258 | 200 |
| After | 163 | 182 | 188 | 150 | 161 | 220 | 235 | 191 |

| | 9 | 10 | 11 | 12 | 13 | 14 | 15 |
|---|---|---|---|---|---|---|---|
| Before | 228 | 246 | 198 | 221 | 236 | 255 | 231 |
| After | 228 | 237 | 209 | 220 | 222 | 268 | 234 |

**AP3.26** b

**AP3.27** c

**AP3.28** d

**AP3.29** a

**AP3.30** b

**AP3.31** *S*: $H_0: \mu_d = 0$ versus $H_a: \mu_d < 0$, where $\mu_d$ is the true mean change in weight (after − before) in pounds for people like these who follow a 5-week crash diet. *P*: Paired *t* test for $\mu_d$. *Random*: Random sample. *10%*: $n_d = 15$ is less than 10% of all dieters. *Normal/Large Sample*: There is no strong skewness or outliers.

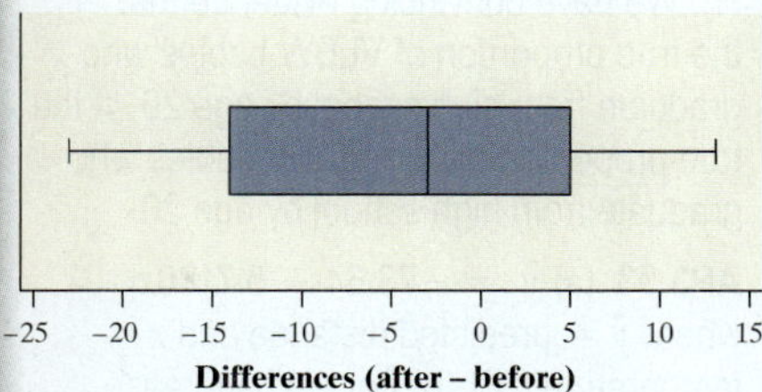


*D*: $\bar{x} = -3.6$ and $s_x = 11.53$. $t = -1.21$. Using df = 14, the *P*-value is between 0.10 and 0.15 (0.1232). *C*: Because the *P*-value of 0.1232 is greater than $\alpha = 0.05$, we fail to reject $H_0$. We do not have convincing evidence that the true mean change in weight (after − before) for people like these who follow a 5-week crash diet < 0.

**AP3.32** Starting in the 1970s, medical technology allowed babies with very low birth weight (VLBW, less than 1500 grams, or about 3.3 pounds) to survive without major handicaps. It was noticed that these children nonetheless had difficulties in school and as adults. A long study has followed 242 randomly selected VLBW babies to age 20 years, along with a control group of 233 randomly selected babies from the same population who had normal birth weight.[49]

(a) Is this an experiment or an observational study? Why?

(b) At age 20, 179 of the VLBW group and 193 of the control group had graduated from high school. Is the graduation rate among the VLBW group significantly lower than for the normal-birth-weight controls? Give appropriate statistical evidence to justify your answer.

**AP3.33** A nuclear power plant releases water into a nearby lake every afternoon at 4:51 P.M. Environmental researchers are concerned that fish are being driven away from the area around the plant. They believe that the temperature of the water discharged may be a factor. The scatterplot below shows the temperature of the water (°C) released by the plant and the measured distance (in meters) from the outflow pipe of the plant to the nearest fish found in the water on eight randomly chosen afternoons.

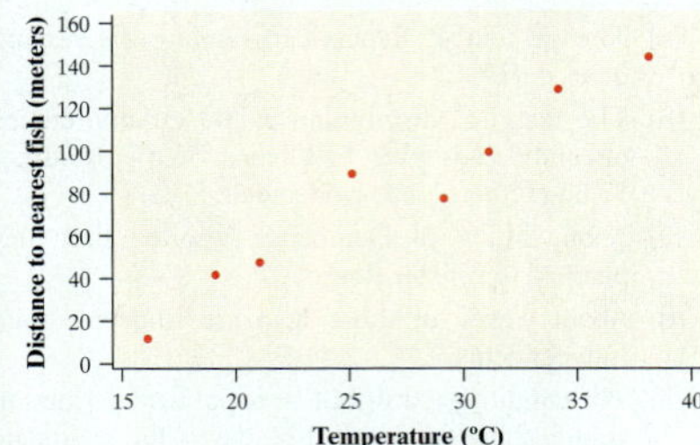


Computer output from a least-squares regression analysis on these data and a residual plot are shown below.

| Predictor | Coef | SE Coef | T | P |
|---|---|---|---|---|
| Constant | -73.64 | 15.48 | -4.76 | 0.003 |
| Temperature | 5.7188 | 0.5612 | -10.19 | 0.000 |

S = 11.4175 R-Sq = 94.5% R-Sq(adj) = 93.6%

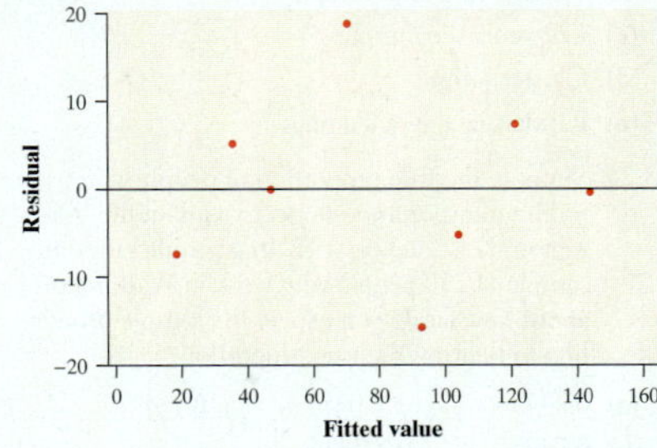


(a) Write the equation of the least-squares regression line. Define any variables you use.

(b) Interpret the slope of the regression line in context.

(c) Is a linear model appropriate for describing the relationship between temperature and distance to the nearest fish? Justify your answer.

(d) Compute the residual for the point (29, 78). Interpret this residual in context.

**AP3.34** The Candy Shoppe assembles gift boxes that contain 8 chocolate truffles and 2 handmade caramel nougats. The truffles have a mean weight of 2 ounces with a standard deviation of 0.5 ounce, and the nougats have a mean weight of 4 ounces with a standard deviation of 1 ounce. The empty boxes weigh 3 ounces with a standard deviation of 0.2 ounce.

**AP3.32 (a)** Observational study. No treatments were imposed on the individuals in the study. **(b)** *S*: $H_0: p_1 - p_2 = 0$ vs. $H_a: p_1 - p_2 < 0$, where $p_1$ is the true proportion of VLBW babies who graduate from high school by age 20 and $p_2$ is the true proportion of non-VLBW babies who graduate from high school by age 20. *P*: Two-sample *z* test for $p_1 - p_2$. *Random*: Independent random samples. *10%*: $n_1 = 242$ is less than 10% of all VLBW babies and $n_2 = 233$ is less than 10% of all non-VLBW babies. *Large Counts*: 179, 63, 193, 40 are all $\geq 10$. *D*: $z = -2.34$ and *P*-value $= 0.0095$. *C*: Because the *P*-value of 0.0095 is less than $\alpha = 0.05$, we reject $H_0$. We have convincing evidence that the true proportion of VLBW babies who graduate from high school by age 20 < the true proportion of non-VLBW babies who graduate from high school by age 20.

**AP3.33 (a)** $\hat{y} = -73.64 + 5.7188x$, where $\hat{y}$ = predicted distance and $x$ = temperature (°C) **(b)** For each increase of 1° (in degrees Celsius) in the water discharge temperature, the predicted distance from the nearest fish to the outflow pipe increases by about 5.7188 meters. **(c)** Yes. The residual plot shows no leftover pattern. **(d)** residual $= 78 - 92.21 = -14.21$ meters. The actual distance on this afternoon was 14.21 meters closer than expected, based on the temperature of the water.

(a) Assuming that the weights of the truffles, nougats, and boxes are independent, what are the mean and standard deviation of the weight of a box of candy?

(b) Assuming that the weights of the truffles, nougats, and boxes are approximately Normally distributed, what is the probability that a randomly selected box of candy will weigh more than 30 ounces?

(c) If five gift boxes are randomly selected, what is the probability that at least one of them will weigh more than 30 ounces?

(d) If five gift boxes are randomly selected, what is the probability that the mean weight of the five boxes will be more than 30 ounces?

**AP3.35** An investor is comparing two stocks, A and B. She wants to know if over the long run, there is a significant difference in the return on investment as measured by the percent increase or decrease in the price of the stock from its date of purchase. The investor takes a random sample of 50 annualized daily returns over the past five years for each stock. The data are summarized below.

| Stock | Mean return | Standard deviation |
|---|---|---|
| A | 11.8% | 12.9% |
| B | 7.1% | 9.6% |

(a) Is there a significant difference in the mean return on investment for the two stocks? Support your answer with appropriate statistical evidence. Use a 5% significance level.

(b) The investor believes that although the return on investment for Stock A usually exceeds that of Stock B, Stock A represents a riskier investment, where the risk is measured by the price volatility of the stock. The standard deviation is a statistical measure of the price volatility and indicates how much an investment's actual performance during a specified period varies from its average performance over a longer period. Do the price fluctuations in Stock A significantly exceed those of Stock B, as measured by their standard deviations? Identify an appropriate set of hypotheses that the investor is interested in testing.

(c) To measure this, we will construct a test statistic defined as

$$F = \frac{\text{large sample variance}}{\text{smaller sample variance}}$$

What value(s) of the statistic would indicate that the price fluctuations in Stock A significantly exceed those of Stock B? Explain.

(d) Calculate the value of the $F$ statistic using the information given in the table.

(e) Two hundred simulated values of this test statistic, $F$, were calculated assuming no difference in the standard deviations of the returns for the two stocks. The results of the simulation are displayed in the following dotplot.

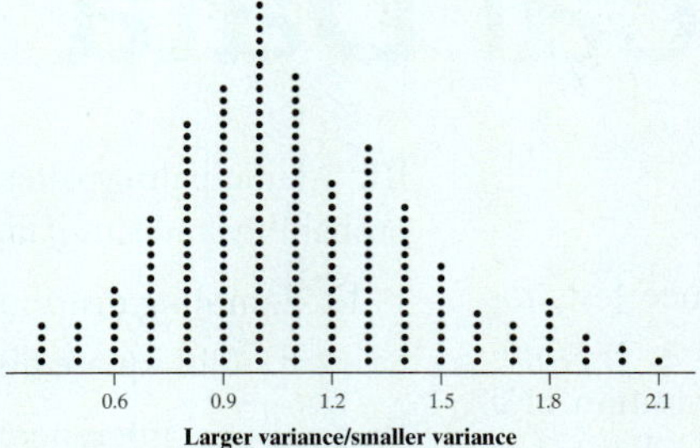


Use these simulated values and the test statistic that you calculated in part (d) to determine whether the observed data provide convincing evidence that Stock A is a riskier investment than Stock B. Explain your reasoning.

**AP3.34 (a)** Define $W$ = the weight of a randomly selected gift box. Then $\mu_W = 8(2) + 2(4) + 3 = 27$ ounces and $\sigma_W = \sqrt{8(0.5^2) + 2(1^2) + 0.2^2} = 2.01$ ounces. **(b)** We want to find $P(W > 30)$ using the $N(27, 2.01)$ distribution. $z = \frac{30 - 27}{2.01} = 1.49$ and $P(W > 30) = 0.0681$. *Using technology*: 0.0678. There is a 0.0678 probability of randomly selecting a box that weighs more than 30 ounces. **(c)** $P$(at least one box > 30 ounces) $= 1 - P$(none of the boxes > 30 ounces) $= 1 - (1 - 0.0678)^5 = 1 - (0.9322)^5 = 0.2960$. **(d)** Because the distribution of $W$ is Normal, the distribution of $\overline{W}$ will also be Normal, with mean $\mu_{\overline{W}} = 27$ ounces and standard deviation $\sigma_{\overline{W}} = \frac{2.01}{\sqrt{5}} = 0.899$. We want to find $P(\overline{W} > 30)$. $z = \frac{30 - 27}{0.899} = 3.34$ and $P(Z > 3.34) = 0.0004$. There is a 0.0004 probability of randomly selecting 5 boxes that have a mean weight of more than 30 ounces.

**AP3.35 (a)** *S*: $H_0: \mu_A - \mu_B = 0$ vs. $H_a: \mu_A - \mu_B \neq 0$, where $\mu_A$ is the true mean annualized return for stock A and $\mu_B$ is the true mean annualized return for stock B. *P*: Two-sample *t* test. *Random*: Independent random samples. *10%*: $n_A = 50$ is less than 10% of all days in the past 5 years and $n_B = 50$ is less than 10% of all days in the past 5 years. *Normal/Large Sample*: $n_A = 50 \geq 30$ and $n_B = 50 \geq 30$. *D*: $t = 2.07$. Using df = 40, the *P*-value is between 0.04 and 0.05. Using df = 90.53, *P*-value = 0.0416. *C*: Because the *P*-value of 0.0416 is less than $\alpha = 0.05$, we reject $H_0$. We have convincing evidence that the true mean annualized return for Stock A is different than the true mean annualized return for Stock B. **(b)** $H_o: \sigma_A - \sigma_B = 0$ vs. $H_a: \sigma_A - \sigma_B > 0$, where $\sigma_A$ is the true standard deviation of returns for Stock A and $\sigma_B$ is the true standard deviation of returns for Stock B. **(c)** When the standard deviation of Stock A > the standard deviation of Stock B, the variance of Stock A will be bigger than the variance of Stock B. Thus, values of $F$ that are significantly greater than 1 would indicate that the price volatility for Stock A is higher than that for Stock B. **(d)** $F = \frac{(12.9)^2}{(9.6)^2} = 1.806$. **(e)** In the simulation, a test statistic of 1.806 or greater occurred in only 6 of the 200 trials. Thus, the approximate *P*-value is 6/200 = 0.03. Because the approximate *P*-value of 0.03 is less than $\alpha = 0.05$, we reject $H_0$. There is convincing evidence that the true standard deviation of returns for Stock A > the true standard deviation of returns for Stock B.

Chapter

11

# Inference for Distributions of Categorical Data

## PD Overview

In this chapter, we learn three different significance tests for distributions of *categorical* data. The chi-square test for goodness of fit is used to determine whether the distribution of a single categorical variable differs from a hypothesized distribution in some population. The chi-square test for homogeneity is used to compare the distribution of a single categorical variable for two or more populations or treatments. Finally, the chi-square test for independence is used to investigate the relationship between two categorical variables in a single population. Students often have a hard time deciding which test to use, so continually remind them of these distinctions.

Chi-square tests differ from the other significance tests we have learned in several ways. First, we usually state the hypotheses in words rather than symbols. Second, there are no one-sided alternative hypotheses. Essentially, the alternative hypothesis is always "the null hypothesis *isn't* true." Students are often happy that they don't have to think as much about how to state the alternative hypothesis or calculate the *P*-value. Also, chi-square tests do not have corresponding confidence intervals like the other significance tests. For example, while a two-sample $z$ interval can accompany a two-sample $z$ test for a difference between two proportions, there is no confidence interval that corresponds to a chi-square test for homogeneity.

The "Inference" special-focus materials on AP® Central are a great resource for helping you understand the logic of significance tests, why we need to check conditions, and how chi-square tests fit in the overall scheme of inference. Another nice resource is a paper written by Dan Teague called "Chi-Square Analyses." If you do an Internet search using his name and the title of the paper, it should come right up.

### Relationship to the AP® Statistics Course Description

Chapter 11 covers the following content as described in the Topic Outline from the AP® Statistics Course Description. Topics marked with a superscript are covered in other chapters as indicated.

III. Anticipating patterns: Exploring random phenomena using probability and simulation

- D. Sampling distributions
  - 8. Chi-square distribution

IV. Statistical inference: Estimating population parameters and testing hypotheses

- B. Tests of significance
  - 6. Chi-square test for goodness of fit, homogeneity of proportions, and independence (one- and two-way tables)

## The Big Ideas

One of the challenges in teaching the AP® Statistics course is keeping students focused on the big picture and not just the details of each section. We outline the big ideas for the chapter here.

### Chapter 11 Introduction

We strongly recommend that you do The Candy Man Can Activity to open this chapter. It is a great way to introduce chi-square tests, and students always enjoy getting a tasty treat in class. Also, we refer back to this Activity in Section 11.1 as we develop the chi-square test for goodness of fit. There are some additional suggestions for this Activity listed on page 678. If you cannot distribute candy at your school, consider using the A Fair Die? Alternate Activity described on page 679 of the *Annotated Teacher's Edition.*

### Section 11.1 Chi-Square Tests for Goodness of Fit

This section begins by developing the chi-square statistic as a way to measure the difference between an observed distribution and a hypothesized distribution of categorical data. Like any other statistic, the chi-square statistic has its own sampling distribution that can be estimated by simulation. Fortunately, when certain conditions are satisfied, we can accurately model the sampling distribution of the chi-square statistic using a density curve, making it easier to calculate *P*-values. Not coincidentally, the density curves come from a family of distributions known as chi-square distributions. This is very similar to what we learned in Chapter 8—the sampling distribution of the *t* statistic can be modeled with a *t* distribution when certain conditions are satisfied.

Chi-square tests for goodness of fit always compare the observed distribution of a categorical variable with a hypothesized distribution. In the M&M'S® example, we compare the observed distribution of colors with the distribution specified by the company. Other examples and exercises test if the distribution of birthdays is uniform or if a proposed genetic model is correct.

Finally, we suggest that students do a follow-up analysis when the results of a chi-square test are significant. This involves looking at which categories of the variable had the largest contributions to the chi-square statistic and whether the observed values in those categories were larger or smaller than expected.

## Section 11.2 Inference for Two-Way Tables

In Chapter 9, we learned how to do significance tests for population means and population proportions. In both cases, we compared the observed value of a statistic to a hypothesized value, such as $p = 0.5$ or $\mu = 10$. In Chapter 10, we compared the means or proportions from two populations or treatments to *each other*, rather than to some hypothesized value. Similarly, while a chi-square test for goodness of fit compares an observed distribution to a hypothesized distribution, a chi-square test for homogeneity compares the observed distributions of a categorical variable to *each other* for two or more populations or treatments. Fortunately, the chi-square statistic and chi-square distributions we use to estimate *P*-values are the same for both tests.

The third and final chi-square test we learn in this chapter is for investigating the relationship between two categorical variables in one population. While there are some important differences in how we state hypotheses and check conditions, many components of the test for independence are the same as those for the test for homogeneity. In fact, many statisticians don't even make a distinction between the two tests. However, because the tests are listed separately on the AP® Statistics Topic Outline, we recommend that your students know the difference. Looking ahead, we will learn how to do significance tests for the relationship between two *quantitative* variables in the next chapter.

# Chapter 11: Resources

## TRM Additional *TPS* Teacher's Resource Materials

The following resources can be found by clicking on the links in the new Teacher's e-Book (TE-book), opening the Teacher's Resource CD (TRCD), or logging onto the book's companion Web site www.whfreeman.com/tps5e (teacher log-in required).

- Printable Chapter 11 Learning Objective Grid (page 680)
- FRAPPY! materials: Question, Model Solution, Scoring Rubric, Sample Responses with commentary (page 730)
- Quizzes: 3 versions for each section; versions A and B are roughly parallel
    - Section 11.1 (page 692)
    - Section 11.2 (page 722)
- Tests: 4 versions; versions A and B are roughly parallel, version D is cumulative (page 731)

## Free-Response Questions from Previous AP® Exams

Questions can be found on the AP® Central Web site: http://apcentral.collegeboard.com/apc/members/exam/exam_questions/8357.html.

Students should be able to answer all the free-response questions listed below with material learned in this chapter. Questions that contain content from this chapter but also require content from later chapters are listed in the last chapter required to complete the entire question. This list will be updated after each AP® exam and will be posted on the book's Web site: www.whfreeman.com/tps5e.

Questions marked with an asterisk are from exams with released multiple-choice questions. You may want to save these questions until the end of the year so you can give your students a complete released exam for practice.

| Year | # | Content |
|---|---|---|
| 2013 | 4 | • Chi-square test for independence |
| 2011B | 4 | • Chi-square test for independence<br>• Type I and Type II errors |
| 2010 | 6 | • Graphing and comparing distributions<br>• Evaluating and using an unfamiliar test statistic |
| 2010B | 5 | • General addition rule<br>• Conditional probability<br>• Independence<br>• Chi-square test for independence |
| 2009 | 1 | • Graphing categorical data<br>• Describing an association between categorical variables<br>• Choosing a correct inference procedure<br>• Stating hypotheses |
| 2008 | 5 | • Chi-square test for goodness of fit<br>• Follow-up analysis |
| 2006 | 6 | • Stating hypotheses<br>• Calculating a test statistic and *P*-value<br>• Rejection regions<br>• Identifying simulated distributions of a test statistic |
| 2004 | 5 | • Chi-square test for independence<br>• Scope of inference |
| 2003 | 5 | • Chi-square test for independence |
| 2003B | 5 | • Multiplication rule for independent events<br>• Expected value<br>• Chi-square test for goodness of fit |
| 2002B | 6 | • Two-sample *t* test<br>• Chi-square test for homogeneity<br>• Comparing distributions using graphs |
| 1999 | 2 | • Chi-square test for independence |
| 1998 | 3 | • Methods of random assignment<br>• Choosing the correct inference procedure |

## TI-Nspire Activities

These Activities are designed to illustrate important statistical concepts using the TI-Nspire. The Activities can be found at www.whfreeman.com/tps5e.

- *Goodness of Fit*: The Nspire is used to complete a chi-square test for goodness of fit for an M&M'S® Activity, including graphs of categorical data.

Chapter

11

# Chapter 11: Pacing Guide, Learning Objectives, and Suggested Assignments

This pacing guide is based on a schedule with 110, 50-minute sessions before the AP® exam. The 110 days include 4 days for midterms and 10 days for AP® exam review. If you have a different number of sessions before the AP® exam, you can modify the pacing guide to suit your needs. If you have additional time, consider incorporating quizzes, released AP® free-response questions, or additional activities. See the Resources section above for suggestions.

The suggested homework assignments list odd-numbered exercises, whenever possible, so students can check their answers against the back-of-the-book answers. If you would rather students not have access to the answers while doing homework, adding 1 to the exercise numbers usually will do the trick, because the homework exercises typically are paired. For example, Exercises 1 and 2 will generally cover the same topics, but in different contexts. You may also choose to include the Recycle and Review questions at the end of each section, which review topics from previous sections or chapters. These questions are denoted with a symbol.

| Day | Topics | **Learning Objectives:** Students will be able to. . . | Suggested Assignment |
|---|---|---|---|
| 1 | Activity: The Candy Man Can; 11.1 Comparing Observed and Expected Counts: The Chi-Square Statistic; The Chi-Square Distributions and *P*-values | • State appropriate hypotheses and compute expected counts for a chi-square test for goodness of fit.<br>• Calculate the chi-square statistic, degrees of freedom, and *P*-value for a chi-square test for goodness of fit. | 1, 3, 5 |
| 2 | 11.1 Carrying Out a Test; Follow-Up Analysis | • Perform a chi-square test for goodness of fit.<br>• Conduct a follow-up analysis when the results of a chi-square test are statistically significant. | 7, 9, 11, 15, 17 |
| 3 | 11.2 Comparing Distributions of a Categorical Variable; Expected Counts and the Chi-Square Statistic; The Chi-Square Test for Homogeneity | • Compare conditional distributions for data in a two-way table.<br>• State appropriate hypotheses and compute expected counts for a chi-square test based on data in a two-way table.<br>• Calculate the chi-square statistic, degrees of freedom, and *P*-value for a chi-square test based on data in a two-way table.<br>• Perform a chi-square test for homogeneity. | 19–22, 27, 29, 31, 33, 35, 37, 39 |
| 4 | 11.2 Relationships between Two Categorical Variables; the Chi-Square Test for Independence; Using Chi-Square Tests Wisely | • Perform a chi-square test for independence.<br>• Choose the appropriate chi-square test. | 41, 43, 45, 47, 49, 51–55 |
| 5 | Chapter 11 Review/FRAPPY! | | Chapter 11 Review Exercises |
| 6 | Chapter 11 Test | | |

**PD Chapter 11 Overview**

Watch the video overview of Chapter 11 (for teachers) by clicking on the link in the TE-book, opening the TRCD, or logging onto the book's companion Web site www.whfreeman.com/tps5e (teacher log-in required).

# Chapter 11

# Inference for Distributions of Categorical Data

case study

## Do Dogs Resemble Their Owners?

Some people look a lot like their pets. Maybe they deliberately choose animals that match their appearance. Or maybe we're perceiving similarities that aren't really there. Researchers at the University of California, San Diego, decided to investigate. They designed an experiment to test whether or not dogs resemble their owners. The researchers believed that resemblance between dog and owner might differ for purebred and mixed-breed dogs.

A random sample of 45 dogs and their owners was photographed separately at three dog parks. Then, researchers "constructed triads of pictures, each consisting of one owner, that owner's dog, and one other dog photographed at the same park." The subjects in the experiment were 28 undergraduate psychology students. Each subject was presented with the individual sets of photographs and asked to identify which dog belonged to the pictured owner. A dog was classified as resembling its owner if more than half of the 28 undergraduate students matched dog to owner.[1]

The table below summarizes the results. There is some support for the researchers' belief that resemblance between dog and owner might differ for purebred and mixed-breed dogs.

| | Breed status | |
|---|---|---|
| **Resemblance?** | **Purebred dogs** | **Mixed-breed dogs** |
| Resemble owner | 16 | 7 |
| Don't resemble | 9 | 13 |

Do these data provide convincing evidence of an association between dogs' breed status and whether or not they resemble their owners? By the end of this chapter, you will have developed the tools you need to answer this question.



### Teaching Tip

Remind students that they have already learned how to analyze two-way tables like this one. Using the methods from Chapter 1, students should be able to calculate, graph, and compare the conditional distributions of resemblance for each type of dog. Also, using the methods from Chapter 5, students should be able to determine if the events "a randomly selected dog looks like its owner" and "a randomly selected dog is a purebred" are independent for the dogs *in the sample*. In this chapter, we go one step further and ask if there is convincing evidence of an association between dogs' breed status and whether or not they resemble their owners for *all dogs in the population*. Making an inference about the relationship between two categorical variables in a population based on data from a sample requires that we do a significance test to account for the variability due to random sampling.

# Introduction

In the previous chapter, we discussed inference procedures for comparing the proportion of successes for two populations or treatments. Sometimes we want to examine the distribution of a single categorical variable in a population. The *chi-square test for goodness of fit* allows us to determine whether a hypothesized distribution seems valid. This method is useful in a field like genetics, where the laws of probability give the expected proportion of outcomes in each category.

We can decide whether the distribution of a categorical variable differs for two or more populations or treatments using a *chi-square test for homogeneity*. In doing so, we will often organize our data in a two-way table. It is also possible to use the information in a two-way table to study the relationship between two categorical variables. The *chi-square test for independence* allows us to determine if there is convincing evidence of an association between the variables in the population at large.

The methods of this chapter help us answer questions such as these:

- Are the birthdays of NHL players evenly distributed throughout the year?
- Does background music influence customer purchases?
- Is there an association between anger level and heart disease?

Of course, we have to do a careful job of describing the data before we proceed to statistical inference. In Chapter 1, we discussed graphical and numerical methods of data analysis for categorical variables. You may want to quickly review Section 1.1 now.

Here's an Activity that gives you a taste (pardon the pun) of what lies ahead.

## ACTIVITY | The Candy Man Can

**MATERIALS:**
Large bag of M&M'S® Milk Chocolate Candies for the class; TI-83/84 or TI-89 for each team of 3 to 4 students

Mars, Incorporated, which is headquartered in McLean, Virginia, makes milk chocolate candies. Here's what the company's Consumer Affairs Department says about the color distribution of its M&M'S Milk Chocolate Candies:

> *On average, M&M'S Milk Chocolate Candies will contain 13 percent of each of browns and reds, 14 percent yellows, 16 percent greens, 20 percent oranges and 24 percent blues.*

The purpose of this activity is to determine whether the company's claim is believable.

1. Your teacher will take a random sample of 60 M&M'S from a large bag and give one or more pieces of candy to each student. As a class, count the number of M&M'S® Chocolate Candies of each color. Make a table on the board that summarizes these *observed counts*.
2. How can you tell if the sample data give convincing evidence to dispute the company's claim? Each team of three or four students should discuss this question and devise a formula for a test statistic that measures the difference between the observed and expected color distributions. The test statistic should yield a

### Teaching Tip

Emphasize that this chapter is all about categorical data by noting that each of the variables in the examples on this page (birth month, type of music, type of purchase, anger level, heart disease status) is categorical.

### Activity Overview

You can watch an overview of this Activity (for teachers) by clicking on the link in the TE-book, opening the TRCD, or logging onto the book's companion Web site www.whfreeman.com/tps5e (teacher log-in required).

### Teaching Tip

Here are some tips for The Candy Man Can Activity:

- Another way to get a sample of 60 M&M'S® is to give out fun-size packs and have each student select 2–3 M&M'S (depending on the number of students you have). Make sure that the total sample size is 60, however.
- If you want to make the Activity more challenging, have them keep their books closed so students can't see the guiding questions in Step 2. However, you may need to make some suggestions if groups are having trouble.
- In Step 2, it is reasonable for students to choose proportions rather than counts and absolute values rather than squared values. When you introduce the chi-square statistic, say that counts are better than proportions because they preserve information about sample size. For example, if the proportion of orange M&M'S is 0.30, we don't know if that is a "big" difference from 0.20. If the sample size is 10, then a difference of 0.10 is no big deal. But, if the sample size is 1000, a difference of 0.10 is a much bigger deal. Also, you can say that squared values are better than absolute values because it is easier to model the sampling distribution when we square the differences. The reason we divide by the expected counts is explained in the Think About It on page 683.
- To sort the values in List 1 as described in Step 6, press the `STAT` button, choose `SortA(`, and execute the command `SortA(L1)`.
- To make the conclusion in Step 9, use the same logic as for other significance tests. If getting a value of the test statistic as large as or larger than the observed test statistic would be unlikely, we have convincing evidence that the distribution of colors is different than what the company claims.
- Note that an example dotplot is given on the bottom of page 679, just below the activity.

single number when the observed and expected values are plugged in. Here are some questions for your team to consider:

- Should we look at the difference between the observed and expected *proportions* in each color category or between the observed and expected *counts* in each category?
- Should we use the differences themselves, the absolute value of the differences, or the square of the differences?
- Should we divide each difference value by the sample size, expected count, or nothing at all?

3. Each team will share its proposed test statistic with the class. Your teacher will then reveal how the *chi-square statistic* $\chi^2$ is calculated.

4. Discuss as a class: If your sample is consistent with the company's claimed distribution of M&M'S® Chocolate Candies colors, will the value of $\chi^2$ be large or small? If your sample is not consistent with the company's claimed color distribution, will the value of $\chi^2$ be large or small?

5. Compute the value of the chi-square test statistic for the class's data. Is this value large or small? To find out, you and your classmates will perform a simulation.

6. Suppose that the company's claimed color distribution is correct. We'll use numerical labels from 1 to 100 to represent the color of a randomly chosen M&M'S Milk Chocolate Candy:

1–13 = brown 14–26 = red 27–40 = yellow 41–56 = green
57–76 = orange 77–100 = blue

Use the calculator command below to simulate choosing a random sample of 60 candies.

TI-83/84: `RandInt(1,100,60) → L1`
TI-89: `tistat.randint(1,100,60) → list1`

Sort the list in ascending order. Then record the observed counts in each color category and compute the value of $\chi^2$ for your simulated sample.

7. Your teacher will draw and label axes for a class dotplot. Plot the result you got in Step 6 on the graph.

8. Repeat Steps 6 and 7 if needed to get a total of at least 40 repetitions of the simulation for your class.

9. Based on the class's simulation results, how surprising would it be to get a $\chi^2$-value as large as or larger than the one you did in Step 5 by chance alone when sampling from the claimed distribution? What conclusion would you draw?

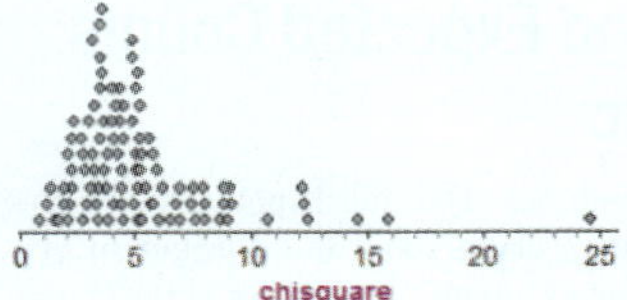


Here is an example of what the class dotplot in the Activity might look like after 100 trials. The graph shows what values of the chi-square statistic are likely to occur by chance alone when sampling from the company's claimed M&M'S® Chocolate Candies color distribution. Where did your class's $\chi^2$-value fall? You will learn more about the sampling distribution of the chi-square statistic shortly.

## ALTERNATE ACTIVITY

*A fair die?*

Before beginning this Activity, make a six-sided die out of paper, clay, wood, or any other material. Have students take turns rolling the die (for a total of 60 rolls) and record how often each number shows up on top. Do the results give convincing evidence that the die isn't fair? Finish the Activity by modifying the instructions for The Candy Man Can Activity in the student text, using $60(1/6) = 10$ as the expected count for each possible outcome.

**PD Section 11.1 Overview**

Watch the video overview of Section 11.1 (for teachers) by clicking on the link in the TE-book, opening the TRCD, or logging onto the book's companion Web site www.whfreeman.com/tps5e (teacher log-in required).

**TRM Learning Objectives Grid**

Remind students about the importance of the Learning Objectives. We include them at the beginning of each section so students know what they are expected to learn. At the end of the chapter, the Learning Objectives are included in a grid, along with references to examples and chapter review exercises. A printable version of this grid is included in the Teacher's Resource Materials.

**TRM Section 11.1 Alternate Examples**

You can find the Alternate Examples for this section in Microsoft Word format by clicking on the link in the TE-book, opening the TRCD, or logging onto the book's companion Web site www.whfreeman.com/tps5e (teacher log-in required).

### Teaching Tip

Ask your students if they can figure out why the table on this page is called a "one-way" table rather than a "two-way" table. It is a one-way table because it displays only the distribution of one categorical variable in one sample. Two-way tables show the distribution of one categorical variable in two or more samples/groups or the relationship between two categorical variables in one sample. To pick the correct chi-square test, students will need to understand these distinctions.

### Teaching Tip

In Chapters 9 and 10, we conducted significance tests about population parameters, such as a population proportion or the difference between two population means. However, in this chapter we are conducting tests about *distributions* of categorical variables.

# 11.1 Chi-Square Tests for Goodness of Fit

**WHAT YOU WILL LEARN** By the end of the section, you should be able to:

- State appropriate hypotheses and compute expected counts for a chi-square test for goodness of fit.
- Calculate the chi-square statistic, degrees of freedom, and *P*-value for a chi-square test for goodness of fit.
- Perform a chi-square test for goodness of fit.
- Conduct a follow-up analysis when the results of a chi-square test are statistically significant.

Jerome's class did the Candy Man Can Activity. The **one-way table** below summarizes the data from the class's sample of M&M'S® Milk Chocolate Candies. In general, one-way tables display the distribution of a single categorical variable for the individuals in a sample.

| Color: | Blue | Orange | Green | Yellow | Red | Brown | Total |
|---|---|---|---|---|---|---|---|
| Count: | 9 | 8 | 12 | 15 | 10 | 6 | 60 |

The sample proportion of blue M&M'S is $\hat{p} = \frac{9}{60} = 0.15$. Because the company claims that 24% of all M&M'S Milk Chocolate Candies are blue, Jerome might believe that something fishy is going on. We could use the one-sample $z$ test for a proportion from Chapter 9 to test the hypotheses

$$H_0: p = 0.24$$
$$H_a: p \neq 0.24$$

Note that the correct alternative hypothesis $H_a$ is two-sided. A sample proportion of blue M&M'S much higher or much lower than 0.24 would give Jerome reason to be suspicious about the company's claim. It's not appropriate to adjust $H_a$ after looking at the sample data!

where $p$ is the true population proportion of blue M&M'S® Chocolate Candies. We could then perform additional significance tests for each of the remaining colors.

Not only would this method be fairly inefficient, it would also lead to the problem of multiple comparisons, which we'll discuss in Section 11.2. More important, this approach wouldn't tell us how likely it is to get a random sample of 60 candies with a color *distribution* that differs as much from the one claimed by the company as the class's sample does, taking all the colors into consideration at one time. For that, we need a new kind of significance test, called a **chi-square test for goodness of fit**.

## Comparing Observed and Expected Counts: The Chi-Square Statistic

As with any test, we begin by stating hypotheses. The null hypothesis in a chi-square test for goodness of fit should state a claim about the distribution of a single categorical variable in the population of interest. In the case of the Candy Man Can Activity, the categorical variable we're measuring is color and the population of interest is all M&M'S® Milk Chocolate Candies. The appropriate null hypothesis is

$H_0$: The company's stated color distribution for all M&M'S Milk Chocolate Candies is correct.

The alternative hypothesis in a chi-square test for goodness of fit is that the categorical variable does not have the specified distribution. For the M&M'S, our alternative hypothesis is

$H_a$: The company's stated color distribution for all M&M'S Milk Chocolate Candies is *not* correct.

THINK ABOUT IT

**Why did we state hypotheses in words for a chi-square test for goodness of fit?** We can also write the hypotheses in symbols as

$$H_0: p_{blue} = 0.24, p_{orange} = 0.20, p_{green} = 0.16, p_{yellow} = 0.14, p_{red} = 0.13, p_{brown} = 0.13,$$

$H_a$: At least two of the $p_i$'s are incorrect

where $p_{color}$ = the true population proportion of M&M'S Milk Chocolate Candies of that color. Why don't we write the alternative hypothesis as $H_a$: At least one of the $p_i$'s is incorrect? If the stated proportion in one category is wrong, then the stated proportion in at least one other category must be wrong because the sum of the $p_i$'s must be 1.

Don't state the alternative hypothesis in a way that suggests that *all* the proportions in the hypothesized distribution are wrong. For instance, it would be *incorrect* to write

$$H_a: p_{blue} \neq 0.24, p_{orange} \neq 0.20, p_{green} \neq 0.16, p_{yellow} \neq 0.14, p_{red} \neq 0.13, p_{brown} \neq 0.13$$

The idea of the chi-square test for goodness of fit is this: we compare the **observed counts** from our sample with the counts that would be expected if $H_0$ is true. (*Remember*: we always assume that $H_0$ is true when performing a significance test.) The more the observed counts differ from the **expected counts,** the more evidence we have against the null hypothesis. In general, the expected counts can be obtained by multiplying the sample size by the proportion in each category according to the null hypothesis. Here's an example that illustrates the process.

## EXAMPLE Return of the M&M'S® Chocolate Candies

### *Computing expected counts*

**PROBLEM:** Jerome's class collected data from a random sample of 60 M&M'S Milk Chocolate Candies. Calculate the expected counts for each color. Show your work.

**SOLUTION:** Assuming that the color distribution stated by Mars, Inc., is true, 24% of all M&M'S Milk Chocolate Candies produced are blue. For random samples of 60 candies, the average number of blue M&M'S should be (60)(0.24) = 14.40. This is our expected count of blue M&M'S® Chocolate Candies. Using this same method, we find the expected counts for the other color categories:

### Teaching Tip

In Chapters 9 and 10, the alternative hypothesis could be one-sided or two-sided, depending on the question we were trying to answer. However, in this chapter, there will be no one-sided tests. The alternative hypothesis will always be "the null hypothesis is *not* correct." In context, of course!

### Teaching Tip

If your students choose to state their hypotheses in symbols, make sure they avoid using $\hat{p}$ to represent a population proportion.

### Teaching Tip

Remind students that the "play" icon at the top of this example (and many others) indicates that they can watch a video presentation of the example. Find the videos at www.whfreeman.com/tps5e. Also, remind them that the icon in the lower right corner indicates a link between this example and Exercise 1. If students are having trouble with the exercise, they can look back to this example for help.

### ALTERNATE EXAMPLE

#### *A fair die?*

Jenny made a six-sided die in her ceramics class and rolled it 60 times to test if each side was equally likely to show up on top.

**PROBLEM:**
Assuming that her die is fair, calculate the expected counts for each possible outcome.

**SOLUTION:**
If the die is fair, each of the six sides has a 1/6 probability of ending up on top. This means that the expected count for each outcome is 60(1/6) = 10.

## AP® Exam Common Error

Some students mistakenly round the expected counts, believing that they must be integers. Students will lose points for this on the AP® exam because it shows a misunderstanding of what expected counts represent—the average number of observations in a given category in many, many random samples.

## Teaching Tip

Remind students that it is generally a good idea to use relative frequency bar graphs to compare two distributions of a categorical variable. However, when the sample sizes are the same, it is OK to use frequency bar graphs as we do on this page. As always, don't forget the labels!

## Teaching Tip

Remind students that the other test statistics we have learned ($z$ and $t$) also measure how far apart the sample value is from the hypothesized value, just on a different scale.

## AP® Exam Formula Sheet

Tell students that the formula for the chi-square statistic is included on the formula sheet that is provided on both sections of the AP® exam.

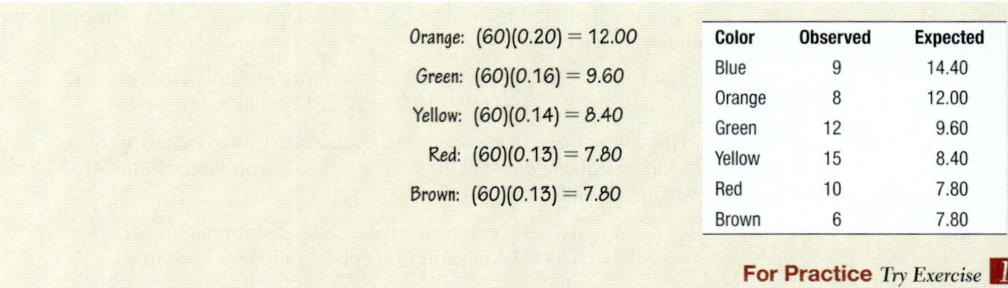

Orange: $(60)(0.20) = 12.00$

Green: $(60)(0.16) = 9.60$

Yellow: $(60)(0.14) = 8.40$

Red: $(60)(0.13) = 7.80$

Brown: $(60)(0.13) = 7.80$

| Color | Observed | Expected |
|---|---|---|
| Blue | 9 | 14.40 |
| Orange | 8 | 12.00 |
| Green | 12 | 9.60 |
| Yellow | 15 | 8.40 |
| Red | 10 | 7.80 |
| Brown | 6 | 7.80 |

**For Practice** *Try Exercise* **1**

Did you notice that the expected count sounds a lot like the expected value of a random variable from Chapter 6? That's no coincidence. The number of M&M'S® Chocolate Candies of a specific color in a random sample of 60 candies is a binomial random variable. Its expected value is $np$, the average number of candies of this color in many samples of 60 M&M'S Milk Chocolate Candies. The expected value is not likely to be a whole number.

To see if the data give convincing evidence for the alternative hypothesis, we compare the observed counts from our sample with the expected counts. If the observed counts are far from the expected counts, that's the evidence we were seeking. The table in the example gives the observed and expected counts for the sample of 60 M&M'S in Jerome's class. Figure 11.1 shows a side-by-side bar graph comparing the observed and expected counts.

**FIGURE 11.1** Bar graph comparing observed and expected counts for Jerome's class sample of 60 M&M'S® Milk Chocolate Candies.

We see some fairly large differences between the observed and expected counts in several color categories. How likely is it that differences this large or larger would occur just by chance in random samples of size 60 from the population distribution claimed by Mars, Inc.? To answer this question, we calculate a statistic that measures how far apart the observed and expected counts are. The statistic we use to make the comparison is the **chi-square statistic**

$$\chi^2 = \sum \frac{(\text{Observed} - \text{Expected})^2}{\text{Expected}}$$

(The symbol $\chi$ is the lowercase Greek letter chi, pronounced "kye.")

**DEFINITION: Chi-square statistic**

The **chi-square statistic** is a measure of how far the observed counts are from the expected counts. The formula for the statistic is

$$\chi^2 = \sum \frac{(\text{Observed} - \text{Expected})^2}{\text{Expected}}$$

where the sum is over all possible values of the categorical variable.

Let's use this formula to compare the observed and expected counts for Jerome's class sample.

## EXAMPLE Return of the M&M'S® Chocolate Candies

### *Calculating the chi-square statistic*

The table shows the observed and expected counts for the random sample of 60 M&M'S Milk Chocolate Candies in Jerome's class.

| Color | Observed | Expected |
|---|---|---|
| Blue | 9 | 14.40 |
| Orange | 8 | 12.00 |
| Green | 12 | 9.60 |
| Yellow | 15 | 8.40 |
| Red | 10 | 7.80 |
| Brown | 6 | 7.80 |

PROBLEM: Calculate the chi-square statistic.

SOLUTION: The formula for the chi-square statistic is

$$\chi^2 = \sum \frac{(\text{Observed} - \text{Expected})^2}{\text{Expected}}$$

For Jerome's data, we add six terms—one for each color category:

$$\chi^2 = \frac{(9-14.40)^2}{14.40} + \frac{(8-12.00)^2}{12.00} + \frac{(12-9.60)^2}{9.60} + \frac{(15-8.40)^2}{8.40} + \frac{(10-7.80)^2}{7.80} + \frac{(6-7.80)^2}{7.80}$$

$$= 2.025 + 1.333 + 0.600 + 5.186 + 0.621 + 0.415 = 10.180$$

**For Practice** *Try Exercise* **3**

THINK ABOUT IT

**Why do we divide by the expected count when calculating the chi-square statistic?** Suppose you obtain a random sample of 60 M&M'S Milk Chocolate Candies. Which would be more surprising: getting 18 blue candies or 12 yellow candies in the sample? Earlier, we computed the expected counts for these two categories as 14.4 and 8.4, respectively. The difference in the observed and expected counts for the two colors would be

Blue: $18 - 14.4 = 3.6$ Yellow: $12 - 8.4 = 3.6$

In both cases, the number of M&M'S® Chocolate Candies in the sample exceeds the expected count by the same amount. But it's much more surprising to be off by 3.6 out of an expected 8.4 yellow candies (almost a 50% discrepancy) than to be off by 3.6 out of an expected 14.4 blue candies (a 25% discrepancy). For that reason, we want the category with a larger *relative* difference to contribute more heavily to the evidence against $H_0$ and in favor of $H_a$ measured by the $\chi^2$ statistic.

If we just computed $(\text{Observed} - \text{Expected})^2$ for each category instead, the contributions of these two color categories would be the same:

Blue: $(18 - 14.40)^2 = 12.96$ Yellow: $(12 - 8.40)^2 = 12.96$

## ALTERNATE EXAMPLE

### *A fair die?*

Here are the results of 60 rolls of Jenny's ceramic die, along with the expected counts:

| Outcome | Observed | Expected |
|---|---|---|
| 1 | 13 | 10 |
| 2 | 11 | 10 |
| 3 | 6 | 10 |
| 4 | 12 | 10 |
| 5 | 10 | 10 |
| 6 | 8 | 10 |
| **Total** | **60** | **60** |

PROBLEM: Calculate the chi-square statistic.

SOLUTION:

$$\chi^2 = \frac{(13-10)^2}{10} + \frac{(11-10)^2}{10} + \frac{(6-10)^2}{10} + \frac{(12-10)^2}{10} + \frac{(10-10)^2}{10} + \frac{(8-10)^2}{10}$$

$$= 0.9 + 0.1 + 1.6 + 0.4 + 0 + 0.4$$

$$= 3.4$$

## Teaching Tip

On the AP® exam, students do not need to show the calculation of every term. Showing something like

$$\chi^2 = \frac{(9-14.4)^2}{14.4} + \frac{(8-12.0)^2}{12.0} + \cdots$$

$= 10.180$ is sufficient.

## AP® Exam Common Error

Some students mistakenly use proportions when calculating the chi-square test statistic. Even though it is possible to state the hypotheses for a chi-square test using proportions, the chi-square test statistic is always calculated using counts.

By using $\frac{(\text{Observed} - \text{Expected})^2}{\text{Expected}}$, we guarantee that the color category with the larger relative difference will contribute more heavily to the total:

$$\text{Blue: } \frac{(18 - 14.40)^2}{14.40} = 0.90 \quad \text{Yellow: } \frac{(12 - 8.40)^2}{8.40} = 1.54$$

Think of $\chi^2$ as a measure of the distance of the observed counts from the expected counts. Like any distance, it is always zero or positive, and it is zero only when the observed counts are exactly equal to the expected counts. Large values of $\chi^2$ are stronger evidence for $H_a$ because they say that the observed counts are far from what we would expect if $H_0$ were true. Small values of $\chi^2$ suggest that the data are consistent with the null hypothesis. Is $\chi^2 = 10.180$ a large value? You know the drill: compare the observed value 10.180 against the sampling distribution that shows how $\chi^2$ would vary in repeated random sampling if the null hypothesis were true.

## CHECK YOUR UNDERSTANDING

Mars, Inc., reports that their M&M'S® Peanut Chocolate Candies are produced according to the following color distribution: 23% each of blue and orange, 15% each of green and yellow, and 12% each of red and brown. Joey bought a randomly selected bag of Peanut Chocolate Candies and counted the colors of the candies in his sample: 12 blue, 7 orange, 13 green, 4 yellow, 8 red, and 2 brown.

1. State appropriate hypotheses for testing the company's claim about the color distribution of M&M'S Peanut Chocolate Candies.
2. Calculate the expected count for each color, assuming that the company's claim is true. Show your work.
3. Calculate the chi-square statistic for Joey's sample. Show your work.

## The Chi-Square Distributions and *P*-Values

We used Fathom software to simulate taking 500 random samples of size 60 from the population distribution of M&M'S Milk Chocolate Candies given by Mars, Inc. Figure 11.2 shows a dotplot of the values of the chi-square statistic for these 500 samples. The blue vertical line is plotted at the value of $\chi^2 = 10.180$ from Jerome's class data.

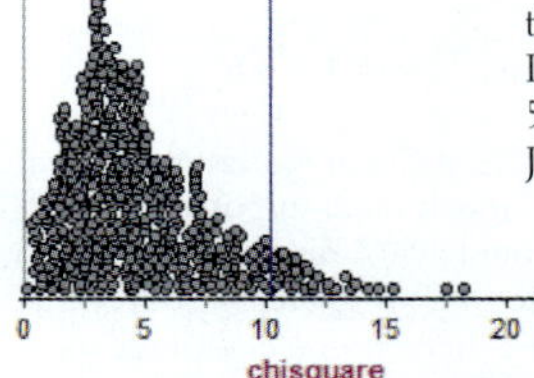


**FIGURE 11.2** Fathom dotplot showing values of the chi-square statistic in 500 simulated samples of size $n = 60$ from the population distribution of M&M'S® Milk Chocolate Candies stated by the company.

Recall that larger values of $\chi^2$ give more convincing evidence against $H_0$ and in favor of $H_a$. According to Fathom, 37 of the 500 simulated samples resulted in a chi-square statistic of 10.180 or higher. Our estimated *P*-value is 37/500 = 0.074. Because the *P*-value exceeds the default $\alpha = 0.05$ significance level, we fail to reject $H_0$. We do not have convincing evidence that the company's claimed color distribution is incorrect.

As Figure 11.2 suggests, the sampling distribution of the chi-square statistic is *not* a Normal distribution. It is a right-skewed distribution that allows only nonnegative values because $\chi^2$ can never be negative.

### Teaching Tip

Emphasize the first paragraph on this page. There are two possible explanations for why we got a chi-square statistic as big as 10.180:

- The claimed distribution is correct, and the differences we observed were due to sampling variability.
- The claimed distribution is not correct.

To decide which of these explanations is more plausible, we need to know how likely it is to get a chi-square statistic at least as big as 10.180 by chance, assuming the claimed distribution is correct. In other words, we need a *P*-value!

### ANSWERS TO CYU

1. $H_0$: The company's claimed color distribution for its Peanut M&M'S® is correct vs. $H_a$: The company's claimed color distribution is not correct.

2. The expected count of both blue and orange candies is 46(0.23) = 10.58, for green and yellow is 46(0.15) = 6.9, and for red and brown is 46(0.12) = 5.52.

3. $$\chi^2 = \frac{(12 - 10.58)^2}{10.58} + \frac{(7 - 10.58)^2}{10.58} + \frac{(13 - 6.9)^2}{6.9} + \frac{(4 - 6.9)^2}{6.9} + \frac{(8 - 5.52)^2}{5.52} + \frac{(2 - 5.52)^2}{5.52} = 11.3724$$

The sampling distribution of $\chi^2$ differs depending on the number of possible values for the categorical variable (that is, on the number of categories).

When the expected counts are all at least 5, the sampling distribution of the $\chi^2$ statistic is modeled well by a **chi-square distribution** with degrees of freedom (df) equal to the number of categories minus 1. As with the $t$ distributions, there is a different chi-square distribution for each possible value of df. Here are the facts.

**THE CHI-SQUARE DISTRIBUTIONS**

The chi-square distributions are a family of density curves that take only non-negative values and are skewed to the right. A particular chi-square distribution is specified by giving its degrees of freedom. The chi-square test for goodness of fit uses the chi-square distribution with degrees of freedom = the number of categories − 1.

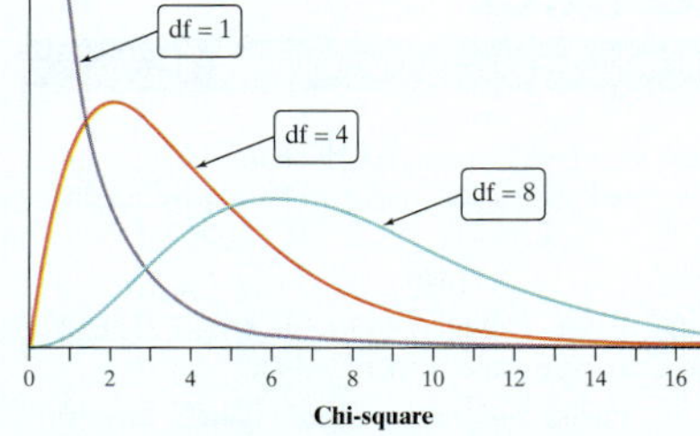


**FIGURE 11.3** The density curves for three members of the chi-square family of distributions.

Figure 11.3 shows the density curves for three members of the chi-square family of distributions. As the degrees of freedom (df) increase, the density curves become less skewed, and larger values become more probable. Here are two other interesting facts about the chi-square distributions:

- The mean of a particular chi-square distribution is equal to its degrees of freedom.
- For df > 2, the mode (peak) of the chi-square density curve is at df − 2.

When df = 8, for example, the chi-square distribution has a mean of 8 and a mode of 6.

To get $P$-values from a chi-square distribution, we can use technology or Table C in the back of the book. The following example shows how to use the table.

## EXAMPLE Return of the M&M'S® Chocolate Candies

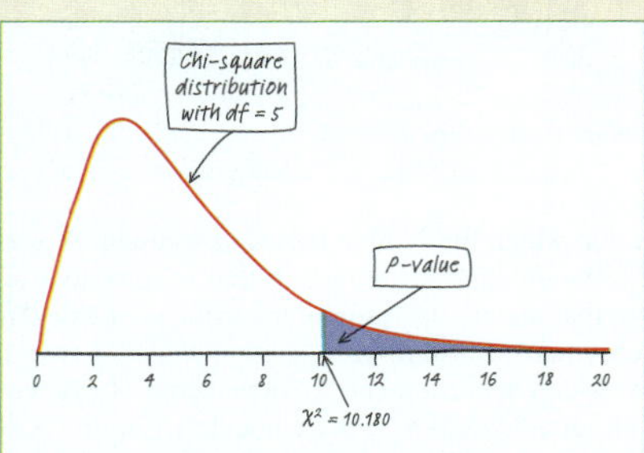


**FIGURE 11.4** The $P$-value for a chi-square test for goodness of fit using Jerome's M&M'S® Chocolate Candies class data.

### *Finding the P-value*

In the last example, we computed the chi-square statistic for the random sample of 60 M&M'S Milk Chocolate Candies in Jerome's class: $\chi^2 = 10.180$. Now let's find the $P$-value. Because all the expected counts are at least 5, the $\chi^2$ statistic will be modeled well by a chi-square distribution when $H_0$ is true. There are 6 color categories for M&M'S Milk Chocolate Candies, so df = 6 − 1 = 5.

The $P$-value is the probability of getting a value of $\chi^2$ as large as or larger than 10.180 when $H_0$ is true. Figure 11.4 shows this probability as an area under the chi-square density curve with 5 degrees of freedom.

## Teaching Tip

Unlike the $t$ distributions, the degrees of freedom for the sampling distribution of the chi-square test statistic do not depend on the sample size. Rather, the degrees of freedom are based on the number of categories.

## Teaching Tip

A chi-square density curve is a good model for the sampling distribution of the chi-square statistic, but only when the sample size is large enough that the expected counts are all at least 5. This is very similar to the Large Counts condition for proportions, where we used a Normal curve to model the distribution of $\hat{p}$ only when the number of successes and failures were at least 10. Here is a histogram showing values of the chi-square statistic in 500 simulated samples of 60 M&M'S® and a chi-square density curve with 5 degrees of freedom:

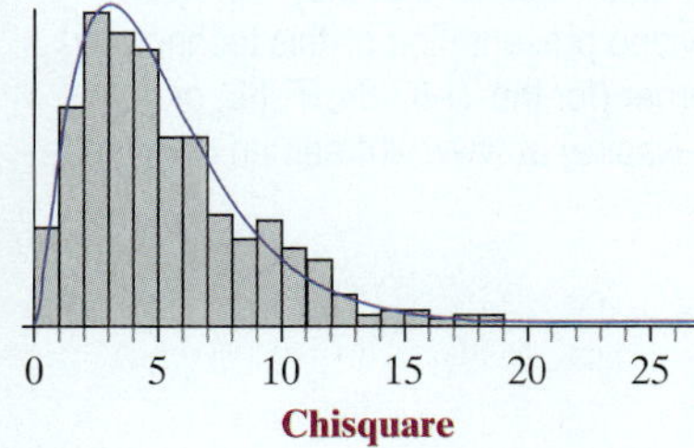


## Teaching Tip

Remind students that the area under a density curve is always equal to 1 and that an area under the curve between two boundaries corresponds to the probability that a randomly selected observation will fall between those two boundaries.

## Teaching Tip

Knowing that the mean of a chi-square distribution is equal to the degrees of freedom can help students get an impression of how unusual a chi-square statistic is. For example, with 10 degrees of freedom, $\chi^2 = 11.2$ is not very unusual, but $\chi^2 = 54.3$ would be very unusual.

## Teaching Tip

Here are two more interesting facts about the chi-square distributions. The variance is 2 × df and the median is approximately df − 2/3 when df > 2.

## ALTERNATE EXAMPLE

### *A fair die?*

When Jenny rolled her ceramic six-sided die 60 times and calculated the chi-square statistic, she got $\chi^2 = 3.4$.

**PROBLEM:**
Using the appropriate degrees of freedom, calculate the *P*-value. What conclusion can you make about Jenny's die?

**SOLUTION:**
Because there are six possible outcomes when rolling her die, degrees of freedom = 6 − 1 = 5. Using Table C, the *P*-value is greater than 0.25 because the $\chi^2$ statistic is smaller than the lowest critical value in the df = 5 row. Because the *P*-value is quite large, we fail to reject $H_0$. We do not have convincing evidence that Jenny's die is unfair.

### TECHNOLOGY TIP

Remind students that they can watch a video presentation of this technology corner (for the TI-83/84, TI-89, or TI-Nspire) at www.whfreeman.com/tps5e.

To find the *P*-value using Table C, look in the df = 5 row. The value $\chi^2 = 10.180$ falls between the critical values 9.24 and 11.07. The corresponding areas in the right tail of the chi-square distribution with 5 degrees of freedom are 0.10 and 0.05. (See the excerpt from Table C on the right.) So the *P*-value for a test based on Jerome's data is between 0.05 and 0.10.

| | *P* | | |
|---|---|---|---|
| df | .15 | .10 | .05 |
| 4 | 6.74 | 7.78 | 9.49 |
| 5 | 8.12 | 9.24 | 11.07 |
| 6 | 9.45 | 10.64 | 12.59 |

Now let's look at how to find the *P*-value with your calculator.

## 25. TECHNOLOGY CORNER — FINDING *P*-VALUES FOR CHI-SQUARE TESTS ON THE CALCULATOR

TI-Nspire instructions in Appendix B; HP Prime instructions on the book's Web site.

To find the *P*-value in the M&M'S® example with your calculator, use the $\chi^2$cdf(Chi-square Cdf on the TI-89) command. We ask for the area between $\chi^2 = 10.180$ and a very large number (we'll use 10,000) under the chi-square density curve with 5 degrees of freedom.

**TI-83/84**

- Press [2nd] [VARS] (DISTR) and choose $\chi^2$`cdf(`.

**OS 2.55 or later:** In the dialog box, enter these values: `lower:10.18`, `upper:10000`, `df:5`, choose `Paste`, and then press [ENTER]. **Older OS:** Complete the command $\chi^2$`cdf(10.180,10000,5)` and press [ENTER].

**TI-89**

- In the Stats/List Editor, Press [F5] `(Distr)` and choose `Chi-square Cdf...`.
- In the dialog box, enter these values: `Lower value:10.18`, `Upper value:10000`, `Deg of Freedom, df:5`, and then choose [ENTER].

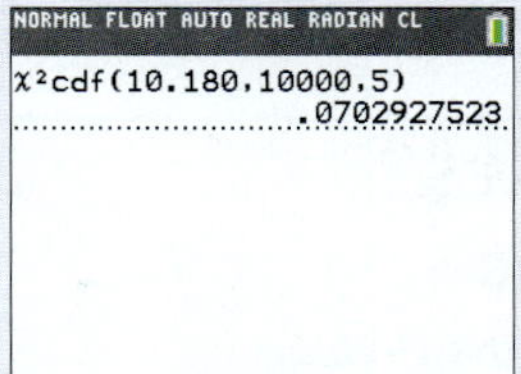

As the calculator screen shots show, this method gives a more precise *P*-value than Table C.

Table C gives us an interval in which the *P*-value falls. The calculator's $\chi^2$`cdf` (`Chi-square Cdf` on the TI-89) command gives a result that is consistent with Table C but more precise. For that reason, we recommend using your calculator to compute *P*-values from a chi-square distribution.

Based on Jerome's sample, what conclusion can we draw about $H_0$: the company's stated color distribution for all M&M'S® Milk Chocolate Candies is correct? Because our *P*-value of 0.07 is greater than $\alpha = 0.05$, we fail to reject $H_0$. We don't have convincing evidence that the company's claimed color distribution is incorrect.

Failing to reject $H_0$ does not mean that the null hypothesis is true! That is, we can't conclude that the color distribution claimed by Mars, Inc., *is* correct. All we can say is that the sample data did not provide convincing evidence to reject $H_0$. 

## AP® Exam Common Error

When stating conclusions for chi-square tests, students often accept the null hypothesis. Points will always be deducted on the AP® exam for this error. In this example, even though the evidence isn't convincing, there *is* evidence that the company's claim is incorrect. Consequently, it would be wrong to conclude that the company's claim is correct. Remind your students about the criminal trial analogy: a verdict of "not guilty" doesn't mean the defendant is innocent!

### CHECK YOUR UNDERSTANDING

Let's continue our analysis of Joey's sample of M&M'S® Peanut Chocolate Candies from the previous Check Your Understanding (page 684).

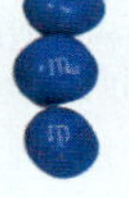

1. Confirm that the expected counts are large enough to use a chi-square distribution. Which distribution (specify the degrees of freedom) should we use?
2. Sketch a graph like Figure 11.4 on page 685 that shows the *P*-value.
3. Use Table C to find the *P*-value. Then use your calculator's $\chi^2$`cdf` command.
4. What conclusion would you draw about the company's claimed color distribution for M&M'S® Peanut Chocolate Candies? Justify your answer.

### ANSWERS TO CYU

1. The expected counts are all at least 5. df = 6 − 1 = 5.

2.

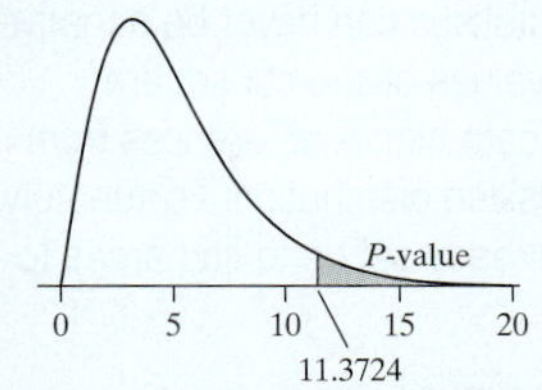


**Chi-square distribution with 5 df**

3. The *P*-value is between 0.025 and 0.05 (0.0445).

4. Because the *P*-value of $0.0445 < \alpha = 0.05$, we reject $H_0$. There is convincing evidence that the color distribution of M&M'S® Peanut Chocolate Candies is different from what the company claims.

## Carrying Out a Test

Like our test for a population proportion, the chi-square test for goodness of fit uses some approximations that become more accurate as we take larger samples. The **Large Counts condition** says that all expected counts must be at least 5. Before performing a test, we must also check that the Random and 10% conditions are met.

### CONDITIONS FOR PERFORMING A CHI-SQUARE TEST FOR GOODNESS OF FIT

- **Random:** The data come from a well-designed random sample or randomized experiment.
  - **10%:** When sampling without replacement, check that $n \leq \frac{1}{10}N$.
- **Large Counts:** All *expected* counts are at least 5.

Before we start using the chi-square test for goodness of fit, we have two important cautions to offer.

1. The chi-square test statistic compares observed and expected *counts*. Don't try to perform calculations with the observed and expected *proportions* in each category. 
2. When checking the Large Counts condition, be sure to examine the *expected* counts, not the observed counts.

## Teaching Tip

If the expected counts are not all at least 5, you can combine categories until the Large Counts condition is satisfied.

## Teaching Tip

The Large Counts condition for this test serves the same purpose as the Large Counts condition for tests for proportions: to ensure that the probability distribution we use to calculate the *P*-value is a good model for the actual sampling distribution of the test statistic we are using. The histogram below shows the sampling distribution of the chi-square statistic for samples of 5 M&M'S®. Notice that the Large Counts condition isn't met and the chi-square density curve doesn't model this distribution nearly as well as the sampling distribution on page 685.

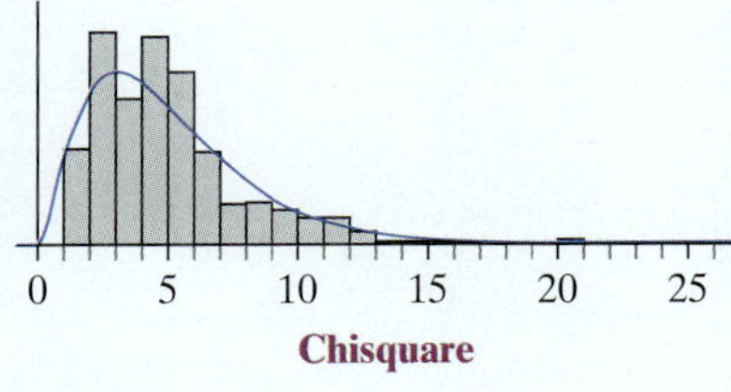

## AP® Exam Common Error

When stating hypotheses, students sometimes refer to the distribution of the sample rather than the distribution of the population. For example, it would be incorrect to state a null hypothesis as "the *observed* distribution of M&M'S® colors is the same as what the company claims."

## Teaching Tip

Unlike the other significance tests we have learned so far, the *P*-value for a chi-square test for goodness of fit is always found by calculating the area *to the right* of the observed chi-square statistic. Because we are squaring the differences between the observed and expected counts, the chi-square statistic can never be negative, and bigger values of the chi-square statistic indicate bigger differences from the hypothesized distribution. Fortunately, Table C is already set up to find areas to the right.

## AP® Exam Common Error

When checking conditions, many students examine the observed counts rather than the expected counts. Also, many students forget to list the expected counts, especially when doing the calculations on their calculators. Simply stating that the expected counts are at least 5 is not sufficient—students must list the expected counts to prove they have actually checked them.

## Teaching Tip

For more information on the "relative age effect" described in the Birthdays and Hockey example, see the article "August's Child is . . . Favoured by Fortune" by Lauren M. Brewer and James J. Cochran in the June 2013 issue of *Significance* magazine, published by the American Statistical Association.

We can also write these hypotheses symbolically using $p_i$ to represent the proportion of individuals in the population that fall in category $i$:

$H_0$: $p_1 =$ ___, $p_2 =$ ___, …, $p_k =$ ___.
$H_a$: At least two of the $p_i$'s are incorrect.

### THE CHI-SQUARE TEST FOR GOODNESS OF FIT

Suppose the conditions are met. To determine whether a categorical variable has a specified distribution in the population of interest, expressed as the proportion of individuals falling into each possible category, perform a test of

$H_0$: The stated distribution of the categorical variable in the population of interest is correct.
$H_a$: The stated distribution of the categorical variable in the population of interest is not correct.

Start by finding the expected count for each category assuming that $H_0$ is true. Then calculate the chi-square statistic

$$\chi^2 = \sum \frac{(\text{Observed} - \text{Expected})^2}{\text{Expected}}$$

where the sum is over the $k$ different categories. The *P*-value is the area to the right of $\chi^2$ under the density curve of the chi-square distribution with $k - 1$ degrees of freedom.

The next example shows the chi-square test for goodness of fit in action. As always, we follow the four-step process when performing inference.

## EXAMPLE Birthdays and Hockey

### *A test for equal proportions*

In his book *Outliers*, Malcolm Gladwell suggests that a hockey player's birth month has a big influence on his chance to make it to the highest levels of the game. Specifically, because January 1 is the cut-off date for youth leagues in Canada (where many National Hockey League (NHL) players come from), players born in January will be competing against players up to 12 months younger. The older players tend to be bigger, stronger, and more coordinated and hence get more playing time, more coaching, and have a better chance of being successful.

To see if birth date is related to success (judged by whether a player makes it into the NHL), a random sample of 80 NHL players from a recent season was selected and their birthdays were recorded. The one-way table below summarizes the data on birthdays for these 80 players:

| Birthday: | Jan–Mar | Apr–Jun | Jul–Sep | Oct–Dec |
|---|---|---|---|---|
| Number of players: | 32 | 20 | 16 | 12 |

Do these data provide convincing evidence that the birthdays of NHL players are not uniformly distributed throughout the year?

## ALTERNATE EXAMPLE

### *Landline surveys*

According to the 2000 census, of all U.S. residents aged 20 and older, 19.1% are in their twenties, 21.5% are in their thirties, 21.1% are in their forties, 15.5% are in their fifties, and 22.8% are 60 and older. The following table gives the age distribution for a sample of U.S. residents aged 20 and older. Members of the sample were chosen by randomly dialing landline telephone numbers.

| Category | Count |
|---|---|
| 20–29 | 141 |
| 30–39 | 186 |
| 40–49 | 224 |
| 50–59 | 211 |
| 60+ | 286 |
| **Total** | **1048** |

Do these data provide convincing evidence that the age distribution of people who answer landline telephone surveys is not the same as the age distribution of all U.S. residents?

STATE: We want to perform a test of the following hypotheses using $\alpha = 0.05$:

The null hypothesis says that NHL players' birthdays are evenly distributed across the four quarters of the year. In that case, all 4 proportions must be 1/4. So we could write the hypotheses in symbols as

$H_0$: $p_{\text{Jan-Mar}} = p_{\text{Apr-Jun}} = p_{\text{Jul-Sep}} = p_{\text{Oct-Dec}} = 1/4$
$H_a$: At least two of the proportions are not 1/4

**STATE:** We want to perform a test of

$H_0$: The birthdays of all NHL players are evenly distributed across the four quarters of the year.

$H_a$: The birthdays of all NHL players are not evenly distributed across the four quarters of the year.

No significance level was specified, so we'll use $\alpha = 0.05$.

**PLAN:** If the conditions are met, we will perform a chi-square test for goodness of fit.

- *Random:* The data came from a random sample of NHL players.
  - *10%:* Because we are sampling without replacement, there must be at least $10(80) = 800$ NHL players. In the season when the data were collected, there were 879 NHL players.
- *Large Counts:* If birthdays are evenly distributed across the four quarters of the year, then the expected counts are all $80(1/4) = 20$. These counts are all at least 5.

**DO:**

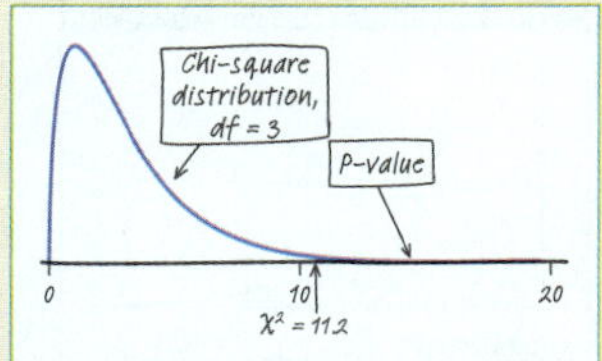


**FIGURE 11.5** The *P*-value for the chi-square test for goodness of fit with $\chi^2 = 11.2$ and df = 3.

- *Test statistic:*

$$\chi^2 = \sum \frac{(\text{Observed} - \text{Expected})^2}{\text{Expected}}$$

$$= \frac{(32-20)^2}{20} + \frac{(20-20)^2}{20} + \frac{(16-20)^2}{20} + \frac{(12-20)^2}{20}$$

$$= 7.2 + 0 + 0.8 + 3.2 = 11.2$$

- *P-value:* Figure 11.5 displays the *P*-value for this test as an area under the chi-square distribution with $4 - 1 = 3$ degrees of freedom.

As the excerpt at right shows, $\chi^2 = 11.2$ corresponds to a *P*-value between 0.01 and 0.02.

| | *p* | | |
|---|---|---|---|
| df | 0.02 | 0.01 | 0.005 |
| 2 | 7.82 | 9.21 | 10.60 |
| 3 | 9.84 | 11.34 | 12.84 |
| 4 | 11.67 | 13.28 | 14.86 |

*Using Technology:* Refer to the Technology Corner that follows the example. The calculator's $\chi^2$ GOF-Test gives $\chi^2 = 11.2$ and *P*-value = 0.011 using df = 3.

**CONCLUDE:** Because the *P*-value, 0.011, is less than $\alpha = 0.05$, we reject $H_0$. We have convincing evidence that the birthdays of NHL players are not evenly distributed across the four quarters of the year.

**For Practice** *Try Exercise* **15**

You can use your calculator to carry out the "Do" step for a chi-square test for goodness of fit. Remember that this comes with potential benefits and risks on the AP® exam.

## 26. TECHNOLOGY CORNER — CHI-SQUARE TEST FOR GOODNESS OF FIT ON THE CALCULATOR

TI-Nspire instructions in Appendix B; HP Prime instructions on the book's Web site.

You can use the TI-83/84 or TI-89 to perform the calculations for a chi-square test for goodness of fit. We'll use the data from the hockey and birthdays example to illustrate the steps.

$H_0$: The age distribution of people who answer landline telephone surveys is the same as the age distribution of all U.S. residents. $H_a$: The age distribution of people who answer landline telephone surveys is not the same as the age distribution of all U.S. residents.

**PLAN:** If conditions are met, we will perform a chi-square test for goodness of fit.

- *Random*: The data came from a random sample of U.S. residents who answer landline telephone surveys.
  - *10%*: 1048 is less than 10% of all U.S. residents aged 20 and older.
- *Large Counts*: The expected counts are $1048(0.191) = 200.2$, $1048(0.215) = 225.3$, $1048(0.211) = 221.1$, $1048(0.155) = 162.4$, $1048(0.228) = 238.9$. All expected counts are at least 5.

**DO:**

- *Test statistic:*

$$\chi^2 = \frac{(141 - 200.2)^2}{200.2} + \cdots = 48.2$$

- *P-value:* Using 5 - 1 = 4 degrees of freedom, *P*-value < 0.0005. *Using technology*: $\chi^2$`cdf(lower:48.2, upper:1000, df:4)` ≈ 0.

**CONCLUDE:** Because the *P*-value of approximately 0 is less than $\alpha = 0.05$, we reject $H_0$. We have convincing evidence that the age distribution of people who answer landline telephone surveys is not the same as the age distribution of all U.S. residents.

### Teaching Tip

Consistently ask your students to think about Type I and Type II errors. It is important that students understand that these errors are possible in any significance test, not just the ones in Chapter 9, where these errors were introduced. In the Hockey example, it is possible we made a Type I error—finding convincing evidence that the birthdays of NHL players are not uniformly distributed when they really are.

### TECHNOLOGY TIP

There are two ways to upgrade the operating system on the TI-84. The easiest is to link an older calculator with an upgraded calculator using a USB or calculator cable. Then press the LINK button on both machines. On the receiving machine, arrow over to RECEIVE and press ENTER. On the sending machine, arrow down to SendOS and press ENTER. The other method is to use free connection software and OS updates from education.ti.com.

### TECHNOLOGY TIP

Students with older calculators are allowed to have a chi-square goodness-of-fit program on their calculators during the AP® exam. In general, students are allowed to have programs that upgrade their calculators to be equivalent to other allowable calculators. However, make sure to check the most recent edition of the AP® Statistics Course Description to see if this policy is still in place.

1. Enter the counts.
- Enter the observed counts in L1/list1. Enter the expected counts in L2/list2.

| Birthday | Observed | Expected |
|---|---|---|
| Jan–Mar | 32 | 20 |
| Apr–Jun | 20 | 20 |
| Jul–Sep | 16 | 20 |
| Oct–Dec | 12 | 20 |

2. Perform a chi-square test for goodness of fit.

*Note: Some older TI-83s and TI-84s don't have this test. TI-84 users can get this functionality by upgrading their operating systems.*

**TI-83/84:** Press STAT, arrow over to TESTS and choose $\chi^2$GOF-Test. . . .

**TI-89:** In the Stats/List Editor APP, press 2nd F1 ([F6]) and choose Chi2GOF. . . .

Enter the inputs shown below. If you choose Calculate, you'll get a screen with the test statistic, *P*-value, and df. If you choose the Draw option, you'll get a picture of the appropriate chi-square distribution with the test statistic marked and shaded area corresponding to the *P*-value.

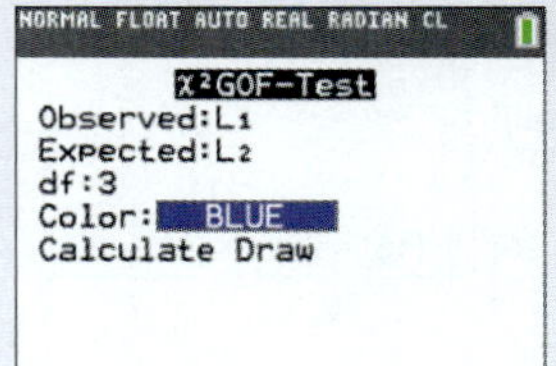


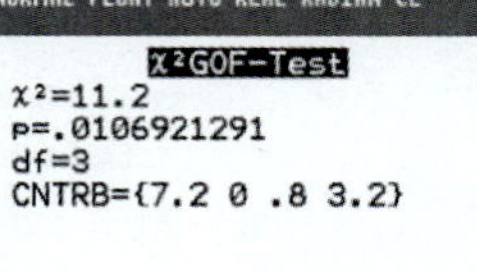


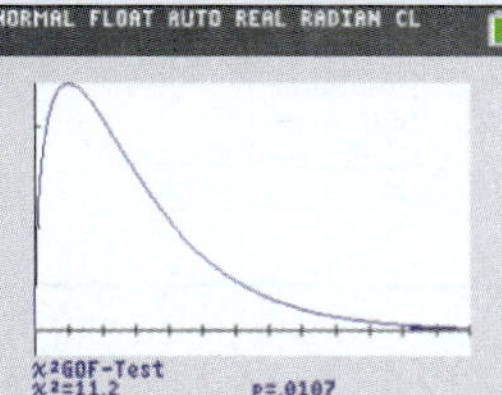

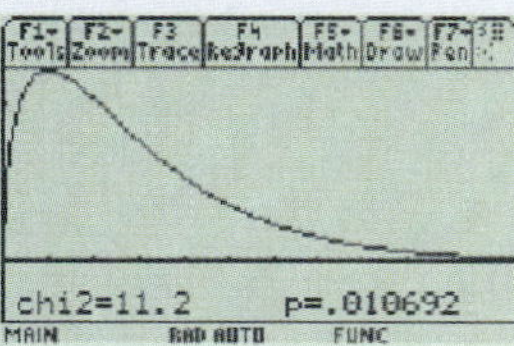


We'll discuss the CNTRB and Comp Lst results shortly.

**AP® EXAM TIP** You can use your calculator to carry out the mechanics of a significance test on the AP® exam. But there's a risk involved. If you just give the calculator answer with no work, and one or more of your values is incorrect, you will probably get no credit for the "Do" step. We recommend writing out the first few terms of the chi-square calculation followed by ". . .". This approach might help you earn partial credit if you enter a number incorrectly. Be sure to name the procedure ($\chi^2$GOF-Test) and to report the test statistic ($\chi^2 = 11.2$), degrees of freedom (df = 3), and *P*-value (0.011).

**Follow-up Analysis** In the chi-square test for goodness of fit, we test the null hypothesis that a categorical variable has a specified distribution in the population of interest. If the sample data lead to a statistically significant result, we can conclude that our variable has a distribution different from the one stated. When this happens, start by examining which categories of the variable show large deviations between the observed and expected counts. Then look at the individual terms $\frac{(\text{Observed} - \text{Expected})^2}{\text{Expected}}$ that are added together to produce the test statistic $\chi^2$. These **components** show which terms contribute most to the chi-square statistic.

## Teaching Tip

Students are not expected to do a follow-up analysis on the AP® exam unless they are specifically asked to do this. See Problem 5 on the 2008 AP® exam for an example of a question that asks students to do a follow-up analysis.

Let's return to the hockey and birthdays example. The table of observed and expected counts for the 80 randomly selected NHL players is repeated below. We have added a column that shows the components of the chi-square test statistic. Looking at the counts, we see that there were many more players born in January through March than expected and far fewer players born in October through December than expected. The component for January to March birthdays made the largest contribution to the chi-square statistic. These results support Malcolm Gladwell's claim that NHL players are more likely to be born early in the year.

| Birthday | Observed | Expected | O−E | $(O-E)^2/E$ |
|---|---|---|---|---|
| Jan–Mar | 32 | 20 | 12 | 7.2 |
| Apr–Jun | 20 | 20 | 0 | 0.0 |
| Jul–Sep | 16 | 20 | −4 | 0.8 |
| Oct–Dec | 12 | 20 | −8 | 3.2 |

*Note:* When we ran the chi-square test for goodness of fit on the calculator, a list of these individual components was stored. On the TI-83/84, the list is called CNTRB (for contribution). On the TI-89, it's called Comp Lst (component list).

## CHECK YOUR UNDERSTANDING

Biologists wish to mate pairs of fruit flies having genetic makeup RrCc, indicating that each has one dominant gene (R) and one recessive gene (r) for eye color, along with one dominant (C) and one recessive (c) gene for wing type. Each offspring will receive one gene for each of the two traits from each parent. The following table, known as a *Punnett square*, shows the possible combinations of genes received by the offspring:

| | Parent 2 passes on: | | | |
|---|---|---|---|---|
| **Parent 1 passes on:** | **RC** | **Rc** | **rC** | **rc** |
| **RC** | RRCC (x) | RRCc (x) | RrCC (x) | RrCc (x) |
| **Rc** | RRCc (x) | RRcc (y) | RrCc (x) | Rrcc (y) |
| **rC** | RrCC (x) | RrCc (x) | rrCC (z) | rrCc (z) |
| **rc** | RrCc (x) | Rrcc (y) | rrCc (z) | rrcc (w) |

Any offspring receiving an R gene will have red eyes, and any offspring receiving a C gene will have straight wings. So based on this Punnett square, the biologists predict a ratio of 9 red-eyed, straight-winged (x):3 red-eyed, curly-winged (y):3 white-eyed, straight-winged (z):1 white-eyed, curly-winged (w) offspring.

To test their hypothesis about the distribution of offspring, the biologists mate a random sample of pairs of fruit flies. Of 200 offspring, 99 had red eyes and straight wings, 42 had red eyes and curly wings, 49 had white eyes and straight wings, and 10 had white eyes and curly wings. Do these data differ significantly from what the biologists have predicted? Carry out a test at the $\alpha = 0.01$ significance level.

## Teaching Tip

When doing a follow-up analysis, don't focus only on the size of the contribution. Also discuss the direction of the difference. For example, saying that the component for January to March is the largest is good, but it is important to also point out that the observed number of players is greater than expected during that time period.

## Teaching Tip

Point out the application to biology in the Check Your Understanding on this page. Students in AP® Biology also learn about the chi-square test for goodness of fit! See Question 1 on the 2013 AP® Biology exam for an example.

## ANSWERS TO CYU

*S*: $H_0$: The distribution of eye color and wing shape is the same as what the biologists predict vs. $H_a$: The distribution of eye color and wing shape is not what the biologists predict. *P*: Chi-square test for goodness of fit. *Random*: Random sample. *10%*: $n = 200 < 10\%$ of all fruit flies. *Large Counts*: 112.5, 37.5, 37.5, 12.5 all $\geq 5$. *D*: $\chi^2 = 6.1867$, df $= 3$, the *P*-value is between 0.10 and 0.15 (0.1029). *C*: Because the *P*-value of $0.1029 > \alpha = 0.01$, we fail to reject $H_0$. We do not have convincing evidence that the distribution of eye color and wing shape is different from what the biologists predict.

## ALTERNATE EXAMPLE

### *Landline surveys*

In the previous Alternate Example, we concluded that the age distribution of people who answer landline telephone surveys is not the same as the age distribution of all U.S. residents. The table shows the observed counts, expected counts, difference in counts (O − E), and chi-square contribution for each age category. The two age groups that contributed the most to the chi-square statistic were the 20- to 29-year-olds (59.2 fewer than expected) and the 50- to 59-year-olds (48.6 more than expected).

| Category | Observed count | Expected count | O − E | $(O-E)^2/E$ |
|---|---|---|---|---|
| 20–29 | 141 | 200.2 | −59.2 | 17.51 |
| 30–39 | 186 | 225.3 | −39.3 | 6.86 |
| 40–49 | 224 | 221.1 | 2.9 | 0.04 |
| 50–59 | 211 | 162.4 | 48.6 | 14.54 |
| 60+ | 286 | 238.9 | 47.1 | 9.29 |

**TRM Section 11.1 Quizzes**

There are three different quizzes available for this section in the Teacher's Resource Materials. Versions A and B are roughly parallel. Find them in the TE-book, or on the TRCD or the book's companion Web site www.whfreeman.com/tps5e (teacher log-in required).

## Section 11.1 Summary

- A **one-way table** is often used to display the distribution of a single categorical variable for a sample of individuals.
- The **chi-square test for goodness of fit** tests the null hypothesis that a categorical variable has a specified distribution in the population of interest.
- This test compares the **observed count** in each category with the counts that would be expected if $H_0$ were true. The **expected count** for any category is found by multiplying the sample size by the proportion in each category according to the null hypothesis.
- The **chi-square statistic** is

$$\chi^2 = \sum \frac{(\text{Observed} - \text{Expected})^2}{\text{Expected}}$$

where the sum is over all possible categories.

- The conditions for performing a chi-square test for goodness of fit are:
  - **Random:** The data were produced by a well-designed random sample or randomized experiment.
    - **10%:** When sampling without replacement, check that the population is at least 10 times as large as the sample.
  - **Large Counts:** All expected counts must be at least 5.
- When the conditions are met, the sampling distribution of the statistic $\chi^2$ can be modeled by a **chi-square distribution**.
- Large values of $\chi^2$ are evidence against $H_0$ and in favor of $H_a$. The *P*-value is the area to the right of $\chi^2$ under the chi-square distribution with degrees of freedom df = number of categories − 1.
- If the test finds a statistically significant result, consider doing a follow-up analysis that compares the observed and expected counts and that looks for the largest **components** of the chi-square statistic.

### 11.1 TECHNOLOGY CORNERS

TI-Nspire Instructions in Appendix B; HP Prime instructions on the book's Web site.

# Section 11.1 Exercises

pg 681 

1. **Aw, nuts!** A company claims that each batch of its deluxe mixed nuts contains 52% cashews, 27% almonds, 13% macadamia nuts, and 8% brazil nuts. To test this claim, a quality-control inspector takes a random sample of 150 nuts from the latest batch. The one-way table below displays the sample data.

| Nut: | Cashew | Almond | Macadamia | Brazil |
|---|---|---|---|---|
| Count: | 83 | 29 | 20 | 18 |

(a) State appropriate hypotheses for performing a test of the company's claim.

(b) Calculate the expected counts for each type of nut. Show your work.

2. **Roulette** Casinos are required to verify that their games operate as advertised. American roulette wheels have 38 slots—18 red, 18 black, and 2 green. In one casino, managers record data from a random sample of 200 spins of one of their American roulette wheels. The one-way table below displays the results.

| Color: | Red | Black | Green |
|---|---|---|---|
| Count: | 85 | 99 | 16 |

(a) State appropriate hypotheses for testing whether these data give convincing evidence that the distribution of outcomes on this wheel is not what it should be.

(b) Calculate the expected counts for each color. Show your work.

pg 683 

3. **Aw, nuts!** Calculate the chi-square statistic for the data in Exercise 1. Show your work.

4. **Roulette** Calculate the chi-square statistic for the data in Exercise 2. Show your work.

5. **Aw, nuts!** Refer to Exercises 1 and 3.

(a) Confirm that the expected counts are large enough to use a chi-square distribution to calculate the *P*-value. What degrees of freedom should you use?

(b) Sketch a graph like Figure 11.4 (page 685) that shows the *P*-value.

(c) Use Table C to find the *P*-value. Then use your calculator's $\chi^2$`cdf` command.

(d) What conclusion would you draw about the company's claimed distribution for its deluxe mixed nuts? Justify your answer.

6. **Roulette** Refer to Exercises 2 and 4.

(a) Confirm that the expected counts are large enough to use a chi-square distribution to calculate the *P*-value. What degrees of freedom should you use?

(b) Sketch a graph like Figure 11.4 (page 685) that shows the *P*-value.

(c) Use Table C to find the *P*-value. Then use your calculator's $\chi^2$`cdf` command.

(d) What conclusion would you draw about whether or not the roulette wheel is operating correctly? Justify your answer.

7. **Birds in the trees** Researchers studied the behavior of birds that were searching for seeds and insects in an Oregon forest. In this forest, 54% of the trees were Douglas firs, 40% were ponderosa pines, and 6% were other types of trees. At a randomly selected time during the day, the researchers observed 156 red-breasted nuthatches: 70 were seen in Douglas firs, 79 in ponderosa pines, and 7 in other types of trees.[2] Do these data provide convincing evidence that nuthatches prefer particular types of trees when they're searching for seeds and insects?

8. **Seagulls by the seashore** Do seagulls show a preference for where they land? To answer this question, biologists conducted a study in an enclosed outdoor space with a piece of shore whose area was made up of 56% sand, 29% mud, and 15% rocks. The biologists chose 200 seagulls at random. Each seagull was released into the outdoor space on its own and observed until it landed somewhere on the piece of shore. In all, 128 seagulls landed on the sand, 61 landed in the mud, and 11 landed on the rocks. Do these data provide convincing evidence that seagulls show a preference for where they land?

9. **No chi-square** A school's principal wants to know if students spend about the same amount of time on homework each night of the week. She asks a random sample of 50 students to keep track of their homework time for a week. The following table displays the average amount of time (in minutes) students reported per night:

| Night: | Sunday | Monday | Tuesday | Wednesday | Thursday | Friday | Saturday |
|---|---|---|---|---|---|---|---|
| Average time: | 130 | 108 | 115 | 104 | 99 | 37 | 62 |

**TRM Full Solutions to Section 11.1 Exercises**

## Answers to Section 11.1 Exercises

**11.1 (a)** $H_0$: The company's claimed distribution for its deluxe mixed nuts is correct vs. $H_a$: The company's claimed distribution is not correct. **(b)** Cashews: $150(0.52) = 78$, almonds: $150(0.27) = 40.5$, macadamia nuts: $150(0.13) = 19.5$, brazil nuts: $150(0.08) = 12$.

**11.2 (a)** $H_0$: The distribution of outcomes is what it should be vs. $H_a$: The distribution of outcomes is not what it should be. **(b)** Red: $200\left(\frac{18}{38}\right) = 94.74$, black: $200\left(\frac{18}{38}\right) = 94.74$, green: $200\left(\frac{2}{38}\right) = 10.53$.

**11.3** $$\chi^2 = \frac{(83-78)^2}{78} + \frac{(29-40.5)^2}{40.5} + \frac{(20-19.5)^2}{19.5} + \frac{(18-12)^2}{12} = 6.599$$

**11.4** $$\chi^2 = \frac{(85-94.74)^2}{94.74} + \frac{(99-94.74)^2}{94.74} + \frac{(16-10.53)^2}{10.53} = 4.034$$

**11.5 (a)** Expected counts are all at least 5 and df = 3.

**(b)**

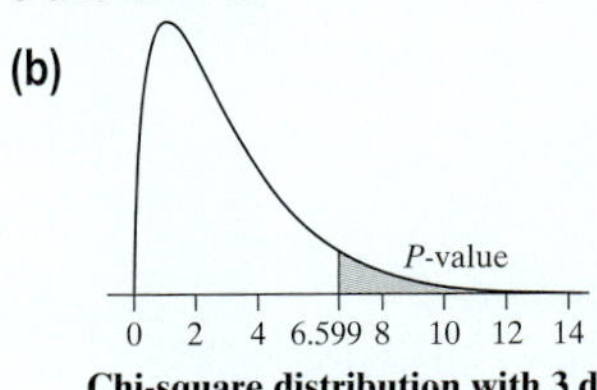

Chi-square distribution with 3 df

**(c)** The *P*-value is between 0.05 and 0.10 (0.0858). **(d)** Because the *P*-value of $0.0858 > \alpha = 0.05$, we fail to reject $H_0$. We do not have convincing evidence that the company's claimed distribution for its deluxe mixed nuts is incorrect.

**11.6 (a)** Expected counts are all at least 5 and df = 2.

**(b)**

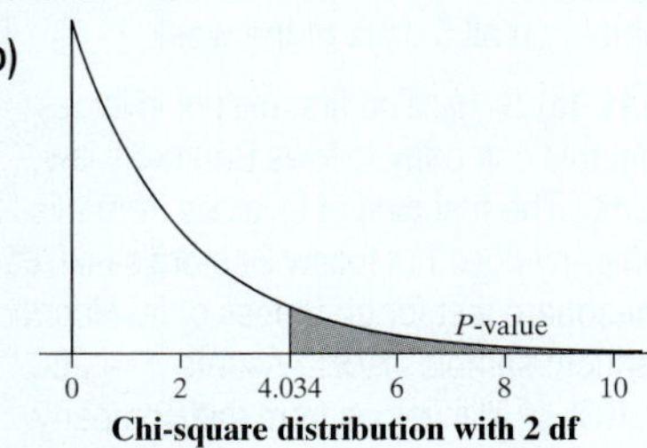

Chi-square distribution with 2 df

**(c)** The *P*-value is between 0.10 and 0.15 (0.133). **(d)** Because the *P*-value of $0.133 > \alpha = 0.05$, we fail to reject $H_0$. We do not have convincing evidence that the distribution of outcomes on the roulette wheel is not what it should be.

**11.7** *S*: $H_0$: Nuthatches do not prefer particular types of trees when searching for seeds and insects vs. $H_a$: Nuthatches do prefer particular types of trees when searching for seeds and insects. *P*: Chi-square test for goodness of fit. *Random*: Random sample. *10%*: $n = 156 < 10\%$ of all nuthatches. *Large Counts*: 84.24, 62.4, 9.36 all $\geq 5$. *D*: $\chi^2 = 7.418$. With df = 2, the *P*-value is between 0.02 and 0.025 (0.0245). *C*: Because the *P*-value of $0.0245 < \alpha = 0.05$, we reject $H_0$. There is convincing evidence that nuthatches prefer particular types of trees when they are searching for seeds and insects.

**11.8** *S*: $H_0$: Seagulls do not have a preference for where they land vs. $H_a$: Seagulls do have a preference for where they land. *P*: Chi-square test for goodness of fit. *Random*: Random sample. *10%*: $n = 200 < 10\%$ of all seagulls. *Large Counts*: 112, 58, 30 all $\geq 5$. *D*: $\chi^2 = 14.474$. With df = 2, the *P*-value is between 0.0005 and 0.001 (0.0007). *C*: Because the *P*-value of $0.0007 < \alpha = 0.05$, we reject $H_0$. There is convincing evidence that seagulls have a preference for the location where they land.

**11.9** Time spent doing homework is quantitative. Chi-square tests for goodness of fit should only be used for distributions of categorical data.

**11.10** The data do not describe the distribution of a single categorical variable. Homework completion status (yes or no) is being recorded for each student in the sample on all 5 days of the week.

**11.11 (a)** *S*: $H_0$: The first digit of invoices from this company follows Benford's law vs. $H_a$: The first digit of invoices from this company does not follow Benford's law. *P*: Chi-square test for goodness of fit. *Random*: Random sample. *10%*: Assume $n = 250 < 10\%$ of all invoices from this company. *Large Counts*: 75.25, 44, 31.25, 24.25, 19.75, 16.75, 14.5, 12.75, 11.5 all $\geq 5$. *D*: $\chi^2 = 21.563$. With df $= 8$, the *P*-value is between 0.005 and 0.01 (0.0058). *C*: Because the *P*-value of $0.0058 < \alpha = 0.05$, we reject $H_0$. There is convincing evidence that the first digit of invoices from this company does not follow Benford's law. *Follow-up analysis*: The largest contributors to the statistic are amounts with the first digit of 3, 4, or 7. There are more invoices that start with 3 or 4 than expected and fewer invoices that start with 7 than expected. **(b)** I: Finding convincing evidence that the company's invoices do not follow Benford's law (suggesting fraud), when in reality they are consistent with Benford's law. A consequence is falsely accusing this company of fraud. II: Not finding convincing evidence that the invoices do not follow Benford's law (suggesting fraud), when in reality they do not. A consequence is allowing this company to continue committing fraud. A Type I error would be more serious for the accountant.

**11.12** *S*: $H_0$: The distribution of race in the complex is the same as the population distribution vs. $H_a$: The distribution of race in the complex is not the same as the population distribution. *P*: Chi-square test for goodness of fit. *Random*: Random sample. *10%*: Assume $n = 800 < 10\%$ of the residents. *Large Counts*: 224, 192, 280, 96, 8 all $\geq 5$. *D*: $\chi^2 = 26.0625$. With df $= 4$, the *P*-value $< 0.0005$ (0.00003). *C*: Because the *P*-value of $0.00003 < \alpha = 0.05$, we reject $H_0$. There is convincing evidence that the distribution of race in the complex is not the same as the population distribution. *Follow-up analysis*: The largest contributor to the statistic, by far, is the other category. There were many more people in the "Other" category than we expected.



Explain carefully why it would not be appropriate to perform a chi-square test for goodness of fit using these data.

10. **No chi-square** The principal in Exercise 9 also asked the random sample of students to record whether they did all of the homework that was assigned on each of the five school days that week. Here are the data:

| School day: | Monday | Tuesday | Wednesday | Thursday | Friday |
|---|---|---|---|---|---|
| No. who did homework: | 34 | 29 | 32 | 28 | 19 |

Explain carefully why it would not be appropriate to perform a chi-square test for goodness of fit using these data.

11. **Benford's law** Faked numbers in tax returns, invoices, or expense account claims often display patterns that aren't present in legitimate records. Some patterns are obvious and easily avoided by a clever crook. Others are more subtle. It is a striking fact that the first digits of numbers in legitimate records often follow a model known as Benford's law.[3] Call the first digit of a randomly chosen record X for short. Benford's law gives this probability model for X (note that a first digit can't be 0):

| First digit: | 1 | 2 | 3 | 4 | 5 | 6 | 7 | 8 | 9 |
|---|---|---|---|---|---|---|---|---|---|
| Probability: | 0.301 | 0.176 | 0.125 | 0.097 | 0.079 | 0.067 | 0.058 | 0.051 | 0.046 |

A forensic accountant who is familiar with Benford's law inspects a random sample of 250 invoices from a company that is accused of committing fraud. The table below displays the sample data.

| First digit: | 1 | 2 | 3 | 4 | 5 | 6 | 7 | 8 | 9 |
|---|---|---|---|---|---|---|---|---|---|
| Count: | 61 | 50 | 43 | 34 | 25 | 16 | 7 | 8 | 6 |

(a) Are these data inconsistent with Benford's law? Carry out an appropriate test at the $\alpha = 0.05$ level to support your answer. If you find a significant result, perform a follow-up analysis.

(b) Describe a Type I error and a Type II error in this setting, and give a possible consequence of each. Which do you think is more serious?

12. **Housing** According to the Census Bureau, the distribution by ethnic background of the New York City population in a recent year was

Hispanic: 28% Black: 24% White: 35%
Asian: 12% Others: 1%

The manager of a large housing complex in the city wonders whether the distribution by race of the complex's residents is consistent with the population distribution. To find out, she records data from a random sample of 800 residents. The table below displays the sample data.[4]

| Race: | Hispanic | Black | White | Asian | Other |
|---|---|---|---|---|---|
| Count: | 212 | 202 | 270 | 94 | 22 |

Are these data significantly different from the city's distribution by race? Carry out an appropriate test at the $\alpha = 0.05$ level to support your answer. If you find a significant result, perform a follow-up analysis.

13. **Skittles** Statistics teacher Jason Molesky contacted Mars, Inc., to ask about the color distribution for Skittles candies. Here is an excerpt from the response he received: "The original flavor blend for the SKITTLES BITE SIZE CANDIES is lemon, lime, orange, strawberry and grape. They were chosen as a result of consumer preference tests we conducted. The flavor blend is 20 percent of each flavor."

(a) State appropriate hypotheses for a significance test of the company's claim.

(b) Find the expected counts for a bag of Skittles with 60 candies.

(c) How large a $\chi^2$ statistic would you need to have significant evidence against the company's claim at the $\alpha = 0.05$ level? At the $\alpha = 0.01$ level?

(d) Create a set of observed counts for a bag with 60 candies that gives a *P*-value between 0.01 and 0.05. Show the calculation of your chi-square statistic.

14. **Is your random number generator working?** Use your calculator's RandInt function to generate 200 digits from 0 to 9 and store them in a list.

(a) State appropriate hypotheses for a chi-square test for goodness of fit to determine whether your calculator's random number generator gives each digit an equal chance to be generated.

(b) Carry out a test at the $\alpha = 0.05$ significance level.

*For parts (c) and (d), assume that the students' random number generators are all working properly.*

(c) What is the probability that a student who does this exercise will make a Type I error?

(d) Suppose that 25 students in an AP Statistics class independently do this exercise for homework. Find the probability that at least one of them makes a Type I error.

15. pg 688  **What's your sign?** The University of Chicago's General Social Survey (GSS) is the nation's most important social science sample survey. For reasons known

**11.13 (a)** $H_0$: The true distribution of flavors for Skittles candies is the same as the company's claim vs. $H_a$: The true distribution of flavors for Skittles candies is not the same as the company's claim. **(b)** Expected counts all $= 12$. **(c)** Using df $= 4$, $\chi^2$ statistics greater than 9.49 would provide significant evidence at the $\alpha = 0.05$ level and $\chi^2$ values greater than 13.28 would provide significant evidence at the $\alpha = 0.01$ level. **(d)** Answers will vary.

**11.14 (a)** $H_0$: Each of the digits 0–9 from my random number generator is equally likely vs. $H_a$: Each of the digits 0–9 from my random number generator is not equally likely. **(b)** *P*: Chi-square test for goodness of fit. *Random*: Random sample. *10%*: Not needed, because we are not sampling without replacement. *Large Counts*: The expected counts are all $= 20$, which are $\geq 5$. *D*: Answers will vary. **(c)** *P*(Type I error) $= \alpha = 0.05$. **(d)** *P*(at least one Type I error) $= 1 -$ *P*(no Type I errors) $= 1 - (0.95)^{25} = 1 - 0.277 = 0.723$.

only to social scientists, the GSS regularly asks a random sample of people their astrological sign. Here are the counts of responses from a recent GSS:

| Sign: | Aries | Taurus | Gemini | Cancer | Leo | Virgo |
|---|---|---|---|---|---|---|
| Count: | 321 | 360 | 367 | 374 | 383 | 402 |
| Sign: | Libra | Scorpio | Sagittarius | Capricorn | Aquarius | Pisces |
| Count: | 392 | 329 | 331 | 354 | 376 | 355 |

If births are spread uniformly across the year, we expect all 12 signs to be equally likely. Do these data provide convincing evidence that all 12 signs are not equally likely? If you find a significant result, perform a follow-up analysis.

16. **Munching Froot Loops** Kellogg's Froot Loops cereal comes in six fruit flavors: orange, lemon, cherry, raspberry, blueberry, and lime. Charise poured out her morning bowl of cereal and methodically counted the number of cereal pieces of each flavor. Here are her data:

| Flavor: | Orange | Lemon | Cherry | Raspberry | Blueberry | Lime |
|---|---|---|---|---|---|---|
| Count: | 28 | 21 | 16 | 25 | 14 | 16 |

Do these data provide convincing evidence that Kellogg's Froot Loops do not contain an equal proportion of each flavor? If you find a significant result, perform a follow-up analysis.

17. **Mendel and the peas** Gregor Mendel (1822–1884), an Austrian monk, is considered the father of genetics. Mendel studied the inheritance of various traits in pea plants. One such trait is whether the pea is smooth or wrinkled. Mendel predicted a ratio of 3 smooth peas for every 1 wrinkled pea. In one experiment, he observed 423 smooth and 133 wrinkled peas. Assume that the conditions for inference were met. Carry out an appropriate test of the genetic model that Mendel predicted. What do you conclude?

18. **You say tomato** The paper "Linkage Studies of the Tomato" (*Transactions of the Canadian Institute*, 1931) reported the following data on phenotypes resulting from crossing tall cut-leaf tomatoes with dwarf potato-leaf tomatoes. We wish to investigate whether the following frequencies are consistent with genetic laws, which state that the phenotypes should occur in the ratio 9:3:3:1.

| Phenotype: | Tall cut | Tall potato | Dwarf cut | Dwarf potato |
|---|---|---|---|---|
| Frequency: | 926 | 288 | 293 | 104 |

Assume that the conditions for inference were met. Carry out an appropriate test of the proposed genetic model. What do you conclude?

*Multiple choice: Select the best answer for Exercises 19 to 22.*

*Exercises 19 to 21 refer to the following setting.* The manager of a high school cafeteria is planning to offer several new types of food for student lunches in the following school year. She wants to know if each type of food will be equally popular so she can start ordering supplies and making other plans. To find out, she selects a random sample of 100 students and asks them, "Which type of food do you prefer: Asian food, Mexican food, pizza, or hamburgers?" Here are her data:

| Type of Food: | Asian | Mexican | Pizza | Hamburgers |
|---|---|---|---|---|
| Count: | 18 | 22 | 39 | 21 |

19. An appropriate null hypothesis to test whether the food choices are equally popular is

(a) $H_0{:}\mu = 25$, where $\mu$ = the mean number of students that prefer each type of food.

(b) $H_0{:}p = 0.25$, where $p$ = the proportion of all students who prefer Asian food.

(c) $H_0{:}n_A = n_M = n_P = n_H = 25$, where $n_A$ is the number of students in the school who would choose Asian food, and so on.

(d) $H_0{:}p_A = p_M = p_P = p_H = 0.25$, where $p_A$ is the proportion of students in the school who would choose Asian food, and so on.

(e) $H_0{:}\hat{p}_A = \hat{p}_M = \hat{p}_P = \hat{p}_H = 0.25$, where $\hat{p}_A$ is the proportion of students in the sample who chose Asian food, and so on.

20. The chi-square statistic is

(a) $\frac{(18-25)^2}{25} + \frac{(22-25)^2}{25} + \frac{(39-25)^2}{25} + \frac{(21-25)^2}{25}$

(b) $\frac{(25-18)^2}{18} + \frac{(25-22)^2}{22} + \frac{(25-39)^2}{39} + \frac{(25-21)^2}{21}$

(c) $\frac{(18-25)}{25} + \frac{(22-25)}{25} + \frac{(39-25)}{25} + \frac{(21-25)}{25}$

(d) $\frac{(18-25)^2}{100} + \frac{(22-25)^2}{100} + \frac{(39-25)^2}{100} + \frac{(21-25)^2}{100}$

(e) $\frac{(0.18-0.25)^2}{0.25} + \frac{(0.22-0.25)^2}{0.25} + \frac{(0.39-0.25)^2}{0.25} + \frac{(0.21-0.25)^2}{0.25}$

21. The $P$-value for a chi-square test for goodness of fit is 0.0129. Which of the following is the most appropriate conclusion?

**11.15** *S*: $H_0$: All 12 astrological signs are equally likely vs. $H_a$: All 12 astrological signs are not equally likely. *P*: Chi-square test for goodness of fit. *Random*: Random sample. *10%*: $n = 4344 < 10\%$ of all people in the United States. *Large Counts*: All expected counts = 362, which are $\geq 5$. *D*: $\chi^2 = 19.76$. With df = 11, the *P*-value is between 0.025 and 0.05 (0.0487). *C*: Because the *P*-value of $0.0487 < \alpha = 0.05$, we reject $H_0$. There is convincing evidence that the 12 astrological signs are not equally likely. *Follow-up analysis*: The largest contributors to the statistic are Aries and Virgo. There are fewer Aries $(321 - 362 = -41)$ and more Virgos $(402 - 362 = 40)$ than we would expect.

**11.16** *S*: $H_0$: Froot Loops contain an equal proportion of each flavor vs. $H_a$: Froot Loops do not contain an equal proportion of each flavor. *P*: Chi-square test for goodness of fit. *Random*: Random sample. *10%*: $n = 120 < 10\%$ of all Froot Loops. *Large Counts*: All expected counts = 20, which are $\geq 5$. *D*: $\chi^2 = 7.9$. With df = 5, the *P*-value is between 0.15 and 0.20 (0.1618). *C*: Because the *P*-value of $0.1618 > \alpha = 0.05$, we fail to reject $H_0$. We do not have convincing evidence that Froot Loops do not contain an equal proportion of each flavor. *Follow-up analysis*: Not necessary.

**11.17** *S*: $H_0$: Mendel's 3:1 genetic model is correct vs. $H_a$: Mendel's 3:1 genetic model is not correct. *P*: Chi-square test for goodness of fit. Conditions are met. *D*: $\chi^2 = 0.3453$. With df = 1, the *P*-value $> 0.25$ (0.5568). *C*: Because the *P*-value of $0.5568 > \alpha = 0.05$, we fail to reject $H_0$. We do not have convincing evidence that Mendel's 3:1 genetic model is wrong.

**11.18** *S*: $H_0$: The proposed 9:3:3:1 genetic model is correct vs. $H_a$: The proposed 9:3:3:1 genetic model is not correct. *P*: Chi-square test for goodness of fit. Conditions are met. *D*: $\chi^2 = 1.469$. Using df = 3, the *P*-value $> 0.25$ (0.6895). *C*: Because the *P*-value of $0.6895 > \alpha = 0.05$, we fail to reject $H_0$. We do not have convincing evidence that the proposed 9:3:3:1 genetic model is wrong.

**11.19** d

**11.20** a

**11.21** c

**11.22** c

**11.23** The distribution of English grades for the heavy readers is skewed to the left, while the distribution of English grades for the light readers is roughly symmetric. The center of the distribution of English grades is greater for the heavy readers than for the light readers. The English grades are more variable for the light readers. There is one low outlier in the heavy reading group but no outliers in the light reading group.

**11.24 (a)** *Random*: Reasonable to consider the two samples independent. *10%*: $n_1 = 47 < 10\%$ of heavy readers and $n_2 = 32 < 10\%$ of light readers at this large school. *Normal/Large Sample*: $n_1 = 47 \geq 30$ and $n_2 = 32 \geq 30$. **(b)** *S*: $\mu_1 =$ the true mean English grade of heavy readers and $\mu_2 =$ the true mean English grade of light readers. *P*: Two-sample *t* interval for $\mu_1 - \mu_2$. *D*: Using df = 30, (0.1163, 0.4517). Using df = 59.46, (0.1197, 0.4483). *C*: We are 95% confident that the interval from 0.1197 to 0.4483 captures the true difference in the mean English grade of heavy and light readers. **(c)** This was an observational study, so no conclusion of cause and effect can be drawn.

**11.25 (a)** For each additional book read, the predicted English GPA increases by about 0.024. The predicted English grade for a student who has read 0 books is about 3.42. **(b)** Residual $= 2.85 - 3.828 = -0.978$. This student's English GPA is 0.978 less than predicted, based on the number of books this student has read. **(c)** Not very strong. On the scatterplot, the points are quite spread out from the line. Also, the value of $r^2$ is 0.083, which means that only 8.3% of the variation in English grades is accounted for by the linear model relating English GPA to number of books read.

**11.26 (a)** There are 6 different possible ways of getting a Yahtzee—one for each number on the die. **(b)** Nassir should not be surprised. $P(\text{Yahtzee}) = 6\left(\frac{1}{6}\right)^5 = 0.000772$. The probability of getting 25 rolls with no Yahtzee is very large: $(0.999228)^{25} = 0.9809$.



(a) Because 0.0129 is less than $\alpha = 0.05$, reject $H_0$. There is convincing evidence that the food choices are equally popular.

(b) Because 0.0129 is less than $\alpha = 0.05$, reject $H_0$. There is not convincing evidence that the food choices are equally popular.

(c) Because 0.0129 is less than $\alpha = 0.05$, reject $H_0$. There is convincing evidence that the food choices are not equally popular.

(d) Because 0.0129 is less than $\alpha = 0.05$, fail to reject $H_0$. There is not convincing evidence that the food choices are equally popular.

(e) Because 0.0129 is less than $\alpha = 0.05$, fail to reject $H_0$. There is convincing evidence that the food choices are equally popular.

22. Which of the following is *false*?

(a) A chi-square distribution with *k* degrees of freedom is more right-skewed than a chi-square distribution with $k + 1$ degrees of freedom.

(b) A chi-square distribution never takes negative values.

(c) The degrees of freedom for a chi-square test is determined by the sample size.

(d) $P(\chi^2 > 10)$ is greater when df $= k + 1$ than when df $= k$.

(e) The area under a chi-square density curve is always equal to 1.

*Exercises 23 through 25 refer to the following setting.* Do students who read more books for pleasure tend to earn higher grades in English? The boxplots below show data from a simple random sample of 79 students at a large high school. Students were classified as light readers if they read fewer than 3 books for pleasure per year. Otherwise, they were classified as heavy readers. Each student's average English grade for the previous two marking periods was converted to a GPA scale where A+ = 4.3, A = 4.0, A− = 3.7, B+ = 3.3, and so on.

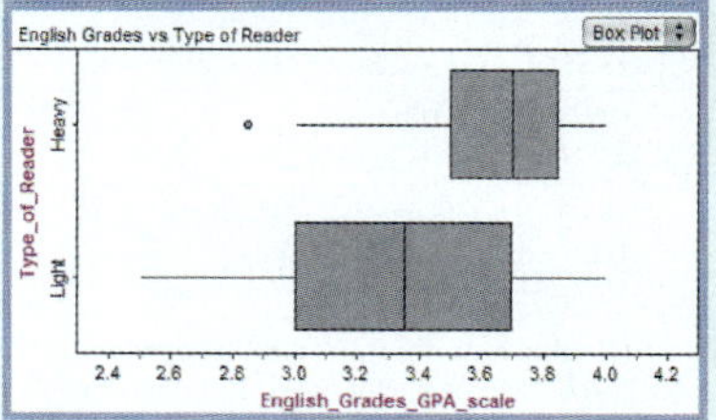


23. **Reading and grades** (1.3) Write a few sentences comparing the distributions of English grades for light and heavy readers.

24. **Reading and grades** (10.2) Summary statistics for the two groups from Minitab are provided below.

| Type of reader | N | Mean | StDev | SE Mean |
|---|---|---|---|---|
| Heavy | 47 | 3.640 | 0.324 | 0.047 |
| Light | 32 | 3.356 | 0.380 | 0.067 |

(a) Explain why it is acceptable to use two-sample *t* procedures in this setting.

(b) Construct and interpret a 95% confidence interval for the difference in the mean English grade for light and heavy readers.

(c) Does the interval in part (b) provide convincing evidence that reading more causes a difference in students' English grades? Justify your answer.

25. **Reading and grades** (3.2) The Fathom scatterplot below shows the number of books read and the English grade for all 79 students in the study. A least-squares regression line has been added to the graph.

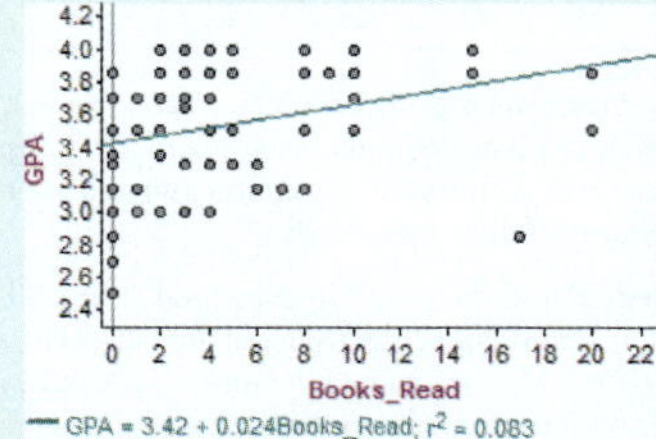


(a) Interpret the meaning of the slope and *y* intercept in context.

(b) The student who reported reading 17 books for pleasure had an English GPA of 2.85. Find this student's residual and interpret this value in context.

(c) How strong is the relationship between English grades and number of books read? Give appropriate evidence to support your answer.

26. **Yahtzee** (5.3, 6.3) In the game of Yahtzee, 5 six-sided dice are rolled simultaneously. To get a Yahtzee, the player must get the same number on all 5 dice.

(a) Luis says that the probability of getting a Yahtzee in one roll of the dice is $\left(\frac{1}{6}\right)^5$. Explain why Luis is wrong.

(b) Nassir decides to keep rolling all 5 dice until he gets a Yahtzee. He is surprised when he still hasn't gotten a Yahtzee after 25 rolls. Should he be? Calculate an appropriate probability to support your answer.

# 11.2 Inference for Two-Way Tables

**WHAT YOU WILL LEARN** By the end of the section, you should be able to:

- Compare conditional distributions for data in a two-way table.
- State appropriate hypotheses and compute expected counts for a chi-square test based on data in a two-way table.
- Calculate the chi-square statistic, degrees of freedom, and *P*-value for a chi-square test based on data in a two-way table.
- Perform a chi-square test for homogeneity.
- Perform a chi-square test for independence.
- Choose the appropriate chi-square test.

The two-sample $z$ procedures of Chapter 10 allow us to compare the proportions of successes in two populations or for two treatments. What if we want to compare more than two samples or groups? More generally, what if we want to compare the distributions of a single categorical variable across several populations or treatments? We need a new statistical test. The new test starts by presenting the data in a two-way table.

Two-way tables have more general uses than comparing distributions of a single categorical variable. As we saw in Section 1.1, they can be used to describe relationships between any two categorical variables. In this section, we will start by developing a test to determine whether the distribution of a categorical variable is the same for each of several populations or treatments. This test is called a **chi-square test for homogeneity**. Then we'll examine a related test to see whether there is convincing evidence of an association between the row and column variables in a two-way table. This test is known as a **chi-square test for independence**.

## Comparing Distributions of a Categorical Variable

We'll start with an example involving a randomized experiment.

**EXAMPLE**

### Does Background Music Influence What Customers Buy?

*Comparing conditional distributions*

Market researchers suspect that background music may affect the mood and buying behavior of customers. One study in a European restaurant compared three randomly assigned treatments: no music, French accordion music, and Italian string music. Under each condition, the researchers recorded the number of

**PD Section 11.2 Overview**

Watch the video overview of Section 11.2 (for teachers) by clicking on the link in the TE-book, opening the TRCD, or logging onto the book's companion Web site www.whfreeman.com/tps5e (teacher log-in required).

**TRM Section 11.2 Alternate Examples**

You can find the Alternate Examples for this section in Microsoft Word format by clicking on the link in the TE-book, opening the TRCD, or logging onto the book's companion Web site www.whfreeman.com/tps5e (teacher log-in required).

**Teaching Tip**

As you begin this section, it might be helpful to briefly review some of the main ideas from Section 1.1, especially the idea of comparing conditional distributions using graphs.

**Teaching Tip**

We can use a chi-square test to compare the distribution of a single categorical variable for 2 or more populations or treatments. To compare the mean of a quantitative variable for more than 2 population or treatments, we use a significance test called ANOVA. This test is not included in the AP® Topic Outline, but it is covered in Chapter 13, available on the book's Web site, www.whfreeman.com/tps5e.

## Teaching Tip

As in Chapters 1 and 5, when we present a two-way table we list the values of the explanatory variable in the columns and the values of the response variables in the rows. However, it isn't wrong to reverse the variables, so don't be surprised if you see a table with this orientation in other places.

**ALTERNATE EXAMPLE**

### *Birthdays*

Has modern technology changed the distribution of birthdays? With more babies being delivered by planned C-section, Mrs. McDonald's statistics class hypothesized that the day-of-the-week distribution for births would be different for people born after 1993 compared to people born before 1980. After all, why would a doctor plan a C-section for the weekend? To investigate, they selected a random sample of people from each of the two age categories and recorded the day of the week on which they were born. The results are shown in the table below. Is there convincing evidence that the distribution of birthdays has changed?

| | Before 1980 | After 1993 | Total |
|---|---|---|---|
| Sunday | 12 | 9 | **21** |
| Monday | 12 | 11 | **23** |
| Tuesday | 14 | 11 | **25** |
| Wednesday | 12 | 10 | **22** |
| Thursday | 7 | 17 | **24** |
| Friday | 9 | 9 | **18** |
| Saturday | 11 | 6 | **17** |
| **Total** | **77** | **73** | **150** |

PROBLEM:

**(a)** Calculate the conditional distribution (in proportions) of the birthday for older people and younger people.

**(b)** Make an appropriate graph for comparing the conditional distributions in part (a).

**(c)** Write a few sentences comparing the distributions of birthdays for each age group.

customers who ordered French, Italian, and other entrees.[5] Here is a table that summarizes the data:

| Entree ordered | Type of Music: None | French | Italian | Total |
|---|---|---|---|---|
| French | 30 | 39 | 30 | **99** |
| Italian | 11 | 1 | 19 | **31** |
| Other | 43 | 35 | 35 | **113** |
| **Total** | **84** | **75** | **84** | **243** |

PROBLEM:

(a) Calculate the conditional distribution (in proportions) of entrees ordered for each treatment.

(b) Make an appropriate graph for comparing the conditional distributions in part (a).

(c) Write a few sentences comparing the distributions of entrees ordered under the three music treatments.

SOLUTION:

(a) When no music was playing, the distribution of entree orders was

$$\text{French: } \frac{30}{84} = 0.357 \quad \text{Italian: } \frac{11}{84} = 0.131 \quad \text{Other: } \frac{43}{84} = 0.512$$

When French accordion music was playing, the distribution of entree orders was

$$\text{French: } \frac{39}{75} = 0.520 \quad \text{Italian: } \frac{1}{75} = 0.013 \quad \text{Other: } \frac{35}{75} = 0.467$$

When Italian string music was playing, the distribution of entree orders was

$$\text{French: } \frac{30}{84} = 0.357 \quad \text{Italian: } \frac{19}{84} = 0.226 \quad \text{Other: } \frac{35}{84} = 0.417$$

(b) The bar graphs in Figure 11.6 compare the distributions of entrees ordered for each of the three music treatments.

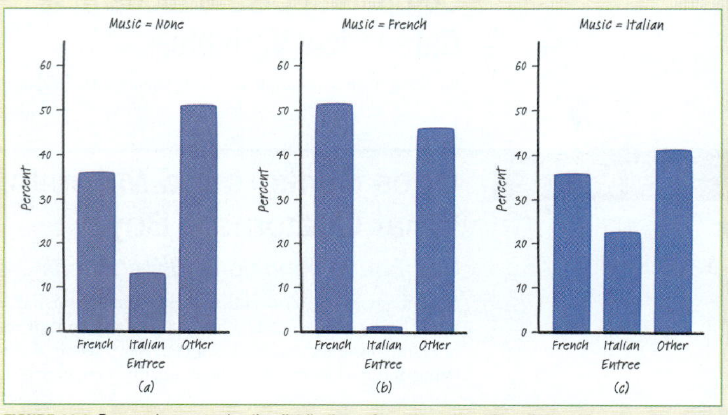


**FIGURE 11.6** Bar graphs comparing the distributions of entrees ordered for different music conditions.

SOLUTION:

(a)

| | Before 1980 | After 1993 |
|---|---|---|
| Sunday | 12/77 = 0.16 | 9/73 = 0.12 |
| Monday | 12/77 = 0.16 | 11/73 = 0.15 |
| Tuesday | 14/77 = 0.18 | 11/73 = 0.15 |
| Wednesday | 12/77 = 0.16 | 10/73 = 0.14 |
| Thursday | 7/77 = 0.09 | 17/73 = 0.23 |
| Friday | 9/77 = 0.12 | 9/73 = 0.12 |
| Saturday | 11/77 = 0.14 | 6/73 = 0.08 |

(b)

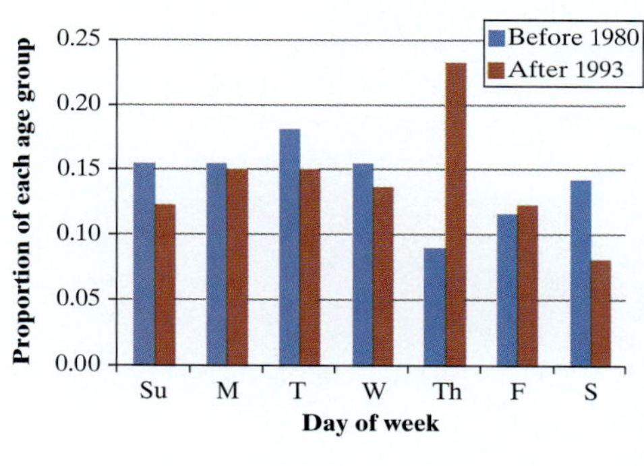

(c) The type of entree that customers order seems to differ considerably across the three music treatments. Orders of Italian entrees are very low (1.3%) when French music is playing but are higher when Italian music (22.6%) or no music (13.1%) is playing. French entrees seem popular in this restaurant, as they are ordered frequently under all music conditions but notably more often when French music is playing. For all three music treatments, the percent of Other entrees ordered was similar.

**For Practice** *Try Exercise* 27

The researchers in the restaurant study expected that music would influence customer orders, so the type of music played is the explanatory variable and the type of entree ordered is the response variable. A good general strategy is to compare the conditional distributions of the response variable for each value of the explanatory variable. That's why we compared the conditional distributions of entrees ordered for each type of music played.

It is common practice to describe a two-way table by its number of rows and columns (not including totals). For instance, the data in the previous example were given in a $3 \times 3$ table. The following Check Your Understanding involves a $3 \times 2$ table.

## CHECK YOUR UNDERSTANDING

The Pennsylvania State University has its main campus in the town of State College and more than 20 smaller "commonwealth campuses" around the state. The Penn State Division of Student Affairs polled separate random samples of undergraduates from the main campus and commonwealth campuses about their use of online social networking. Facebook was the most popular site, with more than 80% of students having an account. Here is a comparison of Facebook use by undergraduates at the main campus and commonwealth campuses who have a Facebook account:[6]

| Use Facebook | Main campus | Commonwealth |
|---|---|---|
| Several times a month or less | 55 | 76 |
| At least once a week | 215 | 157 |
| At least once a day | 640 | 394 |
| **Total Facebook users** | **910** | **627** |

1. Calculate the conditional distribution (in proportions) of Facebook use for each campus setting.
2. Why is it important to compare proportions rather than counts in Question 1?
3. Make a bar graph that compares the two conditional distributions. What are the most important differences in Facebook use between the two campus settings?

### Stating Hypotheses

The null hypothesis in the restaurant example is

$H_0$: There is *no difference* in the true distributions of entrees ordered at this restaurant when no music, French accordion music, or Italian string music is played.

## ANSWERS TO CYU

**1.** Main: 0.060 several times a month or less, 0.236 at least once a week, 0.703 at least once a day. Commonwealth: 0.121 several times a month or less, 0.250 at least once a week, 0.628 at least once a day.

**2.** Because there was such a big difference in the sample size from the two different types of campuses.

**3.** Students on the main campus are more likely to be everyday users of Facebook. Also, those on the commonwealth campuses are more likely to use Facebook several times a month or less.

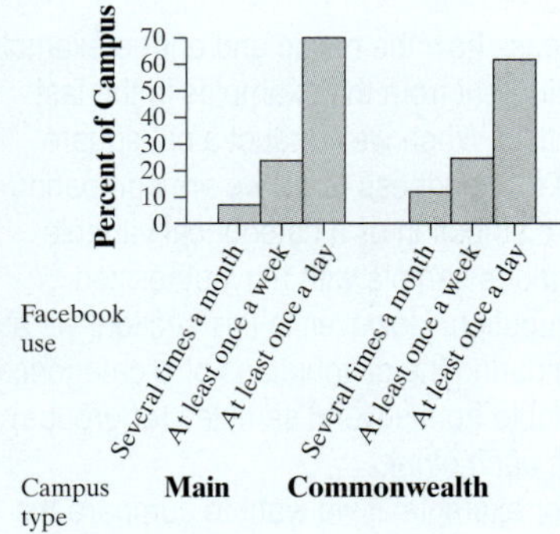


**(c)** There were a higher proportion of older people born on Tuesday, Saturday, and Sunday and a much higher proportion of younger people born on Thursday. The proportions are similar for Monday, Wednesday, and Friday. This supports the idea that doctors plan C-sections for convenient times. There was a smaller proportion of young people born on the weekend and a much bigger proportion born on Thursday, possibly to avoid having their mothers go into labor on the weekend.

## Teaching Tip

As you discuss the hypotheses for the music and entree study, point out the word "true." We use this word to be clear that we are not referring to the observed distributions of entrees ordered at this restaurant. We know the observed distributions are different. We want to know if there is convincing evidence that distributions of entrees would be different if the treatments were re-randomized to different days.

## Teaching Tip

As always, emphasize the big picture when introducing a significance test. There are two possible explanations for why the conditional distributions of purchases are different for the three different treatments.

**1.** People are not influenced by the type of background music, and the differences in the distributions are due to the chance variation in random assignment.

**2.** People are influenced by the type of background music.

We calculate a *P*-value to determine how likely it is to get differences as big as the ones we observed in the experiment due to the chance variation in random assignment only, assuming that the type of background music has no influence on purchases.

## Teaching Tip

Discuss how the music and entree example is different from the examples in the last section. When we conduct a chi-square test for goodness of fit, we are comparing the distribution of a categorical variable from one sample with a hypothesized distribution. However, in this section, we are comparing the distributions of a categorical variable from several samples (or groups) *with each other.*

For example, if we want to compare the distribution of colors of plain M&M'S® to the distribution claimed by the company, we would do a chi-square test for goodness of fit. However, if we want to compare the distributions of colors for plain and peanut M&M'S, we would use the methods of this section.

This is analogous to the distinction between a one-sample z test for a proportion and a two-sample z test for a difference in proportions. In the one-sample case, we compared a sample proportion with a hypothesized proportion, but in the two-sample case, we compared two sample proportions with each other.

If the null hypothesis is true, the observed differences in the distributions of entrees ordered for the three groups are due to the chance involved in the random assignment of treatments. The alternative hypothesis says that there *is* a difference but does not specify the nature of that difference:

$H_a$: There is a difference in the true distributions of entrees ordered at this restaurant when no music, French accordion music, or Italian string music is played.

Any difference among the three true distributions of entrees ordered when no music, French accordion music, or Italian string music is played means that the null hypothesis is false and the alternative hypothesis is true. The alternative hypothesis is not one-sided or two-sided. We might call it "many-sided" because it allows any kind of difference.

With only the methods we already know, we might start by comparing the proportions of French entrees ordered when no music and French accordion music are played. We could similarly compare other pairs of proportions, ending up with many tests and many *P*-values. This is a bad idea. The *P*-values belong to each test separately, not to the collection of all the tests together.

| | Type of Music | | | |
|---|---|---|---|---|
| **Entree ordered** | **None** | **French** | **Italian** | **Total** |
| French | 30 | 39 | 30 | **99** |
| Italian | 11 | 1 | 19 | **31** |
| Other | 43 | 35 | 35 | **113** |
| **Total** | **84** | **75** | **84** | **243** |

Caution

*When we do many individual tests or construct many confidence intervals, the individual P-values and confidence levels don't tell us how confident we can be in all the inferences taken together.* Because of this, it's cheating to pick out one large difference from the two-way table and then perform a significance test as if it were the only comparison we had in mind. For example, the proportions of French entrees ordered under the no music and French accordion music treatments are $30/84 = 0.357$ and $39/75 = 0.520$, respectively. A two-sample $z$ test shows that the difference between the proportions is statistically significant ($z = 2.06$, $P = 0.039$) if we make just this one comparison.

But we could also pick a comparison that is not significant. For example, the proportions of Italian entrees ordered for the no music and Italian string music treatments are $11/84 = 0.131$ and $19/84 = 0.226$, respectively. These two proportions do not differ significantly ($z = 1.61$, $P = 0.107$). Individual comparisons can't tell us whether the three *distributions* of the categorical variable (in this case, type of entree ordered) are significantly different.

The problem of how to do many comparisons at once with an overall measure of confidence in all our conclusions is common in statistics. This is the problem of **multiple comparisons.** Statistical methods for dealing with multiple comparisons usually have two parts:

1. An *overall test* to see if there is convincing evidence of any differences among the parameters that we want to compare.
2. A detailed *follow-up analysis* to decide which of the parameters differ and to estimate how large the differences are.

The overall test uses the familiar chi-square statistic. But in this new setting the test will be used to compare the distribution of a categorical variable for several populations or treatments. The follow-up analysis can be quite elaborate. We will concentrate on the overall test and do a follow-up analysis only when the observed results are statistically significant.

## Expected Counts and the Chi-Square Statistic

A chi-square test for homogeneity begins with the hypotheses

$H_0$: There is no difference in the distribution of a categorical variable for several populations or treatments.

$H_a$: There is a difference in the distribution of a categorical variable for several populations or treatments.

It would also be correct to state the null hypothesis as $H_0$: The distribution of a categorical variable is the same for each of several populations or treatments. We prefer the "no difference" wording because it's more consistent with the language we used in the significance tests of Chapter 10.

To perform a test, we compare the observed counts in a two-way table with the counts we would expect if $H_0$ were true. Finding the expected counts is not that difficult, as the following example illustrates.

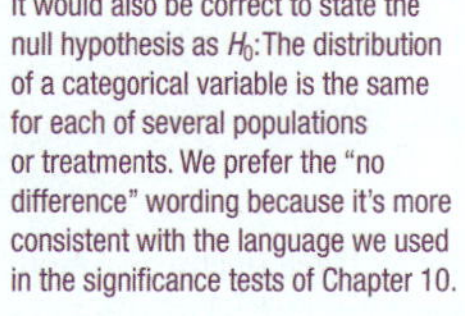

**EXAMPLE**

### Does Background Music Influence What Customers Buy?

*Computing expected counts*

The null hypothesis in the restaurant experiment is that there's no difference in the distribution of entrees ordered when no music, French accordion music, or Italian string music is played. To find the expected counts, we start by assuming that $H_0$ is true. We can see from the two-way table that 99 of the 243 entrees ordered during the study were French.

**Observed Counts**

| | Type of Music | | | |
|---|---|---|---|---|
| Entree ordered | None | French | Italian | Total |
| French | 30 | 39 | 30 | **99** |
| Italian | 11 | 1 | 19 | **31** |
| Other | 43 | 35 | 35 | **113** |
| **Total** | **84** | **75** | **84** | **243** |

If the specific type of music that's playing has no effect on entree orders, the proportion of French entrees ordered under each music condition should be 99/243 = 0.4074. For instance, there were 84 total entrees ordered when no music was playing. We would expect

$$84 \cdot \frac{99}{243} = 84(0.4074) = 34.22$$

**ALTERNATE EXAMPLE**

### *Birthdays*

Here is a summary of the birthday study:

| | Before 1980 | After 1993 | Total |
|---|---|---|---|
| Sunday | 12 | 9 | **21** |
| Monday | 12 | 11 | **23** |
| Tuesday | 14 | 11 | **25** |
| Wednesday | 12 | 10 | **22** |
| Thursday | 7 | 17 | **24** |
| Friday | 9 | 9 | **18** |
| Saturday | 11 | 6 | **17** |
| **Total** | **77** | **73** | **150** |

PROBLEM: Calculate the expected counts for the two age groups, assuming that there is no difference in the distribution of birthdays.

SOLUTION: Because 21/150 = 14% of all respondents were born on Sunday, we expect 14% of older people and 14% of younger people to be born on Sunday.

Older: 77(0.14) = 10.78

Younger: 73(0.14) = 10.22

Here is a table with all the expected counts:

| | Before 1980 | After 1993 |
|---|---|---|
| Sunday | 77(0.14) = 10.78 | 73(0.14) = 10.22 |
| Monday | 77(0.153) = 11.78 | 73(0.153) = 11.17 |
| Tuesday | 77(0.167) = 12.86 | 73(0.167) = 12.19 |
| Wednesday | 77(0.147) = 11.32 | 73(0.147) = 10.73 |
| Thursday | 77(0.16) = 12.32 | 73(0.16) = 11.68 |
| Friday | 77(0.12) = 9.24 | 73(0.12) = 8.76 |
| Saturday | 77(0.113) = 8.70 | 73(0.113) = 8.25 |

### Teaching Tip

Resist the temptation to show your students the formula for expected counts on page 703 right away. It is much better if students can understand why the formula works.

## AP® Exam Common Error

Some students mistakenly round the expected counts, believing that they must be integers. Students will lose points for this on the AP® exam because it shows a misunderstanding of what an expected count represents—the average number of subjects in a treatment group with that response in repeated reassignments of the treatments, assuming the treatments have the same effect.

Although any count of entrees ordered must be a whole number, an expected count need not be. The expected count gives the average number of entrees ordered if $H_0$ is true and the random assignment process is repeated many times.

of those entrees to be French, on average. The expected counts of French entrees ordered under the other two music conditions can be found in a similar way:

French music: 75(0.4074) = 30.56 Italian music: 84(0.4074) = 34.22

We repeat the process to find the expected counts for the other two types of entrees. The overall proportion of Italian entrees ordered during the study was 31/243 = 0.1276. So the expected counts of Italian entrees ordered under each treatment are

No music: 84(0.1276) = 10.72 French music: 75(0.1276) = 9.57

Italian music: 84(0.1276) = 10.72

The overall proportion of Other entrees ordered during the experiment was 113/243 = 0.465. So the expected counts of Other entrees ordered for each treatment are

No music: 84(0.465) = 39.06 French music: 75(0.465) = 34.88

Italian music: 84(0.465) = 39.06

The following table summarizes the expected counts for all three treatments. Note that the values for no music and Italian music are the same because 84 total entrees were ordered under each condition. We can check our work by adding the expected counts to obtain the row and column totals, as in the table. These should be the same as those in the table of observed counts except for small round-off errors, such as 75.01 rather than 75 for the total number of French entrees ordered.

**Expected Counts**

| | Type of Music | | | |
|---|---|---|---|---|
| Entree ordered | None | French | Italian | Total |
| French | 34.22 | 30.56 | 34.22 | **99** |
| Italian | 10.72 | 9.57 | 10.72 | **31** |
| Other | 39.06 | 34.88 | 39.06 | **113** |
| **Total** | **84** | **75** | **84** | **243** |

Let's take a look at the two-way table from the restaurant study one more time. In the example, we found the expected count of French entrees ordered when no music was playing as follows:

$$84 \cdot \frac{99}{243} = 34.22$$

**Observed Counts**

| | Type of Music | | | |
|---|---|---|---|---|
| Entree ordered | None | French | Italian | Total |
| French | 30 | 39 | 30 | **99** |
| Italian | 11 | 1 | 19 | **31** |
| Other | 43 | 35 | 35 | **113** |
| **Total** | **84** | **75** | **84** | **243** |

We marked the three numbers used in this calculation in the table. These values are the row total for French entrees ordered, the column total for entrees ordered when no music was playing, and the table total of entrees ordered during the experiment. We can rewrite the original calculation as

$$\frac{84 \cdot 99}{243} = \frac{99 \cdot 84}{243} = 34.22$$

This suggests a more general formula for the expected count in any cell of a two-way table:

$$\frac{\text{row total} \cdot \text{column total}}{\text{table total}}$$

**FINDING EXPECTED COUNTS**

When $H_0$ is true, the expected count in any cell of a two-way table is

$$\text{expected count} = \frac{\text{row total} \cdot \text{column total}}{\text{table total}}$$

All the expected counts in the restaurant study are at least 5. This satisfies the Large Counts condition. The Random condition is met because the treatments were assigned at random. We don't need to check the 10% condition because the researchers were not sampling without replacement from some population of interest. They just performed an experiment using customers who happened to be in the restaurant at the time.

**CONDITIONS FOR PERFORMING A CHI-SQUARE TEST FOR HOMOGENEITY**

- **Random:** The data come from independent random samples or from the groups in a randomized experiment.
  - **10%:** When sampling without replacement, check that $n \leq \frac{1}{10}N$.
- **Large Counts:** All *expected* counts are at least 5.

Just as we did with the chi-square test for goodness of fit, we compare the observed counts with the expected counts using the statistic

$$\chi^2 = \sum \frac{(\text{Observed} - \text{Expected})^2}{\text{Expected}}$$

This time, the sum is over all cells (not including the totals!) in the two-way table.

### AP® Exam Formula Sheet

The formula for the expected counts is *not* given on the formula sheet that is provided to students during the AP® exam. This is another reason why it is important that students understand why the expected counts are calculated this way.

### Teaching Tip

Emphasize that the 10% and Large Counts conditions for a test of homogeneity are essentially the same as the conditions for a test of goodness of fit. The only difference is the Random condition. For the test of homogeneity, we check that the data come from two or more independent random samples or two or more groups in a randomized experiment.

### AP® Exam Common Error

When checking conditions, many students examine the observed counts rather than the expected counts. Also, many students forget to list the expected counts, especially when doing the calculations on their calculators. Simply stating that the expected counts are at least 5 is not sufficient—students must list the expected counts to prove they have actually checked them.

### AP® Exam Formula Sheet

The formula for the chi-square statistic is given on the formula sheet that is provided to students during the AP® exam. It is exactly the same as the chi-square statistic we used for the test for goodness of fit.

## EXAMPLE Does Background Music Influence What Customers Buy?

### *The chi-square statistic*

PROBLEM: The tables below show the observed and expected counts for the restaurant experiment. Calculate the chi-square statistic. Show your work.

**Observed Counts**

| | Type of Music | | | |
|---|---|---|---|---|
| Entree ordered | None | French | Italian | Total |
| French | 30 | 39 | 30 | **99** |
| Italian | 11 | 1 | 19 | **31** |
| Other | 43 | 35 | 35 | **113** |
| **Total** | **84** | **75** | **84** | **243** |

**Expected Counts**

| | Type of Music | | | |
|---|---|---|---|---|
| Entree ordered | None | French | Italian | Total |
| French | 34.22 | 30.56 | 34.22 | **99** |
| Italian | 10.72 | 9.57 | 10.72 | **31** |
| Other | 39.06 | 34.88 | 39.06 | **113** |
| **Total** | **84** | **75** | **84** | **243** |

SOLUTION: For French entrees with no music, the observed count is 30 orders and the expected count is 34.22. The contribution to the $\chi^2$ statistic for this cell is

$$\frac{(\text{Observed} - \text{Expected})^2}{\text{Expected}} = \frac{(30 - 34.22)^2}{34.22} = 0.52$$

The $\chi^2$ statistic is the sum of nine such terms:

$$\chi^2 = \sum \frac{(\text{Observed} - \text{Expected})^2}{\text{Expected}} = \frac{(30 - 34.22)^2}{34.22} + \frac{(39 - 30.56)^2}{30.56} + \cdots + \frac{(35 - 39.06)^2}{39.06}$$

$$= 0.52 + 2.33 + \cdots + 0.42 = 18.28$$

**AP® EXAM TIP** In the "Do" step, you aren't required to show every term in the chi-square statistic. Writing the first few terms of the sum followed by "..." is considered as "showing work." We suggest that you do this and then let your calculator tackle the computations.

**For Practice** *Try Exercise* **29**

As in the test for goodness of fit, you should think of the chi-square statistic $\chi^2$ as a measure of how much the observed counts deviate from the expected counts. Once again, large values of $\chi^2$ are evidence against $H_0$ and in favor of $H_a$. The *P*-value measures the strength of this evidence. When conditions are met, *P*-values for a chi-square test for homogeneity come from a chi-square distribution with df = (number of rows − 1) × (number of columns − 1).

## EXAMPLE Does Background Music Influence What Customers Buy?

### *P-value and conclusion*

Earlier, we started a significance test of

$H_0$: There is no difference in the true distributions of entrees ordered at this restaurant when no music, French accordion music, or Italian string music is played.

---

**ALTERNATE EXAMPLE**

### *Birthdays*

PROBLEM: The tables below show the observed and expected counts for the birthday study. Calculate the chi-square statistic. Show your work.

| | Before 1980 | After 1993 |
|---|---|---|
| Sunday | 12 | 9 |
| Monday | 12 | 11 |
| Tuesday | 14 | 11 |
| Wednesday | 12 | 10 |
| Thursday | 7 | 17 |
| Friday | 9 | 9 |
| Saturday | 11 | 6 |

| | Before 1980 | After 1993 |
|---|---|---|
| Sunday | 10.78 | 10.22 |
| Monday | 11.78 | 11.17 |
| Tuesday | 12.86 | 12.19 |
| Wednesday | 11.32 | 10.73 |
| Thursday | 12.32 | 11.68 |
| Friday | 9.24 | 8.76 |
| Saturday | 8.70 | 8.25 |

SOLUTION:

$$\chi^2 = \frac{(12 - 10.78)^2}{10.78} + \frac{(9 - 10.22)^2}{10.22} + \ldots$$
$$= 6.55$$

**TECHNOLOGY TIP**

To calculate the chi-square statistic using your calculator, you can follow the instructions on page 706 or you can use lists. Simply enter the observed counts in `L1`, the expected counts in `L2`, and the following formula in `L3: (L1 - L2)^2/L2`. Then find the sum of the values in `L3` using `1-Var Stats L3`.

**ALTERNATE ACTIVITY**

### *Does background music influence what customers buy?*

After calculating $\chi^2 = 18.28$, ask your students how they can tell if this is a big or a small value. Since they do not know what degrees of freedom to use, you might suggest simulating the randomization distribution of $\chi^2$ in this context, assuming that the music does not influence what customers buy. Using 243 index cards, label 99 "French," 31 "Italian," and 113 "Other" to represent the actual purchases. Then shuffle the cards and divide them at random into three piles: one pile of 84 to represent the 84 purchases made with no music playing, 75 to represent the purchases when French music was playing, and 84 to represent the purchases when Italian music was playing. Then calculate the observed number of entrees of each type in each treatment group and calculate the $\chi^2$-value for the simulated experiment. Repeat this process over and over, graphing each $\chi^2$-value on a dotplot. Students should be able to see that $\chi^2 = 18.28$ would be very unlikely to occur simply due to the chance variation in random assignment.

### Teaching Tip

Here is one way to explain why the degrees of freedom for this test are df = (number of rows − 1) × (number of columns − 1). In any two-way table, the values in the last row and last column can be *determined* by the remaining values in the table and the totals. In other words, these values are *not free*. The number of degrees of freedom is equal to the number of cells that are *not* in the last row or last column of the data table. For example, in the following two-way table, once we know the totals and the four values highlighted in yellow, the remaining values are determined.

| | No music | French music | Italian music | Total |
|---|---|---|---|---|
| French entree | 30 | 39 | 30 | **99** |
| Italian entree | 11 | 1 | 19 | **31** |
| Other entree | 43 | 35 | 35 | **113** |
| **Total** | **84** | **75** | **84** | **243** |

$H_a$: There is a difference in the true distributions of entrees ordered at this restaurant when no music, French accordion music, or Italian string music is played.

| Observed Counts | | | | |
|---|---|---|---|---|
| | Type of Music | | | |
| Entree ordered | None | French | Italian | Total |
| French | 30 | 39 | 30 | **99** |
| Italian | 11 | 1 | 19 | **31** |
| Other | 43 | 35 | 35 | **113** |
| **Total** | **84** | **75** | **84** | **243** |

We already checked that the conditions are met. Our calculated test statistic is $\chi^2 = 18.28$.

**PROBLEM:**

(a) Use Table C to find the *P*-value. Then use your calculator's $\chi^2$`cdf` command.

(b) Interpret the *P*-value from the calculator in context.

(c) What conclusion would you draw? Justify your answer.

**SOLUTION:**

(a) Because the two-way table that summarizes the data from the study has three rows and three columns, we use a chi-square distribution with df = (3 − 1)(3 − 1) = 4 to find the *P*-value.

- *Table:* Look at the df = 4 row in Table C. The calculated value $\chi^2 = 18.28$ lies between the critical values 16.42 and 18.47. The corresponding *P*-value is between 0.001 and 0.0025.

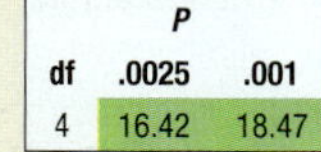

| | *P* | |
|---|---|---|
| df | .0025 | .001 |
| 4 | 16.42 | 18.47 |

- *Calculator:* The command $\chi^2$`cdf(18.28,10000,4)` gives 0.0011.

(b) Assuming that there is no difference in the true distributions of entrees ordered in this restaurant when no music, French accordion music, or Italian string music is played, there is a 0.0011 probability of observing a difference in the distributions of entrees ordered among the three treatment groups as large as or larger than the one in this study.

(c) Because the *P*-value, 0.0011, is less than our default $\alpha = 0.05$ significance level, we reject $H_0$. We have convincing evidence of a difference in the true distributions of entrees ordered at this restaurant when no music, French accordion music, or Italian string music is played. Furthermore, the random assignment allows us to say that the difference is caused by the music that's played.

**For Practice** *Try Exercise*  31

## CHECK YOUR UNDERSTANDING

In the previous Check Your Understanding (page 699), we presented data on the use of Facebook by two randomly selected groups of Penn State students. Here are the data once again.

| Use Facebook | Main campus | Commonwealth |
|---|---|---|
| Several times a month or less | 55 | 76 |
| At least once a week | 215 | 157 |
| At least once a day | 640 | 394 |
| **Total Facebook users** | **910** | **627** |

### ALTERNATE EXAMPLE

#### *Birthdays*

Earlier we started a significance test of $H_0$: There is no difference in the distributions of birthdays for people born before 1980 and people born after 1993. $H_a$: There is a difference in the distributions of birthdays for people born before 1980 and people born after 1993. The value of $\chi^2$ was 6.55.

**PROBLEM:**

**(a)** Verify that the conditions for this test are satisfied.
**(b)** Use Table C to find the *P*-value. Then use your calculator.
**(c)** Interpret the *P*-value from the calculator in context.
**(d)** What conclusion would you draw? Justify your answer.

**SOLUTION:**

**(a)** *Random*: The data came from two independent random samples. *10%*: 77 is less than 10% of all people born before 1980 and 73 is less than 10% of all people born after 1993. *Large Counts*: The expected counts are all at least 5 (the smallest is 8.25).
**(b)** Using df = (7 − 1)(2 − 1) = 6, *P*-value > 0.25. *Using technology*: $\chi^2$ `cdf(lower:6.55, upper:1000,df:6)` = 0.3645.
**(c)** Assuming that the distribution of birthdays is the same, there is a 0.3645 probability of observing a difference in the distributions of birthdays among the two age groups as large or larger than the one in this study.
**(d)** Because the *P*-value of 0.3645 is greater than $\alpha = 0.05$, we fail to reject the null hypothesis. We do not have convincing evidence that there is a difference in the distributions of birthdays for people born before 1980 and people born after 1993.

## ✓ ANSWERS TO CYU

**1.** $H_0$: There is no difference in the distributions of Facebook use among students at the main campus and students at the commonwealth campuses vs. $H_a$: There is a difference. . . .

**2.** $\frac{(131)(910)}{1537} = 77.56$, $\frac{(131)(627)}{1537} = 53.44$, $\frac{(372)(910)}{1537} = 220.25$, $\frac{(372)(627)}{1537} = 151.75$, $\frac{(1034)(910)}{1537} = 612.19$, $\frac{(1034)(627)}{1537} = 421.81$

**3.** $\chi^2 = \frac{(55 - 77.56)^2}{77.56} + \cdots + \frac{(394 - 421.81)^2}{421.81} = 19.49$

**4.** With df = 2, the $P$-value $< 0.0005$ (0.000059).

**5.** Assuming that there is no difference in the distributions of Facebook use between students on Penn State's main campus and students at Penn State's commonwealth campuses, there is a 0.000059 probability of observing samples that show a difference in the distributions of Facebook use among students at the main campus and the commonwealth campuses as large or larger than the one found in this study.

**6.** Because the $P$-value of 0.000059 $< \alpha = 0.05$, we reject $H_0$. There is convincing evidence that the distribution of Facebook use is different among students at Penn State's main campus and students at Penn State's commonwealth campuses.

Do these data provide convincing evidence of a difference in the distributions of Facebook use among students in the two campus settings?

1. State appropriate null and alternative hypotheses for a significance test to help answer this question.
2. Calculate the expected counts. Show your work.
3. Calculate the chi-square statistic. Show your work.
4. Use Table C to find the $P$-value. Then use your calculator's $\chi^2$cdf command.
5. Interpret the $P$-value from the calculator in context.
6. What conclusion would you draw? Justify your answer.

Calculating the expected counts and then the chi-square statistic by hand is a bit time-consuming. As usual, technology saves time and gets the arithmetic right.

## 27. TECHNOLOGY CORNER — CHI-SQUARE TESTS FOR TWO-WAY TABLES ON THE CALCULATOR

TI-Nspire instructions in Appendix B; HP Prime instructions on the book's Web site.

You can use the TI-83/84 or TI-89 to perform calculations for a chi-square test for homogeneity. We'll use the data from the restaurant study to illustrate the process.

1. Enter the observed counts in matrix [A].

**TI-83/84**

- Press [2nd] [x⁻¹] (MATRIX), arrow to EDIT, and choose A.
- Enter the dimensions of the matrix: 3 × 3.

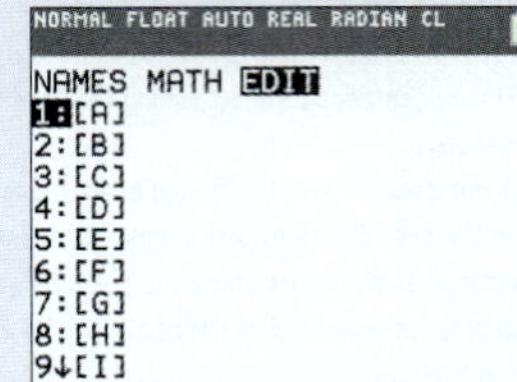


**TI-89**

- Press [APPS], select `Data/Matrix Editor` and then `New. . .`
- Adjust your settings to match those shown.

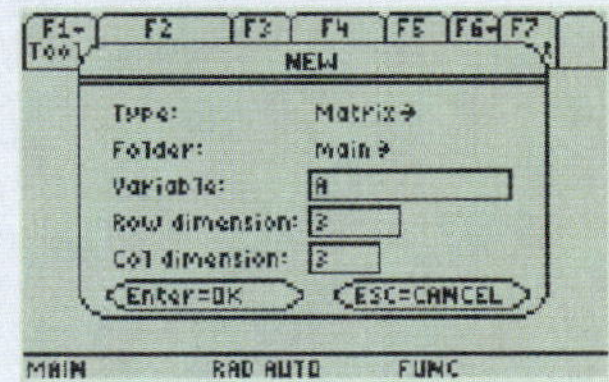


- Enter the observed counts from the two-way table in the same locations in the matrix.

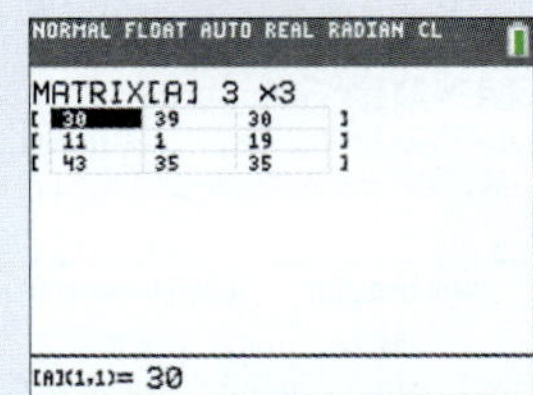


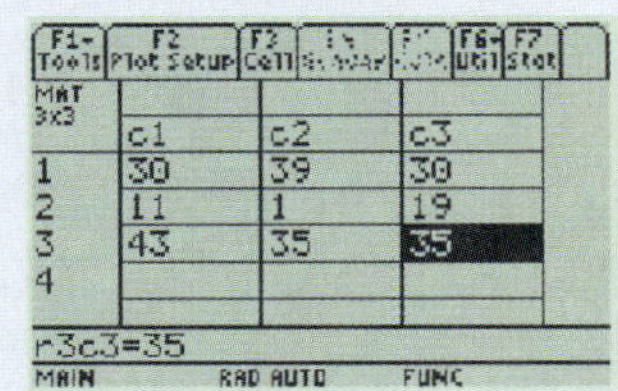


2. Specify the chi-square test, the matrix where the observed counts are found, and the matrix where the expected counts will be stored.

### TECHNOLOGY TIP

Remind students that they can watch a video presentation of this technology corner (for the TI-83/84, TI-89, or TI-Nspire) at the book's companion Web site www.whfreeman.com/tps5e.

### TECHNOLOGY TIP

Using this $\chi^2$ test will not work for a chi-square test for goodness of fit. For this test to run properly, there must be at least two rows and two columns of data. In a test for goodness of fit, there is only one row (or one column) of data.

- Press STAT, arrow to TESTS, and choose $\chi^2$-Test.
- Adjust your settings as shown.

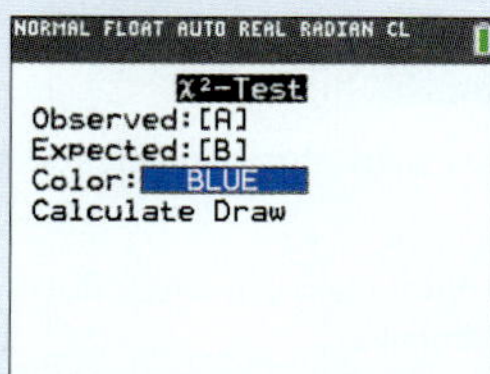


- In the Statistics/List Editor, press 2nd F1 ([F6]), and choose Chi2 2-way....
- Adjust your settings as shown.

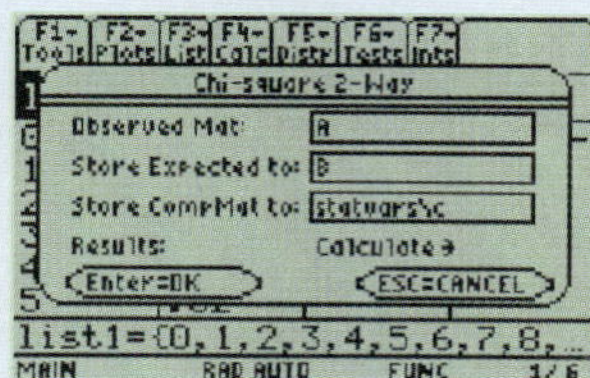


3. Choose "Calculate" or "Draw" to carry out the test. If you choose "Calculate," you should get the test statistic, *P*-value, and df shown below. If you specify "Draw," the chi-square distribution with 4 degrees of freedom will be drawn, the area in the tail will be shaded, and the *P*-value will be displayed.

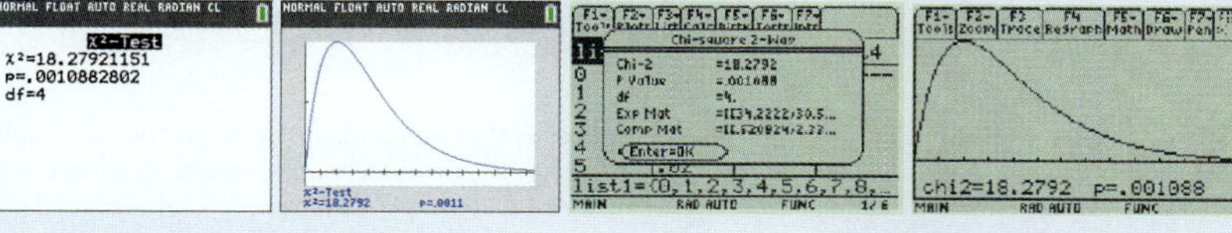


4. To see the expected counts, go to the home screen and ask for a display of the matrix [B].

- Press 2nd x⁻¹ (MATRIX), arrow to EDIT, and choose [B].

- Press 2nd − (Var-LINK) and choose B.

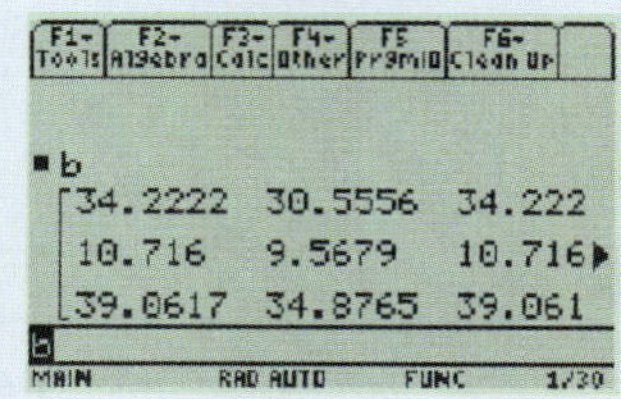


**AP® EXAM TIP** You can use your calculator to carry out the mechanics of a significance test on the AP® exam. But there's a risk involved. If you just give the calculator answer with no work, and one or more of your values is incorrect, you will probably get no credit for the "Do" step. We recommend writing out the first few terms of the chi-square calculation followed by "...". This approach might help you earn partial credit if you enter a number incorrectly. Be sure to name the procedure ($\chi^2$-Test for homogeneity) and to report the test statistic ($\chi^2 = 18.279$), degrees of freedom (df = 4), and *P*-value (0.0011).

## The Chi-Square Test for Homogeneity

In Section 11.1, we used a chi-square test for goodness of fit to test a hypothesized model for the distribution of a categorical variable. Our *P*-values came from a chi-square distribution with df = the number of categories −1. When the Random, 10%, and Large Counts conditions are met, the $\chi^2$ statistic calculated from

### AP® Exam Common Error

When students use their calculator to calculate the chi-square statistic, they often forget to write down the values of the expected counts. This will cause them to lose credit for the required check of conditions.

### Teaching Tip

Many students have trouble deciding which chi-square test to use in a particular context. However, whenever they are analyzing the results of a comparative experiment with a chi-square test, they should always use a test for homogeneity.

a two-way table can be used to perform a test of $H_0$: There is no difference in the distribution of a categorical variable for several populations or treatments. This new procedure is known as a chi-square test for homogeneity.

This test is also known as a chi-square test for homogeneity of proportions. We prefer the simpler name.

#### CHI-SQUARE TEST FOR HOMOGENEITY

Suppose the conditions are met. You can use the **chi-square test for homogeneity** to test

$H_0$: There is no difference in the distribution of a categorical variable for several populations or treatments.

$H_a$: There is a difference in the distribution of a categorical variable for several populations or treatments.

Start by finding the expected counts. Then calculate the chi-square statistic

$$\chi^2 = \sum \frac{(\text{Observed} - \text{Expected})^2}{\text{Expected}}$$

where the sum is over all cells (not including totals) in the two-way table. If $H_0$ is true, the $\chi^2$ statistic has approximately a chi-square distribution with degrees of freedom = (number of rows − 1)(number of columns − 1). The *P*-value is the area to the right of $\chi^2$ under the corresponding chi-square density curve.

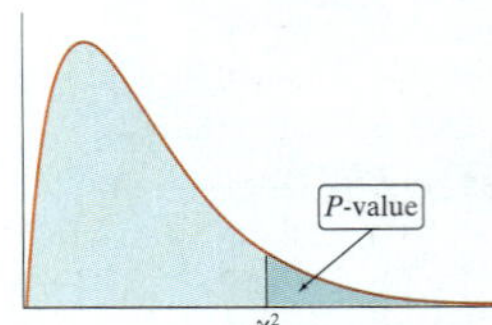


Let's look at an example of a chi-square test for homogeneity from start to finish. As usual, we follow the four-step process when performing a significance test.

EXAMPLE

## Are Cell-Only Telephone Users Different?

*The chi-square test for homogeneity*

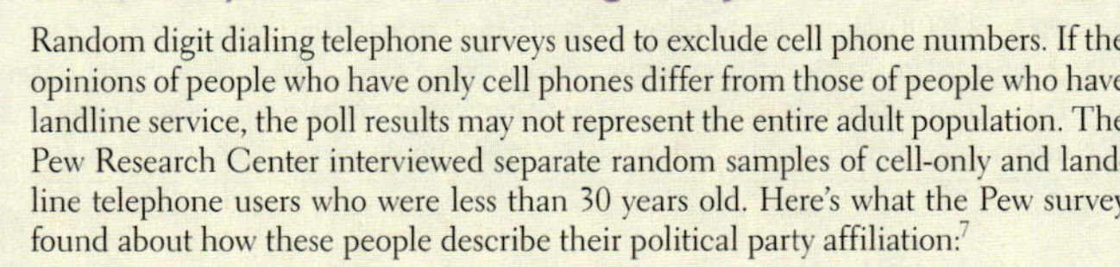

Random digit dialing telephone surveys used to exclude cell phone numbers. If the opinions of people who have only cell phones differ from those of people who have landline service, the poll results may not represent the entire adult population. The Pew Research Center interviewed separate random samples of cell-only and landline telephone users who were less than 30 years old. Here's what the Pew survey found about how these people describe their political party affiliation:[7]

| | Cell-only sample | Landline sample |
|---|---|---|
| Democrat or lean Democratic | 49 | 47 |
| Refuse to lean either way | 15 | 27 |
| Republican or lean Republican | 32 | 30 |
| **Total** | **96** | **104** |

PROBLEM:

(a) Compare the distributions of political party affiliation for cell-only and landline phone users.

(b) Do these data provide convincing evidence at the $\alpha = 0.05$ level that the distribution of party affiliation differs in the under-30 cell-only and landline user populations?

SOLUTION:

(a) Because the sample sizes are different, we should compare the proportions of individuals in each political affiliation category in the two samples. The table below shows the conditional distributions of political party affiliation for cell-only and landline phone users. We made a segmented bar graph to compare these two distributions. Cell-only users appear slightly more likely to declare themselves as Democrats or Republicans than people who have landlines. People with landlines seem much more likely to say they don't lean Democratic or Republican than those who use only cell phones.

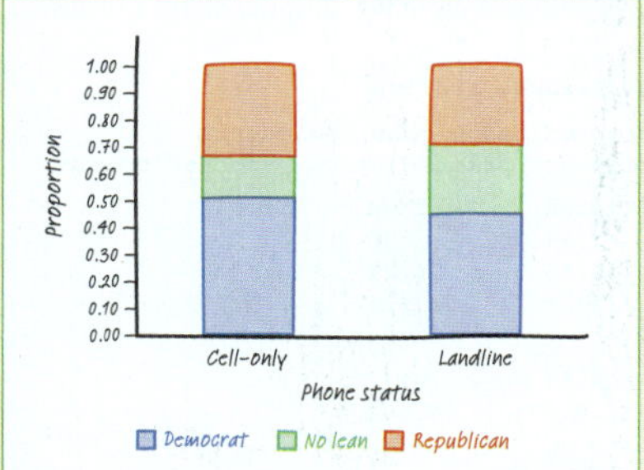


| | Phone Status | |
|---|---|---|
| **Political affiliation** | **Cell only** | **Landline** |
| Democrat | 0.51 | 0.45 |
| No lean | 0.16 | 0.26 |
| Republican | 0.33 | 0.29 |

(b) STATE: We want to perform a test of

$H_0$: There is no difference in the distribution of party affiliation in the under-30 cell-only and landline populations.

$H_a$: There is a difference in the distribution of party affiliation in the under-30 cell-only and landline populations.

at the $\alpha = 0.05$ level.

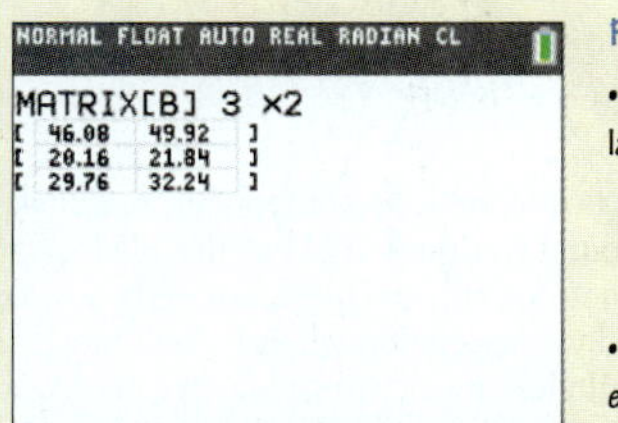


PLAN: If conditions are met, we should use a chi-square test for homogeneity.

- *Random:* The data came from independent random samples of 96 cell-only and 104 landline users.
  - 10%: Sampling without replacement was used, so there need to be at least 10(96) = 960 cell-only users under age 30 and at least 10(104) = 1040 landline users under age 30. This is safe to assume.
- *Large Counts:* We followed the steps in the Technology Corner on page 706 to get the expected counts. The calculator screen shot confirms that all expected counts are at least 5.

DO: A chi-square test on the calculator gave

NORMAL FLOAT AUTO REAL RADIAN CL
χ²-Test χ²=3.2199 p=.1999

- *Test statistic:*

$$\chi^2 = \sum \frac{(\text{Observed} - \text{Expected})^2}{\text{Expected}}$$

$$= \frac{(49 - 46.08)^2}{46.08} + \frac{(47 - 49.92)^2}{49.92} + \cdots = 3.22$$

- *P-value:* Using df = (number of rows − 1)(number of columns − 1) = (3 − 1)(2 − 1) = 2, the *P*-value is 0.1999.

CONCLUDE: Because our *P*-value, 0.1999, is greater than $\alpha = 0.05$, we fail to reject $H_0$. There is not convincing evidence that the distribution of party affiliation differs in the under-30 cell-only and landline user populations.

**For Practice** *Try Exercise* **33**

## ALTERNATE EXAMPLE

### *The color of candy*

Inspired by the Does Background Music Influence What Customers Buy? example, a statistics student decided to investigate other ways to influence a person's behavior. Using 60 volunteers, she randomly assigned 20 volunteers to get the "red" survey, 20 volunteers to get the "blue" survey, and 20 volunteers to get a control survey. The first three questions on each survey were the same, but the fourth and fifth questions were different. For example, the fourth question on the "red" survey was "When you think of the color red, what do you think about?" On the blue survey, the question replaced red with blue. On the control survey, the questions were not about color. As a reward, the student let each volunteer choose a chocolate candy in a red wrapper or a chocolate candy in a blue wrapper. Here are the results:

| | Red survey | Blue survey | Control survey | Total |
|---|---|---|---|---|
| Red candy | 13 | 5 | 8 | **26** |
| Blue candy | 7 | 15 | 12 | **34** |
| **Total** | **20** | **20** | **20** | **60** |

PROBLEM:

**(a)** Compare the distributions of color choice for each of the three treatments.

**(b)** Do these data provide convincing evidence at the $\alpha = 0.05$ level that the true distributions of color choice are different for the three types of surveys?

SOLUTION:

**(a)** As shown in the bar graph below, people with the red survey chose red candies more often and people with the blue surveys chose blue candies more often. People with the control survey chose blue candies slightly more often.

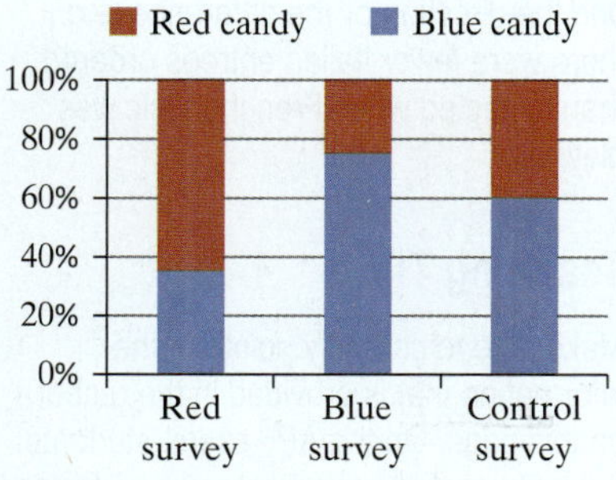


**(b)** STATE: We want to perform a test of the following hypotheses using $\alpha = 0.05$: $H_0$: There is no difference in the true distributions of color choice for people who get the red survey, the blue survey, or the control survey. $H_a$: There is a difference in the true distributions of color choice for people who get the red survey, the blue survey, or the control survey.

PLAN: If conditions are met, we will perform a chi-square test for homogeneity.

- *Random*: The three treatments were assigned at random.
  - *10%*: Because we are not sampling without replacement, we do not need to check the 10% condition.
- *Large Counts*: The expected counts (listed below) are all at least 5.

| | Red survey | Blue survey | Control survey | Total |
|---|---|---|---|---|
| Red candy | 8.67 | 8.67 | 8.67 | **26** |
| Blue candy | 11.33 | 11.33 | 11.33 | **34** |
| **Total** | **20** | **20** | **20** | **60** |

DO:

- *Test statistic:* $\chi^2 = \frac{(13 - 8.67)^2}{8.67} + \cdots = 6.65$
- *P-value*: Using (2 − 1)(3 − 1) = 2 degrees of freedom, the *P*-value is between 0.025 and 0.05. *Using technology*: *P*-value = 0.0359.

CONCLUDE: Because the *P*-value of 0.0359 is less than $\alpha = 0.05$, we reject $H_0$. We have convincing evidence that there is a difference in the true distributions of color choice for people who get the red survey, the blue survey, or the control survey.

**Follow-up Analysis** The chi-square test for homogeneity allows us to compare the distribution of a categorical variable for any number of populations or treatments. If the test allows us to reject the null hypothesis of no difference, we may want to do a follow-up analysis that examines the differences in detail. Start by examining which cells in the two-way table show large deviations between the observed and expected counts. Then look at the individual components $\frac{(\text{Observed} - \text{Expected})^2}{\text{Expected}}$ to see which terms contribute most to the chi-square statistic.

Our earlier restaurant study found significant differences among the true distributions of entrees ordered under each of the three music conditions. We entered the two-way table for the study into Minitab software and requested a chi-square test. The output appears in Figure 11.7. Minitab repeats the two-way table of observed counts and puts the expected count for each cell below the observed count. Finally, the software prints the 9 individual components that contribute to the $\chi^2$ statistic.

```
Chi-Square Test: None, French, Italian

Expected counts are printed below observed counts
Chi-Square contributions are printed below expected counts

          None    French   Italian    Total
1           30        39        30       99
         34.22     30.56     34.22
         0.521     2.334     0.521

2           11         1        19       31
         10.72      9.57     10.72
         0.008     7.672     6.404

3           43        35        35      113
         39.06     34.88     39.06
         0.397     0.000     0.422

Total       84        75        84      243

Chi-Sq = 18.279, DF = 4, P-Value = 0.001
```

**FIGURE 11.7** Minitab output for the two-way table in the restaurant study. The output gives the observed counts, the expected counts, and the individual components of the chi-square statistic.

Looking at the output, we see that just two of the nine components that make up the chi-square statistic contribute about 14 (almost 77%) of the total $\chi^2 = 18.28$. Comparing the observed and expected counts in these two cells, we see that orders of Italian entrees are much below expectation when French music is playing and well above expectation when Italian music is playing. We are led to a specific conclusion: orders of Italian entrees are strongly affected by Italian and French music. More advanced methods provide tests and confidence intervals that make this follow-up analysis more complete.

**What if we want to compare several proportions?** Many studies involve comparing the proportion of successes for each of several populations or treatments. The two-sample $z$ test from Chapter 10 allows us to test the null hypothesis $H_0: p_1 = p_2$, where $p_1$ and $p_2$ are the true proportions of successes for the two populations or treatments. The chi-square test for homogeneity allows us to test $H_0: p_1 = p_2 = \cdots = p_k$. This null hypothesis says that there is no difference in the proportions of successes for the $k$ populations or treatments. The alternative hypothesis is $H_a$: at least two of the $p_i$'s are different. Many students *incorrectly state* $H_a$ as "all the proportions are different." Think about it this way: the opposite of "all the proportions are equal" is "some of the proportions are not equal."

## Teaching Tip

As with the chi-square test for goodness of fit, students are not expected to do a follow-up analysis on the AP® exam unless they are specifically asked to. If they are asked to do a follow-up analysis, students should address the size of the contribution *and* the direction of the difference (e.g., there were *fewer* Italian entrees ordered than expected when French music was playing).

## Teaching Tip

Make sure to carefully point out the information that is provided in the output on this page. On the AP® exam, students will be expected to understand how to read computer output. For computer analyses of two-way tables, there is usually a "key" somewhere in the output. In this case, the second and third lines describe where to find the observed and expected counts and the chi-square contributions.

## Teaching Tip

See Exercises 37 and 38 for examples of studies that compare more than two proportions.

## CHECK YOUR UNDERSTANDING

Canada has universal health care. The United States does not but often offers more elaborate treatment to patients with access. How do the two systems compare in treating heart attacks? Researchers compared random samples of U.S. and Canadian heart attack patients. One key outcome was the patients' own assessment of their quality of life relative to what it had been before the heart attack. Here are the data for the patients who survived a year:

| Quality of life | Canada | United States |
|---|---|---|
| Much better | 75 | 541 |
| Somewhat better | 71 | 498 |
| About the same | 96 | 779 |
| Somewhat worse | 50 | 282 |
| Much worse | 19 | 65 |
| **Total** | **311** | **2165** |

1. Construct an appropriate graph to compare the distributions of opinion about quality of life among heart attack patients in Canada and the United States.
2. Is there a significant difference between the two distributions of quality-of-life ratings? Carry out an appropriate test at the $\alpha = 0.01$ level.

## Relationships between Two Categorical Variables

Two-way tables can arise in several ways. The restaurant experiment compared entrees ordered for three music treatments. The phone use and political party affiliation observational study compared independent random samples from the cell-only and landline user populations. In both cases, we are comparing the distributions of a categorical variable for several populations or treatments. We use the chi-square test for homogeneity to perform inference in such settings.

Another common situation that leads to a two-way table is when a *single* random sample of individuals is chosen from a *single* population and then classified based on two categorical variables. In that case, our goal is to analyze the relationship between the variables. The next example describes a study of this type.

EXAMPLE

### Angry People and Heart Disease

*Relationships between two categorical variables*

A study followed a random sample of 8474 people with normal blood pressure for about four years.[8] All the individuals were free of heart disease at the beginning of the study. Each person took the Spielberger Trait Anger Scale test, which measures how prone a person is to sudden anger. Researchers also recorded whether each individual developed coronary heart disease (CHD). This includes people who had

### ANSWERS TO CYU

1.

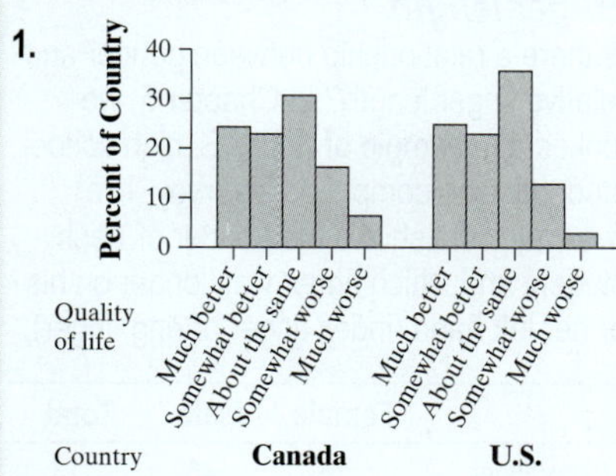


**2.** *S*: $H_0$: There is no difference in the distribution of quality of life for patients who have suffered a heart attack in Canada and the U.S. vs. $H_a$: There is a difference. . . . *P*: Chi-square test for homogeneity. *Random*: Independent random samples. *10%*: $n_1 = 311 < 10\%$ of all Canadian heart attack patients and $n_2 = 2165 < 10\%$ of all U.S. heart attack patients. *Large Counts*: 77.37, 538.63, 71.47, 497.53, 109.91, 765.09, 41.70, 290.30, 10.55, 73.45 all $\geq 5$. *D*: $\chi^2 = 11.725$. With df = 4, the *P*-value is between 0.01 and 0.02 (0.0195). *C*: Because the *P*-value of $0.0195 > \alpha = 0.01$, we fail to reject $H_0$. There is not convincing evidence that there is a difference in the distribution of quality of life for heart attack patients in Canada and the United States.

### Teaching Tip

Let your students know that we will learn how to test for a relationship between two *quantitative* variables in Chapter 12.

ALTERNATE EXAMPLE

### Finger length

Is there a relationship between gender and relative finger length? In Chapter 5, we looked at a sample of 452 U.S. high school students who completed a survey. The two-way table shows the gender of each student and which finger was longer on his or her left hand (index finger or ring finger).

| | Female | Male | Total |
|---|---|---|---|
| Index finger | 78 | 45 | **123** |
| Ring finger | 82 | 152 | **234** |
| Same length | 52 | 43 | **95** |
| **Total** | **212** | **240** | **452** |

PROBLEM:

**(a)** Is this an observational study or an experiment? Justify your answer.
**(b)** Make a well-labeled bar graph that compares the distribution of finger length for females and males. Describe what you see.

SOLUTION:

**(a)** This is an observational study. Gender was not randomly assigned to the members of the sample!
**(b)** A higher proportion of females had longer index fingers compared to males, while a higher proportion of males had longer ring fingers. A slightly higher proportion of females had index fingers and ring fingers of the same length.

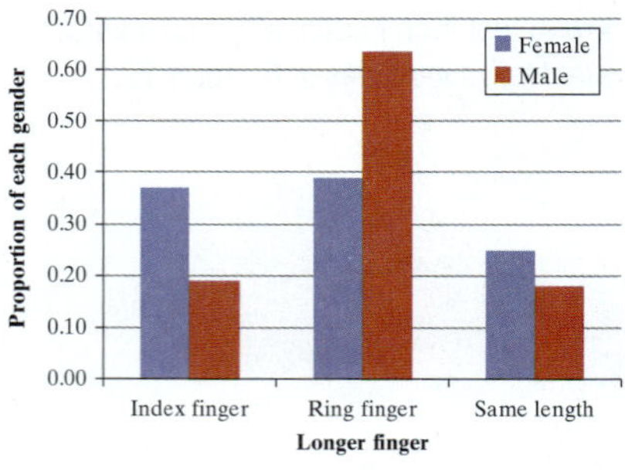


## Teaching Tip

The bars in Figure 11.8 only show the percent with CHD in each anger category. We don't need to include the bars for the percent without CHD because they would provide no additional information—we can calculate the percent without CHD by subtracting the percent with CHD from 100. Also, including the bars for the percent without CHD would make the graph much "taller," making it harder to see the difference in the percent with CHD.

heart attacks and those who needed medical treatment for heart disease. Here is a two-way table that summarizes the data:

| | Low anger | Moderate anger | High anger | Total |
|---|---|---|---|---|
| CHD | 53 | 110 | 27 | **190** |
| No CHD | 3057 | 4621 | 606 | **8284** |
| **Total** | **3110** | **4731** | **633** | **8474** |

PROBLEM:

(a) Is this an observational study or an experiment? Justify your answer.

(b) Make a well-labeled bar graph that compares CHD rates for the different anger levels. Describe what you see.

SOLUTION:

(a) This is an observational study. Researchers did not deliberately impose any treatments. They just recorded data about two variables—anger level and whether or not the person developed CHD—for each randomly chosen individual.

(b) In this setting, anger level is the explanatory variable and whether or not a person gets heart disease is the response variable. So we compare the percents of people who did and did not get heart disease in each of the three anger categories:

| | CHD | no CHD |
|---|---|---|
| Low anger: | $\frac{53}{3110} = 0.0170 = 1.70\%$ | $\frac{3057}{3110} = 0.9830 = 98.30\%$ |
| Moderate anger: | $\frac{110}{4731} = 0.0233 = 2.33\%$ | $\frac{4621}{4731} = 0.9767 = 97.67\%$ |
| High anger: | $\frac{27}{633} = 0.0427 = 4.27\%$ | $\frac{606}{633} = 0.9573 = 95.73\%$ |

The bar graph in Figure 11.8 shows the percent of people in each of the three anger categories who developed CHD. There is a clear trend: as the anger score increases, so does the percent who suffer heart disease. A much higher percent of people in the high anger category developed CHD (4.27%) than in the moderate (2.33%) and low (1.70%) anger categories.

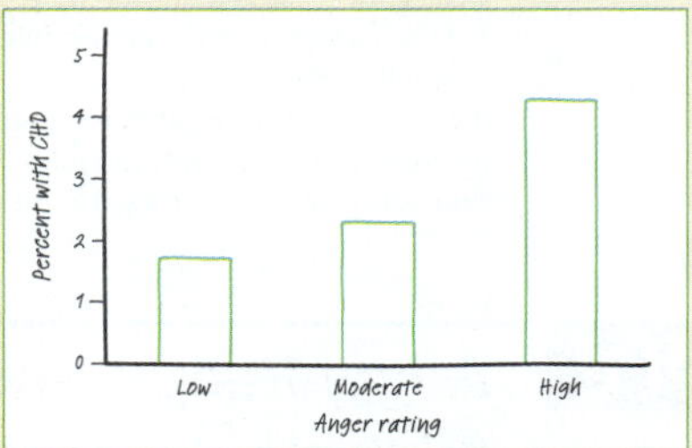


**FIGURE 11.8** Bar graph comparing the percents of people in each anger category who got coronary heart disease (CHD).

**For Practice** *Try Exercise* **41**

Anger rating on the Spielberger scale is a categorical variable that takes three possible values: low, medium, and high. Whether or not someone gets heart disease is also a categorical variable. The two-way table in the example shows the relationship between anger rating and heart disease for a random sample of

8474 people. Do these data provide convincing evidence of an *association* between the variables in the larger population? To answer that question, we work with a new significance test.

## The Chi-Square Test for Independence

We often gather data from a random sample and arrange them in a two-way table to see if two categorical variables are associated. The sample data are easy to investigate: turn them into percents and look for a relationship between the variables. Is the association in the sample evidence of an association between these variables in the entire population? Or could the sample association easily arise just from the luck of random sampling? This is a question for a significance test.

Our null hypothesis is that there is *no association* between the two categorical variables in the population of interest. The alternative hypothesis is that there *is* an association between the variables. For the observational study of anger level and coronary heart disease, we want to test the hypotheses

$H_0$: There is no association between anger level and heart-disease status in the population of people with normal blood pressure.
$H_a$: There is an association between anger level and heart-disease status in the population of people with normal blood pressure.

We could substitute the word "dependent" in place of "not independent" in the alternative hypothesis. We'll avoid this practice, however, because saying that two variables are dependent sounds too much like saying that changes in one variable *cause* changes in the other.

No association between two variables means that knowing the value of one variable does not help us predict the value of the other. That is, the variables are *independent*. An equivalent way to state the hypotheses is therefore

$H_0$: Anger and heart-disease status are independent in the population of people with normal blood pressure.
$H_a$: Anger and heart-disease status are not independent in the population of people with normal blood pressure.

As with the two previous types of chi-square tests, we begin by comparing the observed counts in a two-way table with the expected counts if $H_0$ is true.

### EXAMPLE Angry People and Heart Disease

*Finding expected counts*

The null hypothesis is that there is no association between anger level and heart-disease status in the population of interest. If we assume that $H_0$ is true, then anger level and CHD status are independent. We can find the expected cell counts in the two-way table using the definition of independent events from Chapter 5: $P(A \mid B) = P(A)$. The chance process here is randomly selecting a person and recording his or her anger level and CHD status.

| | Low anger | Moderate anger | High anger | Total |
|---|---|---|---|---|
| CHD | 53 | 110 | 27 | **190** |
| No CHD | 3057 | 4621 | 606 | **8284** |
| **Total** | **3110** | **4731** | **633** | **8474** |

### ALTERNATE EXAMPLE

***Finger length***

The null hypothesis for the previous Alternate Example is that there is no association between gender and relative finger length in the population of U.S. high school students who filled out the CensusAtSchool survey. Here is a two-way table that summarizes the data:

| | Female | Male | Total |
|---|---|---|---|
| Index finger | 78 | 45 | **123** |
| Ring finger | 82 | 152 | **234** |
| Same length | 52 | 43 | **95** |
| **Total** | **212** | **240** | **452** |

If the null hypothesis is true, then *P*(longer index finger | female) = *P*(longer index finger | male) = *P*(longer index finger). Because 123/452 = 27.2% had longer index fingers, the expected count for females is 212(0.272) = 57.7 and the expected count for males is 240(0.272) = 65.3.

Here are all the expected counts:

| | Female | Male |
|---|---|---|
| Index finger | 212(0.272) = 57.7 | 240(0.272) = 65.3 |
| Ring finger | 212(0.518) = 109.8 | 240(0.518) = 124.3 |
| Same length | 212(0.210) = 44.5 | 240(0.210) = 50.4 |

## Teaching Tip

Make sure to discuss the information in the first paragraph on this page. It is a great way to review the concept of independence from Chapter 5. You can also point out that $P(\text{CHD} \mid \text{low anger}) = P(\text{CHD} \mid \text{moderate anger}) = P(\text{CHD} \mid \text{high anger}) = P(\text{CHD})$ if there is no association between anger and heart disease.

## AP® Exam Common Error

Remind students that it is not appropriate to round the expected counts to the nearest integer.

Let's start by considering the events "CHD" and "low anger." We see from the two-way table that 190 of the 8474 people in the study had CHD. If we imagine choosing one of these people at random, $P(\text{CHD}) = 190/8474$. Because anger level and CHD status are independent, knowing that the selected individual is low anger does not change the probability that this person develops CHD. That is to say, $P(\text{CHD} \mid \text{low anger}) = P(\text{CHD}) = 190/8474 = 0.02242$.

Of the 3110 low-anger people in the study, we'd expect

$$3110 \cdot \frac{190}{8474} = 3110(0.02242) = 69.73$$

to get CHD. You can see that the general formula we developed earlier for a test for homogeneity applies in this situation also:

$$\text{expected count} = \frac{\text{row total} \cdot \text{column total}}{\text{table total}} = \frac{190 \cdot 3110}{8474} = 69.73$$

To find the expected count in the "low anger, no CHD" cell, we begin by noting that $P(\text{no CHD}) = 8284/8474 = 0.97758$ for a randomly selected person in the study. Of the 3110 low-anger people in the study, we would expect

$$3110 \cdot \frac{8284}{8474} = 3110(0.97758) = 3040.27$$

to not develop CHD.

We find the expected counts for the remaining cells in the two-way table in a similar way.

| **CHD, Low** | **CHD, Moderate** | **CHD, High** |
|---|---|---|
| 3110(0.02242) = 69.73 | 4731(0.02242) = 106.08 | 633(0.02242) = 14.19 |
| **no CHD, Low** | **no CHD, Moderate** | **no CHD, High** |
| 3110(0.97758) = 3040.27 | 4731(0.97758) = 4624.92 | 633(0.97758) = 618.81 |

The 10% and Large Counts conditions for the chi-square test for independence are the same as for the homogeneity test. There is a slight difference in the Random condition for the two tests: a test for independence uses data from one sample but a test for homogeneity uses data from two or more samples/groups.

### CONDITIONS FOR PERFORMING A CHI-SQUARE TEST FOR INDEPENDENCE

- **Random:** The data come from a well-designed random sample or randomized experiment.
  - **10%:** When sampling without replacement, check that $n \le \frac{1}{10}N$.
- **Large Counts:** All expected counts are at least 5.

If the Random, 10%, and Large Counts conditions are met, the $\chi^2$ statistic calculated from a two-way table can be used to perform a test of $H_0$: There is no association between two categorical variables in the population of interest. $P$-values for this test come from a chi-square distribution with df = (number of rows − 1) × (number of columns − 1). This new procedure is known as a chi-square test for independence.

The chi-square test for independence is also known as the chi-square test for association.

**CHI-SQUARE TEST FOR INDEPENDENCE**

Suppose the conditions are met. You can use the **chi-square test for independence** to test

$H_0$: There is no association between two categorical variables in the population of interest.

$H_a$: There is an association between two categorical variables in the population of interest.

Or, alternatively,

$H_0$: Two categorical variables are independent in the population of interest.

$H_a$: Two categorical variables are not independent in the population of interest.

Start by finding the expected counts. Then calculate the chi-square statistic

$$\chi^2 = \sum \frac{(\text{Observed} - \text{Expected})^2}{\text{Expected}}$$

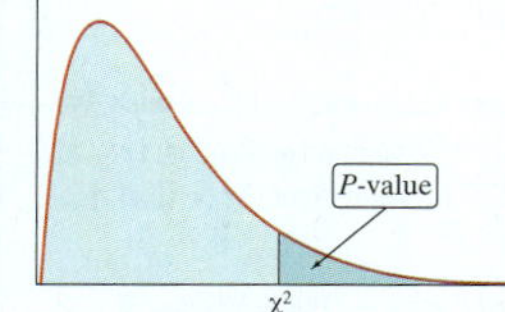


where the sum is over all cells in the two-way table. If $H_0$ is true, the $\chi^2$ statistic has approximately a chi-square distribution with degrees of freedom = (number of rows − 1)(number of columns − 1). The $P$-value is the area to the right of $\chi^2$ under the corresponding chi-square density curve.

Now we're ready to complete the significance test for the anger and heart disease study.

## EXAMPLE Angry People and Heart Disease

### *Chi-square test for independence*

Here is the complete table of observed and expected counts for the CHD and anger study side by side:

| | Observed | | | Expected | | |
|---|---|---|---|---|---|---|
| | **Low** | **Moderate** | **High** | **Low** | **Moderate** | **High** |
| CHD | 53 | 110 | 27 | 69.73 | 106.08 | 14.19 |
| No CHD | 3057 | 4621 | 606 | 3040.27 | 4624.92 | 618.81 |

## Teaching Tip

Although the Do step is the same for a chi-square test for homogeneity and a chi-square test for independence, the State, Plan, and Conclude steps are each a little different. In a test for homogeneity, we are comparing the distribution of one categorical variable in *two or more* populations or groups. However, in a test for independence, we are investigating the relationship between two categorical variables in *one* population. Make sure to highlight these differences as you introduce the test of independence.

## ALTERNATE EXAMPLE

### *Finger length*

Here is a complete table of observed counts and expected counts (in parentheses):

| | Female | Male | Total |
|---|---|---|---|
| Index finger | 78 (57.7) | 45 (65.3) | **123** |
| Ring finger | 82 (109.8) | 152 (124.3) | **234** |
| Same length | 52 (44.5) | 43 (50.4) | **95** |
| **Total** | **212** | **240** | **452** |

Do the data provide convincing evidence of an association between gender and relative finger length for U.S. high school students who filled out the CensusAtSchool survey?

STATE: We want to perform a test of the following hypotheses using $\alpha = 0.05$:

$H_0$: There is no association between gender and relative finger length in the population of U.S. high school students who filled out the CensusAtSchool survey.

$H_a$: There is an association between gender and relative finger length in the population of U.S. high school students who filled out the CensusAtSchool survey.

PLAN: If conditions are met, we will perform a chi-square test for independence.

- *Random*: The sample was randomly selected.
    - *10%:* The sample of 452 students is less than 10% of all U.S. students who filled out the survey.
- *Large Counts*: The expected counts (see table above) are all at least 5.

DO:

- *Test statistic*: $\chi^2 = \frac{(78 - 57.7)^2}{57.7} + \cdots = 29.0$
- *P-value*: Using $(3 - 1)(2 - 1) = 2$ degrees of freedom, the *P*-value is less than 0.0005. *Using technology*: *P*-value $\approx$ 0.

CONCLUDE: Because the *P*-value of approximately 0 is less than $\alpha = 0.05$, we reject $H_0$. We have convincing evidence that there is an association between gender and relative finger length in the population of U.S. high school students who filled out the CensusAtSchool survey.

Do the data provide convincing evidence of an association between anger level and heart disease in the population of interest?

STATE: We want to perform a test of

$H_0$: There is no association between anger level and heart-disease status in the population of people with normal blood pressure.

$H_a$: There is an association between anger level and heart-disease status in the population of people with normal blood pressure.

Because no significance level was stated, we'll use $\alpha = 0.05$.

PLAN: If conditions are met, we should carry out a chi-square test for independence.

- *Random:* The data came from a random sample of 8474 people with normal blood pressure.
    - 10%: Because the researchers sampled without replacement, we need to check that the total number of people in the population with normal blood pressure is at least 10(8474) = 84,740. This seems reasonable to assume.
- *Large Counts:* All the expected counts are at least 5 (the smallest is 14.19), so this condition is met.

DO: We perform calculations assuming $H_0$ is true.

- *Test statistic:*

$$\chi^2 = \sum \frac{(\text{Observed} - \text{Expected})^2}{\text{Expected}}$$

$$= \frac{(53 - 69.73)^2}{69.73} + \frac{(110 - 106.08)^2}{106.08} + \cdots + \frac{(606 - 618.81)^2}{618.81}$$

$$= 4.014 + 0.145 + \cdots + 0.265 = 16.077$$

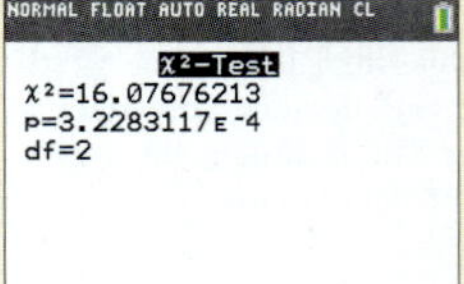


- *P-value:* The two-way table of anger level versus heart disease has 2 rows and 3 columns. We will use the chi-square distribution with df = $(2 - 1)(3 - 1) = 2$ to find the *P*-value. Look at the df = 2 line in Table C. The observed statistic $\chi^2 = 16.077$ is larger than the critical value 15.20 for $\alpha = 0.0005$. So the *P*-value is less than 0.0005.

*Using Technology:* The calculator's $\chi^2$-Test gives $\chi^2 = 16.077$ and *P*-value = 0.00032 using df = 2.

CONCLUDE: Because the *P*-value of 0.00032 is less than $\alpha = 0.05$, we reject $H_0$. We have convincing evidence of an association between anger level and heart-disease status in the population of people with normal blood pressure.

**For Practice** *Try Exercise* **45**

A follow-up analysis reveals that two cells contribute most of the chi-square statistic: Low anger, CHD (4.014) and High anger, CHD (11.564). A much smaller number of low-anger people developed CHD than expected. And a much larger number of high-anger people got CHD than expected.

Caution

Can we conclude that proneness to anger *causes* heart disease? No. The anger and heart-disease study is an observational study, not an experiment. It isn't surprising that some other variables are confounded with anger level. For example, people prone to anger are more likely than others to be men who drink and smoke. We don't know whether the increased rate of heart disease among those with higher anger levels in the study is due to their anger or perhaps to their drinking and smoking or maybe even to gender.

## Teaching Tip

At the end of each significance test, make sure your students think about the scope of inference. Even if the results are statistically significant, we can't necessarily make an inference about a larger population or an inference about cause and effect. The caution on this page illustrates this nicely.

## CHECK YOUR UNDERSTANDING

Many popular businesses are franchises—think of McDonald's. The owner of a local franchise benefits from the brand recognition, national advertising, and detailed guidelines provided by the franchise chain. In return, he or she pays fees to the franchise firm and agrees to follow its policies. The relationship between the local owner and the franchise firm is spelled out in a detailed contract.

One clause that the contract may or may not contain is the entrepreneur's right to an exclusive territory. This means that the new outlet will be the only representative of the franchise in a specified territory and will not have to compete with other outlets of the same chain. How does the presence of an exclusive-territory clause in the contract relate to the survival of the business?

A study designed to address this question collected data from a random sample of 170 new franchise firms. Two categorical variables were measured for each franchisor. First, the franchisor was classified as successful or not based on whether or not it was still offering franchises as of a certain date. Second, the contract each franchisor offered to franchisees was classified according to whether or not there was an exclusive-territory clause. Here are the count data, arranged in a two-way table:[9]

| | Exclusive Territory | | |
|---|---|---|---|
| Success | Yes | No | Total |
| Yes | 108 | 15 | **123** |
| No | 34 | 13 | **47** |
| **Total** | **142** | **28** | **170** |

Do these data provide convincing evidence at the $\alpha = 0.01$ level of an association between an exclusive-territory clause and business survival for new franchise firms?

## ANSWERS TO CYU

*S*: $H_0$: There is no association between an exclusive territory clause and business survival vs. $H_a$: There is an association. . . . *P*: Chi-square test for independence. *Random*: Random sample. *10%*: We assume that $n = 170 < 10\%$ of all new franchise firms. *Large Counts*: 102.74, 20.26, 39.26, 7.74 all $\geq 5$. *D*: $\chi^2 = 5.911$. Using df = 1, the *P*-value is between 0.01 and 0.02 (0.0150). *C*: Because the *P*-value of $0.0150 > \alpha = 0.01$, we fail to reject $H_0$. There is not convincing evidence of an association between exclusive territory clause and business survival.

## Using Chi-Square Tests Wisely

**AP® EXAM TIP** If you have trouble distinguishing the two types of chi-square tests for two-way tables, you're better off just saying "chi-square test" than choosing the wrong type. Better yet, learn to tell the difference!

caution !

Both the chi-square test for homogeneity and the chi-square test for independence start with a two-way table of observed counts. They even calculate the test statistic, degrees of freedom, and *P*-value in the same way. *The questions that these two tests answer are different, however.* A chi-square test for homogeneity tests whether the distribution of a categorical variable is the same for each of several populations or treatments. The chi-square test for independence tests whether two categorical variables are associated in some population of interest.

Unfortunately, it is quite common to see questions asking about association when a test for homogeneity applies and questions asking about differences between proportions or the distribution of a variable when a test of independence applies. Sometimes, people avoid the distinction altogether and pose questions about the "relationship" between two variables.

Instead of focusing on the question asked, it's much easier to look at how the data were produced. If the data come from two or more independent random samples or treatment groups in a randomized experiment, then do a chi-square test for homogeneity. If the data come from a single random sample, with the individuals classified according to two categorical variables, use a chi-square test for independence.

### Teaching Tip

To help your students recognize the difference between the three different chi-square tests, tell them to think about how the data were collected, noting the number of populations (groups) and the number of variables. Here is a summary:

- Goodness of fit: one variable in one population
- Homogeneity: one variable in two or more populations (groups)
- Independence: two variables in one population

You might also list one or two memorable examples of each test along with these descriptions (e.g., M&M'S® for goodness of fit, music and entree choice for homogeneity, and anger and heart disease for independence).

### Teaching Tip

Here is another way to think about the difference between a test for homogeneity and a test for independence. In a test for homogeneity, either the row totals or the column totals are determined by the researcher collecting the data. This allows the researcher to maximize the power of the test by choosing appropriate sample sizes. In a test for independence, both the row totals and the column totals are random.

## EXAMPLE Scary Movies and Fear

### *Choosing the right type of chi-square test*

Are men and women equally likely to suffer lingering fear from watching scary movies as children? Researchers asked a random sample of 117 college students to write narrative accounts of their exposure to scary movies before the age of 13. More than one-fourth of the students said that some of the fright symptoms are still present when they are awake.[10] The following table breaks down these results by gender.

| | Gender | | |
|---|---|---|---|
| Fright symptoms? | Male | Female | Total |
| Yes | 7 | 29 | **36** |
| No | 31 | 50 | **81** |
| **Total** | **38** | **79** | **117** |

Minitab output for a chi-square test using these data is shown below.

```
Chi-Square Test: Male, Female
Expected counts are printed below observed counts
Chi-Square contributions are printed below expected counts
          Male      Female     Total
1            7          29        36
         11.69       24.31
         1.883       0.906

2           31          50        81
         26.31       54.69
         0.837       0.403
Total       38          79       117
Chi-Sq = 4.028, DF = 1, P-Value = 0.045
```

PROBLEM: Assume that the conditions for performing inference are met.

(a) Explain why a chi-square test for independence and not a chi-square test for homogeneity should be used in this setting.

(b) State an appropriate pair of hypotheses for researchers to test in this setting.

(c) Which cell contributes most to the chi-square statistic? In what way does this cell differ from what the null hypothesis suggests?

(d) Interpret the *P*-value in context. What conclusion would you draw at $\alpha = 0.01$?

SOLUTION:

(a) The data were produced using a single random sample of college students, who were then classified by gender and whether or not they had lingering fright symptoms. The chi-square test for homogeneity requires independent random samples from each population.

(b) The null hypothesis is $H_0$: There is no association between gender and ongoing fright symptoms in the population of college students. The alternative hypothesis is $H_a$: There is an association between gender and ongoing fright symptoms in the population of college students.

## ALTERNATE EXAMPLE

### *Online social networking*

An article in the *Arizona Daily Star* (April 9, 2009) included the following table:

| Age (years): | 18–24 | 25–34 | 35–44 | 45–54 | 55–64 | 65+ | Total |
|---|---|---|---|---|---|---|---|
| Use online social networks | 137 | 126 | 61 | 38 | 15 | 9 | **386** |
| Do not use online social networks | 46 | 95 | 143 | 160 | 130 | 124 | **698** |
| **Total** | **183** | **221** | **204** | **198** | **145** | **133** | **1084** |

Suppose that you decide to analyze these data using a chi-square test. However, without any additional information about how the data were collected, it isn't possible to know which chi-square test is appropriate.

PROBLEM:

**(a)** Explain how you know that a test for goodness of fit is *not* appropriate for analyzing these data.

**(b)** Describe how these data could have been collected so that a test for homogeneity is appropriate.

**(c)** Describe how these data could have been collected so that a test for independence is appropriate.

SOLUTION:

**(a)** Because there are either two variables or two or more populations, a test for goodness of fit is not appropriate. Tests for goodness of fit are appropriate only when analyzing the distribution of one variable in one population.

**(b)** To make a test for homogeneity appropriate, we would need to take six independent random samples, one from each age category, and then ask every person whether or not he or she uses online social networks. Or we could take two independent random samples, one of online social network users and one of people who do not use online social networks, and ask every member of each sample how old he or she is.

**(c)** To make a test for independence appropriate, we would take one random sample from the population and ask all members about their

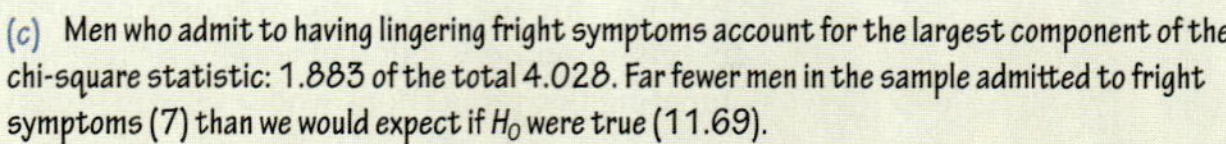

(c) Men who admit to having lingering fright symptoms account for the largest component of the chi-square statistic: 1.883 of the total 4.028. Far fewer men in the sample admitted to fright symptoms (7) than we would expect if $H_0$ were true (11.69).

(d) If gender and ongoing fright symptoms really are independent in the population of interest, there is a 0.045 chance of obtaining a random sample of 117 students that gives a chi-square statistic of 4.028 or higher. Because the *P*-value, 0.045, is greater than 0.01, we would fail to reject $H_0$. We do not have convincing evidence that there is an association between gender and fright symptoms in the population of college students.

**For Practice** *Try Exercise* 47

**What if we want to compare two proportions?** Shopping at secondhand stores is becoming more popular and has even attracted the attention of business schools. A study of customers' attitudes toward secondhand stores interviewed separate random samples of shoppers at two secondhand stores of the same chain in two cities. The two-way table shows the breakdown of respondents by gender.[11]

| | City 1 | City 2 |
|---|---|---|
| Men | 38 | 68 |
| Women | 203 | 150 |
| **Total** | **241** | **218** |

Do the data provide convincing evidence of a difference in the true gender distributions of shoppers at the two stores?

To answer this question, we could perform a chi-square test for homogeneity. Our hypotheses are

$H_0$: There is no difference in the true gender distributions of shoppers at the two stores.

$H_a$: There is a difference in the true gender distributions of shoppers at the two stores.

But a difference in gender distributions would mean that there is a difference in the true proportions of female shoppers at the two stores. So we could also use a two-sample *z* test from Section 10.1 to compare two proportions. The hypotheses for this test are

$$H_0: p_1 - p_2 = 0$$
$$H_a: p_1 - p_2 \neq 0$$

where $p_1$ and $p_2$ are the true proportions of women shoppers at Store 1 and Store 2, respectively.

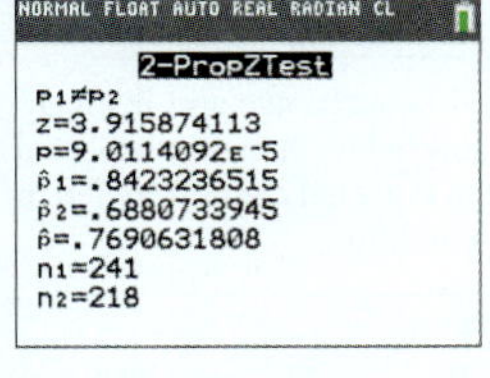


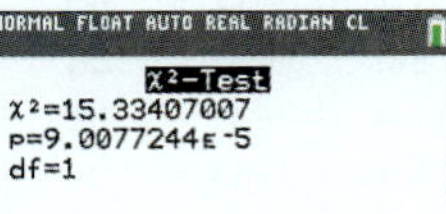


The TI-84 screen shots in the margin show the results from a two-sample *z* test for $p_1 - p_2$ and from a chi-square test for homogeneity. (We checked that the Random, 10%, and Large Counts conditions are met before carrying out the calculations.)

Note that the *P*-values from the two tests are the same except for rounding errors. You can also check that the chi-square statistic is the square of the two-sample *z* statistic: $(3.915\ldots)^2 = 15.334$.

As the previous example suggests, the chi-square test for homogeneity based on a $2 \times 2$ two-way table is equivalent to the two-sample *z* test for $p_1 - p_2$ with a two-sided alternative hypothesis. *We cannot use a chi-square test for a one-sided alternative hypothesis.* The two-sample *z* procedures allow us

age and whether or not they use online social networks. This seems like the most reasonable method for collecting the data, so a test of independence is probably the best choice. But we can't know for sure unless we know how the data were collected.

## Teaching Tip

Any time you conduct a chi-square test for a 2-by-2 table, you can also use a two-sample *z* test for a difference in proportions. Here are some things to keep in mind:

- The chi-square test is always two-sided. That is, it only tests for a difference in the two proportions. If you want to test whether one proportion is larger than the other, use the two-sample *z* test.
- If you want to estimate the difference between two proportions, use a two-sample *z* interval. There are no confidence intervals that correspond to chi-square tests.
- If you are comparing more than two treatments or the response variable has more than two categories, you must use a chi-square test.
- You can also use a chi-square test for goodness of fit in place of a one-sample *z* test for a proportion if the alternative hypothesis is two-sided. The chi-square test will use two categories (success and failure) and have df = 2 − 1 = 1. For example, in the Nonsmokers example on page 562, we found *z* = 2.45 and *P*-value = 0.0142. Using a chi-square goodness of fit test with observed counts of 90 and 60 and expected counts of 75 and 75, we get $\chi^2 = 6.00$ with *P*-value = 0.0143. Note that the *P*-values are essentially the same and $\sqrt{\chi^2} = \sqrt{6} = 2.45 = z$.

## Teaching Tip

Dan Teague has written a brief paper showing the equivalence of chi-square tests and two-sample *z* tests for a difference in proportions. Do an Internet search for Dan Teague, "Chi-Square Analyses."

## Teaching Tip

To illustrate that comparing the distributions can provide more information than comparing means, consider the following two distributions of scores on the AP® Statistics exam. Even though both schools had a mean score of 3, the distributions are very different!

| Score | School A | School B |
|---|---|---|
| 5 | 10 | 1 |
| 4 | 5 | 5 |
| 3 | 1 | 10 |
| 2 | 5 | 5 |
| 1 | 10 | 1 |

However, be careful not to use too few categories when converting a quantitative variable into a categorical variable. For example, knowing the mean score might be more informative than simply reporting the percent of students who scored 3 or higher.

## Teaching Tip

Remind students that we require the expected counts to be at least 5 so that the chi-square probability distribution is a good model for the sampling distribution of the chi-square statistic. A distribution isn't necessarily a good model for the sampling distribution of a statistic just because it has the same name!

to perform one-sided tests and to construct confidence intervals for the difference between proportions. For that reason, we recommend the Chapter 10 methods for comparing two proportions whenever you are given a choice.

**Grouping quantitative data into categories** As we mentioned in Chapter 1, it is possible to convert a quantitative variable to a categorical variable by grouping together intervals of values. Here's an example. Researchers surveyed independent random samples of shoppers at two secondhand stores of the same chain in two cities. The two-way table below summarizes data on the incomes of the shoppers in the two samples.

| Income | City 1 | City 2 |
|---|---|---|
| Under $10,000 | 70 | 62 |
| $10,000 to $19,999 | 52 | 63 |
| $20,000 to $24,999 | 69 | 50 |
| $25,000 to $34,999 | 22 | 19 |
| $35,000 or more | 28 | 24 |

Personal income is a quantitative variable. But by grouping the values of this variable, we create a categorical variable. We could use these data to carry out a chi-square test for homogeneity because the data came from independent random samples of shoppers at the two stores. Comparing the distributions of income for shoppers at the two stores would give more information than simply comparing their mean incomes.

**What can we do if the expected cell counts aren't all at least 5?** Let's look at a situation where this is the case. A sample survey asked a random sample of young adults, "Where do you live now? That is, where do you stay most often?" A two-way table of all 2984 people in the sample (both men and women) classified by their age and by where they lived is shown below.[12] Living arrangement is a categorical variable. Even though age is quantitative, the two-way table treats age as dividing the young adults into four categories. The table gives the observed counts for all 20 combinations of age and living arrangement.

| | Age (years) | | | | |
|---|---|---|---|---|---|
| Living arrangement | 19 | 20 | 21 | 22 | Total |
| Parents' home | 324 | 378 | 337 | 318 | **1357** |
| Another person's home | 37 | 47 | 40 | 38 | **162** |
| Your own place | 116 | 279 | 372 | 487 | **1254** |
| Group quarters | 58 | 60 | 49 | 25 | **192** |
| Other | 5 | 2 | 3 | 9 | **19** |
| **Total** | **540** | **766** | **801** | **877** | **2984** |

Our null hypothesis is $H_0$: There is no association between age and living arrangement in the population of young adults. The table below shows the expected counts assuming $H_0$ is true. We can see that two of the expected counts (circled in red) are less than 5. This violates the Large Counts condition.

| | Age (years) | | | | |
|---|---|---|---|---|---|
| Living arrangement | 19 | 20 | 21 | 22 | Total |
| Parents' home | 245.57 | 348.35 | 364.26 | 398.82 | **1357** |
| Another person's home | 29.32 | 41.59 | 43.49 | 47.61 | **162** |
| Your own place | 226.93 | 321.90 | 336.61 | 368.55 | **1254** |
| Group quarters | 34.75 | 49.29 | 51.54 | 56.43 | **192** |
| Other | 3.44 | 4.88 | 5.10 | 5.58 | **19** |
| **Total** | **540** | **766** | **801** | **877** | **2984** |

A clever strategy is to "collapse" the table by combining two or more rows or columns. In this case, it might make sense to combine the Group quarters and Other living arrangements. Doing so and then running a chi-square test in Minitab gives the following output. Notice that the Large Counts condition is now met.

```
Chi-Square Test: 19, 20, 21, 22

Expected counts are printed below observed counts
Chi-Square contributions are printed below expected counts
            19        20        21        22     Total
1          324       378       337       318      1357
        245.57    348.35    364.26    398.82
        25.049     2.525     2.040    16.379

2           37        47        40        38       162
         29.32     41.59     43.49     47.61
         2.014     0.705     0.279     1.940

3          116       279       372       487      1254
        226.93    321.90    336.61    368.55
        54.226     5.719     3.720    38.068

4           63        62        52        34       211
         38.18     54.16     56.64     62.01
        16.129     1.134     0.380    12.654

Total      540       766       801       877      2984

Chi-Sq = 182.961, DF = 9, P-Value = 0.000
```

case closed

## Do Dogs Resemble Their Owners?

In the chapter-opening Case Study (page 677), we described a study that investigated whether or not dogs resemble their owners. The researchers who conducted the experiment believe that resemblance between dog and owner might differ for purebred and mixed-breed dogs. Here is a two-way table summarizing the results of the experiment:

| | Breed status | |
|---|---|---|
| **Resemblance?** | **Purebred dogs** | **Mixed-breed dogs** |
| Resemble owner | 16 | 7 |
| Don't resemble | 9 | 13 |

**TECHNOLOGY TIP**

Make sure students know how to read the computer output on this page. The "key" is printed at the top of the table and the test statistic, degrees of freedom, and *P*-value are located at the bottom.

## Answers to Case Closed

**1.** The researchers used a random sample of dogs and their owners so they can generalize their results to the population of dogs and their owners.

**2.** We should use a chi-square test for independence because the researchers selected one sample and recorded two variables from each individual: breed status and resemblance. The null hypothesis is $H_0$: There is no association between breed status and whether or not a dog resembles its owner. The alternative hypothesis is $H_a$: There is an association between breed status and whether or not a dog resembles its owner.

**3. (a)** The expected count is $\frac{\text{row total} \times \text{column total}}{\text{table total}} = \frac{23 \times 25}{45} = 12.78$. **(b)** The Large Counts condition is met because all of the expected counts are at least 5.

**4.** $\chi^2 = \frac{(16 - 12.78)^2}{12.78} + \cdots = 3.74$, df $= (2 - 1)(2 - 1) = 1$, $P$-value $= 0.0531$

**5.** Because the $P$-value of 0.0531 is greater than $\alpha = 0.05$, we fail to reject $H_0$. We do not have convincing evidence of an association between breed status and whether or not a dog resembles its owner.

### TRM Section 11.2 Quizzes

There are three different quizzes available for this section in the Teacher's Resource Materials. Versions A and B are roughly parallel. Find them in the TE-book, or on the TRCD or the book's companion Web site www.whfreeman.com/tps5e (teacher log-in required).

1. Why did researchers photograph a random sample of dogs and their owners in this study?

*Do the data from this study provide convincing evidence of an association between dogs' breed status and whether or not they resemble their owners? Questions 2 through 5 address this issue.*

2. Which type of chi-square test should be used to help answer the question of interest? State an appropriate pair of hypotheses for the test you choose.
3. The table shows the expected counts for the appropriate chi-square test in Question 2.

| | Breed status | |
|---|---|---|
| **Resemblance?** | **Purebred dogs** | **Mixed-breed dogs** |
| Resemble owner | 12.78 | 10.22 |
| Don't resemble | 12.22 | 9.78 |

(a) Show how the expected count for the cell "purebred dogs, resemble owner" was computed.
(b) Explain why the Large Counts condition is met.

4. Find the test statistic and $P$-value. Be sure to state the degrees of freedom you are using.
5. What conclusion would you draw?

## Section 11.2 Summary

- We can use a two-way table to summarize data involving two categorical variables. To analyze the data, we compare the conditional distributions of one variable for each value of the other variable. Then we turn to formal inference. Two different ways of producing data for two-way tables lead to two different types of chi-square tests.
- Some studies aim to compare the distribution of a single categorical variable for each of several populations or treatments. In such cases, researchers should take independent random samples from the populations of interest or use the groups in a randomized experiment. The null hypothesis is that there is no difference in the distribution of the categorical variable for each of the populations or treatments. We use the **chi-square test for homogeneity** to test this hypothesis.

- The conditions for performing a chi-square test for homogeneity are:
  - **Random:** The data come from independent random samples or the groups in a randomized experiment.
    - **10%:** When sampling without replacement, check that the population is at least 10 times as large as the sample.
  - **Large Counts:** All expected counts must be at least 5.
- Other studies are designed to investigate the relationship between two categorical variables. In such cases, researchers take a random sample from the population of interest and classify each individual based on the two categorical variables. The **chi-square test for independence** tests the null hypothesis that there is no association between the two categorical variables in the population of interest. Another way to state the null hypothesis is $H_0$: The two categorical variables are independent in the population of interest.
- The conditions for performing a chi-square test for independence are:
  - **Random:** The data come from a well-designed random sample or randomized experiment.
    - **10%:** When sampling without replacement, check that the population is at least 10 times as large as the sample.
  - **Large Counts:** All expected counts must be at least 5.
- The **expected count** in any cell of a two-way table when $H_0$ is true is

$$\text{expected count} = \frac{\text{row total} \cdot \text{column total}}{\text{table total}}$$

- The **chi-square statistic** is

$$\chi^2 = \sum \frac{(\text{Observed} - \text{Expected})^2}{\text{Expected}}$$

where the sum is over all cells in the two-way table.

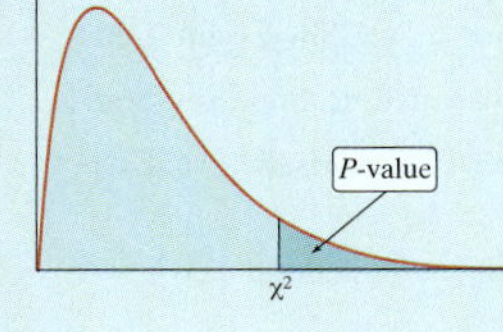


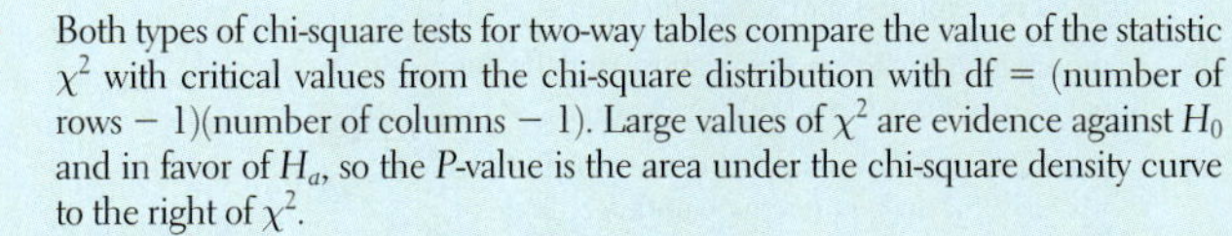

- Both types of chi-square tests for two-way tables compare the value of the statistic $\chi^2$ with critical values from the chi-square distribution with df = (number of rows − 1)(number of columns − 1). Large values of $\chi^2$ are evidence against $H_0$ and in favor of $H_a$, so the *P*-value is the area under the chi-square density curve to the right of $\chi^2$.
- If the test finds a statistically significant result, consider doing a follow-up analysis that compares the observed and expected counts and that looks for the largest components of the chi-square statistic.

## 11.2 TECHNOLOGY CORNER

TI-Nspire Instructions in Appendix B; HP Prime instructions on the book's Web site.

27. Chi-square tests for two-way tables on the calculator page 706

# Section 11.2 Exercises

27. **Why men and women play sports** Do men and women participate in sports for the same reasons? One goal for sports participants is social comparison—the desire to win or to do better than other people. Another is mastery—the desire to improve one's skills or to try one's best. A study on why students participate in sports collected data from independent random samples of 67 male and 67 female undergraduates at a large university.[13] Each student was classified into one of four categories based on his or her responses to a questionnaire about sports goals. The four categories were high social comparison–high mastery (HSC-HM), high social comparison–low mastery (HSC-LM), low social comparison–high mastery (LSC-HM), and low social comparison–low mastery (LSC-LM). One purpose of the study was to compare the goals of male and female students. Here are the data displayed in a two-way table:
pg 697

| | Gender | |
|---|---|---|
| Goal | Female | Male |
| HSC-HM | 14 | 31 |
| HSC-LM | 7 | 18 |
| LSC-HM | 21 | 5 |
| LSC-LM | 25 | 13 |

(a) Calculate the conditional distribution (in proportions) of the reported sports goals for each gender.

(b) Make an appropriate graph for comparing the conditional distributions in part (a).

(c) Write a few sentences comparing the distributions of sports goals for male and female undergraduates.

28. **How are schools doing?** The nonprofit group Public Agenda conducted telephone interviews with three randomly selected groups of parents of high school children. There were 202 black parents, 202 Hispanic parents, and 201 white parents. One question asked was "Are the high schools in your state doing an excellent, good, fair, or poor job, or don't you know enough to say?" Here are the survey results:[14]

| | Black parents | Hispanic parents | White parents |
|---|---|---|---|
| Excellent | 12 | 34 | 22 |
| Good | 69 | 55 | 81 |
| Fair | 75 | 61 | 60 |
| Poor | 24 | 24 | 24 |
| Don't know | 22 | 28 | 14 |
| **Total** | **202** | **202** | **201** |

(a) Calculate the conditional distribution (in proportions) of responses for each group of parents.

(b) Make an appropriate graph for comparing the conditional distributions in part (a).

(c) Write a few sentences comparing the distributions of responses for the three groups of parents.

29. **Why women and men play sports** Refer to Exercise 27. Do the data provide convincing evidence of a difference in the distributions of sports goals for male and female undergraduates at the university?
pg 704

(a) State appropriate null and alternative hypotheses for a significance test to help answer this question.

(b) Calculate the expected counts. Show your work.

(c) Calculate the chi-square statistic. Show your work.

30. **How are schools doing?** Refer to Exercise 28. Do the data provide convincing evidence of a difference in the distributions of opinions about how high schools are doing among black, Hispanic, and white parents?

(a) State appropriate null and alternative hypotheses for a significance test to help answer this question.

(b) Calculate the expected counts. Show your work.

(c) Calculate the chi-square statistic. Show your work.

31. **Why women and men play sports** Refer to Exercises 27 and 29.
pg 704

(a) Check that the conditions for performing the chi-square test are met.

(b) Use Table C to find the *P*-value. Then use your calculator's $\chi^2$cdf command.

(c) Interpret the *P*-value from the calculator in context.

(d) What conclusion would you draw? Justify your answer.

32. **How are schools doing?** Refer to Exercises 28 and 30.

(a) Check that the conditions for performing the chi-square test are met.

(b) Use Table C to find the *P*-value. Then use your calculator's $\chi^2$cdf command.

(c) Interpret the *P*-value from the calculator in context.

(d) What conclusion would you draw? Justify your answer.

**TRM Full Solutions to Section 11.2 Exercises**

## Answers to Section 11.2 Exercises

**11.27 (a)** Female: 0.209, 0.104, 0.313, 0.373. Male: 0.463, 0.269, 0.075, 0.194.

**(b)**

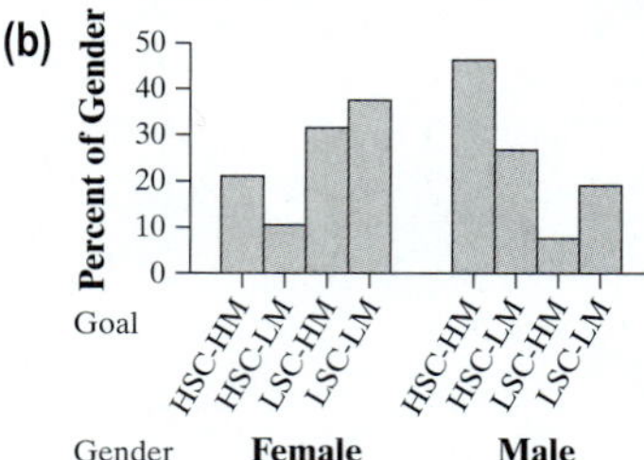


**(c)** In general, it appears that females were classified mostly as low social comparison, whereas males were classified mostly as high social comparison. However, about an equal percentage of males and females were classified as high mastery.

**11.28 (a)** Black: 0.059, 0.342, 0.371, 0.119, 0.109. Hispanic: 0.168, 0.272, 0.302, 0.119, 0.139. White: 0.109, 0.403, 0.299, 0.119, 0.070.

**(b)**

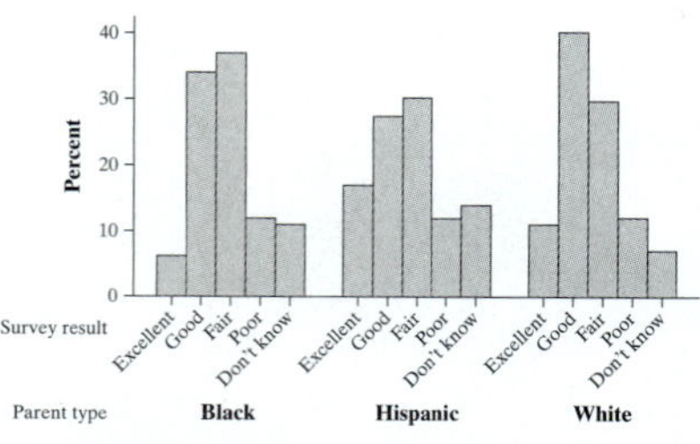


**(c)** In general, Hispanic parents are more likely to think that the high schools are doing an excellent job than white or black parents. White parents are more likely than either of the other two groups to think that the high schools are doing either an excellent or a good job. Black parents are more likely to think that schools are only doing a fair job than either of the other two groups of parents.

**11.29 (a)** $H_0$: There is no difference in the distribution of sports goals for male and female undergraduates at this university vs. $H_a$: There is a difference . . . **(b)** 22.5, 12.5, 13, 19, 22.5, 12.5, 13, 19 **(c)** $\chi^2 = 24.898$

**11.30 (a)** $H_0$: There is no difference in the distribution of opinions about how high schools are doing among black, Hispanic, and white parents vs. $H_a$: There is a difference . . . **(b)** 22.70, 68.45, 65.44, 24.04, 21.37, 22.70, 68.45, 65.44, 24.04, 21.37, 22.59, 68.11, 65.12, 23.92, 21.26. **(c)** $\chi^2 = 22.426$

**11.31 (a)** *Random*: Independent random samples. *10%*: $n_1 = 67 < 10\%$ of all males and $n_2 = 67 < 10\%$ of all females at the university. *Large Counts*: All expected counts $\geq 5$. **(b)** With df $= 3$, the *P*-value $< 0.0005$ (0.000016). **(c)** Assuming that there is no difference in the distributions of goals for playing sports among males and females, there is a 0.000016 probability of observing independent random samples that show a difference in the distributions of goals for playing sports among males and females as large or larger than the one found in this study. **(d)** Because the *P*-value of $0.000016 < \alpha = 0.05$, we reject $H_0$. There is convincing evidence of a difference in the distribution of goals for playing sports among male and female undergraduates at this university.

**11.32 (a)** *Random*: Independent random samples. *10%*: $n_1 = 202 < 10\%$ of all black parents, $n_2 = 202 < 10\%$ of all Hispanic parents, and $n_3 = 201 < 10\%$ of all white parents. *Large Counts*: All expected counts $\geq 5$. **(b)** With df $= 8$, the *P*-value is between 0.0025 and 0.005 (0.0042). **(c)** Assuming that there is no difference in the distributions of opinions about how high schools are doing among black, Hispanic, and white parents, there is a 0.0042 probability of observing samples that show differences in the distributions of opinions as large or larger than the one found in this study. **(d)** Because the *P*-value of $0.0042 < \alpha = 0.05$, we reject $H_0$. There is convincing evidence of a difference in the distribution of opinions about how high schools are doing among black, Hispanic, and white parents.

**33. Python eggs** How is the hatching of water python eggs influenced by the temperature of the snake's nest? Researchers randomly assigned newly laid eggs to one of three water temperatures: hot, neutral, or cold. Hot duplicates the extra warmth provided by the mother python, and cold duplicates the absence of the mother. Here are the data on the number of eggs that hatched and didn't hatch:[15]

pg 708 

| | Water Temperature | | |
|---|---|---|---|
| Hatched? | Cold | Neutral | Hot |
| Yes | 16 | 38 | 75 |
| No | 11 | 18 | 29 |

(a) Compare the distributions of hatching status for the three treatments.

(b) Are the differences between the three groups statistically significant? Give appropriate evidence to support your answer.

**34. Don't do drugs!** Cocaine addicts need cocaine to feel any pleasure, so perhaps giving them an antidepressant drug will help. A three-year study with 72 chronic cocaine users compared an antidepressant drug called desipramine with lithium (a standard drug to treat cocaine addiction) and a placebo. One-third of the subjects were randomly assigned to receive each treatment. Here are the results:[16]

| | Drug administered | | |
|---|---|---|---|
| Relapsed? | Desipramine | Lithium | Placebo |
| Yes | 10 | 18 | 20 |
| No | 14 | 6 | 4 |

(a) Compare the distributions of relapse status for the three treatments.

(b) Are the differences among the three groups statistically significant? Give appropriate evidence to support your answer.

**35. Sorry, no chi-square** How do U.S. residents who travel overseas for leisure differ from those who travel for business? The following is the breakdown by occupation:[17]

| Occupation | Leisure travelers (%) | Business travelers (%) |
|---|---|---|
| Professional/technical | 36 | 39 |
| Manager/executive | 23 | 48 |
| Retired | 14 | 3 |
| Student | 7 | 3 |
| Other | 20 | 7 |
| **Total** | **100** | **100** |

Explain why we can't use a chi-square test to learn whether these two distributions differ significantly.

**36. Going Nuts** The UR Nuts Company sells Deluxe and Premium nut mixes, both of which contain only cashews, brazil nuts, almonds, and peanuts. The Premium nuts are much more expensive than the Deluxe nuts. A consumer group suspects that the two nut mixes are really the same. To find out, the group took separate random samples of 20 pounds of each nut mix and recorded the weights of each type of nut in the sample. Here are the data:[18]

| | Type of mix | |
|---|---|---|
| Type of nut | Premium | Deluxe |
| Cashew | 6 lb | 5 lb |
| Brazil nut | 3 lb | 4 lb |
| Almond | 5 lb | 6 lb |
| Peanut | 6 lb | 5 lb |

Explain why we can't use a chi-square test to determine whether these two distributions differ significantly.

**37. How to quit smoking** It's hard for smokers to quit. Perhaps prescribing a drug to fight depression will work as well as the usual nicotine patch. Perhaps combining the patch and the drug will work better than either treatment alone. Here are data from a randomized, double-blind trial that compared four treatments.[19] A "success" means that the subject did not smoke for a year following the beginning of the study.

| Group | Treatment | Subjects | Successes |
|---|---|---|---|
| 1 | Nicotine patch | 244 | 40 |
| 2 | Drug | 244 | 74 |
| 3 | Patch plus drug | 245 | 87 |
| 4 | Placebo | 160 | 25 |

(a) Summarize these data in a two-way table. Then compare the success rates for the four treatments.

(b) Explain in words what the null hypothesis $H_0$: $p_1 = p_2 = p_3 = p_4$ says about subjects' smoking habits.

(c) Do the data provide convincing evidence of a difference in the effectiveness of the four treatments at the $\alpha = 0.05$ significance level?

**38. Preventing strokes** Aspirin prevents blood from clotting and so helps prevent strokes. The Second European Stroke Prevention Study asked whether adding another anticlotting drug named dipyridamole would be more effective for patients who had already had a stroke. Here are the data on strokes during the two years of the study:[20]

| Group | Treatment | Number of patients | Number of strokes |
|---|---|---|---|
| 1 | Placebo | 1649 | 250 |
| 2 | Aspirin | 1649 | 206 |
| 3 | Dipyridamole | 1654 | 211 |
| 4 | Both | 1650 | 157 |

**11.35** We do not have the actual counts of the travelers in each category. We also do not know if the sample was taken randomly or if the samples are independent.

**11.36** These data are quantitative. A chi-square test for homogeneity requires categorical data.

**11.37 (a)** The data appear in the table below. The best success rate is for the patch plus the drug (0.355), followed by the drug alone (0.303). The patch alone (0.164) is just a little better than the placebo (0.156).

| | Success | Failure | Total |
|---|---|---|---|
| Nicotine patch | 40 | 204 | **244** |
| Drug | 74 | 170 | **244** |
| Patch plus drug | 87 | 158 | **245** |
| Placebo | 25 | 135 | **160** |
| **Total** | **226** | **667** | **893** |

**(b)** Each of the four treatments has the same probability of success for smokers like these. **(c)** *S*: $H_0$: The true proportions of smokers like these who are able to quit for a year are the same for each of the four treatments vs. $H_a$: The true proportions are not the same . . . *P*: Chi-square test for homogeneity. *Random*: 4 groups in a randomized experiment. *Large Counts*: 61.75, 61.75, 62, 40.49, 182.25, 182.25, 183, 119.51 all $\geq 5$. *D*: $\chi^2 = 34.937$. With df $= 3$, the *P*-value $< 0.0005$. *C*: Because the *P*-value of approximately $0 < \alpha = 0.05$, we reject $H_0$. There is convincing evidence that the true proportions of smokers like these who are able to quit for a year are not the same for each of the four treatments.

**11.33 (a)** Cold: 0.593 hatched. Neutral: 0.679 hatched. Hot: 0.721 hatched. As the temperature warms up from cold to neutral to hot, the proportion of eggs that hatch appears to increase. **(b)** *S*: $H_0$: There is no difference in the true proportion of eggs that hatch in cold, neutral, or hot water vs. $H_a$: There is a difference . . . *P*: Chi-square test for homogeneity. *Random*: 3 groups in a randomized experiment. *Large Counts*: 18.63, 38.63, 71.74, 8.37, 17.37, 32.26 all $\geq 5$. *D*: $\chi^2 = 1.703$. With df $= 2$, the *P*-value $> 0.25$ (0.4267). *C*: Because the *P*-value of $0.4267 > \alpha = 0.05$, we fail to reject $H_0$. We do not have convincing evidence that there is a difference in the true proportions of eggs that hatch in cold, neutral, or hot water.

**11.34 (a)** Patients who used desipramine had the lowest relapse rate (0.417), followed by lithium (0.75) and then placebo (0.833). **(b)** *S*: $H_0$: There is no difference in the true proportion of cocaine addicts like these who relapse when using desipramine, lithium, or placebo vs. $H_a$: There is a difference . . . *P*: Chi-square test for homogeneity. *Random*: 3 groups in a randomized experiment. *Large Counts*: 16, 16, 16, 8, 8, 8 all $\geq 5$. *D*: $\chi^2 = 10.5$. With df $= 2$, the *P*-value is between 0.005 and 0.01 (0.0052). *C*: Because the *P*-value of $0.0052 < \alpha = 0.05$, we reject $H_0$. We have convincing evidence that there is a difference in the true proportion of cocaine addicts like these who relapse when using desipramine, lithium, or placebo.

(a) Summarize these data in a two-way table. Then compare the stroke rates for the four treatments.

(b) Explain in words what the null hypothesis $H_0: p_1 = p_2 = p_3 = p_4$ says about the incidence of strokes.

(c) Do the data provide convincing evidence of a difference in the effectiveness of the four treatments at the $\alpha = 0.05$ significance level?

**39. How to quit smoking** Perform a follow-up analysis of the test in Exercise 37 by finding the individual components of the chi-square statistic. Which cell(s) contributed most to the final result and in what direction?

**40. Preventing strokes** Perform a follow-up analysis of the test in Exercise 38 by finding the individual components of the chi-square statistic. Which cell(s) contributed most to the final result and in what direction?

**41. Attitudes toward recycled products** Some people believe recycled products are lower in quality than other products, a fact that makes recycling less practical. Here are data on attitudes toward coffee filters made of recycled paper from a random sample of adults:[21]

| | Recycled coffee filter status | |
|---|---|---|
| Quality rating | Buyers | Nonbuyers |
| Higher | 20 | 29 |
| Same | 7 | 25 |
| Lower | 9 | 43 |

Make a well-labeled bar graph that compares buyers' and nonbuyers' opinions about recycled filters. Describe what you see.

**42. Is astrology scientific?** The General Social Survey asked a random sample of adults their opinion about whether astrology is very scientific, sort of scientific, or not at all scientific. Here is a two-way table of counts for people in the sample who had three levels of higher education:[22]

| | Degree Held | | |
|---|---|---|---|
| | Associate's | Bachelor's | Master's |
| Not at all scientific | 169 | 256 | 114 |
| Very or sort of scientific | 65 | 65 | 18 |

Make a well-labeled bar graph that compares opinions about astrology for the three education categories. Describe what you see.

**43. Attitudes toward recycled products** Refer to Exercise 41.

(a) State appropriate hypotheses for performing a chi-square test of independence in this setting.

(b) Compute the expected counts assuming that $H_0$ is true. Show your work.

(c) Calculate the chi-square statistic, df, and *P*-value.

(d) What conclusion would you draw?

**44. Is astrology scientific?** Refer to Exercise 42.

(a) State appropriate hypotheses for performing a chi-square test of independence in this setting.

(b) Compute the expected counts assuming that $H_0$ is true. Show your work.

(c) Calculate the chi-square statistic, df, and *P*-value.

(d) What conclusion would you draw?

**45. Regulating guns** The National Gun Policy Survey asked a random sample of adults, "Do you think there should be a law that would ban possession of handguns except for the police and other authorized persons?" Here are the responses, broken down by the respondent's level of education:[23]

| | Education | | | | |
|---|---|---|---|---|---|
| | Less than high school | High school grad | Some college | College grad | Postgrad degree |
| Yes | 58 | 84 | 169 | 98 | 77 |
| No | 58 | 129 | 294 | 135 | 99 |

Does the sample provide convincing evidence of an association between education level and opinion about a handgun ban in the adult population?

**46. Market research** Before bringing a new product to market, firms carry out extensive studies to learn how consumers react to the product and how best to advertise its advantages. Here are data from a study of a new laundry detergent.[24] The participants are a random sample of people who don't currently use the established brand that the new product will compete with. Give subjects free samples of both detergents. After they have tried both for a while, ask which they prefer. The answers may depend on other facts about how people do laundry.

| | Laundry Practices | | | |
|---|---|---|---|---|
| | Soft water, warm wash | Soft water, hot wash | Hard water, warm wash | Hard water, hot wash |
| Prefer standard product | 53 | 27 | 42 | 30 |
| Prefer new product | 63 | 29 | 68 | 42 |

**11.38 (a)** The data appear in the table below. The lowest stroke rate is for the subjects who took both aspirin and dipyridamole (0.095). The groups that had either aspirin alone (0.125) or dipyridamole alone (0.128) had about the same stroke rates. The highest stroke rate was for the group taking the placebo (0.152).

| | Stroke | No stroke | Total |
|---|---|---|---|
| Placebo | 250 | 1399 | **1649** |
| Aspirin | 206 | 1443 | **1649** |
| Dipyridamole | 211 | 1443 | **1654** |
| Both | 157 | 1493 | **1650** |
| Total | **824** | **5778** | **6602** |

**(b)** Each of the four treatments has the same stroke rate for patients like these. **(c)** *S*: $H_0$: The true proportions of patients like these who have a stroke are the same for each of the four treatments vs. $H_a$: The true proportions are not the same . . . *P*: Chi-square test for homogeneity. *Random*: 4 groups in a randomized experiment. *Large Counts*: 205.81, 205.81, 206.44, 205.94, 1443.19, 1443.19, 1447.56, 1444.06 all $\geq 5$. *D*: $\chi^2 = 24.243$. With df = 3, the *P*-value < 0.0005. *C*: Because the *P*-value of approximately $0 < \alpha = 0.05$, we reject $H_0$. There is convincing evidence that the true proportions of patients like these who have a stroke are not the same for each of the four treatments.

**11.39** The largest component comes from those who had success using both the patch and the drug (25 more than expected). The next largest component comes from those who had success using just the patch (21.75 less than expected).

**11.40** The largest component comes from those who experienced strokes while taking both aspirin and dipyridamole (48.94 less than expected). The next largest component comes from those who had strokes while on the placebo (44.19 more than expected).

**11.41** Buyers are much more likely to think the quality of recycled coffee filters is higher, while nonbuyers are more likely to think the quality is the same or lower.

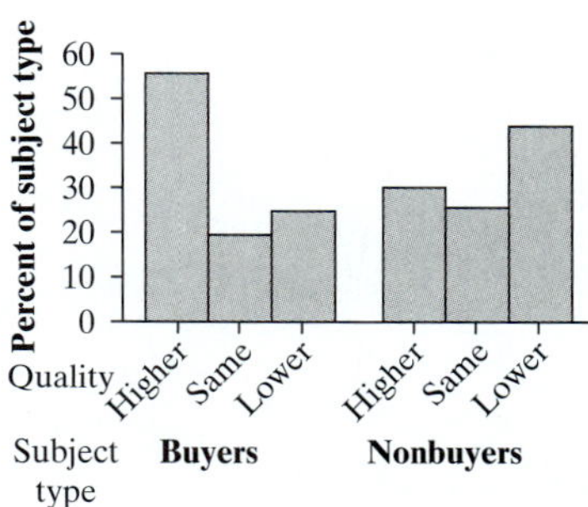


**11.42** The higher the degree earned, the less likely people are to think that astrology is scientific.

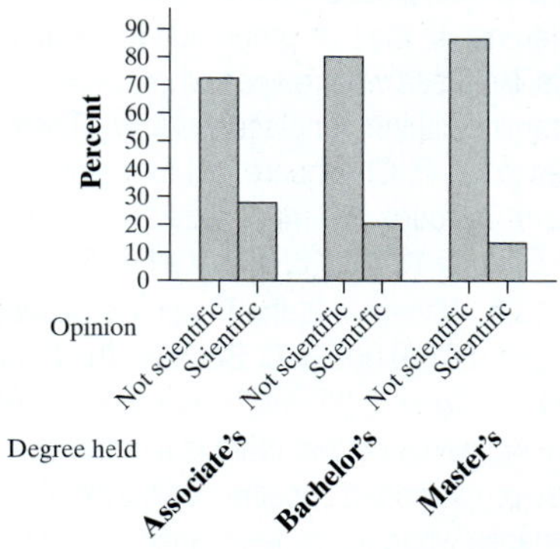


**11.43 (a)** $H_0$: There is no association between beliefs about the quality of recycled products and whether or not a person buys recycled products in the population of adults vs. $H_a$: There is an association. . . . **(b)** 13.26, 35.74, 8.66, 23.34, 14.08, 37.92. **(c)** $\chi^2 = 7.64$. With df = 2, the *P*-value is between 0.02 and 0.025 (0.022). **(d)** Because the *P*-value of $0.022 < \alpha = 0.05$, we reject $H_0$. There is convincing evidence of an association between beliefs about the quality of recycled products and whether or not a person buys recycled products in the population of adults.

**11.44 (a)** $H_0$: There is no association between level of education and opinion about astrology in the population of adults with some higher education vs. $H_a$: There is an association. . . . **(b)** 183.59, 251.85, 103.56, 50.41, 69.15, 28.44. **(c)** $\chi^2 = 10.582$. With df = 2, the *P*-value is between 0.005 and 0.01 (0.0050). **(d)** Because the *P*-value of $0.0050 < \alpha = 0.05$, we reject $H_0$. There is convincing evidence of an association between level of education and

Does the sample provide convincing evidence of an association between laundry practices and product preference in the population of interest?

**47. Where do young adults live?** A survey by the National Institutes of Health asked a random sample of young adults (aged 19 to 25 years), "Where do you live now? That is, where do you stay most often?" Here is the full two-way table (omitting a few who refused to answer and one who claimed to be homeless):[25]

pg 718 

| | Female | Male |
|---|---|---|
| Parents' home | 923 | 986 |
| Another person's home | 144 | 132 |
| Own place | 1294 | 1129 |
| Group quarters | 127 | 119 |

(a) Should we use a chi-square test for homogeneity or a chi-square test for independence in this setting? Justify your answer.

(b) State appropriate hypotheses for performing the type of test you chose in part (a).

Minitab output from a chi-square test is shown below.

```
Chi-Square Test: Female, Male
Expected counts are printed below observed
counts
Chi-Square contributions are printed below
expected counts
         Female      Male    Total
  1         923       986     1909
         978.49    930.51
          3.147     3.309

  2         144       132      276
         141.47    134.53
          0.045     0.048

  3        1294      1129     2423
        1241.95   1181.05
          2.181     2.294

  4         127       119      246
         126.09    119.91
          0.007     0.007

Total      2488      2366     4854
Chi-Sq = 11.038, DF = 3, P-Value = 0.012
```

(c) Check that the conditions for carrying out the test are met.

(d) Interpret the *P*-value in context. What conclusion would you draw?

**48. Students and catalog shopping** What is the most important reason that students buy from catalogs? The answer may differ for different groups of students. Here are results for separate random samples of American and Asian students at a large midwestern university:[26]

| | American | Asian |
|---|---|---|
| Save time | 29 | 10 |
| Easy | 28 | 11 |
| Low price | 17 | 34 |
| Live far from stores | 11 | 4 |
| No pressure to buy | 10 | 3 |

(a) Should we use a chi-square test for homogeneity or a chi-square test for independence in this setting? Justify your answer.

(b) State appropriate hypotheses for performing the type of test you chose in part (a).

Minitab output from a chi-square test is shown below.

```
Chi-Square Test: American, Asian
Expected counts are printed below observed
counts
Chi-Square contributions are printed below
expected counts
        American    Asian    Total
  1          29       10       39
          23.60    15.40
          1.236    1.894

  2          28       11       39
          23.60    15.40
          0.821    1.258

  3          17       34       51
          30.86    20.14
          6.225    9.538

  4          11        4       15
           9.08     5.92
          0.408    0.625

  5          10        3       13
           7.87     5.13
          0.579    0.887

  Total      95       62      157
  Chi-Sq = 23.470, DF = 4, P-Value = 0.0001
```

(c) Check that the conditions for carrying out the test are met.

(d) Interpret the *P*-value in context. What conclusion would you draw?

**49. Treating ulcers** Gastric freezing was once a recommended treatment for ulcers in the upper intestine. Use of gastric freezing stopped after experiments showed it had no effect. One randomized comparative experiment found that 28 of the 82 gastric-freezing patients improved, while 30 of

**11.47 (a)** Independence, because the data come from a single random sample. **(b)** $H_0$: There is no association between gender and where people live in the population of young adults vs. $H_a$: There is an association. . . . **(c)** *Random*: Random sample. *10%*: $n = 4854 < 10\%$ of all young adults. *Large Counts*: The expected counts are all at least 5. **(d)** *P-value*: If there is no association between gender and where people live in the population of young adults, there is a 0.012 probability of getting a random sample of 4854 young adults with an association as strong or even stronger than the one found in this study. *Conclusion*: Because the *P*-value of $0.012 < \alpha = 0.05$, we reject $H_0$. There is convincing evidence of an association between gender and where people live in the population of young adults.

**11.48 (a)** Homogeneity, because there were two independent random samples. **(b)** $H_0$: There is no difference in the distribution of reasons for shopping from catalogs for American and Asian students at this university vs. $H_a$: There is a difference. . . . **(c)** *Random*: Independent random samples. *10%*: $n_1 = 95 < 10\%$ of all American students at this university and $n_2 = 62 <$ 10% of all Asian students at this university. *Large Counts*: The expected counts are all at least 5. **(d)** *P-value*: If the distribution of reasons for shopping from catalogs is the same for both Asian and American students, there is a 0.0001 probability of finding distributions as different or even more different than the observed distributions by chance alone. *Conclusion*: Because the *P*-value of $0.0001 < \alpha = 0.05$, we reject $H_0$. There is convincing evidence of a difference in the distribution of reasons for shopping from catalogs among American and Asian students at this university.

opinion about astrology in the population of adults with some higher education.

**11.45** *S*: $H_0$: There is no association between education level and opinion about a handgun ban in the adult population vs. $H_a$: There is an association. . . . *P*: Chi-square test for independence. *Random*: Random sample. *10%*: $n = 1201 < 10\%$ of all adults. *Large Counts*: 46.94, 86.19, 187.36, 94.29, 71.22, 69.06, 126.81, 275.64, 138.71, 104.78 all $\geq 5$. *D*: $\chi^2 = 8.525$. With df $= 4$, the *P*-value is between 0.05 and 0.10 (0.0741). *C*: Because the *P*-value of 0.0741 $> \alpha = 0.05$, we fail to reject $H_0$. We do not have convincing evidence that there is an association between educational level and opinion about a handgun ban in the adult population.

**11.46** *S*: $H_0$: There is no association between laundry practices and product preference for people who don't currently use the established brand vs. $H_a$: There is an association. . . . *P*: Chi-square test for independence. *Random*: Random sample. *10%*: $n = 354 < 10\%$ of the population who don't currently use the established brand. *Large Counts*: 49.81, 24.05, 47.23, 30.92, 66.19, 31.95, 62.77, 41.08 all $\geq 5$. *D*: $\chi^2 = 2.058$. With df $=$ 3, the *P*-value $> 0.25$ (0.5605). *C*: Because the *P*-value of $0.5605 > \alpha = 0.05$, we fail to reject $H_0$. We do not have convincing evidence that there is an association between laundry practices and product preference among people who don't currently use the established brand.

**11.49** **(a)** *Hypotheses*: $H_0$: There is no difference in the improvement rates for patients like these who receive gastric freezing and those who receive the placebo vs. $H_a$: There is a difference . . . *P-value*: Assuming that there is no difference in the improvement rates between those receiving gastric freezing and those receiving the placebo, there is a 0.570 probability of observing a difference in improvement rates as large or larger than the difference observed in the study by chance alone. *Conclusion*: Because the *P*-value of 0.570 is larger than $\alpha = 0.05$, we fail to reject $H_0$. There is not convincing evidence of a difference in the improvement rates for patients like these who receive gastric freezing and those who receive the placebo. **(b)** The *P*-values are equal and $z^2 = (-0.57)^2 = 0.3249 \approx \chi^2 = 0.322$.

**11.50** **(a)** *Hypotheses*: $H_0$: There is no difference in opinion about the death penalty among all people who only have a high school diploma and all those who have a bachelor's degree vs. $H_a$: There is a difference. . . . *P-value*: Assuming that there is no difference in opinion about the death penalty among all people who only have a high school diploma and all those who have a bachelor's degree, there is a probability of approximately 0 of observing a difference in opinions as large or larger than the difference observed in the study by chance alone. *Conclusion*: Because the *P*-value of approximately $0 < \alpha = 0.05$, we reject $H_0$. There is convincing evidence of a difference in opinion about the death penalty among all people who only have a high school diploma and all those who have a bachelor's degree. **(b)** The *P*-values are equal and $z^2 = (4.19)^2 = 17.556 \approx \chi^2 = 17.590$.

**11.51** d

the 78 patients in the placebo group improved.[27] We can test the hypothesis of "no difference" in the effectiveness of the treatments in two ways: with a two-sample $z$ test or with a chi-square test.

(a) Minitab output for a chi-square test is shown below. State appropriate hypotheses and interpret the *P*-value in context. What conclusion would you draw?

```
Chi-Square Test: Gastric freezing, Placebo
Expected counts are printed below observed
counts
Chi-Square contributions are printed below
expected counts
      Gastric freezing    Placebo    Total
1                   28         30       58
                 29.73      28.27
                 0.100      0.105
2                   54         48      102
                 52.27      49.73
                 0.057      0.060
Total               82         78      160
Chi-Sq = 0.322, DF = 1, P-Value = 0.570
```

(b) Minitab output for a two-sample $z$ test is shown below. Explain how these results are consistent with the test in part (a).

```
Test for Two Proportions
Sample      X      N      Sample p
1          28     82      0.341463
2          30     78      0.384615
Difference = p (1) - p (2)
Estimate for difference: -0.0431520
Test for difference = 0 (vs not = 0):
Z = -0.57 P-Value = 0.570
```

50. **Opinions about the death penalty** The General Social Survey asked separate random samples of people with only a high school degree and people with a bachelor's degree, "Do you favor or oppose the death penalty for persons convicted of murder?" The following table gives the responses of people whose highest education was a high school degree and of people with a bachelor's degree:

| | Highest education level | |
|---|---|---|
| | High school | Bachelor's degree |
| Favor | 1010 | 319 |
| Oppose | 369 | 185 |

We can test the hypothesis of "no difference" in support for the death penalty among people in these educational categories in two ways: with a two-sample $z$ test or with a chi-square test.

(a) Minitab output for a chi-square test is shown below. State appropriate hypotheses and interpret the *P*-value in context. What conclusion would you draw?

```
Chi-Square Test: C1, C2
Expected counts are printed below
observed counts
Chi-Square contributions are printed be-
low expected counts
            C1         C2     Total
1         1010        319      1329
        973.28     355.72
         1.385      3.790
2          369        185       554
        405.72     148.28
         3.323      9.092
Total     1379        504      1883
Chi-Sq = 17.590, DF = 1, P-Value = 0.000
```

(b) Minitab output for a two-sample $z$ test is shown below. Explain how these results are consistent with the test in part (a).

```
Test for Two Proportions
Sample       X       N     Sample p
1         1010    1379     0.732415
2          319     504     0.632937
Difference = p (1) - p (2)
Estimate for difference: 0.0994783
Test for difference = 0 (vs not = 0):
Z = 4.19 P-Value = 0.000
```

*Multiple choice: Select the best answer for Exercises 51 to 56.*

*Exercises 51 to 55 refer to the following setting.* The National Longitudinal Study of Adolescent Health interviewed a random sample of 4877 teens (grades 7 to 12). One question asked was "What do you think are the chances you will be married in the next ten years?" Here is a two-way table of the responses by gender:[28]

| | Female | Male |
|---|---|---|
| Almost no chance | 119 | 103 |
| Some chance, but probably not | 150 | 171 |
| A 50–50 chance | 447 | 512 |
| A good chance | 735 | 710 |
| Almost certain | 1174 | 756 |

51. Which of the following would be the most appropriate type of graph for these data?

(a) A bar chart showing the marginal distribution of opinion about marriage

(b) A bar chart showing the marginal distribution of gender

(c) A bar chart showing the conditional distribution of gender for each opinion about marriage

(d) A bar chart showing the conditional distribution of opinion about marriage for each gender

(e) Dotplots that display the number in each opinion category for each gender

52. The appropriate null hypothesis for performing a chi-square test is that

(a) equal proportions of female and male teenagers are almost certain they will be married in 10 years.

(b) there is no difference between the distributions of female and male teenagers' opinions about marriage in this sample.

(c) there is no difference between the distributions of female and male teenagers' opinions about marriage in the population.

(d) there is no association between gender and opinion about marriage in the sample.

(e) there is no association between gender and opinion about marriage in the population.

53. The expected count of females who respond "almost certain" is

(a) 487.7. (c) 965. (e) 1174.

(b) 525. (d) 1038.8.

54. The degrees of freedom for the chi-square test for this two-way table are

(a) 4. (c) 10. (e) 4876.

(b) 8. (d) 20.

55. For these data, $\chi^2 = 69.8$ with a *P*-value of approximately 0. Assuming that the researchers used a significance level of 0.05, which of the following is true?

(a) A Type I error is possible.

(b) A Type II error is possible.

(c) Both a Type I and a Type II error are possible.

(d) There is no chance of making a Type I or Type II error because the *P*-value is approximately 0.

(e) There is no chance of making a Type I or Type II error because the calculations are correct.

56. When analyzing survey results from a two-way table, the main distinction between a test for independence and a test for homogeneity is

(a) how the degrees of freedom are calculated.

(b) how the expected counts are calculated.

(c) the number of samples obtained.

(d) the number of rows in the two-way table.

(e) the number of columns in the two-way table.

*For Exercises 57 and 58, you may find the inference summary chart inside the back cover helpful.*

57. **Inference recap** (8.1 to 11.2) In each of the following settings, state which inference procedure from Chapter 8, 9, 10, or 11 you would use. Be specific. For example, you might say "two-sample *z* test for the difference between two proportions." You do not need to carry out any procedures.[29]

(a) What is the average voter turnout during an election? A random sample of 38 cities was asked to report the percent of registered voters who actually voted in the most recent election.

(b) Are blondes more likely to have a boyfriend than the rest of the single world? Independent random samples of 300 blondes and 300 nonblondes were asked whether they have a boyfriend.

58. **Inference recap** (8.1 to 11.2) In each of the following settings, state which inference procedure from Chapter 8, 9, 10, or 11 you would use. Be specific. For example, you might say "two-sample *z* test for the difference between two proportions." You do not need to carry out any procedures.[30]

(a) Is there a relationship between attendance at religious services and alcohol consumption? A random sample of 1000 adults was asked whether they regularly attend religious services and whether they drink alcohol daily.

(b) Separate random samples of 75 college students and 75 high school students were asked how much time, on average, they spend watching television each week. We want to estimate the difference in the average amount of TV watched by high school and college students.

*Exercises 59 to 60 refer to the following setting.* For their final project, a group of AP® Statistics students investigated the following question: "Will changing the rating scale on a survey affect how people answer the question?" To find out, the group took an SRS of 50 students from an alphabetical roster of the school's just over 1000 students. The first 22 students chosen were asked to rate the cafeteria food on a scale of 1 (terrible) to 5 (excellent). The remaining 28 students were asked to rate the cafeteria food on a scale of 0 (terrible) to 4 (excellent). Here are the data:

| 1 to 5 scale | | | | | |
|---|---|---|---|---|---|
| Rating | 1 | 2 | 3 | 4 | 5 |
| Frequency | 2 | 3 | 1 | 13 | 3 |

| 0 to 4 scale | | | | | |
|---|---|---|---|---|---|
| Rating | 0 | 1 | 2 | 3 | 4 |
| Frequency | 0 | 0 | 2 | 18 | 8 |

**11.52** e

**11.53** d

**11.54** a

**11.55** a

**11.56** c

**11.57** **(a)** One-sample *t* interval for a mean. **(b)** Two-sample *z* test for the difference between two proportions.

**11.58** **(a)** Chi-square test for independence. **(b)** Two-sample *t* interval for the difference between two means.

59. **Design and analysis** (4.2)

(a) Was this an observational study or an experiment? Justify your answer.

(b) Explain why it would not be appropriate to perform a chi-square test in this setting.

60. **Average ratings** (1.3, 10.2) The students decided to compare the average ratings of the cafeteria food on the two scales.

(a) Find the mean and standard deviation of the ratings for the students who were given the 1-to-5 scale.

(b) For the students who were given the 0-to-4 scale, the ratings have a mean of 3.21 and a standard deviation of 0.568. Since the scales differ by one point, the group decided to add 1 to each of these ratings. What are the mean and standard deviation of the adjusted ratings?

(c) Would it be appropriate to compare the means from parts (a) and (b) using a two-sample $t$ test? Justify your answer.

## FRAPPY! Free Response AP® Problem, Yay!

**The following problem is modeled after actual AP® Statistics exam free response questions. Your task is to generate a complete, concise response in 15 minutes.**

*Directions: Show all your work. Indicate clearly the methods you use, because you will be scored on the correctness of your methods as well as on the accuracy and completeness of your results and explanations.*

Two statistics students wanted to know if including additional information in a survey question would change the distribution of responses. To find out, they randomly selected 30 teenagers and asked them one of the following two questions. Fifteen of the teenagers were randomly assigned to answer Question A, and the other 15 students were assigned to answer Question B.

**A:** When choosing a college, how important is a good athletic program: very important, important, somewhat important, not that important, or not important at all?

**B:** It's sad that some people choose a college based on its athletic program. When choosing a college, how important is a good athletic program: very important, important, somewhat important, not that important, or not important at all?

The table below summarizes the responses to both questions. For these data, the chi-square test statistic is $\chi^2 = 6.12$.

| | Question A | Question B | Total |
|---|---|---|---|
| Very important | 7 | 2 | **9** |
| Important | 4 | 3 | **7** |
| Somewhat important | 2 | 3 | **5** |
| Not that important | 1 | 2 | **3** |
| Not important at all | 1 | 5 | **6** |
| **Total** | **15** | **15** | **30** |

(a) State the hypotheses that the students are interested in testing.

(b) Describe a Type I error and a Type II error in the context of the hypotheses stated in part (a).

(c) For these data, explain why it would *not* be appropriate to use a chi-square distribution to calculate the $P$-value.

(d) To estimate the $P$-value, 100 trials of a simulation were conducted, assuming that the additional information didn't have an effect on the response to the question. In each trial of the simulation, the value of the chi-square statistic was calculated. These simulated chi-square statistics are displayed in the dotplot below.

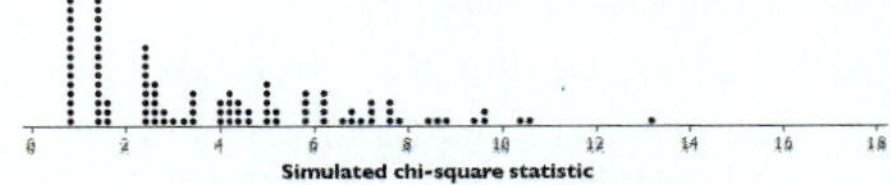


Based on the results of the simulation, what conclusion would you make about the hypotheses stated in part (a)?

**After you finish, you can view two example solutions on the book's Web site (www.whfreeman.com/tps5e). Determine whether you think each solution is "complete," "substantial," "developing," or "minimal." If the solution is not complete, what improvements would you suggest to the student who wrote it? Finally, your teacher will provide you with a scoring rubric. Score your response and note what, if anything, you would do differently to improve your own score.**

**11.59 (a)** Experiment, because a treatment (type of rating scale) was deliberately imposed on the students who took part in the study. **(b)** Several of the expected counts are less than 5.

**11.60 (a)** The mean for the 1–5 scale is 3.545. The standard deviation is 1.184. **(b)** The new mean for the 0–4 scale is $3.21 + 1 = 4.21$. Adding 1 to each of the values does not change the standard deviation, so it remains 0.568. **(c)** No, because the Normal/Large Sample condition is not met. The sample sizes are both less than 30 and there are two low outliers in the distribution of responses for students who were given the 1–5 scale.

**TRM FRAPPY!**

We present a model solution here. Please consult the Teacher's Resource Materials for sample student responses, a scoring rubric, and a black line master of the original question with space for students to write their responses.

**(a)** $H_0$: The true distribution of responses is the same for the two questions *versus* $H_a$: The true distribution of responses is not the same for the two questions.

**(b)** Type I: Finding convincing evidence that the true distribution of responses is different for the two questions, when the true distributions really are the same. Type II: Not finding convincing evidence that the true distribution of responses is different for the two questions, when the true distributions really are not the same.

**(c)** We should not use a chi-square distribution to estimate the $P$-value because the Large Counts condition is not met. All of the expected counts are less than 5:

| | A | B |
|---|---|---|
| Very important | 4.5 | 4.5 |
| Important | 3.5 | 3.5 |
| Somewhat important | 2.5 | 2.5 |
| Not that important | 1.5 | 1.5 |
| Not at all important | 3 | 3 |

**(d)** Because 24 of the 100 trials resulted in a $\chi^2$-value of 6.12 or more, the $P$-value $\approx 0.24$. Because the approximate $P$-value of 0.24 is greater than $\alpha = 0.05$, we fail to reject $H_0$. We do not have convincing evidence that the true distribution of responses is different for the two questions.

# Chapter Review

### Section 11.1: Chi-Square Tests for Goodness of Fit

In this section, you learned the details for performing a chi-square test for goodness of fit. The null hypothesis is that a single categorical variable follows a specified distribution. The alternative hypothesis is that the variable does not follow the specified distribution.

The chi-square statistic measures the difference between the observed distribution of a categorical variable and its hypothesized distribution. To calculate the chi-square statistic, use the following formula that involves the observed and expected counts for each value of the categorical variable:

$$\chi^2 = \sum \frac{(\text{Observed} - \text{Expected})^2}{\text{Expected}}$$

To calculate the expected counts, multiply the total sample size by the hypothesized proportion for each category. Larger values of the chi-square statistic provide more convincing evidence that the categorical variable does not have the hypothesized distribution.

When the Random, 10%, and Large Counts conditions are satisfied, we can accurately model the sampling distribution of a chi-square statistic using a chi-square distribution (density curve). The Random condition says that the data are from a well-designed random sample or a randomized experiment. The 10% condition says that the sample size should be at most 10% of the population size when sampling without replacement. The Large Counts condition says that the *expected* counts for each category must be at least 5. In a test for goodness of fit, use a chi-square distribution with degrees of freedom = number of categories − 1.

When the results of a test for goodness of fit are significant, consider doing a follow-up analysis. Identify which categories of the variable had the largest contributions to the chi-square statistic and whether the observed values in those categories were larger or smaller than expected.

### Section 11.2: Inference for Two-Way Tables

In this section, you learned how to perform inference for categorical data that are summarized in a two-way table. To begin the analysis, compare the conditional distributions of the response variable for each value of the explanatory variable. Displaying these distributions with a bar graph will help you make an effective comparison.

There are two types of chi-square tests that could apply when data are summarized in a two-way table. A test for homogeneity compares the distribution of a single categorical variable for two or more populations or treatments. A test for independence looks for an association between two categorical variables in a single population.

In a chi-square test for homogeneity, the null hypothesis is that there is no difference between the true distributions of a categorical variable for two or more populations or treatments. The alternative hypothesis is that there is a difference in the distributions. The Random condition is that the data come from independent random samples or groups in a randomized experiment. The 10% condition applies when sampling without replacement, but not in experiments. Finally, the Large Counts condition remains the same—the expected counts must be at least 5 in each cell of the two-way table.

To calculate the expected counts for a test for homogeneity, use the following formula:

$$\text{expected count} = \frac{\text{row total} \cdot \text{column total}}{\text{table total}}$$

To calculate the *P*-value, compute the chi-square statistic and use a chi-square distribution with degrees of freedom = (number of rows − 1)(number of columns − 1).

In a chi-square test for independence, the null hypothesis is that there is no association between two categorical variables in one population. The alternative hypothesis is that there is an association between the two variables. For this test, the Random condition says that the data must come from a single random sample. The 10% condition applies when sampling without replacement. The Large Counts condition is still the same—the expected counts must all be at least 5. The method for calculating expected counts, the chi-square statistic, the degrees of freedom, and the *P*-value are exactly the same in a test for independence and a test for homogeneity.

As with tests for goodness of fit, when the results of a test for homogeneity or independence are significant, consider doing a follow-up analysis. Identify which cells in the two-way table had the largest contributions to the chi-square statistic and whether the observed values in those cells were larger or smaller than expected.



**TRM Chapter 11 Tests**

There are four different tests available for this chapter in the Teacher's Resource Materials. Versions A and B are roughly parallel and version D is cumulative. Find them by clicking on the link in the TE-book, opening the TRCD, or logging on to the book's companion Web site www.whfreeman.com/tps5e (teacher log-in required).

**TRM** Learning Objectives Grid

## What Did You Learn?

| Learning Objective | Section | Related Example on Page(s) | Relevant Chapter Review Exercise(s) |
|---|---|---|---|
| State appropriate hypotheses and compute expected counts for a chi-square test for goodness of fit. | 11.1 | 681 | R11.1 |
| Calculate the chi-square statistic, degrees of freedom, and *P*-value for a chi-square test for goodness of fit. | 11.1 | 683, 685 | R11.1 |
| Perform a chi-square test for goodness of fit. | 11.1 | 688 | R11.1 |
| Conduct a follow-up analysis when the results of a chi-square test are statistically significant. | 11.1,11.2 | Discussion on 690–691 | R11.4 |
| Compare conditional distributions for data in a two-way table. | 11.2 | 697, 711 | R11.3, R11.5 |
| State appropriate hypotheses and compute expected counts for a chi-square test based on data in a two-way table. | 11.2 | 701, 713 | R11.2, R11.3, R11.4, R11.5 |
| Calculate the chi-square statistic, degrees of freedom, and *P*-value for a chi-square test based on data in a two-way table. | 11.2 | 704 | R11.3, R11.5 |
| Perform a chi-square test for homogeneity. | 11.2 | 708 | R11.3 |
| Perform a chi-square test for independence. | 11.2 | 715 | R11.5 |
| Choose the appropriate chi-square test. | 11.2 | 718 | R11.4 |

# Chapter 11 Chapter Review Exercises

*These exercises are designed to help you review the important ideas and methods of the chapter.*

**R11.1 Testing a genetic model** Biologists wish to cross pairs of tobacco plants having genetic makeup Gg, indicating that each plant has one dominant gene (G) and one recessive gene (g) for color. Each offspring plant will receive one gene for color from each parent. The Punnett square below shows the possible combinations of genes received by the offspring:

| | | Parent 2 passes on: | |
|---|---|---|---|
| | | G | g |
| Parent 1 passes on: | G | GG | Gg |
| | g | Gg | gg |

The Punnett square suggests that the expected ratio of green (GG) to yellow-green (Gg) to albino (gg) tobacco plants should be 1:2:1. In other words, the biologists predict that 25% of the offspring will be green, 50% will be yellow-green, and 25% will be albino. To test their hypothesis about the distribution of offspring, the biologists mate 84 randomly selected pairs of yellow-green parent plants. Of 84 offspring, 23 plants were green, 50 were yellow-green, and 11 were albino. Do the data provide convincing evidence at the $\alpha = 0.01$ level that the true distribution of offspring is different from what the biologists predict?

**R11.2 Sorry, no chi-square** We would prefer to learn from teachers who know their subject. Perhaps even preschool children are affected by how knowledgeable they think teachers are. Assign 48 three- and four-year-olds at random to be taught the name of a new toy by either an adult who claims to know about the toy or an adult who claims not to know about it. Then ask the children to pick out a picture of the new toy in a set of pictures of other toys and say its name. The response variable is the count of right answers in four tries. Here are the data:[31]

732

**TRM** Full Solutions to Chapter 11 Review Exercises

### Answers to Chapter 11 Review Exercises

**R11.1** *S*: $H_0$: The proposed 1:2:1 genetic model is correct vs. $H_a$: The proposed 1:2:1 genetic model is not correct.
*P*: Chi-square test for goodness of fit. *Random*: Random sample. *10%*: $n = 84 < 10\%$ of all yellow-green parent plants. *Large Counts*: 21, 42, 21 all $\geq 5$.
*D*: $\chi^2 = 6.476$. Using df $= 2$, the *P*-value is between 0.025 and 0.05 (0.0392).
*C*: Because the *P*-value of $0.0392 > \alpha = 0.01$, we fail to reject $H_0$. We do not have convincing evidence that the proposed 1:2:1 genetic model is not correct.

| | Correct Answers | | | | |
|---|---|---|---|---|---|
| | 0 | 1 | 2 | 3 | 4 |
| Knowledgeable teacher | 5 | 1 | 6 | 3 | 9 |
| Ignorant teacher | 20 | 0 | 3 | 0 | 1 |

The researchers report that children taught by the teacher who claimed to be knowledgeable did significantly better ($\chi^2 = 20.24$, $P < 0.05$). Explain why this result isn't valid.

**R11.3 Stress and heart attacks** You read a newspaper article that describes a study of whether stress management can help reduce heart attacks. The 107 subjects all had reduced blood flow to the heart and so were at risk of a heart attack. They were assigned at random to three groups. The article goes on to say:

> One group took a four-month stress management program, another underwent a four-month exercise program, and the third received usual heart care from their personal physicians. In the next three years, only three of the 33 people in the stress management group suffered "cardiac events," defined as a fatal or non-fatal heart attack or a surgical procedure such as a bypass or angioplasty. In the same period, 7 of the 34 people in the exercise group and 12 out of the 40 patients in usual care suffered such events.[32]

(a) Use the information in the news article to make a two-way table that describes the study results.
(b) Compare the success rates of the three treatments in preventing cardiac events.
(c) Do the data provide convincing evidence that the true success rates are not the same for the three treatments?

**R11.4 Sexy magazine ads?** Researchers looked at 1509 full-page ads that show a model. The two-way table below shows the main audience of the magazines in which the ads were found (young men, young women, or young adults in general) and whether or not the ad was "sexual." This was determined based on how the model was dressed (or not dressed).[33]

| | Readers | | |
|---|---|---|---|
| | Men | Women | General |
| Sexual | 105 | 225 | 66 |
| Not sexual | 514 | 351 | 248 |

The following figure displays Minitab output for a chi-square test using these data.

```
Session
              Men    Women  General      All

Sex           105      225       66      396
            16.96    39.06    21.02    26.24
            162.4    151.2     82.4    396.0
           20.312   36.074    3.265        *

notsexy       514      351      248     1113
            83.04    60.94    78.98    73.76
            456.6    424.8    231.6   1113.0
            7.227   12.835    1.162        *

All           619      576      314     1509
           100.00   100.00   100.00   100.00
            619.0    576.0    314.0   1509.0
                *        *        *        *

Cell Contents:        Count
                      % of Column
                      Expected count
                      Contribution to Chi-square

Chi-Square = 80.874,  DF = 2, P-Value = 0.00
```

(a) Describe how these data could have been collected so that a test for homogeneity is appropriate.
(b) Describe how these data could have been collected so that a test for independence is appropriate.
(c) Show how each of the numbers 60.94, 424.8, and 12.835 was obtained for the "notsexy, Women" cell.
(d) Which cell contributes most to the chi-square statistic? How do the observed and expected counts compare for this cell?

**R11.5 Popular kids** Who were the popular kids at your elementary school? Did they get good grades or have good looks? Were they good at sports? A study was performed to examine the factors that determine social status for children in grades 4, 5, and 6. Researchers administered a questionnaire to a random sample of 478 students in these grades. One of the questions they asked was "What would you most like to do at school: make good grades, be good at sports, or be popular?" The two-way table below summarizes the students' responses.[34]

| | Gender | |
|---|---|---|
| Goal | Female | Male |
| Grades | 130 | 117 |
| Popular | 91 | 50 |
| Sports | 30 | 60 |

(a) Construct an appropriate graph to compare male and female responses. Write a few sentences describing the relationship between gender and goals.
(b) Is there convincing evidence at the $\alpha = 0.05$ level of an association between gender and goals for elementary school students?

**R11.2** Several of the expected counts are less than 5.

**R11.3 (a)**

| | Stress management | Exercise | Usual Care | Total |
|---|---|---|---|---|
| Suffered cardiac event | 3 | 7 | 12 | **22** |
| No cardiac event | 30 | 27 | 28 | **85** |
| **Total** | **33** | **34** | **40** | **107** |

**(b)** The success rate was highest for stress management (0.909), followed by exercise (0.794) and usual care (0.70). **(c)** *S*: $H_0$: The true success rates for patients like these are the same for all three treatments vs. $H_a$: The true success rates are not all the same . . . *P*: Chi-square test for homogeneity. *Random*: 3 groups in a randomized experiment. *Large Counts*: 6.79, 6.99, 8.22, 26.21, 27.01, 31.78 all $\geq 5$. *D*: $\chi^2 = 4.840$. With df $= 2$, the *P*-value is between 0.05 and 0.10 (0.0889). *C*: Because the *P*-value $> \alpha = 0.05$, we fail to reject $H_0$. We do not have convincing evidence that the true success rates for patients like these are not the same for all three treatments.

**R11.4 (a)** The data could have been collected from 3 independent random samples—a random sample of ads from magazines aimed at young men, a random sample of ads from magazines aimed at young women, and a random sample of ads aimed at young adults in general. In each sample, the ads would be classified as sexual or not sexual. **(b)** The data could have been collected from a single random sample of ads from magazines aimed at young adults. Then each ad in the sample would be classified as sexual or not sexual and the magazine where the ad appeared would be classified as aimed at young men, young women, or young adults in general.

**(c)** $\frac{351}{576} = 0.6094 = 60.94\%$. $\frac{(1113)(576)}{1509} = 424.8$. $\frac{(351 - 424.8)^2}{424.8} = 12.82$. (The difference is due to rounding error.) **(d)** The "sexual, Women" cell. There were 225 observed ads in this cell, which was 73.8 more than expected.

**R11.5 (a)** For both groups of children, the largest percentage report grades as the goal. But after that, boys were more likely to pick sports, whereas girls were more likely to pick being popular.

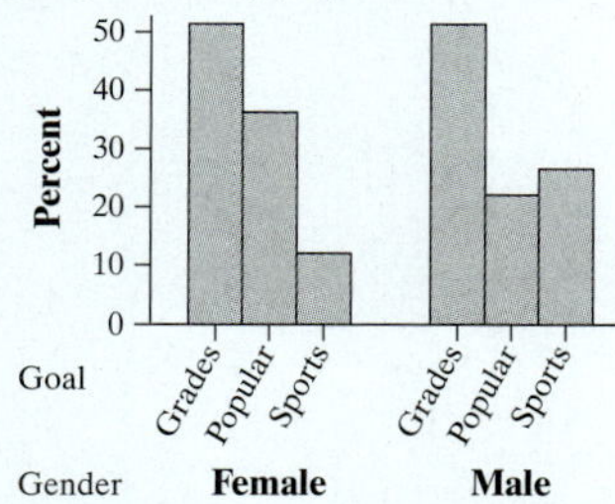


**(b)** *S*: $H_0$: There is no association between gender and goals for 4th, 5th, and 6th grade students vs. $H_a$: There is an association . . . *P*: Chi-square test for independence. *Random*: Random sample. *10%*: $n = 478 < 10\%$ of all 4th, 5th, and 6th grade students. *Large Counts*: 129.70, 117.30, 74.04, 66.96, 47.26, 42.74 all $\geq 5$. *D*: $\chi^2 = 21.455$. With df $= 2$, the *P*-value $< 0.0005$ (0.00002). *C*: Because the *P*-value of $0.00002 < \alpha = 0.05$, we reject $H_0$. There is convincing evidence of an association between gender and goals for 4th, 5th, and 6th grade students.

TRM **Full Solutions to Chapter 11 AP® Statistics Practice Test**

## Answers to Chapter 11 AP® Statistics Practice Test

**T11.1** b
**T11.2** c
**T11.3** e
**T11.4** d
**T11.5** c
**T11.6** c

# Chapter 11 AP® Statistics Practice Test

**Section I: Multiple Choice** *Select the best answer for each question.*

**T11.1** A chi-square test is used to test whether a 0 to 9 spinner is "fair" (that is, the outcomes are all equally likely). The spinner is spun 100 times, and the results are recorded. The degrees of freedom for the test will be

(a) 8. (b) 9. (c) 10. (d) 99. (e) None of these.

*Exercises T11.2 and T11.3 refer to the following setting.* Recent revenue shortfalls in a midwestern state led to a reduction in the state budget for higher education. To offset the reduction, the largest state university proposed a 25% tuition increase. It was determined that such an increase was needed simply to compensate for the lost support from the state. Separate random samples of 50 freshmen, 50 sophomores, 50 juniors, and 50 seniors from the university were asked whether they were strongly opposed to the increase, given that it was the minimum increase necessary to maintain the university's budget at current levels. Here are the results.

| Strongly Opposed? | Year | | | |
|---|---|---|---|---|
| | Freshman | Sophomore | Junior | Senior |
| Yes | 39 | 36 | 29 | 18 |
| No | 11 | 14 | 21 | 32 |

**T11.2** Which hypotheses would be appropriate for performing a chi-square test?

(a) The null hypothesis is that the closer students get to graduation, the less likely they are to be opposed to tuition increases. The alternative is that how close students are to graduation makes no difference in their opinion.
(b) The null hypothesis is that the mean number of students who are strongly opposed is the same for each of the 4 years. The alternative is that the mean is different for at least 2 of the 4 years.
(c) The null hypothesis is that the distribution of student opinion about the proposed tuition increase is the same for each of the 4 years at this university. The alternative is that the distribution is different for at least 2 of the 4 years.
(d) The null hypothesis is that year in school and student opinion about the tuition increase in the sample are independent. The alternative is that these variables are dependent.
(e) The null hypothesis is that there is an association between year in school and opinion about the tuition increase at this university. The alternative hypothesis is that these variables are not associated.

**T11.3** The conditions for carrying out the chi-square test in exercise T11.2 are

I. Independent random samples from the populations of interest.
II. All expected counts are at least 5.
III. The population sizes are at least 10 times the sample sizes.

Which of the conditions is (are) satisfied in this case?

(a) I only (b) II only (c) I and II only (d) II and III only (e) I, II, and III

*Exercises T11.4 to T11.6 refer to the following setting.* A random sample of traffic tickets given to motorists in a large city is examined. The tickets are classified according to the race of the driver. The results are summarized in the following table.

| Race: | White | Black | Hispanic | Other |
|---|---|---|---|---|
| Number of tickets: | 69 | 52 | 18 | 9 |

The proportion of this city's population in each of the racial categories listed above is as follows:

| Race: | White | Black | Hispanic | Other |
|---|---|---|---|---|
| Proportion: | 0.55 | 0.30 | 0.08 | 0.07 |

We wish to test $H_0$: The racial distribution of traffic tickets in the city is the same as the racial distribution of the city's population.

**T11.4** Assuming $H_0$ is true, the expected number of Hispanic drivers who would receive a ticket is

(a) 8. (b) 10.36. (c) 11. (d) 11.84. (e) 12.

**T11.5** We compute the value of the $\chi^2$ statistic to be 6.58. Assuming that the conditions for inference are met, the *P*-value of our test is

(a) greater than 0.20.
(b) between 0.10 and 0.20.
(c) between 0.05 and 0.10.
(d) between 0.01 and 0.05.
(e) less than 0.01.

**T11.6** The category that contributes the largest component to the $\chi^2$ statistic is

(a) White.
(b) Black.
(c) Hispanic.
(d) Other.
(e) The answer cannot be determined because this is only a sample.

*Exercises T11.7 to T11.10 refer to the following setting.* All current-carrying wires produce electromagnetic (EM) radiation, including the electrical wiring running into, through, and out of our homes. High-frequency EM radiation is thought to be a cause of cancer. The lower frequencies associated with household current are generally assumed to be harmless. To investigate the relationship between current configuration and type of cancer, researchers visited the addresses of a random sample of children who had died of some form of cancer (leukemia, lymphoma, or some other type) and classified the wiring configuration outside the dwelling as either a high-current configuration (HCC) or a low-current configuration (LCC). Here are the data:

| | Leukemia | Lymphoma | Other cancers |
|---|---|---|---|
| HCC | 52 | 10 | 17 |
| LCC | 84 | 21 | 31 |

Computer software was used to analyze the data. The output included the value $\chi^2 = 0.435$.

**T11.7** The appropriate degrees of freedom for the $\chi^2$ statistic is

(a) 1. (b) 2. (c) 3. (d) 4. (e) 5.

**T11.8** The expected count of cases with lymphoma in homes with an HCC is

(a) $\frac{79 \cdot 31}{215}$. (b) $\frac{10 \cdot 21}{215}$. (c) $\frac{79 \cdot 31}{10}$. (d) $\frac{136 \cdot 31}{215}$.

(e) None of these.

**T11.9** Which of the following may we conclude, based on the test results?

(a) There is convincing evidence of an association between wiring configuration and the chance that a child will develop some form of cancer.

(b) HCC either causes cancer directly or is a major contributing factor to the development of cancer in children.

(c) Leukemia is the most common type of cancer among children.

(d) There is not convincing evidence of an association between wiring configuration and the type of cancer that caused the deaths of children in the study.

(e) There is convincing evidence that HCC does not cause cancer in children.

**T11.10** A Type I error would occur if we found convincing evidence that

(a) HCC wiring caused cancer when it actually didn't.

(b) HCC wiring didn't cause cancer when it actually did.

(c) there is no association between the type of wiring and the form of cancer when there actually is an association.

(d) there is an association between the type of wiring and the form of cancer when there actually is no association.

(e) the type of wiring and the form of cancer have a positive correlation when they actually don't.

**Section II: Free Response** *Show all your work. Indicate clearly the methods you use, because you will be graded on the correctness of your methods as well as on the accuracy and completeness of your results and explanations.*

**T11.11** A large distributor of gasoline claims that 60% of all cars stopping at their service stations choose regular unleaded gas and that premium and supreme are each selected 20% of the time. To investigate this claim, researchers collected data from a random sample of drivers who put gas in their vehicles at the distributor's service stations in a large city. The results were as follows:

| Gasoline Selected | | |
|---|---|---|
| Regular | Premium | Supreme |
| 261 | 51 | 88 |

Carry out a test of the distributor's claim at the 5% significance level.

**T11.12** A study conducted in Charlotte, North Carolina, tested the effectiveness of three police responses to spouse abuse: (1) advise and possibly separate the couple, (2) issue a citation to the offender, and (3) arrest the offender. Police officers were trained to recognize eligible cases. When presented with an eligible case, a police officer called the dispatcher, who would randomly assign one of the three available treatments to be administered. There were a total of 650 cases in the study. Each case was classified according to whether the abuser was subsequently arrested within six months of the original incident.[35]

| | Treatment | | |
|---|---|---|---|
| Subsequent arrest? | Advise and separate | Citation | Arrest |
| No | 187 | 181 | 175 |
| Yes | 25 | 43 | 39 |

(a) Explain the purpose of the random assignment in the design of this study.

(b) Construct a well-labeled graph that is suitable for comparing the effectiveness of the three treatments.

(c) State an appropriate pair of hypotheses for performing a chi-square test in this setting.

(d) Assume that all the conditions for performing the test in part (b) are met. The test yields $\chi^2 = 5.063$ and a *P*-value of 0.0796. Interpret this *P*-value in context. What conclusion should we draw from the study?

**T11.13** In the United States, there is a strong relationship between education and smoking: well-educated people are less likely to smoke. Does a similar relationship hold in France? To find out, researchers recorded the level of education and smoking status of a random sample of 459 French men aged 20 to 60 years.[36] The two-way table below displays the data.

| | Education | | |
|---|---|---|---|
| Smoking Status | Primary School | Secondary School | University |
| Nonsmoker | 56 | 37 | 53 |
| Former | 54 | 43 | 28 |
| Moderate | 41 | 27 | 36 |
| Heavy | 36 | 32 | 16 |

Is there convincing evidence of an association between smoking status and educational level among French men aged 20 to 60 years?

**T11.7** b

**T11.8** a

**T11.9** d

**T11.10** d

**T11.11** *S*: $H_0$: The distribution of gas types is the same as the distributor's claim vs. $H_a$: The distribution of gas types is not the same as the distributor's claim. *P*: Chi-square test for goodness of fit. *Random*: Random sample. *10%*: $n = 400 < 10\%$ of all customers at this distributor's service stations. *Large Counts*: 240, 80, 80 all $\geq 5$. *D*: $\chi^2 = 13.15$. With df $= 2$, the *P*-value is between 0.001 and 0.0025 (0.0014). *C*: Because the *P*-value of $0.0014 < \alpha = 0.05$, we reject $H_0$. There is convincing evidence that the distribution of gas type is not all the same as the distributor claims.

**T11.12 (a)** Random assignment was used to create three roughly equivalent groups at the beginning of the study.

**(b)**

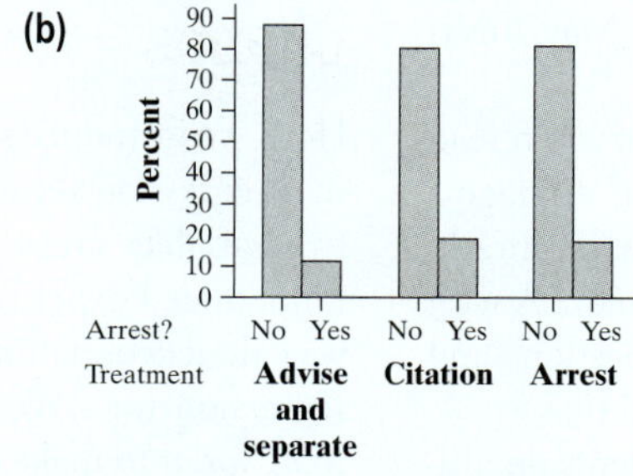


**(c)** $H_0$: The true proportion of spouse abusers like the ones in the study who will be arrested within 6 months is the same for all three police responses vs. $H_a$: The true proportions are not all the same.

**(d)** *P-value*: If the true proportion of spouse abusers like the ones in the study who will be arrested within 6 months is the same for all three police responses, there is a 0.0796 probability of getting differences between the three groups as large as or larger than the ones observed by chance alone. *Conclusion*: Because the *P*-value of 0.0796 is larger than $\alpha = 0.05$, we fail to reject $H_0$.There is not convincing evidence that the true proportion of spouse abusers like the ones in the study who will be arrested within 6 months is not the same for all three police responses.

**T11.13** *S*: $H_0$: There is no association between smoking status and educational level among French men aged 20 to 60 years vs. $H_a$: There is an association. . . . *P*: Chi-square test for independence. *Random*: Random sample. *10%*: $n = 459 < 10\%$ of all French men aged 20 to 60 years. *Large Counts*: 59.48, 44.21, 42.31, 50.93, 37.85, 36.22, 42.37, 31.49, 30.14, 34.22, 25.44, 24.34 all $\geq 5$. *D*: $\chi^2 = 13.305$. With df $= 6$, the *P*-value is between 0.025 and 0.05 (0.0384). *C*: Because the *P*-value of $0.0384 < \alpha = 0.05$, we reject $H_0$. There is convincing evidence of an association between smoking status and educational level among French men aged 20 to 60 years.

Chapter

12

# More about Regression

## PD Overview

Congratulations on making it to the last chapter in the book! In this chapter, we cover two different regression topics. In Section 12.1, we perform inference about the slope of a least-squares regression line. In Section 12.2, we use transformations to achieve linearity when the association between two quantitative variables is nonlinear.

Although the regression setting is different, the types of inference we perform in Section 12.1—confidence intervals and significance tests for the slope of a least-squares regression line—are essentially the same as in previous chapters. Confidence intervals for the slope of a least-squares regression line give a range of plausible values for the slope. Significance tests for a slope allow us to choose between two competing claims about the true value of the slope. We also take the opportunity to review many other topics in regression that your students may have forgotten since Chapter 3.

In Section 12.2, we finally answer a question that was raised in Chapter 3: What do we do when a scatterplot shows a curved relationship? The short answer is that we transform one or both of the variables using logarithms or powers, hoping that we can straighten out the form of the relationship and fit a least-squares regression line. Although we could have included this section at the end of Chapter 3, we decided to move it to the last section of the book for a couple of reasons. First, the mathematics required causes problems for some students. We didn't want to put an obstacle in their way, especially when this material isn't necessary to understand any of the topics that follow Chapter 3. Second, this topic doesn't seem to have the same prominence on the AP® exam as many others. If you are having trouble finishing the curriculum before the AP® exam in early May, this is the best section to shorten.

If you haven't read it recently, we recommend that you reread the section "Preparing Students for the AP® Exam" on page x of the *Annotated Teacher's Edition*. In addition, make sure to spend time reviewing the AP® Exam Tips in Appendix A with your students, along with the inference review questions and flash cards provided in the Teacher's Resource Materials.

We also strongly recommend that you have students do a final project to pull together the four major content areas of the course. Have students design a study, collect the data, analyze the data, and perform inference to draw a conclusion. A project description and sample rubric are included in the Teacher's Resource Materials.

### Relationship to the AP® Statistics Course Description

Chapter 12 covers the following content as described in the Topic Outline from the AP® Statistics Course Description. Topics marked with a superscript are covered in other chapters as indicated.

I. Exploring data: Describing patterns and departures from patterns

- D. Exploring bivariate data
  - 5. Transformations to achieve linearity: Logarithmic and power transformations

IV. Statistical inference: Estimating population parameters and testing hypotheses

- A. Estimation (point estimators and confidence intervals)
  - 8. Confidence interval for the slope of a least-squares regression line
- B. Tests of significance
  - 7. Test for the slope of a least-squares regression line

## The Big Ideas

One of the challenges in teaching the AP® Statistics course is keeping students focused on the big picture and not just the details of each section. We outline the big ideas for the chapter here.

### Chapter 12 Introduction

Here, we introduce students to the idea of performing inference in a regression setting. If there is a linear association between two variables and the data come from a random sample or a randomized experiment, then the least-squares regression line we calculate is actually just an *estimate* of the true least-squares regression line. We need inference to create an interval estimate for or to make decisions about the true value of the slope.

If you have time, we highly recommend that you do The Helicopter Experiment Activity or the Does Seat Location Matter? Alternate Activity with your students. Both are great ways to review regression and the principles of experimental design while simultaneously introducing the ideas of inference for regression.

## Section 12.1 Inference for Linear Regression

In many ways, Section 12.1 parallels Sections 10.1 and 10.2. In Section 10.1, we learned about the sampling distribution of $\hat{p}_1 - \hat{p}_2$ and then used our knowledge of the sampling distribution to create confidence intervals and conduct significance tests for $p_1 - p_2$. In Section 10.2, we learned about the sampling distribution of $\bar{x}_1 - \bar{x}_2$, so we could perform inference about $\mu_1 - \mu_2$. In this section, we learn about the sampling distribution of the sample slope $b$ so we can create confidence intervals and conduct significance tests for the population slope $\beta$. Although it is possible to do inference for other regression parameters, the AP® Statistics Topic Outline includes only inference for slope, so that's what we'll do as well.

As with any other sampling distribution, we are interested in the shape, center, and spread of the sampling distribution of $b$. When certain conditions are met, the sampling distribution of $b$ is approximately Normal and centered at the value of the true slope $\beta$, with a standard deviation of $\sigma_b = \dfrac{\sigma}{\sigma_x\sqrt{n}}$.

Unfortunately, the conditions are more complicated than in previous chapters. Students should know how to check if the relationship is *linear,* if the observations are *independent,* if the residuals are *Normally distributed,* if the residuals have an *equal standard deviation* at each value of $x$, and if the data were produced at *random.* Hopefully, the LINEaR acronym will help students to remember all five conditions!

Once we are satisfied that the conditions have been met, performing inference should be fairly routine. As always, we will expect students to follow the four-step process. Other than the conditions, the remaining steps are very similar to those given in previous chapters.

On previous AP® exams, when students have been asked questions about inference for regression, they have always been provided with computer output. It is very important that your students know how to obtain the equation of the least-squares regression line from the computer output, as well as the standard error of the slope, the test statistic, *P*-value, standard deviation of the residuals, and $r^2$. If you are very short on time, it would be wise to spend the time you have teaching students how to use computer output to perform inference for the slope.

## Section 12.2 Transforming to Achieve Linearity

In Chapter 3, we spent most of our time analyzing linear relationships between two quantitative variables. However, many pairs of quantitative variables have nonlinear relationships, such as the income per person and life expectancy in countries around the world.

There are two basic strategies for modeling associations that are nonlinear:

1. Find a nonlinear function to model the association.
2. Transform the data so a linear model is appropriate.

Both of these strategies are reasonable. However, we will spend our time learning the second strategy. Not only does the AP® Statistics Topic Outline specifically include transformations to achieve linearity, but also understanding how to transform data is a useful skill in other contexts. We have already had experience performing linear transformations of data in Chapter 2 and linear transformations of random variables in Chapter 6.

We begin by using powers and roots to transform data that follow a power model in the form $y = ax^p$. The goal is to make the form of the scatterplot as linear as possible. We know we have found the right transformation when the residual plot from a linear regression analysis has no leftover patterns.

If transformations with powers and roots don't accomplish the goal of straightening out the data, we can also transform one or both variables using logarithms, such as a base-10 logarithm (log) or the natural logarithm (ln). If a graph of log $y$ versus $x$ has a linear form, then the original data follow an exponential model in the form $y = ab^x$. If a graph of log $y$ versus log $x$ has a linear form, then the original data follow a power model in the form $y = ax^p$.

It's easy for us, as math teachers, to get carried away with all the interesting mathematical and statistical connections that can be made in this section. If you have finished the rest of the book and still have time before the AP® exam, explore these if you want. However, if you are running short on time, we recommend that you focus your time on logarithmic transformations. Also, don't worry about re-expressing the linear model in $\hat{y} = ab^x$ or $\hat{y} = ax^p$ form. Students will be expected only to predict values of $y$ for specific values of $x$ using the least-squares regression line and transforming the predicted value back into the original units.

Finally, Dan Teague has written a nice paper on this topic entitled "Re-Expression and Linear Regression." If you do an Internet search with his name and the title of the paper, you are sure to find it.

# Chapter 12: Resources

## TRM Additional *TPS* Teacher's Resource Materials

The following resources can be found by clicking on the links in the new Teacher's e-Book (TE-book), opening the Teacher's

Resource CD (TRCD), or logging onto the book's companion Web site www.whfreeman.com/tps5e (teacher log-in required).

- Helicopter template for The Helicopter Experiment Activity (page 738)
- Printable Chapter 12 Learning Objective Grid (page 739)
- Additional problems involving transformations to achieve linearity (page 784)
    - Baseball Salaries
    - How to Weigh a Gator
- FRAPPY! materials: Question, Model Solution, Scoring Rubric, Sample Responses with commentary (page 793)
- Final project, including rubric (page 794)
- Flash Cards for AP® Exam review (page 794)
- Inference Review (page 794)
- Quizzes: 3 versions for each section; versions A and B are roughly parallel
    - Section 12.1 (page 758)
    - Section 12.2 (page 784)
- Tests: 4 versions; versions A and B are roughly parallel, version D is cumulative (page 794)

## Free-Response Questions from Previous AP® Exams

Questions can be found on the AP® Central Web site: http://apcentral.collegeboard.com/apc/members/exam/exam_questions/8357.html.

Students should be able to answer all the free-response questions listed below with material learned in this chapter. This list will be updated after each AP® exam and will be posted on the book's Web site: www.whfreeman.com/tps5e.

Questions marked with an asterisk are from exams with released multiple-choice questions. You may want to save these questions until the end of the year so you can give your students a complete released exam for practice.

| Year | # | Content |
|---|---|---|
| 2011 | 5 | • Regression output<br>• Interpreting slope<br>• Meaning of $r^2$<br>• $t$ test for slope (conclusion only) |
| 2011B | 6 | • Interpreting slope<br>• Extrapolation<br>• Sampling distribution of $\bar{y}$ in a regression context<br>• Optimal design for estimating slope |
| 2010B | 6 | • Interpreting the slope of a least-squares regression line<br>• Interpreting a residual<br>• Using the residuals to estimate an effect<br>• Testing for a difference between two slopes using a confidence interval<br>• Using two different least-squares regression lines to estimate an effect |
| 2008 | 6 | • Two-sample $t$ test for a difference in means<br>• Stating the equation of a least-squares regression line from computer output<br>• Interpreting the slope of a least-squares regression line<br>• $t$ test for a slope<br>• Comparing inference methods |
| 2007 | 6* | • Interpreting the slope of a least-squares regression line<br>• Using a model with no constant term<br>• $t$ test for slope with $H_0: \beta = 1$<br>• Graphing a multiple regression model with an indicator variable<br>• Interpreting the coefficients of a multiple regression model |
| 2007B | 6 | • Two-sample $z$ test for a difference of proportions<br>• Confidence interval for slope<br>• Using a confidence interval to make a decision<br>• Using transformed data and a least-squares regression line to make predictions |
| 2006 | 2 | • Stating the equation of a least-squares regression line from computer output<br>• Interpreting the standard deviation of the residuals<br>• Interpreting the standard error of the slope |
| 2005B | 5 | • Stating the equation of a least-squares regression line from computer output<br>• Interpreting the slope and $y$ intercept<br>• Confidence interval for slope |
| 2004B | 1 | • Describing a scatterplot<br>• Interpreting $r^2$<br>• Interpreting a residual plot for a least-squares regression line using transformed data |
| 2001 | 6 | • Making graphs and comparing two distributions<br>• $t$ test for slope<br>• Classifying a new observation |
| 1997 | 6* | • Making predictions using least-squares regression lines, including transformed data<br>• Determining if models are appropriate<br>• Creating a better model |

Chapter

# 12

## Applets 

- The *Sampling Regression Lines* applet at www.rossmanchance.com/applets allows students to define a population of bivariate data, take random samples from the population, and record the value of the sample slope and intercept.
- The *Test for a Slope* applet at www.lock5stat.com/statkey allows students to enter bivariate data and simulate the randomization distribution of the slope under the assumption that there is no association between the variables.

## TI-Nspire Activities

These Activities are designed to illustrate important statistical concepts using the TI-Nspire. The Activities can be found at www.whfreeman.com/tps5e.

- *Finding the Best Model:* Using sliders on the Nspire, students will explore models for associations that are nonlinear.
- *Is the Slope Significant?:* Students use simulation to determine if a slope is significant.

## Chapter 12: Pacing Guide, Learning Objectives, and Suggested Assignments

This pacing guide is based on a schedule with 110, 50-minute sessions before the AP® exam. The 110 days include 4 days for midterms and 10 days for AP® exam review. If you have a different number of sessions before the AP® exam, you can modify the pacing guide to suit your needs. If you have additional time, consider incorporating quizzes, released AP® free-response questions, or additional activities. See the Resources section above for suggestions.

The suggested homework assignments list odd-numbered exercises, whenever possible, so students can check their answers against the back-of-the-book answers. If you would rather students not have access to the answers while doing homework, adding 1 to the exercise numbers usually will do the trick, because the homework exercises typically are paired. For example, Exercises 1 and 2 will generally cover the same topics, but in different contexts. You may also choose to include the Recycle and Review questions at the end of each section, which review topics from previous sections or chapters. These questions are denoted with a symbol.

| Day | Topics | **Learning Objectives:** Students will be able to. . . | Suggested Assignment |
|---|---|---|---|
| 1 | Activity: The Helicopter Experiment; 12.1 Sampling Distribution of $b$; Conditions for Regression Inference | • Check the conditions for performing inference about the slope $\beta$ of the population (true) regression line. | 1, 3 |
| 2 | 12.1 Estimating the Parameters; Constructing a Confidence Interval for the Slope | • Interpret the values of $a$, $b$, $s$, $SE_b$, and $r^2$ in context, and determine these values from computer output.<br>• Construct and interpret a confidence interval for the slope $\beta$ of the population (true) regression line. | 5, 7, 9, 11 |
| 3 | 12.1 Performing a Significance Test for the Slope | • Perform a significance test about the slope $\beta$ of the population (true) regression line. | 13, 15, 17 |
| 4 | 12.2 Transforming with Powers and Roots | • Use transformations involving powers and roots to find a power model that describes the relationship between two variables, and use the model to make predictions. | 19–24, 31, 33 |
| 5 | 12.2 Transforming with Logarithms; Putting it all Together: Which Transformation Should We Choose? | • Use transformations involving logarithms to find a power model or an exponential model that describes the relationship between two variables, and use the model to make predictions.<br>• Determine which of several transformations does a better job of producing a linear relationship. | 35, 37, 39, 41, 43, 45, 47–50 |
| 6 | Chapter 12 Review/FRAPPY! | | Chapter 12 Review Exercises |
| 7 | Chapter 12 Test | | Cumulative AP® Practice Test 4 |

**PD Chapter 12 Overview**

Watch the video overview of Chapter 12 (for teachers) by clicking on the link in the Teacher's e-Book (TE-book), opening the Teacher's Resource CD (TRCD), or logging onto the book's companion Web site www.whfreeman.com/tps5e (teacher log-in required).

# Chapter 12

# More about Regression

## case study

## Do Longer Drives Mean Lower Scores on the PGA Tour?

Recent advances in technology have led to golf balls that fly farther, clubs that generate more speed at impact, and swings that have been perfected through computer video analysis. Moreover, today's professional golfers are fitter than ever. The net result is many more players who routinely hit drives traveling 300 yards or more. Does greater distance off the tee translate to better (lower) scores?

We collected data on mean drive distance (in yards) and mean score per round from an SRS of 19 of the 197 players on the Professional Golfers Association (PGA) Tour in a recent year. Figure 12.1 is a scatterplot of the data with results from a least-squares regression analysis added. The graph shows that there is a moderately weak negative linear relationship between mean drive distance and mean score for the 19 players in the sample.[1]

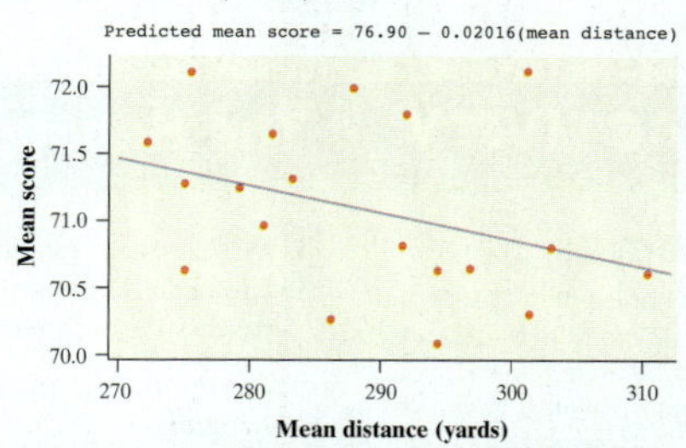


**FIGURE 12.1** Scatterplot and least-squares regression line of mean score versus mean drive distance for a random sample of 19 players on the PGA Tour.

The slope of the least-squares regression line for the sample data is about −0.02. This line predicts a 0.02 decrease in mean score per round for each 1-yard increase in mean driving distance. But a slope of −0.02 is very close to 0. Do these data give convincing evidence that the slope of the true regression line for *all* 197 PGA Tour golfers is negative? By the end of this chapter, you'll have developed the tools you need to answer this question.



### Teaching Tip

Make sure your students understand the two possible explanations for the negative association between mean distance and mean score that we observed in the *sample* of PGA golfers.

- There really is no association between mean distance and mean score in the population of all PGA golfers, and the observed negative association was due to sampling variability.
- There really is a negative association between mean distance and mean score in the population of PGA golfers.

We perform inference to see which explanation is more plausible.

# Introduction

When a scatterplot shows a linear relationship between a quantitative explanatory variable $x$ and a quantitative response variable $y$, we can use the least-squares line calculated from the data to predict $y$ for a given value of $x$. If the data are a random sample from a larger population, we need statistical inference to answer questions like these:

- Is there really a linear relationship between $x$ and $y$ in the population, or could the pattern we see in the scatterplot plausibly happen just by chance?
- In the population, how much will the predicted value of $y$ change for each increase of 1 unit in $x$? What's the margin of error for this estimate?

If the data come from a randomized experiment, the values of the explanatory variable correspond to the levels of some factor that is being manipulated by the researchers. For instance, researchers might want to investigate how temperature affects the life span of mosquitoes. They could set up several tanks at each of several different temperatures and then randomly assign hundreds of mosquitoes to each of the tanks. The response variable of interest is the average time (in days) from hatching to death. Suppose that a scatterplot of average life span versus temperature has a linear form. We need statistical inference to decide if it's plausible that there is no linear relationship between the variables, and that the pattern observed in the graph is due simply to the chance involved in the random assignment.

It is conventional to refer to a scatterplot of the points $(x, y)$ as a graph of $y$ versus $x$. So a scatterplot of life span versus temperature uses life span as the response variable and temperature as the explanatory variable.

In Section 12.1, we will show you how to estimate and test claims about the slope of the population (true) regression line that describes the relationship between two quantitative variables. The following Activity gives you a preview of inference for linear regression.

## ACTIVITY | The Helicopter Experiment

**MATERIALS:**
50 copies of the helicopter template from the *Teacher's Resource Materials*, scissors, tape measures, stopwatches

Is there a linear relationship between the height from which a paper helicopter is released and the time it takes to hit the ground? In this Activity, your class will perform an experiment to investigate this question.[2]

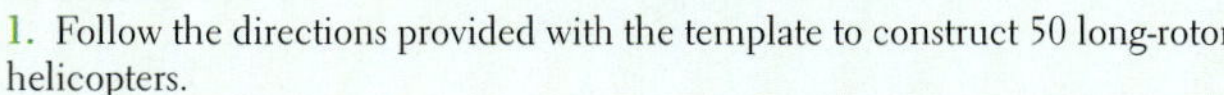

1. Follow the directions provided with the template to construct 50 long-rotor helicopters.

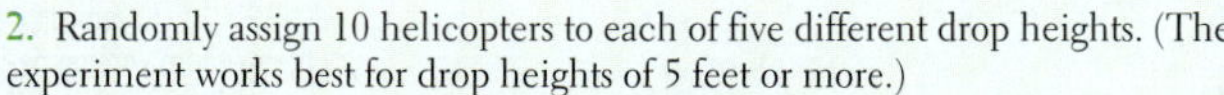

2. Randomly assign 10 helicopters to each of five different drop heights. (The experiment works best for drop heights of 5 feet or more.)

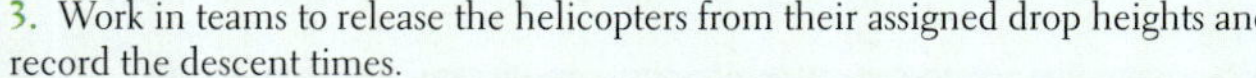

3. Work in teams to release the helicopters from their assigned drop heights and record the descent times.

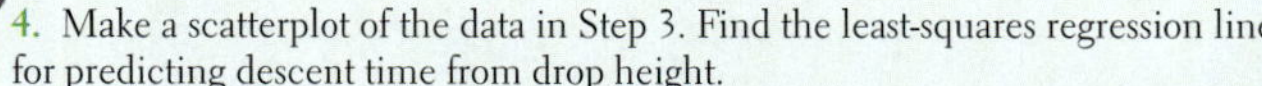

4. Make a scatterplot of the data in Step 3. Find the least-squares regression line for predicting descent time from drop height.
5. Interpret the slope of the regression line from Step 4 in context. What is your best guess for the increase in descent time for each additional foot of drop height?
6. Does it seem plausible that there is really no linear relationship between descent time and drop height and that the observed slope happened just by chance due to the random assignment? Discuss this as a class.

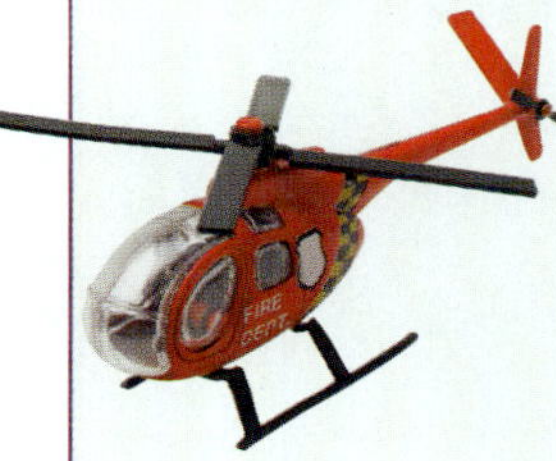

### Activity Overview

You can watch an overview of this Activity (for teachers) by clicking on the link in the TE-book, opening the TRCD, or logging onto the book's companion Web site www.whfreeman.com/tps5e (teacher log-in required).

### TRM Helicopter Activity Template

Here are some tips for the Activity on this page:

- Use this as an opportunity to review the concepts of experimental design. Ask students to identify the explanatory and response variables, the experimental units, and the treatments. Also, ask students what other variables they should be aware of and how they can address them. For example, wind certainly can have an effect on drop times. If possible, the experiment should be conducted indoors so that the wind doesn't become an additional source of variability. If you have to do the experiment outdoors, remind students how the random assignment should balance out the effects of the wind among the different treatments. Another variable to consider is the type of paper used. If the helicopters are made with different types of paper, we are introducing an additional source of variability. To avoid this, we control this variable by using the same type of paper for all helicopters.
- After discussing the questions in Step 6, ask your students how they could use simulation to investigate what values of the slope we could expect to happen by the chance due to the random assignment. Similar to the Distracted Driver Activity in Chapter 4 and the Does Polyester Decay? Activity in Chapter 10, we can write down the time for each drop on a note card, shuffle the cards, randomly assign 10 cards to each of the 5 drop heights, and recalculate the slope. After calculating many values of the slope in this way, check to see if the observed slope is different than what we would expect to happen by chance. See the following Alternate Activity for an example of this process. Also, The *Test for a Slope* applet at www.lock5stat.com/statkey allows students to enter bivariate data and simulate the randomization distribution of the slope under the assumption that there is no association between the variables.

### ALTERNATE ACTIVITY

#### *Does seat location matter?*

Many people believe that students learn better if they sit closer to the front of the classroom. Does sitting closer *cause* higher achievement, or do better students simply choose to sit in the front? To investigate, an AP® Statistics teacher randomly assigned students to seat locations in his classroom for a particular chapter and recorded the test score for each student at the end of the chapter. The explanatory variable in this experiment is which row the student was assigned (Row 1 is closest to the front and Row 7 is the farthest away). Here are the results, including a scatterplot and least-squares regression line:

Row 1: 76, 77, 94, 99
Row 2: 83, 85, 74, 79
Row 3: 90, 88, 68, 78
Row 4: 94, 72, 101, 70, 79
Row 5: 76, 65, 90, 67, 96
Row 6: 88, 79, 90, 83
Row 7: 79, 76, 77, 63

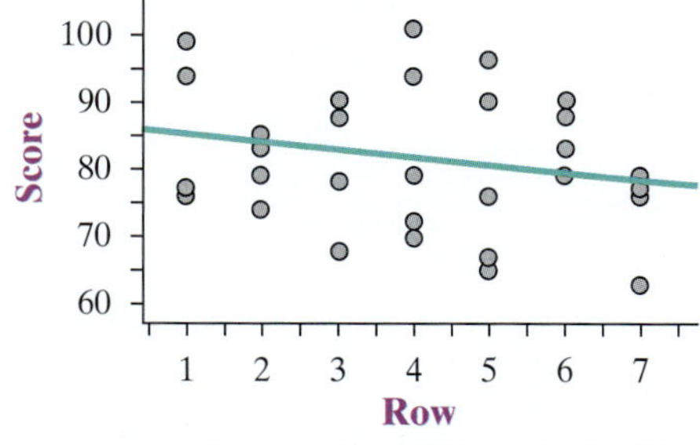

Sometimes a scatterplot reveals that the relationship between two quantitative variables has a strong curved form. One strategy is to transform one or both variables so that the graph shows a linear pattern. Then we can use least-squares regression to fit a linear model to the data. Section 12.2 examines methods of transforming data to achieve linearity.

# 12.1 Inference for Linear Regression

**WHAT YOU WILL LEARN** By the end of the section, you should be able to:

- Check the conditions for performing inference about the slope $\beta$ of the population (true) regression line.
- Interpret the values of $a$, $b$, $s$, $SE_b$, and $r^2$ in context, and determine these values from computer output.
- Construct and interpret a confidence interval for the slope $\beta$ of the population (true) regression line.
- Perform a significance test about the slope $\beta$ of the population (true) regression line.

In Chapter 3, we examined data on eruptions of the Old Faithful geyser. Figure 12.2 is a scatterplot of the duration and interval of time until the next eruption for all 222 recorded eruptions in a single month. The least-squares regression line for this population of data has been added to the graph. Its equation is

$$\text{predicted interval} = 33.97 + 10.36\,(\text{duration})$$

We call this the **population regression line** (or *true regression line*) because it uses all the observations that month.

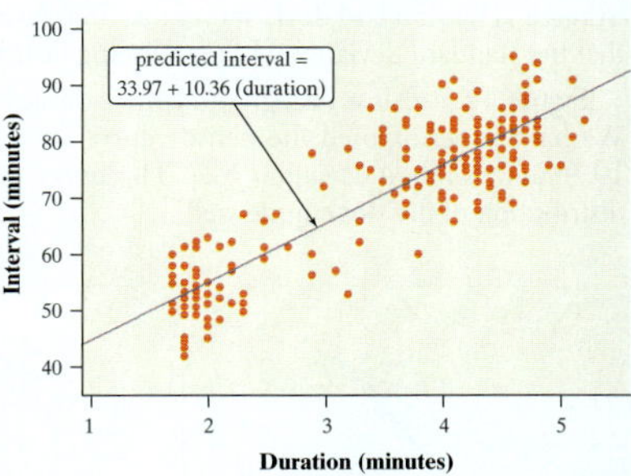


**FIGURE 12.2** Scatterplot of the duration and interval between eruptions of Old Faithful for all 222 eruptions in a single month. The population least-squares line is shown in blue.

The sample regression line is sometimes called the *estimated regression line*.

Suppose we take an SRS of 20 eruptions from the population and calculate the least-squares regression line $\hat{y} = a + bx$ for the sample data. How does the slope $b$ of the **sample regression line** relate to the slope of the population regression line? Figure 12.3 on the next page shows the results of taking three different SRSs of 20 Old Faithful eruptions in this month. Each graph displays the selected points and the least-squares regression line for that sample. Notice that the slopes of the sample regression lines (10.2, 7.7, and 9.5) vary quite a bit from the slope of the population regression line, 10.36. The pattern of variation in the slope $b$ is described by its sampling distribution.

**PD Section 12.1 Overview**

Watch the video overview of Section 12.1 (for teachers) by clicking on the link in the TE-book, opening the TRCD, or logging onto the book's companion Web site www.whfreeman.com/tps5e (teacher log-in required).

**TRM Learning Objectives Grid**

Remind students about the importance of the Learning Objectives. We include them at the beginning of each section so students know what they are expected to learn. At the end of the chapter, the Learning Objectives are included in a grid, along with references to examples and chapter review exercises. A printable version of this grid is included in the Teacher's Resource Materials.

**TRM Section 12.1 Alternate Examples**

You can find the Alternate Examples for this section in Microsoft Word format by clicking on the link in the TE-book, opening the TRCD, or logging onto the book's companion Web site www.whfreeman.com/tps5e (teacher log-in required).

**1.** Interpret the slope of the least-squares regression line in this context.

**2.** Explain why it was important to randomly assign the students to seats rather than letting each student choose his or her own seat.

**3.** Does the negative slope provide convincing evidence that sitting closer causes higher achievement, or is it plausible that the association is due to the chance variation in the random assignment? Let's investigate.

**(a)** Using 30 note cards, label each one with one of the student scores.

**(b)** Shuffle the 30 cards and divide them into seven piles (4 cards for Rows 1, 2, 3, 6, and 7 and 5 cards for Rows 5 and 6).

**(c)** Calculate the slope of the least-squares regression line using the re-randomized data ($x$ = row number and $y$ = test score).

**(d)** Repeat this process over and over, graphing the approximate randomization distribution of the slope on a dotplot.

**(e)** Was the observed slope of −1.12 unusual, or is it likely to get a slope this small due to the chance variation in random assignment? What conclusion should you make based on this study?

*Suggested Answers*:

**1.** For each additional row a student sits from the front, the predicted score decreases by 1.12 points.

**2.** If students are allowed to choose their own seats, the most conscientious students might choose to sit in the front, making it appear that sitting in the front causes higher scores. If we randomly assign the seats, the average ability level of students should be approximately the same in each row.

**3.** Here are the results of 100 trials of this simulation:

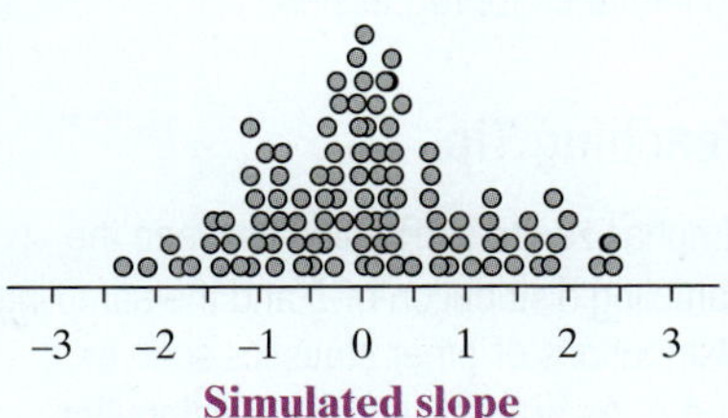


A slope of −1.12 is not very unusual, meaning that a slope this small could have resulted from the chance variation in random assignment. This study does not provide convincing evidence that sitting closer causes higher achievement.

## Teaching Tip

Make sure your students understand the graphs in Figure 12.3. In each graph, a different SRS of size 20 was selected and the sample least-squares regression line was calculated. Emphasize that different samples will produce different lines—and different slopes. This means that the sample slope is a random variable. The sampling distribution of the sample slope describes the possible values of $b$ and how likely they are to occur.

APPLET

To illustrate what is happening in Figures 12.3 and 12.4, consider using the *Sampling Regression Lines* applet at www.rossmanchance.com/applets. This applet allows students to define a population of bivariate data, take random samples from the population, and record the value of the sample slope and intercept to see their approximate sampling distributions. For a nice illustration, set the "pop slope" to 1, "pop intercept" to 0, "$x$ mean" to 4, "$x$ std" to 2, "sigma" to 0.5, and press the "set population" button. Then change the "sample size" to 20 and press the "draw samples" button over and over. The sample regression lines will be recorded on the scatterplot, and the sample slopes and intercepts will be graphed on the dotplots.

## Teaching Tip

As you introduce inference for regression, remind students that we use data from a random sample or a randomized experiment to *estimate* the values of the slope and $y$ intercept, just like we use data from random samples and randomized experiments to estimate means and proportions. Because the slope and $y$ intercept are statistics, we need to know more about their sampling distributions to do inference for regression.

## Teaching Tip

Emphasize the similarities between the sampling distribution of $b$ and the sampling distributions of other statistics such as $\bar{x}$ and $\hat{p}$. As with any sampling distribution, we are interested in the shape, center, and spread of the sampling distribution of $b$. And just as with the sampling distributions of $\bar{x}$ and $\hat{p}$, when certain conditions are met, we can model the sampling distribution of $b$ with a Normal distribution.

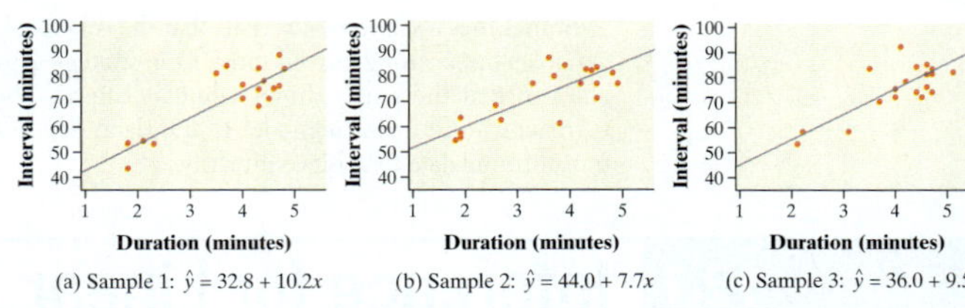


**FIGURE 12.3** Scatterplots and least-squares regression lines for three different SRSs of 20 Old Faithful eruptions.

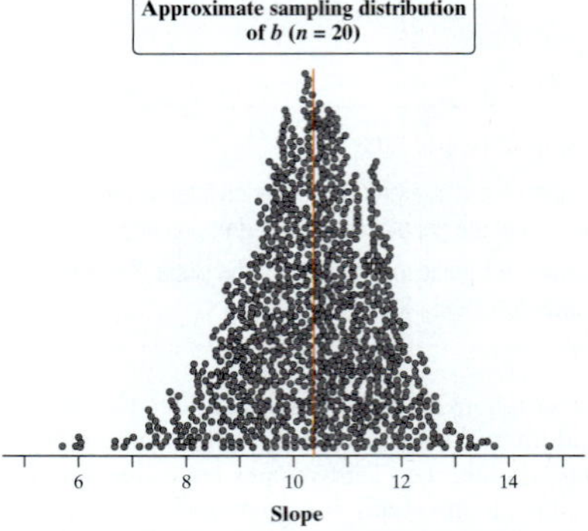


**FIGURE 12.4** Dotplot of the slope $b$ of the least-squares regression line in 1000 simulated SRSs by Fathom software.

## Sampling Distribution of $b$

Confidence intervals and significance tests about the slope of the population regression line are based on the sampling distribution of $b$, the slope of the sample regression line. We used Fathom software to simulate choosing 1000 SRSs of $n = 20$ from the Old Faithful data, each time calculating the equation $\hat{y} = a + bx$ of the least-squares regression line for the sample. Figure 12.4 displays the values of the slope $b$ for the 1000 sample regression lines. We have added a vertical line at 10.36 corresponding to the slope of the population regression line. Let's describe this approximate sampling distribution of $b$.

**Shape:** We can see that the distribution of $b$-values is roughly symmetric and unimodal. Figure 12.5(a) is a Normal probability plot of these sample regression line slopes. The strong linear pattern in the graph tells us that the approximate sampling distribution of $b$ is close to Normal.

**Center:** The mean of the 1000 $b$-values is 10.35. This value is quite close to the slope of the population (true) regression line, 10.36.

**Spread:** The standard deviation of the 1000 $b$-values is 1.29. Soon, we will see that the standard deviation of the sampling distribution of $b$ is actually 1.27.

Figure 12.5(b) is a histogram of the $b$-values from the 1000 simulated SRSs. We have superimposed the density curve for a Normal distribution with mean 10.36 and standard deviation 1.27. This curve models the approximate sampling distribution of the slope quite well.

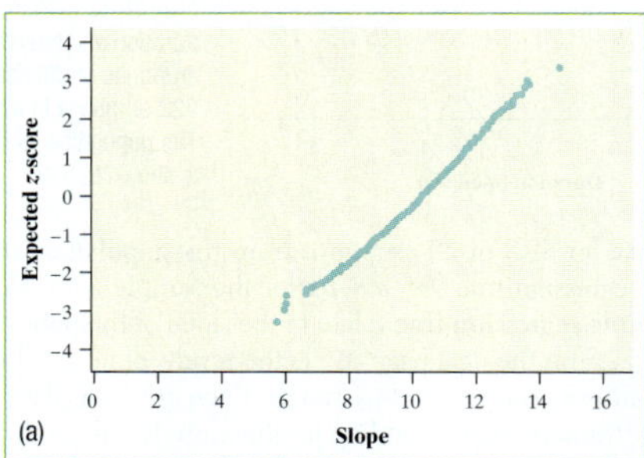


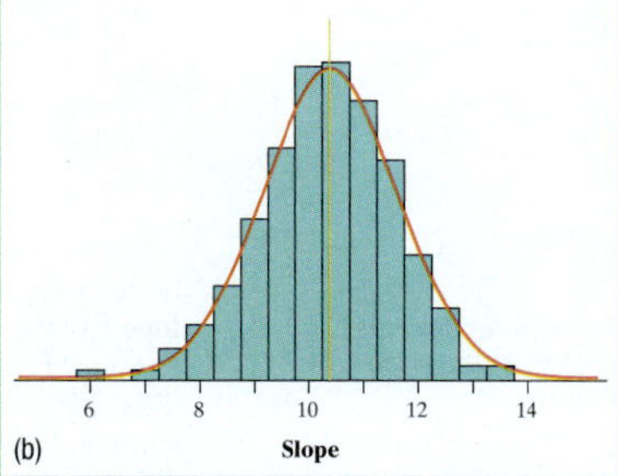


**FIGURE 12.5** (a) Normal probability plot and (b) histogram of the 1000 sample regression line slopes from Figure 12.4. The red density curve in Figure 12.5(b) is for a Normal distribution with mean 10.36 (marked by the yellow line) and standard deviation 1.27.

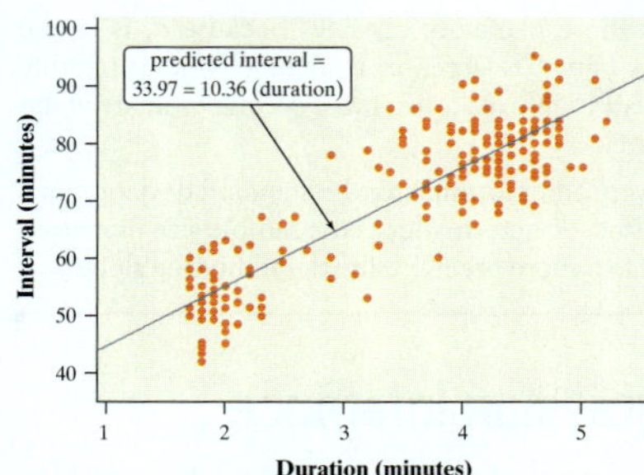


Let's do a quick recap. For all 222 eruptions of Old Faithful in a single month, the population regression line is: predicted interval = 33.97 + 10.36 (duration). We use the symbols $\alpha = 33.97$ and $\beta = 10.36$ to represent the $y$ intercept and slope parameters. The standard deviation of the residuals for this line is the parameter $\sigma = 6.131$.

Figure 12.5(b) shows the approximate sampling distribution of the slope $b$ of the sample regression line for samples of 20 eruptions. If we take *all* possible SRSs of size $n = 20$ from the population, we get the sampling distribution of $b$. Can you guess its shape, center, and spread?

**Shape:** Approximately Normal

**Center:** $\mu_b = \beta = 10.36$ ($b$ is an unbiased estimator of $\beta$)

**Spread:** $\sigma_b = \dfrac{\sigma}{\sigma_x\sqrt{n}} = \dfrac{6.131}{1.0815\sqrt{20}} = 1.27$ where $\sigma_x$ is the standard deviation of the 222 eruption durations.

We interpret $\sigma_b$ just like any other standard deviation: the slopes of the sample regression lines typically differ from the slope of the population regression line by about 1.27.

Here's a summary of the important facts about the sampling distribution of $b$.

Note that the symbols $\alpha$ and $\beta$ here refer to the intercept and slope, respectively, of the population regression line. They are in no way related to Type I and Type II error probabilities, which are sometimes designated by these same symbols.

**SAMPLING DISTRIBUTION OF A SLOPE**

Choose an SRS of $n$ observations $(x, y)$ from a population of size N with least-squares regression line

$$\text{predicted } y = \alpha + \beta x$$

Let $b$ be the slope of the sample regression line. Then:

- The **mean** of the sampling distribution of $b$ is $\mu_b = \beta$.
- The **standard deviation** of the sampling distribution of $b$ is

$$\sigma_b = \frac{\sigma}{\sigma_x\sqrt{n}}$$

as long as the *10% condition* is satisfied: $n \leq \frac{1}{10}N$.

- The sampling distribution of $b$ will be **approximately Normal** if the values of the response variable $y$ follow a Normal distribution for each value of the explanatory variable $x$ (the *Normal condition*).

We'll say more about the Normal condition in a moment.

**What's with that formula for $\sigma_b$?** There are three factors that affect the standard deviation of the sampling distribution of $b$:

- $\sigma$, the standard deviation of the residuals for the population regression line. Because $\sigma$ is in the numerator of the formula, when $\sigma$ is larger, so is $\sigma_b$. When the points are more spread out around the population (true) regression line, we should expect more variability in the slopes $b$ of sample regression lines from repeated random sampling or random assignment.

## Teaching Tip

Remind students that if a statistic is an unbiased estimator of a parameter, the mean of the sampling distribution of the statistic is equal to the value of the parameter. In other words, in repeated sampling, the sampling distribution of the slope will be centered at the value of the population (true) slope.

## Teaching Tip

The guidelines for determining the shape, center, and spread of the *sampling* distribution of $b$ also apply to the *randomization* distribution of $b$. This makes it possible to use the same inference procedures to analyze the results of random samples and randomized experiments.

## Teaching Tip

To illustrate how the standard deviation of $x$ affects the variability of the slope, hold a meter stick with one hand on each end. Then "wiggle" the meter stick by moving your hands up and down by an inch in opposite directions. Ask your students to comment on how much the "slope" of the meter stick changes. Next move your hands toward the middle of the meter stick and wiggle the meter stick by moving your hands up and down by an inch in opposite directions. The "slope" of the meter stick is now much more variable.

## Teaching Tip

Question 6 on the 2011 AP® exam (Form B) led students to discover that the sampling distribution of the slope will be less variable when the values of $x$ are more spread out. In an experiment, researchers can reduce the standard error of the slope (and gain more power) by spreading out the levels of the explanatory variable as much as possible so that the standard deviation of $x$ is larger.

## Teaching Tip

Make sure to spend time carefully explaining Figure 12.6 and how it illustrates why we need the Linear, Normal, and Equal SD conditions.

## Teaching Tip

After discussing the Think About It on this page, ask your students some follow-up questions to check for understanding. For example, ask students for the probability that it takes less than 60 minutes for the next eruption, given that the previous eruption lasted 4 minutes.

When the previous eruption lasts $x = 4$ minutes, the distribution of interval times is approximately Normal with mean $\mu_y = 34 + 10.4(4) = 75.6$ minutes and standard deviation 6 minutes. The $z$-score for an interval of 60 minutes is $z = \frac{60 - 75.6}{6} = -2.60$. Using Table A, the probability that it will take less than 60 minutes is 0.0047—not very likely!

- $\sigma_x$, the standard deviation of the explanatory variable. Because $\sigma_x$ is in the denominator of the formula, when $\sigma_x$ is larger, $\sigma_b$ is smaller. More variability in the values of the explanatory variable leads to a more precise estimate of the slope of the true regression line.
- $n$, the sample size. Just like every other formula for the standard deviation of a statistic, the variability of the statistic gets smaller as the sample size increases. A larger sample size will lead to a more precise estimate of the true slope.

## Conditions for Regression Inference

We can fit a least-squares line to any data relating two quantitative variables, but the results are useful only if the scatterplot shows a linear pattern. Inference about regression involves more detailed conditions. Figure 12.6 shows the regression model when the conditions are met in picture form. The regression model requires that for each possible value of the explanatory variable $x$:

1. The mean value of the response variable $\mu_y$ falls on the population (true) regression line $\mu_y = \alpha + \beta x$.
2. The values of the response variable $y$ follow a Normal distribution with common standard deviation $\sigma$.

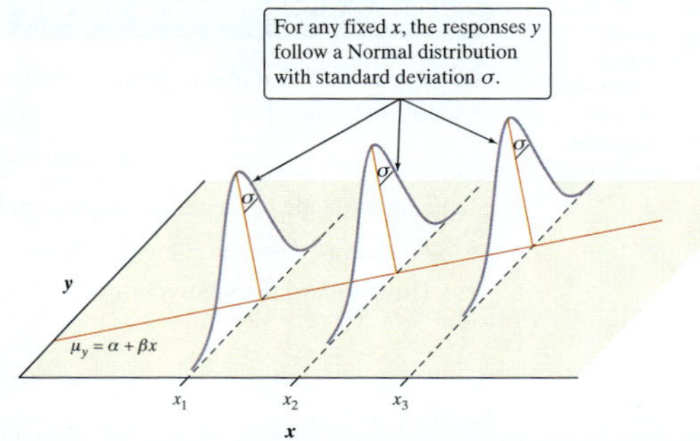


FIGURE 12.6 The regression model when the conditions for inference are met. The line is the population (true) regression line, which shows how the mean response $\mu_y$ changes as the explanatory variable $x$ changes. For any fixed value of $x$, the observed response $y$ varies according to a Normal distribution having mean $\mu_y$ and standard deviation $\sigma$.

**What does the regression model in Figure 12.6 tell us?** Consider the population of all eruptions of the Old Faithful geyser in a given year. For each eruption, let $x$ be the duration (in minutes) and $y$ be the interval of time (in minutes) until the next eruption. Suppose that the conditions for regression inference are met for this data set, that the population regression line is $\mu_y = 34 + 10.4x$, and that the spread around the line is given by $\sigma = 6$. Let's focus on the eruptions that lasted $x = 2$ minutes. For this "subpopulation":

- The average amount of time until the next eruption is $\mu_y = 34 + 10.4(2) = 54.8$ minutes.

- The amounts of time until the next eruption follow a Normal distribution with mean 54.8 minutes and standard deviation 6 minutes.
- For about 95% of these eruptions, the amount of time $y$ until the next eruption is between $54.8 - 2(6) = 42.8$ minutes and $54.8 + 2(6) = 66.8$ minutes. That is, if the previous eruption lasted 2 minutes, 95% of the time the next eruption will occur in 42.8 to 66.8 minutes.

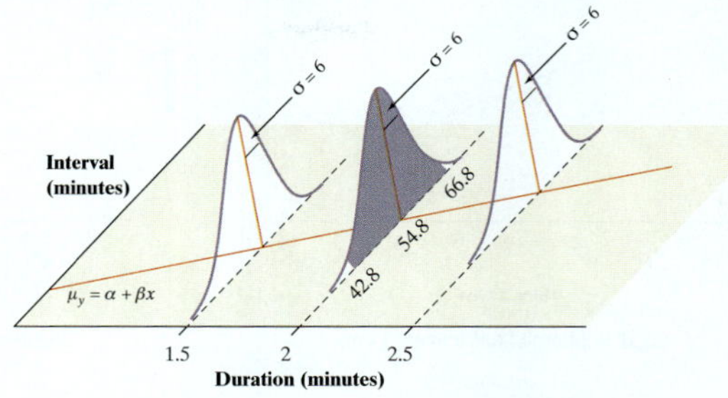


Here are the conditions for performing inference about the linear regression model.

## CONDITIONS FOR REGRESSION INFERENCE

Suppose we have $n$ observations on an explanatory variable $x$ and a response variable $y$. Our goal is to study or predict the behavior of $y$ for given values of $x$.

- **Linear:** The actual relationship between $x$ and $y$ is linear. For any fixed value of $x$, the mean response $\mu_y$ falls on the population (true) regression line $\mu_y = \alpha + \beta x$.
- **Independent:** Individual observations are independent of each other. When sampling without replacement, check the *10% condition.*
- **Normal:** For any fixed value of $x$, the response $y$ varies according to a Normal distribution.
- **Equal SD:** The standard deviation of $y$ (call it $\sigma$) is the same for all values of $x$.
- **Random:** The data come from a well-designed random sample or randomized experiment.

The acronym LINER should help you remember the conditions for inference about regression.

Although the conditions for regression inference are a bit complicated, it is not hard to check for major violations. Most of the conditions involve the population (true) regression line and the deviations of responses from this line. We usually can't observe the population line, but the sample regression line estimates it. The residuals from the sample regression line estimate the deviations from the population line. We can check several of the conditions for regression inference by looking at graphs of the residuals. Start by making a residual plot and a histogram or Normal probability plot of the residuals.

Here's a summary of how to check the conditions one by one.

### Teaching Tip

You could also consider using the acronym LINEaR to help students remember the conditions for regression inference, where the "a" stands for "and." Linear, Independent, Normal, Equal SD, and Random.

### Teaching Tip

Remind students that if the conditions for inference aren't met, then the stated confidence level and significance levels may not be correct. For example, if the conditions aren't met, "95%" confidence intervals for the slope may capture the true slope less than 95% of the time.

## Teaching Tip

Here are some additional comments about checking the conditions for regression inference:

- *Linear*: If there is a leftover pattern in the residual plot (such as a U shape), this condition is not met. One way to address this violation is to transform the data so they become roughly linear. This is the topic of Section 12.2.
- *Independent*: There are methods for doing inference for time-series data, but they are beyond the scope of this book.
- *Normal*: We check the Normal condition in the same manner as in Chapters 8–10. Make a graph of the residuals (not a residual plot) and look for strong skewness or outliers.
- *Equal SD*: Fan-shaped patterns in the residual plot indicate a violation of this condition.
- *Random*: Remind students about the scope of inference that we introduced in Chapter 4. If the data don't come from a well-designed random sample, we shouldn't make an inference about a larger population. Likewise, if the data do not come from a randomized experiment, we shouldn't make inferences about cause and effect.

- **Linear:** Examine the scatterplot to see if the overall pattern is roughly linear. Make sure there are no curved patterns in the residual plot. Check to see that the residuals center on the "residual = 0" line at each $x$-value in the residual plot.

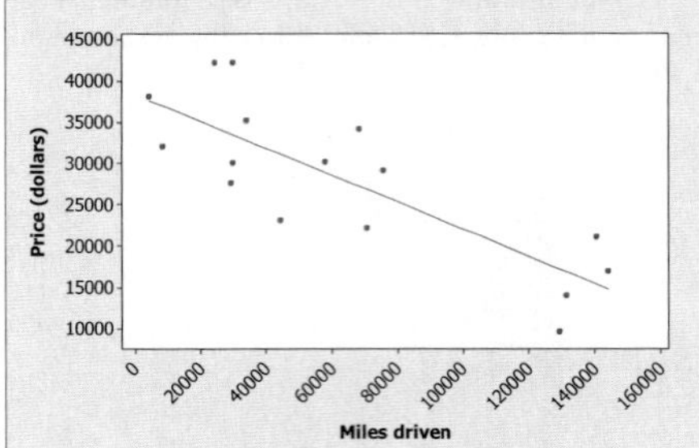


Good: Scatterplot has a linear form.

Bad: Residual plot shows a curved pattern.

- **Independent:** Look at how the data were produced. Random sampling and random assignment help ensure the independence of individual observations. If sampling is done without replacement, remember to check that the population is at least 10 times as large as the sample (*10% condition*). But there are other issues that can lead to a lack of independence. One example is measuring the same variable at intervals over time, yielding what is known as *time-series data*. Knowing that a young girl's height at age 6 is 48 inches would definitely give you additional information about her height at age 7. You should avoid doing inference about the regression model for time-series data.
- **Normal:** Make a stemplot, histogram, or Normal probability plot of the residuals and check for clear skewness or other major departures from Normality. Ideally, we would check the distribution of residuals for Normality at each possible value of $x$. Because we rarely have enough observations at each $x$-value, however, we make one graph of all the residuals to check for Normality.
- **Equal SD:** Look at the scatter of the residuals above and below the "residual = 0" line in the residual plot. The vertical spread of the residuals should be roughly the same from the smallest to the largest $x$-value.

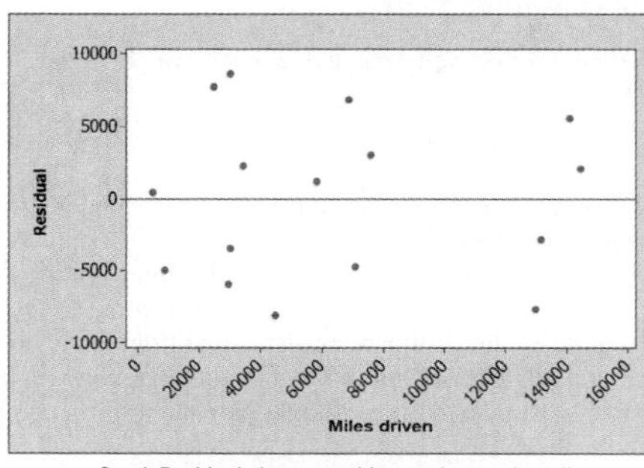


Good: Residuals have roughly equal spread at all $x$-values in the data set.

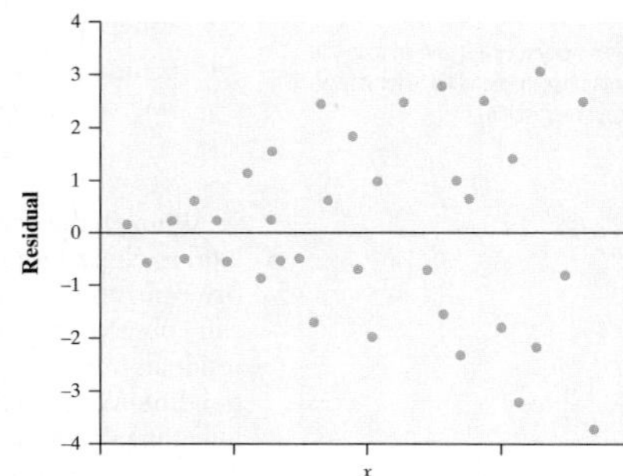


Bad: The response variable $y$ has greater spread for larger values of the explanatory variable $x$.

- **Random:** See if the data came from a well-designed random sample or randomized experiment. If not, we can't make inferences about a larger population or about cause and effect.

Let's look at an example that illustrates the process of checking conditions.

## EXAMPLE The Helicopter Experiment

### *Checking conditions*

Mrs. Barrett's class did a variation of the helicopter experiment on page 738. Students randomly assigned 14 helicopters to each of five drop heights: 152 centimeters (cm), 203 cm, 254 cm, 307 cm, and 442 cm. Teams of students released the 70 helicopters in a predetermined random order and measured the flight times in seconds. The class used Minitab to carry out a least-squares regression analysis for these data. A scatterplot and residual plot, plus a histogram and Normal probability plot of the residuals are shown below.

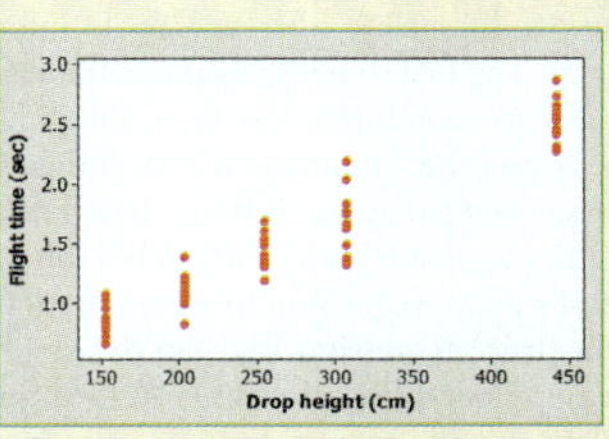


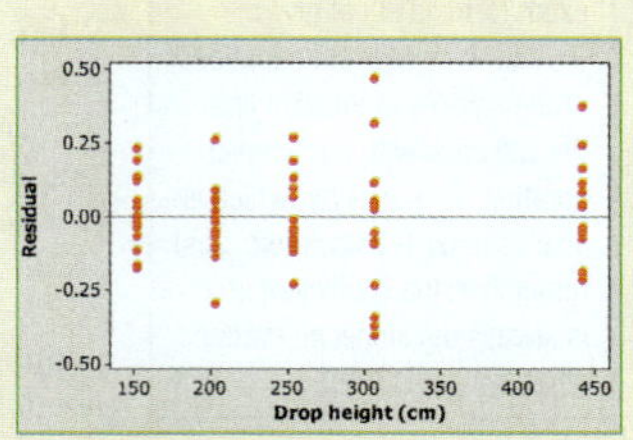


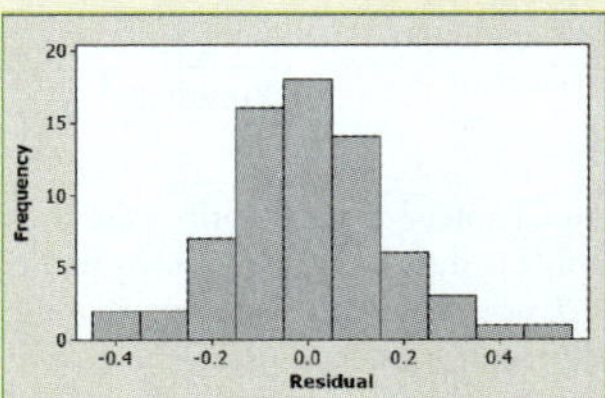


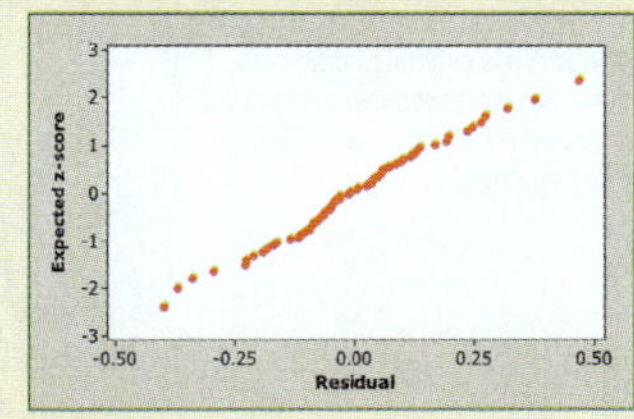


PROBLEM: Check whether the conditions for performing inference about the regression model are met.

SOLUTION: We'll use our LINER acronym!

- *Linear:* The scatterplot shows a clear linear form. The residual plot shows a random scatter about the horizontal line. For each drop height used in the experiment, the residuals are centered on the horizontal line at 0.
- *Independent:* Because the helicopters were released in a random order and no helicopter was used twice, knowing the result of one observation should not help us predict the value of another observation.
- *Normal:* The histogram of the residuals is single-peaked and somewhat bell-shaped. In addition, the Normal probability plot is very close to linear.

Note that we do not have to check the 10% condition here because there was no random sampling.

## Teaching Tip

The residual plot in The Helicopter Experiment Example has five vertical stacks of dots, which some students incorrectly identify as a pattern that violates the linear condition. Remind them that the *x*-values correspond to the five treatments, so we should not be surprised by this "pattern." In examples like this, if the linear condition is met, we should see random scatter about the residual = 0 line for each value of *x*, with roughly half the residuals above the line and half below the line.

## ALTERNATE EXAMPLE

### *Does seat location matter?*

We used Fathom to carry out a least-squares regression analysis for the Does Seat Location Matter? Alternate Activity from page 738. A scatterplot, residual plot, histogram, and Normal probability plot of the residuals are shown.

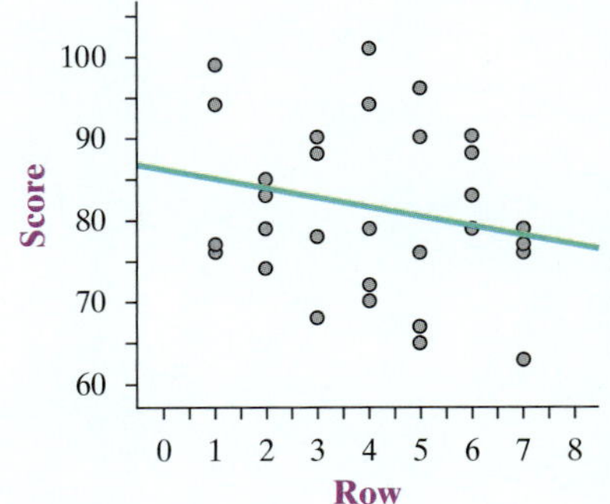


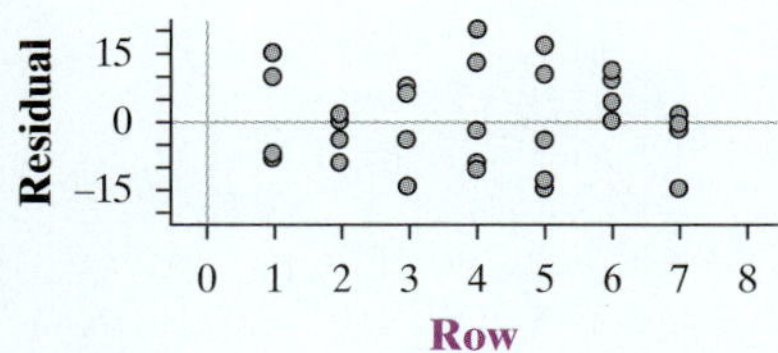


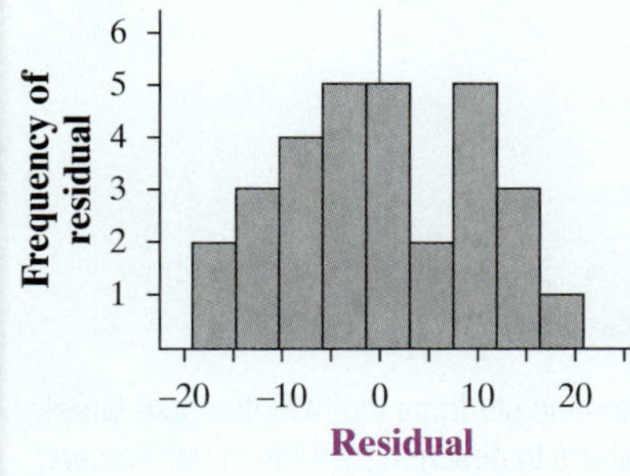


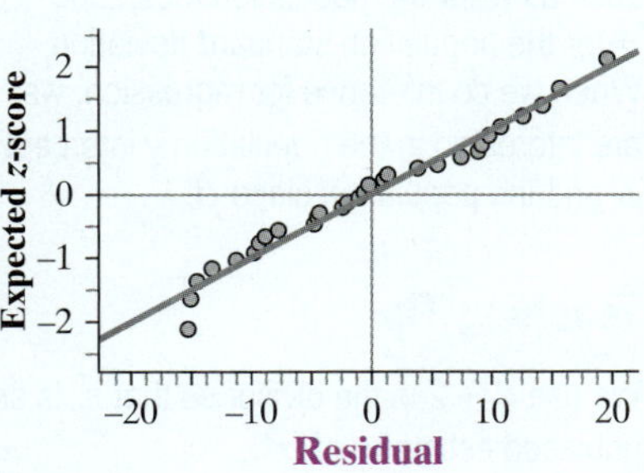


PROBLEM:
Check whether the conditions for performing inference about the regression model are met.

SOLUTION:

- *Linear*: The scatterplot shows a weak linear relationship and the residual plot does not show any obvious leftover patterns.
- *Independent*: Students were randomly assigned to seats and were monitored for cheating, so knowing the score for one student should give no additional information about another student's score.
- *Normal*: The histogram is roughly symmetric and unimodal, and the Normal probability plot is roughly linear.
- *Equal SD*: Although there is a different amount of variability in each row, the differences aren't very large and there is no systematic pattern, such as increasing variability as *x* increases.
- *Random*: The students were assigned to seats at random.

Because there are no serious violations of the conditions, we should be safe performing inference about the regression model in this setting.

- *Equal SD:* The residual plot shows a similar amount of scatter about the residual = 0 line for the 152, 203, 254, and 442 cm drop heights. Flight times (and the corresponding residuals) seem to vary a little more for the helicopters that were dropped from a height of 307 cm.
- *Random:* The helicopters were randomly assigned to the five possible drop heights.

Except for a slight concern about the equal-SD condition, we should be safe performing inference about the regression model in this setting.

**For Practice** *Try Exercise* **3**

You will always see some irregularity when you look for Normality and equal standard deviation in the residuals, especially when you have few observations. Don't overreact to minor issues in the graphs when checking these two conditions.

## Estimating the Parameters

When the conditions are met, we can do inference about the regression model $\mu_y = \alpha + \beta x$. The first step is to estimate the unknown parameters. If we calculate the sample regression line $\hat{y} = a + bx$, the slope $b$ is an unbiased estimator of the true slope $\beta$, and the $y$ intercept $a$ is an unbiased estimator of the true $y$ intercept $\alpha$. The remaining parameter is the standard deviation $\sigma$, which describes the variability of the response $y$ about the population (true) regression line.

The least-squares regression line computed from the sample data estimates the population (true) regression line. So the residuals estimate how much $y$ varies about the population line. Because $\sigma$ is the standard deviation of responses about the population (true) regression line, we estimate it by the standard deviation of the residuals

$$s = \sqrt{\frac{\sum \text{residuals}^2}{n-2}} = \sqrt{\frac{\sum (y_i - \hat{y}_i)^2}{n-2}}$$

Recall from Chapter 3 that $s$ describes the size of a "typical" prediction error.

It is possible to do inference about any of the three parameters in the regression model: $\alpha$, $\beta$, or $\sigma$. However, the slope $\beta$ of the population (true) regression line is usually the most important parameter in a regression problem. So we'll restrict our attention to inference about the slope.

When the conditions are met, the sampling distribution of the slope $b$ is approximately Normal with mean $\mu_b = \beta$ and standard deviation

$$\sigma_b = \frac{\sigma}{\sigma_x \sqrt{n}}$$

In practice, we don't know $\sigma$ for the true regression line. So we estimate it with the standard deviation of the residuals, $s$. We also don't know the standard deviation $\sigma_x$ for the population of $x$-values. For reasons beyond the scope of this text, we replace the denominator with $s_x\sqrt{n-1}$. So we estimate the spread of the sampling distribution of $b$ with the **standard error of the slope**

$$\text{SE}_b = \frac{s}{s_x\sqrt{n-1}}$$

**AP® EXAM TIP** The AP® exam formula sheet gives $\hat{y} = b_0 + b_1x$ for the equation of the sample regression line. We will stick with our simpler notation, $\hat{y} = a + bx$, which is also used by TI calculators. Just remember: the coefficient of $x$ is always the slope, no matter what symbol is used.

Because $s$ is estimated from data, it is sometimes called the *regression standard error* or the *root mean squared error*.

### Teaching Tip

Remind students that we often use Greek letters to describe population parameters, such as $\mu$ for the population mean and $\sigma$ for the population standard deviation. When we do inference for regression, we are interested in the population $y$ intercept $\alpha$ and the population slope $\beta$.

### Teaching Tip

We use $n - 2$ in the divisor so that $s^2$ is an unbiased estimator of $\sigma^2$.

### Teaching Tip

Ask your students to explain why $s$ is sometimes called the "root mean square error." The word "error" is used as a synonym for residual. We "square" the errors so that they don't add up to 0. We find the "mean" by dividing by $n - 2$, and we take the square "root" to return to the original units of the response variable.

### Teaching Tip

The AP® Statistics Topic Outline includes only confidence intervals and significance tests for the slope of a least-squares regression line. It does not include inference for the $y$ intercept, standard deviation of the residuals, or correlation.

### AP® Exam Formula Sheet

The formula for the standard error of the slope is included on the formula sheet that is given to students on both sections of the AP® exam. However, it is in the "Descriptive Statistics" section, not the "Inferential Statistics" section. See the margin note on page 748 for more information.

What happens if we transform the values of $b$ by standardizing? Because the sampling distribution of $b$ is approximately Normal, the statistic

$$z = \frac{b - \beta}{\sigma_b}$$

is modeled well by the standard Normal distribution. Replacing the standard deviation $\sigma_b$ of the sampling distribution with its standard error gives the statistic

$$t = \frac{b - \beta}{\text{SE}_b}$$

which has a $t$ distribution with $n - 2$ degrees of freedom. (The explanation of why df $= n - 2$ is beyond the scope of this book.)

Let's return to the Old Faithful eruption data. Figure 12.7(a) displays the simulated sampling distribution of the slope $b$ from 1000 SRSs of $n = 20$ eruptions. Figure 12.7(b) shows the result of standardizing the $b$-values from these 1000 samples. The superimposed curve is a $t$ distribution with df = 20 – 2 = 18.

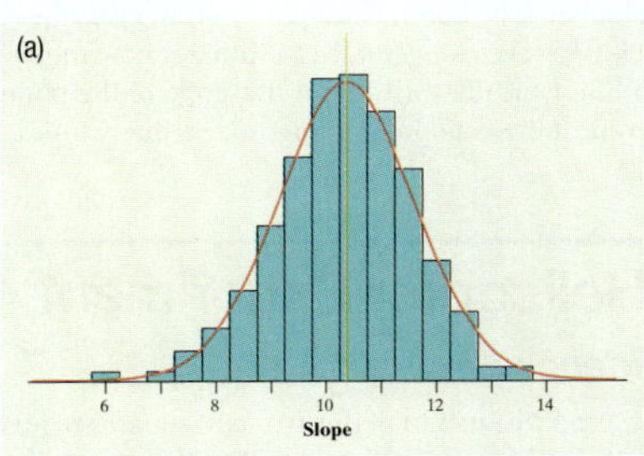


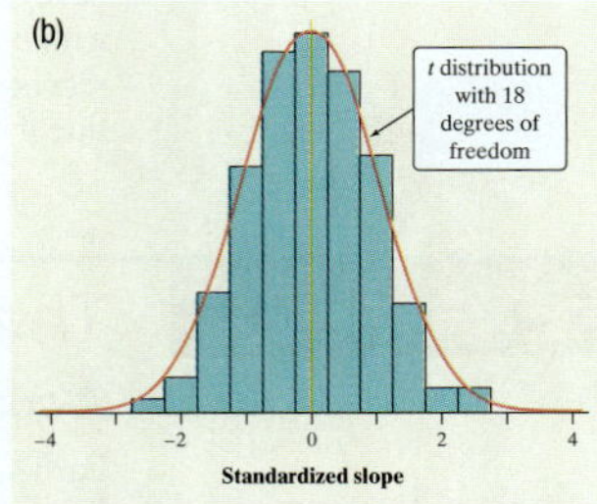


**FIGURE 12.7** (a) The approximate sampling distribution of the slope $b$ for samples of size $n = 20$ eruptions. This distribution has a roughly Normal shape with mean about 10.36 and standard deviation about 1.27. (b) The sampling distribution of the standardized slope values has approximately a $t$ distribution with df $= n - 2$.

## Constructing a Confidence Interval for the Slope

In a regression setting, we often want to estimate the slope $\beta$ of the population (true) regression line. The slope $b$ of the sample regression line is our point estimate for $\beta$. A confidence interval is more useful than the point estimate because it gives a set of plausible values for $\beta$.

The confidence interval for $\beta$ has the familiar form

statistic ± (critical value) · (standard deviation of statistic)

Because we use the statistic $b$ as our point estimate, the confidence interval is

$$b \pm t^* \text{SE}_b$$

### Teaching Tip

The reason we use the $t$ distribution instead of the Normal distribution is the same for slopes as it is for means. Even though $\bar{x}$ and $b$ have approximately Normal distributions when certain conditions are met, when we standardize these statistics, we use an estimate of the standard deviation. Because this estimate is a variable, not a constant, the shape of the distribution of the standardized statistic is no longer Normal.

### Teaching Tip

For more information about why we use $n - 2$, see the Teaching Tip on page 178.

We call this a **$t$ interval for the slope**. Here are the details.

**AP® EXAM TIP** The AP® exam formula sheet gives the formula for the standard error of the slope as

$$s_{b_1} = \frac{\sqrt{\frac{\sum(y_i - \hat{y}_i)^2}{n-2}}}{\sqrt{\sum(x_i - \bar{x})^2}}$$

The numerator is just a fancy way of writing the standard deviation of the residuals $s$. Can you show that the denominator of this formula is the same as ours?

### $t$ INTERVAL FOR THE SLOPE

When the conditions for regression inference are met, a $C\%$ confidence interval for the slope $\beta$ of the population (true) regression line is

$$b \pm t^* \text{SE}_b$$

In this formula, the standard error of the slope is

$$\text{SE}_b = \frac{s}{s_x\sqrt{n-1}}$$

and $t^*$ is the critical value for the $t$ distribution with df $= n - 2$ having $C\%$ of its area between $-t^*$ and $t^*$.

Although we give the formula for the standard error of $b$, you should rarely have to calculate it by hand. Computer output gives the standard error $\text{SE}_b$ along with $b$ itself. However we get it, $\text{SE}_b$ estimates how much the slope of the sample regression line typically varies from the slope of the population (true) regression line if we repeat the data production process many times.

## EXAMPLE The Helicopter Experiment

### *A confidence interval for $\beta$*

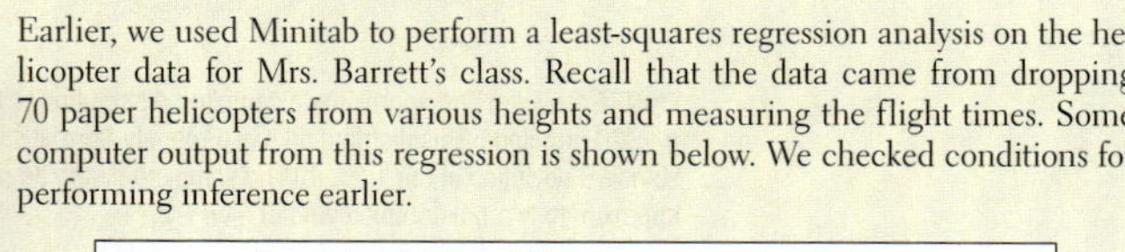

Earlier, we used Minitab to perform a least-squares regression analysis on the helicopter data for Mrs. Barrett's class. Recall that the data came from dropping 70 paper helicopters from various heights and measuring the flight times. Some computer output from this regression is shown below. We checked conditions for performing inference earlier.

Regression Analysis: Flight time versus Drop height

| Predictor | Coef | SE Coef | T | P |
|---|---|---|---|---|
| Constant | -0.03761 | 0.05838 | -0.64 | 0.522 |
| Drop height (cm) | 0.0057244 | 0.0002018 | 28.37 | 0.000 |

S = 0.168181 R-Sq = 92.2% R-Sq(adj) = 92.1%

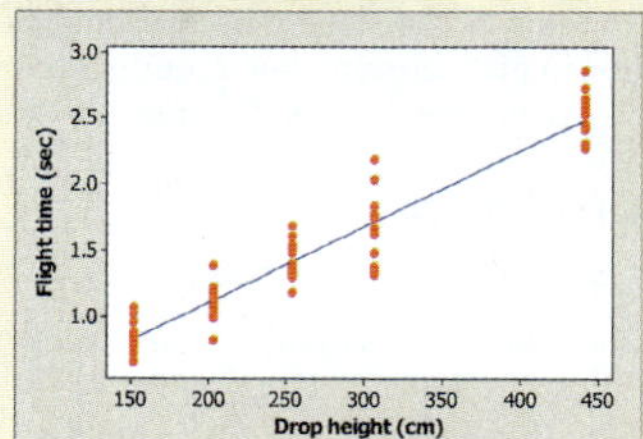


PROBLEM:

(a) Give the standard error of the slope, $\text{SE}_b$. Interpret this value in context.

(b) Find the critical value for a 95% confidence interval for the slope of the true regression line. Then calculate the confidence interval. Show your work.

(c) Interpret the interval from part (b) in context.

(d) Explain the meaning of "95% confident" in context.

## AP® Exam Formula Sheet

The general formula for a confidence interval is included on the formula sheet that is given to students on both parts of the AP® exam. However, the specific formula for a confidence interval for the slope is *not* included on the formula sheet.

## Teaching Tip

Make sure your students know how to interpret the standard error of the slope. Like other standard errors, it measures how far we expect our estimate to be from the truth in repeated random sampling or repeated random assignment.

## Teaching Tip

Point out that we are using the term "true regression line" instead of "population regression line" in this example because we are analyzing the results of an experiment. See the margin note on the next page for more details.

## ALTERNATE EXAMPLE

### *Does seat location matter?*

In a previous alternate example, we looked at the results of an experiment designed to see if sitting closer to the front of a classroom causes higher achievement. We checked the conditions for inference and there were no major violations. Here is a scatterplot of the data and some output from a regression analysis:

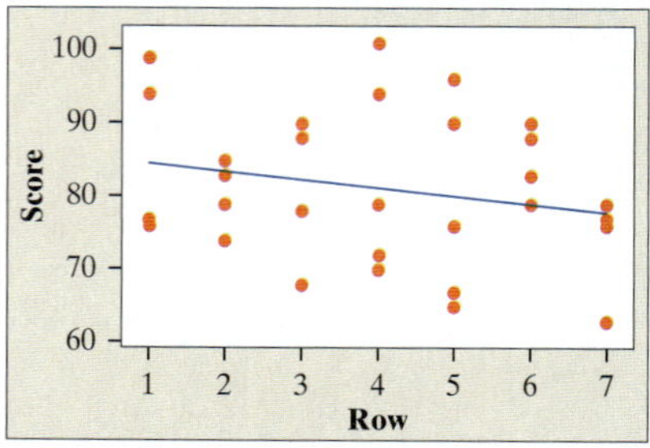


**Regression Analysis: Score versus Row**

| Predictor | Coef | SE Coef | T | P |
|---|---|---|---|---|
| Constant | 85.706 | 4.239 | 20.22 | 0.000 |
| Row | -1.1171 | 0.9472 | -1.18 | 0.248 |

S = 10.0673 R-Sq = 4.7% R-Sq(adj) = 1.3%

PROBLEM:

**(a)** Identify the standard error of the slope $\text{SE}_b$ from the computer output. Interpret this value in context.

**(b)** Calculate the 95% confidence interval for the true slope. Show your work.

**(c)** Interpret the interval from part (b) in context.

**(d)** Based on your interval, is there convincing evidence that seat location affects scores?

SOLUTION:

**(a)** $\text{SE}_b = 0.9472$. If we repeated the random assignment many times, the slope of the estimated regression line would typically vary by about 0.9472 from the slope of the true regression line for predicting test score from row number.

**(b)** Because $n = 30$, df $= 30 - 2 = 28$, and $t^* = 2.048$. The 95% confidence interval is $-1.1171 \pm 2.048(0.9472) = -1.1171 \pm 1.9399 = (-3.0570, 0.8228)$.

**(c)** We are 95% confident that the interval from $-3.0570$ to $0.8228$ captures the slope of the true regression line relating a student's test score $y$ and the student's row number $x$.

**(d)** Because the interval of plausible slopes includes 0, we do not have convincing evidence that seat location affects test scores.

When we compute the least-squares regression line based on a random sample of data, we can think about doing inference for the population regression line. When our least-squares regression line is based on data from a randomized experiment, as in this example, the resulting inference is about the *true* regression line relating the explanatory and response variables. We'll follow this convention from now on.

SOLUTION:

(a) We got the value of the standard error of the slope, 0.0002018, from the "SE Coef" column in the computer output. If we repeated the random assignment many times, the slope of the sample regression line would typically vary by about 0.0002 from the slope of the true regression line for predicting flight time from drop height.

(b) Because the conditions are met, we can calculate a *t* interval for the slope $\beta$ based on a *t* distribution with df $= n - 2 = 70 - 2 = 68$. Using the more conservative df $= 60$ from Table B gives $t^* = 2.000$. The 95% confidence interval is

$$b \pm t^*\text{SE}_b = 0.0057244 \pm 2.000(0.0002018) = 0.0057244 \pm 0.0004036$$
$$= (0.0053208, 0.0061280)$$

*Using technology:* From `invT(.025,68)`, we get $t^* = 1.995$. The resulting 95% confidence interval is

$$0.0057244 \pm 1.995(0.0002018) = 0.0057244 \pm 0.0004026$$
$$= (0.0053218, 0.0061270)$$

This interval is slightly narrower due to the more precise $t^*$ critical value.

(c) We are 95% confident that the interval from 0.0053218 to 0.0061270 seconds per cm captures the slope of the true regression line relating the flight time *y* and drop height *x* of paper helicopters.

(d) If we repeat the experiment many, many times, and use the method in part (b) to construct a confidence interval each time, about 95% of the resulting intervals will capture the slope of the true regression line relating flight time *y* and drop height *x* of paper helicopters.

**For Practice** *Try Exercise* 7

The values of *t* given in the computer regression output are *not* the critical values for a confidence interval. They come from carrying out a significance test about the *y* intercept or slope of the population (true) regression line. We'll discuss tests in more detail shortly.

You can find a confidence interval for the *y* intercept $\alpha$ of the population (true) regression line in the same way, using *a* and $\text{SE}_a$ from the "Constant" row of the Minitab output. However, we are usually interested only in the point estimate for $\alpha$ that's provided in the computer output.

Here is an example using a familiar context that illustrates the four-step process for calculating and interpreting a confidence interval for the slope.

## EXAMPLE How Much Is That Truck Worth?

STEP 4

### *Confidence interval for a slope*

Everyone knows that cars and trucks lose value the more they are driven. Can we predict the price of a used Ford F-150 SuperCrew 4 × 4 if we know how many miles it has on the odometer? A random sample of 16 used Ford F-150 SuperCrew 4 × 4s was selected from among those listed for sale on autotrader.com. The number of miles driven and price (in dollars) were recorded for each of the trucks.[3]

## Teaching Tip

In The Helicopter Experiment example, the *y* intercept *should* be 0, because a helicopter that is dropped 0 cm should have a drop time of 0 seconds. Also, even though we haven't learned about performing significance tests in this context, you can mention that the large *P*-value for the intercept suggests that the null hypothesis of $H_0: \alpha = 0$ should not be rejected. *Note*: Inference for the *y* intercept of a least-squares regression line is not on the AP® Statistics Topic Outline. However, asking your students to do inference for the *y* intercept is a good way to prepare them for the types of questions they might be asked on Question 6 on the free-response section of the exam, called the investigative task. The investigative task often asks students to use concepts they have learned to answer questions slightly beyond the AP® Statistics curriculum. See Question 6 on the 2007 AP® exam for an example where statistical software was used to force the *y* intercept to equal 0.

## AP® Exam Common Error

Make sure to emphasize the caution on this page. On the AP® exam, students often use the *t* test statistic from the computer output as their critical value. The value of *t* in the computer output measures how far the sample slope is from 0 in terms of standard errors. However, the $t^*$ critical value measures how many standard errors we need to go from the mean to have C% of the area under a *t* curve with specified degrees of freedom. To help avoid this error, we use an asterisk to denote the critical value $t^*$.

Here are the data:

| Miles driven: | 70,583 | 129,484 | 29,932 | 29,953 | 24,495 | 75,678 | 8359 | 4447 |
|---|---|---|---|---|---|---|---|---|
| Price (in dollars): | 21,994 | 9500 | 29,875 | 41,995 | 41,995 | 28,986 | 31,891 | 37,991 |

| Miles driven: | 34,077 | 58,023 | 44,447 | 68,474 | 144,162 | 140,776 | 29,397 | 131,385 |
|---|---|---|---|---|---|---|---|---|
| Price (in dollars): | 34,995 | 29,988 | 22,896 | 33,961 | 16,883 | 20,897 | 27,495 | 13,997 |

Minitab output from a least-squares regression analysis for these data is shown below.

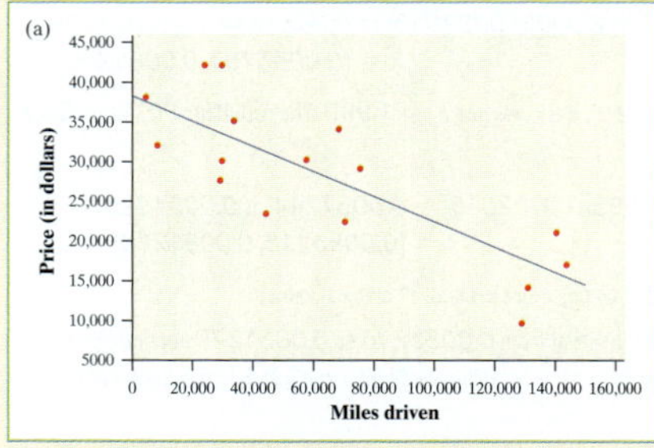


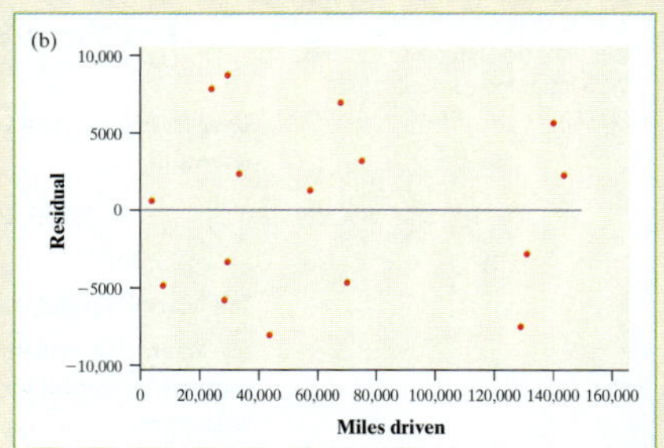


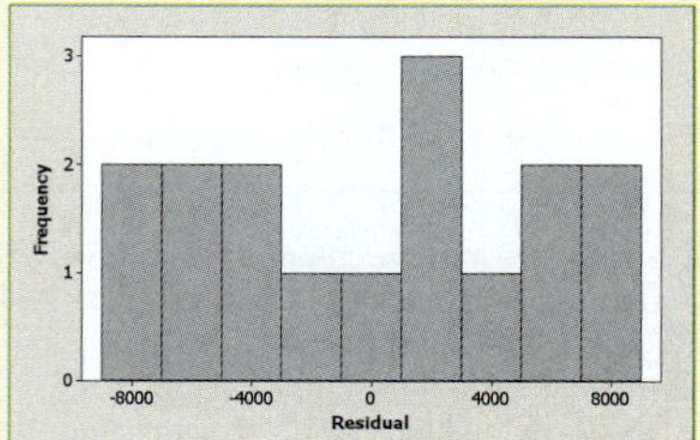


```
Regression Analysis: Price (dollars) versus Miles driven
Predictor      Coef      SE Coef     T       P
Constant       38257     2446        15.64   0.000
Miles driven   -0.16292  0.03096     -5.26   0.000
S = 5740.13    R-Sq = 66.4%    R-Sq(adj) = 64.0%
```

PROBLEM: Construct and interpret a 90% confidence interval for the slope of the population regression line.

SOLUTION: We will follow the familiar four-step process.

STATE: We want to estimate the slope $\beta$ of the population regression line relating miles driven to price with 90% confidence.

PLAN: If the conditions are met, we will use a *t* interval for the slope of a regression line.

- *Linear:* The scatterplot shows a clear linear pattern. Also, the residual plot shows a random scatter of points about the residual = 0 line.
- *Independent:* Because we sampled without replacement to get the data, there have to be at least 10(16) = 160 used Ford F-150 SuperCrew 4 × 4s listed for sale on autotrader.com. This seems reasonable to believe.
- *Normal:* The histogram of the residuals is roughly symmetric and single-peaked, so there are no obvious departures from Normality.
- *Equal SD:* The scatter of points around the residual = 0 line appears to be about the same at all *x*-values.
- *Random:* We randomly selected the 16 pickup trucks in the sample.

## ALTERNATE EXAMPLE

### *More Mentos, more mess?*

When Mentos are dropped into a newly opened bottle of Diet Coke, carbon dioxide is released from the Diet Coke very rapidly, causing the Diet Coke to be expelled from the bottle. Will more Diet Coke be expelled when a larger number of Mentos are dropped in the bottle? How much more? Two statistics students, Brittany and Allie, decided to find out. Using 16-ounce (2-cup) bottles of Diet Coke, they dropped either 2, 3, 4, or 5 Mentos into a randomly selected bottle, waited for the fizzing to die down, and measured the number of cups remaining in the bottle. Then they subtracted this measurement from the original amount in the bottle to calculate the amount of Diet Coke expelled (in cups). Here are the data:

| Mentos | Amount expelled (cups) | Mentos | Amount expelled (cups) |
|---|---|---|---|
| 2 | 1.1250 | 4 | 1.2500 |
| 2 | 1.2500 | 4 | 1.3125 |
| 2 | 1.0625 | 4 | 1.2500 |
| 2 | 1.2500 | 4 | 1.3750 |
| 2 | 1.1250 | 4 | 1.3125 |
| 2 | 1.0625 | 4 | 1.2500 |
| 3 | 1.1875 | 5 | 1.2500 |
| 3 | 1.1250 | 5 | 1.4375 |
| 3 | 1.2500 | 5 | 1.3125 |
| 3 | 1.1875 | 5 | 1.3125 |
| 3 | 1.3125 | 5 | 1.3750 |
| 3 | 1.1875 | 5 | 1.4375 |

Output from a regression analysis is shown below.

```
Predictor   Coef      SE Coef    T      P
Constant    1.00208   0.04511    22.21  0.000
Mentos      0.07083   0.01228    5.77   0.000
S = 0.0672442  R-Sq = 60.2%  R-Sq(adj)
```

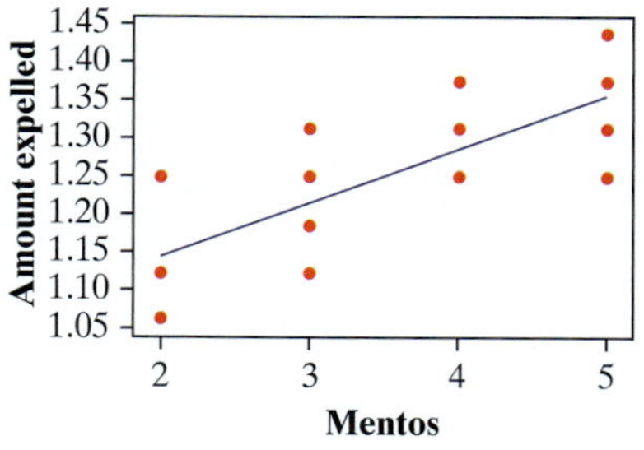


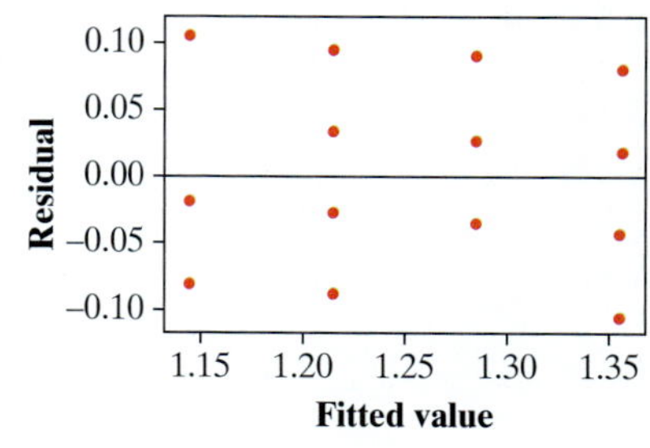


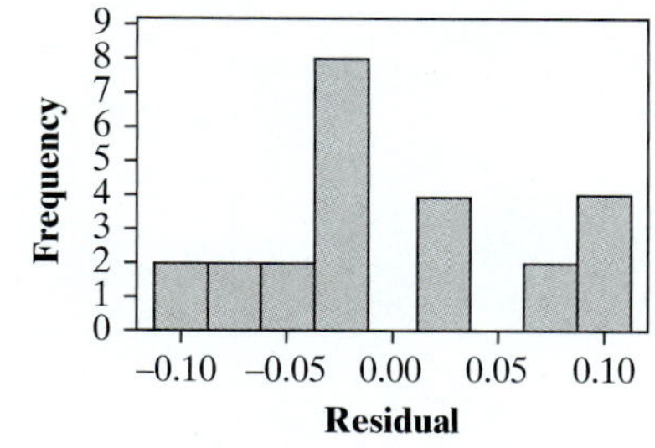


PROBLEM:
Construct and interpret a 99% confidence interval for the slope of the true regression line.

SOLUTION:
STATE: We want to estimate the slope $\beta$ of the true regression line relating amount expelled (in cups) to number of Mentos at the 99% confidence level.

DO: We use the $t$ distribution with $16 - 2 = 14$ degrees of freedom to find the critical value. For a 90% confidence level, the critical value is $t^* = 1.761$. So the 90% confidence interval for $\beta$ is

$$b \pm t^* SE_b = -0.16292 \pm 1.761(0.03096) = -0.16292 \pm 0.05452$$
$$= (-0.21744, -0.10840)$$

*Using technology:* Refer to the Technology Corner that follows the example. The calculator's LinRegTInt gives $(-0.2173, -0.1084)$ using df = 14.

CONCLUDE: We are 90% confident that the interval from $-0.2173$ to $-0.1084$ captures the slope of the population regression line relating price to miles driven for used Ford F-150 SuperCrew 4 × 4s listed for sale on autotrader.com.

**For Practice** *Try Exercise* **9**

The predicted change in price of a used Ford F-150 is quite small for a 1-mile increase in miles driven. What if miles driven increased by 1000 miles? We can just multiply both endpoints of the confidence interval in the example by 1000 to get a 90% confidence interval for the corresponding predicted change in average price. The resulting interval is $(-217.3, -108.4)$. That is, the population regression line predicts a decrease in price of between \$108.40 and \$217.30 for every additional 1000 miles driven.

So far, we have used computer regression output when performing inference about the slope of a population (true) regression line. The TI-83/84, TI-89, and TI-Nspire can do the calculations for inference when the sample data are provided.

## 28. TECHNOLOGY CORNER — CONFIDENCE INTERVAL FOR SLOPE ON THE CALCULATOR

TI-Nspire instructions in Appendix B; HP Prime instructions on the book's Web site.

Let's use the data from the previous example to construct a confidence interval for the slope of a population (true) regression line on the TI-83/84 and TI-89. Enter the $x$-values (miles driven) into L1/list1 and the $y$-values (price) into L2/list2.

**TI-83/84 with recent OS**

- Press STAT, then choose TESTS and LinRegTInt. . . .
- In the LinRegTInt screen, adjust the inputs as shown. Then highlight "Calculate" and press ENTER.

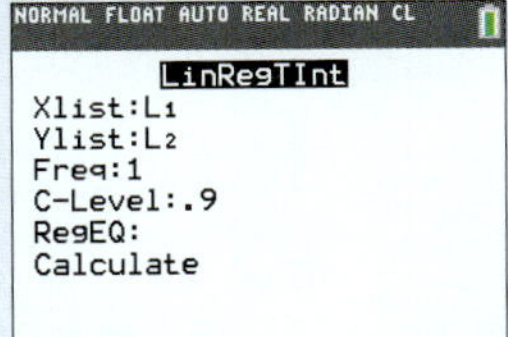


**TI-89**

- Press 2nd F2 ([F7]) and choose LinRegTInt. . . .
- In the LinRegTInt screen, adjust the inputs as shown and press ENTER.

### AP® Exam Common Error

When students interpret the slope, they often forget to include the word "predicted" (or the equivalent) in their response. In the previous example, saying that "the price decreases by 0.16292 dollars for each additional mile driven" implies that price can be predicted exactly. Including the phrase "predicted price" or "expected" acknowledges that our predictions are not guaranteed to be correct.

### TECHNOLOGY TIP

Remind students that they can watch a video presentation of this technology corner (for the TI-83/84, TI-89, or TI-Nspire) at www.whfreeman.com/tps5e.

### TECHNOLOGY TIP

If your students have TI-83s or older TI-84s, they may not have an option for calculating a confidence interval for a slope. However, they can still use their calculators to find the slope $b$ and the standard error of the slope $SE_b$. If they use the LinRegTTest option (see page 756), $b$ = slope and $t = \dfrac{b - 0}{SE_b}$, so $SE_b = \dfrac{b}{t}$.

PLAN: If conditions are met, we will use a $t$ interval for the slope to estimate $\beta$.

- *Linear*: The scatterplot shows a linear pattern, and there is no obvious leftover curvature in the residual plot.
- *Independent*: Knowing the amount expelled for one bottle of Diet Coke shouldn't provide additional information about the amount expelled for any other bottles.
- *Normal*: The histogram of residuals does not show any skewness or outliers.
- *Equal SD*: The variability of the residuals for each number of Mentos is roughly the same, although a little less variability seems to occur when there were 4 Mentos.
- *Random*: The number of Mentos was determined at random for each bottle of Diet Coke.

DO: Using df $= 24 - 2 = 22$, the $t$ critical value is $t^* = 2.819$. Thus, the 99% confidence interval is $0.07083 \pm 2.819(0.01228) = 0.07083 \pm 0.03462 = (0.03621, 0.10545)$.

CONCLUDE: We are 99% confident that the interval from 0.03621 to 0.10545 captures the slope of the true regression line relating the amount of Diet Coke expelled (in cups) to the number of Mentos. Because all of the plausible values for the slope are positive, it appears that more Mentos means more mess!

- The linear regression $t$ interval results are shown below. The TI-84 Plus C fits the results on one screen. The TI-83/84 and TI-89 require you to arrow down to see the rest of the output.

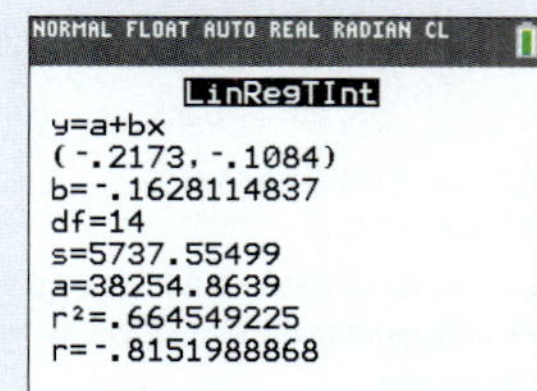

Note that $s$ is the standard deviation of the residuals, *not* the standard error of the slope.

**AP® EXAM TIP** The formula for the $t$ interval for the slope of a population (true) regression line often leads to calculation errors by students. As a result, we recommend using the calculator's `LinRegTInt` feature to compute the confidence interval on the AP® Exam. Be sure to name the procedure ($t$ interval for slope) and to give the interval (−0.217, −0.108) and df (14) as part of the "Do" step.

## CHECK YOUR UNDERSTANDING

Does fidgeting keep you slim? Some people don't gain weight even when they overeat. Perhaps fidgeting and other "nonexercise activity" (NEA) explain why—some people may spontaneously increase nonexercise activity when fed more. Researchers deliberately overfed a random sample of 16 healthy young adults for 8 weeks. They measured fat gain (in kilograms) as the response variable and change in energy use (in calories) from activity other than deliberate exercise—fidgeting, daily living, and the like—as the explanatory variable. Here are the data:[4]

| NEA change (cal): | −94 | −57 | −29 | 135 | 143 | 151 | 245 | 355 |
|---|---|---|---|---|---|---|---|---|
| Fat gain (kg): | 4.2 | 3.0 | 3.7 | 2.7 | 3.2 | 3.6 | 2.4 | 1.3 |
| NEA change (cal): | 392 | 473 | 486 | 535 | 571 | 580 | 620 | 690 |
| Fat gain (kg): | 3.8 | 1.7 | 1.6 | 2.2 | 1.0 | 0.4 | 2.3 | 1.1 |

Minitab output from a least-squares regression analysis for these data is shown below.

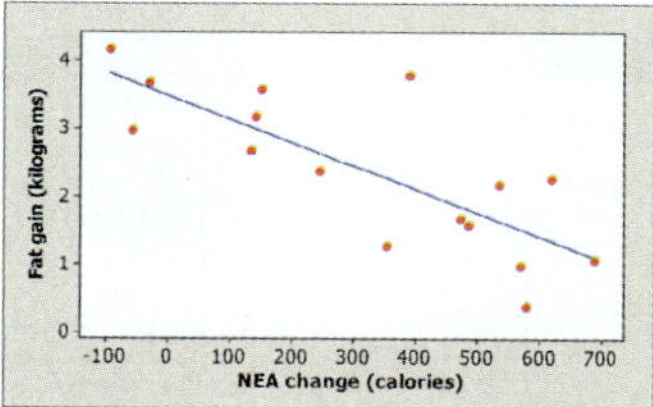


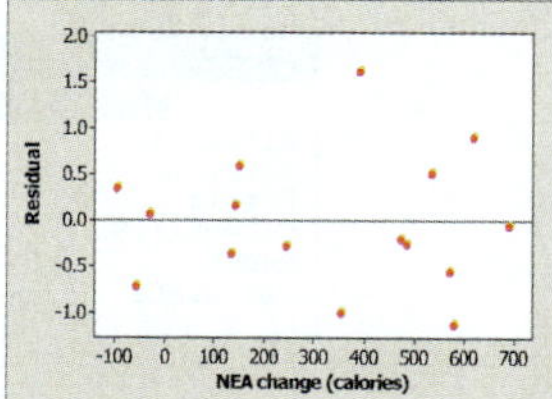


### ANSWERS TO CYU

S: $\beta$ = slope of the population regression line relating fat gain to change in NEA. P: $t$ interval for the slope. *Linear*: There is no leftover pattern in the residual plot. *Independent*: The sample size ($n = 16$) is less than 10% of all healthy young adults. *Normal*: The histogram of the residuals shows no strong skewness or outliers. *Equal SD*: Other than one point with a large positive residual, the residual plot shows roughly equal scatter for all $x$ values. *Random*: Random sample. D: With df = 14, (−0.005032, −0.001852). C: We are 95% confident that the interval from −0.005032 to −0.001852 captures the slope of the population regression line relating fat gain to change in NEA.

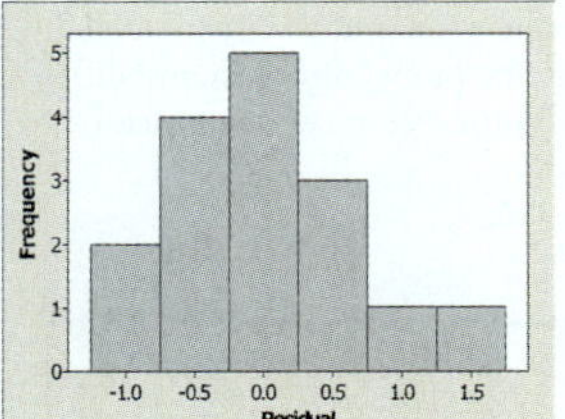


| Regression Analysis: Fat gain versus NEA change | | | | |
|---|---|---|---|---|
| Predictor | Coef | SE Coef | T | P |
| Constant | 3.5051 | 0.03036 | 11.54 | 0.000 |
| NEA change | −0.0034415 | 0.0007414 | −4.64 | 0.000 |
| S = 0.739853 | R-Sq = 60.6% | R-Sq(adj) = 57.8% | | |

Construct and interpret a 95% confidence interval for the slope of the population (true) regression line.

## Performing a Significance Test for the Slope

When the conditions for inference are met, we can use the slope $b$ of the sample regression line to construct a confidence interval for the slope $\beta$ of the population (true) regression line. We can also perform a significance test to determine whether a specified value of $\beta$ is plausible. The null hypothesis has the general form $H_0: \beta = \beta_0$. To do a test, standardize $b$ to get the test statistic:

$$\text{test statistic} = \frac{\text{statistic} - \text{parameter}}{\text{standard deviation of statistic}}$$

$$t = \frac{b - \beta_0}{\text{SE}_b}$$

To find the *P*-value, use a $t$ distribution with $n - 2$ degrees of freedom. Here are the details for the ***t* test for the slope**.

### *t* TEST FOR THE SLOPE

Suppose the conditions for inference are met. To test the hypothesis $H_0: \beta = \beta_0$, compute the test statistic

$$t = \frac{b - \beta_0}{\text{SE}_b}$$

Find the *P*-value by calculating the probability of getting a $t$ statistic this large or larger in the direction specified by the alternative hypothesis $H_a$. Use the $t$ distribution with df $= n - 2$.

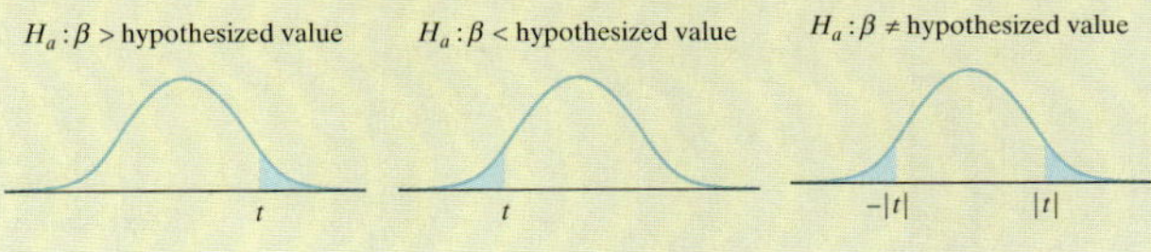


If sample data suggest a linear relationship between two variables, how can we determine whether this happened just by chance or whether there is actually a linear relationship between $x$ and $y$ in the population? By performing a test of $H_0: \beta = 0$. A regression line with slope 0 is horizontal. That is, the mean of $y$ does not change at all when $x$ changes. So $H_0: \beta = 0$ says that there is *no linear relationship* between $x$ and $y$ in the population. Put another way, $H_0$ says that *linear regression of y on x is of no value for predicting y.*

### AP® Exam Formula Sheet

The general formula for a test statistic is included on the formula sheet that is given to students on both parts of the AP® exam. However, the specific formula for the test statistic for the slope is *not* included on the formula sheet.

### Teaching Tip

Make sure that students understand why we perform a significance test for a slope. When data from a random sample or a randomized experiment suggest that there is an association between two variables, there are two possible explanations for why the slope differs from 0. The first explanation is that there really is no association between the variables, and we got a nonzero slope due to sampling variability or the chance variation due to random assignment. The second explanation is that there really is an association between the two variables. We do a significance test to decide which explanation is more plausible.

Most technology will only do a test with $H_0: \beta = 0$.

Regression output from statistical software usually gives $t$ and its two-sided $P$-value for a test of $H_0: \beta = 0$. For a one-sided test in the proper direction, just divide the $P$-value in the output by 2. The following example shows what we mean.

## EXAMPLE Crying and IQ

### Significance test for β

STEP 4

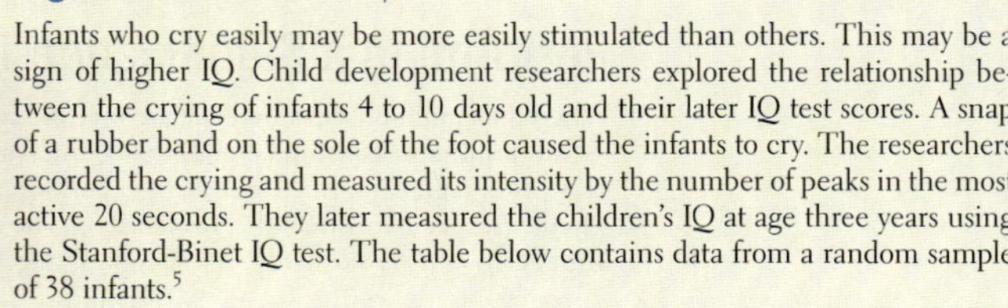

Infants who cry easily may be more easily stimulated than others. This may be a sign of higher IQ. Child development researchers explored the relationship between the crying of infants 4 to 10 days old and their later IQ test scores. A snap of a rubber band on the sole of the foot caused the infants to cry. The researchers recorded the crying and measured its intensity by the number of peaks in the most active 20 seconds. They later measured the children's IQ at age three years using the Stanford-Binet IQ test. The table below contains data from a random sample of 38 infants.[5]

**AP® EXAM TIP** When you see a list of data values on an exam question, don't just start typing the data into your calculator. Read the question first. Often, additional information is provided that makes it unnecessary for you to enter the data at all. This can save you valuable time on the AP® exam.

| Crycount | IQ | Crycount | IQ | Crycount | IQ | Crycount | IQ |
|---|---|---|---|---|---|---|---|
| 10 | 87 | 20 | 90 | 17 | 94 | 12 | 94 |
| 12 | 97 | 16 | 100 | 19 | 103 | 12 | 103 |
| 9 | 103 | 23 | 103 | 13 | 104 | 14 | 106 |
| 16 | 106 | 27 | 108 | 18 | 109 | 10 | 109 |
| 18 | 109 | 15 | 112 | 18 | 112 | 23 | 113 |
| 15 | 114 | 21 | 114 | 16 | 118 | 9 | 119 |
| 12 | 119 | 12 | 120 | 19 | 120 | 16 | 124 |
| 20 | 132 | 15 | 133 | 22 | 135 | 31 | 135 |
| 16 | 136 | 17 | 141 | 30 | 155 | 22 | 157 |
| 33 | 159 | 13 | 162 | | | | |

Some computer output from a least-squares regression analysis on these data is shown below.

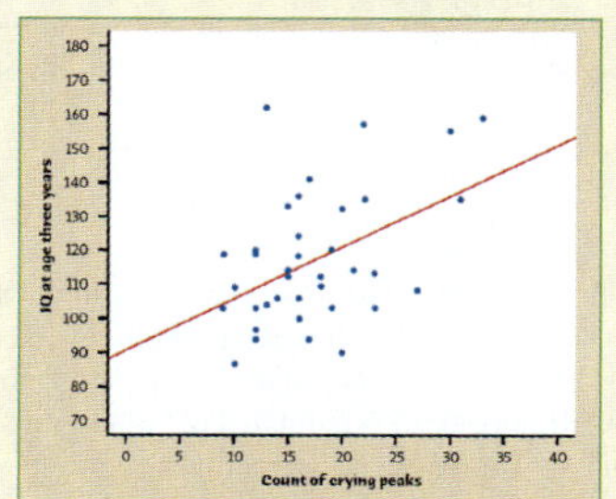


Regression Analysis: IQ versus Crycount

| Predictor | Coef | SE Coef | T | P |
|---|---|---|---|---|
| Constant | 91.268 | 8.934 | 10.22 | 0.000 |
| Crycount | 1.4929 | 0.4870 | 3.07 | 0.004 |

S = 17.50  R-Sq = 20.7%  R-Sq(adj) = 18.5%

## ALTERNATE EXAMPLE

### *Tipping at a buffet*

Do customers who stay longer at buffets give larger tips? Charlotte, an AP® Statistics student who worked at an Asian buffet, decided to investigate this question for her second-semester project. While she was doing her job as a hostess, she obtained a random sample of receipts, which included the length of time (in minutes) a party was at the restaurant and the amount of the tip it left (in dollars). Here are the data, along with some output from a least-squares regression analysis:

| Time (minutes) | Tip (dollars) |
|---|---|
| 23 | 5.00 |
| 39 | 2.75 |
| 44 | 7.75 |
| 55 | 5.00 |
| 61 | 7.00 |
| 65 | 8.88 |
| 67 | 9.01 |
| 70 | 5.00 |
| 74 | 7.29 |
| 85 | 7.50 |
| 90 | 6.00 |
| 99 | 6.50 |

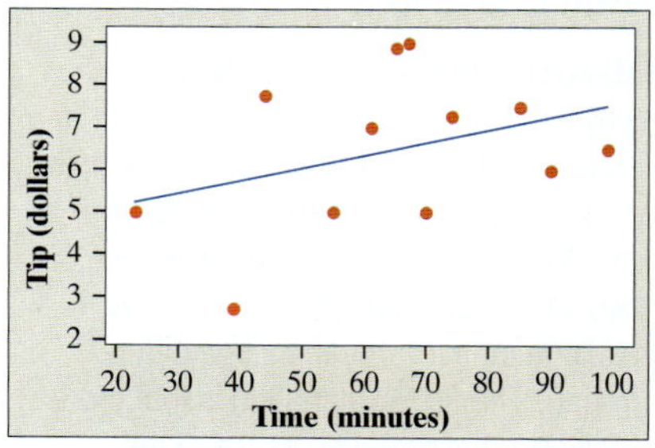


| Predictor | Coef | SE Coef | T | P |
|---|---|---|---|---|
| Constant | 4.535 | 1.657 | 2.74 | 0.021 |
| Time (minutes) | 0.03013 | 0.02448 | 1.23 | 0.247 |

S = 1.77931  R-Sq = 13.2%  R-Sq(adj) = 4.5%

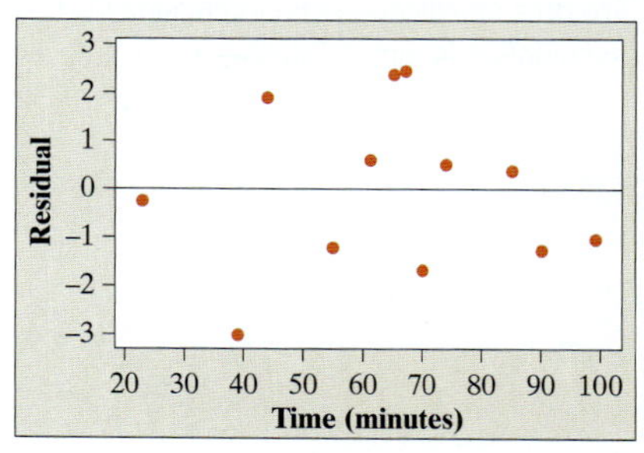


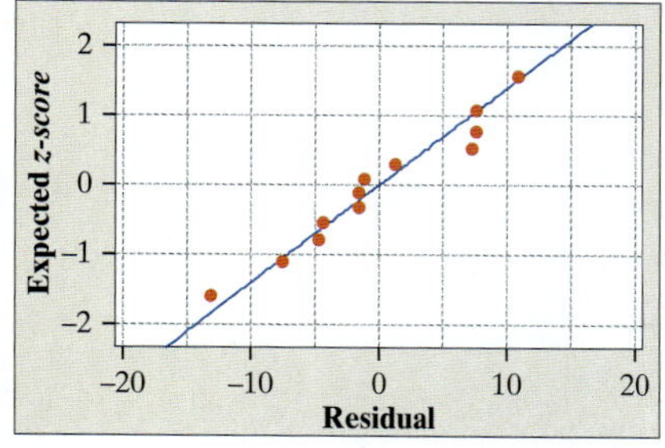


**PROBLEM:**

**(a)** What is the equation of the least-squares regression line for predicting the tip from the amount of time spent at the buffet? Define any variables you use.

**(b)** Interpret the slope of the least-squares regression line.

**(c)** Explain what the value of $r^2$ means in this setting.

**(d)** Do these data provide convincing evidence of a positive linear relationship between the amount of time and amount of tip for customers at this Asian buffet?

**SOLUTION:**

**(a)** $\hat{y} = 4.535 + 0.030x$, where $\hat{y}$ = predicted tip amount (in dollars) and $x$ = length of stay (in minutes).

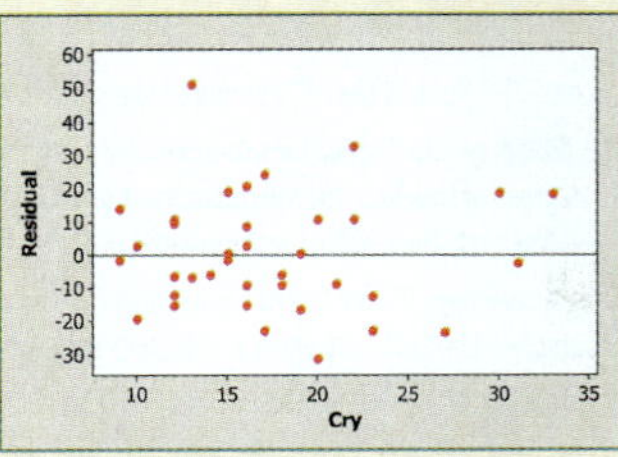
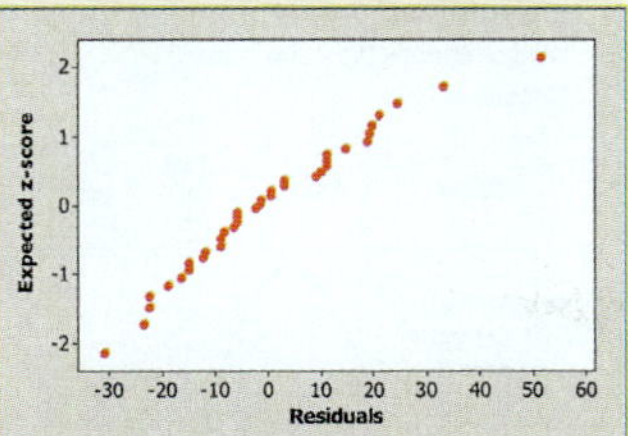

PROBLEM:

(a) What is the equation of the least-squares regression line for predicting IQ at age 3 from the number of crying peaks (crycount)? Interpret the slope and *y* intercept of the regression line in context.

(b) Explain what the value of *s* means in this setting.

(c) Do these data provide convincing evidence of a positive linear relationship between crying counts and IQ in the population of infants?

SOLUTION:

(a) The equation of the least-squares line is

$$\text{predicted IQ score} = 91.268 + 1.4929\,(\text{crycount})$$

Because there were no infants who recorded fewer than 9 crying peaks in their most active 20 seconds, it is a risky extrapolation to use this line to predict the value of *y* when $x = 0$.

*Slope:* For each additional crying peak in the most active 20 seconds, the regression line predicts an increase of about 1.5 IQ points. *y intercept:* The model predicts that an infant who doesn't cry when flicked with a rubber band will have a later IQ score of about 91.

(b) The size of a typical prediction error when using the regression line in part (a) is 17.50 IQ points.

(c) We'll follow the four-step process.

STATE: We want to perform a test of

$$H_0: \beta = 0$$
$$H_a: \beta > 0$$

where $\beta$ is the slope of the population regression line relating crying count to IQ score. No significance level was given, so we'll use $\alpha = 0.05$.

PLAN: If the conditions are met, we will do a *t* test for the slope $\beta$.

- *Linear:* The scatterplot suggests a moderately weak positive linear relationship between crying peaks and IQ. The residual plot shows a random scatter of points about the residual = 0 line.
- *Independent:* Due to sampling without replacement, there have to be at least 10(38) = 380 infants in the population from which these children were selected.
- *Normal:* The Normal probability plot of the residuals shows slight curvature, but no strong skewness or obvious outliers that would prevent use of *t* procedures.
- *Equal SD:* The residual plot shows a fairly equal amount of scatter around the horizontal line at 0 for all *x*-values.
- *Random:* We are told that these 38 infants were randomly selected.

**(b)** For each additional minute that a party stays at the buffet, the least-squares regression line predicts an increase in the tip of \$0.030 (3 cents).
**(c)** 13.2% of the variation in tip amount is accounted for by the least-squares regression line relating $y$ = tip amount to $x$ = amount of time spent at the buffet.
**(d)** STATE: We want to perform a test of $H_0: \beta = 0$ versus $H_a: \beta > 0$, where $\beta$ is the slope of the population regression line relating length of stay $x$ to tip amount $y$. We will use $\alpha = 0.05$.
PLAN: If the conditions are met, we will do a *t* test for the slope $\beta$.

- *Linear*: The scatterplot shows a weak positive linear relationship between length of stay and tip amount. The residual plot shows random scatter about the residual = 0 line.
- *Independent*: Knowing one tip amount shouldn't provide additional information about other tip amounts. Because we are sampling without replacement, we must assume that there are more than 10(12) = 120 receipts in the population from which these receipts were selected.
- *Normal*: The Normal probability plot looks roughly linear.
- *Equal SD*: The residual plot shows a fairly equal amount of scatter around the residual = 0 line.
- *Random*: The receipts were randomly selected.

DO: From the computer output:

- *Test statistic*: $t = 1.23$
- *P-value*: Because the *P*-value in the computer output is for a two-sided test, we must cut it in half for a one-sided test. $P = 0.247/2 = 0.1235$ using df $= 12 - 2 = 10$.

CONCLUDE: Because the *P*-value of 0.1235 is larger than $\alpha = 0.05$, we fail to reject $H_0$. We do not have convincing evidence that parties who stay longer at this Asian buffet leave larger tips.

Our usual formula for the test statistic confirms the value in the computer output:

$$t = \frac{b - \beta_0}{SE_b} = \frac{1.4929 - 0}{0.4870} = 3.07$$

*t* distribution, 36 degrees of freedom

*P*-value = 0.002

Values of *t*

*t* = 3.07

**FIGURE 12.8** The *P*-value for the one-sided test.

DO: We can get the test statistic and *P*-value from the Minitab output.

- *Test statistic:* $t = 3.07$ (look in the "T" column of the computer output across from "Crycount")
- *P-value:* Figure 12.8 displays the *P*-value for this one-sided test as an area under the *t* distribution curve with $38 - 2 = 36$ degrees of freedom. The Minitab output gives $P = 0.004$ as the *P*-value for a two-sided test. The *P*-value for the one-sided test is half of this, $P = 0.002$.

*Using technology:* Refer to the Technology Corner that follows the example. The calculator's LinRegTTest gives $t = 3.065$ and *P*-value $= 0.002$ using df $= 36$.

CONCLUDE: Because the *P*-value, 0.002, is less than $\alpha = 0.05$, we reject $H_0$. There is convincing evidence of a positive linear relationship between intensity of crying and IQ score in the population of infants.

**For Practice** *Try Exercise* 13

## Teaching Tip

Follow-up questions about the scope of inference are common on the AP® exam. Make sure to regularly ask your students if it is reasonable to generalize the results of a test to a larger population or if it is reasonable to make an inference about cause and effect.

Based on the results of the crying and IQ study, should we ask doctors and parents to make infants cry more so that they'll be smarter later in life? Hardly. This observational study gives statistically significant evidence of a positive linear relationship between the two variables. However, we can't conclude that more intense crying as an infant *causes* an increase in IQ. Maybe infants who cry more are more alert to begin with and tend to score higher on intelligence tests.

### TECHNOLOGY TIP

Remind students that they can watch a video presentation of this technology corner (for the TI-83/84, TI-89, or TI-Nspire) at www.whfreeman.com/tps5e.

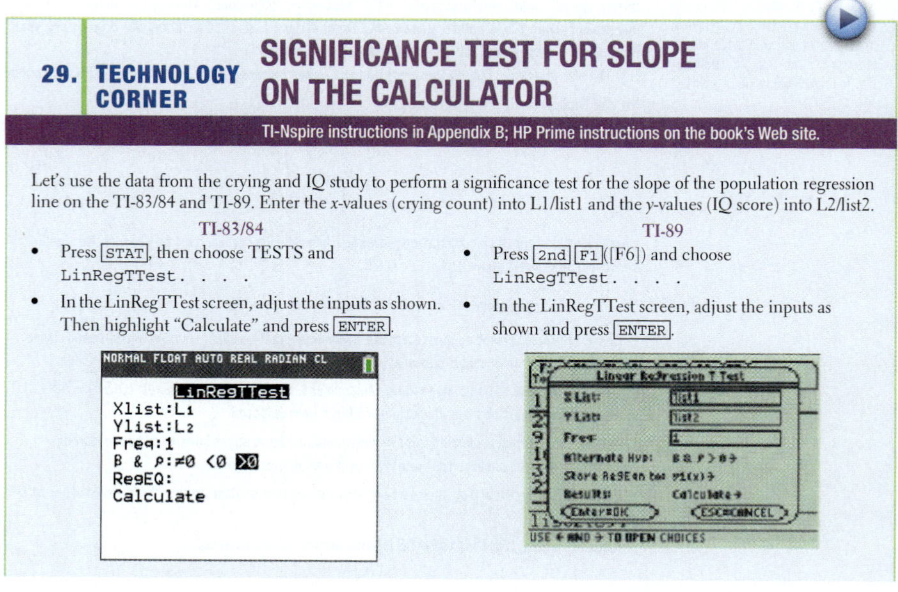

## 29. TECHNOLOGY CORNER — SIGNIFICANCE TEST FOR SLOPE ON THE CALCULATOR

TI-Nspire instructions in Appendix B; HP Prime instructions on the book's Web site.

Let's use the data from the crying and IQ study to perform a significance test for the slope of the population regression line on the TI-83/84 and TI-89. Enter the *x*-values (crying count) into L1/list1 and the *y*-values (IQ score) into L2/list2.

**TI-83/84**

- Press STAT, then choose TESTS and LinRegTTest. . . .
- In the LinRegTTest screen, adjust the inputs as shown. Then highlight "Calculate" and press ENTER.

**TI-89**

- Press 2nd F1 ([F6]) and choose LinRegTTest. . . .
- In the LinRegTTest screen, adjust the inputs as shown and press ENTER.



## Teaching Tip

*Note*: This tip goes beyond the AP® Topic Outline. There is an interesting connection between the *t* test for slope and a two-sample *t* test for a difference of means. For example, on page 645 in Chapter 10, we analyzed the results of an experiment to see if calcium reduced blood pressure better than a placebo. The difference in means was 5.273, the test statistic was 1.604, and the *P*-value was 0.0644. We can also analyze these data using a test for slope if we code the placebo users with $x = 0$ and the calcium users with $x = 1$. Here is a scatterplot showing the same data, along with some regression output:

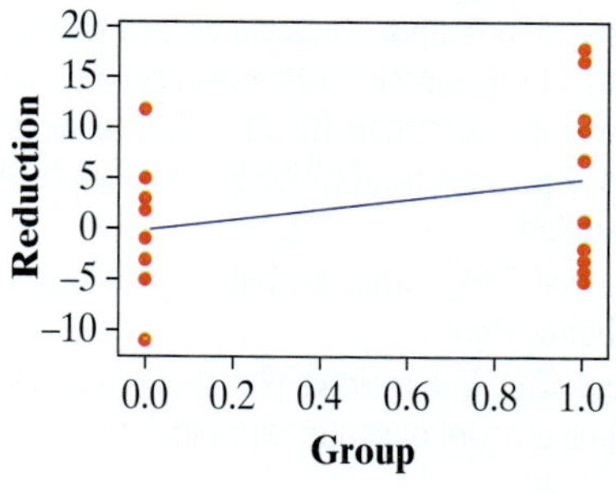


| Predictor | Coef | SE Coef | T | P |
|---|---|---|---|---|
| Constant | -0.273 | 2.227 | -0.12 | 0.904 |
| Group | 5.273 | 3.227 | 1.63 | 0.119 |

S = 7.38484 R-Sq = 12.3% R-Sq(adj) = 7.7%

Notice that slope = 5.273, which is the difference in means. Also, the test statistic is $t = 1.63$, quite close to the previous test statistic of $t = 1.604$. The small difference results because the *t* test for slope pools the data to estimate a common standard deviation (remember the Equal SD condition), while we used the unpooled two-sample *t* test for the difference in means. Had we chosen the pooled two-sample *t* test, the test statistics would be identical. Finally, notice that the *P*-value from the computer output, when cut in half for a one-sided test, would be about the same as the *P*-value in the two-sample *t* test.

- The linear regression $t$ test results take two screens to present. We show only the first screen.

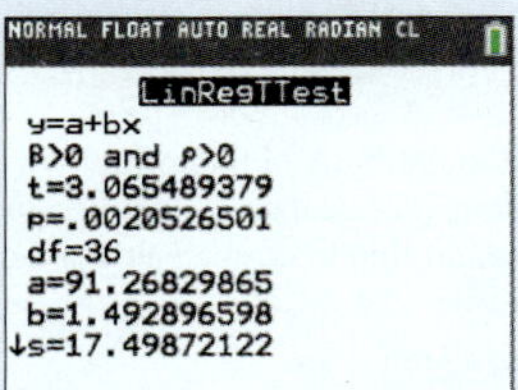


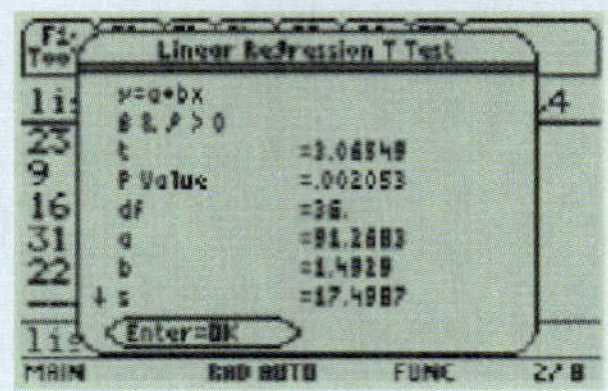


**AP® EXAM TIP** The formula for the test statistic in a $t$ test for the slope of a population (true) regression line often leads to calculation errors by students. As a result, we recommend using the calculator's `LinRegTTest` feature to perform calculations on the AP® exam. Be sure to name the procedure ($t$ test for slope) and to report the test statistic ($t = 3.065$), $P$-value (0.002), and df (36) as part of the "Do" step.

**What's with that $\rho > 0$ in the LinRegTTest screen?** The slope $b$ of the least-squares regression line is closely related to the correlation $r$ between the explanatory and response variables $x$ and $y$. (Recall that $b = r\frac{s_y}{s_x}$). In the same way, the slope $\beta$ of the population regression line is closely related to the correlation $\rho$ (the lowercase Greek letter rho) between $x$ and $y$ in the population. In particular, the slope is 0 when the correlation is 0.

Testing the null hypothesis $H_0: \beta = 0$ is, therefore, exactly the same as testing that there is *no correlation* between $x$ and $y$ in the population from which we drew our data. You can use the test for zero slope to test the hypothesis $H_0: \rho = 0$ of zero correlation between any two quantitative variables. That's a useful trick. Because correlation also makes sense when there is no explanatory-response distinction, it is handy to be able to test correlation without doing regression.

## CHECK YOUR UNDERSTANDING

The previous Check Your Understanding (page 752) described some results from a study of nonexercise activity (NEA) and fat gain. Here, again, is the Minitab output from a least-squares regression analysis for these data.

Regression Analysis: Fat gain versus NEA change

| Predictor | Coef | SE Coef | T | P |
|---|---|---|---|---|
| Constant | 3.5051 | 0.3036 | 11.54 | 0.000 |
| NEA change | −0.0034415 | 0.0007414 | −4.64 | 0.000 |

S = 0.739853   R-Sq = 60.6%   R-Sq(adj) = 57.8%

Do these data provide convincing evidence at the $\alpha = 0.05$ significance level of a negative linear relationship between fat gain and NEA change in the population of healthy young adults? Assume that the conditions for regression inference are met.

## Teaching Tip

If your students are interested, the test statistic for the correlation is $t = \dfrac{r - \rho}{\sqrt{\dfrac{1 - r^2}{n - 2}}}$, where $\rho$ is the hypothesized value of the correlation and df $= n - 2$.

## ANSWERS TO CYU

S: $H_0: \beta = 0$ vs. $H_a: \beta < 0$, where $\beta$ is the slope of the population regression line relating fat gain to NEA change. P: $t$ test for the slope $\beta$. D: $t = -4.64$. $P$-value $\approx 0.000/2 \approx 0$. C: Because the $P$-value of approximately 0 is less than $\alpha = 0.05$, we reject $H_0$. There is convincing evidence that the slope of the population regression line relating fat gain to NEA change is negative.

**TRM Section 12.1 Quizzes**

There are three different quizzes available for this section in the Teacher's Resource Materials. Versions A and B are roughly parallel. Find them in the TE-book, or on the TRCD or the book's companion Web site www.whfreeman.com/tps5 (teacher log-in required).

# Section 12.1 Summary

- Least-squares regression fits a straight line of the form $\hat{y} = a + bx$ to data to predict a response variable $y$ from an explanatory variable $x$. Inference in this setting uses the **sample regression line** to estimate or test a claim about the **population (true) regression line**.
- The conditions for regression inference are
  - **Linear:** The actual relationship between $x$ and $y$ is linear. For any fixed value of $x$, the mean response $\mu_y$ falls on the population (true) regression line $\mu_y = \alpha + \beta x$.
  - **Independent:** Individual observations are independent. When sampling is done without replacement, check the *10% condition*.
  - **Normal:** For any fixed value of $x$, the response $y$ varies according to a Normal distribution.
  - **Equal SD:** The standard deviation of $y$ (call it $\sigma$) is the same for all values of $x$.
  - **Random:** The data are produced from a well-designed random sample or randomized experiment.
- The slope $b$ and intercept $a$ of the sample regression line estimate the slope $\beta$ and intercept $\alpha$ of the population (true) regression line. Use the standard deviation of the residuals, $s$, to estimate $\sigma$.
- Confidence intervals and significance tests for the slope $\beta$ of the population regression line are based on a $t$ distribution with $n - 2$ degrees of freedom.
- The ***t* interval for the slope** $\beta$ has the form $b \pm t^* \mathrm{SE}_b$, where the **standard error of the slope** is $\mathrm{SE}_b = \frac{s}{s_x\sqrt{n-1}}$.
- To test the null hypothesis $H_0: \beta = \beta_0$, carry out a ***t* test for the slope.** This test uses the statistic $t = \frac{b - \beta_0}{\mathrm{SE}_b}$. The most common null hypothesis is $H_0: \beta = 0$, which says that there is no linear relationship between $x$ and $y$ in the population.

**12.1 TECHNOLOGY CORNERS**

TI-Nspire Instructions in Appendix B; HP Prime instructions on the book's Web site.

# Section 12.1 Exercises

1. **Oil and residuals** Exercise 53 on page 194 (Chapter 3) examined data on the depth of small defects in the Trans-Alaska Oil Pipeline. Researchers compared the results of measurements on 100 defects made in the field with measurements of the same defects made in the laboratory.[6] The figure below shows a residual plot for the least-squares regression line based on these data. Explain why the conditions for performing inference about the slope $\beta$ of the population regression line are not met.

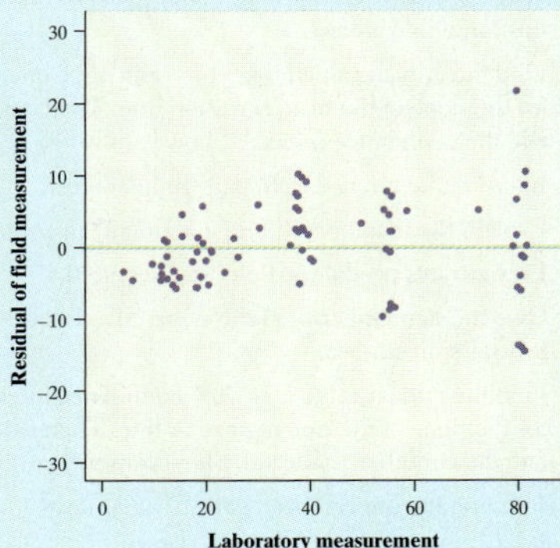


2. **SAT Math scores** In Chapter 3, we examined data on the percent of high school graduates in each state who took the SAT and the state's mean SAT Math score in a recent year. The figure below shows a residual plot for the least-squares regression line based on these data. Explain why the conditions for performing inference about the slope $\beta$ of the population regression line are not met.

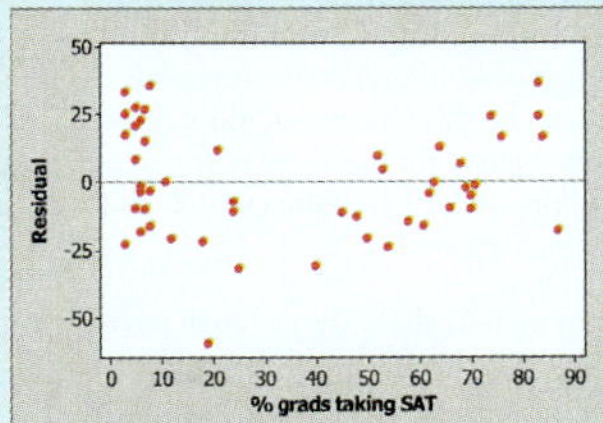


3. pg 745  **Beer and BAC** How well does the number of beers a person drinks predict his or her blood alcohol content (BAC)? Sixteen volunteers aged 21 or older with an initial BAC of 0 took part in a study to find out. Each volunteer drank a randomly assigned number of cans of beer. Thirty minutes later, a police officer measured their BAC. Least-squares regression was performed on the data. A residual plot and a histogram of the residuals are shown below. Check whether the conditions for performing inference about the regression model are met.

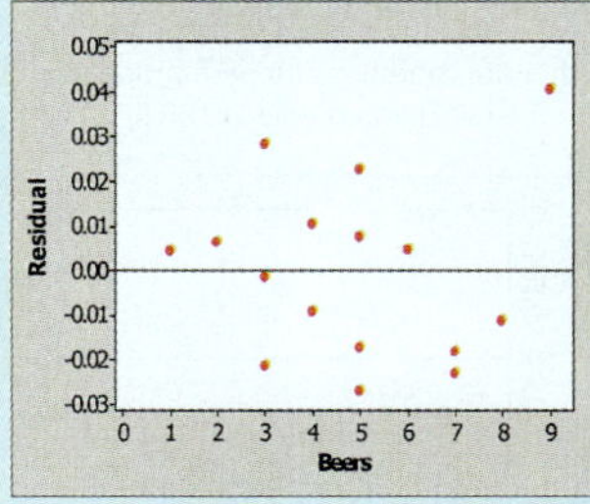


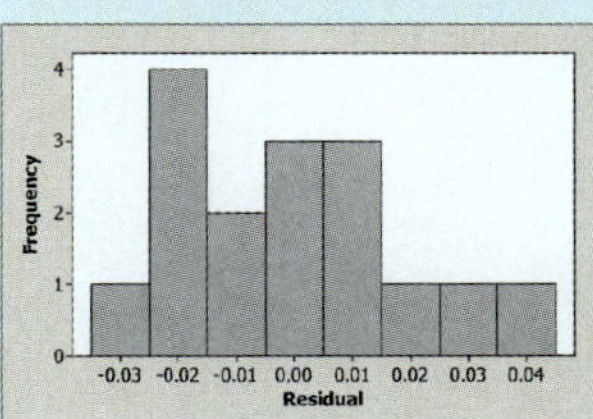


4. **Prey attracts predators** Here is one way in which nature regulates the size of animal populations: high population density attracts predators, which remove a higher proportion of the population than when the density of the prey is low. One study looked at kelp perch and their common predator, the kelp bass. The researcher set up four large circular pens on sandy ocean bottoms off the coast of southern California. He chose young perch at random from a large group and placed 10, 20, 40, and 60 perch in the four pens. Then he dropped the nets protecting the pens, allowing bass to swarm in, and counted the

**TRM Full Solutions to Section 12.1 Exercises**

## Answers to Section 12.1 Exercises

**12.1** The Equal SD condition is not met because the SD of the residuals clearly increases as the laboratory measurement ($x$) increases.

**12.2** The Linear condition is not met because there is clear curvature in the residual plot.

**12.3** *Linear*: There is no leftover pattern in the residual plot. *Independent*: Knowing the BAC for one subject should not help us predict the BAC for another subject. *Normal*: The histogram of the residuals shows no strong skewness or outliers. *Equal SD*: The residual plot shows roughly equal scatter for all $x$ values. *Random*: These data come from a randomized experiment.

**12.4** *Linear*: There is no leftover pattern in the residual plot. *Independent*: Knowing the proportion of perch killed in one pen should not help us predict the proportion of perch killed in another pen. *Normal*: The histogram of the residuals shows no strong skewness or outliers. *Equal SD*: The residual plot shows roughly equal scatter for all *x* values. *Random*: These data come from a randomized experiment.

**12.5** $\alpha$ is the true *y* intercept, which measures the true mean BAC level if no beers had been drunk ($a = -0.012701$). $\beta$ is the true slope, which measures how much the true mean BAC changes with the drinking of one additional beer ($b = 0.018$). Finally, $\sigma$ is the true standard deviation of the residuals, which measures how much the observed values of BAC typically vary from the population regression line ($s = 0.0204$).

**12.6** $\alpha$ is the true *y* intercept, which measures the true average proportion of perch killed if there were 0 fish in the tank to begin with ($a = 0.12049$). $\beta$ is the true slope, which measures how much the true average proportion of fish killed changes for each additional perch in the pen ($b = 0.008569$). Finally, $\sigma$ is the true standard deviation of the residuals, which measures how much the proportion killed values typically vary from the population regression line ($s = 0.1886$).

**12.7** **(a)** $SE_b = 0.0024$. If we repeated the experiment many times, the slope of the sample regression line would typically vary by about 0.0024 from the slope of the true regression line for predicting BAC from the number of beers consumed. **(b)** With df = 14, $0.018 \pm 2.977(0.0024) = (0.011, 0.025)$. **(c)** We are 99% confident that the interval from 0.011 to 0.025 captures the slope of the true regression line for predicting BAC from the number of beers consumed. **(d)** If we repeated the experiment many times and computed a confidence interval for the slope each time, about 99% of the resulting intervals would contain the slope of the true regression line for predicting BAC from the number of beers consumed.

**12.8** **(a)** $SE_b = 0.002456$. If we repeated the experiment many times, the slope of the sample regression line would typically vary by about 0.002456 from the slope of the true regression line for predicting the proportion of perch killed from the number of perch in the pen. **(b)** With df = 14, $0.00857 \pm 1.761(0.002456) = (0.00424, 0.0129)$. **(c)** We are 90% confident that the interval from 0.00424 to 0.0129 captures the slope of the true regression line for predicting the proportion of perch killed from the number of perch in the pen. **(d)** If we repeated the experiment many times and computed a confidence interval for the slope each time, about 90% of the resulting intervals would contain the slope of the true regression line for predicting the proportion of perch killed from the number of perch in the pen.

**12.9** *S*: $\beta$ = the slope of the population regression line relating number of clusters of beetle larvae to number of stumps. *P*: *t* interval for $\beta$. *D*: With df = 21, (8.678, 15.11). *C*: We are 99% confident that the interval from 8.678 to 15.11 captures the slope of the population regression line relating number of clusters of beetle larvae to number of stumps.



perch left after two hours. Here are data on the proportions of perch eaten in four repetitions of this setup:[7]

| Number of Perch | Proportion Killed | | | |
|---|---|---|---|---|
| 10 | 0.0 | 0.1 | 0.3 | 0.3 |
| 20 | 0.2 | 0.3 | 0.3 | 0.6 |
| 40 | 0.075 | 0.3 | 0.6 | 0.725 |
| 60 | 0.517 | 0.55 | 0.7 | 0.817 |

The explanatory variable is the number of perch (the prey) in a confined area. The response variable is the proportion of perch killed by bass (the predator) in two hours when the bass are allowed access to the perch. A scatterplot of the data shows a linear relationship.

We used Minitab software to carry out a least-squares regression analysis for these data. A residual plot and a histogram of the residuals are shown below. Check whether the conditions for performing inference about the regression model are met.

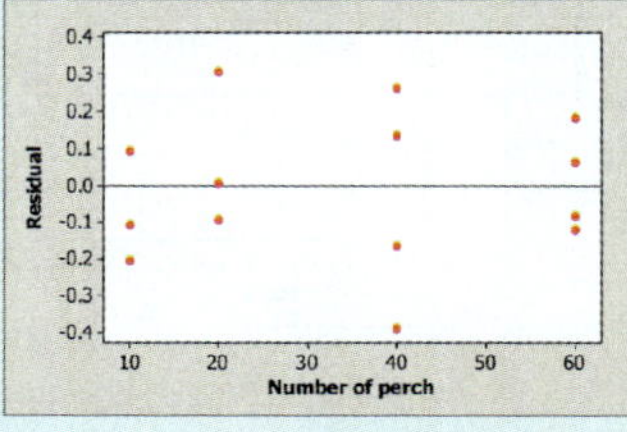


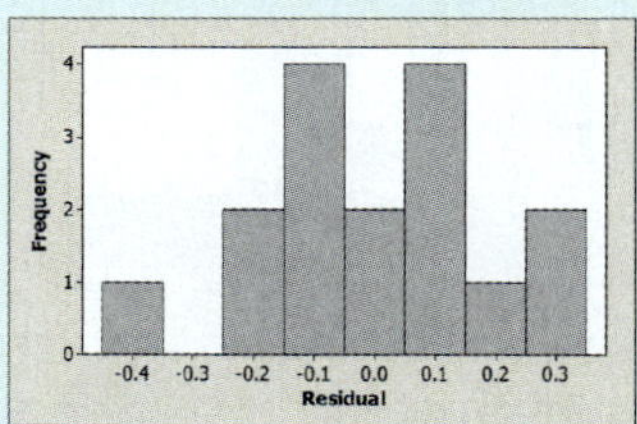


5. **Beer and BAC** Refer to Exercise 3. Computer output from the least-squares regression analysis on the beer and blood alcohol data is shown below.

```
Dependent variable is:   BAC
No Selector
R squared = 80.0% R squared (adjusted) = 78.6%
s = 0.0204 with 16 - 2 = 14 degrees of freedom

Variable  Coefficient s.e. of Coeff t-ratio  prob
Constant  -0.012701    0.0126       -1.00    0.3320
Beers      0.017964    0.0024        7.84    ≤0.0001
```

The model for regression inference has three parameters: $\alpha$, $\beta$, and $\sigma$. Explain what each parameter represents in context. Then provide an estimate for each.

6. **Prey attracts predators** Refer to Exercise 4. Computer output from the least-squares regression analysis on the perch data is shown below.

```
Predictor  Coef       Stdev.     t-ratio  p
Constant   0.12049    0.09269    1.30     0.215
Perch      0.008569   0.002456   3.49     0.004
S = 0.1886    R-Sq = 46.5%    R-Sq(adj) = 42.7%
```

The model for regression inference has three parameters: $\alpha$, $\beta$, and $\sigma$. Explain what each parameter represents in context. Then provide an estimate for each.

7. **Beer and BAC** Refer to Exercise 5.

pg 748

(a) Give the standard error of the slope, $SE_b$. Interpret this value in context.

(b) Find the critical value for a 99% confidence interval for the slope of the true regression line. Then calculate the confidence interval. Show your work.

(c) Interpret the interval from part (b) in context.

(d) Explain the meaning of "99% confident" in context.

8. **Prey attracts predators** Refer to Exercise 6.

(a) Give the standard error of the slope, $SE_b$. Interpret this value in context.

(b) Find the critical value for a 90% confidence interval for the slope of the true regression line. Then calculate the confidence interval. Show your work.

(c) Interpret the interval from part (b) in context.

(d) Explain the meaning of "90% confident" in context.

9. **Beavers and beetles** Do beavers benefit beetles? Researchers laid out 23 circular plots, each 4 meters in diameter, at random in an area where beavers were cutting down cottonwood trees. In each plot, they counted the number of stumps from trees cut by beavers and the number of clusters of beetle larvae. Ecologists think that the new sprouts from stumps are more tender than other cottonwood growth, so that beetles prefer them. If so, more stumps should produce more beetle larvae.[8]

pg 749

Minitab output for a regression analysis on these data is shown below. Construct and interpret a 99% confidence interval for the slope of the population regression line. Assume that the conditions for performing inference are met.

**Regression Analysis: Beetle larvae versus Stumps**

```
Predictor  Coef     SE Coef   T       P
Constant   -1.286   2.853     -0.45   0.657
Stumps     11.894   1.136     10.47   0.000
S = 6.41939   R-Sq = 83.9%   R-Sq(adj) = 83.1%
```

10. **Ideal proportions** The students in Mr. Shenk's class measured the arm spans and heights (in inches) of a random sample of 18 students from their large high school. Some computer output from a least-squares regression analysis on these data is shown below. Construct and interpret a 90% confidence interval for the slope of the population regression line. Assume that the conditions for performing inference are met.

| Predictor | Coef | Stdev | t-ratio | p |
|---|---|---|---|---|
| Constant | 11.547 | 5.600 | 2.06 | 0.056 |
| Armspan | 0.84042 | 0.08091 | 10.39 | 0.000 |

S = 1.613   R-Sq = 87.1%   R-Sq(adj) = 86.3%

11. **Beavers and beetles** Refer to Exercise 9.

(a) How many clusters of beetle larvae would you predict in a circular plot with 5 tree stumps cut by beavers? Show your work.

(b) About how far off do you expect the prediction in part (a) to be from the actual number of clusters of beetle larvae? Justify your answer.

12. **Ideal proportions** Refer to Exercise 10.

(a) What height would you predict for a student with an arm span of 76 inches? Show your work.

(b) About how far off do you expect the prediction in part (a) to be from the student's actual height? Justify your answer.

13. **Weeds among the corn** Lamb's-quarter is a common weed that interferes with the growth of corn. An agriculture researcher planted corn at the same rate in 16 small plots of ground and then weeded the plots by hand to allow a fixed number of lamb's-quarter plants to grow in each meter of corn row. The decision of how many of these plants to leave in each plot was made at random. No other weeds were allowed to grow. Here are the yields of corn (bushels per acre) in each of the plots:[9]

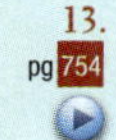


Some computer output from a least-squares regression analysis on these data is shown below.

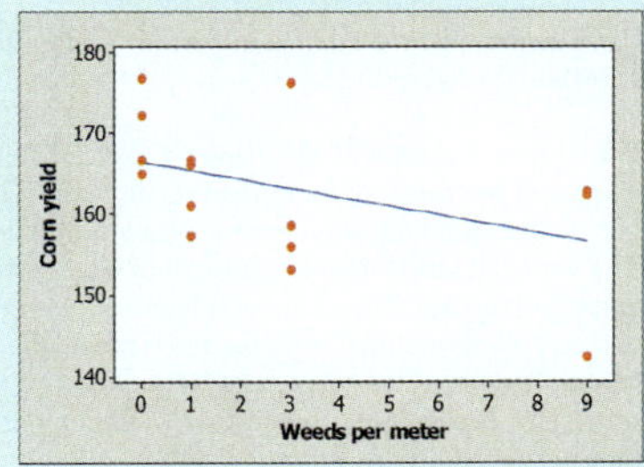


| Predictor | Coef | SE Coef | T | P |
|---|---|---|---|---|
| Constant | 166.483 | 2.725 | 61.11 | 0.000 |
| Weeds per meter | −1.0987 | 0.5712 | −1.92 | 0.075 |

S = 7.97665   R-Sq = 20.9%   R-Sq(adj) = 15.3%

(a) What is the equation of the least-squares regression line for predicting corn yield from the number of lamb's quarter plants per meter? Interpret the slope and *y* intercept of the regression line in context.

(b) Explain what the value of *s* means in this settting.

(c) Do these data provide convincing evidence at the $\alpha = 0.05$ level that more weeds reduce corn yield? Assume that the conditions for performing inference are met.

14. **Time at the table** Does how long young children remain at the lunch table help predict how much they eat? Here are data on a random sample of 20 toddlers observed over several months.[10] "Time" is the average number of minutes a child spent at the table when lunch was served. "Calories" is the average number of calories the child consumed during lunch, calculated from careful observation of what the child ate each day.

Some computer output from a least-squares regression analysis on these data is shown below.

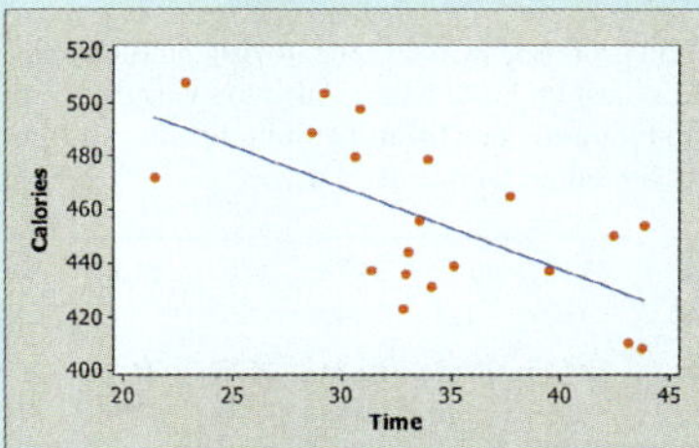


| Predictor | Coef | SE Coef | T | P |
|---|---|---|---|---|
| Constant | 560.65 | 29.37 | 19.09 | 0.000 |
| Time | −3.0771 | 0.8498 | −3.62 | 0.002 |

S = 23.3980   R-Sq = 42.1%   R-Sq(adj) = 38.9%

(a) What is the equation of the least-squares regression line for predicting calories consumed from time at the table? Interpret the slope of the regression line in context. Does it make sense to interpret the *y* intercept in this case? Why or why not?

(b) Explain what the value of *s* means in this setting.

(c) Do these data provide convincing evidence at the $\alpha = 0.01$ level of a linear relationship between time

**12.10** *S*: $\beta$ = the slope of the population regression line relating heights to arm spans for students in this high school. *P*: *t* interval for $\beta$. *D*: With df = 16, (0.6991, 0.9817). *C*: We are 90% confident that the interval from 0.6991 to 0.9817 captures the slope of the population regression line relating heights to arm spans for students in this high school.

**12.11** **(a)** $\hat{y} = -1.286 + 11.894(5) = 58.184$ clusters **(b)** $s = 6.419$, so we would expect our prediction to be off from the actual number of clusters by about 6.419 clusters.

**12.12** **(a)** $\hat{y} = 11.547 + 0.84042(76) = 75.4189$ inches **(b)** $s = 1.613$, so we would expect our prediction to be off from the actual height by about 1.613 inches.

**12.13** **(a)** $\hat{y} = 166.483 - 1.0987x$, where $\hat{y}$ is the predicted corn yield and *x* is the number of weeds per meter. *Slope*: For each additional weed per meter, the predicted corn yield will decrease by about 1.0987 bushels/acre. *y* intercept: If there are no weeds per meter, we would predict a corn yield of 166.483 bushels/acre. **(b)** When using weeds per meter to predict corn yield, the actual yield will typically vary from the predicted yield by about 7.98 bushels/acre. **(c)** *S*: $H_0: \beta = 0$ vs. $H_a: \beta < 0$, where $\beta$ is the slope of the true regression line relating corn yield to weeds per meter. *P*: *t* test for $\beta$. *D*: $t = -1.92$. *P*-value = 0.075/2 = 0.0375. *C*: Because the *P*-value of 0.0375 is less than $\alpha = 0.05$, we reject $H_0$. There is convincing evidence that the slope of the true regression line relating corn yield to weeds per meter is negative.

**12.14** **(a)** $\hat{y} = 560.65 - 3.0771x$, where $\hat{y}$ is the predicted calories consumed and *x* is the time spent at the table. *Slope*: For each additional minute at the table, the predicted number of calories will decrease by about 3.0771 calories. *y* intercept: If there is no time spent at the table, the predicted number of calories consumed would be about 560.65. **(b)** When using time at the table to predict number of calories, the actual number of calories will typically vary from the predicted calories by about 23.3980 calories. **(c)** *S*: $H_0: \beta = 0$ vs. $H_a: \beta \neq 0$, where $\beta$ is the slope of the population regression line relating calorie consumption to time at the table in the population of toddlers. *P*: *t* test for $\beta$. *D*: $t = -3.62$ and *P*-value is 0.002. *C*: Because the *P*-value of 0.002 is less than $\alpha = 0.01$ we reject $H_0$. There is convincing evidence of a linear relationship between time at the table and calorie consumption in the population of toddlers.

at the table and calories consumed in the population of toddlers? Assume that the conditions for performing inference are met.

15. **Is wine good for your heart?** A researcher from the University of California, San Diego, collected data on average per capita wine consumption and heart disease death rate in a random sample of 19 countries for which data were available. The following table displays the data.[11]

| Alcohol from wine (liters/year) | Heart disease death rate (per 100,000) | Alcohol from wine (liters/year) | Heart disease death rate (per 100,000) |
|---|---|---|---|
| 2.5 | 211 | 7.9 | 107 |
| 3.9 | 167 | 1.8 | 167 |
| 2.9 | 131 | 1.9 | 266 |
| 2.4 | 191 | 0.8 | 227 |
| 2.9 | 220 | 6.5 | 86 |
| 0.8 | 297 | 1.6 | 207 |
| 9.1 | 71 | 5.8 | 115 |
| 2.7 | 172 | 1.3 | 285 |
| 0.8 | 211 | 1.2 | 199 |
| 0.7 | 300 | | |

Is there statistically significant evidence of a negative linear relationship between wine consumption and heart disease deaths in the population of countries? Carry out an appropriate significance test at the $\alpha = 0.05$ level.

16. **The professor swims** Here are data on the time (in minutes) Professor Moore takes to swim 2000 yards and his pulse rate (beats per minute) after swimming on a random sample of 23 days:

| | | | | | | |
|---|---|---|---|---|---|---|
| **Time:** | 34.12 | 35.72 | 34.72 | 34.05 | 34.13 | 35.72 |
| **Pulse:** | 152 | 124 | 140 | 152 | 146 | 128 |
| **Time:** | 36.17 | 35.57 | 35.37 | 35.57 | 35.43 | 36.05 |
| **Pulse:** | 136 | 144 | 148 | 144 | 136 | 124 |
| **Time:** | 34.85 | 34.70 | 34.75 | 33.93 | 34.60 | 34.00 |
| **Pulse:** | 148 | 144 | 140 | 156 | 136 | 148 |
| **Time:** | 34.35 | 35.62 | 35.68 | 35.28 | 35.97 | |
| **Pulse:** | 148 | 132 | 124 | 132 | 139 | |

Is there statistically significant evidence of a negative linear relationship between Professor Moore's swim time and his pulse rate in the population of days on which he swims 2000 yards? Carry out an appropriate significance test at the $\alpha = 0.05$ level.

17. **Stats teachers' cars** A random sample of AP® Statistics teachers was asked to report the age (in years) and mileage of their primary vehicles. A scatterplot of the data is shown at top right.

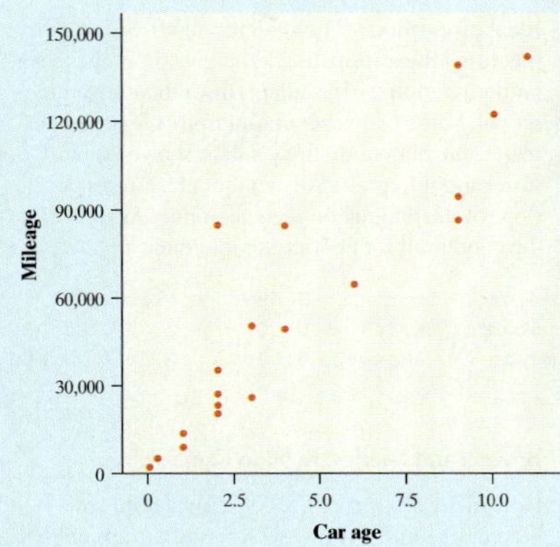


Computer output from a least-squares regression analysis of these data is shown below (df = 19). Assume that the conditions for regression inference are met.

```
Variable Coef        SE Coef  t-ratio  prob
Constant  7288.54    6591     1.11     0.2826
Car age  11630.6     1249     9.31     <0.0001
S = 19280   R-Sq = 82.0%    RSq(adj) = 81.1%
```

(a) Verify that the 95% confidence interval for the slope of the population regression line is (9016.4, 14,244.8).

(b) A national automotive group claims that the typical driver puts 15,000 miles per year on his or her main vehicle. We want to test whether AP® Statistics teachers are typical drivers. Explain why an appropriate pair of hypotheses for this test is $H_0: \beta = 15{,}000$ versus $H_a: \beta \neq 15{,}000$.

(c) Compute the test statistic and P-value for the test in part (b). What conclusion would you draw at the $\alpha = 0.05$ significance level?

(d) Does the confidence interval in part (a) lead to the same conclusion as the test in part (c)? Explain.

18. **Paired tires** Exercise 71 in Chapter 8 (page 529) compared two methods for estimating tire wear. The first method used the amount of weight lost by a tire. The second method used the amount of wear in the grooves of the tire. A random sample of 16 tires was obtained. Both methods were used to estimate the total distance traveled by each tire. The following scatterplot displays the two estimates (in thousands of miles) for each tire.[12]

**12.15** *S*: $H_0: \beta = 0$ vs. $H_a: \beta < 0$, where $\beta$ is the slope of the population regression line relating heart disease death rate to wine consumption in the population of countries. *P*: *t* test for $\beta$. *Linear*: There is no leftover pattern in the residual plot. *Independent*: The sample size ($n = 19$) is less than 10% of all countries. *Normal*: The histogram of residuals shows no strong skewness or outliers. *Equal SD*: The residual plot shows that the standard deviation of the death rates might be a little smaller for large values of wine consumption *x*, but it is hard to tell with so few data values. *Random*: Random sample.

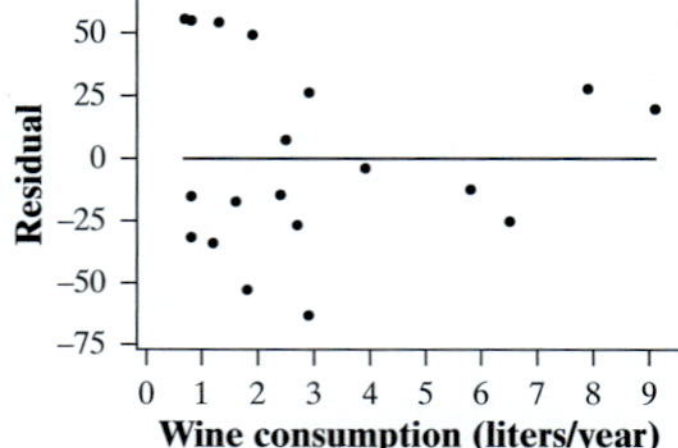


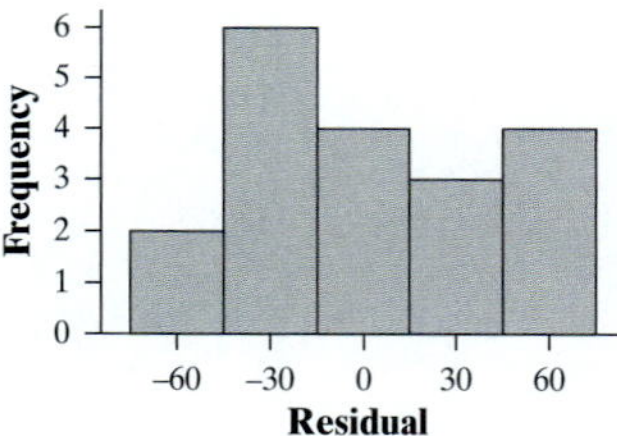


*D*: $t = -6.46$, df = 17, and *P*-value $\approx 0$. *C*: Because the *P*-value of approximately 0 is less than $\alpha = 0.05$, we reject $H_0$. There is convincing evidence of a negative linear relationship between wine consumption and heart disease death rate in the population of countries.

**12.16** *S*: $H_0: \beta = 0$ vs. $H_a: \beta < 0$, where $\beta$ is the slope of the population regression line relating pulse rate to swim time for Professor Moore. *P*: *t* test for $\beta$. *Linear*: There is no leftover pattern in the residual plot. *Independent*: The sample size ($n = 23$) is less than 10% of all days when Professor Moore swims 2000 yards. *Normal*: The histogram of residuals shows no strong skewness or outliers. *Equal SD*: The residual plot shows that the standard deviation of the pulse rates might be a little smaller for small values of *x*, but it is hard to tell with so few data values. *Random*: Random sample.

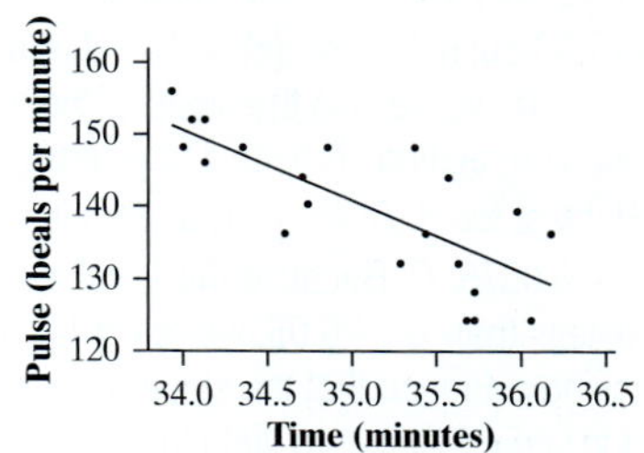


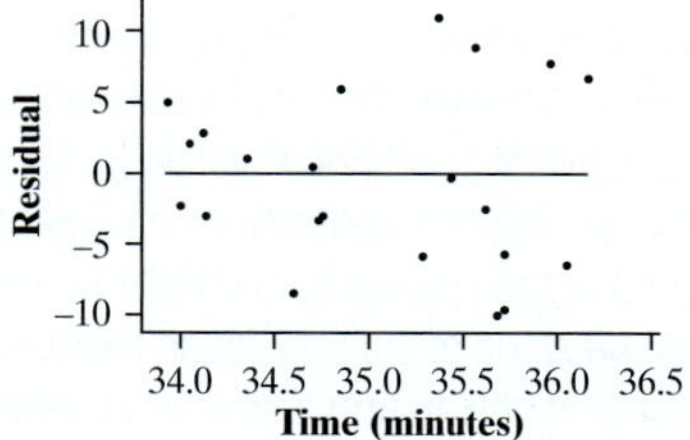


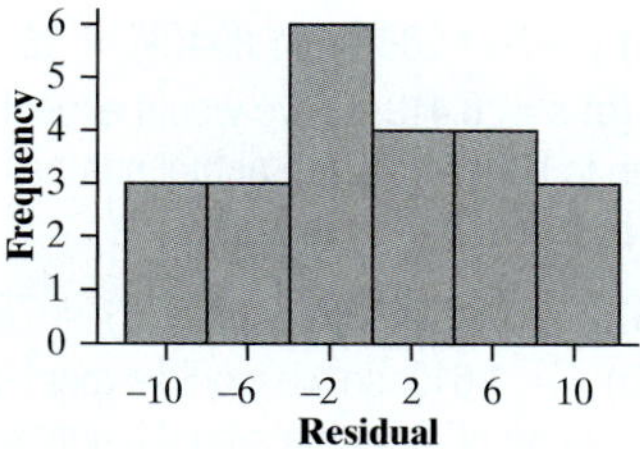

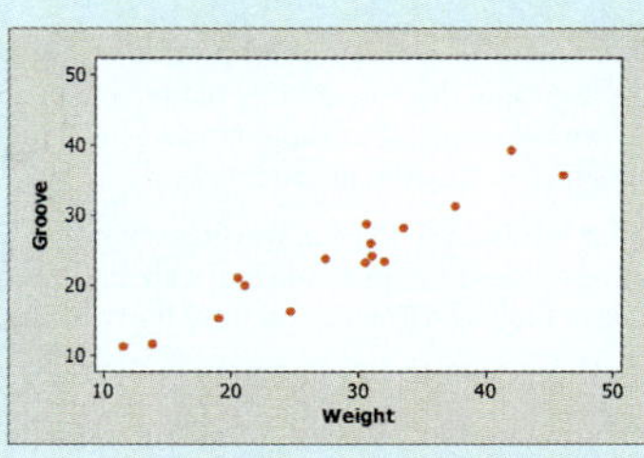


Computer output from a least-squares regression analysis of these data is shown below. Assume that the conditions for regression inference are met.

```
Predictor  Coef     SE Coef   T      P
Constant   1.351    2.105     0.64   0.531
Weight     0.79021  0.07104   11.12  0.000

S = 2.62078  R-Sq = 89.8%  R-Sq(adj) = 89.1%
```

(a) Verify that the 99% confidence interval for the slope of the population regression line is (0.5787, 1.0017).

(b) Researchers want to test whether there is a difference in the two methods of estimating tire wear. Explain why the researchers might think that an appropriate pair of hypotheses for this test is $H_0: \beta = 1$ versus $H_a: \beta \neq 1$.

(c) Compute the test statistic and *P*-value for the test in part (b). What conclusion would you draw at the $\alpha = 0.01$ significance level?

(d) Does the confidence interval in part (a) lead to the same conclusion as the test in part (c)? Explain.

***Multiple choice: Select the best answer for Exercises 19 to 24, which are based on the following information.*** To determine property taxes, Florida reappraises real estate every year, and the county appraiser's Web site lists the current "fair market value" of each piece of property. Property usually sells for somewhat more than the appraised market value. We collected data on the appraised market values *x* and actual selling prices *y* (in thousands of dollars) of a random sample of 16 condominium units in Florida. We checked that the conditions for inference about the slope of the population regression line are met. Here is part of the Minitab output from a least-squares regression analysis using these data.[13]

```
Predictor  Coef    SE Coef  T     P
Constant   127.27  79.49    1.60  0.132
Appraisal  1.0466  0.1126   9.29  0.000

S = 69.7299  R-Sq = 86.1%  R-Sq(adj) = 85.1%
```

19. The equation of the least-squares regression line for predicting selling price from appraised value is

(a) $\widehat{\text{price}} = 79.49 + 0.1126$ (appraised value).

(b) $\widehat{\text{price}} = 0.1126 + 1.0466$ (appraised value).

(c) $\widehat{\text{price}} = 127.27 + 1.0466$ (appraised value).

(d) $\widehat{\text{price}} = 1.0466 + 127.27$ (appraised value).

(e) $\widehat{\text{price}} = 1.0466 + 69.7299$ (appraised value).

20. The slope $\beta$ of the population regression line describes

(a) the exact increase in the selling price of an individual unit when its appraised value increases by \$1000.

(b) the average increase in the appraised value in a population of units when selling price increases by \$1000.

(c) the average increase in selling price in a population of units when appraised value increases by \$1000.

(d) the average increase in the appraised value in the sample of units when selling price increases by \$1000.

(e) the average increase in selling price in the sample of units when the appraised value increases by \$1000.

21. Is there convincing evidence that selling price increases as appraised value increases? To answer this question, test the hypotheses

(a) $H_0: \beta = 0$ versus $H_a: \beta > 0$.

(b) $H_0: \beta = 0$ versus $H_a: \beta < 0$.

(c) $H_0: \beta = 0$ versus $H_a: \beta \neq 0$.

(d) $H_0: \beta > 0$ versus $H_a: \beta = 0$.

(e) $H_0: \beta = 1$ versus $H_a: \beta > 1$.

22. Which of the following is the best interpretation for the value 0.1126 in the computer output?

(a) For each increase of \$1000 in appraised value, the average selling price increases by about 0.1126.

(b) When using this model to predict selling price, the predictions will typically be off by about 0.1126.

(c) 11.26% of the variation in selling price is accounted for by the linear relationship between selling price and appraised value.

(d) There is a weak, positive linear relationship between selling price and appraised value.

(e) In repeated samples of size 16, the sample slope will typically vary from the population slope by about 0.1126.

23. A 95% confidence interval for the population slope $\beta$ is

(a) $1.0466 \pm 1.046$.

(b) $1.0466 \pm 0.2415$.

(c) $1.0466 \pm 0.2387$.

(d) $1.0466 \pm 0.2207$.

(e) $1.0466 \pm 0.2400$.

**12.18 (a)** With df = 14, $0.79021 \pm 2.977(0.07104) = (0.5787, 1.0017)$. **(b)** Both variables are measuring tire wear in the same units. If one method of measuring wear gives an increase in wear of 1 unit, we expect that the other way of measuring wear would also give an increase in wear of 1 unit. **(c)** $t = -2.95$. With df = 14, the *P*-value is between 0.01 and 0.02 (0.0105). Because the *P*-value of 0.0105 is greater than $\alpha = 0.01$, we fail to reject $H_0$. There is not convincing evidence that the slope of the population line relating wear measurements using the groove method to wear measurements using the weight method is different than 1. **(d)** Yes. Because the interval in part (a) includes the value 1, the interval also does not provide convincing evidence that the slope of the population line relating wear measurements using the groove method to wear measurements using the weight method is different than 1.

**12.19** c

**12.20** c

**12.21** a

**12.22** e

**12.23** b

*D*: $t = -5.13$, df = 21, and *P*-value ≈ 0. *C*: Because the *P*-value of approximately 0 is less than $\alpha = 0.05$, we reject $H_0$. There is convincing evidence of a negative linear relationship between swim time and pulse rate in the population of days when Professor Moore swims 2000 yards.

**12.17 (a)** With df = 19, $11{,}630.6 \pm 2.093(1249) = (9016.4, 14{,}244.8)$. **(b)** Because the automotive group claims that people drive 15,000 miles per year, this says that for every increase of 1 year, the mileage would increase by 15,000 miles. **(c)** $t = -2.70$. With df = 19, the *P*-value is between 0.01 and 0.02 (0.0142). Because the *P*-value of 0.0142 is less than $\alpha = 0.05$, we reject $H_0$. We have convincing evidence that the slope of the population regression line relating miles to years is not equal to 15,000. **(d)** Yes. Because the interval in part (a) does not include the value 15,000, the interval also provides convincing evidence that the slope of the population regression line relating miles to years is not equal to 15,000.

**12.24** a

**12.25 (a)** The two treatments (say the color, read the word) were deliberately assigned to the students. **(b)** He used a randomized block design, where each student was a block. He did this to help account for the different abilities of students to read the words or to say the color they were printed in. **(c)** To help average out the effects of the order in which people did the two treatments. If every subject said the color of the printed word first and was frustrated by this task, the times for the second treatment might be worse. Then we wouldn't know the reason the times were longer for the second treatment—because of frustration or because the second method actually takes longer.

**12.26** A boxplot of the differences (color − words) is shown below. The median difference is 6 and the minimum is 0. For all students, the time it took to identify the color was the same or longer than the time needed to read the words. Typically, the students were able to read the words in about 6 seconds less time than it took them to identify the color.

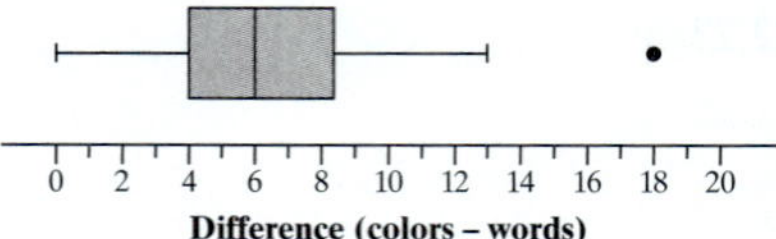


**12.27** There are a small number of differences ($n_d = 16 < 30$) and there is an outlier.

**12.28 (a)** There appears to be a moderately strong, positive linear association between the length of time to read the word and length of time to identify the color.

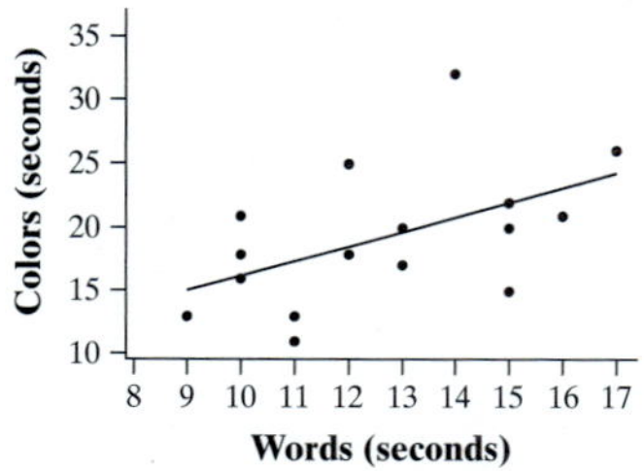


**(b)** $\hat{y} = 4.887 + 1.1321x$, where $\hat{y}$ is the predicted time to identify the color and $x$ is the amount of time to read the words. **(c)** Residual $= 13 - 15.076 = -2.076$ seconds. This student took 2.076 seconds less than expected to identify the colors, based on his or her time to read the words. **(d)** If the true slope of the regression line relating time to identify the colors and time to read the words is 0 and this experiment was repeated many times, there is a 0.0215 probability of getting an observed slope of 1.1321 or larger by chance alone.

**12.29 (a)** (i) $\dfrac{295}{1526} = 0.1933$

(ii) $\dfrac{295 + 77 + 212}{1526} = 0.3827$

(iii) $\dfrac{212}{305} = 0.6951$



24. Which of the following would have resulted in a violation of the conditions for inference?

(a) If the entire sample was selected from one neighborhood

(b) If the sample size was cut in half

(c) If the scatterplot of $x$ = appraised value and $y$ = selling price did not show a perfect linear relationship

(d) If the histogram of selling prices had an outlier

(e) If the standard deviation of appraised values was different from the standard deviation of selling prices

*Exercises 25 to 28 refer to the following setting.* Does the color in which words are printed affect your ability to read them? Do the words themselves affect your ability to name the color in which they are printed? Mr. Starnes designed a study to investigate these questions using the 16 students in his AP® Statistics class as subjects. Each student performed two tasks in a random order while a partner timed: (1) read 32 words aloud as quickly as possible, and (2) say the color in which each of 32 words is printed as quickly as possible. Try both tasks for yourself using the word list below.

| | | | |
|---|---|---|---|
| YELLOW | RED | BLUE | GREEN |
| RED | GREEN | YELLOW | YELLOW |
| GREEN | RED | BLUE | BLUE |
| YELLOW | BLUE | GREEN | RED |
| BLUE | YELLOW | RED | RED |
| RED | BLUE | YELLOW | GREEN |
| BLUE | GREEN | GREEN | BLUE |
| GREEN | YELLOW | RED | YELLOW |

25. **Color words** (4.2) Let's review the design of the study.

(a) Explain why this was an experiment and not an observational study.

(b) Did Mr. Starnes use a completely randomized design or a randomized block design? Why do you think he chose this experimental design?

(c) Explain the purpose of the random assignment in the context of the study.

The data from Mr. Starnes's experiment are shown below. For each subject, the time to perform the two tasks is given to the nearest second.

| Subject | Words | Colors | Subject | Words | Colors |
|---|---|---|---|---|---|
| 1 | 13 | 20 | 9 | 10 | 16 |
| 2 | 10 | 21 | 10 | 9 | 13 |
| 3 | 15 | 22 | 11 | 11 | 11 |
| 4 | 12 | 25 | 12 | 17 | 26 |
| 5 | 13 | 17 | 13 | 15 | 20 |
| 6 | 11 | 13 | 14 | 15 | 15 |
| 7 | 14 | 32 | 15 | 12 | 18 |
| 8 | 16 | 21 | 16 | 10 | 18 |

26. **Color words** (1.3) Do the data provide evidence of a difference in the average time required to perform the two tasks? Include an appropriate graph and numerical summaries in your answer.

27. **Color words** (9.3) Explain why it is not safe to use paired $t$ procedures to do inference about the difference in the mean time to complete the two tasks.

28. **Color words** (3.1, 3.2, 12.1) Can we use a student's word task time to predict his or her color task time?

(a) Make an appropriate scatterplot to help answer this question. Describe what you see.

(b) Use your calculator to find the equation of the least-squares regression line. Define any symbols you use.

(c) Find and interpret the residual for the student who completed the word task in 9 seconds.

(d) Assume that the conditions for performing inference about the slope of the true regression line are met. The $P$-value for a test of $H_0: \beta = 0$ versus $H_a: \beta > 0$ is 0.0215. Explain what this value means in context.

*Note:* John Ridley Stroop is often credited with the discovery in 1935 of the fact that the color in which "color words" are printed interferes with people's ability to identify the color. The so-called Stroop Effect, though, was originally published by German researchers in 1929.

*Exercises 29 and 30 refer to the following setting.* Yellowstone National Park surveyed a random sample of 1526 winter visitors to the park. They asked each person whether he or she owned, rented, or had never used a snowmobile. Respondents were also asked whether they belonged to an environmental organization (like the Sierra Club). The two-way table summarizes the survey responses.

| | Environmental Clubs | | |
|---|---|---|---|
| | No | Yes | Total |
| Never used | 445 | 212 | **657** |
| Snowmobile renter | 497 | 77 | **574** |
| Snowmobile owner | 279 | 16 | **295** |
| **Total** | **1221** | **305** | **1526** |

29. **Snowmobiles** (5.2, 5.3)

(a) If we choose a survey respondent at random, what's the probability that this individual

(i) is a snowmobile owner?

(ii) belongs to an environmental organization or owns a snowmobile?

(iii) has never used a snowmobile given that the person belongs to an environmental organization?

(b) Are the events "is a snowmobile owner" and "belongs to an environmental organization" independent for the members of the sample? Justify your answer.

(c) If we choose two survey respondents at random, what's the probability that

(i) both are snowmobile owners?

(ii) at least one of the two belongs to an environmental organization?

30. **Snowmobiles** (11.2) Do these data provide convincing evidence at the 5% significance level of an association between environmental club membership and snowmobile use for the population of visitors to Yellowstone National Park? Justify your answer.

# 12.2 Transforming to Achieve Linearity

**WHAT YOU WILL LEARN** By the end of the section, you should be able to:

- Use transformations involving powers and roots to find a power model that describes the relationship between two variables, and use the model to make predictions.
- Use transformations involving logarithms to find a power model or an exponential model that describes the relationship between two variables, and use the model to make predictions.
- Determine which of several transformations does a better job of producing a linear relationship.

In Chapter 3, we learned how to analyze relationships between two quantitative variables that showed a linear pattern. When two-variable data show a curved relationship, we must develop new techniques for finding an appropriate model. This section describes several simple *transformations* of data that can straighten a nonlinear pattern. Once the data have been transformed to achieve linearity, we can use least-squares regression to generate a useful model for making predictions. And if the conditions for regression inference are met, we can estimate or test a claim about the slope of the population (true) regression line using the transformed data.

**EXAMPLE** 

## Health and Wealth

*Straightening out a curved pattern*

The Gapminder Web site, www.gapminder.org, provides loads of data on the health and well-being of the world's inhabitants. Figure 12.9 on the next page is a scatterplot of data from Gapminder.[14] The individuals are all the world's nations for which data are available. The explanatory variable is a measure of how rich a country is: income per person. The response variable is life expectancy at birth.

**12.29 (b)** No. The probability that a person is a snowmobile owner ($295/1526 = 0.1933$) is different from the probability that the person is a snowmobile owner given that he or she belongs to an environmental organization ($16/305 = 0.0525$).
**(c)** (i) $P(\text{both are owners}) = \left(\frac{295}{1526}\right)\left(\frac{294}{1525}\right) = 0.0373$.
(ii) $P$(at least one belongs to an environmental organization) $= 1 - P(\text{neither belong})$ $= 1 - \left(\frac{1221}{1526}\right)\left(\frac{1220}{1525}\right) = 0.3599$.

**12.30** *S*: $H_0$: There is no association between environmental club membership and snowmobile use for the population of visitors to Yellowstone National Park vs. $H_a$: There is an association. . . .
*P*: Chi-square test for independence. *Random*: Random sample. *10%*: The sample size ($n = 1526$) is less than 10% of the population of visitors. *Large Counts*: 525.69, 131.31, 459.28, 114.72, 236.04, 58.96 all $\geq 5$. *D*: $\chi^2 = 116.588$. With df $= 2$, the *P*-value $\approx 0$. *C*: Because the *P*-value of approximately 0 is less than $\alpha = 0.05$, we reject $H_0$. There is convincing evidence of an association between environmental club membership and snowmobile use in the population of visitors to Yellowstone National Park.

### PD Section 12.2 Overview

Watch the video overview of Section 12.2 (for teachers) by clicking on the link in the TE-book, opening the TRCD, or logging onto the book's companion Web site www.whfreeman.com/tps5e (teacher log-in required).

### TRM Section 12.2 Alternate Examples

You can find the Alternate Examples for this section in Microsoft Word format by clicking on the link in the TE-book, opening the TRCD, or logging onto the book's companion Web site www.whfreeman.com/tps5e (teacher log-in required).

### Teaching Tip

In addition to the methods we present in this section, there are many other ways to create models for nonlinear associations. Although some of these other methods (including the ladder of powers) might be interesting to your more mathematically inclined students, we suggest that you stick with the methods listed on the AP® Topic Outline: power and logarithmic transformations. If you have extra time, spend it reviewing for the AP® exam.

### Teaching Tip

If you have an extra 20 minutes, there is a fascinating video that demonstrates how the Gapminder Web site can be used to explore data about the world's nations. Do an Internet search for "Debunking Myths about the Third World by Hans Rosling."

**ALTERNATE EXAMPLE**

### More from Gapminder

- Go to www.gapminder.org and select the "GAPMINDER WORLD" tab. A scatterplot of life expectancy versus log(income per person) appears.
- To show students a graph similar to the one on this page, choose "lin" from the menu in the lower-right corner (currently, it should say "log").
- To look at relationships between other variables, click on the variable name on either axis, and a menu with many other variables will pop up. Consider looking at the following relationships that show nonlinear patterns. Transform the variables by using logarithms for one or both variables to see if the relationships can become linear.
  - Income per person *y* versus Children per woman *x*
  - Income per person *y* versus Under-5 mortality rate *x*
  - Income per person *y* versus Literacy rate (adult total) *x*
- To see how these relationships change over time, move the slider at the bottom of the graph to 1800 and press "Play."

We expect people in richer countries to live longer because they have better access to medical care and typically lead healthier lives. The overall pattern of the scatterplot does show this, but the relationship is not linear. Life expectancy rises very quickly as income per person increases and then levels off. People in very rich countries such as the United States live no longer than people in poorer but not extremely poor nations. In some less wealthy countries, people live longer than in the United States.

Four African nations are outliers. Their life expectancies are similar to those of their neighbors, but their income per person is higher. Gabon and Equatorial Guinea produce oil, and South Africa and Botswana produce diamonds. It may be that income from mineral exports goes mainly to a few people and so pulls up income per person without much effect on either the income or the life expectancy of ordinary citizens. That is, income per person is a mean, and we know that mean income can be much higher than median income.

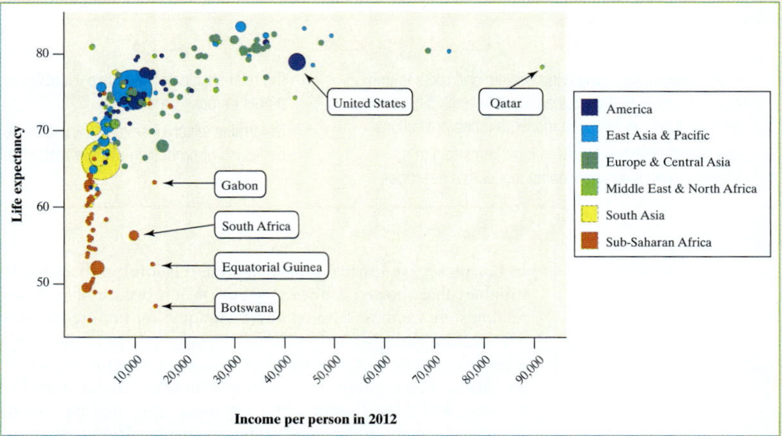


**FIGURE 12.9** Scatterplot of the life expectancy of people in many nations against each nation's income per person. The color of each circle indicates the geographic region in which that country is located. The size of each circle is based on the population of the country—bigger circles indicate larger populations.

The scatterplot in Figure 12.9 shows a curved pattern. We can straighten things out using logarithms. Figure 12.10 (on the facing page) plots the logarithm of income per person against life expectancy for these same countries. The effect is almost magical. This graph has a clear, linear pattern.

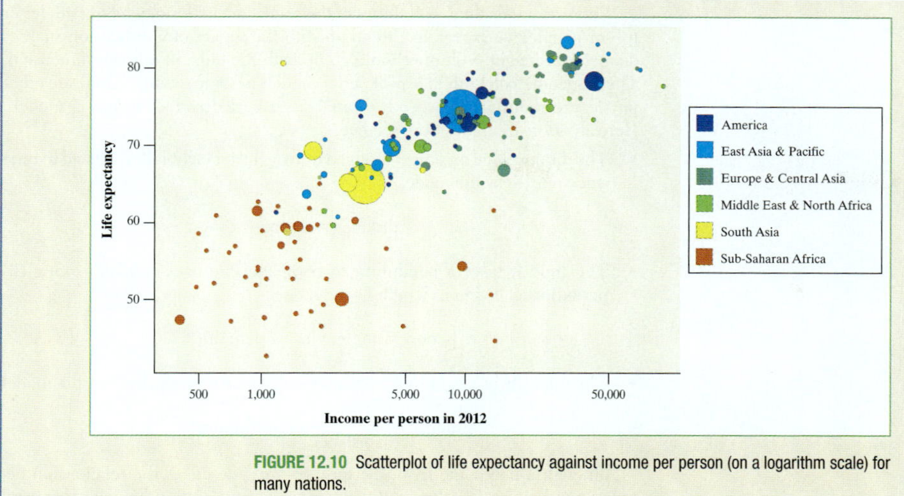


**FIGURE 12.10** Scatterplot of life expectancy against income per person (on a logarithm scale) for many nations.

Applying a function such as the logarithm or square root to a quantitative variable is called **transforming** the data. We will see in this section that understanding how simple functions work helps us choose and use transformations to straighten nonlinear patterns.

Transforming data amounts to changing the scale of measurement that was used when the data were collected. We can choose to measure temperature in degrees Fahrenheit or in degrees Celsius, distance in miles or in kilometers. These changes of units are *linear transformations*, discussed in Chapter 2.

Linear transformations cannot straighten a curved relationship between two variables. To do that, we resort to functions that are not linear. The logarithm function, applied in the "Health and Wealth" example, is a nonlinear function. We'll return to transformations involving logarithms later.

## Transforming with Powers and Roots

When you visit a pizza parlor, you order a pizza by its diameter—say, 10 inches, 12 inches, or 14 inches. But the amount you get to eat depends on the area of the pizza. The area of a circle is $\pi$ times the square of its radius $r$. So the area of a round pizza with diameter $x$ is

$$\text{area} = \pi r^2 = \pi\left(\frac{x}{2}\right)^2 = \pi\left(\frac{x^2}{4}\right) = \frac{\pi}{4}x^2$$

This is a **power model** of the form $y = ax^p$ with $a = \pi/4$ and $p = 2$.

### Teaching Tip

Notice that the power model we present here has only one term. There are other multi-term models that can be used for nonlinear associations, such as quadratic models in the form $y = ax^2 + bx + c$. These more complicated models are not included on the AP® Topic Outline, however.

ALTERNATE EXAMPLE

### *Child mortality and income*

What does a country's mortality rate for children under 5 years of age (per 1000 live births) tell us about the income per person (measured in gross domestic product per person adjusted for differences in purchasing power) for residents of that country? Here are the data and a scatterplot for a random sample of 14 countries in a recent year (data from www.gapminder.org):

| Country | Mortality rate | Income per person |
|---|---|---|
| Switzerland | 4.4 | 38,003.90 |
| Timor-Leste | 56.4 | 2,475.68 |
| Uganda | 127.5 | 1,202.53 |
| Ghana | 68.5 | 1,382.95 |
| Peru | 21.3 | 7,858.97 |
| Cambodia | 87.5 | 1,830.97 |
| Suriname | 26.3 | 8,199.03 |
| Armenia | 21.6 | 4,523.44 |
| Sweden | 2.8 | 32,021.00 |
| Niger | 160.3 | 643.39 |
| Serbia | 7.1 | 10,005.20 |
| Kenya | 84.0 | 1,493.53 |
| Fiji | 17.6 | 4,016.20 |
| Grenada | 14.5 | 8,826.90 |

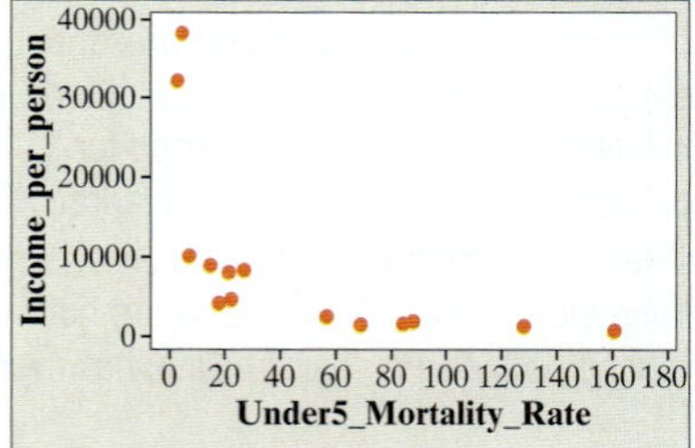


The scatterplot shows a strong negative association that is clearly nonlinear. Because the horizontal and vertical axes look like asymptotes, perhaps there is a reciprocal relationship between the variables, such as

$$\text{income per person} = \frac{a}{\text{child mortality rate}} = a(\text{child mortality rate})^{-1}$$

If we find the reciprocal of child mortality rate and graph income versus 1/(child mortality rate), we find a linear relationship!

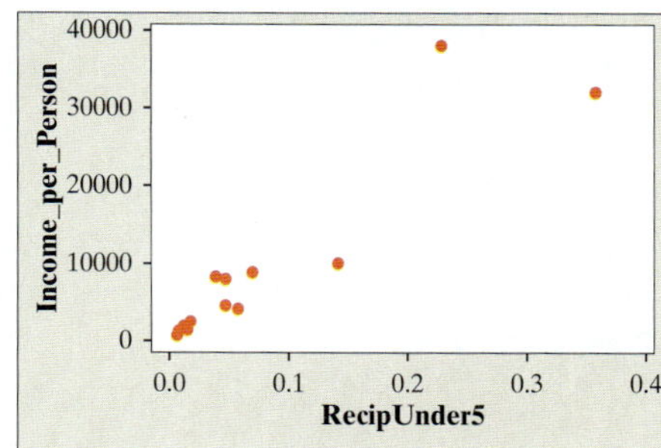


Likewise, we could calculate the reciprocal of the income per person and graph 1/(income per person) versus child mortality rate. Again, we get a linear relationship!

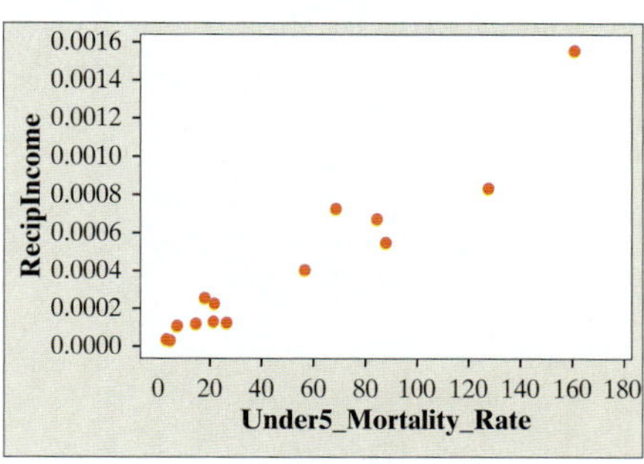


When we are dealing with things of the same general form, whether circles or fish or people, we expect area to go up with the square of a dimension such as diameter or height. Volume should go up with the cube of a linear dimension. That is, geometry tells us to expect power models in some settings. There are other physical relationships between two variables that are described by power models. Here are some examples from science.

- The distance that an object dropped from a given height falls is related to time since release by the model

$$\text{distance} = a(\text{time})^2$$

- The time it takes a pendulum to complete one back-and-forth swing (its period) is related to its length by the model

$$\text{period} = a\sqrt{\text{length}} = a(\text{length})^{1/2}$$

- The intensity of a light bulb is related to distance from the bulb by the model

$$\text{intensity} = \frac{a}{\text{distance}^2} = a(\text{distance})^{-2}$$

Although a power model of the form $y = ax^p$ describes the relationship between $x$ and $y$ in each of these settings, there is a *linear* relationship between $x^p$ and $y$. If we transform the values of the explanatory variable $x$ by raising them to the $p$ power, and graph the points $(x^p, y)$, the scatterplot should have a linear form. The following example shows what we mean.

## EXAMPLE Go Fish!

### *Transforming with powers*

Imagine that you have been put in charge of organizing a fishing tournament in which prizes will be given for the heaviest Atlantic Ocean rockfish caught. You know that many of the fish caught during the tournament will be measured and released. You are also aware that using delicate scales to try to weigh a fish that is flopping around in a moving boat will probably not yield very accurate results. It would be much easier to measure the length of the fish while on the boat. What you need is a way to convert the length of the fish to its weight.

You contact the nearby marine research laboratory, and they provide reference data on the length (in centimeters) and weight (in grams) for Atlantic Ocean rockfish of several sizes.[15]

| **Length:** | 5.2 | 8.5 | 11.5 | 14.3 | 16.8 | 19.2 | 21.3 | 23.3 | 25.0 | 26.7 |
|---|---|---|---|---|---|---|---|---|---|---|
| **Weight:** | 2 | 8 | 21 | 38 | 69 | 117 | 148 | 190 | 264 | 293 |

| **Length:** | 28.2 | 29.6 | 30.8 | 32.0 | 33.0 | 34.0 | 34.9 | 36.4 | 37.1 | 37.7 |
|---|---|---|---|---|---|---|---|---|---|---|
| **Weight:** | 318 | 371 | 455 | 504 | 518 | 537 | 651 | 719 | 726 | 810 |

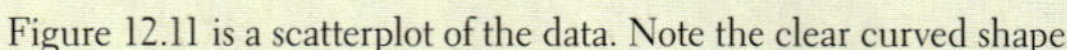

Figure 12.11 is a scatterplot of the data. Note the clear curved shape.

Because length is one-dimensional and weight (like volume) is three-dimensional, a power model of the form weight = $a$ (length)$^3$ should describe the relationship. What happens if we cube the lengths in the data table and then graph weight versus length$^3$? Figure 12.12 gives us the answer. This transformation of the explanatory variable helps us produce a graph that is quite linear.

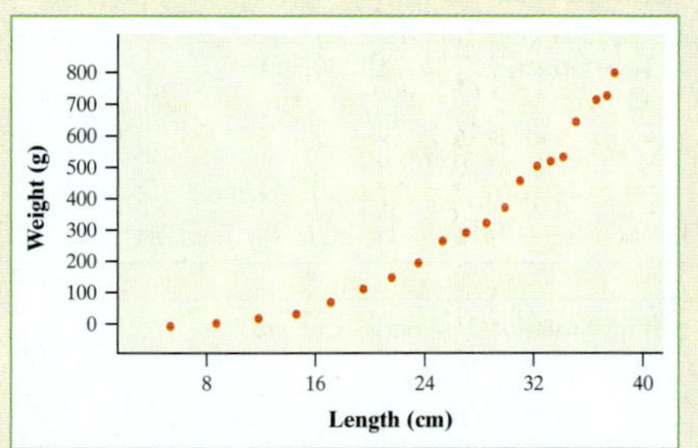


**FIGURE 12.11** Scatterplot of Atlantic Ocean rockfish weight versus length.

**FIGURE 12.12** The scatterplot of weight versus length$^3$ is linear.

There's another way to transform the data in the example to achieve linearity. We can take the cube root of the weight values and graph $\sqrt[3]{\text{weight}}$ versus length. Figure 12.13 shows that the resulting scatterplot has a linear form. Why does this transformation work? Start with weight = $a$(length)$^3$ and take the cube root of both sides of the equation:

$$\sqrt[3]{\text{weight}} = \sqrt[3]{a(\text{length})^3}$$
$$\sqrt[3]{\text{weight}} = \sqrt[3]{a}(\text{length})$$

That is, there is a linear relationship between length and $\sqrt[3]{\text{weight}}$.

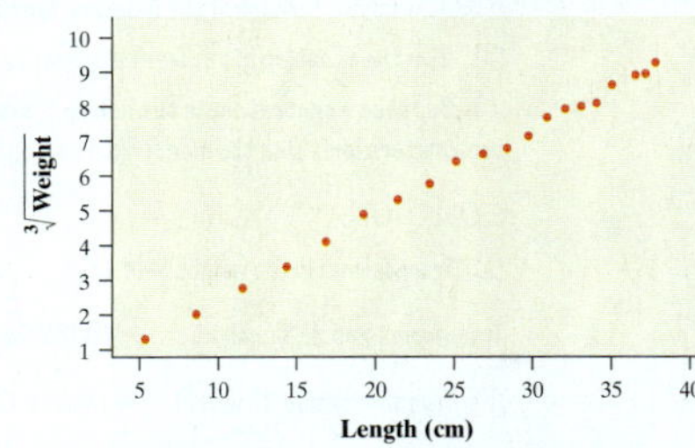


**FIGURE 12.13** The scatterplot of $\sqrt[3]{\text{weight}}$ versus length is linear.

Once we straighten out the curved pattern in the original scatterplot, we fit a least-squares line to the transformed data. This linear model can be used to predict values of the response variable $y$. As in Chapter 3, a residual plot tells us if the linear model is appropriate. The values of $s$ and $r^2$ tell us how well the regression line fits the data.

## Teaching Tip

You may need to help your students see how $\sqrt[3]{\text{weight}} = \sqrt[3]{a}$ (length) is in a linear form. Remind them that an equation is linear if it is in the form: variable = constant + constant(variable). In this case, $\sqrt[3]{\text{weight}}$ is the variable on the left side, 0 is the first constant, $\sqrt[3]{a}$ is the second constant, and length is the second variable.

## Teaching Tip

When reading the example on this page, ask your students why the values of *s* are so different, while the values of $r^2$ are so similar. The standard deviations of the residuals are so different because they are measured in the units of the response variable, which are on different scales because of the cube root transformation in the second model. The $r^2$-values are nearly the same because $r^2$ does not depend on the units the variables are measured in.

## EXAMPLE Go Fish!

### *Transforming with Powers and Roots*

Here is Minitab output from separate regression analyses of the two sets of transformed Atlantic Ocean rockfish data.

Transformation 1: (length$^3$, weight)

| Predictor | Coef | SE Coef | T | P |
|---|---|---|---|---|
| Constant | 4.066 | 6.902 | 0.59 | 0.563 |
| Length^3 | 0.0146774 | 0.0002404 | 61.07 | 0.000 |

S = 18.8412 R-Sq = 99.5% R-Sq(adj) = 99.5%

Transformation 2: (length, $\sqrt[3]{\text{weight}}$)

| Predictor | Coef | SE Coef | T | P |
|---|---|---|---|---|
| Constant | −0.02204 | 0.07762 | −0.28 | 0.780 |
| Length | 0.246616 | 0.002868 | 86.00 | 0.000 |

S = 0.124161 R-Sq = 99.8% R-Sq(adj) = 99.7%

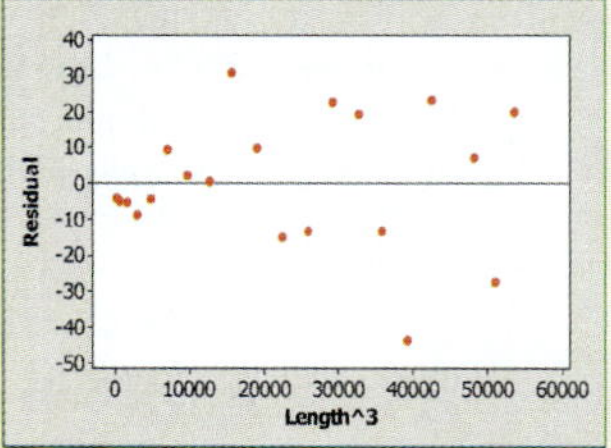


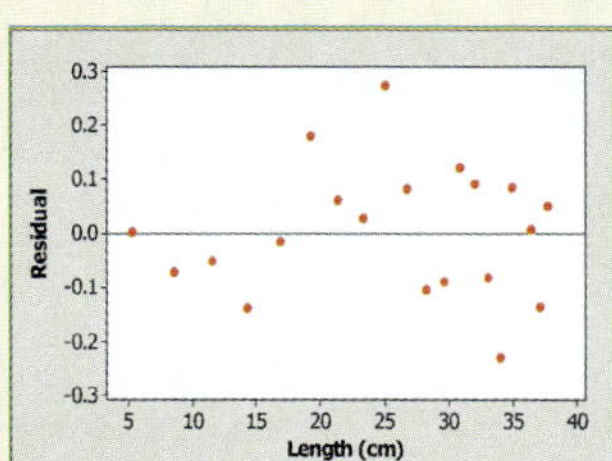


PROBLEM: Do each of the following for *both* transformations.

(a) Give the equation of the least-squares regression line. Define any variables you use.

(b) Suppose a contestant in the fishing tournament catches an Atlantic Ocean rockfish that's 36 centimeters long. Use the model from part (a) to predict the fish's weight. Show your work.

SOLUTION:

(a) Transformation 1: $\widehat{\text{weight}} = 4.066 + 0.0146774(\text{length}^3)$

Transformation 2: $\widehat{\sqrt[3]{\text{weight}}} = -0.02204 + 0.246616(\text{length})$

(b) Transformation 1: $\widehat{\text{weight}} = 4.066 + 0.0146774(36^3) = 688.9$ grams

Transformation 2: $\widehat{\sqrt[3]{\text{weight}}} = -0.02204 + 0.246616(36) = 8.856$

$\widehat{\text{weight}} = 8.856^3 = 694.6$ grams

For Practice Try Exercise 33

## ALTERNATE EXAMPLE

### *Child mortality and income*

Here is Minitab output from separate regression analyses of the two sets of transformed data:

Transformation 1: $\left(\frac{1}{\text{Under5}}, \text{Income}\right)$

| Predictor | Coef | SE Coef | T | P |
|---|---|---|---|---|
| Constant | 869 | 1615 | 0.54 | 0.601 |
| RecipUnder5 | 104868 | 13074 | 8.02 | 0.000 |

S = 4797.65 R-Sq = 84.3% R-Sq(adj) = 83.0%

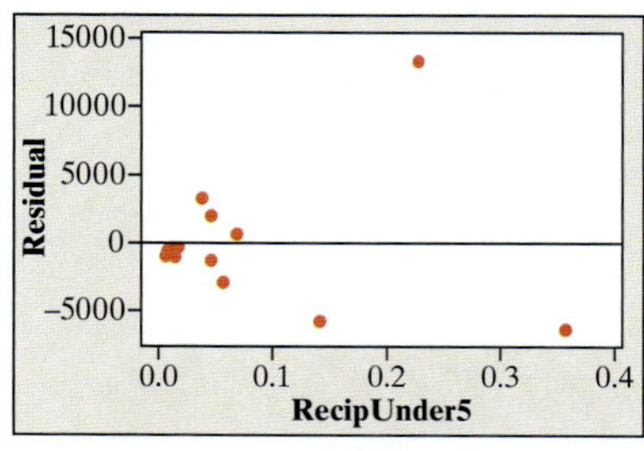


Transformation 2: $\left(\text{Under5}, \frac{1}{\text{Income}}\right)$

| Predictor | Coef | SE Coef | T | P |
|---|---|---|---|---|
| Constant | -0.00000585 | 0.00004849 | -0.12 | 0.906 |
| Under5_Mortality_Rate | 0.00000829 | 0.00000070 | 11.80 | 0.000 |

S = 0.000125175 R-Sq = 92.1% R-Sq(adj) = 91.4%

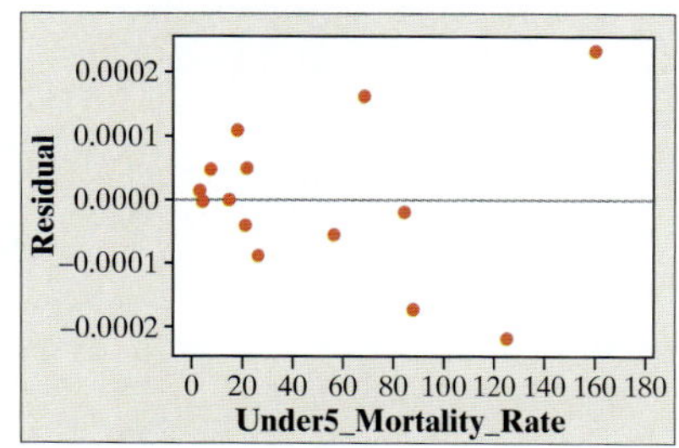

When experience or theory suggests that the relationship between two variables is described by a power model of the form $y = ax^p$, you now have two strategies for transforming the data to achieve linearity.

1. Raise the values of the explanatory variable $x$ to the $p$ power and plot the points $(x^p, y)$.
2. Take the $p$th root of the values of the response variable $y$ and plot the points $(x, \sqrt[p]{y})$

What if you have no idea what power to choose? You could guess and test until you find a transformation that works. Some technology comes with built-in sliders that allow you to dynamically adjust the power and watch the scatterplot change shape as you do.

It turns out that there is a much more efficient method for linearizing a curved pattern in a scatterplot. Instead of transforming with powers and roots, we use logarithms. This more general method works when the data follow an unknown power model or any of several other common mathematical models.

## Transforming with Logarithms

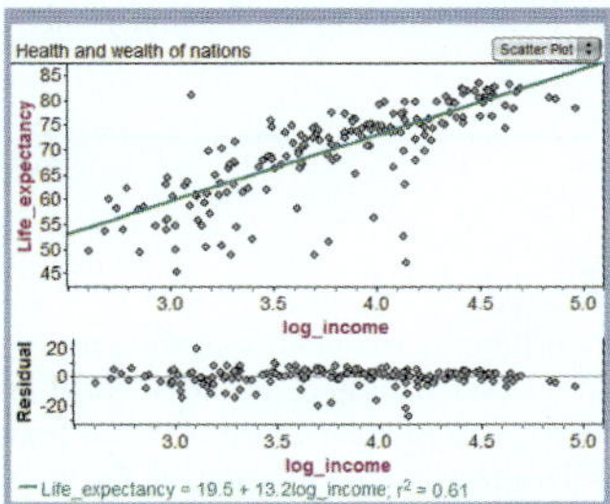


**FIGURE 12.14** Scatterplot with least-squares line added and residual plot from Fathom for the transformed data about the health and wealth of nations.

Not all curved relationships are described by power models. For instance, in the "Health and Wealth" example (page 765), a graph of life expectancy versus the logarithm (base 10) of income per person showed a linear pattern. We used Fathom software to fit a least-squares regression line to the transformed data and to make a residual plot. Figure 12.14 shows the results.

The regression line is

$$\widehat{\text{life expectancy}} = 19.5 + 13.2\log(\text{income})$$

How well does this model fit the data? The residual plot shows a random scatter of prediction errors about the residual = 0 line. Also, because $r^2 = 0.61$, about 61% of the variation in life expectancy is accounted for by the linear model using log(income) as the explanatory variable.

The relationship between life expectancy and income per person is described by a logarithmic model of the form $y = a + b\log x$. We can use this model to predict how long a country's citizens will live from how much money they make. For the United States, which has income per person of \$42,296.20,

$$\widehat{\text{life expectancy}} = 19.5 + 13.2\log(42{,}296.20) = 80.567 \text{ years}$$

The actual U.S. life expectancy in 2012 was 78.80 years.

Taking the logarithm of the income per person values straightened out the curved pattern in the original scatterplot. The logarithm transformation can also help achieve linearity when the relationship between two variables is described by a *power model* or an *exponential model.*

**Power Models** Biologists have found that many characteristics of living things are described quite closely by power models. There are more mice than elephants, and more flies than mice—the abundance of species follows a power model with body weight as the explanatory variable. So do pulse rate, length of life, the number of eggs a bird lays, and so on.

### Teaching Tip

Students do not need to know the algebraic properties of logarithms for the AP® exam. However, students should know how to use their calculators to transform data using logarithms and how to make a prediction using a least-squares regression line created with transformed data.

### Teaching Tip

Make sure your students are comfortable using both the natural log and the base-10 log when transforming data. Both transformations could appear on the AP® exam and we use both throughout this section.

PROBLEM:

Do the following for both transformations:

**(a)** Give the equation of the least-squares regression line. Define any variables you use.

**(b)** Predict the income per person for Turkey, which had a child mortality rate of 20.3.

SOLUTION:

**(a)** Transformation 1: $\widehat{\text{income}} = 869 + \dfrac{104{,}868}{\text{Under5}}$

Transformation 2: $\widehat{\dfrac{1}{\text{income}}} = -0.00000585 + 0.00000829(\text{Under5})$

**(b)** Transformation 1:

$$\widehat{\text{income}} = 869 + \frac{104{,}868}{20.3} = \$6034.91$$

Transformation 2: $\widehat{\dfrac{1}{\text{income}}}$

$= -0.00000585 + 0.00000829(20.3)$

$= 0.000162437$

$$\widehat{\text{income}} = \frac{1}{0.000162437} = \$6156.23$$

Sometimes the powers can be predicted from geometry, but sometimes they are mysterious. Why, for example, does the rate at which animals use energy go up as the 3/4 power of their body weight? Biologists call this relationship Kleiber's law. It has been found to work all the way from bacteria to whales. The search goes on for some physical or geometrical explanation for why life follows power laws.

To achieve linearity from a power model, we apply the logarithm transformation to *both* variables. Here are the details:

1. A power model has the form $y = ax^p$, where $a$ and $p$ are constants.
2. Take the logarithm of both sides of this equation. Using properties of logarithms, we get

$$\log y = \log(ax^p) = \log a + \log(x^p) = \log a + p \log x$$

The equation $\log y = \log a + p \log x$ shows that taking the logarithm of both variables results in a linear relationship between $\log x$ and $\log y$.

3. Look carefully: the *power* $p$ in the power model becomes the *slope* of the straight line that links $\log y$ to $\log x$.

If a power model describes the relationship between two variables, a scatterplot of the logarithms of both variables should produce a linear pattern. Then we can fit a least-squares regression line to the transformed data and use the linear model to make predictions. Here's an example.

## EXAMPLE Go Fish!

### *Transforming with logarithms*

Let's return to the fishing tournament from the previous example. Our goal remains the same: to find a model for predicting the weight of an Atlantic Ocean rockfish from its length. We still expect a power model of the form weight = $a$(length)$^3$ based on geometry. Here once again is a scatterplot of the data from the local marine research lab.

Earlier, we transformed the data in two ways to try to achieve linearity: (1) cubing the length values and (2) taking the cube root of the weight values. This time we'll use logarithms.

We took the logarithm (base 10) of the values for both variables. Some computer output from a linear regression analysis on the transformed data is shown below.

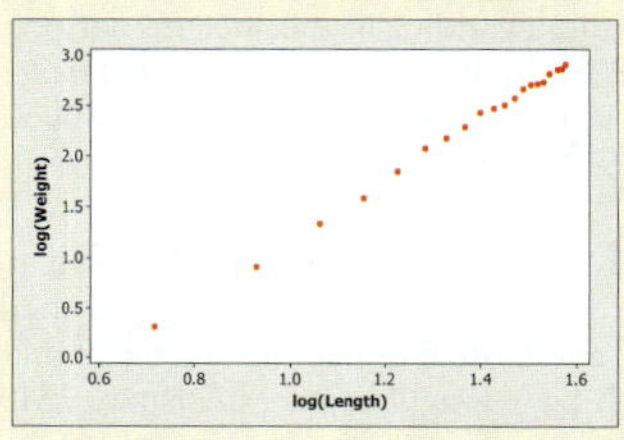


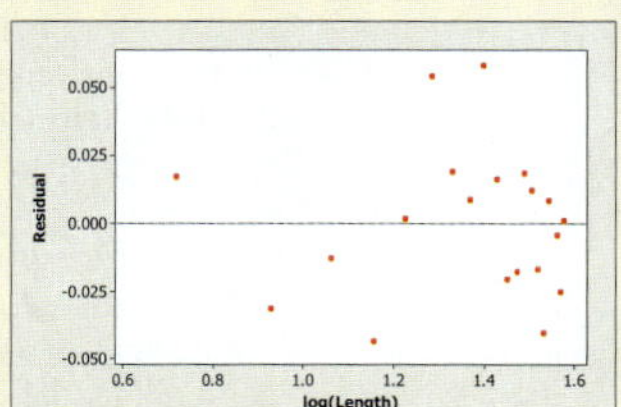


## Teaching Tip

As on page 769, you may need to help your students see how $\log y = \log a + p \log x$ is in a linear form. Remind them that an equation is linear if it is in the form: variable = constant + constant(variable). In this case, $\log y$ is the variable on the left side, $\log a$ is the first constant, $p$ is the second constant, and $\log x$ is the second variable.

## Teaching Tip

Emphasize the paragraph just before the example on this page. Also point out that the converse of the first sentence is also true: if a scatterplot of the logarithm of both variables produces a linear pattern, then a power model describes the relationship between the two variables.

## ALTERNATE EXAMPLE

### *Child mortality and income*

In an earlier Alternate Example, we transformed the variables Under5 child mortality and Income per person using the reciprocal function for a random sample of 14 countries. We can also try transforming these variables using the natural logarithm function. Here is a scatterplot of data:

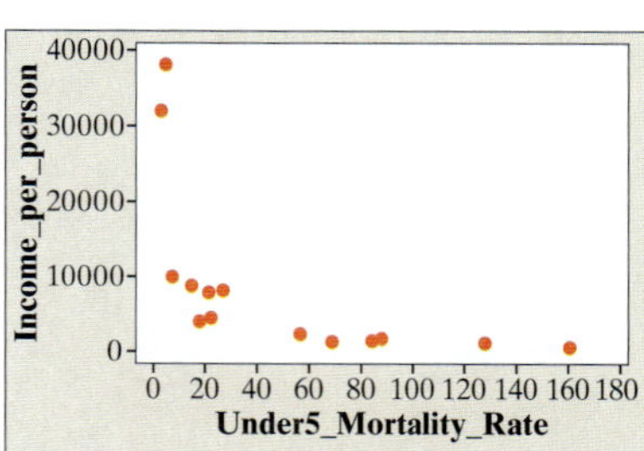


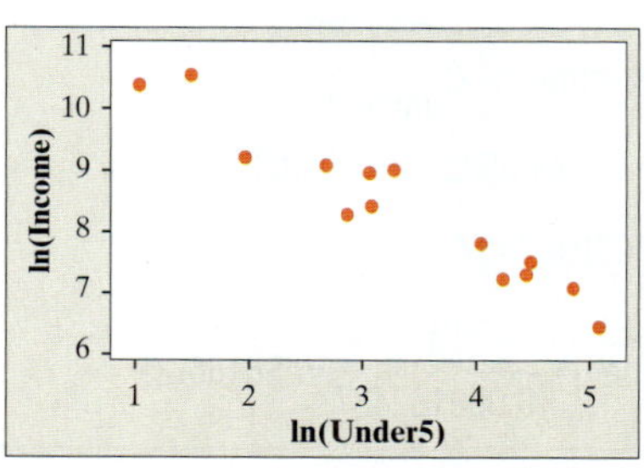


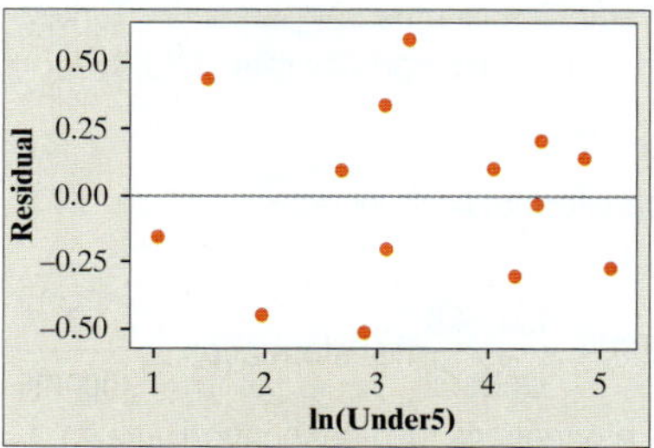


Now, here is a scatterplot and some computer output showing the relationship between the natural logarithm of Income and the natural logarithm of Under5 mortality rate.

Regression Analysis: log(Weight) versus log(Length)

| Predictor | Coef | SE Coef | T | P |
|---|---|---|---|---|
| Constant | −1.89940 | 0.03799 | −49.99 | 0.000 |
| log(Length) | 3.04942 | 0.02764 | 110.31 | 0.000 |

S = 0.0281823 R-Sq = 99.9% R-Sq(adj) = 99.8%

PROBLEM:

(a) Based on the output, explain why it would be reasonable to use a power model to describe the relationship between weight and length for Atlantic Ocean rockfish.

(b) Give the equation of the least-squares regression line. Be sure to define any variables you use.

SOLUTION:

(a) If a power model describes the relationship between two variables *x* and *y*, then a *linear* model should describe the relationship between log *x* and log *y*. The scatterplot of log(weight) versus log(length) has a linear form, and the residual plot shows a fairly random scatter of points about the residual = 0 line. So a power model seems reasonable here.

(b) $\widehat{\log(\text{weight})} = -1.89940 + 3.04942 \log(\text{length})$

**For Practice** *Try Exercise* **35**

On the TI-83/84, you can "undo" the logarithm using the [2nd] function keys. To solve log $y = 2$, press [2nd] [LOG] [2] [ENTER]. To solve ln $y = 2$, press [2nd] [LN] [2] [ENTER].

If we fit a least-squares regression line to the transformed data, we can find the predicted value of the logarithm of *y* for any value of the explanatory variable *x* by substituting our *x*-value into the equation of the line. To obtain the corresponding prediction for the response variable *y*, we have to "undo" the logarithm transformation to return to the original units of measurement. One way of doing this is to use the definition of a logarithm as an exponent:

$$\log_b a = x \Rightarrow b^x = a$$

For instance, if we have log $y = 2$, then

$$\log y = 2 \Rightarrow \log_{10} y = 2 \Rightarrow 10^2 = y \Rightarrow 100 = y$$

If instead we have ln $y = 2$, then

$$\ln y = 2 \Rightarrow \log_e y = 2 \Rightarrow e^2 = y \Rightarrow 7.389 = y$$

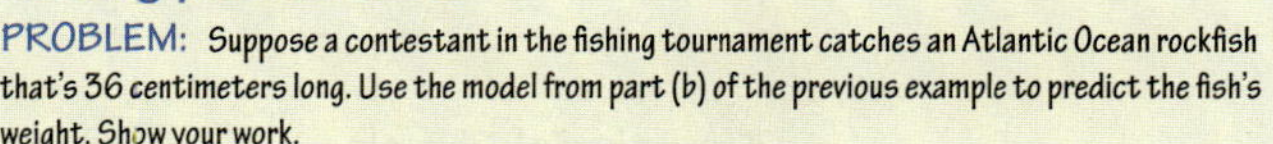

## EXAMPLE Go Fish!

### *Making predictions*

PROBLEM: Suppose a contestant in the fishing tournament catches an Atlantic Ocean rockfish that's 36 centimeters long. Use the model from part (b) of the previous example to predict the fish's weight. Show your work.

SOLUTION: For a length of 36 centimeters, we have

$$\widehat{\log(\text{weight})} = -1.89940 + 3.04942 \log(36) = 2.8464$$

### Teaching Tip

Make sure your students include the "log" when stating the equation of the least-squares regression line.

### Teaching Tip

Although it would be nice if students understood the properties of logarithms, this understanding is not necessary for the AP® exam. As stated in the margin note, students simply need to be able to "undo" a logarithm by pressing `2nd log` or `2nd ln`.

### ALTERNATE EXAMPLE

### *Child mortality and income*

PROBLEM:

Using the model from the previous example, predict the income per person for Turkey, which had a child mortality rate of 20.3.

SOLUTION:

$\widehat{\ln(\text{Income})} = 11.495 - 0.9374 \ln(20.3) = 8.6728$

Thus, $\widehat{\text{Income}} = e^{8.6728} = \$5842$.

Regression Analysis: ln(Income) versus ln(Under5)

| Predictor | Coef | SE Coef | T | P |
|---|---|---|---|---|
| Constant | 11.4950 | 0.2680 | 42.90 | 0.000 |
| ln(Under5) | -0.93740 | 0.07579 | -12.37 | 0.000 |

S = 0.343501 R-Sq = 92.7% R-Sq(adj) = 92.1%

PROBLEM:

**(a)** Based on the output, explain why it would be reasonable to use a power model to describe the relationship between Income and Under5 mortality rate.

**(b)** Give the equation of the least-squares regression line, defining any variables you use.

SOLUTION:

**(a)** The scatterplot of ln(Income) versus ln(Under5) shows a linear form and there are no leftover patterns in the residual plot. Because the relationship between ln(Income) and ln(Under5) is linear, a power model should describe the relationship between Income and Under5.

**(b)** $\widehat{\ln(\text{Income})} = 11.495 - 0.9374 \ln(\text{Under5})$

To find the predicted weight, we use the definition of a logarithm as an exponent:

$$\widehat{\log_{10}(\text{weight})} = 2.8464$$
$$\widehat{\text{weight}} = 10^{2.8464} \approx 702.1$$

This model predicts that a 36-centimeter-long rockfish will weigh about 702 grams.

**For Practice** *Try Exercise* **37**

Your calculator and most statistical software will calculate the logarithms of all the values of a variable with a single command. The important thing to remember is this: if the relationship between two variables is described by a power model, then we can linearize the relationship by taking the logarithm of *both* the explanatory and response variables.

## Teaching Tip

On the AP® exam, students are not expected to work backward to express a power model in $\hat{y} =$ form. However, if you have strong math students, it is fun to solve for $\hat{y}$ and plot the power model with the original data as we did in the Think About It feature on this page.

## Teaching Tip

When using least-squares regression lines in the form $\widehat{\ln y} = a + b(\ln x)$, the slope $b$ corresponds to the exponent in a power model. In the previous Alternate Example, the slope of the least-squares regression line using the transformed data was $-0.9374$. Because this is very close to $-1$, our choice of a reciprocal transformation in the earlier Alternate Example seems justified.

**How do we find the power model for predicting $y$ from $x$?** The least-squares line for the transformed rockfish data is

$$\widehat{\log(\text{weight})} = -1.89940 + 3.04942\log(\text{length})$$

If we use the definition of the logarithm as an exponent, we can rewrite this equation as

$$\widehat{\text{weight}} = 10^{-1.89940 + 3.04942\log(\text{length})}$$

Using properties of exponents, we can simplify this as follows:

$$\widehat{\text{weight}} = 10^{-1.89940} \cdot 10^{3.04942\log(\text{length})} \quad \text{using the fact that } b^m b^n = b^{m+n}$$
$$\widehat{\text{weight}} = 10^{-1.89940} \cdot 10^{\log(\text{length})^{3.04942}} \quad \text{using the fact that } p\log x = \log x^p$$
$$\widehat{\text{weight}} = 0.0126(\text{length})^{3.04942} \quad \text{using the fact that } 10^{\log x} = x$$

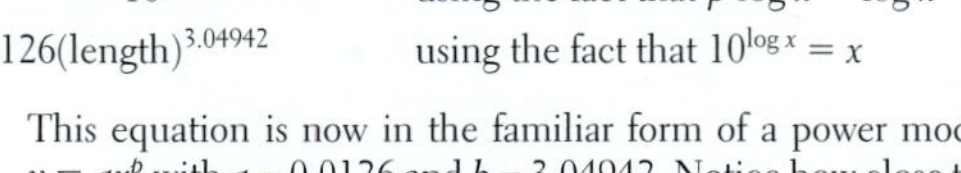

FIGURE 12.15 Rockfish data with power model.

This equation is now in the familiar form of a power model $y = ax^p$ with $a = 0.0126$ and $b = 3.04942$. Notice how close the power is to 3, as expected from geometry.

We could use the power model to predict the weight of a 36-centimeter-long Atlantic Ocean rockfish:

$$\widehat{\text{weight}} = 0.0126(36)^{3.04942} \approx 701.76 \text{ grams}$$

This is the same prediction we got earlier (up to rounding). The scatterplot of the original rockfish data with the power model added appears in Figure 12.15. Note how well this model fits the data!

**Exponential Models** A linear model has the form $y = a + bx$. The value of $y$ increases (or decreases) at a constant rate as $x$ increases. The slope $b$ describes the constant rate of change of a linear model. That is, for each 1 unit increase in $x$, the model predicts an increase of $b$ units in $y$. You can think of a linear model as describing the repeated addition of a constant amount. Sometimes the relationship between $y$ and $x$ is based on repeated *multiplication* by a constant factor. That is, each time $x$ increases by 1 unit, the value of $y$ is multiplied by $b$. An **exponential model** of the form $y = ab^x$ describes such multiplicative growth.

Populations of living things tend to grow exponentially if not restrained by outside limits such as lack of food or space. More pleasantly (unless we're talking about credit card debt!), money also displays exponential growth when interest is compounded each time period. Compounding means that last period's income earns income in the next period.

## EXAMPLE Money, Money, Money

### *Understanding exponential growth*

Suppose that you invest \$100 in a savings account that pays 6% interest compounded annually. After a year, you will have earned \$100(0.06) = \$6.00 in interest. Your new account balance is the initial deposit plus the interest earned: \$100 + (\$100)(0.06), or \$106. We can rewrite this as \$100(1 + 0.06), or more simply as \$100(1.06). That is, 6% annual interest means that any amount on deposit for the entire year is multiplied by 1.06.

If you leave the money invested for a second year, your new balance will be $[\$100(1.06)](1.06) = \$100(1.06)^2 = \$112.36$. Notice that you earn \$6.36 in interest during the second year. That's another \$6 in interest from your initial \$100 deposit plus the interest on your \$6 interest earned for Year 1. After $x$ years, your account balance $y$ is given by the exponential model $y = 100(1.06)^x$.

The table below shows the balance in your savings account at the end of each of the first six years. Figure 12.16 shows the growth in your investment over 100 years. It is characteristic of exponential growth that the increase appears slow for a long period and then seems to explode.

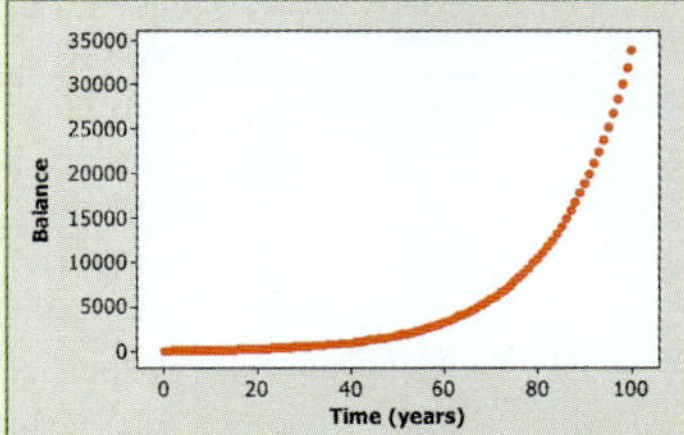


**FIGURE 12.16** Scatterplot of the growth of a \$100 investment in a savings account paying 6% interest, compounded annually.

| Time $x$ (years) | Account balance $y$ |
|---|---|
| 0 | \$100.00 |
| 1 | \$106.00 |
| 2 | \$112.36 |
| 3 | \$119.10 |
| 4 | \$126.25 |
| 5 | \$133.82 |
| 6 | \$141.85 |
| ... | ... |

If an exponential model of the form $y = ab^x$ describes the relationship between $x$ and $y$, we can use logarithms to transform the data to produce a linear relationship. Start by taking the logarithm (we'll use base 10, but the natural logarithm ln

## Teaching Tip

You can also use models in the form $y = ab^x$ to model exponential decay, as in the following Alternate Example. When the value of $y$ decreases as the value of $x$ increases, the base $b$ will be between 0 and 1.

## ALTERNATE EXAMPLE

### *Dating with carbon 14*

When scientists estimate the age of fossils, they often measure the amount of the radioactive isotope carbon 14 in the remains. Carbon 14 has a half-life of approximately 5730 years, meaning that in 5730 years the amount of carbon 14 in organic matter will be half of what it was originally. Suppose that the organic matter in a fossil originally had 1000 milligrams of carbon 14. After 5730 years, the amount of carbon 14 should be $1000(1/2)^1 = 500$. To find $y$ = amount of carbon 14 remaining after $x$ years, use the function $y = 1000\left(\frac{1}{2}\right)^{x/5730}$. Here is a graph of this function for the first 50,000 years:

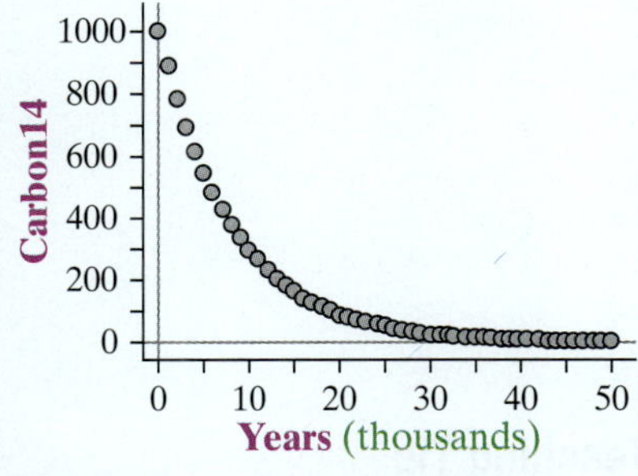

## Teaching Tip

As on page 772, you may need to help your students see how $\log y = \log a + (\log b)x$ is in a linear form. Remind them that an equation is linear if it is in the form: variable = constant + constant(variable). In this case, $\log y$ is the variable on the left side, $\log a$ is the first constant, $\log b$ is the second constant, and $x$ is the second variable.

## Teaching Tip

Emphasize the last sentence before the example on this page. Be sure to point out that the converse of the first sentence is also true: if a scatterplot of $\log y$ versus $x$ produces a linear pattern, then an exponential model describes the relationship between the two variables.

## Teaching Tip

To understand why we use "years since 1970" as the explanatory variable, see page 778 (just under the Think About It feature).

using base $e$ would work just as well). Then use algebraic properties of logarithms to simplify the resulting expressions. Here are the details:

| | |
|---|---|
| $\log y = \log (ab^x)$ | taking the logarithm of both sides |
| $\log y = \log a + \log (b^x)$ | using the property $\log(mn) = \log m + \log n$ |
| $\log y = \log a + x \log b$ | using the property $\log m^p = p \log m$ |

We can then rearrange the final equation as $\log y = \log a + (\log b)x$. Notice that $\log a$ and $\log b$ are constants because $a$ and $b$ are constants. So the equation gives a linear model relating the explanatory variable $x$ to the transformed variable $\log y$. Thus, if the relationship between two variables follows an exponential model, and we plot the logarithm (base 10 or base $e$) of $y$ against $x$, we should observe a *straight-line pattern* in the transformed data.

## EXAMPLE Moore's Law and Computer Chips

### *Logarithm transformations and exponential models*

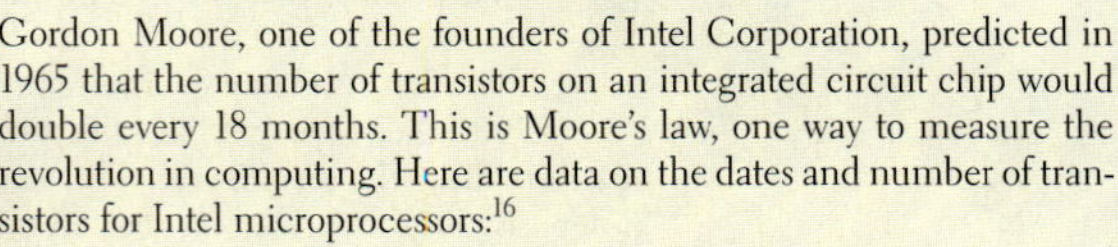

Gordon Moore, one of the founders of Intel Corporation, predicted in 1965 that the number of transistors on an integrated circuit chip would double every 18 months. This is Moore's law, one way to measure the revolution in computing. Here are data on the dates and number of transistors for Intel microprocessors:[16]

| Processor | Date | Transistors |
|---|---|---|
| 4004 | 1971 | 2,250 |
| 8008 | 1972 | 2,500 |
| 8080 | 1974 | 5,000 |
| 8086 | 1978 | 29,000 |
| 286 | 1982 | 120,000 |
| 386 | 1985 | 275,000 |
| 486 DX | 1989 | 1,180,000 |
| Pentium | 1993 | 3,100,000 |
| Pentium II | 1997 | 7,500,000 |
| Pentium III | 1999 | 24,000,000 |
| Pentium 4 | 2000 | 42,000,000 |
| Itanium 2 | 2003 | 220,000,000 |
| Itanium 2 w/9MB cache | 2004 | 592,000,000 |
| Dual-core Itanium 2 | 2006 | 1,700,000,000 |
| Six-core Xeon 7400 | 2008 | 1,900,000,000 |
| 8-core Xeon Nehalem-EX | 2010 | 2,300,000,000 |

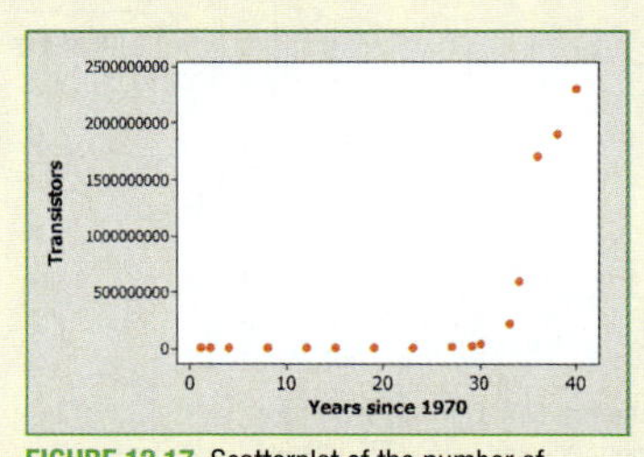


**FIGURE 12.17** Scatterplot of the number of transistors on a computer chip from 1971 to 2010.

Figure 12.17 shows the growth in the number of transistors on a computer chip from 1971 to 2010. Notice that we used "years since 1970" as the explanatory variable. We'll explain this later. If Moore's law is correct, then an exponential model should describe the relationship between the variables.

## ALTERNATE EXAMPLE

### *More M&M'S® chocolate candies*

A student opened a bag of M&M'S, dumped them out, and ate all the ones with the M on top. When he finished, he put the remaining 30 M&M'S back in the bag and repeated the same process over and over until all the M&M'S were gone. Here are a table and scatterplot showing the number of M&M'S remaining at the end of each "course":

| Course | M&M'S remaining |
|---|---|
| 1 | 30 |
| 2 | 13 |
| 3 | 10 |
| 4 | 3 |
| 5 | 2 |
| 6 | 1 |
| 7 | 0 |

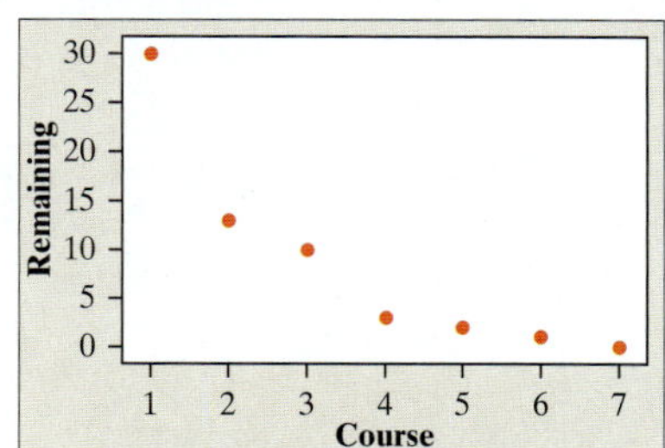


Because the number of M&M'S® should be cut in half after each course, an exponential model should describe the relationship between the variables.

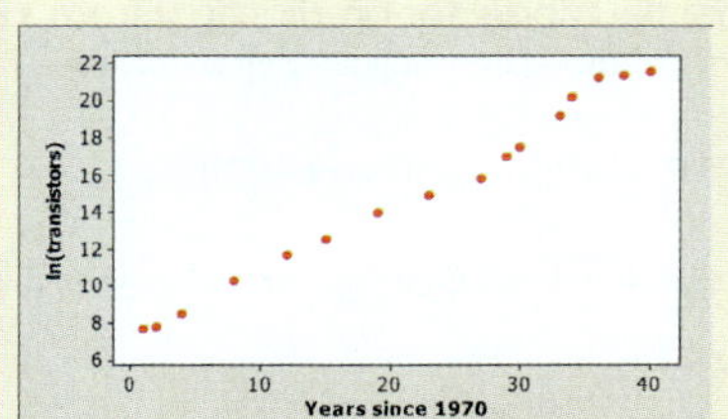


PROBLEM:

(a) A scatterplot of the natural logarithm (log base $e$ or ln) of the number of transistors on a computer chip versus years since 1970 is shown. Based on this graph, explain why it would be reasonable to use an exponential model to describe the relationship between number of transistors and years since 1970.

(b) Minitab output from a linear regression analysis on the transformed data is shown below. Give the equation of the least-squares regression line. Be sure to define any variables you use.

| Predictor | Coef | SE Coef | T | P |
|---|---|---|---|---|
| Constant | 7.0647 | 0.2672 | 26.44 | 0.000 |
| Years since 1970 | 0.36583 | 0.01048 | 34.91 | 0.000 |

S = 0.544467   R-Sq = 98.9%   R-Sq(adj) = 98.8%

(c) Use your model from part (b) to predict the number of transistors on an Intel computer chip in 2020. Show your work.

(d) A residual plot for the linear regression in part (b) is shown at left. Discuss what this graph tells you about the appropriateness of the model.

SOLUTION:

(a) If an exponential model describes the relationship between two variables $x$ and $y$, then we expect a scatterplot of $(x, \ln y)$ to be roughly linear. The scatterplot of ln(transistors) versus years since 1970 has a fairly linear pattern, especially through the year 2000. So an exponential model seems reasonable here.

(b) $\widehat{\ln(\text{transistors})} = 7.0647 + 0.36583(\text{years since } 1970)$

(c) Because 2020 is 50 years since 1970, we have

$$\widehat{\ln(\text{transistors})} = 7.0647 + 0.36583(50) = 25.3562$$

To find the predicted number of transistors, we use the definition of a logarithm as an exponent:

$$\widehat{\ln(\text{transistors})} = 25.3562 \Rightarrow \widehat{\log_e(\text{transistors})} = 25.3562$$

$$\widehat{\text{transistors}} = e^{25.362} \approx 1.028 \cdot 10^{11}$$

This model predicts that an Intel chip made in 2020 will have about 100 billion transistors.

(d) The residual plot shows a distinct pattern, with the residuals going from positive to negative to positive as we move from left to right. But the residuals are small in size relative to the transformed $y$-values. Also, the scatterplot of the transformed data is much more linear than the original scatterplot. We feel reasonably comfortable using this model to make predictions about the number of transistors on a computer chip.

**For Practice** *Try Exercise* **41**

Make sure that you understand the big idea here. The necessary transformation is carried out by taking the logarithm of the response variable. The crucial property of the logarithm for our purposes is that *if a variable grows exponentially, its logarithm grows linearly.*

PROBLEM:

**(a)** A scatterplot of the natural log of the number of M&M'S® remaining versus course number is shown below. The last observation in the table is not included because ln(0) is undefined. Explain why it would be reasonable to use an exponential model to describe the relationship between the number of M&M'S remaining and the course number.

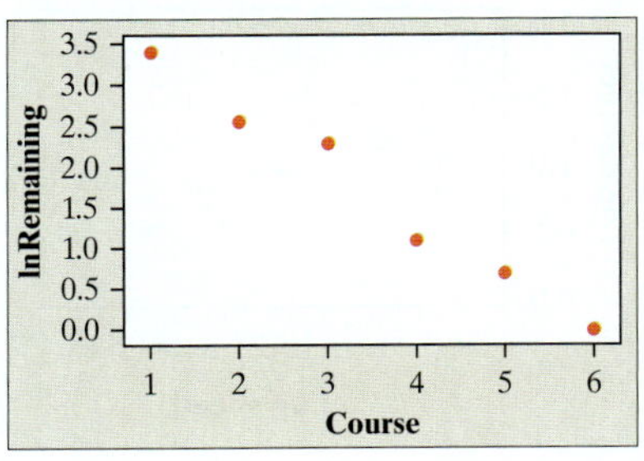


**(b)** Minitab output from a linear regression analysis on the transformed data is shown below. Give the equation of the least-squares regression line. Be sure to define any variables you use.

Regression Analysis: LnRemaining versus Course

| Predictor | Coef | SE Coef | T | P |
|---|---|---|---|---|
| Constant | 4.0593 | 0.1852 | 21.92 | 0.000 |
| Course | -0.68073 | 0.04755 | -14.32 | 0.000 |

S = 0.198897   R-Sq = 98.1%   R-Sq(adj) = 97.6%

**(c)** Use your model from part (b) to predict the original number of M&M'S in the bag.

**(d)** A residual plot of the linear regression in part (b) is shown at top right. Discuss what this graph tells you about the appropriateness of the model.

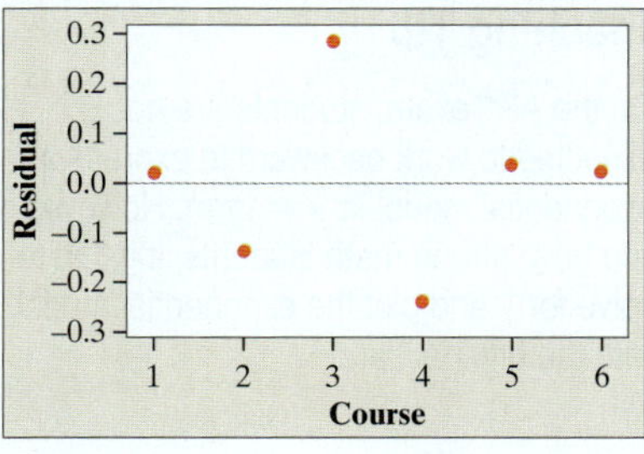


SOLUTION:

**(a)** If there is an exponential relationship between two variables $x$ and $y$, we expect a scatterplot of $(x, \ln y)$ to be roughly linear. Because the scatterplot of ln(remaining) versus course number is roughly linear, an exponential model seems appropriate here.

**(b)** $\widehat{\ln y} = 4.0593 - 0.68073x$, where $\widehat{\ln y}$ = the predicted value of the natural log of the number of M&M'S remaining and $x$ = course number.

**(c)** To estimate the original number of M&M'S, we need to predict the number remaining after Course 0. $\widehat{\ln y} = 4.0593 - 0.68073(0) = 4.0593$. Thus, $\hat{y} = e^{4.0593} = 57.93$ M&M'S®.

**(d)** Because there is no obvious leftover curvature in the residual plot, the model is appropriate for these data.

## Teaching Tip

After discussing the answer to part (d) of the computer chip example, remind students of the quote from George Box: "All models are wrong, but some are useful."

## Teaching Tip

On the AP® exam, students are not expected to work backward to express an exponential model in $\hat{y}$ = form. However, if you have strong math students, it is fun to solve for $\hat{y}$ and plot the exponential model with the original data.

THINK ABOUT IT

**How do we find the exponential model for predicting *y* from *x*?** The least-squares line for the transformed data in the computer chip example is

$$\widehat{\ln(\text{transistors})} = 7.0647 + 0.36583\,(\text{years since 1970})$$

If we use the definition of the logarithm as an exponent, we can rewrite this equation as

$$\widehat{\text{transistors}} = e^{7.0647 + 0.36583(\text{years since 1970})}$$

Using properties of exponents, we can simplify this as follows:

| | |
|---|---|
| $\widehat{\text{transistors}} = e^{7.0647} \cdot e^{0.36583(\text{years since 1970})}$ | using the fact that $b^m b^n = b^{m+n}$ |
| $\widehat{\text{transistors}} = e^{7.0647} \cdot (e^{0.36583})^{(\text{years since 1970})}$ | using the fact that $(b^m)^n = b^{mn}$ |
| $\widehat{\text{transistors}} = 1169.93 \cdot (1.44171)^{(\text{years since 1970})}$ | simplifying |

This equation is now in the familiar form of an exponential model $y = ab^x$ with $a = 1169.93$ and $b = 1.44171$.

We could use the exponential model to predict the number of transistors on an Intel chip in 2020: $\widehat{\text{transistors}} = 1169.93(1.44171)^{50} \approx 1.0281 \cdot 10^{11}$. This is the same prediction we got earlier. How does this compare with the prediction from Moore's law? Suppose the number of transistors on an Intel computer chip doubles every 18 months (1.5 years). Then in the 49 years from 1971 to 2020, the number of transistors would double 49/1.5 = 32.67 times. So the predicted number of transistors on an Intel chip in 2020 would be

$$\widehat{\text{transistors}} = 2250(2)^{32.67} = 1.54 \cdot 10^{13}$$

Moore's law predicts more rapid exponential growth than our model does.

The calculation at the end of the Think about It feature might give you some idea of why we used years since 1970 as the explanatory variable in the example. To make a prediction, we substituted the value $x = 50$ into the equation for the exponential model. This value is the exponent in our calculation. If we had used years as the explanatory variable, our exponent would have been 2020. Such a large exponent can lead to overflow errors on a calculator.

## Putting It All Together: Which Transformation Should We Choose?

Suppose that a scatterplot shows a curved relationship between two quantitative variables *x* and *y*. How can we decide whether a power model or an exponential model better describes the relationship? The following example shows the strategy we should use.

### ALTERNATE EXAMPLE

### *Braking distance*

How is the braking distance for a motorcycle related to the speed the motorcycle was traveling when the brake was applied? Statistics teacher Aaron Waggoner gathered data to answer this question. The table and scatterplot below show the speed (in miles per hour, or mph) and the distance (in feet) needed to come to a complete stop when the brake was applied:

| Speed (mph) | Distance (feet) |
|---|---|
| 6 | 1.42 |
| 9 | 4.92 |
| 19 | 18.00 |
| 30 | 44.75 |
| 32 | 52.08 |
| 40 | 84.00 |
| 48 | 110.33 |

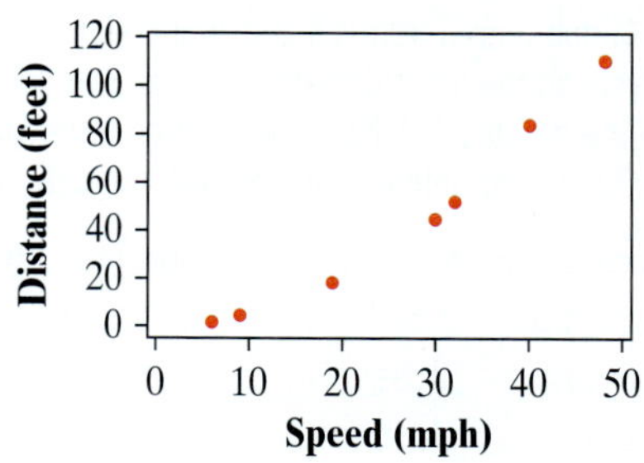


PROBLEM:

The graphs at right show the results of two different transformations of the data. The first graph plots the logarithm of the distance against speed and the second graph plots the logarithm of distance against the logarithm of speed.

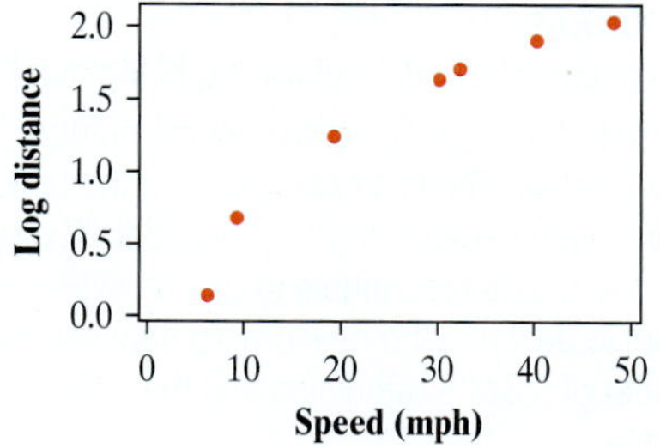


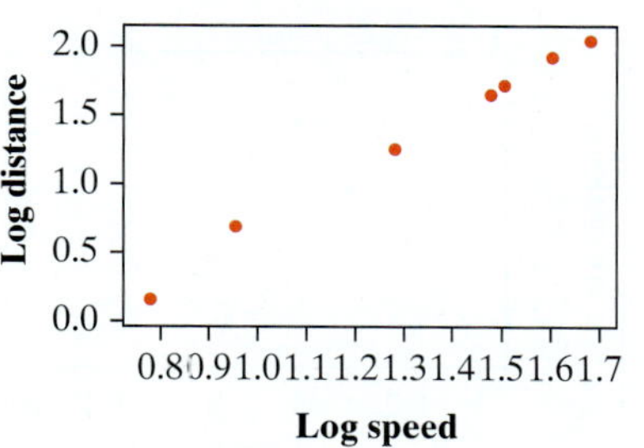

## EXAMPLE What's a Planet, Anyway?

### *Power models and logarithm transformations*

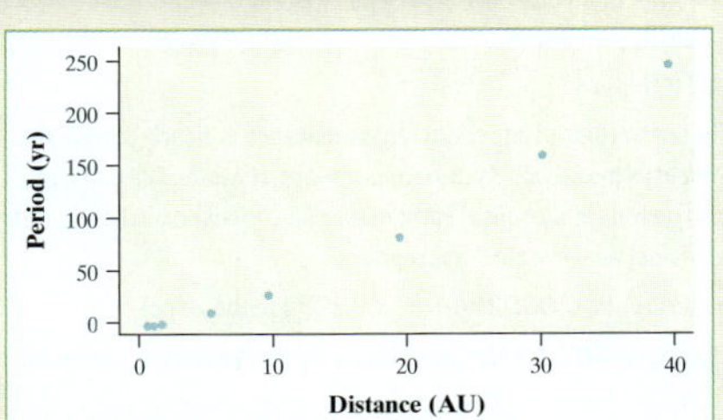

On July 31, 2005, a team of astronomers announced that they had discovered what appeared to be a new planet in our solar system. They had first observed this object almost two years earlier using a telescope at Caltech's Palomar Observatory in California. Originally named UB313, the potential planet is bigger than Pluto and has an average distance of about 9.5 billion miles from the sun. (For reference, Earth is about 93 million miles from the sun.) Could this new astronomical body, now called Eris, be a new planet?

At the time of the discovery, there were nine known planets in our solar system. Here are data on the distance from the sun and period of revolution of those planets. Note that distance is measured in astronomical units (AU), the number of Earth distances the object is from the sun.[17]

| Planet | Distance from sun (astronomical units) | Period of revolution (Earth years) |
|---|---|---|
| Mercury | 0.387 | 0.241 |
| Venus | 0.723 | 0.615 |
| Earth | 1.000 | 1.000 |
| Mars | 1.524 | 1.881 |
| Jupiter | 5.203 | 11.862 |
| Saturn | 9.539 | 29.456 |
| Uranus | 19.191 | 84.070 |
| Neptune | 30.061 | 164.810 |
| Pluto | 39.529 | 248.530 |

**FIGURE 12.18** Scatterplot of planetary distance from the sun and period of revolution.

Figure 12.18 is a scatterplot of the planetary data. There appears to be a strong curved relationship between distance from the sun and period of revolution.

In August 2006, the International Astronomical Union agreed on a new definition of "planet." Both Pluto and Eris were classified as "dwarf planets."

**PROBLEM:** The graphs below show the results of two different transformations of the data. Figure 12.19(a) plots the natural logarithm of period against distance from the sun for all nine planets. Figure 12.19(b) plots the natural logarithm of period against the natural logarithm of distance from the sun for the nine planets.

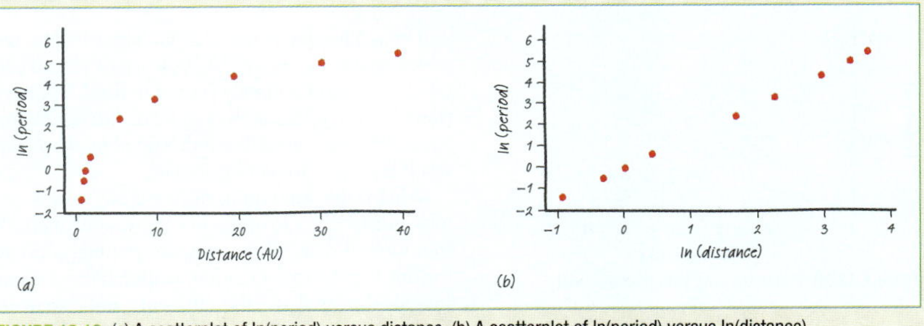


**FIGURE 12.19** (a) A scatterplot of ln(period) versus distance. (b) A scatterplot of ln(period) versus ln(distance).

**(a)** Explain why a power model would provide a more appropriate description of the relationship between braking distance and speed than an exponential model.

**(b)** Minitab output from a linear regression analysis on the second set of transformed data is shown below. Give the equation of the least-squares regression line. Be sure to define any variables you use.

```
Predictor   Coef      SE Coef     T        P
Constant    -1.35063  0.09697     -13.93   0.000
logSpeed    2.03610   0.07118     28.61    0.000
S = 0.0594632   R-Sq = 99.4%   R-Sq(adj) = 99.3%
```

**(c)** Use your model from part (b) to predict the braking distance if the motorcycle was going 55 miles per hour. Show your work.

**(d)** A residual plot for the linear regression in part (b) is shown below. Do you expect your prediction in part (c) to be too high, too low, or just right? Justify your answer.

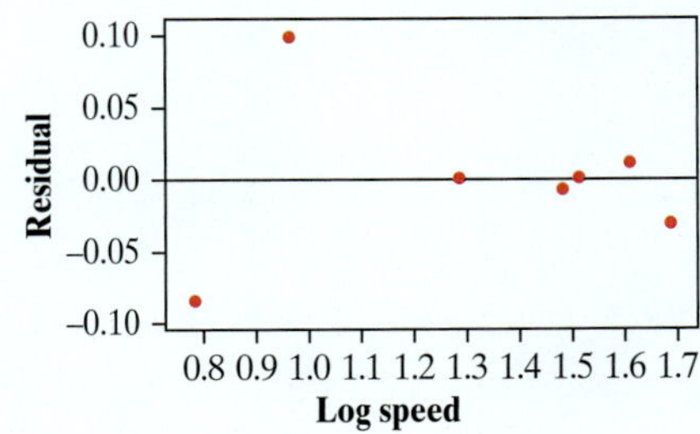


**SOLUTION:**

**(a)** The scatterplot of log(distance) versus speed is clearly curved, so an exponential model would not be appropriate. However, the graph of log(distance) versus log(speed) is roughly linear, indicating that a power model would be more appropriate.

**(b)** $\widehat{\log(\text{distance})} = -1.35063 + 2.03610 \log(\text{speed})$

**(c)** $\widehat{\log(\text{distance})} = -1.35063 + 2.03610 \log(55) = 2.1929$.
Thus, $\widehat{\text{distance}} = 10^{2.1929} = 155.92$ feet.

**(d)** Because there is no leftover pattern in the residual plot, the prediction should be just about right. In other words, we can't tell whether the prediction will be too high or too low.

(a) Explain why a power model would provide a more appropriate description of the relationship between period of revolution and distance from the sun than an exponential model.

(b) Minitab output from a linear regression analysis on the transformed data in Figure 12.19(b) is shown below. Give the equation of the least-squares regression line. Be sure to define any variables you use.

| Predictor | Coef | SE Coef | T | P |
|---|---|---|---|---|
| Constant | 0.0002544 | 0.0001759 | 1.45 | 0.191 |
| ln(distance) | 1.49986 | 0.00008 | 18598.27 | 0.000 |
| S = 0.000393364 | R-Sq = 100.0% | R-Sq(adj) = 100.0% | | |

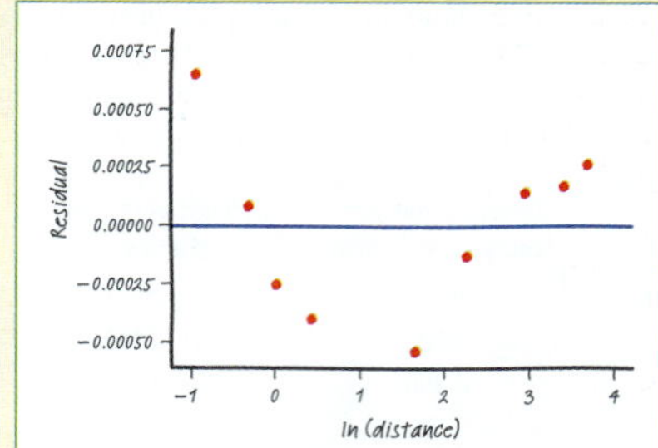


(c) Use your model from part (b) to predict the period of revolution for Eris, which is 9,500,000,000/93,000,000 = 102.15 AU from the sun. Show your work.

(d) A residual plot for the linear regression in part (b) is shown at left. Do you expect your prediction in part (c) to be too high, too low, or about right? Justify your answer.

SOLUTION:

(a) The scatterplot of ln(period) versus distance is clearly curved, so an exponential model would not be appropriate. However, the graph of ln(period) versus ln(distance) has a strong linear pattern, indicating that a power model would be more appropriate.

(b) $\widehat{\ln(\text{period})} = 0.0002544 + 1.49986 \ln(\text{distance})$

(c) Eris's average distance from the sun is 102.15 AU. Using this value for distance in our model from part (b) gives

$$\widehat{\ln(\text{period})} = 0.0002544 + 1.49986 \ln(102.15) = 6.939$$

To predict the period, we have to undo the logarithm transformation:

$$\widehat{\text{period}} = e^{6.939} \approx 1032 \text{ years}$$

We wouldn't want to wait for Eris to make a full revolution to see if our prediction is accurate!

(d) Eris's value for ln(distance) is ln(102.15) = 4.626, which would fall at the far right of the residual plot, where all the residuals are positive. Because residual = actual $y$ − predicted $y$ seems likely to be positive, we would expect our prediction to be too low.

**For Practice** *Try Exercise* **43**

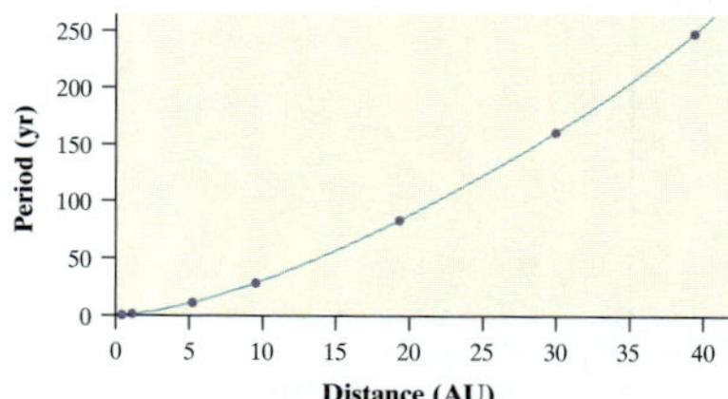


**FIGURE 12.20** Planetary data with power model.

The scatterplot of the original data with the power model added appears in Figure 12.20. It seems remarkable that period of revolution is closely related to the 1.5 power of distance from the sun. Johannes Kepler made this fascinating discovery about 400 years ago without the aid of modern technology—a result known as Kepler's third law.

What if the scatterplots of (log $x$, log $y$) and ($x$, log $y$) *both* look linear? Fit a least-squares regression line to both sets of transformed data. Then compare residual plots and look for the one with the most random scatter. If the residual plots look roughly the same, use the values of $s$ and $r^2$ to decide whether a power model or an exponential model is a better choice.

Here is a scatterplot showing the original braking distance data, along with the power model: $\hat{y} = 0.0446x^{2.0361}$.

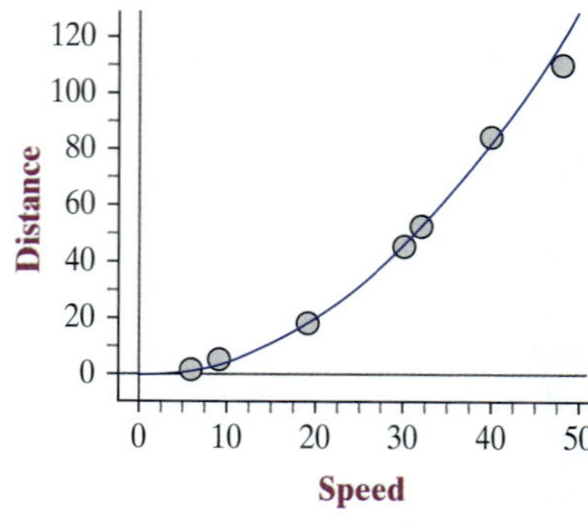

We have used statistical software to do all the transformations and linear regression analysis in this section so far. Now let's look at how the process works on a graphing calculator.

## 30. TECHNOLOGY CORNER — TRANSFORMING TO ACHIEVE LINEARITY ON THE CALCULATOR

TI-Nspire instructions in Appendix B; HP Prime instructions on the book's Web site.

We'll use the planet data to illustrate a general strategy for performing transformations with logarithms on the TI-83/84 and TI-89. A similar approach could be used for transforming data with powers and roots.

TI-83/84 TI-89

- Enter the values of the explanatory variable in L1/list1 and the values of the response variable in L2/list2. Make a scatterplot of $y$ versus $x$ and confirm that there is a curved pattern.

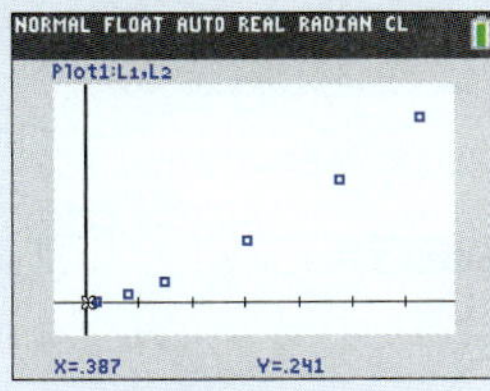


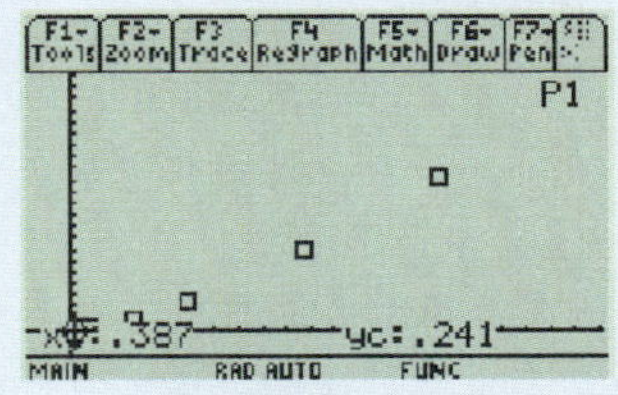


- Define L3/list3 to be the natural logarithm (ln) of L1/list1 and L4/list4 to be the natural logarithm of L2/list2. To see whether a power model fits the original data, make a plot of ln $y$ (L4/list4) versus ln $x$ (L3/list3) and look for linearity. To see whether an exponential model fits the original data, make a plot of ln $y$ (L4/list4) versus $x$ (L1/list1) and look for linearity.

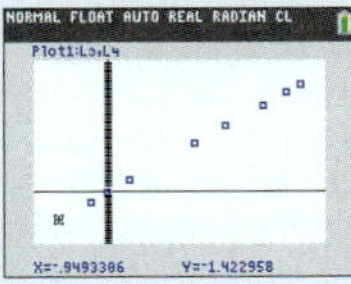
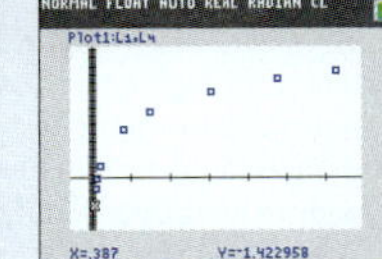
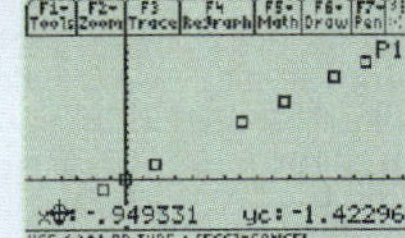

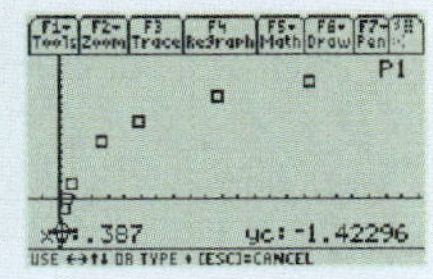


- If a linear pattern is present, calculate the equation of the least-squares regression line and store it in Y1. For the planet data, we executed the command `LinReg(a+bx)L3,L4,Y1`.

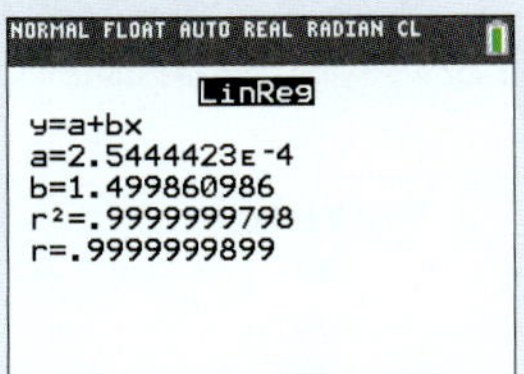


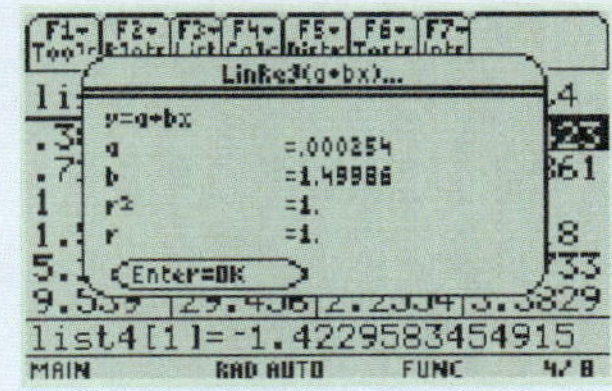


Construct a residual plot to look for any departures from the linear pattern. For Xlist, enter the list you used as the explanatory variable in the linear regression calculation. For Ylist, use the RESID list stored in the calculator. For the planet data, we used L3/list3 as the Xlist.

### TECHNOLOGY TIP

We strongly recommend that you have students use the RESID list on their calculators during this section. Although understanding how to calculate residuals is a valuable skill, there are so many other details in this section that we think it is worthwhile to let the calculator handle this step.

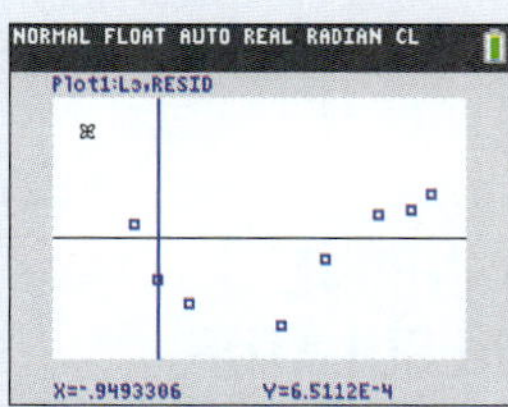


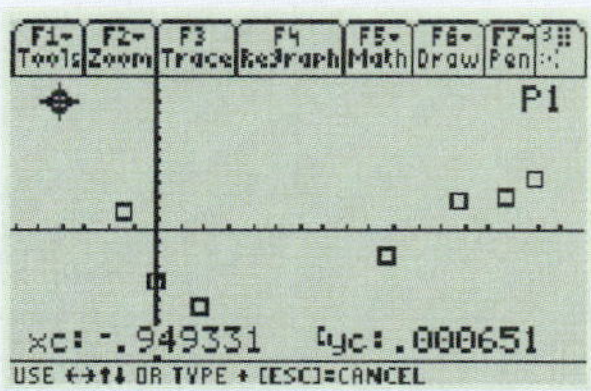


- To make a prediction for a specific value of the explanatory variable, compute log *x* or ln *x*, if appropriate. Then use Y1(*k*) to obtain the predicted value of log *y* or ln *y*. To get the predicted value of *y*, use 10^Ans or *e*^Ans to undo the logarithm transformation. Here's our prediction of the period of revolution for Eris, which is at a distance of 102.15 AU from the sun:

```
NORMAL FLOAT AUTO REAL RADIAN CL
ln(102.15)
                    4.626442321
Y1(Ans)
                    6.939274784
e^(Ans)
                    1032.021505
```

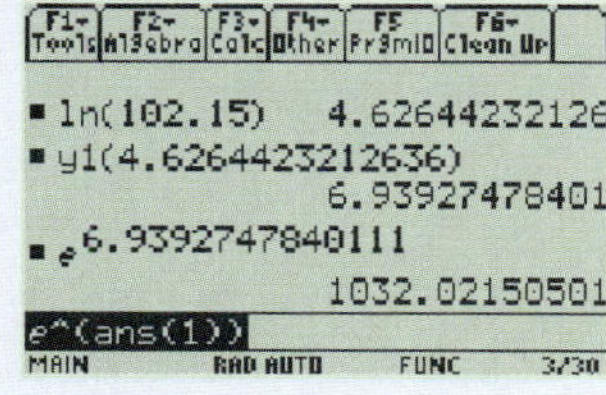


## TECHNOLOGY TIP

If your students aren't that calculator savvy, don't worry about using the Y1(*k*) approach. Simply have them substitute the value of *x* into the regression equation as illustrated in the examples on pages 773, 777, and 780.

## CHECK YOUR UNDERSTANDING

One sad fact about life is that we'll all die someday. Many adults plan ahead for their eventual passing by purchasing life insurance. Many different types of life insurance policies are available. Some provide coverage throughout an individual's life (whole life), while others last only for a specified number of years (term life). The policyholder makes regular payments (premiums) to the insurance company in return for the coverage. When the insured person dies, a payment is made to designated family members or other beneficiaries.

How do insurance companies decide how much to charge for life insurance? They rely on a staff of highly trained actuaries—people with expertise in probability, statistics, and advanced mathematics—to establish premiums. For an individual who wants to buy life insurance, the premium will depend on the type and amount of the policy as well as on personal characteristics like age, sex, and health status.

The table shows monthly premiums for a 10-year term-life insurance policy worth $1,000,000.[18]

| Age (years) | Monthly premium |
|---|---|
| 40 | $29 |
| 45 | $46 |
| 50 | $68 |
| 55 | $106 |
| 60 | $157 |
| 65 | $257 |

The Fathom screen shots below show three possible models for predicting monthly premium from age. Option 1 is based on the original data, while Options 2 and 3 involve transformations of the original data. Each screen shot includes a scatterplot with a least-squares regression line added and a residual plot.

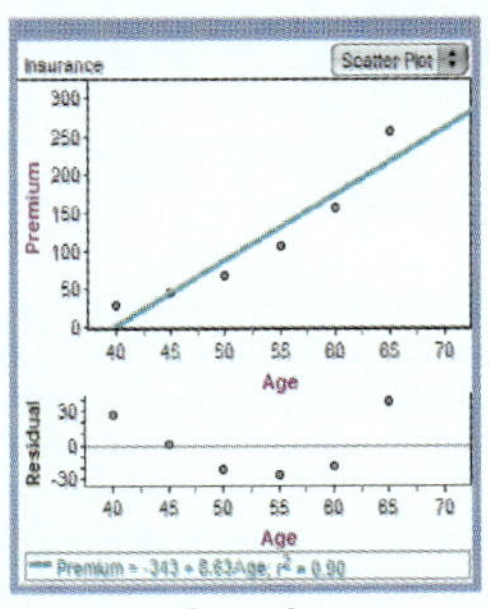


Option 1

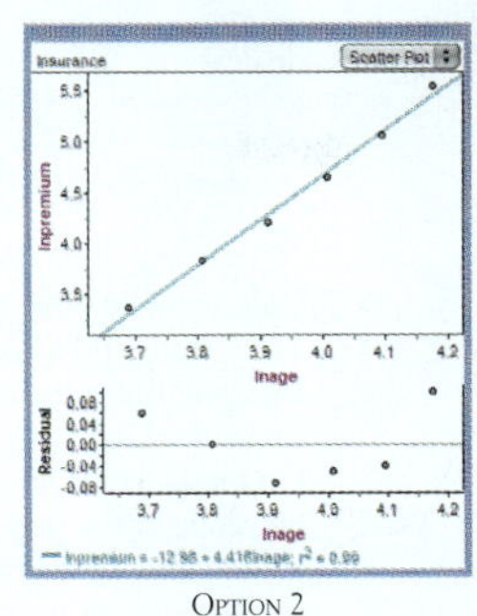


Option 2

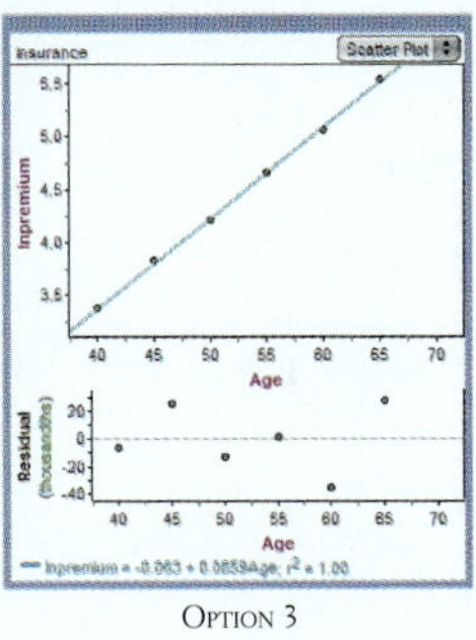


Option 3

1. Use each model to predict how much a 58-year-old would pay for such a policy. Show your work.
2. What type of function—linear, power, or exponential—best describes the relationship between age and monthly premium? Explain.

## ✓ ANSWERS TO CYU

**1.** Option 1: $\widehat{\text{premium}} = -343 + 8.63(58) = \$157.54$

Option 2: $\widehat{\ln(\text{premium})} = -12.98 + 4.416(\ln 58) = 4.9509 \rightarrow \hat{y} = e^{4.9509} = \$141.30$

Option 3: $\widehat{\ln(\text{premium})} = -0.063 + 0.0859(58) = 4.9192 \rightarrow \hat{y} = e^{4.9192} = \$136.89$

**2.** Exponential (Option 3), because the scatterplot showing ln(premium) vs. age was the most linear and this model had the most randomly scattered residual plot.

case closed!

# Do Longer Drives Mean Lower Scores on the PGA Tour?

In the chapter-opening Case Study (page 737), we examined data on the mean drive distance (in yards) and mean score per round for an SRS of 19 of the 197 players on the PGA Tour in a recent year. Here is some Minitab output from a least-squares regression analysis on these data:

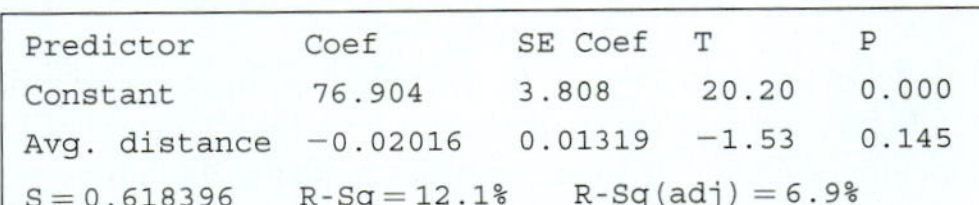

| Predictor | Coef | SE Coef | T | P |
|---|---|---|---|---|
| Constant | 76.904 | 3.808 | 20.20 | 0.000 |
| Avg. distance | −0.02016 | 0.01319 | −1.53 | 0.145 |

S = 0.618396 R-Sq = 12.1% R-Sq(adj) = 6.9%

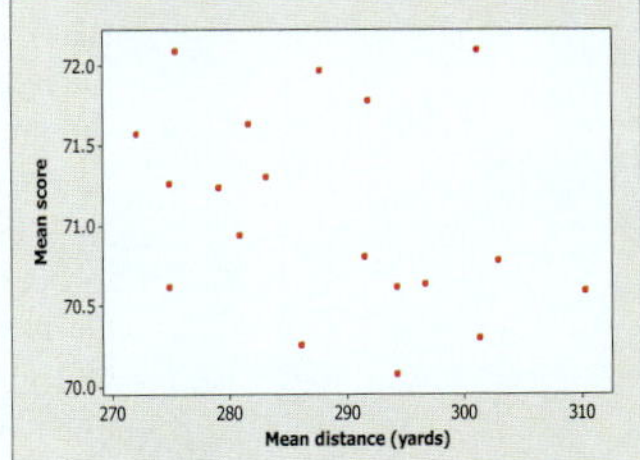

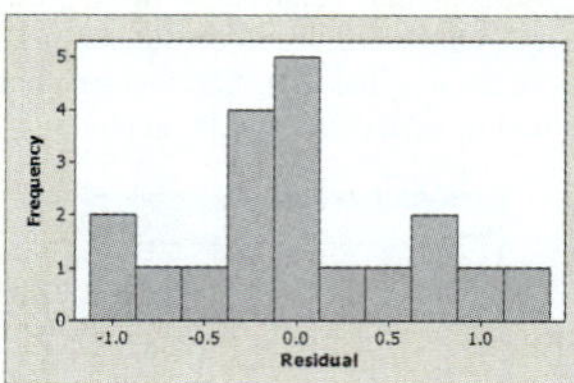


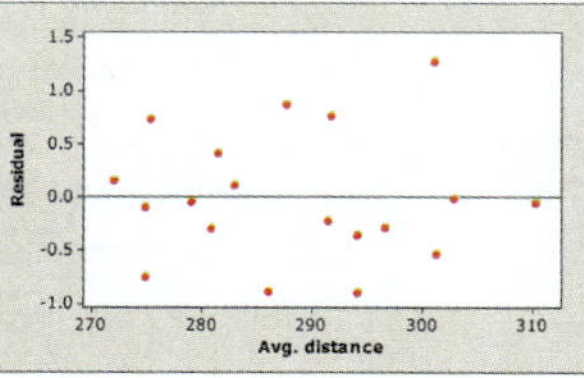


1. Calculate the residual for the player with a mean drive distance of 275.4 yards and a mean score per round of 72.1. Show your work.
2. Interpret the value of $s$ in this setting and explain what parameter $s$ is estimating.
3. Do these data give convincing evidence at the $\alpha = 0.05$ level that the slope of the population regression line is negative?
4. Which kind of mistake—a Type I error or a Type II error—could you have made in Question 3? Justify your answer.

## Section 12.2 Summary

- Curved relationships between two quantitative variables can sometimes be changed into linear relationships by **transforming** one or both of the variables. Once we transform the data to achieve linearity, we can fit a least-squares regression line to the transformed data and use this linear model to make predictions.
- When theory or experience suggests that the relationship between two variables follows a **power model** of the form $y = ax^p$, there are two transformations involving powers and roots that can linearize a curved pattern in a scatterplot. Option 1: Raise the values of the explanatory variable $x$ to the power $p$, then look at a graph of $(x^p, y)$. Option 2: Take the $p$th root of the values of the response variable $y$, then look at a graph of $(x, \sqrt[p]{y})$.
- Another useful strategy for straightening a curved pattern in a scatterplot is to take the **logarithm** of one or both variables. When a power model describes the relationship between two variables, a plot of log $y$ (ln $y$) versus log $x$ (ln $x$) should be linear.
- In a linear model of the form $y = a + bx$, the values of the response variable are predicted to increase by a constant amount $b$ for each increase of 1 unit in the explanatory variable. For an **exponential model** of the form $y = ab^x$, the predicted values of the response variable are multiplied by a factor of $b$ for each increase of 1 unit in the explanatory variable. When an exponential model describes the relationship between two variables, a plot of log $y$ (ln $y$) versus $x$ should be linear.

### Answers to Case Closed

**1.** $\hat{y} = 76.904 - 0.02016(275.4) = 71.35$. Thus, residual $= y - \hat{y} = 72.1 - 71.35 = 0.75$.

**2.** When using $x$ = mean distance to predict $y$ = mean score, the actual values will typically be about 0.618396 from the predicted values. We are using $s$ to estimate $\sigma$, the true standard deviation of the residuals.

**3.** STATE: We want to perform a test of $H_0: \beta = 0$ vs. $H_a: \beta < 0$, where $\beta$ is the slope of the population regression line relating mean distance $x$ to mean score $y$. We will use $\alpha = 0.05$.

PLAN: If the conditions are met, we will do a $t$ test for the slope $\beta$.

- *Linear*: The scatterplot shows a weak negative linear relationship between mean distance and mean score. The residual plot shows random scatter about the residual = 0 line.
- *Independent*: Knowing the mean score for one golfer shouldn't provide additional information about the mean score for other golfers. Also, 19 is less than 10% of 197.
- *Normal*: No strong skewness or outliers appear in the histogram of residuals.
- *Equal SD*: The residual plot shows a fairly equal amount of scatter around the residual = 0 line.
- *Random*: The golfers were randomly selected.

DO: From the computer output:

- *Test statistic*: $t = -1.53$
- *P-value*: Because the $P$-value in the computer output is for a two-sided test, we must divide it in half for a one-sided test. $P$-value $= 0.145/2 = 0.0725$ using df $= 19 - 2 = 17$.

CONCLUDE: Because the $P$-value of 0.0725 is larger than $\alpha = 0.05$, we fail to reject $H_0$. We do not have convincing evidence that there is a negative association between mean distance and mean score for the population of PGA golfers.

**4.** Because we didn't find convincing evidence of a negative association between mean distance and mean score, when it is possible that there really is a negative association, we could have made a Type II error.

### TRM Section 12.2 Quizzes

There are three different quizzes available for this section in the Teacher's Resource Materials. Versions A and B are roughly parallel. Find them in the TE-book, or on the TRCD or the book's companion Web site www.whfreeman.com/tps5e (teacher log-in required).

### TRM Additional Transformation Problems

In the Teacher's Resource Materials, there are two additional problems that you can give your students to help them practice the methods of this section. The first problem asks students to model the growth of baseball salaries over time. The second problem has students develop a model for predicting the weight of an alligator based on its length.

## 12.2 TECHNOLOGY CORNER

TI-Nspire Instructions in Appendix B; HP Prime instructions on the book's Web site.



# Section 12.2 Exercises

**31. The swinging pendulum** Mrs. Hanrahan's precalculus class collected data on the length (in centimeters) of a pendulum and the time (in seconds) the pendulum took to complete one back-and-forth swing (called its period). Here are their data:

| Length (cm) | Period (s) |
|---|---|
| 16.5 | 0.777 |
| 17.5 | 0.839 |
| 19.5 | 0.912 |
| 22.5 | 0.878 |
| 28.5 | 1.004 |
| 31.5 | 1.087 |
| 34.5 | 1.129 |
| 37.5 | 1.111 |
| 43.5 | 1.290 |
| 46.5 | 1.371 |
| 106.5 | 2.115 |

(a) Make a reasonably accurate scatterplot of the data by hand, using length as the explanatory variable. Describe what you see.

(b) The theoretical relationship between a pendulum's length and its period is

$$\text{period} = \frac{2\pi}{\sqrt{g}}\sqrt{\text{length}}$$

where g is a constant representing the acceleration due to gravity (in this case, $g = 980$ cm/s$^2$). Use the following graph to identify the transformation that was used to linearize the curved pattern in part (a).

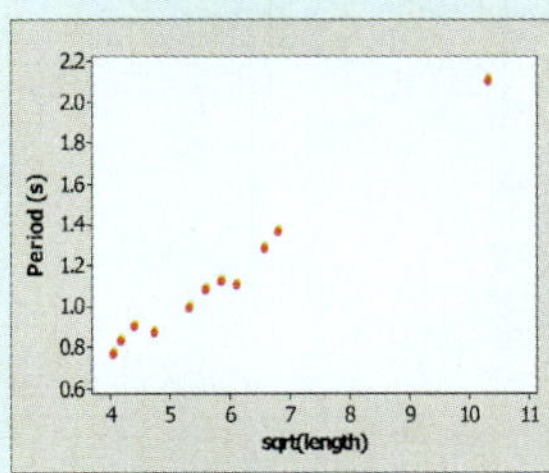


(c) Use the following graph to identify the transformation that was used to linearize the curved pattern in part (a).

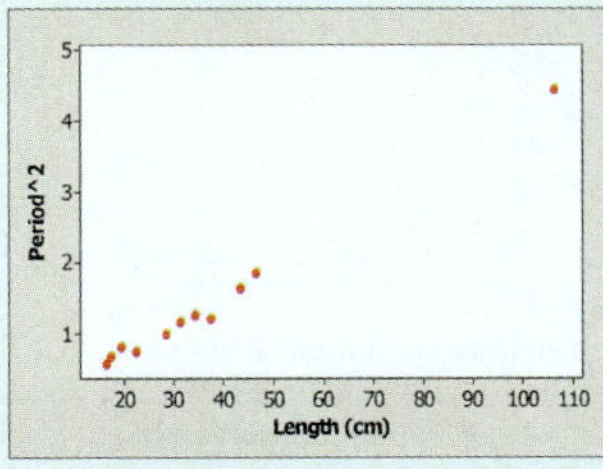


**32. Boyle's law** If you have taken a chemistry or physics class, then you are probably familiar with Boyle's law: for gas in a confined space kept at a constant temperature, pressure times volume is a constant (in symbols, $PV = k$). Students collected the following data on pressure and volume using a syringe and a pressure probe.

**TRM Full Solutions to Section 12.2 Exercises**

## Answers to Section 12.2 Exercises

**12.31 (a)** The scatterplot shows a fairly strong, positive, slightly curved association between length and period with one very unusual point (106.5, 2.115) in the top right corner.

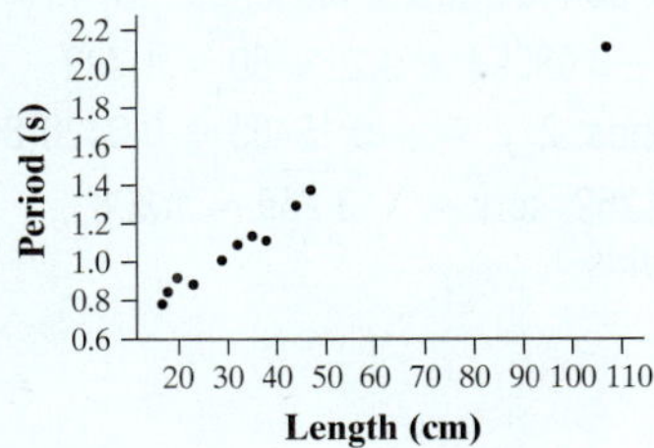


**(b)** The class used the square root of $x$ = length. **(c)** The class used the square of $y$ = period.

**12.32 (a)** The scatterplot shows a strong, negative, curved association between volume and pressure.

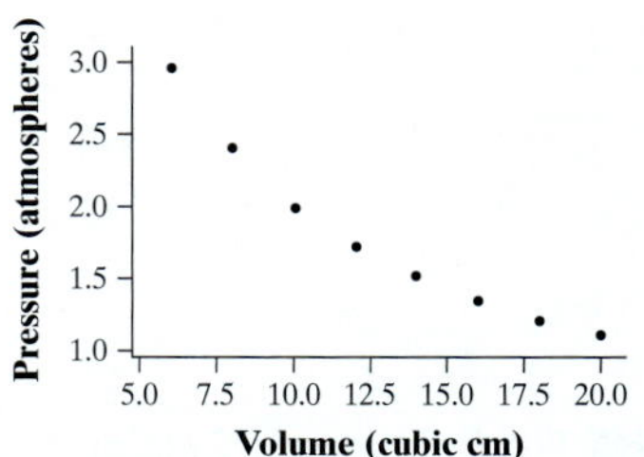


**(b)** The students used the reciprocal of $x$ = volume. **(c)** The students used the reciprocal of $y$ = pressure.

**12.33 (a)** 1: $\hat{y} = -0.08594 + 0.21\sqrt{x}$, where $y$ is the period and $x$ is the length. 2: $\widehat{y^2} = -0.15465 + 0.0428x$, where $y$ is the period and $x$ is the length. **(b)** 1: $\hat{y} = -0.08594 + 0.21\sqrt{80} = 1.792$ seconds. 2: $\widehat{y^2} = -0.15465 + 0.0428(80) = 3.269$, so $\hat{y} = \sqrt{3.269} = 1.808$ seconds.



| Volume (cubic centimeters) | Pressure (atmospheres) |
|---|---|
| 6 | 2.9589 |
| 8 | 2.4073 |
| 10 | 1.9905 |
| 12 | 1.7249 |
| 14 | 1.5288 |
| 16 | 1.3490 |
| 18 | 1.2223 |
| 20 | 1.1201 |

(a) Make a reasonably accurate scatterplot of the data by hand using volume as the explanatory variable. Describe what you see.

(b) If the true relationship between the pressure and volume of the gas is $PV = k$, we can divide both sides of this equation by $V$ to obtain the theoretical model $P = k/V$, or $P = k(1/V)$. Use the graph below to identify the transformation that was used to linearize the curved pattern in part (a).

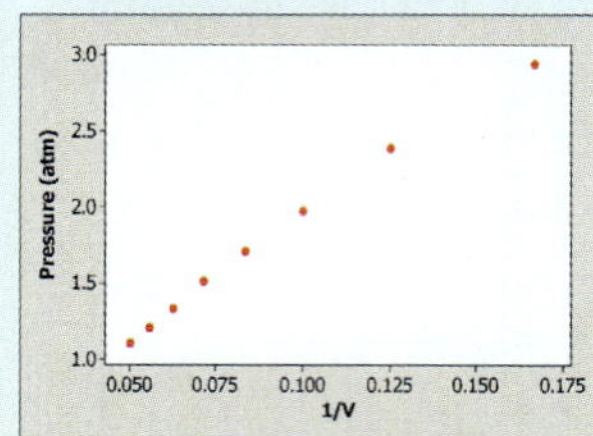


(c) Use the graph below to identify the transformation that was used to linearize the curved pattern in part (a).

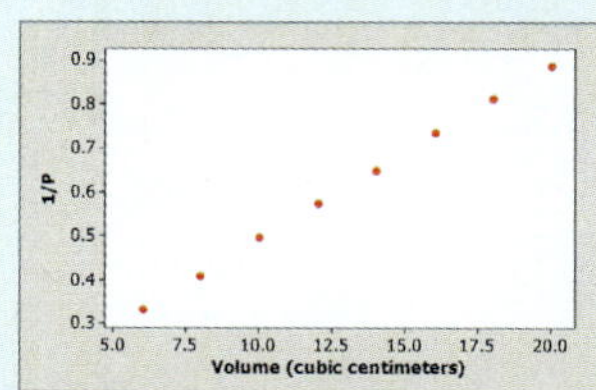


**33. The swinging pendulum** Refer to Exercise 31. Here is Minitab output from separate regression analyses of the two sets of transformed pendulum data:

pg 770

Transformation 1: ($\sqrt{\text{length}}$, period)

```
Predictor Coef      SE Coef  T     P
Constant  -0.08594  0.05046  -1.70 0.123
sqrt      0.209999  0.008322 25.23 0.000
 (length)
S = 0.0464223 R-Sq = 98.6% R-Sq(adj) = 98.5%
```

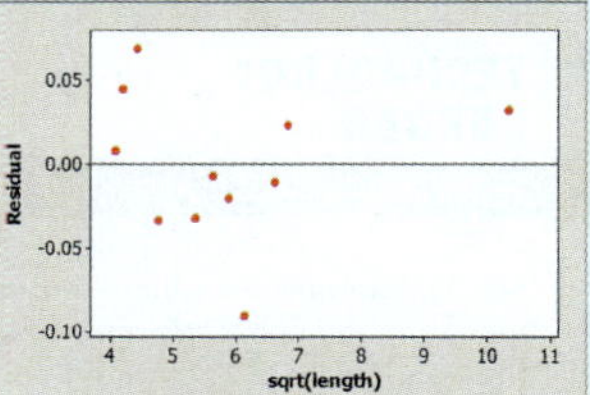


Transformation 2: (length, period²)

```
Predictor    Coef      SE Coef  T     P
Constant     -0.15465  0.05802  -2.67 0.026
Length (cm)  0.042836  0.001320 32.46 0.000
S = 0.105469  R-Sq = 99.2%  R-Sq(adj) = 99.1%
```

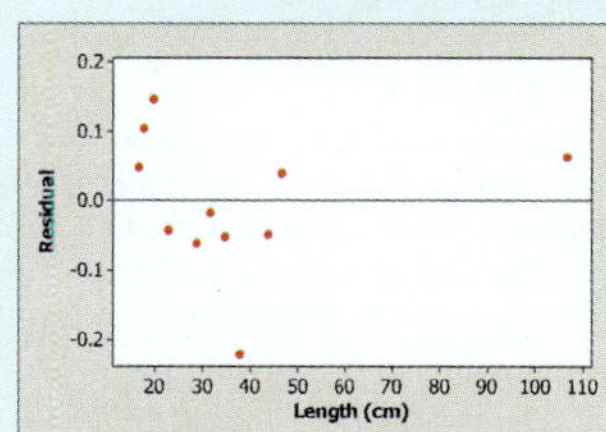


Do each of the following for *both* transformations.

(a) Give the equation of the least-squares regression line. Define any variables you use.

(b) Use the model from part (a) to predict the period of a pendulum with length 80 centimeters. Show your work.

**34. Boyle's law** Refer to Exercise 32. Here is Minitab output from separate regression analyses of the two sets of transformed pressure data:

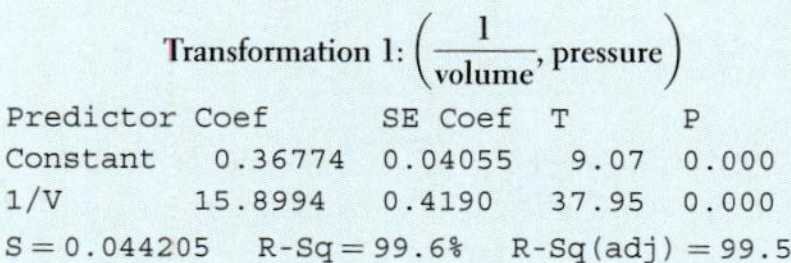

Transformation 1: $\left(\frac{1}{\text{volume}}, \text{pressure}\right)$

```
Predictor Coef     SE Coef  T     P
Constant  0.36774  0.04055  9.07  0.000
1/V       15.8994  0.4190   37.95 0.000
S = 0.044205  R-Sq = 99.6%  R-Sq(adj) = 99.5%
```

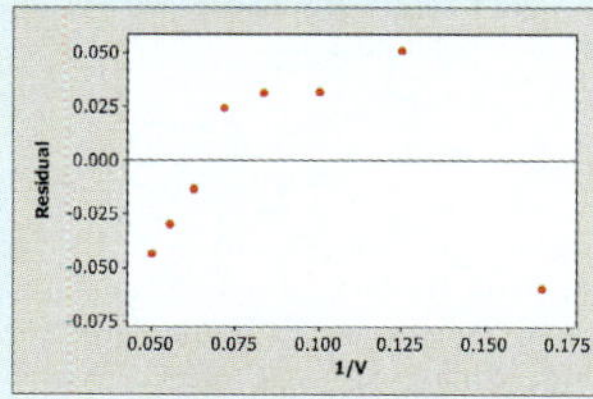

Transformation 2: $\left(\text{volume}, \dfrac{1}{\text{pressure}}\right)$

| Predictor | Coef | SE Coef | T | P |
|---|---|---|---|---|
| Constant | 0.100170 | 0.003779 | 26.51 | 0.000 |
| Volume | 0.0398119 | 0.0002741 | 145.23 | 0.000 |

S = 0.003553 R-Sq = 100.0% R-Sq(adj) = 100.0%

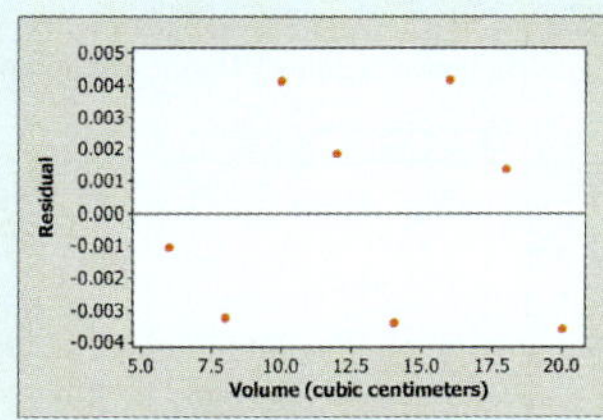


Do each of the following for *both* transformations.

(a) Give the equation of the least-squares regression line. Define any variables you use.

(b) Use the model from part (a) to predict the pressure in the syringe when the volume is 17 cubic centimeters. Show your work.

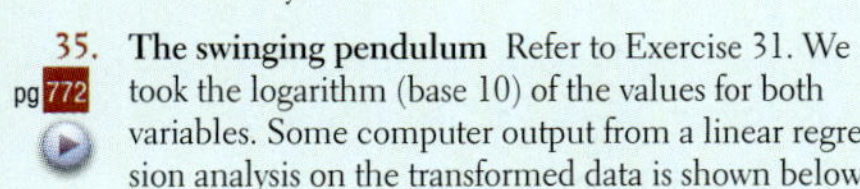

35. **The swinging pendulum** Refer to Exercise 31. We took the logarithm (base 10) of the values for both variables. Some computer output from a linear regression analysis on the transformed data is shown below. pg 772

**Regression Analysis: log(Period) versus log(Length)**

| Predictor | Coef | SE Coef | T | P |
|---|---|---|---|---|
| Constant | −0.73675 | 0.03808 | −19.35 | 0.000 |
| log(Length) | 0.51701 | 0.02511 | 20.59 | 0.000 |

S = 0.0185568 R-Sq = 97.9% R-Sq(adj) = 97.7%

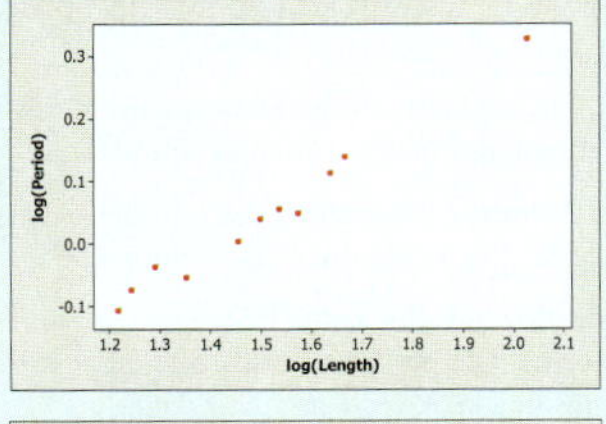


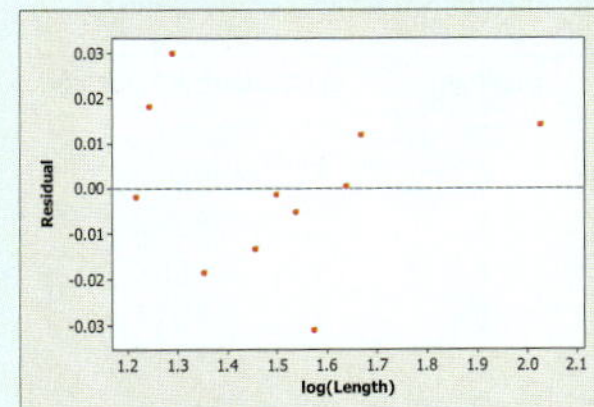


(a) Based on the output, explain why it would be reasonable to use a power model to describe the relationship between the length and period of a pendulum.

(b) Give the equation of the least-squares regression line. Be sure to define any variables you use.

36. **Boyle's law** Refer to Exercise 32. We took the logarithm (base 10) of the values for both variables. Some computer output from a linear regression analysis on the transformed data is shown below.

**Regression Analysis: log(Pressure) versus log(Volume)**

| Predictor | Coef | SE Coef | T | P |
|---|---|---|---|---|
| Constant | 1.11116 | 0.01118 | 99.39 | 0.000 |
| log(Volume) | −0.81344 | 0.01020 | −79.78 | 0.000 |

S = 0.00486926 R-Sq = 99.9% R-Sq(adj) = 99.9%

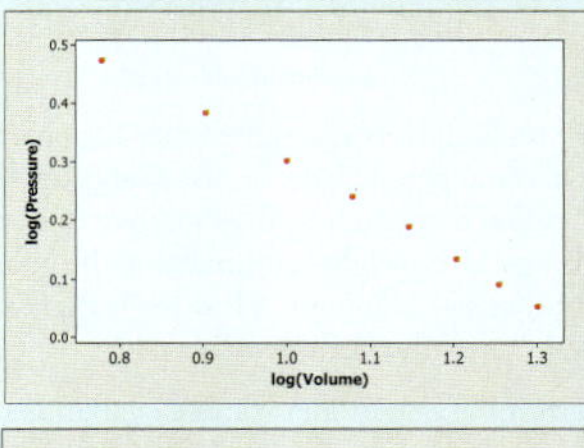


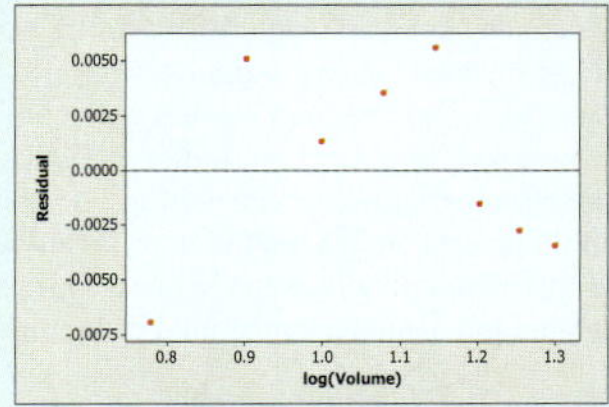


(a) Based on the output, explain why it would be reasonable to use a power model to describe the relationship between pressure and volume.

(b) Give the equation of the least-squares regression line. Be sure to define any variables you use.

37. **The swinging pendulum** Use your model from Exercise 35 to predict the period of a pendulum with length 80 centimeters. Show your work. pg 773

38. **Boyle's law** Use your model from Exercise 36 to predict the pressure in the syringe when the volume is 17 cubic centimeters. Show your work.

39. **Brawn versus brain** How is the weight of an animal's brain related to the weight of its body? Researchers collected data on the brain weight (in grams) and body weight (in kilograms) for 96 species of mammals.[19] The following figure is a scatterplot of

**12.34** **(a)** 1: $\hat{y} = 0.3677 + 15.8994\left(\frac{1}{x}\right)$, where $y$ is the pressure and $x$ is the volume. 2: $\widehat{\frac{1}{y}} = 0.1002 + 0.0398x$, where $y$ is the pressure and $x$ is the volume. **(b)** 1: $\hat{y} = 0.3677 + 15.8994\left(\frac{1}{17}\right) = 1.303$ atmospheres.

2: $\widehat{\frac{1}{y}} = 0.1002 + 0.0398(17) = 0.7768$, so $\hat{y} = \frac{1}{0.7768} = 1.287$ atmospheres.

**12.35** **(a)** The scatterplot of log(period) vs. log(length) is roughly linear and the residual plot shows no obvious leftover patterns. **(b)** $\widehat{\log y} = -0.73675 + 0.51701 \log(x)$, where $y$ is the period and $x$ is the length.

**12.36** **(a)** Even though the residual plot shows a leftover curved pattern, the scatterplot of log(pressure) vs. log (volume) is roughly linear. **(b)** $\widehat{\log y} = 1.11116 - 0.81344 \log(x)$, where $y$ = pressure and $x$ = volume.

**12.37** $\widehat{\log y} = -0.73675 + 0.51701 \log(80) = 0.24717$. Thus, $\hat{y} = 10^{0.24717} = 1.77$ seconds.

**12.38** $\widehat{\log y} = 1.11116 - 0.81344 \log(17) = 0.110264$. Thus, $\hat{y} = 10^{0.110264} = 1.28903$ atmospheres.

**12.39** $\widehat{\log y} = 1.01 + 0.72 \log(127) = 2.525$. Thus, $\hat{y} = 10^{2.525} = 334.97$ grams.

**12.40** $\widehat{\ln y} = -2.00 + 2.42 \ln(30) = 6.231$. Thus, $\hat{y} = e^{6.231} = 508.263$ kg.

**12.41 (a)** The relationship between bacteria count and time is strong, negative, and curved with a possible outlier in the top-left-hand corner.

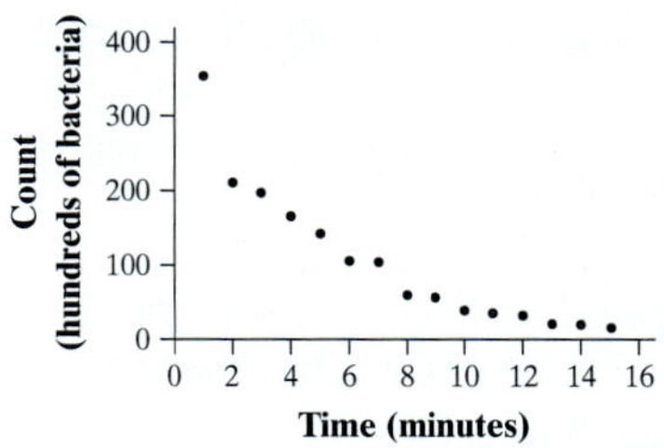


**(b)** Because the scatterplot of ln(count) vs. time is fairly linear.
**(c)** $\widehat{\ln y} = 5.97316 - 0.218425x$, where $y$ is the count of surviving bacteria and $x$ is time in minutes. **(d)** $\widehat{\ln y} = 5.97316 - 0.218425(17) = 2.26$, so $\hat{y} = e^{2.26} = 9.58$ or 958 bacteria.

the logarithm of brain weight against the logarithm of body weight for all 96 species. The least-squares regression line for the transformed data is

$$\widehat{\log y} = 1.01 + 0.72 \log x$$

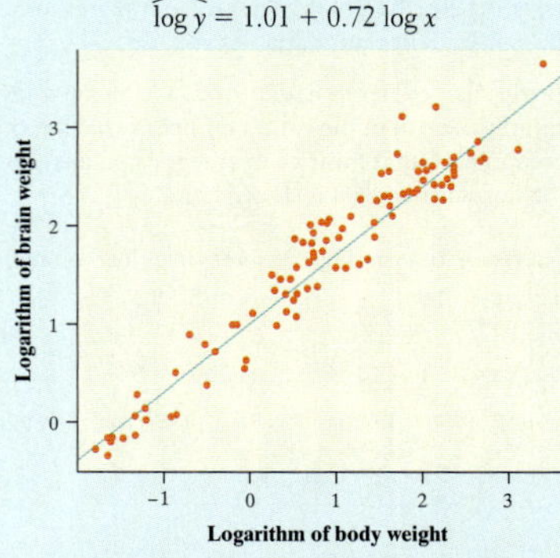


Based on footprints and some other sketchy evidence, some people believe that a large apelike animal, called Sasquatch or Bigfoot, lives in the Pacific Northwest. His weight is estimated to be about 280 pounds, or 127 kilograms. How big is Bigfoot's brain? Show your method clearly.

40. **Determining tree biomass** It is easy to measure the "diameter at breast height" of a tree. It's hard to measure the total "aboveground biomass" of a tree, because to do this you must cut and weigh the tree. The biomass is important for studies of ecology, so ecologists commonly estimate it using a power model. Combining data on 378 trees in tropical rain forests gives this relationship between biomass $y$ measured in kilograms and diameter $x$ measured in centimeters:[20]

$$\widehat{\ln y} = -2.00 + 2.42 \ln x$$

Use this model to estimate the biomass of a tropical tree 30 centimeters in diameter. Show your work.

41. pg 776  **Killing bacteria** Expose marine bacteria to X-rays for time periods from 1 to 15 minutes. Here are the number of surviving bacteria (in hundreds) on a culture plate after each exposure time:[21]

| Time $t$ | Count $y$ | Time $t$ | Count $y$ |
|---|---|---|---|
| 1 | 355 | 9 | 56 |
| 2 | 211 | 10 | 38 |
| 3 | 197 | 11 | 36 |
| 4 | 166 | 12 | 32 |
| 5 | 142 | 13 | 21 |
| 6 | 106 | 14 | 19 |
| 7 | 104 | 15 | 15 |
| 8 | 60 | | |

(a) Make a reasonably accurate scatterplot of the data by hand, using time as the explanatory variable. Describe what you see.

(b) A scatterplot of the natural logarithm of the number of surviving bacteria versus time is shown below. Based on this graph, explain why it would be reasonable to use an exponential model to describe the relationship between count of bacteria and time.

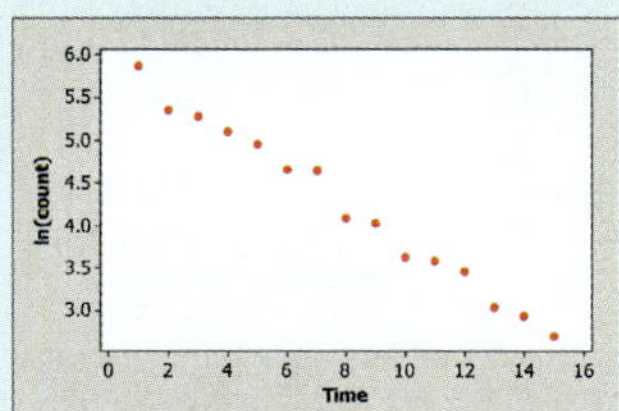


(c) Minitab output from a linear regression analysis on the transformed data is shown below.

```
Predictor  Coef       SE Coef    T       P
Constant    5.97316   0.05978    99.92  0.000
Time       -0.218425  0.006575  -33.22  0.000

S = 0.110016   R-Sq = 98.8%   R-Sq(adj) = 98.7%
```

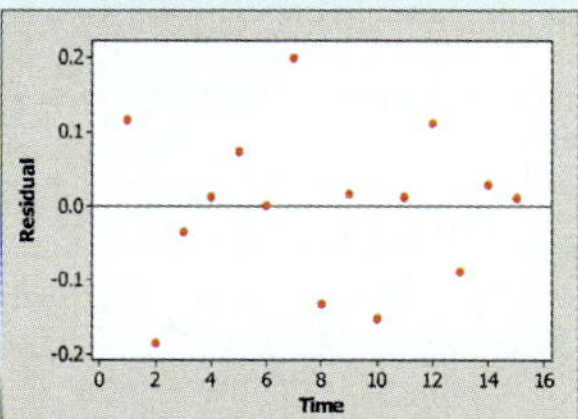


Give the equation of the least-squares regression line. Be sure to define any variables you use.

(d) Use your model to predict the number of surviving bacteria after 17 minutes. Show your work.

42. **Light through the water** Some college students collected data on the intensity of light at various depths in a lake. Here are their data:

| Depth (m) | Light intensity (lumens) |
|---|---|
| 5 | 168.00 |
| 6 | 120.42 |
| 7 | 86.31 |
| 8 | 61.87 |
| 9 | 44.34 |
| 10 | 31.78 |
| 11 | 22.78 |

(a) Make a reasonably accurate scatterplot of the data by hand, using depth as the explanatory variable. Describe what you see.

(b) A scatterplot of the natural logarithm of light intensity versus depth is shown below. Based on this graph, explain why it would be reasonable to use an exponential model to describe the relationship between light intensity and depth.

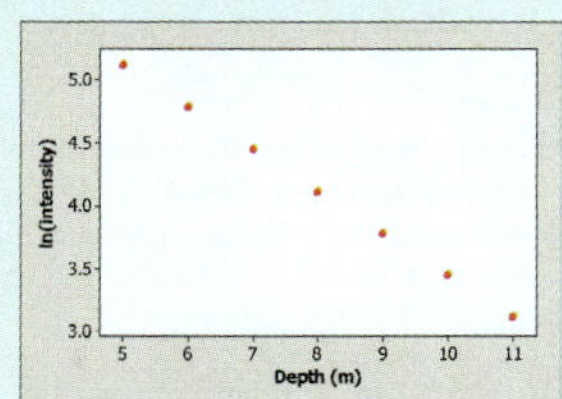


(c) Minitab output from a linear regression analysis on the transformed data is shown below.

```
Predictor  Coef       SE Coef   T          P
Constant   6.78910    0.00009   78575.46   0.000
Depth (m)  -0.333021  0.000010  -31783.44  0.000

S = 0.000055   R-Sq = 100.0%   R-Sq(adj) = 100.0%
```

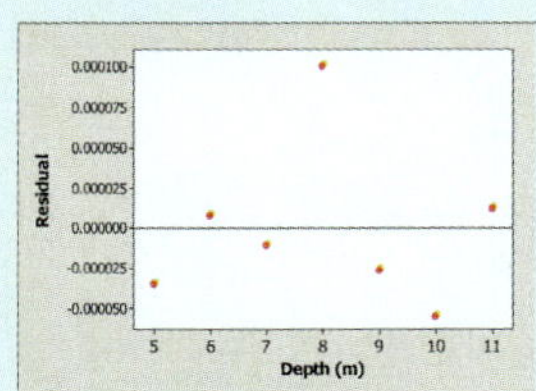


Give the equation of the least-squares regression line. Be sure to define any variables you use.

(d) Use your model to predict the light intensity at a depth of 12 meters. Show your work.

**43. Follow the bouncing ball** Students in Mr. Handford's class dropped a kickball beneath a motion detector. The detector recorded the height of the ball as it bounced up and down several times. Here are the heights of the ball at the highest point on the first five bounces:

pg 779

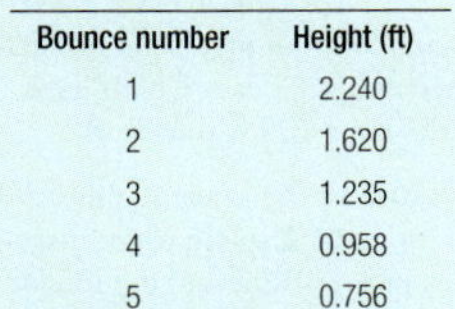

| Bounce number | Height (ft) |
|---|---|
| 1 | 2.240 |
| 2 | 1.620 |
| 3 | 1.235 |
| 4 | 0.958 |
| 5 | 0.756 |

Here is a scatterplot of the data:

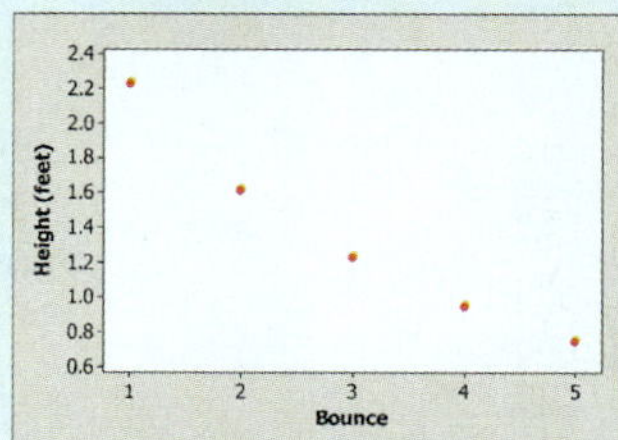


(a) The following graphs show the results of two different transformations of the data. Would an exponential model or a power model provide a better description of the relationship between bounce number and height? Justify your answer.

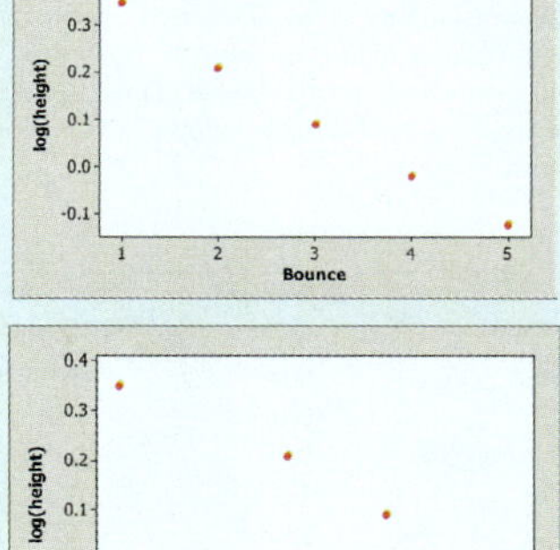


(b) Minitab output from a linear regression analysis on the transformed data of log(height) versus bounce number is shown below. Give the equation of the least-squares regression line. Be sure to define any variables you use.

```
Predictor  Coef       SE Coef   T       P
Constant   0.45374    0.01385   32.76   0.000
Bounce     -0.117160  0.004176  -28.06  0.000

S = 0.0132043   R-Sq = 99.6%   R-Sq(adj) = 99.5%
```

(c) Use your model from part (b) to predict the highest point the ball reaches on its seventh bounce. Show your work.

(d) A residual plot for the linear regression in part (b) is shown on the next page. Do you expect your prediction

**12.42** **(a)** The relationship is strong, negative, and slightly curved with no outliers.

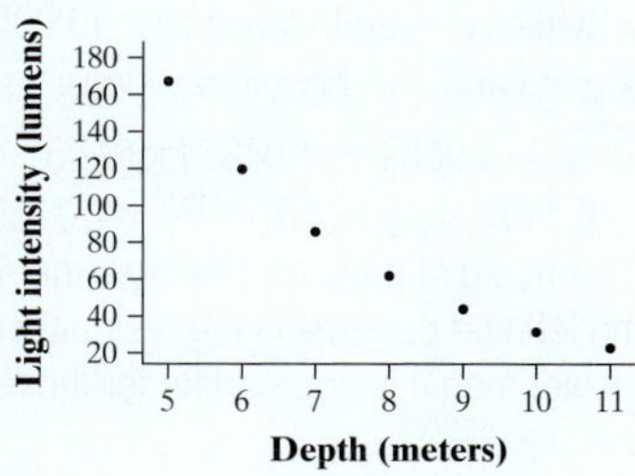


**(b)** Because the scatterplot of ln(intensity) vs. depth is fairly linear. **(c)** $\widehat{\ln y} = 6.789 - 0.333x$, where $y$ is the light intensity (lumens) and $x$ is the depth (meters). **(d)** $\widehat{\ln y} = 6.789 - 0.333(12) = 2.793$, so $\hat{y} = e^{2.793} = 16.33$ lumens.

**12.43** **(a)** Exponential, because the scatterplot of log(height) vs. bounce number is more linear. **(b)** $\widehat{\log y} = 0.45374 - 0.11716x$, where $y$ = height in feet and $x$ = bounce number. **(c)** $\widehat{\log y} = 0.45374 - 0.11716(7) = -0.36638$, so $\hat{y} = 10^{-0.36638} = 0.43$ feet. **(d)** The trend in the residual plot suggests that the residual for $x = 7$ would be positive, meaning that the predicted height will be less than the actual height.

**12.44 (a)** Power, because the scatterplot of log(abundance) vs. log(body mass) is more linear. **(b)** $\widehat{\log y} = 1.9503 - 1.0481 \log x$, where $y$ = abundance (per 10,000 kg of prey) and $x$ = body mass (kg). **(c)** $\widehat{\log y} = 1.9503 - 1.0481 \log(92.5) = -0.1104$, so $\hat{y} = 10^{-0.1104} = 0.7755$ per 10,000 kg of prey. **(d)** Because there are no leftover patterns in the residual plot, the power model is appropriate for these data.

in part (c) to be too high, too low, or about right? Justify your answer.

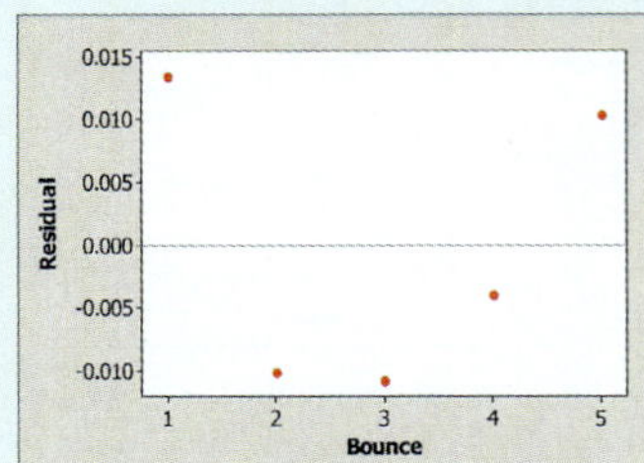


**44. Counting carnivores** Ecologists look at data to learn about nature's patterns. One pattern they have found relates the size of a carnivore (body mass in kilograms) to how many of those carnivores there are in an area. A good measure of "how many" is to count carnivores per 10,000 kilograms (kg) of their prey in the area. The table below gives data for 25 carnivore species.[22]

| Carnivore species | Body mass (kg) | Abundance (per 10,000 kg of prey) |
|---|---|---|
| Least weasel | 0.14 | 1656.49 |
| Ermine | 0.16 | 406.66 |
| Small Indian mongoose | 0.55 | 514.84 |
| Pine marten | 1.3 | 31.84 |
| Kit fox | 2.02 | 15.96 |
| Channel Islands fox | 2.16 | 145.94 |
| Arctic fox | 3.19 | 21.63 |
| Red fox | 4.6 | 32.21 |
| Bobcat | 10.0 | 9.75 |
| Canadian lynx | 11.2 | 4.79 |
| European badger | 13.0 | 7.35 |
| Coyote | 13.0 | 11.65 |
| Ethiopian wolf | 14.5 | 2.70 |
| Eurasian lynx | 20.0 | 0.46 |
| Wild dog | 25.0 | 1.61 |
| Dhole | 25.0 | 0.81 |
| Snow leopard | 40.0 | 1.89 |
| Wolf | 46.0 | 0.62 |
| Leopard | 46.5 | 6.17 |
| Cheetah | 50.0 | 2.29 |
| Puma | 51.9 | 0.94 |
| Spotted hyena | 58.6 | 0.68 |
| Lion | 142.0 | 3.40 |
| Tiger | 181.0 | 0.33 |
| Polar bear | 310.0 | 0.60 |

Here is a scatterplot of the data.

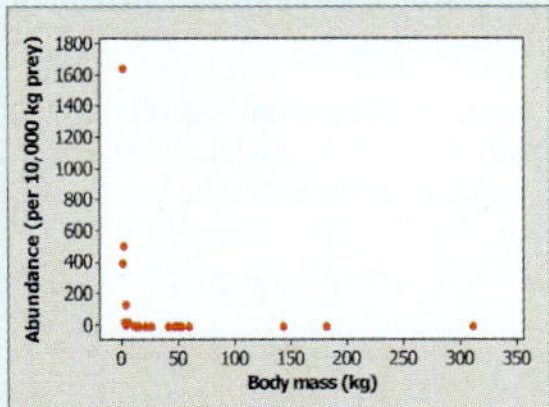


(a) The following graphs show the results of two different transformations of the data. Would an exponential model or a power model provide a better description of the relationship between body mass and abundance? Justify your answer.

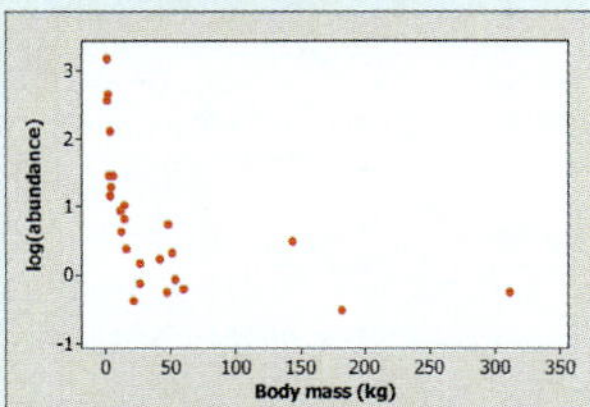


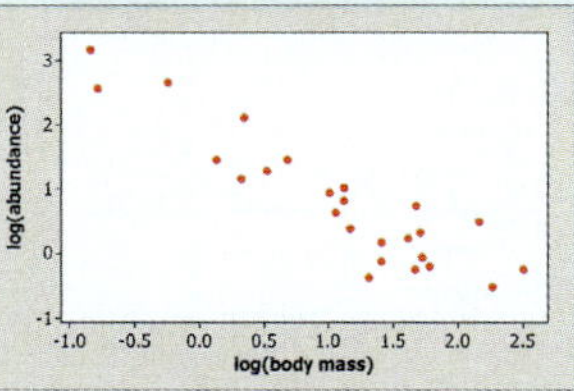


(b) Minitab output from a linear regression analysis on the transformed data of log(abundance) versus log(body mass) is shown below. Give the equation of the least-squares regression line. Be sure to define any variables you use.

```
Predictor  Coef      SE Coef   T       P
Constant    1.9503   0.1342    14.53   0.000
log(body   -1.04811  0.09802  -10.69   0.000
 mass)

S = 0.423352   R-Sq = 83.3%   R-Sq(adj) = 82.5%
```

(c) Use your model from part (b) to predict the abundance of black bears, which have a body mass of 92.5 kilograms. Show your work.

(d) A residual plot for the linear regression in part (b) is shown at top right. Explain what this graph tells you about the appropriateness of the model.

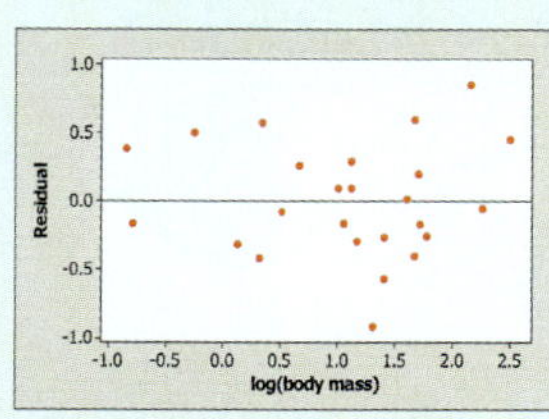


45. **Heart weights of mammals** Here are some data on the hearts of various mammals.[23]

| Mammal | Length of cavity of left ventricle (cm) | Heart weight (g) |
|---|---|---|
| Mouse | 0.55 | 0.13 |
| Rat | 1.0 | 0.64 |
| Rabbit | 2.2 | 5.8 |
| Dog | 4.0 | 102 |
| Sheep | 6.5 | 210 |
| Ox | 12.0 | 2030 |
| Horse | 16.0 | 3900 |

(a) Make an appropriate scatterplot for predicting heart weight from length. Describe what you see.

(b) Use transformations to linearize the relationship. Does the relationship between heart weight and length seem to follow an exponential model or a power model? Justify your answer.

(c) Perform least-squares regression on the transformed data. Give the equation of your regression line. Define any variables you use.

(d) Use your model from part (c) to predict the heart weight of a human who has a left ventricle 6.8 centimeters long. Show your work.

46. **Galileo's experiment** Galileo studied motion by rolling balls down ramps. He rolled a ball down a ramp with a horizontal shelf at the end of it so that the ball was moving horizontally when it started to fall off the shelf. The top of the ramp was placed at different heights above the floor (that is, the length of the ramp varied), and Galileo measured the horizontal distance the ball traveled before it hit the floor. Here are Galileo's data. (We won't try to describe the obsolete seventeenth-century units Galileo used to measure distance and height.)[24]

| Height | Distance |
|---|---|
| 1000 | 1500 |
| 828 | 1340 |
| 800 | 1328 |
| 600 | 1172 |
| 300 | 800 |

(a) Make an appropriate scatterplot for predicting horizontal distance traveled from ramp height. Describe what you see.

(b) Use transformations to linearize the relationship. Does the relationship between distance and height seem to follow an exponential model or a power model? Justify your answer.

(c) Perform least-squares regression on the transformed data. Give the equation of your regression line. Define any variables you use.

(d) Use your model from part (c) to predict the horizontal distance a ball would travel if the ramp height was 700. Show your work.

*Multiple Choice: Select the best answer for Exercises 47 to 50.*

47. Suppose that the relationship between a response variable $y$ and an explanatory variable $x$ is modeled by $y = 2.7(0.316)^x$. Which of the following scatterplots would approximately follow a straight line?

(a) A plot of $y$ against $x$

(b) A plot of $y$ against $\log x$

(c) A plot of $\log y$ against $x$

(d) A plot of $\log y$ against $\log x$

(e) A plot of $\sqrt{y}$ against $x$.

48. Some high school physics students dropped a ball and measured the distance fallen (in centimeters) at various times (in seconds) after its release. If you have studied physics, then you probably know that the theoretical relationship between the variables is distance $= 490(\text{time})^2$. A scatterplot of the students' data showed a clear curved pattern. At 0.68 seconds after release, the ball had fallen 220.4 centimeters. How much more or less did the ball fall than the theoretical model predicts?

(a) More by 226.576 centimeters

(b) More by 6.176 centimeters

(c) No more and no less

(d) Less by 226.576 centimeters

(e) Less by 6.176 centimeters

49. A scatterplot of $x$ = Super Bowl number and $y$ = cost of a 30-second advertisement on the Super Bowl broadcast (in dollars) shows a strong, positive, nonlinear association. A scatterplot of ln(cost) versus Super Bowl number is roughly linear. The least-squares regression line for this association is $\widehat{\ln(\text{cost})} = 10.97 + 0.0971$ (Super Bowl number). Predict the cost of a 30-second advertisement for Super Bowl 40.

**12.45 (a)** There is a strong, positive, curved relationship between heart weight and length of left ventricle for mammals.

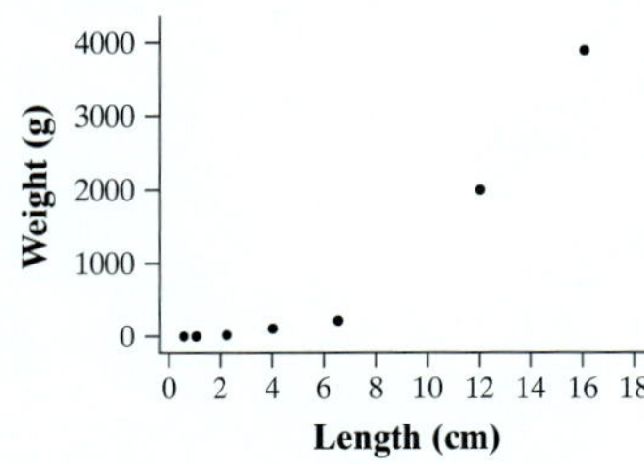


**(b)** Two scatterplots are given at right. Because the relationship between ln(weight) and ln(length) is roughly linear, heart weight and length seem to follow a power model.

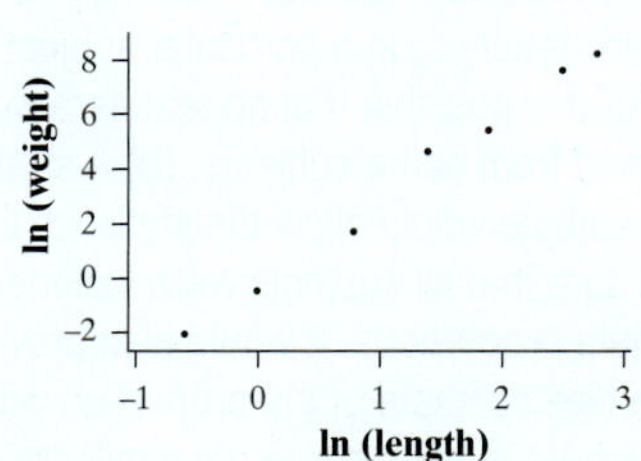


**(c)** $\widehat{\ln y} = -0.314 + 3.1387 \ln x$, where $y$ is the weight of the heart and $x$ is the length of the cavity of the left ventricle. **(d)** $\widehat{\ln y} = -0.314 + 3.1387 \ln(6.8) = 5.703$, so $\hat{y} = e^{5.703} = 299.77$ grams.

**12.46 (a)** There is a strong, positive, slightly curved relationship between height and distance.

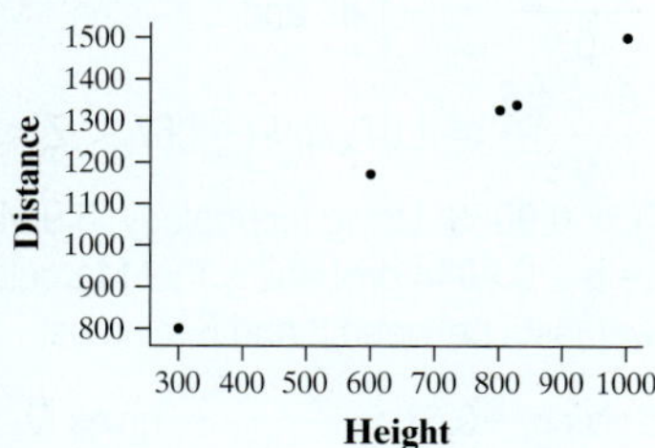


**(b)** Two scatterplots are given below. Because the relationship between ln(distance) and ln(height) is roughly linear, distance and height seem to follow a power model.

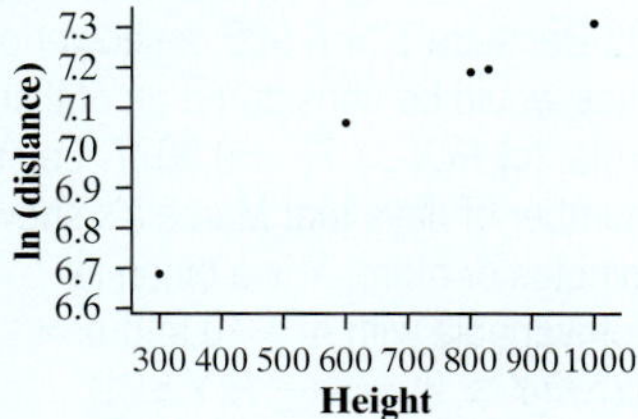


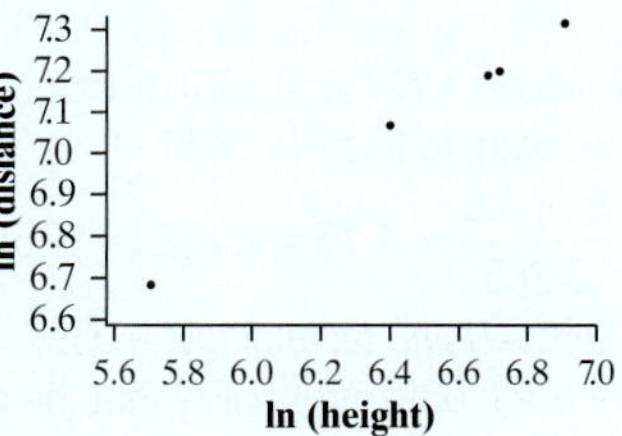


**(c)** The equation is $\widehat{\ln y} = 3.7514 + 0.5152 \ln x$, where $y$ is the distance and $x$ is the height. **(d)** If the ramp height was 700, $\widehat{\ln y} = 3.7514 + 0.5152 \ln(700) = 7.1265$ and $\hat{y} = e^{7.1265} = 1244.51$ units.

**12.47** c

**12.48** e

(a) \$3
(b) \$15
(c) \$58,153
(d) \$83,132
(e) \$2,824,947

50. A scatterplot of $y$ versus $x$ shows a positive, nonlinear association. Two different transformations are attempted to try to linearize the association: using the logarithm of the $y$ values and using the square root of the $y$ values. Two least-squares regression lines are calculated, one that uses $x$ to predict $\log(y)$ and the other that uses $x$ to predict $\sqrt{y}$. Which of the following would be the best reason to prefer the least-squares regression line that uses $x$ to predict $\log(y)$?
    (a) The value of $r^2$ is smaller.
    (b) The standard deviation of the residuals is smaller.
    (c) The slope is greater.
    (d) The residual plot has more random scatter.
    (e) The distribution of residuals is more Normal.

51. **Shower time** (1.3, 2.2, 6.3, 7.3) Marcella takes a shower every morning when she gets up. Her time in the shower varies according to a Normal distribution with mean 4.5 minutes and standard deviation 0.9 minutes.
    (a) Find the probability that Marcella's shower lasts between 3 and 6 minutes on a randomly selected day. Show your work.
    (b) If Marcella took a 7-minute shower, would it be classified as an outlier by the 1.5*IQR* rule? Justify your answer.
    (c) Suppose we choose 10 days at random and record the length of Marcella's shower each day. What's the probability that her shower time is 7 minutes or higher on at least 2 of the days? Show your work.
    (d) Find the probability that the *mean* length of her shower times on these 10 days exceeds 5 minutes. Show your work.

52. **Tattoos** (8.2) What percent of U.S. adults have one or more tattoos? The Harris Poll conducted an online survey of 2302 adults during January 2008. According to the published report, "Respondents for this survey were selected from among those who have agreed to participate in Harris Interactive surveys."[25] The pie chart at top right summarizes the responses from those who were surveyed. Explain why it would *not* be appropriate to use these data to construct a 95% confidence interval for the proportion of all U.S. adults who have tattoos.

**Do you have a tattoo?**

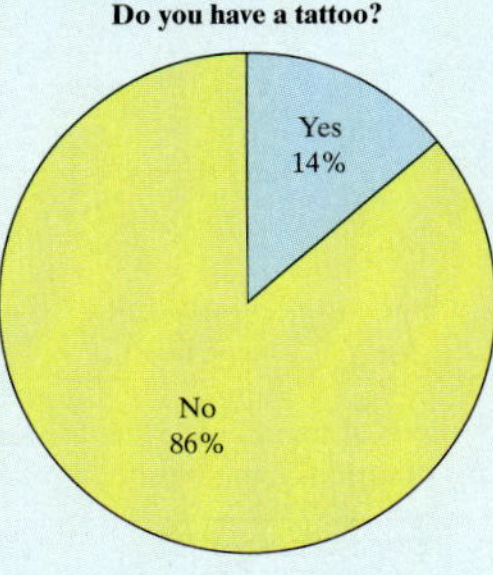


*Exercises 53 and 54 refer to the following setting.* About 1100 high school teachers attended a weeklong summer institute for teaching AP® classes. After hearing about the survey in Exercise 52, the teachers in the AP® Statistics class wondered whether the results of the tattoo survey would be similar for teachers. They designed a survey to find out. The class opted to take a random sample of 100 teachers at the institute. One of the questions on the survey was

Do you have any tattoos on your body?

(Circle one) YES NO

53. **Tattoos** (8.2, 9.2) Of the 98 teachers who responded, 23.5% said that they had one or more tattoos.
    (a) Construct and interpret a 95% confidence interval for the actual proportion of teachers at the AP® institute who would say they had tattoos.
    (b) Does the interval in part (a) provide convincing evidence that the proportion of teachers at the institute with tattoos is not 0.14 (the value cited in the Harris Poll report)? Justify your answer.
    (c) Two of the selected teachers refused to respond to the survey. If both of these teachers had responded, could your answer to part (b) have changed? Justify your answer.

54. **Tattoos** (4.1) One of the first decisions the class had to make was what kind of sampling method to use.
    (a) They knew that a simple random sample was the "preferred" method. With 1100 teachers in 40 different sessions, the class decided not to use an SRS. Give at least two reasons why you think they made this decision.
    (b) The AP® Statistics class believed that there might be systematic differences in the proportions of teachers who had tattoos based on the subject areas that they taught. What sampling method would you recommend to account for this possibility? Explain a statistical advantage of this method over an SRS.

**12.49** e

**12.50** d

**12.51 (a)** For Marcela, $X$ = the length of her shower on a randomly selected day follows a Normal distribution with mean 4.5 minutes and standard deviation 0.9 minutes. We want to find $P(3 < X < 6)$. $z = \frac{3-4.5}{0.9} = -1.67$ and $z = \frac{6-4.5}{0.9} = 1.67$, so $P(-1.67 < Z < 1.67) = 0.9050$. *Using technology*: 0.9044. There is a 0.9044 probability that Marcela's shower lasts between 3 and 6 minutes. **(b)** Solving $-0.67 = \frac{Q_1 - 4.5}{0.9}$ gives $Q_1 =$ 3.897 minutes. Solving $0.67 = \frac{Q_3 - 4.5}{0.9}$ gives $Q_3 = 5.103$ minutes. *Using technology*: $Q_1 = 3.893$ minutes and $Q_3 = 5.107$ minutes. Thus, an outlier is any value above $5.107 + 1.5(5.107 - 3.893) = 6.928$. Because $7 > 6.928$, a shower of 7 minutes would be considered an outlier for Marcela. **(c)** $P(X > 7) = 0.0027$. Let $Y$ = the number of days that Marcela's shower is7 minutes or more. $Y$ is a binomial randomvariable with $n = 10$ and $p = 0.0027$.$P(Y \geq 2) = 1 - P(Y \leq 1)$ = 1 - `binomcdf(trials:10, p:0.0027, x value:1)` = 0.0003. **(d)** $\bar{x}$ follows a $N(4.5, 0.285)$ distribution and we want to find $P(\bar{x} > 5)$. $z = \frac{5-4.5}{0.285} = 1.75$ and $P(Z > 1.75) = 0.0401$. *Using technology*: 0.0397. There is a 0.0397 probability that the mean length of Marcela's showers on these 10 days exceeds 5 minutes.

**12.52** The sample will be representative only of the population of people who have agreed to participate in these polls, not the population of all U.S. adults.

**12.53 (a)** *S*: $p$ = true proportion of all AP® teachers attending this workshop who have tattoos. *P*: One-sample $z$ interval for $p$. *Random*: Random sample. *10%*: The sample size ($n = 98$) is less than 10% of the population of teachers at this workshop (1100). *Large Counts*: 23 and 75 are both $\geq 10$. *D*: (0.151, 0.319). *C*: We are 95% confident that the interval from 0.151 to 0.319 captures the true proportion of AP® teachers at this workshop who have tattoos. **(b)** Yes. Because the value 0.14 is not included in the interval, we have convincing evidence that the true proportion of teachers at the workshop who have a tattoo is not 0.14. **(c)** If we had two more failures, the interval will shift to lower values and might include the value 0.14. However, the new interval is (0.148, 0.312), which does not include the value 0.14. So, the answer would not change if we got responses from the 2 nonresponders.

**12.54 (a)** A possible response would be that (1) it would be hard to find all of the individual teachers in the sample since they were spread out among 40 different sessions and (2) there may not have been many teachers in a particular subject and in an SRS it is possible that no teachers would be selected from some subjects. **(b)** A stratified random sample would allow the statistics teachers to make sure that all subjects were included in appropriate proportions. It would also provide a more precise estimate of the proportion with tattoos if the teachers in each strata are similar to each other and different than teachers in other strata.

## FRAPPY! Free Response AP® Problem, Yay!

The following problem is modeled after actual AP® Statistics exam free response questions. Your task is to generate a complete, concise response in 15 minutes.

*Directions: Show all your work. Indicate clearly the methods you use, because you will be scored on the correctness of your methods as well as on the accuracy and completeness of your results and explanations.*

A random sample of 14 golfers was selected from the 147 players on the Ladies Professional Golf Association (LPGA) tour in a recent year. The total amount of money won during the year (in dollars) and the scoring average for each player in the sample was recorded. Lower scoring averages are better in golf.

The scatterplot below displays the relationship between money and scoring average for these 14 players.

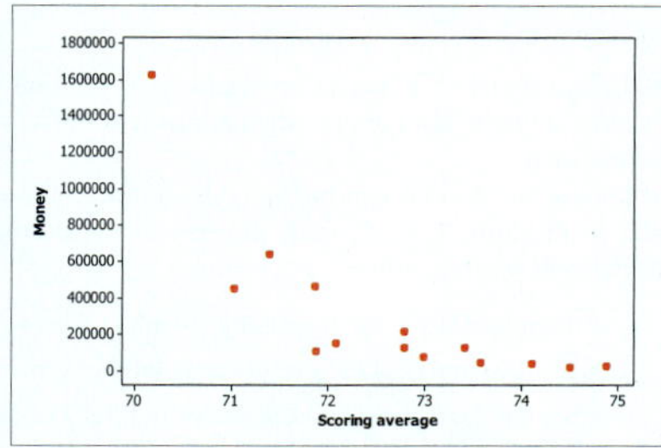


(a) Explain why it would *not* be appropriate to construct a confidence interval for the slope of the least-squares regression line relating money to scoring average.

A scatterplot of the natural logarithm of money versus scoring average is shown at top right along with some computer output for a least-squares regression using the transformed data.

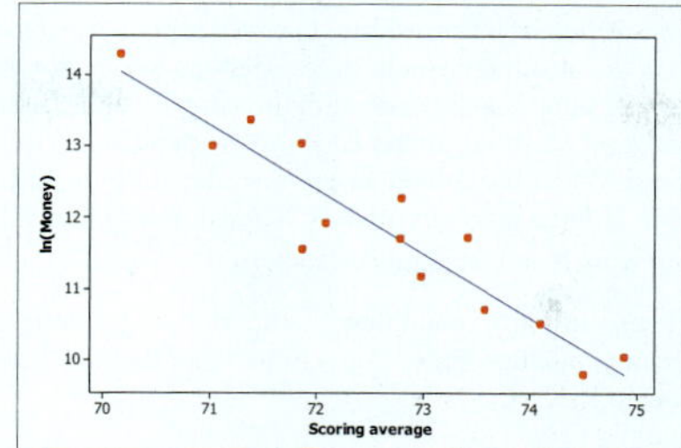


| Predictor | Coef | SE Coef | T | P |
|---|---|---|---|---|
| Constant | 77.537 | 7.035 | 11.02 | 0.000 |
| Scoring average | −0.90470 | 0.09679 | −9.35 | 0.000 |

S = 0.475059  R-Sq = 87.9%  R-Sq(adj) = 86.9%

(b) Predict the amount of money won for an LPGA golfer with a scoring average of 70.

(c) Calculate and interpret a 95% confidence interval for the slope of the least-squares regression line relating ln(money) to scoring average. Assume that the conditions for inference have been met.

After you finish, you can view two example solutions on the book's Web site (www.whfreeman.com/tps5e). Determine whether you think each solution is "complete," "substantial," "developing," or "minimal." If the solution is not complete, what improvements would you suggest to the student who wrote it? Finally, your teacher will provide you with a scoring rubric. Score your response and note what, if anything, you would do differently to improve your own score.

**TRM FRAPPY!**

We present a model solution here. Please consult the Teacher's Resource Materials for sample student responses, a scoring rubric, and a black line master of the original question with space for students to write their responses.

**(a)** It would not be appropriate to calculate a confidence interval for the slope because the Linear condition is not met. The association shown in the scatterplot is clearly curved.

**(b)** $\widehat{\ln(\text{Money})} = 77.537 - 0.90470(70) = 14.208$.

Thus, $\widehat{\text{Money}} = e^{14.208} = \$1{,}480{,}662$.

**(c)** STATE: We want to estimate the slope $\beta$ of the population regression line relating $\ln y$ = natural logarithm of money earned to $x$ = scoring average at the 95% confidence level.

PLAN: We were told the conditions were met, so we will use a $t$ interval for the slope to estimate $\beta$.

DO: Using df = 14 − 2 = 12, the $t$ critical value is $t^* = 2.179$. Thus, the 95% confidence interval is $-0.90470 \pm 2.179(0.09679) = -0.90470 \pm 0.21091 = (-1.11561, -0.69379)$.

CONCLUDE: We are 95% confident that the interval from −1.11561 to −0.69379 captures the slope of the population regression line relating ln(money) to scoring average.

## TRM Chapter 12 Tests

There are four different tests available for this chapter in the Teacher's Resource Materials. Versions A and B are roughly parallel and version D is cumulative. Find them by clicking on the link in the TE-book, opening the TRCD, or logging onto the book's companion Web site www.whfreeman.com/tps5e (teacher log-in required).

## Teaching Tip

As you finish Chapter 12, make sure to reread the section "Preparing Students for the AP® Exam" on page x in the *Annotated Teacher's Edition*. There are many good suggestions, including using the flash cards described below.

## TRM Flash Cards

In the Teacher's Resource Materials, we have provided a set of 60 flash cards that review important topics from the AP® Statistics curriculum. Make a set for each student and encourage them to review the cards every day until the AP® exam. Find them by clicking on the link in the TE-book, opening the TRCD, or logging onto the book's companion Web site www.whfreeman.com/tps5e (teacher log-in required).

## TRM Final Project

A final project is a great way to pull together the important ideas from the course. A description and rubric are provided in the Teacher's Resource Materials. We also included some helpful comments from veteran statistics teacher Floyd Bullard. Find them by clicking on the link in the TE-book, opening the TRCD, or logging onto the book's companion Web site www.whfreeman.com/tps5e (teacher log-in required).

## TRM Inference Review

In the Teacher's Resource Materials, we have provided a short worksheet to help students practice identifying the correct inference procedure. Find it by clicking on the link in the TE-book, opening the TRCD, or logging onto the book's companion Web site www.whfreeman.com/tps5e (teacher log-in required).

# Chapter Review

### Section 12.1: Inference for Linear Regression

In this section, you learned how to conduct inference about the slope of a population (true) least-squares regression line. The sampling distribution of the sample slope $b$ is the foundation for doing inference about the population (true) slope $\beta$. When the conditions are met, the sampling distribution of $b$ has an approximately Normal distribution with mean $\mu_b = \beta$ and standard deviation $\sigma_b = \frac{\sigma}{\sigma_x\sqrt{n}}$.

There are five conditions for performing inference about a population (true) slope. Remember them with the acronym LINER.

- The Linear condition says that the mean value of the response variable $\mu_y$ falls on the population (true) regression line $\mu_y = \alpha + \beta x$. To check the Linear condition, verify that there are no leftover patterns in the residual plot.
- The Independent condition says that individual observations are independent of each other. To check the Independent condition, verify that the sample size is less than 10% of the population size. Also, convince yourself that knowing the response for one individual won't help you predict the response for another individual.
- The Normal condition says that the distribution of $y$ values is approximately Normal for each value of $x$. To check the Normal condition, graph a dotplot, histogram, or Normal probability plot of the residuals and verify that there are no outliers or strong skewness.
- The Equal SD condition says that for each value of $x$, the distribution of $y$ should have the same standard deviation. To check the Equal SD condition, verify that the residuals have roughly the same amount of scatter around the residual = 0 line for each value of $x$ on the residual plot.
- The Random condition says that the data are from a random sample or a randomized experiment. To check the Random condition, verify that randomness was properly used in the data collection process.

To construct and interpret a confidence interval for the slope of the population (true) least-squares regression line, follow the familiar four-step process. The formula for the confidence interval is $b \pm t^*\text{SE}_b$, where $t^*$ is the $t$ critical value with df $= n - 2$. The standard error of the slope $\text{SE}_b$ describes how far the sample slope typically varies from the population (true) slope in repeated random samples or random assignments. The formula for the standard error of the slope is $\text{SE}_b = \frac{s}{s_x\sqrt{n-1}}$. The standard error of the slope is typically provided with standard computer output for least-squares regression.

In most cases, when you conduct a significance test for the slope of the population (true) least-squares regression line, the null hypothesis is $H_0: \beta = 0$. This hypothesis says that a straight-line relationship between $x$ and $y$ is of no value for predicting $y$. To do the calculations, use the test statistic $t = \frac{b - \beta_0}{\text{SE}_b}$ with df $= n - 2$. The value of the test statistic, along with a two-sided $P$-value, is typically provided with standard computer output for least-squares regression.

### Section 12.2: Transforming to Achieve Linearity

When the association between two variables is nonlinear, transforming one or both of the variables can result in a linear association.

If the association between two variables follows a power model in the form $y = ax^p$, there are several transformations that will result in a linear association.

- Raise the values of $x$ to the power of $p$ and plot $y$ versus $x^p$.
- Calculate the $p$th root of the $y$ values and plot $\sqrt[p]{y}$ versus $x$.
- Calculate the logarithms of the $x$ values and the $y$ values and plot $\log(y)$ versus $\log(x)$. You can use base 10 logarithms (log) or base $e$ logarithms (ln).

If the association between two variables follows an exponential model in the form $y = ab^x$, transform the data by computing the logarithms of the $y$ values and plot $\log(y)$ versus $x$ (or $\ln(y)$ versus $x$).

Once you have achieved linearity, calculate the equation of the least-squares regression line using the transformed data. Remember to include the transformed variables when you are writing the equation of the line. Likewise, when using the line to make predictions, make sure that the prediction is in the original units of $y$. If you transformed the $y$ variable, you will need to undo the transformation after using the least-squares regression line.

To decide between two or more transformations, look at the residual plots and choose the one with the most random scatter.

794

## What Did You Learn?

| Learning Objective | Section | Related Example on Page(s) | Relevant Chapter Review Exercise(s) |
|---|---|---|---|
| Check the conditions for performing inference about the slope $\beta$ of the population (true) regression line. | 12.1 | 745 | R12.2, R12.3, R12.4 |
| Interpret the values of *a*, *b*, *s*, $SE_b$, and $r^2$ in context, and determine these values from computer output. | 12.1 | 748, 754 | R12.1 |
| Construct and interpret a confidence interval for the slope $\beta$ of the population (true) regression line. | 12.1 | 749 | R12.3 |
| Perform a significance test about the slope $\beta$ of the population (true) regression line. | 12.1 | 754 | R12.2 |
| Use transformations involving powers and roots to find a power model that describes the relationship between two variables, and use the model to make predictions. | 12.2 | 768, 770 | R12.5 |
| Use transformations involving logarithms to find a power model or an exponential model that describes the relationship between two variables, and use the model to make predictions. | 12.2 | 772, 773, 776 | R12.6 |
| Determine which of several transformations does a better job of producing a linear relationship. | 12.2 | 779 | R12.6 |

TRM Learning Objectives Grid

# Chapter 12 Chapter Review Exercises

*These exercises are designed to help you review the important ideas and methods of the chapter.*

*Exercises R12.1 to R12.3 refer to the following setting.* In the casting of metal parts, molten metal flows through a "gate" into a die that shapes the part. The gate velocity (the speed at which metal is forced through the gate) plays a critical role in die casting. A firm that casts cylindrical aluminum pistons examined a random sample of 12 pistons formed from the same alloy of metal. What is the relationship between the cylinder wall thickness (inches) and the gate velocity (feet per second) chosen by the skilled workers who do the casting? If there is a clear pattern, it can be used to direct new workers or to automate the process. A scatterplot of the data is shown below.[26]

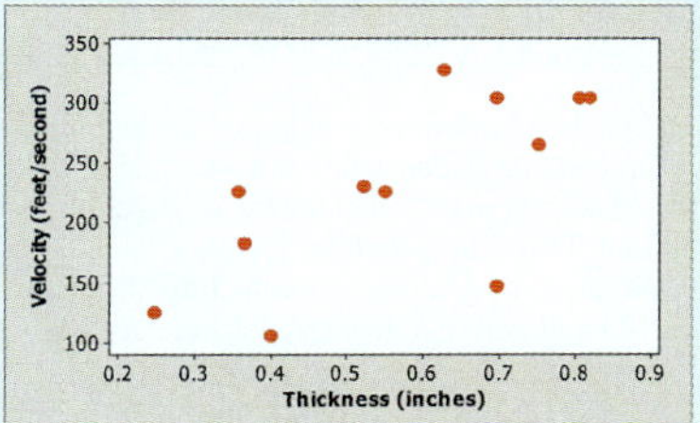


A least-squares regression analysis was performed on the data. Some computer output and a residual plot are shown below. A Normal probability plot of the residuals (not shown) is roughly linear.

```
Predictor   Coef     SE Coef   T      P
Constant    70.44    52.90     1.33   0.213
Thickness   274.78   88.18     ***    ***
S = 56.3641  R-Sq = 49.3%  R-Sq(adj) = 44.2%
```

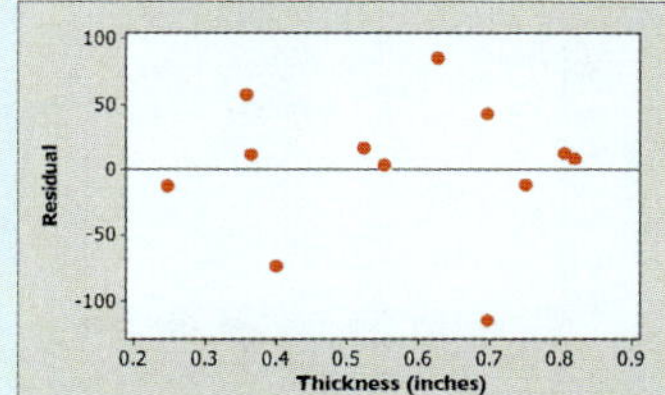

**R12.1 Casting aluminum**

(a) Describe what the scatterplot tells you about the relationship between cylinder wall thickness and gate velocity.

(b) What is the equation of the least-squares regression line? Define any variables you use.

(c) One of the cylinders in the sample had a wall thickness of 0.4 inches. The gate velocity chosen for this cylinder was 104.8 feet per second. Does the regression line in part (b) overpredict or underpredict the gate velocity for this cylinder? By how much? Show your work.

(d) Is a linear model appropriate in this setting? Justify your answer with appropriate evidence.

(e) Interpret each of the following in context:
(i) The slope
(ii) $s$
(iii) $r^2$
(iv) The standard error of the slope

**R12.2 Casting aluminum** Do the data provide convincing evidence at the $\alpha = 0.05$ level of a linear relationship between thickness and gate velocity in the population of pistons formed from this alloy of metal?

**R12.3 Casting aluminum** Construct and interpret a 95% confidence interval for the slope of the population regression line. Explain how this interval is consistent with the results of Exercise R12.2.

**R12.4 SAT essay—is longer better?** Following the debut of the new SAT Writing test in March 2005, Dr. Les Perelman from the Massachusetts Institute of Technology recorded the number of words and score for each essay in a sample provided by the College Board. A least-squares regression analysis was performed on these data. The two graphs at bottom left display the results of that analysis. Explain why the conditions for performing inference are not met in this setting.

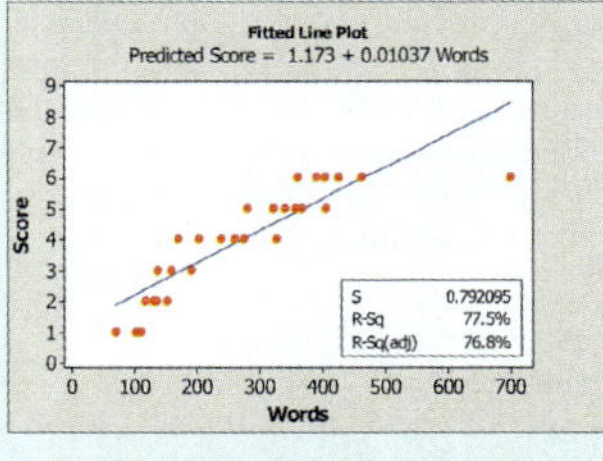


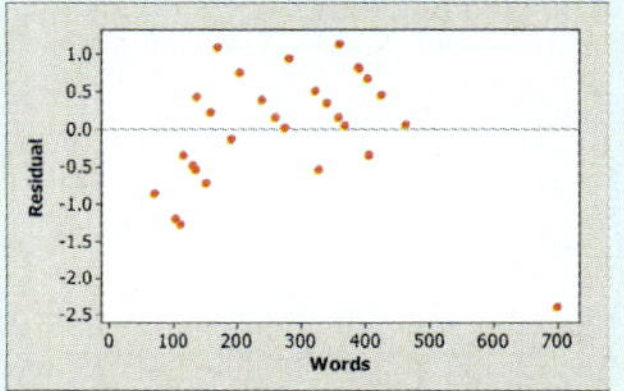


**R12.5 Light intensity** In a physics class, the intensity of a 100-watt lightbulb was measured by a sensor at various distances from the light source. A scatterplot of the data is shown below. Note that a candela is a unit of luminous intensity in the International System of Units.

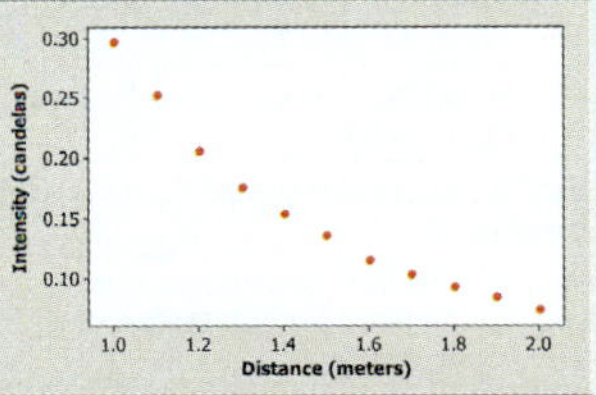


Physics textbooks suggest that the relationship between light intensity $y$ and distance $x$ should follow an "inverse square law," that is, a power law model of the form $y = ax^{-2} = a\frac{1}{x^2}$. We transformed the distance measurements by squaring them and then taking their reciprocals. Some computer output and a residual plot from a least-squares regression analysis on the transformed data are shown below. Note that the horizontal axis on the residual plot displays predicted light intensity.

| Predictor | Coef | SE Coef | T | P |
|---|---|---|---|---|
| Constant | -0.000595 | 0.001821 | -0.33 | 0.751 |
| Distance^(-2) | 0.299624 | 0.003237 | 92.56 | 0.000 |

S = 0.00248369  R-Sq = 99.9%  R-Sq(adj) = 99.9%

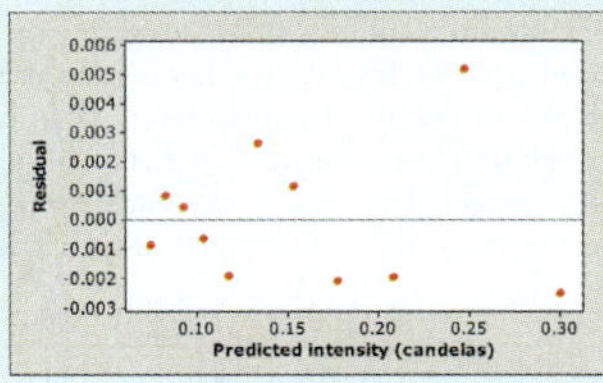


(a) Did this transformation achieve linearity? Give appropriate evidence to justify your answer.

(b) What is the equation of the least-squares regression line? Define any variables you use.

(c) What would you predict for the intensity of a 100-watt bulb at a distance of 2.1 meters? Show your work.

**TRM Full Solutions to Chapter 12 Review Exercises**

## Answers to Chapter 12 Review Exercises

**R12.1 (a)** There is a moderately strong, positive linear relationship between the thickness and the velocity. **(b)** $\hat{y} = 70.44 + 274.78x$, where $y$ is the velocity and $x$ is the thickness. **(c)** Residual $= 104.8 - 180.352 = -75.552$, so the line overpredicts the velocity by 75.552 feet/second. **(d)** The linear model is appropriate. The scatterplot shows a linear relationship and the residual plot has no leftover patterns. **(e)** *Slope*: For each increase of an inch in thickness, the predicted velocity increases by 274.78 feet/second. *s*: When using the least-squares regression line with $x$ = thickness to predict $y$ = velocity, we will typically be off by about 56.36 feet per second. $r^2$: About 49.3% of the variation in velocity is accounted for by the linear relationship relating velocity to thickness. $SE_b$: If we take many different random samples of 12 pistons and compute the least-squares regression line for each sample, the estimated slope will typically vary from the slope of the population regression line for predicting velocity from thickness by about 88.18.

**R12.2** *S*: $H_0: \beta = 0$ vs. $H_a: \beta \neq 0$, where $\beta$ is the slope of the population regression line relating thickness to velocity. *P*: $t$ test for $\beta$. *Linear*: The residual plot shows no leftover patterns. *Independent*: Knowing the velocity for one piston should not help us predict the velocity for another piston. Also, the sample size ($n = 12$) is less than 10% of the pistons in the population. *Normal*: We are told that the Normal probability plot of the residuals is roughly linear. *Equal SD*: The residual plot shows roughly equal scatter for all $x$ values. *Random*: The data come from a random sample. *D*: $t = 3.116$. With df $= 10$, the *P*-value is between 0.01 and 0.02 (0.0109). *C*: Because the *P*-value of 0.0109 is less than $\alpha = 0.05$ we reject $H_0$. There is convincing evidence of a linear relationship between thickness and gate velocity in the population of pistons formed from this alloy of metal.

**R12.3** *D*: With df $= 12 - 2 = 10$, (78.315, 471.245). *C*: We are 95% confident that the interval from 78.315 to 471.245 captures the slope of the population regression line for predicting velocity from thickness for the population of pistons formed from this alloy of metal. Because 0 is not in the interval, we reject 0 as a plausible value for the slope of the population regression line, as in R12.2.

**R12.4** The Linear condition is violated because there is clear curvature to the scatterplot and an obvious curved pattern in the residual plot. The Random condition may not be met because we weren't told if the sample was selected at random.

**R12.5 (a)** Yes, because there is no leftover pattern in the residual plot.

**(b)** $\hat{y} = -0.000595 + 0.3\left(\frac{1}{x^2}\right)$, where $y$ = intensity and $x$ = distance.

**(c)** $\hat{y} = -0.000595 + 0.3\left(\frac{1}{(2.1)^2}\right) = 0.0674$ candelas.

**R12.6** An experiment was conducted to determine the effect of practice time (in seconds) on the percent of unfamiliar words recalled. Here is a Fathom scatterplot of the results with a least-squares regression line superimposed.

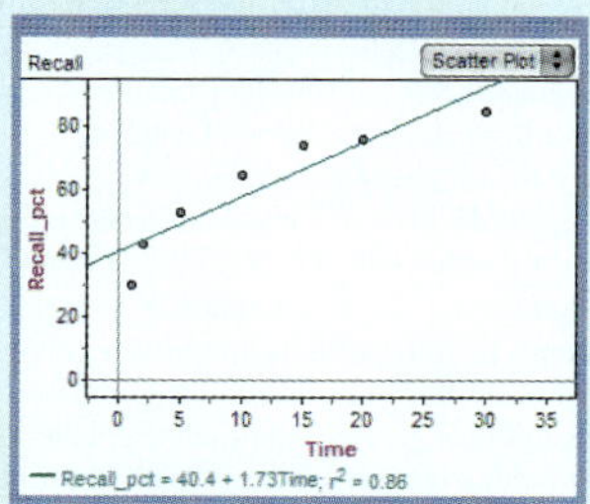


(a) Sketch a residual plot. Be sure to label your axes.

(b) Explain why a linear model is not appropriate for describing the relationship between practice time and percent of words recalled.

(c) We used Fathom to transform the data in hopes of achieving linearity. The screen shots on the right show the results of two different transformations. Would an exponential model or a power model describe the relationship better? Justify your answer.

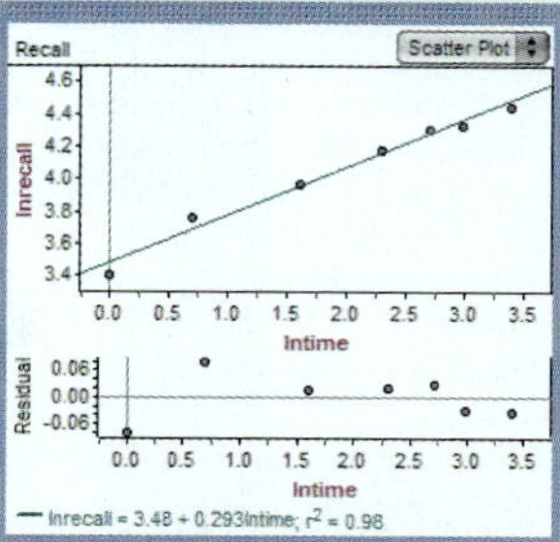


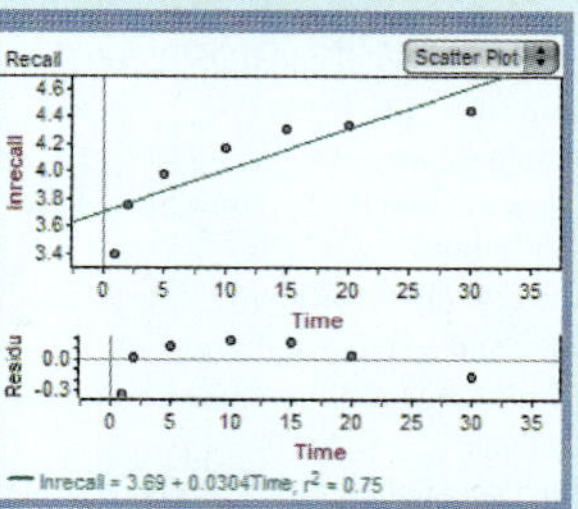


(d) Use each model to predict the word recall for 25 seconds of practice. Show your work. Which prediction do you think will be better?

**R12.6 (a)**

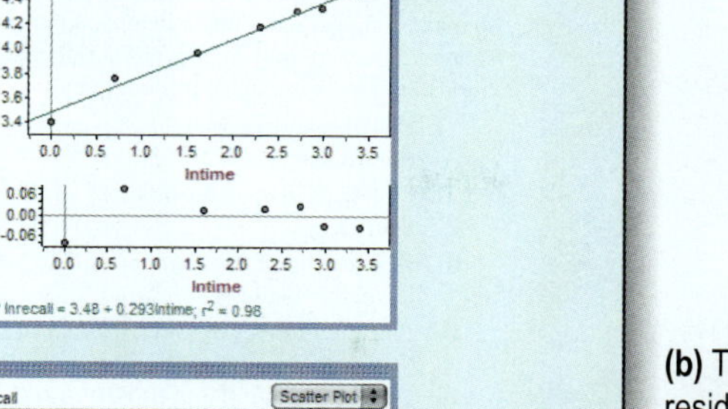


**(b)** There is a leftover pattern in the residual plot, so the relationship between practice time and percent of words recalled is not linear. **(c)** Power, because the scatterplot showing ln(recall) vs. ln(time) is more linear than the scatterplot showing ln(recall) vs. time. **(d)** Power: $\widehat{\ln y} = 3.48 + 0.293\ln(25) = 4.423$ and $\hat{y} = e^{4.423} = 83.35\%$ of words recalled. Exponential: $\widehat{\ln y} = 3.69 + 0.0304(25) = 4.45$ and $\hat{y} = e^{4.45} = 85.63\%$ of words recalled. Based on my answer to part (c), I think the power model will give a better prediction.

# Chapter 12 AP® Statistics Practice Test

**Section I: Multiple Choice** *Select the best answer for each question.*

**T12.1** Which of the following is *not* one of the conditions that must be satisfied in order to perform inference about the slope of a least-squares regression line?

(a) For each value of $x$, the population of $y$-values is Normally distributed.

(b) The standard deviation $\sigma$ of the population of $y$-values corresponding to a particular value of $x$ is always the same, regardless of the specific value of $x$.

(c) The sample size—that is, the number of paired observations $(x, y)$—exceeds 30.

(d) There exists a straight line $y = \alpha + \beta x$ such that, for each value of $x$, the mean $\mu_y$ of the corresponding population of $y$-values lies on that straight line.

(e) The data come from a random sample or a randomized experiment.

**T12.2** Students in a statistics class drew circles of varying diameters and counted how many Cheerios could be placed in the circle. The scatterplot shows the results.

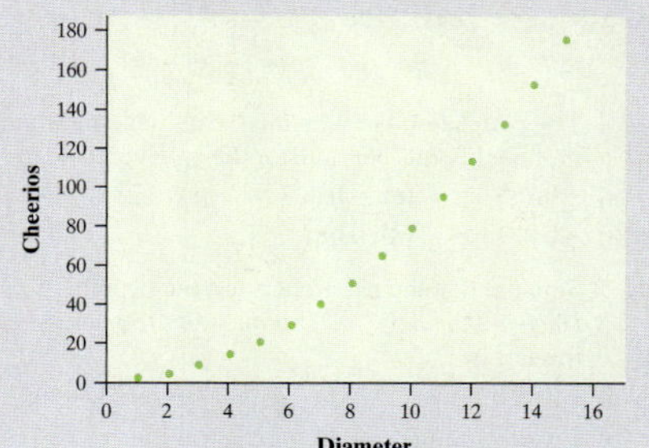


**TRM Full Solutions to Chapter 12 AP® Statistics Practice Test**

## Answers to Chapter 12 AP® Statistics Practice Test

**T12.1** c

T12.2 b

T12.3 d

T12.4 a

T12.5 d

T12.6 d

T12.7 e

T12.8 d



The students want to determine an appropriate equation for the relationship between diameter and the number of Cheerios. The students decide to transform the data to make it appear more linear before computing a least-squares regression line. Which of the following transformations would be reasonable for them to try?

I. Plot the square root of the number of Cheerios against diameter.

II. Plot the cube of the number of Cheerios against diameter.

III. Plot the log of the number of Cheerios against the log of the diameter.

IV. Plot the number of Cheerios against the log of the diameter.

(a) I and II (b) I and III (c) II and III (d) II and IV (e) I and IV

**T12.3** Inference about the slope $\beta$ of a least-squares regression line is based on which of the following distributions?

(a) The $t$ distribution with $n - 1$ degrees of freedom

(b) The standard Normal distribution

(c) The chi-square distribution with $n - 1$ degrees of freedom

(d) The $t$ distribution with $n - 2$ degrees of freedom

(e) The Normal distribution with mean $\mu$ and standard deviation $\sigma$

*Exercises T12.4 through T12.8 refer to the following setting.* An old saying in golf is "You drive for show and you putt for dough." The point is that good putting is more important than long driving for shooting low scores and hence winning money. To see if this is the case, data from a random sample of 69 of the nearly 1000 players on the PGA Tour's world money list are examined. The average number of putts per hole and the player's total winnings for the previous season are recorded. A least-squares regression line was fitted to the data. The following results were obtained from statistical software.

| Predictor | Coef | SE Coef | T | P |
|---|---|---|---|---|
| Constant | 7897179 | 3023782 | 6.86 | 0.000 |
| Avg. Putts | -4139198 | 1698371 | **** | **** |

S = 281777 R-Sq = 8.1% R-Sq(adj) = 7.8%

**T12.4** The correlation between total winnings and average number of putts per hole for these players is

(a) −0.285. (b) −0.081. (c) −0.007. (d) 0.081. (e) 0.285.

**T12.5** Suppose that the researchers test the hypotheses $H_0: \beta = 0$, $H_a: \beta < 0$. The value of the $t$ statistic for this test is

(a) 2.61. (b) 2.44. (c) 0.081. (d) −2.44. (e) −20.24.

**T12.6** The $P$-value for the test in Question T12.5 is 0.0087. A correct interpretation of this result is that

(a) the probability that there is no linear relationship between average number of putts per hole and total winnings for these 69 players is 0.0087.

(b) the probability that there is no linear relationship between average number of putts per hole and total winnings for all players on the PGA Tour's world money list is 0.0087.

(c) if there is no linear relationship between average number of putts per hole and total winnings for the players in the sample, the probability of getting a random sample of 69 players that yields a least-squares regression line with a slope of −4139198 or less is 0.0087.

(d) if there is no linear relationship between average number of putts per hole and total winnings for the players on the PGA Tour's world money list, the probability of getting a random sample of 69 players that yields a least-squares regression line with a slope of −4139198 or less is 0.0087.

(e) the probability of making a Type I error is 0.0087.

**T12.7** A 95% confidence interval for the slope $\beta$ of the population regression line is

(a) 7,897,179 ± 3,023,782.

(b) 7,897,179 ± 6,047,564.

(c) −4,139,198 ± 1,698,371.

(d) −4,139,198 ± 3,328,807.

(e) −4,139,198 ± 3,396,742.

**T12.8** A residual plot from the least-squares regression is shown below. Which of the following statements is supported by the graph?

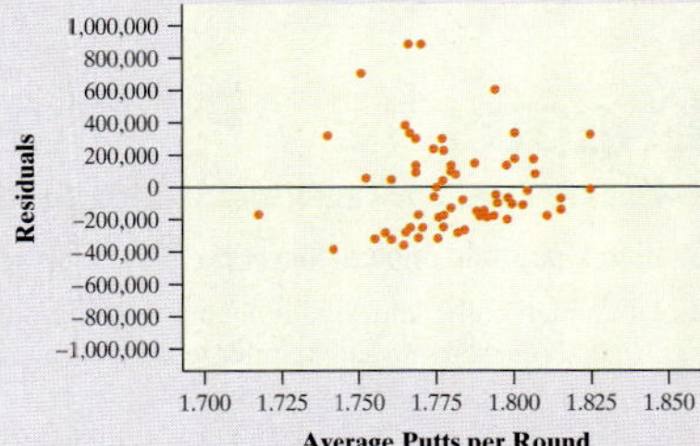


(a) The residual plot contains dramatic evidence that the standard deviation of the response about the population regression line increases as the average number of putts per round increases.

(b) The sum of the residuals is not 0. Obviously, there is a major error present.

(c) Using the regression line to predict a player's total winnings from his average number of putts almost always results in errors of less than $200,000.

(d) For two players, the regression line underpredicts their total winnings by more than $800,000.

(e) The residual plot reveals no correlation between average putts per round and prediction errors from the least-squares line for these players.

**T12.9** Which of the following would provide evidence that a power law model of the form $y = ax^b$, where $b \neq 0$ and $b \neq 1$, describes the relationship between a response variable $y$ and an explanatory variable $x$?

(a) A scatterplot of $y$ versus $x$ looks approximately linear.

(b) A scatterplot of ln $y$ versus $x$ looks approximately linear.

(c) A scatterplot of $y$ versus ln $x$ looks approximately linear.

(d) A scatterplot of ln $y$ versus ln $x$ looks approximately linear.

(e) None of these

**T12.10** We record data on the population of a particular country from 1960 to 2010. A scatterplot reveals a clear curved relationship between population and year. However, a different scatterplot reveals a strong linear relationship between the logarithm (base 10) of the population and the year. The least-squares regression line for the transformed data is

$$\widehat{\log(\text{population})} = -13.5 + 0.01(\text{year})$$

Based on this equation, the population of the country in the year 2020 should be about

(a) 6.7. (b) 812. (c) 5,000,000. (d) 6,700,000. (e) 8,120,000.

**Section II: Free Response** *Show all your work. Indicate clearly the methods you use, because you will be graded on the correctness of your methods as well as on the accuracy and completeness of your results and explanations.*

**T12.11** Growth hormones are often used to increase the weight gain of chickens. In an experiment using 15 chickens, 3 chickens were randomly assigned to each of 5 different doses of growth hormone (0, 0.2, 0.4, 0.8, and 1.0 milligrams). The subsequent weight gain (in ounces) was recorded for each chicken. A researcher plots the data and finds that a linear relationship appears to hold. Computer output from a least-squares regression analysis for these data is shown below. Assume that the conditions for performing inference about the slope $\beta$ of the true regression line are met.

| Predictor | Coef | SE Coef | T | P |
|---|---|---|---|---|
| Constant | 4.5459 | 0.6166 | 7.37 | <0.0001 |
| Dose | 4.8323 | 1.0164 | 4.75 | 0.0004 |

S = 3.135 R-Sq = 38.4% R-Sq(adj) = 37.7%

(a) What is the equation of the least-squares regression line for these data? Define any variables you use.

(b) Interpret each of the following in context:

(i) The slope

(ii) The $y$ intercept

(iii) $s$

(iv) The standard error of the slope

(v) $r^2$

(c) Do the data provide convincing evidence of a linear relationship between dose and weight gain? Carry out a significance test at the $\alpha = 0.05$ level.

(d) Construct and interpret a 95% confidence interval for the slope parameter.

**T12.12** Foresters are interested in predicting the amount of usable lumber they can harvest from various tree species. They collect data on the diameter at breast height (DBH) in inches and the yield in board feet of a random sample of 20 Ponderosa pine trees that have been harvested. (Note that a board foot is defined as a piece of lumber 12 inches by 12 inches by 1 inch.) A scatterplot of the data is shown below.

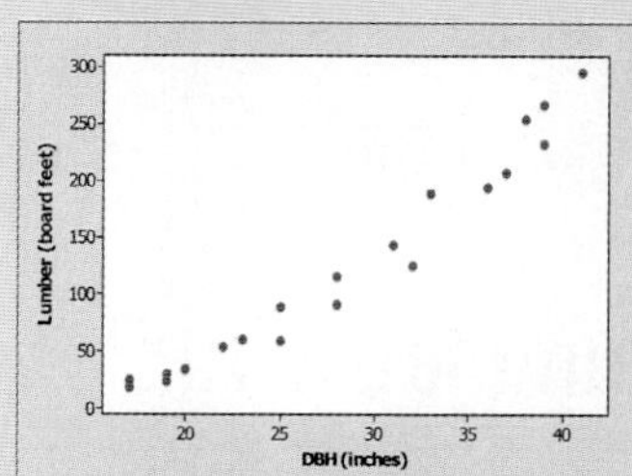


(a) Some computer output and a residual plot from a least-squares regression on these data appear below. Explain why a linear model may not be appropriate in this case.

| Predictor | Coef | SE Coef | T | P |
|---|---|---|---|---|
| Constant | −191.12 | 16.98 | −11.25 | 0.000 |
| DBH (inches) | 11.0413 | 0.5752 | 19.19 | 0.000 |

S = 20.3290 R-Sq = 95.3% R-Sq(adj) = 95.1%

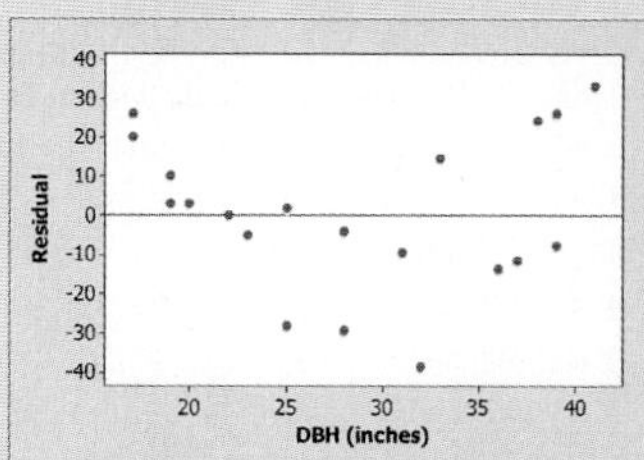


The foresters are considering two possible transformations of the original data: (1) cubing the

**T12.9** d

**T12.10** c

**T12.11** **(a)** $\hat{y} = 4.546 + 4.832x$, where $y$ is the weight gain and $x$ is the dose of growth hormone. **(b)** **(i)** For each 1-mg increase in growth hormone, the predicted weight gain increases by about 4.832 ounces. **(ii)** If a chicken is given no growth hormone ($x = 0$), the predicted weight gain is 4.546 ounces. **(iii)** When using the least-squares regression line with $x$ = dose of growth hormone to predict $y$ = weight gain, we will typically be off by about 3.135 ounces. **(iv)** If we repeated this experiment many times, the sample slope will typically vary by about 1.0164 from the true slope of the least-squares regression line with $y$ = weight gain and $x$ = dose of growth hormone. **(v)** About 38.4% of the variation in weight gain is accounted for by the linear model relating weight gain to the dose of growth hormone. **(c)** *S*: $H_0: \beta = 0$ vs. $H_a: \beta \neq 0$, where $\beta$ is the slope of the true regression line relating $y$ = weight gain to $x$ = dose of growth hormone. *P*: *t* test for $\beta$. *D*: $t = 4.75$, df = 13, and *P*-value = 0.0004. *C*: Because the *P*-value of 0.0004 is less than $\alpha = 0.05$, we reject $H_0$. There is convincing evidence of a linear relationship between the dose of growth hormone and weight gain for chickens like these. **(d)** *D*: With df = 13, (2.6373, 7.0273). *C*: We are 95% confident that the interval from 2.6373 to 7.0273 captures the slope of the true regression line relating $y$ = weight gain to $x$ = dose of growth hormone for chickens like these.

**T12.12 (a)** There is clear curvature evident in both the scatterplot and the residual plot. **(b)** 1: $\hat{y} = 2.078 + 0.0042597(30)^3 = 117.09$ board feet.
2: $\widehat{\ln y} = 1.2319 + 0.113417(30) = 4.63441$ and $\hat{y} = e^{4.63441} = 102.967$ board feet. **(c)** The residual plot for Option 1 is much more scattered, while the plot from Option 2 shows curvature, meaning that the model from Option 1 relating the amount of usable lumber to cube of the diameter is more appropriate.

diameter values or (2) taking the natural logarithm of the yield measurements. After transforming the data, a least-squares regression analysis is performed. Some computer output and a residual plot for each of the two possible regression models follow.

**Option 1:** Cubing the diameter values

| Predictor | Coef | SE Coef | T | P |
|---|---|---|---|---|
| Constant | 2.078 | 5.444 | 0.38 | 0.707 |
| DBH^3 | 0.0042597 | 0.0001549 | 27.50 | 0.000 |

S = 14.3601   R-Sq = 97.7%   R-Sq(adj) = 97.5%

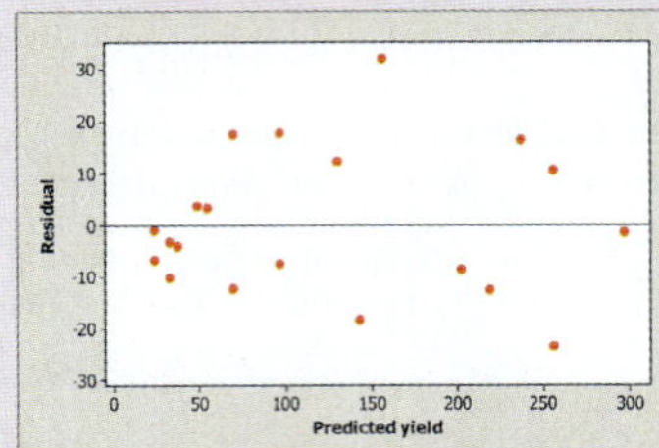


**Option 2:** Taking natural logarithm of yield measurements

| Predictor | Coef | SE Coef | T | P |
|---|---|---|---|---|
| Constant | 1.2319 | 0.1795 | 6.86 | 0.000 |
| DBH (inches) | 0.113417 | 0.006081 | 18.65 | 0.000 |

S = 0.214894   R-Sq = 95.1%   R-Sq(adj) = 94.8%

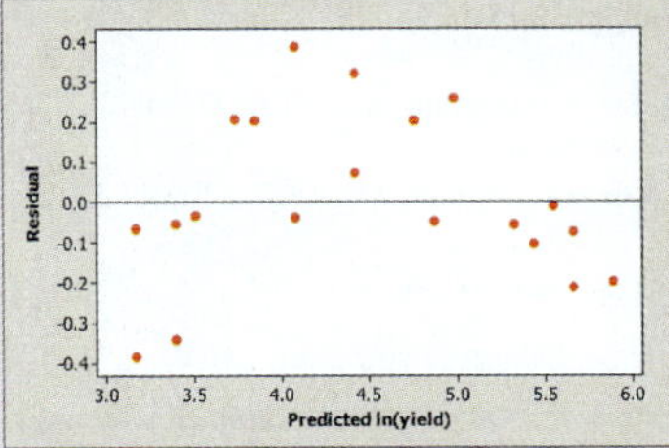


(b) Use both models to predict the amount of usable lumber from a Ponderosa pine with diameter 30 inches. Show your work.

(c) Which of the predictions in part (b) seems more reliable? Give appropriate evidence to support your choice.

**TRM Full Solutions to Cumulative AP® Practice Test 4**

## Answers to Cumulative AP® Practice Test 4

**AP4.1** e

**AP4.2** c

**AP4.3** e

# Cumulative AP® Practice Test 4

**Section I: Multiple Choice** *Choose the best answer.*

**AP4.1** A major agricultural company is testing a new variety of wheat to determine whether it is more resistant to certain insects than is the current wheat variety. The proportion of a current wheat crop lost to insects is 4%. Thus, the company wishes to test the following hypotheses:

$$H_0: p = 0.04$$
$$H_a: p < 0.04$$

Which of the following significance levels and sample sizes would lead to the highest power for this test?

(a) $n = 200$ and $\alpha = 0.01$
(b) $n = 400$ and $\alpha = 0.05$
(c) $n = 400$ and $\alpha = 0.01$
(d) $n = 500$ and $\alpha = 0.01$
(e) $n = 500$ and $\alpha = 0.05$

**AP4.2** If $P(A) = 0.24$ and $P(B) = 0.52$ and events $A$ and $B$ are independent, what is $P(A \text{ or } B)$?

(a) 0.1248 (b) 0.28 (c) 0.6352 (d) 0.76
(e) The answer cannot be determined from the given information.

**AP4.3** As part of a bear population study, data were gathered on a sample of black bears in the western United States to examine the relationship between the bear's neck girth (distance around the neck) and the weight of the bear. A scatterplot of the data reveals a straight-line pattern. The $r^2$-value from a least-squares regression analysis was determined to be $r^2 = 0.963$. Which one of the following is the correct value and corresponding interpretation for the correlation?

(a) The correlation is −0.963, and 96.3% of the variation in a bear's weight can be explained by the least-squares regression line using neck girth as the explanatory variable.
(b) The correlation is 0.963. There is a strong positive linear relationship between a bear's neck girth and its weight.
(c) The correlation is 0.981, and 98.1% of the variation in a bear's weight can be explained by the least-

squares regression line using neck girth as the explanatory variable.

(d) The correlation is −0.981. There is a strong negative linear relationship between a bear's neck girth and its weight.

(e) The correlation is 0.981. There is a strong positive linear relationship between a bear's neck girth and its weight.

**AP4.4** The school board in a certain school district obtained a random sample of 200 residents and asked if they were in favor of raising property taxes to fund the hiring of more statistics teachers. The resulting confidence interval for the true proportion of residents in favor of raising taxes was (0.183, 0.257). The margin of error for this confidence interval is

(a) 0.037. (b) 0.183. (c) 0.220. (d) 0.257. (e) 0.740.

**AP4.5** After a name-brand drug has been sold for several years, the Food and Drug Administration (FDA) will allow other companies to produce a generic equivalent. The FDA will permit the generic drug to be sold as long as there isn't convincing evidence that it is less effective than the name brand drug. For a proposed generic drug intended to lower blood pressure, the following hypotheses will be used:

$$H_0: \mu_G = \mu_N \text{ versus } H_a: \mu_G < \mu_N$$

where

$\mu_G$ = true mean reduction in blood pressure using the generic drug

$\mu_N$ = true mean reduction in blood pressure using the name-brand drug.

In the context of this situation, which of the following describes a Type I error?

(a) The FDA finds convincing evidence that the generic drug is less effective, when in reality it is less effective.

(b) The FDA finds convincing evidence that the generic drug is less effective, when in reality it is equally effective.

(c) The FDA fails to find convincing evidence that the generic drug is less effective, when in reality it is less effective.

(d) The FDA fails to find convincing evidence that the generic drug is less effective, when in reality it is equally effective.

(e) The FDA finds convincing evidence that the generic drug is equally effective, when in reality it is less effective.

**AP4.6** Which of the following sampling plans for estimating the proportion of all adults in a medium-sized town who favor a tax increase to support the local school system does *not* suffer from undercoverage bias?

(a) A random sample of 250 names from the local phone book

(b) A random sample of 200 parents whose children attend one of the local schools

(c) A sample consisting of 500 people from the city who take an online survey about the issue

(d) A random sample of 300 homeowners in the town

(e) A random sample of 100 people from an alphabetical list of all adults who live in the town

**AP4.7** Which of the following is a categorical variable?

(a) The weight of automobiles

(b) The time required to complete the Olympic marathon

(c) The average gas mileage of a hybrid car

(d) The brand of shampoo purchased by shoppers in a grocery store

(e) The average closing price of a particular stock on the New York Stock Exchange

**AP4.8** A large machine is filled with thousands of small pieces of candy, 40% of which are orange. When money is deposited, the machine dispenses 60 randomly selected pieces of candy. The machine will be recalibrated if a group of 60 candies contains fewer than 18 that are orange. What is the approximate probability that this will happen if the machine is working correctly?

(a) $P\left(Z < \frac{0.3 - 0.4}{\sqrt{\frac{(0.4)(0.6)}{60}}}\right)$

(b) $P\left(Z < \frac{0.3 - 0.4}{\sqrt{\frac{(0.3)(0.7)}{60}}}\right)$

(c) $P\left(Z < \frac{0.3 - 0.4}{\frac{\sqrt{(0.4)(0.6)}}{60}}\right)$

(d) $P\left(Z < \frac{0.3 - 0.4}{\frac{(0.4)(0.6)}{\sqrt{60}}}\right)$

(e) $P\left(Z < \frac{0.4 - 0.3}{\frac{\sqrt{(0.3)(0.7)}}{60}}\right)$

**AP4.9** A random sample of 900 students at a very large university was asked which social-networking site they used most often during a typical week. Their responses are shown in the table below.

| Networking site | Male | Female | Total |
|---|---|---|---|
| Facebook | 221 | 283 | **504** |
| Twitter | 42 | 38 | **80** |
| LinkedIn | 108 | 87 | **195** |
| Pinterest | 23 | 26 | **49** |
| MySpace | 29 | 43 | **72** |
| **Total** | **423** | **477** | **900** |

**AP4.4** a
**AP4.5** b
**AP4.6** e
**AP4.7** d
**AP4.8** a

AP4.9 e

AP4.10 a

AP4.11 b

AP4.12 d

AP4.13 e

AP4.14 d



Assuming that gender and preferred networking site are independent, how many females do you expect to choose LinkedIn?

(a) 18.85 (b) 46.11 (c) 87.00 (d) 91.65 (e) 103.35

AP4.10 Insurance adjusters are always vigilant about being overcharged for accident repairs. The adjusters suspect that Repair Shop 1 quotes higher estimates than Repair Shop 2. To check their suspicion, the adjusters randomly select 12 cars that were recently involved in an accident and then take each of the cars to both repair shops to obtain separate estimates of the cost to fix the vehicle. The estimates are given below in hundreds of dollars.

| Car: | 1 | 2 | 3 | 4 | 5 | 6 |
|---|---|---|---|---|---|---|
| Shop 1: | 21.2 | 25.2 | 39.0 | 11.3 | 15.0 | 18.1 |
| Shop 2: | 21.3 | 24.1 | 36.8 | 11.5 | 13.7 | 17.6 |

| Car: | 7 | 8 | 9 | 10 | 11 | 12 |
|---|---|---|---|---|---|---|
| Shop 1: | 25.3 | 23.2 | 12.4 | 42.6 | 27.6 | 12.9 |
| Shop 2: | 24.8 | 21.3 | 12.1 | 42.0 | 26.7 | 12.5 |

Assuming that the conditions for inference are reasonably met, which of the following significance tests could legitimately be used to determine whether the adjusters' suspicion is correct?

(a) A paired $t$ test
(b) A two-sample $t$ test
(c) A $t$ test to see if the slope of the population regression line is 0
(d) A chi-square test for homogeneity
(e) A two-sample $z$ test for comparing two proportions

AP4.11 A survey firm wants to ask a random sample of adults in Ohio if they support an increase in the state sales tax from 5% to 6%, with the additional revenue going to education. Let $\hat{p}$ denote the proportion in the sample who say that they support the increase. Suppose that 40% of all adults in Ohio support the increase. How large a sample would be needed to guarantee that the standard deviation of $\hat{p}$ is no more than 0.01?

(a) 1500 (b) 2400 (c) 2401 (d) 2500 (e) 9220

AP4.12 A set of 10 cards consists of 5 red cards and 5 black cards. The cards are shuffled thoroughly, and you choose one at random, observe its color, and replace it in the set. The cards are thoroughly reshuffled, and you again choose a card at random, observe its color, and replace it in the set. This is done a total of four times. Let X be the number of red cards observed in these four trials. The random variable X has which of the following probability distributions?

(a) The Normal distribution with mean 2 and standard deviation 1
(b) The binomial distribution with $n = 10$ and $p = 0.5$
(c) The binomial distribution with $n = 5$ and $p = 0.5$
(d) The binomial distribution with $n = 4$ and $p = 0.5$
(e) The geometric distribution with $p = 0.5$

AP4.13 A study of road rage asked random samples of 596 men and 523 women about their behavior while driving. Based on their answers, each respondent was assigned a road rage score on a scale of 0 to 20. The respondents were chosen by random digit dialing of telephone numbers. Are the conditions for two-sample $t$ inference satisfied?

(a) Maybe. The data came from independent random samples, but we need to examine the data to check for Normality.
(b) No. Road rage scores in a range between 0 and 20 can't be Normal.
(c) No. A paired $t$ test should be used in this case.
(d) Yes. The large sample sizes guarantee that the corresponding population distributions will be Normal.
(e) Yes. We have two independent random samples and large sample sizes, and the 10% condition is met.

AP4.14 Do hummingbirds prefer store-bought food made from concentrate or a simple mixture of sugar and water? To find out, a researcher obtains 10 identical hummingbird feeders and fills 5, chosen at random, with store-bought food from concentrate and the other 5 with a mixture of sugar and water. The feeders are then randomly assigned to 10 possible hanging locations in the researcher's yard. Which inference procedure should you use to test whether hummingbirds show a preference for store-bought food based on amount consumed?

(a) A one-sample $z$ test for a proportion
(b) A two-sample $z$ test for a difference in proportions
(c) A chi-square test for independence
(d) A two-sample $t$ test
(e) A paired $t$ test

**AP4.15** A Harris Poll found that 54% of American adults don't think that human beings developed from earlier species. The poll's margin of error for 95% confidence was 3%. This means that

(a) there is a 95% chance that the interval (51%, 57%) contains the true percent of American adults who do not think that human beings developed from earlier species.

(b) the poll used a method that provides an estimate within 3% of the truth about the population 95% of the time.

(c) if Harris takes another poll using the same method, the results of the second poll will lie between 51% and 57%.

(d) there is a 3% chance that the interval is correct.

(e) the poll used a method that would result in an interval that contains 54% in 95% of all possible samples of the same size from this population.

**AP4.16** Two six-sided dice are rolled and the sum of the faces showing is recorded after each roll. Let X = the number of rolls required until a sum greater than 7 is obtained. If 100 trials are conducted, which of the following is most likely to be part of the probability distribution of X?

(a)

| Number of rolls *X* | Number of trials |
|---|---|
| 1 | 34 |
| 2 | 20 |
| 3 | 16 |
| 4 | 10 |
| 5 | 6 |
| 6 | 6 |
| 7 | 3 |
| 8 | 2 |
| 9 | 1 |
| 10 | 0 |
| 11 | 1 |
| 12 | 0 |
| 13 | 1 |

(b)

| Number of rolls *X* | Number of trials |
|---|---|
| 0 | 34 |
| 1 | 20 |
| 2 | 16 |
| 3 | 10 |
| 4 | 6 |
| 5 | 6 |
| 6 | 3 |
| 7 | 2 |
| 8 | 1 |
| 9 | 0 |
| 10 | 1 |
| 11 | 0 |
| 12 | 1 |

(c)

| Number of rolls *X* | Number of trials |
|---|---|
| 1 | 18 |
| 2 | 23 |
| 3 | 26 |
| 4 | 15 |
| 5 | 9 |
| 6 | 6 |
| 7 | 1 |
| 8 | 0 |
| 9 | 1 |
| 10 | 0 |
| 11 | 0 |
| 12 | 0 |
| 13 | 1 |

(d)

| Number of rolls *X* | Number of trials |
|---|---|
| 1 | 10 |
| 2 | 9 |
| 3 | 10 |
| 4 | 12 |
| 5 | 7 |
| 6 | 13 |
| 7 | 10 |
| 8 | 7 |
| 9 | 9 |
| 10 | 10 |
| 11 | 2 |
| 12 | 1 |

(e)

| Number of rolls *X* | Number of trials | Number of rolls *X* | Number of trials |
|---|---|---|---|
| 1 | 2 | 8 | 17 |
| 2 | 2 | 9 | 9 |
| 3 | 5 | 10 | 4 |
| 4 | 10 | 11 | 2 |
| 5 | 11 | 12 | 0 |
| 6 | 15 | 13 | 1 |
| 7 | 22 | | |

**AP4.17** Women who are severely overweight suffer economic consequences, a study has shown. They have household incomes that are an average of $6710 lower. The findings are from an eight-year observational study of 10,039 randomly selected women who were 16 to 24 years old when the research began. Does this study give strong evidence that being severely overweight causes a woman to have a lower income?

(a) Yes. The study included both women who were severely overweight and women who were not.

(b) Yes. The subjects in the study were selected at random.

(c) No. The study showed that there is no connection between income and being severely overweight.

(d) No. The study suggests an association between income and being severely overweight, but we can't draw a cause-and-effect conclusion.

(e) There is not enough information to answer this question.

**AP4.15** b
**AP4.16** a
**AP4.17** d

AP4.18 c

AP4.19 a

AP4.20 b

AP4.21 e

AP4.22 e



*Questions AP 4.18 and 4.19 refer to the following situation.* Could mud wrestling be the cause of a rash contracted by University of Washington students? Two physicians at the University of Washington student health center wondered about this when one male and six female students complained of rashes after participating in a mud-wrestling event. Questionnaires were sent to a random sample of students who participated in the event. The results, by gender, are summarized in the following table.

| | Men | Women |
|---|---|---|
| Developed rash | 12 | 12 |
| No rash | 38 | 12 |

Some Minitab output for the previous table is given below. The output includes the observed counts, the expected counts, and the chi-square statistic.

```
Expected counts are printed below observed
counts
                  MEN     WOMEN    Total
Developed rash    12      12       24
                  16.22   7.78
No rash           38      12       50
                  33.78   16.22
Total             50      24       74
ChiSq = 5.002
```

AP4.18 The cell that contributes most to the chi-square statistic is

(a) men who developed a rash.
(b) men who did not develop a rash.
(c) women who developed a rash.
(d) women who did not develop a rash.
(e) both (a) and (d).

AP4.19 From the chi-square test performed in this study, we may conclude that

(a) there is convincing evidence of an association between the gender of an individual participating in the event and development of a rash.
(b) mud wrestling causes a rash, especially for women.
(c) there is absolutely no evidence of any relation between the gender of an individual participating in the event and the subsequent development of a rash.
(d) development of a rash is a real possibility if you participate in mud wrestling, especially if you do so regularly.
(e) the gender of the individual participating in the event and the development of a rash are independent.

AP4.20 Random assignment is part of a well-designed comparative experiment because

(a) it is more fair to the subjects.
(b) it helps create roughly equivalent groups before treatments are imposed on the subjects.
(c) it allows researchers to generalize the results of their experiment to a larger population.
(d) it helps eliminate any possibility of bias in the experiment.
(e) it prevents the placebo effect from occurring.

AP4.21 The following back-to-back stemplots compare the ages of players from two minor-league hockey teams (1|7 = 17 years).

| Team A | | Team B |
|---|---|---|
| 98777 | 1 | 788889 |
| 44333221 | 2 | 00123444 |
| 7766555 | 2 | 556679 |
| 521 | 3 | 023 |
| 86 | 3 | 55 |

Which of the following *cannot* be justified from the plots?

(a) Team A has the same number of players in their 30s as does Team B.
(b) The median age of both teams is the same.
(c) Both age distributions are skewed to the right.
(d) The age ranges of both teams are similar.
(e) There are no outliers by the 1.5*IQR* rule in either distribution.

AP4.22 A distribution that represents the number of cars X parked in a randomly selected residential driveway on any night is given by

| $x_i$: | 0 | 1 | 2 | 3 | 4 |
|---|---|---|---|---|---|
| $p_i$: | 0.1 | 0.2 | 0.35 | 0.25 | 0.15 |

Which of the following statements is correct?

(a) This is a legitimate probability distribution because each of the $p_i$-values is between 0 and 1.
(b) This is a legitimate probability distribution because $\sum x_i$ is exactly 10.
(c) This is a legitimate probability distribution because each of the $p_i$-values is between 0 and 1 and the $\sum x_i$ is exactly 10.
(d) This is not a legitimate probability distribution because $\sum x_i$ is not exactly 10.
(e) This is not a legitimate probability distribution because $\sum p_i$ is not exactly 1.

AP4.23 Which sampling method was used in each of the following settings, in order from I to IV?

I. A student chooses for a survey the first 20 students to arrive at school.

II. The name of each student in a school is written on a card, the cards are well mixed, and 10 names are drawn.

III. A state agency randomly selects 50 people from each of the state's senatorial districts.

IV. A city council randomly selects eight city blocks and then surveys all the voting-age residents of those blocks.

(a) Voluntary response, SRS, stratified, cluster
(b) Convenience, SRS, stratified, cluster
(c) Convenience, cluster, SRS, stratified
(d) Convenience, SRS, cluster, stratified
(e) Cluster, SRS, stratified, convenience

**AP4.24** Western lowland gorillas, whose main habitat is the central African continent, have a mean weight of 275 pounds with a standard deviation of 40 pounds. Capuchin monkeys, whose main habitat is Brazil and a few other parts of Latin America, have a mean weight of 6 pounds with a standard deviation of 1.1 pounds. Both weight distributions are approximately Normally distributed. If a particular western lowland gorilla is known to weigh 345 pounds, approximately how much would a capuchin monkey have to weigh, in pounds, to have the same standardized weight as the lowland gorilla?

(a) 4.08 (c) 7.93
(b) 7.27 (d) 8.20
(e) There is not enough information to determine the weight of a capuchin monkey.

**AP4.25** Suppose that the mean weight of a certain type of pig is 280 pounds with a standard deviation of 80 pounds. The weight distribution of pigs tends to be somewhat skewed to the right. A random sample of 100 pigs is taken. Which of the following statements about the sampling distribution of the sample mean weight $\bar{x}$ is true?

(a) It will be Normally distributed with a mean of 280 pounds and a standard deviation of 80 pounds.
(b) It will be Normally distributed with a mean of 280 pounds and a standard deviation of 8 pounds.
(c) It will be approximately Normally distributed with a mean of 280 pounds and a standard deviation of 80 pounds.
(d) It will be approximately Normally distributed with a mean of 280 pounds and a standard deviation of 8 pounds.
(e) There is not enough information to determine the mean and standard deviation of the sampling distribution.

**AP4.26** Which of the following statements about the $t$ distribution with degrees of freedom df is (are) true?

I. It is symmetric.

II. It has more variability than the $t$ distribution with df + 1 degrees of freedom.

III. As df increases, the $t$ distribution approaches the standard Normal distribution.

(a) I only (c) III only (e) I, II, and III
(b) II only (d) I and III

**AP4.27** A company has been running television commercials for a new children's product on five different family programs during the evening hours in a large city over a one-month period. A random sample of families is taken, and they are asked to indicate which of the five programs they viewed most often and their rating of the advertised product. The results are summarized in the following table.

| | Family program | | | | |
|---|---|---|---|---|---|
| Product rating | A | B | C | D | E |
| Excellent | 23 | 29 | 42 | 48 | 51 |
| Good | 25 | 33 | 44 | 53 | 49 |
| Fair | 31 | 29 | 25 | 16 | 10 |
| Poor | 38 | 32 | 25 | 18 | 12 |

The advertiser decided to use a chi-square test to see if there is a relationship between the family program viewed and the product's rating. What would be the degrees of freedom for this test?

(a) 3 (c) 12 (e) 19
(b) 4 (d) 18

*Questions AP4.28 and AP4.29 refer to the following situation.* Park rangers are interested in estimating the weight of the bears that inhabit their state. The rangers have data on weight (in pounds) and neck girth (distance around the neck in inches) for 10 randomly selected bears. Some regression output for these data is shown below.

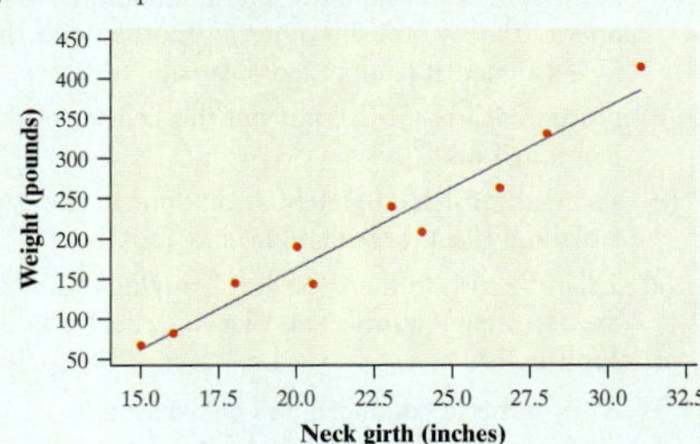


```
Predictor    Coef     SE Coef  T      P
Constant     -241.70  38.57    -6.27  0.000
Neck Girth   20.230   1.695    11.93  0.000
S = 26.7565  R-Sq = 94.7%
```

AP4.23 b
AP4.24 c
AP4.25 d
AP4.26 e
AP4.27 c

AP4.28 c
AP4.29 b
AP4.30 b
AP4.31 d
AP4.32 a
AP4.33 e



**AP4.28** Which of the following represents a 95% confidence interval for the true slope of the least-squares regression line relating the weight of a bear and its neck girth?

(a) $20.230 \pm 1.695$
(b) $20.230 \pm 3.83$
(c) $20.230 \pm 3.91$
(d) $20.230 \pm 20.22$
(e) $26.7565 \pm 3.83$

**AP4.29** A bear was recently captured whose neck girth was 35 inches and whose weight was 466.35 pounds. If this bear were added to the data set given above, what would be the effect on the value of $s$?

(a) It would decrease the value of $s$ because the added point is an outlier.
(b) It would decrease the value of $s$ because the added point lies on the least-squares regression line.
(c) It would increase the value of $s$ because the added point is an outlier.
(d) It would increase the value of $s$ because the added point lies on the least-squares regression line.
(e) It would have no effect on the value of $s$ because the added point lies on the least-squares regression line.

**AP4.30** An experimenter wishes to test whether or not two types of fish food (a standard fish food and a new product) work equally well at producing fish of equal weight after a two-month feeding program. The experimenter has two identical fish tanks (1 and 2) to put fish in and is considering how to assign 40 fish, each of which has a numbered tag, to the tanks. The best way to do this would be to

(a) put all the odd-numbered fish in one tank, the even in the other, and give the standard food type to the odd-numbered ones.
(b) obtain pairs of fish whose weights are roughly equal at the start of the experiment and randomly assign one to Tank 1 and the other to Tank 2, with the feed assigned at random to the tanks.
(c) proceed as in part (b), but put the heavier of the pair into Tank 2.
(d) assign the fish completely at random to the two tanks and give the standard feed to Tank 1.
(e) assign the fish to the tanks using any method that the researcher wants. The placebo effect doesn't apply to fish.

**AP4.31** A city wants to conduct a poll of taxpayers to determine the level of support for constructing a new city-owned baseball stadium. Which of the following is the primary reason for using a large sample size in constructing a confidence interval to estimate the proportion of city taxpayers who would support such a project?

(a) To increase the confidence level
(b) To eliminate any confounding variables
(c) To reduce nonresponse bias
(d) To increase the precision of the estimate
(e) To reduce undercoverage

**AP4.32** A standard deck of playing cards contains 52 cards, of which 4 are aces and 13 are hearts. You are offered a choice of the following two wagers:

I. Draw one card at random from the deck. You win \$10 if the card drawn is an ace. Otherwise, you lose \$1.
II. Draw one card at random from the deck. If the card drawn is a heart, you win \$2. Otherwise, you lose \$1.

Which of the two wagers should you prefer?

(a) Wager 1, because it has a higher expected value
(b) Wager 2, because it has a higher expected value
(c) Wager 1, because it has a higher probability of winning
(d) Wager 2, because it has a higher probability of winning
(e) Both wagers are equally favorable.

**AP4.33** Below are boxplots of SAT Critical Reading and Math scores for a randomly selected group of female juniors at a highly competitive suburban school.

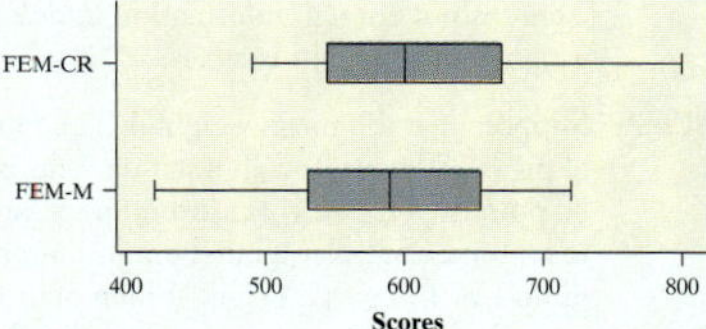


Which of the following *cannot* be justified by the plots shown above?

(a) The maximum Critical Reading score is higher than the maximum Math score.
(b) Critical Reading scores are skewed to the right, whereas Math scores are somewhat skewed to the left.
(c) The median Critical Reading score for females is slightly higher than the median Math score.
(d) There appear to be no outliers in the SAT score distributions.
(e) The mean Critical Reading score and the mean Math score for females are about the same.

**AP4.34** A distribution of exam scores has mean 60 and standard deviation 18. If each score is doubled, and then 5 is subtracted from that result, what will be the mean and standard deviation, respectively, of the new scores?

(a) mean = 115 and standard deviation = 31
(b) mean = 115 and standard deviation = 36
(c) mean = 120 and standard deviation = 6
(d) mean = 120 and standard deviation = 31
(e) mean = 120 and standard deviation = 36

**AP4.35** In a clinical trial, 30 patients with a certain blood disease are randomly assigned to two groups. One group is then randomly assigned the currently marketed medicine, and the other group receives the experimental medicine. Each week, patients report to the clinic where blood tests are conducted. The lab technician is unaware of the kind of medicine the patient is taking, and the patient is also unaware of which medicine he or she has been given. This design can be described as

(a) a double-blind, completely randomized experiment, with the currently marketed medicine and the experimental medicine as the two treatments.
(b) a single-blind, completely randomized experiment, with the currently marketed medicine and the experimental medicine as the two treatments.
(c) a double-blind, matched pairs design, with the currently marketed medicine and the experimental medicine forming a pair.
(d) a double-blind, block design that is not a matched pairs design, with the currently marketed medicine and the experimental medicine as the two blocks.
(e) a double-blind, randomized observational study.

**AP4.36** A local investment club that meets monthly has 200 members ranging in age from 27 to 81. A cumulative relative frequency graph is shown below. Approximately how many members of the club are more than 60 years of age?

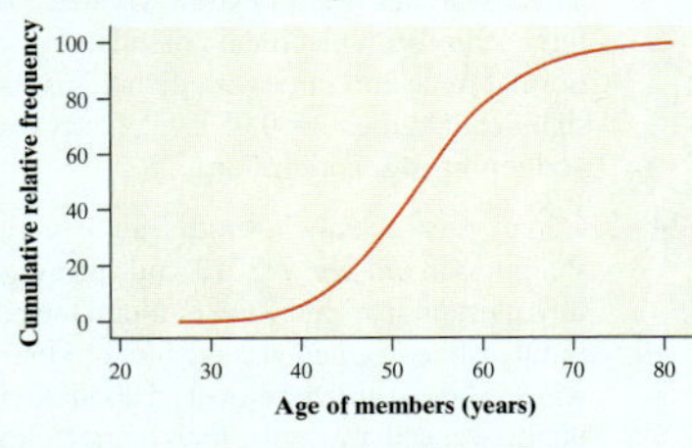


(a) 20
(b) 44
(c) 78
(d) 90
(e) 110

**AP4.37** A manufacturer of electronic components is testing the durability of a newly designed integrated circuit to determine whether its life span is longer than that of the earlier model, which has a mean life span of 58 months. The company takes a simple random sample of 120 integrated circuits and simulates normal use until they stop working. The null and alternative hypotheses used for the significance test are given by $H_0: \mu = 58$ and $H_a: \mu > 58$. The *P*-value for the resulting one-sample *t* test is 0.035. Which of the following best describes what the *P*-value measures?

(a) The probability that the new integrated circuit has the same life span as the current model is 0.035.
(b) The probability that the test correctly rejects the null hypothesis in favor of the alternative hypothesis is 0.035.
(c) The probability that a single new integrated circuit will not last as long as one of the earlier circuits is 0.035.
(d) The probability of getting a sample statistic as far or farther from 58 if there really is no difference between the new and the old circuits is 0.035.
(e) The probability of getting a sample mean for the new integrated circuit that is lower than the mean for the earlier model is 0.035.

*Questions AP4.38 and AP4.39 refer to the following situation.* Do children's fear levels change over time and, if so, in what ways? Little research has been done on the prevalence and persistence of fears in children. Several years ago, two researchers surveyed a randomly selected group of 94 third- and fourth-grade children, asking them to rate their level of fearfulness about a variety of situations. Two years later, the children again completed the same survey. The researchers computed the overall fear rating for each child in both years and were interested in the relationship between these ratings. They then assumed that the true regression line was

$$\mu_{\text{later rating}} = \alpha + \beta\,(\text{initial rating})$$

and that the assumptions for regression inference were satisfied. This model was fitted to the data using least-squares regression. The following results were obtained from statistical software.

| Predictor | Coefficient | St. Dev. |
|---|---|---|
| Constant | 0.877917 | 0.1184 |
| Initial Rating | 0.397911 | 0.0676 |
| S = 0.2374 | R-Sq = 0.274 | |

Here is a scatterplot of the later ratings versus the initial ratings and a plot of the residuals versus the initial ratings.

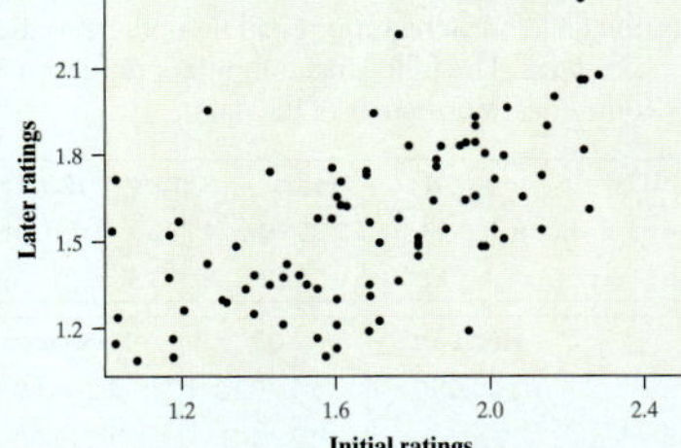


**AP4.34** b
**AP4.35** a
**AP4.36** b
**AP4.37** d

**AP4.38** a

**AP4.39** d

**AP4.40** d

**AP4.41** *S*: $H_0: \mu_1 - \mu_2 = 0$ vs. $H_a: \mu_1 - \mu_2 \neq 0$, where $\mu_1$ = true mean difference in electrical potential for diabetic mice and $\mu_2$ = true mean difference in electrical potential for normal mice. *P*: Two-sample *t* test for $\mu_1 - \mu_2$. *Random*: Independent random samples. *10%*: $n_1 = 24$ is less than 10% of all diabetic mice and $n_2 = 18$ is less than 10% of all normal mice. *Normal/Large Sample:* No outliers or strong skewness. *D*: $t = 2.55$. Using df = 23, the *P*-value is between 0.01 and 0.02. Using df = 38.46, *P*-value = 0.0149. *C*: Because the *P*-value of 0.0149 is less than $\alpha = 0.05$, we reject $H_0$. There is convincing evidence that the true mean difference in electric potential for diabetic mice is different than for normal mice.



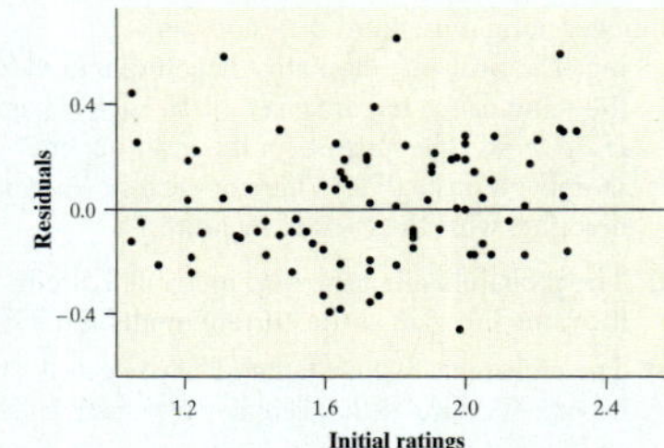


**AP4.38** Which of the following statements is supported by these plots?

(a) There is no striking evidence that the assumptions for regression inference are violated.

(b) The abundance of outliers and influential observations in the plots means that the assumptions for regression are clearly violated.

(c) These plots contain dramatic evidence that the standard deviation of the response about the true regression line is not approximately the same for each *x*-value.

(d) These plots call into question the validity of the assumption that the later ratings vary Normally about the least-squares line for each value of the initial ratings.

(e) A linear model isn't appropriate here because the residual plot shows no association.

**AP4.39** George's initial fear rating was 0.2 higher than Jonny's. What does the model predict about their final fear ratings?

(a) George's will be about 0.96 higher than Jonny's.

(b) George's will be about 0.40 higher than Jonny's.

(c) George's will be about 0.20 higher than Jonny's

(d) George's will be about 0.08 higher than Jonny's.

(e) George's will be about the same as Jonny's.

**AP4.40** The table below provides data on the political affiliation and opinion about the death penalty of 850 randomly selected voters from a congressional district.

| | Favor | Oppose | Total |
|---|---|---|---|
| Republican | 299 | 98 | **397** |
| Democrat | 77 | 171 | **248** |
| Other | 118 | 87 | **205** |
| **Total** | **494** | **356** | **850** |

Which of the following does *not* support the conclusion that being a Republican and favoring the death penalty are not independent?

(a) $\frac{299}{494} \neq \frac{98}{356}$

(b) $\frac{299}{494} \neq \frac{397}{850}$

(c) $\frac{494}{850} \neq \frac{299}{397}$

(d) $\frac{494}{850} \neq \frac{397}{850}$

(e) $\frac{(397)(494)}{850} \neq 299$

**Section II: Free Response** *Show all your work. Indicate clearly the methods you use, because you will be graded on the correctness of your methods as well as on the accuracy and completeness of your results and explanations.*

**AP4.41** The body's natural electrical field helps wounds heal. If diabetes changes this field, it might explain why people with diabetes heal more slowly. A study of this idea compared randomly selected normal mice and randomly selected mice bred to spontaneously develop diabetes. The investigators attached sensors to the right hip and front feet of the mice and measured the difference in electrical potential (in millivolts) between these locations. Graphs of the data for each group reveal no outliers or strong skewness. The following computer output provides numerical summaries of the data.[27]

| Variable | N | Mean | StDev | Minimum |
|---|---|---|---|---|
| Diabetic mice | 24 | 13.090 | 4.839 | 1.050 |
| Normal mice | 18 | 10.022 | 2.915 | 4.950 |

| Q1 | Median | Q3 | Maximum |
|---|---|---|---|
| 10.038 | 12.650 | 17.038 | 22.600 |
| 8.238 | 9.250 | 12.375 | 16.100 |

The researchers want to know whether the difference in mean electrical potentials between normal mice and mice with diabetes is statistically significant at the $\alpha = 0.05$ level. Carry out a test and report your conclusion.

**AP4.42** Can physical activity in youth lead to mental sharpness in old age? A 2010 study investigating this question involved 9344 randomly selected, mostly white women over age 65 from four U.S. states. These women were asked about their levels of physical activity during their teenage years, thirties, fifties, and later years. Those who reported being physically active as teens enjoyed the lowest level of cognitive decline—only 8.5% had cognitive impairment—compared with 16.7% of women who reported not being physically active at that time.

(a) State an appropriate pair of hypotheses that the researchers could use to test whether the proportion of women who suffered a cognitive decline was significantly lower for women who were physically active in their youth than for women who were not physically active at that time. Be sure to define any parameters you use.

(b) Assuming the conditions for performing inference are met, what inference method would you use to test the hypotheses you identified in part (b)? Do *not* carry out the test.

(c) Suppose the test in part (b) shows that the proportion of women who suffered a cognitive decline was significantly lower for women who were physically active in their youth than for women who were not physically active at that time. Can we generalize the results of this study to all women aged 65 and older? Justify your answer.

(d) We cannot conclude that being physically active as a teen *causes* a lower level of cognitive decline for women over 65, due to possible confounding with other variables. Explain the concept of confounding and give an example of a potential confounding variable in this study.

**AP4.43** In a recent poll, randomly selected New York State residents at various fast-food restaurants were asked if they supported or opposed a "fat tax" on nondiet sugared soda. Thirty-one percent said that they were in favor of such a tax and 66% were opposed. But when asked if they would support such a tax if the money raised were used to fund health care given the high incidence of obesity in the United States, 48% said that they were in favor and 49% were opposed.

(a) In this situation, explain how bias may have been introduced based on the way the questions were worded *and* suggest a way that they could have been worded differently in order to avoid this bias.

(b) In this situation, explain how bias may have been introduced based on the way the sample was taken *and* suggest a way that the sample could have been obtained in order to avoid this bias.

(c) This poll was conducted only in New York State. Suppose the pollsters wanted to ensure that estimates for the proportion of people who would support a tax on nondiet sugared soda were available for each state as well as an overall estimate for the nation as a whole. Identify a sampling method that would achieve this goal *and* briefly describe how the sample would be taken.

**AP4.44** Each morning, coffee is brewed in the school workroom by one of three faculty members, depending on who arrives first at work. Mr. Worcester arrives first 10% of the time, Dr. Currier arrives first 50% of the time, and Mr. Legacy arrives first on the remaining mornings. The probability that the coffee is strong when brewed by Dr. Currier is 0.1, while the corresponding probabilities when it is brewed by Mr. Legacy and Mr. Worcester are 0.2 and 0.3, respectively. Mr. Worcester likes strong coffee!

(a) What is the probability that on a randomly selected morning the coffee will be strong? Show your work.

(b) If the coffee is strong on a randomly selected morning, what is the probability that it was brewed by Dr. Currier? Show your work.

**AP4.45** The following table gives data on the mean number of seeds produced in a year by several common tree species and the mean weight (in milligrams) of the seeds produced. Two species appear twice because their seeds were counted in two locations. We might expect that trees with heavy seeds produce fewer of them, but what mathematical model best describes the relationship?[28]

| Tree species | Seed count | Seed weight (mg) |
|---|---|---|
| Paper birch | 27,239 | 0.6 |
| Yellow birch | 12,158 | 1.6 |
| White spruce | 7202 | 2.0 |
| Engelmann spruce | 3671 | 3.3 |
| Red spruce | 5051 | 3.4 |
| Tulip tree | 13,509 | 9.1 |
| Ponderosa pine | 2667 | 37.7 |
| White fir | 5196 | 40.0 |
| Sugar maple | 1751 | 48.0 |
| Sugar pine | 1159 | 216 |
| American beech | 463 | 247 |
| American beech | 1892 | 247 |
| Black oak | 93 | 1851 |
| Scarlet oak | 525 | 1930 |
| Red oak | 411 | 2475 |
| Red oak | 253 | 2475 |
| Pignut hickory | 40 | 3423 |
| White oak | 184 | 3669 |
| Chestnut oak | 107 | 4535 |

(a) Based on the scatterplot below, is a linear model appropriate to describe the relationship between seed count and seed weight? Explain.

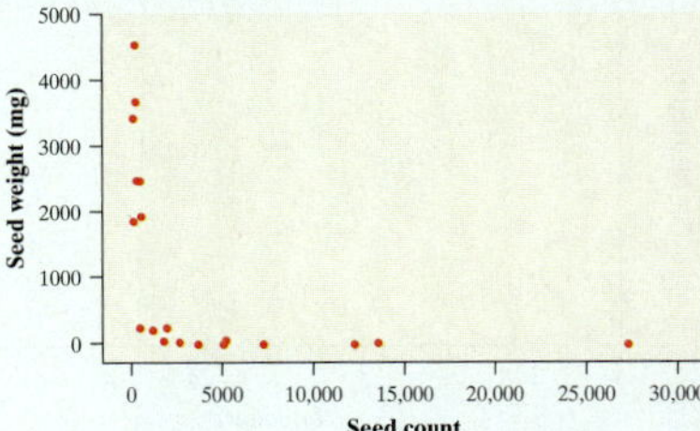


**AP4.42** **(a)** $H_0: p_1 - p_2 = 0$ vs. $H_a: p_1 - p_2 < 0$, where $p_1$ = the true proportion of women like the ones in the study who were physically active as teens that would suffer a cognitive decline and $p_2$ = the true proportion of women like the ones in the study who were not physically active as teens that would suffer a cognitive decline. **(b)** A two-sample *z* test for $p_1 - p_2$. **(c)** No. Because the participants were mostly white women from only four states, the findings may not be generalizable to women in other racial and ethnic groups or who live in other states. **(d)** Two variables are confounded when their effects on the response variable cannot be distinguished from one another. For example, women who were physically active as teens might have also done other things differently as well, such as eating a healthier diet. We would be unable to determine if it was their physically active youth or their healthier diet that slowed their level of cognitive decline.

**AP4.43** **(a)** Because the first question called it a "fat tax," people may have reacted negatively because they believe this is a tax on those who are overweight. The second question provides extra information that gets people thinking about the obesity problem in the U.S. and the increased health care that could be provided as a benefit with the tax money. *Better*: "Would you support or oppose a tax on non-diet sugared soda?" **(b)** This method samples only people at fast-food restaurants. They may go to these restaurants because they like the sugary drinks and wouldn't want to pay a tax on their favorite beverages. Thus, it is likely that the proportion of those who would oppose such a tax will be overestimated with this method. *Better*: Take a random sample of all New York State residents. **(c)** Use a stratified random sampling method in which each state is a stratum.

**AP 4.44** **(a)**

$$P(S) = (0.1)(0.3) + (0.4)(0.2) + (0.5)(0.1) = 0.16$$

**(b)**

$$P(C|S) = \frac{(0.5)(0.1)}{(0.1)(0.3) + (0.4)(0.2) + (0.5)(0.1)} = 0.3125$$

**AP4.45** **(a)** No. The scatterplot exhibits a strong curved pattern.

**AP4.45** **(b)** B, because the scatterplot shows a much more linear pattern and its residual plot shows no leftover patterns. **(c)** $\widehat{\ln(\text{Weight})} = 15.491 - 1.5222 \ln(3700) = 2.984$, thus $\widehat{\text{weight}} = e^{2.984} = 19.77$ mg. **(d)** About 86.3% of the variation in ln(seed weight) is accounted for by the linear model relating ln(seed weight) to ln(seed count).

**AP4.46** **(a)** Let $X$ = diameter of a randomly selected lid. Because $X$ follows a Normal distribution, the sampling distribution of $\bar{x}$ also follows a Normal distribution. $\mu_{\bar{x}} = 4$ inches and $\sigma_{\bar{x}} = \dfrac{0.02}{\sqrt{25}} = 0.004$ inches.

**(b)** We want to find $P(\bar{x} < 3.99$ or $\bar{x} > 4.01)$ using the $N(4, 0.004)$ distribution. $z = \dfrac{3.99 - 4}{0.004} = -2.50$ and $z = \dfrac{4.01 - 4}{0.004} = 2.50$. $P(Z < -2.50$ or $Z > 2.50) = 0.0124$. *Using technology*: 0.0124. Assuming that the machine is working properly, there is a 0.0124 probability that the mean diameter of a sample of 25 lids is less than 3.99 inches or greater than 4.01 inches.



(b) Two alternative models based on transforming the original data are proposed to predict the seed weight from the seed count. Graphs and computer output from a least-squares regression analysis on the transformed data are shown below.

Model A:

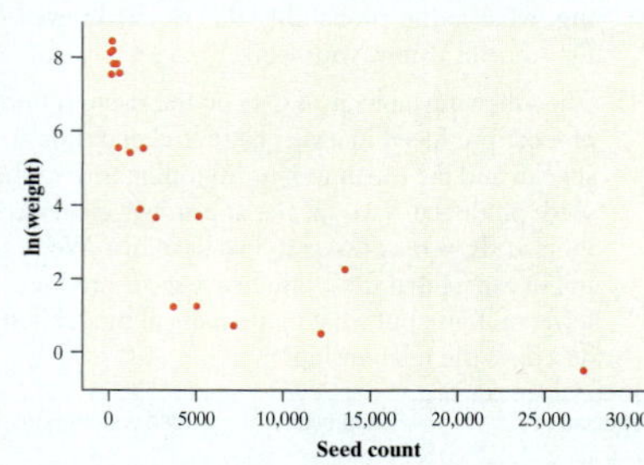


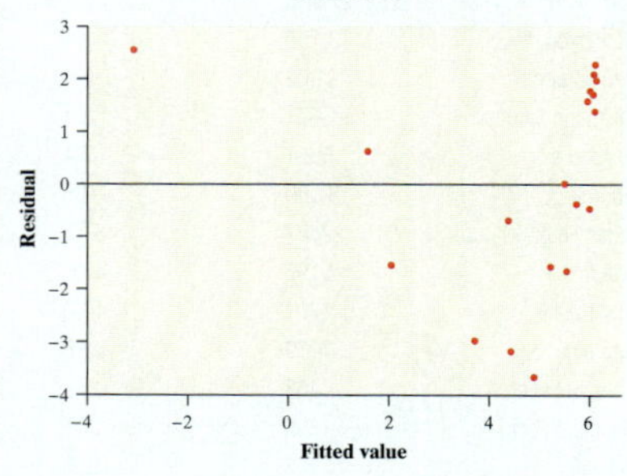


| Predictor | Coef | SE Coef | T | P |
|---|---|---|---|---|
| Constant | 6.1394 | 0.5726 | 10.72 | 0.000 |
| Seed Count | −0.00033869 | 0.00007187 | −4.71 | 0.000 |
| S = 2.08100 | R-Sq = 56.6% | R-Sq(adj) = 54.1% | | |

Model B:

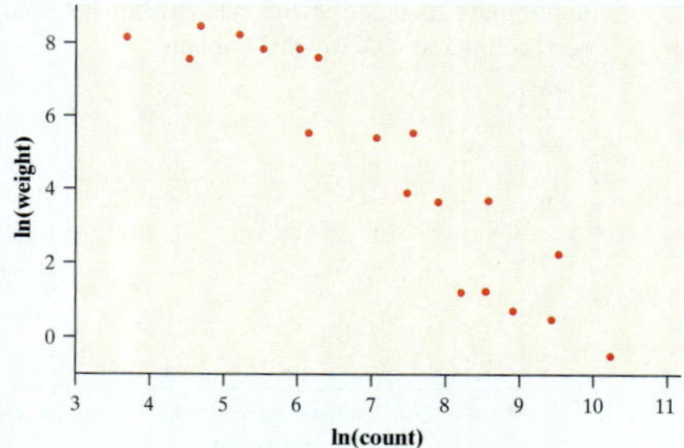


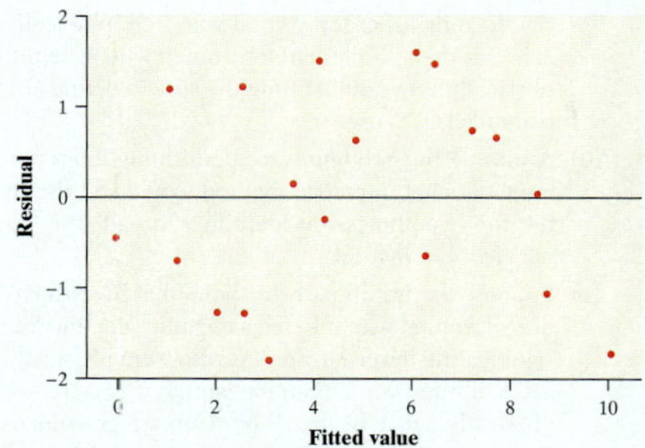


| Predictor | Coef | SE Coef | T | P |
|---|---|---|---|---|
| Constant | 15.491 | 1.081 | 14.33 | 0.000 |
| ln(count) | −1.5222 | 0.1470 | −10.35 | 0.000 |
| S = 1.16932 | R-Sq = 86.3% | R-Sq(adj) = 85.5% | | |

Which model, A or B, is more appropriate for predicting seed weight from seed count? Justify your answer.

(c) Using the model you chose in part (b), predict the seed weight if the seed count is 3700.

(d) Interpret the value of $r^2$ for your model.

**AP4.46** Suppose a company manufactures plastic lids for disposable coffee cups. When the manufacturing process is working correctly, the diameters of the lids are approximately Normally distributed with a mean diameter of 4 inches and a standard deviation of 0.02 inches. To make sure the machine is not producing lids that are too big or too small, each hour a random sample of 25 lids is selected and the sample mean is calculated.

(a) Describe the shape, center, and spread of the sampling distribution of the sample mean diameter, assuming the machine is working properly.

The company decides that it will shut down the machine if the sample mean diameter is less than 3.99 inches or greater than 4.01 inches, because this indicates that some lids will be too small or too large for the cups. If the sample mean is less than 3.99 or greater than 4.01, all the lids from that hour are thrown away because the company does not want to sell bad products.

(b) Assuming that the machine is working properly, what is the probability that a random sample of 25 lids will have a mean diameter less than 3.99 inches or greater than 4.01 inches? Show your work.

Also, to look for any trends, each hour the company records the value of the sample mean on a chart, like the one at top right.

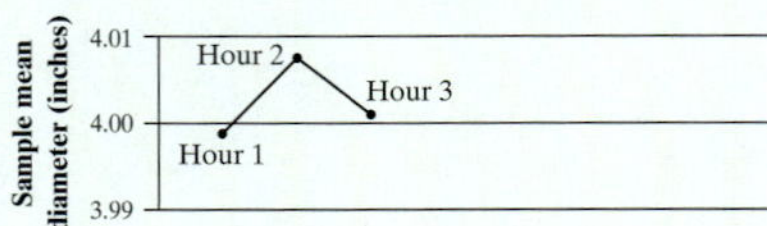


One benefit of using this type of chart is that out-of-control production trends can be noticed before it is too late and lids have to be thrown away. For example, if the sample mean increased in 3 consecutive samples, this would suggest that something might be wrong with the machine. If this trend can be noticed before the sample mean gets larger than 4.01, then the machine can be fixed without having to throw away any lids.

(c) Assuming that the manufacturing process is working correctly, what is the probability that the sample mean diameter will be above the desired mean of 4.00 but below the upper boundary of 4.01? Show your work.

(d) Assuming that the manufacturing process is working correctly, what is the probability that in 5 consecutive samples, 4 or 5 of the sample means will be above the desired mean of 4.00 but below the upper boundary of 4.01? Show your work.

(e) Which of the following results gives more convincing evidence that the machine needs to be shut down? Explain.

1. Getting a single sample mean below 3.99 or above 4.01

*or*

2. Taking 5 consecutive samples and having at least 4 of the sample means be between 4.00 and 4.01.

(f) Suggest a different rule (other than 1 and 2 stated in part (e)) for stopping the machine before it starts producing lids that have to be thrown away. Assuming that the machine is working properly, calculate the probability that the machine will be shut down when using your rule.

**AP4.46** **(c)** We want to find $P(4 < \bar{x} < 4.01)$ using the $N(4, 0.004)$ distribution.

$$z = \frac{4 - 4}{0.004} = 0 \text{ and } z = \frac{4.01 - 4}{0.004} = 2.50.$$

$P(0 < Z < 2.50) = 0.4938$. *Using technology*: 0.4938. Assuming that the machine is working properly, there is a 0.4938 probability that the mean diameter of a sample of 25 lids is between 4.00 and 4.01 inches. **(d)** Let $Y =$ the number of samples (out of 5) in which the sample mean is between 4.00 and 4.01. The random variable $Y$ has a binomial distribution with $n = 5$ and $p = 0.4938$. *Using technology*: $P(X \geq 4) = 1 - P(X \leq 3) =$ `1 - binomcdf (trials:5, p:0.4938, x value:3) =` 0.1798. **(e)** Because the probability found in part (b) is less than the probability found in part (d), getting a sample mean below 3.99 or above 4.01 is more convincing evidence that the machine needs to be shut down. This event is much less likely to happen by chance when the machine is working correctly. **(f)** Answers will vary.

# Photo Credits

Cover: Joseph Devenney/Getty Images
Flamingos used in interior design: © Smileus/istockphoto

p. vi Judy Starnes
p. vi Anne Tabor
p. xxi istockphoto
p. xxi Bloomberg via Getty Images
p. xxii istockphoto
p. xxiii istockphoto
p. xxiii © Leah Warkentin/Design Pics/Corbis

**Chapter 1**

p. xxxii Thinkstock
p. 2 istockphoto
p. 3 Courtesy RSSCSE
p. 5 istockphoto
p. 9 DILBERT: © Scott Adams/Dist. by United Feature Syndicate, Inc.
p. 10 istockphoto
p. 12 istockphoto
p. 15 Thinkstock
p. 19 The Kobal Collection/20th Century Fox/Paramount
p. 25 Chuck Myers/MCT via Getty Images
p. 26 Photo by Bentley Motors via Getty Images
p. 30 istockphoto
p. 31 Thinkstock
p. 33 Thinkstock
p. 50 istockphoto
p. 51 W. H. Freeman and Company
p. 52 istockphoto
p. 57 Photo by Sara Wolfram/Getty Images
p. 62 © 2008 Zits Partnership. Distributed by King Features Syndicate.
p. 65 Getty Images/OJO Images RF
p. 67 Thinkstock

**Chapter 2**

p. 82 PRNewsFoto/Chrysler Group, Diane Bondareff
p. 84 Thinkstock
p. 86 Fotosearch
p. 88 AP Photo/Jae C. Hong
p. 91 istockphoto
p. 95 istockphoto
p. 98 NASA and The Hubble Heritage Team
p. 106 FOXTROT©2005 Bill Amend. Universal Press Syndicate, Inc.
p. 111 istockphoto
p. 115 istockphoto
p. 124 istockphoto
p. 126 PRNewsFoto/Chrysler Group, Diane Bondareff

**Chapter 3**

p. 140 Thinkstock
p. 142 istockphoto
p. 144 istockphoto
p. 145 Chris Graythen/Staff, Getty Images
p. 148 istockphoto/Thinkstock
p. 149 Mother Goose & Grimm-Grimmy, Inc. King Features Syndicate
p. 152 Cal Sport Media via AP Images
p. 156 W. H. Freeman and Company
p. 157 AP Photo/Eric Gay
p. 165 Ford Images
p. 168 Thinkstock
p. 173 Ford Images
p. 180 Sports Illustrated/Getty Images
p. 181 © Fancy/Alamy
p. 183 Thinkstock
p. 185 © Laqhill|Dreamstime.com
p. 190 istockphoto
p. 191 Thinkstock

**Chapter 4**

p. 206 Eric O'Connell/Getty Images
p. 208 © MBI/Alamy
p. 209 istockphoto
p. 210 © Charles O. Cecil/Alamy
p. 212 Thinkstock
p. 214 Cartoons By Landers
p. 215 Urbaniak, G. C., & Plous, S. (2011). Research Randomizer (Version 3.0) [Computer software]. Retrieved on April 22, 2011, from http://www.randomizer.org/
p. 217 istockphoto
p. 218 MPI/Getty Images
p. 219 © Li Ding/Alamy
p. 223 © Richard Levine/Alamy
p. 223 © David Hodges/Alamy
p. 225 istockphoto
p. 226 GSS
p. 227 © 1995 Watterson. Universal Press Syndicate. Reprinted with permission. All rights reserved.
p. 235 istockphoto
p. 237 © Golden Pixels LLC/Alamy
p. 238 istockphoto
p. 239 istockphoto
p. 240 istockphoto
p. 241 © moodboard/Alamy
p. 243 istockphoto
p. 250 istockphoto
p. 266 istockphoto/Thinkstock
p. 269 istockphoto
p. 271 W. H. Freeman and Company
p. 272 Eric O'Connell/Getty Images

**Chapter 5**

p. 286 Hemera/Thinkstock
p. 288 © Jason Stitt|Dreamstime.com
p. 289 Creatas/Thinkstock
p. 293 © Eyebyte/Alamy
p. 294 Mitchell Layton/Getty Images
p. 295 W. H. Freeman and Company
p. 298 Drew Hallowell/Getty Images
p. 306 © Judith Collins/Alamy
p. 308 © Murray Lee/Getty Images
p. 309 © Big Cheese Photo/SuperStock
p. 313 © Flint/Corbis
p. 318 istockphoto
p. 319 istockphoto
p. 320 istockphoto
p. 322 Thomas Starke/Bongarts/Getty Images
p. 324 istockphoto
p. 327 Graeme Harris/Getty Images
p. 330 © 2008 Chris Browne 2010 King Features Syndicate, Inc.
p. 332 Hemera/Thinkstock

**Chapter 6**

p. 344 Comstock/Thinkstock
p. 346 istockphoto
p. 347 istockphoto
p. 349 Thinkstock
p. 351 Thinkstock
p. 352 Scott Adams/Dist. by United Feature Syndicate, Inc.
p. 357 Alamy
p. 365 Alamy
p. 368 istockphoto
p. 372 moodboard/SuperStock
p. 374 *(top)* istockphoto/Thinkstock
p. 374 *(inset)* Alamy
p. 380 istockphoto
p. 381 Thinkstock
p. 383 istockphoto
p. 387 © Alex Clark/Alamy
p. 388 istockphoto
p. 390 Exactostock/SuperStock
p. 396 Thinkstock
p. 399 Dreamstime
p. 401 © Chen Wee Lip|Dreamstime.com
p. 402 istockphoto
p. 403 istockphoto
p. 405 istockphoto/thinkstock
p. 409 Comstock/Thinkstock

**Chapter 7**

p. 420 © Arekmalang|Dreamstime.com
p. 423 istockphoto
p. 425 istockphoto
p. 426 © Kevin Woodrow|Dreamstime.com
p. 430 Alamy
p. 432 CBS/Landov
p. 435 istockphoto
p. 440 FOXTROT©2005 Bill Amend. Universal Press Syndicate.
p. 445 Dreamstime
p. 450 istockphoto
p. 455 Alamy
p. 460 © Arekmalang|Dreamstime.com

**Chapter 8**

p. 474 Brooke Fasani/age fotostock
p. 476 istockphoto
p. 478 istockphoto

p. 480 Garfield ©1999 Paws, Inc. Dist. by Universal Uclick. Reprinted by permission. All rights reserved.
p. 481 Digital Vision/Thinkstock
p. 484 © Jan Haas/dpa/Corbis
p. 485 istockphoto
p. 493 istockphoto
p. 498 istockphoto
p. 500 Alamy
p. 502 istockphoto
p. 509 © Robert Kneschke|Dreamstime.com
p. 519 Alamy
p. 520 istockphoto
p. 522 istockphoto
p. 524 Phototake/Alamy
p. 525 Brooke Fasani/age footstock

## Chapter 9

p. 536 istockphoto
p. 540 Superstock
p. 542 © Otnaydur|Dreamstime.com
p. 543 istockphoto
p. 544 istockphoto
p. 545 istockphoto
p. 546 istockphoto
p. 549 fotosearch
p. 555 fotosearch
p. 559 istockphoto
p. 562 istockphoto
p. 577 istockphoto
p. 580 istockphoto
p. 583 Alamy
p. 586 Alamy
p. 589 Courtesy Judy Starnes
p. 590 istockphoto
p. 592 istockphoto
p. 593 istockphoto

## Chapter 10

p. 608 Jamie Rector/Getty Images
p. 610 © TravelStockCollection—Homer Sykes/Alamy
p. 612 Alamy
p. 615 Envision/Corbis
p. 617 Photo by Peter Macdiarmid/Getty Images
p. 620 Alamy
p. 622 istockphoto
p. 625 Photolibrary
p. 635 iStock
p. 636 Jeff Greenberg/Photo Edit
p. 639 Rich Legg/Getty Images
p. 639 istockphoto
p. 645 Alamy
p. 650 Clerkenwell/Getty Images
p. 651 Jamie Rector/Getty Images
p. 655 istockphoto

## Chapter 11

p. 676 Alex-Takes-Photos/Flickr/Getty
p. 678 R Rivera Moret
p. 681 R Rivera Moret
p. 683 R Rivera Moret
p. 687 R Rivera Moret
p. 688 Christian Petersen/Getty Images
p. 698 RF Pictures/Corbis
p. 701 Alamy
p. 708 Alamy
p. 712 Alamy
p. 715 Alamy
p. 718 Alamy
p. 721 Alex-Takes-Photos/Flickr/Getty

## Chapter 12

p. 736 Mike Ehrmann/Getty Images
p. 738 fotosearch
p. 748 istockphoto
p. 749 Ford Images
p. 754 Fotosearch
p. 767 istockphoto
p. 768 Alamy
p. 773 © RGB Ventures LLC dba SuperStock/Alamy
p. 775 istockphoto
p. 776 © Nandana De Silva/Alamy
p. 779 NASA/JPL-Caltech
p. 783 Mike Ehrmann/Getty Images

# Notes and Data Sources

## Overview: What Is Statistics?

1. Patrizia Frei et al., "Use of mobile phones and risk of brain tumours: update of Danish cohort study," BMJ 2011;343:d6387 (Published 20 October 2011).

2. Nikhil Swaminathan, "Gender Jabber: Do Women Talk More than Men?" *Scientific American*, July 6, 2007.

3. Stephen Moss, "Do women really talk more?" *The Guardian*, November 26, 2006.

4. Data from gapminder.org, accessed October 19, 2013.

5. Data from the Josephson Institute's "2012 Report Card of American Youth," accessed online at charactercounts.org, March 16, 2013.

## Chapter 1

1. These are some of the data from the EESEE story "Stress among Pets and Friends." The study results appear in K. Allen, J. Blascovich, J. Tomaka, and R. M. Kelsey, "Presence of human friends and pet dogs as moderators of autonomic responses to stress in women," *Journal of Personality and Social Psychology*, 83 (1988), pp. 582–589.

2. Roller coaster data for 2009 from www.rcdb.com.

3. Arbitron, *Radio Today*, www.arbitron.com.

4. Arbitron, *Internet and Multimedia*, www.arbitron.com.

5. We got the idea for this example from David Lane's case study "Who is buying iMacs?" which we found online at onlinestatbook.com/case_studies_rvls/.

6. The National Longitudinal Study of Adolescent Health interviewed a stratified random sample of 27,000 adolescents, then reinterviewed many of the subjects six years later, when most were aged 19 to 25. These data are from the Wave III reinterviews in 2000 and 2001, found at the Web site of the Carolina Population Center, www.cpc.unc.edu.

7. The U.K. data were obtained using the Random Sampler tool at www.censusatschool.com. The U.S. data were obtained using the Random Sampler tool at www.amstat.org/censusatschool.

8. 2011 DuPont Automotive Color Survey, at www2.dupont.com/Media_Center/en_US/color_popularity/.

9. Found at spam-filter-review.toptenreviews.com, which claims to have compiled data "from a number of different reputable sources."

10. Centers for Disease Control and Prevention, *Births: Final Data for 2005*, National Vital Statistics Reports, 56, No. 6, 2007, at www.cdc.gov.

11. National Center for Health Statistics, *Deaths: Preliminary Data for 2010*, at www.cdc.gov/nchs.

12. *The Hispanic Population: 2010*, at www.census.gov, based on the data collected from the 2010 U.S. Census.

13. Data for 2005 from the 2007 *Statistical Abstract of the United States* at the Census Bureau Web site, www.census.gov.

14. Arbitron, *Internet and Media*, at www.arbitron.com.

15. *Statistical Abstract of the United States*, 2004–2005, Table 1233.

16. Lien-Ti Bei, "Consumers' purchase behavior toward recycled products: an acquisition-transaction utility theory perspective," MS thesis, Purdue University, 1993.

17. S. V. Zagona (ed.), *Studies and Issues in Smoking Behavior*, University of Arizona Press, 1967, pp. 157–180.

18. From the Web site of the Carolina Population Center, www.cpc.unc.edu.

19. Data on the U.S. Women's National Soccer Team are from the United States Soccer Federation Web site, www.ussoccer.com.

20. *2012 Fuel Economy Guide*, from the U.S. Environmental Protection Agency's Web site at www.fueleconomy.gov.

21. From the American Community Survey, at factfinder2.census.gov.

22. James T. Fleming, "The measurement of children's perception of difficulty in reading materials," *Research in the Teaching of English*, 1 (1967), pp. 136–156.

23. See Note 13.

24. The original paper is T. M. Amabile, "Motivation and creativity: effects of motivational orientation on creative writers," *Journal of Personality and Social Psychology*, 48, No. 2 (February 1985), pp. 393–399. The data for Exercise 43 came from Fred L. Ramsey and Daniel W. Schafer, *The Statistical Sleuth*, 3rd ed., Brooks/Cole Cengage Learning, 2013.

25. The cereal data came from the Data and Story Library, http://lib.stat.cmu.edu/DASL/.

26. From a plot in K. Krishna Kumar et al., "Unraveling the mystery of Indian monsoon failure during El Niño," *Science*, 314 (2006), pp. 115–119.

27. Basketball scores from the *Los Angeles Times*.

28. Monthly stock returns from the Web site of Professor Kenneth French of Dartmouth, mba.tuck.dartmouth.edu/pages/faculty/ken.french. A fine point: the data are the "excess returns" on stocks, the actual returns less the small monthly returns on Treasury bills.

29. C. B. Williams, *Style and Vocabulary: Numerological Studies*, Griffin, 1970.

30. From the American Community Survey, at factfinder2.census.gov.

31. International Energy Agency, *Key World Energy Statistics 2007*, at www.iea.org.

32. Maribeth Cassidy Schmitt, "The effects of an elaborated directed reading activity on the metacomprehension skills of third graders," PhD thesis, Purdue University, 1987.

33. These data were collected by students as a class project.

34. The chest size data came from the Data and Story Library, lib.stat.cmu.edu/DASL/.

35. Advanced Placement exam results from AP Central Web site, apcentral.collegeboard.com.

36. Arbitron, *Internet and Media 2006*, www.arbitron.com.

37. H. Lindberg, H. Roos, and P. Gardsell, "Prevalence of coxarthritis in former soccer players," *Acta Orthopedica Scandinavica*, 64 (1993), pp. 165–167.

38. From the American Community Survey, at the Census Bureau Web site, www.census.gov. The data are a subsample of the 13,194 individuals in the ACS North Carolina sample who had travel times greater than zero.

39. Pets data set from *Guidelines for Assessment and Instruction in Statistics Education: Pre-K–12 Report*, American Statistical Association, 2007.

40. This isn't a mathematical theorem. The mean can be less than the median in right-skewed distributions that take only a few values, many of which lie exactly at the median. The rule almost never fails for distributions taking many values, and most counterexamples don't appear clearly skewed in graphs even though they may be slightly skewed according to technical measures of skewness. See Paul T. von Hippel, "Mean, median, and skew: correcting a textbook rule," *Journal of Statistics Education*, 13, No. 2 (2005), online journal, www.amstat.org/publications/jse/v13n2/vonhippel.html.

41. From the National Association of Realtors, www.realtor.org.

42. Tom Lloyd et al., "Fruit consumption, fitness, and cardiovascular health in female adolescents: the Penn State Young Women's Health Study," *American Journal of Clinical Nutrition*, 67 (1998), pp. 624–630.

43. The daily returns for the year ending November 8, 2007, for the CREF Equity Index Fund and the TIAA Real Estate Fund. Daily price data for these funds are at www.tiaa-cref.org. Returns can be easily calculated because dividends are incorporated in the daily prices rather than given separately.

44. Data from the most recent Annual Demographic Supplement can be found at www.census.gov/cps.

45. Ethan J. Temeles and W. John Kress, "Adaptation in a plant-hummingbird association," *Science*, 300 (2003), pp. 630–633. We thank Ethan J. Temeles for providing the data.

46. 2007 CIRP Freshman Survey; data from the Higher Education Research Institute's report "The American freshman: national norms for fall 2007," published January 2008.

47. From Internet Movie Database, www.imdb.com.

48. Movie ratings data found at informationplease.com.

49. The report "Cell phones key to teens' social lives, 47% can text with eyes closed" is published online by Harris Interactive, July 2008, www.marketingcharts.com.

50. A. Karpinski and A. Duberstein, "A description of Facebook use and academic performance among undergraduate and graduate students," paper presented at the American Educational Research Association annual meeting, April 2009. Thanks to Aryn Karpinski for providing us with some original data from the study.

51. S. M. Stigler, "Do robust estimators work with real data?" *Annals of Statistics*, 5 (1977), pp. 1055–1078.

52. T. Bjerkedal, "Acquisition of resistance in guinea pigs infected with different doses of virulent tubercle bacilli," *American Journal of Hygiene*, 72 (1960), pp. 130–148.

53. Data from the Bureau of Labor Statistics, Annual Demographic Supplement www.bls.census.gov.

54. Data from the report "Is our tuna family-safe?" prepared by Defenders of Wildlife, 2006.

## Chapter 2

1. Will Shortz, "A few words about Sudoku, which has none," *New York Times*, August 28, 2005.

2. Data found online at exploringdata.net.

3. S. M. Stigler, "Do robust estimators work with real data?" *Annals of Statistics*, 5 (1977), pp. 1055–1078.

4. Sheldon Ross, *Introduction to Probability and Statistics for Engineers and Scientists*, 3rd ed., Academic Press, 2004.

5. Stephen Jay Gould, "Entropic homogeneity isn't why no one hits .400 anymore," *Discover*, August 1986, pp. 60–66. Gould does not standardize but gives a speculative discussion instead.

6. Information on bone density in the reference populations was found at www.courses.washington.edu/bonephys/opbmd.html.

7. Data from *USA Today* sports salary data base at www.usatoday.com.

8. Data from Gary Community School Corporation, courtesy of Celeste Foster, Department of Education, Purdue University.

9. Detailed data appear in P. S. Levy et al., *Total Serum Cholesterol Values for Youths 12–17 Years*, Vital and Health Statistics, Series 11, No. 155, National Center for Health Statistics, 1976.

10. Data found at Bureau of Labor Statistics, www.bls.gov.

11. Data provided by Chris Olsen, who found the information in *Scuba News* and *Skin Diver* magazines.

12. See Note 3.

13. The data set was constructed based on information provided in P. D. Wood et al., "Plasma lipoprotein distributions in male and female runners," in P. Milvey (ed.), *The Marathon: Physiological, Medical, Epidemiological, and Psychological Studies*, New York Academy of Sciences, 1977.

14. Data found online at www.earthtrends.wri.org.

15. R. Shine, T. R. L. Madsen, M. J. Elphick, and P. S. Harlow, "The influence of nest temperature and maternal brooding on hatchling phenotypes in water python," *Ecology*, 78 (1997), pp. 1713–1721.

16. We found the information on birth weights of Norwegian children on the National Institute of Environmental Health Sciences Web site. The relevant article can be accessed here: http://www.ncbi.nlm.nih.gov/pubmed/1536353.

17. From a graph in L. Partridge and M. Farquhar, "Sexual activity reduces lifespan of male fruit flies," *Nature*, 294 (1981), pp. 580–582. Provided by Brigitte Baldi.

## Chapter 3

1. SEC football data from espn.go.com.

2. The Florida Boating Accident Statistical Report for each year (at myfwc.com/boating/safety-education/boating-accidents) gives the number of registered vessels. The Florida Wildlife Commission maintains a manatee death data base at research.myfwc.com/manatees/.

3. Michael Winerip, "On education: SAT essay test rewards length and ignores errors," *New York Times*, May 4, 2005.

4. Data from The College Board Web site, www.collegeboard.com.

5. Data provided by Darlene Gordon, Purdue University.

6. From a graph in Bernt-Erik Saether, Steiner Engen, and Erik Mattysen, "Demographic characteristics and population dynamical patterns of solitary birds," *Science*, 295 (2002), pp. 2070–2073.

7. Based on T. N. Lam, "Estimating fuel consumption from engine size," *Journal of Transportation Engineering*, 111 (1985), pp. 339–357. The data for 10 to 50 km/h are measured; those for 60 and higher are calculated from a model given in the paper and are therefore smoothed.

8. From a graph in Christer G. Wiklund, "Food as a mechanism of density-dependent regulation of breeding numbers in the merlin *Falco columbarius*," *Ecology*, 82 (2001), pp. 860–867.

9. From a graph in Naomi I. Eisenberger, Matthew D. Lieberman, and Kipling D. Williams, "Does rejection hurt? An fMRI study of social exclusion," *Science*, 302 (2003), pp. 290–292.

10. M. A. Houck et al., "Allometric scaling in the earliest fossil bird, *Archaeopteryx lithographica*," *Science*, 247 (1990), pp. 195–198.

The authors conclude from a variety of evidence that all specimens represent the same species.

11. *Consumer Reports*, June 1986, pp. 366–367.

12. G. A. Sacher and E. F. Staffelt, "Relation of gestation time to brain weight for placental mammals: implications for the theory of vertebrate growth," *American Naturalist*, 108 (1974), pp. 593–613. We found the data in Fred L. Ramsey and Daniel W. Schafer, *The Statistical Sleuth: A Course in Methods of Data Analysis*, Duxbury, 1997, p. 228.

13. Table 1 of E. Thomassot et al., "Methane-related diamond crystallization in the earth's mantle: stable isotopes evidence from a single diamond-bearing xenolith," *Earth and Planetary Science Letters*, 257 (2007), pp. 362–371.

14. The information for exercise 34 came from libertystreeteconomics.newyorkfed.org/2012/03/grading-student-loans.html

15. Data on used car prices from autotrader.com, September 8, 2012. We searched for F-150 4 x 4's on sale within 50 miles of College Station, Texas.

16. N. R. Draper and J. A. John, "Influential observations and outliers in regression," *Technometrics*, 23(1981), pp. 21–26.

17. Frank J. Anscombe, "Graphs in statistical analysis," *American Statistician*, 27 (1973), pp. 17–21.

18. P. Goldblatt (ed.), *Longitudinal Study: Mortality and Social Organization*, Her Majesty's Stationery Office, 1990. At least, so claims Richard Conniff, in *The Natural History of the Rich*, Norton, 2002, p. 45. We have not been able to access the Goldblatt report.

19. W. M. Lewis and M. C. Grant, "Acid precipitation in the western United States," *Science*, 207 (1980), pp. 176–177.

20. Data from National Institute of Standards and Technology, *Engineering Statistics Handbook*, www.itl.nist.gov/div898/handbook. The analysis there does not comment on the bias of field measurements.

21. We obtained the data for exercise 55 from i.nbcolympics.com/figure-skating/resultsandschedules/event=FSW010000/index.html.

22. Gary Smith, "Do statistics test scores regress toward the mean?" *Chance*, 10, No. 4 (1997), pp. 42–45.

23. We got the data for exercise 66 from www.baseball-reference.com.

24. From a graph in G. D. Martinsen, E. M. Driebe, and T. G. Whitham, "Indirect interactions mediated by changing plant chemistry: beaver browsing benefits beetles," *Ecology*, 79 (1998), pp. 192–200.

25. We got the data for exercise 68 from http://nutrition.mcdonalds.com/getnutrition/nutritionfacts.pdf.

26. The data for exercise 69 came from www.pro-football-reference.com/teams/jax/2011.htm.

27. Debora L. Arsenau, "Comparison of diet management instruction for patients with non–insulin dependent diabetes mellitus: learning activity package vs. group instruction," MS thesis, Purdue University, 1993.

28. A. K. Yousafzai et al., "Comparison of armspan, arm length and tibia length as predictors of actual height of disabled and nondisabled children in Dharavi, Mumbai, India," *European Journal of Clinical Nutrition*, 57 (2003), pp. 1230–1234. In fact, $r^2 = 0.93$.

29. David M. Fergusson and L. John Horwood, "Cannabis use and traffic accidents in a birth cohort of young adults," *Accident Analysis and Prevention*, 33 (2001), pp. 703–711.

30. *The World Almanac and Book of Facts* (2009).

31. G. L. Kooyman et al., "Diving behavior and energetics during foraging cycles in king penguins," *Ecological Monographs*, 62 (1992), pp. 143–163.

32. We found the data on cherry blossoms in the paper "Linear equations and data analysis," which was posted on the North Carolina School of Science and Mathematics Web site, www.ncssm.edu.

33. From a graph in Craig Packer et al., "Ecological change, group territoriality, and population dynamics in Serengeti lions," *Science*, 307 (2005), pp. 390–393.

## Chapter 4

1. Carlos Vallbona et al., "Response of pain to static magnetic fields in postpolio patients, a double blind pilot study," *Archives of Physical Medicine and Rehabilitation*, 78 (1997), pp.1200–1203.

2. "Traffic Safety Facts 2012," published by the National Highway Traffic Safety Administration on their Web site, www.nhtsa.gov.

3. Results from the Cox Communications Teen Online and Wireless Safety Survey, in Partnership with the National Center for Missing and Exploited Children® (NCMEC) and John Walsh, published May 2009, www.cox.com/takecharge/safe_teens_2009.

4. Sheldon Cohen, William J. Doyle, Cuneyt M. Alper, Denise Janicki-Deverts, and Ronald B. Turner, "Sleep habits and susceptibility to the common cold," *Archives of Internal Medicine*, 169, No. 1 (2009), pp. 62–67.

5. Rebecca Smith and Alison Wessner, The Lawrenceville School, "Does listening to music while studying affect performance on learning assessments?" May 2009.

6. Based on a suggested Science Fair project entitled "Do the Eyes Have It?" at the Science Buddies Web site, www.sciencebuddies.org.

7. Frederick Mosteller and David L. Wallace. *Inference and Disputed Authorship: The Federalist*. Addison-Wesley, Reading, MA, 1964.

8. According to the Web site http://en.wikipedia.org/wiki/Federalist_papers.

9. Excerpt obtained from http://www.constitution.org/fed/federa51.htm.

10. For information on the American Community Survey of households (there is a separate sample of group quarters), go to www.census.gov/acs.

11. The Pew press release and the full report "Polls face growing resistance, but still representative" (April 20, 2004) are at people-press.org/reports.

12. Cynthia Crossen, "Margin of error: studies galore support products and positions, but are they reliable?" *Wall Street Journal*, November 14, 1991.

13. Robert C. Parker and Patrick A. Glass, "Preliminary results of double-sample forest inventory of pine and mixed stands with high- and low-density LiDAR," in Kristina F. Connoe (ed.), *Proceedings of the 12th Biennial Southern Silvicultural Research Conference*, U.S. Department of Agriculture, Forest Service, Southern Research Station, 2004. The researchers actually sampled every tenth plot. This is a systematic sample.

14. Gary S. Foster and Craig M. Eckert, "Up from the grave: a socio-historical reconstruction of an African American community from cemetery data in the rural Midwest," *Journal of Black Studies*, 33 (2003), pp. 468–489.

15. Bryan E. Porter and Thomas D. Berry, "A nationwide survey of self-reported red light running: measuring prevalence, predictors, and perceived consequences," *Accident Analysis and Prevention*, 33 (2001), pp. 735–741.

16. Giuliana Coccia, "An overview of non-response in Italian telephone surveys," *Proceedings of the 99th Session of the International Statistics Institute*, 1993, Book 3, pp. 271–272.
17. Mario A. Parada et al., "The validity of self-reported seatbelt use: Hispanic and non-Hispanic drivers in El Paso," *Accident Analysis and Prevention*, 33 (2001), pp. 139–143.
18. These figures are cited by Dr. William Dement at "The Sleep Well" Web site, www.stanford.edu/~dement/.
19. See, for example, Martin Enserink, "The vanishing promises of hormone replacement," *Science*, 297 (2002), pp. 325–326; and Brian Vastag, "Hormone replacement therapy falls out of favor with expert committee," *Journal of the American Medical Association*, 287 (2002), pp. 1923–1924. A National Institutes of Health panel's comprehensive report is *International Position Paper on Women's Health and Menopause*, NIH Publication 02-3284.
20. "Family dinner linked to better grades for teens: survey finds regular meal time yields additional benefits," written by John Mackenzie for ABC News's *World News Tonight*, September 13, 2005.
21. Information found online at http//ssw.unc.edu/about/news/careerstart_1-13-09.
22. Simplified from Arno J. Rethans, John L. Swasy, and Lawrence J. Marks, "Effects of television commercial repetition, receiver knowledge, and commercial length: a test of the two-factor model," *Journal of Marketing Research*, 23 (February 1986), pp. 50–61.
23. Steering Committee of the Physicians' Health Study Research Group, "Final report on the aspirin component of the ongoing Physicians' Health Study," *New England Journal of Medicine*, 321 (1989), pp. 129–135.
24. Martin J. Bergee and Lecia Cecconi-Roberts, "Effects of small-group peer interaction on self-evaluation of music performance," *Journal of Research in Music Education*, 50 (2002), pp. 256–268.
25. The placebo effect examples are from Sandra Blakeslee, "Placebos prove so powerful even experts are surprised," *New York Times*, October 13, 1998.
26. The "three-quarters" estimate is cited by Martin Enserink, "Can the placebo be the cure?" *Science*, 284 (1999), pp. 238–240. An extended treatment is Anne Harrington (ed.), *The Placebo Effect: An Interdisciplinary Exploration*, Harvard University Press, 1997.
27. The flu trial quotation is from Kristin L. Nichol et al., "Effectiveness of live, attenuated intranasal influenza virus vaccine in healthy, working adults," *Journal of the American Medical Association*, 282 (1999), pp. 137–144.
28. From the Electronic Encyclopedia of Statistical Examples and Exercises (EESEE), "Anecdotes of Placebos."
29. David L. Strayer, Frank A. Drews, and William A. Johnston, "Cell phone–induced failures of visual attention during simulated driving," *Journal of Experimental Psychology: Applied*, 9 (2003), pp. 23–32.
30. We obtained the tire pressure loss data from the Consumer Reports Web site: http://news.consumerreports.org/cars/2007/10/nitrogen-tires-.html.
31. J. E. Muscat et al., "Handheld cellular telephone use and risk of brain cancer," *Journal of the American Medical Association*, 284 (2000), pp. 3001–3007.
32. *Early Human Development*, 76, No. 2 (February 2004), pp. 139–145.
33. National Institute of Child Health and Human Development, Study of Early Child Care and Youth Development. The article appears in the July 2003 issue of *Child Development*. The quotation is from the summary on the NICHD Web site, www.nichd.nih.gov.
34. Naomi D. L. Fisher, Meghan Hughes, Marie Gerhard-Herman, and Norman K. Hollenberg, "Flavonol-rich cocoa induces nitric-oxide-dependent vasodilation in healthy humans," *Journal of Hypertension*, 21, No. 12 (2003), pp. 2281–2286.
35. Joel Brockner et al., "Layoffs, equity theory, and work performance: further evidence of the impact of survivor guilt," *Academy of Management Journal*, 29 (1986), pp. 373–384.
36. Details of the Carolina Abecedarian Project, including references to published work, can be found online at abc.fpg.unc.edu.
37. K. B. Suttle, Meredith A. Thomsen, and Mary E. Power, "Species interactions reverse grassland responses to changing climate," *Science*, 315 (2007), pp. 640–642.
38. Christopher Anderson, "Measuring what works in health care," *Science*, 263 (1994), pp. 1080–1082.
39. Esther Duflo, Rema Hanna, and Stephan Ryan, "Monitoring works: getting teachers to come to school," November 21, 2007, at economics.mit.edu/files/2066.
40. See Note 28.
41. Marielle H. Emmelot-Vonk et al., "Effect of testosterone supplementation on functional mobility, cognition, and other parameters in older men," *Journal of the American Medical Association*, 299 (2008), pp. 39–52.
42. Linda Stern et al., "The effects of low-carbohydrate versus conventional weight loss diets in severely obese adults: one-year follow up of a randomized trial," *Annals of Internal Medicine*, 140, No. 10 (May 2004), pp. 778–785.
43. W. E. Paulus et al., "Influence of acupuncture on the pregnancy rate in patients who undergo assisted reproductive therapy," *Fertility and Sterility*, 77, No. 4 (2002), pp. 721–724.
44. Mary O. Mundinger et al., "Primary care outcomes in patients treated by nurse practitioners or physicians," *Journal of the American Medical Association*, 238 (2000), pp. 59–68.
45. Sterling C. Hilton et al., "A randomized controlled experiment to assess technological innovations in the classroom on student outcomes: an overview of a clinical trial in education," manuscript, no date. A brief report is Sterling C. Hilton and Howard B. Christensen, "Evaluating the impact of multimedia lectures on student learning and attitudes," *Proceedings of the 6th International Conference on the Teaching of Statistics*, at www.stat.auckland.ac.nz.
46. Study conducted by cardiologists at Athens Medical School, Greece, and announced at a European cardiology conference in February 2004.
47. Evan H. DeLucia et al., "Net primary production of a forest ecosystem with experimental $CO_2$ enhancement," *Science*, 284 (1999), pp. 1177–1179. The investigators used the block design.
48. Niels Juel-Nielsen, *Individual and Environment: Monozygotic Twins Reared Apart*, International Universities Press, 1980.
49. The sleep deprivation study is described in R. Stickgold, L. James, and J. Hobson, "Visual discrimination learning requires post-training sleep," *Nature Neuroscience*, 2000, pp. 1237–1238. We obtained the data from Allan Rossman, who got it courtesy of the authors.
50. The idea for this chart came from Fred L. Ramsey and Daniel W. Schafer, *The Statistical Sleuth: A Course in Methods of Data Analysis*, 2nd edition, Duxbury Press, 2002.
51. Thanks to Dan Teague for providing this example.
52. Thanks to Josh Tabor for contributing this Activity.
53. *The Health Consequences of Smoking: 1983*, U.S. Health Service, 1983.

54. See the details on the Web site of the Office for Human Research Protections of the Department of Health and Human Services, hhs.gov/ohrp.

55. Charles A. Nelson III et al., "Cognitive recovery in socially deprived young children: the Bucharest Early Intervention Project," *Science*, 318 (2007), pp. 1937–1940.

56. The article describing this study is Nikhil Swaminathan, "Gender jabber: do women talk more than men?" *Scientific American*, July 6, 2007.

57. Marilyn Ellis, "Attending church found factor in longer life," *USA Today*, August 9, 1999.

58. Scott DeCarlo with Michael Schubach and Vladimir Naumovski, "A decade of new issues," *Forbes*, March 5, 2001, www.forbes.com.

59. "Antibiotics no better than placebo for sinus infections" http://in.news.yahoo.com/antibiotics-no-better-placebo-sinus-infections-074240018.html

60. Warren McIsaac and Vivek Goel, "Is access to physician services in Ontario equitable?" Institute for Clinical Evaluative Sciences in Ontario, October 18, 1993.

61. L. E. Moses and F. Mosteller, "Safety of anesthetics," in J. M. Tanur et al. (eds.), *Statistics: A Guide to the Unknown*, 3rd ed., Wadsworth, 1989, pp. 15–24.

62. *Arizona Daily Star* 10-29-2008.

63. R. C. Shelton et al., "Effectiveness of St. John's wort in major depression," *Journal of the American Medical Association*, 285 (2001), pp. 1978–1986.

64. Based on a news item "Bee off with you," *Economist*, November 2, 2002, p. 78.

65. Jay Schreiber and Megan Thee, "Fans concerned about steroid use and believe it's widespread, poll shows," *New York Times*, March 31, 2008.

66. From the Electronic Encyclopedia of Statistical Examples and Exercises (EESEE) case study "Is Caffeine Dependence Real?"

### Addendum for Activity "Sampling sunflowers" (page 219)

**Sampling sunflowers**

| | A | B | C | D | E | F | G | H | I | J |
|---|---|---|---|---|---|---|---|---|---|---|
| **1** | 106 | 104 | 106 | 105 | 105 | 105 | 108 | 108 | 107 | 106 |
| **2** | 103 | 102 | 103 | 104 | 103 | 103 | 103 | 104 | 103 | 105 |
| **3** | 101 | 99 | 102 | 101 | 100 | 100 | 102 | 101 | 102 | 101 |
| **4** | 98 | 99 | 99 | 99 | 100 | 99 | 98 | 101 | 99 | 101 |
| **5** | 98 | 100 | 100 | 100 | 99 | 99 | 99 | 98 | 100 | 98 |
| **6** | 97 | 98 | 98 | 98 | 99 | 99 | 99 | 99 | 97 | 98 |
| **7** | 103 | 102 | 102 | 104 | 101 | 102 | 102 | 102 | 104 | 102 |
| **8** | 104 | 103 | 103 | 103 | 102 | 102 | 103 | 104 | 102 | 102 |
| **9** | 106 | 106 | 104 | 106 | 102 | 107 | 104 | 103 | 106 | 106 |
| **10** | 107 | 108 | 109 | 110 | 106 | 107 | 109 | 107 | 106 | 107 |

## Chapter 5

1. Note that pennies have rims that make spinning more stable. The probability of a head in spinning a coin depends on the type of coin and also on the surface.

2. R. Vallone and A. Tversky, "The hot hand in basketball: on the misperception of random sequences," *Cognitive Psychology*, 17 (1985), pp. 295–314.

3. Gur Yaari and Shmuel Eisenmann, The Hot (Invisible?) Hand: Can Time Sequence Patterns of Success/Failure in Sports Be Modeled as Repeated Random Independent Trials? (2011) PLoS ONE 6(10): e24532. doi:10.1371/journal.pone.0024532

4. The excerpts in Exercise 22 are from http://marilynvossavant.com/game-show-problem/.

5. From the Web site of the Gallup Organization, www.gallup.com. Individual poll reports remain on this site for only a limited time.

6. Information for the 1999–2000 academic year is from the 2003 *Statistical Abstract of the United States*, Table 286.

7. Data for 2006 from the Web site of Statistics Canada, www.statcan.ca.

8. Gail Burrill, "Two-way tables: introducing probability using real data," paper presented at the Mathematics Education into the Twenty-first Century Project, Czech Republic, September 2003. Burrill cites as her source H. Kranendonk, P. Hopfensperger, and R. Scheaffer, *Exploring Probability*, Dale Seymour Publications, 1999.

9. From the EESEE story "Is It Tough to Crawl in March?"

10. Pierre J. Meunier et al., "The effects of strontium ranelate on the risk of vertebral fracture in women with postmenopausal osteoporosis," *New England Journal of Medicine*, 350 (2004), pp. 459–468.

11. The table closely follows the grade distributions for these three schools at the University of New Hampshire in the fall of 2000, found in a self-study document at www.unh.edu/academic-affairs/neasc/. The counts of grades mirror the proportions of UNH students in these schools. The table is simplified to refer to a university with only these three schools.

12. Information about Internet users comes from sample surveys carried out by the Pew Internet and American Life Project, at www.pewinternet.org.

13. Data on Roger Federer's serve percentages from www.atpworldtour.com.

14. Thanks to Michael Legacy for providing these data.

15. This is one of several tests discussed in Bernard M. Branson, "Rapid HIV testing: 2005 update," a presentation by the Centers for Disease Control and Prevention, at www.cdc.gov. The Malawi clinic result is reported by Bernard M. Branson, "Point-of-care rapid tests for HIV antibody," *Journal of Laboratory Medicine*, 27 (2003), pp. 288–295.

16. The National Longitudinal Study of Adolescent Health interviewed a stratified random sample of 27,000 adolescents, then reinterviewed many of the subjects six years later, when most were aged 19 to 25. These data are from the Wave III reinterviews in 2000 and 2001, found at the Web site of the Carolina Population Center, www.cpc.unc.edu.

17. Information about Internet users comes from sample surveys carried out by the Pew Internet and American Life Project, found online at www.pewinternet.org. The music-downloading data were collected in 2003.

18. We got these data from the Energy Information Administration on their Web site at http://www.eia.gov/dnav/pet/pet_sum_mkt_dcu_sct_m.htm.

19. From the National Institutes of Health's National Digestive Diseases Information Clearinghouse, found at http://digestive.niddk.nih.gov/.

20. The probabilities given are realistic, according to the fundraising firm SCM Associates, at scmassoc.com.

21. Probabilities from trials with 2897 people known to be free of HIV antibodies and 673 people known to be infected are reported in J. Richard George, “Alternative specimen sources: methods for confirming positives,” 1998 Conference on the Laboratory Science of HIV, found online at the Centers for Disease Control and Prevention, www.cdc.gov.
22. Robert P. Dellavalle et al., “Going, going, gone: lost Internet references,” *Science*, 302 (2003), pp. 787–788.
23. Margaret A. McDowell et al., “Anthropometric reference data for children and adults: U.S. population, 1999–2002,” *National Center for Health Statistics, Advance Data from Vital and Health Statistics*, No. 361 (2005), at www.cdc.gov/nchs.
24. The General Social Survey exercises in this chapter present tables constructed using the search function at the GSS archive, sda.berkeley.edu/archive.htm.
25. National population estimates for July 1, 2006, at the Census Bureau Web site www.census.gov. The table omits people who consider themselves to belong to more than one race.
26. Data provided by Patricia Heithaus and the Department of Biology at Kenyon College.
27. Thanks to Tim Brown, The Lawrenceville School, for providing the idea for this exercise.

## Chapter 6

1. U.S. Supreme Court, Strauder v. West Virginia, 100 U.S. 303 (1879) https://supreme.justia.com/cases/federal/us/100/303/case.html.
2. U.S. Supreme Court, Berghuis v. Smith, Docket No. 08-1402 http://www.law.cornell.edu/supct/html/08-1402.ZO.html.
3. In most applications, X takes a finite number of possible values. The same ideas, implemented with more advanced mathematics, apply to random variables with an infinite but still countable collection of values.
4. The Apgar score data came from National Center for Health Statistics, *Monthly Vital Statistics Reports*, Vol. 30, No. 1, Supplement, May 6, 1981.
5. Information from www.ncsu.edu.
6. The mean of a continuous random variable X with density function $f(x)$ can be found by integration:

$$\mu_X = \int x f(x)\,dx$$

This integral is a kind of weighted average, analogous to the discrete-case mean

$$\mu_X = \sum x_i p_i$$

The variance of a continuous random variable X is the average squared deviation of the values of X from their mean, found by the integral

$$\sigma_X^2 = \int (x - \mu)^2 f(x)\,dx$$

7. You can find a mathematical explanation of Benford’s law in Ted Hill, “The first-digit phenomenon,” *American Scientist*, 86 (1996), pp. 358–363; and Ted Hill, “The difficulty of faking data,” *Chance*, 12, No. 3 (1999), pp. 27–31. Applications in fraud detection are discussed in the second paper by Hill and in Mark A. Nigrini, “I’ve got your number,” *Journal of Accountancy*, May 1999, available online at http://www.journalofaccountancy.com/issues/1999/may/nigrini.
8. The National Longitudinal Study of Adolescent Health interviewed a stratified random sample of 27,000 adolescents, then reinterviewed many of the subjects six years later, when most were aged 19 to 25. These data are from the Wave III reinterviews in 2000 and 2001, found at the Web site of the Carolina Population Center, www.cpc.unc.edu.
9. Data from the Census Bureau’s 1998 American Housing Survey.
10. Thomas K. Cureton et al., *Endurance of Young Men*, Monographs of the Society for Research in Child Development, Vol. 10, No. 1, 1945.
11. The survey question is reported in Trish Hall, “Shop? Many say ‘Only if I must,’” *New York Times*, November 28, 1990. In fact, 66% (1650 of 2500) in the sample said “Agree.”
12. Office of Technology Assessment, *Scientific Validity of Polygraph Testing: A Research Review and Evaluation*, Government Printing Office, 1983.
13. John Schwartz, “Leisure pursuits of today’s young men,” *New York Times*, March 29, 2004. The source cited is comScore Media Metrix.
14. Population base from the 2006 Labor Force Survey, at www.statistics.gov.uk. Smoking data from Action on Smoking and Health, *Smoking and Health Inequality*, at www.ash.org.uk.

## Chapter 7

1. Data on income and education from the March 2012 CPS supplement, obtained from the Census Bureau Web site.
2. This and similar results of Gallup polls are from the Gallup Organization Web site, www.gallup.com.
3. From a graph in Stan Boutin et al., “Anticipatory reproduction and population growth in seed predators,” *Science*, 314 (2006), pp. 1928–1930.
4. See Note 2.
5. Amanda Lenhart and Mary Madden, “Music downloading, file-sharing and copyright,” Pew Internet and American Life Project, 2003, at www.pewinternet.org.
6. This Activity is suggested in Richard L. Schaeffer, Ann Watkins, Mrudulla Gnanadesikan, and Jeffrey A. Witmer, *Activity-Based Statistics*, Springer, 1996.
7. See Note 1.
8. We found the information on birth weights of Norwegian children on the National Institute of Environmental Health Sciences Web site: The relevant article can be accessed here: http://www.ncbi.nlm.nih.gov/pubmed/1536353.

## Chapter 8

1. Thanks to Floyd Bullard for sharing the idea for this Activity.
2. This and similar results from the Pew Internet and American Life Project can be found at www.pewinternet.org.
3. Michele L. Head, “Examining college students’ ethical values,” Consumer Science and Retailing honors project, Purdue University, 2003.
4. Francisco L. Rivera-Batiz, “Quantitative literacy and the likelihood of employment among young adults,” *Journal of Human Resources*, 27 (1992), pp. 313–328.

5. This and similar results of Gallup polls are from the Gallup Organization Web site, www.gallup.com.

6. Amanda Lenhart and Mary Madden, "Teens, privacy and online social networks," Pew Internet and American Life Project, 2007, at www.pewinternet.org.

7. E. W. Campion, "Editorial: power lines, cancer, and fear," *New England Journal of Medicine*, 337, No. 1 (1997), pp. 44–46. The study report is M. S. Linet et al., "Residential exposure to magnetic fields and acute lymphoblastic leukemia in children," pp. 1–8 in the same issue. See also G. Taubes, "Magnetic field–cancer link: will it rest in peace?" *Science*, 277 (1997), p. 29.

8. Karl Pearson and A. Lee, "On the laws of inheritance in man," *Biometrika*, 2 (1902), p. 357. These data also appear in D. J. Hand et al., *A Handbook of Small Data Sets*, Chapman & Hall, 1994. This book offers more than 500 data sets that can be used in statistical exercises.

9. Data from the College Alcohol Study Web site, www.hsph.harvard.edu/cas/.

10. Linda Lyons, "Teens: sex can wait," December 14, 2004, from the Gallup Poll Web site, www.gallup.com.

11. Eric Sanford et al., "Local selection and latitudinal variation in a marine predator–prey interaction," *Science*, 300 (2003), pp. 1135–1137.

12. See Note 3.

13. Substance Abuse and Mental Health Services Administration, *Results from the 2011 National Survey on Drug Use and Health: Summary of National Findings*, NSDUH Series H-44, HHS Publication No. (SMA) 12-4713. Rockville, MD: Substance Abuse and Mental Health Services Administration, 2012.

14. Wayne J. Camera and Donald Powers, "Coaching and the SAT I," *TIP*, July 1999 (online journal at www.siop.org/tip).

15. "Poll: men, women at odds on sexual equality," Associated Press dispatch appearing in the *Lafayette (Ind.) Journal and Courier*, October 20, 1997.

16. Linda Lyons, "Most teens have in-room entertainment," February 22, 2005, www.gallup.com/poll/14989/Most-Teens-InRoom-Entertainment.aspx.

17. Based on information in "NCAA 2003 national study of collegiate sports wagering and associated health risks," which can be found on the NCAA Web site, www.ncaa.org.

18. T. J. Lorenzen, "Determining statistical characteristics of a vehicle emissions audit procedure," *Technometrics*, 22 (1980), pp. 483–493.

19. Data provided by Drina Iglesia, Purdue University. The data are part of a larger study reported in D. D. S. Iglesia, E. J. Cragoe, Jr., and J. W. Vanable, "Electric field strength and epithelization in the newt (*Notophthalmus viridescens*)," *Journal of Experimental Zoology*, 274 (1996), pp. 56–62.

20. Harry B. Meyers, "Investigations of the life history of the velvetleaf seed beetle, *Althaeus folkertsi* Kingsolver," MS thesis, Purdue University, 1996. The 95% $t$ interval is 1227.9 to 2507.6. A 95% bootstrap BCa interval is 1444 to 2718, confirming that $t$ inference is inaccurate for these data.

21. Lisa M. Baril et al., "Willow-bird relationships on Yellowstone's northern range," *Yellowstone Science*, 17, No. 3 (2009), pp. 19–26.

22. Helen E. Staal and D. C. Donderi, "The effect of sound on visual apparent movement," *American Journal of Psychology*, 96 (1983), pp. 95–105.

23. M. Ann Laskey et al., "Bone changes after 3 mo of lactation: influence of calcium intake, breast-milk output, and vitamin D–receptor genotype," *American Journal of Clinical Nutrition*, 67 (1998), pp. 685–692.

24. TUDA results for 2003 from the National Center for Education Statistics, at nces.ed.gov/nationsreportcard.

25. Chi-Fu Jeffrey Yang, Peter Gray, and Harrison G. Pope, Jr., "Male body image in Taiwan versus the West,"*American Journal of Psychiatry*, 162 (2005), pp. 263–269.

26. Alan S. Banks et al., "Juvenile hallux abducto valgus association with metatarsus adductus," *Journal of the American Podiatric Medical Association*, 84 (1994), pp. 219–224.

27. The vehicle is a 1997 Pontiac Transport Van owned by George P. McCabe.

28. These data are from "Results report on the vitamin C pilot program," prepared by SUSTAIN (Sharing United States Technology to Aid in the Improvement of Nutrition) for the U.S. Agency for International Development. The report was used by the Committee on International Nutrition of the National Academy of Sciences/Institute of Medicine (NAS/IOM) to make recommendations on whether or not the vitamin C content of food commodities used in U.S. food aid programs should be increased. The program was directed by Peter Ranum and Françoise Chomé.

29. R. D. Stichler, G. G. Richey, and J. Mandel, "Measurement of treadware of commercial tires," *Rubber Age*, 73, No. 2 (May 1953).

30. Data from Pennsylvania State University Stat 500 Applied Statistics online course, https://onlinecourses.science.psu.edu/stat500/.

31. Based on interviews in 2000 and 2001 by the National Longitudinal Study of Adolescent Health. Found at the Web site of the Carolina Population Center, www.cpc.unc.edu.

32. Simplified from Sanjay K. Dhar, Claudia Gonzalez-Vallejo, and Dilip Soman, "Modeling the effects of advertised price claims: tensile versus precise pricing," *Marketing Science*, 18 (1999), pp. 154–177.

33. From program 19, "Confidence Intervals," in the *Against All Odds* video series.

34. James T. Fleming, "The measurement of children's perception of difficulty in reading materials," *Research in the Teaching of English*, 1 (1967), pp. 136–156.

35. Bryan E. Porter and Thomas D. Berry, "A nationwide survey of self-reported red light running: measuring prevalence, predictors, and perceived consequences," *Accident Analysis and Prevention*, 33 (2001), pp. 735–741.


## Chapter 9


1. P. A. Mackowiak, S. S. Wasserman, and M. M. Levine, "A critical appraisal of 98.6 degrees F, the upper limit of the normal body temperature, and other legacies of Carl Reinhold August Wunderlich," *Journal of the American Medical Association*, 268 (1992), pp. 1578–1580.

2. Thanks to Josh Tabor for suggesting the idea for this example.

3. R. A. Fisher, "The arrangement of field experiments," *Journal of the Ministry of Agriculture of Great Britain*, 33 (1926), p. 504, quoted in Leonard J. Savage, "On rereading R. A. Fisher," *Annals of Statistics*, 4 (1976), p. 471. Fisher's work is described in a biography by his daughter: Joan Fisher Box, *R. A. Fisher: The Life of a Scientist*, Wiley, 1978.

4. For a discussion of statistical significance in the legal setting, see D. H. Kaye, "Is proof of statistical significance relevant?"

*Washington Law Review*, 61 (1986), pp. 1333–1365. Kaye argues: "Presenting the *P*-value without characterizing the evidence by a significance test is a step in the right direction. Interval estimation, in turn, is an improvement over *P*-values."

5. Julie Ray, "Few teens clash with friends," May 3, 2005, on the Gallup Organization Web site, www.gallup.com.
6. The idea for this exercise was provided by Michael Legacy and Susan McGann.
7. Projections from the 2011 *Digest of Education Statistics*, found online at nces.ed.gov.
8. From the report "Sex and tech: results from a study of teens and young adults," published by the National Campaign to Prevent Teen and Unplanned Pregnancy, www.thenationalcampaign.org/sextech.
9. National Institute for Occupational Safety and Health, *Stress at Work*, 2000, available online at http://www.cdc.gov/niosh/docs/99-101/. Results of this survey were reported in *Restaurant Business*, September 15, 1999, pp. 45–49.
10. See Note 5.
11. Matthew A. Carlton and William D. Stansfield, "Making babies by the flip of a coin?" *The American Statistician*, 59 (2005), pp. 180–182.
12. Dorothy Espelage et al., "Factors associated with bullying behavior in middle school students," *Journal of Early Adolescence*, 19, No. 3 (August 1999), pp. 341–362.
13. Aaron Smith and Joanna Brenner, "Twitter Use 2012," published by the Pew Internet and American Life Project, May 31, 2012.
14. This and similar results of Gallup polls are from the Gallup Organization Web site, www.gallup.com.
15. Linda Lyons, "Teens: sex can wait," December 14, 2004, from the Gallup Organization Web site, www.gallup.com.
16. Michele L. Head, "Examining college students' ethical values," Consumer Science and Retailing honors project, Purdue University, 2003.
17. Based on Stephen A. Woodbury and Robert G. Spiegelman, "Bonuses to workers and employers to reduce unemployment: randomized trials in Illinois," *American Economic Review*, 77 (1987), pp. 513–530.
18. E. C. Strain et al., "Caffeine dependence syndrome: evidence from case histories and experimental evaluation," *Journal of the American Medical Association*, 272 (1994), pp. 1604–1607.
19. Warren E. Leary, "Cell phones: questions but no answers," *New York Times*, October 26, 1999.
20. R. A. Berner and G. P. Landis, "Gas bubbles in fossil amber as possible indicators of the major gas composition of ancient air," *Science*, 239 (1988), pp. 1406–1409. The 95% $t$ confidence interval is 54.78 to 64.40. A bootstrap BCa interval is 55.03 to 62.63. So $t$ is reasonably accurate despite the skew and the small sample.
21. This exercise is based on events that are real. The data and details have been altered to protect the privacy of the individuals involved.
22. From the NIST Web site, www.nist.gov/srd.
23. Data provided by Timothy Sturm.
24. A. R. Hirsch and L. H. Johnston, "Odors and learning," *Journal of Neurological and Orthopedic Medicine and Surgery*, 17 (1996), pp. 119–126.
25. From the story "Friday the 13th," at the Data and Story Library, lib.stat.cmu.edu/DASL.
26. From the Gallup Organization Web site, www.gallup.com. The poll was taken in December 2006.
27. W. S. Gosset, "The probable error of a mean," *Biometrika*, 6 (1908), 1–25.

## Chapter 10

1. This Case Study is based on the story "Drive-Thru Competition" in the Electronic Encyclopedia of Statistical Examples and Exercises (EESEE). Updated data were obtained from the *QSR* magazine Web site: www.qsrmagazine.com/drive-thru.
2. C. P. Cannon et al., "Intensive versus moderate lipid lowering with statins after acute coronary syndromes," *New England Journal of Medicine*, 350 (2004), pp. 1495–1504.
3. This Activity is based on an experiment conducted on the March 9, 2005, episode of the TV show *MythBusters*. You may be able to find a video clip from this program at the Discovery Channel Web site, www.dsc.discovery.com.
4. The study is reported in William Celis III, "Study suggests Head Start helps beyond school," *New York Times*, April 20, 1993. See www.highscope.org.
5. The idea for this exercise was inspired by an example in David M. Lane's *Hyperstat Online* textbook at http://davidmlane.com/hyperstat.
6. Modified from Richard A. Schieber et al., "Risk factors for injuries from in-line skating and the effectiveness of safety gear," *New England Journal of Medicine*, 335 (1996). Internet summary at www.nejm.org.
7. Francisco Lloret et al., "Fire and resprouting in Mediterranean ecosystems: insights from an external biogeographical region, the Mexican shrubland," *American Journal of Botany*, 88 (1999), pp. 1655–1661.
8. Saiyad S. Ahmed, "Effects of microwave drying on checking and mechanical strength of low-moisture baked products," MS thesis, Purdue University, 1994.
9. Aaron Smith and Joanna Brenner, "Twitter Use 2012," Pew Internet and American Life Project, May 31, 2012.
10. From the Web site of the Black Youth Project, blackyouthproject.uchicago.edu.
11. The National Longitudinal Study of Adolescent Health interviewed a stratified random sample of 27,000 adolescents, then reinterviewed many of the subjects six years later, when most were aged 19 to 25. These data are from the Wave III reinterviews in 2000 and 2001, found at the Web site of the Carolina Population Center, www.cpc.unc.edu.
12. Janice Joseph, "Fear of crime among black elderly," *Journal of Black Studies*, 27 (1997), pp. 698–717.
13. National Athletic Trainers Association, press release dated, September 30, 1994.
14. Based on Deborah Roedder John and Ramnath Lakshmi-Ratan, "Age differences in children's choice behavior: the impact of available alternatives," *Journal of Marketing Research*, 29 (1992), pp. 216–226.
15. Louie E. Ross, "Mate selection preferences among African American college students," *Journal of Black Studies*, 27 (1997), pp. 554–569.
16. Martin Enserink, "Fraud and ethics charges hit stroke drug trial," *Science*, 274 (1996), pp. 2004–2005.

17. Kwang Y. Cha, Daniel P. Wirth, and Rogerio A. Lobo, "Does prayer influence the success of *in vitro* fertilization–embryo transfer?" *Journal of Reproductive Medicine*, 46 (2001), pp. 781–787.

18. W. E. Paulus et al., "Influence of acupuncture on the pregnancy rate in patients who undergo assisted reproductive therapy," *Fertility and Sterility*, 77, No. 4 (2002), pp. 721–724.

19. Sapna Aneja, "Biodeterioration of textile fibers in soil," MS thesis, Purdue University, 1994.

20. We obtained the National Health and Nutrition Examination Survey data from the Centers for Disease Control and Prevention Web site at www.cdc.gov/nchs/nhanes.htm.

21. Detailed information about the conservative *t* procedures can be found in Paul Leaverton and John J. Birch, "Small sample power curves for the two sample location problem," *Technometrics*, 11 (1969), pp. 299–307; Henry Scheffe, "Practical solutions of the Behrens-Fisher problem," *Journal of the American Statistical Association*, 65 (1970), pp. 1501–1508; and D. J. Best and J. C. W. Rayner, "Welch's approximate solution for the Behrens-Fisher problem," *Technometrics*, 29 (1987), pp. 205–210.

22. Data for this example from Noel Cressie, *Statistics for Spatial Data*, Wiley, 1993.

23. Based on R. G. Hood, "Results of the prices received by farmers for grain quality assurance project," U.S. Department of Agriculture Report SRB-95-07, 1995.

24. This study is reported in Roseann M. Lyle et al., "Blood pressure and metabolic effects of calcium supplementation in normotensive white and black men," *Journal of the American Medical Association*, 257 (1987), pp. 1772–1776. The data were provided by Dr. Lyle.

25. See Note 19.

26. The idea for this example was provided by Robert Hayden.

27. United Nations data on literacy were found at www.earthtrends.wri.org.

28. Shailija V. Nigdikar et al., "Consumption of red wine polyphenols reduces the susceptibility of low-density lipoproteins to oxidation in vivo," *American Journal of Clinical Nutrition*, 68 (1998), pp. 258–265.

29. Ethan J. Temeles and W. John Kress, "Adaptation in a plant–hummingbird association," *Science*, 300 (2003), pp. 630–633. We thank Ethan J. Temeles for providing the data.

30. Data for 1982, provided by Marvin Schlatter, Division of Financial Aid, Purdue University.

31. Gabriela S. Castellani, "The effect of cultural values on Hispanics' expectations about service quality," MS thesis, Purdue University, 2000.

32. From a graph in Fabrizio Grieco, Arie J. van Noordwijk, and Marcel E. Visser, "Evidence for the effect of learning on timing of reproduction in blue tits," *Science*, 296 (2002), pp. 136–138.

33. This example is loosely based on D. L. Shankland, "Involvement of spinal cord and peripheral nerves in DDT poisoning syndrome in albino rats," *Toxicology and Applied Pharmacology*, 6 (1964), pp. 197–213.

34. Based on W. N. Nelson and C. J. Widule, "Kinematic analysis and efficiency estimate of intercollegiate female rowers," unpublished manuscript, 1983.

35. Adapted from Maribeth Cassidy Schmitt, "The effects of an elaborated directed reading activity on the metacomprehension skills of third graders," PhD dissertation, Purdue University, 1987.

36. M. Ann Laskey et al., "Bone changes after 3 mo of lactation: influence of calcium intake, breast-milk output, and vitamin D–receptor genotype," *American Journal of Clinical Nutrition*, 67 (1998), pp. 685–692.

37. See Note 24 in Chapter 1.

38. The data for this exercise came from Rossman, Cobb, Chance, and Holcomb's National Science Foundation project shared at JMM 2008 in San Diego. Their original source was Robert Stickgold, LaTanya James, and J. Allan Hobson, "Visual discrimination learning requires sleep after training," *Nature Neuroscience*, 3 (2000), pp. 1237–1238.

39. The idea for this exercise was provided by Robert Hayden.

40. Pew Internet Project, "Counting on the Internet," December 29, 2002, www.pewinternet.org.

41. Wayne J. Camera and Donald Powers, "Coaching and the SAT I," *TIP* (online journal at www.siop.org/tip), July 1999.

42. See Note 40.

43. Thanks to Larry Green, Lake Tahoe Community College, for allowing us to use some of the contexts from his "Classifying Statistics Problems" Web site, www.ltcconline.net/greenl/java/Statistics/catStatProb/categorizingStatProblems12.html.

44. JoAnn K. Wells, Allan F. Williams, and Charles M. Farmer, "Seat belt use among African Americans, Hispanics, and whites," *Accident Analysis and Prevention*, 34 (2002), pp. 523–529.

45. Based on Amna Kirmani and Peter Wright, "Money talks: perceived advertising expense and expected product quality," *Journal of Consumer Research*, 16 (1989), pp. 344–353.

46. Francisco L. Rivera-Batiz, "Quantitative literacy and the likelihood of employment among young adults," *Journal of Human Resources*, 27 (1992), pp. 313–328.

47. Clive G. Jones et al., "Chain reactions linking acorns to gypsy moth outbreaks and Lyme disease risk," *Science*, 279 (1998), pp. 1023–1026.

48. Data provided by Warren Page, New York City Technical College, from a study done by John Hudesman.

49. Maureen Hack et al., "Outcomes in young adulthood for very-low-birth-weight infants," *New England Journal of Medicine*, 346 (2002), pp. 149–157. Exercise AP3.32 is simplified, in that the measures reported in this paper have been statistically adjusted for "sociodemographic status."

## Chapter 11

1. M. M. Roy and N. J. S. Christenfeld, "Do dogs resemble their owners?" *Psychological Science*, 15, No. 5 (May 2004), pp. 361–363.

2. R. W. Mannan and E. C. Meslow, "Bird populations and vegetation characteristics in managed and old-growth forests, northwestern Oregon," *Journal of Wildlife Management*, 48 (1984), pp. 1219–1238.

3. You can find a mathematical explanation of Benford's law in Ted Hill, "The first-digit phenomenon," *American Scientist*, 86 (1996), pp. 358–363; and Ted Hill, "The difficulty of faking data," *Chance*, 12, No. 3 (1999), pp. 27–31. Applications in fraud detection are discussed in the second paper by Hill and in Mark A. Nigrini, "I've got your number," *Journal of Accountancy*, May 1999, available online at http://www.journalofaccountancy.com/issues/1999/may/nigrini.

4. The idea for this exercise came from a post to the AP Statistics electronic discussion group by Joshua Zucker.

5. The context of this example was inspired by C. M. Ryan et al., "The effect of in-store music on consumer choice of wine," *Proceedings of the Nutrition Society*, 57 (1998), p. 1069A.

6. Pennsylvania State University Division of Student Affairs, "Net behaviors, November 2006," *Penn State Pulse*, at http://studentaffairs.psu.edu/.

7. Pew Research Center for the People and the Press, "The cell phone challenge to survey research," news release for May 15, 2006, at www.people-press.org.

8. Janice E. Williams et al., "Anger proneness predicts coronary heart disease risk," *Circulation*, 101 (2000), pp. 63–95.

9. P. Azoulay and S. Shane, "Entrepreneurs, contracts, and the failure of young firms," *Management Science*, 47 (2001), pp. 337–358.

10. J. Cantor, "Long-term memories of frightening media often include lingering trauma symptoms," poster paper presented at the Association for Psychological Science Convention, New York, May 26, 2006.

11. William D. Darley, "Store-choice behavior for preowned merchandise," *Journal of Business Research*, 27 (1993), pp. 17–31.

12. The National Longitudinal Study of Adolescent Health interviewed a stratified random sample of 27,000 adolescents, then reinterviewed many of the subjects six years later, when most were aged 19 to 25. These data are from the Wave III reinterviews in 2000 and 2001, found at www.cpc.unc.edu.

13. Joan L. Duda, "The relationship between goal perspectives, persistence, and behavioral intensity among male and female recreational sports participants," *Leisure Sciences*, 10 (1988), pp. 95–106.

14. Data compiled from a table of percents in "Americans view higher education as key to the American dream," press release by the National Center for Public Policy and Higher Education, www.highereducation.org, May 3, 2000.

15. R. Shine et al., "The influence of nest temperatures and maternal brooding on hatchling phenotypes in water pythons," *Ecology*, 78 (1997), pp. 1713–1721.

16. D. M. Barnes, "Breaking the cycle of addiction," *Science*, 241 (1988), pp. 1029–1030.

17. U.S. Department of Commerce, Office of Travel and Tourism Industries, in-flight survey, 2007, at http://tinet.ita.doc.gov.

18. The idea for this exercise came from Bob Hayden.

19. Douglas E. Jorenby et al., "A controlled trial of sustained-release bupropion, a nicotine patch, or both for smoking cessation," *New England Journal of Medicine*, 340 (1990), pp. 685–691.

20. Martin Enserink, "Fraud and ethics charges hit stroke drug trial," *Science*, 274 (1996), pp. 2004–2005.

21. Lien-Ti Bei, "Consumers' purchase behavior toward recycled products: an acquisition-transaction utility theory perspective," MS thesis, Purdue University, 1993.

22. All General Social Survey exercises in this chapter present tables constructed using the search function at the GSS archive, http://sda.berkeley.edu/archive.htm.

23. Based closely on Susan B. Sorenson, "Regulating firearms as a consumer product," *Science*, 286 (1999), pp. 1481–1482. Because the results in the paper were "weighted to the U.S. population," we have changed some counts slightly for consistency.

24. Data produced by Ries and Smith, found in William D. Johnson and Gary G. Koch, "A note on the weighted least squares analysis of the Ries-Smith contingency table data," *Technometrics*, 13 (1971), pp. 438–447.

25. See Note 12.

26. Mei-Hui Chen, "An exploratory comparison of American and Asian consumers' catalog patronage behavior," MS thesis, Purdue University, 1994.

27. Lillian Lin Miao, "Gastric freezing: an example of the evaluation of medical therapy by randomized clinical trials," in John P. Bunker, Benjamin A. Barnes, and Frederick Mosteller (eds.), *Costs, Risks, and Benefits of Surgery*, Oxford University Press, 1977, pp. 198–211.

28. See Note 12.

29. Thanks to Larry Green, Lake Tahoe Community College, for giving us permission to use several of the contexts from his Web site at www.ltcconline.net/greenl/java/Statistics/catStatProb/categorizingStatProblems12.html.

30. See Note 29.

31. Mark A. Sabbagh and Dare A. Baldwin, "Learning words from knowledgeable versus ignorant speakers: links between preschoolers' theory of mind and semantic development," *Child Development*, 72 (2001), pp. 1054–1070. Many statistical software packages offer "exact tests" that are valid even when there are small expected counts.

32. Brenda C. Coleman, "Study: heart attack risk cut 74% by stress management," Associated Press dispatch appearing in the *Lafayette (Ind.) Journal and Courier*, October 20, 1997.

33. Tom Reichert, "The prevalence of sexual imagery in ads targeted to young adults," *Journal of Consumer Affairs*, 37 (2003), pp. 403–412.

34. Based on the EESEE story "What Makes Pre-teens Popular?"

35. Based on the EESEE story "Domestic Violence."

36. Karine Marangon et al., "Diet, antioxidant status, and smoking habits in French men," *American Journal of Clinical Nutrition*, 67 (1998), pp. 231–239.

## Chapter 12

1. Data obtained from the PGA Tour Web site, www.pgatour.com, by Kevin Stevick, The Lawrenceville School.

2. The idea for this Activity came from Gloria Barrett, Floyd Bullard, and Dan Teague at the North Carolina School of Science and Math.

3. See Note 15 from Chapter 3.

4. Data from a plot in James A. Levine, Norman L. Eberhart, and Michael D. Jensen, "Role of non-exercise activity thermogenesis in resistance to fat gains in humans," *Science*, 283 (1999), pp. 212–214.

5. Samuel Karelitz et al., "Relation of crying activity in early infancy to speech and intellectual development at age three years," *Child Development*, 35 (1964), pp. 769–777.

6. Data from National Institute of Standards and Technology, *Engineering Statistics Handbook*, www.itl.nist.gov/div898/handbook. The analysis there does not comment on the bias of field measurements.

7. Todd W. Anderson, "Predator responses, prey refuges, and density-dependent mortality of a marine fish," *Ecology*, 81 (2001), pp. 245–257.

8. Based on a plot in G. D. Martinsen, E. M. Driebe, and T. G. Whitham, "Indirect interactions mediated by changing plant chemistry: beaver browsing benefits beetles," *Ecology*, 79 (1998), pp. 192–200.

9. Data provided by Samuel Phillips, Purdue University.

10. Based on Marion E. Dunshee, "A study of factors affecting the amount and kind of food eaten by nursery school children," *Child Development*, 2 (1931), pp. 163–183. This article gives the means,

standard deviations, and correlation for 37 children but does not give the actual data.

11. M. H. Criqui, University of California, San Diego, reported in the *New York Times*, December 28, 1994.

12. See Note 29 in Chapter 8.

13. Data for the building at 1800 Ben Franklin Drive, Sarasota, Florida, starting in March 2003. From the Web site of the Sarasota County Property Appraiser, http://www.sc-pa.com.

14. Information about the sources used to obtain the data can be found under "Documentation" at the Gapminder Web site, www.gapminder.org.

15. Gordon L. Swartzman and Stephen P. Kaluzny, *Ecological Simulation Primer*, Macmillan, 1987, p. 98.

16. From the Web site, http://en.wikipedia.org/wiki/Transistor_count.

17. Planetary data from http://hyperphysics.phy-astr.gsu.edu/hbase/solar/soldata2.html.

18. Sample Pennsylvania female rates provided by Life Quotes, Inc., in *USA Today*, December 20, 2004.

19. G. A. Sacher and E. F. Staffelt, "Relation of gestation time to brain weight for placental mammals: implications for the theory of vertebrate growth," *American Naturalist*, 108 (1974), pp. 593–613. We found these data in Fred L. Ramsey and Daniel W. Schafer, *The Statistical Sleuth: A Course in Methods of Data Analysis*, Duxbury, 1997.

20. Jérôme Chave, Bernard Riéra, and Marc-A. Dubois, "Estimation of biomass in a neotropical forest of French Guiana: spatial and temporal variability," *Journal of Tropical Ecology*, 17 (2001), pp. 79–96.

21. S. Chatterjee and B. Price, *Regression Analysis by Example*, Wiley, 1977.

22. Chris Carbone and John L. Gittleman, "A common rule for the scaling of carnivore density," *Science*, 295 (2002), pp. 2273–2276.

23. Data originally from A. J. Clark, *Comparative Physiology of the Heart*, Macmillan, 1927, p. 84. Obtained from Frank R. Giordano and Maurice D. Weir, *A First Course in Mathematical Modeling*, Brooks/Cole, 1985, p. 56.

24. Stillman Drake, *Galileo at Work*, University of Chicago Press, 1978. We found these data in D. A. Dickey and J. T. Arnold, "Teaching statistics with data of historic significance," *Journal of Statistics Education*, 3 (1995), www.amstat.org/publications/jse/.

25. Tattoo survey data from the Harris Poll®, No. 15, www.harrisinteractive.com, February 12, 2008.

26. Peter H. Chen, Neftali Herrera, and Darren Christiansen, "Relationships between gate velocity and casting features among aluminum round castings," no date. Provided by Darren Christiansen.

27. Data provided by Corinne Lim, Purdue University, from a student project supervised by Professor Joseph Vanable.

28. Data from many studies compiled in D. F. Greene and E. A. Johnson, "Estimating the mean annual seed production of trees," *Ecology*, 75 (1994), pp. 642–647.

# Appendix A About the AP® Exam

## Chapter 1

• If you learn to distinguish categorical from quantitative variables now, it will pay big rewards later. You will be expected to analyze categorical and quantitative variables correctly on the AP® exam.

• When comparing distributions of quantitative data, it's not enough just to list values for the center and spread of each distribution. You have to explicitly *compare* these values, using words like "greater than," "less than," or "about the same as."

• If you're asked to make a graph on a free-response question, be sure to label and scale your axes. Unless your calculator shows labels and scaling, don't just transfer a calculator screen shot to your paper.

• You may be asked to determine whether a quantitative data set has any outliers. Be prepared to state and use the rule for identifying outliers.

• Use statistical terms carefully and correctly on the AP® exam. Don't say "mean" if you really mean "median." Range is a single number; so are $Q_1$, $Q_3$, and $IQR$. Avoid colloquial use of language such as "the outlier *skews* the mean." Skewed is a shape. If you misuse a term, expect to lose some credit.

## Chapter 2

• Normal probability plots are not included on the AP® Statistics topic outline. However, these graphs are very useful for assessing Normality. You may use them on the AP® exam if you wish—just be sure that you know what you're looking for (a linear pattern).

## Chapter 3

• If you are asked to make a scatterplot for a free-response question, be sure to label and scale both axes. Don't just copy an unlabeled calculator graph directly onto your paper.

• If you're asked to interpret a correlation, start by looking at a scatterplot of the data. Then be sure to address direction, form, strength, and outliers (sound familiar?) and put your answer in context.

• When displaying the equation of a least-squares regression line, the calculator will report the slope and intercept with much more precision than is needed. However, there is no firm rule for how many decimal places to show for answers on the AP® exam. Our advice: Decide how much to round based on the context of the problem you are working on.

• Students often have a hard time interpreting the value of $r^2$ on AP® exam questions. They frequently leave out key words in the definition. Our advice: Treat this as a fill-in-the-blank exercise. Write "________% of the variation in [response variable name] is accounted for by the linear model relating [response variable name] to [explanatory variable name]."

• The formula sheet for the AP® exam uses different notation for these equations: $b_1 = r\frac{s_y}{s_x}$ and $b_0 = \bar{y} - b_1\bar{x}$. That's because the least-squares line is written as $\hat{y} = b_0 + b_1x$. We prefer our simpler versions without the subscripts!

## Chapter 4

• If you're asked to describe how the design of a study leads to bias, you're expected to do two things: (1) identify a problem with the design, and (2) explain how this problem would lead to an underestimate or overestimate. Suppose you were asked, "Explain how using your statistics class as a sample to estimate the proportion of all high school students who own a graphing calculator could result in bias." You might respond, "This is a convenience sample. It would probably include a much higher proportion of students with a graphing calculator than would the population at large because a graphing calculator is required for the statistics class. So this method would probably lead to an overestimate of the actual population proportion."

• If you are asked to identify a possible confounding variable in a given setting, you are expected to explain how the variable you choose (1) is associated with the explanatory variable and (2) affects the response variable.

• If you are asked to describe the design of an experiment on the AP® exam, you won't get full credit for a diagram like Figure 4.5 (page 246). You are expected to describe how the treatments are assigned to the experimental units and to clearly state what will be measured or compared. Some students prefer to start with a diagram and then add a few sentences. Others choose to skip the diagram and put their entire response in narrative form.

• Don't mix the language of experiments and the language of sample surveys or other observational studies. You will lose credit for saying things like "Use a randomized block design to select the sample for this survey" or "This experiment suffers from nonresponse since some subjects dropped out during the study."

## Chapter 5

• On the AP® exam, you may be asked to describe how you will perform a simulation using rows of random digits. If so, provide a clear enough description of your simulation process for the reader to get the same results you did from *only* your written explanation.

• Many probability problems involve simple computations that you can do on your calculator. It may be tempting to write down just your final answer without showing the supporting work. Don't do it! A "naked answer," even if it's correct, will usually earn you no credit on a free-response question.

• You can write statements like $P(B \mid A)$ if events A and B are defined clearly, or you can use a verbal equivalent, such as P(reads *New York Times* | reads *USA Today*). Use the approach that makes the most sense to you.

## Chapter 6

• If the mean of a random variable has a noninteger value but you report it as an integer, your answer will not get full credit.

• When showing your work on a free-response question, you must include more than a calculator command. Writing `normalcdf(68,70,64,2.7)` will *not* earn you full credit for a Normal calculation. At a minimum, you must indicate what each of those calculator inputs represents. Better yet, sketch and label a Normal curve to show what you're finding.

• Don't rely on "calculator speak" when showing your work on free-response questions. Writing `binompdf(5,0.25,3)` = 0.08789 will not earn you full credit for a binomial probability calculation. At the very least, you must indicate what each calculator input represents. For example, "I used `binompdf(trials 5,p:0.25, x value:3)`."

## Chapter 7

• Terminology matters. Don't say "sample distribution" when you mean *sampling distribution*. You will lose credit on free-response questions for misusing statistical terms.

• Notation matters. The symbols $\hat{p}$, $\bar{x}$, $p$, $\mu$, $\sigma$, $\mu_{\hat{p}}$, $\sigma_{\hat{p}}$, $\mu_{\bar{x}}$, and $\sigma_{\bar{x}}$ all have specific and different meanings. Either use notation correctly—or don't use it at all. You can expect to lose credit if you use incorrect notation.

## Chapter 8

• On a given problem, you may be asked to interpret the confidence interval, the confidence level, or both. Be sure you understand the difference: the confidence interval gives a set of plausible values for the parameter and the confidence level describes the long-run capture rate of the method.

• If a free-response question asks you to construct and interpret a confidence interval, you are expected to do the entire four-step process. That includes clearly defining the parameter, identifying the procedure, and checking conditions.

• You may use your calculator to compute a confidence interval on the AP® exam. But there's a risk involved. If you give just the calculator answer with no work, you'll get either full credit for the "Do" step (if the interval is correct) or no credit (if it's wrong). We recommend showing the calculation with the appropriate formula and then checking with your calculator. If you opt for the calculator-only method, be sure to name the procedure (e.g., one-proportion $z$ interval) and to give the interval (e.g., 0.514 to 0.607).

• If a question of the AP® exam asks you to calculate a confidence interval, all the conditions should be met. However, you are still required to state the conditions and show evidence that they are met.

• It is not enough just to make a graph of the data on your calculator when assessing Normality. You must *sketch* the graph on your paper to receive credit. You don't have to draw multiple graphs—any appropriate graph will do.

## Chapter 9

• The conclusion to a significance test should always include three components: (1) an explicit comparison of the P-value to a stated significance level, (2) a decision about the null hypothesis: reject or fail to reject $H_0$, and (3) a statement in the context of the problem about whether or not there is convincing evidence for $H_a$.

• When a significance test leads to a fail to reject $H_0$ decision, be sure to interpret the results as "We don't have enough evidence to conclude $H_a$." Saying anything that sounds like you believe $H_0$ is (or might be) true will lead to a loss of credit. And don't write text-message-type responses, like "FTR the $H_0$."

• You can use your calculator to carry out the mechanics of a significance test on the AP® exam. But there's a risk involved. If you give just the calculator answer with no work, and one or more of your values are incorrect, you will probably get no credit for the "Do" step. We recommend doing the calculation with the appropriate formula and then checking with your calculator. If you opt for the calculator-only method, be sure to name the procedure (one-proportion $z$ test) and to report the test statistic ($z = 1.15$) and P-value (0.1243).

• It is not enough just to make a graph of the data on your calculator when assessing Normality. You must *sketch* the graph on your paper to receive credit. You don't have to draw multiple graphs—any appropriate graph will do.

• Remember: If you give just calculator results with no work and one or more values are wrong, you probably won't get any credit for the "Do" step. If you opt for the calculator-only method, name the procedure ($t$ test) and report the test statistic ($t = -0.94$), degrees of freedom (df = 14), and P-value (0.1809).

## Chapter 10

• The formula for the two-sample $z$ interval for $p_1 - p_2$ often leads to calculation errors by students. As a result, we recommend using the calculator's `2-PropZInt` feature to compute the confidence interval on the AP® exam. Be sure to name the procedure (two-proportion $z$ interval) and to give the interval (0.076, 0.143) as part of the "Do" step.

• The formula for the two-sample $z$ statistic for a test about $p_1 - p_2$ often leads to calculation errors by students. As a result, we recommend using the calculator's `2-PropZTest` feature to perform calculations on the AP® exam. Be sure to name the procedure (two-proportion $z$ test) and to report the test statistic ($z = 1.17$) and P-value (0.2427) as part of the "Do" step.

• The formula for the two-sample $t$ interval for $\mu_1 - \mu_2$ often leads to calculation errors by students. As a result, we recommend using the calculator's `2-SampTInt` feature to compute the confidence interval on the AP® exam. Be sure to name the procedure (two-sample $t$ interval) and to give the interval (3.9362, 17.724) and df (55.728) as part of the "Do" step.

• When checking the Normal condition on an AP® exam question involving inference about means, be sure to include graphs. Don't expect to receive credit for describing a graph that you made on your calculator but didn't put on paper.

• The formula for the two-sample $t$ statistic for $\mu_1 - \mu_2$ often leads to calculation errors by students. As a result, we recommend using the calculator's `2-SampTTest` feature to perform calculations on the AP® exam. Be sure to name the procedure (two-sample $t$ test) and to report the test statistic ($t = 1.60$), $P$-value (0.0644), and df (15.59) as part of the "Do" step.

## Chapter 11

• You can use your calculator to carry out the mechanics of a significance test on the AP® exam. But there's a risk involved. If you give just the calculator answer with no work, and one or more of your values is incorrect, you will probably get no credit for the "Do" step. We recommend writing out the first few terms of the chi-square calculation followed by ". . .". This approach might help you earn partial credit if you enter a number incorrectly. Be sure to name the procedure ($\chi^2$ GOF-Test) and to report the test statistic ($\chi^2 = 11.2$), degrees of freedom (df = 3), and $P$-value (0.011).

• In the "Do" step, you aren't required to show every term in the chi-square statistic. Writing the first few terms of the sum followed by ". . ." is considered as "showing work." We suggest that you do this and then let your calculator tackle the computations.

• You can use your calculator to carry out the mechanics of a significance test on the AP® exam. But there's a risk involved. If you give just the calculator answer with no work and one or more of your values is incorrect, you will probably get no credit for the "Do" step. We recommend writing out the first few terms of the chi-square calculation followed by ". . .". This approach might help you earn partial credit if you enter a number incorrectly. Be sure to name the procedure ($\chi^2$-Test for homogeneity) and to report the test statistic ($\chi^2 = 18.279$), degrees of freedom (df = 4), and $P$-value (0.0011).

• If you have trouble distinguishing the two types of chi-square tests for two-way tables, you're better off just saying "chi-square test" than choosing the wrong type. Better yet, learn to tell the difference!

## Chapter 12

• The AP® exam formula sheet gives $\hat{y} = b_0 + b_1x$ for the equation of the sample (estimated) regression line. We will stick with our simpler notation, $\hat{y} = a + bx$, which is also used by TI calculators. Just remember: The coefficient of $x$ is always the slope, no matter what symbol is used.

• The AP® exam formula sheet gives the formula for the standard error of the slope as

$$s_{b_1} = \frac{\sqrt{\dfrac{\sum(y_i - \hat{y}_i)^2}{n-2}}}{\sqrt{\sum(x_i - \bar{x})^2}}$$

The numerator is just a fancy way of writing the standard deviation of the residuals $s$. Can you show that the denominator of this formula is the same as ours?

• The formula for the $t$ interval for the slope of a population (true) regression line often leads to calculation errors by students. As a result, we recommend using the calculator's `LinRegTInt` feature to compute the confidence interval on the AP® exam. Be sure to name the procedure ($t$ interval for slope) and to give the interval (−0.217, −0.108) and df (14) as part of the "Do" step.

• When you see a list of data values on an exam question, don't just start typing the data into your calculator. Read the question first. Often, additional information is provided that makes it unnecessary for you to enter the data at all. This can save you valuable time on the AP® exam.

• The formula for the test statistic in a $t$ test for the slope of a population (true) regression line often leads to calculation errors by students. As a result, we recommend using the calculator's `LinRegTTest` feature to perform calculations on the AP® exam. Be sure to name the procedure ($t$ test for slope) and to report the test statistic ($t = 3.065$), $P$-value (0.002), and df (36) as part of the "Do" step.

# Appendix B
## TI-Nspire™ Technology Corners

Texas Instruments released the new TI-Nspire CX in March 2011. The new handheld no longer has an interchangeable TI-84 faceplate; however, the body of the Nspire CX is much slimmer and its display is in full color. When you click ctrl menu and arrow down to *Color*, there are several options available: *Line Color*, *Fill Color*, and *Text Color*. If you choose the *Fill Color* option, a color palette will appear. Then you can select the colors you want for your graphs. This feature is quite useful when displaying multiple graphs. When creating a bar graph, the CX even allows you to change the color of each bar!

The keystrokes used for the new CX are the same as for the TI-Nspire Touchpad. The keystrokes for the older Nspire "clickpad" are still different in some ways; therefore they are still shown in parentheses when needed.

Start by updating your device's OS to ensure that your handheld has full capabilities. Go to education.ti.com and search under *Downloads* → *Software, Apps, Operating Systems*… to download the latest version of the OS. If you have the TI-Nspire computer link software, you should be asked automatically to update your handheld's OS.

TOUCHPAD

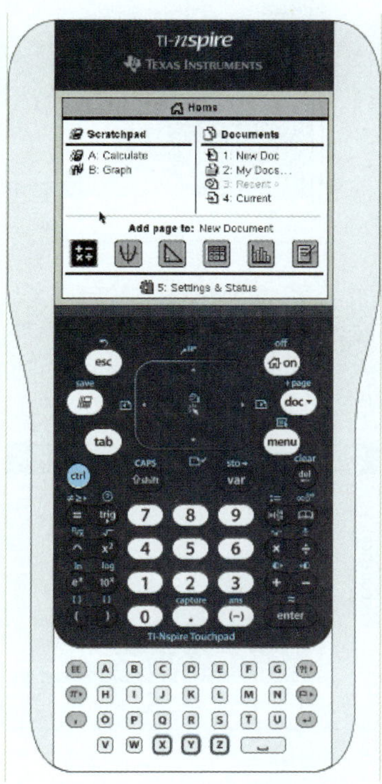


CX

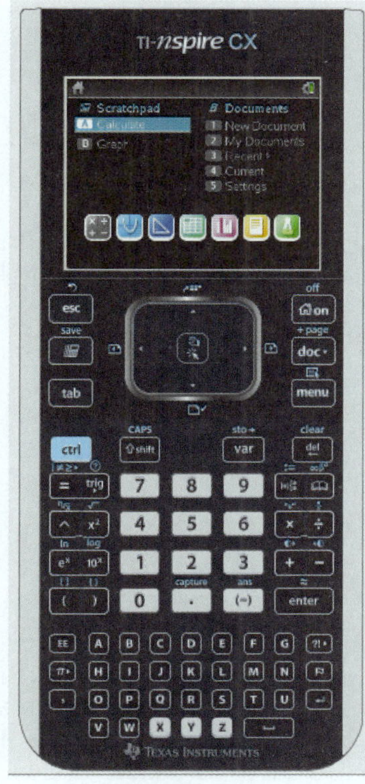

## Chapter 1 TI-Nspire Technology Corners

### 2. Histograms on the calculator

1. Insert a *New Document* by pressing ctrl N.
2. Insert a *Lists & Spreadsheet* page by arrowing down to *Add Lists & Spreadsheet*.
   - Name column A **foreignbrn**.
   - Type the data for the percent of state residents born outside the United States into the list. The data can be found on page 33.

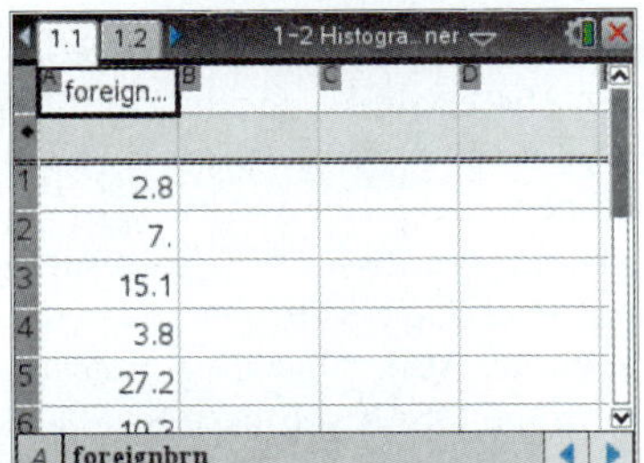


3. Insert a *Data & Statistics* page: press ctrl I, arrow to *Add Data & Statistics*, and press enter.
   - Press tab and *Click to Add Variable* on the horizontal axis will show the variables available. Select **foreignbrn**.

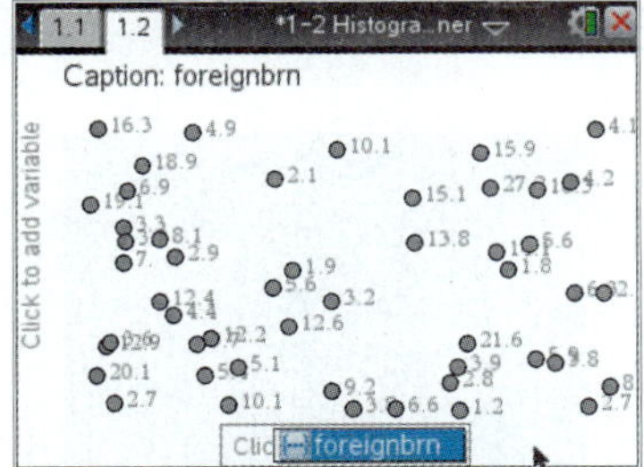


   - The data should now move into a dotplot. Notice the organization of the graph. Even though the data look "lopsided" in some places, you should consider the dots as being directly above each other in each column.

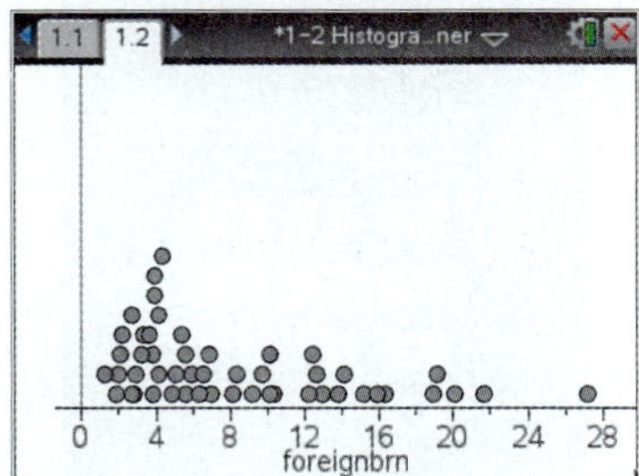


4. To make a better graphical display, let's move the data into a histogram. Use the Navpad to position the pointer in an empty space within the graph. Press ctrl menu and select *Histogram*. You will now see the data move into a histogram.

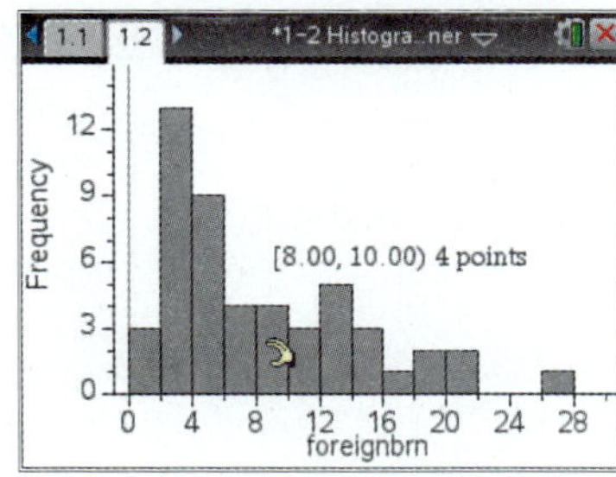


You can now position the pointer over each bar to examine the classes. The pointer will become an open hand and the class size will be displayed along with the number of data values in the class.

5. Adjust the classes to match those in Figure 1.16 (page 34).
   - Arrow into an empty space inside the histogram.
   - Press ctrl menu and select *Bin Settings, Equal Bin Width*. Enter the values shown. tab to OK and press enter. The new histogram should be displayed.

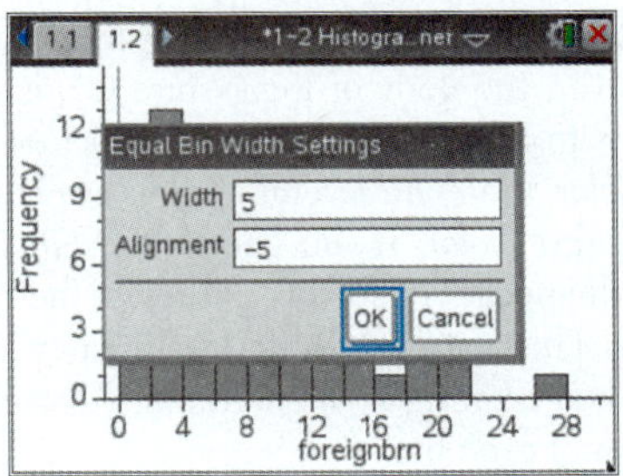


Notice how the first bar is "off the page." To adjust this, arrow over until the pointer becomes the axis-move cursor.

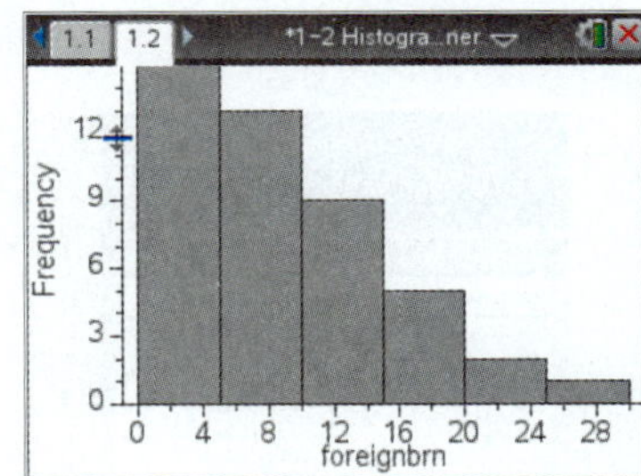


   - Press and hold the click key until the axis-move cursor becomes a grab hand.
   - Use the Navpad and, with the down arrow, "pull" the vertical axis down. Keep arrowing down until the top of the tallest histogram class is visible.

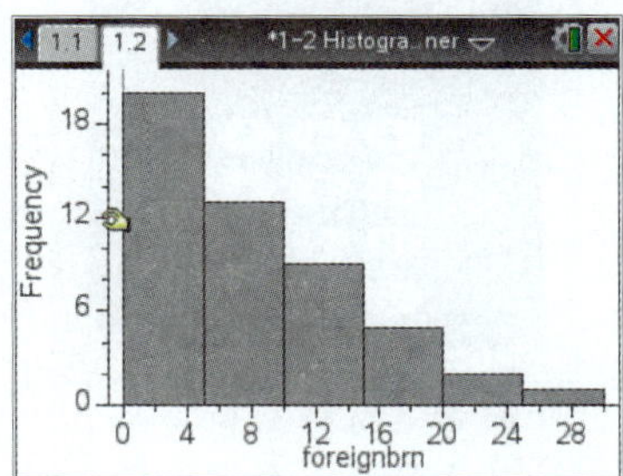


6. See if you can match the histogram in Figure 1.17 (page 35).

### 3. Making calculator boxplots

One of the added benefits of the TI-Nspire is its ability to plot more than three boxplots at a time in the viewing window. Let's use the calculator to make parallel boxplots of the travel data for the samples from North Carolina and New York.

1. Insert a *Lists & Spreadsheet* page: press ctrl I, arrow to *Add Lists & Spreadsheet*, and press enter.
   - Name column A **ncarolina** and column B **newyork**.

- Enter the travel time data from page 52.

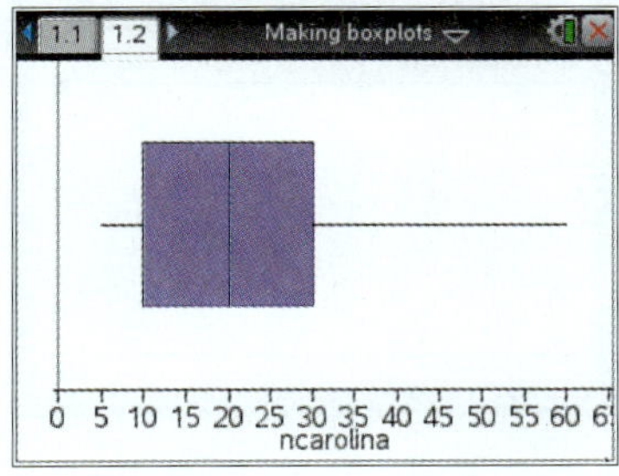


2. Insert a *Data & Statistics* page: press ctrl I, arrow to *Add Data & Statistics*, and press enter.
   - Press tab and *Click to Add Variable* on the horizontal axis will show the variables available. Select **ncarolina**.
   - The data will move into a dotplot. Use the Navpad to position the pointer in an empty space within the graph. Press ctrl menu and select *Box Plot*. You will now see the data move into a boxplot.
   - Using the Navpad, arrow over the plot. You will see the values in the five-number summary display one by one as you move across the boxplot.

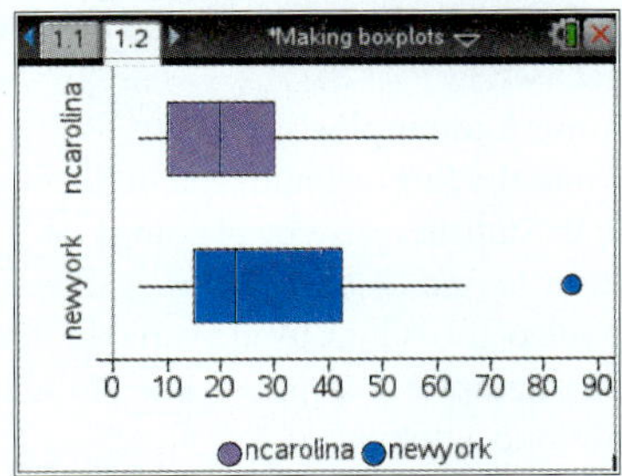


3. To add the boxplot of **newyork** travel times, arrow over "**ncarolina**" on the horizontal axis, press ctrl menu, and choose *Add X Variable*. Select **newyork** and the second boxplot will be added to the page.

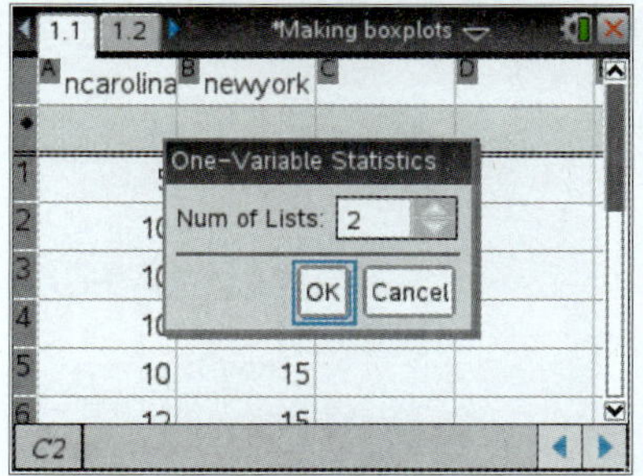


## 4. Computing numerical summaries with technology

Let's find numerical summaries for the travel times of North Carolina and New York workers from the previous Technology Corner. If you haven't already done so, enter the North Carolina and New York data.

- Insert a *Lists & Spreadsheet* page: press ctrl I, arrow to *Add Lists & Spreadsheet*, and press enter. Name column A **ncarolina** and column B **newyork.**
- Arrow down to the first empty cell in column A and type in the data. Repeat the process for **newyork** in column B.

1. The Nspire can calculate one-variable statistics for several lists at the same time (unlike the TI-84 or TI-89).
   - Press menu → *Statistics* → *Stat Calculations* → *One-Variable Statistics*. A dialogue box should appear asking for the number of lists. Press the up arrow (▲) to 2 or type "2." tab to OK and press enter.

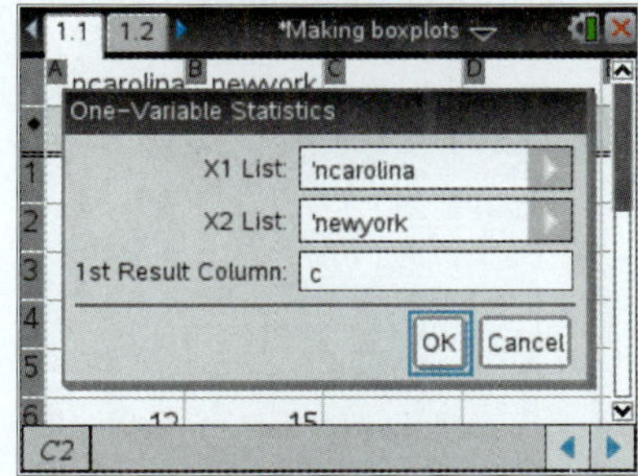


   - Another dialogue box should appear. Select the lists in the drop-down boxes: *X1 list*: **ncarolina** and *X2 list*: **newyork**. tab between the entry boxes to enter the next list and the column where you want the one-variable stats listed: type "c", tab to OK, and press enter. The numerical summaries for both states should now be displayed.

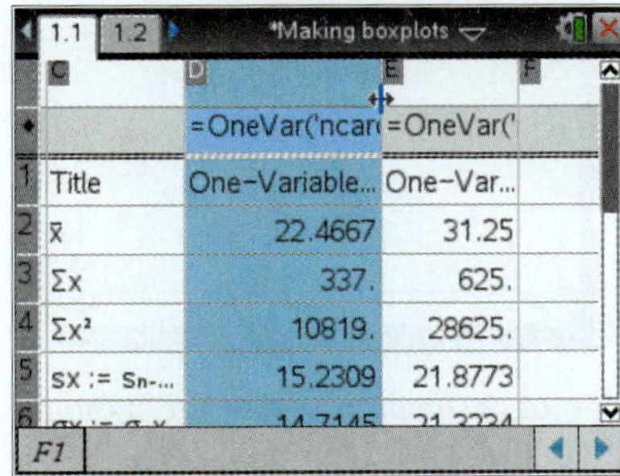


2. You can resize the columns to see which column contains values for which state: D has summary statistics for **ncarolina**, and E has summary statistics for **newyork**.
   - Use the Navpad to place the arrow between the columns. Press and hold the click key until ↔ appears. Use the arrow keys to increase the column width. Press the click key again to release the column.

|  | D | E |
|---|---|---|
|  | =OneVar('ncar | =OneVar(' |
| Title | One-Variable... | One-Var... |
| $\bar{x}$ | 22.4667 | 31.25 |
| Σx | 337. | 625. |
| Σx² | 10819. | 28625. |
| sx := Sn-... | 15.2309 | 21.8773 |

   - Repeat the same process to resize the column with one-variable statistics for **newyork**.

## Chapter 2 TI-Nspire Technology Corners

### 5. From *z*-scores to areas, and vice versa

*Finding areas:* The **normcdf** command on the Nspire can be used to find areas under the Normal curve. The syntax is *normalcdf* (lower bound, upper bound, mean, standard deviation). Let's use this command to confirm our answers to the examples on pages 116–118.

1. On the Home screen, select the *Calculate* scratchpad. This will

take you to a calculator page that is "outside" your current document. Therefore, you do not have to worry about losing/saving a document you are working on or about adding an unneeded page to that document.

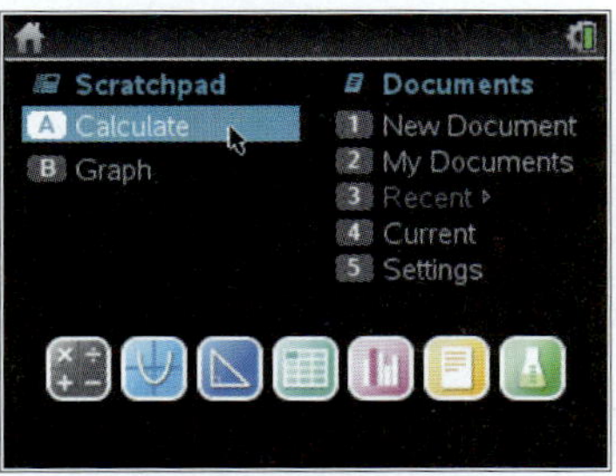


**What proportion of observations from the standard Normal distribution are greater than −1.78?**

Recall that the standard Normal distribution has mean 0 and standard deviation 1.

2. On the *Calculate* scratchpad, press menu → *Statistics* → *Distributions* → *Normal Cdf*. In the dialogue box that appears, type the numbers shown. To move between the drop-down boxes, press tab after typing each number. When the last number is entered, tab to OK and press enter.

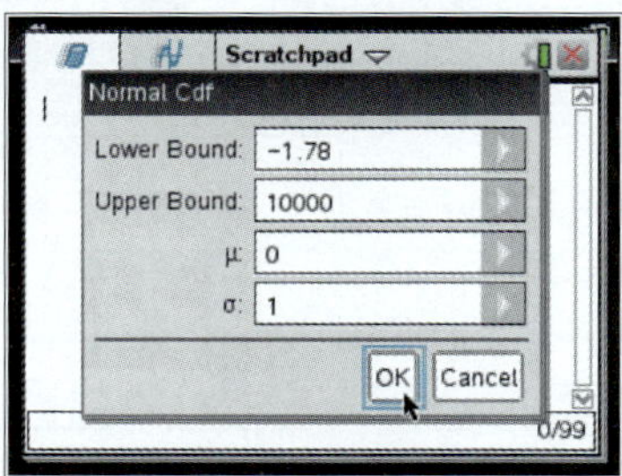


The proportion should now be displayed on the main screen. *Note*: We chose 10,000 as the upper bound because it is many standard deviations above the mean. These results agree with our previous answer using Table A: 0.9625.

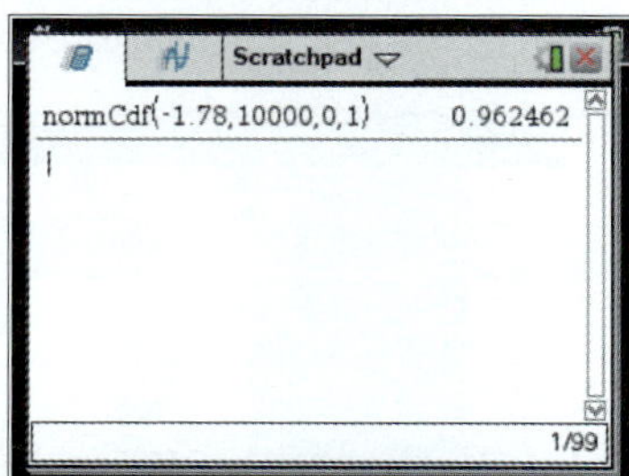


**What proportion of observations from the standard Normal distribution are between −1.25 and 0.81?**

The following screen shot confirms our earlier result of 0.6854 using Table A.

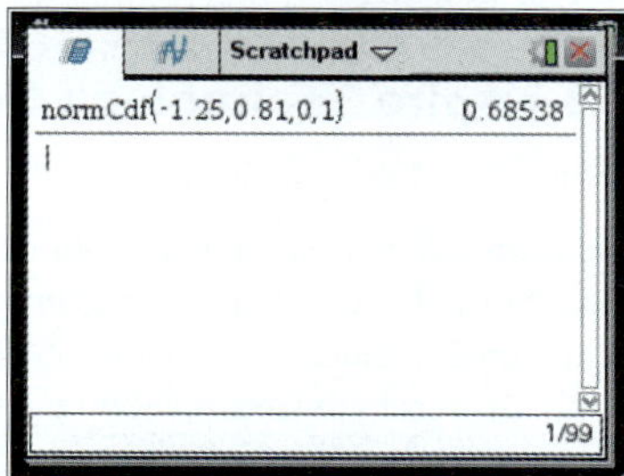


*Working backward*: The Nspire ***invNorm*** function calculates the value corresponding to a given percentile in a Normal distribution. For this command, the syntax is **inv Norm**(*area to the left*, $\mu$, $\sigma$).

3. Let's start with a "clean slate" by clearing the entries on our page. To do this, press menu → *Actions* → *Clear History*. Your scratchpad should now be blank.

**What is the 90th percentile of the standard Normal distribution?**

- Press menu → *Statistics* → *Distributions* → *Inverse Normal*. A dialogue box will appear. Type the numbers in the dialogue box as shown. To enter the numbers, tab between the entry boxes. When the last number is entered, tab to OK and press enter.

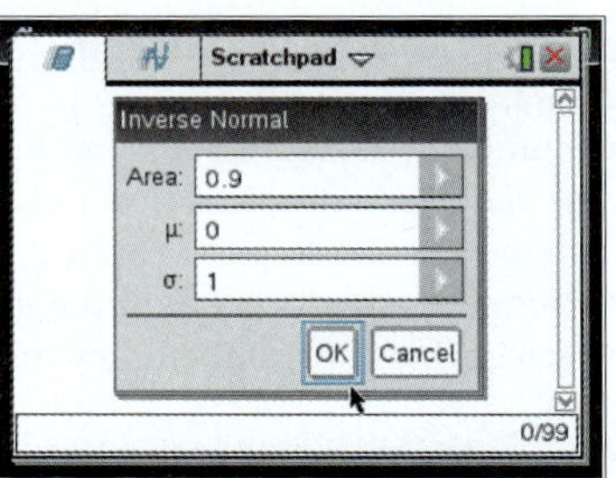


These results match our previous answer using Table A.

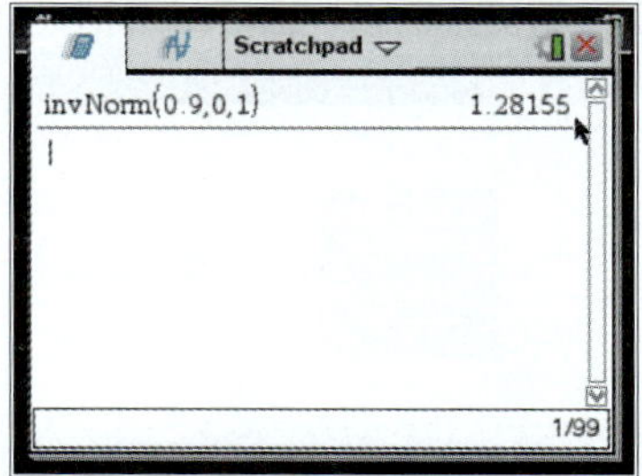


## 6. Normal probability plots

We will use the state unemployment rates data from page 122 to demonstrate how to make a Normal probability plot for a set of quantitative data.

1. Insert a *New Document* by pressing ctrl N.
2. Insert a *Lists & Spreadsheet* page by arrowing down to *Add Lists & Spreadsheet*.
   - Name column A **unemploy**.
   - Arrow down to the first cell and type in the 50 data values.
3. Insert a *Data & Statistics* page by pressing ctrl I and use the Navpad to arrow to *Add Data & Statistics*. Press enter.
4. Press tab to select the "Click to add variable" for the horizontal axis. Arrow to **unemploy** and press enter to select it. The data will now move into a dotplot.
5. Arrow up into an empty region of the dotplot and press ctrl menu. Select *Normal Probability Plot* and press [key].

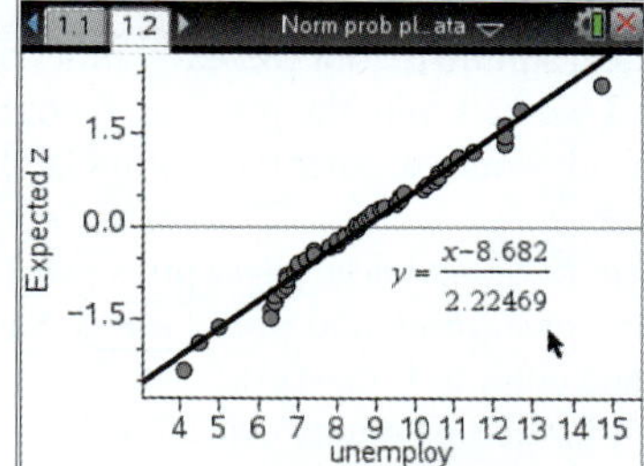


*Interpretation*: The Normal probability plot is quite linear, so it is reasonable to believe that the data follow a Normal distribution.

## Chapter 3 TI-Nspire Technology Corners

### 7. Scatterplots on the calculator

1. Insert a *New Document* by pressing ctrl N.
2. Insert a *Lists & Spreadsheet* page by arrowing down to *Add Lists & Spreadsheet*.
   - Name column A **points** and column B **wins**.
   - Type the corresponding values into each column. The data list follows.

| Points: | 34.8 | 36.8 | 25.7 | 25.5 | 32.0 | 15.8 |
|---|---|---|---|---|---|---|
| Wins: | 12 | 11 | 8 | 7 | 10 | 5 |
| Points: | 35.7 | 16.1 | 25.3 | 30.1 | 20.3 | 26.7 |
| Wins: | 13 | 2 | 7 | 11 | 5 | 6 |

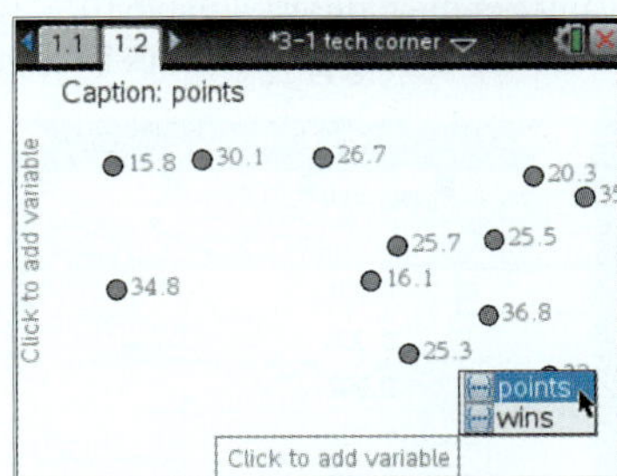

3. Press ctrl I and use the Navpad to arrow to *Add Data & Statistics*. Press enter.
4. Press tab to select the "Click to add variable" for the horizontal axis. Arrow to **points** and press enter to select it.

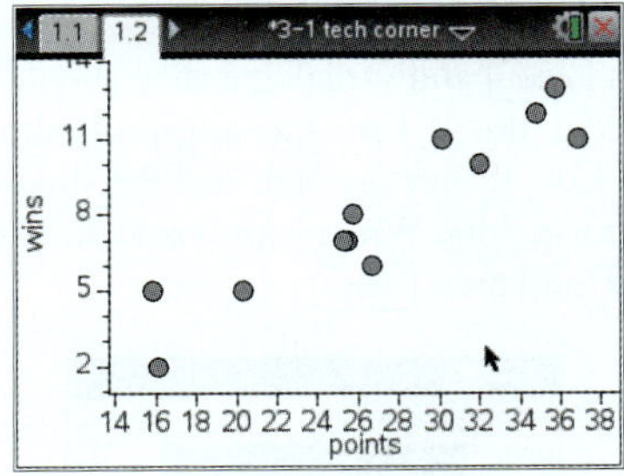

5. Press tab again and the box will move to the vertical axis. Select **wins**. The data will now move into a scatterplot. TI-Nspire labels the $x$ and $y$ axes with the list names, making a well-labeled graph to insert into documents.

### 8. Least-squares regression lines on the calculator

Let's use the Ford F-150 data to show how to find the equation of the least-squares regression line on the TI-Nspire. Here are the data.

| Miles driven | 70,583 | 129,484 | 29,932 | 29,953 | 24,495 | 75,678 | 8359 | 4447 |
|---|---|---|---|---|---|---|---|---|
| Price (in dollars) | 21,994 | 9500 | 29,875 | 41,995 | 41,995 | 28,986 | 31,891 | 37,991 |
| Miles driven | 34,077 | 58,023 | 44,447 | 68,474 | 144,162 | 140,776 | 29,397 | 131,385 |
| Price (in dollars) | 34,995 | 29,988 | 22,896 | 33,961 | 16,883 | 20,897 | 27,495 | 13,997 |

1. Insert a New Document by pressing ctrl N.
2. Insert a *Lists & Spreadsheet* page by arrowing down to *Add Lists & Spreadsheet*.
   - Name column A **miles** and column B **price**.
   - Type the corresponding values into each column.
3. Graph the data in a scatterplot putting **miles** on the horizontal axis and **price** on the vertical axis. Refer to the previous TI-Nspire Technology Corner.
4. To add a least-squares regression line, first ctrl ◀ back to the *Lists & Spreadsheet* page.
5. Press menu, and arrow to Statistics → *Stat Calculations*, *Linear Regression (a + bx)*, enter. You should then see a dialogue box. In the drop-down boxes, arrow down to **miles** for the X *List*:, then press tab and arrow down to **price** for the Y *List*:. tab to OK and press enter.

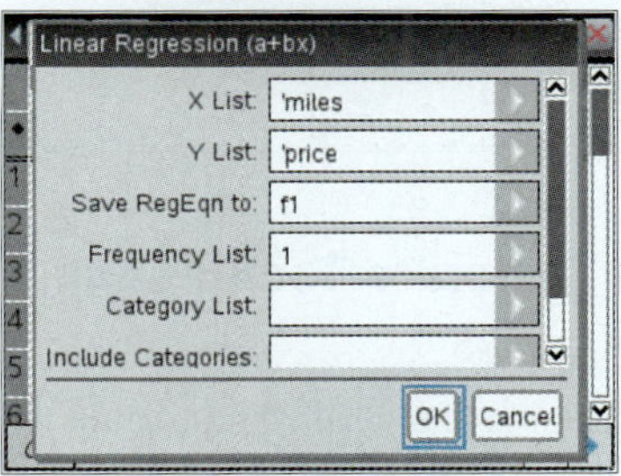

The linear regression information, $a$, $b$, $r^2$, $r$, and *resid* will be displayed in another column within the *Lists & Spreadsheet* page.

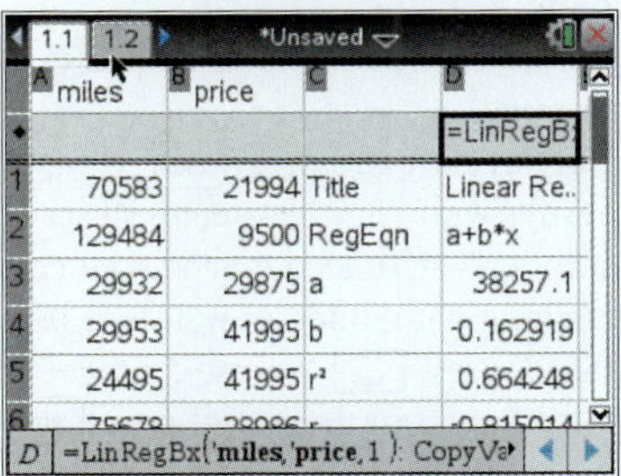

6. Press ctrl ▶ to return to the *Data & Statistics page*. Press menu; arrow to *Analyze* → *Regression* → *Show Linear (a + bx)*, and press enter. The least-squares regression line along with the equation will appear. If you arrow over the equation, the hand will appear. Click and hold. When the hand closes, you can move the equation using the arrow keys.

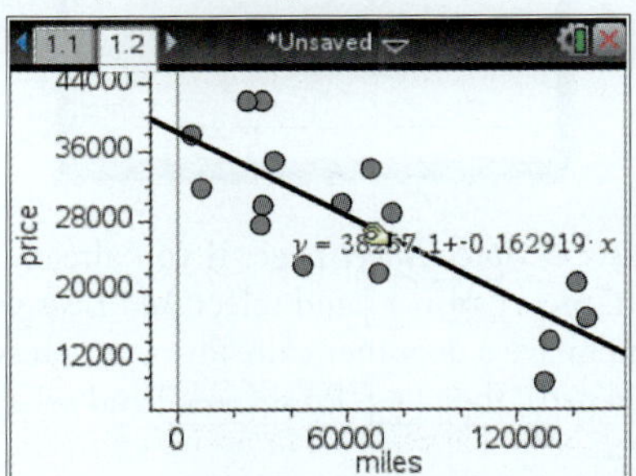

7. Save the document for later use. Press ctrl S. Name your document **Truck prices**.

### 9. Residual plots on the calculator

Let's continue the analysis of the Ford F-150 miles driven and price data from the previous Technology Corner. You should have already made a scatterplot, calculated the equation of the least-squares regression line, and graphed the line on your plot. Now we want to make a residual plot.

1. Open the document **Truck prices**. Press ctrl N, arrow through *My Documents* → *Truck prices* and press enter.
2. Press ctrl ▸ to go to the *Data & Statistics* page.
3. Press menu; arrow to *Analyze* → *Residuals* → *Show Residual Plot*. This will split the screen and the residual plot will be displayed below the graph of the least-squares regression line.

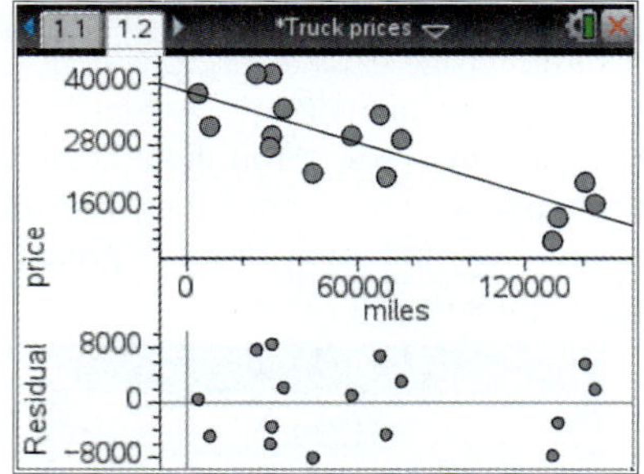


## Chapter 4 TI-Nspire Technology Corner

### 10. Choosing an SRS

The TI-Nspire has a function called **randSamp** that will randomly select individuals for a sample with or without replacement from a population.

1. Check that your calculator's random number generator is working properly.
   - Open the *Calculator Scratchpad* by pressing (scratchpad key) (or (home) Ⓐ on the keypad).
   - Type **randint(1,1750)** and press enter.
   - Compare results with your classmates. If several students have the same number, you'll need to seed your calculator's random number generator with different numbers before you proceed. Type **randSeed**, press ␣ (to insert a space), type < last four digits of your phone number> and press enter. *Done* should appear. Now your calculator is ready to generate numbers that are different from those of your classmates.

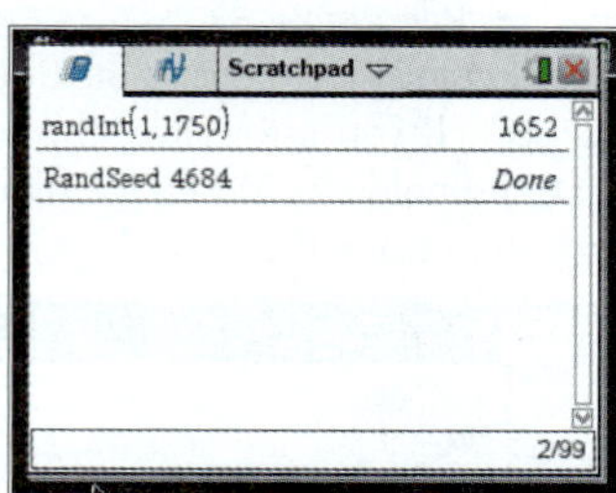


2. Insert a *Lists & Spreadsheet* page. If you already have a document open, press ctrl I and select *Add Lists & Spreadsheet*. If you do not have a document already open, press (home on) ((home) on the clickpad), then 4. Press menu and select *Add Lists & Spreadsheet*.
3. Name column A **students**. Arrow down to the formula cell and press enter. **students:=** should appear. Type **seq(x,x,1,1750)** and enter. This will put the digits 1 through 1750 in this list.
4. Name column B **sampstudents**. Arrow down to the formula cell and press enter. **sampstudents:=** should appear. Type **randSamp(students,10,1)** and press enter. This function will take a random sample of 10 students from the list. "1" lets the function know to do the sampling without replacement.

| | A students | B sampstudents | C | D |
|---|---|---|---|---|
| • | =seq(x,x,1 | =randsamp(stu | | |
| 1 | 1 | 562 | | |
| 2 | 2 | 412 | | |
| 3 | 3 | 1527 | | |
| 4 | 4 | 534 | | |
| 5 | 5 | 1529 | | |

B sampstudents =randsamp(students,1

*Note:* Sampling with replacement is the default setting for this function. You can use 0 as the third input in the **randSamp** command or close the parentheses after the second input.

## Chapter 6 TI-Nspire Technology Corners

### 11. Analyzing random variables on the calculator

Let's explore what the calculator can do using the random variable X = Apgar score of a randomly selected newborn from the example on page 349.

1. Insert a *Lists & Spreadsheet* page. Press ctrl I, arrow to *Add Lists & Spreadsheet*, and press enter.
   - Name column A **apgar** and column B **apgrprob**.
   - Enter the values of the random variable (0 – 10) in the **apgar** list and the corresponding probabilities in **apgrprob**.

| | A apgar | B apgrprob | C | D |
|---|---|---|---|---|
| • | | | | |
| 1 | 0 | 0.001 | | |
| 2 | 1 | 0.006 | | |
| 3 | 2 | 0.007 | | |
| 4 | 3 | 0.008 | | |
| 5 | 4 | 0.012 | | |

A1 0

2. Graph a histogram of the probability distribution.
   - Insert a *Data & Statistics* page. Press ctrl I, arrow to *Add Data & Statistics*, and press enter.
   - Press ctrl menu and select *Add X Variable with Summary List*. Press enter and a dialogue box should appear. **apgar** should be in the *X List* and **apgrprob** should be in the *Summary List*. If they are not, use the drop-down boxes to select your variables. When your box looks like the one here, tab to OK and press enter.

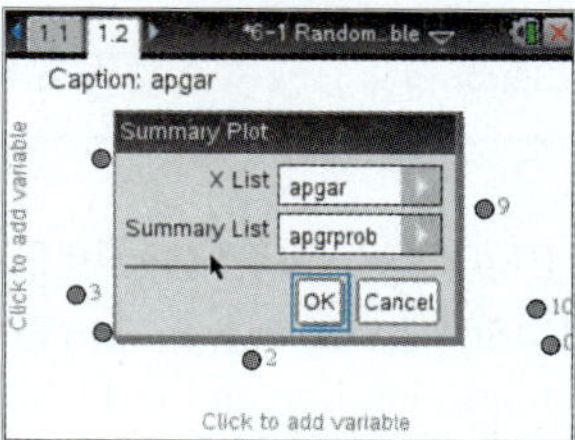

The probability histogram should now be displayed.

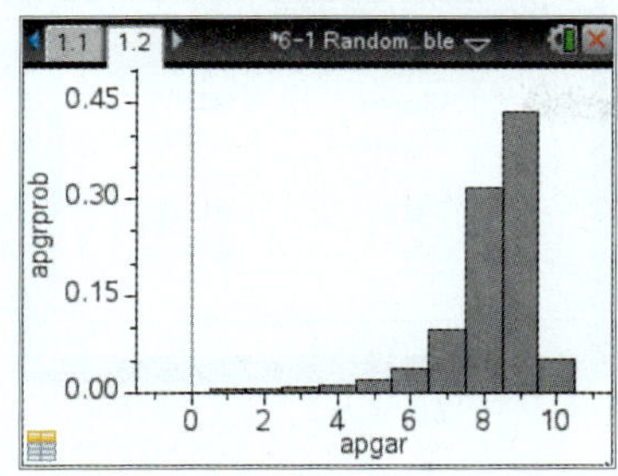


3. To calculate the mean and standard deviation of the random variable, use one-variable statistics with **apgar** as the *Data List* and **apgrprob** as the *Frequency List*.
   - Press ctrl ◀ to go back to the *Lists & Spreadsheet* page.
   - Press menu → *Statistics* → *Stat Calculations* → *One-Variable Statistics*.
   - Make sure your *X1 List*, *Frequency List*, and *1st Result Column* have the variables/values shown (you can press the down arrow in the drop-down boxes to access the variable names and type C for *1st Result Column*). tab to OK and press enter.

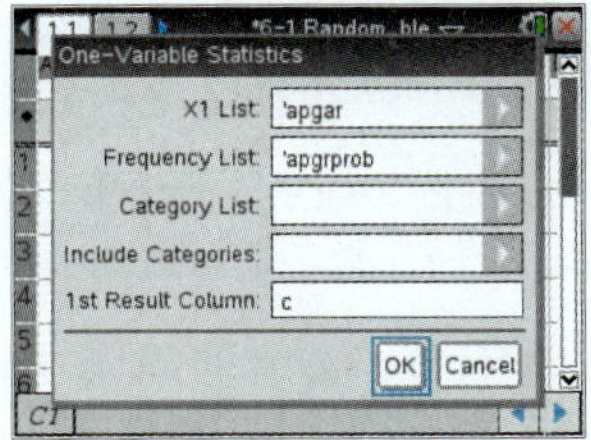


   - The statistics should now be displayed in your *Lists & Spreadsheet* page.

## 12. Binomial coefficients on the calculator

To calculate a binomial coefficient like $\binom{5}{2}$ on the TI-Nspire, proceed as follows. Open the *Calculator Scratchpad* by pressing (or Ⓐ on the clickpad). Press menu → *Probability* → *Combinations*, and then enter. **nCr(** will appear. Complete the command **nCr(5,2)** and press enter.

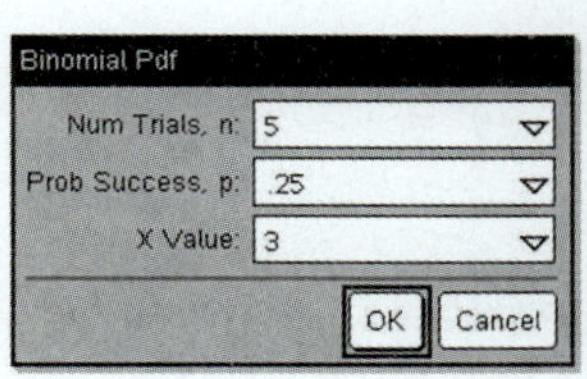


## 13. Binomial probability on the calculator

There are two handy commands on the TI-Nspire for finding binomial probabilities:

`binomPdf(n,p,k)` computes $P(X = k)$

`binomCdf(n,p,k)` computes $P(X \le k)$

You will need to open the *Calculator Scratchpad* (press or Ⓐ on the clickpad). These two commands can be found in the *Distributions* menu within the *Statistics* menu. You can access them by pressing menu → *Statistics* → *Distributions*. A dialogue box will appear. Input $n$ (the number of observations), $p$ (probability of success), and $k$ (number of successes).

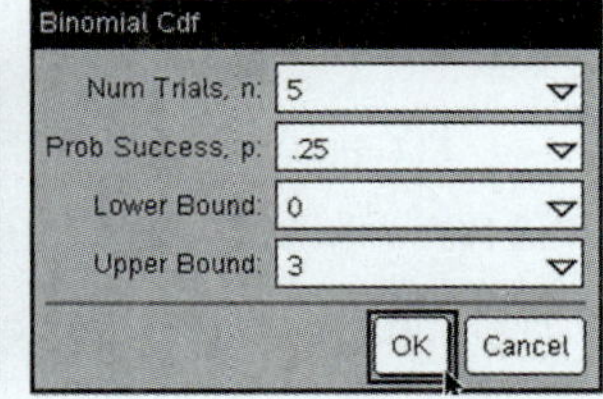


For the parents having $n = 5$ children, each with probability $p = 0.25$ of type O blood:

$$P(X = 3) = \texttt{binomPdf(5,0.25,3)} = 0.08789$$

To find $P(X > 3)$, we used the complement rule:

$$P(X > 3) = 1 - P(X \le 3) = 1 - \texttt{binomCdf(5,0.25,3)} = 0.01563$$

Of course, we could also have done this as

$$P(X > 3) = P(X = 4) + P(X = 5) = \texttt{binomPdf(5,0.25,4)} + \texttt{binomPdf(5,0.25,5)} = 0.01465 + 0.00098 = 0.01563$$

On the TI-Nspire, you can also calculate using

$$P(X > 3) = P(X = 4) + P(X = 5) = \texttt{binomCdf(5,0.25,4,5)} = 0.01563$$

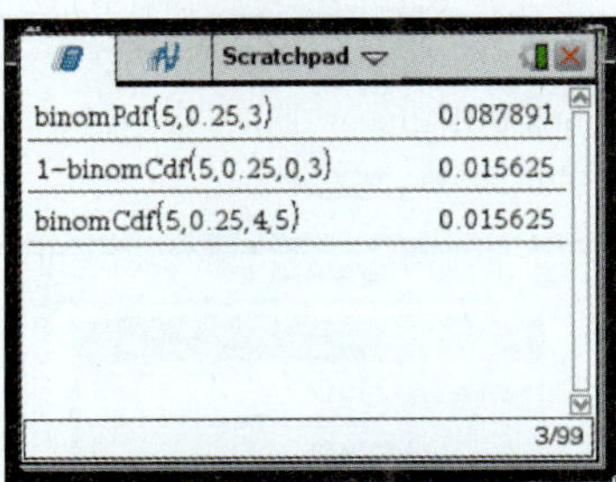


## 14. Geometric probability on the calculator

There are two handy commands on the TI-Nspire for finding geometric probabilities:

`geomPdf(p,k)` computes $P(Y = k)$

`geomCdf(p,k)` computes $P(Y \le k)$

You will need to open the *Calculator Scratchpad* (press or A). These two commands can be found in the *Distributions* menu within the *Statistics* menu. You can access them by pressing menu → *Statistics* → *Distributions*. A dialogue box will appear. Input $p$ (probability of success) and $k$ (number of trials to get the first success).

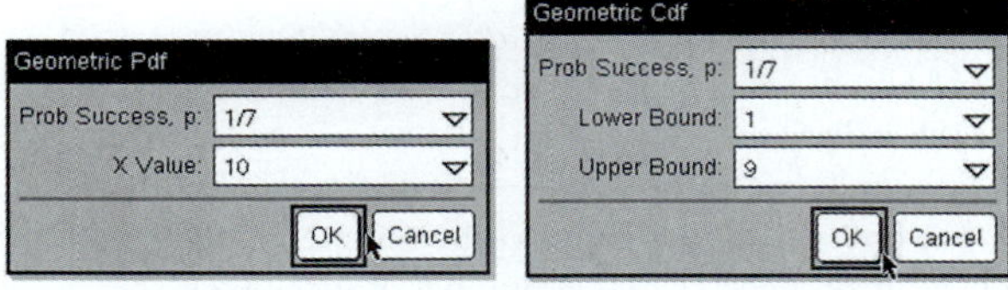


For the Lucky Day Game, with probability of success $p = 1/7$ on each trial,

$$P(Y = 10) = \texttt{geomPdf(1/7,10)} = 0.0357$$

To find $P(Y < 10)$, use `geomCdf`:

$$P(Y < 10) = P(Y \leq 9) = \texttt{geomCdf(1/7,9)} = 0.7503$$

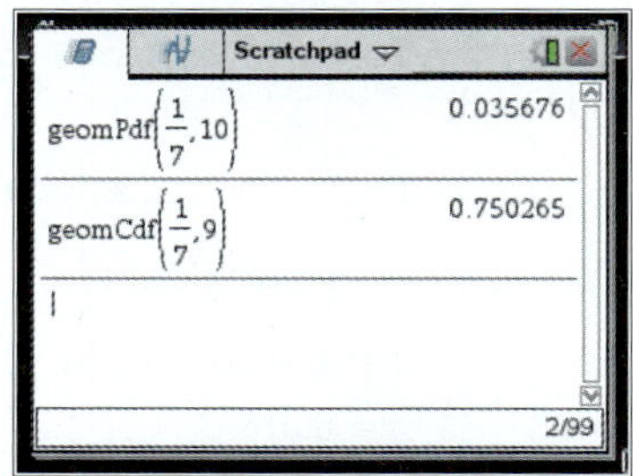


## Chapter 8 TI-Nspire Technology Corners

### 15. Confidence interval for a population proportion

The TI-Nspire can be used to construct a confidence interval for an unknown population proportion. We'll demonstrate using the example on page 500. Of $n = 439$ teens surveyed, $x = 246$ said they thought young people should wait to have sex until after marriage.

To construct a confidence interval:

- Press ( A) to insert a *Calculator Scratchpad*.
- Press menu → *Statistics* → *Confidence Intervals* → *1-Prop z Interval*.
- A dialogue box will appear: Enter the values as shown below. tab to OK and press enter.

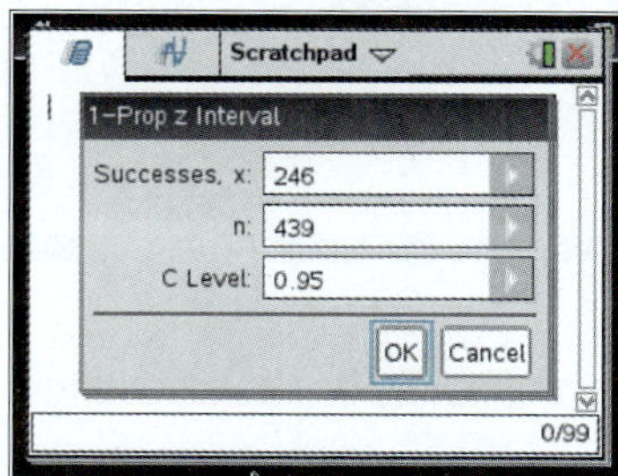


The lower and upper bounds of the confidence interval are reported, along with the sample proportion $\hat{p}$, the margin of error (ME), and the sample size.

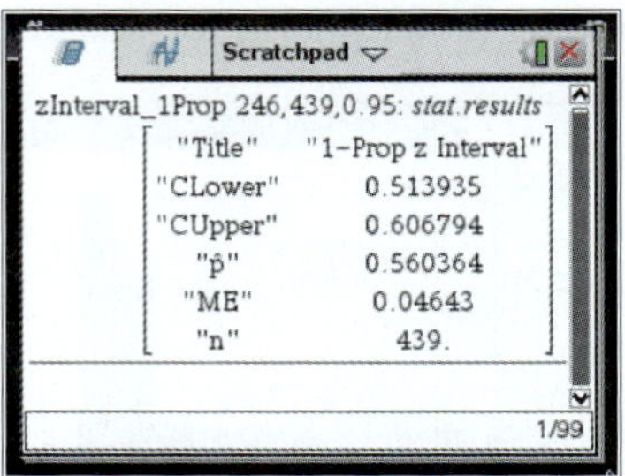


### 16. Inverse *t* on the calculator

The TI-Nspire allows you to find critical values $t^*$ using the inverse $t$ command. As with the calculator's inverse Normal command, you have to enter the *area to the left* of the desired critical value. Let's use the inverse $t$ command to find the critical values for parts (a) and (b) in the example on page 513.

- Press (or A) to insert a *Calculator Scratchpad*.
- Press menu → *Statistics* → *Distributions* → *Inverse t*.
- A dialogue box will appear. For part (a), enter .025 for the *Area* and 11 for the *Deg of Freedom, df*. tab to OK and press enter.

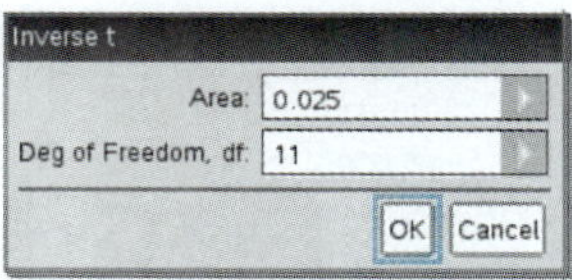


For part (b), enter .05 for the *Area* and 47 for the *Deg of Freedom, df*. tab to OK and press enter.

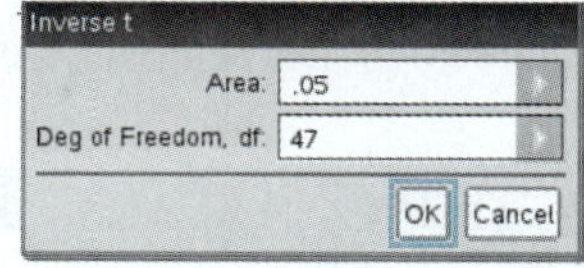


- The critical values $t^*$ should now be displayed.

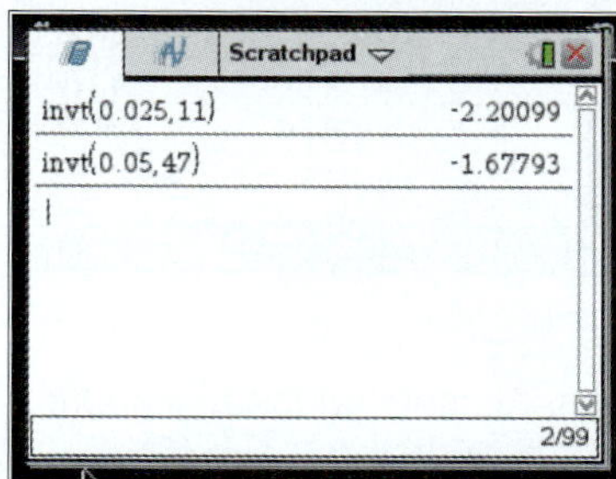


### 17. One-sample *t* intervals for $\mu$ on the calculator

Confidence intervals for a population mean using $t$ procedures can be constructed on the TI-Nspire, thus avoiding the use of Table B. Here is a brief summary of the techniques when you have only numerical summaries and when you have the actual data values.

1. Using summary statistics: Auto pollution example, page 519
   - Insert a *Lists & Spreadsheet* page: Press ctrl I and select *Add Lists & Spreadsheet*.
   - Press menu → *Statistics* → *Confidence Intervals* → *t interval*.

- The first dialogue box that appears asks for *Data* or *Stats* in the drop-down box. Select *Stats*, (tab) to [OK], and press (enter).
- In the next dialogue box, enter the values shown.

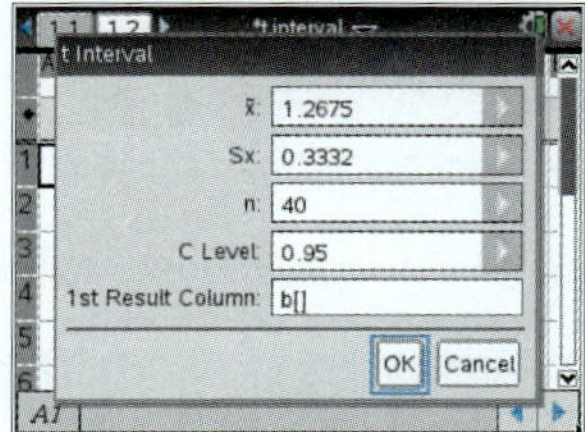


- (tab) to [OK] and press (enter).
- The results should now appear in the spreadsheet.

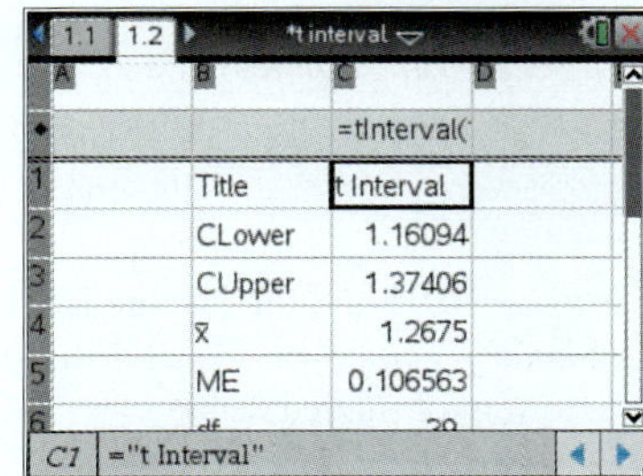


2. Using raw data: Video screen tension example, page 520
Enter the 20 video screen tension readings data using the following procedure.
   - Insert a *Lists & Spreadsheet* page: Press (ctrl) (I) and select *Add Lists & Spreadsheet.*
   - Name the first column **screen**.
   - Arrow down to the first cell and enter the 20 values.

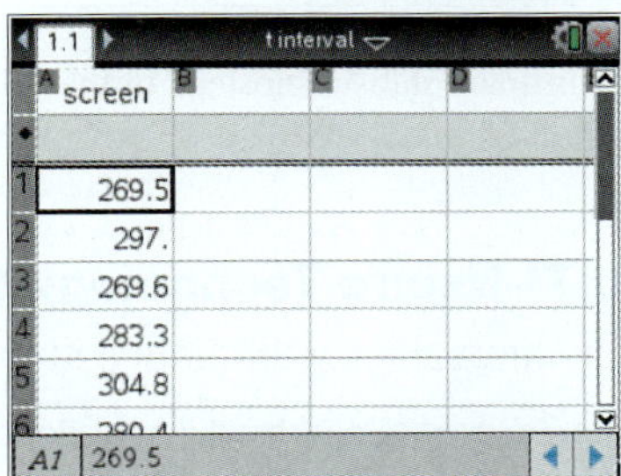


To construct the $t$ interval:

- Press (menu) → *Statistics* → *Confidence Intervals* → *t interval.*
- The first dialogue box that appears asks for *Data* or *Stats* in the drop-down box. Select *Data*, (tab) to [OK], and press (enter).
- In the next dialogue box, select the data list, **screen**, (tab) to [OK], and press (enter).

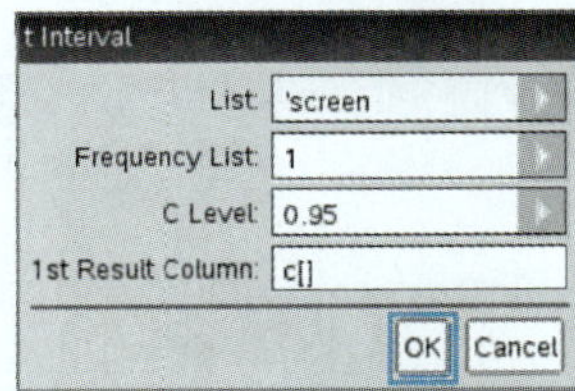


- The results should now appear in the spreadsheet. (You may have to scroll up to see them.)

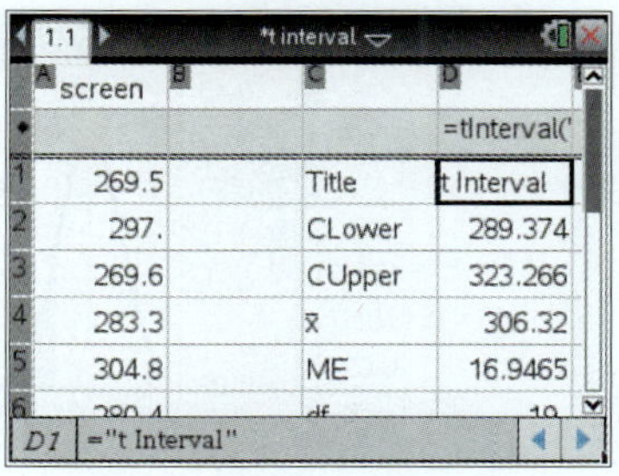


## Chapter 9 TI-Nspire Technology Corners

### 18. One-proportion $z$ test on the calculator

The TI-Nspire can be used to test a claim about a population proportion. We'll demonstrate using the example on page 559. In a random sample of size $n = 500$, the supervisor found $x = 47$ potatoes with blemishes. To perform a significance test:

- Press (calculator) (or (home) (A)) to insert a *Calculator Scratchpad.*
- Press (menu) → *Statistics* → *Stat Tests* → *1-Prop z test.*
- A dialogue box will appear. Enter the values shown: $p_0 = 0.08$, $x = 47$, and $n = 500$. Specify the alternative hypothesis as "$H_a$: prop $> p_0$." (tab) to [OK] and press (enter).

*Note:* $x$ is the number of successes and $n$ is the number of trials. Both must be whole numbers!

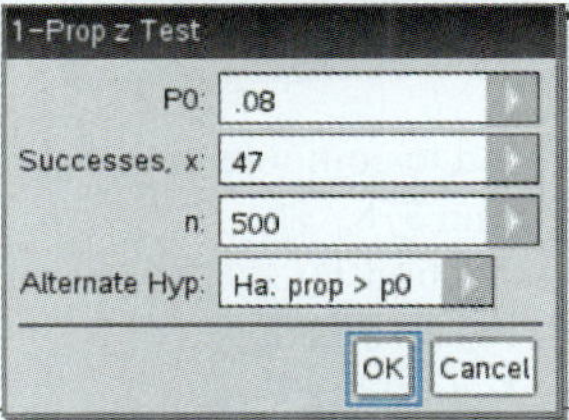


You can see that the test statistic is $z = 1.15392$ and the $P$-value is 0.1243.

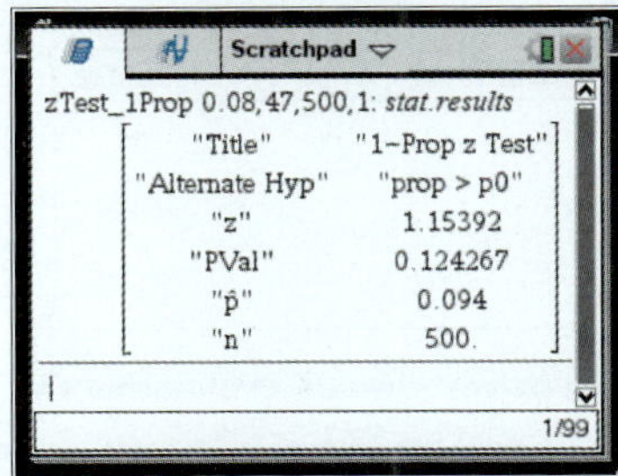


To display the $P$-value as a shaded area under the Normal curve:

- Press (home on) and select the *Lists & Spreadsheet* icon.
- Press (menu) → *Statistics* → *Stat Tests* → *1-Prop z test.*
- A dialogue box will appear: Enter the values shown below. Check the box to *Shade P Value*. (tab) to [OK] and press (enter).

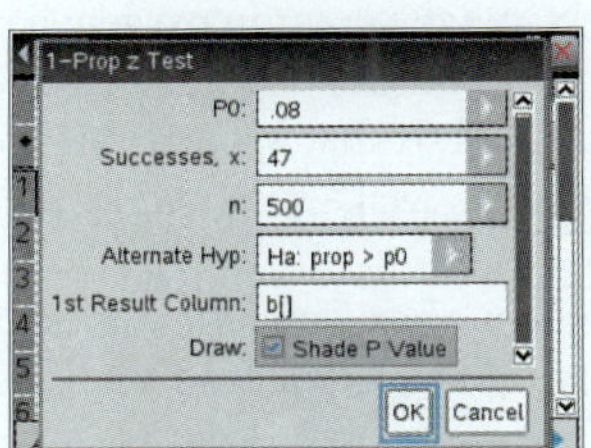

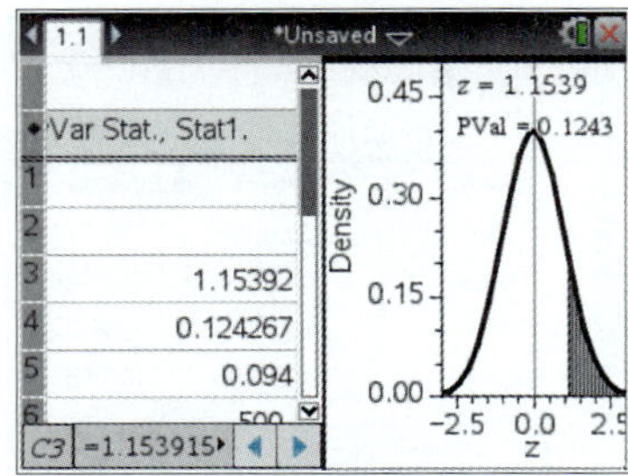


## 19. Computing *P*-values from *t* distributions on the calculator

You can use the *tcdf* command on the TI-Nspire to calculate areas under a *t* distribution curve. The syntax is tcdf(lower bound,upper bound,df). To use this command:

- Press (=▤) (or (⌂) (A)) to insert a *Calculator Scratchpad*.
- Press (menu) → *Statistics* → *Distributions* → *t Cdf*.
- In the dialogue box that appears, enter your lower and upper bound and degrees of freedom.

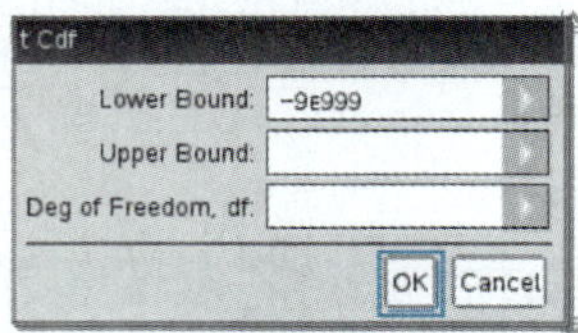


Use the *t Cdf* command to compute the *P*-values from the examples on pages 577 and 578.

- *Better batteries:* To find $P(t \geq 1.54)$, use Lower Bound: 1.54, Upper Bound: 10000, and df:14.
- *Two-sided test:* To find the *P*-value for the two-sided test with df = 36 and $t = -3.17$, execute the command $2 \cdot$ tCdf $(-10000, -3.17, 36)$.

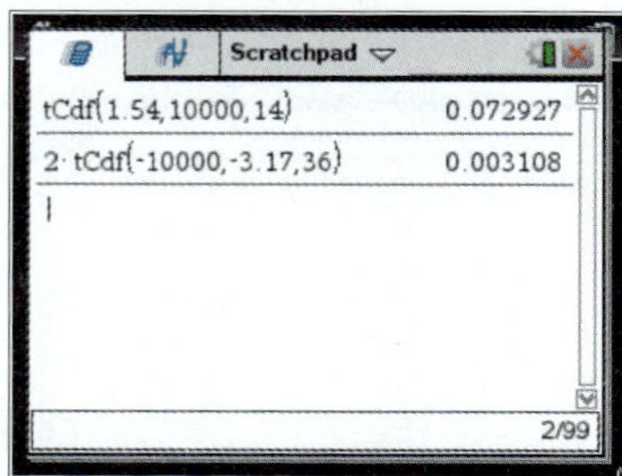


## 20. One-sample *t* test for a mean on the calculator

You can perform a one-sample *t* test using either raw data or summary statistics on the TI-Nspire. Let's use the calculator to carry out the test of $H_0$: $\mu = 5$ versus $H_a$: $\mu < 5$ from the dissolved oxygen example on page 580.

Start by entering the sample data into a column in a *Lists & Spreadsheet* page. Name the column **oxygen**. Then, to do the test:

- Press (menu) → *Statistics* → *Stats Tests* → *t Test*.
- The first dialogue box that appears asks for *Data* or *Stats* in the drop-down box. Make sure *Data* is selected. (tab) to [OK] and press (enter).
- In the next dialogue box, enter the values shown in the following box. To just "calculate," leave the *Shade PValue* option unchecked. Then (tab) to [OK] and press (enter).

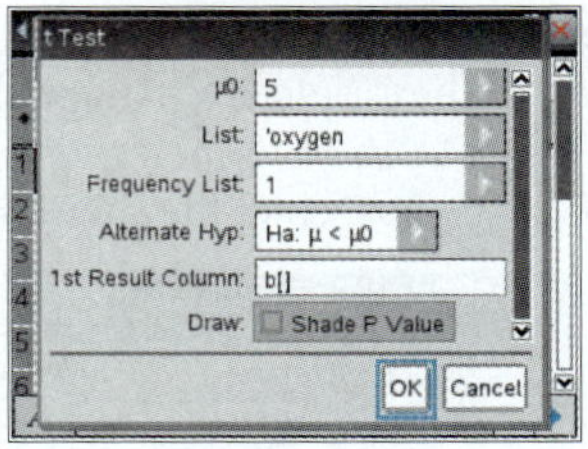


- The results should now appear in the spreadsheet.

| | A oxygen | B | C | D |
|---|---|---|---|---|
| ◆ | | | =tTest(5,'o | |
| 1 | 4.53 | Title | t Test | |
| 2 | 5.04 | Alternate... | μ < μ0 | |
| 3 | 3.29 | t | -0.942556 | |
| 4 | 5.23 | PVal | 0.180945 | |
| 5 | 4.13 | df | 14. | |

The test statistic is $t = -0.94$ and the *P*-value is 0.1809.

If you check *Shade P Value*, you see a *t*-distribution curve (df = 14) with the lower tail shaded.

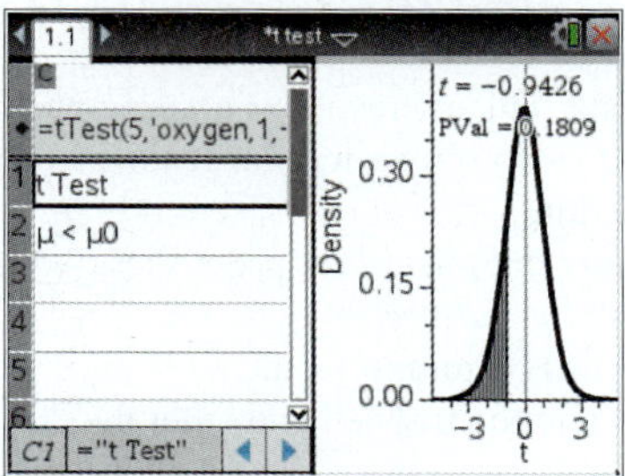


If you are given summary statistics instead of the original data, you would select the "Stats" option in the drop-down box.

# Chapter 10 TI-Nspire Technology Corners

## 21. Confidence interval for a difference in proportions

The TI-Nspire can be used to construct a confidence interval for $p_1 - p_2$. We'll demonstrate using the example on page 617. Of $n_1 = 799$ teens surveyed, $X = 639$ said they used social networking sites. Of $n_2 = 2253$ adults surveyed, $X = 1555$ said they engaged in social networking. To construct a confidence interval:

- Press (=▤) ((⌂) (A)) to insert a *Calculator Scratchpad*.
- Press (menu) → *Statistics* → *Confidence Intervals* → *2-Prop z Interval*.
- A dialogue box will appear. Enter the values shown below. (tab) to [OK] and press (enter).

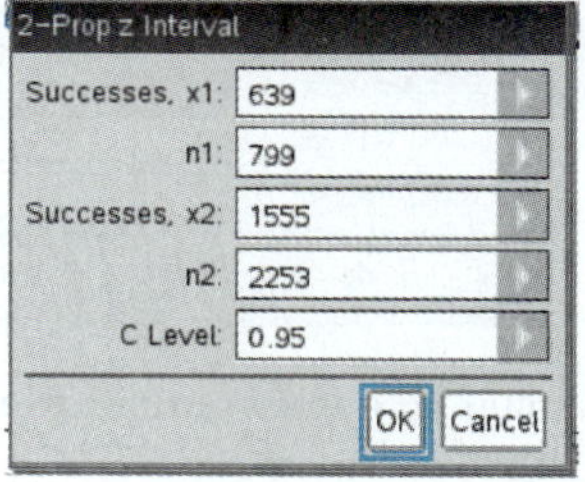

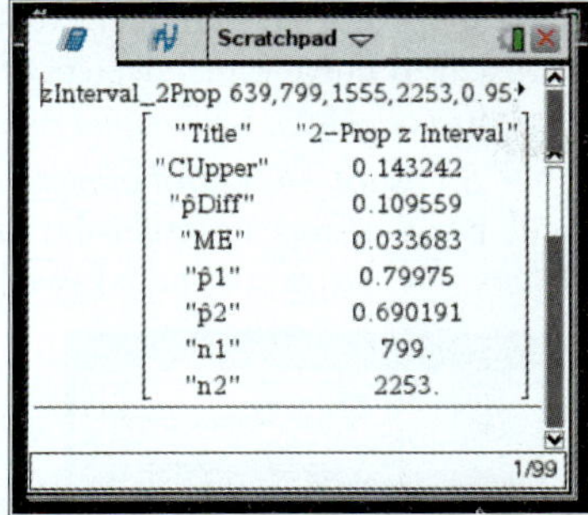


## 22. Significance test for a difference in proportions

The TI-Nspire can be used to perform significance tests for comparing two proportions. Here, we use the data from the Hungry Children example on page 622.

To perform a test of $H_0$: $p_1 - p_2 = 0$:

- Press (or ) to insert a *Calculator Scratchpad*.
- Press menu → *Statistics* → *Stat Tests* → *2-Prop z test*.
- A dialogue box will appear. Enter the values shown: $x_1 = 19$, $n_1 = 80$, $x_2 = 26$, $n_2 = 150$. Specify the alternative hypothesis $H_a$: $p_1 \neq p_2$ as shown.
- tab to OK and press enter.

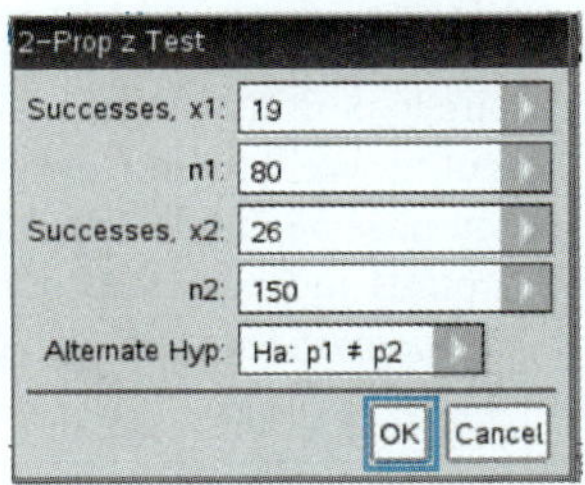


You will see that the $z$ statistic is $z = 1.168$ and the $P$-value is 0.2427, as shown here. Do you see the combined proportion of students who didn't eat breakfast? It's the $\hat{p}$ value, 0.1957.

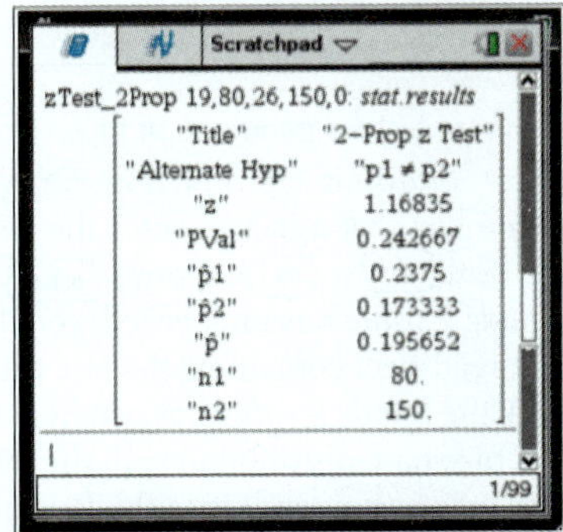


To display the $P$-value as a shaded area under the standard Normal curve:

- Press on and select the *Lists & Spreadsheet* icon.
- Press menu → *Statistics* → *Stat Tests* → *2-Prop z test*. A dialogue box will appear.
- Enter the values shown in the following box. Check the box to *Shade P Value*. tab to OK and press enter.

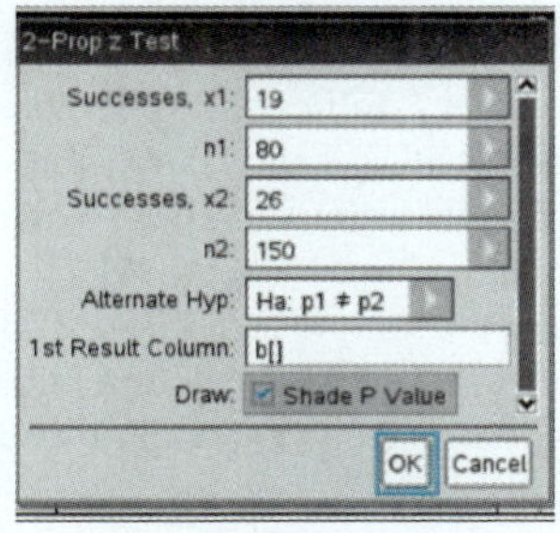


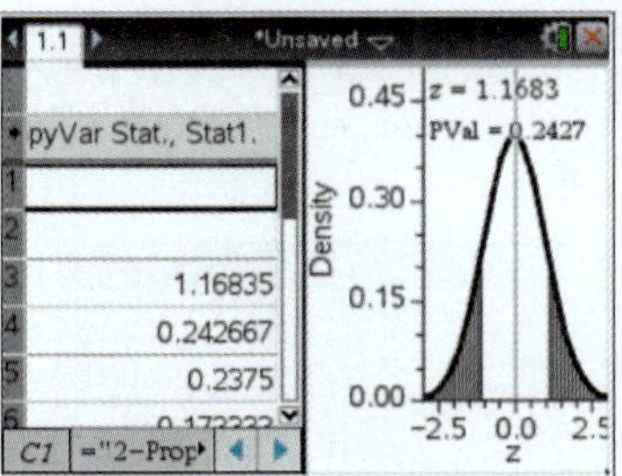


## 23. Two-sample *t* intervals on the calculator

You can use the two-sample *t* interval command on the TI-Nspire to construct a confidence interval for the difference between two means. We'll show you the steps using the summary statistics from the pine trees example on page 641.

- Press (or ) to insert a *Calculator Scratchpad*.
- Press menu → *Statistics* → *Confidence Intervals* → *2-Sample t interval*.
- In the first dialogue box, select *Stats* in the drop-down menu. tab to OK and press enter. Another dialogue box will appear.
- Enter the summary statistics shown:

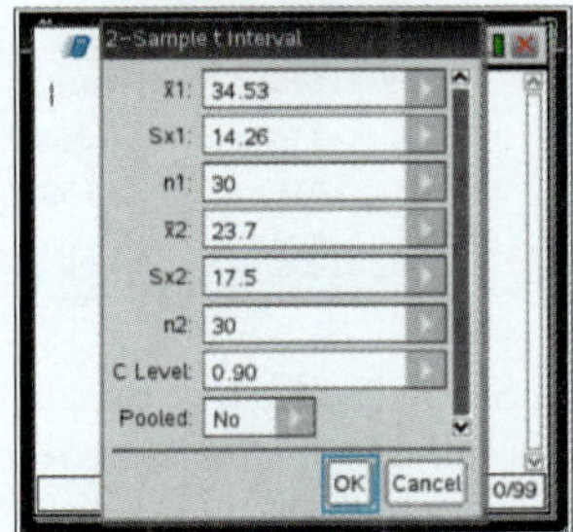


- Enter the confidence level: C level: .90. For pooled: choose "No." (We'll discuss pooling later.) tab to OK and press enter.

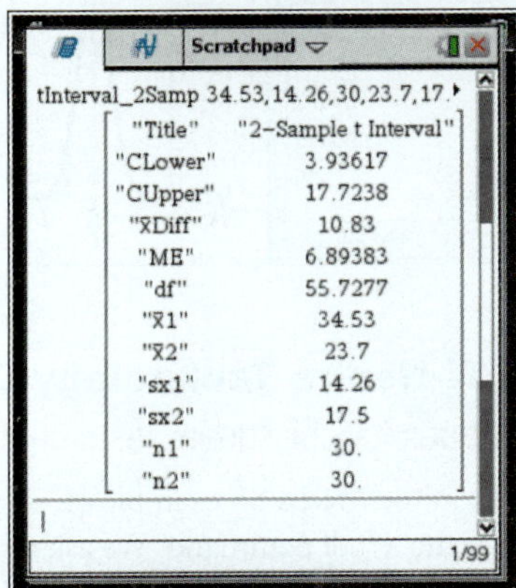

## 24. Two-sample *t* tests on the calculator

Technology gives smaller *P*-values for two-sample *t* tests than the conservative method. That's because calculators and software use the more complicated formula on page 640 to obtain a larger number of degrees of freedom.

Start by entering the sample data into a column in a *Lists & Spreadsheet* page. Name column A **calcium** and enter the Group 1 data. Name column B **placebo** and enter the Group 2 data. Then do the test:

- Press (menu) → *Statistics* → *Stats Tests* → *2-Sample t Test*.
- In the first dialogue box, select *Data* in the drop-down menu. (tab) to [OK] and press (enter).
- In the next dialogue box, enter the values shown, (tab) to [OK], and press (enter).

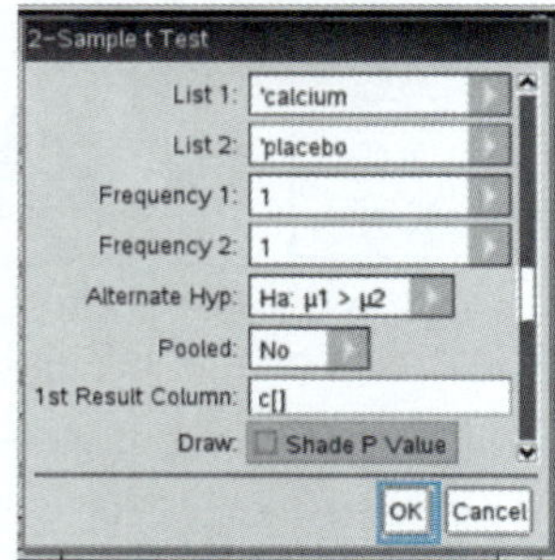


*Note:* To just "calculate," leave the *Shade P value* option unchecked.

- The results should now appear in the spreadsheet.

| | A calcium | B placebo | C | D =tTest_2S |
|---|---|---|---|---|
| 1 | 7 | -1 | Title | 2-Samp... |
| 2 | -4 | 12 | Alternate... | μ1 > μ2 |
| 3 | 18 | -1 | t | 1.60372 |
| 4 | 17 | -3 | PVal | 0.06442 |
| 5 | -3 | 3 | df | 15.5905 |

If you check the *Shade P value* box, the appropriate *t* distribution will also be displayed, showing the same results and the shaded area corresponding to the *P*-value.

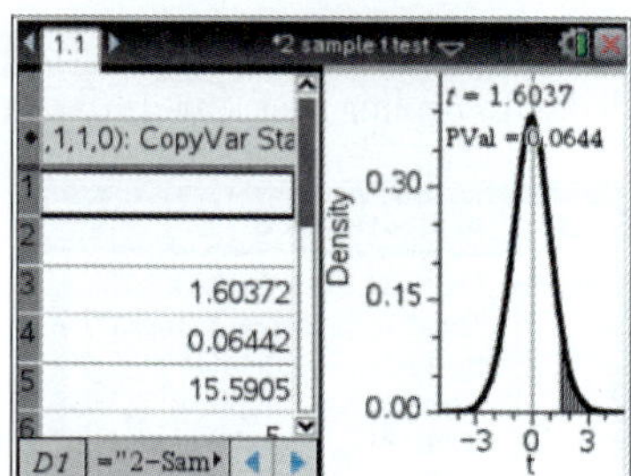


# Chapter 11 TI-Nspire Technology Corners

## 25. Finding *P*-values for chi-square tests on the calculator

To find the *P*-value in the M&M'S® example on page 685 with your calculator, use the $\chi^2$cdf command. We ask for the area between $\chi^2 = 10.180$ and a very large number (we'll use 10,000) under the chi-square density curve with 5 degrees of freedom.

- Press (scratchpad) and the *Calculator Scratchpad* should appear.
- Press (menu) → *Statistics* → *Distributions* → $\chi^2$*Cdf*.
- In the dialogue box that appears, enter the values shown in the following box. (tab) to [OK] and press (enter).

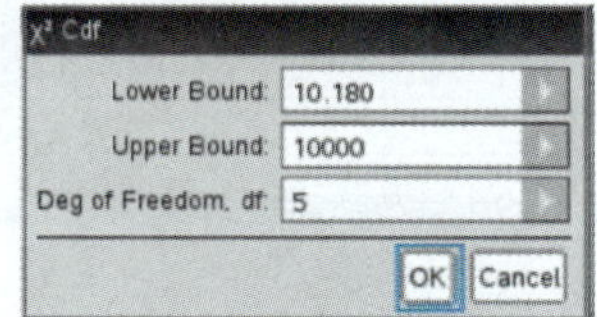


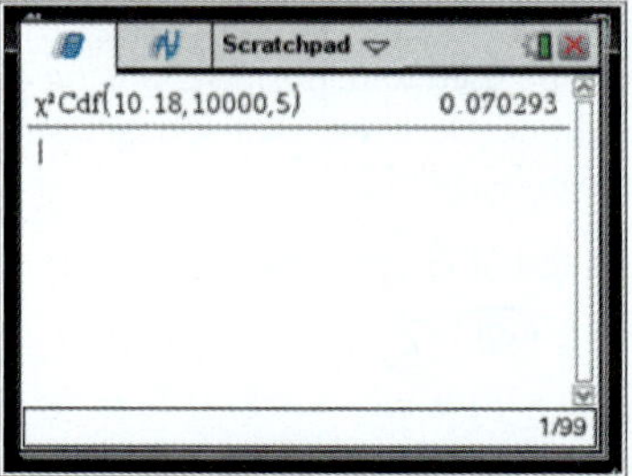


As the calculator screen shot shows, this method gives a more precise *P*-value than Table C.

## 26. Chi-square test for goodness of fit on the calculator

You can use the TI-Nspire to perform the calculations for a chi-square test for goodness of fit. We'll use the data from the hockey and birthdays example on page 688 to illustrate the steps.

1. Enter the observed counts and expected counts in two separate columns in a *Lists & Spreadsheet* page. Name the columns **observed** and **expected**.

| Birthday | Observed | Expected |
|---|---|---|
| Jan-Mar | 32 | 20 |
| Apr-June | 20 | 20 |
| July-Aug | 16 | 20 |
| Sept-Dec | 12 | 20 |

2. Perform a chi-square test for goodness of fit.
   - Press (menu) → *Statistics* → *Stat Tests* → $\chi^2$ GOF.
   - In the dialogue box that appears, enter the values shown in the following box. (tab) to [OK] and press (enter).

If you leave the *Shade P value* box unchecked, you'll get the test results within the spreadsheet containing the test statistic, *P-value*, and df. If you check the *Shade P value* box, you'll get a picture of the appropriate chi-square distribution with the test statistic marked and shaded area corresponding to the *P*-value.

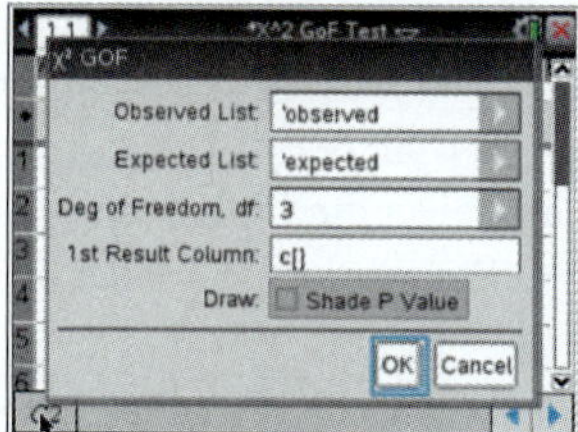

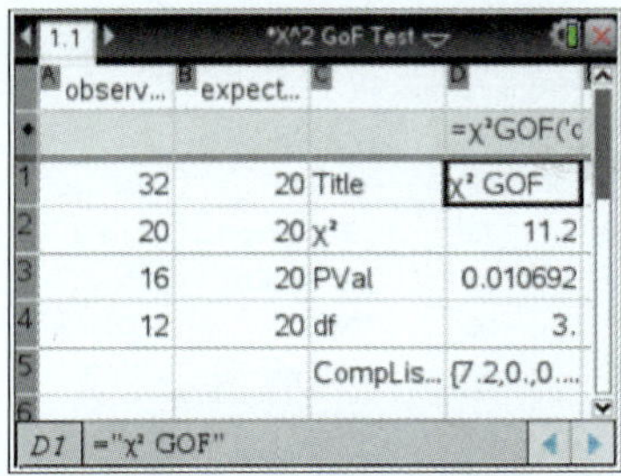


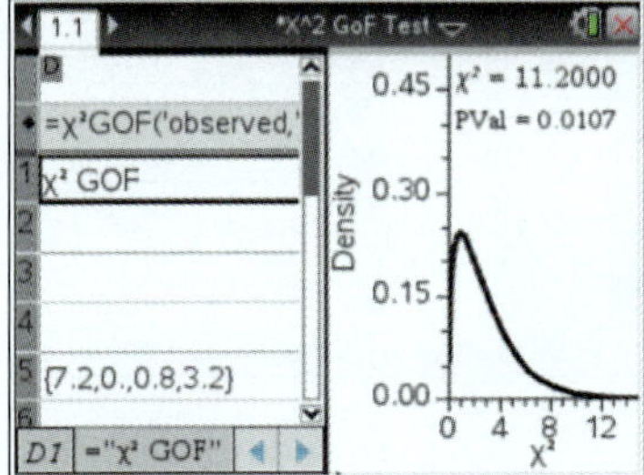


We'll discuss the *Comp List* results later.

## 27. Chi-square tests for two-way tables on the calculator

You can use the TI-Nspire to perform calculations for a chi-square test for homogeneity. We'll use the data from the restaurant study on page 704 to illustrate the process.

1. Press (scratchpad) ((home) (A)) to insert a *Calculator Scratchpad*.
2. Define a matrix by doing the following:
   - Name your matrix by typing **musicinfluence** (ctrl) [:=] (template). A box will appear with different math type options. Select (matrix) and enter "3" for *Number of rows* and "3" for *Number of columns*.

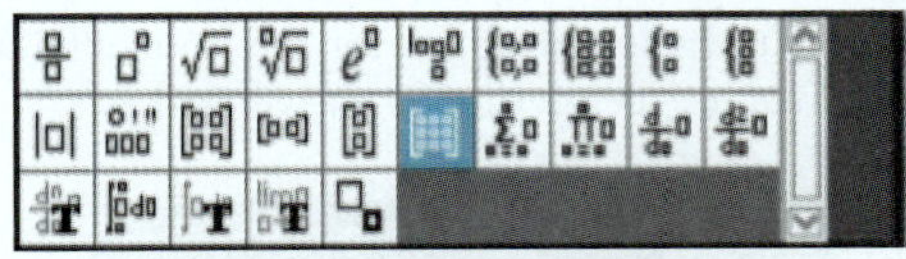

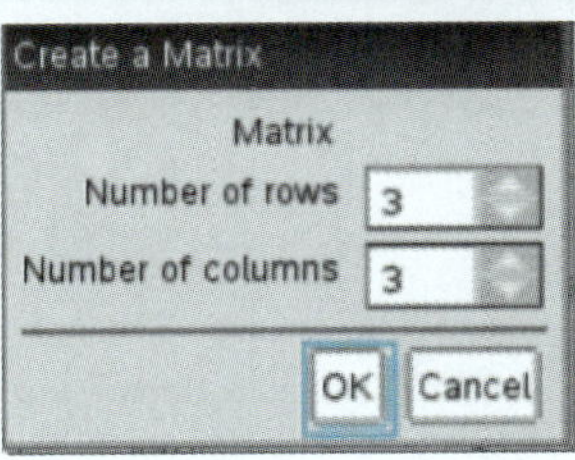


   - Type in the corresponding row data, pressing (tab) between entries. Press (enter) when finished.

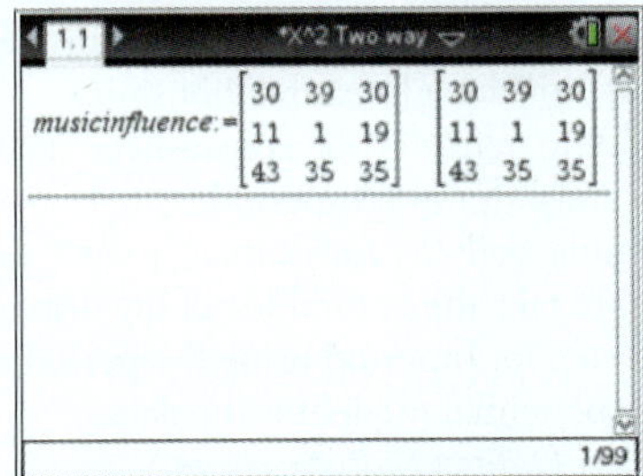


3. To perform the chi-square test, do the following steps:
   - Press (menu) → *Statistics* → *Stat tests* → $\chi^2$ *2-way Test*.
   - Specify the observed matrix, (tab) to (OK), and press (enter).
   - The results will be displayed and the expected matrix and component matrix will be calculated.

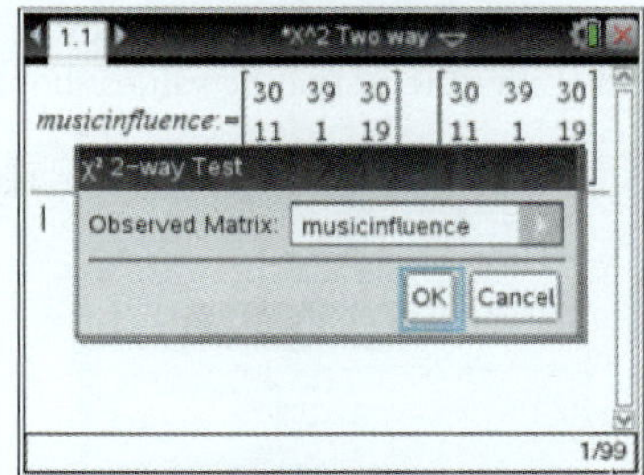


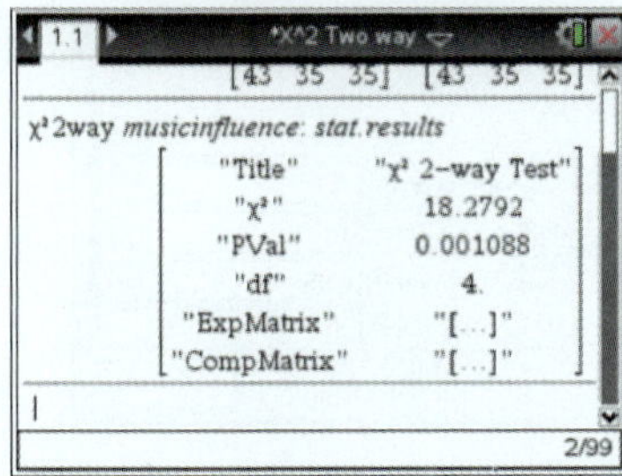


4. To see the expected counts and component matrix, press (var) and select **stat.expmatrix** for the expected matrix or **stat.compmatrix** for the component matrix.

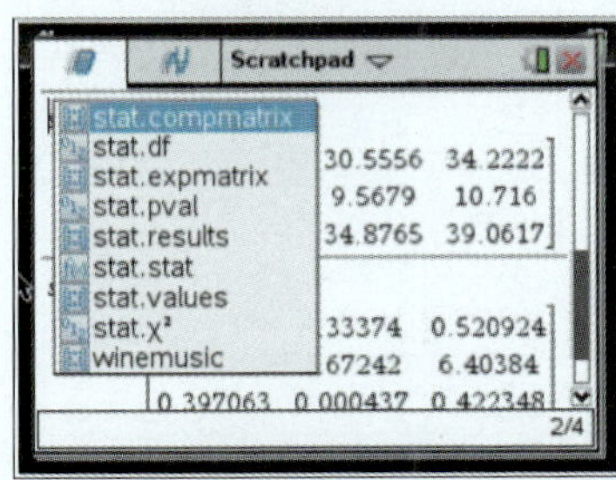


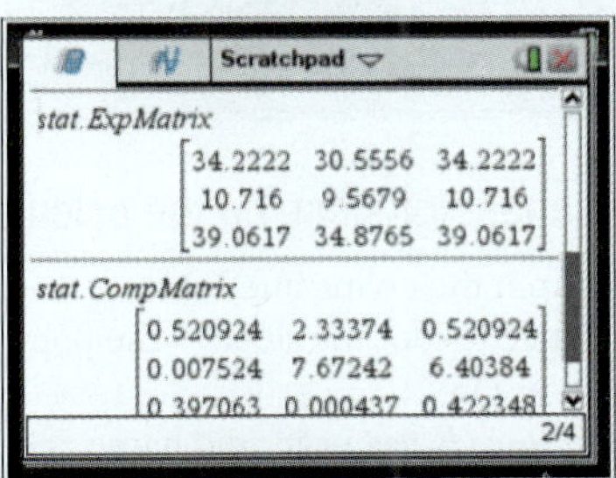


# Chapter 12 TI-Nspire Technology Corners

## 28. Confidence interval for slope on the calculator

Let's use the data from the Ford F-150 truck example on page A-9 to construct a confidence interval for the slope of a population (true) regression line on the TI-Nspire.

1. Insert a *Lists & Spreadsheet* page, and name column A **miles** and column B **price**. Type the corresponding values into each column.
2. To construct a confidence interval:
   - Press (menu) → *Statistics* → *Confidence Intervals* → *Linear Reg t Intervals*.
   - In the first dialogue box, select *Slope*. (tab) to (OK) and press (enter).

- In the next dialogue box, select **miles** for the X *List* and **price** for the Y *List*. Enter the rest of the values shown. (tab) to [OK] and press (enter).

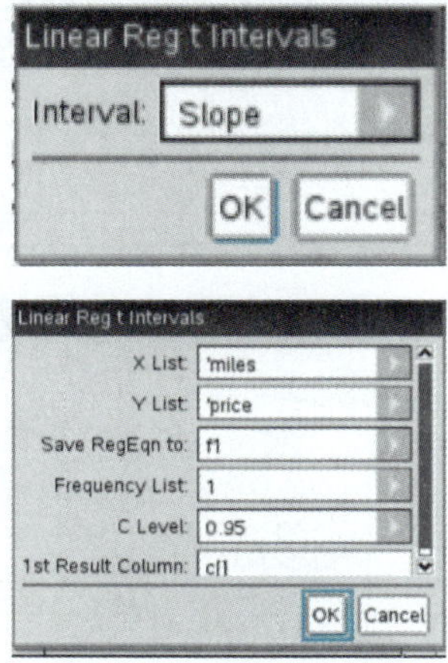


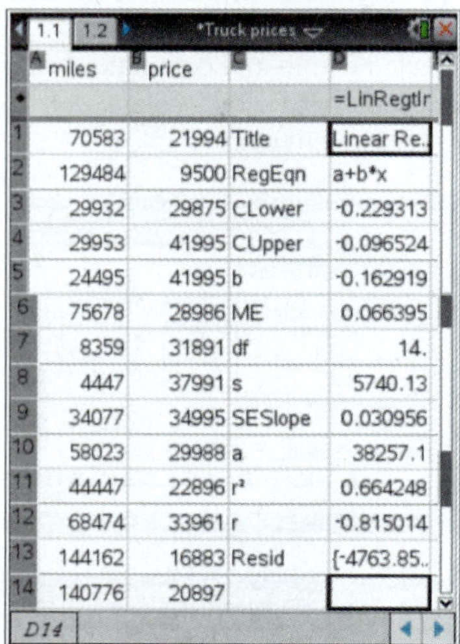

*Truck prices

| | A miles | B price | C | D =LinRegtIr |
|---|---|---|---|---|
| 1 | 70583 | 21994 | Title | Linear Re. |
| 2 | 129484 | 9500 | RegEqn | a+b*x |
| 3 | 29932 | 29875 | CLower | -0.229313 |
| 4 | 29953 | 41995 | CUpper | -0.096524 |
| 5 | 24495 | 41995 | b | -0.162919 |
| 6 | 75678 | 28986 | ME | 0.066395 |
| 7 | 8359 | 31891 | df | 14. |
| 8 | 4447 | 37991 | s | 5740.13 |
| 9 | 34077 | 34995 | SESlope | 0.030956 |
| 10 | 58023 | 29988 | a | 38257.1 |
| 11 | 44447 | 22896 | r² | 0.664248 |
| 12 | 68474 | 33961 | r | -0.815014 |
| 13 | 144162 | 16883 | Resid | {-4763.85.. |
| 14 | 140776 | 20897 | | |

D14

## 29. Significance test for slope on the calculator

Let's use the data from the crying and IQ study on page 754 to perform a significance test for the slope of the population regression line on the TI-Nspire.

1. Insert a *Lists & Spreadsheet* page, and name column A **crycount** and column B **iqscore**. Type the corresponding values into each column.
2. To do a significance test:
   - Press (menu) → *Statistics* → *Stat Tests* → *Linear Reg t Test*.
   - Select **crycount** for the X *List* and **iqscore** for the Y *List*. Enter the rest of the values as shown. (tab) to [OK] and press (enter).

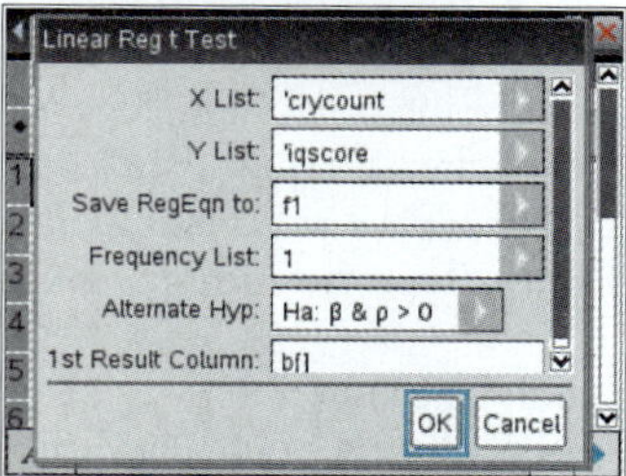


*Crying IQscore

| | A crycount | B iqscore | C | D =LinRegtT |
|---|---|---|---|---|
| 1 | 10 | 87 | Title | Linear Re.. |
| 2 | 12 | 97 | Alternate... | β & ρ ≠ 0... |
| 3 | 9 | 103 | RegEqn | a+b*x |
| 4 | 16 | 106 | t | 3.06549 |
| 5 | 18 | 109 | PVal | 0.004105 |
| 6 | 15 | 114 | df | 36. |
| 7 | 12 | 119 | a | 91.2683 |
| 8 | 20 | 132 | b | 1.4929 |
| 9 | 16 | 136 | s | 17.4987 |
| 10 | 33 | 159 | SESlope | 0.487001 |
| 11 | 20 | 90 | r² | 0.207 |
| 12 | 16 | 100 | r | 0.454973 |
| 13 | 23 | 103 | Resid | {-19.1972.. |
| 14 | 27 | 108 | | |

D =LinRegtTest('crycount,'iqscore,1,0

## 30. Transforming to achieve linearity on the calculator

We'll use the planet data on page 779 to illustrate a general strategy for performing transformations with logarithms on the TI-Nspire. A similar approach could be used for transforming data with powers and roots.

1. Insert a *Lists & Spreadsheet* page, and name column A **distance** and column B **period**. Type the corresponding values into each column.
2. Make a scatterplot of $y$ versus $x$ and confirm that there is a curved pattern.
   - Insert a *Data & Statistics* page. Press (ctrl) (I) and select *Add Data & Statistics*.
   - Press (tab) and select **distance** for the horizontal axis. Press (tab) again and select **period** for the vertical axis.

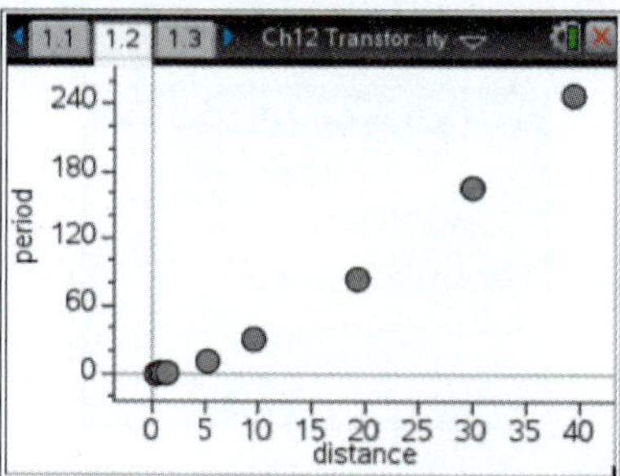


3. To "straighten" the curve (that is, determine the relationship), we can use different models of the explanatory-response data to see which one provides a linear relationship.
   - Press (ctrl) ◀ to return your spreadsheet. Name column c **lndistance** and column d **lnperiod**.
   - In the formula cell for **lndistance**, press (enter) and enter *ln(distance)* to take the natural log of the distance values.
   - Repeat this step for **lnperiod** using the **period** data.
4. To see if an exponential model fits the data:
   - Insert another *Data & Statistics* page.
   - Put **distance** on the horizontal axis and **lnperiod** on the vertical axis. If the relationship looks linear, then an exponential model is appropriate.

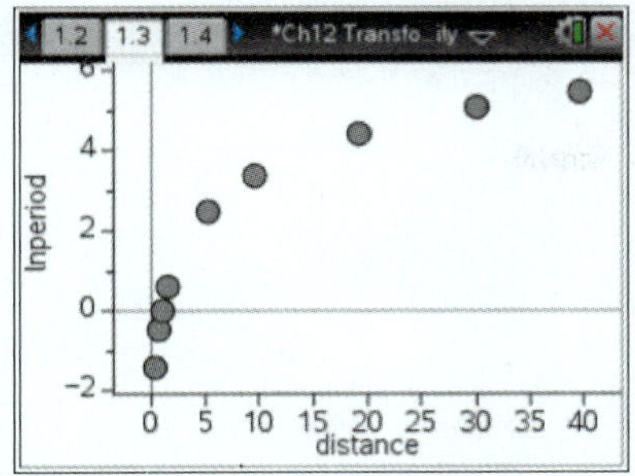


5. To see if a power model fits the data:
   - Using the same *Data & Statistics* page, change the horizontal axis to **lndistance**.
   - If this relationship looks linear, then a power model is appropriate.

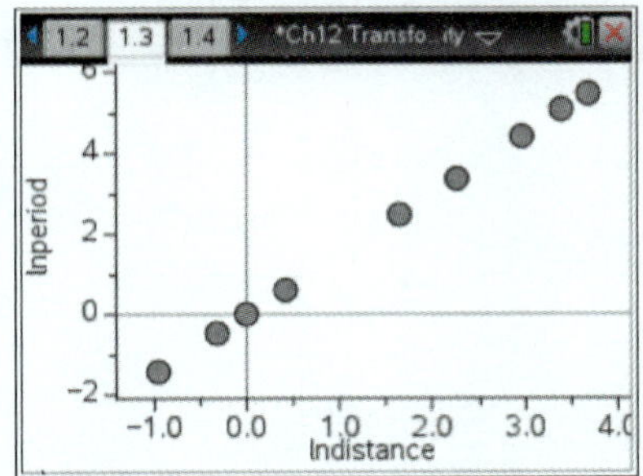


6. If a linear pattern is present, calculate the equation of the least-squares regression line:
   - In the spreadsheet, press (menu) → *Statistics* → *Calculations* → *Linear Regression*$(a + bx)$.
   - In the dialogue box, select **lndistance** for *X List*, **lnperiod** for *Y List*, and enter the rest of the values as shown. (tab) to [OK] and press (enter).

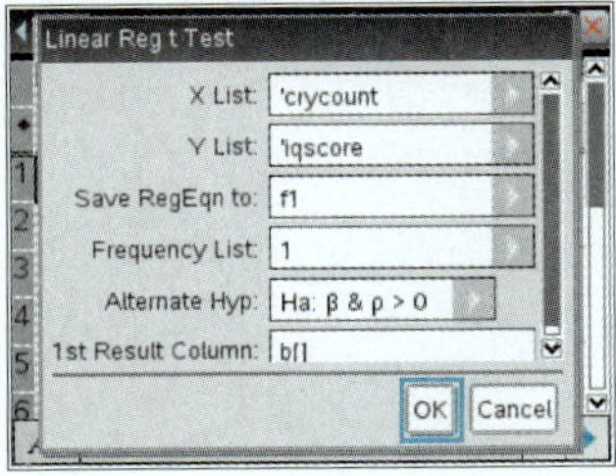


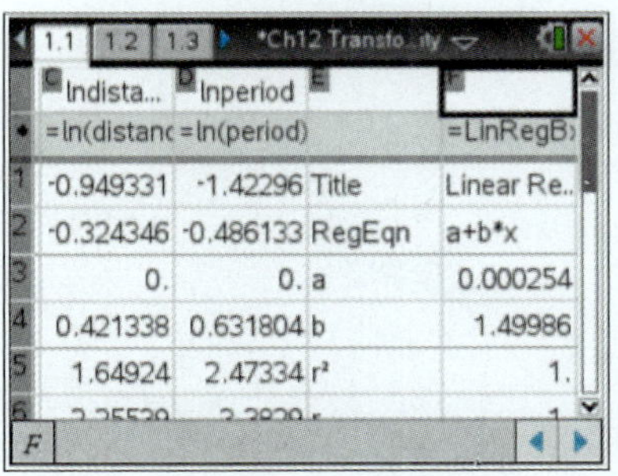


7. Construct a residual plot to look for any departures from the linear pattern.
   - Insert another *Data & Statistics* page.
   - For the horizontal axis select **lndistance**. For *Ylist*, use the **stat.resid** list stored in the calculator.

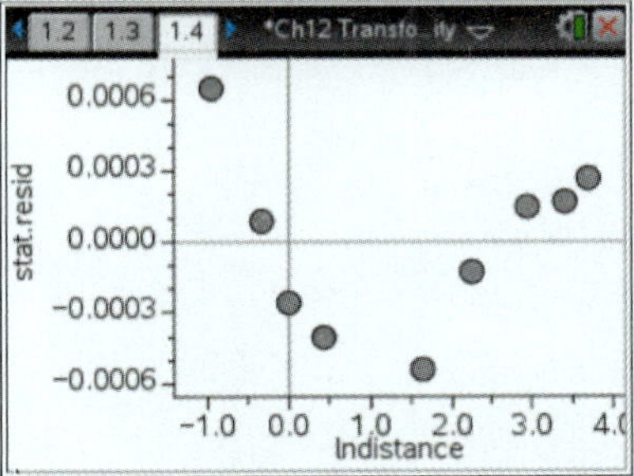


8. To make a prediction for a specific value of the explanatory variable, compute $\log(x)$ or $\ln(x)$, if appropriate. Then do **f1**$(k)$ to obtain the predicted value of $\log y$ or $\ln y$. To get the predicted value of $y$, do 10^Ans or $e$^Ans to undo the logarithm transformation. Here's our prediction of the period of revolution for Eris, which is at a distance of 102.15 AU from the sun.

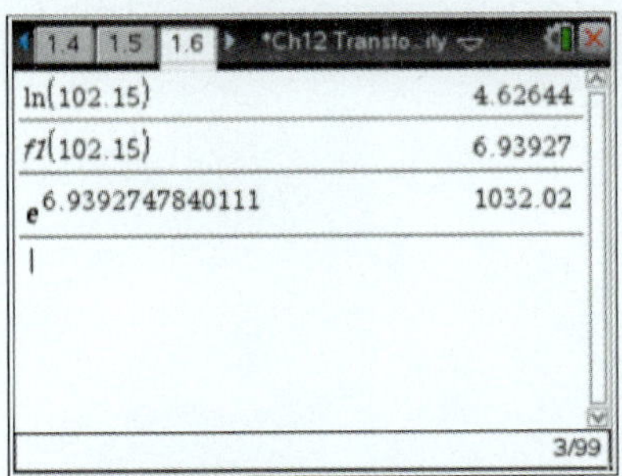

# Formulas for AP® Statistics Exam

Students are provided with the following formulas on both the multiple choice and free-response sections of the AP® Statistics exam.

## I. Descriptive Statistics

$$\bar{x} = \frac{\sum x_i}{n}$$

$$s_x = \sqrt{\frac{1}{n-1}\sum (x_i - \bar{x})^2}$$

$$s_p = \sqrt{\frac{(n_1 - 1)s_1^2 + (n_2 - 1)s_2^2}{(n_1 - 1) + (n_2 - 1)}}$$

$$\hat{y} = b_0 + b_1 x$$

$$b_1 = \frac{\sum (x_i - \bar{x})(y_i - \bar{y})}{\sum (x_i - \bar{x})^2}$$

$$b_0 = \bar{y} - b_1\bar{x}$$

$$r = \frac{1}{n-1}\sum\left(\frac{x_i - \bar{x}}{s_x}\right)\left(\frac{y_i - \bar{y}}{s_y}\right)$$

$$b_1 = r\frac{s_y}{s_x}$$

$$s_{b_1} = \frac{\sqrt{\dfrac{\sum (y_i - \hat{y}_i)^2}{n-2}}}{\sqrt{\sum (x_i - \bar{x})^2}}$$

## II. Probability

$$P(A \cup B) = P(A) + P(B) - P(A \cap B)$$

$$P(A|B) = \frac{P(A \cap B)}{P(B)}$$

$$E(X) = \mu_x = \sum x_i p_i$$

$$\text{Var}(X) = \sigma_x^2 = \sum (x_i - \mu_x)^2 p_i$$

If X has a binomial distribution with parameters $n$ and $p$, then:

$$P(X = k) = \binom{n}{k} p^k (1-p)^{n-k}$$

$$\mu_x = np$$

$$\sigma_x = \sqrt{np(1-p)}$$

$$\mu_{\hat{p}} = p$$

$$\sigma_{\hat{p}} = \sqrt{\frac{p(1-p)}{n}}$$

If $\bar{x}$ is the mean of a random sample of size $n$ from an infinite population with mean $\mu$ and standard deviation $\sigma$, then:

$$\mu_{\bar{x}} = \mu$$

$$\sigma_{\bar{x}} = \frac{\sigma}{\sqrt{n}}$$

## III. Inferential Statistics

Standardized test statistic: $\dfrac{\text{statistic} - \text{parameter}}{\text{standard deviation of statistic}}$

Confidence interval: statistic $\pm$ (critical value) $\cdot$ (std. deviation of statistic)

### Single-Sample

| Statistic | Standard Deviation of Statistic |
|---|---|
| Sample Mean | $\dfrac{\sigma}{\sqrt{n}}$ |
| Sample Proportion | $\sqrt{\dfrac{p(1-p)}{n}}$ |

### Two-Sample

| Statistic | Standard Deviation of Statistic |
|---|---|
| Difference of sample means | $\sqrt{\dfrac{\sigma_1^2}{n_1} + \dfrac{\sigma_2^2}{n_2}}$<br>Special case when $\sigma_1 = \sigma_2$<br>$\sigma\sqrt{\dfrac{1}{n_1} + \dfrac{1}{n_2}}$ |
| Difference of sample proportions | $\sqrt{\dfrac{p_1(1-p_1)}{n_1} + \dfrac{p_2(1-p_2)}{n_2}}$<br>Special case when $p_1 = p_2$<br>$\sqrt{p(1-p)}\sqrt{\dfrac{1}{n_1} + \dfrac{1}{n_2}}$ |

Chi-square test statistic $= \sum \dfrac{(\text{observed} - \text{expected})^2}{\text{expected}}$

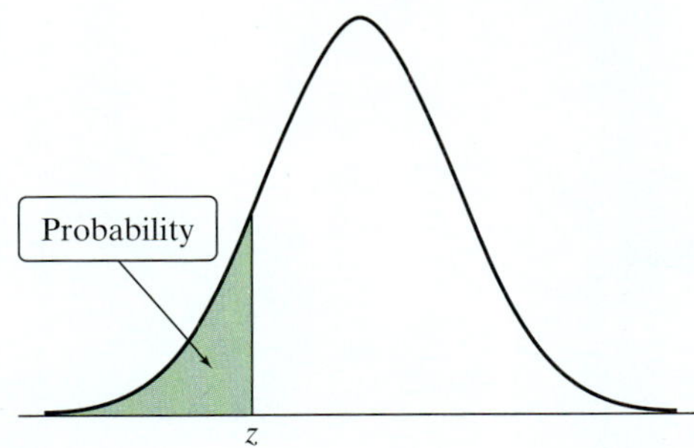


Table entry for $z$ is the area under the standard Normal curve to the left of $z$.

**Table A Standard Normal probabilities**

| $z$ | .00 | .01 | .02 | .03 | .04 | .05 | .06 | .07 | .08 | .09 |
|---|---|---|---|---|---|---|---|---|---|---|
| −3.4 | .0003 | .0003 | .0003 | .0003 | .0003 | .0003 | .0003 | .0003 | .0003 | .0002 |
| −3.3 | .0005 | .0005 | .0005 | .0004 | .0004 | .0004 | .0004 | .0004 | .0004 | .0003 |
| −3.2 | .0007 | .0007 | .0006 | .0006 | .0006 | .0006 | .0006 | .0005 | .0005 | .0005 |
| −3.1 | .0010 | .0009 | .0009 | .0009 | .0008 | .0008 | .0008 | .0008 | .0007 | .0007 |
| −3.0 | .0013 | .0013 | .0013 | .0012 | .0012 | .0011 | .0011 | .0011 | .0010 | .0010 |
| −2.9 | .0019 | .0018 | .0018 | .0017 | .0016 | .0016 | .0015 | .0015 | .0014 | .0014 |
| −2.8 | .0026 | .0025 | .0024 | .0023 | .0023 | .0022 | .0021 | .0021 | .0020 | .0019 |
| −2.7 | .0035 | .0034 | .0033 | .0032 | .0031 | .0030 | .0029 | .0028 | .0027 | .0026 |
| −2.6 | .0047 | .0045 | .0044 | .0043 | .0041 | .0040 | .0039 | .0038 | .0037 | .0036 |
| −2.5 | .0062 | .0060 | .0059 | .0057 | .0055 | .0054 | .0052 | .0051 | .0049 | .0048 |
| −2.4 | .0082 | .0080 | .0078 | .0075 | .0073 | .0071 | .0069 | .0068 | .0066 | .0064 |
| −2.3 | .0107 | .0104 | .0102 | .0099 | .0096 | .0094 | .0091 | .0089 | .0087 | .0084 |
| −2.2 | .0139 | .0136 | .0132 | .0129 | .0125 | .0122 | .0119 | .0116 | .0113 | .0110 |
| −2.1 | .0179 | .0174 | .0170 | .0166 | .0162 | .0158 | .0154 | .0150 | .0146 | .0143 |
| −2.0 | .0228 | .0222 | .0217 | .0212 | .0207 | .0202 | .0197 | .0192 | .0188 | .0183 |
| −1.9 | .0287 | .0281 | .0274 | .0268 | .0262 | .0256 | .0250 | .0244 | .0239 | .0233 |
| −1.8 | .0359 | .0351 | .0344 | .0336 | .0329 | .0322 | .0314 | .0307 | .0301 | .0294 |
| −1.7 | .0446 | .0436 | .0427 | .0418 | .0409 | .0401 | .0392 | .0384 | .0375 | .0367 |
| −1.6 | .0548 | .0537 | .0526 | .0516 | .0505 | .0495 | .0485 | .0475 | .0465 | .0455 |
| −1.5 | .0668 | .0655 | .0643 | .0630 | .0618 | .0606 | .0594 | .0582 | .0571 | .0559 |
| −1.4 | .0808 | .0793 | .0778 | .0764 | .0749 | .0735 | .0721 | .0708 | .0694 | .0681 |
| −1.3 | .0968 | .0951 | .0934 | .0918 | .0901 | .0885 | .0869 | .0853 | .0838 | .0823 |
| −1.2 | .1151 | .1131 | .1112 | .1093 | .1075 | .1056 | .1038 | .1020 | .1003 | .0985 |
| −1.1 | .1357 | .1335 | .1314 | .1292 | .1271 | .1251 | .1230 | .1210 | .1190 | .1170 |
| −1.0 | .1587 | .1562 | .1539 | .1515 | .1492 | .1469 | .1446 | .1423 | .1401 | .1379 |
| −0.9 | .1841 | .1814 | .1788 | .1762 | .1736 | .1711 | .1685 | .1660 | .1635 | .1611 |
| −0.8 | .2119 | .2090 | .2061 | .2033 | .2005 | .1977 | .1949 | .1922 | .1894 | .1867 |
| −0.7 | .2420 | .2389 | .2358 | .2327 | .2296 | .2266 | .2236 | .2206 | .2177 | .2148 |
| −0.6 | .2743 | .2709 | .2676 | .2643 | .2611 | .2578 | .2546 | .2514 | .2483 | .2451 |
| −0.5 | .3085 | .3050 | .3015 | .2981 | .2946 | .2912 | .2877 | .2843 | .2810 | .2776 |
| −0.4 | .3446 | .3409 | .3372 | .3336 | .3300 | .3264 | .3228 | .3192 | .3156 | .3121 |
| −0.3 | .3821 | .3783 | .3745 | .3707 | .3669 | .3632 | .3594 | .3557 | .3520 | .3483 |
| −0.2 | .4207 | .4168 | .4129 | .4090 | .4052 | .4013 | .3974 | .3936 | .3897 | .3859 |
| −0.1 | .4602 | .4562 | .4522 | .4483 | .4443 | .4404 | .4364 | .4325 | .4286 | .4247 |
| −0.0 | .5000 | .4960 | .4920 | .4880 | .4840 | .4801 | .4761 | .4721 | .4681 | .4641 |

*(Continued)*

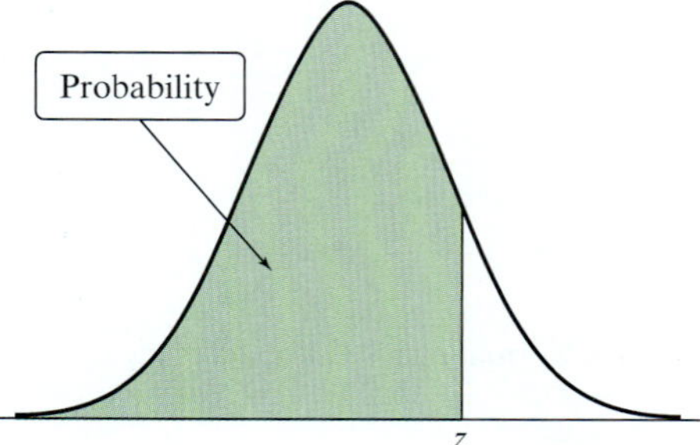


Table entry for $z$ is the area under the standard Normal curve to the left of $z$.

**Table A Standard Normal probabilities (continued)**

| $z$ | .00 | .01 | .02 | .03 | .04 | .05 | .06 | .07 | .08 | .09 |
|---|---|---|---|---|---|---|---|---|---|---|
| 0.0 | .5000 | .5040 | .5080 | .5120 | .5160 | .5199 | .5239 | .5279 | .5319 | .5359 |
| 0.1 | .5398 | .5438 | .5478 | .5517 | .5557 | .5596 | .5636 | .5675 | .5714 | .5753 |
| 0.2 | .5793 | .5832 | .5871 | .5910 | .5948 | .5987 | .6026 | .6064 | .6103 | .6141 |
| 0.3 | .6179 | .6217 | .6255 | .6293 | .6331 | .6368 | .6406 | .6443 | .6480 | .6517 |
| 0.4 | .6554 | .6591 | .6628 | .6664 | .6700 | .6736 | .6772 | .6808 | .6844 | .6879 |
| 0.5 | .6915 | .6950 | .6985 | .7019 | .7054 | .7088 | .7123 | .7157 | .7190 | .7224 |
| 0.6 | .7257 | .7291 | .7324 | .7357 | .7389 | .7422 | .7454 | .7486 | .7517 | .7549 |
| 0.7 | .7580 | .7611 | .7642 | .7673 | .7704 | .7734 | .7764 | .7794 | .7823 | .7852 |
| 0.8 | .7881 | .7910 | .7939 | .7967 | .7995 | .8023 | .8051 | .8078 | .8106 | .8133 |
| 0.9 | .8159 | .8186 | .8212 | .8238 | .8264 | .8289 | .8315 | .8340 | .8365 | .8389 |
| 1.0 | .8413 | .8438 | .8461 | .8485 | .8508 | .8531 | .8554 | .8577 | .8599 | .8621 |
| 1.1 | .8643 | .8665 | .8686 | .8708 | .8729 | .8749 | .8770 | .8790 | .8810 | .8830 |
| 1.2 | .8849 | .8869 | .8888 | .8907 | .8925 | .8944 | .8962 | .8980 | .8997 | .9015 |
| 1.3 | .9032 | .9049 | .9066 | .9082 | .9099 | .9115 | .9131 | .9147 | .9162 | .9177 |
| 1.4 | .9192 | .9207 | .9222 | .9236 | .9251 | .9265 | .9279 | .9292 | .9306 | .9319 |
| 1.5 | .9332 | .9345 | .9357 | .9370 | .9382 | .9394 | .9406 | .9418 | .9429 | .9441 |
| 1.6 | .9452 | .9463 | .9474 | .9484 | .9495 | .9505 | .9515 | .9525 | .9535 | .9545 |
| 1.7 | .9554 | .9564 | .9573 | .9582 | .9591 | .9599 | .9608 | .9616 | .9625 | .9633 |
| 1.8 | .9641 | .9649 | .9656 | .9664 | .9671 | .9678 | .9686 | .9693 | .9699 | .9706 |
| 1.9 | .9713 | .9719 | .9726 | .9732 | .9738 | .9744 | .9750 | .9756 | .9761 | .9767 |
| 2.0 | .9772 | .9778 | .9783 | .9788 | .9793 | .9798 | .9803 | .9808 | .9812 | .9817 |
| 2.1 | .9821 | .9826 | .9830 | .9834 | .9838 | .9842 | .9846 | .9850 | .9854 | .9857 |
| 2.2 | .9861 | .9864 | .9868 | .9871 | .9875 | .9878 | .9881 | .9884 | .9887 | .9890 |
| 2.3 | .9893 | .9896 | .9898 | .9901 | .9904 | .9906 | .9909 | .9911 | .9913 | .9916 |
| 2.4 | .9918 | .9920 | .9922 | .9925 | .9927 | .9929 | .9931 | .9932 | .9934 | .9936 |
| 2.5 | .9938 | .9940 | .9941 | .9943 | .9945 | .9946 | .9948 | .9949 | .9951 | .9952 |
| 2.6 | .9953 | .9955 | .9956 | .9957 | .9959 | .9960 | .9961 | .9962 | .9963 | .9964 |
| 2.7 | .9965 | .9966 | .9967 | .9968 | .9969 | .9970 | .9971 | .9972 | .9973 | .9974 |
| 2.8 | .9974 | .9975 | .9976 | .9977 | .9977 | .9978 | .9979 | .9979 | .9980 | .9981 |
| 2.9 | .9981 | .9982 | .9982 | .9983 | .9984 | .9984 | .9985 | .9985 | .9986 | .9986 |
| 3.0 | .9987 | .9987 | .9987 | .9988 | .9988 | .9989 | .9989 | .9989 | .9990 | .9990 |
| 3.1 | .9990 | .9991 | .9991 | .9991 | .9992 | .9992 | .9992 | .9992 | .9993 | .9993 |
| 3.2 | .9993 | .9993 | .9994 | .9994 | .9994 | .9994 | .9994 | .9995 | .9995 | .9995 |
| 3.3 | .9995 | .9995 | .9995 | .9996 | .9996 | .9996 | .9996 | .9996 | .9996 | .9997 |
| 3.4 | .9997 | .9997 | .9997 | .9997 | .9997 | .9997 | .9997 | .9997 | .9997 | .9998 |

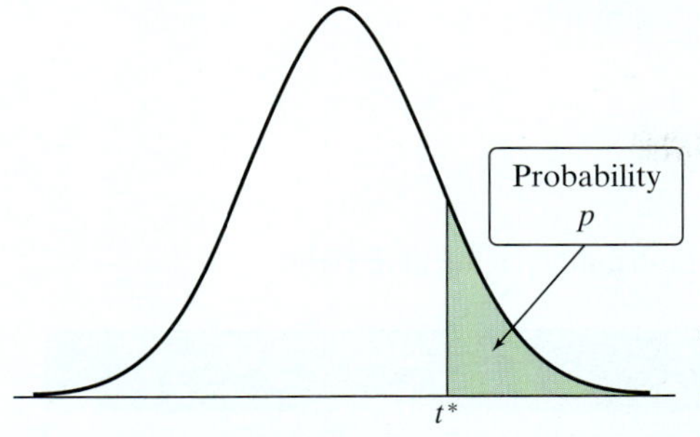


Table entry for $p$ and C is the point $t^*$ with probability $p$ lying to its right and probability C lying between $-t^*$ and $t^*$.

**Table B  *t* distribution critical values**

| | Tail probability *p* | | | | | | | | | | | |
|---|---|---|---|---|---|---|---|---|---|---|---|---|
| **df** | **.25** | **.20** | **.15** | **.10** | **.05** | **.025** | **.02** | **.01** | **.005** | **.0025** | **.001** | **.0005** |
| 1 | 1.000 | 1.376 | 1.963 | 3.078 | 6.314 | 12.71 | 15.89 | 31.82 | 63.66 | 127.3 | 318.3 | 636.6 |
| 2 | 0.816 | 1.061 | 1.386 | 1.886 | 2.920 | 4.303 | 4.849 | 6.965 | 9.925 | 14.09 | 22.33 | 31.60 |
| 3 | 0.765 | 0.978 | 1.250 | 1.638 | 2.353 | 3.182 | 3.482 | 4.541 | 5.841 | 7.453 | 10.21 | 12.92 |
| 4 | 0.741 | 0.941 | 1.190 | 1.533 | 2.132 | 2.776 | 2.999 | 3.747 | 4.604 | 5.598 | 7.173 | 8.610 |
| 5 | 0.727 | 0.920 | 1.156 | 1.476 | 2.015 | 2.571 | 2.757 | 3.365 | 4.032 | 4.773 | 5.893 | 6.869 |
| 6 | 0.718 | 0.906 | 1.134 | 1.440 | 1.943 | 2.447 | 2.612 | 3.143 | 3.707 | 4.317 | 5.208 | 5.959 |
| 7 | 0.711 | 0.896 | 1.119 | 1.415 | 1.895 | 2.365 | 2.517 | 2.998 | 3.499 | 4.029 | 4.785 | 5.408 |
| 8 | 0.706 | 0.889 | 1.108 | 1.397 | 1.860 | 2.306 | 2.449 | 2.896 | 3.355 | 3.833 | 4.501 | 5.041 |
| 9 | 0.703 | 0.883 | 1.100 | 1.383 | 1.833 | 2.262 | 2.398 | 2.821 | 3.250 | 3.690 | 4.297 | 4.781 |
| 10 | 0.700 | 0.879 | 1.093 | 1.372 | 1.812 | 2.228 | 2.359 | 2.764 | 3.169 | 3.581 | 4.144 | 4.587 |
| 11 | 0.697 | 0.876 | 1.088 | 1.363 | 1.796 | 2.201 | 2.328 | 2.718 | 3.106 | 3.497 | 4.025 | 4.437 |
| 12 | 0.695 | 0.873 | 1.083 | 1.356 | 1.782 | 2.179 | 2.303 | 2.681 | 3.055 | 3.428 | 3.930 | 4.318 |
| 13 | 0.694 | 0.870 | 1.079 | 1.350 | 1.771 | 2.160 | 2.282 | 2.650 | 3.012 | 3.372 | 3.852 | 4.221 |
| 14 | 0.692 | 0.868 | 1.076 | 1.345 | 1.761 | 2.145 | 2.264 | 2.624 | 2.977 | 3.326 | 3.787 | 4.140 |
| 15 | 0.691 | 0.866 | 1.074 | 1.341 | 1.753 | 2.131 | 2.249 | 2.602 | 2.947 | 3.286 | 3.733 | 4.073 |
| 16 | 0.690 | 0.865 | 1.071 | 1.337 | 1.746 | 2.120 | 2.235 | 2.583 | 2.921 | 3.252 | 3.686 | 4.015 |
| 17 | 0.689 | 0.863 | 1.069 | 1.333 | 1.740 | 2.110 | 2.224 | 2.567 | 2.898 | 3.222 | 3.646 | 3.965 |
| 18 | 0.688 | 0.862 | 1.067 | 1.330 | 1.734 | 2.101 | 2.214 | 2.552 | 2.878 | 3.197 | 3.611 | 3.922 |
| 19 | 0.688 | 0.861 | 1.066 | 1.328 | 1.729 | 2.093 | 2.205 | 2.539 | 2.861 | 3.174 | 3.579 | 3.883 |
| 20 | 0.687 | 0.860 | 1.064 | 1.325 | 1.725 | 2.086 | 2.197 | 2.528 | 2.845 | 3.153 | 3.552 | 3.850 |
| 21 | 0.686 | 0.859 | 1.063 | 1.323 | 1.721 | 2.080 | 2.189 | 2.518 | 2.831 | 3.135 | 3.527 | 3.819 |
| 22 | 0.686 | 0.858 | 1.061 | 1.321 | 1.717 | 2.074 | 2.183 | 2.508 | 2.819 | 3.119 | 3.505 | 3.792 |
| 23 | 0.685 | 0.858 | 1.060 | 1.319 | 1.714 | 2.069 | 2.177 | 2.500 | 2.807 | 3.104 | 3.485 | 3.768 |
| 24 | 0.685 | 0.857 | 1.059 | 1.318 | 1.711 | 2.064 | 2.172 | 2.492 | 2.797 | 3.091 | 3.467 | 3.745 |
| 25 | 0.684 | 0.856 | 1.058 | 1.316 | 1.708 | 2.060 | 2.167 | 2.485 | 2.787 | 3.078 | 3.450 | 3.725 |
| 26 | 0.684 | 0.856 | 1.058 | 1.315 | 1.706 | 2.056 | 2.162 | 2.479 | 2.779 | 3.067 | 3.435 | 3.707 |
| 27 | 0.684 | 0.855 | 1.057 | 1.314 | 1.703 | 2.052 | 2.158 | 2.473 | 2.771 | 3.057 | 3.421 | 3.690 |
| 28 | 0.683 | 0.855 | 1.056 | 1.313 | 1.701 | 2.048 | 2.154 | 2.467 | 2.763 | 3.047 | 3.408 | 3.674 |
| 29 | 0.683 | 0.854 | 1.055 | 1.311 | 1.699 | 2.045 | 2.150 | 2.462 | 2.756 | 3.038 | 3.396 | 3.659 |
| 30 | 0.683 | 0.854 | 1.055 | 1.310 | 1.697 | 2.042 | 2.147 | 2.457 | 2.750 | 3.030 | 3.385 | 3.646 |
| 40 | 0.681 | 0.851 | 1.050 | 1.303 | 1.684 | 2.021 | 2.123 | 2.423 | 2.704 | 2.971 | 3.307 | 3.551 |
| 50 | 0.679 | 0.849 | 1.047 | 1.299 | 1.676 | 2.009 | 2.109 | 2.403 | 2.678 | 2.937 | 3.261 | 3.496 |
| 60 | 0.679 | 0.848 | 1.045 | 1.296 | 1.671 | 2.000 | 2.099 | 2.390 | 2.660 | 2.915 | 3.232 | 3.460 |
| 80 | 0.678 | 0.846 | 1.043 | 1.292 | 1.664 | 1.990 | 2.088 | 2.374 | 2.639 | 2.887 | 3.195 | 3.416 |
| 100 | 0.677 | 0.845 | 1.042 | 1.290 | 1.660 | 1.984 | 2.081 | 2.364 | 2.626 | 2.871 | 3.174 | 3.390 |
| 1000 | 0.675 | 0.842 | 1.037 | 1.282 | 1.646 | 1.962 | 2.056 | 2.330 | 2.581 | 2.813 | 3.098 | 3.300 |
| $\infty$ | 0.674 | 0.841 | 1.036 | 1.282 | 1.645 | 1.960 | 2.054 | 2.326 | 2.576 | 2.807 | 3.091 | 3.291 |
| | **50%** | **60%** | **70%** | **80%** | **90%** | **95%** | **96%** | **98%** | **99%** | **99.5%** | **99.8%** | **99.9%** |
| | Confidence level *C* | | | | | | | | | | | |

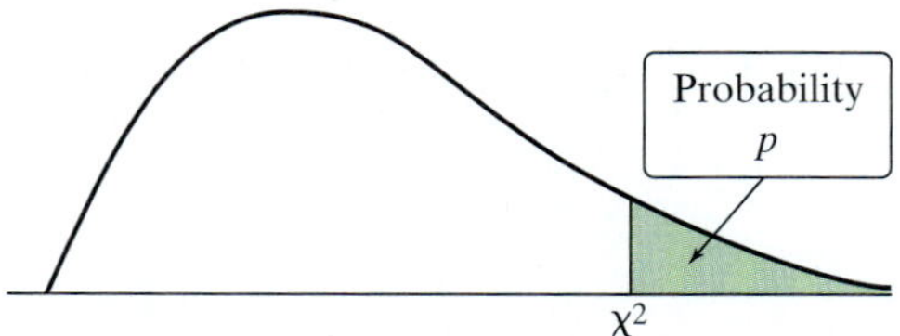


$\chi^2$

Table entry for $p$ is the point $\chi^2$ with probability $p$ lying to its right.

**Table C Chi-square distribution critical values**

| | Tail probability $p$ | | | | | | | | | | | |
|---|---|---|---|---|---|---|---|---|---|---|---|---|
| df | .25 | .20 | .15 | .10 | .05 | .025 | .02 | .01 | .005 | .0025 | .001 | .0005 |
| 1 | 1.32 | 1.64 | 2.07 | 2.71 | 3.84 | 5.02 | 5.41 | 6.63 | 7.88 | 9.14 | 10.83 | 12.12 |
| 2 | 2.77 | 3.22 | 3.79 | 4.61 | 5.99 | 7.38 | 7.82 | 9.21 | 10.60 | 11.98 | 13.82 | 15.20 |
| 3 | 4.11 | 4.64 | 5.32 | 6.25 | 7.81 | 9.35 | 9.84 | 11.34 | 12.84 | 14.32 | 16.27 | 17.73 |
| 4 | 5.39 | 5.99 | 6.74 | 7.78 | 9.49 | 11.14 | 11.67 | 13.28 | 14.86 | 16.42 | 18.47 | 20.00 |
| 5 | 6.63 | 7.29 | 8.12 | 9.24 | 11.07 | 12.83 | 13.39 | 15.09 | 16.75 | 18.39 | 20.51 | 22.11 |
| 6 | 7.84 | 8.56 | 9.45 | 10.64 | 12.59 | 14.45 | 15.03 | 16.81 | 18.55 | 20.25 | 22.46 | 24.10 |
| 7 | 9.04 | 9.80 | 10.75 | 12.02 | 14.07 | 16.01 | 16.62 | 18.48 | 20.28 | 22.04 | 24.32 | 26.02 |
| 8 | 10.22 | 11.03 | 12.03 | 13.36 | 15.51 | 17.53 | 18.17 | 20.09 | 21.95 | 23.77 | 26.12 | 27.87 |
| 9 | 11.39 | 12.24 | 13.29 | 14.68 | 16.92 | 19.02 | 19.68 | 21.67 | 23.59 | 25.46 | 27.88 | 29.67 |
| 10 | 12.55 | 13.44 | 14.53 | 15.99 | 18.31 | 20.48 | 21.16 | 23.21 | 25.19 | 27.11 | 29.59 | 31.42 |
| 11 | 13.70 | 14.63 | 15.77 | 17.28 | 19.68 | 21.92 | 22.62 | 24.72 | 26.76 | 28.73 | 31.26 | 33.14 |
| 12 | 14.85 | 15.81 | 16.99 | 18.55 | 21.03 | 23.34 | 24.05 | 26.22 | 28.30 | 30.32 | 32.91 | 34.82 |
| 13 | 15.98 | 16.98 | 18.20 | 19.81 | 22.36 | 24.74 | 25.47 | 27.69 | 29.82 | 31.88 | 34.53 | 36.48 |
| 14 | 17.12 | 18.15 | 19.41 | 21.06 | 23.68 | 26.12 | 26.87 | 29.14 | 31.32 | 33.43 | 36.12 | 38.11 |
| 15 | 18.25 | 19.31 | 20.60 | 22.31 | 25.00 | 27.49 | 28.26 | 30.58 | 32.80 | 34.95 | 37.70 | 39.72 |
| 16 | 19.37 | 20.47 | 21.79 | 23.54 | 26.30 | 28.85 | 29.63 | 32.00 | 34.27 | 36.46 | 39.25 | 41.31 |
| 17 | 20.49 | 21.61 | 22.98 | 24.77 | 27.59 | 30.19 | 31.00 | 33.41 | 35.72 | 37.95 | 40.79 | 42.88 |
| 18 | 21.60 | 22.76 | 24.16 | 25.99 | 28.87 | 31.53 | 32.35 | 34.81 | 37.16 | 39.42 | 42.31 | 44.43 |
| 19 | 22.72 | 23.90 | 25.33 | 27.20 | 30.14 | 32.85 | 33.69 | 36.19 | 38.58 | 40.88 | 43.82 | 45.97 |
| 20 | 23.83 | 25.04 | 26.50 | 28.41 | 31.41 | 34.17 | 35.02 | 37.57 | 40.00 | 42.34 | 45.31 | 47.50 |
| 21 | 24.93 | 26.17 | 27.66 | 29.62 | 32.67 | 35.48 | 36.34 | 38.93 | 41.40 | 43.78 | 46.80 | 49.01 |
| 22 | 26.04 | 27.30 | 28.82 | 30.81 | 33.92 | 36.78 | 37.66 | 40.29 | 42.80 | 45.20 | 48.27 | 50.51 |
| 23 | 27.14 | 28.43 | 29.98 | 32.01 | 35.17 | 38.08 | 38.97 | 41.64 | 44.18 | 46.62 | 49.73 | 52.00 |
| 24 | 28.24 | 29.55 | 31.13 | 33.20 | 36.42 | 39.36 | 40.27 | 42.98 | 45.56 | 48.03 | 51.18 | 53.48 |
| 25 | 29.34 | 30.68 | 32.28 | 34.38 | 37.65 | 40.65 | 41.57 | 44.31 | 46.93 | 49.44 | 52.62 | 54.95 |
| 26 | 30.43 | 31.79 | 33.43 | 35.56 | 38.89 | 41.92 | 42.86 | 45.64 | 48.29 | 50.83 | 54.05 | 56.41 |
| 27 | 31.53 | 32.91 | 34.57 | 36.74 | 40.11 | 43.19 | 44.14 | 46.96 | 49.64 | 52.22 | 55.48 | 57.86 |
| 28 | 32.62 | 34.03 | 35.71 | 37.92 | 41.34 | 44.46 | 45.42 | 48.28 | 50.99 | 53.59 | 56.89 | 59.30 |
| 29 | 33.71 | 35.14 | 36.85 | 39.09 | 42.56 | 45.72 | 46.69 | 49.59 | 52.34 | 54.97 | 58.30 | 60.73 |
| 30 | 34.80 | 36.25 | 37.99 | 40.26 | 43.77 | 46.98 | 47.96 | 50.89 | 53.67 | 56.33 | 59.70 | 62.16 |
| 40 | 45.62 | 47.27 | 49.24 | 51.81 | 55.76 | 59.34 | 60.44 | 63.69 | 66.77 | 69.70 | 73.40 | 76.09 |
| 50 | 56.33 | 58.16 | 60.35 | 63.17 | 67.50 | 71.42 | 72.61 | 76.15 | 79.49 | 82.66 | 86.66 | 89.56 |
| 60 | 66.98 | 68.97 | 71.34 | 74.40 | 79.08 | 83.30 | 84.58 | 88.38 | 91.95 | 95.34 | 99.61 | 102.7 |
| 80 | 88.13 | 90.41 | 93.11 | 96.58 | 101.9 | 106.6 | 108.1 | 112.3 | 116.3 | 120.1 | 124.8 | 128.3 |
| 100 | 109.1 | 111.7 | 114.7 | 118.5 | 124.3 | 129.6 | 131.1 | 135.8 | 140.2 | 144.3 | 149.4 | 153.2 |

**Table D Random digits**

| Line | | | | | | | | |
|---|---|---|---|---|---|---|---|---|
| 101 | 19223 | 95034 | 05756 | 28713 | 96409 | 12531 | 42544 | 82853 |
| 102 | 73676 | 47150 | 99400 | 01927 | 27754 | 42648 | 82425 | 36290 |
| 103 | 45467 | 71709 | 77558 | 00095 | 32863 | 29485 | 82226 | 90056 |
| 104 | 52711 | 38889 | 93074 | 60227 | 40011 | 85848 | 48767 | 52573 |
| 105 | 95592 | 94007 | 69971 | 91481 | 60779 | 53791 | 17297 | 59335 |
| 106 | 68417 | 35013 | 15529 | 72765 | 85089 | 57067 | 50211 | 47487 |
| 107 | 82739 | 57890 | 20807 | 47511 | 81676 | 55300 | 94383 | 14893 |
| 108 | 60940 | 72024 | 17868 | 24943 | 61790 | 90656 | 87964 | 18883 |
| 109 | 36009 | 19365 | 15412 | 39638 | 85453 | 46816 | 83485 | 41979 |
| 110 | 38448 | 48789 | 18338 | 24697 | 39364 | 42006 | 76688 | 08708 |
| 111 | 81486 | 69487 | 60513 | 09297 | 00412 | 71238 | 27649 | 39950 |
| 112 | 59636 | 88804 | 04634 | 71197 | 19352 | 73089 | 84898 | 45785 |
| 113 | 62568 | 70206 | 40325 | 03699 | 71080 | 22553 | 11486 | 11776 |
| 114 | 45149 | 32992 | 75730 | 66280 | 03819 | 56202 | 02938 | 70915 |
| 115 | 61041 | 77684 | 94322 | 24709 | 73698 | 14526 | 31893 | 32592 |
| 116 | 14459 | 26056 | 31424 | 80371 | 65103 | 62253 | 50490 | 61181 |
| 117 | 38167 | 98532 | 62183 | 70632 | 23417 | 26185 | 41448 | 75532 |
| 118 | 73190 | 32533 | 04470 | 29669 | 84407 | 90785 | 65956 | 86382 |
| 119 | 95857 | 07118 | 87664 | 92099 | 58806 | 66979 | 98624 | 84826 |
| 120 | 35476 | 55972 | 39421 | 65850 | 04266 | 35435 | 43742 | 11937 |
| 121 | 71487 | 09984 | 29077 | 14863 | 61683 | 47052 | 62224 | 51025 |
| 122 | 13873 | 81598 | 95052 | 90908 | 73592 | 75186 | 87136 | 95761 |
| 123 | 54580 | 81507 | 27102 | 56027 | 55892 | 33063 | 41842 | 81868 |
| 124 | 71035 | 09001 | 43367 | 49497 | 72719 | 96758 | 27611 | 91596 |
| 125 | 96746 | 12149 | 37823 | 71868 | 18442 | 35119 | 62103 | 39244 |
| 126 | 96927 | 19931 | 36809 | 74192 | 77567 | 88741 | 48409 | 41903 |
| 127 | 43909 | 99477 | 25330 | 64359 | 40085 | 16925 | 85117 | 36071 |
| 128 | 15689 | 14227 | 06565 | 14374 | 13352 | 49367 | 81982 | 87209 |
| 129 | 36759 | 58984 | 68288 | 22913 | 18638 | 54303 | 00795 | 08727 |
| 130 | 69051 | 64817 | 87174 | 09517 | 84534 | 06489 | 87201 | 97245 |
| 131 | 05007 | 16632 | 81194 | 14873 | 04197 | 85576 | 45195 | 96565 |
| 132 | 68732 | 55259 | 84292 | 08796 | 43165 | 93739 | 31685 | 97150 |
| 133 | 45740 | 41807 | 65561 | 33302 | 07051 | 93623 | 18132 | 09547 |
| 134 | 27816 | 78416 | 18329 | 21337 | 35213 | 37741 | 04312 | 68508 |
| 135 | 66925 | 55658 | 39100 | 78458 | 11206 | 19876 | 87151 | 31260 |
| 136 | 08421 | 44753 | 77377 | 28744 | 75592 | 08563 | 79140 | 92454 |
| 137 | 53645 | 66812 | 61421 | 47836 | 12609 | 15373 | 98481 | 14592 |
| 138 | 66831 | 68908 | 40772 | 21558 | 47781 | 33586 | 79177 | 06928 |
| 139 | 55588 | 99404 | 70708 | 41098 | 43563 | 56934 | 48394 | 51719 |
| 140 | 12975 | 13258 | 13048 | 45144 | 72321 | 81940 | 00360 | 02428 |
| 141 | 96767 | 35964 | 23822 | 96012 | 94591 | 65194 | 50842 | 53372 |
| 142 | 72829 | 50232 | 97892 | 63408 | 77919 | 44575 | 24870 | 04178 |
| 143 | 88565 | 42628 | 17797 | 49376 | 61762 | 16953 | 88604 | 12724 |
| 144 | 62964 | 88145 | 83083 | 69453 | 46109 | 59505 | 69680 | 00900 |
| 145 | 19687 | 12633 | 57857 | 95806 | 09931 | 02150 | 43163 | 58636 |
| 146 | 37609 | 59057 | 66967 | 83401 | 60705 | 02384 | 90597 | 93600 |
| 147 | 54973 | 86278 | 88737 | 74351 | 47500 | 84552 | 19909 | 67181 |
| 148 | 00694 | 05977 | 19664 | 65441 | 20903 | 62371 | 22725 | 53340 |
| 149 | 71546 | 05233 | 53946 | 68743 | 72460 | 27601 | 45403 | 88692 |
| 150 | 07511 | 88915 | 41267 | 16853 | 84569 | 79367 | 32337 | 03316 |

# Index

## D

## TECHNOLOGY CORNERS REFERENCE

TI-Nspire instructions in Appendix B; HP Prime instructions at www.whfreeman.com/tps5e

# Inference Summary

| How to Organize a Statistical Problem: A Four-Step Process | | |
|---|---|---|
| | **Confidence intervals (CIs)** | **Significance tests** |
| STATE: | What *parameter* do you want to estimate, and at what *confidence level*? | What *hypotheses* do you want to test, and at what *significance level*? Define any *parameters* you use. |
| PLAN: | Choose the appropriate inference *method*. Check *conditions*. | Choose the appropriate inference *method*. Check *conditions*. |
| DO: | If the conditions are met, perform *calculations*. | If the conditions are met, perform *calculations*.<br>• Compute the **test statistic**.<br>• Find the ***P*-value**. |
| CONCLUDE: | *Interpret* your interval in the context of the problem. | Make a *decision* about the hypotheses in the context of the problem. |

CI: statistic ± (critical value)·(standard deviation of statistic)

$$\text{Standardized test statistic} = \frac{\text{statistic} - \text{parameter}}{\text{standard deviation of statistic}}$$

| Inference about | Number of samples (groups) | Interval or test | Name of procedure (TI Calculator function) Formula | Conditions |
|---|---|---|---|---|
| **Proportions** | 1 | Interval | One-sample $z$ interval for $p$ (1-PropZInt)<br>$\hat{p} \pm z^*\sqrt{\frac{\hat{p}(1-\hat{p})}{n}}$ | **Random** Data from a random sample or randomized experiment<br>○ **10%:** $n \leq 0.10N$ if sampling without replacement<br>**Large Counts** At least 10 successes and failures; that is, $n\hat{p} \geq 10$ and $n(1-\hat{p}) \geq 10$ |
| | 1 | Test | One-sample $z$ test for $p$ (1-PropZTest)<br>$z = \frac{\hat{p} - p_0}{\sqrt{\frac{p_0(1-p_0)}{n}}}$ | **Random** Data from a random sample or randomized experiment<br>○ **10%:** $n \leq 0.10N$ if sampling without replacement<br>**Large Counts** $np_0 \geq 10$ and $n(1-p_0) \geq 10$ |
| | 2 | Interval | Two-sample $z$ interval for $p_1 - p_2$ (2-PropZInt)<br>$(\hat{p}_1 - \hat{p}_2) \pm z^*\sqrt{\frac{\hat{p}_1(1-\hat{p}_1)}{n_1} + \frac{\hat{p}_2(1-\hat{p}_2)}{n_2}}$ | **Random** Data from independent random samples or randomized experiment<br>○ **10%:** $n_1 \leq 0.10N_1$ and $n_2 \leq 0.10N_2$ if sampling without replacement<br>**Large Counts** At least 10 successes and failures in both samples/groups; that is,<br>$n_1\hat{p}_1 \geq 10$, $n_1(1-\hat{p}_1) \geq 10$,<br>$n_2\hat{p}_2 \geq 10$, $n_2(1-\hat{p}_2) \geq 10$ |
| | 2 | Test | Two-sample $z$ test for $p_1 - p_2$ (2-PropZTest)<br>$z = \frac{(\hat{p}_1 - \hat{p}_2) - 0}{\sqrt{\frac{\hat{p}_C(1-\hat{p}_C)}{n_1} + \frac{\hat{p}_C(1-\hat{p}_C)}{n_2}}}$<br>where<br>$\hat{p}_C = \frac{\text{total successes}}{\text{total sample size}} = \frac{X_1 + X_2}{n_1 + n_2}$ | **Random** Data from independent random samples or randomized experiment<br>○ **10%:** $n_1 \leq 0.10N_1$ and $n_2 \leq 0.10N_2$ if sampling without replacement<br>**Large Counts** At least 10 successes and failures in both samples/groups; that is,<br>$n_1\hat{p}_1 \geq 10$, $n_1(1-\hat{p}_1) \geq 10$,<br>$n_2\hat{p}_2 \geq 10$, $n_2(1-\hat{p}_2) \geq 10$ |